3rd International Conference on Mechatronics and Mechanical Engineering (ICMME 2016)

MATEC Web of Conferences Volume 95 (2017)

Shanghai, China
21 - 23 October 2016

Editors:

H.L. Yuan
R.K. Agarwal
P. Tandon
E.X. Wang

ISBN: 978-1-5108-3692-1

Printed from e-media with permission by:

Curran Associates, Inc.
57 Morehouse Lane
Red Hook, NY 12571

Some format issues inherent in the e-media version may also appear in this print version.

Printed by Curran Associates, Inc. (2017)

For additional information, please contact EDP Sciences – Web of Conferences
at the address below.

EDP Sciences – Web of Conferences
17, Avenue du Hoggar
Parc d'Activité de Courtabœuf
BP 112
F-91944 Les Ulis Cedex A
France

Phone: +33 (0) 1 69 18 75 75
Fax: +33 (0) 1 69 28 84 91

contact-edps@webofconferences.org

Additional copies of this publication are available from:

Curran Associates, Inc.
57 Morehouse Lane
Red Hook, NY 12571 USA
Phone: 845-758-0400
Fax: 845-758-2633
Email: curran@proceedings.com
Web: www.proceedings.com

TABLE OF CONTENTS

MATEC Web of Conferences
Open access conference proceedings in Materials science, Engineering and Chemistry
A flexible, fast and efficient publication service, for an open access recording of your event and a wide dissemination of your proceedings
www.matec-conferences.org
edp open Open Access Publishing from EDP Sciences

Aims & scope

MATEC Web of Conferences is an open-access journal devoted to the publishing of conference proceedings dealing with Materials science, Engineering and Chemistry.

The journal aims to make conference proceedings visible, citable and widely usable by the scientific community. It offers outstanding services for the publication and dissemination of contents in order to ensure long-term archiving and maximum exposure of scientific events.

Coverage includes all engineering disciplines: civil, naval, mechanical, chemical, and electrical engineering as well as nanotechnology and metrology. The journal also deals with the physical-chemical characterization of materials, their implementation, their resistance in their environment, etc. Others subdisciplines of chemistry, such as analytical chemistry, petrochemistry, organic chemistry, etc., and even pharmacology, are also welcome.

Why publish with MATEC Web of Conferences?

- **Open access:** conference papers available worldwide.
- **Fast publication:** less than 3 months.
- **Total flexibility:** no page restrictions, free unlimited use of colour in the online version, any kind of scientific document welcomed (traditional articles, posters, abstracts and slideshows).
- **High visibility:** all documents indexed in main databases and easily found online.
- **Fair author rights:** CC-BY Creative Commons license.
- **Range of additional publishing options:** printed proceedings, CD-ROMs, copy-editing, etc.

Contact: Isaline Augusto
matec@webofconferences.org

ISSN (electronic edition): 2261-236X

www.matec-conferences.org

edp open Open Access Publishing from EDP Sciences

PREFACE

2016 the 3rd International Conference on Mechatronics and Mechanical Engineering was held during October 21-23, 2016 in Shanghai, China.

ICMME aims to provide a professional platform for academic researchers and industrial practitioners for the interchange of information on the latest development and applications in mechatronics and mechanical engineering, and to promote friendship and interdisciplinary research collaborations.

ICMME received a number of 328 papers from universities, research institutes and industries this year. This volume includes 187 selected papers which were presented at the conference and each contributed paper has been strictly peer-reviewed by organizing and advisory committee members as well as external reviewers. The proceedings tend to present to the readers the newest researches results and findings in the field of mechatronics and mechanical engineering, which include 19 chapters, including materials handling methods and performance analysis, material structure and measurement, material structure and measurement, material structure and measurement, material structure and measurement, material structure and measurement, mechanical design-manufacture and automation, robot design and control, vehicle structure design and manufacturing, manufacturing and industrial production, ship and traffic engineering, applied mechanics and dynamics, mechatronics, electronic technology and application, etc.

The organizing committee expresses its entire gratitude to all the authors who presented their works at ICMME 2016 and contributed in this way to the success of this event. Special thanks are due to the authors from abroad for attending the conference and to the reviewers for their support in improving the quality of the papers and finally for the assurance the quality of this volume.

We hope that the contents of this volume will prove useful for mechatronics and mechanical engineering researchers and practitioners in developing and applying new technologies and processes.

Prof. H.L., Yuan
Tongji University, China

Prof. R. K. Agarwal
Washington University in St. Louis, USA

Prof. P. Tandon
Indian Institute of Information Technology, Design and Manufacturing, India

Prof. Everett X. Wang
Guangdong University of Technology, China

COMMITTEE

Advisory Committee Chair

Prof. Lu Li, National University of Singapore, Singapore

Conference Co-Chairs

Prof. Puneet Tandon, Indian Institute of Information Technology, Design and Manufacturing Jabalpur, India

Prof. Wirachman Wisnoe, Universiti Teknologi MARA, Malaysia

Program Chairs

Prof. Ramesh K. Agarwal, Washington University in St. Louis, USA

Prof. Everett X. Wang, Guangdong University of Technology, China

Publication Chair

Dr. Xiande Zhao, Dalian Maritime University, China

Publicity Chair

Dr. H.L., Yuan, Tongji University, China

Technical Committee

Assoc. Prof. Guoliang Lu, Shandong University, China

Assoc. Prof. Dr M. Sivakumar, Amrita Vishwa Vidyapeetham, India

Dr. Alain Kusmoko, University of Wollongong, Australia

Assoc. Prof. Carolina Senabre, Universidad Miguel Hernandez, Spain

Prof. Promod K. Jain, Indian Institute of Technology, India

Dr. Hoe Guan Beh, Universiti Teknologi Petronas, Malaysia

Assoc. Prof. Sergio Valero Verdú, Universidad Miguel Hernandez, Spain

Asst. Prof. Dr. Xili Duan, Memorial University of Newfoundland, Canada

Dr. Najid Ahmad, Dongbei University of Finance and Economics China

Dr. Janice Toh Guat Guan, Tunku Abdul Rahman University College, Malaysia

Dr. Vigneshwaran Krishnamurthy, Indian Institute of Technology, India

Assoc. Prof. Dr. Teik-Cheng LIM, SIM University, Singapore

Dr. Kiattisin Kanjanawanishkul, Mahasarakham University, Thailand

Prof. Dr. Hermann Wellenstein, Brandeis University, USA

Prof. Syed H. Masood, Swinburne University of Technology, Australia

Asst. Prof. Prasanta Kumar Samal, NIE, India

Dr. C. Elanchezhian, Sri Sairam Engg College, India

Dr. Sohaib Khan, Swinburne University of Technology, Australia

Prof. Dr Katherine R. Ilie, Romania Politehnica University of Bucharest, Romania

Asst. Prof. Amir Atrian, Islamic Azad University, Iran

Dr. Xiangdong Li, Shanghai Jiao Tong University, China

Dr. Najid Ahmad, Dongbei University of Finance and Economics, China

Statement of Peer review

In submitting conference proceedings to MATEC Web of Conferences, I certify to the Publisher that I adhere to the **Policy on Publishing Integrity** of the journal in order to safeguard good scientific practice in publishing.

1. All articles have been subjected to peer review administered by the proceedings editors.
2. Reviews have been conducted by expert referees, who have been requested to provide unbiased and constructive comments aimed, whenever possible, at improving the work.
3. Proceedings editors have taken all reasonable steps to ensure the quality of the materials they publish and their decision to accept or reject a paper for publication has been based only on the merits of the work and the relevance to the journal.

Title, date and place of the conference

2016 The 3rd International Conference on Mechatronics and Mechanical Engineering (ICMME 2016)
Shanghai, China
October 21-23, 2016

Proceedings editor(s):

Prof. H.L., Yuan, Tongji University, China
Prof. R. K. Agarwal, Washington University in St. Louis, USA
Prof. P. Tandon, Indian Institute of Information Technology, Design and Manufacturing, India
Prof. Everett X. Wang, Guangdong University of Technology, China

Date and editor's signature

Antimicrobial Property of Hydrocolloid Impression Material Incorporated with Silver Nanoparticles Against *Staphylococcus Aureus*

Norbu Penden Wangchuk[1], Passiri Nisalak[1], Boonyanit Thaweboon [2], Sroisiri Thawboon[2] and Pornrachanee Sawaengkit[1]

[1]Orthodontic Department, Faculty of Dentistry, Mahidol University, Bangkok 10400, Thailand
[2]Oral Microbiology Department, Faculty of Dentistry, Mahidol University, Bangkok 10400, Thailand

Abstract. Dental impressions can easily become contaminated with patient's blood and saliva which are capable of transmitting infectious diseases to dental personnel. The addition of antimicrobial agents into impression materials could be effective in reducing the chances of cross-infection. Silver nanoparticles have been applied in dentistry as a potent antimicrobial agent. This study aims to evaluate the *in vitro* antimicrobial efficacy of silver nanoparticles incorporated to irreversible hydrocolloid impression material against *Staphylococcus aureus*. Silver nanoparticles ($AgZrPO_4$, National Direct Network Company, Thailand) at concentrations of 0.25%, 0.50%, 1.00% and 1.50% w/w were added to powder of impression materials (Kromopan, Lascod, Ilaty). Impression material samples were prepared on sterile plate in accordance with manufacturer's instruction. After setting, a 100 microliter of *S. aureus* ATCC6538 suspension (10^6 cells/mL) were inoculated on the surface of the impression sample and left for 10 minutes. The amount of *S. aureus* on the surface was quantified using imprint technique on Mannitol Salt agar. Impression materials incorporated with $AgZrPO_4$ showed antimicrobial property against *S. aureus* (up to 95% reduction) compared with control (impression material without $AgZrPO_4$). Even though the mechanism of antimicrobial action was not clearly understood, $AgZrPO_4$ incorporated to impression material was demonstrated to possess an inhibitory effect against pathogenic bacteria. Further studies are needed to investigate physical properties of the material and the clinical usage.

1 Introduction

The study of dental biomaterials and its properties are of immense interest in the field of research and innovative technology. One of the most widely accepted and used in dentistry is the hydrocolloid alginate impression material owing to its several advantages over the others [1]. Impression material has been one of the many biomaterials that have captured the attention of many researchers. Firstly in trying to replicate the anatomic details of the oral cavity and then further to addition of antimicrobial agent to render it sterile. The reasons for the purpose of rendering it sterile erupts from the fact that dental impressions become contaminated with patient's saliva and/or blood that can consequently cross-contaminate stone casts poured against them. Subsequent handling puts the dental personnel at high risks of microbial transmission/infection.

The micro flora of the oral cavity has a diverse community of microbes ranging from bacteria to fungi with studies even confirming the presence of viruses [2]. In a three year retrospective study by McCormack *et al*, they reported a large data regarding the recovery of *Staphylococcus aureus* in the oral and perioral cavity [3].

His findings condemned oral rinse as the main specimen from which the *S. aureus* could be isolated. It warns us of the potential cross-infection and dissemination to other body sites. It may be indispensable to the hands of dentists but impression taking is also known for effective cross infection and transmission of microbes and diseases thereof like many other dental practices [4]. As it is known that the oral cavity is a host to many known and unknown microbes [2], researches have confirmed that the oral fluid diffuse into the impression material [5] suggesting a pathway for initiating a cross infection that can be potentially deadly in transmitting infectious microorganisms. Of course it is needless to mention the numerous methods invented to render the impression material sterile in the form of aerosol spraying or immersion in the disinfectant solution or solely rinsing under running tap water[6], [7]. These methods tend to have some shortcomings either in the form of disinfection achieved or the compromise resulting in the physical property of the impression material that is controversial overall [8].

The dawn of nanotechnology, silver nanoparticles to be specific has gained much popularity owing to its potent antimicrobial activity [9]. Its use is well documented especially as a dressing material resulting from injury at times of battle. Dentistry also reports its use in various field including prosthetic dentistry, dental implants and restorative as well as in endodontics in the form of root canal irrigation solution [10].

The purpose of this study is therefore to evaluate the antimicrobial property of hydrocolloid impression material incorporated with silver zirconium phosphate nanoparticles against *S. aureus*.

2 Materials and methods

2.1 Microorganisms

Staphylococcus aureus ATCC 5638 were obtained from the culture collection of the Department of Oral Microbiology, Faculty of Dentistry, Mahidol University, Thailand. They were grown in Mannitol Salt agar at 30 to 35 °C for 7 days. Microbial suspension was prepared in saline solution to approximately 10^8 CFU/ml using Mac Farlands no. 0.5 and further diluted to 10^5 CFU/ml before being used in the antimicrobial testing.

2.2 Irreversible hydrocolloid impression material

The irreversible alginate impression material used for the test was Kromopan class A, type I dust free complying with ISO 1563. Lascod, Italy.

2.3 Silver based nanoparticles

The silver zirconium phosphate nanoparticles ($AgZrPO_4$) were supplied from the National Direct Network Company, Thailand and were mixed with powder of hydrocolloid impression material to the concentrations of 0.25%, 0.50%, 1.00 % and 1.50% w/w.

2.4 Antimicrobial testing

The irreversible hydrocolloid impression material powder incorporated with each concentration of silver nanoparticles (0.25%, 0.50%, 1.00%, and 1.50% w/w) was mixed with water following the manufacturer's instruction and then poured into sterile petri dish. The 100 microliter of the microbial suspension were then spread on the surface of the impression material and left for 10 minutes. A 2×2 cm^2 sterile Whatmans filter paper was then placed on the spread bacteria and left for few minutes. This step simulates the activity of microbes that are harvested during the impression taking.

The filter paper were then removed and placed in the Mannitol Salt agar plate which serves as a selective media for the growth of *S. aureus* and was incubated at 37 °C for 48 hours. This procedure best describes the optimal technique that could give quantitative results by the growth of the *S. aureus,* commonly known as imprint technique. Hydrocolloid impression material without $AgZrPO_4$ served as control. The number of bacterial growth recovered from the media will represent the antibacterial activity.

All the experiments were done in triplicate in three different occasions.

2.5 Statistical analysis

The antimicrobial property was evaluated by the recovery of number of *S. aureus* (CFU/cm^2) on Mannitol Salt agar by imprint technique with non-parametric test using Kruskal-Wallis test. Pairwise comparison of group was done using Mann-Whitney test at a confidence interval set at 95% with the significance level set at α=0.05.

Table 1. The mean number and standard deviation of *S. aureus* detected on impression material (CFU/cm^2) by imprint method.

$AgZrPO_4$ (%w/w)				**no $AgZrPO_4$**
0.25	**0.50**	**1.00**	**1.50**	
1.67±0.4	0.65±0.53	0.31±0.38	0.13±0.28	2.47±0.45

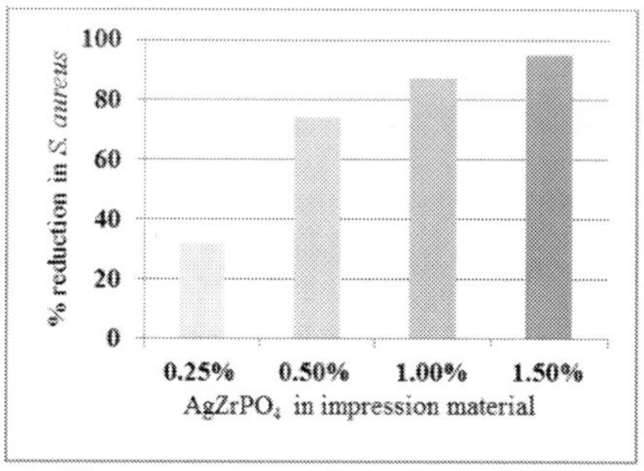

Figure 1. The percentage of *S. aureus* reduction at various concentration of $AgZrPO_4$ containing impression material compared with control.

3 Results

Imprint technique was used for this method, as it is noninvasive easy-to-use that would give representative and quantitative results for samples of specific unit areas. In this study the results were collected as the observed number of bacterial growth in all the media reflecting the different concentrations of $AgZrPO_4$ containing impression material. Table 1 shows the mean and standard deviation of viable bacteria recovered from impression material. It was evident to see that that least number of colonies (0.31±0.38) was observed in the impression material with the highest concentration of $AgZrPO_4$ (1.50%). On the other hand, the highest number of bacteria (1.67±0.48) was recovered from the impression material containing the least $AgZrPO_4$ (0.25%) that was comparable to that of the bacteria recovered from the control group without the $AgZrPO_4$. As seen in the Fig. 1, the percentage reduction of the viable bacteria is seen to increase in a dose dependent manner. The impression material containing (0.25%) $AgZrPO_4$ showed the least percentage reduction of 32%. On the other hand, almost up to 95% reduction in the viable *S. aureus* was seen with impression material containing (1.50%) $AgZrPO_4$ that corresponds to its highest concentration used. Statistical analysis showed no significant difference when the control group was compared to that of the lowest concentration of the $AgZrPO_4$ (0.25%). However, significant difference was seen among the different concentrations of $AgZrPO_4$ containing impression material and that between the control and 0.50%, 1.00% and 1.50%.

4 Discussion

Infection control and prevention is of paramount importance to prevent cross contamination. It has been a huge concern in the medical practice including dentistry. One of the pathways to cross-infection can be demonstrated simply by the use of a common biomaterial by the name, irreversible hydrocolloid impression material while making a negative replica of the patients intra-oral hard and soft tissue. This material owing to their composition and the nature of hydrophilic mechanism of solidification can be easily infected with microorganisms present in the oral cavity and therefore be responsible for cross-infection. Literature has it that the oral fluids diffuse into the impression material up to 540 micrometer establishing the fact that impression material could be contaminated with microorganisms from oral fluid [5], [11]. It is also reported that impression material when not disinfected can be heavily contaminated with microorganisms.

The earliest methods to disinfect the impression material included the rinsing of the impression under running tap water. Many methods have been developed and applied since then in an attempt to disinfect the impression material in the form of either mixing the disinfectant directly with the impression material or immersed into or sprayed in the form of aerosol. Potential disinfectants include iodophors, chlorhexidine, phenolics and inorganic ions such as copper or fluoride. Quaternary ammonium compounds are reported to have limited antimicrobial abilities. Although the above methods and the disinfectants were effective in reducing the number of bacteria, difficulties were faced when complying with the procedure of doing "another step" [12]. Moreover, immersion techniques are known to distort the anatomic details in addition to taking up much of the time before the impression material can be poured. Some of the agents could cause adverse tissue response such as the quaternary ammonium compounds when present in multiple forms. Aerosol spraying has limitations in reaching every nook and corner of the dental impression.

On the other hand, Silver nanoparticles were extensively explored for their antimicrobial activity. The earliest reports of silver as documented were used for wound dressings, catheters and prosthesis[13]. It has been extensively researched and experimented in various dental fields including endodontics, prosthetic dentistry as well as dental implants. And since then, its use evolved giving way to the latest technology of nanoparticles that had the advantage of small-sized particles showing efficient antimicrobial properties due to their larger surface area producing an effective contact with microorganism [14]. Our study demonstrated that $AgZrPO_4$ when used in mixture with the impression material could have antimicrobial effects at concentrations as low as 0.25% against *S. aureus*. This result is in agreement with finding of Saengmee-Anupharb *et al* where silver inorganic materials were evaluated for the antimicrobial activities including silver zeolite (AgZ), silver zirconium phosphate ($AgZrPO_4$) and silver zirconium phosphate silicate (AgZrPSi) [14]. However, this antimicrobial activity exhibited by $AgZrPO_4$ is still an ambiguity. Reliable researches base the whole mechanism of antimicrobial activity firstly to the silver ions interaction with the cell wall, cell membrane and cell envelope of microorganism causing them to rupture and consequently resulting in cell death. It is also shown that silver ions do react with the nucleophilic amino acid residues in proteins that are attached to sulfhydryl, amino, imidazole, phosphate and carbonyl groups of membrane or enzyme proteins [13]. Some studies relate the antimicrobial property to the structural changes induced in the bacterial cell by formation of free radicals and subsequent free-radical induced membrane damage [15].

The results of this study, through the statistical analysis, found no significant difference in bacterial inhibition between the control and the 0.25% $AgZrPO_4$ containing impression material. Significant difference was found between the control and the 0.50%, 1.00% and with 1.50% $AgZrPO_4$ and also between the different concentration of $AgZrPO_4$ containing impression material. With the increase in concentration of the nanoparticles in the impression material, their antimicrobial activity increased vice-versa. From this inference, although a high level of antibacterial activity can be achieved with the use of 1.50% of $AgZrPO_4$ in the impression material, the percentage reduction of *S. aureus* is comparable to that 0.50% $AgZrPO_4$ and therefore recommend the use of 0.50% of silver nanoparticle in impression material that will aim to achieve the maximum antimicrobial benefit while maintaining a low dose level of the disinfectant at the same time.

Few studies have reported the effect disinfection procedures had on the dimensional accuracy and surface quality of the irreversible hydrocolloid impression material. One study reports an increased distortion of irregular areas of the impression when using the immersion technique [8]. While there is a lack of evidence on the effect $AgZrPO_4$ has on the physical property of the impression material, it was interesting to note that Bajrachary *et al*, in their study of Poly methyl methacrylate resin incorporated with silver nanoparticles did not compromise the flexural strength of the Poly methyl methacrylate resin [16]. The findings of Slane *et al*, in acrylic bone cement impregnated with silver nanoparticles authenticate the least effect it has on the materials physical property where it was established that under low concentrations of metallic silver nanoparticles, acrylic bone cements had retained their mechanical and material properties similar to the standard cement [17]. It may not be appropriate to say the physical properties of the impression material incorporated with $AgZrPO_4$ will be least affected but rather be experimented and prove accordingly.

As with the problem of every antibacterial agent, silver nanoparticle has also expressed concern regarding the side effect especially in terms of cytotoxicity and genotoxicity. Although limited studies can be found that are aimed at investigating the genotoxicity of the nanoparticles, it is nevertheless considered complex. One study in particular points out that the most resistant organism to genotoxicity being the human cell cultures [18]. More over the overall time period of contact of the impression material with the oral mucosa and related structures is limited to few minutes. This is even supported by the fact that oral healthcare products do contain silver which is attributed to the broad-spectrum antimicrobial effects and low toxicity to humans [14]. This propels us to further research more regarding the potential ability to produce an impression material equipped with an antibacterial agent least harmful to the patient while exerting the most beneficial antibacterial activity.

Indeed the human oral cavity is extensively populated with a vast variety of microbes out of which *S. aureus* having reported a carriage rate of 24-84% in healthy oral cavity and 48% incidence among the denture-wearing population [19], takes the credit for causing majority of the deleterious clinical diseases involving the hard and soft tissue of the mouth [20]. Since its spread by contact is vividly documented in literature posing a great deal of threat to the dental personnel, this study therefore provides a significantly useful alternative to control the pathologies connected with *S. aureus* and other associated microbes. Dental impressions can be easily contaminated with patient's blood and saliva capable of transmitting infectious diseases to dental personnel. The method adopted in this experimental study was an attempt to produce an impression material with the addition of an antimicrobial agent so as to render the impression material effective against the numerous microbes including the *S. aureus*. The results obtained reflect the antimicrobial property of the $AgZrPO_4$ in the impression material that was seen to be effective in a dose dependent manner.

In conclusion, hydrocolloid impression material containing $AgZrPO_4$ could evidently reduce the number of test bacteria. However, further studies are required to test the accuracy in capturing the anatomic details including the physical properties with this impression material. Nevertheless, this basic research provides useful information on the application of $AgZrPO_4$ in dental product to control or reduce risk of cross-infection. Although the invention of 3-D scanner and printer endanger the practice of using conventional alginate impression material, it is still nevertheless an economic, safe and reliable method of impression taking of the oral cavity in many of the developing countries

Acknowledgement

The authors would like to acknowledge the kind assistance generously offered by Miss Thaniya Maudcheingka including the other staff members of Oral Microbiology Department, Faculty of Dentistry, Mahidol University for trying to make this research a success. We also extend our heartfelt gratitude towards the Faculty of Dentistry, Mahidol University, Thailand and the Ministry of Health, Royal Government of Bhutan for their financial and academic support.

References

1. T.E. Donovan, W.W. Chee, Dent Clin North Am **48,** vi-vii, 445-70. (2004).
2. M. Avila, D.M. Ojcius, Ö. Yilmaz, DNA and Cell Biology **28,** 405-411. (2009).
3. M.G. McCormack, A.J. Smith, A.N. Akram, M. Jackson, D. Robertson, G. Edwards, American Journal of Infection Control **43,** 35-37. (2015).
4. H. Egusa, T. Watamoto, K. Abe, M. Kobayashi, Y. Kaneda, S. Ashida, T. Matsumoto, H. Yatani, Int J Prosthodont **21,** 62-8. (2008).
5. R. Surna, J. Junevicius, E. Rutkauskas, Stomatologija **11,** 129-34. (2009).
6. F.A. Rueggeberg, F.E. Beall, M.T. Kelly, G.S. Schuster, J Prosthet Dent **67,** 628-31. (1992).
7. I.H. Rweyendela, M. Patel, C.P. Owen, Sadj **64,** 208, 210-2. (2009).
8. A. Rentzia, D.C. Coleman, M.J. O'Donnell, A.H. Dowling, M. O'Sullivan, J Dent **39,** 133-40. (2011).
9. K. Chaloupka, Y. Malam, A.M. Seifalian, Trends Biotechnol **28,** 580-8. (2010).
10. J.M. Correa, M. Mori, H.L. Sanches, A.D. da Cruz, E. Poiate, Jr., I.A. Poiate, Int J Biomater **2015,** 485275. (2015).
11. C.P. Owen, R. Goolam, Int J Prosthodont **6,** 480-94. (1993).
12. D.A. Flanagan, C.J. Palenik, J.C. Setcos, C.H. Miller, Dental Materials **14,** 399-404. (1998).
13. J.S. Kim, et al., Nanomedicine: Nanotechnology, Biology and Medicine **3,** 95-101. (2007).
14. S. Saengmee-Anupharb, T. Srikhirin, B. Thaweboon, S. Thaweboon, T. Amornsakchai, S. Dechkunakorn, T. Suddhasthira, Asian Pac J Trop Biomed **3,** 47-52. (2013).
15. M. Chen, Z. Yang, H. Wu, X. Pan, X. Xie, C. Wu, International Journal of Nanomedicine **6,** 2873-2877. (2011).
16. S. Bajrachary, S. Thaweboon, B. Thaweboon, A. Wonglamsam, T. Sirthavaj, In vitro evaluation of candida albicans biofilm formation on denture base PMMA resin incorporated with silver nanoparticles and its effect on flexural strength, 2014, 51-55.
17. J. Slane, J. Vivanco, W. Rose, H.-L. Ploeg, M. Squire, Materials Science and Engineering: C **48,** 188-196. (2015).
18. R. de Lima, A.B. Seabra, N. Duran, J Appl Toxicol **32,** 867-79. (2012).
19. M.S. Jackson, J. Bagg, M.N. Gupta, R.D. Sturrock, Rheumatology (Oxford) **38,** 572-5. (1999).
20. A.J. Singer, J Dent Res **31,** 591-7. (1952).

Antifungal Activity of Type III Dental Gypsum Incorporated with 3-iodo-2-Propynyl-Butylcarbamate

Thitinop Riyatanon[1], Pornrachanee Sawaengkit[1], Sroisiri Thaweboon[2], Boonyanit Thaweboon[2] and Passiri Nisalak[1]

[1]*Orthodontic Department, Faculty of Dentistry, Mahidol University, Bangkok 10400, Thailand*
[2]*Oral Microbiology Department, Faculty of Dentistry, Mahidol University, Bangkok 10400, Thailand*

Abstract. The fungal growth on dental model can damage and affect the physical appearance of the gypsum. Fungi can be transferred among patients and dental personnel. Moreover, they relate to numerous illnesses. Thus, the development of antifungal dental gypsum is required to avoid the fungal growth on dental models. This study evaluated antifungal properties of 3-iodo-2-propynyl-butylcarbamate (IPBC) incorporated into type III dental gypsum. Three types of dental gypsum (Sirius, Ultima, France, 0.005% w/w IPBC and non-IPBC Siam Moulding Plaster, Thailand) were tested according to modified ASTM G 21-96 method with *Penicillium notatum* MI-311, *Aspergillus flavus* MI-321, and *Aspergillus spp.* isolated from orthodontic models. 50 μL of spore suspension of each fungus (10^4CFU/mL) was dropped on the prepared gypsum samples and incubated at room temperature, ≥85% relative humidity for 28 days. Fungal growth was visually scored. No fungal growth was observed on IPBC gypsum while 2 strains of *Aspergillus spp.* could be found on sirius gypsum. Type III dental gypsum incorporated with IPBC shows significant antifungal activity ($p < .001$) compared with non-IPBC and Sirius groups. This developed gypsum with IPBC can be used to fabricate dental models to prevent any damages from fungal growth.

1 Introduction

In orthodontics, type III dental gypsum is traditionally used to fabricate dental models, which required for diagnosis, treatment planning, communication, treatment record, and especially medicolegal purpose. Moreover, The American Board of Orthodontics (ABO) requires dental models to be smoothed and polished in such a manner that tooth and soft tissue details are not destroyed.

The growth of fungi is one of dental model storing problems. It does not only affect the physical appearance of the models but also can damage the gypsum [1], [2]. In addition, fungi associate with numerous health effects such as upper respiratory tract symptoms, cough, wheeze, respiratory infection, bronchitis, asthma and other problems [3], [4].

The isolated fungi from plaster dental models are *Cladophialophora spp.*, *Trichosporon beigelii*, *Aspergillus flavus* and *Aureobasidium pullulans* [5]. The sources of these fungi are from patient's secretion and environment.

The healthy oral cavity presents more than one hundred species of fungi. The most frequently detected oral fungi are *Candida spp.* (75%), followed by *Cladosporium spp.* (65%), *Aureobasidium spp.* (50%), *Saccharomycetales spp.* (50%), *Aspergillus spp.* (35%), *Fusarium spp.* (30%), and *Cryptococcus spp.* (20%) [6]. Furthermore, the predominant fungi isolated from environmental were *Aspergillus spp., Penicillium spp., Stachybotrys spp. and Cladosporium spp* [7], [8].

Using of fungal-resistant construction products that are incorporated with the antimicrobial material is one way to reduce the risk of fungal growth. There are many agents which can be used for this aim, such as copper montmorillonites, zinc montmorillonites, silver montmorillonites [9], silver zeolite, silver zirconium phosphate silicate, silver zirconium phosphate [10], azole compounds [11], and 3-iodo-2-propynyl-butylcarbamate (IPBC) [12].

IPBC has antimicrobial and fungicide properties widely used as a preservation in cosmetics, paint, inks, adhesives, plastics, paper, textiles, emulsions, metalworking fluids, and wood products. It can be used against several types of bacterial and fungi [13], [14]. The effective antimicrobial concentration of IPBC in type III dental gypsum, which has acceptable physical properties, is 0.005% w/w [15]. Thus, the aim of this study was to evaluate the antifungal properties of 3-iodo-2-propynyl-butylcarbamate (IPBC) incorporated into type III dental gypsum.

2 Materials and methods

2.1 Microorganism

Aspergillus spp. isolated from patient's orthodontic dental model and two types of fungi (*Penicillium*

 DOI: 10.1051/matecconf/20179501002

notatum MI-311, *Aspergillus flavus* MI-321) obtained from the Oral Microbiology Department, Faculty of Dentistry, Mahidol University, Bangkok, Thailand were grown on Difco™ Sabouraud Dextrose Agar at 30 to 35°C for 7 days. After that, the tested strains were transferred into RPMI-1640 medium (with glutamine, without bicarbonate, and with phenol red as a pH indicator) (Sigma #R-7755 St. Louis, USA) (Life technologies™) supplemented with 0.2% glucose. Microbial suspensions were adjusted using a spectrophotometer with a tested inoculum in the range $0.4x10^4$ to $5x10^4$ CFU/ml with the optical density (OD) of 0.09 - 0.11 at 530 nm according to CLSI M38-A standard for molds[16].

2.2 Dental gypsum

In this study, type III dental gypsum were divided into 3 groups as follows non-IPBC group (Siam Moulding Plaster Co., Ltd, Thailand without disinfectant), IPBC group (Siam Moulding Plaster Co., Ltd, Thailand with 0.005% w/w IPBC), and Sirius group (Sirius, Ultima, France).

2.3 Antifungal testing

The antifungal testing was done according to the modified ASTM G 21-96 method[17, 18]. All groups of dental gypsums were prepared and poured into petri dishes and then remain undisturbed for 60 minutes to completely set. Fifty µl of each microbial suspension was dropped on the prepared dental gypsum samples. Petri dishes were sealed by laboratory sealing film and were incubated in the container for 28 days at 30 to 35°C and 85 % relative humidity. The density of fungal colonies on gypsum samples were examined under microscope and were scored at day 7, 14, 21 and 28 base on the fungal growth as shown in Table1.

All experiments were done in triplicate with 3 separate occasions. The fungal growth scores were carried out by the same researcher.

Table 1. Fungal growth scoring criteria

Growth	Score
None	0
Traces of growth (Less than 10%)	1
Light growth (10-30%)	2
Medium growth (30-60%)	3
Heavy growth (60% to complete coverage)	4

2.4 Intra-examiner reliability

To evaluate intra-examiner reliability, 10 samples were randomly selected from all tested petri dishes. Each selected sample was scored 3 times with the time interval of 1 week.

2.5 Statistical analysis

The fungal growth scores were analyzed to evaluate the antifungal activity of three groups of type III dental gypsum with the Fisher's exact statistical test at 95 % confidence interval; the fundamental significance level was thus set at $\alpha = 0.05$. The intraclass correlation coefficient (ICC) was performed to evaluate the intra-examiner reliability.

3 Results

Almost perfect intra-examiner reliability was determined (ICC = 0.973). There is only one error of intra-examiner fungal growth scoring from score 1 to score 0.

The median and range of fungal growth scores of each group are demonstrated in the Table 2. No fungal growth was observed on Type III dental gypsum incorporated with IPBC under the experimental conditions. Whereas, the median values of fungal growth scores of *P. notatum*, *A. flavus,* and *Aspergillus spp.* isolated from models in non-IPBC group were 4, 3, and 4 respectively. In Sirius group, the median values of fungal growth scores similar to non-IPBC group's scores except *P. notatum* that showed no growth. Thus, type III dental gypsum incorporated with IPBC exhibited significantly better antifungal activity against all tested fungi than non-IPBC and Sirius groups ($P < .001$).

Table 2. Fungal growth scores on 3 types of dental gypsum.

Fungi	Type III dental gypsum		
	Non-IPBC	IPBC	Sirius
Penicillium notatum	4 (3-4)	0 (0-0)	0 (0-0)
Aspergillus flavus	3 (3-4)	0 (0-0)	3 (3-4)
Aspergillus spp. isolated from model	4 (3-4)	0 (0-0)	4 (3-4)

*Data expressed as median value and (range) of 9 replicates.

4 Discussion

Cross contamination is a major problem in dentistry. Several harmful microorganisms which come from oral cavity and environment [5], [19] can be spread and transferred among patients, dental health personnel, and technicians in dental clinic and laboratory via dental models. For example, orthodontic appliances such as retainer are fabricated on dental models before put on patients. If dental models are contaminated with fungi, they can transfer to dentists and patients during treatment procedures.

Table 3. Fungal growth scores of each sample at time interval.

Fungi	Type III dental gypsum																										
	Non IPBC									IPBC									Sirius								
	A			B			C			A			B			C			A			B			C		
	a	b	c	a	b	c	a	b	c	a	b	c	a	b	c	a	b	c	a	b	c	a	b	c	a	b	c
Penicillium notatum																											
7 d	2	2	2	0	0	0	0	0	0	0	0	0	0	0	0	0	0	0	0	0	0	0	0	0	0	0	0
14d	3	3	3	2	2	2	2	3	2	0	0	0	0	0	0	0	0	0	0	0	0	0	0	0	0	0	0
21d	4	4	4	3	3	3	3	3	4	0	0	0	0	0	0	0	0	0	0	0	0	0	0	0	0	0	0
28d	4	4	4	4	4	3	4	4	4	0	0	0	0	0	0	0	0	0	0	0	0	0	0	0	0	0	0
Aspergillus flavus																											
7 d	3	2	2	1	1	2	0	0	0	0	0	0	0	0	0	0	0	0	3	2	3	0	1	1	1	2	1
14d	3	3	3	2	2	2	1	1	1	0	0	0	0	0	0	0	0	0	4	3	4	1	2	2	2	2	2
21d	4	4	3	3	3	3	2	3	3	0	0	0	0	0	0	0	0	0	4	3	4	2	3	3	2	3	2
28d	4	4	3	3	3	3	3	4	4	0	0	0	0	0	0	0	0	0	4	4	4	3	3	3	3	4	3
Aspergillus spp. from model																											
7 d	3	2	3	3	3	3	3	3	2	0	0	0	0	0	0	0	0	0	1	1	1	3	3	2	0	1	1
14d	3	2	4	3	3	3	3	4	3	0	0	0	0	0	0	0	0	0	2	1	2	3	3	3	1	3	2
21d	4	3	4	4	4	4	4	4	3	0	0	0	0	0	0	0	0	0	3	3	3	4	4	4	2	3	3
28d	4	3	4	4	4	4	4	4	4	0	0	0	0	0	0	0	0	0	4	4	4	4	4	4	3	4	4

d stands for days. a, b and c are three replicas. A, B and C are three occasions.

Furthermore, fungi can damage gypsum of dental models by creating the decay from both physical and chemical actions [20]. Only small decay on dental models can impact the analysis activities such as diagnosis, treatment planning, and treatment evaluation in orthodontics because the analysis of tooth alignment, dental arch perimeter, and occlusion in orthodontics require all details at millimeter-level from dental models. Therefore, dental gypsum containing antimicrobial agents can be one of methods to solve these problems.

3-iodo-2-propynyl-butylcarbamate or IPBC is used as preservative agent in various products. In this study, the incorporation of IPBC into type III dental gypsum demonstrated effective antifungal activity against all tested fungi compared with non-IPBC and Sirius groups of dental gypsum. However, the mechanism of IPBC antimicrobial activity is not clearly understood. IPBC has approximately 1,000 times more toxic to invertebrates and fish than the propargyl butyl carbamate (PBC), primary hydrolysis metabolite without iodine. Thus, a possible mechanism may be related to the iodine [21].

Iodine is a broad-spectrum antimicrobial agent that has simultaneously action against gram positive, gram negative, spores, amoebic cysts, fungi, protozoa and yeasts, and Methicillin-resistant *Staphylococcus aureus* (MRSA). It affects enzymes of respiratory chain, lipid membrane, and nucleic acid function that interferes with the permeability of the cell membranes and denatures protein [22]. Thaweboon's study showed the antimicrobial activity of type III dental gypsum incorporated with IPBC against *Staphylococcus aureus*, *Pseudomonas aeruginosa*, and *Candida albicans* [15]. However, IPBC is degraded by some strains of proteobacteria (Alcaligenes, Enterobacter, Pseudomonas, Ralstonia, etc.) [23].

In term of safety profile, the 2013 Cosmetic Ingredient Review (CIR) Expert Panel concluded that the safety concentration of IPBC for cosmetic is $\leq$ 0.1%. There are low risks of contact sensitization and allergic contact dermatitis at this concentration level. Moreover, IPBC neither has embryotoxic, teratogenic, mutagenic, nor carcinogenic effects [24]. Therefore, the amount of IPBC used in dental gypsum (0.005% w/w) is far below the threshold of safety limit for that used in cosmetics.

From Table 2 and Table 3, dental gypsum in Sirius group also showed some antifungal activity against *P. notatum* and retarded growth of *A. flavus* and *Aspergillus spp.* isolated from models. In accordance with information technique - technical datasheet of Sirius [25], this type of gypsum contains additives to protect absorption of moisture which is one of important factors influencing the growth of fungi on the materials [26].

In dentistry, gypsum models of patients' dentition are one of recognized tools for orthodontic diagnosis and treatment planning as well as medico-legal purpose. Dental models are the traditional three-dimensional patient record for measuring linear changes in the dentition during treatment and, therefore, have to be stored for many years. In our study, antifungal property of dental gypsums was evaluated during the period of 28 days according to the ASTM G 21-96 antifungal testing method for materials. This might be the time limitation of our study.

Concerning other properties of studied dental gypsums, even though the physical properties of IPBC incorporated dental gypsum were not determined in this study compared with others, data from our previous study have shown that this type of dental gypsum has satisfactory physical properties within the ISO standard of dental materials [15]. Further investigation on other contaminated fungi and the sustained period of antimicrobial action of this modified dental gypsum are still required.

As demonstrated in experimental results, Type III dental gypsum incorporated with 3-iodo-2-propynyl-butylcarbamate (IPBC) had no fungal growth whereas other two types of gypsum had medium to heavy fungal growth. Therefore, we can concluded that Type III dental gypsum incorporated with IPBC had antifungal effect on *P. notatum*, *A. flavus* and *Aspergillus spp.* isolated from model at 30 to 35°C, 85 % relative humidity for 28 days. This developed antifungal dental gypsum could be used to fabricate dental models, which can be preserved overtime for the purpose of treatment planning and evaluation. In addition, further study will be conducted with other fungal species and microorganism such as *Cladophialophora spp.*, *Trichosporon beigelii*, *Aureobasidium pullulans,* coagulase-negative *Staphylococcus, Micrococcus spp.*, *Bacillus spp.*, and nonfermenting gram-negative bacillus [5] to investigate the effectiveness of Type III dental gypsum incorporated with IPBC regarding antimicrobial property.

Acknowledgement

The authors would like to acknowledge with deep appreciation the invaluable help of Miss Thaniya Muadcheingka and all staffs of Oral Microbiology Department, Faculty of Dentistry, Mahidol University. We are also very much thankful to Assist. Prof. Chulaluk Komoltri for her precious assistance in the statistical analysis. Finally, we wish to thank the Faculty of Dentistry, Mahidol University, Thailand for financial support. The project could not have been completed without their supports.

References

1. F. Cappitelli, P. Principi, R. Pedrazzani, L. Toniolo, C. Sorlini, Sci. Total. Environ. **385,** 172-81, (2007).
2. J. Lstiburek, T. Brennan, N. Yost, 1-2, (2002).
3. W.J. Fisk, E.A. Eliseeva, M.J. Mendell, Environ. Health. **9**, 72, (2010).
4. W.J. Fisk, Q. Lei-Gomez, M.J. Mendell, Indoor. Air. **17**, 284-96, (2007).
5. S. Gallão, A.C. Pizzolitto, L. Santos-Pinto, A. dos Santos-Pinto, K. Faltin Jr, L.P. Martins, J. World. Fed. Orthod. **2**, e165-e168, (2013).
6. M.A. Ghannoum, R.J. Jurevic, P.K. Mukherjee, F. Cui, M. Sikaroodi, A. Naqvi, P.M. Gillevet, PLoS. Pathog. **6**, e1000713, (2010).
7. A.A.H. Khan, S.M. Karuppayil, Saudi. J. Biol. Sci. **19**, 405-426, (2012).
8. D. Norback, G.H. Cai, J. Environ. Monit. **13**, 2895-903, (2011).
9. K. Malachova, P. Praus, Z. Rybkova, O. Kozak, Appl. Clay. Sci. **53**, 642-645 (2011).
10. S. Saengmee-Anupharb, T. Srikhirin, B. Thaweboon, S. Thaweboon, T. Amornsakchai, S. Dechkunakorn, T. Suddhasthira, Asian. Pac. J. Trop. Biomed. **3**, 47-52, (2013).
11. C.A. Clausen, V.W. Yang, Int. Biodeter. Biodegr. **55**, 99-102, (2005).
12. S. Thaweboon, B. Thaweboon, S. Plang-Ngern, P. Nisalak, R. Kaypetch, Adv. Mater. Res. **898**, 292-295, (2014).
13. S. Badreshia, J.G. Marks, Jr., Am. J. Contact. Dermat. **13**, 77-9, (2002).
14. US EPA Archive Document of 3-Iodo-2-propynyl butylcarbamate (IPBC), 1-6, (1997).
15. S. Thaweboon, P. Nisalak, B. Thaweboon, P. Sawaengkit, S. Plang-Ngern, R. Kaypetch, Adv. Mater. Res. **1052**, 322-326, (2014).
16. CLSI M38-A standard for moulds, The University of Adelaide, http://www.adelaide.edu.au.
17. ASTM Designation: G 21-90 Standard Practice for Determining. Resistance of Synthetic Polymeric Materials to Fungi, 1097-1100, (1990).
18. ASTM G21-96, Standard Practice for Determining Resistance of Synthetic Polymeric Materials to Fungi, ASTM International, West Conshohocken, PA, http://www.astm.org/, (1996).
19. H. Egusa, T. Watamoto, K. Abe, M. Kobayashi, Y. Kaneda, S. Ashida, T. Matsumoto, H. Yatani, Int. J. Prosthodont. **21**, 62-8, (2008).
20. K.L. Garg, K.K. Jain, and A.K. Mishra. Sci. Total Environ. **167**, 255-271, (1995).
21. Canadian environmental quality guidelines (1999).
22. G. Selvaggi, S. Monstrey, K. Van Landuyt, M. Hamdi, P. Blondeel, Acta. Chir. Belg. **103**, 241-247, (2003).
23. L. Reinprecht, *Fungicides for Wood Protection - World Viewpoint and Evaluation/Testing in Slovakia*, InTech, Croatia, 95-122, (2010).
24. R.S. Lanigan, IJT. **17**, 1-37, (1998).
25. Information technique - technical datasheet : Sirius, France.
26. K.F. Nielsen, G. Holm, L.P. Uttrup, P.A. Nielsen, Int. Biodeter. & Biodegr. **54**, 325-336, (2004).

Effect of Bond Layer Properties to Thermo-Mechanical Stresses in Flip Chip Packaging

Y. W. Pok , D. Sujan, M. E. Rahman, S. S. Dol

School of Engineering and Science, Curtin University Sarawak Campus, CDT 250, 98009 Miri, Sarawak, Malaysia

Abstract. The flip chip bonding technology is widely used in electronic packaging as a result of improvements towards mechanical performance of layered structures. However, thermal mismatch shear and peeling stress are often induced by the differences of the material properties and geometries of bond layer during the high temperature change at operating stage. Intrinsically, these thermo-mechanical stresses play a very significant role in the design and reliability of the flip chip package. Therefore, this project aims to develop a methodology to find optimized bonding material thermo-mechanical properties and geometries in relation to the packaging layers in order to eliminate or reduce thermal mismatch stresses that occur in multi-layered structures in electronic packaging. The closed-form solution of thermo-mechanical analysis of bi-material assembly with bond layer is provided .Parametric study will be carried out in order to study the influence of bond layer parameters on interfacial thermal stresses of a flip chip assembly. These parameters include Young modulus, Coefficient of Thermal Expansion (CTE), Poisson's ratio and thickness of the bond layer. It is found that the shearing stresses and peeling stresses decreased considerably at the interface with the increase of bond layer Young Modulus and thickness. On the other side, bond layer CTE and Poisson ratio show almost no significant effect on the interfacial shearing stress and peeling stress along the interface in a bi-material assembly.

1 Introduction

As the dramatic increasing demand of microelectronics over the past decade, flip chip microelectronic packages are increasingly being used in microelectronics system applications such as cellular phones, pagers, laptops, Personal Digital Assistants (PDAs), and watches [1-9].

Original flip chip assembly used the ceramic substrate with CTE that matching with silicon die. However, ceramic substrates are expensive as these materials require high temperature processing, and therefore limits the application of flip chip technology in low cost products. On the other hand, organic substrates are cheaper but the high CTE differences between the substrate and silicon die will induce large thermal stress on the laminate layers during operation of the device. This reliability concern has limited the types of applicable substrate for the flip chip assemblies for many years until the bonding layer was introduced [10-11]. Bonding layer compensates for the difference of material properties between silicon die and the substrate.

However, detrimental thermo-mechanical loadings on flip chip reliability are still a major concern due to CTE mismatch of components in the flip chip assembly even with the introduction of bonding layer [12]. During operation, the multi-material laminated assembly undergoes a significant temperature gradient during cooling. This is mainly due to the heat generated by the circuits and the on-and-off nature of the electronic devices. Besides, the interfaces in multilayered assemblies undergo high stress gradients near the free edge, which will initiates the delamination or micro crack at the free edge. There are two types of major stresses acting between the bonding layer and laminate layers in the flip chip assembly, which is shearing stress and peeling stress.

The importance of such failures is felt more acutely as the power requirement of electronic devices increasing rapidly in recent years. In order to minimize the risk of such structural or functional failures, a detailed and critical understanding of the nature of the interfacial stresses within elastic limit under thermal loading is deemed necessary [13].

Due to limitations of experimental modeling and numerical modeling, analytical modeling had been widely applied in determination of interfacial stresses in the materials and the interfaces. The development of bonded interfacial modeling is inspired by the derivations of Chen and Nelson [14]. Many researchers such as Matthys and Mey [15], Mirman [16], Moore and Jarvis [17], and Ru [18] are contributed in this aspect. Suhir et al. [5] proposed a bi-layered interfacial thermal mismatch model by using integro-differential equation. Sujan et al. [13] had developed another simpler solution using a second order differential equation approach compared to Suhir's solution.

Nevertheless, studies conducted on the effect of bond layer on the interfacial stresses are very few [5, 13-18]. The bond layer acts as interfacial shear stress compliance between the two principal layers. Consequently, it will have some influence on the interfacial stresses in a bi-material assembly. The value of interfacial shear stress compliance for the bond layer at the interface was proposed by Sujan et al. [13] which is given as K_0 as h_0/G_0.A Gold-Tin solder bond is introduced as the bond layer between silicon and diamond layers and they show that the effect of bond layer on the interfacial shearing stress. It is found that the interfacial shearing stress is decreased considerably at the interface with the increase of bond thickness. Besides, it is observed that the effect of linear temperature gradient may influence interfacial stresses considerably. However, to date limited attempt has been made to study the effect of the thermal mismatching interfacial peeling stress with bond subjected to differentially uniform temperature change.

Therefore, this paper aims to present both the interfacial shearing stress and peeling stress analysis of flip chip packaging with consideration of bond layer. Subsequently a parametric study was conducted to examine the influence of thermo-mechanical properties of bond layer on interfacial thermal mismatch stresses. The properties focused in this research including CTE, Young's modulus, Poisson ratio and thickness.

2 Analytical Model

The analytical model used in this project is developed using second order differential equation which does not involved any integro-differential equation .The model is upgraded by Sujan et al. [13] to a generalized form which will account the differential uniform temperatures in the two layers first and later the temperature gradients through the thickness of the layers.

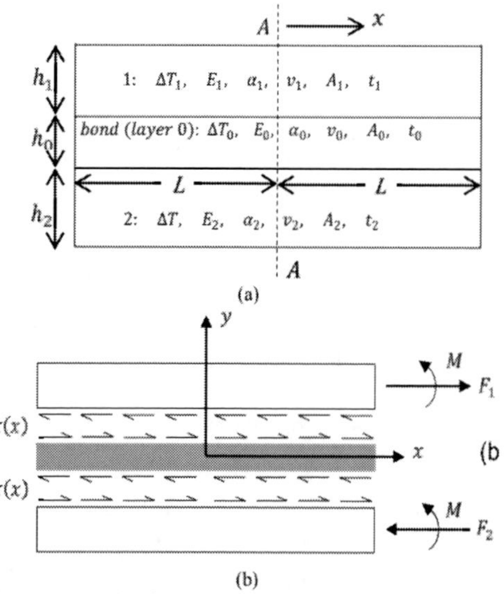

Figure 1. (a) Showing material properties and (b) is the free body diagram of a bi-layered packaging with bond layer at the interface [Adapted from 13].

Figure 1 (a) shows a full length of the model analyzed where A-A represented the center line of the model. The full length of the model is 2L.The model is considered to be a unit of width in a direction perpendicular to the plane of the paper and the forces and moments are defined with respect to the unit width. **Figure 1** (b) showed a free body diagram of the model for a cut at some arbitrary x –location with the two layers designated as 1 and 2, and the bond layer as $0.E, \alpha, v$ and h represented elastic modulus, thermal expansion coefficient, Poisson's ratio and thickness of the $i-th$ layer and is uniform temperature changes in the layers. It is assumed that no external force acting in the layers and axial forces varies along the length, and therefore shear arise at the interface.

For unit width, the force F at any section of layer is given by

$$F_1 = \int_{-L}^{x} \tau dx \tag{1}$$

Here τ represents shear stress between the two bonding surfaces per unit depth.

Considering the effect of bond layer effect, the strain compatibility condition relating to the displacements at the two interfaces can be expressed as:

$$\epsilon_{x(1)} - \epsilon_{x(2)} = K_0 \frac{\partial \tau}{\partial x} \tag{2}$$

Refering to **Figure 1** (b), the resultant moment about z-axis at $x = 0$ and $y = 0$ is given by,

$$M_1 + M_2 - \frac{1}{2}(h_1 + h_0)F_1 - \frac{1}{2}(h_2 + h_0)F_2 = 0 \tag{1}$$

Substituting $M_i = \frac{D_i}{R}$ and $F_1 = F_2 = F$ in eq. (3), result in

$$\frac{1}{R} = \frac{hF}{2D} \tag{2}$$

Where

$$\mathrm{h} = \mathrm{h}_1 + \mathrm{h}_2 + 2\mathrm{h}_0 \tag{3}$$

D is the flexural rigidity given by

$$\sum_{i=1}^{2} \mathrm{D_i} = \frac{\mathrm{E_i h_i^3}}{12(1 - \mathrm{V_i^2})} \tag{4}$$

The axial strains at the interface of the uniformly heated two layerd structure take the form,

$$\epsilon_{x(1)} = \alpha_1 \Delta T_1 + \lambda_1 F_1 + \frac{h_1}{2R} - K_1 \frac{\partial \tau}{\partial x} \tag{7}$$

$$\in_{x(2)} = \alpha_2 \Delta T_2 - \lambda_2 F_2 - \frac{h_2}{2R} + K_2 \frac{\partial \tau}{\partial x} \quad (8)$$

Here,

$\alpha_1 \Delta T$=Strain due to change in Temperature

$\lambda_1 F_1$=Strain due to thermal mismatch axial force, F_1

$\frac{h_1}{2R}$=Strain due to bending

$K_1 \frac{\partial \tau}{\partial x}$=Strain due to shearing force

Where i=1,2

Replacing the components of $\in_{x(1)}$ and $\in_{x(2)}$ from eq. (7) and eq.(8), eq. (2) takes the form as,

$$(\alpha_1 \Delta T_1 - \alpha_2 \Delta T_2) + \lambda F - (K_1 + K_2)\frac{\partial \tau}{\partial x} = K_0 \frac{\partial \tau}{\partial x} \quad (5)$$

Where

$$\lambda = \lambda_1 + \lambda_2 + \frac{h(h_1 + h_2)}{4D} \quad (6)$$

Differentiating eq. (9), results in a 2nd order differential equation as,

$$\frac{\partial^2 \tau}{\partial x^2} - \kappa^2 \tau = 0., \quad (7)$$

Here $\kappa^2 = \frac{\lambda}{K}$ and $K = K_1 + K_2 + K_0$

Where $K_1 = \frac{h_1}{3G_1}$, $K_2 = \frac{h_2}{3G_2}$, and $K_0 = \frac{h_0}{G_0}$

With reference to [13], the solution of eq. (11) can be written as,

$$\tau = C_1 \sinh(\kappa x)\, C_2 \sinh(\kappa x) \quad (8)$$

Applying boundary conditions at $x = 0, \tau = 0$ and at $x = L, F_1 = F_2 = 0$, Eq. (7) and eq.(8), become

$$K \frac{\delta \tau}{\delta x} = \alpha_1 \Delta T_1 - \alpha_2 \Delta T_2, \quad (9)$$

Applying the boundary conditions, the shear stress $\tau(x)$ is given by,

$$\tau = \frac{(\alpha_1 \Delta T_1 - \alpha_2 \Delta T_2)}{K\kappa \cosh(\kappa L)} \sinh(\kappa x) \quad (10)$$

The peeling stress expression can be obtained from the consideration of moment equilibrium of the applied forces combined with the shearing stress given by eq. (14). To develop the peeling stress model similar steps given by Sujan et al. [13] can be applied and the expressions for peeling stress can be obtained as follows:

$$P = \frac{(h_1 D_2 - h_2 D_1)}{2D} \frac{(\alpha_1 \Delta T_1 - \alpha_2 \Delta T_2)}{K \cosh(\kappa L)} \cosh(\kappa x) \quad (11)$$

It is observed that eq. (14) and eq. (15) are of same expressions as equations for the case of bi-material assembly without bond layer effect. The major differences are

$$h = h_1 + h_2 + 2h_0, \lambda_1 + \lambda_2 + \frac{h(h_1 + h_2)}{4D}, \quad (12)$$

and $K = K_1 + K_2 + K_0$ that the quantities h, λ, and K are all redefined. The quantity κ, although given by the same expression in terms of K and λ, is also redefined.

3 Case Study: Silicon-Diamond System

A typical flip chip package, silicon-diamond system was used in this study. **Table 1** below showed the parameters used to calculate the interfacial stresses by using the analytical approach .The length of the assembly is $2L = 0.005m$ .The temperature change, ΔT is taken at 60°C in this computation.

Table 1. Material properties and dimension for case study.

	M_1(Die)	M_2 (Die attach)	M_0(Solder)
Young's Modulus ,$E_i(GPa)$	188	49.7	7.51-62.5
CTE,$\alpha_i(\times 10^{-6}/ ℃)$	3	25	6.5-28.44
Poisson's ratio ,v_i	0.3	0.29	0.25-0.37
Thickness,$h_i(mm)$	0.35	0.15	0.00-0.05

4 Result and Discussion

The shearing stress and peeling stress are calculated using eq. (14) and eq. (15) respectively. The results are plotted in Figure 2 to Figure 9.The interfacial stresses are tabulated from $x/L = 0.7$ to 1, since the stress values and their variations are significantly small when the ratio is lower than 0.7.The maximum shearing stress and peeling stress are recorded at the free end ($x/L = 1.0$) as expected.

4.1 Young's Modulus

Figure 2 and Figure 3 represented shearing and peeling stresses for different values of young modulus along the interface $x/L = 0.7$ to 1.Here the Young modulus of bond layer, E_0 is varied from $5MPa$ with increment of $15MPa$ until $65MPa$. From Figure 2 and Figure 3, it can be observed that the shearing stress is tensile while peeling stress is compressive in nature. The shearing stress and peeling stress decreased due to the compliant (lower young's modulus) bond layer effect at any identical location at the interface. Particularly, near the vicinity of the free end, the differences of thermo-mechanical stress between highest modulus and lowest modulus of bond layer were significant, which is 76.1% for shearing stress and 84.5% for peeling stress. Thus, it indicated that a more compliant bond will results in smaller interfacial stresses compared to stiffer bond.

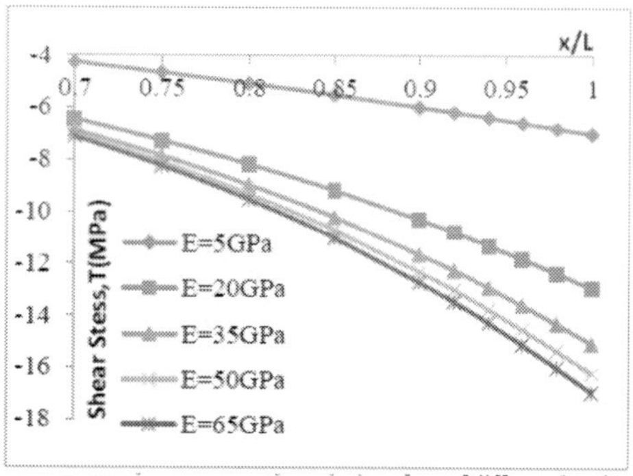

Figure 2. Shear stresses along the interface of different bond layer Young modulus.

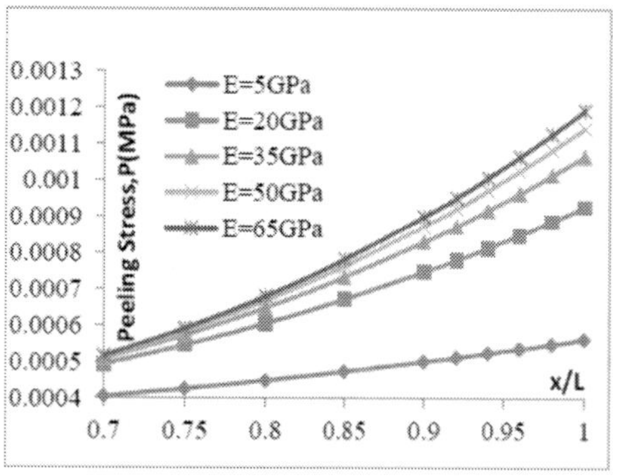

Figure 3. Peeling stresses along the interface of different bond layer Young modulus.

4.2 Coefficient of Thermal Expansion

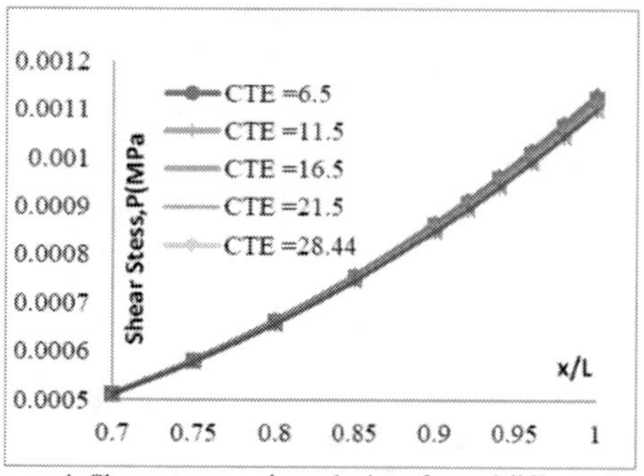

Figure 4. Shear stresses along the interface of different bond layer CTE.

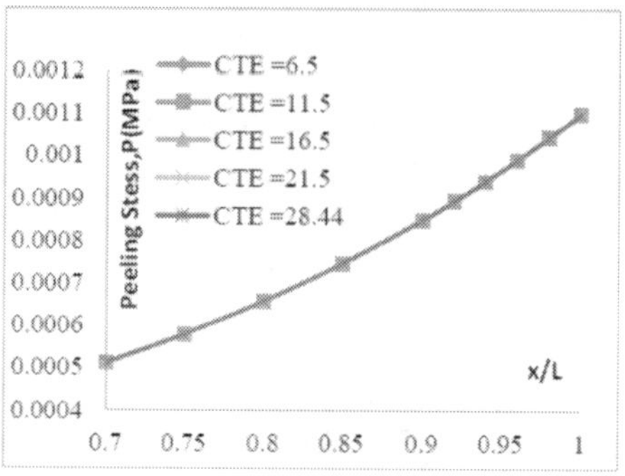

Figure 5. Peeling stresses along the interface of different bond layer CTE.

Figure 4 and Figure 5 represented shearing stress and peeling stress based on various value of CTE. From Figure , it can observed that shearing stress $x/L = 0.7$ for $\alpha_0 = 6.5/°C$,the shearing stress reduced by 0.3MPa compared to $\alpha_0 = 28.44/°C$ .The trends show almost no difference with increment of 5/°C until $\alpha_0 = 28.44/°C$ towards the end of assembly. On the other side referring to Figure , the peeling stress reduced 12% averagely for $\alpha_0 = 6.5/°C$ $\alpha_0 = 28.44/°C$ at any identical location along the interface. Therefore, it is evident that the different level of CTE in bond layer have no significant influence in the shearing and peeling stress development.

4.3 Poisson's Ratio

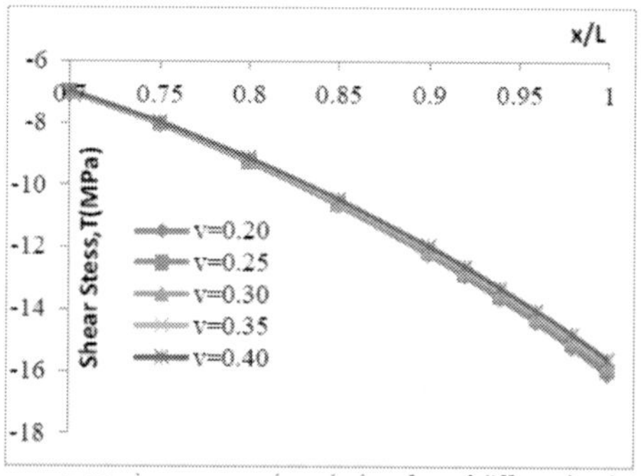

Figure 6. Shear stresses along the interface of different bond layer Poisson ratio.

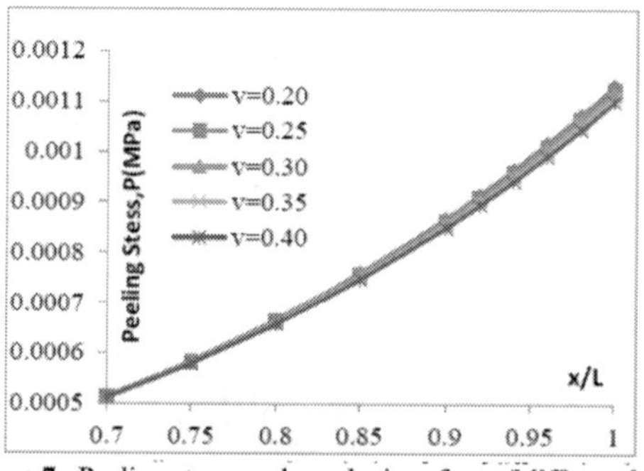

Figure 7. Peeling stresses along the interface of different bond layer Poisson ratio.

Referring to Figure 6 and Figure 7, it was demonstrated that Poisson ratio of bond layer do not play significant role in reducing interfacial stresses in flip chip packaging. For higher Poisson ratio and lower Poisson ratio of bond layer, no significant changes of the shearing stress and peeling stress were observed at any identical location. From the above observation it can be concluded that the effect of Poisson ratio in bond layer may not be essential in predicting stresses development at the interface.

4.4 Thickness

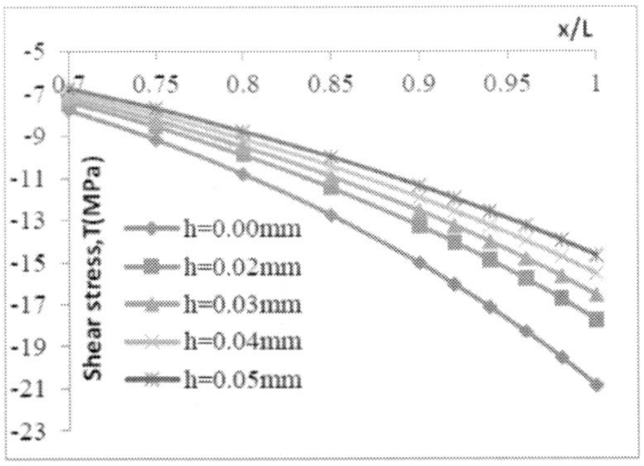

Figure 8. Shear stresses along the interface of different bond layer thickness.

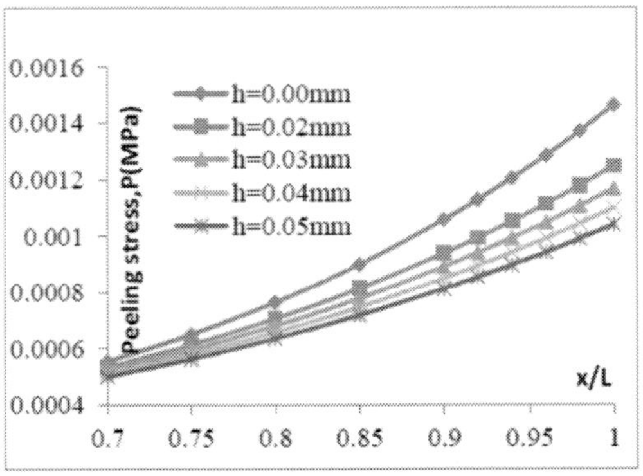

Figure 9. Peeling stresses along the interface of different bond layer thickness.

Figure 8 and Figure 9 showed the graphical plot of analytical results for shearing stress and peeling stress under different thickness of bond layer. The five different thickness values including negligible bond thickness, 0 mm, and increasing bond thickness of 0.02mm, 0.03mm, 0.004mm and 0.05mm. The thermo-mechanical stresses analysis was presented only for the region near the free end of the assembly since it was the region where the stresses generated significantly.

For both interfacial stresses distribution, the shear stress and peeling stress were observed to be reduced along the length when the thickness of bond layer was being increased. The increased bond layer thickness would generated a surrogate layer acted for shear stress absorption along the interface.

Therefore, a higher value of bond layer thickness would be suggested in the flip-chip packaging design. However, the increased bond thickness was seem to be another concern for the increased overall cost of material.

5 Conclusion

The parametric studies on the effect of thermo-mechanical properties and physical properties of bond layer had been carried out .The results obtained from the analysis leads to following conclusion:

- The shearing stresses and peeling stresses decreased considerably at the interface with the increase of bond layer Young Modulus and thickness.
- Bond layer CTE and Poisson ratio show almost no significant effect on the interfacial shearing stress and peeling stress along the interface in a bi-material assembly.

References

1. Tummala, Rao. *Fundamentals of microsystems packaging*. McGraw Hill Professional, 2001.
2. Moore, Gordon E. "Cramming more components onto integrated circuits."
3. Brown, W. D. "Advanced Electronic Packaging with Emphasis on Multichip Modules. ed. WD Brown. 1998."
4. Greig, William. *Integrated circuit packaging, assembly and interconnections: trends and options*. Springer Science & Business Media, 2007.
5. Suhir, Ephraim, Yung-Cheng Lee, and C. P. Wong, eds. *Micro-and Opto-Electronic Materials and Structures: Physics, Mechanics, Design, Reliability, Packaging: Volume I Materials Physics-Materials Mechanics. Volume II Physical Design-Reliability and Packaging*. Vol. 1. Springer Science & Business Media, 2007.
6. Vujošević, Milena. "Thermally induced deformations in die-substrate assembly." *Theoretical and Applied Mechanics* 35, no. 1-3 (2008): 305-322.
7. Gao, Shubo. "New technologies for lead-free flip chip assembly." PhD diss., Department of Electrical and Electronic Engineering, Imperial College, 2005.
8. Chungpaiboonpatana, Surasit, and Frank G. Shi. "Advanced HiCTE flip chip packaging of 90-nm Cu/Low-K chips: Underfill, novel terminal pad structures, and processing optimization." *Journal of electronic materials* 34, no. 7 (2005): 977-993.
9. Schubert, A., R. Dudek, H. Walter, E. Jung, A. Gollhardt, B. Michel, and H. Reichl. "Reliability assessment of flip-chip assemblies with lead-free solder joints." In *Electronic Components and Technology Conference, 2002. Proceedings. 52nd*, pp. 1246-1255. IEEE, 2002.
10. Wong, C. P., Songhua H. Shi, and G. Jefferson. "High performance no-flow underfills for low-cost flip-chip applications: material

 DOI: 10.1051/matecconf/20179501003

characterization."*Components, Packaging, and Manufacturing Technology, Part A, IEEE Transactions on* 21, no. 3 (1998): 450-458.

11. Zhang, Jian. "In-process stress analysis of flip chip assembly and reliability assessment during environmental and power cycling tests." (2003).
12. Zhang, Zhuqing. "Study on the curing process of no-flow and wafer level underfill for flip-chip applications." (2003).
13. Debnath, Sujan, X. Pang, Ekhlasur Rahman, and Mohan Reddy Moola. "Performance study of solder bond on thermal mismatch stresses in electronic packaging assembly." (2012).
14. Chen, W. T., and C. W. Nelson. "Thermal stress in bonded joints." *IBM Journal of Research and Development* 23, no. 2 (1979): 179-188.
15. Matthys, L. and De Mey, G., 1996. An analysis of an engineering model for the thermal mismatch stresses at the interface of a uniformly heated two layer structure. *INTERNATIONAL JOURNAL OF MICROCIRCUITS AND ELECTRONIC PACKAGING*, *19*, pp.323-329.
16. Mirman, Ilya B. "Effects of peeling stresses in bimaterial assembly." *Journal of Electronic Packaging* 113, no. 4 (1991): 431-433.
17. Moore, Thomas D., and John L. Jarvis. "A simple and fundamental design rule for resisting delamination in bimaterial structures." *Microelectronics Reliability* 43, no. 3 (2003): 487-494.
18. Ru, C. Q. "Interfacial thermal stresses in bimaterial elastic beams: modified beam models revisited." *Journal of Electronic Packaging* 124, no. 3 (2002): 141-146.

 DOI: 10.1051/matecconf/20179501004

Performance and Stability Analysis of Sub-stiffening for Mechanical Buckling and Post Buckling: A Selective Study

Hemza Layachi and Yuan-ming Xu

School of Aeronautic Science and Engineering, Beihang University, Beijing 100191, China

Abstract. The recent work is a selective study of different types and shapes of stiffeners and sub-stiffeners. The aim is to get an optimum design model of the sub-stiffened panels. We provide a variety of parameters to enhance the stability performance of sub-stiffened panels. We correspondingly perform the numerical study under ABAQUS 6.13 using the RIKS algorithm in the object of defining the critical loads for both buckling and post-buckling; the ultimate strength load represents the initial structure damage. Comparing to a linear sub-stiffened panel, the simulation results show that the grid and curvilinear sub-stiffening can improve the stability performance.

1 Introduction

Metallic Sub-stiffened panels in aerospace application have to be optimized in manner to remain stable against loading and reduced in mass. Several recent researches have inspected the possibility to reach an optimum model in order to extend the critical load to buckle and save the weight [1]-[3]. It is important to investigate the stability of the sub-stiffened panels against buckling; predictive, parametric, and experimental studies has defined the different buckling modes that can occur due to different loading [4],[5]. The mathematical formulation of buckling issue considers the type of loading which the panel is exposed, the dimensions of the panel and its material proprieties [6]-[8] and [9]. The panel configuration is another important argument in the study of its stability [10]-[12] and [13], the arrangement of the stiffeners and sub-stiffeners is essential. In our study, we will go through the analysis of the stiffened panel's stability. Thus, numerous parameters will be investigated in aim to increase the overall performance.

Buckling of sub-stiffened panels may occur due to different loading types, the instability process passes from local to global. Here under, we briefly introduce the different buckling modes:

1.1 Local buckling

Local buckling is represented when the connection between the plate and the stiffener does not displace vertically. It represents the buckling of the plate or the stiffener without the displacing in the vertical connection.

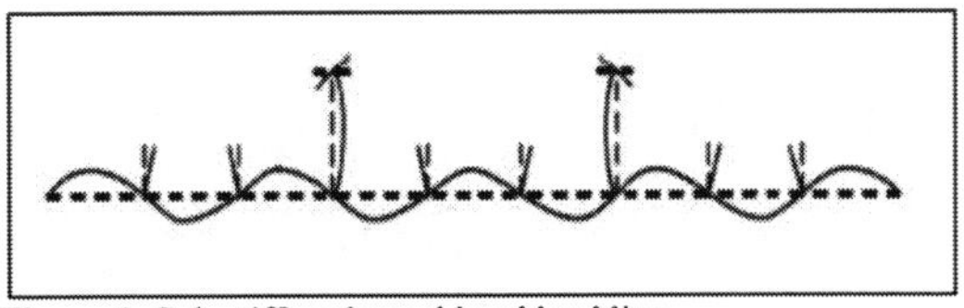

Figure 1. Sub-stiffened panel local buckling.

1.2 Torsional Buckling

Similar to local buckling mode, the difference is that the stiffeners flange are sufficiently stiff to keep that junction straight.

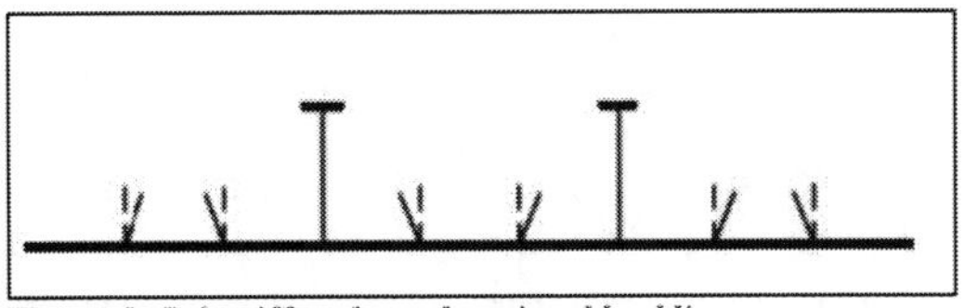

Figure 2. Sub-stiffened panel torsional buckling.

1.3 Partial buckling

Considering the spaces between the stiffeners, if the sub-stiffener remains unbuckled with a torsional buckling mode, the partial buckling mode occurs.

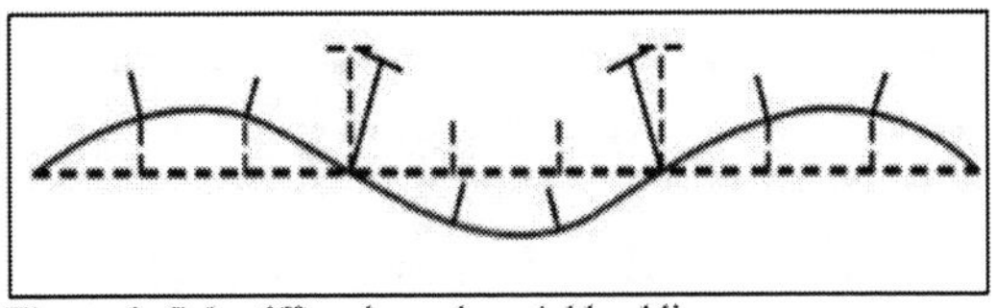

Figure 3. Sub-stiffened panel partial buckling.

DOI: 10.1051/matecconf/20179501004

1.4 Overall buckling

It is the buckling mode that leads to total deformation of the skin and stiffeners at the same time, the local or torsional buckling mode is extended to a complete stiffened panel including several stiffeners.

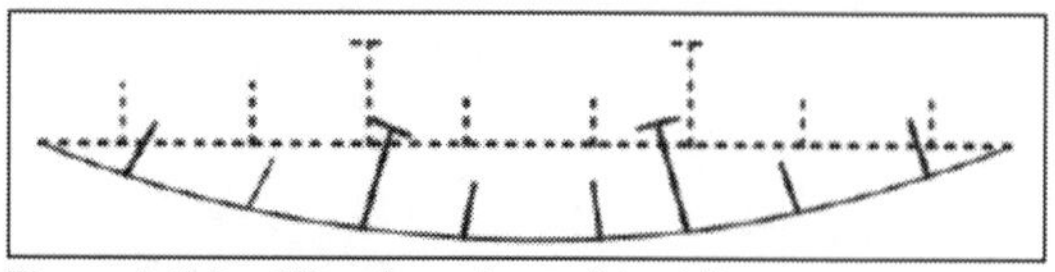

Figure 4. Sub-stiffened panel overall buckling.

The study of post buckling, as the name suggests, describes the behavior of a structure after the critical buckling load[14], [15],. In elastic domain, the stiffened panel can carry additional load when it is locally buckled, and that buckled surface are permissible along the width of the panel. When the whole panel buckles all as a column, it reaches the global buckling, moreover, it cannot carry more loads, it is the limit to failure [16],[17]and [18].

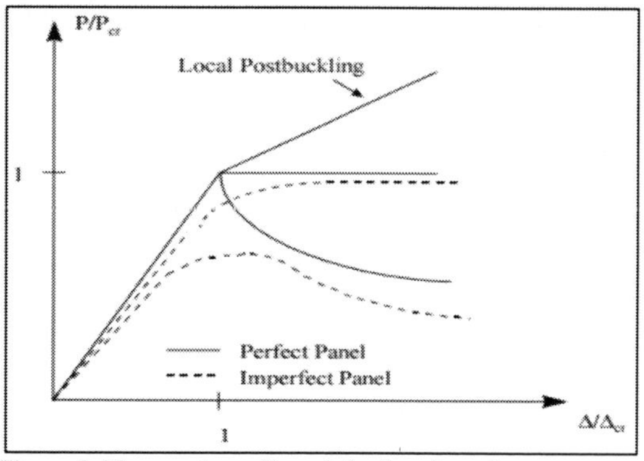

Figure 5. Linear and non-linear behavior of panels.

2 Problem formulation

Assuming that the sub-stiffened panel is exposed to a uniaxial compression in the direction of the stiffeners, and concealed on it transversal boundary conditions, the buckling of the panel will occur at the critical load.

The configuration of the panel and its boundary conditions are shown in the figure bellow,

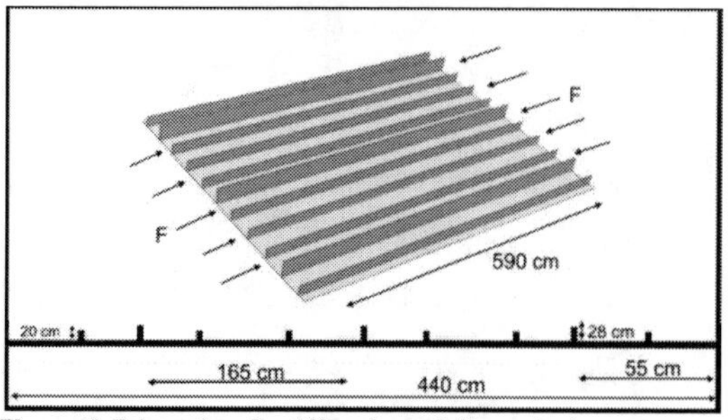

Figure 6. Prismatic sub-stiffened panel dimensioning.

In addition to the dimension given in figure 4, the Table 1 gives more design variables and mechanical proprieties of the manufactured panel in question,

Table 1. Prismatic sub-stiffened panel design variables and mechanical properties.

Design Variables		Mechanical properties	
Prismatic sub-stiffened panel (A)		Aluminum alloy	2024T3
Plate thickness	1.9 mm	Density (Kg/m^3)	2770
Stiffeners thickness	2.8 mm	Young modulus	73.1 GPa
Sub-thickness	2.8 mm	Poisson's ratio	0.33
Distance per stiffener/sub stiffener	55 cm	Expansion module	2.32 10^{-5}
Total mass	2.128 Kg	Yield stress	324 MPa

Buckling analysis: at the loading limit, the structure starts to buckle, which leads to the limits of the elastic range. If the panel buckling response is local to partial as presented in the section before, the structure can carry more load before it buckles on an overall mode (post buckling range).

Post buckling analysis: the panel has not overall buckled, the stability of the stiffened panels will be standing up on the resistance of the stiffeners. Each stiffener will be analyzed individually. The collapse of the entire panel will occur when all the stiffeners can carry no additional load. At this range, the panel will undergo on local plastic deformation.

Collapse behavior: it represents the full plasticization of the structure, it cannot sustain more load, and the damage of the structure is predictable.

3 Methodology

The numerical analysis is based on comparative study of three kinds of sub-stiffened panels. All of them have the same mechanical proprieties and almost equal in mass. The aim is to select the best configuration that guarantee the panel stability under uniaxial compression load and delivers enhanced performance. The stiffeners and the sub-stiffeners are a reinforcement for a better stability of the panel in a mechanical aspect but it is against the concept of weight saving.

The design variable will be the dimensioning of the stiffeners and sub-stiffeners including the number of stiffeners and sub-stiffeners in the panel's cross section, the type of the sub-stiffening and the thickness of the plate section. We give two new concepts of sub-stiffeners that increase the stiffened panels stability along the plan, the first one is a grid sub-stiffened and the second is a curvilinear sub-stiffened. After an overall study, we adopt to use three sub-stiffeners in a cross section for some design and manufacturing reasons. Table 2 represents the two proposed design panels.

Our simulation output are load limit for buckling and post buckling, it is denoted as critical load and bearing ultimate load respectively. We cannot ignore the stresses caused by these loads along the panel σ_s. The gain

 DOI: 10.1051/matecconf/20179501004

percentage is based on ratio limit yield σ_s to maximum mises stress caused by them. The interpolation of the variables in design is taking the prismatic sub-stiffened as a benchmark panel. x_i Are the input design, we list the thickness for stiffeners and sub-stiffeners t_{st}, their height h_{st}, and plate thickness t_p, they should not exceed what the prismatic possess t_{st-max}, h_{st-max}, t_{p-max} . Looking for a better material distribution along the plan, regarding the behavior of the three panels for the uniaxial compression force, we conclude the effect of each parameter listed in Table 2.

$$\begin{cases} \sigma_{max} < \sigma_s \\ t_{st} \le t_{st-max} \\ h_{st} \le h_{st-max} \\ t_p \le t_{p-max} \\ x_{min} < x_i < x_{max}, i = 1, \dots, n \end{cases}$$

Table 2. Proposed design dimensions.

Proposed designs / Design variable	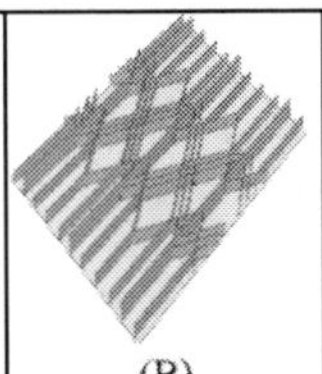 (B)	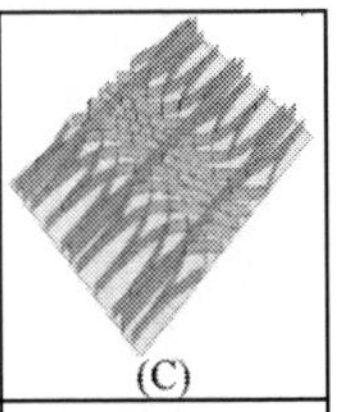 (C)
Plate thickness	1.80 mm	1.90 mm
Distance between stiffeners	165.00 mm	165.00 mm
Initial distance	55.00 mm	55.00 mm
Stiffener thickness	2.80 mm	2.80 mm
High of stiffeners	28.00 mm	28.00 mm
High of sub-stiffeners	16.50 mm	16.32 mm
Distances between sub-stiffeners	55.00 mm	55.00 mm
Thickness of sub-stiffeners	2.32 mm	2.35 mm
Total mass	2.126 Kg	2.133 Kg

4 Results and discussion

4.1 Buckling analysis

The simulation results at the reached eigenvalue show that the structure B provide higher performance ratio (96%) comparing to the other structures, while the structure (C) represents a lesser loss of strength, which is due to the inadequate curving to uniaxial forces in its direction.

Table 3. Buckling results.

Structure panel design	A	B	C
Critical load (Kn)	212.2	222.4	211.8
Equivalent Mises stress (Mpa)	295.5	312.3	278
Buckling mode	Partial	Local	Local
Gain	91%	96%	85%

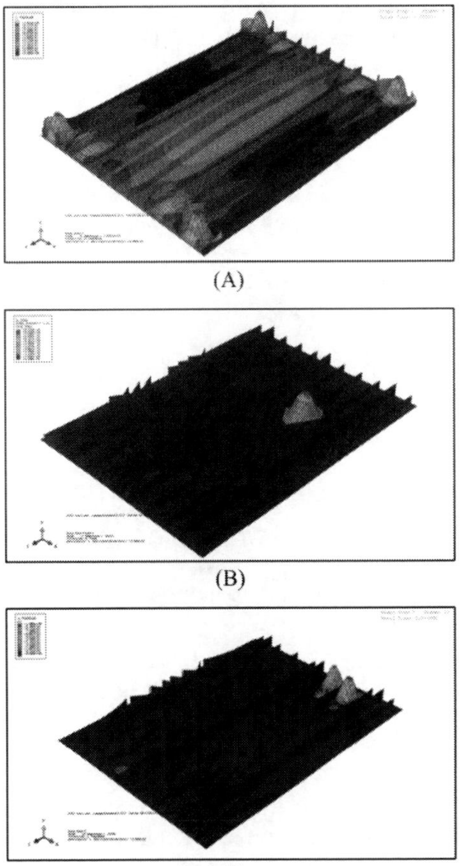

Figure 7. First Buckling modes of sub-stiffened panels (Case A, B and C).

The great advantage of the proposed design (B) and (C) is that, both structures buckle locally in a first step (skin buckling with a great decrease of the half wavelet comparing to the benchmark case that represents a partial mode of buckling, it is just one step closer to the overall buckling (catastrophic).

4.2 Post buckling analysis

Figure 3 shows the numerical results of ultimate loading capacity. After that critical load, the sub-stiffened panels will collapse: the structure B reaches a greater strength capacity with a critical load at 383KN, in addition to the transverse reinforcement that help to avoid the overall buckling also, possess rational continuity along its longitudinal axis in both stiffeners and sub-stiffeners. The design guidelines of such a structure start with ensuring the stability of the panel along the direction of the loading. As in that case, the stiffeners are oriented in the longitudinal axis of the load first, the grid sub-stiffens are a second reinforcement for the basic stiffeners in one hand and, give a better stability along the plan by a reasonable distribution along the plan. The post-buckling range is enhanced in fact that the structure buckles locally in a first stage.

The structures A and C reach its ultimate bearing capacity at 323KN and 337KN respectively, we can realize the importance of the continuity of stiffening and sub-stiffening along the load axis since the structure C represents the lowest strength efficiency, even it is reinforced in its transverse direction. The post buckling range is not great enough to reach the bearing capacity shown by the structure B. The curvilinear sub-stiffeners are not adequate for the uniaxial force: it is due to the non-sufficient continuity along the direction of the load that the structure start to buckle, goes from local to global in a short range.

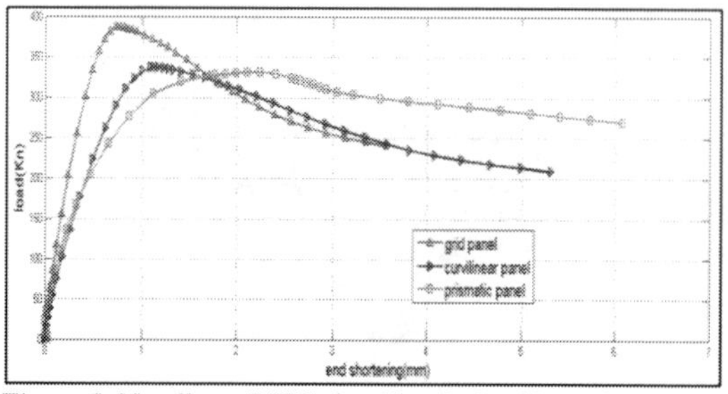

Figure 8. Non-linear RIKS algorithm for load Vs end shortening (Cases A, B and C).

We notice the poor post buckling range of the structural panel design (A), which is due to buckling response. In a first stage, it goes on a partial buckling with an important number of half wavelet.

4.3 predicted collapse mode

Studying the instability process refers to state the ultimate strength capacity which is known as the bearing load, the nonlinear RIKS algorithm performs the behavior of the three proposed design. The damage of the structures is predicted at the bearing critical load, our goal is to try to understand how and when the collapse may happen after the ultimate load:

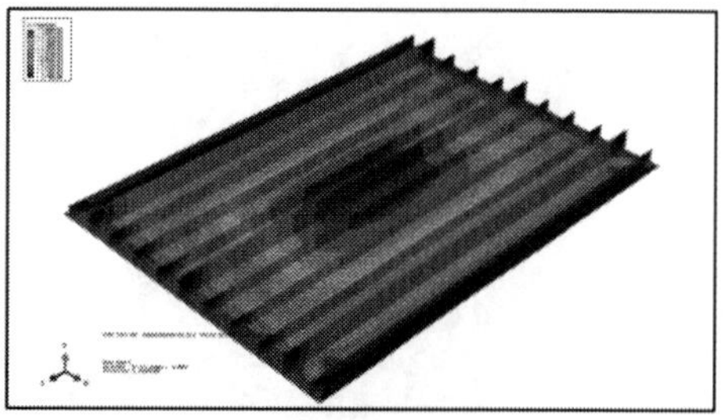

(A)

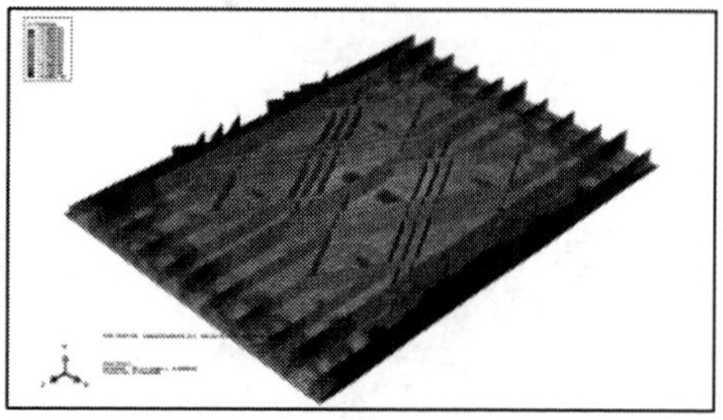

(B)

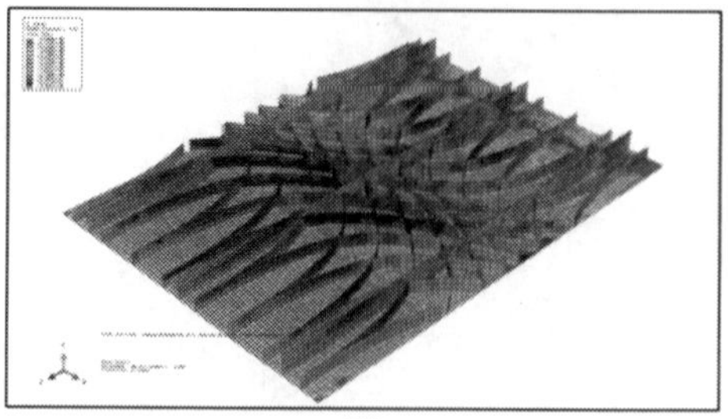

(C)

Figure 9. Predicted collapse modes of the proposed panels (Cases A, B and C).

The structure A, after reaching its ultimate strength capacity will collapse in a mode that both skin, the stiffeners and sub-stiffeners are affected. The partial buckling occur in the first stage affect the stability of the structure, lead it to a speedy instability and goes from partial to over.

The structure B represent a lesser ratio load/displacement, which is due to the, interaction between stiffeners and sub-stiffeners. the stress, is more located of the stiffeners so the collapse of the panel is predicted when the stiffeners breakdown(the sub-stiffeners have no capacity to carry more loads, its designed to enhance the buckling performance by avoiding the overall buckling in first phase.

The structure C represents, a similar behavior shown by the structure A, the stiffeners, sub-stiffeners and the skin plate lose their stability, which goes on an overall buckling. The curvilinear sub-stiffeners cannot locate the stress caused by the axial load along the plan for longer, the panel buckle locally in a first stage, start to lose its stability and end up with failure in all directions.

5 Conclusion

The recent work is a selective study of sub-stiffening of the stiffened panels. An overall study about it leads us to think about the type of sub-stiffening, the dimensioning and how many sub-stiffener needed in a cross section. The compromise loading limits to weight saving should be respected. The results of simulation show that comparing to a normal stiffened panel, sub-stiffening improves the stability performance of the panel against the mechanical charging. :

The type of stiffening and sub-stiffening should take into consideration the type of loading. As for structure B, it may not give such a better performance for complicated loads like coupled compression-torsion.

The sub-stiffening is a second reinforcement for a better stability of the panels, it helps to locate the buckling on the skin and avoid the overall buckling.

The compromised aspect better performance with weight saving should take in care the load direction, that pass by ensuring the reinforcement in its direction first then, looking for a better distribution on the plan. Further numerical studies will focus on the response of such structures to multi-axis loads with design guidelines and outline of topology optimization.

References

1. Khvyiuzov A, Xu YM. Initil buckling of compressed rectangular panels with variable stiffener sizes. *Advanced Materials Research*, 915-916D (2014).
2. Bushnell and C. Rankin, "Optimum design of stiffened panels with sub-stiffeners," *in Proceedings of the forty sixth AIAA/ASME/ASCE/AHS/ASC structures, structural dynamics & materials conference*. Austin, Texas, pp. 18–21, (2005).
3. Wang HM. Additive layer manufacturing technology for high performance of metallic structure, opening new page of defence manufacture. *Defense Manufacture Technology*, (03):003, (2013)
4. Ni Y, Xu YM. Research on the stability of metallic sub-stiffening structure and its optimization under uniaxial loading. *Acta Aeronautica et Astronautica Sinica*. Jan 25, **35** (2014)
5. Xu YM, Li SZ. Approximate Calculation Method of Buckling Load on Intergral Sub-stiffened Panel. *Journal of Beijing University of Aeronautics and Astronautics* (2014)
6. D. Quinn, A. Murphy, W. McEwan, and F. Lemaitre, "Stiffened panel stability behaviour and performance gains with plate prismatic sub-stiffening," *Thin-Walled Struct.*, **47**, no. 12, pp. 1457–1468, (2009).
7. B. W. Schafer, "Local, distortional, and Euler buckling of thin-walled columns," J. *Struct. Eng.*, **128**, no. 3, pp. 289–299, (2002).
8. G. L. Farley, "Selective reinforcement to enhance the structural performance of metallic compression panels," *in Abstract submitted to 45th AIAA/ASME/ASCE/AHS/ASCSDM Conference*, Palm Springs, CA, (2004).
9. G. Grondin, A. . Elwi, and J. J. . Cheng, "Buckling of stiffened steel plates—a parametric study," J. *Constr. Steel Res.*, **50**, no. 2, pp. 151–175, (1999).
10. R. M. F. Paulo, F. Teixeira-Dias, and R. A. F. Valente, "Numerical simulation of aluminium stiffened panels subjected to axial compression: Sensitivity analyses to initial geometrical imperfections and material properties," *Thin-Walled Struct.*, **62**, pp. 65–74, (2013)
11. Mulani SB, Slemp WCH, Kapania RK. EBF3PanelOpt: An optimization framework for curvilinear blade-stiffened panels. *Thin-Walled Structures*, **63**: pp.13-26,(2013)
12. D. G. Stamatelos, G. N. Labeas, and K. I. Tserpes, "Analytical calculation of local buckling and post-buckling behavior of isotropic and orthotropic stiffened panels," *Thin-Walled Struct.*, **49**, no. 3, pp. 422–430, (2011).
13. X. Wang, H. Sun, A. Akiyama, and A. Du, "Buckling and ultimate strength of aluminium plates and stiffened panels in marine structures," *in Fifth International forum on aluminium ships*. Tokyo, Japan, (2005).
14. L. Brubak and J. Hellesland, "Semi-analytical postbuckling and strength analysis of arbitrarily stiffened plates in local and global bending," *Thin-Walled Struct.*, **45**, no. 6, pp. 620–633, (2007).
15. J. K. Paik and B. J. Kim, "Ultimate strength formulations for stiffened panels under combined axial load, in-plane bending and lateral pressure: a benchmark study," *Thin-Walled Struct.*, **40**, no. 1, pp. 45–83, (2002).
16. T. Kubiak, Nonlinear Plate Theory for Postbuckling Behaviour of Thin-Walled Structures Under Static and Dynamic Load. *INTECH Open Access Publisher*, (2012).
17. Barrière, L., Marguet, S., Castanié, B. et Cresta, P..A computational strategy for post-buckling analysis of large stiffened structures. *In28th International Council in Aeronautical Sciences, ICAS2012*, Brisbane, Autralia (2012).
18. Barrière, L., Marguet, S., Castanié, B. et Cresta, P..A. Modelling strategies for post-buckling analysis of large stiffened structures *In 17th International Conference on Composite Structures, ICCS17*, Porto,Portugal(2013).

 DOI: 10.1051/matecconf/20179501005

The Effect of Growth Temperature and V/III Flux Ratio of MOCVD Antimony Based Semiconductors on Growth Rate and Surface Morphology

Ari Handono Ramelan[1], Sayekti Wahyuningsih[2] and Ewa Goldys[3]

[1]Physics Department, Faculty of Mathematics and Natural Sciences, Sebelas Maret University, Jl. Ir. Sutami No, 36A Surakarta, Indonesia.
[2]Chemistry Department, Faculty of Mathematics and Natural Sciences, Sebelas Maret University, Jl. Ir. Sutami No, 36A Surakarta, Indonesia.
[3]Physics Department, Macquarie University, Sydney NSW 2109, Australia.

Abstract. Epitaxial $Al_xGa_{1-x}Sb$ layers on GaSb and GaAs substrates have been grown by atmospheric pressure metalorganic chemical vapor deposition using TMAl, TMGa and TMSb. Nomarski microscope and a profiler were employed to examine the surface morphology and growth rate of the samples. We report the effect of growth temperature and V/III flux ratio on growth rate and surface morphology. Growth temperatures in the range of 520°C and 680°C and V/III ratios from 1 to 5 have been investigated. A growth rate activation energy of 0.73 eV was found. At low growth temperatures between 520 and 540°C, the surface morphology is poor due to antimonide precipitates associated with incomplete decomposition of the TMSb. For layers grown on GaAs at 580°C and 600°C with a V/III ratio of 3 a high quality surface morphology is typical, with a mirror-like surface and good composition control. It was found that a suitable growth temperature and V/III flux ratio was beneficial for producing good AlGaSb layers. Undoped AlGaSb grown at 580oC with a V/III flux ratio of 3 at the rate of 3.5 μm/hour shows p-type conductivity with smooth surface morphology.

1 Introduction

Because of the corresponding wavelengths of the alloys of Gallium antimonide (GaSb) based compound semiconductors cover a wide spectral range from 1.24 μm (AlGaSb or AlGaAsSb) to 4.3 μm (InGaAsSb), they have received increasing attention recently. Consequently, they became promising candidates for applications in long wavelength lasers and photodetectors for fibre optic communication systems [1]. However, due to free-carrier absorption, undoped-GaSb substrates have a strong absorption in the IR wavelength region of interest which is difficult to eliminaten [2]. Furthermore, GaSb substrates are more expensive than GaAs. Thus there is a strong interest in growing superlattices (SLs) on GaAs substrates. So far, molecular beam epitaxy (MBE) has been used as the major growth technology for InAs/GaSb structures [3]-[5]. However, growth of the SLs by metalorganic chemical vapor deposition (MOCVD) is more challenging than that by MBE [6], [7] especially for the growth on GaAs substrates due to the 7.8% lattice mismatch between GaAs and GaSb

It is ideal to grow the AlGaSb films on lattice-matched GaSb substrates. However, due to the lack of semi-insulating GaSb, undoped substrates have a strong free carrier absorption in the mid-IR wavelength region of interest, so that the substrate must be thinned for optimal device operation in the flip-chip mode. For this and commercial reasons, GaAs has been explored as a substrate for GaSb/AlGaSb devices. Unfortunately, a lattice mismatch of approximately 7% exists between GaSb and GaAs. Strain energy can be accommodated by misfit dislocations at the GaAs/GaSb interface. This results in the propagation of threading dislocations through the epilayers [8]. Thus the growth of undoped high quality GaSb/AlGaSb layers on GaAs substrate becomes a necessary process for preparing thin layers.

These films have usually been grown by liquid phase epitaxy or molecular beam epitaxy (MBE) with relatively little data available for growth of AlGaSb by metalorganic chemical vapor deposition (MOCVD) [9]-[13]. AlGaSb has proven to be a much more difficult material to grow by MOCVD than GaSb. The simple reason is the reactivity of Al, which forms strong bonds with carbon [10]-[11]. It is not surprising then that frequently the MOCVD-grown AlGaSb and AlSb have been reported to exhibit poor electronic and optical properties.

In this work, a series of undoped AlGaSb samples are grown on semi-insulating GaAs substrates at different substrate temperatures and V/III flux ratios. There have been no reports on the effect of substrate temperature and V/III flux ratio on growth rate and surface morphology, so we prepared several samples with different V/III flux ratios.The growth rates and surface morphology AlGaSb by metalorganic vapor phase epitaxy were studied as functions partial pressures of precursors.

2 Experimental procedure

2.1 Growth conditions

The growth was carried out in a horizontal MOCVD reactor at atmospheric pressure. The precursors were trimethylaluminum (TMAl), trimethylgallium (TMGa) and trimethylantimony (TMSb) held at a temperature of 18, -9 and 0°C respectively. Source materials were diluted in hydrogen to obtain a carrier gas velocity of 1 sccm in the reactor. The substrates were (100) semiinsulating GaAs. GaAs substrates were prepared by degreasing in trichloroethylene (TCE), rinsing in acetone and methanol. The use of de-ionized water was used for rinsing the samples. The GaAs substrate was etched in a standard solution of H_2SO_4, H_2O_2 and H_2O (5:1:1) for 30 s [14]. The substrates were further rinsed in DI water and rapidly dried under a nitrogen jet. The V/III ratios were varied from 1 to 5.The growths were performed in the range of 520 and 680°C measured by a thermocouple located inside the graphite susceptor. The actual temperature of thesubstrate surface is assumed to be lower due to the cooling effect of the cold carrier gas.

2.2 Characterisation

The surface morphologies of the grown epilayers were observed using Nomarski contrast microscopy. Thicknesses were measured using a Tencor Alpha Step 500 surface profiler. A stylus profilometer was routinely employed to record the surface profile of etched AlGaSb samples. Alpha-Sep can scan the surface at various rates in the range of 0.04-5 μm/sec for a maximum scan distance of 1 cm.The vertical and horizontal resolutions are 5 nm and 12.5 μm radius respectively. For this study, the scan rate and scan distance were selected as 0.2 μm/sec and 400 μm respectively. The etch rate was obtained by dividing the etch depth from the profile by the etching duration. To measure surface textures, Alpha-Step automatically calculated average roughness. However, due to the relatively low resolution, Alpha-Step was only able to determine the surface roughness of some samples. The lowest measurable average roughness is 5 nm.

3 Results and discussion

3.1 Growth Rate

The two growth parameters that most strongly affect the growth rates are the metalorganic sources partial pressure and the growth temperature. Growth rate is defined as the thickness of the epilayer deposited on a substrate during a certain period of growth. The effect of growth temperature, in the range 520 - 680°C, on the growth rate was investigated by keeping others parameters constant. Constant flow rates of TMAl = 5.6 sccm, TMGa = 19.1 sccm and TMSb = 83.6 sccm, or equivalent to V/III = 3 were used.

Fig. 1 shows the temperature dependence of growth rates for $Al_{0.05}Ga_{0.95}Sb$. In the low temperatures range up to 540°C, kinetic processes of alkyl decomposition at the substrate surface limit decomposition. This is known as the *kinetically controlled regime*, where the growth rate depends on temperature (degree of pyrolysis) and very little dependence of growth rate on pressure was observed. A typical exponential relationship between growth rate and growth temperature was observed and the activation energy of 0.73 eV was found. Ungermanns et al. [15] have found an activation energy of 0.80 eV for AlSb growth using dimethylethylaminealane (DMEAAl) as the aluminium source.

The growth rate of AlGaSb increases monotonically from 540 to 600°C, while that of arsenic-containing alloys is almost constant in the same temperature range. The difference in behaviour between arsenic and antimony-containing alloys is explained by taking into account the surface coverage of group V atoms during growth [16]. Sb atoms are excessively absorbed on the growth front because of the low vapour pressure. Since Ga and Al atoms tend to desorb from the excessively Sb absorbed surface, only a few Ga and Al species can contribute to the growth. This is known as the site-blocking effect. Therefore, a higher growth rate, for Sb-containing alloys, will be obtained for lower Sb surface coverage condition by selecting higher growth temperatures and/or lower V/III ratio. However, the low surface coverage of Sb may lead to the formation of Ga droplets. For higher temperatures, in the range of 580°C to 600°C the growth rate is constant. This region is a *mass transport limited* (i.e. the growth rate is controlled predominantly by the concentration of the precursors), because of the complete pyrolysis of the TMGa and TMSb molecules. Further increase in the growth temperatures, leads to a decrease of the growth rate due to desorption of gallium and antimony from the growth surface. The growth rate is not only a function of the availability of atomic species and the rate at which the precursor fragments combine at the surface, but also depends on the carrier gas velocity and the partial pressure of the metalorganic sources [17].

The V/III ratio is another key parameter because both a deficiency and an excess of antimony (Sb) on the semiconductor surface leads to a strong degradation of the material quality. Fig.1 shows the growth rate for the $Al_{0.05}Ga_{0.95}Sb$ films as a function of the V/III ratio. The data indicate that the growth rate is significantly dependent on V/III ratios. At V/III ratios below about 2 the growth rate is low, under $1.87\,\mu m/hour$. As the V/III ratio increases the growth rate increases to reach value of 3.54 µm/min near a V/III ratio of 3. In this region, the availability of Sb appears to be the key factor controlling the growth rate. At higher V/III ratios the growth rate starts to decrease. The strong decrease of the growth rate with increasing TMSb (group V) is caused by excess antimony molecules (due to the low vapour pressure of TMSb) at the substrate surface which leads to

DOI: 10.1051/matecconf/20179501005

a decrease to the number of surface sites available for Al and Ga adsorptions and a resulting absolute decrease in Al and Ga adatoms. This blocking of Al and Ga adatoms by excess Sb results in reduced growth rates. Similar behaviour was reported by Okuno et al. [18] for AlGaSb growth using TMAAl (trimethylaminealane), TMGa and TMSb.

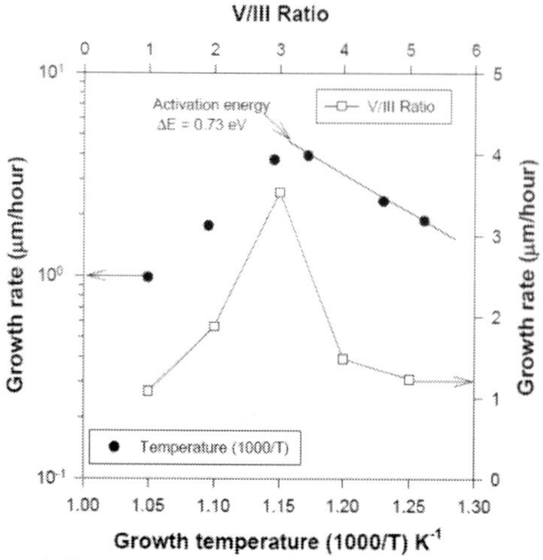

Figure 1. Temperature dependence of the $Al_{0.05}Ga_{0.95}Sb$ growth rates and the growth rate of $Al_{0.05}Ga_{0.95}Sb$ epilayers grown at 580°C as a function of V/III ratio.

3.2 Surface Morphology

Fig. 2 shows the surface morphology of selected $Al_xGa_{1-x}Sb$ compositions for V/III ratio = 3 at 580°C and 600°C respectively. The morphology of $Al_xGa_{1-x}Sb$ on GaAs is mirror-like to the naked eye, but the epilayers exhibit microscopic well defined elongated features. As the Al-content increases, the surface texture changes and the elongated features disappear, to be replaced by featureless texture associated with the highest optical transmission. The interference fringe modulation depths of the epilayers obtained from transmission measurements confirm the high optical quality of the films grown at 580°C. With further increase of Al-content, several vessel-shaped structures appear on the epilayer surface. It has been found for this project that the $Al_xGa_{1-x}Sb$ layers grown at 580°C and 600°C and V/III = 3 with Al composition in the range of 0.5% and 30% are highly regular and accompanied by excellent optical quality, with a very high optical transmission at energies below the bandgap.

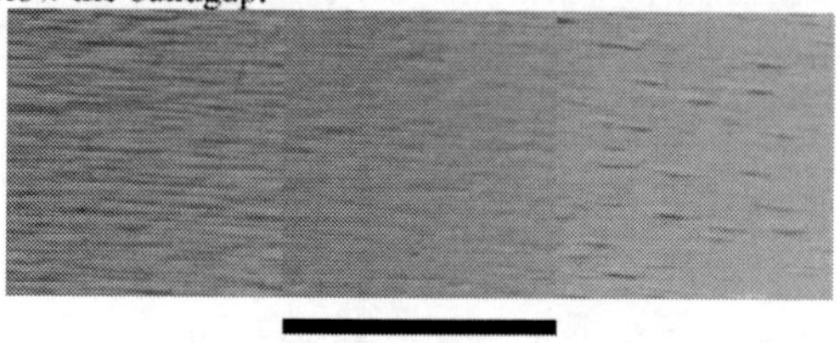

50 μm

(i) x = 10% (ii) x = 20% (iii) x = 25%

(a) Samples grown at 580°C

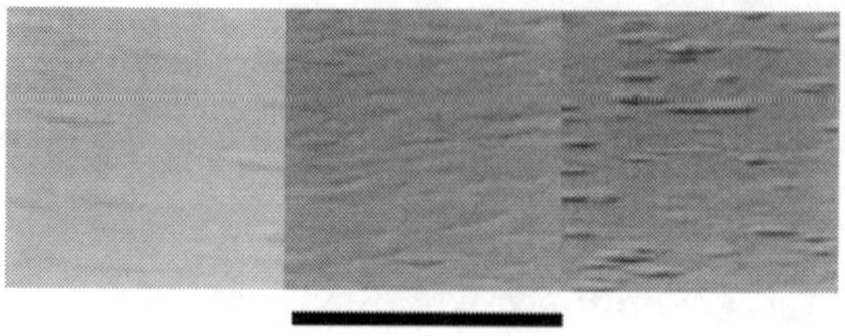

50 μm

(i) x = 15 % (ii) x = 20% (iii) x = 25%

(b) Samples grown at 600°C

Figure 2. Surface morphology of $Al_xGa_{1-x}Sb$ epilayers grown on GaAs with a different Al-content grown at 580°C and 600°C with a V/III ratio of 3.

4 Summary

We have found that the $Al_xGa_{1-x}Sb$ layers grown at 580°C and 600°C and at a V/III ratio of 3 with Al composition in the range of 0.5% and 25% have excellent optical quality with a very high optical transmission at energies below the bandgap. In the low temperature range up to 580°C, the growth rate depends on temperature (degree of pyrolysis of precursors) and the growth rate of AlGaSb increases monotonically. At high temperatures in the range between 580°C and 600°C, the growth rate is constant. Further increasing the growth temperature lead to a decrease of the growth rate due to desorption of gallium and antimony from the growth surface. A growth rate activation energy of 0.73 eV was found.

Acknowledgment

The authors would like to acknowledge the Ministry of Research, Technology and Higher Education (Kemenristek dan Dikti) THE REPUBLIK OF INDONESIA for financial assistance.

References

1. A.G. Milnes and A.Y. Polyakov, *Solid State Electron.*36, 803, (1993).
2. L. Li-Gong, LIU Shu-Man, Luo Shuai, Yang Tao, Wang Li-Jun, Liu Feng-Qi, Ye Xiao-Ling, Xu Bo, Wang Zhan-Guo Li, *Chin.Phys. Lett.* **29**, 076801, (2012).
3. S. Mallick, Banerjee K, Ghosh S, Rodriguez J B and Krishna S., *IEEE Photon. Technol. Lett.*, **19**, 1843, (2007).
4. S.A.Pour, Huang E K, Chen G, Haddadi A, Nguyen B M and Razeghi M, Appl. Phys. Lett., **98**, 143501, (2011).
5. N. Gautam, Naydenkov M, Myers S, Barve A V, Plis E, Rotter T, Dawson L R and Krishna S., *Appl. Phys. Lett.* **98**, 121106, (2011).
6. R.M. Biefeld, *Mater. Sci. Eng.* R **36**, 105, (2002).
7. Aardvark, Mason N J and Walker P J., *Prog. Cryst. Growth Charact.* **35**, 207, (1997).

DOI: 10.1051/matecconf/20179501005

8. D. Lubyshev, J.M. Fastenau, X.M. Fang, Y. Wu, C. Doss, A. Snyder, W.K. Liu, M.S. Lamb, S. Bals, and C. Song, *J. Vac. Sci. Technol.* B, **22**, 1565, (2004).
9. C.B. Cooper, R.P. Saxena and M.J. Ludowise, *Electron. Lett.* **16**, 892, (1980).
10. M. Leroux, A. Tromson-Carli, P. Gibart, C. Verie, C. Bernard and M.C. Schouler, *J.Crystal Growth*, **48**, 367, (1990).
11. A. Tromson-Carli, P. Gibart and C. Bernard, *J. Crystal Growth* **55**, 70, (1981).
12. E.T.R Chidley, S.K. Haywood, R.E. Mallard, N.J. Manson, R.J. Nicholas, P.J. Walker and R.J. Warburton, *J. Crystal Growth* **93**, 70, (1988).
13. F.S. Juang, Y.K. Su, N.Y. Li and K.J. Gan, *J. Appl. Phys.*, **68**, 6383, (1990).
14. T. Koljonen, M. Sopanen, H. Lipsanen and T. Tuomi, *J. Crystal Growth* **169**, 417, (1996).
15. Ch. Ungermanns, H. Hardtdegen, M. Matt, A. Foster, R. Ahe Mvd, R. Carius, R. Schmidt, and Luth, *J. Crystal Growth*, **188**, 32, (1998).
16. H. Asahi, T. Kaneko, Y. Okuno, and S. Gonda, *Jpn. J. Appl. Phys.* **32**, 2786, (1993).
17. T. Koljonen, M. Sopanen, H. Lipsanen and T. Tuomi, *J. Electron. Mat.* **24**, 1691, (1995).
18. Y. Okuno, H. Asahi, X.F. Liu, K. Inoue, Y. Itani, K. Asami and S. Gonda, *J. Crystal Growth* **123**, 143, (1993).

Effect of Different Photovoltaic Materials on Energetic and Exergetic Performance of Photovoltaic Thermal Arrays

C. S. Rajoria[1], Pankaj Kumar Gupta[1], Sanjay Agrawal[2], G. N. Tiwari[3] and D. Singh[1]

[1]Government Engineering College Bikaner, Pugal Road Bikaner-334004
[2]School of Engineering & Technology, IGNOU, New Delhi, India
[3]Centre for Energy Studies, Indian Institute of Technology Delhi, Hauz Khas, New Delhi-110016

Abstract. The study presents the effect of packing factor of Photovoltaic (PV) module on the room temperature, cell temperature and efficiency of a proposed Building Integrated Semi-transparent Photovoltaic Thermal (BiSPVT) and Building Integrated Opaque Photovoltaic Thermal (BiOPVT) systems with duct mounted on the roof of the building. Different PV materials like mono-crystalline silicon (m-Si), amorphous silicon (a-Si), poly-crystalline silicon (p-Si), cadmium telluride (CdTe), copper indium selenide (CIS) and hetero-junction with intrinsic thin layer (HIT) have been considered in the analysis under the cold climatic condition of India. Since Srinagar (India) has the cold climatic condition, therefore, its climatic data has been considered in the present analysis.

1 Introduction

The current energy scenario is gradually shifting towards non-conventional sources of energy and is going to be a main substitute for fossil fuels in the coming years for their clean and renewable nature. The PV technology is gathering fast momentum and would soon penetrate into the mainstream. So far about 85 % of the PV market concerns mono- and poly-crystalline silicon modules (m- or p-Si) [1] and most of the researchers carried out studies on BIPVT systems by considering the same PV modules. An increase in building energy efficiency by using the PV modules integrated on an office building has been thoroughly investigated [2]. Similar research has been conducted showing a significant gain in energy efficiency [3]. The study of temperature profile of photovoltaic (PV) module in a non-steady state condition with respect to time has been investigated by [4]. The overall electrical efficiency of the PV module can be increased by increasing the packing factor (PF) and decreasing the temperature of the PV module. The effect of packing factor on the electrical and thermal efficiencies of a vertically mounted hybrid photovoltaic water heating system has been investigated by various researchers [5, 6, 7].

Mono silicon (m-Si) modules are highly efficient but manufacturing process is slow and requires highly skilled operators. Manufacturing of poly silicon (p-Si) modules is simpler but the modules are less efficient. Shortage and high price of silicon PV modules are the main reasons for many solar cell manufacturers to seek alternative raw materials like thin film and multi junction technologies, [8].The present study extends the research for other type of PV modules (i) m-Si, (ii) p-Si, (iii) amorphous silicon (a-Si), (iv) cadmium telluride (CdTe), (v) copper indium gallium diselenide (CIGS), and (vi) a heterojunction comprised of a thin a-Si PV cell on top of acrystalline silicon (c-Si) cell (HIT), [9,10]. In m-Si and p-Si, packing factor of a PV module can be varied by varying the area of PV cells for a given area of PV module, which has been shown in Figure 1 (a) and (b). In Figure 1 (a), there are 36 cells, each having an area of 0.0139 m^2 for a given area of PV module. On the other side, in Figure 1 (b), each cell area is half of 0.0139 m^2 for the same area of PV module and therefore, increases the non-packing area i.e. decreases packing factor. However, in case of thin film PV, the packing factor can be varied according to the design e.g. as per need the area of thin filmed PV and glass can be arranged in an alternate fashion to allow solar radiation to pass through the glass area, as shown in Figure 1 (c). In this paper different PV materials have been analyzed to find the effect of different packing factor on the performance of BiSPVT system integrated to roof, for with air duct.

(a)

(b)

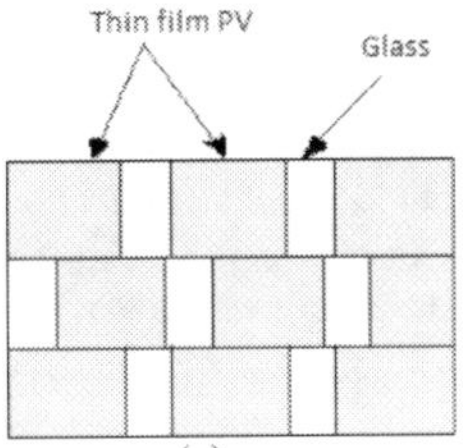

(c)

Figure 1. (a): Photograph of a 75 W, PV module with a cell area of 0.0139 m^2 (b) Photograph of a 37 W, PV module with a cell area of 0.0069 m^2 (c) Proposed design of thin film PV with glass.

2 Identification of problem

In the present study, the proposed design of BiSPVT and BiOPVT system integrated to roof, with the provision of air duct as shown in Figures 2 (a) – (b) respectively has been considered for determining the effect of packing factors (0.42, 0.62, and 0.83) of PV for different PV materials as mentioned in Table 1.

2.1 Thermal modelling

Basic heat transfer equations have been used to write the energy balance of the proposed systems

Energy balance equation for BiSPVT and BiOPVT system integrated to roof with air duct (Fig. 2 (a) and (b))

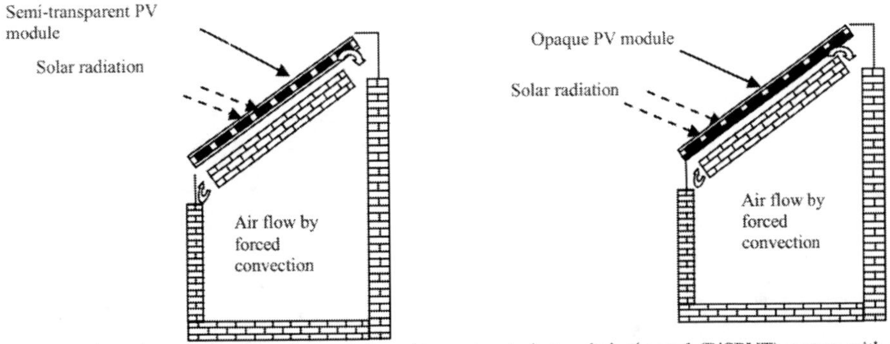

Figure 2. (a): Cross-section view of a building integrated semi-transparent photovoltaic thermal (BiSPVT) system with air duct and integrated to roof a room. (b) Cross-section view of building integrated opaque photovoltaic thermal (BiOPVT) system with air duct and integrated to roof a room.

Table 1. Values of module efficiency and electrical efficiency temperature coefficient for different PV modules.

Type of PV module	Module efficiency STC (%) $\eta_{ref,PV,m}$	Electrical efficiency temp. coefficient % ($\beta_{ref,PV,m}$)
m-Si	13.5	-0.40
a-Si	6.3	-0.26
p-Si	11.6	-0.40
CdTe	6.9	-0.20
CIS	8.2	-0.45
HIT	17.0	-0.33

2.2 For semi-transparent PV module

Using expression given by [11], energy balance equation of solar cell as shown in Fig. 2 (a) is,

$$\alpha_c \tau_g \beta_c I(t) b dx = \left[U_{tc,a}\left(T_c - T_a\right) + U_{Tc,f}\left(T_c - T_f\right)\right] b dx + \eta_{PV,m} I(t) b dx \tag{1}$$

where, α, β, η and τ are=absorptivity, packing factor, efficiency and transmissivity respectively; while subscripts a=ambient; bs=back surface; c=solar cell; eff=effective; f=fluid (air); f_i= inlet fluid; f_o=outgoing fluid; g=glass; L=length; T= tedlar; tc=tedlar to cell the energy balance equation of blackened absorber roof is

$$\alpha_p \left(1 - \beta_c\right) \tau_g^2 I(t) b dx = \left[h_{p,f}\left(T_p - T_f\right) + U_{p,r}\left(T_p - T_r\right)\right] b dx \tag{2}$$

the energy balance equation for air flowing through air duct is

$$\dot{m}_f c_f \frac{dT_f}{dx} dx = [h_{p,f}(T_p - T_f) + U_{Tc,f}(T_c - T_f)] b dx \tag{3}$$

Eqs. (2) and (3) are for BiSPVT and BiOPVT system integrated to roof considered by [12].

Energy balance equation for room air temperature is

$$\dot{m}_f c_f (T_{fo} - T_r) + A_w U_w (\overline{T_f} - T_r) = M_a C_a \frac{dT_r}{dt} + (UA)_t (T_r - T_a) + 0.33 NV (T_r - T_a) \tag{4}$$

Integrating and applying initial conditions at $T_f |_{x=0}$, $T_f = T_{fi}$ and at $T_f |_{x=L}$, $T_f = T_{fo}$, we get

$$\overline{T_f} = \frac{1}{U_{L,G-G}}[(\alpha\tau)_{G-G,eff} I(t) + (U_{L,G-G} - U_T)T_a + U_T T_r](1 - \frac{1 - \exp(\frac{-U_{L,G-G} bL}{\dot{m}_f c_f})}{\frac{U_{L,G-G} bL}{\dot{m}_f c_f}}) + T_r \frac{1 - \exp(\frac{-U_{L,G-G} bL}{\dot{m}_f c_f})}{\frac{U_{L,G-G} bL}{\dot{m}_f c_f}} \tag{5}$$

Using Eqs. (5) and (6) in Eq. (4), we have

$$\frac{dT_r}{dt} + aT_r = f(t) \tag{6}$$

DOI: 10.1051/matecconf/20179501006

where, $a = \frac{[(AF_r + AF_{rw})(U_{L,G-G} - U_T) + (UA)_t + 0.33NV]}{M_a C_a}$

$f(t) = \frac{(AF_r + AF_{rw})(\alpha\tau)_{G-G,eff} I(t)}{M_a C_a} + \frac{[(AF_r + AF_{rw})(U_{L,G-G} - U_T) + (UA)_t + 0.33NV]T_a}{M_a C_a}$

Solution of Eq. (6) is obtained by integrating and applying initial condition at time t=0, $T_r = T_{ri}$

$$T_r = \frac{f(t)}{a}(1 - \exp(-at)) + T_{ri}\exp(-at) \quad (7)$$

The above equation can be computed for any numerical values of $I(t)$, and T_a for a given climatic condition.

3 Results and Discussion

Fig. 4 shows the effect of packing factor on hourly variation of module efficiency, module temperature and room temperature for m-Si module for a day for BiSPVT system integrated to roof with duct and without duct. It is observed that decrease in packing factor from 0.83 to 0.42 decreases the temperature of module from 40.00 °C to 24.00 °C which increases the module efficiency from 12.8 % to 13.6 % and room temperature from 17.60 °C to 26.20 °C.

The observed values of maximum room and cell temperature and efficiency for different packing factors and different PV modules have been tabulated in Table 2.

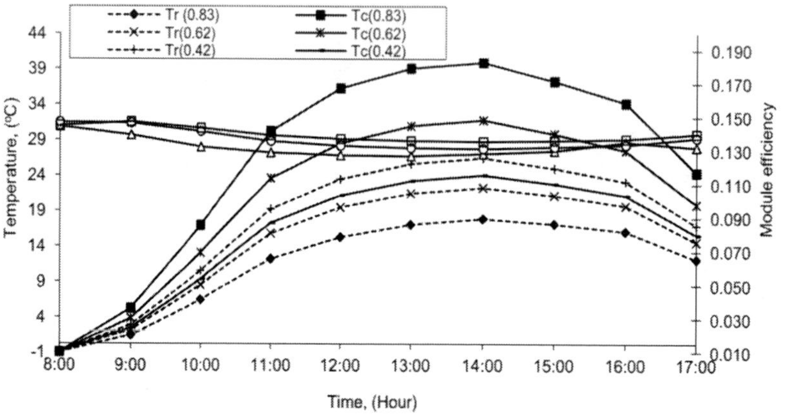

Figure 3. Hourly variation of module efficiency, module temperature and room temperature for different packing factor for m-Si module for a day of month January for BiSPVT system integrated to roof, with duct.

Table 2. Maximum room temperature, module temperature and efficiency for different packing factors and types of PV module (with duct).

PV modules	Packing factor									
	0	**0.42**			**0.62**			**0.83**		
	T_r (°C)	T_c (°C)	η (%)	T_r (°C)	T_c (°C)	η (%)	T_r (°C)	T_c (°C)	η (%)	T_r (°C)
m- Si	35.00	24.00	13.60	26.24	31.56	13.10	22.00	40.00	12.80	17.65
p- Si	35.00	25.00	11.60	26.51	33.00	11.24	22.33	41.00	10.86	18.00
a Si	35.00	28.00	6.20	27.30	36.00	6.10	23.10	44.00	6.00	19.00
CdTe	35.00	28.00	6.90	27.20	35.46	6.80	23.00	43.58	6.60	19.00
CIGS	35.00	27.00	8.10	27.00	34.50	7.80	23.00	43.00	7.50	18.40
HIT	35.00	22.00	17.20	26.00	29.00	16.70	21.55	17.15	16.30	17.15

It is found that the room temperature is maximum in a- Si, CdTe and Copper Indium Gallium Selenide (CIGS) PV modules. As the packing factor decreases from 0.83-0.42, the room temperature increases from 19.00-27.00 °C and 21.00-29.00 °C in with and without duct respectively. The rise in room temperature for both systems is 8.0 °C. It also seems that maximum room temperature is 35.00 and 58.00 °C in zero packing factor (double glazing) in with and without duct respectively. This is valuable only for heating point of view.

4 Conclusions

On the basis of present study, the summary of the study is as follows:

- Forced circulation of air through duct decreases the module temperature in BiSPVT system with duct.
- Indirect heat gain in room due to presence of roof decreases the room temperature in BiSPVT system with duct.
- Decrease in packing factor from 0.83 to 0.42, decreases the module temperature by 16 °C, increase its efficiency by 0.8 % and room temperature by 8.6 °C in BiSPVT system integrated to roof with duct.

References

1. Bloem, J.J., 2008. Evaluation of a PV- integrated building application in a well-controlled outdoor test environment. Building and Environment 43, 205-216.
2. Zogou, O., Stapountzis, H., 2011. Energy analysis of an improved concept of integrated PV panels in an office building in central Greece. Applied Energy 88, 853- 866.
3. Yoon, J.H.,.Song, J.H., Lee, S.J., 2011. Practical application of building integrated photovoltaic (BIPV) system using transparent amorphous silicon thin-film PV module. Solar Energy 85, 723- 733.
4. Jones, A.D., Underwood, C.P., 2001. A thermal model for photovoltaic systems. Solar Energy 70 (4), 349-359.
5. Zondag, H.A., de Vries, D.W., van Helden, W.G.J., van Zolengen, R.J.C., Steenhoven, A.A., 2003. The yield of different combined PV-thermal collector designs. Solar Energy 74 (3), 253-269.
6. Chow, T.T., 2003. Performance analysis of photovoltaic- thermal collector by explicit dynamic model. Solar Energy 75, 143-152.
7. Ji, J., Han, j., Chow, T.T., Yi, H., Lu, J., He, W., Sun, W., 2006. Effect of fluid flow and packing factor on energy performance of a wall- mounted hybrid photovoltaic/water heating collector system. Energy and Buildings 38, 1380-1387.
8. Makrides, G., Zinsser, B., Norton, M., Georghiou, G.E., 2010. Schubert M, Werner JH. Potential of photovoltaic systems in countries with high solar irradiation. Renewable and Sustainable Energy Reviews 14, 754-762.
9. Ordenes, M., Marinoski, D.L., Braun, P., Ruther, R., 2007. The impact of building-integrated photovoltaics on the energy demand of multi-family dwellings in Brazil. Energy and Buildings 39, 629–642.
10. Goetzberger, A., 2002. Photovoltaic material, history, status and outlook. Materials Science and Engineering 40, 1- 46.
11. Dubey, S., Sandhu, G.S., Tiwari, G.N., 2009. Analytical expression for electrical efficiency of PV/ T hybrid air collector. Applied Energy 86, 697-705.
12. Agrawal, B., Tiwari, G.N., 2010. Optimizing the energy and exergy of building integrated photovoltaic thermal (BIPVT) systems under cold climatic conditions. Applied Energy 87, 417- 426.

The Application Research on Waste Clay Brick in the Cement Concrete

Cheng-qin Chen , Yi-lin Zhang, Wei Zhang and Yu-chun Zhang

Northwest University for Nationalities Lanzhou 730124, China
Key Laboratory of New Laboratory of New Building Materials and Building Energy-saving of Gansu Province Lanzhou 730030, China

Abstract. The construction waste has been produced more and more with the rapid development of urbanization and industrialization in China in recent years. The waste clay brick of construction waste named recycled coarse aggregate was mixed with cement concrete mainly by making laboratory experimental study. The clay brick as recycled coarse aggregate was mixed into the concrete at 0%, 20%, 25%, 30 %, 35%, 40%, 50%, 60%, 70%, 80%, 100% and fixed water-cement ratio according to references at home and abroad, which made the C30 standard test blocks, and strictly controlled the variable of the coarse aggregate replacement rate; the slump and 7-day compressive strength were measured, and their performance was compared with that of the standard C30 concrete specimen; the optimum range of incorporation of recycled aggregate would be put forward according to their performance. This result would have a certain significance in waste bricks.

1 Introduction

The demand for construction resources has been rapidly growing with the rapid development of urbanization and industrialization in China in recent years, along with more and more construction waste generated. The current construction waste production has reached about 800 million and 0.6 tons per capita; the amount of waste generated in the construction of the future are predicted that the future construction waste production will reach 10 million tons in 10 years [1]-[5]. The number of construction waste generated in urban construction has accounted for 30% -40% of the total municipal solid waste according to the survey; the vast part of the construction waste is being transported directly into the wild pile or landfill, which will occupy a lot of land, and even the occupation of arable land and farmland. Construction waste will be nowhere in a certain time if they are dealt with in this way a long-term [1]-[5]. While construction debris during transportation will produce dust, gray sand flying and other issues, bring about a lot of pollution to the environment, and lead to people's lives inconvenience. They will also have a small fortune in fees during transportation. It is said that these waste is such as resources that were just put in the wrong place. A lot of research institutions have been conducting research in foreign countries, and made great achievements. Construction waste utilization it is not very common in the domestic, so people can fully draw on advanced foreign technology and processing methods to use construction waste combined with China's actual situation. The foregoing analysis shows that the construction waste recycled has become an urgent problem.

The waste clay brick would be used to replace part of the coarse aggregate in the concrete as recycled coarse aggregate. The technical indicators would be made by crushing test, and to mix design a concrete thickness C30 according to the aggregate value; the recycled coarse aggregate were incorporated into the concrete at 0%, 20%, 25%, 30 %, 35%, 40%, 50%, 60%, 70%, 80%, 100% by a fixed water-cement ratio method, formulating a standard block to measure the compressive strength; the best incorporation of recycled aggregate value range would be obtained.

2 Raw materials and technical indicators

The raw materials contain natural pebbles in Yuzhong County, river sand clay brick construction waste, and naphthalene superplasticizer, P.O 42.5, etc. Their technical indicators were shown in Table 1.

2.1 The main formulas [6]

a. The bulk density of natural coarse aggregate and fine aggregate:

$$\rho_0'=(m_2-m_1)/v_0'*1000 \quad (1)$$

where: m_1—Capacity cartridge mass, kg; m_2—Capacity cartridge and gravel mass, kg

v_0'—Capacity cartridge volume, L

b. The crushing index of natural coarse aggregate:

$$Crushing\ index:=(G_0-G_1)/G_0\times 100\ \% \quad (2)$$

where: G_0—Sample mass, kg; G_1—Sieve allowance, kg

c. The fineness modulus of fine aggregate:

$$Mx=[(A_2+A_3+A_4+A_5+A_6)-5A_1]/(100-A_1) \quad (3)$$

2.2 General technical indexes

The cementitious material is P.O 42.5, and its technical indexes are seen in Table 1. The naphthalene superplasticizer as water reducing agent was used in this experiment, which the water reduction rate of 20% was measured; the appearance took on yellow powder; the solid content was greater than 92%; PH value (5% aqueous solution)was 7-9; sodium content was 16-19%; cement paste fluidity was not less than 200mm; the concrete strength increases by 15-40% at early and night strength for the significant concrete to enhance the effect; it fully improves and enhances the physical and mechanical properties of concrete, and save 10-20% of cement maintaining the strength of the same; it has no corrosion of steel effect.

The recycled aggregate [7], [8] is mainly clay brick used in building demolition wall materials. The collected clay bricks were crushed with jaw crusher; then they were used the standard sieving to sieve and done an adjustments in accordance with the norms [9]. The aggregate Technical indicators were shown in Table 2.

Table 1. The Technical indicators of P.O42.5.

Num.	Test items		Technical indicators	The results	Result verdict
1	Density (kg/m^3)		/	3161	/
2	The specific surface area (m^2/kg)		≮300	342	qualified
3	Standard consistency water (%)		-	27.41	/
4	Setting time (min)	Initial setting	≮45	252	qualified
5		Final setting	≯600	400	qualified
6	Stability		≯5.0	0.6	qualified
7	Flexural strength (MPa)	3d	≮3.5	4.6	qualified
		28d	≮6.5	7.5	qualified
	Compressive strength (MPa)	3d	≮17.0	25.8	qualified
		28d	≮42.5	48.3	qualified
The indicators of P.O42.5 meet the technical requirements in (GB175-2007) "Common Portland cement".					

Table 2. The aggregate technical indicators.

Aggregate Type	Particle size (mm)	Bulk density (kg/m^3)	Crushin g index (%)	M_x
Coarse aggregate	5-31.5	1663	7.95	--
recycled aggregate	5-31.5	952.43	--	
Fine aggregate	0-5	1258	--	3.67

Table 2 showed: the stone is belong to I Class stones by the crushed index value. The sand is I sand area, names the coarse sand.

3 Mix design and production of standard specimens

The trial mixture—C30 concrete was used to make the mixture design, which the mix proportion was that: cement: 328.89kg, sand: 673.09kg, stones: 1250.02kg, water: 148kg, water reducer: 3.29kg. The recycled coarse aggregate was prepared to make cement concrete block; the variable of the coarse aggregate replacement rate was strictly controlled th the test process; recycled coarse aggregate was substituted to make six test groups at respectively 0%, 20%, 25%, 30 %, 35%, 40%, 50%, 60%, 70%, 80%, 100%. The standard test pieces were made with 150mm × 150mm × 150mm, and measured for slump and 7-day compressive strength.

The conservation and test methods for the production of test pieces were mixed strictly in accordance with the standard [10], and the specimens were allowed to stand to form removal after 24h pouring in the laboratory; the blocks were made with conservation trials to 7d after curing under standard conditions.

Since the water absorption of recycled aggregate clay brick was large to bring a lot of trouble, the recycled aggregate was selected to be soaked firstly before making aggregate mixing through experimental analysis; the aggregate fully absorb the moisture, and then dried the aggregate surface, which solves the problem.

4 Analysis of experimental data

The mix proportion used was shown in Table 3.

Coarse aggregate replacement rate: the recycled aggregate volume is equal to the volume of natural aggregate; the bulk density of natural coarse aggregate and recycled coarse aggregate is not the same; the equal volume substitution was taken in the test in order to make the data more reliable.

The concrete slump was measured during the test and the compression testing data in the universal testing machine compression were obtained in Table 4. The incorporation of recycled aggregate concrete had impact on the strength shown in Fig. 1.

Fig. 1 showed: the overall compressive strength firstly trends to increase. Then decrease with the increase in the aggregate amount of renewable incorporated under other factors invariable; the test piece reached to the maximum compressive strength when the amount reaches 25%. Also the recycled aggregate amount was among 20% to 35%, the strength of concrete is greatly improved.

In conclusion, the compressive strength added to the recycled aggregate is higher than that of made of natural aggregate; there are two reasons for primary analysis: firstly, the specific surface area of the recycled aggregate is larger than that of natural aggregate; the contact area increases to enhance the bond strength between the aggregates; secondly, the surface voids of the recycled aggregate is much more than that of natural aggregates, so the grout can fully play its role, and effectively fills into internal porosity of the recycled aggregate, which improves the solidity of the recycled aggregate, making the overall strength of the test piece improved.

Table 3. The amount of material (kg/m3).

Num.	CA replacement rate (%)	W/C	Superplasticizer(g)	Pebble	RA	S	W	C
1	0	0.45	20	1250.02	100	673.09	148	328.89
2	20	0.45	20	1250.02	80	673.09	148	328.89
3	25	0.45	20	1250.02	75	673.09	148	328.89
4	30	0.45	20	1250.02	70	673.09	148	328.89
5	35	0.45	20	1250.02	65	673.09	148	328.89
6	40	0.45	20	1250.02	60	673.09	148	328.89
7	50	0.45	20	1250.02	50	673.09	148	328.89
8	60	0.45	20	1250.02	40	673.09	148	328.89
9	70	0.45	20	1250.02	30	673.09	148	328.89
10	80	0.45	20	1250.02	20	673.09	148	328.89
11	100	0.45	20	1250.02	0	673.09	148	328.89

Table 4. Indexes value of concrete.

Test block Numbering	ressure (kN)	Compressive strength (MPa)	Slump (mm)	Mass (kg)
1	516.033	22.933	72	8.086
2	639.667	28.430	79	7.835
3	788.757	35.056	70	7.920
4	648.000	28.8	75	7.865
5	559.800	24.880	70	7.650
6	521.100	23.16	73	7.486
7	502.425	22.33	75	7.257
8	498.375	22.15	72	7.089
9	481.95	21.42	74	6.881
10	446.625	19.85	75	6.792
11	428.02	19.023	75	6.770

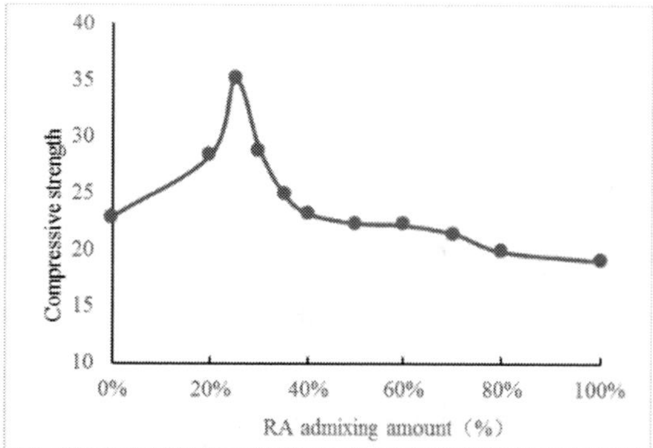

Figure 1. The relations curve between incorporation of recycled aggregate and concrete strength

5 Conclusions

The admixing amount of recycled aggregate as a variable was strictly controlled in the testing process; the strength of the test pieces with recycled aggregate first increased and then decreased. Eventually the recycled aggregate range could be drawn upon incorporation of 20% to 35%; this range could make concrete strength greatly improved. The results showed that the waste clay bricks can not only save resources, but also protect the environment, which will bring great social benefits.

Sources of project funds

The project fund is the professional project funds of the central university basic scientific research business expenses, the project number of which is 31920150071, and the innovative entrepreneurial training projects of college students, the project number of which is 201510742066.

References

1. Q. Zhou, *Research on Force and Permeability Performance of Mixed Recycled Aggregate Concrete*, Xiamen University, a master's degree thesis, (2009)
2. X. Wu, *Application of powder and granular construction waste in cement concrete*, Zhengzhou University, a master's degree thesis, (2004)
3. Y. Li. *Experimental Study on High-Performance Recycled Concrete Produced by Building Waste*, Comprehensive utilization of fly ash, (4):17-20 (2013)
4. S. Jia, J. Wang, et al., *Construction Waste Recycled Concrete Aggregate Performance Study*, Construction Technology, (S1):165-168 (2014)
5. M. Li, *Applied Technique Research of Regeneration Concrete by Use of Construction Waste*, Municipal Engineering Technology, (S1): 254-260 (2010)

6. *Pebbles, Gravel Used in Building*, (GB14685-2001)
7. *Ordinary Concrete Mixed Design Procedures*, (JGJ 55-2011)
8. S. Song, L. Wang. *The Experimental Research on Concrete with Construction Rubbish Recycled Aggregate*, Journal of Wuhan University of Technology (7):56-59 (2009)
9. *Building Pebbles, Gravel Modified Version* (GB14685-2001)
10. *Ordinary Concrete Mixing Performance Test Methods*, (GB/T50080-2002)

The Pavement Performance Research on the Powder Colored Asphalt Mixture

Cheng-qin Chen and Wei Zhang

Northwest University for Nationalities Lanzhou 730124, China
Key Laboratory of New Laboratory of New Building Materials and Building Energy-saving of Gansu Province Lanzhou 730030, China

Abstract. The high-strength powder colored asphalt was studied by sieve analysis, density and cone penetration test in order to study technical performance. Secondly, the mixing process of colored asphalt mixture was put forward through the Marshall Compaction test. The best whetstone of the colored asphalt mixture was determined by Marshall Test design method. Finally, the variation of pavement performance for colored asphalt was analyzed by the rutting test, immersion Marshall Test split and freeze-thaw test and low-temperature bending test. The results showed: the high temperature performance of colored asphalt mixture was outstanding, but the low temperature performance was lower. Therefore the high-strength powder colored asphalt mixture is suitable to be used in high temperature areas.

1 Introduction

The colored asphalt concrete pavement is studied in Europe and other developed countries from the 1950s, and the application is gradually expanded [1], [2]. The colored asphalt paving technology begins with the early 1980s, and the efficiency is rarely, especially the road engineering applications. At present, the colored asphalt concrete pavement is used widely in China, involving more than 20 provinces, and the kind of the colored pavement is also very much, but relatively few high viscosity colored asphalt.

Most southern coastal areas have relatively developed economy, big volume of transportation, more traffic accidents, coupled with sustained high temperatures all year round; above all, the desired material requirements is relatively high, tending to high hardness, high viscosity, which is conducive to resolving early disease pavement occurred. The high-viscosity, high-strength colored asphalt is rarely used in road construction, belonging to the new material; there are no uniform technical indicators. The technical performance of high viscosity color binders will be analyzed by the laboratory tests in this paper based on the colored asphalt from Shenzhen Co., Ltd.

This kind of colored asphalt performance would be studied from high temperature stability, water stability and low temperature crack resistance, comparing with SBS I-C modified asphalt mixture performance.

2 Raw material

2.1 SBS I-C

The modified asphalt—SBS I-C is a superior technical performance of heavy traffic asphalt; the common technical indicators were seen in Table 1.

2.2 The colored asphalt

(1) Sieving. The colored asphalt is a high-strength powder colored asphalt, both having plastic powder particles; it was sieved; the sieving results were shown in Table 2.

Table 2 showed: the particle size distribution range of colored asphalt is more than 2.36mm, most concentrating between 2.36mm ~ 4.75mm and under 0.075mm; the part above 0.075mm accounted for about 70%, and the other under 0.075mm is about 30%. The portions of above 0.075mm were heated, and could be bonded together; this part was the colored asphalt binder. The other did not occur under heating conditions; it would be as filler in the mix, acting as part of slag.

(2) Density. The colored asphalt contains lighter quality plastic particles and powdered substances; the density of the plastic particles is less, floating on the water; the density of the powdered substance is relatively large, sinking to the bottom; therefore, the density of colored asphalt cannot be measured with the ordinary asphalt density method. In addition, it cannot be tested the conventional indicators like ordinary asphalt when the heating temperature of the colored asphalt exceeds 200°C, still in a viscous state. So the colored of asphalt is regarded as cementitious paste for research. It was made possible firstly melted in an oven at 135°C ~ 163°C about to exclude air bubbles, and then re-measured the density of water for measuring the density. The results were shown in Table 3.

Table 1. The common experimental value of SBS I-C.

Number	Test items	Specification Value	Measured value	Test Standard
1	Penetration (25°C, 100g, 5s, 0.1mm)	60~75	63.27	T0604-2011
2	Penetration index PI	≮-0.1	0.87	
3	Ductility (5cm/min, 5°C, cm)	≮35	52.4	T0605-2011
4	Softening Point (°C)	≮80	>90	T0606-2011
5	Kinematic viscosity 135°C (Pa·s)	2~3	2.2	T0625-2011
6	Elastic recovery (25°C, %)	≮80	85.17	T0662-2011
7	Storage stability	≯2.5	0.25	T0661-2011
8	Density(15°C, g/cm3)	The measured records	1.26	T0603-2011
Rolling Thin Film Heating Test (163°C,85min)				T0610-2011
9	Mass loss (%)	≯1.0	0.36	T0609-2011
10	Penetration ratio (25°C, %)	≮65	71.34	T0604-2011
11	Ductility (5°C, cm)	≮25	25.3	T0605-2011

Table 2. The sieving results.

Sieve size	The total amount of the first group (161.55g)				The total amount of the second group(128.89g)				
	Each mesh quality (g)	Single sieve (%)	Accumulated sieve (%)	Passing rate (%)	The mesh quality (%)	Single sieve (%)	Accumulated sieve (%)	Passing rate (%)	Average (%)
4.75	0.00	0.00	0.00	100.00	0.00	0.00	0.00	100.00	100.00
2.36	48.44	30.21	30.21	69.79	40.51	31.53	31.53	68.47	69.13
1.18	0.08	0.05	30.26	69.74	0.06	0.05	31.58	68.42	69.08
0.6	0.08	0.05	30.31	69.69	0.05	0.04	31.62	68.38	69.04
0.3	3.68	2.29	32.60	67.40	2.92	2.27	33.89	66.11	66.75
0.15	22.81	14.23	46.83	53.17	18.22	14.18	48.07	51.93	52.55
0.075	22.54	14.06	60.89	39.11	16.67	12.98	61.05	38.95	39.03
Sieve bottom	62.72	39.11	100.00	0.00	50.04	38.95	100.00	0.00	0
Sum	160.35	-	-	-	128.47	-	-	-	-
Differ.	1.20	-	-	-	0.42	-	-	-	-

Table 3. The determination value of the colored asphalt density.

m_1 (g)	m_2 (g)	m_3 (g)	m_4 (g)	m_5 (g)	Density g/cm^3
47.08	41.27	102.66	147.82	61.06	1.2676
47.71	41.77	105.33	151.98	59.72	1.2202

The equation is:

$$\rho = m_3/[(m_4-m_5)-(m_2-m_1)] \quad (1.1)$$

where: m_1- Cylinder mass; m_2- Cylinder mass in water; m_3- Asphalt mass; m_4-Asphalt mass+ Cylinder mass; m_5-Asphalt mass+ Cylinder mass in water after heating
The density is 1.244g/cm^3 by calculation.

(3) Cone Penetration

a. The test results of cone penetration. The penetration test was made at 25°C, 30°C, 40°C, 50°C, 60°C, 70°C and 5s. Penetration weight in six temperature conditions was uniform. The cone penetration value was made a linear correlation fitting analysis with the temperature, and the cone penetration value took logarithmic; if y=lgP, x=T, y=ax+b was made a linear regression, namely the regression equation. Only 25°C, 50°C, 70°C three temperatures were made a linear regression corresponding cone penetration value; the regression results were shown in Table 4. And the penetration test results were shown in Table 5.

Table 4. The regression result.

Number	The regression equation	R^2
1	*y=0.489x-4.677*	*0.996*
2	*y=0.509x-5.051*	*0.997*
3	*y=0.528x-5.527*	*0.995*

Table 5. The test results of cone penetration.

Test temperature	Cone Penetration 25°C			Cone Penetration 40°C			Cone Penetration 50°C			Cone Penetration 60°C			Cone Penetration 70°C		
Number	1	2	3	1	2	3	1	2	3	1	2	3	1	2	3
Measured value	7.9	8.1	8.0	12.3	12.4	12.6	19	20	20	24	23	24	30	32	32
Average value	8			12.4			19.7			23.7			31		

As can be seen from Table 4 and Table 5: the cone penetration value of colored asphalt is relatively small, indicating that the large hardness. The value of penetration increases with temperature increasing, and the cone penetration value trends to be evident. The correlation coefficient R^2 for the fitting regression equations were above 0.99, indicating that the cone penetration value and temperature changes conform to the form $lgP=AT+K$.

b. Shear Strength τ. Shear strength τ can be measured in terms of cone penetration value according to formula (1.2):

$$\tau = [981Q\cos2(\alpha/2)]/[\pi h2\tan(\alpha/2)] \quad (1.2)$$

where: τ—Shear strength, KPa; Q—Penetration Weight [3], [4], g; h—cone penetration, 0.1mm; α—Cone angle needle number.

The τ was calculated under different temperature conditions according to the Equation (1.1). The results were shown in Fig. 1.

2.3 Physical indicators test of coarse aggregate

The coarse aggregate selected good texture, color uniformity, good acid and alkali diabase. The indicators met the requirements of the literature [5], specified in Table 6.

2.4 Physical indicators test of fine aggregate

The fine aggregate selected the limestone, the basic performances were shown in Table 7.

2.5 Mineral powder

The mineral powder selected limestone filler or other alkaline rocks finely powder; it should look white, and its indicators were shown in Table 8.

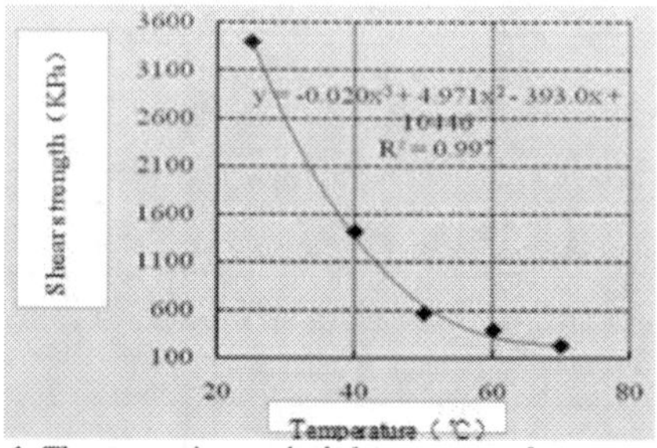

Figure 1. The regression analysis between τ and temperature

Table 6. Technical indicators of the coarse aggregate.

Test item		Unit	Test results	Test standard	Test method
Crushed stone value		%	10.907	≯26	T0316−2005
Los Angeles abrasion loss		%	10.783	≯28	T0317−2005
Gross volume relative density (g/cm³)	13.2~16mm	-	2.833	-	T0304−2005
	9.5~13.2mm	-	2.812		
	4.75~9.5 mm	-	2.784		
Apparent relative density (g/cm³)	13.2~16mm	-	2.898	≮2.60	T0304−2005
	9.5~13.2mm	-	2.890		
	4.75~9.5 mm	-	2.901		
Water absorption	13.2~16mm	%	0.791	≯2.0	T0304−2005
	9.5~13.2mm		0.956		
	4.75~9.5 mm		1.453		
The needle flake content	9.5~13.2mm	%	4.499	≯15	T0312−2005
	4.75~9.5 mm		8.783		

Table 7. Technical specifications of the limestone.

Test item		Unit	Test results	Test method
Density (g/cm³)	2.36~4.75mm	-	2.825	T0304−2005
	1.18~2.36mm	-	2.692	
	0.6~1.18 mm	-	2.746	
	0.3~0.6mm	-	2.740	
	0.15~0.3mm	-	2.771	
	0.075~0.15mm	-	2.703	

Table 8. The indicators of mineral powder.

Test item		Unit	Test results	Test standard	Test method
Apparent relative density(g/cm^3)		-	2.628	≮2.50	T0352-2005
Particle size range	<0.6	%	100	100	T0351-2005
	<0.15	%	98.65	90~100	
	<0.075	%	97.43	75~100	
Hydrophilic coefficient		-	0.794	<1	T0353-2005

3 The mixing process of the colored asphalt

The heating temperature of colored asphalt aggregate was controlled at 220°C ~ 270°C; the colored asphalt mix was added like mineral powder; the mixing temperature is generally between 240°C ~ 270°C; the mixing time was 120s; the compaction temperature was 210°C ~ 230°C. All kinds of mineral aggregate mixture were uniform mixing. If the color material occurred, they would be discarded, the reason why the color material causes: the heating temperature between mineral aggregate and asphalt was insufficient; the material amount of color combination was inadequate; mixing time is too short; he blade was wear due to prolonged use; the fine aggregate amount was the high side; the mix temperature was too high.

4 Pavement performances

Mix Gradation diagram was seen in Fig. 2 accord to [6].

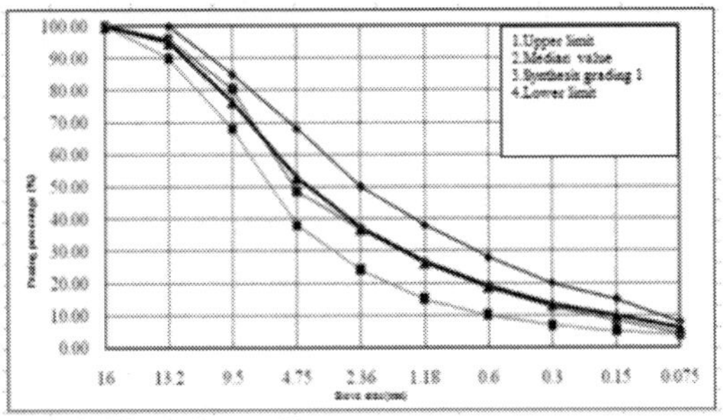

Figure 2. The design grading curve

The Marshall Test of the colored asphalt mixture was made at 8%, 9%, 10%, 11% of asphalt-aggregate ratio to measure the gross volume relative density for *VV*, *VFA*, *VMA*, and the *MS*, *FL* at 60°C after soaking 0.5h. The optimum asphalt-aggregate ratio took 10.4% according to practical experience, and the mixture pavement performance was studied at 10.4% of the asphalt-aggregate ratio.

4.1 High-temperature stability

Table 9 showed: the dynamic stability of the two mixtures was very large. Relative deformation of colored asphalt is 17.8% of the SBS mixture, indicating excellent high temperature performance of colored asphalt.

Table 9. Rutting test results.

Matrix	DS, time/mm	d_{45}, mm	d_{60}, mm	δ (%)
The colored asphalt	14000	0.239	0.284	0.57
SBS I-C	5139	1.481	1.604	3.21

4.2 Water stability

The immersion Marshall Test [7] was made to check the water stability; the method was simple and could better evaluate the water stability of mixture. The test results were shown in Table 10 according to reference [6] about the water stability of ordinary asphalt mixture.

Table 10. Residual stability test results

Matrix	MS(kN)	MS_1(kN)	MS_0(%)
The colored asphalt	65.02	59.78	91.94
SBS I-C	15.32	13.64	89.03

Table 10 showed: the stability value of the colored asphalt mixture is four times of that of SBS mixture; the MS0 of the colored asphalt was larger than that of SBS, namely: colored asphalt> SBS.

4.3 Low temperature anti-cracking performance

The shear strength and deformation of the mixture will affect the low temperature anti-cracking performance [8]. In general, the mixture has high crack resistance, higher strength and deformability is also better. The -10°C Low temperature bending test results of mixture were shown in Table 11 according to the specification [6] the technical requirements of the low temperature bending testing of ordinary asphalt mixture.

Table 11. Low temperature bending test results.

Binder Category	R_B [MPa]	ε_B	S_B, [MPa]	Bending creep compliance [MPa-1]
The colored asphalt	15.06	3358.5	4791.3	233.515
SBS I-C	7.51	2965.1	2534.8	394.508

Table 11 showed: the RB of the colored asphalt was 2.01 times of SBS, showing that the colored asphalt

 DOI: 10.1051/matecconf/20179501008

flexural capacity is strong. The curved tensile strain limits of the two mixtures exceeded the minimum failure strain for cold winter areas, the εB for the colored asphalt is 1.13 times of that of SBS I-C. The SB of the colored asphalt is 1.89 times of that of SBS I-C from the bending stiffness modulus, suggesting that the low temperature resistance of the colored asphalt was poor to deformation.

5 Conclusions

The part above 0.075mm was the binder, and the other 0.075mm below was as part of the filler such as mineral powder for the powder colored asphalt material. It would emit a pungent odor when heated. It is determined by the temperature sensing performance of cone penetration. The powder colored asphalt mixture had good high-temperature stability; the stability value is relatively high in water stability, but the splitting strength is relatively small, indicating that the colored asphalt had poor water stability; the curved tensile strain limits of the mixture exceeded the minimum failure strain for cold winter areas from low temperature crack resistance.

Sources of project funds

The project fund is the professional project funds of the central university basic scientific research business expenses, the project number of which is ZYP2015011.

References

1. H. Cao, Z. Qian, *Color analysis and application of colored asphalt pavement*, Petroleum Asphalt, **21**(5): 1-4). (2007)
2. Y. Chi, W. Liu, G. Yuan. *Application of colour asphalt pavement*, Technology & Economy in Areas of Communications, (1): 18- 26) (2007)
3. M. Liu, C. Yang, J. Liang, *et al.*, *Study on Test about Sclerosis Effect of the Fillings*, Journal of Hebei University of Technology, (5) (2005)
4. L. Liu, P. Hao. *Analysis on Viscosity Character of Asphalt Mortar*, Journal of Hebei University of Technology, **35**(2): 109- 113 (2006)
5. *Highway Engineering Aggregate Testing Procedures*, (JTG E42-2005)
6. *Asphalt Pavement Construction Specifications*, (JTG F40-2004)
7. *Highway Engineering Asphalt and Asphalt Procedures*, (JTJ E20-2011)
8. J. Zhang, *Research on Performance and Construction of Road plus fiber asphalt concrete for bridge deck overlay*, Transportation Enterprise Management, (3) (2007)

MATEC Web of Conferences 95, 01009 (2017) DOI: 10.1051/matecconf/20179501009
ICMME 2016

Investigation into Effects of Scanning Speed on in Vitro Biocompatibility of Selective Laser Melted 316L Stainless Steel Parts

Yitong Shang[1], Yanping Yuan[2], Yongzhi Zhang[2], Dongfang Li[2, 3] and Yansheng Li[1]

[1] College of Mechanical Engineering and Applied Electronics Technology, Beijing University of Technology, Beijing 100124, China
[2] Institute of Laser Engineering, Beijing University of Technology, Beijing 100124, China
[3]Fraunhofer-Institute for Laser Technology, the Federal Republic of Germany

Abstract. In recent years, selective laser melting (SLM) has gained an important place in fabrication due to their strong individualization which cannot be manufactured using conventional processes such as casting or forging. By proper control of the SLM processing parameters, characteristics of the alloy can be optimized. In the present work, 316L stainless steel (SS), as a widely used biomedical material, is investigated in terms of the effects of scanning speed on in vitro biocompatibility during SLM process. Cytotoxicity assay is adopted to assess the in vitro biocompatibility. The results show the scanning speed strongly affects the in vitro biocompatibility of 316L SS parts and with prolongs of incubation time, the cytotoxicity increase and the in vitro biocompatibility gets worse. The optimal parameters are determined as follows: scanning speed of 900 mm/s, laser power of 195 W, hatch spacing of 0.09 mm and layer thickness of 0.02 mm. The processing parameters lead to the change of surface morphology and microstructures of samples, which can affect the amount of toxic ions release, such as Cr, Mo and Co, that can increase risks to patient health and reduce the biocompatibility.

1 Introduction

With an increased demand for fast and less expensive product development, rapidly manufacturing parts from metal powders without moulds becomes more and more desirable [1,2]. Additive manufacturing (AM) has received great attention over the past decade [3],[4],[5].AM by selective laser melting (SLM) is an advanced manufacturing process which uses lasers to melt metal powders one layer at a time to produce components from 3D CAD models, as shown in Fig. 1. The process is suitable to manufacture complex parts which cannot be manufactured using conventional processes such as casting or forging. The process of selective laser melting is a potential manufacturing route for biomedical parts, aero applications, and dental prostheses [6]. Compared with other biomedical metallic materials, 316L stainless steel (SS) has been widely used for biomedical material to make artificial joints, bone plate and bone screw and other medical devices due to its outstanding mechanical properties, corrosion resistance and low price [7,8]. The process has been successfully demonstrated [6] to manufacture 316L SS parts. Some investigation about the mechanical properties and corrosion resistance of 316L SS parts has been reported [9, 10]. However, the release of potential toxic ions such as Cr, Ni and Mo, in 316L parts manufactured via the SLM route, prevents its use for applications in dental prostheses, which also increases risks to patient health. Therefore, during the SLM process, the researchers can determine the input parameters to get the desired product quality in time to meet the needs [11]. The properties of SLM shaped parts are influenced by the process parameters, such as the scanning speed, the laser power and the scanning distance [12, 13]. The influence of processing parameters on the microstructure and mechanical properties of the final parts for different materials has been studied [14–20]. The parameters of SLM stainless steel have been optimized and evaluated in some studies. Riemer etal. [21] have been studied the fatigue crack growth behavior of SLM 316L stainless steel. The feasibility of the preparation of cellular lattice structure of SLM stainless steel has been studied by Yan et al. [22]. Li etal. [23] have studied the optimization of four process parameters: laser power, scanning speed, scanning spacing and layer thickness, to obtain compact specimens. Laser melting and numerical simulation of 316L powder have been used to evaluate the influence of laser power, scanning speed and beam size on the melting zone and the phenomenon of the ball performed by Antony etal. [24]. Hence, effects of scanning speed on in vitro biocompatibility of 316L SS parts manufactured via SLM are experimentally investigated in this study.

DOI: 10.1051/matecconf/20179501009

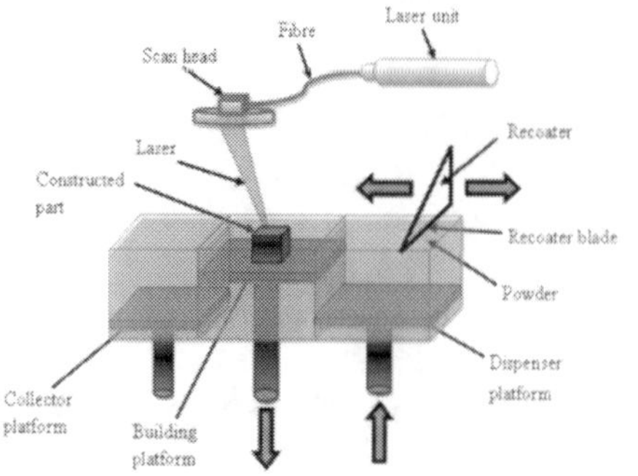

Figure 1. Schematic diagram of SLM process

2 Materials and methods

The material used in our experiments is 316L stainless steel powder (size range from 24μm to 59μm, the mean particle size 38μm), manufactured via gas atomization, as shown in Fig. 2 and Fig. 3. The specification and actual composition of the alloy are shown in Table 1. Based on the information provided by the supplier, this powder is recommended in medical field and has a reported a tensile strength of 590–690 MPa in horizontal direction and 485–595 MPa in vertical direction, a yield strength of 470–590 MPa in horizontal direction and 380–560 MPa in vertical direction, Young's modulus of typical 185 GPa in horizontal direction and 180 GPa in vertical direction, elongation at break of typical 25–55 % in horizontal direction and 30–70 % in vertical direction, and hardness of typical 85 HRB when standard SLM building parameters and strategies are used.In order to avoid the test errors, 15 cylindrically-shaped samples (10 mm diameter and 10 mm thickness, as shown in Fig. 4) are manufactured by EOS M280 (Fig. 5) and divided into five groups. The processing parameters are: scanning speed (v) of 800 mm/s, 900 mm/s, 1050 mm/s, 1100 mm/s, 1200mm/s, laser power (P) of 195 W, hatch spacing (h) of 0.09 mm and layer thickness (d) of 0.02 mm. All samples are cleaned in ultrasonic bath with acetone (15 min), ethanol (15 min) and distilled water (3 min) and then dried in open air. All samples are sterilized in autoclave for 20 min. Five group samples are immersed in 15ml centrifuge tube at a ratio of 3:1 volume of DMEM supplemented with 10% fetal calf serum to surface area of the specimens for 72h at 37 ℃ environment. After 72h, the samples are token out from the DMEM.

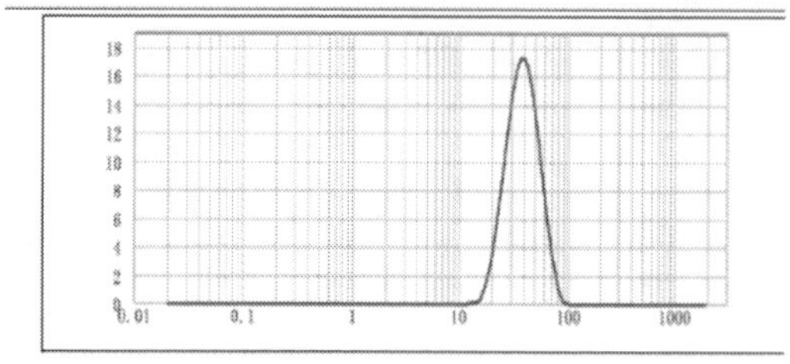

Figure 2. SEM of 316L powder

Figure 3. SEM of 316L powder

Figure 4. Cylindrically-shaped samples

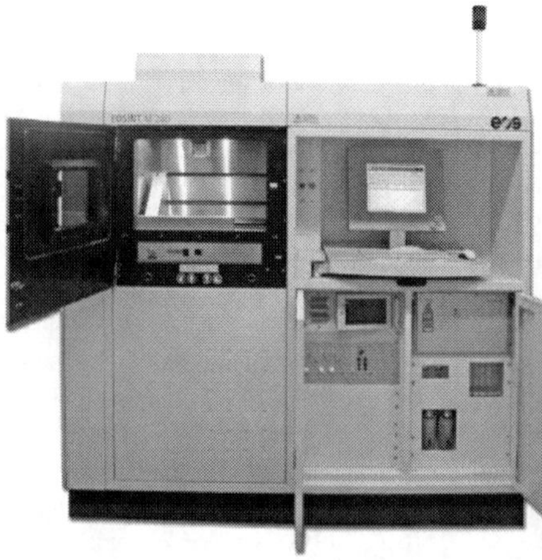
Figure 5.EOS M280

Table 1. Composition of 316L stainless steel powder.

Grade 316L	C	Cr	Ni	Mo	Mn	Si	P	S	Fe
Wt%	0.030	16.97	13.11	2.35	2.00	0.75	0.025	0.010	Balance

In our study, cytotoxicity assay is adopted to assess the in vitro biocompatibility of 316L SS parts. Cytotoxicity assay is performed by using a CCK-8 assay to evaluate the HEK 293T cells. HEK 293T cells are digested with trypsin and seeded at 2000 cells/well into a 96-well culture plate containing the DMEM medium, and cultivated in a 5% CO_2 humidified environment at 37 ℃ for 24h, then the original culture solution is removed when cells are growth attached to the wall. Each well is washed with phosphate buffer solution (PBS), and then the solution of the five group samples are added, and two groups are provided with DMEM culture medium as the control groups. The cell inhibition rate (CIR) is examined using CCK-8 after 24h and 72h of cell culture, which shows a positive correlation with the cytotoxicity of 316L SS parts. Microplate reader is used to test the optical density (OD) of each solution. The CIR is calculated by the following equation:

$$CIR = \left(1 - \frac{E_{OD}}{C_{OD}}\right) \times 100\% \qquad (1)$$

where E_{OD} indicates the OD of experimental groups, C_{OD} indicates the OD of control groups.

3 Results and conclusions

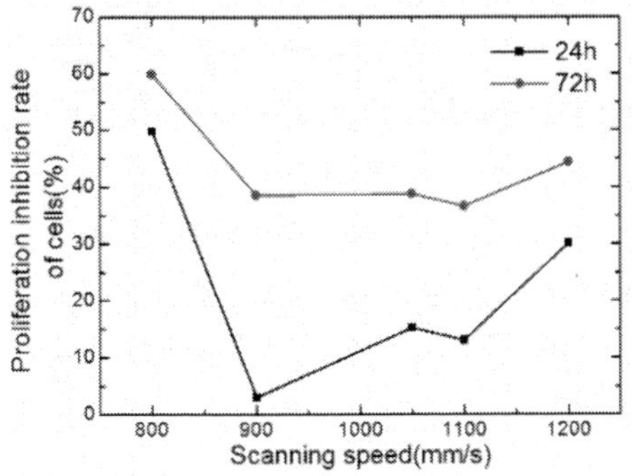

Figure 6. Inhibition rate of HEK 293T cells after 24h and 72h

Fig.6 shows the CIR after 24h and 72h. After 24h, the highest cytotoxicity in the solution (scanning speed of 800 mm/s) is obtained, the CIR is 49.82%. For the scanning speed of 900 mm/s, the CIR of the solution is 3.05%, which is the lowest cytotoxicity of five groups. As the scanning speed increases, the cytotoxicity in corresponding solutions are various, and 15.21%, 13.04%, and 30.11% of the CIR are for the cases of 1050 mm/s, 1100 mm/s and 1200mm/s, respectively. The experimental data shows that the scanning speed strongly affects the cytotoxicity of the 316L SS parts. The inhibition rate of HEK 293T cells in different solutions after 72h is also investigated. The CIR in solutions are 59.92%, 38.53%, 38.86%, 36.60% and 44.38% for the cases of 800 mm/s, 900 mm/s, 1050 mm/s, 1100 mm/s and 1200mm/s. The experiments shows that: 1) after 24h, the lowest cytotoxicity is obtained (scanning speed of 900 mm/s), while the highest value is CIR 49.82% (scanning speed of 800 mm/s); and 2) after 72h, the lowest and highest value of cytotoxicity is CIR 36.60% (1100 mm/s) and 59.92% (800 mm/s). It is found that: 1) with prolongs of incubation time, the cytotoxicity increase. And the in vitro biocompatibility of 316L SS parts manufactured via SLM gets worse. 2) processing parameters: scanning speed of 800 mm/s, laser power of 195 W, hatch spacing of 0.09 mm and layer thickness of 0.02 mm, is not suitable for the manufacture of biomedical parts.

As shown in Fig. 6, the scanning speed has strong effects on the in vitro biocompatibility of 316L SS. The main reasons for the use of 316L SS are its low cost, good machining performance and mechanical properties. The increasing use of this metallic alloy in the medical field has led to an enormous amount of studies related to health. However, clinical experience has revealed that metal ions, which are released from dental restorations, can provoke systemic and local allergic reactions. In addition, these metal ions may also have an adverse effect on adjacent oral soft tissues and nearby alveolar bone. The severe and prolonged processes can lead to failure of the implant. Nickel、Chromium、Molybdenum have been found to be the main elements resulting from dissolution of 316L SS alloys. Yamamoto et al. [25] have been established the sequence from the highest toxic element to the lowest toxic element as being: Ni >Cr>Mo for L929 cells. Particularly, Ni^{3+} undergoes mitochondrial redox metabolism and leads to intermediate reactive oxygen radical formation, which is toxic for the cell. Contrary to a relatively low nickel ion concentration, a higher concentration lead to in vitro cell death and the integrity of a fibroblast monolayer loss, which affect cellular metabolism. In fact, this element can cause allergic sensitization especially in females according to a report that indicate 30% of women had skin allergy to objects containing nickel. The presented in vitro biocompatibility data shows that the scanning speed has strong effects on the in vitro cytotoxicity whereas has no significant effects on the hemolysis. Extensive in vitro studies [26-28] have suggested that metallic dental restorations release metal cations due to corrosion and 316 L SS is always accompanied by corrosion problems. It is reported that microstructure might, in fact, significantly influenced corrosion, that is to say, significantly influenced biocompatibility [29]. Material degradation of stainless steels begins whenever they experience temperature in the range 450-800 °C. After sufficient time at temperature, intermetallic compounds (principally Cr carbides) precipitate in grain boundaries and (later) inside the grains as well. This precipitation creates a condition known as sensitization that ruins corrosion resistance [30]. However, the detailed mechanism is still under investigation. The further study will focus on detection of toxic ion (including ion type and amount of release) and the effects of processing parameters (including power, scanning speed, and scanning strategy) on surface morphology, such as porosity, density and dislocation.

References

1 T.H.C. Childs, C. Hauser, M. Badrossamay, P.I. Mech, Eng. B: J. Eng. 219, 339–357, (2005).
2 K. Osakada, M. Shiomi, Int. J. Mach. Tools Manuf. 46 1188–1193, (2006).

3 G.N. Levy, R. Schindel, J.-P. Kruth, Rapid manufacturing and rapid tooling with layer manufacturing (LM) technologies, state of the art and future perspectives, CIRP Annals, 52/2: 583- 609, (2003).
4 Hopkinson, N., Dickens, P., rapid prototyping for direct manufacture, Rapid Prototyping J., 7/4: 197-202 (2001).
5 J.-P. Kruth, G. Levy, F. Klocke, T.H.C. Childs, Consolidation phenomena in laser and powder-bed based layered manufacturing, CIRP Annals, 56/2: 730-759, (2007).
6 J.A. Cherry, H. M. Davies, S. Mehmood, N.P. Lavery, S. G.R. Brown, J. Sienz. Int. J Adv. Manuf. Techno.176, 869–879, (2015).
7 L. Zhou, K. Nakata, T. Tsumura. Journal of Materials Engineering and Performance 23, 3718-3726, (2014).
8 S.M. Hosseinalipour, A. Ershad-Langroudi, A.N. Hayati. Progress in Organic Coatings 67, 371-374, (2010).
9 B. Zhang, L. Dembinski, C. Coddet. Materials Science and Engineering A 584, 21-31, (2013).
10 Y. Sun, A. Moroz, K. Alrbaey. Journal of materials engineering and performance 23, 518-526, (2014).
11 V.S. Thangarasu, R .Sivasubramanian, HIGH SPEED CNC MACHINING OF AISI 304 STAINLESS STEEL; OPTIMIZATION OF PROCESS PARAMETERS. International Journal of Mechanical Engineering and Robotics Research, 1(9), (2012).
12 G. Casalino, S.L. Campanelli, N. Contuzzi et al. Experimental investigation and statistical optimisation of the selective laser melting process of a maraging steel. Optics & Laser Technology, 65: 151-158, (2015).
13 A. Simchi, Direct laser sintering of metal powders: Mechanism, kinetics and microstructural features. Materials Science and Engineering: A, 428(1): 148-158, (2006).
14 D. Gu, Y. Shen, Balling phenomena in direct laser sintering of stainless steel powder: metallurgical mechanisms and control methods. Materials & Design, 30(8): 2903-2910, (2009).
15 A. Simchi, F. Petzoldt, H. Pohl, On the development of direct metal laser sintering for rapid tooling. Journal of Materials Processing Technology, 141(3): 319-328, (2003).
16 E.O. Olakanmi, Selective laser sintering/melting (SLS/SLM) of pure Al, Al–Mg, and Al–Si powders: Effect of processing conditions and powder properties. Journal of Materials Processing Technology, 213(8): 1387-1405, (2013).
17 E. Sallica-Leva, A.L. Jardini, J.B. Fogagnolo, Microstructure and mechanical behavior of porous Ti–6Al–4V parts obtained by selective laser melting. Journal of the mechanical behavior of biomedical materials, 26: 98-108, (2013).
18 J.P. Kruth, L. Froyen, J. Van Vaerenbergh, et al. Selective laser melting of iron-based powder. Journal of Materials Processing Technology, 149(1): 616-622, (2004).
19 E. Chlebus, B. Kuźnicka, T. Kurzynowski et al. Microstructure and mechanical behaviour of Ti—6Al—7Nb alloy produced by selective laser melting. Materials Characterization, 62(5): 488-495, (2011).
20 R. Wauthle, B. Vrancken, B. Beynaerts et al. Effects of build orientation and heat treatment on the microstructure and mechanical properties of selective laser melted Ti6Al4V lattice structures. Additive Manufacturing, 5: 77-84, (2015).
21 A. Riemer, S. Leuders, M. Thöne et al. On the fatigue crack growth behavior in 316L stainless steel manufactured by selective laser melting. Engineering Fracture Mechanics, 120: 15-25, (2014).
22 C. Yan, L. Hao, A. Hussein et al. Advanced lightweight 316L stainless steel cellular lattice structures fabricated via selective laser melting. Materials & Design, 55: 533-541, (2014).
23 R. Li, Y. Shi, Z. Wang et al. Densification behavior of gas and water atomized 316L stainless steel powder during selective laser melting. Applied Surface Science, 256(13): 4350-4356, (2010).
24 K. Antony, N. Arivazhagan, K. Senthilkumaran. Numerical and experimental investigations on laser melting of stainless steel 316L metal powders. Journal of Manufacturing Processes, 16(3): 345-355, (2014).
25 A. Yamamoto, R. Honma, M. Sumita. Cytotoxicity evaluation of 43 metal salts using murine fibroblasts and osteoblastic cells. Journal of biomedical materials research, 39(2): 331-340, (1998).
26 Y.C. Tang, S. Katsuma, S. Fujimoto et al. Electrochemical study of Type 304 and 316L stainless steels in simulated body fluids and cell cultures. Acta Biomaterialia, 2(6): 709-715, (2006).
27 R. Nishimura, A. Sulaiman, Y. Maeda. Stress corrosion cracking susceptibility of sensitized type 316 stainless steel in sulphuric acid solution. Corrosion Science, 45(2): 465-484, (2003).
28 R.A. Antunes, A.C.D. Rodas, N.B. Lima et al. Study of the corrosion resistance and in vitro biocompatibility of PVD TiCN-coated AISI 316L austenitic stainless steel for orthopedic applications. Surface and Coatings Technology, 205(7): 2074-2081, (2010).
29 X.Z. Xin, N. Xiang, J. Chen et al. In vitro biocompatibility of Co–Cr alloy fabricated by selective laser melting or traditional casting techniques . Materials Letters, 88: 101-103, (2012).
30 D.H. Kumar, A.S. Reddy. STUDY OF MECHANICAL BEHAVIOR IN AUSTENITIC STAINLESS STEEL 316 LN WELDED JOINTS. Int. J. Mech. Eng. Rob. Res, 2: 37-56, (2013).

Dynamic Response Analysis of Elastic-plastic Structures with Component Mode Synthesis

Peng-bo Qian

College of Mechanical Engineering, Nanjing University of Science and Technology, Nanjing, Jiangsu, 210094, China

Abstract. This paper evaluates the validity of component mode synthesis (CMS) technique for dynamic response of elastic-plastic structures. A method incorporating elastic-plastic material behaviour and the fixed interface mode synthesis technology is presented. To do this, the incremental governing equations of elastic-plastic transient problem derived from nonlinear finite element theory are linearized in each time step, the tangent stiffness matrix is used to generate the basis vector of transformation matrix. Using the proposed method, the response of a simply supported beam loaded impulsively is examined. The Numerical results indicate that the proposed method is accurate, convergent and more efficient compared with the full order finite element model.

1 Introduction

The dynamic analysis of elastic-plastic structure is a complicated nonlinear mechanical problem, which is typically solved using numerical method (e.g. finite element method). Because of the nonlinear constitutive relation, the time step for solving should be small enough to ensure the accuracy and reliability of solutions. If plastic deformation occurs during the time step, further iterations are required to obtain convergent results [1]. Thus, the computational cost of performing elastic-plastic dynamic response is high, especially for large or complex structures. In order to overcome this problem, the applications of reduced order models in nonlinear dynamic problems are gaining a lot of attention [2].

A commonly used technique for the dynamic responses analysis of larger or complex structures in linear system is the CMS method. This technique can reduce the size of system by dividing the structure into several subcomponents and neglecting the higher modal frequencies of subcomponents, without significantly degrading the accuracy of the solution. The CMS is originally proposed by Hurty [3], modified by Craig [4], then its theories have been further perfected, and the widespread applications are presented also. For more details, the reader can be refereed to a recent review [5].

However, the applications of CMS of linear system have been presented for many years, but of nonlinear system are still in its infancy [5]. The solutions of contact impact problem with CMS were first proposed by Wu and Haug [6], and then more deep studies are developed in [7, 8]. In certain problems, the nonlinearities can be approximated as an internal force. With the help of this idea the CMS method was used to reduced modelling for geometrical nonlinear problem [9] ,material nonlinear problem[10] and some local nonlinear problem[11]. Recently, some improved CMS for nonlinear system have been a concerned topic [12].

This work is focused on investigating the validity of CMS method for dynamic analysis of elastic-plastic structures. By considering plastic behaviour of structures, one widely used CMS method named fixed-interface mode synthesis technique is enhanced. Firstly, the incremental governing equations of elastic-plastic dynamic problem are derived from nonlinear finite element theory and be linearized in each time step. Then, the tangent stiffness matrix is used to generate the basis vector of transformation matrix for fixed-interface mode synthesis method. Using the presented method, the elastic-plastic dynamic behaviours of simple beams suffered impulsive load are examined to demonstrate the validity, accuracy and efficiency of the proposed method by comparing to some published work.

2 Formulation of CMS for elastic-plastic dynamic problem

2.1 Governing equations of nonlinear problem

Limiting the motion of elastic-plastic structures in the small strain type, with the help of nonlinear finite theory, the general dynamic equilibrium equations of the problem at time can be described as [1]

$$\mathbf{M}\ddot{\mathbf{u}}_t + \mathbf{F}(\mathbf{u}_t) = \mathbf{P}_t \tag{1}$$

Where $\mathbf{M}$, $\mathbf{F}(\mathbf{u}_t)$, $\mathbf{P}_t$, $\mathbf{u}_t$ and $\ddot{\mathbf{u}}_t$ are mass matrix, resisting force vector, applied force vector, nodal displacement vector and nodal acceleration vector of structure, respectively. The $\mathbf{F}(\mathbf{u}_t)$ is calculated by

$$\mathbf{F}(\mathbf{u}_t) = \int_V \mathbf{B}^{\mathrm{T}} \boldsymbol{\sigma}_t \, \mathrm{dv} \tag{2}$$

Where $\mathbf{B}$ is the strain-displacement matrix of element. The $\boldsymbol{\sigma}_t$ stands for the stress vector at time t of structure. The detailed derivation of equilibrium equation can be found in [1]. In Eq.(1), other items such as damping, can be included, but is not pertinent to the topic in this paper, so it is neglected.

After a small time interval Δt, there will be increments on system variables (e.g. stress, displacement, etc.) provided that the applied force is changed slightly at time $t+\Delta t$. And the motion of structure is still governed by Eq.(1), only by replacing variables at time t with time $t+\Delta t$. That is

$$\mathbf{M}\ddot{\mathbf{u}}_{t+\Delta t} + \mathbf{F}(\mathbf{u}_{t+\Delta t}) = \mathbf{P}_{t+\Delta t}\,. \tag{3}$$

Subtracting Eq.(1) from Eq.(3) and considering the unbalance force in Eq.(1), this yields

$$\mathbf{M}\Delta\ddot{\mathbf{u}} + \mathbf{F}(\mathbf{u}_{t+\Delta t}) - \mathbf{F}(\mathbf{u}_t) = \mathbf{P}_{t+\Delta t} - \left(\mathbf{M}\ddot{\mathbf{u}}_t + \mathbf{F}(\mathbf{u}_t)\right) = \Delta\mathbf{Q}\,. \tag{4}$$

This equations will dominate the motion of structure during the time interval $t \sim t+\Delta t$. Typically, the resisting force vector $\mathbf{F}(\mathbf{u}_{t+\Delta t})$ can be differentiable in the neighborhood of deformed shapes $\mathbf{u}_t$, so that

$$\mathbf{F}(\mathbf{u}_{t+\Delta t}) = \mathbf{F}(\mathbf{u}_t) + \frac{\partial \mathbf{F}}{\partial \mathbf{u}}(\mathbf{u}_t)\Delta\mathbf{u} \tag{5}$$

Substituting Eq.(5) into Eq.(4) and defining the tangent stiffness matrix $\frac{\partial \mathbf{F}}{\partial \mathbf{u}}(\mathbf{u}_t) = \mathbf{K}_t$, then

$$\mathbf{M}\Delta\ddot{\mathbf{u}} + \mathbf{K}_t\Delta\mathbf{u} = \Delta\mathbf{Q} \tag{6}$$

The Eq.(6) is the incremental governing equations of motion of elastic-plastic structures. Here, we employed the Newmark-β method in conjunction with MNR iterative technique for solving Eq.(6). By doing so, the Eq.(6) is linearized between time steps. The tangent stiffness matrix of structure at the beginning of each time step is updated and kept unchanged in a time step. To reduce the computational effort in solution of Eq.(6), we consider using CMS method to generate a reduced order model for iterations.

2.2 Substructuring and modal analysis

The essence of the CMS method is the coordinate transformation, where a reduced space is generated by a set of structure modes (e.g. exact eigenmodes, static modes, interface modes, etc.) called component modes for representing the motion of structures. In this paper, we will adopt and improve fixed-interface mode synthesis approach for nonliear analysis.

The dynamic analysis of elastic-plastic structures is typically a strong nonlinear problem, and does not obey the principle of superposition. It is not possible to use CMS method directly. In order to overcome this difficult, we learn from Nickell's idea, which was originally proposed to explore the use of mode superposition method for nonlinear problems [13]. That is, the forced motion of nonlinear system over a small time interval can be represented with the nonlinear (tangent stiffness) frequency spectrum. As a result, we utilize the nonlinear frequency spectrum to generate the reduced modal space for fixed-interface mode synthesis method, and the original method is improved to be able for the nonlinear problem.

By dividing the full model into several substructures, the un-damped incremental governing equations of a substructure during the time interval $t \sim t+\Delta t$, are written in the partitioned form

$$\begin{bmatrix} \mathbf{m}_{ii}^{(s)} & \mathbf{m}_{ib}^{(s)} \\ \mathbf{m}_{bi}^{(s)} & \mathbf{m}_{bb}^{(s)} \end{bmatrix} \begin{bmatrix} \Delta\ddot{\mathbf{u}}_i^{(s)} \\ \Delta\ddot{\mathbf{u}}_b^{(s)} \end{bmatrix} + \begin{bmatrix} (\mathbf{k}_t^{(s)})_{ii} & (\mathbf{k}_t^{(s)})_{ib} \\ (\mathbf{k}_t^{(s)})_{bi} & (\mathbf{k}_t^{(s)})_{bb} \end{bmatrix} \begin{bmatrix} \Delta\mathbf{u}_i^{(s)} \\ \Delta\mathbf{u}_b^{(s)} \end{bmatrix} = \begin{bmatrix} \Delta\mathbf{q}_i^{(s)} \\ \Delta\mathbf{q}_b^{(s)} \end{bmatrix}, \tag{7}$$

Where the subscript i indicates the interior degrees of freedom (DOFs), the subscript b indicates the boundary DOFs. The superscript (s) denotes the serial number of the substructure.

The fixed-interface mode synthesis approach uses two sets of substructure modes to generate transformation matrix, namely, the fixed-interface normal modes $\boldsymbol{\Phi}_N^{(s)}$ and constrain modes $\boldsymbol{\Phi}_C^{(s)}$. Fixed-interface normal modes are defined as the free vibration modes of a restrained substructure, in which all boundary degrees of freedoms are constrained and the displacement are set to zero, i.e., $\Delta\mathbf{u}_b^{(s)} = \mathbf{0}$. Thus, the Eq.(7) is reduced to an eigenvalue problem

$$\left((\mathbf{k}_t^{(s)})_{ii} - w^2\mathbf{m}_{ii}^{(s)}\right)\boldsymbol{\Phi}_{ii}^{(s)} = \mathbf{0} \tag{8}$$

The constraint modes are the static deflections of substructure that result due to successive unit displacements at each boundary degrees of freedoms, all other boundary degrees of freedoms being totally constrained. Thus, the constraint modes of substructure can be calculated by as

$$\boldsymbol{\Phi}_C^{(s)} = \begin{bmatrix} \boldsymbol{\psi}_{ib}^{(s)} \\ \mathbf{I}_{bb}^{(s)} \end{bmatrix} = \begin{bmatrix} -\left((\mathbf{k}_t^{(s)})_{ii}\right)^{-1}\mathbf{k}_{ib}^{(s)} \\ \mathbf{I}_{bb}^{(s)} \end{bmatrix} \tag{9}$$

Remaining the lowest fixed-interface normal modes $\boldsymbol{\Phi}_{ik}^{(s)}$, the coordinate transformation of substructure can be written as

$$\Delta\mathbf{u}^{(s)} = \begin{bmatrix} \boldsymbol{\Phi}_{ik}^{(s)} & \boldsymbol{\psi}_{ib}^{(s)} \\ \mathbf{0} & \mathbf{I}_{bb}^{(s)} \end{bmatrix} \begin{bmatrix} \Delta\mathbf{a}_i^{(s)} \\ \Delta\mathbf{u}_b^{(s)} \end{bmatrix} \tag{10}$$

Where $\Delta\mathbf{a}^{(s)}$ is denoted as the modal displacement vector. The total number of which is equal to the number of retained fixed-interface modes plus the number of boundary DOFs.

2.3 Synthesis of whole system

Finally, all the substructures are assembled to form the full system expressed by modal coordinates, which has a reduced order of DOFs. The process of assembling is similar to finite element analysis. With the consideration of displacement compatibility and force equilibrium of the interface DOFs at the conjunction region of substructures, and eliminating repeated interface DOFs, the dynamic equilibrium equations of full system can be expressed as follows

$$\bar{\mathbf{M}}\Delta\ddot{\mathbf{a}}+\bar{\mathbf{K}}_t\Delta\mathbf{a}=\mathbf{\Phi}^{\mathrm{T}}\Delta\mathbf{P}=\Delta\mathbf{Q} \tag{11}$$

Where $\mathbf{\Phi}$ is the modal matrix of full system assembled from all substructures. The $\bar{\mathbf{M}}=\mathbf{\Phi}^{\mathrm{T}}\mathbf{M}\mathbf{\Phi}$, $\bar{\mathbf{K}}_t=\mathbf{\Phi}^{\mathrm{T}}\mathbf{K}_t\mathbf{\Phi}$ and $\Delta\mathbf{Q}=\mathbf{\Phi}^{\mathrm{T}}\Delta\mathbf{P}$ are modal mass matrix, modal tangent stiffness matrix and modal external force vector of full system, respectively.

3 Numerical example

Considering a simply supported beam under distributed load $p(t)$ as depicted in Fig. 1. This problem has been investigated by many researchers [14-15].The dimensions of the beam are shown as follow: L=0.762 m, h=0.0508 m, w=0.0254 m. The material of this beam is assumed to be ideal plastic with E=206.8 GPa, σ_y=344.74 MPa, ρ = 7800 Kg/ m^3.

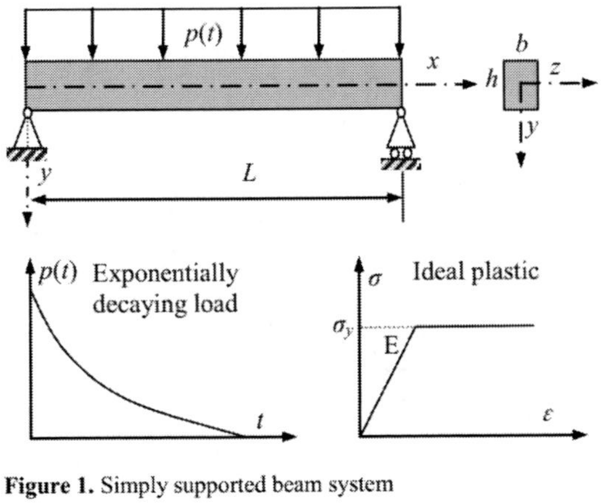

Figure 1. Simply supported beam system

The time variation of the exponentially decaying load is given as $p(t)=np_0\mathrm{e}^{-2t/\mathrm{T}}$. where p_0 is static collapse load defined as the load causing a plastic hinge to occur at the midpoint of the beam. Then n=2.0 and T is the fundamental time period of the beam. The time step of this example equals 1.0e^{-5} sec.

3.1 Convergence and accuracy study

First, a convergence study was developed based on the peak deflection of midpoint of the beam for exponentially decaying load using the proposed method. With the uniform division, each substructure along the beam is discretized into 2 Euler beam elements, and the fixed-interface normal modes of substructure retains 1. The convergent plot of substructures is depicted in Figure 2. The first peak value of midpoint non-dimensional deflection is 8.3 reported in reference [14]. The non-dimensional deflection δ is obtained by normalizing the deflection with respect to Δ, the static elastic deflection caused by the load p_0. As the number of substructures is increased, the accuracy of the solution for the midpoint deflection is improved obviously as shown in Figure 2.

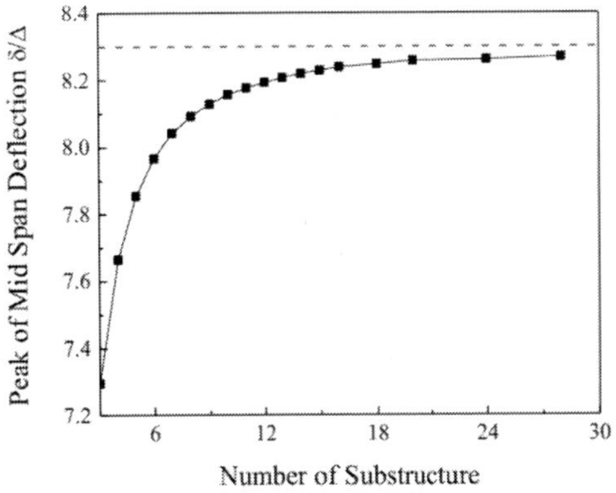

Figure 2. Convergence study of CMS method

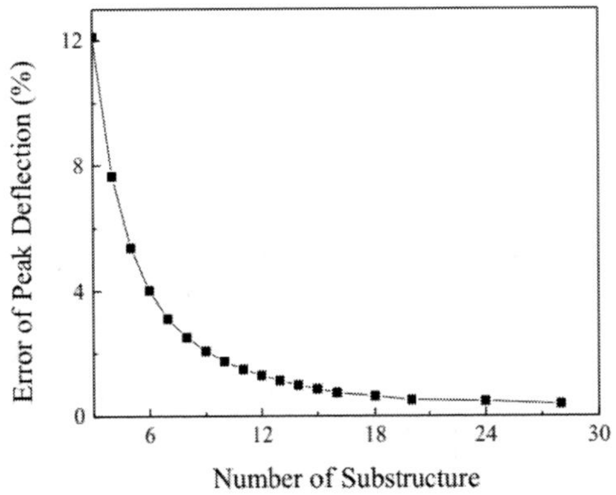

Figure 3. Relative error of the CMS method

The relative error of the solution is calculated by $(\delta_{\mathrm{CMS}}-\delta_{\mathrm{Ref}})/\delta_{\mathrm{Ref}}$, where δ_{CMS} is the numerical solution of the first peak value of deflection from the proposed method, δ_{Ref} is the first value of deflection reported in reference [14]. The relationship between relative error and substructure number is shown in Figure 3. It gives the information that the relative error is decreased from 14 percent with 2 substructures to 2 percent with 8 substructures rapidly, at the begining of the curve. After that, with the increase in the number of substructures, the relative error is reduced significantly slow. For instance, the relative error is 0.378 percent when the substructure number is increased to 28. The Figure 3 also indicates that a solution with <2% relative error can be obtained at the division of 10 substructures.

Then, for the succeeding simulations in this example, the division strategy of 10 substructures with 1 fixed-interface normal modes will be used always. The time history of midpoint deflection of the beam subjected to exponentially decaying load is compared with previous published work in Figure 4. The comparisons show that the result of the proposed method agree well with those of Baron [14], and Nagarajan [15].

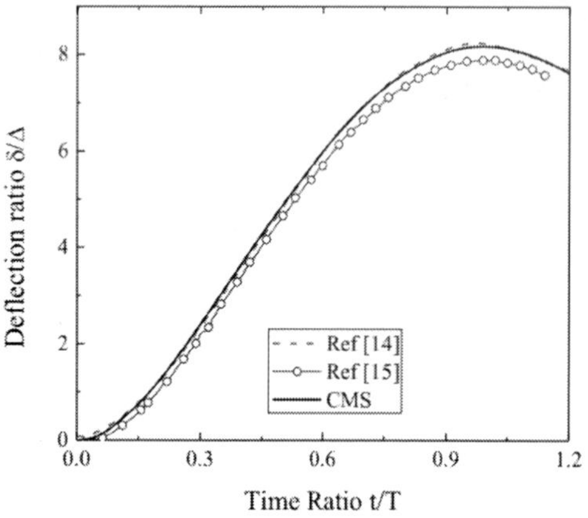

Figure 4. Deflection response of the simply supported beam

3.2 Computational efficiency study

For a concrete elastic-plastic dynamic analysis, it is obviuous that the description with Eq.(11) has much smaller size than the description with Eq.(6). Nevertheless, an additional coordinate transformation is needed to generated Eq.(11) compared with Eq.(6). The time cost of elastic-plastic dynamic analysis should include two parts, the time for solving governing equations and the time for generating governing equations. In order to evaluate the efficiency of the proposed method, we record the time cost of simulating this example with the proposed method and with full order model (description with Eq.(6)), respectively. Although, this example is a simple beam structure, the result still can explain some characteristics of the proposed method.

Table 1. Time cost of simulation.

Method	Discretization	Time cost (s)	Time ratio
Full model	N_e=20	10.72	1.00
CMS	N_s=10	7.15	0.66

The central processing unit (CPU) time for simulation and obtained benefit in different analysis are listed in Table 1. It is clear that the CMS method has higher computational efficiency than full order model. For this example, the computational time cost of proposed method is only about 0.66 times as the full order model.

4 Conclusions

In order to evaluate the validity of CMS for dynamic responses of elastic-plastic structures, a numerical method incorporating elastic-plastic theory and fixed-interface component mode synthesis technique is proposed. To do this, the incrementally governing equations of elastic-plastic dynamic problem are linearized over a small time step, the tangent stiffness matrix is used to generate the transformation matrix . The presented method is demonstrated by the analysis of dynamic elastic-plastic response of simply supported beams. The material of beam is assumed to be ideal plastic and the load is exponentially decaying. Numerical results are in excellent agreement with those obtained by previous researchers. The comparison of the computational time for example also demonstrates that the proposed method has higher computation efficiency compared with finite element method.

Acknowledgment

This research has been supported by Postdoctoral Fund of JiangSu (1402007B), the support is gratefully acknowledged.

References

1. O. C. Zienkiewicz, R.L.T, *The Finite Element Method for Solid and Structural Mechanics*, (2006).
2. A. Corigliano, M. Dossi, S. Mariani, COMPUT METHOD APPL M, **290**, 127-155, (2015).
3. W. C. Hurty, AIAA J, **3** 678-685. (1965).
4. R. R. Craig, M. C. C Bampton, AIAA J, **6**, 1313-1319. (1968)
5. D.d Klerk, D. Rixen, S. Voormeeren, AIAA J, **46**, 1169-1181, (2008).
6. S. C. Wu, E. J. Haug, J MECH DESIGN **112**, 390-398, (1990).
7. A. Guo, S. Batzer, J VIB ACOUST **126**, 126-131, (2004)
8. Y, Shen, X. Yin, APPL MATH MODEL, **40**, 1972-1988, (2016).
9. R. J. Kuether, M. S. Allen, *Dynamics of Coupled Structures*, **1**, 167-178, (2014).
10. J. Bond, T. Khraishi, INT J MECH MATER DES, **5**, 365-380, (2009).
11. D-K, Kim, J-S. Bae, I. Lee, J-H. Han, J SPACECRAFT ROCKETS, **42**, 66-77. (2005)
12. I. P. Krishna, C. Padmanabhan, NONLINEAR DYNAM, **63**, 561-586. (2011)
13. R. E. Nickell, COMPUT METHOD APPL M, **7**, 107-129. (1976)
14. M. L. BARON, H. H. B, P. WEIDLINGER. J ENG MECH, **89**, 23-42. (1961)
15. S. Nagarajan, E. P. Popov, COMPUT STRUCT, **4**, 1117-1134, (1974)

MATEC Web of Conferences 95, 01011 (2017) DOI: 10.1051/matecconf/20179501011
ICMME 2016

The Stress and Stiffness Analysis of Diaphragm

Dongyue Qu , Xiuming Li, Fanbing Liu and Chunhua Zhao

School of Harbin Engineering University, Harbin 150000, China

Abstract. Diaphragm coupling with its simple structure, small size, high reliability, which can compensate for its input and output displacement deviation by its elastic deformation, is widely used in aerospace, marine, and chemical etc. This paper uses the ANSYS software and its APDL language to analysis the stress distribution when the diaphragm under the load of torque, axial deviation, centrifugal force, angular deviation and multiple loads. We find that the value of maximum stress usually appears in the outer or inner transition region and the axial deviation has a greater influence to the distribution of the stress. Based on above, we got three kinds of stiffness for axial, angular and torque, which the stiffness of diaphragm is nearly invariable. The results can be regard as an important reference for design and optimization of diaphragm coupling.

1 Introduction

Diaphragm coupling belongs to metal flexible coupling (fig 1), which is made up of left-right diaphragm, input-output shaft and middle shaft, not only can transfer big torque but also deviate the axial, angular and radial displacement by its elastic deformation of profile.

The working principles of the Diaphragm coupling are described below: the torque is inputted from the input shaft by the circumferentially spaced strength bolts. Then, the torque is transmitted to the middle term of the Diaphragm coupling. Besides, the torque is outputted by the same method at another side, which like the input shaft. What's more, diaphragm is a significant component of the metal flexible coupling. When the coupling works, the profile of diaphragm not only transmits the torque between the input and output but also compensate the displacement deviation by its elastic deformation [1], etc. Usually, the diaphragm coupling suffers centrifugal force, torque, and axial deformation loads. The design of profile plays an important role in design of the diaphragm coupling. From that, by taking the conical diaphragm as an example, we analysis its stress distribution and stiffness, getting its rule of stress distribution at the different loads, which can provide a theoretical basic for the design and optimization of the diaphragm.

For the characteristics of Diaphragm coupling, many scholars at home and abroad made a wide range of research. The research contents mainly involves coupling analysis of static and dynamic, which include the stress distribution under different load, the modal analysis and the analysis of the life [2] etc. From that, they established the formulas of stress and stiffness.

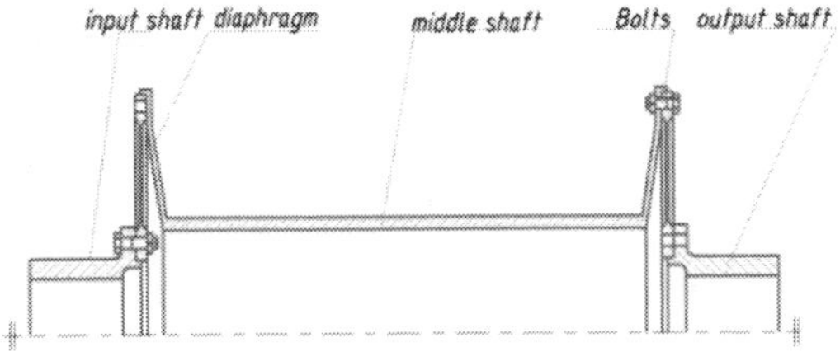

Figure 1. Diaphragm coupling

2 The finite element model

The common thickness calculation formula of diaphragm profile is described below:

$$T = C/R^n \tag{1}$$

The profile of diaphragm is corresponding to linear, conical and hyperbola diaphragm [3,4] when the value of the index is 0, 1, 2 severally. This paper takes the conical diaphragm as research object. In order to facilitate analysis, the model was simplified, which we apply the APDL (ANSYS parametric design language) to create parametric model in ANSYS. Besides, taking the calculation accuracy in consideration, we apply the element type of Solid185 unit to mesh the model and refine the grid in transitional region where the hexahedral element is used as shown in figure 2,3.

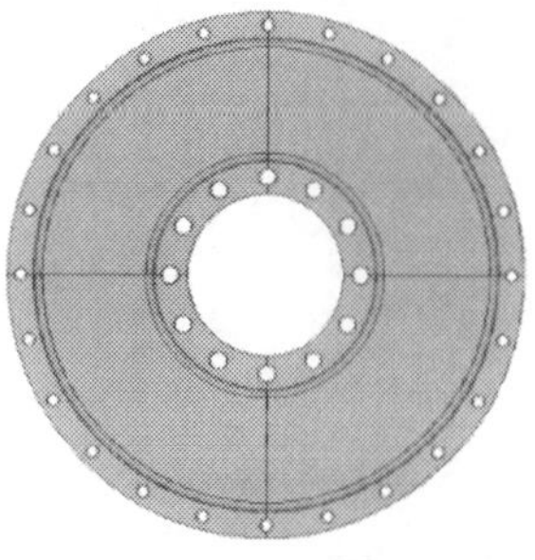

Figure 2. Geometry model of conical diaphragm

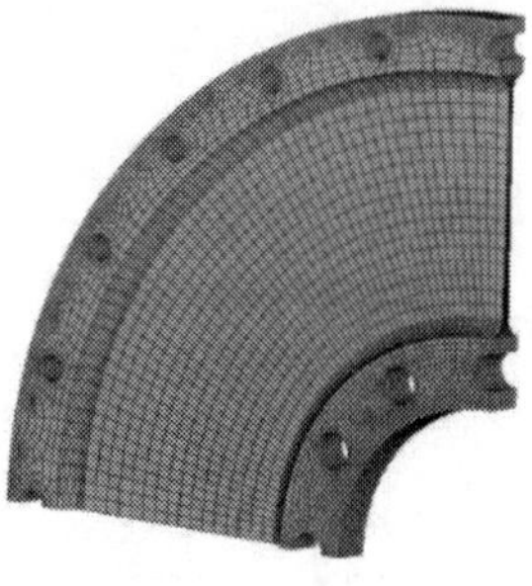

Figure 3. Local grid of conical diaphragm

3 Stress analysis of the diaphragm

3.1 The stress value distribution of diaphragm under the action of torque

Diaphragm suffers the torque that its value is 7958 Nm when it works. Besides, the torque is equally distributed to the hole surface of the bolt after the torque are translated in to the pressure which according to cosine distribution. What's more, taking the action of torque in to consideration only, we fix the radial and axial displacement of the inner holes. Furthermore, all constraints are applied in outer holes. After solving, the stress contours of diaphragm and stress in profile along with the y axis distribution curve are shown.

As shown in figure 4,5, we can know that the value of maximum stress is 247Mpa when the Diaphragm suffer the torque only. The stress value in inner diameter of profile is about 239Mpa. With the value of radius increase, the stress value decreases gradually and changes slowly.

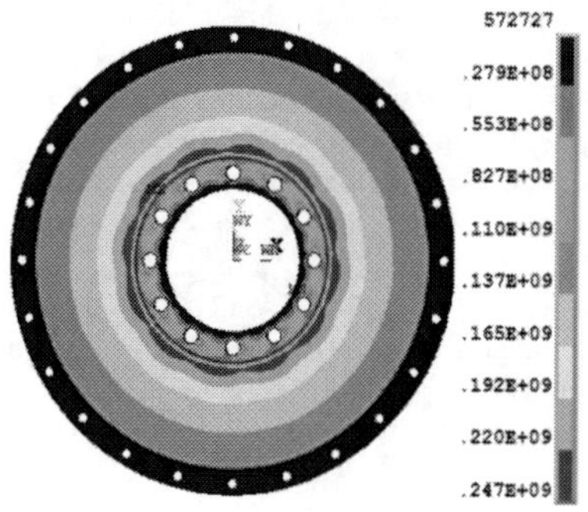

Figure 4. The stress caused by torque

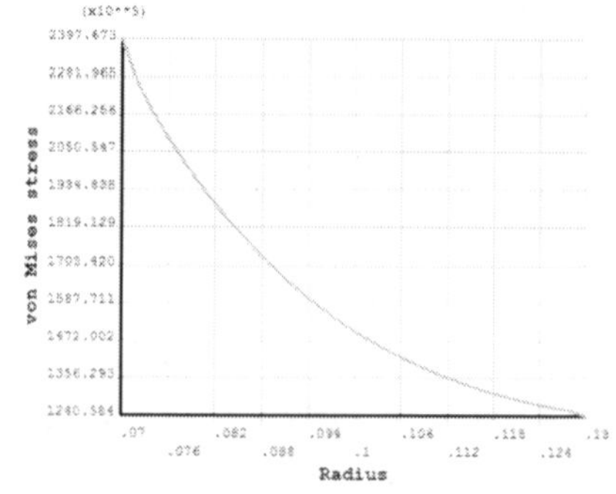

Figure 5. The stress distribution curve of the profile

3.2 Stress value of diaphragm under the action of Centrifugal force

We can know that the rotate speed of Diaphragm is 7200 rpm when Diaphragm works. Besides, we apply the load in the form of angular velocity, meanwhile we fix the angular and axial displacement of the all holes [5]. After solving, we can get the stress contours of diaphragm and stress of profile along with the y axis distribution curve.

As shown in the figure 6,7, we can know that the value of maximum stress is 41.1Mpa, and the stress value in outer diameter of profile is about 30.73Mpa. With the value of the radius increase, the stress showed a trend that decreases firstly and then increase. Compared with the stress of torque load, it has smaller impact to the stress of the diaphragm.

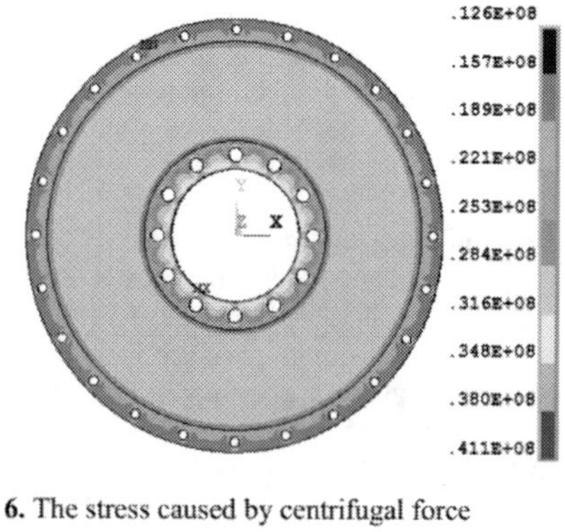

Figure 6. The stress caused by centrifugal force

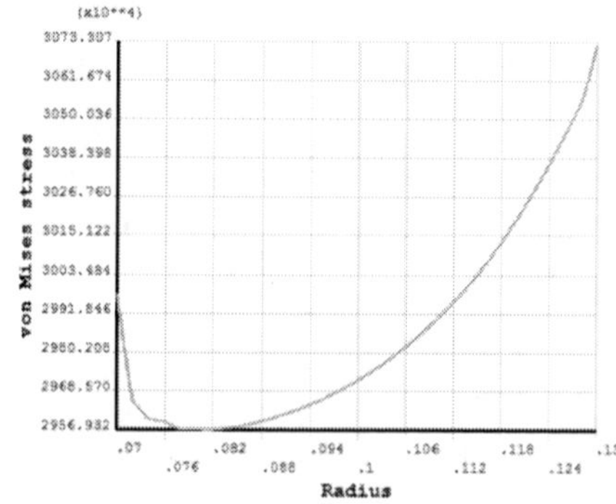

Figure 7. The stress distribution curve of the profile

3.3 Stress value of diaphragm under the action of Axial Deviation

We can know that the axial deviation value of Diaphragm is 3 mm which is applied in inner bolt holes when Diaphragm works. In the process of analysis, we fix all constraints of the all outer holes [5]. After solving, we can get the stress contours of diaphragm and stress of profile along with the y axis distribution curve.

As shown in the figure 8,9, we can know that the value of the maximum stress is 515Mpa when diaphragm suffer axial deviation only. Besides, the stress in inner and outer diameter of profile is about 381Mpa. With the value of the radius increase, the value of stress displays a trend that decreases firstly and then increase. What's more, the stress value in inner and outer radius is almost equal. Compared with the load in the front, the axial deviation has greater impact to the stress value of the diaphragm.

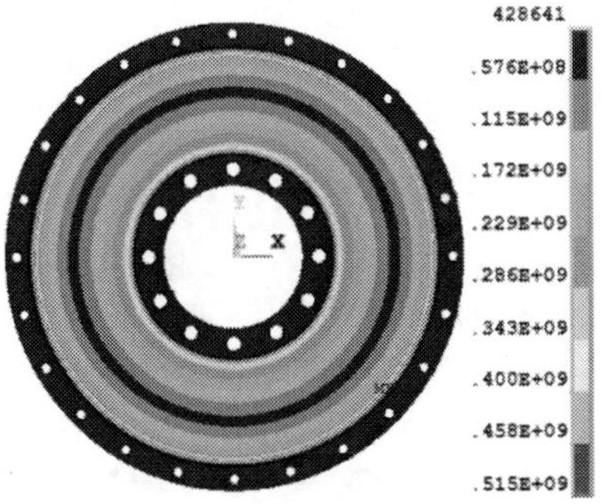

Figure 8. The stress caused by axial deviation

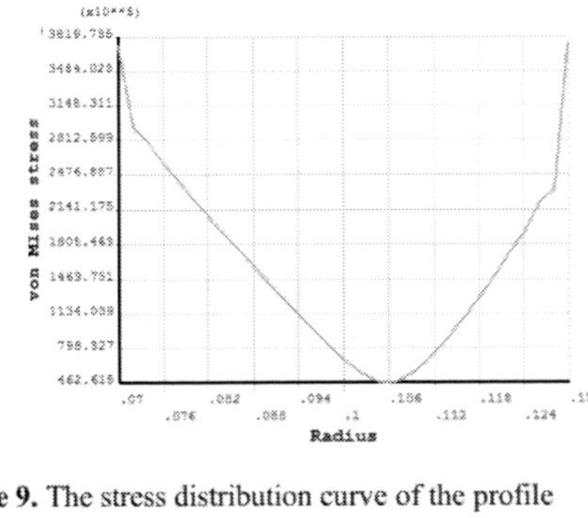

Figure 9. The stress distribution curve of the profile

3.4 Stress value of diaphragm under the action of Angular Deviation

Suppose that the diaphragm withstand value of angular deviation is 0.3°. We transform it to the axial displacement constraint which is applied in the inner bolt holes. All constraints of the outer bolt holes are fixed. After solving, the stress contours of diaphragm and stress of the profile along with the y axis distribution curve are shown

As shown in the figure 10, 11, we can know that the value of maximum stress is 82Mpa when the diaphragm suffer angular deviation only. The stress value in inner diameter of working profile is about 72.23Mpa. With the value of the radius increase, the stress value shows a trend that decreases firstly and then increase.

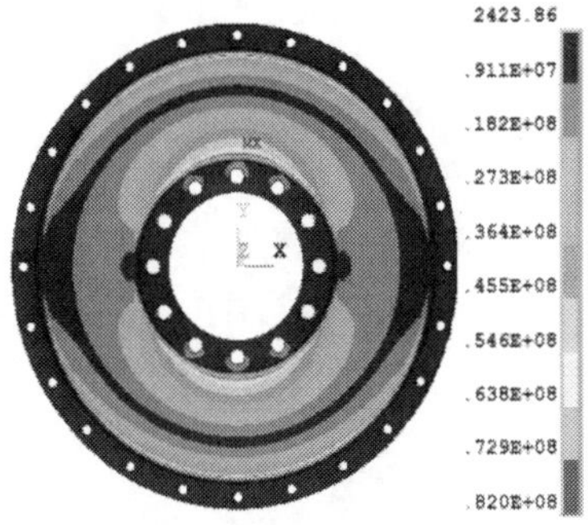

Figure 10. The stress caused by angular deviation

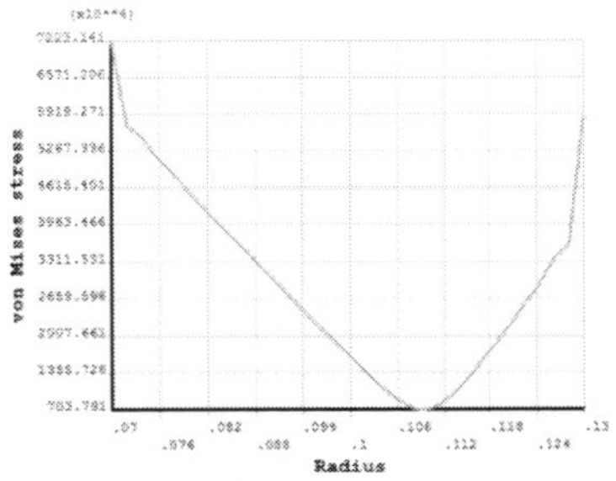

Figure 11. The stress distribution curve of the profile

3.5 Stress value of diaphragm under the action of variety loads

We apply four kinds of aforementioned loads at the same time. By fixing the radial and axial displacement constraint of the outer holes, we can get the stress contours of diaphragm and stress value in working profile along with the Y axis distribution curve after solving.

As shown in the figure 10,11, we can know that the value of the maximum stress is 598Mpa when diaphragm suffer multiple loads. The stress value in inner diameter of working profile is about 510Mpa. With the radius value increase, the stress value showed a trend that decreases firstly and then increase. The distribution form of stress value is the same as the stress caused by axial deviation.

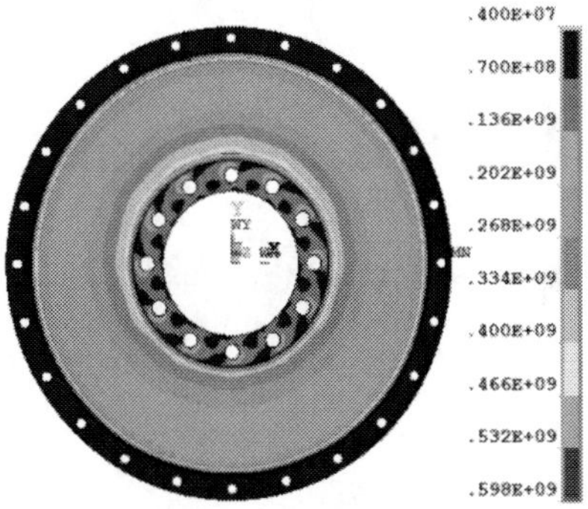

Figure 12. The stress caused by variety loads

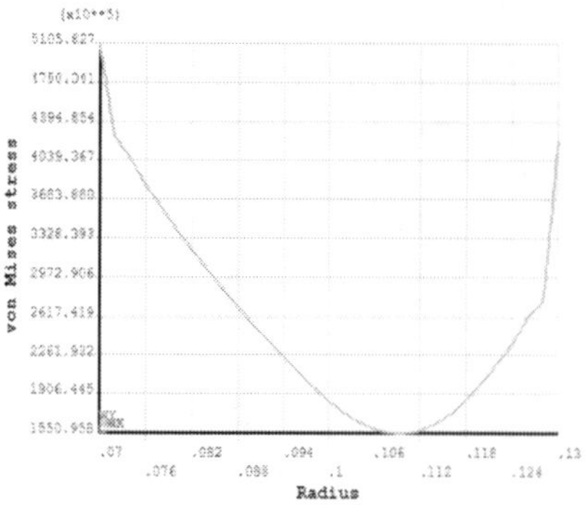

Figure 13. The stress distribution curve of the profile

Based on above, the max stress in profile and its takes place is shown in Table 1.

Table 1. The max stress in profile and its takes place

	torque	Centrifugal force	Axial Deviation	Angular Deviation	variety loads
Stress	247	41.1	515	82	598
place	inside	outside	outside	inside	outside

4 Stiffness analysis of diaphragm

Diaphragm will produce the torsional stiffness when suffer the load of torque, and axial stiffness when suffer axial deviation. Besides, Diaphragm will produce angular stiffness [6,7] when suffer the angular deviation. In this section, we will study on the change of the stiffness when Diaphragm suffers different loads. From that, we can get three kinds of stiffness curve under the different loads are showed in fig 14 to fig 16.

As shown in the figures, we can know that the deflection angle of diaphragm increase linearly and the slope of curve is torsional stiffness, its value is about 7.04 $\times 10^7$ Nm/Rad when the diaphragm suffers the different torque load. Besides, the axial reaction force increases linearly and the value of axial stiffness is about 1.27 $\times 10^6$ N/m when the diaphragm suffers the load of axial deviation [7]. What's more, the torque reaction increases linearly and the value of angular stiffness is about 7435.9Nm/Rad when the diaphragm suffers load of different angular deviation.

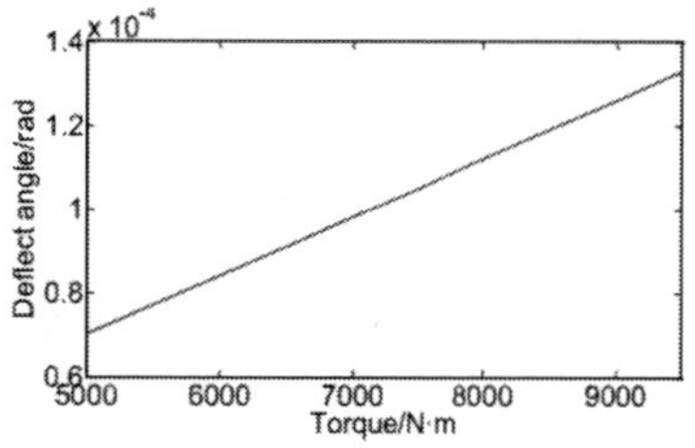

Figure 14. The torsional angle curve of diaphragm

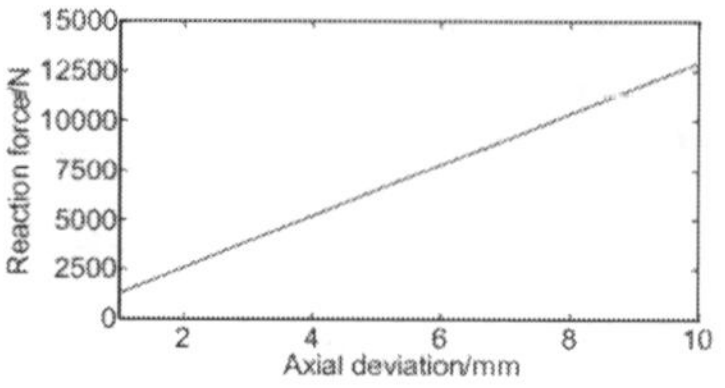

Figure 15. The axial reaction force curve of diaphragm

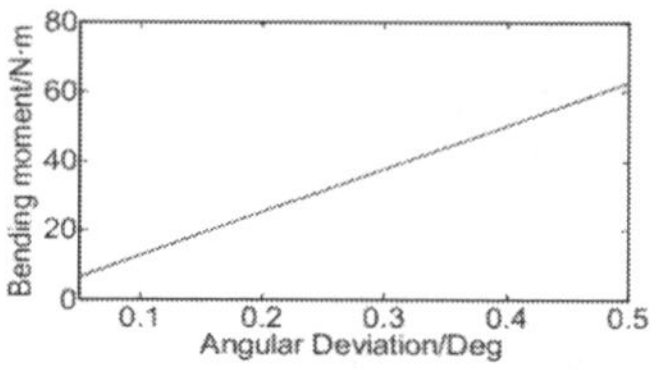

Figure 16. The bending moment curve of diaphragm

5 Conclusion

In this paper, we study on the distribution of the stress and three kinds of stiffness for conical diaphragm when the diaphragm suffers different loads. From that, the results show that the value of maximum stress usually appears in the outer or inner transition region. Compared with the other load, the axial deviation has a greater influence to the distribution of the stress which generates the greatest stress. But the diaphragm coupling usually bear the bigger stress, due to that it usually suffers the several loads that has mentioned above. In the situation that the structure parameters of the diaphragm is constant, the three kinds of stiffness for conical diaphragm is constant when the diaphragm suffer different loads.

References

1. H. Liao, Y.L.He, J. Du, Strength and Fatigue Finite Element Analysis of Diaphragm in Diaphragm Coupling, Machinery & Electronics,**05**,5(2008)
2. Z.G.Qiu, F.H.Zhang, J.H.bai, the design and FEM analysis of flexible diaphragm coupling' web curve, Machinery Design&Manufacture, **7**, 2(2010)
3. Y. Peng, G.L.Deng, L.F.Xing, Diaphragm strength calculation method, Ship Science and Technology, **35**,5(2003)
4. Y. Peng, Y.Zhao, X.X.Liu, Analysis of different type mask plate characteristic, Ship Science and Technology,**35**,5(2013)
5. S. Liu, X.F.Wang, Analysis of axial misalignment based on ANSYS of flexible couplings, Mechanical Research and Application, **16**,2(2003)
6. L.F.Shentu,X.Y.Zhang, Analysis Axial displacement compensation capacity of Joint Flexible Laminated Membrane Coupling (JFLMC),**19**,2(2003)
7. W.P.Chen, Y.M.Ma, analysis axis stiffness of diaphragm coupling, ship engineering, **32**,4(2010)

DOI: 10.1051/matecconf/20179502001

Influences of the Air in Metal Powder High Velocity Compaction

Jun Liu,Haifeng Hu and Xiaolong Luo

Faculty of Mechanical Engineering & Mechanics, Ningbo University, Ningbo 315211, China

Abstract: During the process of metal powder high velocity impact compaction, the air is compressed sharply and portion remains in the compacts. In order to study the Influences, a discrete density volleyball accumulation model for aluminium powder was established with the use of ABAQUS. Study found that the powder porosity air obstruct the pressing process because remaining air reduced strength and density of the compacts in the current high-speed pressing (V≤100m/s). When speed further increased (V≥100m/s), the temperature of the air increased sharply, and was even much higher than the melting point of the material. When aluminium powder was compressed at a speed of 200m/s, temperatures of air could reach 2033 K, far higher than the melting point of 877 K. Increased density of powders was a result of local softening and even melt adhesive while air between particles with high temperature and pressure flowed past.

1 Introduction

Powder metallurgy is a near net shape technology which has a huge advantage on some materials difficult to produce in the traditional way [1]. Powder dynamic repression has been carried out for decades, and pressing speed of 200 meters per second or more, the weight of the hammer also ranged from a few grams to several tons.

During the process of HVC（High velocity impact compaction）, speed not only affects friction state and hardening of powder, but also affects the air to escape from the powder particles gap. In the study of high speed pressing, many scholars consider that the increased density is a result of excluding air in powder after several times of suppression. Wang Jianzhong [2] thinks the density should be higher due to less air remaining after holding pressure in the process of HIP （Hot Isostatic Pressing）. However, some found that air pressure in compacts was lower than the compacting pressure within two orders of magnitude in the process of HVC [3]. Chen Jin [4] found density was lower through experiments which the powder was performed high-speed suppression after pre-pressed. Ma Binbin [5] considered in certain pressing speed and the friction coefficient, powder compaction and sintering of synchronization are achievable for most metal powder.

ABAQUS is one of the most advanced finite element analysis software in the world [6]. Liquid-solid coupling approach was used was used to simulate high-speed aluminium powder pressing process by ABAQUS, and air escaping mode was analysis and discuss the impact of air during the process.

2 Model and parameter settings

There are many kind of particles shape ,and mist method can get a better atomization of spherical particles [7].For convenience ,the powder particle were set as diameter disc with 100 μ m .The model was shown in Fig 1, and the concave edge model was shown in Fig 2.The punch was on the top, and air flow field was established, then divided Euler mesh size and the die cavity Fig 3 equivalents and impart to the flow field at the aperture, so that the impact of the powder particles and air could be simultaneously analyzed at the same time besides the air was depend as an ideal air.

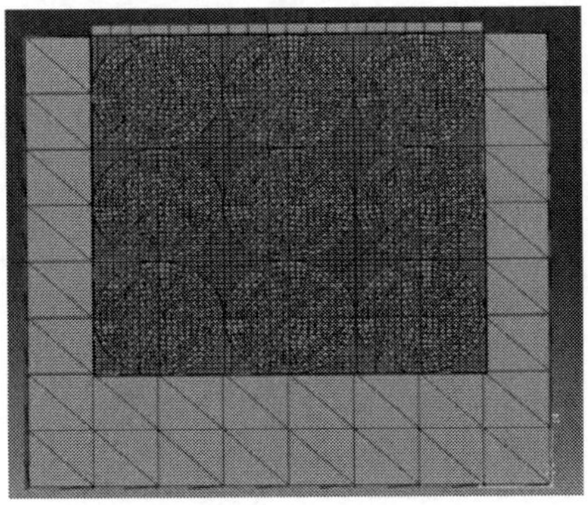

Figure 1. Powder compaction model

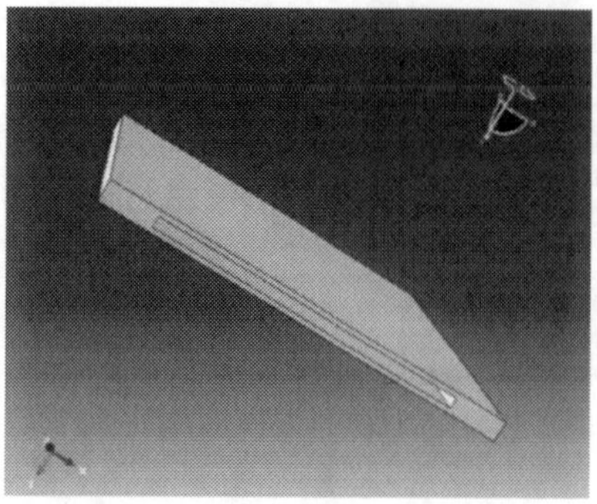

Figure 2. Powder compaction die model

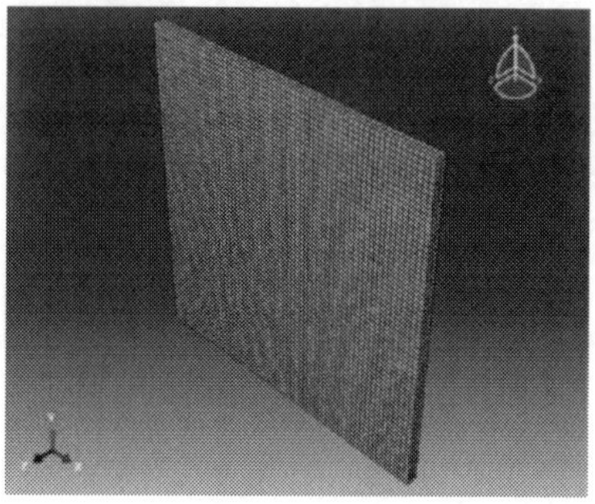

Figure 3. Air flow model of powder compacting

As a commonly kind of material with low melting point, aluminium were selected to be study .Meanwhile Johnson-Cook model can well describe the work hardening of metal materials, the strain rate effect and the softening effect as the constitutive relation of the powder particles[8]. In its general form:

$$\sigma = (A + B\varepsilon^{n}) \cdot (1 + C \ln \dot{\varepsilon}^{*}) \cdot (1 - T^{*}) \quad (2\text{-}1)$$

$$\dot{\varepsilon}^{*} = \frac{\dot{\varepsilon}}{\dot{\varepsilon}_0} \quad (2\text{-}2)$$

$$T^{*} = \frac{T - T_r}{T_m - T_r} \quad (2\text{-}3)$$

A is material in the quasi static yield strength; B is strain hardening modulus; n is strain hardening index; C is the strain rate sensitivity index; m is temperature softening coefficient; ε is equivalent plastic strain; $\dot{\varepsilon}$ is strain rate; $\dot{\varepsilon}_0$ is the reference strain rate; T is temperature material; T_r is room temperature; T_m is melting point materials.

The process was adiabatic due to a very short period of time; Dynamic display method is adopted because the stress field and the temperature field intersect each other in powder compaction .

3 Simulation results and analysis

From Fig 4 to Fig 6, Compaction temperature was 298 K, and pressing speed was 10 m / s, besides round particles represents a metal powder while the white was the air. From the results, the temperature rise of the powder particles were mainly produced by plastic deformation of the particles since the friction was not defined .As a result, the temperature at the grain contacts was higher than in the center, which was correlated with the degree of deformation of the particles. And the air in the models with repression carried out by the punch. Ultimately only a small part of the air was left between the particles and the particles enclosed at the aperture.

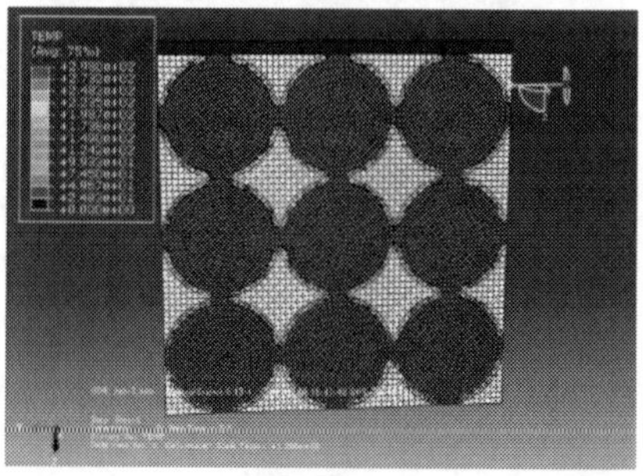

Figure 4. At 0 μs

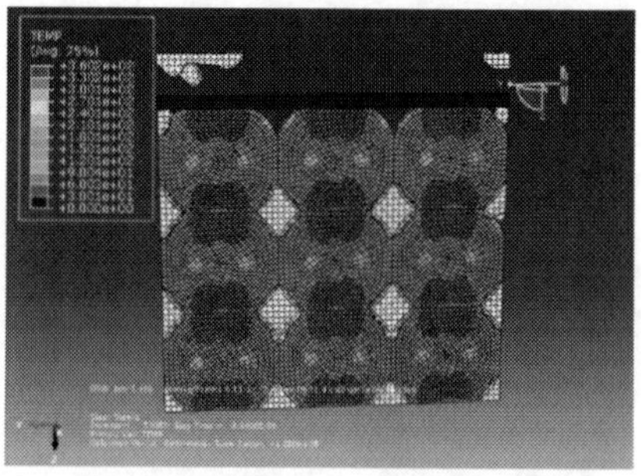

Figure 5. At 4.44 μs

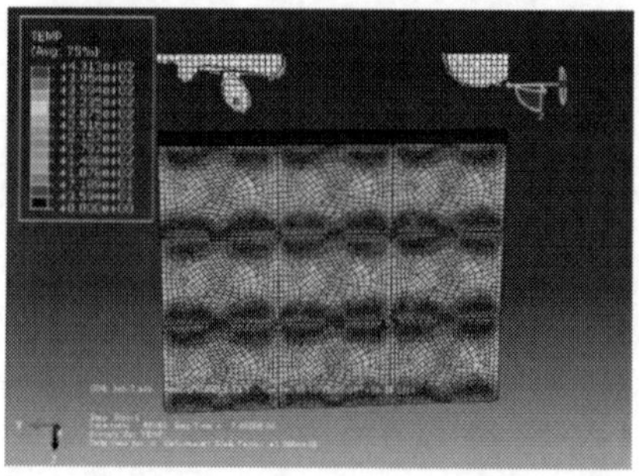

Figure 6. At 7.4 μs

The temperature rise of the air when aluminium powder was pressed at different speeds is shown from Fig 7 to Fig 10. It can be seen the temperature in the air was gradually increased with the pressing speed. At the same time, air temperature rise of inter-particle porosity contrast at 50 m /s was not significantly to 10 m / s, and the temperature was still far below the melting point of the material even at speed of 100m/s.Air temperature would rise sharply when the speed reaches 200 m/s, and the highest temperature reached at 2033 K. Fig 11 is air temperature at different speeds. The air temperature increased dramatically only when the speed was more than 100 m / s. Because the air could not remain in the powders, and most of the air would escape outside the model at a rate of tens of meters per second. The air was too late to escape as a result of abruptly deformed s between particles when a speed more than one hundred meters per second case, and temperature rapid rise with increasing pressure to reduce the volume.

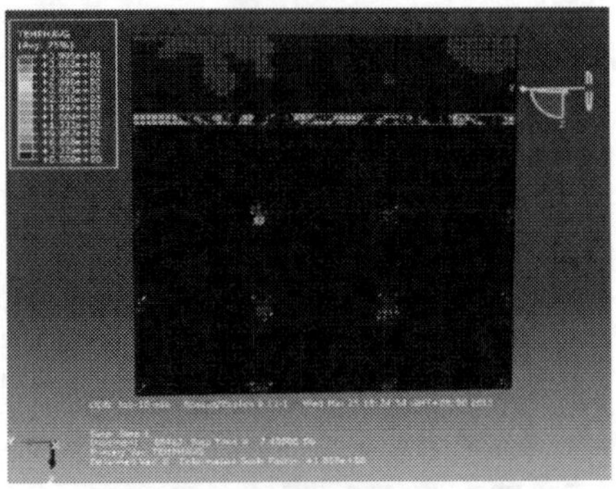

Figure 7. Temperature between the air at **10 m/s**

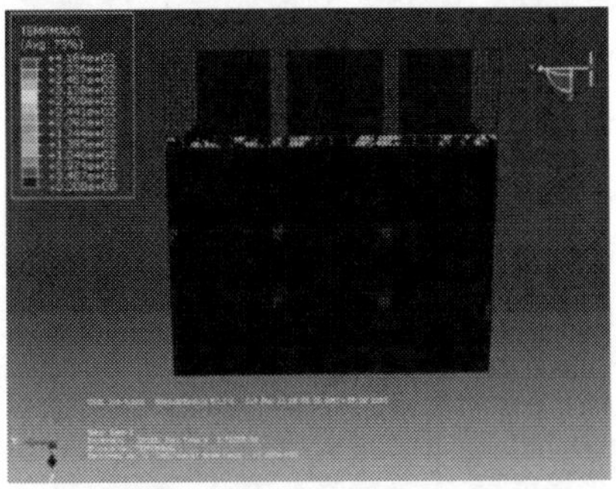

Figure 8. Temperature between the air at **50 m/s**

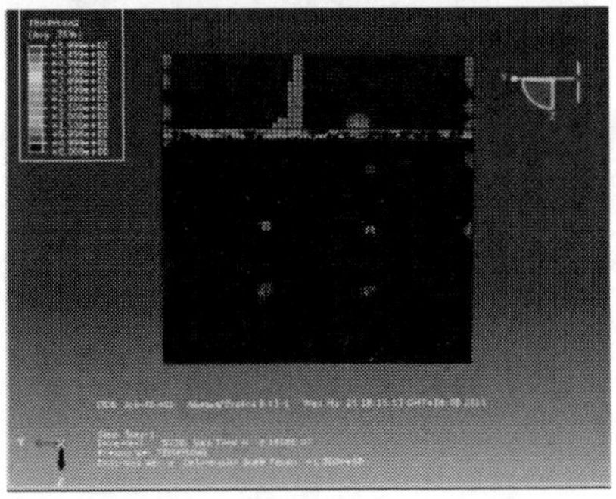

Figure 9. Temperature between the air at **100 m/s**

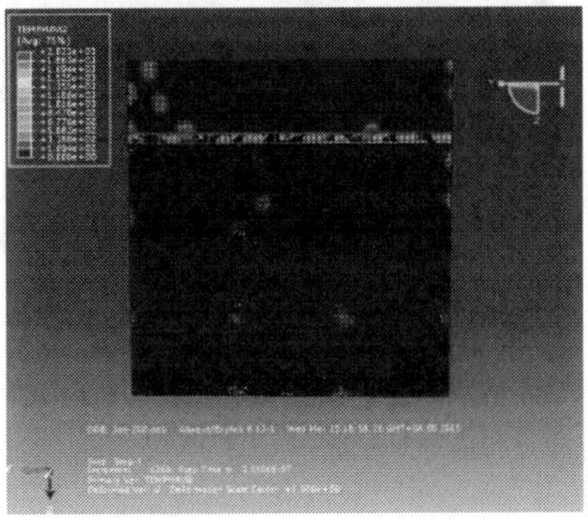

Figure 10. Temperature between the air at **200 m/s**

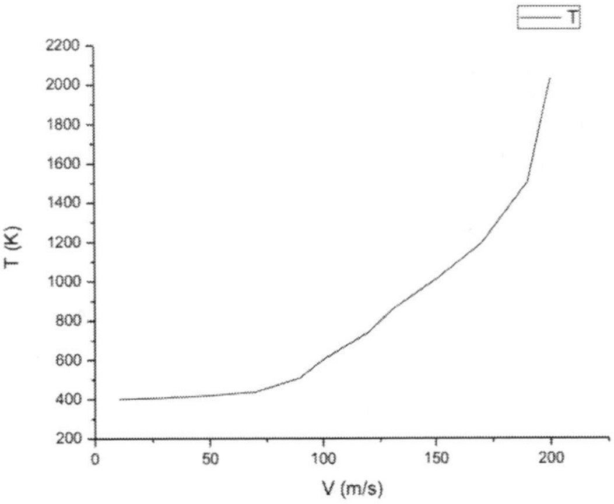

Figure 11. The maximum temperature rise of air at different speeds

The temperature rise rate of air porosity under different initial temperatures of the powder particles at 10m/s is shown in Fig 12 and Fig 13. Pressing temperature had little effect on the initial porosity of the temperature rise, and the final temperature was about 100K.

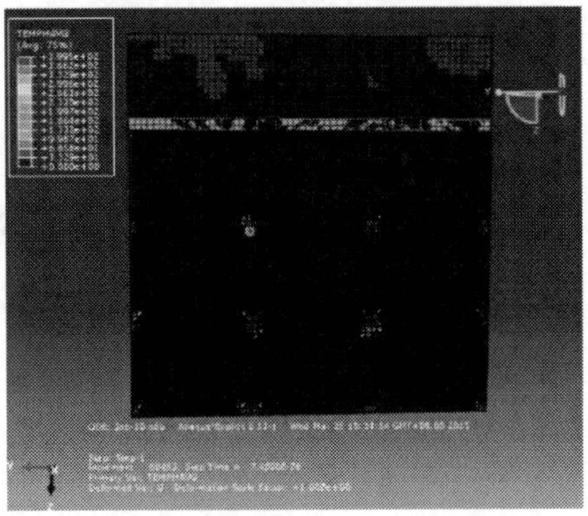

Figure 12. The temperature rise when initial temperatures of air porosity was **298 K**

 DOI: 10.1051/matecconf/20179502001

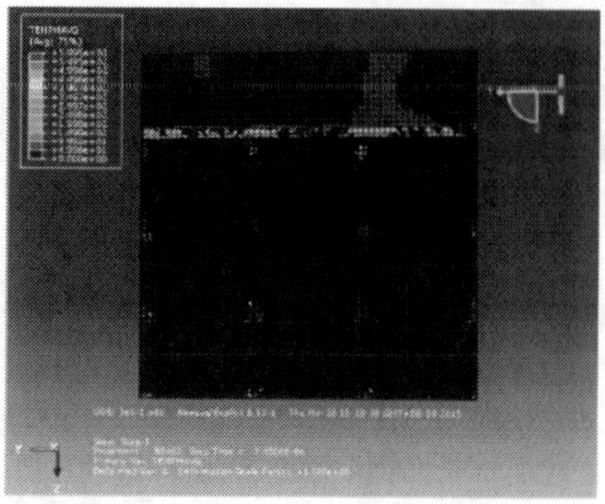

Figure 13. The temperature rise when initial temperatures of air porosity was **498 K**

As can be seen from Figure 14, air pressure finally rises again after the first and then decreased. The reason may be that the air pressure is low in the initial stage of pressing, and Pressing speed is faster than, as a result, the air pressure increases with increasing pressure. The pressure reduced with the rate of air evolution increased. As it can be seen from Fig. 15, the temperature of air rises rapidly in the initial stage, essentially stayed unchanged.

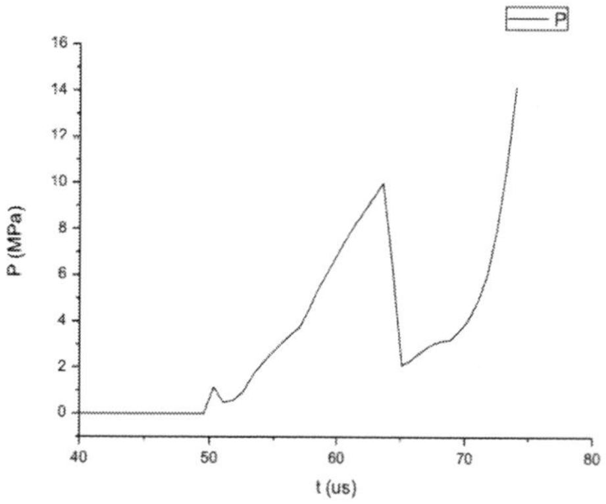

Figure 14. Pressure of air

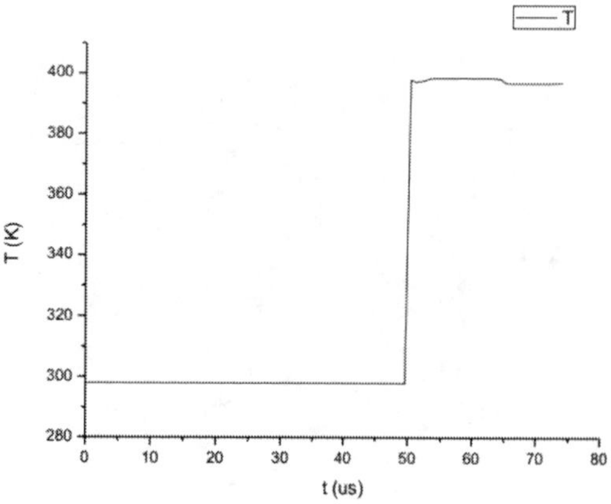

Figure 15. Temperature of air

4 Conclusion

(1) During the process of metal powder high velocity impact compaction, the air temperature rises with the repression though it has little effect on the consolidation of powder particles melt when pressing speed is about tens of meters per second. Meanwhile, holes would form with a fraction of the air remaining at the end. Therefore, air impedes pressing process in the common high-speed.

(2) Temperature of the air rises sharply when the pressing speed is higher than 100 m/s, and is even much higher than the melting point of the material. The surface of the adjacent powder particles would partial soften even melt adhesive as results of further increase in temperature at the interface of high-pressure powder. So air porosity plays a catalytic role during pressing.

(3) The initial temperature has no effect on the temperature rise of the air in a conventional powder pressing since a large amount of air can not remain in the powder because of low speed.

Acknowledgements

This research was financially supported by National Natural Science Foundation of China (No. 11372148).

References

1. Qu X H, Yi H Q. Materials China .J. 29, 45-49(2010).
2. Wang J Z. Research on densification mechanism of powder high velocity compaction (Doctoral dissertation). Beijing:University of Science and Technology Beijing,2009.
3. Sano Y, Miyagi K. Int. J. Powder Metall. Powder Technol.J.**20**,115-118 (1984).
4. Chen J. Investigation on equipment development and densification mechanism of warm powder high velocity compaction (Doctoral dissertation). Guang zhou:South China University of Technology,2011.
5. Ma B B,Hu X P,Liu J.Hot Working Technology .J.01 91-95(2016)
6. Shi Y P, Zhou Y R.ABAQUS Finite Element Analysis [M] .Beijing: China Machine Press,2006.
7. Sun L. Mesoscopic simulation on the compression deformation process of powder particles(Doctoral dissertation). Hefei: Hefei University of Technology (2013)
8. Johnson.G.R, Cook. W.H. A constitutive model and data for metals subjected to large strains, high strain rates and high temperatures[C]//7th International Symposium on Ballistics. The Netherlands: Hague, 1983. 541-547.

Influence of Process Parameters in n-PMEDM of Inconel 800 with Electrode and Coated Electrodes

K.Karunakaran[1] and M.Chandrasekaran [2]

[1] *Assistant Professor, Department of Mechanical Engineering, Vels University, Chennai, Tamil Nadu, India.*
[2] *Professor and Director, Department of Mechanical Engineering, Vels University, Chennai, Tamil Nadu, India.*

Abstract. By considering the unique performances, the Powder Mixed Electrical Discharge Machining (PMEDM) mostly encounter choice for machining the hard materials with job features like intricate shapes with grater accuracy and those materials are difficult to cut in conventional machining processes. This Research aimed to prepare parameter index chart for machining nickel based super alloy - Inconel 800 for various requirements. To achieve the aim, In this experimental study on machining of Inconel 800 in PMEDM with electrolyte copper electrode as well as silver coated electrolyte copper electrode is focused. Taguchi full factorial design derived from MINITAB release-16 software was used to design the experimentation. The factors like pulse off time, Current and Pulse on Time were considered. The responses like amount of Tool Wear, Quantity of material removal, surface finishes were noted against each case. In the same experimental conditions the coated electrode increased the MRR by 36.67% with minor increment of maximum surface roughness to 16.11%.

1 Introduction

The Inconel materials generally have high hardness, thermal strength and low thermal diffusivity and their applications are especially in high pressure and high temperature field such as gas turbine, Nuclear reactors, chemical vessels, electrical power plant equipments etc [1]. The EDM is choice for machining them to meet the definite shape and accuracy requirements because they form build-up-edge and have tendency to stick or weld with tool material and extremely difficult by using traditional processes [2]. The PMEDM extends the cutting efficiency and surface quality by sparking more rapidly and enhances the thermal conductivity surrounded by discharge take place causes the high rate of erosion from the work [3]. The Kuppan *et al.* [4] studied the influence of process parameters in in deep hole EDM drilling of Inconel 718 and developed mathematical model for the responses of depth averaged surface roughness and MRR to interpret process parameters. Rajyalakshmi and Venkata Ramaiah, [5] studied machining of Inconel 825 in Wire EDM. The MuthuKumar et al. [6] studies were on WEDM of Inconel 800 for optimimsing process parameters and later about radial overcut [7]. Mao-yong LIN et al. [8] investigation was parameter optimization in micro milling EDM of Inconel 718. In PMEDM, the Rathi and Mane [9] studied the Effect of Powders of Al_2O_3, SiC and graphite powder for machining of Inconel 718. Tzeng and Lee [10] reported that the density, particle size, Electrical and Thermal Condivity and power concentration in the mix are influencing significantly in Performance of PMEDM. The Syed et al. [11] recommended up to 2g/l of Al Powder mix in Kerosene for Machining of Aluminium-Silicon Carbide Composites. The Shriram et al. [12] used 4g/l of Al powder mix EDM for machining of Tungsten Carbide work. And many authors were recommended Al powder mix in kerosene dielectric. Here the Al powder mixed with kerosene and stirred continuously about 8 to 9 hours in the Magnetic stirrer for reducing the size of particles about to 5 Nm, before they mix in the actual dielectric. Hence this unique case of research may be referred as Nano Powder mixed EDM (n-PMEDM) of Inconel 800.

2 The Experimentation

This research is motivated by industrial requirements. The manufacturer deals various kinds of jobs of inconel 800. The practical issues like, statistical approximation and mathematical models to choose parameters often difficult to interpolate by industrialist. Sometimes the approximate solutions and manual mistakes of interpolating parameter values leads to losses.

Table 1. Factors and responses

Factors	Level 1	Level 2	Level 3
Peak Current (A)	5	10	15
Pulse on time (μs)	6	7	8
Pulse off time (μs)	4	5	6
Electrode	Plain	Coated	-
Responses	MRR, TWR and SR		

Table 2. Taguchi full factorial design of experiments

Expt. No	Current (A)	Pulse On Time (μs)	Pulse Off Time (μs)
1	5	6	3
2	5	6	4
3	5	6	5
4	10	6	3
5	10	6	4
6	10	6	5
7	15	6	3
8	15	6	4
9	15	6	5
10	5	7	3
11	5	7	4
12	5	7	5
13	10	7	3
14	10	7	4
15	10	7	5
16	15	7	3
17	15	7	4
18	15	7	5
19	5	8	3
20	5	8	4
21	5	8	5
22	10	8	3
23	10	8	4
24	10	8	5
25	15	8	3
26	15	8	4
27	15	8	5

Hence the manufacturer demanded that a parameter choice card to select and fix factors for desired requirements of their job for regular use. Hence the research is designed completely experimental. Taguchi general full factorial design was preferred for the experimental design. The MINITAB release 16 software was used for obtaining such design. The factors and their levels and responses were furnished in Table 1. The Experiments mainly classified as PMEDM with Electrolyte copper Electrode and PMEDM with Silver coated Electrolyte copper Electrode.

Table 3. Observations of PMEDM with plain electrode

Expt. No	MRR (g/min)	TWR (g/min)	SR (μm)
1	0.11438	0.00069	0.98
2	0.13967	0.00091	1.04
3	0.15019	0.00127	1.18
4	0.21576	0.00161	1.12
5	0.22548	0.00185	1.21
6	0.25345	0.00224	1.37
7	0.38793	0.00246	1.27
8	0.43584	0.00284	1.34
9	0.46086	0.00175	1.54
10	0.28472	0.00186	1.08
11	0.31560	0.00204	1.18
12	0.33025	0.00219	1.26
13	0.35302	0.00223	1.23
14	0.38582	0.00271	1.35
15	0.43447	0.00313	1.49
16	0.49270	0.00418	1.34
17	0.52210	0.00445	1.54
18	0.58815	0.00328	1.75
19	0.50203	0.00412	1.17
20	0.53702	0.00439	1.26
21	0.55829	0.00475	1.33
22	0.60079	0.00521	1.27
23	0.64298	0.00618	1.38
24	0.67097	0.00644	1.53
25	0.69663	0.00690	1.58
26	0.74722	0.00706	1.63
27	0.79346	0.00745	1.78

Table 3. Observations of PMEDM with coated electrode

Expt. No	MRR (g/min)	TWR (g/min)	SR (μm)
1	0.13086	0.00094	1.13
2	0.16983	0.00124	1.18
3	0.20524	0.00177	1.24
4	0.24762	0.00205	1.28
5	0.26186	0.00254	1.35
6	0.29222	0.00308	1.41
7	0.45716	0.00323	1.47
8	0.52833	0.00368	1.54
9	0.57562	0.00184	1.76
10	0.36433	0.00197	1.21
11	0.37210	0.00216	1.26
12	0.37984	0.00228	1.34
13	0.40252	0.00299	1.37
14	0.41140	0.00324	1.46
15	0.46888	0.00321	1.59
16	0.54627	0.00429	1.54
17	0.61643	0.00489	1.67
18	0.67616	0.00394	1.83
19	0.59742	0.00424	1.24
20	0.61350	0.00469	1.29
21	0.69590	0.00501	1.38
22	0.71494	0.00564	1.42
23	0.73120	0.00637	1.54

24	0.83280	0.00689	1.67
25	0.79329	0.00702	1.64
26	0.81330	0.00745	1.71
27	0.92407	0.00859	1.93

In each case there were 27 experiments. In each experiment, the net observations were three likr TWR, SR and MRR. The experimental design is shown in Table 2 with parameters setting values for each case. The Electronica Machine Tools make Xpert 1model die sinking type CNC EDM machine was employed in this research. Kerosene is dielectric fluid in this machine because of its property of very low viscosity and it gets flushed away easily. Based on the literature review the recommended aluminum nanopowder concentration is 2g/l to 4g/l and fixed for this research as average value of 3g/l. The machining time for each test was 5 minutes exactly. The Taylor Hobson make, Surtronic3+ branded contact type profile-meter was used to measure surface roughness with 0.8mm cut of length and took three sample readings to avoid observational error. The tool and work material weight losses were measured by using laboratory balance of semi micro with accuracy of 0.00001 gm. The observations of MRR, SR and TWR were carried out for each experimental setting as per Table 1. They were recorded and furnished for electrode and coated electrode in Table 3 and Table 4 respectively.

3. Result and Discussion

The Comparative performances of both electrode types in terms of Material Removal Rate, Surface Roughness produced and Tool wear rates were analyzed. In figures the Input current index A, B and C used for experiments with Plain Electrode and 1, 2 and 3 for Silver coated Electrode for 5A, 10A and 15A respectively. The graphs plotted for constant pulse on time 6, 7 and 8μs for MRR, SR and TWR.

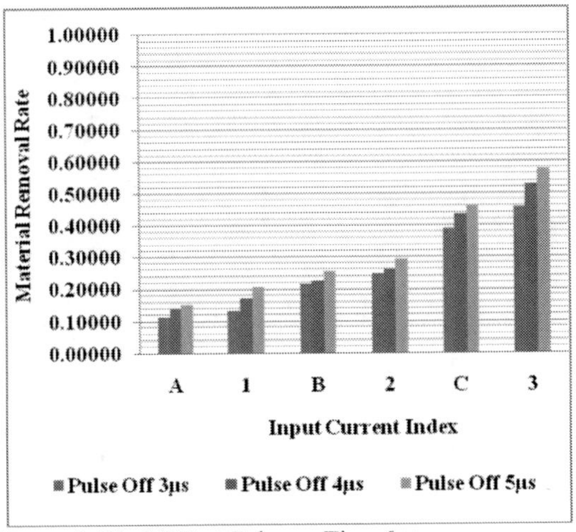

Figure 1. The MRR at Pulse on Time 6μs.

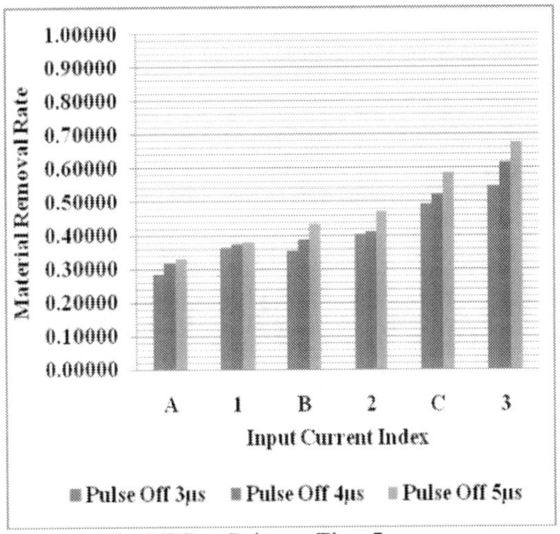

Figure 2. The MRR at Pulse on Time 7μs.

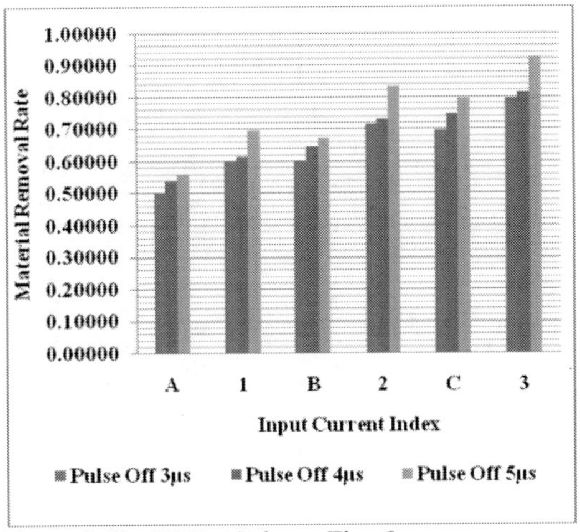

Figure 3. The MRR at Pulse on Time 8μs.

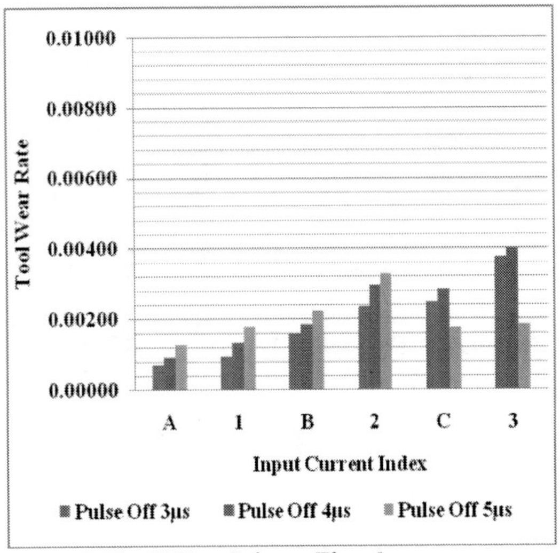

Figure 4. The TWR at Pulse on Time 6μs.

DOI: 10.1051/matecconf/20179502002

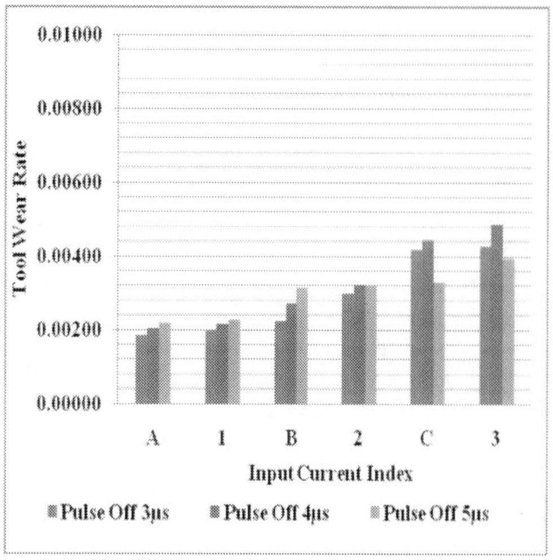

Figure 5. The TWR at Pulse on Time 7μs.

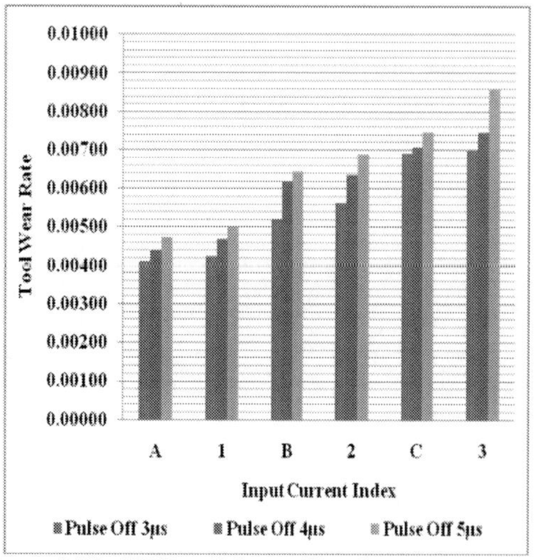

Figure 6. The TWR at Pulse on Time 8μs.

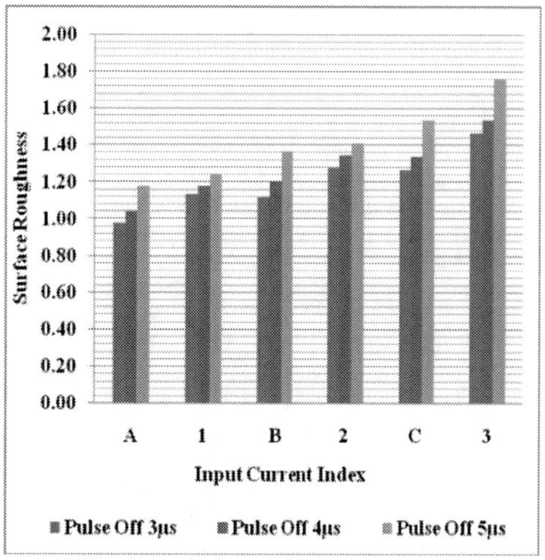

Figure 7. The SR at Pulse on Time 6μs.

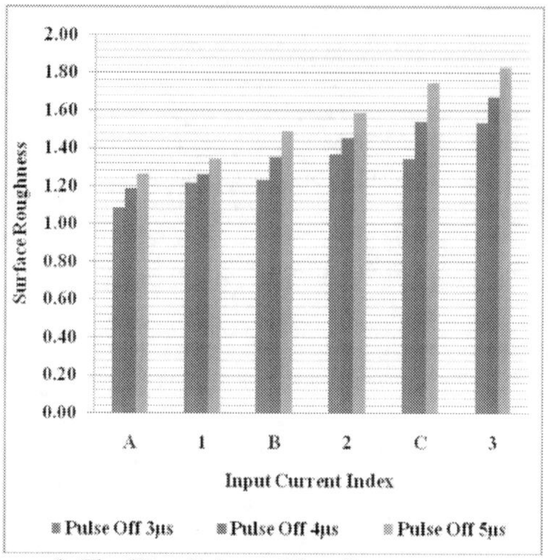

Figure 8. The SR at Pulse on Time 7μs.

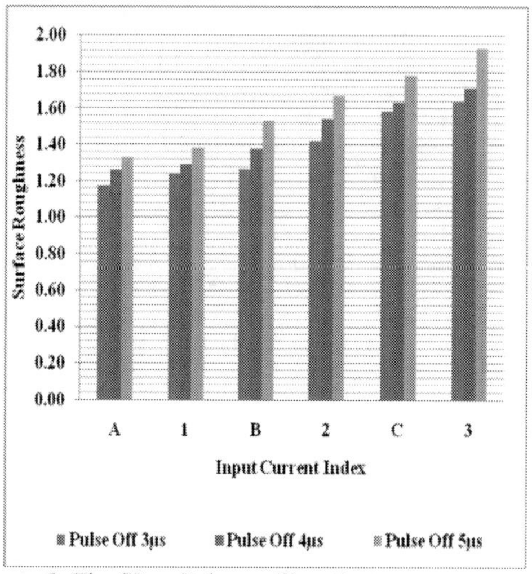

Figure 8. The SR at Pulse on Time 7μs.

The results reveals that the input current influenced more in increasing MRR as well as SR and TWR than Pulse on and pulse off times. But the factors Pulse on time and pulse off time influence are signifiant. The contribution of coated electrode is appréciable.

4. Conclusion

The Al Nano Powder Mixed EDM of Inconel 800 surface machining with plain Electrolyte Copper Electrode and Silver coated Electrolyte Copper Electrode were considered for this study. The reserach concluded that:

1. The maximum MRR can be achieved at higher values of the process parameters especially the input current.
2. The influence of pulse on time and pulse of time along with higher value of input current gives higher MRR.
3. The silver coating on Electrolyte Copper electrode significantly increases the MRR as

DOI: 10.1051/matecconf/20179502002

well as TWR. But lesser percentage of increments were observed in surface finish.

4. The MRR are directly proportional to TWR but inverse to the surface finish.
5. The manufacurer can be used the observation to understand the parameter influences in both cases as well as those table are healpful them to take appropriate decition to fix the parameters based on their requirements for the respective job type.

References

1. Hewidy, M.S., El-Taweel T.A., El-Safty M.F. ''Modelling the machining parameters of wire electrical discharge machining of Inconel 601 using RSM'', Journal of Materials Processing Technology. **169**, pp. 328–336, (2005).
2. Narutaki, N., Yamane, Y. and Y. Lee, 1993. High speed machining of Inconel 718 with ceramic tools. Annals of CIRP 42, 103–106.
3. M. L. Jeswani,. Effect of the addition of Graphite powder to Kerosene Used as the dielectric fluid in Electrical Discharge Machining., Elsevier Sequoia S.A., Lausanne 133-139, (1981).
4. Kuppan, P., Rajadurai, A., Narayanan, S., Influence of EDM process parameters in deep hole drilling of Inconel 718, Int J Adv Manuf Technol, **38**, 74–84, (2008).
5. Rajyalakshmi, G., Venkata Ramaiah, P., Multiple process parameter optimization of wire electrical discharge machining on Inconel 825 using Taguchi grey relational analysis. Int J Adv Manuf Technol, **69**, 1249–1262, (2013).
6. MuthuKumar, V., SureshBabu, A., Venkatasamy, R., Raajenthiren, M., Optimization of the WEDM parameters on machining Incoloy800 Super alloy with multiple quality characteristics, International Journal of Engineering Science and Technology, **2**, 1538-1547, (2010).
7. V. Muthukumar, N. Rajesh, R. Venkatasamy, A. Sureshbabu, N. Senthilkumar, Mathematical Modeling for Radial Overcut on Electrical Discharge Machining of Incoloy 800 by Response Surface Methodology Procedia Materials Science, **6**, 1674 – 1682, (2014).
8. Lin, M.Y., Tsao, C.C., Hsu, C.Y., Chiou, A.H., Huang, P.C., Lin, Y.C., Optimization of micro milling electrical discharge machining of Inconel 718 by Grey-Taguchi method. Trans. Nonferrous Met. Soc. China, **23**, 661−666, (2013).
9. Mahendra G. Rathi, Deepak V. Mane, " Study on Effect of Powder Mixed dielectric in EDM of Inconel 718", International Journal of Scientific and Research Publications, **4**(11), November (2014).
10. Tzeng, Y.F and Lee, C.Y, "Effects of Powder Characteristics on Electro discharge Machining Efficiency", International Journal of Advance Manufacturing Technology, **17**, 586-592, (2001).
11. Khalid Hussain Syed, G. P. Anuraag, Gangarapu Hemanth and Shaik Abdul Subahan, Powder-Mixed EDM Machining of Aluminium-Silicon Carbide Composites. Indian Journal of Science and Technology, **8**(S2), 133–137, January (2015).
12. Shriram Y. Kaldhone, Mukund V. Kavade and Umashankar Rawat, Effect of Powder Mixed Dielectric on Performance Measures of EDM for Tungsten Carbide, International Journal of Innovative Research in Advanced Engineering, **1**(10) pp. 106-111, (2014).

Simulation and Experiment on In-plane Carbon Nanotube Thermoelectric Generator in Parallel

Wenbin Huang , Xiangjun Song, Yaozhou Liu, Wanling Li, Peng Zhang, Xiaopan Liu

Center of Nanoscience and Technology, Institute of Mechanical Technology, Shijiazhuang, Hebei, 050003, China

Abstract. In order to solve the problem of larger internal resistance of thin-film thermoelectric generator (TEG) in series, which could influence the output power and restrict the application, the in-plane carbon nanotube (CNT) TEG in parallel was ingeniously researched. Utilizing the parameters of output power, conversion and energy efficiencies, the ideal and actual models of TEG in parallel were established, respectively. The thermal conduction insulating layer was taken into account, which could provide the theoretical guidance for the experimental test and engineering applications. The CNT films were prepared by the floating-catalyst chemical vapor deposition (CVD), and the experiment and properties of TEG based on CNT films were investigated. The testing circuits of conventional and gas TEGs were designed, and the output powers of the serial and parallel connecting types were tested and compared. The correctness of theoretical model and numerical analysis was proved to be valid. The novel method could effectively enhance the output power, extend the applied range of TEG in MEMS/NEMS and had a fine prospect.

1 Introduction

Along the quick development of micro/nano systems and their extensive applications, the power source supplied by the storage battery cannot be adapted well to the rigorous environments and requests. The size of power sources will restrict the microminiaturization and practicability of micro/nano systems, which cannot be easily replaced by the battery. The power source supplied by the storage battery cannot be adapted well to the rigorous environments and requests. The sustainable self-sufficient micro-power source, which harvests energy from the environment to power the micro/nano systems, is a feasible method of solving the supply problems [1-3]. The thermoelectric generator (TEG) is highly desirable since it can convert the heat energy into the electrical power.

TEG has many advantageous qualities over alternative devices, e.g. comparatively lightweight, completely silent, highly reliable, clean energy, no moving parts and no consumables. The heat energy could be supplied from the solar energy, automobile exhaust, production waste gas or heat dissipation of electronic components [4, 5]. So it may play a key role for green energy production and utilization in the future. However, in order to make the TEG practical, more efficient thermoelectric materials and optimized structures must be developed and researched in recent years.

Since the discovery of CNT in 1991 [6, 7], the correlated research work has increased exponentially over the years owing to the splendid physical, chemical, structural and electrical properties [8,9]. It can generate electricity in the photic, thermic, magnetic or fluidic fields, which can be applied to a more complicated environment [10, 11]. Based on the thermoelectric effects, the thermal gas and solar energy from ambient environment can be converted to electric energy.

Recently, Ref. [11, 12] reported the thermoelectric properties of CNT films and designed an integrated device to directly convert illuminating NIR light into a notable output voltage. The p-type and n-type CNT films were arranged alternatively and linked with each other in series. The output voltage can reach 0.1 V by setting 50 p-type CNT films and 50 n-type CNT films. However, the thin-film TEG has a larger internal resistance due to the smaller cross-sectional area. The maximum power output is directly proportional to the Seebeck coefficient and temperature difference between the hot and cold ends. But it is inversely proportional to the internal resistance. The output electrical properties are greatly influenced by the larger internal resistance. Hence, thermoelectric module (TEM) by connecting diverse p-type and n-type thermoelectric cells in parallel can reduce the internal resistance and enhance the output power. It is significant to the TEG application in diverse fields.

2 Theoretical model

2.1. Ideal theoretical model

In the ideal conditions, suppose the temperatures of the hot and cold ends remains invariable. The convection and radiation heat transfer between the hot and cold ends are

DOI: 10.1051/matecconf/20179502003

neglected. The heat mainly transfers along the couple arms. The electrical properties effected by the electrode resistance, contact resistance and thermal resistance are neglected. When the temperature difference between the hot and cold ends is minor, The Thomson effect could be neglected. The ideal schematic diagram of the membrane thermoelectric module in parallel is shown as Fig. 1(a). Every thermoelectric cell (TEC) includes P-type and N-type semiconductor couple arms and electrode layer. At the low temperature end, the P-type couples are connected on the substrate top surface. The N-type couples are connected on the substrate bottom surface. The membrane thermoelectric module in parallel, abbreviated 'parallel TEM' includes diverse TECs in parallel.

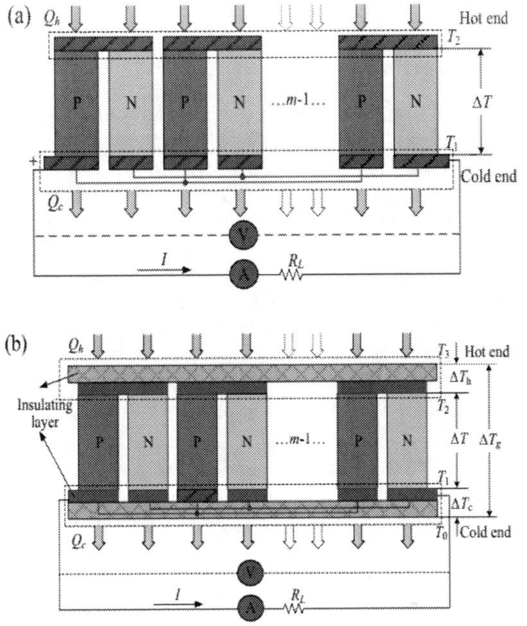

Figure 1. Ideal (a) and actual (b) schematic diagrams of membrane TEM in parallel.

Suppose that the parallel TEM is constituted by m pairs of TECs. Every TEC has the same material properties and temperatures at the cold and hot ends. So the CNT couple arms have the same electrical currents I_i. The current in the main circuit is $I = mI_i$. The heat quantity exchanged between the heat source and hot ends of TEC is approximately equivalent.

The output power expressed by the current in the main circuit is

$$P = Q_h - Q_c = \alpha_{pn} I \Delta T - \frac{I^2 r}{m} \quad (1)$$

The conversion efficiency is

$$\eta = \frac{Q_h - Q_c}{Q_h} = \frac{m\alpha_{pn} I \Delta T - rI^2}{m^2 K \Delta T + m\alpha_{pn} T_2 I - 0.5 I^2 r} \quad (2)$$

2.2. Actual theoretical model

When the thermal conduction insulating layer is considered in the parallel TEM, the actual schematic diagram is shown as Fig. 1(b). The positive-negative electrodes are located at the low-temperature ends. The parallel TEM requires that the P-type couple arms are connected and the N-type couple arms are connected together. So the metal wire will slightly influence the electrical properties, whether the metal wire drills through the thermal conduction insulating layer or substrate material layer. The m pairs TECs in plane substrate are equivalent to the thickness of CNT films increased by m times.

Suppose that the parallel TEM is constituted by m pairs of TECs. Every TEC has the same material properties and temperatures at the cold and hot ends. So the CNT couple arms have the same electrical currents I_i. The current in the main circuit is $I = mI_i$. The heat quantity exchanged between the heat source and hot ends of TEC is approximately equivalent.

The output power expressed by the current in the main circuit is

$$P = [\alpha K_h K_c \Delta T_g I - rK_h K_c I^2/n - 2rK\overline{K}I^2 - \alpha^2 I^2 (K_h T_3 + K_c T_0) + 0.5\alpha r \Delta K_g I^3/n] \times [K_h K_c + 2nK\overline{K} - n\alpha \Delta K_g I - n^2\alpha^2 I^2]^{-1} \quad (3)$$

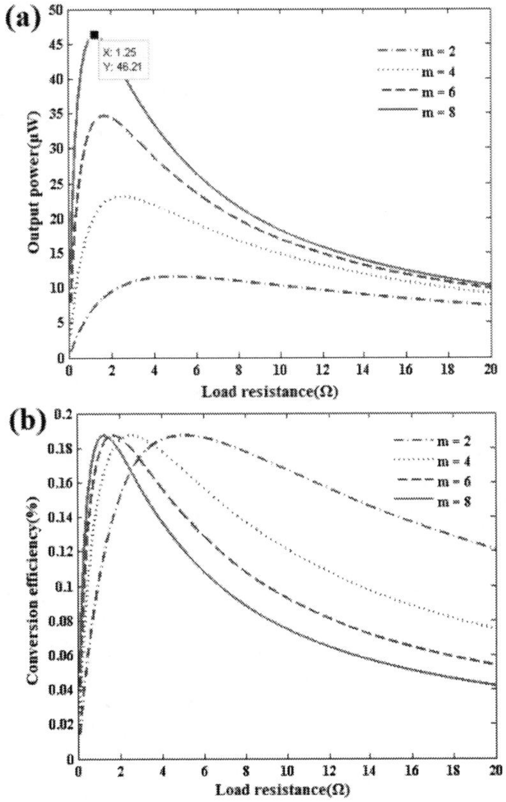

 DOI: 10.1051/matecconf/20179502003

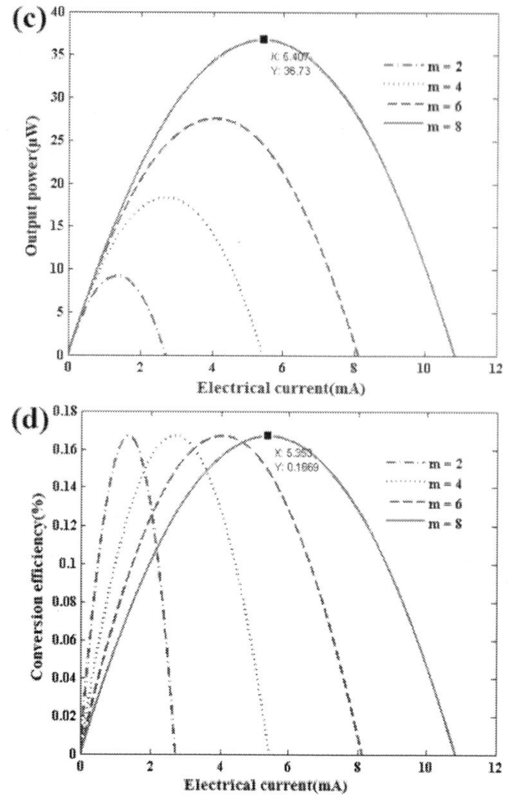

Figure 2. Performance of membrane TEM in parallel, output power (a) and conversion efficiency (b) of ideal model, output power (c) and conversion efficiency (d) of actual model.

The conversion efficiency is expressd as

$$\eta_p = [\alpha K_h K_c \Delta T_g I - 2rK\overline{K}I^2 + 0.5\alpha r \Delta K_g I^3/n] \times K_k^{-1} \times [nKK_c \Delta T_g + \alpha K_c T_3 I - (0.5rK_c/n + rK + \alpha^2 T_3)I^2 + 0.5\alpha r I^3/n]^{-1} \quad (4)$$

The energy efficiency can be got by the equation below

$$\eta_e = \eta T_3/\Delta T_g \quad (5)$$

When the temperature difference is smaller, the performance parameter is propitious to analyze the thermoelectric properties.

3 Simulation and Analysis

Assume the temperature T_1=298 K, T_2=393 K, the internal resistance of TEC r=10 Ω, the performance analysis of membrane TEM in parallel with an ideal model is shown as Fig. 2(a) and (b). Along with the increase of parallel pair m, the whole internal resistance of TEM obviously decreases. The output power can be enhanced by configuring the load resistances. When the internal resistance is equal to the load resistance, the output power reaches the maximal value. The maximal output power of series TEM is equal to the one of parallel TEM when they have the same connection pairs. The maximal value of conversion efficiency or exergy efficiency is independent of the parallel pair, the values are close to the ones of serial TEM.

Assuming the coefficient of thermal conductivity at the hot and cold ends are respectively K_h=0.8mW/K, K_c=0.4mW/K. In order to simplify the calculation, assume the surface heat transfer coefficient of TEM is fixed, K_h and K_c are directly proportional to the parallel pair. The performance analysis of membrane TEM in parallel with an actual model is shown as Fig. 2(c) and (d). The output power and the conversion efficiency could reduce due to the influence of the thermal conduction insulating layer. When m=8, the output power is 36.73 μW, the conversion efficiency is 0.1669%. When the pair of TECs is equal, the output current has larger difference. So we can get the conclusion that the connecting type of TECs is determined by the load power consumption and the electrical current requested.

4 Experiment and verification

The in-plane TEG is composed of p-type and n-type CNT films connected by aurum electrodes. The insulator materials are filled in the gaps between p-type and n-type CNT films, which also have functions of reinforcing and smoothing the material surface. The CNT film is synthesized by a floating catalyst chemical vapor deposition (CVD) method [13, 14]. Its schematic diagram is shown in Fig. 3(a). The experimental setup is a three-section quartz tube mounted in a two-stage furnace system. The quartz tube is first flushed with the argon flow to exhaust the air. Ferrocene as catalyst precursor and sulphur as growth promoter are heated (60~80 ℃) and carried into the reactor (1100 ℃) along with the gaseous mixture of argon and methane. The large-area transparent CNT film will be synthesized in the quartz tube wall after five hours. The SEM morphology of CNT film is shown in Fig. 3(b). By doping n-type semiconductor element or coating PEI[15, 16], the n-type CNT film can be synthesized.

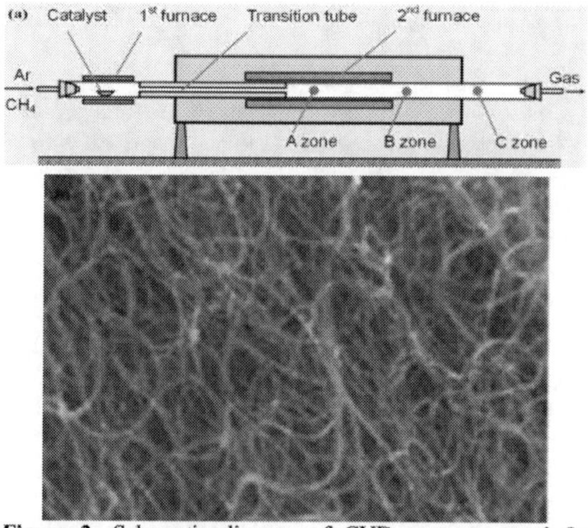

Figure 3. Schematic diagram of CVD apparatus used for synthesizing CNT membrane (a), and SEM morphology of CNT membrane (b).

 DOI: 10.1051/matecconf/20179502003

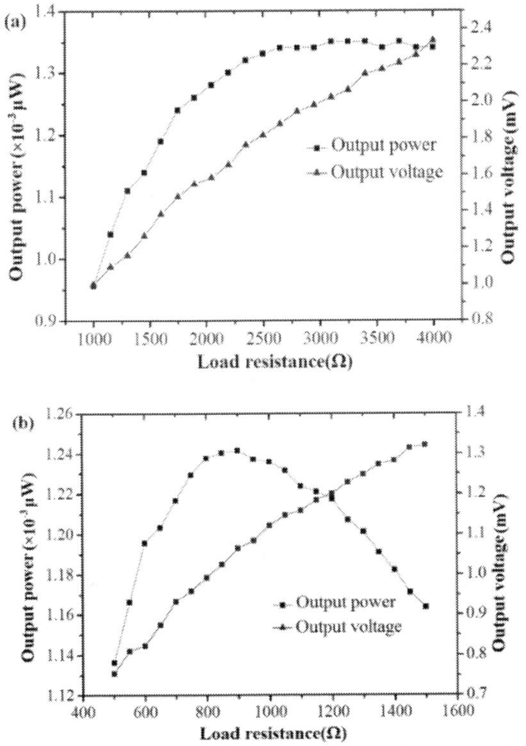

Figure 4. Comparison of thermoelectric properties between the serial (a) and parallel (b) modules.

CNT films are heated by the thermoprobe, the temperature is 348 K at the hot-end and 298 K at the cold-end. The temperature difference is ΔT = 50 K, which is invariable in the experiment. As shown in Fig. 4(a) and (b). The maximum output powers of the TEM in series and parallel are 1.35×10^{-3} μW and 1.24×10^{-3} μW. The output power could be greatly enhanced by increasing the temperature difference and connection pairs. In the experiment, the different maximum output power is attributed to the collocation method of TEC. The parallel TEM includes two TECs are respectively disposed on the top and bottom of substrates. The larger area of electrode layer has bigger contact resistance and thermal resistance, which influence the power output. The parallel TEM can enhance the output power by reducing the internal resistance and increasing the loop current. So it is more significant in the application of thin-film TEG.

5 Summary

In this work, the theoretical model and optimization techniques of conventional TEG based on CNT films are researched. Utilizing the parameters of output power, conversion efficiency and energy efficiency, the ideal model and actual model of TEM in parallel are respectively established. The thermal conduction insulating layer is taken into account, which can provide the theoretical guidance for the experimental test and engineering applications. The experiment and properties of TEG based on CNT films are researched. The CNT films are prepared by CVD method. The testing circuits of conventional and gas TEG are both designed. The output voltage and power are tested in the series-parallel connecting types. The correctness of theoretical model and numerical analysis is proved to be valid. It can achieve the waste heat recovery from the solar energy, automobile exhaust, production waste gas or heat dissipation of electronic components to self-power MEMS/NEMS in the future.

Acknowledgements

The authors sincerely thank National Center for Nanoscience and Technology and Institute of Nanotechnology and Microsystems. This work was supported by National Science Foundation of China (Grant Nos. 10774032, 90921001), Key Knowledge Innovation Project of the Chinese Academy of Sciences on Water Science Research, Instrument Developing Project of the Chinese Academy of Sciences (Grant No. Y2010031).

References

1. Z.L. Wang, J.H. Song: Science, Vol. 312-315(2006), p. 242.
2. X.D. Wang, J.H. Song, J. Liu: Science, Vol. 316-319 (2007), p. 102.
3. R.S. Yang, Y. Qin, C. Li: Nano Lett., Vol. 9-12(2009), p. 1201.
4. D. Kraemer, P. Bed, H.P. Feng: Nat. Mater., Vol. 10-14(2011), p. 532.
5. C.T. Hsu, G.Y. Huang, H.S. Chu: Appl. Energy, Vol. 88-100(2011), p. 1291.
6. S. Iijima: Nature, Vol. 354-356(1991), p. 56.
7. M.E. Itkis, F. Borondics, A.P. Yu: Science, Vol. **312-315** (2006), p. 413.
8. Y.Y. Hsiao, W.C. Chang, S.L. Chen: Energy, Vol. 35-37 (2010), p. 1447.
9. Q. Cao, H.S. Kim, N. Pimparkar: Nature, Vol. 454-**458** (2008), p. 495.
10. M.C. LeMieux, M. Roberts, S. Barman: Science, Vol. **321-325** (2008), p. 101.
11. C. Yu, Y. Ryu, L. Yin: ACS Nano, Vol. 5-7 (2011), p. 1297.
12. C.H. Hu, C.H. Liu, L.Z. Chen, C.Z. Meng, S.S. Fan: ACS Nano, Vol. 4(2010), p. 4701.
13. B.C. St-Antoine, D. Menard, R. Martel: Nano Letters, Vol. **9-12** (2009), p. 3503.
14. M.A. El Khakani, V. Le Borgne, B. Aïssa: Appl. Phys. Lett., Vol. **95**(2009), p. 083114-1.
15. Y.C. Zhao, L. Song and K. Deng: Adv. Mater. Vol. **20-24** (2008), p. 1772.
16. Z. Liu, K.H. Zheng, L.J. Hu: Adv. Mater., Vol. 22-**25** (2010), p. 999.

Finite Element Modeling with Embed Rebar Elements and Steady State Rolling Analysis for Rolling Resistance Test of Pneumatic Tire

Chakrit Suvanjumrat and Ravivat Rugsaj

[1]*Department of Mechanical Engineering, Faculty of Engineering, Mahidol University, Salaya, Nakorn Pathom, 73170, Thailand*
[2]*Laboratory of Computer Mechanics for Design (LCMD), Department of Mechanical Engineering, Faculty of Engineering, Mahidol University, Salaya, Nakorn Pathom, 73170, Thailand*

Abstract. Finite element model of tire rolling resistance test on the drum was developed using 3D steady state rolling analysis coupling with pre-inflation of 2D axisymmetric tire analysis. The complex components of the radial tires composing tread, sidewall, ply layers, steel belts, and lead wires were modeled using rebar elements which were embed into the rubber element using the tying equation. The Mooney-Rivlin hyperelastic constitutive model was employed to describe the large deformation behavior of tread and sidewall, while other components such as plies, steel belts and bead wires were assigned the linear isotropic material. The tire rolling resistance system was modeled by inflation of slick tire and compression on the drum for the footprint analysis regarding the rolling resistance test. The tire's steady state characteristics such as footprint contact pressure, rolling resistance force, and time response characteristic of tires were predicted instead the experiment of the prototype.

1 Introduction

The rolling resistance of a tire had been referred to the ratio between the energy loss per tire per travelling distance due to non-elastic deformation and energy dissipation of tires, while the rolling resistance coefficient was defined by the ratio of rolling resistance and the normal force on the tire [1]. The drum testing is an important method to measure the rolling resistance, which was performed by pressing an inflated tire against a drum with a desirable load. The drum was rotated and the tire's resistance was indirect to any variable such as force, torque, power, and deceleration [2]. The drum testing method had the advantage to reduce various factors from the vehicle suspension system and the pavement structure [3].

The finite element (FE) model of rolling resistance test was simple to run a tire model on a flat plate to reduce complexity caused by the rolling-contact on curved geometry but it was not agreed with the physical experiment caused by the curvature of the drum surface. The different in footprint and rolling characteristic became even more significant as the drum diameter get smaller [4]. In the recent years, FE models of the drum testing had been developed. The rotation system of drum and tire could be modeled by using transient dynamic analysis and steady state rolling (SSR) analysis which was based on the Eulerian and Lagrangian coupling [5]. Tönük [6] used kinematic inversion technique by assigning the drum to move around the fixed tire with same relative motion.

The transient dynamic analysis for the rolling resistance test of the pneumatic tire is preferred to use the explicit finite element method (FEM) to the implicit FEM. It is unnecessary if the SSR characteristic of the tire is considered. The transient dynamic method is disadvantaged because it requires many calculating times and induces cumulative error to solve the high nonlinear complexity of tire rotation and frictional rolling-contact mechanism. The transient dynamic FEM also tends to found limitation in dealing with rigid body rotation, forces users to model the drum with many fine meshes and caused to increase analysis time considerably. Particularly, the transient dynamic analysis will have severe disadvantage for modeling each component of tire for the high prediction accuracy. There was difficult to restore data for analysis due to the high amount of simulation result happened.

The SSR method had more advantages than the transient dynamic analysis because it required less analysis time and resource. The analysis time using may further reduce by refining of element concentration only in the critical contact region between tire and drum. This research aimed to develop a simulation model of the slick tire based on SSR analysis. The tire components were modeled with rebar elements and the tying equation to tie degree of freedom between tire's rubber and component elements. The axisymmetric analysis of the tire inflation was performed and then the results were used for the initial condition of rolling analysis.

2 Hyperelastic material model

The nonlinear large deformation behavior of the rubber material which was used to produce radial tires is governed by the Mooney-Rivlin hyperelastic constitutive model [7] as given by:

$$W = W(\lambda_1, \lambda_2, \lambda_3) \quad (1)$$

where W is the strain energy density function, λ is the stretch ratio.

Strain energy density function may be derived in the form of strain invariant (I) of left Cauchy-Green deformation tensor [8] is written by:

$$W = W(I_1, I_2, I_3) \quad (2)$$

where I_1, I_2 and I_3 are first, second and third strain invariant, respectively. The relation between strain invariant and stretch ratio can be described by:

$$I_1 = \lambda_1^2 + \lambda_2^2 + \lambda_3^2 \quad (3)$$

$$I_2 = \lambda_1^2\lambda_2^2 + \lambda_2^2\lambda_3^2 + \lambda_1^2\lambda_3^2 \quad (4)$$

$$I_3 = \lambda_1^2\lambda_2^2\lambda_3^2 \quad (5)$$

Polynomial hyperelastic model is phenomenological model which based on strain energy in the form of polynomial equation of left Cauchy-Green deformation tensor as shown by:

$$W = \sum_{i+j+k=1}^{N} C_{ijk}(I_1 - 3)^i (I_2 - 3)^j (I_3 - 3)^k \quad (6)$$

where C_{ijk} is material coefficients.

For an incompressible material, it can be simplified to:

$$W = \sum_{i+j=1}^{N} C_{ij}(I_1 - 3)^i (I_2 - 3)^j \quad (7)$$

Mooney-Rivlin hyperelastic model is the 1st order polynomial hyperelastic model or N which equals to 1. Mooney-Rivlin model can be used to reasonably predict stress-strain relationship in a small range of strain. The model can also be used to model nonlinear deformation behavior of tire during rolling. Mooney-Rivlin model may be written as:

$$W = C_{10}(I_1 - 3) + C_{01}(I_2 - 3) \quad (8)$$

where C_{10} and C_{01} is the material coefficient.

3 Finite element modeling

The FE model of tires under the rolling resistance test was performed by using commercial software, MSC. Marc. The slick tire shape like Michelin Agilis 195R14C tire was referred to generate finite element (FE) model for the rolling resistance test. The modelling consisted of two states: (1) the 2D inflation of tire using the axisymmetric analysis and (2) the 3D rolling resistance test analysis.

During the inflation state of the tire, the axisymmetric analysis was used to simplify the process. The slick tire was modeled in the 2D cross section which composed several components. The tread, sidewall, apex and bead wires are modeled by 2D quadratic axisymmetric elements while plies layers and steel belts are modeled by 1D rebar elements with independently assign the thickness in each layer (Fig. 1). Since the 2D and 1D element were meshed independently, nodes of the 2D and 1D element didn't share location with each other. In order to connect both meshes, the tying equation was used to tie degree of freedom between two types of elements. Table 1 describes types and the number of elements which are combined to be the axisymmetric FE model of radial tire.

The 3D radial tire model was then modeled by revolving the 2D axisymmetric model around its rotating axis or x-axis. The 2D axisymmetric elements and 1D rebar elements were expanded to 3D continuum element and 2D rebar elements, respectively. The expanded element was uniformly divided into 20 sectors, which elements in the potential contacted region were divided with finer mesh density. Fig. 2 illustrates the FE model of the radial tire.

The FE model for the rolling resistance test of a radial tire was developed. The drum was modeled by a rigid body surface with a diameter of 1.7068 m. The FE model of the radial tire was placed on a drum model. The drum model is assigned to be contacted with tread elements of a tire model with a friction coefficient of 0.5 (Fig. 3). The Coulomb friction model using nonlinear arctangent smoothing function was assigned to model the contact friction between the radial tire and drum.

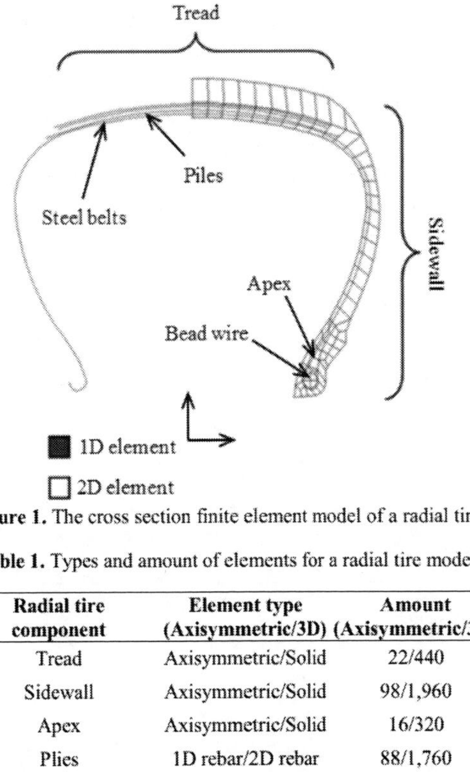

Figure 1. The cross section finite element model of a radial tire.

Table 1. Types and amount of elements for a radial tire model.

Radial tire component	Element type (Axisymmetric/3D)	Amount (Axisymmetric/3D)
Tread	Axisymmetric/Solid	22/440
Sidewall	Axisymmetric/Solid	98/1,960
Apex	Axisymmetric/Solid	16/320
Plies	1D rebar/2D rebar	88/1,760
Steel belts (1st layer)	1D rebar/2D rebar	17/340
Steel belts (2nd layer)	1D rebar/2D rebar	17/340
Cap ply	1D rebar/2D rebar	32/640
Bead wire	1D rebar/2D rebar	8/160

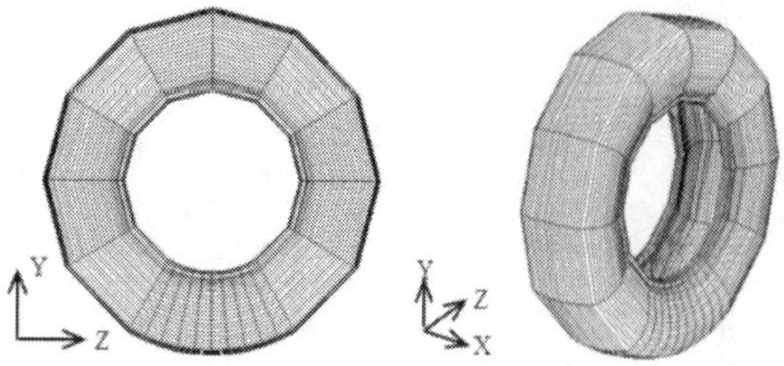

Figure 2. The 3D finite element of a radial tire.

The Mooney-Rivlin hyperelastic constitutive model was employed to tread, apex and sidewall of a tire model, while other components such as plies, steel belts and bead wires were defined as linear isotropic material. The material properties of radial tire components are shown in Table 2. These materials obtained by samples test according to the material testing standard.

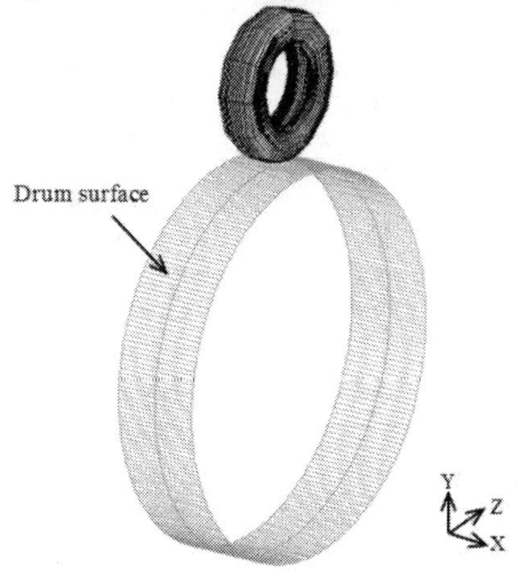

Figure 3. The FE model for the drum rolling resistance test of a radial tire.

Table 2. Mechanical properties of radial tire components.

Radial tire component	Constitutive model	Mechanical property (MPa)			
		C10	C01	E	υ
Tread	Hyperelastic	0.003940	0.862791	-	-
Sidewall	Hyperelastic	0.152081	0.555920	-	-
Apex	Hyperelastic	0.645362	0.873692	-	-
Plies	Linear	-	-	3,800	0.48
Steel belts	Linear	-	-	110,316	0.46
Cap ply	Linear	-	-	3,800	0.48
Bead wire	Linear	-	-	200,000	0.30

Note E is elastic modulus, υ is Poisson's ratio, C10 and C01 is constant.

The 2D axisymmetric model of the tire was then used to analyze tire stress, strain, and deformation during inflation state. The boundary conditions consisted pressure of 0.45 MPa and the fixed constraint were applied to the inside edge of tire element and nodes at the rim location, respectively. The 2D axisymmetric analysis results which consisted of stress, strain and deformation were collected and used to assign the initial condition of the next corresponding 3D rolling resistance analysis. During rolling resistance analysis, the drum was brought to contact with the tire and the load of 807.5 kg was applied to the drum for the footprint and contact pressure prediction. In addition, the drum surface model was assigned to rotate about its center with a rotation velocity of 80 km/hr after load and displacement were successfully assigned. The pressure applied into the tire was held constant during this state.

The FEA of the tire rolling resistance test on the drum was performed using personal computer Core-i5 CPU and 4GB RAM memory. The static and steady state analysis was used to solve for the 2D axisymmetric and 3D model respectively.

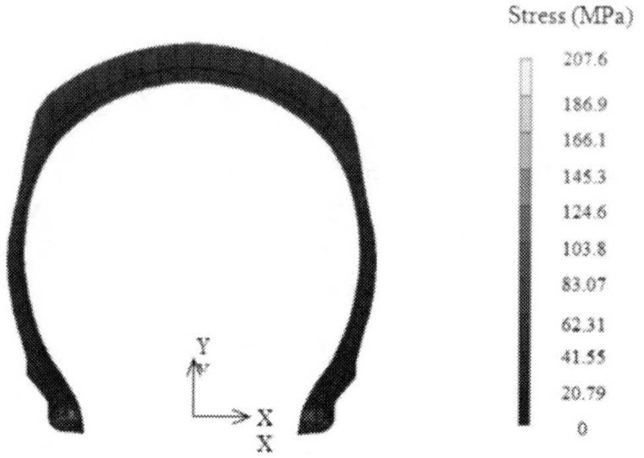

Figure 4. The inflation analysis results of the 2D axisymmetric tire model.

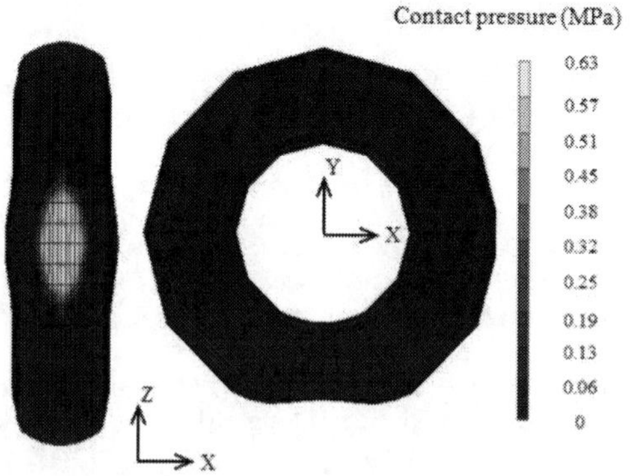

Figure 5. The contact pressure of the footprint analysis results.

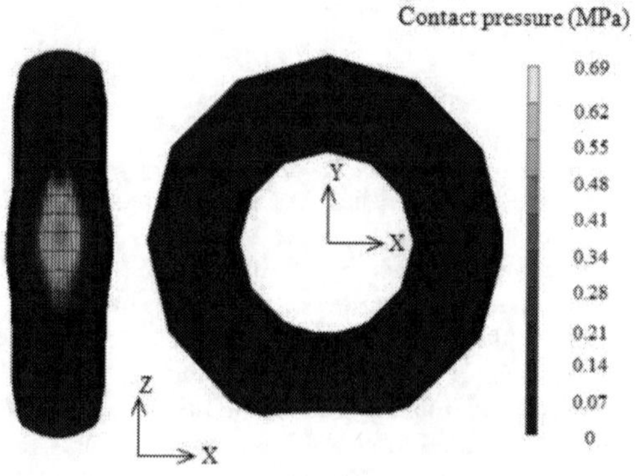

Figure 6. The footprint analysis results during steady state rolling.

4 Results and discussion

The analysis results of the 2D axisymmetric model which compose deformation and stress during tire inflation state are illustrated in Fig. 4. The color contour was shown the maximum result in red and the minimum result in blue. These results were used as initial conditions of the 3D analysis and then the inflated 3D tire model was brought to contact with the drum and the y-direction force of 8,075 N. The 3D footprint analysis results show the contact pressure distribution and deformation at loaded of 8,075 N in Fig. 5. After loading state, the SSR analysis was performed to inflate the 3D tire model for contacting with the drum. The drum surface at the contacted position was assigned to equivalently move with the speed of 22.22 m/sec while the tire was assigned to rotate with spinning velocity of 66.6126 rad/sec, which corresponded to traveling speed at 80 km/hr. The contact pressure distribution and deformation during rolling at steady state are shown in Fig. 6. The sidewall of tire was observed to slightly bulge during footprint analysis and continue to be bulged during rolling. The contact pressure showed an effective area of the tire which contacted on the drum or the road surface in the general usage, therefore, it could be used to indicate how well various tire's parameters such as the tire shape, material, components, pressure, tread pattern and depth effect traction of the vehicle. The rolling resistance force was also calculated from the reaction force between the drum and tire. The rolling resistance force and drum loading are plotted by graphs as shown in Fig. 7. The graphs were divided into three sections upon the time during three states of the 3D analysis which were tire's moving into contact state, drum loading state, and SSR. The force by drum and rolling resistance were satisfied the condition of drum test. The resistance force increased after the inflated tire was loaded with the drum force.

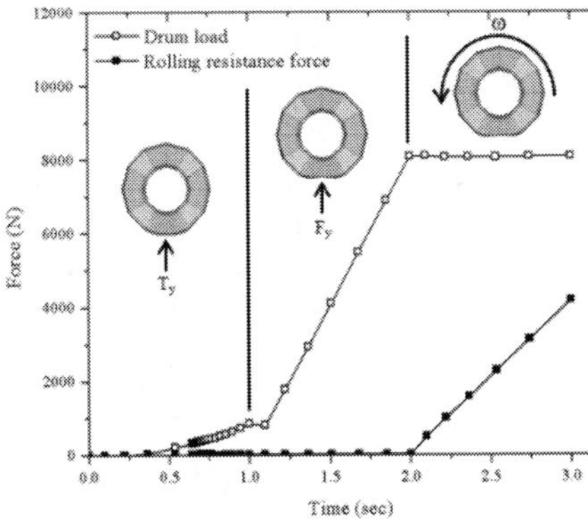

Figure 7. The applied drum load and the rolling resistance force vs. time.

5 Conclusions

The FE model of the tire rolling resistance with drum testing method was developed based on the 2D axisymmetric and the 3D SSR analysis. The Mooney-Rivlin hyperelastic constitutive model was used to describe the nonlinear large deformation behavior of the tire, while other components such as plies, steel belts and bead wires were defined as linear isotropic material. The tire components were modeled with rebar elements and tied to rubber elements which had a different degree of freedom using the tying equation. The 2D axisymmetric elements were expanded to the 3D elements by revolving around the tire axis. The 2D axisymmetric analysis results of the tire during inflation state were used as the initial condition of 3D analysis. The inflated 3D tire model was brought for contacting with the imported drum surface. The drum was loaded and SSR analysis was performed. The analysis results showed that FE model of the tire using the 3D SSR analysis combining with the pre-load of the 2D axisymmetric analysis and the embed rebar element could be used to predict steady state characteristics such as stress, deformation, contact pressure, and rolling resistance force for the drum testing. This technique, there are more advantages, less analysis time and resource requirement than the previous 3D transient dynamic analysis. Finally, this FEM could plentifully give the useful information and understanding for designing and development the radial tires in the near future.

References

1. L.G. Andersen, J.K. Larsen, E.S. Fraser, B. Schmidt, J.C. Dyre, J. Transp. Eng. **141**, 1 (2015)
2. International Organization for Standardization. *ISO 18164:2005 Passenger car, truck, bus and motorcycle tyres-Methods of measuring rolling resistance* (ISO, Geneva, 2005)
3. H.W. Lee, J.R. Cho, W.B. Jeong, K.M. Jeong, K.W. Kim, Int. J. Automot. Techn. **15**, 411 (2014)
4. N. Korunović, M. Trajanović, M. Stojković, D. Mišić, J. Milovanovićet, J. Mech. Eng. **57**, 888 (2011)
5. R. Behnke, M. Kaliske, Int. J. Nonlinear Mech. **68**, 101 (2015)
6. E. Tönük, Y. S. Ünlüsoy. Computers & Structures **79**, 1219 (2001)
7. G.T. Mase, R.E. Smelser, G.E. Mase, *Continuum Mechanics for Engineer* (CRC press, Florida, 2010)
8. G.A. Holzapfel, *Nonlinear Solid Mechanics* (John Wiley & Sons, England, (2000)

 DOI: 10.1051/matecconf/20179502005

Ceramics Vitreous China Produced by Utilizing Sediment Soil from Water Supply Treatment Process

Ubolrat Wangrakdiskul , Jindakarn Wanasbodee and Pornnapa Sansroi

1518 Production Engineering Department, Faculty of Engineering, King Mongkut's University of Technology North Bangkok, Thailand

Abstract. Due to generating the abundant of sediment soil, it makes the high burden of disposal cost to the metropolitan waterworks authority. Enhancing the value of sediment soil has been explored. This research aims to utilize the sediment soil, wastes of water supply treatment process for producing ceramics vitreous china. In this experiment, five types of raw materials are exploited, namely, sediment soil, ball clay, kaolin, feldspar and silica sand. The formulas have been divided into two groups. Sediment soil has been used as substituted material in ball clay for the first group, and substituted in kaolin for the second group. The specimens of each formula are formed by uniaxial pressing at 100 bar of size 50×100x7 mm. Then they have been sintered at two different temperatures, 1200°C and 1250°C, with heating rate 400° C/ hr and soaking for 30 minutes. The result reveals that the suitable formula for ceramics vitreous china is No. 2_4 of Group 2 with sintering temperature 1250°C. Its mixture consists of 0% kaolin, 35% ball clay, 30% feldspar, 20% silica sand and 15% sediment soil. The properties of this formula are 9.4% shrinkage, 9.39 MPa of bending strength, 6.34×10^{-6}/K coefficient of thermal expansion, and 0.66% water absorption.

1 Introduction

Metropolitan Waterworks Authority (MWA) is one of an important state enterprises providing water supply for supporting people in the metropolitan area. With the growing population as well as the change in people's living behaviors nowadays, the demand of water consumption is increasing. Therefore, four main plants of water supply have been constructed for serving the increased demand.

These plants are located at Bangkhaen, Mahasawat, Samsen, and Thonburi area with the capacities 3.6, 0.8, 0.7, 0.17 million cubic meter /day. In addition to producing the quality water, sediment Soil has been generated from the water supply treatment process at the clarification station. They have been considered as wastes of the process and generated more than 0.1 million ton/year. Eliminating these sediment soil lead the high disposal costs to the MWA approximately 2×10^6 THB/ year.

As mentioned above, the problem has been studied and solved by many researchers. Suriyachat et al have reported the application sediment soil from water supply for ceramics products of the SME industry. The research found that ratio of sediment Soil: sand at 70:30 can be used for producing ceramic ware sintered at 800°-900° C [1]. The properties of lighted weight concrete using sediment soil from water supply process as coarse aggregate have been examined by Boonin et al. The results showed that compressive strength of concrete with the ratio 80:20 of plastic clay: sediment soil can achieve the ASTM C338 [2]. Chatsatapattayakul et al have used the chemical sludge from water treatment system with cement in cement mortar and soil cement block. They summarized that soil cement block mixing with 10-30% of chemical sludge met the standards of the Thailand Institute of Scientific and Technological Research (TISTR) for attending a good strength and water resistance [3]. Apart from utilizing sediment soil from water supply treatment process, the similar wastes of the water treatment process from the different sources have been studied. Teixeira et al showed that up to 20% sludge of water treatment plant (silt, sand and clay) can be incorporated into clays used for producing ce ramic bricks [4]. Martínez-García et al have assessed the effect of incorporating waste sludge on the properties and microstructure of clay used for bricks manufacturing. Replacing clay in a ceramic body with different proportions of sludge can reduce the cost due to the utilization of waste. Results have shown that incorporating up to 5 wt% of sludge is beneficial for clay bricks [5]. Investigation utilizing fresh water treatment sludge (WTS) on the physical and mechanical properties, structural parameters as well as mineralogical composition of the ceramic products has been examined by Kizinievic et al. It was found that WTS additive, which is mostly composed from colorific Fe_2O_3, can be utilized as a natural pigment that dyes the ceramic body in darker, more intense red color [6]. Moreover, industrial solid waste generated by a water treatment plant (WTP) at a pulp mill was used for manufacturing construction fired bricks. Wolff et al have proposed that sludge can be used as a substitute for clay in the formulation of clay masses and the mixtures between 50-85% at 850 and 950° C should be tested in the ceramic industry on a pilot scale in order to evaluate their suitability for the production of interior coatings or acoustic bricks [7].

However, the previous researches have not been studied the utilizing of sediment soil for ceramic vitreous china, which can be produced the sanitary ware, table ware, etc. Vitreous china has water absorption less than 1% with coefficient of thermal expansion 4-6x10^{-6}/K [8]. Therefore, this research has focused on studying the effect of sediment soil from water supply treatment process for producing ceramic vitreous china. The benefit of this study can promote the eco-friendly products which utilizing industrial wastes.

Table 1. Chemical analysis of raw materials

% Oxide	Raw Materials				
	Maetan Ball Clay	Ranong Kaolin Clay	Rayong Silica Sand	India Potash Feldspar	Sediment Soil
SiO_2	65.88	45.5	99.16	65.17	58.15
Al_2O_3	20	38	-	18.45	27.84
Fe_2O_3	1.37	0.8	-	0.17	7.17
TiO_2	0.53	0.03	0.03	0.01	0.86
CaO	0.22	0.01	0.03	0.22	0.94
MgO	0.48	0.01	0.02	0.01	1.37
K_2O	1.86	0.9	0.04	12.73	2.76
Na_2O	0.21	0.01	0.09	3.11	0.14
LOI	9.42	13	0.16	0.13	0.77

Table 2. Mixture composition of replacement sediment soil in maetan ball clay and ranong kaolin clay

Raw Materials	% Mixture								
	Group 1.Replacement in Maetan Ball Clay					Group 2. Replacement in Ranong Kaolin Clay			
	No. 1_1	No. 1_2	No. 1_3	No. 1_4	No. 1_5	No. 2_1	No. 2_2	No. 2_3	No. 2_4
Ranong Kaolin Clay	15	15	15	15	15	15	10	5	0
Maetan Ball Clay	35	30	25	20	15	35	35	35	35
India Potash Feldspar	30	30	30	30	30	30	30	30	30
Rayong Silica Sand	20	20	20	20	20	20	20	20	20
Sediment Soil	0	5	10	15	20	0	5	10	15

Note that : No 1_1= No 2_1

2 Materials and methods

2.1 Materials

In this experiment, five types of raw materials are exploited, namely, sediment soil, ball clay, kaolin, feldspar, and silica sand. Wastes material; sediment Soil from water supply treatment process has been utilized for replacement on basic materials. Sediment Soil used in this study has been generated from Mahasawat Plant. The chemical analysis of all materials has been represented in Table 1.

Notice that, all materials are local materials except India Potash Feldspar. The mixture of basic formula has been provided from Ubolrat and Vipa [9]. Considering chemical composition of sediment soil in Table 1, it indicates that main compositions are SiO2 and Al2O3 which similar to Maetan ball clay and Ranong clay. Therefore, the mixture of this experiment has been classified into two groups. Firstly, it is to replace sediment soil in Maetan ball clay. And secondly, sediment soil has been used as replacement material in Ranong kaolin clay. All of formulas have been represented in Table 2. It consists of 8 formulas which No.1_1 is the basic formula.

2.2 Methods

All of materials as mentioned above have been prepared as the procedure illustrated in Fig. 1. It consists of 9 steps. In the 3rd step, formulas of varying the proportion of materials have been constructed which consist of 8 formulas as mentioned in Table 2.

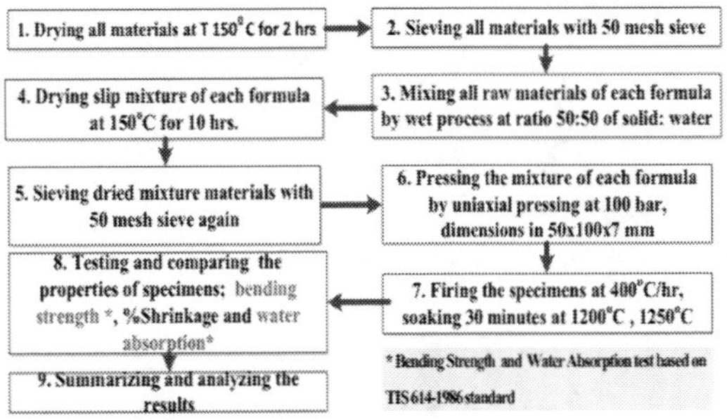

Figure 1. The 9 steps of this experiment.

3 Results and discussion

Specimens of each formula have been tested for analyzing the properties; bending strength, %water absorption, and %shrinkage. They can be described as follows.

3.1 Mechanical and physical properties

Three points bending has been used for testing bending strength of materials. All specimens of two material groups have been represented with their bending strength in Fig. 2 and Fig. 3. Increasing the ratio of sediment soil in the basic formula (No.1_1 or 2_1), the bending strength of them has gradually increased. Similarly, higher sintering temperature also increases the bending strength.

According to Fig. 4 and Fig. 5, both of two material groups have increased shrinkage when adding sediment soil in the basic formula. At the same time, sintering temperature increased also increase %shrinkage of all specimens.

On the contrary, water absorption property is decreased when increasing the ratio of sediment soil in the basic formula (No.1_1 or No.2_1). In addition,

DOI: 10.1051/matecconf/20179502005

increasing sintering temperature also decreases the water absorption of both two formula groups. These have been represented in Fig. 6 and Fig.7.

The results indicated that water absorption can be related to the shrinkage property, with higher shrinkage can provide lower water absorption. Furthermore, lower water absorption can promote the higher bending strength of specimens. These can be summarized that sediment soil has the potential to enhance the beding strength of vitreous china ceramics when comparing with using kaolin clay and ball clay.

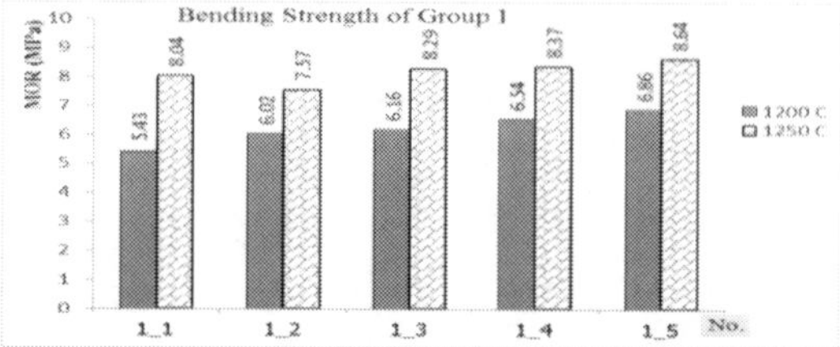

Figure 2. Bending strength of Group 1 at 1200° C, 1250°C

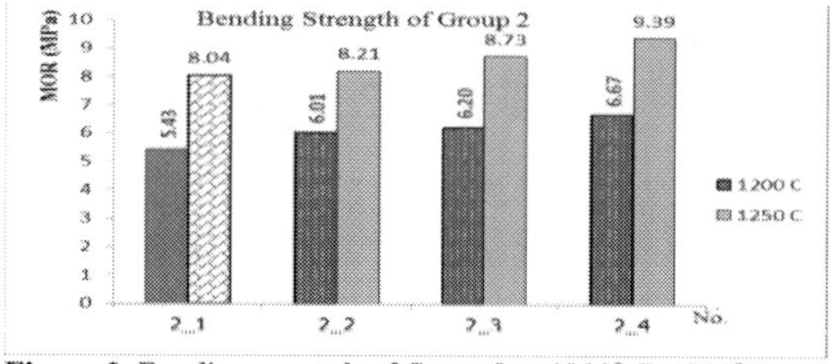

Figure 3. Bending strength of Group 2 at 1200° C, 1250°C

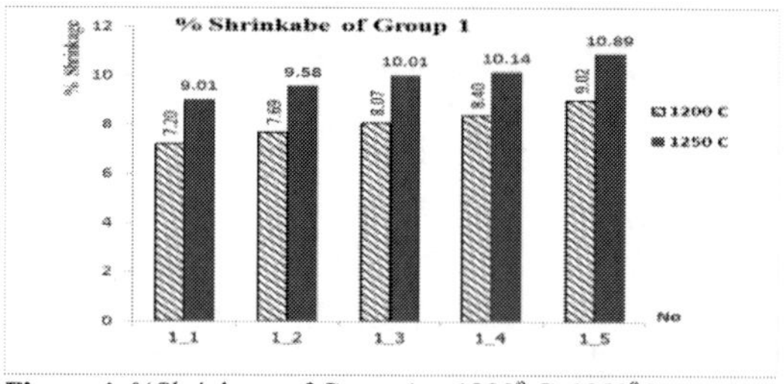

Figure 4. %Shrinkage of Group 1 at 1200° C, 1250°C

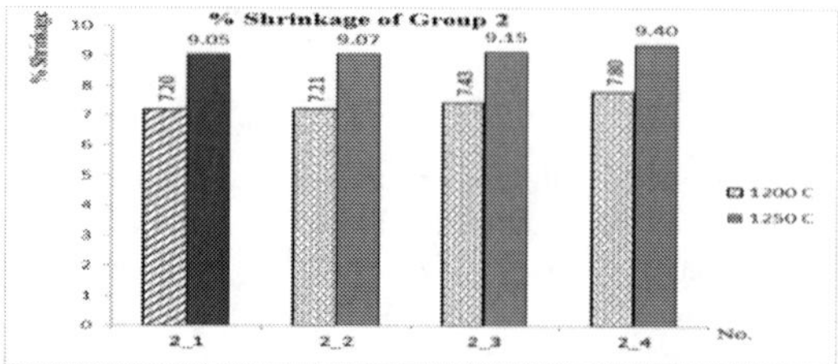

Figure 5 %Shrinkage of Group 2 at 1200° C, 1250°C

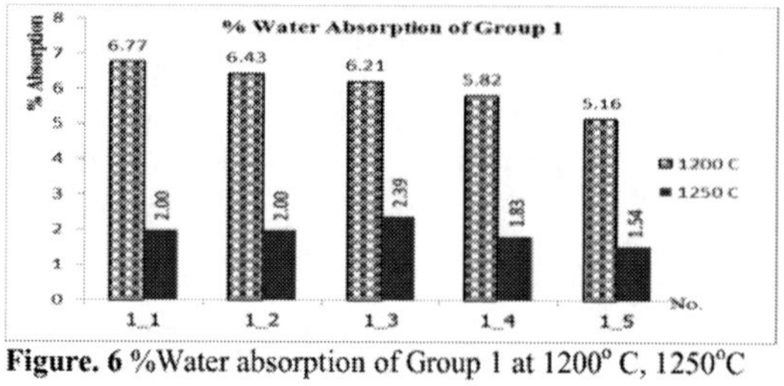

Figure. 6 %Water absorption of Group 1 at 1200° C, 1250°C

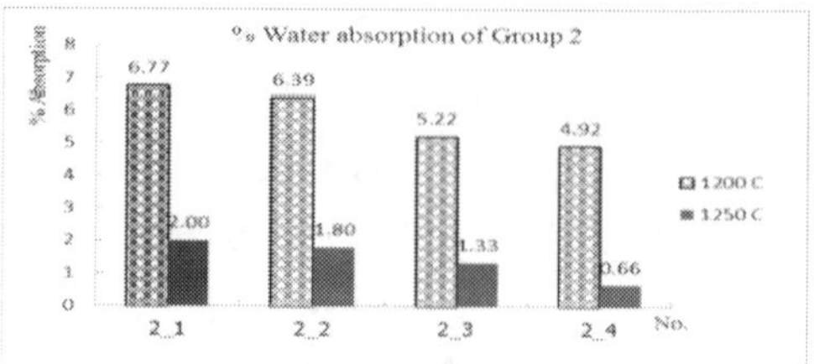

Figure 7. %Water absorption of Group 2 at 1200° C, 1250°C

3.2 Statistical analysis

For approving the results of this study, the experimental design has been conducted [10]. Two way ANOVA has been employed for analyzing the effect of raw material replacement on the properties of each formula. P-values of hypothesis testing with a 0.05 significance level of the experiment are illustrated in Table 3.

As considering Table 3, formula and temperature refer to factors of mixture number and sintering temperature. Their p values are less than 0.05, it means that formula number and temperature have the effects on bending strength, % shrinkage and % water absorption.

Table 3 Two way ANOVA test for 2 factors effect on bending strength, %Shrinkage, and % Water Absorption

P value	Group 1		Group 2	
	Formula	Temp	Formula	Temp
Bending Strength	0.012	0	0.003	0
%Shrinkage	0	0	0.015	0
%Water absorption	0	0	0	0

Two-way ANOVA: MOR versus Formula, Temp

Source	DF	SS	MS	F	P
Formula	4	10.439	2.6097	3.39	0.012
Temp	1	73.803	73.8028	96.00	0.000
Interaction	4	2.713	0.6782	0.88	0.478
Error	90	69.189	0.7688		
Total	99	156.143			

S = 0.8768 R-Sq = 55.69% R-Sq(adj) = 51.26%

Figure 8. Probability plot of bending strength of Group 1

Two-way ANOVA: MOR versus Formula, Temp

Source	DF	SS	MS	F	P
Formula	3	11.237	3.7458	5.06	0.003
Temp	1	97.753	97.7529	132.08	0.000
Interaction	3	8.443	2.8142	3.80	0.014
Error	72	53.288	0.7401		
Total	79	170.721			

S = 0.8603 R-Sq = 68.79% R-Sq(adj) = 65.75%

Figure 9. Probability plot of bending strength of Group 2

Two-way ANOVA: %Sh versus Formula, Temp

Source	DF	SS	MS	F	P
Formula	4	76.790	19.198	49.81	0.000
Temp	1	170.235	170.235	441.68	0.000
Interaction	4	0.226	0.056	0.15	0.964
Error	190	73.231	0.385		
Total	199	320.482			

S = 0.6208 R-Sq = 77.15% R-Sq(adj) = 76.07%

Figure 10. Probability Plot of Shrinkage of Group 1

Details of p value of determined properties for two groups and different temperature have been represented in Fig. 8-9, Fig. 10-11, and Fig. 12-13.

Two-way ANOVA: %Sh versus Formula, Temp

Source	DF	SS	MS	F	P
Formula	3	2.2457	0.7486	3.78	0.015
Temp	1	45.1144	45.1144	227.81	0.000
Interaction	3	1.2993	0.4331	2.19	0.098
Error	64	12.6740	0.1980		
Total	71	61.3334			

S = 0.4450 R-Sq = 79.34% R-Sq(adj) = 77.08%

Figure 11. Probability plot of shrinkage of Group 2

Two-way ANOVA: %Ab versus Formula, Temp

Source	DF	SS	MS	F	P
Formula	4	92.319	23.080	37.66	0.000
Temp	1	277.517	277.517	452.82	0.000
Interaction	4	72.502	18.125	29.57	0.000
Error	90	55.158	0.613		
Total	99	497.496			

S = 0.7829 R-Sq = 88.91% R-Sq(adj) = 87.80%

Figure 12. Probability plot of absorption of Group 1

Two-way ANOVA: %Ab versus Formula,Temp

Source	DF	SS	MS	F	P
Temp	1	383.472	383.472	437.17	0.000
Formula	3	32.327	10.776	12.28	0.000
Interaction	3	2.283	0.761	0.87	0.462
Error	72	63.156	0.877		
Total	79	481.238			

S = 0.9366 R-Sq = 86.88% R-Sq(adj) = 85.60%

Figure 13. Probability plot of absorption of Group 2

4 Conclusion

The chemical composition of sediment soil as shown in Table 1 indicates that its main composition is SiO_2 and Al_2O_3. However, it also contains the fluxing agents, namely, K_2O, CaO, Na_2O which are 2.76%, 0.94%, 0.14%. They take into account for reducing porosity and increasing bending strength of test pieces. Therefore, water absorption and bending strength have showed reducing and increasing of water absorption and bending strength, in Fig. 2,3 and 6,7. Due to reducing porosity, % shrinkage of specimens has been increased when increasing sediment soil, as shown in Fig. 4 and 5.

When recognition on different sintering temperature, it revealed that higher sintering temperature has promoted the higher bending strength, lower water absorption and higher shrinkage of the specimens.

Furthermore, different group of formula materials has given the slightly different properties of test pieces. The effect of replacement sediment soil in Ranong Kaolin clay has promoted the higher bending strength and lower water absorption than replacing in Maethan ball clay. As the effect of Ranong kaolin clay's chemical analysis contains lower fluxing agent than Maethan ball clay and sediment soil.

Because water absorption of vitreous china is less than 1%. The suitable formula of this study is No.2_4 sintering at 1250°C, with less than 1% of water absorption. Its bending strength is higher than the others, which is 9.39 MPa with coefficient of thermal expansion at 6.34×10^{-6} /K at 1000°C. Mixture of this formula is 0% kaolin, 35% ball clay, 30% feldspar, 20% silica sand and 15% sediment soil. It can be concluded that sediment soil can substitue ranong kaolin clay for producing vitreous china ceramics. This means that costs of material producing ceramic products has been decreased (Ranong kaolin clay: 6000 THB/ton) . In addition, the burden of sediment soil diposal of MWA also has been alleviated (20 THB/ton).

However, the future work is to prepare the material for casting the vitreous china and extracting Fe_2O_3 should be performed. If the results of this study can be the commercial products, it may response the customer needs who favors the eco-friendly products.

References

1. D. Suriyachat, P. Vichitamornpun and W. Ruangsumret, Application of Sediment Soils, Metallurgical Technology Group, Bureau of Primary Industries, Department of Primary Industries and Mines, Country Report, Thailand (2004).
2. K. Boonin, S. Sarapol and S. Deesui. Properties of Lighted Weight Concrete used Sediment Soil from Water Supply Process as Coarse Aggregate, Department of Civil Engineering, Faculty of Engineering, Thesis, Khon Kaen University (2012).
3. N. Chatsatapattayakul, M. Wangwiang and P. Pengthamkeerati, Potential use of chemical sludge from water treatment system with cement in cement mortar and soil cement block, Rajabhat Journal of Science, Humanities & Social Sciences. **13**(1) 48-54 (2012)
4. S.R. Teixeira, G.T.A. Santos, A.E. Souza, P. Alessio, S.A. Souza and N.R. Souza, The effect of incorporation of a Brazilian water treatment plant sludge on the properties of ceramic materials, Applied Clay Science, **53**, 561–565 (2011)
5. C. Martínez-García, D. Eliche-Quesada, L. Pérez-Villarejo, F.J. Iglesias-Godino and F.A. Corpas-Iglesias, Sludge valorization from wastewater treatment plant to its application on the ceramic industry, Journal of Environmental Management **95**, S343-S348 (2012)
6. O. Kizinievic, R. Zurauskiene, V. Kizinievic and R. Zurauskas, Utilisation of sludge waste from water treatment for ceramic products, Construction and Building Materials, **41** ,464–473 (2013)
7. E. Wolff , W. Keller Schwabe and S. Vieira Conceiçao, Utilization of water treatment plant sludge in structural ceramics, Journal of Cleaner Production xxx, 1-8 (2014)
8. Silicate ceramics, Information on http://old.vscht.cz/sil/keramika/Ceramic_Technology /SM-Lect-6-A.pdf
9. U. Ounprasertpong and V. Kittidumkeng, Cullet glass as a flux, Material Department, Faculty of Science, Project, Chulalongkorn University (1984)

10. M.J. Anderson, S.L. Kraber, Keys to successful designed experiments. Tech. Rep., Stat-Ease, Inc., Minneapolis, MN (2002). [Online]. Available: http://www.statease.com/pubs/doe-keys.pdf

DOI: 10.1051/matecconf/20179502006

A Surface Plasmon Resonance Histamine Sensor Based on an Electropolymerized Molecularly Imprinted Polymer (E-MIP)

Frances Adeline D. Hubilla, Arvin I.Mabilangan, Rigoberto C. Advincula and Florian R. del Mundo

Institute of Chemistry, University of the Philippines – Diliman, Quezon City, 1101 Philippines

Abstract. In this paper, histamine sensing by surface plasmon resonance (SPR) spectroscopy using an electropolymerized molecularly imprinted polymer (E-MIP) of functionalized polyterthiophene and histamine is reported. The MIP histamine sensor showed good linear relationship (R^2 = 0.995) between the SPR angle shift and the logarithm of histamine concentration, ranging from 15 – 500 μg/mL, with a limit of detection (LOD) of 2.0 μg/mL and a calibration sensitivity of 44.6 m° per log μg/mL histamine. Results also showed higher binding affinity of histamine for the MIP sensor than for the non-imprinted (NIP) sensor indicating excellent specificity. Moreover, the selectivity of the MIP histamine sensor against some of its structural analogues was demonstrated.

1 Introduction

Histamine is a by-product of spoilage and a good indicator of temperature abuse and state of good manufacturing practices adopted in the handling of fish/shrimp harvests. The World Health Organization (WHO) standard for good hygienic practices (GHP) and hazard analysis critical control point (HACCP) as applied to storing fish and shrimp products required that the level of histamine must be lower than 15 ppm (mg/kg) [1].

Conventional methods for histamine detection include HPLC [2], spectrofluorometry [3], colorimetry [4], ELISA [5], and the widely-accepted AOAC fluorometric method [6]. These methods have relatively low limits of detection for histamine, but require expensive instrument and tedious sample pretreatment to eliminate potential interference from other imidazole compounds such as histidine.

Molecular imprinting technology involves the fabrication of materials called molecularly imprinted polymers (MIPs), which are capable of recognizing target molecules with high selectivity and affinity [7]. From the interaction of the functional monomer polymerized around the template molecule, a highly cross-linked polymer complex will be formed. Removing the template molecule will produce an imprint of it on the polymer film and recognition sites for the template molecule will be available.

Surface plasmon resonance spectroscopy is a technique based on the interaction of surface plasmons- a charge density wave which occurs at the interface of a metal and a dielectric, and incident light. The incident electromagnetic wave, when in resonance with the plasmons, leads to collective oscillations, which is observed as a sharp dip in the reflected light intensity of the spectrum (Fig. 1a and 1b).

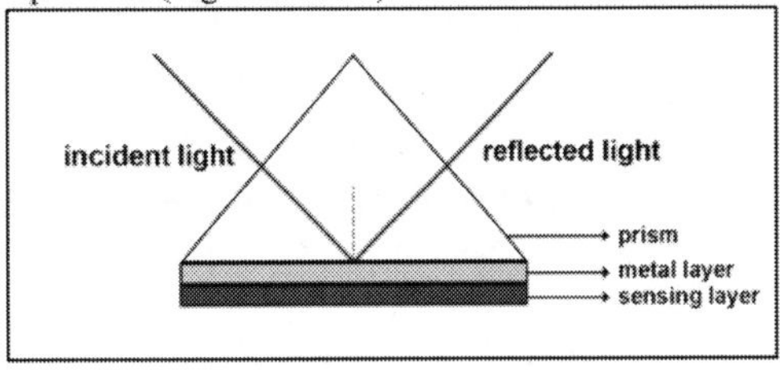

(a)

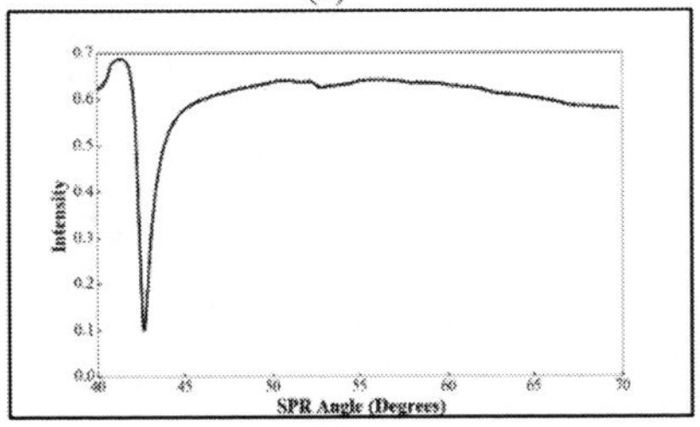

(b)

Figure 1. Kretschman configuration of SPR and **b.** An SPR curve showing the sharp dip caused by excitation of the surface plasmons.

Beyond the metal surface is the evanescent electric field. This field interacts with the area around the metal, which in this study is the area around the immobilized MIP. Every time there is a change in the evanescent region (e.g. presence of analytes interacting with the MIP), the SPR angle changes. This is the basis of SPR for biosensing applications [8].

The development of SPR-based sensors has been applied to a variety of fields over the recent years as they present an alternative in detection as compared to

DOI: 10.1051/matecconf/20179502006

conventional techniques [9]. Not only they can offer fast, sensitive, and on-site analysis, SPR sensors have also made significant contributions in high-throughput screening used in numerous important applications such as medical diagnostics, environmental monitoring, food safety, and security [10].

In this paper, the development of a surface plasmon resonance (SPR) histamine sensor using ultrathin films of electropolymerized molecularly imprinted polymer (E-MIP) of functionalized polyterthiophene and histamine is reported.

2 Experimental

2.1 Chemical reagents

Histamine (freebase), tetrabutylammonioumhexa-fluorophosphate (TBAH) ,cadaverine, putrescine, and L-histidine were purchased from Sigma-Aldrich (Missouri, USA). Acetonitrile (HPLC grade) and sodium hydroxide pellets (ACS reagent) were purchased from J.T. Baker. The terthiophene monomer (Sigma-Aldrich, Missouri) was functionalized with carboxylic acid (G03TCOOH) prior to use [11].

2.2 MIP fabrication

The optimum histamine:G03TCOOH ratio was predetermined using modelling studies via the semi-empirical PM3 calculations of the Orca program. All solids were dissolved with 0.1 M TBAH in 80:20 ACN:H_2O as supporting electrolyte.

The pre-polymer complex solution was electropolymerized by cyclic voltammetry in a standard three-electrode measuring cell using an Ivium Technologies CompactStat potentiostat. The reference electrode is an Ag/AgCl electrode, the counter electrode is a platinum rod electrode and the working electrode is a Bionavis Ltd SPR gold sensor slide (Au-coated BK7 glass slides, 12 x 20 mm). An NIP was also fabricated in the absence of the template molecule to act as the control. The imprinted template molecule was extracted using 0.01 M NaOH in 80:20 ACN:H_2O.

2.3 Performance evaluation of the fabricated E-MIP histamine sensor

SPR measurements were performed with SPR Navi™ 200 (BioNavis Ltd, Finland). The SPR response of the fabricated E-MIP sensor to standard histamine solutions was evaluated from 0-500 µg/mL. All histamine standard solutions were prepared in 75:25 MeOH:H_2O. From the generated standard calibration plot of SPR angle shift versus the logarithm of the histamine concentration (µg/mL), the linear concentration range, calibration sensitivity, LOD, and limit of quantitation (LOQ) were calculated. The MIP SPR responses to histamine and its structural analogues were also compared.

3 Results and discussion

3.1 MIP fabrication

Histamine interacts with G03TCOOH through non-covalent interaction in the aliphatic amine group of histamine. The molecular ratio between histamine and G03TCOOH was optimized through modelling studies and was found to be 1:3 (histamine:G03TCOOH), shown in Fig. 2a and 2b.

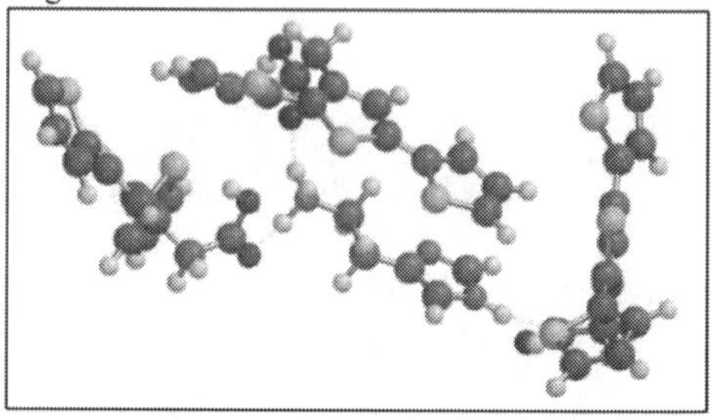

(a)

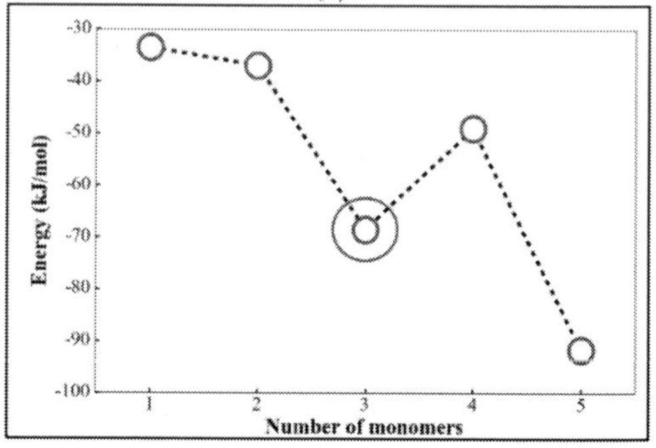

(b)

Figure 2. optimized simulated structure of histamine complexed with three G03TCOOH molecules and **b.** calculated stabilizing interaction in histamine-G03TCOOH complexes. Encircled point indicates the number of G03TCOOH monomers which gives the first and most stable arrangement with histamine.

Fig. 3 illustrates the SPR curves of the MIP at different stages of fabrication (gold slide only, electropolymerization, template removal, rebinding). A noticeable shift in the SPR angle will be observed due to changes in the dielectric constant of the film layer upon interaction of the template and the MIP sensor [12].

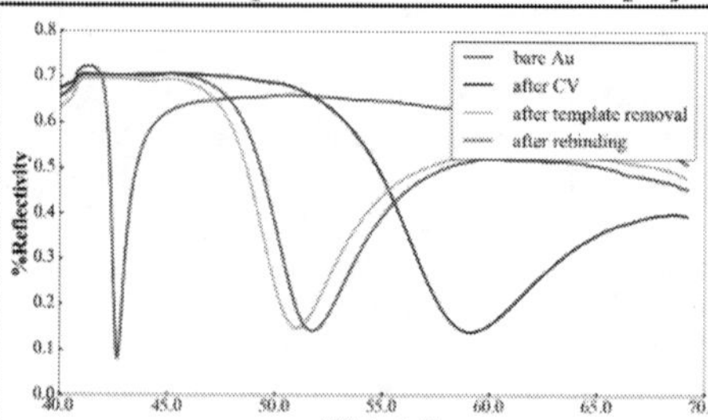

Figure 3. SPR curves of one MIP sensor at different fabrication stages.

Using the pre-determined template monomer ratio of 1:3, the sensor was electropolymerized using cyclic voltammetry. The number of cycles and scan rate were optimized as these affect the thickness of the sensor.

For the extraction of histamine from the electropolymerized polymer complex, a basic solution containing NaOH was used. The base deprotonates the aliphatic amine group of histamine, dissociating the complex and releasing histamine into the solution [13].

SPR was used to monitor the template removal time by obtaining the SPR curves at certain time intervals. A plot of angle shift versus time is shown in Fig. 4.

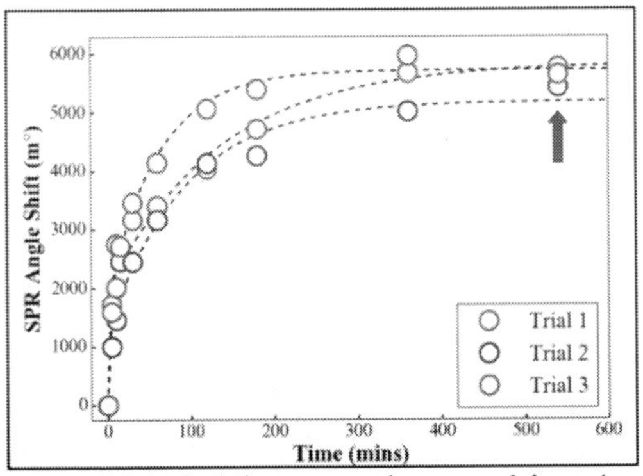

Figure 4. SPR angle shift versus template removal time using 0.01 M NaOH in 80:20 ACN:H_2O for 1mM histamine:3mM terthiophene MIP sensor.

Optimum time was chosen when the SPR angle shift is less than 10 millidegrees per 5 minutes. As shown in Fig. 4, at the optimum time of nine hours, it is assumed that the 1 mM histamine template has been completely extracted from the polymer, hence no changes in the optical thickness (angle shift) can be observed.

Rebinding studies on the extracted histamine were also performed using SPR. The time it takes for the MIP sensor to reach equilibrium response upon rebinding was determined. The sensor was equilibrated with 1mM histamine in 75:25 MeOH:H_2O at different time intervals. SPR readings were taken at the end of each time interval and a plot of SPR angle shift at each time interval was made as shown in Fig. 5.

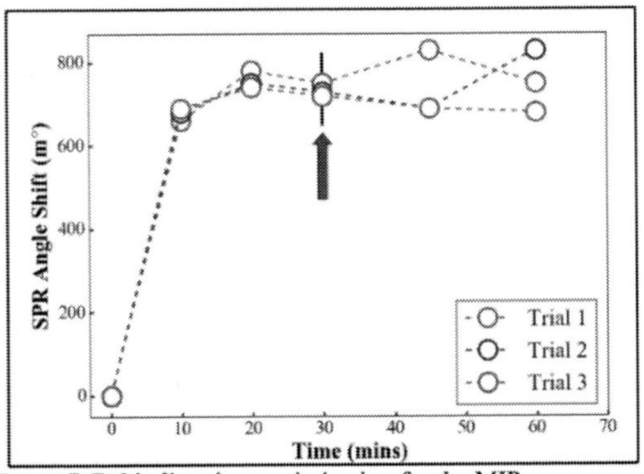

Figure 5. Rebinding time optimization for the MIP sensor. Concentration of the injected histamine solution is 1 mM in 75:25 MeOH:H_2O.

At the plateau portion of the plot, no significant change in the optical thickness (angle shift) was observed. The optimum equilibrium rebinding time of 30 minutes was selected where the SPR angle shift is less than 10 millidegrees per 5 minutes. At this point the MIP sensor has been completely saturated with the injected histamine molecules. This optimum rebinding time was used for the succeeding rebinding experiments, including calibration and selectivity studies.

To check if histamine was indeed imprinted on the MIP sensor, rebinding studies were also performed on the NIP. Injection of histamine in the NIP should show no significant response as it does not have the binding sites for histamine. The plot in Fig. 6 showed that the NIP exhibited no significant response upon injection of histamine at different concentrations compared to the MIP. The slight increase in response in the NIP however, may be attributed to the nonspecific binding of histamine to the NIP.

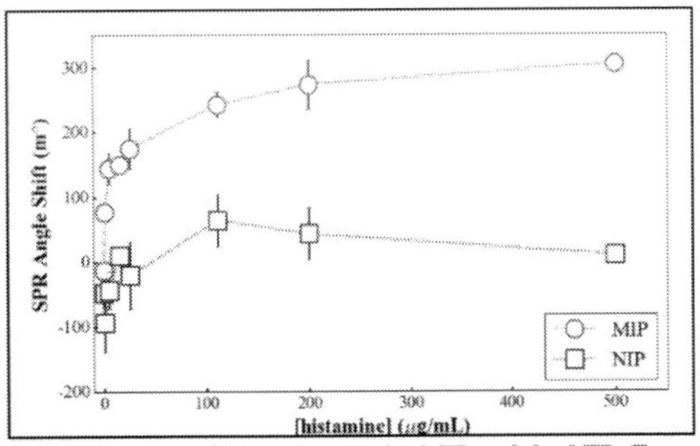

Figire 6. Binding of histamine on the MIP and the NIP. Error bars represent standard deviation of 3 trials.

3.2 Evaluation of the analytical performance of the histamine MIP sensor

The generated response curve of SPR angle shift versus histamine concentration from 0-500 μg/L is shown in Fig. 7.

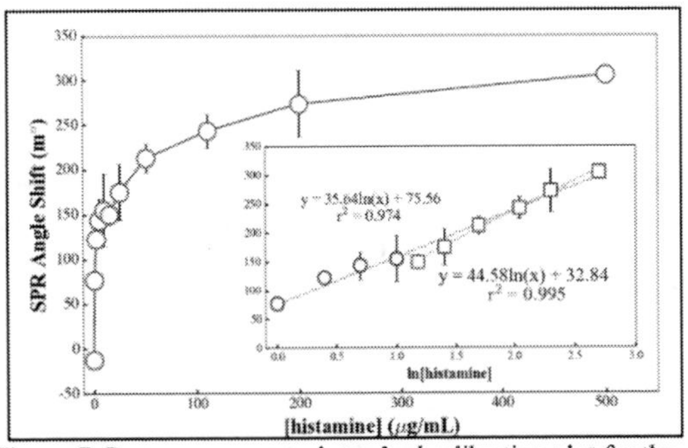

Figure 7. Response curve and standard calibration plot for the histamine MIP sensor. Squares correspond to the linear range.

The response curve is similar to a typical binding isotherm with an initial sharp rise followed by gradual saturation. The standard calibration plot (Fig. 7 inset) of the SPR angle shift versus log histamine concentration (μg/mL) showed good linear relationship (R^2 = 0.995) from 15-500 μg/mL. The LOD and LOQ of the sensor were calculated using the following equations:

$$\text{LOD} = \frac{3\sigma_{\text{blank}}}{m} \quad (1)$$

$$\text{LOQ} = \frac{10\sigma_{\text{blank}}}{m} \quad (2)$$

In equations (1) and (2), σ_{blank} refers to the standard deviation of the blank readings and *m* corresponds to the slope of the standard calibration curve. The obtained LOD and LOQ of the sensor are 2.0 µg/mL and 12 µg/mL, respectively. The calibration sensitivity is equal to 44.6 m° per log µg/mL histamine.

Fig. 8 shows the relative response of the the MIP sensor to histamine structural analogues. The MIP sensor showed no response to L-histidine, and relatively lower responses to cadaverine and putrescine compared to histamine indicating high selectivity of the fabricated MIP sensor to histamine. The possibility that most of the binding sites in the MIP complements the aliphatic amine group of histamine might account for the increasing trend found in the response of the MIP to cadaverine and putrescine, as these two analogues also have aliphatic amines in their structure.

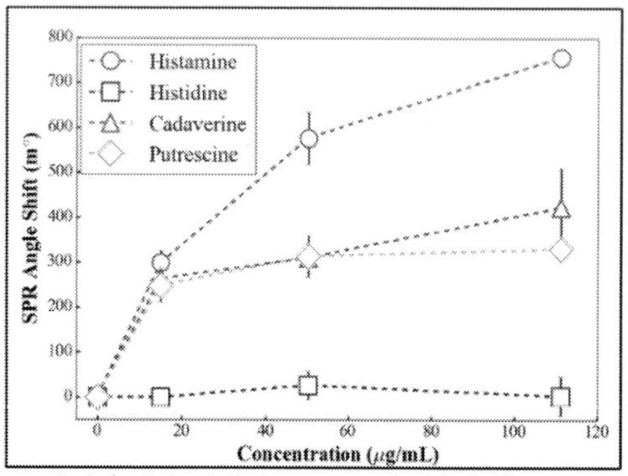

Figure 8. Selectivity (relative response) of the MIP sensor to histamine compared to L-histidine, cadaverine, and putrescine.

4 Conclusions

Detection and sensing of histamine was successfully achieved with the use of SPR spectroscopy and E-MIP of polyterthiophene and histamine. The analytical performance of the fabricated MIP sensor showed a good linear relationship ($R^2 = 0.995$) from 15-500 µg/mL, an LOD of 2.0 µg/mL and a calibration sensitivity equal to 44.6 m° per log µg/mL histamine. The MIP sensor also exhibited high selectivity for histamine versus various histamine analogues like L-histidine, cadaverine, and putrescine. The fabricated MIP histamine sensor has high potential for use as a quality assurance sensor to detect "freshness" of fish and shrimp harvests.

Acknowledgement

This work was supported by the Department of Science and Technology – Philippine Council for Industry, Energy, and Emerging Technology Research and Development (DOST-PCIEERD). The authors also would like to thank Dr. Rigoberto C. Advincula for the donated functionalized terthiophene monomer.

References

1. FAO/WHO (Food and Agriculture Organization of the United Nations/World Health Organization), *Public Health Risks of Histamine and Other Biogenic Amines from Fish and Fishery Products Meeting Report* (2013)
2. A. Sahebnasagh, G. Karimi, and S.A. Mohajeri: Food Anal. Methods, **7** (2014)
3. M.C. Vidal-Carou, M.L. Izquierdo-Pulido, A. Mariné-Font: JAOAC, **72** (1989)
4. T. Kuda, T. Mihara, and T. Yano: Food Control, **18** (2007)
5. A. Marcobal, M.C. Polo, P.J. Martín-Álvarez, and M.V. Moreno-Arribas: Food Res. Int., **38** (2005)
6. AOAC (Association of Official Analytical Chemists), JAOAC, **60** (1977)
7. W. Li and S. Li, Adv. Polym. Sci. **206** (2007)
8. M. Bera and M. Ray: Opt. Eng. **50** (2011)
9. H. Šípová, S. Zhang, A.M. Dudley, D. Galas, K. Wang, and J. Homola: Anal. Chem., **82** (2010)
10. D.C. Apodaca, R.B. Pernites, R. Ponnapati, F.R. del Mundo, and R.C. Advincula: Macromolecules, **44** (2011)
11. M. Piliarik, L. Párová, and J. Homola: Biosens. Bioelectron., **24** (2009)
12. P. Taranekar, T. Fulghum, A. Baba, D. Patton, and R.C. Advincula: Langmuir, **23** (2007).
13. A. Pietrzyk, S. Suriyanarayanan, W. Kutner, R. Chitta, and F. D'Souza, Anal. Chem., **81** (2009)

Micro-Bulges Investigation on Laser Modified Tool Steel Surface

Fazliana Fauzun[1], Syarifah Nur Aqida[1,2], Izwan Ismail[2,3] and Norhafzan Bariman[1]

[1]*Faculty of Mechanical Engineering, Universiti Malaysia Pahang, 26600 Pekan Pahang, Malaysia*
[2]*Automotive Engineering Centre, Universiti Malaysia Pahang, 26600 Pekan Pahang, Malaysia*
[3]*Faculty of Manufacturing Engineering, Universiti Malaysia Pahang, 26600 Pekan, Malaysia*

Abstract. This paper presents micro-bulges investigation on laser modified tool steel. The aim of this study is to understand the effect of laser irradiance and interaction time on surface morphology configuration. An Nd:YAG laser system with TEM_{00} pulse processing mode was used to modify the samples. Metallographic study shows samples were analyzed for focal position effect on melted pool size, angle of peaks geometry and laser modified layer depth. Surface morphology were analyzed for surface roughness. Laser modified layer shows depth ranged between 42.22 and 420.12 μm. Angle of peak bulge was found to be increase with increasing peak power. The maximum roughness, R_a, achieved in modified H13 was 21.10 μm. These findings are significant to enhance surface properties of laser modified steel and cast iron for dies and high wear resistance applications.

1 Introduction

In most engineering cases, the designers are responsible to develop surface that have the ability to perform maximum with longer lifespan at the lowest cost. Common problem in most technology systems found to be wear damage which subjected to elevated temperature during operation [1]. Surface topography defined three characteristics of surface finishing which include surface roughness, lay and waviness [2]. Surface modification importance to most industrial applications which are surface dependent such as adhesion, friction, corrosion resistance and wettability [1], [3], [4]. Surface roughness plays an important role in controlling friction and wear between two contact areas during processing [3], [5], [6]. Excessive friction between contact surfaces after thousands cycles during processing produces abundant heat resulted in reduced die lifecycle [1], [7].

In some cases, surface roughness is crucial to enhance adhesion between coating materials and substrate, holding lubricant in machining process and ease component removal from die surface in forming process [1], [5], [8], [9]. Surface texture affects friction by creating resistance that prevented fluid from escaping from surface contact [3], [5]-[7]. Surface topography and structure influenced friction that between contact surfaces. It became one of key which affect the coefficient of friction (COF) especially during transferring film formation in sliding application [3].

Surface texturing known as one best method using laser application. Specified microstructure on laser treated surface were importance to reduce and control friction and wear in most engineering application. Surface texturing has been applied in mechanical components such as piston rings, thrust rings, sliding bearings, machine tool guide and puncher for friction properties during hot stamping [2].

Various manufacturing process such as machining, grinding, abrasive blasting, milling and electrical discharge machining (EDM) have been introduced to develop surface texture by previous researcher [2], [3], [10], [11]. To date, laser surface texturing became one of the most promising means due to its precision and ability to configure high texture complexity.

Laser melting offers surface properties modification at controlled parameters; energy density, pulse duration, laser spot size and pulse repetition frequency (PRF) [12]-[14]. Intense energy penetration rapidly melted and solidified metal surface for high hardness properties. Most laser processing experimental design prefers pulse mode for a better control the surface integrity. Overlapping pulses, parameters design and material flow behavior affected micro-bulges formation and vary topography of the modified surface [7]. Previous studies found irregularities and mechanical effects resulted from the laser pulses which analyzed by the distribution of residual stress and microhardness on treated surfaces [11]. In order to design most engineering applications, good understanding between mechanical interaction between rough surface become essential [7].

Other than that, laser scanning speed gives crucial relevance in determining surface morphology. Together with high power and high pulse repetition rate, laser scanning speed produced varies machining processes such as texturing, patterning, engraving and structuring could be executed [15]. Nevertheless, the parameters

 DOI: 10.1051/matecconf/20179502007

need to be processed at sufficient energy to avoid surface vaporization which gives burning effect. In this study, laser surface modification was conducted on AISI H13 tool steel at constant scanning speed, PRF and laser energy. Surface morphology, roughness and micro-bulges formation on laser modified surface were analyzed. The effects of peak power to the surface roughness and angle of micro-bulges were investigated.

2 Experimental

Laser surface modification was conducted using a JK300HPS Nd:YAG twin lamp laser source with TEM_{00} mode and 1064 nm wavelength. A flat AISI H13 tool steel plate of 100 x 300 mm dimension was used in as-received condition. The chemical composition of all materials involved is given in Table 1. The samples were processed at controlled parameters of peak power (P_P) from 1.7 kW to 2.5 kW, PRF of 50 Hz and scanning speed (v) of 1000 mm/min. Three responses investigated were depth of modified surface, surface roughness and angle of micro-bulges.

Table 1. Chemical composition of AISI H13 tool steel.

	C	Si	Mn	S
	0.401 - 0.473	0.968 - 1.000	0.327 - 0.343	0.0055 - 0.0069
	Co	Cu	Ti	V
	0.0199 - 0.0207	0.08 - 0.165	0.0094 - 0.0123	0.901 - 0.927
AISI H13 steel	Cr	Mo	Ni	Al
	4.92 - 5.03	1.17 - 1.24	0.073 - 0.107	0.024 - 0.046
	Co	Cu	Ti	V
	0.0199 - 0.0207	0.08 - 0.165	0.0094 - 0.0123	0.901 - 0.927
	W	Nb	Fe	
	0.0252 - 0.0399	0.003	Balance	

Parameters were set to produce significant amount of energy of 2.0 J to melt the sample surface. The samples surface was placed perpendicular to the firing direction of laser beam and was linearly translated by CNC moving control system. The laser processing was assisted by argon gas to avoid surface oxidation. Metallographic study was conducted by IM7000 Series Inverted Optical microscopes. Micro-bulge geometry was designed using computer aided design (CAD) software. Surface roughness was measured by Mitutoyo SURFTEST SJ-410 Roughness Tester.

3 Results and discussion

The surface cross-sectional micrographs in Table 2 indicate presence of laser modified layer on H13 tool steel substrate with corresponding surface morphology. At constant average power of 100 W and 2.0 J laser energy, the surface angle resulted from micro-bulges in micrographs A, B and C increased with increasing peak power. The respective average angle of peak bulge in sample A, B and C was 16° 32° and 36°. Referring to the surface morphology, the micro-bulges direction is determined by linear translation path. The resultant surface roughness was 6.18 μm, sample B with 11.39 μm while sample C with 21.10 μm as peak power increased from 1.7 kW to 2.5 kW.

The depth of modified layer also show the same increment pattern from 272.62 μm for sample A, while 300.71 μm and 338.25 μm for sample B and C respectively. The increase of angle was due to increase of sample surface temperature at high temperature. This can be seen from laser spot marks diameter from A with 0.62 mm while B and C with 0.64 and 0.68 mm respectively from the surface morphology. Different directions of laser linear translation affect the bulges pattern formation as shown in micrograph of sample A, B and C in Table 2.

Though laser spot size used was 0.6 mm, a larger diameter was measured on surface morphology of sample C, as high surface temperature extended the melting zone. The higher surface temperature of samples processed at peak power of 2.0 and 2.5 kW increased molten surface fluidity [12]. High fluidity of molten surface inclined to form micro-bulges when progressing to consequent pulses due to the translation speed of 1000 mm/min.

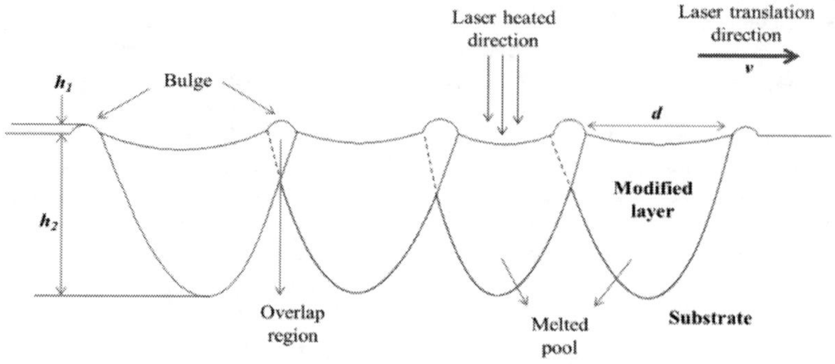

Figure 1. Schematic diagram of the geometrical parameters of laser modified layer.

Formation of micro-bulges is further explained in schematic diagram of Fig. 1. In laser modified layer of Fig. 1, the height of bulge (h_1), height of melted pool h_2, diameter d are determined by laser peak power. The melted pool formation processed on a planar surface starts sequentially from one to the end process along the laser translation direction. Increased peak power gives high impact force by the laser power which influence the depth (h_2) and diameter (d) size of melted pool and formation of bulges [11]. The height of bulges was produced by the high recoil pressure transmitted by the laser radiance [2]. Referring to surface morphology of sample A in Table 2, laser impact direction onto the substrate with high power increased the volumetric force on the surface [2]. This phenomenon resulted in increased bulges height as shown in sample B and C surface morphology based on darker shading at the spot size edges (d) size of melted pool and formation of bulges [11].

The height of bulges was produced by the high recoil pressure transmitted by the laser radiance [2]. Referring to surface morphology of sample A in Table 2, laser impact direction onto the substrate with high power increased the volumetric force on the surface [2]. This phenomenon resulted in increased bulges height as shown in sample B and C surface morphology based on darker shading at the spot size edges. High scanning speed influence the formation of bulges peak angle direction either in x or z-axis.

Surface-laser beam interaction time was essential as it determined the surface geometry of laser modified layer.

The factor which influences the interaction between laser and substrate surface found to be material composition, laser energy, frequency and pulse width [9], [12], [14].

Table 2: Micrographs, surface morphology, surface roughness and depth of modified layer at different peak power settings at 1000 mm/min scanning speed and 50 Hz PRF.

Sample	A	B	C
Angle of peak bulge [°]	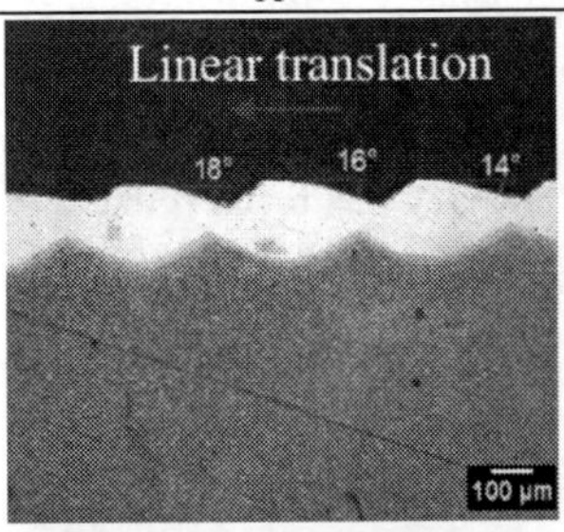	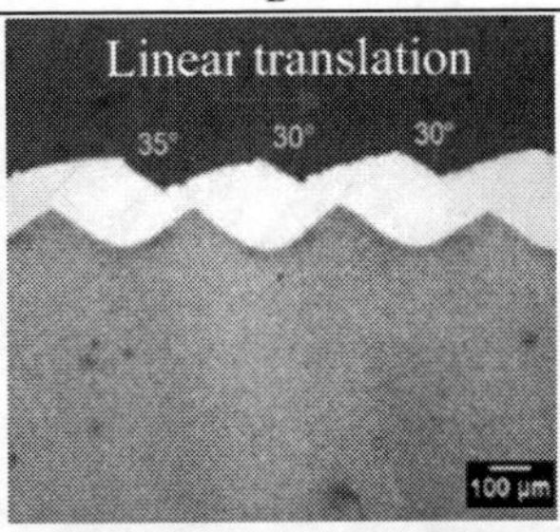	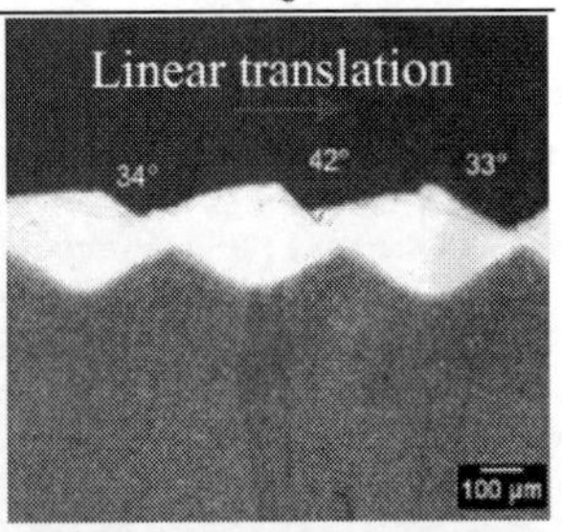
Surface morphology	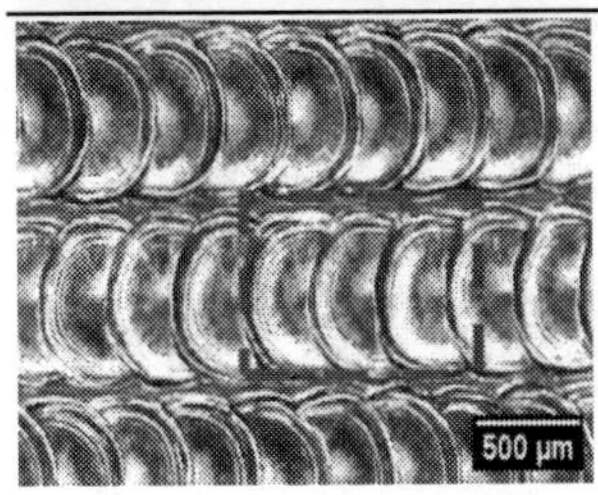	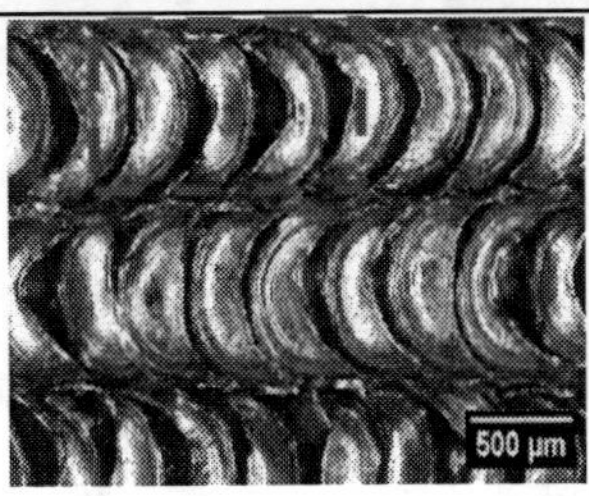	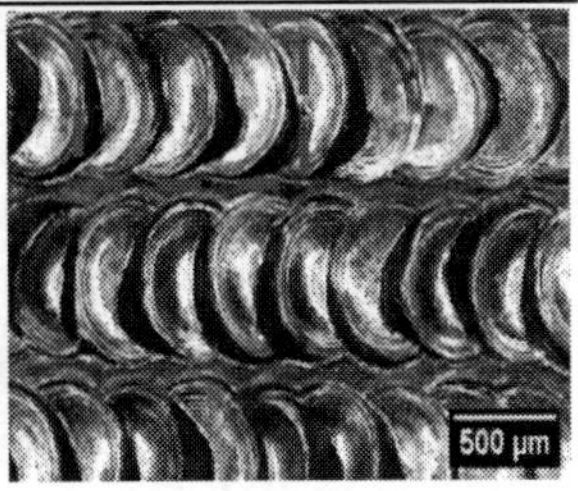
Peak Power [kW]	1.7	2.0	2.5
Surface Roughness [Ra]	6.18	11.39	21.10
Depth of modified layer [µm]	272.62	300.71	338.25

Laser beam generates high strength impact power on the surface thus introduced residual stress. This phenomenon found by previously studies on mechanical effects and regularities due to pulse laser beam [11]. Mechanical response under pulsed laser emitted on processed surface reflected the microscopic surface topography evolution, thus caused the laser frequency to be one of important factor need to be control in order to design surface texture [3], [7], [11]. During laser processing at linear translation, surface material absorbed an amount of energy at from pulse laser that lead to re-melt superficially on overlap region [9]. This is the phenomenon which creates molten pool with bulge on the surface thus influence surface geometry and roughness [2], [7], [13].

4 Conclusion

Experimental analysis of three laser modified sample produced an average surface roughness from 6.11 µm to

21.10 μm with average bulges peak angle ranged from 14° to 42° concluded followings details:

1. Surface morphology of laser modified H13 tool steel comprised micro-bulges at different angle due to peak power intensity.
2. Laser modified layer depth and angle of bulges were increased with increasing peak power due to higher laser irradiance transmitted into substrate.
3. Average peak power ranged from 1.7 kW to 2.5 kW transmitted sufficient amount of energy of 2.0 J to melt the substrate surface.
4. Laser scanning speed influenced direction of bulge peak angle thus produced vary surface morphology for all samples.

These findings signify preliminary effects of parameter on micro-feature geometry formation using laser processing. Laser surface modification introduced one of the best methods of surface texturing due to its high precision processing.

Acknowledgement

This work was financially supported by Universiti Malaysia Pahang (Grant no.: RDU1403150).

References

1. Tomala, A., et al., *Tribological performance of some solid lubricants for hot forming through laboratory simulative tests.* Tribology International. **74**: p. 164-173, (2014).
2. Zhou, J., et al., *Experimental study on laser microstructures using long pulse.* Optics and Lasers in Engineering. **78**: p. 113-120, (2016).
3. Menezes, P.L. and S.V. Kailas, *Role of surface texture and roughness parameters on friction and transfer film formation when UHMWPE sliding against steel.* Biosurface and Biotribology. **2**(1): p. 1-10, (2016).
4. Dunn, A., et al., *Laser surface texturing for high friction contacts.* Applied Surface Science. **357, Part B**: p. 2313-2319, (2015).
5. Vlădescu, S.-C., et al., *Combined friction and wear reduction in a reciprocating contact through laser surface texturing.* Wear. **358–359**: p. 51-61, (2016).
6. Jeong, S.-H., et al., *Effects of solid lubricant and laser surface texturing on frictional performance of pulse electric current sintered Al2O3–ZrO2 composites.* Ceramics International. **42**(6): p. 7830-7836, (2016).
7. Kalin, M., et al., *Comparing surface topography parameters of rough surfaces obtained with spectral moments and deterministic methods.* Tribology International. **93, Part A**: p. 137-141, (2016).
8. Medvedev, A.E., et al., *Effect of bulk microstructure of commercially pure titanium on surface characteristics and fatigue properties after surface modification by sand blasting and acid-etching.* Journal of the Mechanical Behavior of Biomedical Materials. **57**: p. 55-68, (2016).
9. Viana, R., et al., *Laser texturing of substrate of coated tools — Performance during machining and in adhesion tests.* Surface and Coatings Technology. **276**: p. 485-501, (2015).
10. Kurniawan, R., G. Kiswanto, and T.J. Ko, *Micro-dimple pattern process and orthogonal cutting force analysis of elliptical vibration texturing.* International Journal of Machine Tools and Manufacture. **106**: p. 127-140, (2016).
11. Lu, G.X., et al., *Microscopic surface topography of a wrought superalloy processed by laser shock peening.* Vacuum. **130**: p. 25-33, (2016).
12. Aqida, S., D. Brabazon, and S. Naher, *Atomic diffusion in laser surface modified AISI H13 steel.* Applied Physics A. **112**(1): p. 139-142, (2013).
13. Fauzun, F., et al. *Optimisation of Pulsed Nd: YAG Laser Processing of Gray Cast Iron for Enhanced Surface Properties.* in *Advanced Materials Research.* (2014).
14. Fauzun, F., M. Wahab, and S.N. Aqida, *Laser Surface Modification of AISI 1025 Low Carbon Steel Using Pulsed Nd: YAG Laser for Enhance Surface Properties.* Key Engineering Materials. **554**: p. 596-602, (2013).
15. Mincuzzi, G., et al., *Extending ultra-short pulse laser texturing over large area.* Applied Surface Science. **386**: p. 65-71, (2016).

A Master Shape of Bottles for Design under Desirable Geometry and Top Load Test

Piyapat Keawjaroen and Chakrit Suvanjumrat

Department of Mechanical Engineering, Faculty of Engineering, Mahidol University, Salaya, Nakorn Pathom, 73170, Thailand
Laboratory of Computer Mechanics for Design (LCMD), Department of Mechanical Engineering, Faculty of Engineering, Mahidol University, Salaya, Nakorn Pathom, 73170, Thailand

Abstract. Design of plastic bottles had preferred to use computer aided design (CAD) to propose desirable shapes. The strength also was regarded to pass the top load test unless an appearance of plastic bottles. Finite element method (FEM) was employed to analyze and predict the bottle shape which enough to support load under a collapsible regulation. Unfortunately, the redesign of bottle shape always performed when the desirable bottle shape had not passed the test. There was time consumption and loss of opportunity to compete producing of bottles. This research proposed a method to receive a desirable shape of plastic bottles together with top load strength. The master of bottle shape had been created which capable to change any dimensions before generated CAD and performed top load analysis with FEM. The artificial neural network (ANN) was employed to obtain the desirable bottle shape with top load resistance by varying dimension of the master bottle. The plastic bottle design would be performed rapidly with the ANN of master bottle shape. Consequently, the suitable dimension of plastic bottles which achieved by ANN could be used to design a desirable shape of bottles by using CAD and FEM without trial and error.

1 Introduction

Plastic bottles were produced to contain liquid must be had strength to support top load causes a warehousing according to ASTM D2659 [1]. To reduce trial and error by production, computer aided design (CAD) and computer aided engineering (CAE) had been used to design and analyze bottle shape before real production [2]-[4]. Unfortunately, the bottle production will use CAD/CAE method repeatedly until achieves a desirable shape of plastic bottles. The product cost will be increased by a consumption of time and design number. Furthermore, the opportunity to achieve production of plastic bottles from customer will be lost.

Artificial neural network (ANN) is a computational model that mimics human brains to learn relation between inputs and outputs from experiment. Many researchers have applied ANN to design products [5]-[7]. The important feature of ANN is the ability to estimate complex relationships without any prior assumptions about the nature and interrelations of the underlying process. The transfer functions, weight and output are constructed the neural network to recieve input and predict output, respectively [8]. The ANN is trained by input and output data until it can predict output by itsilf [9] [10].

This research has developed the design process of plastic bottles by using ANN. The master shape of 1 liter bottle has created and analyzed the top load resistance. The simulation data is used to train ANN which is the design tool of plastic bottle. The desirable shape of plastic bottle which achieve by ANN will be created CAD model and analyzed with FEM of top load test to produce plastic bottles in the future.

2 Master shape of bottle

2.1 CAD model

The bottle model is created by classify a capacity of 1 liter, no handle and rectangular base from all bottle products of the company. The master bottle shape is defined and generated by SolidWorks software as illustrated by in Fig. 1. The shape of bottle can vary by curvatures which include R_1, R_2, R_3, R_4 and R_5. The curvature of bottle neck (R_5) has many effects to collapse bottles by top load test therefore it is extreamely brought to be a parameter of the master shape. The bottle width is limitted to be constant with a value of 109.51 mm. Fig. 1 shows the 3D-model of the master shape bottle. The bottle neck does not have a screw thread detail because it does not have any effect to simulate the top load test for the bottle shape design. The diameter of bottle neck is limited at 36.55 mm while the height is 232.42 mm. The curvatures of the master model have relation to the bottle geometry therefore it has a dimensional limitation for creating of the 1 liter bottle model to produce the real bottle shape in manufacturing process.

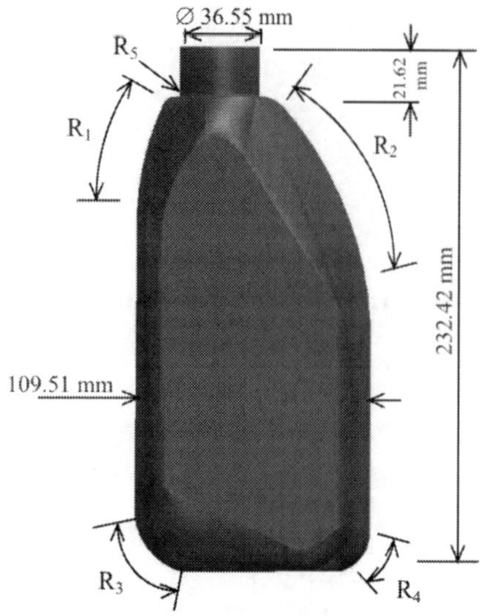

Figure 1. The variable dimension of a 1 liter master shape bottle.

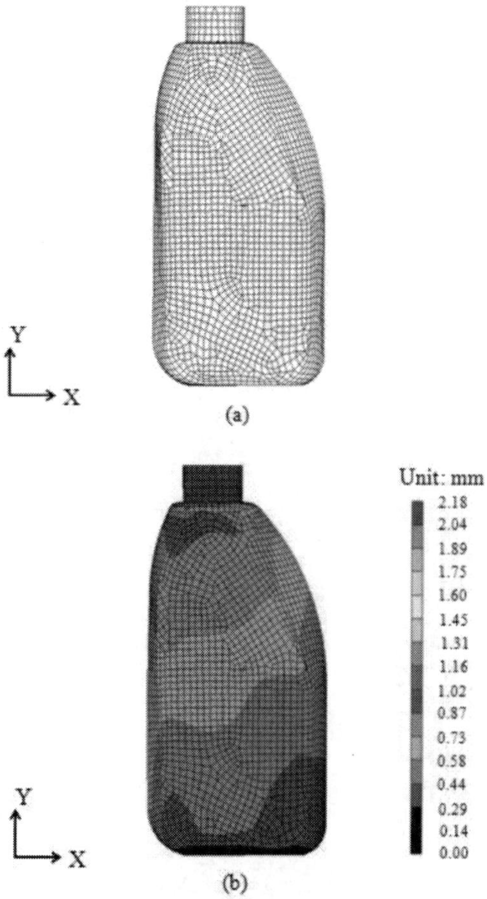

Figure 2. The finite element model of the master shape bottle under top load analysis as a result of: (a) a mesh structure and (b) mesh deformation with a color contour of the vertical displacement.

2.2 Top load analysis

The CAD model of a master shape is transformed into finite element analysis (FEA) software, MSC.Nastran, to simulate top load test with a constant mass of 25 kg. The quadrilateral plate element is used to create the finite element model of the master shape bottle as shown in Fig. 2a. The total plate elements to created finite element model are 4,625 elements. The thickness of plate elements or the bottle wall is assign to be a constant value which is 1 mm. The 1 liter bottle is produced by the high density polyethylene (HDPE) material by the extrusion blow molding process. Therefore, the material properties of the bottle model obtain by the tensile test of HDPE specimens according to ASTM D638 [11]. The Young's modulus and a yield stress of the HDPE material is 522.45 MPa and 27.34 MPa, respectively. The bottle model has been supported the vertical top load on the bottle neck therefore it is collapsed. The deformation of bottle wall is illustrated by color contour which red is the maximum vertical displacement and dark blue is the minimum vertical displacement (Fig. 2b). The maximum vertical displacement of the bottle deformation which varies by the variable shape of the master shape bottle is used to be the output data while curvatures of the master bottle is the input data of ANN, respectively.

3 Artificial neural networks

The architecture of ANN in this research is using the error back-propagation neural network model. The multiple-input neuron includes of the 5 neural units. The multiple-hidden layer includes of 20 tansig, 10 tansig and 5 logsig functions. The output is the vertical deformation of plastic bottles by the top load simulation teat which is one neural unit. The five-layered supervised ANN for this research is described by a diagram as shown in Fig. 3. Each neuron operates by the sum of weighted inputs and passes the result through a transfer function of network which is modeled as:

$$O_{pk} = f_k(net_{pk}) = f_k\left(\sum_n W_{kn}\, f_j\left(\sum_m W_{jm}\, f_i\left(\sum_l W_{il}\, O_{il} + \theta_i\right) + \theta_j\right) + \theta_k\right) \tag{1}$$

where $f_k(net_{pk})$ is a transfer function in k-layer, O_{il} is the output in i-layer, θ_i is the threshold in i-layer, θ_j is the threshold in j-layer, θ_k is the threshold in k-layer, W_{il} is weight of connection in i-layer, W_{jm} is weight of connection in j-layer, W_{kn} is weight of connection in k-layer, O_{pk} is the output in the current layer.

The tansig and logsig transfer function is written by:

$$O_{pi} = \frac{e^{(net_{pi})} - e^{-(net_{pi})}}{e^{(net_{pi})} + e^{-(net_{pi})}} \tag{2}$$

$$O_{pk} = \frac{1}{1 + e^{-(net_{pk})}} \tag{3}$$

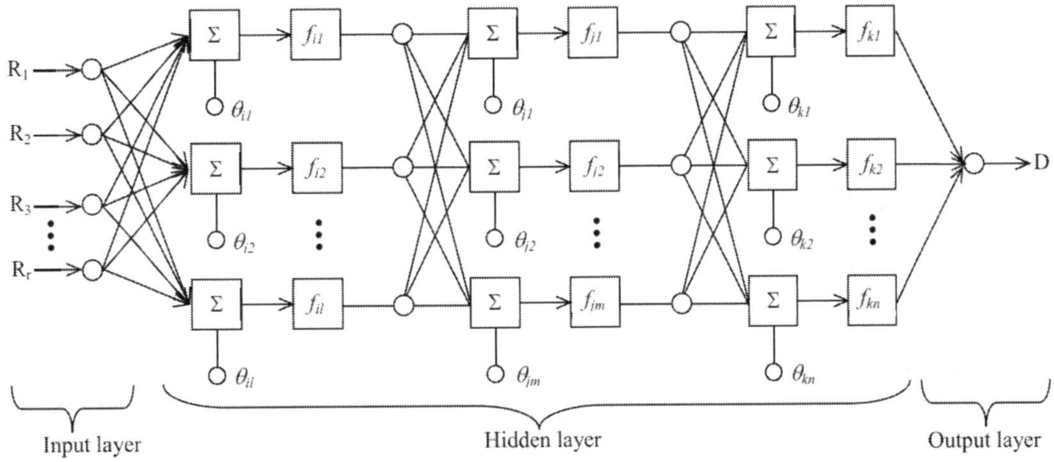

Figure 3. Structure of five-layer ANN.

where O_{pk} is the output of network.

The iterative procedure to adjust weights in each step uses the Levenberg-Marquardt algorithm which is given by:

$$W_{il}(n+1) = W_{il}(n) - e_{il}\frac{J_n^T}{(J_n^T J_n + \mu_n I)} \quad (4)$$

where

$$J = \left[\frac{\partial e_{il}}{\partial W_{il}}\right] \quad (5)$$

$$e_{il} = \begin{bmatrix} T_{i1} - Q_{i1} \\ T_{i2} - Q_{i2} \\ T_{i3} - Q_{i3} \\ \vdots \\ T_{il} - Q_{il} \end{bmatrix} \quad (6)$$

where T_{il} is the target output of i-layer, J is the Jacobian matrix which includes the first order partial derivatives of the network errors with respect to weights, e_{il} is an error vector of i-layer corresponding to the l-data, I is the identity matrix. μ_n is constant between 0 and 1.

4 Results and discussion

The 1 liter bottle was design depending on the desirable geometry by using the master shape bottle. The curvatures of desirable bottle products included R_1, R_2, R_3 R_4 and R_5 are selected following of 140.00, 240.00, 30.00, 20.00 and 0.97 mm, respectively. The desirable bottle was not decorated by ribs or grooves on its wall until there were certain for the corrected shape. Fig. 4 shows model of the bottle which will be predicted the vertical deformation of top load test by ANN. This bottle model achieved the maximum vertical deformation of 2.08 mm which would pass the standard test of the company (less than 4.00 mm).

The bottle model was analyzed the top load test by FEM. The quadrilateral plate element was used to create the finite element model of the desirable bottle. The total plate elements were 4,552 elements. The bottle wall was HDPE material which was assigned a constant thickness of 1 mm. The bottle model had been supported the vertical top load on the bottle neck at 25 kg. The FEA results of the top load test on the desirable shape bottle show in Fig. 5. The maximum vertical deformation of the 1 liter bottle model which analyzed by FEA was 2.09 mm. The FEA results expressed the vertical deformation of the desirable shape bottle less than the company's standard of 1.91 mm therefore it certainly passed the top load test.

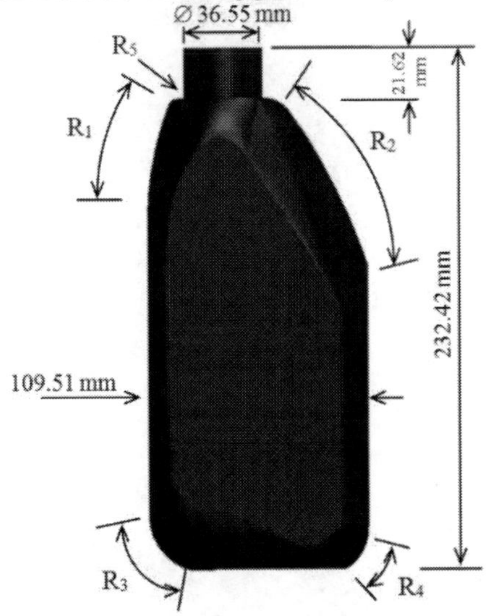

Figure 4. The CAD model of the desirable shape bottle.

The final bottle shape was decorated by groove and sticker area and then it brought to create mold for producing by the extrusion blow molding process. The real 1 liter bottle has been test according to the ASTM D2659 using the universal testing machine (Hounsfield: model H50KS) as shown in Fig. 6. The bottle samples were pressed down on it's neck in the vertical direction until collapsing. Every bottle first collapsed at neck round (R_5) and moved to bottom round R_4 and R_3, respectively. The resistant force was equal to the top load which was recorded by a load cell of the universal testing machine during sample tests. Five bottles give testing results which show by relation graph between top load and vertical deformation in Fig. 7. The average deformation of the bottle test expressed the vertical deformation by the top load at 250 N (25 kg) was 2.40 mm. The vertical deformation of the ANN was different from the FEA and the physical experiment for the top load test was -0.02 and -0.32 mm, respectively.

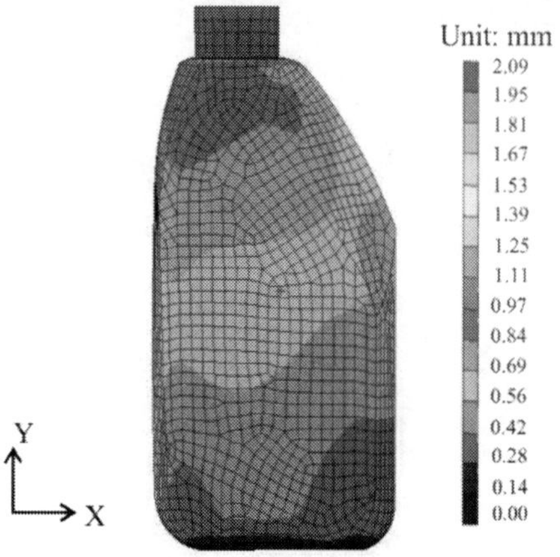

Figure 5. The FEA results of the top load testing under the constant weight of 25 kg.

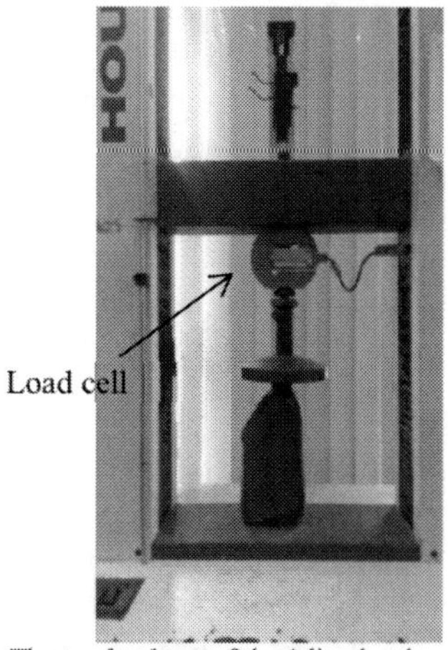

Figure 6. The top load test of the 1 liter bottle using universal testing machine.

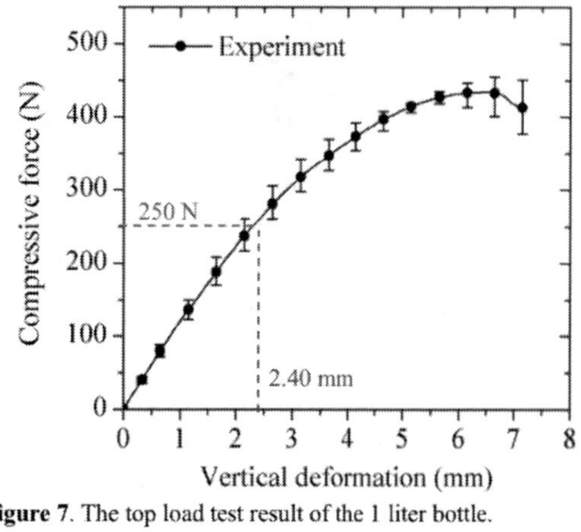

Figure 7. The top load test result of the 1 liter bottle.

5 Conclusions

The master shape of the 1 liter bottle which had the rectangular base and capability to adjust geometry had been created to use for the bottle design. The ANN training had performed to predict the vertical deformation of bottle models under the top load test at 25 kg. The master shape of bottle could used to design shape to pass the top load test by using ANN at the vertical deformation of company specification. Furthermore, the desirable bottle model had been reexamined by FEA before brought to produce. The bottle product had been test by the universal testing machine again and was found the different deformation from the ANN and FEA. The ANN obtained an error less than -13.33% when compared with the experimental data. Nevertheless, the ANN result passed the top load test. Finally, the ANN training of the master shape bottle under the top load test can be used to be the design tool for the 1 liter HDPE bottle of the company in the future.

Acknowledgement

This research was funded by the Thailand Research Fund (TRF) and Panjawattana Plastic Public Company Limited under Research and Researchers for Industries (RRI): Master Degree-grant no. MSD58I0051.

References

1. ASTM International, *D2659-11 standard test methods for column crush properties of blown thermoplastic containers*, ASTM International, PA, (2011)
2. C. Suvanjumrat, E. Chaichanasiri, Kasetsart J. (Nat. Sci.) **48**, 3 (2014)
3. C. Suvanjumrat, T. Puttapitukporn, Kasetsart J. (Nat. Sci.) **45**, 5 (2011)
4. C. Suvanjumrat, T. Puttapitukporn, S. Thusneyapan, Kasetsart J. (Nat. Sci.) **42**, 1 (2008)
5. H. Avci, D. Kumlutas, O. Ozer, M. Ozsen, Int. J. Refrig. **67** (2016)
6. B. Wimarshana, J. Ryu, H.J. Choi, Int. J. Precis. Eng. Man. **15**, 1 (2014)
7. C.C. Zhou, G.F. Yin, X.B. Hu, Mater. Design **30** (2009)
8. M.F. Augusteijn, T.P. Harrington, Neural Compt. & Applic. **13** (2004).
9. D.J. Livingstone, *Artificial Neural Networks: Methods and Applications*, Humana Press, UK, (2009)
10. S. Rajasekaran, G.A. Vijayalakshmi Pai, *Neural Networks, Fuzzy Logic, and Genetic Algorithms: Synthesis and Applications*, PHI, New Delhi, (2008)
11. ASTM International, *D638-14 standard test methods for tensile properties of plastics* (ASTM International, PA, 2014)

 DOI: 10.1051/matecconf/20179502009

Effect of machining parameters on surface finish of Inconel 718 in end milling

Bapi Sarkar , Moola Mohan Reddy and Sujan Debnath

Faculty of Engineering and Science Curtin University, CDT 250 Miri, Sarawak 98009, Malaysia

Abstract. Surface finish is an important criteria in machining process and selection of proper machining parameters is important to obtain good surface finish. In the present work effects of the machining parameters in end milling of Inconel 718 were investigated. Central composite design was used to design the total number of experiments. A Mathematical model for surface roughness has been developed using response surface methodology. In this study, the influence of cutting parameters such as cutting speed, feed rate and depth of cut on surface roughness was analyzed. The study includes individual effect of cutting parameters on surface roughness as well as their interaction. The analysis of variance (ANOVA) was employed to find the validity of the developed model. The results show that depth of cut mostly affected the surface roughness. It is also observed that surface roughness values are comparable in both dry and wet machining conditions.

1 Introduction

Inconel 718 is a well known nickel-based super alloy used in structural materials because of its high strength-to-weight ratio, fracture toughness, fatigue strength, corrosion resistance and ability to operate at high temperatures. Nickel alloys are broadly used in petrochemical, automotive, aerospace, marine, nuclear sectors and biomedical [1, 2]. Although these materials are widely preferred, their machinability is not at the desired level. The reason for this is due to their thermal and mechanical properties, which limit the machinability of Nickel based alloys during the machining process [3, 4]. Surface roughness is significantly influenced by cutting speed, feed rate and depth of cut. Therefore, it is necessary to develop predictive mathematical models in order to determine the surface roughness, in advance (before the actual machining is performed). There are a large number of tools available to develop predictive models for machining processes, namely Response Surface Methodology (RSM), Artificial Neural Network (ANN), Taguchi method, etc [5-9]. Among all predictive model technique, RSM is a widely used tool for designing experiments and analyzing data to predict model and to select the best mathematical model that represents the real life experimental results [10]. ANN model produced more percentage error as compare to RSM model in turning of Al 7050/10/SiCp and Al 7075 composites [11]. Choudhary and Baradie [12] developed response model for tool life surface roughness and cutting force with central composite method using RSM. They found it very useful for assessing the maximum tool life and surface finish. A good design of the experiment and an accurate predictive model need to be developed to minimize the number of experiments optimizing the cutting parameters for desired machinability.

The cutting parameters are an important criterion to obtain good machinability of Nickel-based super alloys [13-15]. Liaoa et al. [16] experimented end milling of Inconel 718 under various cutting speeds by cemented carbide tools for slot and side milling. The results show that cutting speed in the range of 90 to 110 m/min is appropriate for slot milling, and 55 to 135 m/min is appropriate for side milling. Devillez et al. [17] Performed wet and dry turning tests at various cutting speeds using TiCN/Al2O3 coated carbide tool inserts. The results show an acceptable surface quality for a cutting speed of 60 m/min in the dry condition. At the same cutting speed, the result also shows the same order of residual stress and micro hardness for both dry and wet conditions. Zhaopeng et al. [18] studied tool wear pattern in the dry machining of Inconel 718 with PVD coated cemented carbide tool. Their study shows that build up edge (BUE) was formed at a low cutting speed of 20 m/min which resulted in chipping and tool wear. While at a high cutting speed range of 45 m/min to 50 m/min, the wear debris diffused between the tool and the work piece, which accelerated the BUE and peeling wear debris.

Coolants and its influence on machinability have been discussed thoroughly in available literature [19, 20]. Yazid et al. [21] investigated the effect of cutting parameters and machining conditions on surface integrity for turning Inconel 718. They reported that at low speed of 90 m/min and 120 m/min, 50 ml/h of lubricant shows a better surface roughness compared to 100 ml/h lubricant and the dry condition. Kamata and Obikawa [22] applied

MQL while using three different coated carbide tools TiCN/Al2O3/TiN (CVD), TiN/AlN superlattice (PVD) and TiAlN (PVD) for Inconel 718 in turning and reported that air pressure is an important parameter influencing tool life and surface finish. Zhang et al. [23] investigated the effects of different cutting conditions on the tool life and cutting forces in end milling Inconel 718 with a combination of TiAlN and TiN coated tools. They reported a combination of cryogenic compressed air and the microdroplets of biodegradable vegetable oil significantly reduced the tool wear compared to the dry cutting condition. Pusaveca et al. [24] examined the effects of cutting fluid on surface integrity in turning of Inconel 718. The result shows that the combination of cryogenic machining and cutting fluid provides a better surface finish compared to cutting fluid alone.

2 Experimental designs

In all conventional machining process, the most influential machining parameters are cutting speed, feed rate and depth of cut. A proper combination of machining parameters is very important to achieve good machinability. The design of experiments is done by using Central Composite Design (CCD) technique to minimize the number of experiments For CCD, the number of experiments required can be determined from Eq. 1.

$$n = 2^k+2k+C \qquad (1)$$

Where n represents the number of experimentations, k is the total number of parameters considered.

The experimental results are used to develop predictive models by using response surface method for surface roughness. The proposed relationship between the output response and input variables can be represented by Eq. 2

$$R=b_0+b_1\times A+b_2\times B+b_3\times C+b_{11}\times A^2+b_{22}\times B^2+b_{33}\times C^2\times b_{12}\times AB+b_{23}\times BC+b_{13}\times AC \qquad (2)$$

Where R is the response and A, B, and C are the different machining factors. The regression coefficients bi=1, 2, 3 is computed by the regression method using experimental results. The significance of factors and their interactions is computed using statistical analysis. Optimum response model and setting of parameters are produced by using the response model in Eq. 2. A total 20 run or set of experiments were conducted with different cutting parameters on basis of CCD and the surface finish is recorded for each run. Table 1 shows the range of cutting parameters used in the experiments. The adequacy of the quadratic model evaluated using the analysis of variance (ANOVA). Design expert 9.0 was used to establish the design matrix and to analyze the experimental data.

Table 1. Machining parameters.

Cutting parameters	Notation	Unit	Range
Cutting speed	V_c	rpm	1200 to 2500
Feed rate	f	mm/rev	0.1 to 0.5
Depth of cut	D_c	mm	0.1 to 0.5

3 Experimental setup

The work piece material used is Inconel 718. Rectangle block of 202 mm x 102 mm x 12 mm. The general properties are listed on the table for Inconel 718. The experiments are conducted on V30 Horizontal CNC machine (Lead well). The cutting tool was a six flute Mitsubishi milling tool (VC- 8MH). The first test was carried out in dry condition followed by wet condition with a similar experimental setup. According to the experimental setup and cutting parameters the cutting length was 112mm the average interval of one cutting pass is about 0.75 min. After each cutting pass, the surface roughness was checked, and the tool removed to observe tool wear under optical microscope after the observation the tool is again fixed to the tool holder for next set of experiments. Similarly, after each cutting pass the surface roughness was measured by surface roughness tester (Mitutoyo SJ-210), the procedure was repeated until the tool failed for both dry and wet conditions.

4 Experimental result and discussion

To predict the surface roughness in end milling of Inconel 718, the surface roughness value of each run was input and RSM was performed. Table 2 and Table 3 shows the CCD matrix, with variable cutting parameters and experimental surface roughness for a dry condition and wet conditions.

Based on the experimental results second order predictive models have been developed for surface roughness by using Response Surface Method. The obtained equations for the response factors are given below in equation 3. The Fit and summary test suggested that model is significant for dry condition.

$$Ra=0.33303-4.37501\times 10^{-5}\times A+3.06384\times 10^{-3}\times B-0.11178\times C\times 3.07692\times 10^{-7}\times AB-5.70923\times 10^{-5}\times AC+2.0\times 10^{-3}\times BC+1.17619\times 10^{-7}\times A^2\times -1.38276\times 10^{-5}\times B^2+0.40266\times C^2 \qquad (3)$$

Table 2. Cutting parameter of experiments and resultant surface roughness.

Exp. no	Run order	Cutting speed (V_c) *rpm*	Feed rate (f) *mm/rev*	Depth of cut (D_c) *mm*	Surface roughness (Ra) µm	
					Dry Condition	Wet condition
1	5	1200	100	0.5	0.31	0.29

2	7	1200	150	0.5	0.33	0.31
3	6	2500	100	0.5	0.30	0.27
4	20	1850	125	0.3	0.19	0.19
5	8	2500	150	0.5	0.36	0.3
6	15	1850	125	0.3	0.28	0.23
7	18	2500	150	0.3	0.22	0.22
8	4	1850	125	0.1	0.23	0.18
9	13	1850	150	0.1	0.17	0.20
10	17	1850	125	0.3	0.10	0.11
11	14	1850	125	0.5	0.34	0.30
12	19	1850	125	0.3	0.30	0.27
13	16	1850	125	0.3	0.17	0.18
14	2	2500	125	0.1	0.23	0.19
15	3	1200	100	0.1	0.19	0.15
16	12	1850	167	0.3	0.24	0.20
17	10	2943	125	0.3	0.37	0.33
18	11	1850	83	0.3	0.13	0.19
19	1	1200	100	0.1	0.19	0.13
20	9	757	125	0.3	0.33	0.27

Similarly, the obtained equations for the response factors are given below in equation 4. The Fit and summary test suggested that model is significant for wet condition.

$$Ra=0.020757-2.78250\times10^{-4}\times A+6.00110\times10^{-3}\times B-0.23374\times C\times1.53846\times10^{-7}\times AB-.84615\times10^{-3}\times AC+2.50000\times10^{-3}\times BC+7.9276\times10^{-8}\times A^2\times-2.56293\times10^{-5}\times B^2+0.39504\times C^2 \quad (4)$$

The developed model is verified by using ANOVA and results are shown in Table 3 and 4. From the table, The Model F-value of 3.70 implies the model is significant, and the Values of "Prob > F" less than 0.0500 indicate model terms are significant. From the ANOVA result, it is clear that the value of factor A, B, C have effects on surface roughness. The depth of cut influenced more significantly influence on surface roughness compare to cutting speed and feed rate.

Table 3. ANOVA table for surface roughness for dry condition.

Source	Sum of Squares	df	Mean Square	F Value	p-value Prob > F	
Model	0.090	9	9.990E-003	3.41	0.0346	significant
A-Speed	*2.049E-003*	*1*	*2.049E-003*	*0.70*	*0.4224*	
B-Feed rate	*5.142E-003*	*1*	*5.142E-003*	*1.76*	*0.2145*	
C-Depth of cut	*0.041*	*1*	*0.041*	*13.92*	*0.0039*	
AB	*2.000E-004*	*1*	*2.000E-004*	*0.068*	*0.7991*	
AC	*4.500E-004*	*1*	*4.500E-004*	*0.15*	*0.7032*	
BC	*8.000E-004*	*1*	*8.000E-004*	*0.27*	*0.6125*	
A^2	*0.036*	*1*	*0.036*	*12.16*	*0.0059*	
B^2	*1.076E-003*	*1*	*1.076E-003*	*0.37*	*0.5578*	
C^2	*3.739E-003*	*1*	*3.739E-003*	*1.28*	*0.2848*	
Residual	0.029	10	2.927E-003			
Lack of Fit	*2.072E-003*	*5*	*4.145E-004*	*0.076*	*0.9933*	*not significant*
Pure Error	*0.027*	*5*	*5.440E-003*			
Cor Total	0.12	19				

Table 4. ANOVA table for surface roughness for dry condition.

Source	Sum of Squares	df	Mean Square	F Value	p-value Prob > F	
Model	0.065	9	7.232E-003	3.70	0.0267	significant
A-Speed	*2.956E-003*	*1*	*2.956E-003*	*1.51*	*0.2468*	
B-Feed rate	*3.370E-003*	*1*	*3.370E-003*	*1.73*	*0.2183*	
C-Depth of cut	*0.033*	*1*	*0.033*	*16.73*	*0.0022*	
AB	*5.000E-005*	*1*	*5.000E-005*	*0.026*	*0.8761*	

DOI: 10.1051/matecconf/20179502009

AC	*2.000E-004*	*1*	*2.000E-004*	*0.10*	*0.7556*	
BC	*1.250E-003*	*1*	*1.250E-003*	*0.64*	*0.4423*	
A^2	*0.016*	*1*	*0.016*	*8.27*	*0.0165*	
B^2	*3.698E-003*	*1*	*3.698E-003*	*1.89*	*0.1989*	
C^2	*3.598E-003*	*1*	*3.598E-003*	*1.84*	*0.2046*	
Residual	0.020	10	1.953E-003			
Lack of Fit	*4.735E-003*	*5*	*9.470E-004*	*0.32*	*0.8817*	not significant
Pure Error	*0.015*	*5*	*2.960E-003*			
Cor Total	0.085	19				

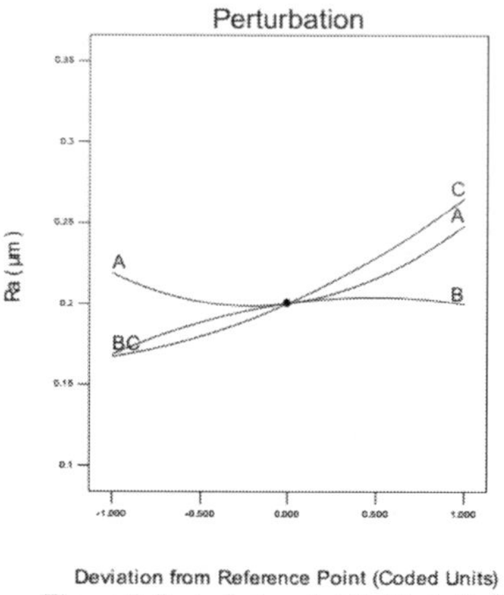

Figure 1. Perturbation plot Ra Vs A, B, C.

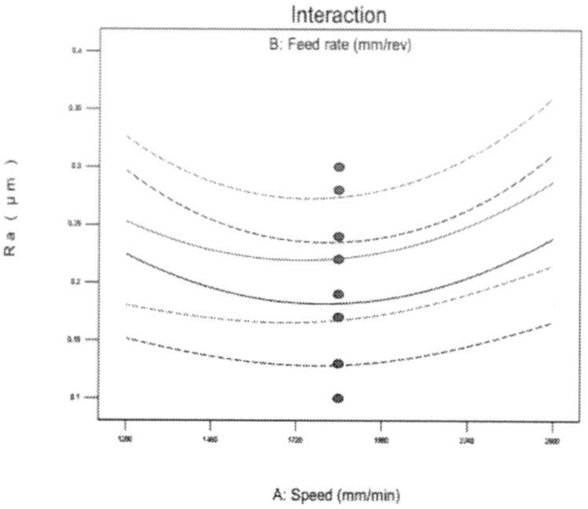

Figure 2. Interaction plot A and B.

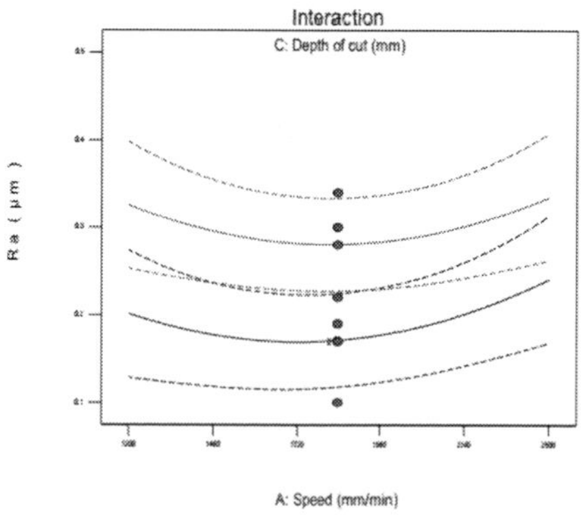

Figure 3. Interaction plot A and C.

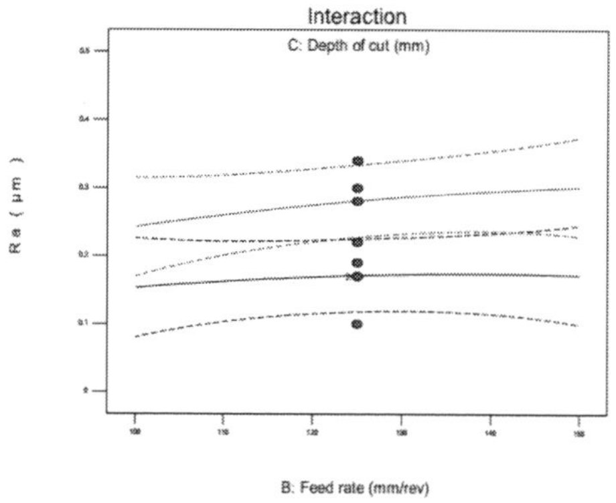

Figure 4. Interaction plot B and C.

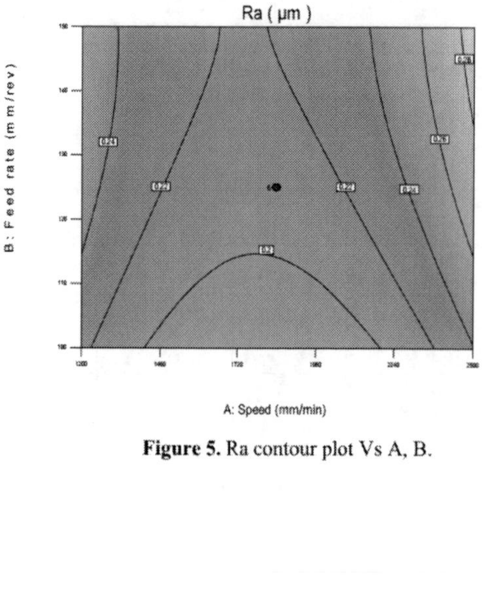

Figure 5. Ra contour plot Vs A, B.

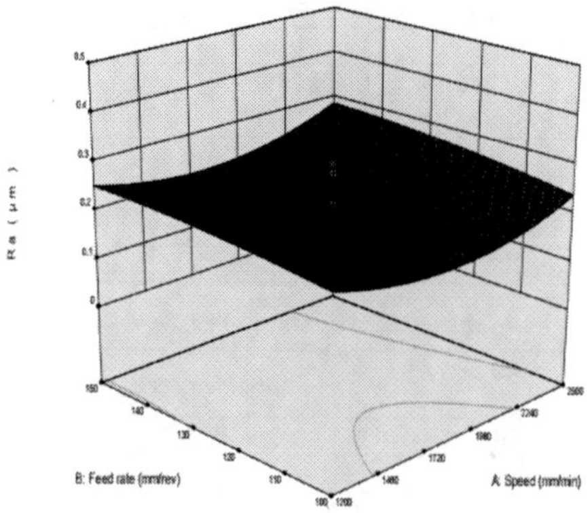

Figure 6. 3D surface plot Vs A, B.

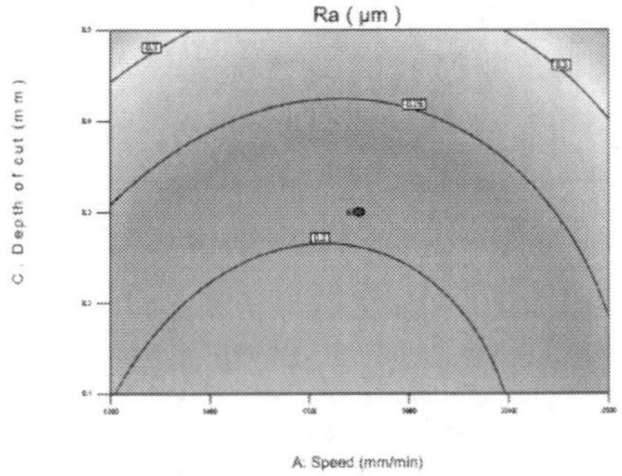

Figure 7. Ra contour plot Vs A, C.

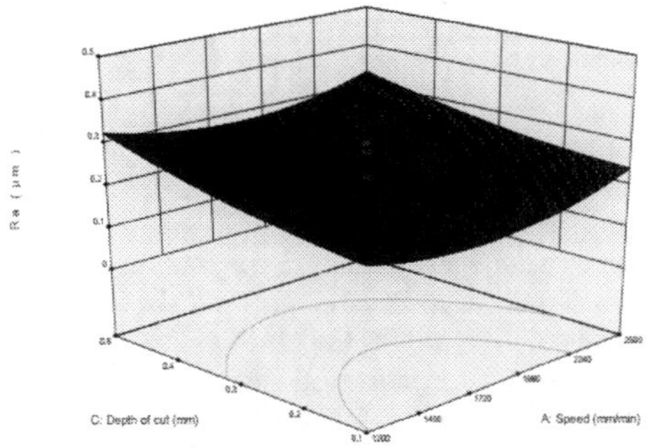

Figure 8. 3D surface plot Vs A, C.

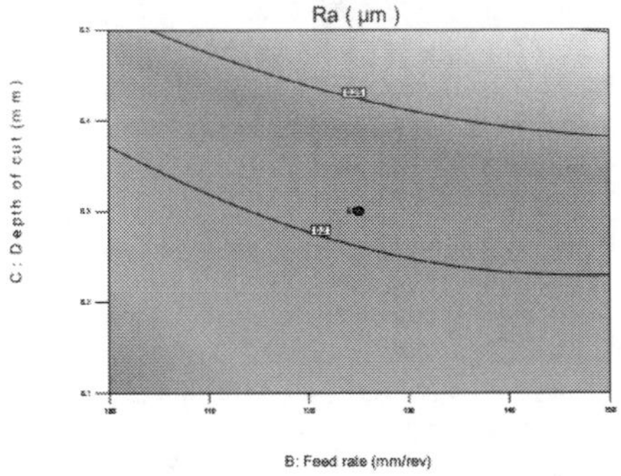

Figure 9. Ra contour plot Vs B, C.

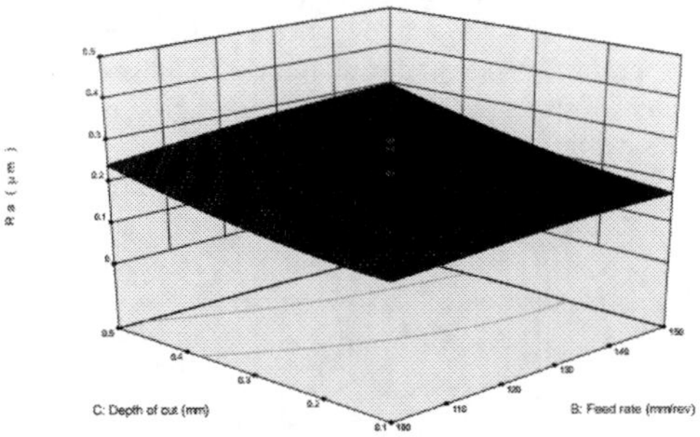

Figure 10. 3D surface plot Vs B, C.

Figure 1 shows the perturbation plot of cutting parameters. The plot indicates that with the increase of each parameter surface roughness increases. But most influential parameter is the depth of cut on surface roughness. Figure 2, 3 and 4 show the plot of the predictive model equation for different cutting parameters and its effect on surface roughness. Figure 2. Indicates that the surface roughness increases with an increase of speed and feed rate. Figure 3. Shows the influence of cutting speed and depth of cut on surface roughness. The trend of the graph clearly indicates that the surface roughness increases with the cutting speed and depth of cut. But surface roughness increase rapidly with increase in depth of cut compare to cutting speed. Figure 4 shows the effect of feed rate and depth of cut on surface roughness. It can be observed that the surface roughness increases with the feed rate and depth. More specifically, the depth of cut has more effect on surface roughness as compared to the feed rate. Figure 5 and 6 shows plot surface roughness Vs A, B at 0.3 depth of cut. Figure 7 and 8 shows plot surface roughness Vs A,C at 125 mm/min feed rate. Surface roughness is low in intermediate cutting speed but is can be observed it increases with increase in depth of cut. Figure 9 and 10 shows plot surface roughness Vs B, C at 1850 rev/min. The trend is simillar when feed rate and depth of cut increases surface roughness also increases.

The surface roughness is highly affected by cutting parameters. For surface roughness, it is observed that depth of cut is the most influential parameter followed by cutting speed and feed rate. The result shows surface roughness is comparable while machining in wet condition and dry machining condition. The lowest surface roughness observed at cutting speed 1850 rev/mm, depth of cut 0.3 µm and feed rate 125 mm/min under wet condition. Whereas lowest surface roughness at cutting speed 1850 rev/mm, depth of cut 0.3 µm and feed rate 85 mm/min under dry condition.

5 Conclusions

In the present work, the effect of machining parameters (cutting speed, feed rate, depth of cut) on surface roughness was experimentally investigated. RSM used to develop the predictive mathematical model for machining of Inconel 718. The validity of the model is verified using ANOVA. The results show low surface roughness at lower cutting parameters in the selected range. The depth of cut has a higher influence on surface roughness compared to other two parameters (cutting speed and feed rate). Experiential results show, obtained surface roughness values are comparable in dry machining with coolant machining.

References

1. S. Pervaiz, A. Rashid, I. Deiab, M. Nicolescu, Influence of Tool Materials on Machinability of Titanium- and Nickel-Based Alloys: A Review, Materials and Manufacturing Processes **29**, 219-252 (2014)
2. D. Ulutan, T. Ozel, Machining induced surface integrity in titanium and nickel alloys: A review,

International Journal of Machine Tools and Manufacture **51**, 250-280 (2011)
3. A. Jawaid, S. Koksal, S. Sharif, Cutting performance and wear characteristics of PVD coated and Carbide tools in face milling Inconel 718 aerospace alloy, journal of materials journal of materials processing technology **116**, 2-9 (2001)
4. J. Zhou, V. Bushlya, R.L. Peng, Z. Chen, S. Johansson, J.E. Stahl, Analysis of Subsurface Microstructure and Residual Stresses in Machined Inconel 718 with PCBN and Al2O3-SiCw Tools, Procedia CIRP **13**, 150-155 (2014)
5. T. Kıvak, Optimization of surface roughness and flank wear using the Taguchi method in milling of Hadfield steel with PVD and CVD coated inserts, Measurement **50**, 19-28 (2014)
6. D. Philip Selvaraj, P. Chandramohan, M. Mohanraj, Optimization of surface roughness, cutting force and tool wear of nitrogen alloyed duplex stainless steel in a dry turning process using Taguchi method, Measurement **49**, 205-215 (2014)
7. M. Sarıkaya, A. Güllü, Taguchi design and response surface methodology based analysis of machining parameters in CNC turning under MQL, Journal of Cleaner Production **65**, 604-616 (2014)
8. P.J. Arrazola, A. Kortabarria, A. Madariaga, J.A. Esnaola, E. Fernandez, C. Cappellini, D. Ulutan, T. Özel, On the machining induced residual stresses in IN718 nickel-based alloy: Experiments and predictions with finite element simulation, SIMPAT **41**, 87-103 (2014)
9. M.S. Kasim, C.H . Che Haron, J.A. Ghani, M.A. Sulaiman, Prediction Surface Roughness in High-Speed Milling of Inconel 718 under Mql Using Rsm Method, Middle-East Journal of Scientific Research **13**, 264-272 (2013)
10. Y. Sahin, A.R. Motorcu, Surface roughness model for machining mild steel with coated carbide tool, Materials & Design **26**, 321-326 (2005)
11. R. Kumar, S. Chauhan, Study on surface roughness measurement for turning of Al 7075/10/SiCp and Al 7075 hybrid composites by using response surface methodology (RSM) and artificial neural networking (ANN), Measurement **65**, 166-180 (2015)
12. M. Alauddin, M.A. Mazid, A.E.B. b, M.S.J. Hashmi, Cutting forces in the end milling of Inconel 718, Journal of Materials Processing Technology **77**, 153–159 (1998)
13. R.M. Arunachalam, M.A. Mannan, A.C. Spowage, Surface integrity when machining age hardened Inconel 718 with coated carbide cutting tools, International Journal of Machine Tools and Manufacture **44**, 1481-1491 (2004)
14. M. Nalbant, A. Altın, H. Gökkaya, The effect of cutting speed and cutting tool geometry on machinability properties of nickel-base Inconel 718 super alloys, Materials & Design **28**, 1334-1338 (2007)
15. S. Amini, M.H. Fatemi, R. Atef, High Speed Turning of Inconel 718 Using Ceramic and Carbide Cutting Tools, Arab J Sci Eng **39**, 2323–2330 (2014)
16. Y.S. Liao, H.M. Lin, J.H. Wang, Behaviors of end milling Inconel 718 superalloy by cemented carbide tools, Journal of Materials Processing Technology **201**, 460-465 (2008)
17. A. Devillez, G. Le Coz, S. Dominiak, D. Dudzinski, Dry machining of Inconel 718, workpiece surface integrity, Journal of Materials Processing Technology **211**, 1590-1598 (2011)
18. Z. Hao, D. Gao, Y. Fan, R. Han, New observations on tool wear mechanism in dry machining Inconel718, International Journal of Machine Tools and Manufacture **51**, 973-979 (2011)
19. R.B. da Silva, Á.R. Machado, E.O. Ezugwu, J. Bonney, W.F. Sales, Tool life and wear mechanisms in high speed machining of Ti–6Al–4V alloy with PCD tools under various coolant pressures, Journal of Materials Processing Technology **213**, 1459-1464 (2013)
20. M. Imran, P.T. Mativenga, A. Gholinia, P.J. Withers, Comparison of tool wear mechanisms and surface integrity for dry and wet micro-drilling of nickel-base superalloys, International Journal of Machine Tools and Manufacture **76**, 49-60 (2014)
21. M.Z.A. Yazid, C.H. CheHaron, J.A. Ghani, G.A. Ibrahim, A.Y.M. Said, Surface integrity of Inconel 718 when finish turning with PVD coated carbide tool under MQL, Procedia Engineering **19**, 396-401 (2011)
22. Y. Kamata, T. Obikawa, High speed MQL finish-turning of Inconel 718 with different coated tools, Journal of Materials Processing Technology **192-193**, 281-286 (2007)
23. S. Zhang, J.F. Li, Y.W. Wang, Tool life and cutting forces in end milling Inconel 718 under dry and minimum quantity cooling lubrication cutting conditions, Journal of Cleaner Production **32**, 81-87 (2012)
24. F. Pusavec, H. Hamdi, J. Kopac, I.S. Jawahir, Surface integrity in cryogenic machining of nickel based alloy—Inconel 718, Journal of Materials Processing Technology **211**, 773-783 (2011)

Thermo-Physical Properties of Kenaf-Filled Acrylonitrile Butadiene Styrene Composites

Siti Nikmatin[1], Achmad Syafiuddin[2], Naresworo Nugroho[3], Widya Utama[2], Agus Sukarto Wismogroho[4]

[1]*Department of Physics, Bogor Agricultural University, 16680 Bogor, Indonesia*
[2]*Department of Geophysics Engineering, Sepuluh November Institute of Technology, 60111 Surabaya, Indonesia*
[3]*Department of Forest Products Technology, Bogor Agricultural University, 16680 Bogor, Indonesia*
[4]*Research Center for Physics, Indonesian Institute of Sciences, 15310 Serpong, Indonesia*

Abstract. Studies on advantageous of natural fillers incorporated into polymer composites on thermo-physical and mechanical properties are still intensively investigated. Several evidences suggest that the natural fillers with small contents combined with polymer increase their composite properties. We thus investigate thermo-physical properties of kenaf-filled acrylonitrile butadiene styrene (ABS) composites. ABS with 5% kenaf microparticle size (ABS/K5), ABS with 5% kenaf short fiber (ABS/KSF5), and recycled ABS with 5% kenaf microparticle size (RABS/K5) were manufactured. Granular composites were manufactured by the twin screw extruder. Composite properties in terms of X-ray diffractions, surface morphologies, and thermal behaviors were investigated. The present work found that ABS/KSF5 has the highest degree of crystallinity compared to others. No significant difference was found in terms of thermal properties of the composites.

1 Introduction

The worldwide availability of natural resources and agro-wastes is the potential to be used as alternative fillers in composites. The rapidly expanding use of composites in automotive and construction leads to develop new sustainable and renewable composites. In addition, the advantages in using natural fillers are low cost, low density, renewable, minimize the environmental pollution caused by the characteristic biodegradability, and may straightly increase socio-economy aspects [1-5]. Several studies reported that the composite materials combined with natural fibers or fillers improved their physical and mechanical properties [1, 6-9]. These improvements were associated with economic advantages, such as low production cost and low resin consumption, which are strongly considered in industrial aspects [10].

Many types of natural fillers, which are used to reinforce polymer composites, were manufactured and investigated [1, 2, 6]. Essabir et al. [1] reported that the reinforcement argan nut shell (ANS) with polypropylene improved intention and torsion of the composites. Essabir et al. [2] noticed that the thermal stability of clay reinforced into high density polyethylene was more thermally stable than the composites produced by high density polyethylene with the oil palm fibers. Alternatively, Nikmatin et al. [6] found that the rattan nanoparticle incorporated into polypropylene increased the thermal stability of the composite. Another reinforcement of natural fibers or fillers of plants or agro-wastes having lower density are essential to be investigated.

Furthermore, the present work aims to investigate thermo-physical properties of kenaf-filled acrylonitrile butadiene styrene composites. Kenaf fiber and kenaf with microparticle size were considered in the present study. This paper is then organized as follows. Materials of the composite are first described in section 2. Then, the procedures for producing kenaf fiber and kenaf with microparticle size are described in detail in the same section. Examination of X-ray diffraction, morphological study, and thermal inspection are also described in section 2. Section 3 provides results and findings that are discussed in detail. This paper ends with section 4 providing conclusions and suggestions for future work.

2 Materials and methods

2.1 Materials

In this study, kenaf stalks (KS) were obtained from farmers under PT Global Agrotek Nusantara, Riau. The KS have about 5 to 6 m in height and 1.5 to 3 cm in diameter. For the matrix, we use original acrylonitrile butadiene styrene (Torray Toyulac Resin 100MPJ 40049689 NLG). For the comparison purpose, recycled ABS supplied by PT MUB Jaya was used. Maleic anhydride from Germany was used as coupling agent. Composition of composites is 5% kenaf (fiber or microparticle), 92% ABS, and 1% coupling agent for all the composites (see Table 1).

Table 1. Composition of composites

Designation	Composition
ABS/K5	ABS with 5% kenaf microparticle
ABS/KSF5	ABS with 5% kenaf short fiber
RABS/K5	Recycled ABS with 5% kenaf microparticle

2.2 Preparation of fillers

In order to obtain kenaf short fibers (KSF), KS were saturated for 7 days. Furthermore, the KS were dried under the sun to reduce KS water content. This is followed by drying process using oven at temperature of 60 °C until the KS have about 8% in water contents. KSF were fabricated using milling machine. It was performed at 25000 rpm in milling speeds for 2 minutes. In order to obtain KSF that are uniform, obtained KSF from milling machine were then sieved with 20 mesh in sieving size. Kenaf microparticle were produced by milling machine with high energy (FOMEC MDY-1000g) using KSF after sieved. The machine was operated with 28000 rpm in milling speeds for 10 minutes with a voltage of 220 V and a power of 2000 watt.

2.3 Preparation of composites

Granular composites were fabricated by the twin-screw extruder (Collin model). Extrusion process was used to produce composites from the combinations of kenaf (fiber or microparticle), ABS, and coupling agent. Type of extrusion applied in this study was direct extrusion. The materials were placed on the hoper machine and a force then is given through barrel using screw. The screw consists of feeding section, compression section, and metering section. The extrusion process was operated at temperature of 170 °C.

2.4 X-ray diffraction analysis

X-ray diagram was recorded by Emma 013B model. It was operated at 35 kV and 28 mA. Radiation wavelength that was used in this study is Cu Kα = 1.540 Å in the range = 10 to 80° with a step of 0.02° at a scan rate of 2° min^{-1}. Weight of all samples that are used in this study is about 10 g.

2.5 Morphological study

Surface morphology of composites was observed by means of Scanning Electron Microscopy (SEM) using JEOL model (JSM-6510LA). It was operated at an accelerating voltage of 20 kV. In order to prevent electrostatic charging and low image resolution during examinations, sputter coated with gold was applied in this examination.

2.6 Thermal inspection

Thermal analysis was carried out by means differential scanning calorimeter (DSC) using JADE model, USA. It was operated in temperature ranges of 30 to 450 °C under nitrogen rate of 20 cc min^{-1} using scan rate of 20° min^{-1}. Weight of all the samples that are used in the current study is about 4 mg.

3 Results and discussion

3.1. X-ray diffraction analysis

X-ray diffraction is one of material examination that is used to identify structure and crystal sizes of materials. Figure 1 shows the X-ray diffraction patterns of ABS with various fillers. Kenaf fiber used in this study has monoclinic structure with a = 7.87, b = 10.31, and c = 10.31; and $\alpha = \gamma = 90°$ and β = 122°. It has the highest peak of 2θ = 22.48° JCPDS. In addition, ABS has consists of phase of polyacrylonitrile ((C3H3N)n), butadiene (C4H6), and polystyrene ((C8H8)n) with dominant phase is polystyrene. It has structure of cubic crystal with $a = b = c$ = 21.95 and $\alpha = \beta = \gamma = 90°$ with the highest peak of 2θ = 21.5°.

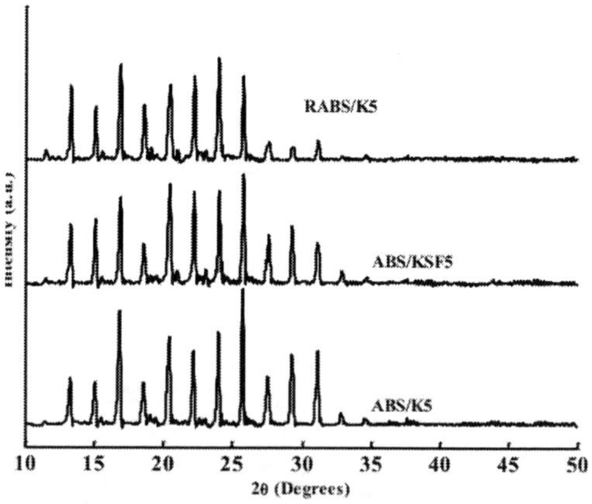

Figure 1. X-ray diffractograms of ABS with various fillers.

Biocomposites investigated in the present study consists of monoclinic and cubic structure. ABS/K5 and ABS/KSF5 have two highest peaks of 2θ = 16.8 and 2θ =25.7. Interestingly, RABS/K5 has the highest peak of 2θ = 16.8 and 2θ = 24. Slightly difference in the second peak may be due to by polymer used in RABS/K5 was from recycled ABS. This is an interesting subject research for future investigating in terms of physical, chemical, and mechanical properties of the composites. Another important finding that was found in this study was that ABS/KSF5 has the highest degree of crystallinity compared to others. The present study found that ABS/K5, ABS/KSF5, and RABS/K5 have degree of crystallinity of 57%, 68%, and 65%, respectively.

3.2 Morphological properties

Observed morphology in scanning electron microscopy (SEM) is formed by using electrons beam. Surface material used must be conductive in order to the electron can be reflected from the surface material. Figures 2(a) – 2(c) show the surface morphology of ABS/K5, ABS/KSF5, and RABS/K5, respectively. We can see that ABS/KSF5 exhibits better interaction between filler and polymer. It is clear that ABS/KSF5 has the highest degree of crystallinity compared to ABS/K5 and RABS/K5. Otherwise, surface morphology of ABS/K5 and RABS/K5 shown that the distribution of kenaf microparticles concentrated in certain part of the composites. These patterns were known a phenomenon defined as agglomeration [11]. These findings suggest for future manufacturing composites using kenaf microparticle incorporated into ABS polymer have to vary another pressure and temperature in TSE machine. These variables used in this study may be appropriate for kenaf short fiber. As discussed, ABS with kenaf short fiber has the highest degree of crystallinity.

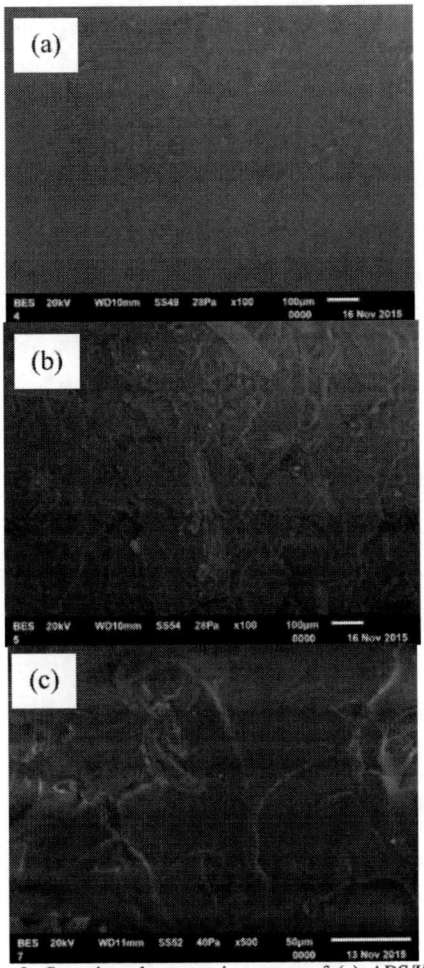

Figure 2. Scanning electron microscopy of (a) ABS/K5, (b) ABS/KSF5, and (c) RABS/K5.

3.3 Thermal properties

Figures 3(a) – 3(c) show the thermal behaviors of ABS/K5, ABS/KSF5, and RABS/K5, respectively. This study found that ABS/K5 has the onset temperature (T_{onset}) at 426.45 °C and the end temperature (T_{end}) at 443.08 °C. In addition, ABS/KSF5 has T_{onset} at 427.09 °C and T_{end} at 443.09 °C and RABS/K5 has T_{onset} at 428.63 °C and T_{end} 442.95 °C. Qiu et al. [12] reported that T_{onset} of composites increases with the rise cellulose content from 0 to 60%. In this study, no significant difference was found in T_{onset} and T_{end} of all composites.

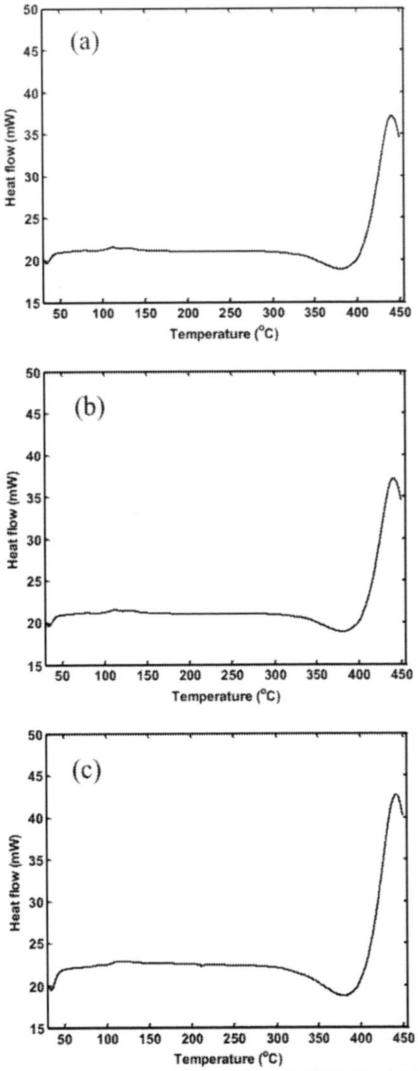

Figure 3. Thermal properties of (a) ABS/K5, (b) ABS/KSF5, and (c) RABS/K5.

4 Conclusion

The present work investigated thermo-physical properties of kenaf-filled acrylonitrile butadiene styrene composites. It was found that ABS/KSF5 has the highest degree of crystallinity compared to RABS/K5 and ABS/K5. These finding were confirmed by the surface morphology of all composites. This study found that the thermal properties of all composites were not significant difference. In general, therefore, it seems that the natural filler used in this study is the potential to be exploited as alternative filler in order to produce a biocomposite materials that can be used as helmet materials, which are parts of our ongoing investigation. The present study makes several noteworthy contributions to materials science especially in exploring recycled matrix and fillers obtained from natural resources. A limitation of this study is that there is no variation in filler contents for recycled or virgin matrix. Further experimental investigations are essentially needed to investigate the composites properties with many variations in filler contents.

References

1. H. Essabir, M.O. Bensalah, D. Rodrigue, R. Bouhfid, A.E.K. Qaiss, Carbohydr. Polym. **143**, (2016)
2. H. Essabir, R. Boujmal, M.O. Bensalah, D. Rodrigue, R. Bouhfid, A.E.K. Qaiss, Mech. Mater. **98**, (2016)
3. I. Naghmouchi, F.X. Espinach, P. Mutjé, S. Boufi, Mater. Des. **65**, (2015)
4. H.M. Akil, M.F. Omar, A.A.M. Mazuki, S. Safiee, Z.A.M. Ishak, A. Abu Bakar, Mater. Des. 32, **8–9** (2011)
5. D.B Dittenber, H.V.S. GangaRao, Composites, Part A 43, **8** (2012)
6. S.Nikmatin, A. Syafiuddin, A.B.H. Kueh, Y.A. Purwanto, Jurnal Teknologi 77, **16** (2015)
7. Y. Ruksakulpiwat, J. Sridee, N. Suppakarn,W. Sutapun, Composites, Part B 40, **7** (2009)
8. S. chi, Mech. Mater. 40, 4–5 (2008)
9. M.S Jamil, I. Ahmad, I. Abdullah, J. Polym. Res. 13, **4** (2006)
10. S.H. Aziz, M.P. Ansell, Compos. Sci. Technol. 64, **9** (2004)
11. M. Inkyo, T. Tahara, T. Iwaki, F. Iskandar, C.J. Hogan Jr, K. Okuyama, J. Colloid Interface Sci. 304, 2 (2006)
12. W. iu, F. Zhang, T. Endo, T. Hirotsu, J. Appl. Polym. Sci. 87, **2** (2003)

Complex Power System for Lithium Batteries and Lead-Acid Batteries

Hendrick , Chen-Chai Tsai, Jyun-Teng Jheng, Zhi-Hao Wang and Gwo-Jia Jong

National Kaohsiung University of Applied Sciences, No.415, Jiangong Rd., Sanmin Dist., Kaohsiung City 807, Taiwan

Abstract. This paper discusses about the power system, and focus on complex power system applications on the public transportation. The objective of this research is to improve power system to achieve green energy applications. Energy depletion is a global problem for people who live on the Earth and we have to face this problem. So before energy depletion occurs, we should develop more energy alternatives. In this paper, we go through two experimental verification of complex power systems. The first part is to establish a platform for static electricity complex experiments. In the second part, using vehicle test platform for dynamic test. Finally, by creating a motor drive mode simulate actual conditions to enhance the accuracy of the experimental results. According to the results we can know the power system after being modified can improve overall system performance.

1 Introduction

Currently, many researchers improve all kinds of energy sources. A lot of energy was limited by natural condition in long time period. It makes some countries difficult to develop efficiently. Some energy are still unable to break through science and technology by using renewable energy but they still get some weakness that need to solve, for example the conversion efficiency of solar panels is too low, the plate is also limited, wind power due to lack of stable wind band, hydroelectric because the terrain cause cannot store enough water to generate electricity and nuclear energy because radioactive material also cannot become applicable energy alternatives. In the case of these resources are unable to successfully use, complex power system is a very important role. It has many advantages, such as there is no the terrain restrictions, accepts multiple power sources, high conversion efficiency, without noise, etc. So complex power system in recent years become the focus of energy alternatives.

2 Research method

In this paper, the experiment method was divided into two parts, which are static test and dynamic test. The following will explain the methods.

2.1 Static test

In the first part of this experiment is focused to test the design of power system. It is really need to do for getting the real condition when this system is used. in the power system it have some elements such as power converter that used to generate the energy resource from power resource1 and power resource N, sensor and a microcontroller as the main element, algorithms to control the energy system then a power supply and load. Beside that giving a light signal is the most impotent thing as the status to indicate complete test already finished. Completely, it will fig out as the following bellow in Fig. 1.

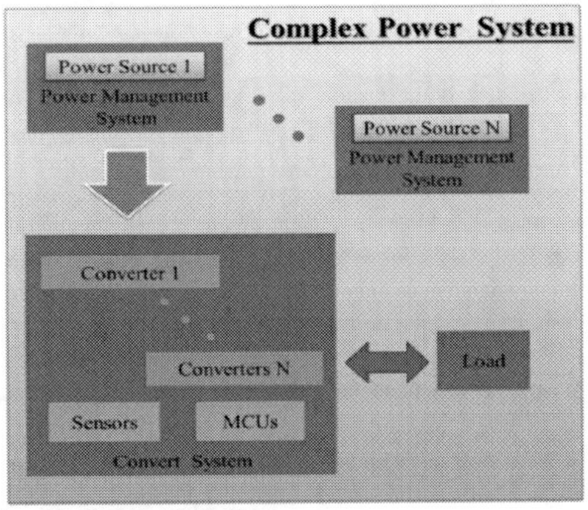

Figure 1. System block diagram

2.2 Dynamic test

The second part of the experiment, we want to know further result when the system that we already create to use for the real application. We use vehicle as testing platform to test complex power system, and with reference to the urban driving mode, simulate the motor load when vehicle driving on the road. We use the experiment by using many kind variations and modes including acceleration driving, deceleration driving, constant speed and slide driving. Apparently, all of these type use different power source consumption. From fig.2 it shows that the red line is the power consumption and the blue one as the speed of vehicle. When the vehicle speed increase smoothly so the power consumption will rise dramatically. Whereas when the speed growth rapidly so the power usage only smoothly increase then when the vehicle have constant speed so power consumption will growth slowly. On the other hands, the consumption of power will be constant when the speed of vehicle is decline. According to this map, consumption of power is depend on speed of vehicle. As power source, it used the Lithium Batteries and Lead-Acid Batteries. So the system has two inputs and one output.

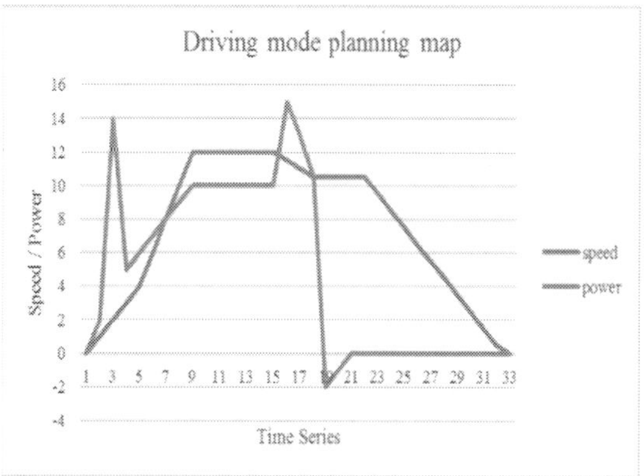

Figure 2. Driving mode planning map.

3 System architecture

The system architecture is divided into part of hardware and algorithms. The following will explain it.

3.1 Lithium-ion batteries

Lithium ion batteries are all secondary batteries among the highest energy density, power density, and also the longest life expectancy. The Battery based on lithium-ion battery cathode materials can be divided into different lithium cobalt batteries, lithium nickel batteries, lithium-nickel-cobalt-manganese and lithium iron phosphate battery. Lithium battery has a stable output voltage curve, so it is more convenient to use the back-end design.

Basically, lithium-ion battery have any characteristic that appropriate as energy resource in our experiment, in Fig. 3 shows the curve of this battery. We can see that the red line is the voltage curve, most of the voltage at 3.7V. When the battery is fully charge the voltage will increase from 3.7 to maximum voltage 4.2V. Which can note characteristic point, which is the transition point with a constant current constant voltage, which it helps algorithm design.

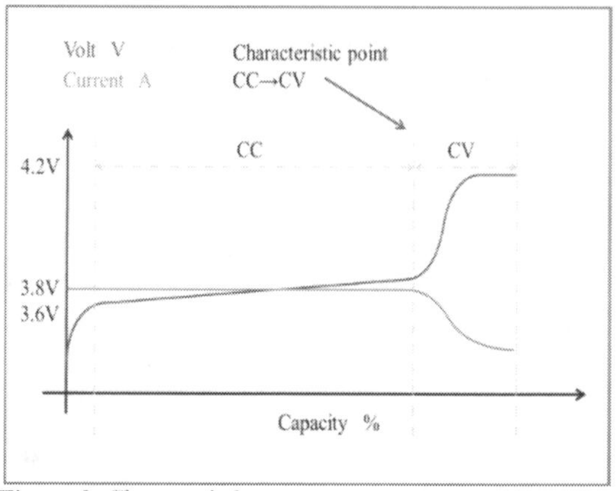

Figure 3. Characteristic curve.

3.2 Lead-acid batteries

In this experiment the use of lead-acid battery voltage is 12V which has 22Ah nominal capacity. The battery is a series consisting of six 2V single cells. In the part of the power output capacity of the battery maximum continuous output current is 330A, and also indicate the maximum continuous output power is 3.96kW. [1]

3.3 Power converter

Power conversion module is mainly composed of half-brick DC converter. Since each battery in different output power will correspond to different output voltages to adjust the final output voltage through the complex power system. The complex power system can arbitrarily set the output voltage and the maximum output current for different needs. At the input of the module, with the use of diode allows converters to be protected.

DC converter used in this experiment is shown in Fig 4, the input and output voltage ranges are 9 ~ 60V and 0 ~ 60V, maximum input and output current is 40A, the maximum module efficiency of up to 97% and an average efficiency of 95% outside. Its output current can be used with an external circuit design to achieve constant current output. The module is also the composite power system module inside the core element, its size and weight to be able to withstand the power of 1.2kW [2].

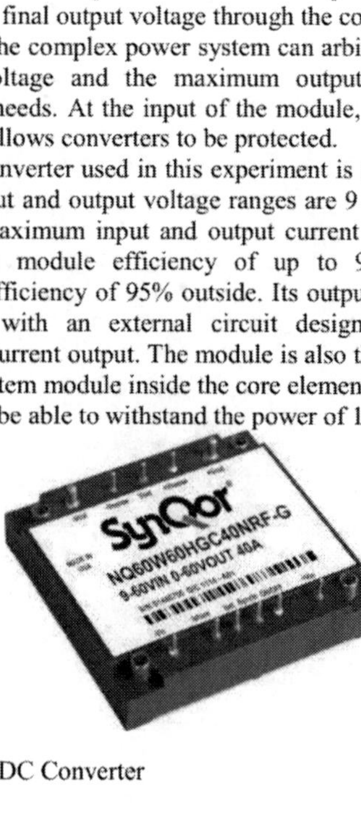

Figure 4. DC Converter

3.4 Microcontrollers

A microcontroller usedin experiment is manufactured by Microchip controlled controller (PIC18F series), which have built in the AD converter, serial communication module and low power consumption, etc.

We use microcontroller as the main control in this system. Microcontroller will determine base on algorithm that we already create. we wrote the algorithm into the controller, which can be set freely to control on different input power based on the user's parameters. Using 12-bit AD converter to precisely measure the amount of electrical signals within the system, provide feedback to the algorithm.

3.5 Sensors

This experiment used two kinds of sensor such as a voltage sensor and a current sensor. Voltage sensor measurement is made by resistance dividing the signal. The resistance value must be a very high value in order to maintain low power consumption and accuracy. Current sensor using a hall sensor, because the amount of non-contact measurement of the loss can be avoided, but must provide a negative supply.

The following equation based experiments converted voltage and current signals.
Where r is AD conversion result.

$$V(R) = 0.0513*R+3.3204 \quad (1)$$

$$I(R)=0.0235*R-1.7977 \quad (2)$$

3.6 Algorithms

In this experiment, the key point is that the algorithms constant voltage and constant current converter. In constant current mode, voltage can be changed. In constant voltage mode, current can be changed. So we have to make lithium-ion batteries operated in two modes to control the output of lead-acid batteries.

Because the characteristics of lithium batteries are not suitable for suddenly output but suitable for stable output, so we will limit its maximum power point, and lead-acid batteries for load changes sharply. We use the complex power system to combined two different characteristics power source.

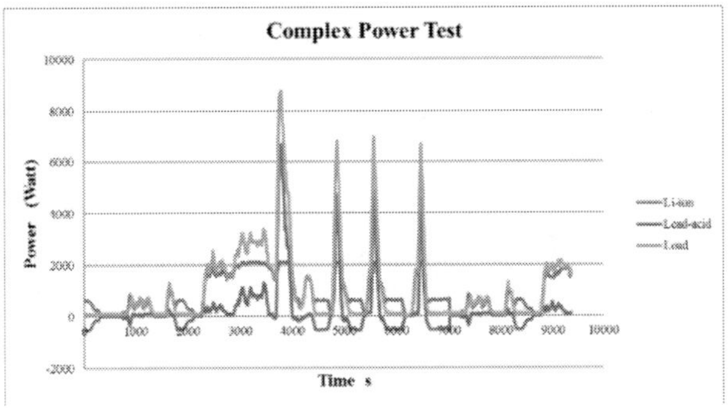

Figure 5. Characteristic curve.

Our system test results are shown in Fig. 5, the blue line represents a lithium-ion battery power. Red is the lead-acid battery power and green is the load power (total power).

4 Conclusion

In the experiment, we established a prototype for complex power system and found a way to control the system. Algorithm limits the maximum output power of lithium batteries, first to protect the lithium battery to maintain the optimum operating point, the second can make overall system to operate in a higher efficiency point and the third is to reach a complex power system performance.

In fact, we can add a variety of other sources of electricity, but after comparison, we chose the lead-acid batteries and lithium batteries as a power source. Both of power sourcenot only can quickly provide a large output power but also has a low cost. The Complex power system can quickly build. The complex power system will be more suitable for public transportation.

References

1. REC22-12, 2015. http://www.yuasa.com.tw/_english/03_vrla/03_detail.php?vd=51.
2. NQ60W60HGC40NRF-G:Half-Brick, http://parts.arrow.com/item/detail/synqor/nq60w60hgc40nrf-g#GGzn
3. Au-Yong-Tien-Tuck, "The Research and Development of Vechicle's Fuel Cell System Control Module with Performances Improved Mechanism", National Kaohsiung University of Applied Sciences, Jul. 2014, Page(s): 61-63.
4. Hong Wen HE, YingQi Zhang, Fan Wan, 2008, "Control Strategies Design for a Fuel Cell Hybrid Electric Vehicle," IEEE Vehicle Power and Propulsion Conference (VPPC).
5. V. H. Johnson, "Battery Performance Models in ADVISOR," Journal of Power Sources, **110**, pp. 321-329, Aug. 2002.
6. D. P. Xu, L. F. Wang, and J. A. Yang, "Research on Li-ion Battery Management System," Proceedings of

DOI: 10.1051/matecconf/20179503002

2010 International Conference on Electrical and Control Engineering, pp. 4106-4109, Jun. 2010.
7. Y. C. Hsieh and C. S. Huang, "Li-ion Battery Charger Based on Digitally Controlled Phase-shifted Full-bridge Converter," IET Power Electron, **4**(2), pp. 242-247, Feb. 2011.

A Prototype of Industrial Waste Water Treatment Using Electrocoagulation

Phonnipha Boriboonsuksri[1] and Natth Jun-krob[2]

[1]SAU- Safety Training centre, Safety Engineering Department, Southeast Asia University, Bangkok, Thailand
[2]Electronics and Computer Engineering Department, Faculty of Engineering, Southeast Asia University, Bangkok, Thailand

Abstract. This paper proposes a construct of electrocoagulation waste water treatment system. The system consists of reactor tank, skimmer, cyclone tank and sediment tank. Waste water is feed into reactor tank. The electrochemical reaction is made emulsification to waste water. The contaminants are removed from waste water and can be divided to two kinds: light weight suspensions be floating up and another be sediment. The flocculants are skim out and the sediments are pumped out to sludge container. An electrical power which feed to electro-coagulation procedure is controlled by microcontroller. The user can be adjusted for suitable with waste water loaded. The input of waste water and output of sediments are controlled by PLC. The results, when operate with industrial waste water, can be treat by 30 m^3/day rates and the controlled parameter value: pH, BOD, Oil & Grease, COD, SS, TDS, and Ni are not exceed than the standard limit. The advantages of this system are consume small area and low power consumption.

1 Introduction

The electrocoagulation wastewater treatment is very interested in the present. Because the system requires less space and lower operating cost, when compared to other systems.[1] Many studies have tested the effectiveness of wastewater treatment by this method in different manners. But mostly confined only in laboratory scale.[2-5] Some work can be done commercially, but must be imported from abroad. That is expensive and lacks of specialists to operate the system. For this reason, the electrocoagulation waste water treatment system has not been used prevalent in Thailand. Our research team is interested in creating a prototype wastewater treatment by this method, which can be used practically in Thailand.

2 The system design

The proposed wastewater treatment process has three major components parts:
(1) The reactor tank (2) Skimmer and (3.) Cyclone and sedimentation tank. Each part-section will have to run the corresponding and appropriate to the flow rate of the treated water with details on Fig. 1. Which is consisting mainly parts of the following functions.
- Sludge pump: Used to pump the waste water from clarifier into the tank.
- Sump tank: Storage and discharge of waste water flowing into the reactor properly.
- Reactor tank: Serves to separate the contaminants in the wastewater into sludge.
- Sludge tank: Serves to storage sludge which formed by the reactor.
- Cyclone tank: This device served to stirring the heavy sediment settling quickly.
- Sediment tank: This tank served to precipitate the lightweight remains to be settling down.
- Power supply Unit: This device controls the electrical power supplied to the reactor. [6]

2.1 Reactor tank

The reaction tank is designed to support the power in series topology in order to reduce the power consumption of electronics devices of the power supply circuits. [6] An electrode, which is an aluminium sheet size 60x120 cm, 19 sheets were placed parallel to each other as a place far away as 56 cm. The total reactions area of 13.68 square meters, as shown in Fig. 2 and the aerated sludge that occurs when the reaction shown in Fig. 3.

2.2 Skimmer

The light weight suspension is separated from waste water to float above level inside the reaction tank. The skimmer consists of sweep mechanism powered by an electric motor which is mounted buried beneath the lid of the reaction tank. All suspensions resulting from these treatments will be swept into a storage tank which is mounted on the side of the tank reactions on both sides. The skimming is controlled by a PLC, which skim automatically every 15 minutes. The skim sediments mechanism is shown in Fig. 4.

DOI: 10.1051/matecconf/20179503003

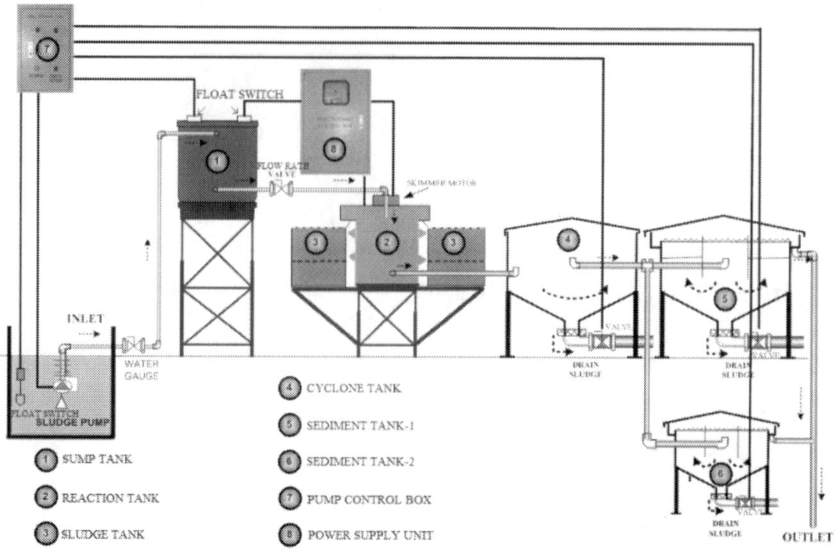

Figure 1. Diagram of the wastewater treatment plant operated with an electrocoagulation

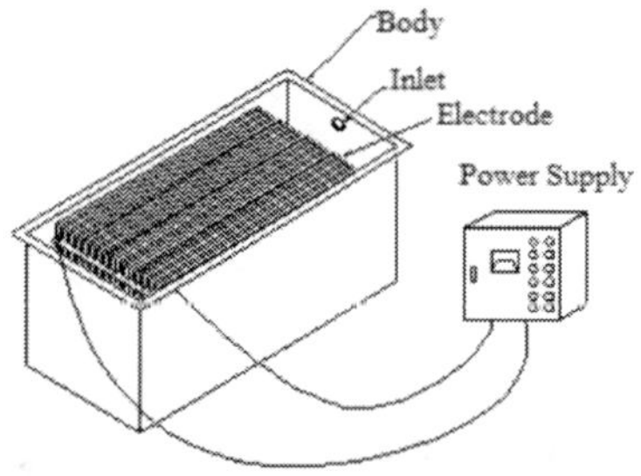

Figure 2. The structure of the reactor and its components

Figure 3. The rapid-settling flocs during test

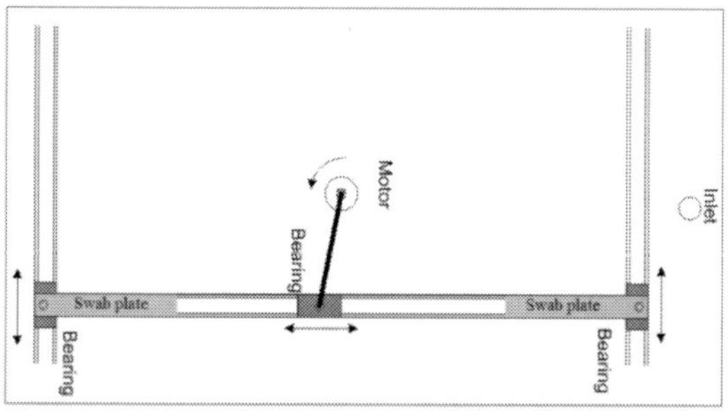

Figure 4. The structure of skimmer and its components

Figure 5. The cyclone tank

2.3 Coagulation

The sediments caused by the extraction of water in the reaction tank have weighs not enough to precipitate. It requires some time to gather the sedimentary. In this design, the system will consist of two parts: a cyclone tank and sedimentation tank.

2.3.1 Cyclone tank

This tank causes sedimentation to weigh enough to precipitate shortly. Which according to the phenomenal movement behaviour of objects in the cyclone tank, as shown in Fig. 5.

2.3.2 Sedimentation tank

This functions to taking sedimentation from the cyclone tank to fall to the bottom of the body. The top of this section constructs with a weir-overflow. And inside them containers with cross flow tube to increase precipitation efficiency as shown in Fig. 6.

2.4 Control sludge management

To store sedimentation, the mechanisms are controlled by a PLC controlled system, as shown in Fig. 7. This

 DOI: 10.1051/matecconf/20179503003

controls the sludge pump which submerged in the cyclone tank, sedimentation tank to carry out the sludge. And control motor skims fluctuate that floats on the surface of the reaction tank. Including controls "on/off" pumps to suction water into the treatment system. Moreover, the system also controls the "on/off" power supply to the reactor tank, which will shut down when there is no treatment wastewater into the system, as shown in Fig. 7.

Figure 6. Sedimentation tank

Figure 7. PLC Control sludge management system

3 The process of systems

Waste water is sucked from clarifier, which has a float switch for on/off the electric pump. In the absence of wastewater in clarifier, an electric pump and power supply will be turned off automatically. The wastewater is pumped into the sump tank to regulate the water pressure which flowing into the system with a fixed rate. The tank has two sets of float switch, the first functions to pumped water into the tank and control "on/off" pump to get the water level in the given point. Another switch functions to disable power supply if the water in the tank runs out. It sends a signal to stop supplying electricity to the system in order to save energy in the absence of wastewater treated. The wastewater will flow from sum tank into the reactor tank. . The flow rate of wastewater is controlled by a ball-valve. Waste water is treated in the reactor by the electrocoagulation function. These reactions reduce the stability and destroy the molecular bonds of the contaminants in the wastewater. As a result, mass spectrometry that can be separated contaminated wastewater, such as proteins, carbohydrates, fats, heavy metals and etc. from waste water. These contaminants are separated as sludge with a bubble floating up into the lighter sludge floating on the surface of the reactor. The reactor has a blade skim to skim floc-sludge to flow into sludge tank, which is placed on the side. The treated water will be delivered to a cyclone tank, which served heavy-sludge falling to the bottom of the cyclone tank rapidly. In the bottom of the tank will be structured as a cone. Its ending is connected to a suction pump to carry heavy sludge to the yard dried automatically controlled by the PLC. After that, the water will be sent to a sedimentation tank to the settling of sediment that slows down again. The sedimentation tank is structured as weir overflow and inside the tank installation cross flow tube to accelerate precipitation. The bottom of the tank installs a pump to carry sludge to the yard to dry, as well as cyclone tank. The water is delivered from the factory as standard for the industry. Fig. 8 shows the overall of a prototype wastewater treatment that have been designed.

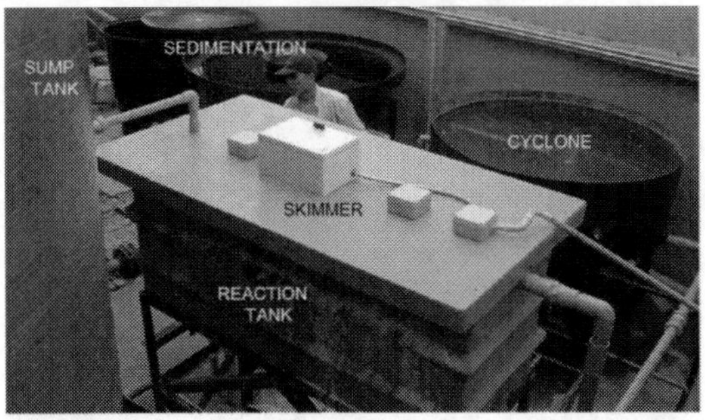

Figure 8 Prototype of industrial wastewater treatment with electrocoagulation

4 The experiment results

4.1 Water quality after treatment

A prototype has been installed to test in actual use of a S.C.I. Electric Manufacturer Co., ltd. This has the load of 500-BOD_5-wastewater from production of electrical wiring equipment parts and factory canteens with the rate of 30 cubic meters per day. The system operates with electric current supplying to the water treatment about 4-6 A. The treatment results showed in Table 1, which the water qualities are within the standard levels of discharged water from the factory that enforced by Thailand's Ministry of Industry.

Table 1. Results of water quality after treated

Parameter	Unit	Analysis Methode	Standard Limit	Result
pH		Electrometric Methode	5.5-9.0	8.16
BOD	mg/L	Axide Modification Methode	20	11.00
Oil and Grease	mg/L	Partition Gravimetric Methode	5	1.00
COD	mg/L	Closed Reflux, Colorimetric Methode	120	56.00
SS	mg/L	In-house Methode LT-01	50	5.60
TDS	mg/L	Dried at 180oC	3000	599.00
Nickel(Ni)	mg/L	ICP	1	0.13

In long-term experiment, electrodes were adhesive with hydroxide and corrosion occurs. When operating over a long period of time, these barriers to the passage of an electric current. As a result, the efficiency of the water treatment is reduced. Fig. 9 shows the formation of hydroxide on the electrode system after use. Fig. 10 shows the erosion of aluminium electrodes in the reaction tank. Fig. 11 the nature of the sludge, which is a liquid cream and leave to dry soil, which can be planting crops.

Figure 9. The hydroxide formation on the electrode

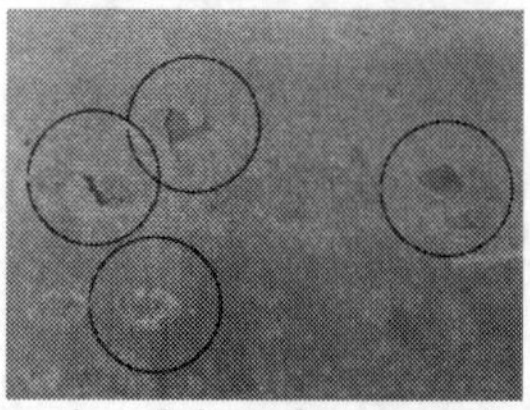

Figure 10. The erosion of electrodes occurs after use

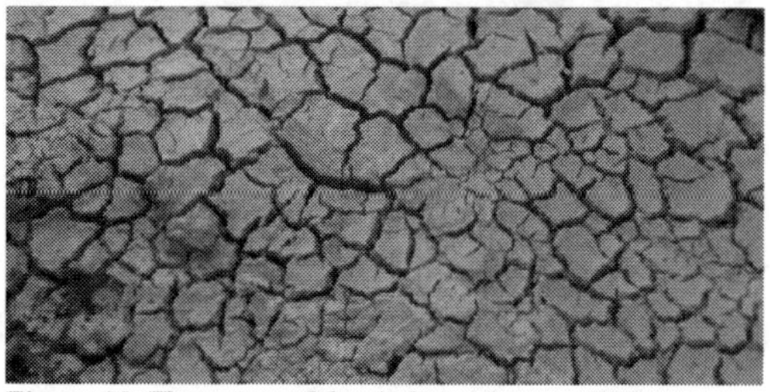

Figure 11. The nature of the sludge on leave until dry

4.2 Energy consumption

The system will consume electric energy in the total average of about 5 A. or about 792 units per month. And the cost of treatment was 2.64 baht per one cubic meter of 500-BOD_5-wastewater.

4.3 Development of system performance

We can see that, the main problem in the system is corrosion and the formation of a hydroxide formed on the electrode. Therefore, it is necessary to regularly to clean the electrode in a reaction tank to provide effective treatment of wastewater is not reduced. That is the burden of maintenance. A solution to these problems can be achieved by creating a mechanical abrasion electrode in a reaction tank, which will be developed in next-generation.

4.4 A Mobile prototype for industrial testing service

An academic service is an important mission of higher education institutions in Thailand. To exploring the extensive testing of waste water to the factory is essential for this work. The mobile unit prototypes for treated wastewater testing service have been designed as shown in Fig. 12. This system is design for compact and high mobility. Make it convenient to use for the service of wastewater treatment testing in the factory of industry in Thailand.

Figure 12. A mobile prototypes for Industrial testing service

5. Conclusion

The created Industrial prototype of the electrocoagulation wastewater treatment is capable up to 30 cubic meters of wastewater treatment per day. It consumables cost per unit for the treatment of about 2.64 baht per cubic meter. It used very little space, only 40 square meters. It has low operating costs compared with other systems. It has flexibility to customize with the volume of wastewater accrued per day. It is easy to use. It is a great solution to wastewater treatment for Thailand's industry in modern times. The problems encountered in operation is to maintenance for not to lose its effectiveness. That is needed to clean the electrode reactions in the tank frequently. And another problem is founded; the sludge that is produced from the electrocoagulation process is a cream that fine sediment. It needs a long time to dry and break up into soil. Therefore, it is necessary to have the location for them to dry sufficiently.

Acknowledgement

This research was funded by Southeast Asia university (ST003/2014) and S.C.I. Electric Manufacturer Co.,Ltd. We so thank you for supporting us in this opportunity.

References

1. S. Krisadawanich, S.E.A. Eng. J. **2**, 11 (2006).
2. C. Barrera-Díaz, B. Frontana-Uribe, B. Bilyeu, Chem.**105**, 160 (2014)
3. V. Khandegar, A. K. Saroha, J. Env. Mgmt. **128**, 949 (2013)
4. C. Ricordel, H. Djelal, J. Env.Chem. Eng. **2**, 1551 (2014)
5. S. Zhao, G. Huang, G. Cheng, Y. Wang, H. Fu, Des.J.**344**, 454 (2014)
6. N. Jun-krob, S. Chean-arom, S. Sonantha, K. Kamonchot, *Conf. ENETT-4th*, 45 (2008)

Mechanical Property of 9 Years Old Thinning of Teak Plantation in Thailand

Somchai Seviset[1], Thanate Piromgran[1], Udomsak Saributr[1], Ratthai Porncharoen[1], Kananke Raerai[1] and Sittichai Charoensettasilp[2]

[1]*Faculty of Industrial Education, King Mongkut's Institute of Technology Ladkrabang, Bangkok 10520, Thailand*
[2]*Faculty of Science, King Mongkut's Institute of Technology Ladkrabang, Bangkok 10520, Thailand*

Abstract. Based on the study of the conditions and problems of thinned teak in Thailand it's with findings that thinned teak or so called thinning for distance clearance being with small stems and can be sold at low price. thinned teak was with mechanical property different with teak from natural forest with age over 30 years old with the results of most of the tests were with findings that the values of the mechanical property of teak from natural forest with age over 30 years were more valuable than the values of the mechanical properties of thinned teak with exception on Standard Test Method for Tensile and Cleavage Test of Timber which thinned teak was with more value than teak from natural forest with age over 30 years old. For guidelines on market value added for thinned teak can be made by producing to use in the context of the design of household goods made of wood as the guide for resolution of problems and creation of artistic work pieces household goods made of wood.

1 Introduction

Teak/teakwood has its origin in Southern areas of India, Burma, Thailand, Laos (the part adjacent to Thailand North Region) and Indonesia. In Thailand teak mostly grows up in Five Kinds Unassorted Forest or Miscellaneous Forest in the North. Teak is fond to grow in mountainous areas or on a plateau conditional to well water drainage and non-flood, which can be a mixture of crumby soil with sand or soil with very much depth especially soil derived from limestone which is broken decayed rotten turning into a depth of crumby soil strata which will help teak to grow very well. Teak usually grows in a cluster of pure teak as a dump of teak or may grow mixing with other miscellaneous woods such as red Pterocarpus, Ormasia Dalbergia, Lagerstroemia etc.[1]

In Thailand teak generally is with scattered growth according to the nature in miscellaneous forest. Teak is wood of high economical value, in the past, it used to be export goods making top income revenue to Thailand second to only rice. Later teak from natural forests in Thailand was with reduced volume and became with higher price. Additionally, there are increasing trends of using teak from forest plantations therefore there is also increasing growth of teak to which Thailand Department of Forest has promoted the private sector and agriculturists/farmers to grow economical wise forest plantations with finding that teak is the kind of wood which the people are fond of growing but the administration in the past growers faced the problems of growing in unsuitable areas, lacking of knowledge and understanding on growing, and bad management, therefore, upon thinning there was a lot of teak scraps to be thinned, thereby, has made some growers turning to grow plants with short-term return instead. [2]

The production of teak for thinning or a large number of teak being thinned for testing of mechanical quality/properties of wood is another variable which is used to specify the quality of wood which wood standard tests are carried out by the Office of Building Control and Inspection, Department of Public Works & Town Plan of Thailand B.E. 2008 which has stipulated the Test Standards as follows.

MorYoPor. 1221-51: Standard Test Method for Compression Perpendicular to Grain;

MorYoPor. 1222-51: Standard Test Method for Compression Parallel to Grain;

MorYoPor. 1223-51: Standard Test Methods for Direct Moisture Content Measurement of Wood;

MorYoPor. 1224-51: Standard Test Method for Static Bending of Timber);

MorYoPor. 1225-51: Standard Test Method for Tensile and Cleavage Test of Timber;

MorYoPor. 1226-51: Standard Test Method for Shear Test of Wood Parallel to Grain;

MorYoPor. 1227-51: Standard Test Method for Specific Gravity to Grain;

The foregoing Standard Tests will make us know the various properties of teak thinned for distance clearance which, based on a collection of research works dealing with teak thinned for distance clearance it's found that the production of teak thinned for distance clearance was still not with sufficient data to be guidelines or guidance for study for exploitation on various aspects and to make added value to growing agriculturists/farmers.

This research article, therefore, would like to make presentation of making added value of teak thinning or teak thinning for distance clearance of 9 years old by testing the mechanical property of the wood as primary information on the development and promotion on usage of thinned teak for production of optimal usage as well as being potentiality development and making added economical value resulting the interest to Thai agriculturists/farmers next.for thinning or a large number of teak being thinned for testing of mechanical.

2 Literature review physical properties of teak in Thailand

In Thailand there is only one kind of Teak but may be called in other names such as Gold/Thong Teak, Stone/Hin Teak, Buffalo Shit/Ki Kwuai Teak following the nature of the wood patterns or colors which are different which may be from factors of different environments such as rain volume, soil type and mineral elements or natures of making use of teak taking into consideration based on the colors of the wood pulp, decoration, hardness and tensility of wood pulp into 5 categories of:

Thong/Gold Teak. The wood pulp is gold brown color, straight grain with ease for furnishing;

Hin/Stone Teak. The wood pulp is brown or pale with ease for furnishing;

Yuak/Soft Teak. The wood pulp is light brown or pale and is with ease for furnishing;

Khai/Egg Teak. The wood pulp is dark brown mixed with yellow, having the content of fat and thus making it difficult for furnishing and coating/painting.

Ki Kwuai/Buffalo Shit Teak. The wood pulp is a mixture of gree with brown, dark brown appearing as blurred coloring.

Thinning. Bunvong Thaiutsa et al. (1992) said that thinning is a necessary forestry practice for forestry management all this irrespective whether thinning for sale will be cost-efficient or not as a by-product of the forestry plantation growing whose genuine philosophy of thinning is for the improvement of the growth of remaining plantation. Thinning is virtually cutting out of partial of wood for teakwood to grow more dense for making the remaining teakwood grow up to larger size to which the size of the wood is normally related direct with its top housing and root housing i.e. if the top housing expands more then the three may grow with larger size. Growing threes must be made with density on the beginning for the trees to compete the height of their stem, straight stem, not much branching saving branch cutting. Teakwood growth may start with distance clearance of 3 x 3 meters. After teak growth for another period teak top housing will start to be close together again, growth will not increase thus necessitating another thinning. Leaving this alone will cause natural competition by teakwood, there is scrambling or competition for water and food causing partial of the trees to die. Part of teakwood will stabilize or stop growing thereby larger size of wood to derive will take longer time and not cost-efficient. Teakwood growth should be with the first thinning upon or when the density starts, top housing scrambling too close together prevents growth not to increase which is depending on the area condition, growing clearance distance, growth duration, maintenance. Generally teakwood will start to be dense at the age of approximately 6–10 years which will necessitate thinning then. Afterwards, let the teakwood grow until their top housings are scrambling close together again and then continue with the 2nd thinning at the age of about 15-22 years. For teakwood growth the first thinning will be with small size of teakwood, low price, most growers of teak plantation forestry normally will not do thinning despite the fact that teakwood trees are very dense and thereby causing the overall growth of teakwood although inability to make sale of the wood cost-efficiently. It may be necessary for the growers of the teak plantation forestry to make thinning by taking the wood for self-usage. In respect of the methods of thinning these are comprises of :

1) Systematic thinning is thinning systematically or thinning one row skip one row or thinning one tree and skip another tree;

2) Selective thinning is selecting to thin larger sized tree or smaller sized tree out. For large scale teak plantations systematic thinning will help facilitation of thinning. For teak plantations with small size area if there is existence of different growth the method of selective thinning should be chosen with the first thinning to choose to screen small sized with bad appearance out leaving teakwood trees with nature of well growth because trees with good growth can grow up to be larger trees betters thereby making the forestry plantation with higher productivity per Rai, cost-efficient on the last thinning round better with teak for sale at good price and mostly needed of the market demand, i.e. with age of more than 15 years old, with circumference at the height of 1.30 meters of more than 60 cm.'s upwards. For some areas which teak growers fail to carry out thinning teakwood will grow densely, scrambling and growth is not even. There are both large and small trees, on the first thinning will then necessitate to choose small trees for thinning and leaving larger trees to remain. Growers' problems on thinning include such as:

1) Market limitation and on the exploitation of the first stage thinning, trees are of small sizes and low price, considered as discarded teakwood.

2) The problems on buyer's selection to buy. Growers let the medium buyers to go in for selection to thin, by selecting to thin only large sized trees out, leaving only small sized unhealthy trees. Their growth will then be slow thereby making the productivity to get in the future low and not cost-efficient

3) Growing agriculturists/farmers are lacking of knowledge and understanding about the relation of the growth of the trees with the density of the tress thereby rendering them not taking the importance of thinning leaving the forestry plantations to grow densely and with competition for nutrients.

3 Mechanical properties of teak test in Thailand

The Department of Public Works and Town Plan has established various standards e.g. standard of material specification, method of test standard, standard of design, construction supervision standard, standard for installation and inspection, standard for maintenance for instances, in order of establishment of safety on lives and properties belonging to the population as well as provision of academic support to local administration organizations, various public offices both centrally and in the regions whose test standards are listed as follows:

1. Standard Test Method for Compression Perpendicular to Grain.

This Standard Test covers the study of behaviors and determination of mechanical property of wood under compression perpendicular to grain to determine required values of Modulus of Elasticity, Elastic Force at Proportional Limit, yield strength of wood at offset distance of 0.05%.

Standard Test Method for Static Bending of Timber wood at offset distance level of 0.05%

This standard uses SI (International System Units) as principle and uses the value of unit conversion of 1 kilogram force equal to 9.806 newton.

Preparation of Sample Pices. The sample preparation of Dressed Timber is required by the wood must be in the good order condition with no cracks of the size 50x50x150 ml. each kind of 3 pieces [3]

2. Standard Test Method for Compression Parallel to Grain.

This standard test covers the study the behaviors and determination of mechanical properties of wood under the pressure parallel to grain to determine the required value of 0.05% and annihilation compression of the wood, as well as the study of the nature of failure of the wood upon reaching the failure point.[4]

This standard uses SI (International System Units) as principle and uses the value of unit conversion of 1 kilogram force equal to 9.806 Newton.

Preparation of Sample. The sample preparation of Dressed Timber of the size 50x50x150 ml. total of 3 pieces is required by the wood pieces must be in the good order condition with no cracks along the length and the cross section profile shall be perpendicular so that the compression scattered on the cross section profile evenly.[5]

3. Standard Test Methods for Direct Moisture Content Measurement of Wood

This standard test covers determination for the volume of humidity of the wood with the method of drying. This Standard test used SI (International System Units) as principle.

The standards referred to in this standard is comprised of the Department of Public Works & Town Plan Standard MorYorPhor. 1104: Woodwork Standard.

Preparation of Sample Pieces. The sample preparation of Dressed Timber, which is uncoated with chemical of total of 3 pieces with cross section profile size of 50x100 ml. with length parallel to grain 25 ml. [6]

4. Standard Test Method for Static Bending of Timber.

This Standard Test Method covers determination of mechanical property of wood with the test of bending resistance force for observation of wood behaviors, nature of wood rupture at point of failure on being acted upon and consideration for values required i.e. Stress in Outer Fiber at Proportional Limit, Modulus of Rupture, Modulus of Elasticity, Maximum Shearing Stress, Modulus of Resilience, Type of Failure.

This standard uses SI (International System Units) as principle and uses the value of unit conversion of 1 kilogram force equal to 9.806 Newton.

Preparation of Sample Pieces. A preparation of Dressed Timber must be made of total of 3 pieces free from buds or defects in good working order condition with no traces of cracks of the size 50x50x760 ml. [7]

5. Standard Test Method for Tensile and Cleavage Test of Timber.

This standard test covers determination of mechanical property of wood upon tensility force perpendicular to grain and failure resistance of the wood.

This standard uses SI (International System Units) as principle and uses the value of unit conversion of 1 kilogram force equal to 9.806 Newton.

Preparation of Sample Pieces. Preparation must be made for Dressed Timber for test of tensility of wood and test on wood failure resistance of 3 pieces of each kind with natures/appearance and size as stipulated in the Form. [8]

6. Standard Test Method for Shear Test of Wood Parallel to Grain

This Standard Test covers determination of ability to receive the shears of wood parallel to grain of wood.

This standard uses SI (International System Units) as principle and uses the value of unit conversion of 1 kilogram force equal to 9.806 Newton.

Preparation of Sample Pieces. Preparation of sample pieces of Dressed Timber of 3 pieces free from buds or defects in good working condition, no traces of crack of 50x50x63 m. size with cross section profile of the size 50x50 ml. [9]

7. Standard Test Methods for Specific Gravity of Wood.

This standard test cover determination of the value of specific gravity of the wood in the normal condition and in the dried condition. This standard test uses SI (International System Units) as primary principle.

Preparation of Sample Pieces. A preparation of sample pieces of Dressed Timber with the size according to the length to parallel grain of 25 ml. of 3 pieces or in case of flat plank sheet to be with space of 75 x 120 ml. (3 x 6 inches) total of 3 pieces.

As for the Structural Elements should be cut of wood pieces to be with minimum length of 450 ml. (18 inches) and then produce for preparation of sample pieces for prevention of loss of humidity at the end of the sample pieces. [10]

4 Research objectives

1. To study the condition and problems of thinned teak (teak thinned for distance clearance of 9 years old) in Thailand;

2. To make an examination of the mechanical properties of thinned teak (teak thinned for distance clearance of 9 years old) in comparison with team from natural forest;

3. As a guidance for making market value added thinned teak (teak thinned for distance clearance of 9 years old).

5 Research hypothesis

Thinned teak (teak thinned for distance clearance of 9 years old) was with mechanical property not different from teak from natural forest.

6 Research methodology

1. Study for Primary Information with the research of data concerning teak, the properties of wood material, quality test, and mechanical properties of thinned teak and other relevant data, relating researches, books, and textbooks for instances by making studies from libraries, various media and theses in order to obtain primary data and basic data of the research.

2. Study Wood Test Standards and methods of tests of mechanical test of wood in preparation for tests on thinned teak;

3. Preparation of teak thinned (teak thinned for clearance of 9 years old) to be in accordance with the forms stipulated ready for the tests;

4. Testing thinned teach and make result analysis;

5. Comparison of mechanical properties of teak thinned with mechanical properties of teal from natural forest;

6. Conclusion and debate of the research results.

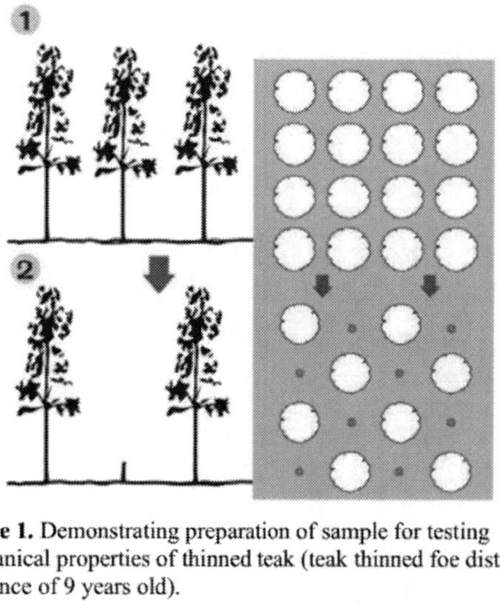

Figrue 1. Demonstrating preparation of sample for testing mechanical properties of thinned teak (teak thinned foe distance clearance of 9 years old).

7 Material and mechanical properties test

When teakwood is during the age of 7-15 years old there must be thinning for distance clearance because teak top housing are closely scrambling thereby rendering growth to be slowed down thus necessitating to make thinning for distance clearance to reduce competition of light and nutrients amongst teakwood trees (Fig. 1) teak of which obtained from thinning for distance clearance is with little value and low price when compared with teak from general natural forest thereby emerging the research concerning thinned teak for making Market Value Add.

On this research the Author has produced teak thinned for preparation as samples for testing of mechanical properties of thinned teak by making preparation according to the instruments used for measurement of the mechanical properties as provided for by each instrument such as on standard test method for compression perpendicular to grain the sample preparation of dressed timber must be made with dressed timber and the wood must be in the good order condition with no cracks of the size 50x50x150 ml. each kind of 3 pieces as shown in Fig. 2.

Figure 2. The sample preparation of dressed timber.

After making preparation of the samples for testing of the mechanical properties then the tests themselves would be carried out including Standard Test Method for Compression Perpendicular to Grain, Standard Test Method for Compression Parallel to Grain, Standard Test Methods for Direct Moisture Content Measurement of Wood, Standard Test Method for Static Bending of Timber, Standard Test Method for Tensile and Cleavage Test of Timber, Standard Test Method for Shear Test of Wood Parallel to Grain, Standard Test Methods for Specific Gravity of Wood.

After the tests the Author took the information obtained from the tests for data analysis with the method of descriptive statistics with determination of the mean of each method of test.

8 Research results

On this research the mechanical properties of teak from natural forest with age over 30 years or more as teak from natural forest for comparison with thinned teak in which the values of figures of each test can be demonstrated in Table 1.

From Table 1. It's found that the test of the mechanical property of thinned teak when the values of data obtained in each test to determine the average (mean) and make comparison with the test value of each test of teak from natural forest with age over 30 years old were with results according to Table 2.

Table 1. Demonstrating the unit (s) and interpretation result of tests of each subject matter.

Test	Unit	Interpretation Result
Compression Perpendicular to Grain Resistance	MPa	Higher (Good)
Compression Parallel to Grain Resistance	MPa	Higher (Good)
Shear Parallel to Grain Resistance	Percentile (%)	Less
Shear Parallel to Grain Resistance	MPa	Higher (Good)
Shear Parallel to Grain Resistance	MPa	Higher (Good)
Shear Parallel to Grain Resistance	MPa	Higher (Good)
Specific Gravity Determination	-	Higher (Good)

Table 2. Demonstrating the test values of each test of thinned teak in comparison with teak from nature forest with age over 30 years old.

Test	Thinned Teak	Teak with Age over 30 years
Compression Perpendicular to Grain Resistance	8.00	22.92
Compression Parallel to Grain Resistance	32.00	35.70
Grain Humidity Determination	0.20	0.20
Static Bending Resistance	7,412	7,996
Tensile and Cleavage Resistance	92.24	63.00
Shear Parallel to Grain Resistance	11.28	13.00
Specific Gravity Determination	0.52	0.60

Note ; Tests with more values are shown in bold letters.

9 Conclusion and discussion

Based on the study of the conditions and problems of thinned teak in Thailand it's with findings that thinned teak or so called thinning for distance clearance being with small stems and can be sold at low price. The application for conversion of their forms are still not varied and inadmissible by the producers and consumers rendering the growers to receive the return on the growth relative low in the first stage where thinning is required, resulting the emergence of this research in seeking for market added value for thinned teak. On the verification of the mechanical property of thinned teak in comparison with teak from natural forest with age of over 30 years, it's found that the research result was with conclusion that it's negative with the hypothesis assumed i.e. thinned teak was with mechanical property different with teak from natural forest with age over 30 years old with the results of most of the tests were with findings that the values of the mechanical property of teak from natural forest with age over 30 years were more valuable than the values of the mechanical properties of thinned teak with exception on Standard Test Method for Tensile and Cleavage Test of Timber which thinned teak was with more value than teak from natural forest with age over 30 years old.

In respect of seeking for guidelines on market value added for thinned teak can be made by producing to use in the context of the design of household goods made of wood as the guide for resolution of problems and creation of artistic work pieces household goods made of wood which is consistent with the test result of thinned teak which is with prominence on tensile resistance and capable of producing for production and creation of household goods made of wood well, capable of helping emergence of economical investment, helping with environments and rural development and for the interest of promotion of teak plantation or forestry growth and resolution of problem on the exploitation from thinned teak as well as the promotion of Thailand teak industry to prosper and progress sustainably and can be competitive with foreign countries. Furthermore, this will also help communities residing in the rural areas to receive contribution and support on the development of artistic and handicraft work pieces and communal products for creation of identities and production of household goods products in the localities and capable of creation of work careers, including the development and creation of knowledge to conserve handicraft work pieces to inherit local/popular wisdoms to the next generations consistent with the national policy without prejudice to the surrounding conditions and environments.

References

1. Forest: Private Sector Forest Plantations, Office of Forest Growth Promotion, (2010)
2. Phusutsawaeng: Plantation and Management Economically for farmers and private sector, (2010), P.5-6.
3. Workman Standard MorYorThor (Thai) 201-2541Standard Test Method for Compression Perpendicular to Grain Department of Public Works Ministry of Interior, (1998).
4. American Society of Testing Materials ASTM D 143:Standard Test Methods for Small Clear Specimens of Timber.
5. Workman Standard MorYorThor (Thai) 202-2541 (Standard Test Method for Compression Parallel to Grain) Department of Public Works Ministry of Interior, (1998).
6. Workman Standard MorYorThor (Thai) 203 - 2541 (Standard Test Methods for Direct Moisture Content Measurement of Wood) Department of Public Works Ministry of Interior, (1998).
7. Workman Standard MorYorThor (Thai) 204-2541(Standard Test Method for Static Bending of

DOI: 10.1051/matecconf/20179503004

Timber) Department of Public Works Ministry of Interior, (1998).
8. Workman Standard MorYorThor (Thai) 205 -2541 Standard Test Method for Tensile and Cleavage Test of Timber Department of Public Works Ministry of Interior, (1998).
9. Thai Industrial Standard TSI. 421-2525 Converted Wood : General Specifications, (1982).
10. Department of Public Works Ministry of Interior 207-2541 Standard Test Method for Specific Gravity to Grain, (1998).

Stress Analysis of Non-Ferrous Metals Welds by Numerical Simulation

Helena Kravarikova

Faculty of Materials Science and Technology in Trnava, Paulinska 16, 917 24 Trnava, Slovakia, Institute of Applied Informatics, Automation and Mechatronics

Abstract. Thermal energy welded material unevenly heated and thus supports the creation of tension. During the fusing process welding transient tensions generated in the welded material. Generation of the transient tensions depends on the thermal expansion and fixed permanently welded parts. Tensions are the result of the interaction of material particles. For welded parts and constructions it is necessary to know the size and direction of application of tensions. The emerging tensions can cause local change or a total deformation of welded materials. Deformations and residual stresses impair the performance of a welded construction, reduces the stability of the parts. To reduce or eliminate of action or a screening direction stresses and strains it is necessary to know the mechanism of their emergence. It is now possible to examine the emergence of tensions numerical experiments on any model using numerical simulation using FEM. Results of numerical experiment is the analysis of stress and deformation course. In the plane the tension it divided into normal σ and τ tangential folders. Decomposition stress on components simplifies the stress analysis. The results obtained from numerical analysis are correct to predict the stress distribution and size. The paper presents the results of numerical experiments stress analysis solutions fillet welds using FEM numerical simulation of welding of non-ferrous metals.

1 Introduction

Internal forces cause stress on the parts of the construction. Internal forces are reaction external load or temperature changes. When welding the material is uneven heating and thermal energy causes the formation tensions. Local or total deformations arise when the size of generated tension is greater than the yield strength or rigidity of the material. Stress and deformation are not separable phenomena in the welding process. Stress and deformations are integral, which dominate tension or deformation. Deformations and residual stresses impair the properties of welded construction, reduces the stability of the parts. Therefore, it is necessary to know the mechanism of their formation and of action. Deformation and stress fields generated during welding due to the introduced thermal energy into the material at the point of weld. It is therefore necessary to examine the progress of the welding temperature cycles.

2 The simulation model of experimental samples fillet welds

For examining the stresses in welding non-ferrous metals was chosen fillet weld. For the numerical experiment was used aluminum welded material having a purity of 99.5%. On the accuracy of the results have affect thermo-physical properties of materials depending on the temperature changes.

Material models include the following thermo - physical properties of materials [1]-[3]:

• Coefficient of thermal conductivity λ [$Wm^{-1}.K^{-1}$]
• Specific heat capacity c [$J.kg^{-1}.K^{-1}$]
• Density ρ [$kg.m^{-3}$]
• Modulus of elasticity E [MPa]
• Linear coefficient of linear expansion α_l [K^{-1}]
• Yield strength Re [MPa]
• Tangent modulus E_t [MPa].

When crossing the yield strength of the material is often used model with linear hardening. When analyzing stress-deformation tasks in ANSYS was used model with kinematic hardening.

Stress, individual parts of the structure, causes the formation of internal forces in response to an external load or temperature change.

Deformation and stress fields are produced by action of thermal cycles of welding. The total of strains in the examined body, the non-uniform temperature field, can be expressed with as the sum of the elastic deformation, depending on stress, the temperature and plastic deformation is [1], [4]-[6]:

$$\varepsilon_x = \varepsilon_{\sigma x} + \varepsilon_T + \varepsilon_x^P = \frac{1}{E}\left[\sigma_x - \mu\left(\sigma_y + \sigma_z\right)\right] + \alpha_T \Delta T + \varepsilon_x^P \tag{1}$$

$$\varepsilon_y = \varepsilon_{\sigma y} + \varepsilon_T + \varepsilon_y^P = \frac{1}{E}\left[\sigma_y - \mu\left(\sigma_x + \sigma_z\right)\right] + \alpha_T \Delta T + \varepsilon_y^P \tag{2}$$

$$\varepsilon_z = \varepsilon_{\sigma z} + \varepsilon_T + \varepsilon_z^P = \frac{1}{E}\left[\sigma_z - \mu\left(\sigma_x + \sigma_y\right)\right] + \alpha_T \Delta T + \varepsilon_z^P \tag{3}$$

$$\gamma_x = \frac{1}{G}\tau_x, \quad \gamma_y = \frac{1}{G}\tau_y, \quad \gamma_z = \frac{1}{G}\tau_z. \tag{4}$$

where:
ε_σ – an elongation by stress in the direction of coordinate axis x, y, z,
ε_T – an elongation of the temperature in the direction of coordinate axis x, y, z,
ε^P – the plastic an elongation in the direction of coordinate axis x, y, z,
γ – shear strain.

Equations 1 to 4 are expressions of Hook's law for linear 3D solid isotropic material in Cartesian coordinates.

Differential equations, in vector notation, for thermo elasticity has the form [1], [4]-[6]:

$$\Delta \mathbf{u} + \frac{1}{1-2\mu}\,\text{grad div}\,\mathbf{u} - \frac{2(1+\mu)}{1-2\mu}\alpha_V\,\text{grad}T = 0 \tag{5}$$

where:
$\mathbf{u}$ [m] – displacement vector,
μ [-] – Poisson´s Ratio,
T [K] – thermodynamic temperature,
α_v [K^{-1}] – coefficient of thermal volumetric expansion.

Cancyho tensor of the mechanical elastic deformation $\varepsilon_{ij}^{m,E}$ and the thermo-elastic deformation tensor $\varepsilon_{ij}^{T,E}$ are defined by the relation [1], [4]-[6]:

$$\varepsilon_{ij}^{m,E} = \frac{1}{2}(u_{i,j} + u_{j,i}) \tag{6}$$

$$\varepsilon_{ij}^{T,E} = \alpha_{i,j}\Delta T \tag{7}$$

Hooke's law generally reflects the dependence of strain and stress in the form [1], [3], [4]:

$$\sigma_{ij} = C_{ijkl}\varepsilon_{kl}^{E} = C_{ijkl}(\varepsilon_{kl}^{m,E} + \varepsilon_{kl}^{T,E}) \tag{8}$$

where:
$\sigma_{i,j}$ [Pa] – tensor elements of stress,
C_{ijkl} [Pa] – elements of the tensor of elasticity.

The overall relative deformation can be expressed as the sum of elastic deformation ε_{ij}^{E} and plastic deformation ε_{ij}^{P} [1], [4]-[6]:

$$\varepsilon_{ij} = \varepsilon_{ij}^{E} + \varepsilon_{ij}^{P} \tag{9}$$

Geometric model shape and size was the same with real experimental sample. The weld was made GMAW method. The stress - strain analysis was solved by numerical simulation using FEM in ANSYS. The solution stress - strain problems of numerical simulation is the thermal analysis. For the calculation of residual stresses and deformation during welding are used the results of the thermal analysis as a load of the model. The task is solved in ANSYS as transient and nonlinear.

3 The results of numerical simulation of structural tasks

Fig. 1 shows the course of the equivalent Mises stress at the time of welding 50 s. At this time, the residual stress is minimum value 0.626 MPa. A small tensions value is in the area of the weld, there is metal in liquid phase. At time 50 s is the maximum value of residual stress 83.3 MPa.

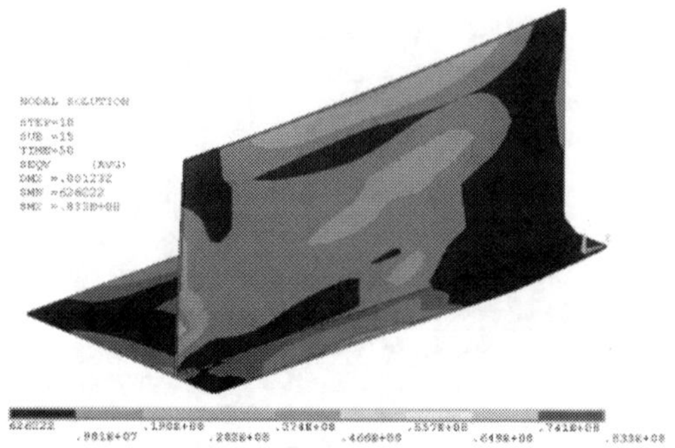

Figure 1. Stress field at the time of welding 50 s

Fig. 2 shows the equivalent Mises stress at the time of welding 130 s. At this time, the minimum stress value is 0.195 MPa and the maximum stress is 24.1 MPa.

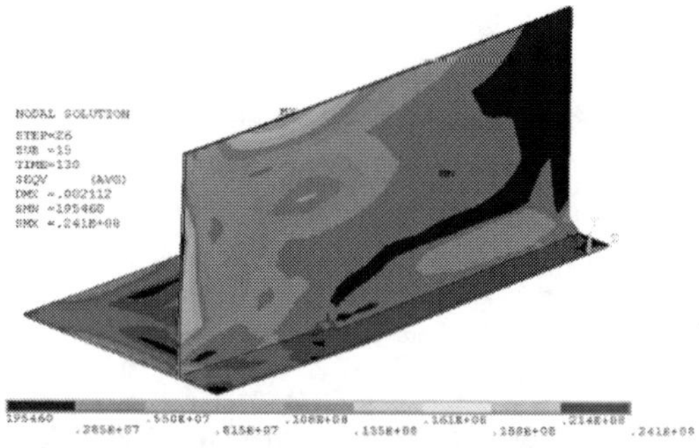

Figure 2. Stress field at the time of welding 130 s

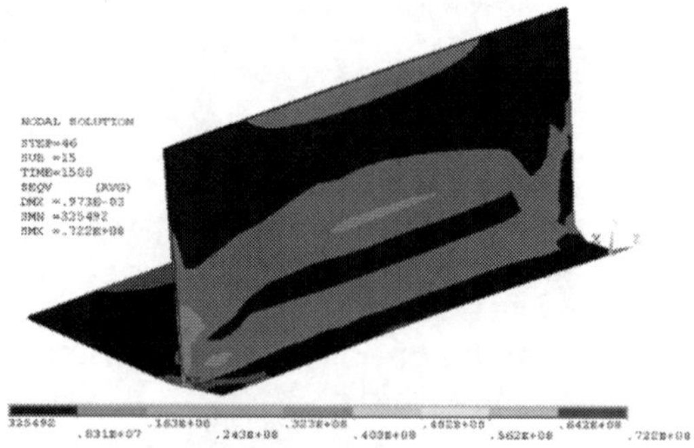

Figure 3. Stress field at the time of welding of 900 s

In Fig. 3 is shown the course of the equivalent Mises stress after cooling to room temperature 22.281 °C at the time of 900 s. At this time, the temperature field in the sample throughout the volume of the balanced and minimum residual stress is 0.325 MPa. The values of residual stress are in the weld between 8.31 MPa and

DOI: 10.1051/matecconf/20179503005

24.3 MPa. The maximum value of the residual stress is 72.2 MPa. The maximum stress is acted simultaneously at the point where the sample has been firmly fixed, and it harvested all degrees of freedom.

Fig. 4 graphically show the results of angular deformation of the sample of numerical experiments.

Figure 4. Graphical representation of angular deformation of numerical experiments

In Fig. 5 is graphically show the results of longitudinal deformation of the sample of numerical experiments.

The angular and longitudinal deformations produced as a result of supply of heat energy in the process of welding experimental samples.

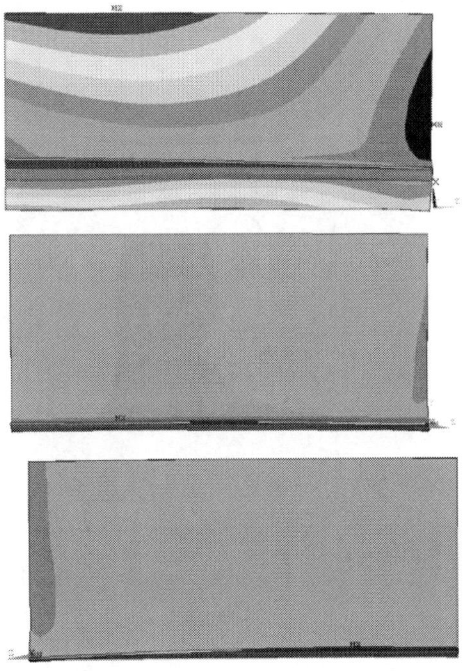

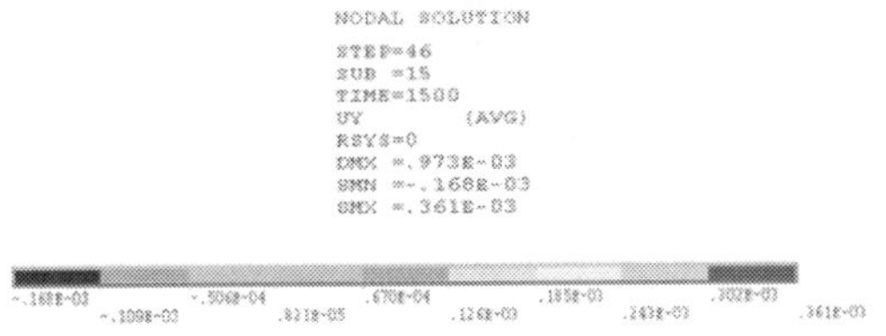

Figure 5. Graphical representation of longitudinal deformations of numerical experiments

4 Evaluation of the results

Angular weld is deformation of numerical experiments with shape deformation identical to the real experimental samples.

In Fig. 6 is graphically show the angular deformation of the sample of real experiments. Fig. 7 graphically show the longitudinal deformation of the sample of real experiments.

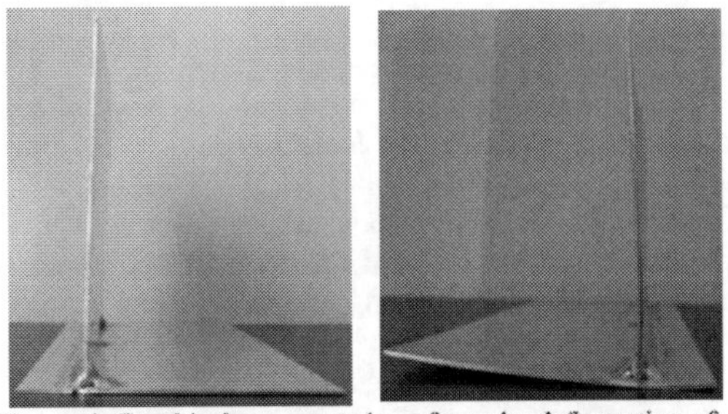

Figure 6. Graphical representation of angular deformation of real experiments

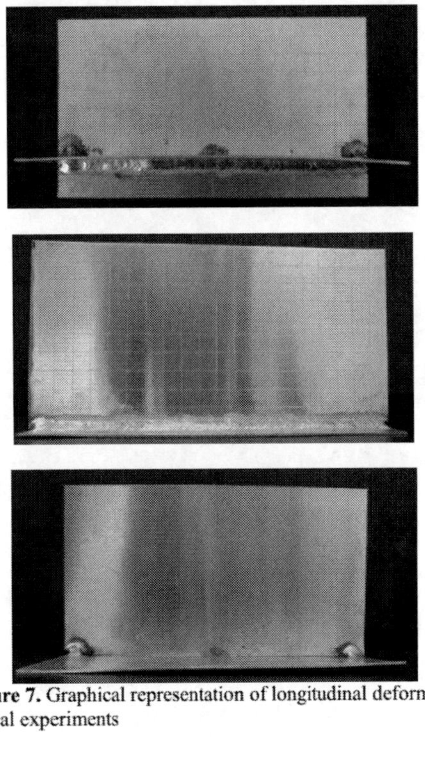

Figure 7. Graphical representation of longitudinal deformation of real experiments

In the Fig. 4 to Fig. 7 it is seen that the deformed shape of the experimental sample in the real experiment is identical to the shape of the sample graphical display of the numerical simulation of the all of the above views.

On the basis of the results obtained deformations of numerical and real experiments it can be concluded that the results of deformation for mating. Therefore, it is possible to replace the real experiment numerical simulation.

The accuracy of results obtained numerical experiments depend on many factors. The most important factors affecting the accuracy of the results are thermo-physical properties of welded materials depending on temperature change, method of simulating the heat flow in the welding site and its surroundings, initial and boundary conditions and welding method. Subject investigation the accuracy of the results of the numerical experiment might be the chemical composition of the welded materials and their dependency, temperature changes. Deviations that can arise are caused by inaccurate by subtracting thermo-physical properties of materials from the chart. The simulation is always contemplated welding speed uniformly. It is also contemplated steady movement of volumetric heat source. In manual welding it is not possible to meet uniformly speed and also moving volumetric heat source. On the accuracy of the results of numerical simulation has also implications welding parameters, the depth of fusion penetration and the size of the heat affected zone. The accuracy of the results achieved in part affects the human factor.

5 Summary

Experimental problem solving in different areas of engineering technology is very often replaced by computer simulation and numerical solution. Substitution of experimental solutions using numerical simulation has significant scientific and economic benefits. Further examination might be those of other advanced techniques such as welding special welding technology, as well as examination of the impact of the welding parameters on the accuracy of the numerical simulation. Numerical experiment is a flexible solution of stress and thermal deformation process produced during welding.

Acknowledgement

This publication is the result of implementation of the project: "University Scientific Park: Campus Mtf Stu - Cambo" (ITMS: 26220220179) supported by the Research & Development Operational Program funded by the EFRR and of the project VEGA 1/0481/14, 1/1010/16.

References

1. Hrivňák: Theory of weld ability of metals and alloys Elsevier, USA, (1992)
2. Kolektív autorov: Konštrukcia a navrhovanie, Zeross, Ostrava, (2001)
3. H. F. Wang, D. W. Zuo, H. Miao, H. J. Wang, "Numerical Simulation of the Prestressed Laser Welding of 7075—T7451 Aluminum Alloy Sheet", Key Engineering Materials, Vols. 431-432, pp. 13-16, (2010)
4. M. H. Sadd: Elasticity, Theory, Applications and Numerics, 2. Edt., Elsevier, USA, (2009)
5. F. Trebuňa, F. Šimčák, V. Jurica: Pružnosť a pevnosť I, Vienala, Košice, (2000)
6. M. S. Węglowski, P. Sedek, C. Hamilton, "Experimental and Numerical Analysis of Residual Stress in Cast Aluminum Alloy after FSP Process", Key Engineering Materials, Vols. 651-653, pp. 1563-1568, (2015)

MATEC Web of Conferences 95, 03006 (2017)
ICMME 2016 DOI: 10.1051/matecconf/20179503006

Numerical Investigation of the Air Flow Effects on Amorphous Alloy Ribbon Formation

Yongpeng Li, Cuiping Wang and Fengyin Wang
Energy Engineering Institute, Qingdao University, Qingdao 266071, China

Abstract: During the amorphous ribbon production process, the mechanisms that cause uneven thickness and herringbone on two surfaces of the ribbon are very complicated. The high velocity rotating roller causes periodic air eddies near the surface and air flow around the contact zone. The air flow might be the major reason that causes the non-uniformity thickness and the ripples. In this paper, a two-dimension model was developed to simulate the air flow eddies and a three-dimensional simulation was performed for the air flow field around the puddle zone. Simulation results show that the higher velocity of air flow around the puddle edges is closely related to thicker ribbon rim. The continuous air eddies along the roller surface and the air flow fluctuations along the roller width are the major reasons causing non-uniform convective heat transfer during ribbon solidification. The frequency of air eddy matches well with the nominal ripple frequency along the ribbon length, and the air flow fluctuation coincides with the herringbone wavelength along the ribbon width. The influences of the air flow on cooling, ribbon thickness, and herringbone phenomena are all discussed here.

1 Introduction

The amorphous alloy ribbon is a function material with the advantages of excellent magnetism, anti-corrosion, high mechanical strength and toughness, high electrical resistivity and conductivity. It has been widely used as the core material for electric motors, transformers, sensors, etc.

The rapid solidification process, shown in Fig. 1 [1], is a popular method for amorphous ribbon production nowadays, the melt Fe-based alloy liquid flows out of the crucible nozzle, impinges on the surface of the high speed rotating roller surface, spreads out rapidly forming thin ribbon. Similar approach has been presented in works [2]-[3]. The high pressure and high speed circulating cooling water inside the rotating copper roller takes the main solidification heat away. The width of the jet nozzle (B) and the gap (G) between the nozzle and the roller surface are only about 0.3 mm, as shown in Fig. 1(b). The alloy liquid solidifies into ribbon in a milliseconds to 0.5 second.

The heat transfer coefficient is a key factor to the solidification process, which has been studied by some researchers. Hui[4] calculated the cooling rates under different wheel velocities. The modeling results agree well with the experimental results except the region near the up-meniscus and down-meniscus. The simulation [3] showed that the heat transfer from the molten liquid to the air by convection is negligible, while the majority of heat (Qw, approx. 2.7 MW) is conducted from the molten liquid to the roller, about $(1.8\sim2)*10^6$ W/m^{2}K[5] .

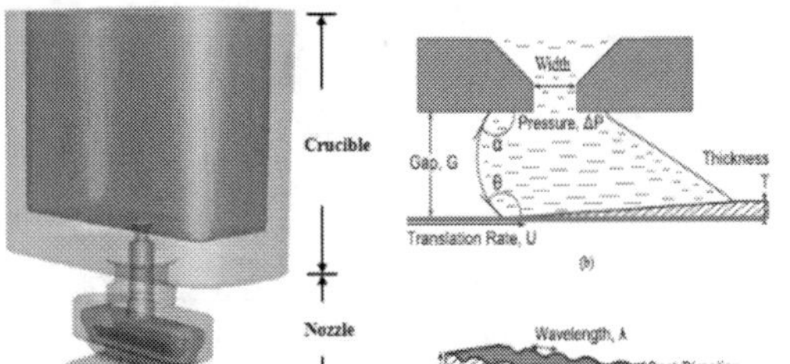

Figure 1. The ribbon production and the herringbone formation. (a) the amorphous ribbon production device; (b) the contact puddle zone; (c) ribbon cross section

So the size and shape of contact zone are very important to the cooling rate, which in turn influences the dendritic crystal growth and ripple wavelength formation [6]. Fig. 1(c) shows the ripple or herringbone formed on both ribbon surfaces during the solidification process. The pattern of ripple or herringbone is λf_λ = U [1]. The ripple or herringbone is desired for ribbon used as electrical resistivity and conductivity material because it reduces the energy lose when the electric motor no load working. Fig. 2 shows the ribbon produced by YUNLU Ltd. The Herringbone on the wheel-side is shorter in wavelength and shallower than those on the air-side. The study conducted [7] showed that no ribbon was formed under vacuum. So the ripple is clearly related to the air

 DOI: 10.1051/matecconf/20179503006

flow. The ripple width along the rotating direction is about 1 mm in the ribbon center and 2~3 mm near the edge.

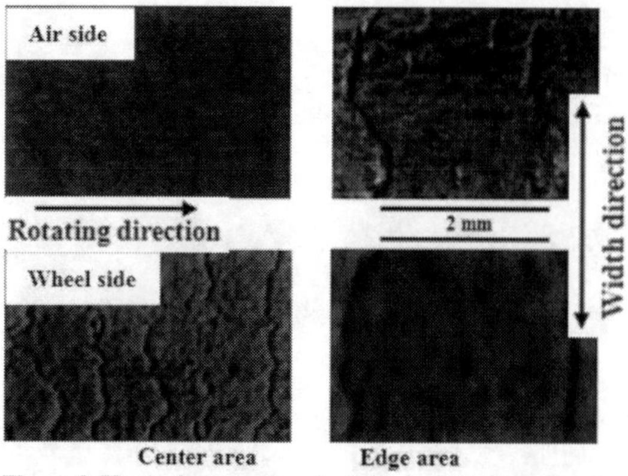

Figure 2. Herringbone on the wheel side and Ripple on the air side surface of ribbon

The other distinct feature is the ribbon thickness. Previous studies show the thickness is related with the nozzle-wheel gap [8], wheel speed [2], [4], liquid viscosity [9], and thermal conductivity [5]. The thickness along the ribbon width is also uneven, which is thicker in the center and edge zones, as shown in Fig. 3. For the ribbon produced by Yunlu here, the variation in ribbon thickness is about 2~3 μm.

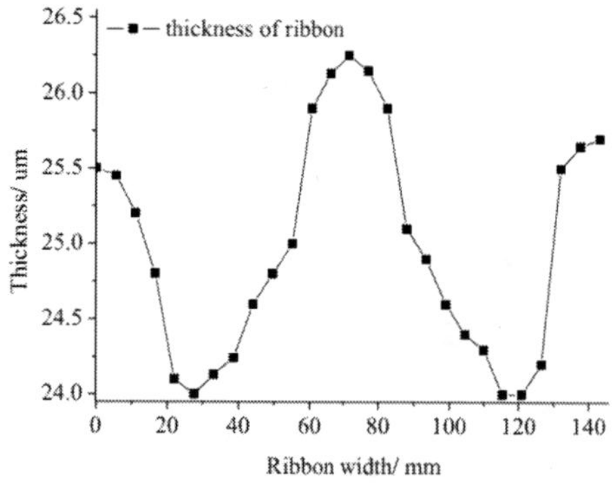

Figure 3. The ribbon thickness variation along the width.

The air flow over the roller played a key role, like the turbulence flow around a two-dimensional hump[10] and to fluctuated cool the alloy liquid. The nozzle effect and geometry are also the important factors[11]-[12]. In this paper, numerical simulation of air flow around the contact zone was carried out, including the air flow field over the rotating roller, the flow air scouring the contact zone, and the air flow around the nozzle. The influences of air flow on the convective heat transfer, ripple formation, and ripple thickness were discussed. The conductive heat transfer to the copper roller was regarded as uniform and the solidification process is neglected in this paper.

2 Numerical simulation models

The model developed in this paper is based on the rapid cooling device of the amorphous ribbon production line in Qingdao Yunlu New Energy Technology Co., Ltd. The study focuses on the rapid solidification area, including the nozzle and slot, the contact zone, the wheel-nozzle gap, and the rotating roller surface. For the ribbon width and length are all larger several degrees than the nozzle and puddle zone, a two-dimensional model was developed to simulate and focus on the shape of the molten puddle under the nozzle and the air flow regime along the roller surface with the roller swirling. Based on the inlet and outlet parameters obtained in this 2D simulation, setting the corresponding parameter values for a simplified 3D model, then the air flow in the gap and the entrainment and turbulent of flow around air was simulated three dimensional.

2.1 Two-dimensional simulation of the molten puddle shape

Fig. 4 shows the two-dimensional model developed for the molten puddle zone. The molten liquid flows out of the vertical nozzle slot (0.3 mm in width), spreads on roller surface forming thin layer due to the rotating roller drag force. The data from YUNLU Co. were used as the boundary conditions for all simulations, which were listed in Table 1, including the roughness of the aluminum oxynitride refractory and the roller copper surfaces, the mean liquid velocity of the nozzle inlet, the dynamic pressure outside the contact zone, and the velocity of the roller surface. The main liquid iron physical properties at 1300℃ used in all simulations were also listed. The $k-\varepsilon$ turbulent model and VOF two-phase flow model were adopted for both the air flow and molten iron liquid. Since the length of the contact surface is much longer than the gap and nozzle slot, the roller surface was treated as a flat plane.

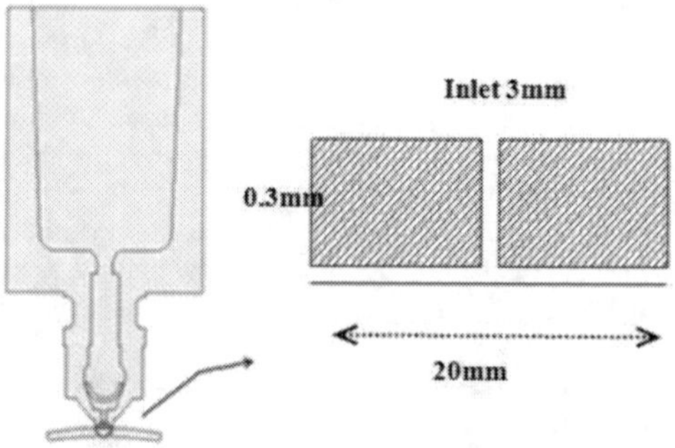

Figure 4. Two-dimensional model of the molten puddle zone

Table 1. The material physical properties in simulations

Material	ρ/(kg/m^3)	Cp/(J/kg·K)	viscosity/(Pa·s)	λ/(W/m·K)	roughness/(um)
liquid iron	7000	462	0.00225	54	/
refractory	2500	680	/	11	3.2
copper roller	8978	381	/	388	0.8

2.2 Two-dimensional simulation of the air flow in the roller surface zone

The diameter of the rotating roller is 800 mm, several order of magnitudes larger than the nozzle and the gap between the roller and the nozzle. In this study, the region 30° before and after the nozzle was studied, as shown in Fig. 5. The height of simulated zone is 25 mm. The roller rotating velocity is 580 r/min, which is the line speed of 24.3 m/s, the $k-\varepsilon$ turbulent model was also mainly used.

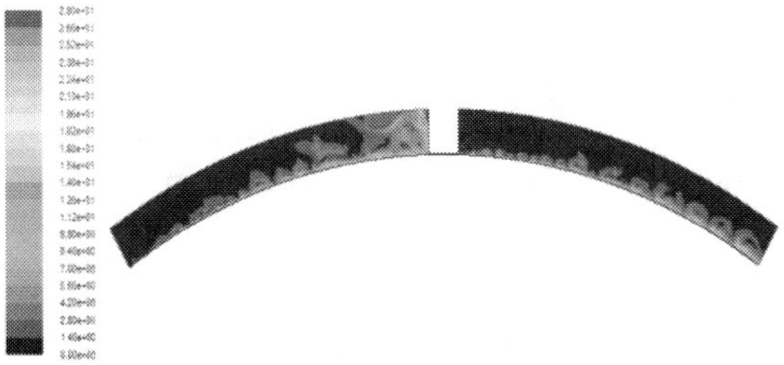

Figure 5. Two dimensional simulation zone and air flow velocity.

2.3 Three dimensional model of the air flow around the nozzle

A three-dimensional flow simulation was carried out for the analysis of air flow along the ribbon width. Fig. 6 shows the dimensions of the simulation region. The actual roller width is 165 mm, the jet slot length is 142 mm, shorter than the nozzle refractory solid, and shorter than the roller width, as shown in Fig. 6(a). The bottom is the roller surface, and the height of air flow zone is 25 mm. The simulation was carried out using the Ansys Fluent 14.5 software. The inlet pressure (1.35bar) and outlet pressure (1bar) were used as boundary conditions for both two dimensional and three dimensional models. The rotating roller surface was treated as the moving-wall. The simulation was transient and the time step was set as variable in order to avoid the courant number exceeding the default limit.

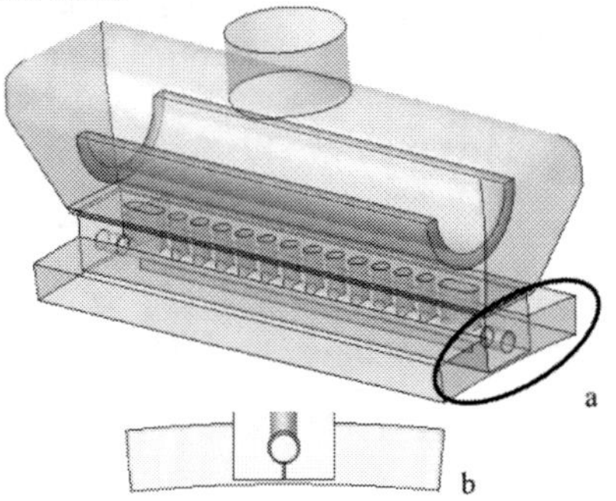

Figure 6. The three dimensional simulation region of air flow (b is the side view of simulation area).

3 Results and discussion

3.1 The molten puddle shape

Since the liquid iron is of weak wettability with the nozzle refractory materials, the iron liquid flows out without adhesion to the nozzle slot. As shown in Figure 7, the liquid iron was spread and layered due to the drag force of the high speed rotating roller surface[13]. Due to the fast roller rotation speed, very thin iron liquid layer (25um) was ultimately formed on the roller surface, by rapid cooling.

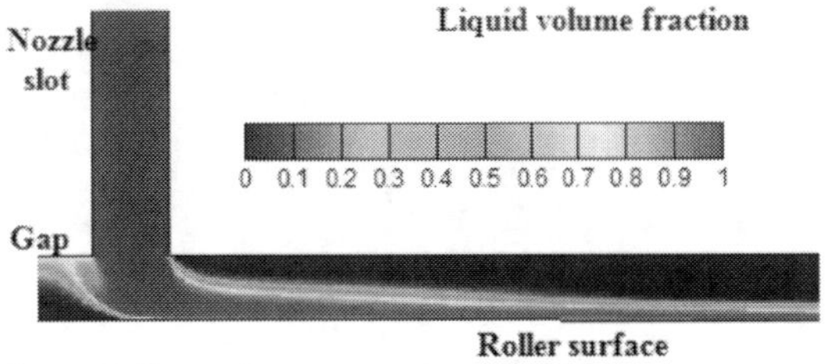

Figure 7. Shape of contact zone (molten puddle)

3.2 Eddies on the roller surface

Compared from the velocity vectors variation of two neighbored eddies shown in Fig. 7, the air flow has higher initial velocity tangential to the local roller surface, causing lower pressure between the roller and tangential air flow. When leaving the curve surface, the air flow slows down rapidly. Some air could be drawn back to form eddy flow, and some could blend with the follow up air to form another eddy. Some slower air flows into the back eddies, then the air is accumulated and the eddies become bigger and stronger till they hit the nozzle body. The air flow along with the rotating roller is blocked by the nozzle, forming large vortex, as shown in Fig. 8.

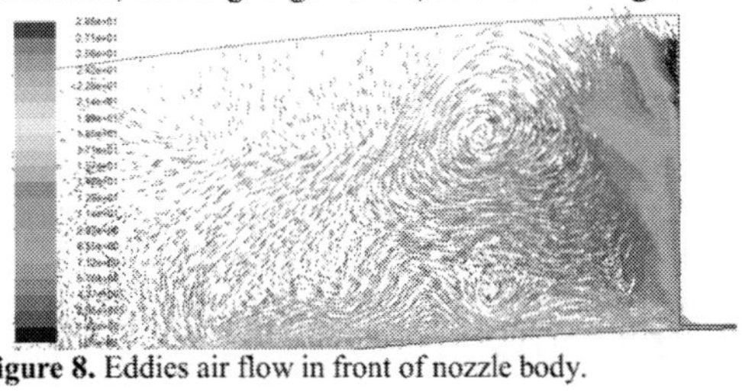

Figure 8. Eddies air flow in front of nozzle body.

Only a very thin layer of air very close to the roller flows into the gap under the nozzle. The air flow in gap is compressed and disturbed by the periodically reflected eddies, which causes air flow vibration and the subsequent periodic change in convective heat transfer between air and the molten iron liquid. The air in the gap after flowing around the molten puddle, ultimately flows out over the surface of the roller and the molten iron liquid, shown in Fig.9. there are also serious eddies along the rotating roller surface. Because the quantity of air flow through the gap is small, the initially formed eddies are small, which were later enhanced due to the air entranced from the surrounding area. So the ribbon would pass through several air eddies after peeling off the roller.

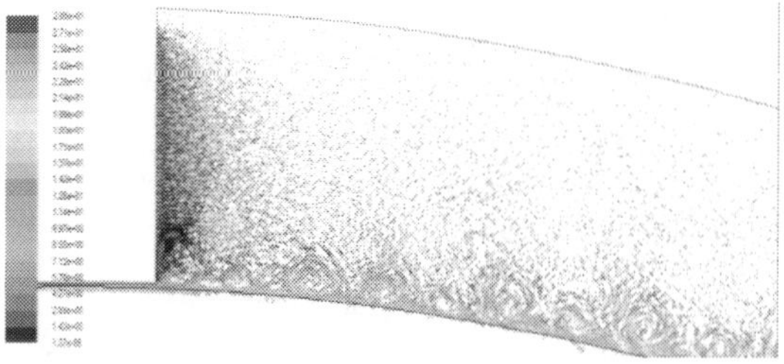

Figure 9. The eddy development of air flow at the back of nozzle.

The air eddies near the nozzle body could affect the cooling to the ribbon, cause periodic changes in the convective heat transfer and even the dynamic pressure on the ribbon surface. The eddy formed out of the solidification area exhibits the similar size to the ribbon ripple wavelength [1], [10], the formation mechanism of ripple on the air-side could be due to the periodic air flow eddies. When the upstream air flows in the gap approaching the liquid film, shown in Fig.10, the velocity is lower near center with regularly vibration; the amplitude of flow vibration is at least 3 m/s, which could contribute to the ripple formation from both solidification rate and air flow impact vibration.

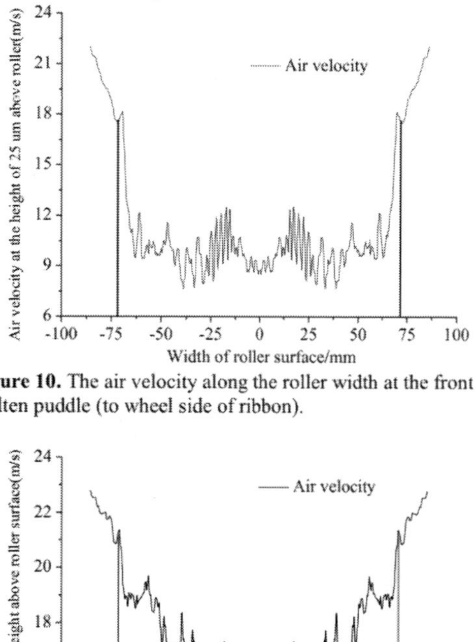

Figure 10. The air velocity along the roller width at the front of molten puddle (to wheel side of ribbon).

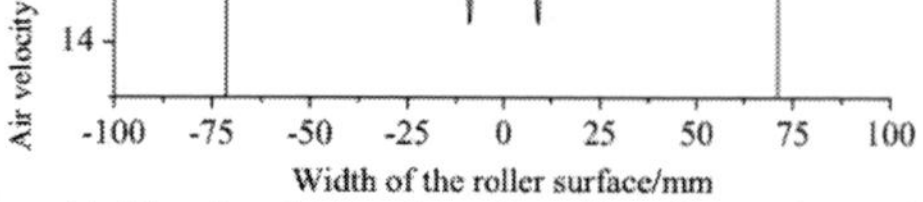

Figure 11. The air velocity along the roller width at the back of molten puddle (to air side of ribbon).

Fig. 11 shows the air velocity on the line of 1 mm downstream of the liquid film. The liquid is directly to be the air-side of the ribbon. The wavelength judged from the curve is shorter in the center and become longer on both sides, which match the feature of ripple as shown in Fig.2. The herringbones formed on the wheel-side surface is about 1~3 mm in wavelength, similar to the velocity vibration frequency.

For the air flow at the back of liquid film, the velocity is also similarly symmetric in Fig. 11, which is higher in velocity and more severe in vibration at the center area, the velocity increases gradually to its peak value and reduces a bit to a new valley, then increases to the highest value near the edges. The air velocity on the ribbon side is always higher than 15 m/s, and the velocity on the wheel side is about 10 m/s. As a consequence, the wave length is clearer and matches the air-side ripple size, causing deeper herringbones on the air side surface.

As seen, the ripples on the two sides are different due to the difference in air flow field. On the wheel-side, although the conductive heat transfer on the roller copper surface is uniform and is the main source taking away the solidification heat, the periodic vibration of air flow and the dynamic pressure shock on the contact zone cause difference in heat transfer and thus formed herringbones. Since the solidification of wheel side liquid is faster, the ripple is shallow and compressed to form herringbones. On the air side, the mixture air and periodic eddies influence the ribbon solidification and cause wave ripple formation.

4 Conclusions

(1) The non-uniform air flow and the corresponding influences on convective heat transfer could be the major reason for the two common features observed during the ribbon production. The air flow is induced by the fast rotating roller, and the air flow velocity in the gap is very high with regular fluctuation.

(2) The uneven ribbon thickness along the width is influenced by the air flow velocity variation and the high velocity air flow around the contact zone edges.

(3) The wave ripples formation is related with the periodic vibration along the width and the air eddies distribution along the rotation direction. The wheel-side surface undergoes more rapid cooling and compress, causing shallow herringbones. The continuous air eddies in the front of the contact zone influence the cooling and solidification, causing deeper wave ripple lines.

Acknowledgment

The financial support was received from the Natural Science Foundation of Shandong Province (Project No.ZR2015EM004) and National Natural Science Foundation of China (Project No.51676102) . Thanks for the Taishan Scholars Construction Engineering of Shandong province.

References

1. BL Cox, PH Steen. J. Mater. Process. Tech. **213**,1743–1752 (2013).

DOI: 10.1051/matecconf/20179503006

2. DR Li, JH Zhuang, TC Liu, ZC Lu, SX Zhou, J. Mater. Process. Tech. **211**, 1764–1767(2011).
3. M Sowjanya, TKK Reddy. J. Mater. Process. Tech. **214**, 1861–1870(2014).
4. XD Hui, YS Yang, XM Chen, ZQ Hu, Sci. Technol. Adv. Mater. **2**, 265-270(2001).
5. MBussmanna, J Mostaghimi, DW Kirk, JW Graydon., Int. J. Heat Mass Tran. **45**, 3997–4010(2002).
6. C J. Byrne, A M. Kueck, S P. Baker, P H. Steen. Mat. Sci. Eng. A-Struct. **459**, 172–181(2007).
7. R Seino, YC Sato. J. Alloy Comp. 586(2014),S150–S152.
8. D Raybould, M Meola, R Bye, SK Das. Mat. Sci. Eng. **A 241**,191–201(1998).
9. L.A.Anestiev. Mat. Sci. Eng. A **110**,131-138(1989).
10. Yakeno, S Kawai, T Nonomura, K.Fujii. Int. J. Heat Fluid Fl. **22**,235 - 240(2015).
11. ZG Pei, CW Yu. J. Fluid Struct. **27(1)**, 121-133(2011).
12. Y. Otobe, H. Kashimura, S. Matsuo, T. Setoguchi, H.-D. Kim. J. Fluid Struct. **24(2)**, 281-293(2008).
13. CW Chen and WS wang. App. Math. Model. **19**,704-712(1995).

Numerical Simulation Analysis of Seismic of Frame Structure on Hill Terrain

Weisu Weng , Xupeng Zhao and Zhanxue Zhou

Hebei University of Architecture, 075000 Zhangjiakou, Hebei, China

Abstract. In recent year, Wenchuan,Ya'an,Yushu and other areas in china occur a series of high earthquake, however areas of earthquake is similar as mountainous terrain, building structure of seismic increasingly aroused our concern, and the research that hill topography affected building structure seismic in shallow mountain. The research content mainly includes: through modelling was built by the ANSYS software, the cooperative effects of a ten layer of frame structure- hill system were calculation. First, simple comparative dynamic characteristics analysis of soil - structure interaction and the rigid foundation assumption conditions; Second, put Hill-Soil-Structure Interaction(referred to as HSSI) and Soil - Structure - Interaction(referred to as SSI) further analysis of the dynamic response, including: including structural modal analysis (vibration mode, cycle), the time history analysis (such as displacement, internal force and acceleration) and so on. Through Hill-Soil-Structure Interaction research, taking each factor in consideration, giving structure seismic key technology measures about shallow mountain to provide reference for such structure theory research.

1 Introduction

Our country is a vast land, but two-thirds of the area is mountain. With the growth of economy, urban land used by built construction does not meet the requirements of people to pursue a more healthy and comfortable environment, so the expansion of the city is imminent, therefore the development of the town is at its time. At the same time, with the adjoining hill land becoming more precious, the development of it is becoming more and more urgent. In engineering design, seismic research hill on building structures is lack of mature theory, so it may cause some unexpected risks of structural safety. For example: Many building structure is facing such a problem, is adjacent to mountain according to the design of the rigid ground assumption, which means there is no real consideration about hill and the role of its soil, simplified calculation program for analysis did not consider this part of the factors that could cause the internal forces of structure to reverse the sudden increase, these problems are not conducive to seismic design. Meanwhile, China is located between the Eurasian seismic belt and the Pacific Rim seismic belt, where seismic activity is so wide, and China is also an more and more seismic country, so the seismic design of frame structures for shallow mountain terrain should arouse our attention .So the analysis on such structures overall seismic is the most important issue for seismic areas.

Nowadays, the research of soil - structure interaction has made some achievements 1, but the application in practice is relatively less, and the analysis of hill - soil - structure is even fewer. Through theoretical study of HSSI issues, we will recognize the difference of the dynamic characteristics of the entire system and a single soil - structural; when the hill - soil - structure analysis is used as a whole, which is more different from the actual design of the structure, which is separated of the upper structure and rigid foundation. Therefore, the research fills up the deficiency of the theory, which provides us a reference for the future and the experience of the actual project.

2 Principle of analysis model

2.1The soil structure

Soil structure is divided into two parts, one part is the soil which is at the bottom of the upper structure, the other part is the soil which is next to the upper structure of the hill. When often encountered or rarely met earthquake seismic action, the soil meets into the plastic zone. Considering the practical calculation of the analysis, this article mainly studies the effect of the hill and the structure, and reduce the process of calculation at the same time. Integrated all factors, the two parts of the soil both use ideal elasticity. So the results of the analysis is ideal, and we save much time for the calculation. Parameters of materials are as follows: elastic modulus of soil, poisson's ratio, density.

2.2 The upper structure

For the analysis of the upper structure, we adopt 2-D truss - layer mechanical model, which is built by the basic analysis unit, and to simplify the beams and columns for the neutral axis without mass, so the quality will be focused on each node, the DOF will be coherency by certain assumption, and the dynamic degree of freedom will be reduced, and the dynamic analysis will be carried out with in layer model. Duing to the greatness of the upper structure stiffness, compared with soil stiffness, the elastic analysis will be used.

2.3 The overall model

The soil structure model and the structure model are combined to form the whole stiffness matrix, mass matrix, and join Rayleigh damping, and to form the whole matrix. First, the step is going to analyse about static analysis; then, which are applied seismic action for the modal analysis and transient analysis.

Modal analysis is the analysis on the characteristic of structural natural vibration is used to determine the natural frequency and vibration mode of structure, the structure as the basis of transient analysis. Also can through the modal analysis preliminary judgment of the structure of the weak link, provide reference for further in-depth analysis.

And the time history analysis method is also known as direct dynamic method (or transient dynamics analysis), the modal analysis is further added, which is based on the dynamics equation of motion. Through the seismic time history curve as incentives, direct integral operation, to solve the dynamic response of each moment, including: displacement, stress and strain, acceleration, force, etc.

Any time of the whole model under earthquake action balance equation, the following formula:

$$Ku(t)+C\dot{u}(t)+M\ddot{u}(t)=F(t) \quad (1)$$

$$F(t)=-M\ddot{u}(t) \quad (2)$$

Type: K, C, M, respectively represent the whole stiffness matrix, damping matrix of the whole, the overall quality of matrix; F(t) represent dynamic loading, $\ddot{u}$(t) represent he earthquake acceleration.

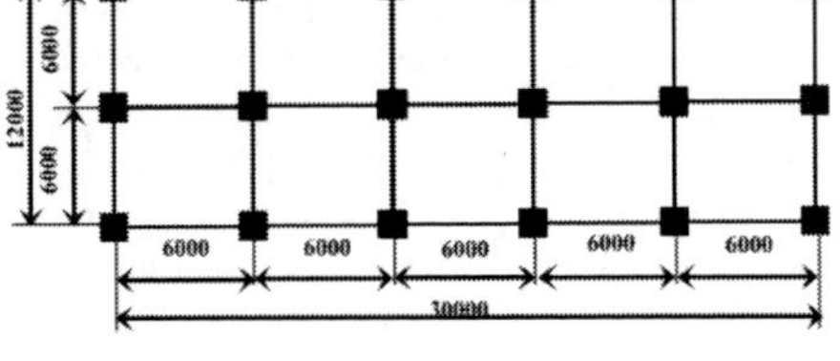

Figure 1. building plan

3 Project summary

A building structure system for cast-in-situ concrete frame structure, building standards plan is shown in Fig. 1. A total of ten layers structure, average height of 3m, span is 6m. Site casting thickness: 120mm, beam cross section: 250mmx550mm, column section: 600 mmx600mm, concrete strength: C30, reinforced models: HRB400. Site category belongs to three classes, considering the seismic fortification intensity of 7 degrees often met earthquake (0.15 g).

3.1 Model parameters

Whole structure is simplified, Relevant data of the upper structure includes beams and columns using Beam23 unit, the elastic modulus E = 3.15×10^{10}N/m^2, poisson's ratio ν =0.3, the damping ratio of concrete ξ = 0.05, concrete density ρ = 2700 kg/m^3.

Soil of structure bottom: using Plane42, hill is 50m deep.

Related introduction is as follows: BEAM23 unit is 2D plastic beam element, can have some property (tension- compression and bending bending capabilities), each node has three degrees of freedom, namely the translational motion displacement along the node coordinates X, Y direction and rotation displacement around the Z axis.PLANE42 entity unit for a 2D structure, which is simulated entity structure plane. The element is used for graphic unit (plane strain and plane stress), and can be used for axisymmetric element. The unit has four nodes, each node has two degrees of freedom [2].

3.2 The artificial boundary and damping

In the soil-structure dynamic finite element time history, it is very important to select the appropriate finite soil on behalf of infinite field. In order to ignore the influence of the boundary of soil to structure, it needs to select part of the limited computing area in infinite field and choose the right size artificially. The soil boundary of the project model is greater than 10 times the length of one frame, which implements the scattering of soil boundary. This example uses the fixed constraints at the bottom of artificial boundary and restricts the vertical displacement at right and left, which meets the precision requirement.

In structural dynamic analysis, it is a key factor of damping selection in the process of the seismic capacity dissipation which caused by damping of the whole structure in the time course. ANSYS provides five kinds of damping. The results show that several kinds of damping effect on the structure is the same after comparing several damping. In order to simplify the calculation of structure form, this article chooses Rayleigh damping which has simple parameters setting. Firstly, extract the former two-class frequency based on modal analysis, which computes as:

$$C=A\ M+B\ K$$

where α is Alpha damper, also called mass damper;β is Beta damper, also called stiffness and damping coefficients.

3.3 Input seismic wave

When the dynamic time history analysis is carried out on the building structure model, a reasonable choice of seismic wave analysis is a key factor to success. when choosing different seismic waves, the dynamic responses of the structure is very different; coupled with the randomness of the earthquake, so input a reasonable seismic wave is important. Referring to many research results show, seismic wave selected according to the site classification, through the use of China's code for seismic design of buildings GB50011-2010, an acceleration peak value change processing, simplified calculation formula and also meet the accuracy requirements, reflecting the real dynamic response of the structure .Using Ning he-Tianjin's seismic waves have been record, he peak acceleration is: =104cm/ ,The time interval is:0.01s,The duration time is 19.2s. Input Seismic wave in the X direction, Acceleration adjustment formula is as follows:

$$a' = \frac{A'_{max}}{A_{max}} a(t). \quad (3)$$

4 Interaction analysis of model

The soil - structure interaction seismic calculation and model history analysis method of the calculation process is as follows [4]: (1) consider the weight of structure, the static analysis (ANTYPE, STATIC);(2) under the condition of without considering gravity, modal analysis (ANTYPE, MODAL);(3) the last transient analysis (ANTYPE, TRANS).Applying acceleration for the elastic analysis in static analysis; In modal analysis, extracting the model cycle to transformation period for the angular frequency, then prepare the way for applying damping at the back of the time history analysis; On the basis of the first two steps, the last time history analysis, after the input amplitude modulation of earthquake acceleration, structural damping parameters and Rayleigh damping. Applying level X vibration direction, analysis of three different calculation models respectively design calculation, specific parameters are shown in Table 1 below, model shown in figure.

Table 1. Different parameters of calculation models

Content / Group	Model 1	Model 2	Model 3
Model type	Rigid foundation assumption	Soil- structure	Hill-soil-structure
Difference	The bottom of the column uses rigid connection	Infrastructure is a deep soil (deep50 m *wide 300 m)	There were 10 m away from the structure, and high 20 m *long 100 m hill

4.1 Cycle analysis of structure model

Through the modal analysis, the first 5 orders of the three models are listed, as follows Table 2. Observed under the same vibration type, different model simulations are more accurate and more close to the engineering, then longer period of the structure; At the same time by two cycles ratio index (T_2/T_1, T_3/T_2) shows, the cycle of model 1 and model 2 significantly increased due to considering the impact of soil; and model 2 and model 3 compared to the former increased slightly because of considering the hill. On the whole, the period of the hill-soil structure is much larger than that of the rigid ground hypothesis. The main reason is that relative to the upper structure, the foundation soil and hill has a trend of mutual movement , while compared to the rigid foundation model, the overall stiffness and damping of the overall analysis model became larger , and thus the cycle becomes longer.

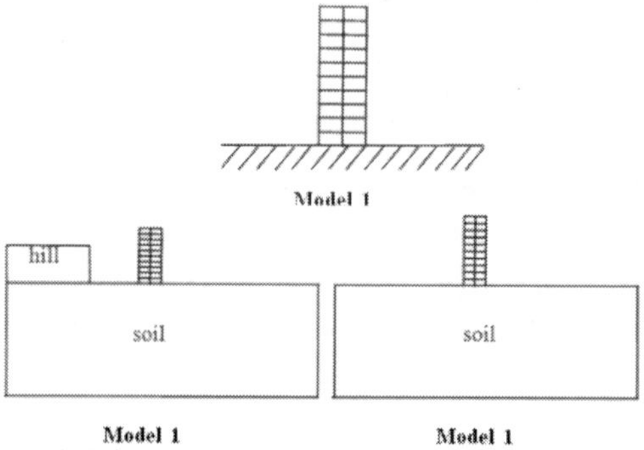

Figure 2. Calculation's model

Table 2. Three model cycle

Mode of vibration	1	2	3	4
T_1 (s) (Model 1)	4.6594	1.3727	0.6773	0.3957
T_2 (s) (Model 2)	6.0194	1.6743	0.9140	0.7990
T_3 (s) (Model 3)	6.0194	1.6752	1.2906	0.8452
$(T_2-T_1)/T_1$	0.2914	0.2197	0.3494	1.0180
$(T_3-T_2)/T_2$	0.0000	0.0006	0.4120	0.0578
$(T_3-T_1)/T_1$	0.29188	0.2204	0.9054	1.1363

4.2 The acceleration and displacement analysis of top structure

Fig. 3 shows three acceleration curves of top-level structure system , the largest acceleration of the rigid foundation (Model 1) is 0.497m/s^2 ,and the maximum acceleration of soil-structure system for 0.519m/s^2, while the hill-soil-structure (with the model of the closest actual situation) maximum acceleration is 0.534m/s2 ; the maximum acceleration of model 2 is more than that of model 1 by 5.633%, and less then that of model 3 by 1.685% The acceleration curve of the rigid foundation is assumed to be slower than the other two, and the main factor is the existence of soil damping. The Through the modal analysis, the first 5 orders of the three models amplitude variation of acceleration time history curve of model 2 and model 3 is basically the same, which shows that the influence of soil on structure can not be ignored in practical engineering.

Fig. 4 shows three top-level displacement curves of these models, rigid foundation during the whole time history analysis, which compared to displacement trend of model 1 and model 2 in the most of the time,corresponding to the maximum, The datum indicates that soil of stiffness and damping matrix has decreased the top displacement of the structure, the soil can not be ignored influence of the top displacement of the upper structure .The maximum displacement ratio of model 3 is larger than that of model 2 by 16.43%, which shows that compared to the soil structure model, the existence of the hill cause the trend of the large displacement. In the whole history, the rigidity foundation reach the peak of the maximum displacement is relatively fast, and the existence of hill is the main reason that model 3 is slightly slower than the model 2.

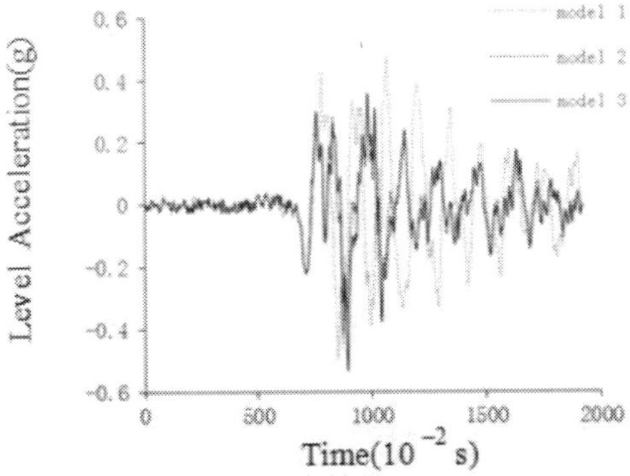

Figure 3. The acceleration curve of top-level

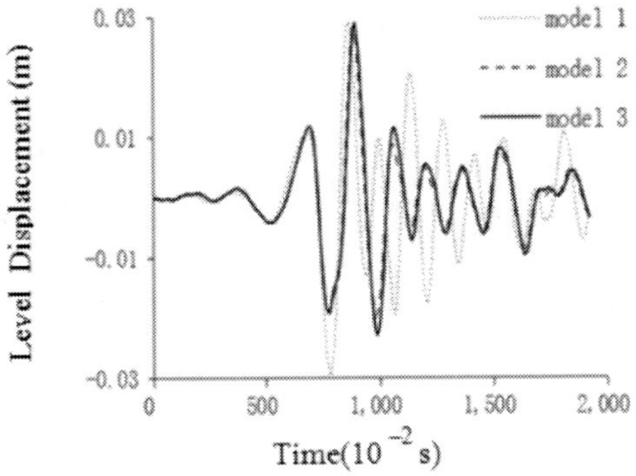

Figure 4. The displacement curve of top-level

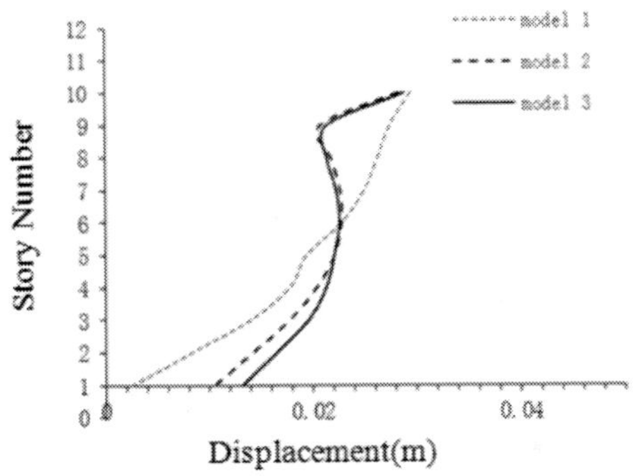

Figure 5. Layer displacement

Layer displacement of three models as shown in Fig. 5, it can be seen that the existence of the soil and scarp from three curves roughly changes, and the maximum layer displacement of 6th-10th floor have mutation, which is less than that of corresponding layers compared to the rigid foundation; structure of the layers below, layer displacement spurt, top layer shift basis unchanged to the present.

5 Conclusions

In seismic performance analysis of building structure, the whole model response more realistic the stress characteristics of the structure , embodies an interaction between one upper,lower structure and the surrounding environment.In this paper,three different models under the action of vertical load and horizontal dynamic load make the numerical simulation , analyzes the influence on the upper structure under the existence of hill.Through analysis, compare, and summarize, the conclusions as follows:

(1) Through the comparative analysis of three model, the cycles of model one, two, three are gradually increasing, the main reason is the involvement of the hill and soil, compared with the rigid ground assumption ,model stiffness decreases, lead to the overall structure become soft,which is in favor of the seismic energy dissipation.

(2) Considering the hill - soil - structure dynamic time history analysis, because of the existence of scarp and flexible foundation layer, top displacement and displacement of the upper structure acceleration has a different degree of influence.Under the overall modeling analysis,there is a difference with rigid foundation assumption ,actual hill, soil influence on the structure cannot be ignored.

(3) Considering the scarp - soil - structure interaction is a very complex tedious study, as the limitation of the time and research conditions, this case is simply analyzed the influence of the hill, provide a reference for engineering design, as well as the subsequent researchers.

References

1. Zhan-xue. Zhou, Li-jun. Wang, Yan-tao, Li, Hai-yang. Feng, JOURNAL OF HEBEI INSTITUTE OF ARCHITECTURAL SCIENCE & TECHNOLOGY, Seismic,responses on str-ucture-basements-piles-soil interaction,**40**,4(2005)
2. Hong-nan. Li, guo-xing. Chen, Earthquake Engin eering: China Machine PRESS (2013)
3. GB50011-2010.Code for Seismic Design of Buildings: China Building Industry Press (2010)
4. Xin-min. Wang, Numerical Analysis of ANSYS Engineering Structure: China Communications Pr-ess (2007).

MATEC Web of Conferences **95**, 03008 (2017)
ICMME 2016
DOI: 10.1051/matecconf/20179503008

Effect of Geometrical Parameters on Deflection of Different Chassis Component Sections

Sumedh Kousadikar , Mangesh Yadav, Dattaprasad Lomate and Manoj Ukhande

Kalyani Centre for Technology & Innovation, Bharat Forge Ltd. Pune, India.

Abstract. Many of the chassis and suspension components are made from variety of sections. Most of them are the robust components assembled critically with other chassis part of the vehicle. The chassis components are used to carry the weight of the vehicle as well as some of them facilitates steering and absorb shocks due to road surface variations. Present competitive market of automotive industry demands for low cost and light weight components. Therefore more emphasis is towards design optimization for reducing weight while maintaining the structural stiffness unchanged. This paper more focuses on effect of geometrical parameters of different sections of chassis components on deflection characteristics by using a hybrid approach of statistical Taguchi method based on orthogonal array and Finite elements analysis (FEA). Using this method the most influencing parameters for deflection are identified through sensitivity analysis. The results obtained are used as a design guideline for generating optimized components with improved deflection performance.

1 Introduction

The present competitive market of automotive industry demands for low cost and light weight components. As emission norms are becoming more and more stringent, the subsequent challenges in reducing the greenhouse gas emissions are forcing the vehicle manufactures to develop light weight vehicles without compromising its strength and stiffness criteria. Recent research in the US shows that on average across all vehicle models every 100kg weight reduction will achieve a reduction in fuel consumption of 0.39 L/100km for cars and 0.48 L/100km for light trucks [1]. So many of the OEM's are concentrating on the redesigning the chassis/suspension system to smaller and lighter extent for getting substantial reduction in the total vehicle weight.

In chassis applications the opportunities to reduce weight have two major benefits. Any contribution to fuel economy reduction is important, but for many chassis and suspension parts, the reduction of unsprung weight is even more significant, offering improvement in handling performance and drive dynamics. Many of the chassis and suspension components e.g. axle beams, body frames are made from variety of sections. The automotive axle beam is one of the robust components assembled critically with other chassis part of the vehicle. The strength and stiffness of these components is an important aspect to avoid sudden failure. To solve these problems, most of the researchers have been focused on improving material strength, replacement of alloy grades and geometry weight optimization, etc. These techniques will definitely reduce the weight of the component but adversely affects the deflection characteristics. However to balance the weight and stresses within controlled deflection for such critically loaded components, the authors believe that, combination of statistical approach and Finite elements analysis (FEA) are an efficient way to optimize the critically loaded components. As most of the chassis components are in simply supported and cantilever loading conditions, instead of taking entire component; the study on different geometrical sections is an effective and quick way of optimization.

The objective of this study is to design stiffer sections with improved moment of inertia (I) which resulted into reduced weight and deflection at cantilever loading condition. The deflection of simple beam at cantilever loading condition is as given (1)

$$Deflection= PL3/3EI \quad (1)$$

where P is applied load, l is length of the beam, E is the young's modulus and I is the moment of inertia.

This paper studied the geometrical parameter effect of one chassis component e.g. axle beam which consist of different section mostly I and T shapes. The optimization is carried out using Taguchi orthogonal array and Finite element analysis. These optimized sections generated guidelines for designing the component with improved deflection performance [2].

2 Experimental works

A multi-disciplinary approach of Taguchi analysis in conjunction with FEA was used for the optimization. A

Taguchi orthogonal array method is used as a statistical approach for optimization. The Taguchi method involves reducing the variation in a process through robust design of experiments. Taguchi method is used to investigate how different parameters affect the mean and variance of a process performance characteristic that defines how well the process is functioning. The experimental design proposed by Taguchi involves orthogonal arrays to organize the parameters affecting the process and the levels at which they will be varies.

2. 1. Statistical methodology

There are many geometrical parameters to affect the deflection of the considered beam geometry. In this study, we have chosen three sections as control factors A. Start section, B. Middle section and C. End section as shown in Fig. 1. The levels of the factors are listed in Table I.

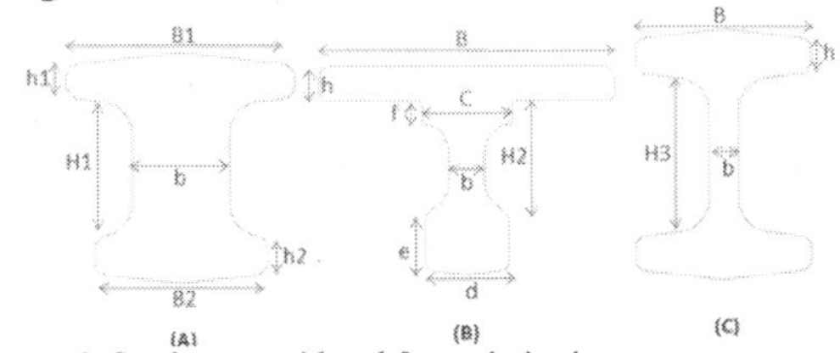

Figure 1. Sections considered for optimisation

Table 1 Levels of the control factors.

Label	Factor Name	Level1	Level 2
A	Start Section	Thicker	Slimmer
B	Middle section	Thicker	Slimmer
C	End section	Thicker	Slimmer

Each section consists of four to five geometrical parameters. The parameters notations are described in Table II.

Table 2 Parameter notations for sections

Start Section A	
h1	Upper flange thickness
B1	Upper flange width
h2	Lower flange thickness
B2	Lower flange width
b	Web Thickness
H1	Web height
Middle section B	
h	Upper flange thickness
B	Upper flange width
C	Resting support width
f	Resting support height
b	Web Thickness
H2	Pad height
e	Lower flange height
End section C	
h1	Flange thickness
B	Flange width
b	Web Thickness
H3	Flange height

The experiments are designed based on Taguchi 2X2 orthogonal array with 16 numbers of experiments for each section. The parameters are varied within two levels as targeted. The Table 3 shows Taguchi 2X2 orthogonal array with combination of levels of factors. [3] These factors 1, 2, 3, etc are replaced with original geometrical parameter dimensions considering levels and targets.

Table 3 Taguchi 2X2 orthogonal array

	Parameter1	Parameter2	Parameter3	Parameter4	Parameter5	Parameter6
EXP 1	1	1	1	1	1	1
EXP 2	1	1	1	2	1	2
EXP 3	1	1	2	1	2	1
EXP 4	1	1	2	2	2	2
EXP 5	1	2	1	1	2	2
EXP 6	1	2	1	2	2	1
EXP 7	1	2	2	1	1	2
EXP 8	1	2	2	2	1	1
EXP 9	2	1	1	1	2	2
EXP 10	2	1	1	2	2	1
EXP 11	2	1	2	1	1	2
EXP 12	2	1	2	2	1	1
EXP 13	2	2	1	1	1	1
EXP 14	2	2	1	2	1	2
EXP 15	2	2	2	1	2	1
EXP 16	2	2	2	2	2	2

Based on the decided 16 number of experiments, 16 geometries of section have been modeled using 3D CAD packages and FE analysis has been carried out with the help analysis tools. Based on these experiments, a set of optimized parameters has been identified. These parameters are referred as Taguchi optimized parameters. The evaluated set further analyzed for sensitivity analysis and most influencing parameter on deflection is evaluated.

2.2 FE analysis

The FE analysis is carried out at cantilever loading condition. The commercial software packages ANSYS workbench and UG NX is used for analysis and 3D CAD modeling. The section geometry is kept as similar length of 200mm for all analysis. The sections are finely meshed with the mesh size of 1mm before analysis. One end of the section is fixed with frictionless support and at other end vertical load of 41.4KN is applied. The analysis flow is as shown in Fig. 2. The applied load in this example is considered based on maximum load acting on the respective beam section after assembly and loading condition. The analysis methodology for all the experiments is kept similar. Based on the analysis, the results of deflection, equivalent stresses and weight is calculated. Series of analysis is conducted for other two sections to obtain the optimum weight for allowable.

2.3 Sensitivity analysis

The sensitivity analysis is carried out using S/N (Signal to noise) ratio which is useful to check the robustness of the control factors. Noises are unwanted or uncontrollable factors given from the surroundings of the system. The response functions are weight, equivalent stresses and deflection. There are three categories of quality characteristic in the analysis of the S/N ratio, (i) the-lower-the-better, (ii) the-higher-the-better and (iii) the-nominal-the better. Regardless of the category of the quality characteristic, process parameter settings with the highest S/N ratio always yield the optimum quality with minimum variance. The category the-lower-the-better was used to calculate the S/N ratio for deflection [4].

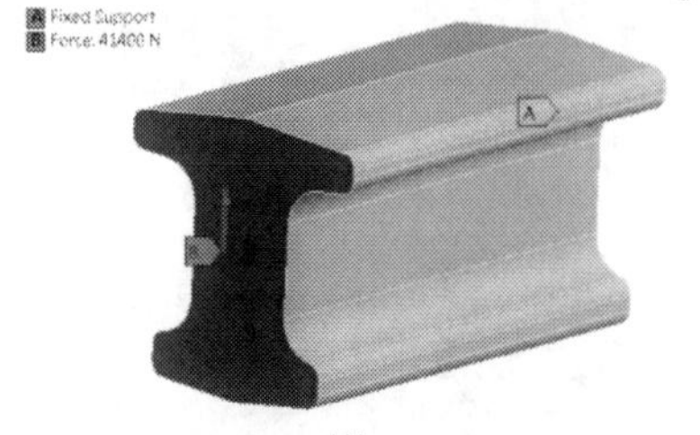

(A)

(B)

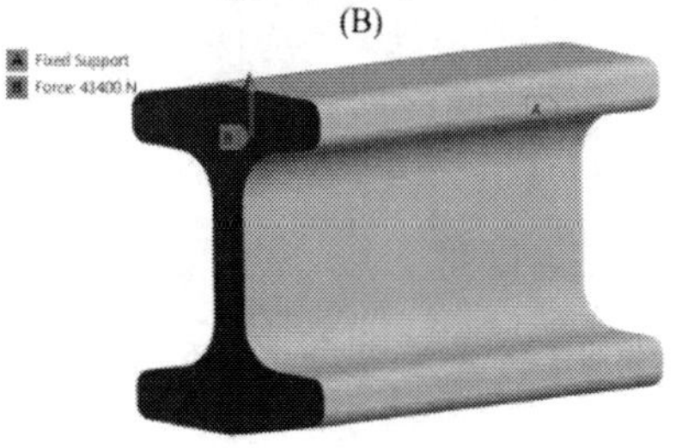

(C)

Figure 2. FE analysis of the sections

2.4 Results and discussions

The equivalent stresses and deflection are measured for each set of parameter using FEA in Ansys. The results are as shown in Fig. 3, and the results of FEA are analyzed for sensitivity using Minitab [5].

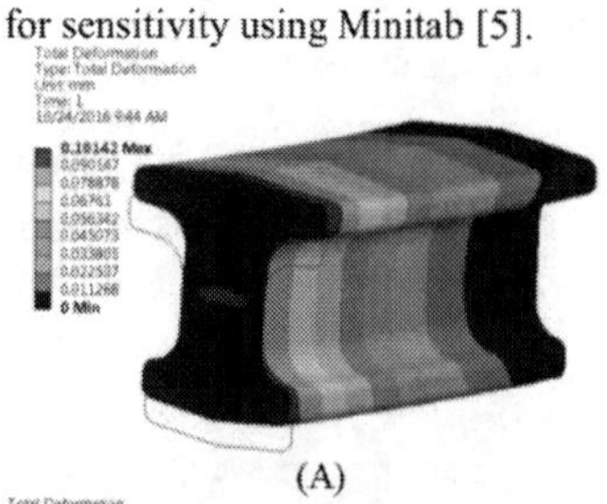

(A)

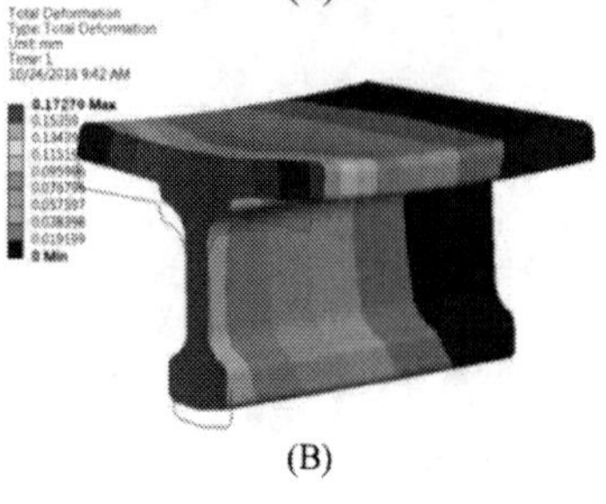

(B)

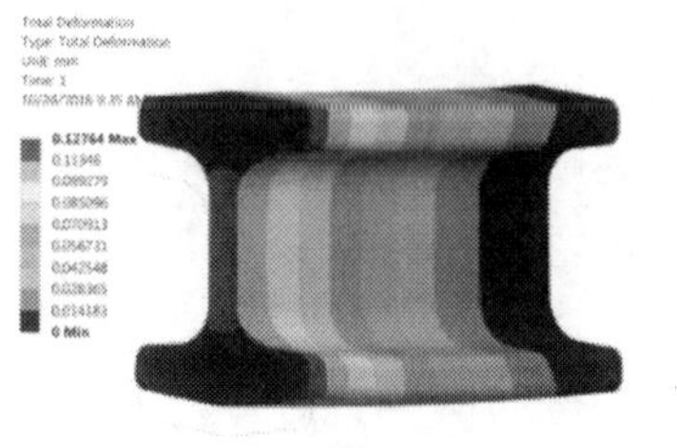

(C)

Figure 3. FE analysis results of the sections under cantilever loading condition.

The S/N ratio for minimum deflection is coming under "Smaller is-better" characteristic, which can be calculated as logarithmic transformation of the loss function [4]. Taguchi design the experiments using especially constructed tables known as "orthogonal arrays" (OA). The use of these tables makes the design of experiments very easy and consistent. The results of the sensitivity analysis for section A is as shown in Fig. 4.

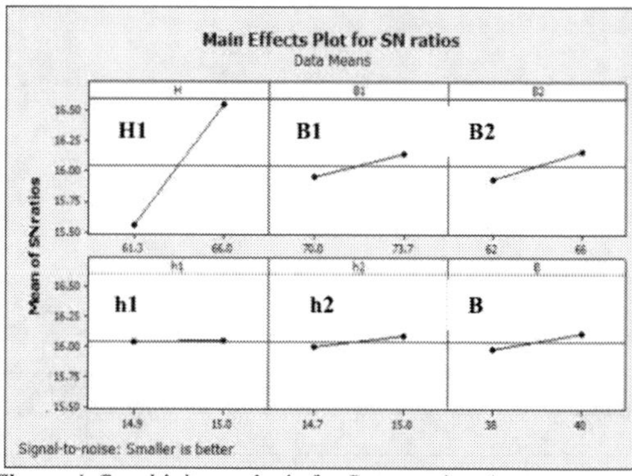

Figure 4. Sensitivity analysis for Start section A

Looking into the results of start section A, It is found that the level 2 of parameter web height (H1) is good choice in sensitivity and plays value addition role in minimizing deflection. Increasing the web height (H1) as shown in Fig. 5 is beneficial to reduce the deflection of the section.

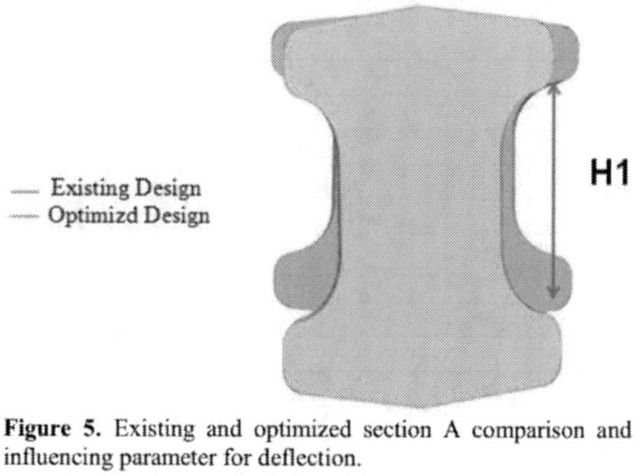

Figure 5. Existing and optimized section A comparison and influencing parameter for deflection.

The results of the sensitivity analysis for middle section B is as shown in Fig. 6. Looking into the results of middle section B, It is found that the level 2 of

DOI: 10.1051/matecconf/20179503008

parameters pad height (H2), web thickness (b) and lower flange height (e) are good choice in sensitivity and plays value addition role in minimizing deflection. Increasing the pad height (H2) and reducing web thickness (b) as well as lower flange height (e) as shown in Fig. 7 is beneficial to reduce the deflection of the section with optimum weight.

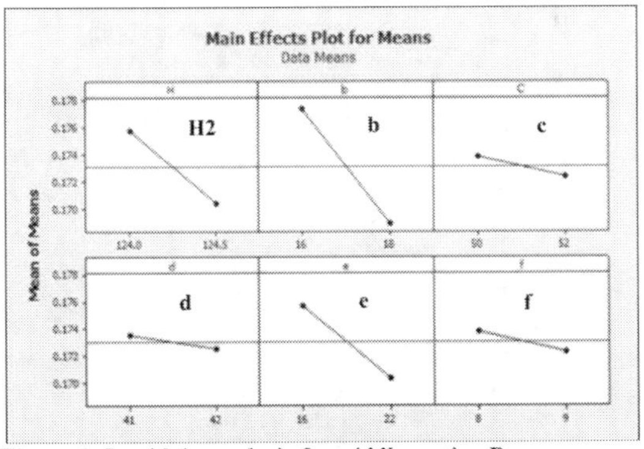

Figure 6. Sensitivity analysis for middle section B

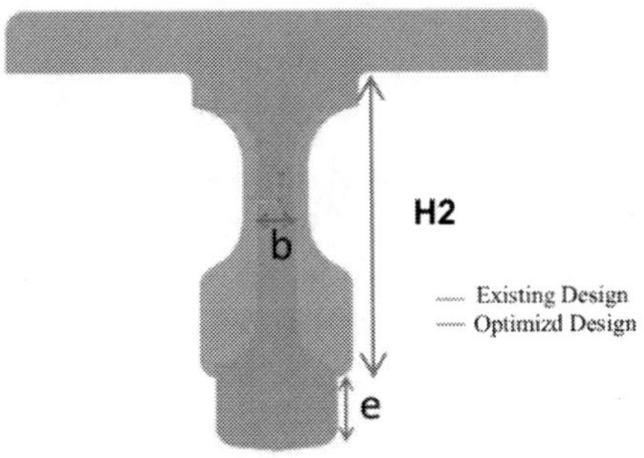

Figure 7. Existing and optimized section B comparison and influencing parameters for deflection.

The results of the sensitivity analysis for end section C is as shown in Fig. 8. Looking into the results of end section C, It is found that the level 2 of parameters flange height (H3) and flange thickness (h1) are good choice in sensitivity and plays value addition role in minimizing deflection.

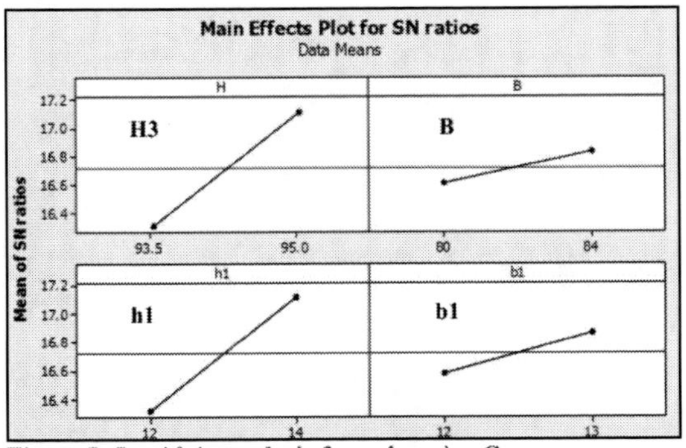

Figure 8. Sensitivity analysis for end section C

Increasing the flange height (H3) and reducing the flange thickness (h1) as shown in Fig. 9 is beneficial to reduce the deflection of the section with optimum weight.

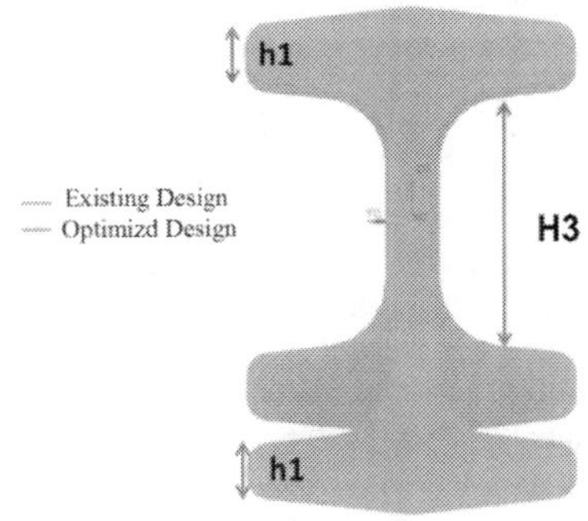

Figure 9. Existing and optimized section C comparison and influencing parameters for deflection.

It is also observed that, moment of inertia of each optimized sections is about 2 to 3% more than the existing section in all three directions [6]. In the investigation, it has been found that by optimizing the influencing parameters for respective sections will reduce the deflection as well as weight [7]. The stresses observed are consistent and in line with the existing design.

3 Conclusions

In this paper, the geometric parameter effect on deflection is studied using statistical Taguchi method. The S/N ratios are analyzed with the deflection results by means of FE analysis. As a result, following conclusions are arrived at from this investigation.

1. The FEM-based Taguchi methods have effectively decreased the time and efforts required for evaluating and optimizing the design variables of chassis component.
2. The optimal parameter combination for the minimum weight with permissible value of deflection is obtained by using the analysis of S/N ratio. According to the results slimmer section parameters are the optimal parameters for permissible deflection.
3. The most influencing parameter for section A is web height (H1), by increasing this parameter deflection of this section under cantilever loading condition deceases to appreciable extent.
4. The most influencing parameter for section B is pad height (H2), web thickness (b) and lower flange height (e), by increasing pad height (H2) and decreasing web thickness (b) and lower flange height (e) the deflection of this section under cantilever loading condition decreases to appreciable extent.
5. The most influencing parameter for section C is web height (H3), flange thickness (h1) by increasing web height (H3) and reducing flange thickness (h1) the deflection of this section under cantilever loading condition deceases to appreciable extent.
6. The deflection performances of chassis components are greatly related to geometry of the structure, so by optimization of these sections substantially reduces the weight of the components.

Acknowledgement

The authors gratefully acknowledge the extended support provided to this work by KCTI (Kalyani Centre for Technology and Innovation) for providing financial funding, laboratory and library facilities. The authors also acknowledge the support provided by Bharat Forge Ltd, Pune and DSIR (Department of Scientific and Industrial Research), Govt. of India. Finally, the authors would like to express special thanks and gratitude to review committee and top management of Bharat Forge Ltd for granting the permission to publish/present the research work. Thanks to our departmental colleagues and others who helped us directly or indirectly.

References

1. IHS Automotive, sectorial report on "*Weight Reduction in Automotive*", pp. 30-56, (2014).
2. Patel, H., Panchal, K. C., Jadav, C. S. Structural Analysis of Truck Chassis Frame and Design, *International Journal of Engineering and Advanced Technology (IJEAT)*, **2**;4, p. 665-668, (2013).
3. A. Yasar, D.A. Bircan, Design, Analysis and Optimization of Heavy Vehicle Chassis Using Finite Element Analysis, *International Journal of Scientific and Technological Research, ISSN 2422-8702, Vol 1, No.6*, (2015).
4. R.K Roy Design of Experiments Using The Taguchi Approach, *Willy publication*, ISBN-0-471-36101-1, pp. 95-171, (2001).
5. Wen-Hsien Hsua, Ching-Kong Chaoa, Hsi-Ching Hsub, Jinn Lin b, Ching-Chi Hsua, Parametric study on the interface pullout strength of the vertebral body replacement cage using FEM-based Taguchi methods, *Medical Engineering & Physics* **31**, 287–294, (2009).
6. S. Tiwari, S. Kumar, V.Pattabiraman "A statistical approach towards connecting rod optimization" , *SAE paper 2006-32-0068, Japan*, (2006).
7. M. Ramgir, S. Neelkanth, I. Poonawala, "Design Optimization of Four Wheel Drive Tractor Front Axle Housing", *International Journal of Engineering Research & Technology (IJERT), ISSN: 2278-0181, Vol. 4 Issue 08, August-*(2015).

DOI: 10.1051/matecconf/20179503009

The Effect of Nano-Aluminumpowder on the Characteristic of RDX based Aluminized Explosives Underwater Close-Filed Explosion

Yin Junting, Yuan Baohui, Zhou Tao, Li Gang and Ren Xinlian

Xi'an Modern Chemistry Research Institute, Xi'an, China

Abstract. In order to investigate the effect of nano-aluminum powder on the characteristic of RDX based aluminized explosives underwater closed-filed explosions, the scanning photographs along the radial of the charges were gained by a high speed scanning camera. The photographs of two different aluminized explosives underwater explosion have been analyzed, the shock wave curves and expand curves of detonation products were obtained, furthermore the change rules of shock waves propagation velocity, shock front pressure and expansion of detonation products of two aluminized explosives were investigated, and also the parameters of two aluminized explosives were contrasted. The results show that the aluminized explosive which with nano-aluminum whose initial shock waves pressure propagation velocity, shock front pressure are smaller than the aluminized explosive without nano-aluminum and has lower decrease rate attenuation of energy.

1 Introduction

Aluminized explosives is a kind of high-density, high-power high-explosive heat of mixing explosives, damaging effects on the target can be increased, which has important applications in underwater weapons, air weapons and armor charge [1], especially in the underwater explosion the aluminized explosive shock wave and bubble can significantly higher than non-aluminized explosives. The aluminized explosives detonation is not ideal detonation process, for the aluminum involved in the reaction [2], the content of aluminum powder, particle size, shape and activity will effect on the detonation process and produces a large charge influence [3]-[4], therefore, such as the impact of ultrafine particles of aluminum nanopowders of aluminized explosive properties, becoming one of hot research. The effect of Nano-aluminum on aluminized explosives near field underwater explosion characteristics can provide important reference for the water weapons design and the power assessment.

Much research has been carried out on the impact of nano-aluminum aluminized explosive properties and the characteristicsof aluminized explosives for underwater explosions predecessors. Huang Hui, etc.through studied, found that thenano-aluminum composite explosiveshave stronger ability of metal acceleration, and the reaction time is more short [5]; Shen Fei, etc.have studied different aluminum-containing explosive shock wave propagation in water near field, and the results were compared with TNT, it showed the pressure decay process, the reaction of aluminum makes shock wave pressure decay rate has been reduced [6]; Zhou Lin, etc. By measuring and analyzing different types of aluminized explosives in underwater explosion energy output parameters, given the different types of aluminized explosives underwater explosion energy output characteristics [7]. However, there is little literature available on the nano-aluminum aluminized explosives underwater explosion characteristics.

In this study, a series experimental investigation were done to characterize the effects of nano-aluminum powder on the shack wave pressure decay and bubble expansion speed of aluminized explosives .

Table 1. The samples composition of aluminized explosives

Serial number	RDX	Micron Al	Nano Al	Binder	Density/(g/cm3)
ES-1	65%	30%	—	5%	1.80
ES-2	65%	20%	10%	5%	1.80

DOI: 10.1051/matecconf/20179503009

2 Experiment

2.1 Experimental samples

The experiment used ES-1, ES-2 two aluminized explosives formulation as shown in Table 1, wherein the nano-aluminum powder particle size: 100-200nm, chargewas collected press-fit shape, with section Φ30mm × 30mm, the main charge by the six charge bonded together, the total charge length of 180mm. TheΦ25mm × 25mm, crimping density of 1.67g / cm^3 of JH-14 used as a Booster, Booster paste in the end position of the center of the main charge.

2.2 Experimental testing principle

Strong shock waves generated by underwater explosives detonation, which will produce severe compression of its aqueous medium at a spread, so that the density of the water medium increases, resulting in greatly reduced transmittance. Thus, the shock wave went, it will form a dark layer in an aqueous medium. An argon bomb explosion,which Placed in the rear of the tank, as a light source, in the shock wave went dark layer blocking the light which emitted from the light source, while the track movement of the dark layer is recorded by the high-speed scanning camera[8].

2.3 Experimental device and layout

The cube tank used for the experiment with 400mm side length, between the charge andthe tank cover is connected by wire, adjusting the length of the wire, so that the charge is parallel to the bottom surface of the tank. Employed SJZ-15-type high-speed rotating mirror scanning cameras which scanning speed set to 3mm / μs, the charge axis and high-speed cameras, slits are perpendicular to the optical axis. Φ200mm × 300mm argon gas bombs used as an experimental light source, argon bomb optical axis coincides with the axis of the camera and is located behind the water tank. Experimental layout and photo shown in Fig. 1.

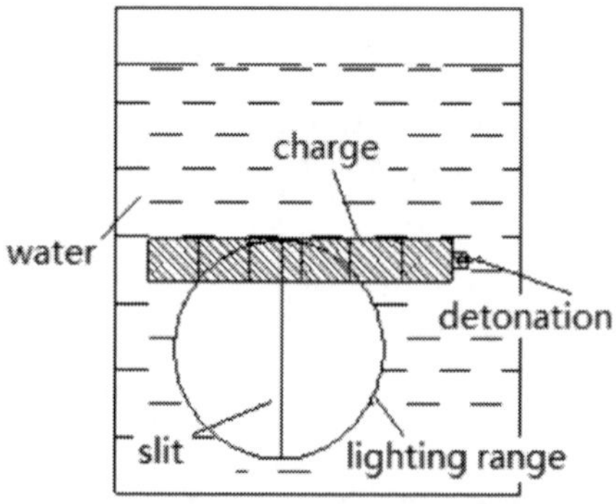

(a) Experimental layout sketch map

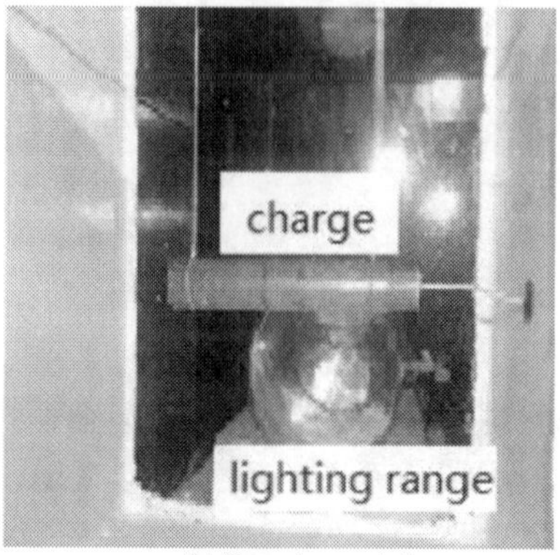

(b) Experimental site

Figure 1. Layout of the experiment

3 Experimental results and discussion

3.1 Experimental data processing

The typical scanning trajectory as shown in Fig. 2, as the figure shows, the two moving trajectory obtained, the trajectory of the shock wave spread was faster, while the detonation products bubble boundary spread slower. Combined zoom ratio and camera scanning speed image scanning negatives interpretation, described a series of two traces obtained data points (Fig. 3). As can be seen from the figure, ES-1 shock wave propagation slightly faster than the ES-2, but the expansion curve of the product bubble slower than the ES-2.

Figure2. Typical scanning trajectory (ES-1)

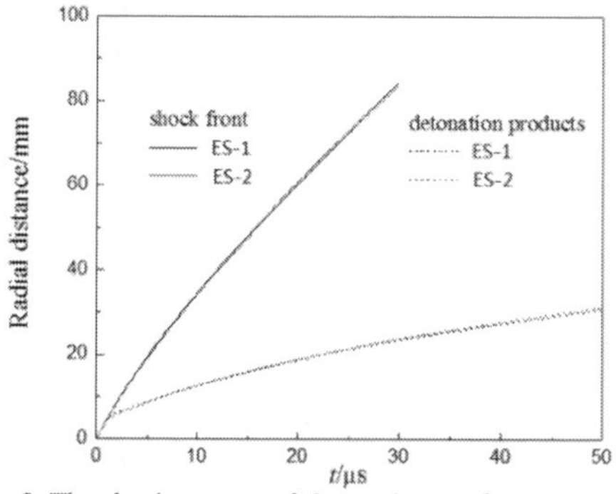

Figure 3. The shock waves and detonation products curves of radial distance-time history

3.2 Shockwave trace analysis

The nonlinear function Can be used to fit the experimental data of the shock wave propagation trace [9], the specific form of the function as follows:

DOI: 10.1051/matecconf/20179503009

$$y_1 = A_1[1-\exp(-B_1 t)] + A_2[1-\exp(-B_2 t)] + c_0 t \quad (1)$$

wherein, y_1 represents the propagation distance of the shock wave along the axial direction of the charge, mm; c_0 =1.483mm/μs, represents the velocity of sound in the water; trepresents time, μs; A_1, A_2, B_1, B_2 on behalf of the fitting parameters. Equation (1) can better reflect the law of shock waves propagation in the water, Table 2 shows the fitting parameters.

Table 2. Fitting coefficients

explosives	A_1/mm	B_1/(1/μs)	A_2/mm	B_2/(1/μs)
ES-1	8.426	2.141×10-1	83.059	1.619×10-2
ES-2	8.269	2.089×10-1	82.689	1.596×10-2

In order to obtain the propagation velocity of the shock wave, the equation (1) derivative of time t:

$$u_s = \mathrm{d}y_1/\mathrm{d}t = A_1 B_1 \exp(-B_1 t) + A_2 B_2 \exp(-B_2 t) + c_0 \quad (2)$$

The Pressure of shock wave front p_s get according to the impact of water adiabatic equation and momentum conservation relations,

$$u_s = c_0 + 25.306\lg(1 + u_p/5.19) \quad (3)$$

$$p_s = \rho_{w0} u_s u_p \quad (4)$$

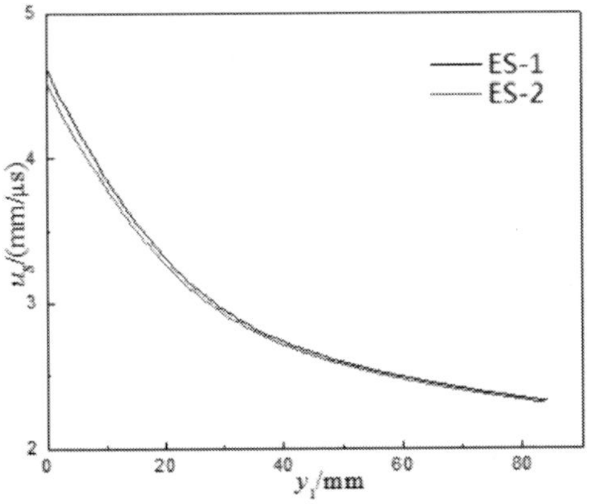

Figure 4. Velocity-distance history of the shock waves

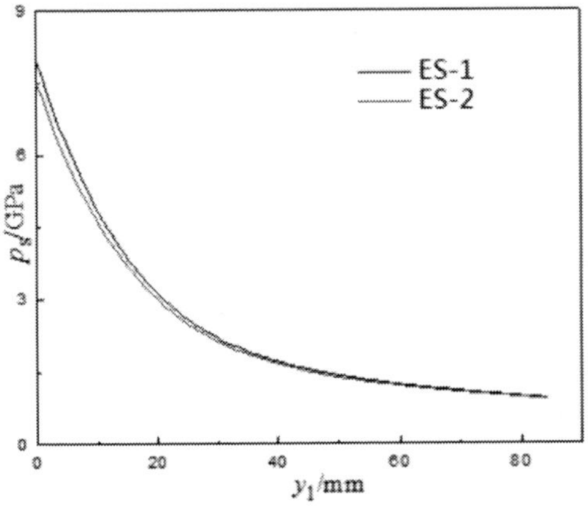

Figure 5. Pressure-distance history of the radial shock waves

wherein, u_p indicates the particle velocity, mm / μs; ρ_{w0} indicates the density of the water, which value is 1.0g / cm3. Using the formula (2) - (4) to calculate the shock wave propagation velocity u_s and pressure p_s, where along the charge radial in the near field underwater explosion, with the variation of the propagation distance, the graphs shown in Fig. 4 and Fig. 5, at the same time, the initial values of these parameters are shown in Table 3 below.

Table 3 Initial parameters of radial shock waves

explosives	u_{s0}/(mm/μs)	u_{p0}/(mm/μs)	p_{s0}/GPa
ES-1	4.63	1.72	7.97
ES-2	4.53	1.66	7.51

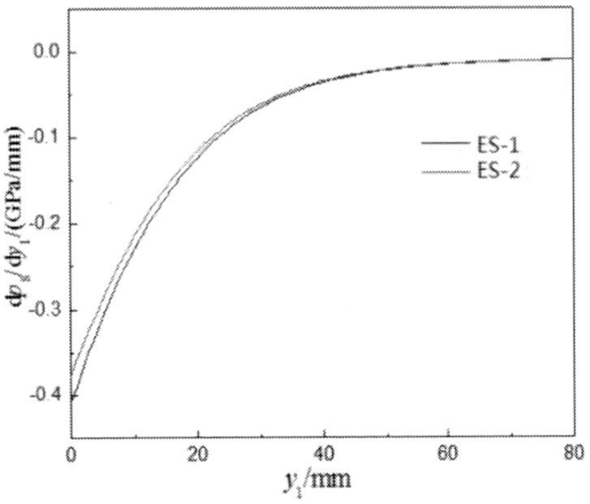

Figure 6. Pressure attenuation rate of radial shock waves

The initial shock of the pressure front is mainly affected by the explosion pressure of the explosives, propagation velocity is mainly determined by the value of the surface pressure, since the ES-2 containing nano-aluminum, its detonation velocity and pressure has declined compared ES-1 Therefore at the beginning of underwater explosion, ES-2's shock wave front propagation velocity and pressure were slightly lower than the ES-1. However, there are accompanied by a shock wave propagation wavefront energy dissipation, the main part of the kinetic energy of the shock wave irreversible converted into medium heat, pressure and shock wave front, the higher the greater the energy dissipation. As can be seen from Fig. 4 and Fig. 5, ES-1, although the curve value higher than the initial period ES-2, but the energy loss is higher than the ES-2, after a short propagation distance, it is very close to the value of the two curves .Front pressure decay rate curve shown in Fig. 6 also shows that, when the propagation distance is about 40mm, front pressure decay rate will be almost identical.

4 Conclusion

The shock wave propagation traces observed byHigh-speed scanning camera from the radial of the cylindrical charge can more accurately calculate the propagation velocity and attenuation of the shock wave pressure in the explosion near field. The ES-2 explosive, which Contained nano-aluminum powder ,whose initial propagation velocity and pressure of the shock wave are smaller compared to the ES-1explosive, but its decay rate of energy and pressure is relatively low, resulting in the propagation distance of about 40mm when two explosive fronts pressure will be more similar.

According to the experimental data, the explosive containing nano-aluminum powder, the energy and pressure decrease with increasing distance, but the attenuation rate is lower than that without aluminum explosive, so the underwater explosion operations used in explosives, explosives should be determined according to the actual needs of the aluminum content. Within 40mm, such as the need for higher pressure, it should be used without nano-aluminum powder, such as the need to slow the pressure of explosives, it is necessary to use nano-aluminum containing explosives. When the distance is greater than 40mm in the working environment, whether or not the aluminized explosives aluminized explosive, its effect is basically the same.

In military applications, especially large-scale torpedo warhead, because the explosive contains nano-aluminum can increase the loading density of explosive in warhead, and increase the heat generation after the explosive reaction, when the warhead in thin-walled shell structures such as ships, can give full play to the power of aluminized explosives.

References

1. Li Yuan-yuan, Wang xiao-feng, Nan Hai, et al. Effects of Aluminized ParticleSize on HMX-based Aluminized Casting PBX Explosives Detonation Heats. Chinese Journal of Science Technology and Engineering, **14**(10):206-209, (2014).
2. Guiruis R H, et al. Time-Dependent Equation of Sate for Aluminized Underwater Explosive. Tenth Symposium on Detonation.675-682. (1993).
3. Miao Qin-shu, Xu Geng-guang, Wang Ting-zeng. Mechanism Analysis of the Influense of Shape and Size on the Detonation Properties of Aluminized Explosives. Chinese Journal of Explosive and Propellants, **2**:4-8, (2002)
4. Feng Xiao-jun, Wang Xiao-feng, Li Yuan-yuan, et al. Effect of Aluminum Particle Size and Explosion Atmosphere on the Energy of Explosion of Aluminized Explosive. Chinese Journal of Explosive and Propellants, **36**(6): 24-26, (2013)
5. Huang Hui, Huang Heng-jian Huang Yong, et al. The influence of aluminum particle size and oxidizer morphology in RDX-based aluminized explosives on their ability to accelerate metals. Chinese Journal of Explosion and Shock Waves, **26**(1):7-11, (2006).
6. Shen Fei, Wang Hui, Yu Ran, et al. Propagation characteristics of close-field shock wave for two aluminized explosives by underwater explosion. Chinese Journal of Explosive Materials, **43**(5):26-29, (2014).
7. Zhou, Lin, Xu Geng-guang. Configuration of Underwater Energy Output for aluminized Explosive Mixtures. Chinese Journal of Explosives and Propellants, **26**(1):30-36, (2003).
8. Sun Cheng-wei, et al. Applications Detonation Physics. Beijing: Defense Industry Press, (2000).
9. Itoh S, Suzuki O, Nagano S, et al. Investigations on fundamental properties of underwater shock waves by high speed photography. Journal of Materials Processing Technology, **85**(3): 226-230, (1999).

Analysis of Dynamic Mechanical Properties for Simply Supported Beam Including ESP Glass

Guoliang Zhang and Jishi Du

Institute of Electronic Engineering, China Academy of Engineering Physics, Mianyang, Sichuan, China

Abstract. There is growing awareness of the need to protect system circuit from the abnormal impact loading environment. Impact mechanical sensor can immediately hold back system circuit in the extreme environment. Especially, mechanical response component of it must possess the characteristics of high fracture strength and high Weibull modulus simultaneously. Engineered stress profiles glass (ESP glass) is the best choice for achieving these two performances in all kinds of material. In many cases, it is desirable that once the impact mechanical sensor has served its function that ESP glass then fractures into very small fragments. ESP glass with controlled static fracture properties so far has been carried out theoretically and experimentally, whereas impact properties of it are rarely performed. In order to understand mechanical properties of ESP glass under impact loading environment, the mechanical behavior of the simply supported beam including ESP glass is investigated by theoretical formulae and finite element method (FEM) simulation in this paper. The theoretical computational results of the dynamic flexural strength of ESP glass are consistent with the simulation results of it. Results gained show that the relation between static mechanical properties and dynamic mechanical properties of ESP glass can be established by dynamic coefficient.

1 Introduction

With growing awareness of the need to protect system circuit from the abnormal impact loading environment, mechanical response properties of impact mechanical sensor have been a topic of growing interest gradually [1], [2]. As the key part, mechanical response component of the sensor must possess the characteristics of high fracture strength and high Weibull modulus simultaneously. With controlled fracture properties, ESP glass has very high reliability strength of four to five times that of regular glass. At same time, visible warning cracks appeared on the surface of ESP glass are generated before disastrous failure occurs. At this level, the advantages of a Weibull modulus are shown in Fig. 1. It assume that the ESP glass must sustain 80% of the average failure stress. At a given stress is increased significantly with higher Weibull modulus of 60, only a one in one million chance of failing will occur for this ESP glass. In contrast, regular annealed glass with lower Weibull modulus of 5 has a 30% failure probability [3]. Although regular glass does not have a microstructure, ESP glass has been designed a built-in stress profile that would produce R-curve (increasing crack resistance) behavior. Based on this design idea, there is a high compressive stress just below the glass surface rather than at the surface (Fig. 2). This devise can also make glass strength controlled by the size of the worst defect.

Therefore, ESP glass is the best choice for achieving these two performances in all kinds of material. In many cases, it is desirable that once the impact mechanical sensor has served its function that ESP glass then fractures into very small fragments [4]. Under impact loading environment, it will interfere with the functioning of system circuit of a device drastically. ESP glass with controlled static fracture properties so far has been carried out theoretically and experimentally [5]-[7], whereas impact properties of it are rarely performed. Meanwhile, analysis of dynamic load and dynamic coefficient of regular elastic and plastic material had been researched many years [8]-[10]. Although the material properties of ESP glass vary considerably from regular elastic and plastic material, these investigations can be referred to studying impact properties of ESP glass.

In order to understand mechanical properties of ESP glass under impact loading environment, the mechanical behavior of the simply supported beam including ESP glass is investigated by theoretical formulae and finite element method (FEM) simulation in this paper. Hence, theoretical formulae and FEM simulation are simultaneously used to know the relation between static mechanical properties and dynamic mechanical properties of ESP glass. It also can be used to understand mechanical properties of ESP glass under impact loading environment. Moreover, the results gained can be utilized to improve the structure design of impact mechanical

DOI: 10.1051/matecconf/20179503010

sensor and adequately exert advantages of ESP glass with controlled fracture properties.

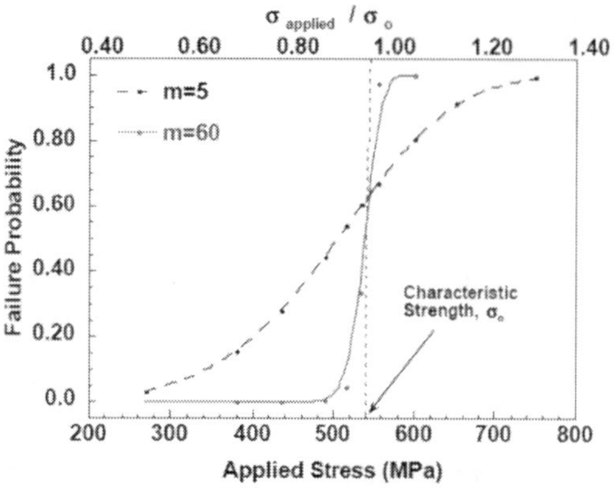

Figure 1. The relationship between failure probability and various Weibull moduli (m).

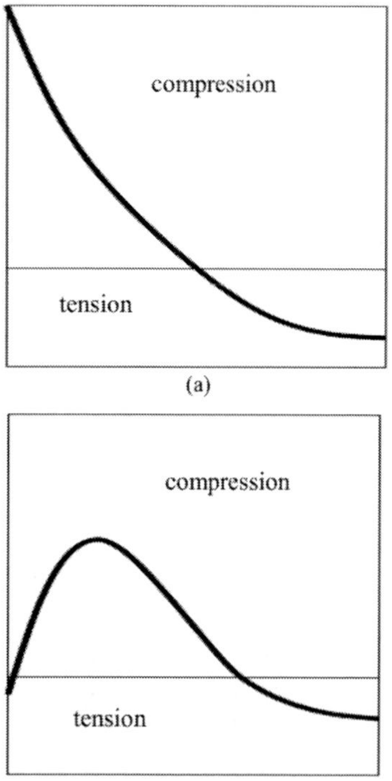

(b)

Figure 2. Stress profiles of two kinds of glass: (a) regular annealed glass, (b) ESP glass.

2 Theoretical calculation of a dynamic stress in a simply supported beam

A simply supported beam is loaded with different types of force in Fig. 3. Theoretical calculation of a dynamic stress in this beam can be acquired from the energy equilibrium condition [11]. Kinematics energy of a mass of block is equal to a work done during a dislocation with dynamic deformation and free falling height. The mass of block C is m. The length of beam is L. In Fig. 3(a), a static force F_j=mg is acted on the middle of the beam whose bending deformation is calculated from the equation (1) as follows:

$$t_j = \frac{F_j L^3}{48EI_z} \tag{1}$$

where t_j indicates the static deformation, E is the elastic modulus, I_z is the cross-section moment of inertia for Z axis, EI_z is the bending stiffness and its value is constant. In Fig. 3(b), the dynamic force F_d will be generated as the block C is falling from the height h and impacting on the middle of the beam. Its bending deformation is calculated from the equation (2) as follows:

$$t_d = \frac{F_d L^3}{48EI_z} \tag{2}$$

where t_d indicates the dynamic deformation whose value is larger than the static deformation t_j.

Combined with equation (1) and equation (2), the dynamic force F_d is given as:

$$F_d = \frac{t_d}{t_j} F_j \tag{3}$$

in which the ratio of the dynamic deformation t_d to the static deformation t_j denotes the dynamic coefficient K_d. It is expressed as:

$$K_d = \frac{t_d}{t_j} \tag{4}$$

The dynamic coefficient K_d can be calculated from the energy equilibrium condition which is read as follows:

$$F_j(h + t_d) = \frac{1}{2} F_d t_d \tag{5}$$

Combined with equation (3) and equation (5), the dynamic coefficient K_d becomes:

$$K_d = 1 + \sqrt{1 + \frac{2h}{t_j}} \tag{6}$$

The free falling height h can be deduced from the energy equilibrium condition as follows:

$$h = \frac{(at)^2}{2g} \tag{7}$$

where a is impact acceleration, g is gravity acceleration, t is impact time.

The static flexural strength of brittle materials is given as:

$$\sigma_j = \frac{8F_j L}{\pi d^3} \tag{8}$$

where d is diameter of the beam. Associated with equation (1) ~ (7), the dynamic flexural strength σ_d of ESP glass within the simply supported beam could be reckoned.

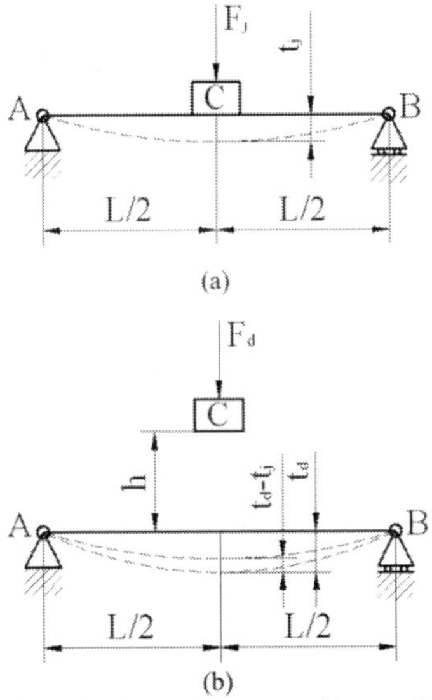

Figure 3. Schematic of a simply supported beam with different types of force: (a) static force, (b) dynamic force.

3 Finite element analysis

Under impact loading environment, the dynamic stress on ESP glass will be greater than the fracture strength on it. For validating theoretical calculation of a dynamic stress in a simply supported beam including ESP glass, a three-dimensional finite element model of it is established in ANSYS simulator (Fig. 4). Non-uniform grid is introduced for reducing the amount of meshing as well as receiving the dynamic stress distributions in the middle concerned region. Hexahedron grids are introduced in the entire computational region for enhancing the accuracy of simulation.

In Fig. 4, the beam is ESP glass, the mass of block is stainless steel, the foundation is stainless steel, and the support post is stainless steel. The geometric and mechanic characteristics of main components are summarized in Table 1. The condition of first row is for the beam. The condition of second row is for mass of block. Thereinto, L is the length, D is the diameter, E is elastic modulus, v is Poisson's ratio, ρ is density, σ_f=400MPa is fracture strength of ESP glass, a is impact acceleration, t is impact time.

The assumptions for computational region of the simply supported beam are given as follows: 1) The mass of block, the foundation and the support post are set as rigid body; 2) The beam, the foundation and the support post are tied up together; 3) The friction generated on the dynamic contact surfaces is ignored.

Figure 4. Proposed finite element composite model.

4 Results and discussions

4.1 Theoretical calculation

Be based upon Table 1 and equation (7), the free falling height h is about 31m. From equation (1), the static bending deformation t_j is 9.8E-5mm and the static force F_j is 0.013N. Then, according to equation (3) and equation (4), the dynamic coefficient K_d is near 4E3 and the dynamic force F_d is close to 52N. Therefore, in the light of the dynamic flexural strength of brittle materials, the dynamic flexural strength σ_d of ESP glass is about 415MPa. It can be seen that the dynamic stress on the ESP glass has been greater than the fracture strength on it under the given impact loadings from Table 1.

Table 1. Geometric and mechanic characteristics of simply supported beam and initial conditions.

L (mm)	D (mm)	ρ (T/mm^3)	E (MPa)	v	a (m/s^2)	t (ms)
25	2	2.2E-9	5.5E4	0.25	0	0
6	6	7.8E-9	2E5	0.3	5E4	0.5

On the basis of the above analysis, the dynamic load resulting from an impact load is bigger than a static load loaded by the same mass. In other words, the same mass will cause very different loads on the same beam as this mass falls from 0 to the specifically height.

Properties of all kinds of material are also significant for theoretical analysis. Impulse loads maybe alter the attributes of the material.

4.2 Simulation

Fig. 5 shows the response of impact acceleration during the mass of block is freely falling from height h and impacting on the beam. The response of impact acceleration produced on the beam fast changes from the maximum 5E4m/s^2 to zero. It is consistent with the load conditions from Table 1. The kinematics energy of the mass of block transforms into a potential energy of an elastic deformation for the beam. Meanwhile, the response of impact acceleration appeared on the foundation fast changes from the maximum 4E4m/s^2 to zero. It should be noted that the response of impact acceleration will be attenuated as it is transferring from the beam to the foundation. Combined the energy equilibrium condition, the more realistic dynamic stress distributions can be gotten.

Fig. 6 depicts dynamic stress contour on the ESP glass as the response of impact acceleration of the beam reaches its maximum. The maximum dynamic stress is located at the middle of the beam with different lengths. As the length of the beam is 25mm, this maximum dynamic stress of 428 MPa agrees well with the theoretical computational results of it. When the length of the beam becomes 13mm, the maximum dynamic stress will increase up to 624 MPa under the greater response of impact acceleration of 8E4m/s^2. In summarized, ESP glass has fractured under the given impact load conditions.

DOI: 10.1051/matecconf/20179503010

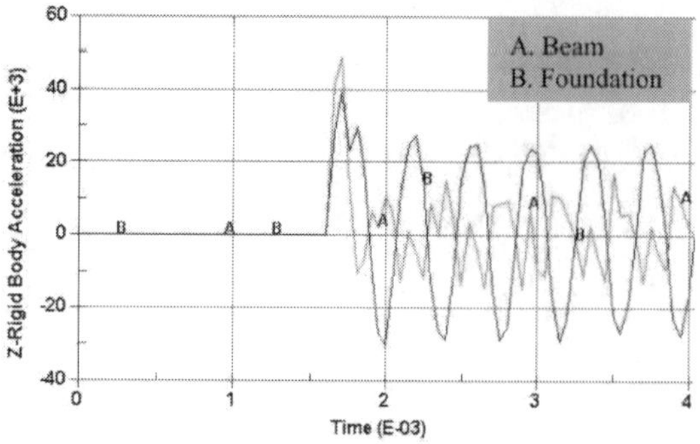

Figure 5. Response of impact acceleration.

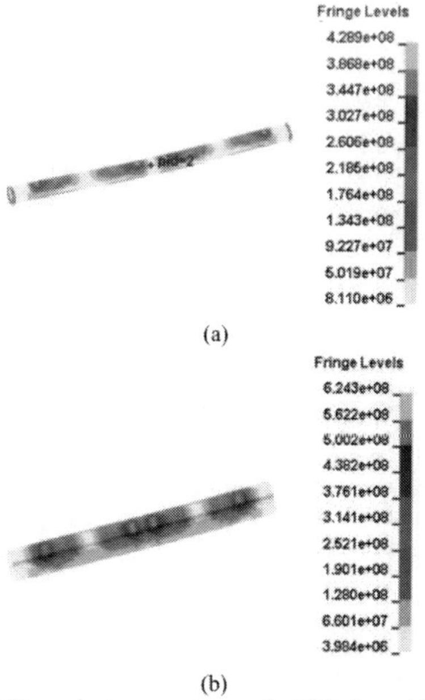

Figure 6. Dynamic stress contour on the ESP glass with different length: (a) L of the beam is 25mm, (b) L of the beam is 13mm.

5 Conclusions

Impact mechanical sensor has been a topic of growing interest with the growing awareness of the need to protect system circuit from the abnormal impact loading environment. As the key part of the impact mechanical sensor, ESP glass is the best choice for achieving these two performances of the characteristics of high fracture strength and high Weibull modulus in all kinds of material.

In order to understand mechanical properties of ESP glass under impact loading environment, the mechanical behavior of the simply supported beam including ESP glass is investigated by theoretical formulae and FEM simulation in this paper. Moreover, the relation between static mechanical properties and dynamic mechanical properties of ESP glass has been established by dynamic coefficient. Based on the energy equilibrium condition, the better agreement of the dynamic stress produced on the ESP glass between theoretical calculation and simulation has been obtained. These results show that research of the dynamic impulse for an impact mechanical sensor probably discovers the value of the dynamic stress as well as a dynamic coefficient. It is notable that impact loads may radically change the mechanical properties of the material. For improving the structure design, calculated protection coefficients for an impact mechanical sensor containing a bending beam should be higher than the ones used during designing. ESP glass and stainless steel mentioned in this article should be assumed as the introduction to the further theoretical analysis as well as practical verification.

Acknowledgement

The authors are very grateful to our colleagues for their discussion and data processing. This work is supported financially by the Science and Technique Innovation Foundation of Institute of Electronic Engineering of China Academy of Engineering Physics, China (Grant No. S20140806).

References

1. B. Denkena, F.L. Hackeloer, *2010 IEEE Sensors* 2518-2524 (2010)
2. A. Stevan, *2012 Mediterranean Conference on Embedded Computing* 50-53 (2012)
3. D.J. Green, R. Tandon, V.M. Sglavo, Science. **283,** 1295 (1999)
4. D.J. Green, J Non-Cryst Solids. **316,** 35 (2003)
5. V.M. Sglavo, A. Prezzi, D.J. Green, Eng Fract Mech. **74,** 1383 (2007)
6. S. Blanka, S. Jarmila, N. Pavla, J Non-Cryst Solids. **356,** 1509 (2010)
7. P.K. Kreski, A.K. Varshneya, A.N Cormack, J Non-Cryst Solids. **358,** 3539 (2012)
8. P. Ratko, K. Predrag, R. Predrag, Journal of Sound and Vibration. **301**, 690 (2007)
9. E. Wlodarczyk, M. Zielenkiewicz, Journal of Theoretical and Applied Mechanics. **49,** 457 (2011)
10. B. Rosnawati, M.R. Munzilah, E.A. Mohd, Applied Mechanics and Materials. **405**, 1900 (2013)
11. Z.X. Gong, Z.Z. Li, *Materials and Mechanics*, Science Press, Beijing, (1999) (in Chinese)

DOI: 10.1051/matecconf/20179503011

Thermal and Hydraulic Performances of Nanofluids Flow in Microchannel Heat Sink with Multiple Zigzag Flow Channels

Weerapun Duangthongsuk

Department of Mechanical Engineering, Faculty of Engineering, Southeast Asia University, Bangkok, Thailand, 10160

Abstract. This article presents an experimental investigation on the heat transfer performance and pressure drop characteristic of two types of nanofluids flowing through microchannel heat sink with multiple zigzag flow channel structures (MZMCHS). SiO_2 nanoparticles dispersed in DI water with concentrations of 0.3 and 0.6 vol.% were used as working fluid. MZMCHS made from copper material with dimension of 28 x 33 mm. Hydraulic diameter of MZMCHs is designed at 1 mm, 7 number of flow channels and heat transfer area is about 1,238 mm^2. Effects of particle concentration and flow rate on the thermal and hydraulic performances are determined and then compare with the common base fluid. The results indicated that the heat transfer coefficient of nanofluids was higher than that of the water and increased with increasing particle concentration as well as Reynolds number. For pressure drop, the particle concentrations have no significant effect on the pressure drop across the test section.

1 Introduction

A decade ago, thermal performance of the air-cooled was quite small due to limitation of the thermal performance. Later, liquid cooled system such as water or other liquids were used to instead of the air-cooled system. Because of the thermal performance of liquid-cooled system are very high compared to the air-cooled system. Now a day, the microscale and nanoscale heat transfer devices became the crucial tool in many common applications such as integrated circuits (IC), VLSI devices, laser-diode arrays, micro heat exchanger and computer chips. Thus, the heat transfer area was limited and unprecedented high load was also generated. Improving the thermal performance of the coolants and innovative cooling equipment with special geometric structures were required. Tuckerman and Pease [1] who first introduced the concept of the microchannel heat sink (MCHS) to increase the thermal performance of the cooling system. Later, Masuda and colleges [2] was first to improve the thermal performance of the coolants by using very small solid particles dispersed in the common fluids. In 1995, Choi [3] was first named the common heat transfer fluids with nanoparticle suspension that "Nanofluids". Many researchers reported that the thermal performance of nanofluids was higher than the common heat transfer fluids and little penalty drop in pressure [4]-[11]. As a result, nanofluid-cooled heat sink was estimated to be a very effective cooling system for removing high heat densities generated by the modern electronic devices.

The benefits of nanofluid-cooled heat sinks have been summarized by Daungthongsuk and Wongwises [9]-[11]. Thus, the target of the present work is to evaluate the heat transfer performance of nanofluids flowing through MCHS with multiple zigzag flow channel structures (MZMCHS). Effect of particle concentration and flow rate on the heat transfer coefficient and pressure drop are reported.

2 Experimental apparatus

The experimental system is constructed in the present study to evaluate the thermal performance and flow characteristic of nanofluids flowing through MZMCHS. Schematic diagram of the experimental apparatus used in the present work is demonstrated as shown in Fig. 1. It mainly consists of the test section, a pump with speed controller, two storage tanks, and a receiver tank. The configuration of the MZMCHS used in this work is shown in Fig. 2. Copper material is used to make the MZMCHS with dimension of 28 x 33 mm. It also consist of 7 number of flow channels, heat transfer area of 1,238 mm^2 and hydraulic diameter of 1 mm. The heat sink cover plate is made from acrylic material and attached to the top of the heat sink to completely seal the flow passages. To simulate heat load, a 95 W electric heater is used and is attached at the below side of the test section. Two storage tanks with capacity of 15 liters are made from stainless steel. For the storage tank No.1, it consist of a cooling coil with cooling capacity of 3.5 kW and a 2 kW electric heater. In order to adjust the fluid temperature, a thermostat is used to control the temperature at the desired value. To meet the steady state condition, the storage tank No.2 is used to decrease the fluid temperature leaving from the MZMCHS to the temperature of the tank No. 1. It consist of a 3.5 kW cooling coil and a thermostat. To adjust the flow rate of the working fluid, a pump with speed controller is used. T-type thermocouples are installed at the both ends on the test section to measure the bulk fluid temperature. To evaluate the surface temperature of the test section, two thermocouples are placed at 5 mm and 15 mm from the top surface of the test section for measuring the temperature gradient and then used to estimate the surface temperature. For pressure drop measurement, two digital pressure gauges are located at the inlet and exit of the test section. The receiver tank is used to measure the mass flow rate by the time taken for a given volume of the working fluid to be discharged. All instruments are calibrated to ensure the accuracy and reliability of the measured data.

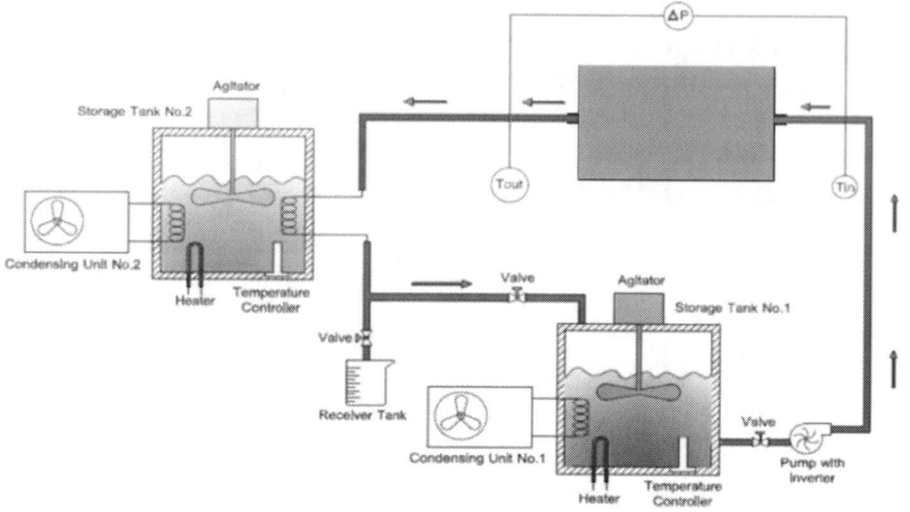

Figure 1. Schematic diagram of the experimental system

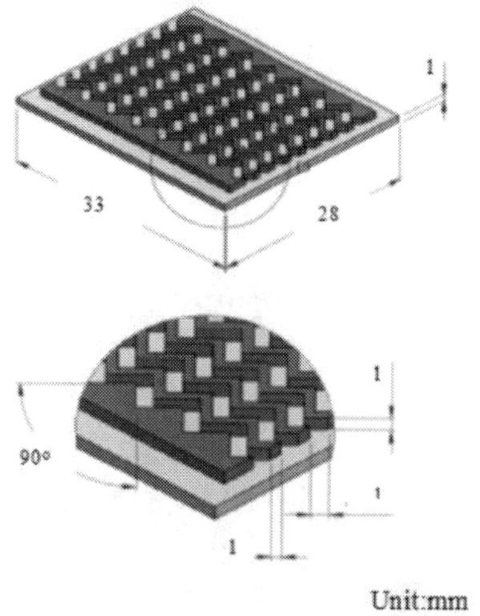

Figure 2. MZMCHS used in this work

To evaluate the thermal performance and pressure drop across the test section, the inlet and exit temperatures of the working fluid, surface temperatures, mass flow rates, and pressure drop across the test section. However, all data are measured after steady state condition reached.

3 Sample preparation

The 2-step method was used to prepare the nanofluids with the desired particle volume fractions. SiO_2 nanoparticles with an average diameter of 15 nm were used and dispersed in the DI water at particle volume fractions of 0.3 and 0.6 vol.% . Firstly, SiO_2 nanoparticles were dispersed and stirred in the DI water at the required particle concentration. Secondly, a 500 W ultrasonic vibrator was continuously sonicated the nanofluid for 2 hours in order to break down the agglomeration of the ultrafine nanoparticle. The thermophysical properties of each nanoparticle are expressed in Table 1.

Table 1. Thermophysical properties of SiO_2 nanoparticle.

Properties	Value
Thermal conductivity (W/m^{2}K)	1.37
Specific heat (kJ/kgK)	0.742
Density (kg/m^3)	2,648

4 Data reduction

In the present study, SiO_2-water nanofluids with particle fractions of 0.3 and 0.6 vol.% are tested and compared with the data for water-cooled heat sink. The thermal performance can be calculated by the following equations.

The heat transfer rate of the nanofluids is computed from:

$$Q_{nf} = \dot{m}_{nf} Cp_{nf} (T_{out} - T_{in})_{nf} \quad (1)$$

in which, Q is the heat transfer rate, $\dot{m}$ is the mass flow rate, T is the temperature and subscript nf, in and out are nanofluid, inlet and exit, respectively.

The Nusselt numbers of the nanofluids are defined as follows:

$$Nu_{nf} = \frac{h_{nf} D_H}{k_{nf}} = \frac{Q_{nf} D_H}{A_S (T_S - T_{nf}) k_{nf}} \tag{2}$$

where h is the heat transfer coefficient, k is the thermal conductivity. T_S and A_S are the surface temperature and the heat transfer area, Nu is the Nusselt number and D_H is the hydraulic diameter based on the each flow channel.

The Reynolds number based on hydraulic diameter are calculated from:

$$\mathrm{Re}_{D_H} = \frac{\rho_{nf} u_{ch} D_H}{\mu_{nf}} = \frac{\rho_{nf} u_{ch}}{\mu_{nf}} \frac{4A_{ch}}{p_{ch}} \tag{3}$$

where ρ is the density, μ is the viscosity, u_{ch} and p_{ch} are the mean velocity and the perimeter of the each flow channel and A_{ch} is the cross-sectional area of the each flow channel.

The Pak and Cho [12] correlations are used to calculate the specific heat and density of the nanofluids which are expressed as follows:

$$Cp_{nf} = \phi Cp_p + (1-\phi) Cp_w \tag{4}$$

$$\rho_{nf} = \phi \rho_p + (1-\phi) \rho_w \tag{5}$$

where Cp is the specific heat, ϕ is the particle volume fraction, and subscript nf, p and w are nanofluid, nanoparticles and water, respectively.

The Hamilton and Crosser model [13] and Einstein model [14], are used to calculate the thermal conductivity (k) and viscosity (μ) of nanofluids which are expressed as follows:

For thermal conductivity;

$$k_{nf} = \left[\frac{k_p + (n-1)k_w - (n-1)\phi(k_w - k_p)}{k_p + (n-1)k_w + \phi(k_w - k_p)} \right] k_w \tag{6}$$

$$n = 3/\psi \tag{7}$$

For viscosity of the nanofluids;

$$\mu_{nf} = (1 + 2.5\phi)\mu_w \tag{8}$$

where ψ is the sphericity which is defined as the ratio of the surface area of a sphere to the surface area of the particle. The sphericity is 1 and 0.5 for the spherical and cylindrical shapes, respectively. n is the empirical shape factor.

5 Results and discussion

The variation of the surface temperature with heat load as a function of particle concentration was shown in Fig. 3. The measured data indicated that the surface temperatures of MZMCHS increased with increasing heat load and decrease as particle concentration increased. This mean that nanofluid-cooled MCHS gave better thermal performance than that of the water-cooled MCHS which led to increase in the heat removal rate from the system. This is due to the fact the nanofluid have a large thermal property compared to the conventional pure water. As a result, the surface temperature of the test section was lower than usual.

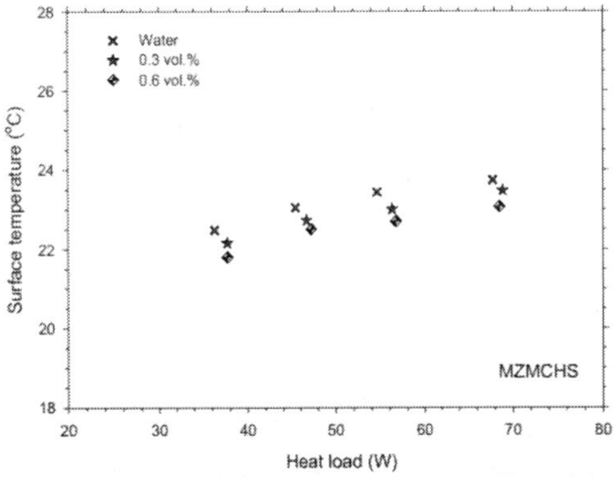

Figure 3. Variation of the surface temperature of heat sink as a function of heat load

Comparison of the measured Nusselt number (Nu) between nanofluids-cooled and water-cooled MCHS was shown in Fig. 4. The measured data illustrated that Nu increases with increasing Reynolds number as well as particle concentrations. Moreover, it evidently seen that the nanofluid-cooled heat sink gave higher heat transfer performance than that of the water-cooled heat sink about 8%. This means that the use of nanofluids as coolant gave better transfer performance than that of the common pure water and can be suitable for used in the case of high heat load dissipation

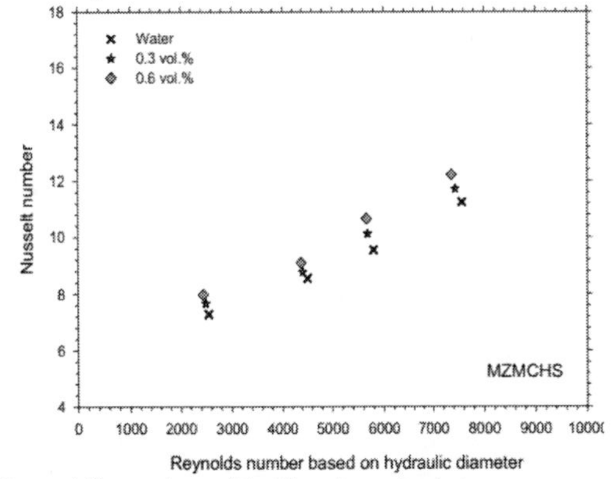

Figure 4 Comparison of the Nusselt number between nanofluids-cooled and water-cooled heat sink

Fig. 5 shows the comparisons between the pressure drops across the MZMCHS obtained from the nanofluids and common pure water. The results indicated that the pressure drop across the test section increases with Reynolds number increased. Furthermore, it clearly seen that the use of nanofluids had no significant effect on the pressure drop. Thus, it may concluded that nanofluids have high potential to enhance the thermal performance of the cooling system without penalty drop in pressure.

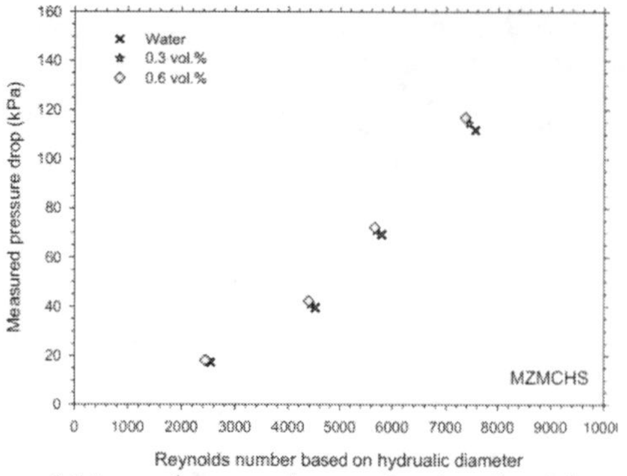

Figure 5. Measured pressure drop as a function of particle concentration and Reynolds number

6 Conclusions

The heat transfer performance and pressure drop characteristic of nanofluids-cooled heat sink were experimentally studied. Microchannel heat sink with multiple zigzag flow channe (MZMCHS) was used. SiO_2-water nanofluids with particle volume concentration of 0.3 and 0.6 vol.% were used as coolant and then compared to the common water. The effect of particle concentrations and the Reynolds number on the Nusselt number and pressure drop were examined. Important conclusions have been described as follows:

• The Nusselt number increased with increasing particle concentration as well as Reynolds number. And, nanofluids-cooled heat sink gave lager thermal performance than that of the water-cooled heat sink by approximately 8% (average at particle concentration of 0.6 vol.%).

• The use of nanofluids as coolant had no significant effect on the pressure drop.

The authors would like to express their appreciation to the office of research of the Southeast Asia University, the Thailand Research Fund (TRF), and the Office of the Higher Education Commission to provide financial support under Grant No. 5880269.

References

1. D.B.Tuckerman, R.F.W. Pease, IEEE Electr. Device L. **2(5)**, 126 (1981)
2. H. Masuda, A. Ebata, K. Teramae, N. Hishinuma, Netsu Bussei (Japan) **7(4)**, 227 (1993)
3. S.U.S. Choi, ASME Fluids Eng Division **231**, 99 (1995)
4. W. Duangthongsuk, S. Wongwises, Int. J. Heat Mass Trans. **22**, 2059 (2009)
5. W. Duangthongsuk, S. Wongwises, Exp. Therm. Fluid Sci. **33**, 706 (2009)
6. W. Duangthongsuk, S. Wongwises, Int .J. Heat Mass Trans. **52,** 2059 (2009)
7. W. Duangthongsuk, S. Wongwises, Int .J. Heat Mass Trans. **55,** 3138 (2012)
8. W. Duangthongsuk, S. Wongwises, Curr. Nanosci. **8 (3)**, 317 (2012)
9. W. Duangthongsuk, S. Wongwises, Exp. Therm. Fluid Sci. **66,** 28 (2015)
10. W. Duangthongsuk, S. Wongwises, Adv. Mat. Res. **1105,** 253 (2015)
11. W. Duangthongsuk, S. Wongwises, Exp. Therm. Fluid Sci. **69,** 111 (2015)
12. B.C. Pak, Y.I. Cho, Exp. Heat Transfer **11**, 151 (1998)
13. R.L. Hamilton, O.K. Crosser, Ind. Eng. Chem. Fund. **1 (3)**, 187 (1962)
14. D.A. Drew, S.L. Passman, *Theory of multi component fluids* (Springer, Berlin, 1999)

MATEC Web of Conferences 95, 03012 (2017)
ICMME 2016
DOI: 10.1051/matecconf/20179503012

Seismic Response of Building Structure Near Tunnel

Zhanxue Zhou and Hongliang Hu

School of Civil Engineering, Hebei University of Architecture, 075000Zhangjiakou, China

Abstract. To research the influence of seismic response of building by the underground tunnel, a high-rise frame structure on the uniform field was analysed by using finite element method (FEM).The results showed that when the tunnel was located below the building the biggest influence on structural seismic response will been emerged, and made the structural displacement and natural vibration period increase and bending moment decrease. Tunnel located elsewhere on the influence of the structure response was not simply decreases with increasing of the distance, but to show some volatility. The tunnel will have little impact on seismic response of the underground part of the structure.

1 Introduction

With the rapid development of economy and the advancement of urbanization in our country, the development and utilization of underground space such as subway and shelter is expanding on a large scale, a lot of buildings are built near the underground tunnels, and the phenomena that tunnels through the structure closely increased obviously. Previous studies mostly focused on the seismic response of the tunnel structure itself, while there are less studies on the tunnel structure's influences to the seismic performance of buildings near the ground [1]. Studies show that seismic waves will interact with the tunnels in the process of transmission, so that any point on the tunnel becomes a new wave source, which send out secondary waves in all directions, forming the seismic wave scattering, and then it will impact the seismic response of surface buildings [2]-[6].

The paper studies the seismic response of surface high-rise buildings near the tunnel by the finite element method, to analyse how the distance between the tunnel and the overground construction impact on the seismic response of the building structure.

2 Structure calculation model and related parameters

The soil parameters of the calculation model have no difference, to simulate and analyze the soil and the structure by plane strain unit, the left and right boundaries are viscoelastic artificial boundary, the bottom is bed rock. The analytic site is uniform field, whose length is 150 meters, depth is 30 meters, elasticity modulus is 40×10^8Pa, poisson ratio is 0.3, density is 2049kg/m^3. The surface structures is reinforced concrete frame structure, it has two floors underground and ten floors above, the span is 10 meters, the floor height is 3 meters. Taking a common double span frame to calculate, whose elasticity modulus is 4.83×10^{10}Pa, poisson ratio is 0.2 and density is 2500 kg/m^3. The diameter of tunnel is 10 meters. It locates 15 meters under the ground. Taking five working situations in consideration, project status one is with no tunnel near the building structure, project status two, three, four, five respectively are the horizontal distance between the tunnel center and the axis of surface building is 0, 10, 20, 30meters. Finite element meshing is shown in the Fig. 1, according to the shear wave velocity to determine the grid size, both sides are viscoelastic artificial boundary, simulating by using spring-damper elements, the bottom soil is fixed [7].

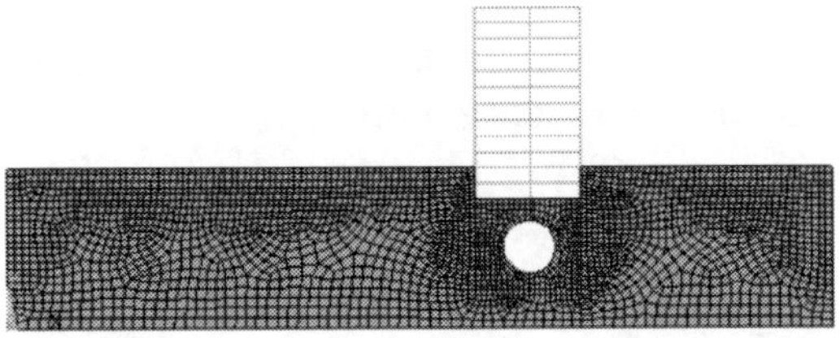

Figure 1. Finite element model of tunnel and structure system

The input seismic waves is EI-Centro waves in north-south, inputing horizontally from the bedrock. In order to compare commodiously, we adjust the peak acceleration of EI-Centro wave to 0.1g, as shown in the Fig. 2.

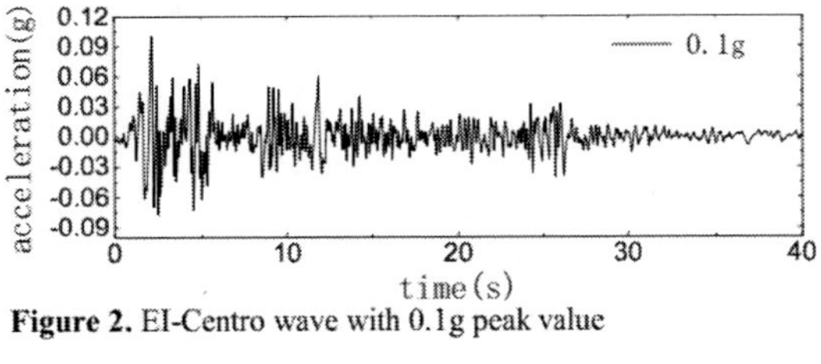

Figure 2. EI-Centro wave with 0.1g peak value

3 Tunnel's influence to building structural frequency

The calculation results of each order mode of vibration frequency of the frame structure under Five kinds of project status as shown in the Table 1 and Fig. 3.

Table 1. Each order modal frequency unit:Hz

mode of vibration	project status 1 (no tunnel)	project status 2 (location 1)	project status 3 (location 2)	project status 4 (location 3)	project status 5 (location 4)
1	2.076	0.62136	1.7966	2.0679	1.7983
2	6.313	1.932	5.7279	6.3043	5.7435
3	7.431	3.4395	7.2652	7.3478	7.2547
4	8.88	4.7793	8.6743	8.7486	8.6892
5	10.155	5.1738	10.856	9.9179	10.978
6	11.618	6.4794	10.991	11.589	11.634
7	12.369	7.1274	12.103	12.081	12.101
8	12.964	7.3324	12.754	12.742	12.886
9	13.483	7.6613	13.424	13.422	13.211
10	13.977	7.7867	13.875	13.625	13.668

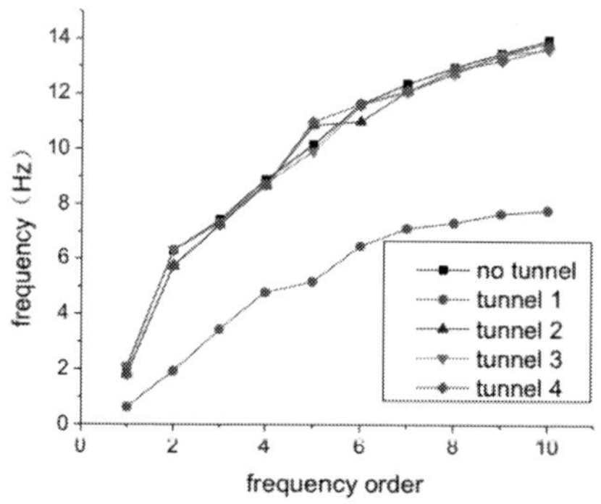

Figure 3. Frequency of structure

It can be seen that when the tunnel under the structure system, that is to say structure frequency decreases more when it is in the project status 2, while other cases are close to the case 1 (without tunnel). compared with the case with no tunnel, the natural frequency of project status 2 dropped by 70% at most.

4 Tunnel's influence to building structural displacement

The calculation results of each floor peak displacement of the structure under five kinds of project status as shown in the Fig. 4, the displacement history response as shown in the Fig. 5.

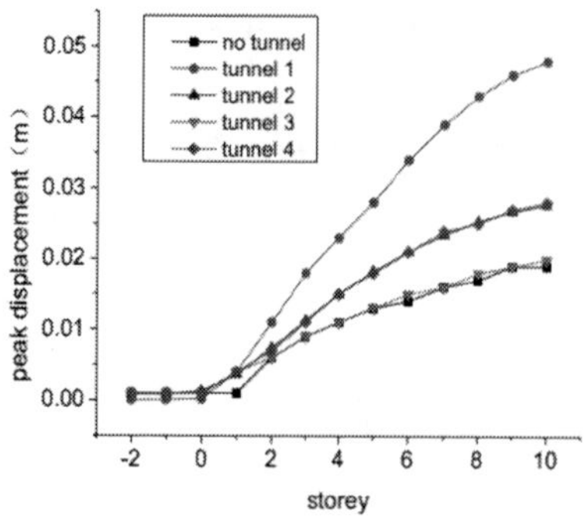

Figure 4. Displacement of the structure

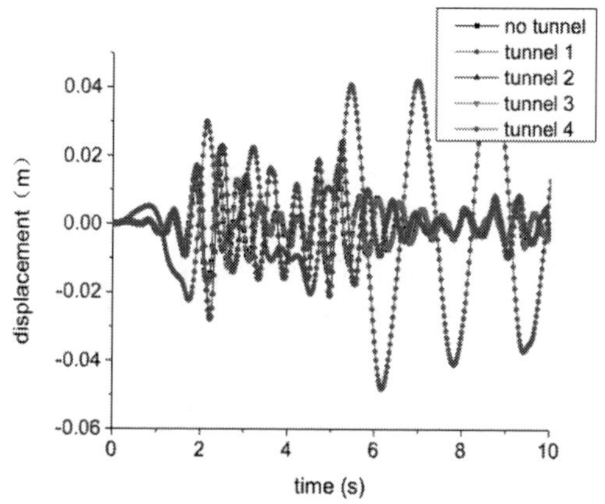

Figure 5. The top floor displacement time history curve

Fig. 4 is the peak displacement curve of the second member across the left beam for each floor of the structure with different project status , Figure 5 is the displacement history curve of the second member across the left beam for top floor of the structure with different project status, the history curve cut out a period of 1 to 10 seconds.

As shown in the figures above, Tunnel's influence on the structural displacement can be roughly divided into three groups similar curve: project status 2(tunnel under the structure center)is the first group, it is 90% bigger than that of project status 1; project status3 and project status 5 are the second group, its average value is 50% bigger than that of project status 1; project status 4 is the third group, it is 5% bigger than that of project status 1. It is clear that the horizontal distance between the tunnel and the surface building structure has a big influence on the structure displacement response, especially, the influence is bigger than others when the tunnel under the structure. However, we should notice that the structure displacement is not simply decreasing with the increase of the distance, but showing some hop. What's more, positions of the tunnel have little impact on the displacement response of the underground part of structure. From the Fig. 5, the peak displacement response of top floor mainly appear on the later part of the earthquake when the tunnel under the structure.

5 Tunnel's influence to building structural internal force

The calculation results of peak bending moment of the second member across the left beam for each floor of the structure under five kinds of project status as shown in the Fig. 6, the bending moment history response of the second member across the left beam for top floor as shown in the Fig. 7.

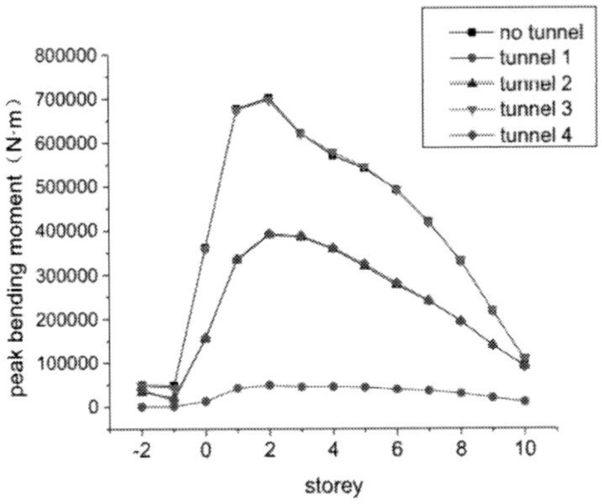

Figure 6. Floor bending moment

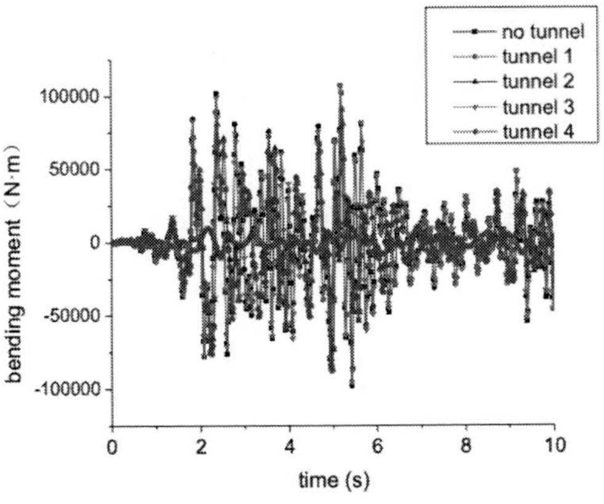

Figure 7. The top floor bending moment time history curve

Fig. 6 is the peak bending moment curve of the second member across the left beam for each floor of the structure with different project status, Fig. 7 is the moment history curve of the second member across the left beam for top floor of the structure with different project status.

As shown in the figures above, Tunnel's influence on the structural bending moment can also be roughly divided into three groups similar curve: project status 2(tunnel under the structure center)is the first group, its value reaches the maximum compared with project status 1, both the status differ by 93% in average;project status3 and project status 5 are the second group, which has larger difference with project status 1, both the status differ by 41% in average;project status 4 is the third group, it differs by 1% in average with project status 1. It is clear that the horizontal position of the tunnel has a big influence on the bending moment in the midpoint of the structure, the bending moment reaches the minimum when the tunnel under the structure, while the value of bending moment in other position is relatively large. However, we should notice that the distance between the tunnel and the building is equal to a diameter of the tunnel which is closed to the case without tunnel. The change of bending moment value shows some volatility. What's more, the tunnel's location has little impact on the bending moment of the underground part of structure.

6 The influence of different seismic waves to building structure

In order to consider the influence of different seismic wave input to architectural structure system, we choose Tianjin wave to excite horizontally from the bedrock, and adjusting the peak acceleration of Tianjin wave to 0.1g. The calculation results of each floor peak displacement of the structure under five kinds of project status as shown in the Fig. 8.

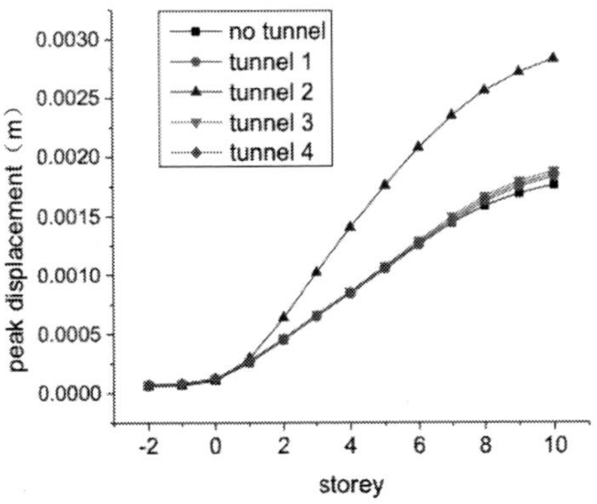

Figure 8. Displacement of the structure when tianjin wave

From Fig. 8, we can see the tunnel's influence to structural displacement when inputting Tianjin wave: project status 3(the tunnel locate in corner of the structure) is 61% bigger than that of project status 1; other project statuses is similar to the project status 1. It is clear that although inputting different earthquake waves, overall, the tunnel's influence to structural displacement response reaches the maximum when it under the building, structural seismic response is relatively small when tunnel located in other status. What's more, the tunnel's location also has little impact on the displacement response of the underground part of structure.

7 Conclusions

1. Tunnel's influence to structural seismic response reaches the maximum when it under the building. At this point, the interaction between tunnel and building structure is the most intense, the natural vibration period of the whole structure increases obviously, it makes the structure displacement increasing and internal force decreasing.
2. Tunnel's influence to structural seismic response is relatively small when tunnel located in other status, but the impact on structural response is not simply decreasing with the increase of distance, it shows some volatility.
3. The tunnel's location has little impact on seismic response of the underground part of structure.

References

1. Li Fang-jie, Zhao Feng-xin, Zhang Yu-shan, You Hong-bing. *Influence of ground building's relative position on seismic response of subway structure.* Earthquake research in China, **26**(2):201 ~ 209, (2010)
2. Liang Jian-wen, JI Xiao-dong, Vincent W Lee. *Effects of an underground lined tunnel on ground motion (II): numerical results.* Rock and Soil Mechanics, **26**(5):687~692, (2005)
3. Chen Guo-xing, Zhuang Hai-yang, Xu Ye. *A study on influence of excavated shallow tunnel on design parameters of ground motion in the soft site.* Chinese Journal of Geotechnical Engineering, **26**(6):739 ~ 744, (2004)
4. Zhu Feng-bin, Yang Ping. *Study on interaction and influence of one on another during construction of urban tunnel and adjacent underground structure.* Railway Engineering, **(2)**: 48~51, (2007)
5. He Wei , Chen Jian-yun, YU Pin-qing. *A Study on Influence of Underground Structure on Ground Surface Response Spectra.* Chinese Journal of Underground Space and Engineering, **5**(6)**:**1099 ~ 1114, (2009)
6. Maria I. Todorovska, Ph.D., Mihailo D. Trifunac, Ph.D., and Vincent W. Lee, Ph.D. *Synthetic Earthquake Ground Motions for the Design of Long Structures[J].*International Efforts in Lifeline Earthquake Engineering, 592~599, (2013)
7. Wang Zhen-yu, Liu Jing-bo. *Study on wave motion input and artificial boundary for problem of nonlinear wave motion in layered soil.* Chinese Journal of Rock Mechanics and Engineering, **23**(7):1169~117(2004)

Solar Wear Tattoo: New Technology Through the use of Different Material Behaviours

Vanessa da Graça Mendes[1] and Mario S. Ming Kong[2]
[1]Faculty of architecture - University of Lisbon - FA-UL 1349-055 Polo Universitário - Alto da Ajuda, Lisboa, Portugal
[2]Faculty of architecture - University of Lisbon Researchers and CIAUD-permanent member of the Research Centre for Architecture, Urban Planning and Design FA-UL, Portugal

Abstract. This study comes within the framework of a PhD course in Design, which is the continuation of a Masters Degree research in Fashion Design, searching for wardrobe technology and innovation through material behaviour. The intent of this project is to present solutions that promote a new and versatile product allowing for a different kind of aesthetic. This will be achieved through different materials with different behaviours that when put together create an innovative technology thus creating a new practice of body art, through the use of fashionable beachwear created with this technology. This technology allows the transformation of the suntan marks left by female swimwear. This will result in the creation of designs and drawings on the body, using shapes inspired by the Oriental Zodiac Signs, as well as colours drawn from the five elements of Feng Shui, which are also present in the Western Astrological Signs.

1 Introduction

Material behaviour and its potential for creating unique functions and new types of objects when combined, and its ability for transforming the suntan marks left by the usual bikini or swimsuit, has been one of the main themes and subjects of this research. This allows for new creative boundaries to emerge in body art through textile technology.

2 Body art and new practices through beachwear

2.1 Body art

Everyday people produce art on the body in a almost unconscious way, working hair, nails and makeup, because these physical interventions aren't usually associated or even called as a from of body art, although all these actions are a part of a body transformation process, metamorphosing the projected exterior appearance, communicating a set of feelings and values as a part of mans natural social instincts and behaviours. [1]

Focusing on a specific type of body art, tattoo, especially temporary tattoo, it already exists in the form of a practice called *Mehdi*. [2] This practice is essentially the art of painting a body with *henna* (*Lawsonia Inermis* in Latin [3]), an Egyptian plant with a strong characteristic smell. Its leafs and flowers, after dried up, are grinded into dust, mixed with water, lemon juice and some drops of kerosene to create a paste used to make a natural skin dye. [2]

2.2 New practice through beachwear

Beachwear originally existed to allow water-sport fans to go to the beach to watch this kind of competitions. [4] Last century, in the last mid 20's, Patou was one of the big fashion creators to conceive such outfits suitable for beach practice. These cloths were made for sportive woman. Mainly, these outfits consisted in short dresses made from different materials, with soft coats to match, which were highly in *vogue* at that decade.

The main public figure iconic enough to transform fashion beachwear into a body art practice/body cult was Coco Chanel. In the same decade that Patou fashion was in *vogue*, Chanel practiced outdoor spots often, becoming one of the first women to appear in public with a suntan. [5]

This information makes clear that, since then, suntans and beach practice became a relatively recent, new form of body art in Western fashion.

3 Harmonious beachwear through *Feng Shui* logic

3.1 Observation of the natural elements

Pushing the envelope of new body art forms is not easy and, in this case, it would not have been possible without applying *Feng Shui* logic.

Western people call it intuition; in Eastern culture that intuition is defined by guide lines based on the logic of nature, which can be acquired by observation of the natural interactions between the various natural elements.

That natural observation is the key. For that reason, *Feng Shui* is applied as a main theme. It is the concept that gave birth to the hole, extending its principles not only to the choice of materials, but to colour choice and combinations as well, so that it is possible for the users and observers to have a functional and harmonious beachwear experience. [4]

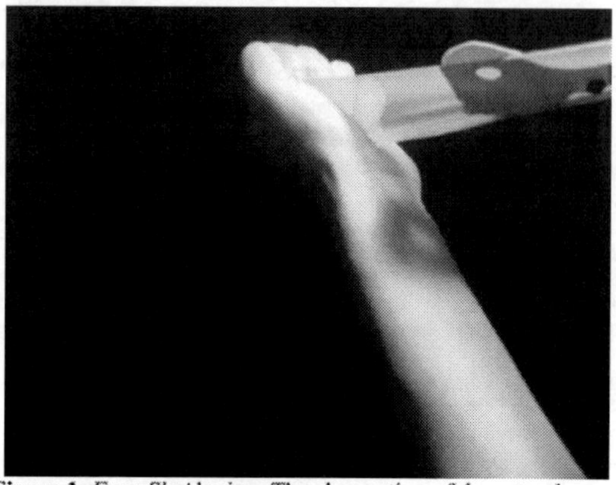

Figure 1. *Feng Shui* logic. - The observation of the natural elements interacting with materials. - Authors Photography. [4]

The harmonious aspect of *Feng Shui* application does not only come in form of colour, because the technology that results out of the observations of the natural elements, comes with the benefit of providing a natural tattoo freed of the black ink and dyes that artificial *henna* tattoos use. These artificial dyes can become dangerous to the user in case of occurrence of an allergic skin reaction, due to scarification risk. [6]

3.2 Colour in harmony

Colour is an important factor in fashion design. "At the start of each year since the dawn of the new century, Pantone has selected a Colour of the Year. This is the shade that best encapsulates the spirit of the time, and that can be expected to pop up everywhere throughout the coming months, both in fashion and in other areas of design and daily life." [7] This is more commonly known as fashion tendencies or trends. [8]

Whatever the trend of the year is, each user has his very personal colour perception and preferences that can be defined through his psychology or physiological sensations. Heller studied relations between colour and emotion as a psychological result of cultural influence. [9] Silva clearly describes physiological reactions to colour as "colour sensations". [10]

Around the world, colours describe emotional characteristics. Melancholy is usually associated to blue, green to envy and yellow to cowardice. Yet, for the Chinese, colour has a complete different significance and application in order to achieve and acquire harmony and balance. [11]

Conceiving healthy and harmonious beachwear as a goal and applying the Oriental theme as a hole, studies about colour and their impact to the user, will be associated and respectively distributed on the Five *Feng Shui* Elements.

As a result, since each element is never associated with a single colour alone, a group of gradients are needed to form a sensation of the presence of a *Feng Shui* element through colour. The tests made with various materials (in order to obtain the desired results of suntan manipulation) showcase the *Feng Shui* elements through colour.

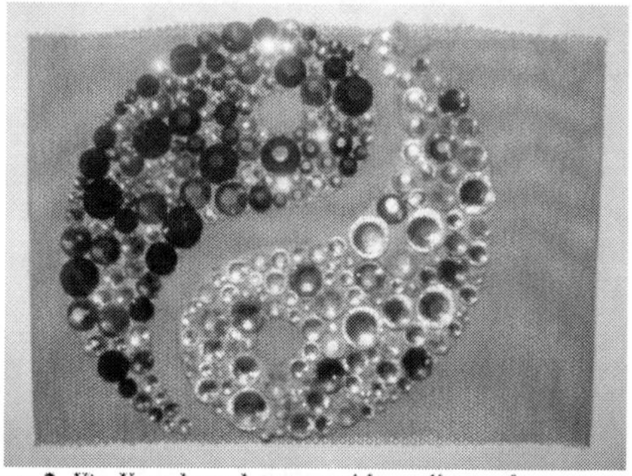

Figure 2. *Yin-Yang* bracelete test with gradient color interpertation. - Authors Photography. [4]

Following these principals, a bracelet was made to test the behaviour of the materials when exposed to the sun, expecting that a natural solar tattoo could come out of these material combos. The following materials were tested: Polyamid, lycra and silicone. [4] These materials not only are eco-friendly, they have proven to be functional for body art purposes.

3.3 Environemental harmony

With the best regards for the environment, the components from which the bracelet is made of are separable; which is a good thing when talking about recycling since the recycling process becomes much easier when components can easily be separated from each other. [12]

According to Wang, fashion products made out of multi-material composite system with detachable connection prove to be balanced in maters of recycling process and durability of the components.

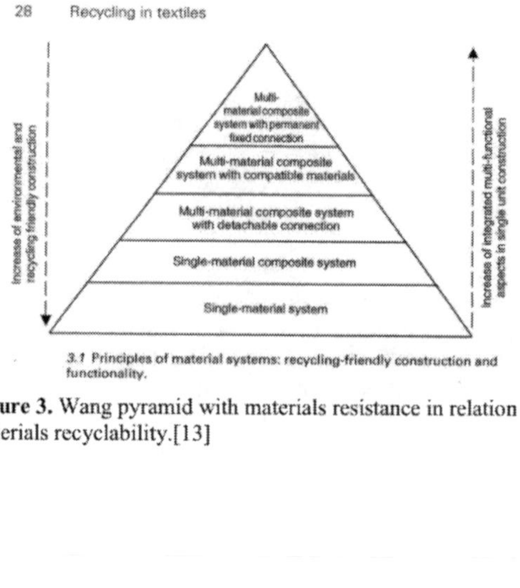

Figure 3. Wang pyramid with materials resistance in relation to materials recyclability.[13]

3.4 First results

After a week of sun exposure, a natural tattoo appeared on skin, [4] as a result of material behaviour between the body and the sun, perhaps, making a new body art practice possible for the user. [14]

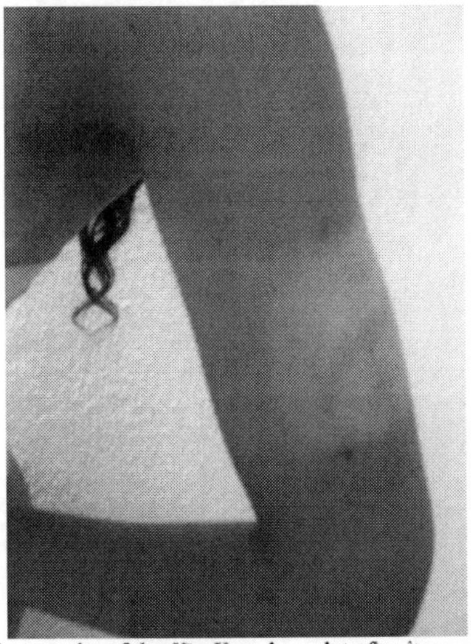

Figure 4. Test results of the *Yin-Yang* bracelet after its use in sun exposure.- Authors Photography. [14]

The present investigation is also dedicated to test other material behaviours that can have the same outcome in the clasp of the beachwear. Inspired by Chinese seals, this clasp is meant to be tested in acrylic.

3D prints are already made and ready to come alive in other materials through mould extraction, being also possible to test with other materials that are test ready after 3D the printing process is over, without the need to extract moulds in order to obtain the final material intended.

Figure 5. - 3D print of Chinese Seal Clasp.- Authors Photography. [4]

For both clasp and textile to become functional, the need of an application of a layer of silicone, on the side that comes into contact with the user's body is going to be considered and studied, in order to provide a greater adherence to the skin, allowing for a more precise tattooing of the skin by holding it in place.

It is a matter of function, and concern with the user's comfort, thus creating a product that is ergonomic and has the wearer's best interests in mind.

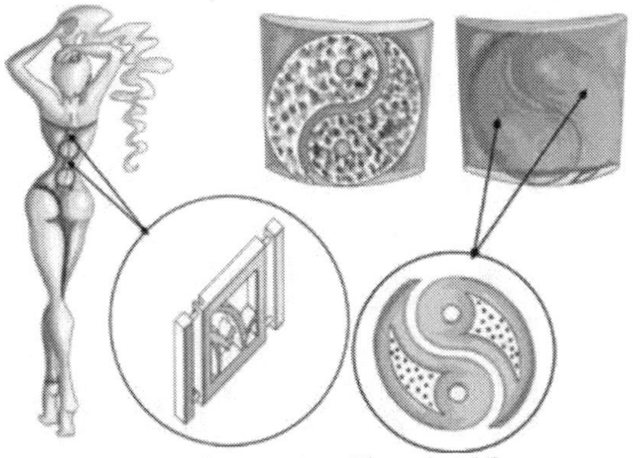

Figure 6. Silicone applications on textile and Chinese clasp.- Authors Illustration. [4]

References

1. N. Thomas: *Body Art,* Thames & Hudson world of art, London, p. 12, 34, 126, (2014).
2. The Unesco Courier: A window open to the world. A lost metropolis of 4,000 years ago discovered in Syria. **2**, p. 19, (1977).
3. C. Fabius: *Mehndi: The Art of Henna Body Painting* Thames & Hudson world of art, London, (2014).
4. V. Mendes: *Solar Wear Tattoo: Harmonia do Design de Moda, no Beachwear como Tatuagem - Fusão Cultural,* Faculty of Architecture - University of Lisbon, **55**, pp. 9-11, 26-27, 2015.
5. G. Lehnert: *História da Moda do século XX* Konemann, Germany, p. 28, (2001).
6. S. Benomar, L. Ismaili, L. Benzekri, B. Hassam: *Tatouagem au henné noir: au-del'de l'eczéma de contact.* Elsevier B.V., Critica Françêsa de Alergologia - Fato clinico, (2009).
7. L. Eiseman, E. P. Cutler: *Pantone® on Fashion: A Century of Colour in Design,* Chronicle Books LLC, San Frncisco, p. 109, (2014).
8. M. Sabino: *Dicionario da Moda,* Elsevier Editora Ltd, Rio de Janeiro, p. 574, (2007).

9. E. Heller: *A psicologia das cores,* Editorial Gustavo Gili, SL, Barcelona, (2009).
10. F. M. Silva: *A Colour Space. Its quality management in architecture. The colour/space unity as an unity of visual communication,* Ph. D. Thesis, UK, (1999).
11. M. S. M. Kong: *Harmonia e Proporção. Um olhar sobre o desenho arquitetónico no Ocidente e no Oriente,* Inside city, Lisboa, p. 117, (2012).
12. C. Morais: *A Sustentabilidade no Design de Vestuário,* Ph. D. Thesis, Lisboa, (2013).
13. Y. Wang: *Recycling in textiles.,* Woodhead Publishing Limited & Cr Press LLC, England, (2006).
14. V. Mendes: *Solar Wear Tattoo* (to be published).

Prediction of Vortex-Induced Vibration of Bare Cylinder and Cylinder Fitted with Helical Strakes

KeeQuen Lee[1], Aminudin Abu[1], Pauziah Muhamad[1], Asnizah Sahekhaini[2], Litken Tan[1] and HooiSiang Kang[3]

[1]Malaysia – Japan International Institute of Technology (MJIIT), University Teknologi Malaysia Kuala Lumpur, Malaysia
[2]Politeknik Sultan Salahudin Abdul Aziz Shah, 40150 Shah Alam, Malaysia
[3]Institute for Vehicle Systems & Engineering (IVeSE), Faculty of Mechanical Engineering, Universiti Teknologi Malaysia, Johor Bahru, Malaysia

Abstract. In the study, the vortex-induced vibration (VIV) of a cylinder fitted with and without helical strakes is investigated using fluid-structure interaction (FSI) software. The purpose is to predict the VIV characteristic of cylinder fitted with and without helical strakes. Fluid and structural solvers in commercial computational fluid dynamic (CFD) software were implemented as two-way FSI solvers to develop the simulation. A flexible circular cylinder of diameter, D = 0.018 m was tested. The pitch and height of the helical strakes were 10D and 0.10D, respectively. Existing experiment outputs of bare cylinder were used as benchmark to verify the simulated results. In overall, the simulation was able to predict the trend of the amplitude response and the mean drag coefficient. However, a slight over-prediction was noticed. It was also found that the helical strakes was able to reduce up to 60% of the VIV in water.

1 Introduction

Marine riser, which is one of the important components in the offshore structure, play a crucial role in extracting the hydrocarbon products from undersea drilling site to the oil platform. Due to the unpredictable sea current, risers face several problems such as creeping, buckling, expansion and fatigue generated by vortex-induced vibration (VIV). Among these problems, VIV, which can be explained as the vibration stimulated by the external fluid flow vortices when interacting with the structure, is one of the critical design issue of the riser. Currently most of the investigations of the VIV of riser are conducted experimentally [1-4]. Nevertheless, due to the high cost in constructing the experimental facilities and large aspect ratio of the riser, several factors are usually being constrained in the model testing. This initiates the exploration of fluid-structure interaction (FSI) software in predicting the VIV characteristic of a riser. Although FSI software is widely used in several applications such as combustion system [5,6] and arterial blood flow [7,8], its involvement on flow pass a circular riser/cylinder in three-dimensional is still rare. Several studies using two-dimensional (2D) FSI simulation are found. Pereira Gomes et al. [9] conducted a 2D FSI simulation on an aluminium cylinder in a laminar flow. The outputs were acceptable, but deficiencies were observed due to the constraint of damping coefficient. Facchinetti et al. [10] utilized classical van der Pol equation with equation of motion to examine VIV of a rigid cylinder in 2D. The authors recommended that the van der Pol wake oscillator was suitable in 2D simulation and was in high potential to be used in 3D analysis. Gustafsson [11] was one of the few who conducted 3D FSI simulation to analyse the VIV of a short riser. The author pointed that time step was an important factor to decide the simulation's accuracy. Besides, he explained that FSI simulation was much more time consuming compare to the pure computational fluid dynamic (CFD) simulation. Therefore, high level of mesh refining was unrealistic as it increases the computational time.

On the other hand, the simulation of the application of helical strakes on a circular cylinder has not been widely explored. Constantinides and Oakley [12] were the very few researchers that investigate the flow around a cylinder fitted with helical strakes. The authors observed that the VIV suppression mechanism was lack of spanwise correlation, and found that the vortices formation was farther downstream.

To the author knowledge, there is still no published work using FSI simulation to predict the response of cylinder fitted with helical strakes. Since the reviews have shown that FSI study on riser is very limited and preliminary, we hope to explore its capability in predicting the VIV characteristic of cylinder fitted with and without helical strakes in three-dimension. The existing experimental results of bare cylinder – cylinder without fitted with helical strakes [13] will be used as benchmark to validate the simulation outputs.

2 Computational study

 DOI: 10.1051/matecconf/20179504001

2.1. Fluid equation

Fluid-structure interaction, a combination of fluid and structural solvers, is implemented in the study. The unsteady Reynolds-averaged Navier-Stokes equations (RANS) is used as the governing equation for the fluid. For an incompressible Newtonian fluid, the equation is expressed as below:

$$\frac{\partial U_i}{\partial x_i} = 0 \quad (1)$$

$$\rho \frac{\partial U_i}{\partial t} + \rho U_j \frac{\partial U_i}{\partial x_j} = -\frac{\partial P}{\partial x_i} + \frac{\partial}{\partial x_j}(2\mu S_{ji} - \rho \overline{u_j' u_i'}) \quad (2)$$

where U is the averaged velocity, u' as the fluctuating velocity, P denotes pressure, ρ represents density, and Sji and μ are rate of strain tensor and molecular viscosity, respectively. Reynolds-stress tensor is represented by the term . At this stage, the system is not closed because of the number of equations are less than the number of unknown components. Hence, additional equations are needed. In the study, k-ω based Shear Stress Transport (SST) eddy viscosity turbulence model is chosen as this model considers the turbulent shear stress, which is not accounted in other turbulence model. Therefore, it is able to estimate the separation flow under adverse pressure gradient more precisely [19].

2.2 Structural equation

To couple with fluid response, equation of motion is utilized as structural dynamic system for a cylinder in the study. A spring-damper-mass system with external fluid forces for a cylinder is described as:

$$m_s\ddot{\chi} + c_s\dot{\chi} + k_s\chi = F_{fluid}(t) \quad (3)$$

where ms denotes the structure mass, cs as the damping coefficient, ks as the spring stiffness and Ffluid(t) represents the forces created by fluid on the cylinder body. The fluid force is obtained from the pressure and body viscous friction based on Navier-Stokes equation solutions. χ = Xi +Yj + Zk, where X, Y and Z is the instantaneous displacement of the cylinder in x, y and z direction, respectively. The first and second order-partial derivative of the displacement are denoted as $\dot{\chi}$and $\ddot{\chi}$, respectively.

2.3 Coupling for fluid and structure

The concept of FSI is to coupling the equation of fluid and structure, as shown in Figure 1. The system coupling is a two-way coupling. Partitioned approach is implemented as suitable solver because it allows the selection of each domain based on the requirement. The concept of the coupling is as following: Firstly, to achieve converged solution, the fluid calculation is executed. The calculated fluid forces of the fluid interface are then inserted to the structural domain. In the structural domain, the hydrodynamic forces are used as the boundary conditions to further compute the deformation of structural mesh. The deformation output is the cylinder displacement. Then, the displacements information is sent back to the fluid domain to recalculate the change of hydrodynamic forces. The process is repeated until the predetermined limits are reached for both displacement and force values.

3 Boundary condition

3.1 Structural domain

The cylinder is modelled to be similar to a pre-tensioned flexible cylinder, where both ends of the cylinder are pinned, and one the cylinder's end is allowed to move in axial direction. To achieve that, at the ends of the cylinder, all of the displacements and rotation in z direction are set to be zero. However, rotation is allowed in x and y directions. For the cylinder end where axial movement is allowed, instead of applying zero in z direction, an axial force of 147 N is used to represent the pre-tensioned effect. The gravity effect of the cylinder is represented by applying a standard earth gravity in -Y direction. Function of Fluid Solid Interface is defined at the cylinder surface to allow the integration of force and displacement between structural and fluid domains.

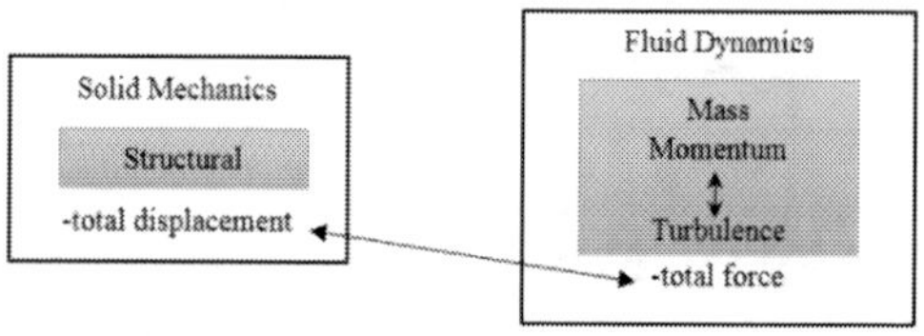

Figure 1. Coupling between structural and fluid equations.

3.2 Fluid domain

The fluid domain is consists of seven boundaries, namely upstream and downstream boundaries, left and right boundaries, upper and lower boundaries and the cylinder wall. In deciding the size of the computational domain, the distance between cylinder and boundaries is very important to avoid the effect of side boundaries [14]. According to Borch and Rodi [15], the minimum distance should be 4.5D to evade the influence of pressure and velocity, where D denotes the diameter of the cylinder. The authors recommended that boundary distance of 10D was appropriate to eliminate the effect. Hence, distance of 10D from the cylinder is set for upstream, lower and upper boundaries, as shown in Figure 2. To ensure the vortices of the cylinder can be captured successfully without any disturbance from the downstream boundary, distance of 20D is used between the cylinder and downstream boundary.

3.3 Grid independence study

Three different level of meshing sizes, namely coarse, medium and fine meshing are tested to check the

 DOI: 10.1051/matecconf/20179504001

sensitivity of the mesh towards the output of the simulation. The mesh size should be appropriate to avoid mesh dependency outputs. The grid independence study is tested on a fix end short cylinder. The outputs are presented in terms of cross-flow (CF) amplitude response, as shown in Table 1.

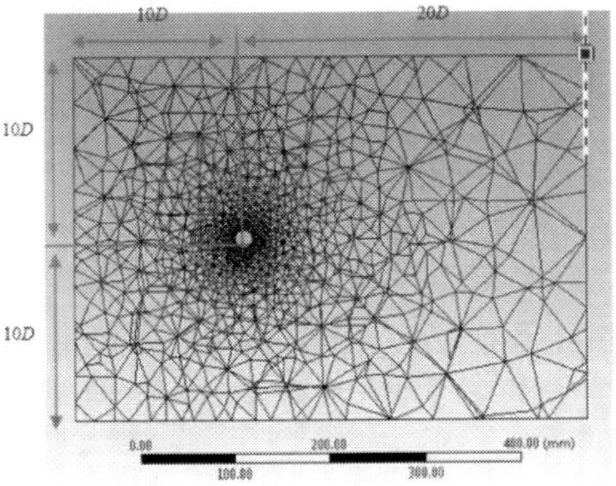

Figure 2. Cylinder mesh in fluid domain.

Table 1. Amplitude response of different mesh size.

Meshing size	Number of elements ($x10^5$)	Amplitude response (m)	Discrepancy of amplitude (standard deviation)
Coarse	11.6	0.0162	13%
Medium	11.7	0.0143	2.5%
Fine	11.8	0.0140	-

Based on table 1, the discrepancy of amplitude in term of standard deviation is relatively minor in overall. The difference between the coarse and fine mesh is around 13%, while only 2.5% of changes in amplitude is found between the medium and fine mesh. This indicates that the mesh sensitivity especially for medium and fine meshing is adequate and independent from the change of mesh size. To minimize the total computational time, medium mesh size is selected for the rest of the simulations. Table 2 shows the dimensions and the summary of computational meshes used for bare cylinder and cylinder fitted with helical strakes. Only strakes with pitch (p) of 10D and height (h) of 0.10D is tested in the study.

4 Result and discussion

The purpose of conducting the simulation is to estimate the ability of the FSI software in predicting the response of a flexible cylinder. Only lock-in region is investigated in the study as it is the most critical region that required more attention. Lock-in is defined as the occurrence of large amplitude oscillation due to the synchronization of structural vibration frequencies and the vortex shedding [16]. The simulation is run for cylinder fitted with and without helical strakes. The existing experimental results of bare cylinder with similar configuration [13] are used as for comparison.

Table 2. Summary of computational meshes used.

Type of cylinder	Length, L (m)	Diameter , D (m)	Number of elements ($x10^5$)
Bare	2.92	0.018	11.7
Helical strakes ($p = 10D$; $h = 0.10D$)	2.92	0.018	13.2

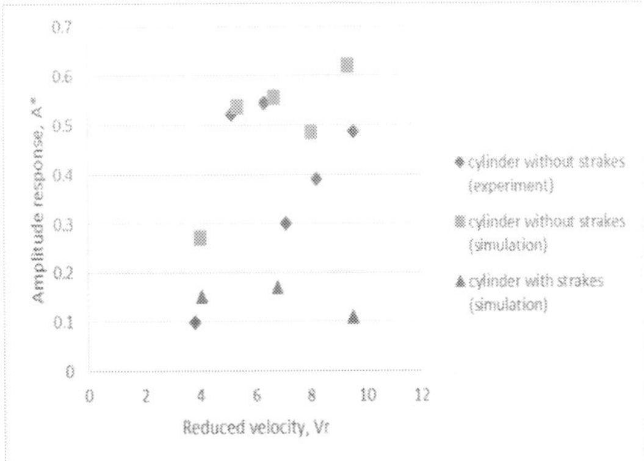

Figure 3. Amplitude response of cylinder fitted with and without helical strakes.

Figure 3 indicates the amplitude response of the simulated and experimental results. Five cases of different reduced velocity are tested for bare cylinder in simulation. The reduced velocity is defined as

$$V_r = \frac{U}{f_n D} \tag{4}$$

where D is the diameter of cylinder, fn is the natural frequency and U as the flow velocity. In overall, the simulated bare cylinder results follow the trend of the experimental results. Nevertheless, the values are slightly overvalued. Another three runs of simulations on cylinder fitted with helical strakes show apparent amplitude reduction from 0.5 in average to less than 0.2. This shows that the application of helical strakes is able to reduce the amplitude and hence the vibration of the cylinder. The VIV suppression efficiency ratio of the helical strakes of the present study is 82.5 percent, which can calculated by:

$$efficiency\ ratio = \frac{(\frac{A}{D})_{max\_bare} - (\frac{A}{D})_{max\_strakes}}{(\frac{A}{D})_{max\_bare}} x100\% \tag{5}$$

where (A/D)max indicates the maximum amplitude ratio. Percentage as high as 93.1% is reported by Boubenider et al. [17]. The output indicates that FSI is able to capture the response of a cylinder fitted with and without helical strakes.

Figure 4 shows the mean drag coefficient of the simulated and experimental results. Similar to the amplitude response, the mean drag coefficient of the simulated bare cylinder is slightly overestimated compare to the experimental output. However, the graph trend is the same, where the mean drag coefficient increases gradually until Vr ≈ 5.5 and then slowly decreases to lower values. The mean drag coefficient of the simulated cylinder fitted with helical strakes, on the other hand, is found to be higher than the bare cylinder. The same condition is reported by Gomez et al. [18] where the installation of helical strakes increases the drag force of a cylinder. The mean drag coefficient of helical strakes remains around value of 3 without influence of the reduced velocity. It is reasonable because the vortex behind the cylinder has been disrupted by the helical strakes, hence the change of the reduced velocity does not create different patterns of vortices that can affect the drag coefficient. The constant values of drag coefficient is also as reported by Boubenider et al. [17] in their study.

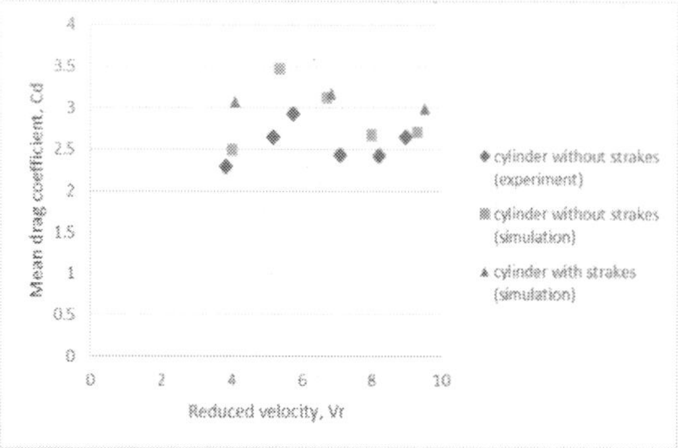

Figure 4. Mean drag coefficient of cylinder fitted with and without helical strakes.

5 Conclusion

The purpose of the study, which is to test the capability of FSI simulation in predicting the VIV characteristic of cylinder fitted with and without helical strakes, is done successfully. Fluid and structural solvers are used as two-way FSI solvers to develop the simulation. In general, minor over-prediction is found when comparing the simulated outputs with the experimental data. However, the graph trends are similar. Significant amplitude reduction and slightly increase of mean drag coefficient are observed after the application of helical strakes on the cylinder. The findings are in agreement with previous research works. Further work should be conducted to improve the over-prediction of the outputs.

Acknowledgements

This research was financially supported by Universiti Teknologi Malaysia under Potential Academic Staff research grant (Q.K130000.2743.02K26).

References

1. Y. Gao, S. Fu, T. Ren, Y. Xiong, L. Song, Appl. Ocean Res. **52**, 102 (2015)
2. S. Huang, A. Sworn 2013. Appl. Ocean Res. **43**, 21 (2013)
3. F.J. Huera-Huarte, Z.A. Bangash, L.M. Gonzalez J. Fluid Struct **48**, 81 (2014)
4. J.R. Chaplin, P.W. Bearman, F.J. Huara Huarte, R.J. Pattenden. J. Fluids Struct. 21, 3 (2005)
5. Z. Khatir, A.K. Pozarlik, R.K. Cooper, J. W. Watterson. Int. J. Numer. Method Fluid, 1 (2006)
6. M. Shahi, J.B.W. Kok, P.R. Alemela. Proceedings ASME Turbo Expo. Copenhagen, Denmark (2012)
7. C.A. Figueroa, I.E. Vignon-Clementel, K.E. Jansen, T.J.R. Hughes, C.A Taylor. Comput. Method Appl. M. **195**, 5685 (2006)
8. J.F. Gerbeau, M. Vidrascu, P. Frey. Comput. Struct. **83**, 155 (2005)
9. J. Pereira Gomes, S. Yigit, H. Lienhart, M. Schafer. J. Fluid Struct **27**, 43 (2011)
10. M.L. Facchinetti, E. Langre, F.. J. Fluid Struct. **19**, 123 (2004)
11. A. Gustafsson. Master thesis. Chalmers University of Technology, Sweden (2012)
12. Y. Constantinides and O.W. Oakley. Proceedings of OMAE, 25th International Conference on Offshore Mechanics and Arctic Engineering, Germany (2006)
13. L.K. Quen, A. Abu, N. Kato. Appl. Ocean Res. **44**, 82 (2014)
14. T. Farrant, M. Tan, W.G. Price. Comput. Fluid **30**, 211 (2001)
15. G. Bosch and W. Rodi. Int. J. for Numer. Method Fluid **28**, 601 (1998)
16. R. Bourguet, G.E. Karniadakis, M.S. Triantafyllou. J. Fluid Struct **27**, 838 (2011)
17. R. Boubenider, K. Alptunaer, P. Fourchy,, J.J. Wilde. Offshore Technology Conference in Houston, Texas (2008)
18. F. Gómez, J.H. Quesada, R. Gómex, V. Theofilis, B. Carmo, J. Meneghini. 43rd AIAA Fluid Dynamics Conference, San Diego, CA. (2013)
19. F.R. Menter,. AIAA J. **32**, 8, 1598 (1994)

DOI: 10.1051/matecconf/20179504002

A Method for Generating Helical Surface of the VPA Hob on a CNC Worm Grinding Machine

Van-The Tran[1, 2], Ruei-Hung Hsu[3], Shih-Sheng Chen[1] and Yu-Ren Wu[1]

[1]Department of Mechanical Engineering, National Central University, 300, Jhongda Rd., Jungli City, Taoyuan 320, Taiwan
[2]Department of Mechanical Engineering, Hung Yen University of Technology and Education, 5 Rd., Hung Yen City, Vietnam
[3]Bachelor's Program in Precision System Design, Feng Chia University, 100 Wenhwa Rd., Seatwen, Taichung 40724, Taiwan

Abstract. Variable pressure angle (VPA) hob is an important cutting tool for generating the twist-free tooth flanks on a crowned involute helical gear used in the high precision gear boxes. However, an appropriate method for generating the VPA hob surface has not proposed yet. Therefore, this paper presents a worm grinding method for generating the helical surface of VPA hob on a CNC worm grinding machine. A mathematical model of the proposed method is established based on a standard hob by controlling rotational motion of the hob axis in the worm grinding process. A numerical example is implemented to verify the merits of the proposed method.

1 Introduction

In batch production of small external gears and consideration of more efficient process, the generating grinding method is usually adopted for the hard finishing of gears instead of the form grinding method. The tooth profile of a cylindrical grinding worm equates a rack profile in the normal section. The involute is generated by the continuous and relative rotation motion between grinding worm and work-piece based on the envelope process. Besides, in the generating grinding process, many points of the grinding worm are always in contact with the gear workpiece and the number of contact points change continuously during the tool rotation. Thus the stock removal rate in the generating grinding of gear is very high. In the past researches, the worm grinding process has been applied for some kind of the gears. Therefore, in this study, we proposed a worm grinding method for grinding the VPA hob on a CNC grinding machine.

The basic geometric design, tooth surface envelope theory, processing method and tool design of various worms are published by Litvin and Fuentes [1]. Also, some related literatures about the worm grinding process are published. Wirz [2]-[4] patented a method, tool and device for profiling the grinding worm that applies in the continuous gear grinding. Besides, a process for dressing a single and multiple-grooved worm by a dressing tool is also proposed. Then a grinding worm that can be profiled on the grinding machine is introduced by Thyssen [5]. Jankowski [6], [7] provides a grinding worm with a geometry and face topology to obtain a high time-cutting volume and a tooth face modifications on the micro-meter level of the work gear. Stadtfeld and Mundt [8] patented a dressing method for the threaded grinding wheel that can obtain the twist-free tooth flanks of gear workpiece in the gear grinding process. The profile and lead modifications of a helical gear is produced by using the threaded wheel and profile grinding [9]. Klocke et al. [10], [11] introduced a research method for analyzing the gear generating grinding process based on the manufacturing simulation as well as an analogy trial and the cutting forces of grinding are also determined depending on the process parameters. Denkena et al. [12] presented an approach for analyzing the tool engagement in gear grinding and calculating the material removal rate. Ophey and Reimann [13] introduced a thermo-mechanical model for describing influence on the surface zone in the gear generating grinding process. Recently, Tran et al. [14], [15] proposed a novel hobbing method to free twist tooth flanks of the longitudinal crowned helical gear by using variable pressure angle hob cutter on CNC hobbing machine. Besides, the topologies, contact bearings and transmission errors of the double-crowned work gear pairs generated by a modified hob in the gear-hobbing process are investigated.

A method for grinding a standard hob as a VPA one is proposed by using the grinding worm on a CNC grinding machine. A mathematical model of grinding process is also established. Numeral results show that the proposed VPA hob surface generated by grinding worm is close to the theoretical VPA hob surface.

2 Generation of theoretical VPA hob surface

Basically, the standard hob's profile is the same as the standard helical gear's profile that can be generated by a standard rack cutter. To obtain a twist-free tooth flank of helical gears, the pressure angle of standard rack cutter's profile is modified as a variable function in its longitudinal direction, as shown in Fig. 1. Accordingly, the position vector and unit normal vector of the rack cutter's right-hand side profile can be expressed in the coordinate system $S_r(x_r, y_r, z_r)$ as follows:

$$\mathbf{r}_r(u_1, v_1) = [x_r, y_r, z_r, 1]^T = [u_1 \cos\alpha_{on}, u_1(-\sin\alpha_{on} + bv_1), v_1, 1]^T, \quad (1)$$

$$\mathbf{n}_r(u_1, v_1) = [n_{x_r}, n_{y_r}, n_{z_r}]^T = [\sin\alpha_{on} - bv_1, \cos\alpha_{on}, -bu_1\cos\alpha_{on}]^T, \quad (2)$$

where u_1 and v_1 are the rack cutter surface parameters, α_{on} is the pressure angle of the rack cutter, and b is VPA coefficient of rack cutter. The dotted profile of the standard rack cutter are determined by setting the coefficient, b, to zero, as shown in Fig. 1.

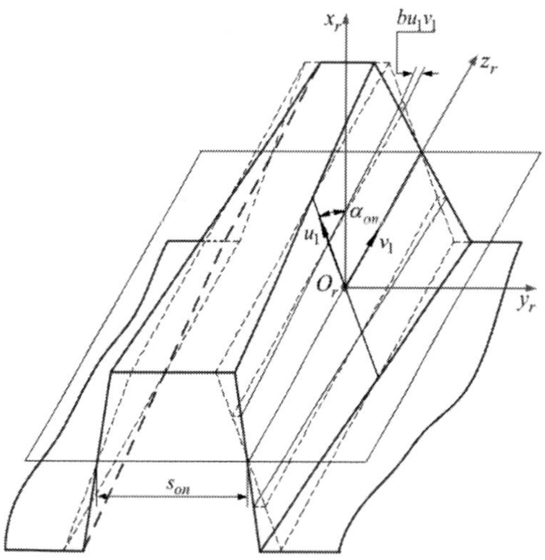

Figure 1. Surface parameters of the VPA rack cutter.

The schematic generation mechanism of the hob is shown in Fig. 2, where coordinate systems $S_r(x_r, y_r, z_r)$, $S_h(x_h, y_h, z_h)$, and $S_4(x_4, y_4, z_4)$ are rigidly connected to the rack cutter, hob, and frame, respectively. By applying the homogenous coordinate transformation matrix equation, the locus and unit normal vectors of the left-hand side rack cutter surface represented in the coordinate system $S_h(x_h, y_h, z_h)$ are attained by

$$\mathbf{r}_h(u_1, v_1, \varphi_1) = \mathbf{M}_{hr}(\varphi_1)\cdot\mathbf{r}_r(u_1, v_1)$$
$$= \begin{bmatrix} \cos\varphi_1 & -\cos\beta_{o1}\sin\varphi_1 & -\sin\beta_{o1}\sin\varphi_1 & r_{o1}(\cos\varphi_1 + \varphi_1\sin\varphi_1) \\ \sin\varphi_1 & \cos\beta_{o1}\cos\varphi_1 & -\sin\beta_{o1}\cos\varphi_1 & r_{o1}(\sin\varphi_1 + \varphi_1\cos\varphi_1) \\ 0 & -\sin\beta_{o1} & \cos\beta_{o1} & 0 \\ 0 & 0 & 0 & 1 \end{bmatrix} \cdot \begin{bmatrix} x_r \\ y_r \\ z_r \\ 1 \end{bmatrix}, \quad (3)$$

$$\mathbf{n}_h(u_1, v_1, \varphi_1) = \mathbf{L}_{hr}(\varphi_1)\cdot\mathbf{n}_r(u_1, v_1), \quad (4)$$

where $\mathbf{L}_{hr}(\varphi_1)$ is the upper-left (3×3) sub-matrix of the (4×4) homogeneous coordinate transformation matrix, $\mathbf{M}_{hr}(\varphi_1)$. And φ_1 is the rotation angle of hob, β_{o1} is the helix angle of rack cutter and r_{o1} is the operating radius of hob.

According to the theory of gearing, the equation of meshing between the right-hand side rack cutter surface and left-hand side hob surface can be obtained by

$$f_1(u_1, v_1, \varphi_1) = \mathbf{n}_h^T \cdot \partial_{\varphi_1}[x_h, y_h, z_h] = 0 \quad (5)$$

Then, the locus of tooth profile and its unit normal of the hob are obtained by simultaneously considering Eqs. (3), (4) and (5).

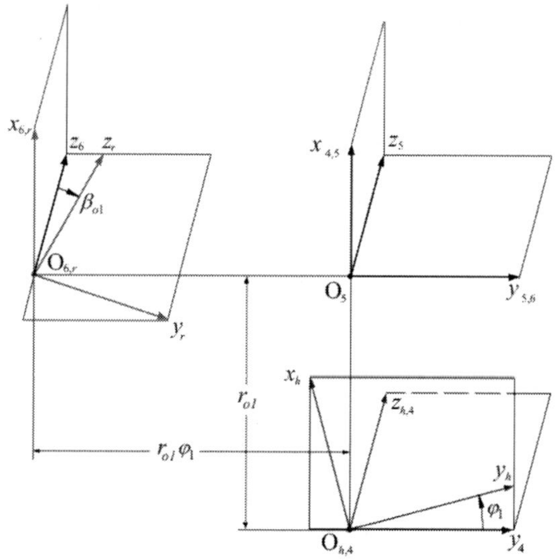

Figure 2. Coordinate systems for generation surface of the VPA hob.

3 General mathematical model for generating the surface of grinding worm on a universal machine

Theoretically, for grinding the surface of VPA hob, a general coordinate systems is established for obtaining a corresponding grinding worm from the VPA hob as shown in Fig. 3. Wherein the coordinate systems $S_w(x_w, y_w, z_w)$ and $S_h(x_h, y_h, z_h)$ are rigidly connected to the grinding worm and hob, respectively, while the coordinate system $S_d(x_d, y_d, z_d)$ is rigidly connected to the frame of grinding machine, and $S_a(x_a, y_a, z_a)$, $S_b(x_b, y_b, z_b)$ and $S_c(x_c, y_c, z_c)$ are auxiliary coordinate systems for simplifying the coordinate transformation. The crossed angle γ_o of the rotaing axes of hob and grinding worm is usually a machine-tool setting. According to Fig. 3, by applying the homogeneous coordinate transformation from S_h to S_w, the locus and unit normal vector of grinding worm surface can be represented in coordinate system S_w as follows:

$$\mathbf{r}_w(u_1,v_1,\varphi_1,\phi_h,l_z)=\mathbf{M}_{wh}(\phi_h,l_z)\cdot\mathbf{r}_h(u_1,v_1,\varphi_1), \qquad (6)$$

$$\mathbf{n}_w(u_1,v_1,\varphi_1,\phi_h,l_z)=\mathbf{L}_{wh}(\phi_h,l_z)\cdot\mathbf{n}_h(u_1,v_1,\varphi_1), \qquad (7)$$

where

$$\mathbf{M}_{wh}(\phi_h,l_z)=\mathbf{M}_{wc}\cdot\mathbf{M}_{ch}(\phi_h,l_z)=$$
$$\begin{bmatrix}-\cos\phi_w & \sin\phi_w\cos\gamma_o & \sin\phi_w\sin\gamma_o & 0\\ -\sin\phi_w & -\cos\gamma_o\cos\phi_w & -\cos\phi_w\sin\gamma_o & 0\\ 0 & -\sin\gamma_o & \cos\gamma_o & 0\\ 0 & 0 & 0 & 1\end{bmatrix}\cdot\begin{bmatrix}\cos\phi_h & -\sin\phi_h & 0 & -E_o\\ \sin\phi_h & \cos\phi_h & 0 & 0\\ 0 & 0 & 1 & -l_z\\ 0 & 0 & 0 & 1\end{bmatrix}. \qquad (8)$$

and the transformation matrix $\mathbf{L}_{wh}$ is the sub-matrix of $\mathbf{M}_{wh}$ by deleting the last column and row, E_o is the center distance between grinding worm and hob, l_z is axis feed of the grinding worm, ϕ_h and ϕ_w are rotation angles of hob and grinding worm, respectively. The relation between rotation angle of the grinding worm, ϕ_w, and hob, ϕ_h, is defined as

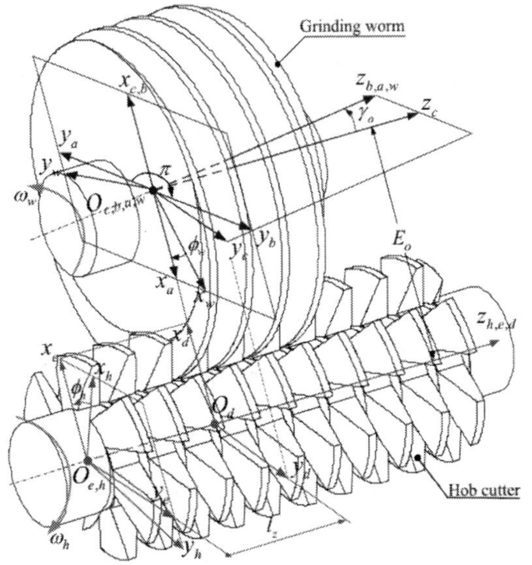

Figure 3. General coordinate systems for grinding the hob cutter with a grinding worm.

$$\phi_w=\frac{N_2}{N_1}\phi_h, \qquad (9)$$

where symbols N_2 and N_1 indicate the number of teeth of the hob and grinding worm, respectively.

Then, the tooth surface and unit normal vector of grinding worm can be determined by using Eqs. (6) and (7) combined with the following equations of meshing. There are two independent kinematic parameters l_z and ϕ_h needed for grinding the helical surface of hob. Therefore, there are two equations of meshing between the hob and the grinding worm as follows:

$$f_2(u_1,v_1,\varphi_1,\phi_h,l_z)=\mathbf{n}_w^{\ T}\cdot\partial_{\varphi_h}\left[x_w,y_w,z_w\right]=0, \qquad (10)$$

$$f_3(u_1,v_1,\varphi_1,\phi_h,l_z)=\mathbf{n}_w^{\ T}\cdot\partial_{l_h}\left[x_w,y_w,z_w\right]=0, \qquad (11)$$

4 Mathematical model for the generation of VPA hob on a CNC worm grinding machine

Fig. 4 presents the 3-D model and seven motion axes of a worm grinding machine. Wherein four axes A_1, X_1, Y_1 and Z_2, are used for setting tool: A_1 is the rotary axis of the worm-head that uses for setting crossed angle between grinding worm and hob, X_1 is the horizontal axis respect to the VPA hob that uses for setting center distance between grinding worm and hob, Y_1 is the vertical axis for setting position of grinding worm and Z_2 for setting position of grinding hob. And three axes B_1, B_2 and Z_1 are are used for generating tooth surfaces of the VPA hob: B_1 is the rotary axis of the grinding worm spindle, B_2 is the rotary axis of workpiece spindle and Z_1 is the tangential axis of griding worm. Fig. 5 shows coordinate systems of the CNC worm grinding machine. By applying the homogeneous coordinate transformation matrix equation, transforming from coordinate system S_1 to S_2, the locus of the grinding worm, $\mathbf{r}_1$, and the unit normal vector, $\mathbf{n}_1$, in the coordinate system S_1, can be represented in coordinate system S_2 as follows:

$$\mathbf{r}_2(C_x,C_y,C_z,\phi_1,\phi_2)=\mathbf{M}_{21}(C_x,C_y,C_z,\phi_1,\phi_2)\cdot\mathbf{r}_1, \qquad (12)$$

$$\mathbf{n}_2(C_x,C_y,C_z,\phi_1,\phi_2)=\mathbf{L}_{21}(C_x,C_y,C_z,\phi_1,\phi_2)\cdot\mathbf{n}_1, \qquad (13)$$

$$\mathbf{M}_{21}(C_x,C_y,C_z,\phi_1,\phi_2)=\mathbf{M}_{2m}(C_x,C_y,\phi_2).\mathbf{M}_{m1}(C_z,\phi_1). \qquad (14)$$

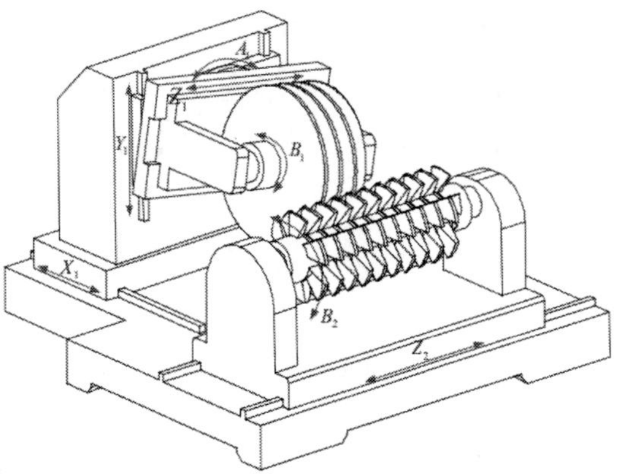

Figure 4. 3-D model of a CNC worm grinding machine.

 DOI: 10.1051/matecconf/20179504002

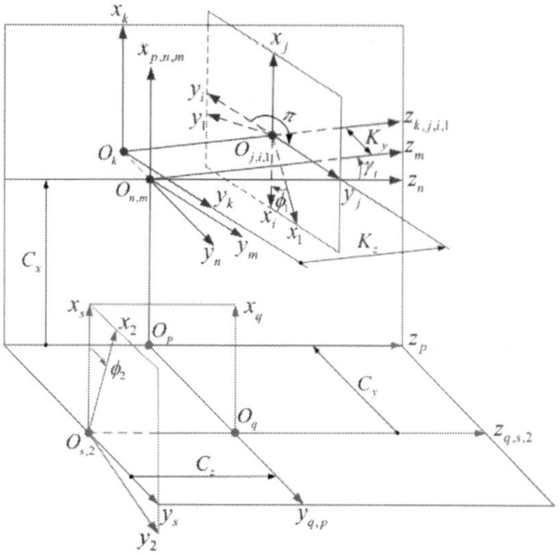

Figure 5. Coordinate system of a CNC worm grinding machine.

where

$$\mathbf{M}_{2m}=\begin{bmatrix}\cos\phi_2 & \sin\phi_2 & 0 & C_x\cos\phi_2-C_y\sin\phi_2\\ -\sin\phi_2 & \cos\phi_2 & 0 & -C_y\cos\phi_2-C_x\sin\phi_2\\ 0 & 0 & 1 & K_z\\ 0 & 0 & 0 & 1\end{bmatrix},$$

$$\mathbf{M}_{m1}=\begin{bmatrix}-\cos\phi_1 & -\sin\phi_1 & 0 & 0\\ \cos\gamma_t\sin\phi_1 & -\cos\gamma_t\cos\phi_1 & -\sin\gamma_t & -K_y\cos\gamma_t-C_z\sin\gamma_t\\ \sin\phi_1\sin\gamma_t & -\cos\phi_1\sin\gamma_t & \cos\gamma_t & C_z\cos\gamma_t-K_y\sin\gamma_t\\ 0 & 0 & 0 & 1\end{bmatrix},$$

and $\mathbf{L}_{21}$ is the upper-left (3×3) sub-matrix of the (4×4) homogeneous coordinate transformation matrix $\mathbf{M}_{21}$.

The spatial relative position of the grinding worm with respect to the VPA hob should be the same whether the VPA hob is ground on an universal or a CNC worm grinding machine. Therefore, the Eq. (8) and Eq. (14) should be satisfied the following relation:

$$\mathbf{M}_{21}(C_x,C_y,C_z,\phi_1,\phi_2)=\mathbf{M}_{hw}(\phi_h,l_z)=\mathbf{M}_{wh}^{-1}(\phi_h,l_z). \quad (15)$$

Assuming that the rotational angle of the grinding worm and hob on the universal machine is the same as on the CNC machine as follows:

$$\begin{cases}\phi_1(\phi_h,l_z)=\dfrac{N_2}{N_1}\phi_h\\ \phi_2(\phi_h,l_z)=\phi_h\\ \gamma_t(\phi_h,l_z)=\gamma_o\end{cases}. \quad (16)$$

Combining the translation vectors in Eq. (17) with the above results yields the following solution for the translational positions:

$$\begin{cases}C_x(\phi_h,l_z)=E_o\\ C_y(\phi_h,l_z)=-K_y\cos\gamma_o-K_z\sin\gamma_o\\ C_z(\phi_h,l_z)=l_z-K_y\sin\gamma_o-K_z\cos\gamma_o\end{cases}. \quad (17)$$

The rotating angle of hob, ϕ_2, is set as a function of grinding worm's rotation angle, ϕ_1, and grinding worm's traverse feed, C_z as follows:

$$\phi_2=\frac{N_1}{N_2}\phi_1+\frac{\cot\lambda_{o2}}{r_{o2}}C_z+a\phi_1C_z. \quad (18)$$

where λ_{o2} is the lead angle of hob, $\lambda_{o2}=\pi/2-\beta_{o2}$ and a is the control coefficient for rotational axis of hob.

5 Numerical example

The purpose of this example is to verify the merits of proposed worm grinding method. The basic data of the hob and the grinding worm are given in Table 1. The setup data of grinding machine is calculated according to the basic meshing conditions as illustrated in Ref. [1].

Table 1. Basic data of the hob and the grinding worm.

Items	Value
Hob data	
Number teeth of hob (N_2)	1
Normal module (m_{pn})	2
Normal circular-tooth thickness of hob (s_{on})	3.14 mm
Normal pressure angle (α_{on})	20.0°
Helix angle of hob (β_{o2})	87.8° R.H.
Lead angle of hob (λ_{o2})	2.2°
Hob operating radius (r_{o2})	26.05 mm
Hob form radius (r_{f2})	24.21 mm
Hob outer radius (r_{a2})	28.05 mm
Hob length (F_{w2})	100 mm
Grinding worm data	
Number of starts of grinding worm (N_1)	1
Number of threads of grinding worm (N_T)	3
Grinding worm operating radius (r_{o1})	41.0 mm
Helix angle of grinding worm (β_{o1})	88.8° R.H.
Lead angle of grinding worm (λ_{o1})	1.2° L.H.
Grinding process data	
Center distance between worm and hob (E_o)	87.05 mm
Crossed angle between worm and hob (γ_o)	1.0°
Control coefficient (a)	-4.5×10^{-5}
VPA coefficient (b)	5.0×10^{-6}
Machine constant (K_y)	5.0 mm
Machine constant (K_z)	15.0 mm

DOI: 10.1051/matecconf/20179504002

The Topography of the theoretical VPA hob surface and proposed VPA hob surface generated by grinding worm are shown in Figs. 6-7, respectively. The maximum deviation of the tooth flank of the proposed VPA hob surface, $M_d = 3.26\mu m$, is so small and it is approach to that of the theoretical VPA hob surface, $M_d = 1.35\mu m$. And the ratio of deviation of the proposed VPA hob surface, $R_d = 0.82$, is also approach to that of the theoretical VPA hob surface, $R_d = 0.91$. It reveals that the proposed VPA hob surface generated by grinding worm fits well with the theoretical VPA hob surface, as shown in Figs. 6-7.

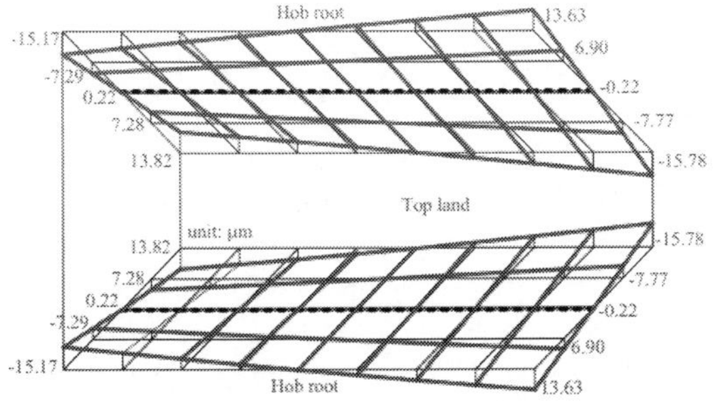

Figure 6. Topography of the theoretical VPA hob surface (maximum deviation of the tooth flank $M_d = 1.35\mu m$ and ratio of deviation $R_d = 0.91$).

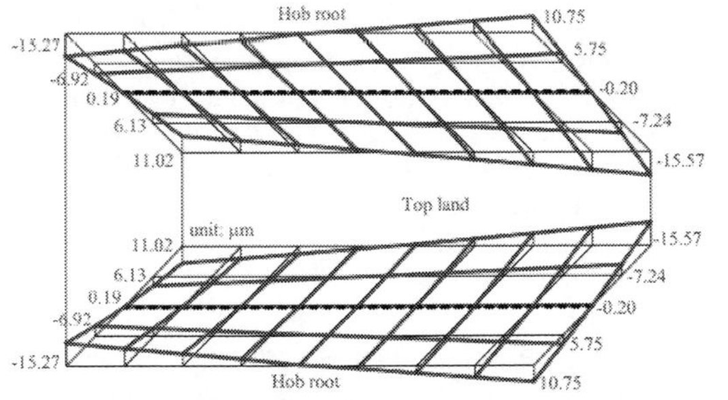

Figure 7. Topography of the proposed VPA hob surface generated by grinding worm (maximum deviation of the tooth flank $M_d = 3.26\mu m$ and ratio of deviation $R_d = 0.82$).

6 Conclusion

In this paper, a new method for grinding the VPA hob with a grinding worm on a CNC machine is proposed by controlling rotational motion of hob axis (Eq. (20)). The mathematical models for generating helical surfaces of the theoretical VPA hob and the proposed VPA hob are established. The numerical result shows that the proposed VPA hob surface generated by the grinding worm can fits well with the theoretical VPA hob surface.

Acknowledgment

The authors are grateful to the Ministry of Science and Technology in Taiwan (R.O.C.) for financial supports from two projects (project no.: MOST 104-2221-E-008-007 and MOST 104-2221-E-035-087).

References

1. F.L. Litvin, A. Fuentes, *Gear Geometry and Applied Theory*, 2nd ed., Cambridge University Press, Cambridge, UK (2004)
2. W. Wirz, Process for machining a worm-type workpiece with a worm-type tool, Patent No. US 4475319 A (1984)
3. W. Wirz, Method, tool and device for the profiling of grinding worms for continuous gear grinding, Patent No. US5954568 A (1999)
4. W. Wirz, Topological profiling of grinding worms for continuous generating grinding of gear teeth, Patent No. US6386953 B1 (2002)
5. W. Thyssen, Grinding worm for the continuous generating grinding of gear wheels, Patent No. US 6257963 B1 (2000)
6. R. Jankowski, Profiling methods for generation of modified grinding worms, Patent No. US 6077150 A (2000)
7. R. Jankowski, Method for flexible profiling of grinding worms, Patent No. US 6012972 A (2000)
8. H.J. Stadtfeld, E.G. Mundt, Threaded grinding wheel and method of dressing, Patent No. US 6217409 B1 (2001)
9. A. Türich, Producing profile and lead modifications in threaded wheel and profile grinding, Gear Technology, **27**(1), 54–62 (2010)
10. F. Klocke, M. Brumm, J. Reimann, Continuous Generating Gear Grinding: A report on new possibilities in the process design and analysis of continuous generating gear grinding, Gear Solutions, 31-44 (2012)
11. F. Klocke, M. Brumm, J. Reimann, Modeling of Surface Zone Influences in Generating Gear Grinding, Procedia CIRP, **8**, 21-26 (2013)
12. B. Denkena, J. Köhler, A. Schindler, S. Woiwode, Continuous Generating Gear Grinding-Material engagement in gear tooth root machining, Mechanism and Machine Theory, **81**, 11–20 (2014)
13. M. Ophey, J. Reimann, Prediction of Surface Zone Changes in Generating Gear Grinding, Gear Technology, 52-59 (2015)
14. V.T. Tran, R.H. Hsu, C.B. Tsay, A Novel Finish Hobbing Methodology for Longitudinal Crowning of a Helical Gear with Twist-Free Tooth Flanks by Using Dual-Lead Hob Cutters, ASME International Mechanical Engineering Congress & Exposition, Montreal, Quebec, Canada. Paper No. IMECE2014-36149, 1-9 (2014)
15. V.T. Tran, R.H. Hsu, C.B. Tsay, Tooth Contact Analysis for a Double-Crowned Involute Helical Gear with Twist-Free Tooth Flanks Generated by Dual-Lead Hob Cutters, ASME Journal of Mechanical Design, **137**, 1–11 (2015)

 DOI: 10.1051/matecconf/20179504003

Fracture Failure Analysis of Fuel Pump Transmission Shaft of Dual-Fuel Engine

Pei-hong Chen, Jun-jie Zou and Li Fan

Marine Engineering Department, Nantong Shipping College, 226010, Nantong Jiangsu, China

Abstract. NTS6Z_LCz-129 dual-fuel turbocharged and intercooled engine durability test at 1000h, fuel pump shaft fractured. Fracture analysis, chemical analysis, microstructure examination and finite element stress analysis were carried out on the fractured shaft. The analysis result showed that the shaft fracture cause is forging fold. By improving the forging process, the forging fold was solved, and the durability test can be carried out smoothly.

1 Introduction

High pressure pump is driven by the fuel pump transmission shaft. NTS6ZLCz-129 dual-fuel turbocharged intercooled engine, in 1000h durability test, fuel pump transmission shaft fractured suddenly. Fuel pump could not supply oil, the engine stopped. The fractured transmission shaft is shown in Fig. 1. The transmission shaft is made of 40Cr, fracture diameter is 30mm, and manufacturing process consists of cutting, forging, roughing, quenching and finishing.

To identify shaft fracture causes, fracture analysis, chemical analysis, microstructure examination and finite element stress analysis were carried out on the fractured shaft.

Figure 1. Fractured fuel pump transmission shaft

2 Inspection and analysis

2.1 Fracture macroscopic analysis

Fracture surface rust obviously, certain areas have bumps. Specimens were cut from fracture surface, fracture surface observed after ultrasonic cleaning with alcohol. Macro picture of fracture is shown in Fig. 2. The fracture surface was rough, around the bumps and final crack area had obvious plastic deformation, the other parts had no significant deformation. Fracture in most regions showed obvious macroscopic fatigue Bay Ridge. According to the shape of the distribution of fatigue Bay ridges, the path of crack extension, the crack source faults and final fracture zone can be determined. The source region of the crack and the final fracture zone are shown in Fig. 2. The crack source area is located in a small longitudinal cross section. This section is close to the keyway site and nearly vertical to step surface and the main fracture. Because of the crack source area had obvious impact on the abrasion, obvious macroscopic defects failed to be observed. The final fracture zone is located on the circumference, opposite to the source region of crack, and has a smaller shear lip. Since the fracture is not well protected, no obvious rapid tear streaks are observed in the final fracture zone except in the cut lip.

Figure 2. Macro picture of fracture

Take the area near the final fracture as the rapid tear streaks area before the final fracture. This area has no obvious macroscopic fatigue Bay ridges. This area is very small, and takes a small proportion of the total area of the fracture. From the above examination and fracture analysis, it can be determined that the fracture of the shaft is caused by low stress and high cycle fatigue in the keyway.

2.2 Metallographic examination

Three metallographic specimens are cut from the fracture surface area, which is vertical to the main fracture surface through the source region of the crack. Etchant is 4% Nital.

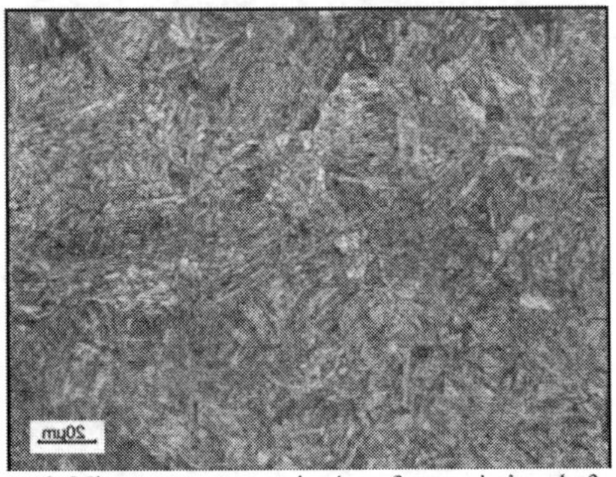

Figure 3. Microstructure organization of transmission shaft (Magnification of 500, 4% Nital)

Fig. 3 illustrates the microstructure of fractured transmission shaft is tempered sorbite. Grain-size number is 8 according to GB / T 6394-2002, and it is the normal quenching and tempering heat treatment state organization.

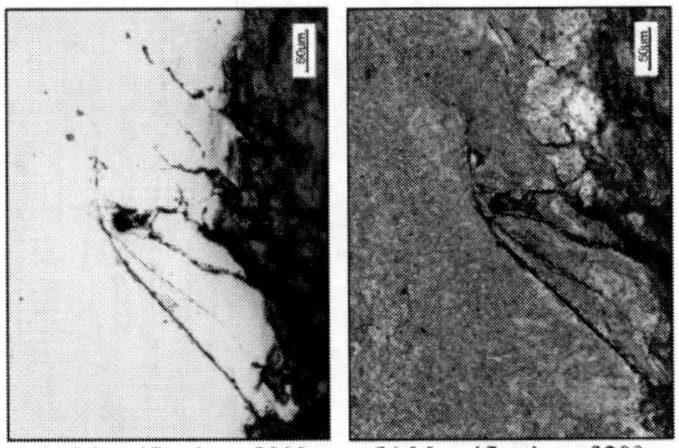

(a) Magnification of 200 No etchant (b) Magnification of 200 4% Nital

Figure 4. Surface defects and microstructure of the fracture surface in the source region of the crack

Fig. 4 illustrates there are cracked defects oblique stretched into internal in the crack source region. Oxide inclusion exists in the fracture surface, inner and surrounding areas of defects. Nearby the fracture surface and cracked defects, there is obvious decarbonization phenomenon. The defect has forging or rolling folding defect characteristics.

Fig. 5 illustrates there are cracked defects oblique and smooth stretched into internal. Oxide inclusion exists in the fracture surface and inner area of defects. There are continuous tiny dots and granular inclusions on both sides near the crack. The defect has forging or rolling folding defect characteristics.

Fig. 6 is a partially enlarged view of Fig. 5. Fig. 6 illustrates oxide inclusion exists in the fracture surface and inner area of defects. Continuous tiny dots and granular inclusions on both sides near the crack are clearly visible.

Figure 5. Surface defects of the fracture surface in the source region of the crack (Magnification of 500, no etchant)

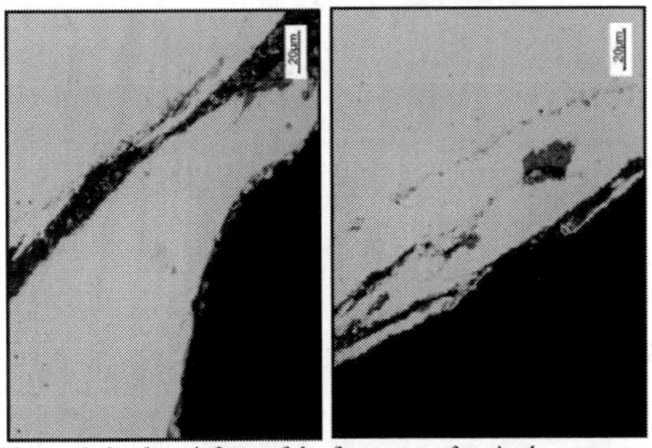

Figure 6. Surface defects of the fracture surface in the source region of the crack (Magnification of 500, no etchant)

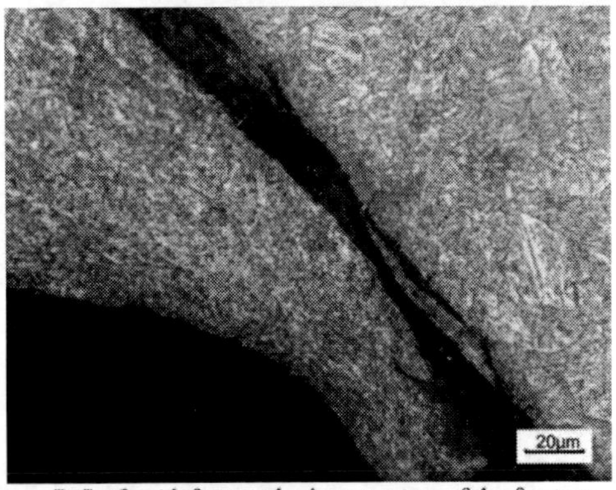

Figure 7. Surface defects and microstructure of the fracture surface in the source region of the crack (Magnification of 500, 4% Nital)

 DOI: 10.1051/matecconf/20179504003

The microstructure in Fig. 7 obtained after etching in 4% Nital. There are decarbonization in the surface and both sides of the crack.

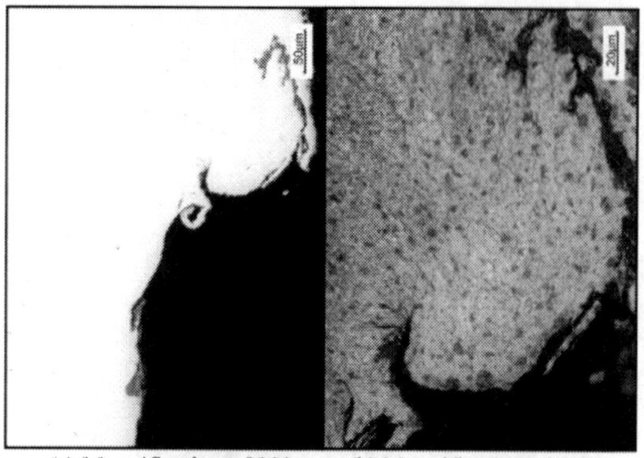

(a) Magnification of 200 (b) Magnification of 500
No etchant 4% Nital

Figure 8. Surface defects and microstructure of the fracture surface in the source region of the crack

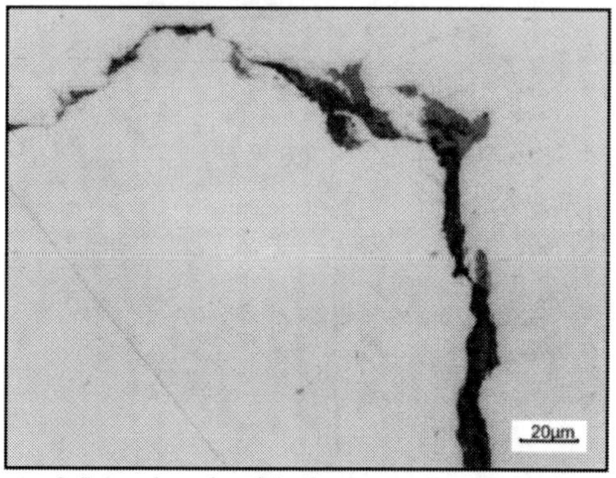

Figure 9. Internal cracks of the fracture surface nearby the source region of the crack (Magnification of 50, no etchant)

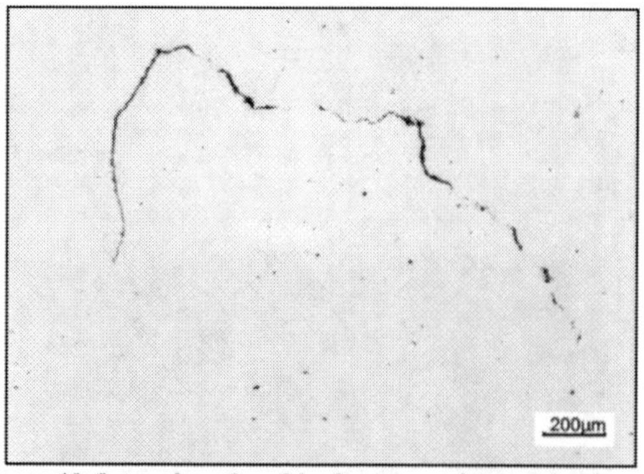

Figure 10. Internal cracks of the fracture surface nearby the source region of the crack (Magnification of 50, no etchant)

Fig. 8 illustrates there are curling defects and oxides in the crack source region. Obvious decarbonization exists in the surface organization, thus shows that this defect has forged or rolled folding defect characteristics.

Fig. 9 illustrates there are internal cracks nearby the source region of the crack. The crack snaked stretches and has a tapering tail. There are oxide inclusions inside the crack.

Fig. 10 is a partially enlarged view of Figure. Internal crack is filled with oxide, it can be seen clearly.

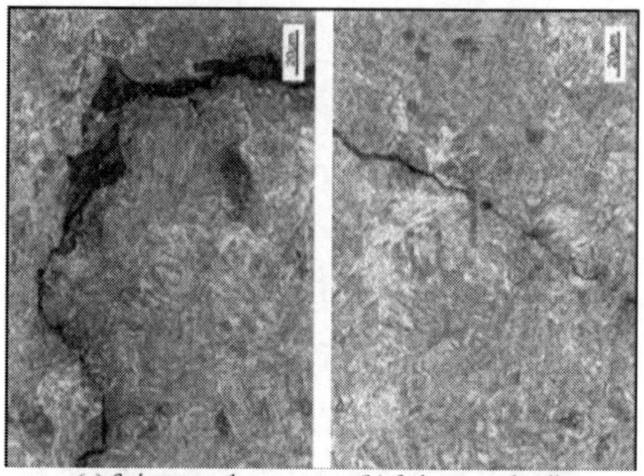

(a) fatigue crack (b) fatigue crack tail

Figure 11. Internal cracks and microstructure of the fracture surface nearby the source region of the crack (Magnification of 500, 4% Nital)

Fig. 11(a) is microstructure of fatigue crack, and Fig. 11(b) is the microstructure of fatigue crack tail. Except the wider parts of the crack, the crack is mainly dominated by the extension of transgranular crack. This crack may be an extension of the original crack in heat treatment, and has a tapering tail. The crack should be fatigue extension during service of the transmission shaft.

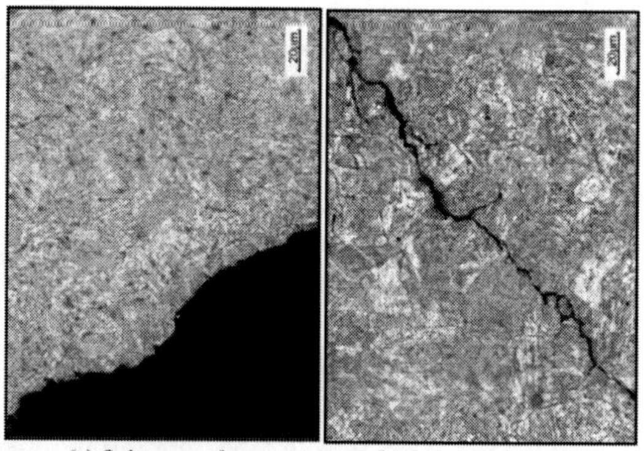

(a) fatigue crack (b) fatigue crack tail

Figure 12. Fatigue Crack Growth Area and organization (Magnification of 500, 4% Nital)

Fig. 12 (a) is microstructure of fatigue crack, and Fig. 12 (b) is the morphology and microstructure of fatigue crack tail. Fig. 12 illustrates the fatigue crack is mainly transgranular cracks. The microstructure is tempered sorbite, which belongs to the normal quenching and tempering heat treatment microstructure.

2.3 Analysis of the chemical composition

Specimens were cut from fracture surface, and SPECTRO MAXx optical direct-reading spectrometer was used to analysis chemical composition, as observed in Table 1. It was shown that chemical composition of the transmission shaft met the requirements of GB/T 3077-1999 for 40Cr.

Table 1. Chemical composition test results (%)

Item	Measured value	Technical specifications project
C	0.40	0.37~0.44
Si	0.24	0.17~0.37
Mn	0.60	0.50~0.80
Cr	0.99	0.80~1.10
P	0.009	≤0.035
S	0.008	≤0.035

3 Finite element analysis

The CREO 3D software of PTC Company can be used to test, study and improve the product design in the virtual condition. [1] In this study, CREO are used for strength check of fuel pump transmission shaft.

3.1 Boundary conditions and loads

With any FEA package, accurate loadestimation depends on the quality of the modelbuilt by the analyst. [2] Using CERO, transmission shaft 3D model containing axial keyway and shoulder were built as observed in Fig. 13. Canters of the two key-ways are used as the two fixed ends. Transfer torque is 380N.m, the diameter is 30mm and the material is 40Cr.

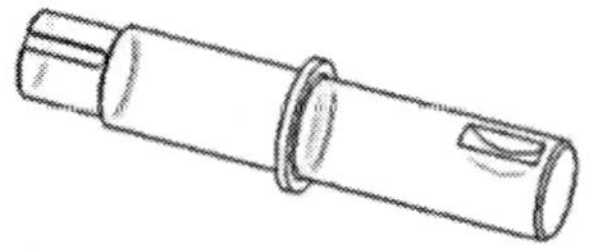

Figure 13. Three-dimensional model of transmission shaft

3.2 Static analysis

The static analysis is mainly used to analyze the maximum stress of the transmission shaft. [3] Fig. 14 shows the shear stress distribution and the maximum stress of transmission shaft. The maximum stress is 198.1MPa at the flat key, and the allowable shearing stress for 40Cr is 227 MPa. Obviously, the maximum stress is less than the allowable shearing stress. Thus the strength of the fuel pump transmission shaft meets the requirements.

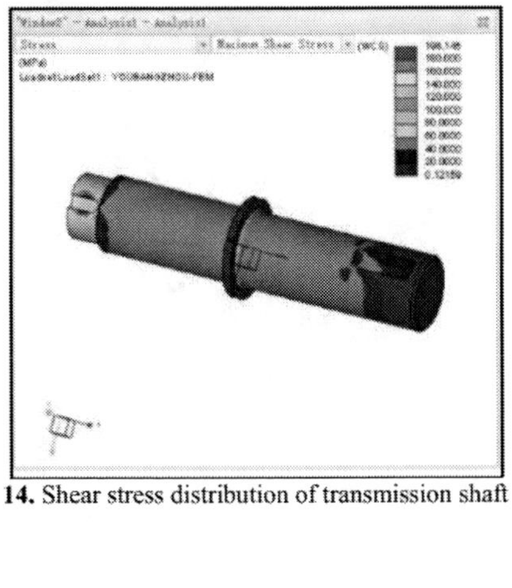

Figure 14. Shear stress distribution of transmission shaft

4 Results and improvement

Fracture analysis, chemical analysis, microstructure examination and finite element stress analysis were carried out on the fractured shaft. The analysis results show that the microstructure of transmission shaft is the normal quenching and tempering heat treatment state organization and the chemical composition of the shaft meets the requirements of GB/T 3077-1999 for 40Cr. In the crack source region, there is a tail round blunt crack type defect. This defect is at a certain angle with the external surface of parts and stretches into internal. Internal crack is filled with oxide inclusions. And there are obvious decarbonization oxidation and in the surface and both sides of the crack. This type defect is a typical forging or rolling folding defect. Shaft fracture is a type of low stress and high cycle fatigue fracture which is caused by a forging or folding defect on the outer circle near keyway. [4]The fracture mechanism is that there is a forging defect on the outer circle near keyway under long-term alternative working stress. The folding defect tip expands and leads to fatigue crack. With the increase of the number of stress, the crack continues to expand, when it expands to critical dimension, the actual bearing area is reduced. The remaining cross-sectional area cannot withstand the load, eventually led to the shaft fracture immediately.

The main causes of forging and folding of shaft parts are as follows:

(1) Due to the difference in the diameter of the metal at the step, the metal flow is poor. The forgings rotate during forging process, most metals flow along the tooling thus form a step, the rest of the metals are bent and pasted on the surface of forging because of the thin thickness. Thus they are pressed into the metal blank and formed folding in the subsequent forging forming process.

(2) The existence of the chamfer can reduce the stress concentration and make the flow smooth. However, the chamfer will produce a step gap and lead to the formation of folding easily.

(3) During the end stretching process, when the feed amount is very small and the press-fitting is large, the local deformation of the upper and lower ends of the metal is easy to form a folding.

According to the fuel pump shaft fracture position, it is preliminarily concluded that the shaft end stretching and small amount of feed lead to the folding defects. [5] Then, shaft forging process checked, and feed amount found indeed too small when stretching the shaft end. The pushing amount is too large, feed and unilateral compression ratio L / (△h / 2) is less than 1, thus leads to a large metal deformation on surface layer and the end concave into a fold. Increase the amount of feed, when the feed unilateral compression ratio reached 1.2, the transmission shaft was reprocessed. By ultrasonic inspection, no folding defects were found. The shaft was installed and the durability test was carried out, then the fuel pump transmission shaft worked properly.

5 Conclusions

DOI: 10.1051/matecconf/20179504003

The main conclusions can be drawn as follows:

(1) The finite element analysis shows that the strength of the fuel pump transmission shaft meets the requirements.

(2) Chemical analysis shows that the chemical composition of the shaft meets the requirements of GB/T 3077-1999 for 40Cr.

(3) Metallographic analysis shows that there are cracked defects oblique stretched into internal in the crack source region. Internal crack is filled with oxide inclusions. And there are obvious decarbonization oxidation and in the surface and both sides of the crack. This type defect is a typical forging or rolling folding defect.

(4) Forging process analysis shows that inappropriate parameters of forging process lead to folded forging defects on the outer circle near Keyway. Thus the fuel pump transmission shaft occurs low stress and high cycle fatigue fracture. By changing forging parameters, the problem of fractured shaft was effectively solved.

Acknowledgment

This work is supported by Nantong Shipping College school level key project "Fishing LNG / diesel dual fuel engine NOX Emissions Research" (project number: HYKJ2015 / A02).

References

1. H.R. Sheng, Jixie Gongchengshi/Mechanical Engineers **2**, 144 (2015)
2. C.S. Baviskar, R. M. Tayade, G. Vinay, Int. J. Mech. Eng. & Rob. Res **1**(3), 22 (2012)
3. M. Shafaee, E. Abbas, F. Hamed, B. Jebraiel, Int. J. Mech. Eng. & Rob. Res **5**(2), 133 (2016)
4. Q.R. Xiang, Keji xiangdao/Technology Guide **2**, 51 (2011)
5. J.C. Huang, Qiye Jishu Kaifa /Enterprise Technology Development **32**, 104 (2014)

Concept Design of Movable Beam of Hydraulic Press

Li Yancong[1] Wan Shengmou[2] Liu Yule[1] Song Xin[1]

[1]Tianjin Agricultural University, 300384, Tianjin China,
[2] China Lianhe Credit Rating Co., 100101, Beijing China

Abstract. The hydraulic press movable beam is one of the key components of the hydraulic press; its design quality impacts the accuracy of the workpiece that the press suppressed. In this paper, first, with maximum deflection and material strength as constraints, mechanical model of the movable beam is established; next, the concept design model of the moveable beam structure is established; the relationship among the force of the side cylinder, the thickness of the inclined plate, outer plate is established also. Taking movable beam of the 100MN type THP10-10000 isothermal forging hydraulic press as an example, the conceptual design result is given. This concept design method mentoned in the paper has general meaning and can apply to other similar product design.

1 Introduction

Conceptual design situated at the beginning of product design, is most important stage in the entire design process. A good concept design result can improve the quality of follow-up technical design, detailed design, reasonable technology and processing, as well as reduced the cost [1,2,3] fundamentally. Conceptual design stage is the most creative stage in the process of product design also. At present, in traditional hydraulic machine design process, case based reasoning method as conceptual design solution is commonly used mostly, which exists problem of hydraulic press designer rely on the personal experience mostly and design results can not actively and accurately controlled. How to design from the product function demands and how to solve the mapping mechanism from the demand of the product function to the structure are the difficulty and direction in the research of the hydraulic machine's conceptual design. Therefore, this paper puts forward a hydraulic press movable beam concept design method [4]. Hydraulic pressure has four main parts: upper beam, lower beam, upright column and movable beam. So the movable beam is key component of pressure. The movable beam situated at the nearest place to the workpiece, so its design quality direct influence the precision of workpiece. In this article, a new concept design method is put forwarded out, and apply this method to the design of movable beam. The concept design method has universal significance in mechanical design, and can be used in other similar mechanical part's concept design.

2 Conceptual design of movable beam

2.1 Mechanical model of movable beam

Mechanical model of the hydraulic press components adopted the following basic assumptions: continuous uniform hypothesis ; isotropic assumption; small deformation hypothesis ; deform of the geometry and size of the object is very small compared with the original size; the moving beams of the hydraulic press is regarded as beam.

There are many structural type of movable beam,while Fig. 1 is a conventional movable beam drawing[5]. Its outer contour profile is very big,which made its weight is heavy. In order to reduce weight, it is made with box shape. stiffened plate is added in it ; at load bearing place, the stiffened plate arranged intense so as to increase ragid and reduce stress, at other place, the stiffened plate arranged sparse. Rational arrengement of stiffened plate made the movable beam weight reducd, at the same time made the movable beam has sufficient strength and uniform stiffness. The movable beam has three contour profile dimention, its length dimention depend on work cylinders arrengement and the center distance of the columns, narrow side size usually is smaller than length side, and high dimention diffinition depend on the strength of material.

In the ideal condition, the beam has no constraint in the right and left, front and back direction; at upper side, the beam contacts with pistons, bearing the load from the piston; at lower side, the beam contacts with mould, bearing the load from the mold ; therefore the movable beam subjected to bending, shear and compression, and then produced the corresponding deformations.

DOI: 10.1051/matecconf/20179504004

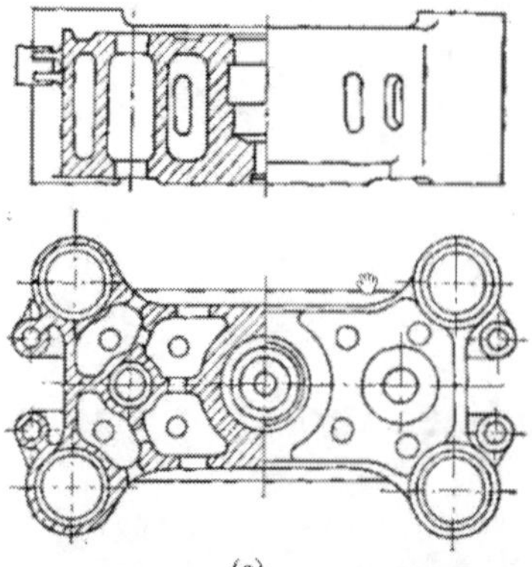

(a)

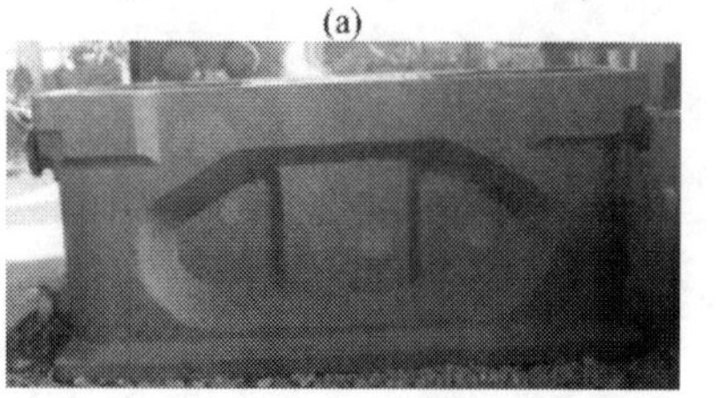

(b)

Figure 1. Movable beam

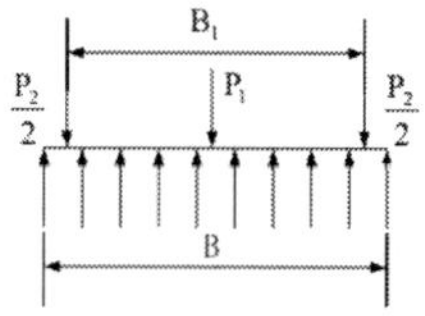

(a)

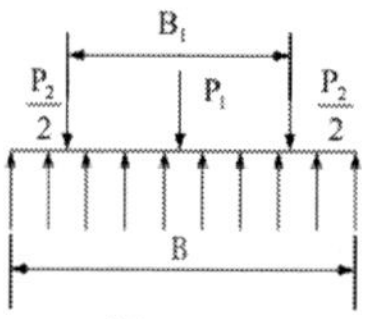

(b)

Figure 2. Force analysis of movable beam

For single cylinder hydraulic press, usually check the extrusion stress on the bearing surface when designing. In this paper, taking three cylinder hydraulic press as an example. According to the relationship between the size of the movable beam and the mould size, the force of the movable beam has two forms. One type is the size of the movable beam and mould nearly equal, which makes area of force exert on the movable beam from the mould and piston very closer, then the movable beam mainly bear extrusion force, consequently occurs compressive deformation; at this situation, bending deformation by the action of bending moment is not considered; From Fig..1, we extract force model using material mechanics method. force exert on the movable slide is shown in Figure 2(a), considering the mold stiffness is much smaller than the movable beam, lower side of the movable beam can be considered as bearing uniform load; considering horizontal plate thick is much smaller than the hight of vertical plate, so we ignore the effect of horizontal plate, just take into account the vertical plate. The vertical plate of movable beam meet the extrusion conditions as

$$\frac{P_1 + P_2}{A_{sb\min}} \leq [\sigma_{sb}] \qquad (1)$$

Where, P_1—the master cylinder pressure, N;

P_2—Side cylinder pressure, N;

$[\sigma_{sb}]$—allowable extrusion stress, MPa.

Another situation is the range of mould exert force on the movable beam is not the same with the area of force that the plunger on movable range, and a lot of difference existed; considering mold stiffness much less than the movable beam, and movable beam lower side subjected to uniform load, beam will occur bend deformation, and then the movable beam bearing bending moment. the movable beam bearing force is shown in Figure 2(b).

As for the Figure 2(b) shown, the maximum bending moment occure in the middle of beam. Through a set of calculation, the middle section bending moment is as following

$$M = \frac{1}{8}P_1 B + \frac{1}{8}P_2(B - 2B_1)$$

The maximum bending deflection in the middle is

$$f_{sb} = \frac{3P_1B^3}{384EJ} + \frac{3P_2}{384EJ}\left[3B^2 - 4B_1^2(3B - B_1)\right] \qquad (2)$$

Where, f_{sb} ——Center deformation of the movable beam, mm;

B—— center distance of both sides limit block, mm;

B_1——center distance of two side cylinder, mm.

the condition of shear stress strength must meet the requirement : $\tau = \frac{1.5Q}{A} \leq [\tau]$. While stress $Q = P/2$, Change the above formula, the shear area isas following :

$$A \geq \frac{1.5P}{2[\tau]} \qquad (3)$$

Where, $[\tau]$——the allowable shear stress, Mpa ;

P——the cylinder pressure, N.

According to the formula of the tensile strength, the moment of inertia of the cross section is satisfied.

$$I_Z \geq \frac{\left[\frac{1}{8}P_1B + \frac{1}{8}P_2(B - 2B_1)\right]y}{[\sigma]} \qquad (4)$$

Where, $[\sigma]$——the allowable bending stress, MPa;

y——The distance from the neutral axis of the cross section.

From Formula(1), we can see that when the nominal load is determined, movable beam structure must enture that the movable beam deflection does not exceed the maximum allowable value; at the same time, when designing the movable beam section, the cross-sectional area and moment of inertia must also satisfy requirement of Formula (2) and Formula (3).

2.2 Conceptual design model of movable beam

Apply the above concept design method to the design of movable beam. the movable beam with cylinders as an example; its stress type is shown in Figure 2(a). Typical structure with cylinder block is shown in Figure 3. The master cylinder seated in the center, the side cylinders distributed separately around it. Refering to the oil cylinder design method, the design of the main cylinder is given, while design the thickness of the beam vertical platea based on the side cylinder pressure. According to Figure 2, we considered that the four piston apply totally forces on the corresponding four vertical plates, and the contact length is the same as the four piston diameter. Considering the material extrusion stress, the relationship between the force of the side cylinder and the thickness of the vertical plate is established as following:

$$t \geq \frac{P_2}{8d[\sigma_{sb}]} \tag{4}$$

where, d —the piston diameter of the side cylinder, mm;
$[\sigma_{sb}]$—material extrusion stress, MPa;
t —the thickness of vertical plate, mm.

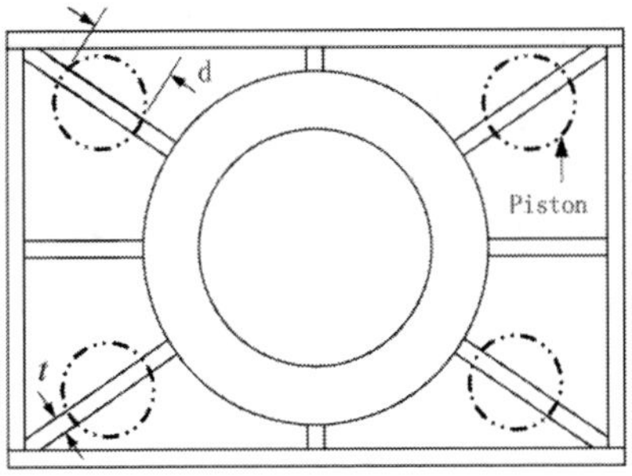

Figure 3. Typical structure of movable beam with cylinder block

Taking movable beam on the THP10-10000 type 100MN isothermal forging hydraulic press as an example, the length of movable beam is 3180mm, the width is 3500mm ; the master cylinder is piston driven, which provide the maximum load of 60Mn; the four other auxiliary hydraulic cylinder are piston driven, whicn providing the maximum load of 40Mn totally. The inner diameter of the master cylinder is 1060mm, and the diameter of the auxiliary cylinder is 795mm. Define the inner and around virtical plate are the same thickness. For material of Q235 steel, the allowable extrusion stress is $[\sigma_{sb}]=175\text{MPa}$. The above value is brought into the Formula (4), can obtaine the following result :

$$t \geq 80\text{mm}$$

From the material saving and the weight reducing considering, the minimum value is selected. So we determined the thickness of theinner and around vertical plate is 80mm.

Apply the above concept design method to the struture design of movable beam(Figure 2(b)). At this type, deformation and stress must be calculated considered, and satify the deformation and stress requirement simultimosly. design steps is the same as the tredition beam design method, so there is not described detailed there.

2.3 Prototype design of movable beam

By establishing the mechanical model of movable beam, the moment of inertia of the component section is calculated, and the initial size of the equivalent section is determined. On the basis of conceptual design prototype, the prototype design of the hydraulic press's movable beam can be further carried out. As to the conceptual prototype design, the structural prototype design needs to be designed according to the concept prototype; this phrase usually needs to combine with the previous case experience to complete.

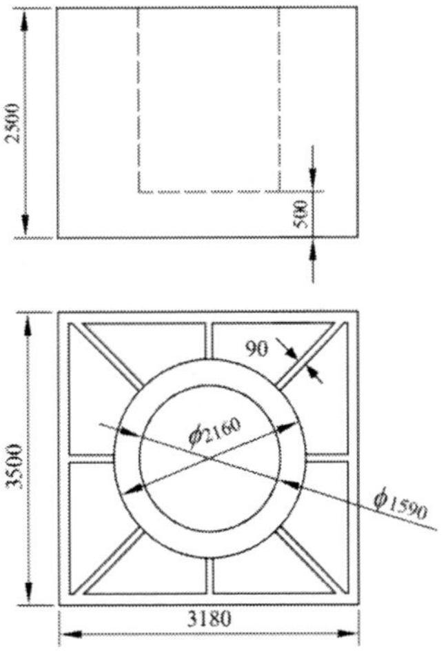

Figure 4. The prototype of the active beam structure

In the concept design phase, hydraulic pressure's movable beam geometry information is incomplete, so when establishing the preliminary structural design of the prototype, a certain amount of geometric information must added to prototype [6], such as: distance between longitudinal and transverse ribs, rib plate size and so on. Taking the movable beam on the 100MN type THP10-10000 isothermal forging hydraulic press as an example, taking into account the above fact, the structure prototype of the movable beam of the hydraulic press is shown in Fig. 4. This design result is the initial scheme comparing with the optimization design of the movable beam of the hydraulic press. According to this, the structure design, detailed design and optimization design can be carried out successively.

The structure design includes function and structure mapping, preliminary structure design, scheme layout. Detailed design is the structure of the specific products, process details. When conducting constrained structure optimization, the unimportant detail structure needs to ignore. Based on this, movable beam initial structure is shown on Fig. 5.

2.4 Case study

In Fig. 5, the size of the middle cylinder can be calculated according to the strength of the cylinders, not as a variable parameter. Besides that, there are still much variable structures, which have great influence on the movable beam strength.

Combined with the specific structure of the movable beam, the thickness of the outer plate of the moving beam and the thickness of the inclined plate thickness of the moving beam are selected as the control variables. Determine the key variables that affecting the performance of the active beam structure is the thickness of the outer plate x_1 and the thickness of the inclined plate x_2.

By calculating the volume of moving beam, the mass of moving beam is described as

$$m_{mb} = (1755636648 + 7567400x_1 + 2503642x_2) \times 7.8 \times 10^{-9} \quad (5)$$

where, x_1 —the outer plate, mm;

x_2 —inclined plate, mm;

m_{mb} —the mass of moving beam, kg.

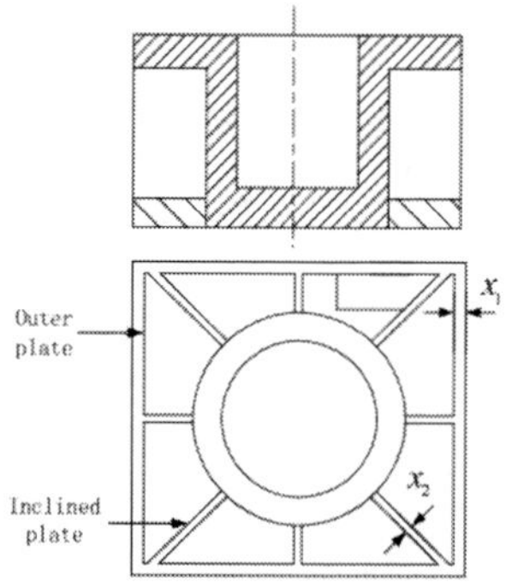

Figure 5. Movable beam parameters

Tradition methods of design and calculation of press are based on old experience, according to the original condition and data, or referring to the same type of press[7,8]. When conducting design, some parameters are chosen as the initial conditions, and give the design result. Next, the strength and stiffness of the moving beam are checked.

The design result using this method is only one feasible plan, not an optimal result. So there using the optimization design method to conduct the active beam design. There must add other equation: mass equation of upper beam, lower beam and column, deformation equation of upper beam, lower beam and column, the range of each variable, constraint conditions of deformation. The minimum mass and minimum deformation as the optimization goal, the structure optimization of the hydraulic press is conducted. Obtained that the outer plate thickness x_1 is 95 mm, the thickness of the inclined plate x_2 is 85 mm.

3 Conclusions

Taken material maximum allowable stress as constraint, constructed the movable beam mechanical model, established relational formula between sectional area and moment of inertia. The relationship between the force of the side cylinder and the thickness of the inclined plate and outer plate is established, and then the concept design model of the movable beam is obtained. The conceptual design method is used to the 100MN type THP10-10000 isothermal forging hydraulic press, and the concept design results of moving beam is obtained. This concept design method has general meaning and can apply to other similar product design.

References

1. Deng Jiat. Product concept design [M]. Beijing: China Machine Press, 2002. (in chinese)
2. Morgan Swink, Technological Innovativeness as a Moderator of new Product Design Integration and Top Mangement Support ,J PROD INNOV MANAG 2000,17: 208~220
3. Tang Lin. Basic principles and methods of product conceptual design[M]. Beijing: National Defence Industry Press, 2006. (in chinese)
4. li Yancong, Zhang Lianhong, Liu Zhanwen, etc. Multi-objective Optimization Design of Heavy Forge Framed Hydraulic Press Driven by Stiffness and Mass[J]. Journal of Mechanical Engineering, 2010, 46(l):140-146. (in chinese)
5. Yu Xinglu. design and application of hydraulic press[M] . Beijing : China Machine Press, 2007. (in chinese)
6. Chen Yongliang, Geng Wenxuan, Man Jia, et c. Adaptable design of machine tool structure based on quantification of structural configuration and performance improvement [J]. China Mechanical Engineering, 2009, 20 (9) :1029 - 1032. (in chinese)
7. Oblak M, Harl B, Butinar B. Optimal design of hydraulic support. Structural and Multidisciplinary Optimization, 2000, 10: 76~82.
8. He Baiyan, Zhang Lianhong, Wang Shuxin, etc. Finite element analysis on the main frame and component of model THP37-150A hydraulic press. Journal of basic science and engineering, 2009, 17(3):438-444. (in chinese)

MATEC Web of Conferences 95, 04005 (2017) DOI: 10.1051/matecconf/20179504005
ICMME 2016

Study on Installation Angle of the Milling Wheel Accurate Estimation and Compensation

ZongXuemei[1], Guo Qiang[2], Kang Kaixuan[1], JiaHaibo[1] and He bing[1]

[1]Jiangsu Xuzhou Construction Machinery Research Institute, Xuzhou 221004, Jiangsu, China
[2]School of Mechanical and Power Engineering, Henan Polytechnic University, Jiaozuo, 454000, China

Abstract: An accurate estimation and compensation algorithm is proposed to calculate the manufacture deviation of the milling wheel based on optical measurement. Firstly, the spatial position of the cutting tool tip and the installation angle of the cutting tool are calculated based on measured data from hand held coordinate measuring instrument. Secondly, the installation angle field between the measurement model and CAD modelis calculated and analyzed by statistics methods. Thirdly, the deviation compensation strategy is given to correct the manufacturing process base on the results of the analysis. The feasibility of the proposed method has been validated by an illustrated experiment at last. The comparison results show that the deviation of the installation angle can be effectively constrained and the proposed method is helpful for the improvement of quality and accuracy in milling wheel.

1. Introduce

As the key part of the milling machine, the milling wheel plays an important role in the working performance. The road material is cut down in turn by the cutting tool, and concentrated in the middle of the wheel by the spiral blades which are arranged in the milling wheel. The road material is transferred into belt conveyor by cast material plate. The quality of the milling wheel greatly affects the service life and construction effect of the milling machine. During the manufacture, whether or not the installation angle reaches the designed accuracy is a problem[1, 2]. The tool apron is fixed on the milling drum by welding. Because of the deviation of the welding fixture, welding heat deformation, welding surrounding and welding process, there is a certain installation angle deviation between the design and manufacture.

With the high speed development of the optical technology, the optical measurement technology is widely used to measure the geometric deviation [1,2]. The traditional method to measure the welding heat deformation is mainly applied to regulation components and parts [3, 4]. The optical measurement that is applied to irregular components and parts is rare. For that reason, an accurate estimation and compensation algorithm is proposed to calculate the manufacture deviation of the milling wheel based on optical measurement in this paper. Firstly, the spatial position of the cutting tool tip and the installation angle of the cutting tool are calculated based on measured data from handheld coordinate measuring instrument. Secondly, the installation angle field between the measurement model and CAD model is calculated and analyzed by statistics methods. Thirdly, the deviation compensation strategy is given to correct the manufacturing process base on the results of the analysis. The method has realized the accurate estimation of each tool deformation of the milling wheel, and it can offer correct guidance for the amendment of the design and manufacture.

2. Installation angle analysis

A reasonable installation angle has important implication for the service life and construction effect and surrounding. An unreasonable installation angle may lead to abnormal wear of the cutting tool, vibration of the machine and safety of the workers[5, 6].

The spatial position of the cutting tool is determined jointly by the spatial position of the cutting tool tip and the installation angle of the cutting tool. As shown in the Fig.1, Cartesian coordinate system **OXYZ** set on the end surface of the milling drum. The rotational axis of the milling drum is represented by a vector $\boldsymbol{Z}(u)$. The cutting tool coordinate is defined as the part coordinate system $\mathbf{O_1X_1Y_1Z_1}$ that is obtained by translating the **OXYZ** a distance along the **Z** axis and a radius of the cutting tool tip circle along the **X** axis and rotating the **OXYZ** a circumferential angle θ around the **Z** axis.

The installation angle consists of the cutting angle φ and roll angle δ, as show in the Fig.1, the cutting angle φ is defined as the included angle between the rotational axis $\boldsymbol{H}(u)$ of the cutting tool and the center line $\boldsymbol{M}(u)$ of the cutting tip circle, which is the axis $\mathbf{X_1}$ in the part coordinate system. The roll angle δ is defined as the included angle between the rotational axis of the cutting tool and the forward direction of the cutting tool, and is also described as the included angle between the end face OXY of the milling drum and the roll face ***P***(u) which is determined by the rotational axis ***H***(u)of the cutting tool and the axis $\mathbf{X_1}$.

Define:

- α, β and γ are the included angle between the rotational axis of the cutting tool and axis X, Y and Z in coordinate system OXYZ, respectively.
- α_1, β_1 and γ_1 are the included angle between the rotational axis of the cutting tool and axis X_1, Y_1 and Z_1 in the part coordinate system $\mathbf{O_1X_1Y_1Z_1}$,respectively.

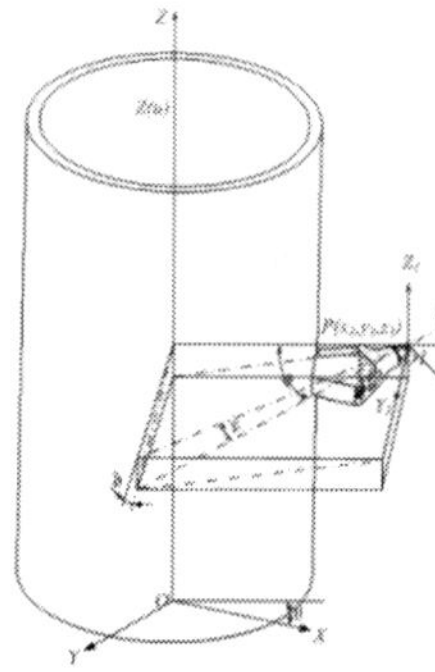

Figure 1. The installation angle model

As shown in the Fig.1, the α_1, $\beta_1$andγ_1 can be given as follows:

$$\begin{cases}\alpha_1 = \varphi \\ cos\beta_1 = cos\delta \cdot sin\varphi \\ cos^2\alpha_1 + cos^2\beta_1 + cos^2\gamma_1 = 1\end{cases} \quad (1)$$

Then Eq. (1) can be simplified as

$$\begin{cases}\alpha_1 = \varphi \\ \beta_1 = arc[cos\delta \cdot sin\varphi] \\ \gamma_1 = arc[sin\varphi \cdot sin\delta]\end{cases} \quad (2)$$

Variable ε is represented the included angle between the projection of the rotational axis of the cutting tool in $\mathbf{O_1X_1Y_1}$ surface and the axis $\mathbf{X_1}$. Thus, the (90°-ε) is represented the included angle between the projection of the rotational axis of the cutting tool in $\mathbf{O_1X_1Y_1}$ surface and the axis $\mathbf{Y_1}$.

$$\begin{cases}\alpha_1 = \varphi \\ cos\beta_1 = cos\delta \cdot sin\varphi \\ cos^2\alpha_1 + cos^2\beta_1 + cos^2\gamma_1 = 1\end{cases}$$

According to the geometry model, the α_1, $\beta_1$andγ_1 can also be given as follows:

$$\begin{cases}cos\alpha_1 = cos\varepsilon \cdot cos(90° - \gamma_1) \\ cos\beta_1 = cos(90° - \varepsilon) \cdot cos(90° - \gamma_1)\end{cases} \quad (3)$$

Then the variable ε is calculated as

$$\varepsilon = arccos\left(\frac{cos\alpha_1}{sin\gamma_1}\right) \quad (4)$$

Base on the same principle, ε-θ and 90°-(ε-θ) are represented the included angle between the projection of the rotational axis of the cutting tool in **OXY** surface and the axis **X** and **Y**. The relations between α, β and γ and the variable ε can be given as

$$\begin{cases}cos\alpha = cos(\varepsilon - \theta) \cdot cos(90° - \gamma_1) \\ cos\beta = cos(90° - \varepsilon + \theta) \cdot cos(90° - \gamma_1) \\ \gamma = \gamma_1\end{cases} \quad (5)$$

Furthermore, the rotational axis of the cutting tool can be given as

$$\frac{X - x_1}{cos\alpha} = \frac{Y - y_1}{cos\beta} = \frac{Z - z_1}{cos\gamma} \quad (6)$$

From Eq. (6), it can be seen that the rotational axis of the cutting tool has relationship with the six variables (X, Y, Z, α, β, γ). However, from Eq. (2) to Eq. (5), it can be seen that α, β and γ *can be* represented byvariableφ *and* δ. Thus, the rotational axis of the cutting tool can be determined by (X, Y, Z, φ, δ).

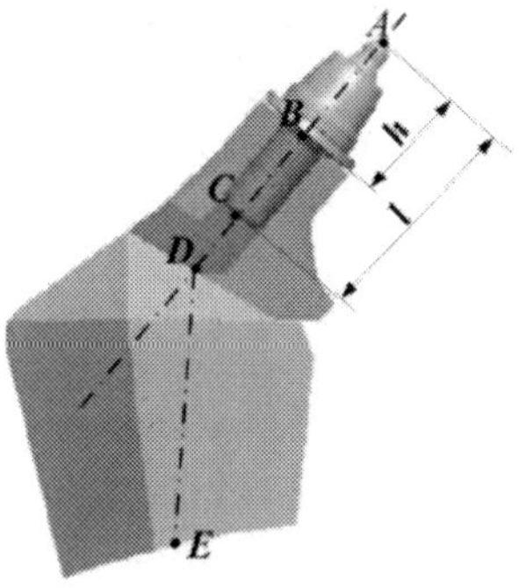

Figure 2. The tool apron model

If l represents the length of the cutting tool, then the spatial position of the cutting tool root, as shown in Fig. 2, is obtained as

$$(x_1 - lcos\alpha, y_1 - lcos\beta, z_1 - lcos\gamma) \quad (7)$$

3. Deviation measurement and estimation

3.1 Elementary parameter measurement

The elementary parameter of the milling wheel is measured by handheld coordinate measuring instrument that is based on laser triangulation method in this paper. As is shown in Fig.3, a laser, which is sent by the handheld coordinate measuring instrument, irradiate to the surface of an object to be measured, then an imaging will present on the double CCD sensor through reflection. With the displacement change of the surface of the object to be measured, the position of the imaging change. The position of the object to be measured is calculated through geometry model at last.

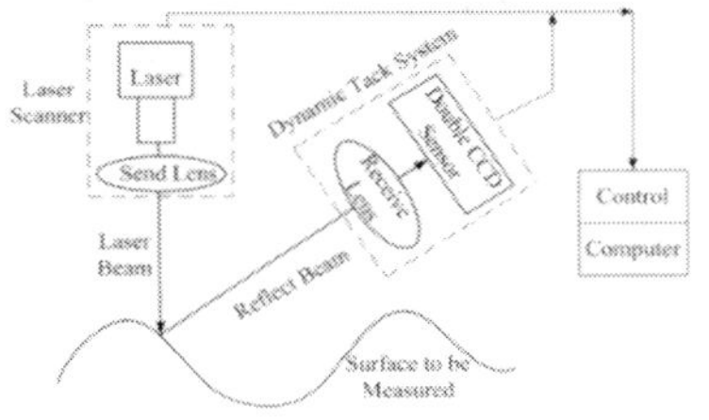

Figure 3. Laser triangulation method

As is shown in Fig.4, the rotational axis $\boldsymbol{Z(u)}$ of the milling drum and the end surface **XOY** of the milling drum are measured by reference to the original design firstly. The coordinate system **OXYZ** is built according to the measured data. Then location parameter that is consist of side face of the tool apron, end face and internal face of the cutting tool installation hole is obtained in the coordinate system **OXYZ**. The center point $\boldsymbol{C}(u)$ of the cutting tool installation hole and the rotational axis $\boldsymbol{H}(u)$of the cutting tool are calculated through the measuring software.

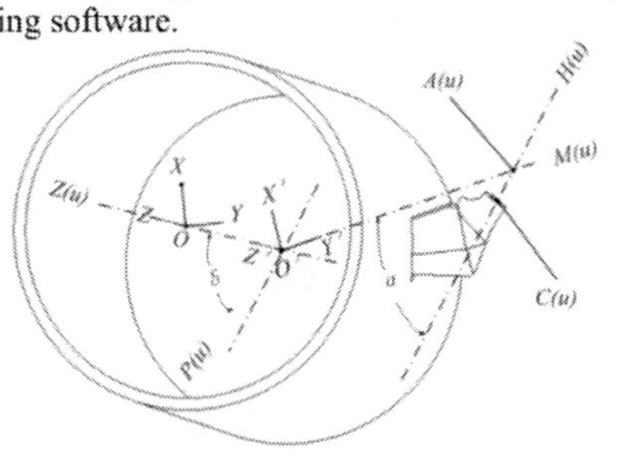

Figure 4. The measurement scheme

3.2 Installation angle solution

Based on the transformational relation of the elementary parameter and the installation angle, the transformational algorithm is built to calculate practical installation angle of the milling wheel. The detailed algorithm is shown as follows:

As is shown in Fig.4, the cutting tool tip $\boldsymbol{A}(u)$ and the center point $\boldsymbol{C}$(u) of the cutting tool installation hole are fall on the rotational axis $\boldsymbol{H}(u)$ of the cutting tool, thus the cutting tool tip $\boldsymbol{A}(u)$ can be written as,

$$A(u) = C(u) + |h|H(u) \tag{8}$$

Where $|h|$ represents the length of the cutting tool bit.

The cutting angle φ is the included angle between the rotational axis $\boldsymbol{H}(u)$ of the cutting tool and the center line $\boldsymbol{M}(u)$ of the cutting tip circle, the Eq. (9) is given as

$$\begin{cases} M(u) = A(u) - O(u) \\ H(u) \cdot M(u) = |M(u)||H(u)| \cdot cos\varphi \end{cases} \tag{9}$$

Where $\boldsymbol{O}(u)$ represents the gyration center of the cutting tool tip $\boldsymbol{A}(u)$, then the cutting angle φ is obtained as

$$\varphi = cos^{-1}\frac{H(u) \cdot (A(u)-O(u))}{|H(u)||A(u)-O(u)|} \tag{10}$$

The roll angle δ is the included angle between the end face $\boldsymbol{Z}(u)$ of the milling drum and the roll face $\boldsymbol{P}(u)$, the Eq. (9) is given as

$$\begin{cases} M(u) = A(u) - O(u) \\ P(u) = H(u) \times M(u) \\ Z(u) \cdot P(u) = |Z(u)||P(u)| \cdot cos\delta \end{cases} \tag{11}$$

Then the roll angle δ is obtained as

$$\delta = cos^{-1}\frac{Z(u) \cdot \left(H(u) \times (A(u)-O(u))\right)}{|H(u) \times (A(u)-O(u))|} \tag{12}$$

3.3 Deviation analysis and compensation

Through the comparison for the original date and the practical date, the deviation of the installation angle is solved. The normality, the mean value and variance of the cutting angle and roll angle are calculated through *Minitab* software. The distribution of the installation angle field is analyzed.

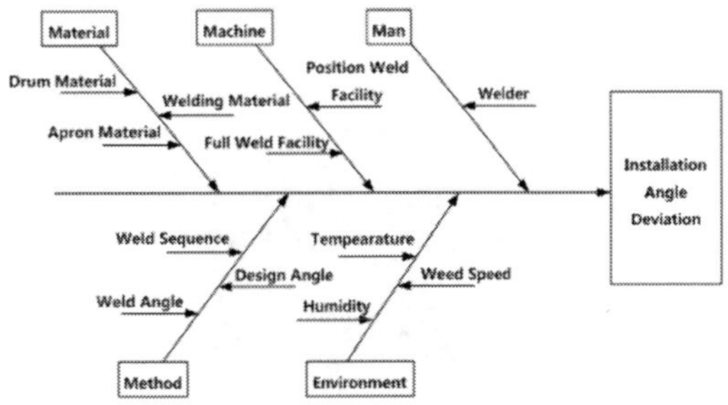

Figure 5. Fishbone diagram

Combined to the manufacture process of the milling wheel, the fishbone diagram method, as shown in Fig.5, is proposed to analyze the reason of the deviation from five respects, such as man, machine, materials, method and environment.

It can be found that the mean value deviation because of the welding deformation of the tool apron at the full weld stage, the main reason of the variance deviation is that the initial point and direction is inconsistent.

As you saw earlier, the rotational axis of the cutting tool can be determined by (X, Y, Z, φ, δ). The (X, Y, Z) is the spatial position of any point of the cutting tool axis. As is shown in Fig.2, the relative position of the five points (A B C D E) is fixed. Combined to the manufacture process of the milling wheel, the welding deformation, which is very small, can be neglected at the position welding stage. At the full weld stage, the large welding deformation of the tool apron causes the deviation of installation angle. However, the position of the center point of the tool apron at the bottom is still changeless. Therefore, it can take the angle reversible deformation method to compensate the mean value deviation of installation angle.

Meanwhile, improving the process capability of the manufacture and ensuring the consistent of the initial point and direction is important. It can reduce the variance deviation in order to guarantee the quality of the milling wheel.

4. Experimental verification

In the section, an illustrated experiment is given to evaluate the feasibility and applicability of the proposed measurement and compensation algorithm. The CAD model and the finished product of the same milling wheel, which are shown in Fig.6 and Fig.7, are used to validate the algorithm. About 88 tool aprons are set on the milling wheel regularly.

Figure 6. The CAD model of the milling wheel

Figure 7. The finished product of the milling wheel

The comparison installation angle results of the CAD model and the finished product are given in Fig.8 and Fig.9. From Fig.8, one can see that the mean value and variance of the cutting angle both exceed the bounding scope. The mean value is designed as 40 degree, while the mean value of manufacture is about 39.38 degree. Meanwhile, the value of the *Cp*, which is represented the process capability of the manufacture, is only 0.87. It is less than the standard value 1.33.

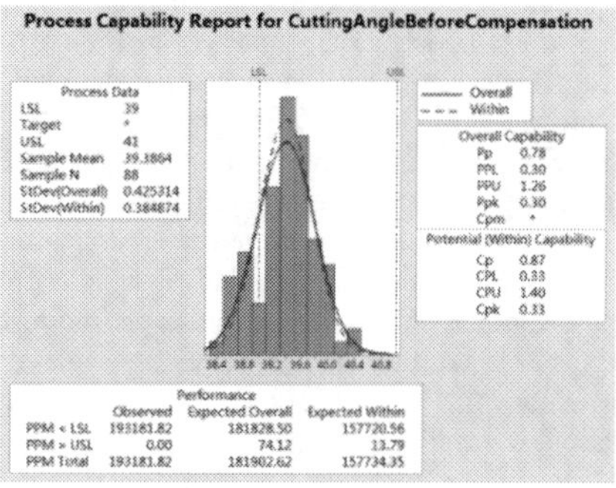

Figure 8. Process Capability Report for Cutting Angle Before Compensation

From figure 9, it can be seen that the mean value of the roll angle, which is 7.46 degree, meet the design requirement which is designed as 40 degree. But the value of the *Cp* is just 0.63. Therefore, it has a strong impact on the product quality.

The installation angle result of milling wheel that is optimized is shown in Fig.10 and Fig.11. It can be seen that the mean value and variance of the cutting angle both meet the design requirement.

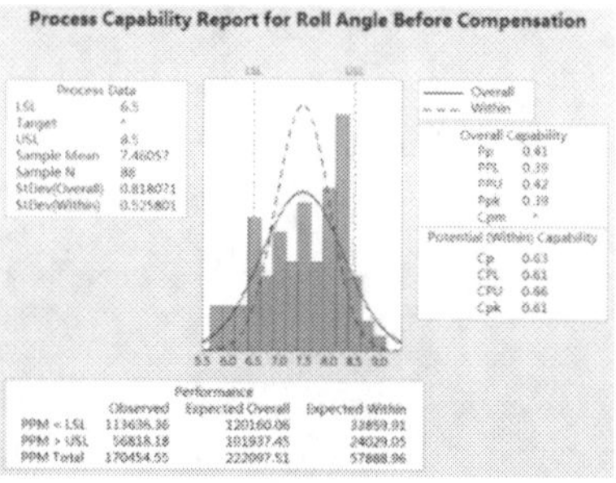

Figure 9. Process capability Report for roll angle before compensation

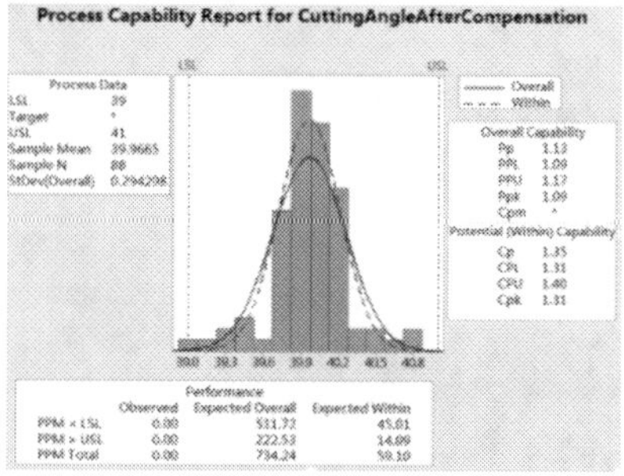

Figure 10. Process Capability Report for Cutting Angle After Compensation

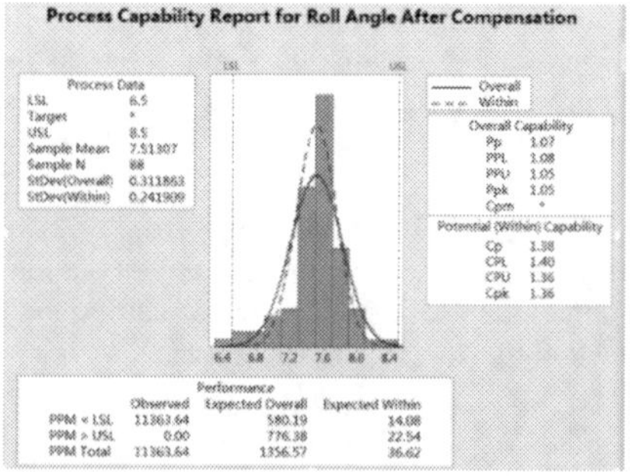

Figure 11. Process Capability Report for Roll Angle After Compensation

5. Conclusions

In this study, an accurate estimation and compensation algorithm was proposed to calculate the manufacture deviation of the milling wheel based on optical measurement. The scheduling method mainly concentrates on the assurance of manufacture accuracy and working performance of the milling wheel. Firstly,

 DOI: 10.1051/matecconf/20179504005

the elementary parameter of the milling wheel is measured by handheld coordinate measuring instrument. Then, the practical installation angle of the milling wheel is calculated and the process capability of the manufacture is analyzed. Furthermore, combine to the manufacture process of the milling wheel, the reversible deformation and process capability control method is proposed to compensate the deviation. The feasibility of the proposed method has been validated by an illustrated experiment. The comparison results show that the deviation of the installation angle can be effectively constrained and the proposed is helpful for the improvement of quality and accuracy in milling wheel.

Acknowledgement

This research is supported by Special Research Found for the Doctoral Program of Higher Education (Num. 20114116110002), Doctoral Fund of Henan Polytechnic University (Num. B2014-034), Foundation of He'nan Education Committee (No. 15A460025)

References

1. Jiang S.R., Zhang D.H., Wang W.H., Bu K., Cheng Y.W., Estimation of Displacement Field for Turbine Blade Profile Based on Reverse Engineering in Investment Casting. Special Casting & Nonferrous Alloys, 29(1) (2009):13-15.
2. Gong Z.H., Zhong Z.H., Estimation of the Springback of Auto-Body PanelsBased on Reverse Engineering. Journal of Hunan University (Natural Sciences), 34(1) (2007):24-27, 55.
3. He H.Q., Zhao H.Y., Niu W.C., Wang P., A method to measure welding deformation of plate by three dimensional laser scanner. Transactions of the china welding institution, 32(12)(2011):9-12
4. Li L., Study on the rapid three-dimension optical measurement for welding deformation of sheet structure .Optical Technique, 36 (6)(2011):825- 828.
5. Tian J.Y., Xiang H.R., Analysis on parameter influence laws and milling resistances of pavement miller. Journal of Jiangsu University (Natural Science Edition), 25(5)(2004):380-384.
6. Gu H.R., Jiao S.J., Xiao C.Y., Analysis and Test of Milling Load Characteristics in Milling Machine. Chinese Journal of Highway and Transport, 25(3)(2012):154-156.

MATEC Web of Conferences 95, 04006 (2017) DOI: 10.1051/matecconf/20179504006
ICMME 2016

Synthesis of Two Four-bar in Series for Body Guidance

Wen-Yeuan Chung

Department of Mechanical Engineering, Chinese Culture University, Taipei, Taiwan

Abstract. Using planar two-degree-of-freedom mechanism for the task of body guidance is evaluated and presented. This mechanism consists of two four-bar in series. For some applications of simple planar movement, planar robot with three actuators can be replaced by this mechanism to save cost and reduce complexity. This work reveals that at most seventeen precision positions can be prescribed theoretically. Based on synthetic consideration, each four-bar is proposed to be in charge of some prescribed positions and thus can be designed separately. Decomposing the prescribed precision positions into the assignment of body guidance for each constituent four-bar is developed. Both four-bar are then assembled to successfully execute the task of at most nine precision positions. Numerical example is given to illustrate the synthesis process. Problems of order defect, branch defect and ratio of link lengths are all considered to ensure the performance and applicability of the mechanism.

1 Introduction

Body guidance is one of major applications for various low-cost linkages. The well-known planar four-bar can be synthesized by prescribing at most five positions theoretically [1]-[2]. The six-bar linkages [3] and a special cam-linkage [4] mechanism were synthesized and introduced for similar tasks. Eleven given positions are expected to be visited by using a complicated eight-bar linkage [5]. The concepts of multi-phase and adjustable link length are also proposed to improve the capability of planar four-bar [6]-[8].

The RRRRR five-link is a two-degree-of-freedom linkage, and the joints at both ends are fixed on the frame and driven by two actuators [9]-[10]. The type of some joints can be prismatic [11]-[12], and the PRRRP parallel mechanism actuated by linear actuators was also proposed [13]. For the purpose of improving performances, such as payload capability and stiffness, the mechanism with seven links including a redundant dyad was suggested [14]-[16]. However, the feature of these linkages is that the middle moving pivot can reach any points within the workspace, and path generation is thus the only application.

An expensive planar robot usually has three DOF (degrees of freedom) and is equipped with three actuators [17]. Hence, its end-effector can be moved to any positions and orientations. The present work focuses on the evaluation and synthesis of mechanisms with two DOF. Body guidance is the major task and the precision positions that can be prescribed should be more than five. Moreover, the synthesized mechanism should have no branch defect and order defect.

2 Description of mechanism

Fig. 1 shows the proposed mechanism that mainly consists of two four-bar, A_oABB_o and C_oCDD_o, in series. The driving links are A_oA and C_oC. Link CEFD is the guided body. Two frames X-Y and x_b-y_b are respectively fixed on the base and attached to the guided body. The parameter ${}^b\mathbf{C}$ represents the coordinate of a point C, being on the guided body, with respect to the frame x_b-y_b. For the guided body CEFD at the ith prescribed position, the angle between axes X and x_b is denoted by α_i and the vector between both origins is denoted by $\mathbf{V}_i=\overrightarrow{OO_b}$. Furthermore, points A, B, C, C_o, D, and D_o are represented by A_i, B_i, C_i, C_{oi}, D_i, and D_{oi} when the guided body is at the ith prescribed position. The parameter $\mathbf{C}_i$ that is the coordinate of point C_i with respect to the base frame can thus be derived as

$$\mathbf{C}_i=\mathbf{R}_i\cdot{}^b\mathbf{C}+\mathbf{V}_i \quad (1)$$

where

$$\mathbf{R}_i(\alpha_i)=\begin{bmatrix}\cos\alpha_i & -\sin\alpha_i\\ \sin\alpha_i & \cos\alpha_i\end{bmatrix} \quad (2)$$

The moving frame x_t-y_t is attached to the link AC_oD_oB. Its origin is coincident with point A and the x_t axis is along the vector $\overrightarrow{AB}$. The angle between axes x_t and X is denoted by β_i when the linkage is at the ith

position. Unknown variables $\mathbf{A}_o$, $\mathbf{B}_o$, ${}^b\mathbf{C}$, ${}^b\mathbf{D}$, ${}^t\mathbf{C}_o$, ${}^t\mathbf{D}_o$, x_t component of ${}^t\mathbf{B}$, $\mathbf{A}_i$ and β_i $(i=1\sim k+1)$ are assumed, and the total number is $2\cdot 6+1+(2+1)\cdot(k+1)$. All $\mathbf{D}_i$ can then be derived by using equation like (1). Furthermore, any $\mathbf{B}_i$ can be obtained by

$$\mathbf{B}_i=\mathbf{R}_i(\beta_i)\cdot{}^t\mathbf{B}+\mathbf{A}_i \quad i=1\sim k+1 \tag{3}$$

All $\mathbf{C}_{oi}$ and $\mathbf{D}_{oi}$ can be derived similarly. After deriving coordinates of all points corresponding to the ith prescribed position, the conditions that lengths of four links A_oA, B_oB, C_oC, and D_oD remain constant lead to constraint equations. For example, the equation for the link length $|A_oA|$ is

$$(\mathbf{A}_o-\mathbf{A}_i)^T\cdot(\mathbf{A}_o-\mathbf{A}_i)=(\mathbf{A}_o-\mathbf{A}_1)^T\cdot(\mathbf{A}_o-\mathbf{A}_1)$$
$$i=2\sim k+1 \tag{4}$$

Therefore, there are $4\cdot k$ constraints like Eq. 4. The maximum value of k can be calculated from the equality

$$4\cdot k=2\cdot 6+1+(2+1)\cdot(k+1) \tag{5}$$

The value of k is 16, and this implies that at most 17 positions can be prescribed by using this mechanism.

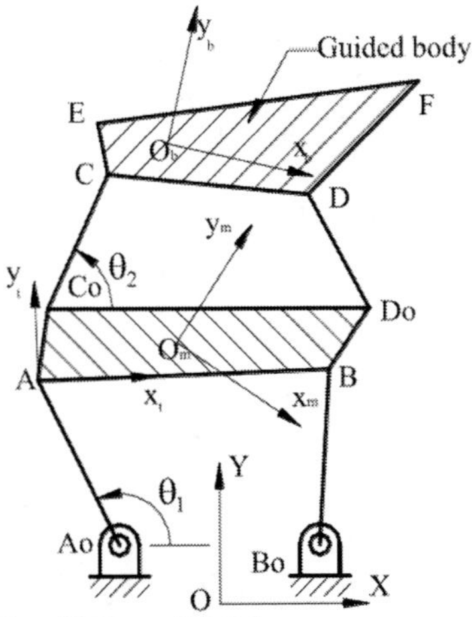

Figure 1. Type III: Seven-link linkage

3 Synthesis of mechanism

This mechanism can be synthesized by theoretically solving 16 nonlinear equations after seventeen positions are prescribed, but this task is definitely cumbersome. Even though the solutions can be obtained, the linkage might have order defect and branch defect and is thus useless.

An alternate strategy for applying or synthesizing this mechanism is to let each four-bar be in charge of some prescribed positions separately. For example, if the jth prescribed position is under the charge of four-bar A_oABB_o, only A_oA is actuated when the guided body is moved from the (j-1)[th] position to the jth position. Consequently, each four-bar can be synthesized independently as explained in the following.

The coordinates of point E at the 1st and i[th] positions can be related by a displacement matrix ${}^i_1\mathbf{T}$ as

$$\mathbf{E}_i={}^i_1\mathbf{T}\cdot\mathbf{E}_1 \tag{6}$$

and

$${}^i_1\mathbf{T}=\begin{bmatrix}\mathbf{R}_i & \mathbf{V}_i\\ \mathbf{0} & 1\end{bmatrix} \tag{7}$$

Both $\mathbf{E}_1$ and $\mathbf{E}_i$ are 3 by 1 vectors and in the form of $[E_x\ E_y\ 1]^T$ in Eq. 6 and the following derivation. The third moving frame x_m-y_m is attached to the coupler link AC_oD_oB as shown in Fig. 1. For the ith prescribed position, this frame and the fixed X-Y frame can be related by a matrix ${}^f_m\mathbf{T}_i$ and thus

$$\mathbf{E}_i={}^f_m\mathbf{T}_i\cdot{}^m\mathbf{E}_i \tag{8}$$

The displacement matrix ${}^j_i\mathbf{N}$ is introduced to describe the movement of four-bar A_oABB_o between the ith and j[th] positions. For example, both $\mathbf{B}_i$ and $\mathbf{B}_j$ are related by

$$\mathbf{B}_j={}^j_i\mathbf{N}\cdot\mathbf{B}_i \tag{9}$$

Similarly, the matrix ${}^j_i\mathbf{M}$ is used to describe the motion of four-bar C_oCDD_o with respect to the frame x_m-y_m. Hence, both ${}^m\mathbf{E}_i$ and ${}^m\mathbf{E}_j$ can be related by

$${}^m\mathbf{E}_j={}^j_i\mathbf{M}\cdot{}^m\mathbf{E}_i \tag{10}$$

These displacement matrices, such as ${}^j_i\mathbf{M}$, ${}^j_i\mathbf{N}$, or ${}^j_i\mathbf{T}$, can be multiplied together to obtain a single matrix that relates the first position to the nth position. For example, ${}^n_1\mathbf{M}$ can be expressed as

$${}^n_1\mathbf{M}={}^n_{n-1}\mathbf{M}\cdot{}^{n-1}_{n-2}\mathbf{M}\ldots{}^3_2\mathbf{M}\cdot{}^2_1\mathbf{M} \tag{11}$$

If all ${}^i_1\mathbf{N}$ and ${}^i_1\mathbf{M}$ can be derived from prescribed motion or ${}^i_1\mathbf{T}$, the technique of synthesizing a four-bar as discussed in [1, 2] can be followed to synthesize four-bar A_oABB_o and C_oCDD_o.

Let the movement of the body from the (j-1)[th] position to the jth position be charged by the four-bar A_oABB_o. The four-bar C_oCDD_o thus remains stationary and

$${}^{\ j}_{j-1}\mathbf{M}=\mathbf{I} \tag{12}$$

The parameter **I** represents an identity matrix. The matrix ${}^{\ j}_{j-1}\mathbf{N}$ is the same as ${}^{\ j}_{j-1}\mathbf{T}$ that can be derived from

DOI: 10.1051/matecconf/20179504006

$$ {}^{j}_{j-1}\mathbf{T} = {}^{j}_{1}\mathbf{T}\cdot\left({}^{j-1}_{1}\mathbf{T}\right)^{-1} \quad (13)$$

Both matrixes ${}^{j}_{1}\mathbf{N}$ and ${}^{f}_{m}\mathbf{T}_{j}$ can then be obtained as

$$ {}^{j}_{1}\mathbf{N} = {}^{j}_{j-1}\mathbf{N}\cdot{}^{j-1}_{1}\mathbf{N} = {}^{j}_{j-1}\mathbf{T}\cdot{}^{j-1}_{1}\mathbf{N} = {}^{j}_{1}\mathbf{T}\cdot\left({}^{j-1}_{1}\mathbf{T}\right)^{-1}\cdot{}^{j-1}_{1}\mathbf{N} \quad (14)$$

$$ {}^{f}_{m}\mathbf{T}_{j} = {}^{j}_{j-1}\mathbf{N}\cdot{}^{f}_{m}\mathbf{T}_{j-1} \quad (15)$$

For the case that the four-bar C_oCDD_o is actuated and four-bar A_oABB_o keeps stationary, we have

$$ {}^{j}_{j-1}\mathbf{N} = \mathbf{I} \quad (16)$$

$$ {}^{f}_{m}\mathbf{T}_{j} = {}^{f}_{m}\mathbf{T}_{j-1} \quad (17)$$

Based on Eq. 6, Eq. 8, and Eq. 10, the term $\mathbf{E}_j$ can be derived respectively as

$$ \mathbf{E}_j = {}^{j}_{1}\mathbf{T}\cdot\mathbf{E}_1 = {}^{j}_{1}\mathbf{T}\cdot{}^{f}_{m}\mathbf{T}_1\cdot{}^{m}\mathbf{E}_1 \quad (18)$$

$$ \mathbf{E}_j = {}^{f}_{m}\mathbf{T}_j\cdot{}^{m}\mathbf{E}_j = {}^{f}_{m}\mathbf{T}_j\cdot{}^{j}_{1}\mathbf{M}\cdot{}^{m}\mathbf{E}_1 \quad (19)$$

By relating both Eq. 18 and Eq. 19, the matrix ${}^{j}_{1}\mathbf{M}$ is derived as

$$ {}^{j}_{1}\mathbf{M} = \left({}^{f}_{m}\mathbf{T}_j\right)^{-1}\cdot{}^{j}_{1}\mathbf{T}\cdot{}^{f}_{m}\mathbf{T}_1 \quad (20)$$

The term ${}^{f}_{m}\mathbf{T}_1$ that relates $x_m - y_m$ frame and X-Y frame for the linkage at the first position is given or can be specified freely. With the responsible four-bar for each prescribed movement being assigned, all matrixes ${}^{j}_{1}\mathbf{M}$, ${}^{j}_{1}\mathbf{N}$, and ${}^{f}_{m}\mathbf{T}_1$ can be derived successively by using Eq. 12, Eq. 14, and Eq. 15 or Eq. 16, Eq. 17, and Eq. 20. Both four-bar are synthesized separately and then assembled with ${}^{f}_{m}\mathbf{T}_1$.

4 Order defect and branch defect

If a linkage is used to execute body guidance or path generation, problems of branch defect and order defect should be considered and avoided. A four-bar has two configurations for a given input as shown in Fig. 2. If link A_oA is the driving, both configurations can be distinguished from the sign of the criterion Δ_i.

$$ \Delta_i = \left(\left(\mathbf{B}_i - \mathbf{A}_i\right)\times\left(\mathbf{B}_o - \mathbf{B}_i\right)\right)\cdot\hat{\mathbf{k}} \quad (21)$$

Eq. 21 is a three dimensional vector operation and $\hat{\mathbf{k}}$ is a unit vector normal to the plane. The linkage synthesized should be branch defect-free and associate with the same configuration for all prescribed positions. In other words, the signs of all Δ_i must be the same. On the other hand, the order defect-free implies the satisfaction of the following inequality

$$ (\theta_{1j} - \theta_{1i})\cdot(\theta_{1k} - \theta_{1j}) > 0 \quad (22)$$

where θ_{1j} represents the value of input angle θ_1 corresponding to the jth position, and j is between i and k.

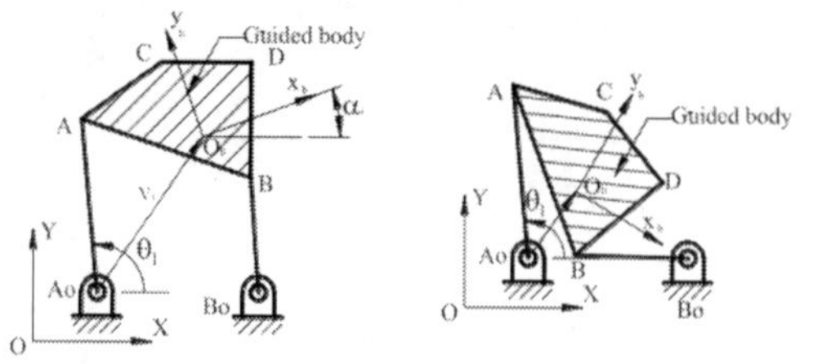

Figure 2. Planar four-bar at two batches

5 Application and example

With the assignment of prescribed positions for the mechanism, the results described in Section 3 are used to derive ${}^{j}_{1}\mathbf{N}$ and ${}^{j}_{1}\mathbf{M}$ for synthesizing four-bar A_oABB_o and C_oCDD_o respectively. The synthetic techniques are inversion, circle point curve, center point curve, and Burmester theory. Any linkages that have order defect or branch defect are discarded based on the criteria described in Section 4.

Example: A mechanism as shown in Fig. 1 is to be designed to execute body guidance. There are eleven prescribed positions as listed in Table 1. Two points $E_1:(-2,0)$ and $F_1:(-1,0)$ on the coupler link CD are used to verify the movement of body guidance. The precribed movements of points E and F or coupler link CD are shown in Fig. 3.

The data of prescribed positions were proposed in [5], and a linkage with eight links and one degree of freedom was synthesized. However, neither order defect nor branch defect was considered.

Table 1. The data of eleven positions

	1	2	3	4	5	6
α_i	0	-12.64	-34.97	-59.80	-79.97	-88.29
x_i	0	0.45	1.32	2.29	2.88	2.95
y_i	0	0.24	0.42	0.17	-0.43	-0.97

	7	8	9	10	11
α_i	-79.97	-59.80	-34.97	-12.64	0
x_i	2.60	1.93	1.36	1.06	1.00
y_i	-1.26	-1.57	-2.06	-2.63	-3.00

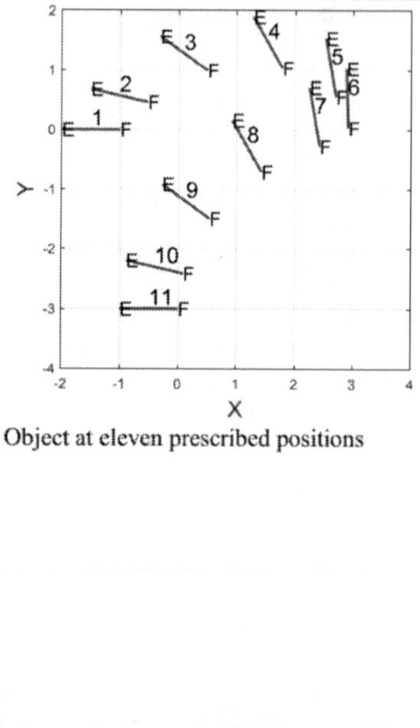

Figure 3. Object at eleven prescribed positions

Sol: 1. By observing the feature of the prescribed positions, the four-bar A_0ABB_0 is assigned to guide the link CEFD through the 1st to the 6th positions and four-bar C_0CDD_0 guides through the other positions. Therefore, each four-bar has to match six prescribed positions

2. Although a four-bar can match precisely at most five prescribed positions based on Burmester theory, the linkages obtained usually have branch defect or even order defect. Therefore, only four of six prescribed positions are chosen to derive the circle point curve, and the others become approximate positions. A linear spaced array covering a specified interval is schemed, and the elements are used for the x coordinate of pivot A_1 or B_1. For each x coordinate, at most three solutions are obtained for pivot A_1 or B_1. Numerous four-bar are then synthesized by combining any A_1 and B_1. Both branch defect and ratio of link length are checked to discard unacceptable linkages. The others are compared by evaluating the deviations corresponding to both approximate positions. For the ith approximate position, the ideal or desired coordinate of pivot A or B can be derived as $\mathbf{A}_{id}={}_1^i\mathbf{N}\cdot\mathbf{A}_1$ or $\mathbf{B}_{id}={}_1^i\mathbf{N}\cdot\mathbf{B}_1$. The deviations are defined by the following functions

$$Err_{ia}=\left|\left|A_oA_{id}\right|-a\right|/a \tag{23}$$

$$Err_{ib}=\left|\left|B_oB_{id}\right|-b\right|/b \tag{24}$$

where a is the length of input link $|A_oA_1|$ and $b=|B_oB_1|$. Four deviations are obtained since there are two approximate positions, and the maximum one is used as the index for comparison. The linkage with the minimum value of index is thus selected. For four-bar A_oABB_o, the 1st, 4th, 5th, and 6th positions listed in Table 1 are used to derive circle point curve. The data of the best linkage, after comparing the deviations or index, are listed in Table 2.

Table 2. The data of the first four-bar at the first position

$\mathbf{A}_o$	$\mathbf{A}_1$
(1.4373, -1.7443)	(2.0000, 1.4667)
$\mathbf{B}_1$	$\mathbf{B}_o$
(1.3333, -1.9073)	(6.0457, -4.8063)

3. The four-bar C_oCDD_o is then to be synthesized. Both x_m-y_m frame and X-Y frame are assumed parallel with translation vector as (-1.5, 2.5) when the linkage is at the 1st position, and ${}_m^f\mathbf{T}_1$ is thus defined. The matrix ${}_1^j\mathbf{M}$ is just an identity matrix for $j\le 6$ since this four-bar remains still. For $j>7$, Eq. 20 has to be used to find ${}_1^j\mathbf{M}$. The data like those in Table 1 are obtained and listed in Table 3.

By choosing the 6th, 7th, 9th, and 11th as the precision positions, this four-bar is synthesized as in Step 2. The coordinates of four pivots with respect to the moving frame x_m-y_m are listed in Table 4.

Table 3. The data of six positions for the second four-bar

	6	7	8	9	10	11
α_i	0	8.32	28.49	53.32	75.65	88.29
x_i	0	-0.066	-0.442	-0.359	0.309	0.927
y_i	0	-0.602	-2.056	-3.831	-5.272	-5.934

Table 4. The data of the second four-bar at the first position

${}^m\mathbf{C}_o$	${}^m\mathbf{C}_1$
(3.8714, -2.9097)	(3.3333, -4.9840)
${}^m\mathbf{D}_1$	${}^m\mathbf{D}_o$
(3.3333, -2.5859)	(4.1963, -0.4869)

4. The motion of each four-bar can be verified by using several animation tools. Both four-bar are then assembled with ${}_m^f\mathbf{T}_1$ to finish the design. The configurations of the linkage at some positions are shown in Fig. 4. The four-bar C_oCDD_o is drawn by dashed lines. The ideal or desired position of segment EF is drawn by bold and solid lines.

5. Fig. 5 shows the stuctural error at eleven prescribed positions. The values are the distances between the actual positions of point E or F and the ideal ones. Evidently, the link CEFD can be sequentially guided to the desired positions precisely except for trivial error at the 2nd, 3rd, 8th, and 10th positions. The maimum error is less that 0.05.

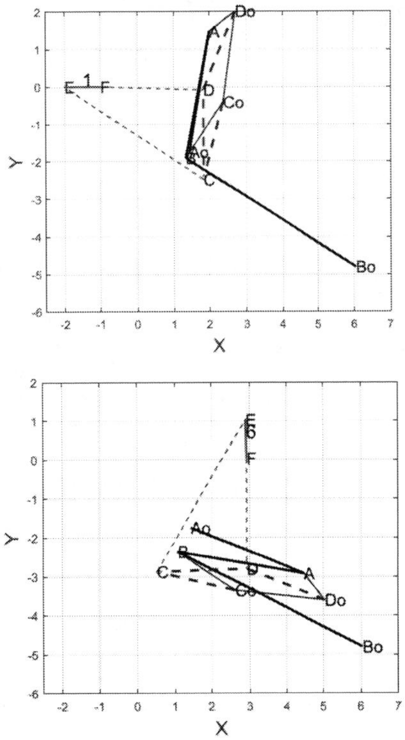

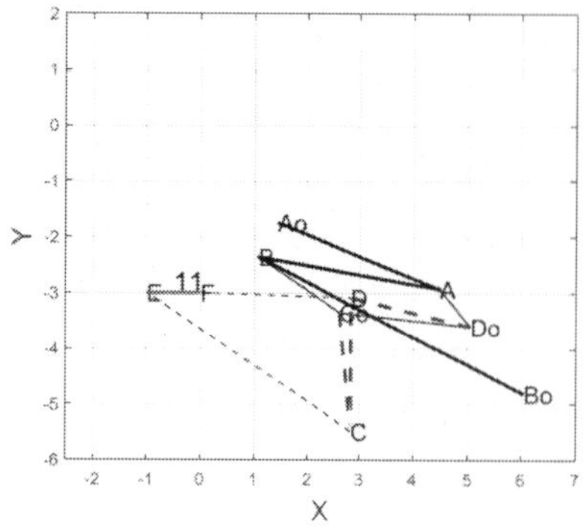

Figure 4. Seven-link mechanism at 1st, 6th, 11th positions

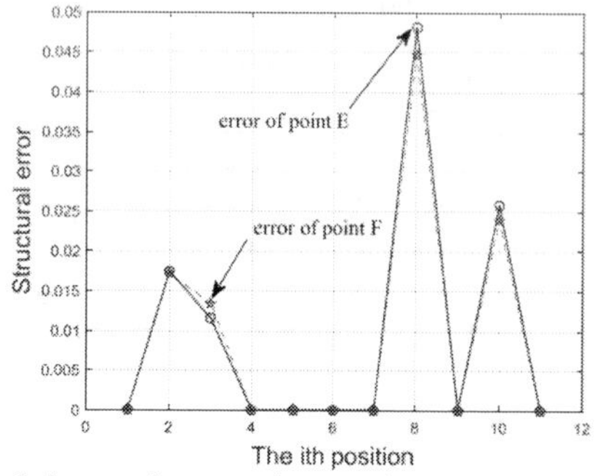

Figure 5. Structural error at eleven positions

6 Conclusion

A mechanism with two four-bar in series and with two DOF is proposed for the task of body guidance. Its application and synthesis is presented for the first time. This mechanism can be synthesized by theoretically solving numerous nonlinear equations and at most seventeen positions can be prescribed.

An example illustrates the effective design of this mechanism by mainly actuating each four-bar in sequence. The well-developed techniques for synthesizing four-bar can thus be used. Although the mechanism can guide the body through nine positions precisely, there is few choices left to match other requirements. Therefore, the mechanism synthesized in this example passes through only seven positions precisely. The problems of order defect, branch defect and ratio of link lengths are all considered, and the errors for approximate positions are trivial. The techniques used in this example can certainly be integrated with optimization search so that more positions can be prescribed and more consideration can be included. Evidently, this mechanism can replace complicated and expensive planar robots in some ways.

References

1. C.S. Suh and C.W. Radcliffe, *Kinematics and Mechanisms Design* (1978)
2. C.H. Chiang, *Kinematics and Design of Planar Mechanisms*, Krieger Publishing Company (2000)
3. G.S. Soh and J.M. McCarthy, "The synthesis of six-bar linkages as constrained planar 3R chains," Mechanism and Machine Theory, **45**, pp. 160-170 (2008)
4. G. Gatti and D. Mundo, "Optimal synthesis of six-bar cammed-linkages for exact rigid-body guidance," Mechanism and Machine Theory, **42**, pp. 1069–1081 (2007)
5. C. Chen and J. Angeles, "A novel family of linkages for advanced motion synthesis," Mechanism and Machine Theory, **43**, pp. 882–890 (2008)
6. A. Ahmad and K.J. Waldron, "Synthesis of adjustable planar 4-bar mechanisms", Mechanism and Machine Theory, **14**, pp. 405–411 (1979)
7. H.Z. Edmund and H.M. Cheung, "Adjustable four-bar linkages for multi-phase motion generation", Mechanism and Machine Theory, **39**, pp. 261–279 (2004)
8. K. Russell and R.S. Sodhi, "On the design of slider-crank mechanisms Part I: multi-phase motion generation", Mechanism and Machine Theory, **40**, pp. 285–299 (2005)
9. F. Gao; X.Q. Zhang, Y. S. Zhao, and H. R. Wang, "A physical model of the solution space and the atlas of the reachable workspace for 2-DOF parallel planar manipulators, " Mechanism and Machine Theory, **31**, pp. 173-184 (1996)
10. F. Gao, X.Q. Zhang, Y.S. Zhao, and W.B. Zu, "Distribution of some properties in physical model of the solution space of 2-DOF parallel planar manipulators," Mechanism and Machine Theory, **30**, pp. 811-817 (1995)
11. J.J. Cervantes-Sánchez and J. G. Rendón-Sánchez, "A simplified approach for obtaining the workspace of a class of 2-dof planar parallel manipulators," Mechanism and Machine Theory, **34**, pp. 1057-1073 (1999)
12. T. Huang, M. Li, Z. Li, D.G. Chetwynd, and D.J. Whitehouse, "Optimal kinematic design of 2-DOF parallel manipulators with well-shaped workspace bounded by a specified conditioning index," Transactions on Robotics and Automation, **20**, pp. 538-543 (2004)
13. X. J. Liu, J. Wang, and G. Pritschow, "On the optimal kinematic design of the PRRRP 2-DoF parallel mechanism," Mechanism and Machine Theory, **41**, pp. 1111-1130 (2006)
14. Y.X. Zhang, S. Cong, W.W. Shang, Z.X. Li, and S.L. Jiang, "Modeling, Identification and Control of a Redundant Planar 2-DOF Parallel Manipulator," International Journal of Control, Automation, and Systems, **5**, pp. 559-569 (2007)
15. W.W. Shang, S. Cong, and Y.X. Zhang, "Nonlinear friction compensation of a 2-DOF planar parallel manipulator," Mechatronics, **18**, pp. 340–346 (2008)
16. J. Wu, J.S. Wang, and L.P. Wang, "A comparison study of two planar 2-DOF parallel mechanisms: one with 2-RRR and the other with 3-RRR structures," Robotica, **28**, pp. 937-942 (2010)
17. J.J. Craig, *Introduction to Robotics: Mechanics and Control*, Pearson Education (2005)

DOI: 10.1051/matecconf/20179504007

Research on Thermal-Mechanical Coupling Modeling and Simulation of the Spindle Feed System of Machine Tool

Jianhua Xu, Wenhua Ye , Yang Feng and Chunyu Li

College of Mechanical and Electrical Engineering, Nanjing University of Aeronautics and Astronautics, Nanjing, Jiangsu 210016, China

Abstract. Performance of spindle feed system affects the accuracy of machine tools directly. Aiming at the problem that most research works focused on mechanical characteristics or thermal characteristics of the feed system so far, the thermal-mechanical coupling characteristic is studied in this paper. The coupling mechanism and theoretical model are established for a machine tool feed system, its coupling modal and harmonic response are simulated and analyzed by the FEM software. By comparing with the mechanical characteristics, thermal-mechanical coupling characteristics has directly influence on the dynamic performance of spindle feed system, its displacement amplitude is significantly weakened.

1 Introduction

As a key component of machine tool, the dynamic performance of spindle feed system (SFS) has a non-negligible impact on machining accuracy. Many studies concerning the influence of mechanical properties on the SFS have been reported [1], [2]. And some others concerned the influence of thermal characteristics [3], [4]. However, the dynamic performance of the SFS is not only influenced by the mechanical properties or the thermal characteristics but also the interaction of them. It is necessary to study the thermal-mechanical coupling influence on the dynamic performance of the SFS. Research on the influence of thermal-mechanical coupling [5], [6] is quite few. Most of the researches have not made a comparison with the influence of the mechanical properties or the thermal characteristics.

Based on a five-axis linkage gantry machine tool SFS, the thermal-mechanical coupling model were established in this paper. Then the influence of the coupling effect on the dynamic performance of SFS was obtained by comparing the analysis results between thermal-mechanical coupling simulation and structure field simulation.

2 Thermal-mechanical coupling modeling

In this paper, the SFS of five-axis linkage gantry machine tool is taken as study object. Its mechanical structure is shown in Fig. 1.

In the SFS working process, displacement and velocity are time-varying. The internal force equilibrium equation turns into:

$$\sigma_{ij,i} + F_{Ni} + F_{Ti} - \rho u_{i,tt} - \mu u_{i,t} = 0 \quad (1)$$

where $\sigma_{ij,i}$ is internal stress component, F_{Ni} and F_{Ti} are body forces caused by external force and temperature rise, $u_{i,tt}$ is the second order derivative to time of the displacement component (i.e. acceleration), $u_{i,t}$ is the first derivative of it (i.e. velocity), ρ is density, μ is damping coefficient.

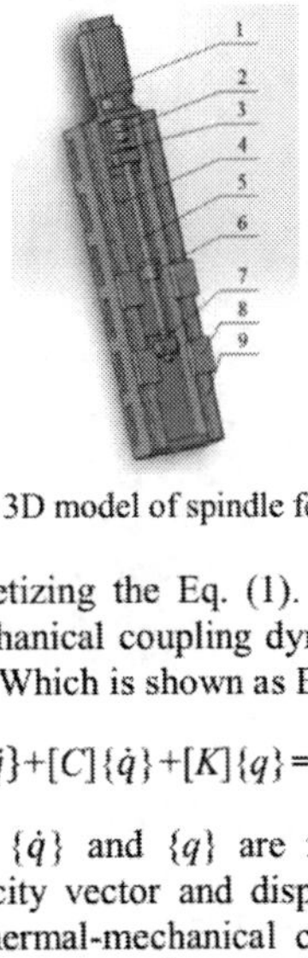

Figure 1. The 3D model of spindle feed system

By discretizing the Eq. (1). with finite element, the thermal-mechanical coupling dynamic model of SFS can be obtained. Which is shown as Eq. (2).

$$[M]\{\ddot{q}\}+[C]\{\dot{q}\}+[K]\{q\}=\{F_N\}+\{F_T\} \quad (2)$$

where $\{\ddot{q}\}$, $\{\dot{q}\}$ and $\{q\}$ are respectively acceleration vector, velocity vector and displacement vector of SFS under the thermal-mechanical coupling effect, $[M]$, $[K]$ and $[C]$ are the quality matrix, stiffness matrix and damping matrix of the SFS, respectively. $\{F_N\}$ and $\{F_T\}$ are node load vectors caused by external force and temperature rise.

For Eq. (2), its characteristic equation is Eq. (3), and it meets Eq. (4).

$$[K][A]=[M][A][\Lambda] \quad (3)$$

$$\begin{cases} [\phi]^T[M][A]=I \\ [\kappa]^T[K][A]=[\Lambda]. \end{cases} \quad (4)$$

where $[A]=diag(\omega_1^2,\omega_2^2,...,\omega_n^2)$ is the natural frequencies matrix for SFS, I is unit matrix, $[A]=[A_1,A_2...A_n]$ is matrix of modal shape of SFS. The natural frequency and vibration mode under the effect of thermal-mechanical coupling of the SFS can be obtained by the above equations.

According to Eq. (2), modal parameter of the SFS is related to the mass matrix, the stiffness matrix, the damping matrix of the system, the force it suffered and the thermal stress caused by temperature rise. The temperature has influence on the modal parameters of the SFS. The temperature field will affect the structure field. While the influence of structure field on temperature field is relatively small [7]. So the thermal-mechanical coupling of SFS is an one-way coupling.

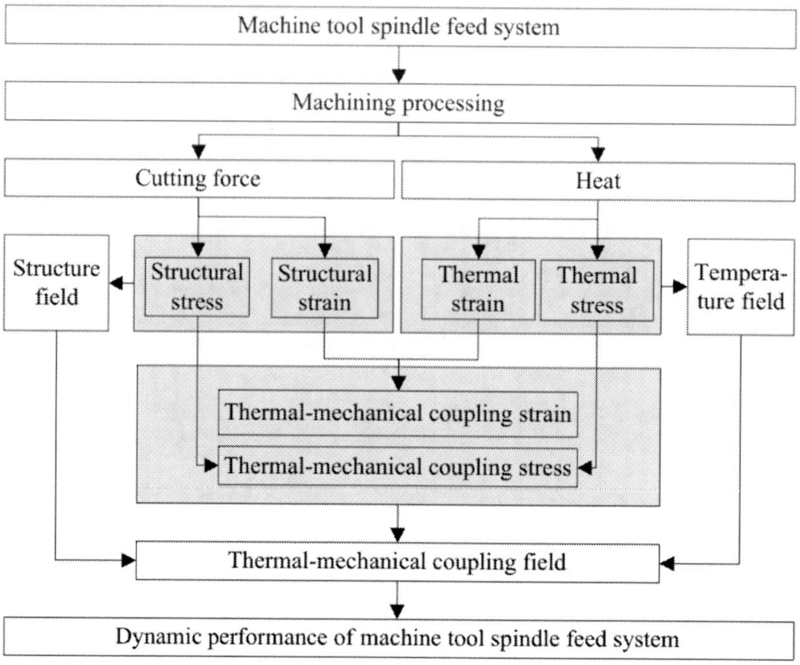

Figure 2. Thermal-mechanical coupling relationship

3 Simulation and analysis of thermal-mechanical coupling characteristics

3.1 Simulation method

The thermal-mechanical coupling problem can be divided into three sub-problems: dynamic characteristics of structure, thermal effects and thermal-mechanical coupling effect. Their coupling relationship is illustrated in Fig. 2.

To compare and verify the influence on SFS dynamic performance by thermal-mechanical coupling effect, structure field simulation and thermal-mechanical coupling simulation are carried out. For the latter, modal analysis and harmonic response analysis are conducted respectively.

3.2 Dynamic performance simulation

The modal analysis is carried out by simulation with a finite element analysis software. The first six order natural frequencies are gained in coupling field and structure field respectively. The results are summarized in Table 1.

Table 1. Contrast of the first six natural frequencies

Modal order	Structure field[Hz]	Coupling field[Hz]	Frequency deviation[%]
1	168.53	167.16	0.81%
2	178.79	177.99	0.45%
3	234.03	233.07	0.41%
4	388.99	387.25	0.45%
5	430.57	427.33	0.75%
6	481.36	479.21	0.44%

3.3 Contrast and analysis of modal parameters

The contrast of modal parameters between the thermal-mechanical coupling field and the structure field is shown in Table 1. It shows that each order of natural frequencies makes no great difference. Among them, the first natural frequency is most significant (0.81%), the fifth natural frequency takes second place (0.75%), other are all within 0.5%. Deviation of each order natural frequency are within 4 Hz. Therefore, the internal thermal stress of SFS has a relatively small influence on the modal parameters.

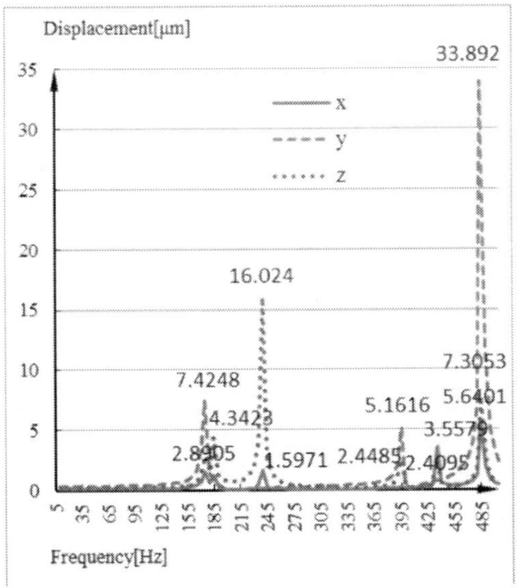

Figure 3. Displacement-frequency curves in the structure field

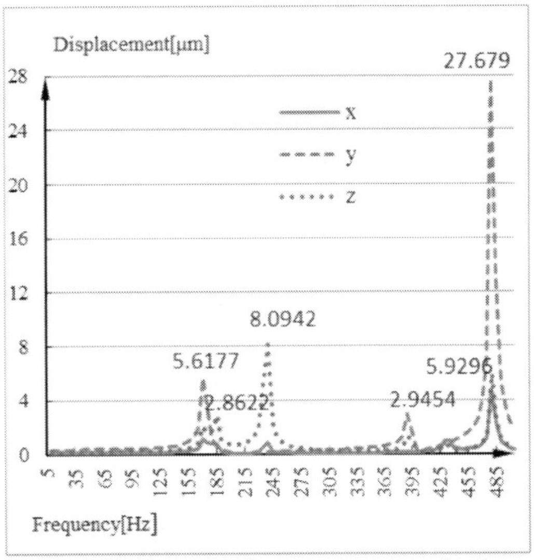

Figure 4. Displacement- frequency curves in the thermal-mechanical coupling field

Table 2. Summary and contrast of displacement amplitudes

Amplitude in different directions		**Frequency**					
		170 Hz	**180 Hz**	**240 Hz**	**390 Hz**	**430 Hz**	**480 Hz**
X	Structure field[μm]	1.5226	1.267	1.5971		3.5579	5.6401
	Coupling field[μm]	1.1336	0.797	0.83468		1.0037	4.7087
	Deviation[%]	-25.55%	-37.10%	-47.63%		-71.79%	-16.51%
Y	Structure field[μm]		7.4248		5.1616		33.892
	Coupling field[μm]		5.6177		2.9454		27.679
	Deviation[%]		-24.34%		-42.94%		-18.33%
Z	Structure field[μm]	2.8905	4.3423	16.024	2.4485	2.4095	7.3053
	Coupling field[μm]	2.0539	2.8622	8.0942	1.2624	0.71359	5.9296
	Deviation[%]	-28.94%	-34.09%	-49.49%	-48.44%	-70.38%	-18.83%

3.4 Harmonic response analysis and contrast

By contrasting the displacement-frequency curves in Fig. 3 and Fig. 4, it can be found that their changing trend are similar, but the amplitudes in coupling field are less than that of the structure field.

A comparison of displacement amplitudes in three directions are listed in Table 2. For X, the maximum reduction in displacement is 71.79% at frequency 430Hz. For Y, the maximum reduction is 42.94% at 390 Hz. For Z, the maximum reduction is 70.38% at 430Hz.

According to the analysis above, the displacement of the SFS is reduced under the coupling field. That is, the thermal-mechanical coupling effect plays a weakening role in the displacement.

4 Conclusions

In this paper, the thermal-mechanical coupling mechanism and its theoretical coupling model are established for the SFS of machine tools. The coupling modal and harmonic response are simulated and analyzed by FEM. The simulation results show that the thermal stress caused by the temperature rise has little influence on the modal parameters, but the displacement amplitude has been significantly weakened under thermal-mechanical coupling effect.

Acknowledgment

 DOI: 10.1051/matecconf/20179504007

The work is supported by the National Natural Science Foundation of China (Grant No. 51575272).

References

1. Y. Altintas, A. Verl, C. Brecher, et al., CIRP Ann-Manuf Techn, **60**, 779-796 (2011)
2. N. Tounsi, T. Bailey, M.A. Elbestawi, Int J Mach Tool Manu, **43**, 440-451 (2003)
3. H. Li, Y.C. Shin, Int J Mach Tool Manu, **44**, 347-364 (2004)
4. J. Xia, B. Wu, Y. Hu, et al., Precision Engineering, **34**, 357-368 (2010)
5. S.H. Liu, W.H. Ye, J China Soc Mech Eng, **34**, 177-185 (2013)
6. J.L. Tian, J. Hong, Y.S. Zhu, et al., Journal of Xi'an Jiaotong University, **46**, 63-68 (2012)
7. J.S. Dhupia, B. Powalka, A.G. Ulsoy, et al., J Manu Sci E, **129**, 943-950 (2007)

Stochastic Stability and Bifurcation of A High Speed Rotor-Bearing System with Random Excitation

Jiarong Lu[1], Jiangang Zhang[1], Wenju Du[2] and Hongwei Luo[1]

[1]*Department of Mathematics, Lanzhou Jiaotong University, 730070, Anning, Lanzhou, China*
[2]*School of Traffic and Transportation, Lanzhou Jiaotong University, 730070, Anning, Lanzhou, China*

Abstract. In this paper, a high speed rotor-bearing system with stochastic excitation is considered, and study the stability and Hopf bifurcation of the system by using quasi-nonintegrable Hamiltonian system theory. Then, the conditions of local and global stability of system are obtained by largest Lyapunov exponent and boundary category. Finally, the solution of FPK equation can be got, which is stationary probability density function and jointly stationary probability density function, and then, by simulating its graph to illustrate the results.

1 Introduction

The rotor bearing system involves many nonlinear problems in rotating machinery, and its dynamic behaviors are very complex. Hu and Xian [1] established a mathematical model of an asymmetric double-disc rotor-bearing system and use nonlinear dynamics theory to consider nonlinear oil-film force and rub-impact force. And then, Ninoslav Truhar et al. [2] considered a mechanical system excited by external force with the development of nonlinear dynamics theory, nonlinear dynamics connects with rotor dynamics tightly, which makes the research on rotor dynamics present a new outlook. In the past decade, many researches about the nonlinear stochastic dynamics and controls which have been shifted from Lagrange system to Hamilton system, and the nonlinear stochastic dynamical system is expressed as dissipative Hamilton system with stochastic excitation. Finally, according to the intergrability and resonance of Hamilton system, a theory based on Hamilton system with stochastic excitation is presented and developed, which is used to study mechanical system with stochastic excitation. Xu et al. [3] discussed the stochastic stability and bifurcation of multiwalled carbon nanotubes-absorbing hydrogen atoms subjected to thermal perturbation. They studied the dynamic behavior by useing the quasi-nonintegrable Hamiltonian system theory and the stochastic parametric excitation is regarded as Gauss white noise. Li [4] also studies stochastic stability of quasi-integrable and resonant (nonresonant) Hamiltonian systems under parametric excitations of combined Gaussian and Poisson white noises. In this study, the high speed rotor-bearing system with stochastic parametric excitation is presented in section 2. Then, the largest Lyapunov exponent [5] of Ito stochastic differential equation can be obtained by using stochastic averaging method in section 3. In the section 4, the stochastic Hopf bifurcation of the system study is studied. Finally, some examples and numerical simulations are given in section 5.

2 System model

A high speed rotor-bearing system [6] is considered. Due to the external random factors and the internal random vibration, then regarding all stochastic excitations as Guass white noise. The equation of motion of the non rotating damped system when the global assembling of the matrices is done is given by

$$M\ddot{q}(t)+C\dot{q}(t)+Kq=fq\xi(t) \tag{1}$$

where $q(t)$ is a vector containing the generalized coordinates, referred to an inertial frame, M is the symmetric mass, C is the symmetric damping coefficient, K is the symmetric stiffness coefficient, and f is a function of p and q , $\xi(t)$ is the Gauss white noise whose mean is zero and intensity is 2D.

Let $p=\dot{q}$, $c=C/M$, $k=K/M$, then the Eq. (1) becomes

$$\begin{cases}\dot{q}=p\\ \dot{p}=-cp-kq+fq\xi(t)\end{cases} \tag{2}$$

The Hamiltonian function of Eq. (2) can be shown as follows:

$$H=\frac{1}{2}p^2+\frac{k}{2}q^2 \tag{3}$$

According to the quasi-nonintegrable Hamiltonian system theory, the Hamiltonian function $H(t)$ converges weakly in probability to a one-dimensional Ito diffusion process [7]. The averaged Ito equation about the Hamiltonian function can be shown as follows:

$$dH = m(H)dt + \sigma(H)dB(t) \tag{4}$$

where $B(t)$ is standard Wiener process, $m(H)$ and $\sigma(H)$ are drift and diffusion coefficients of Ito stochastic process can be obtained in stochastic averaging method [8].

$$m(H) = -cH + \frac{2Df^2}{3c}H^2 \tag{5}$$

$$\sigma^2(H) = \frac{4Df^2}{3c}H^2 \tag{6}$$

3 Stochastic stability

The local stability of the stochastic system will be discussed in the following, and the largest Lyapunov exponent of Ito stochastic differential equation can be defined as

$$\lambda = \lim_{t\to\infty}\frac{1}{t}\ln\|Z(t,z_0)\| \tag{7}$$

then the largest Lyapunov exponent is

$$\begin{aligned}\lambda &= \lim_{t\to\infty}\frac{1}{t}\ln H^{1/2}(q,p) = \left[\dot{m}(0) + (\dot{\sigma}(0))^2/2\right]/2 \\ &= -\frac{c}{2}\end{aligned} \tag{8}$$

By Oselesec multiplicative ergodic theorem [9], the condition that the trivial solution of system keeps asymptotically stable with probability 1 can be got:

Lemma 3.1. The conditions to keep the system local stable can be obtained: the largest Lyapunov exponent is less than zero, which means $c > 0$.

It is remarkable that that largest Lyapunov exponent can judge local stability of the trivial solution but is helpless for global stability. Therefore, using the boundary category of Eq. (4) to study global stability. Then the two boundaries will be mainly discussed in the following, which is the left boundary $H \to 0$ and right boundary $H \to \infty$.

Because of $\sigma^2(H)=0$ and $m(H)=0$ when $H \to 0$, then the left boundary $H \to 0$ belong to the first singular boundary and the drift and diffusion coefficients converge asymptotically to the two equations

$$m(H) = -cH + O(H),\ \sigma^2(H) = \frac{4Df^2}{3c}H^2 + O(H^2) \tag{9}$$

then, the diffusion exponent α_l, drift exponent β_l and character value c_l of left boundary can be obtained

$$\alpha_l = 2,\ \beta_l = 1$$

$$\begin{aligned}c_l &= \lim_{H\to 0^+}\frac{2m(H)H^{\alpha_l-\beta_l}}{\sigma^2(H)} \\ &= \frac{3c^2}{2Df^2}\end{aligned} \tag{10}$$

all the solution curves in state space of system is converging asymptotically to left boundary $H \to 0$ when left boundary $H \to 0$ is attractive to nature, which means $c_l < 1$.

Similarly, it can be get that $\sigma^2(H) = \infty$ and $m(H) = \infty$ when $H \to \infty$, then the right boundary $H \to \infty$ belong to the second singular boundary and the drift and diffusion coefficients converge asymptotically to the two equations

$$m(H) \to O(H^2),\ \sigma^2(H) \to O(H^2) \tag{11}$$

then, the diffusion exponent α_r, drift exponent β_r and character value c_r of right boundary can be obtained

$\alpha_r = 2,\ \beta_r = 2$

$$c_r = \lim_{H\to\infty}\frac{2m(H)|H|^{\alpha_r-\beta_r}}{\sigma^2(H)} \tag{12}$$

Since $\beta_r > \alpha_r - 1$ and $m(+\infty) < 0$, we can get the right boundary $H \to \infty$ is entering boundary.

Lemma 3.2. The conditions to keep system global stable can be obtained: the left boundary $H \to 0$ is attractive to nature and the right boundary $H \to \infty$ is entering boundary, which means $3c^2/2Df^2 < 1$

4 Stochastic hopf bifurcation

There are two kinds of stochastic bifurcations, one is dynamic bifurcation (D-bifurcation), which is used to study the stationary probability density of system response and determined by the value of lyapunov exponent. Another is phenomenology bifurcation (P-bifurcation), which is used to study the joint probability density, and determined by the number, place and shape of peak.

The averaging FPK equation [10] of Eq. (4) is

$$\frac{\partial f}{\partial t} = -\frac{\partial}{\partial H}\left[m(H)f\right] + \frac{1}{2}\frac{\partial^2\left[\sigma^2(H)f\right]}{\partial H^2} \tag{13}$$

then the solution of Eq. (13) is

$$f(H) = C\exp\left[-\int_0^h \frac{\frac{d\sigma^2(t)}{dt} - 2m(t)}{\sigma^2(t)}dt\right] \tag{14}$$

then the stationary probability density function can be got

$$f(H) = CH^{\eta}\exp[-H] \tag{15}$$

where $\eta = \dfrac{3c^2}{2Df^2} - 2$, C is normalized constant.

Then, the jointly stationary probability density function also can be got

$$f(q,p) = C\left[\frac{1}{2}p^2 + \frac{k}{2}q^2\right]^{\eta}\exp\left[-\left(\frac{1}{2}p^2 + \frac{k}{2}q^2\right)\right] \tag{16}$$

The $f(H)$ is a Delta function when $\eta < -1$, which is a monotone decreasing function and gets maximum value when $-1 < \eta < 0$, and the first bifurcation will be appeared. The $f(H)$ will still get the maximum value when $\eta > 0$ but the maximum value deviates from the origin, and the second bifurcation will be appeared. Then we have the following conclusion from above analysis. The system (12) will undergoes stochastic D-bifurcation when $\eta = -1$, and will undergoes stochastic P-bifurcation when $\eta = 0$, finally, the Hopf bifurcation of the system is constituted by these two bifurcations.

5 Numerical example

We choice c as the bifurcation parameter, the qualitative variation of the system response with the change of the parameter c was observed, and mainly discuss the Hopf bifurcation at $\eta = 0$, and observe the numerical results of stationary probability density function and jointly stationary probability density function with variation of parameter c. We consider the system with the following coefficients: k=1, $D = 0.5$, $f = 0.5$.

The images of stationary probability density function and jointly stationary probability density function show in Fig 1-4, when c varies near $c = 0.7348$. The stationary probability density changed from Monotone decreasing function to single peak function and Hopf bifurcation appeared. Similarly, the jointly stationary probability density changed from single peak shape to crater-shape and Hopf bifurcation appeared when $c = 0.7348$.

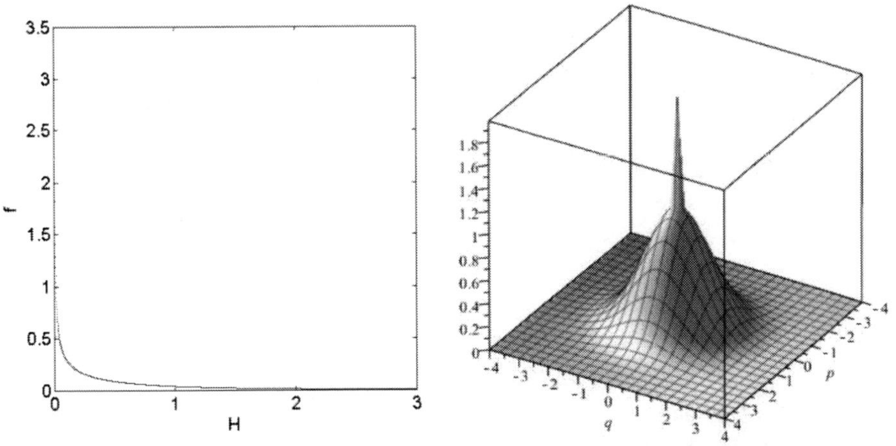

Figure 1. Stationary probability density and jointly stationary probability density when $c = 0.5$, $\eta = -0.67$.

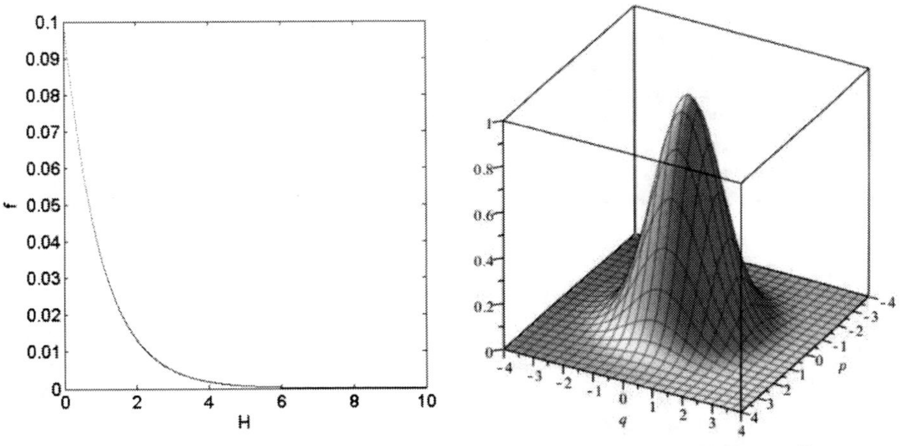

Figure 2. Stationary probability density and jointly stationary probability density when $c = 0.7348$, $\eta = 0$.

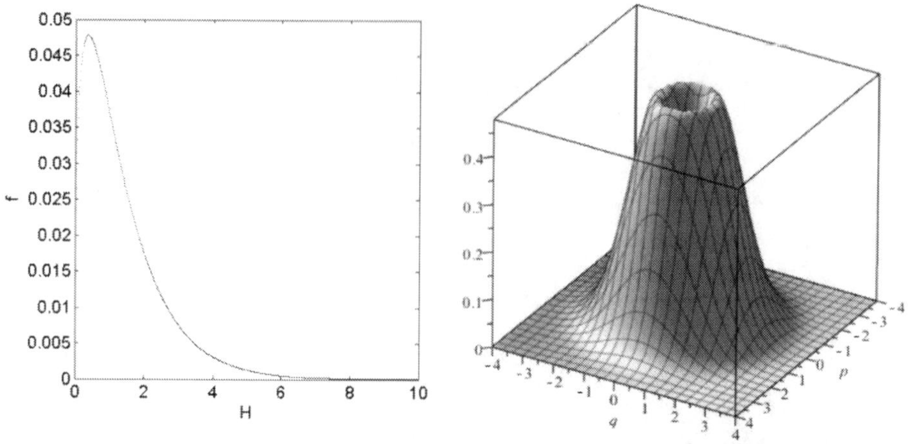

Figure 3. Stationary probability density and jointly stationary probability density when $c = 0.8$, $\eta = 0.37$.

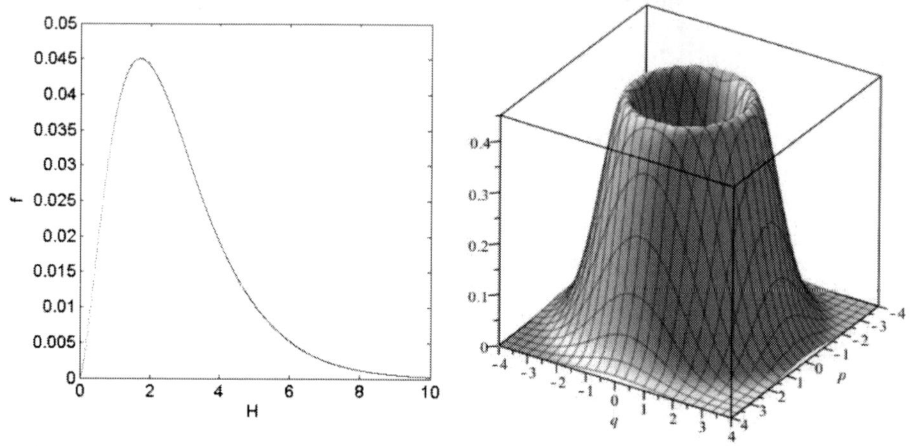

Figure 4. Stationary probability density and jointly stationary probability density when $c = 1$, $\eta = 1.7$.

6 Conclusion

In this paper, the stochastic stability and Hopf bifurcation of nonlinear damping in a high speed rotor-bearing system with random excitation are discussed by using the method of quasi-nonintegrable Hamiltonian system theory. The local stability of the system can be judged by largest Lyapunov exponent, then, we get the parameter conditions when Hopf bifurcation appears. Finally, some numerical results are given to illustrate the results. In the following study, the stochastic averaging method will be used to transform the Hamilton function into a one-dimensional Ito diffusion process, and will establish the reliability function and first-passage time to study the effect of first-passage phenomenon on system characteristics.

Acknowledgment

This The authors gratefully acknowledge the support from the National Natural Science Foundation (No. 61364001) and Science and Technology Program of Gansu Province (No. 144GKCA018), China.

References

1. L. Xiang, A. J. Hu, L. L. Hou, Nonlinear coupled dynamics of an asymmetric double-disc rotor-bearing system under rub-impact and oil-film forces, Applied Mathematical Modelling 40 (2016) 4505-4523.
2. N. Truhar, Z. Tomljanovic, K. Veselic, Damping optimization in mechanical systems with external force, Applied Mathematics and Computation 250 (2015) 270-279.
3. J.Z. Xu, D.X. He, X.L. Zhao, Status and prospects of Chinese wind energy, Energy2010;35:4439e44.
4. W.Y. Liu, W.Q. Zhu, Stochastic stability of quasi-integrable and resonant Hamiltonian systems under parametric excitations of combined Gaussian and Poisson white noises, International Journal of Non-Linear Mechanics 67 (2014) 52-62.
5. X.D. Gu, W.Q. Zhu, A stochastic averaging method for analyzing vibro-impact systems under Gaussian white noise excitations. JS. Vibration. Vol. 333, No. 9, 2632-2642 (2014).

6. M.H. Jalali, M. Ghayour, S. Ziaei-Rad, B. Shahriari, Dynamic analysis of a high speed rotor-bearing system, Measurement 53 (2014) 1-9.
7. W.Q. Zhu, Z.L. Huang, Y.Q. Yang, Stochastic averaging of quasi integrable Hamiltonian systems. J. Appl. Mech. Vol. 64, 157-64 (1997).
8. W.T. Jia, W.Q. Zhu, Y. Xu, Stochastic averaging of quasi-non-integrable Hamiltonian systems under combined Gaussian and Poisson white noise excitations, International Journal of Non-Linear Mechanics 51 (2013) 45-53.
9. Pavel Chigansky, An ergodic theorem for filtering with applications to stability, Volume 55, Issue 11, November 2006, Pages 908-917.
10. W.Q. Zhu, Z.L. Huang, Stochastic Hopf bifurcation of quasi nonintegrable Hamiltonian systems, Int J Nonlin Mech 1999; 34: 437-47.

MATEC Web of Conferences 95, 04009 (2017) DOI: 10.1051/matecconf/20179504009
ICMME 2016

Design of Gear Churning Power Loss Measurement Device

Bin Wang[1], Ya Jie Zhou[2] and Ping Wang[1]

[1]*Yancheng Institute of Technology, School of Mechanical Engineering, 224051 Yancheng, China*

[2]*Jiangsu University, School of Mechanical Engineering, 212013 Zhenjiang, China*

Abstract. To explore the impacts of gear churning power losses, a research was conducted to achieve the internal causes of power losses of churning gear by designing a gear churning power losses measurement device. The gear churning power losses could be influenced by different gear modules, the number of teeth and the axial position of gear. Finally, the impacts of gear churning power losses were discussed by comparing experimental data and theoretical data.

1 Introduction

At present, there is a complex relationship between gear churning power losses and some factors which include gear speed, the level of lubrication oil, lubrication oil viscosity and temperature. In addition, it is difficult to directly determine by theoretical analysis in which most of them are empirical formula. Thus, it is usually confirmed by carrying out experiments.

Research methods of load-independent power consumption were described by Seetharaman et al [4]. Furthermore, Seetharaman et al state that most of the previous calculation formulas are derived from the semi-empirical formula or the basis of the disk. Moreover, Seetharaman et al [5] claim that gear speed, the level of lubrication oil and tooth thickness can mostly influence churning power losses as a result of the method of gear churning experiment on a pair of spur gears which are submerged in oil bath. Also, they established the theoretical model and calculation formula of the churning losses of a pair of spur gears. The overview of classical literature on the power consumption of gear transmission was conducted by Valerity Stavytskyy et al [6]. They listed calculation formulas of spur gear, helical gear and bevel gear churning losses and windage losses, which were established by scholars during nearly half a century, so as to provide reference for future generations to understand and master the research content and direction of the load-independent power consumption. Yasutsune et al [7] conducted a research by using a flash frequency detector which aims to analyze and discuss the characteristics of churning losses. They indicated that the churning losses mainly consists of four parts: 1) Power losses are generated by the lubricating oil which is in accelerating gear; 2) Power losses are generated by lubricating oil which move in the meshing gear cavity; 3) Power losses are generated by lubricating oil splash in the pool; 4) When the lubricating oil come into meshing gear position, power losses are generated by bending lubricating oil. Changenet et al [8, 9] pointed out that most of the current gear churning losses formulas can be more precise calculate the churning losses that were generated in meshing gear or single disk, Petry-Johnson et al [10] obtained same calculation results by analysing the experimental data which is measured under different conditions. Kahraman et al. [5] claim that the churning losses is mainly divided into two parts: 1) When the gear is rotated, the power losses were caused by the movement of the lubricating oil stuck on the gear (This part includes three aspects: power losses were generated by interaction of the periphery of gear and lubricating oil; power losses were generated by interaction of the face of gear and lubricating oil; Scroll power losses were generated by lubricating oil in gear cavity); 2) When the gear is meshed, the gap between the two gears is changed and makes the oil being sucked and extruded which cause power losses. At present, the researchers are only conducted about the power consumption of single-stage gear or disk [11]. Boness [12, 13] has measured the drag torque of different size disks which stir in different level of the lubricating oil, compared to Karman's formula [14, 15] for calculating the churning losses and summed up the formula and method for calculating the drag torque of churning losses. In recent years, the churning power losses are concentrated by foreign countries in the helical gears, bevel gears and circular arc gears in the high speed transmission gear box [16, 17]. There is no relevant literature at home and abroad for the problem of churning losses of spur gears and helical gears in medium and low speed drives. The churning losses of gear transmission [18, 19, 20] mainly includes friction power losses, meshing gear power losses, friction power losses of bearing, lubricating oil power losses and so on. They are interacted which makes it impossible to draw a separate assessment of the churning losses. Churning losses are also known as the load-independent losses. Because of the uncertainty of the reasons for the churning losses, it is difficult to discuss it.

In response to these problems, a measuring device for churning losses will be designed and tested to infer the defect of the theory research. In addition, the churning losses can be measured. The aim of this research is to explicit the relationship between the churning losses and the gear modulus, the number of teeth and the axial distance of the gear and find ways to reduce the power consumption which in turn improve the energy efficiency.

2 Overall design

In order to discuss the influence of internal spur gear on the churning power losses, a measuring device for churning power losses is designed. As shown in Fig.1, 1 is oil tank, 1-3 temperature sensor, 1-4 oil inlet, 1-5 oil outlet, 2 churning gear, 3 single-axis actuator, including ball, guide bar, bearing block which fixes gear shaft, bearing block displacement monitor and rotation wheel. 4 gear shaft (1), 5 gear shaft (2), 6 coupling (1), 7 drive motor, 8 coupling (2), 9 motor load, 10 speed and torque sensor (1), 11 speed and torque sensor (2), 12-1 drip tube, 12-2 valve, 12-3 drums, 13-1 stents, 13-2 ball screw (2), 13-3 guide bar (2), 13-4 slider (1), 13-5 rotation wheel (2), 14-1 oil-dripping device, 15 cross-shaped mouth.

The rotational speed of drive motor is adjusted and controlled by a controller. A coupling is used to connect the drive motor to speed and torque sensor. It measures speed and torque of the drive gear by speed and torque sensor, a coupling is used to connected the load motor (9) to speed and torque sensor, it calculates the power losses by the parameters of the input and output speed and torque sensor. Under the same conditions of other parameters, through single-axis actuator changes the axial distance of two gear, it calculates the churning losses by the parameters of speed and torque sensor and bearing block displacement monitor to research the gear churning power losses on different axial distance.

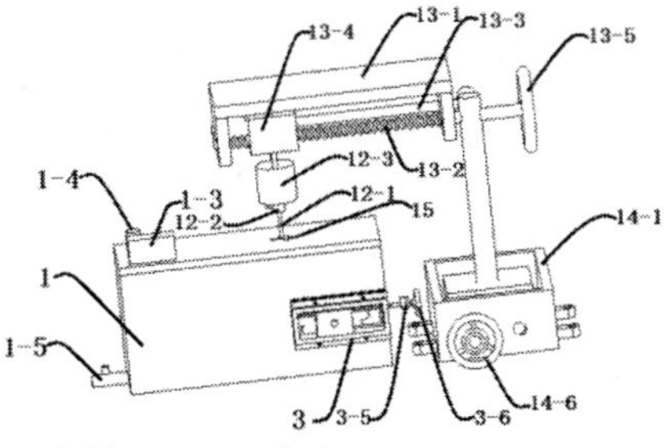

Figure 1. Measurement device.

3 Design of the tank

This gear churning power losses measurement device is mainly composed of gear, box, single axis actuator. As shown in Figure 1, the device comprises tank (1), two meshing churning gears (2) and a single-axis actuator (3) which is fixed on the side of rectangular track grooves in tank, One of the churning gear (2) is fixed on gear shaft (4) which is fixed on bearing block in single-axis actuator, another churning gear (2) is fixed on the tank (1) by gear shaft (5) and bearing block. A gear shaft (4) is connected to the output of drive motor (7) through coupling (6), the gear shaft (5) is connected to the output load motor (9) by coupling (8), while coupling (6) and coupling (8) respectively provided with speed and torque sensor (10), speed and torque sensor (11), it installs locking device inside of the rotation wheel on the ball to make that ball screw is not rotating while churning gear is rotating, drive motor (7) installs in the carriage which is moved with the moving of bearing block. A temperature sensor is fixed on the upside of the tank, which extends deeply into the oil of the tank. Oil port 1-4 is also fixed on the upside of the tank, and there is a port 1-5 on one side of the box, which is controlled by the valve. It sets a transparent surface on side of the tank (1) to observe the movement of the two churning gear. Single-axis actuator (3) is installed in the rectangular slot which is provided in the transparent surface of the tank (1) by four nuts.

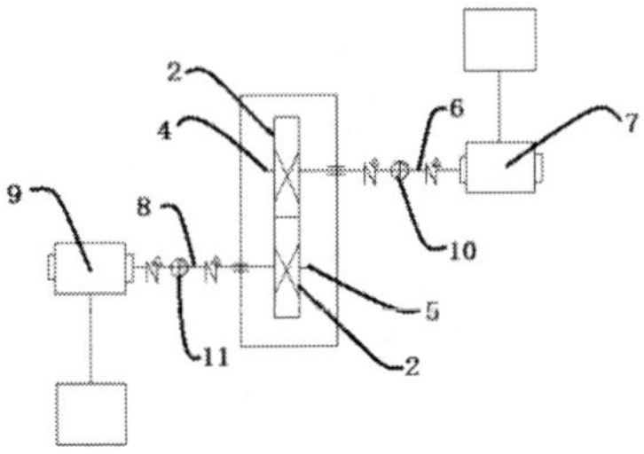

Figure 2. Overall device.

Temperature sensor measures the changing values of churning gear temperature, it injects lubricating oil though the oil inlet on the upside of the tank. The level of lubricating oil is adjusted by oil outlet on the side. It is easy to adjust the position of the oil pipe by the cross-shaped mouth which is installed on the upside of tank. Effect of axial distance on power consumption of gear can be measured by applying a single-axis actuator to adjust two axis distance of gear, measuring its change distance and calculating the power consumption of multiple sets of center distance of the meshing gear pair.

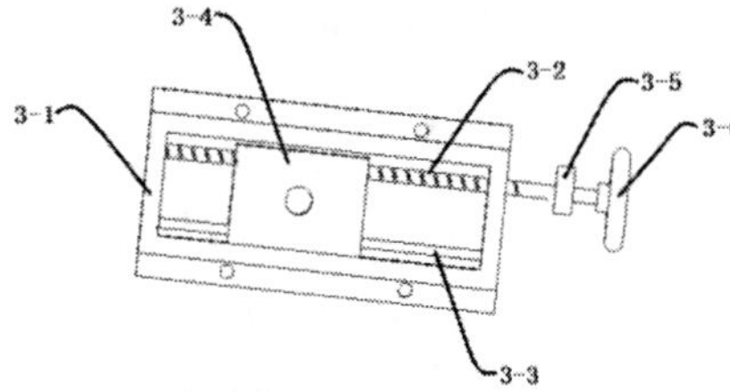

Figure 3. Single shaft drive.

4 Design of oil-Dripping device

Oil-dripping device is installed above the meshing gears, the oil-dripping device includes a drip tube 12-1 which controls the speed of drip oil by a valve 12-2, the drum 12-3 is connected to the top of drip tube, the X, Y direction moving device adjust the position of drip tube 12-1 with the change of the meshing gear position. At the same time, the position of drip tube is adjusted with the change of the gear meshing points though the cross-shaped mouth which was installed on the top of the tank. the X-direction moving device comprises a stent 13-1 which includes ball 13-2 and guide bar 13-3, the slider were provided on the ball 13-2 and guide bar 13-3, on the side of ball 13-2 was installed a rotation wheel 13-5 while a slider 13-4 is fixedly connected to the top of the tank 12-3; The Y-direction moving device includes a base, ball and guide bar which are installed in the bottom, the slider (2) is installed on the ball screw (3) and guide bar (3), a pillar is vertically installed on slider (2), one side of the ball (3) installs a rotation wheel (3) while pillar and ball 13-2 are connecting to guide bar 13-3, the base is installed on the table by four nuts.

Gear meshing position is changed when measuring different gear modulus. Common churning device can not adjust the position of the drip tube. The device changes the X, Y direction position which is correspond with the change of meshing gear position by two drive devices, it reduces friction in meshing gear to reduce churning power losses.

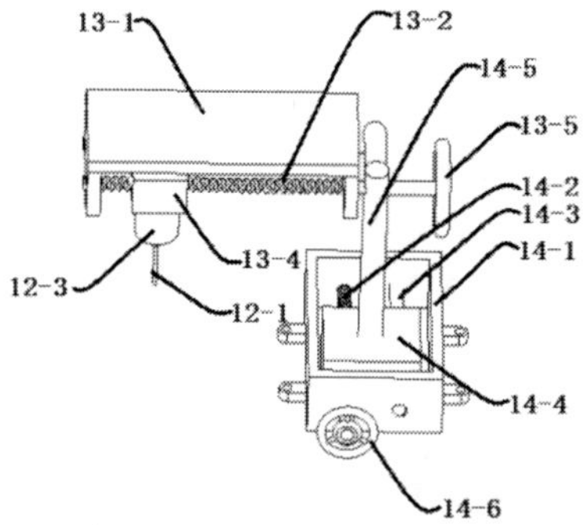

Figure 4. Oil-dripping device.

5 Testing principle of gear churning power losses measurement device

5.1 Experimental test methods

Controller adjusts the drive motor torque, motor load is adjusted by the speed controller.

The specific steps are as follows:

Power test points are respectively setting as $p_1 \ldots p_i \ldots p_j$, the speed of drive motor is setting respectively as $n_1 \ldots n_i \ldots n_j$;

At measurement point p_1, the speed of drive motor is setting as n_1, it maintains the drive motor which runs for some time at this speed to make that the input power of drive motor and the output power of the dynamometer motor keep stability. The input torque of drive motor is Tin-1, the output torque of dynamometer is Tout-1; then testing the input torque and output torque of the next point p_2, until all testing points p_1 ... p_i ... p_j are measured. Then, the input torque of drive motor is Tin-1 ... Tin-i ... Tin-j, the output torque of the dynamometer motor is Tout-1 ... Tout-i ... Tout-j;

Finally, gear churning losses of the test point p_i is:

$$P_i = \frac{(T_{out-i} - T_{in-i}) \times n_i}{9550} \quad (i=1\ldots m) \tag{1}$$

5.2 Theoretical calculation

The oil-churning power losses P_C consist of three parts: P_{c1} expresses stirring loss related to the outer diameter of the optical axis; P_{c2} expresses the churning losses with the two side of gear; P_{c3} expresses the churning losses related to the tooth surface (the outer surface of the two gears). It can be expressed by the following formula.

$$P_C = P_{C1} + P_{C2} + P_{C3} \tag{2}$$

$$P_{C1} = \frac{7.37 f_g v n_i^{\,3} D_o^{4.7} L}{A_g \times 10^{26}} \tag{3}$$

$$P_{c2} = \frac{1.474 f_g v n_i^{\,3} D_O^{5.7}}{A_g \times 10^{26}} \tag{4}$$

$$P_{C3} = \frac{7.37 f_g v n_i^{\,3} D_o^{4.7} B(\frac{R_f}{\sqrt{\tan\beta}})}{A_g \times 10^{26}} \tag{5}$$

where f_g is gear submergence coefficient (under oil-injecting condition, f_g is set as 0; under full immersion condition, f_g is set as 1; under other circumstances, f_g has a value between 0~1 according to the height of immersion.); v is the kinematic viscosity of lubricating oil under working temperatures; n_i is the rotational speed of the ith gear; D_o is the size of the outer diameter of the ith gear; L is the length of shaft; A_g is a constant; B is the width of big gear or small gear; R_f is roughness factor, it is related to tooth size, it can be obtained by approximate formula:

$$R_f = 7.93 - \frac{4.648}{m_t} \tag{6}$$

where Mt is Gear transverse module; β is helical gear spiral Angle, when $\beta \le 10^{\circ}$, $\beta = 10^{\circ}$.

5.3 Experimental data analysis method

- When the meshing gears are of same modulus, different tooth thickness and the other working conditions (working temperature, the level of lubricating oil, etc.) keep unchanged, the power losses of oil churning can be calculated by the following process:

The parameters used for calculating gear power losses P_i can be attained from speed and torque sensor (1), speed and torque sensor (2) and temperature sensor. Thus the value of gear power losses P_i can be easily calculated by Formula 1 and 2.

- The modulus of the meshing gears are different while the other parameters are equal.

Under this condition, the equation of centre distance of the meshing gear pair can be written as:

$$a = \frac{(d_1 + d_2)}{2} = (z_1 + z_2) \times \frac{m}{2} \quad (7)$$

where d_1 and d_2 respectively represent circle diameter of gear 1 and 2, z_1 and z_2 represent the number of teeth on the gear 1 and 2, m is gear modulus, the centre distance of two gears can be obtained by formula 7. Through Single-Axis Actuator adjusts the position of meshing gear, which is corresponding with the position of oil-dripping device, through the parameters of the sensor, it respectively calculates the churning power losses P_i by Formula 1 and 2.

- It changes the axial distance between two gears while the other parameters are equal. It measures the change parameters in different axial distance by speed and torque sensor and respectively calculates the churning power losses Pi by Formula 1 and 2.

6. Conclusions

The gear churning power consumption measuring device can easily measure the gear churning power losses in different modulus, different numbers of teeth and different gear axial distance. It respectively obtains experimental data and theoretical data by Formula 1 and 2. It obtains key factors which impact gear churning power losses by comparing and analysing these parameters. Then, it explores methods to reduce gear churning power losses.

Acknowledgment

The authors gratefully acknowledge the financial support of Jiangsu Provincial Research Foundation for Basic Research (Jiangsu Provincial Natural Science Foundation, China) under Grant No.BK20131221 and Jiangsu Provincial Science and Technology Program under Grant No. BY2015057-38.

References

1. W.H. Liang, K.Liu, Y.H. Cui, Engineering Mechanics. J. 05, 220-226 (2016)
2. A. Kahraman, D.R Hilty, A. Singh, Mechanism & Machine Theory. J. 86(1), 48 - 61 (2015)
3. A. Kahraman, D.R Hilty, A. Singh. Internation Gear Conference. C. 1150 - 1161 (2014)
4. S. Seetharaman, A. Kahraman, Columbus: Ohio State University. D. 33 - 42 (2009)
5. S. Seetharaman, A. Kahraman, M.D. Moorhead, et al. Journal of Tribology. J. 131(2), 022202 - 022211 (2009)
6. S. Valeriy, N. Pavel, F. Pavel, et al. A Review. TEKA Kom. Mot. I Energ. Roln. - OLPAN. J. 10B, 205-213 (2010)
7. Y. Ariura, T. Ueno, T. Sunaga, et al. Transactions of the Japan Society of Mechanical Engineers. J. 38, 2403-2413 (1973)
8. R.W. Mann and C.H. Marston, ASME Journal of Basic Engineering. J. 83 (4), 719 – 723 (1961)
9. C. Changenet, P. Velex, ASME Journal of Mechanical Design. J. 129 (1), 128-133 (2007)
10. T.T. Petry-Johnson, A. Kahraman, N.E . Anderson, et al. Journal of Mechanical Design. J. 130(6), 373-8 (2007)
11. P.B. Pawar, A.A. Utpat, Materials Today Proceedings, J. 2 (4 - 5), 2968-2974 (2015,)
12. R.J. Boness, Proc.ASME Int. Power Trans. Gearing Conf,Chicago, J. 1, 355-359 (1989)
13. R.J. Bones, Proceedings of the ASM 5th International Power Transmission and Gearing Conference, Chicago, J. 355-359 (1989)
14. T. Von Karman, National Advisory Committee for Aeronautics, R. (1946)
15. B. Wang, X.B. Chen, Journal of Tongji University. Nature Science, J. 42 (12), 1904 - 1911 (2014)
16. B. Wang, Z.J. Lu, X.B. Chen, et al. International Journal of Automotive Technology, J. 16 (3), 417-426 (2015)
17. Y. Diab, F. Ville, H. Houjoh, P. Sainsot, P. Velex, Proceedings of the Institution of Mechanical Engineers, Part C, Mechanical Engineering Science, J. 219, 785-800 (2005,)
18. Z.X. Li, Y. Jiang, H. Chao, et al. A Review. Measurement, J. 90, 4 - 19 (2016)
19. H. Blok, Proceedin of the Institution of Mechanical Engineers, J. 2, 222-235 (1957)
20. B. Wang, Z.J. Lu, X.B. Chen, et al. International Journal of Automotive Technology, J. 16(3), 417-426. (2015)

Surface polish of PLA parts in FDM using dichloromethane vapour

Yifan Jin, Yi Wan and Zhanqiang Liu

Key Laboratory of High Efficiency and Clean Manufacturing, School of Mechanical Engineering, Shandong University, Jinan, China

Abstract. Fused deposition modelling has become one of the most diffused rapid prototyping techniques, which is widely used to fabricate prototypes. However, further application of this technology is severely limited by poor surface roughness. Thus it is necessary to adopt some operations to improve surface quality. Chemical finishing is typically employed to finish parts in fused deposition modelling (FDM). The purpose of this paper is to decrease the surface roughness for polylactic acid (PLA) parts in FDM. The chemical reaction mechanism during the treating process is analysed. Then NaOH solution and dichloromethane vapour are used to treat FDM specimens respectively. A 3D laser microscope has been applied to assess the effects in terms of surface topography and roughness. The experimental results show that treatment using dichloromethane vapour performs much better than NaOH solution. Compared with the untreated group, surface roughness obtained through vapour treatment decreases by 88 per cent. This research has been conducted to provide a better method to treat PLA parts using chemical reagents.

1 Introduction

Fused deposition modelling (FDM) is the most diffused rapid prototyping (RP) technique in the world [1]. The main advantages of this technology include: low maintenance cost, toxic free materials, quick production of complexity parts, and high utilization rate of raw materials. These aspects make it suitable for a wide range of applications in aerospace, automotive, biomedical, customer product industry, design and tooling [2-4]. Nevertheless, the main disadvantages are anisotropic property, requirement of support structures for overhangs and high surface roughness [5]. Particularly, the poor surface limits the further application.

Several researches have been investigated to reduce surface roughness of FDM parts by optimizing fabrication parameters. A study conducted by Thrimurthulu [6] et al. aimed to obtain an optimum part deposition orientation by using a real coded genetic algorithm. Ahn [7,8] et al. proposed a model to represent average roughness by assuming the filament profile as an elliptical curve, and regarded cross-sectional shape, surface angle, layer thickness and overlap between adjacent layers as the main factors affecting surface quality. An approach for adaptive slicing that applied the image processing technique to determine appropriate thickness and to recommend slicing positions on a 3D model was proposed, then it was implemented on the LabVIEW platform and compared with other approaches to reduce build time while maintaining surface quality [9]. Mahapatra and Sood[10] proposed use of artificial neural network model to determine the relationship between five input parameters, subsequence bayesian regularization was adopted for selection of optimum network architecture to obtain reasonably good surface quality.

Whereas, roughness on the vertical orientation caused by staircase effect cannot be avoided through optimizing fabrication parameters. Hence, post-treatment operations to improve surface quality is needed.

At present, two primary approaches are used to achieve smooth surfaces on parts: chemical and mechanical smoothing [11]. A chemical post-treatment was adopted after studying the influence of FDM machining parameters on acrylonitrile butadiene styrene (ABS) prototypes surface finish, subsequently, surface finish of products had been measured, which showed a significant improvement in surface finish at the expense of a negligible change in the dimensional accuracy[12]. Application of surface coatings is another approach to achieve a desired surface finish, in addition to adding strength to a finished part. Such coatings may include primers, paints or metallic electroplated coatings [13].

However, chemical polish of polylactic acid (PLA) has been rarely investigated, not to mention using different chemical reagents.

In this paper, two distinct chemical reagents are adopted to reduce surface roughness of PLA parts respectively, with one method using NaOH solution, and the other using dichloromethane (CH_2CL_2) vapour. Subsequently, the surface topography and roughness are observed by a 3D laser scanning microscope to verify the mechanism of chemical reaction analysed.

2 Theory analysis

2.1. Effects of fabrication parameters

Many fabrication parameters have impacts on the surface quality of FDM parts, such as layer thickness, nozzle diameter, printing speed, raster angle, shell thickness and infill overlap etc., which are shown in **Figure 1**. Moreover, different factors make diverse influences on the surface quality. The major ones are layer thickness and nozzle diameter. The thin layer produces a smoother surface than the thick layer, whether it is measured vertically or horizontally [14]. Surface roughness and part deposition time are always contradicted with each other. It is the same with nozzle diameter. However, in any case the staircase effect on the vertical orientation cannot be eliminated by optimizing fabrication parameters.

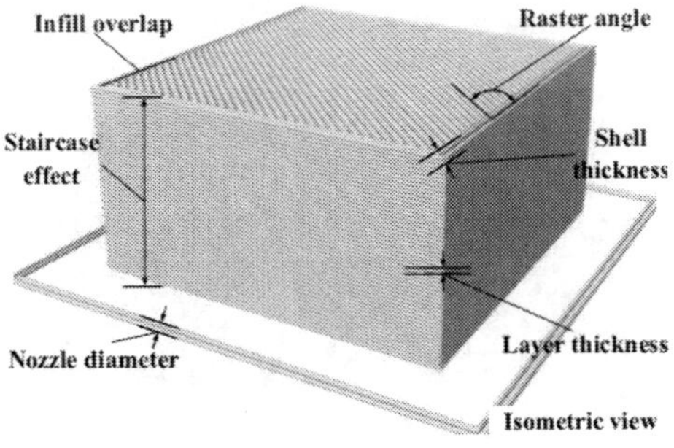

Figure 1. Illustration of partial fabrication factors affecting surface quality

2.1. Effects of fabrication parameters

Opposite to the polycondensation of esters, there are three methods during the hydrolysis process of PLA: (i) occurring on the main chain to generate monomers or oligomers, (ii) hydrolysis on the side chain to generate soluble main chain polymer and (iii) hydrolysis on the linkage chain to generate linear polymer [15]. And the principal way is (i), as shown in **Figure 2**. The hydrolysis of PLA begins with the absorption of water, then small molecules of water move to the surface of the sample, spreading around the ester bond or hydrophilic group. Especially in the medium of acid or alkali, the ester bond in macromolecular main chain is broken continually. Ultimately samples dissolve when the molecule reduces to a certain extent. Furthermore, the characteristic of this process is that the hydrolysis happens at the interface simultaneously, and small molecules generated leave the model surface to diffuse in aqueous solution.

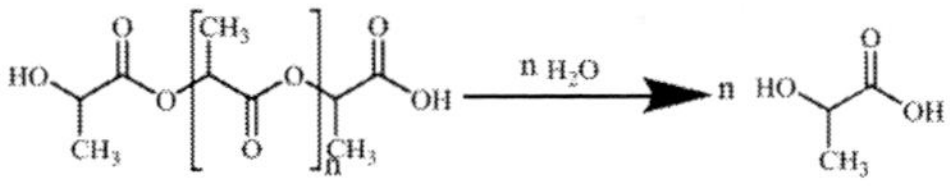

$$(C_3H_4O_2)_n + n\,H_2O \longrightarrow n\,C_3H_6O_3$$

Figure 2. Illustration of hydrolysis of PLA

2.3. Dissolution of PLA in organic solvent

For FDM parts made up of PLA, there exists a range of corrugated structures on the surface due to the staircase effect, which is supposed to be eliminated by using vapour treatment of easy volatile solvent such as chloroform, dichloromethane and tetrahydrofuran etc. Ahn [16] et al. has proposed a magnified cross-sectional shape of the deposited filaments, which is close to an elliptic curve. Appear on model surface, peaks and valleys become droplets through absorbing the volatile solvent. Then, under the effects of liquid surface tension and gravity, these two kinds of droplets blend continually, resulting in lower peak and filled valley. As it goes on, a dynamic balance will be achieved. Finally, a smoother surface can be obtained. The process of dissolving and flow on the profile of PLA part is shown in **Figure 3**. Laser Raman spectra shows that there exists no dichloromethane in the dried product, and it has been volatilized completely during this processing [17]. The main distinction compared with hydrolysis of PLA is that the dissolved molecules attach to the model surface during the progress.

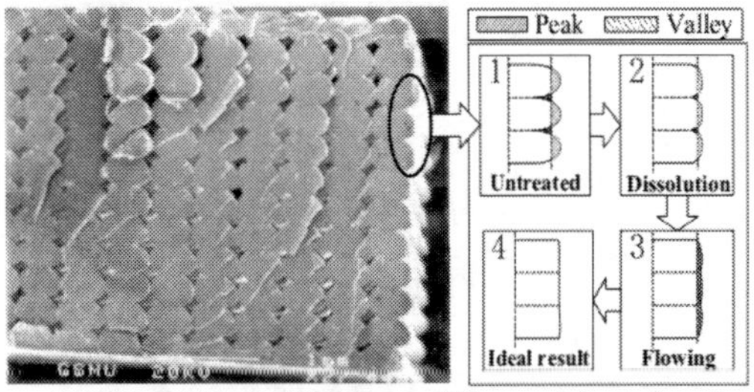

(a)Shape of deposited filaments (b)Process of chemical reaction

Figure 3. Process of dissolution and flow on the profile of PLA part. (a) Shape of deposited filaments [16] and (b) process of chemical reaction.

3 Experimental

In this paper, two chemical methods for enhancing the surface finish of specimens are investigated. The NaOH solution is adopted in order to treat the surface of specimens in the first method, and dichloromethane is used as the organic solvent in the second method. The progress is carried out over three steps. **Figure 4** displays the workflow of the experiment.

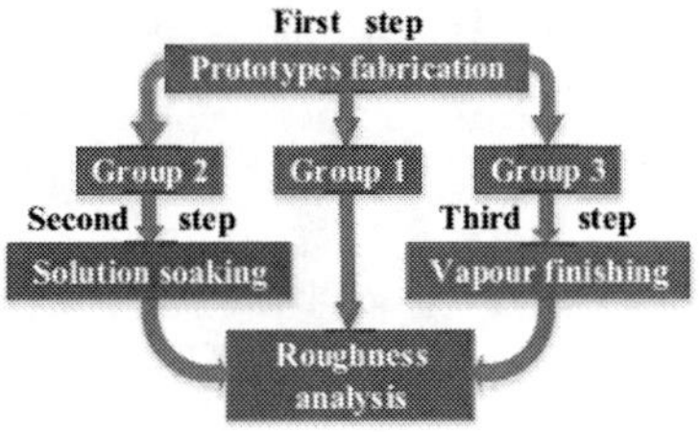

Figure 4. Experimental workflow

3.1. First step –specimens manufacture

A cuboid test specimen is designed and its dimension is shown in **Figure 5** with the unit used is millimetre. All specimens for groups of three are fabricated using a Prusa

DOI: 10.1051/matecconf/20179505001

Mendel desktop 3D printer, and the filament feedstock is PLA. Printing parameters are adopted by empirical values in order to eliminate the manufacture error. Table 1 shows the main printing factors above-mentioned in Chapter 2.1. Other factors are maintained at their default levels.

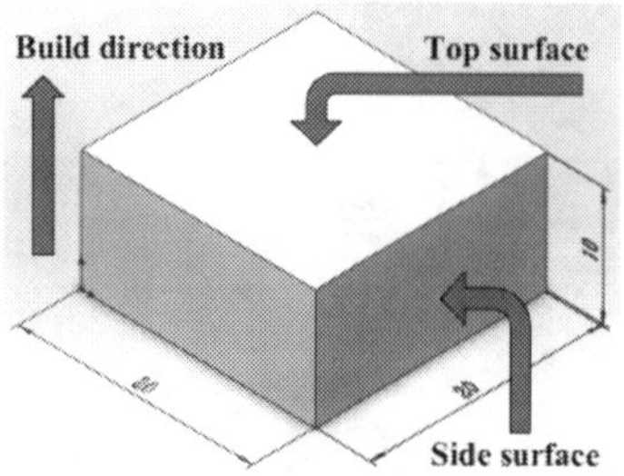

Figure 5. Shape and dimension of specimens

Table 1. Printing factors and their levels

Factor	Level	Unit
Layer thickness	0.2	mm
Raster angle	45	°
Nozzle diameter	1.0	mm
Printing speed	30	mm/s
Shell thickness	1.0	mm
Infill overlap	10%	-

3.2. Second step - treatment with NaOH solution

In the second step of the experimentation, the specimens of group two were subjected to treatment of soaking in NaOH solution. Heating by an alcohol lamp is added for the reason that reaction performs quite slow at room temperature. The process was carried out by immersing in the 5 wt. % NaOH boiling solution for half an hour. Subsequently, the roughness of side and top surfaces of all the specimens were measured by the laser microscope for 3D & profile measurement, in order to evaluate any differences happened. Three specimens were assessed to avoid the accidental error.

3.3. Third step – treatment with dichloromethane vapour

In the third step, the specimens of group three were subjected to an organic chemical treatment. The dichloromethane was selected according to its low cost, very low toxicity and extremely high diffusion. Before the test, a processing time trial was made to decide the suitable treatment time. The process was undertaken in a glass bottle full of dichloromethane vapour for 300s. The operation was conducted in room temperature at 25℃ and the pressure of vapour using is at normal atmospheric pressure. The surface roughness of three specimens was evaluated using the same laser microscope and compared with the above two groups in order to evaluate the improvement of surface finish.

4 Results and discussion

4.1. Analysis of surface roughness

Surface roughness were measured by a laser scanning microscope (KEYENCE VK-X200). Images were taken over a 1.414*1mm area of the side surface. Surface roughness was analysed by the VK Analyzer 3.3 software (KEYENCE, JAPAN).

In **Figure 6**, the 3D surface topography and its line roughness of side surfaces are shown. Three colourful lines show the position of the line roughness, while the corresponding curves below represent the line roughness measured. Surface defects near the edges were avoided.

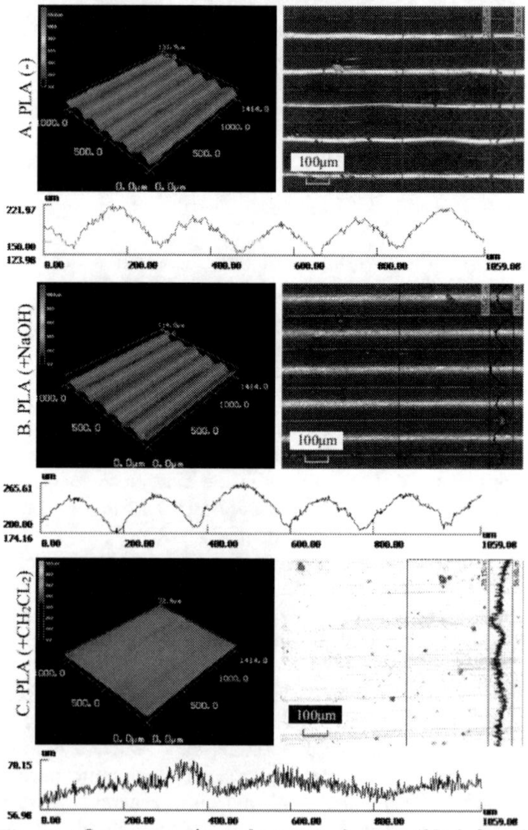

Figure 6. Laser scanning microscope images (3D-left and 2D-right) of the side surface showing before (-) and after (+) chemical treatments. (A) Untreated PLA, (B) treated PLA (+NaOH), (C) treated PLA (+CH_2CL_2). All images are at the same scale.

As shown in **Figure 6**, the roughness of the specimens in group three have been greatly reduced by using the dichloromethane vapour treatment described above. Roughness before and after chemical treatments is reported in Table 2. Compared with group one, the surface roughness of group two reduced by 13 per cent. However, the surface treated by NaOH solution loses its lustre which can be obviously seen in the middle of **Figure 6**. Further investigation on the nano-level needs to be conducted. An upgrade reflection effect of light is obtained in group three. Meanwhile, the surface line between neighbour layers cannot be observed by the naked eye. A markedly roughness reduction is obtained with a lowering of 88 per cent than group one, and 86 per cent than group two, which indicates the treatment used in group three performs much better than that used in group two. In other words, the method using dichloromethane vapour almost eliminates the influence of staircase effect. As to the top surface, the results show that there is poorly difference amongst the three groups. The roughness of top surface is decided by many factors except the staircase effect. Hence, the measured results have been elided in this paper.

Table 2. Roughness on side surface of three groups

	Group number	Specimens number	Ra (μ m)	Average Ra(μ m)
Before treatment	I	1	18.84	20.46
		2	22.33	
		3	20.20	
After NaOH treatment	II	4	17.34	17.79
		5	19.15	
		6	16.92	
After CH_2CL_2 treatment	III	7	1.90	2.46
		8	2.47	
		9	3.00	

In Boschetto and Bottini [18], through barrel finishing, the Ra decreases from 16.54 to 4.23 μ m in 960 minutes for vertical walls of ABS FDM parts. In this paper, the results show that after vapour smoothing for just 30 seconds the peak and valley of the surface cannot be distinguished any more, with average Ra passes from 20.46 to 2.46 μ m. However, there still exists a tiny uneven area as shown at the bottom of **Figure 6**. The reason for this phenomenon is unknown. Better surfaces may be produced if a regularly vibrating treatment applied on the specimens is adopted simultaneously with vapour treatment, which may promote the flowing droplets to homogenize.

4.2. Case study

The investigation of this paper has proposed a chemical method to improve the surface finish, which can be used to fabricate parts. To provide a case of treatment on free form surfaces, a cartoon character (**Figure 7** – left side) was manufactured using the vapour post-treatment. Then the prototype was immersed in the chemical vapour for thirty seconds and its roughness was measured on a 1.414*1mm area considered on the front part. The roughness reduces from an average Ra equal to 18.24μ m to 3.77μ m. The 3D map and 2D surface topography are shown in **Figure 7** – right side. It can be observed that the surface quality has been greatly improved.

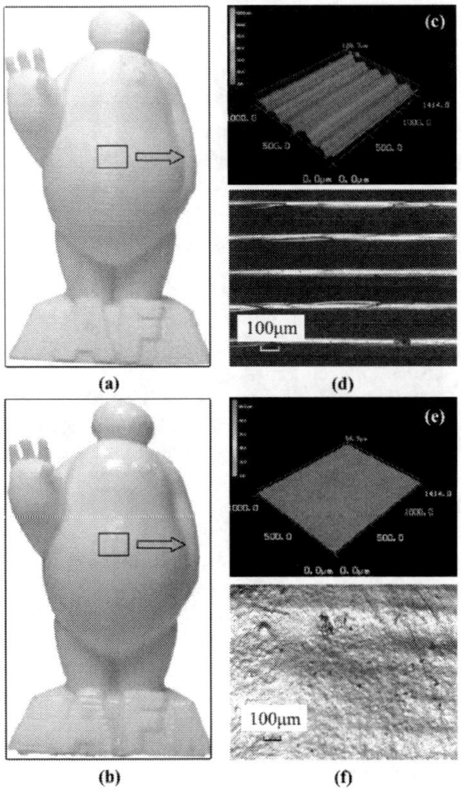

Figure 7. Images of free form surface for (a) untreated and (b) treated by dichloromethane vapour

5 Conclusion

In this paper, the chemical reaction process of FDM prototypes made up of PLA is analysed originally. Using different chemical reagents, two methods of improving the surface finish are investigated, with vapour treatment yielding a significant enhancement of surface roughness. Compared with untreated specimens, average Ra is found to decrease by 88 per cent using dichloromethane vapour and near mirror quality surfaces can be obtained regardless of staircase effect. While the specimens treated by NaOH solution can also decrease average Ra in some degree. The proposed chemical treatment is economic, fast and easy to use. With simple treatment and little time, the experimental findings provide the operator a method to polish PLA parts, which can reduce instrument and labour cost in the post-processing operation. Further studies need to be conducted on the mechanical property after vapour treatment, particularly on the vertical orientation.

Acknowledgements

The authors would like to acknowledge the financial support from National Natural Science Foundation of China (51175306 and 51575320), Tai Shan Scholar Foundation and Fundamental Research Funds for the Central Universities (2014JC020).

References

1. Wohlers, T. (2012), "Executive Summary", Wohlers Report, Wohlers Associates, Fort Collins, CO. Wohlers, T. (2012), "Executive Summary", Wohlers Report, Wohlers Associates, Fort Collins, CO.
2. Chua, C.K., Leong, K.F. and Lim, C.S. (2010), Rapid Prototyping: Principles and Applications, World Scientific, River Edge, NJ.
3. Ingole, D.S., Kuthe, A.M., Thakare, S.B. and Talankar, A.S. (2009), "Rapid prototyping – a technology transfer approach for development of rapid tooling", Rapid Prototyping Journal, Vol. 15 No. 4, pp. 280-290.
4. Ivanova, O., Williams, C. and Campbell, T. (2013), "Additive manufacturing (AM) and nanotechnology: promises and challenges", Rapid Prototyping Journal, Vol. 19 No. 5, pp. 353-364.
5. Brooks, H., Lupeanu, M.E. and Piorkowski, B. (2013), "Research towards high speed extrusion freeforming", International Journal of Rapid Manufacturing, Vol. 3 Nos 2/3, pp. 154-171.
6. Thrimurthulu, K., Pandey, P.M. and Reddy, N.V. (2004), "Optimum part deposition orientation in fused deposition modeling", International Journal of Machine Tools and Manufacture, Vol. 44 No. 6, pp. 585-594.
7. Ahn, D.K., Kim, H. and Lee, S. (2009a), "Surface roughness prediction using measured data and interpolation in layered manufacturing", Journal of Materials Processing Technology, Vol. 209 No. 2, pp. 664-671.
8. Ahn, D.K., Kweon, J.H., Kwon, S., Song, J. and Lee, S. (2009b), "Representation of surface roughness in fused deposition modelling", Journal of Materials Processing Technology, Vol. 209 Nos 15/16, pp. 5593-5600.
9. Rianmora, S. and Koomsap, P. (2010), "Recommended slicing positions for adaptive direct slicing by image processing technique", The International Journal of Advanced Manufacturing Technology, Vol. 46 Nos 9/10/11/12, pp. 1021-1033.
10. Mahapatra, S.S. and Sood, A.K. (2012), "Bayesian regularization-based Levenberg–Marquardt neural model combined with BFOA for improving surface finish of FDM processed part", The International Journal of Advanced Manufacturing Technology, Vol. 60 Nos 9/10/11/12, pp. 1223-1235.
11. Gibson, I., Rosen, D.W. and Stucker, B. (2010), Additive manufacturing technologies: rapid prototyping to direct digital manufacturing, Springer, New York, NY.
12. Galantucci, L., Lavecchia, F. and Percoco, G. (2009), "Experimental study aiming to enhance the surface finish of fused deposition modeled parts", CIRP Annals-Manufacturing Technology, Vol. 58 No. 1, pp. 189-192.
13. Turner, B.N., Strong, R. and Gold, S.A. (2014), "A review of melt extrusion additive manufacturing processes: I. Process design and modeling", Rapid Prototyping Journal, Vol. 20 No. 3, pp. 192-204.
14. Bakar, N.S.A., Alkahari, M.R. and Boejang, H. (2010), "Analysis on fused deposition modelling performance", Journal of Zhejiang University Science A, Vol. 11 No. 12, pp. 972-977.
15. Siparsky, G.L., Voorhees, K.J. and Miao, F. (1998), "Hydrolysis of polylactic acid (PLA) and polycaprolactone (PCL) in aqueous acetonitrile solutions: autocatalysis", Journal of environmental polymer degradation, Vol. 6 No. 1, pp. 31-41.
16. Ahn, S.-H., Montero, M., Odell, D., Roundy, S. and Wright, P.K. (2002), "Anisotropic material properties of fused deposition modeling ABS", Rapid Prototyping Journal, Vol. 8 No. 4, pp. 248-257.
17. Kasuga, T., Fujikawa, H. and Abe, Y. (1999), "Preparation of polylactic acid composites containing β–Ca (PO3)2 fibers", Journal of materials research, Vol. 14 No. 2, pp. 418-424.
18. Boschetto, A. and Bottini, L. (2015), "Surface improvement of fused deposition modeling parts by barrel finishing", Rapid Prototyping Journal, Vol. 21 No. 6, pp. 686-696.

Laser Welding Process Parameters Optimization Using Variable-Fidelity Metamodel and NSGA-II

Chaochao Wang, Qi Zhou, Ping Jiang , Leshi Shu and Xiongbin Li

The State Key Lab of Digital Manufacturing Equipment and Technology, School of Mechanical Science and Engineering, Huazhong University of Science and Technology, Wuhan 430074, China

Abstract. An optimization methodology based on variable-fidelity (VF) metamodels and nondominated sorting genetic algorithm II (NSGA-II) for laser bead-on-plate welding of stainless steel 316L is presented. The relationships between input process parameters (laser power, welding speed and laser focal position) and output responses (weld width and weld depth) are constructed by VF metamodels. In VF metamodels, the information from two levels fidelity models are integrated, in which the low-fidelity model (LF) is finite element simulation model that is used to capture the general trend of the metamodels, and high-fidelity (HF) model which from physical experiments is used to ensure the accuracy of metamodels. The accuracy of the VF metamodel is verified by actual experiments. To slove the optimization problem, NSGA-II is used to search for multi-objective Pareto optimal solutions. The results of verification experiments show that the obtained optimal parameters are effective and reliable.

1 Introduction

Laser welding has been widely used in a variety of industrial applications due to its significant advantages including deep penetration, narrow heat-affected zone. It is very important to determine process parameters that are highly relevant to impact welding bead profile. Generally, a promising strategy is to utilize metamodels fitting the relationships between the input process parameters and the output responses for the sake of optimal welding bead in the laser welding process parameters optimization.

In the broadest sense, metamodel assisted process parameters optimization methods can be divided into two distinct types: physical experiment-based optimization approach and simulation-based optimization approach. Physical experiment-based optimization approach obtains the data for constructing the metamodel by conducting laser welding experiments [1-4]. The main shortcoming of this method is that it requires a significant time and economic costs to ensure the accuracy of the metamodel. Simulation-based optimization approach, on the other hand, obtains the bead profile by running computational simulation models [5-7]. Compared with the physical experiment-based optimization approach, the most obvious advantage is that it can greatly shorten the design period and reduce the overall cost. However, this approach is difficult to ensure the accuracy of metamodel. This issue, to some extent, has limited their capability of guiding the actual laser welding processing.

To overcome the above mentioned shortcomings, this paper proposes an optimization methodology based on variable-fidelity (VF) metamodel [8, 9], which is able to combine the information from two different levels fidelity models, low-fidelity (LF) simulation model and high-fidelity (HF) physical experiment, for laser welding process parameter optimization. In the proposed VF metamodels approach, the LF simulation model is a three-dimensional thermal finite element (FE) model that is used to capture the general trend of the metamodels and the HF model which from physical experiments is used to ensure the accuracy of metamodels. The two levels fidelity models are integrated into VF metamodels using the method of Hierarchical-Kriging (HK) [10]. The constructed VF metamodels will be used to descript the relationship between process parameters and welding bead profile. In optimization process, the nondominated sorting genetic algorithm II (NSGA-II) is adopted to facilitate design space exploration and multi-objective Pareto optimal search [11]. The proposed methodology is applied to obtain optimum process parameters of the laser bead-on-plate welding for stainless steel 316L.

2 Optimization methodology

In this paper, the optimization methodology is proposed where HK metamodel is used to establish the VF metamodels between the process parameters and welding bead profiles; NSGA-II is used to facilitate design space exploration and global optimum search. The overall optimization process based on VF metamodels is depicted in the flowchart of Figure 1. The details steps are as follows:

DOI: 10.1051/matecconf/20179505002

Step 1 Determine the laser welding process parameters optimization problem including the target function, the amount of parameters and parameter ranges.
Step 2 Construct LF simulation model and HF physical experiment model to investigate the relationship between process parameters and welding bead profile.
Step 3 Construct VF metamodels using the method of HK metamodel.
Step 4 Check whether the accuracy of the constructed VFM. If yes, go to next step, otherwise, adjust the LF model.
Step 5 Implement NSGA-II to obtain the optimum process parameters according to the prediction function of the constructed VF metamodels.
Step 6 Verify optimal results through experiments.

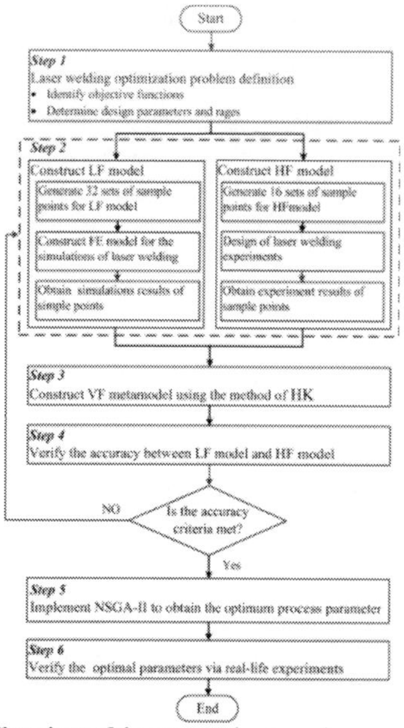

Figure 1. Flowchart of the proposed approach.

The goal of optimization problem is to solve the nonlinear task of the form as blow:

$$\begin{aligned} \text{Objective: } & \text{Minimize BW} \\ & \text{Maximize DP} \\ \text{Subject to: } & 2.0 \le LP \le 3.5 \\ & 2.5 \le WS \le 3.5 \\ & -2.0 \le LF \le 0 \end{aligned} \quad (1)$$

where the input variables include laser power (LW), welding speed (WS) and laser focal position (LF). The output parameters are depth of penetration (DP) and bead width (BW), as shown in Figure 2. The objective is to obtain the maximized DP and minimized BW. The two objectives and the optimization problem in equation (1) are trade-offs for a multi-objective optimization problem.

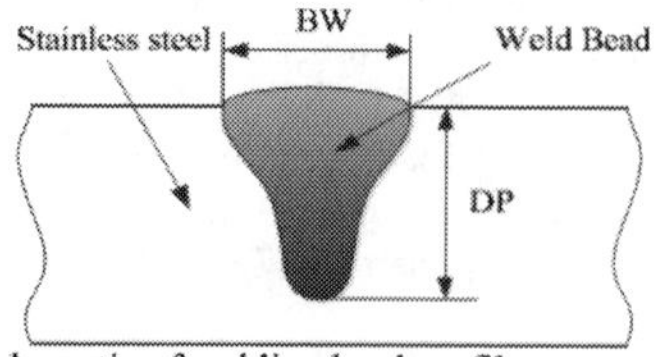

Figure 2. schematic of welding bead profile.

3 VF metamodel construction

3.1 Construct LF simulation model

In this study, a three-dimensional thermal FE model is developed to simulate the bead profile in laser welding process. Due to complex changes in the actual welding process, simplifying assumptions are made as follows [12, 13]:

- The initial temperature of the weldment is 300K.
- The finite element model is heat conduction model, which the formation of keyhole is not considered.
- The thermal transfer of weldment is of solid-solid, and there is no chemical reaction, agitation or oxidation in the melt pool.

The half of weldment by the centerline of welding bead is taken to construct FE model using ICEM CFD 14.0 software. The computational domain of model has a dimension of 6*4*3(mm). The size of mesh is 0.1mm and consists of 66,729 elements.

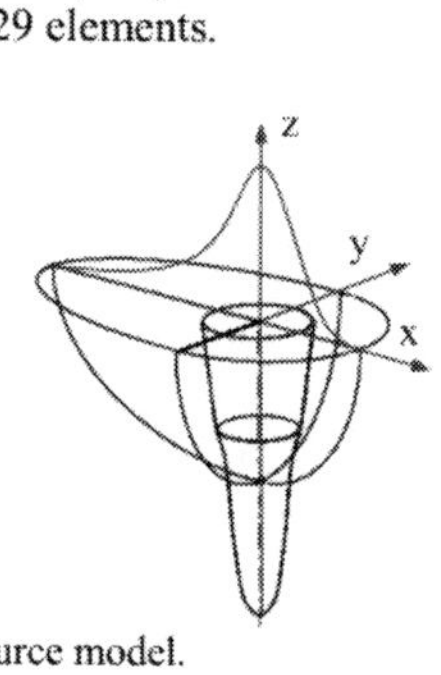

Figure 3. Heat source model.

The most commonly used heat sources of laser welding simulation have a Gaussian distribution [14]. In this study, the body heat source model combined by double-ellipsoids and rotating-Gaussian is developed for FE simulation, shown in Figure 3.

Table 1. Thermal properties of stainless steel 316L.

Thermal properties	Value
Density (g / cm^3)	7.91
Solid temperature (K)	1616
Liquid temperature (K)	1718

Table 2. Temperature dependent thermal properties of stainless steel 316L.

Temperature (K)	Conductivity ($W/m \cdot K$)	Specific Heat ($J/G \cdot K$)
300	15.91	0.45
800	22.46	0.66
1300	29.01	0.65
1616	33.15	0.79
1660	33.67	1.12
1718	30.62	0.78
2000	35.07	0.79
2500	42.95	0.81
3000	50.83	0.83

The physical properties parameters of the stainless steel 316L are shown in Table 1. The specific heat capacity and thermal conductivity change with temperature, calculated by JMatPro software, are shown in Table 2.

The FE simulation for laser welding process is calculated by FLUENT 13.0 software which about 16.5 hours was totally spent for computation. 32 sets of sample points were obtained by the method of optimal Latin hypercube [15], and these sample points were employed to obtain the simulation results of welding bead profile, namely the LF model data. Figure 4 shows the transient temperature contours of laser welding process at 5ms. Figure 4(a) shows the whole three-dimensional transient temperature contours by mirror symmetry; Figure 4(b) shows whole isothermal surface with temperature above the fusion temperature (T>1616K) by mirror symmetry; Figure 4(c) shows isothermal surface with all temperatures of half model; Figure 4(d) shows the temperature contours at the cross section of welding bead.

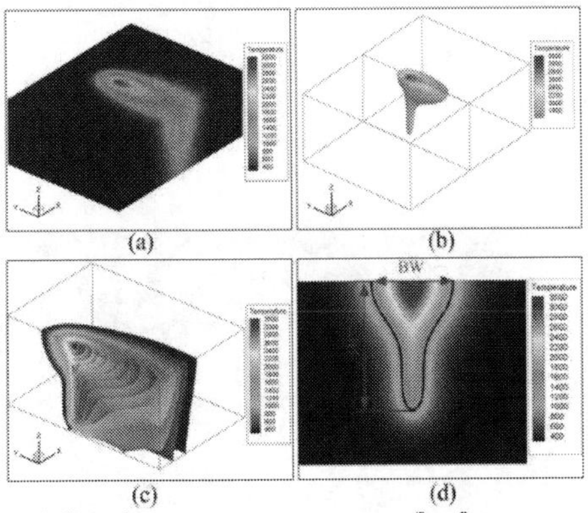

Figure 4. Transient temperature contours after 5ms.

3.2 Design HF physical experiment

In this paper, the data of HF model is from laser welding physical experiments. Figure 5 demonstrates the laser welding equipment used in this study. The laser welder is IPG YLR-4000 ytterbium-doped fiber laser. Laser is delivered through the optical fiber to the laser welder head. The laser welding header is installed on the robot ABB IRB4400.

Figure 5. Laser welding equipment.

16 set of sample points are generated by the method of optimal Latin hypercube. Then, the experiment results of sample points are obtained through the designed experiments. As shown in Figure 6, the two outputs of welding bead profile were observed on optical microscope and measured with measurement software CSM1 according to the transverse section of each weldment.

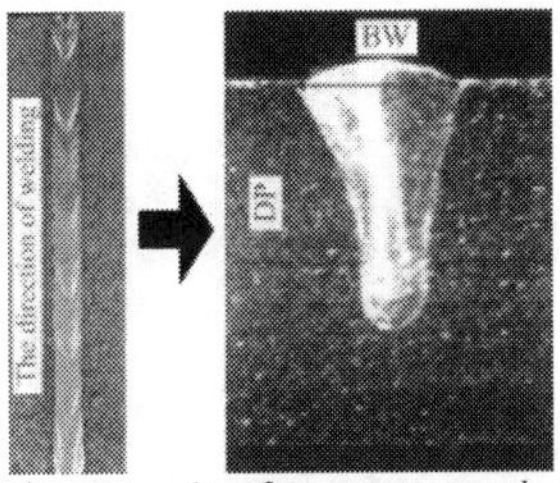

Figure 6. Experiment results after post processing.

3.3 Construct HK metamodel

In this paper, the VF metamodels are constructed by the method of HK. We can get the HK predictor that can be written in a form as:

$$\hat{y}(x) = \beta_0 \hat{y}_{lf}(x) + r^T(x) \underbrace{R^{-1}(y_s - \beta_0 F)}_{=V_{HK}} \tag{2}$$

where $\hat{y}_{lf}(x)$ is low-fidelity model. β_0 is a scaling factor indicating how much the low-fidelity and high-fidelity functions are correlated to each other, and it is calculated only by the initial sample points.

The LF models and HF models are integrated into VF metamodels using the HK metamodel. The constructed VF metamodels can descript the relationship between process parameters and welding bead profile. As shown in Figure 7, the constructed VF metmodels for BW and DP are demostrated.

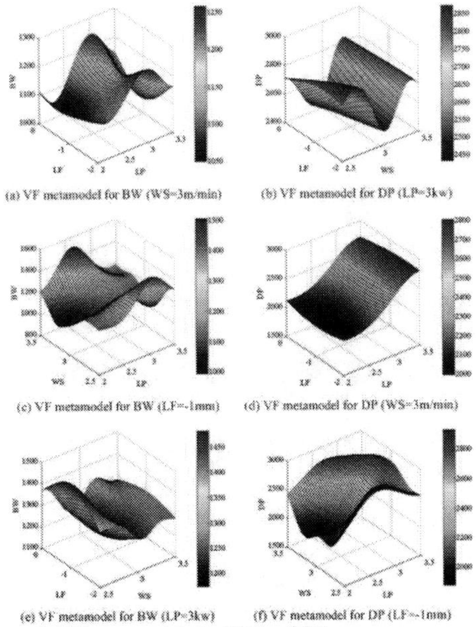

Figure 7. VF metamodels for BW and DP.

3.4 Verify the variable-fidelity metamodel

DOI: 10.1051/matecconf/20179505002

6 sets of sample points (NO.1-6) are randomly selected to verify the accuracy of VF metamodels by laser welding experiments. Comparison of predicted values and simulated values is shown in Figure 8. There has a good agreement between predicted values and experiment values. Table 3 demonstrates the predicted results of VF metamodels. The average relative error of VF metamodels for BW and DP are 9.11% and 7.00%. Therefore, the constructed VF metamodels are reliable and can be used for predicting welding bead profile in optimization section.

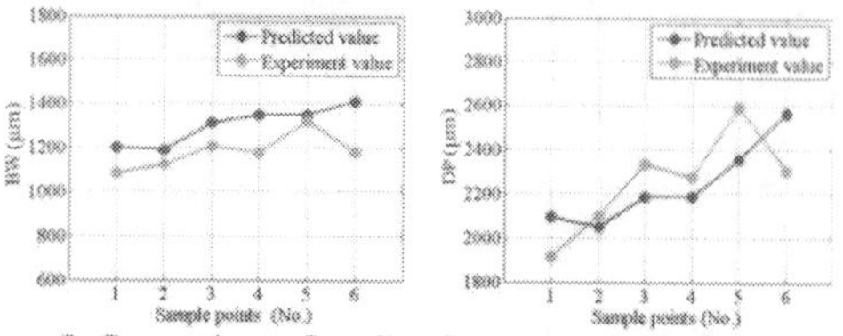

Figure 8. Comparison of predicted values and experiment values.

Table 3. Predicted results of VF metamodels.

No.	LP	WS	LF	Relative error (%)	
				BW	DP
1	2.10	3.00	-1	9.86	8.56
2	2.10	3.00	-2	5.42	2.45
3	2.18	2.68	-2	8.16	6.78
4	2.40	2.83	-2	12.80	4.04
5	2.63	3.34	-1	2.15	10.06
6	2.58	2.88	-2	16.25	10.13
			Average error:	***9.11***	***7.00***

4 Optimization of laser welding process parameters

4.1 Multi-objective optimization using NSGA-II

NSGA-II is implemented to solve the laser welding process parameters optimization problem and obtain the Pareto-optimal front. The settings for the NSGA-II in this two objective optimization problem are listed in Table 4. Fitness values of BW and DP in NSGA-II optimization process are predicted by VF metamodels constructed in the previous steps.

Table 4. Settings of NSGA-II parameters.

Parameter	Value
Population size	40
Maximum iterations	300
Elite count	2
Pareto fraction	0.4
Crossover fraction	0.8
Mutation fraction	0.2
Function tolerance	10E-06
Scaling fitness function	Rank

Figure 9 plots the obtained Pareto-optimal front for BW and DP. Each point illustrates a specific optimal solution. It can be concluded from Figure 9 that all optima are non-inferior with each other because there are trade-offs among the BW and DP.

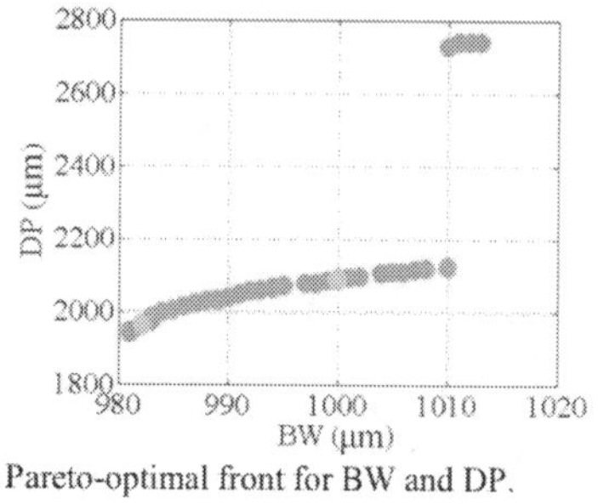

Figure 9. Pareto-optimal front for BW and DP.

4.2. Verification experiments of optimal solation

Two solutions in the Pareto-optimal front are selected and carried out to verify the effectiveness of the optimal results. Figure 10 shows the welding bead profile and dimension of these three verification experiments. The optimal parameters and corresponding outputs responses are listed in Table 5. As illustrated in Table 5, the maximum relative error of BW is 14.16%. For DP, the maximum relative error is 3.13%. It can indicate that the prediction accuracy of proposed approach can meet the requirement of process parameters optimization.

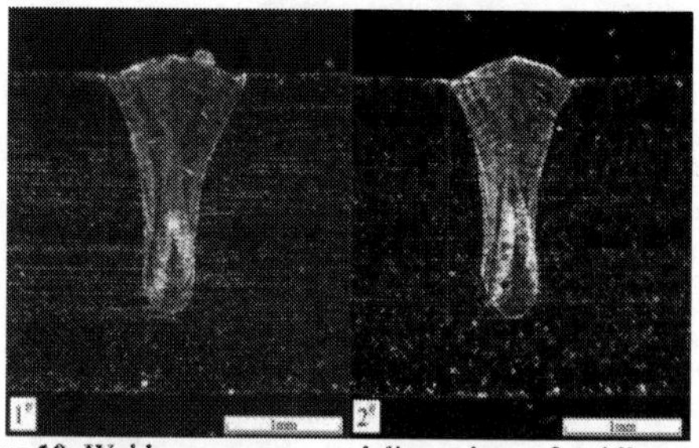

Figure 10. Weld appearances and dimensions of welding cross-sections of optimal results.

Table 5. Results of the verification experiments.

No.	LP	WS	LF		BW	DP
1	3.50	2.91	-1	Optimized (μm)	1000	2088
				Experimental (μm)	1165	2024
				Relative error (%)	***14.16***	***3.07***
2	2.30	3.12	0	Optimized (μm)	982	1952
				Experimental (μm)	1060	2013
				Relative error (%)	***7.36***	***3.13***

In terms of the weld appearances, it can be observed in Figure 10 that the weld appearances of two optimal results are beautiful and few defects exist in the welding bead.

5 Conclusion

In this paper, an optimization methodology using VF metamodel and NSGA-II is proposed for process parameters optimization in laser bead-on-plate welding for stainless steel 316L. The proposed approach could obtain the desired welding bead profile which can improves the quality of the welding bead remarkably. In addition, the verification experiments for VF metamodel and optimal results have been analyzed. It is obvious that the approach can provide a reliable guidance for laser welding experiments. Following conclusions can be drawn from the above investigation:

1. The predicted values of VF metamodels have agreement with experiment results. The VF metamodels could predict the welding bead profile with small error.
2. Verification experiments of obtained optimal process parameters have proved that the optimization results is in good agreement with experiments results.

Acknowledgements

This work was supported by the National Natural Science Foundation of China (NSFC) under Grant No.51505163, No.51421062 and No.51323009, National Basic Research Program (973 Program) of China under grant No.2014CB046703. The authors also would like to thank the anonymous referees for their valuable comments.

References

1. Y Rong, Z Zhang, G Zhang, C Yue, Y Gu, Y Huang, et al, *Optics and Lasers in Engineering,* **67**, 94-104 (2015)
2. P Jiang, L Cao, Q Zhou, Z Gao, Y Rong, X Shao, *IJMAT*,(to be published) 2016
3. Q Zhou, P Jiang, X Shao, Z Gao, L Cao, C Yue, X Li, *Metallurgical and Materials Transactions B*, (to be published) 2016
4. Q Zhou, Y Rong, X Shao, P Jiang, Z Gao, L Cao, *Journal of Intelligent Manufacturing*, (to be published) 2016
5. X Wang, H Chen, H Liu, P Li, Z Yan, C Huang, et al. *Optics and Lasers in Engineering*, **51(11):**1245-1254 (2013)
6. M.H. Shojaeefard, M Akbari, P. Asadi, *International Journal of Precision Engineering and Manufacturing*, **15(11)**: 2351-2356 (2014)
7. M. Islam, A. Buijk, M. Rais-Rohani, K. Motoyama, *Advances in Engineering Software*, **79**: 127-136 (2015)
8. Q Zhou, X Shao, P Jiang, L Cao, H Zhou, L Shu, *CMES*, **106(5)**: 323-355 (2015)
9. G. Sun, G. Li, M. Stone, Q. Li, *Computational Materials Science*, **49(3)**: 500-511 (2010)
10. Z-H Han, R Zimmermann, S Görtz, AIAA Paper, 1225: 2010(2010)
11. K. Deb, A. Pratap, S. Agarwal, T. Meyarivan, IEEE Transactions on, **6(2):** 182-197 (2002)
12. M.R.N. Esfahani, J. Coupland, S. Marimuthu, *Journal of Materials Processing Technology*, **224**: 135-142 (2015)
13. R. Wang, Y. Lei, Y. Shi, *Optics & Laser Technology*, **43(4)**: 870-873 (2011)
14. J. Goldak, A. Chakravarti, M. Bibby, *Metallurgical transactions B*, **15(2)**: 299-305 (1984)
15. J.-S. Park, *Journal of statistical planning and inference*, **39(1)**: 95-111 (1994)

Rapid Modelling and Grinding of Workpieces' Inner-surface by Robot with Impedance Model Based Fuzzy Force Control Algorithm

Wang Fei , Luo Zhong and Liu Hongyi

School of Mechanical Engineering & Automation, Northeastern University, 110819, Shenyang, China

Abstract. To achieve rapid automatic grinding of workpieces' inner-surface by industrial robot, a rapid translational detection strategy of workpieces' inner-surface and fuzzy force control algorithm of grinding are proposed in this paper. The rapid translational detection strategy introduces a way to establish an inner-surface's model quickly by recording key points of the axial section contour which reflects big curvature changes of the contour. The established model is feasible but imprecision. The force control algorithm is based on impedance model. To promote adaptability to the imprecision of the established inner-surface's model, a fuzzy adjusting strategy is introduced in the force control algorithm. By adopting an adjusting factor, which determined by force response and a fuzzy logic, the strategy can adjust the reference trajectory of impedance model in time. Taking advantage of proposed detection and force control method, grinding experiments shows that the contact normal force maintains approximately constant, the relative mean error is within 6.5%, and the material removal thickness of the inner-surface is approximately consistent. The proposed strategy's feasibility is verified.

1 Introduction

Industrial robot is a tendency in the manufacturing process in INDUSTRIE 4.0[1]. When robot contact with a workpiece, such like grinding, deburring and polishing, compliance is necessary for the robot[2]. Moreover, the precision of workpieces' model and workpieces' location in its jig are both important to the constraint machining process.

A. Robertsson etc. established a robot platform with external fast open sensor integration system, and implemented force-controlled grinding, finishing and deburring experiments with the platform[3]. In the experiments, an off-line programming including path generation based on work-pieces' CAD model is introduced in the hybrid force/position control strategies.

Considering the position and orientation errors of the workpiece, Hee-Chan Song and Jae-Bok Song developed a tool-path modification method based on a CAD model and direct teaching. By setting a virtual wall, the force control based on impedance model is used to applying an accurate and stable contact force[4].

Accuracy of workpieces affects the performance of force control intensively. To obtain accurate model of the workpiece, an off-line pre-programming is needed in those researches. Pre-programming is obviously time-consuming but beneficial to force control. In order to shorten the whole work time, a less-accurate model of workpieces is acceptable. Meanwhile, adaptability to the imprecision model is put forward to the force control algorithm.

To solve the above mentioned problem, a rapid translational detection strategy and fuzzy force control algorithm, which can establish a workpiece's model quickly and adapt the imprecision of the model, are proposed in this paper. Modelling and grinding experiments are described in the end of the paper.

2 Model of the grinding robot system

When grinding a cavity's inner-surface with a constant material removal thickness, the sketch map of the grind robot system is shown in figure1. The cavity is assumed to be a rotary shell. The robot has 5-DOF to grinding the inner-surface of the shell, including a translational joint, 2 rotational joints and a tool and a shell's rotational movement[5].

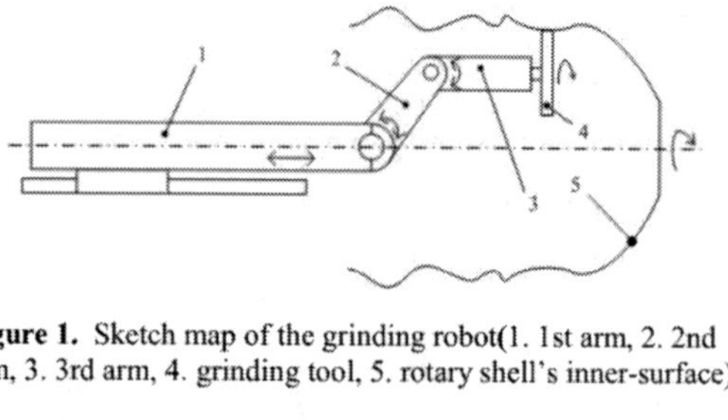

Figure 1. Sketch map of the grinding robot(1. 1st arm, 2. 2nd arm, 3. 3rd arm, 4. grinding tool, 5. rotary shell's inner-surface)

2.1 Model of the grinding robot

The dynamic model of an n-link rigid-joint manipulator in joint coordinate is described by

$$M(q)\ddot{q}+C(q,\dot{q})\dot{q}+G(q)=\tau+J^{T}F_{e} \quad (1)$$

where $q,\dot{q},\ddot{q}$ correspond to the joints' position, velocity and acceleration vectors, respectively. $M(q)$ is the n×n positive-definite inertia matrix. $C(q,\dot{q})$ is the n × n centripetal and Coriolis matrix. $G(q)$ is the n×1 vector of gravitational terms in workspace. τ is the n×1 joint input torque. J represents the n×6 Jacobian matrix. The robot dynamic model in the Cartesian space can be derived from formula(1):

$$M_{x}(x)\ddot{x}+C_{x}(x,\dot{x})\dot{x}+G_{x}(x)=J^{-T}\tau+F_{e} \quad (2)$$

where $M_{x}(B)=J^{-T}MJ^{-1}$, $C_{x}(x,\dot{x})=J^{-T}(C-MJ^{-1}\dot{J})J^{-1}$ and $G_{x}(x)=J^{-T}G$.

2.2 Contour detecting and modelling of the inner-surface

Usually, the contour of the shell's inner-surface has great difference with its CAD model, cause the shell's crafting process. And position errors inevitably occur when a workpiece is placed in the jig, shown in figure2. The CAD model can not be used directly in robot's grinding process. Therefore, the inner-surface's model needs to be established to guarantee the accurate grinding process.

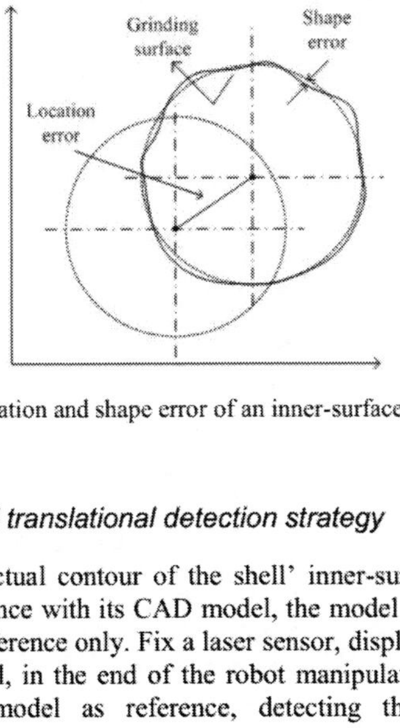

Figure 2. Location and shape error of an inner-surface's contour

2.2.1 Rapid translational detection strategy

When the actual contour of the shell' inner-surface has great difference with its CAD model, the model could be used as a reference only. Fix a laser sensor, displacing the grinding tool, in the end of the robot manipulator. With the CAD model as reference, detecting the inner-surface's axial contour, the detection could be conducted according to the following steps.

1. Actuating the manipulator into the cavity, let the end of the manipulator locate approximately in the middle of the cavity. The serial number of this cavity's axial section represents by $j(j=0,1,\cdots,m)$.

2. Taking the CAD model as reference, let the laser beam be perpendicular to the suppositional inner-surface. Set the orientation of the end-effector as the initial orientation. Actuate the manipulator in the initial orientation until the laser sensor got a valid data equalled to D_{d} , which is the median of the laser sensor's measuring range. Calculate this first detection point's coordination value P_{0}^{j} in the base coordinate system according to the robot's kinematics model.

3. Maintaining the end-effector's gesture, let the parallel motion of laser beam be along with one axial direction of the cavity until the sensor's output D_{out} changed out of range Δ_{f} . When sensor's output changed out of rang, which means $|D_{out}-D_{d}|\geq\Delta_{f}$, the point detected can be recorded as $P_{i}^{j}(i=1,2,\cdots t)$.

4. On the basis of maintaining the laser beam sticking on detection point P_{i}^{j} , adjusting end-effector's position and gesture, let laser sensor's output be equalled to D_{d} and laser beam be perpendicular to the line formed by P_{i-1}^{j} and P_{i}^{j} .

5. Repeating step 3 and 4 with direction $P_{i}^{j}-P_{i-1}^{j}$ until laser sensor got no valid output, a series $[P_{0}^{j}\ P_{1}^{j}\ \cdots\ P_{t}^{j}]$ will be recorded.

6. Actuating the manipulator return to the status of the first point P_{0}^{j} , and repeating step 3 to 5 in the opposite direction, series $[P_{t+1}^{j}\ P_{t+2}^{j}\ \cdots\ P_{n}^{j}]$ will be recorded.

7. Resettling and combining the series $[P_{0}^{j}\ P_{1}^{j}\ \cdots\ P_{t}^{j}]$ and $[P_{t+1}^{j}\ P_{t+2}^{j}\ \cdots\ P_{n}^{j}]$ along with the cavity's axial direction, the resettled series $[P_{0}^{j}\ P_{1}^{j}\ \cdots\ P_{n}^{j}]$ will sketch an axial contour of the cavity's inner-surface, shown as figure3.

8. Rotating the cavity in a small degree and repeating step 1 to 7, series $[P_{0}^{j+1}\ P_{1}^{j+1}\ \cdots\ P_{k}^{j+1}]$ will be recorded. Where, *k*, representing the quantity of recorded points, is determined by complexity of the axial contour.

9. Repeating step 1 to 8, after the whole axial contours being detected, key points of the cavity's inner-surface will form a bivector as $P_{i}^{j}(i=1,2,\cdots k,\ j=1,2,\cdots m)$.

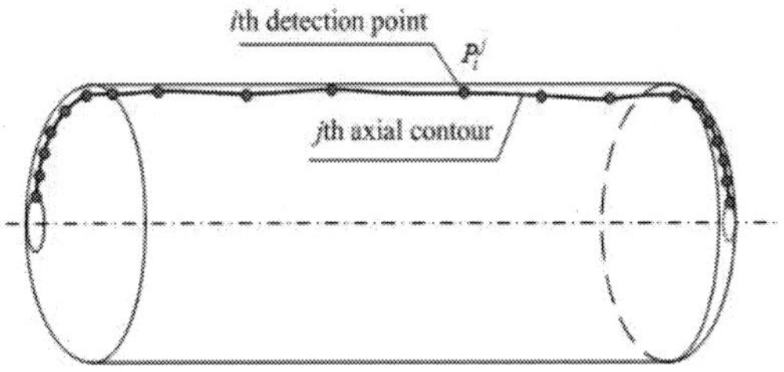

Figure 3. Detection points series of a cavity's inner-surface contour

2.2.2 Simulation of translational detection strategy

DOI: 10.1051/matecconf/20179505003

Setting Δ_f as 0.5mm and 2mm, the detection simulation results are shown in Fig. 4 and Fig. 5.

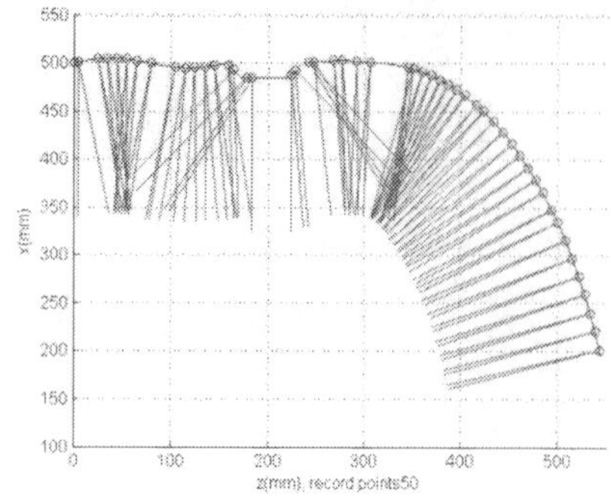

Figure 4. Detection with $\Delta_f = 0.5\text{mm}$

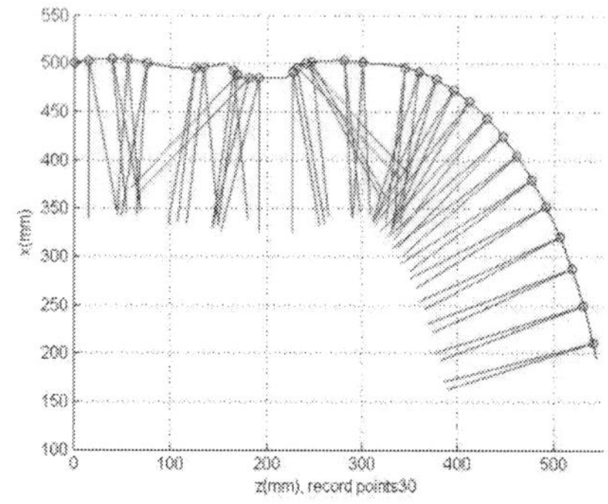

Figure 5. Detection with $\Delta_f = 2\text{mm}$

The simulation results shown that the translational detection strategy proposed above is feasible. Moreover, the smaller Δ_f is the more accurate the model goanna be. However, the model is not detailed enough. It only includes several key points that can reflect big curvature changes of the contour. Therefore, the model established by proposed rapid translational detection strategy is feasible but imprecision. An intelligent force control algorithm which can be adaptive the imprecision is needed.

3 Force control algorithm

Since the existing imprecision of both contour model and location of the inner-surface, the force control algorithm should be adaptive to the mentioned error. A fuzzy force control strategy based on impedance control model is proposed.

3.1 Impedance control model

Using the following impedance mathematic model [6]:

$$\boldsymbol{M}_{\mathrm{d}}(\ddot{\boldsymbol{x}}_{\mathrm{d}} - \ddot{\boldsymbol{x}}) + \boldsymbol{D}_{\mathrm{d}}(\dot{\boldsymbol{x}}_{\mathrm{d}} - \dot{\boldsymbol{x}}) + \boldsymbol{K}_{\mathrm{d}}(\boldsymbol{x}_{\mathrm{d}} - \boldsymbol{x}) = \boldsymbol{F}_{\mathrm{d}} - \boldsymbol{F}_{\mathrm{e}} \quad (3)$$

where $\boldsymbol{M}_{\mathrm{d}}$, $\boldsymbol{D}_{\mathrm{d}}$ and $\boldsymbol{K}_{\mathrm{d}}$ are respectively the desired inertia, damping and stiffness matrices of the manipulator-environment system. $\boldsymbol{F}_{\mathrm{d}}$ is the desired contact force between manipulator's end-effector and the constraint surface. $\boldsymbol{x}$, $\dot{\boldsymbol{x}}$ and $\ddot{\boldsymbol{x}}$ are the actual position, velocity and acceleration of the robot manipulator, respectively. $\boldsymbol{x}_{\mathrm{d}}$, $\dot{\boldsymbol{x}}_{\mathrm{d}}$ and $\ddot{\boldsymbol{x}}_{\mathrm{d}}$ is respectively the desired trajectory, velocity and acceleration.

3.2 Reference trajectory's fuzzy adjusting strategy

Assuming x_{e} and K_{e} to be the actual position and stiffness of the constraint surface. x_{d} can be represents as:

$$x_{\mathrm{d}} = x_{\mathrm{e}} + \frac{F_{\mathrm{d}}}{K_{\mathrm{e}}} \quad (4)$$

Usually, the accurate x_{e} and K_{e} are unavailable. Let the inaccurate value x_{ec} and K_{ec} replace the x_{e} and K_{e} in formula(4). And the approximate K_{ec} is usually experiential.

According to the approximately rough function, the reference trajectory should be:

$$x_{\mathrm{r}} = x_{\mathrm{ec}} + \frac{F_{\mathrm{d}}}{K_{\mathrm{ec}}} \quad (5)$$

To promote adaptability of the control algorithm, an intelligent prediction and adjusting of the reference trajectory is introduced into the impedance model [7].

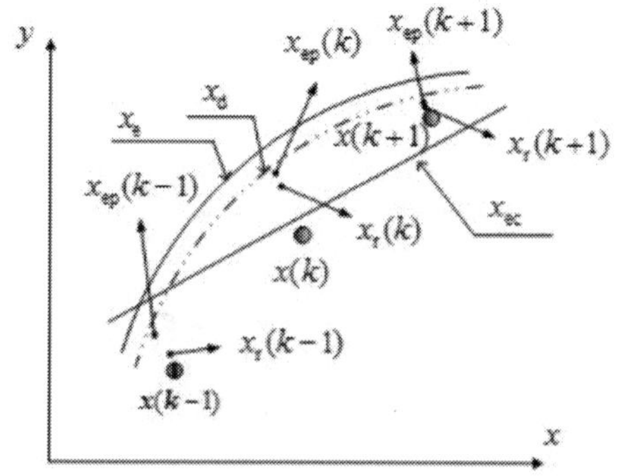

Figure 6. Reference trajectory adjusting

Establish the reference trajectory intelligent adjusting law in discrete form as:

$$\boldsymbol{x}_{\mathrm{r}}(k+1) = \boldsymbol{x}_{\mathrm{r}}(k) + \Delta x_{\mathrm{ec}}(k) \quad (6)$$

where k is defined as the current moment. $\boldsymbol{x}_{\mathrm{r}}(k+1)$ is the predictive position of reference trajectories in time $\boldsymbol{k}+1$. $\boldsymbol{x}_{\mathrm{r}}(k)$ is the reference trajectory in time $\boldsymbol{k}$. The term $\Delta x_{\mathrm{ec}}(k)$ is an adjustment based on the predicted location and geometric error:

$$\Delta x_{ec}(k)=x_{ec}(k+1)-x_{ec}(k)+k_c\cdot e_{xp}(k) \tag{7}$$

where, $e_{xp}(k)=\Delta x_{ep}(k)-\Delta x_{ep}(k-1)$ represents the tendency of the adjusting error. Setting $x_{ep}(k)=x(k)-\dfrac{F_e(k)}{K_{ec}}$, represents the approximately actual position of the constraint surface, $\Delta x_{ep}(k)=x_{ep}(k)-x_{ec}(k)$ approximately reflects the distance from the reference trajectories to the actual surface. k_c is the adapting scale factor determined by a fuzzy controller.

The input of the fuzzy controller is $|e_{xp}(k)|$ .The linguistic variables assigned to e_{xp} and k_c are EX and KC. The corresponding fuzzy subsets are A and B. The set A comprises VB, B, M, S and VS, which respectively denote very big, big, middle, small and very small. The set B comprises KB, BK, KM, SK and KS, which denote very big, big, middle, small and very small respectively. The fuzzy rules for input EX and output KC are shown in Table I.

Table 1. Fuzzy Control Rules for The Output KC

	EXP				
	VB	B	M	S	VS
KC	KB	BK	KM	SK	KS

Replacing the $\boldsymbol{x}_d$, $\dot{\boldsymbol{x}}_d$ and $\ddot{\boldsymbol{x}}_d$ with $\boldsymbol{x}_r$, $\dot{\boldsymbol{x}}_r$ and $\ddot{\boldsymbol{x}}_r$ in formula(3), define $\boldsymbol{u}\equiv\ddot{\boldsymbol{x}}$ as the control rule. It could be educed as:

$$\boldsymbol{u}=\ddot{\boldsymbol{x}}_d+\boldsymbol{M}_d^{-1}(\boldsymbol{D}_d\dot{\boldsymbol{e}}_r+\boldsymbol{K}_d\boldsymbol{e}_r-\boldsymbol{F}_d+\boldsymbol{F}_e) \tag{8}$$

where, $\boldsymbol{e}_r=\boldsymbol{x}_r-\boldsymbol{x}$, $\dot{\boldsymbol{e}}_r=\dot{\boldsymbol{x}}_r-\dot{\boldsymbol{x}}$ and $\ddot{\boldsymbol{e}}_r=\ddot{\boldsymbol{x}}_r-\ddot{\boldsymbol{x}}$.

Combining formula(1) and (8), the join torque control law of the force control model with reference trajectory online intelligent adjusting could be:

$$\boldsymbol{\tau}=\boldsymbol{J}^T[\boldsymbol{M}_x\boldsymbol{u}+\boldsymbol{C}_x\dot{\boldsymbol{x}}+\boldsymbol{G}_x-\boldsymbol{F}_e] \tag{9}$$

3.3 Impedance model based force control

The control law is shown in Fig. 3. When robot's in a constraint motion, the impedance based intelligent force controller will be implemented to perform a constant force between the robot end-effector and constraint surfece.

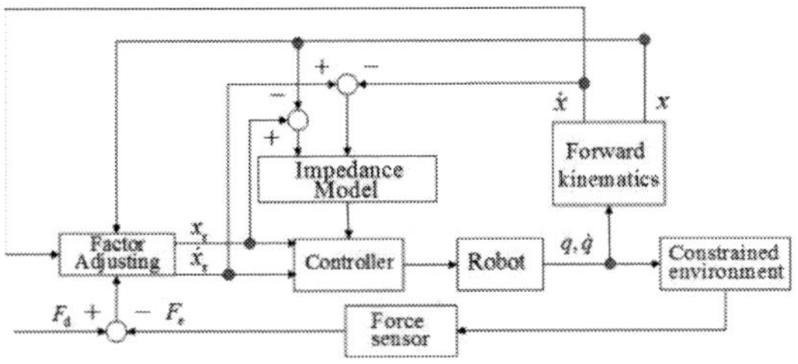

Figure 7. Scheme of the impedance model based force control

4 Detecting and grinding experiment

The grinding robot system is shown in Fig. 8. The CAD model of cavity's inner-surface is available but has a great difference with the actual object. Firstly detecting the inner-surface of the shell with a laser sensor, the 3D model of the inner-surface could be built as figure 9. Then executing the grinding process based on the proposed fuzzy force control algorithm, with an expected contact force 5N, the force response is shown in figure 10.

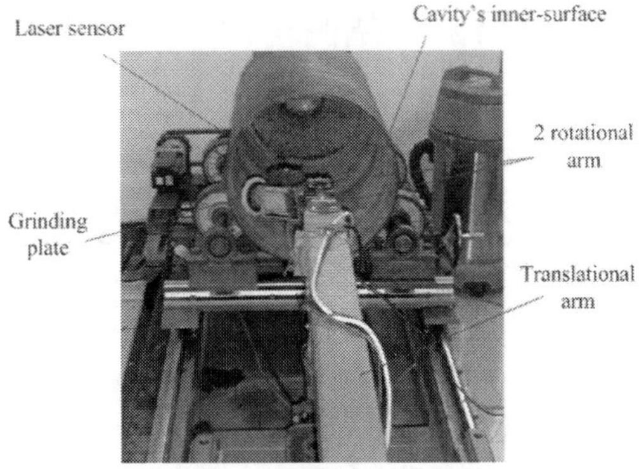

Figure 8. Grinding experiment of a rotary shell's inner-surface

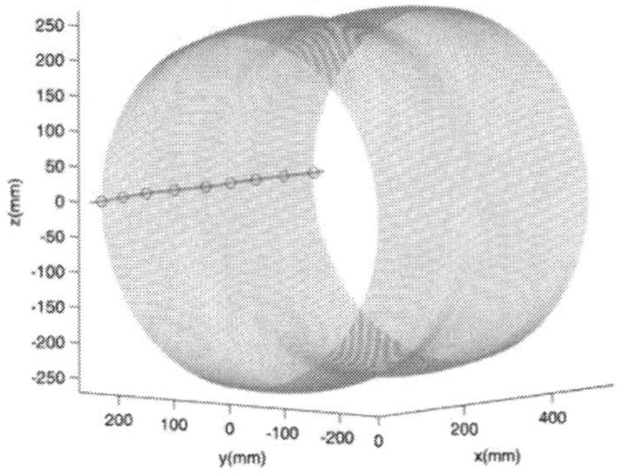

Figure 9. Reconstructed 3D model of the rotary shell's inner-surface

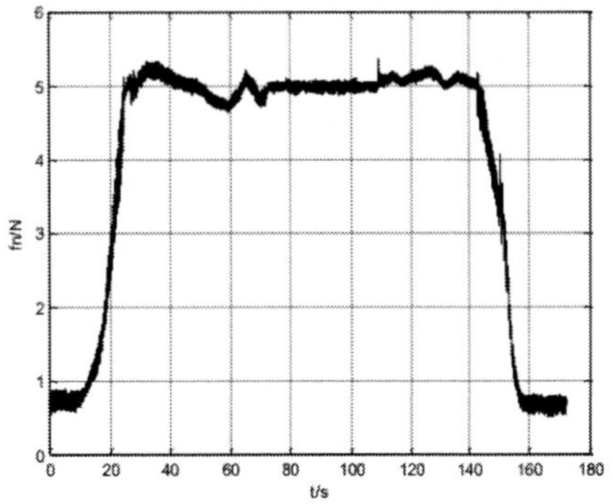

Figure 10. Normal force response with expected 5N

Fig. 10 shows that the contact normal force between manipulator's end-effector and the constrained inner-surface is controlled to be around its expected value. The fluctuation in the response is caused by inaccurate model of the inner-surface and impact of the manipulator's movement from free movement to a restraint movement in the very beginning of the grinding process. Moreover, when grinding process is in time 50~70s, there's fluctuation because that the curvature of the constrained surface is very small and changed drastically. Taking advantage of proposed fuzzy force control strategy, error of contact normal force fluctuate in [-0.3, 0.2]N, with relative error within 5%. Overall, in the whole grinding process, the relative mean error of normal contact force from the expected value is within 6.5%. The material removal thickness of the inner-surface is approximately consistent.

5 Conclusion

To grind a cavity's inner-surface automatically with robot, a rapid translational detection strategy and fuzzy force control algorithm are proposed in this paper. A compendious model could be established quickly by following the rapid translational detection strategy which developed to record key points of the inner-surface's contour. Taking advantage of the fuzzy force control algorithm based on impedance control algorithm, the material removal thickness in grinding process can be controlled to be approximately consistent. The force control algorithm is adaptive to the contour model's imprecision, and maintain the contact force be approximately constant equal to the expect value with relative mean error 6.5%. The experience of this research provides an important reference for the practical application.

References

1. J. F. Wan, H. Cai, K. L. Zhou. 2015 ICIT, 135(2015)
2. Y. Shiote, J. Akiyama, H. Yoshida, Y. Harada. Robotics and Automation, IEEE International Conference on Robotics and Automation, **1**, 788(1995)
3. A. Robertsson, T. Olsson, R. Johansson, etc.. Proceedings of the 2006 IEEE/RSJ International Conference on Intelligent Robots and Systems, 2743(2006)
4. H. C. Song, J. B. Song. International Journal of Precision Engineering & Manufacturing, **14(1)**, 85(2013)
5. F. Wang,H. Y. Liu, Z. Luo. Optics and Precision Engineering, **21(6)**,1479(2013)
6. M. K. Vukobratovi, A. G. Rodic. Journal of Intelligent and Robotic Systems, **18(2)**, 191(1997)
7. F. Wang, Z. Luo, H. Y. Liu, L. Wang. 2010ROBIO, 1555(2010)

 DOI: 10.1051/matecconf/20179505004

Improvement of EDM Machining Speed by Using Magnetic/Piezoelectric Hybrid Drive Actuator

Xiaoyou Zhang and Kouki Uchiyama

Department of Mechanical Engineering, Nippon Institute of Technology, 4-1, Gakuendai, Miyashiro-cho, Minamisaitama-gun, Saitama, Japan

Abstract. The objective of this research is to achieve high-speed electrical discharge machining (EDM) of small holes using a high-speed, high-precision, 3-DOF controlled, magnetic/piezoelectric hybrid drive actuator. In this paper, the proposed actuator was attached to the conventional electrical discharge machine, and the increase in the machining speed using the actuator was experimentally confirmed. The relationships between the machining speed, the amplitude and the frequency of the electrode vibration were discussed. Experimental results show that the use of the proposed actuator can speedily adjust the distance between the electrode and the workpiece, and the machining speed was increased by 138% compared with the conventional EDM. Moreover, the machining speed using the actuator also was increased as the changes of the amplitude and the frequency of the electrode vibration, and it was increased by 76% compared to without the electrode vibration.

1 Introduction

Electrical discharge machining (EDM) is a non-conventional material removal processes based on the thermo-electric energy created between an electrically conductive workpiece and an electrode submerged in a machining fluid. EDM has several advantages such as the capability of machining of all conductive materials, regardless of their hardness, and the ability to deal with complex shapes[1]. However, the machining speed and accuracy of conventional EDM are limited by probability and efficiency of the electrical discharges. To obtain a stable electrical discharge, the electrode needs to be speedily re-positioned in order to maintain a suitable distance from the workpiece, and the debris around the electrode due to EDM has to be removed immediately.

To improve the positioning response and the positioning accuracy, a combination of a conventional electrical discharge machine and a wide-bandwidth, high-precision, multi-degree- of-freedom local actuator is required. As the local actuators, a piezoelectric actuator[2], a voice coil motor[3], and a magnetic bearing[4] have been used to improve the positioning response of the electrode.

Moreover, to immediately remove the debris around the electrode, in the conventional EDM, the cylindrical electrode is usually rotated or jumped. The fresh machining fluid can be introduced into the machined hole and the debris can be removed. However, in the rectangular hole machining or 3D surface machining, the electrode cannot be rotated, and adhesion of the workpiece and electrode due to the improper flow of the debris is a problem. To increase the flushing effect during the EDM, an ultrasonic vibration has been superposed on the normal electrode movement[5]. Furthermore, to flush out debris from the machined hole, a vibration-assisted machining also has been introduced to micro-EDM using piezoelectric[6]

In this paper, the developed 3-DOF controlled, high-speed, high-precision, local actuator is attached to an electrical discharge machine to experimentally confirm the increase in the machining speed. Then, the relationships between the machining speed, the amplitude and the frequency of the electrode vibration are discussed.

2 Magnetic/piezoelectric hybrid drive actuator

Fig. 1 shows the principle of the proposed 3-DOF controlled, magnetic/piezoelectric hybrid drive actuator. The actuator primarily consists of a thrust magnetic bearing, a pair of elastic hinges, and an electrode vibration mechanism in the radial directions. The motion of the spindle attached the electrode in the Z direction is controlled by the voice coil motor type of thrust magnetic bearing consisting of an air-core coil and two permanent magnet rings, as shown in Fig. 1(a). The thrust magnetic bearing with a positioning stroke of a few millimeters can realize speedy position of the electrode in order to maintain a suitable distance from the workpiece, and the fast jump theelectrode to introduce fresh machining fluid into the hole and wash away debris. The motions of the spindle in X, Y, Θ, Φ and Ψ directions are constrained by upper and lower elastic hinges shown in Fig. 1(b).

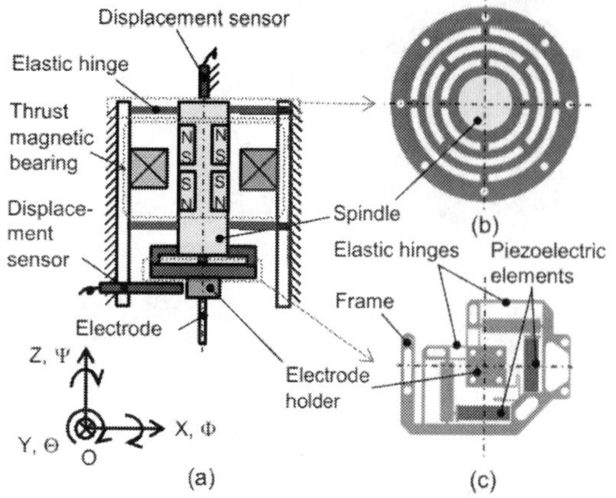

Figure 1. Configuration of 3-DOF controlled actuator

The electrode vibration mechanism shown in Fig. 1(c) is attached to the lower end of the spindle. The main components comprise elastic hinges, two piezoelectric elements, and an electrode holder. The electrode used to machining is fixed by the electrode holder. The motions of the electrode in X and Y directions are driven by two piezoelectric elements, to achieve the vibration of the electrode to remove the debris. To realize a requested actuation stroke at higher operating frequencies, a displacement magnification mechanism for piezoelectric elements are applied.

The height of the actuator is 97mm, the diameter is 110mm, and the mass is 4.7kg. The actuator possesses a positioning resolution of 1 m, a bandwidth of 120Hz and a positioning stroke of 2mm in the thrust direction, and a positioning resolution of 0.05 m, a positioning stroke of 20 m at 1000Hz in the radial directions.

3 Electrical discharge machining system using the local actuator

In order to improve the machining speed of EDM, the developed actuator was attached to an electrical discharge machine (EA8PV, Mitsubishi Electric Corp.), as shown in Fig. 2. In the experiments, the conventional electrical discharge machine was used only to move the electrode to the initial position and to supply the voltage to the electrode. The electrode was controlled solely by the local actuator.

A cylindrical shaped copper electrode with a diameter of 2 mm was used and was not rotated or retracted during EDM. The machining fluid was EDF-K (Nippon Oil Corp.). The EDM experiments were carried out, until a 0.5 mm thick SUS304 steel plate was penetrated. The discharge circuit was a transistor circuit with a peak current of 29.0A, a pulse on-time of 70.8μsec, and an off-time of 115.2μsec.

To maintain a stable electrical discharge, the voltage between the electrode and the workpiece was kept at a constant value by rapidly changing the distance between the electrode and the workpiece. A block diagram of the EDM system is shown in Fig. 3. The gap voltage V, is amplified and averaged by a low-pass filter, and the average voltage, V_{fb}, is used as a feedback signal. The gain of the amplifier is 0.03 and the cutoff frequency of the low-pass filter is 330 Hz. The deviation between the reference voltage, V_r, and the feedback voltage, V_{fb}, is sent to a PI controller to calculate the value V_{ar}, which is used to position the electrode with respect to the workpiece. The reference $\boldsymbol{R}$ is the motion path of the electrode. To give a stable electrical discharge, the reference voltage V_r, was set to 1.6 V, and the values of and in the PI controller were set to $1.0\text{x}10^{-6}$ and $1.0\text{x}10^{-4}$, respectively, by trial and error.

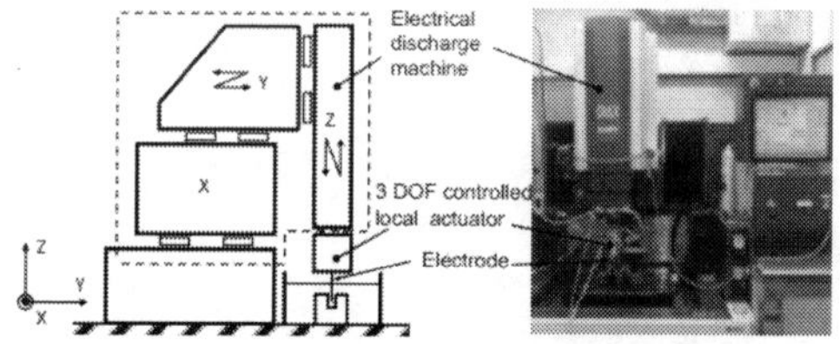

Figure 2. Electrical discharge machine with local actuator

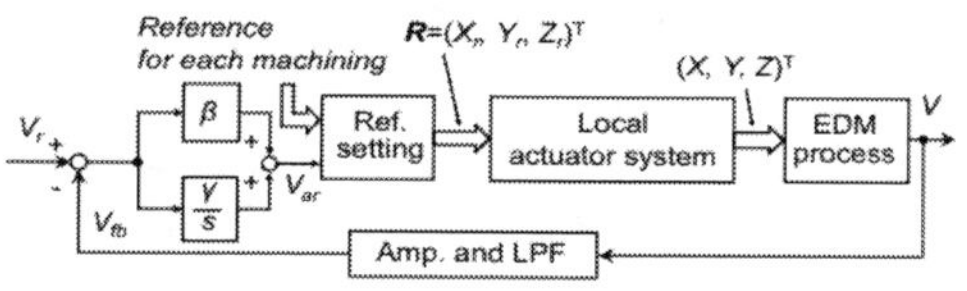

Figure 3. Block diagram of EDM system

4 Electrical discharge machining experiments

4.1. Hole machining without electrode vibration

4.1.1 EMD method

In order to evaluate the effectiveness of the proposed actuator for improvement of EDM machining speed, straight holes were machined in the thrust direction. In EDM experiments, the electrode position in this direction is controlled by the voltage across the gap between the electrode and the workpiece, and the electrode vibration in the radial directions is not carried out. In this case the Z_r of the input reference $\boldsymbol{R}$ of the actuator system was set to V_{ar}, and X_r, Y_r were set to zero. That is,

$$R=(X_r\ Y_r\ Z_r)^T=(0\ 0\ V_{ar})^T \tag{1}$$

For comparison, an EDM experiment was also carried out using the conventional electrical discharge machine only. The parameters of the PI controller and the machining conditions were identical to those used for the actuator.

4.1.2 Machining results

Fig. 4 shows photographs of the machined holes without and with the proposed actuator. Using the proposed actuator, the diameter of the hole was 2.13mm. On the other hand, using conventional EDM only, the diameter of the hole was 2.14mm, and the diameters were almost the same. The feed displacements of the electrode in the thrust direction were measured to evaluate the motion of the electrode during EDM as shown in Fig. 5. By using the proposed actuator, the machining time was reduced from 124.8s to 50.8s.

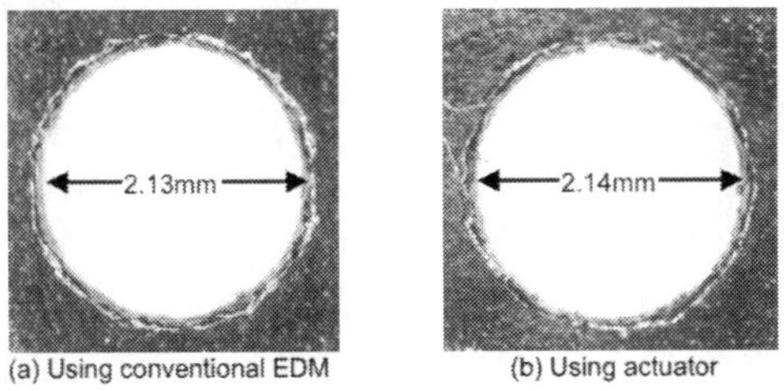

Figure 4. Machined holes

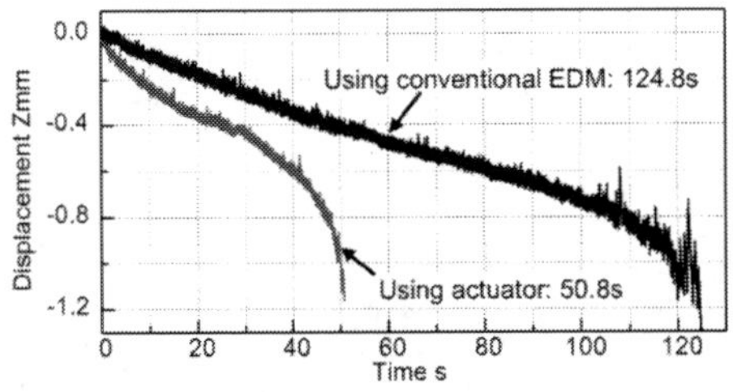

Figure 5. Feed displacements of electrode

Moreover, the feedback voltage V_{fb}, was monitored by a digital signal processor (DS1103, dSPACE Corp., Resolution, 16bit) at a sampling frequency of 10 kHz, as shown in Fig. 6. Comparing Fig. 6(a) with Fig. 6(b), it can be seen that the response of the electrode to maintain a stable electrical discharge is enhanced when using the proposed actuator.

For each EDM experiment, several holes were machined under the same conditions. Five holes except the holes influenced by large abnormal electrical discharges were used for the sample data. Defining the average machining speed as the ratio of thick of the workpiece to averaged machining time, the average machining speeds using the maglev local actuator and using only a conventional EDM were 9.59 m/s and 4.03 m/s, respectively. Thus, the average machining speed was improved by 138%.

4.2 Hole machining with electrode vibration

4.2.1 EMD method

To evaluate the relationships between the machining speed, the amplitude and the frequency of the electrode vibration, the amplitude and the frequency of the electrode in radial directions were tuned respectively during the EDM experiments. In this case, the electrode is driven in a circular path in the radial directions, and ***R*** is given by Eq. (2).

$$R=(X_r\ Y_r\ Z_r)^T=(\frac{A}{2}\cos 2\pi f t\ \frac{A}{2}\sin 2\pi f t\ V_{ar})^T \qquad (2)$$

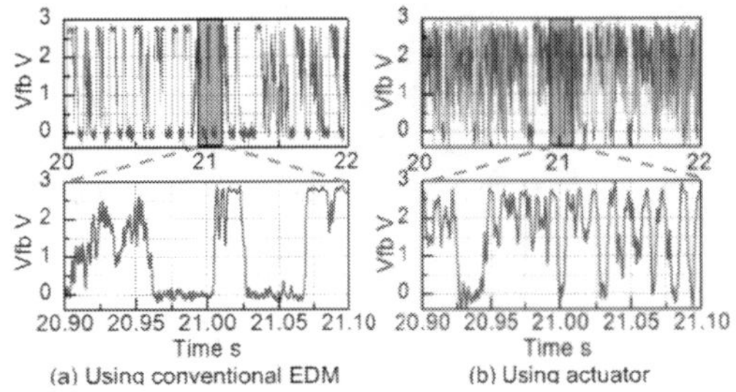

Figure 6. Feedback voltage (V_{fb}) during EDM

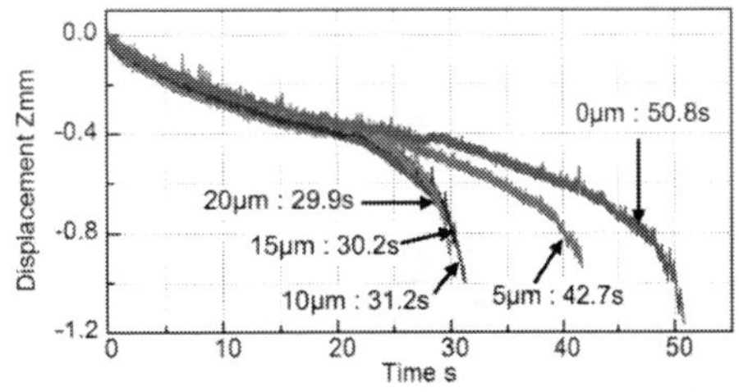

Figure 7. Feed displacements of electrode on 1000Hz

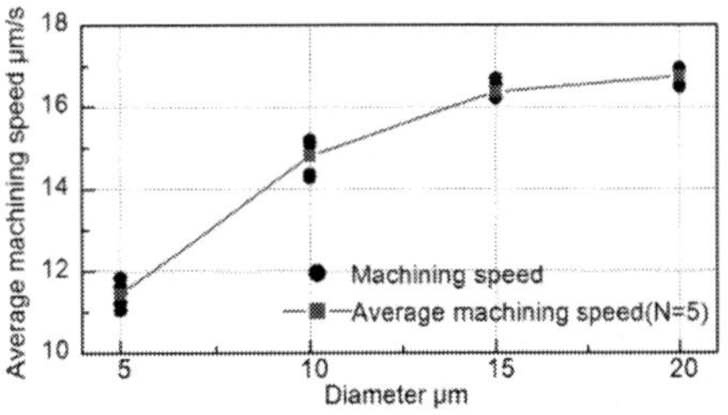

Figure 8. Average machining speed on 1000Hz

Where A and f are the diameter and frequency of the circular path, respectively. The frequency of the circular path was set to 200Hz ~ 1200Hz, and the diameter was set variously to 5μm ~ 20μm.

4.2.2 Machining results

Firstly, the frequency of the circular path was set to 1000Hz, and the diameter was tuned to 5 m, 10 m, 15 m, 20 m in EDM experiments. Fig. 7 shows the feed displacements of the electrode in the thrust direction. From these results, it can be seen that as the amplitude of the electrode vibration was increased, the machining time can be reduced to a saturated value.

For each EDM experiment, Five holes were also machined under the same conditions. The relationships between average machining speed and the amplitude of the electrode vibration are shown in Fig. 8. As the amplitude of the electrode vibration was increased, the average machining speed also increased. Therefore, to realize high speed machining, it is necessary to select a large the amplitude for the electrode vibration. However, for the accurate hole machining, it is necessary to consider the influence of the amplitude of the electrode vibration.

Secondly, the diameter of the circular path was set to 15μm, and the frequency was tuned to 200Hz, 350Hz, 600Hz, 800Hz, 1000Hz and 1200Hz. Five holes were also machined under the same conditions, and the calculated average machining speed is shown in Fig. 9. When the frequency of the electrode vibration is smaller than 1000Hz, the average machining speed gradually increases.

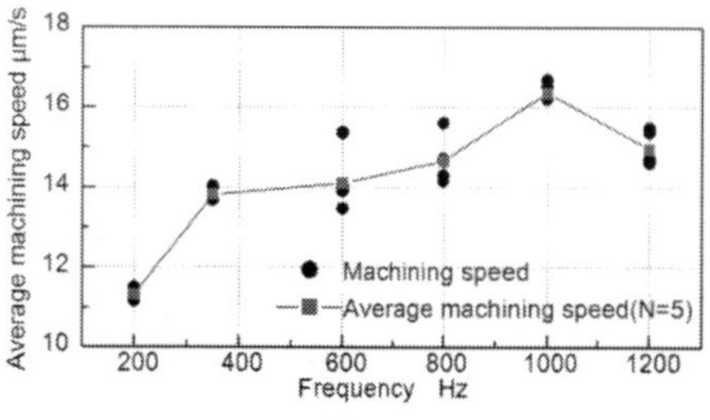

Figure 9. Average machining speed on 15μm

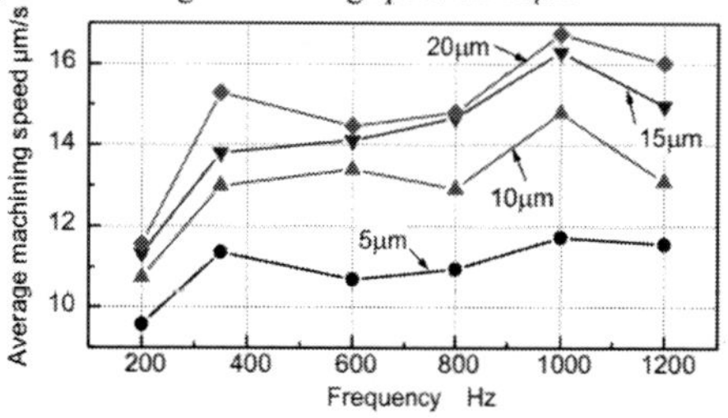

Figure 10. Relationships between average machining speed, amplitude and frequency of electrode vibration

The relationships between the average machining speed, the amplitude and the frequency of the electrode vibration are shown in Fig. 10. When the diameter of the circular path was set to 20 m, and the frequency was set to 1000Hz, the machining speed becomes the maximun, and it was 16.7 m/s. The average machining speed was increased by 76 % compared to without the electrode vibration, and 315% compared to conventional EDM.

5 Conclusions

In this paper, to evaluate the effectiveness of the proposed actuator for improvement of EDM machining speed, straight holes were machined in the thrust direction by using the proposed actuator. Experimental results show that the use of the proposed actuator can speedily adjust the distance between the electrode and the workpiece, and the machining speed was increased by 138% compared with the conventional EDM.

Moreover, the relationships between the machining speed, the amplitude and the frequency of the electrode vibration were discussed. The machining speed using the actuator also was increased as the changes of the amplitude and the frequency of the electrode vibration, and it was increased by 76% compared to without the electrode vibration, and 315% compared to conventional EDM.

Acknowledgement

This study was supported in part by JSPS KAKENHI Grant Number 15K05768 and SUZUKI Foundation.

References

1. T. Masuzawa: State of Art of Micromachining. Annals of the CIRP, Vol. 49 (2000), No.2, p. 473.
2. Y. Imai, A. Satake, A. Taneda and K. Kobayashi: Improvement of EDM Machining Speed by Using High Frequency Response Actuator. Int. J. Electr. Mach., (1996), No.1, p. 21.
3. T. Masuzawa, K. Tanaka and M. Fujino: Study on the High Speed Machining by EDM Using a Moving Coil Head Type Feed Control. In: Proceedings of the 19th International MTDR Conference, (1978), p. 543.
4. X. Zhang, T. Shinshi, T. Fukuoka and T. Nakai: Development of a 2-DOF Controlled Magnetic Drive Actuator for Laser Beam Cutting, Journal of Advanced Mechanical Design, Systems, and Manufacturing, JSME, Vol. 7 (2013), No. 3, p. 448.
5. H. Huang, H. Zhang, L. Zhou and H. Y. Zheng: Ultrasonic Vibration Assisted Electro-discharge Machining of Microholes in Nitinol. Journal of Micromechanics and Microengineering, Vol. 13 (2003), No.5, p. 693.
6. T. Endoa, T. Tsujimoto and K. Mitsui: Study of Vibration-assisted Micro-EDM. -The Effect of Vibration on Machining Time and Stability of Discharge. Precision Engineering, Vol. 32 (2008), No.4, p. 269.

The Tool Path Planning for Ring Torus Optical Surface Diamond Turning with Parallel 2DOF Fast Tool Servo

Aiyun Hao and Sixiang Zhang

Hebei university of technology, School of mechanical engineering 5340 Xiping Road, Beichen District, P.R.China

Abstract. FTS (Fast Tool Servo) has always been an important method for manufacturing non-axisymmetric optical surface. In this paper, a novel tool path planning method is presented which plans tool path in two coordinate directions of a parallel structure 2 DOFs FTS simultaneously. Comparing with single DOF FTS, this method has significantly improved the ability of producing non-axisymmetric optical surface, such as ring torus surface.

1 Introduction

It is a effective and flexible method that turning optical surface using single point diamond tool. High precise geometric tolerance and nano scale surface roughness can be gained by holding workpiece only one time. One of approaches for turning non-axisymmetric optical surface is using diamond turning incorporate with FTS. FTS provide high frequency reciprocate motion along x or (and) z axis of turning lathe synchronize with slow speed feed motion of diamond turning.

Large stroke and high bandwidth are always be pursued and they are main directions which researchers has worked on to improve. Recently, approaches for improving these two performances are centralized in how to innovate or refine driving method, for example, driving by piezoelectric stacks [1][2], magnetostrictive FTS[3], normal stress electromagnetically driven motor[4][5]. The capability of single DOF FTSs introduced above is always been limited by the stroke of actuators. There is another way to improve the performance of FTS. Xiao-qin Zhou from Jilin university provided a new approach that turning non-axisymmetric optical surface by FTS with multi-DOF[6]. In this paper, we present a method that magnificent improve FTS's manufacturing ability by planning tooltip trajectory in both x and z axis of a parallel structure 2DOF FTS simultaneously, achieving high frequency motion in both directions.

In the paper, ring torus optical surface is taken for example. By comparing the tool path planning of 2DOF FTS and single DOF FTS, Conclusion is drawn that 2DOF FTS has remarkable ability of performance improving than single DOF FTS.

2 The tool path planning for ring torus surface turning

2.1 The equation of ring torus surface

Equation 1 describes ring torus surface in a cylindrical coordinates system. The origin of the coordinates system is coincident with the vertex of the ring torus surface; the z axis is collinear with spindle axis of lathe and the direction is point to cross curve origin. The polar angle's reference direction is collinear with x axis of lathe. θ is polar angle and ρ is radial coordinate

$$z = R_b - \sqrt{\left(R_b - R_c + \sqrt{R_c^{\ 2} - r^2 \sin^2 \theta}\right)^2 - r^2 \cos^2 \theta} \quad (1)$$

where: $R_b = b + a$, $R_c = a$, a is radius of base curve of ring torus, b is radius of cross curve of ring torus.

2.2 The tool path planning for ring torus surface using 2DOF FTS

Ring torus optical surface can be created by using two slow-speed feed motions of diamond turning lathe along x and z axes, and two fast-speed motions of 2DOF FTS along the same axes as well. These motion can be described as equation 2:

$$\begin{cases} r = r_s(t) + r_f(r_s, \theta) \\ z = z_s(r_s) + z_f(r_s, \theta) \end{cases} \quad (2)$$

where subscript s means slow-speed of turning lathe, f means fast-speed of 2DOF FTS.

Polar coordinate equation 3 is designed for description the x axis motion of turning lathe and the motion along workpiece radius of 2DOF FTS.

DOI: 10.1051/matecconf/20179505005

$$\begin{cases} \theta = \omega t \\ r_s = R - vt \\ r_f = A\dfrac{x_{max}}{2}\cos(2\theta) \end{cases} \tag{3}$$

where:

$$A = \begin{cases} 1 & r_s > R - S \\ r_s/(R-S) & r_s \le R - S \end{cases}$$

θ is polar angle, ω is angle velocity of lathe spindle, R is the maximum radius of workpiece, r is the distance between tooltip and origin of polar coordinate system, S is the maximum stroke of 2DOF FTS along one axis, x_{max} is the maximum stroke along radius.

The equation that describes the motion of 2DOF FTS along z axis is constructed as equation 4

$$z_s = a + b - \left[\sqrt{(b+a)^2 - (r_s + AS)^2} + b + \sqrt{a^2 - (r_s - AS)^2}\right]/2 \quad \ldots\ldots(4)$$

$$\begin{aligned} z_f &= z - z_s \\ &= \left[\sqrt{(b+a)^2 - (r_s + AS)^2} + b + \sqrt{a^2 - (r_s - AS)^2}\right]/2 \\ &- \sqrt{\left[b + \sqrt{a^2 - (r_s + AS\cos 2\theta)^2 \sin^2\theta}\right]^2 - (r_s + AS\cos 2\theta)^2 \cos^2\theta} \end{aligned} \quad \ldots\ldots(5)$$

The simulation of trajectory of tooltip on the 2DOF FTS that constructed in this paper is shown in Fig.1. Ring torus geometry is shown in Fig. 2 and the combination of ring torus geometry and tooltip trajectory curve is shown in Fig. 3. It's obviously that the geometry surface and the curve is matched perfectly.

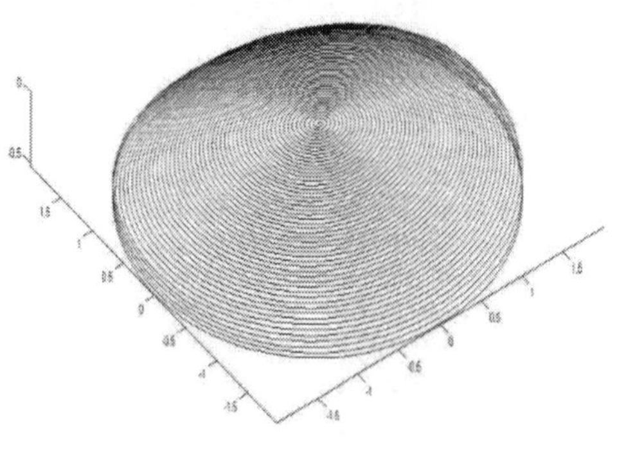

Figure 1. Tool path plan for turning ring torus.

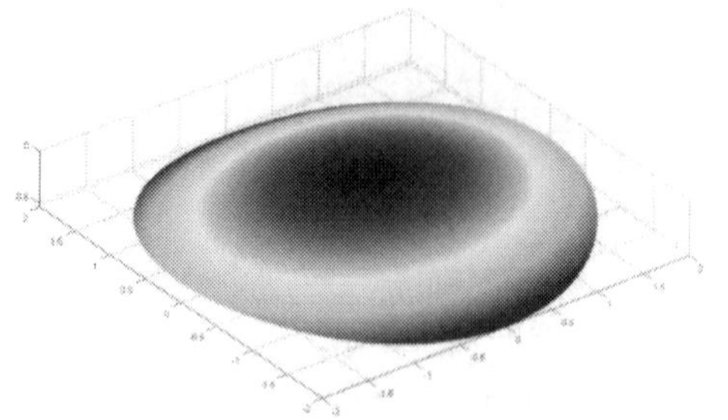

Figure 2. Ring torus geometry.

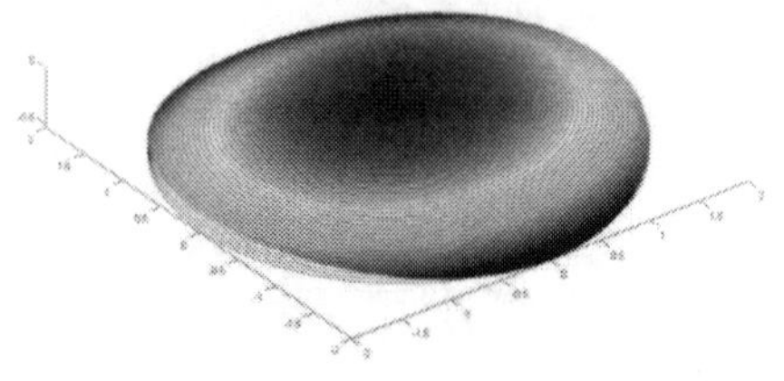

Figure 3. match of ring torus geometry and tool path.

3 Performance comparison between single DOF FTS and 2DOF FTS

3.1 Qualitative performance comparison between single DOF FTS and 2DOF FTS

For Qualitative comparison between single DOF FTS and 2DOF FTS, the curves of 2DOF FTS motions along x and z axis are drawn in Fig. 4 and Fig. 5 separately. In these graphs, horizontal coordinate is the angle of turning lathe spindle and vertical coordinate is x or z displacement.

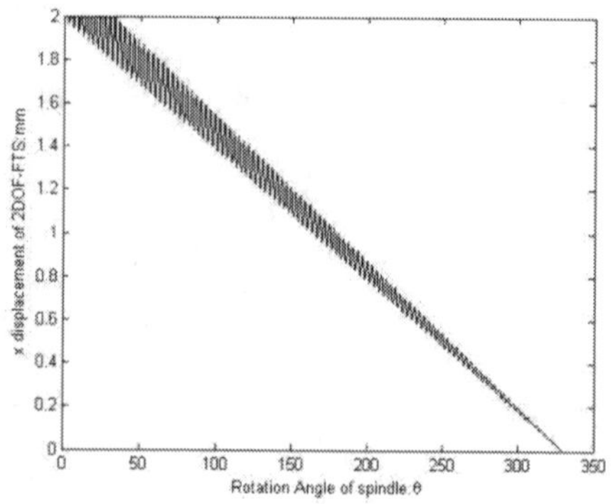

Figure 4. x displacement by angle of spindle.

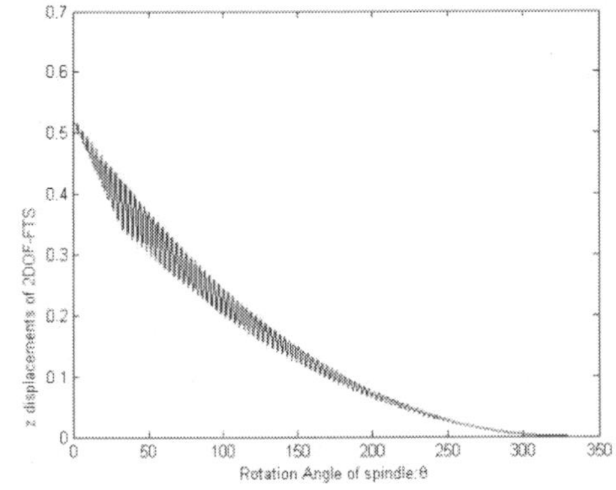

Figure 5. z displacement by angle of spindle.

The high frequency motion component shown in Fig. 4 and 5 are accomplished by 2DOF FTS along x and z axis separately meanwhile turning lathe's x and z slider fulfill the low frequency motion components.

According the Fig, 4 and Fig. 5, the max strokes along x and z axis are not needed at first period of cutting process but after several ones. If we turn the same workpiece by single DOF FTS, the max stroke always happened at the first period as the light curve shown in Fig. 6. The reason is that the tooltip is not always keep in touch with workpiece before the period of max strokes happened if the ring torus fabricated by 2DOF FTS. Before the period during which max stroke happens, the cutting process is not continuous to avoid collision between tooltip and workpiece by swing tooltip out of workpiece outer cycle along x axis. Once tooltip swing away from the workpiece, the motion of 2DOF FTS along z axis will be hold. That is the key why smaller strokes needed by using 2DOF FTS than single DOF FTS. The method is described clearly by the trajectory shown in Fig. 1, Fig. 3 and equation 5.

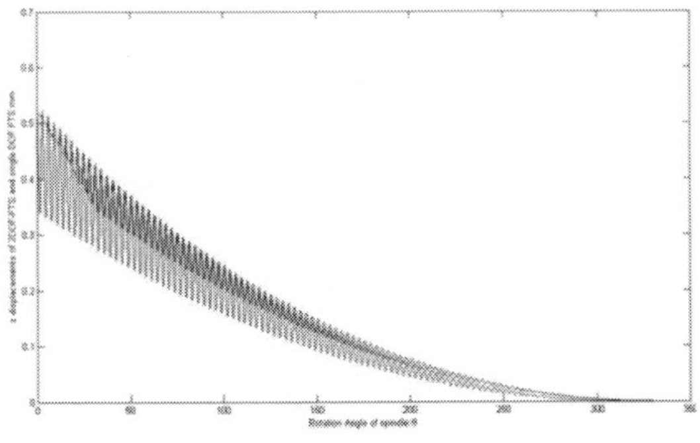

Figure 6. Two kinds of FTS tooltip motion curves along z axis

In order to illustrate the comparing more clearly, both the single DOF FTS's and the 2DOF FTS's tooltip motion curve along z axis of turning lathe is drawn in the same graph as shown in Fig. 6. In Fig. 6, the light curve is single DOF FTS motion trajectory, and the dark curve is 2DOF FTS motion curve. Obviously, about 100% more stroke is required using single DOF FTS than using 2 DOF FTS for fabricating the same ring torus surface in this example.

3.2 Quantitative performance comparison between single DOF FTS and 2DOF FTS

According to equation 5, the maximum displacement value of 2 DOF FTS along z axis arise while the angle of turning lathe spindle $\theta = 0$ and the minimum value arise while $\theta = \pi/2$. The maximum and minimum value are expressed as equation 6.

$$\begin{cases} z_0 = a + b - \sqrt{(a+b)^2 - R^2} \\ z_{\pi/2} = a + b - \left[b + \sqrt{a^2 - (R-2S)^2} \right] \end{cases} \quad (6)$$

where z_0 means the displacement along z axis while $\theta = 0$ and $z_{\pi/2}$ means the displacement along z axis while $\theta = \pi/2$.

The maximum z displacement can be obtained from equation 6 as equation 7:

$$\Delta z_{max} = \sqrt{(a+b)^2 - R^2} - \sqrt{a^2 - (R-2S)^2} - b \,....(7)$$

Equation 7 presents the limit of single DOF FTS displacement.

The 2 DOF FTS is going to its limit of displacement when $\Delta z_{max} = S$. Substitute $\Delta z_{max} = S$ into the equation 7, an one-Variable Quadratic inequality of S is obtained. Arranging the result, S is expressed as equation 8.

$$S = \frac{m - \sqrt{m^2 - 5n}}{5} \quad (8)$$

where $m = \sqrt{(a+b)^2 - R^2} - b + 2R$, $n = \left[\sqrt{(a+b)^2 - R^2} - b \right]^2 + R^2 - a^2$

Equation 8 describes the relationship between ring torus with parameters a, b, R and the manufacturing ability of 2DOF FTS with determined stroke S. The maximum stroke required for ring torus with parameters a, b, R using 2 DOF FTS is expressed as equation 9.

$$\Delta z'_{max} = \sqrt{(a+b)^2 - R^2} - \sqrt{a^2 - R^2} - b \quad (9)$$

Let $\alpha = b/a, \beta = R/a$ and substitute α, β in equation 9, equation 10 is obtained.

$$P_r = \frac{\sqrt{1-(\beta - 2S/\alpha)^2} - \sqrt{1-\beta^2}}{S/\alpha} \quad (10)$$

The parameters a, b and R in equation 9 is the absolute dimensions of ring torus surface. However, α and β are the opposite, they indicate the degree of non-axisymmetric without relationship of the dimensions of ring torus. Hence, equation 10 has more meaning that is

DOI: 10.1051/matecconf/20179505005

general. The function graph of equation 10 is shown in Fig. 7.

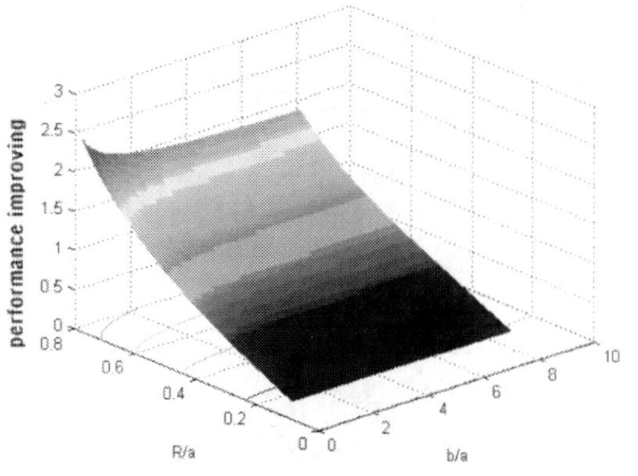

Figure 7. Performance improving with 2DOF FTS.

According to the figure 7, as β increases, the performance of 2DOF FTS is significent improved respect to single DOF FTS. As α decreases, the performance improving of 2DOF FTS is not obvious. during the value of α change from 0.1 to 0.8 and β change from 0.1 to 8, P_r is going to change from 0.19 to 2.53 continuously. In other word, the manufacturing ability of 2 DOF FTS is improved about 20% to 250% respect to single DOF FTS. In figure 7, the value range of α and β are wide. In practical application, these two parameters is located in the middle of surface in figure 7. For example, Radius of curvature of Astigmatism correcting lens is between 50mm to 130mm. Correspondingly, let $b = 100mm$, $a = 60mm$ and $R = 30mm$, the results of equation 7 and equation 9 are $\Delta z_{max} = 2.6mm$ and $\Delta z'_{max} = 5.2mm$. That means, the lens can be fabricated using 2 DOF FTS with both stroke of 2.6mm instead of single DOF FTS with stroke of 5.2mm. In this example, the manufacture ability of 2 DOF FTS is 200% times than single DOF FTS.

4. Conclusion

In this paper, a novel tool path planning method is presented. The method plans tool path along both x and z axis of single point diamond turning lathe using parallel structure 2 DOF FTS. Take the fabrication of ring torus optical surface for example, the equation of tooltip motion trajectory is given. It is proved that the fabricating ability of FTS will be significant improved using the method presented in this paper by analysing fabrication limits of both single DOF FTS and 2 DOF FTS.

References

1. Kim.H.S, Kim.I.L, Lee.K.M. Fabrication of free-form surfaces using a long-stroke fast tool servo and corrective figuring with on-machine measurement. International Journal of Machine Tools & Manufacture, pages 991-997, 2009.
2. Altintas Y, Woronko A. A piezo tool actuator for precision turning of hardened shafts[J]. CIRP Annals - Manufacturing Technology, 51(1): 303-306, 2002.
3. D.Liu. Surface texture improvement in the turning process via application of a magnetostrictively actuatored tool holder. ASME Journal of DSMC, 120, 1998.
4. Xiaodong Lu and David L. Trumper. Electromagnetically driven ultrafast tool servo. Proceedings from ASPE 2004 annual meeting, pages 269-272, 2004.
5. Dan Wu, Xiaodan Xie, and shunyan Zhou. Design of a Normal Stress Electromagnetic Fast Linear Actuator. IEEE TRANSACTIONS ON MAGNETICS, 46(4): 1007-1014, 2010.
6. Q Liu, S Pan, H Yan, X Zhou, R Wangln. Situ measurement and error compensation of optical freeform surfaces based on a two DOF fast tool servo. International Journal of Advanced Manufacturing Technology, 1-6, 2015.

DOI: 10.1051/matecconf/20179505006

Tool Wear Analysis due to Machining In Super Austenitic Stainless Steel

Ashwin Polishetty, Mohanad Alabdullah and Guy Littlefair

School of Engineering, Deakin University, 3216 Waurn Ponds, VIC, Australia

Abstract. This paper presents tool wear study when a machinability test was applied using milling on Super Austenitic Stainless Steel AL6XN alloy. Eight milling trials were performed under two cutting speeds, 100 m/min and 150 m/min, combined with two feed rates at 0.1mm/tooth and 0.15 mm/tooth and two depth of cuts at 2 mm and 3 mm. An Alicona 3D optical surface profilometer was used to scan cutting inserts flank and rake face areas for wear. Readings such as maximum and minimum deviations were extracted and used to analyse the outcomes. Results showed various types of wear were generated on the tool rake and flank faces. The common formed wear was the crater wear. The formation of the build-up edge was observed on the rake face of the cutting tool.

1 Introduction

Super austenitic stainless steel (SASS) alloy-AL6XN from now on referred as AL6XN is a type of austenite stainless steel distinguishable by its high alloying contents. The presence of alloying elements such as chromium, nickel, molybdenum and nitrogen enhance the resistance of the alloy to the stress corrosion cracking (SCC), crevice and pitting corrosion. AL6XN alloy is applied in various engineering applications such as in offshore structures, chemical waste processing, food containers, transformer cases and pump parts [1]-[3].

Contrary, the alloying elements render the alloy to be classified as a difficult to machine material. Machinability of a material is a criterion used to define the ease with which the material can be machined. In general, machinability is often judged by; the magnitude of cutting forces, tool wear/tool life, surface finish and surface integrity; and chips formation and morphology [4]. The machinability process is sensitive to the properties of the machined material, such as high level of strength, ductility, work hardening and decreased thermal conductivity [5]-[9]. Machinability studies have been conducted by researchers on different Austenitic Stainless Steel (ASS) grades.

The authors of this work conducted a recent study on machinability of AL6XN SASS alloy [10]. The cutting forces, surface microhardness and roughness were investigated when the alloy was machined under up end milling operation. A correlation between the machining aspects and the cutting parameters was explained.

Research study conducted by Nordin et.al says that flank, crater and chipping wear was observed during milling of 316 AISI austenite stainless steel [11]. Similarly, Abou-El-Hossein, K. and Z. Yahya [12] studied the machinability behaviour of AISI 304. Coated carbide inserts were used in an end milling operation. The main objective of the research was to evaluate the wear and life of the inserts. When high feed rate was used, the tool wear significantly decreased. However, the large value for the tool wear was acquired when the cutting speed increased. In addition, high feed rate and moderate cutting speed caused the formation of the BUE on the cutting inserts. V. P. Astakhov [13] studied the influence of the depth of cut and feed rate on tool wear. Low carbon austenite stainless steel of 18.5 % chromium and 15.5% nickel has been machined in this study. Results convey that the depth of cut did not affect the rate of the tool wear during the cutting process. S.-T. Kim and Y.-S. Park [14] performed a dry turning experiment to machine the high performance austenite stainless steel. This type of steel has a chromium content of 21%. Flank and notch wear were observed on the cutting insert at a cutting speed of 100 m/min. Furthermore, high wear and short tool life was observed on the cutting insert at high cutting speed. Ciftici evaluated the cutting tool wear during dry machining of AISA 316 and AISI 304 austenite stainless steel using carbide tools [15]. Xavior MA, Adithan M studied the effects of the coolant on the tool wear during turning of austenite stainless steel. The outcomes revealed that using low feed rate during machining reduced the tool wear and improved the quality of the machined surfaces [16]. The correlations between the tool wear with the cutting speed and feed rate were also studied by Z. Tekıner and S. Yeşılyurt [17]. In most of the reviewed research studies, Scanning Electron Microscope (SEM) was used to inspect the tool wear and BUE formation on the cutting tool faces. However, Królczyk G, Legutko S, Raos P used Alicona Infinite Focus (AIF) 3D optical profilometer technique to establish the topography of the cutting tool surface [18].

The outcomes showed the wear on the tool edge increased when the cutting speed increased.

Maximum and minimum deviation calculations were used to determine the propagation of the wear and build-up edge on the cutting faces. Crater wear and BUE on the rake face can be precisely calculated when (AIF) profilometer is used. From the reviewed literature, it can be seen that machinability studies have been conducted on common austenite stainless steel grades. According to the literature, a research gap exists in literature regards to gradual progress of machining ASS to SASS. As AL6XN has been a classified material so far with limited research conducted on it, this paper aims to continue machining investigation on AL6XN in order to study and analyse the generated wear and surface topography of the cutting edge and BUE formation using Alicona Infinite Focus 3D optical profilometer.

2 Experimental design

The chemical composition of the AL6XN alloy has been evaluated using spectrometry. The elements present in the alloy are listed in Table 1. A total of 8 milling trials were executed under a combination of cutting parameters which included cutting speed of 100 and 150 m/min; feed rates of 0.1 and 0.15 mm/tooth; and depth of cuts of 2 and 3 mm. Coolant made of phenol-water mixture (1:10) was applied during machining.

Table 2 lists the eight cutting trials that represent the predefined cutting parameters. The selection and the arrangement of the cutting trials were designated according to the recently published study by the authors on machinability of the AL6XN SASS alloy.

Table 1. Chemical composition of AL6XN SASS

C	Si	Mn	P	S	Cr	Mo	Ni
0.026	0.36	0.37	0.044	0.002	21	6.066	24

Table 2. List of cutting trials.

Trial	Cutting speed (m/min)	Feed/tooth (mm/tooth)	Cutting depth (mm)
1	100	0.1	2
2	100	0.1	3
3	100	0.15	2
4	100	0.15	3
5	150	0.1	2
6	150	0.1	3
7	150	0.15	2
8	150	0.15	3

Machining trials were conducted on 5-Axis CNC milling machine (SPINNER U620) having specification of spindle power, 15 kW and a table diameter of 650 mm. Insert type cutting tool-ISCAR HELIDO 490-09 grade IC830, TiCN, TiN coated inserts were used for the trials. The designation of the cutting insert according to the manufacturer standards is H490 ANCX 090416PDR. The cutting insert has 4 cutting edges and a 0.8 mm nose radius. This type of insert is used in rough and hard machining applications.

The cutting tool holder was designed to clamp two cutting inserts. Each cutting insert has four cutting edges of 0.8 mm radius per edge. A new cutting edge was used for each cutting trial in order to maintain reference (zero) tool wear condition. The workpiece of a rectangular block configuration was fixed on the milling dynamometer as shown in Fig. 1. The cutting time was constant and equal to 20 seconds per trial during the cutting process. Side by side cutting process was executed and eight machined bays were formed on the mounted alloy.

Figure 1. Experimental setup.

2.1 Tool wear measurements using Alicona Infinite Focus 3D optical profilometer

The worn cutting inserts were scanned using Alicona Infinite Focus (AIF) 3D optical surface profilometer. The cutting insert was fixed on to a customised tool holder which can scan sample at inclination angles from 0° to 35°as shown in Fig. 2. The best scan quality was obtained at 25°. A full scan was performed on a new cutting edge which was later used as a reference edge. Later, the worn edge was scanned using same elected scanning coordinates of the new cutting edge. Using difference analysis available on IF-Measure Suite 5.1 software, the new and worn cutting edges were superimposed to evaluate the tool wear (blue region as shown in Fig. 2).

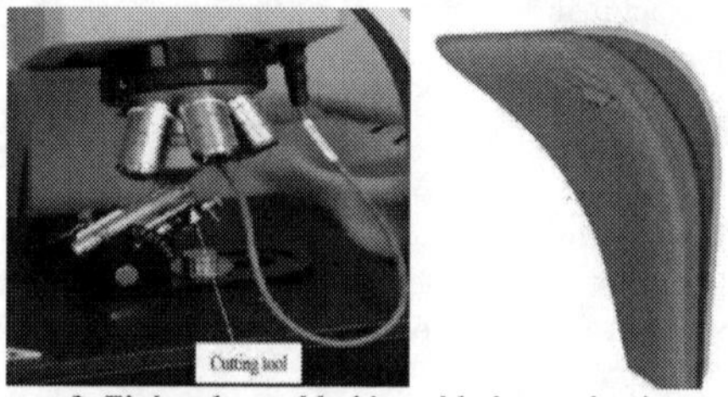

Figure 2. Fixing the tool holder with the cutting inserts on the stage of the AIF device and difference analysis of cutting tool wear using Alicona optical surface profilometry.

3 Results and discussion

Four wear types were identified on the edges of the cutting tool when the AIF scanning was run on the cutting inserts as shown in Fig. 3.

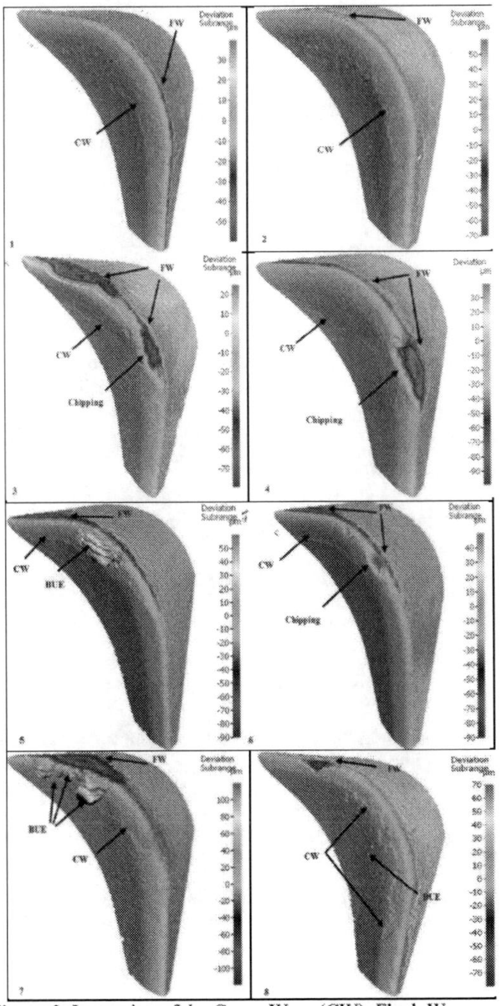

Figure 3. Inspection of the Crater Wear (CW), Flank Wear (FW), Chipping and BUE on the cutting edges of the 8 cutting inserts.

3.1. Flank wear

Flank Wear (FW) was observed in all the cutting trials but with different values as shown in Fig. 3. Flank wear was found to increase when the feed rate was increased from 0.1 mm/tooth in trial 2 to 0.15 mm/tooth in trial 3 and trial 7. The effect of the depth of cut on the flank wear was also observed in trials 2 and 3. Reducing the depth of cut from 3 mm to 2 mm led to increase in flank wear as in trial 3. In trial 7, changing the depth of cut from 3 mm to 2 mm also caused the flank wear on the insert edge. Generally, when the cutting speed was elevated from 100 m/min to 150 m/min the flank wear was reduced.

3.2 Crater wear

The rake faces of the inserts suffered from the presence of the crater wear. In Fig. 3, the Crater Wear (CW) was observed in all the trials but with different values. Figure 7 shows that applying a cutting speed of 100 m/min while increasing the feed from 0.1 to 0.15 mm/ tooth increased the CW in trials 3 and 4. Such an increase was found to be larger than the CW in trials 1 and 2. Changing the speed to 150 m/min caused a significant wear on the rake faces of the inserts in the last four trials. In other words, high speed, as well as high feed rate, used in trials 7 and 8 produced a large crater wear.

3.3 Chipping wear

Chipping wear is the separation of small parts from the cutting tool edge. This type of wear occurred when the material was machined under trials 3 and 4 cutting conditions, as shown in Fig. 3, due to the use of a high feed with a low cutting speed.

3.4 Build-Up Edge (BUE)

The formation of BUE was detected when the cutting edges were scanned. Fig. 3 shows the presence of the BUE in trials 5, 7 and 8. At a 150 m/min cutting speed, the BUE was formed on the rake face of the cutting insert used in trial 5. Keeping the cutting speed and the depth of cut constants and increasing the feed rate to 0.15 mm/tooth in the trial 7 led to increase the BUE formation. In trial 8, a feed rate of 0.15 mm/tooth and a depth of cut of 3 mm lowered and limited the formation of the BUE on the rake face as shown in Fig. 3.

3.5 Tool wears statistical results obtained by the optical profilometer

The statistics obtained from AIF can be presented by the minimum and maximum deviation. Minimum deviation reveals the magnitude of the formed BUE whereas the maximum deviation is an indicator of how much the cutting edge is lowered [18]. The inserts of the cutting trials 5 and 7 have the BUE formed and their minimum deviations magnitudes were equal to 368 µm and 152 µm respectively.

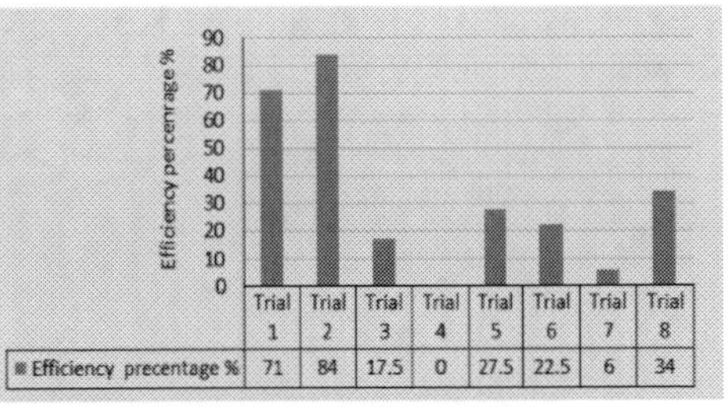

Figure 4. Efficiency percentages of the 8 cutting edges.

The BUE was measured and Table 3 lists the minimum and maximum deviations of the eight cutting edges. It can be seen from table 3 that the cutting parameters used in trials 1, 3, 6 and 7 lowered the cutting edges more in comparison to the other cutting trials. Such lowered cutting edge causes tool failure and as a consequence, shortens the tool life. Maximum value of the maximum deviation was placed in the cutting trial number 4 and nominated as the lowest effective edge during machining. This value was used as a reference to estimate the efficiency percentages of the cutting edges. Fig. 4 shows the efficiency percentage of each cutting trial calculated regarding to trial 4.

Table 3. BUE and the lowered cutting edge measurements using AIF statistics.

Trial	1	2	3	4	5	6	7	8
Maximum deviation (μm)	35	20	99	120	87	93	113	79

The maximum efficiency percentage of the cutting edge was achieved in trial 2 where the low speed, feed rate were applied. The efficiency was reduced to 17.5 % in trial 3, and the minimum efficiency was recorded in trial 4 due to the occurrence of the chipping wear. When the cutting speed increased to 150 m/min, the efficiencies of the cutting edges were reduced in trials 5 and 6 and the minimum efficiency value at high cutting speed was equal to 6 % in trial 7 due to the existence of the BUE.

4 Conclusion

Super Austenitic Stainless Steel alloy-AL6XN is hard to be machined material. Machinability test was conducted under a combination of various cutting parameters. The cutting inserts used were inspected for tool wear using Alicona Infinite Focus 3D optical profilometer.

1. Flank wear was obtained with low amounts on the flank faces of the cutting inserts.
2. Crater wear was located on the rake faces of all the cutting inserts with a varied length and depth. Maximum crater wear length of 2624 μm and maximum depth of 380 μm was obtained in trial 1 and 4 respectively. The length and the depth of the crater wear decreased up to 58% and 71% respectively in trial 2 where the low speed and feed with high depth of cut were applied.
3. Significant BUE was located on the rake faces of the cutting inserts in trials 5 and 7 in which high cutting speed, depth of cut and low feed rate were applied.
4. Chipping wear, was found in the cutting trials 3, 4 and 6.
5. Minimum and maximum deviations received from the alignment process of the new and worn cutting edges present the magnitude of the BUE and the lowered cutting edges respectively. Maximum deviation values were used to calculate the efficiency percentages of the cutting tools. Maximum efficiency of 84 % was gained at low cutting speed and feed rate in trial 2 whereas minimum efficiency was gathered at cutting trial 4 due to the presence of the chipping wear. At high speed and feed rate in trial 7, the formed BUE lowered the efficiency of the cutting edge to 6%.

References

1. F.H. Abed, G.Z. Voyiadjis, Int. J. Plast. **21**(8): p. 1618-1639, (2005).
2. S. Kalnaus, F. Fan, A.K. Vasudevan, Y. Jiang, Eng. Fract. Mech. **75**(8): p. 2002-2019, (2008).
3. S. Nemat-Nasser, W.-G. Guo, D.P. Kihl, J. Mech. Phys. Solids. **49**(8): p. 1823-1846, (2001).
4. S. Kalpakjian, S.R. Schmid, *Manufacturing engineering and technology*, (2014).
5. M.P. Groover, *Fundamentals of modern manufacturing: materials processes, and systems*, (2007).
6. M.M.C.A.P. Handbook, Inc., Sweden, (1997).
7. T. Akasawa, H. Sakurai, M. Nakamura, T. Tanaka, K. Takano, J. Mater. Process. Technol. **143**: p. 66-71, (2003).
8. T. Kosa, R. Ney, ASM Handbook. **16**: p. 681-707, (1989).
9. J. Kopač, S. Šali, Journal of Materials Processing Technology. **113**(1–3): p. 312-316, (2001).
10. A. Polishetty, M.F.A. Alabdullah, N. Pillay, G. Littlefair. *ASME Proceedings Advanced Manufacturing* (2015).
11. M. Nordin, R. Sundström, T.I. Selinder, S. Hogmark, Surf. Coat. Technol. **133–134**: p. 240-246, (2000).
12. K. Abou-El-Hossein, Z. Yahya, J. Mater. Process. Technol. **162**: p. 596-602, (2005).
13. V.P. Astakhov, Int. J. Adv. Manuf. Technol. **34**(7-8): p. 631-640, (2007).
14. S.-T. Kim, Y.-S. Park, Met. Mater. Int. **15**(2): p. 221-230, (2009).
15. I. Ciftci, Tribol. Int. **39**(6): p. 565-569, (2006).
16. M.A. Xavior, M. Adithan, J. Mater. Process. Technol. **209**(2): p. 900-909, (2009).
17. Z. Tekıner, S. Yeşılyurt, Mater. Des. **25**(6): p. 507-513, (2004).
18. G. Królczyk, S. Legutko, P. Raos, Tehn. vjesn. Tech. Gaze. **20**(3): p. 413-418, (2013).

 DOI: 10.1051/matecconf/20179505007

Development of Inner-Grooved Tubes by Dieless Micro-Forming with Laser-as-Sisted Air Pressure

Ching-Min Cheng[1], Ming-Zheng Tu[1], Tun-Ping Teng[2], Shang-Pang Yu[1], Cheng-Yi Yu[2] and Yu-Tai Chang[1]

[1]Department of Mechatronic Engineering, National Taiwan Normal University, Taiwan
[2]Department of Industrial Education, National Taiwan Normal University, Taiwan

Abstract. Grooved tubes are widely used in electronic cooling modules. They are usuallt drawn, a term referring to molding. However, this process has limitations, especially in the production of smaller tubes. Therefore, this study examines the potential of an alternative process, dieless forming, in which the grooves on stainless steel are generated by pulsed laser-assisted air pressure. To begin with, the stainless steel tube in ambient air was heated using a pulsed laser. Then, groove samples were sliced for observation using an optical microscope (OM). Finally, the samples were analyzed using Scanning Electronic Microscope (SEM). The results demonstrate that grooves can be formed utilizing dieless forming with pulsed laser-assisted air pressure. The average observed depth of the grooves 208.5 um at 10-watt power and 0.01 Mpa gas pressure. In conclusion, grooves at a micro scale can be produced effectively by dieless forming for a range of industrial applications.

1 Introduction

Moore's Law [1] holds that, all electronic devices become smaller and smaller over time. This trend extends to consumer electronics, whose requirements for tiny components are constantly increasing. Rapid development of both the software and hardware components of smartphones, for instance, implies that there will be increasing demand for advanced heat-dissipation solutions. In recent years, such solutions have included micro heat pipes. [2]

The diameter of micro heat pipes can affect the efficiency of heat dissipation, because increasing surface area reduces air-flow resistance and elevates thermal conductivity. As compared to heat pipes without internal structures, the thermal conductivity of heat pipes with internal tread increases by 30 % to 40 %. In the early stages of their development, the usual diameter of heat pipes was 9.52 mm. this was subsequently reduced to 7 mm, and from 1995, to 5 mm.

Tube forming refers to changing the shape of a tube by external force using a mold or die. Recently, there has been a marked downward shift in the size of tube forming. In seeking to obtain an ideal size of tube while maintaining the quality of the material, mold processing is of crucial concern. [3]

There are three types of heat pipes: 1) grooved, 2) screen, and 3) sintered. This paper will focus on the grooved type, which is generally produced by roll forming. The roll-forming process begins with the feeding of a flat metal sheet into a series of calendering machines in order to produce a tube with an inner spiral ribbed surface. [4] After the tube has been formed, its size will be reduced using multiple rollers rotating in a cycle around the tube while it is drawn in the axial direction. This process has the limitation that, once the tube is reduced, the geometric section of its inner groove will deform. [5], [6] The present research attempts to solve this problem by forming inner-grooved tubes via dieless forming instead of roll forming.

2 Experimental

This experiment was performed using a ROFIN starweld 40 W Nd:YAG pulse-laser welding machine with a FANUC Robot LR Mate200i six-axis robot arm. (Fig. 1) The forming process as show in Fig. 2 took place at standard atmospheric pressure without preheating or noble gas. A precision pressure regulator and air filter regulator were chosen to maintain the pressure. Because an air lubricator would have changed the accuracy of the precision pressure regulator, it was excluded from this experiment.

Prior studies have noted that the forming pressure of tubes at ambient temperature is generally beyond 10 Mpa; and at low pressure of around 1 Mpa, several attempts may be required for forming. In this study, the forming pressure was much lower due to the use of laser heating. The pressure parameters should be in keeping with the parameters of the laser if fractures are to be avoided. The main laser parameters are power, pulse width, pulse frequency, robot-arm movement speed, and overlap rate. Among all five of these parameters, power and pressure are the most important. Pulse width affects the geometric

 DOI: 10.1051/matecconf/20179505007

size of the groove, while pulse frequency and the speed of the robot arm control the overlap rate. In this research, SUS 301 φ2 mm x 0.2 mm stainless steel was used as a specimen. The related parameters were set as follows: power, 8-12 W; pulse width, 3-5 ms; pulse frequency, 4-5 Hz; robot-arm movement speed, 1 mm/sec; and pressure, 0.007-0. 015 Mpa in Table 1.

The experiment results were observed using SEM and an OM.

Table 1. Experiment parameters

material	SUS301
power (W)	8-12
Pulse Width (ms)	3-5
Pulse frequency (Hz)	4-5
speed (mm/sec)	1
Pressure (Mpa)	0.007-0. 015

Figure 1. Experimental equipment, (a) FANUC Robot LR Mate200i six-axis robot arm (b) Nd:YAG pulse-laser welding machine.

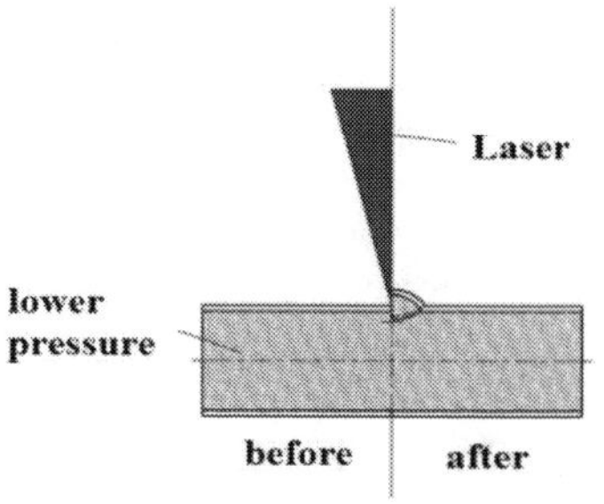

Figure 2. Approach forming

3 Results and discussion

The specimens were analyzed using SEM, and a significant result was found regarding atmospheric pressure. The related parameters were as follows: power 10 W, pulse width 4ms, and pulse frequency 4.5 Hz. While the specimens' surfaces were being heated by the laser, the groove was not generated, since there were no pressure differences. However, the melted inner wall was found to be lower than the surface due to effect of gravity, as shown in Fig. 3. If the forming pressure was increased to 0.007 MPa, the groove could be shaped, with a width of 79 um. There was no sharp difference in wall thickness, as shown in Fig. 4. When the forming pressure was augmented to 0.01 MPa, the groove was completed, and its average width was 208.5 um. the wall thichness was thinner than it was originally, as shown in Fig. 5. If the forming pressure was increased to 0.015 MPa, random fractures would occur, as shown in Fig. 6. The relationship between pressure and the average depth of inner grooves is presented in Fig. 7.

Figure 3. Inner groove was not formed since there were no pressure differences at atmospheric condition. (a) Outside of the tube (b) inside of the tube.

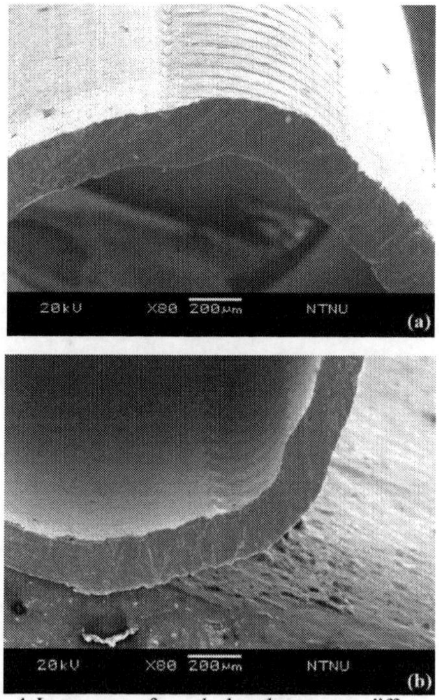

Figure 4. Inner groove formed when the pressure differences reach to 0.007 Mpa, and the average depth was 79 um. (a) Outside of the tube (b) inside of the tube.

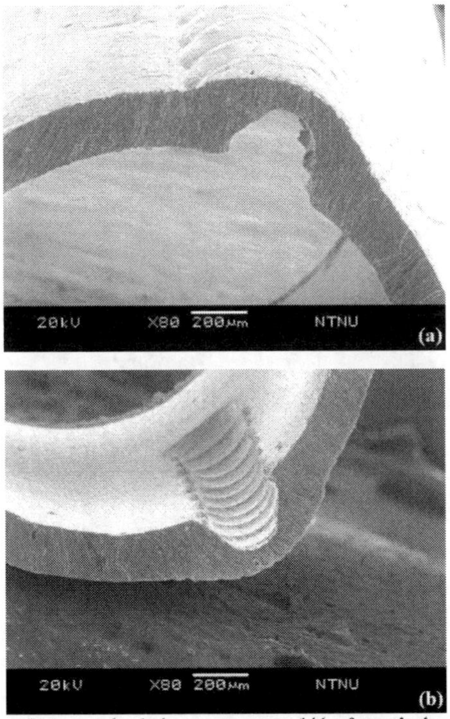

Figure 5. A completely inner groove could be formed when the pressure differences reach to 0.01 Mpa, and the average depth was 208.5 um. (a) Outside of the tube (b) inside of the tube.

Figure 6. Random fracture would happen when the pressure differences reach to 0.015 Mpa. (a) Outside of the tube (b) inside of the tube.

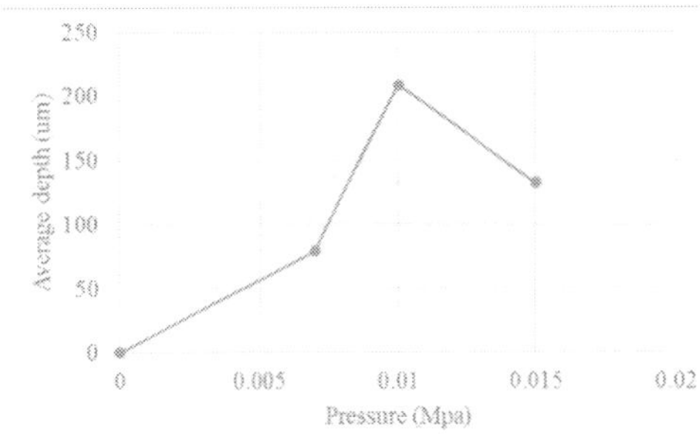

Figure 7. The relationship between pressure and average depth.

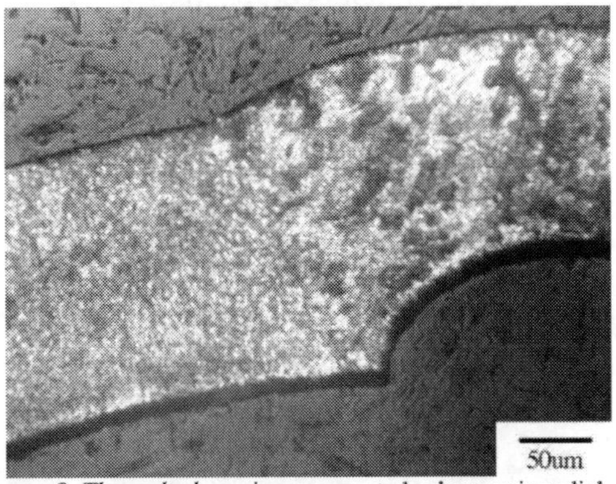

Figure 8. The melted specimen presented columnar in radial direction after being pressed.

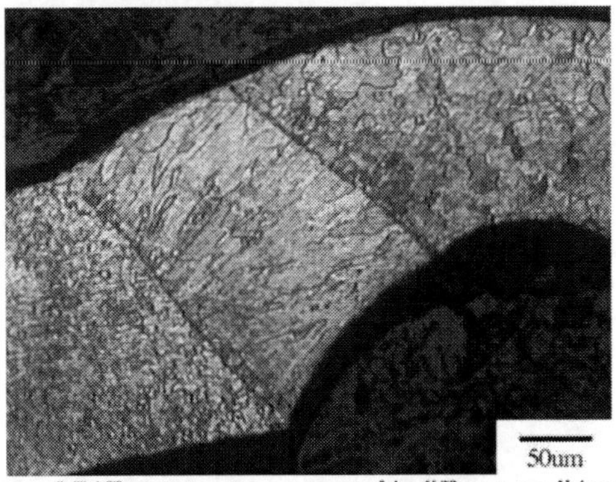

Figure 9 Different structures generated in different conditions

The metallography results imply that the setting of parameters impacts considerably on the structure of the groove. Two conditions were selected, and are depicted in Fig. 8&9. In Fig. 8, the groove was formed due to inner pressure while the specimen was melted. The structure in the radial direction was columnar, and that in the axial direction tool the form of dots. In Fig. 9, the inner pressure also affected the forming when at duplex.

4 Conclusion

This research on the feasibility of creating inner-grooved tubes by dieless forming utilized. Stainless steel specimens heated by laser in standard atmospheric conditions. It was found that:

1. Inner tube structure can be formed successfully using this dieless process.
2. The average depth of the inner grooves obtained was 208.5 um.
3. Inner tubes can be formed dielessly at lower pressure (0.01 MPa), which is safer and less costly than hydroforming, which requires pressure of 2 MPa.
4. The diameter of the hole was controllable.
5. The forming position and length were unrestricted.

References

1. G. E. Moore, Cramming more components onto integrated circuits, Electronics, **38**, p. 114, (1965).
2. H.-T. Lim, S.-H. Kim, H.-D. Im, K.-H. Oh, and S.-H. Jeong, Fabrication and evaluation of a copper flat micro heat pipe working under adverse-gravity orientation, Journal of Micromechanics and Microengineering, Vol. 18, Number 10, (2008)
3. G. Ngaile, J. Lowrie: Micro-tube hydroforming system based on floating die assembly, 60 Excellent Inventions in Metal Forming p.427-432, (2015).
4. L.-S. Lu, Y. Tang, W.-Q. Fang, J. Cheng, Pipe reduction of miniature inner grooved copper tubes through rotary swaging process, Trans. Nonferrous Met. Soc. China **23,** p.377−384, (2013)
5. L. Lu, D. Yuan, Y. Tang, J. Cheng, Slave rotation analysis of miniature inner grooved copper tube through rotary swaging process, Int J Adv Manuf Technol, **61**, p. 185–193, (2012)
6. T. Tangsri, S. Norasethasopon, Influences of total reduction of area on drawing stress and tube dimension in inner spiral ribbed copper tube sinking, Int J Adv Manuf Technol, **77**, p. 2269–2276, (2015)

An Analytical and Experimental Investigation of Average Laser Power and Angular Scanning Speed Effects on Laser Tube Bending Process

Khalil Ibraheem Imhan[1,3] B.T.H.T. Baharudin [1], Azmi Zakaria [2], Mohd Idris Shah b. Ismail [1], Nasser Mahdi Hadi Alsabti [3] and Ahmad Kamal Ahmad[4]

[1]Department of Mechanical and Manufacturing Engineering, Faculty of Engineering, University Putra Malaysia, 43400 UPM Serdang-Selangor Darul Ehsan-Malaysia
[2]Faculty of Science, University Putra Malaysia, 43400 UPM Serdang-Selangor Darul Ehsan-Malaysia
[3]Laser research center, Ministry of science and Technology, Baghdad, Iraq
[4]Faculty of Science, Al-Nahrain University, Baghdad, Iraq

Abstract. Laser tube bending is a new technique of laser material forming to produce a complex and accurate shape due to its flexibility and high controllability. Moreover, the defects during conventional tube forming such as thinning, wrinkling, spring back and ovalization can be avoided in laser tube bending process, because there is no external force used. In this paper an analytical investigation has been conducted to analyses the effects of average laser power and laser scanning speed on laser tube bending process, the analytical results have been verified experimentally. The model used in this study is in the same trend of the experiment. The results show that the bending angle increased with the increasing of average laser power and decreased with the increasing of angular scanning speed.

1 Introduction

The laser forming has taken place during the last three decades due to its ability to produce accurate and complicated shapes with perfect controllability [1]. Nowadays, researches have been conducted and focused on laser tube bending due to its importance to different applications such as heat exchangers, automotive, aerospace, and medical field. Moreover, different aspects have been studied to understand the process behavior and the effect of the parameters especially laser and laser beam delivery. Hao, N. [2] proposed an equation for the bending angle taking into account plastic mechanics and validated by an experimental investigation. He reported that the bending angle is as a consequence of average laser power, laser angular scanning speed, tube measurements, and material properties. The average laser power and the scanning speed, in general, have the strong correlation with each other; hence, the produced bending angle mainly depends on these two factors and their effects on the rest aspects of the laser tube bending. Theoretically, the bending angle is directly proportional to the average laser power and inversely proportional to the angular scanning speed [3]. Moreover, in laser tube bending the ratio between the laser beam diameter and the tube thickness is considered high, and the angular scanning speed is low to have a close homogenous temperature distribution via the wall thickness direction for the duration of laser irradiation. Hsieh and Lin [4] solved the uncoupled thermal–mechanical problem in laser tube bending using a three-dimensional finite element method. Zhang et al. [5] investigated a numerical simulation for tube bending, where there are four lasers scanning schemes, involving point-source circumferential scanning, pulsed line-source axial procession, and line-source axial scanning without and with water cooling. It is clear that scanning schemes extremely impact laser tube bending. In another sentence, the laser beam diameter is also an influential factor in laser tube bending. Lawrence et al. [6] mentioned that the laser beam diameter is one of the parameters which have to be taken into account in the process design to avoid melting of the material surface. It also affects clearly on laser process mechanism. The general analytical expressions for tube bending are mostly circumferential scanning schemes assuming a two-dimensional temperature distribution and temperature independent material properties. In addition, there are other types of scanning such as axial and alternate circumferential [7]. Furthermore, Jie Zhang et al. [5] investigated a numerical simulation for tube bending, where there are four lasers scanning schemes, involving point-source circumferential scanning, pulsed line-source axial procession, and line-source axial scanning without and with water cooling. It is clear that scanning schemes extremely impact laser tube bending. Comparing the axial scanning scheme to the circumferential scanning scheme the angle produced is twice times; hence the number of scanning would be lesser. The mechanical method is used to bend tubes with a thin wall, but it is unable to bend large tube diameter more than 2 inch, due to difficulties of producing the templates and its high cost. Hence, the extrados portion suffers stretching and the intrados portion suffers from wrinkling rather than the distortion of the cross section [8]. Zhang et al. [9] suggested laser light to do such task and bend large tube diameter 6 inches with a thin wall, they refer to wall factor > 80 which can define as the ratio between OD and wall thickness. Many techniques have been used to assist the laser forming process and overcome the challenges. Where plugs inserted inside the free end of the tube to avoid ovalization and active cooling was used to increase the process speed. Moreover, the tube surface is coated to enhance the absorptivity as well as the rectangular laser beam is utilized to produce a uniform thermal profile. Tube diameter over 12 inches has bent to 5 degrees after 200 scans with 20 minutes in 200 mm length. In this paper an analytical investigation has conducted to study the effects of average laser power and laser angular scanning speed on laser tube bending process, the analytical results have verified experimentally.

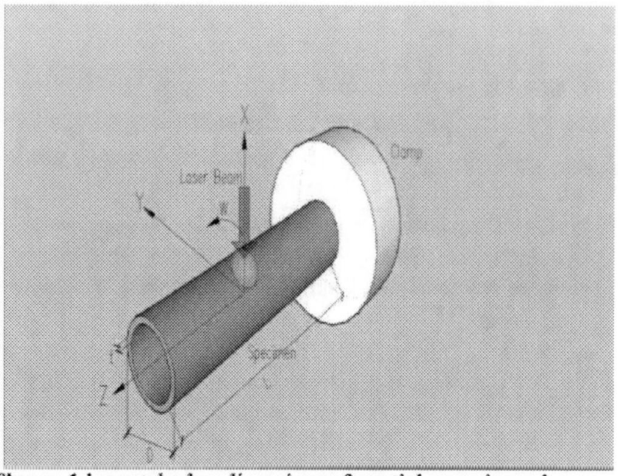

Figure 1 laser tube bending circumferential scanning scheme

2 Analytical approaches

Laser tube bending is considered a heat treatment process in a high level of temperature, but still the material surface temperature is less than the melting point. In order to understand the nonlinear phenomena accompany the process Hao and Li [10] established a new analytical model to determine the bending angle and studied the effects of significant parameters. Circular tube with circumferential scanning was used as shown in Fig. 1, the tube rotates on its axial axis around 180°, where the laser light is defocused on the tube surface with laser beam diameter much bigger than tube thickness. The heated area suffered wall thickening and compressive plastic deformation because of the thermal expansion limitation by the unheated material. When the laser source is turning off a rapid cooling, occur, with material shrinkable on the heated surface. As a result, the shortening of irradiated material in the axial of the tube direction forced it to bend towards of laser source. During bending process, heat is generated because of strain energy, but it is very small compared to input laser beam energy so it can be neglected. The importance of the analytical solution and mathematical representation is obvious in understanding the laser tube bending process behavior and diagnosis the parameters effects. Hao and Li [10], [11] Explicates the laser tube bending and create an analytical model that can involve the effects of laser, material, and geometry parameters to find the bending angle θ, as well as they, used numerical analysis to simulate the process. The laser is heating the material surface and induced plastic compressive strain, from Fig. 2 there are two Force's F exerts inversely to each other between the heated and unheated material where the laser scanning is only on the upper portion of the tube circumference and in the axial direction.

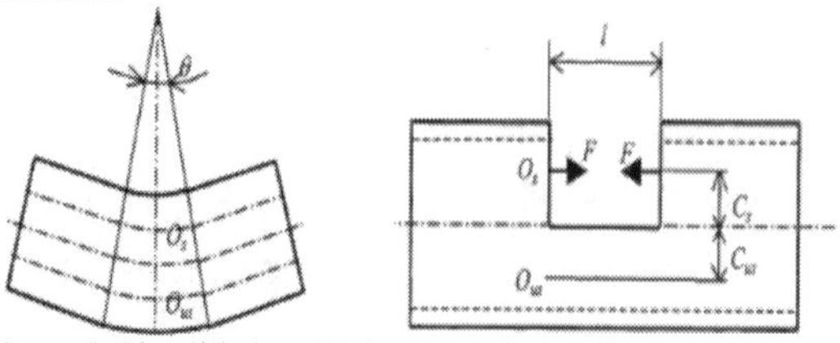

Figure 2. Simplified model for laser tube bending

Strain on scanning area is:

$$\varepsilon_s = -\Delta T\alpha/2 + F/SE \qquad (1)$$

where S is the area of the section of the half tube. E is the Yong's modulus of the material.

The strain of unscanned portion is

$$\varepsilon_{us} = -F/SE \qquad (2)$$

The angle bend can be found from the below:

$$\theta = (\pi l/4D)[\Delta T\alpha - 2(\sqrt{2}-1)\sigma_y/E] \qquad (3)$$

where l is the length of heated region and can be considered the laser beam diameter, D is a tube diameter, and σ_y is a yield stress of the material.

The heat energy generated by laser for bending is:

$$E = P\eta\upsilon t' \qquad (4)$$

where P is a laser power, η is the absorption coefficient υ is the ratio of heating power to input power, and t' is a time period of laser scan.

From all the above the laser tube bend angle can be written as:

$$\theta = (\pi l/2D)[(P\eta\upsilon\alpha/\omega Dlt\rho c)-(\sqrt{2}-1)\sigma y/E] \qquad (5)$$

where ω is an angular scanning speed, ρ is a density of the material, t tube thickness, and c is a specific heat of the material.

3 Laser parameters

The laser parameters are the most important among other parameters in laser tube bending process due to its being flexible and can be controlled in many aspects of its properties. All the literature mentioned that the effect of laser power is linearly proportional to the value of angle produced, whilst the scanning speed has inverse proportionality. In order to be more specified, one has to study the laser beam and scanning characteristics[12]. The general analytical expressions for tube bending are mostly circumferential scanning schemes assuming a two-dimensional temperature distribution and temperature independent material properties.

DOI: 10.1051/matecconf/20179505008

4 Experiment setup

A high power pulsed Nd-Yag laser JK300HPS of Maximum Average Power Laser 300 (W) emitting at 1064 nm and fiber coupled has been used to irradiate stainless steel 304 (SS304) tubes of diameter 12.7 mm, 0.6 mm thickness and 60mm in length. Moreover, computerized RSA Series Motorized Rotation Stage with its controller (from Zolix Company) is used for holds and rotates the tube as illustrated in Fig. 3. Hence, the tube is irradiated with different average laser power range from 0 to 200 w of laser beam diameter 7.5 mm; the angular scanning speed is 20 to 40 deg/sec and rotated angle 180°. The deviation of the tube around the x-axis is measured by dial gauge to determine the bending angle θ. Furthermore, the following specification of the SS404 has used, the yield stress of 205 MPa, density 7930 kg/m3, Young's modulus of 193 GPa, the coefficient of thermal expansion of 17.5x10-6 k^{-1} , the specific heat of 500 J/Kg.K, the absorption coefficient η supposed to be 0.6 [13]. On the other hand, the ratio of heating power to input power υ depends on many conditions such as the heat losses due to convection and radiation which is a function of the material, scanning speed and geometry specification. In this study, υ is set 0.8 and extreme care is paid to avoid surface melting of the SS 304 during the bending processes.

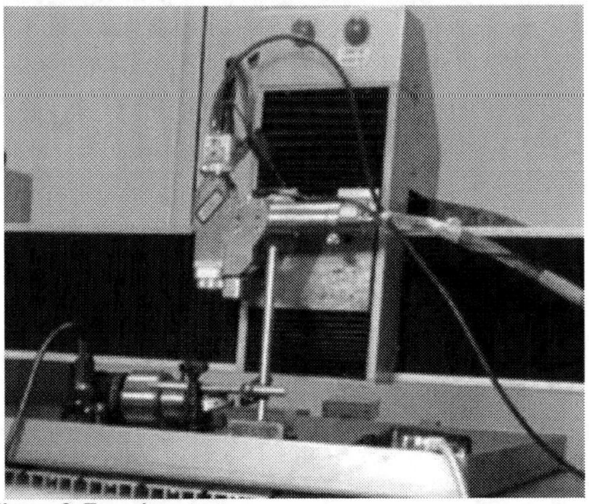

Figure 3. Experiment setup

5 Results and discussions

From the relationship between the laser power and the bending angle in both cases, the bending angle increases with the increasing of average laser power as illustrated in Fig. 4. The bending angle in analytical results is higher than the experimental results, especially at the low average laser power amount. This is due to the nonlinear absorption of the material to the laser power, but it is increased with the increasing of the laser power and the gap become very small when the laser power reaches 200 W. With temperature increasing, absorption coefficient increases also other temperature dependency properties such as thermal expansion coefficient and specific heat, hence more of laser energy is penetrated inside the material [14].

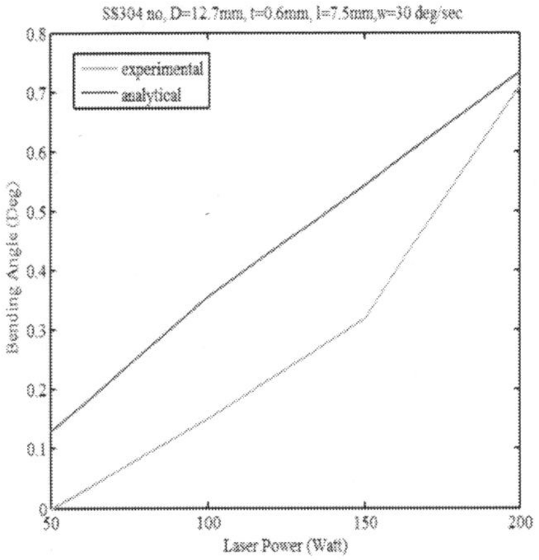

Figure 4 Relationship between average laser power and bending angle

The maximum bending angle is become close to 0.7 degree once the analytical and the experimental using average laser power 200 W. the analytical and the experimental results are in the same trend in this level of average laser power and when the angular scanning speed hold steady on 30 deg/sec.

On the other hand, Fig.5 shows the relationship between the angular scanning speed and the bending angle of analytical and experimental studies which in the same trend but a different slope. As the angular scanning speed decreases as the bending angle increased [3], analytically the bending angle can reach to 3° if the scanning speed is reduced to 5deg/sec. However, experimentally that cannot be achieved because the material temperature increased and reach to the melting point and the tube collapses. Furthermore, when the angular scanning speed decreased the time for energy absorption is increased by the material which enhances the process. In Fig. 5 the average laser power 200 W was used, when the scanning speed has reduced to 20 deg/ sec the experiment bending angle is about 30% bigger than the analytical bending angle. Hence, the experiment results are higher than analytical due to the increase of absorption coefficient.

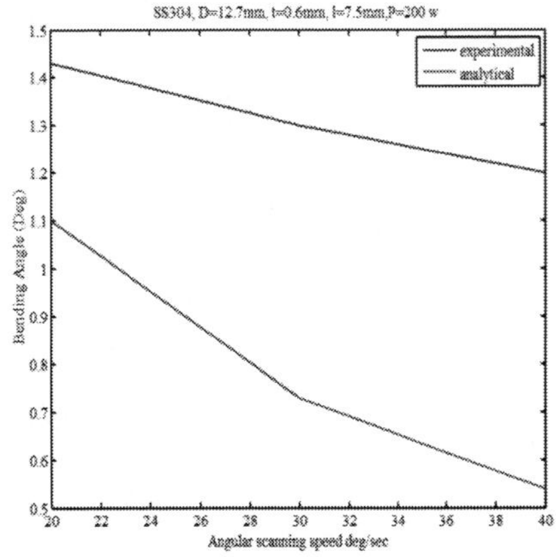

Figure 5. relationship between angular scanning speed and bending angle

DOI: 10.1051/matecconf/20179505008

6 Conclusions

Analytical and experimental investigations were conducted to study the effect of the average laser power and angular scanning speed on the laser tube bending process. Moreover, the analytical model depends mainly on the average laser power and angular scanning speed because those parameters can be changed during the process in spite the effects of other parameters are present. The analytical results are verified by sets of experiments and hence the conclusions can be listed as below:

• The angle bending θ is proportional directly to the average laser power.

• The angular bending θ is proportional reversely to the angular scanning speed.

• The analytical results are in acceptable agreement with the experiments results.

• The average laser power has to match with the angular scanning speed to avoid the scraping effect on the specimen surface.

References

1. Kannatey-Asibu Jr, E., *Principles of laser materials processing*. Vol. 4: John Wiley & Sons. (2009).
2. Hao, N. *On the process parameter of laser tube bending*. in *2010 International Conference on.Mechanic Automation and Control Engineering (MACE)*, (2010)
3. Safari, M. and M. Farzin, *A study on laser bending of tailor machined blanks with various irradiating schemes*. Journal of Materials Processing Technology,. **214**(1): p. 112-122, (2014).
4. Hsieh, H.-S. and J. Lin, *Study of the buckling mechanism in laser tube forming*. Optics & Laser Technology. **37**(5): p. 402-409, (2005).
5. Zhang, J., et al., *Effects of scanning schemes on laser tube bending*. Journal of manufacturing science and engineering. **128**(1): p. 20-33, (2006).
6. Lawrence, J.R., *Advances in laser materials processing: Technology, research and application.*: Elsevier, (2010).
7. Safdar, S., et al., *Finite element simulation of laser tube bending: Effect of scanning schemes on bending angle, distortions and stress distribution*. Optics & Laser Technology. **39**(6): p. 1101-1110, (2007).
8. He, Y., et al., *Advances and trends on tube bending forming technologies*. Chinese Journal of Aeronautics. **25**(1): p. 1-12, (2012).
9. Zhang, W., et al. *Large diameter and thin wall laser tube bending*. in *24th International Congress on Applications of Lasers and Electro-Optics, ICALEO*. (2005).
10. Hao, N. and L. Li, *An analytical model for laser tube bending*. Applied surface science. **208**: p. 432-436, (2003).
11. Hao, N. and L. Li, *Finite element analysis of laser tube bending process*. Applied surface science. **208**: p. 437-441, (2003).
12. Jamil, M.C., M. Sheikh, and L. Li, *A study of the effect of laser beam geometries on laser bending of sheet metal by buckling mechanism*. Optics & Laser Technology. **43**(1): p. 183-193, (2011).
13. Davis, J.R., *Stainless steels*. 1994: ASM international.
14. Li, W. and Y.L. Yao, *Laser bending of tubes: mechanism, analysis, and prediction*. Journal of manufacturing science and engineering. **123**(4): p. 674-681, (2001).

Research on Mass Imbalance Fault of Wind Turbine Based on Virtual Prototype

Ping Zhao , Xiaoguang Li and Liu Yang

CRRC Zhuzhou Electric Locomotive Research Institute, Zhuzhou, 412001

Abstract. Wind turbine virtual prototype was set up with SIMPACK software. The influence of different mass in different blades was researched by changing size and position of imbalance mass. Results showed that the period of imbalanced moment inducted by different size of imbalance mass imposing blades at different position was the same as the wind turbine rotor, but the amplitude and phase was different. Electricity generation during starting was associated with not only the size of mass but also the position. After starting, wind turbine would keep the stable state in simulation, power and revolution speed and pitch angle curves would wave periodically, the periods were all the same as the wind turbine rotor, the size of imbalance masses were bigger, the fluctuation amplitude of curves were greater.

1 Introduction

In recent years, wind energy, as a new, safe and reliable, clean energy, has been given more and more attentions and large-scale developed by international which possess abundant resources[1-2]. Developing wind electricity has also been an important measure to carry out the sustainable energy development strategy of China. As the capacity of single wind turbine has become larger and larger, the impeller diameter has become longer and longer, the flexibility of wind turbine has become stronger, and influence of impeller imbalanced faults has received more and more attentions, such as the increasing of huge vibration, drive train vibration, and parts fatigue stress of whole wind turbine caused by it. And the sand abrasion, blade icing, loosen of the inner filling material will lead to mass unbalance of impeller.

Many researchers have carried out relevant researches aim at impeller imbalance faults of wind turbine generator set. German research institution ISET found that [3] it could be gotten the mass imbalance faults of wind turbine through testing electrical signal of generator, however, they didn't offer any theoretical explanation of this diagnostic technique. RONNY and the others [4-5] from university of linz in Austria analyzed vibration characteristics wind turbine. Identify the size and direction of impeller imbalance mass through using reverse engineering research methods. JIANG [6-7] carried out experiments aimed at three malfunction of impeller including mass imbalance, asymmetric aerodynamic force, yaw misalignment, using way which is based on vibration spectrum analysis to research on main shaft vibration of impeller caused by mass imbalance. Yang Tao [8-9] conducted simulation for impeller imbalance faults including mass imbalance, asymmetric aerodynamic force malfunction, they pointed out under the state of mass imbalance faults, wave of electric power was resulted from the wave of impeller output torque caused by gravity of imbalance mass, and under the state of asymmetric aerodynamic force faults, because of the vibration of tower leads to the wave of aerodynamic force on impeller, then leads to wave of torque, and at last leads to the wave of electric power. In other words, aerodynamic force model and tower vibration model are coupled. Guo Yuanchao [10] compared the changing situation of equivalent fatigue loading at all important point of load of wind turbine generator in different condition of maximum mass eccentricity, and it explained that mass imbalance of wind wheel makes serious influence on fatigue loads.

The research methods above are mainly based on two class methods, one of it is a fault diagnosis method based on vibration, and this method need to install a large number of sensors, low reliability, but high costing. Another one is based on simulation analytical calculation, modeling and simulation of faults can make a more profound understanding of malfunction formulation mechanism and influences, and it will be easier to proceed calculation and comparison for different malfunction signal, however, this kind of simulation is based on physical parameters used of wind turbine without considering dynamical behavior of the wind turbine, and neither do consider the relevant connected relations and coupling of impeller, transmission chain and tower-silo, so that this simulation can only aim at one physical quantity of electric power.

This paper infers the influence that the size of impeller imbalance mass moment of wind turbine do on electric power in theory, and wind turbine real simulation was set up with many-body dynamics software of SIMPACK, makes simulation on location distribution on wind turbine of imbalance mass and mass of different size to load on the same direction of same impeller to research on the influences it makes on generator power, rotate speed and blade variable pitch respectively.

2 Wind turbine mass imbalance moment

Wind turbine generator of three-vanned is showed in Figure 1, the impeller is rotates clockwise at speed of ω, suppose the initial position of blade 1 is right above, when impeller turn a angle of wt at speed of ω, the mass moment of impeller T_g is:

$$T_g = G_1r_1\sin(wt) + G_2r_2\sin(wt + \frac{2\pi}{3}) + G_3r_3\sin(wt - \frac{2\pi}{3}) \quad (1)$$

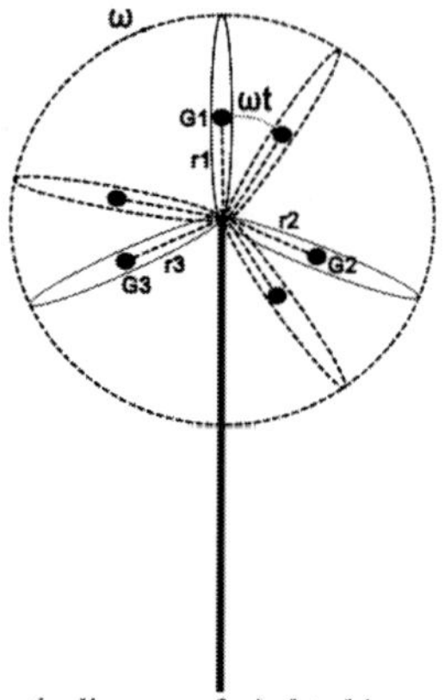

Figure 1. Schematic diagram of wind turbine mass distribution.

Unfold the formula (1), it can be inferred as follows:

$$T_g = (G_1r_1 - \frac{1}{2}G_2r_2 - \frac{1}{2}G_3r_3)\sin(wt) + \frac{\sqrt{3}}{2}(G_2r_2 - G_3r_3)\cos(wt) \quad (2)$$

Let

$$k_1 = G_1r_1 - \frac{1}{2}G_2r_2 - \frac{1}{2}G_3r_3, k_2 = \frac{\sqrt{3}}{2}(G_2r_2 - G_3r_3),$$

So that formula (2) can be turned into formula (3).

$$T_g = k_1\sin(wt) + k_2\cos(wt) \quad (3)$$

Suppose

$$\sin\varphi = \frac{k_2}{\sqrt{k_1^2 + k_2^2}}, \cos\varphi = \frac{k_1}{\sqrt{k_1^2 + k_2^2}},$$

It can be inferred from formula (8) as follow:

$$T_g = \sqrt{k_1^2 + k_2^2}\sin(wt + \varphi) \quad (4)$$

From the deduction above, it means all the mass moment that is generated from impeller mass imbalance can be expressed by formula (4). Therefore, moment change period from impeller mass imbalance in different conditions is the same with impeller period of rotation, but here are some differences in amplitude and phase position of mass, and we can infer backwards the mass imbalance situation according to size of amplitude and value of phase position. Suppose that the impeller receives Ta aerodynamic moment, we can get the total moment received by impeller and electric power of generator respectively:

$$T = T_a + T_g \quad (5)$$

$$P_e = (T_a + T_g)\omega/k \quad (6)$$

where, P_e represents output power of generator, ω represents geostationary angular spin rate, k represents transmission ratio of gearbox. From formula (6), we can see that ω and k are constant, we suppose impeller aerodynamic moment in rotation process T_a is also a constant value, so, the change of output power is proportional to impeller imbalance mass moment, change period is the same as imbalance mass moment period.

3 Virtual prototype model

Based on the analysis above, this paper use software of SIMPACK to set up wind turbine generator virtual prototype model, topological structure is showed in Figure 2. The model includes impeller, hub, spindle, gearbox model and generator model, we can research on the influences of wind turbine generator caused by different sizes of imbalance mass on different blades through changing the size and position of imbalance mass. Force of wind turbine take way of applying wind load, number 241 of aerodynamic element in SIMPACK software is interface of Aerodyn aerodynamic software. It will calculate aerodynamic through associating simulation form of Aerodyn software at run time. We can achieve simulation of pitch control through force element FE243（Wind Controller Interface）. Add virtual hub of zero mass and virtual variable pitch of zero mass respectively between blade and hub, allowing blade rotate in axial, so that SIMPACK software can call DLL file of dynamic depot in control model directly to achieve variable pitch control when calculate it. For achieving variable pitch control, SIMPACK software usually consider rotate speed of engine rotor and electric power as output, and then, put it into control procedure, variable pitch angle, as the output of control procedure, will be delivered back into dynamic model as the input of SIMPACK, so that to achieve joint simulation.

Hub and spindle use the power force of FE13 Spring-Damp Rot Meas Inp Cmp) to represent the bolted connection between them. Spindle connect with rack through main bearing, represented by FE41 (Spring-Damp Matrix Cmp). Guide the whole gearbox into the whole model as a substructure. Force element FE5 (Spring-Damp Parallel Cmp) represent flexible support of

 DOI: 10.1051/matecconf/20179506001

which connect gearbox and rack. The connection between coupling, generator, front and back gearbox is permanent connection, release the rotate DOF in axial between individual couplings, torsional rigidity and damping reflect in force element FE13. There are bearings and electromagnetic force between generator rotor and stator, use FE43 (Bushing Cmp) and FE50 (Force Torque Expression Cmp) to represent them respectively. The same as force element FE5 (Spring-Damp Parallel Cmp) to represent flexible support of which connect generator and rack. A virtual rack of zero mass is built in the bottle of the rack to connect with tower-silo, and the whole wind turbine connects with ground through tower-silo, tower-silo simulates with fully flexible form. Model is proceeded meshing and built main node in ANSYS software at first, and then, guide it into SIMPACK model. Damping d calculations of flexible support and coupling are showed here:

Flexible support: $d_1 = 2D_1\sqrt{K_1 m_{eq} / n}$,

D_1 represents damping coefficient, K_1 represents flexible support rigidity, n represents flexible support quantity, meq represents equivalent mass, $m_{eq}=I/r^2$, r represents radius of rotary arm, I represents rotary inertia for direction of rotation.

Rotary inertia calculation of gearbox flexible support is expressed as follows: $I=I_{bades}+I_{mainshaft}+I_{hub}+I_{housing}$, I_{bades} is rotary inertia of whole impeller relative to spindle, $I_{mainshaft}$ represents rotary inertia of spindle rotate in axial, I_{hub} represents the moment of inertia of the wheel around the axial rotation, $I_{housing}$ represents the moment of inertia of the wheel around the axial rotation.

Rotary inertia calculation of generator flexible support is expressed as follows: $I=I_{rotor}+I_{stator}$, I_{rotor} is rotary inertia of spindle when generator rotor rotates around spindle, I_{stator} is rotary inertia of spindle when generator stator and frame rotate around spindle.

Damping calculation formula of coupling is: $d_2 = 2D_2\sqrt{K_2 I_2}$, D_2 is damping coefficient, K_2 is torsional rigidity, calculation of rotary inertia I are here below:

$$I=I_{coupling2}+ I_{coupling3}+ I_{coupling4}+I_{rotor}$$

where, $I_{coupling2}$, $I_{coupling3}$, $I_{coupling4}$ are the rotary inertia of spindle when the second, third, fourth section of coupling rotate around spindle.

This simulation of this paper aims at low-velocity wind turbine, the main simulation parameters are showed below:

Table 1. Options for simulation parameters

parameter	value
Rated power/MW	2
Rated wind speed /m.s-1	9.5
Radius of impeller /m	55
Speed increasing ratio of gearbox	91
Wind speed/m.s-1	12
Impeller rated rotate speed /rpm	13.3

4 Mass imbalance of impeller

4.1 Simulation of applying imbalance mass to different blade

The model achieves its impeller imbalance through applying different mass on blade. We set five simulation working conditions and show it in Table 2 below to research on the influence caused by the direction distribution of impeller mass imbalance moment on wind turbine. The first working condition is the mass of three blades are totally equal, impeller is in the idea mass balance condition; and the second, third, fourth working condition correspond to put 200 kilogram mass on the same section of first, second, and third blade, respectively; the fifth working condition is corresponding to put 400kg and 600kg mass on the second and third blade respectively to form comparison with the four working condition above. The initial position of calculation blade for all the working condition correspond to three position of blade in Figure 1, the position of the first blade is zero position, with vertical direction. Wind wheel rotate clockwise, the 1,2,3,4,5 here in this chapter correspond to working condition of 1,2,3,4,5, respectively.

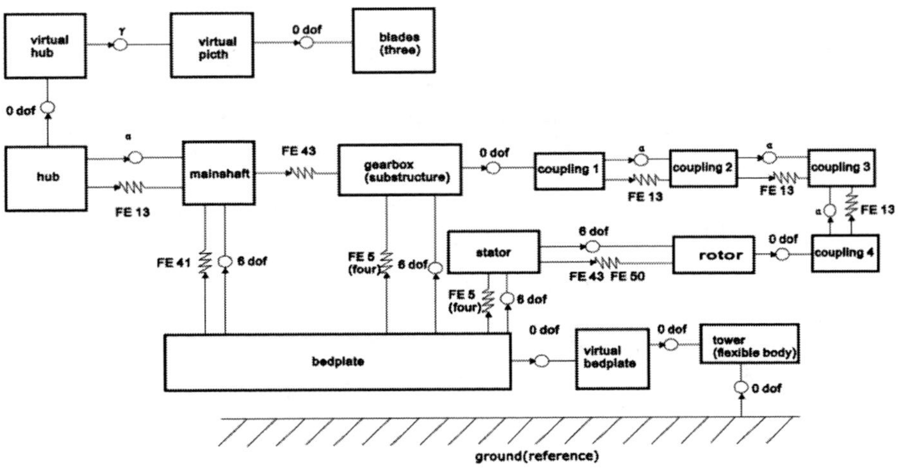

Figure 2. Structure picture of wind turbine topology.

Under this five working conditions, in the time of simulation, the changing relationship of generator output power variation with time is showed in Figure 3, we can see that there exist larger differences of generator characteristic curve under different working condition. Figure 4 correspond to the characteristic curve when power of generator tends to be stable, working condition of 2,3,4,5 under imbalance condition, its value of power fluctuation range is obviously larger than the balanced state 1, the same imbalance mass 2, 3, 4 have basically the same value of power fluctuation rang, the same as the change period, but exist some difference of phase, it's completely consistent with the analysis conclusion of the first chapter, so it also verify the correctness of multi-body dynamic simulation model. The generating capacity of wind turbine generator during one period time is equal to the square surrounded by power curve and timer shaft, move forward from Figure 5, we can get during the period of starting wind turbine, the rank of generating capacity is 2≈3>1>5>4, we can know that not only the generating capacity is related to the size of imbalance mass, but also to its distribution position during the period of starting wind turbine. Because of the three blades which make up the impeller is unequal, as the impeller rotates, wind turbine will generate forced vibration due to suffering from alternating loads. Under the vibration, relatively incoming flow wind speed of impeller will have a period change to lead to wave of aerodynamic force. Wave of aerodynamic force will lead to the wave of impeller output torque, finally make influence on generator end electric power. The fluctuation range of electric power rank is 5>2≈3≈4>1, we can get the size of fluctuation range is relevant with imbalance mass moment, as the imbalance mass moment turns bigger, the corresponding the electric power waves turn bigger. When apply equal imbalance mass moment to different blades, there is hardly change of electric power wave. When impeller is in equilibrium state, the corresponding electric power wave is very small. Period of wave average on 4.55s, and convert it into frequency is 0.22Hz. It is close to stable impeller rotate speed of 13.3rpm when the corresponding frequency is 0.222Hz, imbalance mass moment will lead to electric power wave when they have the same frequency. Wave frequency has nothing to do with the size and position of imbalance mass moment.

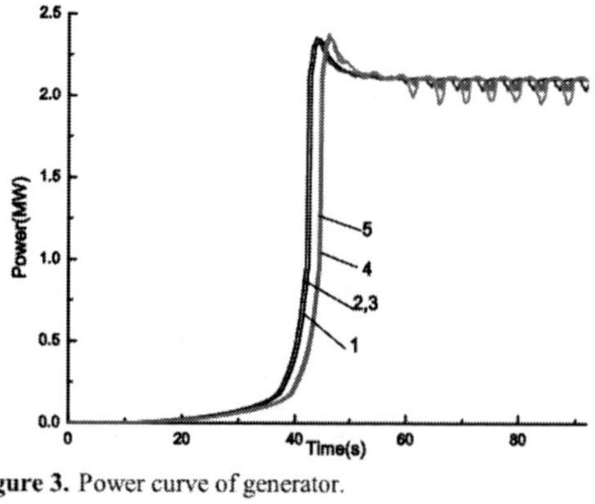

Figure 3. Power curve of generator.

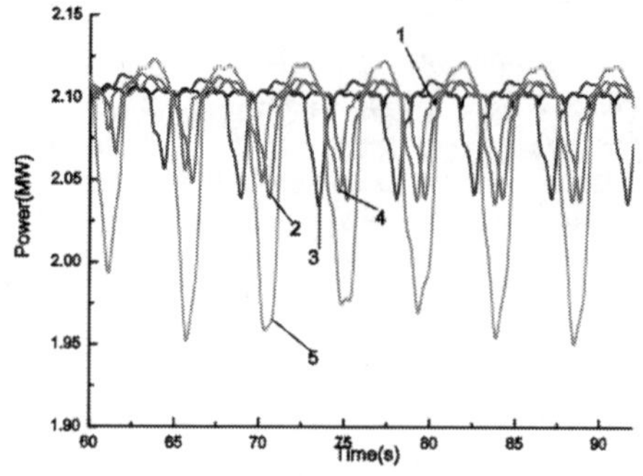

Figure 4. Electric power curve of stabled generator.

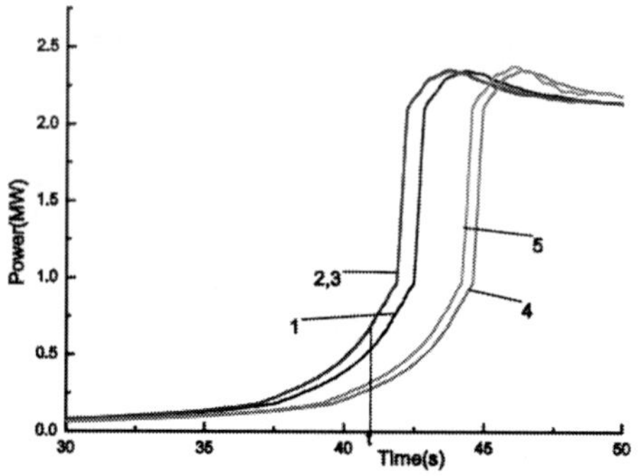

Figure 5. Generator power curve of the starting process.

4.2 Apply different mass to the same blade

Figure 6 represents the relevant curve between electric power fluctuation range of stabled generator and the size of imbalance mass moment in five working condition. We can see the amplitude of electric power wave has an approximate linear relation with imbalance mass moment. As the imbalance mass moment enlarge, the electric power fluctuation range will increase in proportion.

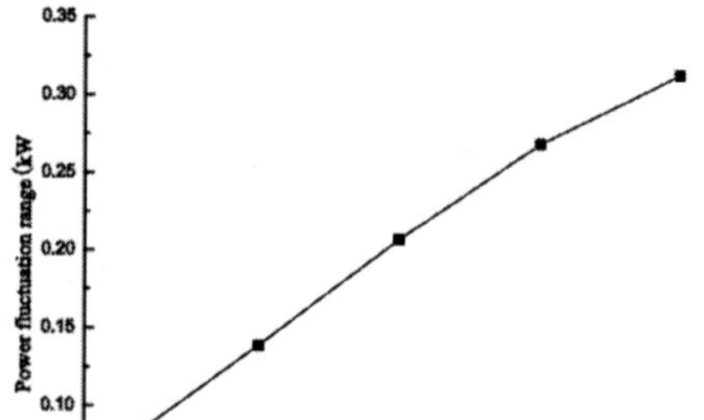

Figure 6. Relevant curve between electric power fluctuation range of generator and imbalance mass moment.

We set five simulation working condition in Table 3 below to research on the influences caused by imbalance mass moment of impeller on wind turbine generator, from working condition 1 to condition 5 correspond to apply 200kg, 400kg, 600kg, 800kg, 1000kg to the same direction of the third blade respectively. The initial position of calculation blade for all the working condition correspond to three position of blade in Figure 1, the position of the first blade is zero position, with vertical direction. Wind wheel rotate clockwise, the 1,2,3,4,5 here in this chapter correspond to working condition of 1,2,3,4,5,respectively.

The relationship of generator output power history under five different mass moments is showed in the Figure 7 below, we can see that there are differences in generator characteristic curve of different working conditions. Figure 8 represents the characteristic curve when generator tends to be stable, and as the imbalance mass moment enlarge, the electric power fluctuation range increase along. The generating capacity of wind turbine generator during one period time is equal to the square surrounded by power curve and timer shaft, move forward from Figure 7, we can get during the period of starting wind turbine, the rank of generating capacity is 1>2>3>4>5, we can see that the generating capacity is related to the size of imbalance mass, the bigger imbalance mass moment is, the smaller generating capacity during starting process will be, it means mass imbalance of wind turbine makes direct influence on generating capacity during starting process. Period of wave average on 4.55s, and convert it into frequency is 0.22Hz. It is close to stable impeller rotate speed of 13.3rpm when the corresponding frequency is 0.222Hz, imbalance mass moment will lead to electric power wave when they have the same frequency. Wave frequency has nothing to do with the size of imbalance mass moment. In addition, there exist differences on phase position of electric power wave because of the imbalance mass will change the aerodynamic force to lead to phase deviation.

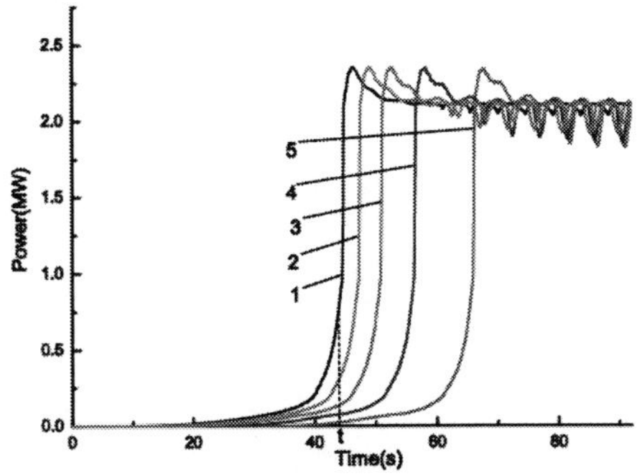

Figure 7. Electric power curve of generator.

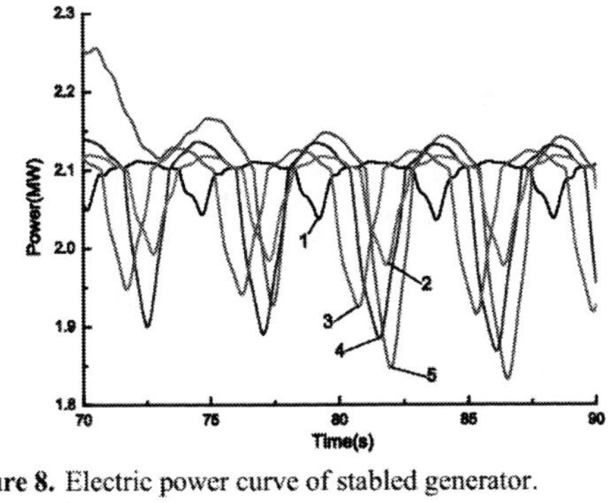

Figure 8. Electric power curve of stabled generator.

5 Conclusions

(1)This paper analyze calculation of wind turbine impeller mass moment in five different conditions, period of imbalanced moment inducted by different size of imbalance mass imposing blades at different position was the same as the wind turbine rotor, but the amplitude and phase was different, and we can infer backwards the mass imbalance situation according to size of amplitude and value of phase position. And we further infer the variation of output power is in proportion to imbalance mass moment of impeller, its change period is the same with the change period of imbalance mass moment

(2) This paper use software of SIMPACK to set up wind turbine generator virtual prototype model. The model includes impeller, hub, spindle, gearbox model and generator model, we can research on the influences of wind turbine generator caused by different sizes of imbalance mass on different blades through changing the size and position of imbalance mass. We verify the veracity of this model through testing that the simulation result of electric power variation period is the same as the inferred one.

(3) We analyze the generating capacity, rotate speed, variable pitch angle through model calculating of simulation. Not only generating capacity is related to the size of imbalance mass, but also to its distribution position during the period of starting wind turbine. Generator speed curve and variable pitch angle curve in stable condition have the same change rule with power curve. Both of them have the same wave frequency with impeller rotate frequency, the bigger imbalance mass moment is, the bigger fluctuation range will be.

References

1. Saidur R., Rahim N. A., Islam M. R., Solangi K.H. Environmental impact of wind energy[J]. Renewable and Sustainable Energy Reviews, 2011, **15**(5): 2423–2430.
2. E Jia-qiang, LI Guang-ming, ZHANG Bin, DONG Jiang-Dong, ZHU Hao. Fatigue Simulation and Analysis on Planetary Gear of Megawatt Wind Power Yaw Reducer[J]. Advanced Materials Research, 2012, **455-456**: 187-193.
3. Caselitz P, Giebhardt J. Rotor condition monitoring for improved operational safety of offshore wind energy converters[J]. Journal of Solar Energy Engineering, 2005, **127**(2): 253-261.
4. Ronny R, Jenny N. Imbalance estimation without test masses for wind turbines[J]. Journal of Solar Energy Engineering, 2009, **131**(1): 1-7.
5. Jenny N, Ronny R, Thien T N. Mass and aerodynamic imbalance estimates of wind turbines[J].Energies, 2010, **3**(4): 696-710.
6. Jiang Dongxiang. Theoretical and experimental study on wind wheel unbalance for a wind turbine[C]//2009 World Non-Grid-Connected Wind Power and Energy Conference, 24-26 Sept. 2009, Nanjing, China, 2009：10975727.1-10975727.5.

DOI: 10.1051/matecconf/20179506001

7. Zhao Minghao, Jiang Dongxiang, Li Shaohua. Research on fault mechanism of icing of wind turbine blades[C]//2009 World Non-Grid-Connected Wind Power and Energy Conference, 24-26 Sept. 2009, Nanjing , China, 2009: 10975712.1-10975712.4.
8. Yang Tao, Ren Yong, Liu Xia, et al. Research on modeling and simulation of wind turbine rotor imbalance fault[J]. Journal of Mechanical Engineering, 2012, **48**(6):130-135.
9. Ren Yong. Modeling and Simulation of the unbalance of wind turbine blade[D].Wuhan: Huazhong University of Science and Technology, 2012.
10. Guo Yuanchao, Xu Yiqing,Wang Fan, et al. Analysis of the influence of wind turbine mass unbalance on the load of wind turbine [J]. Wind Energy, **2012**(2): 70-72.

 DOI: 10.1051/matecconf/20179506002

Five Types of Ruled Helical Surfaces for Helical Conveyers, Support Anchors and Screws

S. N. Krivoshapko and Marina Rynkovskaya

RUDN University, Engineering Faculty, 117198, Miklukho-Maklaya Str., 6, Moscow, Russia

Abstract. The paper provides an overview of five types of ruled helical surfaces which can be used for helical conveyers, support anchors and screws. The geometry of helical surfaces has been well studied. Several available methods of strength analysis of helicoidal shells give one a choice in solving one-dimensional or two-dimensional problems. The basic problems considered in this paper include geometrical research and static analysis of right and developable helicoids. Several works dealing with stress-strain state of oblique, pseudo-developable helicoids, and pseudo-developable helicoid of general type are given in References.

1 Introduction

Geometry of helical surfaces has started to be investigated in the 18th century. An important contribution to the theory of helical surfaces was made by L. Euler (Switzerland), G. Monge (France), J. Meusnier, E. Catalan (France), G. Darboux, D. Hilbert (Germany), and many others.

The peak of publications on analytical methods of linear analysis of helicoidal shells was in the 50s. Professor S.N. Krivoshapko [1] described some theoretical methods of static analysis of engineering structures in the form of right and oblique helicoids. Later, in a paper [2] with 181 references, the detailed analysis of the existing methods for calculating of the helicoidal thin shells was presented, as well as the issues of geometric design and profiling of the helical surfaces. The study of the existing scientific literature showed that currently the engineers know only right and oblique helicoids, although an encyclopedia [3] provides information about the five types of ruled helical surfaces (Fig. 1).

Mechanical engineering is the main sphere of application of helicoidal shells.

2 Methods of geometrical research

Right helicoid is a minimal surface and its Gaussian curvature is negative (Fig. 1, a).

Oblique helicoid is the second well-known helical surface having rectilinear generatrixes inclined with respect to the axis of the helicoid at an angle of $\alpha < \pi/2$ and moves remaining parallel to the generatrixes of a director cone's surface (Fig 1, *b*):

$$x(u,v) = u \sin\alpha \cos v,$$
$$y(u,v) = u \sin\alpha \sin v, \qquad (1)$$
$$z(v) = bv + u\cos\alpha,$$

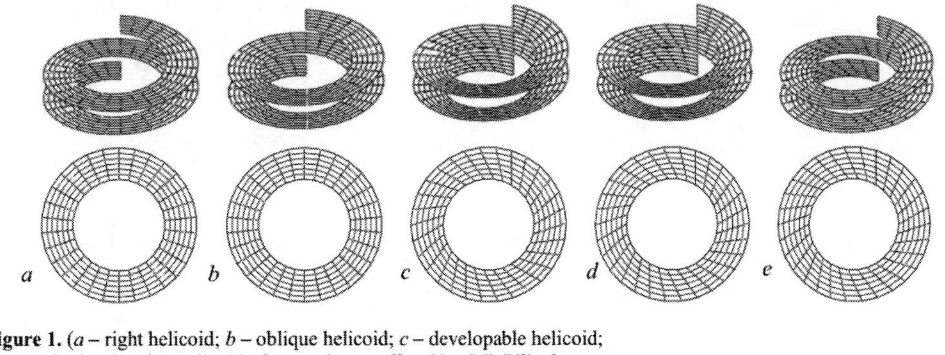

Figure 1. (*a* – right helicoid; *b* – oblique helicoid; *c* – developable helicoid; *d*– pseudo-developable helicoid of general type offered by S.F. Pilipaka; *e* – pseudo-developable helicoid).

where $|u|$ is the distance the axis from the selected point taken along the straight inclined generatrix.

The third ruled helical surface is called a developable helicoid (Fig. 1, *c*) and it has the parametric equations in the form:

$$\begin{aligned} x &= x(u,v) = a\cos v - \frac{au\sin v}{m}, \\ y &= y(u,v) = a\sin v + \frac{au\cos v}{m}, \\ z &= z(u,v) = bv + \frac{bu}{m}, \\ m &= \sqrt{a^2+b^2}. \end{aligned} \tag{2}$$

So the curvilinear coordinate u (or v = const) is a rectilinear generatrix, tangent to the helical cuspidal edge, but lines v (u = const) are helixes. The curvilinear coordinates u, v are parametric non-orthogonal conjugate curves. If we take a plane circular area with inside radius a_0 and cut it along a tangent to the inside contour then this circle area can be transformed into any open helicoid with a helical cuspidal edge lying on a cylinder with radius a, where $a = a_0\cos^2\varphi$, here φ is the helix angle.

S. N. Krivoshapko proposed a new form of presentation of the parametric equations of developable helicoid:

$$\begin{aligned} x &= x(u,s) = a_0\cos^2\varphi(\cos\frac{s}{m} - \frac{u}{m}\sin\frac{s}{m}), \\ y &= y(u,s) = a_0\cos^2\varphi(\sin\frac{s}{m} + \frac{u}{m}\cos\frac{s}{m}), \\ z &= z(u,s) = (s+u)\sin\varphi,\ m = a_0\cos\varphi, \end{aligned} \tag{3}$$

u, s are curve non-orthogonal conjugate coordinates on the surface. The arc length s of a helical cuspidal edge can be determined by the formula $s = mv$.

For the fourth helical surface (pseudo - developable helicoid of general type), parametric equations were obtained by S.F. Pilipaka [4]. However, the authors believe that this surface is better to set in the following form (Fig. 1, *d*):

$$\begin{aligned} x &= a\cos v - u\cos\varepsilon\sin v, \\ y &= a\sin v + u\cos\varepsilon\cos v, \\ z &= bv + u\sin\varepsilon, \end{aligned} \tag{4}$$

where ε is the angle of the the straight generatrix of the surface with the horizontal plane.

Fifth helical ruled surface which is defined by parametric equations (Fig. 1, *e*):

$$\begin{aligned} x(u,v) &= a\cos v - u\sin v, \\ y(u,v) &= a\sin v + u\cos v, \\ z(v) &= bv \end{aligned} \tag{5}$$

can be considered to be a special case of the surface (4) when $\varepsilon = 0$. The surface (5) is a helical trace of uniform motion of horizontal rectilinear generatrix on the given helical guideline. Moreover, the straight generatrixes on the plane will coincide with the tangents to the helical guideline as well as in the case of the developable helicoid.

All these helical ruled surfaces are used in screw conveyors, the full length of which can reach 76 m, screw support anchors, and in screws of different purposes.

3 Methods of strength analysis

It should be noted that geometry of these helical surfaces is studied quite well, but the analysis of the stress-strain state of thin shells with middle surfaces in the form of these ruled helicoids still cause some difficulties.

The first calculations of right helicoidal shells were presented by L.I. Solomon, V.G. Rekach and J.W. Cohen [5]. At present time, strength analysis is studied by M. Rynkovskaya [6], who found a number of inaccuracies in the method of V.G. Rekach and suggested the modified method for stress-strain analysis of long right helicoidal shells.

A. R. Yaroshenko [7] studied the axisymmetrical loading of an oblique helicoid given by formulas

$$x(r,\varphi) = r\cos\varphi;\ y(r,\varphi) = r\sin\varphi;$$
$$z = r\operatorname{ctg}\alpha + b\varphi,$$

where $b = l/(2\pi)$ and l is the constant pitch of the helicoid. Substituting $\varphi = \psi - \operatorname{ctg}\alpha \cdot \operatorname{arctg}(r/b)$, he went on to an orthogonal non-conjugate system of curvilinear coordinates.

The coordinate lines in the accepted coordinate system are helixes and their orthogonal trajectories. A system of equilibrium equations was reduced to a system of ordinary differential equations. The sixth equation of equilibrium is satisfied identically.

At present, the calculations for determination of the stress-strain state of oblique helicoids are conducted by E. Tupikova [8].

One-dimensional problem for shallow developable helicoid with two fixed helical sides was firstly solved by G.Ch. Bajoria [9], who took the shell material with a Poisson's ratio equal to zero ($\nu = 0$) in order to simplify the computation. His calculation is based on the system of three ordinary differential eighth-order equations with three unknown displacement parameters. The problem was solved numerically. Later, S.N. Krivoshapko and Kumudini Jayawardena [10] examined the one-dimensional problem for a non-shallow thin shell in the form of long developable helicoid with several loops having arbitrary Poisson's ratio.

It was determined that the difference between the numerical results of calculations for $\nu = 0.17$ and $\nu = 0$ was, on average, about 10%, but for the normal internal forces it was increased to 20%.

Formulas (3) define a surface of the developable helicoid in non-orthogonal conjugate curvilinear coordinates. That is why, for the calculation of this type of helicoids it is necessary to use only the system of governing equations (six equilibrium equations, six geometrical and eight physical equations) of A.L. Gol'denveizer. But his equilibrium equations contain internal pseudo-forces and pseudo-moments, that is why, S.N. Krivoshapko suggested to work with the system of governing equations containing normal and shearing forces, bending and twisting moments reckoned per unit of curvilinear coordinates' length which are generally used in engineering practice:

Equilibrium equations:

$$\frac{\partial}{\partial v}(AS_v)+\frac{N_u-N_v}{\sin\chi}\left(\frac{\partial B}{\partial u}-\frac{\partial A}{\partial v}\cos\chi\right)+\frac{\partial A}{\partial v}S_u+$$
$$+B\frac{\partial S_u}{\partial u}\cos\chi+B\frac{\partial N_u}{\partial u}\sin\chi-\frac{AB}{R_u}Q_u+ABX\sin\chi=0,$$

$$\frac{\partial}{\partial v}(AN_v)+\frac{S_u+S_v}{\sin\chi}\left(\frac{\partial B}{\partial u}-\frac{\partial A}{\partial v}\cos\chi\right)-\frac{\partial A}{\partial v}N_u+$$
$$+B\frac{\partial S_u}{\partial u}\sin\chi-B\frac{\partial N_u}{\partial u}\cos\chi-\frac{AB}{\sin\chi}\left(\frac{Q_v}{R_v}-\frac{Q_u}{R_u}\cos\chi\right)+$$
$$+ABY\sin\chi=0,$$

$$\frac{N_u}{R_u\sin\chi}+\frac{N_v}{R_v\sin\chi}+\frac{1}{AB}\left[\frac{\partial}{\partial u}(BQ_u)+\frac{\partial}{\partial v}(AQ_v)\right]-$$
$$-Z\sin\chi=0,$$

$$-\frac{\partial}{\partial v}(AM_v)+\frac{M_{uv}+M_{vu}}{\sin\chi}\left(\frac{\partial B}{\partial u}-\frac{\partial A}{\partial v}\cos\chi\right)+\frac{\partial A}{\partial v}M_u+$$
$$+B\frac{\partial M_u}{\partial v}\cos\chi+B\frac{\partial M_{uv}}{\partial u}\sin\chi+ABQ_v\sin\chi=0,$$

$$\frac{\partial}{\partial v}(AM_{vu})+\frac{M_v-M_u}{\sin\chi}\left(\frac{\partial B}{\partial u}-\frac{\partial A}{\partial v}\cos\chi\right)+\frac{\partial A}{\partial v}M_{uv}$$
$$-B\frac{\partial M_u}{\partial u}\sin\chi+B\frac{\partial M_{uv}}{\partial u}\cos\chi+AB(Q_u+Q_v\cos\chi)=0,$$
$$AB(Q_u+Q_v\cos\chi)=0,$$
$$(S_u-S_v)\sin\chi+(N_v-N_u)\cos\chi+$$
$$+\frac{M_{uv}}{R_u\sin\chi}-\frac{M_{vu}}{R_v\sin\chi}=0, \qquad (6)$$

geometrical equations derived by A.L. Gol'denveizer were taken by Krivoshapko without any modification; physical equations:

$$N_v=\frac{Eh}{1-\nu^2}(\varepsilon_v-\varepsilon_{uv}\mathrm{ctg}\chi+\nu\varepsilon_u),$$
$$N_u=\frac{Eh}{1-\nu^2}(\varepsilon_u-\varepsilon_{uv}\mathrm{ctg}\chi+\nu\varepsilon_v),$$
$$S_v=\frac{1-\nu}{2}C[\varepsilon_{uv}+(\varepsilon_v-\varepsilon_u)\mathrm{ctg}\chi],$$
$$S_u=\frac{1-\nu}{2}C[\varepsilon_{uv}+(\varepsilon_u-\varepsilon_v)\mathrm{ctg}\chi],$$
$$M_{vu}=\frac{Eh^3}{12(1+\nu)}(\kappa_{vu}-\kappa_u\cos\chi),$$
$$M_{uv}=\frac{Eh^3}{12(1+\nu)}(\kappa_{vu}-\kappa_v\cos\chi), \qquad (7)$$
$$M_v=-\frac{Eh^3}{12(1-\nu^2)}\left[\frac{\kappa_u+\kappa_v}{\sin\chi}-(1-\nu)(\kappa_u\sin\chi+\kappa_{uv}\mathrm{ctg}\chi)\right],$$
$$M_u=-\frac{Eh^3}{12(1-\nu^2)}\left[\frac{\kappa_u+\kappa_v}{\sin\chi}-(1-\nu)(\kappa_v\sin\chi+\kappa_{uv}\mathrm{ctg}\chi)\right].$$

The positive direction of the forces and moments are shown in Fig. 2.

The hypothesis of non-interaction between layers of the shell which are "parallel" to the middle surface in the direction of its normal was used during the derivation of physical equations (7). The hypothesis show the fact that the normal component of the stress acting on the sections which are parallel to the middle surface is equal to zero.

Systems of equations (6), (7) in conjunction with geometrical equations of A.L. Gol'denveizer were used repeatedly by S.N. Krivoshapko and his disciples for the calculation of developable thin helicoidal shells loaded by axially symmetric load such as own weight [11].

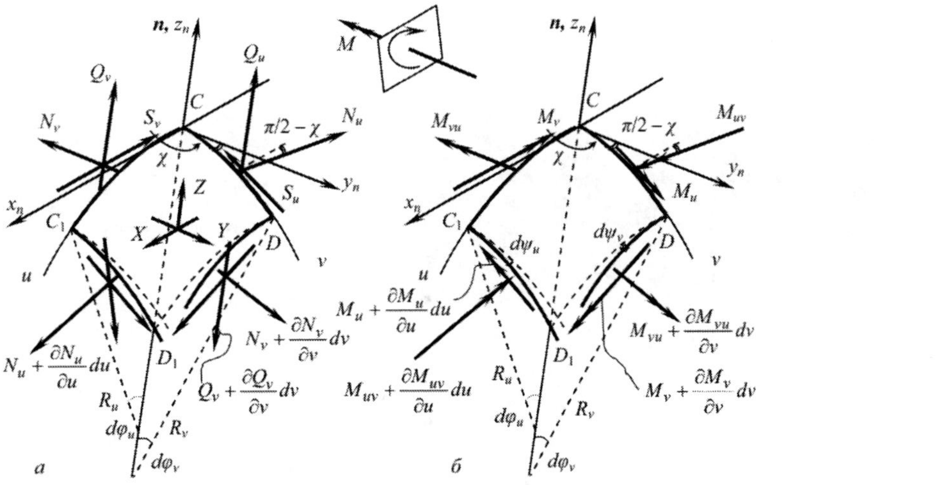

Figure 2. The positive direction of the inner forces and moments.

For example, with the help of the asymptotic method of a small parameter, the proposed system of equations can be reduced to a system of three ordinary differential equations containing three unknown parameters of displacements that were presented in the series

$$\begin{aligned} u_1 &= u_1(\alpha,\mu) = \sum_{i=0}^{\infty} H_i(\alpha)\mu^i, \\ u_2 &= u_2(\alpha,\mu) = \sum_{i=0}^{\infty} V_i(\alpha)\mu^i, \\ w &= w(\alpha,\mu) = \sum_{i=0}^{\infty} W_i(\alpha)\mu^i. \end{aligned} \tag{8}$$

For the determination of the first members of the series (8), generating equations are used upon condition that the small parameter is equal to zero ($\mu = 0$). These generating equations are easily solved analytically. Solutions for the second and subsequent members of the series can be found similarly [11].

However, there were found some problems in analytical calculations of several integrals which could not be directly solved. The approximation of these integrals led to complicated calculations. M. Rynkovskaya's suggestion to use Bernoulli's numbers [12] for these integrals simplified the calculations and increased the practicability of the method.

The governing equations of Gol'denveizer will coincide with the equations (6, 7) of Krivoshapko if one uses the following relations:

$$N_v = N_v^* \sin\chi, \quad N_u = N_u^* \sin\chi,$$
$$S_v = -S_v^* + N_v^* \cos\chi, \quad S_u = S_u^* + N_u^* \cos\chi,$$
$$M_{vu} = M_{vu}^* \sin\chi, \quad M_{uv} = -M_{uv}^* \sin\chi,$$
$$M_v = -M_v^* + M_{vu}^* \cos\chi, \quad M_u = -M_u^* - M_{uv}^* \cos\chi.$$

These relations connect internal forces and moments with pseudo-forces and pseudo moments denoted by the letters with stars, χ is the angle between the coordinate lines.

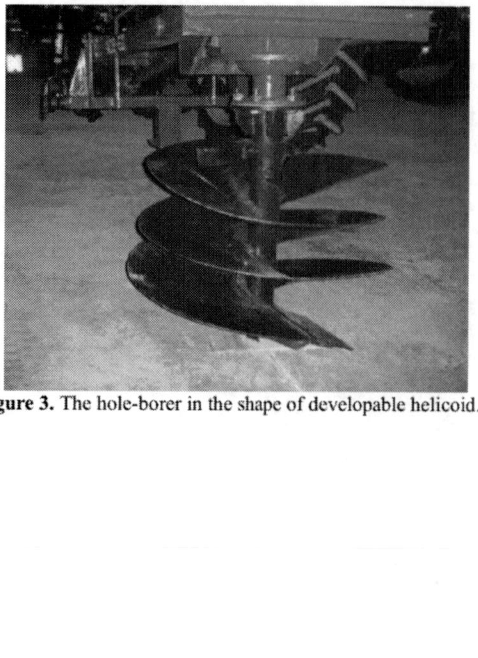

Figure 3. The hole-borer in the shape of developable helicoid.

Analytical solutions for the pseudo-developable helicoids of general form (Fig. 1, *d*) are not yet available, so in practice, it is necessary to carry out experimental research for each specific case.

A pseudo-developable helicoid was analyzed by S.M. Halabi [13], who solved some one-dimensional problems for this type of helicoid.

Various problems on the application of ruled helical surfaces, mostly right helicoids, are presented in [14-20]. The Fig. 3 shows the use of the developable helicoid as the hole-borer.

4 Results and discussion

The authors adduced some formulas for the examination of two types of ruled helicoidal thin shells that were used by them and by some other researchers for the determination of stress-strain state of long helicoids [21]. In engineering, right, oblique, developable and pseudo-developable helicoids are widely used; however, their strength is determined in most cases on the basis of experiments. The monograph of D. Taylor [22] and article of D.Y. Panov [23] were the first two works devoted to the calculation of the strength of the right helicoidal shells in relation to parking structures. Right helicoids which are used in mechanical engineering are usually called twisted plate. This term describes the configuration better to most people, dealing with the calculation of compressor or turbine blades.

The example of using of the formulas presented in the paper can be demonstrated for the developable helicoids obtained by bending of the same annulus (Fig. 4).

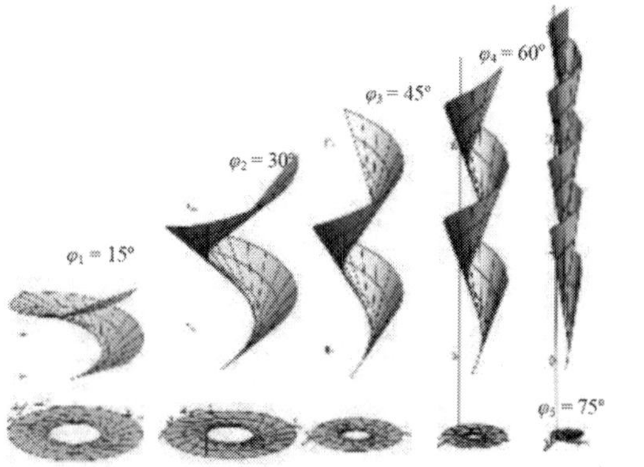

a_1 = 1,871 m; a_2 = 1,504m; a_3 =1,002 m; a_4 =0,501 m; a_5=0,134m;
b_1 = 0,501 m; b_2 = 0,868 m; b_3 =1,002 m; b_4 =0,868 m; b_5= 0,501

Figure 4. Developable helicoids obtained from the same annulus.

The mathematical apparatus of process of bending of a given annular plate into the helicoids with a given angle φ of inclination of the rectilinear generatrixes was derived by S.N. Krivoshapko.

In the Fig. 5, the results of calculations of the reinforced concrete helicoidal shells with the middle surfaces presented in Fig. 4 are shown. The helicoids are rigidly fixed along the contour helical edges and are loaded by uniformly spread load:

$$X = -\frac{bg}{m}, \; Y = 0, \; Z = \frac{ag}{m},$$

where g is a dead (own) weight of 1 m^2 of the shell with the thickness of 10 cm; Poison ratio is 0.17; a and b are geometrical parameters of the middle surfaces shown in Fig. 4.

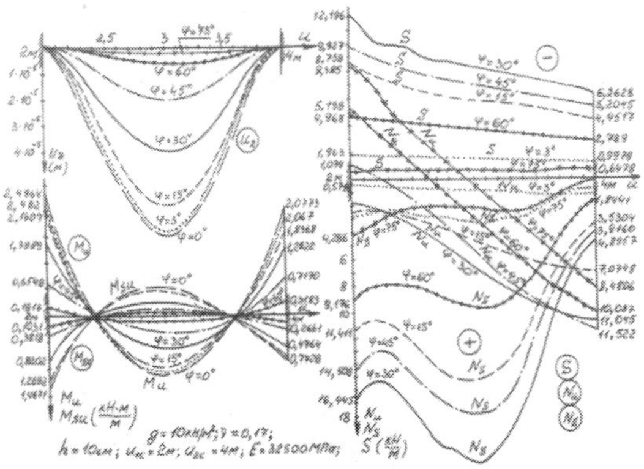

Figure 5. Results for the developed helicoids with various angles of generatrix inclination.

5 Conclusion

Although the subject of the geometry of ruled helical surfaces had its beginning in the eighteenth century, it still retains its attractiveness for an engineer.

Moreover, geometrical research on these surfaces is still in progress. In this paper, the authors described only analytical methods, numerical methods they used for comparison of results obtained by the both methods or for solution of the system of eight ordinary differential equations of the first order. It has been shown above that elastic behavior of right and developable helicoidal shells submitted to axisymmetric loading is well represented by various analytical methods.

References

1. Krivoshapko S.N. Helicoidal shells. Problems in theory and practice of engineering research, p.132-136. (1998).
2. Krivoshapko S.N. Geometry and strength of general helicoidal shells, Appl. Mech. Rev., **52**, No 5, p.161-175 (1999).
3. Krivoshapko S.N., Ivanov V.N. Encyclopedia of Analytical Surfaces, Springer, 752 p. (2015).
4. Pylypaka S.F. Control of bending of ruled surfaces on an example of a screw conoid. Prikl. Geom. ta Inzh. Grafika, **70**, p. 180-186 (2002).
5. Cohen J.W. On stress calculations in helicoidal shells and propeller blades. Holland, Walman, 100 p. (1955).
6. M.I. Rynkovskaya. Analytical Methods of Analysis of Right and Open Helicoids. Northeast Amer. Society of Eng. Educ. Conf., P.1-5 (2009).
7. Yaroshenko, A.R. Axe-symmetrical deformation of helicoidal shell with rectangular profile. Din. i Prochn. Mashin, **12**, p. 3-9 (1971).
8. Tupikova E.M. Analysis of the thin elastic shells in the form of long oblique helicoid, Struct. Mech. of Eng. Constr. and Build., № 3, p. 23-27 (2015).
9. Badjoriya G.Ch. Analysis of long developable helicoid on moment theory in displacements. Struct. Mech. and Calcul. of Structures, 3, p.22-24 (1985).
10. Krivoshapko S.N., Kumudini D. Analysis of thin elastic shells in the form of Archimedes screw with the same angles of inclination of its generatrixes. Math. Probl. in Phys. and Techn., MFTI, p. 96-103 (1992).
11. Krivoshapko S.N. Geometry of linear surfaces with the return rib and linear theory for torso-helicoid calculation, RUDN, p.357 (2009).
12. M. Rynkovskaya, On application of Bernoulli's numbers for thin torso-helicoids calculated by asymptotic method of small parameter, MOO Spatial Structures, **12**, p. 59-64 (2009).
13. Halabi, S.M. Moment theory of thin helicoidal pseudo-developable shells. Struct. Mech. of Eng. Constr. and Build., **10**, p. 61-67 (2001).
14. Roberts A.W., Manjunath K.S., Mcbride W. The mechanics of screw feeder performance for bulk solids flow control. Nat. Conf. Publ., Inst. Eng., **92/7**, P. 333-338 (1992).
15. Dabrowski Otton, Sapian Czeslaw. Loading of a central screw chute in a coal storage container. Pr. nauk. Inst. beed. pwrocl., **51**, P. 125-130 (1987).
16. Teraoka Ats. Analisis del husillo de alta plastificacion para el moldeo por inyeccion. Rev. plast. mod., **46**, № 463, P. 55-64 (1995).
17. Iura Masashi and Hirashima Masaharu. Fourier analysis of shallow right helicoidal shells. Trans. Jap. Soc. Civ. Eng., **14**, P. 55-59 (1984).
18. Knabel J., Lewinski T. Selected equilibrium problem of thin elastic helicoidal shells. Arch. Civil Eng., **42(2),** P. 245-257 (1999).
19. Kielb R.E., Leissa A.W., MacBain J.C. Vibrations of twisted cantilever plates - a comparison of theoretical results. Int. J. for NumericaL Meth. in Engineering, **21**, P. 1365-1380 (1985).
20. Hoyashita Shigeru. Calculation method for profile of helical gear with tooth-trace modification finished by worm-shaped tool or form grinding wheel, Jap. Soc. Mech. Eng. C., **61**, №583, P. 1149-1156 (1995).
21. Krivoshapko S.N., Gbaguidi A.G. Two methods of analysis of thin elastic open helicoidal shells.Int. Journ. of Res. and Rev.in Appl. Sc., **12**, Iss.3, p.382-390 (2012).
22. Taylor D.W. The Speed and Power of Ships. – Ransdell Incorporated, Washington D.C., 130 p. (1933).
23. Panov D.Yu. Wind screw calculation [Raschet vozdushnogo vinta], Tr. TzAGI, **288** (1937).

Effect of Injection Location and Multi-Hole Nozzle on Mixing Performance in a CNG-Fuelled Engine with Port Gas Injection, Using CFD Analyses

Tianbo Wang and Siqin Chang

School of Mechanical Engineering, Nanjing University of Science and Technology, Nanjing 210094, China

Abstract. A compressed natural gas injection device with moving-coil electromagnetic linear actuator and mushroom type poppet valve was designed to supply a large-bore intake port injection type engine with sufficient fuel timely. The transient engine CFD model combined with the poppet valve's motion was established to research the effect of injection conditions on the mixing homogeneity in the intake port and cylinder. Computed penetration distances of impinging jet with different k-ε models and different wall functions are compared with measured results in the literature to validate the model. It was observed that the gas injection location and number of nozzle holes have a significant effect on mixing performance both in the intake port and cylinder.

1 Introduction

Compressed natural gas (CNG), one of the most promising alternative fuels, is widely used in engines due to its rich resource and cheap price [1]. Although the direct injection technology has been adopted to some gas fuel engine models [2], the multi-point port fuel injection (PFI) is still an important developing direction [3, 4].

The gas injection device (GID) is the ultimate component of engine fuel supply system, whose controllability and injection characteristics have a great influence on the engine performance. Currently the driver of GID is mostly solenoid, and the executing component is mostly in the form of spherical or needle valve [4]. Such kind of GIDs are small and can be installed in the engine conveniently, however, its mass flow rate is quite low. More importantly, its control accuracy, working reliability will be seriously dropped by the solenoid [5]. On the contrary, the moving-coil electromagnetic linear actuator has higher power density and better controllability, so it can be used as the driver of GID [6]. The mushroom type poppet valve can be used to deliver gas fuel to the large-bore engine because they have better sealing performance and higher mass flow rate.

The turbulent mixing in the port and inside the engine cylinder is a crucial issue for the investigation of the performance of a spark-ignition engine. The fuel-air mixture from liquid fuel is largely dependent on atomization and evaporation of the injected fuel droplets, while gaseous injection does not involve the complex phenomena. Mixing of gaseous fuel with ambient air is usually slower than that of liquid fuel due to the lower mass and momentum [7].

The CNG injection process in engine is similar to the transverse flow in the scramjet engine. Using flow visualization techniques, Fric and Roshko [8] found four different types of vortices between the jet and the cross-flow. They are the following: (i) the jet shear-layer vortices; (ii) the system of horseshoe vortices; (iii) the counter-rotating vortex pair (CVP); and (iv) the wake vortices.

The wall impingement jet is also involved with injection process both of liquid and gas fuel. The effect of cross-flow on gasoline spray impingement with port fuel injection was studied to conclude that the cross-flow decreases the axial velocity of impinging droplets [9]. The wall impingement effect on mixture preparation in a direct injection hydrogen-fuelled engine was studied using CFD and high-speed schlieren imaging technique [10].

The injection timing, location and number of nozzle holes of the GID have a significant impact on the in-cylinder mixture formation. Yamato, et al. [11] used CFD calculation and PIV measurement to see the effect of injection position on in-cylinder fuel distribution with an experiment engine. Scarcelli, et al. [12] used the commercial code Fluent to study the influence of single- and multi-hole nozzles on mixture preparation in a direct injection hydrogen-fuelled engine. However, all of the above studies must set the flow pattern and outlet mass flow before CFD simulation because they did not take into consideration of the internal structure of GID, which will affect the accuracy of simulation seriously.

In the present study, the GID using moving-coil electromagnetic linear actuator and mushroom type poppet valve is designed. Detailed computational fluid dynamics (CFD) simulations combined with the poppet valve's motion have been performed to analyse the effect of injection location and number of nozzle holes on the mixing performance in the intake port and cylinder. The

DOI: 10.1051/matecconf/20179506003

cross-flow and wall impingement within intake port are emphasized.

2 Working principle and structure of the injection device

The moving-coil electromagnetic linear actuator, with higher power density and better controllability than the solenoid, is taken as driver of GID. Figure 1 shows the structure and prototype of GID. The driver consists of a moving coil, permanent magnets, an inner core, and an outer core. When the coil is energized, the Lorentz force produced in the coil actuates the valve to move. The electromagnetic force can be bidirectional as the direction of the current changes [6].

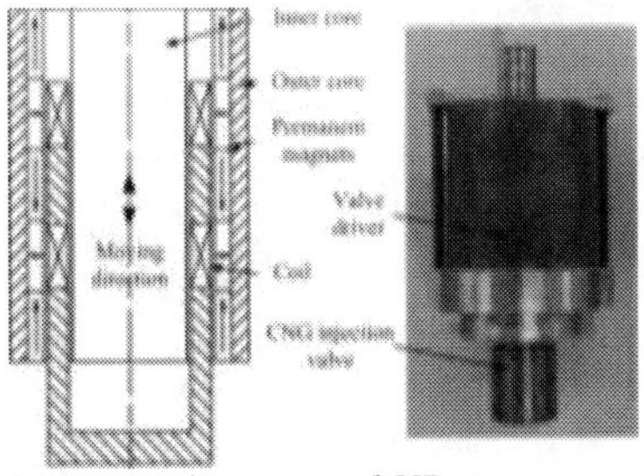

Figure 1. Structure and prototype of GID.

The large-bore CNG engine operating conditions and main dimensions of GID are listed in Table 1 and Table 2 respectively. The engine is fixed at 1900rpm in this study. According to the theoretical fuel consumption given in Equation(1)-(3), the steady CFD model keeping GID at maximum lift was established to validate that the GID can supply enough fuel to the engine at any operating condition within 7.89ms corresponding to 90 crank angle (CA) at 1900rpm.

Table 1. Specifications of engine.

Parameter	Value
Bore(mm)×Stroke(mm)	131×155
Displacement volume(L)	12.53
Compression ratio	11.5
Rated power(kW)/speed(rpm)	255/1900
IVO/IVC(CA)	30°BTDC/46°ABDC
EVO/EVC(CA)	78°BBDC/30°ATDC

Table 2. Specification of gas injection device.

Injection device parameter	Value
Outlet diameter(mm)	7
Valve lift(mm)	1
Injection pressure(MPa)	0.7
Transition time(ms)	1.5
Injection duration(ms)	7.89

Under a certain engine operating condition with output power P_e, the fuel consumption is given in Equation (1).

$$m_g = P_e \cdot \rho / \eta_e h_\mu \tag{1}$$

And the fuel needed per cycle is given in Equation (2).

$$m_{cyl} = 120 m_g / n_e \tag{2}$$

Eventually the steady mass flow of the injection device is calculated as given in Equation (3).

$$m_G = 720 m_g / (\phi_g - 6 n_e \Delta t) \tag{3}$$

where P_e, is the engine power; η_e, is the engine efficiency; h_μ, is the low heating value of gas fuel; ρ, is the density; n_e, is the engine speed; Φ_g, is the injection duration in CA; Δt, is the transition time of the driver.

3 Numerical model and validation

3.1 Turbulence model validation

An impinging jet can be divided into three regions: the free jet region, the impingement region and the wall jet region. Earlier numerical studies [13] have found some difficulties to correctly simulate simple impinging jets with standard k-ε turbulent model.

This study follows a RANS approach combined with a k-ε model to describe in-cylinder turbulence. The quality of the RANS results is not comparable to more detailed approach such as LES or DNS. However, the latter are not computationally affordable yet for engine applications. Turbulence is modelled using a standard k-ε model, a RNG k-ε model and a realizable k-ε model. Standard wall functions and non-equilibrium wall functions are used to model momentum and heat fluxes at the walls respectively.

Experimental measurements made by Fujimoto et al. [14] show motion and air-entrainment characteristics of round impinging gas jets. The measurements were obtained by injecting acetylene gas, C_2H_2, through a nozzle of 0.16 cm diameter into air where the pressure was 101.3kPa and the temperature was 293K. The injection duration was 26 ms with a mean injection pressure of 2.9kPa. The nozzle orifice placed at 1.5 cm above the wall, at L/D ratios of 9.4 is picked here to validate the turbulence model. The highest resolution in the radial direction is near the axis. A uniform grid size of 0.025 cm is used in the axial direction. For the free jet computations, the chamber height is 150 mm.

The diffusion characteristic of wall-impinging jet is mostly indicated by the penetration distance. In this paper, the penetration of the transient wall-impinging jet after impingement is defined as the summation of the impinging distance, L, and the radial distance, X, to which the jet has penetrated along the wall. The computed penetration for impinging jets is defined as the impinging distance plus the maximum radial distance where the fuel mass fraction reaches a cut-off value of 0.01 [13].

As shown in Figure 2(a)(b), compared to the standard k-ε model and realizable k-ε model, the RNG model has been shown to be in better agreement with experiments in predicting gaseous jet penetration history for free jet and wall-impinging jet. And it can be seen from Figure 2(c) that non-equilibrium wall functions seem to be more precise than standard wall functions.

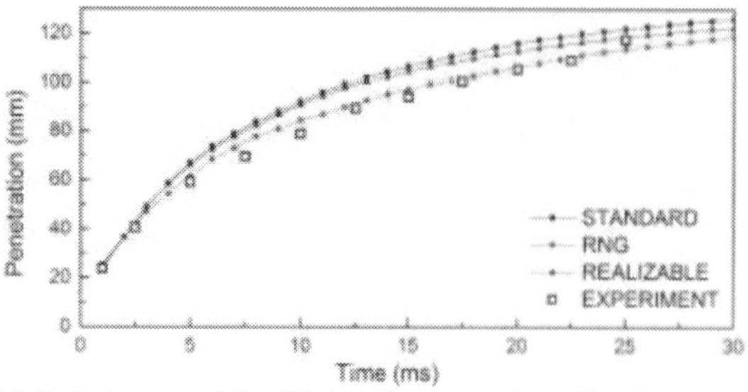

(a) Turbulence model's effect on the penetration of free jet

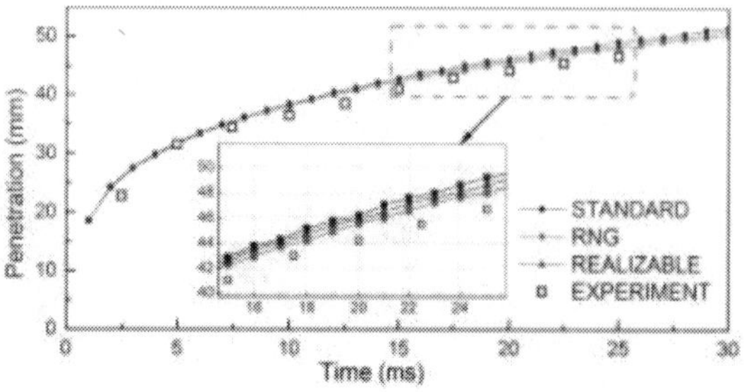

(b) Turbulence model's effect on the penetration of wall jet

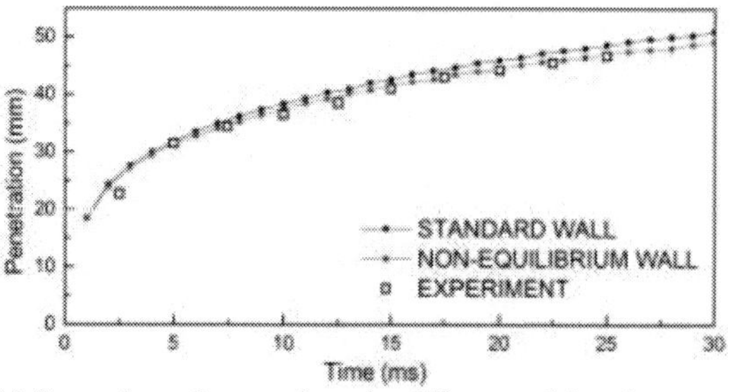

(c) Comparison of penetration using different wall functions

Figure 2. Comparison of penetration with Fujimoto's experiments.

3.2 Grid independence study

One of the most important variables affecting the numerical model ability to predict the flow correctly is the number of grid points across the GID exit. Due to the complexity of GID geometry, it is initially necessary to fix the poppet valve position during the computations, instead of implementing the valve lift with time. And to exclude the influence of intake port wall on the gas jet pattern, the jet export zone was set as a cylinder-shaped chamber with diameter of 70mm, height of 150mm.

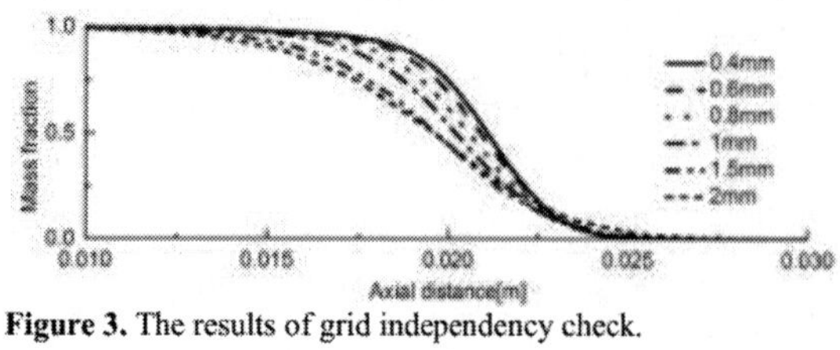

Figure 3. The results of grid independency check.

The results for mass fraction distribution along the exit axis of the cylinder-shaped chamber at time 0.1ms after injection start have been plotted against size of grid cells across the exit in Figure 3. It could be seen that mass fraction tends to be stable with decrease in size of grid cells. But decreasing the size of grid cells to smaller than 1mm seems to have little effect on the mass fraction. This suggests 1mm of the grid cells size is appropriate for grid-independent numerical simulation of CNG injection.

3.3 Case studies

As mentioned in the earlier section, injection location and number of nozzle holes are considered in the case studies. Different cases according to injection location and number of nozzle holes are shown in Figure 4. For case1 the GID is installed on the up side of the intake port, perpendicular to the axis of intake port. Based on case1, one short tube, with the same diameter of the GID outlet, was added to the outlet nozzle for case2. The length of the tube is 30mm. Case 2 shortens the impinging distance L with respective to case1. Case3 has 5 holes of the same diameter, and the total cross area of these 5 holes is the same as case1. At last, the gas jet is oriented to the helical intake and tangential intake port separately for case4 and case5. It should be noted that all the above cases are compared by keeping all the other parameters, such as injection timing, injection duration, engine speed and load, identical.

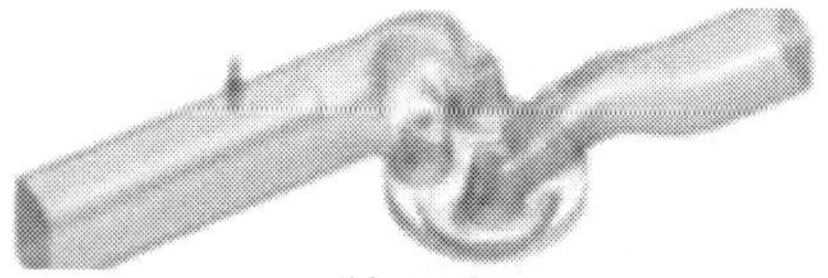

(a) case1

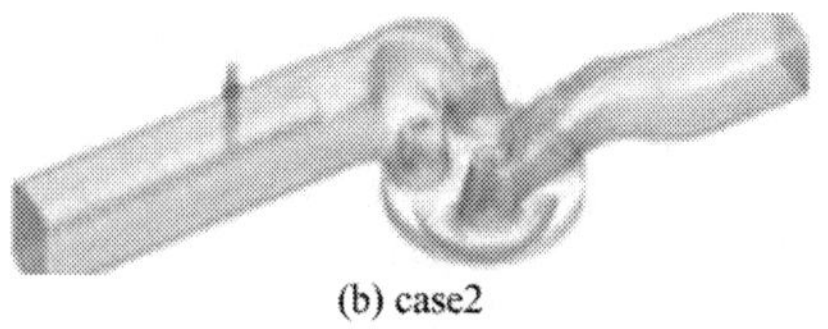

(b) case2

(c) case3

(d) case4

(e) case5

Figure 4. Case studies.

3.4 Meshing and boundary conditions

According to the results of the grid independence study, for single-hole (case1, 2, 4, 5), the grid near GID outlet is shown in figure 5(a), and for 5-hole (case3), the grid is shown in figure 5(b). The valve regions (near the poppet valve and engine valve) involve finer discretization, and the other regions are discretized with coarser grid.

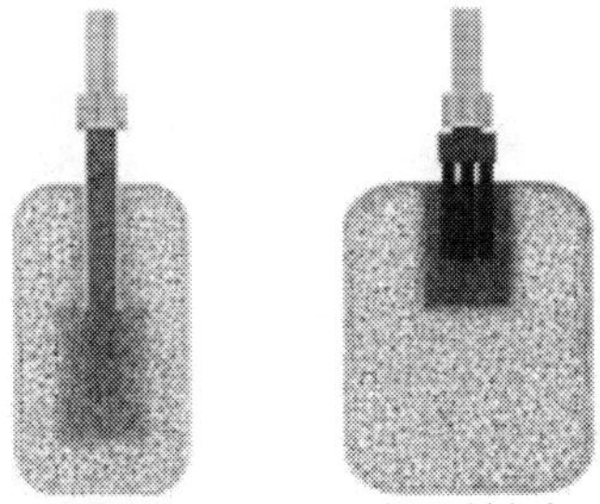

(a) single-hole (b)multi-hole

Figure 5. Computational grids.

The intake boundary condition of engine is set as 2bar (turbo-charged), 353K, and the exhaust condition is set as 1.8bar, 773K. The whole domain is assumed to be initially quiescent. Calculation starts at 30° BTDC when the intake valve opening (IVO). The RNG k-ε turbulence model with non-equilibrium wall functions is used for closure. The program is based on the pressure-correction method and uses the PISO algorithm. The first order upwind differencing scheme is used for the momentum, energy and turbulence equations. The run time for each case is typically about 70 hours on Intel core i7 computer.

4 Results and discussion

The gas fuel mass fraction contour of seven isometric cross sections after the installation location of GID and iso-surface under these conditions are shown in figure 6. The comparisons are carried out at CA430 when GID begins to close. The iso-surface value level in the intake port is 0.2, and the in-cylinder value level is 0.08.

It can be observed that the CVP dominates the mixing process in the port for case1 and fuel mass fraction contour on every cross section has a heart shaped distribution. The gas fuel gets into cylinder through both the helical and tangential intake port, and then impacts on the piston. After the impact, the flow direction is changed to along with the groove at the top of piston. The mixing performance in intake port for case2 is quite similar to that of case1, except that the regular heart shaped distribution of mass fraction is broken by the down side wall of the intake port. For case3 with 5 nozzle holes, two fuel jets are compelled to the up side of the intake port, while the other three jets still take the form of CVP under the effect of air flow. The in-cylinder iso-surface for case1, case2 and case3 is quite analogical with each other. For case4, the gas jet will flow to the cylinder centre under the guide of intake valve in helical port, and impinge on the cylinder wall and piston top. On the other side, for case 5, the gas jet will flow along the cylinder wall under the guide of intake valve in tangential port.

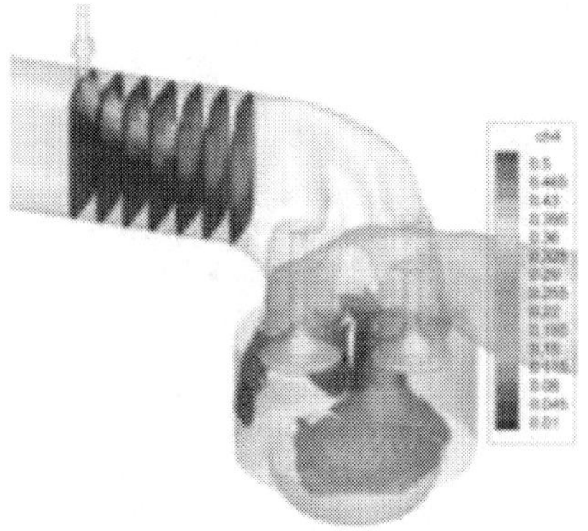

(a) case1

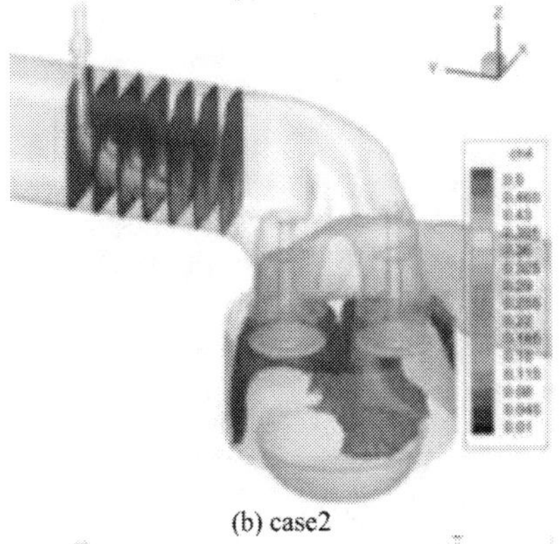

(b) case2

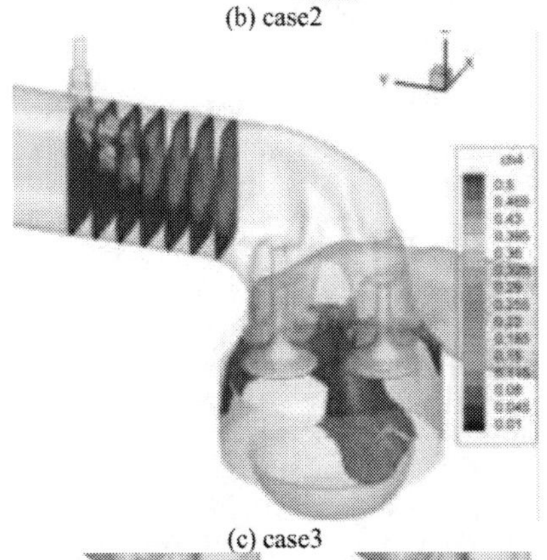

(c) case3

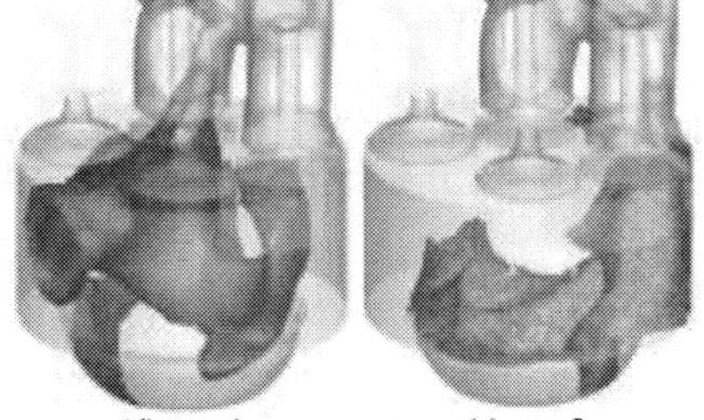

(d) case4 (e) case5

Figure 6. The fuel mass fraction contour and iso-surface at CA430.

In order to analyse the mixing degree in the intake port of case1-3, the standard deviation of fuel mass fraction in the cross section 100mm downstream installation position is plotted in figure 7 because the mixture formation of all these cross sections are quite similar to each other as shown in Figure 6. General speaking, the standard deviation for case2 is larger than case1 and case3 during the injection process (CA340-CA447), that is, the mixing performance is worse. On the contrary, the standard deviation for case3 is smaller. The plane vertical to the axis of cylinder, near the spark plug (3mm apart from the cylinder top) is taken to discuss in-cylinder mixing performance. The standard deviation of the in-cylinder cross section is also plotted in Figure 7. It is obvious that the mixing performance of intake port agrees with that of cylinder.

However, the effect of different injection conditions on in-cylinder standard deviation will be weakened when the intake valve is closed. It is meaningless to use standard deviation as evaluation criterion at firing TDC because the standard deviation is less than 0.003 for all cases by this time. So the probability distribution frequency (PDF) of different mass fraction domains is compared further.

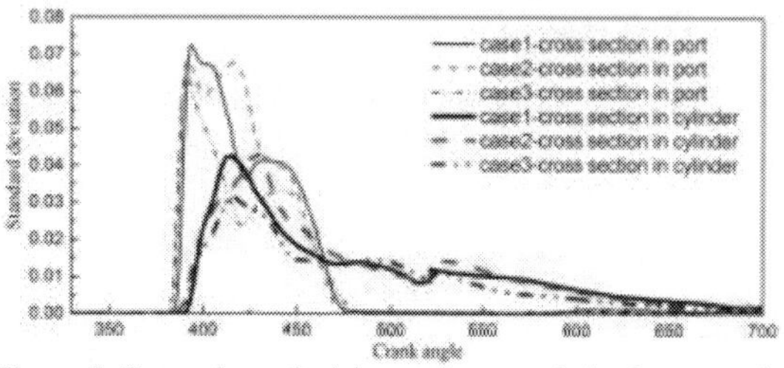

Figure 7. Comparison of mixing performance in intake port and cylinder for case1-3.

The cylinder zone is divided into eight parts according to in-cylinder fuel mass fraction, from lean to rich: 0-0.01, 0.01-0.015, 0.015-0.02, 0.02-0.025, 0.025-0.03, 0.03-0.035, 0.035-0.04, 0.04-0.045. The PDF of these parts are shown in Figure 8. According to total in-cylinder air and fuel mass when the intake valve is closed, the mean fuel mass fraction is in the region of 0.025-0.03 for case1-2, and 0.02-0.025 for case3-5. These two regions are defined as the best mixture concentration region (BMCR) for corresponding injection conditions. It can be assumed that with the larger of PDF in the BMCR the better is in-cylinder mixing performance.

For case2 the PDF of the BMCR is smaller, and there is more lean and rich mixture in cylinder at sparking time than for case1. So it can be concluded that the mixing performance both in intake port and cylinder would be worse if the impinging distance L is shortened. The PDF of the BMCR at the sparking time is about 73.7% for case3, while this value for case1 is about 63.3%. What's more, for case3 there is almost no mixture in the concentration region of 0.03-0.045 and 0-0.02 at TDC. The multi-hole nozzle is beneficial for improving the mixing degree both in intake port and cylinder. For case4, the PDF of the BMCR is 90.2%, larger than other cases. This can be regarded as complete mixing. On the contrary, for case5, because the fuel jet flows along the cylinder wall, the PDF of each concentration region changes very little and the PDF of the BMCR at sparking time is about 41.5%, which is the worst compared with the other cases.

The mixing performance will be improved if the gas fuel is oriented to the helical port alone, while the mixing performance will be weakened greatly if fuel is oriented to the tangential port alone.

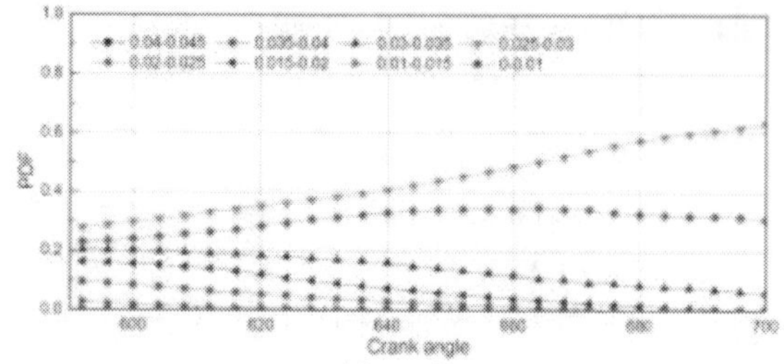

(a) case1

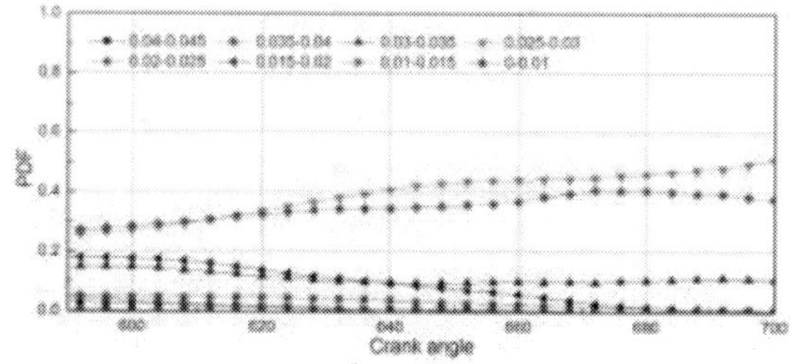

(b) case2

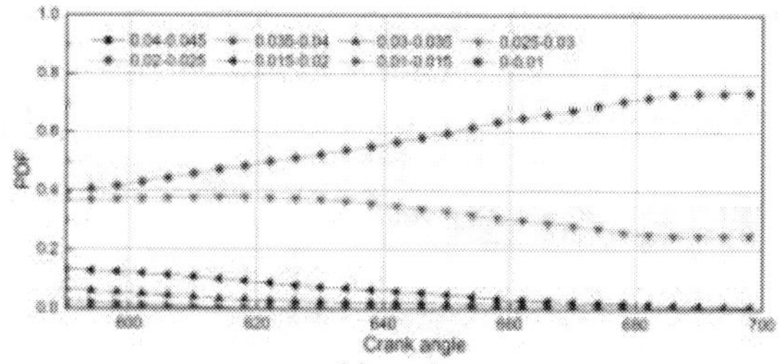

(c) case3

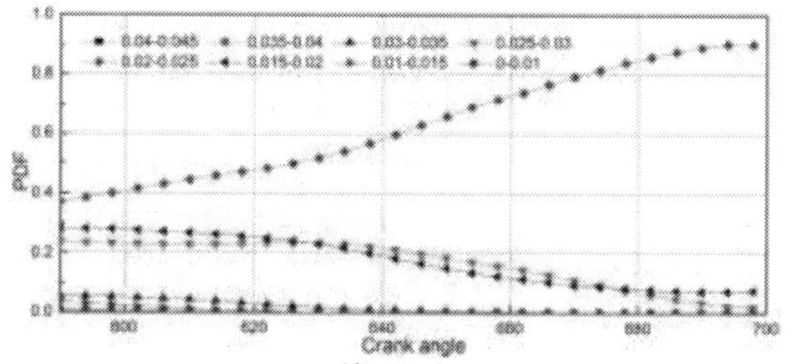

(d) case4

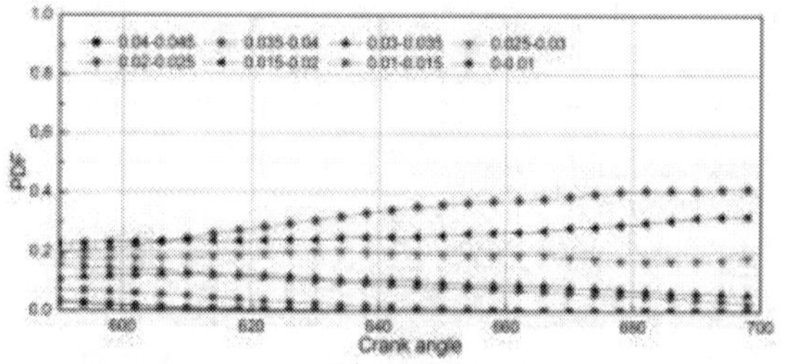

(e) case5

Figure 8. PDF of all mass fraction regions for different injection conditions.

5 Conclusion

The main conclusions are drawn as given below:

1) The GID using the moving-coil electromagnetic linear actuator and mushroom type poppet valve was projected.

2) The RNG k-ε model and non-equilibrium wall functions are more accurate in predicting the gas jet. Different k-ε models are employed to represent turbulence, and two kinds of wall functions are employed to model momentum fluxes at the walls. The computed penetration distances of free jet and impinging jet are compared with measured results in the literature.

3) The mixing performance both in intake port and cylinder would be worse if the impinging distance L is shortened.

4) Number of nozzle holes has a significant effect on the mixture formation in the port. The multi-hole nozzle is beneficial for improving the mixing performance both in intake port and cylinder.

5) The mixing performance will be improved if the gas fuel is oriented to the helical port alone, while the mixing performance will be weakened greatly if fuel is oriented to the tangential port alone.

6) Injection location and number of nozzle holes do not seem to have an obvious effect on the in-cylinder gas movement intensity.

Acknowledgement

This work was supported by the National Natural Science Foundation of China [grant number 51306090]; and the Natural Science Foundation of Jiangsu Province, China [grant number BK20130762].

References

1. Y. K. Gebre-Mariam, Testing for unit roots, causality, cointegration, and efficiency: The case of the northwest US natural gas market, Energy **36**,5, 3489−3500(2011).
2. M. H. Li, Q. Zhang, G. X. Li, and S. D. Shao, Experimental investigation on performance and heat release analysis of a pilot ignited direct injection natural gas engine, Energy **1**,10(2015).
3. M. A. Soberanis Escalante and A. M. Fernandez, A review on the technical adaptations for internal combustion engines to operate with gas/hydrogen mixtures, Int. J. Hydrogen Energy **35**, 21, 12134−12140(2010).
4. R. Bircann, Y. Kazour, K. Dauer, M. Fujita, A. Wells, D. F. Kabasin, and H. Husted, Cold performance challenges with cng pfi injectors, SAE Paper No.2013-01-0863(2013).
5. H. Glasmachers, J. Melbert, and A. Koch, Sensorless movement control of solenoid fuel injectors, SAE Paper No.2006-01-0407(2006).
6. L. Liu, and S. Q. Chang, Motion control of an electromagnetic valve actuator based on the inverse system method, Proc. IMechE Part D:J. Automobile Engineering, **226**,1, 85-93(2011).
7. J. Abraham, V. Magi, J. MacInnes, and F. V. Bracco, Gas versus spray injection: which mixes faster, SAE Paper No.940895(1995).
8. T. Fric, and A. Roshko, Vortical structure in the wake of a transverse jet, J. Fluid Mech, **279**, 1-47(1994).
9. M. R. Panao, A. L. N. Moreira, and D. F. G. Durao, Effect of a cross-flow on spray impingement with port fuel injection systems for HCCI engines, Fuel **106**, 249-257(2013).
10. R. Scarcelli, T. Wallner, N. Matthias, V. Salazar, and S. Kaiser, Numerical and optical evolution of gaseous jets in direct injection hydrogen engines, SAE paper no. 2011-01-0675(2011).
11. T. Yamato, H. Sekino, T. Ninomiya and M. Hayashida, Stratification of in-cylinder mixture distributions by tuned port injection in a 4-valve SI gas engine, SAE Paper No.2001-01-0610(2001).
12. R. Scarcelli, T. Wallner, N. Matthias, V. Salazar, and S. Kaiser, Mixture formation in direct injection hydrogen engines: CFD and optical analysis of single- and multi-hole nozzles, SAE paper no. 2011-24-0096(2011).
13. S. M. Hosseinalipour and A. S. Mujumdar, Comparative evaluation of different turbulence models for confined impinging and opposing jet flows, Numer. Heat Transf., **28**,6,647-666(1995).
14. H. Fujimoto, G.-S. Hyun, M. Nogami, K. Hirakawa, T. Asai, and J. Senda, Characteristics of free and impinging gas jets by means of image processing, SAE paper no. 970045(1997).

Design and Pitch Angle Optimisation of Horizontal Axis Hydrokinetic Turbine with Constant Tip Speed Ratio

Suyash Nigam[a], Shubham Bansal, Tanmay Nema, Vansh Sharma and Raj Kumar Singh

Department of Mechanical, Automotive and Production Engineering, Delhi Technological University, Delhi, India

Abstract. Booming population and associated energy demands, looming threat of exhaustion of conventional sources of energy and the severe environmental repercussions of the same call for alternate sources of clean energy. Hydrokinetic turbine is one such developing technology which harnesses zero-head free flow of water and affects hydrological ecology minimally. This paper discusses the optimisation of Horizontal Axis Hydrokinetic Turbine (HAHkT) blade chord length and twist angle using blade element momentum (BEM) theory to achieve a constant optimal angle of attack (AoA), thus maximising the power output. To achieve this while maintaining robustness at the hub end and eliminate cavitation, two different hydrofoils (S832 and E817) are selected. S832 is simulated using ANSYS 14.0 at low (0^0) and high (15^0) angles of attack and compared against more widely used NACA 4412 to study flow separation characteristics. This is followed by calculating angles of relative flow, ratios of chord length and subsequently twist angles for each blade element using MATLAB simulations. A blade model is thus developed for visualisation using computer aided designing after obtaining optimal chord lengths and pitch angles.

1 Introduction

Today's world sees an unprecedented demand for clean and economic sources of energy. Globally, 1.6 billion do not have access to electricity. [1] This calls for a greater shift in focus towards the renewable sources of energy for an energy-secure future.

According to IEA report in 2014, hydropower constitutes about 16.2% of our total energy demands and thus presents a great alternative. [2] However there are several problems associated with the creation of dams and embankments on water bodies to create power, such as sedimentation in the water bodies, significant damage to the aquatic ecosystem and socio-political concerns over displacement of multitudes of people.

Hydrokinetic turbine is a technology which harnesses zero-head free flow of water and presents a low emission way to satisfy the increasing energy demand while minimally affecting the hydrological ecology. This makes these turbines a lucrative alternative in a developing country like India which has a high hydropower potential.

These hydrokinetic turbines are advantageous as compared to wind power because of more predictability of the medium of flow, higher level of energy extraction due to near surface placement, minimal infrastructural investment, reduced environmental hazards and lower noise problems.

The primary classification of these turbines is done on the basis of axis of rotation relative to the flow- horizontal and vertical hydrokinetic turbines. Vertical axis hydrokinetic turbines are subject to some inherent disadvantages such as high starting torque requirements, lower power output at same flow velocity and higher vibrational losses. However Horizontal Axis Hydrokinetic Turbines (HAHkTs) improve upon these problems.

2 Literature review

Hydrokinetic turbines have seen a lot of research work over the past 2 decades. Ducted horizontal hydrokinetic turbines such as two-bladed SeaGen project turbines in the UK manufactured by Marine Current Turbines Limited are axial flow rotors whose mechanism is similar to that of a wind turbine with a capacity of 300kW-1.2MW. [3-5] Three-bladed Hammerfest Strom turbine (Norway) incorporates arrays of multi rotor underwater turbines whose energy is transported using subsea cables running on a maximum capacity of 300kW. Verdant Power has developed two projects- Roosevelt Island Tidal Project (RITE, New York East River, US) [6-8] and Cornwall Ontario River Energy Project (CORE, St. Lawrence River, Canada) [9] projects. It installs turbines which self-rotate from prevailing currents with optimally designed blades. These operate at 35kW-1MW capacity. Un-ducted hydrokinetic turbines comprise those such as multi-bladed Lunar Energy Rotate Tidal Turbine, housed in a converging-diverging duct so as to create a venturi action, and multi-bladed Underwater Electric Kite (UEK) Corp. [10] who now operate a fully functional plant at Eagle, Alaska.

[a] Corresponding author: suyashnigam@dtu.ac.in

From studying the functional projects, potency and viability of these hydrokinetic turbines can be established. Since HAHkT offer various advantages over the vertical ones, the design optimisation of the same will be the primary objective of this research.

2.1 Selection of hydrofoil geometry

A hydrofoil is a two dimensional surface which operates in water, producing lift force. The following parameters were found to be desirable:

- Glide ratio: This is a measure of the force of lift vs drag. It is important because lift is the primary source of power generation in horizontal axis turbines.
- Stall angle: The foil must produce lift until a high AoA as the velocity of flow varies greatly. Foils with high stall angles must be used towards the tip of the rotor, where the linear velocity is maximum.
- Leading edge area: The leading edge surface area must be small so as to reduce skin friction drag.
- Aft-loading: The trailing edge geometry must permit high values of aft-loading to counter cavitation.

2.2 Series of hydrofoils studied

Many organisations have worked in developing novel foils and studying the viability of already popular foils which are employed for wind turbines and commercial aeronautics.

a) NACA 4 series profiles: The NACA 4412 foil was used extensively for horizontal axis wind turbines (HAWT) purposes owing to large glide ratio values.
b) NREL S Series: National Renewable Energy Laboratory designed several geometries specifically for HAWT. Although they produce lesser coefficient of power than the NACA 4 series, they can be designed to operate at various tip speed ratios. [11]
c) NACA 5 Series: They have a high value of maximum coefficient of lift and can be used at operation near stall angle and off design conditions.
d) Eppler foils: Eppler E817 foil produces close to zero cavitation till 700 pounds per square inch aft loading. [12] This is achieved by using under cambered trailing edges.
e) Delft University Foils: These foils have a specific S-curve in trailing edge [13] to increase permissible aft-loading and a small upper surface area to decrease skin friction.

Several parameters were discussed upon for the optimisation process, namely solidity, chord length, pitch angle, hub design, number of blades, etc. Among these the optimisation of chord length and pitch angle was chosen as the objective of our work because the angle of relative velocity changes at every blade element. To maintain a high coefficient of lift, an optimal AoA is desirable and hence the pitch angles at every blade element have to be optimised accordingly.

3 Methodology

The methodology selected for developing a computational model of the HAHkT involves the selection of hydrofoil using k- ω turbulence model using Ansys Fluent, optimisation of pitch angle and chord length using MATLAB simulations and development of the optimised blade profile using computer-aided modelling for visual representation through Solidworks.

3.1 Procedure for ANSYS simulation

We decided to do a computational analysis of NACA 4412 foil and the NREL S832.

ANSYS Fluent was deployed to index the performance of hydrofoil under various spatial arrangements. For the 2 dimensional computational analysis a standard foil geometry was modeled using ANSYS design modeler and a C-mesh was constructed by employing ANSYS mesh module.

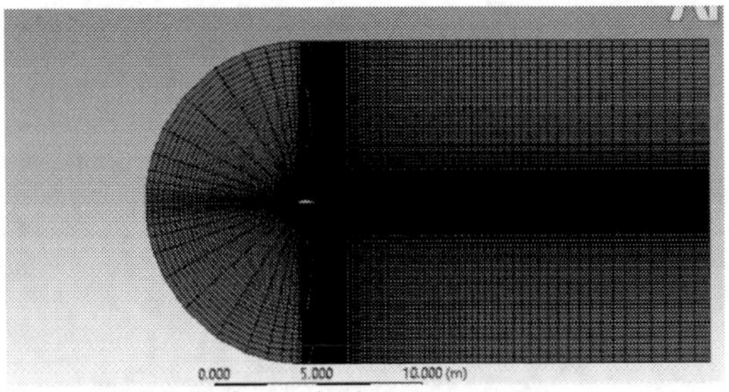

Figure 1. C-Mesh of control volume.

The number of nodes are in the 20,000 range with detailed meshing near the foil boundary so as to accurately capture the physics of boundary-layer flow. Mapped meshing operation was selected and the resultant mesh is displayed in Figure 1.

Simulation parameters:-
Inlet: Velocity inlet
Outlet: Pressure Outlet
Fluid density: 999.7 kg/m^3
Inlet velocity (U): 2m/s
Number of iterations: 500

Turbulence model: The k-ω [14] model was chosen by the virtue of its proven robustness in accurately predicting turbulent flow without high computational requirements.

It's a 2 equation model where 'k' stands for turbulent kinetic energy and 'ω' the specific dissipation rate. 'k' and 'ω' are computed by empirical relations [15]:

$$k = \frac{3}{2}(U * I)^2 \quad (1)$$

$$\omega = C\mu^{\frac{-1}{4}} * \frac{\sqrt{k}}{l} \quad (2)$$

where 'l' is 0.02 meters and denotes average Prandtl-mixing length typically 10% of chord length of 0.2m. 'I'

 DOI: 10.1051/matecconf/20179506004

is the ratio of fluctuation component of velocity to the root mean square velocity of the flow, called turbulence intensity. The value of 'I' for a rotating turbomachinery component is around 10%. 'μ' is the viscosity of flow.

3.1.1 Simulation results

The computation was carried out for NREL S832 and NACA 4412 at 0 degree AoA and a high AoA of 15 degrees to gain insight in flow phenomenon at varying conditions.

At higher AoA the flow separates in case of NACA 4412 while the flow is largely attached in the case of NREL S832. This implies S832 is more robust against varying angular positions.

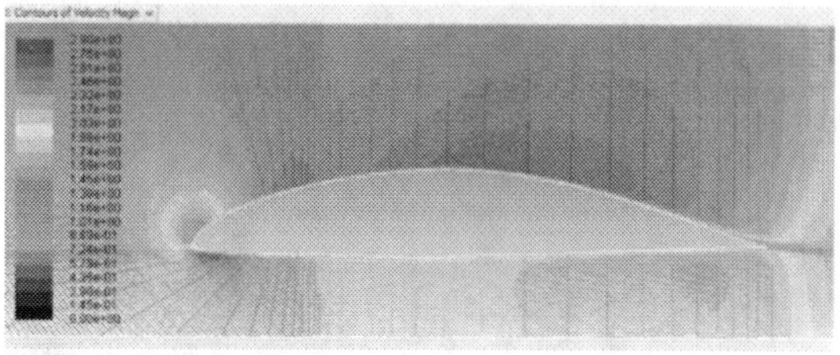

Figure 2. S832 at 0 degree AoA.

Figure 3. S832 at 15 degree AoA.

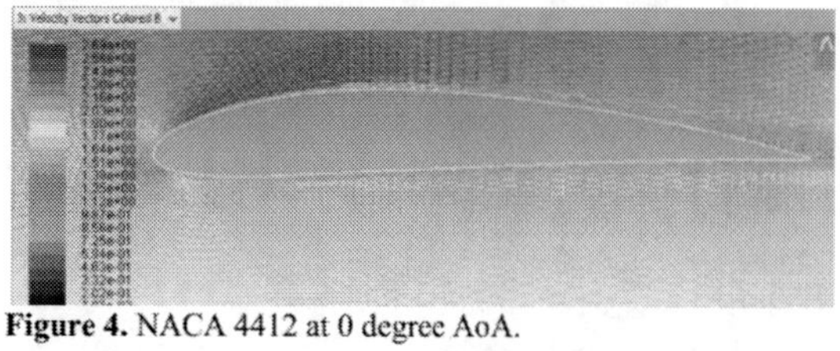

Figure 4. NACA 4412 at 0 degree AoA.

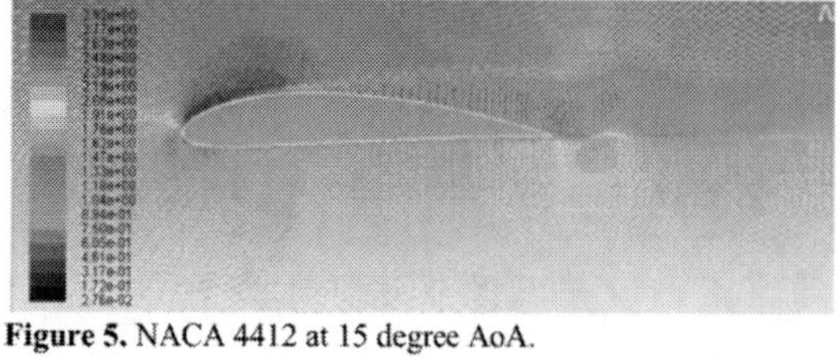

Figure 5. NACA 4412 at 15 degree AoA.

3.2 Optimisation of blade chord length and pitch angle

Two of the most important parameters in deciding the hydrofoil profile and orientation for the turbine blade so as to achieve maximum power output are: ratio of the area occupied by blade with respect to the total area covered by the blade (solidity) and the angle of relative velocity.

Simulation of blade profile data to optimise the chord length and the element twist angle along the length of the blade are done to achieve maximum power output. The variable pitch is important so as to operate at fixed optimal AoA for the respective hydrofoil geometry.

The following assumptions were made for calculations:

- The blade is considered to be a combination of discrete elements of equal length instead of a long single unit.
- The tip speed ratio (TSR) at each element varies linearly along the length of the blade.
- The induced velocity at each element is determined by performing the momentum balance for an annular control volume containing the blade element.
- Tip losses and hub losses and vortices are ignored.

For a specific AoA lift coefficient must be maximum, drag should be minimum while neither flow separation nor cavitation should occur. Optimal pitch angle enables the system to operate at higher relative water velocities. Hydrokinetic turbines are subjected to high thrust and torsional loads due to high density of water which causes high bending moment at the blade root. So, thicker blade sections are preferred near the hub.

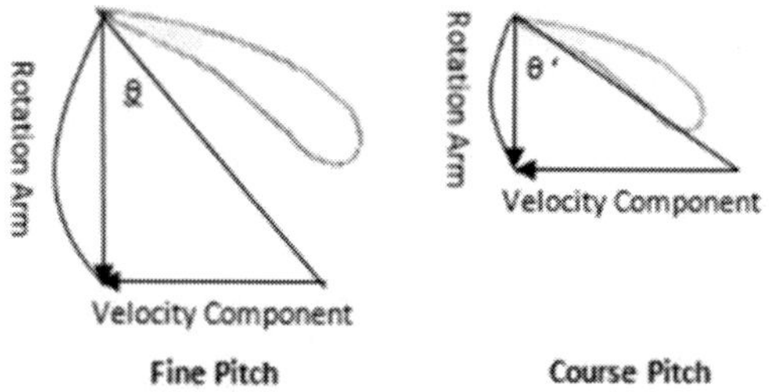

Figure 6. Variation of Relative Velocity Angle with pitch.

At a given water speed, the speed of different blade sections differ depending upon the distance of the section from the hub i.e the rotation arm. We observe that this results in a very low AoA for the sections closer to hub. This needs to be rectified. This is achieved by going for a coarse pitch for the closer section and a fine pitch for sections as we move away from hub.

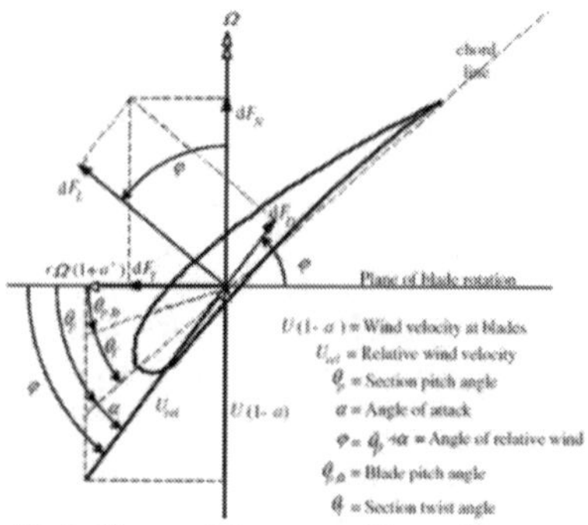

Figure 7. Blade Element Momentum Theory Parameters.

DOI: 10.1051/matecconf/20179506004

3.2.1 MATLAB simulation

The following parameter values are used:

Number of blades = B = 3
Tip Speed Ratio = TSR = 3
Rated Water Speed (m/s) = 2
Estimated efficiency of system = 0.45
Rotor diameter (m) = (d) = 2.2
Number of blade elements = (ne) = 10
AoA for max L/D ratio for S832 (α1) = 9.75
AoA for max L/D ratio for E817 (α2) = 7.5
Lift coefficient for S832 (Cl1) = 1.1207
Lift coefficient for E817 (Cl2) = 0.9083
Drag coefficient for S832 (Cd1) = 0.0622
Drag coefficient for E817 (Cd2) = 0.04476

An element data matrix is prepared in MATLAB after successive iterations of the calculations on initial data. The hydrofoils E817 and S832 are used in the ratio of length as 1:3.

Length of each element = (L) = (d/2) / ne
Distance from hub = (R) = L/2 + (i-1)*L
Tip Speed Ratio = (TSR) = ΩR/V
Local Tip Speed Ratio = (LSR) = TSR * (R/ (d/2))
Angle of relative velocity = $\varphi = (2/3) * \tan^{-1}(1 / LSR)$
Chord Length for S832 = (8*π*L*(1-φ) / (B*Cl1)
Chord Length for E817 = (8*π*L*(1-φ) / (B*Cl2)
Pitch angle (in radians) = θ = (φ – α)
Solidity Ratio = σ = B*c/ (2πr)

3.2.2 Results and inference

The twist angle/pitch angle output obtained from the optimisation code is mentioned in Table 1. This is obtained by using the following relation: θ = (φ – α)

The first blade element is considered at the hub end and the successive blades are positioned radially outwards.

Table 1: Element Data Parameters from MATLAB Code.

Element No.	Radial Distance Ratio (R/ (d/2))	LSR	Angle of Relative Velocity (°)	Ratio of Chord with radial distance	Pitch Angle (°)
1	0.05	0.15	54.313	0.192	46.813
2	0.15	0.45	43.848	0.386	36.348
3	0.25	0.75	35.420	0.346	25.670
4	0.35	1.05	29.069	0.330	19.319
5	0.45	1.35	24.353	0.299	14.603
6	0.55	1.65	20.812	0.268	11.062
7	0.65	1.95	18.100	0.240	8.350
8	0.75	2.25	15.975	0.217	6.225
9	0.85	2.55	14.275	0.196	4.525
10	0.95	2.85	12.890	0.179	3.140

For blade elements upto one-fourth of the length of the blade, the AoA is chosen for S832 blade profile i.e. α1 and for the remaining three-fourth length, that of E817, i.e. α2.

a) Chord Length vs Ratio of Radial Distance: Figure 8 shows the graph. The maxima of the chord length is at a value of ratio of radial distance close to 0.15, i.e. between the first and second element. Beyond that, it gradually decreases, narrowing upto the tip. This is in line with the aim to model blades with thicker sections towards the hub so as to bear the high bending moment at the blade root.
b) Angle φ vs Ratio of Radial Distance: Figure 9 shows the graph. It shows a continuous decreasing trend of angle of relative velocity with respect to the radial distance. Decreasing angle of relative velocity result in a lesser lift force and thus a lower TSR towards the root of the blade i.e. the inner elements. Higher relative velocity angles also indicate a higher value of twist angles towards the root of the blade because of the constant AoA for a given blade profile.

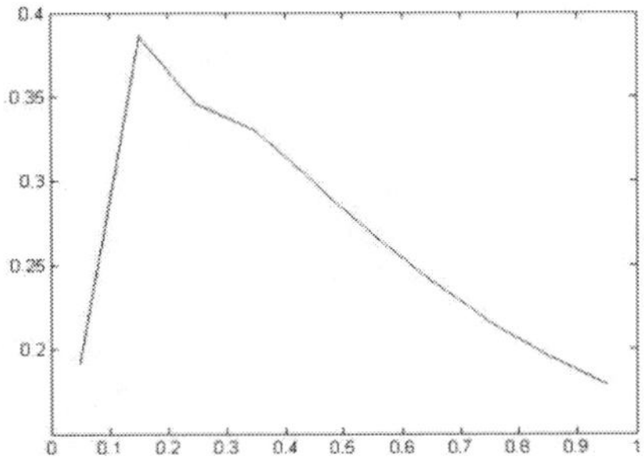

Figure 8. Graphical representation of Chord Length vs Ratio of Radial Distance.

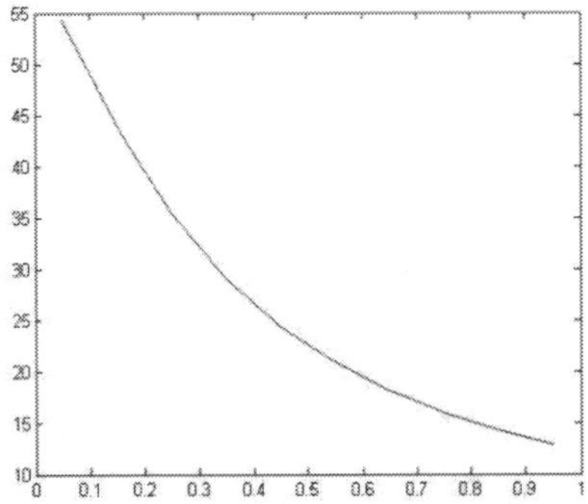

Figure 9. Graphical representation of Angle φ vs Ratio of Radial Distance.

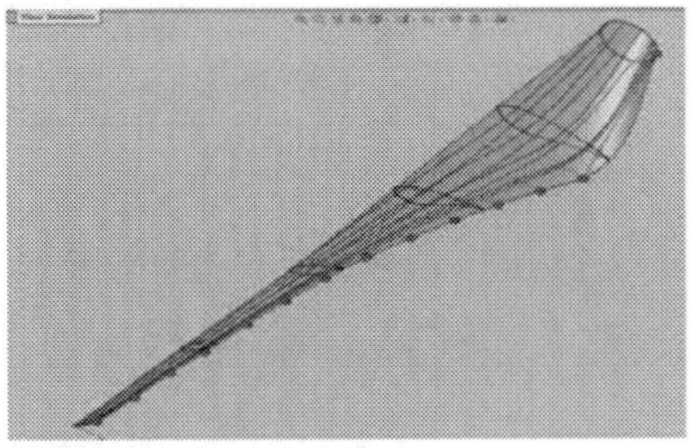

Figure 10. Optimised blade model.

3.3 Computer aided design of blade model

The different profiles are generated by the obtained phase angle and chord length values. Higher the number of blade elements, more accurate blade geometry results will be obtained.

The Fig. 10 depicts the development stages of the blade model.

The primary function of the hub body is to transfer the load from the blades to the low speed shaft. The hub diameter of the turbine is kept 40% of the blade length. The hub design is in accordance with the work of J. Cotrell. [17]

4 Conclusion

The HAHkT blade is formed of two hydrofoil geometries which are twisted about an axis so as to maintain the optimum AoA. The blade uses a combination of E817 and S832 foils. The distribution of the profiles is as follows: E817 constitutes of 25% of the total blade length on the hub end to reduce stresses and avoid cavitation, and the remaining 75% is S832 for no-flow separation even at higher AoA. The hydrofoils are arranged along a common axis to form a blade model which consists of three uniformly distributed blades as shown in Fig. 11.

This HAHkT blade model improves upon the structural stability and power output characteristics of traditional single hydrofoil constant twist angle profiles and thus holds promise as an efficient model.

Figure 11. CAD prototype of turbine.

References

1. *BP: Statistical Review of World Energy*, Workbook (xlsx), London, (2013)
2. http://www.iea.org/publications/freepublications/
3. SeaGen Environmental Monitoring Programme Final Report, as available on http://seageneration.co.uk/files/SeaGen-Environmental-Monitoring-Programme-Final-Report.pdf, accessed on October 24, 2015
4. C.A. Douglas, G.P. Harrison, J.P. Chick, Life cycle assessment of the Seagen marine current turbine. *Proceedings of the Institution of Mechanical Engineers*
5. A. Røkke, R. Nilssen, Marine Current Turbines and Generator preference. A technology review, *International Conference on Renewable Energies and Power Quality (ICREPQ'13)* Bilbao (Spain)(2013)
6. J. Colby. Multi-scale Hydrodynamic analysis of kinetic hydropower arrays. *NortekUSA Technical Symposium* (2011)
7. Verdant Power, Inc, Final Report - GO18168 Improved Structure and Fabrication of Large, High-Power KHPS Rotors, Award No. DE-FG36-08GO18168
8. D. Sale, J. Jonkman, W. Musial, A Hydrodynamic Optimization Method and Design Code for Stall-Regulated Hydrokinetic Turbine Rotors (2009)
9. http://www.power-technology.com/projects/cornwall-ontario-river-energy-project/, as accessed on October 27, 2015.
10. P. Vauthier. The underwater electric kite. East river deployment, UEK Corporation.
11. http://www.nrel.gov/docs/legosti/old/7109.pdf
12. http://www.foils.org/hysecdes.pdf
13. https://gcep.stanford.edu/pdfs/energy_workshops_04_04/wind_van_rooij.pdf
14. http://www.cfd-online.com/Wiki/K-omega_models
15. http://www.cfd-online.com/W/index.php?title=Turbulence_free-stream_boundary_conditions&oldid=22693
16. J. Cotrell, The Mechanical Design, Analysis, and Testing of a Two-Bladed Wind Turbine Hub (1995)

DOI: 10.1051/matecconf/20179506005

Design of Radial Inflow Turbine for 30 kW Microturbine

Thanate Sangsawangmatum and Udomkiat Nontakaew

Faculty of Engineering, Department of Mechanical and Aerospace Engineering, KMUTNB, Thailand

Abstract. Microturbines are small gas turbines that have the capacity range of 25-300 kW. The main components of microturbine are compressor, turbine, combustor and recuperator. This research paper focuses on the design of radial inflow turbine that operates in 30 kW microturbine. In order to operate the 30 kW microturbine with the back work ratio of 0.5, the radial inflow turbine should be designed to produce power at 60 kW. With the help of theory of turbo-machinery and the analytical methods, the design parameters are derived. The design results are constructed in 3D geometry. The 3D fluid-geometry is validated by computational fluid dynamics (CFD) simulation. The simulation results show the airflow path, the temperature distribution, the pressure distribution and Mach number. According to the simulation results, there is no flow blockage between vanes and no shock flow occurs in the designed turbine.

1 Introduction

Microturbines are small gas turbines that have the capacity range of 25-300 kW. They offer several advantages compared to other technologies including compact size, small number of moving parts, low emissions and lead to opportunities to utilize waste fuels. [1]

The main components of microturbine are compressor, turbine, combustor and recuperator. This research paper focuses on the 30 kW microturbine with an operating speed is about 40,000 rpm. The design and analysis of a radial inflow turbine that is a specific kind of turbine is suitable for microturbine application. In order to investigate the appropriate geometry of radial inflow turbine for 30 kW microturbine applications, the theory of turbo machinery is applied and the analytical design result is validated by the methods of computational fluid dynamics (CFD).

The method of computational fluid dynamics (CFD) is a branch of fluid mechanics that applies numerical analysis and algorithm to analyse the fluid flow problem and has been widely used in several research papers as shown [2]. This research paper analyses an intermediate pressure steam turbine using CFD. The domain of the steam turbine bladed region is modelled discretized in three zones; stationary blades, moving blades and labyrinth seals, and governing equations are solved. The CFD results show the pressure contour plot for each stage. It can be seen that the method of computational fluid dynamics is very useful for visualize and understanding the flow in the domain of complex geometries.

The ideal cycle, Brayton cycle, for 30 kW microturbine is shown in Fig. 1.

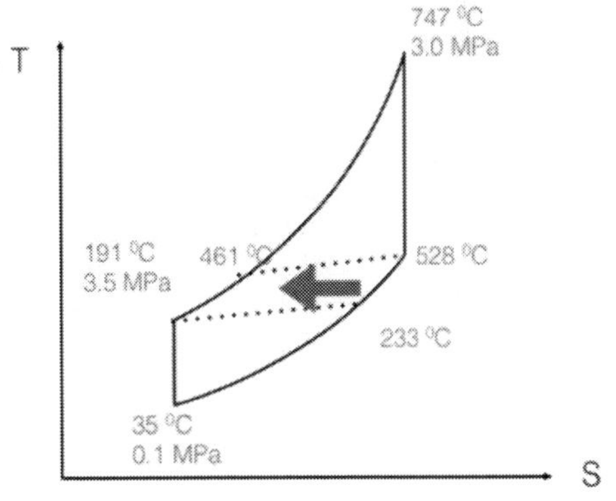

Figure 1. The Brayton cycle for 30 kW microturbine.

The assumptions of ideal conditions made in the design process are as follows: The compression and expansion processes are adiabatic reversible or isentropic. The changes of kinetic energy of the working fluid between inlet and outlet of each component are neglect. There are no pressure losses in the inlet ducting, combustion chambers, heat exchangers, exhaust ducting, and ducts connecting the components.

The working fluid has the same composition throughout the cycle and is a perfect gas with constant specific heats. The mass flow rate of gas is constant throughout the cycle. Heat transfer in a counterflow heat exchanger (recuperator) is complete so that the temperature rise on the cold side is the maximum as possible and can be considered as exactly equal to the temperature drop on the hot side.

DOI: 10.1051/matecconf/20179506005

2 Radial inflow turbine design process

The radial-inflow turbine stage geometry is illustrated in [3]. The flow enters the stage through an inlet volute and the flow area is constant in the circumferential direction before passes through a vaneless annular passage into a nozzle blade row.

The flow passes through another vaneless annular passage to enter the rotor row. An exhaust diffuser is next after the rotor row to convert some of the rotor exit velocity to static pressure. Hence, the basic stage components to be analysed are the inlet volute, nozzle, rotor, and vaneless annular passages and exhaust diffuser. In this research paper, the radial-inflow turbines to be analysed are single-stage units. The theoretical analysis and studies of the boundary layer, the boundary layer loss coefficient, the inlet volute, the nozzle row, the rotor, the vaneless annular passage, and the exhaust diffuser are also summarized. The basic analysis for each stage of radial flow turbines is performed in [4] and [5].

In the radial inflow turbine design process, air is considered as the working fluid with a flow rate of 0.28 kg/s. Turbine inlet temperature and turbine inlet pressure are 747 K and 3 bar, respectively.

The specific speed (N_s) and the specific diameter (D_s) can be calculated as

$$N_s = N\,(Q_3)^{1/2} \,/\, (\Delta H_{ad})^{3/4} \qquad (1)$$

$$D_s = D\,(\Delta H_{ad})^{1/4} \,/\, (Q_3)^{1/2} \qquad (2)$$

where N is the impeller rotation speed (rpm), Q_3 is the turbine exhaust air flow rate (m^3/s), ΔH_{ad} is an isentropic head expand in turbine (m) and D is the diameter of the impeller (m). The air velocity equivalent to the isentropic enthalpy drop, c_o , can be calculated as, see Fig. 2,

$$c_o = \{2C_p\,[T_{o1} - T_{o3}]\}^{1/2} \qquad (3)$$

where C_p is the specific heat at constant pressure (kJ/kg·K), T_{o1} is the stagnation temperature at turbine inlet (K), γ is the pressure ratio and k is the ratio of specific heat at constant pressure and specific heat at constant volume and is equal to a constant of 1.4.

Substitute the value of C_p, T_{o1}, γ and k into equation (3), the air velocity which is equivalent to the isentropic enthalpy drop, c_o ,can be calculated as

$$c_o = \{2(1004)[722 - 528]\}^{1/2} = 624\ m/s \qquad (4)$$

According to the previous researches and studies, the best efficiency point of radial inflow turbine can be determined by the quotient of the air velocity at the turbine outlet (U_2) to the velocity which equivalent to the isentropic enthalpy drop (c_o) and is found between 0.68 and 0.71.

$$0.68 < U_2/c_o < 0.71 \qquad (4)$$

Turbine mass flow rate is equal to 0.32 m^3/s. Gas density at the exit condition is equal to 0.44 kg/m^3. The volume flow rate at the turbine exit can be determined as

$$Q_3 = 0.32 \,/\, 0.44 = 0.73\ m^3/s = 25.78\ ft^3/s$$

$$W = h_{01} - h_{03} = C_p\,(T_{01} - T_{03}) \qquad (5)$$

$$H_{ad} = W \,/\, g = C_p\,(T_{01} - T_{03})\,/g \qquad (6)$$

$$H_{ad} = 1004(722\text{-}528)\,/9.81 = 19{,}854.84\ m$$

$$= 65{,}140.55\ ft$$

From equation (1),

$$N_s = 47{,}000\ (15.78)^{1/2} \,/\, (65{,}140.55)^{3/4} \qquad (7)$$

$$N_s = 58.53$$

According to the specific speed chart, the specific diameter is equal to 1.8. Then

$$D = D_s\,(Q_3)^{1/2} \,/\, (\Delta H_{ad})^{1/4}$$

$$D = 1.8\ (25.78)^{1/2} \,/\, (65{,}140.55)^{1/4}$$

$$D = 0.57\ ft = 174\ mm.$$

$$U = \Pi ND/\ 60 \qquad (8)$$

$$U = \Pi\cdot 47{,}000\cdot 174 \,/\, (60\cdot 1000) = 427.98\ m/s$$

From S.L.Dixon, the optimum ratios of the tip diameter to the turbine diameter and the hub diameter to the tip diameter for microturbine are

$$D_t/D_2 < 0.7 \quad and \quad D_h/D_t > 0.35 \qquad (9)$$

This research paper uses

$$D_t/D_2 = 0.65 \quad and \quad D_h/D_t = 0.35$$

Therefore, the D_t and D_h are equal to 113.1 mm and 40 mm, respectively.

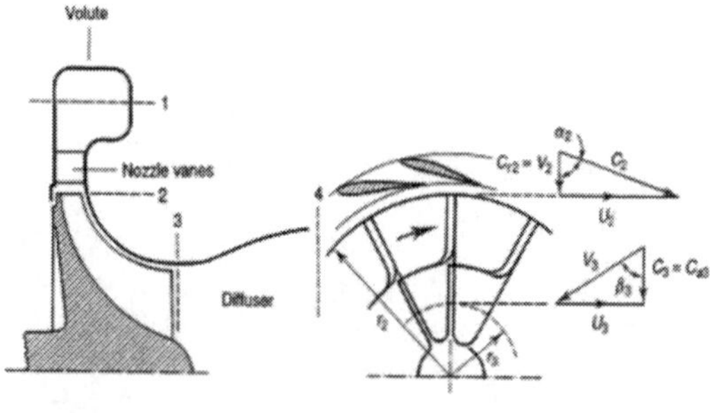

Figure 2. Turbine velocity diagram.

Assume that the outlet gas flows are uniform,

$$U_{3,mean} = (2\Pi N/\ 60)\ r_{3,mean} \qquad (10)$$

$$U_{3,mean} = (2\cdot\Pi\cdot 47{,}000/60)\cdot 38.25/1000 = 188.65\ m/s$$

The turbine outlet area are then calculated, the blade thickness is negligible,

$$A_3 = \Pi(D_h^{\,2} - D_t^{\,2})\,/4 = 8.77x10^{-3}\ m^2$$

The axial outlet gas velocity are calculated as

$$C_3 = Q_3 \,/\, A_3 = 0.73 \,/\, (8.77x10^{-3}) = 83.26\ m/s$$

 DOI: 10.1051/matecconf/20179506005

Then the rotor outlet angle ($\beta_{2,mean}$) can be calculated as

$$\beta_{3,mean} = tan^{-1}\ (U_{3,mean}/C_3) = tan^{-1}\ (188.65/83.26) = 66.19°$$

3 Computational fluid dynamics (CFD)

The control volume is separated into two parts with the interface surface is between. The first part is a stationery part: the nozzle vane, as shown in Fig. 3. This nozzle vane composed of 11 air foils NACA with an exit angle of 58°.

Figure 3. Computational domain of the nozzle vane.

The second part is a moving part: the turbine impeller, as shown in Fig. 4. The designed parameter is based on the value from the previous section. The turbine impeller is also composed of 11 blades. With the reason of the strength of material, the blade thickness at the hub and the tip are designed to be equal to 5 mm and 2 mm, respectively.

Figure 4. Computational domain of turbine impeller.

Fig. 5 shows the computational grid of both two zones. Each grid is in a shape of tetrahedral grid and hexahedral grid in the near wall area (the boundary zone). The computational domain of nozzle vane is divided into 226,923 elements and the computational turbine impeller is divided into 327,076 elements. The grid independent process ensure that the number of grid in both two domains are appropriate.

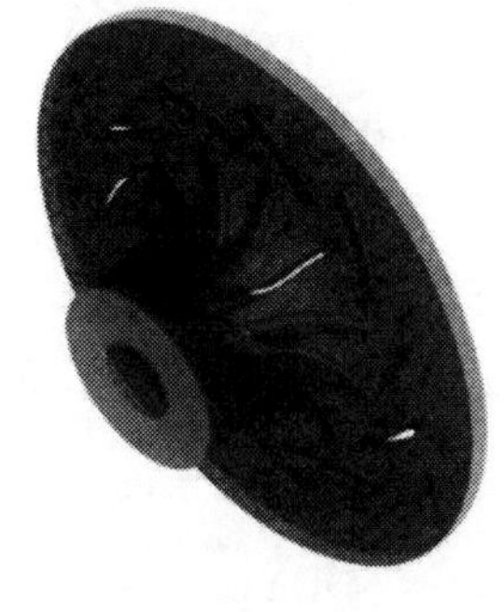

Figure 5. Computational grid.

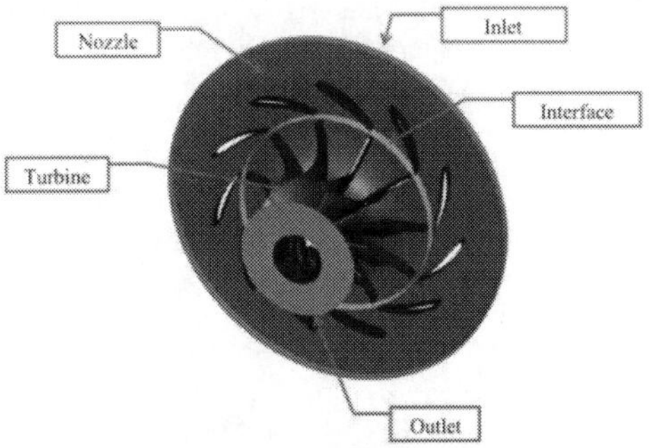

Figure 6. Boundary condition.

The boundary condition surface, Fig. 6, for numerical method is set up as shown in table 1 and the physical set up is shown in Table 2.

Table 1. Boundary condition setup

Position	Boundary condition
Inlet	Pressure ratio = 3 and 747°c
Outlet	Pressure outlet at 0.5 bars and 528°c
Blade and hub	Wall sliding mesh at 47,000 rpm, adiabatic wall
Nozzle	Stationary wall, adiabatic wall
Side shroud	Stationary wall, adiabatic wall

Table 2. Physical setup

Solver setting	Pressure-based
Discretization	Monotone upstream – Centered schemes for conservation law (MUSCL)
Interpolation methods	Green-Gauss Node-Based
Interpolation methods for face pressure	PRESTO
Pressure velocity coupling	Pressure-Implicit with splitting of operators (PISO)
Convergence residuals	Continuity equation = 10^{-5} Momentum equation = 10^{-5} Energy equation = 10^{-7}

DOI: 10.1051/matecconf/20179506005

4 Results and Discussions

The numerical solution from a time-dependent calculation at the time step is equal to 1×10^{-6} sec. The residual solution for continuity and momentum equation are below 10^{-5} and for the energy equation is below 10^{-7}. Flow time is equal to 0.01 sec. The solutions obtained from the numerical method are presents as shown in Fig. 7 to Fig. 11.

Fig. 7 shows the variation of air density. The maximum density occurs at the inlet and gradually decreases from the inlet to the outlet. Figure 8 shows that the Mach number is below the supersonic flow where the physical properties of fluid are rapidly changes. In other words, the flow is acceptable.

The pressure distribution, in Fig. 9, is corresponding to the density distribution. The maximum pressure occurs at the inlet. The temperature is about 540°c at the outlet. The specified temperature at the inlet and outlet are distributed as shown in figure 10. It can be seen that there is no heat accumulate within the domain. From the above CFD results, there is no unusual phenomenon. The airflow path in Fig. 11 emphasize that there is no separation flow in the nozzle blade and impeller blade which is the main reason of the flow to be blocked and also decrease the flow channel. This leads to the decrease of mass flow rate and turbine efficiency.

Figure 7. Density distribution.

Figure 8. Mach number.

Figure 9. Pressure distribution.

Figure 10. Temperature distribution.

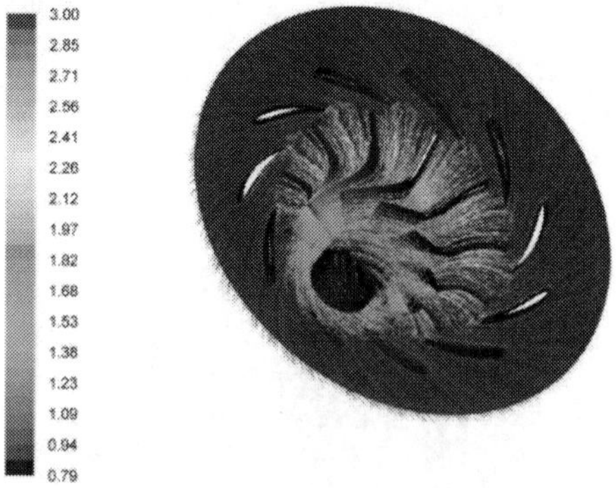

Figure 11. Air flow path depend on pressure.

5 Conclusions

This research paper focuses on the design of radial inflow turbine that operates in 30 kW microturbine. In order to operate the 30 kW microturbine with the back work ratio of 0.5, the radial inflow turbine should be designed to produce power at 60 kW. The radial inflow turbine is designed based on the turbo machinery theory. With the help of the analytical methods, the design parameters are derived. The results are constructed in 3D geometry and the 3D fluid-geometry is validated by computational fluid dynamics (CFD) simulation.

According to the CFD investigation, it shows that the temperature distribution and the pressure distribution at the surface of the nozzle and turbine impeller are simultaneous. The simulation results also show the smoothness of the airflow path in the designed radial inflow turbine.

As can be seen from the figures, there is no flow blockage between vanes. The Mach number value is between 0 and 0.8, which is acceptable. This means that there is no shock flow occurs. In an engineering point of view, it can be concluded that the designed radial inflow turbine works well with the 30 kW microturbine.

References

1. C.F. McDonald, Applied Thermal Engineering, Vol.20(2000) 471-497.
2. S.Vasmate, Int.J.Mech.Eng.& Rob, Vol.3, No.4 (2014)
3. R.H. Aungier, Turbine Aerodynamics: Axial-Flow and Radial-Inflow Turbine Design and Analysis (ASME Press, 2006)
4. D. Japikse and N.C.Baines. Introduction to Turbomachinery (Oxford University Press, 1994)
5. S.L.Dixon. Fluid Mechanics, Thermodynamics of Turbomachinery (Pergamon Press, 1978)

MATEC Web of Conferences **95**, 06006 (2017)
ICMME 2016
DOI: 10.1051/matecconf/20179506006

Implementation of Multi-Body Dynamics Simulation for the Conveyor Chain Drive System

Chakrit Suvanjumrat , Wasan Suwannahong and Sommart Thongkom

[1]*Department of Mechanical Engineering, Faculty of Engineering, Mahidol University, Salaya, Nakorn Pathom, 73170, Thailand*
[2]*Laboratory of Computer Mechanics for Design (LCMD), Department of Mechanical Engineering, Faculty of Engineering, Mahidol University, Salaya, Nakorn Pathom, 73170, Thailand*

Abstract. The conveyor chain drive system which included four sprockets, four flat bars (chain guides) and one conveyor chain had been assembled into commercial software, MSC.ADAMS. The conveyor chain was connected with rollers for running on the chain guides. The multi-body dynamics (MBD) simulation was implemented to study the large displacement of chain components which happened during rotations of the symmetric model of the conveyor chain drive system. The physical experiment had obtained by construction of the conveyor chain drive system regarding the MBD model. The speeds of chain drive were adjusted by a gear motor with a converter. The trajectory of conveyor chain links which respectively meshed and ran on sprockets and chain guides was recorded by a high speed camera. Furthermore, impact-contact forces by a collision of components in the conveyor chain drive system during rotation were analysis. The comparison between MBD simulation and physical experiment of the conveyor chain drive system was performed for validation of simulation models. The MBD simulation results were in good agreement with the experimental data which obtained an average error of 3.95%.

1 Introduction

The conveyor chain has used in a long distance of producing lines to transport products or parts which have a heavy load. This conveyor chain has many components which build up to move and hold products. The contact-impact between conveyor chain and other components in producing line always happen which induces noise and fatigue damage of the conveyor chain system. The conveyor chain in producing line is an enormous system, therefore, complex to analyze with simulation method such as finite element analysis (FEA). The MBD simulation is widely used to model the different body motion for each component of the large system [1], [2]. The constraints which relate with bodies are used to limit motions of the system. The contact-impact constraint is a major condition to detect the precise of impact and to achieve the contact-impact force by MBD simulation [3]. The dynamic characteristics of movable components in systems can be investigated. The dynamic path, force and moment are the whole results which achieve by the MBD simulation to improve rigid components of the dynamic system [4], [5].

In this research, MBD simulation for the conveyor chain system had performed by using MSC.ADAMS software. The trajectory of MBD simulation was validated with the physical experiment. The impact-contact force which happened between contact components was studied. The simulation of the conveyor chain drive system will be used to design and analyze the larger conveyor chain drive system of the rubber glove production line in the further works.

2 Theory of multi-body dynamics

Multi-body dynamics system is a whole system composed of multiple bodies which are combined by the series of geometry constraints to calculate an expected motion. The bodies in multi-body dynamics system concern rigid bodies to form equation of motion by the Lagrangian equation [6]. The dynamic equation of each body of the system is given by:

$$\frac{d}{dt}\left(\frac{\partial L}{\partial \dot{x}_k}\right) - \frac{\partial L}{\partial x_k} = F_k;\ k = 1, 2, 3, \cdots, n \quad (1)$$

where

$$L = T - U \quad (2)$$

where x_k is the generalized coordinate of the k-th body, F_k is the generalized force on k-th body, n is number of bodies, L is the Lagrangian function, T is kinetic energy, U is potential energy.

The constraint equation relates with bodies is given to assemble bodies by:

$$C_i(x, t) = 0;\ i = 1, 2, 3, \cdots, m \quad (3)$$

where m is the number of constraints.

The dynamic equation of system is expressed in the matrix form by:

$$\frac{d}{dt}\left(\frac{\partial T}{\partial \dot{x}_k}\right) + Kx_k = F_k \quad (4)$$

where $\boldsymbol{K}$ is the rigidity matrix. $\boldsymbol{F}_k$ is the force vector.

The bodies are assembled to a system through constraints. The differential equation of the system is generated by the Lagrange multiplier method which written by matrix form as:

$$M\ddot{x}_k + Kx_k + J_C^T\lambda = F_k \quad (5)$$

where $\boldsymbol{M}$ is the mass matrix. $\boldsymbol{J}_C$ is the Jacobian matrix of constraints. This dynamic governing equation of the system will solve by the numerical technique through MSC.ADAMS in this research.

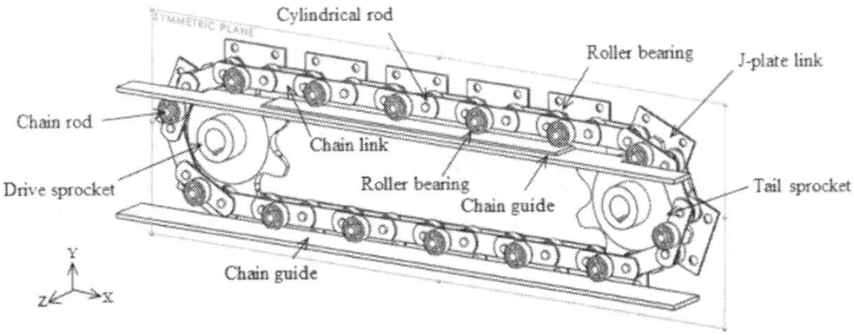

Figure 1. The MBD model of the conveyor chain drive system.

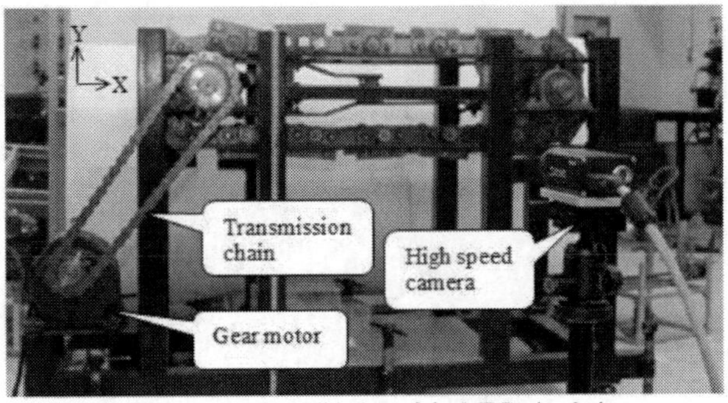

Figure 2. Apparatus of the conveyor chain drive system for validation of the MBD simulation.

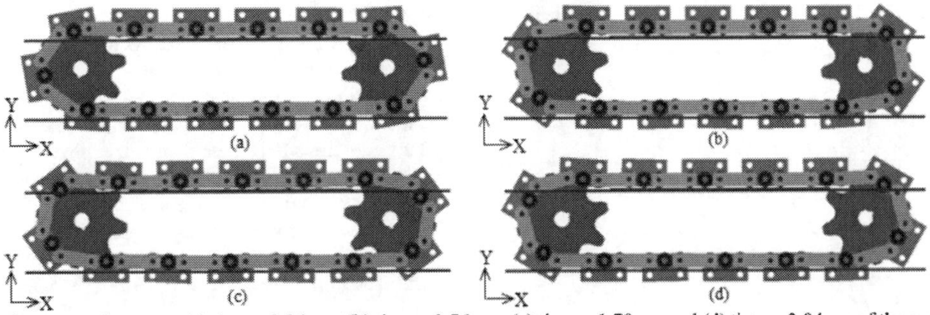

Figure 3. The sequent image at: (a) time = 0.06 sec, (b) time = 0.76 sec, (c) time = 1.78 sec and (d) time = 3.04 sec of the conveyor chain drive system rotating with the rotational speed of 3.40 rad/s.

3 Multi-Body dynamics simulation

The computer aided design (CAD) model of the conveyor chain drive system was created by SolidWorks software which comprises of sprockets, conveyor chain and chain guides. This CAD model was imported into MSC.ADAMS to create the MBD model for the first step of the modeling process. Fig. 1 shows the MBD model of the conveyor chain drive system. The symmetric plane was used to model the conveyor chain drive system to reduce the processing time. The conveyor chain components were assembled including roller bearings, cylindrical rods, chain links and J-plate links. Total components of the symmetric conveyor chain model which wrap around two sprockets were 126 pieces. The conveyor chain which was modeled for the MBD simulation composed of 28 links. The roller bearings which wore chain rods would roll on the chain guide. The contact-impact constraint between components such as roller bearings and chain guides, cylindrical rods and J-plates, roller bearings and sprocket tooth and chain links and roller bearings were carefully built to prevent the software to automatically detect and remove any redundant constraints. The input of the MBD model was the rotational speed of the driving sprocket. The 8-teeth sprocket in front of the conveyor chain drive system drove the chain in the clockwise direction with the angular velocity of 1.46, 2.33 and 3.40 rad/s. The component of the conveyor chain drive system had been used material density to specify weight by multiplying with the component volume and the gravitational acceleration. The personal computer with core-i5 CPU and RAM memory of 4 GB was used to perform the MBD simulation. The roller bearing which meshed sprocket tooth was calculated the trajectory to compare with the experimental data. The trajectory of roller bearing which rolled on chain guides also was calculated and corrected to compare with the physical experiment. The contact-impact force which immediately happened when bearings contacted and impacted with chain guides and sprocket tooth was gathered for calculation to analyze the dynamic appearance of the conveyor chain drive system.

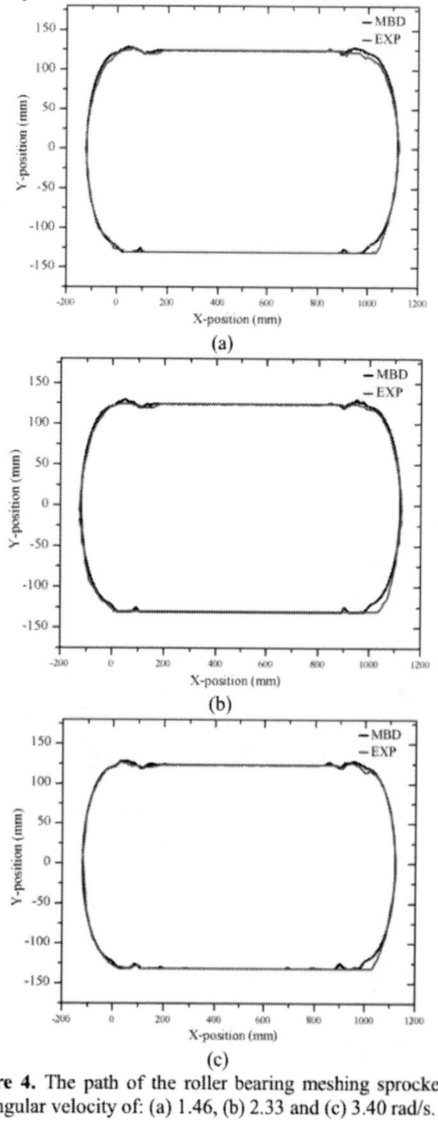

Figure 4. The path of the roller bearing meshing sprockets at the angular velocity of: (a) 1.46, (b) 2.33 and (c) 3.40 rad/s.

4 Experiment

The conveyor chain drive system is constructed and shown in Fig. 2. The eight-leg frame which built by the 4″×4″ square steel tube was supported fully components of the conveyor chain drive system. The driving sprocket was driven by a gear motor through a small chain. The motor speed was controlled by a motor converter which installed in a control box. The motor speed had been adjusted to rotate drive sprockets according to the angular velocity of the MDB simulation of the conveyor chain drive system for investigating the variable dynamic behavior of the conveyor chain. The high speed camera (BASLER Scout scA640-70fm) was set up to record the conveyor chain driving by using a frame rate and image resolution of 71 fps and 659×494 pixels, respectively.

5 Results and discussion

The conveyor chain drive system was controlled to rotate with the different rotational speed. Fig. 3 shows sequent images of the conveyor chain drive system rotating with the rotational speed of 1.23 rad/s. The path of a center position of roller bearings which are meshed sprockets has been plotted to compare with experimental data as shown in Fig. 4. The X-Y path of roller bearings which run on chain guides is plotted by comparison graphs between MBD simulation and physical experiment as in Fig. 5. The oscillation of roller bearing was increased following the angular velocities. The comparison shows the MBD simulation result was in a good agreement with the experimental data. The path of bearings is used to calculate distance from a center point of bearings to a center point of the driving sprocket by:

$$D_c = \sqrt{X^2 + Y^2} \tag{6}$$

where X is the distance of bearings from the center point of the driving sprocket in x-direction, Y is the distance of bearings from the center point of the driving sprocket in y-direction. The error of the MBD simulation was used the comparison of the distance of bearing movement between the MBD model and the physical experiment. The roller bearing of MBD model which meshed sprockets obtained an average error of 4.73, 3.47 and 3.11% while roller bearing which ran on chain guides obtained an average error of 2.50, 5.70 and 4.22% under the rotational speed of 1.46, 2.33 and 3.40 rad/s, respectively.

The maximum contact-impact force occurred on the bearings which was connected to chain links and meshed with the sprocket tooth. The magnitude of contact-impact force which varies along the rotational time of the conveyor chain drive system can express by the graph as shown in Fig. 6. The magnitude of contact-impact force was increased while the time period was reduced when increased angular velocities of drive sprockets. The contact-impact forces on chain guides were very less than on sprockets. Fig. 7 shows contact-impact forces of bearings which roll on the chain guides.

DOI: 10.1051/matecconf/20179506006

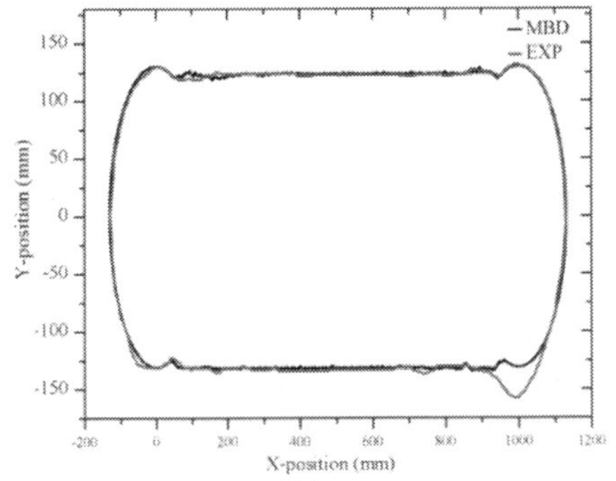

(a)

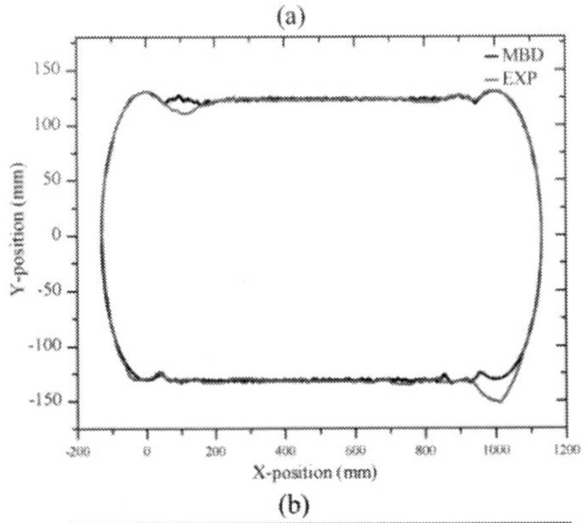

(b)

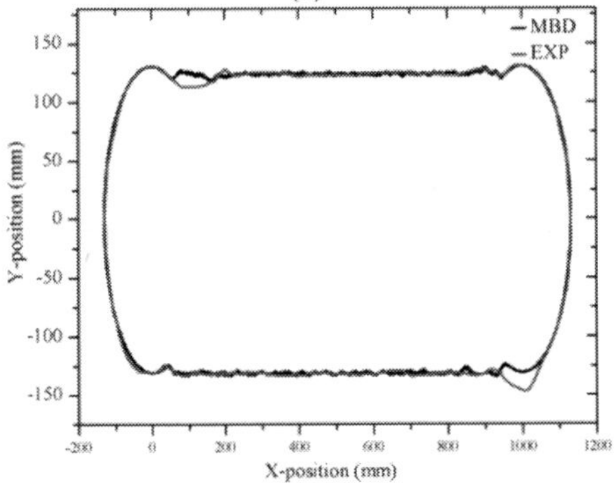

(c)

Figure 5. The path of the roller bearing running on chain guides at the angular velocity of: (a) 1.46, (b) 2.33 and (c) 3.40 rad/s.

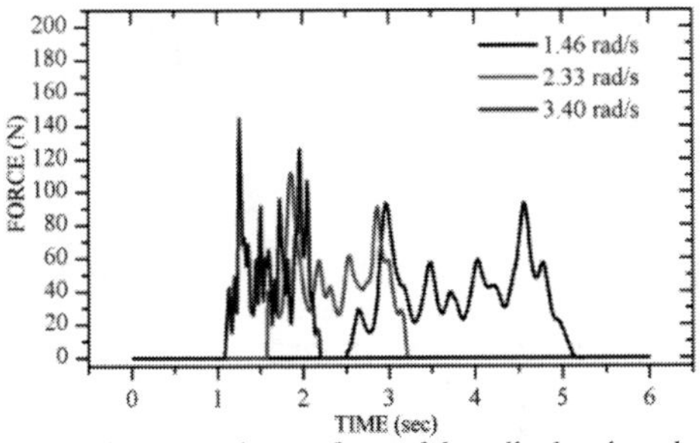

Figure 6. The contact-impact force of the roller bearings during meshing with the sprocket tooth.

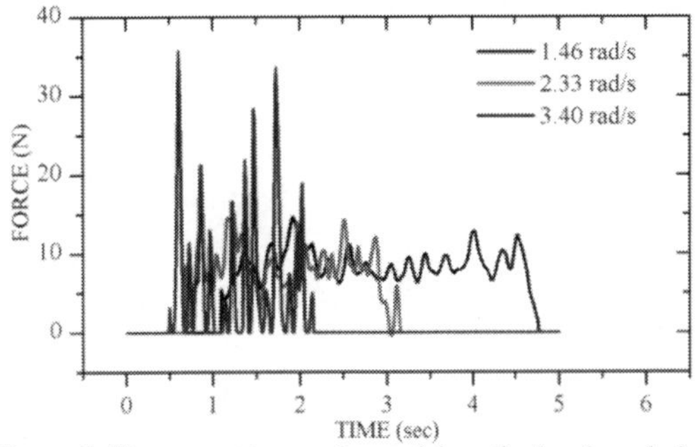

Figure 7. The contact-impact force of the roller bearings during running on the chain guides

6 Conclusions

The MBD simulation of the conveyor chain drive system was performed by using the commercial software, MSC.ADAMS. The physical experiment of the conveyor chain drive system had been constructed to validate the MBD simulation results. The path of the bearings which was the component of the system had been used to validate by comparison between simulation results and experimental data. The MBD simulation results were in a good agreement with experimental data and obtain an average error less than 3.95%. The simulation results shown the oscillation of components increased by increasing of the angular velocities of the drive sprockets. The MBD simulation not only calculated path of components but also the contact-impact force between components could also achieve. The large contact-impact force happened on sprockets of the conveyor chain drive system. The MBD simulation would be useful and advance to design and calculate the strength of the conveyor chain drive system for the producing line in a further work.

References

1. C.J. Park, G. Gschwendtner, J. Mech. Sci. Technol. **29**, 7 (2015)
2. Y. Li, A.M. Castro, T. Sinokrot, W. Prescott, P.M. Carrica, Renew. Energ. **76** (2015)
3. P. Flores, J. Ambrosio, Multibody Syst. Dyn. **24** (2010)
4. M. Schwaze, C. Hurschler, F. Seehaus, S. Oehler, B. Welke, J. Biomech. **46** (2013)
5. E. Abele, J. Bauer, T. Hemker, R. Laurischkat, H. Meier, S. Reese, O. von Stryk,CIRP-JMST. **4** (2011)
6. F.E. Udwad, A.D. Schutte, Acta. Mech. **213** (2010)

DOI: 10.1051/matecconf/20179506007

Analysis of Roller Chain Drive System with Multi-Flexible Body Dynamics Methodology

Wasan Suwannahong and Chakrit Suvanjumrat

Department of Mechanical Engineering, Faculty of Engineering, Mahidol University, Salaya, Nakorn Pathom, 73170, Thailand
Laboratory of Computer Mechanics for Design (LCMD), Department of Mechanical Engineering, Faculty of Engineering, Mahidol University, Salaya, Nakorn Pathom, 73170, Thailand

Abstract. Finite element method (FEM) was employed to analyze the roller chain drive system which complicated to determine various design factors. The multi-flexible body dynamics (MFBD) method was proposed to model and analysis the roller chain drive system. The roller chain which composed of plates, pins and rollers was meshed with solid elements under the convergence test. The eight tooth sprocket was modeled by rigid body to roll and contact with the roller chain. The dynamic load distribution of the roller chain on the rigid sprocket was compared with the analytical solution. The MFBD simulation results had a good agreement with analytic results which obtained an average error of 12.32%.

1 Introduction

The roller chain has an important role as to transport heavy load of products in along distance of producing line that make wear in chain and sprocket, fracture in chain and drive shaft, chain climbing on sprocket, loud noise chain and elongation of chain. These problems are difficult to predict for design and preventive maintenance of the roller chain to protect the breakdown of the producing line. Many factors consist of load, chain tension, contact force between chain and sprocket, useful life, fatigue, alignment and pretension were considered for the chain design [1]. Unfortunately, the roller chain drive system has a lot of components and large movements therefore the design of roller chain was difficult and complicate by using analytical method.

In recent years, finite element method (FEM) was conducted to analyze the roller chain of many transmission applications [2]-[4]. There was the simple methodology to calculate strength of chain components by static analysis but dynamic effects were still indistinct to determine by FEM. The stress and strain which accurred by the element model were deformed, were the finite element simulation results. The multi body simulation (MBS) had been used for dynamic simulation of the chain drive system. The MBS result obtained chain velocity, impulsive and meshing area because the rigid body of chain components was employed for simulation [5], [6]. To calculate load and strength on the dynamic behavior chain simultaneously, the integration between FEM and MBS which known as the multi-flexible body dynamic (MFBD) had developed [7]-[9]. This research had apllied MFBD to analyze the roller chain drive system. The MFBD methodology would be useful to design and predict lift time of the roller chain for the producing line in a further work.

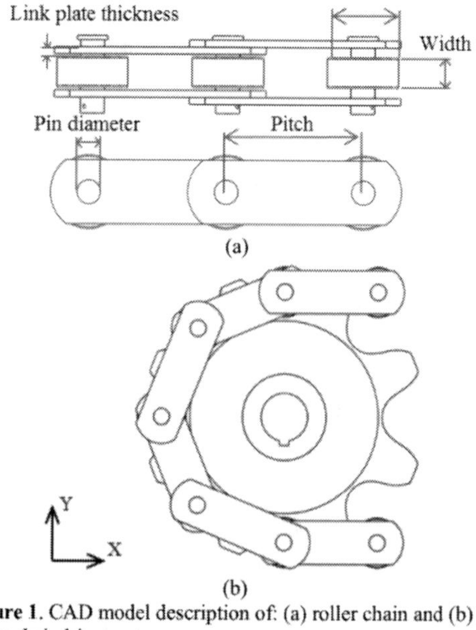

Figure 1. CAD model description of: (a) roller chain and (b) roller chaindrive system

2 Analytical method

The dynamic load distribution in the roller chain was calculated following assumptions: (a) the pitch of the roller chain was equal to the pitch of sprocket; (b) all

DOI: 10.1051/matecconf/20179506007

rollers contacting was seated on the sprocket tooth; (c) no mechanical tolerances were present; and (d) the chain weight was negligible [10]. The component of the roller chain which was included in analytical method composed of cylindrical pins, link plates and rollers. The roller chain had pitch, link plate thickness, roller diameter, pin diameter and width of 100.00, 5.00, 52.00, 17.00 and 22.00 mm respectively (Fig. 1a). The sprocket tooth profile was modeled by parameters which comprised of a pitch diameter of 100 mm, a tooth thickness of 43 mm, a dedendum of 26.13 mm, an addendum of 20.05 mm and seating curve diameter of 52 mm [11]. The roller chain which engaged with an eight-tooth sprocket was illustrated by computer aided design (CAD) model in Fig. 1b. The roller and sprocket of the roller chain drive system had been depicted by the free body diagram as shown in Fig. 2. The sprocket was rotated in a clockwise, therefore the tension T_t and T_s was the tight and slack side of chain tensions (T_i) respectively. The rollers were in the seated position and contacted the tight flank of the sprocket tooth, therefore tooth loads (P_i) were acted on rollers and passed through center of rollers. The chain rollers were numbered i started from the tight side to the slack side or the counterclockwise direction. The pressure angle (θ) and articulation angle (α) also were described in Fig. 2. The link of the tight and slack strand made an angle ξ and ε respectively. The effect of friction force (F_i) had changed the pressure angle by a friction angle (δ). The free body diagram of chain rollers (Fig. 3) under equilibrium consideration had relative equations between tooth load and chain tension assuming as following:

$$P_i = \frac{T_{i-1} - T_i cos\alpha}{cos(\theta_i - \delta_i)}, i = 1,2,3,\dots,n \quad (1)$$

$$P_i = \frac{T_i - T_{i+1} cos\alpha}{cos(\theta_{i+1} - \delta_{i+1})}, i = 1,2,3,\dots,n \quad (2)$$

$$P_i = \frac{T_i sin\alpha}{sin(\theta_i - \delta_i)}, i = 1,2,3,\dots,n \quad (3)$$

$$P_{i+1} = \frac{T_{i+1} sin\alpha}{sin(\theta_{i+1} - \delta_{i+1})}, i = 1,2,3,\dots,n \quad (4)$$

$$T_0 = T_t \frac{sin(\theta - \delta_1)}{sin(\theta + \xi - \delta_1)} \quad (5)$$

$$T_s = T_n \frac{sin(\theta - \delta_1)}{sin(\theta + \varepsilon - \delta_1)} \quad (6)$$

$$\delta = tan^{-1}(\mu_v) \quad (7)$$

where μ_v is the dynamic coefficient.

3 Multi-Flexible body dynamics method

The CAD model of roller chain components had been meshed with solid elements by using commercial software, MSC.Patran. The flexible roller chain under the convergence test was obtained 173,298 elements. The material properties of roller chain composed of elastic modulus, poisson ratio and density at 210 GPa, 0.3 and 7,800kg/m^3 respectively. The Craig-Bampton modal synthesis method in MSC.Nastran was performed to obtain model neural file (MNF), a binary file relating to a flexible body, of chain components. A sprocket surface was defined by a rigid body and assembled to MNF of roller chain components in MSC.Adams. The roller chain drive system model which was generated in MSC.Adams shown in Fig. 4. The four flexible chian links were meshed to an eight-tooth sprocket on the initial position that made angle between the tight chain link and slack chain link at 0 degree.

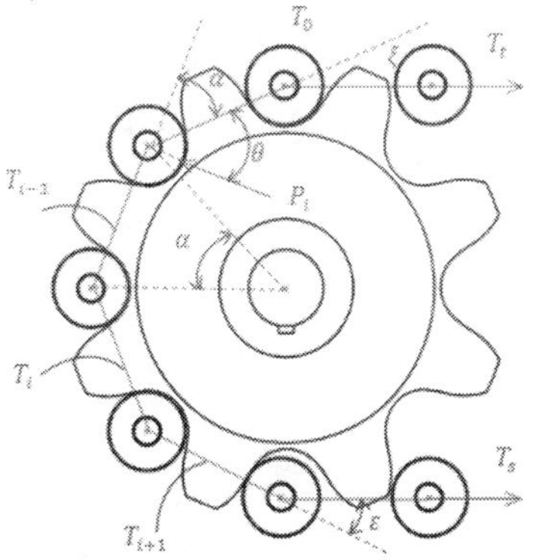

Figure 2. Free body diagram of chain rollers seating on a sprocket.

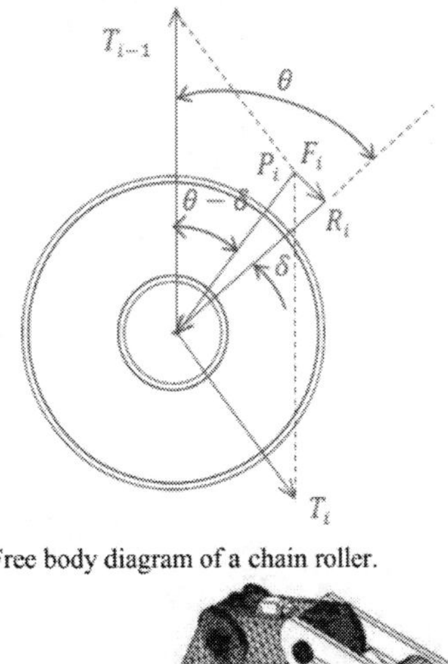

Figure 3. Free body diagram of a chain roller.

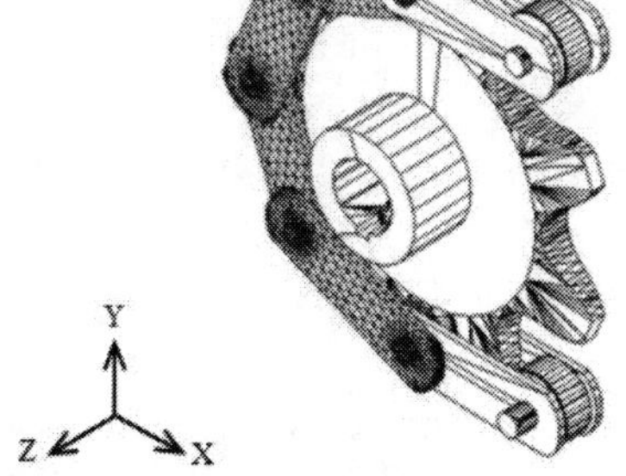

Figure 4. FE model of roller chain engaing with CAD model of a sprocket.

The tight and slack chain link still were the rigid body because they were specified velocity and tension force (assumption of chain weight) respectively. This study case shown that the tight chain link had velocity of 400 mm/s and the slack chain link had load of 2,000 N. The contact parameter was defined on contact of 1 x 10^5 N/mm, a force exponent of 2.2, a damping of 10 Ns/mm and a penetration depth of 0.1 mm. The Gear Stiff Integrator (GSTIFF) which was solver in MSC.Adams [12] employed to calculate dynamic response of the roller chain drive system.

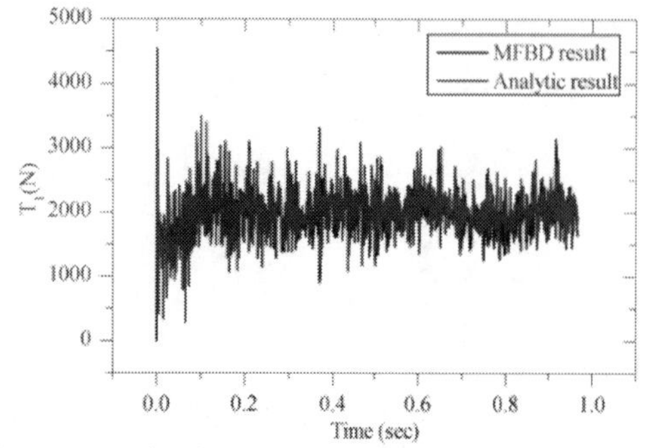

Figure 5. Tension force of chain link no. 1.

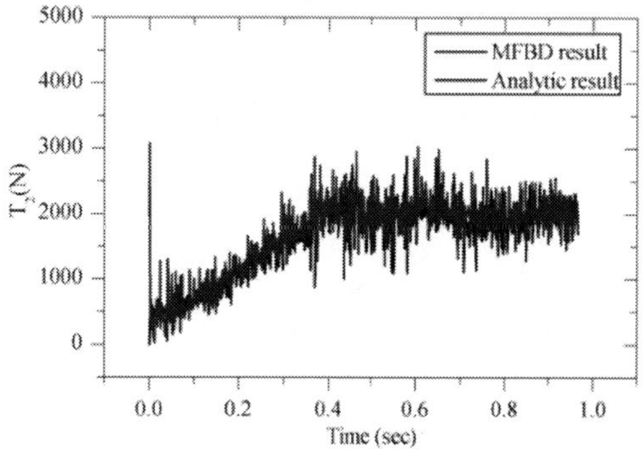

Figure 6. Tension force of chain link no. 2.

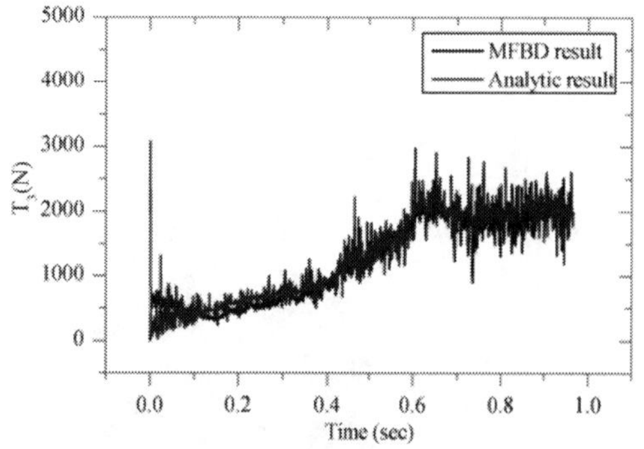

Figure 7. Tension force of chain link no. 3.

4 Results and discussion

The dynamic tension force result by using the MFBD simulation was compared with the analytical result. The comparison of tension forces T_1 to T_4 (starting from flexible link of tight side) were shown in Fig. 5 to Fig. 8 respectively. The tension force of the first flexible chain link was started about 2,000 N from the initial time because of the initial condition calculation. The second chain link had the maximum tension force led the third flexible chain link of 0.2 sec. The fouth link supported the maximum force of 2,000 N when the rotating time was 0.85 sec. All of chain links supported the maximum tension force when rotated at twelve o'clock of a sprocket. The magnitude of simulated tension force T_1, T_2, T_3 and T_4 which happened at the same time of the analytic solution had an average error of 9.99, 15.87, 13.62 and 9.82% respectively. The errors might be that the assumption of analytic method was negligible the chain weight therefore it had value less than the simulation result. The chain moving related with the sprocket rotating shown by sequent images in Fig. 9. The results of the MFBD simulation could be illustrated by stress contour on a flexible roller chain by red; maximum von mises stress and blue; minimum von mises stress that occurred when the flexible chain was deformed by tension and impact force. The maximum stress always occurred on chain links and rollers when it rotated at twelve o'clock position of a sprocket. This position happened the maximum tension force on chain links. The area where near the pin holes of chain links occured red color when rotated to twelve o'clock position of a sprocket. The von mises stress had a peak value and reduced after chain links left of twelve o'clock position. This phenomenon that happened on the chain links was induced the cyclic stress and strain on the chain and the sprocket system.

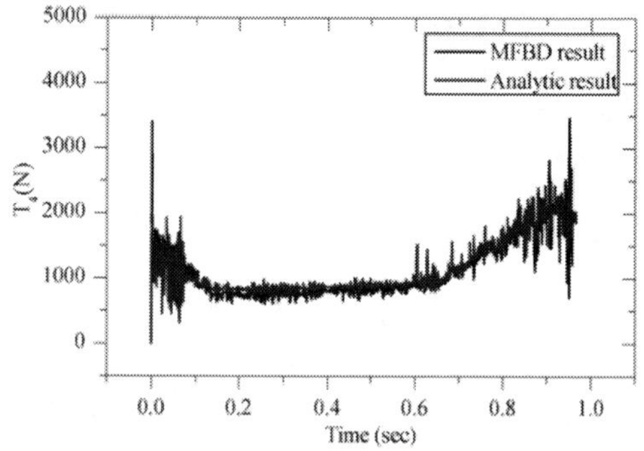

Figure 8. Tension force of chain link no. 4.

The chain roller impacted sprocket tooth during a sprocket drove a roller chain. The impact force between rollers of chain and sprocket tooth which occurred on the second roller of the tight side were indicated by P_1. The third roller to fifth roller were indicated by P_2 to P_4, respectively. The dynamic force P_1, P_2, P_3 and P_4 during rotation of the sprocket shown by a graph in Fig. 10. The impact force was periodic and increased to the maximum value of 2,337.45 N when rollers fully seated on the sprocket tooth and decreased almost to zero when rollers left the sprocket. The speed of a roller chain increased the impact force which was more than the drag load on the slack chain link of 1.17 times. The graph of impact force of roller no. 5 (Fig. 10) shown the cyclic form distincly because it was started from six o'clock to twelve o'clock position of the sprocket for the MFBD simulation. This impact force was a good result for leading to calculate strength of sprocket tooth.

DOI: 10.1051/matecconf/20179506007

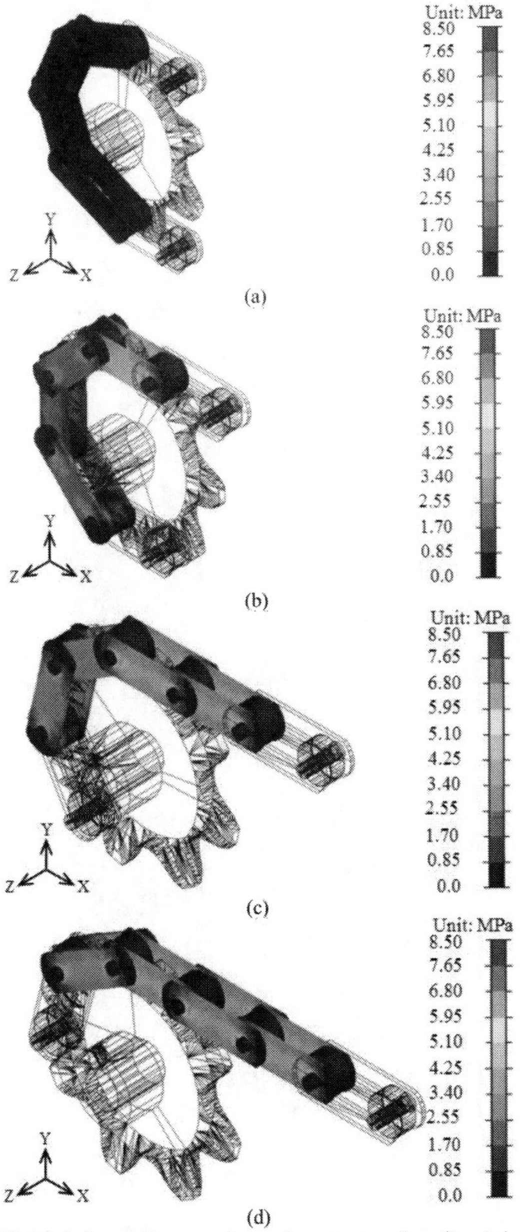

Figure 9. Sequent images of von mises stress on the roller chain driving at: (a) time = 0.0000 sec, (b) time = 0.1200 sec, (c) time = 0.4450 sec and (d) time = 0.7050 sec.

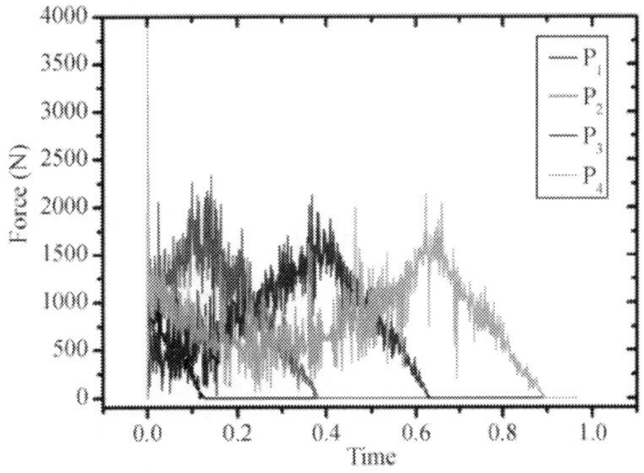

Figure 10. Impact force P_1, P_2, P_3, and P_4 of roller no. 2, 3, 4 and 5, respectively.

5 Conclusions

The dynamic load distribution in a roller chain drive system was simulated with integration of FEM and MBS. The MFBD method result had been compared with analytical method and obtained an average error of 12.32%. The impact force and von mises stress were the advance result of the MFBD method which could be used to analyze chain strength under dynamic simulation. The magnitude of maximum force and the failure limit of the roller chain were investigated obviously. The MFBD simulation would be useful and advance to design and predict strength of the roller chain drive system for the producing line in a further work.

References

1. J.L. Wright, *Standard Handbook of Chains*, second ed. (CRC Press, Florida, 2005)
2. Ç. Özes, M. Demirsoy, Compos. Struct. **69** (2005)
3. A. Bravo, D. Koffi, L. Toubal, F. Erchiqui, Eng. Fail. Anal. **58** (2015)
4. C. Yabing, G. Haitao, F. Zhenming, W. Nen, L. Lei, W. Yang, Chin. J. Mech. Eng. **28**, 1 (2015)
5. H. Junzhou, Y. Shiqiang, Y. Jing, L. Tao, The Open Mechanical Engineering J. **7** (2013)
6. C.J. Park, G. Gschwendtner, J. Mech. Sci. Technol. **29**, 7 (2015)
7. F. Lin, D. Changlong, G. Kuidong, Adv. Mat. Res. **338** (2011)
8. G. Liu, B. Xu, X. Jiao, T. Zhen, Chinese J. Aronaut. **27**, 5 (2014)
9. O. Wallrapp, S. Wiedemann, Nonliner Dynam. **34** (2003)
10. R.N. Mohammad, M.M. Kurt, Mech. Mach. Theory. **18**, 5 (1983)
11. R.G. Budynas, J.K. Nisbett, *Shigley's Mechanical Engineering Design*, ninth ed. (McGraw-Hill, New York, 2011)
12. C.W. Gear, Commun. Acm.**14**, 3 (1971)

DOI: 10.1051/matecconf/20179506008

Optical Display of Transverse Jet Flow Field Under Supersonic Inflow

Qian LI , Ji-fei Ye, Dian-kai Wang

State Key Laboratory of Laser Propulsion & Application, Equipment Academy, Beijing 101416, China

Abstract. Optical display and measurement of transverse jet flow field under supersonic inflow is performed with continuous xenon lamp color schlieren, strobo light color schlieren and strobo light monochrome schlieren. Characteristic of fuel transverse jet in scramjet combustion chamber is studied. Results indicate that wave structure can be displayed through images obtained from long time exposure of schlieren, exposure time in dozens of nanoseconds is suitable for the display of turbulent structure, and comparing with color schlieren, amounts of information obtained from monochrome schlieren are few, whose sensitivity is much lower, either.

1 Introduction

Ignition, flame holding and fuel mixing are primary problems in supersonic combustion. When vehicle flights in supersonic/hypersonic speed, velocity of gas entering into the combustion chamber is high, and the retention time is very short, so the promotion of propulsion system performance is restricted. Therefore, high performance propulsion system needs promoting mixture degree of fuel and air. Less drag loss is better with fuel jet, and favorable fuel distribution should be formed on cross section of combustion chamber.

Combustion efficiency of scramjet is directly influenced by transverse jet characteristic of fuel[1]. Study on the characteristic of transverse jet flow under supersonic condition is helpful to the structure designing of combustion chamber, confirming the igniting position and igniting mode of scramjet[2]. As a non-contacting optical display and measurement technology, schlieren would not interfere the flow field, and can reflect density variation of the fluid directly, so it is widely applied in the measurement of high-velocity flow field. Color schlieren is one of the schlieren techniques which is paid much attention to. A. K. Agrawal et al. [3-8], G. E. Elsinga et al. [9] and DLR Aerospace Center[10] are well known of their research on color schlieren technique.

For studying the distribution of fuel after it is injected into the combustion chamber and the interaction between fuel and supersonic inflow, optical display and measurement of transverse jet flow field under supersonic inflow is performed with continuous xenon lamp color schlieren, strobo light color schlieren and strobo light monochrome schlieren, so as to choose the appropriate optical display and measurement technique according to different requirements in further study.

2 Experimental system and measurement

2.1 Experimental table

Experimental table is composed of supersonic flow table and transverse jet, whose photo is shown in Figure1.

Figure 1. Photo of experimental table

Mach number of the supersonic flow is 2.5, mass flow is 0.5kg/s, test time is 1s, and the total pressure of transverse jet is 1MPa. Cross section of the transverse jet is circular, whose exit diameter is Φ2mm.

2.2 Schlieren technique

Photo of schlieren measurement system is shown in Figure 2.

Color schlieren and monochrome schlieren are both adopted in the experiment. In order to display different wave structure system, three schlieren techniques are used.

Figure 2. Photo of schlieren measurement system

3 Three schlieren techniques and optical display

3.1 Continuous xenon lamp color schlieren

For displaying the structure of shock wave, continuous xenon lamp color schlieren is used in the measurement. The exposure time is 2ms, and the experimental photos are shown in Figure 3.

(a)original image of scale

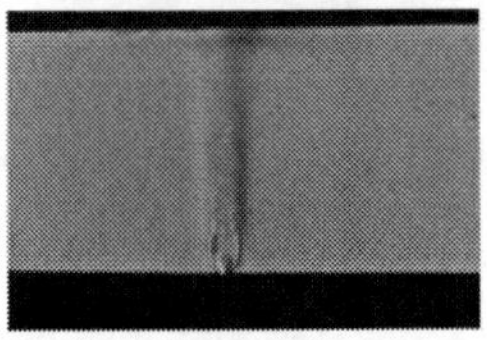

(b)original image of transverse jet

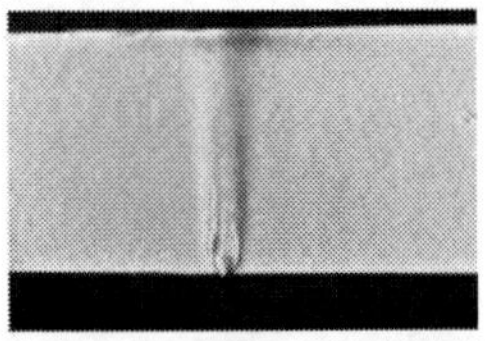

(c)image of transverse jet with hue disposal

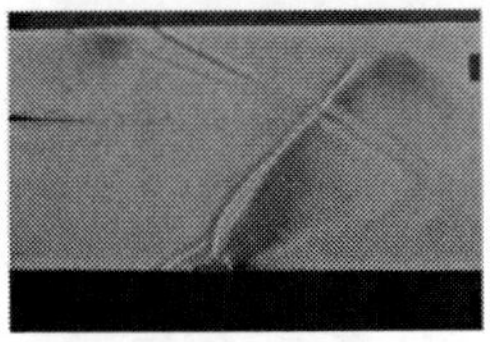

(d)original image with inflow

(e)image with greasy dirt disposal

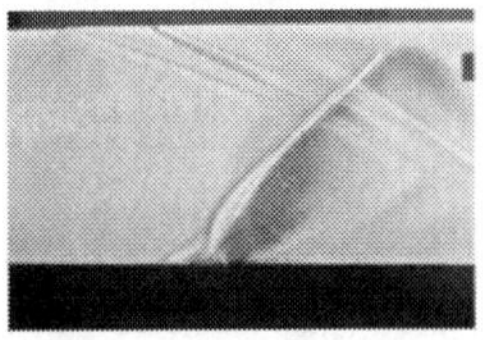

(f)image with hue and greasy dirt disposal

Figure 3. Photos taken by continuous xenon lamp color schlieren

3.2 Strobo light color schlieren

For displaying structure of turbulence, strobo light color schlieren is used in the measurement. The exposure time is 100ns, and the single flash time of light source is 20ns. Schlieren photo of transverse jet is shown in Figure 4. The photos is hue disposed. From the amplifying photo, it can be seen that the structures of turbulence and shock wave are both obvious.

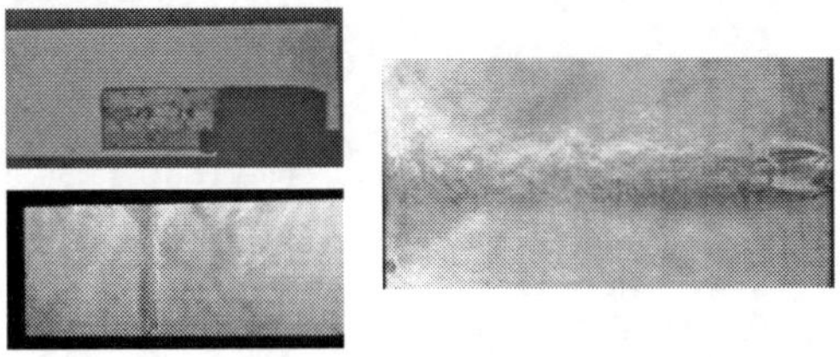

Figue 4. Schlieren photo of transverse jet

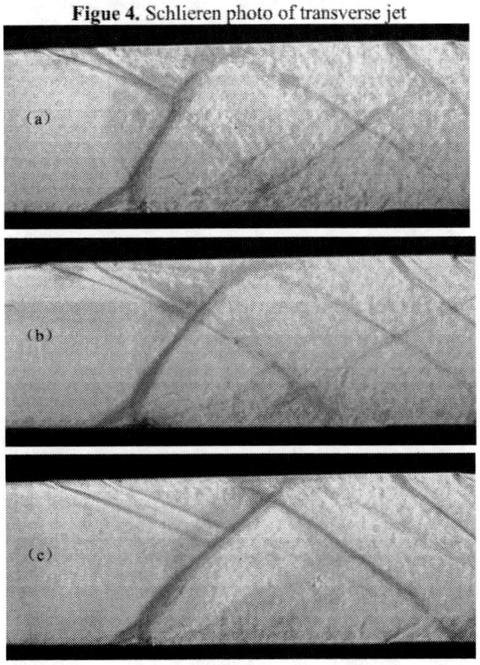

Figure 5. Schlieren photos under different inflow speeds with the start-up of wind tunnel

Schlieren photos under different inflow speeds with the start-up of wind tunnel are shown in Figure 5. From the oblique angle of the second shock wave on the upper wall and the turbulent structures after the shock, it can be concluded that the inflow speed is becoming faster from Figure 5(a) to Figure 5(c). The situation of total start-up of the wind tunnel is shown in Figure 5(c), and the mach number is 2.5. There are shock structures in Figure 5(a) and Figure 5(b), so the inflow speed is supersonic too. It is worth to notice that in Figure 5(a) and Figure 5(b), when the inflow speed is relatively low, mach disk, bulk shock and large scale turbulent structures are more obvious.

3.3 Strobo light monochrome schlieren

In measurements with strobo light monochrome schlieren, the exposure time is 20ns and high speed enhancing camera is adopted. Typical result obtained by strobo light monochrome schlieren with greasy dirt disposal is shown in Fig .6.

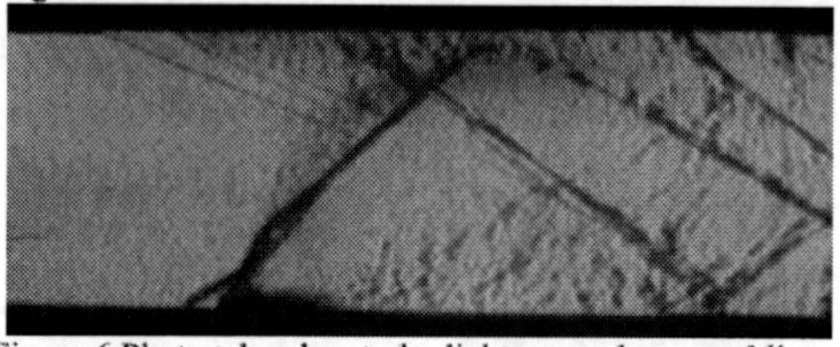

Figure 6 Photo taken by strobo light monochrome schlieren

It can be seen that there is bulk shock in Figure 6, but not obvious, and large scale vortexes are in disorder. Information obtained from monochrome schlieren is less than that of color schlieren. Comparison between typical result in this paper and that of references is shown in Figure 7.

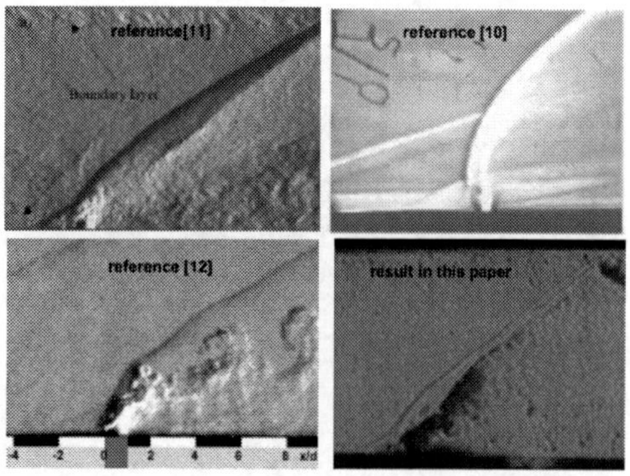

Figure 7 Comparison between our result and references

Table 1. Parameters of the experimental table in reference[11] and ours

parameters of experimental table	reference [11]	our experimental table
total pressure of free inflow	279kPa	400kPa
boundary layer thickness of free inflow	3.2mm	≈3mm
mach number of free inflow	2.25	2.5
total pressure of transverse jet	378kPa	1MPa
momentum ratio between transverse jet and free inflow	1.7	3.7
exit mach number of transverse jet	1	1
exit diameter of transverse jet	4.8mm	2mm

It should be noticed that cold flow tunnel is used in reference [11], but in reference[12] high enthalpy expanding tube is adopted(total enthalpy is 4MJ/kg and inflow mach number is 3.5), and in reference [10], exit speed of transverse jet is supersonic, not sonic. Parameters of the experimental table in reference[11] and ours are shown in Table 1.

4 Conclusion

1)Wave structures, such as bow shock, separate shock, bulk shock and mach disk, can be displayed by schlieren technique with long time exposure. But for mixed layer boundary, compressive wave, boundary layer and trail, it is unfavorable.

2)Dozens of nanoseconds exposure time of schlieren is suitable for the display of turbulent structure, especially measured by color schlieren, disordered state of turbulence can be clearly observed.

3)Compared with color schlieren, information obtained from monochrome schlieren is much less, and the sensitivity is much lower too. However, if the knife edges is more enough, theoretically to say, bulk shock structure and the mixed area can also be observed. Otherwise, difference between bow shock and mixed layer boundary can be recognized in monochrome schlieren which is difficult to be distinguished in color schlieren because of its high sensitivity.

Acknowledgement

This research was supported by the Natural Science Foundation of China (11372356, 11372357).The authors would like to acknowledge all discussions and technical help provided by the former and current members of the State Key Laboratory of Laser Propulsion and Application.

References

1. T. H. Chen, C. R. Smith and D. G. Schommer. Multi-zone behavior of transverse liquid jet in high-speed flow[R]. AIAA 1993-0453.
2. Y. H. Tong. Injection characteristic and breakup process of transverse liquid jet in crossflow[D]. Changsha: National University of Defense Technology, 2012.
3. Agrawal, K. Alammar, and S. Gollahalli. Application of rainbow schlieren deflectometry to measure temperature and oxygen concentration in a laminar gas-jet diffusion flame[J]. Experiments in Fluids, 2002, 32(6): 689-691.
4. S. Yildirim and A. Agrawal. Full-field measurements of self-excited oscillations in momentum-dominated helium jets[J]. Experiments in Fluids, 2005, 38(2): 161-173.
5. T. Wong and A. Agrawal. Quantitative measurements in an unsteady flame using high-speed rainbow schlieren deflectometry[J] Measurement Science and Technology, 2006, 17(6): 1503-1510.
6. P. Kolhe and A. Agrawal. Role of buoyancy on

DOI: 10.1051/matecconf/20179506008

instabilities and structure of transitional gas jet diffusion flames[J] Flow Turbulence and Combustion, 2007, 79(4): 343-360.

7. R. Satti, P. Kolhe, S. Olcmen, and A. Agrawal. Miniature rainbow schlieren deflectometry system for quantitative measurements in microjets and flames[J]. Applied Optics, 2007, 46(15): 2954–2962.
8. P. Kolhe and A. Agrawal. Density measurements in a supersonic microjet using miniature rainbow schlieren deflectometry[J]. AIAA Journal , 2009, 47(4): 830-838.
9. G. E. Elsinga, B. W. van Oudheusden, F. Scarano and D. W. Watt. Assessment and application of quantitative schlieren methods with bi-directional sensitivity: CCS and BOS[C]. Proceedings of PSFVIP-4 June 3-5, 2003, Chamonix, France.
10. D. Herrmann, W. W. Koschel. Experimental investigation of the internal compression of a hypersonic intake[C], 38th AIAA/ASME/SAE/ASEE Joint Propulsion Conference & Exhibit 7-10 July 2002, Indianapolis, Indiana.
11. L. Eli, G. Elliott, N. Glumac. Energy deposition applied to a transverse jet in a supersonic crossflow[R]. AIAA 2009-1534.
12. A. Ben-Yakar. Experimental investigation of mixing and ignition of transverse jets in supersonic crossflows[D]. Stanford University, 2001.

Transient Analysis of Air/Oil Two-Phase Flow in Bearing Chamber under the Periodic Boundary Conditions

Wang Fei, Guoding Chen , Yang Fu and Wang Tao

School of Mechanical Engineering, Northwestern Polytechnical University, Xi'an 710072, China

Abstract. The air/oil two-phase flow is extremely complex in the aero-engine bearing chamber, so the comprehensive understanding of the flow phenomenon is helpful to improve the design of the lubrication and heat exchange system. In this paper, transient CFD simulation is developed to analyse the air/oil two-phase flow in the bearing chamber. The velocity and temperature distributions of air and oil with time and space are discussed. The results show that under the periodic conditions, the velocity and temperature of the air and oil in the bearing chamber present quasi periodic characteristics. And the velocity and temperature distribution of air/oil two-phase flow in bearing chamber is independent with time in the bearing chamber. The research work will contribute to understand the nonlinear characteristics of air/oil two-phase flow in the bearing chamber.

1 Introduction

Design of aero-engine lubrication system and secondary air flow system depends on the understanding of complicate air/oil two-phase flow in the bearing chamber. But lubricating oil and sealing air in the bearing chamber formed a complex air/oil two-phase flow, which results the design of the lubrication system is very difficult [1]. However, scholars have carried out a lot of research on the air/oil two-phase flow and the influence of different operating conditions in the bearing chamber.

Gorse et al. [2] measure the air phase velocity field in the bearing chamber by Laser Doppler Anemometry (LDA), analyse the effect of the bearing chamber air phase swirl on the flow field distribution, and discuss the influence of different operating conditions on air phase velocity distribution. Aroussi et al. [3] measure the velocity of oil droplets and air in the bearing chamber by PIV technology, and analyse the effect of rotor speed on the movement velocity of oil droplets. Gorse et al. [4] use three-dimensional particle image velocimetry (SPIV) technology in the bearing chamber test rig to reveal the oil droplet generation mechanism of the cylindrical roller bearing and carry out experiment research, probes into the influence of bearing support structure of oil droplets flow model, and analysis the influence of rotor speed, fuel supply and flow rate on the oil droplet velocity. Rodkey et al. [5] describe the flow characteristics of oil film on the wall of the bearing chamber, and analyse the effect of rotor speed and flow rate on the oil film thickness. Chen Guoding et al. [6] discuss the effect of the structure parameters of bearing chamber and working conditions on flow patterns, reveal the pressure and velocity distribution laws under different rotating speed. Ao Wenwei et al. [7] utilize CFD software to simulate the air/oil two-phase flow in the bearing chamber, and analyses the influence of operating conditions on the flow state of the lubricating medium.

But these research works mainly focus on the constant air and oil supply, did not consider the influence of varying air and oil supply on the velocity and temperature of air/oil two-phase flow in the bearing chamber. However, in practical engineering, the air and oil supply is not constant, but change with time for special purpose. Therefore, the periodic air and oil supply with time was considered in this paper. Air/oil two-phase flow phenomena in the bearing chamber was simulated. Firstly, the geometric model of air/oil two-phase flow was built and then meshed. Secondly, the transient CFD simulation was completed by setting period boundary conditions of air and oil supply. Then the velocity and temperature field distribution of air/oil two-phase flow was analysed. Finally, the change of velocity and temperature of air/oil two-phase flow with time and space position was discussed. This work contributes to understand the nonlinear problem of air/oil two-phase flow in the bearing chamber.

2 Bearing chamber structure and air/oil physical properties

Figure 1 shows the basic structure of the bearing chamber, and table 1 lists its main structural parameters. The lubricating oil is thrown into the bearing chamber by the bearing retainer, then oil film is formed on the surface of the wall. Sealing air flows into the bearing chamber through the seal clearance. Lubricating oil and sealing air

 DOI: 10.1051/matecconf/20179506009

in the bearing chamber form a complex air/oil two-phase flow.

The physical properties of lubricating oil are described as follow: oil density ρ_L=954kg/m^3, dynamic viscosity $\mu_L=9.5\times10^{-3}$Pa·s, specific heat capacity C_L=2000 J/(kg·K), thermal conductivity rate λ_L=0.15 W/(m·K), and the surface tension coefficient σ_L=0.035 N/m; The physical properties of seal air are described as follow: air density ρ_g=2.92kg/m3, dynamic viscosity $\mu_g=1.84\times10^{-5}$Pa·S, specific heat capacity C_g=1004.4 J/(kg·K), thermal conductivity rate λ_g=0.0261W/(m·K).

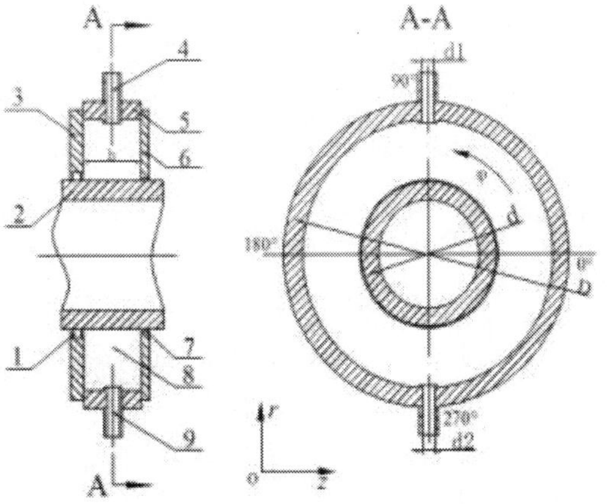

1-bearing rolling body; 2-rotator; 3-bearing seat; 4-vent pipe; 5-chamber wall; 6-end cover; 7-seal clearance; 8-bearing chamber; 9-scavenge pipe

Figure 1. Schematic of bearing chamber.

Table 1. The main structural parameters of the bearing chamber.

Parameter	value
Bearing chamber diameter D (mm)	180
Bottom Rotor diameter d (mm)	124
Bearing chamber width b (mm)	20
Vent pipe diameter d1(mm)	15
Scavenge pipe diameter d2 (mm)	15

2 Mathematical model

Lubricating Oil deposits on the wall of the bearing chamber, and the air/oil two-phase flow states of the air and oil film is formed. The equations of continuity, momentum and energy conservation equations are used to describe the air/oil two-phase flow state in the bearing chamber, and the expressions of the 3 governing equations are as follows:

$$\frac{\partial(\gamma_\alpha\rho_\alpha)}{\partial t}+\nabla\cdot(\gamma_\alpha\rho_\alpha U_\alpha)=0 \tag{1}$$

$$\frac{\partial(\gamma_\alpha\rho_\alpha U_\alpha)}{\partial t}+\nabla\cdot(\gamma_\alpha\rho_\alpha U_\alpha\otimes U_\alpha)=-\gamma_\alpha\nabla \mathrm{p}_\alpha$$
$$+\nabla\cdot(\gamma_\alpha\mu_\alpha(\nabla U_\alpha+(\nabla U_\alpha)^{\mathrm{T}}))+S_{M\alpha}+M_\alpha \tag{2}$$

$$\frac{\partial(\gamma_\alpha\rho_\alpha H_\alpha)}{\partial t}-\gamma_\alpha\frac{\partial p}{\partial t}+\nabla\cdot(\gamma_\alpha\rho_\alpha U_\alpha H_\alpha-\gamma_\alpha\lambda_\alpha\nabla T_\alpha)$$
$$-\nabla\cdot(\gamma_\alpha\mu_\alpha(\nabla U_\alpha+(\nabla U_\alpha)^{\mathrm{T}}-\frac{2}{3}\nabla U_\alpha\delta)U_\alpha)=S_\alpha+Q_\alpha \tag{3}$$

where ρ , μ and λ is respectively for the density, viscosity and thermal conductivity of lubricating oil or sealing air; γ is volume fraction; U is the fluid velocity vector; T is the fluid temperature; t is time; p is pressure; S_M is momentum source terms; S is heat source; M is interphase drag force; Q is interphase heat exchange; H is total energy; the subscript α respectively denotes the lubrication oil when α equals to L, and sealing air when α equals to g. The units of varitions in the formula adopt internationally standard units.

3 Numerical model

3.1 Geometry model and mesh

According to the structure of the bearing chamber as shown in Figure 1, the three dimensional geometry model is built, and the ICEM software is used to mesh the geometric model. Hexahedron element is employed to divide the computational domain. Considering air swirl flow near the wall, the region grids are refined. The total number of grids is 324948. The quality of the grid is judged by the quality criterion. The whole mesh quality is more than 0.3, which meets the requirements of the calculation.

3.2 Boundary and initial conditions

In order to achieve the transient CFD simulation of air/oil two-phase flow in the bearing chamber, the applied boundary conditions include inlet and outlet boundary conditions and wall surface conditions. The the boundary conditions set as follows:

(1) inlet boundary conditions: the periodic flow rate with time. The sealing air varies in the range of 2.5~3.5g/s, time period is 3 seconds; the lubricating oil rate varies in the range of 0.63~3.14L/min, time period is 3 seconds; the temperature of sealing air is 450K, where the temperature of lubricating oil is 330K; the function expressions of sealing air mass flow rate and lubricating oil volume flow rate were written by CEL language, then implemented on the vent port and the scavenge port. The operate temperature is 330K.

(2) outlet boundary conditions: the relative pressure of the vent port and the scavenge port is 0; the environment temperature of the vent port is 540K;

DOI: 10.1051/matecconf/20179506009

(3) wall boundary conditions: no slip wall boundary condition implements on the rotor surface, and the rotor speed is 5000 rpm; bearing chamber wall surface sets no slip stationary wall, and initial wall oil film thickness is 1mm.

3.3 Computing method

The Simple algorithm is used to solve the continuity equation, momentum equation and energy conservation equation and turbulent equation. The calculation step sets 120, and the convergence criterion is 1×10^{-4}.

4 Results and discussion

4.1 The change of the velocity and temperature of air/oil two-phase flow with time

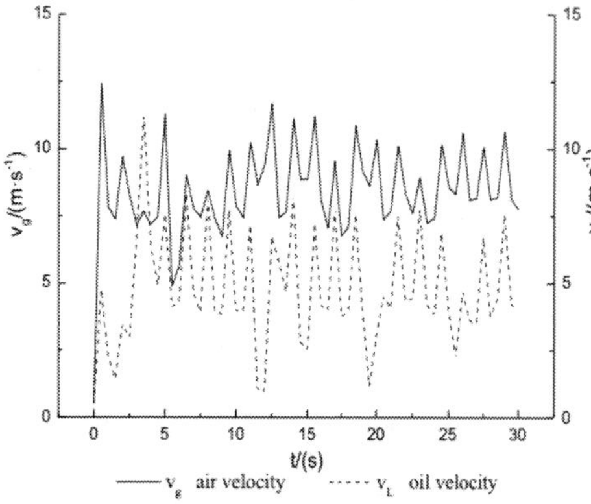

(a). The change of air and oil velocity with time

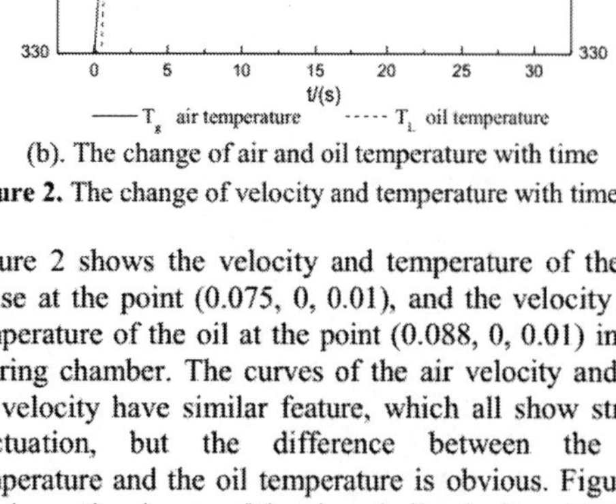

(b). The change of air and oil temperature with time

Figure 2. The change of velocity and temperature with time.

Figure 2 shows the velocity and temperature of the air phase at the point (0.075, 0, 0.01), and the velocity and temperature of the oil at the point (0.088, 0, 0.01) in the bearing chamber. The curves of the air velocity and the oil velocity have similar feature, which all show strong fluctuation, but the difference between the air temperature and the oil temperature is obvious. Figure 2 (a) shows the change of the air and oil velocity with time. During the initial period, the variation of air velocity is larger than the variation of oil velocity, and the distributions of air and oil velocity did not present period changes. But with time increasing, the air/oil two-phase flow in the bearing chamber tends to stable, and the velocity distributions of air and oil shows a quasi-periodic change. The reason is that periodic air mass flow rate and periodic oil volume rate were given as initial conditions, but a large number of oil droplets and the secondary oil droplets exist in the bearing chamber, which results that the velocity distributions of air and oil change with time at the same position. Figure 2 shows the change of the air and oil temperature with time. In the initial time, because the flow of air and oil in the bearing chamber has not reached a steady state, the air and oil temperature fluctuation is fierce. But with time increasing, the fluctuation of air and oil temperature tends to stable in a certain temperature range. Due to the smaller oil volume fraction, so the temperature range of oil is larger than air, the period feature of oil is not obvious. In general, the temperature and velocity of air and oil in the bearing chamber appear a periodic variation with time due to periodic air mass flow and oil volume flow.

4.2 The change of the velocity and temperature of air/oil two-phase flow with space

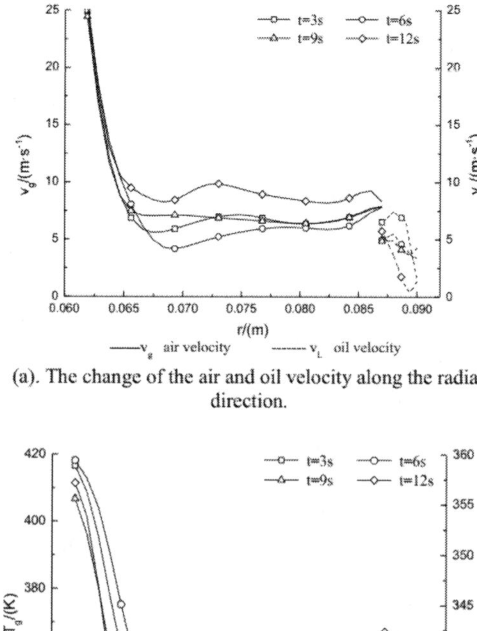

(a). The change of the air and oil velocity along the radial direction.

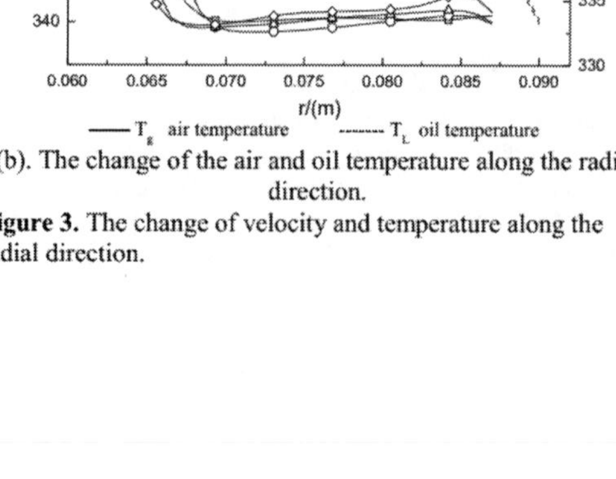

(b). The change of the air and oil temperature along the radial direction.

Figure 3. The change of velocity and temperature along the radial direction.

Figure 3 shows the change curves of velocity and temperature of air and oil at four different times along the radial direction in the bearing chamber. Influenced by rotor speed, the velocity and temperature gradient near the rotor surface is larger, but the velocity and temperature gradient of the air and oil has little change at different locations of the bearing chamber core area. Figure 3 (a) presents the change of the air phase and oil phase velocity at different times along the radial direction.

It shows that the air and oil phase velocity did not appear periodicity in the radial direction of the bearing chamber. In the vicinity of rotor surface, the air velocity phase is higher than other region, and the velocity gradient is larger, but the velocity gradient is not great in the bearing chamber core area. The oil film of bearing chamber wall is affected by the boundary layer, so the velocity gradually decreases along the radial direction. Fig. 3 (b) gives the variation curve of the air phase and oil phase temperature at four different times along the radial direction in the bearing chamber. The air and oil temperature in the radial direction did not show a period change. In the radial direction, the air temperature near the rotor surface is higher than the core area of bearing chamber. In the core area, the temperature distribution of air phase is relatively uniform, and at the different times the changes are tiny. And the oil temperature changes in a certain range at the different times. The reason is that period oil supply makes the heat transfer of oil volatile.

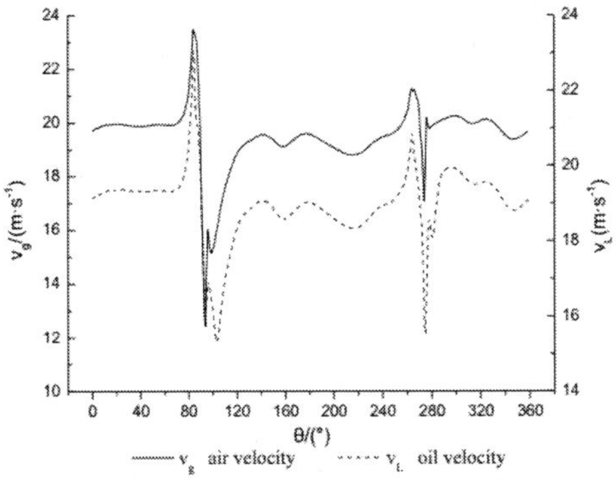

(a). The change of the air and oil velocity along the circumferential direction.

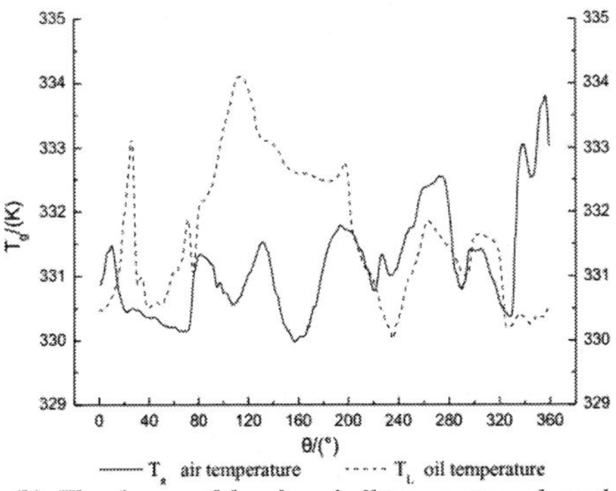

(b). The change of the air and oil temperature along the circumferential direction.

Figure 4. The change of velocity and temperature along the circumferential direction.

Figure 4 shows the velocity and temperature of the air at t=12s along the circumferential direction r=0.75m, and the velocity and temperature of the oil along the r=0.88m circumferential direction. As can be seen from the figure, the velocity field of air and oil in the circumferential direction of the bearing chamber is similar to the velocity field. The temperature field is more complicated than the velocity field. Figure 4 (a) shows the air and oil velocity along the circumferential direction at t=12s. Due to the existence of the vent port and the scavenge port, the velocity of the air and oil leaps in these regions, but the velocity is relatively stable in the other regions of bearing chamber. The air velocity distribution is similar with oil velocity distribution. Figure 4 (b) gives the change of air and oil temperature in the bearing chamber along the circumferential direction at t=12s. As seen in the figure the change of air and oil temperature along the circumferential direction is more complex. The highest air temperature is near the vent. Because environment temperature near the vent is large. Influence by the wall shear stress, the lubricating oil accumulates at the lower right part in bearing chamber, so temperature is higher.

5 Conclusions

With the velocity and temperature distributions of the air/oil two-phase flow in the bearing chamber show an approximate periodicity with time, but the velocity and temperature distribution of the air/oil two-phase flow in the bearing chamber did not show obvious periodicity with space. The research results in this paper contribute to the understand and study the periodic and nonlinear problems of air/oil two-phase flow in the bearing chamber.

Funding

This study is supported by the National Natural Science Foundation of China (grant number 51275411).

References

1. C. Guoding, C. Bo, W. Jun, C. Chen, J. Mech. Eng, **139** (2011).
2. P. Gorse, K. Willenborg, S. Busam, J. Ebner, K. Dullenkopf, S. Wittig, Proc. ASME T. Expo, **16** (2003).
3. A. Aroussi, G. Ishaq, M. Menacer, Proc. ASME Joint Fluids Eng. Conf, **6** (2003).
4. P. Gorse, K. Dullenkopf, H. J. Bauer, S. Wittig, Proc. ASME T. Expo, **9** (2008).
5. S. C. Rodkey, Heister S D, S. H. Collicott, 43rd AIAA/ASME/SAE/ASEE Joint Prop. Conf, **8**(2007).
6. C. Guoding, C. Bo, L. Yajun, C. Chen. J. Nor. Pol. Uni, **62** (2011).
7. A. Wenwei, S. Hua, C. Guoding. Mech. Sci. Tech. Aero. Eng, **1064** (2009).

The Characteristic of Thermoelastic Waves in Transversely Isotropic Finite Cylinders

Hao Bai

Lakehead University, Ontario, Canada, P7B5E1

Abstract. A theoretical as well as a numerical investigation of the propagation of thermoelastic waves and vibration of transversely isotropic cylinders of finite length is discussed. Lord-Shulman theory is adopted here to model the thermoelastic deformation of cylinders. A semi analytical finite element (SAFE) method is employed to study dispersion of thermoelastic waves and natural frequencies of vibration of finite cylinders with traction free curved surfaces having both ends insulated and constrained by frictionless rigid walls. Numerical results obtained by the SAFE method for the frequencies of vibration of a sapphire rod are found to be in excellent agreement with published results. Natural frequencies of vibration for first three axisymmetric and asymmetric modes are presented graphically for a silicon nitride thermoelastic cylinder. Also, numerical results showing dispersion of both propagating and evanescent circumferential waves in infinite and finite cylinders are presented also.

1 Introduction

Isothermal vibration of finite–length cylinders has been studied by many investigators. Since exact analytical solutions for a finite-length vibrating cylinder cannot be obtained in general, approximate solutions have been proposed to estimate natural frequencies of finite isotropic circular cylinders by [1]-[5]. If the material of the cylinder is anisotropic, the solution of the free vibration problem becomes more complex. [6] used separation of variables technique presented by [7], who derived the dispersion equations governing compressional wave propagation along the axis of a transversely isotropic cylinder. In their study [6], they considered the axisymmetric vibrations of finite – length transversely isotropic cylinders and presented experimental results for sapphire rods. [8] estimated natural frequencies of axisymmetric modes of finite length transversely isotropic cylinders by applying the Ritz method. [9] and [10] studied axisymmetric and asymmetric vibrations of a finite transversely isotropic cylinder using potential functions and compared their results with the experimental observations of [6]. [11] studied free vibrations of finite orthotropic thin cylindrical shells by employing shell theory.

In this paper, thermoelastic circumferential wave propagation in a finite – length transversely isotropic cylinder has been investigated. [12] studied analytically the two-dimension problem of isothermal guided circumferential waves in a circular annulus. To our knowledge, propagating and evanescent modes of circumferential waves in a finite-length transversely isotropic cylinder have not been studied in the past. Since finite - length cylinders are common structural components in many practical applications, the safety of these structural components must be assessed regularly for their continual usage or for their repair. In order to monitor their health, it is necessary to assess their material properties and detect critical cracks or defects in them. Guided ultrasonic waves have proven to be very efficient for structural health monitoring (SHM). In recent years, laser based ultrasonic techniques for material property measurements and defect characterizations have proven to be very reliable. The coupled theory of thermoelasticity must be used to find complete solutions for the guided waves generated by pulsed lasers.

Recently the thermoelastic wave in a generalized cross section cylinder immersed in a fluid is studied by using the Fourier expansion collocation method by [13]. [14] studied the propagation of thermoelastic circumferential waves in orthotropic cylindrical curved plates using the Green–Naghdi (GN) generalized thermoelastic theory (without energy dissipation). It is shown that thermal properties have no effect on the circumerential SH wave and the thermoelasticity has very limited effects on the other circumferential modes. In 2011, a work on the randomness of thermoelasticity wave in thick hollowed cylinder was reported [15]. They analytically studied the thermal and mechanical waves using theory of coupled thermoelasticity without energy dissipationbased on Green–Naghdi model.

Propagation of thermoelastic waves in a circular annulus and finite length circular cylindrical shells have been studied here in the context of LS [16] generalized theory of thermoelasticity. A semi-analytical finite element (SAFE) method [17-19, and references therein] is employed to study characteristics of guided wave modes in the cylinder. The dispersion equation governing the modes is obtained in the form of an eigenvalue problem. Solutions obtained by the SAFE formulation are validated by comparing the numerical results for finite length transversely isotropic elastic cylinders (isothermal case) with those published by [9], [10]. Frequency spectra obtained for the (plane strain) circumferential wave propagation problem are shown to agree well with those published by [12] for an isotropic (isothermal) case. Numerical results are presented here for both propagating and evanescent modes in a transversely isotropic silicon nitride cylinder.

2 Equations of motion

Consider a finite - length transversely isotropic circular cylinder of inner radius r_i and outer radius r_o in the cylindrical coordinate system (r, θ, z) shown in Fig. 1. The cylinder is initially in the undisturbed state with uniform temperature T_0. The inner and outer surfaces of the cylinder are traction free and the heat flux vanishes at inner and outer surfaces in the radial direction. The ends of the cylinder are insulated thermally and also constrained by frictionless rigid walls.

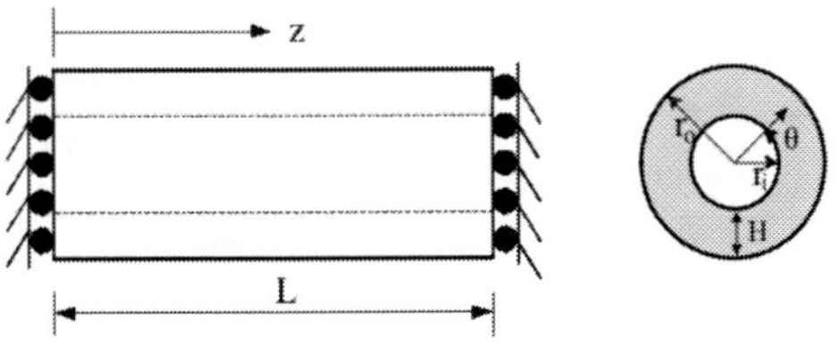

Figure 1. Geometry of the cylinder

The generalized LS governing equations of thermoelasticity, in the absence of body force and heat source, are given by [20] as,

$$\sigma_{ij,j} = \rho \ddot{u}_i \quad (1)$$

$$T_0 \rho \dot{\eta} = -q_{i,i} \quad (2)$$

$$\sigma_{ij} = C_{ijkl}\varepsilon_{kl} - \beta_{ij}T \quad (3)$$

$$\rho \dot{\eta} = \beta_{ij}\dot{\varepsilon}_{ij} + \frac{\rho c_E}{T_0}\dot{T} \quad (4)$$

$$q_i + \tau_0 \dot{q}_i = -k_{ij}T_{,j} \quad (5)$$

here σ_{ij}, the components of the stress tensor; u_i, the components of the displacement; ε_{ij}, the components of the strain tensor; C_{ijkl}, the elastic constants; q_i, components of the heat flux vector; ρ, the mass density; T_0, the reference temperature; η, the entropy density; T, the temperature change; c_E, the specific heat at constant deformation; τ_0, the thermal relaxation time; β_{ij}, the thermal coefficients; k_{ij}, the coefficients of thermal conductivity. In the above equations, the superposed dot indicates the derivative with respect to time.

A normalization procedure is needed to find the solution of the problem. A careful identified set of basic variables was choosen and see detail in [21]. It may be noted that the non-dimensional scheme yields all dimensionless equations in the same form as their dimensional counterparts. Therefore, this normalization can be used to solve multilayer cylinders as well. The final equations of motion in terms of displacements and temperature are given by

$$c_{ijkl}u_{k,lj} - \beta_{ij}T_{,j} = \rho\, \ddot{u}_i \quad (6)$$

$$\beta_{ij}T_0\,(1+\tau_0 \frac{\partial}{\partial t})\dot{u}_{i,j} - k_{ij}T_{,ij} + \rho\, c_E(1+\tau_0 \frac{\partial}{\partial t})\dot{T} = 0 \quad (7)$$

It's worth to mention that the above quantities are in their respective non-dimensional form.

3 Semi analytical FE formulation

In this method, for a finite cylinder of length L, the radial dependence of the displacement, $\mathbf{u}$, and temperature, T, is approximated by one dimensional isoparametric finite elements. The total thickness of the cylinder, H, is composed of cylindrical layers and each layer can have distinct thermoelastic properties and thickness. By using the SAFE method, thickness of the cylinder is discretized into N laminas. Quadratic interpolation polynomials are used to approximate the displacement and temperature field over each lamina in the radial direction.

The displacements and temperature of the kth lamina are expressed as

$$\mathbf{u}(r,\theta,z,t) = Z\mathbf{N}_1(\zeta)\mathbf{u}^e(\theta,t)$$
$$\mathbf{T}(r,\theta,z,t) = c\mathbf{N}_2(\zeta)\mathbf{T}^e(\theta,t) \quad (8)$$

and

$$r = \mathbf{N}_2(\zeta)\mathbf{R}^e \quad (9)$$

here

$$\mathbf{N}_1(\zeta) = \begin{bmatrix} n_1 & 0 & 0 & n_2 & 0 & 0 & n_3 & 0 & 0 \\ 0 & n_1 & 0 & 0 & n_2 & 0 & 0 & n_3 & 0 \\ 0 & 0 & n_1 & 0 & 0 & n_2 & 0 & 0 & n_3 \end{bmatrix},$$

$$\mathbf{N}_2(\zeta) = [n_1 \quad n_2 \quad n_3], \quad \mathbf{Z} = \begin{bmatrix} c & 0 & 0 \\ 0 & c & 0 \\ 0 & 0 & s \end{bmatrix}$$

$$\mathbf{u}^e = (u_{r1}, u_{\theta 1}, u_{z1}, u_{r2}, u_{\theta 2}, u_{z2}, u_{r3}, u_{\theta 3}, u_{z3})^T,$$

$$\mathbf{T}^e = (T_1, T_2, T_3)^T, \quad \mathbf{R}^e = (R_b, R_m, R_f)^T$$

where $s = \sin(pz)$, $c = \cos(pz)$, $p = {m\pi}/{L}$ and L is the length of the cylinder and, m is the axial wave number. The shape functions are

$$n_1 = \tfrac{1}{2}\zeta(\zeta-1),$$
$$n_2 = 1-\zeta^2, \qquad (10)$$
$$n_3 = \tfrac{1}{2}\zeta(\zeta+1), \quad -1 \le \zeta \le 1$$

where

$$\zeta = \frac{2r - R_b - R_f}{R_f - R_b} \qquad (R_b \le r \le R_f)$$

The nodal displacements $u_{rj}, u_{\theta j}, u_{zj}$, and temperature T_j, where j = 1, 2, 3, are taken at the inner surface R_b, middle layer R_m and the outer surface R_f of the kth lamina.

By employing the variational principle of thermoelasticity [20], the equations of motion can be written as, in the matrix form,

$$\mathbf{E}_1\ddot{\mathbf{V}} + \mathbf{E}_2\ddot{\mathbf{V}}_{,\theta} + \mathbf{E}_3\dot{\mathbf{V}} + \mathbf{E}_4\dot{\mathbf{V}}_{,\theta} + \mathbf{E}_5\mathbf{V} + \mathbf{E}_6\mathbf{V}_{,\theta} + \mathbf{E}_7\mathbf{V}_{,\theta\theta} = 0 \quad (11)$$

where $\mathbf{E}_i\ (i = 1,2,...,7)$ are the global matrices, see details in [20], $\mathbf{V}$ is the global nodal displacement and temperature vector.

Traction free boundary conditions on surfaces of cylinder require that the stresses at inner and outer surfaces of the cylinder are zero

$$\sigma_{rr} = \sigma_{r\theta} = \sigma_{rz} = 0 \quad \text{at} \quad r = r_i \text{ and } r = r_o$$

End boundary conditions of the cylinder are

$$u_z = \sigma_{rz} = \sigma_{\theta z} = 0 \quad \text{at} \quad z = 0, L$$

Thermal boundary conditions are considered as

$$T_{,z} = 0 \quad \text{at} \quad z = 0, L$$

$$T_{,r} = 0 \quad \text{at} \quad r = r_i \text{ and } r = r_o$$

The solution of Eq. (11) is assumed as

$$\mathbf{V} = \mathbf{V}_0 e^{i(n\theta - \omega t)} \qquad (12)$$

where, $\mathbf{V}_0$ is the nodal amplitude vector and ω is the nondimensional circular frequency. For waves propagating in the θ -direction, $e^{in\theta}$ is defined as $e^{in\theta} = e^{i(kr_o)\theta}$ where k is the wave number defined as $k = \omega / c(r_o)$. $c(r)$ is the phase velocity of the circumferential wave travelling along the surface at a distance r. Substitution of Eq. (12) into (11) results in a set of linear homogeneous equations. For integer values of m, two eigenvalue problems can be obtained depending on whether n or ω is chosen as the eigenvalue. If n serves as the eigenvalue with assigned values for ω, it will give a quadratic eigenvalue problem:

$$\left(-n^2\mathbf{E}_7 + n\mathbf{E}_B + \mathbf{E}_A\right)\mathbf{V}_0 = 0 \qquad (13)$$

where
$$\mathbf{E}_A = \left(-\omega^2\mathbf{E}_1 - i\omega\mathbf{E}_3 + \mathbf{E}_5\right)$$
$$\mathbf{E}_B = \left(-\omega^2 i\mathbf{E}_2 + \omega\mathbf{E}_4 + i\mathbf{E}_6\right).$$

Equation (13) can be converted into a first order equation in the form

$$\begin{bmatrix} 0 & \mathbf{I} \\ \mathbf{E}_A & \mathbf{E}_B \end{bmatrix} \begin{Bmatrix} \mathbf{V}_0 \\ n\mathbf{V}_0 \end{Bmatrix} = n \begin{bmatrix} \mathbf{I} & 0 \\ 0 & \mathbf{E}_7 \end{bmatrix} \begin{Bmatrix} \mathbf{V}_0 \\ n\mathbf{V}_0 \end{Bmatrix} \qquad (14)$$

Solution of the generalized eigenvalue problem (14) gives the dispersion relation for guided circumferential thermoelastic waves in the finite cylinder. On the other hand, for given integer values of m and n, i twill generate the frequencies of free thermoelastic vibration of the shell, see detail in [21].

4 Numerical results and discussion

In order to validate the SAFE formulation, a couple of examples were studied and results were compared with published literatures. The first example was guided circumferential waves for the isothermal plane strain problem, it was reported in [12]. The comparison between SAFE results and those obtained by Liu and Qu [12] is shown in below in Fig. 2. The agreement is excellent. The second example was the vibration of a finite transversely isotropic cylinder and it was compared with the results in [9, 10], the comparison does not show here to save space. They all showed excellent agreement.

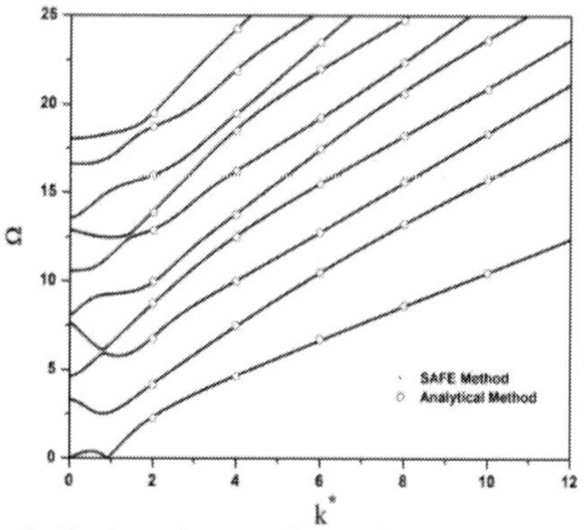

Figure 2. 2D view of propagating modes for $\eta_s = 0.1$ ([12], Fig. 2(a))

A finite cylindrical tube of length 15mm with inner and outer radii of 1.5 mm and 2.5 mm, respectively, is considered here. Thus, the nondimensional length is 15 and, nondimensional inner and outer radii are, respectively, $r_i = 1.5$ and $r_o = 2.5$. The material of the tube is taken to be transversely isotropic silicon nitride (Si_3N_4) [20]. The symmetry axis of the material is aligned with the axis of the cylinder. For the purpose of nondimensionalization, normalizing velocity $\overline{V}$ is chosen to be 0.25 km/s. Furthermore, the normalized density ρ^* is taken to be 1. Then, the nondimensional material properties of the tube are given by

$$T_0^* = 1.0,\ \ \tau_0^* = 1.081\times10^{-7},\ \ \beta_{rr}^* = \beta_{\theta\theta}^* = 70.03,\ \ \beta_{zz}^* = 83.21$$

$$k_{rr}^* = k_{\theta\theta}^* = 0.0785,\ \ k_{zz}^* = 0.10,\ \ c_E^* = 967.50,\ \ c_{11}^* = \frac{c_{11}}{\overline{c}} = 2165,$$

$$c_{12}^* = \frac{c_{12}}{\overline{c}} = 975,\ \ c_{13}^* = \frac{c_{13}}{\overline{c}} = 635,\ \ c_{33}^* = \frac{c_{33}}{\overline{c}} = 2870,\ \ c_{44}^* = \frac{c_{44}}{\overline{c}} = 540$$

DOI: 10.1051/matecconf/20179506010

The dispersion curves for thermoelastic waves in a cylinder of finite length were evaluated for different value of m – the wave number in the axial direction. Group velocities of the circumferential waves are computed at the outer surface of the cylinder as well. Due to the limitation, only the results for m = 1 are presented here.

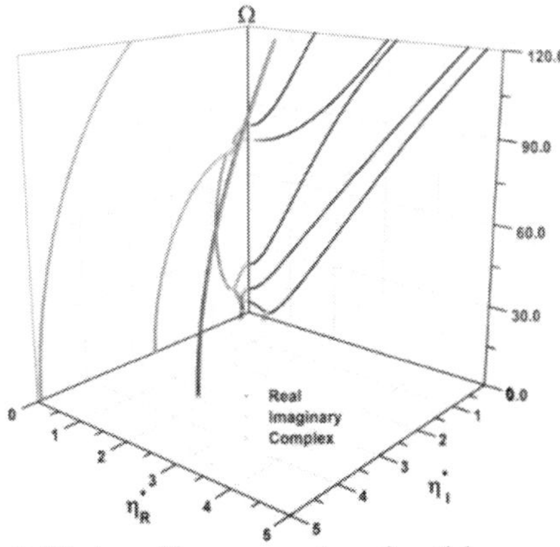

Figure 3. 3D view of frequency – circumferential wave number plots of a silicon nitride cylinder when axial wave number, m = 1

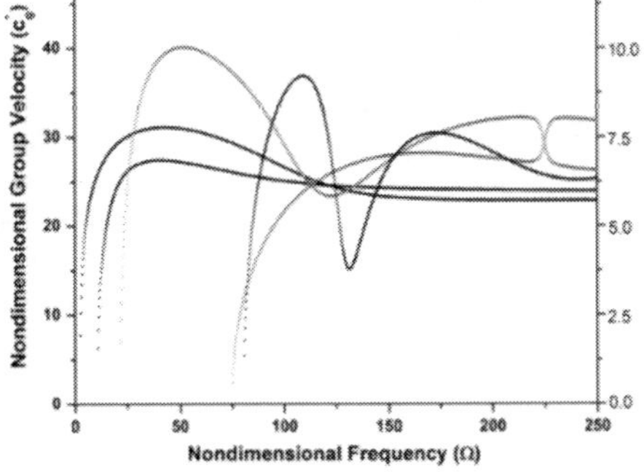

Figure 4. Group velocity of circumferential propagating waves of a silicon nitride cylinder when axial wave number, m = 1

From the results it is observed that first two modes are almost non-dispersive at high frequencies and approach the Rayleigh wave velocity. It is found that elastic modes are not much affected by the thermal effects within the frequency range considered here. Thermal modes are found to have very high attenuation when compared with elastic modes. Higher order thermal modes originate with high imaginary values of the wave number when $\omega^* = 0$. It is also found that the higher order thermal modes approach the first thermal mode as the frequency increases.

The dependence of natural frequencies of the finite cylinder on geometry ratio was studied as well by considering different geometry ratio $(L^* = L/H)$ and the material properties of the cylinder. The results will be shown here due to the limitation.

5 Conclusion

The natural frequencies of vibration of finite length cylinders have been studied in the context of the L-S generalized theory of thermoelasticity. The proposed SAFE formulation is validated by comparing predicted numerical results with those obtained in earlier investigations. Thermal modes have a much higher attenuation than the elastic modes and the influence of thermal effects on elastic modes is not significant. The guided waves studied in this paper can be used for detecting cracks predominantly in the axial direction in finite cylinders.

References

1. G.W. McMahon, J. of the Acou. Soc. Am., **36**, 85–92 (1964)
2. G.W. McMahon, J. of the Acou. Soc. Am., **48**, 307–312 (1970)
3. Hutchinson, J.R., J. Acou. Soc. Am., **51**, 233–240(1972)
4. G.M.L. Gladwell, and U. C. Tahbildar, J. Sound & Vib, **22,** 143–157 (1972)
5. G.M.L. Gladwell, and D.K. Vijay, J. Sound & Vib, **42,** 387–397 (1975)
6. C.P. Lusher, and W.N. Hardy, ASME J. Appl. Mech., **55**, 855–862 (1988)
7. R.W. Morse, J. Acou. Soc. Am., **26**, 1018–1021 (1954)
8. P. R. Heyliger, J. Sound & Vib, **148**, 507–520 (1991)
9. K.T. Chau, ASME J. Appl. Mech., **61**, 964–970 (1994)
10. F. Honarvar, E. Enjilela, and A.N. Sinclair, Acous. Phy., **55**, 708–714 (2009)
11. W. Chen, J. Cai, G. Ye, and H. Ding, J. Sound & Vib, **195**, 117–135 (1996)
12. G. Liu, and J. Qu, ASME J. Appl. Mech., **65**, 424 – 430 (1998)
13. M. Venkatesan, and P. Ponnusamy Int. J. Mech. Sci., **49**, 741-751 (2007)
14. J. Yu, B. Wu, and C. He, Ultrasonics, **50**, 416-423 (2010)
15. S.M. Hosseini, and F. Shahabian, Appl. Math. Modelling, **35**, 4697-4714 (2011)
16. H.W. Lord, and Y. Shulman, J. Mech. Phy. Solids, **15**, 299-309 (1967)
17. R.B. Nelson, S.B. Dong, and R.D. Kalra, J. Sound & Vib, **18**, 429–444 (1971)
18. K.H. Huang, and S.B. Dong, J. Sound & Vib, **96**(3), 363-379 (1984)
19. S.K. Datta, and A.H. Shah, Elastic Waves in Composite Media and Structures: With Applications to Ultrasonic Nondestructive Evaluation, CRC Press, Boca Raton, (2009)
20. H.M. Al-Qahtani, and S.K. Datta, J. Appl. Phys., **96**, 3645-3658 (2004)
21. R. Chitikireddy, Ph.D. Dissertation, University of Manitoba, Winnipeg, MB, Canada, (2012)

Random Fatigue Analysis of Drill-Strings in Frequency Domain

JiaHao Zheng[1], Jianming Yang[1], Ping Yang[2]

[1]*Department of Mechanical Engineering, Memorial University of Newfoundland, A1B 3X5, St John's Canada*
[2]*School of Mechanical Engineering, Jiangsu University, 212013, Zhenjiang, Jiangsu, P.R.China,*

Abstract. Predicting fatigue damage induced by vibrations is of benefit to oil and gas industry. In this paper a spectral method is developed to estimate the damage of vibration, including both deterministic and random vibration, on the fatigue life of drillstrings. The drill-string is first modeled by the continuous parameter method with two vibration modes: axial and torsional. After the vibration response is obtained, the equivalent stress spectra are calculated for any position along the drill-string, based on maximum shear stress fatigue failure criterion. The fatigue life of the drill-string at arbitrary position is then estimated by the developed spectral methods

1 Introduction

A drill-string is a slender structure used in the oil and gas industry. Fatigue of drill-string is a critical issue faced by the industry, and rich research works have dealt this research topic. These works can be roughly divided into two categories: experimental and theoretical. Bachman[1] investigated the fatigue of full-size drill-pipes experimentally with the drill-pipes clamped as cantilevered beams and subjected to rotational bending. However, he did not consider the axial load to the drill-pipes, and the test was conducted only in the condition of air. Morgan and Robin [2] reported their testing results from small-scale coupons cut from as-produced seamless hot rolled drill pipes. Axial load was applied to the specimens to gain more realistic data. In 1991, Grondin and Kulak [3] conducted experiments using full-size drill pipes subjected to axial load, fluctuating bending and rotating in both air and corrosive media. Their tests provided the most comprehensive data so far for drill-string fatigue. In addition to the experimental works, researchers developed various theoretical methods for drill-string fatigue calculation. The theoretical work has been mainly conducted along two lines. One is based on the fracture mechanics theory [4]. These methods assume micro cracks exist in the drill pipe materials. These micro cracks propagate in the drill pipe body along certain directions under cyclic loadings, and eventually develop into fractures [5] which lead to fatigue failures. The other combines the fatigue S-N relation with the linear damage cumulative law to compute an equivalent damage. One of the most representative work along this line was by Lubinski [6], in which the fatigue damage induced by the dog-legs was calculated. Thereafter, modifications were made to the S-N relation and boundary conditions by Hansford and Lubinski[7],Wu [8,9] and other researchers to improve the accuracy of the prediction. In real drilling operation, vibration is a important factor causing fatigue. Further the vibration is more likely random. However, in the existing literature, the vibration effect was rarely considered. Patel and Vaz's work [10] seems the only one accounting for the contribution of vibration to the fatigue of drill-string. In this work, the vibration was considered as periodical and the random nature was not considered. This paper, using spectral method [12], investigates the effect of axial and torsional vibrations on the fatigue LIFE of drill strings. The vibrations are examined with consideration of the randomness. It is organized as follows: section 1 gives a background review on the research of drill-string fatigue. In section 2, continuous vibration models of the drill-string are developed. Section 3 presents the stress transformation and the determination stress spectrum. Following that spectral method for fatigue prediction is presented in section 4. Simulation examples and result analysis are given in the following two sections. The last section concludes the whole paper.

2 Vibration model

Given the similarity between the axial and torsional vibrations, only axial mode is presented in detail below. The torsional part can be easily derived following the steps in the axial vibration. A model of the drill-string under axial vibration is shown in Fig. 1. Let x denote the axial position. And x = 0 is set to be at the bottom of the drill-string. The axial vibration of the drill-string is expressed as:

$$A_i E \frac{\partial^2 \bar{u}_i(x,t)}{\partial x^2} - \gamma_i \frac{\partial \bar{u}_i(x,t)}{\partial t} - \rho A_i \frac{\partial^2 \bar{u}_i(x,t)}{\partial t^2} - (\rho - \rho_m) g A_i = 0 (1)$$

where ρ; E are the material's density and Young's modulus. γ_i denotes the viscous damping coefficient and A_i represents the cross section area. i = 1, 2 stands for the drill-pipe and the drill-collar section respectively.

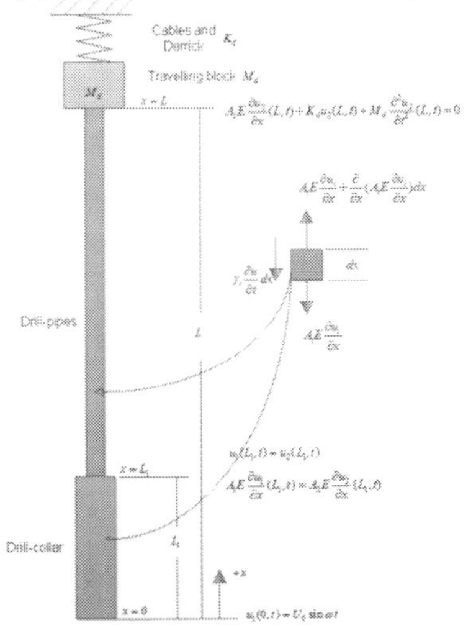

Figure 1. Schematic of a drill-string.

$\tilde{u}_i(x,t)$ denotes the elastic displacement in axial direction, which includes static and dynamic parts: $\tilde{u}_i(x,t) = u_{is}(x) + u_i(x,t)$. Substituting this equation into Eq. 1 reduces it into

$$A_i E \frac{\partial^2 u_i(x,t)}{\partial x^2} - \gamma_i \frac{\partial u_i(x,t)}{\partial t} - \rho A_i \frac{\partial^2 u_i(x,t)}{\partial t^2} = 0 \quad (2)$$

The excitation is assumed from the base motion, represented by $u_b(t)$, at the bottom of the drill-string caused by the unevenness of the bottom. Then Eq. 2 becomes

$$A_i E \frac{\partial^2}{\partial x^2}[u_i(x,t) - u_b(t)] - \gamma_i \frac{\partial u_i(x,t)}{\partial t} - \rho A_i \frac{\partial^2 u_i(x,t)}{\partial t^2} = 0 \quad (3)$$

By defining the axial displacement relative to the base as $v(x,t) = u(x,t) - u_b(t)$, we get:

$$\frac{\partial u}{\partial t} = \frac{\partial v}{\partial t} + \frac{\partial u_b}{\partial t} \quad (4)$$

$$\frac{\partial^2 u}{\partial t^2} = \frac{\partial^2 v}{\partial t^2} + \frac{\partial^2 u_b}{\partial t^2} \quad (5)$$

Inserting the above two equations into Eq. 3 leads to

$$\rho A_i \frac{\partial^2 v_i}{\partial t^2} + \gamma_i \frac{\partial v_i}{\partial t} - A_i E \frac{\partial^2 v_i}{\partial x^2} = -\gamma_i \frac{\partial u_b}{\partial t} - A_i \frac{\partial^2 u_b}{\partial t^2} \quad (6)$$

The right-hand side of this equation represents excitation induced by the base motion. It can be expressed as

$$u_b(t) = p(t) + w(t) \quad (7)$$

where p(t) and w(t) represent the deterministic base motion and random base motion, respectively.

Solution to the equation can be written in the following form using the Galerkinmethod[14]:

$$v(x,t) \approx \sum_{n=1}^{R} X_n(x) p_n(t) \quad (8)$$

where R stands for the order to be retained in the discretization.pn(t) are the generalized coordinates. Xn(x) are the appropriatebasis functions satisfying the specific boundary conditions.The boundary conditions at the top, the interface betweendrill-pipe and drill-collar and the bottom are presented as (SeeFig. 1):

The Top: $2E\frac{\partial u_2}{\partial x}(L,t) + K_d u_2(L,t) + M_d \frac{\partial^2 u_2}{\partial t^2} = 0$

The interface: $A_1 E \frac{\partial u_1}{\partial x}(L_1,t) = A_2 E \frac{\partial u_2}{\partial x}(L_1,t)$;

$$u_1(L_1,t) = u_2(L_1,t)$$

The bottom: $u_1(0,t) = 0$

where K_d, M_d are stiffness of the derrick cable and mass oftraveling block. L_1 is the drill-collar section length and L is thelength of the drill-pipe section.

The mode shapes are solved from the free vibration equation by the variable separation technique which can be found in most vibration texts. Then substituting Eq.(8) into (6) can reduce the partial differential equation into the following ordinary differential equations.

$$\ddot{\mathbf{p}}(\mathbf{t}) + \boldsymbol{\Xi}\dot{\mathbf{p}}(\mathbf{t}) + \boldsymbol{\Omega}^2\mathbf{p}(\mathbf{t}) = \tau_1 \frac{\partial u_b}{\partial t} + \tau_2 \frac{\partial^2 u_b}{\partial t^2} \quad (9)$$

where **p(t)**, is the column vector, and **Ξ, Ω** are matrices with the entries as below.

$$\Xi_{kj} = -\frac{\gamma}{\rho A}\delta_{kj}; \quad \Omega_{kj}^2 = \frac{E\int_0^L X_k(x) X_j''(x)dx}{\rho \int_0^L X_k^2(x)dx}$$

$$\tau_{k1} = -\frac{\gamma \int_0^L X_k(x)dx}{\rho A \int_0^L X_k^2(x)dx}; \quad \tau_{2k} = -\frac{\int_0^L X_k(x)dx}{\int_0^L X_k^2(x)dx}$$

The orthogonality of the mode shapes have been used in in reaching the above. The torsional vibration can also be derived in a similar way following the steps above.

3 Stress spectrums

In general the drill string is under a plane stress status with the largest shear stress at the most outside part under the axial and torsional vibration. However, we start our discussion with the 3 dimensions for generality by assuming stresses exist on all three planes as shown in Fig. 2. Thenthe stress tensor is represented as.

$$\sigma(x,t) = \begin{bmatrix} \sigma_{xx} & \sigma_{xy} & \sigma_{xz} \\ \sigma_{yx} & \sigma_{yy} & \sigma_{yz} \\ \sigma_{zx} & \sigma_{zy} & \sigma_{zz} \end{bmatrix} \quad (10)$$

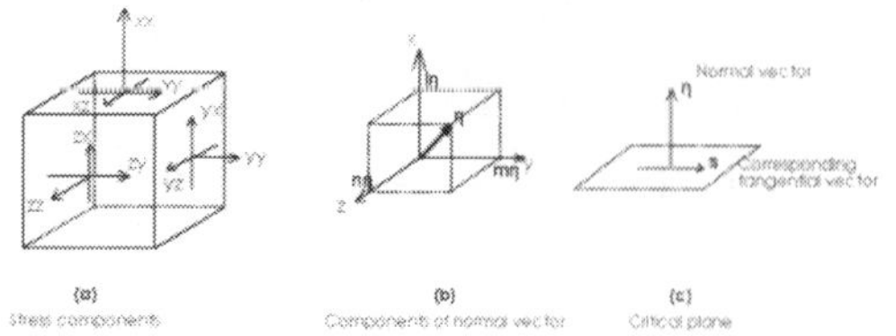

Figure 2. Stress components

For steel drill strings, the above matrix is symmetric; therefore only 6 components are independent which are $\sigma_{xx}, \sigma_{yy}, \sigma_{zz}, \sigma_{xy}, \sigma_{yz}, \sigma_{zx}$. Then the equivalent maximum shear stress is computed through the following equation.

$$\sigma_{eq} = (l_\eta^2 - l_s^2)\sigma_{xx} + (m_\eta^2 - m_s^2)\sigma_{yy} + (n_\eta^2 - n_s^2)\sigma_{zz} + 2(l_\eta m_\eta - l_\eta m_s)\sigma_{xy} + 2(l_\eta n_\eta - l_\eta n_s)\sigma_{xz} + 2(m_\eta n_\eta - m_\eta n_s)\sigma_{yz} \quad (11)$$

Where l_η, m_η and n_η are the directional cosines of the vector normal to the critical plane and l_s, m_s and n_s are those of the vector tangential to it as shown in Fig.2 (b) and (c). The tangential vectors $\vec{s}$ will be found by simulation.

If the PSD of the randominput is known, then the PSD of the general axial coordinates (p(t)) can be computed through the transfer function of Eq. 8.Then the PSD of the physical coordinates (u(t)and Φ(t)) can be obtained as well.Then based on $\sigma_{axial} = E(\partial u/\partial x)$; $\sigma_{shear} = GR(\partial\Phi/\partial x)$, the PSD of the stresses can be found which will be used for fatigue prediction in following sections.

4 Fatigue predictions

First, the S-N curve of the drill string material is represented as

$$N(s)s^b = c \quad (12)$$

where N and s are the cycle to failure and the stress amplitude. B and c are positive constants estimated from material experimental test.

The fatigue life is defined as

$$T = \frac{1}{E[D]} \quad (13)$$

where T and E(D) are the fatigue life of the drill string and the fatigue damage intensity defined in [17] and given below.

$$E[D] = \frac{v_p}{c} m_0^{\frac{b}{2}}[D_1 Q^b \Gamma(1+b)] + 2^{\frac{b}{2}}\Gamma\left(1+\frac{b}{2}\right)(D_2|R|^b + D_3 \quad (14)$$

The parameters used in this equation are as below.

$$v_p = \frac{1}{2\pi}\sqrt{\frac{m_4}{m_2}};\ D_1 = \frac{2(x_m - \alpha_2^2)}{1+\alpha_2^2};$$

$$D_2 = \frac{1-\alpha_2 - D_1 + D_1^2}{1-R};\ D_3 = 1 - D_1 - D_2$$

$$\alpha_2 = \frac{m_2}{\sqrt{m_0 m_4}};\ Q = \frac{1.25(\alpha_2 - D_3 - D_2 R}{D_1};$$

In those equations, m_n are the so-called spectrum moment and Γ is the Gamma function.

5 Simulation examples

For a drill string with the parameters are given in table 1 and the excitations are given in table 2.The S-N curve is taken from [3] as

$$log_{10}N = 14.8 - 3.47 log_{10}S_r - 1.65 \times 10^5 S_m^2 \quad (15)$$

The S-N curves for different S_m is shown in Fig.3 and the S_m for different positions is shown in Fig.4.

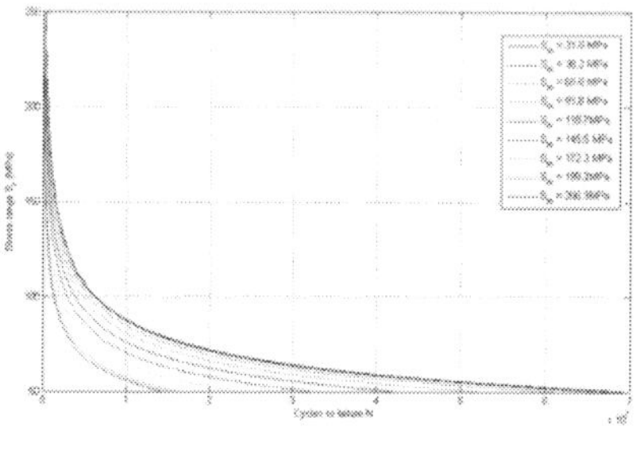

Figure 3. S-N curves with different mean stress

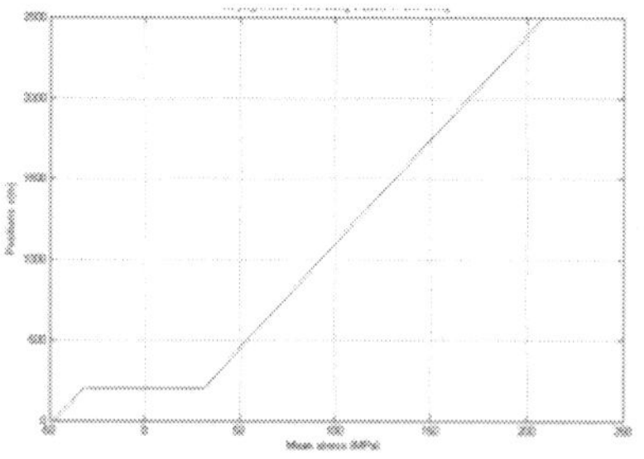

Figure 4. Varying means stress

Table 1. Parameters of simulation

Parameter	Collar O.D	Collar I.D	Pipe O.D
values	0.1651 [m]	0.0572 [m]	0.114 [m]
Parameter	L_1	L	N
values	200 [m]	2500 [m]	1000 [rpm]
Parameter	F_h	E	f
values	7×10^5 [N]	$2.07 \times 10^{10} [N/m^2]$	1000 [Hz]
Parameter	Pipe I.D	A_1	A_2
values	0.0972 [m]	0.0188 $[m^2]$	0.0028 $[m^2]$
Parameter	W_s	ρ	ρ_m
values	1×10^5 [N]	$7.89 \times 10^3\ kg/m^3$	$1.5 \times 10^3 [kg/m^3]$
Parameter	γ_1	γ_2	g
values	239.4 $[Ns/m^2]$	33.5 $[Ns/m^2]$	9.81$[m/s^2]$

Table 2. Excitation parameters

Parameter			$A_d\,(A)$	$A_d\,(T)$	$A_w\,(A)$	$A_w\,(T)$
No.1			3	150	0	0
No.2	3	150	0.3	15		
No.3	3	150	0.6	30		
No.4	0	0	0.3	15		

The No.1 excitation contains only deterministic input; while No.4 contains only random. The simulation results for these two cases are shown in Fig.5 and 6.

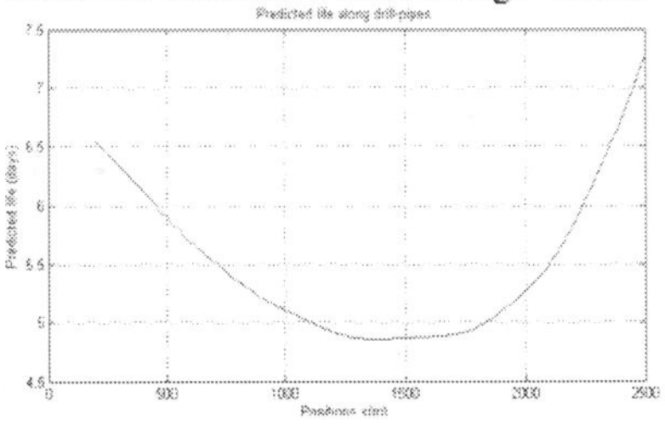

Figure 5. Predicted life under excitation(No.1)

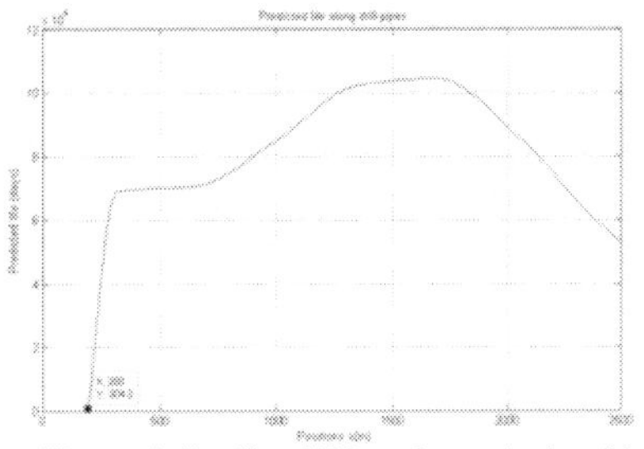

Figure 6. Predicted life under excitation (No.4)

It is noticed that the minimum life happens aboutx = 1500m in Fig.5. If we refer to the vibration model used in Fig. 1,thisis reasonable. In this model the upper end and the bottom endure modelled as simply supported, so the elastic deformation at both ends is zero and the maximum is somewhere in between the two ends. Thus the stresses caused by the vibration reaches the maximum as well somewhere in between the two ends. Also from Fig.6 it can be seen that the most noticeable effect of the random excitation is on the part near the drill collar(the most left part of the curve). To understand the random effect we computed the PSD of the equivalent stress at 9 different positions and shown in Fig.7. Clearly first position (x = 200m) has the largestvalue of the first frequency peak which means the most energy transferred to this part of the drill string. If we compare the effect of different intensities of the random excitation on the fatigue, Excitations No.1, No.2 and No.3 have the same level of deterministic excitation, but with different random excitations. The fatigue life predicted is shown in Fig. 8. It is clearthat the stronger the random excitation, the shorter the fatigue life of the drill-string at any position. This makes sense because stronger random excitation causes more severe vibrationand does more harm to the fatigue life of the drill-string.

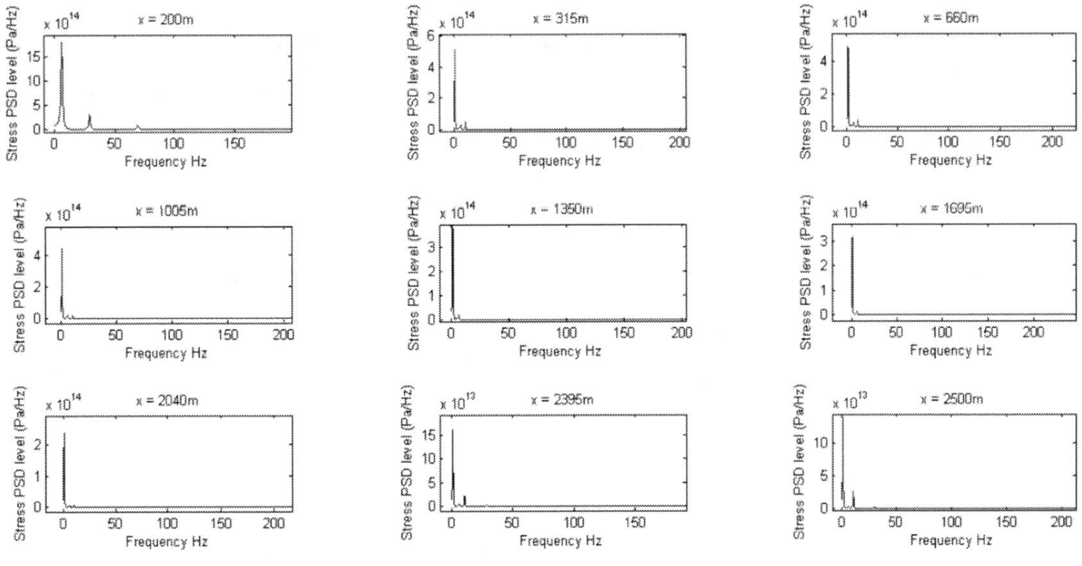

Figure7. PSD of equivalent stresses

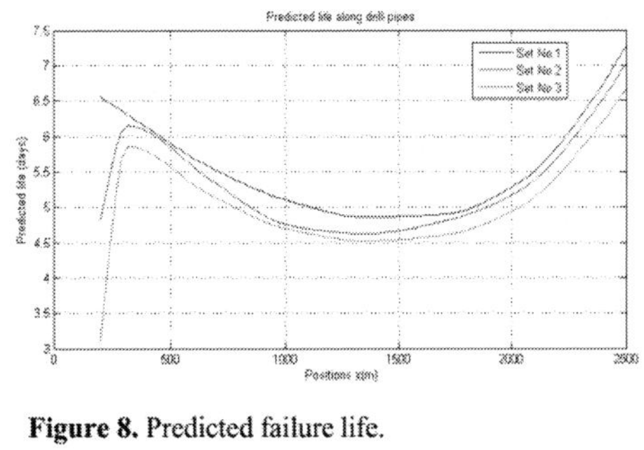

Figure 8. Predicted failure life.

6 Conclusions

A vibration model considering the axial and torsional vibrations of a drill-string is first developed. Then a combined deterministic and random excitation at the bit-rock interaction is assumed and the vibration response is obtained. Based on the vibration model, a spectral method is applied to estimate the effect of the vibration of the drill-string, including both deterministic and random. The ultimate purpose of this paper is to develop a feasible spectral method of fatigue life estimation to account for the vibration's influence on the fatigue of the drill-string.

References

1. W. Bachman, World Oil.132, 104 (1951)
2. R. P. Morgan, M. J. Robin, A method for the investigation of fatigue strength in seamless drillpipe. (1969)
3. G. Y. Grondin, G. L. Kulak,Fatigue of drill pipe. Department of Civil Engineering, University of Alberta (Technical report) (1991)
4. K. Sobczyk, Jr, B. Spencer, Random fatigue: from data to theory(Academic Press, 1992)
5. O. Vaisberg, O. Vincke, G. Perrin, J. Sarda, Oil & Gas Science and Technology, 57, 7 (2002).
6. A. Lubinski, J. of Petrol. Tech.13(2),175(1961)
7. J. E. Hansford, A. Lubinski, J. Petro. Tech., 18, 359(1966)
8. J. Wu, SPE Eastern Regional Meeting,SPE 37353,195,(1996).
9. J. Wu, Oil and Gas Journal, 95, 5(1997).
10. M. Patel, M. Vaz, Proceedings of the Institution of Mechanical Engineers, Part E: Journal of Process Mechanical Engineering, 209(1), 17(1995).
11. J. Grzelak, T. Lagoda, E. Macha Materialwissenscha ftund Werks to fftechnik, 22(3), 85 (1991)
12. Dirlik, T. "Application of computers in fatigue analysis" PhD thesis, University of Warwick,(1985)
13. S. S. Rao, Vibration of continuous systems(John Wiley & Sons2007).
14. M. Vaz, M. Patel, Engineering structures, 17(6), 437(1995)
15. L. D .Lutes, S. Sarkani, Random vibrations: analysis of structural and mechanical systems(Butterworth-Heinemann2004)
16. F. Deily, D. Dareing, G. Paff, J. Ortloff, J. Manufacturing Science and Engineering, 90(2), 217(1968)
17. A. Nieslonyand, E. Macha, Spectral method in multiaxial random fatigue, Springer, (2007)
18. D. Benasciutti, R. Tovo, International Journal of fatigue, 27,8, 867-877(2005).
19. D. Benasciutti, Fatigue analysis of random loadings. PhD University of Ferrara, Department of Engineering(2004)
20. J. W. Miles, Journal of the Aeronautical Sciences, 21, 11, 753-762 (2012).

Dynamic Characteristics Analysis of a Micro Grinding Machine Tool

Wei Li , BeizhiLi and Jianguo Yang

College of Mechanical Engineering, Donghua University, Shanghai, 201620, China

Abstract. Developing miniaturizedultraprecision machine tools (MUMTs) is a rational approach to manufacture micro/mesoscale mechanical components. The dynamic characteristics of MUMTs can be different from that of conventional machine tools, because of its miniaturized structure, downsized components and more flexible assembly. Their dynamics behaviorneeds to be studied. In this paper, the dynamic characteristics of a verticalultraprecisionmicro grinding machine tool were analyzed. An analysis model was established based on finite element method(FEM).Modal analysis was then conducted to obtain vibration mode characteristics of the machine tool, and then the harmonic response analysis was applied to machine tool for verifying its mechanical behaviour. This work helps MUMT developers make a better product at the early design stage with lower cost and development time

1 Introduction

Currently, most of the micro/mesoscale mechanical components are machined by conventional ultra-precision machine tools and MEMS manufacturing techniques. A rational approach for micro/mesoscale manufacturing is to develop miniaturized ultra-precision machine tools(MUMTs). The advantages of MUMTs include: small footprint, low cost of energy, low operational costs, quick response to demands,etc.[1].

Dynamic characteristics of machine tools structure have an important influence on cutting efficiency and machining accuracy. They are important performance indicator of machine tool and have been widely studied for design and optimization of conventional machine tools to enhance their machining performance[2-5]. Dynamic characteristics of miniaturized ultra-precision machine tools also need to be analyzed for structure optimization and performance enhancement. In this paper, the dynamic characteristics of a verticalultra-precisionmicro grinding machine tool were researched based on finite element method(FEM).The natural frequency, vibration modes and displacements of dynamic response were analyzed.

2 Theory of dynamic analysis

Dynamiccharacteristics analysis of the machine tools incl udes modal analysis and harmonic response analysis.

2.1 Theory of modal analysis

Modal analysis is used to determine the vibration characteristics of the structure, it is also the basis for the analysis of the harmonic response [6]. The Eigen frequency equation for modal analysis is given by:

$$\left(K - M\omega_i^2\right)\varphi_i = 0 \tag{1}$$

where K is symmetric stiffness matrix of the structure, $M$is symmetric mass matrix, ω is the natural frequency, and φ is the vibration mode.$i$is the number of modes, and it is equal to the number of degrees of freedom.

2.2 Theory of harmonic response analysis

Harmonic response analysis predicts the response of a structure subjected to continuous harmonic excitation. There are many forced vibration in engineering applications, the vibration is caused by disturbance force or displacement. The dynamic equations of motion can be expressed as:

$$M\ddot{x} + I - P = 0 \tag{2}$$

where M is symmetric mass matrix of the structure, I is internal forces, P is external forces, $\ddot{x}$ is acceleration. By solving equation (2),the steady state response of the structure generated by excitation can be obtained.

3 FEM model

The finite element method (FEM) is a numerical technique for finding approximate solutions to boundary value problems for partial differential equations. FEM cuts a structure into several elements (pieces of the

DOI: 10.1051/matecconf/20179507002

structure), then reconnects elements at "nodes", as if nodes were pins or drops of glue that hold elements together. This process results in a set of simultaneous algebraic equations. The simple equations that model these finite elements are then assembled into a larger system of equations that models the entire problem. FEM then uses variational methods from the calculus of variations to approximate a solution by minimizing an associated error function. Abaqus is one of the outstanding commercial FEM softwares applied to a wide variety of engineering problems. Its accuracy of calculation results have been verified by many experiments.

The main components of the vertical micro grinding machine tool are machine bed, column, spindle, stages. All the stages are set in their initial position. The 3D model of the machine tool was established in Solidworks, and then imported into Abaqus to get the finite element model. In the FEM simulations, tetrahedron elements defined by 10 nodes are suitable for irregular geometry. Meshing were used to mesh the virtual machine structure. The bottom surface of the 3D structure was constrained during the whole computations. Most of the components are connected by bolts, such as the connection between machine bed and column, spindle and stage in z direction. These connection are simulated by bolt load model in Abaqus. The relationship between carriage and base of the stage was simulated with linear spring elements and an example is shown in Fig. 1.

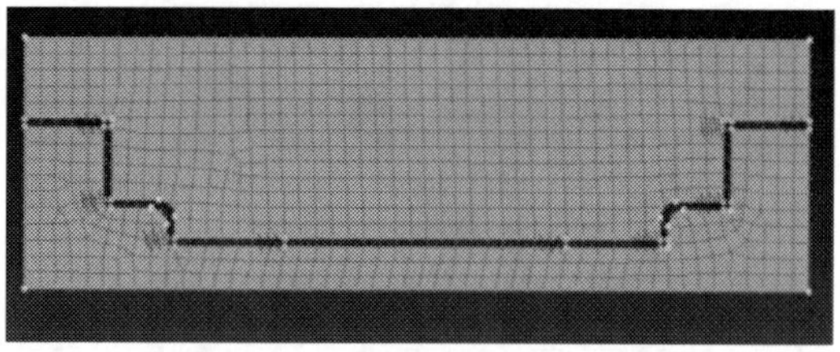

Figure 1. Linear spring element for stage.

The main steps in the whole procedure of FEM analysis are shown in Fig. 2.

1) Modeling the 3D model of all the important component of the machine;
2) Define the materials properties;
3) Mesh the model with C3D10M element (a 10 nodes quadratic tetrahedron);
4) Define load and boundary condition;
5) Solve, and read the results;
6) Improve the structure if necessary.

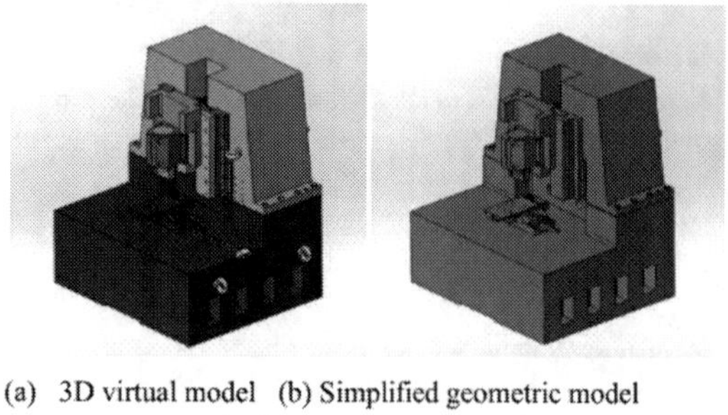

(a) 3D virtual model (b) Simplified geometric model

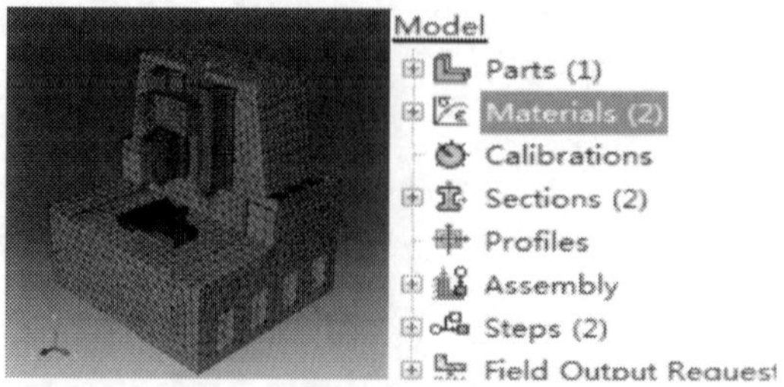

(c) FEM model (d) Define analysis step

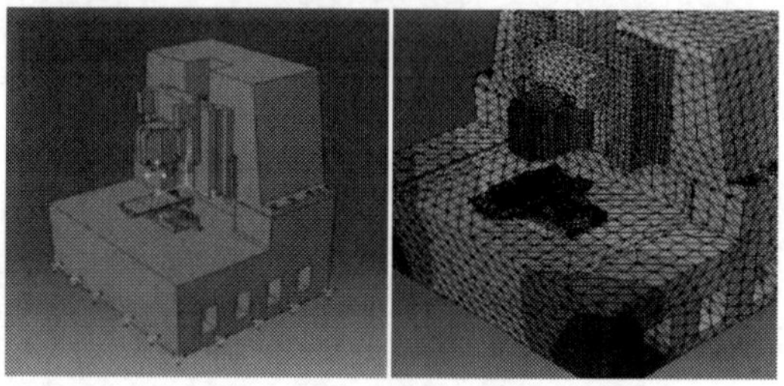

(e) Boundary and load condition (f) Result of analysis

Figure 2. The main procedure for FEM analysis.

4 Modal analysis

Modal analysis can obtain the fundamental vibration mode shapes and corresponding frequencies of the machine tool. Lanzosalgorithm in Abaqus for solving eigenvalue has the most general capabilities and high speed. In this paper, lanczos algorithm is selected to extract the first 60 natural frequencies of the machine tool.

The total effective mass of the model is 1.165T, the excited effective mass in x, y, z directions of the first 60 modes are 1.0625T, 1.0587T, and 1.0733T respectively. The proportion of the effective mass of x, y, z direction in total effective mass is 91.2%, 90.8%, 92.1%. It can be considered enough modal frequencies are extracted [7]. The first 60 natural frequencies of the machine tool are listed as below:

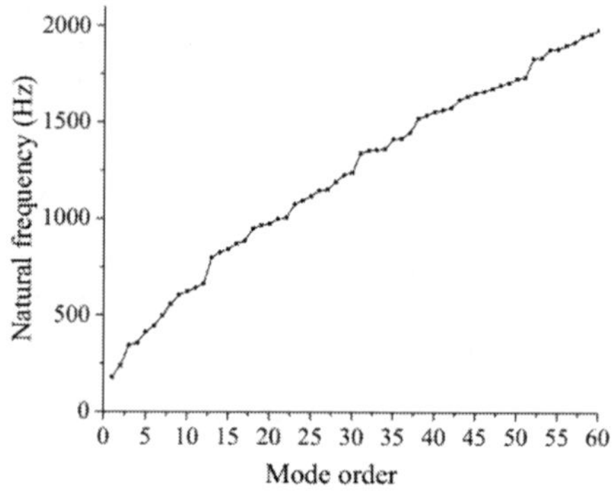

Figure 3. First 60 mode of the machine tool

Table 1 lists the first 10 natural frequencies and the description of the mode for the micro grinding machine tool. Higher order vibration shapes (7th to 10th mode) are considered as combined shapes and are hard to describe.

Table 1. First 10 natural frequencies of the micro grinding machine tool

No.	Mode	Frequency(Hz)
1	Column bending	179
2	Column rolling	239
3	Column twisting	344
4	Z slide bending	356
5	Z slide rolling	413
6	Z slide pitch	444
7	Complex vibration shapes	497
8	Complex vibration shapes	557
9	Complex vibration shapes	604
10	Complex vibration shapes	623

Detailed pictorial descriptions of first 6 natural frequencies and vibration modes are showed in Fig. 3.

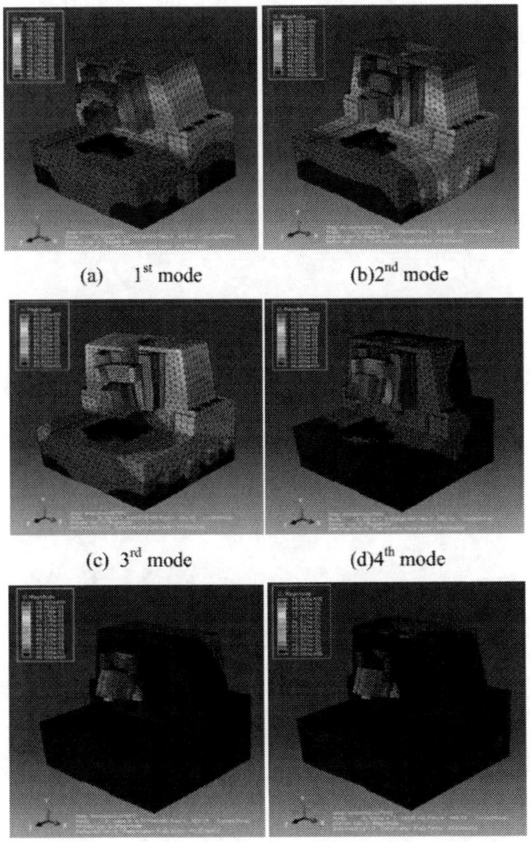

(a) 1st mode (b)2nd mode

(c) 3rd mode (d)4th mode

(e)5th mode (f)6th mode

Figure 3. First 6 natural frequencies and their mode.

The excited effectivemass of each mode is listed in Fig. 4.

It can be found the 2nd,5th,10thand 13th mode are main vibration mode ,the excited mass which corresponds to these mode is much more than other modals excited.

The 2nd mode has the most extracted mass, its corresponding frequency is 239Hz.The 5th mode has the second most extracted mass, its corresponding frequency is 413Hz. The mass extracted by the other mode are much less than the 2nd and 5th mode. It can be considered they have the most influence on dynamic behavior. Figure 3 (b) shows the 2nd mode of the machine tool, and figure 3 (e) shows the 5th mode of the machine tool. With the deformation of column, the displacement of spindle isobviously large, and therefore the column is the weak part in the machine tool configuration. It should be paid more attention to improve the structure of column, which could be achieved by increasing the thickness of the column or mounting bolts in the back of the column.

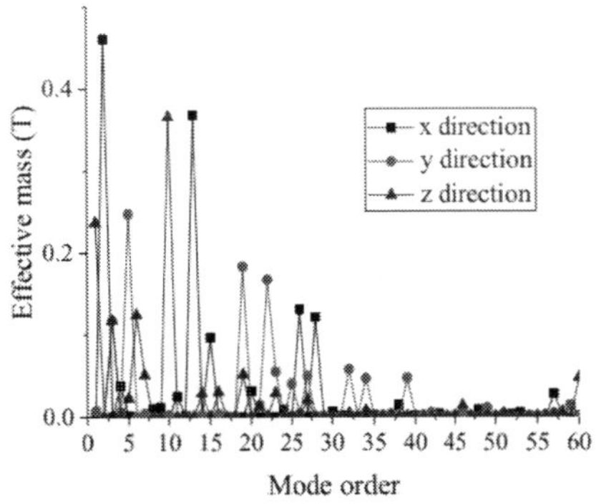

Figure 4. Excited effective mass.

The first natural frequency has been recognized it should be higher than the machine operation frequency in the design of conventional precision machine, for example, the spindle rotational frequency. However, this principle cannot be used in the design of miniaturized ultra-precision micro grinding machine tool, because the maximum spindle speed of the micro grinding machine tool can be up to 2000 Hz (120,000 rpm). Modal analysis results provide a frequency selection guide for machine operation rather than a threshold for maximum spindle speed[8].

5 Harmonic response analysis

When the natural frequencies and their modes of vibration are obtained, harmonic response analyses were performed to determine the steady-state response of the machine to loads that vary harmonically with time. The harmonic response analysis verify whether or not the structure is able to overcome resonance and harmful effects caused by forced vibrations.

In this analysis, the machine structure was excited by a series of harmonic forces acting between work piece and cutting tool. To analysis the dynamic response of the cutting tool under heavy load, 5N force in 3 directions is applied to cutting tool. Frequency sweep range is set to 25-2100Hzwith 25 Hz intervals, it covers the main frequency range of the structure.Figure5.(a)-(c) show the harmonic response of relative displacement in 3 directions respectively.

 DOI: 10.1051/matecconf/20179507002

The maximum dynamic response displacement of about 2.8μm(y direction) occurs at 350Hz, which corresponds to a dynamic loop stiffness of 1.78N/μm. In the actual operation frequency range(1200- 2000 Hz), the dynamic displacement in 3 directions is much less than 0.5μm, and the dynamic loop stiffness is higher than 10N/μm. It can be considered the machine occurs few reasonce when machining.

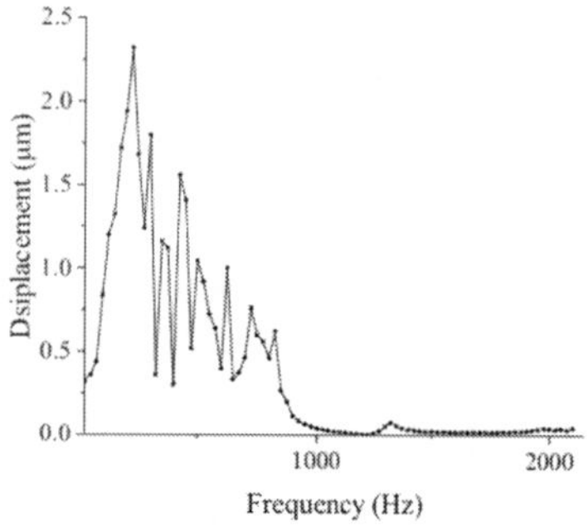

(a)x direction

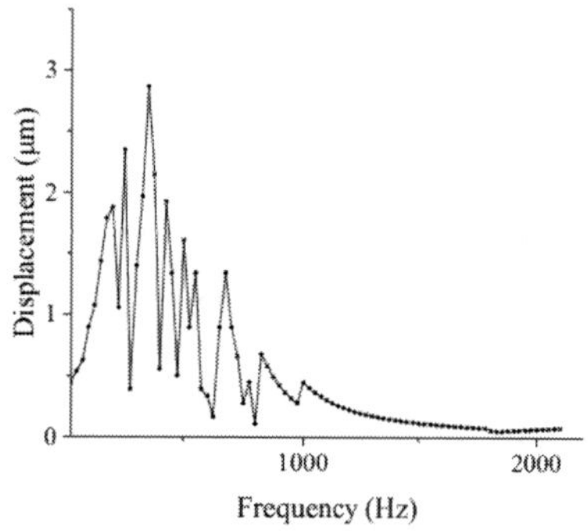

(b)y direction

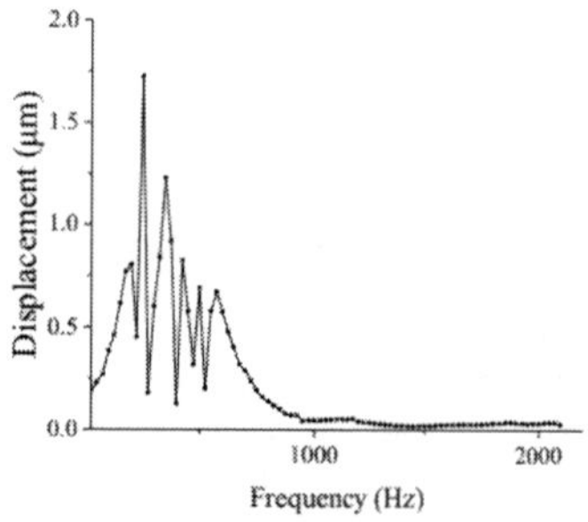

(c) z direction

Figure 5. Harmonic response of micro grinding machine tool

6 Conclusion

A series of simulations were performed on the miniaturized ultra-precision micro grinding machine tool, including modal analysis and harmonic analyses. The simulations based on FEM are used as a powerful and effective tool for supporting the full design process in an iterative manner, which also enable the machine design to be optimized efficiently and effectively.

1) The first 60 mode were extracted, the excited effective mass in 3 directions are obtained, and the main vibration shape in x, y,z directions were found.
2) The cutting tool was applied with force to simulate the harmonic response of displacement, and the response curve was obtained. The max displacement is 2.38μm which occurs at 350Hz.In the actual operation frequency range, the machine tool is hardly to occur reasonce.

This work contributes to the development of a vertical miniaturized ultra-precision micro grinding machine tool. Simulation method is mainly used in this work. Therefore, the physical experiments are needed to verify the results after building the prototype machine.

Acknowledgement

This research was supported byNational Science andTech nology Major Project of the Ministry of Science and Tech nology of China (2013ZX04001-141).

References

1. D. H. Huo, K. Cheng, F. Wardle, Int. J. Adv. Manuf. Technol, 47, **867**(2010)
2. M. Langballe, E. Aasen, T. Mellem, Computers & Structures,4,**149**(1974)
3. R. A. Mahdavinejad,Int. J. Mach. Tools Manuf, 45, **753** (2005)
4. A.V. Pradeep, K. Ramprasad, T. V. Babu, Int. J. Eng Res Appl, 2, **67**(2012)
5. Y. C. Shin, K. F. Eman, S. M. Wu, J. Eng.Ind,111,**116** (1989)
6. W. Heylen, S. Lammens, P. Sas, *Modal Analysis Theory and Testing*(KUL, Leuven, Belgium, 1997)
7. Y.P.Shi, Y.R.Zhou, *ABAQUS example explanation* (CMP, Beijing, China, 2006)
8. D. H. Huo, K. Cheng, F. Wardle, Int. J. Adv. Manuf. Technol, 47, **879** (2010).

DOI: 10.1051/matecconf/20179507003

Research on the development of the centrifugal spinning

Zhang Zhiming , Sun Jun

School of Electrical & Mechanical Engineering, Wuhan Textile University, 430073 Wuhan City, China

Abstract. Centrifugal spinning is a new and efficient method to produce nanofibers quickly. It makes use of the centrifugal force instead of high voltage to produce the nanofibers. The centrifugal spinning has many advantages such as no high voltage, high yield, simple structure, no pollution and can be applied to high polymer material, ceramic and metal material. In order to have more understand about this novel nanofibers formation method, this paper introduces the method of centrifugal spinning and the effect of rotation speed, the properties of material such as viscosity and solvent evaporation, collector distance which have an impact on nanofibers morphology and diameter were also analyzed.

1 Introduction

Nanofibers are widely used in a lot of fields such as textile industry, medicine, energy and polymer transistors and have attracted much attention because of the high surface area to volume and superior mechanical performance[1-2]. Besides, nanofibers have another character that they have shown great mechanical and thermal properties[3-5]. There are a great deal of ways to produce nanofibers, such as stretching method and electrospinning. Electrospinning is a traditional method to generate nanofibers due to the simple operation and the relatively high production efficiency. Electrospinning is a method in which the solution of the material is forced though a spinneret which is charged at high voltage[6]. Electrostatic force makes the material stretched, when the surface tension is overcome, a jet is ejected from the spinneret, then the jet undergoes a spiral trajectory and finally the jet arrives the collector. Although there are many virtues to the electrospinning process, the low fiber production rates, high voltage (5kV~60kV), solvent limitations and so on make it unable to large scale commercial use[7-10]. In recent years, some progresses have been made to increase the production rate of electrospinning. For instance, modied single-needle, multi-needle, needleless systems have been improved to increase nanofiber production rate. But from some previous research, applying the centrifugal force may result in a great increase in the production rate of nanofibers. Some researchers started to use centrifugal forces from a rotating disk to fabricate nano-scale fibers.

So a new way of nanofibers production which is called centrifugal spinning came up, it has overcome many weaknesses of the electrospinning process and has proved to be a better alternative way for nanofibers formation[11]. Unlike the process of electrospinning, the centrifugal spinning utilizes centrifugal force to produce nanofibers. The principle of the centrifugal spinning is shown in Fig.1.

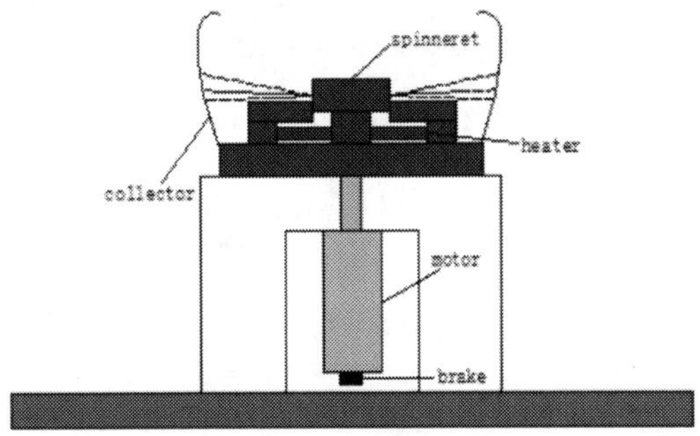

Figure 1. The illustration of the centrifugal spinning

2 The method of centrifugal spinning

The equipment of the centrifugal spinning consists of a spinneret containing material which is in solution form such as polymer solutions, then the motor drives the spinneret to spin about its axis of symmetry[12]. When the angular speed is larger than the limitation which is determined by the balance between surface tension and centrifugal force, the material is then expelled out as a jet from a small orifice of the spinneret[13]. If the forces exerted on the jet are too high, the surface tension of the polymer solution will cause the jet to break up and produce beads. This may occur in certain conditions where the angular velocity of the spinneret is low because the centrifugal forces is low and the viscosity of the polymer solution is not too high. As the solvent evaporates, the material goes along a curved way. At this moment, the centrifugal force drives the material. Then the material continues to move and finally it reaches the wall of the collector. After the jet expelled out, it forced by the air resistance, so the jet follows a curved trajectory. Air resistance has significant effects on the movement of the liquid jet. An effect of air resistance is the deflection of the jet and causes the jet to progressively bend up. Another effect of air resistance is that it will enhance the surface instabilities. Exposing the jet to the high velocity airflow causes the jet to lose its solvent rapidly and as a consequence, the extension of the jet becomes more difficult. It will also result in thicker nanofibers as the other effect of air resistance. Finally, air resistance could lead to fabricate beaded nanofibers[14]. But it is also said that if both the spinneret and the collector rotate at the same time and direction, air stream will rotates as a rigid body. Therefore, the effects which caused by the air resistance on the jet are negligible. So this method is variable and can be divided into three stages.

2.1 The first stage

The first stage of the centrifugal spinning is injecting the material into the spinneret. As mentioned above, there is a motor which drives the spinneret rotating. When the material rotates with the spinneret, the material is forced by many forces, such as friction, gravity, surface tension and centrifugal force, but the surface tension and the centrifugal force take the leading part, others forces are negligible. This stage is shown in Fig.2. As the spinneret rotates, a small solution droplet is seen to originate from the orifice due to the centrifugal forces experienced on the solution within the spinneret. Once the solution droplet reached the lower edge of the orifice, the second process started[15].

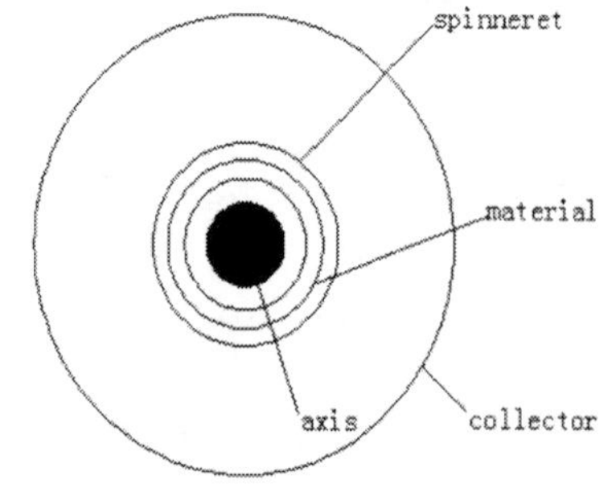

Figure 2. The first stage of the centrifugal spinning

2.2 The second stage

The second stage of the centrifugal spinning is that the material is expelled out from the orifice when the centrifugal force is larger than the surface tension[16-17]. Because of the centrifugal force, the polymer solution sustained by its surface tension is explled outward through the orifice.

The centrifugal force of the material can be expressed as equation(1).

$$F = mr\omega^2 \tag{1}$$

Where m is the mass of the material, r is the radius of the spinneret, ω is the speed of the spinneret.

This equation shows that there is a critical angular velocity exists that produces a jet.

After this stage, the material will go into the third stage.

2.3 The third stage

The third stage of the centrifugal spinning is the stage that the nanofibers form. After the material expelled out from the spinneret, it will go along a curve way with the solvent is evaporating, at the same time, the inertia force will make the material stretch. So the diameter of the material will become smaller and smaller, and finally when it reaches the collector, it forms the nanofibers. The third stage is shown in Fig.3.

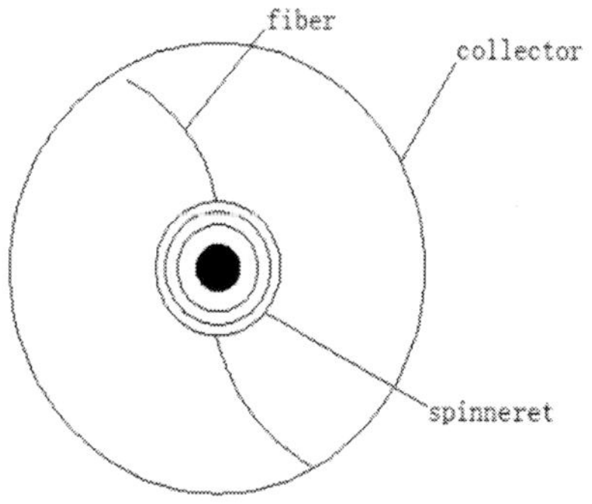

Figure 3. The third stage of the centrifugal spinning

This great new method of nanofibers production relies on the centrifugal force, so it can be used to other new materials that are impossible with the electrospinning. The production efficiency of the centrifugal spinning is much higher than electrospinning. By summarizing, fiber formation requires a critical angular velocity because only if it is reached, a jet can be produced. These conditions have been established in the fiber formation part. Afterwards, another angular velocity must be established at which the fiber reaches what will from now on be described as the trajectory of fiber. During trajectory, the fiber reaches a stable position during which the fiber continues to be stretched without a significant change in its distance to the collector. It is during this period that the fiber diameter experiences the most reduction. It is also during this time that the fiber experiences vibration in the form. After several spinneret rotations, the fiber diameter is reduced considerably and begins to once again expand outwards to eventually reach the collector. Finally, a collector is used to gather the produced fibers.

3 Influence of material parameters to nanofibers morphology

Understanding the three stages of this method for nanofibers formation, it will provide the proper choose of material properties such as viscosity and surface tension, operating parameters such as angular velocity, temperature, machine devices such as spinnerets and collectors. When using centrifugal spinning to produce nanofibers, the material must be in solution or melt form. And these characters of the solution are important in determining the final nanofibers diameter and morphology[18]. There are many parameters that affect the quality of nanofibers. At the process of nanofibers production, two significant material parameters which are the surface tension and viscosity should be considered. Besides, there are some others parameters that should also be considered, such as the speed of the spinneret, the distance of the collector and the diameter of the orifice[19-20]. The other parameter such as initial polymer concentration, solution density, perturbation frequency and solvent vapour pressure has been found to be moderate, so these parameters can be neglected.

3.1 Surface tension

In the centrifugal spinning, the surface tension of the material is a key parameter in nanofibers production. The centrifugal force and the surface tension together affect the formation of nanofibers[21-22]. At the second stage of the centrifugal spinning, the surface tension must be overcome by centrifugal force so the solution of material can form the Taylor cone at the orifice. The surface tension will make surface energy of the solution less and this will lead to the formation of droplets rather than the nanofibers, but the centrifugal force can connect the droplets together and make sure that the nanofibers will not break, so the centrifugal force is the key in the process of nanofibers formation. When other parameters are proper, decreasing the surface tension of the material will help to form the nanofibers.

3.2 Viscosity of solution

Another parameter that must be controlled for good fiber formation is the viscosity of the polymer melt or solution. Due to the large viscosity, the forces being used to draw the fiber may not be strong enough to create a jet. On the other hand, the viscosity is low, the jet may break up and produce beads instead of fibers. For the polymer solution, if the viscosity of solution is too large, the gravitation between molecules will be large and it will make the molecules tangle together, the external force cannot make the solution expel out from the spinneret and the nanofibers cannot be produced too[23]. If the viscosity of solution is too small, the droplet will form or the material will break in the third stage of the centrifugal spinning. The number of the viscosity of solution is related to the temperature and the properties of solution. Besides, the viscosity of solution will affect the velocity of the solution and the morphology of the nanofibers. When the velocity of the solution increases, the diameter of nanofibers will decrease.

3.3 The speed of the spinneret

The speed of the spinneret is also an important parameter that will determine the nanofibers production, because the solution is forced by the surface tension and centrifugal force, when the speed of the spinneret is small, the centrifugal force can not overcome the surface tension, the solution will be stuck in the spinneret. So the speed of the spinneret should be proper. When the speed is proper, the quality and morphology of the nanofibers are good[24]. So the centrifugal force plays a leading role in the process of nanofibers formation, when the speed of the spinneret increase in a right range, the diameter of the nanofibers will decrease. Some experiments show that the 14000 rpm is the best speed to produce nanofibers. It is said that when the speed of the spinneret is increased and the trajectory is expanded, the fiber diameters are reduced, and the fibers produced become more uniform in size.

3.4 The distance of the collector

The distance of the collector means the distance between the spinneret and the collector. When only the distance of the collector is changed, the diameter of the nanofibers does not change much, but there are some changes in the morphology of the nanofibers: the nanofibers are more twisted and broken[25-26]. When the speed of the spinneret increases, the trajectory of the material will expand, so the distance of the collector should also increase. When the viscosity of the solution increases, the speed of the nanofibers will decrease, so the distance of the collector should decrease. But if the distance of the collector is too short, the fiber will not be allowed to stretch enough and then the diameter of the produced fiber will be large. How to decide the most appropriate distance is only depended on how long it will take for the solvent to evaporate, the distance required to allow the fiber to begin to be expelled outwards and the model of collection method used.

3.5 The evaporation rate

The evaporation rate of the polymer solution jet means much in centrifugal spinning. As the jet is expelled out from the spinneret and went toward the collector, the fiber would go on a rapid evaporation which also decreased the diameter of the fiber. If the evaporation rate of the solution is too low, then the fibers may be converted to a thin film as the layers build up on the collector. If the evaporation rate is too high, then the third process of the fiber formation is disturbed and finally it will produce fibers with large diameters[27]. This is dependent of a lot of factors such as the rate of heat supply, boiling point, surface tension of liquid, vapor pressure, interaction between solvent and solute molecules, and air movement on the solution surface.

3.6 The diameter of the orifice

The diameter of the orifice directly determines the flow of the solution and the initial diameter of the fiber. When the orifice is too large, it will result in droplet forming[28]. When the diameter of the orifice is adequately small, it is easier to produce the nanofibers. Some researches indicated that decreasing the orifice diameter decreases the final diameter of the collected fibers. Studying the effects of the various parameters in centrifugal spinning, it was found that the two most influential parameters are the angular velocity of the spinneret and the viscosity of the solution.

4 Conclusions

This paper introduces the method of centrifugal spinning and the three stages of it and also briefly describes the advantages of the centrifugal spinning. And in this paper, the centrifugal spinning and the electrospinning are compared. The centrifugal spinning is a more proper way to produce nanofibers, although there are some achievements, this method also has a lot of problems that need to be studied. For example, the parameters that affect the nanofibers production and the morphology of nanofibers are already analyzed, but due to the complexity of these factors, the mathematical model that can describe the relationship between these parameters and the diameter of nanofibers is not established, and there is no effective way to solve the problems that are relation to the effects of solvent evaporation and temperature on fiber diameter and morphology. But with the development of science and technology, there will be more and more researches on these problems and the centrifugal spinning will finally be widely used.

References

1. R. Nayak, R. Padhye, I. L. Kyratzis,Textile Research Journal,82,129(2012)
2. S.Ramakrishna,K.Fujihara,W.E.Teo,World Scientific Publishing Company,(2006)
3. Y. Ito, H. Hasuda, M. Kamitakahara, Journal Of Bioscience and Bioengineering,100,43(2005)
4. J. Yoon, Y. S. Jung, J. M. Kim, Advanced Functional Materials,19,209(2009)
5. S.Ramakrishna, K.Fujihara, V.K.Ganesh, American Scientific Publisher, 7(2006)
6. A. Baji, Y. Mai, S. C. Wong,Composites Science and Technology,70,703(2010)
7. K. Sarkar, C. Gomez, S. Zambrano, Materials Today, 13,11(2010)
8. Z. Mceachin, K. Lozano, Journal of Applied Polymer Science,126,473(2012)
9. T. Senthilram, L. A. Mary, J. R.Venugopal,Materials Today,14,226(2011)
10. X. W. Zhang, Y. Lu, Polymer Reviews,54,677(2014)
11. E. Kolk, Texas Center for Educational Research,23, 429(2005)
12. Z. M. Zhang, Q. Xu, S. Q. Mei. Polymer Bulletin, 5, 29(2013)
13. P. Mellado, H. A. Mcilwee, Applied Physics Letters, 99,20(2011)
14. J L. Cummings, D P. Howell, Journal of Fluid Mecha nics, 4,361(1999)
15. A. Valipouri, S.A.H. Ravandi, A.R. Pishevar, Fibe rs and Polymers,14(6),941-949(2013)
16. S. P. Decent, A. C. King, I. M. Wallwork, Journal of Engineering Mathematics, 42,265(2002)
17. I. M. Wallwork, S. P. Decent, A. C. King, Applied M athematical Modelling,33,4283(2002)
18. S. Padron, D. Caruntu, K. Lozano, Asme Internation al Mechanical Engineering Congress & Exposition, 821(2011)
19. S. A. Theron, A. L. Yarin, E. Zussman, Polymer,46, 2889(2005)
20. D. H. Reneker, A. L. Yarin, H. Fong, Journal of Applied Physics, 87,4531(2000)
21. H. Z. Xu, H. H. Chen, X. G. Li, Technical Textiles,1, 25(2016)
22. S.Ramakrishna, T.C.Lim, R.Inai, Mechanics of Adv-anced Materials & Structures, 13(1),77-81(2006)
23. T. D. Song, X. H. Li, Y. Liu, China Textile Leader,1, 61(2014)
24. N. E. Zander, Journal Of Applied Polymer Science, 132,2(2015)
25. Y. Lu, Y. Li, S. Zhang, Polymer Journal, 49,3834 (20 13)
26. X. Li, H. Liu, J. Wang, Polymer,53,248(2012)
27. S. Padron, A. Fuentes, D. Caruntu, K. Lozano, Journ al of Applied Physics, 113(2),24318(2013)
28. S. Parden, A. Fuentes, Journal of Applied Physics, 113,1(2013)

Correlations between Axial and Oblique Loaded Column

Salwani Mohd Salleh[1] , Barkawi Sahari[2] , Aidy Ali[3], Mas Ayu Hassan[1], and Rosdi Daud[1]

[1] Faculty of Mechanical Engineering, Universiti Malaysia Pahang, 26600 Pekan, Pahang.
[2] Department of Mechanical and Manufacturing Engineering, Faculty of Engineering, Universiti Putra Malaysia, 43400 Serdang,Selangor
[3] Department of Mechanical Engineering, Faculty of Engineering, Universiti Pertahanan Nasional Malaysia, Kem Sungai Besi, 57000 Kuala Lumpur.

Abstract. In an impact, a structure is rarely subjected to pure axially loading. Load endured by an axially loaded column has been analytically derived by previous researchers, whereas it is very limited works on predicting the obliquely loaded column performance. In this study, the force response curve of the obliquely loaded empty aluminum column is analyzed and compared to its axially loaded column. Theoretical formula to calculate the axial mean force has been applied to obliquely loaded column, and it has yields a largely deviated value. Thus, a modified equation to cater for obliquely loaded column has been proposed.

1 Introduction

The recent trend in automobile design is aimed at improving crash safety. For the crash safety, energy absorbing members have to absorb sufficient collision energy. Automotive structures are designed to sustain impact loading in diverse crash directions such as frontal, lateral and rear impact. Above all crash events, frontal collision was identified as the most common accidents on the road and gives rise to high portion of death [1-2]. Thus, it is vital to have an efficient energy absorbing structure on the front side of an automobile. The main energy absorber on the front side of automotive body is crash box and longitudinal members that commonly represented by column.

The threat to passenger safety arouses the need for the crash test. In accordance with the safety requirement, Federal Motor Vehicle Safety Standard (FMVSS) No. 208 has set a procedure for a frontal rigid barrier test of up to 48 kmph, at angles from the perpendicular (90 degrees) to the line of travel of up to 30 degrees [3]. It has been a current practice to initiate crash testing with simulation to avoid expensive investment on repeated physical testing. Due to the rapid development in the computational capabilities, simulation is well accepted by the as a way to reduce cost associated with physical test of preliminary design. Numerous research works has been presented on structure subjected to axial and bending [4-7], whereas a very limited study has been conducted on theoretical equation for oblique loading. This study will propose improvised axially loaded equation to predict the mean force of the obliquely loaded column based on simulation.

2 Theoretical Background

Crashworthiness parameters are used to assess the performance of energy absorber structure. A good crashworthiness structure can absorb high energy in a controlled manner and yet the forces experiences during collapse do not exceed the allowable range. The primary data usually extracted from a crushing test is the load-displacement curve. Typical load-displacement curve for a thin-walled structure is shown in Fig. 1.

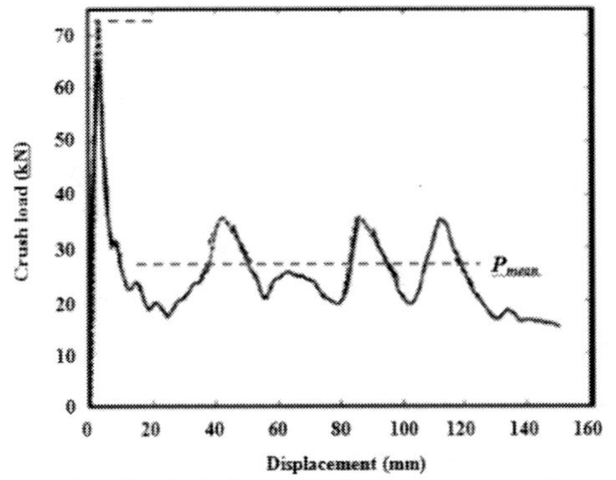

Figure 1. Typical force-displacement curve for axial loading

The mean load, P_{mean} is a parameter commonly used to calculate the energy absorbing capability of a structure. Generally, it is defined as [8]:

$$P_{mean} = \frac{E_a}{\delta} \quad (1)$$

where is the total energy absorbed during collapse and represents displacement. According to Wierzbicki and Abramowicz [7], the mean crushing force of axially loaded square tube can be calculated by

$$P_{mean,A} = 13.06\sigma_0 \cdot w_m{}^{1/3}t^{5/3} \quad (2)$$

where σ_0 is the characteristics stress taken as

$$\sigma_0 = \frac{\sigma_{0.2}+\sigma_u}{2} \quad (3)$$

$\sigma_{0.2}$ is the stress at 0.2% plastic strain, is the ultimate stress and is the mean width defined as

$$w_m = w - t \quad (4)$$

t is the column thickness and w is the outer column width.

2 Methodology

2.1 Validation

Fig. 2 shows the deformation pattern for both experimental and simulation. The simulated crushed empty column is deformed in the same manner as in the experiment.

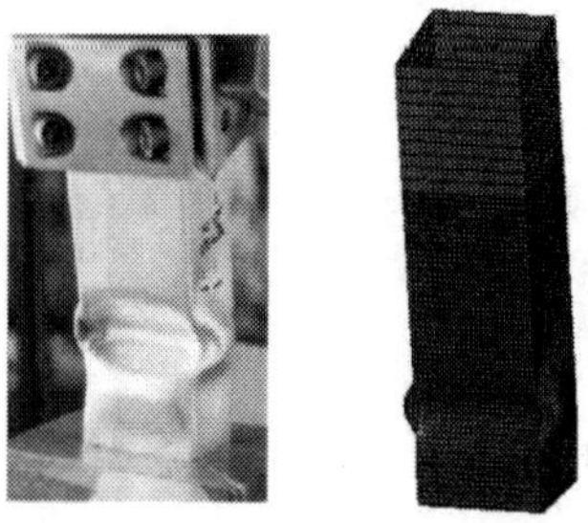

(a) (b)

Figure 2. Deformation of empty column subjected to 5 degrees loading angle (a) experimental [9] and (b) simulation

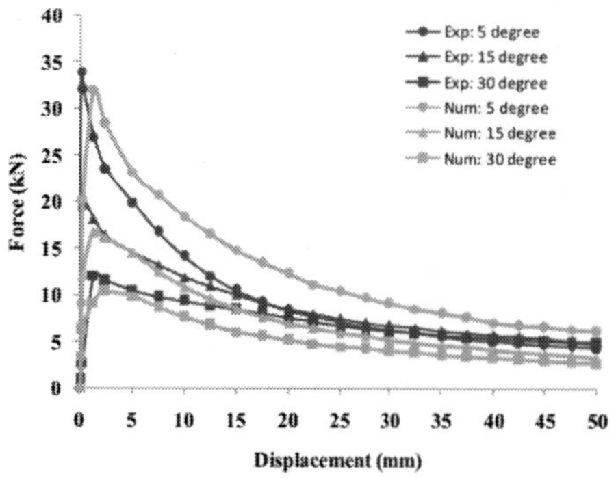

Figure 3. Force-displacement curve comparison between experimental and simulation

Fig. 3 compares quasi-static simulation of aluminum column subjected to oblique loading to the experimental work done by previous researcher. Good agreement between the simulation and experimental result is obtained.

2.2 Finite element analysis

In this study, the model is drawn using the CATIA and meshed by using Hypermesh. The keywords file is then imported into Ls-Dyna. Fig. 4 shows mesh of the square column and direction of load during impact.

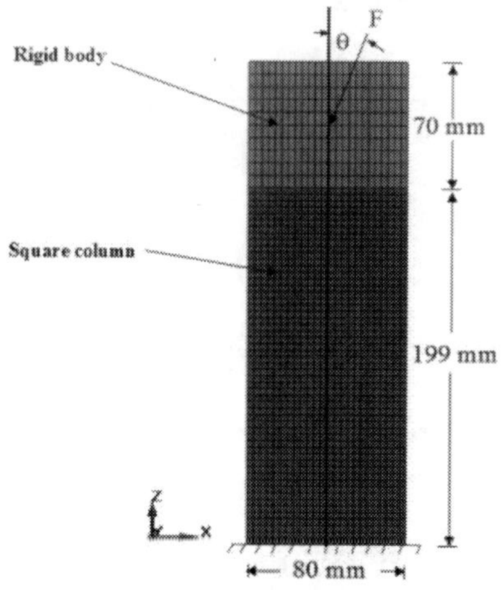

Figure 4. direction of load during impact

The square column and the automotive side member are modeled using shell elements in LS-Dyna. There is a broad library of material models available in LS-Dyna. The material model type 24 [PIECEWISE_LINEAR_PLASTICITY] is used to model the behavior of aluminum alloy under impact. A stress versus strain curve is defined for this elasto-plastic aluminum alloy. Aluminum alloy used in this study is AA6060, temper T4. Table1 summarizes the mechanical properties of AA6060 T4. The important properties are the density, ρ; Young's modulus, E; Poisson ratio, ν; yield stress, σ_y; ultimate tensile stress, σ_u and fracture strain, εf.

Table 1. Mechanical properties of AA6060 T4 [10]

Material	ρ (kg/m^3)	E (N/mm^2)	ν	σ_y (N/mm^2)	σ_u (N/mm^2)	ε_f
AA6060	2700	66820	0.28	91	175	0.22

In the dynamic simulation, the square column is subjected to oblique loading for an angle between 5 to 30 degrees as shown in figure 3.2. A velocity of 48 km/hr as indicated in the frontal barrier crash test in fmvss no. 208 is prescribed to the rigid body.

3 Result and Discussions

Fig. 5 plots the correlation between the numerical and theoretical results. The straight line indicates the unity between the numerical and theoretical values. Theoretical values are calculated based on equation (2). It can be seen that the theoretical model over-predicted the Pmean values.

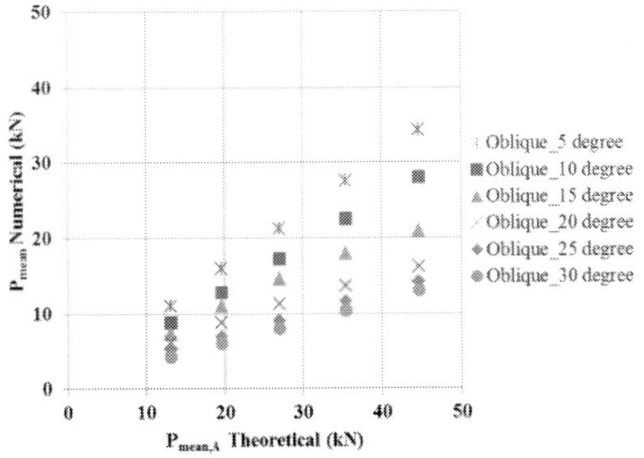

Figure 5. Correlation between results obtained from numerical and theoretical

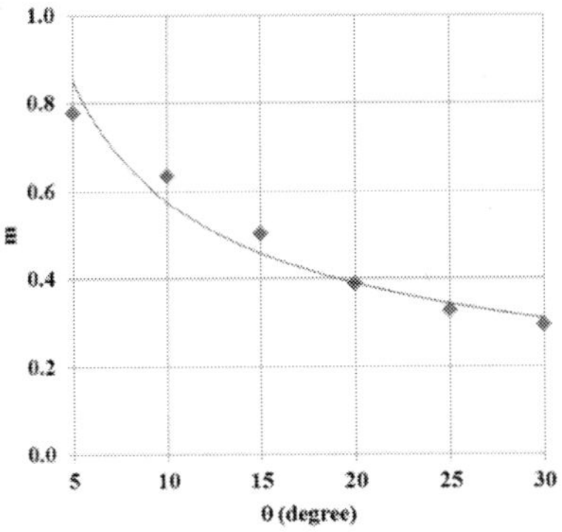

Figure 6. Slopes of various loading angles

It is understandable since equation (2) is meant for axially loaded column. Axially loaded column commonly failed under progressive buckling, absorbing more energy, Ea and yielding a higher Pmean values. Whereas, obliquely loaded column is commonly associated with local bending, reducing the energy absorption capability, Ea and causing a lower Pmean values.

Fig. 5 suggests that Pmean,A for theoretical of different loading angle can be shifted to its numerical values by introducing an equivalent factor, Ef.

By setting an intercept of zero, fitting lines are drawn for each loading angles, θ in Figure 6. The slope, m, obtained from Fig. 4 is plotted in Fig. 6.

Thus, the power law equivalent factor, Ef generated from Fig.6 is

$$E_f = 2.106 \cdot \theta^{-0.56} \quad (5)$$

where θ is the loading angle. In order to better predict the mean force for obliquely loaded column, this study proposes a modified equation of the axial mean force, Pmean,A to incorporate the loading angle associated with oblique loading. The oblique mean load is defined as

$$P_{mean,O} = 13.06 \cdot E_f \cdot \sigma_0 \cdot t^2 \cdot \left(\frac{W_m}{t}\right)^{1/3} \quad (6)$$

Fig. 7 plots the correlation between the numerical and theoretical results. The theoretical values are calculated based on equation (6). It can be seen that the modified equation improves the prediction of obliquely loaded column.

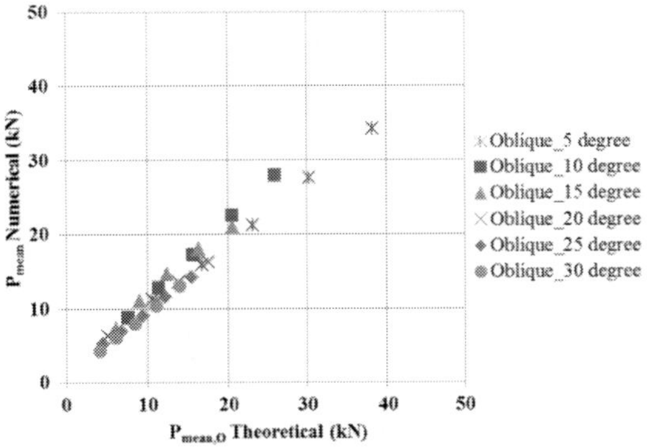

Figure 7. Correlation between results obtained from numerical and theoretical

4 Conclusion

Theoretical formula to calculate the axial mean force has been applied to obliquely loaded column, and it has yields a largely deviated value. Thus, a modified equation to cater for obliquely loaded column has been proposed as:

$$P_{mean,O} = 13.06 \cdot E_f \cdot \sigma_0 \cdot t^2 \cdot \left(\frac{W_m}{t}\right)^{1/3}$$

Acknowledgement

The authors would like to thank Universiti Malaysia Pahang (UMP) for financial support on publication under RDU1303114.

References

1. Kokkula, S., Hopperstad, O. S., Lademo, O. G., Berstad, T., and Langseth, M. (2006). Offset impact behaviour of bumper beamâ€"longitudinal systems:

numerical simulations. International Journal of Crashworthiness, 11(4), 317 - 336.

2. Kokkula, S., Langseth, M., Hopperstad, O. S., and Lademo, O. G. (2006). Offset impact behaviour of bumper beamâ€"longitudinal systems: experimental investigations. International Journal of Crashworthiness, 11(4), 299 - 316.
3. Hollowell, W. T., Gabler, H. C., Stucki, S. L., Summers, S., and James R. Hackney, N. (1999). Updated review of Potential Test Procedures for FMVSS No. 208: National Highway Traffic Safety Administration (NHTSA)o. Document Number)X.
4. M. S. Salwani, B. B. Sahari, Aidy Ali and A. A. Nuraini, The effect of automotive side member filling on car frontal impact performance, Journal of Mechanical Engineering and Sciences, 6 (2014) 873-880.
5. Zhang, Z. Wen, H. Zhang, Axial crushing and optimal design of square tubes with graded thickness, Thin Wall Struct, 84 (2014), pp. 263–274.
6. X. Zhang, H. Zhang, Axial crushing of circular multi-cell columns, Int J Impact Eng, 65 (2014), pp. 110–125.
7. R. Mohammad Javad and J. Ali, Effect of initiator, design, and material on crashworthiness performance of thin-walled cylindrical tubes: A primary multi-criteria analysis in lightweight design, Thin-Walled Structures 96 (2015) 169–182.
8. W. Abramowicz and N. Jones, Dynamic progressive buckling of circular and square tubes. International Journal of Impact Engineering, 4(4) (1986) 243-270.
9. Reyes, M. Langseth and O.S. Hopperstad, Crashworthiness of aluminum extrusions subjected to oblique loading: experiments and numerical analyses. International Journal of Mechanical Sciences, 44(9) (2002) 1965-1984.
10. Reyes, M. Langseth and O.S. Hopperstad, Square aluminum tubes subjected to oblique loading. International Journal of Impact Engineering, 28(10) (2003) 1077-1106.

 DOI: 10.1051/matecconf/20179507005

An Experimental Investigation of Cutting Temperature and Tool Wear in 2 Dimensional Ultrasonic Vibrations Assisted Micro-Milling

Mohd Rasidi Ibrahim[1], Zulafif Rahim[1], Erween Rahim[1], Latif Tobi[1], Kai Cheng[2], Hui Ding[3]

[1]Faculty of Mechanical & Manufacturing Engineering, Universiti Tun Hussein Onn Malaysia, Parit Raja, Johor, Malaysia
[2]Advance Manufacturing Engineering Enterprise, School of Engineering and Design, Brunel University, West London UB8 3PH,
[3]School of Mechatronics Engineering, Harbin Institute of Technology, Harbin 150001, China

Abstract. Two dimensional Ultrasonic vibration assisted milling (2D UVAM) well knows process that involved in high tech system to generate ultra range of frequency applied to the milling process. More industries nowadays become aware taking this opportunity to improve their productivity without decreasing their product accuracies. This paper investigate a comparative machining between UVAM and conventional machining (CM) in tool wear and cutting temperature in milling process. Micro amplitude and sine wave frequency will be generate into the workpiece jig by piezo-actuator. Thus, creating a micro gap that allow heat remove effectively with the chip produces. A more complex tool trajectory mechanics of 2D UVAM has been found during this research. The approaching the tool tip into the workpiece surfaces is affected by the amplitude displacement along the frequency applied. It is found that the tool wear was reduce and surface roughness improvement by applying the 2D UVAM compared to the CM when choosing the optimum amplitude and appropriate frequency.

1 Introduction

The using of advance and high machine technology for manufacturing production has become increasingly year by year. The high demand become seriously affected by the rapid innovation and new technology especially in telecommunication, military, aerospace, electronic gadget etc. Moreover, the development of new high specification and high strength material is a new challenge to the researcher and industries to fabricate for making a mould, tools, jig, die and so on.

Common conventional milling is a traditional method that needs to be improving along the technology development. Thus, 2D UVAM is the one machining system that attracted researcher worldwide for improving time to time. Lian and Guo found the surface roughness on the machining surfaces of Al6061 has smaller roughness that machining by ordinary micro-milling [1]. They conclude the UVAM has addition tool trajectory making the surface better than CM. It has been proved by their mathematical modeling by using Matlab software to simulate the trajectory of the cutter tooth in Figure 1.

Coating technology has one of the solutions for solving the wear problem in machining. The coatings are about 3μm - 5μm layer applied to the cutting tool especially in end mill on the tool tip. It is no doubt the coated tool proven hardened the endmill and approximately up to 80 to 100GPA and thermal stability 1200°C [2]. However, the cost to coat the endmill to its application is below the expectation to some of the small scale industries. The experiment has been carried out to investigate machinability improvement between CM and 2D UVAM focused on the tool wear and cutting temperature looking deeply on the tool UVAM locus trajectory.

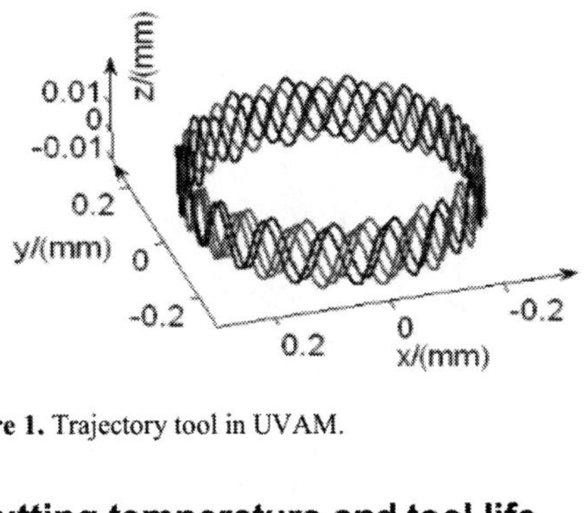

Figure 1. Trajectory tool in UVAM.

2 Cutting temperature and tool life

Rubbing and friction during ploughing and shearing during milling process create exponential energy. As a consequence the energy will convert to heat and spread between the tool and the workpiece. The temperature on the rake face angle has a strong effect on tool life [3]. When the tool tip temperature increases, the changes of graphitization molecules will that soften the tool rapidly occur. The increasing of temperature will affect the tool life because of:

- Raising of cutting force/speed
- Constant stress between tool and workpiece
- Inappropriate cooling medium
- Tool becoming blunt

Intermittent cutting in VAM creates a gap between the tool and the workpiece. This can be expressed:

$$\begin{aligned} y &= a \sin \omega t \\ x &= b \cos \omega t \end{aligned} \qquad (1)$$

Where y, x and ω are the displacement, amplitude acting in X, Y axis and angular velocity of tool tip respectively. In conventional machining, the tool is always kept in contact with the workpiece so the heat will increase as long as it is still in contact. The tool cutter will soften more than half of their hardness when reaching high temperatures.

In 2D UVAM, this will not happen as the temperature remains constant so as to maintain the tool's hardness. Figure 2 shows the mechanism in of 2D UVAM towards workpiece. It also shows how the intermittent mechanism, by amplitude and frequency pulse impacts on tool separation from the chip, promotes aerodynamic lubrication increases. Consequently, shows the tool path mechanism of conventional machining. In addition, the clearance angle and rake angle are free from cutting force and release the stress of pressure from the rubbing and ploughing process.

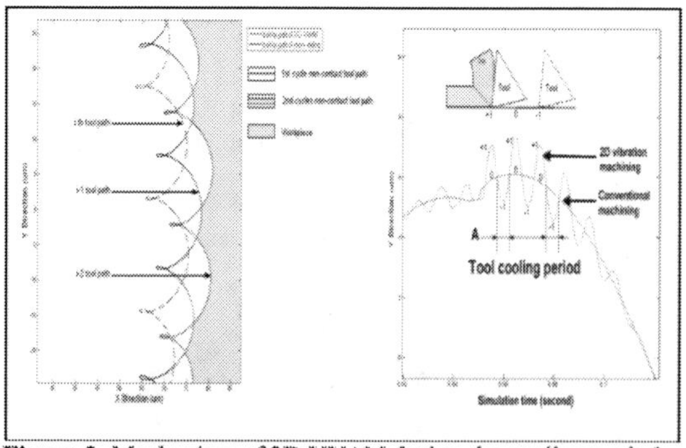

Figure 2. Mechanism of 2D UVAM during the cooling period.

The displacement of tool vibration in the cutting operation at a frequency in the period of A. The digging/ploughing between point +1 and -1 is to remove material causes chips to form. The retraction tool and workpiece between points +1 to -1 promoted dynamic force for the chip to leave the cutting area efficiently. Thus, this beneficial strongly agrees with Shaw theory that 90% of the heat generated goes into chips, 5% into the workpiece and 5% into the tool [4].

Figure 3 shows a simulation created from Matlab platform showing the cooling gap that has been created during the VAM cutting process. With the parameters a is 3 µm, b is 1 µm, frequency X and Y are 1,000 Hz, feedrate is 2 µm per tooth and spindle speed is 2,000 rpm. The clearance or gap during alternating motion allows heat transfer to work efficiently toward air or coolant, cooling down the tool and surface of the workpiece, which prevents the cutting temperature on the tool and workpiece rising especially, in very hard and very soft materials. It also reduces the time available for tools to change their molecular structure or for other types of wear.

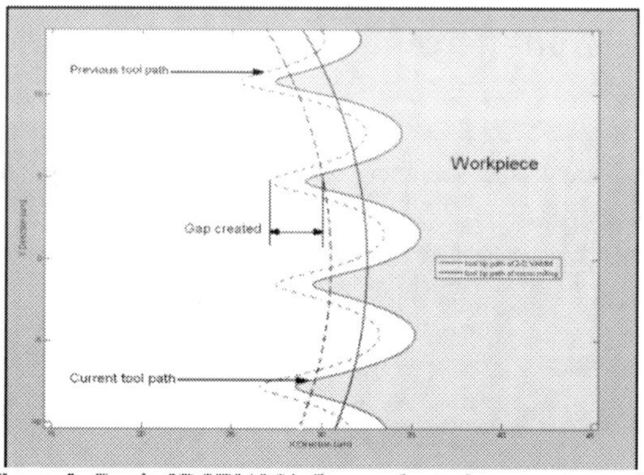

Figure 3. Gap in 2D UVAM influence the tool temperature and tool life.

It is found, during the intermittent cutting, that VAM creates a small energy instantaneous mechanics to lower down tool force affect the decreasing the local stress on the tool. It has been reported that the tendency of cutting force reduces about 40% to 50% by alternating cycles by frequency and amplitude.

3 Surface roughness

The surface produced by conventional milling depends on the result of geometric and kinematic reproduction of the tool point shape factor [4]. Through the experimental studies, the concavity and axial relief angle in the end mill has a big effect on the surface produced, especially on the peak to valley measurement. Figure 4 shows the simulation of mechanics cutting by conventional machining and 2D UVAM from side view.

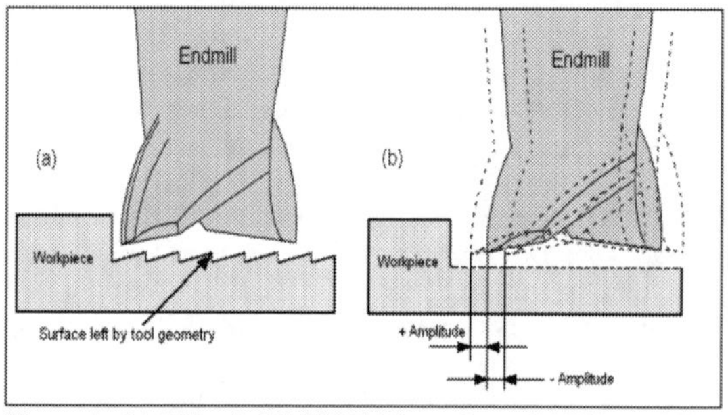

Figure 4. (a) Effect in 2D UVAM in surfaces, (b) Effect of CM in surfaces.

The circular motion nature of 2D UVAM allows the tool cutting point to traverse the cut area from the preceding cycle and remove the material peaks and valleys left over from the last time. During the cutting circular motion of 2D UVAM, the smaller force will deduce the average force and promote to surface accuracy and surface finish by compensating the relative vibration between cutter and workpiece. In addition from that point the reduced tool cutting force also insist chatter on the surface produced.

So, by investigating the Figure 4, the peak to valley of the surface has been reduced by the cutter via an overlapped process. The higher frequency of cutting, the more chances there is of the tool cutter in one particular area reducing the peak and valley to obtain a good surface result. Rasidi and Ding found the 2D UVAM was helping to cut the crest of the surfaces into smaller pieces while opposing the frequency and amplitude thus, eliminated by the alternating cycles phenomena [5]. From this, we observe that the good surface is made by cutting the crests and peak on the surface with the vibratory tool.

4 Experimental case study

Figure 5 shows the comparison of 2 tools that have been cut without 2D UVAM (a) and with aid of 2D UVAM (b). The distance of slot cutting is extended to 90mm to 90mm to observe the tool wear. Figure a clearly shown the wear on the rake angle surface due to abrasive wear. Abrasive wear cause by two common reasons, first the cutting speed increase affect the tool surface sliding in a short time period. Second, rapid increasing a temperature in the cutting area. The hardness of the tool material wills drops directly proportional with temperature. These phenomena called a thermal softening behaviour. Figure (b) shows the abrasive wear has take part as same position as Figure (a), but there is approximately 60% reduction of the worn area. It is because the 2D UVAM create an "instantaneous gap" helping the chip flush away efficiently thus reduced the temperature. In result from that, 2D UVAM chip produced much smaller and shorter chip instead large and continuous chip in CM. [6]. So the temperature in 2D UVAM is reduced and keep constant thus prevent the tool changes the hardness properties. It will keep the tool last longer during the cutting process.

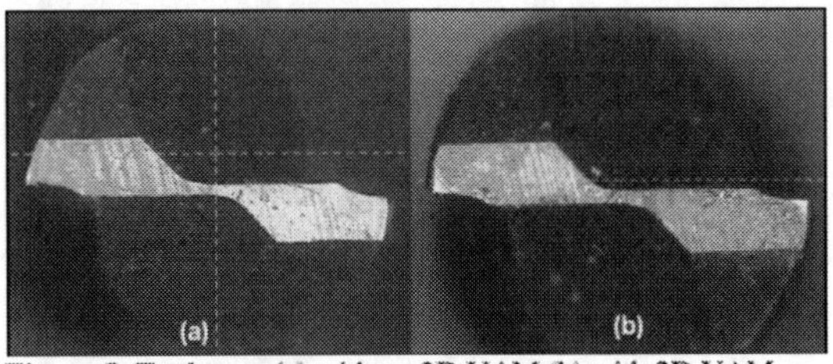

Figure 5. Tool wear (a) without 2D VAM (b) with 2D VAM.

5 Conclusion

From this paper we can conclude the 2DUVAM have distinct advantages compared with conventional machining methods summaries on Table 1 below.

Table 1. Summary camparison between CM and 2DUVAM.

	Conventional Milling	Vibration assisted milling	Comments
Cutting temperature	• No gap to dissipate heat energy. • Tool always making contact with workpiece.	• Alternating cycles (gap) promotes heat dissipate between cutting tool and workpiece, therefore 40% to 50% heat reduced. • Less time contact between tool and workpiece providing chances of tool and workpiece to release heat energy. • Lubricating flowing resulting form periodic tool-work separation. • the work-tool contact area varies at several points on the elliptical path.	• 100% cutting time tool point keep in contact on conv. machining with the workpiece, but in 2D VAM, the clearance distance has been created when the tool point retracted from the workpiece.
Tool life	• Temperature rising is accumulating to soften the tool. • Tool always in contact and always in stress and strain condition.	• Constant temperatures encourage tool cutter have an extra hardness and strength ability. (Thermal softening theory). • 40% to 50% time contact reduced between rake angle and clearance angle and workpiece chances of tool to release stress force. • Intermittent contact between tool and work material in VAM reduces the time available for graphitization and other wear reactions.	• Tool life in conv. machining is reduced cause by thermal softening theory cause by high temperature, but in 2D UVAM, the temperature will constant and reduced 30%-40%. • Tool failure ratio from stress and strain is low.

Acknowledgement

The authors would like to thank Registrar Office, ORICC UTHM for facilitate in this research, Precision Machining Research Centre (Premach) UTHM and lastly the Cedrat Technology SA for providing the piezo-actuator system.

References

1. H. Lian., Z. Guo,Z. Huang, Y. Tang, J. Song : Experimental Research of AL6061 on Ultrasonic Vibration Assisted Micro-Milling, The Seventeenth CIRP Conference on Electro Physical Chemical Machining (ISEM) volume 6, (2013)
2. Information on http:// http://www.endmill.com/pages/coatings.html
3. Negeishi, N. Elliptical vibration-assisted machining with single crystal diamond tools" MS thesis, North Carolina State University, 2003.
4. **Shaw, C.** "Energy conversion in cutting and grinding", Annals of the CIRP, 1996, 101-104.
5. I. Rasidi, N. H Rafai, E. A. Rahim, S.A. Kamaruddin, H. Ding and K. Cheng: An investigation of Cutting Mechanics in 2 Dimensional Ultrasonic Vibration Assisted Milling Toward Chip Thickness and Chip Formation, 3rd International Conference of Mechanical Engineering Research, IOP Conference Series: Material Science volume 100 (2015).
6. I, Rasidi, N.H. Rafaai, E. A. Rahim, S. A. Kamruddin, H. Ding and K. Cheng: An investigation of Cutting Mechanics in 2 Dimensional Ultrasonic Vibration Assisted Milling toward Chip Thickness and Chip Formation, 3rd International Conference of Mechanical Engineering Research (ICMER 2015), IOP Conf. Series, Material Science and Engineering 100, 2015.
7. Brehl, D., and Dow, A.: Review of vibration-assisted machining. Precision Engineering, 32, 2008, 153-172.
8. Yusuf, A. "Manufacturing Automation: Metal Cutting Mechanics, Machine Tool Vibrations and CNC Design. Cambridge University Press, 2000.
9. I, Rasidi, N.H Rafai, E.A. Rahim, S. A. Kamaruddin, H. Ding and K. Cheng : An investigation of cutting mechanics in 2 dimensional ultrasonic vibration assisted milling toward chip thickness and chip formation
10. Ahmed, N., Mitrofanov, A., Babitsky, V., and Silberschmidt, V. "Analysis of forces in ultrasonic assisted turning". Journal of Sound and Vibration, 2007, 308, 845-854.

 DOI: 10.1051/matecconf/20179507006

Differential Quadrature Method Based Study of Vibrational Behaviour of Inclined Edge Cracked Beams

Shivani Srivastava and Raju Sethuraman

Department of Mechanical Engineering, IIT Madras, India

Abstract The study of vibration behaviour of cracked system is an important area of research. In the present work we present a mathematical model to study the effect of inclination, location and size of the crack on the vibrational behavior of beam with different boundary conditions. The model is based on the assumption that the equivalent flexible rigidity of the cracked beam can be written in terms of the flexible rigidity of the uncracked beam, based on the energy approach as proposed by earlier researchers. In the present work the Differential Quadrature Method (DQM) is used to solve equation of motion derived by using Euler's beam theory. The primary interest of the paper is to study the effect of inclined crack on natural frequency. We have also studied the beam vibration with and without vertical edge crack as a special case to validate the model. The DQM results for the natural frequencies of cracked beams agree well with other literature values and ANSYS solutions.

1 Introduction

The behavior of structures containing cracks is the interesting area of research in the light of potential developments in automatic monitoring of structure quality. The study of influence on eigen-frequencies and modes shapes of the structure due to crack is important in many aspects. A number of research has been reported their work in this area. A crack introduces a local flexibility in a system which is a function of crack depth. The dynamical behavior of the system and its stability characteristics changes due to the flexibility. Here, in this work, we have taken the beam structure, specifically the Bernoulli-Euler beam is of our interest with appropriate boundary conditions. In this paper, the assumption and formulation of the model has been discussed in Section 2. Section 3 deals with the methodology used to solve the governing equation. Case studies are reported in Section 4, and the concluding remarks are given in Section 5.

2 Model

The variation of the equivalent bending stiffness and depth (along the beam length) for a cracked beam are obtained using an energy-based model as proposed by Yang et.al. [1] to investigate the influence of cracks on structural dynamic characteristics during the vibration of a beam with open crack. transverse vibration are obtained for a rectangular beam containing cracks. Here, we have extended the model for inclined crack (Figure 1).

2.1 Assumptions for the model

- At the location of the crack the local stiffness got reduced due to crack.
- The change in strain energy due to crack under constant load assumption is computed using energy balance approach.
- The equivalent bending stiffness and equivalent depth of the beam is obtained by modeling strain energy variation along the beam length.
- Crack is always open during vibration.

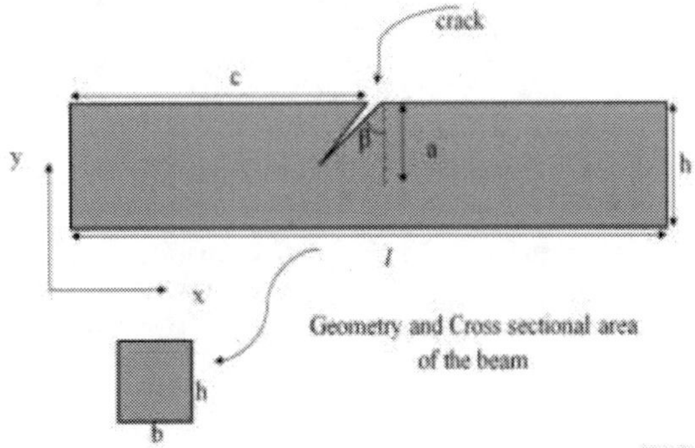

Figure 1. Geometry of the inclined edge cracked beam.

For an uncracked beam the strain energy is given by

$$U = \frac{1}{2}\int_0^l \frac{M^2}{EI}dx \tag{2.1}$$

Energy needed for a crack growth of length 'a' is given by

$$E_c = \int_0^a G dA = \int_0^a Gb da \quad (2.2)$$

where, G is the strain energy release rate.

2.1.1 Vertical edge crack

For vertical edge crack G is given as,

$$G = \frac{K_1^2}{E} \quad (2.3)$$

where, K_I is the stress intensity factor of the first mode and is given by,

$$K_1 = K_{I(0)} = \frac{6M\sqrt{\pi a}}{bh^2} f(a) \quad (2.4)$$

where, M is the bending moment of the beam, and

$$f(a) = 1.12 - 1.4\left(\frac{a}{h}\right) + 7.33\left(\frac{a}{h}\right)^2 - 13.8\left(\frac{a}{h}\right)^3 + 14\left(\frac{a}{h}\right)^4 \quad (2.5)$$

where, $a/h < 0.6$

2.1.2 Inclined edge crack

Let the kink angle is defined by α and crack is assumed to be inclined on an angle β [2],

$$G = \frac{k_1^2(\alpha) + k_2^2(\alpha)}{E} \quad (2.6)$$

where,

$$\begin{aligned} k_1(\alpha) &= C_{11}K_I + C_{12}K_{II} \\ k_2(\alpha) &= C_{21}K_I + C_{22}K_{II} \end{aligned} \quad (2.7)$$

where, k_1 and k_2 are the local stress intensity factor at the tip of the kink and K_I and K_{II} are the stress intensity factors for the main tilted crack given by equation (2.8-2.9), (figure(2))

$$\begin{aligned} K_I &= K_{I(0)} \cos^2 \beta \\ K_{II} &= K_{I(0)} \cos\beta \sin\beta \end{aligned} \quad (2.8)$$

The coefficients C_{ij} are defined by,

$$\begin{aligned} C_{11} &= \frac{3}{4}\cos\left(\frac{\alpha}{2}\right) + \frac{1}{4}\cos\left(\frac{3\alpha}{2}\right) \\ C_{12} &= -\frac{3}{4}\left[\sin\left(\frac{\alpha}{2}\right) + \sin\left(\frac{3\alpha}{2}\right)\right] \\ C_{21} &= \frac{1}{4}\left[\sin\left(\frac{\alpha}{2}\right) + \sin\left(\frac{3\alpha}{2}\right)\right] \\ C_{22} &= \frac{1}{4}\cos\left(\frac{\alpha}{2}\right) + \frac{3}{4}\cos\left(\frac{3\alpha}{2}\right) \end{aligned} \quad (2.9)$$

Further, if EI_c is the bending stiffness of the cracked beam, strain energy in the cracked beam can be written as:

$$U_c = \frac{1}{2}\int_0^1 \frac{M^2}{EI_c} dx \quad (2.10)$$

For the transverse vibration of the long beam, the crack is mainly subjected to the direct bending stresses and the shear stresses can be neglected; therefore, only the first mode crack exists. While the inclined crack includes both Mode I and Mode II (mixed type) crack. The model defined in section 2 can be defined for vertical edge crack by using the strain energy release rate as shown in equation (2.3). For inclined crack the equivalent stored energy G is the function of both K_I and K_{II} (equation(2.6)). Equation 2.16-2.19 gives the model of equivalent approach for vertical edge crack.

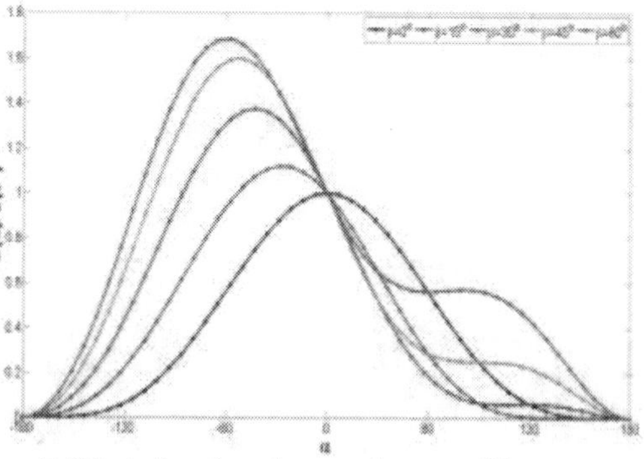

Figure 2. Effect of crack angle on strain energy [2].

A similar model has been used for the inclined edge crack with the modification in strain energy release rate as discussed above.

2.2 Energy expression

The stresses/strains are highly concentrated around the crack tip, and reach the nominal stress at a location far away from the crack. So it can be assumed that the increase of strain energy due to crack growth, under constant applied moment, is concentrated mainly around the crack region. The energy consumed for crack growth along the beam defined by [3-4], equation1, (figure (3)).

$$E_c = A(a,c)\exp\{-|B(x)/C(a)|\} \quad (2.11)$$

for the continuity of the function let, $|B(x)/C(a)| = y^2$ where, $y^2 = ((x-c)/k(a))^2$, which gives,

$$E_c = A(a,c)\exp\{-((x-c)/k(a))^2\} \quad (2.12)$$

on expanding the exponential series, and neglecting the $O(h^2)$ terms, we get the final energy function form as,

$$E_c = A(a,c)/\{1+((x-c)/k(a))^2\} \quad (2.13)$$

where, the terms $A(a,c)$ and $k(a)$ are determined over the beam length for which

$$E_c = \int_0^l A(a,c)/\{1+((x-c)/k(a))^2\}\,dx \quad (2.14)$$

where, c is location of the crack from one end of the beam. Now with the help of equations 2.2-2.5 and 2.14, we can obtain,

$$A(a,c) = \frac{D(a)M^2}{k(a)\left[\tan^{-1}\left(\frac{l-c}{\{k(a)\}}\right)+\tan^{-1}\left(\frac{c}{\{k(a)\}}\right)\right]} \quad (2.15)$$

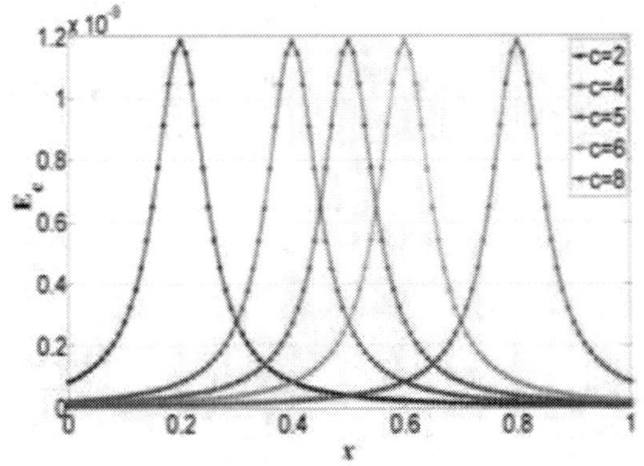

Figure 3. Variation of energy function with crack location for $a=0.8$.

and by using 2.1, 2.10 and 2.14, together with the assumption that the final strain energy in the cracked beam is $U_c = U + E_c$, we get the following relations:

Modified bending stiffness and height (figures 4and 5) of the beam is given by:

$$EI_c = \frac{EI}{1+\frac{EIR(a,c)}{\left[1+\left(\frac{x-c}{\{k(a)\}}\right)^2\right]}} \quad (2.16)$$

$$R(a,c) = \frac{2D(a)}{k(a)\left[\tan^{-1}\left(\frac{1-c}{\{k(a)\}}\right)+\tan^{-1}\left(\frac{c}{\{k(a)\}}\right)\right]} \quad (2.17)$$

$$D(a) = \frac{18\pi[f(a)^2]a^2}{Ebh^4} \quad (2.18)$$

$$k(a) = \frac{3\pi[f(a)^2](h-a)^3 a^2}{[h^3-(h-a)^3]h} \quad (2.19)$$

Also, at the position when $x=c$, we have the following relation:

$$\frac{EI}{EI_c} = \frac{h^3}{(h-a)^3} \quad (2.20)$$

2.3 Equation of motion

For an Euler beam the transverse vibration equation of cracked beam is given by,

$$\frac{\partial^2}{\partial x^2}\left[EI_c\frac{\partial^2 w}{\partial x^2}\right] + \rho A\frac{\partial^2 w}{\partial t^2} = 0 \quad (2.21)$$

where, ρ is density and A is the cross-sectional area of the beam.

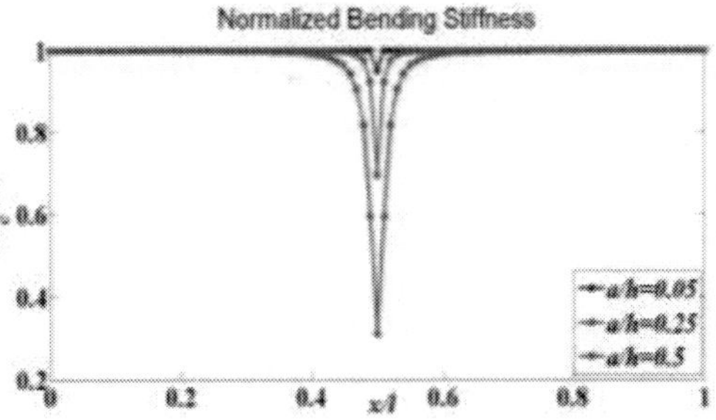

Figure 4. Equivalent bending stiffness *for c/l=0.5, l=10, b=1,h=1.*

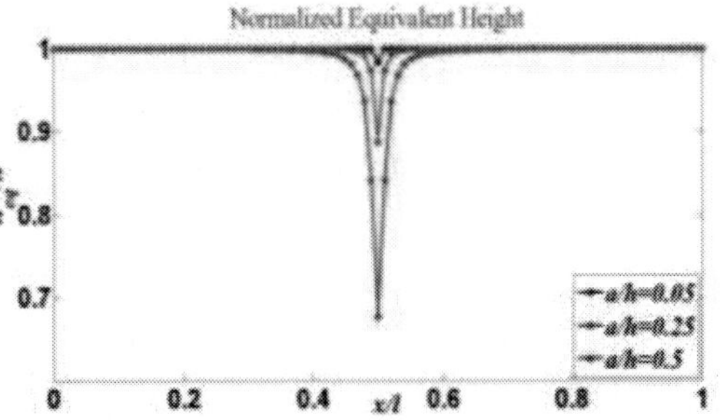

Figure 5. Equivalent height variation for *c/l=0.5 l=10, b=1,h=1.*

Let, $w(x,t)=W(x)exp(i\omega t)$, equation (2.21) becomes

$$\frac{d^2}{dx^2}\left[EI_c\frac{d^2W}{dx^2}\right]-\rho A\omega^2 W=0 \qquad (2.22)$$

where, EI_c is given by (2.16). Equation (2.22) can be rewritten as,

$$EI_c\frac{d^4W}{dx^4}+2\left(\frac{dEI_c}{dx}\right)\frac{d^3W}{dx^3}+\left(\frac{d^2EI_c}{dx^2}\right)\frac{d^2W}{dx^2}-\rho A\omega^2 W=0 \quad (2.23)$$

3 Methodology (DQM)

In this section a brief overview of DQM has been discussed [5]. It is assumed that a function $W(x)$ is sufficiently smooth over the whole domain. The nth order derivative of the function $W(x)$ with respect to x at m number of grid points x_i, is approximated by a linear sum of all the functional values in the whole domain, that is,

$$W_x^{(n)}(x_i)=\sum_{j=1}^{m}c_{ij}^{(n)}W(x_j),\ i=1,23,...,m; \qquad (3.1)$$

where, c_{ij} represent the weighting coefficients, and n is the number of grid points in the whole domain. Equation 1 is called differential quadrature (DQ). It should be noted that the weighting coefficients c_{ij} are different at different locations of x_i. The weighting coefficients required in DQ method as shown in equation(1) are defined recursively by equations(3.2-3.5).

$$c_{ij}^{(n)}=n\left(c_{ii}^{(n-1)}c_{ij}^{(n-1)}-\frac{c_{ij}^{n-1}}{x_i-x_j}\right) \text{ for } i,j=1,2,3,...,m;$$
$$j\neq i;\ n=2,3,4. \qquad (3.2)$$

$$c_{ii}^{(n)}=-\sum_{\substack{j=1\\ j\neq i}}^{m}c_{ij}^{(n)},\ i=1,2,3,...,m; \qquad (3.3)$$

$$c_{ij}^{(1)}=\frac{M^{(1)}(x_i)}{(x_i-x_j)M^{(1)}(x_j)},\quad i,j=1,2,...,m;\ i\neq j \qquad (3.4)$$

$$M^{(1)}(x_i)=\prod_{\substack{j=1\\ j\neq i}}^{m}(x_i-x_j) \qquad (3.5)$$

3.1 Grid point Distribution

The selection of number and type of grid points has a significant effect on the accuracy of the DQM results. It is found that the optimal selection of the sampling points in the vibration problems is the normalized Chebyshev-Gauss-Lobatto points [5].

4 Case Studies

The Model discussed in this paper has been used for different cracks and boundary conditions. The governing equation of motion (equation(2.23)) has been transformed into a discrete eigen value problem with the help of DQM as given in equations 3.1-3.5. Also the ANSYS models has been used to compare the solution. Figure 6 shows an ansys model of actual vertical edge cracked beam, while figure 7 depicts the equivalent cracked model of the beam. The inclined edge cracked model of the beam is shown in Figure 8. Table 1-2 shows the numerical results obtained for uncracked beams by using our model as a special case. While in table 3 the effect of crack location on frequency has been shown. Tables 4 and 5 compares the effect of crack angle on frequency parameters for clamped and fixed boundary conditions. In table 6 the effect of orientation of crack has been shown for clamped beam.

Figure 6. Vertical U-notch crack in beam ANSYS.

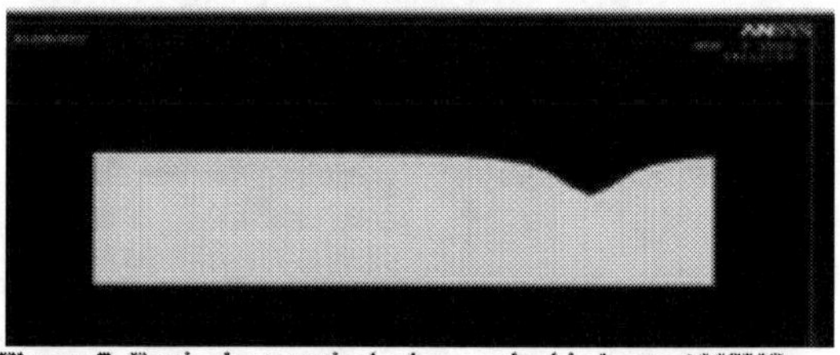

Figure 7. Equivalent vertical edge cracked in beam ANSYS.

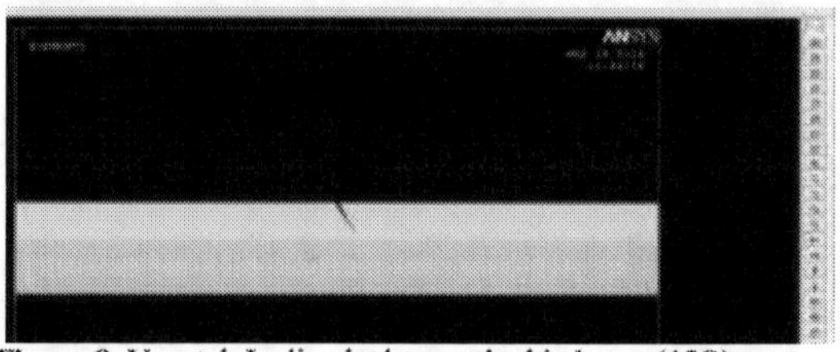

Figure 8. V-notch Inclined edge cracked in beam (45O) ANSYS.

Table 1. Comparison of frequency parameter for uncracked simply-supported beam for $E=2.16GPa$, $\rho=7650kg/m^3$, $l=0.4m$, $b=h=0.01m$.

Frequency (Hz)	Experimental	ANSYS				DQM
	Ref.[6]	20 Node Brick Element	Beam Element 2D	Plane Stress with Thickness	Plane stress with υ=0.3	15 Grid Point
I	151.5	150.44	150.55	150.46	157.69	150.593
II	602.5	599.92	601.75	600.26	628.69	602.373

Table 2. Comparison of frequency parameter for uncracked cantilever beam for *E=70GPa, ρ=2710kg/m³, l=800mm, b=60mm, h=6mm*

Frequency (Hz)	Experimental	ANSYS				DQM
	Ref.[7]	20 Node Brick Element	Beam Element 2D	Plane Stress with Thickness	Plane stress with υ=0.0	10 Grid Point
I	8.0217	7.7486	7.6969	7.7003	7.6966	7.6969
II	50.256	48.539	48.232	48.245	48.223	48.3753
III	141.13	135.92	135.04	135.03	134.98	139.4364

Table 3. Effect of crack location on vertically edge cracked simply-supported beam for *E=2.16GPa, ρ=7650kg/m³, l=0.4m, b=h=0.01m.*

Frequency (Hz)	*a/h*=0.1	Experimental Ref.[6]	DQM 26 pts
I	*c/l*=0.2	151.2	150.49
II		599.8	597.42
I	*c/l*=0.3	150.5	150.59
II		604.2	601.03
I	*c/l*=0.5	149.2	147.17
II		599.8	599.21

Table 4. Comparison of frequency parameter for inclined (30°) edge cracked cantilever beam for *E=70GPa, ρ=2710kg/m³, l=800mm, b=60mm h=6mm, c/l=0.40, a/h=0.20.*

Mode	ANSYS		Experimental [7]	FEA	DQM
	υ=0.0	υ=0.346			
I	7.6799	7.6795	7.8200	7.6925	7.5137
II	47.282	47.061	49.5900	48.542	51.5731
III	128.25	126.87	138.41	135.68	136.371

Table 5. Effect of crack inclination on frequency parameter for fixed beam for *E=2GPa, ρ=7850kg/m³, l=10m, b=h=1m, c=5.0, a=0.4 υ=0.3.*

Angle	Actual Crack Model ANSYS			Equivalent Crack Model ANSYS			DQM Model		
	I	II	III	I	II	III	I	II	III
30°	7.8140	42.168	129.60	8.0250	47.193	129.89	7.9591	43.4381	122.493
45°	7.7651	42.004	120.63	7.8393	41.862	119.66	6.7952	41.9609	118.747
60°	7.7110	41.526	129.28	8.1002	48.903	129.91	7.3828	45.8426	129.004

Table 6. Effect of crack orientation about the crack tip on frequency in fixed beam for *E=2GPa, ρ=2700kg/m³, l=10m, b=h=1m, c/l=0.5, a/h=0.4 υ=0.3.*

Angle	Left side inclination			Right side inclination		
	I	II	III	I	II	III
0°	4.1918	22.803	69.921	4.1918	22.803	69.921
15°	4.1895	22.761	69.909	4.1889	22.758	69.915
30°	4.1892	22.737	69.882	4.1873	22.736	69.898
45°	4.1870	22.649	69.822	4.1806	22.646	69.850
60°	4.1297	20.664	67.677	4.1578	22.391	69.710
75°	4.0963	22.525	67.156	4.0518	21.359	68.808

5 Conclusion

The model discussed in this paper has been used to find the effect of inclined crack on beam vibrations. The governing equation of motion is solved by using Differential Quadrature Method using Chebyshev's collocation points. It has been observed that the model gives optimum results for different types of boundary conditions. The applicability of the model has been obtained by the comparing the results with ANSYS actual crack model and ANSYS equivalent crack model for two types of crack (i) inclined crack and (ii) vertical crack. The results are also compared with the uncracked beam. The comparative study justify and validate the model.

References

1. X. F. Yang, A. S. J. Swamidas, R. Seshadri, Crack Identification in Vibrating Beams using The Energy Method. *J Sound Vib.*, **244**, 339-357 (2001).
2. T.L. Anderson, Fracture Mechanics Fundamental and Application, Third edition, CRC Press, Taylor and Francis Group, (2013).
3. S. Christides, A.D.S. Barr, One-Dimensional Theory of Cracked Bernoulli-Euler Beams. Int. J. Mech. Sci., **26(11/12)**, 639-648 (1984).
4. M.I. Friswell, J.E.T. Penny, Crack Modeling for Structural Health Monitoring. Struct Heal Monit., **1(2)**, 139-148 (2002).
5. C. Shu, Differential Quadrature and Its Applicationin Engineering, Springer, London, (2000).
6. H. Yoon, I. N. Son, S. J. Ahn, Free Vibration Analysis of Euler-Bernoulli Beam with Double Cracks, Journal of mechanical science and technology, **21**, 476-485 (2007).
7. R.K. Behera, A. Pandey, D.R. Parhi, Numerical and Experimental Verification of a Method for Prognosis of Inclined Edge Crack in Cantilever Beam Based on Synthesis of Mode Shapes. *Procedia Technol*, **14**, 67-74 (2014).

Three-Dimensional Assembly Tolerance Analysis Based on the Jacobian-Torsor Statistical Model

Heping Peng [1] and Wenlong Lu [2]

[1] *School of Electromechanical & Architectural Engineering, Jianghan University, Wuhan 430056, China*
[2] *School of Mechanical Science & Engineering, Huazhong University of Science and Technology, Wuhan 430074, China*

Abstract. The unified Jacobian-Torsor model has been developed for deterministic (worst case) tolerance analysis. This paper presents a comprehensive model for performing statistical tolerance analysis by integrating the unified Jacobian-Torsor model and Monte Carlo simulation. In this model, an assembly is sub-divided into surfaces, the Small Displacements Torsor (SDT) parameters are used to express the relative position between any two surfaces of the assembly. Then, 3D dimension-chain can be created by using a surface graph of the assembly and the unified Jacobian-Torsor model is developed based on the effect of each functional element on the whole functional requirements of products. Finally, Monte Carlo simulation is implemented for the statistical tolerance analysis. A numerical example is given to demonstrate the capability of the proposed method in handling three-dimensional assembly tolerance analysis.

1 Introduction

With the release of new generation GPS tolerancing standards, geometric tolerances are generally accepted as industry practices [1]. Geometric tolerances and dimensional tolerances, as well as the interaction between them in the tolerance zone can be taken into consideration by 3D tolerance analysis methods. Desrochers et al. [2] proposed the unified Jacobian-Torsor model for deterministic tolerance analysis by the unifying of two existing models: the torsor model and the Jacobian matrix model in which the torsor model is used for tolerance representation and the Jacobian matrix for tolerance propagation. Currently, the research on combinations of 3D tolerance analysis technology and statistical analysis method is attracting the attention of many researchers [3, 4]. This paper focuses on the development of an analysis method to realize the statistical tolerance analysis of a mechanical assembly by integrating the unified Jacobian-Torsor model with Monte Carlo simulation.

2 Unified Jacobian-Torsor model

2.1 Surface graph of an assembly

The surface model concept is a basic concept within the new generation GPS standards system and is an abstract model of the physical interface between a workpiece and its environment [5]. Teissandier et al. [6] presented the method to describe an assembly using surface model. For example, the assembly shown in Fig.1 comprises seven parts, and the parts can be subdivided into ideal surfaces. The designation of a surface is made up of two numbers separated by a ",": (1, 1) means surface 1 of part 1.

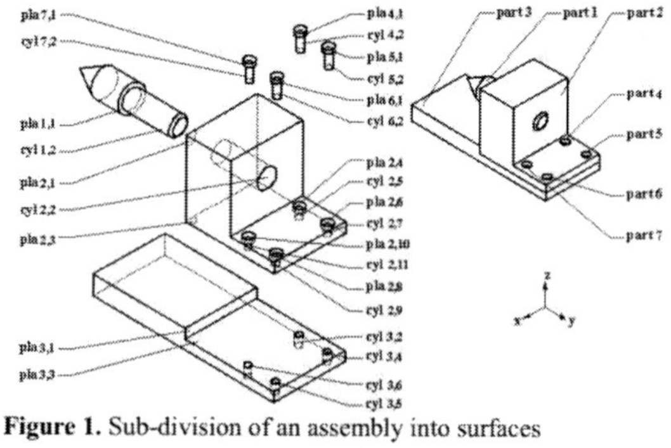

Figure 1. Sub-division of an assembly into surfaces

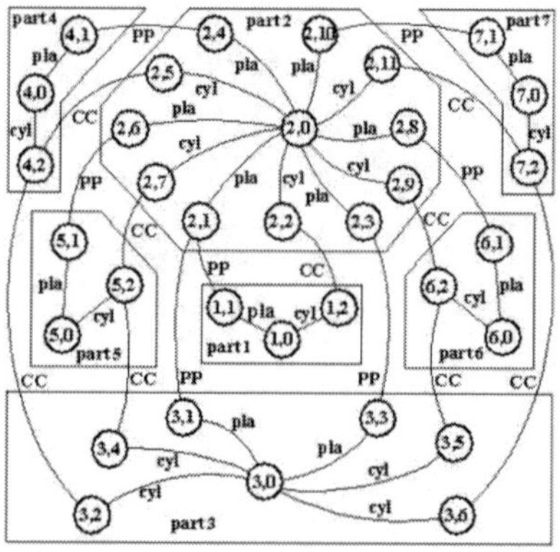

pla: tolerance zone of plane surface CC: joint between two coaxial cylinders
cyl: tolerance zone of cylindrical surface PP: joint between two merged planes

Figure 2. Surface graph of the assembly (Fig. 1)

The part 1 (see Fig. 1) is divided into two surfaces: plane (1, 1) and cylinder (1, 2). The vertex (1, 0) corresponds to nominal surfaces of part 1. Any edge between vertices $(i, 0)$ and (i, j) symbolizes the tolerance zone associated with surface (i, j). Any edge between vertices (i, j) and (k, l) with $i \neq k$; $j \neq 0$ and $l \neq 0$ symbolizes a joint (see Fig. 2). The surface graph of the assembly can be created as shown in Fig. 2.

2.2 Creation of a 3D dimension-chain

In assembly tolerance analysis, all variations should be quantitatively described including dimensional and geometric feature variations of each *individual* part in the assembly and kinematic variations occurring in the contacts or joints between mating parts. After an assembly is subdivided into surfaces, the relative position between any two surfaces of an assembly is determined with the aid of a set of SDT parameters. That is, it is possible to express the limits of relative small displacements between an associated surface and a nominal surface inside a tolerance zone, and the limits of relative displacements between two associated surfaces of two different parts in a joint. In order to determine the torsor between any two surfaces, two governing rules are defined: the union and the intersection. Based on the effect of each functional element on the whole functional requirements of products, a 3D dimension-chain can be created.

2.3 Jacobian-Torsor model of tolerance analysis

Based on the combination of the Jacobian model and the torsor based representation of tolerance zones, the unified Jacobian-Torsor model proposed by Desrochers et al [2] is represented as:

$$\begin{bmatrix}[\alpha^-,\alpha^+]\\ [\beta^-,\beta^+]\\ [\gamma^-,\gamma^+]\\ [u^-,u^+]\\ [v^-,v^+]\\ [w^-,w^+]\end{bmatrix}_{FR} = [[J_1\ J_2\ J_3\ J_4\ J_5\ J_6]_{FE1} \cdots [J_1\ J_2\ J_3\ J_4\ J_5\ J_6]_{FEk} \cdots [J_1\ J_2\ J_3\ J_4\ J_5\ J_6]_{FEn}] \cdot \begin{bmatrix}\begin{bmatrix}[\alpha^-,\alpha^+]\\ [\beta^-,\beta^+]\\ [\gamma^-,\gamma^+]\\ [u^-,u^+]\\ [v^-,v^+]\\ [w^-,w^+]\end{bmatrix}_{FE1} \cdots \begin{bmatrix}[\alpha^-,\alpha^+]\\ [\beta^-,\beta^+]\\ [\gamma^-,\gamma^+]\\ [u^-,u^+]\\ [v^-,v^+]\\ [w^-,w^+]\end{bmatrix}_{FEk} \cdots \begin{bmatrix}[\alpha^-,\alpha^+]\\ [\beta^-,\beta^+]\\ [\gamma^-,\gamma^+]\\ [u^-,u^+]\\ [v^-,v^+]\\ [w^-,w^+]\end{bmatrix}_{FEn}\end{bmatrix}^T \quad (1)$$

where: $\begin{bmatrix}[\alpha^-,\alpha^+]\\ [\beta^-,\beta^+]\\ [\gamma^-,\gamma^+]\\ [u^-,u^+]\\ [v^-,v^+]\\ [w^-,w^+]\end{bmatrix}_{FR}$ is SDT matrix associated to assembly functional requirement (FR); $\begin{bmatrix}[\alpha^-,\alpha^+]\\ [\beta^-,\beta^+]\\ [\gamma^-,\gamma^+]\\ [u^-,u^+]\\ [v^-,v^+]\\ [w^-,w^+]\end{bmatrix}_{FEk}$ is SDT matrix of the kth ($k = 1, 2, \cdots, n$) functional element pair; n is the number of torsors in the tolerancing chain; α^-, β^-, γ^-, u^-, v^-, w^- and α^+, β^+, γ^+, u^+, v^+, w^+ are lower and upper limits of α, β, γ, u, v, w respectively; $[J_1\ J_2\ J_3\ J_4\ J_5\ J_6]_{FEk}$ is the Jacobian matrix expressing a geometrical relation between the assembly functional requirement and the kth functional element pair, and $J_1\ J_2\ J_3$ are three rotation components; $J_4\ J_5\ J_6$ are three translation components; The multiplication operation in Eq.(1) uses the interval arithmetic [2].

The Jacobian matrix is formulated as:

$$[J_1\ J_2 \cdots J_6]_{FEk} = \begin{bmatrix}[R_0^k]_{3\times3}\cdot[R_{PT}]_{3\times3} & [W_k^n]_{3\times3}\cdot[R_0^k]_{3\times3}\cdot[R_{PT}]_{3\times3}\\ [\,0\,]_{3\times3} & [R_0^k]_{3\times3}\cdot[R_{PT}]_{3\times3}\end{bmatrix}$$

where $[R_0^k]_{3\times3}$ is the orientation matrix of reference frame k relative to 0, and reference frame 0 is the global reference frame; $[R_{PT}]_{3\times3}$ is the projection matrix whose purpose is to establish the projection of a traditional torsor along the direction of analysis of the tolerancing chain for the tilted tolerance zone; $[W_k^n]_{3\times3}$ is a Skew-symmetric matrix allowing the representation of vector $[\vec{d}_n - \vec{d}_k]$ with $dx_k^n = dx_n - dx_k$, $dy_k^n = dy_n - dy_k$, $dz_k^n = dz_n - dz_k$, and $\vec{d}_k = [dx_k\ dy_k\ dz_k]^T$ is the position vector defining the position of the origin of reference frame k in 0. It can be defined as follows:

$$[W_k^n]_{3\times3} = \begin{bmatrix}0 & -dz_k^n & dy_k^n\\ dz_k^n & 0 & -dx_k^n\\ -dy_k^n & dx_k^n & 0\end{bmatrix}$$

3 Statistical tolerance analysis

The aim of statistical tolerance analysis is to get the statistical limits of the assembly functional requirement (*FR*) in the desired direction. Yet, the model in Eq. (1) is deterministic tolerance analysis model. In order to perform statistical tolerance analysis, a Monte Carlo simulation will be used to iteratively evaluate the deterministic model using sets of random numbers as inputs. Once after a great many of random numbers are used as inputs, the deterministic model will become a stochastic model.

It is well known that the functional requirements for a complex mechanical assembly are normally evaluated by several assembly critical dimensions. These critical dimensions are simultaneously controlled within certain variation ranges for the best working performances [7]. One of the assembly response functions is given by:

$$z = g(y_1, y_2, \cdots, y_n) \quad (2)$$

where y_k ($k = 1, 2, \cdots, n$) is component design dimension. Using Taylor series expansion and neglecting the higher-order terms beyond the first order, Eq. (2) can be approximated as:

$$\Delta z = \sum_{k=1}^{n}\left(\left.\frac{\partial g}{\partial y_k}\right|_{\bar{y}} \cdot \Delta y_k\right) \tag{3}$$

where Δz is the variation of assembly critical dimension. Δy_k is the deviation of y_k from its mean value $\bar{y}_k$, $\Delta y_k = y_k - \bar{y}_k$. In the bilateral tolerance system, the nominal value of the dimensions is just the mean value of the dimensions.

Statistical tolerance analysis makes use of the relationship expressed by Eq. (3), in which the random variables are assigned as input variables. Random number generators are used to generate a sample of numbers Δy_1, Δy_2, …, Δy_n. The value of Δz corresponding to this sample can be computed. This process is replicated a large number of samples, we will gain a random sample for Δz. The mean and standard deviation of the output Δz (that is *FR*) can be determined through statistical analysis of this sample.

This process comprises the following steps:

- Creation of the unified Jacobian-Torsor model for the assembly tolerance analysis. According to Eq. (1), SDT matrixes of all functional element pairs and the corresponding Jacobian matrixes will be determined in this step.
- Generate real random values for each SDT component. Several software packages can generate random variables, such as we can use the function Rand ( ) in Microsoft Excel. It's important to note that the random values should be within the bounds of the intervals defined by each SDT.
- Compute real FR values for each set of random values. Simulation was stopped as the number of iterations increases past the limited number, and all these simulated data are recorded in a data structure of the program.
- Analyze the results and determine mean and standard deviation of *FR* in the direction of tolerance analysis.

4 Numerical example

In this section, a numerical example is given to illustrate the application of the proposed tolerance analysis model. The assembly consists of two blocks with inclined surfaces (parts 1 and 2) and a space (part 3) as shown in Fig. 3. The functional requirement of this assembly is the contact gap, shown on the rightmost of the assembly, which is the critical characteristic of this assembly wish to control.

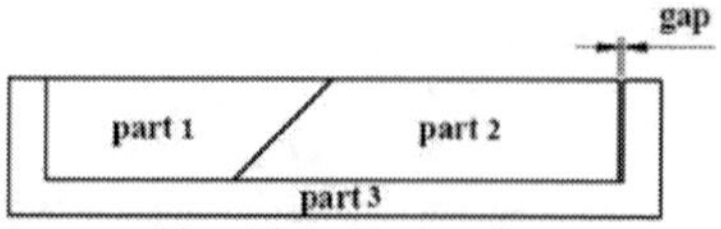

Figure 3. An assembly with inclined surfaces

The relevant dimensions and tolerances of the three parts are shown in Fig. 4.

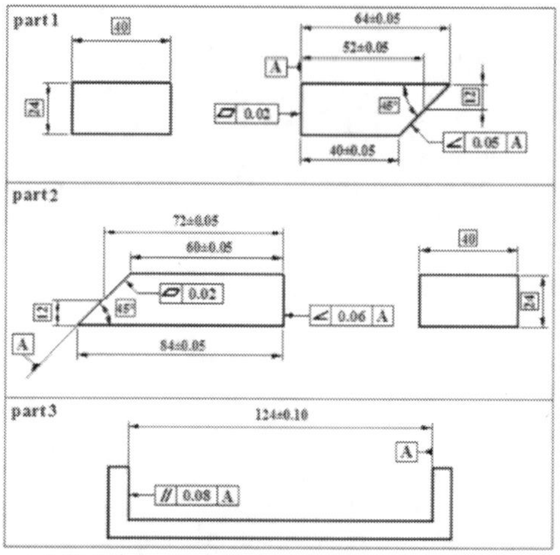

Figure 4. Detail drawing of part 1-3

In order to determine the tolerance value of the contact gap (Fig. 3), the effective feature surfaces between two parts are identified. There are six effective feature surfaces in this assembly as shown in Fig. 5. The designation of a surface is made up of two number separated by a "," e.g., $S_{1,1}$ means surface 1 of part 1. The geometric variations of these features can be calculated based proposed modeling method at first, and then these variations can be accumulated to derive the required tolerance value.

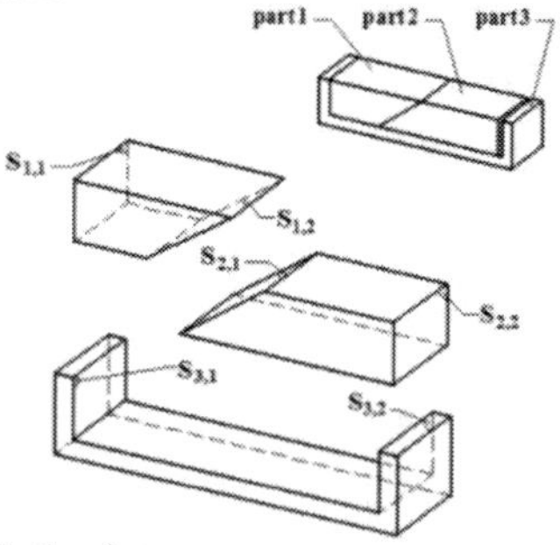

Figure 5. Effective features

The local coordinate systems are established on the effective features as shown in Fig. 6.

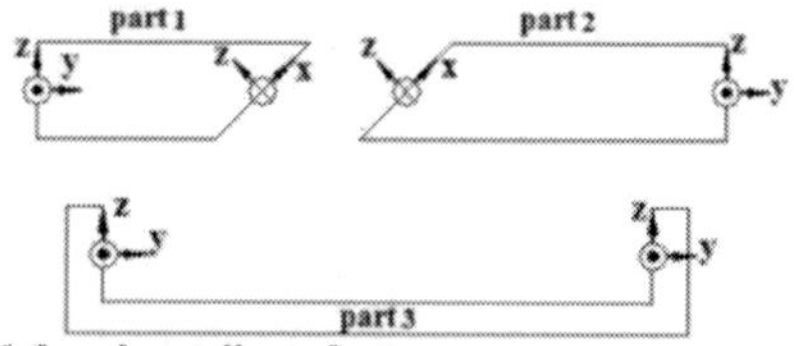

Figure 6. Local coordinate frames

After identifying all features and setting of local coordinate frames, the surface graph of the assembly can be constructed. Considering the Union rule of torsors, torsor $\boldsymbol{\tau}_{2,2/3,2}$ associated to the functional requirement consists of torsors $\boldsymbol{\tau}_{2,2/2,1}$, $\boldsymbol{\tau}_{2,1}$, $\boldsymbol{\tau}_{2,1/1,2}$, $\boldsymbol{\tau}_{1,2/1,1}$, $\boldsymbol{\tau}_{1,1}$, $\boldsymbol{\tau}_{1,1/3,1}$ and $\boldsymbol{\tau}_{3,1/3,2}$. The two kinematic torsors $\boldsymbol{\tau}_{2,1/1,2}$ and $\boldsymbol{\tau}_{1,1/3,1}$ are considered null because the contact between the two planes is assumed perfect.

According to Eq. (1), we can obtain a unified Jacobian-Torsor model about the FR in this assembly which can be written as follows, where the variables C and S correspond respectively to $\cos 45^{\circ}$ and $\sin 45^{\circ}$.

$$\begin{bmatrix}[\alpha^-,\alpha^+]\\ [\beta^-,\beta^+]\\ [\gamma^-,\gamma^+]\\ [u^-,u^+]\\ [v^-,v^+]\\ [w^-,w^+]\end{bmatrix}_{FR} = \left[\begin{bmatrix}1&0&0&0&0&0\\0&1&0&0&0&0\\0&0&1&0&0&0\\0&0&0&1&0&0\\0&0&0&0&1&0\\0&0&0&0&0&1\end{bmatrix} \begin{bmatrix}0&-1&0&0&0&0\\C&0&-CS-C^2&0&0&0\\S&0&-S^2+SC&0&0&0\\72S&0&-72S^2+72SC&0&-1&0\\0&0&0&C&0&-CS-C^2\\0&72&0&S&0&-S^2+SC\end{bmatrix} \begin{bmatrix}0&-1&0&0&0&0\\C&0&-CS-C^2&0&0&0\\S&0&-S^2+SC&0&0&0\\72S&0&-72S^2+72SC&0&-1&0\\0&0&0&C&0&-CS-C^2\\0&72&0&S&0&-S^2+SC\end{bmatrix} \begin{bmatrix}1&0&0&0&0&0\\0&1&0&0&0&0\\0&0&1&0&0&0\\0&0&124&1&0&0\\0&0&0&0&1&0\\-124&0&0&0&0&1\end{bmatrix} \begin{bmatrix}1&0&0&0&0&0\\0&1&0&0&0&0\\0&0&1&0&0&0\\0&0&124&1&0&0\\0&0&0&0&1&0\\-124&0&0&0&0&1\end{bmatrix}\right]$$

$$\bullet \left[\begin{bmatrix}[-0.0025,+0.0025]\\ [0,0]\\ [-0.0015,+0.0015\\ [0,0]\\ [-0.05,+0.05]\\ [0,0]\end{bmatrix} \begin{bmatrix}[-0.00050,+0.00050]\\ [-0.00059,+0.00059]\\ [0,0]\\ [-0.01,+0.01]\\ [0,0]\\ [-0.01414,+0.01414]\end{bmatrix} \begin{bmatrix}[-0.00125,+0.00125]\\ [-0.00147,+0.00147]\\ [0,0]\\ [-0.03536,+0.03536]\\ [0,0]\\ [-0.05,+0.05]\end{bmatrix} \begin{bmatrix}[-0.00083,+0.00083]\\ [0,0]\\ [-0.00050,+0.00050]\\ [0,0]\\ [-0.01,+0.01]\\ [0,0]\end{bmatrix} \begin{bmatrix}[-0.00333,+0.00333]\\ [0,0]\\ [-0.002,+0.002]\\ [0,0]\\ [-0.10,+0.10]\\ [0,0]\end{bmatrix}\right]^{T}$$

Using the interval algorithm, we draw the conclusion that the functional requirement or contact gap of this assembly in the y direction must lie in an interval of [-0.25621, +0.25621].

In the statistical approach, statistical torsor parameters of functional element pairs are listed in Table 1.

Table 1. Statistical torsor parameters of functional element pairs

Torsors	Components of Torsors	Specified distribution/ Rejected	Statistical parameters
$\tau_{2,2/2,1}$	$\begin{Bmatrix}[\alpha^-,\alpha^+]\\ [\beta^-,\beta^+]\\ [\gamma^-,\gamma^+]\end{Bmatrix}=\begin{Bmatrix}[-0.0025,+0.0025]\\ [0,0]\\ [-0.0015,+0.0015]\end{Bmatrix}$ $\begin{Bmatrix}[u^-,u^+]\\ [v^-,v^+]\\ [w^-,w^+]\end{Bmatrix}=\begin{Bmatrix}[0,0]\\ [-0.05,+0.05]\\ [0,0]\end{Bmatrix}$	Normal/ 5%	μ=0.00,σ=0.00128 μ=0.00,σ=0.00 μ=0.00,σ=0.00077 μ=0.00,σ=0.00 μ=0.00,σ=0.02551 μ=0.00,σ=0.00
$\tau_{2,1}$	$\begin{Bmatrix}[\alpha^-,\alpha^+]\\ [\beta^-,\beta^+]\\ [\gamma^-,\gamma^+]\end{Bmatrix}=\begin{Bmatrix}[-0.00050,+0.00050]\\ [-0.00059,+0.00059]\\ [0,0]\end{Bmatrix}$ $\begin{Bmatrix}[u^-,u^+]\\ [v^-,v^+]\\ [w^-,w^+]\end{Bmatrix}=\begin{Bmatrix}[-0.01,+0.01]\\ [0,0]\\ [-0.01414,+0.01414]\end{Bmatrix}$	Normal/ 5%	μ=0.00,σ=0.00026 μ=0.00,σ=0.00030 μ=0.00,σ=0.00 μ=0.00,σ=0.00510 μ=0.00,σ=0.00 μ=0.00,σ=0.00721
$\tau_{1,2/1,1}$	$\begin{Bmatrix}[\alpha^-,\alpha^+]\\ [\beta^-,\beta^+]\\ [\gamma^-,\gamma^+]\end{Bmatrix}=\begin{Bmatrix}[-0.00125,+0.00125]\\ [-0.00147,+0.00147]\\ [0,0]\end{Bmatrix}$ $\begin{Bmatrix}[u^-,u^+]\\ [v^-,v^+]\\ [w^-,w^+]\end{Bmatrix}=\begin{Bmatrix}[-0.03536,+0.03536]\\ [0,0]\\ [-0.05,+0.05]\end{Bmatrix}$	Normal/ 5%	μ=0.00,σ=0.00064 μ=0.00,σ=0.00075 μ=0.00,σ=0.00 μ=0.00,σ=0.01804 μ=0.00,σ=0.00 μ=0.00,σ=0.02551
$\tau_{1,1}$	$\begin{Bmatrix}[\alpha^-,\alpha^+]\\ [\beta^-,\beta^+]\\ [\gamma^-,\gamma^+]\end{Bmatrix}=\begin{Bmatrix}[-0.00083,+0.00083]\\ [0,0]\\ [-0.00050,+0.00050]\end{Bmatrix}$ $\begin{Bmatrix}[u^-,u^+]\\ [v^-,v^+]\\ [w^-,w^+]\end{Bmatrix}=\begin{Bmatrix}[0,0]\\ [-0.01,+0.01]\\ [0,0]\end{Bmatrix}$	Normal/ 5%	μ=0.00,σ=0.00042 μ=0.00,σ=0.00 μ=0.00,σ=0.00026 μ=0.00,σ=0.00 μ=0.00,σ=0.00510 μ=0.00,σ=0.00
$\tau_{3,1/3,2}$	$\begin{Bmatrix}[\alpha^-,\alpha^+]\\ [\beta^-,\beta^+]\\ [\gamma^-,\gamma^+]\end{Bmatrix}=\begin{Bmatrix}[-0.00333,+0.00333]\\ [0,0]\\ [-0.002,+0.002]\end{Bmatrix}$ $\begin{Bmatrix}[u^-,u^+]\\ [v^-,v^+]\\ [w^-,w^+]\end{Bmatrix}=\begin{Bmatrix}[0,0]\\ [-0.10,+0.10]\\ [0,0]\end{Bmatrix}$	Uniform	α^-=-0.00333,α^+=+0.0033 β^-=0.00, β^+=0.00 γ^-=-0.002,γ^+=+0.002 u^-=0.00, u^+=0.00 v^-=-0.10,v^+=+0.10 w^-=0.00, w^+=0.00

Simulation was stopped after 1000 iterations (can obtain 1000 values for each of the six torsor component values for FR). According to the functional requirement of this assembly, it is necessary to perform a statistical analysis for the v torsor component values for FR. The histogram of FR along the direction of analysis is shown in Fig. 7; the figure shows the FR data distribution is normal distribution. Thus, the functional requirement of this assembly in the direction of analysis determined by statistical analysis is: [-0.13880, +0.13880], the corresponding rejection percentage is 5%.

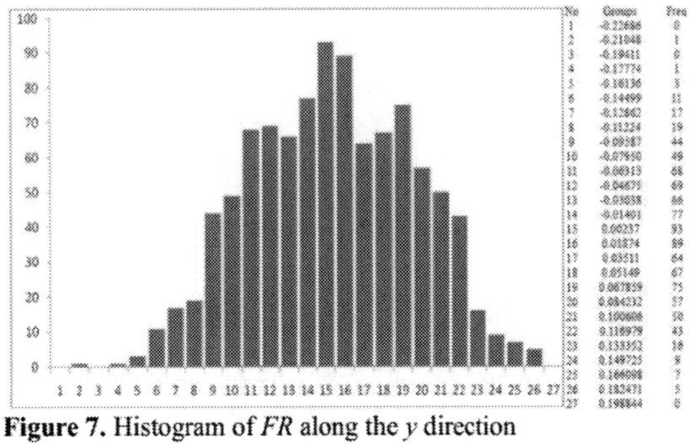

Figure 7. Histogram of *FR* along the *y* direction

5 Conclusions

The assembly tolerance analysis provides an effective evaluation tool for estimating the effects of variations caused by the dimensional and geometrical feature variations of the parts in an assembly on assembly functional requirement. This paper has described an analytical model integrated with the Monte Carlo simulation tool to obtain an algorithm which performs the statistical tolerance analysis of a mechanical assembly. The proposed method can be easily implemented in CAT systems and will provide a more effective manner for design and manufacturing engineers to deal with the statistical tolerance analysis.

Acknowledgment

The authors would like to acknowledge the financial support of the National Natural Science Foundation of the People's Republic of China (No. 51575235).

References

1. ISO 1101: Geometric Product Specifications (GPS)-Geometrical tolerancing–tolerancing of form, orientation, location and run-out, (2004)
2. A. Desrochers, W. Ghie and L. Laperrière, J. Comput. Inf. Sci. Eng. **3**, 1 (2003)
3. W. Ghie, Int. J. Multiphys. **3**, 1 (2009)
4. W. Ghie, L. Laperrière and A. Desrochers, Int. J. Prod. Res. **48**, 15 (2010)
5. F. Charpentier, A. Ballu and J. Pailhès, *Proceedings of the 12th CIRP conference on computer aided design,* (2012).
6. D. Teissandier, Y. Couétard and A. Gérard, Comput. Aided Des., 31 (1999)
7. H. P. Peng, X. Q. Jiang and X. J. Liu, Int. J. Prod. Res. **46**, 24 (2008)

MATEC Web of Conferences 95, 07008 (2017) DOI: 10.1051/matecconf/20179507008
ICMME 2016

Research on Optimization of Formula SAE Truss-Frame

Yu Wang[1], Jianwei Liu[2], Yuhan Li[1] and Nianjiong Yang[3]

[1]*Guilin University of Electronic Technology, School of Mechanical and Electrical Engineering, 541000 Guilin, China*
[2]*Guilin University of Electronic Technology, Department of Practice Teaching, 541000 Guilin, China*
[3]*Guangxi University of Science and Technology, Guangxi key laboratory of Automobile Components and Vehicle technology, 545006 Liuzhou, China*

Abstract. The frame, as one of the basic components of a car, plays an important role of connecting components and withstanding various loads. The truss-frame is the commonest form of Formula SAE car in that it has many advantages, such as low cost, simple manufacture, high rigidity and high strength. The development of frame is reviewed. First, the key performance indicators and the mechanical state of the frame under different conditions are discussed. Second, a variety of optimization methods used in the design of frame are described. Finally, the test method of frame are introduced, with the experimental data and the finite element analysis of results being compared. Through analyzing and summarizing the development of the design, optimization and testing technology of the Formula SAE truss-frame, forecasting the future trends, the great theoretical reference is provided for the design and research of the follow-up frame.

1 Introduction

Formula SAE (FSAE) was originally initiated by the American Society of Automotive Engineers for college students in a formula car race. The FSAE event was organized by China in 2010, with 20 college teams participating. In 2015, the number of college teams had reached 67.

The frame of Formula Student China carries the engine, body and various ancillary facilities. The structure strength of frame greatly affects the car's safety, power, comfort, handling stability and other performance indicators [1]. To ensure the normal functioning of the frame, the FSAE car frame must have sufficient rigidity, strength. And the ergonomic requirements are taken into account. The light-weight design of frame is an important goal in that the lighter frame will help improve acceleration performance and fuel efficiency. Thus, the high strength, high stiffness and light weight of the frame to improve race performance has a significant impact by the college team's attention.

There are two common frame types, one is the monocoque structure, the other is the truss structure. The monocoque is a shell frame made of carbon fiber, which has light weight, high strength and so on. However, the expensive price, complex production process and other issues limit its use in college teams. Truss-frame is inexpensive and simply manufactured, having now been adopted by most college teams. Therefore, this paper details the truss-frame's mechanical properties, optimization and testing methods.

2 Mechanical analysis of frame

Mechanics analysis is the basis of structural design. According to the load condition of the frame, an accurate equivalent mechanical model is proposed, which is helpful to understand the mechanism of the force and deformation of the frame. The favorable theoretical foundation is provided for optimization of the frame.

2.1 Working conditions of frame

From bending conditions, reversing the condition, a combination of bending and torsion conditions, lateral acceleration conditions, longitudinal acceleration condition and a combination of horizontal and vertical acceleration conditions, the FSAE car frame is analyzed.

The bending condition is the condition of bending deformation of the frame, which is caused by the gravity of the driver and other parts of the vehicle. The torsional condition is the condition of the torsional deformation of the vehicle frame, which is caused by the uneven stress of the four tires. A single torsion condition is not exist. Because the frame torsion deformation, it is bound to bear the gravity of the driver and the parts. Therefore, it could only be carried out on the frame of the other five kinds of working conditions, with ergonomics being considered. In the course of driving, the main forms of motion include uniform speed, acceleration, braking and turning. According to these forms, the common condition of the vehicle frame can be divided into three kinds, that is, the bending condition, the accelerating condition and

DOI: 10.1051/matecconf/20179507008

the turning condition. Dynamic load coefficient was added in the analysis of the strength of the frame under each condition. By this method, accurate results could be obtained, with the efficiency of the analysis being improved.

2.2 Kinetics analysis

The static analysis can not meet the design requirements of the frame. Dynamic analysis and optimization of dynamic performance are of great significance to improve the performance of the frame.

Resonance can be caused, reducing the comfort of the vehicle and increasing the load of the frame when the natural frequency of the frame is close to the excitation frequency of road surface or the engine. The natural frequencies and vibration modes of the frame were obtained by the modal analysis. The frame structure was optimized, so that the natural frequency of the frame to avoided the road and engine vibration frequency. To further analyze the performance of the frame, the frequency of the engine and suspension assembly on the frame were analyzed. This frequency needed to avoid tire rotation frequency of the normal speed.

The vehicle collision test is an important way to study the safety of vehicles, but most of the college team does not have the test conditions. Therefore, it is an ideal way to study the safety of the vehicle by using the computer simulation of vehicle collision. The crash simulation of 40% offset frontal was carried out to study the change of acceleration and energy during the collision and the deformation and damage of frame structure after impact.

2.3 Torsional stiffness

Torsional stiffness and weight are the two most important quantifiable aspects to the frame of any race car. Thus, the first aim of this project is to design a frame of which the trade-off between high torsional stiffness and low weight is balanced to achieve high vehicle performance across the various competition events [1].

As a method for analyzing and balancing the frame, the frame was divided into three main section, the front bulkhead, the driver compartment and the rear section. A schematic diagram is shown in Fig. 1. The torsional stiffness of each part was analyzed separately. The total stiffness of the frame can be expressed by equation (1). The K is the total stiffness of the frame. The kf is the stiffness of the front bulkhead. The kd is the stiffness of the cockpit. The kr is the stiffness of engine compartment.

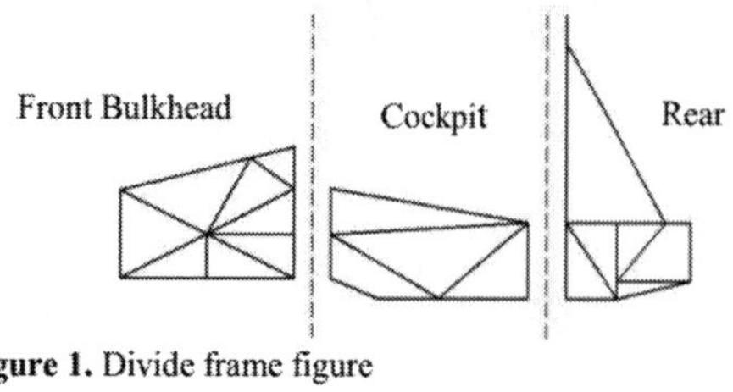

Figure 1. Divide frame figure

$$\frac{1}{K} = \frac{1}{k_f} + \frac{1}{k_d} + \frac{1}{k_r} \tag{1}$$

The equivalent mechanical model was established, as shown in Fig. 2. Further analysis of the various parts of the frame, it could be found that the cockpit is one of the important factors affect the level of the vehicle stiffness. By increasing the cockpit section, the torsional rigidity of the frame could be improved effectively. The smaller front is helpful to reduce the torque. Engine parts such as stiffness is bigger, reasonable installation can effectively increase the nacelle stiffness. So, the reasonable frame structure should be compact front, wide of the cockpit and compact engine compartment.

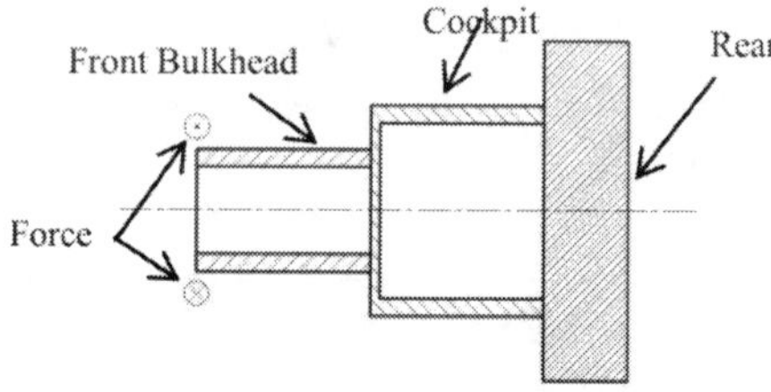

Figure 2. The equivalent mechanical model

Analysis the stress and the deformation law of frame is the basis of optimized frame. The most basic requirements is that frame can work in a variety of conditions. Torsional stiffness of the frame is an important performance index. High torsional stiffness and light weight are the goal of optimizing the frame. According to the research of Jacob I. Salter, William B. Riley and Albert R. George et al. [2], [3], a small cabin could effectively reduce the torsional moment of the frame. The torsional performance of frame could be increased by the large cockpit, with engine and other major components increasing the stiffness of the rear frame stiffness. The research of W. Billygolfer and K. Kcarney et al. [4] shows that the engine compartment can be omitted, in order to further reduce the weight of the frame and improve the performance of the frame. This structure is shown in Fig. 3. This space layout provides a good reference for the frame structure design.

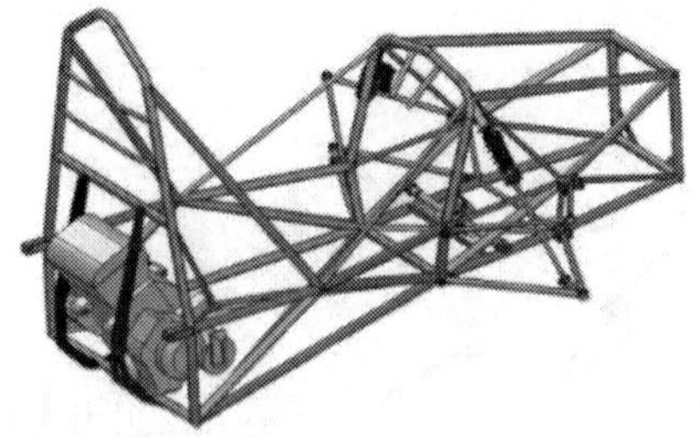

Figure 3. The structure of frame

3 Optimization method

The truss-frame is made up of steel tubes. The spatial position, thickness and many other factors of the steel tubes have great influence on the strength, rigidity and weight of the frame. Optimization algorithm has become an important method to optimize the frame.

3.1 Topology optimization

Topology optimization is a mathematical method to optimize the distribution of materials in a given region, according to the given load condition, constraint condition and performance index. This method provides designers with an ideal structure for reference.

Finite element model was established by using shell element. The topology optimization was carried out under six conditions. Considering the force of the frame, the structure of the final frame was designed. This method can provide a variety of references for the conceptual design of the frame, with allowing a more rational arrangement of steel pipe and reducing the quality of the frame.

3.2 Multi-objective optimization

The truss frame is made up of many different sizes of steel tubes, with the parameters of each steel tube affecting the performance of the frame. It is difficult to find the appropriate value in many of the design parameters. Therefore, the multi-objective optimization has become one of the most widely used method for frame optimization.

The section size of steel tube was set as the optimization variable. The stress and deformation were set as constraint conditions. The stiffness, vibration frequency and quality of frame were set as the objective function. The sensitivity analysis of frame was carried out, with the correlation between size of steel tube and frame stiffness, vibration frequency and mass being obtained. The appropriate size parameters were obtained by response surface analysis of some steel tubes.

There are two main difficulties in the design of the truss frame. One is the rational design of space structure. The other is the choice of the size of the pipe. The new spatial layout and optimum material distribution program can be provided by the topology optimization. Multi-objective optimization can find the suitable values in a large number of design parameters. It provides an effective way to optimize the structure size. According to the research of Fu Yuanfang and Wang Shuting [5], [6], using the optimization algorithm is an effective method to obtain the excellent performance of the frame.

4 Test method of the frame

Using the finite element software to analyze the frame is an important method to get the performance parameters of the frame. It is also one of the most popular methods to study the frame. However, the results obtained from the finite element software have a certain error with the actual performance. Therefore, testing the physical frame is of great significance in that it can provide more accurate data.

The common torsional stiffness measurement method is measured by the sensor. Some pressure sensors are installed on the four test points of front and rear suspensions. Two test points of rear suspension are fixed on the test table, with a pair of forced displacement being applied on two test points of the front suspension. By measuring the force and torsion angle, the actual torsional rigidity of the frame is calculated. A simple measurement method is that two rear suspension test points and right front suspension test points were fixed. The load would be applied by attaching a steel rod through the front bulkhead of the chassis, with the second rear rod being used to counter acts the moments created by the first and thus stabilized the rig. A horizontal bar was set up in front of the frame. Weight would then be added to both rods, with the angle between the center line of the front bulkhead and the horizontal bar being measured. According to this angle, the torsional stiffness was calculated. The schematic diagram of the experiment is shown in Fig. 4:

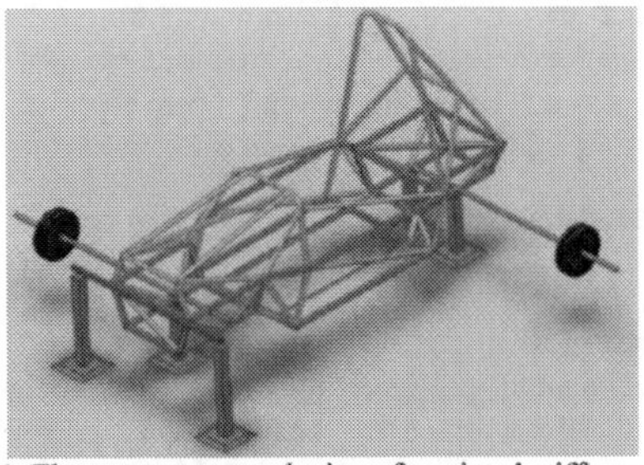

Figure 4. The measurement device of torsional stiffness

The angle of wheel is an important parameter to ensure the smooth running of the vehicle. When the frame occur deformation, the relative position of the fixed point of the suspension is changed, with the angle of wheel being changing. The ride comfort of vehicle also deteriorates. A new test method is that the suspension is mounted on the frame, with an aluminum plate being installed in the column. The sensor was mounted on the aluminum plate. Restraint and load was applied on the aluminum plate. The stiffness of frame was measured. At the same time, the change law of the tire angle was obtained. The comprehensive data reference of frame optimization was provided. The experimental apparatus is shown in Fig. 5:

Figure 5. The experimental apparatus

Due to the influence of positioning error and welding distortion, the actual stress, strain and the design value of the frame have certain error. The measurement method proposed by Jacob I. Salter and Thompson et al. [3], [7] could accurately measure the stiffness and strength of the frame. William B. Riley and Albert R. George et al. [2] put forward a new method. This method of applying the load through the suspension to measure not only can measure the rigidity of the frame, but also can measure the variation of the working angle of the tire, which provides accurate data for the analysis of vehicle performance. The accurate data for the strength analysis of the frame is provided.

5 The development trend

After reasonable design, the truss-frame can be used to meet the requirements. Thanks to its simple and easy implementation, truss-frame has become the mainstream of the frame form the team of University. Through the analysis, several development of the truss-frame are listed. First, the core goal is lightweight. Second, the front bulkhead is more compact, more spacious cockpit. Third, optimization method is applied more widely. Last, the test equipment of Frame will be improved.

6 Conclusions

With the promotion of the FSAE tournament, the research of the truss frame has been developed, but there are still a number of deficiencies. The vehicle collision test is an important part of vehicle safety research. Due to limited experimental conditions, the University team is difficult to carry out relevant tests. Using the computer to simulate the vehicle collision is the best way to study this kind of problem, but there is little research on the FSAE racing crash in China. In view of the test equipment of the FSAE truss-frame is less, design of the test equipment for the University team to use is an important problem to be solved.

The competition of Formula Student China start time is late, compared with the developed countries, so the technology and management of China are backward. At present, the design and manufacturing technology of Chinese automotive are weak. Through the competition of FSAE, the students' engineering practice ability is tempered. In 2010, the first FSAE competition was held. After six years of development, the number of college teams is more, with the car being getting better and better.

The frame of Formula Student China carries the engine, body and various ancillary facilities. The frame designers all hope that frame has high strength and light weight. Therefore, it is significantly important for the development of car to research frame design and optimization.

Acknowledgment

The authors are thankful to the Guangxi Natural Science Foundation (Grant no: 2013GXNSFBA019245), Science and technology research project of Guangxi University (Grant no: KY2015YB095), the Opening Project of Guangxi Key Laboratory of Automobile Components and Vehicle Technology, Guangxi University of Science and Technology (No. 2014KFMS05), and the graduate student scientific research innovation projects of Guilin University of electronic technology (Grant no: YJCXS201504)

References

1. X. Zhang, B. Gao, Y. Wang, The finite element analysis of FSC car under different conditions, Manufacturing automation, **37**, 10, 44-46 (2015)
2. W. Riley, A. George, Design, Analysis and Testing of a Formula SAE car chassis, SAE technical paper series, 2002-01-3300 (2002)
3. J. Salter, Design, Analysis and Manufacture of 2011 REV FSAE Vehicle Chassis, The University of Western Australia (2011)
4. W. Billygolfer, K. Kcarney, J. Jleith, A. Antonk, Design and Optimization of a Formula SAE race car, Worcester Polytechnic Institute (2012)
5. S. Wang, X. Liu, Y. Wu, G. Jin, Lightweight and fatigue life estimation of frame based on sensitivity analysis, China Mechanical Engineering, **22**, 16, 2001-2006 (2011)
6. Y. Fu, D. Jin, Y. Qiao, Optimum design method for frame structure of mini electric vehicle, Journal of Mechanical Engineering, **45**, 9, 210-213 (2009)
7. L.L. Thompson, J.K. Lampert, and E.H. Law, Design of a Twist Fixture to Measure Torsional the Torsional Stiffness of a Cup Chassis, SAE Paper, 983054

Application Research for Performance of Refrigeration Units with Flash Tank Economizer

Qin Huanghui[1] and Fan Li[1,2]

[1]*Department of Marine Engineering, Nantong Shipping College, Nantong, China*
[2]*College of Ocean Science and Engineering, Shanghai Maritime University, Shanghai, China*

Abstract. This paper studies the influence of flash economic system on performance of refrigeration unit using Refrigerant R290. Theoretical analyses of refrigeration system with flash type economizer are carried out firstly, which provide theoretical basis for working parameters. Under different evaporation temperature, influences of supply pressure on the refrigerating capacity and COP are studied and verified through experimental. Experimental results show that, using the economizer can effectively improve the cooling capacity and COP, and achieve the effect of saving energy. With the increase of supply pressure, refrigeration capacity of refrigerating unit and shaft power decreases gradually and the refrigeration coefficient rise first and then drop, to determine the best economic position.

1 Introduction

With the improvement of people's living standard and the development of industry, refrigerating units are widely used in the food, drinks, dairy products, petrochemical industry, coal, medicine, textile, paper making, tobacco, printing and other industries. The cooling temperature generally is between -80°C~0°C. For example, there are cooling, cold storage, frozen food, ultra-low temperature frozen in the process of frozen food, in detail , 0°C for the cooling room, -18°C for storage room, -35°C for refrigerating room, -30°C~-80°C for ultra-low temperature frozen. The lower the evaporation temperature of refrigerating unit, the greater the compression ratio, the sharp fall in the efficiency of compression. When compression ratio is greater than 8 to 10, two-stage refrigeration system should be used [1]. Refrigeration unit with economizer is a simplified two-stage compression refrigeration cycle system, the refrigerating capacity and refrigeration coefficient can be improved. When the evaporation temperature is lower than -30°C~-40°C and the compression ratio is greater than 8 or 10, the effect is obvious. At present, there generally two types of economizer, they are flash type and sub-cooled type. The latter is safe and reliable, but the system is high cost and complex. And it is achieved by direct expansion. The former is simple in structure, easy to control and low in cost, so is mostly used in refrigeration unit [2].

When screw compressor works, the refrigerating capacity and the screw refrigeration coefficient can be increased by adding a new supply port at the rotor to draw in gas from the economizer [3], [4]. Yang Li, etc. from the Shanghai Jiao Tong University studied the performance of screw refrigeration compressor using R134a and R22 by numerical simulation, and compared the performance difference between economizer units and common refrigeration units [5]. Sun Zhen, etc. from the Shanghai Maritime University studied performance of screw refrigeration unit using R134a with ejector and economizer in different operating conditions, and optimized injection port working with economizer [6]. Sun Chao etc. from Huazhong University of Science and Technology took numerical analysis and study on R134a applying in gas supply on refrigeration system of screw compressor [7]. This paper studies the gas supply process of the refrigeration unit system with flash type economizer. The refrigerant is propane (R290), a kind of natural refrigerants [8].

Under different evaporation temperature, influences of supply pressure on the refrigerating capacity and COP are studied in this paper, and verified through experimental. This research has an important guiding significance for the design and application of refrigerating units with flash type economizer.

2 Theoretical analysis

Refrigeration system with flash type economizer is shown in Fig. 1. The LCV throttles the liquid refrigerant, some of the liquid refrigerant change into gas after passing this valve, and flow into the flash type economizer.

The opening of regulating valve-4 is controlled by adjusting the liquid level in the condenser. Pressure control valve-6 maintains the supply pressure in the flash type economizer. The flashed refrigerant gas enters the economizer port of the compressor, while the saturated liquid through the economizer piping flows into the low pressure evaporator by the second throttling valve. In the economizer, the flashed refrigerant gas enters into the primitive volume chamber (the suction chamber) after passing through the pressure control valve and the gas supply port in the screw compressor.

The flashed refrigerant gas and the gas in the primitive volume chamber are compressed together, so the compressor power will increase.

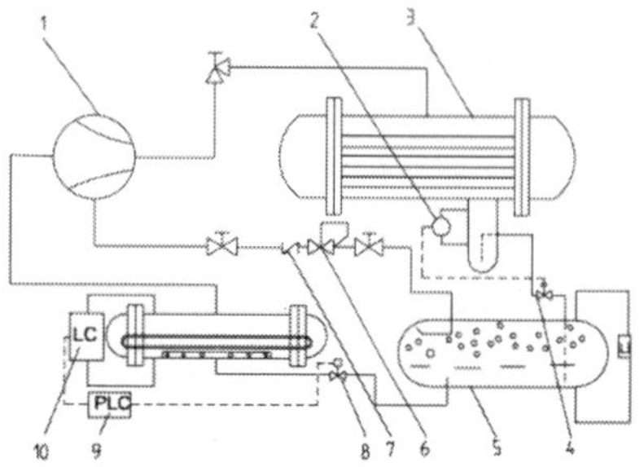

1. Screw Comp. 2.LCV 3.Condenser 4.Control Valve
5. Flash Eco. 6.PCV 7.CV 8.SV 9.PLC 10.ELC

Figure 1. Flow chart of refrigeration system

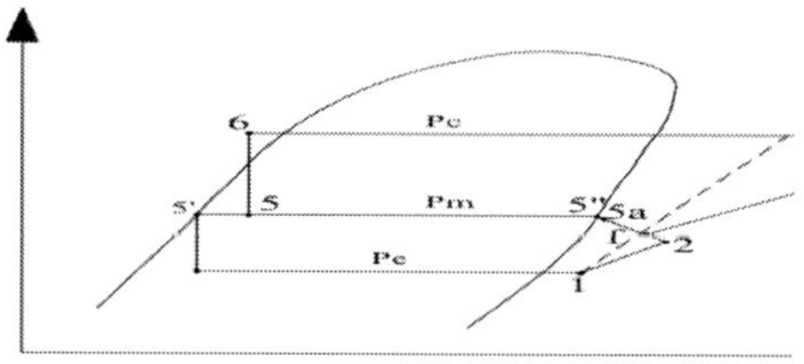

Figure 2. P-h of flash type economizer system

The screw compressor with gas supply port can be considered as a two-stage system compression process, and the P-h shown in Fig. 2. The compressor draws in the gas first. Then, the gas suction chamber in the compressor is separated from the suction port at the end of suction process, and the primitive volume chamber becomes an enclosed space [9]. After a short period of compression, the primitive volume chamber immediately connects with the gas supply port. At this position, the gas from the flash type economizer at pressure P_m enters the compressor immediately. The opening of the gas supply port will increase gradually as the rotation of the screw, and more gas is drawn in, thus higher pressure is produced in the chamber. With the increase of the pressure in the chamber, the differential pressure and the amount of supply gas decreases. At the end of gas supply process, the pressure at the primitive volume chamber is P_f, which is less than P_m. In the screw compressor, the compression process is a continuous process. The gas in the gas supply chamber is at a relatively stable state, so the gas supply process is a rotating pressurization and adiabatic gas mixed combination process. When the gas supply port is separated from the suction port the second stage compression starts. In the beginning of compression, oil is sprayed into the compression chamber. Because the temperature of oil is higher than that of gas, so gas is heated by oil. At the end of compression, the temperature of gas is higher than that of oil, thus the compression process is a polytrophic compression process. On the p-h graph, the slope of f-3 should be greater than the adiabatic compression process (f-3'). If the screw compressor internal pressure ratio is less than the external the pressure ratio, that is to say exhaust pressure (P3) is less than condensing pressure (Pc), the exhaust process is approx isovolumic compression process. If the compressor internal pressure ratio is greater than the external pressure ratio, that is P3>Pc, the exhaust process is constant volume expansion [10]. In short, the position of gas supply port and the oil temperature will influence the indicating efficiency of the second stage compression, change refrigerating capacity, compressor power and COP in a certain extent [11], [12].

2.1 Determination of the parameters of primary compression process

The supply port is added in the compressor rotor on the direction of the length of the screw, and the location of this port determines volume ratio. The pressure and temperature at the end of the primary compression are calculated by formula (1), (2) [13].

$$P_2 = P_1(\eta_{v1}\varepsilon_{v1})^n \tag{1}$$

$$T_2 = T_1\left[\frac{P_2}{P_1}\right]^{\frac{n-1}{n}} \tag{2}$$

Among them, P_1 : compressor suction pressure, Pa.

P_2: Pressure at the end of the primary compression, Pa.

T_1 : Compressor suction temperature, K.

T_2 : Temperature at the end of the primary compression, K.

η_{v1} : Volume efficiency of the primary compression.

ε_{v1} : Volume ratio of the primary compression.

2.2 Mass and energy balance of economizer

For the flash type economizer, the gas mass and energy balance equation in the supply process are as follows:

$$q_c = q_e + q_f \tag{3}$$

$$q_c h_5 = q_e h_{5'} + q_f h_{5''} \tag{4}$$

Among them, q_c : Mass flow of the compressor exhaust, kg s^{-1}.

q_e : Mass flow of evaporator, kg s^{-1}.

q_f : Mass flow of the economizer supply gas, kg s^{-1}.

$h_{5'}$: Specific enthalpy of saturated liquid refrigerant in the flash type economizer, kJ kg^{-1}.

$h_{5''}$: Specific enthalpy of saturated refrigerant gas in the flash type economizer, kJ kg^{-1}.

h_5 : Specific enthalpy of sub-cooled liquid, kJ kg^{-1}.

2.3 Determination of the parameters for the supply process

Supply process is an approximate isentropic and adiabatic process. Gas supply equation of per refrigerant suction volume, the energy equation of gas supply process, the specific volume equation and the specific entropy equation are shown as formula (5), (6), (7), (8) [11]:

$$\alpha = \frac{q_f}{q_e} \tag{5}$$

$$h_2 + \alpha \cdot h_{5"} = (1+\alpha) h_f \tag{6}$$

$$v_f = v_2 / (1+\alpha) \tag{7}$$

$$(1+\alpha) \cdot s_f = \alpha \cdot s_{5"} + s_2 \tag{8}$$

Among them,

α : Gas supply ratio of flash type economizer refrigerant system;

v_2 : Specific volume at the end of the primary compression, $m^3\ kg^{-1}$.

v_f : Specific volume at the end of the supply process, $m^3\ kg^{-1}$.

s_f : Specific entropy at the end of the supply process, kJ $(kg{\cdot}K)^{-1}$.

s_2 : Specific entropy at the end of the primary compression, kJ $(kg{\cdot}K)^{-1}$.

$s_{5"}$: Specific entropy of saturated gas at the economizer pressure, kJ $(kg{\cdot}K)^{-1}$.

2.4 Exhaust temperature of compressor

The compression process is a changeable compression process after the supply process, the exhaust temperature T_3 is calculated as follows:

$$T_3 = T_f \left[\frac{P_c}{P_f} \right]^{\frac{n-1}{n}} \tag{9}$$

2.5 Refrigerating capacity and compression power

The refrigerating capacity is calculated as follows:

$$Q_L = q_e (h_1 - h_6) \tag{10}$$

For the compression power calculation:

$$W = \frac{q_e \left[(h_3 - h_1) + \alpha \cdot (h_3 - h_{5"}) \right]}{\eta_e \eta_d} \tag{11}$$

Among them, η_e : Adiabatic efficiency of the compressor.

η_d : Motor efficiency (including the transfer efficiency between the motor and the compressor).

In this paper, the theoretical displacement of the compressor is $650m^3\ h^{-1}$. The refrigerating capacity, COP value of refrigeration unit, and compressor power are showed in Fig. 4. The curve is the theoretical calculation results at different economizer pressure when the suction pressure is 135 kpa and exhaust pressure is 1370 kpa .

3 Experimental

Refrigerant unit experimental is completed in environmental simulation laboratory. The experimental devices include brine system and refrigerant system. Technical parameters and precision of experiment instruments are shown in Table 1.

Table 1. Experimental instruments and equipment.

The serial number	Name of equipment	Technical parameters	Precision
1	Current transformer	0~600A	0.2 level
2	Digital power meter	5.0~500V/ 0.01~40A	0.5 level
3	Electromagnetic flowmeter	$0\sim100m^3\ h^{-1}$	0.50%
4	PT temperature sensor	-60~60°C	A level
5	Pressure sensor	-50~2000kPa	0.10%

The flash type economizer is applied in the refrigerating unit, using R290 as refrigerant, with the open type twin-screw compressor, the water-cooled shell and tube condenser, and half bundle evaporator.

The rotor diameter of the screw is 163.2mm. The length diameter ratio is 1.65. The theoretical displacement of compressor is $640.7m^3\ h^{-1}$. The throttling valve adopts Fisher LCV to control the refrigerant level of condenser. The pressure control valve controls the pressure of the economizer. Using PID, it can achieve the stability of the economizer pressure and liquid level at different load and pressure.

4 Experimental data analysis

At different supply pressure, the theoretical and experimental results of the unit capacity, power and COP are shown in Fig. 3. The experimental conditions are: evaporate at -35±1°C, condensate at 40±1°C, superheat 1°C at the 100% load. The Fig shows that the unit capacity decreases with the increase of the economizer supply pressure, the same is true for compressor power. The Cal. capacity is about 96.7% to that of the Exp. capacity, the Cal. power is about 101.2%, and the Cal. COP is about 95.5%. So for the refrigeration unit with flash type economizer, the theoretical calculation results are basically consistent with the experimental data. Both to theoretical and experimental data, the refrigerant system COP is parabola distributed with the increase of gas supply pressure. COP increases first, then decreases, because of the reducing rate of refrigerating capacity is less than the rate of power when the supply pressure is closer to the suction pressure. And the opposite occurs when the supply pressure is closer to the exhaust pressure. As can be seen from Fig. 3, refrigerating capacity curve is approximate linear curve, and the power curve is about quadratic curve. So there must be an optimal supply pressure for flash type economizer system. Theoretical calculated COP reaches a peak at the pressure of 402.9 kPa, and the experimental test is about 412 kPa. This shows that when the gas supply pressure is 2.98 ~ 3.05 times of the compressor suction pressure, COP is the highest, and the gas supply port position is reasonable.

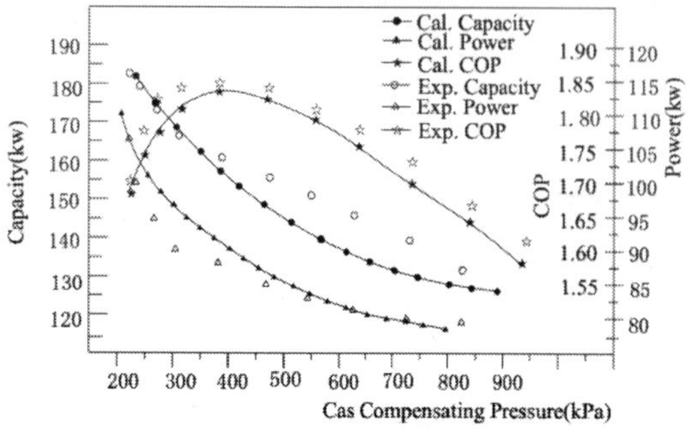

Figure 3. T&E curve in different supply pressure

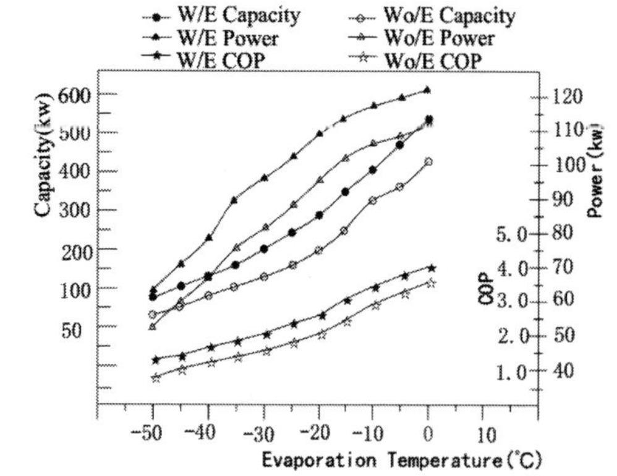

Figure 4. The thermal performance of different evaporation temperature

Flash type economizer has great impact on the performance of refrigeration unit. In two different working conditions (flash type economizer put into use and fall out of use), when the condensing temperature remains at 40°C, the thermal performance (such as refrigerating capacity, shaft power and COP) at different evaporation temperature is shown in Fig. 4. Regardless of whether the economizer is put into use, with the increase of the evaporation temperature, the cooling capacity and the compressor shaft power will increase, and both are basic linear change. When evaporate at -5°C, the refrigerating capacity of the unit with flash type economizer increased by 30% than without the economizer, shaft power increased by 10%, and the cop increased by 17.1%. And when evaporate at - 50°C, the refrigerating capacity of the unit with flash type economizer increased by 72%, shaft power increased by 18%, and the COP increased by 45.5%. Experimental data show that the capacity and shaft power are greatly improved with the use of economizer. With the decrease of the evaporation temperature, the change rate of refrigerating capacity is greater than that of power. Because the COP of the unit with economizer is higher than without economizer, refrigeration performance of flash type economizer unit is obviously improved. In lower evaporation temperature, that is to say, compression ratio is higher, the greater the COP is. Therefore, in low-temperature refrigeration occasions, refrigerating units using flash type economizer can greatly improve the refrigerating capacity and cooling efficiency of the unit.

The change trend of refrigerating capacity and shaft power can be explained by theoretical analysis of refrigeration unit. With the change of air supply pressure, air supply flow is different. When the air supply port is more close to the suction chamber, supply pressure decreases, the gas supply flow increases, and the refrigeration capacity increases greatly. And the second stage compression section of the compressor will extend, the increasing amplitude of shaft power becomes large. When the air supply port is more close to the discharged port, supply pressure increases, the gas supply flow decreases, and the refrigeration capacity decrease greatly. And it will shorten the second stage compression section of the compressor, causing the increasing amplitude of shaft power becomes small. So the change trend of refrigerating capacity with supply pressure is the same as shaft power. With the increase of the supply pressure, the actual growth rate of polytrophic compression power will become large. Because of the injection of higher pressure gas into the compression chamber, more power will be consumed than the equivalent low pressure gas, thus cause the compressor shaft power growth rate becomes larger. However, with the gas flow becomes smaller the growth rate of theoretical adiabatic power will become lower. So there must exist an optimal supply pressure makes the most efficiency of the compressor. That is, the greatest ratio of theoretical adiabatic power to the actual polytrophic compression power.

The above experimental and theoretical analyses show that the low temperature cooling capacity and efficiency of refrigerating unit are improved obviously with the use of flash type economizer. With the use of flash type economizer system, the scope of low temperature refrigeration has been effectively expanded, and the operation reliability of refrigeration unit greatly improved.

5 Conclusions

This paper makes an experimental in thermo performance of refrigeration unit with flash type economizer using Refrigerant R290. The data show that, using the economizer can effectively improve the unit's cooling capacity and COP, so as to achieve the effect of saving energy. And when the compression ratio increases, the energy saving effect will be more significant.

The experimental data of the effect of supply pressure on the compressor refrigeration efficiency is consistent with theoretical analysis. For the refrigerating unit, there is an optimal point of economizer supply pressure. The actual supply pressure is slightly higher than this point.

For the application of environmental protection refrigerant R290 in refrigerating units, the technology is still in research stage, especially in the flash type economizer. In this paper, theoretical analysis have an important guiding significance for the design of refrigerating units with flash type economizer, and provide theoretical basis and practical support for improving the refrigerating unit's performance at low evaporation temperature in long-term reliable operation.

Acknowledgment

We gratefully acknowledge the support of this effort from the sponsors of the Refrigeration Technologies and Applications Laboratory of Nantong Shipping College.

References

1. X. Xu, Y. Hwang, R. Radermacher, Transient and Steady-State Experimental Investigation of Flash Tank Vapor Injection Heat Pump Cycle Control Strategy, Int. J. Refrig. **34**,1922-1933 (2011)
2. J.H. Qiu, Design Analysis of Flash Type Economizer System for Ammonia as a Refrigerant, China Journal of Refrigeration, **32**(2), 29-32 (2013)
3. Z.W. Xing, Screw Compressor - Theory, Design and Application, China Mechanical Industry Press, (2008)
4. J.J. Tian, Z. Zhang, Experimental Study on Air-Cooled Heat Pump on the Frosting Conditions, Refrigeration and Air Conditioning (China Sichuan), **22** (1): 88-90 (2008)
5. L. Yang, W. Wang, Y.F. Bai, Influence of Economizer on Compression Refrigeration Cycle, Journal of Refrigeration (China), **4**, 35-38+56 (2010)
6. Z. Sun, W. Chen, R. Yang, et al. Research of Refrigeration System with Ejector and Economizer, China Journal of Refrigeration, **1**, 15-20 (2010)
7. C. Sun, Numerical Analysis and Study on R134a Applying in Gas Supply on Refrigeration System of Screw Compressor, Compressor Technology (China), **1**, 33-34 (2012)
8. Analysis of Vapour Compression Refrigeration System with R404A, R407C and R410A, International Journal of Mechanical Engineering and Robitics Research, **2**(1), 25-36
9. H.G. Wu, X.Y. Peng, Z.W. Xing, P.C. Su, Experimental Study On the Effects of Superfeed Pressure of Ecomizer on Twin-Screw Refrigeration Compressor Performance, Journal of Refrigeration (China), **4**, 10-13 (2003)
10. L. Fu, G.L. Ding, C.L. Zhang, Z.J. Su, Development and Application of Screw-type Water Chiller Steady-state Simulation Model, HVAC, **35**(5), 71-75 (2005)
11. H.H. Qin, Experimental Study on Performance of Air Cooled Heat Pump Device with Flashing System Economics, Journal of refrigeration (China), **34**(5), 55-58 (2013)
12. Y.C. Li, Z.H. Cai, Y.F. He, et al. Study on Performance Characteristic of Heat Pump with Economizer, Journal of Refrigeration(China), **32**(6), 17-19, 23, (2011)
13. Q.H. Chai, G.Y. Ma, Y. Jiang, J.J. Xia, et al. Thermodynamic Analysis of Scroll Compressor Refrigeration Cycle with Economizer, Journal of Tsinghua University (Natural Science Edition), **43**(10), 1401-1404 (2003)

A Hybrid Prediction Method of Thermal Extension Error for Boring Machine Based on PCA and LS-SVM

QiangCheng[1] Yanwei Yu[2], Guangpeng Li[1] , WeishuoLi[1] Bingwei Sun[1] and Ligang Cai[1]

[1]*College of Mechanical Engineering and Applied Electronics Technology,Beijing University of Technology, Beijing 100124, China*
[2]*Henan Mechanical and electrical Vocational college ,Zhengzhou 451191,China*

Abstract: Thermal extension error of boring bar in z-axis is one of the key factors that have a bad influence on the machining accuracy of boring machine, so how to exactly establish the relationship between the thermal extension length and temperature and predict the changing rule of thermal error are the premise of thermal extension error compensation. In this paper, a prediction method of thermal extension length of boring bar in boring machine is proposed based on principal component analysis (PCA) and least squares support vector machine (LS-SVM) model. In order to avoid the multiple correlation and coupling among the great amount temperature input variables, firstly, PCA is introduced to extract the principal components of temperature data samples. Then, LS-SVM is used to predict the changing tendency of the thermally induced thermal extension error of boring bar. Finally, experiments are conducted on a boring machine, the application results show that Boring bar axial thermal elongation error residual value dropped below 5 μm and minimum residual error is only 0.5 μm. This method not only effectively improve the efficiency of the temperature data acquisition and analysis, and improve the modeling accuracy and robustness.

1 Introduction

In recent years, with the continuous development of equipment manufacturing industry, higher and higher requirement of precision is presented to machine tools [1]. For boring machine, thermal error of the boring bar is one of the key error sources [2-3], and research shows that the thermal error of boring machine can accounts for 40% ~ 70% of total errors[4-5].In general, the methods to reduce the thermal errors can be classified into error prevent method and error compensation method [6]. Error prevention refers that the thermally induced errors could be reduced with structure improvement of the machine tool itself through design and manufacturing technology. In the process of implementation, however, there are many physical limits, cannot completely overcome design difficulties [7-8]. As a result, the thermal error compensation technology is an effective and economical method of reduce the thermal error of machine tool [9-10].

Generally speaking, in order to get a high precision and good robustness of thermal error model, it is important to obtain the temperature data of each site and the corresponding heat deformation in the processing. However, the boring machine temperature field is nonlinear, degeneration and the extremely complex distribution, which requires layout of a large number of temperature sensors to obtain more accurate temperature field on boring machine. Arrangement of much temperature sensors can lead to the problem of dealing with the coupling mong a large amount of data, and at the same time much wiring will also affect the machining process. Therefore, many scholars carried on the thorough research in terms of temperature measuring point optimization and achieved some results. Yang et al. proposed a method based on thermal error sensitivity figure, however this method needs to select more measuring points to ensure accuracy and otherwise the efficiency of modeling will be reduced [11]. Wang et al. proposedPSO optimization K mean clustering method to optimize the temperature measurement points. The method can improve the speed of modeling, but the random initial values may lead to different clustering results, and even no solution [12]. Luo et al. offered a method based on the stepwise linear regression, which provides a new way for thermal error modeling [13]. According to the thermal error and the temperature field of the dynamic, Ma optimized the temperature measurement point [14]. Lee of South Korea has been reduced temperature measurement point by using correlation coefficient and linear regression method. Some results have been obtained [15], but the nonlinear of temperature data is not considered adequately. Creighton et al. established the model of temperature rise and thermal deformation by using finite element method[16].The temperature rise and thermal deformation of the key points were obtained by using thermocouple and capacitance meter. The optimized process is relatively straightforward, but the boundary conditions of finite element method are not uniform. Han et al. proposed a correlation coefficient analysis and fuzzy C means clustering method to reduce the number of temperature sensors, the number of sensors from 32 to 5 [17]. However, this method has the following disadvantages: in order to select the appropriate temperature measurement point must collect a large number of temperature data, the available data must be typical distribution, such as normal (Gaussian) distribution.

It can be seen that in the process of the thermal error modeling and compensation, one of the key is to reduce the large number of temperature measuring points. Therefore, PCA method is proposed to reduce temperature measuring point in this paper, and then LS-SVM modeling method is used to predict the thermal error of boring machine. In order to optimize temperature measuring points and realize accurate prediction of thermal extension error, this paper is arranged as following. The section 2 introduces the reduction method of PCA. Section 3 introduces the relation prediction of the error and temperature based on LS-SVM method. In Section4,a CNC boring machine is selected as an example to verify the method. The conclusions are presented in Section 5.

2 Reduction the dimension of temperature data based on PCA

Theoretically, thermal extension error of boring baris mainly influenced by the whole boring machine temperature changes. In order to establish the accurate thermal extension error of boring bar prediction model, it is required that all parts on the boring machine are installed a large number of temperature sensors to measure real-time temperature during boring machining. Generally speaking, the temperature measuring point selecting from several to several hundreds [18]. However, too much temperature measuring point not only increase the workload, but also increase the mutual coupling between the measured data, which may affect the accuracy of prediction. In order to reduce the amount of input and improve the speed and accuracy of modeling, therefore, in this section a boring machine showed in Fig. 1 is selected as an illustrative object to apply PCA to reduce the temperature measuring points.

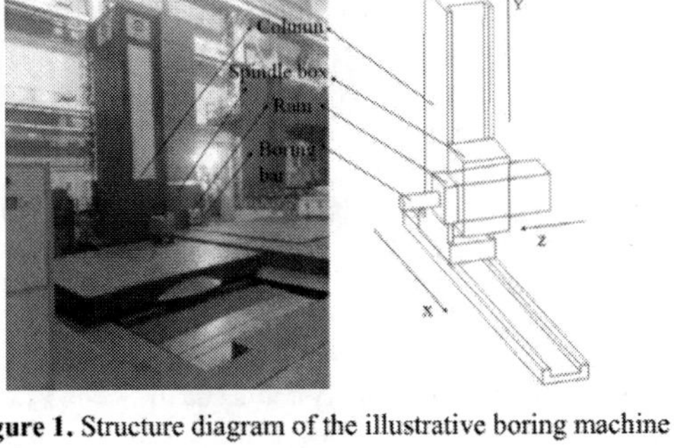

Figure 1. Structure diagram of the illustrative boring machine

PCA aims to use the ideas of dimension reduction to make the majority of indicators transform into a few comprehensive indexes [19-20].The main idea is to find the spindle direction of the data and constitute a new coordinate system spindle, the dimension of the new coordinate system can be lower than that of the original dimension, then the data are mapped to the new coordinate system from the original coordinate system. This mapping process is the process of feature data dimension reduction. So, PCA is used to reduce the dimension of the temperature data, the specific steps are as follows:

(1) Firstly, sort out a lot of boring machine temperature data to construct matrix $\mathbf{T}_{mn}$ Where, m is the number of temperature data collection sample, n is the number of temperature sensors installed on the boring machine.

$$T_{m\times n}=\begin{bmatrix} t_{11} & t_{12} & \cdots & t_{1n} \\ t_{21} & t_{22} & \cdots & t_{2n} \\ \vdots & \vdots & & \vdots \\ t_{m1} & t_{m2} & \cdots & t_{mn} \end{bmatrix}$$

(2) Secondly, standardize the elements in temperature matrix according to Eq.(2).

$$\begin{cases} T_{ij}^{*}=\left(T_{ij}-\overline{T_j}\right)/\,std\left(T_j\right) \\ \overline{T_j}=\dfrac{1}{M}\sum_{i=1}^{M}T_{ij} \\ Std\left(T_j\right)=\sqrt{\dfrac{1}{M}\sum_{i=1}^{M}\left(T_{ij}-\overline{T_j}\right)^2} \end{cases} \tag{2}$$

In which, $i=1,2,\cdots,m,\ j=1,2,\cdots n$; $\overline{T_j}$ is the mean value of the j column temperature, and $Std(T_j)$ is the standard deviation of the j column temperature.

(3)Thirdly, calculate the covariance matrix (S)of the standardization data F_{ij}^{*} by Eq.(3).

$$s_{jk}=\frac{1}{M}\sum_{i=1}^{M}\left(T_{ij}^{*}-\overline{T}_j\right)\left(T_{ik}^{*}-\overline{T}_k\right) \tag{3}$$

$$j=1,2,\cdots,n;\ k=1,2,\cdots,n;$$

(4)Fourthly, calculate the eigenvalues of the covariance matrix S and order by size, $\lambda_1\geq\lambda_2\geq\lambda_3\geq\cdots\lambda_n$;The eigenvalue of the corresponding eigenvector are $\alpha_1,\alpha_2,\alpha_3,\cdots\alpha_n$.

(5)Finally, by the eigenvalues of the covariance matrix S, compute variance contribution rate β_k and the cumulative variance contribution rate $\beta(k)$ is obtained as:

$$\begin{cases} \beta_k=100\%\lambda_k\Big/\sum_{k=1}^{M}\lambda_k \\ \beta(k)=\sum_{q=1}^{k}\lambda_q\Big/\sum_{q=1}^{M}\lambda_q \end{cases}$$

The number of principal components which needed to be extracted is determined by the principle that the cumulative variance contribution rate is more than 85%, and the principal component of $\boldsymbol{A_i}$ can be expressed as following:

$$A_i = T^*_{ij}\alpha_i$$

In which, $i=1,2,\cdots,q$; q is the number of the principal components.

3 Relation prediction of the error and temperature

In machine learning field, SVM is a supervised learning model, which is usually used for pattern recognition, classification, and regression analysis [21].The basic idea is to map a given sample set $T=\{(x_i,y_i)|x_i\in R^n, y_i\in(-1,+1)\}_{i=1}^n$ to a high dimensional feature space by nonlinear kernel function that meets Mercer condition, and construct an optimal classification hyper-plane, which can separate two classes of data points as far as possible.

Training sample (x_i,y_i) is given, in which, $i=1,2,\cdots,n, x\in R^d$, $y\in R$. x_i is training sample input for n dimension , y_i is training sample output for n dimension .Decision function is $f(x,w)=w\phi(x)+b$. So, the training of the model is transformed into the structural risk minimization. The mapping function of the original space sample to high dimensional feature space is $\phi(t)$, W is the weight vector, $x=(T_1,T_2,T_3,\cdots,T_n)$ is composed of N boring machine temperature vector, and b is a threshold. Based on the principle of structural risk minimization, w and b are obtained by minimizing the structural risk function $J=\|w\|^2/2+sE$;Where, s is the regularization parameter, E is error loss function.LS-SVM is developed on the basis of SVM, and the estimation of the model function can be described as the optimization of J function:

$$\begin{cases} \min J(w,\xi)=\frac{1}{2}w^T w+\frac{1}{2}s\sum_{i=1}^n \xi_i^2 \\ s.t.\quad y_i=w^T\varphi(x_i)+b+\xi_i \\ \qquad \xi_i\ge 0 \quad i=1,2,\cdots,n \end{cases} \tag{4}$$

In order to convert the Eq. (4) to the dual problem, the Lagrange function is introduced as follows:

$$L(w,b,\xi,\alpha)=\frac{1}{2}w^T w+\frac{1}{2}s\sum_{i=1}^n \xi_i^2 - \sum_{i=1}^n \alpha_i\left[w^T\varphi(t_i)+b+\xi_i-y_i\right] \tag{5}$$

In the formula, $\alpha_i=(\alpha_1,\alpha_2,\cdots,\alpha_3)$ is the Lagrange multiplier; w,b is model parameters; ξ is training set prediction error vector. According to the necessary condition of extreme value, partial derivative of the Eq. (5) is calculated as follows:

$$\begin{cases} \frac{\partial L}{\partial w}=0 \rightarrow w=\sum_{i=1}^n \alpha_i\varphi(x_i) \\ \frac{\partial L}{\partial b}=0 \rightarrow \sum_{i=1}^n \alpha_i=0 \\ \frac{\partial L}{\partial \xi_i}=0 \rightarrow \alpha_i=2s\xi_i \\ \frac{\partial L}{\partial \alpha_i}=0 \rightarrow w^T\cdot\varphi(x_i)+b+\xi_i-y_i=0 \end{cases}$$

The optimization problem is transformed into solving linear equation Group(6).

$$\begin{cases} \begin{pmatrix} 0 & 1 & \cdots & 1 \\ 1 & A & \cdots & B \\ \vdots & \vdots & & \vdots \\ 1 & B & \cdots & A \end{pmatrix} \begin{pmatrix} b \\ \alpha_1 \\ \vdots \\ \alpha_i \end{pmatrix} = \begin{pmatrix} 0 \\ y_1 \\ \vdots \\ y_i \end{pmatrix} \\ \begin{cases} A=\varphi(x_1)\varphi(x_i)+\frac{1}{s} \\ B=\varphi(x_i)\varphi(x_1) \end{cases} \end{cases} \tag{6}$$

According to the Mercer conditions, the Gauss kernel function $K(x,x_i)$ is selected to as kernel function. So, LS-SVM regression modeling can be expressed as follows:

$$f(x)=\sum_{i=1}^n \alpha_i K(x,x_i)+b$$

The kernel function can be chosen to satisfy the symmetric function of Mercer condition, commonly kernel function has linear kernel function and polynomial kernel function and Gaussian kernel function. In this paper, RBF core $K(x,x_i)=\exp\left[-\|x-x_i\|^2/(2\sigma^2)\right]$ is selected; So that the least squares support vector machine estimation algorithm contains two sensitive parameters: regularization parameter s and kernel function parameter σ, and these two parameters can be obtained by cross identification (CV)[22-23].

4 Experimental analysis and model- ing verification

4.1 Experimental data acquisition

The boring bar system structure of machine tool is as shown in the Fig. 2, in which the spindle box is mainly composed of three parts, including a static pressure radial bearing, a gear box, and a spindle motor. The gear box was divided into two grades, the higher speed was 2000r/min, and the lower onewas600r/min. Experiment was divided into two parts. Firstly, the gear box was set to be 600r/min, and the purpose was to make the natural extension of boring bar to reach the thermal equilibrium state, and the spindle was stopped in order to make the boring bar cool naturally until the length of the boring bar was no longer significantly reduced. In Second part, the gear box was set to be 2000r/min, and the above process was repeated. The sensors were mounted as shown in Fig. 3 and Table 1, and the temperature values were read by the numerical control system.

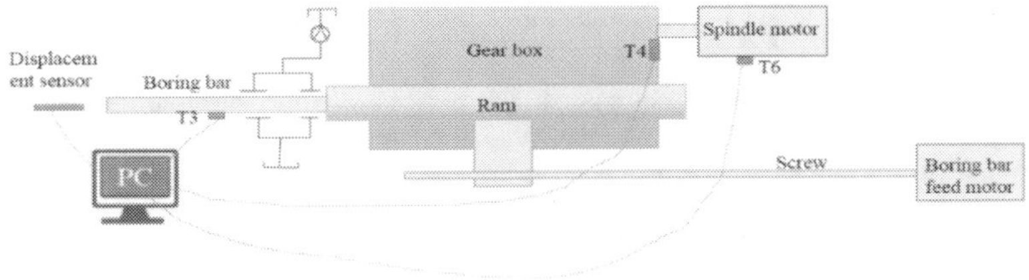

Figure 2. The boring bar structure diagram

(a) Temperature sensor of T5

(b)Temperature sensor of T7 and T8

Figure 3. Installation of instrument

Table 1. The arrangement of 11 sensor locations

T2, T5	The boring bar hydrostatic oil temperature and hydrostatic pump location
T3, T4	Front of the boring bar, gear box position
T9, T10, T11	Up and down of boring bar bearing end and after boring bar bearing position
T7, T8, T6	Boring bar static pressure oil outlet and inlet position and the main shaft motor position
T1	environmental temperature

Table 2. Partial temperature data

T1	T2	T3	T4	T5	T6	T7	T8	T9	T10	T11
23.8	22.2	23.7	26	39	27	20.9	18.7	35.6	36.9	38.4
23.3	21.3	24.4	25.3	40	27	22.6	18.8	33.6	35.6	37.1
23.2	21.2	23.7	26.6	42	75	22.2	17.1	34.5	36.1	38.1
23	21.1	23.3	29	43	81	21.2	16.2	31.8	33.6	35.1
22.8	21.1	23.8	33.2	44	90	21.3	21.3	29.7	31.1	34.6
22.7	21.1	24.3	36.5	44	96	21.9	22	29.1	30.1	34.2
22.6	21	24.6	38	45	101	22.7	21.7	31.4	31.6	35.1
22.5	20.9	23.8	40.5	46	104	22.1	17	35.1	35.6	36.4
22.4	20.9	23.2	41.6	47	106	21.1	16.1	35.9	35.8	36.2
22.2	20.8	23.4	43.3	47	108	20.9	20.8	29.4	30.8	33.3
22.1	20.8	23.8	44	46	109	21.4	21.5	29.3	30.2	33.4
22	20.7	24	45.3	47	110	22.3	19.8	31.7	32.2	35.1
22	20.7	23.2	45.7	47	111	21.6	16.4	35.3	35.6	36
22	20.7	22.8	46.2	47	112	20.7	20.2	31.1	31.5	33.1
22	20.6	23.6	46.8	47	113	21.2	21.3	29.7	30.6	35.6
22	20.6	24	47.2	47	113	22.2	21.9	30.1	30.4	33.3
22	20.6	23.2	47.5	47	113	21.8	16.6	35.2	35.6	36.1
22	20.6	22.7	47.9	48	114	20.9	18.6	31.8	32.4	33.2
22.1	20.5	23.4	48.2	47	114	21.2	21.1	30.1	30.8	33.5
22.1	20.5	23.7	48.4	47	114	21.9	22	29.6	30.7	33.7
22.2	20.5	23.4	48.7	48	115	22	17.1	35	35.6	36.7
22.2	20.5	23	48.8	48	114	21.1	16	36.3	36.1	36.3
22.3	20.5	23.2	49.1	48	115	21.1	21	30.7	31.4	33.6

DOI: 10.1051/matecconf/20179507010

In this experiment, stick out length of boring bar was set as 60 cm, a dial indicator was installed at the front of the boring bar to measure the boring bar elongation. The inlet oil temperature and the oil return temperature of the static pressure radial bearing and the temperature of the sensor were measured by the temperature sensor Pt100. Temperature and the corresponding thermal elongation error data is recorded at every five minutes. The changing curves of temperature data gotten at each measuring point are shown in Fig. 4, and part data are shown in Table 2.

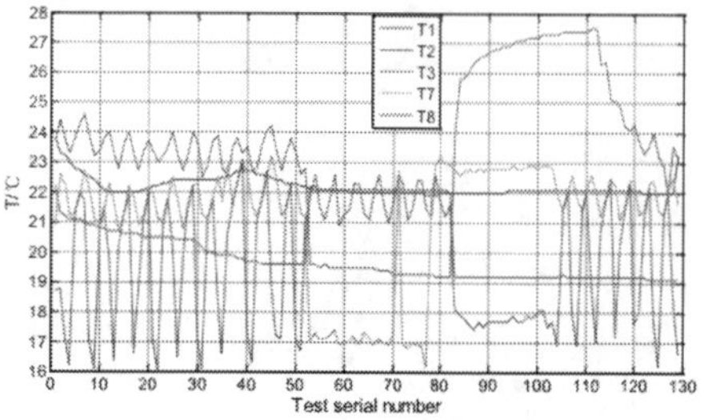

(a) Measured data curve of T1, T2, T3, T7, T8

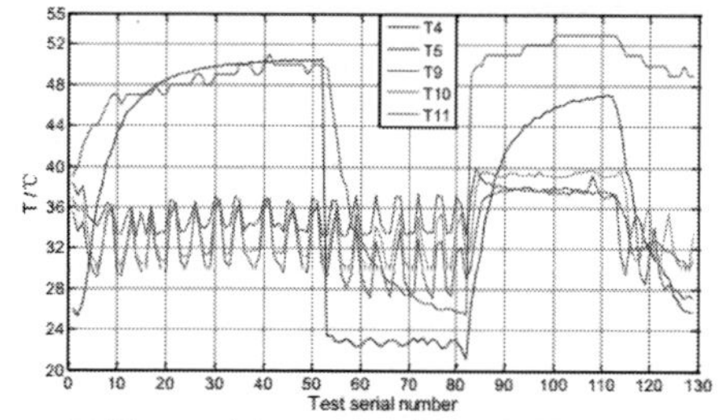

(b) Measured data curve of T4, T5, T9, T10, T11

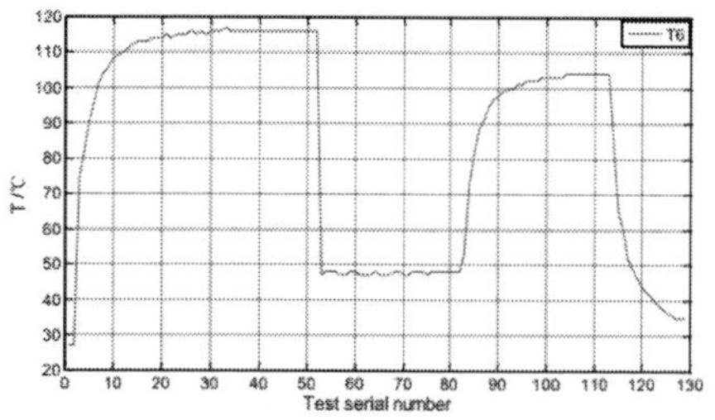

(c) Measured data curve of T6

Figure 4. Measured data of all temperature sensor

4.2 Extraction the principal component of the temperature data by PCA

130 sets of data were collected at every five minutes during the experiment, and so a 130*11 temperature matrix can be established. By the method in Section 2, 11 principal component contribution rate are extracted and the cumulative contribution rates are shown Table 3. From the Table 3, it can be seen that the cumulative contribution rates of the first 4 principal components are reached 88.7835%, so the number of main components is determined as 4.

Table 3. Principal component calculation

Principal component	Variance contribution rate%	Cumulative variance contribution rate
1	53.2409	53.2409
2	16.7726	70.0135
3	10.6200	80.6335
4	8.1500	88.7835
5	4.7514	93.5349
6	3.2069	96.7418
7	0.9729	97.7147
8	0.7847	98.4994
9	0.7600	99.2594
10	0.5768	99.8362
11	0.1638	100.0000

According to the calculation of the Table 3, the feature vector $C=(\alpha_1,\alpha_2,\alpha_3,\alpha_4)$ of the principal component $X=(t_1,t_2,t_3,t_4))$ is obtained, which is shown in Table 4.

Table 4. Feature vector of principal components

a_1	a_2	a_3	a_4
0.0253	-0.1065	0.7499	0.5480
-0.3338	0.2409	0.3576	-0.1725
0.3777	-0.0665	-0.1703	0.0813
0.3232	-0.2946	0.2553	-0.3712
0.3496	-0.2782	-0.1306	0.0714
0.2962	-0.2049	0.3673	-0.4936
0.3099	-0.2516	-0.0517	0.2529
-0.2711	-0.1956	-0.0246	-0.4152
0.3321	0.3911	0.0623	-0.0867
0.3116	0.4248	-0.1255	0.0547
0.2323	0.5339	0.2017	-0.1727

The extracted principal components can be expressed as follows:

$$t_1=0.0253T1-0.3338T2+0.3777T3+0.3232T4+0.3496T5+0.2962T6+0.3099T7-0.2711T8+0.3321T9+0.3116T10+0.2323T11$$

$$t_2=-0.1065T1+0.2409T2-0.0665T3-0.2946T4-0.2782T5-0.2049T6-0.2516T7-0.1956T8+0.3911T9+0.4248T10+0.5339T11$$

$$t_3=0.7499T1+0.3576T2-0.1703T3+0.2553T4-0.1306T5+0.3673T6-0.0517T7-0.0246T8+0.0623T9-0.1255T10+0.2017T11$$

$$t_4=0.5480T1-0.1725T2+0.0813T3-0.3712T4+0.0714T5-0.4936T6+0.2529T7-0.4152T8-0.0867T9+0.0547T10-0.01727T11$$

4.3 Prediction of thermal extension error by LS-SVM

As written as in Section 3, the LS-SVM model can be expressed as

$$f(x)=\sum_{i=1}^{n}\alpha_i K(x,x_i)+b$$

The four temperature sensitive point is used as the input of the model.

$$x=[T_3,T_4,T_6,T_9]$$

The temperature matrix corresponding to the i

$$x_i=[T_3^i,T_4^i,T_6^i,T_9^i]$$

Among them , $j=(1,2,\cdots,l)$, $l=130$ .Before using the Gauss radial basis function as the kernel function of LS-SVM to carry on the training model, the value of the kernel parameter and the regularization parameter s need to be determined. For the choice of these two parameters, directly determines the accuracy of the model, and they are determined to belong to the optimal model selection problem, so far there is no unified principle. In paper, by continuously adjusting the value of the two parameters, the combination of the minimum error of cross validation is selected as the best value of the parameters. Then, it can be obtained that $\sigma=20.3$, $s=0.085$, $b=0.239$, $l=130$ And, the prediction results of LS-SVM thermal error model are shown in Fig. 5.

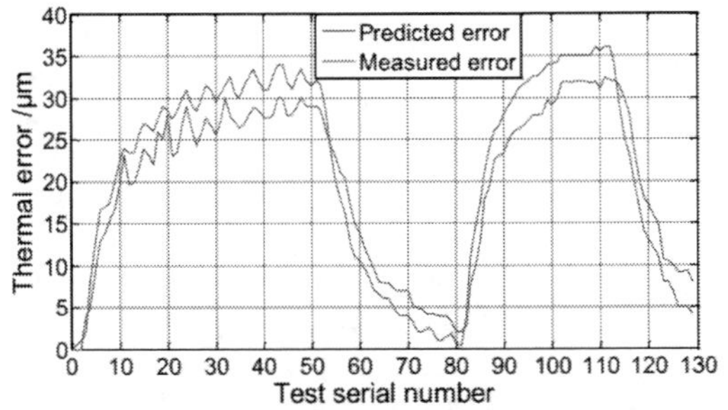

(a) Comparison of measured error and predicted error

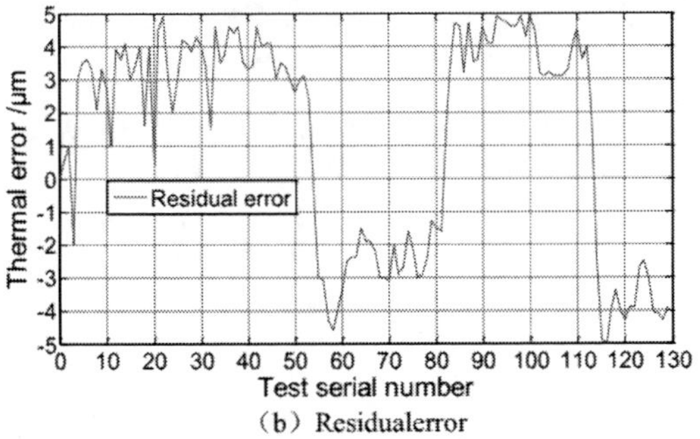

(b) Residualerror

Figure 5. Prediction results of LS-SVM method

By the method proposed in this paper, the results show that the residual value of the model is small, and the maximum error is only 4.9μm just as shown in Fig. 5. However, the maximum residual error gotten by the multiple linear regression model is 7.5 μm, which is shown in Fig. 6. Therefore, it is feasible to optimize the temperature measuring points by using the PCA and the LS-SVM method, and the accuracy and robustness of the model are remarkable.

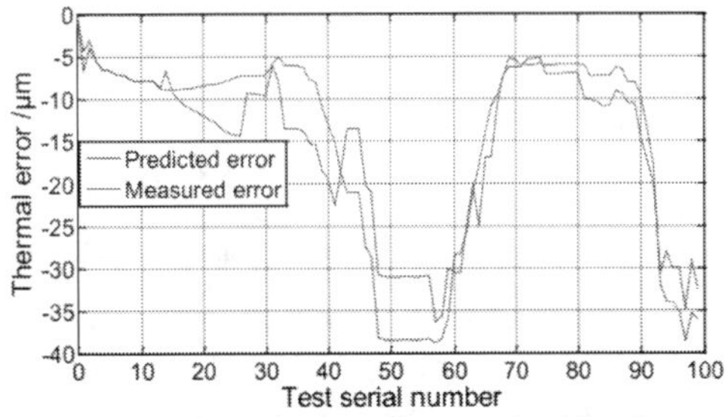

(a) Comparison of measured error and predicted error

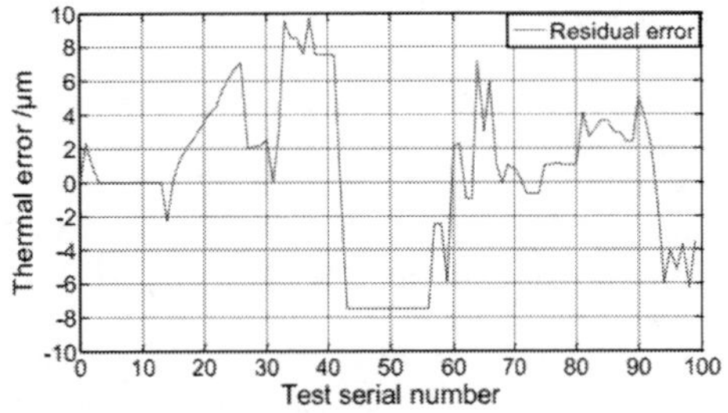

(b) Residualerror

Figure 6. Prediction results of multiple linear regression method

5 Conclusions

Thermal errors are the major contributor to the dimensional errors of a work-piece in precision machining. Error compensation technique is a cost-effective way to reduce thermal errors, but accurate modeling of errors is a prerequisite of error compensation. To this aim, in this paper, a method of PCA is proposed to select the optimization data of temperature measurement points, and SVM is introduced to predict thermally induced extension error of boring machine. Conclusions of this research are summarized as follows.

(1) By the method of PCA the dimension of temperature data from 11 down to 4.

(2) LS-SVM are used to predict the thermally induced thermal extension error of boring bar, and the prediction accuracy of LS-SVM is superior to that of traditional model after comparison. Boring bar residual errors of thermal elongation down to 5 μm or less, and the prediction can reduce the influence of unpredictable noises.

This method can effectively reduce the dimension of the temperature measuring points data and improve the robustness of the error needed in compensation model. Next, how to realize the real-time thermal error compensation efficiently will be the future research focus.

Acknowledgement

The authors are most grateful to National Natural Science Foundation of China(51575010), The Leading Talent Project of Guangdong Province, Beijing Nova Program (Z151100000- 3150138), Shantou Light Industry Equipment Research Institute of Science and Technology Correspondent Station (2013B0909- 00008), Open Research Fund of Key Laboratory of High Performance Complex Manufacturing, Central South University (Kfkt2014-09),Basic Research Foundation of Beijing University of Technology, which they support the research presented in this paper.

References

1 Cheng Q, Luo R, Gu P, et al. Load Induced Error Identification and Camber Curve Design of a Large-Span Crossbeam[J]. Advances in Mechanical Engineering, 2013, 5: 847194.
2 Mehrabi M G, O'Neal G, Min B K, et al. Improving machining accuracy in precision line boring[J]. Journal of Intelligent Manu- facturing, 2002, 13(5): 379-389.
3 Bryan J. International status of thermal error research (1990)[J]. CIRP AnnalsManufacturing Technology, 1990, 39(2): 645-656.
4 Mayr J, Jedrzejewski J, Uhlmann E, et al. Thermal issues in machine tools[J]. CIRP Annals-Manufacturing Technology, 2012, 61(2): 771-791.
5 Postlethwaite S R, Allen J P, Ford D G. The use of thermal imaging, temperature and distortion models for machine tool thermal error reduction[J]. Proceedings of the Institution of Mechanical Engineers, Part B: Journal of Engineering Manufacture, 1998, 212(8): 671-679.
6 Ni J. CNC machine accuracy enhancement through real-time error compensation[J]. Journal of manufacturing science and engineering, 1997, 119(4B): 717-7
7 Yang S, Yuan J, Ni J. The improvement of thermal error modeling and compensation on machine tools by CMAC neural network[J]. International Journal of Machine Tools and Manufacture, 1996, 36(4): 527-537.
8 Kim H S, Jeong K S. Design and manufacture of a three-axis ultra-precision CNC grinding machine[J]. Journal of Materials Processing Technology, 1997, 71(2): 258-266.
9 Ramesh R, Mannan M A, Poo A N. Error compensation in machine tools—a review: Part II: thermal errors[J]. International Journal of Machine Tools and Manufacture, 2000, 40(9): 1257-1284.
10 Mayr J, Jedrzejewski J, Uhlmann E, et al. Thermal issues in machine tools[J]. CIRP Annals-Manufacturing Technology, 2012, 61(2): 771-791.
11 Zhao HT, et al. Selection method of temperature critical point based on thermal error sensitivity chart. Journal of Shanghai Jiaotong University, 2015,49(005): 725-729.
12 Wang XL, et al. Thermal error modeling of machine tool spindle based on PSO clustering and ELM neural network [J]. Combination machine tools and automatic processing technology, 2015 (7): 69-73.
13 Luo FJ,Song DL, et al.Optimization of the Temperature Measuring PointsBased on LSR. Modular Machine Tool &Automatic Manufacturing Technique ,2015,03(3):56-58.
14 Ma Y. Sensor placement optimization for thermal error compensation on machine tools[M]. University of Michigan., 2001.
15 Lee J H, Yang S H. Statistical optimization and assessment of a thermal error model for CNC machine tools[J]. International Journal of Machine Tools and Manufacture, 2002, 42(1): 147-155.
16 Creighton E, Honegger A, Tulsian A, et al. Analysis of thermal errors in a high-speed micro-milling spindle[J]. International Journal of Machine Tools and Manufacture, 2010, 50(4): 386-393.
17 Han J, Wang L, Cheng N, et al. Thermal error modeling of machine tool based on fuzzy c-means cluster analysis and minimal-resource allocating networks[J]. The International Journal of Advanced Manufacturing Technology, 2012, 60(5-8): 463-472.
18 Cheng Q, Qi Z, Zhang G, et al. Robust modelling and prediction of thermally induced positional error based on grey rough set theory and neural networks[J]. The International Journal of Advanced Manufacturing Technology, 2015: 1-12.
19 Zhao C, Gao F, Wang F. Nonlinear Batch Process Monitoring Using Phase-Based Kernel-Independent Component Analysis– Principal Component Analysis (KICA– PCA)[J]. Industrial & engineering chemistry research, 2009, 48(20): 9163-9174.
20 Zhao C H, Wang F L, Lu N Y, et al. Stage-based Soft-transition Multiple PCA Modeling and on-line Monitoring Strategy for Batch Processes [J]. Journal of Process Control, 2007, 17(9) : 728 −741.
21 Suykens J A K, Vandewalle J. Recurrent least squares support vector machines[J] . IEEE transactions on circuits and systems-I : Fundamental theory and applications, 2000, 47(7): 1109-1114.
22 Duan K, Keerthi S S, Poo A N. Evaluation of simple performance measures for tuning SVM hyper-parameters[J]. Neurocomputing, 2003, 51: 41-59.
23 Miao E M, Niu P C, Fei Y T, et al. Selecting temperature-sensitive points and modeling thermal errors of machine tools[J]. Journal of the Chinese Society of Mechanical Engineers, 2011, 32(6): 559-565.

A Method to Optimize Geometric Errors of Machine Tool based on SNR Quality Loss Function and Correlation Analysis

Ligang Cai[1], Jiaying Li[1], Qiang Cheng[1,2], Bingwei Sun[1] and Yan Wang[1]

[1]Beijing Key Laboratory of Advanced Manufacturing Technology, Beijing University of Technology, Beijing 100124, China
[2]Key Laboratory of High Performance Complex Manufacturing, Central South University, Changsha, Hunan, 4310083,China

Abstract. Instead improving the accuracy of machine tool by increasing the precision of key components level blindly in the production process, the method of combination of SNR quality loss function and machine tool geometric error correlation analysis to optimize five-axis machine tool geometric errors will be adopted. Firstly, the homogeneous transformation matrix method will be used to build five-axis machine tool geometric error modeling. Secondly, the SNR quality loss function will be used for cost modeling. And then, machine tool accuracy optimal objective function will be established based on the correlation analysis. Finally, ISIGHT combined with MATLAB will be applied to optimize each error. The results show that this method is reasonable and appropriate to relax the range of tolerance values, so as to reduce the manufacturing cost of machine tools.

1 Introduction

With the development of advanced manufacturing technology, the requirement of machine tool is increasing, which is not only reflected in the machining accuracy of the machine tool, but also reducing costs that can meet the demand of the market. Although the machining accuracy is an important index to measure the performance of the machine, reducing the production cost of CNC machine tools also has important significance in the premise of ensuring the processing precision.

In order to facilitate the research of the relationship between machining accuracy and production cost, the error modeling is the most important task. Domestic and foreign scholars has carried out many researches in the field of machine tool error modeling and proposed many modeling method, such as kinematics model method [1], mechanism method [2], neural network method [3-4], homogeneous transformation matrix method [5], screw theory method [6], etc. These methods can accurately establish the accuracy model of the machine tool, and it is the basis of the research of all aspects of the machine tool. In this paper, the method of homogeneous transformation matrix is used to model the error model of five axis machine tool.

After the machine tool error model is established, the research between the total cost and precision machine tool will be carried out. At first, the tolerance optimization model based on the SNR quality loss function will be established, serving as cost basis to achieve optimal allocation of each parts tolerance. At present, some domestic scholars have some improvement on the quality loss function. Feng et al. [7] proposed an improved multivariate quality loss function, which can be used for all the quality characteristics, and fully considered the correlation between the corresponding variables. Ai et al. [8] combined multivariate quality loss function and genetic algorithm to optimize the assembly quality. Duan et al. [9] improved the multivariate quality loss function, the incremental model of multivariate quality loss function is formed, and the feasibility of the model is proved by an example. Chang et al. [10] consider the trade-off problem between quality and cost for the case when the input quality characteristic is normally distributed. According to the research of Chen et al. [11], the management can adjust the investment on prevention and appraisal costs on quality improvement that enhances process capability, reduces product defect rate and, as a result, generates remarkable financial return. Ram Ganeshan et al. [12] established a model which determines the optimal levels of inventory, and the production lot-size that minimizes the sum of inventory and quality-related costs. Noel Artiles-León [13] described a methodology for optimizing several quality characteristics simultaneously. It was assumed that some type of experimentation had been carried out to determine a relationship between each quality characteristic and the experimental factors. Ramesh Kumar et al. [14] applied quality loss theory to allocate tolerance economically and precisely. Balamurugan, C. et al [15] finished the optimal allocation of geometric and process tolerances based on the present worth of quality loss. Pradeep Kumar Gupta et al [16] enhanced machine accuracy by economic analysis.

Optimization model based on the correlation analysis of the geometric error of machine tool is regarded as

 DOI: 10.1051/matecconf/20179507011

another geometric error basis for the optimal allocation of each parts tolerance. There are also some researches in this respect at present, the commonly used correlation analysis method are Two step cluster [17], K - Means cluster [18], and Hierarchical cluster [19]. Hierarchical cluster analysis is a multivariate statistical analysis method, which is widely used to establish a classification. This method in the SPSS will be used to cluster and analyze the 37 geometric errors of five-axis machine tool.

In this paper, based on the combination of the SNR quality loss function and the geometric error correlation analysis, the tolerance distribution model of the machine tool can be obtained, and the tolerance of the machine tool will be optimized. This method can ensure the machining accuracy of the machine, and appropriately increase geometric errors, so that it can reduce the cost of machine tools, which has a positive effect on the development of the machine tool market.

2 Error modeling of five-axis machine tool based on homogeneous transformation matrix

In this paper, the XKH800 five-axis linkage blade machining center is used to develop the geometric error model. Three-dimensional model of this XHK800 five-axis machining center is shown in Figure 1. There are 37 geometric errors including positioning errors, straightness errors, angular errors, etc., which are shown in Table 1. The topological structure of this machining center is shown in Figure 2. The homogeneous transformation matrices of the machine tool between the moving parts are shown in Table 2.

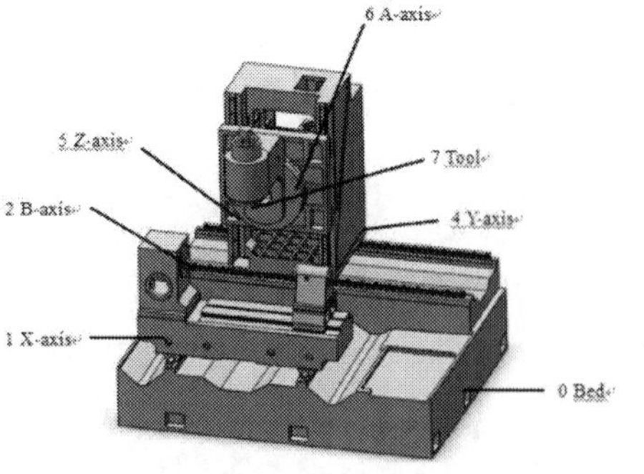

Figure 1. The three-dimensional model of the XHK800 five-axis machining center.

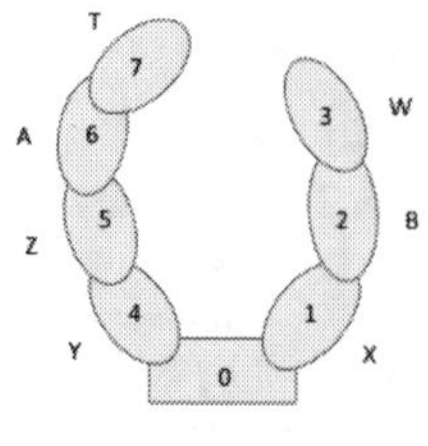

Figure 2. The topological structure of the XHK800 five-axis machining center.

Suppose that the tool forming point coordinate in the tool coordinate system is:

$$P_t = \begin{pmatrix} p_{tx} & p_{ty} & p_{tz} & 1 \end{pmatrix}^T \tag{1}$$

The work-piece forming point coordinate in the work-piece coordinate system is:

$$P_w = \begin{pmatrix} p_{wx} & p_{wy} & p_{wz} & 1 \end{pmatrix}^T \tag{2}$$

When the machine tool moves in ideal case, the machine tool is without error, that means the tool forming point and work-piece forming point will overlap together, which can be described as:

$$T_{07}P_t = T_{03}P_{wideal} \tag{3}$$

The ideal tool forming point in the work-piece coordinate system can be obtained according to Eq. (3):

$$P_{wideal} = (T_{01}^P T_{01}^S T_{12}^P T_{12}^S T_{23}^P T_{23}^S)^{-1} T_{04}^P T_{04}^S T_{45}^P T_{45}^S T_{56}^P T_{56}^S T_{67}^P T_{67}^S P_t \tag{4}$$

In Eq. (4), P and S mean static and dynamic respectively, so $T_{J-1,J}^P$ refers to the ideal static homogenous transformation matrix of the adjacent body and $T_{J-1,J}^S$ refers to the ideal dynamic homogenous transformation matrix as shown in Table 3. In the actual machining process, the position of actual tool forming point will inevitably deviate from the ideal. Therefore, the actual tool forming point in the work-piece coordinate system is:

$$P_{wactrual} = ({}^E T_{03})^{-1}\, {}^E T_{07} P_t \tag{5}$$

And thereupon, geometric errors of five-axis machine tool caused by the gap between actual and ideal forming point can be described as:

$$E = {}^E T_{03} P_{wideal} - {}^E T_{07} P_t \tag{6}$$

The values of the expressions in the Eq. (4)-Eq. (6) can be checked from Table 2. E represents geometric errors of this machining center including E_x, E_y, and E_z, which is:

$$E = \begin{bmatrix} E_x & E_y & E_z & 1 \end{bmatrix}^T \tag{7}$$

3 Tolerance optimal allocation modeling based on the total cost of the SNR quality loss function

This model consists of two parts: manufacturing cost function and SNR quality loss function. Manufacturing cost function control the precondition which is meeting the requirements of product performance and the lowest manufacturing cost, and optimize the parts tolerance. The basic purpose of quality loss function is to evaluate the loss caused by low quality in a quantitative way [7].

3.1 Manufacturing cost

In this article, the tolerance and accuracy of the 37 machine tool geometric errors are supposed to be T_i and A_i separately. Suppose that the half tolerance bandwidth of machine tool geometric accuracy is t_i ($t_i = T_i/2$; i=1, 2…37). As the implementation cost of the machine tool geometric accuracy is different, the cost weight coefficient h_i and the cost characteristic index r_i are introduced. The cost of the machine tool can be expressed as:

$$C_\Sigma(t_i) = \sum_{i=1}^{37} (h_i t_i)^{-r_i} \tag{8}$$

Table 1. Description and representation of the five-axis machine tool geometric errors.

Identification mark	Geometrical significance	Identification mark	Geometrical significance
Δx_x	positioning error	Δx_A	X direction run-out error
Δy_x	Y direction of straightness error	Δy_A	Y direction run-out error
Δz_x	Z direction of straightness error	Δz_A	Z direction run-out error
$\Delta \alpha_x$	Roll error	$\Delta \alpha_A$	Turning error
$\Delta \beta_x$	Pitch error	$\Delta \beta_A$	Around the Y-axis turning error
$\Delta \gamma_x$	Yaw error	$\Delta \gamma_A$	Around the Z-axis turning error
Δx_y	X direction of straightness error	Δx_B	X direction run-out error
Δy_y	positioning error	Δy_B	Y direction run-out error
Δz_y	Z direction of straightness error	Δz_B	Z direction run-out error
$\Delta \alpha_y$	Pitch error	$\Delta \alpha_B$	Around the X-axis turning error
$\Delta \beta_y$	Roll error	$\Delta \beta_B$	Turning error
$\Delta \gamma_y$	Yaw error	$\Delta \gamma_B$	Around the Z-axis turning error
Δx_z	X direction of straightness error	$\Delta \gamma_{xy}$	X,Y-axis perpendicularity error
Δy_z	Y direction of straightness error	$\Delta \beta_{xz}$	X,Z-axis perpendicularity error
Δz_z	positioning error	$\Delta \alpha_{yz}$	Y,Z-axis perpendicularity error
$\Delta \alpha_z$	Pitch error	$\Delta \gamma_{xB}$	B-axis parallelism error in YZ plane
$\Delta \beta_z$	Yaw error	$\Delta \alpha_{zB}$	B-axis parallelism error in XY plane
$\Delta \gamma_z$	Roll error	$\Delta \gamma_{yA}$	A-axis parallelism error in XZ plane
		$\Delta \beta_{zA}$	A-axis parallelism error in XY plane

Table 2. Ideal and error homogeneous transformation matrixes of the five-axis machine tool.

Adjacent body	Body ideal static, motion HTMs ($T^P_{J-1,J}$, $T^S_{J-1,J}$)	Body ideal static, motion error HTMs (${}^eT^P_{J-1,J}$, ${}^eT^S_{J-1,J}$)
0-1 X-axis	$T^P_{01} = I_{4\times4}$ $T^S_{01} = \begin{pmatrix} 1 & 0 & 0 & x \\ 0 & 1 & 0 & 0 \\ 0 & 0 & 1 & 0 \\ 0 & 0 & 0 & 1 \end{pmatrix}$	${}^eT^P_{01} = I_{4\times4}$ ${}^eT^S_{01} = \begin{pmatrix} 1 & -\Delta\gamma_x & \Delta\beta_x & \Delta x_x \\ \Delta\gamma_x & 1 & -\Delta\alpha_x & \Delta y_x \\ -\Delta\beta_x & \Delta\alpha_x & 1 & \Delta z_x \\ 0 & 0 & 0 & 1 \end{pmatrix}$

 DOI: 10.1051/matecconf/20179507011

1-2 B-axis	$T_{12}^P = I_{4\times4}$ $T_{12}^S = \begin{pmatrix} \cos B & 0 & \sin B & 0 \\ 0 & 1 & 0 & 0 \\ -\sin B & 0 & \cos B & 0 \\ 0 & 0 & 0 & 1 \end{pmatrix}$	${}^eT_{12}^P = \begin{pmatrix} 1 & -\Delta\gamma_{xB} & 0 & 0 \\ \Delta\gamma_{xB} & 1 & -\Delta\alpha_{zB} & 0 \\ 0 & \Delta\alpha_{zB} & 1 & 0 \\ 0 & 0 & 0 & 1 \end{pmatrix}$ ${}^eT_{12}^S = \begin{pmatrix} 1 & -\Delta\gamma_B & \Delta\beta_B & \Delta x_B \\ \Delta\gamma_B & 1 & -\Delta\alpha_B & \Delta y_B \\ -\Delta\beta_B & \Delta\alpha_B & 1 & \Delta z_B \\ 0 & 0 & 0 & 1 \end{pmatrix}$
2-3 Work piece	$T_{23}^P = \begin{pmatrix} 1 & 0 & 0 & x_{wd} \\ 0 & 1 & 0 & y_{wd} \\ 0 & 0 & 1 & z_{wd} \\ 0 & 0 & 0 & 1 \end{pmatrix}$ $T_{23}^S = I_{4\times4}$	${}^eT_{23}^P = \begin{pmatrix} 1 & -\Delta\gamma_{wd} & \Delta\beta_{wd} & \Delta x_{wd} \\ \Delta\gamma_{wd} & 1 & -\Delta\alpha_{wd} & \Delta y_{wd} \\ -\Delta\beta_{wd} & \Delta\alpha_{wd} & 1 & \Delta z_{wd} \\ 0 & 0 & 0 & 1 \end{pmatrix}$ ${}^eT_{23}^S = I_{4\times4}$
0-4 Y-axis	$T_{04}^P = I_{4\times4}$ $T_{04}^S = \begin{pmatrix} 1 & 0 & 0 & 0 \\ 0 & 1 & 0 & y \\ 0 & 0 & 1 & 0 \\ 0 & 0 & 0 & 1 \end{pmatrix}$	${}^eT_{04}^P = \begin{pmatrix} 1 & -\Delta\gamma_{xy} & 0 & 0 \\ \Delta\gamma_{xy} & 1 & 0 & 0 \\ 0 & 0 & 1 & 0 \\ 0 & 0 & 0 & 1 \end{pmatrix}$ ${}^eT_{04}^S = \begin{pmatrix} 1 & -\Delta\gamma_y & \Delta\beta_y & \Delta x_y \\ \Delta\gamma_y & 1 & -\Delta\alpha_y & \Delta y_y \\ -\Delta\beta_y & \Delta\alpha_y & 1 & \Delta z_y \\ 0 & 0 & 0 & 1 \end{pmatrix}$
4-5 Z-axis	$T_{45}^P = I_{4\times4}$ $T_{45}^S = \begin{pmatrix} 1 & 0 & 0 & 0 \\ 0 & 1 & 0 & 0 \\ 0 & 0 & 1 & z \\ 0 & 0 & 0 & 1 \end{pmatrix}$	${}^eT_{45}^P = \begin{pmatrix} 1 & 0 & \Delta\beta_{xz} & 0 \\ 0 & 1 & -\Delta\alpha_{yz} & 0 \\ -\Delta\beta_{xz} & \Delta\alpha_{yz} & 1 & 0 \\ 0 & 0 & 0 & 1 \end{pmatrix}$ ${}^eT_{45}^S = \begin{pmatrix} 1 & -\Delta\gamma_z & \Delta\beta_z & \Delta x_z \\ \Delta\gamma_z & 1 & -\Delta\alpha_z & \Delta y_z \\ -\Delta\beta_z & \Delta\alpha_z & 1 & \Delta z_z \\ 0 & 0 & 0 & 1 \end{pmatrix}$
5-6 A-axis	$T_{56}^P = I_{4\times4}$ $T_{56}^S = \begin{pmatrix} 1 & 0 & 0 & 0 \\ 0 & \cos A & -\sin A & 0 \\ 0 & \sin A & \cos A & 0 \\ 0 & 0 & 0 & 1 \end{pmatrix}$	${}^eT_{56}^P = \begin{pmatrix} 1 & -\Delta\gamma_{yA} & \Delta\beta_{zA} & 0 \\ \Delta\gamma_{yA} & 1 & 0 & 0 \\ -\Delta\beta_{zA} & 0 & 1 & 0 \\ 0 & 0 & 0 & 1 \end{pmatrix}$ ${}^eT_{56}^S = \begin{pmatrix} 1 & -\Delta\gamma_A & \Delta\beta_A & \Delta x_A \\ \Delta\gamma_A & 1 & -\Delta\alpha_A & \Delta y_A \\ -\Delta\beta_A & \Delta\alpha_A & 1 & \Delta z_A \\ 0 & 0 & 0 & 1 \end{pmatrix}$
6-7 Tool	$T_{67}^P = \begin{pmatrix} 1 & 0 & 0 & x_{td} \\ 0 & 1 & 0 & y_{td} \\ 0 & 0 & 1 & z_{td} \\ 0 & 0 & 0 & 1 \end{pmatrix}$ $T_{67}^S = I_{4\times4}$	${}^eT_{67}^P = \begin{pmatrix} 1 & -\Delta\gamma_{td} & \Delta\beta_{td} & \Delta x_{td} \\ \Delta\gamma_{td} & 1 & -\Delta\alpha_{td} & \Delta y_{td} \\ -\Delta\beta_{td} & \Delta\alpha_{td} & 1 & \Delta z_{td} \\ 0 & 0 & 0 & 1 \end{pmatrix}$ ${}^eT_{67}^P = I_{4\times4}$

In order to determine the cost weight coefficient, the 37 geometric accuracies of machine tool are classified according to the categories, and it is considered that the cost weight coefficient of the same type of machine tool's geometric accuracy is the same. Linear displacement positioning accuracy is regarded as a benchmark, of which the cost weight coefficient is 1. And introducing 3 proportionality coefficients h_p, h_a, and h_d, which severally means the ratio of the cost weight coefficient of positional accuracy, angular displacement positioning accuracy and non-positioning accuracy to the cost weight coefficient of linear displacement positioning accuracy, and they are shown in Table 3. According to the specific structure of the machine tool chosen by this paper and the reference [20] [21], This paper set up $h_p = 3$, $h_a = 4$, $h_d = 5$, $r_i = 2$.

Table 3. The classification and cost weight coefficient of machine tool geometric accuracy.

Category		The symbol of geometric accuracy	Cost weight coefficient
Positional accuracy		A_{31} A_{32} A_{33} A_{34} A_{35} A_{36} A_{37}	h_p
Kinematic accuracy	Linear displacement positioning accuracy	A_1 A_8 A_{15}	1
	Angle position precision	A_{22} A_{29}	h_a
	Linear displacement non-positioning accuracy	A_2 A_3 A_7 A_9 A_{13} A_{14} A_{19} A_{20} A_{21} A_{25} A_{26} A_{27}	h_d
	Angle non-position precision	A_4 A_5 A_6 A_{10} A_{11} A_{12} A_{16} A_{17} A_{18} A_{23} A_{24} A_{28} A_{30}	h_a h_d

3.2 Cost of quality loss based on signal-noise ratio

The smaller-the-better characteristic refers to the closer to zero the better quality characteristics.
In the actual calculation, the SNR of the smaller-the-better characteristic is:

$$\eta_S = 10\log\frac{1}{\sigma^2+\mu^2} \tag{9}$$

Its estimated value is:

$$\eta_S = 10\log\frac{1}{\sigma^2+\mu^2} = -10\log\left(\frac{1}{n}\sum_{i=1}^{n} y_i^2\right) \tag{10}$$

The greater the value of η is, the product quality will be more stable, and the loss smaller [22]. According to this conclusion, the size of the contribution of the quality indicators to the volatility will be determined, and different weights can be assigned:

$$\lambda_i = \frac{\frac{1}{\eta_i}}{\sum_{j=1}^{n}\frac{1}{\eta_j}} \tag{11}$$

In this paper, 37 geometric errors can be regards as the quality characteristics of the machine tool, and they are the smaller-the-better characteristic. The multivariate quality loss function of the smaller-the-better characteristic can be expressed as:

$$L(y_1, y_2 \ldots y_n) = \sum_{i=1}^{n}\lambda_i L_S(y_i) = \sum_{i=1}^{n}\lambda_i y_i^2 \tag{12}$$

Calculating SNR η_i of the 37 quality indexes t_i by Eq. (10), and replacing them into Eq. (11) and Eq. (12) in turn, the quality loss function of 37 geometric errors will be obtained:

$$L_\Sigma(t_i) = \sum_{i=1}^{37}\lambda_i t_i^2 \tag{13}$$

3.3 Total cost optimization model

The sum of the cost of integrated manufacturing and quality lose reaches the minimum, which is the evidence of tolerance optimization allocation of various parts [23]. According to the manufacturing cost and quality loss cost mentioned above, the objective function of the machine total cost can be expressed as:

$$M1_{\min}\left[C_\Sigma(t_i) + L_\Sigma(t_i)\right] = \min\left[\sum_{i=1}^{37}(\mathrm{h}_i \mathrm{t}_i)^{-r_i} + \sum_{i=1}^{37}\lambda_i t_i^2\right] \tag{14}$$

Constraint conditions are:

$$\begin{aligned} \Delta x_{\max} &\le def(\Delta x) \\ \Delta y_{\max} &\le def(\Delta y) \\ \Delta z_{\max} &\le def(\Delta z) \end{aligned} \tag{15}$$

In Eq. (15), $def(\Delta x)$——Maximum allowable error in X direction; $def(\Delta y)$——Maximum allowable error in Y direction; $def(\Delta z)$——Maximum allowable error in Z direction.

4 Optimal distribution model based on machine tool geometric error correlation analysis

4.1 Machine tool geometric error correlation statistics analysis and classification

In this article hierarchical clustering method in the SPSS (Statistical Product and Service Solutions) is applied to count the correlationship between each error term, and the error terms which ones have stronger relationship are divided into a category. According to the subsequent requirements of modeling, geometric error terms will be divided into four categories. The first kind contains $\Delta\alpha_x$、$\Delta\beta_x$、$\Delta\gamma_x$、$\Delta\alpha_y$、$\Delta\beta_y$、$\Delta\gamma_y$、$\Delta\alpha_z$、$\Delta\beta_z$、$\Delta\gamma_z$、$\Delta\alpha_A$、$\Delta\beta_A$、$\Delta\alpha_B$、$\Delta\beta_B$、$\Delta\gamma_B$、$\Delta\gamma_{xy}$、$\Delta\beta_{xz}$、$\Delta\alpha_{yz}$、$\Delta\gamma_{xB}$、$\Delta\alpha_{zB}$、$\Delta\gamma_{yA}$、$\Delta\beta_{zA}$. The second kind contains Δz_y. The third kind contains Δx_A、Δy_A、Δz_A、$\Delta\gamma_A$、Δx_B、Δy_B、Δz_B. And the forth kind contains Δx_x、Δy_x、Δz_x、Δx_y、Δy_y、Δx_z、Δy_z、Δz_z.

4.2 The optimal distribution model based on precision grade of machine tool key components

There is a certain mutual influence between the geometric errors of machine tools and the geometric precision of components. According to the relationship between the components precision parameters and the geometric errors, the precision parameters of components are regarded as design variables, and the following objective function can be constructed.

$$M2_{\max} = \max\sqrt{\sum_{i=1}^{3}\Delta s_i^{\,2}(x)+\sum_{j=1}^{3}\Delta s_j^{\,2}(y)+\sum_{k=1}^{3}\Delta s_k^{\,2}(z)+\sum_{m=1}^{3}\Delta s_m^{\,2}(A)+\sum_{l=1}^{3}\Delta s_l^{\,2}(B)+S^2(x)+S^2(y)+S^2(z)+S^2(A)+S^2(B)} \tag{16}$$
$$= \max\sqrt{t_1^2+t_2^2+t_3^2+t_4^2+t_7^2+t_8^2+t_9^2+t_{11}^2+t_{13}^2+t_{14}^2+t_{15}^2+t_{18}^2+t_{19}^2+t_{20}^2+t_{21}^2+t_{22}^2+t_{25}^2+t_{26}^2+t_{27}^2+t_{29}^2}$$

Combined with the results of the previous error terms correlation analysis and classification, the formula (16) is changed into the following form and the optimal matching model is obtained based on the correlation analysis of the machine tool geometric errors.

$$M2_{\max} = \max\left(\sqrt{t_1^2+t_2^2+t_3^2+t_7^2+t_8^2+t_{13}^2+t_{14}^2+t_{15}^2}+\sqrt{t_4^2+t_{11}^2+t_{18}^2+t_{22}^2+t_{29}^2}+\sqrt{t_9^2}+\sqrt{t_{19}^2+t_{20}^2+t_{21}^2+t_{25}^2+t_{26}^2+t_{27}^2}\right) \tag{17}$$

The constraint conditions are as follows:

$$\begin{aligned}
&\Delta x_{\max} \le def(\Delta x)\\
&\Delta y_{\max} \le def(\Delta y)\\
&\Delta z_{\max} \le def(\Delta z)\\
&lb(\Delta s_i(u)) \le \Delta s_i(u) \le ub(\Delta s_i(u))\\
&(i=1,2,3; u=x,y,z)\\
&lb(S(u)) \le S(u) \le ub(S(u))\\
&(u=x,y,z)
\end{aligned} \tag{18}$$

In the constraint conditions, the last two inequalities limit the upper and lower bounds of the component accuracy parameters. Meanwhile, the implied condition is that the difference between the precision parameters of the same component is not too large, and the optimized variable value should be greater than the standard value.

4.3 Tolerance optimal distribution model of machine tool

From the foregoing, the optimization problem of this paper can be understood as the goal of minimizing the total cost and the maximization of the Euclidean norm of error terms in the constraint condition of ensuring the machining accuracy of the machine tool. The optimization model is as follows:

$$M1_{\min}\left[C_\Sigma(t_i)+L_\Sigma(t_i)\right] = \min\left[\sum_{i=1}^{37}(\mathrm{h}_i\mathrm{t}_i)^{-r_i}+\sum_{i=1}^{37}\lambda_i t_i^2\right] \tag{19}$$
$$M2_{\max} = \max\left(\sqrt{t_1^2+t_2^2+t_3^2+t_7^2+t_8^2+t_{13}^2+t_{14}^2+t_{15}^2}+\sqrt{t_4^2+t_{11}^2+t_{18}^2+t_{22}^2+t_{29}^2}+\sqrt{t_9^2}+\sqrt{t_{19}^2+t_{20}^2+t_{21}^2+t_{25}^2+t_{26}^2+t_{27}^2}\right)$$

The constraint conditions are as follows:

$$\begin{aligned}
&\Delta x_{\max} \le def(\Delta x)\\
&\Delta y_{\max} \le def(\Delta y)\\
&\Delta z_{\max} \le def(\Delta z)\\
&lb(\Delta s_i(u)) \le \Delta s_i(u) \le ub(\Delta s_i(u))\\
&(i=1,2,3; u=x,y,z,A,B)\\
&lb(S_{300P}(u)) \le S_{300P}(u) \le ub(S_{300P}(u))\\
&(u=x,y,z,A,B)
\end{aligned} \tag{20}$$

In Eq. (20) M_1 represents total cost function; M_2 represents Euclidean norm function of all the parameters variable. According to the requirement of user's processing precision, the maximum allowable error in X, Y, Z direction of the five axis machining center are:

$$def(\Delta x) = -0.015,\ def(\Delta y) = 0.015,\ def(\Delta z) = 0.01 \tag{21}$$

At the same time, in order to achieve the optimization objective, the optimized error should not be less than standard value, and accuracy parameters of the same parts should not be too much difference. Therefore, the standard value is taken as the lower bound of the variable, and the 3 times of the standard value is taken as the upper bound of the variable.

5 Accuracy optimization allocation based on NSGA-II

The ISIGHT and MATLAB integration method will be applied to optimize the above multi-objective optimization problem. The above two objective functions are as the target function, and after the variables constraints range of the 37 geometric errors and requirements of machine tool precision are set, NSGA-II can be used to optimize the 37 geometric errors through combining with MATLAB. Optimization results are shown in Table 4.

Table 4. The optimization results of the five-axis machining center 37 geometric errors.

Error terms	**Standard values** (mm)	**Optimal values** (mm)	**Error terms**	**Standard values** (mm)	**Optimal values** (mm)
Δx_x	0.005	0.0150	$\Delta \beta_x$	0.003125/1000	0.009342/1000
Δy_x	0.005	0.0150	$\Delta \gamma_x$	0.003125/1000	0.003215/1000
Δz_x	0.005	0.0150	$\Delta \alpha_y$	0.003125/1000	0.009323/1000
Δx_y	0.005	0.0100	$\Delta \beta_y$	0.003125/1000	0.009331/1000
Δy_y	0.005	0.0150	$\Delta \gamma_y$	0.003125/1000	0.003126/1000
Δz_y	0.005	0.0150	$\Delta \alpha_z$	0.003125/1000	0.003126/1000
Δx_z	0.005	0.0150	$\Delta \beta_z$	0.003125/1000	0.003281/1000
Δy_z	0.005	0.00866	$\Delta \gamma_z$	0.003125/1000	0.003138/1000
Δz_z	0.005	0.00827	$\Delta \alpha_A$	0.04/1000	0.007509/1000
Δx_A	0.0075	0.0231	$\Delta \beta_A$	0.0075/1000	0.04026/1000
Δy_A	0.0075	0.0206	$\Delta \gamma_A$	0.0075/1000	0.008015/1000
Δz_A	0.0075	0.00801	$\Delta \alpha_B$	0.0075/1000	0.007531/1000
Δx_B	0.0075	0.0225	$\Delta \beta_B$	0.04/1000	0.04001/1000
Δy_B	0.0075	0.0218	$\Delta \gamma_B$	0.0075/1000	0.009955/1000
Δz_B	0.0075	0.00790	$\Delta \gamma_{xB}$	0.0125/300	0.01251/300
$\Delta \gamma_{xy}$	0.005/500	0.0108/500	$\Delta \alpha_{zB}$	0.0125/300	0.01503/300
$\Delta \beta_{xz}$	0.005/500	0.0050/500	$\Delta \gamma_{yA}$	0.0125/300	0.01263/1000
$\Delta \alpha_{yz}$	0.005/500	0.0050/500	$\Delta \beta_{zA}$	0.0125/300	0.01486/1000
$\Delta \alpha_x$	0.003125/1000	0.009374/1000			

6 Conclusions

Table 4 shows that all the variables optimized are less than the standard value. At the same time, the value of e_x, e_y, e_z are at the maximum allowable error range. It shows that the geometric errors are appropriate amplification in the premise of ensuring the processing precision of machine tool, so that it can avoid improving the precision grade of the key components blindly in the production process in order to improve the precision of the machine tool. It is found that through the optimization method in this paper, it was able to appropriately enlarge the geometric errors, which can reduce the manufacturing cost of the machine tool, in the premise of ensuring the machining accuracy of the machine tool.

Acknowledgement

The authors are most grateful to National Natural Science Foundation of China (51575010), Beijing Nova Program (Z151100003150138), The Leading Talent Project of Guangdong Province, Open Research Fund of Key Laboratory of High Performance Complex Manufacturing, Central South University (Kfkt2014-09), Shantou Light Industry Equipment Research Institute of science and technology Correspondent Station

(2013B090900008), which support the research presented in this paper.

References

1. M.Anjanapva, D.K.Anand, J.A.Kirk. Error correction methodologies and control strategies for numerical control machines. *Control Method for Manufacturing Process*. **7**:41-49 (1988).
2. P. D. Lin, K. F. Ehmann. Direct volumetric error evaluation for multi-axis machines. *International Journal of Machine Tools & Manufacture*. **33**, 5: 675-693 (1993).
3. V.Kiridena，P.M.Ferreira. Mapping the effects of positioning errors on the volumetric accuracy of five-axis CNC machine tools. *International Jounal of Machine Tools & Manufacture*, **33**, 3: 417-437 (1993).
4. John C., Ziegert, Prashant Kalle. Error compensation in machine tools: a neural network approach. *Journal of Intelligent Manufacturing*, **5**: 143-151 (1994).
5. Mahbubur Rahman, Jouko Heikkala, Kauko Lappalainen. Modeling measurement and error compensation of multi-axis machine tools. Part l: theory. *International Journal of Machine Tools & Manufacture*, **40**: 1535-1546 (2000).
6. Moon S K, Moon Y M, Kota S, et al. Screw theory based metrology for design and error compensation of machine tools. *Proceedings of DETC*, 1: 697-707 (2001).
7. Yizhong Ma, Shaohua Cheng, Yanjun Li. Improvement of multivariate quality loss function and its empirical analysis. *Systems Engineering*, **20**, 4: 54-58 (2002).
8. Tingting Ai, Jing Tian, Hongchi Gao. A research on the technology of assembly quality optimization based on multiple quality loss and genetic algorithm. *China Mechanical Engineering*, 10: 1161-1164 (2012).
9. Saisai Duan, Shuhai Fan, Tinghua Huang, Qinghong Xie, Nanyun Jiang. A research on incremental model based on multivariate quality loss function. *Machinery Design and Manufacture*, 7: 253-256 (2013).
10. Chang, Y.-C., Hung, W.-L. Linex loss functions with applications to determining the optimum process parameters. *Qual. Quant*, **41**: 291-301 (2007).
11. Jen-Ming Chen, Jia-Chi Tsou. An optimal design for process quality improvement: modeling and application. *Production Planning & Control*, **14**, 7: 603-612 (2003).
12. Ram Ganeshan, Shailesh Kulkarni, Tonya Boone. Production economics and process quality: A Taguchi perspective. *Production Economics*, 71: 343-350 (2001).
13. Noel Artiles-León. A pragmatic approach to multiple-response problems using loss functions. *Quality Engineering*, **9**, 2: 213-220 (1996).
14. Ramesh Kumar, L., Padmanaban, K.P., Balamurugan, C. Least cost-tolerance allocation based on Lagrange multiplier. Concurrent Engineering Research and Applications, 24, 2: 164-177 (2016).
15. Balamurugan, C., Saravanan, A., Dinesh Babu, P., Jagan, S., Ranga Narasimman, S. Concurrent optimal allocation of geometric and process tolerances based on the present worth of quality loss using evolutionary optimisation techniques. Research in Engineering Design, 1-18 (2016).
16. Pradeep Kumar Gupta1, Kunal Sharma, and Vibhu Tripathi. Economical Analysis and Up Gradation of Existing Machine Tools through Retrofitting. International Journal of Mechanical Engineering and Robotics Research, 2, 4: 52-58(2013).
17. Manikandan, R., Swaminathan, P., Sujitha, R. Unimodular hypergraph based clustering approaches for VLSI circuit partitioning. International Journal of Engineering and Technology, 5, 3: 2755-2758 (2013).
18. Shijin Kumar, P.S., Dharun, V.S. Hybrid brain MRI segmentation algorithm based on K-means clustering and texture pattern matrix. International Journal of Applied Engineering Research, 11, 6: 4343-4348 (2016).
19. Daie, P., Li, S. Matrix-based hierarchical clustering for developing product architecture. Concurrent Engineering Research and Applications, 24, 2: 139-152 (2016).
20. Jinyuan Tang,Yanfeng Lu, Chao Zhou. Contact analysis of spiral bevel gear transmission with error. *Chinese Journal of Mechanical Engineering*, **44**, 7: 16-23 (2008).
21. Fang Kang, Jinwei Fan. Optimal allocation method of NC machine tool manufacturing precision. Mechanical Science and Technology, **27**, 5: 588-591 (2008).
22. Lloyd S. Nelson. Some notes on variation. *Journal of Quality Technology*, **31**, 4: 459-462 (1991).
23. Wen Wu, Weishe Zhang. Tolerance optimal allocation method based on manufacturing and quality loss cost. *Journal of Chang 'an University (Natural Science Edition)*, **25**, 3 (2005).

Research on the Frequency Aliasing of Resistance Acceleration Guidance for Reentry Flight

HAN Pengxin[1] , XU Jiangtao[2] ,Li Shang[2] and YAN Zixiang[2]

[1]China Academy of Launch Vehicle Technology, R&D Center, 100076 Beijing, China
[2]Harbin Engineering University, Department of Aerospace Engineering, 150001 Harbin, China

Abstract: According to the special response of resistance acceleration during hypersonic reentry flight, different guidance frequency will result to very different flight and control response. The analysis model for the response of resistance acceleration to the attack angle and dynamic press is put forward respectively in this paper. And the frequency aliasing phenomenon of guidance is revealed. The simulation results to the same vehicle sufficiently substantiate the frequency aliasing of resistance acceleration during reentry guidance.

1 Summary

During the process of large lift-drag ratio flight, Lift-body is usually controlled by given flight profile. Under this background, one of the most classical guidance schemes is space flight reentry guidance [1]. In the guidance scheme of the space shuttle, the resistance acceleration signal is used as the core to control the reentry guidance and the successful implementation of the project is carried out. However, the test results show that there has some defect of using resistance acceleration signal as the controlled quantity. After the opening of the space shuttle test data, a lot of research focused on reentry guidance, which also shows that there is a certain lack in acceleration signal guidance[2].

The traditional guidance design is based on the dynamic equation of a particle. In this method, the assumption of the guidance command and the centroid motion of the aircraft tend to be too idealistic, while the dynamic characteristics of the guidance loop are ignored. It leads to the problem that the guidance loop does not match the control loop, which brings difficulties to the integrated design of the guidance and control. In this paper, the frequency aliasing effect has been considered, and the lack of using drag acceleration signal as the control signal is studied. The study found that high frequency drag acceleration signal is sampled to produce the low frequency interference signal, and this low frequency interference and the low frequency of the guidance are superposed together. It is difficult to eliminate through the filtering process, which will cause serious pollution to the signal and then affect the terminal guidance performance.

2. Frequency aliasing

The frequency aliasing effect can be explained by the following example: a sinusoidal signal with a certain frequency after sampling in different frequencies, the reconstructed signal is directly related to the sampling frequency [3,5]. As shown in Fig.1, the sampling frequency is obviously reduced when the sampling frequency is less than two times of the signal, that is, the low speed rate of the high frequency signal sampling will produce a new low frequency signal. This low frequency signal will seriously affect the performance of the system.

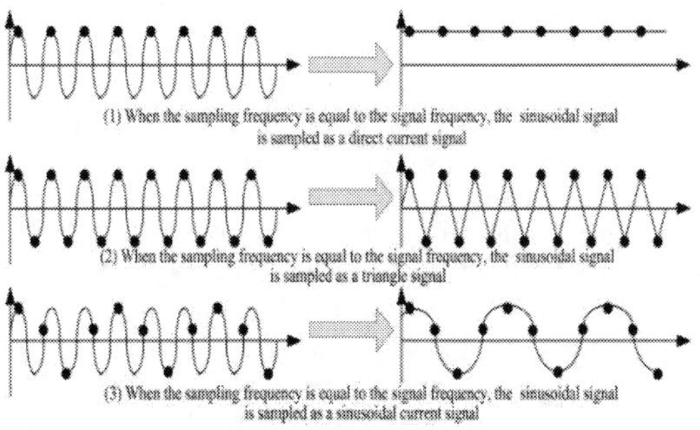

Figure 1. Frequency aliasing

3 Modal analysis of resistance acceleration

According to the dynamic characteristics of reentry flight, the drag acceleration is greatly influenced by dynamic pressure and attack angle. The response of the drag

Supported by National Natural Science Foundation of China (11372080)

 DOI: 10.1051/matecconf/20179507012

acceleration to the attack angle signal is formed by the superposition of two types of oscillation mode, and the response characteristics are relatively complicated. The attack angle and the dynamic pressure are two key factors of the drag acceleration.

As shown in equation (1), $\overline{q}$ is Dynamic pressure, S is Aircraft reference area. C_D refers to Drag coefficient, while it is generally determined by the attack angle and Maher number. D_a is Drag acceleration.

$$D_a = \frac{\overline{q}SC_D(\alpha, mach)}{m} \tag{1}$$

The root locus and modal characteristics indicate that: The transfer function of Drag acceleration-Attack angle has two pairs of conjugate complex roots, which are shown as long-period and short-period mode respectively. Guidance loop design process shows that: The short period motion modes can be improved obviously by the guidance law, but the long period motion mode is attributed to a low frequency response, which influences the guidance characteristics. Fig. 2 is the schematic diagram of open loop root locus.

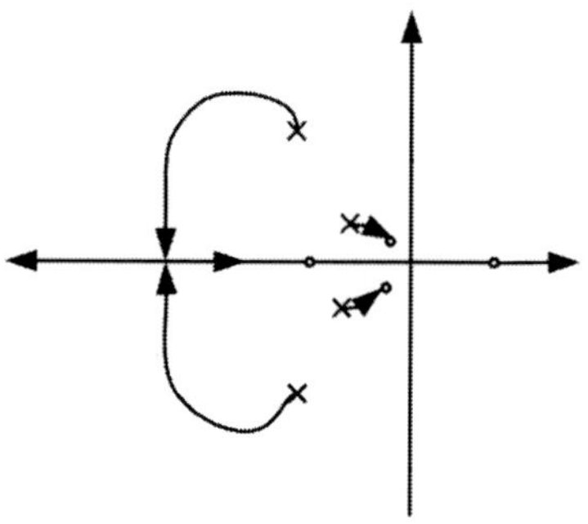

Figure 2. The root locus of drag acceleration and attack angle under typical flight

It is learned from formula (1) that, the drag acceleration is determined by the drag coefficient and dynamic pressure, in which the drag coefficient is greatly affected by the attack angle. It will cause the drag acceleration response to be a short-period motion at initial time. The dynamic pressure is mainly determined by the velocity and altitude, and the response of the velocity to the attack angle is a long-period motion. In the equilibrium state, the moment can be recovered in a relatively short time when the flight vehicle is disturbed by the angle of attack, but the angle of attack caused a negative trajectory angle, leading to the decline of the aircraft, which is gradually accelerated in the tangential direction along the trajectory of gravity force, the velocity increases cause the dynamic pressure increases, and the lift force is also increased. When the lift is greater than gravity, trajectory bend up, and the component of gravity will make airplane slow down, dynamic pressure decreased. When the lift is less than gravity, the trajectory is bent downward, and the dynamic pressure increases with the increase of the velocity. The essence of such alternation is the transformation of the kinetic energy and potential energy, it shows the oscillation motion of the dynamic pressure and the trajectory angle, therefore, dynamic pressure is the physical quantity of long period motion[5].

Fig.3 shows the above discussion about this motion. Above all, when instantaneous changes of attack angle occur, the response of D_a is leaded by C_D at initial time, performing as short period fast oscillation attenuation. Thereafter D_a is mainly influenced by $\overline{q}$, the long-period motion characteristics of $\overline{q}$ makes D_a performance as long-period oscillation.

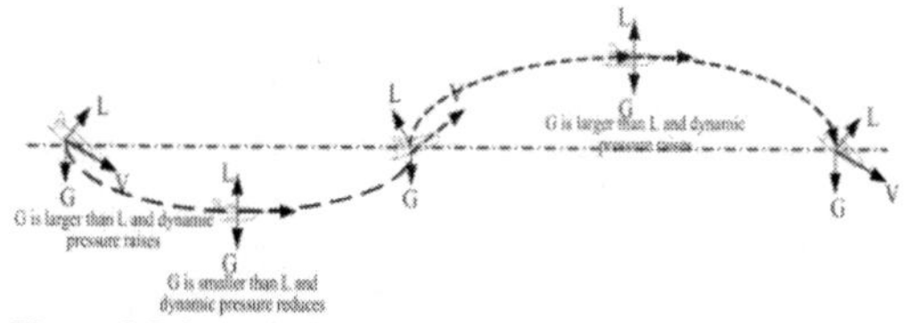

Figure 3. Schematic diagram of the dynamic pressure and long period motion

4 Response characteristic of drag acceleration

4.1 Effect of attack angle on drag acceleration

In the process of high altitude and high speed reentry, the dynamic pressure changes significantly, the sensitivity of the drag acceleration to the angle of attack is different in different dynamic pressure. Fig.5 and Fig.4 shows the response of the drag acceleration during the whole flight when under the condition that, the roll angle is fixed and the attack angle deviation is in $\pm 2^\circ$.

Analysis of simulation shows that, under the same attack angle, when the dynamic pressure is small, the resistance acceleration error is very small, about 5%, with the increase of the dynamic pressure, the drag acceleration error is significantly larger, about 20%. It shows that the influence of the attack angle on drag acceleration is related to the dynamic pressure, the greater the dynamic pressure, the more sensitive to the attack angle.

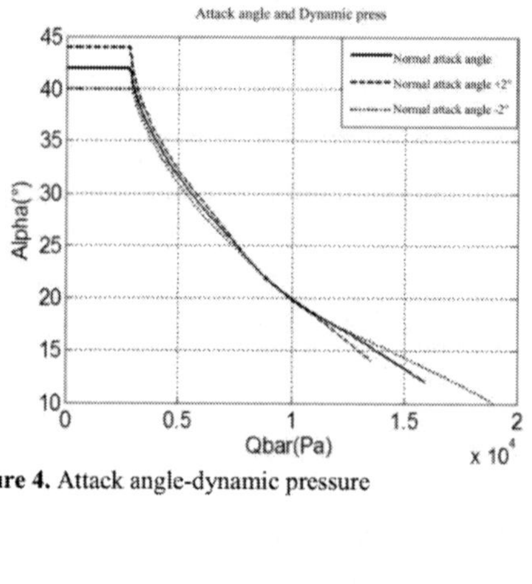

Figure 4. Attack angle-dynamic pressure

DOI: 10.1051/matecconf/20179507012

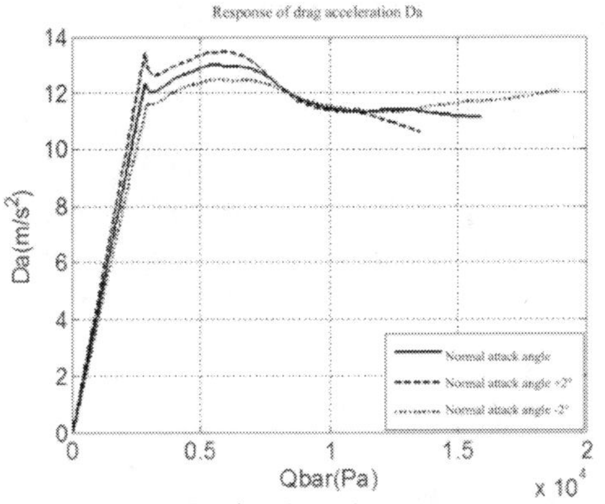

Figure 5. Drag acceleration-dynamic pressure

4.2 Effect of dynamic pressure on drag acceleration

In the guidance loop, the attack angle is the output of the guidance command, but in the control loop, the attack angle is the angular motion. The dynamic characteristics between the attack angle command and the actual attack angle can affect the integrated design of the guidance and control. In order to study the influence of the dynamic characteristic of the attack angle on the drag acceleration, the dynamic characteristics of the attack angle is characterized by a second order system.

$$\frac{\alpha}{\alpha_c} = \frac{\omega_n^2}{s^2 + 2\xi\omega_n s + \omega_n^2} \tag{2}$$

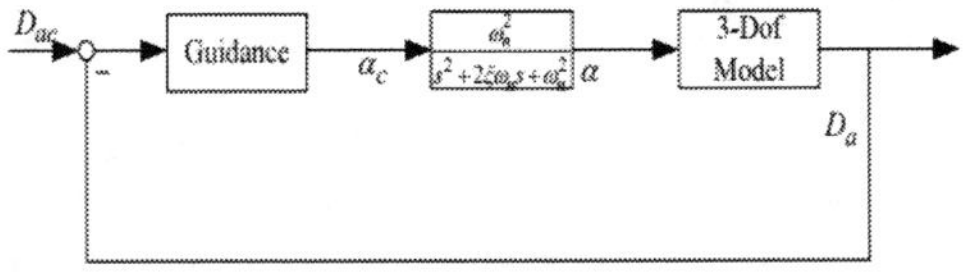

Figure 6. The guidance loop considering the dynamic characteristic of the attack angle

From Fig. 7 to Fig. 10 is the response of the attack angle and drag acceleration under different dynamic link. Considering the dynamic characteristics of the attack angle, as the guidance actual angle command input model, the attack angle command has a certain response time and overshoot. The dynamic characteristic of the attack angle has an influence on the step response of the drag acceleration. When the dynamic pressure is relatively small, due to the dynamic characteristics of the attack angle is poor, the attack angle has a larger overshoot, and the impact of the drag acceleration is also larger. While the dynamic pressure is large, the dynamic characteristic of the attack angle is better, and the influence of the drag acceleration is small.

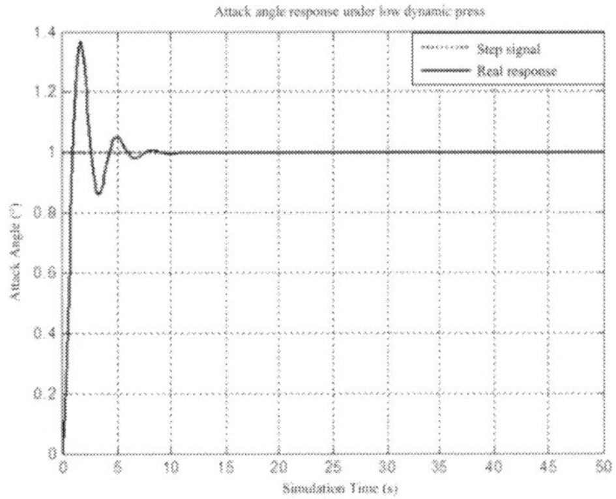

Figure 7. Dynamic characteristics of the attack at low dynamic pressure

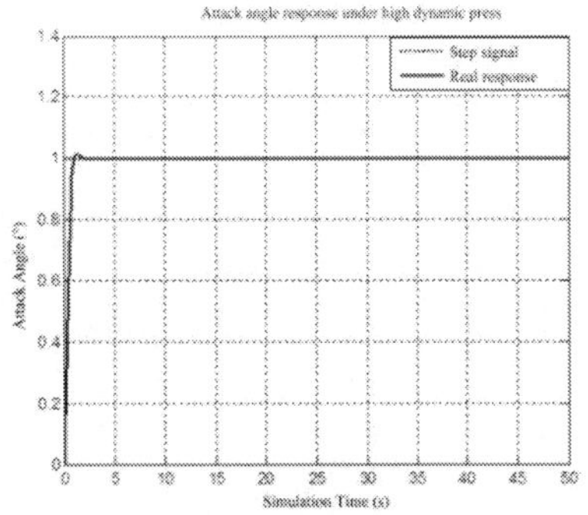

Figure 8. Dynamic characteristics of the attack at high dynamic pressure

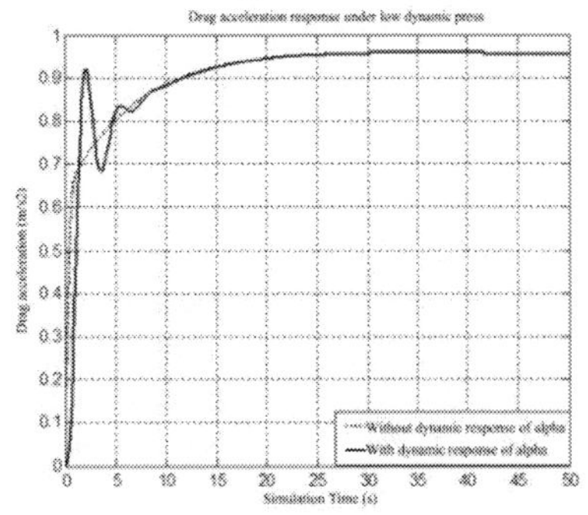

Figure 9. Drag acceleration response at low dynamic pressure

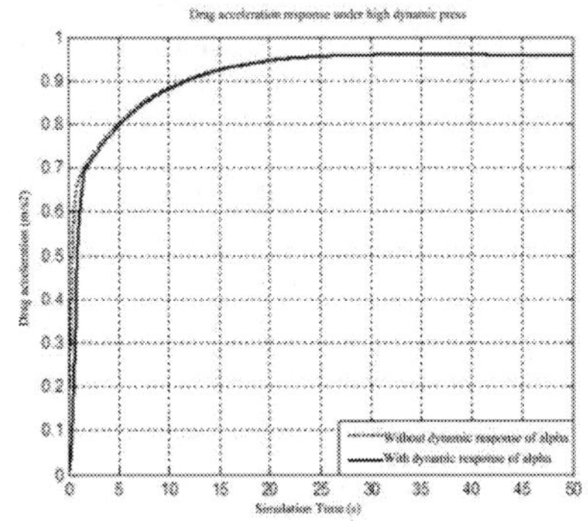

Figure 10. Drag acceleration response at high dynamic pressure

5 Aliasing phenomenon of drag acceleration

The traditional guidance design is based on the dynamic equation of a particle. In this method, the assumption of the guidance command and the centroid motion of the aircraft tend to be too idealistic, while the dynamic characteristics of the guidance loop are ignored. It leads to the problem that the guidance loop does not match the control loop, which brings difficulties to the integrated design of the guidance and control. In fact, due to the short-period motion of the drag acceleration signal being dominated by the attack angle, the presence of aliasing will produce a new low-frequency drag acceleration signal when using drag acceleration guidance. This low-frequency signal is mixed with the real sensor signal that can be used as an instruction input guidance loop, which affects the precision of the guidance.

The signal of drag acceleration is affected by attack angle and dynamic pressure. The analysis of drag acceleration motion shows that in a short time, the attack angle plays a leading role in the drag acceleration response. Taking into account the engineering practice, the whole reentry attack angle is used as a kind of angular motion, and the attack angle is frequent jitter interference. This is equivalent to a high frequency oscillation signal at the attack angle. With the increase of the dynamic pressure, the shake of attack angle is more obvious. The drag acceleration signal is very sensitive to the high frequency disturbance of attack angle, high frequency drag acceleration response is produced. The guidance process is a long period, and the sampling frequency of the guidance command is low. It equals to sampling high frequency drag acceleration signal with low frequency guidance commands. Drag acceleration signal in sampling process existing aliasing phenomenon, drag acceleration signal, which is sampled after mixed with low drag acceleration signal, resulting in resistance acceleration signal polluted thus affecting the performance of guidance.

In order to verify the aliasing phenomenon after the drag acceleration signal is disturbed by the attack, in this paper, a sinusoidal signal with a frequency of $5\,rad/s$ is chosen to simulate the high frequency interference of the attack angle. In the case of considering the dynamic characteristic of the attack angle, the response of the drag acceleration and the dynamic pressure to the high frequency interference of the attack angle is reflected. Fig.11 is schematic diagram of guidance structure. Fig.12 is Signal of attack angle after adding interference.

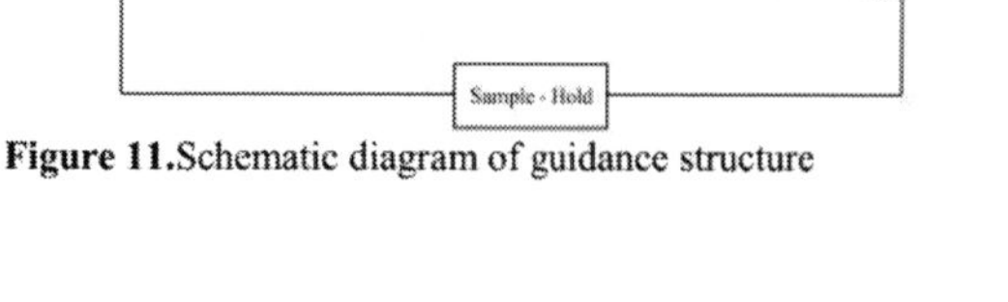

Figure 11. Schematic diagram of guidance structure

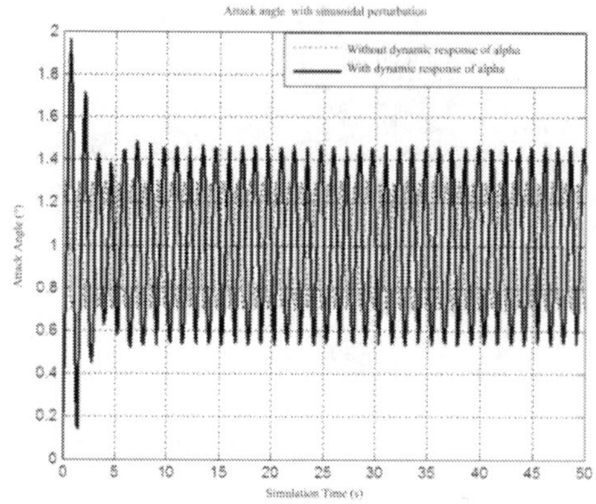

Figure 12. Signal of attack angle after adding interference

The simulation results show that when the sampling frequency is 10Hz, the frequency of the drag acceleration signal after sampling is basically the same as that of the angle of attack. Both of them are shown as high frequency oscillation signals. This is equivalent to the high frequency of attack angle disturbance causes the high frequency drag acceleration of the vibration signal, the drag acceleration of the high frequency oscillation signal can filter out through the high frequency filter. But when the sampling frequency is 1Hz, the low frequency sampling of the high-frequency signal occurs. A new low frequency drag acceleration signal is produced by the aliasing phenomenon and this oscillation signal has a large amplitude. The low frequency drag acceleration signal generated by the high frequency drag of the attack angle can be superimposed on the real resistance acceleration signal, which will cause pollution of the signal. It is difficult to be effectively filtered, because the frequency of new signal is low. In order to avoid this problem, the dynamic pressure can be chosen as the core of the guidance law. Compared with the drag acceleration signal, the dynamic pressure is a long-period change signal, which has little effect by disturbance of attack angle.Fig.13 is the response of 10Hz and 1Hz in the sampling frequency for the drag acceleration. Fig.14 is dynamic pressure response.

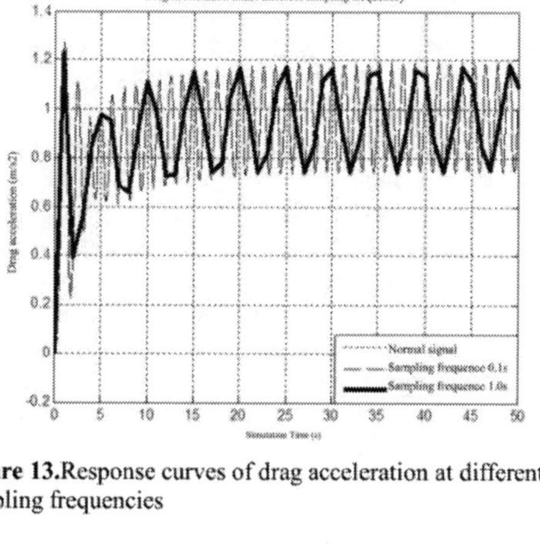

Figure 13. Response curves of drag acceleration at different sampling frequencies

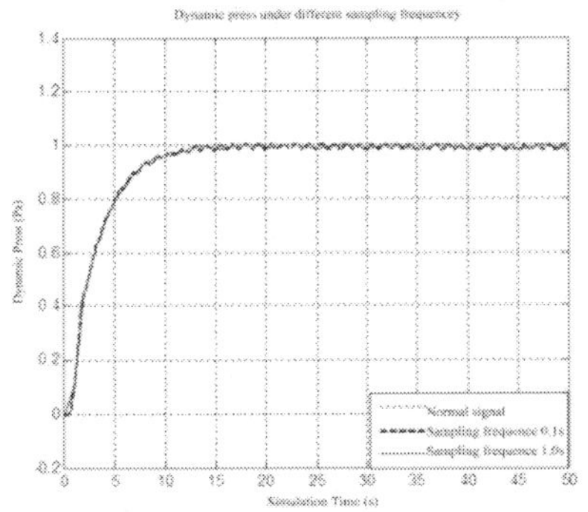

Figure 14. Response curves of dynamic pressure at different sampling frequencies

Fig.13 shows that, due to the drag acceleration being sensitive to the angle of attack, the jitter of the angular signal will result in a larger amplitude response to the drag acceleration. After reducing the sampling frequency, the frequency response of the acceleration signal of the high frequency resistance is mixed, and the new low frequency interference signal is generated. Because the frequency of the guidance loop is low, it is difficult to eliminate the low frequency interference signal through the filter, which will affect the performance of the guidance.

Fig.14 indicates that, dynamic pressure is not sensitive to the attack angle. This is caused by the long period characteristic of dynamic pressure. The dynamic pressure being used as the controlled quantity of the longitudinal direct guidance is relatively well. Even if the sampling frequency is decreased, the dynamic pressure response is very small.

From the above analysis, using drag acceleration as the reentry guidance command has the following deficiencies:

(1) The response performance of the resistance acceleration to the angle of attack is two kinds of motion modes, which are long and short period motion. The initial time of the response is dominated by the angle of attack, which performed short period motion. Then, the dynamic pressure is the main performance for the long period motion, and the response process is complex.

(2) The sensitivity of resistance acceleration to the angle of attack is related to the dynamic pressure, the greater the dynamic pressure, the more sensitive to the angle of attack, and have influence on the integrated design of the guidance and control.

(3) The dynamic characteristic of the angle of attack has great influence on the acceleration of resistance. Resistance acceleration aliasing will be caused after the angle of attack high frequency disturbances. Low frequency resistance acceleration interference is generated by high frequency resistance acceleration signal after aliasing, which can pollution guidance loop and affect the precision of guidance.

6 Conclusion

According to the kinetic law and response characteristics of drag acceleration for high speed reentry vehicle, this paper studies the modal characteristics of the drag acceleration and the response of the guidance command to the attack angle dynamic pressure, It points out the reason that the drag acceleration response is sensitive to the attack angle and objective fact that the frequency aliasing of the guidance commands due to the high frequency attack angle or sampling. Finally, the results of the modal analysis are confirmed by simulation, and it is pointed out that the sampling dynamic pressure guidance can reduce the sensitivity to the attack angle.

References

1. Zhao Hanyuan. *Reentry dynamics and guidance of aircraft*. Changsha: National University of Defence Technology press, 1997:34~45.
2. Peng Xiaobo. Design and implementation of the American space shuttle, Beijing, China Aerospace publishing house, 2015:5~26.
3. Lu Zuliang. *Measurement strategy of periodic signal sampling*. Electrical measurement and instrumentation press, 2008, 45 (10): 1-6.
4. Leng Jianhua. *Fourier transform*. Tsinghua University press, 2004.
5. Zhang Minglian, Yang Yidong. *Flight control system*, Beijing aviation industry press, 1994:41~78.

MATEC Web of Conferences **95**, 07013 (2017) DOI: 10.1051/matecconf/20179507013
ICMME 2016

Design and Testing of a 2-DOF Flexure-based Compliant Stage

Yunsong Du[1], Tiemin Li[1], Wei Ji[2] and Yao Jiang[1]

[1]*Mechanical Engineering Department, Tsinghua University, 100084, Beijing, China*
[2]*Foreign Language Department, Hebei Normal University for Nationalities, Chengde, China*

Abstract. This paper presents the mechanical design and testing of a 2-DOF flexure-based compliant stage. The prismatic joints are first discussed. To obtain high resonance frequency and decoupled property, the stage composed of four parallel chains with symmetrical configuration should be employed. Finally, an experimental platform is set up to test the cooperative tracking performance. The experimental trajectory is in good agreement with the desired trajectory under various input frequencies. It shows that the maximum tracking error exhibits an increase from 0.8% to 2.7%. The experiment results demonstrate that the proposed compliant stage possess good tracking performance under 100 Hz, which also validates the reasonability of the stage.

1 Introduction

Flexure-based compliant mechanisms have many advantages, such as no clearance, no friction losses, no need for lubrication, compactness, and ease of fabrication [1-3]. Considering these advantages, they have been widely used in many industrial applications, such as scanning probe microscopy [4-6], micro-grippers [7-8], and biological science [9-10].

In the past decades, various flexure-based compliant stages have been developed. Awtar [11] presented a parallel XY flexure mechanism designs based on systematic constraint patterns. The advantages of the parallel flexure structure were discussed, and the constraint arrangement was given directly. Qin [12] designed a 2-degree of freedom (DOF) decoupled flexure-based mechanism which was composed of statically indeterminate symmetric flexure structures in parallel. The symmetric design helped to eliminate the thermal error and obtain decoupled property. The experiment results demonstrated that the proposed mechanism had good decoupled ability and load capability, and wide potentials in precision positioning applications. Yong [13] introduced a flexure-based XY stage for fast nanoscale positioning. The design of the flexure-based nanopositioning stage was based on the concept of flexible mechanisms. The stage was composed of two main parts: the outer section that consisted of amplification levers and circular flexures, and the inner section that included a stage and beam flexures. Li [14] designed a totally decoupled flexure-based XY parallel micromanipulator. The double compound parallelogram flexures and a compact displacement amplifier were employed, and they were simple and symmetric. Both finite element analysis and experiments could validate the decoupled property.

There have been many compliant stages with various good performances, such as decoupled ability, high resonance frequency and large workspace. Nevertheless, the design of these existing mechanisms is not systematic. Therefore, the present study addresses the design and testing of a 2-DOF flexure-based mechanism. The mechanical design is first introduced, and the experiments on cooperative tracking performance are tested. The remaining sections of this paper are organized as follows. In section 2, the mechanical design of the stage is derived. In section 3, a prototype of the stage is manufactured an experimental platform is set up. The circular tracking performance is subsequently tested under various input frequencies (from 20 to 100 Hz).

2 Mechanical design

Flexure-based mechanisms can be classed into two types: planar and space mechanisms. The planar flexure-based mechanisms have been applied most widely, and they are usually composed of several prismatic joints, such as linear, parallelogram and amplifying prismatic joints. A linear prismatic joint is illustrated in Fig. 1, and it is composed of flexure hinges, rigid beams and one output platform. The structure is symmetric about the output platform and both ends are fixed. Being symmetric, the output platform would rotate around the center point O when a moment M_z is applied. In addition, the output platform can translate along y-axis if the force F_y is applied to point O. The parasitic motions in x-axis and the parasitic rotations about z-axis can be attenuated by the symmetrical configuration. By contrast, it could remain static in conjunction with consideration of the high stiffness along x-axis when the force F_x is applied to point O.

DOI: 10.1051/matecconf/20179507013

Figure 1. The linear prismatic joint

The parallelogram and double parallelogram prismatic joints are the common joints which can provide translational motion. To derive the decoupled property, flexure-based mechanisms are usually composed of double parallelogram rather than parallelogram. As shown in Fig. 2, the double parallelogram is composed of four limbs, and both sides of the joint are fixed.

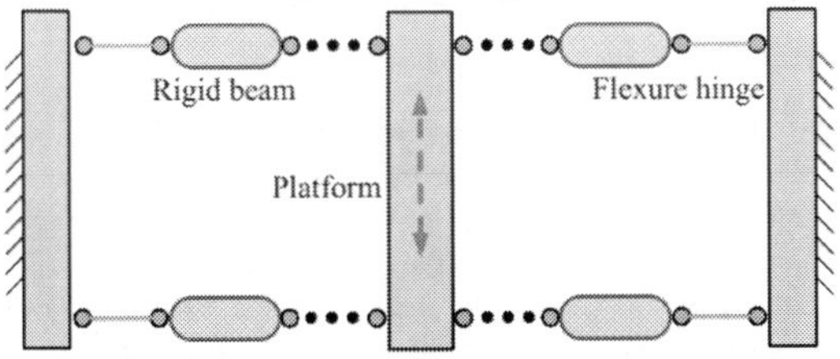

Figure 2. The double parallelogram prismatic joint

For a notch flexure hinge, there are strong couplings between the rotational and translational motions. Furthermore, the rotation center of a flexure hinge drifts whenever the hinge works, resulting in motion errors. To obtain decoupling properties, multiple flexure hinges can be combined to form certain structures. As illustrated in Fig. 3, the parallelogram prismatic joint can keep the orientation of the output platform invariant when the platform is actuated by the force F_x along x-axis. However, undesired cross-coupling error e_y along y-axis is generated at the same time. Compared with the parallelogram joint on one side, with a symmetrical configuration about the actuating axis, the double parallelogram prismatic joint can eliminate the coupling error completely, and thus it possesses output decouple property. In practice, the parallelogram joints are usually used as prismatic joints.

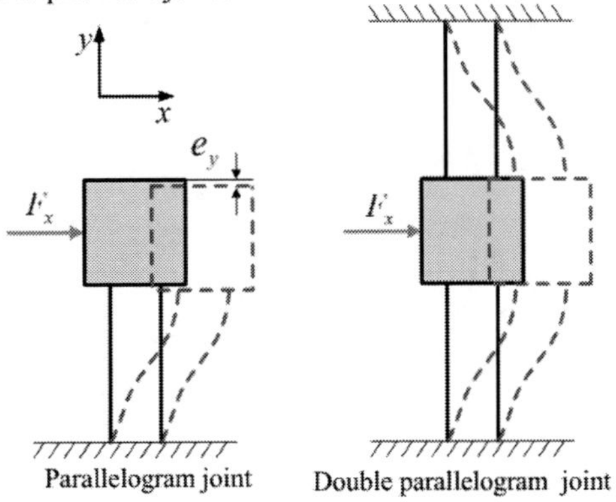

Figure 3. Parallelogram and double parallelogram joint

The amplifying prismatic joint is similar to the bridge-type displacement amplification mechanism. As shown in Fig. 4, surface D is fixed. There is a displacement output on surface C when imposing a force on surfaces A and B. Using this joint, the input displacement can be amplified by more than ten times. Nevertheless, it is only used for compliant mechanisms with low resonant frequency.

In general, flexure-based mechanisms can be classified into two categories according to the kinematic structure: the serial and parallel mechanisms. Serial mechanisms are easier to design, and they have substantially decoupled degree-of-freedom. By contrast, parallel mechanisms with closed-loop kinematic features possess many advantages, such as high load carrying capacity, rigidity and accuracy.

According to the kinematic structure, the flexure-based micro-motion stages can be classified into two categories: the serial structure and the parallel structure. The parallel structure which possesses high rigidity, high load carrying capacity, and high accuracy could be selected. To derive high resonant frequency, the parallel structure should be employed, and the stages should be rigid and compact.

Flexure-based parallel mechanisms are usually built on 2-*PP*、3-*RRR*、3-*PPP* and 6-*SPS* parallel structure, where *P*, *R*, *U*, and *S* denote the revolute, prismatic, universal and spherical joints, respectively. To achieve 2-DOF motion, the 2-*PP* parallel structure could be selected. In addition, to achieve high precise motion, magnetic levitation motors, electromagnetic actuators, voice coil motors and piezoelectric actuators (PZTs) are usually employed. The PZTs are selected to drive the flexure-based compliant stage because they can provide fast and ultra-precision motions.

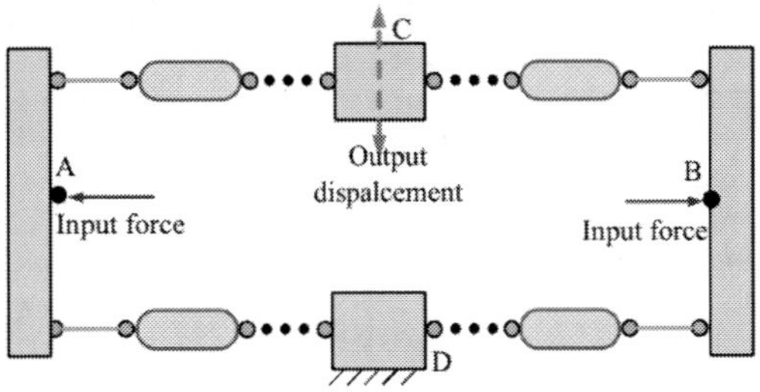

Figure 4. The amplifying prismatic joint

Meanwhile, to obtain decoupled property, the stage with symmetrical configuration should be employed, and thus it can be depicted as Fig. 5. It is composed of four parallel chains, and each chain consists of two prismatic joints. One joint P_1 is used to provide the forces actuated by the PZTs to the stage. The other joint P_2 is employed to transmit the forces and deformations of the flexure hinges to the output platform.

According to the analysis above, to obtain compact structure, P_1 and P_2 can be the linear and parallelogram prismatic joints rather than the amplifying prismatic joints. In addition, the parallelogram is virtually composed of several linear prismatic joints. Therefore, only the linear prismatic joints could be employed to form the stage.

DOI: 10.1051/matecconf/20179507013

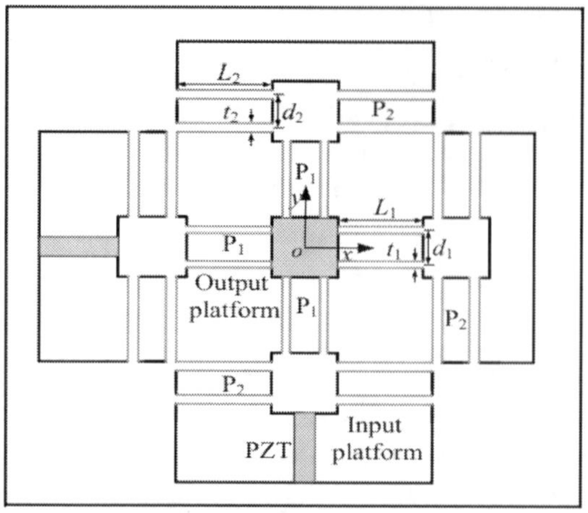

Figure 5. The flexure-based compliant stage

Notch flexure hinges can be manufactured by drilling or milling to form a circular, corner-filled or constant rectangular cutout. A flexure hinge is actually a complex spring element that can respond to and transmit both rotation and translation. The constant rectangular cross-section flexure hinge is selected because it is simple and easy to design. The prismatic joints P_1 and P_2 are composed of several flexure hinges, and their performances are easily influenced by their geometrical parameters. Thus, the parameters of the stage should be optimized.

The objective of the optimization is to maximize the resonance frequency. Meanwhile, the parameters are optimized to obtain the workspace of the stage which is above 18 μm. The thickness of the stage is chosen as 10 mm to guarantee the stiffness of the stage in z-axis. Meanwhile, the length and width of the output platform could be selected as 20 mm× 20 mm to install sensors and accessories. Considering the size of the PZTs, the input platform is chosen as 20 mm× 20 mm to connect to the mechanism and PZTs. The flexure hinge thickness t can be divided into n parts. According to the analysis in our earlier paper [15], the divided parts n are selected 2 to maximize the resonance frequency, and the optimized parameters are listed as follows: L_1=25 mm, L_2=28 mm, t_1=2.5 mm, t_2=3 mm, d_1=11.5 mm and d_2=11 mm.

3 Experiments

An experimental platform is set up. It includes a dSPACE-DS1107 board, two catalog-length gauges (MT 1281), two PZTs (PSt 150/7/40 VS12 from XMT Harbin, Inc.), a piezo amplifier and drive (XE500/E05 from XMT Harbin, Inc.), and the proposed flexure-based compliant stage. As illustrated in Fig. 6, the material of the stage is aluminum 7075. The two preloaded PZTs are actuated by the driver under control of the dSPACE board, with a voltage range of 0-120V. To realize the PI control, the displacements of the output platform are measured by two catalog-length guages (MT 1281 from Heidenhain, Inc.). To reduce vibrations, the air spring optical platform (1209M-100T) made in South Korea is employed.

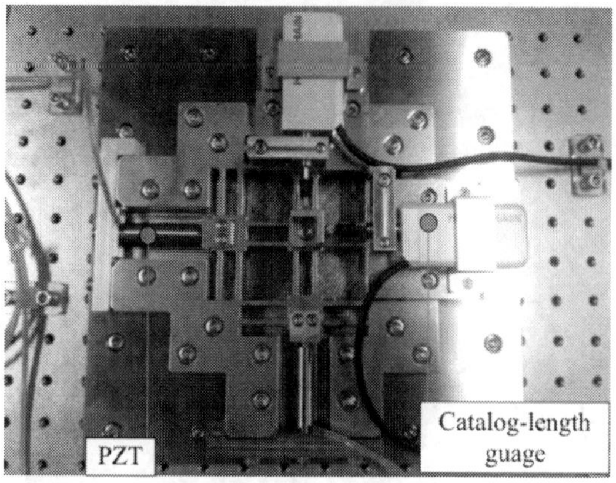

Figure 6. The experiment platform

The dSPACE board can be applied for receiving displacement feedback from the catalog-length gauges, calculating the control, and then sending the it to the amplifier and drive. The 16-bit analog to digital converters is used to translate the analog voltage positional signal to digital signals used by the PI controller. The amplifier and drive can be empoyed to receive command signals from the PI controller and provide amplified voltage to drive the preloaded PZTs.

To verify the proposed mechanical design of the compliant stage, the cooperative tracking performance is tested. A circular trajectory test is performed under different frequencies (from 20Hz to 100Hz), and the input circular radius is 5μm. The compliant stage can be actuated by driving it in x-axis and y-axis with the appropriate amplitudes and phase quadrature (shifted in phase by 90 degrees). Without loss of generality, only the circular tracking performance test under input frequency 80 Hz is illustrated in Fig. 7. It can be seen that the actual trajectory derived from experiments is in good agreement with the desired trajectory. The tacking error is less than 100 nm, and the maximum error is 2% in terms of the circular radius 5μm.

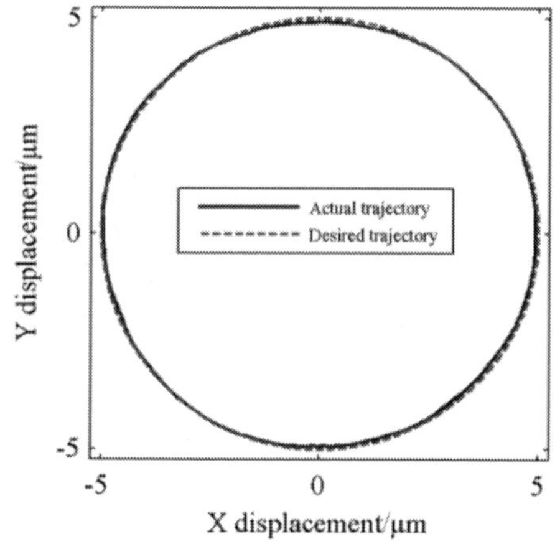

(a) Tracking results

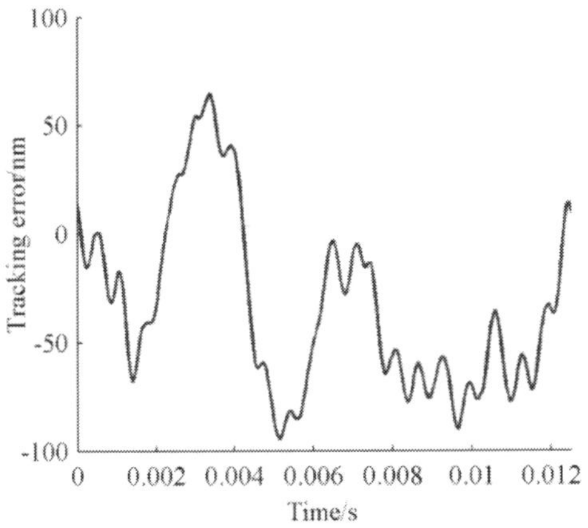

(b) Tracking errors

Figure 7. Circular tracking performance under 80 Hz

The circular tracking error under various input frequencies is studied, as demonstrated in Fig. 8. The tracking error exhibits an increase with increasing the input frequency. To be more specific, the maximum tracking exhibits an increase from 0.04 μm to 0.135 μm. Considering that the input circular radius is 5μm, the maximum tracking error would be 2.7% under input frequency 100 Hz, and the minimum tracking error would be 0.8% under input frequency 20 Hz. Especially, the minimum tracking error would be 0.8% under input frequency 60 Hz. The experiment results indicate that the flexure-based compliant stage possess good cooperative tracking performances.

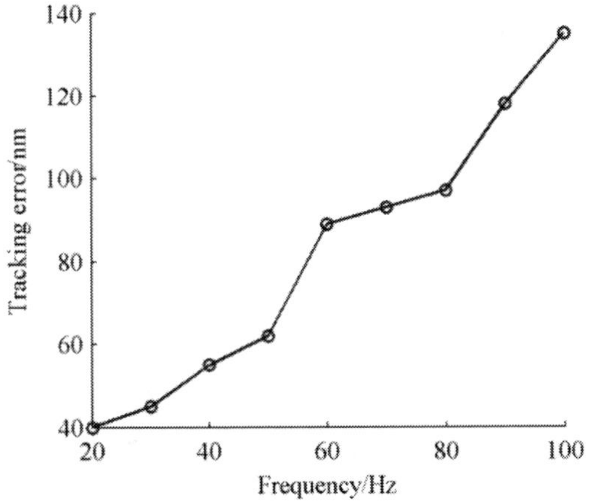

Figure 8. Tracking error under various input frequencies

4 Conclusions

This paper presented the mechanical design and testing of a flexure-based compliant stage with symmetrical configuration. The stage was designed based on the prismatic joints to obtain high resonance frequency and good decoupled property. The flexure parallel structure was selected, and it was composed of four parallel chains. Each chain consisted of two prismatic joints which were composed of the constant rectangular cross-section flexure hinges. The optimization is carried on to maximize the first resonance frequency. Finally, a prototype of the stage is manufactured and an experimental platform is set up. And the circular tracking performance was subsequently tested. The experiment results indicated that the proposed flexure-based compliant stage possessed good tracking performance (from 20 to 100 Hz), which validates the reasonability of the design.

Acknowledgement

This work was supported by the National Basic Research Program of China (Grant No.2011CB302400), the National Science Foundation of China (Grant No.51275260), and the National Science and Technology Major Project of China (Grant No.2015ZX04001002).

References

[1] N. Lobontiu . Compliant mechanisms: design of flexure hinges. CRC press, 2010, pp.2-15.
[2] W. Widhiada. T. G. T. Nindhia, N. Budiarsa. Robust control for the motion five fingered robot gripper. . Int. J. Mech. Eng. And Rob. Res, **4** (2015), 226-232.
[3] Y. Tian, B. Shirinzadeh, D. Zhang. Three flexure hinges for compliant mechanism designs based on dimensionless graph analysis. Precision Engineering, **34** (2010), 92-100.
[4] G. Schitter, K. J. Åström, B. E. DeMartini, P. J. Thurner, K. L. Turner, P. K. Hansma. Design and modeling of a high-speed AFM-scanner. IEEE. T. CONTR. SYST. T, **15** (2007), 906-915.
[5] B Sollapur Shrishail, P Deshmukh Suhas. XY scanning mechanism: a dynamic approach. Int. J. Mech. Eng. And Rob. Res, **3** (2014), 140-154.
[6] Leang, K. K., and Fleming, A. J.: High - speed serial - kinematic SPM scanner: design and drive considerations, Asian. J. Control, **11** (2009), 144-153.
[7] M. N. M. Zubir, B. Shirinzadeh, Y. Tian. Development of a novel flexure-based microgripper for high precision micro-object manipulation. Sensors and Actuators A: Physical, **150** (2009), 257-266.
[8] Q. Xu. Design and development of a flexure-based dual-stage nanopositioning system with minimum interference behavior. IEEE. T. Autom. Sci. Eng, **9** (2012), 554-563.
[9] T. Ando, N. Kodera, D. Maruyama, E. Takai, K. Saito, A. Toda. A high-speed atomic force microscope for studying biological macromolecules in action, Jpn. J. Appl. Phys, 41 (2002), 4851.
[10] H. Y. Kim, D. H. Ahn, D. G. Gweon. Development of a novel 3 degrees of freedom flexure based positioning system, Rev. Sci. Instrum, 83 (2012), 055114.
[11] S. Awtar, A. H. Slocum. Constraint-based design of parallel kinematic XY flexure mechanisms. J. Mech. Des, 129 (2007), 816-830.
[12] Y. Qin, Y. Tian, D. Zhang. Design and dynamic modeling of a 2-DOF decoupled flexure-based mechanism. Chin. J. Mech. Eng, **25** (2012), 688-696.
[13] Y. K. Yong, S. S. Aphale, S. O. R. Moheimani. Design, identification, and control of a flexure-based XY stage for fast nanoscale positioning. IEEE. T. Nanotechnol, **8** (2009), 46-54.

[14] Li Y, Xu Q. Design and analysis of a totally decoupled flexure-based XY parallel micromanipulator. IEEE. T. Robot, **25** (2009), 645-657.

[15] Y. S. Du, T. M. Li, Y. Jiang, H. T. Wang. Design and analysis of a 2-degree-of-freedom flexure-based micro-motion stage. Adv. Mech. Eng. **8** (2016), 1-13.

 DOI: 10.1051/matecconf/20179507014

Wire Rupture Optimization in Wire Electrical Discharge Machining using Taguchi Approach

Ibrahem Maher , Ahmed A. D. Sarhan and Houriyeh Marashi

Center of Advanced Manufacturing and Materials Processing, Department of Mechanical Engineering, University of Malaya, Kuala Lumpur 58100, Malaysia

Abstract. Wire electrical discharge machining (WEDM) is one of the most important nontraditional machining process that is well-known for cutting difficult to machine materials. The wire electrode along with machining parameters control the WEDM process. This research work focuses on optimizing WEDM parameters using Taguchi technique to minimize wire rupture. Experiments have been done using the L18 orthogonal array. Each experiment is repeated three times to ensure accurate readings of the wire rupture. The statistical methods of signal to noise ratio (S/N ratio) is applied to study effects of peak current, pulse width, charging time, wire speed, and wire tension on wire rupture. As a results, the peak current, pulse width, and wire tension have the most significant effect on wire rupture followed by charging time and wire speed. The developed analysis can be used in the metal cutting field to identify the optimum machining parameters for less wire rupture.

1 Introduction

Wire electrical discharge machining (WEDM) is one of the most commonly known and applied nonconventional machining processes in today's industrial practice. WEDM is used to cut any electrically conductive materials. WEDM also facilitates machining complicated shapes that cannot otherwise be achieved with conventional machining processes like turning, milling, and grinding [1]. WEDM is an electro-thermal production process, in which a thin, single-strand metal wire in conjunction with deionized water cuts through metal using heat from electrical sparks. Wire electrode is one of the most important components of cutting process in WEDM. WEDM Machining performance depends on the electrical, mechanical, physical, and geometrical properties of the wire electrode. Besides, other WEDM factors including the mechanical theory, machine intelligence, type of pulse generator, and dielectric flushing system affect machining performance.

The wire electrode is the central component of a WEDM system for the process of producing steady electrical discharge. Therefore, WEDM performance is largely affected by the wire electrode [2]. Wire rupture is a critical phenomenon that affects quality and productivity in the WEDM process. It requires some prior knowledge about the wire electrode and workpiece characteristics in the cutting process in order to avoid such ruptures for continuous process operation. Different factors lead to wire breakage, such as high wire tension, thermal load, electrical discharge impact, and poor flushing. When the developed stresses in the wire are greater than the wire strength, the wire will rupture. The stresses developing in the wire increase with changing wire properties and characteristics, reducing the cross section and increasing the wire temperature. High temperature, varying workpiece thickness, and process parameters influence the wire strength and consequently, wire rupture [3]. Lin et al. [4] developed a control strategy based on fuzzy logic to improve the machining accuracy and concentrate the sparking in the corner areas without affecting the cutting feed rates. During WEDM of pure titanium, wire breakage occurs when the wire encounters non-conducting particles. Many research works have been done on WEDM in contrast to very few investigations on the wire breakage frequency of pure titanium. Kinoshita et al. [5] analyzed the various types of wire breakage. To prevent wire breakage, they developed a control system with a means of monitoring the pulse frequency.

The majority of efforts in past decades have been directed to increasing machining speed without reducing machining accuracy while considering various process-related constraints like wire breakage. Following a comprehensive review of published research works in this field, so far, no research has been introduced that identifies and predict the wire rupture size. Hence, the aim of this work is to obtain the best cutting parameters (peak current, pulse width, charging time, wire speed, and wire tension) for less wire rupture in WEDM process using Taguchi approach.

DOI: 10.1051/matecconf/20179507014

2 Methodology

The effects of machining parameters related to power supply, wire electrode were investigated on wire rupture or wire loss (*WL*). The parameters were peak current (*IP*), pulse width (*Ton*), charging time (*Toff*), wire speed (*WS*), and wire tension (*WT*). The machining parameter levels (Table 1) were selected from prior studies, with the working range and WEDM process parameter levels without wire breakage following the one-factor-at-a-time method.

Table 1. Levels of machining parameters

Machining parameter	Symbol	Units	Levels		
			1	2	3
Peak current	*IP*	A	16	17	-
Pulse on time	*Ton*	μs	0.2	0.3	0.4
Pulse off time	*Toff*	μs	0.5	0.9	1.3
Wire speed	*WS*	m/min	3	7	11
Wire tension	*WT*	N	3.5	6	10.5

The other machining parameters were kept constant during the experiments in order to optimize the process. The machining parameters were selected based on the literature review, experience, and machine maker's recommendations. The Taguchi L18 orthogonal array was selected because of the four machining variables with three levels each and one machining variable with two levels, as shown in Table 1.

The experiments were performed using a computer numerical control WEDM machine tool. The machine consists of four systems that are illustrated in [6]. A coated brass wire has a core of 65%Cu 35%Zn coated with a high zinc concentration layer (40%Cu 60%Zn by weight) was used to cut The workpiece material (Ti6Al4V). The wire electrode tensile strength of 875 N/mm^2, Elongation of 2%, and electric conductivity of 20% IACS. The electrical resistivity and thermal conductivity of Titanium alloy grade 5 (Ti6Al4V) were 17.8x10-5 Ω·cm and 6.7 W/(m·K) respectively.

A workpiece plate 100 x 100 x 20 mm was machined into 5 x 5 x 20 mm specimens. The wire mass loss per meter (*WL*) was used to calculate the wire rupture. The wire mass loss per meter was calculated by weighting 10 meters of the wire electrode before and after the cutting process using a precise balancer. Then the difference in wire mass was divided by ten meters to get the mass loss per meter.

3 Results analysis and discussion

The average wire loss was calculated from three data recorded under the same conditions, as summarized in Table 2. The Taguchi design (L18) was selected with four machining variables with three levels and one machining variable with two levels, as shown in Table 1. The goal of the experiment is to optimize the WEDM variables to obtain low wire loss values using the smaller-is-better characteristic criteria as in the following Equation.

$$S/N = -10\ log(1/n(\sum y^2))$$

where n is the number of experiments, and y is the observed data.

Table 2 introduce the real data which used to perform this approach and show signal to noise ratios of wire loss. Table 3 present the mean *S/N* ratios for each level of wire loss. These data were plotted as shown in Fig. 1. In this study, the *S/N* ratio should have a maximum value to obtain optimum cutting conditions, according to the Taguchi method. Thus, the optimum cutting condition was found as -6.56 *S/N* ratio in Table 2. The optimum cutting conditions, which were the peak current of 16 A, the pulse width of 0.2 μs, the charging time of 1.3 μs, the wire speed of 11 m/min and the wire tension of 10.5 N were obtained for the wire rupture.

Table 3 and Fig. 1 indicate that peak current, pulse width, and wire tension are more significant to the mean *S/N* ratios to the wire rupture and charging time and wire speed have low effect on wire rupture. In addition, it is seen in Fig. 1 and Table 3 that the first level of peak current, pulse width, and charging time factors, the second level of wire speed factor, and the third level of wire tension factor are higher. And hence, the optimum machining settings determined under the same conditions for the experiments to be conducted will be 16 A for the peak current, 0.2 μs for the pulse width, 0.5 μs for the charging time, 7 m/min for the wire speed, and 10.5 N for the wire tension.

Table 2. Measured *WL* at different machining conditions

No.	Machining parameters					*WL* (mg/m)				*S/N* ratios
	IP (A)	*Ton* (μs)	*Toff* (μs)	*WS* (m/min)	*WT* (N)	1	2	3	Avg.	
1	16	0.2	0.5	3	3.5	2.98	2.97	3.09	3.01	-9.58
2	16	0.2	0.9	7	6	2.28	2.3	2.16	2.25	-7.03
3	16	0.2	1.3	11	10.5	2.23	2.09	2.06	2.13	-6.56
4	16	0.3	0.5	7	3.5	2.38	2.34	2.42	2.38	-7.53
5	16	0.3	0.9	11	6	2.58	2.7	2.79	2.69	-8.60
6	16	0.3	1.3	3	10.5	4	4.01	4.22	4.08	-12.21
7	16	0.4	0.5	3	6	4.81	4.75	4.57	4.71	-13.46
8	16	0.4	0.9	7	10.5	3.76	3.85	3.89	3.83	-11.67
9	16	0.4	1.3	11	3.5	3.46	3.55	3.52	3.51	-10.91
10	17	0.2	0.5	11	10.5	2.73	2.84	2.89	2.82	-9.01
11	17	0.2	0.9	3	3.5	2.92	3.08	3.03	3.01	-9.57
12	17	0.2	1.3	7	6	2.66	2.53	2.35	2.51	-8.02

13	17	0.3	0.5	11	6	2.09	2.25	2.23	2.19	-6.81
14	17	0.3	0.9	3	10.5	4.31	4.16	4.25	4.24	-12.55
15	17	0.3	1.3	7	3.5	3.62	3.69	3.79	3.70	-11.37
16	17	0.4	0.5	7	10.5	8.64	8.49	8.61	8.58	-18.67
17	17	0.4	0.9	11	3.5	7.99	7.83	7.94	7.92	-17.97
18	17	0.4	1.3	3	6	11.77	11.86	12.01	11.88	-21.50

Table 3. Response table for S/N ratios

Response table for *S/N* ratios of Wire loss (Smaller is better)					
Level	*IP*	*Ton*	*Toff*	*WS*	*WT*
1	-9.729	-8.295	-10.845	-11.156	-13.145
2	-12.830	-9.845	-11.234	-10.904	-10.715
3		-15.697	-11.759	-11.778	-9.977
Delta	3.101	7.402	0.914	0.874	3.168
Rank	3	1	4	5	2

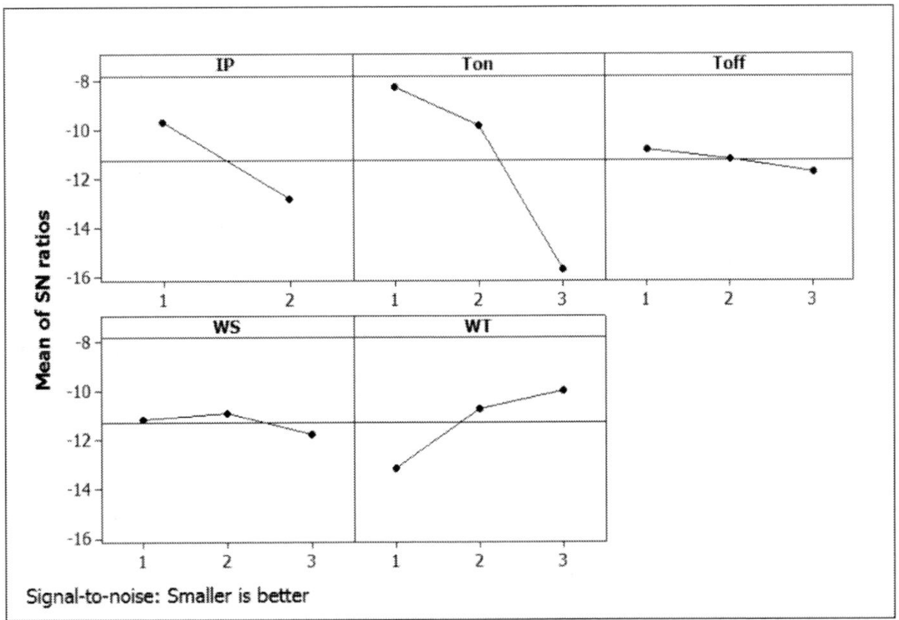

Figure 1. Effect of process parameters on *WL*

4 Conclusion

The Taguchi approach was used optimize machining parameters on wire electrical discharge machining process. Taguchi's L18 orthogonal array was selected for three different levels of pulse width, charging time, wire speed, and wire tension and two levels of peak current. Wire rupture' S/N ratios was found as a result of experiments conducted according to the orthogonal array. The maximum value was found by using the S/N ratio equation of "the smaller-the better," the maximum S/N ratio yielded optimum cutting parameters. Optimum cutting conditions-which correspond to maximum -6.56 S/N value of the smaller WL value for the smaller wire rupture in WEDM operation were found to be 16 A for the peak current, the pulse width of 0.2 µs, the charging time of 1.3 µs, the wire speed of 11 m/min and the wire tension of 10.5 N. The obtained results introduce a good knowledge that may be used in future academic and industrial research.

Acknowledgement

The authors acknowledge the financial support of The University of Malaya Postgraduate Research Grant (PPP) Program No. PG020-2013B.

References

1. I. Maher, A.A.D. Sarhan, M.M. Barzani, M. Hamdi, Increasing the productivity of the wire-cut electrical discharge machine associated with sustainable production, Journal of Cleaner Production, **108**, 247-255, (2015).
2. I. Maher, A.A.D. Sarhan, M. Hamdi, Review of improvements in wire electrode properties for longer

working time and utilization in wire EDM machining, The International Journal of Advanced Manufacturing Technology, **76**, 329-351, (2015).

3. Y.F. Luo, Rupture failure and mechanical strength of the electrode wire used in wire EDM, Journal of Materials Processing Technology, **94**, 208-215 (1999).
4. C. T. Lin, I.F. Chung, S.-Y. Huang, Improvement of machining accuracy by fuzzy logic at corner parts for wire-EDM, Fuzzy Sets and Systems, **122**, 499-511, (2001).
5. N. Kinoshita, M. Fukui, G. Gamo, Control of wire-EDM preventing electrode from breaking, CIRP Annals - Manufacturing Technology, **31**, 111-114, (1982).
6. I. Maher, A.A.D. Sarhan, H. Marashi, Effect of Electrical Discharge Energy on White Layer Thickness of WEDM Process, in: Reference Module in Materials Science and Materials Engineering, Elsevier, (2016).

Simulation of the Carton Erection for the Rubber Glove Packing Machine Using Finite Element Method

Kawin Jewsuwun and Chakrit Suvanjumrat

[1]*Department of Mechanical Engineering, Faculty of Engineering, Mahidol University, Salaya, Nakorn Pathom, 73170, Thailand*
[2]*Laboratory of Computer Mechanics for Design (LCMD), Department of Mechanical Engineering, Faculty of Engineering, Mahidol University, Salaya, Nakorn Pathom, 73170, Thailand*

Abstract. The rubber glove packing machine had been designed an important function which worked with folding carton. Each folded paper carton would be pulled to be erected by vacuum cups. Some carton could not completely form because of an unsuitable design of the erector. Cartons were collapsed or buckling while pulled by vacuum cups that cause to sudden stop the packing process and affect to number and cost of rubber glove production. This research aimed to use simulation method to erect the folded carton. Finite element (FE) model of the rubber glove carton was created with shell elements. The orthotropic material properties were employed to specify FE model for analysis erection behavior of the folding carton. Vacuum cups number, positions and rotation points were simulated until obtained a good situation of the folding carton erector. Subsequently, finite element analysis results will be used to fabricate erector of the rubber glove packing machine in a further work.

1 Introduction

The competition to increase productivity has effected to the rubber glove industry. The production line of rubber gloves is divided into the rubber glove production and packaging. The packaging process required a lot of manpower starting from removing rubber gloves from formers, counting of gloves, folding of cartons and packaging of gloves into carton. Currently, the packaging machine becomes another alternative to replace workers and effectively reduce the process time and cost. Unfortunately, the packaging machine is encountered several issues during carton erection process. The important problem may be buckling of the carton during the transition between flattened and erected states.

The carton which is used in glove industry generally, made from paperboard. The crease on carton board happens by pressing die which makes a groove line on one side and a protuberant line on the other side. Recently, several works have been performed to study the material behavior of paperboard. Carlsson et al. [1] investigated crease stiffness and bending strength of crease zones. A parallel beam model was used to describe bending behavior of locally delaminated zones, which were creased zones of paperboard. Nagasawa et al. [2] investigated creasing characteristic by varying effect of crease depth and crease deviation. The 3-D analytical estimation and 2-D finite element method consisted of carton panel and the crease springs have been developed to study the stiffness characteristic of creases and panels of a carton. The resistive moment resulting from carton crease and force required for folding cartons was investigated [3-6]. Recently, the carton manufacturers have forced to use lighter-weighted carton in order to reduce wasted and production cost. This packaging legislation had decreased carton stiffness which may lead to be failure of the folding process. In order to tackle these problems, the FE model of carton processing is developed to simulate the behavior of folded carton during the complex packing operations. The orthotropic constitutive model was used to model nonlinear behavior of carton wall. The linear elastic model is insufficient since stress of the carton wall in thickness directions being low mechanical resistant and tend to cause buckling of the individual plies and may result in unacceptable damage of the inter-laminar bonds [7].

This research aimed to develop finite element model of carton unfolding process using packaging machine. The orthotropic constitutive model is then used to describe carton behavior, while the multiple point constraint is applied to model simplified crease between each panel. The model is then analyzed to observe important characteristic of folding cartons such as carton deformation and mechanical response.

2 Finite element method

The carton is composed of four rectangular panels with L×W×H of 220×120×70 mm. The erector arm has L-shape with a rotating point to rotate up and down. The height of the erector arm (B) can be adjusted regarding to the installation position. The length of erector arm for

DOI: 10.1051/matecconf/20179507015

installation vacuum cup is 232 mm. The vacuum cups which is a diameter of 19.2 mm, are installed on the erector arms. The number of vacuum cups can be selected to install on the erector arm. Fig. 1 shows schematic of the carton and the erector arm that bring to analyze with finite element method (FEM). Fig. 2 shows FE model to simulate the carton erection by using commercial finite element software, MSC. Patran. The quadratic shell elements are used to create skillet model. The skillet is simplified by modeling with only four panels. The FE model of vacuum cups is also used the quadratic shell elements. The erector arms are used the tetrahedron elements. Description of elements which used to create FE model show in Table 1.

Table 1. Types and amount of elements using for the FE model.

Part	Element type	Dimension	Amount
Carton	Quadratic	2-D	1,219
Vacuum cup	Quadratic	2-D	111
Elector arm I (B = 52.5, C = 30.0 mm)	Tetrahedron	3-D	1,046
Elector arm II (B = 52.5, C = 40.0 mm)	Tetrahedron	3-D	1,180
Elector arm III (B = 67.5, C = 30.0 mm)	Tetrahedron	3-D	1,092
Elector arm IV (B = 70.0, C = 40.0 mm)	Tetrahedron	3-D	1,185

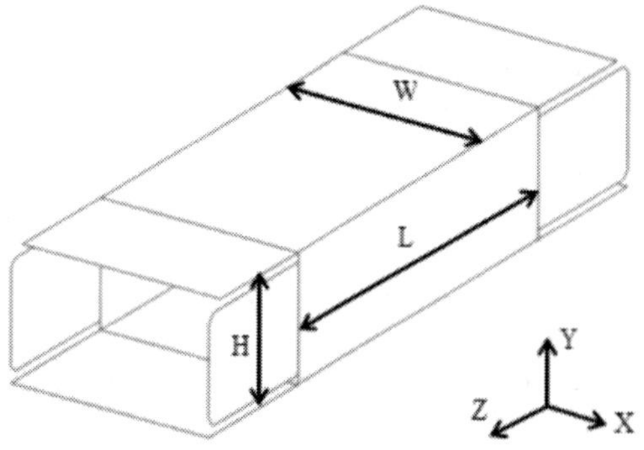

(a)

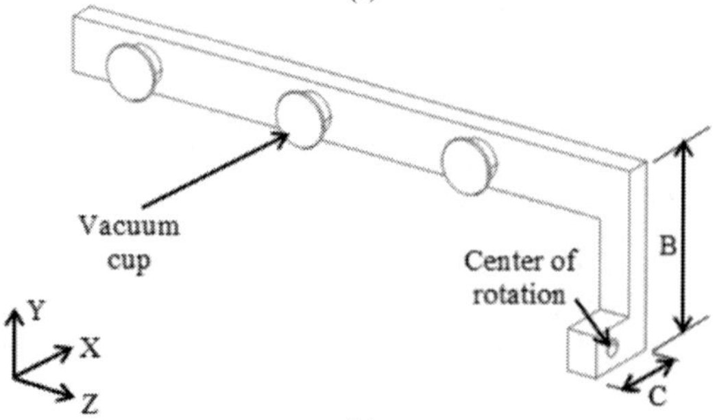

(b)

Figure 1. Schemetric of: (a) a carton and (b) an elector arm

Boundary conditions of the FE model of carton erection were defined to analyze deformation of cartons with different positions of erector arms and number of vacuum cups. Multiple point constraint (MPC) was used to model the crease and connect each panels of a carton together. The fixed constraint was applied to simulate the attached vacuum cups under the bottom of the skillet as shown in Fig. 2c.

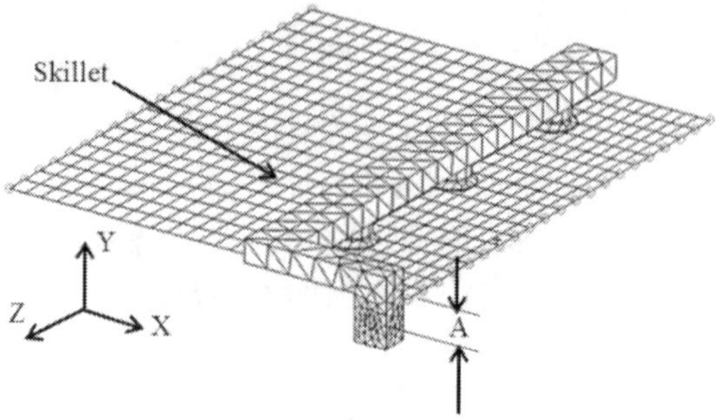

(a)

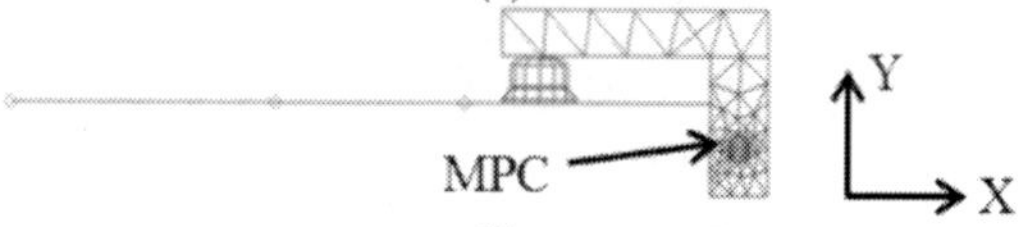

(b)

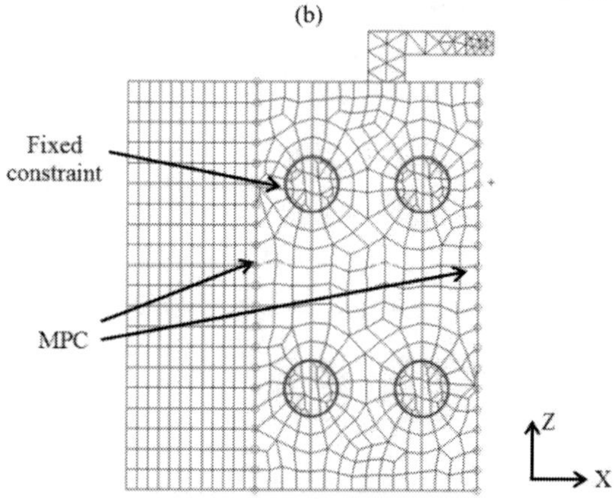

(c)

Figure 2. The FE model of the carton erection in: (a) isomatric view, (b) front view and (c) bottom view.

The MPC was also used to model a center of rotation of the erector arm. The vacuum cups were assigned to contact with skillet element. The contact conditions were applied to be "glued after contact" to simulate suction of the vacuum cups. The FE model of a skillet, vacuum cups and erector arm is assembled in the different position. Fig. 3 shows FE models which are created to analyze a suitable position of the carton erections. The center of rotation of erection arm I and III was on the level of skillet while the center of rotation of erection arm II and IV was lower than skillet of 10 mm. The FE model of carton erections was analyzed by using the FE software, MSC.Nastran. Six conditions of the carton erection were simulated in this research.

The orthotropic constitutive model has been used to describe the nonlinear behavior of carton wall during unfolding process. The carton board can be considered orthotropic since most of its fibers have been reoriented during manufacturing, resulted in different mechanical strength between oriented direction and the others. The oriented direction is commonly known as machine direction (MD), while the perpendicular direction is known as cross direction (CD). The MD and CD direction are laid on the carton panels. In addition, the mechanical properties in the thickness direction are also different from the other two directions due to laminar construction, this thickness direction may be abbreviated as ZD. The element thickness of the vacuum cup was 1 mm while carton board was 0.8 mm. The orthotropic material model can be written in the form of compliance matrix as:

$$\begin{Bmatrix} \varepsilon_{11} \\ \varepsilon_{22} \\ \varepsilon_{33} \\ \gamma_{12} \\ \gamma_{13} \\ \gamma_{23} \end{Bmatrix} = \begin{bmatrix} \frac{1}{E_1} & \frac{-v_{21}}{E_2} & \frac{-v_{31}}{E_3} & 0 & 0 & 0 \\ \frac{-v_{12}}{E_1} & \frac{1}{E_2} & \frac{-v_{32}}{E_3} & 0 & 0 & 0 \\ \frac{-v_{13}}{E_1} & \frac{-v_{23}}{E_2} & \frac{1}{E_3} & 0 & 0 & 0 \\ 0 & 0 & 0 & \frac{1}{G_{12}} & 0 & 0 \\ 0 & 0 & 0 & 0 & \frac{1}{G_{13}} & 0 \\ 0 & 0 & 0 & 0 & 0 & \frac{1}{G_{23}} \end{bmatrix} \begin{Bmatrix} \sigma_{11} \\ \sigma_{22} \\ \sigma_{33} \\ \sigma_{12} \\ \sigma_{13} \\ \sigma_{23} \end{Bmatrix} \quad (1)$$

where E_{ij}, G_{ij} and v_{ij} is Young's modulus, shear stress and Poisson's ratio in i and j direction, respectively. The first direction of Young's modulus (E_1) which is corresponding to MD was aligned to the principle x-axis of the model while the second direction (E_2) is corresponding to CD in z direction. The third direction (E_3) was normal to shell elements. The mechanical properties of orthotropic material which are obtained from [8] shown in Table 2.

Table 2. Mechanical properties of the caton board.

Property	Value
E_1(MPa)	4460
E_2(MPa)	2385
E_3(MPa)	22
G_{12}(MPa)	1272
G_{13}(MPa)	81
G_{23}(MPa)	68
v_{12}	0.15
v_{13}	0.08
v_{23}	0.21
ρ (kg/m^3)	615

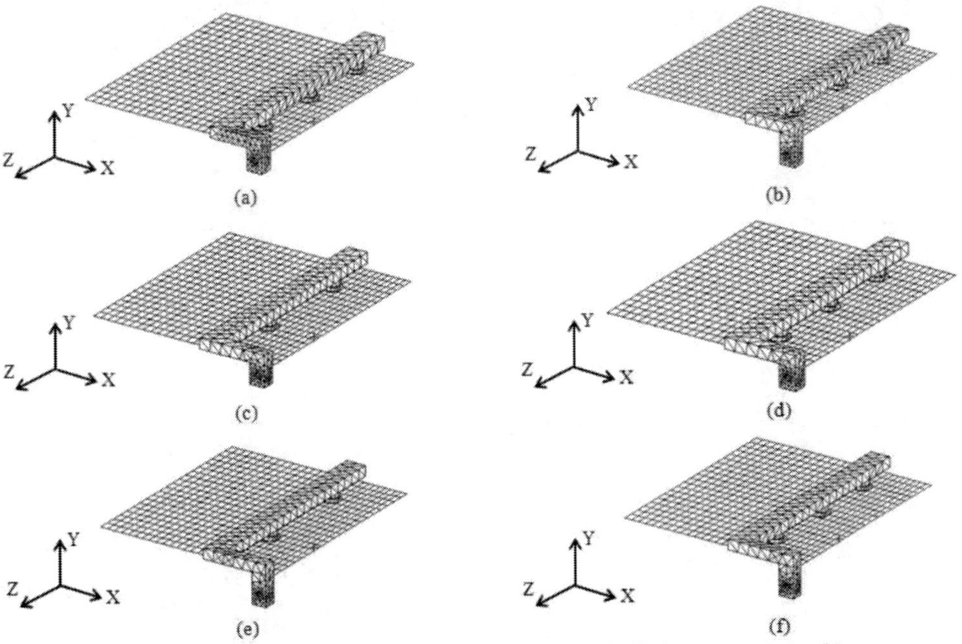

Figure 3. The FE models of the carton erections by variation of: (a) an elector arm I with three vacuum cups, (b) an erector arm II with three vacuum cups, (c) an elector arm III with two vacuum cups, (d) an erector arm III with three vacuum cups, (e) an elector arm IV with two vacuum cups and (f) an erector arm IV with three vacuum cups.

3 Results and discussion

The finite element analysis results of carton erection process with different erector arm positions are shown in Fig. 4. The stress and deformation of erected carton which assembled with an erector arm were illustrated. The red portion of the model indicated maximum stress, while the minimum was indicated by blue color of the contour. The erector arm I and II with three vacuum cups (Fig. 4a and 4b) and an erector arm III, IV with two vacuum cups (Fig 4c and 4e) is shown carton erection at the angle of 45° from the initial skillet position. The erected carton with an erector arm III, IV with three vacuum cups (Fig. 4d and 4f) is shown erection at the angle of 90°. The comparison between analyzed results of the different erector arm models shows the carton model with an erector arm I and II yield more buckling than an erector arm III and IV. The over buckling effect may lead to warpage of carton board near the crease corner, which may be divergence of the simulation results. Two vacuum cups which installed on the erector arm III forced bending of the carton panel. The large concentrated stress happened on the panel by two vacuum cups were more than three vacuum cups on the erector arm. The erector arm IV did not obtain the divergent result but it made carton panel delaminate. The advantage of 3-D orthotropic material is distinct wrinkle of the panel. Consequently, the erector arm III is appropriate to unfold this carton.

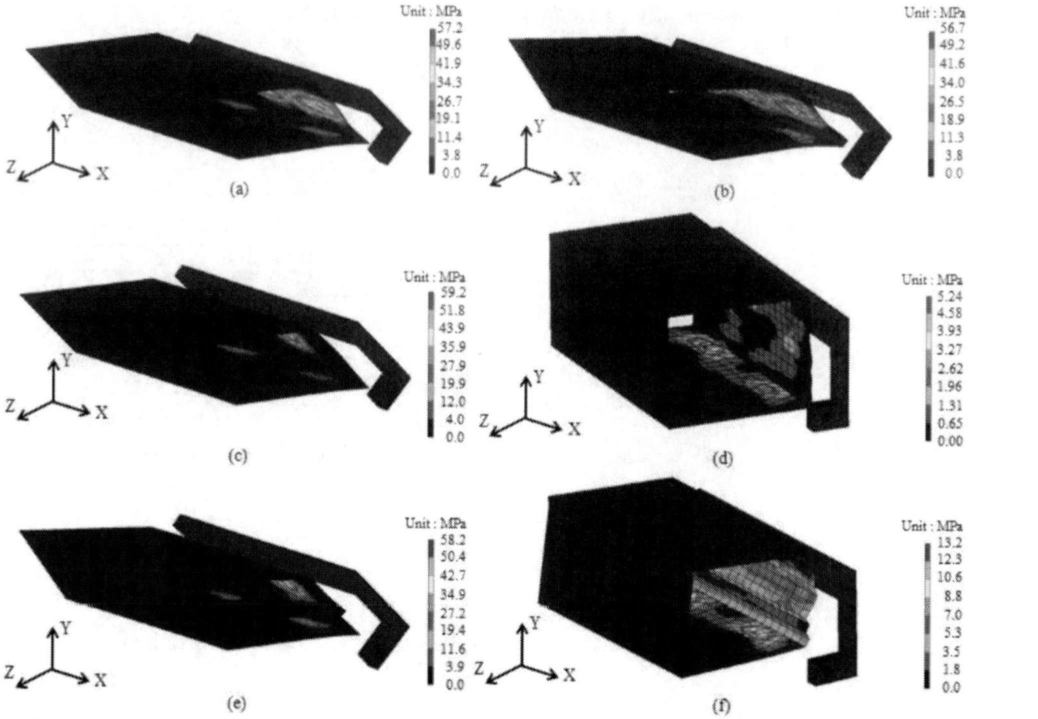

Figure 4. The FE simulation results of the carton erections by variation of: (a) an elector arm I with three vacuum cups, (b) an erector arm II with three vacuum cups, (c) an elector arm III with two vacuum cups, (d) an erector arm III with three vacuum cups, (e) an elector arm IV with two vacuum cups and (f) an erector arm IV with three vacuum cups.

4 Conclusions

The finite element models of carton erection process were developed. The model consisted of skillet, erector arms and vacuum cups. The skillet and vacuum cups were modeled with 2-D shell elements while erector arm was modeled with 3-D solid elements. The skillet was simplified by modeling with four panels. The orthotropic material model was used to describe the nonlinear behavior of cartons, while the erector arm was defined as linear isotropic material. The vacuum cups were assigned to glued contact with the skillet and the erector arm was rotated 90° from initial skillet position. The finite element models of six different carton erections were analyzed to study the effects of machine design. Buckling and distortion effects occurred on carton wall during erection process could be also predicted by FE results. The FE models could be used to study for appropriate geometry and position of the carton erection mechanism and give useful information and understanding for designing and development of the glove packaging machine in the near future.

References

1. L. Carlsson, A.D. Ruvo, C. Fellers, J. Mater. Sci. **18** (1983)
2. S. Nagasawa, Y. Fukuzawa, T. Yamaguchi, S. Tsukatani, I. Katayama, J. Mater. Process. Tech. **140** (2003)
3. F. Cannella, J.S. Dai, J. Mech. Eng. Sci. **220** (2006)
4. A. Giampieri, U. Perego, R. Borsari, Int. J. Solids Struct. **48** (2011)
5. L. Mentrasti, F. Cannella, M. Pubilli, J.S. Dai, Int. J. Solids Struct. **50** (2013)
6. H. Huang, A. Hagman, M. Nygards, Mech. Mater. **69** (2014)
7. D.M. Sirkett, B.J. Hicks, C. Berry, G. Mullineux, A.J. Medland, J. Mech. Eng. Sci. **220** (2006)
8. D.M. Sirkett, B.J. Hicks, C. Berry, G. Mullineux, A.J. Medland, J. Mech. Eng. Sci. **221** (2007).

Acoustic Emission Technique Applied in Textiles Mechanical Characterization

Carlos Rolando Rios-Soberanis[1], Jose Rodriguez-Laviada[1], Emilio Perez-Pacheco[2], Takenobu Sakai[3] and Shuichi Wakayama[4]

[1]*Centro de Investigación Científica de Yucatán, Mérida, Yucatán, México*
[2]*Instituto Tecnológico Superior en el Estado de Campeche (ITESCAM), Calkini, Campeche, México*
[3]*Graduate School of Science and Engineering, Saitama University, Saitama, Japan*
[4]*Department of Mechanical Engineering, Tokyo Metropolitan University, Japan*

Abstract. The common textile architecture/geometry are woven, braided, knitted, stitch boded, and Z-pinned. Fibres in textile form exhibit good out-of-plane properties and good fatigue and impact resistance, additionally, they have better dimensional stability and conformability. Besides the nature of the textile, the architecture has a great role in the mechanical behaviour and mechanisms of damage in textiles, therefore damage mechanisms and mechanical performance in structural applications textiles have been a major concern. Mechanical damage occurs to a large extent during the service lifetime consequently it is vital to understand the material mechanical behaviour by identifying its mechanisms of failure such as onset of damage, crack generation and propagation. In this work, textiles of different architecture were used to manufacture epoxy based composites in order to study failure events under tensile load by using acoustic emission technique which is a powerful characterization tool due to its link between AE data and fracture mechanics, which makes this relation a very useful from the engineering point of view.

1 Introduction

In order to predict the mechanical behavior of a composite during its service life, it is important to study the initiation and development of cracks and its effects induced by degradation. The onset of damage is related to the structural integrity of the component and its fatigue life. For this, among other reasons, non-destructive techniques have been widely used nowadays in composite materials characterization such as acoustic emission. Acoustic Emission (AE) analysis is an extremely powerful technology that can be deployed within a wide range of application of non-destructive testing: metal pressure vessels, piping systems, reactors and similar [1]-[4]. However lately, AE technique has been used extensively in materials characterization, especially on composite materials research by investigating properties of materials, breakdown mechanisms and damage behavior [5]-[7]. AE technique is the most powerful method since it correlates AE signals detected intimately to materials fracture mechanics, which makes this relation a very useful tool from the engineering point of view [7]. In this context, AE can also monitor the damage onset and the damage evolution as well as its profile in real time data, which can assess the structural integrity of the component under loading. Damage mechanisms up to failure are rather complex in composite laminates [8]. One reason is that several damages phenomena (matrix cracking, delamination, fibre breakage, fibre/matrix debonding, fibre pull-out, etc.) are acting alone or coupled. The identification and evaluation of damage or fracture stages of polymer composites are extremely complex due to a composite material exhibiting heterogeneous nature, lacking of uniformity in composition owing to its components and, therefore, the assignment may be controversial [9], [10]. The new advanced technologies in textile industry has allowed the emerge of novel fabric geometries used for composites reinforcement in addition to those traditionally produced. When textiles are used as reinforcements in composite materials, they exhibit very heterogeneous structures As a consequence; the stress and strain fields within it depend greatly on the constituents repartition, on their size and on their spatial arrangement [11]-[13]. In other words, it depends on the textile internal geometry. Hence, a characterization procedure must be able to deal with this geometrical heterogeneity and this mechanical anisotropy. This paper focuses in AE damage analysis in diverse textiles of different architecture/geometry, therefore, damage sequence and its dependence is discussed.

2 Case 1. Damage analysis in multiaxial textile reinforced composites

A multi-axial E-glass textile exhibiting a mass per unit area of 972 ±5% g/m2 and a [0°,+45°,90°,-45°] stacking

DOI: 10.1051/matecconf/20179507016

sequence was used as a reinforcement in epoxy resin matrix composite. The layers have relative mass fractions as indicated in Table 1 and are stitched together with a polyester multifil binding yarn. The unidirectional fiber bundles in the four planes are held together by a fine polyester thread in both sides. In the outer layer where fibers are running at 0°, the polyester (PES) thread is knitted in a zig-zag pattern and in the other side (outer layer at -45°) is knitted at 0° (Fig. 1).

Table 1. Technical specification of the textile.

Layer orientation	Weight g/m²	Composition
0°	354	E- Glass
+45°	200	E- Glass
90°	207	E- Glass
-45°	200	E- Glass
Knit yarn	11	Polyester
Total	972 ± 5%	

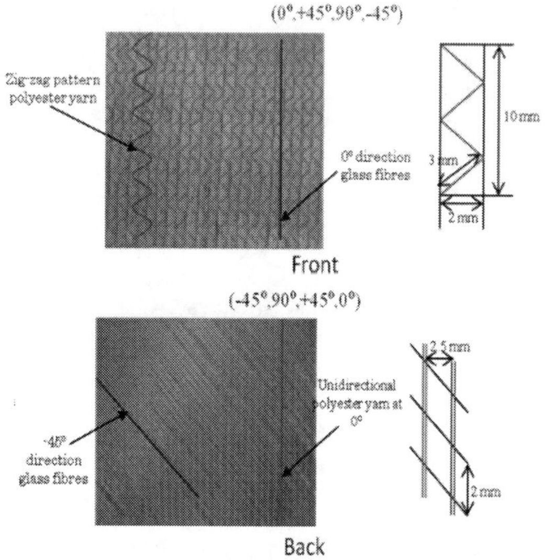

Figure 1. Textiles interweaving, front and back.

Fig. 2 illustrates both sides of the Z-pinned multiaxial textile. Epoxy system chosen was completely transparent when cured allowing visualization of the textiles architecture. Using one layer of textile and epoxy resin system as matrix, a composite with a fibre volume fraction of about 18% was manufactured by wet lay-up process. Laminas were elaborated taking in account the two principal direction layers that conform the multiaxial textile (0°, and 90°).

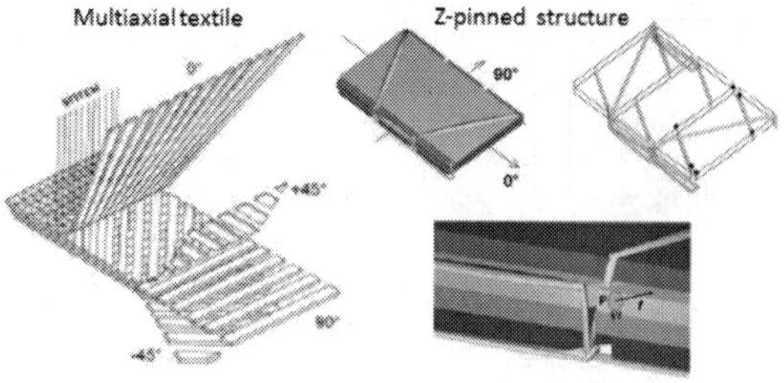

Figure 2. Z-pinned multiaxial textile architecture/geometry.

A Shimadzu universal machine was used to carry out mechanical tests in tension. The textile direction in question was parallel to load direction in which the samples were loaded to progressively higher strain at a cross-head speed of 0.5 mm/min. Acoustic emission sensors were attached to the sample as shown in Fig. 3A to obtain the signal in order to be related to de damage development and mechanisms of fracture.

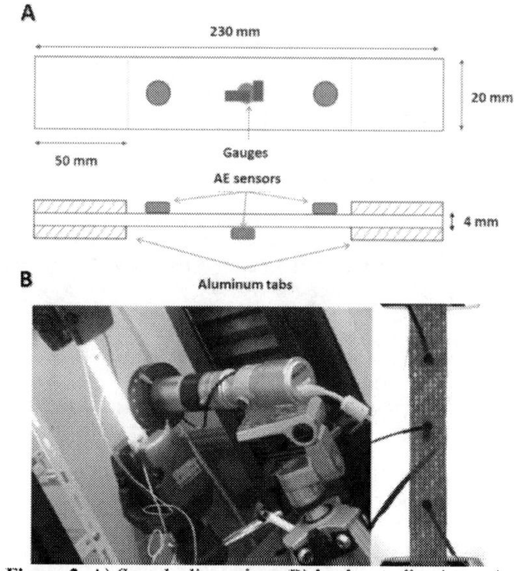

Figure 3. A) Sample dimensions, B) load recording in tension.

During tensile loading, in each stress-strain curves, several discontinuities that are related to significant cracks developed transversally to the sample were observed. At a higher strain discontinuities were more frequent, the higher the load, the more cracks are formed. Samples at 90° exhibited slightly higher crack density (cracks per area). Apparently, textile geometry (stacking sequence) does not affect significantly the mechanical parameters values; however, damage development and propagation along samples depending on the loading direction are influenced by the PES knit yarn structure (Fig. 4).

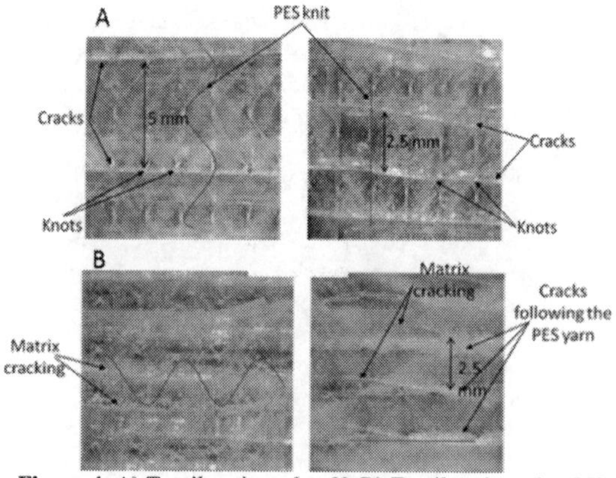

Figure 4. A) Textile oriented at 0°, B) Textile oriented at 90°.

The load (N)-time (s)-amplitude curves for samples when the textile is oriented at 0° is presented in Fig. 5A.

AE signal detected over 280 s are intimately related to the first deflection in the curve which is also correlated to the appearance of the first significant damage that emits high energy.

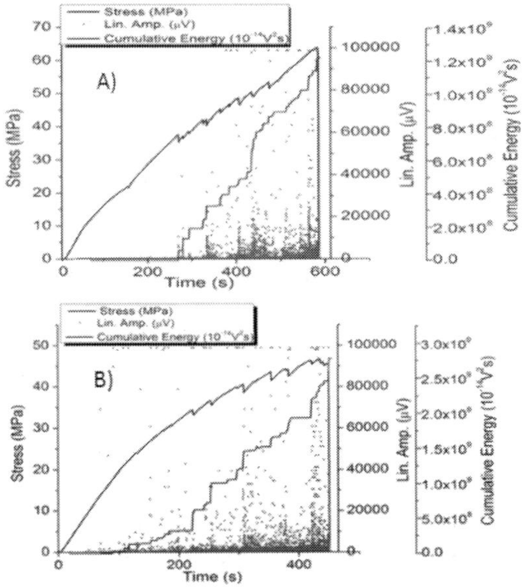

Figure 5. A) Textile oriented at 0°, B) Textile oriented at 90°.

This damage is associated to matrix cracking in the composite. At 90° direction (Fig. 5B), is possible to observe more small discontinuities related with the AE signals identified as matrix cracking. Energy is augmented rapidly as the more cracking damage occurs. In all the samples tested in tension (0°, and 90°), no fiber breakage was observed or AE detected until total fracture happened. Samples at 90° exhibited an early damage development in comparison with 0°. Lower damage initiation at 90° samples apparently was due to the major amount of unidirectional glass fibres at 0° layer (denser) that is positioned perpendicular to the applied stress and allows more transversal matrix cracks to be developed along the coupon.

3 Case 2.- Mechanical properties of geotextiles: effect of architecture

Two geotextiles of different geometry were used to manufacture a polymeric matrix based composite material. First one is made of polyethylene-terephthalate (PET) fibers constituted by unidirectional threads (woven) oriented in longitudinal and transversal directions. Transversal yarns are wider presenting higher amount of filaments than in longitudinal direction. Second one is a polypropylene (PP) based geotextile constituted by a diagonal longitudinal "tape like" and transversal yarns having an oval shape in the intersections (Fig. 6). It is observed that this textile exhibits imperfections in the surface such a microcracks. In order to observe the onset and development of damage to identify the fracture progression, the textiles were placed in epoxy matrix to manufacture a composite material.

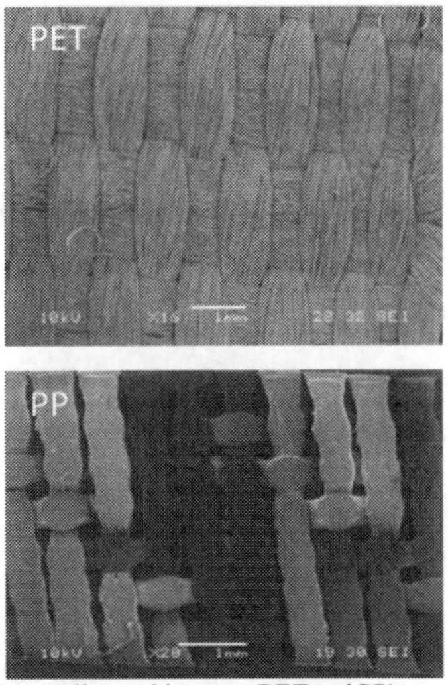

Figure 6. Geotextiles architecture (PET and PP).

Geotextiles were mechanically characterized individually in tension mode in both longitudinal (0°), and transversal direction (90°), in order to compare their mechanical behavior. Composite materials were tensile tested in a Universal Machine Shimadzu using 5 kN load cell at 0.1 mm/min in order to observe the crack progression along the sample. Acoustic Emission sensors were attached to the composite samples in order to monitor damage in "real time "as seen in Fig. 3.

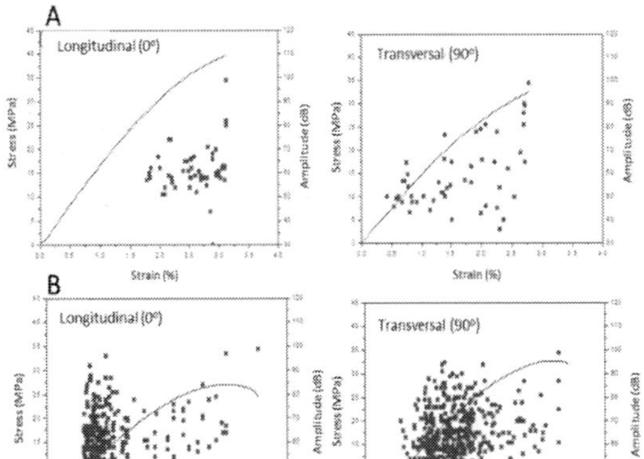

Figure 7. A) Epoxy/PET geotextile composites, B) Epoxy/PP geotextile composites.

Results of mechanical behavior in tensile test coupled to acoustic emission analysis for epoxy/geotextile composites are shown in Fig. 7. Fig. 7A shows the stress-strain behavior of epoxy/PET geotextile composite tested in longitudinal and transversal direction respectively. In longitudinal direction, first EA signals are sensed at about 1.75% strain with high amount of signals with high amplitude (50-80 dB) less disperse. Such signals can be correlated to interfacial damage and matrix crack progression. On the other hand, stress-strain curve when epoxy/PET geotextile is tested in transversal direction,

AE signals were detected significantly sooner compared with the longitudinal sample. Stress-strain curve for epoxy/PP geotextile composite tested in longitudinal and transverse directions are presented in Fig. 7B. In composites tested longitudinally, AE signals showed up at lower strains (0.6%) after linearity but with higher amplitude. Higher cracking density was found between 0.7 and 1.2 that may be related to the extensive crack progression generated by the complex textile geometry. Mechanical tendency observed for epoxy/PP geotextile composites tested in transversal direction is somehow similar to longitudinal samples, however, transversal samples showed higher density of AE signals of significant amplitude. This effect may be related to the fact that this direction presented major stress resistance, therefore it requires more load and consequently, the damage generated in the composite can be larger.

4 Conclusions

In both cases, it has been demonstrated that architecture/geometry of the textiles are an essential factor that influences the mechanical behavior of a composite material and that it is dependent on the loading orientation. Damage cracking initiation, development and progression are governed by the way in which fiber bundles are interweaved. In Case 1, damage cracking development was affected by the polyester thread (zig-zag and unidirectional). It was demonstrated the influence of the textile stitched geometry in the initiation sites and the damage development of the composite. Zig-zag polyester knit yarn was observed to govern the transversal crack development while unidirectional (0°) polyester thread only influenced damage by the stitch knots and when it was oriented transversally to load direction. In the 90° direction, transversal flaws were formed due to fibre/matrix debondings that coalesce at higher strain to generate the significant cracks. These visible cracks have a nearly equidistant spacing and are equally distributed in the +45° and -45° plies. First AE signals detected during tensile tests were identified as damage induced in the material at the fibre/matrix interphase and demonstrated how damage initiation is affected by NCF orientation. In Case 2, mechanical properties were studied in geotextiles made of polyethylene terephthalate (PET) and polypropylene (PP), both textiles with different architecture/geometry of waving. Acoustic emission signals in epoxy/PP geotextile composite exhibited higher density in comparison to epoxy/PET geotextile composite. This effect indicates that epoxy/PP geotextile composite generates more damage events during tensile tests which are intimately related to the complexity of threads architecture/geometry that originates active concentrations sites that release energy when having internal damage. Finally, it was shown that Acoustic Emission technique is a powerful characterization tool for understanding materials mechanical response under any type of external load due to its link between AE data and fracture mechanics, which makes this relation a very useful for identification of damage sequence.

Acknowledgement

The author is grateful to the Mexican Council for Science and Technology (CONACYT) for the provision of grant 60204/ CM0042.

References

1. G. Du, J. Li, W.K. Wang, C. Jiang, S.Z. Song, Corrosion Science, **53**, Issue 9, 2918–2926, (2011).
2. P. Nivesrangsan, J. A. Steel, R.L. Reuben, Mechanical Systems and Signal Processing, **21**, Issue 2, 1103–1114, (2007)
3. G.C. McLaskey, S.D. Glaser, C.U. Grosse, Journal of Sound and Vibration, **329**, no. 12, 2384–2394, (2010)
4. D.G. Aggelis, S. Verbruggen, E. Tsangouri, T. Tysmans, D. Van Hemelrijck, Construction and Building Materials, **47**, 1037–1045, (2013)
5. C.R. Ramirez-Jimenez, N Papadakis, N Reynolds, T.H Gan, P Purnell, M Pharaoh, Composites Science and Technology, **64**, no. 12, 1819–1827 (2004)
6. M. Wevers, NDT & E International, **30**, no. 2, 99-106, (1997)
7. R. Gutkin, C.J. Green, S. Vangrattanachai, S.T. Pinho, P. Robinson, P.T. Curtis, Mechanical Systems and Signal Processing, **25**, Issue 4, 1393–1407, (2011)
8. O.Y. Andreykiv, M.V. Lysak, O.M. Serhiyenko, V. R. Skalsky, Engineering Fracture Mechanics, **68**, Issue 11, 1317–1333, (2001)
9. I.M. De Rosa, C. Santulli, F. Sarasini, Composites Part A: Applied Science and Manufacturing, **40**, no. 9, 1456–1469, (2009)
10. H. Ku, H. Wang, N. Pattarachaiyakoop, M. Trada, Composites Part B: Engineering, **42**, no. 4, 856–873, (2011)
11. Y.H. Yu, J.H. Choi, J.H. Kweon, D.H. Kim, Composite Structures, **75**, Issues 1–4, 163–169, (2006)
12. N. Kosmann, J.M. Karsten, M. Schuett, K. Schulte, B. Fiedler, Composites Part B: Engineering, **70**, 184–188, (2015)
13. L. Lia, S.V. Lomov, X. Yan, Composite Structures, **123**, 45–53, (2015)

MATEC Web of Conferences 95, 07017 (2017) DOI: 10.1051/matecconf/20179507017
ICMME 2016

Design and Analysis of an Abbé Free Coplanar Stage

Tien-Tung Chung , Teng-Hui Tseng, and Ching-Siong Lim Chan

Dept. of Mechanical Engineering, National Taiwan University, Taipei 10617, Taiwan.

Abstract. Design and analysis of a new Abbé free coplanar xy stage are presented in this paper. The xy stage is formed as conventional xy stages which combine with two stacked up linear guides. The x-guide is on the bottom and the y-guide is on top of the x-guide. The travel range of this xy stage is 300mm x 300mm, which fits dimensions of 12-inch wafers. A special mechanism is designed such that the z-surface of the y-guide has the same height as the z-surface of x-guide, and the Abbé error and cumulative error of this coplanar stage can be reduced. The concept of symmetric structure design is also considered to eliminate the structural deformation due to driving force of two guides. For this long travel range precision stage, the finite element method (FEM) is applied to analyze the structural deformation and vibration natural frequencies. The requirement of structural deformation due to self-weight load is limited to 1.5μm, and first natural frequency is limited to over 100 Hz.

1 Introduction

Recently, precision positioning technique is important in the development of science and technology. The positioning stages with high accuracy is required in various fields such as semiconductor manufacturing, precision machining, precision measurements, etc. XY stages with long travel range are normally combined with two stacked up linear stages [1]. The traditional configuration of stacked up type stages has disadvantages of large assembly errors and geometry errors, such as Abbé error. Therefore, stacked up type stages reduce the motion stability and accuracy for positioning. This paper proposes a design of the coplanar XY stage to reduce the accumulation of geometric errors and Abbé error. The coplanar stage developed in this paper contains several advanced design methods, including Abbé free concept and coplanar concept.

2 Coplanar XY stage design

Linear encoder is usually used to measure displacement in direct drive linear stage. If the work piece is put on the moving stage, the work piece and the linear encoder are on the different z-plane [2]. When the moving stage is unstable, the Abbé offset will appear, as shown in Fig. 1. To eliminate the Abbé offset in the direct driver linear stage, laser interferometer is considered, as shown in Fig.2. Reflection mirror, moving stage and laser interferometer locate at the same height. The laser beam will be emitted from the laser interferometer to the reflection mirror and will be reflected back into the receptacle on the laser interferometer, the displacement of the work piece will be measured by the laser interferometer. The laser interferometric measurement is free from Abbé errors of the first order in all the three xyz coordinates [3].

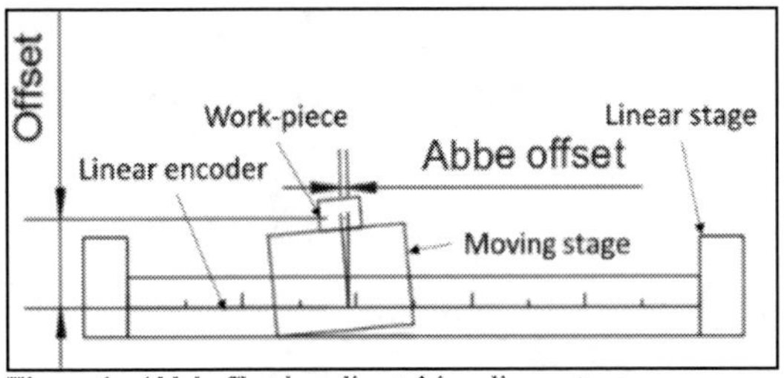

Figure 1. Abbé offset in a direct driver linear stage

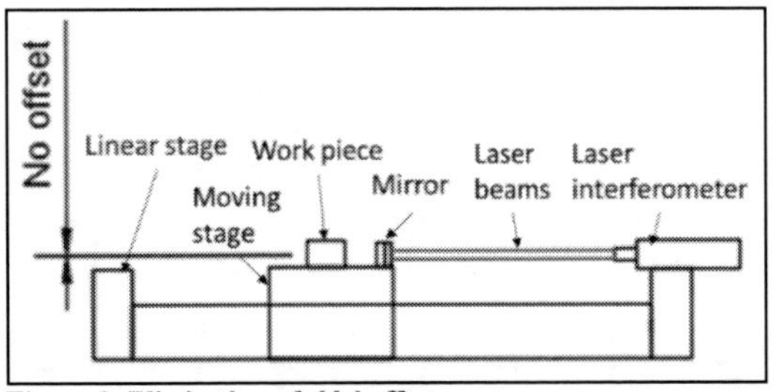

Figure 2. Elimination of abbé offset

Conventional XY stage is stacked up by two linear guides. The Abbé error of the lower guide is high because of the vertical distance from work piece to the bottom of the stage. The concept of coplanar stage is proposed to eliminate the Abbé error. As shown in Fig. 3, point 1 is X-direction laser interferometer, point 2 is X-direction reflection mirror, point 3 is Y-direction laser interferometer and point 4 is Y-direction reflection mirror.

The Abbé error in a two-dimension space can be eliminated when the point 1, point 2, point 3, and point 4 are of the same z-plane heights and the line of point 1 and point 2 intersect the line of point 3 and point 4.

The coplanar stage is proposed in this paper, as shown in Fig. 4. The Y-plate is supported by two precision guideways that are mounted on the X-plate. The X-plate is supported by another two precision guideways that are mounted on the granite. According to the coplanar design and Abbé free design, component 1, component 2, component 3, component 4, component 5, and component 6 have the same z-plane heights. Component 1, component 2 and component 3 are in one straight line in X-direction. Component 4, component 5 and component 6 are in one straight line in Y-direction. The line in X-direction intersects the line in Y-direction, located at the measurement point on the surface of the work piece to be measured [4]. This is a combined concept of coplanar design and Abbé free design.

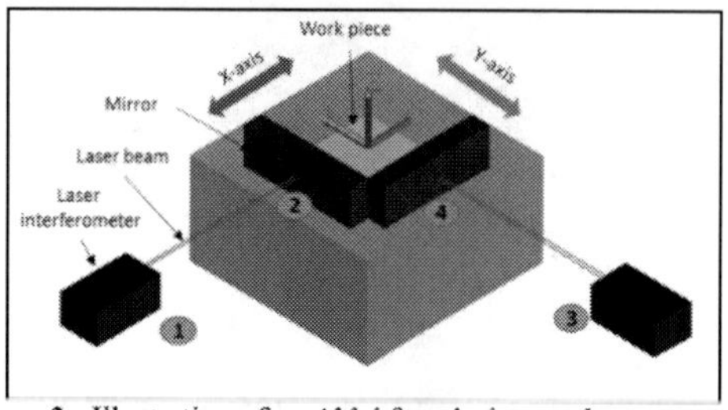

Figure 3. Illustration of an Abbé free design coplanar stage

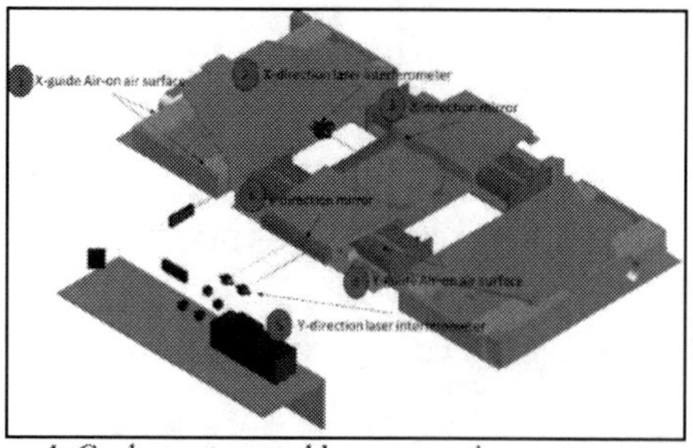

Figuer 4. Coplanar stage and laser measuring system

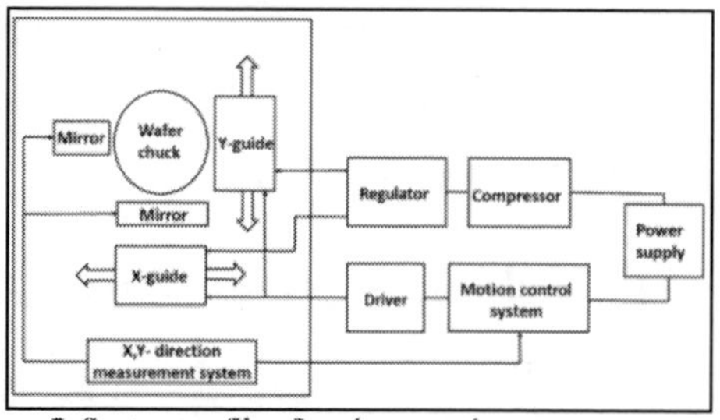

Figure 5. System profile of coplanar motion stage

The overall system configuration of the coplanar motion stage is shown in Fig. 5. The coplanar stage is controlled by the motion control system through the driver connected to a computer. This positioning system is designed for wafer inspection. The exploded view of the coplanar stage is shown in Fig. 6. The Y-stage incorporates a wafer chuck, two mirrors, a Y-plate, and two Y-guides, as shown in Fig. 7. The X-stage incorporates X-plate, two air bearings, and two X-guides, as shown in Fig. 8. Fig. 9 shows the box-in-box concept, which makes the guideways in X-direction and Y-direction are of the same height.As a result, the Abbé error and the cumulative error can be reduced [5] .

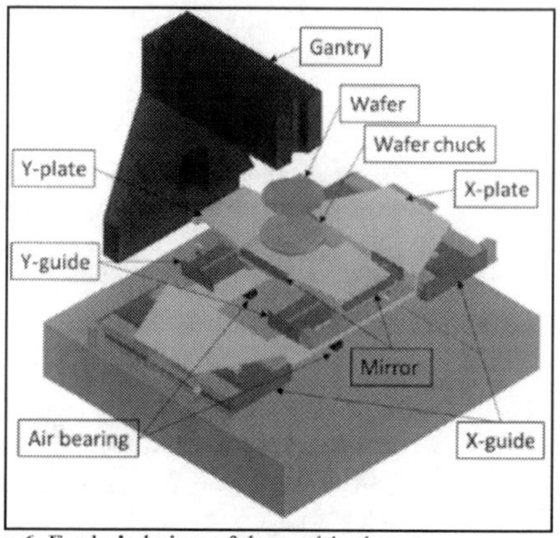

Figure 6. Exploded view of the positioning stage

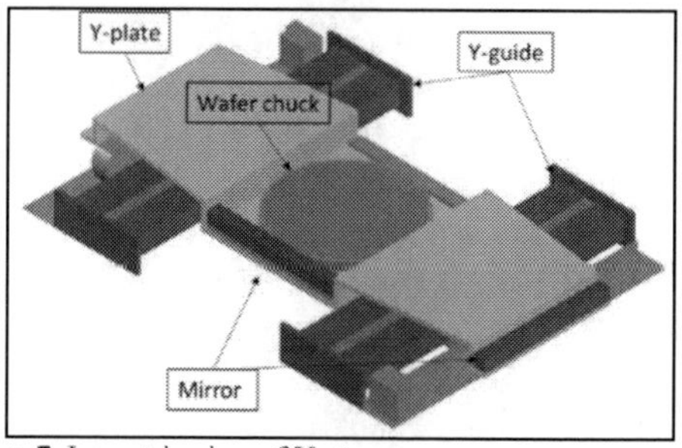

Figure 7. Isometric view of Y-stage

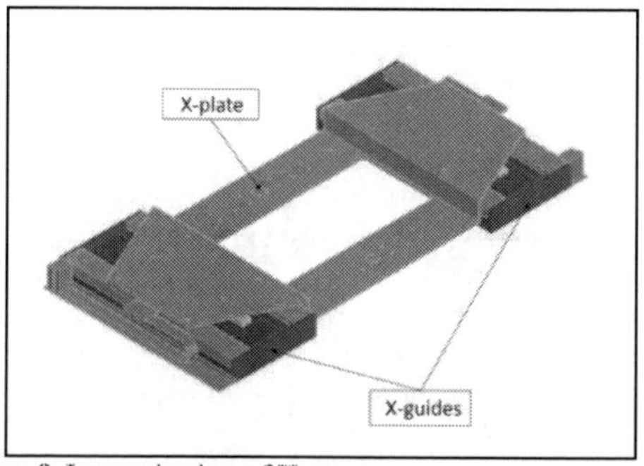

Figure 8. Isometric view of X-stage

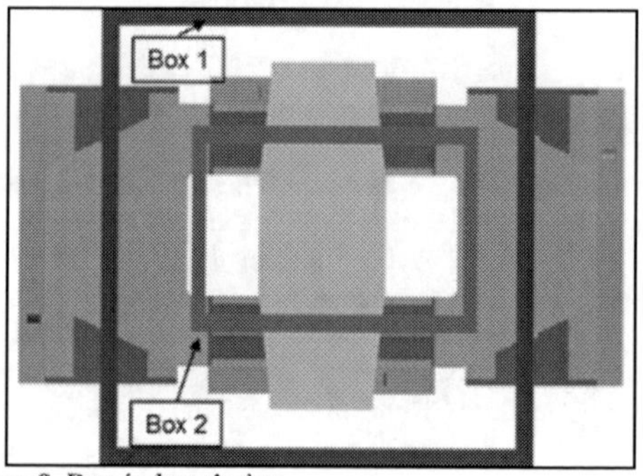

Figure 9. Box-in-box design

3 Finite element analysis

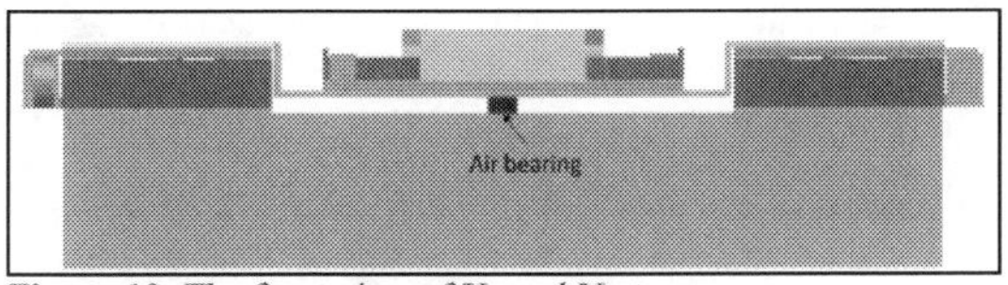

Figure 10. The front view of X- and Y-stage

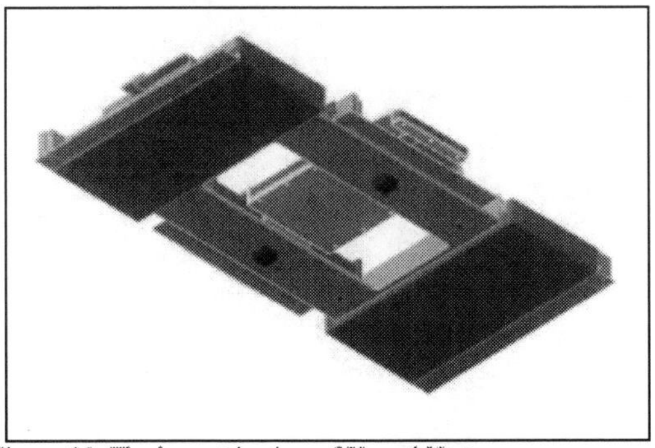

Figure 11. The isometric view of X- and Y-stage

Structure deformation is important in this coplanar stage. Positioning accuracy will be influenced when the deformation is too large. To reduce self-weight deformation, two air bearings are mounted under the X-stage, as shown in Fig. 10. Isometric view of X- and Y-stage with two air bearings are shown in Fig. 11. Then the self-weight deformation of the X- and Y- stages can be analyzed by using finite element analysis software ANSYS. The deformation of the X- and Y- stages is 1.469μm, as shown in Fig. 12. The deformation of the gantry is 1.459μm, as shown in Fig. 13.

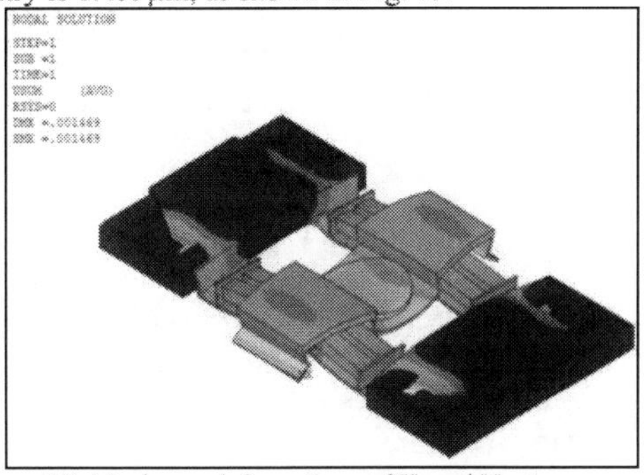

Figure 12. Maximum deformations of X- and Y- stages

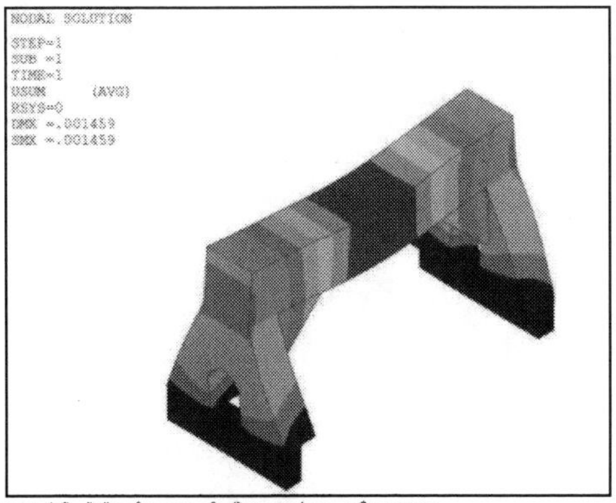

Figure 13. Maximum deformation of gantry

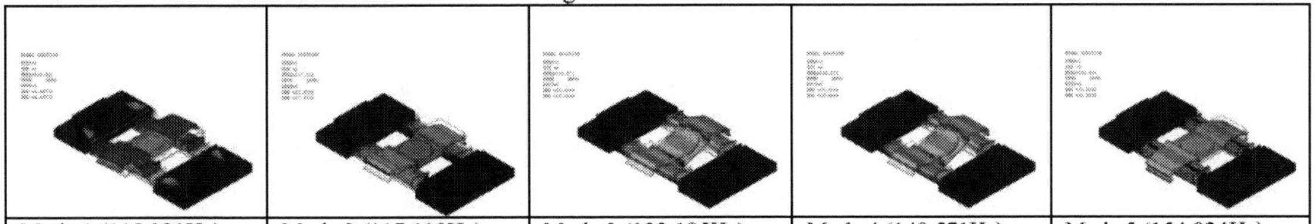

Mode 1 (115.321Hz)	Mode 2 (117.118Hz)	Mode 3 (122.195Hz)	Mode 4 (148.571Hz)	Mode 5 (154.924Hz)

Figure 14. Mode shapes and natural frequencies of X-and Y-plate

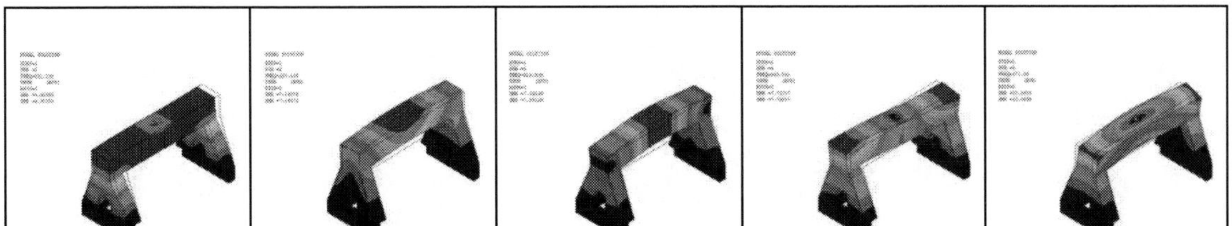

Mode 1 (181.238Hz)	Mode 2 (223.138Hz)	Mode 3 (319.508Hz)	Mode 4 (448.041Hz)	Mode 5(571.86Hz)

Figure 15. Mode shapes and natural frequencies of gantry

The design of the high accuracy positioning stage needs to consider not only the structure deformation but also the natural frequency of the stage structure. The coplanar stage must have the operating frequency far from its structure natural frequency in order to prevent the resonance of the structure, which will influence the accuracy of the stage seriously. Hence, it is necessary to analyze the natural frequency and mode shape of coplanar stages and gantry. For a high accuracy requirement, the operating frequency of the controlling system should be different from the structure operating frequency.

The results of the first five modes and mode shapes from 1st to 5th of X- and Y- stages are shown in Fig. 14. The 1th mode is a translation motion in X-direction, the 2th mode is rotation motion in X-direction, the 3th mode is translation motion in Z-direction, the 4th mode is rotation motion about X-direction, and the 5th mode is

DOI: 10.1051/matecconf/20179507017

translation motion about Z-direction. From the Fig. 14, the frequencies of first five modes are between 100Hz and 160Hz. The natural frequency will be changed due to different mechanism design. To avoid resonance, the mechanism design must be altered as the frequencies of the coplanar motion stage are measured.

The results of the first five modes and mode shape from 1st to 5th of gantry are shown in Fig. 15. The 1st mode is a translation motion in Y-direction, the 2nd mode is translation motion in X-direction, the 3rd mode is translation motion in Z-direction, the 4th mode is rotation motion about Z-direction, and the 5th mode is rotation motion about X-direction.

4 Summary

This paper proposed a new coplanar XY stage used for wafer inspection. This coplanar stage with special mechanism design is Abbé free in both X- and Y-directions. The travel range of this coplanar motion stage is 300mm x 300mm, which makes it possible for a 12-inch wafer inspection. The self-weight analysis and modal analysis are carried out by using finite element analysis software ANSYS. The structural characteristic of this coplanar stage satisfies the design requirements of given design specifications.

References

1. Information on http://www.hiwin.com/pdf/linear_motors.pdf.
2. Osita D. I. Nwokah, and Yildirim Hurmuzlu, CRC press LLC, pp.160-162, (2001).
3. Gunter Wilkening and Ludger Koenders, *Nanoscale Calibration Standards and Methods: Dimensional and Related Measurements in the Micro- and Nanometer Range*, Wiley-VCH, pp. 47-59, (2005).
4. Bryan, J.B., Precision Engineering, pp.129-132,(1979).
5. W. L. Wang, Y. T. Fei, and K. C. Fan, *Proceedings of the 1st IEEE International Conference on Nano/Micro Engineered and Molecular Systems*, Zhuhai, China, pp. 18–21, (2006).

Determination of the Number of Fixture Locating Points for Sheet Metal By Grey Model

Bo Yang , Zhongqi Wang, Yuan Yang, Zewang Jing and Yonggang Kang

The Ministry of Education Key Laboratory of Contemporary Design and Integrated Manufacturing Technology, Northwestern Polytechnical University, No. 127, Youyi Road (West), Xi'an 710072, China

Abstract. In the process of the traditional fixture design for sheet metal part based on the "N-2-1" locating principle, the number of fixture locating points is determined by trial and error or the experience of the designer. To that end, a new design method based on grey theory is proposed to determine the number of sheet metal fixture locating points in this paper. Firstly, the training sample set is generated by Latin hypercube sampling (LHS) and finite element analysis (FEA). Secondly, the GM(1, 1) grey model is constructed based on the established training sample set to approximate the mapping relationship between the number of fixture locating points and the concerned sheet metal maximum deformation. Thirdly, the final number of fixture locating points for sheet metal can be inversely calculated under the allowable maximum deformation. Finally, a sheet metal case is conducted and the results indicate that the proposed approach is effective and efficient in determining the number of fixture locating points for sheet metal.

1 Introduction

In order to not only determine the spatial location but also constrain the excessive deformation of sheet metal part, Cai et al. proposed the "N-2-1" (N>3) locating principle for sheet metal fixture design and at the same time pointed out that the "N-2-1" locating principle was more suitable for sheet metal part than "3-2-1" principle [1]. Hence, one of the keys is to find the minimum number of locators their optimal positions to keep the sheet metal part under the allowable maximum deformation and reduce the maximum sheet metal deformation as greatly as possible in fixture design based on the "N-2-1" locating principle.

To deal with the problem above, many scholars and technicians carried out lots of research. Krishnakumar et al. [2] combined genetic algorithm (GA) with finite element analysis (FEA) to search the optimum fixture layout to minimize the part deformation. Liu et al. [3] determined the initial number and positions of the locators by adding locators on the datum plane at the position with the maximum deformation repeatedly until the deformation was reduced within the range of milling accuracy. Prabhaharan et al. [4, 5] presented the fixture layout optimization methods that used GA and ant colony algorithm (ACA) separately combined with FEA to reduce the dimensional and form errors of the deformable workpiece. Through employing FEA to compute the workpiece deformation for a given fixture layout, Dou et al. [6] conducted the applications and comparisons of GA, improved GA, particle swarm optimization (PSO) and improved PSO for the fixture layout locating optimization to minimize the elastic deformation of the workpiece. Xing et al. [7] proposed a new method to optimize fixture layout by a non-domination sorting social radiation algorithm (NSSRA) combined with FEA to satisfy the requirements of the assembly tolerance. What's more, in order to reduce the computational cost and improve the optimization efficiency, approximation or surrogate models [8] such as response surface methodology (RSM), back propagation neural network (BPNN) and radial basis function neural network (RBFNN) have been introduced and applied to computationally expensive fixture layout optimization for controlling the undesirable deformation of the flexible or deformable workpiece and improving the manufacturing accuracy in recent years [9~11]. However, most studies only optimized the sheet metal fixture locating layout for a given number of fixture locating points and few research have been carried out on the design of the number of fixture locating points.

The grey system theory was first proposed and established by Julong Deng [12, 13] in 1982 to study of the problems involving small samples and poor information. This theory states that for a grey system just some information is known and some is uncertain or fully unknown. However, there is a necessary intrinsic relation between them, through generating and exploiting the limited known information, the system behavior can be described and controlled. At present, grey system theory has been applied to predict and analyze the practical problems in different fields. Wang et al. [14] presented a systematic methodology for the thermal-error correction of a machine tool and used the grey system model to

DOI: 10.1051/matecconf/20179507018

predict and compensate for the thermal error. Li et al. [15] studied the optimization of the thermal sensors' placement on machine tools based on grey system theory and reduced the temperature variables in the thermal error model significantly.

In this paper, a new approach based on grey system theory is presented to obtain the final number of fixture locating points for sheet metal part. With this section included, this paper is made up of six sections. The remainder of the paper is organized as follows: in section 2, the design model of the number of fixture locating points for sheet metal part is constructed. In section 3, the bases of the GM (1, 1) grey model are introduced. Section 4 depicts the flowchart of the proposed method. In section 5, a sheet metal case study is presented to verify the effectiveness and efficiency of the method. Finally, this paper is concluded in section 6.

2 Problem formulation

In order supply more reinforcements to avoid the excessive deformation of the workpiece at machining, assembly and measuring stages throughout the whole manufacturing process, sheet metal part is often kept under an over-constraint condition based on the so-called "N-2-1" locating principle. The principle holds that there are "N" (N>3) locating points on the first datum plane of sheet metal part, and "2" and "1" on the second and third datum plane respectively. Figure 1 illustrates a typical "4-2-1" principle, where 4 locators are required in order to support sheet metal on the primary datum plane to avoid the excessive deflection of the part. Apparently, the determination of the locator number "N", which is always more than three, is one of the main tasks to minimize the sheet metal part deformation in computer aided fixture design.

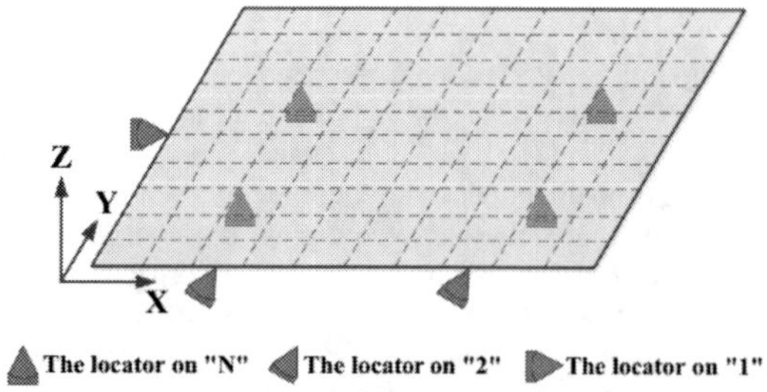

Figure 1. The "N-2-1" locating principle.

The design model for the number of fixture locating points for sheet metal part can be formulated as:

$$\begin{cases} \text{Find: } N \text{ for } L = [l_1, l_2, \cdots, l_N] \\ F = \text{Max}\{\varepsilon_i\}, i = 1, 2, \cdots, \text{K} \\ \text{s.t. Minimum } F \le \delta \end{cases} \tag{1}$$

where N is the design variable standing for the number of fixture locating points; $L = [l_1, l_2, \cdots, l_N]$ is the N fixture locating points; ε_i is the normal deflection of the i-th mesh node; K is the total number of the mesh nodes of the finite element model of sheet metal part; δ is the allowable maximum deformation in engineering.

3 Grey prediction model

Grey means uncertain. The aim of grey system theory is to provide theory, techniques and methods for resolving (analyzing) latent and intricate grey systems which lack concerned information. As one of the most popular grey models, GM (1, 1) model has been studied deeply and applied in many fields since first proposed by Deng [12].

Let $X^{(0)}$ denote the original series: $X^{(0)} = \{x^{(0)}(1), x^{(0)}(2), \cdots, x^{(0)}(n)\}$, where $x^{(0)}(i)$, $(i = 1, 2, \cdots, n)$ is non-negative.

In order to reduce the noise and eliminate the disorderliness of the raw series data, the original data is converted into a monotonically increasing data series through accumulated generating operation (AGO):

$$X^{(1)} = \{x^{(1)}(1), x^{(1)}(2), \cdots, x^{(1)}(n)\} \tag{2}$$

where $x^{(1)}(k) = \sum_{i=1}^{k} x^{(0)}(i)$, $(k = 1, 2, \cdots, n)$.

The new series $X^{(1)}$ is monotonic increase sequence and follows an approximate exponential law supposing the original series is smooth enough. The mean sequence can be generated by:

$$z^{(1)}(k) = \frac{x^{(1)}(k) + x^{(1)}(k-1)}{2}, k = 2, 3, 4 \cdots \tag{3}$$

The definition type of GM (1, 1) model is given as:

$$x^{(0)}(k) + az^{(1)}(k) = b \tag{4}$$

where a is called the development coefficient and represents the development state of the prediction value, b is called the grey action quantity and represents change contained in the data. To build the GM (1, 1) grey model, only a few data are needed to distinguish it. What's more, to estimate the coefficients a and b, least square method is adopted here.

Let's denote $B = \begin{bmatrix} -z^{(1)}(2) & 1 \\ -z^{(1)}(3) & 1 \\ \vdots & \vdots \\ -z^{(1)}(n) & 1 \end{bmatrix}$, $Y = \begin{bmatrix} x^{(0)}(2) \\ x^{(0)}(3) \\ \vdots \\ x^{(0)}(n) \end{bmatrix}$,

$u = \begin{bmatrix} a \\ b \end{bmatrix}$, then based on the principle of least square method, we have:

$$\tilde{u} = \left(B^T B\right)^{-1} B^T Y \tag{5}$$

Therefore, using a and b, we can establish the following response equation of GM (1, 1) for future prediction:

$$\tilde{x}^{(1)}(k+1)=\left(x^{(0)}(1)-\frac{b}{a}\right)e^{-ak}+\frac{b}{a}, k=1,2,\cdots,n-1 \quad (6)$$

Thus the simulative value of $X^{(0)}$ can be calculated through

$$\tilde{x}^{(0)}(k+1)=\tilde{x}^{(1)}(k+1)-\tilde{x}^{(1)}(k) \quad (7)$$

At last, the Q test of relative residual should be conducted on the grey model. The smaller the relative error Q is, the more accurate the prediction model is. The relative error is defined as:

$$Q=\frac{1}{n}\sum_{k=1}^{n}\Delta_k \quad (8)$$

where n is the total number of the original data, and $\Delta_k=\frac{\tilde{x}^{(0)}(k)-x^{(0)}(k)}{x^{(0)}(k)}$ is the relative error of the k-th original data.

4 Method

In this paper, a new approach to determining the number of sheet metal fixture locating points based on the "N-2-1" locating principle by grey model is proposed. The flow chart shown in Fig. 2 explains the procedure of the proposed method.

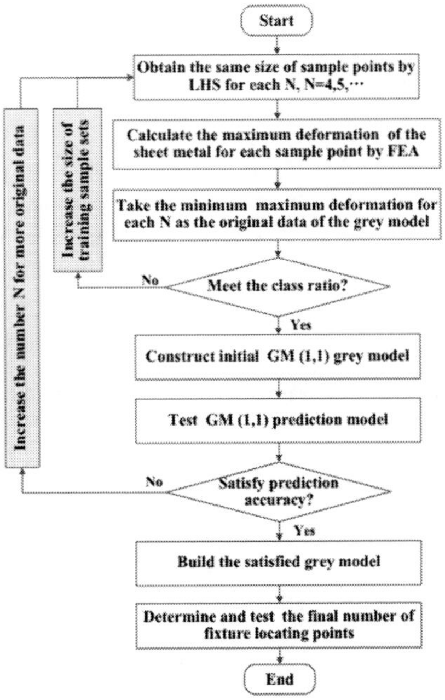

Figure 2. The flowchart of the determination of the number of fixture locating points for sheet metal by grey model.

The locator number "N" is treated as the design variable while the minimum maximum deformation of sheet metal under its dead weight is treated as the evaluation function. The finite element model of sheet metal locating is established to compute the evaluation function value, the training data set is constructed through Latin hypercube sampling (LHS) [16] and FEA and the established GM (1, 1) model is applied to predict the number of sheet metal fixture locating points. Taking the allowable maximum deformation as the restrictive condition, the final design variable can be inversely calculated for the appropriate maximum deformation.

5 Case study

In this section, the determination of the number of fixture locating points for sheet metal by grey model is demonstrated by an aluminium alloy case. The fixture locating scheme of the case based on the "N-2-1" locating principle is shown in Figure 3. The sheet metal has dimensions of 100 mm×100 mm×1 mm and the physical material properties are listed in Table 1. The "N" locating points on the first datum plane are l_1, l_2, ···, l_N, and the "2" locating points on the second datum plane are Fl_1 and Fl_2 while the "1" locating point on the third datum plane is $Fl3$. The coordinates of the fixed locating points Fl_1, Fl_2 and Fl_3 are set as (33, 0), (67, 0) and (0, 50). The locating points to be optimized are l_1, l_2, ···, l_N and their corresponding position coordinates are denoted by (x_1, y_1), (x_2, y_2), ···, (x_N, y_N) respectively. In the finite element model, the grid size is set as 1mm×1mm and the mesh type is S4R. The allowable maximum deformation is presumably set as δ=0.0005 mm.

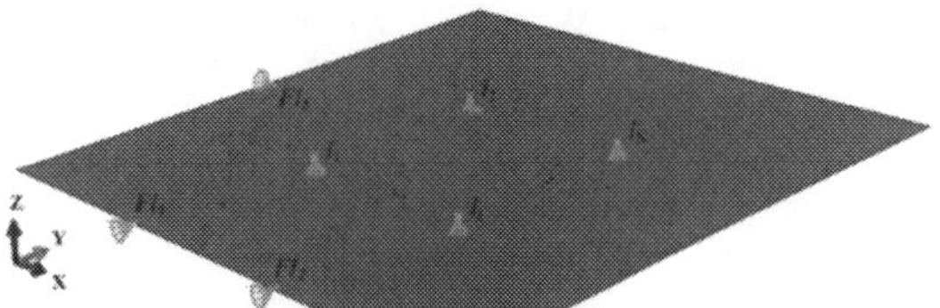

Figure 3. The initial fixture locating scheme of the aluminum alloy sheet metal part.

Table 1. The physical properties of material.

	Value
Mass density	2.8×10^{-3} g/mm^3
Young's modulus	7.3×10^4 MPa
Poisson ratio	0.33

The training data set is generated by LHS and FEA. In order to calculate the evaluation values of the training data set, the commercial finite element software ABAQUS^{TM} [17] is employed to compute the maximum deformation of the sheet metal part under its deadweight because it is always hard or even impossible to derive the explicit and closed-form function expression between the fixture locating layout and the concerned sheet metal deformation. In our implementation, after several trials, the original data set for the grey model, as shown in

Table 2, is obtained from eight different training data sets of the same size of 20 selected by LHS.

Table 2. The original data for GM (1, 1).

N	4	5	6	7
F (mm)	0.001788 41	0.001595 15	0.0013054 1	0.0011525 7
N	8	9	10	11
F (mm)	0.001046 66	0.000901 08	0.0008331 11	0.0006895 32

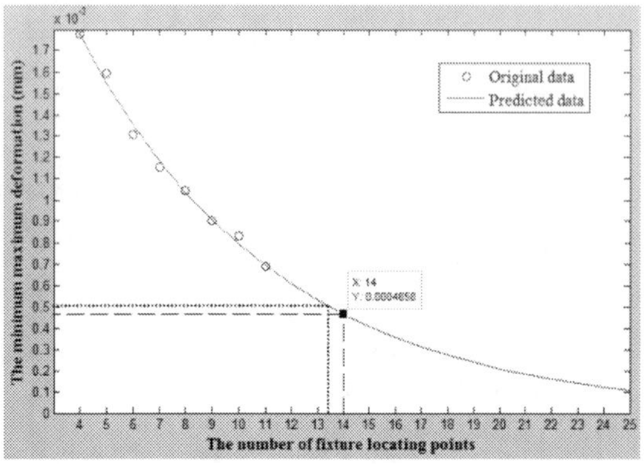

Figure 4. Prediction curve for the number of fixture locating points by grey model

In summary, referring to the flowchart of the determination of the number of fixture locating points for sheet metal by grey model, the GM (1, 1) grey model is established on the platform of MATLAB™ [18, 19]. Therefore, the prediction curve describing the mapping relationship between the number of fixture locating points and the concerned sheet metal maximum deformation can be established, as shown in Figure 4. The predicted data of the original data is listed in Table 3 and the tested data is listed in Table 4. What's more, the Q test of relative residuals is 0.15%, and its accuracy reaches the first grade (good).

Table 3. The predicted data of the original data.

N	4	5	6	7
F (mm)	0.001778	0.0015452 6	0.0013525	0.0011837 8
N	8	9	10	11
F (mm)	0.001036 1	0.0009068 5	0.0007937 2	0.0006947 1

Table 4. The predicted data for testing.

N	15	20	25
F by grey model (mm)	0.0004077	0.00020941	0.00010756
F by FEA (mm)	0.000416886	0.00020959	0.000107342
Relative error	2.20%	0.09%	0.20%

Referring to the prediction curve, the final number of fixture locating points and the corresponding maximum deformation by GM (1, 1) are shown in Table 5. For further comparative analysis, the maximum deformation of the sheet metal part with the final number of fixture locating points of 14 is also calculated by FEA. The results are also listed in Table 5. It can be seen from the table that the result obtained by the proposed approach shows fine agreements with that by FEA and that the final relative error (0.21%) is within the precision requirement (5%).

Table 5. The relative errors of the prediction model.

F by grey model (mm)	F by FEA (mm)	The relative error
0.000466	0.000467	0.21%

6 Conclusion

In order to inversely determine of the minimum number of fixture locating points for sheet metal part under the allowable maximum deformation, this paper establishes a GM (1, 1) based grey prediction model to describe the mapping relationship between the number of fixture locating points and the corresponding sheet metal maximum deformation. The major contributions of this paper include:

(1) A prediction method based on GM (1, 1) grey model for the number of sheet metal fixture locating points is developed, and the prediction precision meets the general requirement at the industry level.

(2) After the determination of the final number of sheet metal fixture locating points, the subsequent sheet metal fixture locating layout design and optimization can be conducted. In addition, the approach by grey prediction model can be further extended and applied to the fixture design subject to more complex loading conditions.

Acknowledgements

This work has been supported by National Natural Science Foundation of China (Grant No. 51375396).

References

1. W. Cai, S. J. Hu, J. X. Yuan, Deformable Sheet Metal Fixturing: Principles, Algorithms, and Simulations. ASME J. Manuf. Sci. Eng., **118**, 3:318-324 (1996)
2. K. Krishnakumar, S. N. Melkote, Machining fixture layout optimization using the genetic algorithm. Int. J. Mach. Tools Manuf., **40**, 4:579-598 (2000)
3. S. G. Liu, L. Zheng, Z. H. Zhang, Z. Z. Li, D. C. Liu, Optimization of the number and positions of fixture locators in the peripheral milling of a low-rigidity workpiece. Int. J. Adv. Manuf. Technol., **33**, 7-8: 668-676 (2007)
4. G. Prabhaharan, K. P. Padmanaban, R. Krishnakumar, Machining fixture layout

optimization using FEM and evolutionary techniques. Int. J. Adv. Manuf. Technol., **32**, 11-12:1090-1103 (2007)
5. K. P. Padmanaban, K. P. Arulshri, G. Prabhakaran, Machining fixture layout design using ant colony algorithm based continuous optimization method. Int. J. Adv. Manuf. Technol., **45**, 9:922-934 (2009)
6. J. P. Dou, X. S. Wang, L. Wang, Machining fixture layout optimisation under dynamic conditions based on evolutionary techniques. Int. J. Prod. Res., **50**, 15:4294-4315 (2012)
7. Y. F. Xing, M. Hu, H. Zeng, Y. S. Wang, Fixture layout optimisation based on a non-domination sorting social radiation algorithm for auto-body parts. Int. J. Prod. Res., **53**, 11:3475-3490 (2015)
8. T. W. Simpson, A. J. Booker, D. Ghosh, A. A. Giunta, P. N. Koch, R. J. Yang, Approximation methods in multidisciplinary analysis and optimization: a panel discussion. Struct. Multidiscip. O., **27**, 5:302-313 (2004)
9. M. Hamedi, Intelligent fixture design through a hybrid system of artificial neural network and genetic algorithm. Artif. Intell. Rev., **23**, 3: 295-311 (2005)
10. C. Lu, H. W. Zhao, Fixture layout optimization for deformable sheet metal workpiece. Int. J. Adv. Manuf. Technol., **78**, 1-4: 85-98 (2015)
11. Z. Q. Wang, B. Yang, Y. G. Kang, Y. Yang, Development of a prediction model based on RBF neural network for sheet metal fixture locating layout design and optimization, Computational Intelligence and Neuroscience, **2016**: 1-6 (2016)
12. J. L. Deng, Introduction to grey system theory. J. Grey Syst., **1**, 1: 1-24 (1989)
13. J. L. Deng, *Grey system theory*, Huazhong University of Science and Technology Press, Wuhan, (2002) (in Chinese)
14. Y. Wang, G. Zhang, K. S. Moon, J. W. Sutherland, Compensation for the thermal error of a multi-axis machining center. J. Mater. Process. Tech, **75**, 1: 45-53 (1998)
15. Y. X. Li, J. G. Yang, T. Gelvis, Y. Y. Li, Optimization of measuring points for machine tool thermal error based on grey system theory. Int. J. Adv. Manuf. Technol., **35**, 7-8: 745-750 (2008)
16. M. D. Mckay, R. J. Beckman, W. J. Conover, A comparison of three methods for selecting values of input variables in the analysis of output form a computer code. Technometrics, **42**, 1: 55-61 (2000)
17. D. S. Simulia, Abaqus 6.12 documentation, Providence, Rhode Island, US, (2012)
18. J. W. Zhuo, *Application of MATLAB Software in Mathematical Modeling*, Beihang University PRESS, Beijing, (2011) (in Chinese)
19. Matlab 8.0, http://www.mathworks.com/, (2012)

 DOI: 10.1051/matecconf/20179507019

The Effect of Orifice Diameter to the Acoustic Signals Captured at the Cold Part of a Ranque-Hilsch Vortex Tube

Wirachman Wisnoe[1] , Khairil Muhaimin Abd Rahman[1], Yusman Istihat[2] and Valliyappan David Natarajan[1]

[1]*Faculty of Mechanical Engineering, Universiti Teknologi MARA (UiTM), 40450 Shah Alam, Malaysia*

[2]*Faculty of Engineering, Universiti Selangor (UNISEL), 45600 Bestari Jaya, Malaysia*

Abstract. In this study, acoustic analysis and thermofluid performance of a Ranque-Hilsch Vortex Tube (RHVT) is experimentally investigated under different orifice diameters at its cold tube. The orifice diameters used are 2mm, 3 mm, 4 mm, 5 mm and 6 mm. The inlet pressure (gage) is set at 10 psi, 15 psi, 20 psi and 25 psi for each orifice diameter. The sound produced by the tube is recorded using a microphone located outside the cold tube. The acoustic signal is processed using Fast-Fourier Transform (FFT) to obtain the frequency representation. Main frequencies are then extracted to constitute the signature of the signal for that specific configuration. It is observed that different orifice diameters give different signatures. These signatures are then associated with the thermofluid performance of the device to obtain the relation among the parameters.

1 Introduction

Ranque-Hilsch Vortex Tube (RHVT) (see figure 1) is an industrial device that is able to produce hot and cold temperatures of air flow from compressed air. Compressed air supplied at the inlet of the device enters several nozzles at a swirl generator to create swirl motion. This swirl flow moves close to the tube wall towards the hot outlet and exits through the conical valve. The centre part of the flow in the tube is bounced back creating a secondary flow towards the orifice and exits through the cold outlet of the device. The flow of air leaving the conical valve has a higher temperature relative to the inlet temperature, while the flow leaving the orifice has a lower temperature relative to the inlet temperature.

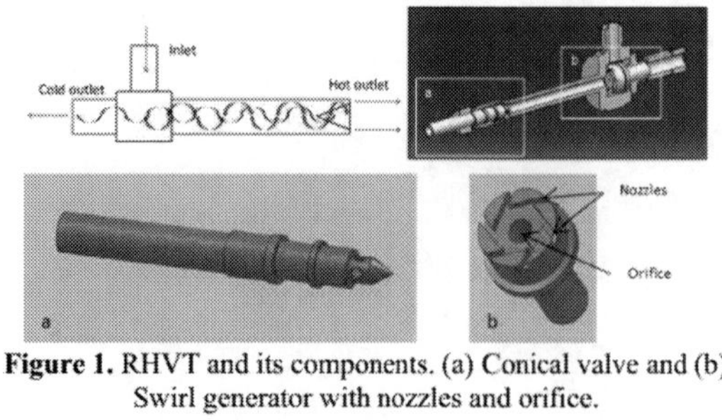

Figure 1. RHVT and its components. (a) Conical valve and (b) Swirl generator with nozzles and orifice.

Despite its simplicity of mechanism, the phenomenon is not well understood. Throughout the years, many researches have been done in order to explain the phenomena. Universiti Teknologi MARA (UiTM) Malaysia has started the research on RHVT since 2011. The study covered experimental, CFD simulation, visualization and acoustic analysis. The study conducted by Ismail et al. [1] revealed the influence of parameters to the overall performance of an RHVT. They conducted experiments by varying the inlet pressure, nozzle depth and orifice diameter. They found that the increase of inlet pressure and nozzle depth would improve the performance of the device. However, the performance of the device was not in linear relation with the orifice diameter. This experimental study is also done by other researchers such as Thakera [2] and Nimbalkar [3]. The visualisation of the device was done by Istihat et al. [4] using a transparent tube with bigger scale from the actual RHVT. Through this visualisation they could observe the swirl motion inside the tube. Wisnoe et al. [5] conducted experiments to associate the effect of different nozzles depth to the acoustic signals produced by the RHVT, as also suggested by Kurosaka [6] about the acoustic streaming of the device. They observed that increasing the inlet pressure did not displace the signal frequencies significantly when the same swirl generator was used within the range of inlet pressures studied. The magnitude of the frequencies, however, changed significantly when different swirl generator was used. Different swirl generator produced different set of frequencies. The set of frequencies-magnitudes represent unique signature of the RHVT for that specific configuration and performance of the RHVT. Abd Rahman et al. [7] further investigated the acoustic signatures produced by the RHVT by studying the effect of orifice diameter to the acoustic signature at the hot tube. It was observed that the acoustic signature remained unchanged when different orifice diameters were used. In other study, Istihat et al. [8] presented the wavelet transform of the acoustic signals recorded at the outlet of cold and hot tubes. The wavelet analysis allowed obtaining the time-frequency representation of the signals.

This paper is focusing on the effect of the orifice diameter to the acoustic signal produced at the cold side of the tube and to its thermofluid performances.

2 Methodology

Fig. 2 shows the schematic diagram of the experiment. This experiment setting is to ensure the thermofluid data (mass flow rate, pressure and temperature) at the inlet and both outlets are measured properly and the sound produced at the cold tube of the RHVT is captured. An "anechoic" box (acoustic box) was used to isolate the sound produced by the tube from the surrounding noise. The sound from tube was captured by a B&K microphone connected to a multichannel spectrum analyzer.

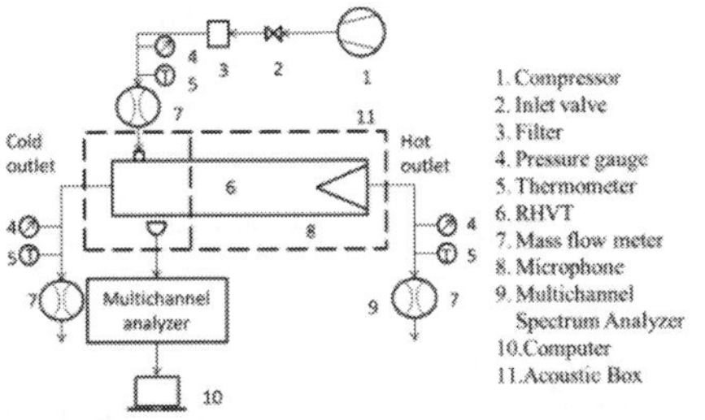

Figure 2. Schematic of the experimental setup

The air pressure is controlled by a valve placed before the inlet. The pressures were set at 10 psi, 15 psi, 20 psi and 25 psi. This experiment was held in a thermodynamic laboratory at around 101kPa of atmospheric pressure.

The "anechoic" box is divided into two chambers to separate the hot tube and the cold tube. This is to ensure that only the sound produced on each side is recorded.

Fig. 3 shows the swirl generator used in this experiment. Four orifice diameters were applied: 2 mm, 3 mm, 4 mm, 5 mm and 6 mm.

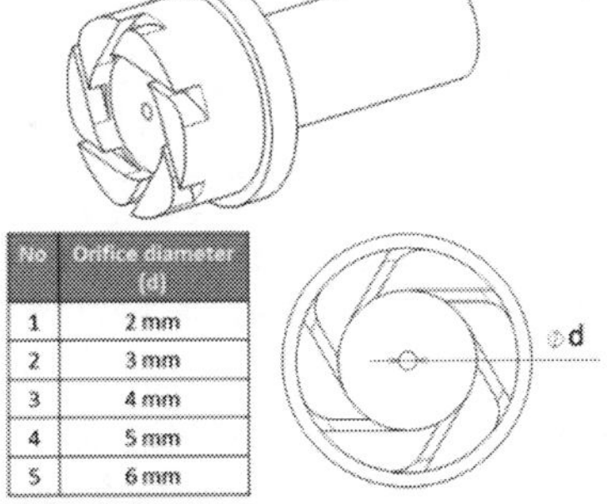

Figure 3. Drawing of the swirl generator

ΔT_c is the temperature difference between the air at the inlet (T_i) and the cold air exiting from the cold tube (T_c) (Eq. 1). This difference of temperatures represents the cooling capability of the device.

$$\Delta T_c = T_i - T_c \tag{1}$$

Mass fraction is the ratio between the outlet mass flow rate ($\dot{m}_c$) and the inlet mass flow rate ($\dot{m}_i$) (Eq. 2). It represents the portion of air flowing through the cold outlet.

$$\mu_c = \frac{\dot{m}_c}{\dot{m}_i} \tag{2}$$

Isentropic efficiency (η_{isen}) is calculated using the principle of adiabatic expansion of an ideal gas (Eq. 3), where γ is the specific heat ratio (for air $\gamma = 1.4$), P_a and P_i represent the atmospheric pressure and the inlet pressure of the air respectively.

$$\eta_{isen} = \frac{T_i - T_c}{T_i\left(1 - \left(\frac{P_a}{P_i}\right)^{\frac{\gamma-1}{\gamma}}\right)} \tag{3}$$

The recorded sounds were processed using Fast-Fourier Transform (FFT) to obtain the frequency spectrum of the signals. Main frequencies are then extracted and replaced by bars representing the frequency location and magnitude.

3 Result And Discussion

3.1 Thermofluid analysis

Fig. 4 shows the effect of orifice diameter to the isentropic efficiency RHVT. The isentropic efficiency gradually increases from orifice diameter of 2 mm to 4 mm then it decreases at higher diameters. Among five orifice diameters used, the diameter of 4 cm produces maximum efficiency (around 16% - 18%) for all four pressures applied at the inlet. The highest isentropic efficiency achieved is 17.88% for orifice diameter 4 mm at 15 psi of inlet pressure. It is also observed that at 2 mm orifice diameter, the value of efficiency is almost the same for all inlet pressures.

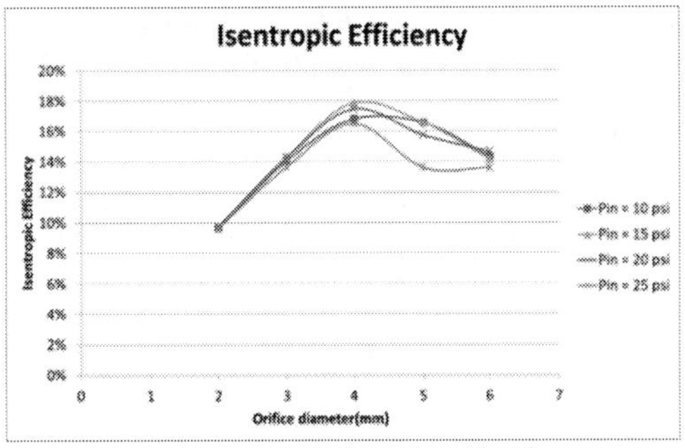

Figure 4. Isentropic efficiency vs orifice diameter.

Fig. 5 shows the variation of difference of temperature between the inlet and the cold outlet when the orifice diameter is varied. Four gage pressures were applied at

DOI: 10.1051/matecconf/20179507019

the inlet. All curves give almost the same trend, it climbs from orifice diameter 2 mm to 4 mm then it descends at bigger diameters. Higher inlet pressure gives higher temperature difference for all diameters. The maximum temperature difference is obtained when the orifice diameter equals 4 mm (12°C at 25 psi inlet pressure).

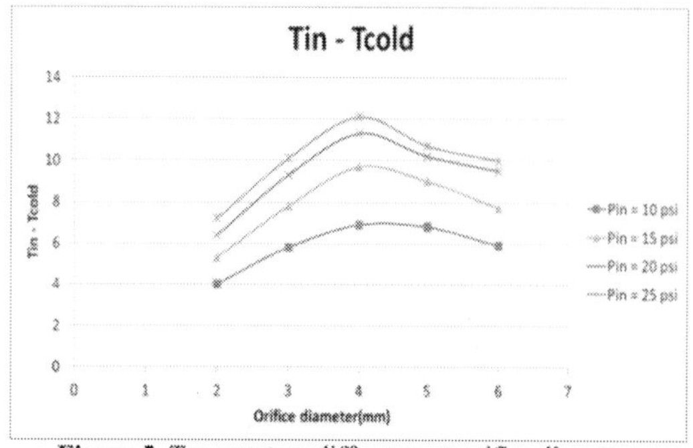

Figure 5. Temperature difference vs orifice diameter.

Fig. 6 represents the variation of cold mass fraction when the inlet pressure is varied at different orifice diameter. The results show that the cold mass fraction is almost constant at each orifice diameter. The mass fraction is higher when the orifice diameter is bigger, as it allows larger amount of air passing through it. The average cold mass fractions produced are 0.28, 0.43, 0.53, 0.58 and 0.61 at orifice diameters of 2 mm, 3 mm, 4 mm, 5 mm and 6 mm respectively.

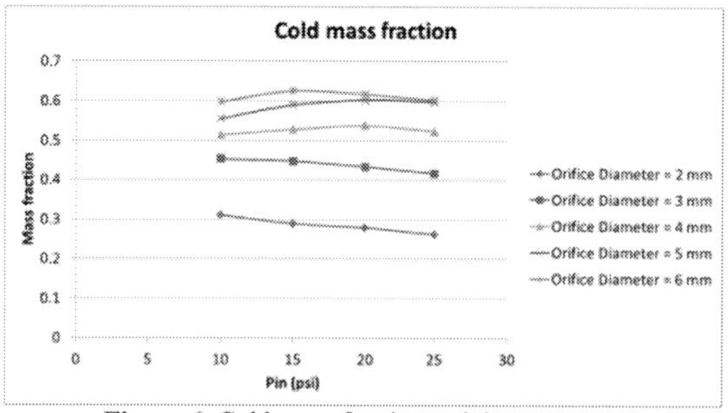

Figure 6. Cold mass fraction vs inlet pressure.

3.2 Acoustic analysis

Fig. 7 represents the sound signals produced at the cold tube. The sound is recorded for 80 ms using 2 mm orifice diameter (1st column), 3 mm diameter (2nd column), 4 mm diameter (3rd column) and 5 mm diameter (4th column) at inlet pressures of 10 psi (1st row), 15 psi (2nd row), 20 psi (3rd row) and 25 psi (4th row). The signal produced by orifice diameter 6 mm is not included in this acoustic analysis as the frequency produced is very weak.

The signal produced at the cold tube mostly contains two categories of frequencies, i.e. low and high frequencies. The low frequency is produced by the surrounding noise, such as from the air-conditioning vent (around 1200 Hz), compressor (around 2500 Hz), etc. The use of "anechoic" box eliminates the noise coming from the air-conditioning system. However, as the compressor is always connected to the device through the inlet pipe, the noise coming from the compressor (when it is running) is more difficult to eliminate. During experiments, the compressor ran intermittently. The presence of the compressor noise can be noticed in figure 7 (c), (h), (j), (k), (l), (m), (n) and (o).

The magnitude of the Fourier Transform of the signals in Fig. 7 is presented in Fig. 8. It is obtained using FFT algorithm to show the frequency spectrum of the signals. The unit of the vertical axis is Pa representing the sound pressure. From the FFT results, some peaks present at different locations representing significant frequencies produced by the device captured at the cold tube for different orifice diameters. It is observed that when the inlet pressure increases (from top to bottom) the magnitude of the peaks increases. The position (the frequency) of the peaks does not seem to displace significantly when the same orifice diameter is used. For orifice diameter equals 2 mm one significant frequency appears at around 5920 Hz which is produced by the orifice. For larger diameters, more peaks appear. The highest magnitude still comes from the orifice and the others come from the right side of the orifice (the swirl generator and the hot tube) which are captured by the microphone located at the cold tube. This is confirmed by the acoustic analysis of the hot tube [7] which captured the same frequencies but with higher magnitudes (see Fig. 9). Except the frequency of the sound created by the orifice, other frequencies show higher in magnitude at the hot tube side. This indicates these sounds are produced inside the hot tube or closer to the hot tube. Returning back to Fig. 8, it is observed that for orifice diameter equals 3 mm, 4 mm and 5 mm, the frequencies produced by the orifice are around 3630 Hz, 3300 Hz and 3130 Hz respectively, which are decreasing following the increase of the orifice diameter.

If the peaks that are present in the FFT results (Fig. 8) are replaced by bars with the same order of magnitude (after removing the low frequencies coming from the surrounding noise), the frequency representation becomes the one presented in Figure 10. This made the analysis clearer. Each acoustic signal at different inlet pressure with different orifice diameter is now represented by a set of bars located at specific frequencies carrying different magnitude. The term "signature" can be used for each representation. These cold tube signatures represent the thermofluid performance shown in Figures 4, 5 and 6 from the cold tube point of view. For example, the signature from 2 mm orifice diameter containing frequencies around 5920 Hz and 19600 Hz with magnitudes of 0.0036 Pa and 0.0009 Pa respectively (Fig. 10 (i)) represents the isentropic efficiency of 10% (see Fig. 4), temperature difference of 6.4°C (see Fig. 5) and a cold mass fraction of 0.279 (see Fig. 6). Another example, the signature of 4 mm orifice diameter at 20 psi containing frequencies around 3312 Hz, 7750 Hz, 16362 Hz, 18512Hz and 19950Hz with magnitudes of 0.00163 Pa, 0.00067 Pa, 0.00087 Pa, 0.00083 Pa and 0.00088Pa (Figure 10 (k)) represents the isentropic efficiency of 17.47%, temperature difference of 11.3°C and a cold mass fraction of 0.538. For other signatures, the thermofluid performance can be consulted directly in

DOI: 10.1051/matecconf/20179507019

Figure 10. These cold tube signatures complement the hot tube signatures from the reference [7].

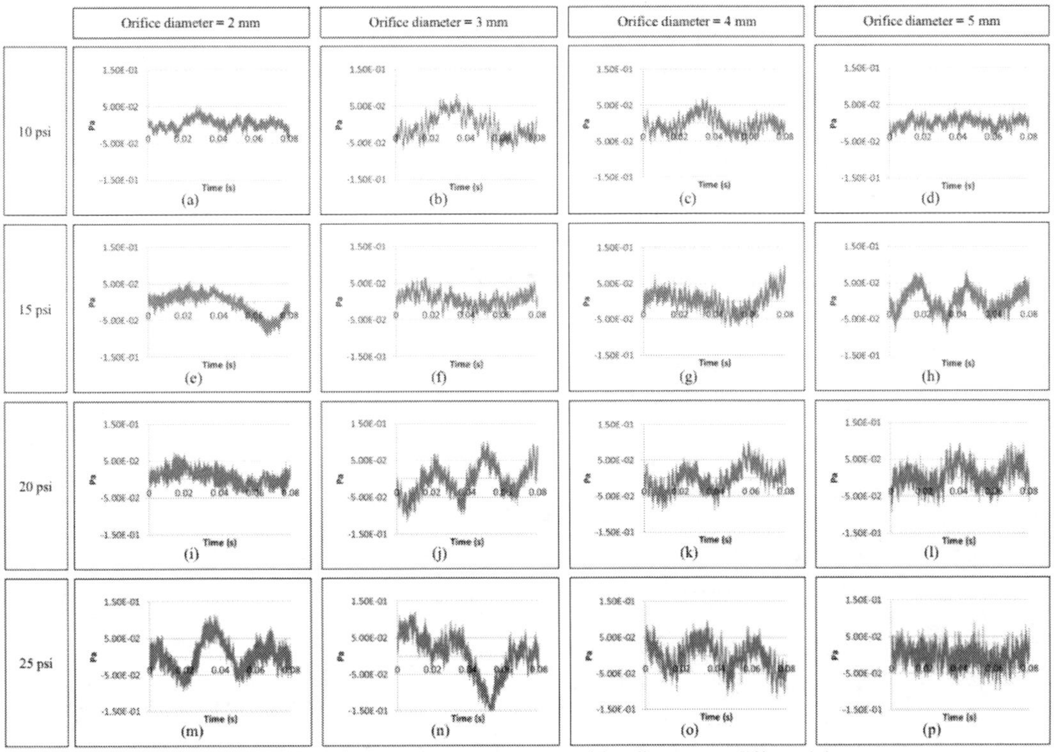

Figure 7. Signal produce by the RHVT with different orifice diameters at different inlet pressures.

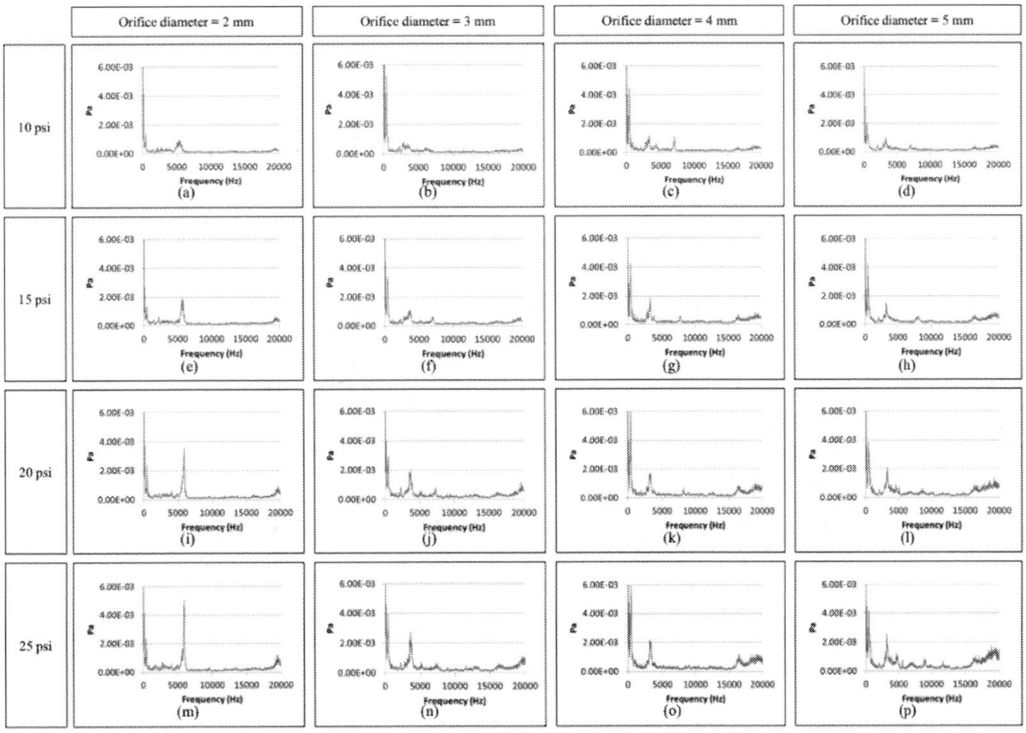

Figure 8. Magnitude of FFT of the signals in Figure 7.

 DOI: 10.1051/matecconf/20179507019

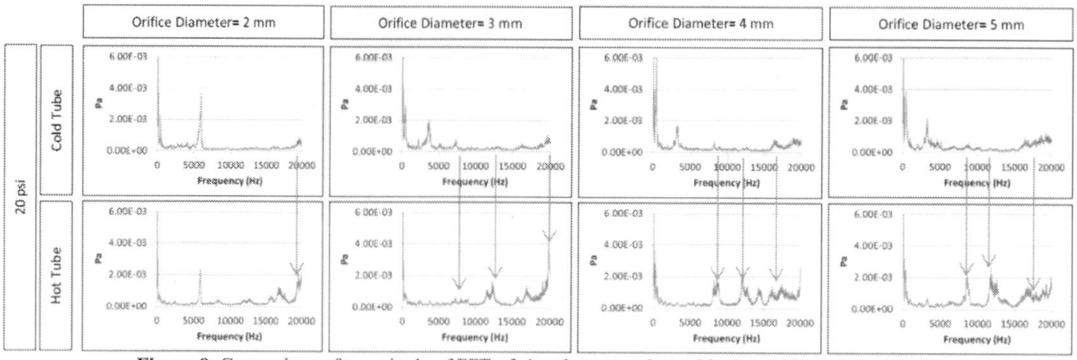

Figure 9. Comparison of magnitude of FFT of signals captured at cold tube and hot tube at 20 psi.

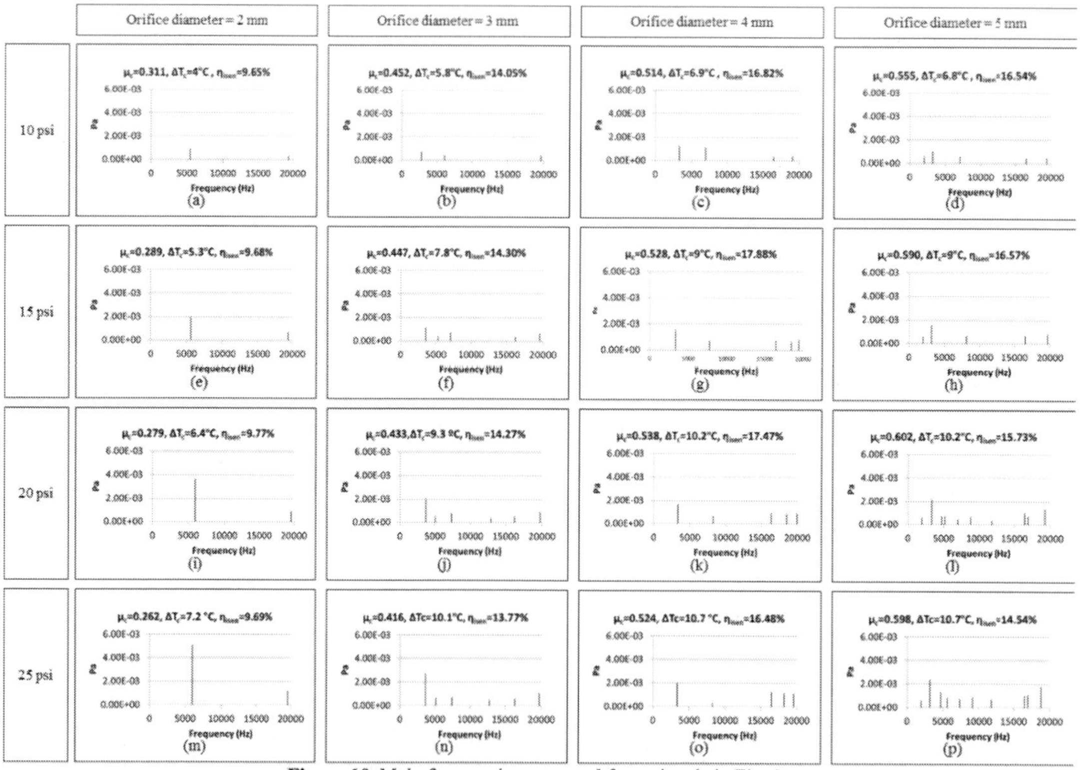

Figure 10. Main frequencies extracted from signals in Fig. 8.

4 Conclusions

In this study, the effect of orifice diameter on the thermofluid performance of the RHVT and the acoustic signal produced on the cold side of the tube is investigated. The investigation has been conducted experimentally and simple signal processing using FFT has been applied. It is observed that different orifice diameter produces different set of acoustic frequencies. This set of frequencies forms a signature of the signal from the cold tube point of view that can be associated to the thermofluid performance of the device. These cold tube signatures complement the hot tube signatures, so that the performance of the RHVT can be represented by the acoustic signatures (both cold and hot tubes) produced by the device.

Acknowledgements

Authors would like to thank the Malaysian Ministry of Higher Education (MOHE), the Research Management Centre (RMC) and the Faculty of Mechanical Engineering, Universiti Teknologi MARA (UiTM) for providing fund and support for this research under Fundamental Research Grant Scheme (FRGS), File No.: 600-RMI/FRGS 5/3 (74/2014).

References

1. N. Ismail, W. Wisnoe, M. F. Remeli, Experimental Investigation of Orifice Diameter, Swirl Generator and Conical Valve Shape to the Cooling Performance of Ranque-Hilsch Vortex Tube, *Applied Mechanics and Materials* **510**, 174-178 (2014).
2. H. R. Thakare, A. Monde, A. D. Parekh, Experimental, computational and optimization studies of temperature separation and flow physics of vortex tube: A review, *Renewable and Sustainable Energy Reviews* **52**,1043-1071 (2015).
3. S. U. Nimbalkar, M. R. Muller, An experimental investigation of the optimum geometry for the cold end orifice of a vortex tube, *Applied Thermal Engineering,* **29(2-3)**, 509-514, (2009).
4. Y. Istihat, M. H. Sabri, W. Wisnoe K. M. Abd Rahman, Flow Visualization in a Ranque-Hilsch Vortex Tube, (submitted to Journal of Mechanical Engineering).
5. W. Wisnoe, K.M. Abd Rahman, Y. Istihat, D. V. Natarajan, Thermofluid-Acoustic Analysis of a Ranque-Hilsch Vortex Tube, *3rd International Conference on System-integrated Intelligence: New Challenges for Product and Production Engineering, SysInt, Procedia Technology*, (2016)
6. M. Kurosaka, Acoustic streaming in a swirling flow and the Ranque-Hilsch (vortex tube) effect, *Journal of Fluid Mechanics,* **124**, 139-172, (1982).
7. K. M. Abd Rahman, W. Wisnoe, Y. Istihat, D. V. Natarajan, Acoustic analysis on different orifice diameter of a Ranque-Hilsch Vortex Tube (hot side), *Mechanical Engineering & Science Postgraduate International Conference,* (2016).
8. Y. Istihat, W. Wisnoe, "Wavelet Transform of Acoustic Signal From A Ranque-Hilsch Vortex Tube, *IOP Conference Series: Materials Science and Engineering*, **88**, (2015).
9. N. Ismail, W. Wisnoe, M. F. Remeli, Experimental Investigation on the Effect of Orifice Diameter and Inlet Pressure to the Ranque-Hilsch Vortex Tube Performance. *Applied Mechanics and Materials,* **465-466**, 515-519 (2014)

MATEC Web of Conferences 95, 07020 (2017) DOI: 10.1051/matecconf/20179507020
ICMME 2016

Topological Optimization of Continuum Structure based on ANSYS

Li Xue-ping[1] , Zhao Lian-yu [1] and Liu Zheng-zhong[1]

[1]*School of Mechanical Engineering, Tianjin Key Laboratory of the Design and Intelligent Control of the Advanced Mechanical System, Tianjin University of Technology, Tianjin 300384,China*

Abstract. Topology optimization is at the phase of structural concept design and the result of it is foundation for succeeding design, therefore, structural topology optimization is more important to engineering design. in this thesis, in order to seek the optimal structure shape of the winch's mounting bracket of ROV simulator, topology optimization design of it by finite element analysis software ANSYS was carried out. the results show that the topology optimization method is an effective optimization method and indicate that the method is correct and effective, it has a certain engineering application prospect.

1 Introduction

According to the type of design variables and degree of difficulty, the structural optimization of continuum structure can be divided into size optimization, shape optimization and topology optimization, corresponding to different stages of product design, namely the three stages of detailed design, basic design and concepts design. Compared to the first two optimizations, topology optimization can fundamentally change the structure of the object, the result is the basis of all the subsequent optimization of the design, so it has a great significance for engineering design; Though topology optimization has good prospects, it is recognized as the most difficult task in the field of structural optimization design.

In this study, we will use topology optimization module of ANSYS to optimize the design the winch' mounting bracket of ROV simulator and achieve the improvement of the virtual design of the winch' mounting bracket of ROV simulator, avoiding the design of blindness and improving the feasibility and effectiveness of the design, thus shortening the design cycle. We would verify the optimization results through the situation in real use. It provides an idea of simple principle taking advantage of topology optimization for design geometry element.

2 Overview of topology optimization

2.1 Method for topology optimization

In 1854, Maxwell made the basic topological analysis of the truss of minimum weight under stress constraints, laying the foundation for the study of topology optimization. In 1904, Michell put forward the theory of truss Michell [1], which veritably started the topological optimization. In 1964, Dorn and some others introduced the value theory to the field of topology optimization and proposed the Ground structural method [2] to establish a collection of nodes constituted by the structural node, the point of load application and point of support, connected with each other by member bars, and formed the so-called "ground structure" In the single load case, we would take the stress constraint into consideration, with the internal forces as design variables to build the liner programming for solution. In 1988 Bendsoe and some others proposed the homogenization method [3],which realizes the topology optimization of continuum structure by dividing the number of different micro-hole structural unit within the design area. Chinese scholars Sui Yunkang et al [4] raised the optimization model approach based on ICM (independent, continuous mapping variable) to study the topology optimization under global stress constraints, greatly reduced the number of constraints and avoided the local optimization problem. At the University of Sydney J.Lin et al [5]studied the topological optimization model of compliant mechani sms combining the advantages of SIM and PP.

Structural topology optimization theory has been proposed for more than 100 years, and it is in recent decades that it has been in rapid development, with new theoretical and computational methods emerging, being applied in various fields of mechanical, of aerospace, marine engineering and shipbuilding, etc. the topology optimization of structure is one of the effective method for the precise design of the large parts [6].

2.2 Theory of topology optimization of Ansys

Topology optimization theory means to seek the maximum or minimum objective function under certain constraints. The principle of topology optimization Ansys is to minimize the structural compliance meeting the constraints of structural volume reduction. Minimization of structural compliance is in fact a seeking of maximization of structural stiffness.

Ansys topology optimization module changes the topology optimization problem to a problem of shape optimization under a special form. The goal is to seek the maximum stiffness and the minimum volume of the structure under certain constraints, with the function of material distribution as optimization parameters. ANSYS topology optimization can be used to find the maximum stiffness, the minimum volume and maximum natural frequency of the structure. The objective function of the topology optimization reduces the structural strain energy on the condition of meeting the structural constraints. To reduce the structural strain energy means to increase the stiffness of the structure. This technology can be reached by giving density to each unit of finite element by means of design variables (η_i).

Design variables assigned to each finite element (i). pseudo-density (η_i). Pseudo-density value changes from 0 to 1, $\eta_i \approx 0$ represents the remove of the material, $\eta_i \approx 1$ represents the preservation of the material. The total volume is the sum of volume of all the units. $V = \sum_i \eta_i V_i$ the pseudo-density affects the volume and elasticity tensor of each unit, $[E_i] = [E(\eta_i)]$. In the conventional linear elastic equation, the elastic tensor equals the stress and strain vector, in the formula $[\sigma_i] = [E_i][\varepsilon_i]$, $[\sigma_i]$ is the stress vector of the unit i, while $[\varepsilon_i]$ is its strain vector.

Taking the calculating the maximum static structural stiffness for example, under the conditions of the volume constraints, the maximum structural stiffness is to seek the minimum statistic strain energy given a certain load, which means to calculate the minimum compliance. In this case, the formula is:

$$\begin{cases} u_c = \min F(\eta_i) \\ \text{s.t} \int_\Omega \eta_i \, d\Omega \le aV_2 \end{cases} \tag{1}$$

where u_c is the compliance;η_i means the pseudo density of the unit number i, $0 \le \eta_i \le 1$; a is the ratio of volume reduction;V_2 is the volume before the optimization of the structure.

When the structure receives different loading conditions, the maximum stiffness can be expressed as:

$$F(U_C^1, U_C^2, \cdots U_C^n) = \sum_{i=1}^{n} W_i U_C^i, W_i \ge 0 \tag{2}$$

where U_c^i is the compliance in the loading condition number i, W_i is the component of loading conditions under the minimum stain energy.

3 Optimization design of the winch's mounting bracket of ROV simulator

3.1 Description of the problem

ROV (remotely operated underwater vehicle) is an important underwater tool of marine engineering, but the complex underwater environment leads to greater job difficulty. Taking as object a certain ROV of type open-frame for observation, our research group guided by author's tutor developed 'simulator for dedicated channel of the underwater production system integration test' realizing the ROV underwater training functions. The pilot can simulate the underwater workflow and pre-evaluate the work program, thus reducing the risk of ROV in the actual job.

In the process of the design of ROV simulator, in order to achieve the control of front and rear pitching position, the author needs to design a mounting bracket of winch machine for fixing the winch. This paper solves this problem by means of Ansys topology optimization .

3.2 Preprocessor before the Ansys topology optimization

• Establish the basic structure of the topology optimization. The basic structure is the initial structure prior to the optimization. The structure should possess the features of the stress, strain etc to facilitate optimization calculations. After a comprehensive analysis of each case ultimately, the final optimal structure is determined as cuboid with a length of 1800mm, width of 700 mm, and thickness 200mm. According to preliminary relative calculation, in extreme working conditions, winch's mounting bracket load bearing condition is shown in Fig.1, which is a schematic diagram of basic structure and stress for the topology optimization of winch mounting bracket, each of the four arrows along the Y axis represents horizontal shear stresses of the same value along the Y axis F_N= 15000N, while the two arrows indicate the equivalent clockwise torque of X axis, T_N=1500N.M.

• Determine the type of unit. The type of initial structure unit should be 10-node Solid 185.

• Define material properties. The selected material is304 stainless steel, elastic modulus of 1.9e+11Pa, Poisson's ratio of 0.29.

• Meshing and selection of the unit 1, control the overall unit size to 0.003 m, and then choose Quad / Free for the entire rectangular meshing.

• Apply a load to the basic structure of the model topology optimization. The finite element model of the topology optimization after the first treatment is shown in Fig. 2.

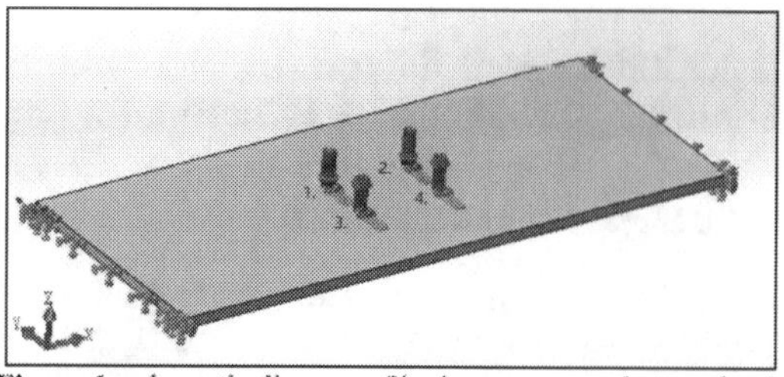

Figure 1. schematic diagram of basic structure and stress for the topology optimization

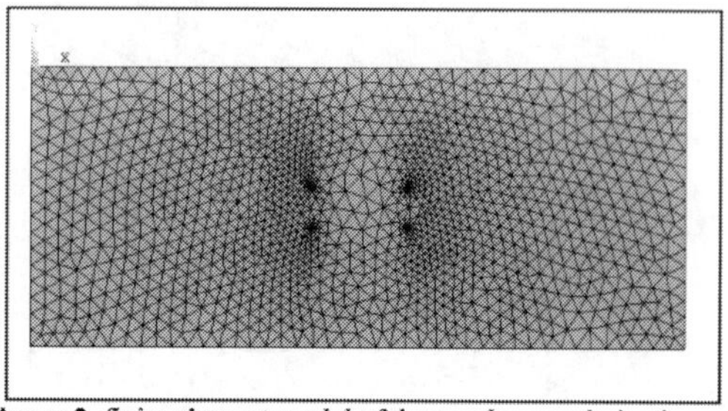

Figure 2. finite element model of the topology optimization

3.3 Solution of topology optimization of Ansys

• Define a single load case as the topology optimization function, the function name is Scomp.

• Define the topology optimization function Scomp as the objective function.

• Define the volume function VOLUME as the constraint function, set the reduction of volume to 60%.

• Choose the method of OC(optimality criteria)for topology optimization solution, the convergence accuracy is set to 0.0001, while the number of iterations is set to 30.

• Run optimization program and display results

3.4 Topology optimization results and analysis

ANSYS topology optimization result is output as density cloud map, in fig. 3, the red section(where labeled 1) represents reserved material area, blue section (where labeled 2) represents the area of removing material.

In Fig.3, below the density of the cloud is given the contrast scale of the topology optimization design variable density value, the position of density value of 1 means the red area(where labeled 1) indicates that the structure should be arranged in structural design; while the density value of 0.001 corresponding to the blue area(where labeled 2) of density cloud map, showing that this area does not need arrangement in structural design. These are the two limiting cases, there are some color regions(where labeled 3) between the two values that need to be necessarily treated in the design process, the basic structure of the volume was significantly reduced weight of the structure is also reduced.

The objective function curves plotted course. Course objective function curve shown in Fig.4, as shown in Fig4.From the calculation process, it can be seen that structural compliance decreases with the increase of iteration, and iterative results are in stable condition after the 16th step and calculation results converge. it reflects the iterative process of structural changes in the stain energy, in an iterative process, the stain can be decreased, and gradually increases structural rigidity.

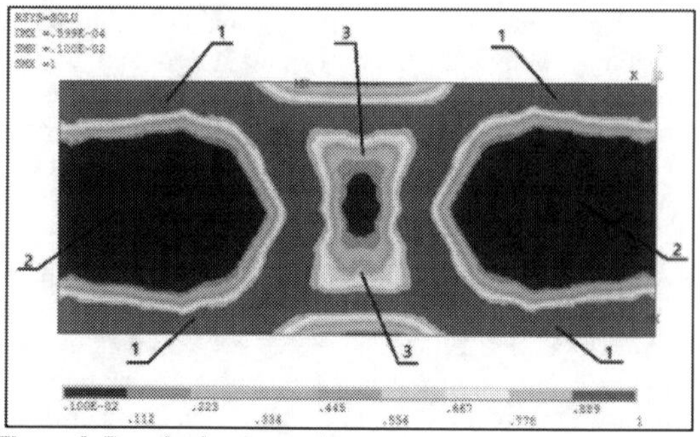

Figure 3. Pseudo-density nephogram.

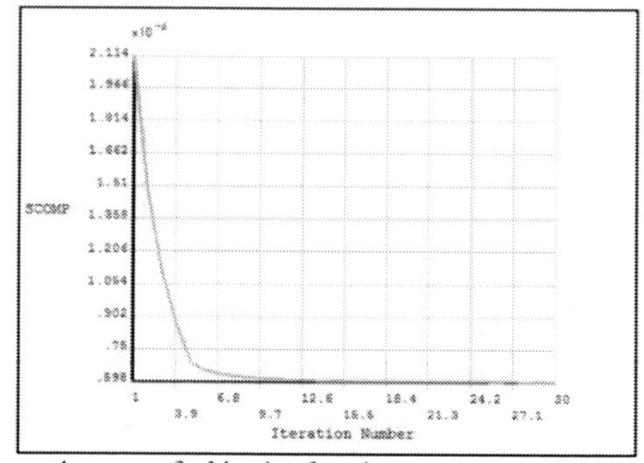

Figure 4.curves of objective function

3.5 Design scheme of the winch's mounting bracket of ROV simulator & Finite element Analysis

Corresponding to pseudo density structure shown in Fig.3, core structure of the mounting bracket of winch , central area is shown in Fig.3(where labeled 1,3). This part of structure, which has greatest impact on mounting bracket's stiffness, however the optimized result in conceptual phase of design. upon most occasions it could guide the shape design of the structure, but there is still a great difference between with realistic engineering design problem the Result cannot be applied to engineering practice directly. Combine the results with the author's experience in engineering design this paper puts forward the design scheme of the winch's mounting bracket of ROV simulator as shown in Fig.5.

The core structure of the design is composed of hollow square steel tube whose cross section is 80mm×80mm, thickness is 4mm and material is 304 stainless steel,stiffened-plates are set in specific area of the structure to improve structure's stress performance. the thickness of stiffened-plate is 6mm and material is 304 stainless steel.

Build the finite element model of the mounting bracket of winch with ANSYS; add material properties; apply load and boundary conditions on model; run static analysis program; display results as shown in fig.6, fig.7.

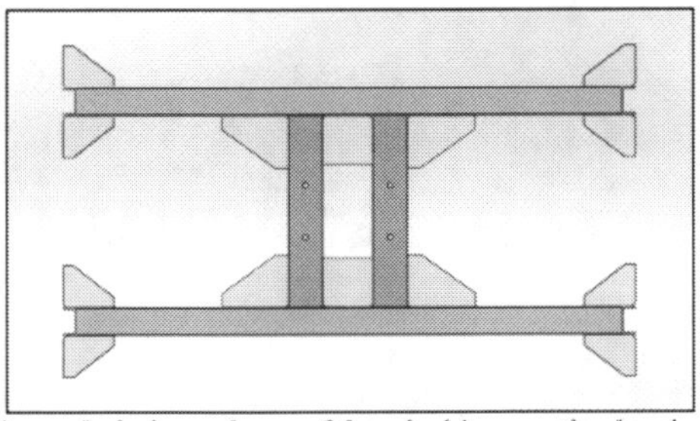

Figure 5. design scheme of the winch's mounting bracket

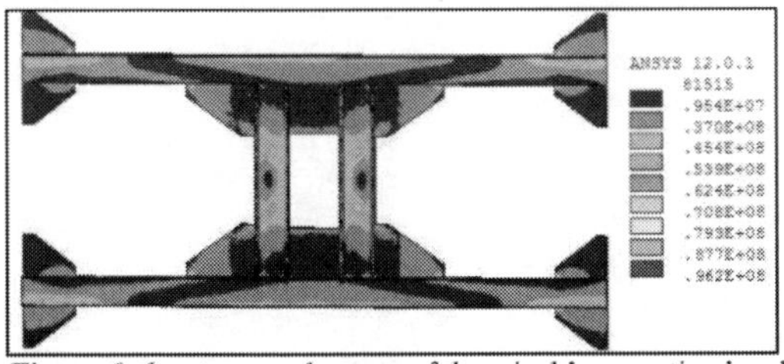

Figure 6. the stress nephogram of the winch's mounting bracket under operating limits

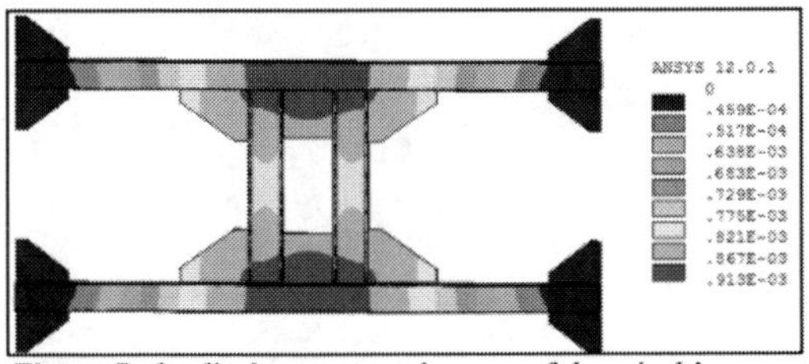

Figure 7. the displacement nephogram of the winch's mounting bracket under operating limits.

Fig. 6 shows that the maximum stress of the structure is 92.6MPa, while the yield strength of the 304 stainless steel is 207MPa ,set the safety parameter as1.5, the permissible stress of the selected material is: 138MPa. From the comparison between the maximum stress of the structure and the permissible stress of the selected material, it is concluded that the maximal stress of this design is under the material's allowable stress.

As shown in fig. 7 the maximum displacement of the structure is 0.913mm,which meet the demand of industrial design. the winch's mounting bracket of ROV simulator is produced based on topology optimization as shown in fig.8 it's function is reliable and perfect through practical testing.

It is concluded that the design of the winch's mounting bracket of ROV simulator is a great success and proved reasonable and safe.

4 Conclusions

• This paper expounds the mathematical processing of topological optimization of continuum structure in theory,then using the ANSYS software platform for optimization calculation. in order to realize the conceptual design.it helps to achieve stiffness and strength condition, as well as to reduce the weight of the structure reasonably and to satisfy requirement of reliability of the winch's mounting bracket of ROV simulator .

• From The results of finite element analysis and the test of winch's mounting bracket in practice, it is shown that the topology optimization model is correct and the load, the boundary constraint and the objective function is reasonable.

• Compared with traditional structural design, the structural topology optimization design method not only guarantees the reliability of the design, but also obviously shortens the design cycle, and avoid blindness and enhance the effectiveness; In addition, cost is reduced, and the topology design result has been tested in engineering application as shown in fig.8 It is of great significance to apply the method in practice.

Figure 8. The winch's mounting bracket of ROV simulator produced based on topology optimization

References

1. Michell A G. The limits of economy of materials in frame structures [J]. Philosophical Magazine, **8(47)**,589-597(1904)
2. Dorn W S, Gomory R E, Greenberg H J. Automatic design of optimal structures[J]. J de Mechnique, **3(1)**,25- 52(1964)
3. Bendsoe M P,Kikuchi N. Generating optimal typologies in structural design using a homogenization method[J].Comp. Applied Mech. Engrg,**71**,197-224(1988)
4. SUI YK,YEHL,PENG XR. Topological optimization of continuum structure with global stress constraints based on ICM method[J].Computational Methods, **PTS1-2**,1003-1014(2006)
5. Lin Jiang-zi,Luo Zhen,Tong Li-yong. A new mufti-objective programming scheme for topology optimization of compliant mechanisms[J].Structural and Mufti-disciplinary Optimization,**40(1-6)**,241-25(2010)
6. Zhu Deng-lin,Chen Jun-wei, Yu Jie etc. Developments of Structural Topology Design and its Application in Structural Design [J]. Machine Building and Automation,**34(6)**,7-11(2005)

 DOI: 10.1051/matecconf/20179507021

Research on Winding System of Twisting Machine Based on Multi-parameters

Zhang Zhiming , Sun Jun, Wu Xiaoqing

School of Electrical & Mechanical Engineering, Wuhan Textile University, 430073 Wuhan City, China

Abstract. Winding movement of twisting machine is closely related to the package build, package density and yarn quality. The requirements are that the package build is good shaped and the winding tension is proper technically. Namely the angle of winding is fine and there is no overlapping and convex, hard edge on the package. The winding density is same and reasonable on the whole package build. And the fluctuation of winding tension is small. It is becoming development trend that driving of multiple-motors is used on twisting machine. There are many parameters to be controlled such as degree of twist, direction of twist, winding angle and spindle speed and so on. Therefore it is essential to set up reasonable mathematical model of yarn for winding movement.

1 Introduction

Twisting machine mostly adopted the single motor driving system home and abroad during the past years[1]. Degree of twist is gained by changing the transmission ratio of twist gears. Direction of twist is changed by the rotation direction of the main motor. This can gain the S twist and Z twist. The winding movement and the reciprocating yarn guiding movement are realized by the gear wheel and the stepped cam transmission. In this way, twisting movement of the spindle, winding movement of rolls and reciprocating yarn guiding movement are realized on twisting machine[3]. This transmitting is long and the structure is complex. Nowadays, with the development of mechanical and electrical integration technology and demand of saving energy and reducing consumption, multi-motor driving has become the development trend of the multi twisting machine[3-5]. According to the characteristics of traditional twisting machine driving system, this paper puts forward a new twisting machine winding based on multi-parameter controlling system, which can satisfy the whole technological requirements of twisting machine.

2 Mathematical model of yarn winding movement

There are several forms such as cylindrical, conical and conical at both ends for the yarn package build. But the principle of these winding movements is same from kinematic of machinery[6].

The yarn winding movement on the twisting machine is the same as the cheese winding machine, spinning frame and roving frame, which are made up of the reciprocating yarn guiding motion and winding rotary movement[7]. In this paper, taking the winding conical package build as an example, the movement of the conical winding package build is shown in Fig. 1.

According to the winding theory of twisting machine, the package build is rotated through friction by the winding roller. According to the basic theory of yarn winding, the relationship between the moving speed V_T of winding point T and speed V_H of yarn guiding point H is shown as equation (1).

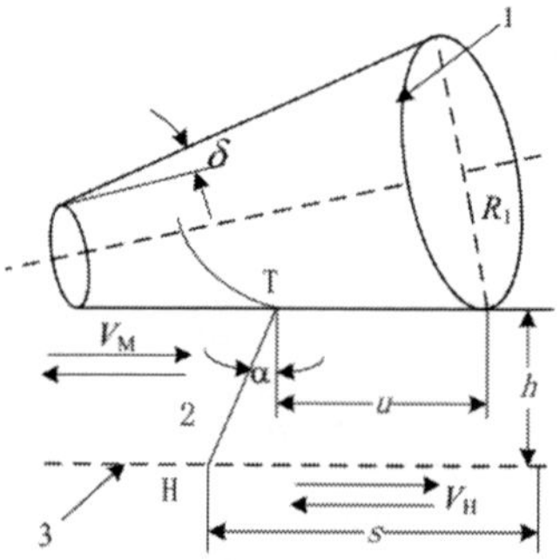

1- package build 2-yarn 3-guiding rod
Figure 1. winding movement of yarn

$$\frac{ds}{dt}=\frac{du}{dt}+\frac{h}{\omega(R_1+u\sin\delta)}[\frac{d^2u}{dt^2}-\frac{\sin\delta}{R_1-(l-u)\sin\delta}\left(\frac{du}{dt}\right)^2] \quad (1)$$

$$V_H = ds/dt \quad (2)$$

$$V_T = du/dt \quad (3)$$

 DOI: 10.1051/matecconf/20179507021

Where u is displacement of winding point (m), s is distance of guiding rod (m), h is normal distance of winding point to guiding rod (m), ω is rotational angular velocity of the package build (rad/s), R_l is big radius of cone package build (m), l is the length of cone package build(m), δ is half cone angle of conical package (rad), t is winding time (s).

The winding movement curve is shown as Fig. 2.

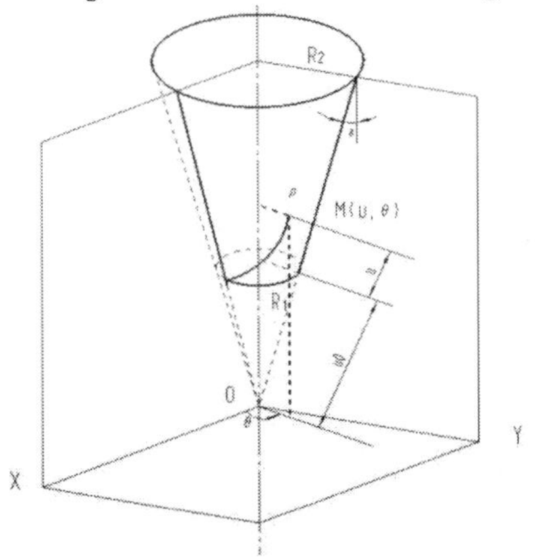

Figure 2. winding movement curve

3 Mathematical model of yarn guiding movement

When the winding roller is rotated, the circumferential speed of the points along the bus bar the surface of the roller is not the same at all. It is difficult to achieve the same tension and same density package build at the same time. Intensive analyses show that the tension fluctuation and the density fluctuation are small. And the shape is good by using the appropriate variable acceleration through yarn guiding rule. The practice shows that when the cone angle is small and the speed of guiding rod is not very large. The good winding can also be got with constant velocity yarn guiding rule[8].

At present, with the development and wide application of mechanical and electrical integration technology, the development trend of the high speed and precision winding drive technology is to directly drive guiding rod and yarn guiding parts to realize yarn guiding movement rule by controlling the motor through software. The main advantages are that it abandons the complex cam box and the machine has the advantages of simple structure. It is also very convenient to replace the twist and roll form and very easy to realize high precision winding.

The curve of yarn guiding movement is improved after the transition curve at re-entry point. In Fig 3, the *AB* and *CD* segments are the transition curve of the re-entry point. *BC* segment is the middle part of the movement curve. The sine acceleration of the law is used in the vicinity of the two direction-changing points. It uses law of constant velocity to transit in the middle. The reciprocating yarn curve, velocity curve and acceleration curve are shown in Fig.3.

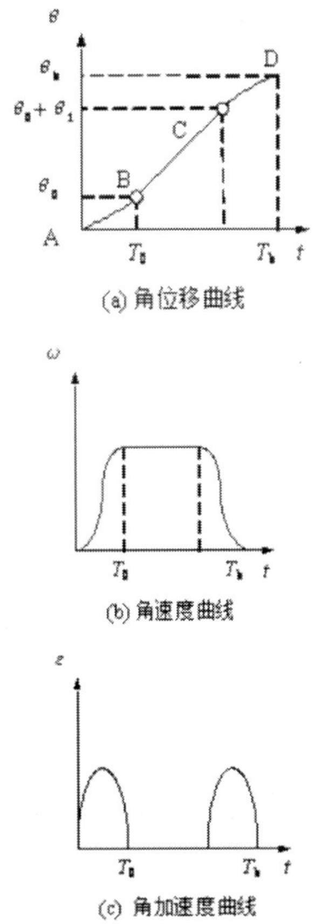

Figure 3. curve of improved yarn guiding movement

The angular displacement of the package build is θ (rad), the displacement of the guiding rod is S(m),then

$$S = \frac{\theta}{2\pi} L \tag{4}$$

Where L is the lead of guiding rod (m).

Therefore, the vertical coordinate of the curve is respectively angular displacement (θ), angular velocity (ω), angular acceleration (ε) of the guiding rod. The horizontal coordinate is the time of movement.

The guiding rod comes and goes both once in a cycle yarn guiding movement. The law of come-and-go movement is basically same, but in the opposite direction. Guiding rod moves forward and reverse correspondingly according to the same curve. Therefore, the movement curve can be studied only in half of the yarn guiding cycle.

From Fig.3, in the travel of yarn guiding movement, the velocity and acceleration at the two re-entry points are zero. The speed of guiding rod reduced to zero and then the guiding rod returned. It can avoid the inertial impact at re-entry point. The maximum acceleration in the whole trip can also be suppressed.

In Fig.3, the time interval for variable motion near the two re-entry points is T_0. Displacement is θ_0. Half cycle is T_b. The maximum angular displacement is θ_b. Then *AB*, *BC*, *CD* segment displacement, velocity and acceleration equations are shown as (3)-(11).

AB segment: $0 \le t \le T_0$

$$\theta_J = \theta_0\left(\frac{t}{T_0} - \frac{1}{\pi}\sin\frac{\pi}{T_0}t\right) \tag{5}$$

$$\omega_J = \frac{\theta_0}{T_0}\left(1 - \cos\frac{\pi}{T_0}t\right) \tag{6}$$

$$\varepsilon_J = \frac{\pi\theta_0}{T_0^2}\sin\frac{\pi}{T_0}t \tag{7}$$

BC segment: $T_0 \le t \le T_b - T_0$

$$\theta_J = \theta_0 + \frac{\theta_1(t - T_0)}{T_b - 2T_0} \tag{8}$$

$$\omega_J = \frac{\theta_1}{T_b - 2T_0} \tag{9}$$

$$\varepsilon_J = 0 \tag{10}$$

CD segment: $T_b - T_0 \le t \le T_b$

$$\theta_J = -\theta_0\left[\frac{T_b - t}{T_0} - \frac{1}{\pi}\sin\frac{\pi}{T_0}(T_b - t)\right] + \theta_b \tag{11}$$

$$\omega_J = \frac{\theta_0}{T_0}\left[1 - \cos\frac{\pi}{T_0}(T_b - t)\right] \tag{12}$$

$$\varepsilon_J = -\frac{\pi\theta_0}{T_0^2}\sin\frac{\pi}{T_0}(T_b - t) \tag{13}$$

where θ、ω、and ε with the subscript "$_J$" is the basic yarn guiding rule.

There should be proper relation in order to ensure the continuous speed curve.

$$\frac{\theta_1}{\theta_0} = \frac{2T_b}{T_0} - 4 \tag{14}$$

Or:

$$\frac{T_b}{T_0} = \frac{\theta_1}{2\theta_2} + 2 \tag{15}$$

Where T_0、T_1、θ_1、θ_0 are determined by different process conditions and requirements. For example, when $\frac{\theta_1}{\theta_0} = 4$, $\frac{T_b}{T_0} = 4$, then $\theta_b = \theta_1 + 2\theta_0 = 6\theta_0$, when $\frac{T_b}{T_0} = 6$, $\frac{\theta_1}{\theta_0} = 8$, then $\theta_b = \theta_1 + 2\theta_0 = 10\theta_0$.

4 Improvement and anti-aliasing of yarn coiler motion

4.1 winding tension improvement

From the previous analysis, it is known that it is easy to implement equal tension and equal density when winding cylindrical roll, but not for winding cone roll. When winding, line speed of small end is small (less than line speed of the friction roller) and line speed of big end is big. In order to make the winding tension uniform, the winding speed should be as stable as possible. Therefore, in the small end of the yarn bobbin, the guide speed should be larger, in the big end of the yarn bobbin, guide speed should be smaller. There are usually two ways, one is the yarn guiding speed gradually decreases when the yarn is from the small end to big end and it gradually increases when the yarn is from the big end to small end. Another way is to make the time of the reciprocating yarn guiding movement stroke different. The time of big end is a little bigger than the time of small end. In this yarn winding movement, the winding density of small end of the bobbin will be slightly less than that of the big end, but it does not affect the handling and unwinding, so it can still be used.

4.2 ribbon breaking mechanism of yarn

There are the following measures to realize ribbon breaking mechanism[9-10]. One way is changing the speed of friction roller periodically to make the roller fast and slow down intermittently. Under the inertial force, the change of bobbin speed always lags behind the change of the friction roller speed, it makes the some regularity transmission relationship between the two is destroyed. So the circle number of yarn winding is not a continuous integer, making it can avoid overlapping.

Another way is changing the yarn guiding speed periodically, making the yarn winding speed change periodically. That also can make the yarn winding cycle not consecutive integer.

In addition, due to the impact of free yarn, the hardness of the bobbin at both ends is usually larger than that at the middle, resulting in a hard edge and a convex edge, also resulting in uneven dyeing, edge collapse.

The main measures to solve uneven dyeing and selvage collapse are differential yarn guiding. In traditional yarn guiding, a low frequency movement can be attached, which can make the winding position of yarn guiding change slowly and periodically.

Above all, the winding movement can be improved as follows.

(1) The improvement of yarn guiding movement

$$\theta(t) = \theta_J(t) + k_1\theta_0\sin\frac{\pi}{k_2T_0}t \tag{16}$$

Where k_1, k_2 are adjusting coefficients, refer to experience. $\theta_J(t)$ is the basic yarn guiding law determined by the equation (3) to (11). The difference of time between the go and return stroke is also considered, which can make $T_b(small \to big) > T_b(big \to small)$.

(2) Changing package build speed (ω_g, rad/s)

$$\omega_g = \omega_g^0(1 + k_3 \sin\frac{\pi}{k_2 T_0} t) \quad (17)$$

Where k_3 is adjusting coefficients of the roller speed, refer to experience, ω_g^0 is the basic angular velocity of roller (rad/s).

According to the movement rules established by the above equations, it can realize high speed and precision yarn winding in twisting machine. Similarly, in the multi motor drive system in this paper, the improvement of the yarn guiding movement and the change of friction roller speed can simply realized by changing the control program of yarn guiding movement drive motor and winding rotary movement drive motor. It can be very flexible and accurate.

5 conclusions

This paper introduces differential equation of yarn winding movement on all kinds of spinning machine. It is the theoretical basis for further research on the various kinds of problem of winding movement. From the angle of mechanics, the faults are analysed about the law of yarn guiding movement of traditional cylindrical cam yarn delivery device and a feasible yarn guiding movement basic law can avoid the inertia impaction at the two re-entry point. The issues of winding tension, winding density, yarn anti-ribboning are also analysed and the method of improving the winding movement is derived. These mathematical models are the theoretical basis of the multi motor driving system of twisting machine.

Acknowledgments

This work was jointly supported by the key project of the Science and Technology Project of Hubei Educational Committee, China (Q20151603).

References

1. S. A. Fang. Journal of Donghua University(natural science edition), 21. 95 (1995)
2. Z. M. Zhang, S. Q. Mei. W.B.Xiao. Microcomputer Information,26,21(2010)
3. Z. M. Zhang, S. Q. Mei. Cotton Textile Technology. 34, Oct. 9(2006)
4. F. Xiao, J. S. Mei. Journal of Dandong Textile College. 11, 18(2004)
5. J. P. Wei, L. L.Cai, G.Li, Journal of Textile Research, 28,12(2007)
6. S.Q.Mei, Development and Innovation of Machinery and Electrical Products,17,2(2004)
7. H.Rong, Cotton Textile Technology.30,5(2002)
8. S. E. Fang, Journal of China Textile University, 21,2(1995)
9. Y. H. Zhou. Journal of Textile Research. 23, 18(2002)
10. J. L. Liu, Gao P. Xiao, Z. X. Zhang. Textile Machinery. 6, 17(2009)

MATEC Web of Conferences 95, 07022 (2017) DOI: 10.1051/matecconf/20179507022
ICMME 2016

Calibration of an Indexing Table Using a High Precision Angle Comparator

Ming Huang[1,2], PinKuan Liu[1], Yangqiu Xia[2,3], Qiu Hu[2], Mengyang Li[2], Qiang Tang[2,3]

[1]*State Key Laboratory of Mechanical System and Vibration, Shanghai Jiao Tong University, Shanghai, China*
[2]*Institute of Machinery Manufacturing Technology, China Academy of Engineering Physics, Mianyang, China*
[3]*National Machine Tool Production Quality Supervision Testing Center(Sichuan), Mianyang, China*

Abstract. The calibration of indexing table is important to ensure high accuracy and reliability in angle metrology. In this paper, we present calibrations of an indexing table and uncertainty analysis to evaluate the performance of the high precision angle comparator. The configuration of the angle comparator is explained. In the calibration, a complete closure method is applied to obtain the angle comparator's position error and the indexing table's angular deviation simultaneously based on the least square method. The position error of the angle comparator is evaluated to be about 0.08". The calibration result of an indexing table is compatible with that of the manufacturer and National Institute of Metrology of China within the expanded uncertainty of calibration, 0.04".

1 Introduction

Accurate angle measure technologies are required in various precision engineering applications, e.g., manufacturing, robotics [1,2]. Indexing tables generate accurate angular positions, and their calibrations are important to ensure high accuracy and reliability in angle metrology [3].

A multi-tooth indexing table is a circular indexing table based on the mechanical indexing principle. A multi-tooth indexing table is composed of two face gears with the same diameter, tooth shape and tooth number, and a similar structure of a multi-tooth face clutch.

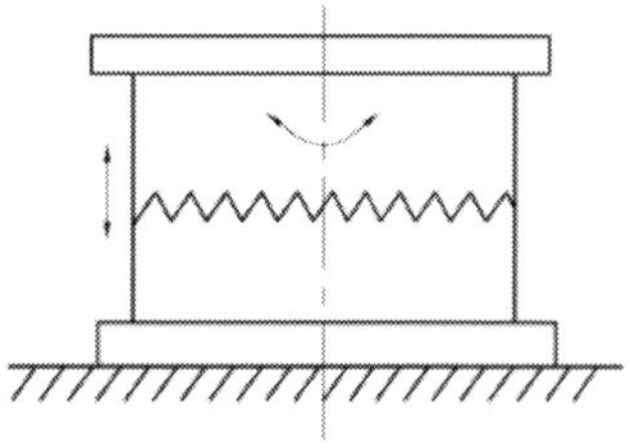

Figure 1. Indexing table.

As shown in Figure 1, when the upper gear disc along the axial direction rises to a certain height, gear discs remove meshing. After a turn of several teeth relative to the fixed lower gear disc, the upper disc moves down again and meshes with the lower gear disc with a certain axial pressure, thus generating. a certain angle. Obviously, a multi-tooth indexing table can generate or calibrate constant angles. Since all the teeth are involved in meshing positioning, the average effect and elastic overconstrained principle of multi-tooth meshing make indexing table positioning accuracy much higher than gear disc machining accuracy. 0.1" accuracy can be reached. Because on any concentric circle, the thickness and slot width remain the same, the indexing accuracy is not affected by the direction of rotation. The repeatability is also ensured by the centrality of teeth.

The angle calibration has been widely studied by many researchers. Yandayan [4] has calibrated high-resolution electronic autocollimators with demanded low uncertainties using single reading head angle encoders. At the Physikalisch-Technische Bundesanstalt (PTB) [5], high-resolution electronic autocollimators have been calibrated with the aid of the WMT 220 angle comparator to ensure their direct traceability to the SI unit of plane angle, the radian (rad). The Korea Research Institute of Standards and Science (KRISS) also developed a high precision angle generator [6, 7], and Jong-Ahn Kim [8] calibrated several angle artifacts and instruments using the high precision angle generator to demonstrate the performance of the angle generator. Huang [9] has compared angle interferometer with angle encoder during calibration of autocollimator.

In this paper, we analyze the performance of the angle comparator, and adopt the complete closure method to calibrate an indexing table. In the calibration of the indexing table, the performance of the angle comparator was evaluated and the measurement uncertainty was analyzed.

2 High precision angle comparator

The high precision angle comparator consists of a radial aerostatic bearing, a thrust aerostatic bearing, four ultrasonic motors, a precision angle encoder, and a

 DOI: 10.1051/matecconf/20179507022

controller. The thrust aerostatic bearing is preloaded with vacuum to stabilize the air clearance and improve the stiffness. The thrust stiffness is 400N/μ m and the load capacity is 80kg. There are four ultrasonic motors placed around the rotating plate. The motor tips drive the plate running at fine resolution less than 0.002″ . The radial run-out of the comparator is less than 0.05μ m.

Figure 3 shows the block diagram of the angle comparator control system, which is composed of a SPiiPlus universal drive interface (UDIhp4422N01N, ACS™), a PC based real-time motion controller (SPiiPlusSC HP0800004BAY5A, ACS™), a motor driver (AB2, Nanomotion™), four Ultrasonic motors (HR2, Nanomotion™) and a feedback encoder with two encoder heads (ERP880, Heidenhain™). The four ultrasonic motors are connected in parallel to synchronize the motion of each motor. The position data of the two heads are averaged to obtain the readout of the angle generator by using a virtual gantry control mode. The sampling and update rates of the servo loop are both 20 kHz.

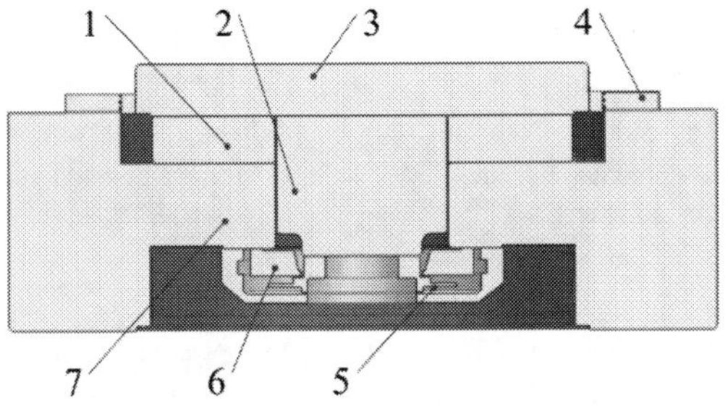

Figure 2. Schematic view of the angle comparator ZT 320: (1) air-bearing casing; (2) air-bearing rotor; (3) table; (4) ultrasonic motor; (5) divided circle; (6) reading heads; (7) supports.

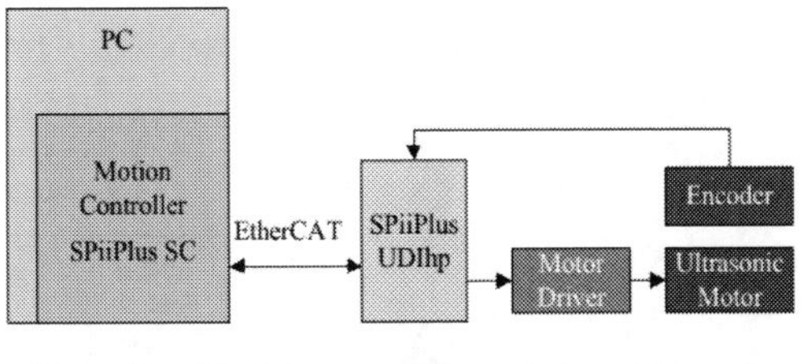

Figure 3. Block diagram of the controller of the angle comparator.

3 Calibration of indexing tables with the angle comparator

Figure 4 shows the measurement arrangement for the calibration of a 1440 precision indexing table (±0.1" of arc accuracy) on the angle comparator. The measurement system is composed of the optical polygon, the indexing table, the angle comparator, and the autocollimator. The autocollimator (TA US 300-57 from TRIOPTICS GMBH) has a measurement range of 3000"×2200" and a resolution of 0.005". According to manufacturer specifications, the accuracy amounts to ±0.05" over any 10" range and ±0.10" over any 20" range.

The axes of the angle comparator, the indexing table and the polygon should coincide with each other. Centred with respect to the comparator's axis of rotation, the autocollimator is standing on an adjustable plate on the granite base. With the aid of the adjustable plate, the autocollimator is aligned such that the horizontal measuring axis and the optical axis of the autocollimator lie in the comparator's measuring plane, and aimed at the reflecting surface of the polygon. The polygon and the indexing table lie in the comparator's axis of rotation to avoid any lateral displacements of the mirror in the autocollimator aperture. The flatness deviations of the mirror have nearly no influence on the calibration result. The air path of 20mm between the autocollimator and the reflecting surface of the polygon does not need to be specially shielded since the constant temperature and the laminar airflow in the clean room laboratory do not exert any significant disturbing influences.

Figure 4. Measuring arrangement for the calibration of an indexing table on the angle comparator.

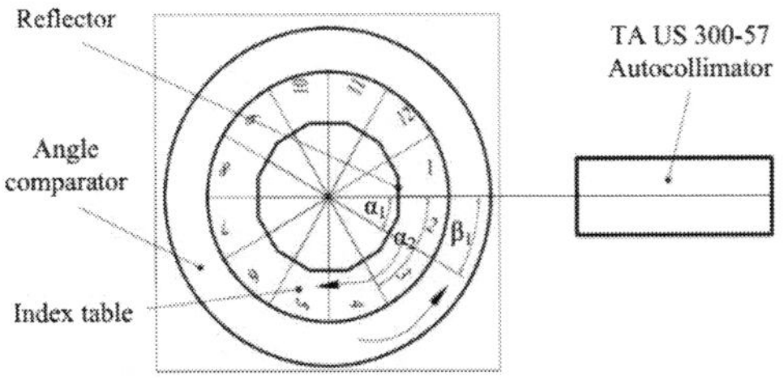

Figure 5. The diagram of the complete closure method.

Calibration is performed by the complete closure method (Figure 5). The method is that a group of constant angles, which are proper, not precise but must be stable, are chosen and used to compare with the measured angles in sequence, respectively and independently in a closed circle. And it can eliminate the influence of the angular deviation of the constant angles and calculate the measured angular deviation at equal weight based on the least square method. If we choose the same angles as the indexing table from the angle comparator, the angular deviations of the indexing table and the comparator will be obtained simultaneously.

During calibration, by rotating the angle comparator clockwise and the indexing table counter-clockwise, by the same nominal angle(30°), the autocollimator is always aimed at the same reflecting surface of the polygon, the angular deviation will be measured by the autocollimator. The measurement circle will be repeated 12 times in one cycle. The indexing table will be shifted by one nominal angle relative to the angle comparator. After that, the indexing table will rotate from next angle step counter-clockwise to ensure that the autocollimator is always aimed at the identical reflecting surface of the polygon.

The method will eliminate the influence of the constant angle's deviation on the measurement and give both the angle comparator's position error and the indexing table's angular deviation simultaneously based on the least square method.

n (number of the constant angles) is 12, and the nominal values of the constant angles can be expressed as

$$\varphi_i = i\frac{360^\circ}{n}, \quad i=1,2,3,\cdots,n-1 \tag{1}$$

When the constant angles have been defined, according to the definition of error, the autocollimator reading value can be expressed by

$$\theta_i = \alpha_i - \beta_i = (\varphi_i + \delta\alpha_i) - (\varphi_i + \delta\beta_i) = \delta\alpha_i - \delta\beta_i \tag{2}$$

where α_i is the rotation angle of the indexing table, β_i is the angle of the angle comparator rotation, $\delta\alpha_i$ is the deviation of angle α_i to the constant angles, $\delta\beta_i$ is the deviation of angle β_i to the constant angles.

For the first measurement cycle, the indexing table will be rotated in sequence to compare with the angle comparator rotation and a group of measurement numbers will be obtained:

$$\begin{cases} (\delta\alpha_1 - \delta\alpha_0) - \delta\beta_1 = \theta_{0,1} \\ (\delta\alpha_2 - \delta\alpha_0) - \delta\beta_2 = \theta_{0,2} \\ \vdots \\ (\delta\alpha_{11} - \delta\alpha_0) - \delta\beta_{11} = \theta_{0,11} \end{cases} \tag{3}$$

For the i-th measurement circle,

$$\begin{cases} (\delta\alpha_{i \bmod 12} - \delta\alpha_{i-1}) - \delta\beta_1 = \theta_{i-1, i \bmod 12} \\ (\delta\alpha_{(i+1) \bmod 12} - \delta\alpha_{i-1}) - \delta\beta_2 = \theta_{i-1, (i+1) \bmod 12} \\ \vdots \\ (\delta\alpha_{(i+10) \bmod 12} - \delta\alpha_{i-1}) - \delta\beta_{11} = \theta_{i-1, (i+10) \bmod 12} \end{cases} \tag{4}$$

For the last measurement circle,

$$\begin{cases} (\delta\alpha_0 - \delta\alpha_{11}) - \delta\beta_1 = \theta_{11,0} \\ (\delta\alpha_1 - \delta\alpha_{11}) - \delta\beta_2 = \theta_{11,1} \\ \vdots \\ (\delta\alpha_{10} - \delta\alpha_{11}) - \delta\beta_{11} = \theta_{11,10} \end{cases} \tag{5}$$

Combine the above equations,

$$\begin{cases} \sum_{i=1}^{12}(\delta\alpha_{i \bmod 12} - \delta\alpha_{i-1}) - 12\delta\beta_1 = \sum_{i=1}^{12}\theta_{i-1, i \bmod 12} \\ \sum_{i=1}^{12}(\delta\alpha_{(i+1) \bmod 12} - \delta\alpha_{i-1}) - 12\delta\beta_2 = \sum_{i=1}^{12}\theta_{i-1, (i+1) \bmod 12} \\ \vdots \\ \sum_{i=1}^{12}(\delta\alpha_{(i+10) \bmod 12} - \delta\alpha_{i-1}) - 12\delta\beta_{11} = \sum_{i=1}^{12}\theta_{i-1, (i+10) \bmod 12} \end{cases} \tag{6}$$

and

$$\delta\beta_j = -\frac{1}{12}\sum_{i=1}^{12}\theta_{i-1, (i+j-1) \bmod 12}, \quad j=1,2,\cdots,11 \tag{7}$$

There are totally 132 measurements and 11 constant angles will be obtained. Through Eq. (7), all the deviation of the constant angles generated by the indexing table will be calculated out. In every measurement round, the error equations containing the residuals V can be listed :

$$\begin{cases} (\delta\alpha_{i \bmod 12} - \delta\alpha_{i-1}) - \delta\beta_1 - \theta_{i-1, i \bmod 12} = V_{i-1, i \bmod 12} \\ (\delta\alpha_{(i+1) \bmod 12} - \delta\alpha_{i-1}) - \delta\beta_2 - \theta_{i-1, (i+1) \bmod 12} = V_{i-1, (i+1) \bmod 12} \\ \vdots \\ (\delta\alpha_{(i+10) \bmod 12} - \delta\alpha_{i-1}) - \delta\beta_{11} - \theta_{i-1, (i+10) \bmod 12} = V_{i-1, (i+9) \bmod 12} \end{cases} \tag{8}$$

11 unknowns $\delta\alpha_i$ are to be solved from the above 132 equations, and it is obviously indefinite. In order to get the best value of the unknown $\delta\alpha_i$, the equation that conforms to the least square principle need to be established,

$$Q = \sum_{i=0}^{11}\sum_{j=0}^{11} V_{i,j}^{\ 2} = \min \quad , i \neq j \tag{9}$$

If the partial derivative of the function Q to the unknown is 0,

$$\frac{\partial Q}{\partial \delta\alpha_i} = 0, \quad i=1,2,\cdots,11 \tag{10}$$

The deviation of the angle comparator's angles can be calculated by the following equations

$$\delta\alpha_i = \frac{t_i - t_0}{2*12} \tag{11}$$

where,

$$t_i = q_i - p_i$$

$$q_i = \sum_{j=0}^{11}\theta_{j,i}, \quad i \neq j$$

$$p_i = \sum_{j=0}^{11}\theta_{i,j}, \quad i \neq j$$

4 Results

4.1 Experiment results

The deviations of nominal angles of a 1440 precision indexing table were measured using the angle comparator, the optical mirror and the calibrated autocollimator aforementioned in Sec. 3. In the experiment, we applied the complete closure method to obtain the deviation angle of the indexing table and the error of the angle comparator separately.

DOI: 10.1051/matecconf/20179507022

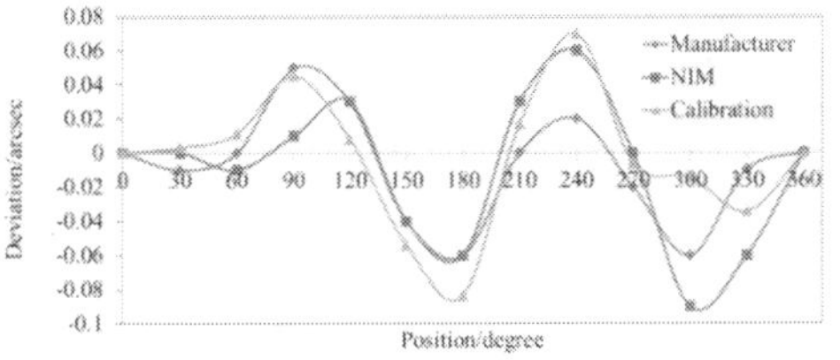

Figure 6. Comparison of the calibration results of the indexing table.

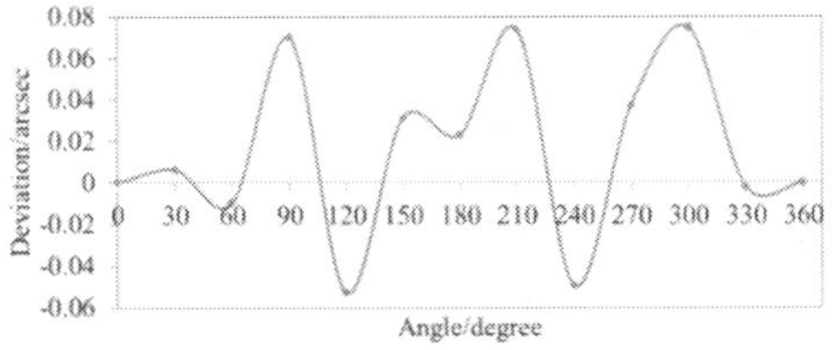

Figure 7. The position error of the angle comparator by the complete closure method.

The measurement result of the indexing table was compared with the calibration result provided by the manufacturer and the National Institute of Metrology of China (NIM). The comparison data curve is shown in Figure 6. The expanded uncertainty (k=2) reported in the calibration certificate of the manufacturer is 0.05", and that of the NIM is also 0.05". These calibration results show good agreement considering the magnitude of each uncertainty.

Figure 7 shows the position error of the angle comparator by the complete closure method, the error of the angle comparator was evaluated to be about 0.08".

4.2 Uncertainty analysis

The combined measurement uncertainty u_c, attributed to the indexing table calibration is determined in accordance with the Guide to the Expression of Uncertainty in Measurement (GUM) [10] by the addition of several single standard uncertainties that are important for the measuring process. According to the Eq. (7), the propagation formula can be deduced as:

$$u_c^2 = \sum_{i=1}^{m} c_i^2 u_i^2 \tag{12}$$

where the sensitive coefficient of every component c_i is 1, u_i is a relevant standard uncertainty.

Table 1 shows a list of all relevant uncertainty components for the calibration of indexing table using a high precision angle comparator and states the estimates, the distribution function and the resulting uncertainty value as standard uncertainly. Measurement uncertainty contributions are explained in following.

Table 1. Uncertainty budget for the calibration of indexing table.

Type	Uncertainty component	Estimate /arcsec	Distribution function	sensitive coeff. c_i	Uncertainty contribution/ arcsec u_i
A	Standard deviation of the mean value		Normal	1	0.017
B	Measurement repeatability of the angle comparator	0.02	Normal		0.0067
B	Resolution of the angle comparator	0.0017	Rectangular		0.0005
B	Resolution of the autocollimator	0.005	Rectangular		0.0014
B	Repeatability of the test indexing table	0.02	Rectangular		0.0058
B	Uncertainty of the angular reflector	0.012	Rectangular		0.007
Combined standard measurement uncertainty u_c = 0.02 arcsec					
Expanded measurement uncertainty $U(k=2)$ = 0.04 arcsec					

Table 2. Calculating data of measurement repeatability.

Position	0°	30°	60°	90°	120°	150°	180°	210°	240°	270°	300°	330°
a_i	0.00	0.00	0.11	0.08	0.01	0.12	0.13	0.13	0.09	0.12	0.15	0.08
a_i'	0.00	0.00	0.10	0.11	0.07	0.18	0.16	0.17	0.12	0.15	0.18	0.13

4.2.1 The uncertainty contribution of autocollimator standard deviation of the mean value u1

During the calibration, standard deviation of the mean value produced from 10 series of repeat measurements and the maximal standard measurement deviation of 10 measurements is 0.055". According to the principle of the standard values merge, the uncertainty u1 is 0.017".

4.2.2 The uncertainty contribution of measurement repeatability of the angle comparator u_2

The measurement repeatability of the angle comparator is obtained by double-observation method, that is, measurement is performed at 12 positions evenly distributed in the whole circumference. Table 2 shows the calculating data of measurement repeatability.

 DOI: 10.1051/matecconf/20179507022

The measurement repeatability is

$$u_2 = \frac{1}{3}s_r = \frac{1}{3}\sqrt{\frac{\sum_{i=1}^{n}\left(\alpha_i - \alpha_i'\right)^2}{2n}} = 0.0067'' \quad (13)$$

4.2.3 The uncertainty contribution of resolution of the angle comparator u_3

The resolution of the angle comparator is 0.0017" resulting in $u_3 = 0.0017/\left(2\times\sqrt{3}\right)$=0.0005" uncertainty contribution.

4.2.4 The uncertainty contribution of resolution of the autocollimator u_4

The resolution of the autocollimator is 0.005" resulting in $u_4 = 0.005/\left(2\times\sqrt{3}\right)$=0.0014" uncertainty contribution.

4.2.5 The uncertainty contribution of repeatability of the test indexing table u5

The repeatability of the test indexing table provided by the manufacturer is 0.02 resulting in $u_5 = 0.02/\left(2\times\sqrt{3}\right)$=0.0058" uncertainty contribution.

4.2.6 The uncertainty contribution of the angular reflector u6

During the calibration, the optical polygon, whose working surface is used as an angular reflector, and autocollimator are used together, the flatness of the working surface is 50 nm in its central area, the dimension of the working surface is 12mm×15mm. When the autocollimator aims at different position of the reflector, the flatness will have influence on the angle measurement.

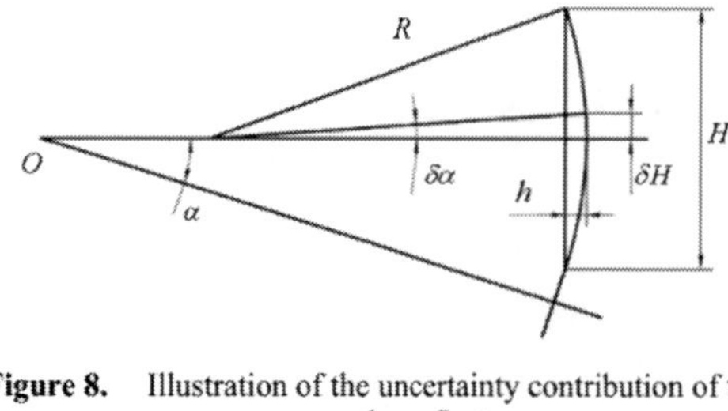

Figure 8. Illustration of the uncertainty contribution of the angular reflector.

Assuming that the surface of the reflector is approximately arc shape, the curvature radius of the reflector is

$$R = \frac{4h^2 + H^2}{8h} \approx \frac{H^2}{8h} \quad (14)$$

where h is the flatness of the reflector, H is the width of the reflector.

When the autocollimator optical axis deviates from working face center at a distance of δH, it will cause the normal angle of working face to produce angle deviation $\delta\alpha$,

$$\delta\alpha = \frac{\delta H}{R}\rho = \frac{8h\delta H}{H^2}\rho \quad (15)$$

where δH is 0.02mm, ρ is 206265"/rad.

So, the uncertainty contribution of the angular reflector is

$$u_6 = \frac{\delta\alpha}{\sqrt{3}} = \frac{8h\delta H}{\sqrt{3}H^2}\rho = 0.007'' \quad (16)$$

4.2.7 The expanded measurement uncertainty U

The expanded measurement uncertainty U is calculated by multiplying the combined standard uncertainty, uc , by the coverage factor k = 2. The value of the calibration results thus lie with a probability of 95% within the attributed interval of values.

$$U = k\sqrt{\sum_{i=1}^{m} c_i^2 u_i^2} = 0.04''$$

5 Conclusions

A high precision angle comparator is developed in this research. It uses four ultrasonic motors and a double read-heads encoder adopting a virtual gantry control mode to produce fine driving resolution and high accuracy with simple configuration. To calibrate the indexing table, a measurement system including the optical polygon, the angle comparator, and the autocollimator was built up, and the mathematical model of complete closure method is established. Applying the complete closure method to the calibration of the indexing table, both the angle comparator's position error and the indexing table's angular deviation are obtained separately based on the least square method. To verify the performance of the angle generator, the indexing table is measured and the experimental results are analyzed. Through the calibration, the position error of the angle comparator is evaluated to be about 0.08". The calibration result of the indexing table agreed with the calibration data from both the manufacturer and National Institute of Metrology of China within the calibration uncertainty, 0.04" . Through the experiment, the performance of the angle generator is proved.

Acknowledgment

This work was supported by Science and Technology Major Project of China Academy of Engineering Physics: (9120602), Calibration and Testing Lab Project of China Academy of Engineering Physics: (16S006).

References

DOI: 10.1051/matecconf/20179507022

1. A. Just, M. Krause, R. Probst, et al. Comparison of angle standards with the aid of high-resolution angle encoder. Precision Engineering, 33, 4(2009): 530-533
2. Ralf D Geckeler, Andreas Fricke, Clemens Elster, Calibration of angle encoders using transfer functions. Measurement Science and Technology, 17, 10(2006): 2811-2818
3. W.T. Estler, Y.H. Queen, J. Bryan. Advanced angle metrology system. CIRP Annals, 42, 1(1993): 573-576
4. Tanfer Yandayan, S Asli Akgoz, Muharrem Asar. Calibration of high-resolution electronic autocollimators with demanded low uncertainties using single reading head angle encoders. Measurement Science and Technology, 25, 1(2014): 015010
5. A Just, M Krause, R Probst , et al. Calibration of high-resolution electronic autocollimators against an angle comparator. Metrologia, 40, 5(2003): 288–294
6. Jong-Ahn Kim, Jae Wan Kim, Chu-Shik Kang, et al. Precision angle comparator using self-calibration of scale errors based on the equal-division-averaged method. Proc. of MacroScale, (2011)
7. Jong-Ahn Kim, Jae Wan Kim, Chu-Shik Kang, et al. High precision angle generator using multiple ultrasonic motors and a self-calibratable encoder. Review of Scientific Instruments, 82, 11(2011): 116108-1-116108-3
8. Jong-Ahn Kim, Jae Wan Kim, Chu-Shik Kang, et al. Calibration of angle artifacts and instruments using a high precision angle generator. International Journal Of Precision Engineering And Manufacturing, 14, 3(2013): 367-371
9. Yao Huang, Zi Xue, Heyan Wang. Comparison between angle interferometer and angle encoder during calibration of autocollimator. Ninth International Symposium on Precision Engineering Measurement and Instrumentation. 9446, (2015): 944624
10. ISO/IEC. Uncertainty of measurement: part 3. Guide to the Expression of Uncertainty in Measurement. ISO/IEC Guide 98-3, (2008)

MATEC Web of Conferences 95, 07023 (2017) DOI: 10.1051/matecconf/20179507023
ICMME 2016

Submersible Unmanned Aerial Vehicle: Configuration Design and Analysis Based on Computational Fluid Dynamics

Wang Qinyang[1], Wu Songping[1], Hong Weijiang[1], Zhuang Wenzheng[1] and Weiyihao[2]

[1]*School of Aeronautic Science and Engineering, Beihang University, Beijing, China*
[2]*Large Advanced Aircraft Training Center, Beihang University, Beijing, China*

Abstract. Submersible aerial vehicle is capable of both flying in the air and submerging in the water. Advanced Research Project Agency (DARPA) outlined a challenging set of requirements for a submersible aircraft and solicited innovative research proposals on submersible aircraft since 2008. In this paper, a conceptual configuration design scheme of submersible unmanned aerial vehicle is proposed. This submersible UAV lands on the surface of water, then adjusts its own density to entry water. On the contrary, it emerges from water by adjusting its own density and then takes off from the surface of water. Wing of the UAV is whirling wing. It is set along aircraft's fuselage while submerging for lift reduction. We analysis aerodynamic and hydrodynamic performance of this UAV by CFD method, especially compare the hydrodynamic performance of the whirling wing configuration and normal configuration. It turns out that whirling wing is beneficial for submerging. This result proves that the configuration design scheme proposed in this paper is feasible and suitable for a submersible unmanned aerial vehicle.

1 Introduction

A vehicle called submersible aerial vehicle that is flight-capable and submersible is wildly concerned around the world for its great advantages that it is capable of submerging to avoid missiles and taking off from water surface to attack enemy. It should combine the key capabilities of three different platforms: 1) the speed and range of an aircraft; 2) the loiter capabilities of a boat; 3) the stealth of a submarine [1]. Submersible aerial vehicle is a kind of aerial weapons which maybe used to enhance tactical ability of armies. Several interesting conceptual schemes and principle prototypes are proposed to verify the feasibility of a hybrid vehicle that combines an aircraft and a submarine together.

The first concept design named "flying submarine" was shown by Dzerzhinskiy Ushakov in 1934, which is a weapon that submerges and attacks enemy's ships underwater and then escapes by taking off from the surface of water. In 1962, a flight-capable submarine was presented by Convair. The submarine was designed in detail and tried to build. However, the plan could not accomplished due to some technical problems. In 2008, a set of requirements for a submersible aircraft was outlined by Defense Advanced Research Project Agency (DARPA) [2]. DAPRA was calling for innovative research proposals on submersible aircraft from 2008 to 2011, but few feasible proposals were submitted, which leads to the project be canceled in 2011. These three submersible aerial vehicles need complex life supporting system to keep crews safe. Compared with these manned vehicles, unmanned submersible aerial vehicles without life supporting system are easier to come true [3].

Actually, a submersible UAV called "cormorant" is developed by US, which is launched from submarine. Then sailing in water pushed by a rocket engine. After through water surface, its air-breathing engine works. What's more, a submersible UAV named "fly-boat" that imitates flying fish was proposed and built up by a research team in Beihang University. Their submersible UAV is similar to a biological flying fish [4]. It adopts soft landing type to land on water surface and also takes off from water surface. This UAV adopts sweptback wing. So the wing provides enough lift in the air for vehicle to fly and folds back to reduce the resistance underwater.

Although some concept designs are made to provide feasible proposals on the submersible aircraft, there is still a large gap between these schemes and products for practical utilization. Accordingly, it is still very valuable and meaningful to research the complex technological problems on the submersible UAV persistently.

2 Technological problems

To satisfy the requirements that both flying in the air and diving under water several technological problems have to be solved. Firstly, flying requires a large wing area to provide enough lift but large wing area will cause large resistance and excess lift under water. Secondly, in one hand, for flying the density of UVA should be as light as better, in the other hand, diving under water requires that

 DOI: 10.1051/matecconf/20179507023

the density of UVA is similar to it of water. Thirdly, a flying UVA are made up by light structures. However, shell structures of a submarine are thick to resist the high pressure caused by water. Fourthly, power system is also special in a submersible vehicle. Fuel engines can't be used underwater. And the shape of propellers for air propulsion and water propulsion are different. Fifthly, the fuselage of UVA will not cause large drag in the air, but the resistance caused by fuselage should not be ignored while sailing underwater. In Table 1, the contradictions are enumerated.

Table 1. Contradictions for flying and submerging.

Parameter	Flying	Submerging
Wing area	Bigger enough to provide lift	Smaller enough for resistance reduction and avoiding excess lift
Density	Much smaller than water	Similar to water
Structure	Light shells and beams	Thick shells
Propelling system	Fuel engine	Electric motor
Fuselage	Contributes a little drag	Causes much drag
Water discharge	Small	About 100%

Except for the contradictions mentioned above, how to transfer modes between water and air should be considered seriously. Based on the previous studies, there are two main feasible modes to transfer between different media. The first mode is plunge-diving and vertical take-off mode. This mode simulates prey process of aquatic birds, which was firstly public aroused by Hawkes on 2010 [5]. After that, some researchers, coming from Beihang University carried out lots of pioneering experiments to study the mechanism of gannet's plunge-diving process [6], [7]. The second is soft landing and taxiing take-off mode. The researchers, coming from Beihang University created a submersible unmanned flying boat named Flying Fish and verified this mode by "Flying Fish" first time in the world [4]. They made further experiments to research this kind of submersible UAV, which already verify the second mode is feasible, but they do not give a perfect explanation for all the technological problems caused by this mode.

3 Concept design of submersible UAV

Because the submersible UAV is mainly used in the air, so traditional method about aircraft design is suitable for determining variables of submersible UAV. However, there are two special problems should be considered. One is about the density, another one is about the Reynolds number.

3.1 Density adjustment

In order to suitable for flying, the density of UAV should be small. On the contrary, the density should be adjust similar to water for sailing underwater. Some parameters are shown in Table 2 to state this problem.

In Table 2, $V_{structure}$ means volume of all the solid objects in the UAV. So, in the fact, ρ_{air_model} is the average density of the UAV while flying in the air. While sailing in the water, the submersible UAV is capable of keeping water away from its inner space. So V_{whole} stands for the whole volume of the submersible UAV.

Table 2. Parameters about density adjustment.

Parameter	symbol	Parameter	symbol
Cross weight	W_{c_w}	Water tanks cubage	C_{tank}
Density in the air	ρ_{air_mode}	Volume of solid parts in the UAV	V_{solid}
Density underwater	ρ_{under_mode}	Water density	ρ_{water}
Density control coefficient	η	Volume of the whole UAV	V_{whole}

Two inequalities are used to introduce principles of density adjustment:

$$\rho_{air\_mode} = \frac{W_{c\_w}}{V_{structure}} \leq \eta \cdot \rho_{water} \tag{1}$$

$$\rho_{under\_mode} = \frac{2C_{tank} \cdot \rho_{water} + W_{c\_w}}{V_{whole}} \geq \rho_{water} \tag{2}$$

From inequality (2), we get inequality (3):

$$C_{tank} \geq \frac{\rho_{water} \cdot V_{whole} - W_{c\_w}}{2\rho_{water}} \tag{3}$$

Inequality (2) and (3) are two constraints conditions for the design of submersible UAV.

3.2 Relationships between cruising speed and Reynolds number

Based on the flow similarity principle, Reynolds numbers in the air and underwater should be the same:

$$\frac{\rho_{air} \cdot V_{c\_air} \cdot L}{\mu_{air}} = \frac{\rho_{water} \cdot V_{c\_underwater} \cdot L}{\mu_{water}} \tag{4}$$

From (4), we get:

$$\frac{V_{c\_air}}{V_{c\_underwater}} = \frac{\rho_{water}}{\rho_{air}} \cdot \frac{\mu_{air}}{\mu_{water}} \tag{5}$$

In equations above, V_{c_air} and $V_{c\_underwater}$ indicate the cruising velocity. The reference length is indicated by L. μ_{air} and μ_{water} indicate the dynamic coefficient of viscosity.

Equation (4), which shows the relationship between cruising velocities, is important for the design.

DOI: 10.1051/matecconf/20179507023

3.3 Configuration of the submersible UAV

A new concept design scheme of unmanned submersible aerial vehicle is proposed in this paper in order to suitable flying and diving requirements better.

According to the previous studies, solutions are proposed in this paper to realize technological problems shown in part 2. Whirling wing is designed to keep large wing area in flying mode and reduce the area of wing which generates excess lift while submerging. Two float bowls that contain water tanks inside are installed beside the fuselage in order to adjust the density of UVA by injecting water into water tanks or pumping water out of water tanks. Two different power systems are used to drive the UAV. One system combines fuel engine and air propeller to drive the UAV in the air. The other system combines electric motor and water propeller together to drive the UAV under water. And the structure is designed enough strong to resist pressure caused by water while the UAV sailing underwater.

In this paper, our conceptual submersible UAV will choose the second mode which is introduced in part 2 to transfer media. Differently, we add two float bowls on both sides of the UAV, which are capable to stretch out and pick up. When the submersible UAV is going to land on the water surface, float bowls will be stretched out to provide drag to contribute to reduce velocity. And stability is increased by stretching float bowls out to resist waves while floating. On the contrary, the float bowls will be picked up while flying in the air or diving under water to reduce resistance. In addition, water tanks are designed inside the float bowls in order to adjust the density of UAV. A solenoid value is installed in the head of the float bowls. When it opens, water will be injected into the water tanks. A pump is installed in the water tank and links with media outside by pipeline. When the pump works, water will be driven out from the water tanks. In order to keep the air pressure balance between the water tanks and outside, air duct is used to link water tanks to the air hole.

Fuselage of the submersible UAV is an important component. In order to design a perfect shape for the fuselage, two aspects of things have to be considered. For one thing, the density of water is larger than air, so reducing resistance becomes very important. The shape of fuselage is similar to a submarine that just like a drop of water. For another thing, take-off mode of the submersible UAV proposed in this paper is taxiing take-off from water surface. While taxiing on the water, the fuselage of submersible UAV will cause large resistance. If the taxiing process is ignored in the design, the UAV will fail to take off. Inspired by amphibious planes, the shape of fuselage is modified for drag reduction.

Wing and tail wing are also key parts of an aircraft. Airfoil effects wing's performance. The software named "Profili" is used to choose airfoils for the UAV. In the end, CLARK-Y is picked for the wing and NACA4412 is picked for the tail wing.

The principal design parameters of the submersible UAV is presented in Table 3. The configuration of the submersible aircraft is presented in Fig. 1, Fig. 2 and Fig. 3.

Table 3. Principal design parameters of the submersible UAV.

Parameter	Value	Parameter	Value
Cross weight	45kg	Body length	3.0m
Wing span	4.1m	Aspect ratio	8.0
Wing area	2.05m	Flight-speed	216km/h
Sail speed	15km/h	Conventional cruising depth	3m
Conventional flight attitude	20~600m	Load capacity	8kg

As shown in Table 3, values of some key parameters are given.

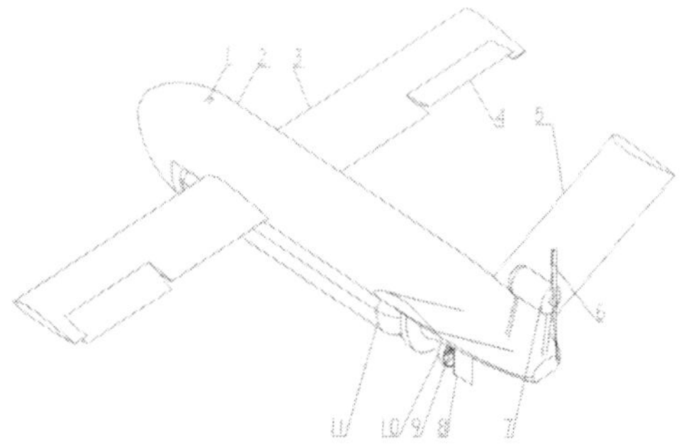

Figure 1. Configuration (In the air).

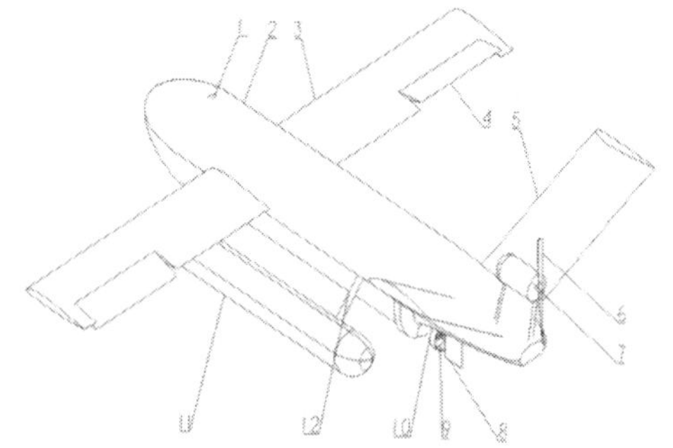

Figure 2. Configuration (On the surface of water).

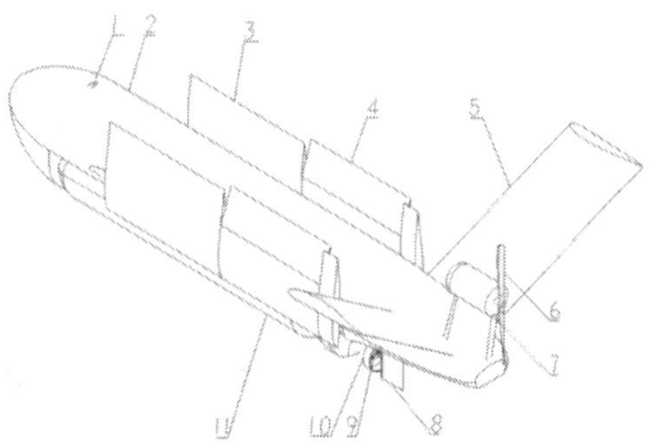

Figure 3. Configuration (Underwater).

Fig. 1 indicates the mode in the air. Fig. 2 shows the mode on the surface of water. Fig. 3 shows the mode

 DOI: 10.1051/matecconf/20179507023

underwater. Numbers in these picture means: 1. spiracle (contact with float bowls); 2. fuselage; 3. whirling wing; 4. aileron; 5. empennage; 6. air propeller; 7. diesel engine; 8. rudder; 9. underwater propeller; 10. electromotor; 11. float bowls; 12. connecting rod.

4 Configuration analysis based on CFD

4.1 Working conditions

In order to prove that the configuration of submersible UAV proposed in this paper is feasible and the scheme of whirling wing is effective, CFD method is used to analysis aerodynamic and hydrodynamic performance of the submersible UAV. 3D model is built by Catia. ICEM is used to repair geometrical model and generate mesh. Fluent is the solver.

Whirling wing is a key component. The wing keeps in the original place while flying in the air. In this mode, the UAV just like a fixed wing aircraft. While submerging underwater, the wing is set along the fuselage in order to reduce lift. If the lift is too large, the submersible UAV will not dive into water. Whirling wing's performance in different modes is mainly concerned, so we simply the model that is used in CFD analysis in order to improve computational efficiency.

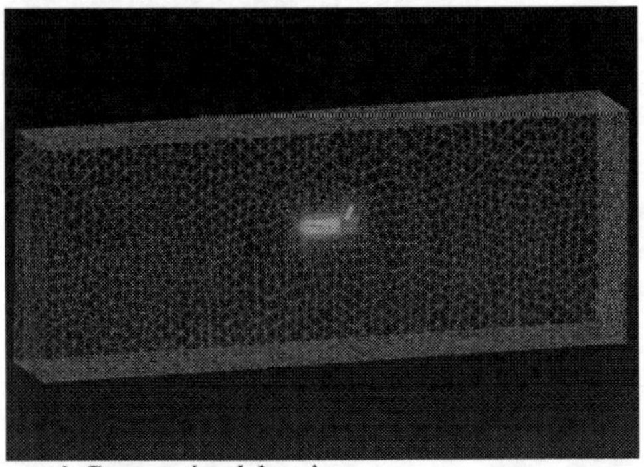

Figure 4. Computational domain.

As shown in Fig. 4, computational domain is 10 times larger than the UAV model.

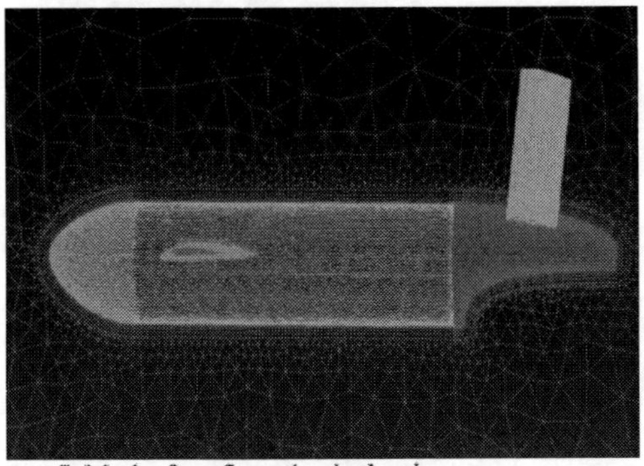

Figure 5. Mesh of configuration in the air.

Mode 1 of the submersible UAV is suitable for flying in the air and mode 2 of the submersible UAV is suitable for sailing underwater. Mesh for different modes are presented in Fig. 5 and Fig. 6.

Figure 6. Mesh of configuration underwater.

The boundary conditions: surface of the UAV is set as "Wall" boundary condition, the far field is set as "Wall" boundary condition. Inlet of the computational domain is set as "Inlet". Outlet of the computational domain is set as "Outlet" boundary condition.

Import geometry model of the submersible UAV in Fluent. Based on pressure method, turbulence model is S-A model in the air and $k-\varepsilon$ model underwater. Set velocity values and attack angle values based on working conditions of the UAV. Choose second-order upwind scheme and coupled method. Set the monitors to get lift coefficient and drag coefficient. Reference area of wing is $3.5m^2$, the length of average aerodynamic chord is 0.5m. Do not stop iterate until the results converged. After this, we can get data of the submersible UVA about its aerodynamic and hydrodynamic performance. The attack angle of working conditions is from 0 degree to 30 degree. The conceptual scheme proposed in this paper is designed for low speed, so we set the speed cases as 45 m/s and 60 m/s in the air, 2 m/s and 4 m/s underwater.

4.2 Analysis of aerodynamic and hydrodynamic performance based on CFD

As introduced before, to satisfy flying and submerging requirements, the submersible UAV has two different configurations. The configuration in the air is mode 1, the configuration underwater is mode 2.

By CFD method, in this paper, a serials of data are got about mode 1 and mode 2. Lift coefficient, drag coefficient and lift-drag ratio about the UAV in different working conditions will be shown as follows.

4.2.1 Aerodynamic performance

Pressure distribution in the case of "15 degree attack angle and 60m/s in the air" is presented in Fig. 7. Aerodynamic properties curves are presented in Fig. 8, Fig. 9 and Fig. 10.

According to the pressure distribution shown in Fig. 7, we can find the high pressure part and low pressure part, which is a visual display of lift and drag.

 DOI: 10.1051/matecconf/20179507023

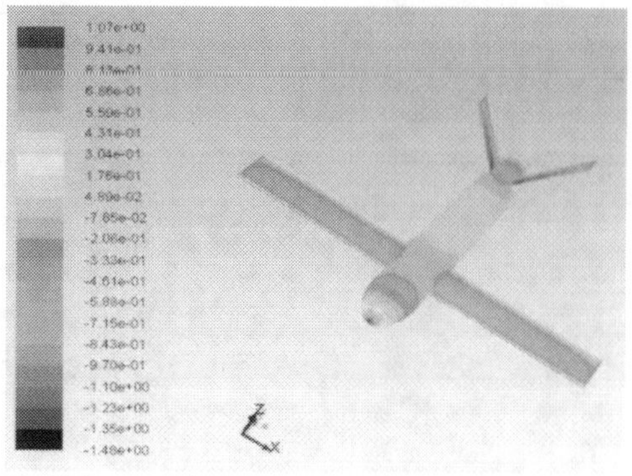

Figure 7. Pressure distribution of mode 1.

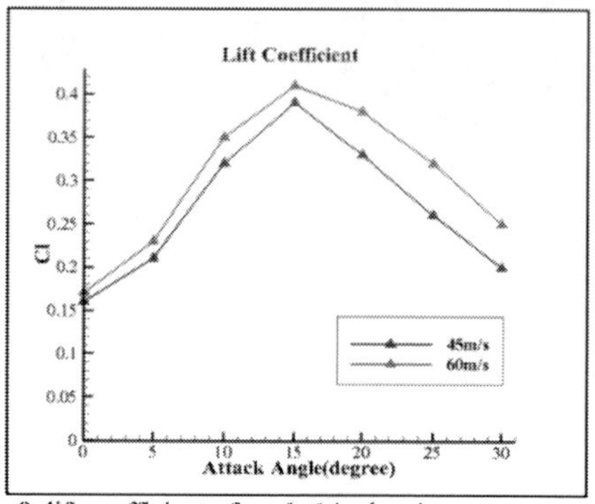

Figure 8. lift coefficient of mode 1 in the air.

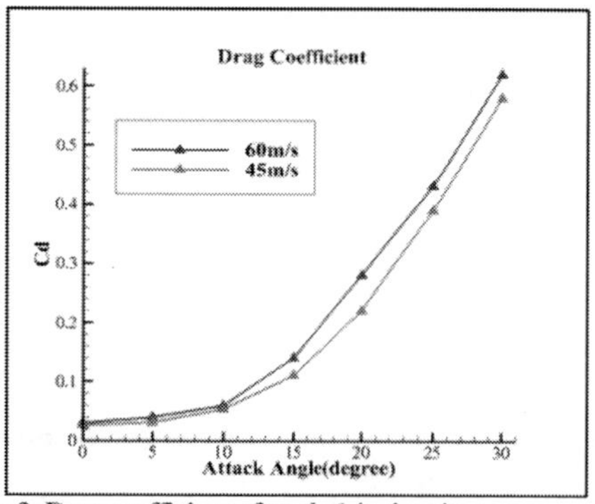

Figure 9. Drag coefficient of mode 1 in the air

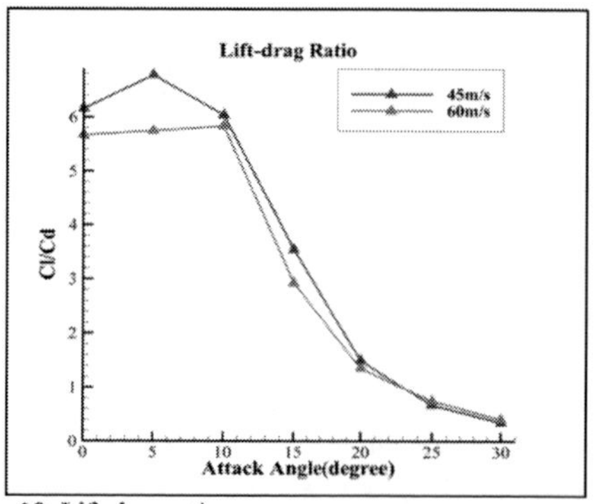

Figure 10. Lift-drag ratio.

From the aerodynamic properties curves, we know that the lift coefficient doesn't stop increasing with the attack angle's increase until 15 degree. After 15 degree, the lift coefficient is decreasing with attack angle's increase. Peak value of the lift coefficient appears in 15 degree. Drag coefficient is increasing with the attack angle's increase. In the whole process of attack angle's increase, the drag coefficient increases ceaselessly. The lift-drag ratio is an important parameter. Its peak value appears before 10 degree. Before 10 degree, the ratio nearly increases with attack angle's increase. After 10 degree, the ratio is decreasing with attack angel's increase.

Accordingly, mode 1 of the submersible UAV is feasible for its flying.

4.2.2 Hydrodynamic performance

Computational domain of hydrodynamic performance analysis is as same as it of aerodynamic performance analysis. And the mesh built in the former case can still be used in the new analysis. Of course, the boundary conditions are different. In the underwater simulation, turbulence mode should be set as $k-\varepsilon$ turbulence model [8]. And a bigger relaxing factor is set to improve the velocity of convergence.

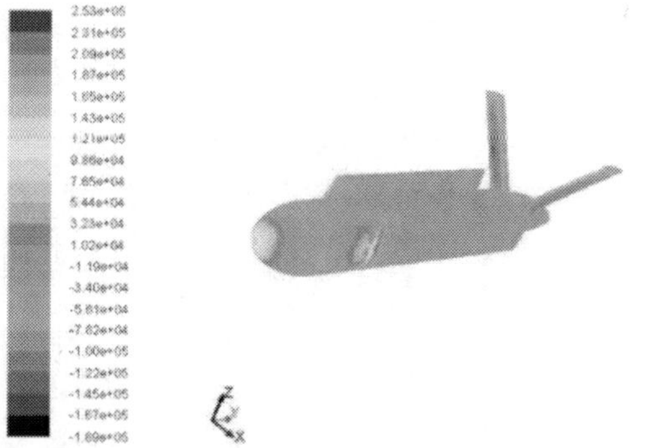

Figure 11. Pressure distribution of mode 2

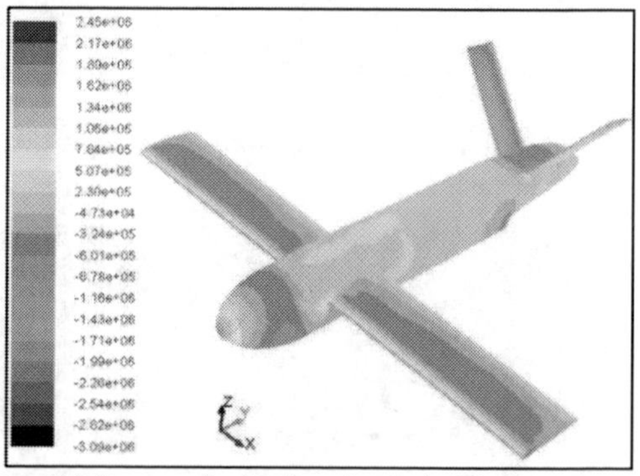

Figure 12. Pressure distribution of mode 1 underwater

In the first, we use mode 2 of the submersible UAV to make a hydrodynamic performance analysis. Fig. 11 shows the pressure distribution of mode 2 that the wing is already be whirled. The working condition is 4m/s with 15 degree attack angle underwater.

In order to prove that our scheme is effective, we also analyze the hydrodynamic performance of mode 1. Fig.

DOI: 10.1051/matecconf/20179507023

12 shows the pressure distribution of mode 1. The working condition is 4m/s with 15 degree attack angle underwater.

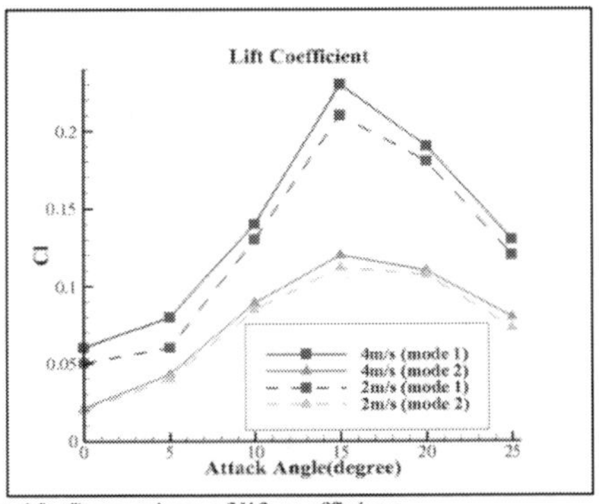

Figure 13. Comparison of lift coefficient

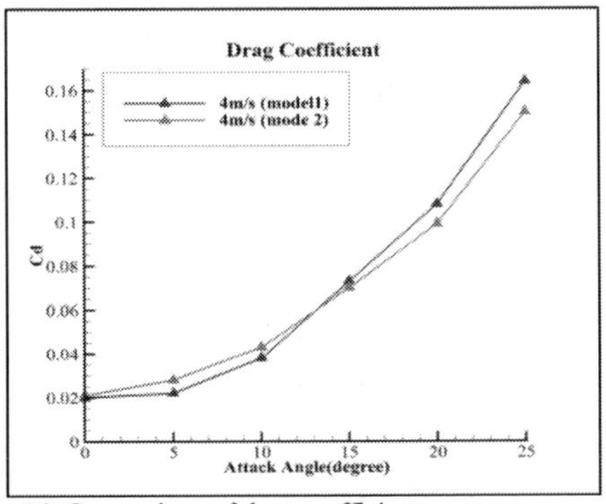

Figure 14. Comparison of drag coefficient

We get relevant data of the two modes and plot curves to compare their hydrodynamic performance. Fig. 13 shows the comparison of lift coefficient. Fig. 14 shows the comparison of drag coefficient. Fig. 13 contains 2 two working conditions of each mode. Fig. 14 contains only one working condition of each mode. The sailing velocity of the first case is 4m/s. And the velocity of the second case is 2m/s.

According to the curves shown in pictures above, lift coefficient increases before 15 degree and decreases after it. If velocity increases, lift coefficient will increase. Lift coefficient of mode 1 is much bigger than it of mode 2. Drag coefficient increases with the increase of attack angel. There is not a big gap between the drag coefficient of mode 1 and mode 2.

The analysis above means our design is effective. Whirling wing reduces the lift coefficient of submersible UAV underwater, which helps the UAV submerging underwater. And this design doesn't cause much increase of drag.

5. Conclusion

Some technological problems of submersible UAV are discussed in this paper. And a conceptual design scheme of submersible UAV that adopts whirling wing is proposed. The submersible UAV has two modes. By CFD analysis, we get a serials of data about aerodynamic performance and hydrodynamic performance of the UAV. Based on these data and property curves that plotted by these data, we conclude:

• Aerodynamic and hydrodynamic performance analysis of submersible unmanned aerial vehicle is useful for engineers to build feasible configurations of submersible UAV during the phase of scheme design.

• CFD analysis method is an excellent method to get lift coefficient, drag coefficient and pressure distribution of a UAV, which helps engineers to improve the performance of a UAV.

• Whirling wing configuration is capable of providing enough lift for flying and reducing lift for submerging by changing the mode of wing. This configuration is feasible for a submersible UAV.

Acknowledgment

This work was supported by the 2015 postgraduate students innovation fund, which provided by advanced aviation and spaceflight innovation organization in Beihang University. (YCSJ-01-2015-12)

References

1. L. David, R. Paul, Nature, **293**, 293-294, (1981)
2. W. Andrew, Z. Gregory, R. Robert, S. Tom, *45th AIAA Aerospace Sciences Meeting and Exhibit,* Reno, Nevada, (2007)
3. Y. Xingbang, L. Jianhong, *2013 Aviation Technology, Integration, and Operations Conference*, Los Angeles, US, (2013)
4. G. Yao, L. Jianhong, *2014 IEEE International Conference on Robotics and Biomimetics*, Bali, Indonesia, (2014)
5. X. Wenzhong, J. Qin, Journal of ship and guidance, **35**, 113-117, (2015)
6. J. Liping, S. Kang, Computer Aided Engineering, **20**, 97-101, (2011)
7. Y. Ming, G. Dongyuan, Y. Xiang, Mechanical design and manufacturing engineering, **10**, 21-26, (2014)
8. Z. Jiaqiang, F. Jinfu, X Qifeng, Acta Aerodynamica Sinica, **30**, 685-691, (2012)

Reactive Ion Etching Process of Micro Mechanical Pendulum

Wei Zhang , Xiaoyan Duan, Jin Qiao and Ping Yue

Sensing Technology Center of Beijing Information Science and Technology University, Beijing 100101, China

Abstract. This paper describes the reactive ion etching (RIE) technique of micro mechanical pendulum chip. Micro mechanical pendulum chip processed by the RIE has excellent performances and surface of the chip is smoother than the chip by wet etching. Properties of chip are closely associated with the process parameters. Reactive ion etching technique improves the problems such as intricate lateral etching and surface deficiency of the wet etching. On the basis of method and many experiments of the ion etching, relation between etching rate, etching uniformity, sidewall profiles and process parameters was analyzed and a set process parameters with greater etching effect was obtained. The set process parameters includes etching depth 55μm, etching uniformity (U) 0.63% and selectivity (P) 90:1, etching rate (V) 5.75μm/min, verticality of sidewall 90°±1°.

1 Introduction

Nowadays, micro mechanical system (MEMS) is an important direction of the developing-fast semiconductor industry, combining micro electronic technique with micro fabrication technique, mixing electronic and machine together. Miniaturization, high sensitivity, low cost, low energy consumption and so on are advantages of MEMS. MEMS is widely used in automobile industry, biological medicine, communications, aerospace, military and other fields [1].

Micro mechanical pendulum chip is used for sensing the inertial attitude of moving carrier and composed of silicon pendulum mass, frame and elastic torsion beam connecting silicon pendulum mass and frame [2]. Micro mechanical pendulum chip was processed by KOH wet etching with some disadvantages. There is a (111) surface on the silicon side forming an angle (54.7°) with (100) surface. Thus beam structure and sensitivity of micro mechanical pendulum are greatly influenced with increasing of etching depth. Lateral etching is serious in the process of wet etching and results in the low accuracy of size. In addition, reaction products of wet etching lead to more surface defects; temperature of wet etching liquor is difficult to remain stable and affects the consistency of etching rate; cost of handling waste liquid is high. Dry etching is known as another process of micro electronic technique. There are some outstanding advantages: high etching rate, high selectivity, good anisotropy, good uniformity in chip and better etching quality of profile [3]. In order to transfer patterning on mask layer to substrate with higher precision, we explore to process micro mechanical pendulum chip by reaction ion etching.

This study pays attention to analyzing relation between etching rate, uniformity, sidewall profiles and process parameters according to method and experiments, aiming to obtain ideal parameters of reaction ion etching and improve process accuracy, quality uniformity and reliability.

2 RIE process of micro mechanical pendulum chip

Ion etching adopts ASE™ process investigated by Bosch. Passivating and etching are conducted alternatively in the ASE™ [4]. Firstly depositing a layer of polymer keeps silicon sidewall from etching. Next polymer and Si are etched simultaneously. At last, keeping balance of depositing and etching realizes anisotropic etching.

Etching is carried out on the silicon substrate. Etching gas is SF6 and passivating gas is C4F8. C4F8 is passed into the chamber firstly, ionized and transformed into CF_x^+, CF_x^- and F^- in high frequency magnetic field and induced electric field. These free radicals diffuse to surface of wafer and polymer passivation coating is formed. With SF6 introduced, there is more F- reacting with Si and producing SiFx, as shown in following equations [5]:

$$C_4F_8 + e^- \rightarrow CF_x^- + CF_x^+ + F^- \tag{1}$$

$$CF_x^- \rightarrow (CF_2)_n \tag{2}$$

$$SF_6 + e^- \rightarrow S_xF_y^+ + S_xF_y^- + F^- \tag{3}$$

$$nCF_2 + F^- \rightarrow CF_x^- \rightarrow CF_2\uparrow \tag{4}$$

$$Si + F^- \rightarrow SiF_x\uparrow \tag{5}$$

In the experiments, same batch of double-sided polishing N100 silicon wafer (4 inches, thickness 375μm) are used for research. Every silicon wafer is composed of 4x3 micro mechanical pendulum chip unit and lithography is performed on every unit as chrome mask shown in Fig. 1 (colored for the required etching parts). Etching opening area is less than 40% and etching depth is 55μm. Conducting etching is based on DSE200E ion etching machine made by the North microelectronics company. Etching depth and shape of the window are measured by step instrument and scanning electron microscope.

Reaction chamber pressure, RF power, gas flow, ratio, etching time and passivating time are main parameters of process. These parameters determine the etching rate and uniformity. Select five points on the silicon wafer upper, middle, lower, left and right five positions as the test points and try to change process parameters and analysis the relation between these parameters and etching rate, uniformity, sidewall profile.

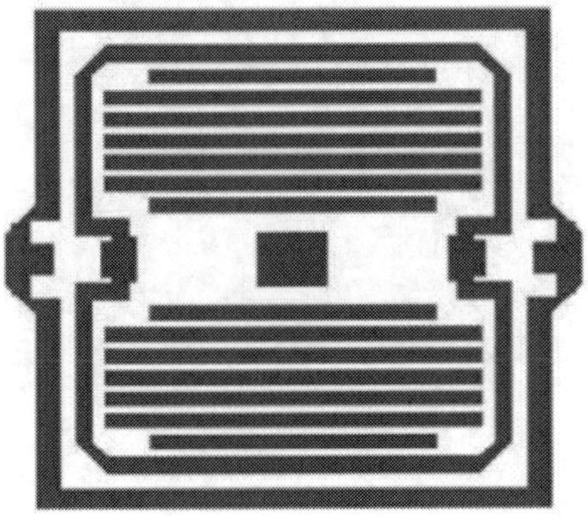

Figure 1. Chrome mask of lithography

Etching process is shown in Fig. 2. First, the silicon wafer was cleaned. Then a layer of SiO_2 was formed as masking layer through oxidizing silicon wafer in the high temperature. SiO_2 of every chip has same density, refractive index and other characteristics. Next, silicon wafer was etched. The process of lithography mainly includes spin-coating (speed is 2500r/min), prebaking (90°C), exposing, developing and postbaking (120°C). When etching was finished, SiO_2 masking layer exposed was needed to remove. Finally, photoresist was removed and reaction ion etching was conducted on silicon.

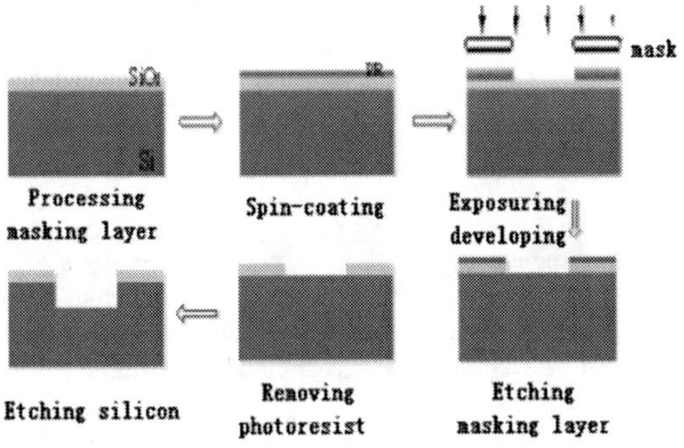

Figure 2. Etching process of micro mechanical pendulum chip

3 Methods and results

3.1 Etching rate

$$V = \overline{D} / T \tag{6}$$

$\overline{D}$ is the average value of etching depth, T is the total time of etching.

To study relation between ion etching rate and process parameters, etching rates were measured with different RF power shown in Fig. 3. It is presented etching rate increases with increasing of power. Increase of power results in the increase of ionization probability and more reactive particles. At the same time, there are more electrons absorbed by ungrounded RF source substrate, leading to higher negative potential, thus reaction rate is intensive and etching rate is larger. However, when the power is too high, bombarding energy will be larger and surface roughness of silicon wafer will be impacted.

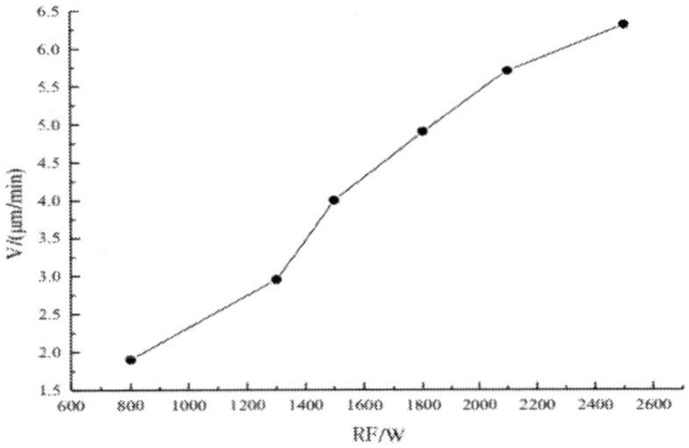

Figure 3. Relation between etching rate and RF power

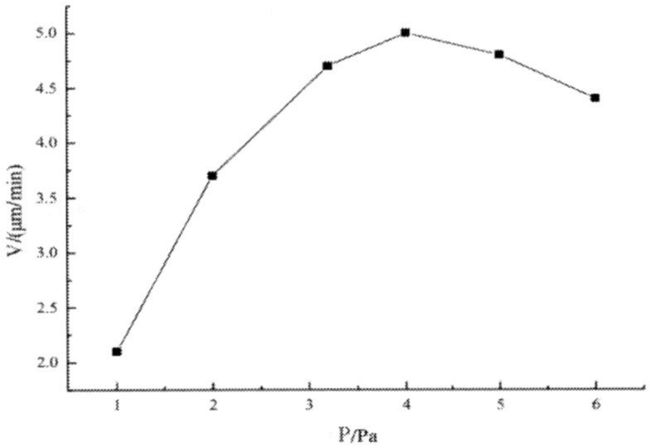

Figure 4. Relation between etching rate and chamber pressure

Etching rate is also related to pressure of process chamber, as shown in Fig. 4. If pressure is smaller, etching rate will increase with pressure rising. When pressure increases to a certain value, etching rate will dropped with pressure increasing. The reason is that increase of gas pressure results in increase of reaction gas concentration in process chamber. Consequently, chemical reaction is more intense and etching rate is quicker. With pressure increasing to a certain value, free radicals, ions and atoms participating in reaction reach saturation, and hence etching rate can not be enhanced with pressure rising. On the contrary, it shows a

DOI: 10.1051/matecconf/20179507024

decreasing trend with energy loss caused by colliding between reaction particles. What's more, flux of reactive gas has an effect on etching rate, which is similar to the effect derived from pressure of reaction chamber.

3.2 Etching uniformity

Etching uniformity directly affects quality consistency of chips. Typically, etching uniformity is calculated as following equation:

$$U = \left(D\max - D\min\right) \div \left(2\overline{D}\right) \times 100\% \quad (7)$$

D_{max} and D_{min} are the maximum and minimum of etching depth of five test points.

Silicon wafers were etched in different pressure and RF power. Etching rate of five test points are shown in Fig. 5. By calculating, it is reached that when the pressure increases, uniformity of etching will be deteriorated. Etching rate of the middle test point is slower than other test points and the difference from edge to center becomes more obvious with higher pressure. This is because the entire etching process consists of hundreds of etching cycles and etching gas is introduced in each cycle. At the moment of etching gas was passed into the chamber, test point in the middle was under the max pressure and other points were under small pressure, which caused the pressure of chamber is nonuniform. With gas in the middle flowing to around, difference of etching rate from center to edge is increasing. Therefore, etching uniformity is deteriorated.

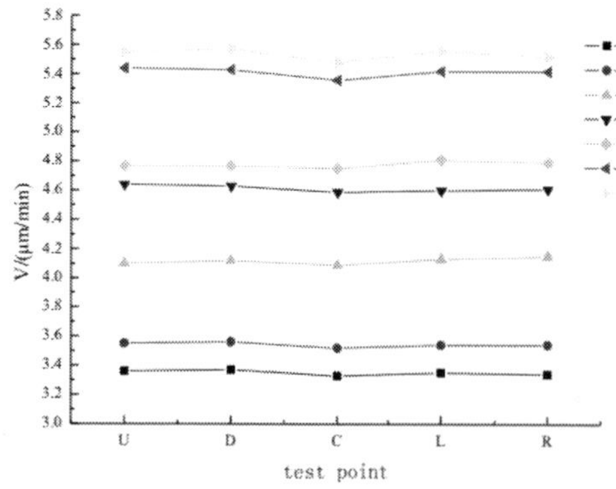

Figure 5. Etching depth of five points (U-upper, D-lower, C-middle, L-left, R-right)

3.3 Vertically of sidewall

Good sidewall profile is the most notable advantage compared with wet etching. Sidewall profile of chip processed in the higher electric power is shown in Fig. 6. Etching depth is larger and etching width is larger, that is to say, vertically of sidewall is over 90°. More reaction particles are excited as RF power increasing, but passivating gas is comparatively few and can not protect sidewall. Therefore, etching depth is larger and etching area is larger. Sidewall profile is shown in Fig. 7 with larger flow rate of C_4F_8. Sidewall inclination is smaller and effect of etching is close to the wet etching. As flow rate of the passivating gas increases, more polymers are produced and enhance the protection of sidewall. As a result, reaction is weak and etching area is smaller with increasing of etching depth.

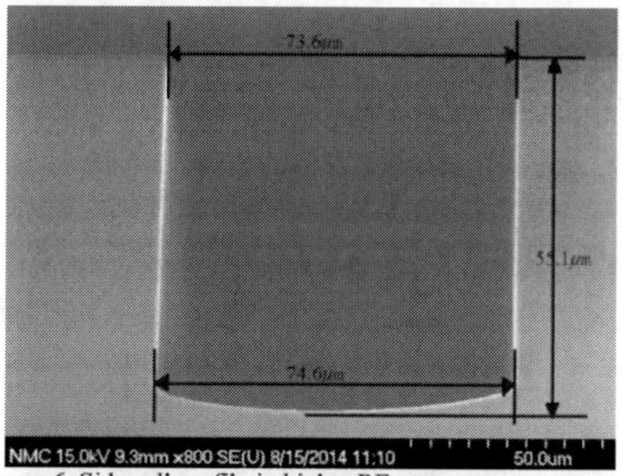

Figure 6. Sidewall profile in higher RF power

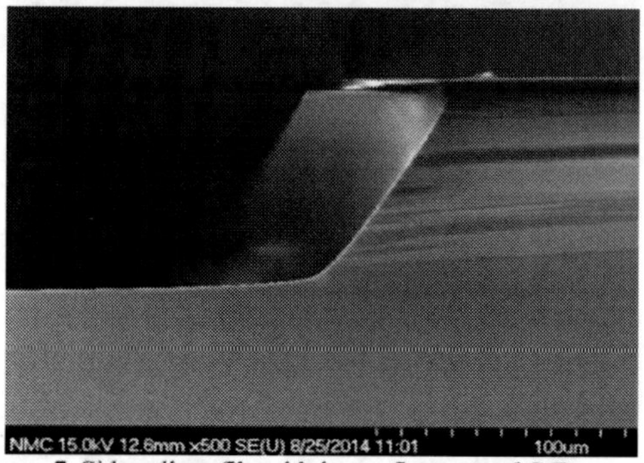

Figure 7. Sidewall profile with larger flow rate of C_4F_8

3.4 Analysis of ion etching

Considering influence of parameters on etching effect and development trend, a set process parameters with good etching effect is selected in the light of experiment and shown in Table 1.

Table 1. Parameters of ion etching process

Process parameters	Depositing	Etching
Pressure/Pa	4	4
C_4F_8/500sccm	170	0
SF_6/500sccm	0	300
Loading power/W	30	30
RF/W	1600	2000
Cycles/times	120	120
Time/s	1.5	2.2

Parameters of five points were measured by scanning electron microscope (SEM) and shown in Table 2. It can be obtained etching rate (V) is 5.75μm/min from Table 2 and Eq. (6). Uniformity (U) is 0.63%, according to Eq. (7) and Table 2.

Table 2. Measuring results of five points by SEM

	Etching depth/μm	Etching width of top edge/μm	Etching width of lower edge/μm	Angle/°	V /(μm/min)
Upper	55.0	75.7	75.9	89.90	5.75
Lower	55.1	75.2	75.1	89.95	5.76
Middle	54.6	75.7	76.0	89.84	5.70
Left	54.9	74.8	74.8	90.00	5.74
Right	55.3	75.2	75.9	89.64	5.79
Maximum difference	0.70	0.90	1.20	0.36	—
Average value	54.98	75.32	75.54	89.87	5.75

As parameters of Table 1, the ion etching was performed and the measuring results shown in Table 2 indicate that angle of Si side section is up to 90°±1° and vertically is far larger than the wet etching. Patterning on mask layer is preferably duplicated. Width of lateral etching is smaller than the wet etching, shown as Fig. 8 and Fig. 9. The black area on both sides of the damping strip is inclined plane produced by lateral etching.

Surface of micro mechanical pendulum chip is shown as Fig. 10. It is obvious that the surface is more smooth and no pits, no reaction residues comparing with the wet etching. Fig. 11 is the photo of silicon wafer surface by wet etching.

Before the ion etching, the surface of the wafer should be checked. Developing time should be accurately grasped during the etching process. The application of ion etching improves the quality of the chip, rate of finished products and reduces the problem of waste liquid treatment. However, it is needed that the equipments of ion etching are maintained regularly.

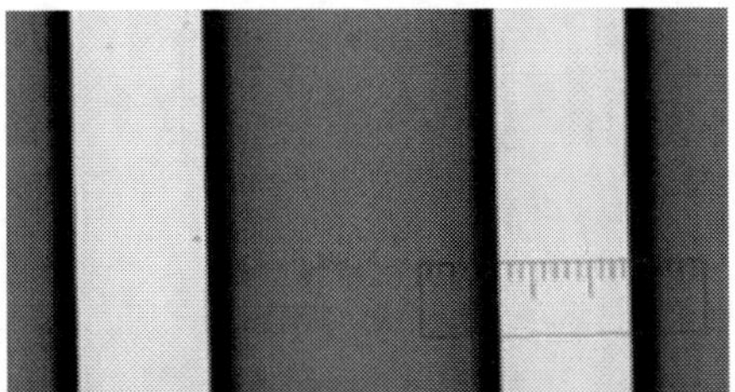

Figure 8. Damping strip of ion etching

Figure 9. Damping strip of wet etching

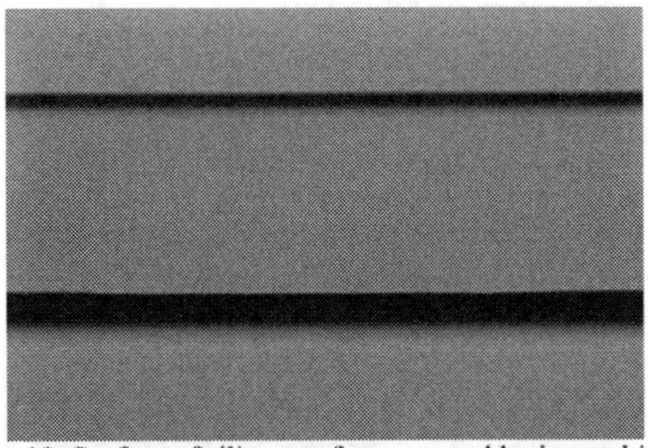

Figure 10. Surface of silicon wafer processed by ion etching

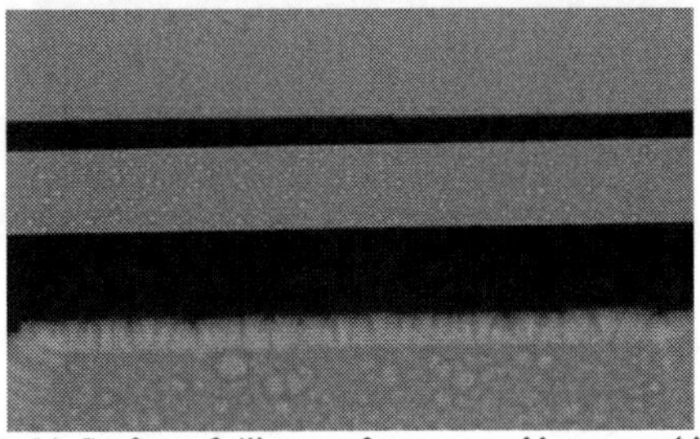

Figure 11. Surface of silicon wafer processed by wet etching

4 Conclusion

This study is about exploring to product micro mechanical pendulum chip by reaction ion etching. Combining experiment and method, a set parameters of etching process has been presented and it is concluded that etching rate, uniformity, sidewall profile are relevant to RF and pressure of chamber.

(1) Etching rate increases with increasing of RF and it will not increase when RF reaches a certain value.

(2) As chamber pressure is larger, etching rate increases first and then decreases.

(3) Increasing of chamber pressure causes uniformity is worse.

(4) In certain conditions, etching rate (V) is up to 5.75μm/min and is about four times as large as the rate of wet etching (1.4μm/min) [2], thus taping out time of micro mechanical pendulum is shorter. Vertically of sidewall (90°±1°) and uniformity (0.63%) are good. By contrast to the wet etching, surface of micro mechanical pendulum chip is smoother and no harmful gas is

produced during processing. In summary, reactive ion etching process provides experimental basis for improving quality consistency and reliability of the micro mechanical pendulum chip.

Acknowledgment

This research was supported by the National Nature Science Foundation of China subsidization project (61372016) and Beijing key laboratory of sensors open project fund project (2015XXX).

References

1. Z.P. Zhao. New developments in typical MEMS and wearable sensing technologies J. *Micro/Nano Electronics*. **1**, 1-4 (2015)
2. F.X. Zhang, W. Zhang. Micro mechanical gyroscope without driving structure and its application. Beijing: *National Defense Industry Press*. 302-306 (2013)
3. T. Mishima et al. Crystalline anisotropic dry etching *J. MEMS 2011*. January 23-27. Cancun. MEXICO. 221-224 (2011)
4. X.C. Jing et al. Optimization of silica etching parameters of the dry etching *J. Manufacturing technology*. **6**(1), 37-40 (2005)
5. Y. G. Li, S. J. Chen. Optimization of dry etching parameters for Si based on high aspect ratio J. *MEMS device and technology*. **12**(46), 750-754 (2009)

 DOI: 10.1051/matecconf/20179507025

Evaluate of Chill Vent Performance for High Pressure Die-Casting Production and Simulation of Motorcycle Fuel Caps

Piyanut Meethum[1] and Chakrit Suvanjumrat[2,3]

[1]Department of Mechanical Engineering, Faculty of Engineering, Siam University, Bangkok, 10163, Thailand
[2]Department of Mechanical Engineering, Faculty of Engineering, Mahidol University, Salaya, Nakorn Pathom, 73170, Thailand
[3]Laboratory of Computer Mechanics for Design (LCMD), Department of Mechanical Engineering, Faculty of Engineering, Mahidol University, Salaya, Nakorn Pathom, 73170, Thailand

Abstract. Inappropriate air vent design always induced porosities in thin and complex shape of high pressure die-casting (HPDC) products. Chill block installation was preferred to improve permanent molds for reducing porosities. There were not regular to obtain an appropriated chill vent from this method unless an experience of mold makers. This research was performed to investigate and evaluate the chill block performance in the HPDC mold of motorcycle fuel caps. The HPDC mold of the thin and complex shape product without chill vents was improved by inserting of chill blocks. The motorcycle fuel caps of molds with and without chill vents were compared to verify size and position of porosity defects. Furthermore the HPDC simulation using the finite element analysis (FEA) had been performed according to the physical experiment. The FEA results were compared and in agreement with the experimental data. Therefore an improvement of the HPDC mold should be used FEA to obtain the best way of the air vent design.

1 Introduction

The gas entrapment during high pressure die casting (HPDC) processes always induces porosity occurring in many products [1]-[3]. The thin and complex shape of products had more occasion to meet porosities on its surface [4]. The causes ofporosity might be formed by unsuitable gate, runner, overflow and vents design [5, 6]. Reduction of porosities in parts could start with an original design of runner geometry [7]. Chill block installation is the additional air- or gas-vent for permanent HPDC molds after usage and porosity detecting [8]. The efficiently exhausting residual air from the inside to the outside of mold cavity is mean by chill vent features. Unfortunately, there are destitute technique papers to use chill blocks then the mold maker has to use experience for the chill block selection.

Nowadays, computer-aided engineering (CAE) is an advance technology to design the HPDC product. The porosity in HPDC products could be assumed from the gas entrapment phenomenon using finite element analysis (FEA) [9]-[12]. Porosity could be reduced by optimization shapes of runner, gating systems or process parameters [13]-[15]. The FEA not only uses for design but also advances to improve permanent molds. Effects of the chill block installation could be simulated to determine a suitable shape and size of chill blocks.

This research attempted to use FEA to specify the chill vent performance. The simulated results were compared with experimental data and evaluated accuracy of simulation techniques. The procedure from simulation of the HPDC process will be used to improve further HPDC parts which have thin and complex shape for better surface quality.

2 HPDC experiment

Mold cavity of motorcycle fuel caps had been design with the important feature comprised of gate, runner and vent system. The chill block was inserted to increase zigzag profile chill vents inside an original mold with width, length, height, pitch and gap thickness of 45, 95, 8, 9 and 1 mm, respectively (Fig. 1). The die-casting machine model TOYO-350T was employed to cast fuel caps with ACD12 alloy. The HPDC process was setup an accumulator pressure of 12 MPa, an injection high speed of 2.3 m/s, an aluminum temperature of 660 °C and a mold temperature of 120 °C.

Sampling of HPDC parts which is removed from the original mold cavity is shown in Fig. 2. Some features of the HPDC part such as vents were broken while removing from its mold therefore the HPDC part not had completely components as same as mold cavity. Small porosities were observed distinctly on surface of motorcycle fuel caps. The interesting porosities on surface of a left side cap of a symmetry HPDC part was closed by circles (A-, B- and C-circle). The scanning electron microscope (SEM) was employed to reveal characteristic of porosities. Micrographs of porosities in

A-, B- and C-circle are shown in Fig. 3. Defects were deep then it was seen as a dark hole. The maximum length of porosity in A-, B- and C-circle was measured which obtained 0.278, 0.374 and 0.188 mm, respectively. The HPDC part of the improved mold is shown in Fig. 4. Porosities still appeared on same positions of the original mold but the characteristic of porosities had changed. The porosity on an improved part was shallow more than an unimproved part and could not specify the maximum porosity length in A-, B- and C-circle. Fig. 5 shows the micrograph of porosities on a HPDC part of the improved mold.

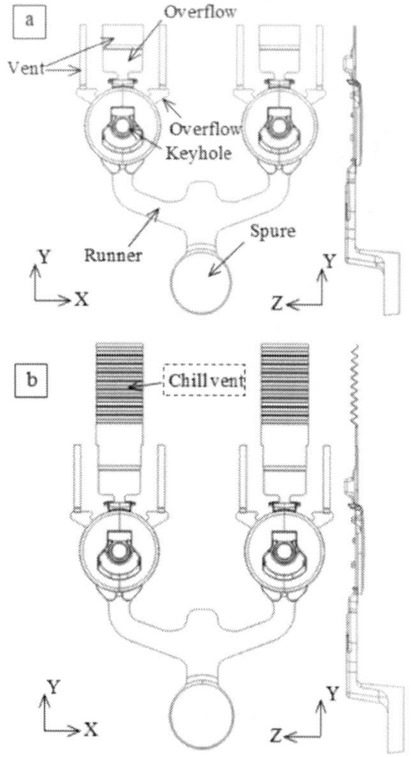

Figure 1. Motorcycle fuel caps of: (a) an original mold cavity and (b) an improved mold cavity.

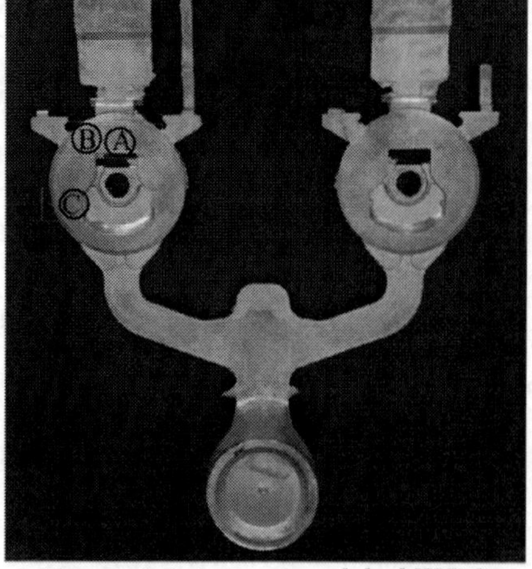

Figure 2. Motorcycle fuel caps of an original HPDC part.

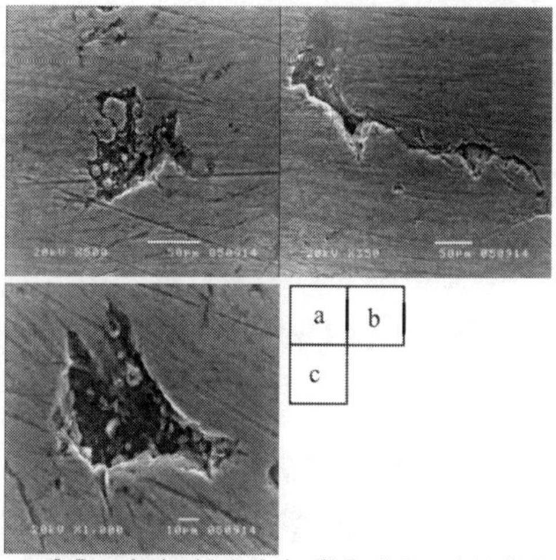

Figure 3. Porosity in: (a) A-circle, (b) B-circle and (c) C-circle on an original HPDC part.

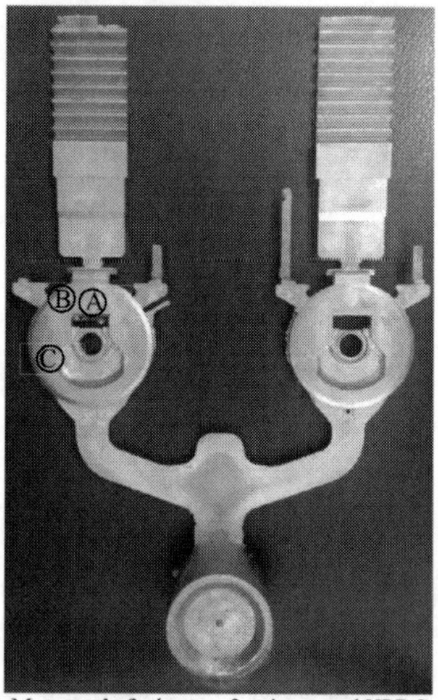

Figure 4. Motorcycle fuel caps of an improved HPDC part.

3 HPDC simulation

Mold cavity for producing motorcycle fuel caps in the previous section was modeled by computer aided design (CAD) software, SolidWorks version 2011. Subsequently, the mold cavity model was exported for the standard file, .stp file, to use in the CAE software, CAST-DESIGNER software, to create finite element (FE) model. The FE model was test convergence of results of mesh quantity. The 413,615 and 469,320 hexahedral elements are used to divide the cavity model and the FE model is shown in Fig. 6. The maximum size of elements was 1.0 mm which far away from an interesting area. The smallest element which had size about 0.7 mm was controlled to locate around the keyhole of the motorcycle fuel cap. The material property of molten aluminum was

DOI: 10.1051/matecconf/20179507025

defined into the FE model as same as the experiment data. Boundary conditions such as the injection high speed and the aluminum temperature were specified on the FE model also according to the HPDC process parameters. The personal computer with Core-i5 CPU and RAM memory of 4 GB was used to perform the HPDC simulation.

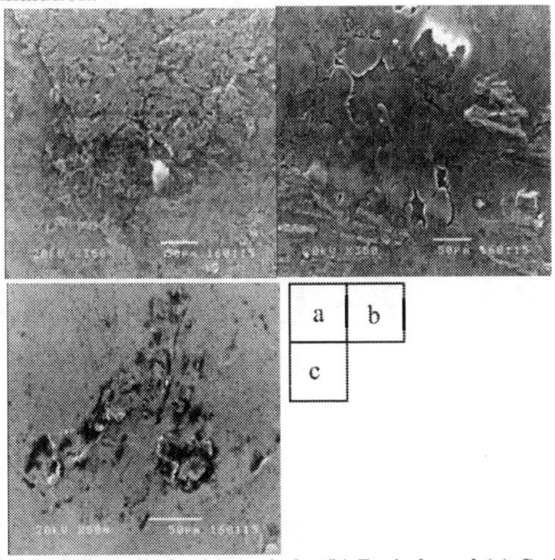

Figure 5. Porosity in: (a) A-circle, (b) B-circle and (c) C-circle on an improved HPDC part.

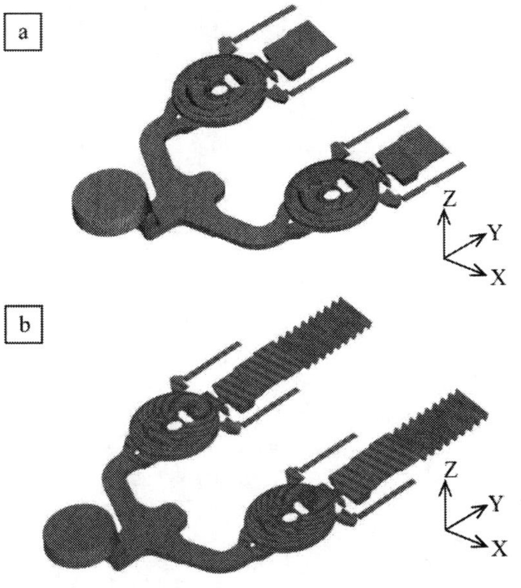

Figure 6. The finite element model of: (a) an original mold cavity and (b) an improved mold cavity of motorcycle fuel caps.

4 Results and discussion

The sequent images of molten aluminum filling into mold cavities of the original and improved mold are shown in Fig. 7 and Fig. 8, respectively. The molten aluminum velocity was illustrated by a color contour. The maximum velocity was signified by red and minimum velocity was blue. The unfilled space occurred at twelve o'clock position of a left and right cap cavity caused molten aluminum velocity reduced to zero around key hole. This area had many ribs to obstruct molten aluminum flow therefore was risk to occur an air entrapment phenomenon which induced porosities happened on surface of the motorcycle fuel cap. Porosity defects occurred after motorcycle fuel caps fully filled with molten aluminum in the filling time of 0.03 sec. Final points of porosity in mold cavities are indicated by shape of black volumes as shown in Fig. 9. The simulation results were distinct porosity location of porosities typically occurred on a left and right cap of the original and improved mold cavity of FE model. The maximum size of porosities in A-, B- and C-circle on the original and improved mold cavity are regarded to compare with the experiment data. The maximum size of porosities in circles A, B and C reduced from 0.383, 0.499 and 0.465 to 0.358, 0.469 and 0.433 mm, respectively. Even though the porosity was not eliminated from fuel caps by improvement of the HPDC mold, it could be easy to remove by polishing process.

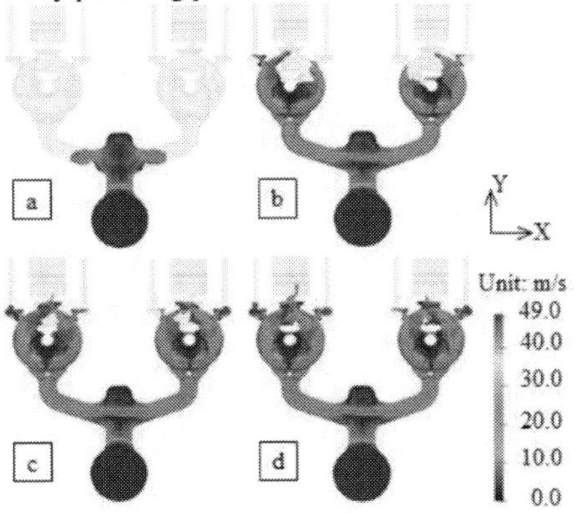

Figure 7. Sequent images of molten aluminum filling into an original mold cavity at: (a) 0.014, (b) 0.020, (c) 0.022 and (d) 0.024 sec.

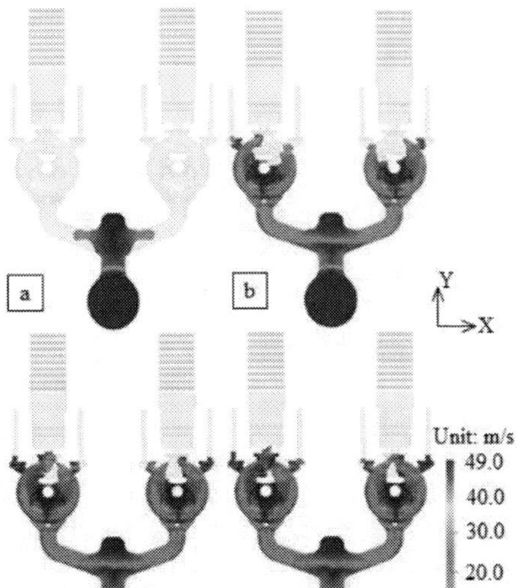

Figure 8. Sequent images of molten aluminum filling into an improved mold cavity at: (a) 0.014, (b) 0.020, (c) 0.022 and (d) 0.024 sec.

5 Conclusions

The HPDC mold had unsuitable air vent was used to produce aluminum fuel caps therefore porosities occurred on surface nearly the keyhole. CAST-DESIGNER which was FEA software was used to simulate the HPDC process. The FEA results illustrated the keyhole of fuel cap obstructed flowing path of the molten aluminum therefore it induced porosities on the motorcycle fuel cap surface where the molten aluminum could not flow. The porosity defects of FEA results had locations in agreement with the experiment data. The maximum size of simulated porosities was compared with experiments obtained the average different length of 0.169 mm on the original HPDC part. Porosity was reduced at an average size of 6.47% when chill blocks were inserted in the original mold. The comparison confirmed an accuracy of the HPDC software therefore the setting parameters of the simulation will be useful for the improvement of fuel cap molds. The appropriate shape of fuel caps under a comfortable flow of molten aluminum will be considered with design in further works.

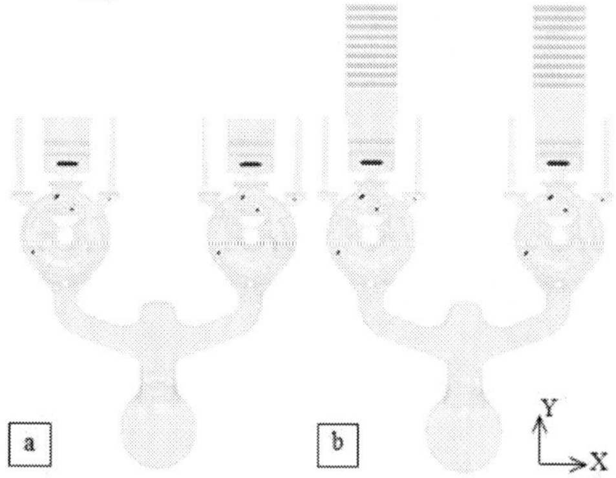

Figure 9. Images of porosity on surface of: (a) an original and (b) improved HPDC part.

Acknowledgement

This research was funded by The Thailand Research Fund (TRF) and Micro Precision Co., Ltd under Research and Researchers for Industries (RRI): Master Degree-grant no. MSD57I0137. The CAST-DESIGNER software had permitted for using a license by 4D Corporation Ltd.

References

1. Y. Hangai, S. Kitahara, Mater. Design **30** (2009)
2. J.J. Hernandez-Ortega, R. Zamora, J. Lopez, F. Faura, Int. J. Adv. Manuf. Technol. **67** (2013)
3. A. Wilczek, P. Dlugosz, M. Hebda, J. Nondestruct. Eval. **34** (2015)
4. E. Aghion, N. Moscovith, A. Arnon, Mat. Sci. Eng. A 447, 341 (2007)
5. S.G. Lee, A.M. Gokhale, G.R. Petal, M. Evans, Mat. Sci. Eng. A **427**, 99 (2006)
6. L. Wang, P. Turnley, G. Savage, J. Mater. Process. Tech. **211**, 1510 (2011)
7. R. Malterer, US Patent 0016710A1. (2005)
8. D.R. Gunasegaram, M. Givord, R.G. O'Donnell, B.R. Finnin, Mat. Sci. Eng. A **559**, 276 (2013)
9. H.D. Zhao, F. Wang, Y.Y. Li, W. Xia, J. Mater. Process. Tech. **209** (2009)
10. H.D. Zhao, Y.F. Bai, X.X. Ouyang, P.Y. Dong, T. Nonferr. Metal. Soc. **20**, 2064 (2010)
11. C.K. Jin, C.H. Jang, C.G. Kang, J. Mech. Sci. Technol. **27**, 10 (2013)
12. O. Boydak, M. Savas, B. Ekici, Int. J. Metalcasting **10**, 1 (2016)
13. C. C. Tai, J.C Lin, J. Mat. Process. Tech. **84** (1998)
14. R. Singh, R. Kapoor, J. Inst. Eng. India Ser. C **94**, 2 (2013)
15. B.V. Ramnath, C. Elanchezhian, V. Chandrasekhar, A.A. Kumar, S.M. Asif, G.R. Mohamed, D.V. Raj, C.S. Kumar, P. Mat. Sci. **6**, 1312 (2014)

Experimental Studies of Composite Sandwich Columns with Face/Core Debond under Axial Compression

Chen Yue, Zhu Xi , Zhu Zi-xu and Li hua-dong

Dept. of Naval Architecture Engineering, Naval Univ. of Engineering, Wuhan, China

Abstract: Experimental studies of buckling failure and debond propagation were carried out on composite sandwich columns with face/core debond. The edgewise compression tests were conducted on two type specimens. Namely, one containin throughthe-width rectangle face/core debond, and the other with embedded circular interfacial debond. High-speed photography technology has been used to record the debond propagating paths. Two main failure modes of collapse were recorded in the tests: core shear buckling and mixed buckling.

1 Introduction

Sandwich columns basically consist of two thin and stiff load carrying face sheets, which are adhered to the opposite sides of the core. The core serves as a spacer that keeps the face sheets apart, giving the sandwich columns a high bending stiffness to weight ratio. However, the debond located at face/core interface is one of the most common flaws in the composite sandwich structures. It might cause a significant reduction in stiffness and strength, and lead to the buckling of the structure.

Thus, investigation and analysis of mechanical behaviors for the composite sandwich structures with interfacial debond are of great interest to engineers, and many valuable papers have been published [1]-[3]. Theoretical prediction formulas of global buckling and local buckling for sandwich structures can be found in Allen's book [4]. Woo-Min [5] analyzed the debond buckling and growth of one-dimensional beam/plate by analytical methods. Singh S B [6] summarized the post-buckling behavior and progressive failure of rectangular laminates under in-plane shear loads. Hassan M [7] performed analytical and finite element calculations to investigate the effect of core density, and face/core debonds on the overall buckling behavior of the sandwich composites. Wang S[8]developed a more accurate analytical theory to express the post-buckling mode shape, then used the method to calculate the strain energy release rate and predicted the debond propagation behavior of one-dimensional debond include through-width debond in beams. Moslemian R [9] investigated sandwich columns with implanted through-width face/core debond both experimentally and numerically, and the numerical results were in overall agreement with experimental results. Jeyakrishnan R P [10] carried out experimental, theoretical and FEA analysis on the buckling characteristic of honeycomb sandwich panel.

The purpose of this work is to study the influence of the debond buckling and compressive properties of composite sandwich columns with embedded rectangle or circular face/core debonds. Experimental studies have been carried out to study the failure modes, debond growth path and ultimate load capability.

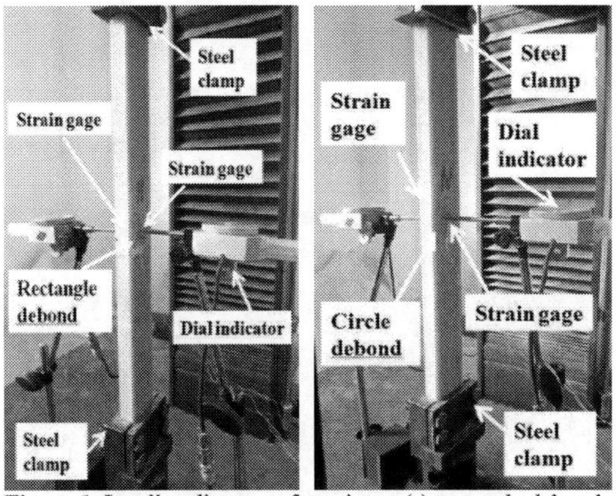

Figure 1. Loading diagram of specimens(a) rectangle debond (b) circular debond

2 Experimental investigation

2.1 Test specimens

Sandwich composite columns were manufactured by vacuum assisted resin infusion molding (VARIM) process. The sandwich composite columns consisted of core materials and pre-impregnated fiber material as face sheets. All layers of sandwich composite columns were laid along the direction of 0°. The fabric was plain weave

SW220 glass fibers. The core was high-performance buoyancy foam Divinycell H100. Primarily two types of sandwich columns with rectangle F/C debond and with circular F/C debond were prepared and tested. The cores were bonded to the face sheets with epoxy adhesive. The debonds were defined by implanting non-stick red teflon cloth in the face/core interface before processing. Only one of the face/core interfaces contain the implanted debond. As seen in Fig. 1, the debonded side of the column was marked as N, and the perfect side was marked as S.

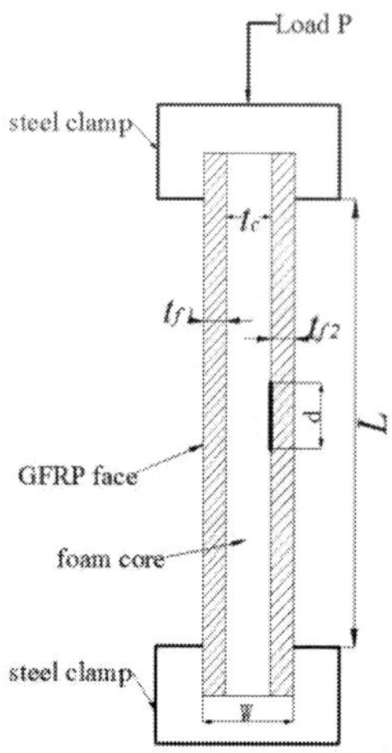

Figure 2. Geometric parameters of the specimen

Two types of specimens, namely containing rectangular face/core debond and circular interfacial debond, were fabricated to study the load capability and failure modes under edgewise compression. Three sets of both topologies sandwich composite columns with different length L, width W, face sheet thickness t_f,core thickness t_c and debond sizes, as shown in Fig. 2, were studied in the damage propagation investigation.

2.2 Test set-up

In order to assess the influence of face/core debond, edgewise compression tests were performed on sandwich composite columns. Loading was applied by a electro-mechanical actuator with an axial capacity of 50kN. Both ends of the specimens are clamped tightly to restrict the rotation. The compression tests were conducted with a displacement rate of 0.3mm/min. A high-speed camera was used to track the debond propagation at a frame rate of 4000 images per second. To record the in-plane strains, strain gages were bonded in the centre of both sides. These polyimide backing strain gauges with the resistance 120.0±0.1Ω were produced by Shanxi Hanzhong Precise Electric Apparatus Co., LTD. The sensitive grid size of the strain gage is 5.0×1.9mm, and the backing size is 8.7×3.6mm.

3 Compressive responses of sandwich composite columns

3.1 Failure modes

The buckling modes of debond sandwich composite columns could be global, local or mixed types. In the compression test, core shear buckling, core failure, face crushing, face wrinkling and debond propagating were found. Table 1 shows the dimensions of all sandwich composite columns, as well as the observed failure modes. The debond size 50 mm covers the entire width of the specimens

Table 1. List of sandwich composite specimens with face/core debond

Specimens type	Specimens number	Length L/mm	Width W/mm	Skin thickness $t_{f1}=t_{f2}$/mm	Core thickness tc/mm	Debond size /mm	Failure modes
Rectangle debond	R1	400	50	2	16	10×50	CSB+CF+ FC
	R2	400	50	2	16	20×50	MB+CF+FC
	R3	400	50	2	16	40×50	MB+CF+FC
Circular debond	C1	400	50	2	16	R=5	CSB+CF+ FC
	C2	400	50	2	16	R=10	CSB+CF+ FC
	C3	400	50	2	16	R=20	MB+CF+FC

Note: core shear buckling, core failure, mixed buckling and face crushing were abbreviated as CSB, CF, MB, FC, respectively.

As shown in Table 1, core shear buckling and mixed buckling were the dominant elastic macro-buckling modes for sandwich columns. Fig. 3(a) shows the failure mode of specimen R1 with a 10 mm rectangle face/core debond. Both Euler buckling and core shear buckling are global buckling modes, which often occurred in specimens containing relatively small debond. The main feature of the core shear buckling is that the core does not bent together with the face-sheets under compression, but the core shear occurred instead. The core shear buckling observed can be attributed to the weak shear stiffness of core.

The deformed mixed buckling configurations of the specimen C3 were showed in Fig. 3(b). Mixed buckling modes is a combination of global buckling and local buckling, which involved out-of-plane deflections for both face-sheets of the column. And we found that once the debonded face-sheet buckled, the debond tips propagated rapidly towards the ends of the specimen. Besides, crack kinking into the core was observed resulting in the crack propagating path changing from the interface to the core side. Typically and as a thumb rule,

 DOI: 10.1051/matecconf/20179507026

core shear failure happens at 45 degrees relative to the face sheets, whereas kinking happens under a 60 degree kinking angle, as seen in Fig. 3.

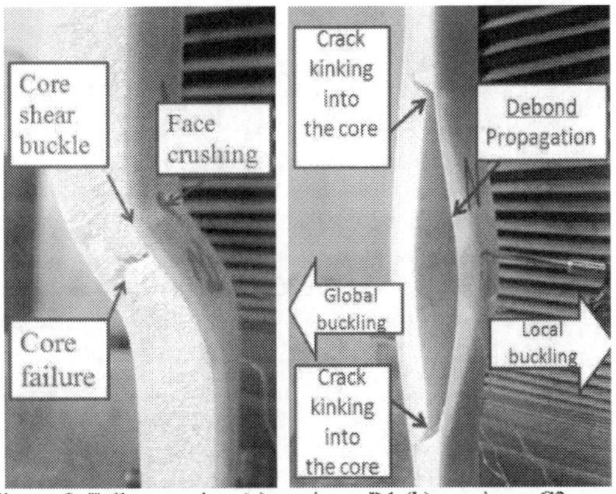

Figure 3. Failure modes: (a)specimen R1 (b) specimen C3

3.2 Debond propagating

The composite sandwich column R3 failed due to the crack propagation and face crushing. Representative photographs at various stages were taken by a high-speed camera during the testing. As shown in Fig. 4(a), under the compression load, firstly the structure became bended slightly, and the failure initiated in the form of debond on the interface. Subsequently, as the debonded zone of the face/core interface expanded, the debonded face-sheets were prone to buckle away from the core under compressive load. As we can see from Fig. 4(b), over the debonded zone, the face sheet buckled. Then, the debond may propagate along the interface or kink into the core. We can see clearly from the Fig. 4(c) that the debond were propagated mainly on the face/core interface. Only a small amount of foam was peeled and adhered to the debonded face-sheet in Fig. 4(d). The debond may kinking into the core when the following inequality is satisfied:

$$\left(\frac{\max G}{G_{\mathrm{I},cr}}\right)_{core} \succ \left(\frac{G}{G_{cr}(\gamma)}\right)_{int}$$

The $G_{\mathrm{I},cr}$ is the critical strain energy release rate for the core material in mode I, and $G_{cr}(\gamma)$ is the critical interfacial strain energy release rate. Slight crack kinked into the core resulting in the crack propagation path changed to the core side. With the crack propagation, out-plane bending tension strain and shearing strain may occur in the tips of the debond zone. And these interaction may cause shear failure of the foam. Eventually, the entire sandwich specimen collapsed in fiber compression failure and lost the capability of load bearing.

As shown in Fig. 4, for sandwich columns R3 with through-the-width face/core debond, both fronts of the debond propagated towards the specimen ends rapidly. While for specimens with embedded circular face/core debonds, the debond propagated perpendicularly along the compression direction until form a through-the-width debond. Then, the debond propagated along the compression direction. After propagation, both two types of specimens collapsed by face-sheet compression failure and core shear failure.

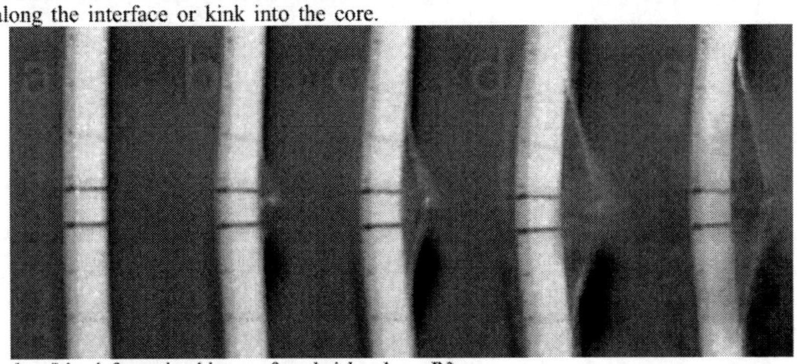

Figure 4. Photographs of the deformation history of sandwich column R3

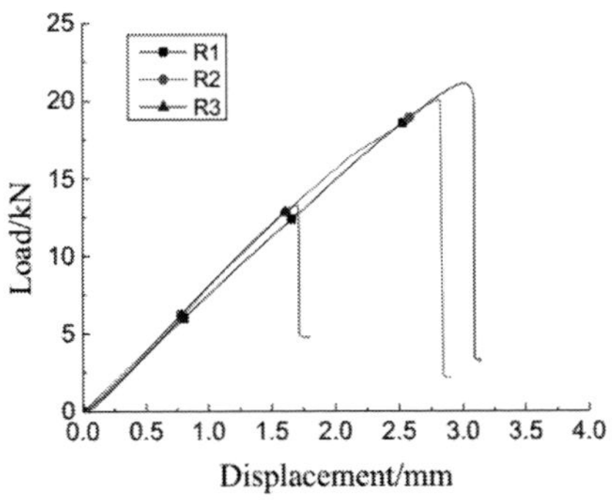

Figure 5. End shorting-Limit load curves of specimens R1-R3

3.3 Load capability

The load-shortening curves of all specimens are depicted in Fig. 5 and Fig. 6. The curves are almost linear after the initial stiffening region, and a reduction in the overall stiffness is observed at the final stage of the curve. As we all know, the overall stiffness was estimated as the tangent to the curves. It is evident that the increase of debond size weakened the ultimate load capability of the structure. The responses of the columns at this stage of loading were governed by a combination of buckling and debonding propagation. As the compressive load reached the critical load, core shear crushing and fiber compression failure

 DOI: 10.1051/matecconf/20179507026

were found. Simultaneously, the overall stiffness decreased, the resisting force of the column decreased sharply.

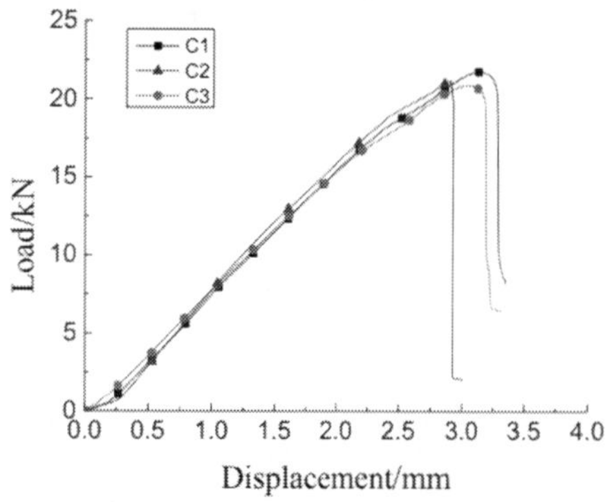

Figure 6. End shorting-Limit load curves of specimens C1-C3

4 Concluding remarks

The compressive response and failure mechanism of composite sandwich columns with rectangle or circular face/core debonds under edgewise compression were investigated in this paper. The main results can be outlined as the following:

During the experimental study, several modes of buckling and failure modes, namely, core shear buckling, mixed buckling, interfacial debonding, core shear failure and fiber compression failure, have been observed. Most specimens failed due to the delaminated buckling followed by rapid debond propagation. And the debond size has a significant effect on the buckling failure modes. Generally, as the increase of the debond size, the buckling load was on the decline. When the debond size exceeded a certain value, the failure mode changed from global buckling to mixed buckling.

References

1. Libove C, Butdorf S B. A general small-deflection theory for flat sandwich plates[M]. NASATN 1526, (1948).
2. Zenkert D. An introduction to sandwich construction[M]. EMAS, London, (1997).
3. Plantema F J. Sandwich construction: the bending and buckling of sandwich beams, plates and shells[M]. New York: John Wiley and Sons, 1966.
4. Allen H G. Analysis and Design of Structural Sandwich Panels[M]. Pergamon Press, London, (1969).
5. Woo M K., Chun G K. Debond buckling and growth of composite laminated plates with transverse shear deformation[J]. Composite Material, **29**:2047 –2068, (1995).
6. Singh S B,Kumar A. Postbuckling response and failure of symmetric laminates under in-plane shear[J]. Composite Sciencen & Technology, **58**(12):1949–1960, (1988).
7. Hassan M, Syful I, Mrinal S, et al.Buckling of sandwich composites:effects of core–skin debonding and core density [J].Applied Composite Materials, **12**(2):73–91, (2005).
8. Wang S, Harvey C M, Wang B, WatsonA. Post-local buckling-driven debond in bilayer composite beams[J]. Composite Structures, **133**: 1058–1066, (2015).
9. Moslemian, R., Berggreen, C., Carlsson, L. A. and Avilés, F., "Failure Investigation of Debonded Sandwich Columns: An Experimental and Numerical Study", Journal of Mechanics of Materials and Structures, **4**(7-8):1469–1487, (2009).
10. Jeyakrishnan P R, Chockalingam K K S K, Nara yanasamy R. Studies on buckling behavior of honeycomb sandwich panel[J]. International Journal of Advanced Manufacturing Technology, **65**(5-8):803-815, (2013)

The Vibration Impact Determination of the Helicopter Structural Components

Zeinab Khaksar[1], Sreenatha Anavatti[1], Krishna Shankar[1], Alessandro Ceruti[2] and Heath Pratt[1]

[1]University of New South Wales, the School of Engineering and Information Technology, Canberra, Australia
[2]University of Bologna, the School of Engineering, Bologna, Italy

Abstract. This paper presents the determination of the vibration impact of the helicopter structural components and skin repairs in terms of frequency characteristics. To address this issue, a 3D Finite Element Method (FEM) model of 349 Gazelle helicopter has been developed in ABAQUS and the frequency analysis is conducted. The results on the natural frequencies of the full structure reasonably match with the literature giving confidence in the baseline model. The main advantage of this FEM model is that, it can be used to predict the natural frequencies of the full structure, precisely. In addition, the material properties and conditions of the components can be updated based on the applied conditions during the repair and maintenance period. Thus, the model gives a comprehensive design tool for analysing the frequencies of the helicopter with differing components. The effective variations in the frequency changes due to repair are predicted numerically. The discussion of these results helps in developing leads to improved selection of replacement materials and their properties.

1 Introduction

One of the critical facts about the rotorcraft is that it has more complexity in both design and maintenance phases in comparison with the fixed-wing aircraft [1], [2].This is due to the purpose for which the rotorcraft has been inherently designed, for example violent, aggressive and rescue operations [3].

One of the most significant barriers on the development of the rotorcraft is the vibration which decelerates the advances in reliability, safety, occupant's comfort and increasing the maximum forward speed [4]. Moreover, the vibration in helicopters makes difficulties in reading the instruments, weapon operations, depreciation and failure in structural and non-structural components. Therefore, helicopters need to be monitored periodically and faulty components have to be replaced with the new ones [5]. By changing the faulty component, especially structural components focused in this research; with a new one (which could not be exactly the same as its original counterpart) there will be changes in the vibration characteristics of the system.

By 1990s, the vibration had not been considered in design phase of a helicopter, but some passive approach to capture it [5]. Nowadays, in order to minimize the costs, the vibration is being taken into account at design stage [6]. With this aim, there is a need to evaluate the vibration primarily. Therefore, obtaining a model of the structure as precise as possible, an aerodynamic model to simulate forces acting on the main rotor head, and the coupling between the fuselage and the main rotor are completely essential [7]. In this research, the impact of structural components on free vibration of the helicopter will be determined. During repair and maintenance of a helicopter, it is necessary to fix the cracks and damages on the fuselage or on the helicopter skin. But making new patches on the helicopter skin will lead to changes in vibration characteristics of the helicopter structure. This paper intends to determine the vibration impact of the helicopter structural components and skin repairs in terms of frequency characteristics.

1.2 The structural model of the helicopter

Yeo and Chopra (2001) consider the Finite Element Method (FEM) modelling as the first step of systematic modelling and analysis technique of the helicopter. In their research, the efforts of four helicopter companies (Bell, Boeing, the former McDonnell Douglas and Sirkorsky) to develop FEM models of the airframe and running their ground test, has been summarized. At the first they started with an FEM model of AH-1G helicopter which has 15 degrees of freedom (DoF), and totally is made of 39 elastic beam elements. In the mid-1970s, they developed a 3D FEM model of that helicopter in NASTRAN software which consisted of 2965 elements including rods, bars, scalar springs, etc. which brought structural parts into modelling.

Gmurczyk et al. (2011) indicated that due to the complexity of rotorcraft structure, the FEM model should be developed in a hierarchy manner. They modelled the structure components of Mil Mi-24 like stringers and frames using bar and beams and then assigned the material properties to them, but about other structural components which does not have a great role in transferring the loads they considered their weights. Since the computer model of that helicopter was not available, a photogrammetric measurement was done and a model of fuselage with 3920 points was received. Based on the geometrical model, an FEM model of Mi-24 with 30000 elements was created in MSC.Patran. [8]. In a similar research done by Annett and Polanco (2010), an integrated FEM analysis of MD-500 was investigated. In a hierarchical approach, in the same manner of automotive industry which models the whole car system, its seats and even the occupants, the whole helicopter structure, the seats and pilots were modelled [9].

In another research carried under a cooperation between National aeronautics and space administration (NASA) and French ministry of defence, the vibration characteristics of a French helicopter called SA349/2 Gazelle was determined. In this NASA report the free and forced vibration of that helicopter was analysed and the relations between the main rotor frequency, fuselage vibration and the vibration at the pilot's seat were demonstrated. In this research two FEM models are developed with the same basic bars, beams, quadrangles and triangles. The updated model included 10 substructures, the total number of 730 nodes and 1063 elements. Also 51 concentrated masses are considered in order to simulate the overall weight of the helicopter [7].

There are two significant points on the presented literature. The first one is the extracting of the helicopter geometry through 3D photogrammetric technique [8] [9], modelling the helicopter by CAM/CAD method and finally producing the FEM model, which is quite expensive, needs especial instruments that can only be provided by giant industries or big research centres with huge budgets. The second point is about those FEM models created through elastic beam elements. The definite numbers of points connected through elastic beams do not model the real shape, cross sections and properties of the structural components precisely.

This paper describes the structural modelling of the helicopter and the effective variations in frequency changes due to repair through the methodology is described in section 2. The numerical prediction of natural frequencies for the basic and full structure and repaired structure are summarized in section 3. The paper concludes in section 4.

2 Methodology

The following paragraphs discuss the development of the 3D ABAQUS model for the 349 Gazelle helicopter. The characteristics of 349 Gazelle helicopter are summarized in Table 1.

Table 1. 349 gazelle characteristics

Rotor radius (m)	5.25
Number of blades	3
Rotor solidity	0.0637
Design aircraft mass (kg)	2000
Design main rotor rotational speed (rpm)	387

Some steps have been taken towards simplifying the modelling in ABAQUS software. The first step is extracting the basic geometry of the Gazelle helicopter and generating the basic structure of the helicopter including the bulkheads, main cantilever beams, tail boom ribs and the tail boom skin which connects all the ribs together. Also, the curved beams are used to build the cockpit top section and the ones in which the windscreen part (forward cockpit section) sits are designed. In this way a precise model of basic frame with regards to the material properties and the cross sections is established. The need to precise modelling of the basic frame or the basic structure is that it contributes to helicopter stiffness characteristics.

The second step is to add the secondary structural components that have a critical role in the vibration characteristics of the model. So, the fuselage skin, the cabin floor, the windscreen glass and doors are created with their assigned properties and attached to the model.

In the next step, the weight of the non-structural components including main rotor head, tail rotor head, gearboxes, engine, fuel tank, landing skid, swash plate, propulsion subsystem, flight boosted control and flight manual control have been calculated and added to certain points of the model as lumped mass. These weights have been calculated based on the formula [Boeing Vertol, Research and Technology Laboratories (RTL) of the US army] available for "up to 12,000 lb gross weight class" [10].

During the damage repair, it is normal to fix the cracks by patching the particular part of the skin with an equivalent skin. In this study, we have looked at the skin patch of size 20*20 cm at the helicopter undercarriage. Variations in the density of the skin material are used to study the frequency variations. The material properties and the considered scenarios for the patched plate are as shown in Table 2 and Table 3 respectively.

Table 2. Material characteristics of the designed model in ABAQUS

Material	Density (kg/m3)	Poisson's ratio	Young's modulus (GPa)
Aluminium 7075	2810	0.33	71.7
Plexiglass	1162.5	0.35	2.4

Table 3. Scenarios for changing the density of the patched plate

	+10%	+20%	+30%	+40%	+50%	-10%	-20%	-30%	-40%	-50%
Density	3091	3372	3653	3934	4215	2529	2248	1967	1686	1405

3 Results and discussions

The basic structure modelled and meshed in ABAQUS, with total number of 12826 elements, 24002 nodes, and the total weight of 251.97 kg is shown in Fig. 1.

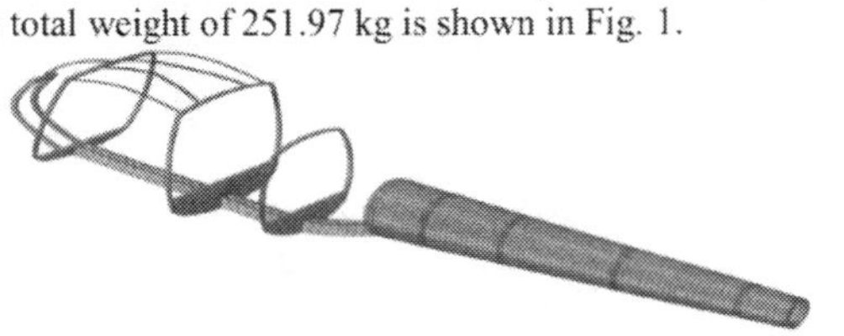

Figure 1. Basic structure of gazelle helicopter developed in ABAQUS software

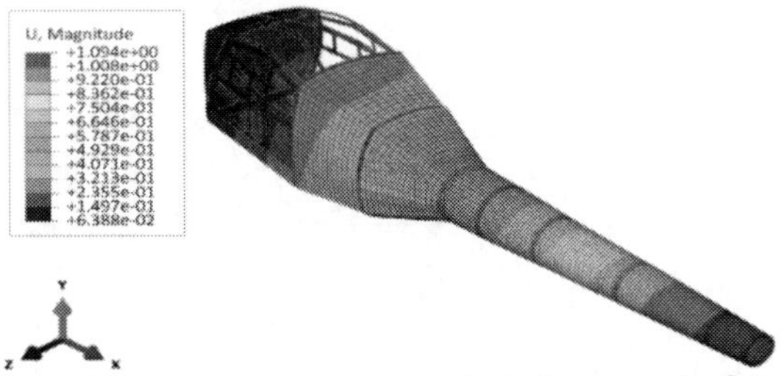

Figure 2. Completed model of gazelle helicopter at its first mode

A free vibration analysis was carried out to extract the natural frequencies of the basic frame up to its 40 modes, which is 76.4 Hz. To summarize, the results have been reported by mode number up to 20 which is around 37.1 Hz, as shown in Table IV (the second column). As it can be seen from the results, the first six modes are coupled rigid modes. The other structural parts of the helicopter including the fuselage, the doors, the windscreen glass, and floors are added to the model to monitor the frequency changes due to their addition. The completed model with the total mass of 375.45 kg at its first mode is shown in Fig. 2.

Table 4. Natural frequencies of the basic frame (the second column) and the full structure (the third column)

Mode	The frequency of basic frame (Hz)	The frequency of full structure (Hz)
1,2,3,4,5,6	0.0	0.0
7	8.2	3.4
8	9.2	4.1
9	11.2	9.1
10	14.1	9.9
11	14.8	10.5
12	20.52	14.1
13	20.4	17.1
14	22.8	18.9
15	25.1	20.1
16	26.7	23.3
17	29.8	25.6
18	30.6	26.3
19	33.0	27.3
20	37.1	30.4

The effects of adding the fuselage shells, the wind screen glass and doors are summarized in Table 4 (in the third column). As it can be seen from the results of simulations, adding shells and secondary components to the basic structure reduces the frequency considerably. For example, it is almost 59% reduction on the 7th mode and 18% on the 20th mode. It is well known that the shells, doors and wind screen are not contributing to the stiffness of the model as much as their total weights. The results are summarized as given in Table 5, as the full weighted model (the 2nd column). Since the exciting frequency of the rotor head is almost 6.45 Hz (387 rpm), the first harmonic of the main rotor head (regarding the three bladed main rotor) will be 19.35 Hz. So, the mode number 17 and 18 will be critical modes and have to be considered carefully in design stage in order to prevent resonance on the structure.

Obviously by adding the non-structural masses the frequency response experienced another reduction which is strictly related to increasing the weights. The frequency range reasonably matches with the results available in literature [7]. This establishes the confidence in the baseline model developed here.

To monitor the changes of repairing the cracks on the fuselage through patching new plates, a plate is patched to the undercarriage of the helicopter with the size of 20*20 cm. Obviously, the number of places which are likely to be patched and their locations (e.g. on the fuselage, tail boom, etc.) will be important. However the aim of this research is to show the ability of the 3D model in investigation of different structural repairs on the vibration characteristics of the helicopter. So the problem has been simplified only by considering one patched plate at the particular location. It can be seen from Table V, a patch even with the same material as its original at the undercarriage of the helicopter will decrease the natural frequencies, especially in higher modes.

This phenomenon will intensify if the number of patched places increases. It is possible to run further scenarios on the model beforehand.

Using patched plates which do not have the same material property as the original will lead to changes in the modes (19 and 20) which are close to those natural frequencies (17 and 18) that can lead to resonance of the structure and catastrophic failures. With increasing the severity of the difference in material which has been used, the impact of patched plate on the free vibration response of the system will magnify, especially when the plate has higher density compare to the original value.

4 Conclusion

A precise FEM model of a helicopter can provide a broad insight in the design stage of the helicopter. The results on the natural frequencies of the full structure reasonably match with the literature giving confidence in the baseline model. Thus, the model gives a comprehensive design tool for analysing the frequencies of the helicopter with differing components. The free vibration analysis done in this research shows higher natural frequencies for the basic frame than the full structure of the helicopter. It shows the significant effects of secondary components on structural modelling of the helicopter. Based on the FEM model and the helicopter rotor speed, modes number 17 and 18 are identified as the critical modes. These modes have to be focused on to avoid resonance. The applied repair scenario of patching a new plate (withconsidered ±50% deviation from its original density) to fix undercarriage section shows effects on mode 19 and 20 which are close to the critical modes. So, it is strongly suggested to test more patched plates in different locations of the helicopter structure to monitor the changes, precisely. It is also recommended to have arranged scenarios collected during repair and maintenance period of the helicopter structure and then applied them to developed model in future research.

Table 5. Natural frequencies of the full weighted structure (2nd column), with a patched plate in different scenarios

Mode	Freq. of full weighted structure	Patched plate with original material	Density scenario of the patched plate									
			+10%	+20%	+30%	+40%	+50%	-10%	-20%	-30%	-40%	-50%
1,2,3,4,5,6	0.0	0.0	0.00	0.00	0.00	0.00	0.00	0.00	0.00	0.00	0.00	0.00
7	3.39	3.39	3.39	3.39	3.39	3.39	3.39	3.39	3.39	3.39	3.39	3.39
8	3.77	3.77	3.77	3.77	3.77	3.77	3.77	3.77	3.77	3.77	3.77	3.77
9	4.20	4.20	4.20	4.20	4.20	4.20	4.20	4.20	4.20	4.20	4.20	4.20
10	8.29	8.28	8.28	8.28	8.28	8.28	8.28	8.28	8.28	8.28	8.28	8.28
11	9.00	8.99	8.99	8.99	8.99	8.99	8.99	8.99	8.99	8.99	8.99	8.99
12	9.40	9.40	9.40	9.40	9.40	9.40	9.40	9.40	9.40	9.40	9.40	9.40
13	10.16	10.16	10.16	10.16	10.16	10.16	10.16	10.16	10.16	10.16	10.16	10.16
14	13.37	13.37	13.37	13.37	13.37	13.37	13.37	13.37	13.37	13.37	13.37	13.37
15	15.85	15.82	15.82	15.82	15.82	15.82	15.82	15.82	15.82	15.82	15.82	15.82
16	16.91	16.91	16.91	16.91	16.91	16.91	16.91	16.91	16.19	16.91	16.91	16.91
17	18.72	18.71	18.71	18.71	18.71	18.71	18.71	18.71	18.17	18.71	18.71	18.71
18	20.62	20.61	20.61	20.61	20.61	20.61	20.61	20.61	20.61	20.61	20.61	20.61
19	24.99	24.64	24.52	24.28	24.08	23.88	23.68	24.48	97	24.99	25.00	25.00
20	25.96	24.96	24.96	24.93	24.88	24.80	24.69	24.95	25.11	25.30	25.50	25.70

References

1. M. Giglio, A. Manes, U. Mariani, R. Molinaro, W. Matta, Helicopter fuselage crack monitoring and prognosis through on-board sensor network, *Proc. of CM* (2009)
2. S. Castillo-Rivera, M.Tomas-Rodriguez, Hover Flight Helicopter Modelling and Vibrations Analysis, XXXVI Jornadas de Automática, (2015)
3. O. A.Bauchau, J. Rodriguez, S. Y Chen, Coupled rotor fuselage analysis with finite motions using component mode synthesis. Journal of the American Helicopter Society, **49** (2004)
4. Tamer, V. Muscarello, P. Masarati, G. Quaranta, Y. Yaman, Helicopter vibratory loads and vibrations reduction using higher-harmonic control, *39th European Rotorcraft Forum* (2014)
5. S. Stupar, A. Simonovic, M. Jovanovic, Measurement and analysis of vibrations on the helicopter structure in order to detect defects of operating elements, Scientific Technical Review, **62** (2012)
6. H. Yeo, I. Chopra, Coupled rotor/fuselage vibration analysis using detailed 3-D airframe models, Mathematical and computer modelling, **33** (2001)
7. R. Heffernan, D. Precetti, W. Johnson, NASA report: Vibration analysis of the SA349/2 Helicopter (1991).
8. G. Gmurczyk, P. Reymer, M. Kurdelski, Global FEM model of combat helicopter, Journal of Kones, **18** (2011).
9. M. S. Annett, M. A. Polanco, System-integrated finite element analysis of a full-scale helicopter crash test with deployable energy absorbers, *American helicopter society 66th annual forum* (2010).
10. W. A Stepniewski, Comparative Study of Soviet VS Western Helicopters.-Pt. 2: Evaluation of Weight, Maintainability and Design Aspects of Major Components, NASA (1983).

Empirical Model and Validation of Bar Formation in Sand Bed Channel

Duratul Ain Tholibon[1], Junaidah Ariffin[2], Jazuri Abdullah[2], Ainamardia Nazarudin[1], Nur` Ain Mat Yusof[1] and Dulina Tholibon[3]

[1]*Faculty of Civil Engineering,Universiti Teknologi MARA (UiTM), 26400 Jengka, Malaysia*
[2]*Faculty of Civil Engineering,Universiti Teknologi MARA (UiTM), 40000 Shah Alam, Malaysia*
[3]*Faculty of Mechanical Engineering, Universiti Kebangsaan Malaysia, 43600 Bangi, Malaysia*

Abstract. Both present experimental and previous historical data were used to develop the empirical model. Generally, the aim of this study to establish an empirical model of bar formation from the present experimental data and selected historical data. Statistical techniques using Multiple Linear Regression Analysis were employed for empirical model development and the selection of bar formation parameters were done based on the works of previous investigators. Validation of the newly developed empirical model was done by using a different set of historical data from selected laboratory studies. Model development involved selection of parameters through review of established models, dimensional analysis to check on the homogeneity of the model and statistical analysis. Derived empirical model has been validated using a different set of data from previous studies. Analysis confirmed that the empirical model derived using linear regression technique depicts the highest discrepancy ratio accuracy of 90% with $\frac{D}{d_s}$ and $\frac{B}{D}$ as the most significant parameters that promote bar height formation.

1 Introduction

Bars formation are present in most sand and gravel bed rivers and shows strong influence on flow and sediment transport processes and consequently on channel morphology [1] and as a fundamental sedimentary feature of braided rivers [2]. It is being observed in a field study that simple lobate unit bars occur with distinct downstream margins that may have avalanche faces, especially in the sand bed channel. Many data have been collected on the occurrence of bars in laboratory study to develop the empirical criteria for alternate bar formations and predictors for the equilibrium length and height of bars [3] [5]. Theoretically, a large number of linear studies that seek the conditions for incipient bar formation and linear growth rate under steady flow conditions has been developed.

Generally, the aim of this study to establish an empirical model of bar formation from the present experimental data and selected historical data. Statistical techniques using Multiple Linear Regression Analysis were employed for empirical model development and the selection of bar formation parameters were done based on the works of previous investigators. Validation of the newly developed empirical model was done by using a different set of historical data from selected laboratory studies.

2 Multiple linear regression analysis

In multiple linear correlation and regression, the additional independent variables denoted by $X_1, X_2 \ldots X_k$ help to better explain or predict the dependent variables (Y). The general linear model (First order multiple regression model) are as follows;

$$Y = \beta o + \beta_1 X_1 + \beta_2 X_2 + \beta_3 X_3 + \ldots + \beta_k X_k + \varepsilon \quad (1)$$

or

$$Y = A + B_1 X_1 + B_2 X_2 + B_3 X_3 + ..B_k X_k + \varepsilon \quad (2)$$

where;
Y = dependent variable
$X_1, X_2, X_3, \ldots, X_K$ = dependent variables
β_0/A = intercept, the value of Y when all the X's are zero
β_j/B_j = amount by which Y changes when that particular X_j increase by one unit, with the values of all other independent variables held constant.
ε =random error term

If the multiple linear regression is estimated using sample data, the estimated regression equation is written as

$$\hat{Y} = \hat{\beta}_o + \hat{\beta}_1 X_1 + \hat{\beta}_2 X_2 + \hat{\beta}_3 X_3 + \ldots + \hat{\beta}_k X_k \quad (3)$$

or

 DOI: 10.1051/matecconf/20179507028

$$\widehat{Y} = a + b_1X_1 + b_2X_2 + b_3X_3 + ... + b_kX_k \quad (4)$$

where $\hat{Y}$ is the estimated value for Y.

The assumptions of the multiple linear regression method are [6].

(a) The mean of the probability distribution of ε is zero, that is E (ε) = 0

(b) The errors must be normally distributed and have a constant variance, σ_ε^2

(c) The independent variables are not linearly related, but they can have a nonlinear relationship. High correlation among independent variables would result to the existence of multicollinearity.

(d) There is no linear association between the random error term, ε and each independent variable.

3 Methodology

Selections of the bar formation parameters are based on the review done by previous researcher and through dimensionless analysis. In this analysis, six dimensionless independent parameters namely ratio of grain size ($\frac{d_s}{RS}$), ratio of velocity to shear velocity ($\frac{V}{U*}$), ratio of channel width to grain size ($\frac{B}{d_s}$), ratio of channel depth to grain size ($\frac{D}{d_s}$), Shields stress (τ^*), and ratio of width to depth ($\frac{B}{D}$) were chosen and analyzed on their correlations with the dependent variable, ratio of bar height to grain size ($\frac{H_B}{d_s}$). Multiple linear and nonlinear regression technique used to investigate the relationship between the dependent variable and independent variables.

The statistical technique was employed in the analysis, namely multiple linear regression. The new empirical model will be developed using data from the present experimental study and selected data from previous researchers. The newly developed empirical model will be validated using a different set of data from previous investigators. The procedure involved in the analyses are as follows;

(a) Selection of the data from the previous researchers.

(b) Derivation of the empirical models through a various combination of the variables using multiple linear regression techniques.

(c) Validation of the newly developed empirical model using selected data from previous researchers.

There were a total of ten experimental works regarding bar height formation, including the present study that gives a total of 189 data. These data were being classified into two groups; data that collected during the experimental works and data of independent parameters that computed using the collected data. Selection of data to be use in model development and model validation were depend on the range of flow rates. For the data of present study, the recorded flow rates were the initial flow rates imposed to the channel while the recorded data for velocity, channel width and flow depth were for final condition. The grains size and slope of the channel were assumed constant for all experimental run. Velocity recorded is below the critical velocity, as this is due to resistance of bed configuration of bar formation at the final measurement. The data were tabulated in Table 1.

Table 1. Range of hydraulics and sediment characteristics data used in the analyses

No.	Researcher	No of Data	Independent Parameters						Dependent Parameter
			$\frac{d_S}{RS}$	$\frac{V}{U_*}$	$\frac{B}{ds}$	$\frac{D}{ds}$	τ_*	$\frac{B}{D}$	$\frac{H_B}{d_S}$
1	Present study	12	7.49-37.95	5.05-9.1	173.913-347.826	4.35-26.09	0.016-0.08	3.571-21.053	66.09-106.96
2	Kinoshita (1961)	26	2.06-10.6	4.54-7.93	77.647-347.368	3.76-14.47	0.06-0.29	8.2-33.85	3.53-36.84
3	Yoshino (1967)	38	1.45-13.66	5.46-10.74	65.22-688.89	5.35-46.67	0.04-0.42	3.53-30.1	2.43-55.78
4	Fujita (1980)	6	2.93-10.85	7.7-9.32	555.556	12.73-25.45	0.06-0.2	21.83-43.65	26.57-45.05
5	Ikeda (1984)	29	6.73-12.29	6.36-9.24	246.15	7.54-27.54	0.05-0.09	8.94-32.65	5-21.54
6	Fujita & Muramoto (1985)	7	4.25-13.65	7.95-11.08	340.91-2272.73	14.32-49.09	0.04-0.14	20-48.19	15.25-83.52
7	Tubino (1991)	6	6.71-7.67	8.76-9.09	545.45	20-23.09	0.08-0.09	23.62-27.27	13.64-35.18
8	Garcia & Nino (1993)	37	4.92-12.76	8.05-10.62	754.72	14.53-43.77	0.05-0.12	8.62-26.08	22.64-89.30
9	Rahman et al. (1996)	17	14.94-20.31	7.03-9.86	98.59	7.04-24.65	0.03-0.04	4-14	3.76-26.25
10	Lanzoni (2000a)	11	17.63-44.54	8.3-10.5	312.5	6.88-17.29	0.014-0.034	18.07-45.45	8.33-17.71

4 Results and discussions

A total of 41 linear and nonlinear regression equations were derived. Accuracy of the equations is measured based on 3 elements, (i) discrepancy ratio, (ii) statistical test analysis and (iii) graphical analysis. Discrepancy ratio is the ratio of the predicted values to the measured values and these values are deemed accurate if they lie within 0.5-2.0 limit. The statistical test is used to further quantify the accuracy of the model. A model is deemed significant if the collinearity statistics give values of tolerance higher than 0.1 and VIF lower than 10. Visual analysis through graphs representation is used to confirm the trends and accuracy of the prediction. The accuracy can be evaluated from the distribution pattern of the data. Homoscedastic scatter with high positive correlation signify good agreement of the predicted to observed values.

Based on these three elements, linear equation is seemed to perform better than nonlinear models for all sets of data from the present study, Kinoshita, 1961; Yoshino, 1967; Fujita, 1980; Ikeda, 1984; Lanzoni, 2000a. [7]-[9], [3], [10].The derived model using linear regression have established two dimensionless parameters namely $\frac{D}{d_s}$ and $\frac{B}{D}$ with the other parameters of $\frac{d_s}{RS}, \frac{V}{U^*}, \frac{B}{d_s}$ and τ^* excluded. The prediction accuracy for this linear regression model is 90% on data from the present experimental study and previous researchers. Summary of the performance of derived empirical models using linear and nonlinear regression method with the respective value of R^2 and discrepancy ratio are shown in Table 2.

Table 2. Performance of derived models in this study

Method	Equation	R^2	D.R (0.5– 2.0) (%)
Linear	$\frac{H_B}{d_s} = -3.35 + 1.63\frac{D}{d_s} + 0.17\frac{B}{D}$	0.17	90
Non linear	$\frac{H_B}{d_s} = 0.429\left(\frac{B}{d_s}\right)^{0.71}\left(\tau^*\right)^{-0.035}$	0.50	88

4.2 Validation of empirical models

An attempt was also made to validate all the derived models for other sets of data from previous studies [5; 11; 12; and 13] and the results are shown in Table 3. Analysis confirms that newly derived empirical model prediction gives satisfactory performance when validated on data from previous researchers. The results from [5] indicates the most accurate prediction as the value of discrepancy ratio of 100%. This is evident from the distribution pattern of the data of each researcher that lies within the acceptable limit as shown in Fig. 1.

Table 3. Validation of derived models on data by previous studies

Method	Discrepancy ratio (0.5 – 2.0) (%)				
	[12] [6 data]	[5] [7 data]	[13] [37 data]	[11] [17 data]	Combine [67 data]
Linear	67	100	92	88	90
Non linear	67	100	95	76	88

Non-linear equation gives significant contribution that shows the highest R^2 value of 0.5 as compared to linear regression method. However, there is clear indication that distribution pattern of data in which the scatter plot shown in Fig. 1 shows well distribution pattern for linear Equation as compared to nonlinear. Thus linear equation derived using linear regression technique is the best derived empirical model with 90% accuracy with $\frac{D}{d_s}$ and $\frac{B}{D}$ as the most significant parameters that promote bar height formation. This indicates the findings of the present study corroborate the previous work done by [3].

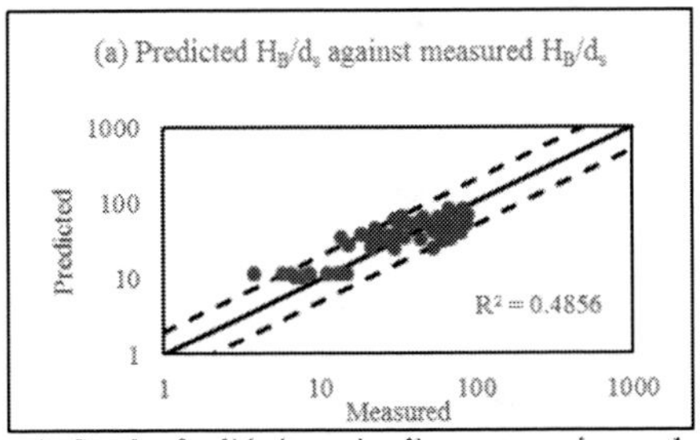

Figure 1. Graph of validation using linear regression method.

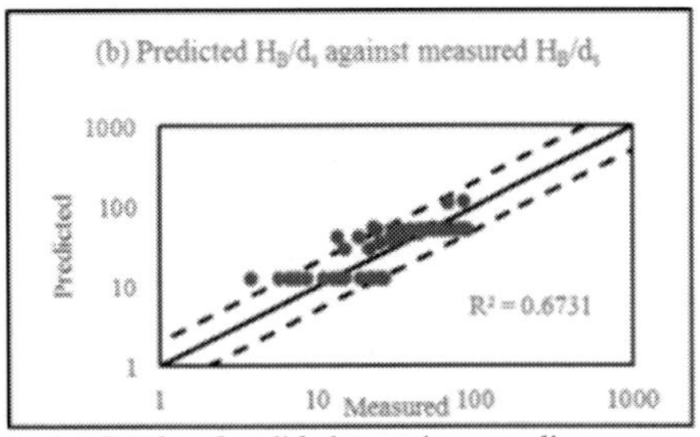

Figure 2. Graph of validation using non-linear regression method.

5 Conclusion

New empirical model to predict the bar height formation in a sand bed channel have been developed using experimental data from the present study and selected sets of historical data from previous investigators. Techniques used in developing the equation are the Multiple Linear Regression and Nonlinear Regression. Derived empirical model has been validated using a different set of data from previous studies. Analysis confirmed that the empirical model derived using linear regression method gave the best prediction accuracy of

90% with $\frac{D}{d_s}$ and $\frac{B}{D}$ as the most significant parameters that promote bar height formation.

Acknowledgement

The authors gratefully acknowledge Faculty of Civil Engineering, Universiti Teknologi MARA, Shah Alam for the facility provided. This research grant was provided by the Ministry of Higher Education Malaysia (600-RMI/RAGS 5/3 (176/2014)) and the Research Management Institute (RMI) of Universiti Teknologi MARA, Shah Alam, Malaysia.

References

1. P. J. Whiting and W. E. Dietrich. Convective accelerations and boundary shear stress over a channel bar. Water Resources Research. **27**, 783–796. (1991).
2. P. Ashmore. Morphology and dynamics of braided rivers. *Treatise on Geomorphology*, **9**, 289-312. (2013).
3. S. Ikeda. Prediction of alternate bar wavelength and height. *Journal Hydraulic Engineering*, **110**, 371-386. (1984).
4. M.N.R. Jaeggi. Formation and effects of alternate bars. *Journal Hydraulics Engineer*, **110**, 142–156. (1984).
5. Y. Fujita and Y. Muramoto. Studies on the process of development of alternate bars. *Bulletin of the Disaster Prevention Research Institute*, **35**, 55-86. (1985).
6. S. Landau and B.S. Everitt. *A handbook of statistical analyses using SPSS.* Chapman & Hall/CRC Press LLC. (2004).
7. R. Kinoshita. Investigation of channel deformation in Ishikari River. *Report of Bureau of Resources, Department of Science and Technology, Japan,* 1-174. (in Japanese). (1961).
8. F. Yoshino. Study on bed forms. *Collected Papers, Department of Civil Engineering, University of Tokyo, Tokyo,* Japan, **4**, 165-176. (in Japanese). (1967).
9. Y. Fujita. *Fundamental study on channel changes in alluvial rivers.* Kyoto University, Japan, 1-210. (in Japanese). (1980).
10. S. Lanzoni. Experiments on bar formation in a straight flume: 1. Uniform sediment. *Water Resources Research*, **36**, 3337-3349. (2000).
11. M. M. Rahman, N. Nagata, T. Hosoda, and Y. Muramoto. Experimental study on morphological process of meandering channel with bank erosion. *Annual Journal of Hydraulic Engineering*, **40**, 947-952. (1996).
12. M. Tubino. Growth of alternate bars in unsteady flow. *Water Resources Research*, **27**, 37-52. (1991).
13. M. Garcia and Y. Nino. Dynamics of sediment bars in straight and meandering channels: experiments on the resonance phenomenon. *Journal Hydraulics Research*, **31**, 739–761. (1993).

Economic Floating Waste Detectionfor Surface Cleaning Robots

Jakkrit Sumroengrit and Niramon Ruangpayoongsak

Production Engineering Department, Faculty of Engineering, King Mongkut's University of Technology North Bangkok, Thailand

Abstract. Removing waste out of water surface is a routine task and can be operated by using autonomous surface cleaning robots. This paper presents amethodoflaser-based floating waste detection for surface robot guidance when waste positions are unknown beforehand. Basing on concept of refraction and reflection of laser ray, the proposed laser-based technique is proven to be applicable on floating waste detection. The economic waste detector is constructed and mounted on the robot. Five DOF equations of motion are formulated for calculation of waste position incorporating distance measured by the laser and also the robot motion caused by external wind force as well as water surface tension. Experiments were conducted on a pond with calm water and results show that the presented economic waste detection successfully identify and locate position of plastic bottles floating on water surface within the range of 5 meters.

1 Introducion

In developing countries, accumulation of floating waste such as plastic scraps, foam scraps or tree leaves on city canals or ponds can block water drainage and also cause pollutions. Cleaning water surface is therefore an essential routine task.

Surface cleaning robots on lake, pond, or reservoirs gained interests from researchers and issues were addressed such as motion control and autonomous driving. Motion of the robot was represented by hydrodynamic model and maneuver ability performance was predictable [1]. The asymptotic stabilizaion control was also theoretically validated for motion control of the surface robot [2]. As for autonomous driving, bank following algorithm with obstacle avoidance was implemented by using ultrasonic sensors [3]. Later, a low cost robot structure design was presented and simulation of multi robot system for lake cleaning was proposed [4]. The strategy of waste removal in a large area was presented whereas waste is assumed to be randomly distributed but waste position is known. On the other hand, the problem of waste detectionwhen waste position is unknown had not yet been investigated.

Laser range finder showed high performance on ocean navigation of unmanned surface vehicles. LIDAR was utilized for obstacle detection and avoidance of obstacle above ocean level for autonomous driving [5-6].The laser scanner is also used for 3D surface reconstruction in which both below and above water level structure were scanned[7].Along with the performance, cost of the laser scanner is high.

This work presents a method for identifying and locating waste floating on calm water surface by using a low cost laser system before the cleaning robot starts navigation. A Pan-Tilt laser unit is utilized as a waste detector heading into water surface and scanning for floating waste. When the waste detector is mounted on the robot floating on water surface, external wind force causes three dimensional rotations. Accordingly, waste position can be obtained from five degree of freedom equations of motion.

This paper is organized as follows: In section 2, the Floating Waste Removal Robot and the waste detector are introduced. Next, the method to identify floating waste is explained and equations for waste position are formulated in section 3. Section 4 presents experimental results. Finally, conclusions and future works are given.

2 Floating waste removal robot

2.1 The robot structure

As shown in Fig. 1, the floating waste removal robot was developed for cleaning on calm water surface such as on city canal, lake, pond, pool, etc.The robot structure is apontoondriven by two paddle wheels and a rudder. The robot has two types of waste collectors and a waste container. The front waste collector is a flight conveyor with half-cylindrical fin developed in [8] and side collectors are net basket lifted by arms. Performances of the waste collectors were evaluated in [9]. Plastic bottles were used in experimentsas floating waste and the best waste collecting performance was at 1.71 kg/minute.

The robot is controlled by Arduinos interfacing with sensors and motor driver modules. The GPS and IMU sensorsprovide information about robot position and inertia. Encodersmeasurerobot driving speed and

 DOI: 10.1051/matecconf/20179508001

conveyor belt speed.The robot is tele-operated control by using joystick and keyboard whereasan operator can manually control the robot speed andthe flight conveyor belt speed. Thoughthe operator cansee floating waste from video taken via a wireless camera onboard, it is hard to estimate distance to the waste floating.

Figure 1. The floating waste removal robot and the waste detector mounted on the top.

Figure 2. The wastedetector.

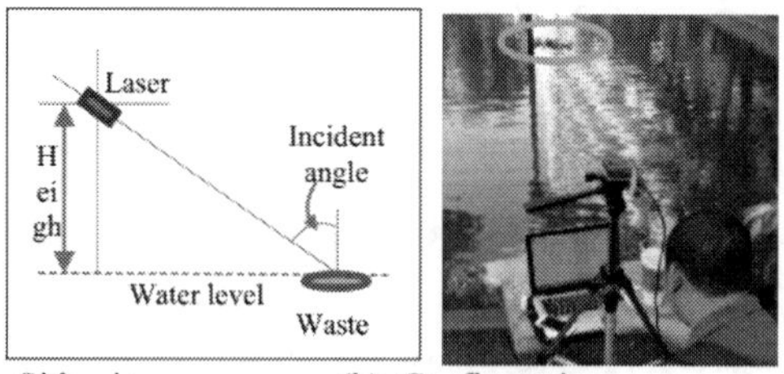

Side view (b) Configuration

Figure 3. Test of laser tilt range.

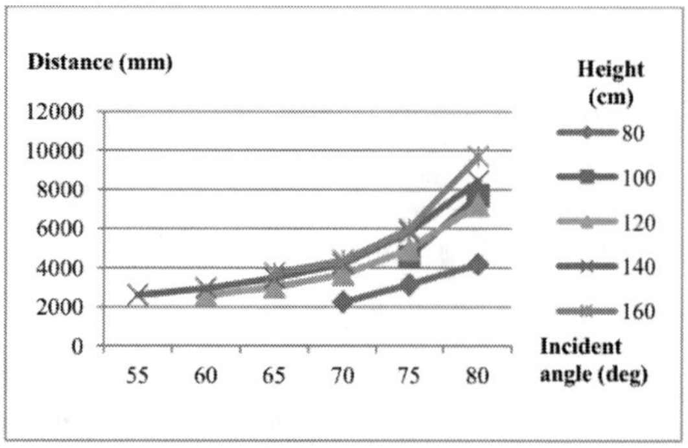

Figure 4. Tilt range test results.

2.2 The waste detector

The waste detector is mounted on the top of the robot as shown in Fig. 1 marked as a red circle.Fig. 2 shows the waste detectorconsisting of a Pan-Tiltlaser unitanda digital distance meter UNI-T-UT-390B mounted on the top. The meter is selected since it is a low cost laser device. Arduino Mega 2560interfaces with the digital distance meter and two Turnigy620DMG servo motors controlling pan and tilt angle. In order to detect the robot motion,CMPS10 sensoris also mounted at the center of robot's base for measuring roll, pitch, and yaw angle.

3 Laser based waste detection

3.1 Background

Laser system has been recently developed for water level measurement [10]. A pulse laser rangefinder sensor using phase method was pointing perpendicularly to calm and sloshing water surface for water level detection.

When laser ray is non-perpendicular to water surface, refraction occurs at interface between air and water where the refractive index depends on density of medium [11]. The refracted ray always point to the denser medium. In this case, the refracted ray points into water. As a result, distance to water surface is not measurable. On the other hand, when there is waste floating such as plastic bottles, the incident ray is reflected on the waste back to the digital meter and distance to the waste is measurable.

A simple test isdone to provethe conceptabove by holding the digital meterabove water level andreadvaluedisplayed on the meter. By several trials, the sameresultisobtained. That iswhen the meterpoints non-perpendicularlyinto water,measured value is not obtainedfrom the digital distance metersince the distance meter shows error message.Whenthe meter points perpendicularlyinto water, the meter shows water level.

3.2 Range of laser scan

Before mounting the waste detector on the robot, several tests were conductedin order to obtain appropriate mounting configuration.

The first step isfinding the range of maximum and minimum incident anglethat can identify waste on water surface. As shown in Fig. 3a, the height and the incident angle are adjustable. During the test, the waste detector is mounted on a camera stand as shown in Fig. 3b andplastic bottles are marked as an orange oval. By varying the height of the waste detector from 80 to 160 cm above water level and varying the incident angle from 50 to 85 degrees, the distance to the waste is measured by laser. Fig. 4 shows the measured distance at each height and incident angle.It is shown that different range of incident angle is obtained for each height level. At 160 cm above water level and 80 degree of incident angle, the waste can be detected at nearly 10 meters.The widest range of incident angle, from55 to80 degrees,is obtained at 140 cm.

The secondstep is to ensurethe appropriate height for mounting the waste detector on the robot. The plastic bottlesare fixed at 4 meters away from the camera stand. For each height, the incident angle is varied until the waste is detected. As shown in Table 1, when the laser unit is higher, the incident angle is smaller.At 160 cmheight, the error is large.Therefore, the appropriate height for mounting on the robot is up to 140 cm above water level. This complies with the results in Fig. 4 showing that the 120 cm and 140 cm height yields wide range of measurement.

Table 1. Distance tothe floating plastic bottles (mm)

Height (cm) \ Incident angle (degree)	80	100	120	140	160
80	4180	-	-	-	-
75	-	4099	-	-	-
72	-	-	4091	4188	
70	-	-	-	-	4495

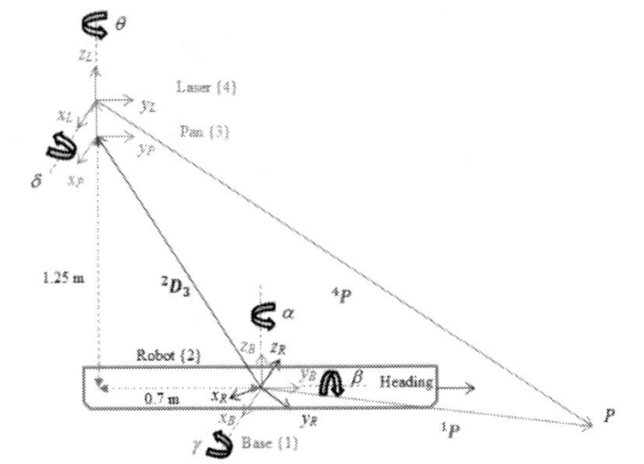

Figure 5. Sideview of the robot illustratingrelationshipamong frames.

3.3 Calculation for waste position

Four reference frames are defined as the Base frame, the Robot frame, the Pan frame, and the Laser frame, respectively.Assume the robot is stand still and represented by the Base frame as in Fig. 5. The robot is heading along y-axis. The waste detector motion is represented as the Pan and the Laser frame on the top.Based on the composite transformation of frames [12], neglecting linear motion, rotation from frame 2 to frame 1 is the rotation of robot caused by external wind force and surface tension can be written as

$$ {}^{1}\boldsymbol{R}_2(\boldsymbol{\alpha},\boldsymbol{\beta},\boldsymbol{\gamma}) = \begin{bmatrix} cos\alpha & -sin\alpha & 0 \\ sin\alpha & cos\alpha & 0 \\ 0 & 0 & 1 \end{bmatrix} \begin{bmatrix} cos\beta & 0 & sin\beta \\ 0 & 1 & 0 \\ -sin\beta & 0 & cos\beta \end{bmatrix} \begin{bmatrix} 1 & 0 & 0 \\ 0 & cos\gamma & -sin\gamma \\ 0 & sin\gamma & cos\gamma \end{bmatrix}. \quad (1) $$

where α, β, and γ, are roll, pitch, and yaw angle, respectively. There is no translation from frame 2 to frame 1, therefore the translation vector ${}^{1}\boldsymbol{D}_2$ is zero.The transformation matrix from frame 2 to frame 1 can be written as

$$ {}^{1}\boldsymbol{T}_2 = \left[\begin{array}{c:c} {}^{1}\boldsymbol{R}_2(\boldsymbol{\alpha},\boldsymbol{\beta},\boldsymbol{\gamma}) & {}^{1}\boldsymbol{D}_2 \\ \hdashline 0 & 1 \end{array}\right]. \quad (2) $$

Frame 3 is rotated by the pan angle, θ. The rotation matrix from frame 3 to frame 2 is

$$ {}^{2}\boldsymbol{R}_3(\boldsymbol{\theta}) = \begin{bmatrix} cos\theta & -sin\theta & 0 \\ sin\theta & cos\theta & 0 \\ 0 & 0 & 1 \end{bmatrix}. \quad (3) $$

Frame 3 is 1.25 m above and 0.7 m behind Frame 2. Thus, the translation vector from frame 3 to frame 2 is

$$ {}^{2}\boldsymbol{D}_3 = [0 - 0.7\ 1.25]^T. \quad (4) $$

According to Eq. (3) and (4), the transformation matrix from frame 3 to frame 2 is

$$ {}^{2}\boldsymbol{T}_3 = \left[\begin{array}{c:c} {}^{2}\boldsymbol{R}_3(\boldsymbol{\theta}) & {}^{2}\boldsymbol{D}_3 \\ \hdashline 0 & 1 \end{array}\right]. \quad (5) $$

Frame 4 is rotated by the tilt angle, δ. The rotation matrix from frame 4 to frame 3 is

$$ {}^{3}\boldsymbol{R}_4(\boldsymbol{\delta}) = \begin{bmatrix} 1 & 0 & 0 \\ 0 & cos\delta & -sin\delta \\ 0 & sin\delta & cos\delta \end{bmatrix}. \quad (6) $$

Frame 4 is 4.25 cm above frame 3. Thus, the translation vector from frame 4 to frame 3 is

$$ {}^{3}\boldsymbol{D}_4 = [0\quad 0\quad 0.00425]^T. \quad (7) $$

According to Eqs. (6) and (7), the transformation matrix from frame4 to frame3 is

$$ {}^{3}\boldsymbol{T}_4 = \left[\begin{array}{c:c} {}^{3}\boldsymbol{R}_4(\boldsymbol{\delta}) & {}^{3}\boldsymbol{D}_4 \\ \hdashline 0 & 1 \end{array}\right]. \quad (8) $$

Finally, the waste position P referred to frame 1 can be calculated by

$$ {}^{1}\boldsymbol{P} = {}^{1}\boldsymbol{T}_2\, {}^{2}\boldsymbol{T}_3\, {}^{3}\boldsymbol{T}_4\, {}^{4}\boldsymbol{P}. \quad (9) $$

The vector ${}^{1}\boldsymbol{P}$ is a vector to point P on frame 1, $[x\ y\ z\ 1]^T$. The vector ${}^{4}\boldsymbol{P}$ is a vector to point P on frame 4, $[0\ d\ 0\ 1]^T$ and d is distance from frame 4 to point P and is measured by the digital meter.

4 Experimental results

The waste detection istested on a pond under no sway, no surge, and no heave. During the experiment, the robot is standstill and the water is calm. Due to the external wind force and surface tension, the robot motion is slightly rotated in the roll, pitch, and yaw direction. Plastic bottles are used as floating waste since it is often found waste in city canals and river.

Figure 6. Test environment.

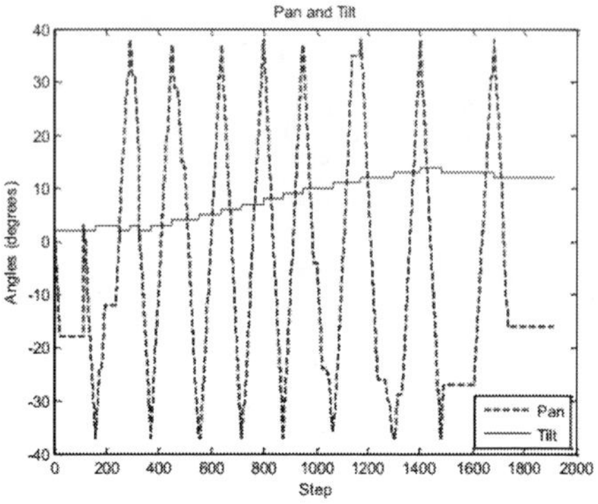

Figure 7. Pan and tilt angles.

The coordinate of the robot is at (0,0). As shown in Fig. 6, in the front of the robot are plastic bottles floating on water surface at coordinate(250,400) referred to the robot position. The size of the plastic bottles is 0.5 m radius.

During the waste detection, the laser scan is stepped by servomotor command as shown in Fig. 7. The results are analysed by using MATLAB. Range of the pan angle is from -38 to +38 degrees and range of the tilt angle is from 2 to 14 degrees. For each tilt angle step, the pan angle is varied from the minimum to the maximum angle. The scan stops when both the maximum pan and tilt angle are reached. At each scan step, the roll, pitch, and yaw angle, as well as the distance are measured.

Fig. 8 shows the yaw angle at each step during the scan. At the beginning, the yaw angle swings in the range of 141.5 to 145 degrees due to surface tension and external wind force. After a while, the yaw angle lies between 141.5 to 142 degrees. Meanwhile, both the roll and the pitch angle are quite still. Both of them lie between 0 and 1 degree.

The laser measurement returns two types of output, zero and distance value. When there is no waste found on water surface, measured distance is set to zero as the incident ray is refracted into water and the meter returns no value. When the waste is found, measured distance is obtained as the laser ray reflects on plastic bottles back to the meter and the meter returns a valid distance value. The raw data of measured distance is plotted in Fig. 9. From the beginning, between steps 1 to 400, the distance values are zero. Later, between steps 400 and 505, some distance values are slightly obtained. After that the distance value is again zero in the range of steps 505 to 1015. Between steps 1015 and 1085, distance values are densely obtained since the plastic bottles are foundand the distance is at approximately480 cmaway. After step 1085, the distance values are again zero with some values until the last step.

In total there are 1915 scanned points, in which only 85 points contains distance value. Fig. 10 shows the plot of waste positions on xy plane using the distance value of those 85 points and eq. (9). Out of 85 points, 61 points are the plastic bottles and 24 points are unknown objects. The density of those 61 points verifies that the plastic bottles are successfully detected. The center of the detected waste points is closed to (250,400).

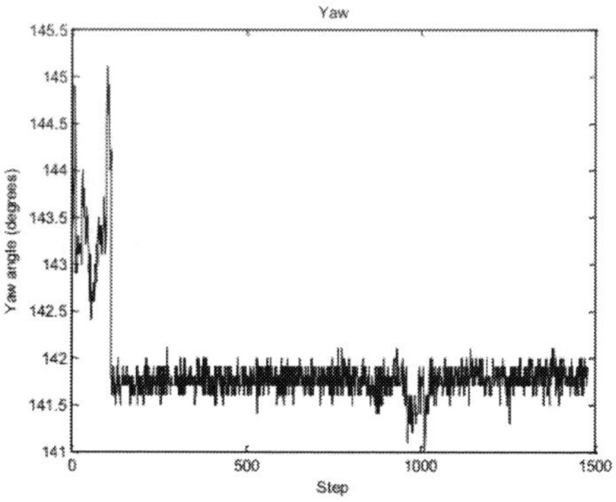

Figure 8. Yaw angle on the Base frame.

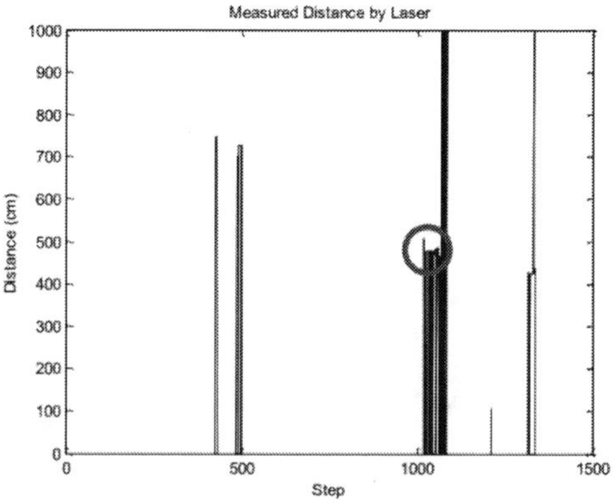

Figure 9. Distance measured by laser.

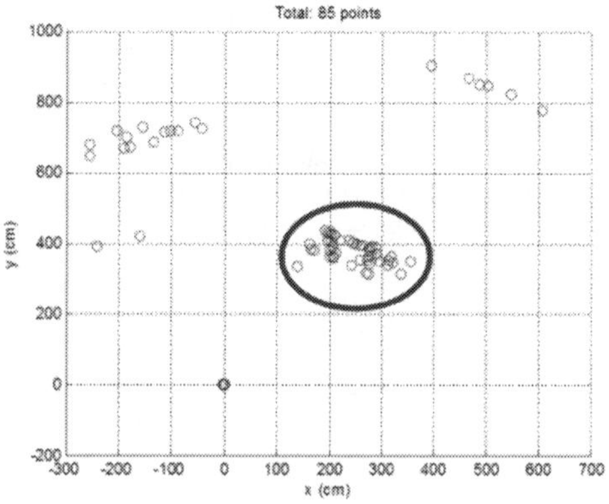

Figure 10. Waste detection result.

Please note that the floating waste in Fig. 6 is also marked in Figs. 9 and 10 accordingly. From the experiments, it is found that brightness of sunlight on water surface can cause error on distance measurement of the laser. When the sun light is too bright such as at noon or too dark at night, the digital distance meter returns no value.

5 Conclusions and future works

The presentedlaser-basedwastedetection technique wassuccessfullyvalidated on a pond.The experimentalresultsshowedthatthe constructedlow-cost waste detector is able to identify and locate position of plastic bottlesfloatingon calm water within the range of 5 metersawayfrom the surface robot. This technique is applicable for autonomousrobot guidance in unknownenvironmentunder medium sunlight illuminationwherewaste position ispriori unknown.

Future workisimprovement ofthe wastedetectionperformance and integrationwith GPS data in order tolocatewaste position in global coordinate.

Acknowledgements

The authorswouldlike to thank Faculty of Engineering and King Mongkut'sUniversity of TechnologyNorth Bangkok for grants and thank Mr. MonthianLeanglum for technical support.

References

1. Z. Wang and Y. Liu, "Estimating hydrodynamic parameters of a lake surface cleaning robot using numerical methods,"in Proc. 2008 IEEE Int. Conf. on Robotics and Biomimetics (2009), **1146 - 1151**.
2. Z. Wang and Y. Liu, "Global AsymptoticStabilization Control of a Lake Surface Cleaning Robot,"in Proc. 2009 IEEE Int. Conf. on Information and Automation (2009), **271-276**.
3. S. Chen, D. Wang, T. Liu, W. Ren, "An AutonomousShip for Cleaning the GarbageFloating on a Lake,"in 2nd Int.Conf. on Intelligent Computation Technology and Automation, (2009), **471 – 474**.
4. P. Agrawal, B. Bhattacharya, "Aquatic Multi-Robot System For Lake Cleaning,"*Nature-Inspired Mobile Robotics* (World Scientific,Singapore, 2013), **171-178**.
5. R. Halterman, M. Bruch, "Velodyne HDL-64E LIDAR for Unmanned Surface Vehicle Obstacle Detection, " in *Proc. SPIE 7692*, UnmannedSystemsTechnology XII, 76920D (May 07, 2010)
6. M. Dunbabin, A. Grinham, J. Udy, "An Autonomous Surface Vehicle for Water Quality Monitoring, " Australasian Conf. on Robotics and Automation (2009), **1-6**.
7. G. Papadopoulos, H. Kurniawati, A. S. B. M. Shariff, "3D-Surface Reconstruction for Partially Submerged Marine Structures Using an Autonomous Surface Vehicle," 2011 IEEE/RSJ International Conference on Intelligent Robots and Systems, **3551 - 3557**.
8. M. Leanglum, N. Ruangpayoongsak, A. Phunopas, "Design and Development of the Waste Collector for a Surface Robot Cleaner," Industrial Engineering Network Conference (2015), **900-905**.
9. J.Sumroengrit and N. Ruangpayoongsak, "Performance of Waste Collectors for Water Surface Cleaning," Int. Conf. on Engineering Innovation (2016, in press).
10. M. Liu, Z. Gai, J. Zhao, X. Cui, L. Yang, S. Chu, J. Yang, "Development of Laser Water Level Measuring System Without Cooperative Target," Symp. on Photonics and Optoelectronics (2012), **1 - 3**.
11. S. D. Gupta, N. Ghosh, A. Banerjee, *WaveOptics: Basic Concepts and Contemporary Trends* (CRC Press, Florida, 2016)
12. R.K. Mittal, I.J. Nagrath, *Robotics and Control* (Tata McGraw-Hill, New Delhi, 2003)

DOI: 10.1051/matecconf/20179508002

ADAMS-MATLAB Co-Simulation of A Serial Manipulator

Tejaswin Parthasarathy, Vignesh Srinivasaragavan and Soundarapandian Santhanakrishnan

Department of Mechanical Engineering, Indian Institute of Technology Madras, Chennai 600036, India

Abstract. This paper presents the dynamic modelling and simulation of a now redundant robot, Mitsubishi RM-501, and proposes a general algorithm for experimental simulation in kinematics, dynamics and control analysis to any such robot. Through reverse engineering, a model as accurate as the real robot was developed in SolidWorks.The simulations of the same were performed in ADAMS (dynamicmodeling software offered by MSC Software Corp)along with MATLAB for motion studies and control dynamics. Finally, with a user-input path the accuracy and precision of the simulator was verified.

1 Introduction

The institute of robotics in United States of America (USA) gives the definition for a robot as "a reprogrammable, multifunctional manipulator design to move material, parts, tools, or specialized devices through various programmed motions for the performance of variety of tasks." Nowadays, robots in industries have a wide spectrum of applications. The most notable ones are the welding and painting robots in car plants [1], electronic board assembly, repairing nuclear installations [2] etc. Moreover, many academic and research laboratories have been trying and developing new methods (motion planning, manipulationplanning, grasp planning) and algorithms (position and force control algorithms) in order to extend its usage to diverse applications. High amount of cost and time is involved in testing and validation of robots pave the way to use simulated models, as a cost–effective approach and areextensively demanded in research labs.

Simulation and analysis of PUMA 560 [3] and Stabuli TX-4.0 [4] have previously been performed. However, inverse kinematics, dynamic analysis and implementation of a control algorithm have not been done so far. Thus, enough thrust has to be given holistically to satisfy the present needs. This is important in an economic, research and environmental point of view, and the present work is aimed to achieve the said goals.

This paper presents a detailed procedure for virtual testing offive degrees of freedom (5-DOF) type model based on the Mitsubishi RM-501 robot through simulations. All procedures followed are presented in this paper in detail such a way that the reader can model any serial robot manipulators in the given software environments.

The main highlight of the present work is to demonstrate easy ways of co-simulation, accurate controlling of system design as an insight into trajectory planning. After the construction of a three-dimensional (3D) model of the robot manipulator in DDS SolidWorks, the model is then exported to ADAMS dynamic modeling software for simulating kinematics and dynamic behavior of robotic manipulator with an acceptable error (compared to the real models). For obtaining an exact model, inertial and geometric parameters are accurately measured and recorded in the software database, for simulating control algorithms, simulation between ADAMS and MATLAB is performed.

2 Modelling

The robot itself is made up of 5 links namely: base, body, shoulder, elbow and wrist. Each of this entity was designed in SolidWorks. The sketches and dimensions were made using the data extracted from [5] and the model was made accordingly. Then these parts were assembled in the 'Assembly' environment, using appropriate constraints as deemed fit. The process of defining all the constraints was started with fixing the base of the manipulator (to which the universal frame was attached). Then the 'Coincidence' and 'Concentric' commands were used to couple the other parts as deemed fit. The revolute joints were used to couple the links in RM-501, thus simplifying the construction of the model and its assembly.

Once the assembly is done, the "Motion Study" module in SolidWorks was usedto verify the robots predetermined DOF. The "Motion Study" also provides an option to export the geometric model to the ADAMS environment for further processing.

2.1 Importing into ADAMS

ADAMS software was used for the dynamic analysis of the robot model. The basic pre-requisites includes geometric correctness and the validity of constraints were done in SolidWorks. The material of the robot parts (as this determines the mass and other basic dynamic parameters) is equally significant as observed during reverse engineering (RE). Once this is done in SolidWorks the "Motion Study" itself is capable of exporting these details along with the geometric model to ADAMS. Verification of the coherence in inertial parameters can be confirmed in both ADAMS and SolidWorks platforms (Table 1).

Table 1. Inertial parameters of RM-501

Link	**m (kg)**	**I_{xx} x 10^{-3} (kg.m^2)**	**I_{yy} x 10^{-3} (kg.m^2)**	**I_{zz} x 10^{-3} (kg.m^2)**
Base	NA	NA	NA	NA
Body	2.627	22.218	69.890	61.674
Shoulder	4.875	551.141	123.225	73.646
Elbow	1.839	7.152	7.152	5.399
Wrist	1.202	0.834	0.887	0.952
End Effector	0.185	0.078	0.078	0.033

*m is the mass of each link
**I is the Moment of Inertia about the parts Centre of Mass (CoM)
*** *x,y,z*are defined according to the local (not fixed) coordinate system attached to each and every part

The motion in ADAMS can be given in two different ways. The first one involves providing the direct motion of the joint (in terms of angular displacement/velocity/acceleration) in SolidWorks Motion Manager and importing it to ADAMS. This ensures that the same motion is replicated in ADAMS and later it can be changed according to the needs. The other way is fundamentally cumbersome but more practical, which involves the user giving the necessary joint torques so that the end effector follows the desired path. This involves using the V_TORQUE option in ADAMS in each and every joint, which essentially creates a 3D torque vector. Non-zero values were then filled in for the Z-direction component in each V_TORQUE, while the other two components were made zero as practical actuation was done by using individual servo motors for a joint. The servo motors can provide torque only in one direction. It was noted that chain drives and gear systems should also be incorporated into the model. It was done manually, while defining the parameters of the individual motors. This paper illustrates the second approach, which involves computing the torque (and hence power) required.

3 Kinematic Analysis

Kinematic analysis in a broad sense involves both forward and inverse kinematics. Emphasis should be given equally for both as they directly determines the joint torques and other related parameters in order to make the end effector of the robot to precisely follow a pre-defined path. The coordinate's representation [6] and the modified Denavit-Hartenberg parameters are shown in Figure 1 and Table 2, respectively.

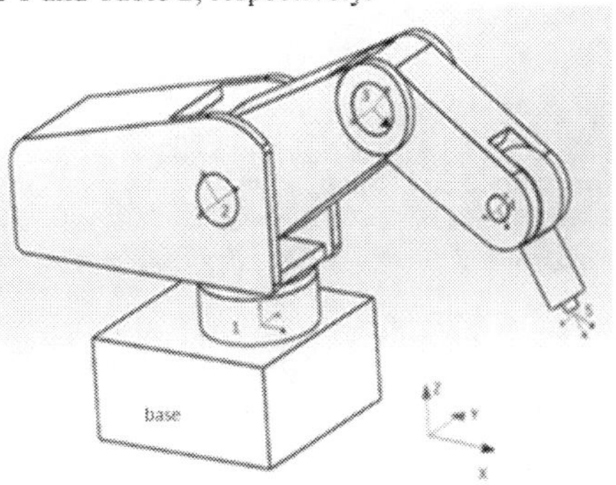

Figure 1. The assembled model of the Mitsubishi RM 501, with the attached frames and coordinate systems

Table 2. Denavit-hartenberg parameters of RM-501

Link (i)	**α_{i-1}**	**a_{i-1}**	**d_{i-1}**	**θ_{i+1}**
Base	0	0	0	θ_1
Body	90	0	0	θ_2
Shoulder	0	220	0	θ_3
Elbow	0	160	0	90+ θ_4
Wrist	90	0	0	θ_5
End Effector	0	0	137	0

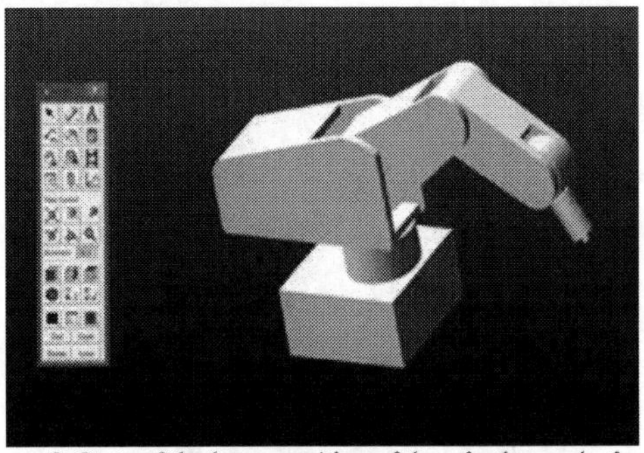

Figure 2. State of the home position of the robotic arm in the ADAMS environment

Care must be taken while importing the CAD model of the robot into ADAMS. The position of joints must eithercorrespond with the default 'Home' position (Figure 2) or a state where all the initial joint parameters, namely the five joint angles are known. Throughout this paper the positive direction of rotation is compatible with the right handed direction of Z axis.

Quintessential to any kinematic calculations are transformation matrices and there are six of them (one for each degree of freedom and one for the end-effector orientation).In this context, one each from the previous coordinate system to the next. Transformation matrices provide complete information about the frames. The information includes the orientation and position of frames, along with information on any vector of a particular frame, if that vector is viewed from some other frame. Transformation matrices for RM 501 are elaborated in next section.

3.1 Forward kinematics

Given the set of joint angles (θ_i) at any instance, the forward kinematics is simple, as the basic equations of forward kinematics easily determines the end-effector position and orientation. The modified Denavit - Hartenberg (DH) convention and methodology was used to derive the kinematics.

The following convention was followed: (x_0, y_0, z_0) to (x_4, y_4, z_4) represent the local coordinate frames at the five joints respectively, (x_5, y_5, z_5) represented the local coordinate frame at the end-effector, α, γ and θ are the rotation angles about x, y, and z axis respectively. The base local coordinate frame (x_0, y_0, z_0) overlapped with the global coordinate system (X, Y, Z) as it was attached to the base frame.

Based on the DH convention, the transformation

$$ {}^{i-1}_{i}T = \begin{bmatrix} c\theta_i & -s\theta_i & 0 & a_{i-1} \\ s\theta_i c\alpha_{i-1} & c\theta_i c\alpha_{i-1} & -s\alpha_{i-1} & -s\alpha_{i-1}d_i \\ s\theta_i & c\theta_i s\alpha_{i-1} & c\alpha_{i-1} & c\alpha_{i-1}d_i \\ 0 & 0 & 0 & 1 \end{bmatrix} \quad (1) $$

matrix from joint n to joint n + 1 is given by:

, where θ and α are DH parameters and cθi, sθi represent cos (θi) and sin (θi), respectively.

Another mathematical entity which is frequently used for describing the frame motion is a 3×3 rotation matrix(${}^{i-1}_{i}R$) that is essentially the first 3 rows and 3 columns of the transformation matrix (${}^{i-1}_{i}T$). The transformation matrices for all the frames are listed here. Hence, the position and orientation of the end-effector can be calculated if all the joint angles are given. Once the joint angles are found, other kinematic variables can be calculated through forward kinematics using the following equations.

$$ {}^{i+1}_{i+1}\omega = {}^{i+1}_{i}R\,{}^{i}_{i}\omega + \dot{\theta}_{i+1}\,{}^{i+1}_{i+1}\hat{Z} \quad (2) $$

$$ {}^{i+1}_{i+1}\dot{\omega} = {}^{i+1}_{i}R\,{}^{i}_{i}\dot{\omega} + {}^{i+1}_{i}R\,{}^{i}_{i}\omega \times \dot{\theta}_{i+1}\,{}^{i+1}_{i+1}\hat{Z} + \ddot{\theta}_{i+1}\,{}^{i+1}_{i+1}\hat{Z} \quad (3) $$

$$ {}^{i+1}_{i+1}\dot{v} = {}^{i+1}_{i}R\left({}^{i}_{i}\dot{\omega} \times {}_{i+1}^{i}P + {}^{i}_{i}\omega \times \left({}^{i}_{i}\omega \times {}_{i+1}^{i}P\right) + {}^{i}_{i}\dot{v}\right) \quad (4) $$

$$ {}^{i+1}_{i+1}\dot{v}_c = {}^{i+1}_{i+1}\dot{\omega} \times {}_{i+1}^{i}P_c + {}^{i+1}_{i+1}\omega \times \left({}^{i+1}_{i+1}\omega \times {}_{i+1}^{i}P_c\right) + {}^{i+1}_{i+1}\dot{v} \quad (5) $$

Here, 'i' varies from 0 to 5. Furthermore ω is the angular velocity of a link, $\dot{\omega}$ is the angular acceleration of a link, $\dot{\theta}$ is the angular velocity of a joint motor, $\ddot{\theta}$ is the angular acceleration of a joint motor, v is the velocity of a link, $\dot{v}$ is the acceleration of a link and $P$is the distance from the joint. A subscript c indicates the reference to the center of mass of a link.${}^{i+1}_{i+1}\hat{Z}$is the unit vector along Z-axis of the frame i+1 . All the above mentioned quantities are 3×1 column vectors (for x, y and z axis).

3.2 Inverse kinematics

Inverse kinematics, as the name suggests compliments the forward kinematics. It deals with identifying the joint angles when the position and orientation of end effector is given. This is very much useful if the robot is to follow a predetermined path.

The non-linearity involved in forward kinematicsmakes the solution to the inverse kinematics not only difficult but also non-unique. A single strategy to tackle the challenging of inverse problem is by using translation Jacobian matrices. This approach isused in this work. The results obtained using Jacobian matrices are in close agreement with the actual angles, provided that the motion imparted to the robot is small between two successive measuring positions.

Let X represent a 6×1 column vector containing the end positions and orientation of the end effector tip.Then the Jacobian (J) for the base-tool transformation matrix ${}^{0}T_6$ is defined as:

$$ J = \left[\frac{\partial X}{\partial \theta_1}\frac{\partial X}{\partial \theta_2}\frac{\partial X}{\partial \theta_3}\frac{\partial X}{\partial \theta_4}\frac{\partial X}{\partial \theta_5}\right] \quad (6) $$

where θ_i represents the corresponding D-H parameter. The forward kinematic equation is:

$$ dX = J(\theta_1, \theta_2, \theta_3, \theta_4, \theta_5)d\theta \quad (7) $$

where θ represents 5×1 the column vector containing the relevant D-H parameters. As both sides are deterministic, it is possible to invert the Jacobian using erstwhile θ_i and obtain the value of $d\Theta$ and subsequently the next set of θ_i, according to the equation,

$$ d\Theta = J^{-1}(\theta_1, \theta_2, \theta_3, \theta_4, \theta_5)dX \quad (8) $$

$$ \Theta_{new} = \Theta + d\Theta \quad (9) $$

It is clearly seen that the Jacobian is a matrix of partial derivate of the position and orientation with respect to the five joint angles. Inverse solutions for the velocity and acceleration of each link with respect to its frame i.e. $\dot{\theta}$ and $\ddot{\theta}$ can be obtained by directly differentiating the obtained values of θ. If the simulation is done in discrete steps, it suffices to differentiate using a first order scheme to obtain the required derivatives.

3.3 Trajectory planning

Our simulation requires, the path (trajectory) as closely spaced discrete points in ADAMS environment 3D space and whose coordinates are stored in a database (Microsoft Excel Spreadsheet), using a specially-constructed macro routine. These points are taken as guides in planning the path to be taken by the robot using the command of inbuilt database read. The interpolation between any pair of these consecutive points is given as a cubic polynomial:

$$ \theta(t) = A_i + B_i t + C_i t^2 + D_i t^3 \quad (10) $$

Here i is the storing indexed positions (points) in the

database. The curve is constructed with two basic constraints. Firstly, the velocity at the starting point and the ending point are zero. Secondly, the velocity at the intermediate points are unique. The coefficients of the polynomials is solved using a Tri-Diagonal Matrix Algorithm (TDMA).

The only disadvantage of a cubic interpolation is the jerks in $\theta(t)$. As we can clearly see the second derivative of $\theta(t)$ yields a linear function which will change between a pair of consecutive input points. Though this could have been avoided by a quadratic polynomial, it would render the equation solution-less due to the constraints involved.

4 Dynamic analysis

Dynamic analysis is the study of motion with regard to the causal entities (forces and torques). Dynamic modeling is vital for simulation and implementation of control algorithm. The Newton-Euler methodology used here is based on Newton's second law,in order to determine inverse dynamics that is used in this paper [6]. This equation can describe the behavior of a robot manipulator link-by-link and joint-by-joint from base to end-effector, called forward recursion and also transfer the essential information from end-effector to base frame, called backward recursion.

The forward recursion dynamic equations for the force and net moment acting on a joint are:

$$^{i+1}_{i+1}F = m_{i+1}\,^{i+1}_{i+1}\dot{v}_c \tag{11}$$

$$^{i+1}_{i+1}N = {}^{C_{i+1}}_{i+1}I\,^{i+1}_{i+1}\dot{\omega} + {}^{i+1}_{i+1}\omega \times {}^{C_{i+1}}_{i+1}I\,^{i+1}_{i+1}\omega \tag{12}$$

where F is the force acting on a link, N is the moment acting on a link, m is the mass of a link, $\omega$is the angular velocity of a link, $\dot{\omega}$ is the angular acceleration of a link, $\dot{v}$ is the acceleration of a link and^c_iI is the moment of inertia of a link 'i' about its center of mass. $^c_iI$is a 3×3 matrix and function of mass of the body and geometry. The rest of the quantities are 3×1 column vectors (for *x*, *y* and *z* axis).

The inverse recursion dynamic equations for calculating the joint forces and joint torques are:

$$^i_if = {}_{i+1}^{i}R\,^{i+1}_{i+1}f + {}^i_iF \tag{13}$$

$$^i_in = {}^i_iN + {}_{i+1}^{i}R\,^{i+1}_{i+1}n + {}^i_iP_c \times {}^i_iF + {}_{i+1}^{i}P \times {}_{i+1}^{i}R\,^{i+1}_{i+1}f \tag{14}$$

$$\tau_i = {}^i_in^T\,{}^i_i\hat{Z} \tag{15}$$

where$f$ is the force acting on a joint, n is the moment acting on a joint and τ is the component of n along the local rotational axis. These quantities are 3×1 column vectors (for *x*, *y* and *z* axis). A subscript *c* indicates the reference to the center of mass of a link.

Based on the same, required forces and joint torques can be calculated easily. It is noted that a pseudo-force is applied on each and every link that simulates gravity in real life situations by explicitly defining the acceleration of the base to be equal to g in the upward direction. Also the maximum force and torque that the end effector experiences are clearly specified for any simulation.

5 Control systems

Robotic control is considered to be the spine of robotics. It involves making a robot manipulator to perform operations automatically and precisely with the help of controllers. Hence it is a vital part in all modern robots. Typically, the controllers take the form of an equation or an algorithm which is realized via specialized computer programs.

Present industrial systems use a combination of proportional, integral and derivative (PID) control. This particular method has a wide range of applicability, ease of implementation and is very robust, while sacrificing some accuracy in comparison to non-linear control methods.The values of K_p, K_d and $K_i$in the PID control loop can be fixed based on the response pattern. With any control system, the key parameters are rise time, overshoot, settling time, and the steady state error for a step input. Each of the three parameters (K_p, K_d, K_i) have distinct effects on the four output parameters above, and fine tuning is necessary to obtain the best possible response. MATLAB also offers a PID tuning algorithm that can help in deciding the values of K_p, K_d and K_i after an iterative process. In the given example, the value of K_p, K_d and K_i are approximately 10 each. The values are obtained with the help of MATLAB toolbox mentioned above. In this simulation a simple PID control system was implemented for each individual angle outputs.

6 Simulation and results

Co-simulation between ADAMS and MATLAB is of paramount importance due to their efficient simulation platform. An acceptable format for the inputs and outputs of each program is required to execute the co-simulation. The objective of co-simulation is to establish a connection so that any change in one of the programs is reflected and further affects the other one, thus utilizing the benefits of both the program modules [7]. To provide a simulation that enables ADAMS to recognize the exported output from MATLAB, there is a need for activating the 'Control' plug-in in the 'Plug-in Manager' of ADAMS and defining the robot as a plant. After activation, a new window appears for the determination of plant and its inputs and outputs.

To call the generated plant in MATLAB, the plant name should be entered in the 'MATLAB command window' such that the computer recognizes and displays the input and output information. By executing the 'Adams_sys' command, the block containing the information about the dynamic model is created and loaded in the SIMULINK environment. The 'Adams-sub' block created in the SIMULINK should be configured as shown in Figure3. This is a vital component, as this block is included in the master simulation and control model, which is again defined in SIMULINK as a separate entity.

DOI: 10.1051/matecconf/20179508002

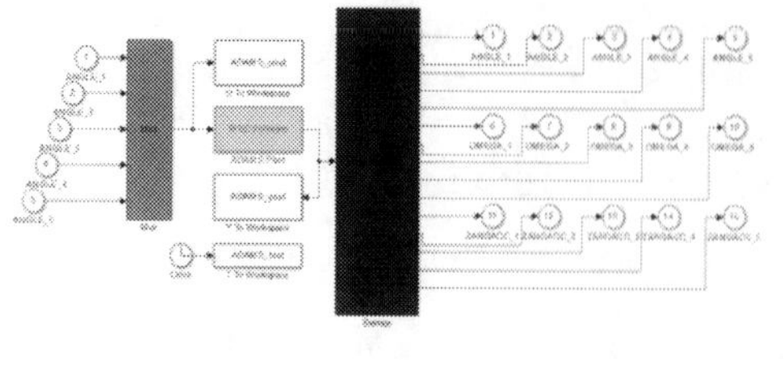

Figure 3. Configured Adams-sub block

The master model, referred to as Master here onwards contains all the modules of the previous sections, which again uses all the concepts (transformation matrices, inverse kinematics) mentioned earlier. The Master contains the trajectory planning module, inverse dynamics module, mainframe model of the robot and its associated control systems. The modules are made of snippets of MATLAB codes hence the blocks are also

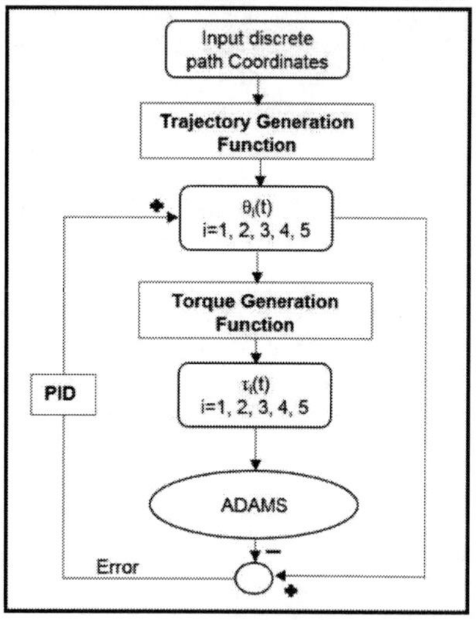

Figure 4. Algorithm for MATLAB/SIMULINK – ADAMS co-simulation

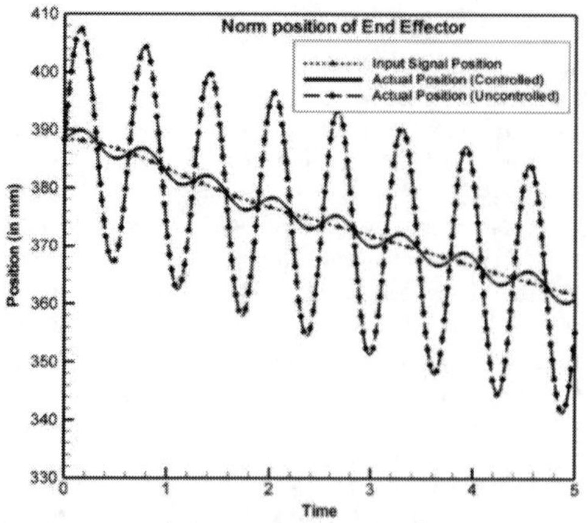

Figure 5. 2nd – Degree Norm Position* of End-Effector (with and without control systems)
*2nd Degree Norm Position is $\sqrt{x^2 + y^2 + z^2}$

user-defined and embedded MATLAB functions. Outputs of the dynamics block represent the plant inputs of the robot in ADAMS, namely the joint torques. By starting the simulation in SIMULINK, the signals denoting the joint torques enter the plant and hence runs ADAMS simultaneously which continues to reach the definite final state. The system is also capable of adjusting the size of the discrete steps for each calculation. The generated trajectory functions for each motor yields the end effector to track the desired position and orientation. The flow diagram indicating path generation and inverse dynamics are shown in Figure 4.

By the end of the simulation, the end effector position in the work space is depicted in a scope of MATLAB-SIMULINK. During the simulation, real-time interactive process was shown on the ADAMS interface [8]. By comparing the results with the forward kinematics solution, it can easily be inferred that the simulation runs with the desired accuracy and precision.The graphs were generated in MATLAB-SIMULINK using the 'Scope' module. From simulation results (Figure 5), it can be seen that the links can track the given trajectory curve well with smallerror.

7 Conclusions

In this paper, the virtual development of a 5-DOF robot has been done using SolidWorks and the simulation, testing and validation were carried out using ADAMS and MATLAB-SIMULINK. The validity of the control loop was put to test by imparting a sinusoidal disturbance to the joint torques.Please do note that this observed error was because of the purposeful sinusoidal disturbance. Furthermore, loading the .res file into ADAMS post-processor, it was possible to cross-check all the inertia forces, gravity, centrifugal forces, Coriolis forces and instantaneous power of the driver motorwith the results that were obtained from calculation in MATLAB.

References

1. K.T. Park, Y.J. Shin, C.H. Park, Y.S. Mo, and D.C. Jeong,ICCAS, *"Robot application for assembly process of engine part,"*(2008)
2. D. D. Ray and M. Singh, CARPI,*"Development of a force reflecting Tele-robot for remote handling in nuclear installations,"* (2010)
3. B. Armstrong, O. Khatib, and J. Burdick, ICRA, *"The explicit dynamic model and inertial parameters of the PUMA 560 arm,"* (1986).
4. F. Cheraghpour, M. Vaezi, H. E. ShooriJazeh, and S. A. A. Moosavian, IEEE-ICM *"Dynamic modeling and kinematic simulation of Stäubli© TX40 robot using MATLAB/ADAMS co-simulation,"* (2011).
5. Mitsubishi Electric Corporation Tokyo,*"Movemaster II RM-501 model instruction manual"*, (1984).
6. J. J. Craig, *Introduction to Robotics: Mechanics and Control, 3rd ed. United Kingdom: Prentice Hall,* (2003).

7. Z. Yi, X. Min-min, Q. Jin-yi, and Z. Hu, ICCASM,*"Research on co-simulation using ADAMS and MATLAB for automobile active suspension system,"* (2010)
8. R. M. Inigo and J. S. Morton, IEEE Trans. Educ., *"Simulation of the dynamics of an industrial robot,"* **vol. 34, no. 1**, pp. 89–99, (1991)

DOI: 10.1051/matecconf/20179508003

Research on the Obstacle Negotiation Strategy for the Heavy-duty Six-legged Robot based on Force Control

Mantian Li, Enbo Cong, Pengfei Wang and Wei Guo

State Key Laboratory of Robotics and Systems,Harbin Institute of Technology, Harbin, 150080, China

Abstract. To make heavy-duty six-legged robots without environment reconstruction system negotiate obstacles after the earthquake successfully, an obstacle negotiation strategy is described in this paper. The reflection strategy is generated by the information of plantar force sensors and Bezier Curve is used to plan trajectory. As the heavy-duty six-legged robot has a large inertia, force controller is necessary to ensure the robot not to lose stability while negotiating obstacles. Impedance control is applied to reduce the impact of collision and active force control is applied to adjust the pose of the robot. The robot can walk through zones that are filled with obstacles automatically because of force control. Finally, the algorithm is verified in a simulation environment.

1 Introduction

Compared with wheeled robots and tracked robots, legged robots can choose non-continuous foothold, which make them adapt to the rough terrain with obstacles easily [1-2]. Heavy-duty six-legged robots are the most suitable legged robots for transporting supplies because of their characteristic of high stability and strong loading capacity [3]. For heavy-duty robots, the inertia is large, force control is extremely necessary. If the robot lose stability while contacting obstacles, it will be difficult to adjust back. So the pose of the robot need to be adjusted in real time.

At present, there are two methods for muti-legged robots to negotiate obstacles. One is recognising obstacles by environment reconstruction devices. For instance, the Little Dog built a 3D terrain model and picked up obstacle information by vision sensor, then they planned an appropriate path to negotiate obstacles [4]. But this method will be restricted by the environmental factors. Sandstorm and rain-snow environment will affect the performance of the vision sensor, laser radar or other external environment sensors. So this method can't be applied to the robot working in the severe environment. The other method is receiving contact information by plantar force sensors. Typical examples include Tekken designed by Japanese [5] and DLR-Crawler designed by German [6]. These robots replanned the trajectory after contacting obstacles. As soon as plantar force sensors receive signals that any foot has contacted obstacles, robots adjust swing track quickly. Central pattern generator(CPG) controller is used in Tekken and the distributed artificial neural network controller WALKNET is used in DLR-Crawler. All the two intelligent control technology need large computational quantity and not handy for real-time control.

This paper describes a strategy for the heavy-duty six-legged robot to negotiate obstacles in real-time in the severe environment. Reflection generation strategy and force control strategy are presented in detail in this paper.

The rest sections are organized as follows. In Section 2, the model is established. Sections 3 and 4 describe the two subtasks of obstacle negotiation strategy for heavy-duty six-legged robot: the reflection strategy and the force control strategy. Section 5 presents results of simulation. Finally, Section 6 summarizes the work and presents the conclusions that can be drawn from it.

2 Model

A model in Adams is established to simulate the real environment, as we can see in Figure 1.

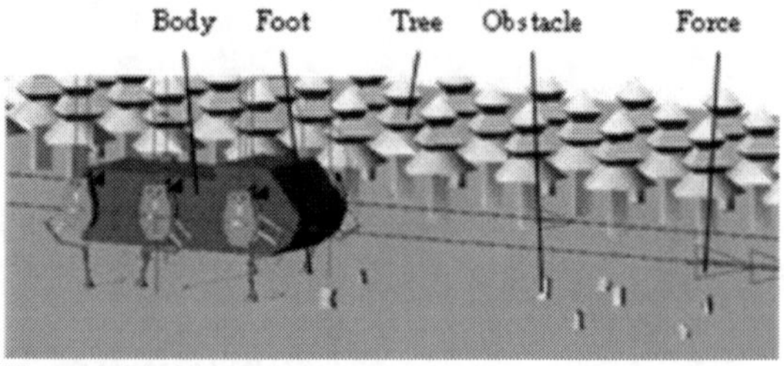

Figure 1. Model in Adams.

The model's parameters are shown in Table 1,where M represents the weight of the heavy-duty six-legged robot, H represents the height from bottom of the body of the robot to ground. L represents the length of the robot. d represents the width of the robot. f represents friction factor between feet of the robot and ground. n represents joints number in one leg. h_{max} represents the vertical height of the highest obstacle.

Table 1. Parameters of the model.

M	H	L	d	f	n	h_{max}
3000 kg	1000 mm	4000 mm	2200 mm	0.5	3	350 mm

3 Reflection Strategy

3.1 Establishment of reflection rule

Legs may contact obstacles at any time in a period time, different reflection rules are made for different collision time. We can insert 4 time points A~E during the swing phase time, which divide the swing phase into 3 stages, as shown in Figure 2.

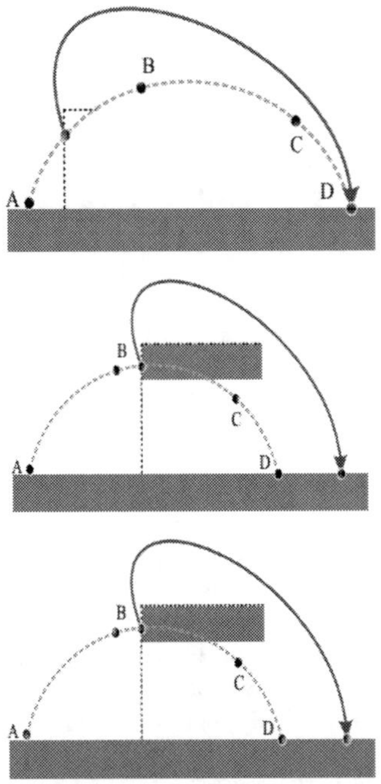

Figure 2. Different reflection rule in different situations.

In Figure 2, AB is the early stage of the swing motion. If the foot contacts obstacles at this stage, reflection motion will be trigged. The foot will be retreated and the height of the new trajectory is higher than the original trajectory.

BC is the later stage, and the reflection rule is roughly identical to BC. The difference is that the foothold of BC is in front of the foothold of AB.

CD is the final stage of the swing motion. The foot will be retreated to a position near the obstacles if touching obstacles because the time left is not enough for the foot to accomplish negotiating obstacles.

As stated above, reflection rule is established based on collision time, then it can ensure the foot to have enough time to negotiate obstacles.

3.2 The curve of reflection Trajectory

For the heavy-duty six-legged robot, the trajectory need to be smooth and compliant. So Bezier Curve is applied to plan the trajectory, because the main advantage of this kind of curve is smooth and compliant. Parametric equation of n order Bessel curve is described as $Q(t)$ in Eq.(1).

$$Q(t) = \sum_{i=0}^{n} P_i B_{i,n}(t) \quad t \in [0,1] \tag{1}$$

Where P_i represents position vector of n+1 control points. $B_{i,n}(t)$ represents Bernstein polynomial, which can be described in Eq.(2).

$$B_{i,n}(t) = \frac{n!}{i!(n-i)!} t^i (1-t)^{n-i} \tag{2}$$

If t_{swing} represents the total time of swing phase, $t_{collision}$ represents the collision time point, the terminal velocity v_{sf} of the foot in swing phase can be calculated by Eq.(1) and Eq.(2), as shown in Eq.(3).

$$v_{sf} = \dot{Q}(1) = n(P_n - P_{n-1}) \frac{1}{t_{swing} - t_{collision}} \tag{3}$$

In Eq.(3), if appropriate values are assigned to P_n and P_{n-1}, the terminal velocity of foot in swing phase will equal to the velocity of foot in stance phase. In this way, the sudden change of velocity can be eliminated. Then the trajectory will be smooth enough for heave-duty robot to negotiate obstacles.

4 Force Control Strategy to Guarantee Stability

To guarantee stability of the heavy-duty robot, force control strategy is applied. Impedance control strategy is applied to swing legs, and active force control strategy is applied to stance legs. These two strategies are described as follow.

4.1 Impedance control strategy for swing legs

Compared with light-duty robots, heavy-duty robots have large inertia. The impact force caused by the collision between robot and convex obstacles is large enough to make the robot overturn. So in order to reduce the impact of collision, position-based impedance control method is applied to swing legs.

Impedance control is a method to adjust force and position dynamically. The input of impedance controller is the deviation ΔF between actual contact force F_γ measured by plantar force sensors and target force F'. It's obvious F'=0 in the swing phase. The output of impedance controller is the deviation ΔP, which can adjust the current position P_γ to a position P where F'=0, as shown in Figure 3.

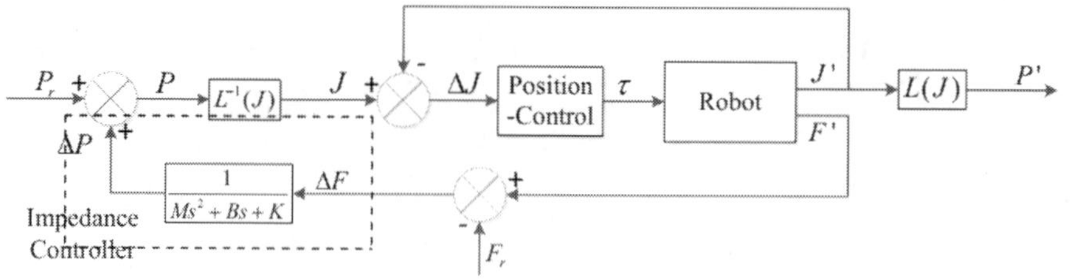

Figure 3. Impedance control system for swing legs.

Impedance controller can be described as Eq.(4).

$$\Delta F = M_d \Delta \ddot{P} + C_d \Delta \dot{P} + K_d \Delta P \tag{4}$$

Where M_d represents inertia coefficient, C_d represents damping coefficient and K_d represents stiffness coefficient.

By impedance controller, contact force and position of the foot are dynamically adjusted. If appropriate M_d, C_d and K_d are assigned to the controller, the foot can get away from obstacles quickly.

4.2 Active force control strategy for stance legs

When the robot contact obstacles, there will be an error between actual pose and target pose of the body caused by collision. The greater the impact force is, the larger the error is. Large error may cause the robot lose its stability. So the pose is adjusted through allocating appropriate force to every stance foot.

The pose of the robot is adjusted by virtual suspension model, which is an imaginary spring-damping system. As shown in Figure 4.

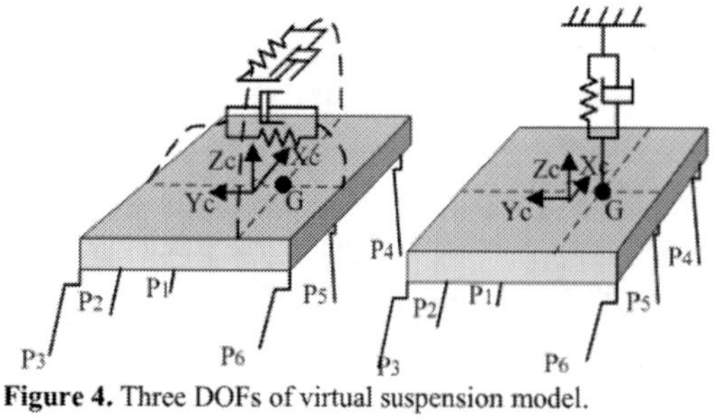

Figure 4. Three DOFs of virtual suspension model.

In Figure 4, virtual suspension model is used to adjust pitch angle, roll angle and vertical height of the robot. The pose of robot is measured by inertial navigation devices in real-time. When the robot contacts obstacles, there will be deviations between actual pose and target pose. If the deviation of pitch angle is represented by $\Delta\beta$, the deviation of roll angle is represented by $\Delta\gamma$, and the deviation of vertical height is represented by Δd, there will be corresponding virtual generalized force ΔM_β, ΔM_γ, ΔM_d provided by stance legs.

The virtual generalized force required to eliminate deviation is shown as Eq.(5). Stiffness coefficients k_β, k_γ, k_d and impedance coefficients c_β , c_γ, c_d are parameters to correct the deviations.

$$\begin{aligned} \Delta M_\beta &= k_\beta \Delta\beta - c_\beta \Delta\dot{\beta} \\ \Delta M_\gamma &= k_\gamma \Delta\gamma - c_\gamma \Delta\dot{\gamma} \\ \Delta F_d &= k_d \Delta d - c_d \Delta\dot{d} \end{aligned} \tag{5}$$

In fact, k_β, k_γ, k_d are related to control stiffness in z direction of stance legs, which is represented by K_{iz}. Their mathematical relationship is shown as Eq.(6).

$$\begin{aligned} k_\beta &= \frac{\sum_{i=1}^{6} K_{iz}({}^C P_{ix} - {}^C G_x)\sin\Delta\beta({}^C P_{ix} - {}^C G_x)}{\Delta\beta} \\ &\approx \sum_{i=1}^{6} K_{iz}({}^C P_{ix} - {}^C G_x)^2 \\ k_\gamma &= \frac{\sum_{i=1}^{6} K_{iz}({}^C P_{iy} - {}^C G_y)\sin\Delta\gamma({}^C P_{iy} - {}^C G_y)}{\Delta\gamma} \\ &\approx \sum_{i=1}^{6} K_{iz}({}^C P_{iy} - {}^C G_y)^2 \\ k_d &= \frac{\sum_{i=1}^{6} K_{iz}\Delta d}{\Delta d} = \sum_{i=1}^{6} K_{iz} \end{aligned} \tag{6}$$

Where i =1,2,···, 6. ${}^C P_{ix}$ represents ith foot position along x coordinate in the body coordinate system, ${}^C P_{iy}$ represents ith foot position along y coordinate in the body coordinate system. ${}^C G_x$ represents centre of gravity position along x coordinate in the body coordinate system. ${}^C G_{iy}$ represents centre of gravity position along y coordinate in the body coordinate system.

By Eq.(5) and Eq.(6) and coordinate transformation equation, we can calculate position variation of every foot relative to the body, as shown in Eq.(7).

$$\Delta {}^C P_{iz} = -{}^C P_{ix}\frac{\Delta M_\beta}{k_\beta} + {}^C P_{iy}\frac{\Delta M_\gamma}{k_\gamma} + \frac{\Delta F_d}{k_d} \tag{7}$$

Then the force ${}^C F_{iz}$(i=1,2,3,4,5,6) allocated to every foot to eliminate the variation can be calculated by Eq.(8).

$${}^C F_{iz} = -(K_{iz}\,{}^C P_{ix})\frac{\Delta M_\beta}{k_\beta} + (K_{iz}\,{}^C P_{iy})\frac{\Delta M_\gamma}{k_\gamma} + (K_{iz})\frac{\Delta F_d}{k_d} \tag{8}$$

By allocating force , the robot can walk stably when it contacts convex obstacles.

DOI: 10.1051/matecconf/20179508003

5 Simulation and Analysis

As trajectory planning and force control for negotiating obstacles have been completed, the algorithm is verified by joint simulation of Simulink and Adams, the sample period of which is 10*ms*, as shown in Figure 5.

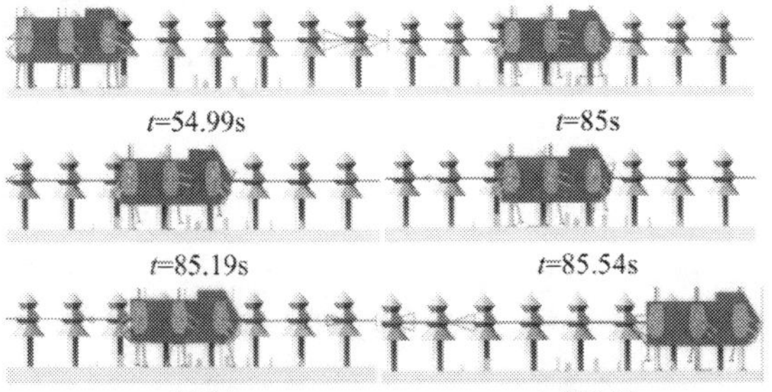

Figure 5. Process of obstacle negotiation in simulation.

In Figure 5, the robot is able to pass through terrain with plenty of convex obstacles after thefiftieth second. Figure 6 is the trajectory when a foot contact obstacles 3 times in a swing period. And the trajectory generated by Bezier Curves is smooth enough to ensure the robot to negotiate obstacles.

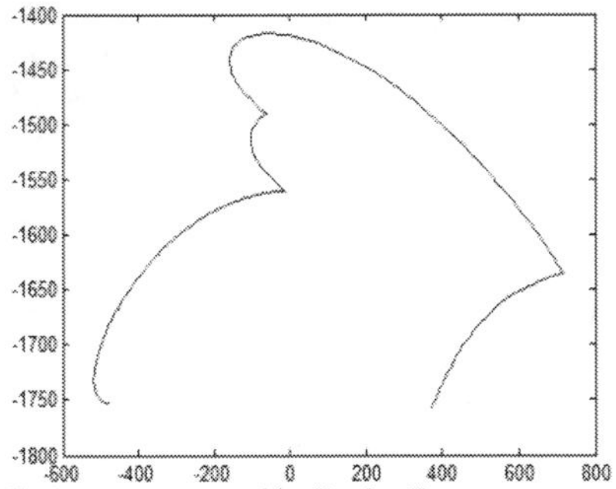

Figure 6. Trajectory generated by Bezier Curves.

As force control strategy has been applied, the roll angle, pitch angle and height of the robot can be maintained around target values, which is shown in Figure 7. The deviations between target values and actual values are so small that can be ignored.

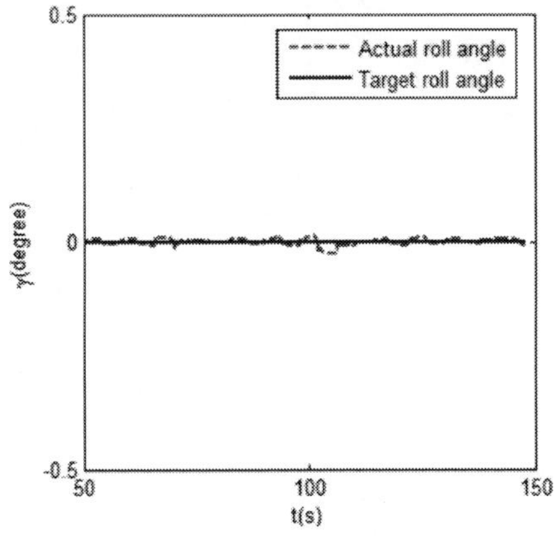

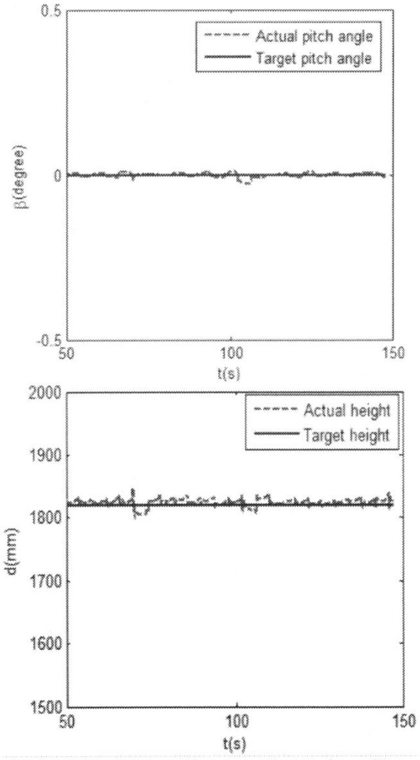

Figure 7. Results of maintaining pose during the collision.

6 Conclusion

An obstacle negotiation strategy for heavy-duty six-legged robot without environment reconstruction system is described in this paper. A trajectory planning strategy is established by the use of plantar force sensors and Bezier Curves is used to plan trajectory. In order to prevent heavy-duty six-legged robots with large inertia from overturning, impedance control strategy is applied to swing leg, and active force control is applied to stance legs. Finally, the algorithm is verified to be correct and effective by joint simulation.

References

1. R.B. McGhee, G.I. Iswandhi, IEEE Trans Syst Man Cybern A **9**,176(1979)
2. Q.J.Huang,K.Nonami, ASME T MECH **13**,773(2003)
3. E. García Armada, M. A. Jiménez, G.D.S Pablo, IEEE Trans Rob Autom **14**,90(2007).
4. V. Dürr, J. Schmitz, H. Cruse, Arthropod Struct DEV **33**, 37(2004)
5. S. M. Song, K. J. Waldron, Nasa Sti/Recon Technical Report A, **89**, 7 (1989)
6. J. Z. Kolter, M. P. Rodgers, A. Y. Ng., *International Conference on Robotics and Automation*(IEEE, New York, 2015)

Implementation of a Mobile Robot Platform Navigating in Dynamic Environment

Hadjira Belaidi[1], Hamid Bentarzi[1] and Mohamed Belaidi[2]

[1] Signals and Systems Laboratory (SisyLab), Institute of Electrical and Electronic Engineering (IGEE), University M'hamed Bougara of Boumerdes (UMBB), 35000 Boumerdes, Algeria
[2] Infotronics Departement, Science Faculty, M'hamed Bougara University of Boumerdes (UMBB), Algeria

Abstract. Currently, problems of autonomous wheeled mobile robots in unknown environments are great challenge. Obstacle avoidance and path planning are the back bone of autonomous control as it makes robot able to reach its destination without collision. Dodging obstacles in dynamic and uncertain environment is the most complex part of obstacle avoidance and path planning tasks. This work deals with the implementation of an easy approach of static and dynamic obstacles avoidance. The robot starts by executing a free optimal path loaded into its controller; then, it uses its sensors to avoid the unexpected obstacles which may occur in that path during navigation.

1 Introduction

Robotics helps human in difficult, repetitive or tedious tasks. Moreover, it is the dream of substituting the machine to man in these tasks. Currently, the faculties of robots perception and reasoning are progressing every day and more in the future, they will play an increasingly important role in our lives.

Navigation of mobile robots in dynamic environments still represents a challenge for real world applications. The robot should be able to gain its goal position navigating safely among moving people or vehicles, facing the implicit uncertainty of the surrounding world and the limits of its perception system.

The problem of autonomous navigation has been deeply studied in literature and several techniques have been developed. These techniques use probabilistic and complex algorithms which are hard to implement and requires specific sets of hardware and software [1], [2].

In this paper a technique that is easy to be implemented using low cost materials is introduced. First, the robot is supposed to follow a predefined optimal free path, generated using the method given by [3], to navigate safely in static environment. Then, a strategy enabling the robot to navigate in dynamic environment considering the uncertain obstacles is proposed. Thus sonar sensors are used for this purpose.

2 Path generation and obstacle avoidance in static environment

First, and prior to spraying the case of dynamic environment, the robot needs to generate a navigation path in a static environment. After that, this trajectory will be saved in the internal memory of a robot and run repeatedly; and at the same time avoiding unexpected obstacles (and dynamic ones).

2.1 Path planning approach

A path planning and obstacle avoidance approach using parametric curves was used for the generation of the predetermined path that the robot should follow [4]. It is assumed that the current position coordinates of the robot and of the target are known using exteroceptive sensors. The robot can measure the distance to nearby obstacles that are in the range of sensors and avoid them to achieve the goal by following an optimal smoothed path by running the algorithm proposed in [4], [5].

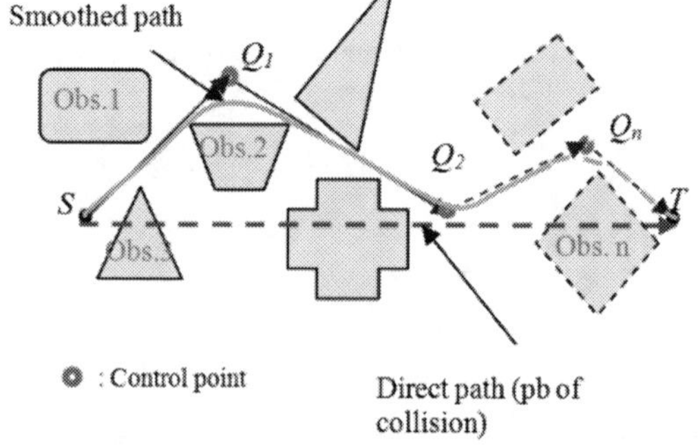

Figure 1. Path planning using parametric curves and control points insertion

In [5], a linear parametric curve is used for path planning, and then this curve is smoothed around the control point. The linear path connecting the initial position (S) to the target (T) is considered first. If the trajectory hits an obstacle, a control point is introduced between the initial position and the target and an intermediate connection point is created as shown in Fig. 1. Then, the collision between the line (QT), connecting the control point Q to the target T, and the obstacle is tested. In case of collision, another control point must be added, and so on (see Fig. 1).

Non-uniform rational B-spline (NURBS) curves were used to smooth the path, as they preserve the C^2 continuity of the trajectory which is very important for the robot dynamic. Furthermore, the addition of control points to avoid obstacles does not affect the whole trajectory. The smoothed curve is shown in green in Fig. 1.

2.2 Static trajectory data extraction

There are several ways to generate a path to follow for a robot such as: Arc trajectory [4], which is easy to be generated; but the combination of obstacles avoidance with other factor changes in a dynamic environment is difficult because of the circular arc. Thus, the trajectory generation using Bezier curves is introduced [6], which is characterized by defining the orientation and direction of the robot. However, the major drawback of this method is the mathematical complexity of Bezier curves, which makes it difficult to create paths a robot can follow with high speed and the same result is presented in straight line guide where the robot must stop walking to turn.

To reduce these problems, in our work, the robot must follow the trajectory generated using NURBS. First, the following parameters must be extracted from the NURBS curve generated using Visual Basic instruction [5]:

- The tangent from the initial position, which defines the initial orientation of the robot.
- The distance (di) between each two control point which defines the distance that the robot has to move and which can be calculated using the following formula:

$$d_i = \sqrt{\left(Q_{(i+1)x} - Q_{i_x}\right)^2 + \left(Q_{(i+1)y} - Q_{i_y}\right)^2} \quad (1)$$

- The tangent at each control point Qi to define the orientation α_i of the robot (see Fig. 2(a)).
- The speed of the robot is defined by the derivative of the NURBS curve at each control point. Between the control points, the robot has to move with a constant speed.

Initially, NURBS is presented as a database containing the coordinates (x_i, y_i) of each control point Q_i. Next, the distances d_i and the orientations α_i are computed as explained above and as shown in Fig. 2(b). After extracting these data, they will be recorded as an Excel database (* .xls) to send them to the platform of the robot via USB (universal serial bus) port, which allows series data transfers from the PC to the robot microcontroller.

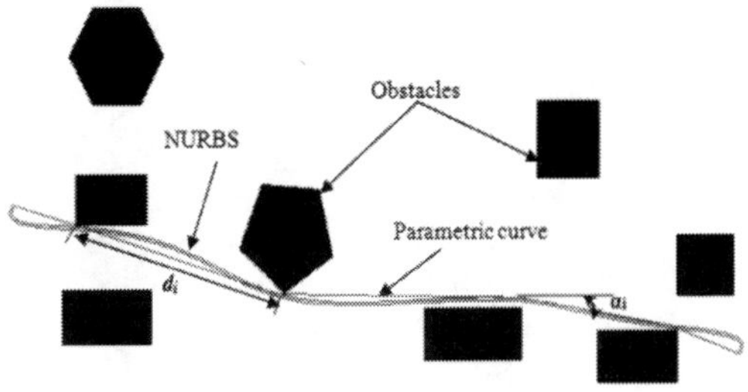

(a)

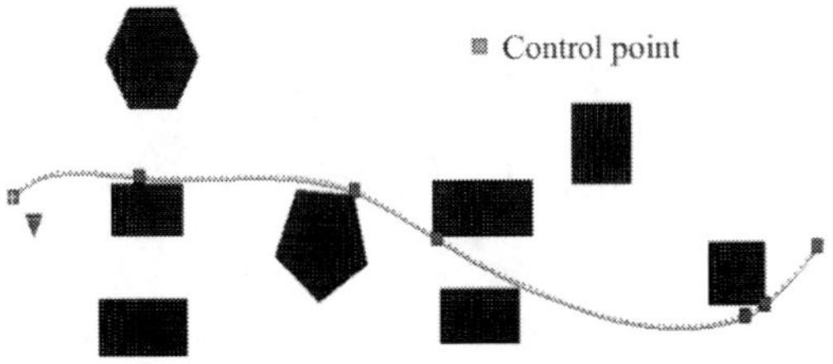

(b)

Figure 2. AutoCAD simulation results. (a) Path planning in static environment. (b) Static trajectory data extraction

Table 1 is an illustration of database containing the coordinates (x_i, y_i) of control points Q_i constituting the free path, and Table 2 illustrates the database that contains the translation of these data to distances and orientations.

Table 1. Illustration of database extraction from static free path (the control points (x, y) coordinates).

$x_i(mm)$	$y_i(mm)$
892,8212	1228,2581
1451,0873	1300,5873
2413,4103	1248,5391
4179,7141	795,9584
2791,8635	1075,9884
4270,0362	838,2637
4506,3260	1043,2394

Table 2. Translation of Q_i coordinates to distances and orientations.

d_i	α_i
1518,4688	30,252
1948,6358	0
2717,2410	-35,456
4254,8278	-45,001
2992,0315	10,983
4351,5394	20,980
4625,5078	45,032

This free path data will be executed by the robot platform. If the robot encounters an unexpected obstacle so it sends an error message to the microcontroller to notify the existence of an obstacle not point in the map of the environment.

3 Obstacles avoidance in dynamic environment

A function called "dynamic" is created to support the dynamic obstacle avoidance. This function helps the robot to navigate in an uncertain environment with four (04) ultrasonic sensors that will collect the environmental information, locate the obstacles and avoid them.
The principle of this function is as follows:

- Read the data of the front sensor;
- If there is an obstacle, the robot moves backward for a distance that allows it to avoid the obstacle without collision by a test of the rear sensor;
- Read the data of the right sensor;
- If there is no obstacle the robot turns right and reads the right sensor data until it exceeds the obstacle then it recovers the initial trajectory by the inverse performance of data travelled;
- If there is an obstacle the robot tests the left sensor, otherwise it avoids the obstacle from the left.
- If both sensors indicate that the path is not free an error message will be sent to the controller to assign that "the way is occupied";
- If a moving obstacle moves toward the robot, the robot will move backward with sending a message that contains the value # DANGER#. The flowchart in Fig. 3 summarizes the proposed approach.

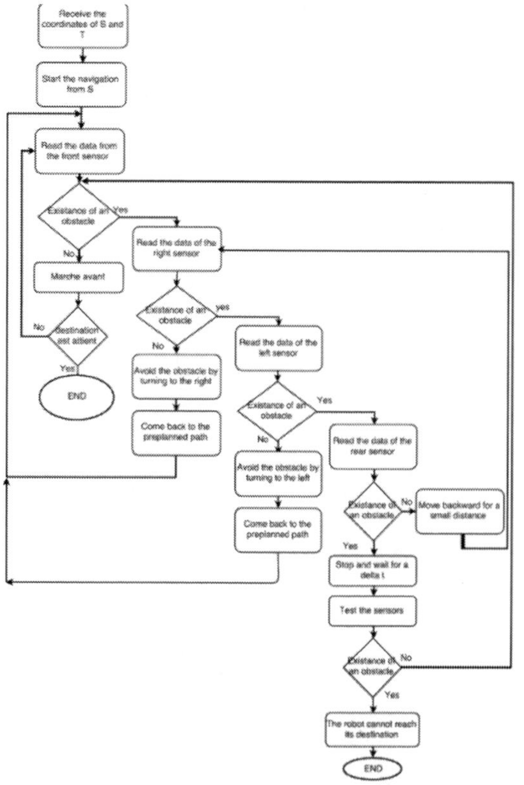

Figure 3. Flowchart simurizing the approach of obstacles aviodance in dynamic environment.

4 The robot platform

Fig. 4 shows the general electronic circuit connecting the various components constituting the robot platform. The denotation of each component is given below.

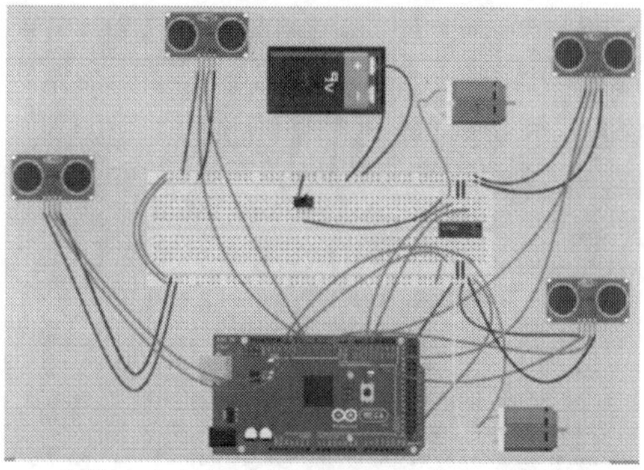

Figure 4. The robot circuit Platform. 1: Arduino MEGA2560 card, it contains the program that allows the robot to navigate in its environment; 2: Ultrasonic sensors used to calculate the distance between the robot and the obstacles; 3: 6V battery energy source; 4: DC-motors; 5: On / off switch; 6: L293D used to drive the DC motors

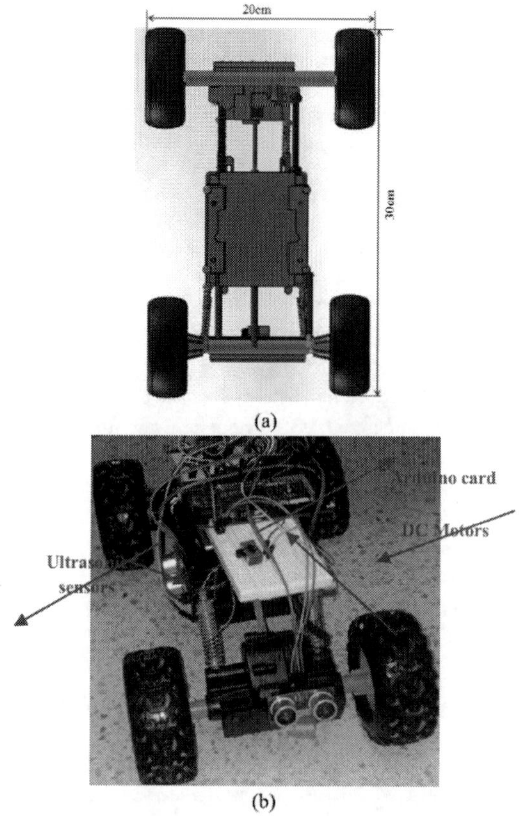

Figure 5. The used Autonomous mobile robot platform. (a) Platform geometry of the used autonomous mobile robot. (b)The image of the robot

The prototype represents an autonomous mobile robot platform which moves in three dimensional space. The platform has two rear driving wheels and two front orientation wheels. The wheels are supplied by two independent DC motors. The geometry of the platform is depicted in Fig. 5(a). Hence an image of the robot platform is given in Fig. 5(b).

5 Experimental results

To validate our approach, it has been tested in several scenarios.

In the scenario shown in Fig. 6, the robot follows its path planned previously, and loaded into its controller; once an unexpected obstacle appears, the robot stops, go back (for a small distance which permits it to avoid the obstacle) then turns right and avoids the obstacle. If no other obstacle occurs, the robot turns left and come back to its initial path.

Figure 6. Obstacle avoidance scenario

Before turning, the robot tests the two side sensors. Hence, the robot avoids the obstacle by the side free of obstacles, and it will not come back to its initial path until the sensor face of the obstacle be free.

One important case to notice is when the robot avoids the obstacles to a specific side and detects another obstacle in that way. The robot automatically goes back and tries the other side. If the two sides are blocked, the robot backs off in a straight line and avoids it again.

When the robot is driven out of its path by many obstacles, it always tries to rejoin its path.

5.1 Results and discussion

This robot is equipped by ultrasonic sensors which make it able to detect transparent objects, and navigate in an indoor environment. So this robot can be used in our dignified house, offices, supermarkets etc .... For example the robot can detect glass sliding doors.

In addition, the robot can navigate in outdoor environments where the surface can be rough, because of its chassis with suspension (see Fig. 5) and the approach that allow it to navigate in uncertain environments.

However, there are some critical cases that this robot can encounter and which are not taken into consideration in our work. The following section discusses some of these cases.

5.2 Some critical cases

Our robot encounters great difficulties to detect circular objects and obstacles with a width lower than 5cm due to the characteristics of ultrasonic sensors when sending sound waves. These waves will hit the obstacle and change direction and they do not return to the sensor, so the obstacle avoidance is not very accurate in this case.

In the case 1 of Fig. 7, the distance measurement is accurate because the obstacle is perpendicular and in front of the sensor, however in case 2 it will generate an inaccurate measurement since the inclined position sensor. Among the critical cases also, the positions which generate a measuring antagonism as shown in Fig. 7 case 3.

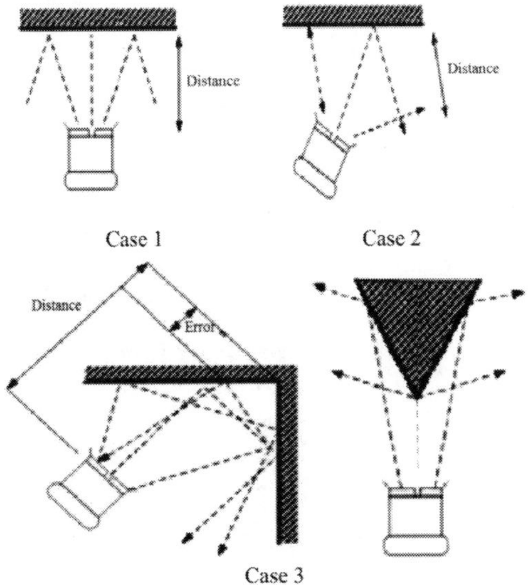

Figure 7. Some critical cases which cannot be avoided accurately by the robot

Apparently, the robot will not detect objects that have a vacuum of more than 10cm high and in the case where the object has a substantial length. This length may be a function of speed. This situation can create a robot damage (crushing) Fig. 8.

Figure 8. Objects having a vacuum of more than 10cm high.

Another drawback to notice is that high-speed obstacles cannot be avoided. The obstacle velocity should be less or equal to the robot's velocity.

 DOI: 10.1051/matecconf/20179508004

6 Conclusion

In this work, the design, simulation and real time testing of an autonomous mobile robot in an unknown environment have been successfully carried out. We were able to come up with a strategy of navigation and try its effectiveness in a real time changing indoor environment.

Though some defects, the results were satisfying and the robot was able to navigate safely in dynamic environment.

Hence, in the future work, the robot's performance can be enhanced by installing a GPS module to localize the robot at any given moment. This will simplify the program and drive the robot to any goal with high precision.

Moreover, the ultrasonic sensors can be replaced by laser sensors which have a wider range to recognize the exact shape and position of obstacles. We can go further in customizing with adding up a Radar sensor to work in outdoor environment (rough environment) since the ultrasonic sensors performance is limited.

References

1. H. Cho, et al., "A Multi-Sensor Fusion System for Moving Object Detection and Tracking in Urban Driving Environments," 2014 IEEE International Conference on Robotics & Automation (ICRA), Hong Kong Convention and Exhibition Center, Hong Kong, China, (May 31-June 7, 2014)
2. C. Saranya, et al., "Path Planning Through Partially Occupied Cells using Modified D* Algorithm," ISTE-ACEEE Int. J. in Computer Science, **1**(1), (March 2014)
3. B.J. Lee, S.O. Lee and G.T. Park, "Trajectory Generation and Motion Tracking Control for the Robot Soccer Game," Proceedings of the 1999 IEEE/RSJ International Conference on Intelligent Robots and Systems, (1999)
4. H. Belaidi, "Optimal control and strategy of autonomous mobile robot 2D/3D simulation," Magister thesis, Electrical and Electronic Institution, UMBB, (June 2009)
5. H. Belaidi, A. Hentout, B. Bouzouia , H. Bentarzi, A. Belaidi, "NURBs Trajectory Generation and Following by an Autonomous Mobile Robot Navigating in 3D Environment," 4th Annual International Conference on Cyber Technology in Automation, Control, and Intelligent Systems (CYBER), 2014 IEEE, Hong Kong, (4-7 June 2014), pp. 168–173.
6. J.C. Wolf, "Fast Autonomous Mobile Robot Platform," A report submitted to the University of Plymouth In partial fulfillment for the degree in B Eng (Hons) Robotics and Automated Systems, (2003-2004)

Implementation of a map route analysis robot: combining an Android smart device and differential-drive robotic platform

Chi-Hung Tseng[1], Ching-Chang Wong[1] and Wen-Hsuan Kuan[2]

[1]*Department of Electrical Engineering, TamKang University, 151, Yingzhuan Rd., New Taipei City, Taiwan*
[2]*Department of Applied Physics and Chemistry, University of Taipei, 1, Ai-Kuo W. Rd., Taipei, Taiwan*

Abstract. This paper proposes an easy-to-implement and relatively low-cost robotic platform with capability to realize image identification, object tracking, and Google Map route planning and navigation. Based on the JAVA and Bluetooth communication architectures, the system demonstrates the integration of Android smart devices and a differential-drive robotic platform.

1 Introduction

Robotic applications have attracted a great deal of attention in four areas: image identification and object tracking (IDOT) [1-2], map positioning and navigation (MPN) [3-4], web communication (WC) [5-6], and vocal control for speech recognition (VCR) [7-8]. In IDOT, feature-based or appearance-based recognition algorithms are usually adopted for facial recognition, road marking, and vehicle tracking. For better accuracy and reliance, the combination of autonomous self-deployment sensor networks with mobile robots [9] allows for the integration and incorporation of map route analysis, obstacle detection and avoidance. Parallel to the visual control of robots, vocal-sensitive robots designed to adaptively characterize the fundamental frequency and signal envelope of a sound [10] may greatly support surveillance and care services. Conducted by machine learning, recent advances in VCR for semantic analysis [11-12] and emotion analysis have achieved more realistic intelligence in human-computer dialogue.

These applications indicate that big data information [13] and WC technologies form the bedrock upon which innovative solutions of the Internet of Things [14] are being built, and are pushing forward the development and service of robotics. However, the aforementioned robotic applications also indicate that robots are able to be well placed in highly dynamic environments and can simultaneously feedback rich information. The cost of the huge computational efforts is, however, prohibitive, let alone extending them to outdoor or large-scale scenarios [15] where expensive beacon and differential-GPS [16] are commonly used at present. Therefore, the use of heterogeneous technologies to improve their performance, or at first at least to improve their affordability, has become an emergent and practical issue.

In practice, the strategy of installing the above four sensors or extended modules on robotic platforms is known for its lack of a price advantage. Meanwhile, there is too great a loading on the robot's on-board controller for it to support multidisciplinary control, especially for image processing. The realization of real-time data acquisition and analysis on a 16-bit processor would be tough. As a comparison, the state-of-the-art microelectronic technologies have been built to easily resolve 20 fps ultrafast and 640 ×480 2D real-time video streaming by the quad-core 64-bit processor with 1.5 Ghz clock speed of a smart phone. Nevertheless, the robotic microcontroller utilities remain significant in serving as middleware to bridge I/O pines, joints and communication protocols in the control of sensors and actuators via a smart phone. We expect that this short description has provided, either from an economic or performance perspective, a persuasive and broad motivation to investigate the combination of a mobile driving robotic platform such as the LEGO or Arduino suites with smart phones.

Due to the space limitation, in this report we focus specifically on introducing some studies on IDOT and MPN. We used the LEGO NXT robotic platform to build the main body of a mobile robot. A JAVA-based object-oriented algorithm was used to develop the programs on Android smart phones. The machine-human interaction was mediated by Bluetooth transmissions.

Figure 1. Two LEGO NXT mobile robotic architectures.

 DOI: 10.1051/matecconf/20179508005

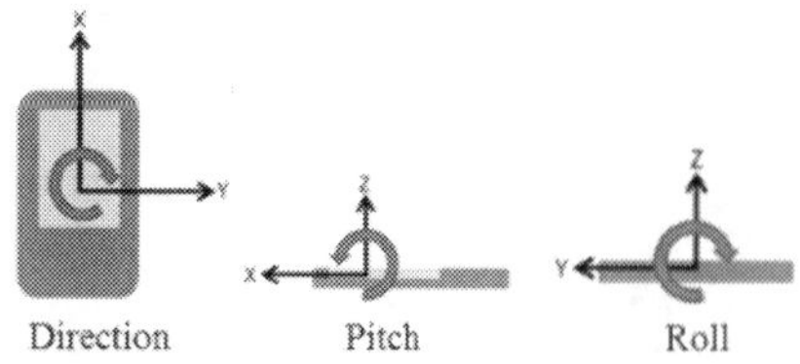

Figure 2. Orientation control of a smart phone.

2 The Architecture

While traditionally using differential two-wheeled robots encountered limitations of movement and direct control, these shortcomings were overcome by Ribeiro et al. who used a robot with three Omni-directional wheels [17]. The transverse drift induced by the axial-movement on that platform however leads to a reduction in the precision required in this work. Hence, a differential-driving mobile robot was adopted, considering the practical points of reducing the mechanical complicity and extensive accessibility. Translational displacement and the speed signals of the robot were provided via fast-response DC encoder servomotors. Two LEGO NXT mobile robotic architectures, as shown in Figure 1, were adopted in our tests. While Newtonian kinematics assumes that optimal rolling may occur without slipping, destructive pure rotations were found in some tests to irregularly induce a translational mismatch using the two-wheeled robot (a). Therefore, a tracked robot (b) with higher resistance contact was also implemented as a comparison. The environment errors were statistically reduced via multiple measurements and featured the one with better performance.

Due to its resource advantages, NXT Direct Commands provide Bluetooth protocols in direct communication with the robot [18]. The robot can interpret the commands embedded in the Direct Commands from the smart phone and then execute the corresponding action. This brings great convenience in robotic control to developers who can pay attention to programs for smart phones rather than for the robot. The communication platform is supported by leJOS NXJ, a JAVA virtual machine executed on NXT, for which many application objects can be freely accessed and developed on Eclipse [19]. Based on these specifications, a JAVA-based control kernel is appropriate for bridging the smart device of the Android core and the LEGO robot.

The operation of the robot was tested in two ways: (i) manual and (ii) autonomous. In case (i), the action control on the speed of the wheels was realized via an orientation sensor by detecting the variation in the orientation of the smart phone with pitch and roll (Figure 2), and via the displacement of the touch point (Figure 3). The control of the differential-drive is depicted in the right 4-quadrant circle, by assuming 4 definite movements for the left and right wheels as (v,v), (-v,-v), (v,-v), and (-v,v). In this manner, the speed of the individual wheels (v_R or v_L) is definitely returned whenever the (x,y) coordinates of a specific (yellow) point are acquired. On the other hand, in case (ii), the robot is completely directed by the route planning accomplished with the importation of the Google Earth-supplied KML (Keyhole Markup Language) into the Google map [20].

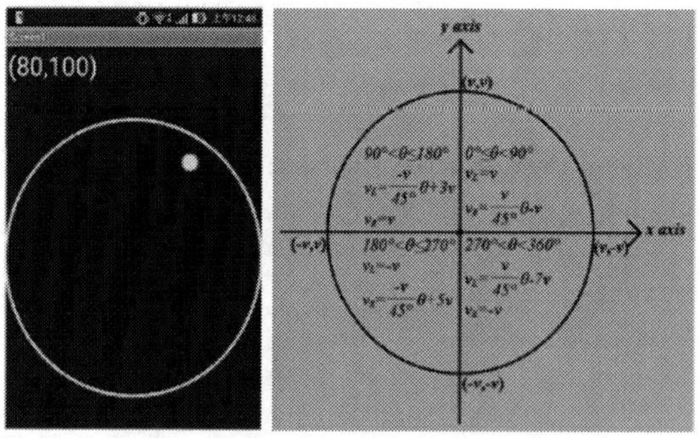

Figure 3. Touch control and 4-quadrant speed conversion.

```
import lejos.nxt.*;
import lejos.robotics.RegulatedMotor;
import lejos.robotics.subsumption.*;
import lejos.robotics.navigation.DifferentialPilot;
import lejos.robotics.navigation.RotateMoveController;
import lejos.util.PilotProps;
public class LineFollower {
public static void main (String[] aArg)
throws Exception
{
  PilotProps pp = new PilotProps();
  pp.loadPersistentValues();
  float wheelDiameter = Float.parseFloat(pp.getProperty(PilotProps.KEY WHEELDIAMETER, "4.96"));
  float trackWidth = Float.parseFloat(pp.getProperty(PilotProps.KEY TRACKWIDTH, "13.0"));
  RegulatedMotor leftMotor = PilotProps.getMotor(pp.getProperty(PilotProps.KEY LEFTMOTOR, "B"));
  RegulatedMotor rightMotor = PilotProps.getMotor(pp.getProperty(PilotProps.KEY RIGHTMOTOR, "C"));
  boolean reverse = Boolean.parseBoolean(pp.getProperty(PilotProps.KEY REVERSE,"false"));
  Behavior DriveForward = new Behavior()
 {
   public boolean takeControl() {return light.readValue() <= 40;}
   public void suppress() {  pilot.stop(); }
   public void action() {
        pilot.forward();
        while(light.readValue() <= 40) Thread.yield(); //action complete when not on line  }  };
```

Figure 4. JAVA-based leJOS protocol for the LEGO NXT robotic platform, a partial demonstration of object tracking.

3 Image Identification & Object Tracking

In this section we demonstrate the results of the light tracking robot. A part of the JAVA-based leJOS program is shown in Figure 4, and three screenshots from a video are shown in Figure 5. IDOT on obstacle and marking detections was implemented via an on-board library, openCV [21], a very powerful open source for image identification. As the coordinates of the object are read, the smart device is able to estimate the relative position between the object and the robot. This allows the robot to adjust its movement. Besides, the captured image provides the information of the object size by which the relative distance between two subjects can also be estimated and conducted to determine the correspondent moving speed. The combination of image identification and dynamic fine tuning directly leads to successful improvement in obstacle avoidance in the Google map routing.

Figure 5. Image identification and light avoiding realized via openCV calculation.

In Figure 5, the robot was designed to avoid illumination from a flashlight. The idea was realized by openCV calculation, incorporating algorithms of image binarization and multiscale edge detection. The theoretical evaluation of the coordinates and the size of the light return the information of the direction and distance to the following movement of the robot.

In our experiment, it was found that the homogeneity of the illumination has a direct influence on the accuracy of the image identification. This problem may possibly be fixed by adopting a Kalman filter or particle filter algorithm to speed up the reaction of the smart phone to external light sources. Further verification of this reasoning and extended investigation will be reported elsewhere in the near future.

4 Map Positioning & Navigation

In this section we demonstrate the integration of Google map navigation and mobile robot control via the Android platform. As shown in Figure 6, the experiment was performed by first choosing the starting point and the destination on the Google map. The GPS returned latitude B and longitude λ of the objects in terms of [22]

$$B = \cos^{-1}\sqrt{\frac{b^2 - Z^2}{b^2 + Z^2 e'^2}} \qquad \lambda = \cos^{-1}\frac{X\sqrt{1-e^2\sin^2 B}}{a\cos B} \qquad (1)$$

on the reference of the geodesic coordinates, in which $a(b)$ is the semi-major(minor) axis of the Earth, XYZ are Cartesian coordinates, and $e=\sqrt{a^2-b^2}/a$ and $e'=\sqrt{a^2-b^2}/b$ are the first and second eccentricities of the spheroid. The arc length d that the robot has travelled was evaluated via the Haversine formula [22]:

$$d = 2R \times \sin^{-1}\left(\sqrt{\sin^2\left(\frac{B_2 - B_1}{2}\right) + \cos B_1 \cos B_2 \sin^2\left(\frac{\lambda_2 - \lambda_1}{2}\right)}\right) \qquad (2)$$

regarding R the radius of the right sphere Earth, B_{12} and λ_{12} represent the geographical latitude and longitudeof points 1 and 2. Then the corresponding displacement and rotation angle of the robot can be calculated by applying the commands bearingTo() and distanceTo() classified in the Location of Android APIs (application programming interface).

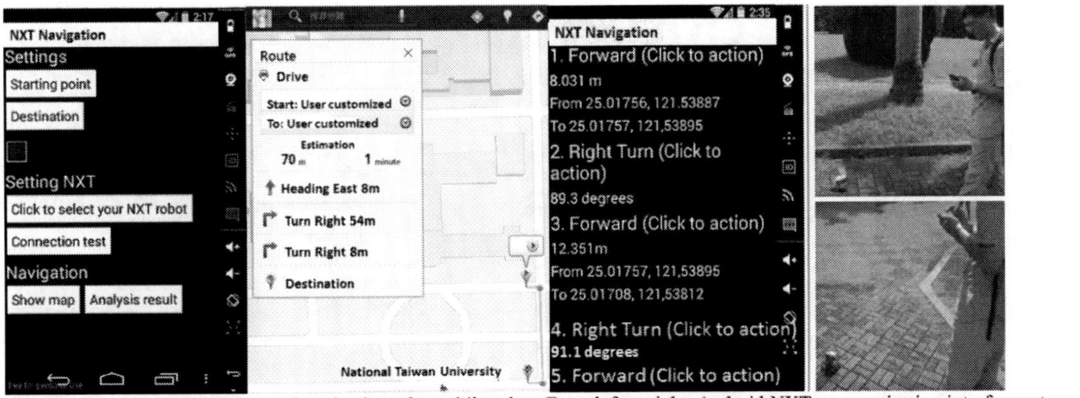

Figure 6. Google map route planning and navigation of a mobile robot. From left to right: Android-NXT communication interface set-up display, route planning information, and moving instruction.

To reach a precise measurement, an infinitesimal displacement within an extremely slow movement, although inconvenient, is established. For practical use, inevitable zig-zag overshooting and the cumulative error have to be considered. We applied a compass sensor with PID controller (proportional–integral–derivative) for stepwise error correction. Eventually an excellent refinement was revealed in the dynamic control.

Furthermore, we also investigated the data transmission between the smart phone and the robot. The

 DOI: 10.1051/matecconf/20179508005

experiment was designed by transferring the detection readings of a supersonic sensor carried by the robot to the Android smart phone, and displaying the data in terms of various kinds of Google Charts. Figure 7 demonstrates a radar chart generated from a fixed point rotation, in which data acquisition was implemented every 45°. The path on which the chart can be accessed is given byhttps://chart.googleapis.com/chart?cht=rs&chs=500x500&chd=t:**77,66,15,0,31,48,100,77**&chco=FF0000,FF9900&chls=2.0,4.0,0.0%7C2.0,4.0,0.0&chxt=x&chxl=*0:%7C0%7C45%7C90%7C135%7C180%7C225%7C270%7C315*&chxr=0,0.0,360.0&chg=25.0,25.0,4.0,4.0&chm=B,FF000080,0,1.0,5.0%7CB,FF990080,1,1.0,5.0%27, in which the numbers in bold type are the sample data and those in italics denote the spacing.

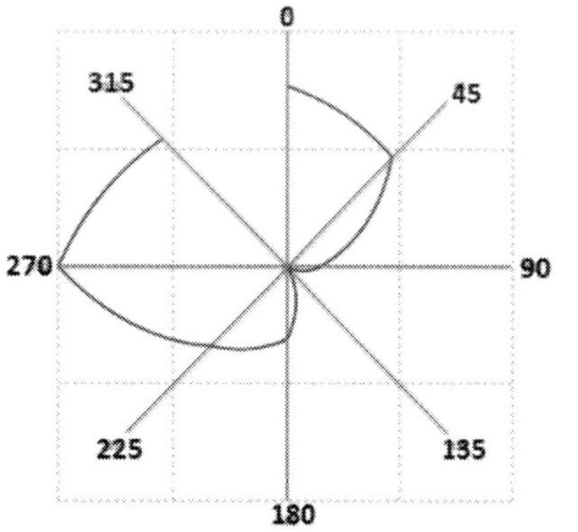

Figure 7. A radar chart. Environment data detected by supersonic sensor are shown by the red curve.

In our experiment, it was found that the web condition plays a primary role in the signal distortion and data reliability. Unfortunately, the signal transmission took roughly 0.3~0.4 seconds by Bluetooth communication, and another few seconds are required to the Cloud. Therefore, an optimal sampling and trigger has to be seriously considered. Moreover, the 45° sampling was found to reach the best recognition in our tests.

However, the impact of disturbance can be statistically reduced in long-time or multi-term measurements. Therefore, it is still worth noting that since real-time data storage is likely to soon be available on both the Cloud and smart phones, Google Charts will then provide fast and multivariate interfaces for big data processing and display. On the other hand, the combination of a smart phone reveals the power of synchronized remote control of the mobile robot and server end, bringing the potential of intensive applications to the industrial Internet of Things.

5 Conclusion

In this project we demonstrate image identification, object tracking, and Google Map route planning and navigation with an easy-to-implement and relatively low-cost robotic LEGO NXT platform. The robotic kinematics was under autonomous control based on the JAVA and Bluetooth communication architectures. The systematic connections on the leJOS, openCV and Android APIs applications lead to the realization of an integration of Android smart devices and a differential-drive robotics platform. Finally, we have demonstrated that environment detection by a mobile robot can be easily achieved. The combination of smart phone and Google Charts made it easy to realize big data storage and characteristic distribution. The use of smart phones for robotic control achieves a powerful integration of control, monitoring, communication, sensing, and calculation on a personal mobile device. Undoubtedly, strong connections between smart phones, the Cloud, and instruments must greatly inspire extensive and intensive creativity and imagination regarding innovative applications of cyber-physical integration.

Acknowledgement

This work is partially supported by the Ministry of Science and Technology of Taiwan under MOST 104-2511-S-845 -009 -MY3. The authors would like to thank C. H. You, C. M. Yang and C. S. Chang for their help with the graphic editing.

References

1. Ying Lan, Gangfeng Yan, and Zhiyun Lin, IEEE Transactions on Automatic Control, **56**, 1170 (2011).
2. F. Heidari and R. Fotouhi, J. Mechanisms Robotics, **7**, 041025 (2015)
3. Cheng Chen, Wennan Chai, and Hubert Roth, J Intell. Robot Syst., **80**, 365 (2015)
4. Haibin Yan, Marcelo H. Ang Jr., Aun Neow Poo, Int. J. Soc. Robot, **6**, 85 (2014)
5. Rafael Soca, Sebastián Dormido, Raquel Dormido and Ernesto Fabregas, Sensors **15**, 30076 (2015).
6. Jong-Hoon Kim, Gokarna Sharma, Noureddine Boudriga, S. Iyengar and Nagarajan Prabakar, EURASIP Journal on Wireless Communications and Networking 2015:262 (2015).
7. Naouel Ayari, Abdelghani Chibani, Yacine Amirat, and Eric Matson, Robotics and Autonomous Systems 75, 17 (2016).
8. J. Fernandes, J. Martinez-de-Dios, I. Maza, F. Fabresse, and A.Ollero, Int. J. Adv. Robot. Syst., 12, 70 (2015)
9. V. Kumar, D. Rus, S. Singh, IEEE Pervasive Comput. **3**, 24 (2004).
10. M. Kitani, T. Hara, H. Hanada, and H. Sawada, 3rd Conference on Human System Interactions, 728 (2013).
11. Miguel Oliveira, et. al., IEEE/RSJ International Conference on Intelligent Robots and Systems, 2216 (2014).
12. Li Zhang, Ming Jiang, Dewan Farid, and M.A. Hossain, Expert Systems with Applications **40**, 5160 (2013).
13. Ming Jiang and Li Zhang, Procedia Computer Science **53**, 141 (2015).

14. Luigi Atzori, Antonio Iera, Giacomo Morabito, Computer Networks **54**, 2787 (2010).
15. Christian Siagian, Chin-Kai Chang, and Laurent Itti, 2013 IEEE International Conference on Robotics and Automation, 564 (2013).
16. D. B. Wolf, C.L. Judy, E. J. Haukkala, and D. J. Godfrey, OCEANS 2000 MTS/IEEE Conference and Exhibition 1, 79 (2000).
17. F. Ribeiro, I. Moutinho, P. Silva, C. Fraga, N. Pereira, Control 2004, University of bath, UK, September, 2004.
18. http://www.mindstorms.rwth-aachen.de/documents/downloads/doc/version-4.04/direct_commands.html
19. https://eclipse.org/
20. https://developers.google.com/kml/
21. R.W. Sinnott, Sky and Telescope, **68**, 159 (1984).
22. Hoffmann-Wellenhof, B. H. Lichtenegger, and J. Collins, *GPS: Theory and Practice*. 3rd ed. New York: Springer-Verlag (1994).

DOI: 10.1051/matecconf/20179508006

Mobile Robot Positioning by using Low-Cost Visual Tracking System

Niramon Ruangpayoongsak

Production Engineering Department, Faculty of Engineering, King Mongut's University of Technology North Bangkok, Thailand

Abstract. This paper presents an application of visual tracking system on mobile robot positioning. The proposed method is verified on a constructed low-cost tracking system consisting of 2 DOF pan-tilt unit, web camera and distance sensor. The motion of pan-tilt joints is realized and controlled by using LQR controller running on microcontroller. Without needs of camera calibration, robot trajectory is tracked by Kalman filter integrating distance information and joint positions. The experimental results demonstrate validity of the proposed positioning technique and the obtained mobile robot trajectory is benchmarked against laser rangefinder positioning. The implemented system can successfully track a mobile robot driving at 14 cm/s.

1 Introduction

Various techniques can be applied for mobile robot positioning [1]. For the relative position, odometry or intertial sensor detect and estimate robot motion based on kinematic or dynamic robot model. One drawback is that when robot motion is unsmooth, especially in rough terrain, obtained trajectory often contains uncertainty. Visual based positioning is a possible solution for this problem.

Visual servo control has been developed since last two decades [2]. The problem of eye-in-hand configuration on robotic system running on high speed controller [3] was implemented as image based visual servoing (IBVS) [4-5]. As for moving object tracking, the target tracking was addressed for robot manipulators in [6-7]. Later, solutions without needs of camera calibration are proposed. Different methods were presented as camera parameter independent or uncalibrated camera [8-9]. Further improvement of motion control by estimating depth of feature in image was proposed [10]. For pan-tilt mechanism, the target tracking was also addressed such as controller design for AC-servomotor visual tracking system [11] and real time position estimation of a constant speed moving object [12].

In this paper, IBVS is applied for mobile robot tracking. The idea is to reduce complexity of acquiring image data and depth information. The camera is utilized without calibration. Also, instead of esimating depth of feature in image as in [10], the distance between camera and the mobile robot is measured directly by using laser sensor mounting above the camera. As robot usually drives with constant speed, our problem is similar to [12] except that the pan-tilt unit is constructed using low-cost hardware and controller is implemented on Arduino.

This paper is organized as follows: In section 2, the hardware and software of target tracking system is introduced. After that the motion control and estimation of robot trajectory is explained. Section 3 presents experimental results. Finally, conclusion is given.

2 Mobile Robot Positioning

2.1 Target Tracking System

In this work, a camera system is built and the state space model is identified. To validate the system, a standard LQR controller is selected for motion control.

2.1.1 Hardware components

As shown in Fig. 1, the camera system consists of

- Two 12V 200 rpm DC gear motors with 1:40 worm gears.
- Two 1000 ppr optical rotary encoders with 15:62 pulleys. The resolution of joint position is 0.09 degree/pulse.
- C920 Logitech web camera with 1280x720 pixels resolution and 30 fps.
- LIDAR-Lite Laser Rangefinder (PulsedLight) measuring distance to object up to 40 meters with 2.5 cm accuracy.
- Arduino M Zero Pro having a 32-bit ARM Cortex microprocessor with 48 MHz clock speed running at 3.3V logic level.
- Intel core i7-4710Q CPU Labtop upto 3.50 GHz with 4 GB RAM running on Windows 8.1 64 bit.

The system is mounted on a general camera stand for convenience of testing configuration. Cost of parts is listed in Table 1. The total cost is approximately 349 US Dollars.

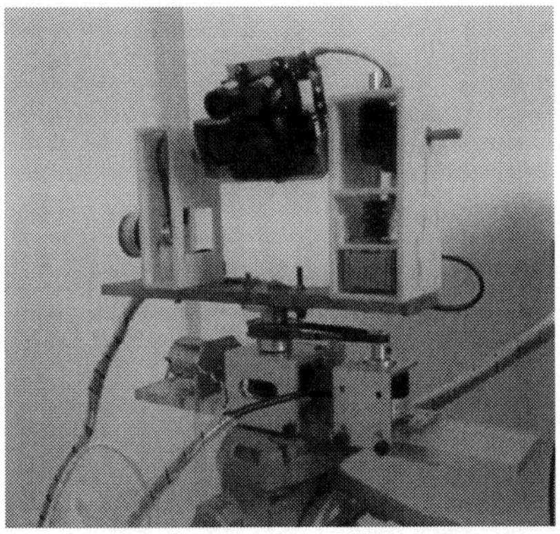

Figure 1. The constructed low cost camera system for mobile robot positioning.

Table 1. Cost of the constructed camera system.

Parts	Cost (US Dollars)
Motors	24
Worm gears & pulleys	59
C920 Logitech web camera	94
LIDAR-Lite Rangefinder	90
Electronics	82
Total	349

2.1.2 Software

The image is implemented by using OpenCV and Codeblocks. The HSV color-based recognition yields background subtraction on white-black image. Then, position of object is taken as center of foreground object. The position of object in image frame is sent to Arduino and sensor data is obtained at every 80 ms approximately. In parallel, LQR controller is running on Arduino.

2.2 Motion control

2.2.1 Kinematic model

According to Denavit-Hartenberg notation [13], parameter are listed in Table 2 for the link transformation matrix according to Fig. 1.

Table 2. Denavit-Hartenberg Parameters.

Link	a_i	α_i	d_i	θ_i
1	0	$-\pi/2$	0.10	θ_1
2	0.03	0	0	θ_2

$$^0T_1 = \begin{bmatrix} \cos\theta_1 & 0 & -sin\theta_1 & 0 \\ sin\theta_1 & 0 & cos\theta_1 & 0 \\ 0 & -1 & 0 & 0.1 \\ 0 & 0 & 0 & 1 \end{bmatrix}, \quad (1)$$

$$^1T_2 = \begin{bmatrix} \cos\theta_2 & -sin\theta_2 & 0 & 0.03cos\theta_2 \\ sin\theta_2 & cos\theta_2 & 0 & 0.03sin\theta_2 \\ 0 & 0 & 1 & 0 \\ 0 & 0 & 0 & 1 \end{bmatrix}, \quad (2)$$

$$^0T_2 = {}^0T_1 \cdot {}^1T_2 , \quad (3)$$

$$^0T_2 = \begin{bmatrix} C_1C_2 & -C_1S_2 & -S_1 & 0.03(C_1C_2) \\ C_2S_1 & -S_1S_2 & C_1 & 0.03(C_2S_1) \\ -S_2 & -C_2 & 0 & 0.1-0.03S_2 \\ 0 & 0 & 0 & 1 \end{bmatrix}, \quad (4)$$

where $C_1 = \cos\theta_1$, $C_2 = \cos\theta_2$, $S_1 = \sin\theta_1$ *and* $S_2 = \sin\theta 2$.

2.2.2 State space model

The dynamic equations of *n*-DOF manipulator [13] are derived from Lagrangian equation, the Lagrange-Euler dynamic is formulated as

$$\tau_i = \frac{d}{dt}\left(\frac{\delta\mathcal{L}}{\delta\dot{q}_i}\right) - \frac{\delta\mathcal{L}}{\delta q_i} \quad (5)$$

and is derived into vector matrix notation:

$$\tau(t) = \boldsymbol{M}(\boldsymbol{q}(t))\ddot{q}(t) + \boldsymbol{H}(\dot{\boldsymbol{q}}(t), \boldsymbol{q}(t)) + \boldsymbol{G}(\boldsymbol{q}(t)) \quad (6)$$

where $\boldsymbol{q}(t)$, $\dot{\boldsymbol{q}}(t)$,and $\ddot{\boldsymbol{q}}(t)$ are $n \times 1$ vector of generalized joint variables (displacements), velocities, and accelerations for joints *i*, respectively. $\tau(t)$ is $n \times 1$ vector of generalized torques applied at joints *i*. $\boldsymbol{M}(\boldsymbol{q})$is $\boldsymbol{n} \times \boldsymbol{n}$ inertial acceleration symmetric matrix whose elements are M_{ij}. $\boldsymbol{H}(\dot{\boldsymbol{q}}, \boldsymbol{q})$ is $n \times 1$ nonlinear Coriolis and centrifugal force vector. $\boldsymbol{G}(\boldsymbol{q})$ is $n \times 1$ gravity loading force vector. Eq. (6) is the state space equation because the term $\boldsymbol{H}(\dot{\boldsymbol{q}}, \boldsymbol{q})$ has both position and velocity dependence.

The state space model of the pan-tilt unit is

$$\dot{x} = Ax + Bu \quad (7)$$

where $x = [\theta_1 \quad \theta_2 \quad \dot{\theta}_1 \quad \dot{\theta}_2]^T$. By model linearization, the transition matrix and the input matrix are:

$$A = \begin{bmatrix} 0 & 0 & 1 & 0 \\ 0 & 0 & 0 & 1 \\ 328.6710 & 43.8228 & 0 & 0 \\ 92.9991 & 0 & 0 & 0 \end{bmatrix},$$

$$B = \begin{bmatrix} 0 & 0 \\ 0 & 0 \\ -9.0528 & 0 \\ 0 & 19.2114 \end{bmatrix}.$$

DOI: 10.1051/matecconf/20179508006

2.2.3 LQR controller

The feedback control law of LQR controller is

$$u = -Kx(t) \tag{8}$$

that minimize the performance index [14]

$$J = \int_0^{\infty}(x^*Qx + u^*Ru)\,dt. \tag{9}$$

When $(A\text{-}BK)$ is a stable matrix, there exists positive-definite matrix P that is a solution of

$$(A - BK)^*P + P(A - BK) = -(Q + K^*RK). \tag{10}$$

The minimization of J with respect to K gives

$$K = R^{-1}B^*P. \tag{11}$$

The feedback gain K can be obtained by choosing

$$Q = \begin{bmatrix} 1000 & 0 & 0 & 0 \\ 0 & 1000 & 0 & 0 \\ 0 & 0 & 1 & 0 \\ 0 & 0 & 0 & 1 \end{bmatrix},\ R = \begin{bmatrix} 1\mathrm{e}-6 & 0 \\ 0 & 1\mathrm{e}-6 \end{bmatrix}.$$

The different between object and center of image frame is pixel error in image. This error is used as a set point for the LQR controller as shown in Fig. 2. When tracked object stays at the center of image, the set point is zero.

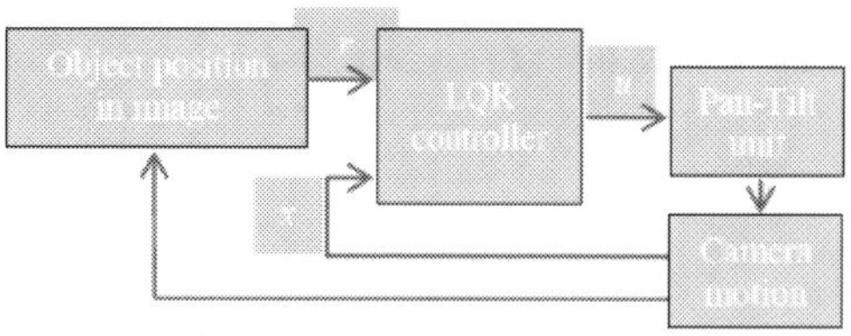

Figure 2. The diagram of implemented feedback control.

2.3 Trajectory estimation

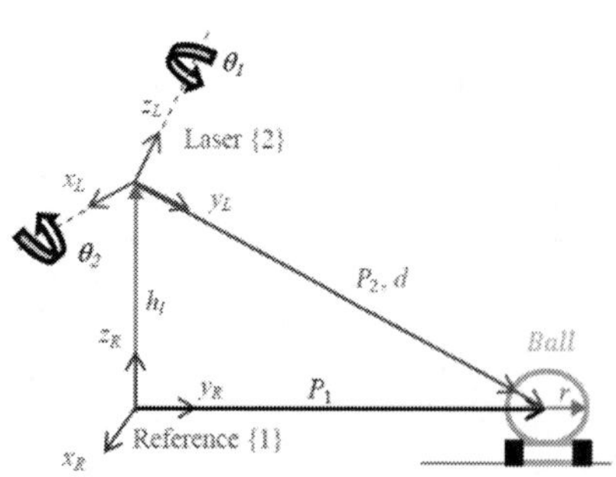

Figure 3. Relationship among the Base frame, the Laser frame and the robot position.

Based on the proposed tracking method [15], the state variables are estimated by motion of joints and distance to the ball measured by laser mounted on the pan-tilt unit above the web camera. Laser frame rotates in pan and tilt directions as shown in Fig. 3. The reference frame is above ground level and a ball is mounted on the mobile robot. The laser frame is above the reference frame at height h_l. The distance d is measured along y_L to the ball having radius r. Note that the ball is use to represent robot position and simplify image processing.

The vector to the ball on the reference frame, P_1, can be calculated by using composite transformation from frame 2 to frame 1 as follows.

$$P_1 = {}^1\boldsymbol{T}_2\, P_2. \tag{12}$$

The vector P_2 to the ball center in frame 2 is

$$P_2 = [0 \quad d+r \quad 0 \quad 1]^T \tag{13}$$

and transformation matrix between the base and the laser frame is

$${}^1\boldsymbol{T}_2 = \left[\begin{array}{ccc:c} & \boldsymbol{R} & & 0 \\ & & & 0 \\ & & & h_l \\ \hdashline 0 & 0 & 0 & 1 \end{array}\right], \tag{14}$$

$$\boldsymbol{R} = \begin{bmatrix} 1 & 0 & 0 \\ 0 & cos\theta_2 & -sin\theta_2 \\ 0 & sin\theta_2 & cos\theta_2 \end{bmatrix} \begin{bmatrix} cos\theta_1 & -sin\theta_1 & 0 \\ sin\theta_1 & cos\theta_1 & 0 \\ 0 & 0 & 1 \end{bmatrix}. \tag{15}$$

According to Eq. (12) - (15), the coordinate of the robot is

$$x = -(d + r)sin\theta_1, \tag{16}$$

$$y = (d + r)cos\theta_1 cos\theta_2, \tag{17}$$

$$z = h_l + (d + r)cos\theta_1 sin\theta_2. \tag{18}$$

3 Experimental results

Figure 4. Experiment setup.

The mobile robot positioning is tested by setting up experiment as shown in Fig. 4. A 18 cm diameter ball is mounted on a line tracking mobile robot driving with constant speed of 14 cm/s. The test field is made of white paper with black tape as a rectangle of 210 cm x 190 cm size with round corner.

The web camera is at 120 cm height pointing downward to the ball and the distance sensor is 3 cm above the camera. The ±30 mm accuracy Hokuyo URG-04LX-UG01 is put at the same height as the ball center and tracked robot position is referred as actual trajectory. From beginning, the robot drives straight on, turns right at corners and returns to starting point.

Fig. 5 illustrates robot trajectory obtained from measured data, estimated trajectory, and the actual trajectory. The robot starts at (0,0) and returns to this point. Due to delay time in communication between the laptop and Arduino and noisy measurement as well as property of LQR controller, even the estimated trajectory is close to the actual trajectory, it has some deviations. However, with respect to low-cost components, the result is acceptable.

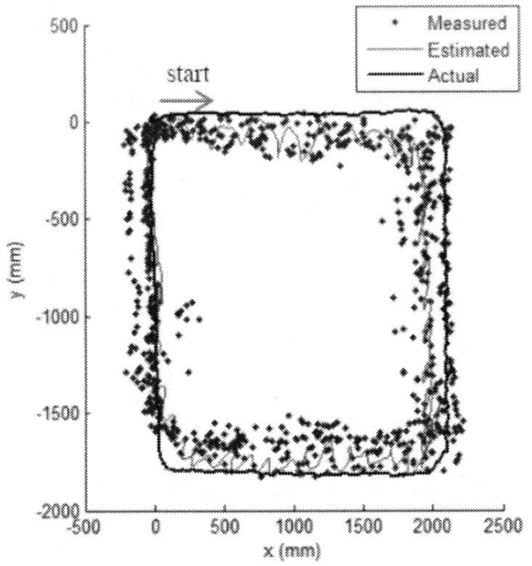

Figure 5. Mobile robot trajectory.

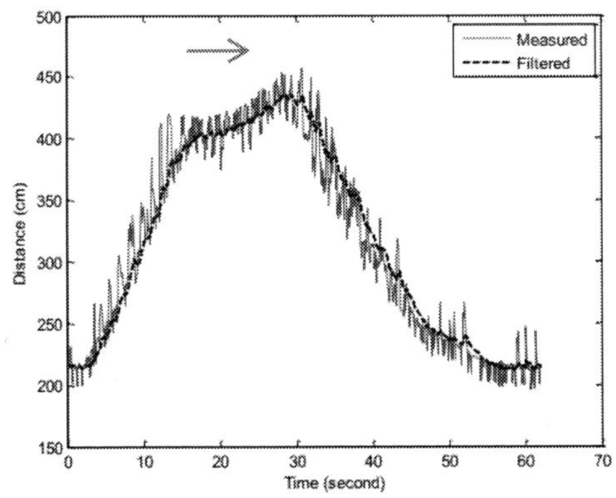

Figure 6. Measured and filtered distance *d*.

Fig. 6 shows noisy measured value obtained from laser distance sensor and its filtered value. The range of measured distance in the experiment is from 2 to 4.5 meters.

Fig. 7 shows position of pan and tilt joint. Starting from zero degree, the peak pan joint position is almost 37 degrees on the third corner before turning back to the starting point. The maximum tilt joint position is 14 degrees when robot is driving far along x-axis.

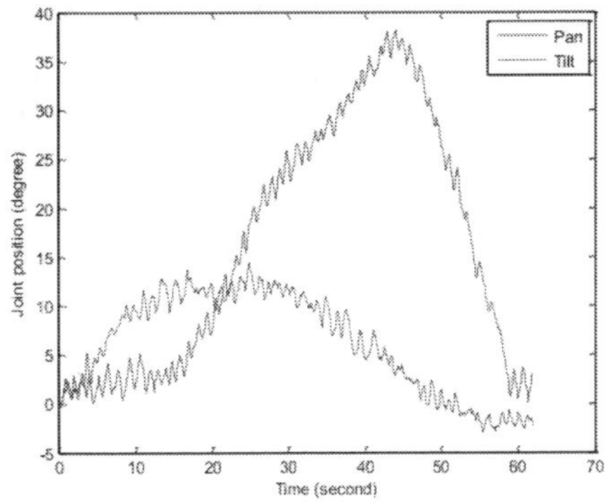

Figure 7. Pan-tilt joint position.

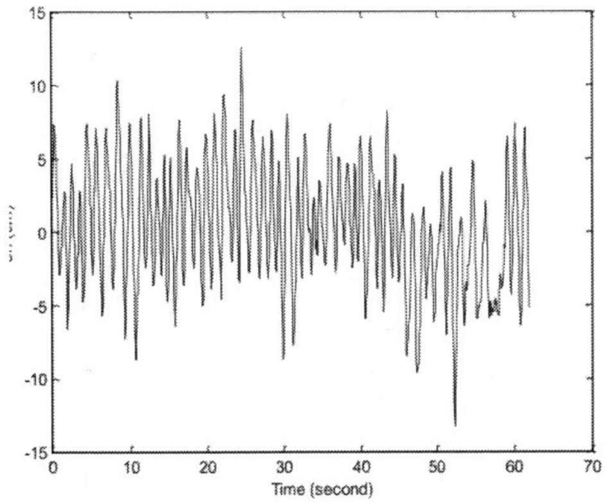

Figure 8. Error in horizontal direction.

Fig. 8 shows horizontal pixel error e_h and Fig. 9 shows vertical pixel error e_v between object and center of image frame. In the image, the tracked object stays at center in both directions with standard deviation of 4.16 and 4.61 cm in horizontal (pan) and vertical (tilt) direction, respectively.

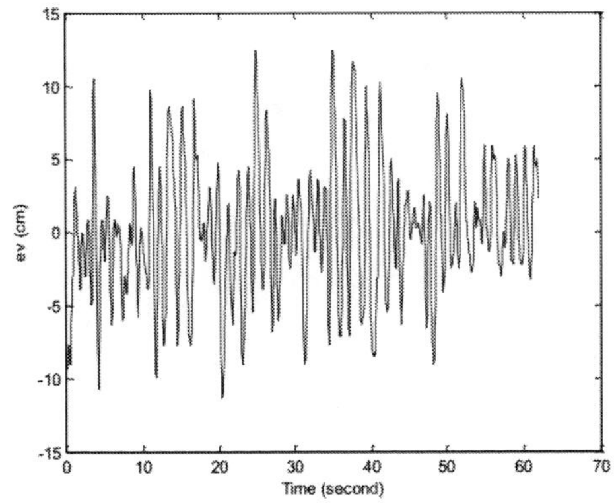

Figure 9. Error in vertical direction.

4 Conclusions

The low-cost target tracking system for mobile robot positioning was proposed, constructed and successfully implemented. The experimental results demonstrate tracked trajectory of a mobile robot driving on ground at constant speed of 14 cm/s. The system can keep tracking object at the center of image frame all the time.

This method can be applied for tracking ground and also flying mobile robots. The maximum tracking range is adjustable by camera height configuration. Future work is to improve pan-tilt motion control performance and object recognition for mobile robot.

Acknowledgement

This research was funded by King Mongut's University of Technology North Bangkok. Contract no. KMUTNB-GOV-58-31.

References

1. J. Borenstein, H.R. Everett, L. Feng, D. Wehe, Mobile Robot Positioning – Sensors and Techniques, J. Robotic Systems, **14**, 4 (1997), 231-249.
2. N. P. Papanikolopoulos and P.K. Khosla, Real-Time LQG Robotic Visual Tracking, Robotic System, **10**, (1992), 305-312.
3. P. Corke and M. Goods, "Controller design for high performance visual servoing," in Proc. 12th World Congress IFAC (1993), 395–398.
4. F. Chaumette and S. Hutchinson, Visual servo control. I. Basic approaches, IEEE Robotics Automation Magazine, **13**, 4 (1993), 82–90.
5. S. Hutchinson, G. Hager, and P. Corke, A tutorial on visual servo control, IEEE Trans. on Robotics and Automation, **12**, 5 (1996), 651 –670.
6. F. Bensalah and F. Chaumette, "Compensation of abrupt motion changes in target tracking by visual servoing," in IEEE/RSJ Int. Conf. on Intelligent Robots and Systems (1995), 181–187.
7. F. Chaumette and S. Hutchinson, Visual servo control. II. Advanced approaches [tutorial], IEEE Robotics Automation Magazine, **14**, 1(2007), 109–118.
8. J. A. Piepmeier, "Experimental Results for Uncalibrated Eye-in-Hand Visual Servoing," Proc. the 35th Southeastern Symposium on System Theory (2003), 335-339.
9. H. Wang, Y.H. Liu, D. Zhou, Adaptive Visual Servoing Using Point and Line Features With an Uncalibrated Eye-in-Hand Camera, IEEE Trans. ROBOTICS, **24**, 4 (2008), 843-857.
10. N. Shahriari, S. Fantasia, F. Flacco, G. Oriolo, "Robotic Visual Servoing of Moving Targets," IEEE/RSJ Int. Conf. on Intelligent Robots and Systems (IROS), (2013), 77-82.
11. M-C. Tsai, K-Y. Chen, M-Y. Cheng and K.C. Lin, Implementation of a real-time moving object tracking system using visual servoing, Robotica, **21**, 6 (2003), 615 – 625.
12. B. Torkaman, M. Farrokhi, A Kalman-Filter-Based Method for Real-Time Visual Tracking of a Moving Object Using Pan and Tilt Platform, Int. J. Sci. & Eng. Research, **3**, 8 (2012), 1-6.
13. R.K. Mittal, I.J. Nagrath, *Robotics and Control* (Tata McGraw-Hill, New Delhi, 2003)
14. K. Ogata, *Modern Control Engineering* (Prentice Hall, 2010)
15. N. Ruangpayoongsak, Object Trajectory Tracking on a Plane Using Camera and Infrared Sensor, The Journal of KMUTNB, **24**, 2 (2014), 247-256.

DOI: 10.1051/matecconf/20179508007

Dimension Synthesis of a Spatial Parallel Kinematic Manipulator Based on S-shaped Test Specimen

Yuzhe Liu, Jun Wu , Liping Wang and Jinsong Wang

State Key Laboratory of Tribology and Institute of Manufacturing Engineering, Department of Mechanical Engineering, Tsinghua University, 100084 Beijing, China
Beijing Key Lab of Precision/Ultra-precision Manufacturing Equipments and Control, 100084 Beijing, China

Abstract. With the development of the aerospace industry, the hybrid machine tool for machining the complicated curved surface is becoming a research hotspot. The S-shaped test specimen proposed by Chengdu Aircraft Industrial Company is an effective test method to verify the ability of machining the complicated curved surface for a machine tool. Considering the spatial parallel kinematic manipulator in a hybrid machine tool provides the complicated orientation motion and its design of structure dimension affects the accuracy of the machine tool, this paper investigates the issue about dimension synthesis of the manipulator based on S-shaped test specimen. By establishing models of the S-shaped test specimen and parallel kinematic manipulator, the motion trajectory of the parallel kinematic manipulator for machining S-shaped test specimen is obtained. According to the motion trajectory and expected performance indices, the result of dimension synthesis for the manipulator is plotted and a set of suitable dimensions is achieved. With the analysis result in this paper, it is helpful to improving the accuracy of machining the S-shaped test specimen.

1 Introduction

With the rapid development of the aerospace industry, a 5-axis machine tool with high accuracy, high speed and high orientational capability is becoming a research hotspot [1-4], which leads to two important issues: how to detect the machining performance of a 5-axis machine tool and which kind of machine tool is suitable.

For the first issue, many investigations have been addressed on it. In general, the accuracy detection test specimens are usually used as the detecting method. The most used one may be the ISO international standard test specimen, which usually is adopted to detect the translational axis accuracy of a 3-axis machine tool. However, this one cannot reflect the motion accuracy of multiple axes. Anther accuracy detection test specimen used a lot in practice is the NAS979 proposed by NASA in 1969. This one is used widely in 5-axis machine tools for detecting the motion accuracy of multiple axes. However, the most outstanding problem about this test specimen is that it cannot reflect the orientation accuracy of a 5-axis machine tool. Recently, in order to completely test the motion accuracy of multiple axes and satisfy the machining demand of the aerospace industry, a novel test specimen, named as the S-shaped test specimen and covering a mass of complicated machining poses, was proposed by Chengdu Aircraft Industrial Company, which has been applied successfully in a lot of factories and brought into ISO 10791-7 criterion group as an additional test specimen in the 78th ISO conference in 2015 [5-6]. Hence, it is necessary to investigate this new standard test specimen.

For the second issue, it is mainly to find a kind of 5-axis machine tool which owns the performances of high accuracy, high speed and high orientational capability. 5-axis machine tools usually can be divided into two categories: serial machine tools and serial-parallel (or hybrid) machine tools. It is well known that the serial machine tools are characterized by large workspace and simple kinematics but are subject to high inertia and error accumulation. The hybrid machine tool, combining the advantages of the parallel kinematic machines and the serial kinematic machines, is recently proposed and designed, which is more suitable for the aerospace industry [7-8]. In the past successful applications of 5-axis hybrid machine tools, considering the structural features of serial and parallel mechanisms and the demands of the machining performance, a 3-Degree-Of-Freedom (3-DOF) tool head with parallel kinematics and a 2-DOF translational table with serial kinematics are always adopted as the parallel module and the serial module in the 5-axis hybrid machine tool. Among them, the most successful application is the combination of a 3PRS parallel kinematic manipulator and a 2-DOF serial platform [9]. The 3PRS parallel kinematic manipulator is a spatial parallel kinematic manipulator and provides the orientation of the end effector of the 5-axis hybrid machine tool, which is the key component to guarantee the accuracy of machining aeronautical parts.

DOI: 10.1051/matecconf/20179508007

Considering the special machining demand of the aerospace industry, such as the S-shaped test specimen covering a mass of complicated machining poses, and in order to improve the machining accuracy of this kind of 5-axis hybrid machine tool which is used for the aerospace industry, it is necessary to study the relation between the 3PRS parallel kinematic manipulator and the S-shaped test specimen and deal with the dimension synthesis of the 3PRS parallel kinematic manipulator based on S-shaped test specimen. In order to achieve this goal, the most widely used method in the field of parallel kinematic manipulators, namely Performance-Chart based Design Methodology (PCbDM), is employed to investigate the dimension synthesis problem [10-11].

2 Model of S-shaped test specimen

The S-shaped test specimen consists of two similar ruled surfaces. For one ruled surface, it consists of a family of straight lines and is defined by two B-spline curves, which is shown in Fig.1. Its expression is written as:

$$S(u,v)=(1-v)\times C_1(u)+v\times C_2(u) \qquad (1)$$

where $S(u, v)$ represents the function of the ruled surface. $C_1(u)$ represents the function of Curve 1. $C_2(u)$ represents the function of Curve 2. The parameters u and v are the variables of $S(u, v)$ and both vary in [0, 1].

One may see that the function of B-spline curve totally affects the shape of ruled surface. Hence, Chengdu Aircraft Industrial Company uses the De Boor-Cox method to completely define the function of B-spline curve in order to guarantee the uniqueness of the S-shaped test specimen. The function of B-spline curve is expressed as:

$$C(u)=\sum_{i=0}^{11}\boldsymbol{T}(u)\boldsymbol{M}_i\boldsymbol{P}_i \qquad (2)$$

where $C(u)$ represents the function of B-spline curve. $\boldsymbol{T}(u)\boldsymbol{M}_i$ defines the base function. $\boldsymbol{P}_i$ is the control points.

$\boldsymbol{T}(u)$ determines the order of the base function. As $C(u)$ is a three-order uniform rational B-spline curve, it leads to:

$$\boldsymbol{T}(u)=\begin{bmatrix}1 & u & u^2 & u^3\end{bmatrix} \qquad (3)$$

$\boldsymbol{M}_i$ determines the parameters of the base function. According to the De Boor-Cox method, $\boldsymbol{M}_i$ is determined in Table 1. $\boldsymbol{P}_i$ is given by Chengdu Aircraft Industrial Company and ignored in this paper. According to the Eqs (1) and (2), the model of S-shaped test specimen can be obtained and one of the ruled surfaces in S-shaped test specimen is plotted in Fig. 2. At this point, any orientations for machining the S-shaped test specimen can be obtained by the model.

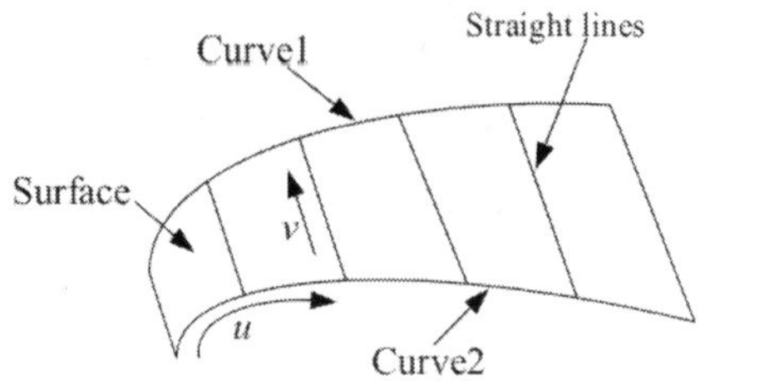

Figure 1. The ruled surface diagram

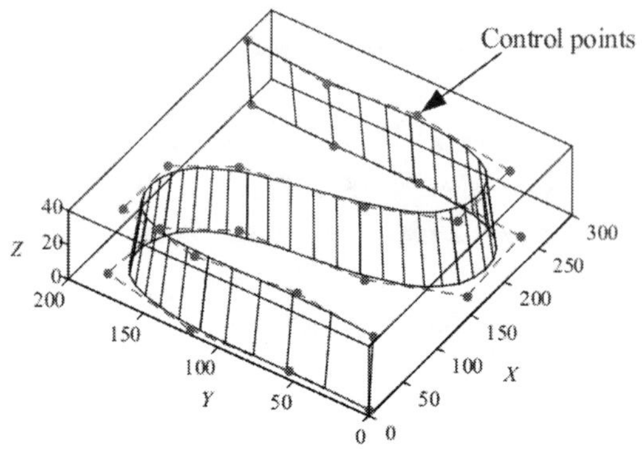

Figure 2. One of the ruled surface in S-shaped test specimen

Table 1. Definition of $\boldsymbol{M}_i$

i	$\boldsymbol{M}_i$
0	$\begin{bmatrix}1 & 0 & 0 & 0\\ -3 & 3 & 0 & 0\\ 3 & -9/2 & 3/2 & 0\\ -1 & 7/4 & -11/12 & 1/6\end{bmatrix}$
1	$\begin{bmatrix}1/4 & 7/12 & 1/6 & 0\\ -3/4 & 1/4 & 1/2 & 0\\ 3/4 & -5/4 & 1/2 & 0\\ -1/4 & 7/12 & -1/2 & 1/6\end{bmatrix}$
2~6	$\begin{bmatrix}1/6 & 2/3 & 1/6 & 0\\ -1/2 & 0 & 1/2 & 0\\ 1/2 & -1 & 1/2 & 0\\ -1/6 & 1/2 & -1/2 & 1/6\end{bmatrix}$
7	$\begin{bmatrix}1/6 & 2/3 & 1/6 & 0\\ -1/2 & 0 & 1/2 & 0\\ 1/2 & -1 & 1/2 & 0\\ -1/6 & 1/2 & -7/12 & 1/4\end{bmatrix}$
8	$\begin{bmatrix}1/6 & 7/12 & 1/4 & 0\\ -1/2 & -1/4 & 3/4 & 0\\ 1/2 & -5/4 & 3/4 & 0\\ -1/6 & 11/12 & -7/4 & 1\end{bmatrix}$

3 Error model of the 3PRS parallel kinematic manipulator

The 3PRS parallel kinematic manipulator is shown in Fig. 3. It is composed of a moving platform, a base platform and three supporting legs with identical kinematic structure. Each PRS leg contains one P joint, one R joint and one S joint. A fixed Cartesian reference coordinate system O-XYZ is fixed at the center of the base platform

$B_1B_2B_3$. A moving Cartesian reference coordinate system *N-UVW* is fixed at the center of the moving platform $A_1A_2A_3$. For simplicity and without losing the generality, let the *X*-axis point along the direction of vector OB_1 and the *U*-axis point along the vector OA_1. B_iC_i on the base platform for *i*=1, 2 and 3 represents the guide rail of each leg. The guide rail of each leg is perpendicular to the base platform. Each P joint moves along B_iC_i. P joint connects the base platform to R joint. R joint connects the P joint to the C_iA_i link. S joint (attached to the moving platform) connects the moving platform to C_iA_i link. Furthermore, all the joints attached to the base and moving platform are symmetrically distributed at vertices of the equilateral triangles.

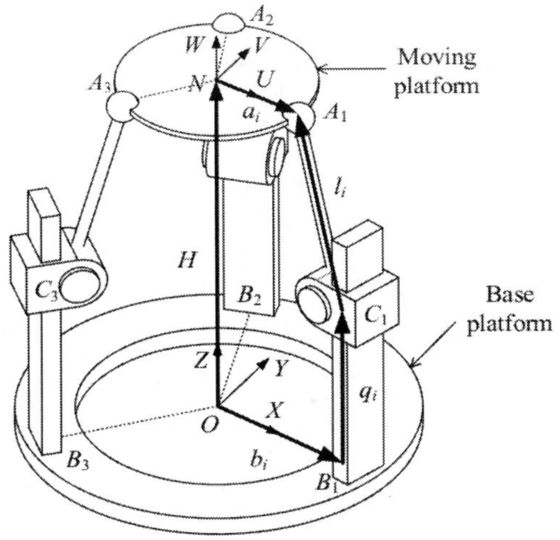

Figure 3. Structure of 3PRS parallel kinematic manipulator

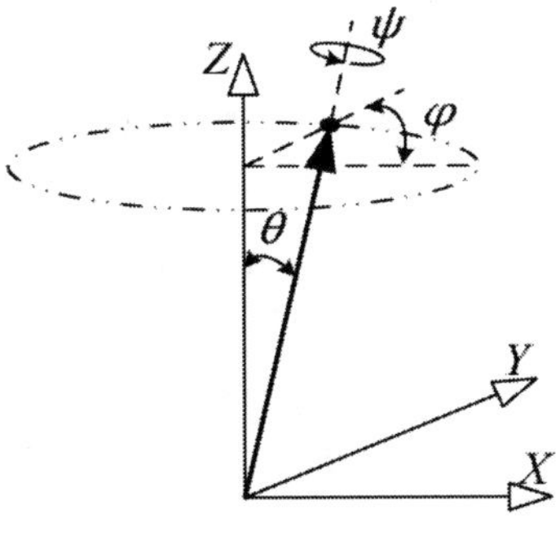

Figure 4. TT angle

In order to simplify the orientation expression, TT rotation angle in Fig. 4 is introduced to describe the orientation of the moving platform. The rotational matrix can be expressed as:

$$\boldsymbol{R}_{\mathrm{TT}}(\varphi,\theta,\psi)=\begin{bmatrix}\cos\varphi & -\sin\varphi & 0\\ \sin\varphi & \cos\varphi & 0\\ 0 & 0 & 1\end{bmatrix}\begin{bmatrix}\cos\theta & 0 & \sin\theta\\ 0 & 1 & 0\\ -\sin\theta & 0 & \cos\theta\end{bmatrix}\begin{bmatrix}\cos\varphi & \sin\varphi & 0\\ -\sin\varphi & \cos\varphi & 0\\ 0 & 0 & 1\end{bmatrix}\begin{bmatrix}\cos\psi & -\sin\psi & 0\\ \sin\psi & \cos\psi & 0\\ 0 & 0 & 1\end{bmatrix}$$

The angle ψ rotates around the normal direction of the moving platform and it is zero in an ideal situation. Furthermore, the accuracy of the angle ψ is not necessary to guarantee when the machine tool machines a part. Hence, ψ can be ignored and only two angles are kept in the rotational matrix.

Based on the vector loop $OB_iC_iA_iN$, the kinematic model can be written as:

$$\boldsymbol{H}+\boldsymbol{R}_{\mathrm{TT}}\boldsymbol{R}_i\boldsymbol{a}_i=\boldsymbol{R}_i\boldsymbol{b}_i+\boldsymbol{R}_i\boldsymbol{R}_{Bi}\boldsymbol{q}_i+\boldsymbol{R}_i\boldsymbol{R}_{Bi}\boldsymbol{R}_{Ci}\boldsymbol{l}_i \tag{4}$$

where $\boldsymbol{R}_i$ represents the rotation matrix of each leg. Furthermore, $\boldsymbol{a}_i$, $\boldsymbol{b}_i$ and $\boldsymbol{l}_i$ represent the position vectors of NA_i, OB_i and C_iA_i. $\boldsymbol{H}$ represents the position vector of the moving platform. $\boldsymbol{q}_i$ represents the input vector of each P joint. The orientation of the P joint and the orientation of the R joint are represented by matrices $\boldsymbol{R}_{Bi}$ and $\boldsymbol{R}_{Ci}$, respectively.

Based on the normal kinematic model above, the error model can be obtained by taking perturbations on both side of Eq. (4):

$$\begin{aligned}&\boldsymbol{H}+\Delta\boldsymbol{H}+(\boldsymbol{E}+\Delta\boldsymbol{\alpha}\times)\boldsymbol{R}_{\mathrm{TT}}\boldsymbol{R}_i(\boldsymbol{a}_i+\Delta\boldsymbol{a}_i)\\&=\boldsymbol{R}_i(\boldsymbol{b}_i+\Delta\boldsymbol{b}_i)+\boldsymbol{R}_i(\boldsymbol{E}+\Delta\boldsymbol{\theta}_{Bi}\times)\boldsymbol{R}_{Bi}\boldsymbol{e}_{zi}(q_i+\Delta q_i)\\&+\boldsymbol{R}_i(\boldsymbol{E}+\Delta\boldsymbol{\theta}_{Bi}\times)\boldsymbol{R}_{Bi}(\boldsymbol{E}+\Delta\boldsymbol{\theta}_{Ci}\times)\boldsymbol{R}_{Ci}\boldsymbol{\omega}_i(l_i+\Delta l_i)\end{aligned} \tag{5}$$

where the parameters with a triangular symbol represent the corresponding error parameters in the error model.

4 Determining the parameter design space and evaluation indices

4.1 Determining the parameter design space

In order to keep the isotropy of the 3PRS parallel kinematic manipulator, each of $|a_i|$, $|b_i|$ and $|l_i|$ is kept in the same so that there are 3 parameters to design it, namely *a*, *b* and *l*. Let

$$D=(a+b+l)/3 \tag{6}$$

Then

$$d_1=a/D \tag{7}$$

$$d_2=b/D \tag{8}$$

$$d_3=l/D \tag{9}$$

Hence

$$d_1+d_2+d_3=3 \tag{10}$$

According to the working mode of the manipulator, two geometric constraint equations of the parameters can be derived as:

$$d_3\ge d_2-d_1 \tag{11}$$

$$d_2\ge d_1 \tag{12}$$

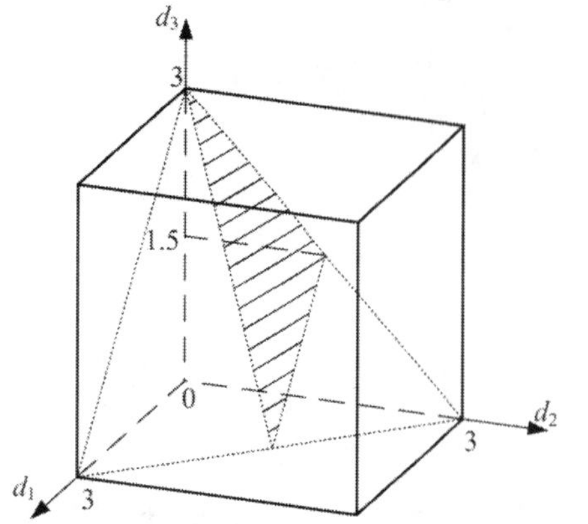

Figure 5. Parameter design space

Using Eqs. (10)-(12), a planar region can be obtain in Fig. 5, which is called the parameter design space. For the convenience of illustration, this region is transformed into a planar system defined by s and t, as shown in Fig. 6. The mapping function of the related variables can be expressed as:

$$t = d_3 \tag{13}$$

$$s = \frac{2d_2}{\sqrt{3}} + \frac{d_3}{\sqrt{3}} \tag{14}$$

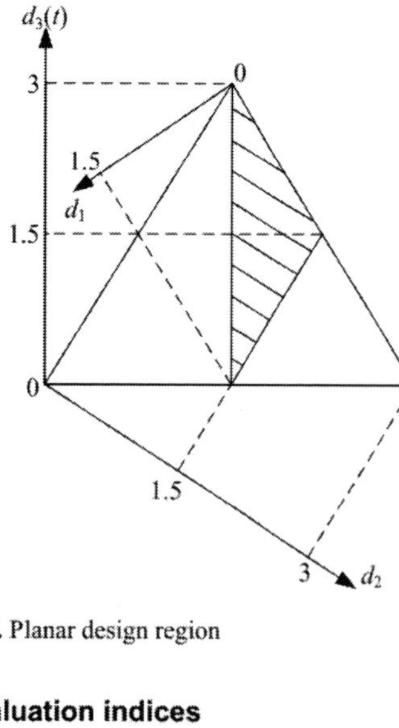

Figure 6. Planar design region

4.2 Evaluation indices

In order to improve the machining accuracy of this kind of 5-axis hybrid machine tool which is used for the aerospace industry, the workspace, position and orientation error sensitivity based on the machining trajectory of S-shaped test specimen are taken as the performance indices of concern to investigate.

For the same input range (assumed as [-q, q] for each leg) and considering the isotropy of the manipulator, the performance index of workspace can be expressed as:

$$w = \sin\{\theta[\mathit{forward\_solution}(\boldsymbol{q})]\} \tag{15}$$

where this index w represents the sine function value of θ solved by forward solution based on the input $\boldsymbol{q}=[q, q, -q]$.

A determination of a global position error sensitivity based on the machining trajectory of S-shaped test specimen is put forward in the following:

$$\sigma_{\mathrm{H}m} = \frac{\int_V \sqrt{\left[\frac{\partial(\Delta H_x)}{\partial(\Delta P_{\mathrm{H}m})}\right]^2 + \left[\frac{\partial(\Delta H_y)}{\partial(\Delta P_{\mathrm{H}m})}\right]^2 + \left[\frac{\partial(\Delta H_z)}{\partial(\Delta P_{\mathrm{H}m})}\right]^2}\, dV}{V} \tag{16}$$

where V denotes the machining trajectory of S-shaped test specimen. ΔH_x, ΔH_y and ΔH_z respectively represent elements of the position error vector of the end effector. ΔP_{H} represents an element of the position error vector of a joint. m is the number of position error parameters of joints.

Then

$$\sigma_{\mathrm{H}} = max(\sigma_{\mathrm{H}m}) \tag{17}$$

where σ_{H} is the maximum $\sigma_{\mathrm{H}m}$, namely the index of position error sensitivity.

Similarly, a determination of a global orientation error sensitivity based on the machining trajectory of S-shaped test specimen is put forward in the following:

$$\sigma_{\alpha m} = \frac{\int_V \sqrt{\left[\frac{\partial(\Delta \alpha_x)}{\partial(\Delta P_{\alpha m})}\right]^2 + \left[\frac{\partial(\Delta \alpha_y)}{\partial(\Delta P_{\alpha m})}\right]^2 + \left[\frac{\partial(\Delta \alpha_z)}{\partial(\Delta P_{\alpha m})}\right]^2}\, dV}{V} \tag{18}$$

$$\sigma_{\alpha} = max(\sigma_{\alpha m}) \tag{19}$$

where σ_{α} is the index of orientation error sensitivity.

It should be noted that the above trajectory based on S-shaped test specimen just denotes the orientation trajectory because the orientation error is the main error source and the position error is easy to guarantee in the machine tool.

5 Dimension synthesis

Before dimension synthesis, the ranges of evaluation indices should be determined. Considering the small error sensitivities are helpful for improving the accuracy and the large workspace is helpful for satisfying the machining demand, the ranges of evaluation indices are determined as: w>0.8, σ_{H}<0.72 and σ_{α}<20. According to the value requirements, the result of dimension synthesis is plotted with the shadow area in Fig. 7. The dimension satisfied with the evaluation indices has been covered by the shadow area. A solution can be selected and the corresponding dimensions can be obtained. In this paper, it is assumed that D = 302 mm and the selected point is (s, t) = (1.92, 1.5). Then, a=177 mm, b=276 mm and l=453 mm.

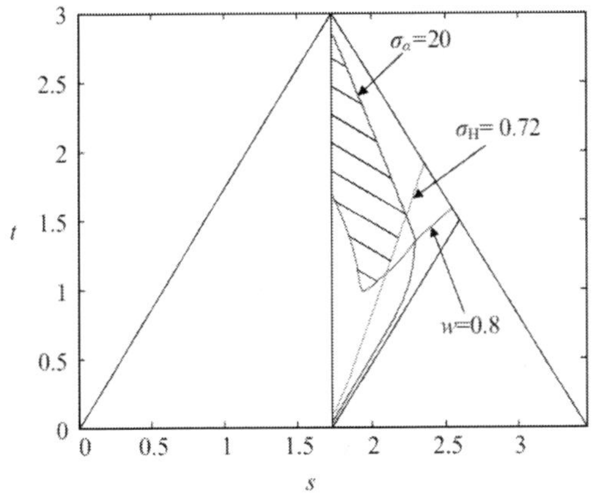

Figure 7. Optimum region for the manipulator

6 Conclusions

Based on the machining trajectory of S-shaped test specimen, this paper investigates the dimension synthesis of the 3PRS parallel kinematic manipulator used in a hybrid machine tool. By analyzing the mathematical characteristics of a ruled surface and using the data provided by Chengdu Aircraft Industrial Company, the model of S-shaped test specimen is established and the corresponding orientations for machining the S-shaped test specimen are obtained. In order to investigate the relation between the machining trajectory accuracy of S-shaped test specimen and the 3PRS parallel kinematic manipulator, the kinematic model and error model of the manipulator are established, respectively. According to the parameter design space determined by the working mode of the manipulator and the expected performance indices, the result of dimension synthesis is plotted. Within the dimension synthesis solutions, a set of suitable dimensions of the 3PRS parallel kinematic manipulator, namely a=177 mm, b=276 mm and l=453 mm, is achieved. With this set of dimensions, it is helpful to improving the accuracy of machining the S-shaped test specimen.

Acknowledgements

This work is supported by the National Natural Science Foundation of China (Grant No. 51575307 and 51225503), the Science and Technology Major Project-Advanced NC Machine Tools & Basic Manufacturing Equipments (2013ZX04004021 and 2014ZX04002051), and Top-Notch Young Talents Program of China.

References

1. J. Wu, J. Wang, L. Wang, T. Li. Dynamics and control of a planar 3-DOF parallel manipulator with actuation redundancy. Mech. Mach. Theory **44**(4), 835-849 (2009)
2. R. Ramesh, M.A. Mannan, A.N. Poo. Error compensation in machine tools—a review: Part I: geometric, cutting-force induced and fixture-dependent errors. Int. J. Mach. Tools Manuf. **40**(9), 1235-1256 (2000)
3. P.S. Rao, N.M. Rao. Position analysis of spatial 3-RPS parallel manipulator. Int. J. Mech. Eng. Rob. Res. **2**(2), 80-90 (2013)
4. V.V. Kamesh, A.B.S. Rao, K.M. Rao. A computer based technique for structural comparison of in parallel robotic manipulators applicable for higher pairs. Int. J. Mech. Eng. Rob. Res. **1**(1), 107-119 (2012)
5. Z. Su, L. Wang. Latest development of a new standard for the testing of five-axis machine tools using an S-shaped test piece. Proc. Inst. Mech. Eng. Part B-J. Eng. Manuf. **229**(7), 1221-1228 (2015)
6. Z. Jiang, J. Ding, Z. Song, L. Du, W. Wang. Modeling and simulation of surface morphology abnormality of 'S'test piece machined by five-axis CNC machine tool. Int. J. Adv. Manuf. Technol. 1-15 (2015)
7. F. Xie, X.J. Liu, T. Li. A comparison study on the orientation capability and parasitic motions of two novel articulated tool heads with parallel kinematics. Adv. Mech. Eng. **5**, 249103 (2013)
8. T. Huang, M. Li, X.M. Zhao, J.P. Mei, D. G. Chetwynd, S.J. Hu. Conceptual design and dimensional synthesis for a 3-DOF module of the TriVariant-a novel 5-DOF reconfigurable hybrid robot. IEEE Trans. Robot. **21**(3), 449-456 (2005)
9. P. Huang, J. Wang, L. Wang, R. Yao. Kinematical calibration of a hybrid machine tool with Regularization method. Int. J. Mach. Tools Manuf. **51**(3), 210-220 (2011)
10. X.J. Liu, J. Wang. A new methodology for optimal kinematic design of parallel mechanisms. Mech. Mach. Theory **42**(9), 1210-1224 (2007)
11. F. Xie, X.J. Liu, J. Wang. A 3-DOF parallel manufacturing module and its kinematic optimization. Robot. Comput.-Integr. Manuf. **28**(3), 334-343 (2012)

Two Error Models for Calibrating SCARA Robots based on the MDH Model

Xiaolong Li[1], Hong Hu[1] and Wenhua Ding [2]

[1]*Harbin Institute of Technology Shenzhen Graduate School, China*
[2]*Shenzhen Colibri Technologies Co. Ltd., China*

Abstract. This paper describes the process of using two error models for calibrating Selective Compliance Assembly Robot Arm (SCARA) robots based on the modified Denavit-Hartenberg(MDH) model, with the aim of improving the robot's accuracy. One of the error models is the position error model, which uses robot position errors with respect to an accurate robot base frame built before the measurement commenced. The other model is the distance error model, which uses only the robot moving distance to calculate errors. Because calibration requires the end-effector to be accurately measured, a laser tracker was used to measure the robot position and distance errors. After calibrating the robot and, the end-effector locations were measured again compensating the error models' parameters obtained from the calibration. The finding is that the robot's accuracy improved greatly after compensating the calibrated parameters.

1 Introduction

Industrial robots usually work from one taught point to another taught point, which makes them highly repeatable although their accuracy may be poor due mainly to geometric errors.

The good news is that the accuracy can be improved by calibrating the robots. The two most important parts of robot calibration are building an accurate model and a valid method to measure the robot's position. Although some researchers, such as Marko Švaco[1],have calibrated robots using the Denavit-Hartenberg (DH) model, it is not good enough for some robots that have parallel links. So Hayati[2]proposed a model called the MDH model, which many researchers now use to calibrate robots. Some researchers have proposed methods for third level calibrations, such as Albert Nubiola[3, 4] who introduced the temperature influence. Wang Zhen Hua [5] proposed a method to calibrate robots without building a base frame by using the distance errors of moving robots. Zhou Jian[6] proposed a pose selection algorithm that allows one to select a given number of optimal poses out of a large set of previously measured poses to avoid local convergence.

Some researchers concentrated on the measure methods, such as: laser tracker[7], coordinate measuring machine (CMM)[8, 9], telescoping ballbar[4], and the vision system[10]. There are also some researchers who use novel methods such as a probe and a ball[11] to measure the robot's moving distance, and XiNing[12]used a laser and a PSD to calibrate robots.

In this paper, two error models are used to calibrate a SCARA robot. One is a position error model and the other is a distance error model, both of which are based on the MDH model.

The paper is organized into the following sections: section 2 proposes a kinematic model of the industrial robot called MDH model; section 3 presents the two calibration error models; section 4 presents the experiments and results; and section 5 is the conclusion.

2 MDH Model

A robot model must be built before the robot can be calibrated. The DH model is the most popular model as its four parameters can express two link positions very well. But the DH model will results in singularities with respect to two adjacent parallel links, even though a very small parallel error will cause a large deviation. So a MDH model developed by Hayati[2] was used, involving a new parameter termedβ as the angle link i rotates around the y axis, written as $Rot(y,\beta)$. The MDH model can be expressed as follows:

$$
{}^{i-1}_{i}A = {}^{i-1}_{i}T \cdot R_y(\beta) =
\begin{bmatrix}
\cos\theta\cos\beta & -\sin\theta & \sin\beta\cos\theta & a \\
\sin\alpha\sin\beta + \cos\alpha\cos\beta\sin\theta & \cos\alpha\cos\theta & \cos\alpha\sin\beta\sin\theta - \cos\beta\sin\alpha & -d\sin\alpha \\
\cos\beta\sin\alpha\sin\theta - \cos\alpha\sin\theta & \sin\alpha\cos\theta & \cos\alpha\cos\beta + \sin\alpha\sin\beta\sin\theta & d\cos\alpha \\
0 & 0 & 0 & 1
\end{bmatrix}
\tag{1}
$$

3 Error Model

Assume that the model errors are very small and use A_i to express the nominal transformation matrix from joint *i-1* to joint *i*; A_i'then expresses the actual transformation matrix as:

$$\begin{aligned} A_i' = R_X(\alpha_{i-1}+\delta\alpha_{i-1})\cdot D_X(a_{i-1}+\delta a_{i-1}) \\ \cdot R_Z(\theta_i+\delta\theta)\cdot D_Z(d_i+\delta d_i)\cdot R_Y(\beta_i+\delta\beta_i) \end{aligned} \quad (2)$$

The error of one joint is equal to the sum of errors caused by every model error[5]:

$$\begin{aligned} dA_i = \frac{\partial A_i}{\partial \alpha_{i-1}}\delta\alpha_{i-1} + \frac{\partial A_i}{\partial a_i}\delta a_{i-1} + \frac{\partial A_i}{\partial \theta_i}\delta\theta_i \\ + \frac{\partial A_i}{\partial d_i}\delta d_i + \frac{\partial A_i}{\partial \beta_i}\delta\beta_i \end{aligned} \quad (3)$$

Just let,

$$Q_a = A^{-1}\frac{\partial A_i}{\partial \alpha_{i-1}} = \begin{bmatrix} 0 & 0 & 0 & c\theta c\beta \\ 0 & 0 & 0 & -s\theta \\ 0 & 0 & 0 & c\theta s\beta \\ 0 & 0 & 0 & 0 \end{bmatrix} \quad (4)$$

$$Q_a = \begin{bmatrix} 0 & -c\theta s\beta & -s\theta & -ds\theta c\beta \\ c\theta s\beta & 0 & -c\theta c\beta & -dc\theta \\ s\theta & c\theta c\beta & 0 & -ds\theta s\beta \\ 0 & 0 & 0 & 0 \end{bmatrix} \quad (5)$$

$$Q_d = \begin{bmatrix} 0 & 0 & 0 & -s\beta \\ 0 & 0 & 0 & 0 \\ 0 & 0 & 0 & c\beta \\ 0 & 0 & 0 & 0 \end{bmatrix} \quad (6)$$

$$Q_\theta = \begin{bmatrix} 0 & -c\beta & 0 & 0 \\ c\beta & 0 & s\beta & 0 \\ 0 & -s\beta & 0 & 0 \\ 0 & 0 & 0 & 0 \end{bmatrix} \quad (7)$$

$$Q_\beta = \begin{bmatrix} 0 & 0 & 1 & 0 \\ 0 & 0 & 0 & 0 \\ -1 & 0 & 0 & 0 \\ 0 & 0 & 0 & 0 \end{bmatrix} \quad (8)$$

Therefore the error of MDH model is:

$$dA_i = A_i\left(Q_{a_{i-1}}\delta a_{i-1} + Q_{\alpha_{i-1}}\delta\alpha_{i-1} + Q_{d_i}\delta d_i + Q_{\theta_i}\delta\theta_i + Q_{\beta_i}\delta\beta_i\right) \quad (9)$$

3.1 SCARA Position Error Model

For the SCARA robot, the final error model is:

$${}_0^4A + d\,{}_0^4A = \prod_{i=1}^{n}\left(A_i + A_iQ_i\delta_i\right) \quad (10)$$

Since the geometric errors are very small, the higher-order terms can be neglected.

$$d\,{}_0^4A = A_1Q_1\delta_1A_2A_3A_4 + \cdots + A_1A_2A_3A_4Q_4\delta_4 \quad (11)$$

When the position errors or distance errors are considered, only the first three rows of the forth columns ofd_0^4Ashould be considered; in other words, the focus should be on the positions of the robot. Therefore, $B_i$is defined as the coefficient matrix of the MDH model errors.

When robot error is measured, there may be errors in the measuring tools. If the tool coordinates parallel to the base frame are defined, they will always exist as δx,δy, δz with respect to the measuring tools and never change. So the relationship of position errors with respect to the model error is:

$$\begin{bmatrix} dx \\ dy \\ dz \end{bmatrix} = \sum_{i=1}^{4} B_i\delta_i + B_{tool}\begin{bmatrix} \delta x \\ \delta y \\ \delta z \end{bmatrix} \quad (12)$$

$$\delta_i = \begin{bmatrix} \delta a_{i-1} & \delta\alpha_{i-1} & \delta d_i & \delta\theta_i & \delta\beta_i \end{bmatrix}^T$$

Whereδ_i is the MDH model errors of the link *i-1* to link i *and*B_i is the coefficient matrix of δ_i.
Equation(12) can be written as:

$$\begin{bmatrix} dx & dy & dz \end{bmatrix}^T = B\delta \quad (13)$$

3.2 Distance error model

Unlike the difficulties involved in building accurate base frame when using the position error model to calibrate the robot, the distance error model allows distance to be measured easily and accurately without the need for a pre-determined base frame. Accordingly, the distance error model is used here to calibrate the robot.

The measured distance when the robot is moving is l_R, the command distance of the robot is l_C, and $\Delta l = l_R - l_C$ is the distance error between the measured distance and the command distance. Therefore:

$$\begin{aligned} (\Delta l + l_R)^2 = & \left(x_R(j) - x_R(i) + \Delta x(j) - \Delta x(i)\right)^2 \\ & + \left(y_R(j) - y_R(i) + \Delta y(j) - \Delta y(i)\right)^2 \\ & + \left(z_R(j) - z_R(i) + \Delta z(j) - \Delta z(i)\right)^2 \end{aligned} \quad (14)$$

So the final distance error model can be written as:

$$\begin{aligned} \Delta l = \left(\frac{x_R(j) - x_R(i)}{l_R(i,j)}, \frac{y_R(j) - y_R(i)}{l_R(i,j)}, \frac{z_R(j) - z_R(i)}{l_R(i,j)}\right) \\ \times\left(B_j - B_i\right)\Delta q \end{aligned} \quad (15)$$

4 Experiment and results

4.1 Experiment set-up

The first step is to build the SCARA MDH model. The

DOI: 10.1051/matecconf/20179508008

SCARA is a very simple robot with all of its links parallel and three of its four joints rotational. The process involved in building the MDH model is depicted in Figure 1 and Table 1.

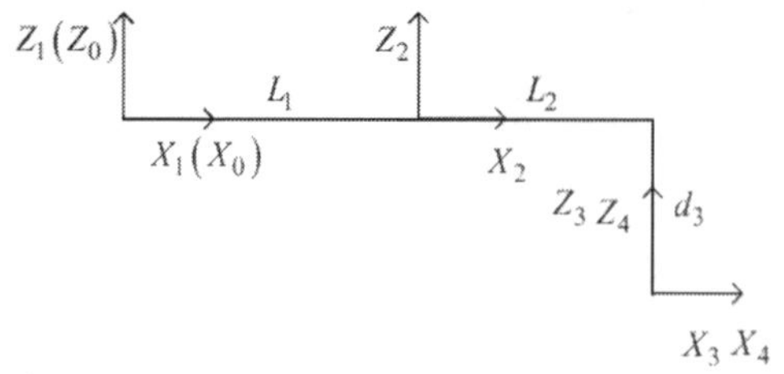

Figure 1. SCARA MDH model

Table 1.Nominal kinematic parameters

Joint	a_{i-1}	α_{i-1}	d_i	θ_i	β_i
1	0	0	0	θ_1	0
2	180	0	0	θ_2	0
3	180	0	0	0	0
4	0	0	0	θ_4	0

The end-effector could only be measured after the MDH model had been built. In this experiment, a FARO laser tracker with an accuracy of 10*um* +2.5*um/m* was used for measuring the errors of the SCARA robot. In order to measure robot position error, a robot base frame was built to allow the robot to move in the XY plane and was fitted so that the XY plane normal vector is the Z axis. The base frame was completed by allowing the robot to move along the X axis so that the axis could be identified as such.

Figure 2.Experiment field

Two groups of positions of the robot end-effector were measured. The calibrated group was used to calibrate the robot, while the check group was to check the accuracy of the robot after calibration. A total of 88 positions were measured for each group.

Finally, the measured data are substituted into equation(13) and equation(15), resulting in a set of equations that are all linear. The least-square method can then be used to solve the equations.

4.2 Position error model results

The lease-square equation for the position error model was built according to equation (13) using the measured data:

$$\boldsymbol{b}_p = C_p \Delta \boldsymbol{q}_p \tag{16}$$

Where, $\boldsymbol{b_p}$ is the position errors vector between the measured position and command position of the robot end-effector, C_p is the coefficient matrix consisting of Jacobian matrix B_i, and$\Delta\boldsymbol{q_p}$is the MDH model errors.

Table 2.Calibration results of position error

i	δa_{i-1}	$\delta\alpha_{i-1}$	δd_i	$\delta\theta_i$	$\delta\beta_i$
1	0	0.0002	0	-0.0065	0.0027
2	-0.8340	0.0011	0	0.0105	0.0018
3	-0.2226	0.0003	0	0	0
tool	δx = -0.8438 δy = -1.8113 δz = -18.5336				

Table 2 is the MDH model errors after equation(16) had been calibrated. The absolute error of the end-effector before calibration and after calibration of the calibrated group pointsis shown in Figure 3.

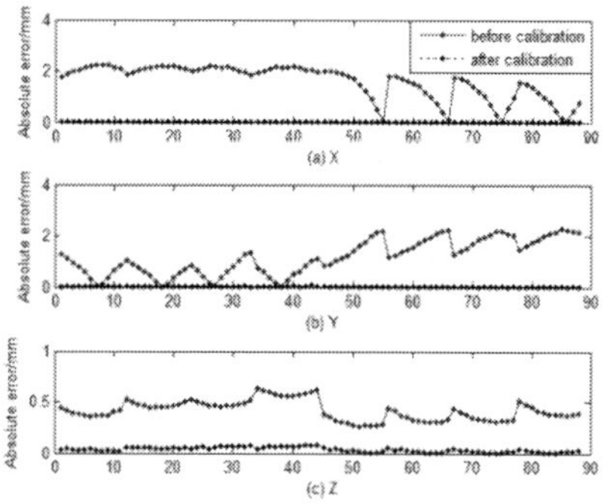

Figure 3.Position error of calibrated group

Because the model errors were calculated with the calibrated group points only, the improvement of accuracy may only be effective for the calibrated groups. To compensate for this another 88 points were measured as a check group in different positions. Their position errors after calibration are shown in Figure 4.

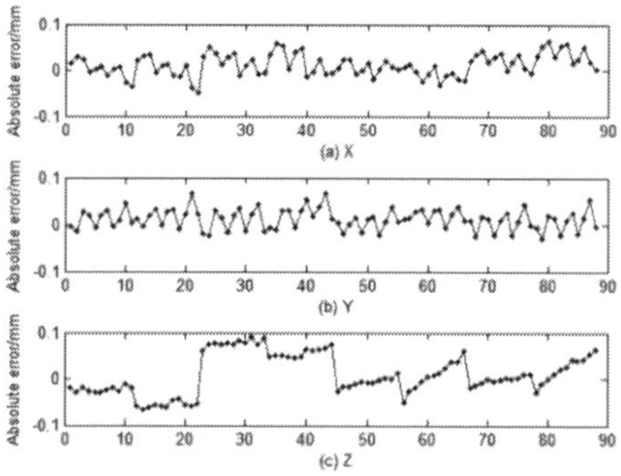

Figure 4.Position error of check group

Table 3.Position error analysis

Error		Calibrated Group Error (mm)			Check Group Error (mm)		
		X	Y	Z	X	Y	Z
BeforeCalibration	Mean	1.6101	1.1801	0.4173	NONE		
	SD	0.6438	0.6931	0.0943			
After Calibration	Mean	0.0173	0.0181	0.0361	0.0214	0.0204	0.0359
	SD	0.0117	0.0131	0.0218	0.0162	0.0139	0.0258

Once the data was obtained, the results were analyzed and the position error expressed respectively in the X, Y and Z axes. From Table3 it can be seen that for the calibrated group points after the calibration the mean position errors could be decreased from 1.6101mm to 0.0173mm in the X axis; 1.1801mm to 0.0181mm in the Y axis; and 0.4173mm to 0.0361mm in the Z axis. The standard deviation (SD) could be improved from 0.6438 to 0.0117 in the X axis; 0.6931 to 0.0131 in the Y axis; and 0.0943 to 0.0218 in the Z axis. For the check group points, the mean position errors are 0.0214 mm in the X axis; 0.0204 mm in the Y axis; and 0.0359 mm in the Z axis. The SD in three axes respectively is 0.0162 mm, 0.0139 mm, and 0.0258 mm. These data show that this method substantially improves robot positioning accuracy. However, the method requires that before measuring position errors it is necessary to build a robot base frame and confirm the relationship between the base frame and the measurement system. If the relationship is not accurate, it may cause other errors. To overcome this difficulty, a distance error model [5, 11]is proposed in the next section.

4.3 Distance error model results

When using the laser tracker as the measurement system, it is easy to confirm the base frame's accuracy because of the laser's high measurement accuracy. However, when it is not possible to get an accurate base frame, a distance error model can make it easier to calibrate the robot.

After the measurements are obtained, they can be put into equation(15), so that:

$$\Delta \boldsymbol{l}_d = D_d \Delta \boldsymbol{q}_d \tag{17}$$

Where, $\Delta \boldsymbol{l}_d$ is the distance errors vector between the measured positions and command positions; D_d is the coefficient matrix consisting of B_j - B_i, and$\Delta \boldsymbol{q}_d$is the MDH model errors.

After the robot is calibrated with the distance error, the results are as shown in Table 4.

Table 4.Results of distance error model

i	δa_{i-1}	$\delta \alpha_{i-1}$	δd_i	$\delta \theta_i$	$\delta \beta_i$
1	0	0	0	0	-0.0022
2	-0.9140	-0.0015	0	0.0104	-0.0042
3	-0.3299	0.0014	0	0	0

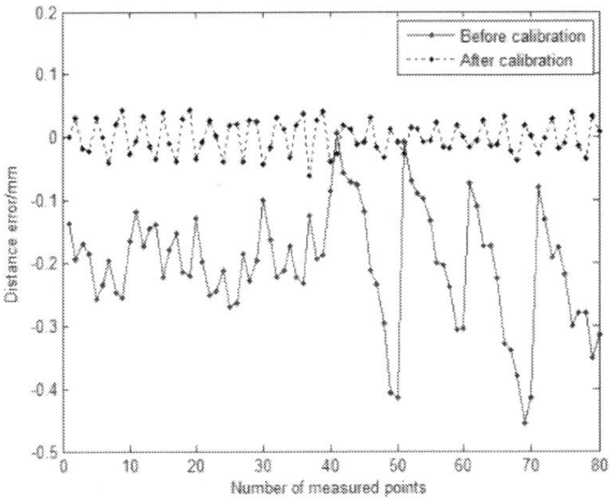

Figure 5.Distance error of calibrated group

The robot can be corrected for distance errors with the calibration parameters: the distance errors before calibration and after calibration can be seen in Fig. 5. Just as with the position error, the calibration may only improve the accuracy of the calibrated group, so it is necessary to measure distance errors of other points that are different to the calibrated group points. Therefore, 88 points, which is 80 distance errors, were measured and are shown in Fig. 6.

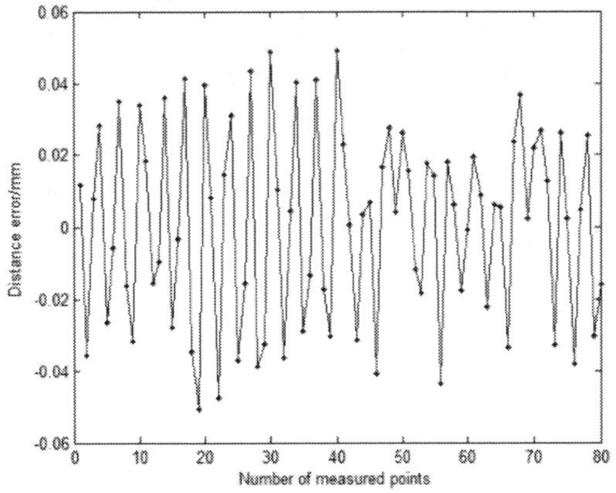

Figure 6.Distance error of check group

Table 5.Distance error analysis

Error		Calibrated Group	Check Group
BeforeCalibration	Mean	0.2031	None
	SD	0.0928	
After Calibration	Mean	0.0225	0.0230
	SD	0.0260	0.0268

From Figure 6 and Table 5 it can be seen that for the calibrated group points after the calibration, the mean distance errors decreased from 0.2031mm to 0.0225mm, the standard deviation (SD) improved from 0.0928 mm to 0.0260mm, and the check group points' mean distance error and SD are 0.0230 mm and 0.0268 mm respectively.

From these data it is clear that the distance error model works very well and can be used as a method when it is not possible to construct an accurate base frame.

5 Conclusion

This paper described the building of a SCARA robot model based on the MDH model, and proposed two error models for calibrating the robot. After the robot was calibrated, the parameters were used to correct the robot, which greatly improved the robot's accuracy. When it is not possible to build an accurate robot base frame and establish the relationship of the base to the measurement system, it is suggested to use the distance error model to achieve accuracy. Both error models proved to be effective for calibrating the SCARA robot based on an accurate robot base frame.

Acknowledge

The authors gratefully acknowledge the support from the National Nature Science Foundation of China under the NSFC Grant 51375118, and the support from Shenzhen Government Founding GJHZ20140419142750948.

References

1 M. Švaco, B. Šekoranja, F. Šuligoj, et al., Calibration of an Industrial Robot Using a Stereo Vision System, Procedia Engineering, **69**, 459-463 (2014)
2 S. A. Hayati, Robot arm geometric link parameter estimation, Decision and Control, 1983. The 22nd IEEE Conference on, 1477-1483 (1983)
3 A. Nubiola, I. A. Bonev, Absolute calibration of an ABB IRB 1600 robot using a laser tracker, Robotics and Computer-Integrated Manufacturing, **29**, 236-245 (2013)
4 A. Nubiola, I. A. Bonev, Absolute robot calibration with a single telescoping ballbar, Precision Engineering, **38**, 472-480 (2014)
5 W. Zhenhua, X. Hui, C. Guodong, et al., A distance error based industrial robot kinematic calibration method, Industrial Robot: An International Journal, **41**, 439-446 (2014)
6 J. Zhou, H. N. Nguyen, H. J. Kang, Selecting Optimal Measurement Poses for Kinematic Calibration of Industrial Robots, Advances in Mechanical Engineering, **6**, 291389-291389 (2015)
7 J. Santolaria, J.-A. Yagüe, R. Jiménez, et al., Calibration-based thermal error model for articulated arm coordinate measuring machines, Precision Engineering, **33**, 476-485 (2009)
8 A. Nubiola, M. Slamani, A. Joubair, et al., Comparison of two calibration methods for a small industrial robot based on an optical CMM and a laser tracker, Robotica, **32**, 447-466 (2013)
9 A. Joubair, M. Slamani, I. A. Bonev, Kinematic calibration of a five-bar planar parallel robot using all working modes, Robotics and Computer-Integrated Manufacturing, **29**, 15-25 (2013)
10 G. Du, P. Zhang, Online robot calibration based on vision measurement, Robotics and Computer-Integrated Manufacturing, **29**, 484-492 (2013)
11 A. Joubair, I. A. Bonev, Kinematic calibration of a six-axis serial robot using distance and sphere constraints, The International Journal of Advanced Manufacturing Technology, **77**, 515-523 (2014)
12 N. Xi, Y. Liu, Low-cost and automated calibration method for joint offset of industrial robot using single-point constraint, Industrial Robot, **38**, 577-584 (2011)

Type Synthesis of Parallel 2R1T Remote Center of Motion Mechanisms Based on Screw Theory

Long Huang, Chenhan Guang, Yang Yang and Peng Su

School of Mechanical Engineering and Automation, Beihang University, 100191 Beijing, China

Abstract. The remote center of motion (RCM) mechanism is an important part of a minimally invasive surgery (MIS) robot. As a practical type of RCM mechanisms, 2R1T RCM mechanisms are synthesized in this paper using a modified screw theory method. Differing from the conventional screw theory method, the modified method brings forward the assessment of different wrench systems, and proposes several general subchains to satisfy each geometrical condition of the wrench system, which can ensure the motion continuity to a great degree. First, the motion pattern and the twist system of 2R1T RCM mechanisms are presented. According to the reciprocal rules of screw theory, three corresponding wrench systems, along with the geometrical conditions for them, are enumerated. After assessing these wrench systems by the comparison of their geometrical conditions, the optimum wrench system is selected for synthesis, and then it is decomposed into sub-wrench-systems. To satisfy each condition of the sub-wrench-systems, several general subchains are proposed, and consequently some preferred general chains are constructed by the serial assembly of the corresponding subchains. These general chains are applied to the leg synthesis. Finally, practical examples of 2R1T RCM mechanisms are constructed by assembling the legs.

1 Introduction

Minimally invasive surgery (MIS) robots have attracted considerable attention in recent years [1-2]. In a MIS robot, the surgical tool should be capable of 3-DOF rotation around the incision and 1-DOF translation along the axis of the instrument. The remote center of motion (RCM) mechanism is particularly well suited for this. RCM mechanisms can be divided into six categories by DOF: 1-DOF rotation (1R), 2-DOF rotation (2R), 3-DOF rotation (3R), 1-DOF rotation and 1-DOF translation (1R1T), 2-DOF rotation and 1-DOF translation (2R1T), 3-DOF rotation and 1-DOF translation (3R1T). The multi-DOF RCM mechanism can be constructed by the serial assembly of RCM mechanisms with fewer DOFs. Taylor's group introduced the RCM mechanism to a laparoscopic robot [3-4]. Afterwards various kinds of RCM mechanisms have been reported, such as spherical mechanisms [5], arc-prismatic-joint RCM mechanisms [6], and parallelogram-based RCM mechanisms [7]. Also, several methods, such as the virtual chain method and the module combination method, are proposed for the type synthesis of some specific kinds of RCM mechanisms [8-10]. However, for multi-DOF mechanisms, a widely used type synthesis method is the screw theory method. The major disadvantage of this method is that it can only synthesize instantaneous motion mechanisms [11-14].

As a practical type of RCM mechanisms, 2R1T RCM mechanisms are synthesized in this paper using a modified screw theory method. First, the twist system of 2R1T RCM mechanisms are presented, and the corresponding wrench system with the simplest conditions are selected and decomposed into sub-wrench-systems. Section 5 concentrates on the leg synthesis. In this section, the conditions for each wrench are analyzed in detail, and each condition can be satisfied by a serial of newly-constructed subchains; by assembling the corresponding subchains, some general chains which satisfy all the conditions are obtained. On this basis, legs with each sub-wrench-system are synthesized. Finally in Section 6, the examples of 2R1T RCM mechanisms are constructed by assembling the corresponding legs. Compared with the conventional screw theory based synthesis processes, this paper brings forward the assessment of different wrench systems, and proposes several general subchains for each geometrical condition, which can ensure the motion continuity of the mechanisms to a large degree.

2 Type synthesis method based on screw theory

Based on the line geometry, screw theory can be used to express the motion and constraint of a mechanism, and these two aspects of a mechanism are connected through the reciprocal rules. In screw theory, the motion of a mechanism is represented by a twist system which is composed of several twists, and the constraint for a mechanism is represented by a wrench system which is composed of several wrenches. The details about screw theory can be seen in [11-13].

Assuming that a twist system (or wrench system) contains *n* independent non-redundant twists (or wrenches), then the corresponding wrench system (or twist system) contains (6-*n*) independent non-redundant wrenches (or twists). Furthermore, the general reciprocal rules and type synthesis processes in screw theory can be found in [11-12]. However, owing to the particularity of the RCM motion pattern, the processes should be adapted for the specific synthesis problems.

In this paper, twists and wrenches of pitch h are represented by $\$^h$ and $\hat{\$}^h$ respectively. Twist systems of order n formed by zero pitch twists (or infinite pitch twists) are denoted by $n-\$^0$ (or $n-\$^\infty$), and wrench systems of order n formed by zero pitch wrenches (or infinite pitch wrenches) are denoted by $n-\hat{\$}^0$ (or $n-\hat{\$}^\infty$). Twist systems and wrench systems can be illustrated in graph [12]: zero pitch twists and wrenches are represented by lines, while infinite pitch twists and wrenches are represented by arrows.

3 Preparation for type synthesis

The output link of the 2R1T RCM mechanism can tilt and roll around the incision point, and translate along the axis of the instrument. Assuming the instrument is mounted on the moving platform of the mechanism with the axis perpendicular to the plane of the moving platform. The base frame $\{F_b\}$ and the moving platform frame $\{F_m\}$ are shown in Fig. 1. Then the moving platform of the 2R1T RCM mechanism can rotate about the origin of $\{F_b\}$ and translate along the z_m-axis in $\{F_m\}$.

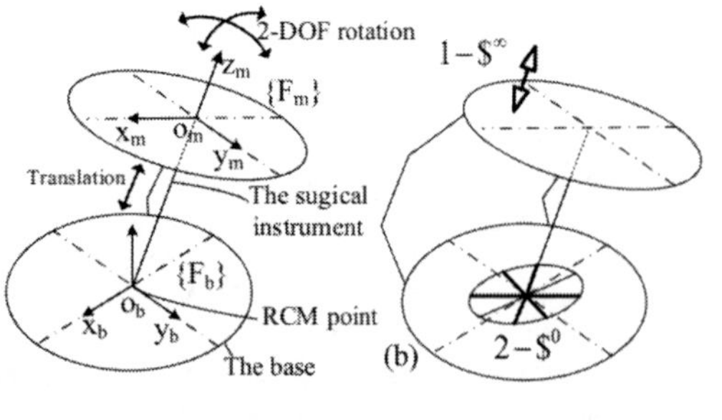

Figure 1. (a) Coordinate frame definition of the motion pattern; (b) the line graph representation of twist system.

According to the motion pattern of the mechanism, the required twist system for it is derived to be $2-\$^0-1-\$^\infty$. To realize the required motion pattern, $2-\$^0-1-\$^\infty$ system should satisfy the following conditions at each configuration of the given workspace: (1) The axis of $\$^\infty$ should be parallel to the z_m-axis; (2) The center of $2-\$^0$ should be the origin of $\{F_b\}$, and the plane of $2-\$^0$ should not be coplanar with the z_m-axis. The line graph representation of the twist system is shown in Fig. 1(b).

4 The wrench systems of the 2R1T RCM mechanism

According to the reciprocal rules, the wrench system of the mechanism is a 3-system, which is generated by three independent wrenches. Three wrench systems corresponding to the twist system are shown in Fig. 2.

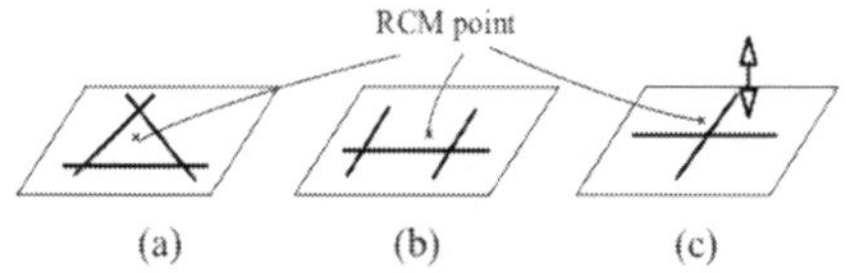

Figure 2. The line graph representation of three wrench systems

Wrench systems in Fig. 2a and Fig. 2b are both $3-\hat{\$}^0$ systems. According to the geometrical conditions for the twist system and the reciprocal rules, the wrenches in these systems should fulfill the following conditions.

Condition (a1): three zero pitch wrenches should be coplanar; they either form a triangle on a plane, or two of them are parallel and intersect the rest wrench.
Condition (a2): the plane generated by the wrenches should be parallel to the plane of the moving platform.
Condition (a3): The RCM point should lie on the plane generated by the wrenches.

Wrench system in Fig. 2c is a $2-\hat{\$}^0-1-\hat{\$}^\infty$ system. The wrenches in this system should fulfill the following conditions.

Condition (c1): two zero pitch wrenches should intersect. The direction of the infinite pitch wrench $\hat{\$}^\infty$ should not be parallel to the plane generated by $2-\hat{\$}^0$.

Condition (c2): the plane generated by $2-\hat{\$}^0$ should be parallel to $x_m y_m$-plane.
Condition (c3): The RCM point should lie on the plane generated by $2-\hat{\$}^0$.

For each one of these wrench systems, if all the listed conditions are satisfied at any configuration in a certain area, then the corresponding mechanisms can maintain the motion pattern in this area. However, not every condition can be satisfied easily. Through the comparison of the conditions for these wrench systems, we can find the optimum wrench system for synthesis. First, each one of these three wrench systems contains two zero pitch wrenches, and the conditions for them are basically the same. And evidently coplanarity (Condition (a3)) is a much stronger condition than non-parallelism (Condition (c2)). If we choose wrench systems $3-\hat{\$}^0$, it is likely that we will obtain 2R1T RCM mechanisms with bifurcated motion or parasitic motion [9]. Hence, we choose wrench system $2-\hat{\$}^0-1-\hat{\$}^\infty$ to synthesize 2R1T RCM mechanisms. This wrench system can be decomposed in many ways, and each part of it corresponds to a single leg. All possible parts of the wrench system can be listed as follows: $2-\hat{\$}^0-1-\hat{\$}^\infty$, $1-\hat{\$}^0-1-\hat{\$}^\infty$, $2-\hat{\$}^0$, $1-\hat{\$}^0$ and $1-\hat{\$}^\infty$. The following paragraphs deal with the synthesis of legs corresponding to each of these sub-wrench-systems.

5 Synthesis of legs

Unlike the conventional screw theory method, the modified method regards the processes of leg synthesis as the key to the guarantee of motion continuity. Based on the condition analysis of each wrench, we can translate each condition into a class of subchains, and furthermore assemble the subchains into general chains. For clarity of explanation, we adopt the visual line graph to represent the wrench system, instead of mathematical equations.

5.1 Condition analysis

Condition (c1) claims that the two zero pitch wrenches should intersect. For convenience of leg synthesis, the intersecting point is designated as the RCM point. In order to ensure the zero pitch wrenches of a leg passing through a fixed point at every configuration, an effective way is to install a subchain consisting of two or more revolute joints intersecting at a fixed point in the leg (the subchain is denoted by $(R...R)_S$ or $\underline{R}...\underline{R}$). Since there are only three independent twists (or wrenches) among all of the concurrent twists (or wrenches) in space [11], the required non-redundant subchains can be of two types: (1) $(RR)_S$, and (2) $(R...R)_S$, as Fig. 3a and 3b shows; furthermore, to keep the intersecting point fixed while the configuration changes, $(RR)_S$ -subchain or $(R...R)_S$ -subchain in the leg should be mounted as close to the base as possible. In addition, there are a class of chains containing two intersecting revolute joints, yet these joints are separated by other joints, for instance, the $\underline{R}RR\underline{R}$ -subchain shown in Fig. 3c. In order to keep the axes of the first joint and the last joint intersecting at a fixed point, there should be some additional constraint generated by other legs, which will be discussed later.

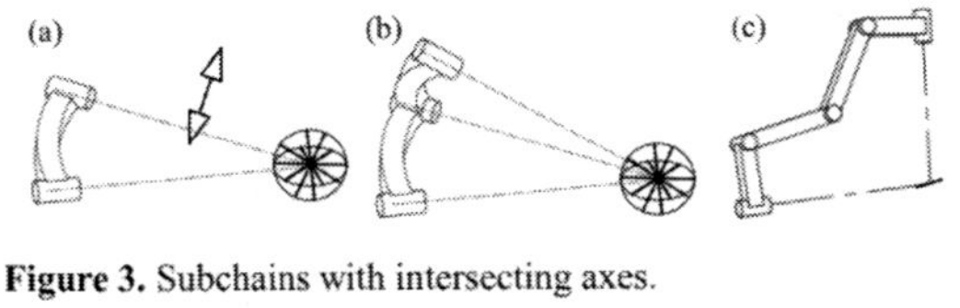

Figure 3. Subchains with intersecting axes.

Condition (c2) says each zero pitch wrench should be parallel to the $x_m y_m$-plane. There are two methods for this condition according to the reciprocal rules: (1) the leg includes a prismatic joint with its axis perpendicular to the $x_m y_m$-plane; (2) the leg includes a subchain consisting of two or more parallel revolute joints whose axes are parallel to the $x_m y_m$-plane in the leg (this subchain is denoted by $(R...R)_E$ or $\bar{R}...\bar{R}$). Since there are only three independent parallel twists (or wrenches) among all of the spatial parallel twists (or wrenches), the required non-redundant parallel subchains can be of two types: (1) $(RR)_E$, and (2) $(RRR)_E$. One thing to note is that each one of these two types has a class of variants. Revolute joints in $(R...R)_E$ -subchain can be replaced by prismatic joints partially without influencing the wrench system, as long as the subchain is still a planar mechanism. Fig. 4 shows some examples of $(RR)_E$ -type and $(RRR)_E$ -type. Furthermore, in order to guarantee the parallel relationship while the configuration changes, the prismatic joint or $(R...R)_E$ -subchain in the leg should be mounted as close to the moving platform as possible.

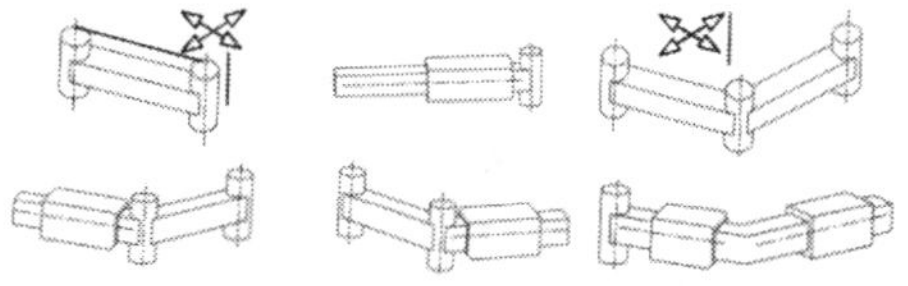

Figure 4. Subchains of $(RR)_E$ -type and $(RRR)_E$ -type

Condition (c3) says that the direction of the infinite pitch wrench $\hat{\$}^{\infty}$ is not parallel to the plane generated by $2-\hat{\$}^0$, which is automatically satisfied on almost every occasion. Therefore condition (c3) can be ignored during the leg synthesis.

In conclusion, during the leg synthesis process, we give preference to the following general serial chains: $(R...R)_S...(R...R)_E$, $(R...R)_S...P$ and $\underline{R}...(R...R)_E\underline{R}$. First, we apply the preferred chains to the specific situation so that we can obtain chains corresponding to the required wrench system without any geometric conditions. Then, in terms of all geometric conditions for the wrench system, the qualified chains are selected and the assembly conditions for them are obtained.

5.2 Synthesis of legs corresponding to sub-wrench-system $2-\hat{\$}^0-1-\hat{\$}^{\infty}$

For a non-redundant serial kinematic chain corresponding to sub-wrench-system $2-\hat{\$}^0-1-\hat{\$}^{\infty}$, the order of it is three. Therefore, the serial chain has three joints according to the reciprocal rules. In fact, the corresponding chain is actually a serial 2R1T RCM mechanism.

Applying the preferred chains to the situation, we only obtain the $(RR)_S P$-chain (Fig. 5). The serial chain shown in Fig. 5a is the simplest chain with the required sub-wrench-system. The kinematic chain shown in Fig. 5b is constructed through serial connection of a revolute joint, an arc prismatic joint and a prismatic joint. It can also be considered as a $(RR)_S P$-chain. According to the reciprocal rules, the rest of the assembly conditions for the joints are: (1) the axes of the first two joints should intersect at the appointed RCM point; (2) the axis of the prismatic joint should be parallel to the z_m-axis.

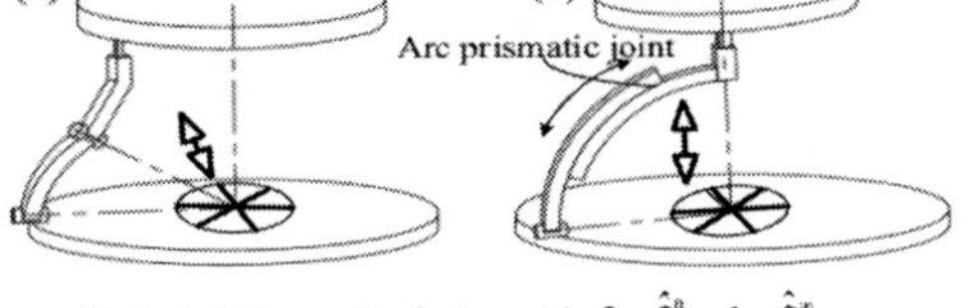

Figure 5. Serial kinematic chains with $2-\hat{\$}^0-1-\hat{\$}^{\infty}$

5.3 Synthesis of legs corresponding to sub-wrench-system $1-\hat{\$}^0-1-\hat{\$}^\infty$

A non-redundant serial chain with sub-wrench-system $1-\hat{\$}^0-1-\hat{\$}^\infty$ has four joints. Applying the preferred chains to the situation, we obtain two possible types: $(RR)_S(RR)_E$-type and $R(RRR)_E$-type.

For $(RR)_S(RR)_E$-type, the sub-wrench-system requires that the axes of revolute joints in $(RR)_E$-subchain be parallel to the plane generated by the axes of $(RR)_S$. Therefore, the possible serial chains include: $(R\bar{R})_S\bar{R}\bar{R}$, $(R\bar{R})_S\bar{P}\bar{R}$, $(R\bar{R})_S\bar{R}\bar{P}$ and $(R\bar{R})_S\bar{P}\bar{P}$. Other than the internal geometrical relationships, these chains should also satisfy the following conditions: (1) the intersecting point of $(RR)_S$-subchain should be the RCM point, and (2) the axis of $\bar{R}\bar{R}\bar{R}$-subchain should be parallel to the $x_m y_m$-plane.

For $R(RRR)_E$-type, there are following possible chains: $R(RRR)_E$, $R(RPR)_E$, $R(PRR)_E$, $R(RRP)_E$, $R(RPP)_E$, $R(PRP)_E$, $R(PPR)_E$. The additional assembly conditions for them are: (1) the axis of the first revolute joint should pass the RCM point; (2) the axis of $(RRR)_E$-subchain should be parallel to the $x_m y_m$-plane; (3) the adjacent axes of the two subchains should be set skewed. Fig. 6 shows two examples of $(RR)_S(RR)_E$-type and $R(RRR)_E$-type.

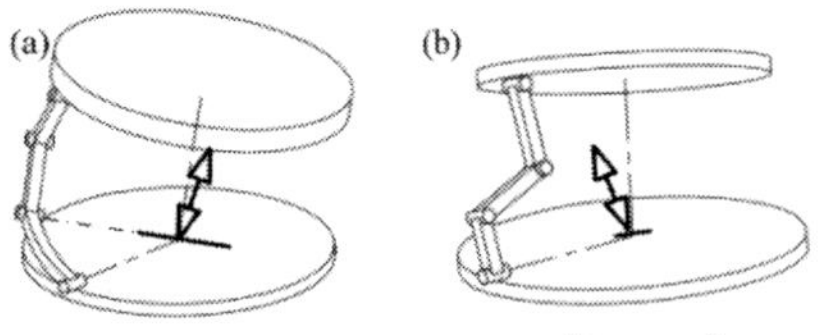

Figure 6. Two legs corresponding to $1-\hat{\$}^0-1-\hat{\$}^\infty$

5.4 Synthesis of legs corresponding to sub-wrench-system $2-\hat{\$}^0$

A non-redundant serial chain corresponding to sub-wrench-system $2-\hat{\$}^0$ also contains four joints. Since the geometrical conditions focus on the two zero pitch wrenches, there are few qualified serial chains. After verifying each one of the preferred general chains, we only obtain the $(RRR)_S P$ chain. Two examples are shown in Fig. 7. Each leg consists of three revolute joints intersecting at the RCM point and a prismatic joint perpendicular to the platform.

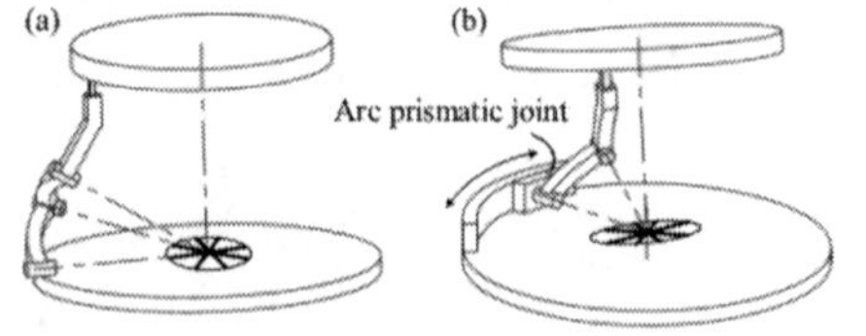

Figure 7. Two examples of $(RRR)_S P$-chain

5.5 Synthesis of legs corresponding to sub-wrench-system $1-\hat{\$}^0$

The conditions for sub-wrench-system $1-\hat{\$}^0$ are listed as follows: (1) the wrench should be parallel to the $x_m y_m$-plane, and (2) it should pass the RCM point. A non-redundant serial chain corresponding to sub-wrench-system $1-\hat{\$}^0$ contains five joints. Applying the preferred serial chains to the situation, we can obtain three following types which satisfy the geometric conditions for the sub-wrench-system: $(RRR)_S(RR)_E$-type, $(RR)_S(RRR)_E$-type and $\underline{R}(RRR)_E\underline{R}$-type.

For $(RRR)_S(RR)_E$-type, there are following possible chains: $(RRR)_S(RR)_E$, $(RRR)_S(PR)_E$, $(RRR)_S(RP)_E$, and $(RRR)_S(PP)_E$. For $(RR)_S(RRR)_E$-type, there are $(RR)_S(RRR)_E$, $(RR)_S(RRP)_E$, $(RR)_S(RPR)_E$, $(RR)_S(PRR)_E$, $(RR)_S(RPP)_E$, $(RR)_S(PRP)_E$ and $(RR)_S(PPR)_E$. The additional conditions for them are: (1) the intersecting point of $(R...R)_S$-subchain should be the RCM point; (2) the axis of $(R...R)_E$-type subchain should be parallel to the $x_m y_m$-plane; (3) the adjacent axes of the two subchains should be set skewed.

For $\underline{R}(RRR)_E\underline{R}$-type, there are following possible chains: $\underline{R}(RRR)_E\underline{R}$, $\underline{R}(RPR)_E\underline{R}$, $\underline{R}(RRP)_E\underline{R}$, $\underline{R}(PRR)_E\underline{R}$, $\underline{R}(PPR)_E\underline{R}$, $\underline{R}(RPP)_E\underline{R}$ and $\underline{R}(PRP)_E\underline{R}$. Fig. 8 shows four examples of these three types.

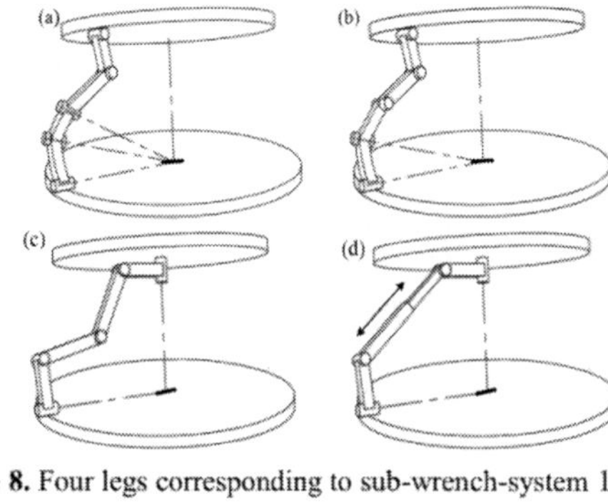

Figure 8. Four legs corresponding to sub-wrench-system $1-\hat{\$}^0$: (a) $(RRR)_S(RR)_E$-type, (b) $(RR)_S(RRR)_E$-type, (c) $\underline{R}(RRR)_E\underline{R}$ and (d) $\underline{R}(RPR)_E\underline{R}$.

5.6 Synthesis of legs corresponding to sub-wrench-system $1-\hat{\$}^\infty$

Condition (c3) brings in little constraint on the wrenches. Hence, there is no need to apply the preferred serial chain in this section. Two typical types are $(RRR)_E(RR)_E$-type and $(RR)_E(RRR)_E$-type [11]. For both types, the assembly conditions are listed as follows: (1) the adjacent axes of the two subchains should be set skewed, and (2) the axes of both subchains should not be perpendicular to the $x_m y_m$-plane.

6 Synthesis of 2R1T RCM mechanisms

On the basis of leg synthesis, several practical mechanisms with required motion pattern can be obtained. As mentioned before, only 3-legged parallel mechanisms are considered. For convenience of expression, these three legs are denoted by leg I, leg II and leg III. Using combinatorial method, the possible types of non-redundant 2R1T RCM mechanisms are listed in Table 1.

Table 1. The combination of three legs corresponding to the given sub-wrench-systems.

	Leg I	Leg II	Leg III
Type (i)	$2-\hat{\$}^0-1-\hat{\$}^\infty$	Ø	Ø
Type (ii)	$1-\hat{\$}^0-1-\hat{\$}^\infty$	$1-\hat{\$}^0$	Ø
Type (iii)	$2-\hat{\$}^0$	$1-\hat{\$}^\infty$	Ø
Type (iv)	$1-\hat{\$}^\infty$	$1-\hat{\$}^\infty$	$1-\hat{\$}^\infty$

The concrete mechanisms of these types can be constructed by simply assembling the corresponding legs, and some examples are shown in Table 2. It is easy to verify the motion continuity of these mechanisms.

Table 2. Examples of the four types and a practical prototype.

	Examples of 2R1T RCM mechanism
Type (i)	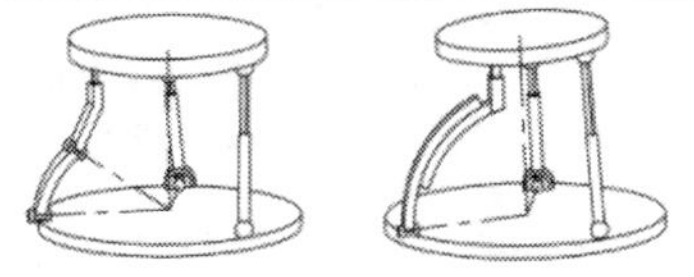
Type (ii)	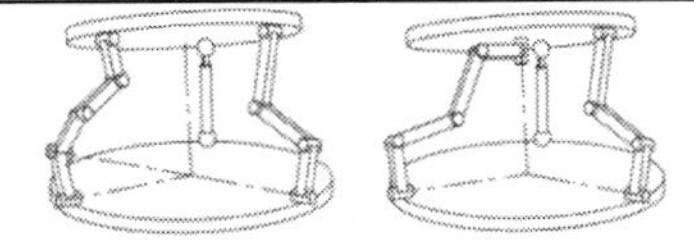
Type (iii)	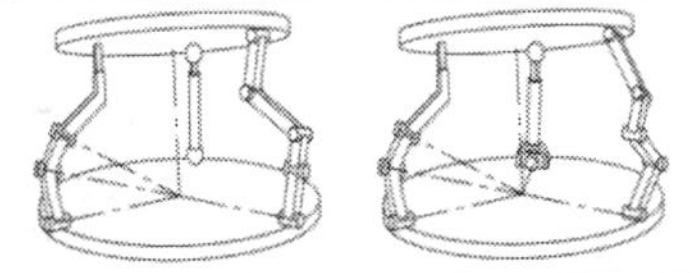
Type (iv)	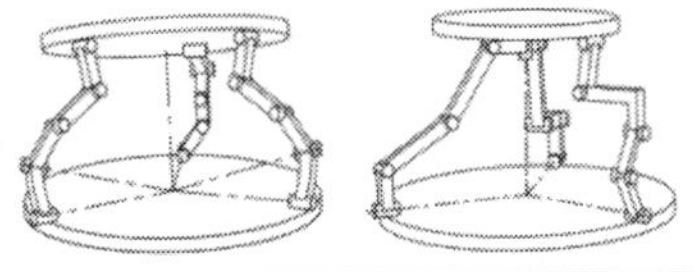
Prototype	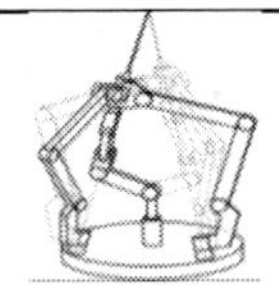

7 Conclusions

As an important kind of RCM mechanisms, the 2R1T RCM mechanisms are synthesized using a modified screw theory method. By following standard procedures, the twist system and several wrench systems of 2R1T RCM mechanisms are obtained. Then, this paper brings forward discussion of the synthesis difficulties, which contributes to the selection of the optimum wrench system. Afterwards, the wrench system is decomposed and synthesized respectively. During the processes, some preferred serial chains are proposed to meet the geometrical conditions, which can ensure the motion continuity of mechanisms to a great extent. Finally, several practical examples of 2R1T RCM mechanisms are constructed by assembling the corresponding legs. Some of them can be translated into practical prototypes.

References

1. A. Gijbels, E.B. Vander Poorten, P. Stalmans, H. Van Brussel, Design of a teleoperated robotic system for retinal surgery, *IEEE ICRA*, 2357-2363 (2014).
2. G.S. Guthart, J.K. Salisbury, The Intuitive telesurgery system: overview and application, *IEEE ICRA*, 618 - 621 (2000).
3. R.H. Taylor, J. Funda, B. Eldridge, S. Gomory, A telerobotic assistant for laparoscopic surgery. *IEEE Eng. Med. Biol. Mag.* **14**, 279-288 (1995).
4. R.H. Taylor, D. Stoianovici, Medical robotics in computer-integrated surgery. *IEEE Trans. on Rob. Autom* **19**, 765-781(2003).
5. L. Huang, Y. Yang, P Su, J Xiao, Type Synthesis of 1R1T Remote Center of Motion Mechanisms. *J. Mech. Eng* **13**, 131-136 (2015). (in Chinese)
6. T. Ueta, Y. Yamaguchi, Y. Shirakawa, T. Nakano, R. Ideta, Y. Noda, Robot-assisted vitreoretinal surgery: development of a prototype and feasibility studies in an animal model. *Ophthalmology* **116**, 1-2 (2009).
7. M.J.H. Lum, D.C.W. Friedman, H. King, The RAVEN: Design and Validation of a Telesurgery System. *Int. J. Robot. Res.* **28** (2009) 1183-1197.
8. G. Zong, X. Pei, J. Yu, S. Bi, Classification and type synthesis of 1-DOF remote center of motion mechanisms. *Mech. Mach. Theory*. **43**, 1585-1595 (2008).
9. J. Li, G. Zhang, A. Müller, S. Wang, A Family of Remote Center of Motion Mechanisms Based on Intersecting Motion Planes. *J. Mech. Design* **135** , 091009-1-10 (2013).
10. X. Kong, C.M. Gosselin, Type synthesis of 4-DOF SP-equivalent parallel manipulators: A virtual chain approach. *Mech. Mach. Theory* **41** 1306-1319 (2006).
11. Z. Huang, Q. Li, H. Ding, Theory of Parallel Mechanisms. *Mech. Mach. Science*. **6**, 653-4 (2013).
12. F. Xie, X.J. Liu, T. Li, Type Synthesis and Typical Application of 1T2R-Type Parallel Robotic Mechanisms. *Math. Probl. Eng.* **6**, 497-504 (2013).
13. N. Kumar, O. Piccin, B. Bayle, A task-based type synthesis of novel 2T2R parallel mechanisms. *Mech. Mach. Theory*. **77**, 59-72 (2014).
14 P. S. Rao, N. M. Rao Position Analysis of Spatial 3-RPS Parallel Manipulator. *Int. J. Mech. Eng. & Rob. Res.* **2**, 80-90 (2013).

The Effect of Foot Structure on Locomotion of a Small Biped Robot

Tinh Nguyen , Linh Tao and Hiroshi Hasegawa

Graduate School of Engineering and Science, Shibaura Institute of Technology, 307 Fukasaku, Minuma-ku, Saitama-shi, Saitama, 337-8570, Japan.

Abstract. This paper is a presentation of a work that consists of considering a novel foot structure for biped robot inspired by human foot. The specific objective is to develop a foot mechanism with human-like toes for a small biped robot. The chosen architecture to present the biped includes ten degrees of freedom (DoF) on ten articulations between eleven links. Our study considers the effect of varying foot structure on a walking process of the robot in simulation by ADAMS (MSC software, USA) through gait generation method. In toe mechanism, aiming to reduce the energy consumption, the passive joint was selected as the toe joint. The center of gravity (CoG) point trajectories of the robot with varying toe is compared with each other in normal motion on flat terrain to determine the most consistent toe mechanism. The result shows that the selected foot structure enables the robot to walk stably and naturally.

1 Introduction

Research on the biped robots constitutes one of the fascinating axes in robotics. Over the past several decades, several anthropomorphic robots have constructed. Some of them became well known even for non-specialists [1]-[5]. The prior target of researchers carried out in the field of biped robot attempts to solve the following problem: How to build a robot prototype able to walk as the humans are doing.

In the motion, one of the characteristics of the human walk is heel-contact and toe-off motions in steady walking. To implement adaptive walking, a foot is one of the most important regions of the human body in bipedal locomotion because it is a unique one that directly physically interacts with the environment. The human foot is the complicated structure which consists of many toes and joints. On the human walk, this structure not only makes ground reaction force smoothly change in the toe-off period but also equip foot with good adaption on uneven terrain.

From human foot inspiration, there have been some papers mentioned on the foot structure. For instance, Yu Ogura et al. have proposed the new foot mechanism by implement one passive joint for bending toe motion of Wabian-2R. However, in this study, the number of robot's DoF is reduced due to the predetermination is complemented by waist rolling motion [6]. Katsu Yamane et al. have investigated feet composed of curved surfaces at toe and heel and a flat section for a simple planar biped robot [7]. Shotaro Mamiya et al. have proposed a foot structure with point contact links to adapt to the ground surface with various geometry and hardness [8]. Sebastian Lohmeier et al. have designed a humanoid robot LOLA with actively driven toe joints [9]. Kenji Hashimoto et al. have developed a human-like foot mechanism mimicking the medial longitudinal arch to clarify the function of the foot arch structure [10]. However, above mentioned papers mainly focus on the humanoid robot whose dimension is similar to human's one. The human-size robots are very convenient for designing structure and integrating actuator on foot.

By the contrary, a small robot has difficulty in building foot structure caused by limited dimension. Wooseok Choi et al. have developed a foot structure composed of a new toe mechanism with variable stiffness implemented using a leaf spring and rubber balls for a child-size humanoid robot. This structure can provide functionality and adaptability to humanoids locomotion on uneven terrains [11]. However, the spring stiffness range is still small. It does not reach the range of toe stiffness variation which is similar to human toes. Krissana Nerakae and Hiroshi Hasegawa have presented the foot mechanism with the big toe and tiptoe for a 10 DoF small biped robot. Nevertheless, in this work, the trajectories of corresponded joints on both legs are generated by seven isolated gait functions which make a gait pattern generation become complicated. Besides, the robot cannot walk naturally in comparison to human motion [12].

This study proposes a novel foot structure with toes for a small biped robot to cultivate its walking behavior on flat ground in the toe-off period. That enables the robot to walk stably and naturally. In toe mechanism, considering a reduction in the energy consumption, thus, the passive joint using torsion spring was selected as the toe joint. It can be said that its walking style, in comparison to those of the other small biped robots, is more similar to that of a human.

The rest of this paper is organized in the following manner. Mechanical description of the robot is presented in Section 2. The principle of gait pattern generation is in Section 3. Section 4 shows the results of the development of the robot by dynamic simulation on ADAMS. Finally, Section 5 includes some brief conclusions and future works.

2 Experimental robot models

2.1 Overview of structural design

In this study, the proposed model is built based on the robot KHR-3HV of Kondo Kagaku Company which is the third generation of humanoid robots developed by this company. The robot KHV-3HV has the weight of 1.5kg, the height of 401.05mm and up to 22 DoF with 17 actual servos and 5 dummy servos. However, in the work, we concentrate on robot legs. Thus, upper body joints are fixed and lower body have 10 controlled joints for the legs as shown in Fig. 1.

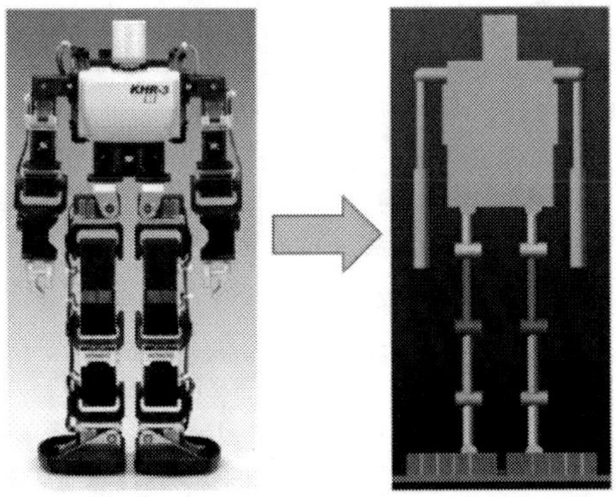

Figure 1. Real robot and proposed robot

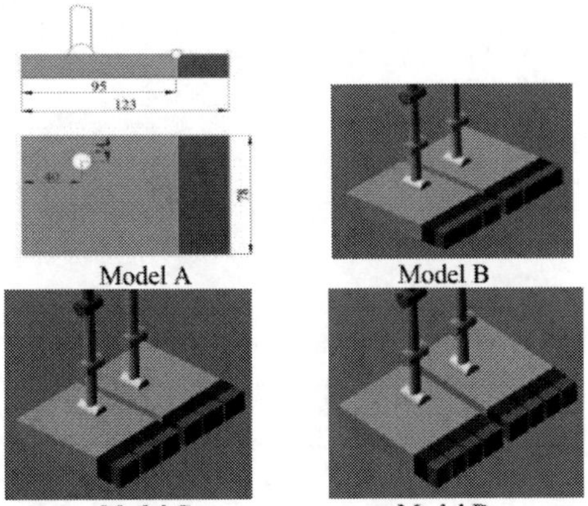

Model A Model B
Model C Model D

Model A: Foot mechanism with tiptoe
Model B: Foot mechanism with two toes
Model C: Foot mechanism with three toes
Model D: Foot mechanism with four toes

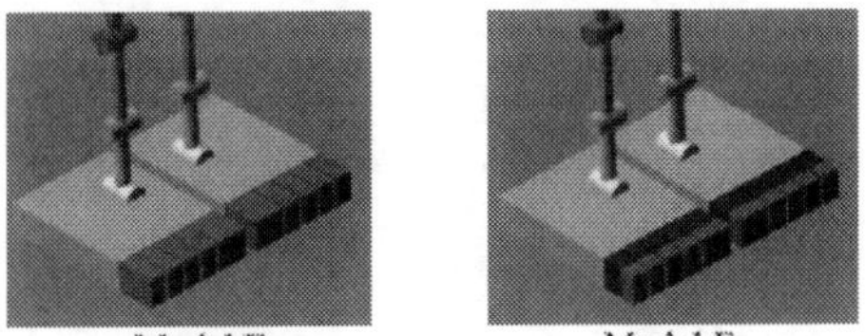

Model E Model F

Figure 2. Foot mechanism with toe
Model E: Foot mechanism with five rigid toes
Model F: Foot mechanism with five arthrogenous toes

2.2 Foot mechanism

During a stance phase of the human walk, support area continuously varies on the sole of the foot. At the end of toe-off, there is a phase switch from stance to swing. Toes have an important role in this process, it makes the phase switch smoother and more stable due to a decrease of the effect of the ground reaction force varied unexpectedly.

By this idea, we study the effect of the novel foot structure with toes on the locomotion of a small biped robot. The six toe mechanisms are designed for simulation experiments as depicted in Fig. 2.

In Fig. 2, model A shows the basic dimensions of the robot foot. In [13], Chockalingam et al. have proved that average ratio between the foot length and heel varies from 1.196 to 1.426. In this model, the length of heel and foot are 95mm and 123mm, respectively. Hence, the length of toes will be 28mm. The ankle joint position is determined based on the real robot. These dimensions are used to all remain models.

As proven in [12], the biped robot whose big toe width ratio per foot equals 0.28, has the longest walking distance when big toe length is fixed and this ratio is similar to the ratio of the human foot. Thus, the width of big toe and foot were designed of 22mm and 78mm. This ratio is applied to model B, C, D, E and F. Spring stiffness coefficients is respectively set of 0.52N.mm/deg, 0.26N.mm/deg to big toe joint and remain toe joints. For making the foot structure, two different types of material are used for heel and toes. Coefficients of static and kinetic friction are 0.5, 0.417 for heel and ground, 0.17, 0.155 for toes and ground, respectively.

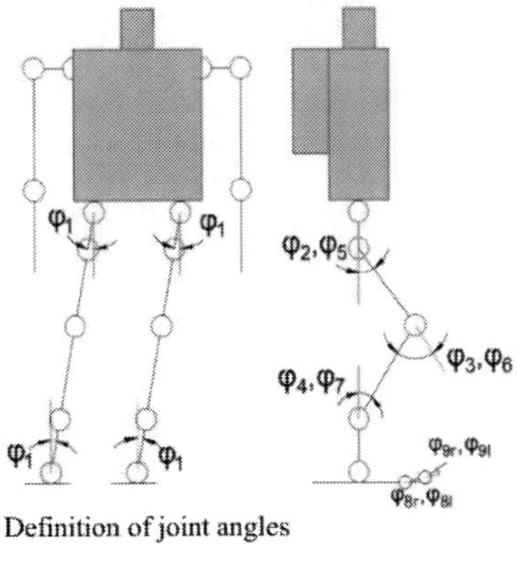

Figure 3. Definition of joint angles

3 Gait pattern generation

3.1 Definition of joint angle

The joint angles are defined as depicted in Fig. 3.

In Fig. 3, the hip, knee and ankle joints are active joints whose motion is supplied by the actuator. Toe joints designed by using torsion springs are passive joints. For model A and E, φ_{9r} and φ_{9l} are not used.

3.2 Gait function

Basing on the human walking pattern as depicted in [14], we assumed the robot control data was generated by the gait function as trigonometric function shown in Equation (1). By changing a, b, c, d coefficients, the gait functions will be created to allocate to each joint of the biped robot.

$$\varphi_i(t) = a_i + b_i.\cos(\omega t) + c_i.\sin(\omega t) + b_i.\cos(\omega t) \quad (1)$$

where a, b, c, d are coefficients, t is time, ω is angular velocity and i is index of joint.

Gait functions are assigned to all joints as shown in Equation (2-8).

$$\varphi_1 = \begin{cases} 0, \ t = 0 \text{ or } t = 3.6 \\ \pm 1.5, \ t = 0.3 \text{ and } t = 3.3 \\ \varphi_1(t), \ 0.3 < t < 3.3 \end{cases} \quad (2)$$

$$\varphi_2 = \begin{cases} 0, \ t \le 0.3 \text{ or } t \ge 3.6 \\ \varphi_2(t+0.6), \ 0.3 < t < 3.3 \\ 15, \ t = 3.3 \end{cases} \quad (3)$$

$$\varphi_3 = \begin{cases} 0, \ t \le 0.3 \text{ or } t \ge 3.6 \\ \varphi_3(t+0.6), \ 0.3 < t < 3.3 \\ 30, \ t = 3.3 \end{cases} \quad (4)$$

$$\varphi_4 = \begin{cases} 0, \ t \le 0.3 \text{ or } t \ge 3.6 \\ \varphi_4(t+0.6), \ 0.3 < t < 3.3 \\ 15, \ t = 3.3 \end{cases} \quad (5)$$

$$\varphi_5 = \begin{cases} 0, \ t = 0 \text{ or } t \ge 3.3 \\ 15, \ t = 0.3 \\ \varphi_2(t), \ 0.3 < t < 3.3 \end{cases} \quad (6)$$

$$\varphi_6 = \begin{cases} 0, \ t = 0 \text{ or } t \ge 3.3 \\ 30, \ t = 0.3 \\ \varphi_3(t), \ 0.3 < t < 3.3 \end{cases} \quad (7)$$

$$\varphi_7 = \begin{cases} 0, \ t = 0 \text{ or } t \ge 3.3 \\ 15, \ t = 0.3 \\ \varphi_4(t), \ 0.3 < t < 3.3 \end{cases} \quad (8)$$

In toe mechanism, due to considering a reduction in the energy consumption of the robot, the passive joint was selected as toe joint. Consequently, φ_{8r}, φ_{8l}, φ_{9r} and φ_{9l} have values in the range from 0° to 30°; their values depend on the robot geometric posture as well as impact force when the robot walks.

4 Simulation result

In our experiments, the robot motion is simulated in three cycles. One cycle was set up to 1.2 seconds. Thus, three cycles spent on 3.6 seconds. Next, 1.2 seconds was used for checking robot stability. In this simulation, one step took 0.02 second, so the total number of steps was 240 steps. The simulation result is shown in Table 1.

Where X_f, Z_f, R_f are respectively sided distance, walking distance and angle of rotation at the final position of the robot.

The trajectory of the CoG point is depicted as in Fig. 4.

Table 1. Simulation result

Model	Distance		Rotation
	X_f(mm)	Z_f(mm)	R_f(deg)
A	14.91	193.91	-5.867
B	-28.187	197.76	-9.557
C	-0.632	192.77	-2.795
D	9.350	195.31	1.309
E	0.384	195.19	2.998
F	-0.007	177.94	-0.091

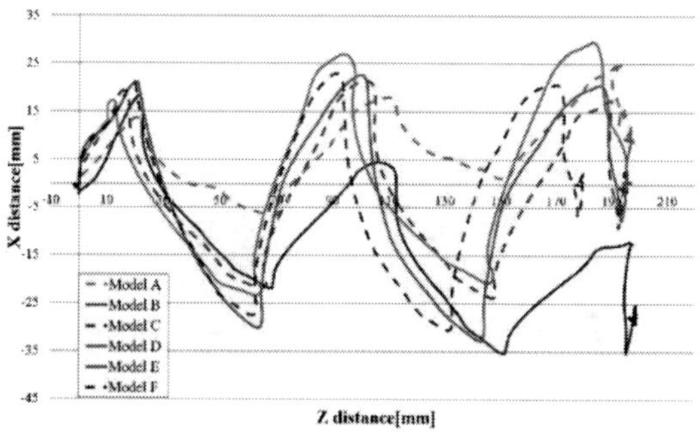

Figure 4. Robot's CoG point trajectory

In Table 1 and Fig. 4, all the trajectory of the robot's CoG point is approximately the waveform of the circular function which is similar to that of the human's CoG point. Maximum walking distance is 197.76mm; side distance is -28.187mm in the case of the model B. This side distance is quite large. Thus, consideration of straight walking and stability, model E with X_f, Y_f, R_f of 0.384mm, 195.19mm and 2.998° is the best model to do the research further.

The waveform of the gait functions allocated to all joints of the biped robot is shown in Fig. 5.

5 Conclusion

A novel foot structure with toes for the small biped robot is proposed in this paper. The locomotion of the robot is simulated on a flat plane by multi-body dynamics simulation software, ADAMS. In this work, we concentrate on the toe-off period of the biped robot in walk. As a result, all the models can walk stably with the bending of the toe which is similar to that of a human; model E obtains the best performance. Therefore, this model is selected for next research stages.

DOI: 10.1051/matecconf/20179508010

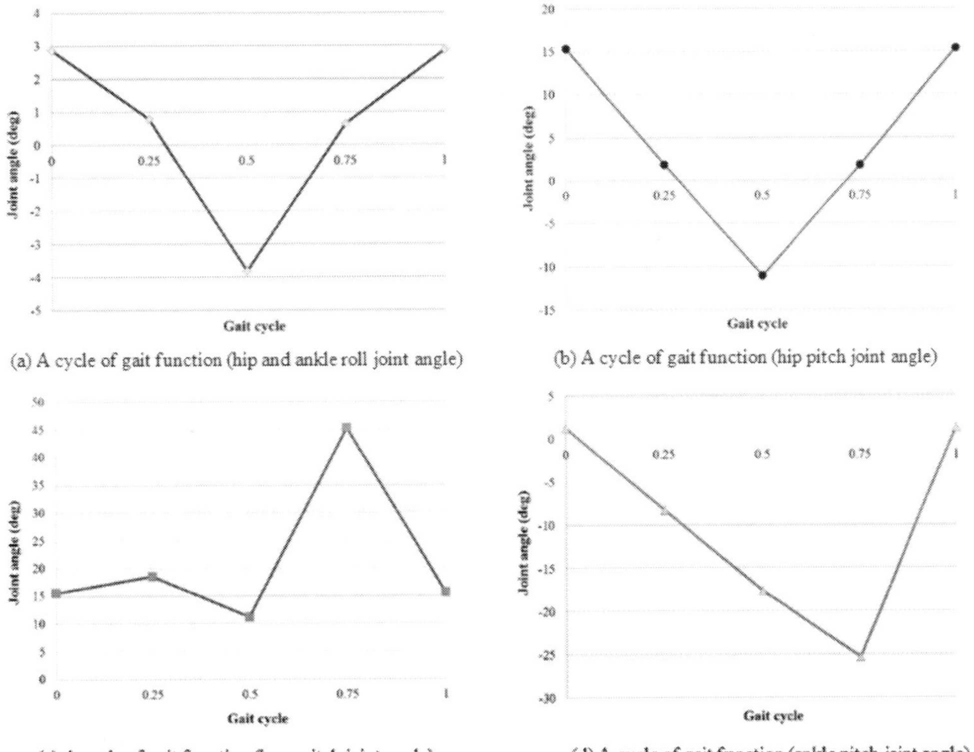

(a) A cycle of gait function (hip and ankle roll joint angle)

(b) A cycle of gait function (hip pitch joint angle)

(c) A cycle of gait function (knee pitch joint angle)

(d) A cycle of gait function (ankle pitch joint angle)

Figure 5. Waveform of the gait function

As for future work, we will try to consider the locomotion of the robot on rough ground, adjust spring stiffness to learn more. Besides, the foot structure of the model E will be also made and applied to the real robot.

References

1. Y. Sakagami, R. Watanabe, C. Aoyama, S. Matsunaga, N. Higaki, K. Fujimura, The intelligent ASIMO, System overview and integration. *Proc. of the IEEE/RSJ Int. Conf. on Intelligent Robots and Systems*, Switzerland, pp 2478-83 (2002)
2. S. Lohmeier, K. Loffler, M. Gienger, H. Ulbrich, F. Pfeiffler, Computer system and control of biped "Johnnie". *Proc. of IEEE Int. Conf. on Robotics and Automation*, New Orleans, LA, **4** pp 4222-27 (2004)
3. Y. Ogura, H. Aikawa, K. Shimomura, A. Morishima, H. O. Lim, A. Takanishi, Development of a new humanoid robot WABIAN-2. *Proc. of IEEE Int. Conf. on Robotics and Automation*, pp. 830-835, Orlando, Florida (2006)
4. K. Kaneko, F. Kanehiro, S. Kajita, K. Yokoyama, K. Akachi, T. Kawasaki, S. Ota, T. Isozumi, Design of Prototype Humanoid Robotics Platform for HRP, *Proc. of the IEEE/RSJ Int. Conf. on Intelligent Robots and Systems*, pp. 2431–2436 (2002)
5. T. Ishida, Development of a Small Biped Entertainment Robot QRIO. *Proc. of The Fourth Symposium Micro-Nano mechatronics for Information - Based Society*, pp.23-28 (2004)
6. Y. Ogura, H. Aikawa, K. Shimomura, A. Morishima, H. O. Lim, A. Takanishi, Development of a new humanoid robot WABIAN-2. *Proc. of IEEE Int. Conf. on Robotics and Automation*, pp. 830-835, Orlando, Florida (2006)
7. K. Yamane, L. Trutoiu, Effect of Foot Shape on Locomotion of Active Biped Robots, a study on effect of two-arch structure of foot for biped robots, *Proc. of 9th IEEE-RAS Int. Conf. on Humanoid Robots*, pp.2230-2236, Paris, France (2009)
8. S. Mamiya, S. Sano, N. Uchiyama, A New Foot Structure with Springs and Flat Soles for Biped Robot Walking, *Proc. of IEEE 13 international workshop on advanced motion control*, pp. 144-148, Yokohama, Japan (2014)
9. S. Lohmeier, T. Buschmann, H. Ulbrich, F. Pfeiffer, Modular joint design for performance enhanced humanoid robot LOLA, *Proc of IEEE international conference on robotics and automation*, pp. 88-93, Florida, USA (2006)
10. K. Hashimoto, Y. Takezaki, H. Motohashi, T. Otani, T. Kishi, H-o. Lim, A. Takanishi, Biped Walking Stabilization Biped on Gait Analysis, *Proc. of IEEE Int. Conf. on Robotics and Automation*, pp. 154-159, Minnesota, USA (2012)
11. W. Choi, G. A. Medrano-Cerda, D. G. Caldwell, N. G. Tsagarakis, Design of Variable Compliant Humanoid Foot with a New Toe Mechanism, *Proc. of IEEE Int. conf. on Robotics and Automation*, pp. 642-647, Stockholm, Sweden (2016)

 DOI: 10.1051/matecconf/20179508010

12. K. Nerakae, H. Hasegawa, Big toe sizing design of small biped robot by using gait generation method. Applied Mechanics and Materials, **541-542** pp. 1079-86 (2014)
13. N. Chockalingam, R. Ashford, Selected foot length ratios in a non-pathological sample, Revista Internacional de Ciencias Podológicas, **1** No. 2 pp. 25-30 (2007)
14. M. W. Whittle, *An Introduction to Gait Analysis*. 4th ed., Oxford, p 59 (2007)

The Motion Path Study of Measuring Robot Based on Variable Universe Fuzzy Control

Ma Guoqing , Yu Zhenglin, Cao Guohua, Zhang Ruoyan and Zheng Yanbin

School of Mechatronical Engineering, Changchun University of Science and Technology, Changchun 130022

Abstract. For the problem of measuring robot requires a higher positioning, firstly learning about the error overview of the system, analysised the influence of attitude, speed and other factors on systematic errors. Then collected and analyzed the systematic error curve in the track to complete the planning process. The last adding fuzzy control in both cases, by comparing with the original system, can found that the method based on fuzzy control system can significantly reduce the error during the motion.

1 Introduction

Robots are complex mechatronic system and easily influenced by external and self structure [1]. Modern control technology for solving robot self adaptation, self-learning ability provides a powerful help. At present, the common robot control technology includes adaptive control, robust control, optimal control, neural network control and sliding mode control.

Chang-Min Ko take a heavy industrial robot as the research object, PI control and fuzzy PD control are studied respectively. The result proves that the fuzzy PD control is more effective [2]. Yu mingliang designed the fuzzy PD control and separate integral control for the mobile robot, simulation results show that the tracking error of the robot becomes very small [3]. The fuzzy control was applied to a three freedom parallel robot by Jingjun Zhang [4]. To prove the control precision of the robot is improved by using fuzzy control, ADAMS and MATLAB joint simulation were used; Gao yanfeng proved that the predictive fuzzy control is more advantage than traditional fuzzy controller by arc seam tracking experiments [5].

2 Establish a model of measuring robot

Robot kinematics study the relationship between the the each joint variable and the end position, in order to describe movement or rotation relationship between the joint and the end. Denavit and Hartenberg put forward DH parameter method. This section establishing the robot joint coordinates (as shown in Fig. 1) based on DH parameter method. Its corresponding DH parameters as shown in Table 1.

θ is Joint angle, which represents angle between the the x_{i-1} and x_i axis about the z_{i-1} axis;

d is link offset, which represents the distance from the origin of frame $i-1$ to the x_i axis along the z_{i-1} axis;

α is link twist, which represents the angle from z_{i-1} axis to the z_i axis about the x_i axis;

a is link length, which represents the distance from the the z_{i-1} axis and the z_{i-1} axis along the x_i axis, for intersecting axis is parallel to $z_{i-1} \times z_i$;

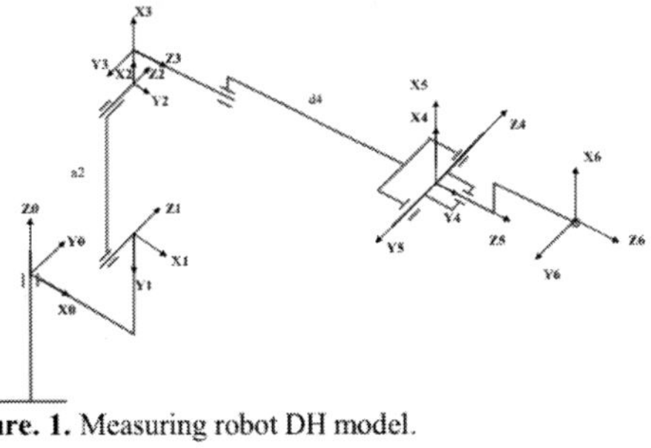

Figure. 1. Measuring robot DH model.

Table 1. DH parameters.

Axis	α (°)	a (m)	θ (°)	d (mm)
1	-90	0.150	0	0.173
2	0	0.600	-90	0.001
3	-90	0.125	0	0.000
4	90	0.000	0	0.640
5	-90	0.000	0	0.000
6	0	0.090	0	0.266

3 PID parameters optimization based on variable universe fuzzy theory

For a low stiffness series system such as a measurement robot, with a PID parameters cannot meet the control precision of the system obviously. Controller can adjust PID parameters according to the system overshoot and time, steady-state error, and so on, in order to achieve the purpose of optimal control. The detailed process is as follows: the object of fuzzy control is proportional gain K_p, integral gain K_i and differential gain K_d .For the velocity K_{vff} and acceleration feed-forward coefficient K_{aff} .After initial position on the robot manually setting will no longer changed.

3.1 The design of fuzzy controller

3.1.1 Language and language variable value

Fuzzy language is used to describe an object by fuzzy control, fuzzy control objects has 5. Set the language of error variable is E , linguistic variable error rate of change of E_c, the proportional gain correction amount of linguistic variables is ΔK_p , integral gain correction amount of linguistic variables is ΔK_i, differential gain correction amount of linguistic variables is ΔK_d . They are the domain $\{-3,-2,-1,0,1,2,3\}$ the corresponding language is {NB NM NS ZO PS PM PB}, expressed great negative, negative, small medium large zero, forward small medium, large forward great.

3.1.2 Language membership function values

Membership function of fuzzy control is mainly used for object describe the extent of the value, the language is related to the actual situation of this system and the objects understanding and judgment of operator. Membership function in compliance with objective reality as the standard,the membership function have many form, commonly used trigonometric functions, Gaussian function and so on, using the trigonometric functions in this project as shown in Fig. 2, the membership function of the specific value show in Table 2:

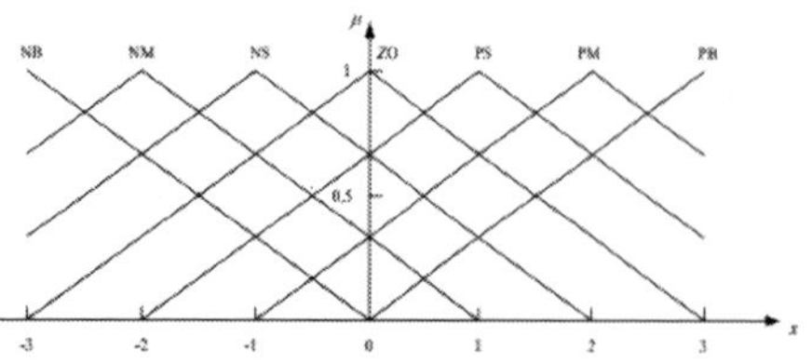

Figure. 2. Triangular membership function

Table 2. Membership function value

input / μ_E / E	-3	-2	-1	0	1	2	3
NB	1	0.65	0.3	0	0	0	0
NM	0.7	1	0.65	0.35	0	0	0
NS	0.4	0.7	1	0.65	0.35	0	0
ZO	0	0.3	0.65	1	0.65	0.3	0
PS	0	0	0.35	0.65	1	0.7	0.4
PM	0	0	0	0.35	0.65	1	0.7
PB	0	0	0	0	0.3	0.65	1

3.1.3 Language and language variable value

Fuzzy rules is composed of a series of fuzzy conditional statement, each statement on behalf of a relationship, such as when the error and error change rate is great, The output of the ΔK_p negative is great, This statement can be represented as IF E = PB and E_c = PB THEN=NB. Fuzzy rules to a large extent affected the effect of fuzzy control. Here according to the experience of the engineering practice, set up ΔK_p ΔK_i and ΔK_d , fuzzy rule shown in Table 3.

Table 3. Fuzzy rule table of ΔK_p

Ec / E	NB	NM	NS	ZO	PS	PM	PB
NB	PB	PB	PM	PM	PS	ZO	ZO
NM	PB	PB	PM	PS	PS	ZO	NS
NS	PM	PM	PM	PS	ZO	NS	NS
ZO	PM	PM	PS	ZO	NS	NM	NM
PS	PS	PS	ZO	NS	NS	NM	NM
PM	PS	ZO	NS	NM	NM	NM	NB
PB	ZO	ZO	NM	NM	NM	NB	NB

3.1.4 Defuzzification

The result is a collection of fuzzy reasoning, the process of transforming the fuzzy set to the output is called fuzzy solution, fuzzy methods include weighted average method and the maximum membership degree method.Maximum membership degree method, just as its name implies is to maximum membership degree in fuzzy set value corresponding linguistic variables as output.Set the output fields to A, then the output value of the maximum membership degree method is:

$$U=\{e^* \mid \mu(e^*)>\mu(e) \quad e,e^* \in A\} \tag{1}$$

In the actual situation of Chinese style e^* may have more than one, then you can take the average or maximum minimum value, this article USES the method of minimum maximum, the description is as follows:

$$U=Min\{e^* \mid \mu(e^*)>\mu(e) \quad e,e^* \in A\} \tag{2}$$

3.1.5 Variable universe factor

In fuzzy control, as the error decreases, error corresponding language variable values can also reduce accordingly. This will affect the precision of the fuzzy control, but the introduction of the theory of domain factor can contribute to real-time expansion theory domain.Improve the fine degree of control. In the literature [6] puts forward a new theory of variable domain factor,not only few parameters and is also small amount of calculation, front design, with reference to the literatur [7] to be considered at the same time , designed variable domain factors as follows:

$$\tau(x)=3^{-\lambda}|x|^{\lambda}+\varepsilon \quad (0<\lambda<1) \tag{3}$$

in the type: λ - scaling factor parameters, this article take 0.5 ε - the smallest positive number, this article take 0.001; Literature pointed out that the theory of variable domain factor in theory domain $U=[3\ \ 3]$ need to meet the dual character, avoid zero, monotonicity, coordination and normality. For type (3) the variable domain factor to fully meet the above requirements.

4 The fuzzy optimization experiments of measuring robot path

4.1 The design of fuzzy controller

In Table 4 for the experiment of four posture of the robot, under the initial posture has completed the system manual adjustment of PID parameters, take 1 shaft, the step response as shown in Fig. 3 (a). Now by robot step response under different attitude to observe attitude of the influence of system error,this article selects three kinds of typical gesture,concrete numerical value shown in table 6.Tests found only 1 shaft has obvious change under the three attitude dynamic performance.

Fig. 3(b)/3(c)/3(d) in addition to position 3, the remaining 1 shaft dynamic performance under two kinds of gestures are getting worse, step response show obvious fluctuation.Contrast three kinds of attitude discovery, all dynamic performance with poor posture.

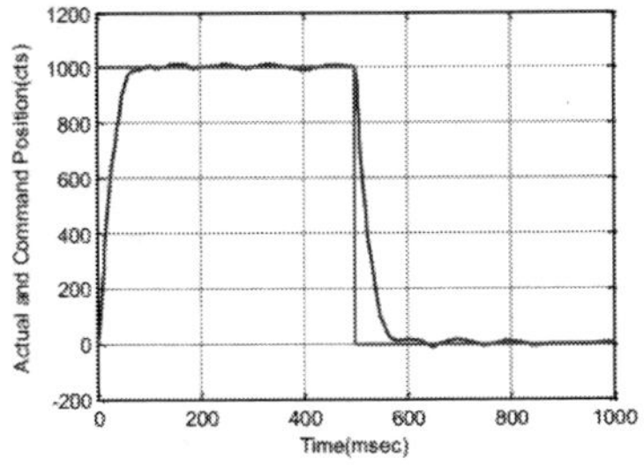

(a) step response of initial posture

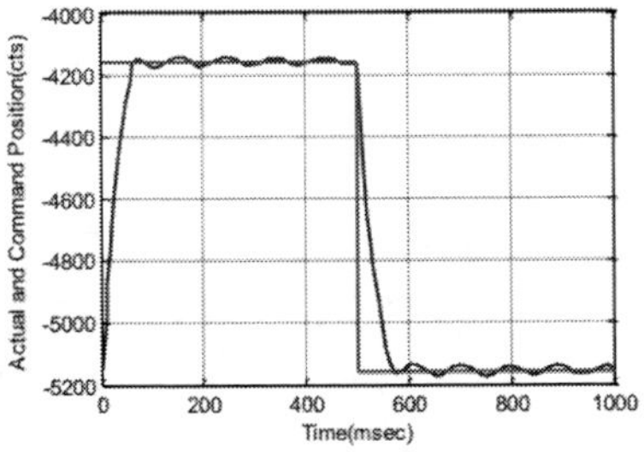

(b) step response of pose1

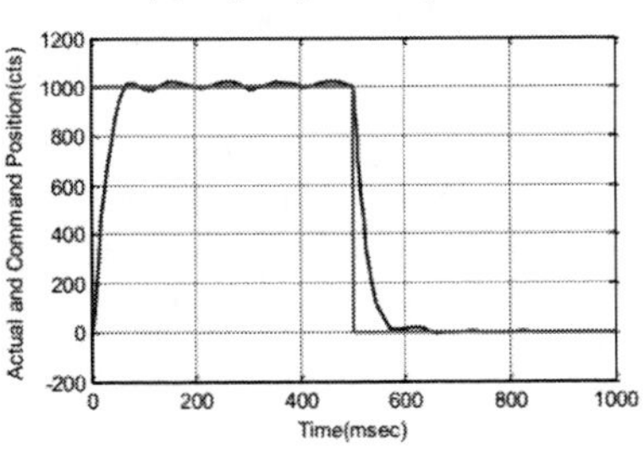

(c) step response of pose2

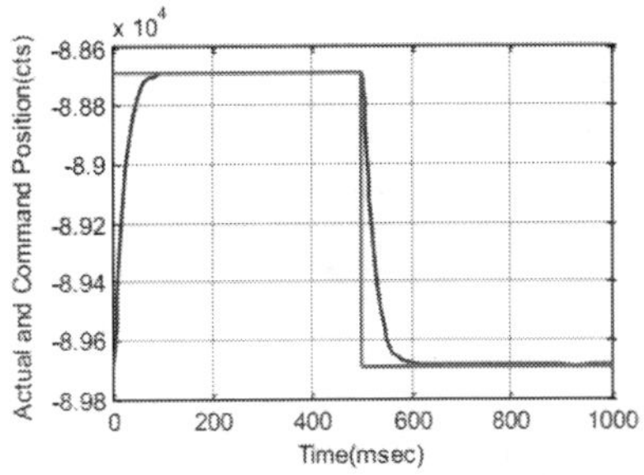

(d) step response of pose3

Figure. 3. The step response of 1 axis in different posture

Table 4. Pose data of robot(uint (°)).

pose	A1	A2	A3	A4	A5	A6
pose1	-1.83	18.95	-12.86	-18.95	-14.40	-15.32
pose2	0	26.68	20.57	18.95	24.00	38.30
pose3	-31.81	-18.95	2.57	-12.63	9.60	7.66

4.2. The change of system error in the motion

From the previous section, the big arm forward 1 shaft dynamic performance is poorer, so when planning robot trajectory. Ensure the protraction of the big arm as far as possible. This paper shows a way to teach the planning of a straight line (two points determine a straight line) and a circular arc (three points to determine a circular arc). Concrete numerical value shown in table 5 and table 6.The speed of the robot is 90 mm/s and 70 mm/s in the finish line and arc trajectory.

Table 5. Trajectory data of line(uint (°))

Axis / Point	A1	A2	A3	A4	A5	A6
origin	0	0	0	0	0	0
Start	-43.24	26.22	15.21	-31.01	-23.79	42.27
end	-57.73	55.91	33.20	19.48	-39.36	-42.28

Table 6. Trajectory data of circle(uint (°))

Axis / Point	A1	A2	A3	A4	A5	A6
origin	0	0	0	0	0	0
Apoint	0.05	34.12	27.83	0.01	-48.67	0.02
Bpoint	-20.50	25.56	14.09	16.15	-46.63	-28.32
Cpoint	-11.46	-3.81	-20.97	10.52	-38.89	-17.66

Fig. 4 and Fig. 5 is spindle motor error variation of 1 shaft in the motion of robot, can see from the diagram, whether a straight line trajectory or circular trajectory, and can match with velocity changes.In the process of robot movement, the following error is the dominance in error sources,and the error is relatively large. Linear and circular arc in the process of movement error range respectively is [285.83, 2443.26] and [226.42, 361.73], unit is pulse.

Now as an example, analysis the change of the error in a straight line motion, for the convenience of description. Set 1 shaft motor corresponding to the origin, the position of the starting point and end point of A, B, C respectively. As shown in Fig. 5, when the motor 1 received instructions from A to B, motor acceleration

DOI: 10.1051/matecconf/20179508011

instantaneous added to a larger values, making speed up in a short time. The resulting is follow error also rising sharply, when the speed reaches a certain value when the basic remain unchanged.At this point, the following error is similar to the horizontal line, will soon arrived at point B ,speed quickly dropped to zero speed error caused by suddenly disappeared. 1 is the above process in Fig. 7, 2 is 1 shaft spindle motor curves in the process of the robot in a straight line (from B to C), the motor is reverse. The speed is negative. There for when the error is negative, due to the robot slower (90 mm/s). At this time of the error value relative to the previous inhibits PTP (A to B) are smaller. The formation of 3 and 1 are similar, just the distance of the robot walk further， speed is relatively large, so the error value also rose slightly.

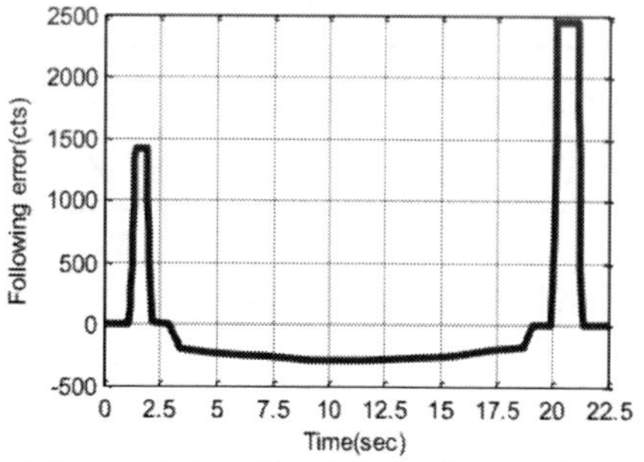

Figure 4. Error variation of 1 axis during linear motion

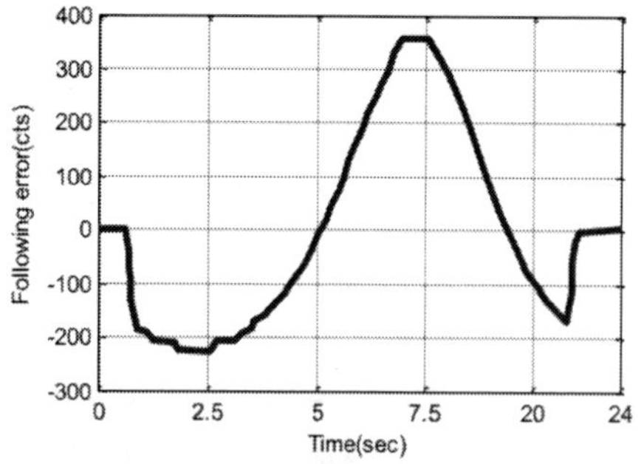

Figure 5. Error variation of 1 axis during arc motion

Through the above analysis that can draw the following conclusion, the following error in the process of movement is the main error; he greater the velocity and acceleration, the greater the error also; the plus or minus of error depend on the direction of speed.

4.3. Experiment and analysis

To make up for the defects, this article into the fuzzy adaptive control. Through the optimization of PID parameters to further reduce the error.

Consider safety factors, in practical applications， the PID parameters, set the upper and lower limits， effect of the fuzzy control is shown in Fig. 6 and Fig. 7, you can see that the feedforward, fuzzy control effect is more obvious.

In linear motion, for example, when the error suddenly rises, fuzzy control quickly down the overshoot (1) in Fig. 6, Once to set the upper and lower limit of PID parameters, PID parameters will not change, the error of line approximation (2 in Fig. 6). After movement has stopped, the speed of error caused by rapidly disappearing.At this time of the error is mainly composed of noise, friction factors, such as its value in the range of 3 cts change. Due to the change range is relatively small.Approximation error line is a horizontal line (3) in Fig. 7.

Here it should be pointed out that, the robot in the movement process and motion stops, the main factors causing error are not the same. So use movement at the end of the optimization of the PID parameters to adjust the system error of resting, the effect is not very ideal.This topic research the repositioning precision of the robot in between 0.06mm, the PID value of manual setting in the system when the speed suddenly 0 means that the error of the effect is satisfying, so when the velocity is zero, PID value is given the system initialization.

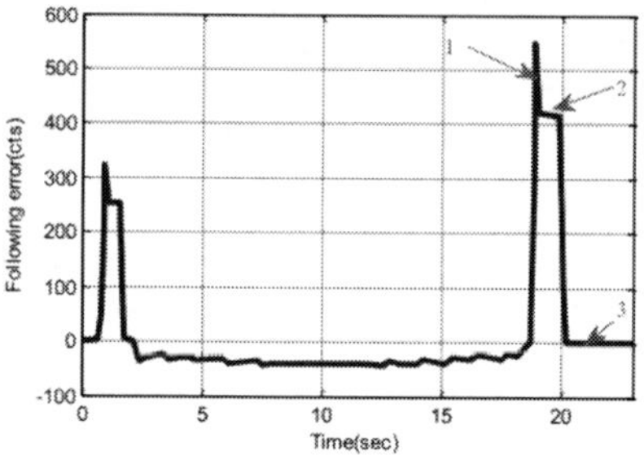

Figure 6. Linear motion error

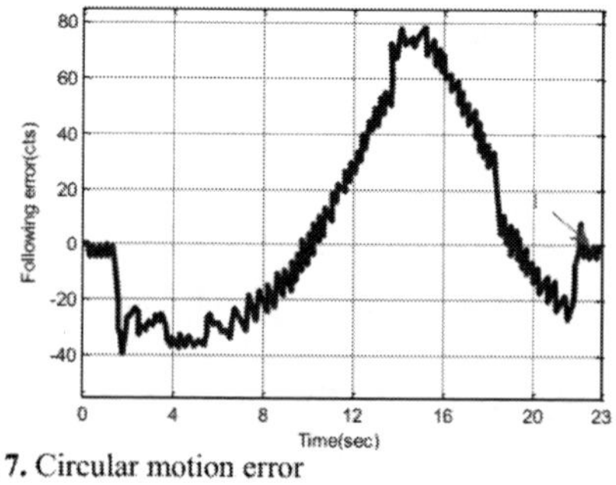

Figure 7. Circular motion error

Mentioned earlier, the PID value is set the upper and lower, many times experiments show that the PID adjustable range is wide, error drop, the more obvious. But if the PID adjustment range is too wide, PID values has far deviated from initial value when speed is zero, if once the speed is zero at this time, PID suddenly give initial value, when system occur large overshoot and oscillation, it will be stabilized at about 20cts even after the shock elimination of error value. This is deadly, because it will greatly influence the positioning precision of the system repeated, so the PID upper and lower's setting needs to be appropriate.After many experiments, can found,when the quantitative factor was set to 0.1, the scaling factor is set to ,change PID parameters in between, and the tracking accuracy and repeat positioning accuracy are in good level (see Table 9 and Table 10).As shown in figure1,because suddenly initialise cause slight overshoot, the super tone value for 9 cts, but it can be seen that overshoot is almost just a moment, then the system back to the ideal level (3 cts).Visible is reasonable setting in front of the upper and lower limits. Table 9 and Table 10 is the optimization effect compared with the before.

Table 7. Set of quantization and scaling factor

factor	E	E_c	ΔK_p	ΔK_i	ΔK_d
vale	10	10	1	1	1

Table 8. PID parameter range of motor 1

parameters	K_p	K_d	K_i
initial	290	380	410
range	[220,360]	[310,450]	[340,480]

Table 9. Effect comparison before and after optimization (Line)

scheme / max error(cts) / process	primeval	fuzzy control
origin->Start	1426.22	254.19
Start->end	-286.18	-32.22
end->Start	2443.26	418.27
Speed of 0	2.99	2.96

Table 10. Effect comparison before and after optimization (circle)

scheme / max error(cts) / process	primeval	fuzzy control
forward direction	362.33	77.62
negative direction	-226.64	-37.36
Speed of 0	2.96	2.96

5 Summary

Analysised the influence of attitude, speed and other factors on systematic errors, planed the two trajectories (lines and circles), collected and analyzed the systematic error curve in the track to complete the planning process. added fuzzy control in both cases, by comparing with the original system, can found that the method based on fuzzy control system can significantly reduce the error during the motion path. This provide a basis for conduct robot motion path planning and optimization based on fuzzy adaptive control in the future

References

1. Matveev A S, Hoy M, Katupitiya J, et al. Nonlinear sliding mode control of an unmanned agricultural tractor in the presence of sliding and control saturation. Robotics and Autonomous Systems, **61**(9): 973-987, (2013).
2. Chang-Min Ko, Seung-Kyu Park, Tae-SungYoon, et al. Comparative Study of Fuzzy PD Control and PI Control for Heavy Duty Robot. International Conference on Industrial Technology, (2009).
3. Yuming Liang, Lihong Xu, Ruihua Wei, et al. Adaptive Fuzzy Control for Trajectory Tracking of Mobile Robot. International Conference onIntelligent Robots and Systems, Taipei, Taiwan: 4755-4760, (2010).
4. Jingjun Zhang, Chaoyang Lian, Ruizhen Gao, et al. 3-Degree-of-freedom Parallel Robot Control Based Fuzzy Theory. International Conference on Intelligent Human-Machine Systems and Cybernetics: 221-224, (2010).
5. Gao yanfeng, Zhang hua, Mao zhiwei, et al.Predictive Fuzzy Control for a Mobile Welding RobotSeam Tracking. World Congress on Intelligent Control and Automation: 2271-2276, (2008).
6. Cheng Shao, Dong Xiwen, Wang Xiaofang, etc. Variable Universe Fuzzy Controller Selection Factor telescopic. Information and Control, **39**(5): 536-541, (2010).
7. Li Hongxing variable universe adaptive fuzzy controller. Science in China: Series E., **29**(1): 32-42, (1999).

DOI: 10.1051/matecconf/20179508012

Reduced Attitude Control of a Robotic Underwater Vehicle

Lukáš Bláha

Department of Cybernetics, University of West Bohemia, Univerzitní 8, 30614, Pilsen, Czech Republic

Abstract. This paper deals with stabilization and reduced attitude control of a robotic underwater vehicle. The vehicle is assumed to be able to perform a full stable rotations around all axes in underwater space, that is why the standard bottom-heavy structure is not used. The system preferably uses a vectored-thrust arrangement and is built as an over-actuated system, which enables to gain a better robustness and guarantees a stable controlled motion even if some thruster suddenly stop working. Because the heading angle cannot be measured, the reduced attitude control strategy is designed and the stability of reduced state of the system is proved using perturbation method.

1 Introduction

In the last few decades, the research of small underwater vehicles has been in the focus of the robotic and offshore engineering community. Especially in recent years the underwater unmanned vehicles have received a new attention that comes from advanced technologies and growing industry [1]. The vehicles for visual and ultrasonic inspection are specific part of that research [2], which very differ from traditional remotely operated vehicle (ROV) [3]. Compared to traditional submarines, they are differently shaped and equipped [4]. These properties have direct connection to accurate motion control. Due to inspection purposes, they has to be able to perform a full attitude maneuvers and therefore the standard auto-stable bottom-heavy structure [5] is not used. On the contrary, full attitude control is suitable for this kind of inspection systems. Research of rigid-body attitude control is therefore motivated not only from space or flight applications.

An accurate dynamic model is a crucial part of designing a stable, controllable and autonomous robotic system. However, the modeling of submerged system dynamics is full of nonlinearities and uncertainties [6]. Moreover the hydrodynamic effects are often coupled between individual degrees of freedom, and therefore it seems very natural to design a submerged robotic vehicle as symmetric as possible and use an over-actuated control structure, see figure 1. This ensures that the model parameters appear mainly as a diagonal terms of related matrices and the system can be more robust to parametric uncertainties.

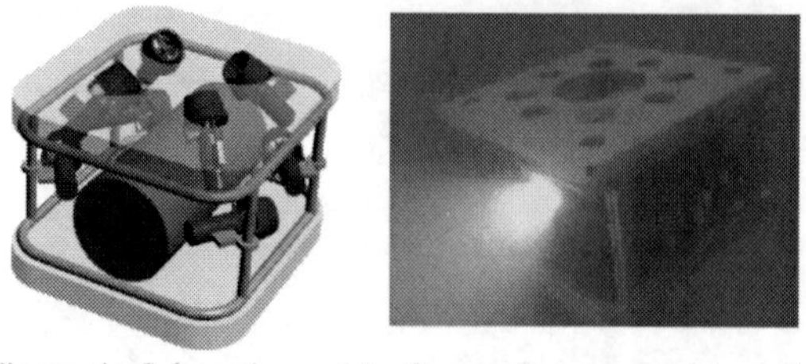

Figure 1: Schematic model of an underwater vehicle with vectored-thrust arrangement and corresponding real prototype.

This paper deals with modeling and attitude control of a robotic underwater vehicle. The key part of modeling is to get a qualitatively good model, respecting all the major aspects of kinematics and dynamics, important to design of a stable attitude control. The identification techniques of model parameters are detailed described in [7]. Attitude control is based on nonlinear controller with PID structure. It is shown using the perturbation method [8] that for some unknown measurement, the stability of reduced system is ensured.

The paper is organized as follows. Section 2 provides some notations and tools that are used to derive rigid body rotations. Then the vehicle dynamic model is derived and some useful simplifications are described. The benefits of over-actuated vehicle are presented. The parametric model is then used for design of a stable motion controller for a real prototype. Section 4 and 5 deals with attitude control and reduced attitude control of given system with emphasis on the stability and robustness in the entire or reduced working range. The stability for given controller is proved using perturbation method and Lyapunov indirect method. Last section contains some concluding remarks and acknowledgments.

2 Rigid body kinematics

This section defines some notation and gives some mathematical background used in the following sections.

Note that the time dependence is left out for compactness through all paper.

2.1 SO(3), matrix of rotation, quaternions

The attitude of rigid body can be represented by rotation matrix $R \in SO(3)$ representing a rotation from body fixed C-frame to inertial O-frame and fulfilling the properties of special orthogonal group

$$SO(3)=\{R \in \mathbb{R}^{3\times3} : R^T R = I,\ det\, R = 1\}. \qquad (1)$$

Euler angles, together with any standard set of intrinsic rotations, can be used for specification of rotation matrix $R(\boldsymbol{\theta})$, where the vector of Euler angles is defined as

$$\boldsymbol{\theta}=[\phi, \theta, \psi]^T. \qquad (2)$$

This specific sequence of 3-2-1 intrinsic rotation is chosen according to limits on measurement of Euler angles, which are then used for a reduced attitude controller, as detailed in Section 4 . Consequently the rotation matrix is as follows

$$R(\boldsymbol{\theta})=\begin{bmatrix} c\theta c\psi & s\phi s\theta c\psi - c\phi s\psi & c\phi s\theta s\psi + s\phi s\psi \\ c\theta s\psi & s\phi s\theta s\psi + c\phi c\psi & c\phi s\theta s\psi - s\phi c\psi \\ -s\theta & s\phi c\theta & c\phi c\theta \end{bmatrix}$$

where $s\cdot=\sin(\cdot)$, $c\cdot=\cos(\cdot)$, $t\cdot=\tan(\cdot)$.

The body angular velocity vector $\boldsymbol{\omega}=[\omega_1, \omega_2, \omega_3]^T$ and Euler rate vector $[\dot{\phi}, \dot{\theta}, \dot{\psi}]^T$ are related through a transformation $\dot{\boldsymbol{\theta}}=S(\boldsymbol{\theta})\boldsymbol{\omega}$, expanding as

$$\begin{bmatrix} \dot{\phi} \\ \dot{\theta} \\ \dot{\psi} \end{bmatrix} = \begin{bmatrix} 1 & s\phi t\theta & c\phi t\theta \\ 0 & c\phi & -s\phi \\ 0 & s\phi/c\theta & c\phi/c\theta \end{bmatrix} \begin{bmatrix} \omega_1 \\ \omega_2 \\ \omega_3 \end{bmatrix},\ c\theta \neq 0. \qquad (3)$$

The transformation is undefined for a pitch angle $\theta=\pm\pi/2$. This is known as a gimbal lock. Note that the matrix $S(\boldsymbol{\theta}) \notin SO(3)$ is not orthogonal.

Other representation of rigid body rotation, such as quaternions, Rodrigues parameters, axis-angle or other parametrization of *R*, are also possible. Their advantages and drawbacks are described for example in [9], [10].

3 Vehicle dynamics

The underwater vehicle is a highly nonlinear complex system. The standard general form of mathematical model of submerged rigid body is defined using Kirchhoff equations with external forces representing the thrust of actuators and the hydrodynamic drags

$$\begin{aligned} & M\dot{v} - D\dot{\omega} - M v \times \omega + D\omega \times \omega = \sum_{j=1}^{8} T_j + R^T f_g + f_h \\ & J\dot{\omega} + D\dot{v} + \omega \times J\omega + \omega \times D v + v \times M v + D\omega \times v = \\ & = r_g \times R^T m_b g e_z + \sum_{j=1}^{8} \rho_j \times T_j + \tau_h \\ & \dot{\Gamma} = \Gamma \times \omega \end{aligned} \qquad (4)$$

where v, ω indicates the linear and angular velocity of coordinated body *C*-frame, situated in the center of buoyancy *CB*, *M* is general mass matrix representing the body and added mass $M=m_b I + M_{ad}$, $J=J_b+J_{ad}$ is a tensor of inertia given by body inertia matrix J_b and added inertia matrix J_{ad} produced by fluid, $D=m_b \hat{r}_g$, where $\hat{}$ is skew-symmetry operator applied on position vector r_g oriented from *C*-frame to center of gravity, $\Gamma=R^T e_z$, f_g is the force produced by gravity and buoyancy. Hydrodynamic effects f_h, τ_h are modeled as an external forces, in general nonlinear form, acting on the vehicle and T_j is a vector of thrust force, produced by j-th thruster, mounted on the position ρ_j, where $j=1,2,...,8$.

3.1 Useful simplifications

The vehicle is built as a neutrally buoyant rigid body. Moreover the center of gravity is coincident[1] with the center of buoyancy, which results to

$$\begin{aligned} & (i)\ f_g=0, \\ & (ii)\ r_g=0 \Rightarrow D=0. \end{aligned} \qquad (5)$$

From operating purposes it is assumed that the vehicle velocities v, ω are low enough making the vortex shedding not dominant and viscous effects can be modeled as a drag in form of linear and/or quadratic damping [6]

$$f_h=-\beta_{tr} v,\ \tau_h=-\beta_{rot}\omega-\gamma_{rot}\omega|\omega|. \qquad (6)$$

where $\{\beta_{tr}, \beta_{rot}, \gamma_{rot}\} \in \mathbb{R}^3$ are positive definite diagonal matrices in form

$$\begin{aligned} & \beta_{tr}=diag\{b_{tr1}, b_{tr2}, b_{tr3}\}, \\ & \beta_{rot}=diag\{b_{rot1}, b_{rot2}, b_{rot3}\}, \\ & \gamma_{rot}=diag\{c_{rot1}, c_{rot2}, c_{rot3}\}. \end{aligned} \qquad (7)$$

The added mass and added inertia matrices are hard to identify in whole complexity. Considering the vehicle prototype is designed to have almost rounded cubic shape symmetrical about all principal axes, the added matrices can be simplified to the diagonal form

$$\begin{aligned} & M=(m_b+m_{ad})I, \\ & J=J_b+diag\{n_1, n_2, n_3\}=diag\{J_1, J_2, J_3\}, \end{aligned} \qquad (8)$$

and their elements can be identified using experimental identification process [7]. It is reasonable to identify only these diagonal terms of added mass and added inertia matrices, because the stabilizing control law will then be less dependent on generally poorly known parameters.

3.2 Equation of motion

Considering only the rotational part of equations (4) and taking the above simplifications into account, the mathematical model of submerged vehicle is in the form

$$J\dot{\omega}+\omega\times J\omega+\beta_{rot}\omega+\gamma_{rot}\omega|\omega|=\sum_{j=1}^{8}\rho_j\times T_j. \qquad (9)$$

The equations (9) together with kinematics (3) represent the attitude equations of the vehicle.

[1] It means almost coincident, but the neglecting is reasonable and the inaccuracies are covered by a robust controller.

3.3 Over-actuated control

The vehicle prototype is designed to be over-actuated, which means that a number of linearly independent columns of any input matrix B has to be greater than the number of rows of B.

Consider input vector $\tau=[T_1,\dots,T_8]^T$. The vehicle uses the vectored-thrust arrangement, from which the matrix B: $B\tau=\sum_{j=1}^{8}\rho_j\times T_j$ is generated as

$$B=\begin{bmatrix} \frac{k_6}{J_1} & \frac{k_6}{J_1} & \frac{-k_6}{J_1} & \frac{-k_6}{J_1} & \frac{k_8}{J_1} & \frac{k_8}{J_1} & \frac{-k_8}{J_1} & \frac{-k_8}{J_1} \\ \frac{-k_7}{J_2} & \frac{k_7}{J_2} & \frac{k_7}{J_2} & \frac{-k_7}{J_2} & \frac{-k_9}{J_2} & \frac{k_9}{J_2} & \frac{k_9}{J_2} & \frac{-k_9}{J_2} \\ \frac{k_{10}}{J_3} & \frac{-k_{10}}{J_3} & \frac{k_{10}}{J_3} & \frac{-k_{10}}{J_3} & 0 & 0 & 0 & 0 \end{bmatrix}. \quad (10)$$

The matrix is composed of 5 constant parameters $k_6,\dots,k_{10}$, which define the positions and orientations of mounted thrusters [2].

Taking all possible combinations of arbitrarily chosen 7, 6, 5 and 4 thrusters, it can be shown that B still has full rank, and so τ can affect all degrees of freedom. In case that only arbitrarily 3 thrusters remain operational, the rank of input matrix respects this inequality

$$rank(B)\geq 2 \quad (11)$$

and moreover the input matrix B still has full rank with probability equals to 0.785,

$$P(rank(B)\equiv 2)=3/14, \quad P(rank(B)\equiv 3)=11/14. \quad (12)$$

The pseudo-inverse is used for dealing with a number of inputs and a number of states

$$BB^{+}=I \quad (13)$$

where B^{+} is right inverse of input matrix (10). Using (13) the input $\tau=B^{+}u$ is modified to get the system in form

$$J\dot{\omega}=-\omega\times J\,\omega-\beta_{rot}\omega-\gamma_{rot}\omega|\omega|+u \quad \dot{\theta}=S(\theta)\omega, \quad (14)$$

where $u: dim(u)=3$ is a new input vector.

4 Attitude control

4.1 Attitude measurement

The measurement of necessary variables is supported by Inertial Measurement Unit (IMU) together with implemented Kalman bias estimating and filtering algorithm. The description of that algorithm is behind the scope of this paper. See [10] or upcoming papers for details.

Let's define an attitude error as

$$e_\theta=\theta-\theta^*, \quad (15)$$

where $\theta^*(t)=\theta^*(0)$ is a constant setpoint. Similarly the angular velocity error is defined as

$$e_\omega=\omega-\omega^*, \quad (16)$$

where $\omega^*(t)=\omega^*(0)$ is a constant setpoint, which will be further specified. Next the attitude error integral is given as

$$e_\varphi=\int_0^t \theta-\theta^*\,d\tau=\int_0^t e_\theta\,d\tau, \quad (17)$$

4.2 Attitude control

Assume we want to stabilize the vehicle attitude around an equilibrium $[\theta^{*T},\omega^{*T}]^T, \omega^*=0$. The standard attitude controller uses proportional-derivative feedback, and for large angle maneuvers is in the form

$$u=-D_\omega\omega-S^T(\theta)P_\theta e_\theta,\ P_\omega>0,\ P_\theta>0. \quad (18)$$

The stability of closed loop system can be proved by Lyapunov direct method.

Proof of Stability: Consider a Lyapunov function candidate as

$$V=\frac{1}{2}\omega^T J\,\omega+h(e_\theta), \quad (19)$$

where $h(e_\theta)\geq 0$ is any positive definite function depending on the attitude error. By the differentiation and using of (14), the change of V can be expressed as

$$\dot{V}=\omega^T J\,\dot{\omega}+\dot{e}_\theta^T\frac{\partial h}{\partial e_\theta} =\omega^T[-\beta_{rot}\omega-\gamma_{rot}\omega|\omega|+u+S^T(\theta)\frac{\partial h}{\partial e_\theta}]\leq 0. \quad (20)$$

Provided $h(e_\theta)=\frac{1}{2}e_\theta^T P_\theta e_\theta, P_\theta>0$, then one control law satisfying Lyapunov function candidate and stability condition can be specified as in the (18) which results to

$$\dot{V}=\omega^T[-\beta_{rot}\omega-\gamma_{rot}\omega|\omega|-D_\omega\omega]\leq 0. \quad (21)$$

The control law (18) is a standard type of attitude rigid body controller, used for example in [11]. Modification for SO(3) and quaternion domain is presented in [9],[12],[13] and references therein.

Nevertheless the given controller is not suitable for underwater vehicle, where the imperfections in shape, inertia tensors and balancing can very affect the dynamic behavior of system, even the stability.

For improving the robustness of the closed-loop system, and for suppression of attitude errors arising from imperfect balancing of device, it is appropriate to modify the feedback to the form of nonlinear PID type. Then, for example, acting of static gravitational force, arising from imperfect balancing, can be suppressed with desired speed of convergence.

Without loss of generality, assume the nonzero position vector r_g is aligned with e_3 axis of C-frame. Then the static force, produced by nonzero distance of CG and CB, is in the form

$$r:\ r=r_g\times R^T m_b g\,e_z=\begin{bmatrix}-m_b g\,r_{g3}\cos\theta\sin\phi\\ -m_b g\,r_{g3}\sin\theta\\ 0\end{bmatrix} \quad (22)$$

The next section describes the proposed type of controller, including the properties of control structure and the proof of stability of the reduced closed loop feedback.

5 Reduced Attitude control

The operating conditions of vehicle prototype does not allow to reliably use any heading sensor. Heading is therefore manually commanded by operator, but the reduced attitude stability and full command control has to be fulfilled. Hence is necessary to prove the stability of the reduced controlled system.

5.1 Reduced attitude kinematics

As mentioned above the unmeasurable (even unobservable) heading variable means that Euler angle ψ is not known. The kinematic matrix $S(\boldsymbol{\theta})$, used in control loop, depends on parametrization of rotation matrix R and therefore is necessary to find any sequence of intrinsic rotations, which produces the $S(\boldsymbol{\theta})$ independent of ψ. One such a parametrization is the sequence of 3-2-1 intrinsic rotation, known as Cardanian angles, which is presented in section 2.1 .

5.2 Structure and design of reduced controller

Assuming the balance imperfection in form of (22) and a reduced attitude controller in the form

$$\boldsymbol{u}=-D_\omega\boldsymbol{\omega}-S^T(\boldsymbol{\theta})P_\theta\boldsymbol{e}_\theta-S^T(\boldsymbol{\theta})I_\varphi\boldsymbol{e}_\varphi, \qquad (23)$$

where $\{D_\omega, P_\theta, I_\varphi\}\in\mathbb{R}^3$ are diagonal matrices in form

$$\begin{aligned} D_\omega &= diag\{D_{\omega_1}, D_{\omega_2}, D_{\omega_3}\},\\ P_\theta &= diag\{P_{\theta_1}, P_{\theta_2}, 0\},\\ I_\varphi &= diag\{I_{\varphi_1}, I_{\varphi_2}, 0\}. \end{aligned} \qquad (24)$$

The zero diagonal terms reflect the unavailability of heading measurement. Then the two states, used for integral part of controller, are added to equations of motion (14), resulting to expression

$$\begin{aligned} J\dot{\boldsymbol{\omega}} &= -\boldsymbol{\omega}\times J\,\boldsymbol{\omega}-\beta_{rot}\boldsymbol{\omega}-\gamma_{rot}\boldsymbol{\omega}|\boldsymbol{\omega}|+\boldsymbol{r}+\boldsymbol{u}\\ \dot{\boldsymbol{\theta}} &= S(\boldsymbol{\theta})\boldsymbol{\omega}\\ \dot{e}_{\varphi_1} &= \phi-\Phi\\ \dot{e}_{\varphi_2} &= \theta-\Theta, \end{aligned} \qquad (25)$$

where the Φ,Θ are the given fixed roll and pitch angles. The perturbation analysis [8] is used to prove the stability of reduced attitude controlled system.

Proof of Stability: Because the angle ψ cannot be controlled, let's assume the vehicle spins around the $\boldsymbol{e}_z$ axis with a constant angular velocity $\dot{\psi}=\Omega_\psi$ and is oriented to given fixed roll and pitch angles Φ,Θ. The corresponding constant angular velocities $\boldsymbol{\Omega}=[\Omega_1,\Omega_2,\Omega_3]^T$ are given from (3) as

$$\begin{bmatrix}\Omega_1\\ \Omega_2\\ \Omega_3\end{bmatrix}=\begin{bmatrix}-\Omega_\psi\sin\Theta\\ -\Omega_\psi\cos\Theta\sin\Phi\\ \Omega_\psi\cos\Theta\cos\Phi\end{bmatrix} \qquad (26)$$

Consider a perturbation of system state $\boldsymbol{\omega}=\boldsymbol{\Omega}+\boldsymbol{\delta}_\omega$, $\boldsymbol{\theta}=[\Phi+\delta_\Phi,\Theta+\delta_\Theta,\Omega_\psi t+\psi_0]^T$,

$$\boldsymbol{e}_\varphi=\Big[\int_0^t\delta_\Phi d\tau,\int_0^t\delta_\Theta d\tau\Big]^T.$$

The differentiation implies $\dot{\boldsymbol{\omega}}=\dot{\boldsymbol{\delta}}_\omega$, $\dot{\boldsymbol{\theta}}=[\dot{\delta}_\Phi,\dot{\delta}_\Theta,\Omega_\psi]^T$, $\dot{\boldsymbol{e}}_\varphi=[\delta_\Phi,\delta_\Theta]^T$ and the equations of perturbed system arise from substitution to (25), together with (22), (23),

$$\begin{aligned} J\dot{\boldsymbol{\delta}}_\omega &= -\boldsymbol{\omega}\times J\,\boldsymbol{\omega}-\beta_{rot}\boldsymbol{\omega}-\gamma_{rot}\boldsymbol{\omega}|\boldsymbol{\omega}|+\boldsymbol{r}+\boldsymbol{u}\\ \big[\dot{\delta}_\Phi\ \ \dot{\delta}_\Theta\ \ \text{-}\big]^T &= S(\Phi+\delta_\Phi,\Theta+\delta_\Theta)\boldsymbol{\delta}_\omega\\ \dot{\boldsymbol{e}}_\varphi &= [\delta_\Phi,\delta_\Theta]^T \end{aligned} \qquad (27)$$

By omitting the heading state and by linearization around the constant trajectory

$$x_{traj}=[\Omega_1,\Omega_2,\Omega_3,\Phi,\Theta,\Phi,\Theta]^T, \qquad (28)$$

the dynamics of reduced closed-loop perturbed system can be expressed as

$$\begin{bmatrix}\dot{\omega}_1\\ \dot{\omega}_2\\ \dot{\omega}_3\\ \dot{\phi}\\ \dot{\theta}\\ \dot{e}_{\varphi_1}\\ \dot{e}_{\varphi_2}\end{bmatrix}=\boldsymbol{x}_{traj}+A(\Phi,\Theta,\Omega_\psi)\begin{bmatrix}\delta_{\omega_1}\\ \delta_{\omega_2}\\ \delta_{\omega_3}\\ \delta_\Phi\\ \delta_\Theta\\ \delta_\Phi\\ \delta_\Theta\end{bmatrix}, \qquad (29)$$

where the constant matrix A is parametrized among other model parameters by $\{\Phi,\Theta,\Omega_\psi\}$.

The stability of perturbed system along the trajectory can be checked by Lyapunov indirect method. It is possible to find appropriate controller parameters, guaranteeing that all eigenvalues of A have negative real part and therefore the given controller stabilizes the reduced system, except the neighborhood of gimbal-lock. Figure 2 and 3 show the dependence of maximal real part of eigenvalues for the attitude range $\Phi\in\langle-\pi,\pi\rangle$, $\Theta\in(-\pi/2,\pi/2)$, with given spin Ω_ψ and with respect to model parameters and specific $\{D_\omega, P_\theta, I_\varphi\}$ setting. Note that the limit stable spin Ω_ψ is about *4 rad/s*.

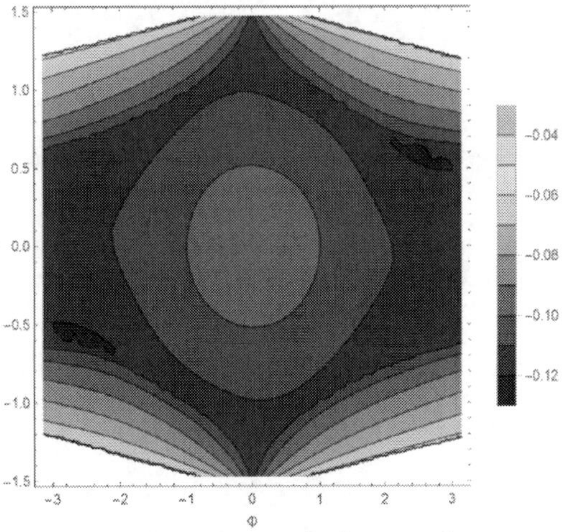

Figure 2: Contour plot of $max(\mathrm{Re}[\lambda_1,\lambda_2,..\lambda_7])$ for given pitch and roll parameters Φ,Θ and $\Omega_\psi=0$.

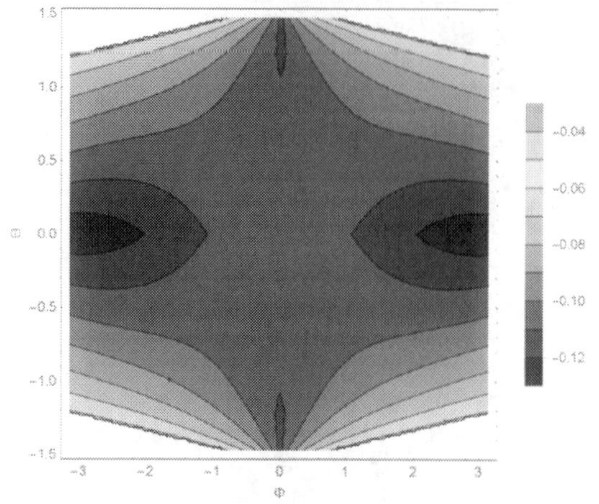

Figure 3: Contour plot of $max(\mathrm{Re}[\lambda_1,\lambda_2,..\lambda_7])$ for given pitch and roll parameters Φ,Θ and $\Omega_\psi = 1$.

6 Experimental results

The controllers were designed to fulfill the stability property of closed loop system as derived above. The parameters of each controller were tuned to get the monotonic response with maximal suitable gain and phase margin on whole attitude range. Figure 4 shows a sample record of attitude control of the real prototype.

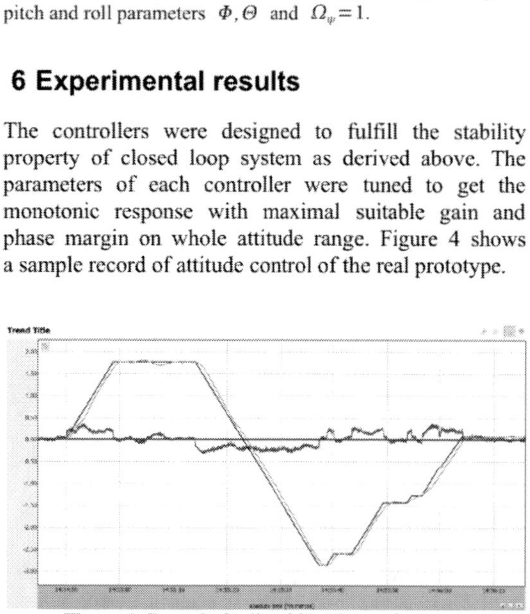

Figure 4: Record of roll stabilization and control.

7 Conclusions

In this paper, the mathematical modeling and attitude control of underwater vehicle is presented. On contrary to traditional submarines the robotic inspection vehicles are very differently shaped and equipped, which very affect the required complex motion and stability. Nevertheless, even for full attitude control, some useful simplifications are reasonable, together with over-actuation of system. Their benefits are analyzed. Real applications showed that the model is rich enough to be relevant yet simple to be useful for control design. Attitude control is based on nonlinear controller with PID structure. It is shown that for some unknown measurement, the stability of reduced system is ensured, provided a proper choice of parametrization of rotational matrix and structure of controller. The stability of reduced system was proved using perturbations and Lyapunov indirect method, with domain of attraction in entire reduced working range, except the gimbal-lock. Avoiding this singularity using reduced controller, developed directly on SO(3), or using quaternions, is a subject of the current research.

Acknowledgment

This work was supported by the university research program Postdoc, by the project LO1506 of the Czech Ministry of Education, Youth and Sports, and the Technology Agency of the Czech Republic under Grant TE02000103.

References

1. A.V. Inzartsev, *Underwater Vehicles*. In-Tech, Austria. (2009).
2. L. Bláha, M. Schlegel and J. Königsmarková, Design and control of underwater vehicle for NDT inspections, *2014 IEEE/OES Autonomous Underwater Vehicles (AUV)*, Oxford, MS, (2014).
3. T.I. Fossen, *Guidance and control of ocean vehicles*. **Vol. 199.** No. 4. New York: Wiley, (1994).
4. S. Mohan, J. Kim, Coordinated motion control in task space of an autonomous underwater vehicle–manipulator system,*Ocean Engineering*, **104**, (2015).
5. E. N. Leonard, Stability of a bottom-heavy underwater vehicle, *Automatica* **33.3** (1997).
6. W. Wang, *Autonomous control of a differential thrust micro rov*. (2007).
7. M. Langmajer, L. Bláha, Structural Parameter Identification of a Small Robotic Underwater Vehicle, *Mechatronics and Robotics Engineering for Advanced and Intelligent Manufacturing*, Springer, (2016).
8. H. K. Khalil, *Nonlinear systems*, Prentice-Hall, (1996).
9. N. A. Chaturvedi, A. K. Sanyal and N. H. McClamroch, Rigid-Body Attitude Control, in *IEEE Control Systems*, **31**, no. 3, (2011).
10. F. L. Markley, J. L. Crassidis, *Fundamentals of Spacecraft Attitude Determination and Control*, Springer, (2015).
11. T.I.Fossen, *Mathematical models for control of aircraft and satellites*, (2011).
12. Y. Yu, S. Yang, M. Wang, C. Li and Z. Li, High performance full attitude control of a quadrotor on SO(3), *IEEE International Conference on Robotics and Automation (ICRA)*, (2015).
13. T. Lee, M. Leoky and N. H. McClamroch, Geometric tracking control of a quadrotor UAV on SE(3), *49th IEEE Conference on Decision and Control*, (2010).

MATEC Web of Conferences 95, 08013 (2017) DOI: 10.1051/matecconf/20179508013
ICMME 2016

Dynamic Obstacle Detection Based on Background Compensation in Robot's Movement Space

Tao Ni[1], Qidong Li[2], Le Sun[2], and Lingtao Huang[1]

[1]School of Mechanical Science and Engineering, Jilin University, Changchun City, Jilin Province, China, 130022
[2]Master Degree Candidate, School of Mechanical Science and Engineering, Jilin University, Changchun City, Jilin Province, China, 130022

Abstract. In the safety area of industrial production, robots are required to have the ability of quickly detecting the dynamic obstacles which appear in the working space of robots when robots move along a predefined route. In order to solve the problem of dynamic obstacles detection more effectively, the paper proposes a quick detection method based on background compensation for dynamic obstacles in robot movement space, which makes extraction of feature points on dynamic obstacles in robot movement space, establishes the background model of optical flow velocity field between adjacent frames and makes background compensation for images with the use of block-matching algorithm in order to realize quick and accurate dynamic obstacles detection in dynamic environment. The experiment results show that the obstacles detection method can effectively eliminate the background movement caused by the movement of camera at the end of robot and can extract a more complete target object. Moreover, it also has good robustness.

1 Introduction

Dynamic obstacle detection based on background compensation in the robot movement space mainly includes static scene and dynamic scene. The background is relatively fixed in the static scene because camera at the end of robot remains still. So in the static scene, the study of dynamic obstacle detection becomes relatively simple, but existing main problems include illumination changes and so on. Whereas, the background moves as the robot moves in dynamic environment. Besides facing with various disadvantageous factors in static scene, differentiating background with foreground is also essential in dynamic environment [1]. What's more, dynamic compensation of background and complex movement of camera add difficulties to obstacle detection.

Moving target detection under complex background mainly includes the following three ways: the background model method [2], inter-frame difference method [3] and the optical flow field method [4]. Background model method is mainly used in static or quasi-static video surveillance applications, which can acquire very comprehensive feature data of moving objects, but it is very sensitive to changes in light and other changes in external dynamic scene, and need to use a lot of frames data, which is not suitable for real-time implementation. Inter-difference method, realizing time-domain difference by subtracting images between continuous two or three frames, which is very effective to dynamic scene, but which can not extract all characteristics pixels of a moving target generally, and the premise of which is that background itself is relatively static. The optical flow field method, with strong versatility, can detect target objects effectively when camera makes some complex movements, but along with the limitation of complex computation, therefore it is rather difficult to be implemented in real time.

In this paper, synthesizing three target detection methods above, a method is provided for the detection of dynamic obstacles based on background compensation of robot movement space. As shown in Figure 1, we start from the extraction of moving velocity of feature pixel in grayscale images [5], calculate the optical flow velocity in pixel points using the method based on block matching algorithm, and then establish an optical flow velocity field background model for two frames front and behind, and then compensate for background movement in the moving space, finally accomplish a moving obstacle detection. Experiment shows that the prediction results can effectively detect the spatial mobile obstacles, using optical flow velocity field based on block matching algorithm to accurately describe the movement of background, it also demonstrates that dynamic compensation can eliminate the effects caused by background movement for spatial dynamic obstacle detection effectively. Not only that, the algorithm adopted in the paper has a strong adaptability of dynamic jitter, noise and some other factors like that.

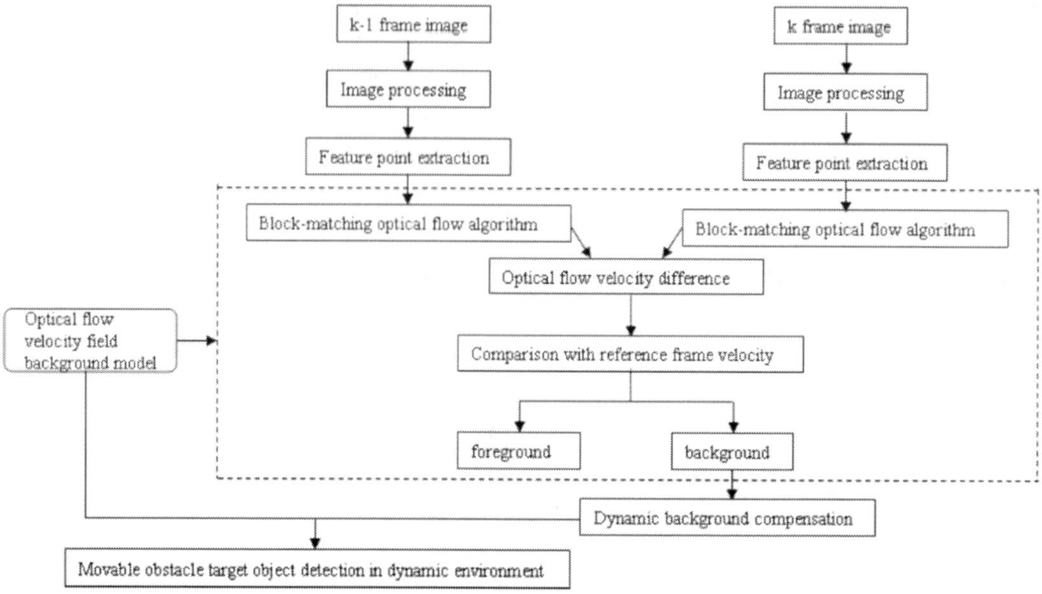

Figure 1. Background modeling based on block-matching optical flow velocity field

2 Optical flow velocity field background modeling based on block-matching algorithm

It is hard to differentiate background from dynamic obstacles effectively in robot's movement space because both dynamic obstacles and robot move at the same time. For the phenomena above, the paper computes the optical flow velocity field background model with the use of block-matching optical flow method. The method adopted to build background model is to calculate optical flow velocity field background model for feature points extracted from m × n non-overlapping bits with the same algorithm, which can improve the accuracy of matching between feature points effectively as well as decrease calculation time.

2.1 Optical flow algorithm based on block matching

The main thought of the optical flow algorithm based on block matching is to divide an image into m × n non-overlapping bits, supposing that all the pixel points in every image bit have the same displacements, that is every image bit is regarded as a movable object. Presuming that the image sequence at time t corresponds to the image at the frame of k and the image sequence at time $t+\tau$ corresponds to the image at the frame of k+1. And then we search for the most similar image bit with that at the frame of k in the image at the frame of k+1 according to certain matching criterion and search algorithm, and we think that the image bit moves from the position of k frame image to the position of k+1 frame image after moving. As for the matching criterions of block-matching algorithm, because the Mean Absolute Differences (abbreviated as MAD algorithm) doesn't need to take the multiplication, which is used widely with easy and convenient implementation. In real calculation, unnecessary division is able to be removed. Usually, MAD will be turned into SAD[6], that is, Sum of Absolute Difference, its definition is as follows:

$$SAD(i,j)=\sum_{i=1}^{M}\sum_{j=1}^{N}\left|I_k(i,j)-I_{k-1}(i+u,j+v)\right| \quad (1)$$

The matching criterion not only reflects the similar relationship between two bits, but also the accuracy of moving state of the object achieved by the matching criterion will influence the authenticity and reliability of final optical flow field image directly. On the condition that illumination variation, zooming and rotational motion of the object exist, the optical flow field got by SAD criterion will have obvious discretization and will not exactly reflect the real movement condition of the object. On the basis of assumption of movement conservation, a method is put forward in the paper, which combines SAD matching criterion with gradient-based block-matching criterion to meet the requirements of moving state detection of the object.

Block-matching criterion is combined by SAD matching criterion and gradient-based block-matching criterion. According to the block-matching criterion and the impacts weighting coefficient and gradient have on the images, we can get the following equation:

$$E=\sum_{i=1}^{M}\sum_{j=1}^{N}\left\{\begin{array}{l}\left|K_\rho\times[I_1(i,j)-I_0(i+u,j+v)]\right|+\\ \gamma\left|K_\rho\times[\nabla I_1(i,j)-\nabla I_0(i+u,j+v)]\right|\end{array}\right\}(\gamma=0.4) \quad (2)$$

At the same time, the equation of optical flow field of the image is as follows:

$$(u,v)=\arg\min\sum_{i=1}^{M}\sum_{j=1}^{N}\left\{\begin{aligned}&\left|K_{\rho}\times\left[I_1(i,j)-I_0(i+u,j+v)\right]\right|+\\&\gamma\left|K_{\rho}\times\left[\nabla I_1(i,j)-\nabla I_0(i+u,j+v)\right]\right|\end{aligned}\right\} \quad (3)$$

After combining SAD matching criterion with gradient-based block-matching criterion, the differences between matched blocks and blocks to be matched depend on gradient information of the image instead of grayscale information, which is correspond to gradient conservation assumption when the object moves. At the same time, due to the existence of gradient information of the image, the criterion no longer considers the grayscale conservation between different blocks in two frame images, but asks for the gradient conservation between matching block and original block, which makes the algorithm more accurate to deal with the condition of illumination variation.

2.2 Construction of the optical flow velocity field background model

In the process of constructing the optical flow velocity field model, as for block-matching optical flow algorithm, at first, we calculate the optical flow velocity vector v_1 at every feature points for results got by block-matching in current frame at time t. And then we calculate the optical flow velocity vector v_2 at every feature points in current frame at time t-1. The specific calculation method is as follows: 1) we preread one frame image (Template Image) and regard it as a reference frame for calculating optical flow velocity. Supposing that v,u,w are the spatial coordinate information at feature points for every frames respectively. We initialize $v_{\min}$, $diff(v_{\min})$, $u_{\min}$, $diff(u_{\min})$, $w_{\min}$, $diff(w_{\min})$ into the smallest integer, initialize $v_{\max}$, $diff(v_{\max})$, $u_{\max}$, $diff(u_{\max})$, $w_{\max}$, $diff(w_{\max})$ into the largest integer; 2) we calculate the optical flow velocity v,u,w based on previous frame image, and meanwhile the optical flow velocity got above are subtracted with the optical flow velocity in previous frame image to get $diff(v)$, $diff(u)$, $diff(w)$ for every frame images; 3) we compare v , u , w , $diff(v)$, $diff(u)$, $diff(w)$ with the initialized values of $v_{\min}$, $u_{\min}$, $w_{\min}$, $diff(v_{\min})$, $diff(u_{\min})$, $diff(w_{\min})$, if $v < v_{\min}$, then we suppose $v_{\min} = v$, afterwards, update $u_{\min}$, $w_{\min}$, $diff(v_{\min})$, $diff(u_{\min})$, $diff(w_{\min})$, successively according to the method above, after that, we compare v , u , w , $diff(v)$, $diff(u)$, $diff(w)$ with the initialized values of $v_{\max}$, $u_{\max}$, $w_{\max}$, $diff(v_{\max})$, $diff(u_{\max})$, $diff(w_{\max})$, if $v > v_{\max}$,then we suppose $v_{\max} = v$, in a similar way, update $v_{\max}$, $u_{\max}$, $w_{\max}$, $diff(v_{\max})$, $diff(u_{\max})$, $diff(w_{\max})$ successively. According to this algorithm, build the optical flow velocity filed model for velocity relations corresponding to every matching block in every frame and its previous frame. We describe background movement with the use of the optical flow velocity field model. As shown in Figure 2, it is obvious that mobile background is mistaken as an obstacle to be detected when moving state of background changes. Hence, in order to eliminate the influence of background movement caused by robot's movement on dynamic obstacle detection, we need to make dynamic compensation for background of the movement.

Figure 2. Extracting images from optical flow velocity field

3 Movement background compensation and dynamic obstacle detection

When the condition is that the camera at the end of robot is moving, assuming that there are no obstacles in moving space, so we can think that the reason of causing image background to change is completely due to movement of robot's space. At this time, we make global movement compensation [7] for the image extracted between frames to make the scene in current frame consistent with the scene in compensated frame, which can effectively eliminate background movement caused by robot's spatial movement. As for background movement compensation, the paper adopted the adaptive random point distance block motion estimation method. The basic thought of the algorithm is to take the movement vector information of the image in previous frame as the initial estimate value of searching in current frame and to do match searching in order to increase speed of the algorithm. The specific measures are: 1) inputting the image in current frame and the image in previous frame, and judge the existence and validity of block movement estimation in previous frame; 2) for every block in current frame, establishing two big rhombus searching areas centering on (0, 0) and movement estimation value in previous frame to look for the best matching point; 3) if the best matching point is one point among the other points beyond central area of movement estimation big rhombus in previous frame, we will search this area with the adaptive point distance searching method; 4) if the best matching point falls into the central area of big rhombus searching range centering on movement estimation in previous frame, then we establish a small rhombus area to search centering on this point; 5) according to rhombus searching algorithm, we stop cycling until searching the central part of small rhombus and take the point as the result of movement estimation.

After getting block movement estimation value, we are able to calculate parameter values of the 6 parameters model [8] of the camera. We calculate approximation function numbers for the parametric model with the least square method based on curve-fitting (or data-fitting) in the paper. For the six parameters model:

$$\begin{cases} u = a_0 + a_1 x_1 + a_2 x_2 \\ v = a_3 + a_4 x_1 + a_5 x_2 \end{cases} \quad (4)$$

Six parameters are estimated in the model with the method of least square estimation to make the sum of errors (denoted by E) the minimum in the following:

$$E = \sum_i \sigma_i^2 = \sum_N \begin{bmatrix} (u - a_0 - a_1 x_1 - a_2 x_2)^2 + \\ (v - a_3 - a_4 x_1 - a_5 x_2)^2 \end{bmatrix} \quad (5)$$

In the Eq. (5), N represents all the points used in the calculation.

As for the errors caused by the block estimation values partly or overall belonging to the movement target area in the process of estimation calculation of six parameters, every estimation result is assessed with the six parameters fitted. If the bias is bigger than a certain threshold value, then the movement estimation value of the point will be removed and six parameters will be fitted again. Putting all the qualified block movement estimation results into the six parameters model, six parameters are fitted with the least square method. Putting the six parameters model fitted into the centric coordinate of every block, and the motion vectors of the six parameters model are calculated. Because the calculation results may not be integer, pixel value is not obtained for the digitized image. And then we adopt an image interpolation method of closure - bilinear interpolation to receive the statistical properties of both the motion vectors obtained by calculating all the qualified blocks and movement estimation results.

The good background compensation is obtained in the dynamic scene, and dynamic obstacles are detected with the optical flow method in static scene by analyzing and resolving robot's camera six parameters model and combining with block matching optical flow velocity field model.

4 Experimental results and analysis

The experiment adopts three-dimensional software to build the virtual situation [9] and control the robot to realize the dynamic situation where movable obstacles appear in the path, along which robot approaches the target object little by little (as is shown in Fig. 3) with the first perspective, at the same time, the experiment makes use of the block-matching optical flow method and background compensation algorithm to detect dynamic obstacles.

Figure 3. Dynamic movement situations with the robot's first perspective

We calculate the optical flow velocity vectors in the feature points of every frame after extracting the feature points in every frame image, and then build the background model of the optical flow velocity field by comparing the velocity difference in every frame. After building the background model, the detection of target objects in different frames is respectively conducted with the optical flow method based on the background model. The detection results are shown in Fig. 4.

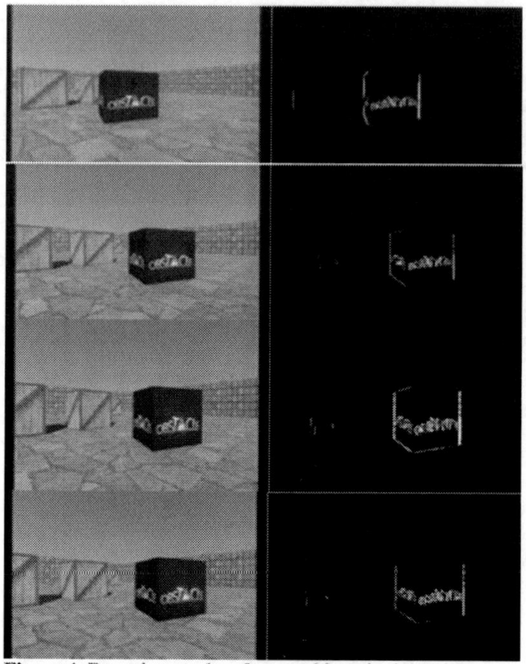

Figure 4. Detection results of target objects in dynamic background

The experimental results show that dynamic background compensation of the optical flow velocity field has received good detection effects for dynamic obstacles in robot's motion space. Influenced by noises and other factors, there is some phenomena such as vague and distortion in the background model, so we need to make further improvement in post production; in the process of dynamic obstacle detection, the detection for single mobile object has better feasibility and higher accuracy in dynamic background with the block-matching optical flow method, at the same time, under the influence of the color of background, target objects and surrounding reference objects, integrity of the objects under detection is far from satisfactory, so it requires more improvement about algorithm and image processing. On the whole, the experimental results reach the anticipative design goal, verify the detection algorithm for dynamic obstacles in robot's movement space based on background compensation adequately and demonstrate the feasibility of the algorithm fully.

5 Conclusions

The paper provides the dynamic obstacle detection algorithm in robot's movement space based on background compensation. Moreover, the paper introduces a kind of novel method about how to establishing the background model of the optical flow velocity field, which simplifies the dynamic modeling in the calculation of the camera affine motion model greatly by means of the background model established. The method is also a very effective method to detect dynamic obstacles when the camera is mobile, because it brings optical flow algorithm's superiority in target detection into full play. The drawback of the algorithm is relatively complex calculation and lower precision, so the algorithm remains to be further optimized.

Acknowledgement

The authors would like to thank the financial support from the National Natural Science Foundation of China (nos. 51305153 and nos. 51575219).

References

1. L. Li, W. Huang, Y.H. Gu, Q. Tian, "Statistical Modeling of Complex Backgrounds for Foreground Object Detection" IEEE TRANSACTIONS ON IMAGE PROCESSING, **13**, 1459-1472 (2004).
2. S. Denman, V. Chandran, S. Sridharan, "An adaptive optical flow technique for person tracking systems" Pattern Recogn, 28, 1232-1239 (2007).
3. W. H. Warren, D. J. Hannon, "Direction of self-motion is perceived from optical flow" Nature, **336**, 162-163 (1988).
4. S. Baker, D. Scharstein, J.P. Lewis, S. Roth, M. J. Black, R. Szeliski, "A Database and Evaluation Methodology for Optical Flow" INT J COMPUT VISION, **92**, 1-31 (2007).
5. D. Decarlo, D. Metaxas, "The Integration of Optical Flow and Deformable Models with Applications to Human Face Shape and Motion Estimation" *Conference on Computer Vision and Pattern Recognition* (San Francisco, CA, USA, 1996).
6. D. Honegger, L. Meier, P. Tanskanen, M. Pollefeys, "An Open Source and Open Hardware Embedded Metric Optical Flow CMOS Camera for Indoor and Outdoor Applications" *IEEE International Conference on Robotics and Automation* (Karlsruhe, Germany, 2013).
7. K. Steinkraus, L. P. Kaelbling, "Optical Flow for Obstacle Detection in Mobile Robots" Artificial Intelligence Laboratory Mit (2001).
8. Y. P. Tan, S. R. Kulkarni, P. J. Ramadge, "A New Method for Camera Motion Parameter Estimation" International Conference on Image Processing, 1, 406 (1995).
9. N. Johansson, F. Williamsson, "Human Machine Interface Visualization Enhancement of an ABB Quality Control System" Umea University Department of Computing Science (2009).

Optimal Design for PID Controller Based on DE Algorithm in Omnidirectional Mobile Robot

Peizhang Wu[1], Kai Wang[2], Juzhong Zhang[1] and Qi Zhang[3]

[1]*Automation Engineering, The 713th Institute of China Shipbuilding Industry Corporation, 450000 Zhengzhou City, Henan Province, China*
[2]*Mechatronics, Naval Representative Office in Zhengzhou Region, 450000 Zhengzhou City, Henan Province, China*
[3]*Railway Engineering Railway Bureau of Zhengzhou, 450000 Zhengzhou City, Henan Province, China*

Abstract. This paper introduces a omnidirectional mobile robot based on Mecanum wheel, which is used for conveying heavy load in a small space of the automatic warehousing logistics center. Then analyzes and establishes the omnidirectional chassis inverse and forward kinematic model. In order to improve the performance of motion, the paper proposes the optimal PID controller based on differential evolution algorithm. Finally, through MATLAB simulation, the results show that the kinematic model of mobile robot chassis is correct, further more the controller optimized by the DE algorithm working better than the traditional Z-N PID tuned. So the optimal scheme is reasonable and feasible, which has a value for engineering applications.

1 Introduction

Mobile robots are increasingly used in automated industrial environments [1]. The mobile robot technology includes the latest research results of many subjects, such as machinery, sensors, electronic, computer, information processing, automatic control and the artificial intelligence, it represents the highest achievement of electromechanical integration [2].

The Mecanum wheel shown in Fig. 1 which is widely used in omnidirectional mobile mechanism is a kind of the mature technology. It first proposed in 1973 by a Swedish engineer called Bengt Ilon [3]. The Mecanum wheel bearing capacity is strong, the relative position between wheel and suspension on the omni-directional mobile mechanism is fixed, it turns without independent steering mechanism, just depends on coordinating 4 wheels velocity and direction that can realize the omnidirectional movement, and mechanical structure of the system is simple, reliable and easy to control [4].

Figure 1. Structure of the Mecanum wheel.

2 Kinematics analysis

The structure of the mobile robot motion platform which described in this paper is shown in Fig. 2.

Figure 2. Structure of the mobile robot platform.

In order to establish and do the research on the kinematics model, we construct the reference coordinate system shown in Fig. 3. The coordinate system origin OXY is based on the center of the mechanical structure of the mobile platform, and the coordinate system origin $O_iX_iY_i$ is based on the center of the mechanical structure of each Mecanum wheel, any of the coordinate system conforms to the right-hand rule.

In particular, the installation of the wheel body has deferent spin around (the left forward and right rear wheel is left-spin, the right forward and left rear wheel is right-spin, specifically, the angle between wheel axis and roller axis is 45°. it's worth noting that the figure 1-a is observed from top to down of the omnidirectional mobile robot platform, as well as the oblique dashed lines drawn in each wheel is the contact line between roller surface and road surface. In Fig. 3, L is the distance that along Y-axis between the center of wheel and chassis, l is the distance that along X-axis between the center of wheel and chassis, and we assume that this mobile platform is running with the velocity (V_x V_y W_z) , at the same time, the wheel i is running with the line velocity V_i (angular velocity is W_i), α_i is the offset angular of wheel i, R is the radius of each Mecanum wheel.

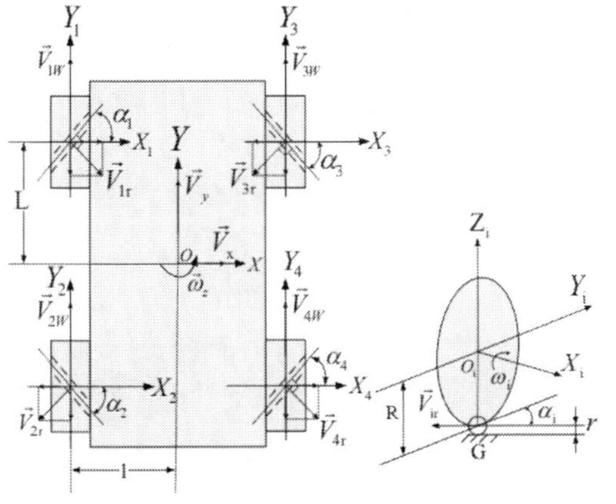

Figure 3. Moving principle of the omnidirectional mobile robot platform.

According to the reference paper [5], it deduced the general kinematic mathematical model of the 4 Mecanum wheels omnidiretional mobile platform.

$$[\omega_1 \ \ \omega_2 \ \ \omega_3 \ \ \omega_4]^{\mathrm{T}} = K_i\left[V_x \ \ V_y \ \ \omega_z\right]^{\mathrm{T}} \quad (1)$$

in formula (1), the inverse kinematic matrix K_i is:

$$K_i = \frac{1}{R}\begin{bmatrix} cot\alpha & 1 & -(Lcot\alpha + l) \\ -cot\alpha & 1 & -(Lcot\alpha + l) \\ -cot\alpha & 1 & Lcot\alpha + l \\ cot\alpha & 1 & Lcot\alpha + l \end{bmatrix} \quad (2)$$

Thus, we transform the matrix formula (1), then get the forward kinematic model in formula (3).

$$\left[V_x \ \ V_y \ \ \omega_z\right]^{\mathrm{T}} = K_f\left[\omega_1 \ \ \omega_2 \ \ \omega_3 \ \ \omega_4\right]^{\mathrm{T}} \quad (3)$$

in formula (3), the forward kinematic matrix K_f is:

$$K_f = \frac{\mathrm{R}}{4}\begin{bmatrix} tan\alpha & -tan\alpha & -tan\alpha & tan\alpha \\ 1 & 1 & 1 & 1 \\ \frac{-1}{Lcot\alpha + l} & \frac{-1}{Lcot\alpha + l} & \frac{1}{Lcot\alpha + l} & \frac{1}{Lcot\alpha + l} \end{bmatrix} \quad (4)$$

From the above formulas, it shows that the velocity of omnidirectional mobile robot is decided by the cooperation from revolving velocity of the 4 Mecanum wheels. The detailed moving direction by different cooperation wheel speed of the chassis is shown in Fig. 4.

3 Principle of DE

Differential Evolution (DE) Algorithm is a simple and efficient method to solve the problem of global optimization, it is also a kind of stochastic search algorithm based on population [6]. The DE Algorithm first proposed by Rain Storn et al. who came from California State University at Berkeley[7]. In view of the advantage of efficient, DE has drawn great attention of the relative scholars all over the world, which quickly became a hot research topic of optimizing computation.

Figure 4. Diagram of omnidirectional moving of chassis.

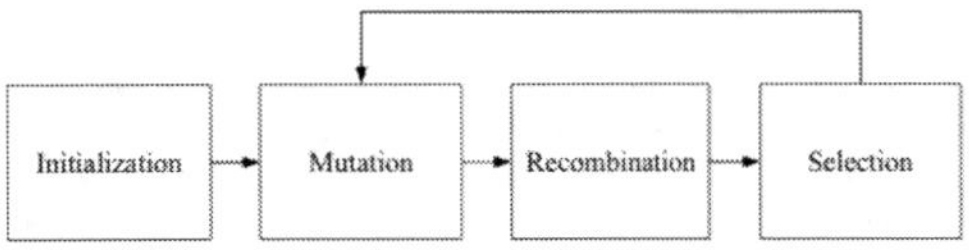

Figure 5. Operation flow of DE algorithm.

3.1 Initialization

The initial population is generated a size of Np randomly which include D dimension vector in each population, for example :

$$X_{i,G} = \{X_{i,G}^1, X_{i,G}^2, ..., X_{i,G}^D\}, i = 1, 2, ..., Np \quad (5)$$

In formula (5), the Np is a constant, every individual need make mutation, recombination and selection in turn, generations of population to be made evolution until computing over.

3.2 Mutation

The mutant vector individual is computed by the follow fomula.

$$V_{i,G} = X_{r1,G} + F(X_{r2,G} + X_{r3,G}), r1, r2, r3 \in \{1, 2, ..., Np\} \quad (6)$$

In formula (6), $i \neq r1 \neq r2 \neq r3$, F is a real constant factor to control the differential variation and F>0.The $\mathbf{V_{i,G}}$ is prepared for the next generation mutation vector, $\mathbf{X_{ri,G}}$ is the current generation vector from different population.

3.3 Recombination

The recombination operation cross over each pair of the target vector $\mathbf{X_{r1,G}}$ and its corresponding mutant vector $\mathbf{V_{i,G}}$ for generating a trial vector $\mathbf{U_{i,G}}$

$$U_{i,G} = \{U^1_{i,G}, U^2_{i,G}, ..., U^D_{i,G}\}, i = 1,2,...,Np \quad (7)$$

In the basic version, DE employs the crossover defined as follows:

$$U^j_{i,G} = \begin{cases} V^j_{i,G} & if(rand_j[0,1] \le CR) \quad or \quad j = j_{rand} \\ X^j_{i,G} & otherwise \end{cases}, j = 1,2,...,D \quad (8)$$

In formula (8), CR controls the fraction of parameter values copied from the mutant vector, CR is a user-defined real constant and randj is a random real constant, they are both between the range [0,1].

3.4 Selection

The DE algorithm uses greedy method that the best individuals are selected to replace parent individuals to generate new individual in general evolutionary strategies. The selection operation can be expressed as follows:

$$X_{i,G+1} = \begin{cases} U_{i,G} & if(f(U_{i,G}) \le f(X_{i,G})) \\ X_{i,G} & otherwise \end{cases} \quad (9)$$

In formula (9), the fitness function f(x) is to evalutate the best result(individual).

4 Optimal controller design

As we know, PID controller has been widely used in most industry control system, optimizing the PID parameters has been a considerable topic for the people [8]-[10]. In this paper, we use DE algorithm for optimizing and tuning parameters of PID controller in omnidirectional mobile robot. The principle of optimal PID controller designing based on DE algorithm is shown in Fig. 6.

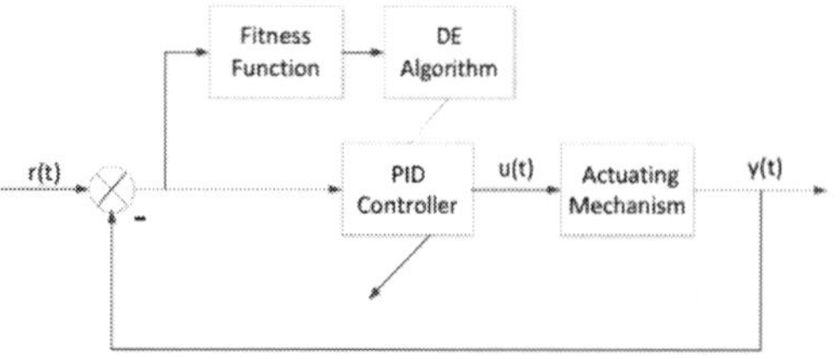

Figure 6. Principle of optimal PID controller design based on DE algorithm

4.1 Encoding

In the PID controller, there are 3 parameters(K_p, K_i, K_d) to be optimized. DE algorithm using floating-point encoding method so that the 3 parameters could be directly served as a vector of individual. Then DE algorithm searching space is worked from the results of Ziegler-Nichols (Z-N) tuning method, thus making full use of the reasonable PID value of Z-N method.Assuming that K^*_p, K^*_i, K^*_d are the tuned value by Z-N method, α, β are the scaling factors,we can define the boundry of searching by formula (10) :

$$\begin{cases} \alpha K^*_p \le K_p \le \beta K^*_p \\ \alpha K^*_i \le K_i \le \beta K^*_i \\ \alpha K^*_d \le K_d \le \beta K^*_d \end{cases} \quad (10)$$

4.2 Fitness function

The target of optimizing PID controller is to improve the key performance indicators, such as stability, precision and highly response, so we indroduce overshoot(δ), rise time(t_r) and accumulative error($\Sigma |e_k|$) into the fitness function F as follows:

$$F = w_1\delta + w_2 t_r + w_3 \sum_{k=1}^{n} |e_k| \quad (11)$$

in formula (11), the constant weight coefficient w_1, w_2, w_3 are used for adjusting the influence of corresponding variables in system.

4.3 Optimizing flow

Step1: Define searching boundaries for PID parameters (K_p, K_i, K_d) by formula (10).

Step 2: Define the population size Np, crossover probability CR, maximum number of iteration G_{max}, then randomly generating initial population in searching boundaries.

Step 3: According to the formula (6) make mutation for individuals in population.

Step 4: According to the formula (8) make recombination for mutation individuals.

Step 5: According to the formula (11) Compute the fitness values of every individual vector.

Step 6: According to the formula (9) make selection for the next generation individuals.

Step 7: If the number of iterations less than G_{max}, repeat Step 2-7, otherwise end.

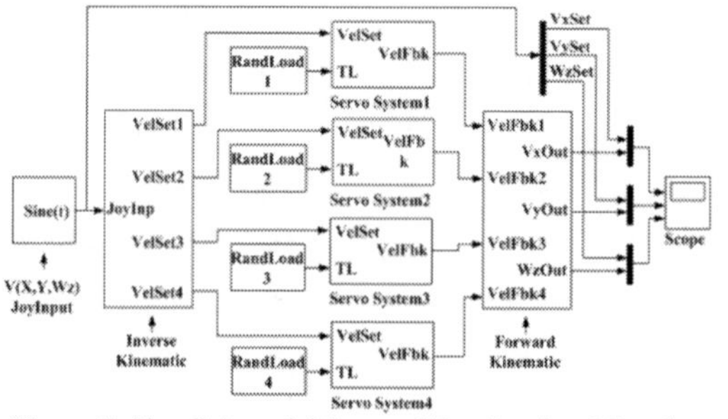

Figure 7. Simulink model for omnidirectional mobile robot

5 Simulation and analysis

5.1 Simulation model

In order to verify the correctness of the forward and inverse kinematics model,as well as the effectiveness of the optimized PID controller by DE algorithm, based on the actual conditions we set up an omnidirectional robot motion control platform by MATLAB simulink, please see Fig. 7.

In Fig. 7, the 'JoyInput' model is a 3*1 column vector with sine signal, each element respectively corresponds to the input of direction *X*, *Y*, W_z; the 'Inverse Kinematic' model represents the matrix K_i in formula(2); 'Random Load' model generates the 3Nm torque random disturbance to the wheels; 'Servo system' model is the servo driver with space-vector PWM technique, each unit has a double-loop control system, furtherly the PID controller of speed loop is optimized by DE algorithm; the 'Forward Kinematic' model represents the matrix K_f in formula(4).

5.2 Simulation parameters

a) Mobile robot chassis parameters: L=1.0m, l=0.5m, transmission ratio i=100;

b) Mecanum wheel parameters: roller offset angle α =45°, wheel radius R=0.25m;

c) PMSM parameters: rated speed w_n=3000, torque T_n=20Nm;

d) The fitness function weight coefficient: w_1=10,w_2=2,w_3 =5;

e) The DE algorthm population size Np=30, crossover probability CR=0.8, iterations max G_{max}=50.

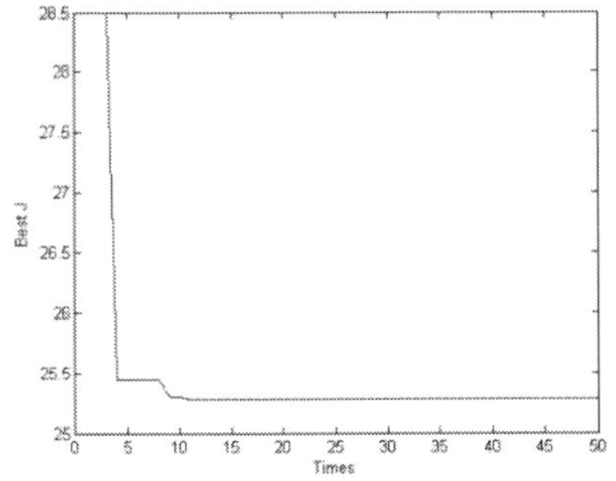

Figure 8. Simulink model for omnidirectional mobile robot

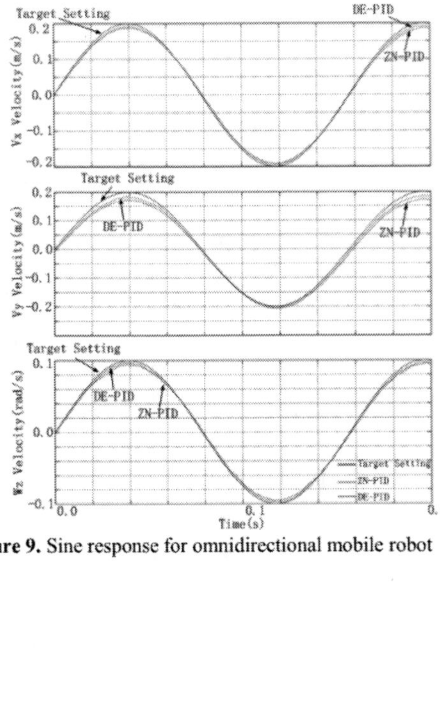

Figure 9. Sine response for omnidirectional mobile robot

In Fig. 8, the convergence curve is the processing of DE algorithm. The Z-N PID parameters are K_p=31.2, K_i= 5.3, K_d =1.2, after optimized by DE algorithm, the PID parameters are K_p=42.4, K_i= 6.3, K_d =0.92.

When we set the target sine signal with angular frequency ω=40rad/s, amplitude of three DOF V_x =*0.2m/s*, V_y =*0.2m/s*, W_z =*0.2rad/s*, the simulation results are shown as following Fig. 9.

5.3 Simulation results

Given that any velocity loop PID controller of the 4 wheels is working in the same way, so we show the optimized results by an example here.

From the simulation results of sine response, at the three DOF direction, the optimized PID controller by DE algorithm works with following error much less than the PID controller by Z-N tuned. It indicates the optimal controller design method is effective and control system has strong robustness, the results of simulation also show that the proposed inverse and forward kinematic are both correct and feasible.

6 Conclusion

In this paper, we analyzed the inverse and forward kinematics of omnidirectional mobile robot based on Mecanum wheel, then put forward the DE algorithm for optimizing PID controller of speed loop in servo system.Through MATLAB simulation platform, we bulit the simulaiton model of system, the results showed that the inverse and forward kinematics model are reasonable, DE algorithm optimization for the PID controller is better than traditional Z-N tuned. The DE algorithm can be implemented with ease and few parameters need to be tuned, which could be widely used in engineering practice.

References

1. Bashra K. O. ChaborAlwawi and Hubert Roth, etc, Mobile Robot Motion Planning and Multi Objective Optimization Using Improved Approach[J], International Journal of Mechanical Engineering and Robotic Research, **4**.(Oct. 2015)
2. Yaonan Wang, *Robot intelligent control engineering*.8.(Science Press, Beijing, 2004)
3. Ilon BE, Directionally statble self propelled vehicle, P.3746112.(U.S., 1973)
4. Dickerson SL, Lapin BD, Control of an omnidirectional robotic vehicle witch Mecanum wheels, *IEEE proceedings of the Telesystems Conference*, 323.(Atlanta, 1991)
5. Zhou Liu, Hongtao Wu, Omnidirectional moiton analysis and simulation of Mobile mechanism with Mecanum wheel, Machine Design and Manufacture Engeering, 43-44, **40(05)**. (SEU Press, Nanjing, 2011)
6. K. Price, Differential evolution a fast and simple numerical optimizer, *1996 Biennial Conference of*

DOI: 10.1051/matecconf/20179508014

the North American Fuzzy Information Processing Society, 524-527.(New York, 1996)
7. R. Storn, K. Price, Differential evolution a simple and efficient adaptive scheme for global optimization over continuous spaces[C], *Technical report International Computer Science Institute*, 41-47.(Berkley, 1995)
8. Jinkun Liu, *advanced PID control and MATLAB Simulink*, 136. Electronics Industry Press, Beijing, (2002)
9. Peng, Wang, et al. PID Parameter Tuning Based on Improved DE Algorithm and the Application, Journal of Chinese Society of Power Engineering. (2015)
10. Youxin Luo, and Xiaoyi Che, Tuning PID control parameters on hydraulic servo control system based on differential evolution algorithm, *Logistics Systems and Intelligent Management 2010 International Conference on IEEE*, 1861-1864.(2010)

Real Time FPGA-Based Ethernet Control Communication for Robotic Arm

Mohamad Khairi Ishak, Ho Kar Hwai and Lee Kong Lam

School of Electrical and Electronics Engineering, Universiti Sains Malaysia, Engineering Campus, 14300, Penang, Malaysia.

Abstract. In this paper, an approach for real time control communication using Ethernet is proposed. The strategy to support this at the network level and include Field Programmable Gate Array (FPGA) implementation on the Ethernet platform for robotic arm. An embedded Ethernet controller is designed to send data packet via Ethernet Local Area Network (LAN). The transferring data also employs Arduino Mega as the medium of communication between FPGA board and the robotic arm. It is used as the receiver to receive data packet from FPGA board with the interface of Arduino Ethernet shield. The control operation on the robotic arm is performed once the desired data packet length is reached to the Arduino Mega. SolidWorks and MATLAB software are used to design the robotic arm and simulate the robotic arm working flexibility in real world respectively. The result of the average data packet delay between FPGA boards is lower in comparison to Arduiono board. The data packet can send successfully in through the network to test the robotic arm.

1 Introduction

Real time system is a system that produces instant output response within a specific amount of time without delay after receiving input signal [1]. It has the advantages over the non-real time system because non real time systems do not produce immediate response and do not have accountability [2]. Embedded communication network real time system is used in applications which require a set of synchronized data to be delivered in time and high response speed. Many field of application especially safety critical application use embedded communication network control to interact with real world due to flexible of designed system and less wiring complexity [3, 4]. Examples of safety critical application are military, robotics, medical, transportation system and avionics. Various types of communication control protocols which developed for network embedded real time system are used as a solution for communication problems. Altera DE2 and DE2-115 boards are used in this project as the embedded real time Ethernet controller system.

Currently, there are high demands from various industries to use different types of robot to perform different tasks in the workplace. The controller chosen to manipulate the robot is very important. Most of the digital control systems for conventional robots still employ microcontroller as the brain of the robot in which cannot keep pace with the new applications that require higher control performance especially when the robots are used for high speed manufacturing and medical operations [5]. Although microcontroller such as Arduino is cheaper compared to FPGA but it cannot provide parallel processing at the same time which is required for most of the industrial robots control nowadays [6,7,8]. Hence, the development of real time FPGA controller is necessary to achieve higher control performance on the robotic arm.

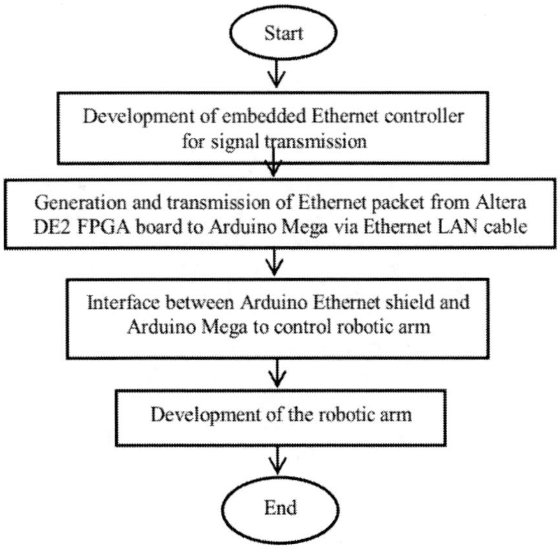

Figure 1. Flow chart of the project development

This project consists of four major parts. The first part is the development of embedded Ethernet controller for signal transmission. The second part is the generation and transmission of Ethernet packet from Altera DE2 FPGA board to Arduino Mega via Ethernet LAN cable. The third part is the interface between Arduino Ethernet shield and Arduino Mega to control robotic arm. The fourth part is the development of the robotic arm. The overall flow chart of this project is shown in Figure 1.

2 System design development

In this section, the step by step methods used to complete this project are highlighted and explained in details. The explanations are important in complete the objectives of the project. Both software and hardware are used in this project.

2.1 Development of embedded ethernet controller for signal transmission affiliations

Ethernet is the main control communication protocol as this protocol is a low cost, robustness, easy to available, expendability and non-proprietary network communication solution. In order to control the Ethernet chip DM9000A in Altera DE2 FPGA board, a digital hardware system that contain components such as internal processor, RAM memory, parallel input/output interfaces, serial communication protocol, and Ethernet interface protocol will be built using SOPC builder tools in Quartus II software. It is used to run application program that handle the data transmitted to or received from the Ethernet controller chip. DM9000A from Davicom is a fully integrated and low pin count Fast Ethernet controller chip with a general processor interface, a media access control (MAC) unit, a 10/100Mbps PHY transceiver and 16K bytes SRAM. Ethernet protocol associated with DM9000A chip included in this digital hardware system works as a Media Access Controller which work along with the Nios II processor and external PHY chip. Besides, the interface for communication between MAC and external PHY chip is chosen as Media Independent Interface (MII) since this Ethernet controller only support data transfer rate of 10/100Mbps.

After Verilog code associated with the components of digital hardware system is generated using SOPC builder, this system need to be instantiated in our design's top level module in order for the components this system to perform its function. The main function of this system is to generate Ethernet packet in Altera DE2 board. The block diagram of digital hardware system is as shown in Figure 2.

2.2 Generation and transmission of ethernet packet from altera DE2 FPGA board to arduino mega via ethernet

Nios II processor is used to send the Ethernet packet from FPGA board to Internet by connecting Ethernet LAN cable through PHY layer via serial interface. The end point of the Ethernet packet will be stored at either server or computer. In this project, the Ethernet packet will send to the IP address of the Arduino Ethernet shield from Altera DE2 FPGA board when a push button is pressed. After this digital hardware system is created, an application program is written in C/C++ code through Nios II IDE software build tool. This application program is implemented in order to initialize the DM9000A Ethernet controller chip, generate Ethernet packets, and drive the Ethernet controller to send the packets over built network through external PHY chip of physical layer.

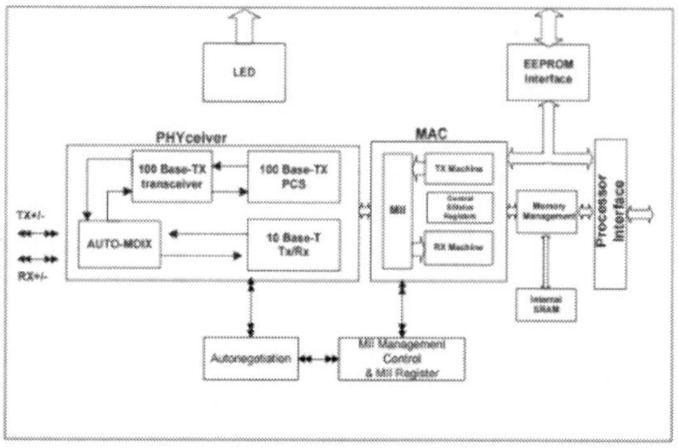

Figure 2: Block diagram of Ethernet controller DM9000A

2.3 Interface between arduino ethernet shield and arduino mega to control robotic arm

Arduino Ethernet shield is interfaced with Arduino Mega to perform control operation on robotic arm. Arduino Ethernet shield is used to enable Arduino Mega to act as the medium of communication between FPGA board and robotic arm so that signal can be transferred via (RJ45) Ethernet LAN cable. Correct IP address need to be obtained from Arduino Mega after Ethernet shield has been connected onto it.

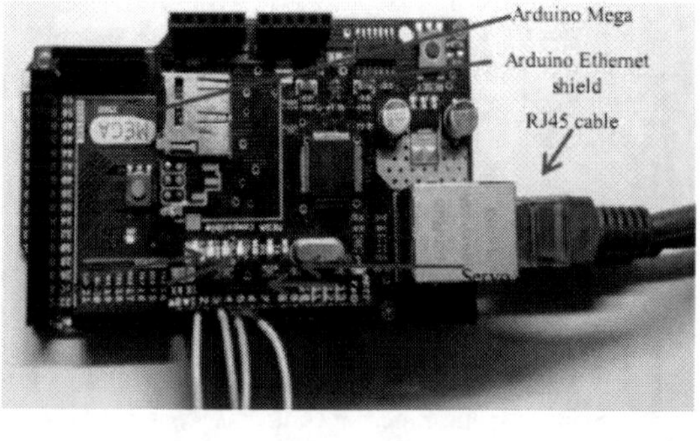

Figure 3: Hardware interface between Arduino Mega and Arduino Ethernet shield

Once the Ethernet packet has reach to the correct IP address of the Arduino Ethernet shield, the Arduino Ethernet shield will prompt the Arduino Mega to perform control operation on robotic arm. The hardware interface is shown in Figure 3.

2.4 Development of the robotic arm

Robotic arm with 3 degree of freedom and the end effector is a gripper is designed by using of SolidWorks software and simulated with MATLAB software. The robotic arm is capable of doing the task of pick and place of objects. All 3 links and joints are revolute. This robot is configured as an articulated robot. Estimation is being made for design material specification, theory to calculate the torque of the robotic arm is applied and its relationship towards rotational speed.

Servo motor is a rotary actuator that allows for precise control of angular position [9]. In this project, type of analog servo motors HD-1501MG are used. Servo motor is the most suitable motor in this project because it is low cost and has a lot of advantage compared to the other motor such as stepper motor, brushless motor, ac motor, etc. The robotic arm is printed with 3D printer so that precise dimension guaranteed. Servo motors are used on the robotic arm.

By using SolidWorks software, the robotic arm is designed. After designing, simulation process must be carried out to make sure there is zero interference and to test on the robotic arm joints and links feasibility as in real world [10].

2.5 Development of real-time communication system between two FPGA devices

In this digital hardware system, Triple Speed Ethernet MegaCore is chosen as the Ethernet Controller instead of DM9000A. This is because the FPGA boards use to build this communication system is DE2-115 series and the Ethernet PHY chip (Marvell 8E1111) of this board can only support this Ethernet controller. Besides, a timer will be used to determine the transmission time and reception time of Ethernet packets at transmitter and receiver. This timer is implemented by using a 32-bit counter and its timing behavior is based on the internal clock frequency of DE2-115 FPGA board.

Practically, the Ethernet communication system is designed using two FPGA boards which communicate to each other based on their unique MAC address and Internet Protocol (IP) address. The experimental setup for this Ethernet real-time communication system is presented in Figure 4. One of the FPGA board will be setup as transmitter and the other FPGA board will be setup as receiver. The process of data transmission will be begun with transmitter send data with data bytes limited by user to receiver. During the process of data transmission, transmitter is allow to send data packets to receiver only by addressing transmitter to one of the receiver's unique IP address. The data packets transmitting time and data packets receiving time are determined from transmitter and receiver by triggering timer in NIOS II software build tool. These timers use counters with internal clock frequency of 10MHz to calculate both data packets transmitting time and receiving time. The obtained timing results will be recorded on workstations which connected to the FPGA boards.

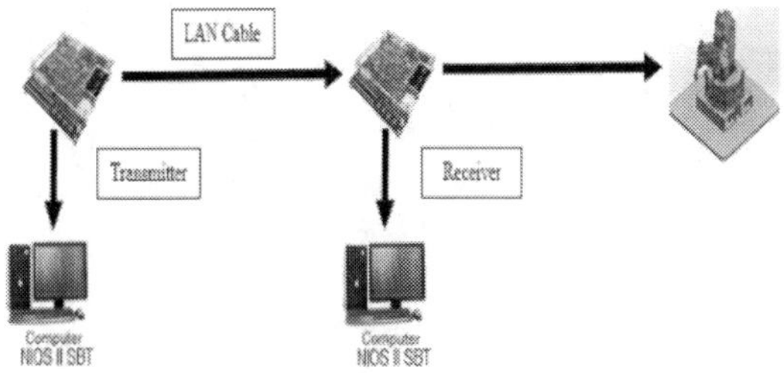

Figure 4: Experimental setup for FPGA based communication system

3 Results and discussion

The Altera DE2 FPGA board transmits Ethernet data packet of different sizes to the Arduino Mega. If the data packet of greater or equal to 1200 bytes is received by the Arduino Ethernet shield, the Arduino Mega will perform control operation of pick and place object on two points. The object to be picked up is a sponge.The combine (Front, Left, Top and Trimetric) view of the robotic arm is shown in Figure 5.

Figure 5: Combine view of the robotic arm

Once the static IP address of the Arduino Ethernet shield is obtained via Ethernet switch, the program of receiving Ethernet data packet from FPGA board and prompt robotic arm operation is uploaded into the Arduino Mega. The serial monitor in the Arduino software is open to observe when the data packet is reached to the Arduino. If the data packet size is greater or equal to 1200 bytes, the robotic arm is activated. The robotic arm picks and place operation is shown in Figure 6.

For the communication setup which used FPGA and Arduino board, the average transmission delay of each data packet is 16ms. From the data measurements, transmission delay of each data packet can be calculated by using the formula:

$$\textit{Transmission Delay} = \frac{\textit{Total Data Reception Time}}{\textit{Total Data Packets Received}} - \frac{\textit{Total Data Transmission Time}}{\textit{Total Data Packets Transmitted}}$$

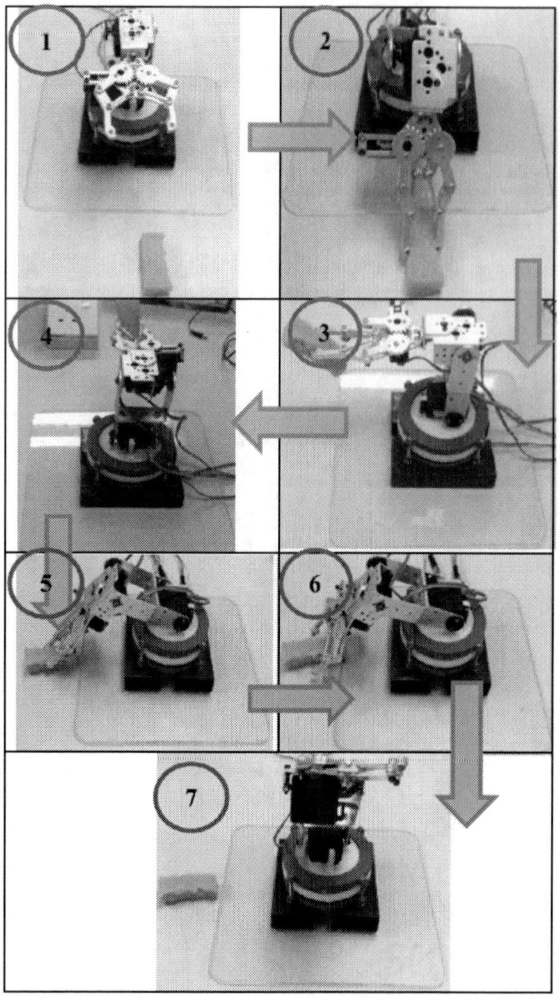

Figure 6: The robotic arm picks and place operation

The transmission delay for each data packet size is calculated and recorded in Table 1. According to Table 1, the average data packet transmission delay for FPGA-FPGA based communication system is about 1.0642ms and this delay is much smaller than the transmission delay of FPGA-Arduino communication system (16ms). This significant delay reduction is because FPGA development boards possess better features compared to Arduino board. These features include high processing speed, high precision, parallel processing, short development cycle and perfect combination of standardized programming structure of Verilog HDL.

Table 1: Data transmission delay for communication between two FPGA devices

Data Packet Size (Bytes)	Total Data Packets Transmitted (Packets)	Total Data Packets Received (Packets)	Total Data Transmission Time (s)	Total Data Reception Time (s)	Transmission Delay for Each Data Packet (millisecond)
500	100000	32103	40.2693	40.2683	0.8517
1000	50000	16107	22.4752	23.4742	1.0079
1200	41667	12334	19.3421	18.3411	1.0228
1468	34060	8039	18.3781	15.3861	1.3744
Average					**1.0642**

4 Conclusion

In conclusion, the control communication is successfully developed at high data transfer for real-time control communication to communicate with other controller device for robotic arm with Arduino board and FPGA board via Ethernet. Through this implementation, the ability to communicate safety data directly over an Ethernet/IP network is achieved. The practical test results have shown that the delay is reduced by 93.3% after the receiver in a communication system is changed from Arduino board to FPGA board.

References

1. Joseph, M. and P. Pandya, *Finding Response Times in a Real-Time System.* The Computer Journal, 1986. 29(5): p. 390-395
2. Wong, T.C., J.W. Mark, and K.C. Chua. *Approximate performance analysis of resource allocation for real-time and non-real-time multiclass services in cellular systems*. in *Communication Systems, 2002.*
3. Lee, K.C. and S.Lee, *Performance evaluation of switched ethernet for real-time industrial communications.* Computer Standards and Interfaces, 2002. p. 411-423.
4. Venkatramani, C. and T.-c. Chiueh, *Design, implementation,and evaluation of a software-based real-time ethernet protocol.* SIGCOMM Comput. Commun. Rev., October 1995.: p. 27-37.
5. Ishak, M.K., G. Herrmann, and M. Pearson. *Reducing delay and jitter for real-time control communication in Ethernet*. in *Advanced Communication Technology (ICACT), 2013 15th International Conference on*. 2013.
6. Kadir, W.M.H.W., R.E. Samin, and B.S.K. Ibrahim, *Internet Controlled Robotic Arm.* Procedia Engineering, 2012. p. 1065-1071.
7. Sam, R., K. Arrifin, and N. Buniyamin. *Simulation of pick and place robotics system using Solidworks Softmotion.* in *System Engineering and Technology (ICSET), 2012 International Conference on*. 2012.
8. C. A. Lara-Niño, C. Torres-Huitzil and J. H. Barron-Zambrano, "Versatile educational and research robotic platform based on reconfigurable hardware," *2014 International Conference on ReConFigurable Computing and FPGAs (ReConFig14)*, Cancun, 2014, pp. 1-6.
9. N. K. Anish, B. Kowshick and S. Moorthi, "Ethernet based industry automation using FPGA," *AFRICON, 2013*, Pointe-Aux-Piments, 2013, pp. 1-4.
10. M. Engin, "*Embedded and real time system design: A case study fire fighting robot*," *2016 in 5th Mediterranean Conference on Embedded Computing (MECO)*,Bar,2016.

Control Strategy for Automatic Transmission under the Condition of Rough Road Feature Recognition

Yulong Lei, Qingkai Wei, Xingzhong Li, Xuesong Zheng and Yao Fu

State Key Laboratory of Automotive Simulation and Control, Jilin University, 130022, Changchun, China

Abstract. To solve the cycle shifting problem caused by double-parameter shifting schedule on the rough road, it is need to recognize rough road and then take corresponding control strategy. Firstly, vehicle longitunal dynamic on rough road is analyzed, which verified the possibility using longitudinal acceleration to recognize rough road. Then longitudinal acceleration expressions on the rough road are analyzed, which proves the turbine speed change rate is effective on rough road recognition, and time domain analysis of turbine speed change rate is done. With the above analysis, a method combined with frequency of turbine speed change rate passing through the limits and receding horizon is used to recognize rough road. Finally, the fixed method of shift schedule on rough road is proposed based on the above passing through frequency, and control strategy of automatic transmission under the condition of road recognition is also given. Road experiments show that road roughness is effectively recognized and cycle shifting problem can be effectively solved.

1 Introduction

When vehicle runs on rough road, rough road excitation will cause signal interference of vehicle speed sensor. Thus traditional dual-parameter shift schedule is based on the driving state on the fine level road, which may cause the cycle shifting on the rough road. So it is significant to recognize the rough road and then change the control strategy.

The real time estimations of the rough road feature and shift schedule design are two key factors. Various approaches to identify the rough road are developed which are based on wheel acceleration sensors of Anti-lock Braking System (ABS) [1-3] or vehicle body vertical acceleration [4-5]. Toshiaki Tsuyam [1] analyzed the impact of the rough road on the angular acceleration and wheel acceleration and proposed the recognition method of rough road based on wheel acceleration. Kong propose "energy specific weight method" and "account the times crossing method" to identify road based on the time-frequency analysis of wheel speed acceleration, and then designed a real-time algorithm for road roughness feature recognition. Cui [6] analyzed road surface power spectral density and proposed a method based on BP neural network to detect road roughness. But these methods cannot solve the problem of rough road recognition for automatic transmission.

Various researches have done on fixing the shift schedule to avoid cycle shifting problem. Wang proposed the accelerator passivated control strategy; the accelerator signal was treated with step by step method. Liu Hongbo [7] adopted the strategy of limiting the highest gear. Li Runze [8] fixed the shift schedule of tracked vehicle based on engine speed change rate.

This paper proposes a novel rough road feature recognition method for automatic transmission. Based on the feature, control strategy on the rough road is proposed and applied to an off road vehicle.

2 Vehicle longitudinal dynamics on rough road

To analyse the rough road's effect on powertrain system, we need to analyse the rough road's effect on speed signal from the view of vehicle longitudinal dynamics. Rough road's feature can be described as road surface with changing slope angle and rolling resistance coefficient. As Figure 1 shown, the rate of slope angle change should be in a small range, changing direction, randomly and unavoidable. Rolling resistance coefficient is determined by road roughness, coefficient of road adhesion and other factors.

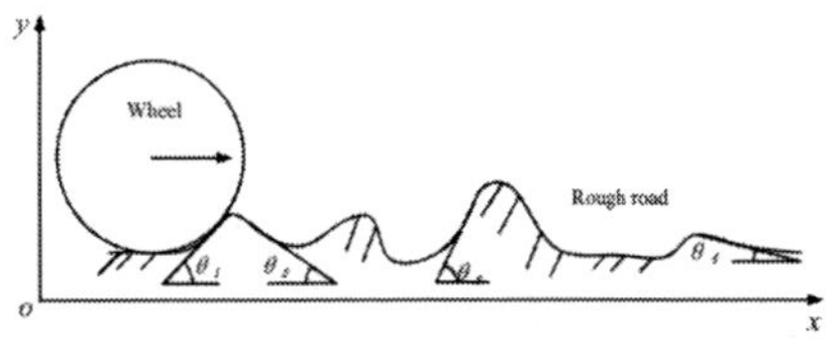

Figure 1. Schematic diagram of vehicle on rough road.

From the automotive running equations, acceleration can be right as the follow form:

$$\frac{du}{dt}=\frac{1}{\delta m}\left(\frac{T_t i_g i_0 \eta_T}{r}-mg\tilde{f}\sin\tilde{\theta}-mg\cos\tilde{\theta}-\frac{C_D A u_a^2}{21.15}\right) \quad (1)$$

In the formula:

$\tilde{f}$ — changing rolling resistance coefficient on rough road;

$\tilde{\theta}$ — changing slope angle on rough road;
u — vehicle speed;
m — vehicle mass;
δ — rotating mass conversion factor;
T_t — turbine torque;
i_g, i_0 — gear ratio of transmission and main reducer;
η_T— transmission efficiency;
A —air resistance coefficient;
C_D — automotive frontal area;
r — tyre rolling radius.

On the rough road, two factors mainly impacts on change of driving resistance: one is the change of slop angle which leads to the change of rolling resistance and climbing resistance; the other is the change of wheel's contact area with road surface and road adhesion coefficient which lead to the change of rolling resistance. If the shift process is not started, turbine torque will not have obviously fluctuation, so the changes of driving resistance will finally lead to the fluctuant acceleration.

3 Rough road feature recognition method

3.1 Decision of recognition signal

According to the above analysis, longitudinal acceleration can be used to identify rough road. For automatic transmission, longitudinal acceleration's manifestation includes turbine speed change rate and vehicle speed change rate.

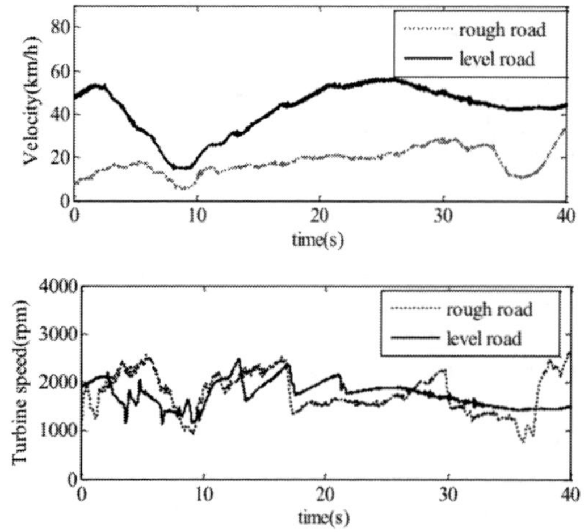

Figure 2. Comparison of vehicle speed and turbine speed on rough road and level road.

For vehicle speed change rate, vehicle speed should be got firstly. Vehicle speed can be calculated by output speed sensor. Vehicle speed on rough road is usually lower than that on level road, as vehicle speed change rate is the speed difference of current time and previous time. If vehicle speed is high, speed difference will be large; otherwise speed difference is small. So vehicle speed is not effective on recognizing acceleration caused by road surface, that is to say different speed disturbs the recognition of vehicle speed change rate.

For turbine speed change rate, value range of turbine speed is smaller, namely different speed will not disturb the recognition of turbine speed change rate. Besides, small obstacles, such as small stone, will not disturb turbine speed obviously. Figure 2 shows the comparison of vehicle speed and turbine speed on rough road and level road.

3.2 Time-domain signal analysis

Based on the above analysis, turbine speed is more suitable for recognizing rough road than vehicle speed. Figure 3 is the comparison of turbine speed change rate on rough road and level road.

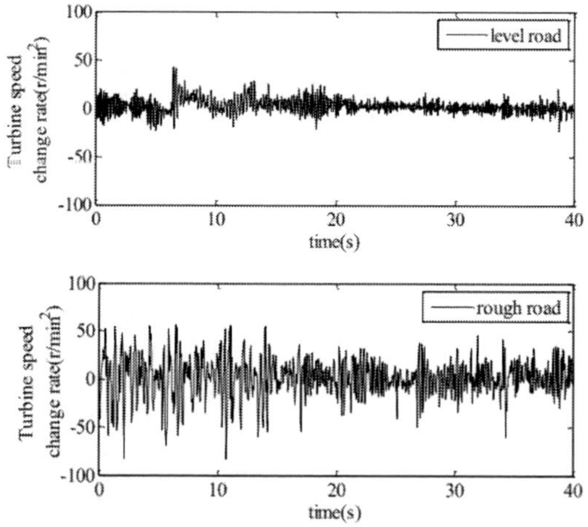

Figure 3. Comparison of turbine speed change rate on rough road and level road.

It can be seen from Figure 3 that turbine speed change rate is larger than change rate on level road. Table 1 is the statistical analysis of these two change rates.

Table 1. Statistical analysis of turbine speed change rate on rough road and level road.

Road surface	Level road	Rough road
Maximum Value	41.2	59.3
Minimum Value	-21.9	-85.6
Average absolute value	16.8	37.6
Standard deviation	7.8965	21.3527

As Table 1 shown, fluctuation range of turbine speed change rate on rough road is greater than that on level road. So we can assign a limit value of turbine speed change rate. Then the times passing through the limit value is recorded in a count cycle and the times is defined as passing through frequency. The passing through frequency is used as numerical feature of turbine speed change rate in time domain. The statistical law of passing through frequency after 10-cycle experiments is as Figure 4 and Table 2 shown. So a threshold value is assigned, when passing through frequency is larger than the threshold value, this road is determined to be rough road; otherwise, this road is level road. Furthermore, a larger passing through frequency means that the vehicle is on a rougher road.

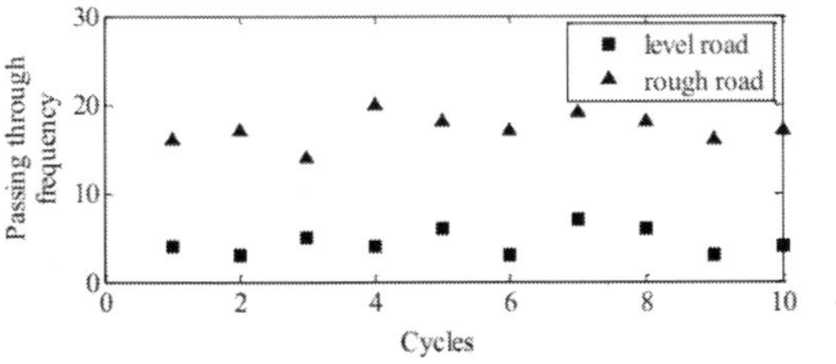

Figure 4. Passing through frequency of 10-cycle.

Table 2. Statistical analysis of passing through frequency.

Road surface	Level road	Rough road
Maximum Value	3	14
Minimum Value	7	20
Average absolute value	4.5	17.38
Standard deviation	1.4337	1.6865

3.3 Feature recognition method based on receding horizon

From the above analysis, statistical law of passing through frequency agrees with the actual road condition, it can be seen that this method is effective to recognize rough road. But if the count cycle is set short, a larger statistical error may occur; while a long period affects its real-time quality. So a receding horizon is designed, which is based passing through times receding time window instead of the current period. Figure 5 is the flowchart of real-time road surface recognition algorithm.

Real-time algorithm introduction is as follows:

(1) Set the initial value of turbine speed change rate limit value $\pm\alpha$, threshold of passing through frequency β, flag of rough road *RFlag*, receding window size *K*, count cycle *T*, road roughness sum counter *R*, $R_{count}[i]=0$, $i=1, 2, \ldots, K$.

(2) Caculate $R_{count}[i]$. $R_{count}[i] = R_{count}[i+1]$, $i=1, 2, \ldots, K-1$, $R=\sum R_{count}[i]$, $i=1, 2, \ldots, K-1$.

(3) When turbine speed change rate Δn_t exceeds $\pm\alpha$ in a count cycle, counter $R_{count}[K]= R_{count}[K]+1$.

(4) When a count cycle finish, $R= R+ R_{count}[K]$.

(5) If roughness counter *R* is greater than β, road surface in this count cycle is considered to be rough road, set *RFlag*=1; otherwise is level road, set *RFlag*=0.

(6) Reset counter *R*, R_{count}, enter the next count cycle. Go to step (2).

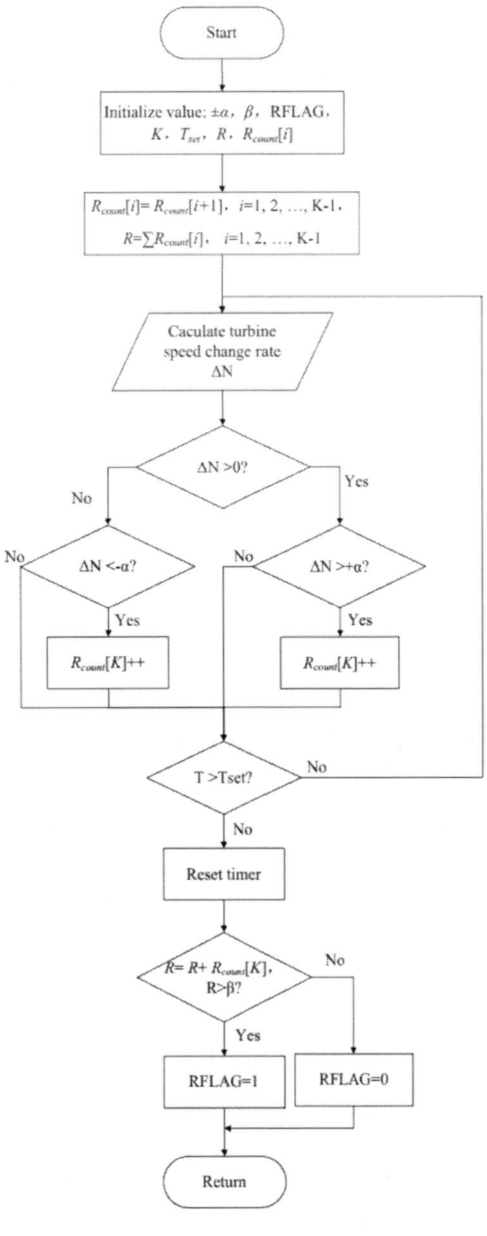

Figure 5. Flowchart of algorithm based on receding horizon.

4 Control strategies under the condition of road recognition

4.1 Shift schedule on rough road

Traditional shift schedule is based on vehicle speed and accelerator pedal position as control parameters and established on level road. But on the rough road, driver's pedal is usually in a fluctuant position, thus this fluctuation is not driver's intension, as figure 6 shows, so accelerator pedal position is not suitable for shifting schedule's control parameter.

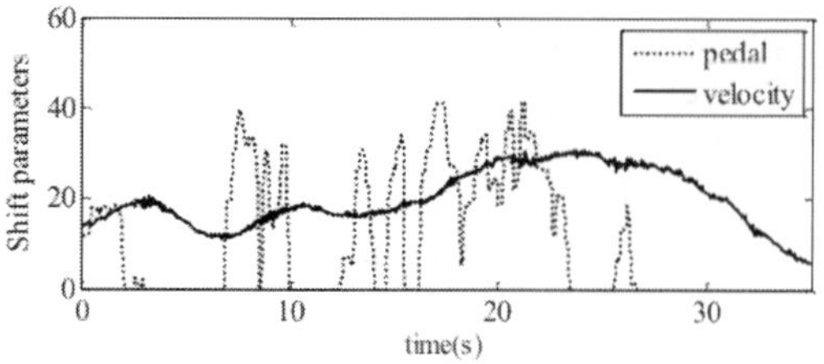

Figure 6. Vehicle speed and acceleration pedal on rough road.

Based on the above analysis, this article use turbine speed change rate passing through frequency and vehicle speed as shift schedule's control parameter, meanwhile limits the highest gear according to this passing through frequency. For an off-road vehicle with 6-speed automatic transmission in this research, the fixed shift schedule is as Figure 7 shown.

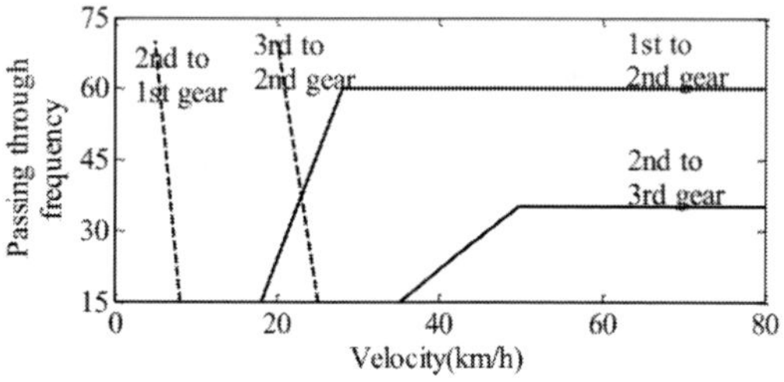

Figure 7. Shift schedule based on vehicle speed and passing through frequency.

If the passing through frequency is less than 30Hz, the current road surface is seen as a common rough road, maximum allowable gear is 3rd gear; otherwise if the passing through frequency is less than 60Hz, the current road surface is seen as a highly rough road; if the passing frequency is greater than 60Hz, upshift is forbidden. At the same time, if the passing through frequency is higher, upshift point of vehicle speed is higher but downshift point is lower, the reason is to avoid cycle shifting lead by vehicle speed fluctuation.

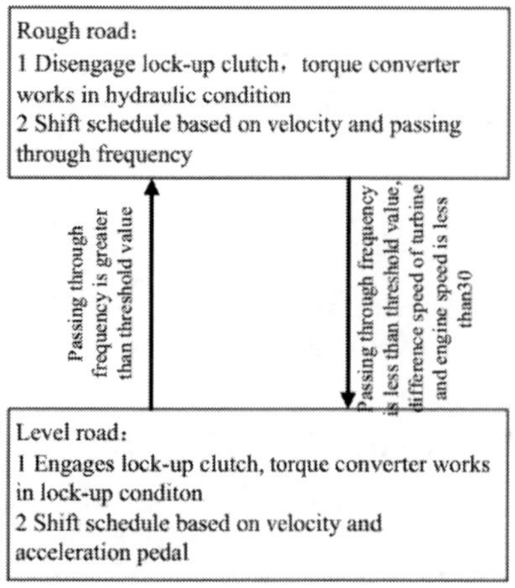

Figure 8. Control strategy under the condition of road feature recognition.

4.2 Architecture of control strategy under the road condition recognition

When rough road is recognized, transmission control unit (TCU) needs to adopt corresponding control strategy to ensure the vehicle riding comfort, meanwhile recognize whether the vehicle leaves rough road so as to quit the control strategy on rough road. The flow chart is as Figure 8 shown.

When vehicle travels on rough road, besides changing the shift schedule, lock-up clutch should be disengaged to make the torque converter working in hydraulic state, so as to improve the vehicle ride performance. When TCU recognize that passing through frequency is less than threshold value and difference speed of engine speed and turbine speed is less than set value, the vehicle is considered to leave rough road.

5 Analysis of the vehicle tests results

5.1 Rough road recognition method tests

Apply the recognition method on actual road surface recognition, vehicle tests results is as Figure 9 shown.

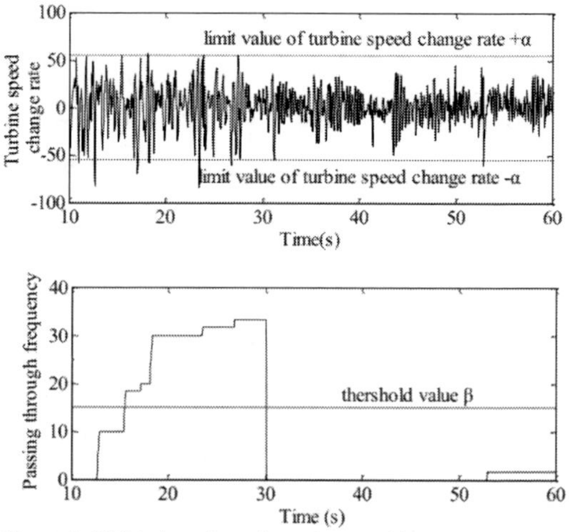

Figure 9. Validation of rough road recognition.

It can be seen from figure 9 that passing through frequency exceeds threshold value from 15s that is to say vehicle begins to run on rough road. The vehicle quits from 30s as the passing through frequency is less than threshold value. The recognition result meets the real road surface.

5.2 Control strategy on rough road tests

Figure 10 is the gear situation comparison of tradition shift schedule and proposed shift schedule.

In the traditional shift schedule, vehicle and pedal acceleration reaches upshift point at about 2s, transmission shifts to 3rd gear. But as the impact of rough road transmission suddenly starts down-shift process, which is lead to cycle shift problem. But the proposed method avoids this problem, as the upshift point of vehicle speed is higher. Vehicle tests prove the effectiveness of the propose shift schedule.

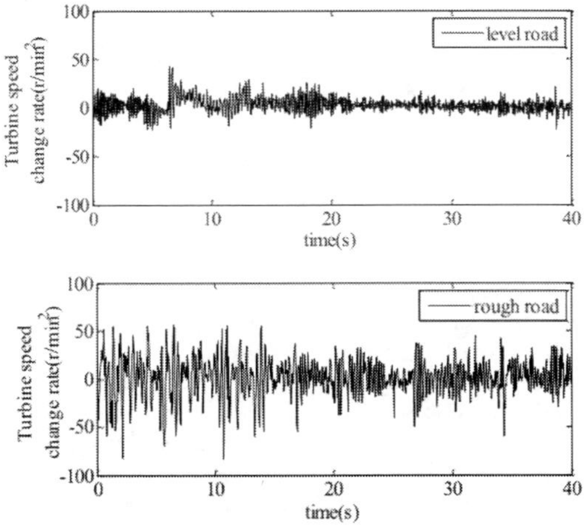

Figure 10. Comparison of turbine speed change rate on rough road and level road.

6 Conclusions

(1) With the dynamic analysis of vehicle on rough road, longitudinal acceleration is effective on rough surface recognition. Furtherly, turbine speed change rate is a kind of more effective longitudinal acceleration signal.

(2) By time-domain analysis of turbine speed change rate, passing through frequency method is proposed. Combined with receding horizon, this method is effective on rough road recognition.

(3) Shift schedule fixed method is researched. Both considering vehicle's passing ability and avoiding cycle shifting, shift schedule based on passing through frequency and vehicle speed is proposed. This method is well performed on the road test.

Acknowledgements

This study was supported by the National Natural Science Foundation of China (Grant No. 51575220).

References

1. T. Tsuyama, K. Nobumoto, K. Sotoyama, T. Onaka, F. Kageyama, & H. Okazaki. EP US5117934A (1992).
2. H. Zhang, B. Chen, T. Zhang, D. Liu, Automotive Engineering, **24** (2002)
3. L. Kong, J. Song, Y. Yan, J. Shen, CHIN J MECH ENG, **43** (2007).
4. R. Wang, X. J, G. Wang, Automotive Engineering, **30** (2008).
5. D. Cui, C. Zhang, D. Han, Computer Simulation, **31** (2014)
6. H. Wang, H. Liu, W. Zhang, H. Chen, J B INST TECHNO, **29** (2009)
7. H.B. Liu. *Changchun: Jilin University*(2012).
8. R.Z. Li. *Beijing: Beijing Institute of Technology* (2015).

Design and test of urea hydrolysis reactor for vehicle

Qi Zhanfeng[1] , Li Shusen[1] and Guo Xiuli[2]

[1]*Northeast Forestry University, 150040, harbin, China*
[2]*Dalian University, 116622, Dalian, China*

Abstract. Ammonia production technology of urea-SCR system for vehicle is mainly used in pyrolysis. The reaction is complex, and there are some side effects. So a kind of urea hydrolysis device for vehicle is designed. Based on the in-depth analysis of ammonia production technology's mechanism of urea hydrolysis for vehicle, the modified extended UNIQUAC equation and PHS equation of state were used to solve the thermodynamic model, and the experimental verification was carried out on the JX493ZLQ3 diesel engine. The results show that the design of urea hydrolysis reactor is in agreement with the experimental results. In the environment of urea hydrolysis, the higher the hydrolysis temperature, the faster the urea hydrolysis and the faster the production of ammonia. Under the same conditions, the hydrolysis rate of urea was the same, the higher the initial quality, the more ammonia produced. The method is suitable for the ammonia production technology of urea hydrolysis for vehicle and can be used for the formation of urea hydrolysis SCR system.

1 Introduction

Urea-SCR technology can effectively reduce the emissions of nitrogen oxides in diesel engines, so it is widely used. The preparation method of reducing agent NH3 in the technology has the urea pyrolysis and urea hydrolysis. The current vehicle SCR system is based on the development of urea pyrolysis technology. The core technology of urea pyrolysis is to inject the urea solution into the high temperature medium. The pyrolysis reaction of urea was decomposed to generate NH_3, CO_2 and H_2O [1]. But because of the vehicle body and SCR volume limit, SCR system cannot achieve complete pyrolysis reaction. So the research focus is mainly on the control strategy of injection. And urea in the pyrolysis process, there will be a certain byproduct, such as isocyanate, melamine, etc.. These reduce the utilization rate of urea and the catalytic effect, and produce a lot of side effects such as urea crystallization.

Urea hydrolysis process is a completely different technical route from the pyrolysis process. The hydrolysis reaction of urea is a side reaction in the process of urea production. In the field of chemical fertilizer, the hydrolysis reaction belongs to the category of deep hydrolysis. It is usually used in the form of a series of cascade. The process flow is complex and the equipment is huge. In the field of denitrification in power plant, the hydrolysis of urea belongs to the category of shallow hydrolysis. It is usually carried out using a pressure vessel. Hydrolysis of urea solution with high concentration was used to generate NH_3, CO_2 and H_2O. The urea hydrolysis process for vehicle is used for the preparation of ammonia in the vehicle condition, which is more convenient and economical. Compared with deep hydrolysis in the chemical fertilizer and hydrolysis denitrification in power plant, its operating conditions and process routes were significant difference. Therefore, the research of urea hydrolysis technology of the vehicle will be a brand-new topic.

2 Design of urea hydrolysis reactor for vehicle

The structure diagram of the urea hydrolysis device is shown in Figure 1. The reactor utilizes the waste gas heat of diesel engine as an external heat source to carry out the hydrolysis of urea. NH_3 is produced and then discharged into the tail gas.

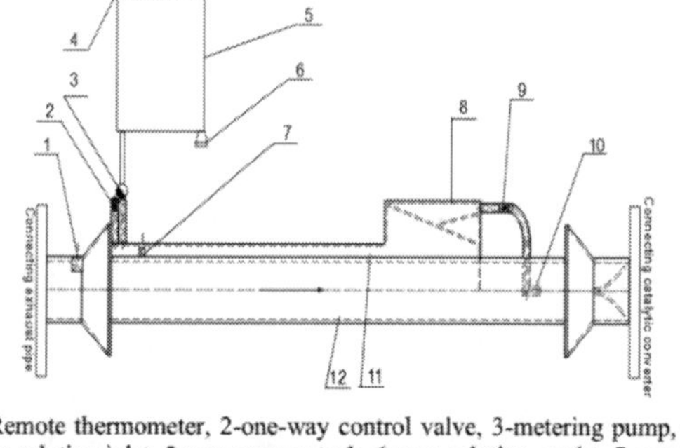

1-Remote thermometer, 2-one-way control valve, 3-metering pump, 4-urea solution inlet, 5-urea storage tank, 6-urea solution outlet, 7-remote liquidometer, 8-gas-liquid separation chamber (plenum chamber), 9-ammonia metering valve, 10-product gas outlet, 11-hydrolysis chamber, 12-exhaust pipe

Figure 1. Schematic diagram of the structure of urea SCR hydrolysis reactor for vehicle.

The hydrolysis system is mainly composed of urea storage tank 11, hydrolysis chamber 5, metering pump 3, gas-liquid separation chamber 8, ammonia metering valve 9, product gas outlet 10, one-way control valve 2 and the corresponding connection pipe.Urea solution inlet 4 and urea solution outlet 6 on the urea storage tank. The whole reactor can be connected with the exhaust pipe and the vehicle SCR catalytic converter. Product gas outlet is right at the center of the exhaust pipe 12. The reactor is installed remote liquidometer 7 and remote thermometer 1. The urea solution hydrolysis reaction system which is heated by tail gas is used in the vehicle SCR system. The urea solution complete heating, hydrolysis reaction, gas product release, gas-liquid separation, steady flow in the hydrolysis reactor in a series of physical and chemical change.

The test proved [2] that there is oxygen in urea ammonium carbamate solution, make chemical passivation of stainless steel. An oxide film formed on the surface of the stainless steel, especially when used in chrome and nickel stainless steel. They can cause the surface to form a dense oxide film, the corrosion rate is greatly reduced. In order to maintain the formation of the oxide film and make the stainless steel passivation, the oxygen content in the solution is the lowest limit of 10×10^{-6}. So it is necessary to add air to the raw materials. The urea hydrolysis reactor is made of 316L double phase stainless steel. And in the design of a one-way control valve to increase the supply of oxygen.

3 Study on the ammonia production mechanism of urea hydrolysis for vehicle

The urea hydrolysis system involves NH_3-CO_2-H_2O-Urea(Ammonia-Carbon dioxide-Water-Urea, ACWU) multicomponent solution system. The thermodynamic equilibrium calculation of the system should take into account both the vapor liquid equilibrium, the ionization equilibrium and the equilibrium of chemical reaction.

Due to high mass fraction of urea hydrolysis process in urea solution of vehicle, ionization can be neglected in liquid phase weak electrolyte. The liquid phase can be reasonably simplified as 4 components of NH_3, CO_2, H_2O and urea. And the gas phase contains 3 components, including H_2O, CO_2 and NH_3. According to the ACWU gas liquid equilibrium system composed of the urea hydrolysis reaction in the reactor, the thermodynamic model and dynamic model are established, and the reaction mechanism is described.

Urea hydrolysis reaction can be expressed as follows. It reflects the relationship between the chemical equilibrium constant and the component.

$$K_r = \frac{x_{CO2}\cdot x_{NH3}^2}{x_{Urea}\cdot x_{H2O}}\frac{\gamma_{CO2}(T,X)\cdot\gamma_{NH3}^2(T,X)}{\gamma_{Urea}(T,X)\cdot\gamma_{H2O}(T,X)} \quad (1)$$

In the formula, x_i is the mole fraction of the component, and γ_i is the activity coefficient of the component, and the K_r is the equilibrium constant of the reaction.

The calculation of activity coefficients of each component was calculated using the extended UNIQUAC equation [3].

In the calculation of vapor liquid equilibrium of ACWU system, according to the properties of H_2O, NH_3 and CO_2 and their content in liquid phase, two kinds of different relations of Henry or Raoult are used in the description of the vapor liquid equilibrium of the system. In the liquid phase, a large number of components of H_2O are found in the liquid phase, and the modified Raoult model is adopted. Component of NH_3 and CO_2 are easy to be volatile and less in liquid phase. Their reference state of the infinite dilute solution is selected in the water. Their gas phase is subject to the modified relation of Henry's law [4].

Under a certain temperature and time condition the urea hydrolysis reactor is, which causes the urea in urea solution hydrolysis to be ammonia and carbon dioxide. The ammonia gas is discharged into the exhaust pipe through a pressure stabilizing cavity. As the use of exhaust heat as a heat source, temperature and flow rate of the exhaust are not fixed. Therefore, the amount of ammonia obtained from the hydrolysis reaction is not fixed, it must be calculated accurately. The core calculation of urea hydrolysis system is mainly based on the model of urea hydrolysis reactor. The calculation model is combined with the least square method and the Newton method in order to obtain a more accurate approximation and the stability of the convergence. And the above analysis was developed to calculate the program with matlab.

4 Experiment and result analysis

The experiment was carried out to verify the thermodynamic and dynamic simulation of the system by means of the test device shown in Figure 2.

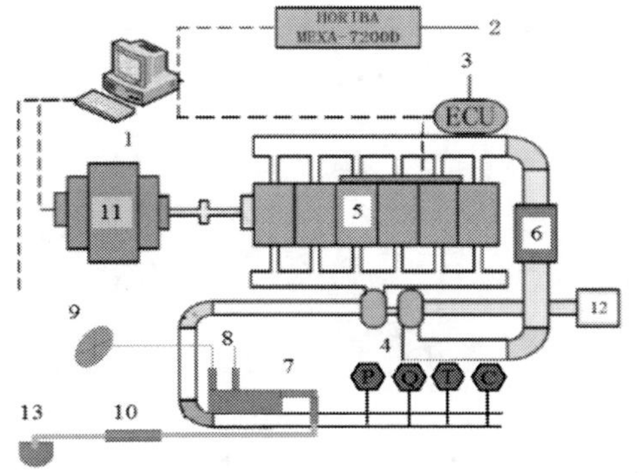

1-calibration computer, 2-gas analyzer, 3-engine ECU, 4-booster, 5-diesel engine, 6-cooler, 7-hydrolysis device, 8-urea solution inlet, 9-blower, 10-water cooled condenser, 11-dynamometer, 12-air filter, 13-boric acid solution, P-pressure sensor, Q-air flow meter, T-temperature sensor, C-NO_x sensor

Figure 2. Schematic of engine bench test system.

When the engine reaches a steady state, certain quantity of urea solution is injected into the hydrolysis reactor through urea solution inlet of hydrolysis reactor. With the blower through the air inlet to the hydrolysis

 DOI: 10.1051/matecconf/20179509002

reactor blowing into the air, gas is generated from the air outlet out. The generated gas through a water condenseris absorbed by a test acid solution that is placed below the water condenser. By changing the reaction conditions of urea hydrolysis, including the hydrolysis temperature and the initial mass, the flow and heat were measured and calculated. The hydrolysis of urea under different conditions was studied. The diesel engine model is JX493ZLQ3, and the basic parameters are shown in Table 1.

Table 1. Basic parameters of diesel engine.

Parameter name / unit	values
engine type	Water cooled, four stroke, inline, turbocharged inter-cooling, common rail
Cylinder	4
cylinder stroke	102mm
bore	93 mm
displacement	2.771 L
rated revolutions	3600 r/min
maximum power	85 k W

When the influence of the exhaust temperature on the urea hydrolysis, the control of the engine is controlled in 5 different stable conditions. 20ml urea solution was added to the reactor, and then through the air, urea hydrolysis experiments started. The hydrolysis reaction time was 2min, and then the samples were analyzed to study the hydrolysis of urea. When the operating conditions change with the exhaust temperature changes, the comparison of the hydrolysis rate and the theoretical value of the urea hydrolysis reactor is shown in Figure 3.

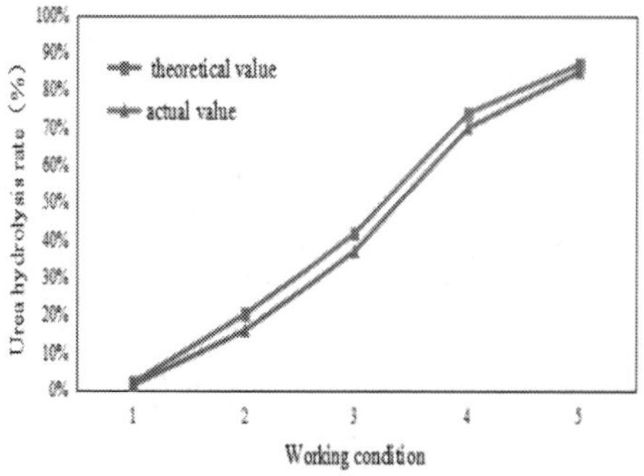

Figure 3. Comparison between theoretical value and experimental value of the rate of urea hydrolysis.

Under the same reaction time, with the increase of the hydrolysis temperature, the hydrolysis rate of urea is increasing, and the hydrolysis rate of urea is positively correlated with the hydrolysis temperature. It can be seen from the figure that the experimental data is always smaller than the theoretical data, and the lower the temperature, the greater the deviation. This is because on the one hand, impurities in the urea solution, will lead to a high value of theoretical calculation. On the other hand, when the temperature is low, the amount of ammonia produced is small, and it is easy to produce large measurement error.

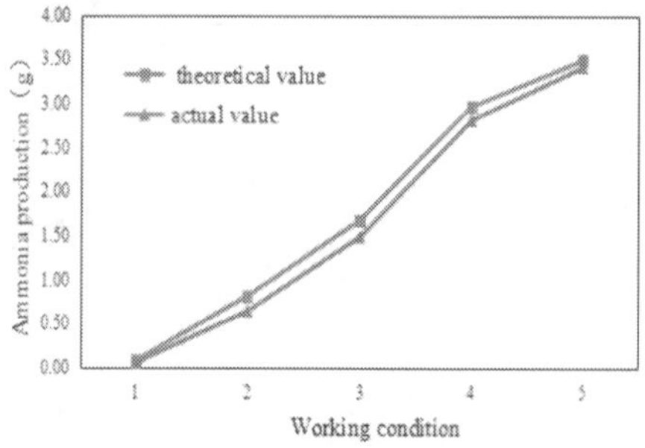

Figure 4. Comparison between theoretical value and experimental value of ammonia production.

As can be seen from Figure 4, the yield of urea hydrolysis and reaction temperature also showed a positive correlation, the higher the hydrolysis temperature, the faster the hydrolysis of urea, the faster the production of ammonia.

When the initial mass of urea hydrolysis is affected by the experiment, it is to control the engine to run in a stable condition. The urea solution of 30ml, 20ml, 40ml and 10ml were added into the hydrolysate, and then the air was added into the reactor to carry out the urea hydrolysis experiment. The hydrolysis reaction time was 2min, and then the samples were analyzed to study the hydrolysis of urea. Under different conditions, the comparison of the theoretical values is shown in Figure 5.

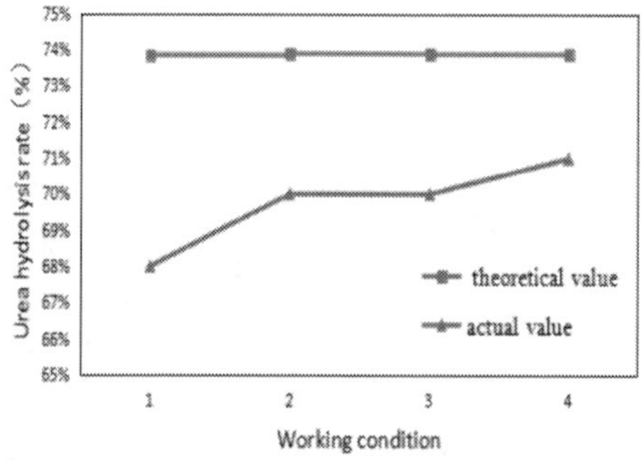

Figure 5. Comparison between theoretical value and experimental value of the rate of urea hydrolysis.

As can be seen from the figure 5, different initial mass of urea solution, in the same condition, the theoretical urea hydrolysis rate.is the same in the theory. This is because in the case of constant concentration, the urea hydrolysis rate is only related to the temperature, as shown in the formula of reaction rate constant [5], which has nothing to do with other factors. The urea hydrolysis rate of the experimental data was slightly lower than that of the theoretical data, and the value of the experimental data was slightly fluctuation. This is also caused by impurities in the urea and measuring errors in two ways.

 DOI: 10.1051/matecconf/20179509002

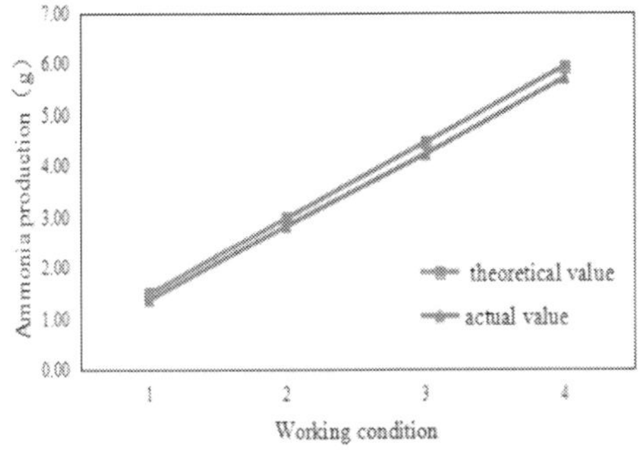

Figure 6. Comparison between theoretical value and experimental value of ammonia production.

As can be seen from the diagram, when the initial mass of the urea solution is increased, the yield of urea hydrolysis in the same condition is also increasing. After the 2min hydrolysis, the ammonia yield of 10ml urea solution was 1.37g. With the continuous improvement of the quality of 20ml, 30ml, 40ml, in the case of constant hydrolysis time, the amount of ammonia produced by urea hydrolysis reached 2.81g, 4.22g and 5.7g. Compared with the yield of 10ml urea solution, the amount of ammonia production increased by 2 times, 3 times and 4 times, respectively. In Figure 5 and figure 6, we can find that the higher the initial mass, the higher the initial quality, the higher the rate of urea hydrolysis. Therefore, under the condition of lower exhaust temperature, the urea solution can not be completely hydrolyzed, but the amount of ammonia required to complete hydrolysis can be achieved by increasing the initial mass of the urea solution.

5 Conclusion

(1)The ammonia reaction system of urea hydrolysis for vehicle belongs to the ACWU vapor liquid equilibrium system, and its thermodynamic simulation can be solved by extended UNIQUAC equation, modified Raoul equation, Henry equation and PHS equation of state. The calculated results of the method are in agreement with the experimental results, which can be used for the hydrolysis process of urea for vehicle.

(2)In the environment of urea hydrolysis for vehicle, the higher the hydrolysis temperature, the faster the hydrolysis of urea, and the faster the production of ammonia. Under the same conditions, the hydrolysis rate of urea is the same, the higher the initial quality, the more ammonia is produced.

(3)The reason of the error of the theory and the experiment mainly lies in the two aspects of the impurity and the measuring error in the urea.

Acknowledgements

This paper is supported by the National Natural Science Foundation of China(51306025) and the Fundamental Research Funds for the Central Universities(2572015AB13).

References

1. Zheng, Y., Luss, D., and Harold, M. SAE Int. J. Engines. **7**, 3 (2014)
2. Sahu J N, Gangadharan P, Patwardhan A V, et al. Ind. Eng. Che. Res. **48**, 3 (2009)
3. Xiang K. Therm. Pwr. Gen. **40**, 6 (2011)
4. Wei S., Yang N., Fu X., et al. Che. Eng. **39**, 11 (2011)
5. Zhang S., Shao F., Wang N. Comput. Appl. Che. **30**, 7 (2013)

Simulating Study on Drive System Performance for Hybrid Electric Bus Based on ADVISOR

Xingxing Wang, Hongjun Ni , Yu Zhu, Shuaishuai Lv, Mingyu Huang and Zheng Zhang

School of Mechanical Engineering, Nantong University, Nantong, 226019, China

Abstract. Hybrid electric bus has a number of advantages when compared with ordinary passenger cars, but in the dynamic matching and the vehicle performance are difficult to detect, thus limits its development process. In this paper, combined with the actual models, the hybrid electric bus module parameters were modified in the software of ADVISOR (Advanced Vehicle Simulator), main including: module of the vehicle, the wheel module, motor module, a battery module and engine module, three kinds of bus models for A, B and C were established, and the related performance that need to be analyzed was set up, such as acceleration, gradability, emissions and energy utilization and so on, in order to ensure the vehicle running in the same environment and convenient for comparison, a fixed vehicle driving cycles was chose, then the simulation results was analyzed, and the various performance was compared with the dynamic indicators and economic indicators which determined by referencing of traditional city bus standard and each other, and finally, the performance optimal model of B was chose out which can meet the demand, its related performance parameters of the simulation results are as follows: the best gradability is 26%, maximum speed is 72.7km/h, maximum acceleration is 1.7m/s2, 0~50km/h acceleration time is 9.5s and fuel consumption is 25L/km.

1 Introduction

Along with the development of the economy, rising energy demand and the increasingly serious environmental problems are not only the two major challenges facing the world today, the sustainable development of the automobile industry is also necessary to overcome the difficulties. Therefore, R&D personnel related to the development of electric vehicles, hybrid vehicles and fuel cell vehicles [1], [2]. City bus, is one of the urban residents travel tools, can not only use the limited road resources, for people to travel to bring convenience, but also reduce the traffic congestion, resource consumption and air pollution problems, therefore, it is one of the key research object, the application of hybrid technology on city bus is good for the driving condition that complex and changeable characteristics [3]-[5].

ADVISOR software is under the environment of MATLAB/Simulink software development of the advanced vehicle simulation software platform, it can predict, analyze and evaluate the performance, fuel economy and emissions of the vehicle and brake performance. Based on ADVISOR software own hybrid bus models, according to the known parameters of three kinds of hybrid bus model of components, and in particular vehicle driving cycle simulation, according to the simulation results of vehicle fuel economy and acceleration performance analysis, finally get a better comprehensive performance of the models [6]-[8].

2 The establishment of simulation model

Hybrid city bus should be level with the same length of the traditional city bus is quite dynamic. Therefore, similar to the design principle of the transmission system and the traditional city buses are judged the maximum speed, climbing ability, the ability to accelerate according to these three indicators. Reference of traditional city bus standard, determine the hybrid electric bus and the dynamic indicators and economic indicators of the following: a top speed of 70km/h, or 0 to 50 km/h speed up 30 s or less time, maximum gradability is more than or equal to 20% [1], [7].

The establishment of simulation model mainly comes with through the model of the software—the secondary development of hybrid electric vehicle (HEV), the object development including: module of the vehicle, the wheel module, motor module, a battery module and engine module. Development method is using MATLAB software to open the corresponding module of the script file, which is called "m" file. Modifying existing data script file according to the specific data of the target models in the corresponding module [9], [10].

2.1 Target vehicle parameters

For identify easily, the three kinds of object models named as A, B and C in this paper, the relevant parameters are listed in the following tables, where the Table 1 for the main parameters of the vehicle model, Table 2 for the main parameters of the engine model, Table 3 for the main parameters of the motor model, Table 4 for the main parameters of the battery model.

In addition, the rolling resistance coefficient and air resistance coefficient are 0.013 and 0.6 respectively, which are based on the concrete or asphalt pavement urban road sand vehicle types [11], [12].

Table 1. The main bodywork model parameters of A, B and C.

Title	A	B	C
Length (mm)	11980	11990	12000
Width (mm)	2540	2500	2550
Height (mm)	3180	3150	3250
Wheel base (mm)	5980	5875	6000
Front wheel base (mm)	2096	2096	2020
Back wheel base (mm)	1860	1836	1860
Windward area (m^2)	8.08	7.88	8.29
Shipping mass (kg)	12300	12500	12930
Gross mass (kg)	17800	18000	18000
Fuel type	diesel	diesel	diesel

Table 2. Engine parameter list.

Title	A	B	C
Delivery capacity	6.5L	6.7L	6.5L
Rated power	147kW /2500rpm	136kW /2500rpm	140kW /2500rpm
Maximum torque	730Nm /1400rpm	700Nm /1400rpm	700Nm /1400rpm
Maximum speed	2500rpm	2500rpm	2500rpm

Table 3. Motor parameter.

Title	A	B	C
Model	YHD280M-4	GHS280M-4	GHS280M-4
Mass (kg)	500	500	500
Voltage (V)	220	275	275
Electricity (A)	280	230	230
Rated power (kW)	120	155	155
Maximum torque (N·m)	2200	1950	1950

Table 4. Battery parameter.

Title	A	B	C
Voltage (V)	384	384	384
Capacity (F)	41.25	41.25	41.25

2.2 Second development of simulation model

According to the parameters of each mentioned model, modify the vehicle module, the wheel module, the motor module, the battery module and the engine module of the selected software. When using the selected vehicle top model of ADVISOR (Fig. 1), it will automatically call these modified modules. In this way, the simulation model of the target vehicle is established.

3 Simulation and results discussion

3.1 The simulation setup

Open the ADVISOR, through the vehicle input interface, choose the modified hybrid electric bus model, click the Continue button in the lower-right corner of the window interface, enter the parameters set interface, select condition CYC_CSHVR (composite driving cycle) as the simulation running road conditions, using the same kind of driving cycle in the process of simulation to facilitate comparing the performance of the three kinds of models, through checking the constant road grade option in the parameters set interface to validate vehicle climbing performance, and then set the climbing gradient for 10%, 15%, 20%, 25%, 20%, interval is 5%, in determining the optimal gradability and narrow interval value to 1%, the same method to determine the optimal gradability; for vehicle Acceleration Test, checking Acceleration Test option in the parameters set interface, in the pop-up interface set the shift delay time is 0.2s, initial battery capacity is 0.7, the current quality, 0 to 50 km Acceleration time parameters to Test the performance. The speed, elevation and running time of CYC_CSHVR conditions as shown in Fig. 2. The Acceleration parameter Settings interface as shown in Fig. 3 [13]-[16].

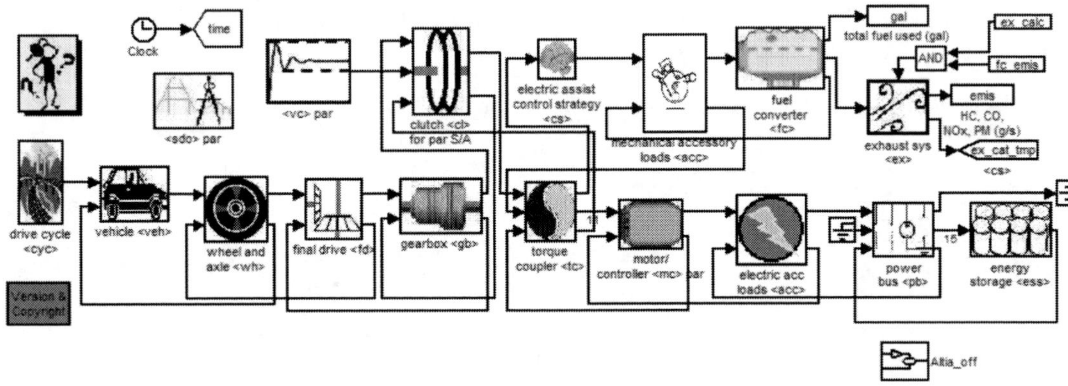

Figure 1. The top model of selected models

DOI: 10.1051/matecconf/20179509003

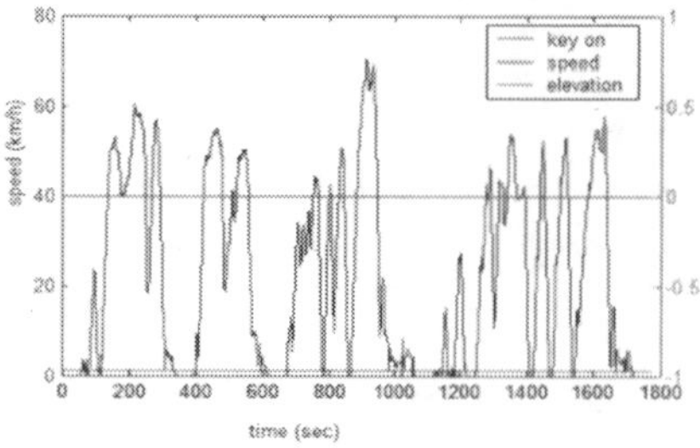

Figure 2. Parameters of CYC_CSHVR

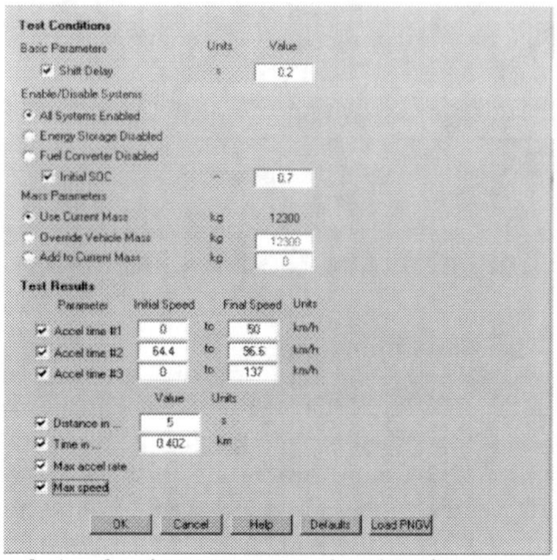

Figure 3. Acceleration parameter settings interface

3.2 The results of simulation

Simulated under CYC_CSHVR conditions, than got A, B and C three kinds of car models of operation conditions. The speed and storage battery charged state relationship with time as shown in Fig. 4, the Fig. 4(a), (b) and (c) respectively corresponding to the three kinds of bus models for A, B and C. From the Fig. 2 and Fig. 3 got that: A, B and C three models of the speed is in good condition and can meet the requirements of working condition, battery charged state (SOC) from 0.7 to 0.6, also in the reasonable scope of charged.

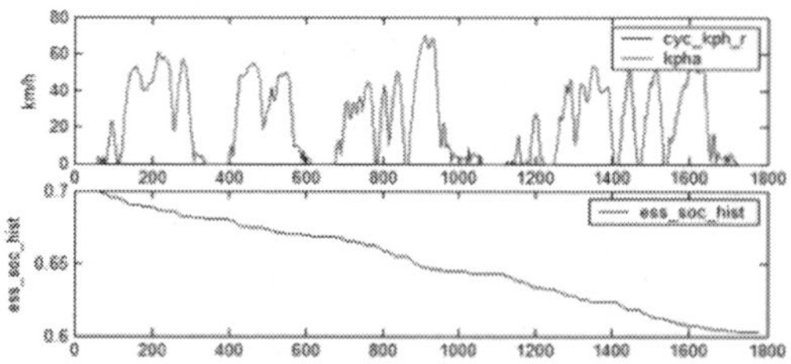

(a) Result of model A

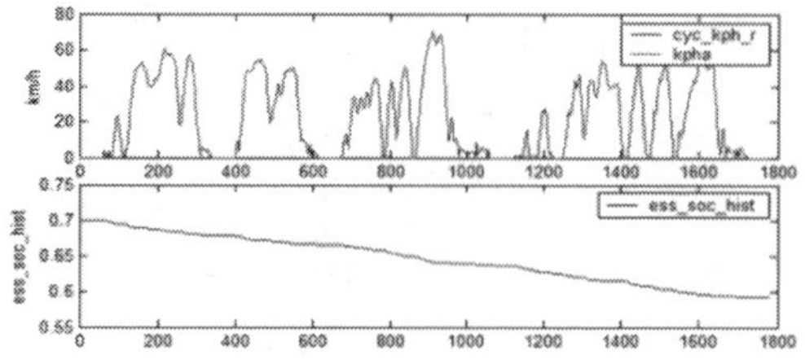

(b) Result of model B

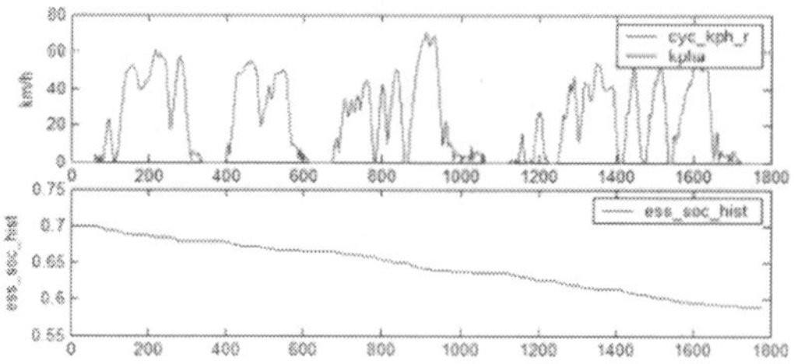

(c) Result of model C

Figure 4. Relationship between speed, battery charged and time

For vehicle climbing performance simulation, it is concluded that the best gradability of A, B and C is 24%, 26% and 20%.

For fuel consumption rate, acceleration, top speed and maximum acceleration of the simulation results are shown in Table 5.

Table 5. Part of the simulation results.

Title	A	B	C
Fuel consumption (L/km)	27.2	25	25.7
0~50km/h acceleration time (s)	10.6	9.5	9.2
Maximum speed (km/h)	74.5	72.7	69.8
Maximum acceleration (m/s2)	1.4	1.7	1.13

The simulation results of exhaust emission situation as shown in Fig. 5, the Fig. 5(a), (b) and (c) respectively corresponding to the three kinds of bus models for A, B and C, the specific discharge conditions are shown in Table 6.

The motor working characteristic curve as shown in Fig. 6, the Fig. 6(a), (b) and (c) respectively corresponding to the three kinds of bus models for A, B and C.

Simulation about vehicle interior parts, mainly involves the efficiency of energy utilization, details are shown in the Table 7.

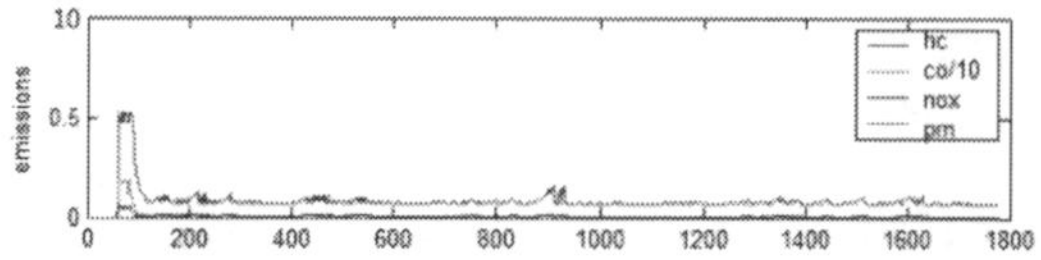

(a) Result of model A

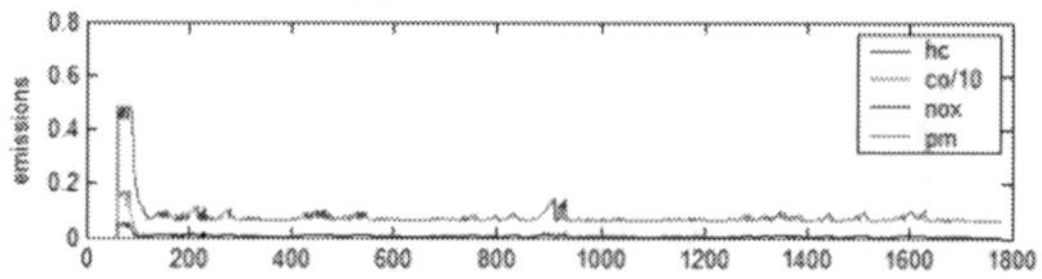

(b) Result of model B

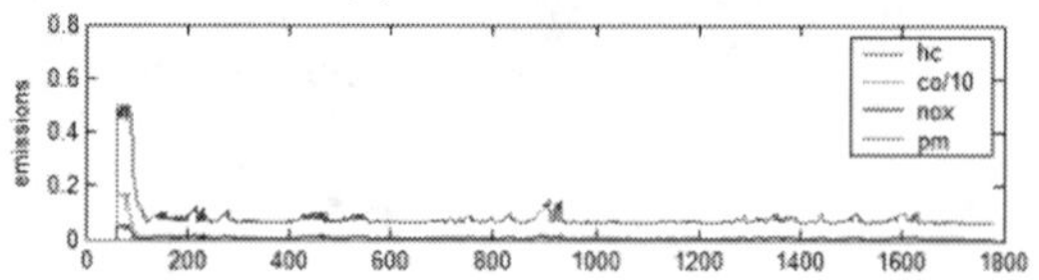

(c) Result of model C

Figure 5. Exhaust emission of A, B and C

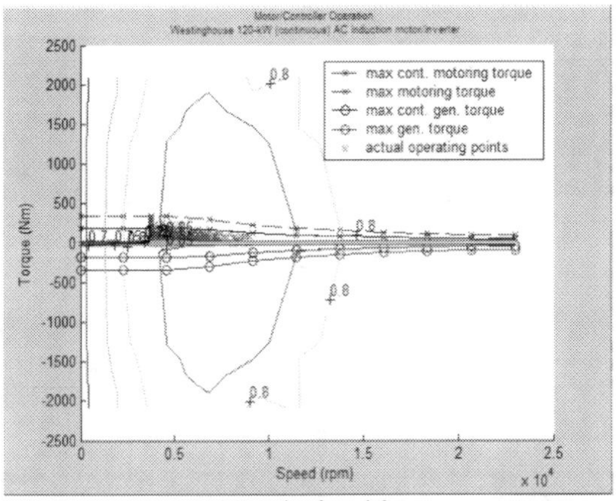

(a) Result of model A

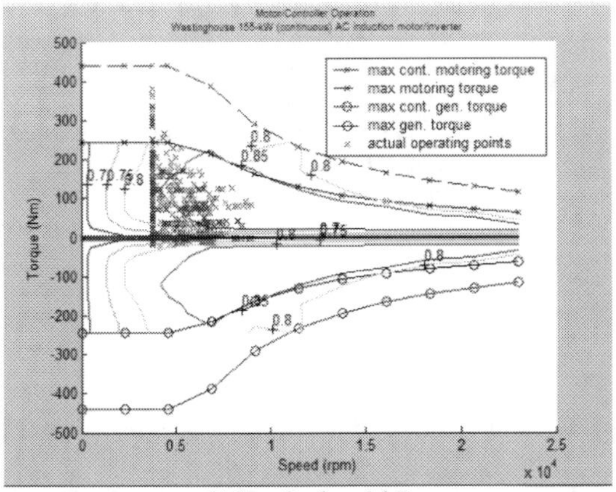

(b) Result of model B

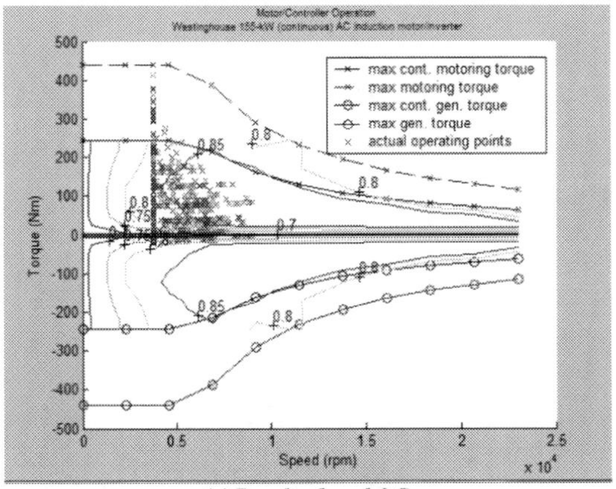

(c) Result of model C

Figure 6. The motor working characteristic curve

Table 6. Exhaust emission.

Title	A	B	C
HC (g/km)	13.925	9.713	13.227
CO (g/km)	18.124	13.876	17.116
NOX (g/km)	1.089	0.809	1.039

Table 7. The simulation results of energy utilization efficiency.

Title	A	B	C
Engine	0.28	0.3	0.3
Storage battery	0.68	0.88	0.88
Electromotor	0.83	0.81	0.81
Transmission	0.95	0.95	0.95
Wheel	0.95	0.95	0.95

3.3 Contrast and analyze

Based on speed and battery charged state relationship with time figure and Table 5 shows that A and B models can not only meet the requirements of the driving cycles—the maximum speed and acceleration to the corresponding values, but also achieve the above certain hybrid city bus performance standards, on the contrary, C bus are not able to meet the two conditions; For A and B models, through the analysis of Table 5: in terms of fuel consumption and accelerate, B is better than A, and in terms of exhaust emissions and energy use efficiency, B is also better than A ,the rest of the module are quite similar. Comprehensive the all above aspects of the comparison results, known B more able to meet demand.

4 Conclusion

In this paper, on the basis of the original hybrid bus, by modifying the module parameters, three kinds of bus models for A, B and C are established according to the existing models of the simulation model, and through the simulation and analysis, the relatively optimal models of B was determined, the related performance parameters of the simulation results are as follows: the best gradability is 26%, maximum speed is 72.7km/h, maximum acceleration is 1.7m/s^2, 0~50km/h acceleration time is 9.5s and fuel consumption is 25L/km.

(1) Using the method of modifying parameter on a existing car model in ADVISOR, establish a simulation model, convenient for using of ADVISOR software to develop more models and simulation in subsequent;

(2) Analyzed the vehicle's dynamic performance, energy efficiency, emissions and so on, easy to determine whether the vehicle's power performance meet demand and the energy matching is reasonable and exhaust emission is in line with the standard, for vehicle performance and other aspects of the design and optimization has the guiding effect.

Acknowledgment

This work is supported by National Science and Technology Support Program of China (2011BAG02B10), A Project Funded by the Priority Academic Program Development of Jiangsu Higher Education Institutions (PAPD), Jiangsu province policy guidance class plan (industry university research cooperation) project (BY2015047-02), Frontier and Key Technical Innovation Program of Nantong City (MS22015028).

References

1. C.J. Yan, J.W. Xia, Y.Y. Cui, *Keche Jishu Yu Yanjiu/ Bus & Coach Technology and Research* **36**(3), 8 (2014)
2. C. Yang, X.H. Jiao, L. Li, S.X. You, B.J. Yan, Z.L. Ma, *Jixie Gongcheng Xuebao/ Journal of Mechanical Engineering* **51**(22), 111 (2015)
3. J.P. Gao, J.Z. Ge, J.B. Zhao, H.B. Qiao, J.G. Xi, *Tuxue Xuebao/ Journal of Graphics* **35**(6), 905 (2014)
4. J. Wang, Q.N. Wang, P.Y. Wang, *Proc. Inst. Mech. Eng. Part C J. Mech. Eng. Sci.* **229**(15), 2760 (2015)
5. T. Ba, Y.H. Gao, Q.N. Wang, X.H. Zeng, *Qiche Gongcheng/ Automotive Engineering* **37**(7), 835 (2015)
6. Y.X. Chen, Q. Ma, X.X. Wang, *Zhonguo Keji Lunwen/ China Sciencepaper* **10**(19), 2261 (2015)
7. Y. Sun, L. Li, B.J. Yan, C. Yang, C.Y. Tang, *Mech Syst Signal Process* **68**, 416 (2015)
8. Y.M. Li, F. He, X.S. Jing, L. Yang, *Zhizaoye Zidonghua/ Manufacturing Automation* **36**(9), 68 (2014)
9. X.S. Hu, J.C. Jiang, B. Egardt, D.P. Cao, *IEEE Trans Ind Electron* **62**(12), 7847 (2015)
10. J. Wang, Q.N. Wang, X.H. Zeng, P.Y. Wang, *Proc. Inst. Mech. Eng. Part D J. Automob. Eng.* **229**(11), 1529 (2015)
11. X.H. Zeng, T. Ba, H. Tian, K. Wu, Q.N. Wang, *Jilin Daxue Xuebao (Gongxueban)/Journal of Jilin University (Engineering and Technology Edition)* **43**(2), 278 (2013)
12. F. Hu, Z.G. Zhao, Z.C. Sun, *Qiche Gongcheng/ Automotive Engineering* **34**(4), 283 (2012)
13. M.L. Dou, C. Shi, Z.L. Chu, G. Wu, Z.J. Ding, S.D. Chen, L.P. Xiong, Y. Hong, J.A. Wang, *Kongzhi Gongcheng/Control Engineering of China* **19**(4), 610 (2012)
14. X.H. Zeng, N.N. Yang, J.N. Wang, D.F. Song, N. Zhang, M.Q. Shang, *Mech Syst Signal Process* **60-61**, 785 (2015)
15. M.Y. Huang, X.B. Feng, D.T. Li, H.J. Ni, X.X. Wang, *Nantong Daxue Xuebao(Ziran Kexueban)/Journal of Nantong University (Natural Science Edition)* **12**(4), 29 (2013)
16. Y. Kim, K. Nam, H. Fujimoto, Y. Hori, Y. Hori, *Int. J. Mech. Eng. & Rob. Res* **4**(1), 249 (2015)

MATEC Web of Conferences **95**, 09004 (2017) DOI: 10.1051/matecconf/20179509004
ICMME 2016

Design of a Maglev Vibration Test Platform for the Research of Maglev Vehicle-girder Coupled Vibration Problem

Danfeng Zhou , Lianchun Wang, Jie Li and Peichang Yu

College of Mechatronics Engineering and Automation, National University of Defense Technology, 410073 Changsha, P.R. China

Abstract. The maglev vehicle-girder coupled vibration problem has been encountered in many maglev test or commercial lines, which significantly degrade the performance of the maglev train. In previous research on the principle of the coupled vibration problem, it has been discovered that the fundamental model of the maglev girder can be simplified as a series of mass-spring resonators of different but related resonance frequencies, and that the stability of the vehicle-girder coupled system can be investigated by separately examining the stability of each mass-spring resonator – electromagnet coupled system. Based on this conclusion, a maglev test platform, which includes a single electromagnetic suspension control system, is built for experimental study of the coupled vibration problem. The guideway of the test platform is supported by a number of springs so as to change its flexibility. The mass of the guideway can also be changed by adjusting extra weights attached to it. By changing the flexibility and mass of the guideway, the rules of the maglev vehicle-girder coupled vibration problem are to be examined through experiments, and related theory on the vehicle-girder self-excited vibration proposed in previous research is also testified.

1 Introduction

As the development of the maglev transport around the world, the maglev vehicle-girder coupled vibration problem has been discovered, and it gains more and more attention as more and more maglev test or commercial lines are constructed [1-4]. Recent test run of the Changsha urban maglev line revealed that the resonance is more easily to be encountered when the maglev train is passing through some steel girders at low speed. Similar problem has also been encountered in the Tangshan maglev test line. In [5], the coupled vibration problems in the EMS (Electromagnetic Suspension) maglev train are reviewed. In [6], the principle of the maglev vehicle-girder self-excited vibration problem was investigated in the frequency domain, and some important conclusions were drawn, but without experimental validation.

On a commercial maglev line, the parameters of the girders, such as the mass and the rigidness, are fixed and they are difficult to change in the presence of maglev vehicle-girder self-excited vibration problem; therefore, experimental study on a commercial line are quite limited to the given conditions. To overcome this problem, based on the maglev vehicle-girder coupled model developed in previous research, a maglev vibration test platform is built for experimental study of the self-excited vibration problem. The test platform equipped with a single electromagnet levitation control system and a flexible guideway. To best imitate the theoretical model of the maglev vehicle-girder coupled system, the levitation control system is equipped with the same sensors as a real maglev train does. As an approximation of the flexible girder in a maglev line, the guideway of the test platform is designed as a rigid beam supported by elastic springs. The mass of the rigid beam, as well as the elasticity of the springs, can be adjusted so as to study the stability of the coupled system under different girder conditions. This configuration provides a wide range of girder parameters which are benefit for the study of the conditions when the maglev vehicle-girder self-excited vibration problem may occur, and it also provides an approach to seek control algorithms that is useful to suppress the maglev vehicle-girder self-excited vibration through experiments.

2 Modelling of the maglev vehicle-girder coupled system

As shown in Fig. 1, the maglev vehicle-girder self-excited vibration generally occurs when a maglev vehicle is passing through an elevated girder at a very low speed, in which condition many levitation control units are capable of sensing the vertical displacement of the girder and adjusting their levitation currents thereafter. For some reason, the adjustments of the levitation currents form a consultant force which enlarges the displacement of the girder, resulting in an increasing vibration of the girder, which is called self-excited vibration (or resonance in some literature). It is believed that the self-excited vibration is partly caused by the flexibility of the girder, hence the displacement of the girder subject to external forces are investigated first.

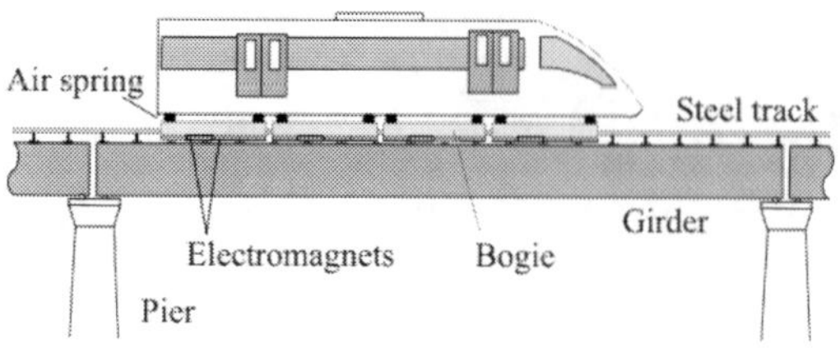

Figure 1. Schematic of a maglev vehicle passing through an elevated girder.

Given the span length of the girder, L, the mass per unit length and the bending stiffness of the bridge, ρ and EI, respectively, the displacement of the girder, $y(x,t)$, can be described by the following equation:

$$EI\frac{\partial^4 y(x,t)}{\partial x^4}+\rho\frac{\partial^2 y(x,t)}{\partial t^2}=f(x,t), \tag{1}$$

where $f(x,t)$ is the distributed force acting on the girder. For a maglev girder, it can be treated as a Bernoulli-Euler beam, and the vibratory motion of the beam can be decomposed into the superposition of all mode shapes; namely,

$$y(x,t)=\sum_{n=1}^{\infty}y_n(t)\phi_n(x), \tag{2}$$

where $\phi_n(x)$ is the n-th order modal shape, which is a combination of $\sin(\lambda_n x)$, $\cos(\lambda_n x)$, $\sinh(\lambda_n x)$ and $\cos(\lambda_n x)$; $y_n(t)$ is the generalized displacement of the n-th order modal shape. The coefficients and λ_n in $\phi_n(x)$ are determined by the boundary conditions of the beam. Substituting Eq. (2) into Eq. (1), multiplying both sides of the resultant equation by $\phi_n(x)$, and integrating both sides along the beam, gives:

$$\lambda_n^4 y_n(t)\frac{EI}{\rho}\int_0^L\phi_n^2(x)\mathrm{d}x+\ddot{y}_n(t)\int_0^L\phi_n^2(x)\mathrm{d}x$$
$$=\frac{1}{\rho}\int_0^L f(x,t)\phi_n(x)\mathrm{d}x. \tag{3}$$

Note that the mode orthogonality condition for linear beams has been applied in the computation. Eq. (3) can be rewritten in another form:

$$\ddot{y}_n(t)+\omega_n^2 y_n(t)=\frac{\bar{F}_n(t)}{M_n}, \tag{4}$$

where

$$M_n=\rho\int_0^L\phi_n^2(x)\mathrm{d}x \tag{5}$$

$$\bar{F}_n(t)=\int_0^L f(x,t)\phi_n(x)\mathrm{d}x \tag{6}$$

$$\omega_n^2=\frac{EI\lambda_n^4}{\rho} \tag{7}$$

Eq.(4) reveals that the dynamic response of the beam under external force excitations is equivalent to the superposition of the responses of a series of mass-spring systems under generalized force excitations. Compared with the length of the elevated girder, the length of the electromagnetic forces can be neglected, and they can be approximately treated as point contact forces. For the k-th electromagnetic force, F_k, we have

$$f_k(x,t)=F_k(t)\delta(x-x_k), \tag{8}$$

where $\delta(\cdot)$ is the Dirac delta function; and x_k corresponds to the location of the k-th electromagnetic force. Consequently,

$$\bar{F}_{n,k}(t)=\int_0^L f(x,t)\phi_n(x)\mathrm{d}x=F_k(t)\phi_n(x_k). \tag{9}$$

Substituting Eq.(10) into Eq.(5), the following equation can be obtained:

$$\ddot{y}_n(t)+\omega_n^2 y_n(t)=\frac{F_k(t)\phi_n(x_k)}{M_n}. \tag{10}$$

It may be noticed that the dynamic equation for a mass-spring resonator is

$$\ddot{y}(t)+\omega^2 y(t)=\frac{F(t)}{m}, \tag{11}$$

where m is the mass of the resonator, and ω is the natural frequency of the resonator. Comparing Eq.(11) with Eq.(10), it can be seen that the dynamic behavior of the elevated girder subject to an electromagnetic force is equivalent to the superposition of the responses of infinitely many mass-spring resonators under the excitation of the same force, as demonstrated in Figure 2.

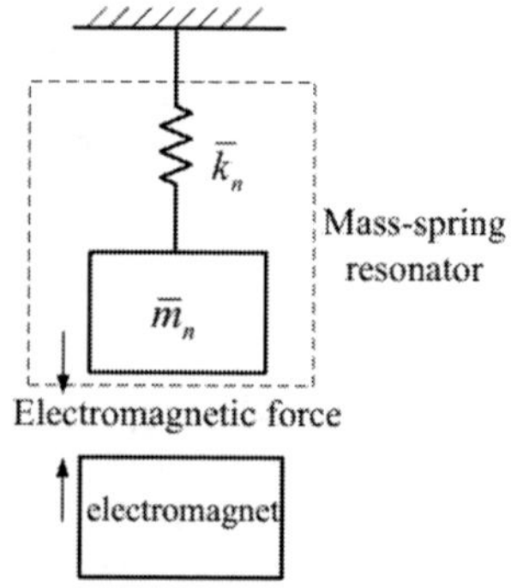

Figure 2. Model of a mass-spring resonator.

The equivalent mass and rigidness of the n-th resonator can be given by

$$\bar{m}_n=\frac{M_n}{\phi_n(x_k)}, \tag{12}$$

$$\bar{k}_n=\frac{M_n\omega_n^2}{\phi_n(x_k)}. \tag{13}$$

Since $\phi_n(x_k)$ is not a constant, Eqs.(12) and (13) gives a hint that the mechanical impedance of the girder is not a constant, either, but a variable depending on the location of the electromagnetic force. For example, the displacement of a simply supported beam can be described as a superposition of sinusoidal functions:

$$y(x,t)=\sum_{n=1}^{\infty} y_n(t)\sin\frac{n\pi x}{L}. \quad (14)$$

The equivalent mass of the n-th mass-spring resonator is:

$$\bar{m}_n=\frac{\rho L}{2}\csc\frac{n\pi x_0}{L}. \quad (15)$$

Here, csc(·) is the cosecant function. Consequently, the equivalent stiffness of the SDOF resonator can also be obtained, which is:

$$\bar{k}_n=\frac{EI}{2}\left(\frac{n\pi}{L}\right)^4\csc\frac{n\pi x_0}{L}. \quad (16)$$

By analogy to the SDOF system, the mechanical impedance of the n-th vibration mode of the girder can be defined as:

$$Z_n=\sqrt{\bar{k}_n\bar{m}_n}=\frac{\sqrt{EI\rho L}}{2}\left(\frac{n\pi}{L}\right)^2\csc\frac{n\pi x_0}{L}. \quad (17)$$

Eq.(17) shows that the mechanical impedance of the girder has a lower bound, which corresponds to the situation when the electromagnetic force is at the location where the displacement of the mode shape reaches its maximal value. Furthermore, the mechanical impedance of the girder is proportional to n^2, which indicates that the higher order the vibration mode is, the larger becomes the mechanical impedance, and the less likely the coupled vibration will occur. This explains why the maglev vehicle–girder self-excited vibration generally occurs at the fundamental order mode of the girder. For a full scale maglev train, the electromagnetic forces are distributed at different locations along the girder, thus the mechanical impedance for each electromagnet is different from others. In this way, the stability of the vehicle-girder coupled system can be deemed as a result of separated electromagnetic forces acting on different size of mass-spring resonators. If most of the electromagnet-resonator coupled systems are stable, the corresponding vehicle-girder coupled system would be stable; otherwise, the vehicle-girder coupled system would be unstable.

As a basic unit of the vehicle-girder coupled model, the investigation of a single electromagnet-resonator coupled system is essentially important to the theoretical analysis of the stability of the vehicle-girder coupled system. This work has been reported in previous research [6], yet has not been verified experimentally due to the limit of conditions on a real maglev test line. Therefore, the purpose of this paper is to validate the theoretical conclusions by setting up a small scale experimental platform which includes a levitation control system and a mass-spring resonator corresponding to a lower vibration mode of the girder. Compared with the full-scale maglev train, the small scale platform is convenient to change the track and levitation parameters, and to obtain a wider range of parameters, which will benefit for the validation of the theoretical analysis under some extreme conditions.

3 Design of the maglev vibration test platform

As discussed in the previous section, the maglev vehicle-guideway coupled test platform should include a mass-spring resonator and a levitation control system. The schematic, which demonstrates the basic structure of the platform, is shown in Fig. 3, in which it can be seen that this platform consists of an electromagnet and a segment of track which is supported by springs. To keep balance, the track is supported by two springs located at each terminal of the beam; therefore, the mass of the track and the flexibility of the springs make up of a mass-spring resonator in vertical direction. Since the rigidness of the beam is much larger compared with the rigidness of the springs, the flexibility of the beam can be neglected, and the natural frequency of the mass-spring resonator can be approximated calculated by

$$\omega_0=\sqrt{\frac{2k_s}{m}}, \quad (18)$$

where k_s is the rigidness of a single spring, m is the mass of the track.

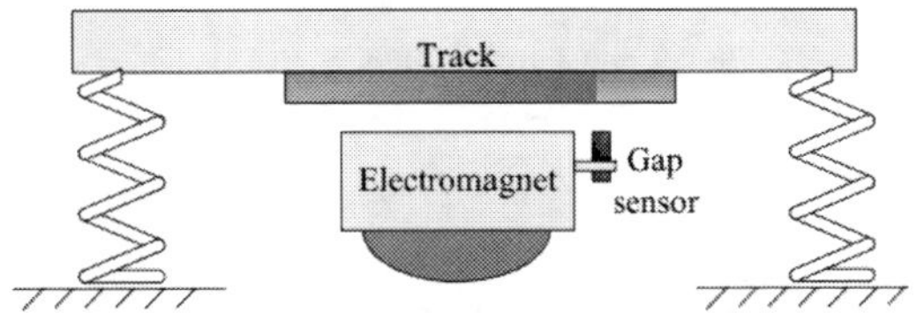

Figure 3. Schematic of the vehicle-guideway coupled vibration test platform.

It may be noticed that the structural of the track system in Fig. 3 cannot keep balance since the rail may swing horizontally. To solve this problem, a guiding gear in vertical direction is needed. Moreover, a guiding and supporting mechanism is also required to guide the movement of the electromagnet, and to support the electromagnet when the equipment is not suspending. A photo of the designed platform is shown in Figure 4, in which the guiding system and the supporting system are considered.

For a real maglev system, the damping of the girder plays an important role on the stability of the vehicle-girder coupled system, hence the damping of the track in the test platform should also be considered. To meet this requirement, several small dampers, whose damping can be adjusted by a screw, are fixed to the track.

DOI: 10.1051/matecconf/20179509004

Figure 4. Produced test platform.

To adjust the stiffness of the mass-spring resonator, more springs can be applied to the track to enhance the stiffness of the resonator in vertical direction. Similarly, to adjust the mass of the mass-spring resonator, extra weights (steel blocks) can be attached to the track. Increasing the mass of the track and the stiffness of the springs simultaneously would increase the mechanical impedance of the track. These adjustments can also guarantee that the frequency of the resonator is able to cover a rather wide range so as to simulate different guideway conditions.

A single electromagnet is designed for the levitation control system, and the material of the iron core is the same as that used in a real maglev train. But the material of the coil is enamelled copper wire, rather than aluminium wires used in a real maglev train. Some key parameters of the test platform are listed in Table 1.

Table 1. Key parameters of the designed platform.

Parameter	Value	Unit
Mass of track	5.4	kg
Mass of suspension system	8.02	kg
Mass of small weight	0.33	kg
Mass of big weight	0.99	kg
Area of electromagnetic pole	0.001	m^2
Number of coil	843	-

To achieve stable levitation, a gap sensor is essential to measure the air gap between the rail and the electromagnetic poles; therefore, an eddy-current gap sensor is mounted to one side of the electromagnet and it moves along with the electromagnet. Besides, an accelerometer, which is used to measure the acceleration of the electromagnet in vertical direction, is also mounted to the other side of the electromagnet. The electromagnet is driven by a PWM chopper and is controlled by a digital micro-processor. To study the influence of the control voltage on the stability of the coupled system, the voltage of the chopper is provided by an adjustable power supply with 60V maximum output.

4 Test result of the platform

4.1 Frequency range test of the track

In a real maglev system, the fundamental frequency of the elevated girder is generally below 16Hz, and the fundamental frequency of the concrete girder is generally lower than that of the steel girder. For example, the fundamental frequencies of the girders in the Tangshan maglev test line have been measured, and it is found that the frequency of the concrete girder is around 9Hz for the 18-m long concrete girder, while it is over 13Hz for the steel girder of the turnover. To be close to reality, the frequency range of the track in the test platform should cover the frequency range of the real girders. Since the frequency of the track can be lower down by attaching weights onto it, the highest natural frequency should be considered first. Fig. 5(a) shows the free vibration of the track by knocking it with a hammer, which appears that the attenuation ratio of the vibration resembles that of a concrete girder. Figure 5(b) shows the power spectrum density (PSD) of the gap signal when the track is repeatedly knocked. It can be seen that natural frequency of the track without additional weight is 18.4Hz, which is higher than the fundamental frequency of the real steel girders. If higher natural frequency is required, additional springs can be applied to increase the stiffness of the track.

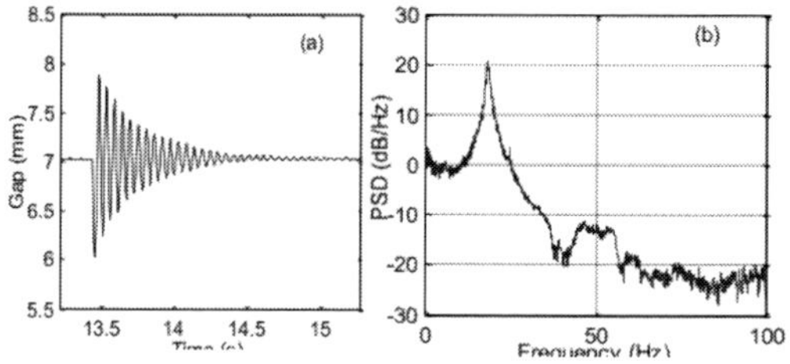

Figure 5. Free vibration and PSD of the track.

The natural frequency of the track can also be decrease by attaching additional weights to the rail. Field tests show that the natural frequency of the track decreases to 16.4 Hz when 3 small weights are attached to each side of the track, and the frequency drops down to 12.0 Hz when 4 big weights are added to each side of the track. Lower frequency can also be achieved by adding more weights to the track.

To investigate the differences of the steel girder and the concrete girder, the dampers in the test platform can be adjusted so as to obtain required damping ratio of the track.

4.2 Stability test of the coupled system

In the previous research on the principle of the maglev vehicle-girder self-excited vibration problem [6], the frequency characteristics of the levitation system and the girder were separately analyzed, and the stability of the coupled system was investigated by the open–loop system which joins these two subsystems together. The

Nyquist stability theory was employed as the basic stability criterion. The investigation drew a conclusion that if the self-excited vibration occurs, the frequency of the vibration is determined by both the unstable mode frequency of the girder and the critical frequency of the levitation system. The latter, which was defined in the literature [6], corresponds to the frequency when the phase response of the levitation system crosses the −180° line downwards. Theoretically, the frequency of the self-excited vibration is higher than either of these two frequencies mentioned above (conclusion 1).

Another conclusion had also been drawn in [6] that when the mode frequency of the girder is lower than the critical frequency of the levitation system, the self-excited vibration is unlikely to occur; otherwise, the self-excited vibration will sure to occur when the damping of the track is neglected. However, when the damping of the track is taken into account, the self-excited vibration only occurs when the mode frequency of the girder is close to the critical frequency of the levitation system (conclusion 2).

In this section, these conclusions will be tested through a series of experiments in the platform. It has been tested in the previous section that the natural frequency of the test platform covers the interested frequency range for a real maglev girder, thus the experiments will be organized by changing the natural frequency of the track at each time. The experiments starts under the condition when the track is free of extra weights, and followed by the decrease of the track natural frequency by adding some weights on the track at each time. The test results will then be listed in a table for comparison, so as to testify the conclusions mentioned above.

Fig. 6 shows the test result of the case when no extra weights are added. It can be shown that after a short period of time, the levitation begins to oscillate, and self-excited vibration occurs afterwards. Investigation of the track indicates that under such a circumstance, the natural frequency of the track is 18.4 Hz, which is quite high compared with the critical frequency of an ordinary levitation system, thus self-excited vibration tends to occur. Fig. 6(b) shows the corresponding PSD of the levitation gap, which shows that the frequency of the self-excited vibration reaches 20.7 Hz, which is higher than the natural frequency of the track. This agrees with the conclusion 1 mentioned above.

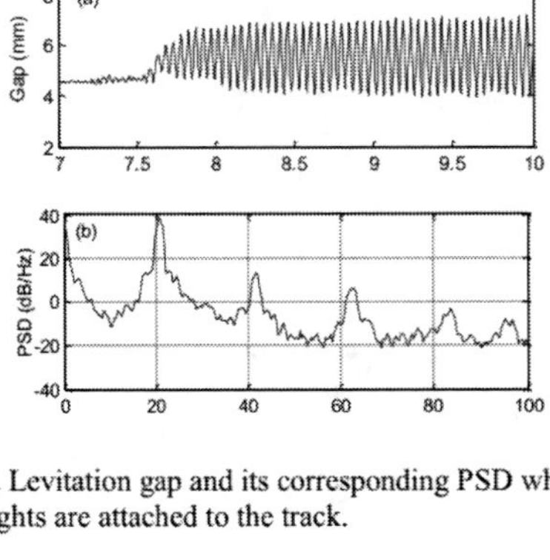

Figure 6. Levitation gap and its corresponding PSD when no extra weights are attached to the track.

Other tests under different track conditions are listed in Table 2, and the stability of these tests are also presented therein. It should be noticed that the values of the vibration frequency are obtained by power spectrum analysis of the levitation gap, regardless of the stability of the coupled system. The vibration frequency can be identified even for the stable cases, which may arise from external disturbances which cause vibratory movements.

It can be seen that as the frequency of the track lowers down, the coupled system becomes stable, this may agree with conclusion 2 presented above. It can also be seen that all the vibration frequencies are higher than the natural frequencies of the track. Therefore, conclusion 1 is also validated.

Table 2. Test results under different track conditions.

Track freq. (Hz)	k_p	k_d	Vibration freq. (Hz)	Stability
12.0	40	50	13.7	Stable
12.0	60	50	15.2	Stable
12.0	80	60	15.6	Stable
13.2	60	60	15.7	Stable
15.8	50	50	17.1	Stable
15.8	70	70	17.8	Stable
16.4	60	60	19.0	Stable
18.4	60	60	20.7	Unstable

Table 2 also gives a hint that the control parameters also affects the frequency of the self-excited vibration, which is a topic to be discussed in further study.

5 Conclusions

The maglev vehicle-girder coupled vibration problem has been encountered in many maglev test or commercial lines, which significantly degrade the performance of the maglev train. This paper presents a methodology to study the principle of the coupled vibration problem by designing a small test platform. The results show that the equipment can well be used to study the fundamental model the vehicle-guideway coupled system. Using this platform, some theoretical results proposed previously are also testified, which shows that the designed platform is a good supplement for the study of the maglev vehicle-girder coupled vibration problem.

Acknowledgement

This work was supported by the National Key Technology R&D Program of the 12th Five-year plan (No. 2012BAG07B01-13), and National Nature and Science Foundation of China (No. 11302252).

References

1. H. Wang, X. Zhong, G. Shen, Journal of Vibration and Control. **21**, 408–416 (2015)
2. J.S. Lee, S.D. Kwon, M.Y. Kim, et al., Journal of Sound and Vibration. **328**, 301–317 (2009)

DOI: 10.1051/matecconf/20179509004

3. B.M. Jin, I.G. Kim, Y.J. Kim, et al., *International Conference on Electrical Machines and Systems*; Seoul, Korea. 1959–1962 (2007)
4. K.J. Kim, J.B.Han, H.S. Han, et al., Vehicle System Dynamics.**53**, 587–601 (2015)
5. D.F. Zhou, C. H. Hansen, J. Li, et al., International Journal of Acoustics and Vibration. **15**, 10–23 (2010)
6. D.F. Zhou, C.H. Hansen, J. Li, Journal of Sound and Vibration. **330**, 883–901 (2011)

 DOI: 10.1051/matecconf/20179509005

Design and Analysis of the AlNiCo Hybrid Magnet in EMS Maglev Vehicles

LV Chao and LI Jie

College of Mechatronics Engineering and Automation, National University of Defense Technology, Changsha, Hunan 410073，China

Abstract: In order to solve the problem of hybrid electromagnet lock orbit, we design a new type of AlNiCo-NdFeB hybrid levitation electromagnet. The theoretical analysis has be carried on and mathematical model is established for AlNiCo-NdFeB hybrid levitation electromagnet. Through two dimensional simulation, the electromagnetic characteristics of the suspended electromagnet are analyzed in the 3 typical operating conditions , which are in heavy load at gap 8mm, in full load at gap 16mm and in no-load at gap 3mm. And it's compared with the traditional electromagnetic magnet and NdFeB hybrid electromagnet. Calculation and analysis show that the new hybrid levitation electromagnet can effectively solve the problems of the electromagnet lock orbit, at the same time, have a good dynamic performance and suspension regulation performance.

1 Introduction

EMS maglev train suspended on the track in a given gap by the attraction of the electromagnet, with the advantages of small bending radius, strong gradeability, low noise, comfort and safety, etc. It gained more and more attention and favor, and it has been hailed as urban internal and inter-city new traffic system in the 21st century [1, 2].

The EMS maglev train electromagnet species are the electromagnetic suspension, electromagnetic-permanent magnet hybrid levitation, superconducting suspension, etc [3]. The representation of the EMS maglev train is the HSST maglev train in Japan and TR08 maglev train in Germany. They have adopted conventional electromagnetic suspension electromagnet structure, matured in technology, realized commercial operation respectively in Japan and shanghai china [4]. And conventional electromagnetic levitation maglev trains provided all electromagnet levitation force, in order to meet the design requirements of bearing capacity, witch need large suspended current, and that lead to the coil current density sharp fever beyond its critical value.Under the condition of full load, although the conventional electromagnetic levitation electromagnetic ferromagnetic are unsaturated, saturated. But due to the current saturation, carrying capacity cannot be improved.

To solve the defects existed in the conventional electromagnetic, the new suspension system was designed. The permanent magnet was added to the electromagnet, then electromagnetic permanent magnet hybrid suspension system formed. Under the condition that the carrying capacity is not reduced, the suspension current is reduced, and the current density of the coil is controlled in the range of no sharp heating of the conductor. At present, the electromagnetic permanent magnet hybrid electromagnet structure are series the permanent magnets in the magnetic yoke, series coil magnetic potential [5]. After adding permanent magnets, suspended current decreased, and suspended power consumption decreased. Coil winding of the current density can be greatly reduced, the coil heating problems are solved, and suspended sensor working environment is improved.

However, the existing electromagnetic permanent magnet hybrid suspension system has inherent defects. It usually adopts zero current suspension. That is to say, in the equilibrium position, electromagnet provides most of the levitation force. The square of the electromagnetic force of permanent magnet and suspension gap is inverse proportion. So in the case of levitation gap smaller, suction increases quickly, easy to lose control, adsorption on the orbit, cause harm to the train.

Take the advantage of energy-saving of electromagnetic permanent magnet hybrid system, at the same time, improv the ability to prevent the adsorption orbit. To improve the existing electromagnetic permanent magnet hybrid electromagnet structure. The two kind of permanent magnet NdFeB and AlNiCo are seried into the yoke, instead of traditional permanent magnet part of hybrid electromagnet. NdFeB permanent magnet is used as the permanent magnetic circuit, which is the main source of levitation force, clinging to reduce power consumption. At the same time, AlNiCo permanent magnet is used as the reversible magnetic circuit, while the electromagnetic coil is designed around the alnico magnets, constitute the modules to prevent lock orbit.

In a new hybrid levitation electromagnet, when loading a positive dc pulse, AlNiCo magnets was magnetized, produced the same pole direction as the

 DOI: 10.1051/matecconf/20179509005

NdFeB magnetic, thus magnet magnetic force is strengthened; When loading reverse dc pulse, AlNiCo magnets was reverse magnetized, so magnet magnetic force reductions and suction less than body gravity, thus the body fall down from orbit, achieve the function of prevent lock orbit.

This article mainly carries on the finite element electromagnetic simulation, and contrastive analysis with the electromagnetic magnet, traditional electromagnetic permanent magnet hybrid electromagnet. Demonstrate the performance and feasibility of it [6]. For applied in maglev train, lay the foundation for the appliment of new electromagnetic permanent magnet hybrid levitation electromagnet structure , which have abilities of prevent lock orbit and good dynamic performance, in maglev train.

2 New hybrid electromagnet structure model

Compared with the traditional hybrid electromagnet, on the basis of the original structure, this new hybrid levitation electromagnet series AlNiCo permanent magnet in the yoke. The 3 kinds of electromagnet coil has 360 turns. Materials of new hybrid electromagnet is the same as traditional electromagnet. Specific size is shown in Fig. 1 . In (a), the thickness of NdFeB is 26 mm, and it' s height is 90 mm. In (b), the thickness of NdFeB is 26 mm, and it's height is 90 mm; the thickness of AlNiCo is 26 mm, and it's height is 90 mm. Orbital size design adopted with existing rail F the corresponding specifications.

As shown in Fig. 1, from left to right, followed by NdFeB magnet, AlNiCo NdFeB magnet blend, electromagnetic magnet. The grey part of the figure is NdFeB permanent magnets. Upper to orbit, the white part of the lower is permeability core.

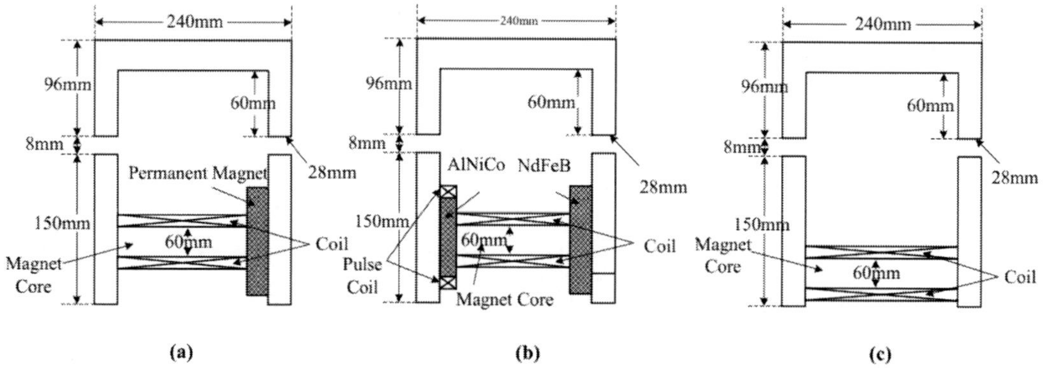

Figure 1. 3 kinds of electromagnet structure model.

3 Calculation of electromagnetic force

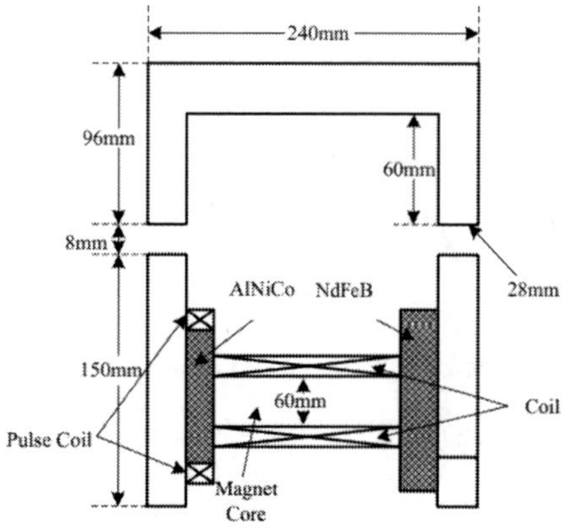

Figure 2. AlNiCo hybrid electromagnet structure model.

By the ampere loop theorem:

$$\oint H.dl=NI \tag{1}$$

Thus:

$$H=NI/l \tag{2}$$

Among them, N is defined as the coil number of turns, I is defined as the coil current. Set core of absolute permeability as , thus the magnetic induction intensity :

$$B=\mu H=\frac{\mu NI}{l} \tag{3}$$

The core flux:

$$\phi=BS=\frac{\mu NIS}{l} \tag{4}$$

Magnetic circuit magnetic potential:

$$Em=\oint H.dl=NI \tag{5}$$

By the magnetic circuit ohm's law can be core magnetic resistance of magnetic circuit.

 DOI: 10.1051/matecconf/20179509005

$$Rm = \frac{Em}{\phi} = \frac{l}{\mu S} \quad (6)$$

The magnetic circuit equation is established:

$$R_{fe}\phi + R_{co}\phi + R_{pm}\phi + R_{air}\phi = NI + E_{pm} + E_{co} \quad (7)$$

Thus:

$$\phi = \frac{NI + E_{co} + E_{pm}}{R_{air} + R_{fe} + R_{co} + R_{pm}} \quad (8)$$

$$B = \frac{\phi}{A} = \frac{NI + E_{co} + E_{pm}}{(R_{air} + R_{fe} + R_{co} + R_{pm})A} \quad (9)$$

Electromagnetic force:

$$\mathrm{F} = \frac{dW_{air}}{dx} = BHA = \frac{B^2 A}{\mu_0} = \frac{\phi^2}{\mu_0 A} = \frac{1}{\mu_0 A}\left(\frac{NI + E_{co} + E_{pm}}{R_{air} + R_{fe} + R_{co} + R_{pm}}\right)^2 \quad (10)$$

The magnetic resistance :

$$R_{air} = 2x / \mu_0 A_{air} \quad (11)$$

$$R_{fe} = L_1 / \mu_{fe} A_{fe} \quad (12)$$

$$R_{pm} = L_2 / \mu_{pm} A_{pm} \quad (13)$$

$$R_{co} = L_3 / \mu_{co} A_{co} \quad (14)$$

4 Under the typical levitation gap, analysis electromagnetic force

4.1. The selection of simulation conditions

Maglev train has three typical load conditions, the no load in 23 t, heavy load in 31. 5 t, full load in 35. 5 t, and whole vehicle has 40 suspension electromagnet. A Maglev train has a total of 40 suspended electromagnet. Corresponding to each electromagnetic module, no load in 0. 575 t, heavy load in 0. 7875t, full load in 0. 8875t. The electromagnetic characteristics of the suspended electromagnet in 3 typical operating conditions are analyzed, which are suspended in heavy load at gap 8mm, full load at gap 16mm, and no load at gap 3mm. Through the simulation of the rated suspension gap and floating gap, the two typical working conditions of the new type electromagnet are compared and analyzed. When the gap is relatively small, the traditional hybrid electromagnet in the suspension, will be adsorbed on the track, after power off, the electromagnet can not drop the case. So in the manufacture of mixed suspension electromagnet, two U-shaped iron core pole installed anti locking-track copper, in order to ensure minimum suspension air gap of 3 mm. And in the case of no load, it is more likely to happen adsorption orbit fault. In the case of the gap 3mm adsorption orbit, the aluminum nickel and cobalt reverse magnetization can be safely separated from the adsorption orbit state, to test whether it has the function of preventing the adsorption orbit.

4.2 Simulation and analysis of electromagnetic force in suspension

4.2.1 Heavy load suspended at gap 8mm

In the gap 8mm, the current from 0A to 30A, the electromagnetic force vary with the change of the current situation, as shown in Fig. 3.

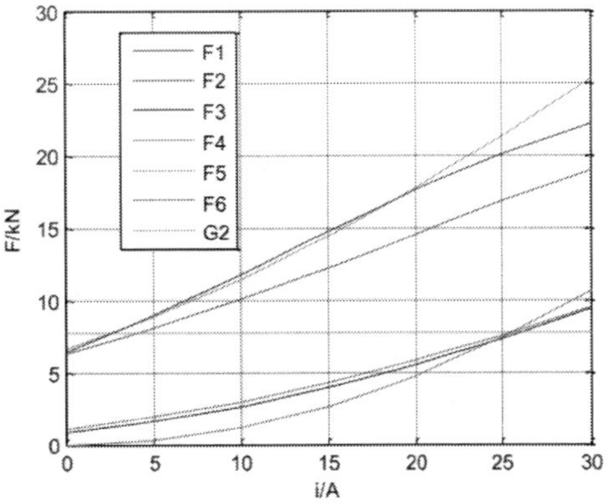

Figure 3. The electromagnetic force changes with the current situation at gap 8mm.

Notes: F1 is simulation value of electromagnetic force of NdFeB hybrid electromagnet; F2 is simulation value of electromagnetic force of AlNiCo (positive) hybrid electromagnet; F3 is simulation value of electromagnetic force of AlNiCo (negative) hybrid electromagnet; F4 is simulation value of electromagnetic force of electromagnetic magnet; F5 is theoretical calculation of electromagnetic force of AlNiCo (positive) mixed electromagnet; F6 is theoretical calculation of electromagnetic force of AlNiCo (negative) mixed electromagnet; G2 is the gravity of the electromagnet is under heavy load;

Can be seen from the figure,suspended in the gap 8mm with heavy load, AlNiCo (positive) hybrid electromagnet suspension current is 3A or so, slightly smaller than NdFeB hybrid electromagnet. From the electromagnetic force curve with the current changes, we can know, AlNiCo (positive) hybrid electromagnet and traditional NdFeB hybrid electromagnet has similar characteristics of suspension and curve slope for higher, performance is better than the traditional mixed electromagnet. From the graph, we can see that the curve of the simulation curve and the theoretical calculation of AlNiCo (positive) hybrid electromagnet is basically consistent with the curve. AlNiCo (positive) hybrid electromagnet has the same energy saving effect as the traditional mixed suspension electromagnet, and the control performance is better.

 DOI: 10.1051/matecconf/20179509005

4.2.2 Full load suspended in gap 16mm

Similarly in the 16mm gap, to establish a current range from 0A to 80A, the levitation force with the change of the current situation. As shown in Fig. 4.

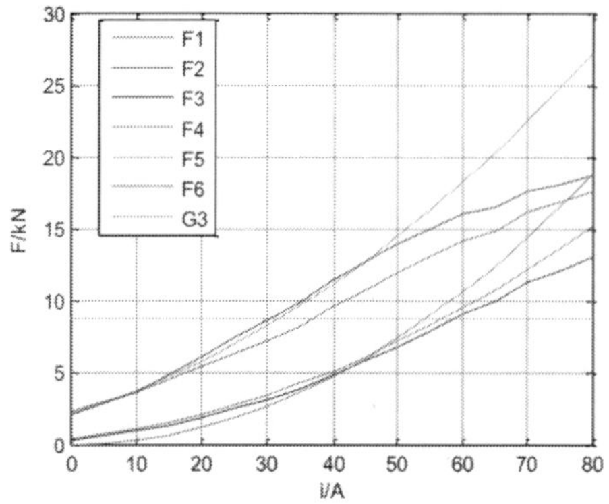

Figure 4. The electromagnetic force changes with the current situation at gap 16mm.

Notes: G3 is the gravity of the electromagnet is under full load;

When the full load of 16mm is suspended, the electromagnetic electromagnet current reaches 52A, which almost reaches the saturation state. The traditional neodymium iron boron hybrid electromagnet current is 36A, and the floating current is relatively small. And AlNiCo (positive) hybrid electromagnet current is only 30A, the current is relatively low, and there is a lot of magnetic saturation allowance. It is in the middle of the electromagnetic force curve, the slope is big, the dynamic control performance is very good. Therefore, the aluminum nickel and cobalt mixed electromagnet is relatively small, the current load capacity of the circuit device can be relatively low, which is beneficial to simplify the power supply system and reduce the cost.

4.2.3 No load suspended in gap 3mm

Also at the gap 3mm, to establish a current range from 0A to 30A, the levitation force with the change of the current situation. As shown in Fig. 5.

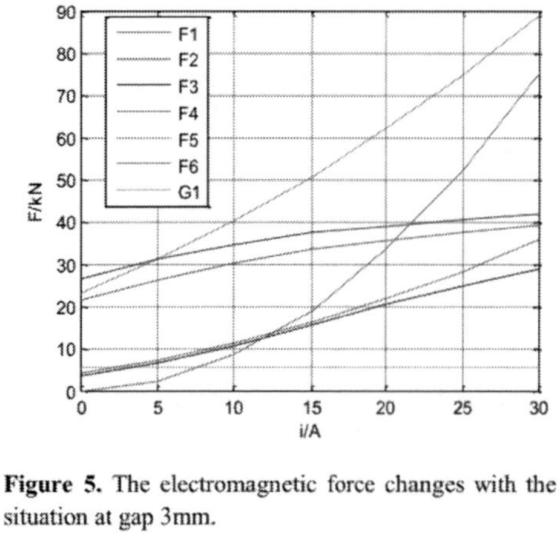

Figure 5. The electromagnetic force changes with the current situation at gap 3mm.

Notes: G1 is the gravity of the electromagnet is under no load;

No load at gap 3mm is the position of the hybrid electromagnet which is easy to be adsorbed on the track. Usually, if the hybrid electromagnet is adsorbed on the track, then the reverse current is large, so that it falls off the track. But if this is the case, disconnect the power supply, hybrid electromagnet will lock rail, may damage the body. Therefore, at the gap 3mm, the investigation of the electromagnet in the 0 current situation, whether it has the ability to prevent the death of the dead, it is very important.

Can be obtained from Fig. 5, when the current is zero, NdFeB hybrid electromagnet electromagnet force about 22kN, AlNiCo (positive) hybrid electromagnet electromagnetic force is about 26 kN, and the electromagnet in the no-load under the force of gravity for 5. 635kN, electromagnetic force is much larger than the gravity. So the hybrid electromagnets would lock the track. While AlNiCo (positive) hybrid electromagnet is adsorbed on the track, the pulse power supply through the reverse current around the aluminum nickel cobalt , so that the magnetic reverse, obtained AlNiCo (negative) hybrid electromagnet. And the excitation current is zero, the electromagnetic force of AlNiCo (negative) hybrid electromagnet is 3. 44kN, which is much less than the weight of the bearing. The electromagnet track will not be locked. Therefore, it has a good ability to prevent locking track. Parallel hybrid magnet and solenoid type electromagnet is almost as electromagnetic properties, not like NdFeB hybrid electromagnet that locking track.

5 Conclusion

In this paper, the theoretical derivation and Simulation of the new type of AlNiCo-NdFeB hybrid electromagnet are carried out. Confirmed the this structure can not only improve the electromagnetic force but also reducing the working current, and the aluminum nickel cobalt are reversely magnetized, thus achieving the goal of anti lock track.

Under the condition of heavy load at gap 8mm, excitation current of AlNiCo hybrid electromagnet, compared to solenoid type electromagnet, reduced about 750%, that effectively reduces the excitation current of the electromagnet, so as to achieve the purpose of energy saving. At the rated operating gap, the slope of the current-electromagnetic force curve of AlNiCo hybrid electromagnet is larger than that of NdFeB hybrid electromagnet and the electromagnetic electromagnet, and the dynamic performance of AlNiCo hybrid electromagnet is optimal.

And under the condition of heavy load at gap 16mm, operating current of AlNiCo hybrid electromagnet is only about 83% of NdFeB hybrid electromagnet, and is about 57% of the electromagnetic magnet. AlNiCo hybrid electromagnet can greatly reduce the float current, and effectively reduces the power consumption of the system and installed capacity. Electromagnet volume, weight and the heat generation are also reduced. Thus the low-speed maglev train system cost and operation cost are reduced.

At the same time, AlNiCo hybrid electromagnet can prevent locking track, and effectively overcome the security risks of traditional hybrid electromagnet locking rail. The new hybrid levitation electromagnet can effectively solve the problems of the electromagnet lock orbit, at the same time, have a good dynamic performance and suspension regulation performance. It has a very good application prospect.

References

1. Zhu Z Q, Chen Y S, Howe D. Maximising the flux-weakening capability of permanent magnet brushless AC machine and drives [C]. The Third International Power Electronics and Motion Control Conference, Beijing, China, (2000)
2. Morimoto S, Sanada M, Takeda Y. Performance of PM-assisted synchronous reluctance motor for high-efficiency and wide constant-power operation[J]. IEEE Transactions on Industry Applications, (2001)
3. Wu Xiangming. Maglev train [M]. Shanghai: Shanghai Science and Technology Press(in Chinese), (2003)
4. Thornton R. The M3 Urban Transportation System[R]. MagneMotion Inc. America：Federal Transit Administration as Part ofFFA Project MA-26-7077. (2003)
5. Xie Yunde. Modeling and simulation on EMS maglev vehicles [D]. Changsha: Natinal University of Defense Technology(in Chinese), (1998)
6. Ostovic V. Pole-changing permanent-magnet machines[J]. IEEE Trans. on Industry Applications, (2002)

DOI: 10.1051/matecconf/20179509006

Formation keeping of unmanned ground vehicles

Kamonwan Muangmin[1] and Thanapat Wanichanon[1,a]

[1] *Department of Mechanical Engineering, Mahidol University, 25/25 Puttamonthon NakornPathom 73170, Thailand*

Abstract. Controlling motions of an unmanned ground vehicle becomes more popular in real world practices. Its application is useful for household chores, military services, medical purposes, and industrial revolutions, etc. An analysis of motions by using the Fundamental Equations of Constrained Motion (FECM) is one effective tool to determine the motions. Its conceptualization is done in three-step procedure as follows: (I) Determining an unconstrained motion (II) Assigning constraint equations and (III) Computing a constrained motion. The equations of motion obtained are expressed as liner functions of acceleration. Then other kinematical information of the unmanned ground vehicles can be obtained by integration its acceleration. In this work, the FECM is used as a tool to analyze motions of a group of unmanned ground vehicles in various forms. The simulation results show that control forces obtained from the approach can regulate motions of unmanned ground vehicles to maneuver in desired formations.

1 Introduction

Currently, researches in tracking of the Unmanned Ground Vehicle (UGV) have widely attracted attention both in academics and industries because of their important roles. For example, an application in a dangerous environment where is unsafe for human. Study of UGV is a broad area in scientific researches, such as transportation, space detection, defense, rescue, industry and other fields. Like many other mechanical systems, UGVs are often required an assign destination or formation pattern. The formations of UGVs use a reference position of a leader to provide patterns [1]. For the complicate formation, UGVs are provided series of formation maneuver patterns by points and use relative dynamic to control [2].

In general, Newton's laws of motion, a theory that predicates displacement, velocity, and acceleration of movements is a standard tool in obtaining motions of objects. However, using this law of motion to study movements of multi-body UGVs is not an easy task. Recently, another concept of analyzing motions was introduced by considering the energy [3] and was developed base on the idea of Gauss's principle [4] and Lagrange's equation [5] called the Fundamental Equation. It now becomes a widely use and effective tool to obtain the control of any complex multi-body movements.

The aim in this paper is to analyze the formation of UGVs by using the concept of the FECM to create the control force needed that satisfies constraint requirements of formations. The constrained equation of motion obtained provides precision analysis since it can be analyzed without linearization or approximation. It also could define multiple constraints simultaneously. The computation used by the proposed conceptualization is done in three steps and it is not complicate. The constrained formation of UGVs can be chosen as continuous or piecewise continuous paths. The simulation results show that motions of the multi-body UGVs can be determined by the closed form constrained equation and the results are obtained with high accuracy.

2 On the Dynamics of the Unmanned Ground Vehicle

Let's begin by introducing the description of the Unmanned Ground Vehicle system. It is useful to conceptualize its description in three-step procedure [4]-[6], which is done in the following way:

First, the uncontrolled system is described in which the coordinates are all assumed independent of each other. The equation of motion of this system is given, using Lagrange's equation, by

$$M(q,t)\ddot{q} = Q(q,\dot{q},t) \tag{1}$$

with the initial conditions

$$q(t=0) = q_0,\ \dot{q}(t=0) = \dot{q}_0 \tag{2}$$

where q is the generalized coordinate n-vector, $M > 0$ in the n by n mass matrix which is a function of q and t, and Q is an n-vector, called the 'given' force, which is a known function of $q, \dot{q}$ and t. From Eq. (1) the acceleration of the uncontrolled system given by

$$a := M^{-1}(q,t)\, Q(q,\dot{q},t) \tag{3}$$

Second, on this uncontrolled system, a set of control requirements is imposed as constraints. The uncontrolled

[a] Corresponding author: thanapat.wan@mahidol.ac.th

system is now subjected to the m sufficiently smooth control requirement given by [5]

$$\varphi_i(q,\dot{q},t)=0, \quad i=1,2,...,m \tag{4}$$

where $r \leq m$ equations in the equation set Eq. (4) are functionally independent. The control constraints described by Eq. (4) include all the usual varieties of holonomic and/or nonholonomic constraints. The presence of the control requirements does not permit all the components of the n-vector q_0 and $\dot{q}_0$ to be independently assigned. The initial conditions Eq. (2) are assumed to satisfy in the m control requirements. (If not, the control constraints can be expressed in an alternative form so that they are asymptotically satisfied [7])

Differentiating the control requirements Eq. (4) with respect to time t, the constraint relation is obtained as [8]

$$A(q,\dot{q},t)\ddot{q}=b(q,\dot{q},t) \tag{5}$$

where A is an m by n matrix whose rank is r, and b is an m-vector. It is noted that each row of A arises by appropriately differentiating one of the m control requirements in the set given in Eq. (4).

In the third step, the equation of constrained motion of the system is given by

$$M(q,t)\ddot{q}=Q(q,\dot{q},t)+Q^c(q,\dot{q},t) \tag{6}$$

where Q^c is the control force n-vector that arises to ensure that the control requirements Eq. (5) are satisfied. The explicit equation of motion of the system is thus given by the FECM [5], [8]

$$M\ddot{q}=Q+A^T(AM^{-1}A^T)^+(b-Aa) \tag{7}$$

wherein the various quantities have been defined in the previous two steps and the superscript "+" denotes the Moore-Penrose (MP) inverse of a matrix [9]. In the above equation, and in what follows the arguments of the various quantities are suppressed unless required for clarity.

The control force that the uncontrolled system is subjects to, because of the presence of the control requirements Eq. (4), can be explicitly expressed as

$$Q^c(t):=Q^c(q(t),\dot{q}(t),t)=A^T(AM^{-1}A^T)^+(b-Aa). \tag{8}$$

The control force given in Eq. (8) is optimized in the sense that it minimizes the control cost $Q^{cT}M^{-1}Q^c$ at each instant of time [7, 8]. The weighting matrix in the control cost has been chosen to be M^{-1}, although other positive definite matrices can be easily chosen [10]. Pre-multiplying both side of Eq. (7) with M^{-1}, the acceleration of system that satisfies the constraint Eq. (4) can be expressed as

$$\ddot{q}=a+M^{-1}A^T(AM^{-1}A^T)^+(b-Aa):=a+M^{-1}Q^c(t) \tag{9}$$

3 Numerical Example

To demonstrate the applicability of the control methodology, an example of a UGVs system will be introduced. This example will also be continued all the way through this paper.

Consider the UGVs system consisting of three vehicles m_1, m_2 and m_3 which are independent to each other. The UGVs system is moving in the XY-plane.

For simplicity, in this work masses are considered as point masses and in indiscriminate size of the UGVs. The force that used to propel UGVs to move forward is required in the X-axis. The three-step approach described in the last sub-section is now illustrated. Let's begin by writing the equation of the uncontrolled system Eq. (1) using the generalized coordinate 6-vector $q(t)=[x_1,y_1,x_2,y_2,x_3,y_3]^T$ whose components, in the absence of any constraints are independent of one another. Lagrange's equations then yield the relation

$$M\ddot{q}=Q \tag{10}$$

where the element of the 6 by 6 symmetric matrix M and the elements of the 6-vector Q are given by

$$M\ddot{q}:=\begin{bmatrix} m_1 & 0 & 0 & 0 & 0 & 0 \\ 0 & m_1 & 0 & 0 & 0 & 0 \\ 0 & 0 & m_2 & 0 & 0 & 0 \\ 0 & 0 & 0 & m_2 & 0 & 0 \\ 0 & 0 & 0 & 0 & m_3 & 0 \\ 0 & 0 & 0 & 0 & 0 & m_3 \end{bmatrix}\begin{bmatrix} \ddot{x}_1 \\ \ddot{y}_1 \\ \ddot{x}_2 \\ \ddot{y}_2 \\ \ddot{x}_3 \\ \ddot{y}_3 \end{bmatrix}=\begin{bmatrix} F_{x_1} \\ F_{y_1} \\ F_{x_2} \\ F_{y_2} \\ F_{x_3} \\ F_{y_3} \end{bmatrix}:=Q \tag{11}$$

For the second step, in this work, two types of constraint equations (continuous and piecewise continuous desired paths) are chosen to demonstrate the ability of the proposed controller.

3.1 Continuous Path

For the aim of formation, the constraints are given for the purpose of keeping a triangular formation of all three UGVs at all time. The two UGVs aside move for protecting the leader UGV in front while the leader moves forward for guiding directions. The sinusoidal pattern is chosen to imitate behavior for avoiding obstacles while operating missions (see Figure 1). The distance in X-axis between the second and third UGV are assigned to be 10 meters and the distance in Y-axis between each other is chosen to be 1 meter. Thus the constraints are

$$y_2-y_1=1, y_3-y_2=1, x_3-x_1=0, x_2-x_3=10, y_2=Psin(\omega x_2) \tag{12}$$

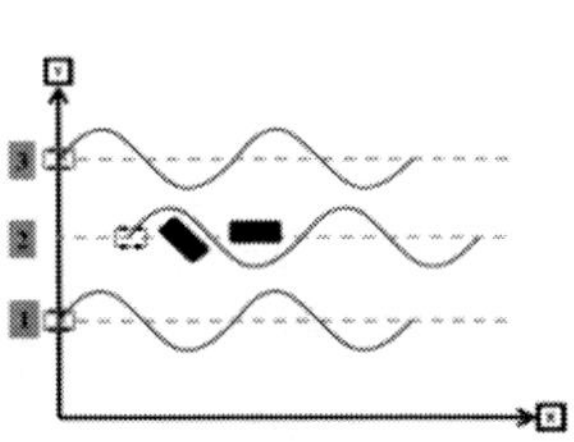

Figure 1 Continuous Path Formation

Since the system may not initially satisfy these control requirements, the trajectory stabilization relation [6] is used to modify the constraints. From Eq. (12) the control requirement is thus obtained as

 DOI: 10.1051/matecconf/20179509006

$$A\ddot{q} := \begin{bmatrix} 0 & 1 & 0 & -1 & 0 & 0 \\ 0 & 0 & 0 & 1 & 0 & -1 \\ 1 & 0 & 0 & 0 & -1 & 0 \\ 0 & 0 & -1 & 0 & 1 & 0 \\ 0 & 0 & P\omega cos(\omega x_2) & -1 & 0 & 0 \end{bmatrix} \begin{bmatrix} \ddot{x}_1 \\ \ddot{y}_1 \\ \ddot{x}_2 \\ \ddot{y}_2 \\ \ddot{x}_3 \\ \ddot{y}_3 \end{bmatrix} = \quad (13)$$

$$\begin{bmatrix} \alpha(\dot{y}_2 - \dot{y}_1) + \beta \\ \alpha(\dot{y}_3 - \dot{y}_2) + \beta \\ \alpha(\dot{x}_3 - \dot{x}_1) + \beta \\ \alpha(\dot{x}_2 - \dot{x}_3) + \beta \\ P\omega^2 \dot{x}_2^{\,2} sin(\omega x_2) + \alpha(\dot{y}_2 - P\omega x_2 cos(\omega x_2)) + \beta \end{bmatrix} := b$$

where $\alpha(t), \beta(t) \geq 0$ is a positive function.

For the final step using the information from Eqs. (11) and (13) in Eq. (9), the desired constrained motions of the UGVs are thus obtained.

3.2 Piecewise Continuous Path

For this scenario, it aims to simulate the formation of UGVs that has changed patterns with time. The purpose of formation is for imitating on attack behavior by beginning with the three UGVs move equally forward in the X-axis and keep equally 10 meters distance away with each other in the Y-axis for a while then the first and third move away from the second UGV (see Figure 2). Thus the constraints are

$$x_2 - x_1 = 0, x_3 - x_2 = 0, y_2 - y_1 = 10, y_3 - y_2 = 10. \quad (14)$$

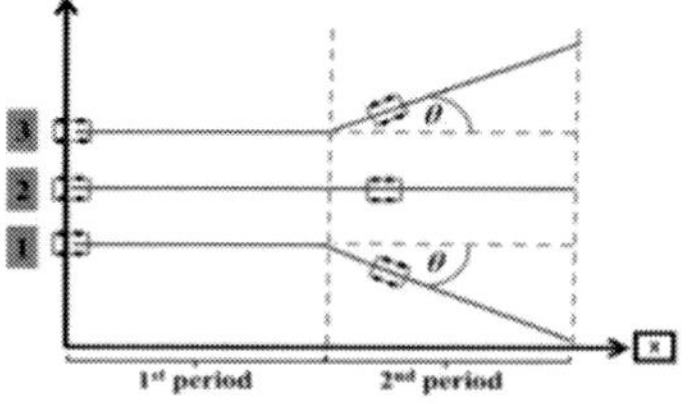

Figure 2 Piecewise Continuous Path Formation

At 10 seconds, the first and third UGVs are driving apart from the second by θ degree, while the second UGV maintains its direction and each of UGVs move equally in the X-axis. Thus in the second period, the constraints are

$$y_1 = -(tan\theta)x_1, y_3 = (tan\theta)x_3, x_2 - x_1 = 0, x_3 - x_2 = 0. (15)$$

Since the system may not initially satisfy these control requirements, the trajectory stabilization relation [6] is again used to modify the constraints. From Eq. (14) and Eq. (15) the control requirements will thus obtained respectively as

$$A\ddot{q} := \begin{bmatrix} 1 & 0 & -1 & 0 & 0 & 0 \\ 0 & 0 & 1 & 0 & -1 & 0 \\ 0 & 1 & 0 & -1 & 0 & 0 \\ 0 & 0 & 0 & 1 & 0 & -1 \end{bmatrix} \begin{bmatrix} \ddot{x}_1 \\ \ddot{y}_1 \\ \ddot{x}_2 \\ \ddot{y}_2 \\ \ddot{x}_3 \\ \ddot{y}_3 \end{bmatrix} = \begin{bmatrix} \gamma(\dot{x}_2 - \dot{x}_1) + \eta \\ \gamma(\dot{x}_3 - \dot{x}_2) + \eta \\ \gamma(\dot{y}_2 - \dot{y}_1) + \eta \\ \gamma(\dot{y}_3 - \dot{y}_2) + \eta \end{bmatrix} := b \quad (16)$$

and

$$A\ddot{q} := \begin{bmatrix} -tan\theta & -1 & 0 & 0 & 0 & 0 \\ 0 & 0 & 0 & 0 & tan\theta & -1 \\ 1 & 0 & -1 & 0 & 0 & 0 \\ 0 & 0 & 1 & 0 & -1 & 0 \end{bmatrix} \begin{bmatrix} \ddot{x}_1 \\ \ddot{y}_1 \\ \ddot{x}_2 \\ \ddot{y}_2 \\ \ddot{x}_3 \\ \ddot{y}_3 \end{bmatrix} = \begin{bmatrix} \gamma(\dot{y}_1 + tan\theta \dot{x}_1) + \eta \\ \gamma(\dot{y}_3 - tan\theta \dot{x}_3) + \eta \\ \gamma(\dot{x}_2 - \dot{x}_1) + \eta \\ \gamma(\dot{x}_3 - \dot{x}_2) + \eta \end{bmatrix} := b \, (17)$$

where $\gamma(t), \eta(t) \geq 0$ is a positive function.

Similarly, for the final step to obtain the equations of constrained motion the information from Eqs. (11), (16) and (17) is used in the constrained equation Eq. (9).

4 Simulation Result

The masses and the 'given' forces in the X-axis and the Y-axis of each vehicles are assumed as $m_1, m_2, m_3 = 1\,\text{kg}$, $F_{x_1}, F_{x_2}, F_{x_3} = 1\,\text{N}$ and $F_{y_1}, F_{y_2}, F_{y_3} = 0\,\text{N}$.

4.1 Continuous Function

The initial displacements and initial velocity are assumed as

$$\begin{aligned} &x_1(0) = -10, \quad y_1(0) = -1, \quad \dot{x}_1(0) = 0, \quad \dot{y}_1(0) = 0 \\ &x_2(0) = 0, \quad y_2(0) = 0, \quad \dot{x}_2(0) = 0, \quad \dot{y}_2(0) = 0 \quad (18) \\ &x_3(0) = -10, \quad y_3(0) = 1, \quad \dot{x}_3(0) = 0, \quad \dot{y}_3(0) = 0 \end{aligned}$$

The amplitude and frequency of sinusoidal function are given by $P = 0.5$ and $\omega = 0.2$ respectively. Figure 3 plots the trajectories of the three UGVs in the XY-plane for 10 seconds. The three UGVs move forward in sinusoidal pattern while keeping triangle form at all time. Figure 4 shows errors in displacement in order to satisfy the constraints Eq. (12), $E1 = y_2 - y_1 - 1$, $E2 = y_3 - y_2 - 1$, $E3 = x_3 - x_1$, $E4 = x_2 - x_3 - 10$ and $E5 = y_2 - Psin(\omega x_2)$. In figure 5, the forces used to make UGVs move in order to satisfy constraints and keeping formation are shown.

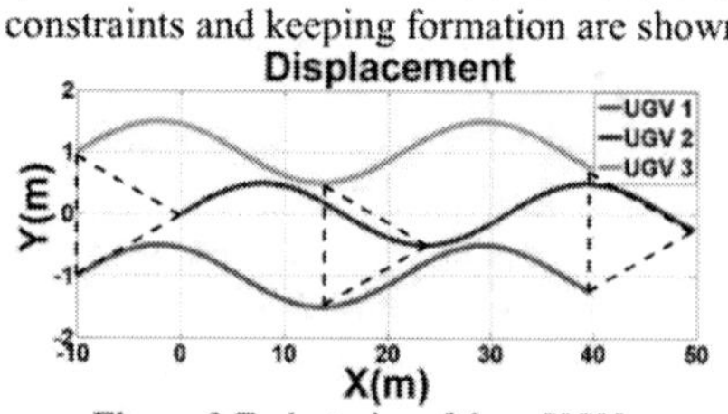

Figure 3 Trajectories of three UGVs

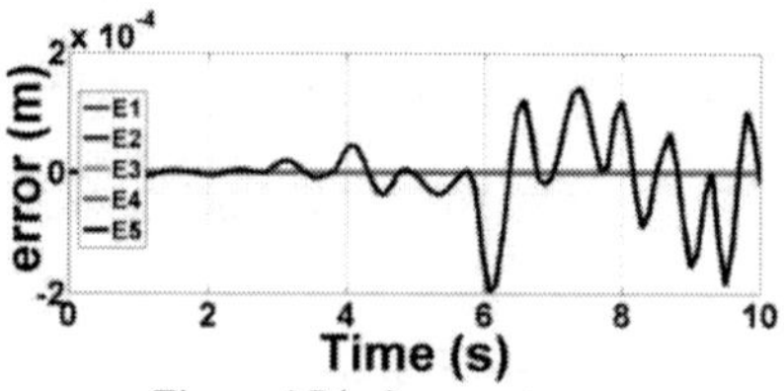

Figure 4 Displacement errors

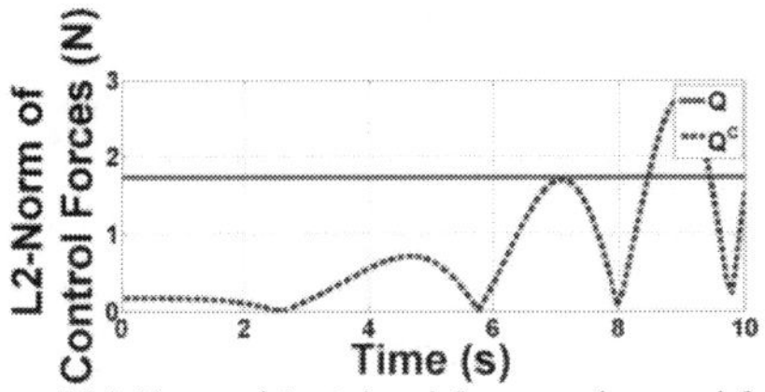

Figure 5 L2-Norm of the 'given' forces and control forces

4.2 Piecewise Continuous Function

The initial displacements and initial velocities of the first period are assumed as

$$\begin{aligned} &x_1(0)=0, \quad y_1(0)=-10, \quad \dot{x}_1(0)=0, \quad \dot{y}_1(0)=0 \\ &x_2(0)=0, \quad y_2(0)=0, \quad \dot{x}_2(0)=0, \quad \dot{y}_2(0)=0 \\ &x_3(0)=0, \quad y_3(0)=10, \quad \dot{x}_3(0)=0, \quad \dot{y}_3(0)=0. \end{aligned} \tag{19}$$

For the next period, the initial displacements of $x_1, y_1, x_2, y_2, x_3, y_3$ are continuous with the last positions from the first period and the initial velocities are chosen as

$$\begin{aligned} &\dot{x}_1(10)=1, \quad \dot{y}_1(10)=-tan\theta \\ &\dot{x}_2(10)=1, \quad \dot{y}_2(10)=0 \\ &\dot{x}_3(10)=1, \quad \dot{y}_3(10)=tan\theta \end{aligned} \tag{20}$$

where angle of the formation is assumed $\theta = 30^\circ$. Figure 6 plots the trajectories of the three UGVs in the XY-plane for 20 seconds. In this figure, the three UGVs equally move forward in X-axis while keeping 10 meters distance in Y-axis to the alongside one. After 10 seconds, the first and third UGVs move away from the second one while all of them have the same movement in X-axis. Figure 7 shows errors in satisfying the constraints, $\mathrm{E1} = x_3 - x_2$, $\mathrm{E2} = x_2 - x_1$, $\mathrm{E3} = y_3 - y_2 - 10$ and $\mathrm{E4} = y_2 - y_1 - 10$ for the first period 10 seconds and for second period, since continuous information from the first period is used, so the errors are $\mathrm{E1} = y_1 + (tan\theta)x_1$, $\mathrm{E2} = y_3 - (tan\theta)x_3$, $\mathrm{E3} = x_2 - x_1$ and $\mathrm{E4} = x_3 - x_2$. In Figure 8, L2-norm of the given forces and control forces are shown. This system uses small portion of control forces to make UGV move to satisfy the constraints.

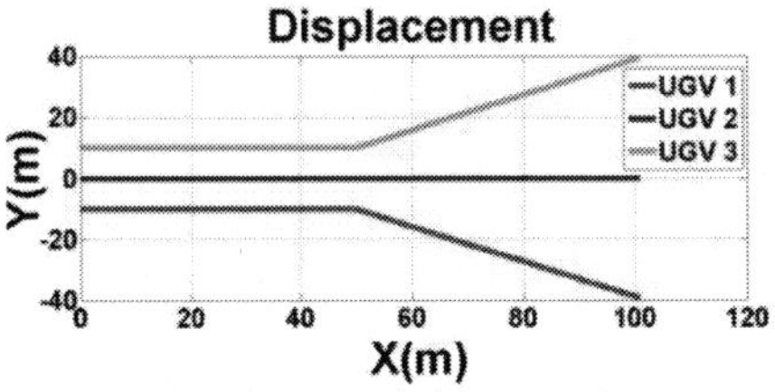

Figure 6 Trajectories of three UGVs

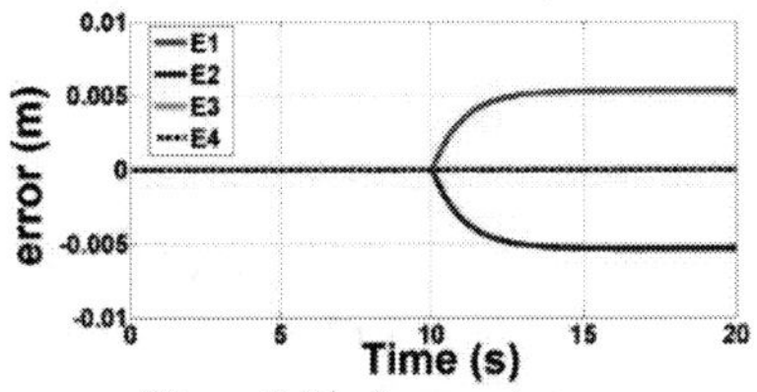

Figure 7 Displacement errors

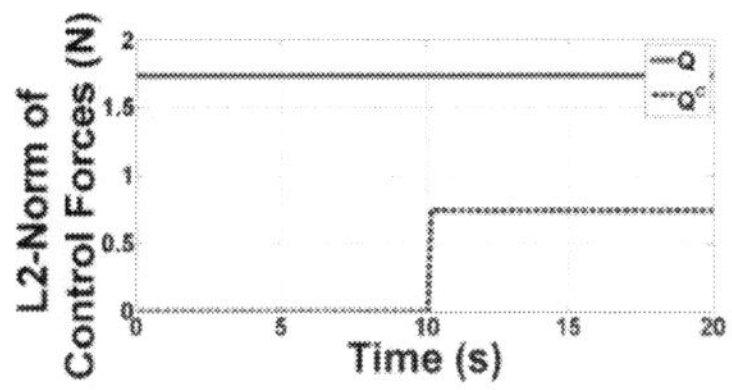

Figure 8 L2-Norm of the 'given' forces and control forces

5 Conclusion

The FECM can be used to analyze movements of multi-UGVs. The general form of the equation is easy to use. It is suitable for both continuous and piecewise continuous constraint paths. Its equation is considered in each axis. So, this equation is regardless of the direction of the axes. This work uses the FECM to define patterns of the UGVs in various paths. The simulation results verify that the UGVs move by following the desired constraint formations.

References

1. H. Lim, Y. Kang, J. Kim and C. Kim, Formation control of leader following unmanned ground vehicles using nonlinear model predictive control, *Advanced Intelligent Mechatronics*, 945-950, (2009).
2. J. R. Lawton, R. W. Beard and B. J. Young, A decentralized approach to formation maneuvers, *Robotics and Automation,* **19**,933-941, (2003).
3. F.E. Udwadia, P. Phohomsiri, Explicit equations of motion for constrained mechanical systems with singular mass matrices and applications to multi-body dynamics, *Proc. Soc. London A*, **462**, 2097-2117, (2006).
4. R.E. Kalaba and F.E. Udwadia, Equations of Motion for Nonholonomic, Constrained Dynamical Systems via Guss's Principle, *J. Appl. Mech.*, **60**, 662-668, (1993).
5. F.E. Udwadia, Nonideal Constraints and Lagrangian Dynamics, *J. Aerosp. Eng.*, **13**, 17-22, (2000)
6. F.E. Udwadia, Equations of Motion for Mechanical Systems: A Unified Approach, *Int. J. Nonlin. Mech.*, **31**, 951-958, (1996).
7. F.E. Udwadia, A New Perspective on the Tracking Control of Nonlinear Structural and Mechanical Systems, *Proc. Soc. London* A, **459**, 1783-1800, (2003).
8. F.E. Udwadia, Equations of Motion for Constrained Multi-body Systems and their Control, *J. Optimiz. Theory App.*, **127**, 627-638, (2005).
9. Prasad, K. Manjunatha, and R.B. Bapat, The generalized Moore-Penrose inverse, *Linear Algebra and its Applications*, **165**, 59-69, (1992).
10. F.E. Udwadia, Optimal Tracking Control of Nonlinear Dynamical Systems, *Proceedings of the Royal Society of London, Series A: Mathematical and Physical Seciences*, **464**, 2341-2363, (2008).

Design, Modeling and Virtual Validation of Mobility platform of Tracked Vehicle

Syed Talat Shabbir and Chen Hui-yan

School of Mechanical and Vehicular Engineering Beijing Institute of Technology, Beijing, 100081, China

Abstract. Hull design of a tracked vehicle plays a critical role in the vehicle mobility, crew protection and combat performance. This work is intended to design a hull of 50 Ton tracked vehicle using conventional layout of suspension and running gear components. Driver compartment of hull is designed based upon driver height of 1.69m. Critical hull parameters such as angle of departure, location of 1st road wheel w.r.t idler wheel, location of last road wheel w.r.t sprocket and distribution of middle road wheels are calculated. Finally, designed hull is modeled in Pro/Engineer software and Recurdyn software is used to ascertain vehicle performance in various prototype tests. In the wake of scarcity of papers on this subject, this paper not only outlines the process of hull design for conventional tracked vehicles but also for unconventional ones such as autonomous tracked vehicles.

1 Introduction

The mobile platform of the tracked vehicle, which can carry the turret with or without gun, is called as the hull of a tracked vehicle. This platform or hull ensures high mobility and protection in combat condition. Three main characteristic of the tracked vehicle are fire power, mobility and protection [1]. The hull of tracked vehicle has very high relation to the mobility and reasonable relation to the protection. The purpose of this study is to design a mobility platform of a tracked vehicle and to investigate its performance prior to field testing. In the course of design of high speed tracked vehicle, number of field tests are required to analyse the performance of designed vehicle. Development cycle based upon field testing is long and cost intensive. Therefore, mathematical design and simulation process is adopted for initial design calculations of mobility platform of 50 Ton tracked vehicle. Designed mobility platform is investigated by using multibody dynamic simulation in Recurdyn program.

2 Hull design

2.1 Height of the hull

In Asian standard a driver's compartment with a height of 900 - 1000 mm and a suitably designed seat a man can be easily accommodated in the compartment. This information is from the ergonomic study of a man and experience [2]. We cannot play much with this height or increase it to our desire since it effects the overall height of the tracked vehicle.

2.2 Length of the hull.

Again the ergonomic studies on tracked vehicles and practical experience indicates that the driver should have some suitable distance at his disposal for his body and controls. Here the height of the man considered is 1.69 m tall. Including the protection thickness of the front nose gives the length from front to the driver's compartment.

2.3 Width of the hull

The width of the tracked vehicle would be restricted by the railway carriages and the transporters to be used for the tracked vehicles. The performance or steerability would also be effected by too wide or too narrow tracked vehicle. As an experience the hull width is generally kept up to 2 meters. This would end up with the tracked vehicle width up to 3.5m.

2.4 Design of hull profile.

Following steps indicate the procedure for hull profile design. First we draw a line to indicate the ground. Next draw a reference line 470mm higher from the ground line to represent ground clearance. Now draw a line at a height of 990 mm to represent driver's compartment. Draw another horizontal line in the front at a height of 0.85 meters which is the required vertical obstacle crossing capability of the tracked vehicle. Now place the nose tip at the 0.85 m high line. If it is lower the tracked vehicle will have difficulty in negotiating vertical obstacles. Idler wheel position is determined on the basis of the required angle of approach for the tracked vehicle. Draw a line at 8° from the tip backward to mark the profile of the top plate cutting the 990-mm height. Angle of 8° is the angle with the nose armour. If it is reduced the blind zone of the driver would increase.

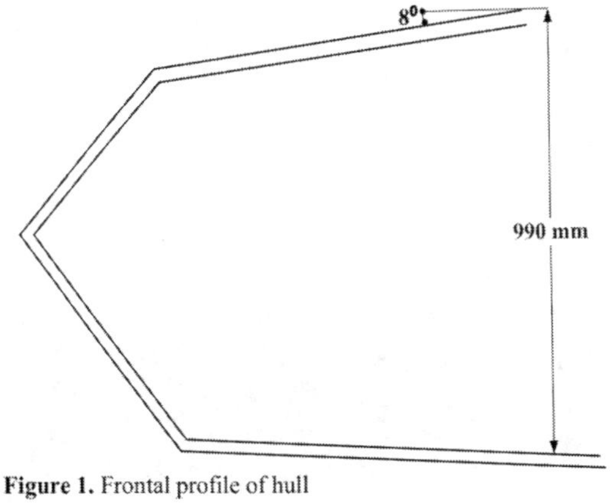

Figure 1. Frontal profile of hull

2.5 Layout of chassis/Hull.

In designing running gear arrangement of a tracked vehicle, the critical parameters considered are location of 1st road wheel w.r.t idler wheel, location of 6th road wheel w.r.t sprocket, angle of approach, angle of departure and distribution of middle road wheels. The angle of approach is generally kept between 30-40° for modern tracked vehicles. Therefore, the hull is designed to negotiate a slope of 60% (31°) and vertical obstacle crossing of 0.85 meter. Angle of departure or angle between ground and rear can only be found after determining the position of the sprocket and the last road wheel. The location of the sprocket basically depends on the output of the engine through transmission and final drive. This gives us the starting point and is marked as O_1 in Fig. 2 where sprocket radius is R_z , R_w is road wheel radius and R_b is radius of balance arm.[3], [4].

$$O_1O_3 = A_1 - R_b \tag{1}$$

Refer Fig. 2 to determine distance between center of final drive and torsion bar at sixth road wheel (A)

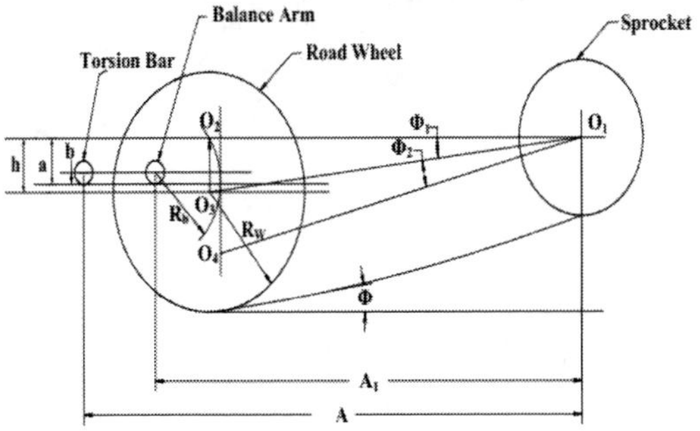

Figure 2. Angle of departure

$$O_1O_2 = R_1 + R_Z + \delta \tag{2}$$

R_1 = vertical distance of shock absorber from bottom;

δ = shock absorber and sprocket clearance (7~10mm)

$$O_2O_5 = \sqrt{(O_1O_2)^2 - (O_1O_5)^2} \tag{3}$$

$$A = O_2O_5 + a_1 \tag{4}$$

a_1 = distance of shock absorber center from torsion bar

$$\phi_1 = \tan^{-1}\frac{h}{O_1O_2} \tag{5}$$

Now (Fig. 2)

$$\phi_2 = \sin^{-1}\left(\frac{Rw - Rz}{\sqrt{(O_1O_2)^2 + h^2}}\right) \tag{6}$$

Finally, angle of departure is as under

$$\phi = \phi_1 + \phi_2 \tag{7}$$

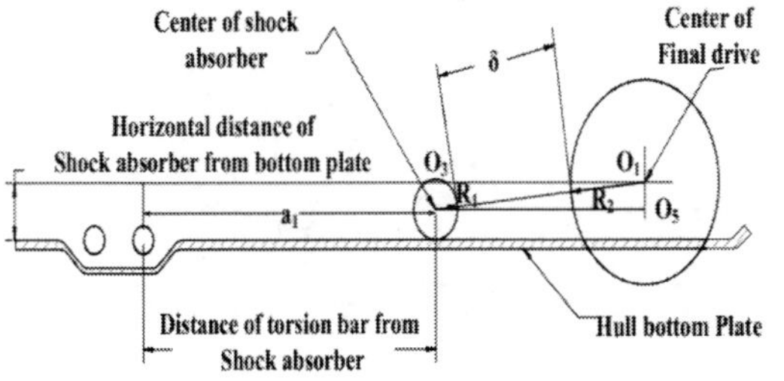

Figure 3. Layout of torsion bar, shock absorber and Final drive

Location of the first road wheel is calculated with respect to idler wheel by keeping the angle of approach as 35° as shown in Fig. 4.

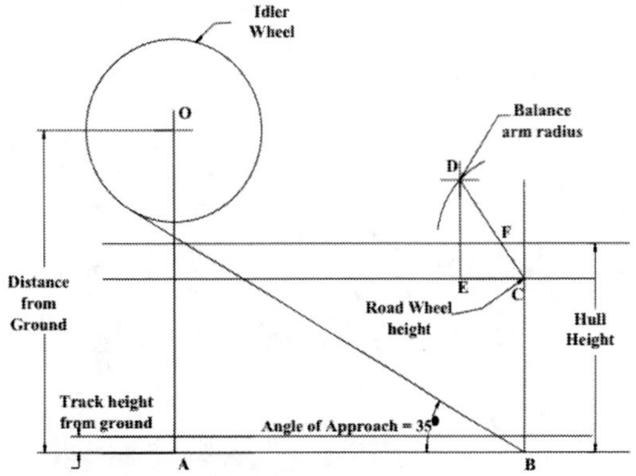

Figure 4. Location of 1st road wheel from Idler

Between the sixth and the first road wheel we can now distribute the other 4 road wheels. Before doing the distribution, it is essential to determine the distance between the two torsion bars. The procedure to be adopted is for worst condition at maximum travel of

 DOI: 10.1051/matecconf/20179509007

wheel no 6 in upward direction. D_1D_2 is the distance between two road wheels which is calculated from the dimensions given in Fig. 5. D_1D_2 is also the minimum between other road wheels for which allowance is added according to the space requirement for suspension system.

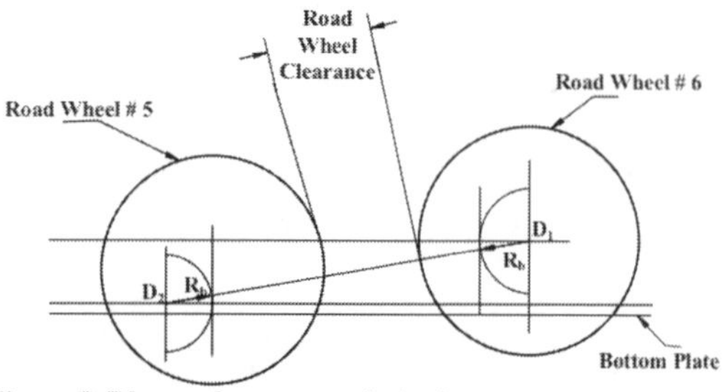

Figure 5. Distance between road wheels

Calculation of layout parameters of hull are summarized in Table-1.

Table 1. Summary of Hull critical parameters.

S/No	Critical Hull Parameter	
1	Distance b/w Wheels 1-2	850
2.	2-3	890
3.	3-4	890
4	4-5	890
5	5-6	882
6.	Location of 1st Rd wheel w.r.t Idler wheel (mm)	600
7.	Location of 6th rd wheel w.r.t Final drive (mm)	1038
8	Angle of approach	35°
9.	Angle of departure	22.2°

2.6 Hull modeling

As per the design process outlined in previous section, hull is modeled in Pro/Engineer software [5, 6] which is best suited for parametric modeling. Designed hull exhibits 3 rotational and 3 translation DOF. Ride (pitch and bounce) and handling (yaw, sides slip and roll) characteristics of tracked vehicle are highly dependent on the design layout hull of tracked vehicle. If these motions are controlled the platform is considered to be stabilized. 3D CAD model of designed hull is shown in Fig. 6.

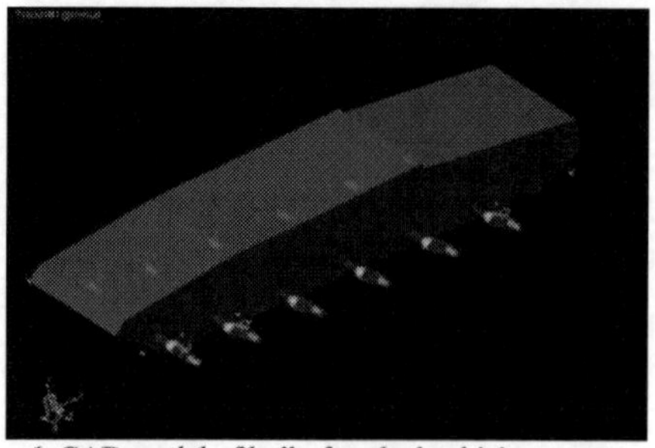
Figure 6. CAD model of hull of tracked vehicle

3. Virtual prototyping

Virtual prototype of tracked vehicle [7,8] is built to perform field test prior to design finalization. These tests are costly and often dangerous when conducted on real prototype. In order to construct a virtual model of tracked vehicle we used two software namely Recurdyn and Matlab. 3D designed CAD model is imported in Recurdyn to add suspension and running gear components. Matlab is used to calculate design parameters such as suspension stiffness, damping ratio, etc. Random road profile is also generated in Matlab which is later used in Recurdyn to construct 3D road surface as shown in Fig. 7.

Figure 7. Random road surface

Balance arms, road wheels, sprocket, idler wheels are added to hull using revolute joints. A real damper curve is added to represent non-linear behaviour of damper. A "sphere-cylinder" connection is also used [9] to represent hydraulic limiter to restrain vertical motion of road wheel.

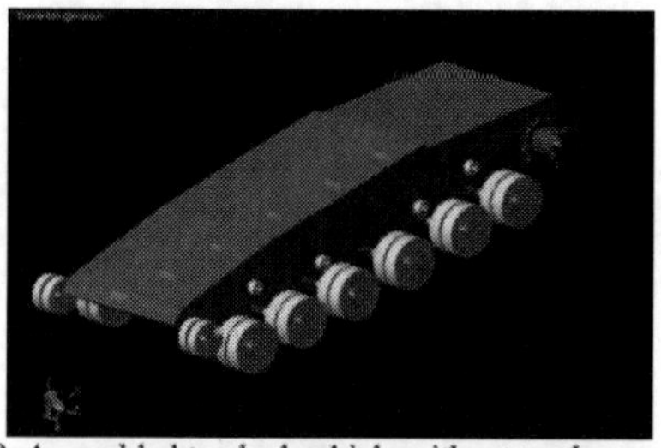
Figure 8. Assembled tracked vehicle without track

Figure 9. Assembled tracked vehicle with track

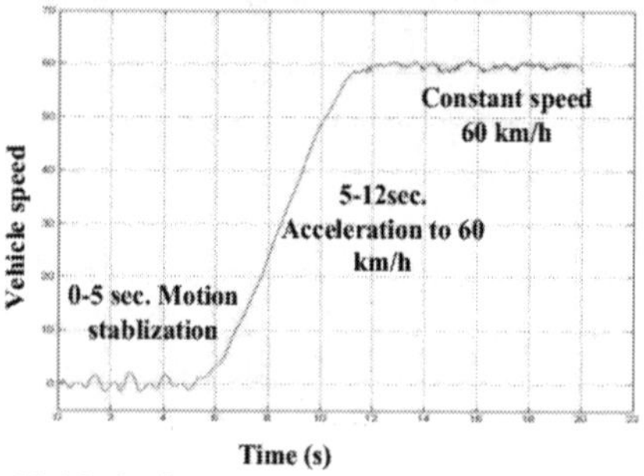

Figure 10. Motion Input curve

3.1 Motion input

Motion input is given at revolute joints provided on sprockets. Step input relation is used to represent tracked vehicle acceleration from 0 to 60 km/h in 12 seconds. Vehicle speed curve is shown in Fig. 10.

4 Results

Various real field tests were conducted in Recurdyn software to ascertain the performance of designed hull for tracked vehicle. These tests include ditch crossing, slope negotiation and high speed run on a road surface having random profile. All these tests completed successfully and vehicle validated the design process prior to assembling a real prototype.

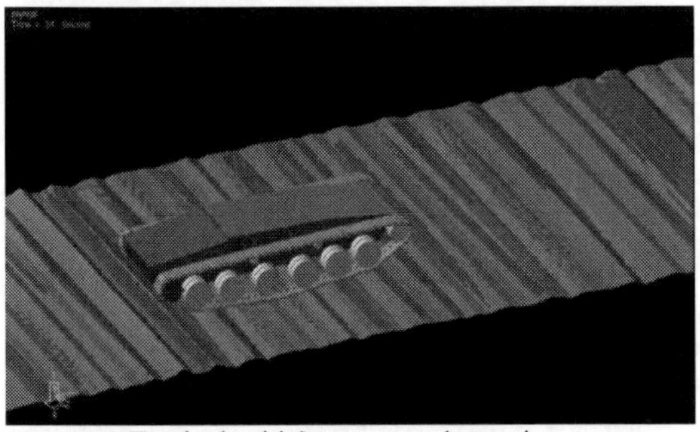

Figure 11. Tracked vehicle over rough terrain

5 Conclusion

In the wake of scarcity of papers on the design of tracked vehicle, a complete design process for the hull of tracked vehicle is outlined. Virtual testing of prototype validated the design process and it is inferred that simulation completed successfully and designed tracked vehicle passed prototype testing such as high speed run, trench and obstacle crossing. Design process can be further extended for selection and optimization of suspension and running gear components for different hull layouts.

Acknowledgement

We gratefully acknowledge to NORINCO China (HIT-NORINCO PhD Scholarship 2013-17) and Vehicle Laboratory of Beijing Institute of technology, Beijing for provisioning the technical support and bearing the financial effects of our whole research and in particular for this work.

References

1. National Research Council Canada, Armour Combat Vehicle Mobility Development Study, (Phase 2 Report, 5 July 1999), 29-49.
2. Richard Ogorkiewicz, Technology of Tracked vehicles II, Jane's Information Group Ltd, (1991)
3. Christian Guenther and Cornelius Leondes. Synthesis of a high-speed tracked vehicle suspension system - part I: Problem statement, suspension structure, and decomposition. IEEE. **22**: 158-165, April (1977).
4. Murakami M, Watanabe K and Kitano. A mathematical model for spatial motion of tracked vehicles on soft ground. Journal of Terramechanics, **29**(1) :71-81, (1992).
5. Zi-Yue, Ya-Dong GAO *et al.* Modeling and Simulation of tracked vehicle based on Pro/E and RecurDyn. 5th International Conference on Advanced Design and Manufacturing Engineering, (2015).
6. ChangLiang Cai, Simulation of the tracked duct cleaning robot based on Pro/E and RecurDyn[J], Manufacturing automation, (11):53-56, (2014).
7. Yin Huabing, Hu Licheng, Zhang Zhongsheng, Huang Yingxin. Modeling and simulation of virtual prototype for tracked vehicle with hydro-pneumatic suspension system. 6th Int. Conf. on System Simulation and Scientific Computing (Beijing). Oct. 24-27. (2005)
8. C. Q. Ma, X. X. Liu, B. Q. Dai, "Research and Design of Mechanical System Virtual Prototyping", Applied Mechanics and Materials, **52-54**, pp. 624-629, (2011)
9. RuiQiang and Wang Hongyan. Multi-body dynamic simulation and experimental validation of high-speed tracked vehicle. Proceedings of the 7th ICFDM 2006 International Conference on Frontiers of Design and Manufacturing, Guangzhou, China, pp: 31-34, June 19-22, (2006).

DOI: 10.1051/matecconf/20179509008

Innovative Design and Manufacture of "S" Type Carbon-Free Cars

Jianwei Liu[1] , Yuhan Li[2] and Nianjiong Yang[3]

[1]Engineering Training Center of Guilin University of Electronic Technology, Guilin 541004, China
[2]Electromechanical Engineering College of Guilin University of Electronic Technology, Guilin 541004, China
[3]Guangxi University of Science and Technology, Guangxi key laboratory of Automobile Components and Vehicle technology, 545006 Liuzhou, China

Abstract. Based on the new rules of the 4th national college students' engineering and comprehensive training ability competition, established three-dimensional model using software UG NX, designed a kind of "S" type carbon-free,it can adapt to various poles' distance and easy assembling and debugging.Focus on variable pitch mechanism and steering mechanism's designing, and by motion simulation verify its rationality, the simulation analysis showed the car trajectory accords with a requirement. Finally,processed parts,assembled and debugged of the cars.Practice has proved that the design of the car conform to the requirements of the game, is reasonable, assembling outfit is convenient, easy to debug, can meet the requirements of a variety of stem from, smooth finish and get good grades.

1 Introduction

Carbon-free car, a kind of three-wheel cars, is driven by given gravitational potential energy, which has the function of continuous obstacle avoidance. The "S"-shaped circuit requires the car to be able to automatically bypass the obstacles on the track when the car is moving. Obstacle for the diameter of 20mm, high 200mm round bar, along the track line and other distance placed. It is different from the previous game rules that the obstacle space to expand the scope of 700mm~1300mm and require to the car on-site dismantling debugging before the game. In the past, the carbon free cars [1]-[4] are only one kind of transmission ratio, can not meet the current requirements of the game. Lu Jian [5] designed a carbon-free car which can adjust the transmission ratio, although it can adapt to the "S"-shaped path of different rod distance, but it is difficult to change the transmission ratio due to the complexity of the structure. Therefore, it is urgent to design a kind of variable distance mechanism which is convenient to adjust many kinds of transmission ratio, and it is convenient to disassemble and adjust the steering mechanism.

2 Integrated design of the carbon-free car

According to the rules of the three wheeled structure and to achieve the "S" shape of the line shape, the overall design of the car as shown in Fig. 1, including the frame, the transmission part, the steering part, the fine-tuning part of the 4 parts.

The principle is that lifting the weight to a specified height, then weight falls automatically due to the gravity.The flexible cord in the weight will change direction of tension by the installation of pulley in the pulley seat, and the other end of the rope will wind on the driving mechanism. The movement is transferred to the rear wheel shaft, which drives the rear wheel to rotate, thereby completing the movement of the trolley.The front wheel rotates to the left or to the right according to the sine period to rotate through the steering mechanism, which causes the car to walk out of the S shape path and avoid the obstacle rod which is arranged at a certain distance.

In addition to the standard thread parts and bearings of standard parts, the rest of the car are required students to manufacture their own, so must first consider the car material. According to the previous experience in design and processing [6], select aluminum body frame.

3 Design of transmission part

As shown in Fig. 1, the transmission part by pulley, wire wheel, wheel, gear and small gear adjustment and back axle, etc. Flexible rope through the pulley lead the line wheel to rotate, and the coaxial with the big gear to rotate together, so as to drive the meshed with the adjusting gear, rotate the pinion gear. The pinion drives the rear axle to rotate. So that the mounting in the rear ends of the two rear wheels rotate, the car forward.

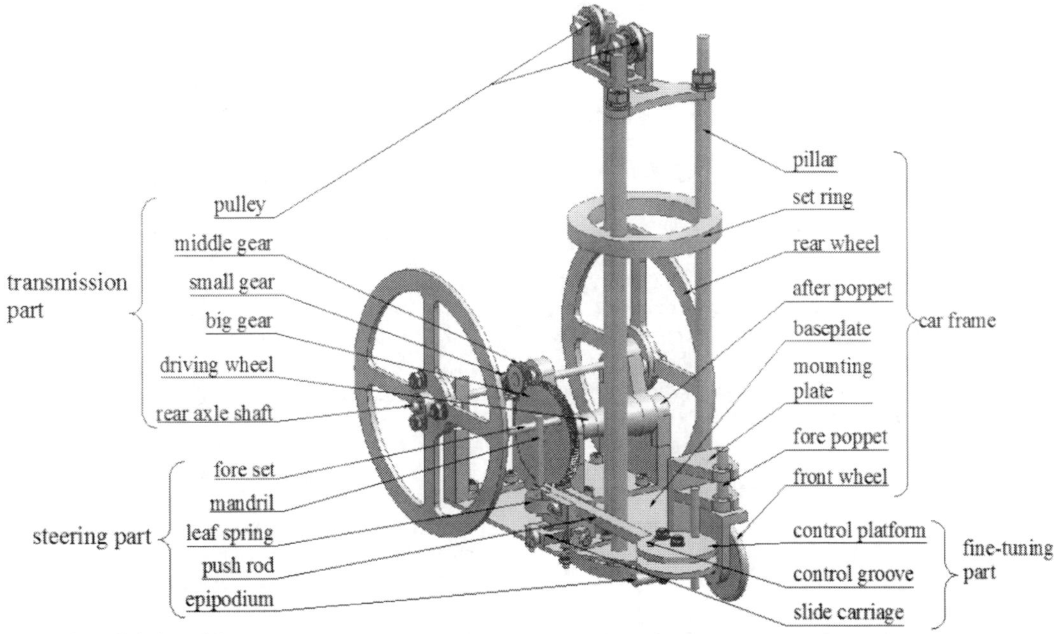

Figure 1. Overall design of the car

3.1 Calculation of transmission ratio

The game provides a random sample of obstacles in the range of 700mm~1300mm. According to the debugging experience, the car can meet the biggest obstacle distance of 200mm by adjusting its own steering angle. So the range is divided into three groups: 700mm~900mm, 900mm~1100mm, 1100mm~1300mm, the median values of each group were 800mm, 1000mm and 1200mm, respectively, and The optimal amplitude of the "S"-shape trajectory is 350mm. Using CAD and other drawing software for spline curve fitting trolley walking route, as shown in Fig. 2.

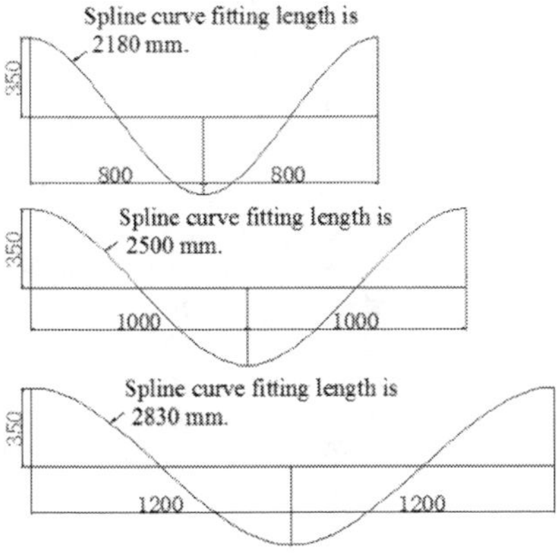

Figure 2. Movement track diagram of the car

Measuring the length of the spline curve can be obtained from the three space under the car a cycle of the theoretical distance respectively 2180mm, 2500mm, 2830mm.

The relationship between the transmission ratio and the distance of a cycle:

$$S = i \times \pi D \tag{1}$$

where S is the distance, 2180mm,2500mm and 2830mm, respectively; i is the transmission ratio; D is the rear wheel diameter.

By formula (1) can calculate transmission ratio i_1=5.3, i_2=6.1, i_3=6.9, when the distance between 800mm, 1000mm, 1200mm, respectively. The number of gear teeth is 120, and the number of small gear teeth is 23, 20 and 17.

3.2 Design of variable pitch mechanism

In order to adapt to the different distance, the design of the car with a variable pitch mechanism, as shown in Fig. 3. It only need to replace a small gear, and adjust the position of the gear to achieve the three gears meshing. Compared with the multi group gear meshing method [2], the operation is simple, no need to adjust the position of the gear support to make the gear mesh.

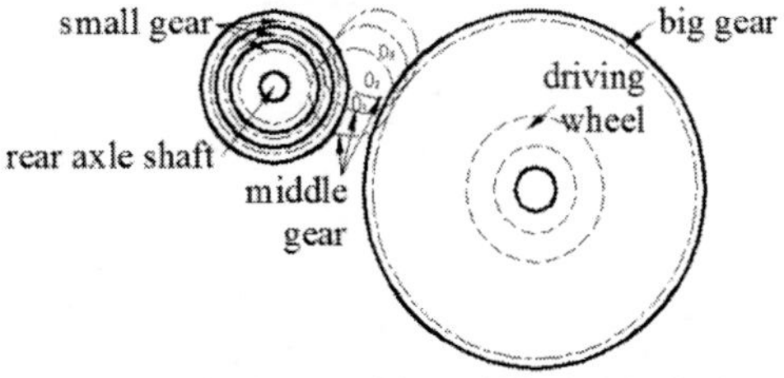

Figure 3. Principle diagram of the variable pitch institutions

4 Design of steering part

The "S"-shaped carbon free car commonly used in the steering mechanism has cam mechanism [7], incomplete gear mechanism, crank rocker mechanism, etc. This car using the sine mechanism, as shown in Fig. 1, steering part is composed of the top of the column, a push rod, a push rod, a spring piece and a front axle. The kinematic diagram [8] of the steering part as shown in Fig. 4.

When the big gear rotates, eccentric top column which fixed on the big gear followed to make circular motion. The front and back straight line motion of the push rod is driven by the eccentric top post, which drives the push rod to slide. The push rod is arranged in the slider and the two ends are respectively connected with the push rod and the front axle is in contact with *C* and *D*. The top rod transfer to the front, so that the front axle to point e for the center for small angle swing back and forth, so as to realize the front left and right steering.

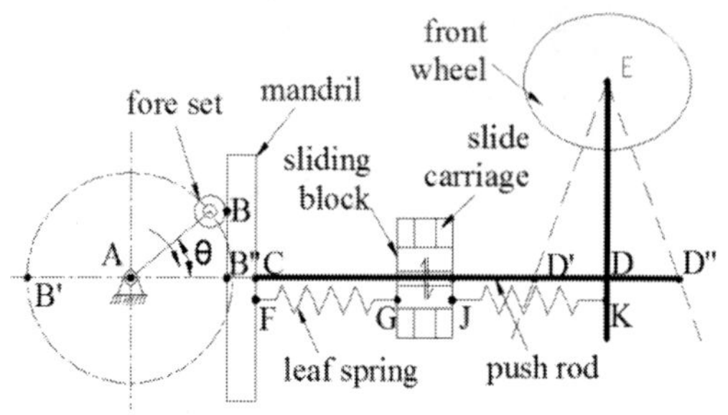

Figure 4. Steering mechanism kinematic sketch

As shown in Fig. 4, When the rocker *AB* moves to *B'* and *B"* position, the push rod has just arrived at the closest point *D'* and the farthest point *D"*. Assume that the length of the crank *AB* is l_1, the *D'D"* is $2l_1$. Assuming the length *DE* is l_2, and the slider is in the middle position of the sliding plate, the front wheel rotation angle as

$$\angle D^{'}ED^{''} = 2arc\tan\frac{l_1}{l_2} \quad (2)$$

The maximum distance that the slider can be adjusted up and down is l_3, then the maximum ψ and minimum values φ for the front wheel angle as:

$$\psi = \angle D^{'}ED^{''}_{\max} = 2arc\tan\frac{l_1 - l_3}{l_2} \quad (3)$$

$$\varphi = \angle D^{'}ED^{''}_{\min} = 2arc\tan\frac{l_1 + l_3}{l_2} \quad (4)$$

Different from the previous design of the steering mechanism, the car designed the compression spring [8] on the basis of the sine mechanism. Due to the effects of the front axle, front push rod spring and the front wheel shaft to maintain contact, and the rear ejector rod is in contact with an eccentric top post because of the action of the push rod spring. So it can keep gapless between the push rod with the front axle, also can maintain gapless between the push rod and the top pillar, making a smooth shift, and keeps the advantages of the sine mechanism. In addition, in order to overcome the uneven force of the spring plate, the two material and the same size spring piece are selected.

In ensuring the accuracy of processing and assembly accuracy, the car can be driven by the standard "S"-shaped track, and do not need to adjust to find a balance position.

5 Design of fine tuning part

In order to make the car to avoid obstacles more accurate, fine adjustment mechanism [9] is needed to adjust the front wheel swing angle range. As shown in Fig. 5, tuning part is composed of sliding board, sliding block, regulating tank, adjusting plate and the like. The sliding block is arranged on the sliding plate. Using screw rod device as device for regulating horizontal movement of slide plate. The trimming principle is by turning screw-steering mechanism of lateral movement, change the top bar and the position of the front wheel axis to change the front wheel angular size, adjusting the amplitude of the trolley routes, and achieve precise fine-tuning of trolley routes. Meanwhile, in regulating trimmer slot is opened on, putting the front-end and adjusting connection point is located in the fine tuning of the disc slot.

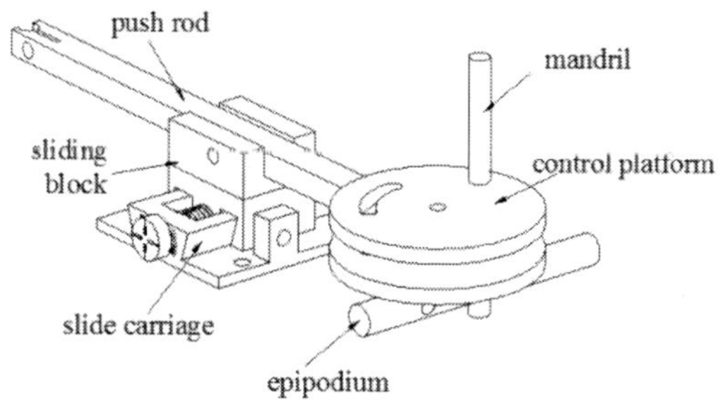

Figure 5. Fine adjustment mechanism

6 Conclusions

The overall innovative design for Carbon-free car is achieved, and the analysis of its structure and movement aimed at the parts of transmission, steering and fine tune is also completed in this study. For the pitch mechanism, it is equipped with many advantages, such as the transmission ratio adjusted easily, handling simply and disassembling and debugging conveniently. The precise control of track route is realized by the micro-adjusting mechanism providing two different measures, i.e. the horizontal adjusting mechanism and fine-tuning slots.

The practice proved that the car has the characteristics of simple structure, adjusting easily and the adaptation for multi-spacing, and provide a certain reference for the design of Carbon-free car under the new rules.

Acknowledgment

The authors are thankful to the Guangxi Natural Science Foundation (Grant no: 2013GXNSFBA019245), Science and technology research project of Guangxi University (Grant no: KY2015YB095), the Opening Project of

Guangxi Key Laboratory of Automobile Components and Vehicle Technology, Guangxi University of Science and Technology (No.2014KFMS05), and the graduate student scientific research innovation projects of guilin university of electronic technology(Grant no: YJCXS201504)

References

1. Ji Yuan-jin, Ren Li-hui. Journal of Machine Design. Innovationa lstructure designo fcarbon-freecar byusing center-changed geart echnique. **14**, 9 (2013)
2. Zou Guang-ming, Yang Xiu-guang. Research on the potential energy-driven car with energy stored by vortex coil spring. **7**, 32-35, (2012)
3. Wang Bin, Wang Yan, Li Run-lian.Innovational design of carbon-free car. **2**, 59-62, (2012).
4. Zhou Cheng, Gao Kun-shan, Ju Chen-ming, Zhang Ya-li. Dynamic Simulation of Carbon-Free Caron ADAMS. **8**, 197-200, (2015).
5. Lu Jian, Wei De-qiang, Sui Xin-meng. An S-shaped path walking shift variable pitch carbon-free car:China, 201410839319.3[P]. (2015)
6. Chen Xiao-dong, Shi Yan-nan, Zhang Li-li. The Design, Manufacture and Innovative Practice of Carbon-free Car. **2**, 92-295, (2013).
7. Zhang Chun, Zheng Ying-bin, Ma Yong-chang. Wide-angle turning performance optimization of gravitational potential energy driven three-wheeled car. **10**, 50-55, (2014).
8. Liu Run, Zhu Xian –yong, Li Zhi -dong.Carbon -free car steering mechanism modeling and simulation based on spring constraint. **9**, 24-27, (2013).
9. Li Yu-han, Liu Jian-wei, Ye Zi-lon. A kind of can adapt to different stem from carbon-free car S-shaped path:China, 201510175306.5[P], (2015)

Genetic Algorithm Combined with Gradient Information for Flexible Job-shop Scheduling Problem with Different Varieties and Small Batches

Ming Chen[1], Jie-Lin Li[2]

[1]*Sino-German College of Applied Sciences, Tongji University, Shanghai 201804, China*
[2]*School of Mechanical Engineering, Tongji University, Shanghai 201804, China*

Abstract. To solve the Flexible Job-shop Scheduling Problem (FJSP) with different varieties and small batches, a modified meta-heuristic algorithm based on Genetic Algorithm (GA) is proposed in which gene encoding is divided into process encoding and machine encoding, and according to the encoding mode, the machine gene fragment is connected with the process gene fragment and can be changed with the alteration of process genes. In order to get the global optimal solutions, the crossover and mutation operation of the process gene fragment and machine gene fragment are carried out respectively. In the initialization operation, the machines with shorter manufacturing time are more likely to be chosen to accelerate the convergence speed and then the tournament selection strategy is applied due to the minimum optimization objective. Meanwhile, a judgment condition of the crossover point quantity is introduced to speed up the population evolution and as an important interaction bridge between the current machine and alternative machines in the incidence matrix, a novel mutation operation of machine genes is proposed to achieve the replacement of manufacturing machines. The benchmark test shows the correctness of proposed algorithm and the case simulation proves the proposed algorithm has better performance compared with existing algorithms.

1 Introduction

With the transformation from manufacturing industry into service industry, the production process under the intelligent manufacturing environment should satisfy the customization and individuation demands of the customer, which has brought the requirement of scheduling problem with different varieties and small batches [1-2]. Compared with traditional Job-shop Scheduling Problem (JSP), it is much harder to achieve the global optimal solutions of FJSP because there exist some redundancies of the manufacturing resources, which causes the rapid increase of search space with the expansion of the problem scale [3-6].

As a kind of Non-deterministic Polynomial-hard (NP-hard) problems, there are mainly two directions of solution methods: exact methods and approximate methods [7]. The exact methods such as mathematical programming [8], branch and bound algorithm [9-10] can obtain the final theoretical optimal solutions of FJSP with small scale. For the large-scale FJSP, this kind of exact methods is not suitable because the expansion of problem scale brings exponential explosion, which takes much more time cost using such enumeration algorithms [11]. To solve this kind of large-scale scheduling problems with lower time cost, corresponding to the exact methods, the approximate algorithms such as GA [12-14], Particle Swarm Optimization (PSO) [15-18], Binary Differential Evolution Harmony Search [19] and Simulated Anneal algorithm [20] have been integrated. These meta-heuristic algorithms maybe cannot get the theoretical global optimal solutions of large-scale FJSP, but the obtained solutions can meet the actual production requirement. In these meta-heuristic algorithms, GA is adapted to solve such highly discrete problem like FJSP very well since the encoding and decoding mode of GA can reflect the discrete solutions. But in the evolutionary process of GA, the new-generated populations may not satisfy the constraint requirement owing to the gradient information loss between the father population and child population. To improve the controllability of evolutionary process and ensure the convergence speed, a GA-based algorithm is proposed with high solving efficiency and optimization performance.

2 Mathematical model

FJSP with different varieties and small batches can be described as follows:

In order to produce N kinds of products with job sum J , the i thworkpiece of k th assortment should be processed according to the process order O_{kj} on the selected machine, in which $1 \le i \le J$, and $1 \le j \le n_i$, n_i means the process amount of the i thworkpiece. Meanwhile, $J = \sum_{k=1}^{N} P_k$, P_k denotes the batch amount of the k th assortment and the operation time T_{ijl} on a concrete machine l is already defined. So the incidence matrix J_o between J and process order O as well as the time matrix T can be obtained. Based on these two matrixes, the optimization model can be established.

The whole production process with different varieties and small batches ends once the last process on a concrete machine is finished, so the minimum value of max make span is utilized as the optimization objective to dispatch the manufacturing resources. So the fitness function in optimization model can be summarized in following Eq. (1):

$$f(x) = max(T_m) \quad (1)$$

in which $1 \le m \le M$, M is the sum number of alternative machines.
Problem constraints are as follows:
- The next process of any job can start only the current process is finished;
- One machine can only process one workpiece at the same time;
- Once the process starts it cannot be suspended until it is finished;
- One workpiece can only be processed on one machine at the same time;
- The workpieces of same kind have the same process order;
- The situation of machine fault is out of consideration.

3 GA-based algorithm design

Fig. 1 depicts the computation framework of proposed algorithm based on GA. Different from traditional GA, the encoding genes are divided into process gene fragment and machine gene fragment which have close relationship with each other, and the following evolutionary operations are all based on these two kinds of genes. The specific operations are described in detail in the following sections.

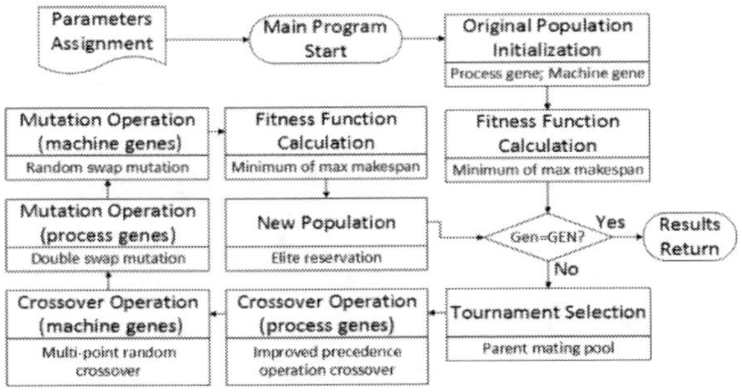

Figure 1. Tournament selection and elite reservation

3.1 Population initialization

Aiming at solving double decision-making problems of the job processes and machines under production demand of different varieties and small batches, the chromosome is composed of two kinds of gene fragments: the process genes and machine genes. The first process gene fragment means the process order on the alternative machines according to the manufacturing procedures of the given job. The latter machine gene fragment means the machine number in the alternative machine sets of the corresponding process order. In order to maintain the consistency with the solutions, decimal coding is applied to compose the population chromosome.

(1) For the process genes, a gene fragment composed of J genes is initialized, and the value of each gene means the process order of the i thwork piece.

(2) For the machine gene, an initialization strategy is applied to accelerate the algorithm convergence speed because the machines with shorter manufacturing time are more likely to been chosen. The max time difference of every non-empty positions in the time matrix T is calculated and sorted in ascending order. The machine genes of the first partare initialized randomly within the range of corresponding positions in the incidence matrix J_o, and for the latter part, the machines with minimum manufacturing time are chosen as the machine genes.

3.2 Tournament selection and elite reservation

The classic roulette selection is not suitable because the optimization objective is to minimize the max make span of a given job and some information of chromosome may

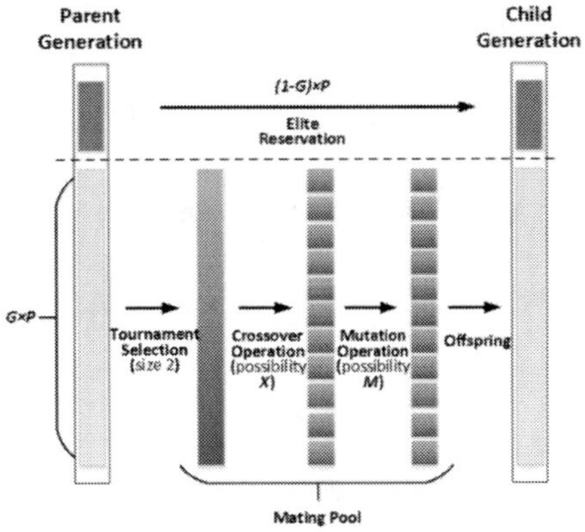

Figure 2. Tournament selection and elite reservation

lose in the transmission operation. Therefore, tournament selection strategy is utilized to select the superior offspring. At a time two individuals are chosen randomly from the parent generation, and the individual with a smaller fitness function is put into the mating pool.

In order to ensure the superior offspring can continue the genetic procedure instead of elimination, an elite reservation strategy using generation gap mechanism is proposed. For every evolutionary iteration, the mating pool consists of parent chromosomes selected by tournament strategy with quantity $G \times P$, in which G means the generation gap coefficient, and P means the population quantity. Fig. 2 shows the important role of tournament selection and elite reservation in the pretreatment phase of the evolutionary process. From Fig. 2, the parent chromosomes with the proportion of $(1-G)$ are selected to be the offspring directly.

3.3 Crossover operation

3.3.1 Crossover operation of process genes

The Improved Precedence Operation Crossover (IPOX) developed by Zhang CY [21] is applied to achieve the crossover operation of process genes.More detail information can be found in reference [21] and will not be described here.From the population initialization process, we can see the process genes and machine genes have a very close connection between each other and the loci of the machine genes should follow the change of the process genes, so after the crossover operation of the process genes, the corresponding machine genes must execute the gene loci based on the change of the process genes.

3.3.2 Crossover operation of machine genes

Different from single-point crossover, a multi-point crossover strategy is introduced to achieve the crossover operation of machine genes and to accelerate the evolutionary speed, the quantity of crossover points is related with the difference degree between machine gene fragments in different parent individuals directly. When the machine genes at the same gene locus of process gene fragment are different, the corresponding genes should be interchanged if the judgment condition $rand(0,1) > 0.5$ is satisfied.

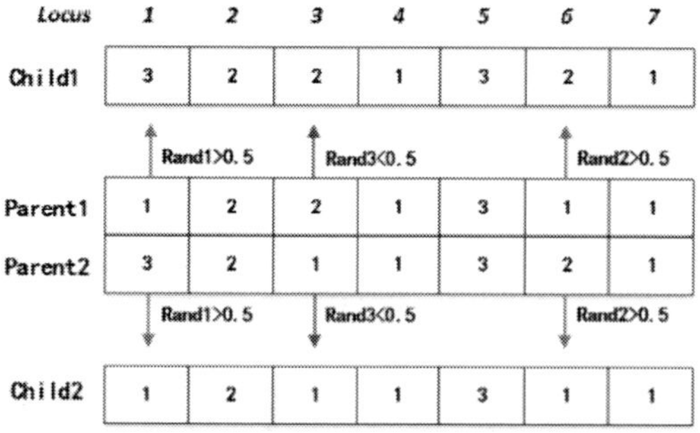

Figure 3. Multi-point random crossover operation.

Fig. 3 shows a simple crossover executionexample of a concrete machine gene fragment. There are three gene loci with different genes: locus 1, 3 and 6. The genes at loci 1 and 6 are interchanged with each other between the two parent chromosomes to achieve the crossover operation because the random condition is qualified but the genes at locus 3 of both parent chromosomes should remain without any change.

3.4 Mutation operation

3.4.1 Mutation operation of process genes

The genetic procedure depends much on the gradient evolutionary information because the arbitrary change between two elements may generate invalid child individuals which cannot satisfy the constraint conditions. Therefore, the double mutation operation should be limited under a certain condition. Supposing double selected gene loci are $L1$ and $L2$, the genes of these double loci can be interchanged only if the genes are different and there is no same gene as that of $L1$ or $L2$ in the gene fragment between these two gene loci. This limitation can avoid destroying the process order of any workpiece so that the validity of child populations is guaranteed.

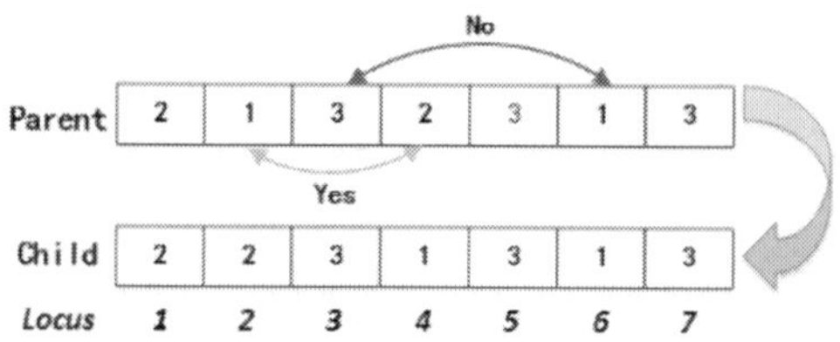

Figure 4. Double swap mutation operation.

Fig. 4 shows a simple mutation example of the process genes. The genes at loci 3 and 6 cannot be interchanged because there exists a same process number 3 between these two loci, and the genes at loci 2 and 4 can carry out the mutation operation because the mutation condition is satisfied.

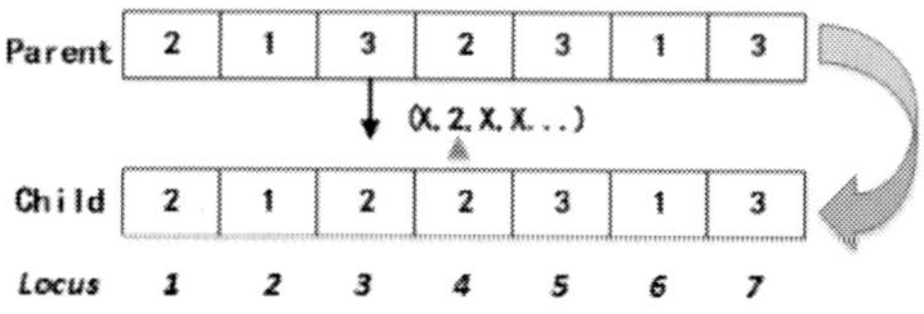

Figure 5. Random swap mutation operation

3.4.2 Mutation operation of machine genes

The mutation operation of machine genes is very important because this operation builds up the interchange relationship between the current machine with the alternative machines prepared in the incidence matrix J_O. For a randomly selected machine gene in the gene fragment, a random machine in the alternative machine set of corresponding position in J_O is picked out to replace the current machine gene.

Fig. 5 depicts a mutation example of machine genes. The gene at locus 3 is randomly selected to carry out the mutation operation and then the machine with number 2 is picked out from the alternative machine set in the incidence matrix J_O. The child chromosome is generated as shown from the parent generation finally.

3.5 Self-adaptive strategy

Compared with the crossover operation of the gene fragments between the parent chromosomes, the mutation operations of process genes and machine genes are miniature, so a self-adaptive mechanism is introduced to adjust the mutation times. Hereon, self-adaptive operator $floor\left(log10(G_{EN}-G_{en})\right)+1$ is utilized to control the mutation procedure, in which $floor$ means the floor function, G_{en} means the current population generation and G_{EN} means the termination criteria. At the beginning of the evolutionary process, this parameter can control the mutation times in the every generation process of offspring, which can improve the chances of better offspring and accelerate the evolutionary rate. With the evolutionary process, this parameter helps the population converge to the optimal results.

4 Experiments and discussion

The simulation experiments are carried out on Windows 7 platform with the hardware parameters as follows: CPU Intel E8400 3.00GHz, RAM 2GB. The programming environment is MatLab R2013a.According to the implement procedure in Fig. 1, the parameters are assigned as follows: population size $P = 100$, termination criteria $G_{EN} = 200$, generation coefficient $G = 0.9$, crossover possibility $X = 0.8$ and mutation possibility $M = 0.6$.

4.1 Benchmark test

The global optimal solutions of scheduling case with different batch scales are as shown in Table 1, and the detail parameters of benchmark test can be found in reference [22]. In this case, there are four kinds of product demands with different batch sizes, and the theoretical solutions as follows. The benchmark test shows that the proposed algorithm can also obtain the global results, which proves the correctness of proposed algorithm.

Table 1. Test results of proposed algorithm

Problem Size	P_1	P_2	P_3	P_4	Optimal Solutions	Obtained Solutions
1x4	1	1	1	1	17	**17**
2x4	2	2	2	2	34	**34**
3x4	3	3	3	3	51	**51**
5x4	5	5	5	5	85	**85**
10x4	10	10	10	10	170	**170**

Additionally, the optimized makespan of case 4 in reference [22] is 33, but the result obtained by proposed algorithm 32, which shows the priority of proposed algorithm applied in the scheduling optimization problems. The concrete scheduling scheme can be seen in Fig. 6.

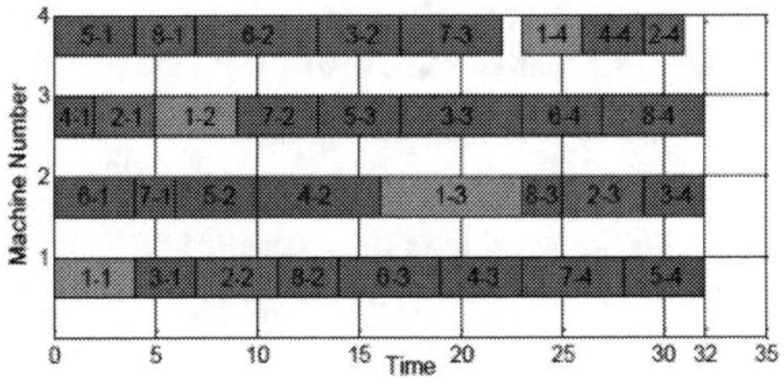

Figure 6. Gantt diagram of optimized scheduling result

4.2 Case study

For the case of FJSP with different varieties and small batches presented by Fu WP, the final optimized solution is 121 [23], but the solution got by the proposed algorithm is 113, which shows the superiority clearly. The convergence procedure is as shown in Fig. 7. It can be seen that the fitness function value converges from 181 to 113 eventually, which is a new lower bound compared with the result in reference [23]. Meanwhile, the proposed algorithm can obtain the solution which can satisfy the industrial scheduling requirements with an acceptable computation cost with 86 evolutionary times.

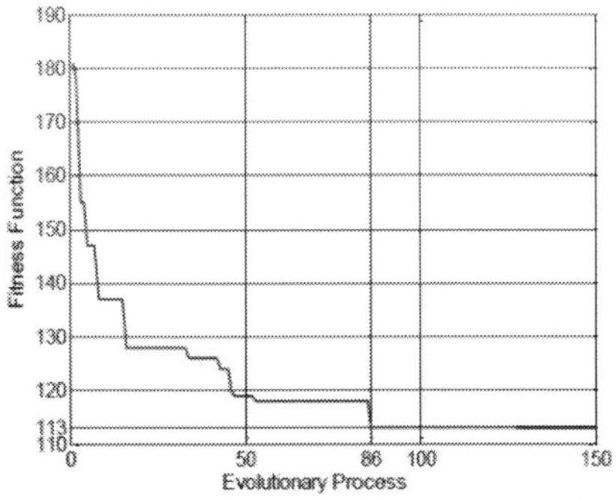

Figure 7. Fitness function with respect to evolutionary process.

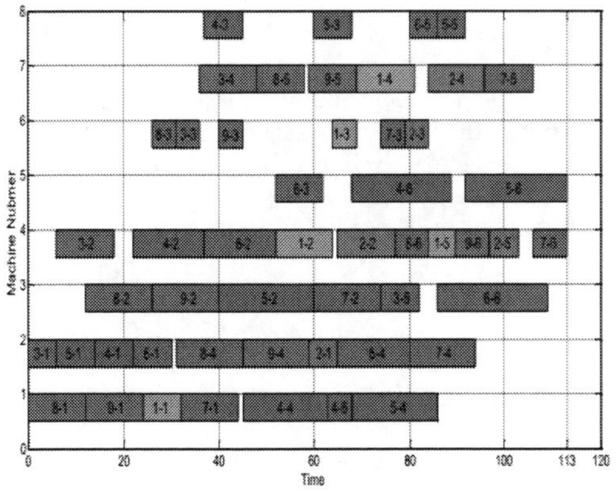

Figure 8. Gantt diagram of optimized scheduling result

The corresponding Gantt diagram in Fig. 8 shows the scheduling of manufacturing machines according to the job requirements with different varieties and small batches. The global optimal solution of this case has not been found until now, and the results with different scheduling schemes got by proposed algorithm are the minimum solutions for the moment. The next research step focuses on finding out the theoretical global solution or updating a new lower bound again.

5 Conclusions

A modified scheduling algorithm based on GA is proposed to solve the FJSP with different varieties and small batches. The initialization operation associated with time matrix is introduced to accelerate the convergence speed and the generation gap coefficient is applied to guarantee the survival rate of superior offspring. To maintain the population diversity and avoid falling into local optimal solutions, IPOX and multi-point random crossover operation are utilized to achieve the crossover of chromosomes, and double swap mutation operation of the process genes and random swap mutation operation of the machine genes are introduced to realize the variation evolution. The benchmark test shows the correctness of proposed algorithm and the case simulation proves the proposed algorithm has better performance compared with existing algorithms. The benchmark tests and simulation results show the proposed algorithm can obtain the global solutions and update the new lower bound of FJSP with different varieties and small batches got by existing algorithms.

References

1. Lee J, Kao HA, Yang S (2014) Service innovation and smart analytics for industry 4.0 and big data environment. Procedia CIRP, 16: 3-8
2. Brettel M, Friederichsen N, Keller M et al (2014) How virtualization, decentralization and network building change the manufacturing landscape: An Industry 4.0 Perspective. International Journal of Mechanical, Industrial Science and Engineering, 8(1): 37-44
3. Wang Y (2012) A new hybrid genetic algorithm for job shop scheduling problem. Computers & Operations Research, 39(10): 2291-2299
4. Nouri HE, Driss OB, Ghédira K (2015) Aholonicmultiagent model based on a combined genetic algorithm-tabu search for the flexible job shop scheduling problem. In Highlights of Practical Applications of Agents, Multi-Agent Systems, and Sustainability-The PAAMS Collection, pp 43-54
5. Wang L, Wang S, Xu Y et al (2012) A bi-population based estimation of distribution algorithm for the flexible job-shop scheduling problem. Computers & Industrial Engineering, 62(4): 917-926
6. Wang J, Chu K (2012) An application of genetic algorithms for the flexible job-shop scheduling problem. International Journal of Advancements in Computing Technology, 4(3)
7. Li J, Pan Q, Xie S (2012) An effective shuffled frog-leaping algorithm for multi-objective flexible job shop scheduling problems. Applied Mathematics and Computation, 218(18): 9353-9371
8. Baptiste P, Le PC, Nuijten W (2012) Constraint-based scheduling: applying constraint programming to scheduling problems (Vol. 39). Springer Science & Business Media
9. Brucker P, Burke EK, Groenemeyer S (2012) A branch and bound algorithm for the cyclic job-shop problem with transportation. Computers & Operations Research, 39(12): 3200-3214
10. Liaw, CF (2013) An improved branch-and-bound algorithm for the preemptive open shop total completion time scheduling problem. Journal of Industrial and Production Engineering, 30(5): 327-335
11. Fribourg L, Soulat R, Lesens D et al (2012) Robustness analysis for scheduling problems using the inverse method. In Temporal Representation and Reasoning (TIME), 2012 19th International Symposium on pp 73-80
12. Chen JC, Wu CC, Chen CW et al (2012) Flexible job shop scheduling with parallel machines using Genetic Algorithm and Grouping Genetic Algorithm. Expert Systems with Applications, 39(11): 10016-10021
13. Chirici L, Wang KS (2014) Tackling the storage problem through genetic algorithms. Advances in Manufacturing, 2(3): 203-211
14. Nouri HE, Belkahla DO, Ghedira K (2015) Genetic algorithm combined with tabu search in a holonicmultiagent model for flexible job shop scheduling problem. In 17th International Conference on Enterprise Information Systems (ICEIS)
15. Tavakkoli-Moghaddam R, Azarkish M, Sadeghnejad-Barkousaraie A (2011) A new hybrid multi-objective Pareto archive PSO algorithm for a bi-objective job shop scheduling problem. Expert Systems with Applications, 38(9): 10812-10821
16. [16] Yu Q, Wang K (2014) A hybrid point cloud alignment method combining particle swarm optimization and iterative closest point method. Advances in Manufacturing, 2(1): 32-38
17. Moslehi G, Mahnam M (2011) A Pareto approach to multi-objective flexible job-shop scheduling problem using particle swarm optimization and local search. International Journal of Production Economics, 129(1): 14-22
18. Wang KS (2013) Towards zero-defect manufacturing (ZDM)—a data mining approach. Advances in Manufacturing, 1(1): 62-74
19. Wang L, An L, Ni HQ et al (2016) Pareto-based multi-objective node placement of industrial wireless sensor networks using binary differential evolution harmony search. Advances in Manufacturing, 4(1): 66-78
20. Elmi A, Solimanpur M, Topaloglu S et al (2011) A simulated annealing algorithm for the job shop cell scheduling problem with intercellular moves and reentrant parts. Computers & industrial engineering, 61(1): 171-178
21. Zhang CY, Dong X, Wang XJ et al (2010) Improved NSGA-II for the multi-objective flexible job-shop scheduling problem. Journal of Mechanical Engineering, 46(11): 156-164
22. Zhang XX, Zhang WL (2010) FJSP based on timed Petri Net and GA. Systems Engineering, (11): 86-94
23. Fu WP, Liu DM, Lai CW et al. (2011) Polychromatic-sets-based improved genetic algorithm for solving multi-species FJSP. Computer Integrated Manufacturing Systems, 17(5): 1004-1010

MATEC Web of Conferences 95, 10002 (2017) DOI: 10.1051/matecconf/20179510002
ICMME 2016

Research on Management System of Mold Manufacturing Enterprise Based on RFID Technology

Bin Wang[1], Defang Liu[1] and Mingkun Ji[1,2]

[1]*Yancheng Institute of Technology, School of Mechanical Engineering, 224051 Yancheng, China*
[2]*Anhui University of Science and Technology, School of Mechanical Engineering, 232001 Huainan, China*

Abstract. To promote manufacturing management efficiency and decrease error rate in production process, this paper proposed a management system framework of mold manufacturing enterprises based on RFID technology. The system consists of four subsystems, which are warehouse management, production management, distribution and customer. The structure of each subsystem is presented in detail based on the analysis of overall design process of mold management system. Finally, a testing table of RFID-based mold manufacturing system was established to illustrate the feasibility of the system.

1 Introduction

In recent years, Radio Frequency Identification (RFID) has been widely used in product manufacturing field as an automatic identification technology, which depends on remotely storing and retrieving data based on tags and readers [1, 2]. Many manufacturing enterprises have already followed the trend of the next industrial revolution under the heading of Industrie 4.0 by investing in hardware, software, and networking systems to build the Internet of Things (IoT) [3, 4]. Reyes et al. [5] identified the determinants of RFID adoption stage and explored the perceived benefits from RFID adoption and found that RFID adoption had a significant benefits to perceived customer service and productively.

To gain insights for the factors that influence the RFID adoption, Bhattacharya [6] conducted a Delphi-based research to explore opinions of experts. Munilla et al. [7] had shown that protocol falls short of its security objectives, and describe attacks that allow: (a) an eavesdropper to trace a tag, (b) the previous owner to obtain the private information that the tag shares with the new owner, and (c) an adversary that has access to the data stored on a tag to link this tag to previous interrogations (violating forward-secrecy) and had analysed the security proof and found that while the first two cases can be addressed with a more careful design, strong privacy remains an open problem for lightweight RFID applications. Tsai et al. [8] proposed a protocol named as Smart Trend-Traversal (STT) to simplify the collecting process of tag IDs, which is regarded as a time-consuming process for most manufacturing enterprises.

To develop single node multi-parameter sensing chipless RFID sensors in versatile RF sensing applications, Amin et al. [9] proposed a novel chipless RFID sensor platform for wireless sensing of tagged objects, which need not require any on-board micro-chip and power source. Xie et al. [10] focused on RFID data management and described the research work on algorithm, protocol and performance evaluation. In his work, he introduced the research progress in RFID with anti-collision algorithm, authentication and privacy protection protocols. Zhang et al. [11] combined IOT operation principle and working mechanism with EPC global standard and established a structural model of IOT and gave the aware-layer, network-layer and application-layer works in this model.

Many scholars have researched on the RFID-based manufacturing management system and achieved successes in this field. For example, Zhang et al. [12] proposed a new job-shop real-time data collecting system based on the analysis of the application demands of RFID technology in manufacturing process. His system combined multi-threads technology and database technology, which performed the functions of the real-time monitoring, the timely transfer of the materials and reasonable scheduling for production plan. Staring from the RFID application requirements analysis of enterprise management, Ni et al. [13] established an automobile manufacturing line vertically distributed application model system framework and gave the system process of each function modules, which finally was testified solving data loss and omissions in information collection and the low efficiency caused by artificial operation in the production process. In order to provide reference for RFID system and promote the optimization of manufacturing system used in discrete manufacturing enterprises, Li et al. [14] analysed RFID adoption process for discrete manufacturing enterprises and summarized adoption influencing factors. The case of automobile manufacturing enterprise in Chongqing was employed to testify the feasibility and effectiveness of the operation mode.

Above research achievements have greatly improved the development of RFID-based manufacturing management. While mold manufacturing enterprises has their own speciality, such multi-variety, multi-process, and small amount, a customized IoT-based management system should be studied and applied in mold manufacturing field. The aim of this study is to improve the management efficiency and decrease the error ratio in mold manufacturing process.

2 Overall design process of IoT-based mold management system

A complete mold manufacturing system includes 4 modules such as warehouse management, production, distribution and customer communication. Fig.1 illustrates the complete IoT system of mold manufacturing.

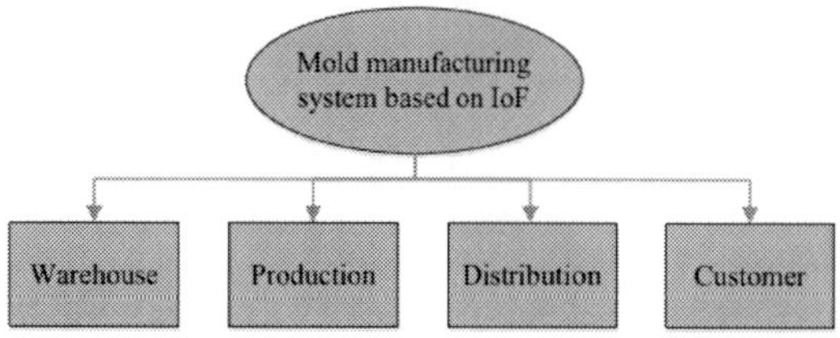

Figure 1. Main modules of mold manufacturing system based on IoF.

2.1 Subsystem of warehouse management

When a mold manufacturing enterprise has received an order, the subsystem of warehouse will automatically inquire the category and quantity of the raw materials, cost of manufacturing, and period of production that will be needed in this production process. The subsystem of warehouse will output an appraisement result for the possible profit of the manufacturing order, and send it managers for making a decision whether the order will be accepted. The structure of the subsystem of warehouse management is illustrated in Fig.2.

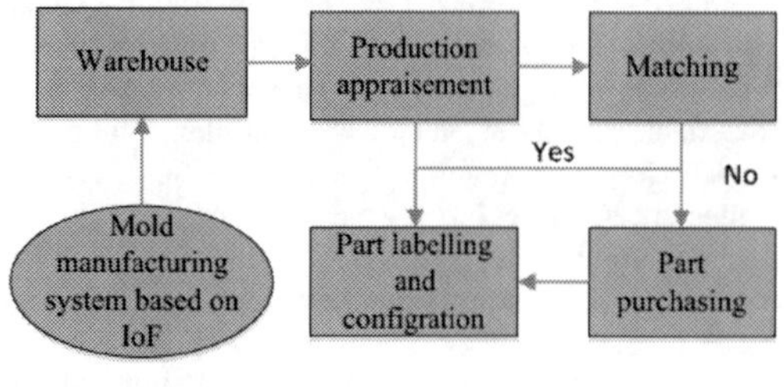

Figure 2. Subsystem of warehouse management.

If the manufacturing order is accepted, the purchasing department will make a further assessment on the required raw materials based on the output results of the subsystem of warehouse. After consulting with part suppliers, the purchasing department will submit the data of names of part suppliers, costs and types of part purchasing, quantity needed to the mold manufacturing system. The design department will design the mold meeting the requirements of the order. The overall 3D model and 2D detail drawing will be finished and submitted to the system by the design department for the reuse of other manufacturing relative departments. For example, the process department will make out a detail manufacturing process of the mold product according the 3D and 2D models and send the manufacturing tasks to the manufacturing plants via the mold manufacturing system.

2.2 Subsystem of production management

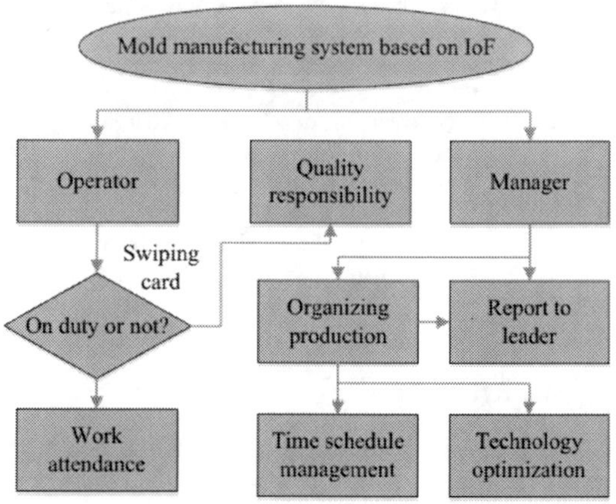

Figure 3. Information flow in production subsystem.

RFID technology employed on manufacturing process helps provide the information, performance and using states of the manufacturing equipment. With RFID, the effectiveness of overall equipment is greatly improved. There are much data needed to be collected by RFID during the mold manufacturing process, such as punching times, temperature, pressure, working condition, noise, speed, weight, and lubricating status. These data are very important for monitoring and predicting the working condition of the mold manufacturing equipment. Through punching time, we can easily deduce the next time for mold maintenance and when a new mold should be prepared in advance to keep the continuity of the production. The temperature data collected during the initial, processing, and finished time of manufacturing process can be effectively used to keep the quality the products, because product quality always has a close relationship with temperature. Pressure data collected are also used to control quality of the manufacturing process. Noise is monitored in real time to keep the safety of operators of the manufacturing equipment. The checking data of casual samples are transmitted to production subsystem and compared with normal data every a fixed span of time. If the dimensions of samples do not in the range of normal size, the mold manufacturing equipment will be forced to stop for an overhaul. Two kinds of persons are involved in the production subsystem, one is the direct operator of mold manufacturing equipments, and the other is the managers of manufacturing plants. The information flow between the two kinds of persons and subsystem is illustrated in Fig.3.

All kinds of sensors are employed in the process of manufacturing to monitor the production state on the site. Through sensors, the production data of temperature, pressure, noise etc. are collected and saved in the system for production management. After the manufacturing of molds, the data of assembling and testing are collected and stored in the system base for future's reuse.

2.3 Subsystem of distribution

RFID technology has been widely used in distribution field. In a mold manufacturing enterprise, the RFID labels are automatically input basic production information, which mainly includes delivery time, production process, number of product, and relative equipment.

The key of solution of mold product distribution is allocation and traceability of the products. The subsystem of product distribution ensure the parts to be sent to the right person. And some troubles such as parts lost and wrong distribution are avoided as well as other hardworking such as parts selection, and checking.

In RFID distribution system, each task needing to be finished should depends on high-efficient computers, which include the support of hardware, software, and database. To realize automatic tracing on materials, each cycle of RFID system should be equipped with an advanced computer, and the security of the system should also be ensured. A software system is more complicated one, which should own the abilities to identify, store and process the information collected by readers and to compare the newly collected information with stored data in the database. A powerful database is the basis of a RFID system, which can support the work of each cycle and improve the efficiency and accuracy of the system.

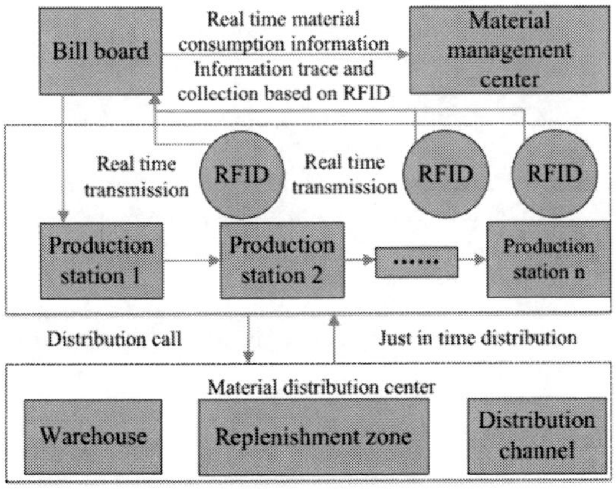

Figure 4. Information flow in production subsystem.

As shown in Fig.4, RFID is used in each production station to collect the material consumption information data, and the material distribution center acquire the real time material consumption data and prepare the materials for next distribution in advance. Thus the continuity of the production is ensured by the subsystem. Storekeepers are asked to label the RFID chips with some basic information of mold materials that will be sent the each working station. The basic information includes delivery time, production process, and product number. A GPRS location system will be encapsulated in the package of the mold. During the process of delivery, the managers can scan the transmission condition in real time to ensure the security the products.

2.4 Subsystem of customer

When the product is sold to customers, customer can inquire the whole manufacturing process of the product by entering the ID of RFID label. Thus the information of production, delivery position, and predicted time of arrival of the mold can be easily acquired by customers. If the some information need to be changed, customer may communicate with mold manufacturing enterprise via the mold management platform to decrease economic loss on both sides. Through the system, customers can feedback their requirements to the manufacturing enterprises in time and improve the efficiency of cooperation between them.

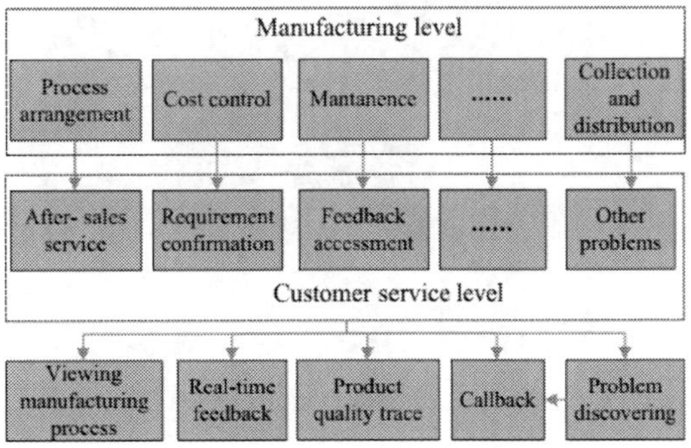

Figure 5. Information flow of customer subsystem.

From Fig.5, we can find that the information of each behaviour of customers will be collected and feedback to Cthe mold manufacturing process, which means any mold product is a comprehensive product joint-designed by customers. The subsystem of customer permits customers participating the whole manufacturing process of mold products, and thus the qualities that meets the requirements of customers are ensured.

3 Example analysis

An example of RFID based mold manufacturing process is employed to illustrate the working principle provided in this paper. The hardware of the system is illustrated in Fig.6, which is mainly made up of controller, mold, data displaying screen, power supply, time configurator, working table, etc.

During the product manufacturing process, terrace die are mainly relied to punch the part and form the final product, which is always doing telescopic movement. A controller is a main component which are used to ensure moving back and forth freely of terrace die. A circular Hall sensor is fixed on the head zone of the mold. When the mold is moving back and forth, the Hall sensor acts accordingly and transfers the collected information to liquid crystal display, which shows the current acting times in real time. Each action of mold will lead to an automatic time counting of the system, which will be shown on the LCD screen. In the hardware system, red lamp is indicator light of power supply, and green lamp is the indicator light of working condition. RFID electric labels are assembled inside the data screen mould, which can express the uniqueness of the mold. Through the unique number of the RFID label, the effective information of the mold storing in the database of the mold management system can be easily inquired. After careful processing, the data of mold manufacturing can help managers of enterprise realize information adjustment and control in real time, and achieve optimization of working efficiency of manufacturing process of the enterprise.

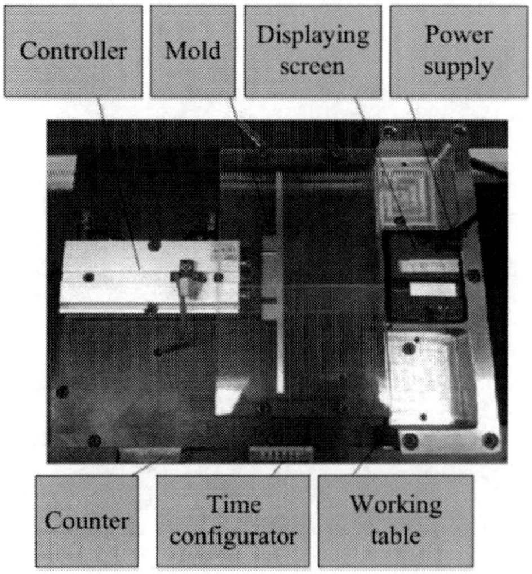

Figure 5. Testing table of RFID-based mold manufacturing system.

4 Conclusions

Mold enterprises are typical order-oriented discrete manufacturing enterprises. Traditional mold management mode cannot meet the requirements of practical operation because of random order and production process certainty. The research achievement of this paper helps promote the transformation and escalation of traditional mold manufacturing enterprises. RFID technology is an effective way to solve real manufacturing problems of mold factory, such as large amount of production data, frequent turnover of employees, incomplete managing records, and communication shortage among different departments, etc. RFID technology effectively accelerates the revolution of mold manufacturing pattern and promotes informatization and intellectualization of mold manufacturing enterprises. The work of this paper will display remarkable instruction for traditional mold manufacturing enterprises.

Acknowledgements

The authors gratefully acknowledge the financial support of the National Key Technology Support Program of China under Grant No. 2014BAF07B01.

References

1. X. Song, X. Li, W. Tang, W. Zhang. Information Fusion, 31 (2016), pp. 76-86.
2. S.H. Cho, K.H. Kim, S. Hong, J. Comput. Sci. Technol., 29 (1) (2014), pp. 4-20.
3. D.M. Segura Velandia, N. Kaur, W.G. Whittow, P.P. Conway, A.A. West. Robotics and Computer-Integrated Manufacturing, 41 (2016), pp. 66-77.
4. H. Kagermann, W. Wahlster, J. Helbig, Recommendations for implementing the strategic initiative INDUSTRIE 4.0 Final report of the Industrie 4.0 Working Group, 2013.
5. P.M. Reyes, S.H. Li, J.K. Visich, European Journal of Operational Research, 254 (3) (2016), pp.801-812.
6. M. Bhattacharya, Business Process Management, 21 (3) (2015), pp.517-540.
7. J. Munilla, M. Burmester, A. Peinado, Computer Communications, 88 (15) (2016), pp.84-88.
8. S.C. Tsai, Y.M. Hu, C.H. Chai, J.S. Li, Computer Communications, 88 (15) (2016), pp.73-83.
9. E.M. Amin, N.C. Karmakar, B.W. Jensen, Sensors and Actuators A: Physical, 17 (7) (2016) (accepted) DOI:http://dx.doi.org/doi:10.1016/j.sna.2016.06.014
10. L. Xie, Y.F. Yin, X. Chen, S.L. Lu, D.X. Chen, Chinese Journal of Computers, 36 (3) (2013), pp.457-470.
11. H.D. Zhang, L. Zhu, Computer Technology and Development, 21 (5) (2011), pp.56-59.
12. Y.Y. Zhang, J. Chen, Journal of Jiangnan University (Natural Science Edition), 12 (1) (2013) pp.54-58.
13. L. Ni, H. Zhong, C. Duan, Computer Engineering, 38 (4) (2012), pp.224-226.
14. W.C. Li, Y. Jing, X. Wang, Computer Integrated Manufacturing Systems, 19 (7) (2013), pp.1684-1691.

MATEC Web of Conferences **95**, 10003 (2017) DOI: 10.1051/matecconf/20179510003
ICMME 2016

Redesigning of Shopping Cart for Cost Reduction Using DFMA

C. D. Naiju, Pranav. V. Warrier and V. Jayakrishnan

School of Mechanical Engineering, VIT University, Vellore, India

Abstract. Product Design is a way of creating new designs for various products to be sold by a business to customers. It is very efficient as certain new designs and ideas can lead to products with less cost. The two methodologies of design for manufacture and assembly (DFMA) include design for manufacture (DFM) which is concentrated on reducing the number of redundant parts for the ease of manufacturing and also design for assembly (DFA) which is concentrated on reducing the assembly time of the parts included in the design of the product. It also helps to reduce the inefficiency and waste in the product design. This paper includes a case study of manufacture and assembly costs estimation of a shopping cart which is widely used in supermarkets, department stores all around the world. The main aim of this paper is to reduce the overall manufacturing cost of the shopping cart. The CAD model of the different parts of the shopping cart was generated using solid works and then analysed using the DFMA software. The outcomes were studied and the re-design is done. The final assembly cost of the re-designed product was compared to that of the existing design and the results are discussed.

1 Introduction

The demand for consumer products increases day by day and the technological advancements and development in the industries increases the demand of the products in market. To meet this demand, speed and the production capability must be increased. The ease in manufacturing provided by DFMA was studied by Andrew et al [1] where the end result was the detailed design of manufacturing assembly line. To improve in the product design process, design for manufacture and design for assembly were combined to form as design for manufacture and assembly (DFMA). The effectiveness of the new method was tested on a heavy duty stapler.

A study showed that a complete DFMA analysis helped give a categorical approach to the cost to be infused with each division within the firm. This enables the product designers to use all resources to the most optimized capabilities. As a result of which, more efficient product with less cost can be developed and thus improving sales [2]. According to Dewhurst et al [3], early cost estimation in product design is intended to form the basis for the design analysis for product design for efficient manufacture.

The methodology of DFMA can be used effectively to reduce the overall manufacturing costs by reducing the number of parts in the product and also by redesigning some parts to reduce the material used. [4].

Recent studies also showed that material selection process can help the product designers to find out what kind of aspects of materials are significant in their selection [5].

Design for Assembly evaluation is a systematic approach in determining how well a product is designed from assembly point of view. A study discusses a framework for automating DFA evaluation procedure, which is a key step in achieving a concurrent design environment. The modules required for this are assembly sequence generation, assembly features extraction, assembly code and other necessary operations generation, data manipulation and computation and re-design suggestion [6]. Recent studies about DFMA is related to the products likediesel engine, washing machine, refrigerators which has turned to be more effective as they are used in daily lives which makes them quite important. The abundant number of products around the world is also another reason to take consumer products under the DFMA technology [7].

2 Shopping cart – A case study

A shopping cart or buggy as used in some parts in United States and Canada, is a trolleywhich is supplied by a shop, especially supermarkets, department stores, for use by the customers inside the store to transport the merchandises from the shopping are to the checkout counter. Customers can also transport it to the parking lots outside the stores. Customers can take or leave the shopping cart outside the premise of the supermarket. This is very common in our day to day life and it is an improvement over the old shopping basket where customers had to carry it in their arms. This shopping basket became impractical. A large amount of merchandises could be transported using the shopping

 DOI: 10.1051/matecconf/20179510003

cart. This was possible with the help of wheels at the bottom of the cart. Carts in different designs are used all over around the world like trolleys in airports which helps passengers move heavy luggage easily from drop off area to check in as well as from the luggage carousel to the arrival area. The existing design available by all the leading department stores is given in the Fig. 1.

Figure 1. Existing shopping cart model.

Main parts in the shopping cart are castor wheels, frame, trim, mesh, handle and under carriage. These parts form the main parts, but there are a total of 26 parts in the existing design taken up for study. The different parts and their CAD models are given in Table 1. The processes included in the manufacturing of the shopping car include cutting (Advanced), automated bending, jig loading, spot welding, trimming, die punching, metal cleaning and finally the parts assembly.

In order to implement DFMA methodology to find the cost estimate included in the manufacturing of the shopping cart, the CAD models were then analyzed to find the manufacturing cost of the parts using DFM. Then the cost analysis for the final assembly of the cart is carried out using DFA. An alternative design is suggested and re-estimation is carried out using the DFMA software. The re-designed shopping cart is shown in Fig. 2.

Figure 2. Re-designed shopping cart model.

Table 1. Part list and CAD model.

Part Name	Major parts of shopping cart			
Trim (Part-1-4)	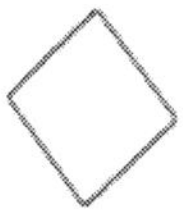			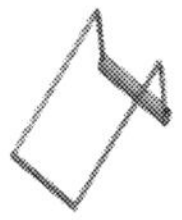
Frame (Part-5-7)	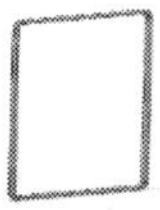	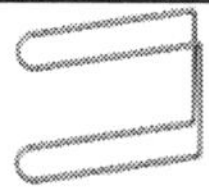		
Main Carriage and Castor Assembly (Part8-26)				

3 DFM, DFA and concurrent costing of shopping cart assembly

To estimate the costing analysis of the shopping cart, each part was first modeled using*Solidworks*: CAD software and then the individual cost per part is found out, which includes the DFM procedure in the DFMA methodology. The analysis is done for a product life volume of 10,000. DFM cost analysis of the final assembly with details of the individual partsincluded in the shopping cart is given in the table 2 and DFA cost analysis of the final assembly is given in the Table 3.

4 Redesign and analysis of shopping cart

DFMA analysis was carried out for the alternative design of shopping cart with changes in a few parts. The redesigned cart had the same number count as that of the existing design. The analysis showed that there was no much change in cost while selecting the existing manufacturing process. Again analysis was carried out with an alternative design, which helped to reduce the cost from $ 327 to $ 314. The DFMA analysis of redesigned cart for time saving and percentage reduction in assembly time is shown in Table 4.

Table 2. DFM cost analysis of the final assembly.

Part Designation	Total Count	Material	Total Item Cost($)
Part 1-4	1	Ferritic Stainless Steel	50.04
Part 5-7	1	Ferritic Stainless Steel	46.45
Part 8-13	16	Ferritic Stainless Steel	81.30
Part 14-26	55	Ferritic Stainless Steel	116.76
Total			294.55

Table 3. DFA of final assembly.

Per Product data	Entities (Including repeats)	Labour Time (s)	Labour Cost ($)
Component Parts	240	3208.29	31.45
Subassemblies Partially or Fully Analyzed	5	99.11	0.97
Subassemblies not to be Analyzed (excluded)	0	0.00	0.00
Standard and Library Operations	0	0.00	0.00
Totals	245	3307.40	32.43

Table 4. DFMA analysis of redesigned product.

Part	Quantity	Time Savings(s)	Percentage Reduction (%)	Part	Quantity	Time Savings (s)	Percentage Reduction (%)
Part 1	1	22.01	0.67	Part 15	11	218.50	6.61
Part 2	1	4.01	0.12	Part 16	15	296.90	8.98
Part 4	1	22.01	0.67	Part 17	2	42.10	1.27
Part 5	1	22.01	0.67	Part 18	7	140.10	4.24
Part 7	1	19.11`	0.58	Part 19	7	140.10	4.24
Part 8	16	120.16	3.63	Part 20	7	132.75	4.01
Part 9	15	51.75	1.56	Part 21	2	42.10	1.27
Part 10	16	106.80	3.23	Part 22	2	40.00	1.21
Part 11	32	128.32	3.88	Part 23	2	42.10	1.27
Part 12	16	36.04	1.09	Part 24	11	218.50	6.61
Part 13	4	30.04	0.91	Part 25	11	218.50	6.61
Part 14	55	1080.90	32.68	Part 26	1	4.01	0.12
Total	**Total Time Savings(s)**		**3178.82**	**Total Percentage Reduction (%)**			**3.89**

5 Conclusion

The principle of DFMA is at least more than ten decades old, by which they provide a huge potential savings in design and manufacture of a product. Total time savings was found to be 3178.82s and the percentage reduction in assembly time was reduced by 3.89 %. By using the DFMA software working in accordance with the Boothroyd principles, it is observed that the price of the existing product using the specified manufacturing process is $327.

The alternative design to the existing product which is manufactured using commercially available weaved meshes of required gauge instead of casting the product itself has given a reduced price of $ 314. This reduction in cost from $327 to $314 is due to the re-design of some parts which the resulted in ease of assembly and also elimination of welding at each and every joint which has led to great reduction in the manufacturing and assembly times. This shows that the modified product is better optimized than the existing original product.

 DOI: 10.1051/matecconf/20179510003

Acknowledgement

The authors wish to acknowledge the School of Mechanical Engineering (SMEC) and Product Design Laboratory, VIT University for supporting the project and for carrying out the study.

References

1. K. Asar, J. A. Day: A Knowledge Based Design Methodology for Manufacturing Assembly Lines, Computers and Industrial Engineering **41**, no. 4, 441-467 (2002)
2. M. Oakley, Product Design and Development in Small Firms, Operations Management Group, University of Aston Management Centre, vol. 3, no. 1, 5-10 (1982)
3. P. Dewhurst, G. Boothroyd, Early Cost Estimation in Product Design, Journal of Manufacturing Systems **26**, no. 7, 505-520(1994)
4. F. H. Ramy, N. Sahmrani, DFMA, A Quantitative DFMA Methodology, Computer Aided Design and Application **7**, no. 5, 701-709 (2010)
5. K. Elvin, H. Paul, K. Prabhu, Material consideration and Product Design: A Survey on Crucial Material Aspects used by Product Designers, Materials and Design **28**, no. 6, 1081-1089 (2008)
6. K. Rong, H. Cheng-Long, A Framework for Automatic DFA System Development, Computers and Industrial Engineering **22**, no. 4, 403-413(1992)
7. K. Annamalai, C. D. Naiju, S. Karthik, M. Mohan Prashanth: Early Cost Estimate of Product During Design Stage Using Design for Manufacturing and Assembly (DFMA) Principles, Advanced Material Research **622-623**, 540-544(2013)

Development of a Low Cost MQL Setup for Turning Operations

Sumaiya Islam , Noman Khandoker, Mohamad Izham, Tengku Azizi and Sujan Debnath

Department of Mechanical Engineering, Faculty of Engineering and Science, Curtin University Sarawak CDT 250, 98009 Miri, Sarawak, Malaysia

Abstract. In this study, the effect of MQL application on the tool wear, surface roughness and chip formation in turning Aluminum alloy 6061 is investigated. Experiments were carried out by plain turning of an Aluminum bar with varying depth of cut, cutting speeds (spindle speed) and cutting environments (Dry, Wet and MQL). A newly designed, cost efficient and portable MQL setup was developed for this study. For each experimental trial, five passes were carried out in order to promote the formation of tool wear. After each pass, the tool surface was examined under a microscope and the surface roughness was measured using a stylus based surface tester. The extent of tool wear from each pass is measured by using *ImageJ* software. Chips were collected after the five passes and were physically examined. The newly designed MQL delivery system was successful in delivering a fine mist of lubricant at the cutting point. MQL lubrication provided lower surface roughness and tool wear values compared to wet and dry conditions due to effective temperature reduction and improved lubrication penetration of cutting zones, as well as better chip flushing. Chip formed under MQL conditions had reduced chip thickness due to reduced temperatures. MQL was found to have the greatest positive impact on tool wear and surface roughness parameters based on ANOVA results.

1 Introduction

In manufacturing industries, machining processes such as turning is commonly used to obtain the desired size, shape, and surface of a material through the removal of material from a workpiece. During the turning operation, significant amount of heat is generated at the tool-chip interface due to friction and stresses generated from the shearing of chips. The heat generated at the cutting zone contributes to the tool wear of the cutting tool and has a negative effect on the surface roughness and surface finish. Chip formation is a machining criterion that is important in order to assess the cutting performance. In order to lower the temperatures at the cutting zone, lubricants or cutting fluids are widely used in the manufacturing industry for cooling and lubricating the tool-workpiece interface. The most common method of applying cutting fluids in turning process is by flood cooling. However, the usage of flood coolant has several harmful effects, namely environmental pollution if mishandled and high costs associated with flood machining. It is stated that "the coolants and lubricants used for machining represents 16% to 20% of the manufacturing costs" [1]. Thus, it is vital to find a way to manufacture products using more sustainable techniques, which would minimize the use of cutting fluids and promote a healthy and safe working environment. Minimum Quantity Lubrication (MQL), is an efficient cooling method that is more environmentally friendly compared to flood coolant while retaining the positive benefits of conventional cooling [2]. Aluminium alloys are commonly used in automobile and aerospace industries due to its excellent material properties, which include its high strength to weight ratio, excellent corrosion resistance and excellent thermal conductivity [3].

In this study, a newly designed cost effective MQL setup was developed for the turning operation. The effects of different process parameters (depth of cut and cutting speeds) and cutting environments (Dry, Wet and MQL) on the tool wear, surface roughness and chip formation in turning Aluminum alloy 6061 were investigated.

2 Literature review

Machining can be defined as a group of process of removing material from a workpiece in the form of chips using a cutting tool to obtain the desired shape, size and surface finish. According to Khan et al. [4], machinability is evaluated by cutting temperature (influences product quality and cutting tool performance), mode of chip formation, cutting forces (affects power requirements), dimensional accuracy, surface finish and tool wear and tool life. During machining with high cutting speeds and feed rates, high temperatures are generated at the cutting zone. This causes reduction in tool life by tool wear which will eventually lead to deviation of dimensional accuracy of the final product. The surface integrity is also affected due to tensile residual stresses, surface micro cracks and rapid oxidation present on the workpiece surface. According to Dhar et al. [5], it is stated that in machining of ductile materials, heat is generated at three regions. Firstly, heat is generated due to shear and plastic deformation at the primary deformation zone. The second region is at the chip–tool interface due to secondary deformation and sliding. The third region is at the work–tool interfaces due to rubbing. These three regions create a combined effect to produce maximum temperatures at the chip-tool interface, which have an adverse effect on tool wear and surface roughness.

Tool wear is defined as the gradual failure of the cutting tool due to regular operation. Tool wear occurs due to the increase in temperature in the cutting zone during turning. There are many types of tool wear, such as flank wear, crater wear and localized wear (chipping of cutting edge). Tool wear results in the formation of an adhesive layer and a built up edge (BUE), which causes a deterioration of the quality of the workpiece surface. Theoretically, built up edge occurs on at low cutting speeds and should not be encountered on coated cutting tools, however if the coating is scratched, built up edge is formed [6]. This is shown in Sharman et al. [7] research, where during the machining of Inconel 718, it was reported that the main wear mechanism was adhesion and built up edge formation, which eventually led to removal of the tool coating. This is because the coating substrate has a higher chemical affinity to the workpiece, and with time the coating peeled off. The tungsten carbide substrate was then exposed and rapid wear of the tool was visible.

At low cutting speeds, the cutting temperatures are relatively low and abrasion wear dominates when there is only wear due to sliding friction. In these conditions, thermal softening and cutting forces are important. Built up edge can occur, and will increase the rake angle which will ease chip formation. At higher cutting speeds, built up layer (BUL) will form at the contact zone, adhesion wear will dominate and it will facilitate the sliding of the chips produced. However, in these conditions, the temperatures will be greater, and chemical diffusion between the contact pieces will facilitate the failure of the tool [8]. From List et al. [8] research on the dry machining of Aluminium alloy with cemented carbide tools, it is stated that when soft materials such as Aluminum are machined, the workpiece material adheres to the cutting tool mainly in the form of built up edge. Due to built up edge formation on the cutting tool, the geometry of the tool is changed to be more inferior with the elongation of the flank face. Inferior tool geometry leads to decreased tool machining performance as well as tool life.

During turning, the most significant type of wear which must be controlled is the flank wear. Flank wear is described as the criteria or predominant parameter in determining the tool life of a cutting tool [9]. In Paul et al. [10] study in turning AISI 1060 steel while applying cryogenic cooling, it is stated that the average auxiliary flank wear is a vital parameter to be controlled as it determines the quality of the surface finish and dimensional accuracy of the final product. Dhar et al. stated that principal flank wear (V_B) is the most predominant wear as it raises cutting forces.

The surface roughness is very important as it defines the geometry of the surface to be machined and its surface texture. Surface quality is a parameter used to monitor the productivity of machine tools in producing machined components. Surface roughness is described as a vital quality indicator for machined surfaces and it has correlation to other properties such as wear resistance, fatigue strength, coefficient of friction, lubrication, wear rate and corrosion resistance of machined parts.

MQL provides reduction in tool wear and surface roughness when applied with the right conditions. Dhar et al. [5] used MQL machining technique in turning AISI 4340 steel with uncoated carbide tool. Under the same process parameters with dry and wet turning, MQL caused a significant reduction in tool wear. The reason for the reduction in tool wear was due to the reduction in flank wear (VB), which can be attributed to the reduction in flank temperature by MQL. MQL reduced the abrasive wear by retaining the tool hardness, while also reducing other temperature sensitive wear types such as adhesion and diffusion wear. Due to the reduced average auxiliary flank wear and notch wear on auxiliary cutting edge, surface roughness was also lowered when machining with MQL. The lower surface roughness with MQL was due to better control of auxiliary cutting edge deformation by abrasion, chipping and built up edge formation. Dry machining displayed higher surface roughness due to higher temperatures and stresses at tool tip.

The form, thickness and shape of chips produced when machining ductile materials depend upon the mechanical properties of the work material, the tool geometry, the cutting parameters, chip-tool interaction and cutting environment. Hence, a lot of information regarding the machining performance can be obtained from visually examining the chips formed. Machining with MQL condition is able to produce favorable chip formation by reducing chip-tool temperatures. Khan et al. [4] studied the effect of MQL on the surface roughness, tool wear and chip formation during turning AISI 9310 low alloy steel by SNMG carbide insert. In their research, it was found that the chip reduction coefficient is one of the most important machining index. Chip reduction coefficient can be defined as the degree of chip thickening and is influenced by all the machining parameters. From these results, it was found that the values of chip reduction coefficient were reduced for MQL conditions compared to dry and wet conditions. He attributed this to reduced temperatures generated at the chip tool interface, retained tool geometry from MQL application and lower built up edge formation with higher cutting speeds.

A review of various machining characteristics during turning was carried out. From the review, it is found that machinability is evaluated by cutting temperature, chip formation, cutting forces, dimensional accuracy, surface finish and tool wear. Hence, in this investigation, the three machinability criteria, tool wear, surface roughness and chip formation, will be studied in turning. Besides that, a review on the comparison of different lubrication modes in turning and its effect on machinability indicated that MQL provides positive machinability characteristics, such as reduced tool wear and reduced surface roughness and more favourable chip formation, when compared to dry and wet machining. MQL is also less of an environmental hazard compared to flood coolant, due to the reduced amount of lubricant used.

Aluminum alloys are widely used in manufacturing industries due to its excellent material properties. However, Aluminum alloys are difficult to machine, due to its high ductility and low melting point, which creates problems due to its affinity for bonding of work material on the cutting edges of the tool in the form of built up edge. Formation of built up edge on tool edges causes increased tool wear and eventually higher measured surface roughness values of workpiece during machining [3]. Thus, it is essential to study the effects of MQL machining with Aluminum alloys.

3 methodology

In this study, the MQL setup was designed to be cost-effective, portable and able to function well on both turning and milling machines. The MQL system had to be designed and developed before the experiment could be carried out. The developed MQL setup is shown in the Figure 1.

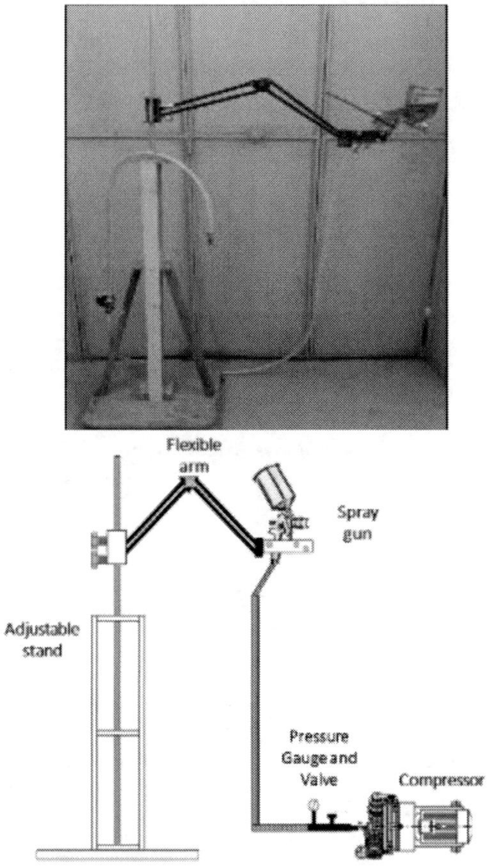

Figure 1. Developed MQL setup.

The main principle of this design is the usage of a paint spray gun, which is able to provide a fine mist of oil-air mixture easily. The spray gun used in this study is the Wynn Spray Gun WF75G, which is gravity feed spray gun. The advantage of using gravity feed gun compared to a conventional spray gun is that less pressure is required for it to operate. This provides other benefits, such as less overspray which means less amount of lubricant consumed from spraying, less wastage and a greater control of the spray. This type of spray gun also provides the option of fine tuning the spray in order to obtain a consistent spray.

The spray gun is connected to the compressor via a high pressure hose. A stand was designed to support the spray gun. The spray gun is sandwiched between two steel plates, and is clamped by tightening the bolts. The two steel plates are connected to a flexible arm, which allows for three axis movement. The flexible arm is connected to the main stand, which is also adjustable by tightening the bolts. The main stand is supported by wooden supports, to prevent the cylindrical Aluminum rod in the centre from swaying side to side. Schematic diagram of the setup is shown below

Based on the cost analysis of the design, the total cost for the materials and items purchased for the MQL setup is RM 130.21. This cost is considered low compared to the cost of a setup which would include the usage of a pump and several pressure gauges and valves, as reviewed in the literature. Hence, the MQL design is cost effective, portable (due to the ease of assembly and disassembly) and also easy to develop as the items used were readily available from the hardware shops.

The experiment was carried out by plain turning an Aluminium bar on a lathe at two different depth of cuts and three different cutting speeds under dry, wet and MQL machining environment. This was performed to study the effect of varying conditions on the machinability characteristics of the workpiece in regards to tool wear, surface roughness and chip formation. The experimental conditions are shown below in Table 1.

Table 1. Experimental Conditions.

Experimental Conditions		
Machine tool	Lathe Machine (Dalian Universal Lathe)	
Work specimen	**Material**	Aluminium Alloy 6061
	Diameter	50.9 mm
	Length	200 mm
Cutting Tool	**Manufacturer**	Kyocera Tooling Japan Co.
	Material	CVD Titanium coated carbide insert
	Model	TJR01856
Process parameters	**Cutting Speed (Spindle Speed), n**	700 rpm, 420 rpm and 300 rpm
	Feed rate, So	0.15 mm/rev (constant)
	Depth of cut, t	1.0 mm and 1.5 mm
	Cutting	140 mm

	Length	
Environment	Dry, Wet and MQL	
MQL Supply	**Air pressure**	3 bar
	Lubricant	Micro-emulsion cutting fluid (Yushiro MIC2500)
	Flow rate	100 ml/hr

A Leica DM 2500M light microscope was used to capture images of the tool cutting edge. The magnification was set to a default 5x. The tool wear images captured were the flank face, the rake face and the auxiliary flank face. Once the image is captured and a scale has been burned on to the image, the extent/length of the tool wear can be measured using Image J software. The tool wear image was captured after each pass in order to study the wear growth/development. Colour images of the tool surface after each experiment was completed was taken using the Nikon Eclipse LV100 POL microscope. A stylus type surface tester (Surfcom SJ- 310), manufactured by Mitutoyo was used to measure the average surface roughness of the work piece. The stylus has a diamond tip with a 90° cone angle and a 4μm tip radius. Each work piece was measured 3 times at three locations along the workpiece.

The chip thickness was measured using an electronic digital slide callipers manufactured by SENATOR. The digital callipers has a measuring range of 150mm/6", a resolution of 0.01mm/0.0005" and an accuracy of +/- 0.03mm. The chips were collected after each experiment and put into containers, where they are labelled for easy identification. Chip thickness were measured for each experiment, in order to obtain the chip reduction coefficient, ζ (ratio of chip thickness and depth of cut), which is an important index of machinability.

ANOVA procedure was used to determine the process parameters that were statistically significant. The purpose of ANOVA analysis is to investigate which factors affect the experiment data significantly. This technique will be used to quantitatively estimate the relative contribution, which each controlled parameter makes to the overall measured response and expressing it as a percentage.

4 Results and discussion

Under MQL condition, the chip reduction coefficient (*Figure 1 and 2*) is significantly lower compared to the dry and wet conditions, when the cutting speed is increased. Thus, the chips produced in dry and wet condition were thicker compared to MQL condition for all experiments. This is because with MQL, there is a reduction of friction and cutting temperature generated at the cutting zone due to the MQL lubrication and cooling effect, which reduced deterioration of the effective rake angle due to built-up edge formation and wear at cutting edges. This is consistent with Khan et al. [4] study and it was added that the chip reduction coefficient decreases with an increase in cutting speed due to plasticization and shrinkage of the shear zone. At higher cutting speed, there is too much softening of the chip material due to the increased temperature and as a result the chip sliding speed is increased.

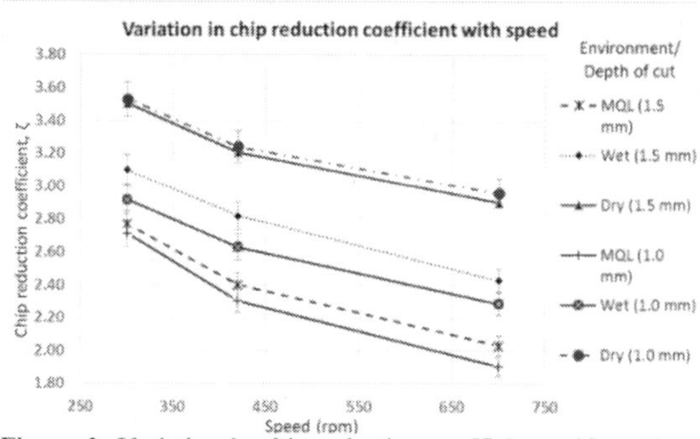

Figure 2. Variation in chip reduction coefficient with cutting speed.

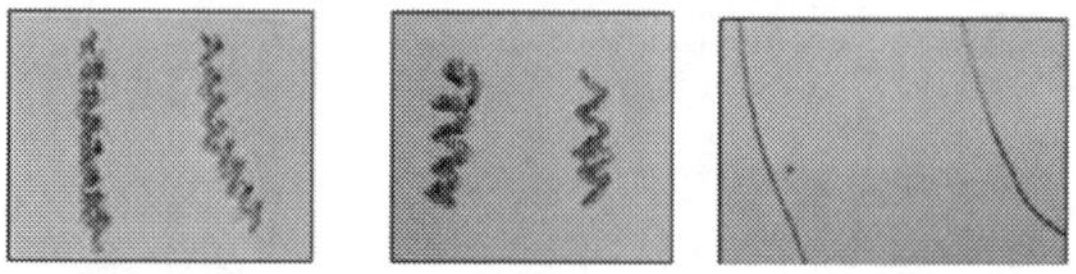

a) MOL b) Dry c) Wet

Figure 3. Chips produced under MQL, Dry and Wet under same parameters.

From the experimental findings, it is observed that increased cutting speed led to an increase in average principal and auxiliary flank wear for all cutting environments (*Figure 3*). As the machining speed is increased, quantity of work material adhered to the tool also increases. Besides that, the adhesion between the tool and chip also increased due to thermal softening of the chip with increased cutting speeds. Hence, increased material and chip adhesion over the tool rake, flank and clearance surfaces will lead to an increase likelihood of the tool material to break off when the chip deforms, causing the tool wear to increase with higher cutting speeds [11]. In addition, increased cutting speeds affect coated cutting tools by weakening the cutting edges as the crater wear is increased. This is the result of lower adhesion strength of the coating with the substrate at higher speeds. Thus, tool wear phenomenon is elevated by increasing speeds.

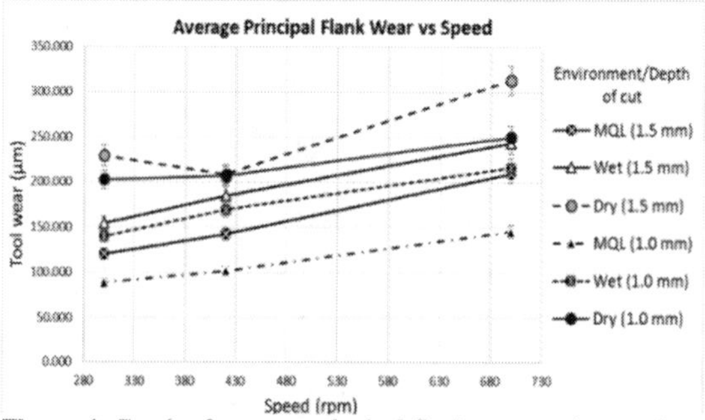

Figure 4. Graph of average principal flank wear against cutting speed.

An increase in depth of cut led to an increase in tool flank wear, based on the plot in *Figure 3 and 4*. This is because by increasing the depth of cut, cutting forces are also raised and this results in an increase in mechanical and thermal stresses on the flank area of the tool. This is also aggravated by the increase in tool contact length between the cutting edge and workpiece which results in higher amount of friction generated between the tool and workpiece. Thus, the flank wear area is extended and this leads to an increase in the flank wear width [12]. The observation that higher depths of cut increases tool flank wear is consistent with Phafat et al. [13] research and he added that at larger depths of cut, the heat and forces generated are larger and this results in an increase amount of adhered workpiece material onto the tool flank face, which accelerates the tool wear.

The flank wear values can be seen to be generally lower for MQL compared to dry and wet conditions from the plot in *Figure 4*. The reason for the reduction of flank wear observed under MQL condition could be due to the decreased flank temperature with MQL machining, which aid in reducing the abrasion wear by maintaining the tool hardness and also in reducing the adhesion and diffusion wear modes. The reduction in flank wear values under MQL is also reported by Hadad et al. [14] and he provided the theory that MQL reduces cutting forces which facilitates improved chip-tool interaction and maintains the cutting edge integrity. The oil droplets from the MQL reduced the friction at the sliding region as there is thinner clad material on the tool. As a result, MQL provides superior cutting performance due to the penetration of coolant/lubricant in the cutting area.

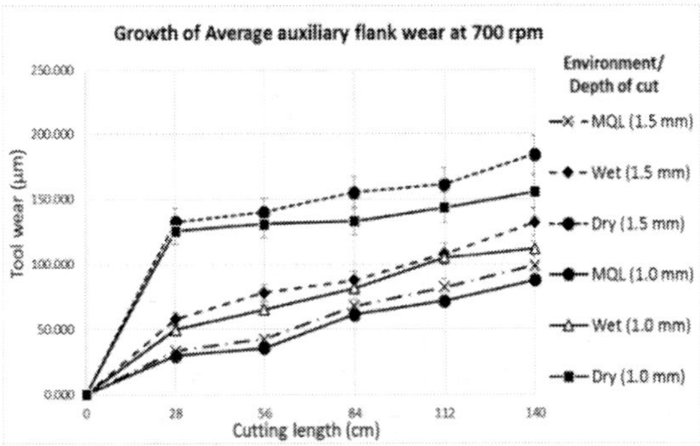

Figure 5. Graph of growth of average auxiliary flank wear at 700 rpm.

In addition, MQL condition also provided lower average auxiliary flank wear values compared to dry and wet condition based on the graph in *Figure 5*. Under MQL conditions, less abrasive wear, chipping and built up edge formation can be observed from the tool auxiliary flank face. The reduced auxiliary flank wear under MQL conditions is as a result of the high pressure jet of MQL behaving similar to a chip breaker and this improves the chip flushing at chip-tool-workpiece area, which provides the benefit of less work material adhered on the tool and less chip contact on the tool surfaces.

Referring to graph of surface roughness against cutting speed (*Figure 5*), it is observed that for MQL and wet condition, the surface roughness values decreased with increased cutting speeds. This is due to the elimination of built up edge (BUE) and reduced material adhered on tool surface at high speed machining and also the effect of MQL acting as a chip breaker. The increase in cutting speed led to the decreased formation of built up edge as there is increased heat generated at higher speeds, which causes the BUE to break off, making the cutting edge sharp again, thus leading to lower surface roughness values. This is comparable to Yoshimura et al. [15] findings, and he theorized that the surface roughness improved when the amount of adhesion decreased, as the repeated formation and breaking of adhered metal degraded the surface of the workpiece at higher speeds. For Aluminum alloys, the amount of adhered material is reduced as cutting speed is increased.

There also appears to be relationship between the tool flank wear and the surface roughness values. Based on the results, it is observed that flank wear increased with increasing cutting speeds and increased cutting speeds led to a decreased surface roughness for MQL condition. This is consistent with Shahabi et al. [16] findings and he mentioned that flank wear is the predominant wear type at the initial stages of tool wear, however at the second stage; other types of wear such as crater wear, nose wear and notch wear have a greater influence on surface roughness.

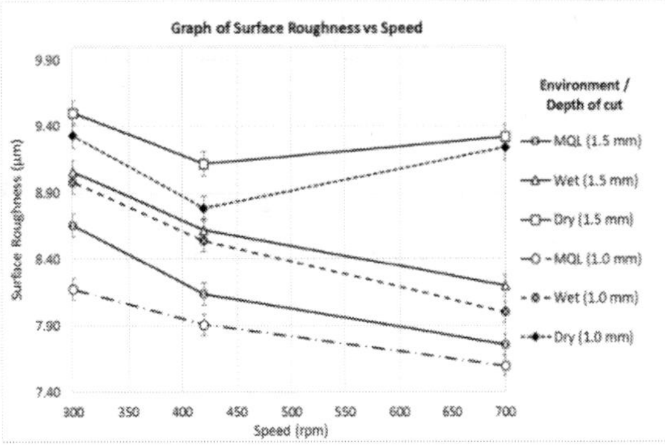

Figure 6. Graph of surface roughness against cutting speed.

In contrast, under dry conditions, the surface roughness (*Figure 5*) did not show a noticeable trend with cutting speeds. However, the surface roughness appeared to be highest at lower cutting speeds (300 rpm), which is a similar observation for MQL conditions. This could be attributed to the type of chip produced during the machining operation, as it has a significant effect on the surface finish. At lower cutting speeds, chip thickness is increased due to the altered shear plane angle, which results in increased cutting forces and vibration that have a detrimental effect on surface finish. As a result, increased surface roughness is obtained at lower speeds. At 700 rpm under dry condition, the surface roughness value appeared to be higher compared to 420 rpm. This could be an effect of the increased material adhered on the workpiece surface at higher cutting speeds. At high speeds, chip adhered on the tool would have continuously scraped and ploughed the workpiece. When the adhered material becomes unstable, it will detach from the tool and adhere on the workpiece, which would increase the surface roughness of the workpiece [17].

For all cutting environments, an increase in depth of cut led to higher surface roughness values (*Figure 6*). This can be explained by the effect of increased tool wear towards surface roughness parameter. At higher depth of cut, the tool wear at both flank faces are increased, as there is an increased in the tool-workpiece contact length and tool-chip interaction. This results in increased friction generated and more material from both the tool and chips to be adhered to the workpiece surface, causing a detrimental impact on the surface roughness values. Increased depth of cut also reduced the lubrication penetration efficiency of MQL to the cutting zone due to increased plastic flow at the tool-chip contact region which contributed to higher surface roughness values.

Values of surface roughness appear to be lower for MQL condition compared to dry and wet conditions. This is due to the reduced principal flank wear with MQL, which caused the auxiliary flank wear to decrease as well. The auxiliary flank wear has a major influence on the surface finish and dimensional accuracy of the machined surface. Thus, reduced auxiliary flank wears result in reduced surface roughness values. Another factor could be due the high velocity jet of MQL forced cutting fluid particles to penetrate the tool-chip interface, resulting in reduced friction. The micro-channels present on the tool chip interface during machining serves as capillaries and through capillary action, cutting fluid can enter the cutting zone. As the MQL used is water soluble and contains a higher ratio of water to oil, heat generated through machining is evaporated. Evaporation is a more efficient mode of heat transfer compared to conduction and this contributes to lower temperatures at the chip tool interface and retained sharpness of tool cutting edge. As a result, the surface roughness is improved [18].

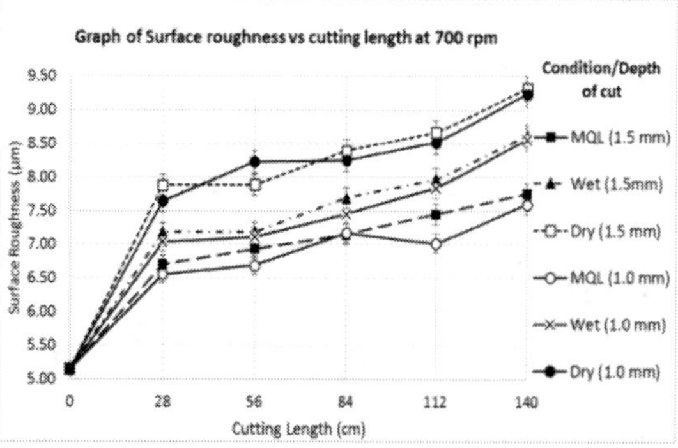

Figure 7. Graph of surface roughness against cutting length at 700 rpm.

The growth of notch wear with machining caused a change to the tool shape to a less flattened shape, which reduced the sharpness of the cutting edge (*Figure 7*). As the tool nose shape has a significant effect on cutting tool performance, hence there is a correlation between the tool notch wear and surface roughness values. It can be inferred that reduced notch and crater wear present under MQL condition also contributed to lower surface roughness values compared to the wet and dry conditions.

The results obtained from the ANOVA analysis (in Table – 2) for tool wear (principal flank wear) showed that the cutting environment (60.66 %) has the highest contribution to tool wear. Cutting speed (26.11 %) is the second most significant factor to tool wear and was followed by the depth of cut (9.90 %).

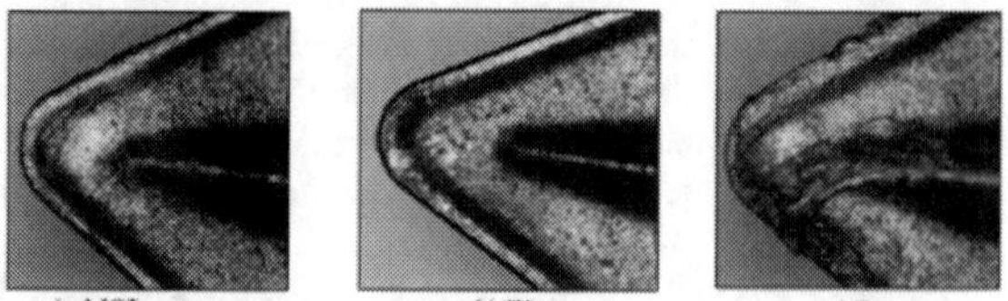

a) MQL b) Wet c) Dry

Figure 8. Comparison of tool rake face under same cutting parameters.

Table 2. ANOVA for tool wear.

No.	Factor	Degree of Freedom	Sum of Squares	Mean Squares	F-Value	Percentage Contribution (%)
1	Environment	1	29168.444	29168.444	72.735445	60.66
2	Cutting Speed	2	12553.469	6276.735	15.651883	26.11
3	Depth of cut	1	4760.721	4760.721	11.871499	9.90
4	Error	4	1604.084	401.021	-	3.34
5	Total	8	48086.718	-	-	100.00

Cutting condition has a large influence on flank wear as MQL provides lubrication, cooling effect and chip flushing/breaking effect. Machining with MQL resulted in shorter tool-chip contact length, less work and chip material adhered on the rake face, lower cutting temperatures due to lower friction generated, lower cutting forces and less possibility for built up edge formation, which ultimately resulted in lower tool wear on the flank surfaces compared to dry condition. Cutting speed also affects the MQL effectiveness. At lower speeds, MQL lubricating ability is more essential while at higher speeds, MQL cooling properties are much more significant due to increased heat generated at higher speeds. As the cutting fluid used in this study was water soluble, hence at higher speeds, the MQL effectiveness should be greater due to increased vaporized liquid at higher speeds, which would reduce cutting temperatures involved and consequently lesser tool wear [19].

Table 3. ANOVA for surface roughness.

No.	Factor	Degree of Freedom	Sum of Squares	Mean Squares	F-Value	Percentage Contribution (%)
1	Environment	1	3.077	3.077	204.25174	67.64
2	Cutting Speed	2	1.091	0.545	36.194951	23.97
3	Depth of cut	1	0.321	0.321	21.321758	7.06
4	Error	4	0.060	0.015	-	1.32

5	Total	8	4.54 9	-	-	100.00

The ANOVA analysis (in Table – 3) for surface roughness also described cutting environment (67.64 %) as the highest contributor to surface roughness. The second most significant factor is the cutting speed (23.97 %) followed by depth of cut (7.06 %). Thus, cutting environment has the greatest influence in the surface roughness and tool wear parameter, with MQL providing improved tool wear and surface roughness values.

5 Conclusion

The MQL setup developed was successful in delivering a steady jet of air-oil lubrication at the cutting point. The minimum quantity lubrication (MQL) technique ensured lower surface roughness and tool wear (flank wear) values due to lower cutting temperatures, increased lubrication penetration and better chip flushing. Chips formed under MQL technique had decreased chip thickness compared to wet and dry conditions, due to reduced cutting temperature, improved chip-tool interaction and reduced built up edge (BUE) formation. Flank wear and notch wear were found to be influential to the surface roughness when machining. Besides that, MQL reduced the growth of principal flank wear and notch wear, resulting in lower auxiliary flank wear and lower surface roughness values compared to wet and dry conditions. The ANOVA results for tool wear and surface roughness revealed that cutting environment has the greatest influence on the tool wear and surface roughness parameters, with MQL providing improved tool wear and surface roughness values.

References

1. D. J. Paulo, P. S. Sreejith, and J. Silva. Materials and Manufacturing Processes, Vol.22 (2007), p. 45-50.
2. Y. K. Hwang, and C. M. Lee, Journal of Mechanical Science and Technology, Vol. 24, (2010) p.1669
3. M. Nouari, G. List, F. Girot, and D. Coupard. Wear, Vol.255, (2003), p.1359
4. M. M. A.Khan, M. A. H. Mithu, and N. R. Dhar. Journal of Materials Processing Technology (2009), Vol.209, p.5573
5. N. R. Dhar, M. Kamruzzaman, and M.Ahmed. Journal of Materials Processing Technology, Vol.172, (2006), p.299
6. P. Zoltán, and G. Csernák. Journal of Sound and Vibration, Vol. 332,(2013), p.2057
7. A.Sharman, R. C. Dewes, and D. K. Aspinwall. Journal of Materials Processing Technology, Vol.118, (2001), p.29
8. G. List, , M. Nouari, D. Géhin, S. Gomez, J. P. Manaud, Y. Le Petitcorps, and F. Girot. Wear Vol.259, (2005), p.1177
9. M. A. Xavior, and M. Adithan. Journal of Materials Processing Technology, Vol.209, (2009), p.900
10. S. Paul, N. R. Dhar, and A. B. Chattopadhyay. Journal of Materials Processing Technology, Vol.116, (2001) p.44

[1] P. S. Sreejith, Materials Letters, (2008) Vol.62, (2008), p.276

[2] S. Hamed, S.P. Khanghah, and R. Teimouri. The International Journal of Advanced Manufacturing Technology, Vol.76,(2015), p.2099

11. N. G. Phafat, R. R. Deshmukh, S. D. Deshmukh, Applied Mechanics and Materials, Vol. 393, (2013), p.240
12. H. Mohammadjafar, and B. Sadeghi, Journal of Cleaner Production, Vol.54, (2013), p.332
13. Y. Hiromi, T. Moriwaki, N. Ohmae, T. Nakai, T. Shibasaka, H. Kinoshita, M.Matsui, and M. Shimizu. JSME International Journal Series C Mechanical Systems, Machine Elements and Manufacturing, Vol.49, (2006), p.83
14. H. H. Shahabi, and M. M. Ratnam. The International Journal of Advanced Manufacturing Technology,(2009), Vol. 40, p.1148
15. K. Ihsan, M. Kasap, I. Ciftci, and U. Seker. Materials & Design, Vol.25, (2004),p.303
16. D. G,Thakur, B Ramamoorthy, and L Vijayaraghavan. International Journal of Materials and Structural Integrity, Vol.5, (2011),p.1
17. V. S. Sharma, M. Dogra, and N. M. Suri. International Journal of Machine Tools and Manufacture, Vol.49 (2009), p.435

MATEC Web of Conferences **95**, 10005 (2017) DOI: 10.1051/matecconf/20179510005
ICMME 2016

3D-Printed Linear Positioner with Micrometer Accuracy

Yin-Yen Kuo, Chiao-Hua Cheng and Shao-Kang Hung

National Chiao Tung University, Department of Mechanical Engineering, 30010 Hsinchu City, Taiwan

Abstract. This article presents a positioner, whose flexure main body is made by a commercial 3D printer. Using this method, manufacturing a positioner can be cost efficient and much easier to customize. Integrating a laser displacement sensor, an electromagnetic actuator, and a feedback controller, this positioning system has 100 micron translational stroke with 1 micron resolution. Experiments also demonstrate sinusoidal motions at different frequencies. Using the method developed by this article, micro-positioners with customized specifications can be implemented rapidly, iteratively, and cost-effectively.

1 Introduction

Precision positioners are typical mechatronic devices which are widely used in automatic manufacturing [1], [2] and measurement. In the field of atomic level topography, they can be equipped in scanning tunneling microscopes [3], [4] and atomic force microscope [5]-[7]. Using piezoelectric elements in devices for positioning is a common method to achieve the goal of precision [8]. The advantages of piezoelectric are light-weight, high resolution, and fast response. Despite of all the advantages mentioned, a crucial shortcoming, short traveling range, will limit the applications. To find the solution of long traveling range, there are many researches of piezoelectric positioning [9]. One of the positioning solutions is apply electromagnetic actuators, which do not need to contact mechanisms and possess the ability to achieve a larger displacement.

The cooperation of electromagnetic and flexure hinge mechanism becomes popular nowadays. Utilizing the flexure hinge mechanism [8] has many advantages, such as no backlash, smoothly continuous displacement and no friction between mechanisms. Nevertheless, most of the flexure hinge mechanisms are manufactured by electrical discharge machining (EDM) process [10]-[12] with metal materials, which often spends lots of effort and cost. In order to improve the shortcomings, we decided to manufacture flexure hinge mechanism by 3D-printing [13] with polylactic acid (PLA) material. In this case, not only the cost of producing is reduced, but also making it easier to implement or to be improved iteratively.

In this paper, a 3D-printed flexure hinge positioner cooperate with electromagnetic actuator is presented. Moreover, experiment results reveal the satisfactory performance.

2 System design

Due to the rapid development of 3D-printing, also known as additive manufacturing, we can customize many complicated structures easily [14], [15]. Fig. 1(a) shows the top view of the experimental setup including a permanent magnet, a coil, a moving stage, and a laser displacement sensor.

(a)

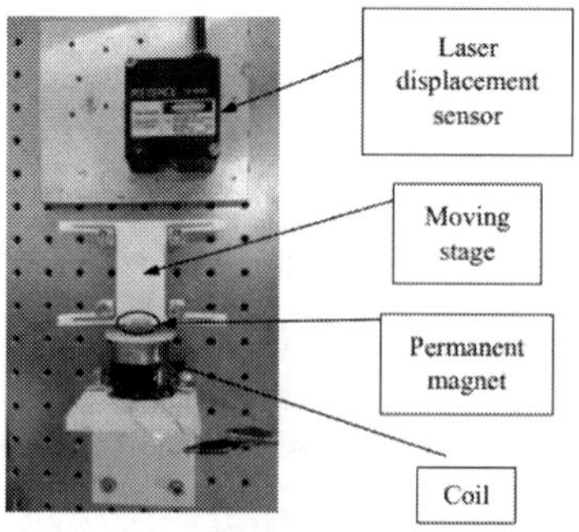

(b)

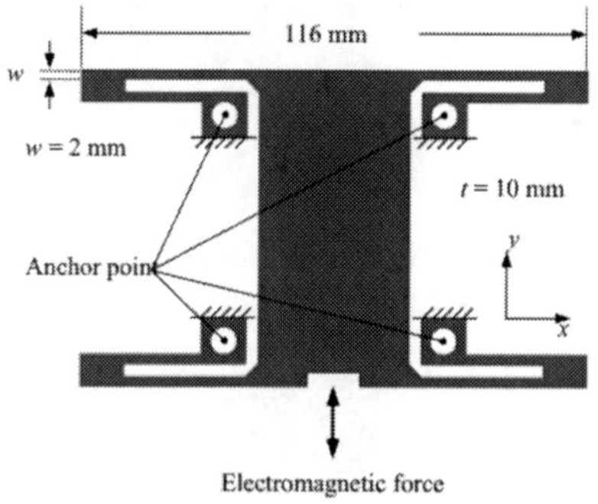

Figure 1. (a) The experimental setup and (b) the detailed structure of moving stage.

A permanent magnet is fixed at one end of the moving stage for interacting with the coil. A laser displacement sensor (LK-H020, KEYENCE) is fixed facing the other end of the moving stage for measuring displacement. The detailed design of the moving stage is presented in Fig. 1(b). The thickness t of the moving stage is 10 mm. The width w is designed to 2 mm because of the limitation of printing quality of the commercial 3D printer. With the flexure structure, the moving stage can move smoothly in y direction.

Fig. 2 shows the schematic block diagram. For the purpose of achieving high-precision positioning, an appropriate feedback controller is necessary. Using proportion-integral-derivative control scheme is a common method, and it can be easily applied in many systems. However, in this paper, a proportion-integral (PI) control scheme is enough. We utilize PI closed-loop control to overcome the hysteresis problem and reduce the steady stare error of the mechanical motion. Furthermore, an optical table is used to stabilize the whole system in order to decrease vibration from external environment.

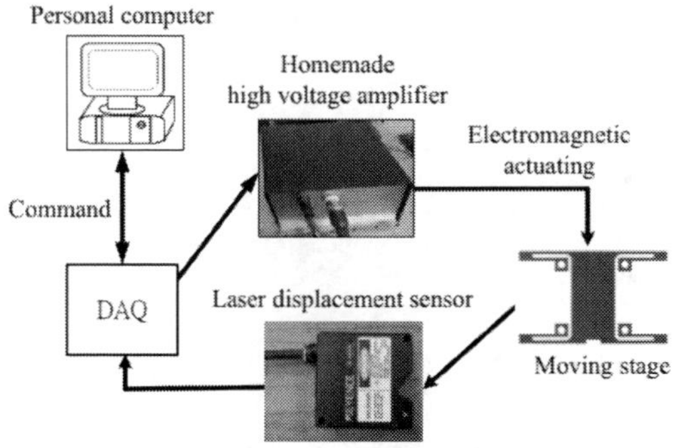

Figure 2. The block diagram of the proposed design.

National Instrument USB-6341 Data Acquisition (DAQ) card is selected to integrate the whole system. Also, a homemade high voltage amplifier is used to magnify the output voltage of the DAQ signal by 10 times. The maximum range can reach from -30 V to +30 V.

In order to prevent from the electromagnetic interference, the contactless optical sensor is chosen in the experiments. Having the active range from -3 mm to +3 mm and 10k Hz sampling frequency, the laser displacement sensor is utilized as the feedback sensor. The optical sensor has another advantage of avoiding the impact of the unevenness surface, which is especially significant in the 3D-printed surface relative to the EDM metal surface.

The 3D printer used in this article is AURORA 3D printer F1, shown in Fig. 3. The product specification of AURORA 3D printer F1 is indicated in Table 1. The build envelope is 168 mm x 168 mm x 168 mm, which is sufficient to print the moving stage in this article. PLA is biodegradable, and has lower melting point compared to Acrylonitrile Butadiene Styrene (ABS). Also, Unlike ABS, PLA is possible to print successfully without heating the printing bed first. Moreover, When PLA is heated, it will become much more liquid, which leads to demonstrate much sharper details. With all the advantages mentioned, the AURORA 3D printer F1 and PLA material are chosen.

Figure 3. AURORA 3D printer F1

Table 1. AURORA 3D printer F1 product specifications

Overall dimensions	450 mm* 420 mm* 460 mm
Weight	19 kg
Build envelope	168 mm* 168 mm* 168 mm
Minimum layer thickness	0.1 mm
Material	PLA Ø1.75mm±0.1mm
Nozzle diameter	0.4mm

3 Experiment result

3.1. Large step response

The full stroke capability of the proposed design mentioned in this paper can achieve from -50 μm to +50 μm. Fig. 4 shows the large step response ranged from 0 μm to 50 μm. In the configuration, we can see that the overshoot is about 3% and the settling time is about 0.3 second.

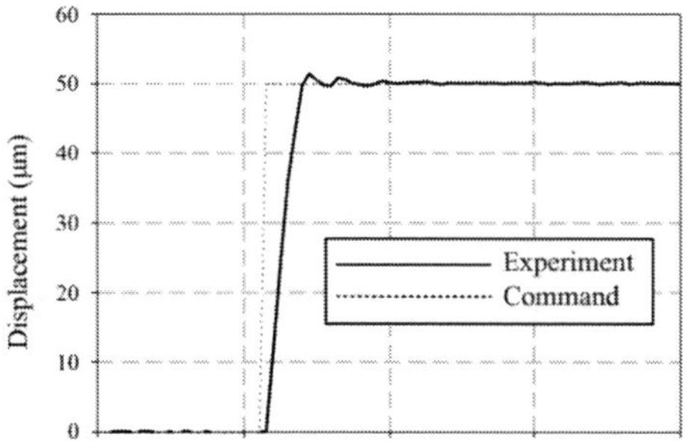

Figure 4. The large step response.

3.2 Step-train response

While Step-train motion is one of the most common operations of the positioning system, we make some experiments of continuing stepping response. In Fig. 5(a), the moving stage moves between -50 μm and +50 μm with each step equal to 5 μm. And in Fig. 5(b), the

 DOI: 10.1051/matecconf/20179510005

moving stage moves between -5 μm and +5 μm with each step equal to 1 μm.

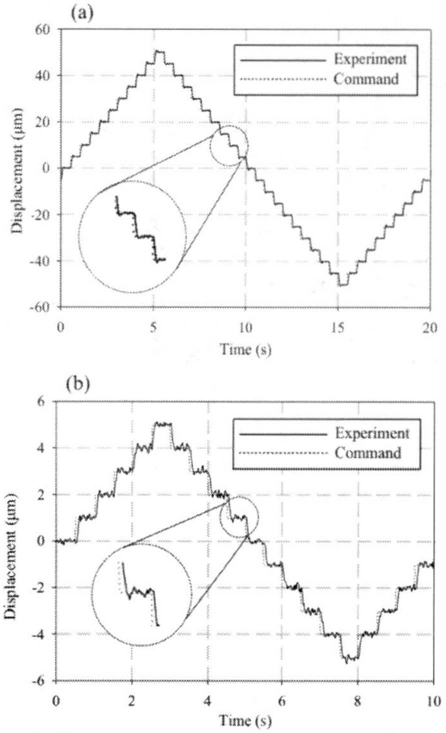

Figure 5. The step-train motion with range (a) ±50 μm with 5 μm each step and (b) ±5 μm with 1 μm each step.

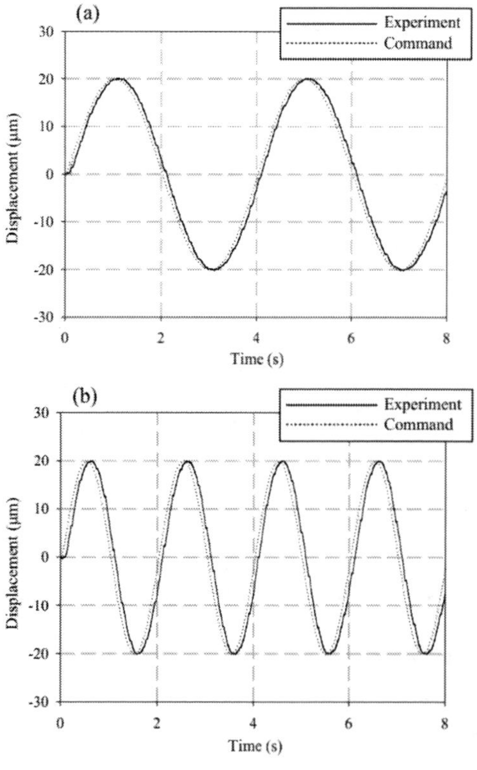

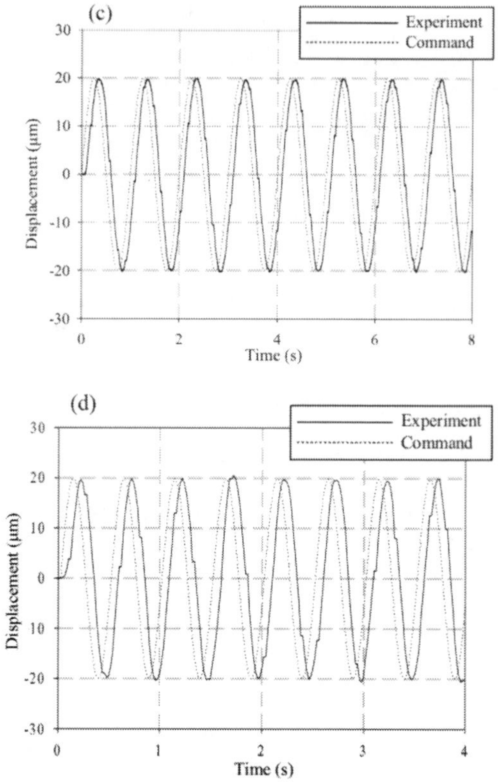

Figure 6. The sinusoidal motion, (a) 0.25 Hz, (b) 0.5 Hz, (c) 1 Hz and (d) 2 Hz.

3.3 Sinusoidal response

The result of the sinusoidal response is illustrated for testing the tracking capability of the proposed design. With the amplitude of 20 μm, the frequency at 0.25 Hz, 0.5 Hz, 1Hz and 2 Hz are presented in Fig. 6(a)-(d), respectively. As the frequency exceeds 1 Hz, the performance becomes unsatisfactory.

4 Conclusion and outlook

A novel approach of manufacturing micro-positioner is demonstrated in this article. With 3D-printing technique, fabricating customized positioners costs less effort, time, and budget. The proposed design has 100 μm traveling, 1 μm resolution, and sinusoidal motion tracking ability. This design will be soon improved iteratively in the future. More advanced control schemes are planned to enhance the dynamical performance of our next generation 3D-printed positioner.

References

1. T. Hu, W. J. Kim, IEEE/ASME Trans. Mech., **11**, 682 (2006)
2. I. J. Busch-Vishniac, IEEE Circuits Devices Mag. **7**, 32 (1991)

3. G. Binnig, H. Rohrer, Surf. Sci., **126**, 236-244(1983)
4. K. Uozumi, K. Nakamoto, K. Fujioka, Jpn. J. Appl. Phys., **27**, L123-L126 (1988)
5. G. Binnig, C. F. Quate, Ch. Gerber, Phys. Rev. Lett., **56**, 9 (1986)
6. S. K. Hung, L. C. Fu, IEEE Trans. Instrum. Meas., **55**, 5 (2006)
7. S. Hwang, S. K. Hung, L.C. Fu, M. Y. Lin - US Patent 7249494 (2007)
8. H. R. Lin, C. H. Cheng, S. K. Hung, Mechatronics, **31**, 180-188 (2015)
9. W. Dong, L. N. Sun, Z. J. Du, Sens. Actuator A-Phys., **135**, 250-256 (2007)
10. E. C. Jameson, *Electrical Discharge Machining* (Society of Manufacturing Engineers, 2001)
11. Y. C. Lin, C. C. Tsao, C. Y. Hsu, S. K. Hung, D.C. Wen, Int. J. Adv. Manuf. Technol., **62**, 1013-1023 (2012)
12. M. Y. Chen, H. H. Huang, S. K. Hung, IEEE Trans. Ind. Electron., **57**, 1 (2010)
13. I. Gibson, D. Rosen, B. Stucker, *Additive Manufacturing Technologies* (Springer, 2010)
14. R. R. Ma, LU. Odhner, A. M. Dollar, Int. Conf. Robot. Autom., 2737-2743 (2013)
15. R. MacCurdy, R. Katzschmann, Y. Kim, D. Rus, 2016 IEEE Int. Conf. Robot. Autom., 3878-3885 (2016)

MATEC Web of Conferences **95**, 10006 (2017) DOI: 10.1051/matecconf/20179510006
ICMME 2016

Effect of Parameters of Isopipe on the Quality of Glass Sheet Produced from Overflow Fusion Process by Numerical Simulation

Yansheng Hou[1], Jinshu Cheng[1,2], Junfeng Kang[1], Jingjing Cui[1], Xinjian Xue[3] and Dong Zuo[3]

[1]*State Key Laboratory of Silicate Materials for Architectures, Wuhan University of Technology, Wuhan, 430070, China*
[2]*Hebei Shahe Glass Technology Research Institute, Shahe, 054100, China*
[3]*Shaanxi Caihong Electronic Glass Company Limited, Xianyang, 712000, China*

Abstract. A numerical model for simulation of molten glass flow with semi-implicit method for pressure-linked equations (SIMPLE) and Volume of fluid (VOF) method during the overflow fusion process was carried out by using FLUENT software. The effect of the geometry parameters of the isopipe and flow rate of molten glass on the flow patterns during overflow was investigated. The results showed that the overflow trough depth only had an effect on the flow rate distribution of glass melt around the inlet point. The tilt angle at the top of the trough had a significant effect on flow rate distribution along the length of the trough, and the desired mass distribution was achieved at the tilt angle 4°~6°. The flow rate had a large influence on the overflow of the molten glass, which had a directive significance to the overflow down draw sheet manufacturing process.

1 Introduction

Currently, the glass used as LCD substrates are produced by two different processes: floating process [1], and overflow fusion process [2]-[5]. One of the major advantages of the overflow fusion process is that the surfaces of the resulting glass sheet are pristine without any contact with refractory material during forming. As a result, the overflow fusion process has become a particular importance method in the production of the glass substrates used in the manufacture of LCDs.

To gain a better understanding of molten glass flow during the overflow fusion process, the physical phenomena should be well studied. Therefore, in this study, FLUENT software was used to simulate the 3D flow of the molten glass in the isopipe. A finite different volume of fluid (VOF) approach for tracking the waving of the surface of the molten glass was proposed. The effect of the geometry of the isopipe on the flow patterns during overflow was investigated to obtain the relationships among flow pattern and geometry of the isopipe. Then, an optimized design of isopipe could be achieved. In addition, flow rate of molten glass on the flow patterns during overflow were also studied, which had a certain guiding role for actual production.

2 Physical model

The molten glass was conducted through an inlet pipe to the sheet forming apparatus, which was formed by a refractory material known as an "isopipe". The schematic diagram of the isopipe is shown in Fig. 1. After filling the trough of the isopipe, the molten glass flowed over both side walls of the trough (called weirs), down along the side surfaces of the isopipe as two glass ribbons. The two sheets met at the root of the isopipe, where they fused together into a single sheet. The single sheet was then fed to a drawing equipment, which controlled the thickness of the sheet by the rate at which the sheet was drawn away from the root.

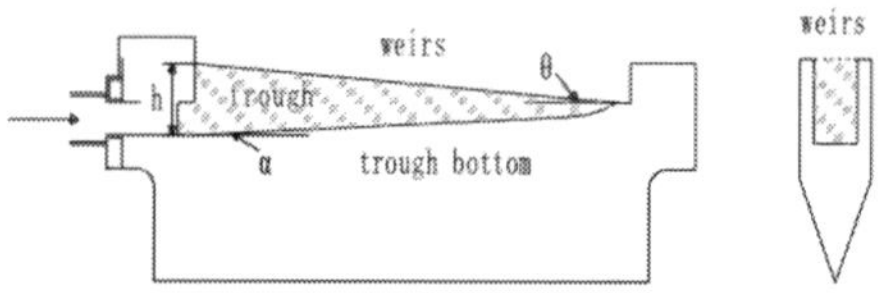

Figure 1. The schematic diagram of the isopipe for the overflow process.

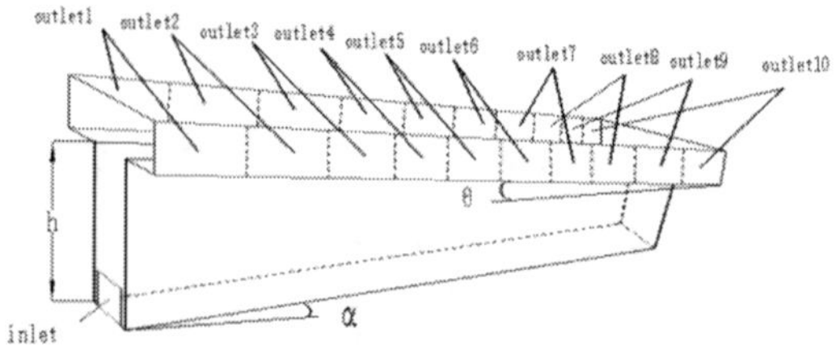

Figure 2. Simulation model of the trough.

The process of molten glass flowing in the isopipe included three parts. The molten glass was conducted through the inlet pipe to the isopipe (Y direction), and then the molten glass rose in vertical direction (Z direction). At last the molten glass overflowed from the weir surfaces (X). Due to the stresses from X, Y, and Z three directions, so the geometry was meshed using 3D GAMBIT software with hexahedral cells. In order to make the analysis easier, each side of overflow weirs was average divided into 10 outlets (shown as Fig. 2). Therefore, the glass melt flow distribution at top of the trough along the length of the isopipe could be calculated respectively.

3 Simulation details

In this paper, the powerful FLUENT software was used to simulate the overflow fusion process. SIMPLE algorithm was selected in the calculation process. The surface tracking of the molten glass-air interface was studied by VOF model with open channel method [6].

In order to simulate the molten glass flowing in the trough, some assumptions were proposed [7]:

1. The molten glass with high viscosity had low Reynolds numbers, assuming that the fluid flow was a laminar flow motion.

2. The molten glass was stable without component unevenness, chemical reaction, bubbles, stones and so on.

3. The molten glass was a uniform incompressible Newtonian fluid.

4. The temperature of the molten glass was considered as a constant without heat transfer, that is, the thermal conductivity C_p was 0.

Some physical properties of the molten glass were approximately constant, including density, viscosity, coefficient of expansion.

3.1 Governing equations

The molten glass was approximatively considered as an incompressible viscous fluid. Therefore the density was equal to a constant in the incompressible Navier-Stokes equation. The continuity equation [8]:

$$\nabla \mathbf{V} = 0 \tag{1}$$

Suppose the viscosity was equal to a constant in the entire flow process. The momentum equations:

X direction momentum equation:

$$\rho \frac{\mathrm{D}u}{\mathrm{D}t} = -\frac{\partial p}{\partial x} + \eta \nabla^2 u + \rho f_x \tag{2}$$

Y direction momentum equation :

$$\rho \frac{Dv}{Dt} = -\frac{\partial p}{\partial y} + \eta \nabla^2 v + \rho f_y \tag{3}$$

Z direction momentum equation:

$$\rho \frac{\mathrm{D}w}{\mathrm{D}t} = -\frac{\partial p}{\partial z} + \eta \nabla^2 w + \rho f_z \tag{4}$$

Assuming the temperature T was a constant, the viscosity η was derived as a constant (because it is a function of T). The four equations(1)~(4) are closed, which have four equations and four unknowns u, v, w, and p. Assuming the ρ=constant、η=constant, the energy equation is decoupled. Therefore the continuity equation and momentum equation were used to solve the problems of molten glass flowing process.

3.2. Treatments for the free surface flow

VOF model can be used to model two or more immiscible fluids. This is done by solving a single set of momentum equations and by tracking the volume fraction of both the liquids-air throughout the domain.

The volume fraction equation [9]:

$$\frac{1}{\rho}\left[\frac{\partial}{\partial t}(\alpha_q \rho_q) + \nabla(\alpha_q \rho_q \mathbf{v}_\mathbf{p}) = S_{\alpha_q} + \sum_{p=1}^{n}(m_{pq} - m_{qp})\right] \tag{5}$$

Here ρ_q is the density, $\mathbf{v}_\mathbf{p}$ is the velocity and α_q is the volume fraction of the q phase. m_{pq} and m_{qp} are mass flow rates of phases p to q and q to p respectively. The volume fraction of secondary phases is computed from the equation below:

$$\sum_{q=1}^{n} \alpha_q = 1 \tag{6}$$

3.3 Parameters of the model materials

The various material properties are defined (as shown in Table 1) and then suitable boundary (as shown in Table 2) conditions are applied. The parameters of glass melt in simulation process were determined by practical production process. The three most important geometric parameters of the isopipe were the depth (h) of the inlet point, the bottom tilt angle (α) of the trough and the tilt angle (θ) of the overflow weirs. Different flow rates of molten glass on the flow patterns during overflow were also investigated. The detail parameters used in this study are shown in Table 3.

Table 1. Materials properties.

Material	Physical property	Value
Molten glass	Density	2500 kg/m^3
	Viscosity	1750 Pa·s
	Surface tension	300 dyne/cm
Air	Density	1.225 kg/m^3
	Viscosity	1.784×10^{-5} Pa·s

Table 2. Boundary conditions.

Contents	Specification
Inlet	Velocity Inlet
	Molten glass volume fraction=1
Outlet	Pressure Outlet
	Gauge pressure=0 Pa
	Backflow volume fraction=0
	Open channel flow specification

Table 3. Parameters for simulation.

	Height of trough at inlet end (cm)	Tilt angle of trough bottom α(°)	Tilt angle of weir surface θ(°)	Flow velocity (kg/h)
Model 1	15,20,25,30, 35,40,50	0	0	300
Model 2	40	1,2,3,4,5,6	0	300
Model 3	40	0	1,2,3,4,5,6,7	300

DOI: 10.1051/matecconf/20179510006

4. Simulation results and discussion

4.1 The effect of geometric parameters of overflow trough

Due to the importance of stability in production process, the numerical studies were carried out under steady conditions, and then obtained the steady state of fluid at different initial conditions. Each model was calculated iteratively under the initial conditions until the residual standard error remained unchanged (the difference in mass between inlet and outlet was less than 0.01%). The flow of the overflow glass melts was obtained by calculating that of the ten outlet (evenly distributed),then the mass distribution of three models were obtained, as shown in Figs. 3, 4 and 5.

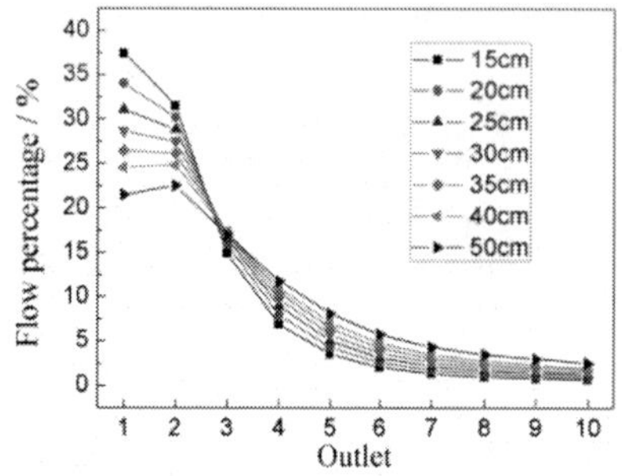

Figure 3. Mass distribution of glass melt with different height values of trough.

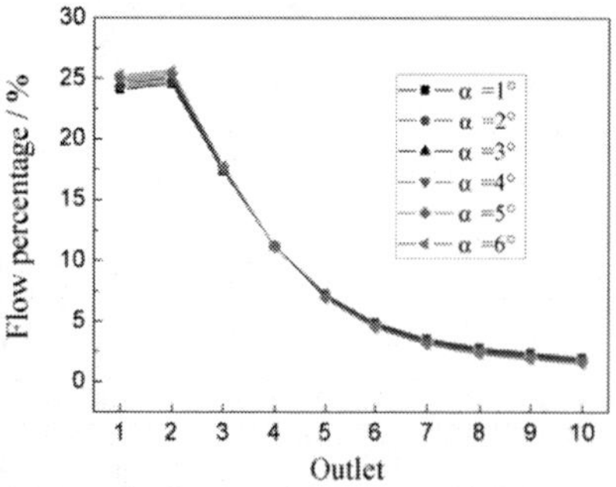

Figure 4. Mass distribution of glass melt with different tilt angle of trough bottom.

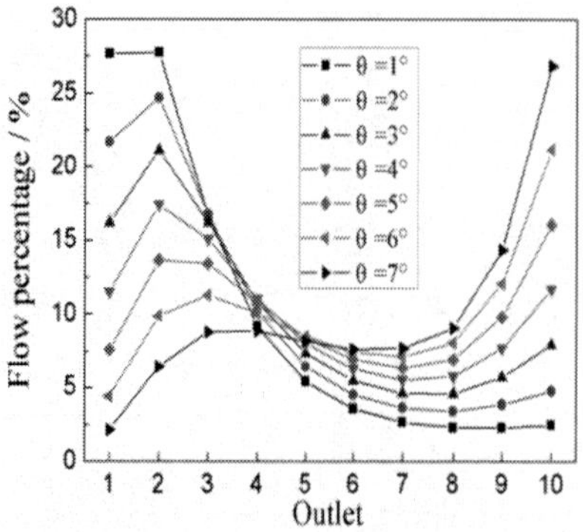

Figure 5. Mass distribution of glass melt with different tilt angle of trough surface.

The flow of the overflow molten glass with different values of h and α was shown in Fig. 3 and Fig. 4. The results showed that the flow of molten glass at inlet point was greater than that of the far end. As shown in Fig. 3, the depth of overflow trough (h) influenced molten glass flow at inlet point greatly. The molten glass flow (within 1/5 of the length of the inlet point) decreased rapidly when h increased, but h had negligible impact to the rest part. The flow distribution curves of molten glasses with different values of α were almost overlapped (as shown in Fig. 4), which indicated that the change of bottom tilt angle of trough did not make a significant difference in the flow distribution of molten glass.

As shown in Fig. 5, with the overflow surface tilt angle (θ) gradually increased, the flow distribution of molten glass in the overflow surface changed significantly. With the value of θ increasing, the flow of molten glass at the inlet point reduced, and reduction of the level height of the overflow surface along the extent of the trough provided the possibility of low potential outflow at the far end. With the tilt angle θ increasing from 4° to 6°, the flow rate distribution curve close to a smooth level, particularly between outlet 2 and outlet 8. Although the molten glass mass distribution curve is still not very even, the distribution that the large overflow in the both ends and small overflow in the middle can meet the actual production needs. Because when the glass overflow down, the non-flowing width of the edge of the glass requires a stretching rolls. In order to ensure the strength of the edges, the thickness of the edges should be increased, so it is necessary to increase the flow in the both edges.

4.2 The effect of flow rate

According to the above results, the optimal geometry model is that: the bottom tilt angle is 0, overflow surface tilt angle is 5°, the depth of trough at the inlet point is 0.4m, and the length of trough is 2 m. The effect of velocity on the mass distribution of the molten glass was investigated, shown in Fig. 6.

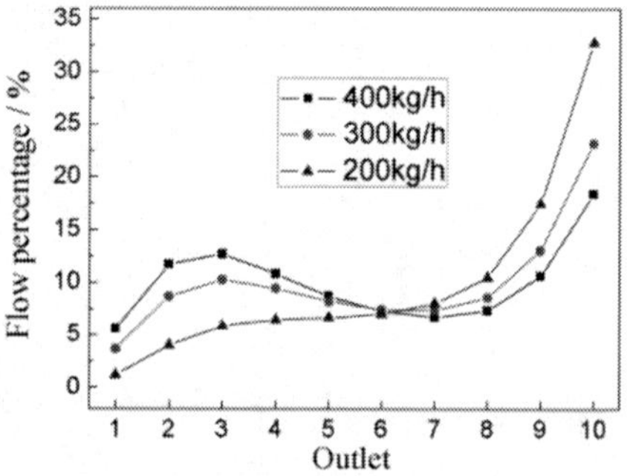

Figure 6. Mass distribution of glass melt with different flow rate.

In model 4, the initial speed of glass melt has greatly influenced on the mass distribution overflow from the weir surfaces, as showed in Fig. 6. The most even flow rate distribution in middle area of trough was obtained with the initial speed of 300kg/h. In addition, since the

DOI: 10.1051/matecconf/20179510006

mass distribution of glass melt was dependent on the flow rate, during practical production, when optimum process had been determined, the viscosity and speed could not be changed casually. If different thickness of glass is wanted (e.g. 0.3mm, 0.4mm, and 0.5mm), the only factor that can be adjusted is the pulling speed, not the flow rate of glass melt, for fast speed could get thinner glass. This method of control the thickness of glass is a basic adjustment way during overflow down-draw process of glass industry, under the condition of a steady flow rate.

5. Summary

After the above modeling, calculation and analysis, some conclusions were obtained. The tilt angle of weir surface θ was the most important factor of the geometry parameters of the isopipe, which had a significant effect on flow rate distribution of the glass melt along the length of the trough, and the desired mass distribution was achieved at the tilt angle 4°～6°. Height of trough at inlet end and tilt angle of trough bottom had little effect on the glass overflow mass distribution, but the groove depth only had some influence on mass distribution around the inlet point. With the increase of inlet velocity, the flow rate of the glass melt decreased at the inlet point, and increased at the distal end.

Acknowledgement

This work was supported by the National Key Research and Development Program of China (2016YFB0303702).

References

1. C.Y. Wang, Q. Lu, Glass & Enamel **3**, 44 (2016).
2. R.P. William, P.R. Kamjula, Chinese Patent Number CN 101925545 A, September 12, (2010).
3. S.M. Milillo, R.L. Rhoads, United States Patent Number US8973402, October 24, (2011).
4. K.W. Aniolek, S.R. Burdette, L.R. de Paor, et al. United States Patent Number US8393178, October 18, 2011.
5. R. Delia, European Patent Number EP 2253598 B1, May 21, 2010.
6. C.W. Hirt, B.D. Nichols, J. Comput. Phys. **39**, 1 (1981).
7. H.J. Lin, F.Y. Hsu, W.K. Chang, Adv. Mater. Res. **39-40** (2008).
8. J. Abbott, 26th Annual Workshop on Mathematical Problems in Industry Rensselear Polytechnic Institute (2010).
9. A. Tiwari, S.S. Mondal, P. Kumar, et al, Adv. Mech. Eng. **1**, 516 (2011).

Factors that Influenced the Quality Inspection on the Production Line in Manufacturing Industry

Yunos Ngadiman[1], Burairah Hussin [2], Abdul Talib Bon[3] and Nor Aziati Abdul Hamid[1]

[1]*Faculty of Technology Management and Business, Universiti Tun Hussein Onn Malaysia, Parit Raja, Batu Pahat, Johor*
[2]*Faculty of Information Technology and Communacation, Universiti Teknikal Malaysia Melaka, Durian Tunggal, Melaka*
[3] *Faculty of Technology Management and Business, Universiti Tun Hussein Onn Malaysia, Parit Raja, Batu Pahat, Johor*

Abstract. The quality of a product is the main concern for customers. Hence, the objectives of this research are to determine factor that affect the inline quality inspection and improvement solutions of the quality inspection standard in the production process. Meanwhile, the scopes of the research are studying of all the factors that influence inline quality inspection and techniques to improve quality inspection standard. Hence, interviews had been carried out towards the respondents who work under the Quality Assurance Department and in charge of inline quality inspection. Through analyzing the chosen products which were cot sheet and latex cap, it can be said that worker, material, and machine are the most common factors that cause defects on the products. Therefore, the result of this research shows that the worker's training, maintenance and checking procedure of incoming raw materials are the main influence factors to overcome the problems occurred.

1 Introduction

This study is to determination of the factors that influence in-process quality inspection standard for manufacturer of water sports and leisure swimming products at Rubber Leisure Private Limited. Inspection is the most common method attaining standardization, uniformity and quality of workmanship. Inspection activity play a crucial role to ensure the products meet the specific requirements and produce good quality of products to fulfill the customer needs and wants. There are three types of quality inspection which are incoming inspection, in-process inspection and final inspection. This case study focus only in-process quality inspection in the manufacturing industry.

The inspection methods that used by this company are sampling inspection and 100% inspection. Sampling inspection is conduct on the semi-finished product or work-in-process products. Meanwhile, 100% inspection is conduct for all the products in the company and ended with segregate out the good products and non-conformance products.

The factors that influence in-process quality inspection are being analyzed for its causes and effects. The research is done for determine the improvement solutions for quality inspection standard by conducting more literature review to search for journals, books and any other source on quality inspection method. The research methodology continues by doing the case study of the company.

2 Problem statement

Rubber Leisure Pte. Ltd. had implemented sampling inspection and 100% inspection to ensure the products meet the specific requirements and identify problems as they arise during the production process, thus allowing corrective action to be taken at an early stage.

Sampling inspection [1] either for the grading or for checking the correctness of the grading should be done most accurately, because unless the sample is truly representative of the entire lot, correct grading cannot be done. The general characteristics of the commodity should be preliminarily inspected before drawing samples from the lot. Meanwhile, 100% inspection determining whether a commodity in a particular lot complies with the standard specifications applicable to it or any other specification stipulated in the trade contract.

In Rubber Leisure Pte. Ltd., there are actually some problems occur and affect the end product's quality. Normally, this is due to the mistakes that made on the in-process quality inspection [2]. One of the problem is the product looks similar between bad product and good product, it can't be differentiate by the eye visual inspection. Then, the customer loyalty may drop. There are several factors that may influence the in-process quality inspection in this company. Therefore, improvement solutions will be proposed to the company

DOI: 10.1051/matecconf/20179510007

and may implement a proper management procedures in monitoring defective products [3].

3 Data analysis

According to Table 1, it showed the quality policy, objectives and target for the year 2014 and 2015. Based on the major products reject rate, latex cap and the cot sheet both products need to achieve below 1.0 % of reject rate. If exceed 1.0 % of reject rate, the quality assurance's officer will act to necessary action .

Table 1. Quality policy, objective and target

Quality Objectives	Target
Customer complaint	< 0.5 %
ETD fulfillment	> 99%
Customer satisfaction index	= 100%
Number of new products	> 4 products
Fulfillment of training pelan	> 60%
Major products reject rates: - Latex cap/Latex stock/Cot Sheet/ Hand Built Cap/ Silicone Cap/Baby Bath Mat - Place Mat/Baby Bib	Below 1.0 % Below 2.0 %

4 Discussion

This research discussed regarding the factors that influence quality inspection during production process in Rubber Leisure Pte. Ltd.. There are also search to identify the improvement solutions of the quality inspection standard in the the production process [5].

4.1 Implication of men factor

Base on the research's result, training is an important criteria to ensure that the workers have sufficient knowledge and skills to complete tasks during the production process [6]. According to the interviews transcript for this company, when there is new worker start working, the company will apply on job training toward them. Throughout the time, the employer will provide them a lot in-house training. The benefits of doing on job training is it can provides a variety of job experiences for the workers [7]. Besides, it can help the worker to understand all the processes activities in manufacturing rubber products [8].

Besides, there is a method that triggered worker's performance which is applying job appraisal [9]. It is used by this company to ensure high performance of employee [10]. This actually quite effective way to attract it's workers intention and motivate them to work effectively. An appraisal may provides an insight into the work being done and the workers who are doing it.

In addition, the surrounding or environment in the workplace also plays a crucial role to ensure the workers can complete their tasks safely and comfortably [11]. Physical environmental issues like distracting noise, bad lightning, rubber smell, uneven temperature, cleanliness and others factors may influence the influence feeling and emotion of a worker to carry out work [11]. Hence, it might indirectly affect men's performance during production process.

4.2 Implication of machine factor

Machine factor is also plays an important role to ensure quality products being produce and complete task in scheduled time. Normally, the machine problems are due to the downtime or machine spoilt and cannot run [12]. Even if the machine can run, but the product quality will be affected.

Hence, preventive maintenance had applied by this company in order to overcome the problem. It is because preventive maintenance [13] can provide systematic inspection, detection and correction of incipient failures either before problems occur or before problems develop into major defects. If there is no application of proactive action, it might cause higher downtime and lower productivity rate in the production process. Besides, it is necessary to conduct routine inspection before machine spoilt like servicing the machine within the time schedule [14].

4.3 Implication of material factor

Material is the main concern of producing rubber products [15]. Hence, the incoming quality control (IQC) plays an important role to check the incoming raw material before being processing to end products [16].

There are several reasons that may affect the rubber product's quality due to the material in this company. Firstly, improper control when purchasing raw material from the suppliers. Actually, it is need to ensure the raw material comply to the incoming material specifications [17]. Besides that, a lab test will be conducted on certain material like rubber and color of materials.

Secondly, this company never simply change the supplier. If inconsistent of supplier, the elements can affect the characteristics of a compound. Hence, the rubber product's quality will be affected.

5 Recommendation

Base on the analysis that had been carried out, researcher had recognized the three main factors that influence quality inspection in the production process which are incompetence of men power, low performance of machines and wrong mixing compounds of materials. Therefore, researcher did gave some suggestions towards the Rubber Leisure Pte. Ltd.. based on the three factors.

5.1 Improvement of men's performance

A competence, knowledgeable and skilled men power is important to a manufacturing company [6]. By improve the competency of workers, it can help the company to

DOI: 10.1051/matecconf/20179510007

increase its productivity and greater revenue. Actually, it can be done by coaching and mentoring the workers [4].

It is an on-going process designed to help the employee gain greater competence and overcome obstacles to improve performance. Through coaching, it can prepare employees for advancement and additional responsibility [8]. It can be done by identify the performance problem, setting a goal or objective and develop a performance improvement plan to let the employees follow [18]. Supervisor will need to concern the worker's performance and give the feedback towards them. Two way communication is an effective one way to determine both positive and negative performance efforts. In addition, it can help motivate the workers through the consistent guidance and boost their work performance [19].

5.2 Prevention of low performance machines

Machinery performance is the key factor that always overlooked by the management and this can cause looses which reduce the yield. Improper maintenance or low service will result in low standard of products. Therefore, a teamwork maintenance with engineering and production can be formed to fix machinery problems more regularly and precisely. It is advisable to put a group of knowledgeable and skilled engineers together in which each member know various things about the machines and processes. Hence, they able solve the problems immediately and prevent the machine from downtime [20].

Furthermore, if there is error occurred during production process, the machine shall be stop immediately and check the machine's condition. It is to prevent greater loss in revenue. Therefore, it is necessary to reduce machine's operating hours since it has certain life span limits [21].

5.3 Prevention of low quality material

Base on the research and analysis, it is being proved that mixing wrong compound is the major problem that influence the quality of rubber products. It is important to check the incoming raw materials at the first place. Raw material shall be stored in the correct storage conditions either in normal temperature or cold storage area [1].

When comes to the mixing compound, the raw materials have to be checked to ensure no changes have occurred during storage [22]. Then, all ingredients have to be weighed with high accuracy according to the types of products and then mixed together according to specified process parameters like temperature, timer and pressure. The mixing compound process is the most sensitive part, so extra concentration and careful need to put into it. If mistake is made here, it is too late to correct it later on.

6 Conclusion

As a conclusion, this research is being done to give a clear illustration regarding the factors that influence quality inspection during production process [3] in the rubber manufacturing industry. This research able to help the relevant companies to determine the root causes of influencing in-process quality inspection especially Rubber Leisure Pte. Ltd.. Then, the suitable solutions being recommended to improve the quality inspection standard in the production process [16]. Overall, researcher hope that this research paper able to give advantage to the rubber manufacturing industry to improve quality inspection standard through discover the root causes of influencing inspection standard and solve the problems successfully [2].

Acknowledgement

The researchers sincerely have been appreciating and gratitude an invaluable pronounced to the Research, Innovation and Commercialization Centre of Universiti Tun Hussein Onn Malaysia under a research fund and the Registrar Department that provides for a sufficient time to enable the researchers were successfully attending to present this paper.

References

1. Anjoran, R.. 15 Ways to Avoid Problems With Your Suppliers (2011)
2. Vivek, N. *Quality Management System Handbook for Product Development Companies* (pp. 1–22). New York: CRC Press (2005)
3. Hessel Schuurman. Quality Management and Competitiveness The Diffusion of The ISO 9000 Standards in Latin America and Recommendations for Government Strategies (1977)
4. Kao, T. Human factors in quality inspection process. 1-5 (2003)
5. Davis S. B. G. D.. *Introduction to Total Quality Management for Production, Processing, and Services* (3rd ed.). United States: Prentice Hall (2000)
6. Wu, H. Y., Chuang, C. L., Kung, Y. S., & Lin, R. H.. Determination of optimal inspection sequence within misclassification error bound. *Expert Systems with Applications*, **38**(5), 5366–5372 (2011)
7. Vienna. The Influence of The Workplace On Perceived Quality (2011)
8. Nzonzo, J. C. Training and Development Practices In An Organisation : An Intervention To Enhance Organisational Effectiveness, **2**(4), 187–198 (2011)
9. Yu, H.-F., Yu, W. C., & Wu, W. P. A mixed inspection policy for CSP-1 and precise inspection under inspection errors and return cost. *Computers & Industrial Engineering*, *57*(3), 652–659 (2009)
10. Kao, T. (2003). Human factors in quality inspection processes, 1–5 (2003)
11. Hessel Schuuman. Quality Management and competitiveness The Diffusion of The ISO 9000 Standards in Latin America and Recommendations for Government Strategies (1997)

12. Dhillon, B. S., & Liu, Y. Human error in maintenance: a review. *Journal of Quality in Maintenance Engineering*, **12**(1), 21–36 (2006)
13. Su, L.-H., & Tsai, H.-L. Flexible preventive maintenance planning for two parallel machines problem to minimize makespan. *Journal of Quality in Maintenance Engineering*, **16**(3), 288–302 (2010)
14. Sondalini, M. 6 Ways to Improve Maintenance and Production, 1–5 (2002)
15. Shin, Sangman, Jung Tae Kim, and Yun Su Choi.. "Rubber Composition for Tire Tread and Tire Manufactured by Using the Same." U.S. Patent No. 20,150,299,436 (2015)
16. Besterfield, D. H. *Quality Control* . Eight Edit., pp. 77–109). New Jersey: Southern Illinois University (2009)
17. Vrat. Prem. "Material Specification, Codification, and Standardization."*Materials Management*. Springer India. 211-224 (2014)
18. Bamford, D. R., & Greatbanks, R. W. The use of quality management tools and techniques: a study of application in everyday situations. *International Journal of Quality & Reliability Management*, **22**(4), 376–392 (2005)
19. Henley, Amy J., et al. "Quantifying efficacy of workplace reinforcers: An application of behavioral economic demand to evaluate hypothetical work performance." *Translational Issues in Psychological Science* 2.2: 174 (2016)
20. Sultana, F., Science, C., Road, P., Razive, N. I., Tower, G., & Azeem, A. Implementation of Statistical Process Control (SPC) For Production Data Control Chart Pareto Chart Pivot Chart Data Analysis Maintenance Communication Electrical Findings, *M*(1), 15–21 (2009)
21. Way, F. Improve your machines and manufacturing processes: Milo, Christian, and Werner Grosch. "Changes in the odorants of boiled salmon and cod as affected by the storage of the raw material." *Journal of Agricultural and Food Chemistry* **44**.8: 2366-2371 (2009)
22. Gu, Ben J., et al. "A quantitative method for measuring innate phagocytosis by human monocytes using real-time flow cytometry." *Cytometry Part A* 85.4: 313-321 (2014)

DOI: 10.1051/matecconf/20179510008

The Design of Drawing Process of Cylindrical Cup with Oval Bottom Using Computer Simulation

Mária Kapustová and Róbert Sobota

Slovak University of Technology in Bratislava, Faculty of Materials Science and Technology in Trnava, Institute of Production Technologies, J. Bottu 25, 91724 Trnava, Slovakia

Abstract. Based on technological calculations, technology of one-operation drawing without blankholder was designed for production of a large-dimensional drawpiece with shape of welded on concave bottom. The necessary drawing force was also defined by calculation as maximum drawing force for trimming of drawpiece bottom. This contribution points out to the importance of FEM simulation of drawing process in order to verify the design of production technology of drawpiece with shape of welded on concave bottom on the basis of technological calculations. Simulation programme Deform verified the construction design of drawing tool determined for drawpiece production by first draw, and also determined the drawing force actually consumed during plastic deformation of drawpiece shape. Used simulation software confirmed correctness of drawing tool construction, as correct material plastic flow in drawing tool was monitored and production of drawpiece was faultless.

1 Introduction

Seamless steel piping parts (pipe fittings) are widely used in various technical equipment as e.g. steam boilers, pressure vessels, containers, as well as steam, gas and petroleum piping, etc., which are applied at concurrent effect of temperature and pressure. One of essential properties of pipe fittings is condition of their weldability. These pipe fittings are represented also by welded on concave bottoms that are used for ending of piping or as a part of exterior casing of small and large pressure vessels, containers, cisterns, etc. Welded on concave bottoms with various diameters and wall thickness are produced by cold drawing using hydraulic presses. Production of drawpieces with required dimensions may be realized by first or more draws within drawing with or without blank holder.

The drawing tool consists of a die and a punch which are made of alloy tool steel and produced by chip machining with subsequent heat treatment. Mentioned heat treatment comprises of hardening and tempering at low temperatures. At construction of drawing tool a drawing clearance between the die and the punch is important, whereby the dimension of this clearance depends on thickness of drawn metal sheet [1]-[5].

2 Description of drawing material and construction of draw piece shape

The selection of appropriate material for drawpieces determined for steel piping parts (pipe fittings) is important for correct, faultless production of drawpieces.

The most common material used for production of welded on concave bottoms is high-temperature strong steel sheet hot rolled which has guaranteed ductility. From the aspect of shape the designed drawpiece belongs to cylindrical large-dimensional drawpieces with oval bottom, i. e. with simple shapes. Actual drawpiece with welded on concave bottom shape illustrated on Fig. 1 will be made of steel 13CrMo4-5. The mentioned steel belongs to a group of chrome-molybden high-temperature strong steels which are appropriate to use at increased temperatures up to 560°C. Therefore these steels are used mainly for production of seamless tubes, component parts for steam boilers and steam turbines.

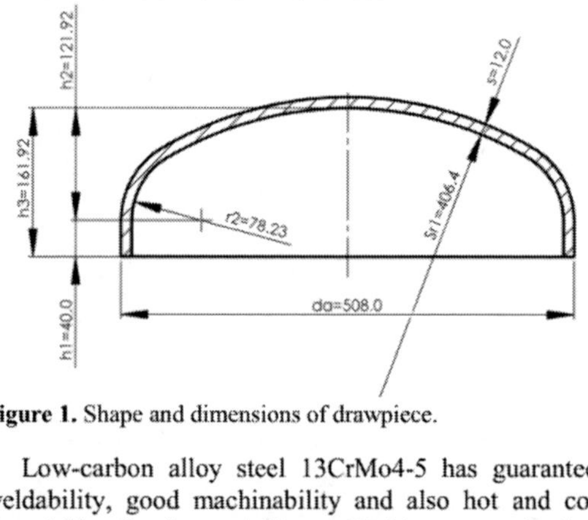

Figure 1. Shape and dimensions of drawpiece.

Low-carbon alloy steel 13CrMo4-5 has guaranteed weldability, good machinability and also hot and cold formability. Steel material is supplied in a state after heat treatment, i. e. normalizing at 890-950°C and tempering at 630-730°C. Mechanical properties of steel are as follows: yield point $R_{p0,2}$ = min. 300 MPa, tensile strength R_m = 450-600 MPa and percent elongation A = min. 19 %. Chemical composition of steel 13CrMo4-5 is described in Table. 1.

Table 1. Chemical composition of steel 13CrMo4-5 (wt %).

Element		C	Mn	Si	Cr	Cu
Composition	min	0,08	0,40	0,35	0,70	0,30
	max	1,18	1,00		1,15	
		Mo	N	P	S	
	min	0,40	0,012	0,025	0,010	
	max	0,60				

3 Calculation of the draw process parameters

In order to produce the cylindrical cup with oval bottom according to dimensions as stated in Fig. 1, first it is necessary to determine the dimension of workpiece (blank) by calculation. The workpiece prepared out of the metal sheet will be circular-shaped with wall thickness s = 12.0 mm and its size (diameter) will be defined by following relation:

$$D = \sqrt{d_a^2 + 4.\left(h_2^2 + d_a.h_1\right)} + w = 667.5\ \text{mm} \qquad (1)$$

where: d_a - external diameter of drawpiece [mm], h_1 - height of cylindrical part [mm], h_2 - height of oval bottom [mm], w – allowance for trimming [mm], w = (2 ÷ 5) . s = 3.12 = 36 mm.

Diameter dimension of circular-shaped workpiece is D = 670.0 mm.

In order to determine number of draws necessary for production of this drawpiece the calculation of total drawing ratio is essential. The production of drawpiece with oval bottom will be one-operational as the real value of total drawing ratio is smaller than the allowable value for the first draw. For consideration of the blankholder necessity the relation according to the author Šofman [6] is used. This relation recommends drawing without blankholder if the following condition is fulfilled:

$$\begin{aligned} D - d_a &\leq 18.s \\ 670 - 508 &\leq 18.12 \end{aligned} \qquad (2)$$

For the condition 162 ≤ 216 the blankholder is not necessary.

For definition of the drawing force needed for one-operational production of drawpiece with oval bottom the relation for calculation of maximum drawing force is commonly used:

$$F_{tmax} = S.\ R_m.\ K_t \qquad (3)$$

where: R_m - tensile strength of steel 13CrMo4-5 (R_m = 450 MPa), K_t - coefficient depending on the draw ration for the first draw (K_t = 1.0), S - perforated drawpiece area (S = 18698.7 mm^2)

$$F_{tmax} = 18699.\ 450.\ 1 = 8414550\ \text{N} \approx 8414\ \text{kN} \qquad (4)$$

The calculation of drawing force through the above-mentioned Eq. (3) is not accurate (only informative), as calculated force value is maximal, needed for damage of drawpiece, i. e. for tearing of its bottom. Therefore it is necessary to verify by computer simulation the correctness of the technology design of one-operational drawing of particular drawpiece and determine the real value of the drawing force needed for drawpiece production.

4 Computer simulation of material plastic flow in drawing tool using programme deform

Computer simulation of drawing process regarding steel alloy 13CrMo4-5 was performed through the programme DEFORM-3D (Design Environment for FORMing). It is a high-performance FEM simulation system designed for three-dimensional (3D) analysis of material flow within broad scope of forming processes. It is a practical and effective tool for prediction of plastic material flow in forming tools for the purpose of reduction of production costs and acceleration of the whole production process.

It is used for FEM simulation of drawing, die forging and smith forging processes, rolling, extruding, etc. This programme is suitable for simulation of forming processes with regard to hot, warm and also cold forming [5]-[9].

User environment of programme Deform contains two dialog windows illustrated in Fig. 2.

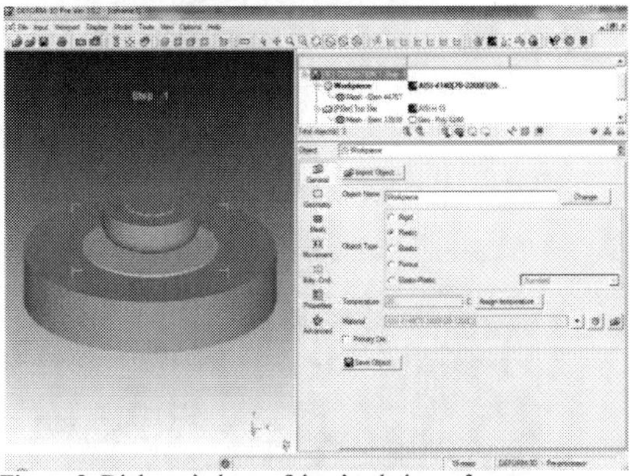

Figure 2. Dialog windows of the simulation software DEFORM 3D.

After definition of input conditions e.g. process type (drawing), number of objects, objects type, import geometry, temperature, mesh, material type, boundary conditions, movement (hydraulic press), contact (Coulomb coefficient of friction 0.2) and simulation controls a database of information necessary for initiation of drawing process simulation was generated [10]-[12]. Numeric simulation verified the correctness of design of technological conditions concerning drawing process and the correctness of drawing tool construction. Based on Fig. 3 the correct material plastic flow is evident and the drawpiece shape was faultless.

DOI: 10.1051/matecconf/20179510008

Figure 3. Simulation of plastic deformation of the drawpiece with shape of welded on concave bottom.

The results of output phase, i.e. simulation postprocessing, are the courses of effective strain and temperature fields in a final shape drawpiece shown on Fig. 4, Fig. 5.

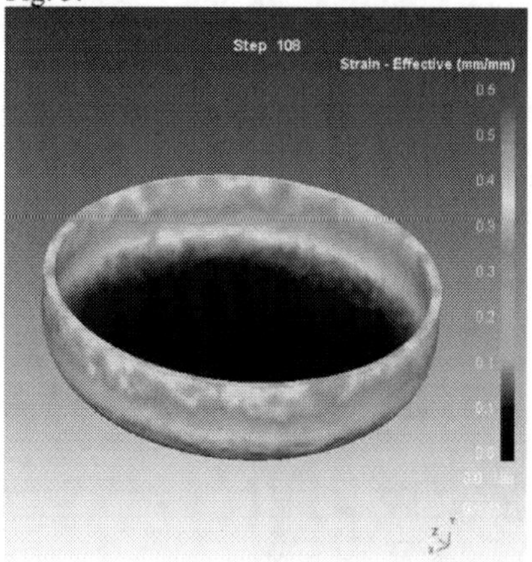

Figure 4. Course of effective strain in final shape drawpiece.

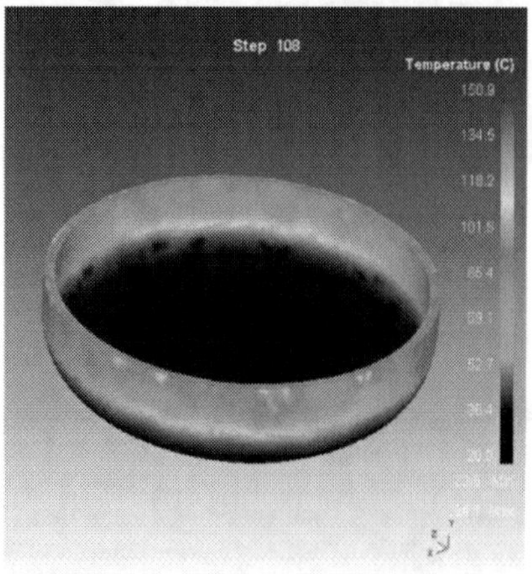

Figure 5. Temperature fields in final shape drawpiece.

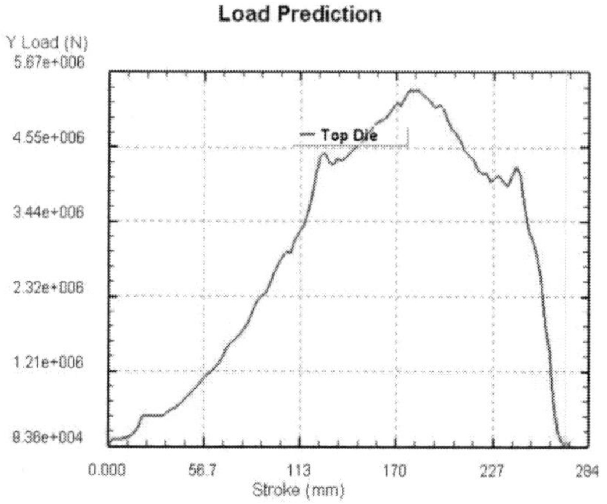

Figure 6. Graphic course of drawing force during a press stroke.

The course of drawing force illustrated on Fig. 6 is an important result of simulation which is necessary for the selection of appropriate hydraulic press for the production of given drawpiece.

5 Summary

Achieved results of FEM simulation of drawing process regarging the drawpiece with oval bottom show correctness of production technology design and geometry of drawing tool without blankholder. Material plastic flow was optimal and final shape of drawpiece did not show any visible defects on its surface. From the course of drawing force based on the drawing process simulation it results that maximum force value is approx. 5500 kN. This force value was achieved at about 65 % of hydraulic press stroke and represents real value compared to the calculated force value of 8400 kN. The course of drawing force detected by numeric simulation will be a valuable assistance in selection of the appropriate press for production of the given drawpiece under practical conditions.

Acknowledgment

This article has been created with the support of projects VEGA No. 1/0669/15 and Form Tool MANUNET-2014-11283 and was supported also by the Slovak Research and Development Agency under the Contract No.APVV-15-0319.

References

1. W.F. Hosford, R.M. Caddell, *Metal Forming: Mechanics and Metallurgy*, 4. ed., Cambridge University Press, New York, p. 331, (2011).
2. I. Görögová, *Technological heredity in bearing production*, 1st ed., Scientific Monographs, Aleš Čeněk, Plzeň, p. 69, (2016).
3. G. Gantar, A. Sterzing, Robust design of forming processes, Journal of Mechanical Engineering, **54**, p. 249-257, (2008).
4. M. Ridzoň, *The Effect of Technological Parameters Influencing the Properties of Seamless Cold-Drawn Tubes*, 1st ed., Scientific Monographs, Hochschule Anhalt, Köthen, p. 89, (2012).
5. E. Spišák, *Mathematical modeling and simulation of technological processes - Drawing*, TYPO Press, Košice, p. 156, (2000).
6. M. Forejt, M. Píška, *Theory of forming, machining and tools*, CERM, Brno, p. 225, (2006).
7. P.M. Dixit, U.S. Dixit, *Modeling of Metal Forming and Machining Processes by Finite Element and Soft Computing Methods*, Springer Verlag, London, p. 590, (2008).
8. S. Kobayashi, S.I. Oh, T. Altan, *Metalforming and the Finite-Element Method*, Oxford University Press, London, p. 402, (1989).
9. Edward R. Champion, Jr., *Finite Element Analysis in Manufacturing Engineering*, McGraw-Hill, New York, p. 310, (1992).
10. S. Natarajan, S. Venkataswamy, P. Bagavathiperumal, A note on deep drawing process: numerical simulation and experimental validation, Journal of Material Processing Technology, **127**(1), p. 64-67, (2002).
11. D.M. Neto, M.C. Oliveira, J.L. Alves, L.F. Menezes, Influence of the plastic anisotropy modelling in the reverse deep drawing process simulation, Materials and Design, **60**, p. 368-379, (2014).
12. G. Ferran, A.C. de Moura, L.P. Moreira, Computer aided development of a bending–drawing test for thin metallic sheets, Journal of Material Processing Technology, **80-81**, p. 531-537, (1998).

Applying Triz for Production Quality Improvement

Nikalus Shu Luing Swee[1], Guat Guan Toh[2], Mum Wai Yip[3], Chee Sheng Keong[4] and See Chew Tai[5]

[1]Tunku Abdul Rahman University College, Department of Computer Science and Mathematics, Malaysia
[2]Tunku Abdul Rahman University College, Deputy Branch Campus Head, Malaysia
[3]Tunku Abdul Rahman University College, Department of Mechanical Engineering, Malaysia
[4]Tunku Abdul Rahman University College, Centre of Pre-University Studies, Malaysia
[5]Tunku Abdul Rahman University College, Department of Electronic Engineering, Malaysia

Abstract. This paper aims to provide a thorough analysis on the application of TRIZ in improving the quality of canned food production. TRIZ tools such as engineering systems analysis, function analysis, cause and effect chain analysis, By-separation model and 40 Inventive Principles are applied in order to discover some feasible and elegant solutions to alleviate the problem. Findings revealed that the rejected canned products on the conveyor belt will be isolated or picked up with other good condition canned products which are lined up very closely to the rejected cans; though the visioning system is able detect the fault printing on the canned product. The main root cause is that the rejected canned product is picked up with other canned products in good condition because all cans are lined up on the belt and are very close to each other or having no gaps between the cans. Conversely, all cans on the conveyor belts are required to be very close to each other to avoid collisions that may damage the cans. The root cause is solved by applying function analysis, By-separation tool and Inventive Principles. Therefore, it can be concluded that TRIZ is a powerful tool in inventive problem solving.

1 Introduction

Production lines comprise of a set of sequential processes whereby materials are out through a refining operation to produce an end-product that is suitable for consumer market; or components that are assembled to make finished product [1]. It also includes inspecting, labeling, and packing the final product before the product is released.[2] Equipments, materials and human resources influence the reliability and capacity of the production operation. Therefore, identifying bottleneck resources and employing an effective scheduling structure for the production process were critical. [2]. Production line commonly is equipped with a belt conveyor system. A belt conveyor system consists of more than 2 pulleys, with endless loop of transporting products or materials. With respect to the forecasts and plans, the expected multitude of options are being developed in order to meet the demand of an industry requiring ever more complex devices which exhibit higher reliability, speedy production and lower cost [3-4]. There are challenges in handling production process. The production technologies need to be aware of the issues of downstream handling, especially loading and unloading of goods, and transportation by all channels [4]. The production applications commonly include vision-guidance of high-speed robots in product packing with simultaneous rejection of substandard items, robotic fish-filleting, color-sorting and grading of capsicum peppers, counting glass anesthesia vials in a tray, and detecting the tracks on the circuit board in a medical device. Many of the applications concerned traceability: for example, checking the presence and readability of date codes and bar codes on product packaging, and checking the labeling and etc. [5].

TRIZ is a Russian acronym for "Teoriya Resheniya Izobreatatelskikh Zadatch", equivalent to "Theory of Inventive Problem Solving" in English. TRIZ methodology was founded in 1940's by Genrich Altshuller and his team. He was an inventor, a writer and a patent engineer who studied intellectual property contained in approximately 200,000 patents [6]. He discovered and organized his study of 40,000 patents according to innovative patterns of design as well as the inventive principles in these innovative solutions. Findings revealed that problems and solutions, patterns of technical evolution were repeated across industries and sciences, and innovations used scientific effects outside the field where they were developed. Therefore, Genrich Altshuller derived 40 inventive principles [6-13]. TRIZ comprises of several essential tools such as engineering systems analysis, function analysis, cause and effect chain analysis, trimming, engineering contradiction, by-separation, substance-field model, Trends of Engineering System Evolutions, ARIZ, and etc. TRIZ uses 40 inventive principles and 39 parameters to help inventors to derive many solutions [6-13].

2 Problem statement

Direct observation revealed that the rejected canned products on the conveyor belt will be picked up with other good condition canned products which are lined up very closely to the rejected cans; though the visioning system is able detect the fault printing on the canned product. If the rejected cans are lined up with other cans without gaps (adjacent to each other very closely), the good condition canned products before and after the rejected canned product will be picked up together.

There are a few constraints to be taken into account such as the speed of the conveyor and the level of the conveyor cannot be altered, and the material of the belt cannot be changed. No budget for sophisticated solutions such as robotic arms. The following are the details on how TRIZ is applied in this case study.

3 Triz models and tools

TRIZ flow process is shown in Figure 1. First and foremost, research and brainstorming help identify the original problem to resolve, and this is followed by function analysis, cause and effect chain analysis, and Physical contradiction. Finally, By-Separation strategy is applied to derive specific inventive principles and specific solution(s) [6-7].

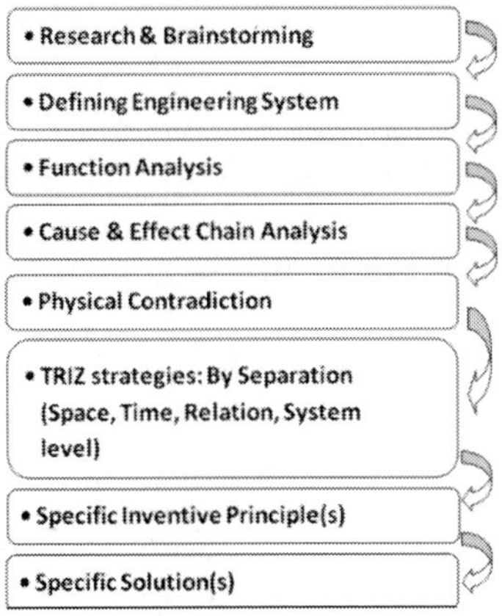

Figure 1. Flow of TRIZ processes.

3.1. Engineering system definition

An Engineering System consists of several components that are interacted among each other. These components are commonly accepted as system components (subsystems) that are listed in Table 1. Along with subsystems, there are also interactions between engineering system and external entities called supersystems. Supersystems are not designed as part of the Engineering System; however, they can influence the Engineering System [6-7].

Table 1. Engineering system.

Sub/System Components	visioning system, canned product, conveyor belt, guided track, conveyor frame, conveyor tripod, motor, pulley
Supersystems	air particles/dust, workers, freight, Humidity, Ambient Thermal

3.2 Function analysis

Function analysis shows the interactions between two or more systems/subsystems (Engineering System components) which are listed in Figure 2. These interactions are called functions. Functions are simply actions between two components, i.e., a subject and an object in which the subject acts upon and modifies a parameter(s) of the object [6]. Two main types of functions are useful function and harmful function. As for useful function, it comprises of "normal", "insufficient", and "excessive" functions. [6-7].

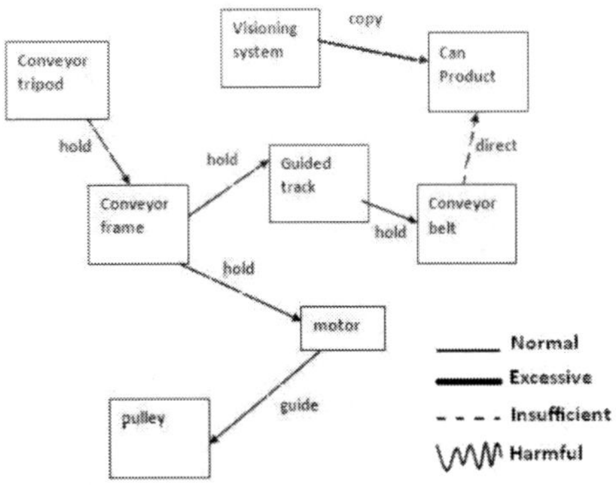

Figure 2. Function model.

3.3 Cause and effect chain analysis (CECA)

The following stage is Cause and Effect Chain Analysis(CECA). CECA is a crucial tool in the TRIZ methodology. It helps identify the right root cause(s) pertaining to the problem in Figure 3. If the wrong root cause is derived, it leads to ineffective solution. Fundamentally, CECA is very similar to "5 Whys". We prompt for causes continuously for the problem from high level causes to low level causes by asking "the question "why?" [6].

From the CECA, the root cause identified is that the canned goods are lined up very closely or even without gaps on the conveyor to prevent collisions among the cans. However, the adjacent good condition canned goods before and after the rejected canned goods will be picked up together.

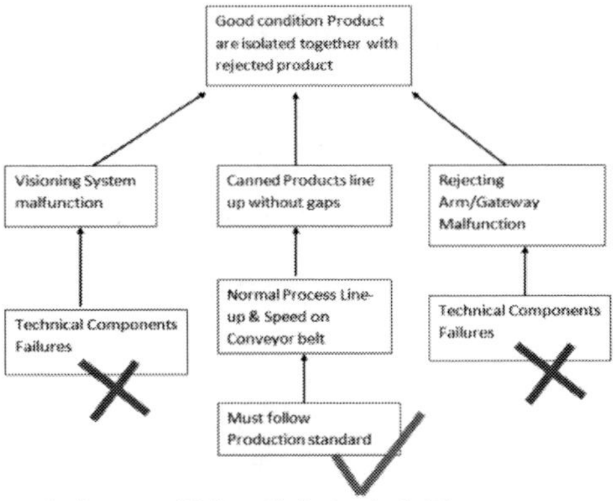

Figure 3. Cause and Effect Chain Analysis Diagram.

3.4 Physical contradiction

Physical Contradiction of an engineering system refers to the presence of a contradiction in an engineering system involving a single parameter (e.g. area, length, volume, etc.) which has a contradiction at two different values. [6-7]. For example, an airplane wing needs to have large surface for easy takeoff and small surface for fast speeds. Airplane landing gear needs to be present during landing but does not need to be present during flight to reduce wind resistance. The essence of the Physical Contradiction is to find the controlling characteristic. The Physical Contradiction can be solved through methods such By-Separation in (Space, Time, Relation, System level) and By–Satisfaction, and Bypass. In this case study, we employ By-Separation model.

Based on the case study, the following Physical Contradiction is proposed.

- **Canned goods need to be separated (with noticeable gaps) in order to be picked up individually if found containing printing error.**

- **Canned goods need to be very close to each other (with no gaps) to avoid collisions which may damage the cans.**

3.5 TRIZ tools: By Separation (space, time, relation, system level)

To determine whether the Separation can be done in Space, in Time, or in Relation, the following scenarios should be established [6]:

- Separation in Space – **Where** do I need condition A? **Where** do I need condition -A?
- Separation in Time – **When** do I need condition A? When do I need condition -A?
- Separation in Relation – I need condition A **IF** ? I need condition -A **IF**?
- Separation in System level – The Physical Contradiction's controlling parameter has a value at the system level but has an opposite value at the component level or a controlling parameter exists at the system level but not at the component level.

During operation time analysis, let's define the following variables: T1(Time during visioning system), T2 (Time during rejecting process), T3 (Time after rejecting process). During T1, all cans move in a normal speed and are adjacent to each other without gaps towards the visioning system to detect any printing errors on the cans and to avoid collisions. The visioning system is able to detect the all printing errors effectively.

During T2, after leaving the visioning system section and reach the rejecting arm/gateway, all cans should be separated (with noticeable gaps) to allow rejection of individual defective can accurately without pulling in any good condition cans.

During T3, all "Pass-QA" canned goods will continue to drift the next production section.

Hence, based on the operation time analysis, Separation In Time strategy is suitable for the problem. There are many recommendations of Inventive principles in Separation in Time [6], only Inventive principles #10 Prior Action, Inventive principles , #15 Dynamization, Inventive principles #16 Partial and Excessive Action, Inventive principles #19 Periodic Action are deemed suitable for the problem after thorough discussion with production engineers.

4 TRIZ solutions and discussion

4.1 Solution derived by Inventive Principle #10 Prior action and #15 Dynamization

Referring to the proposed inventive principles above, Inventive Principle #10 Prior action suggests pre-arranging objects such that they can come into action from the most convenient place without losing time. Inventive principles #15 Dynamization suggests to divide an object into parts capable of movement relative to each other. Based on the suggested Inventive principles, production engineers can install rubbers with spring rods on the conveyor before the visioning system and also before the rejecting arm to separate the cans. The rubber rods/arms act as a diverter gate/arm to separate the cans with acceptable range of gap to avoid serious collisions among the cans 14].

4.2 Solution derived by inventive principle #19 periodic action and #16 partial or excessive action

Inventive principles # 16 Partial or Excessive Action suggests that if the finished product is hard to achieve using a given solution method then, use "slightly less" or "slightly more" of the similar method. Inventive Principle #19 Periodic Action suggests to use pulsating actions to change the situation. Based on suggested Inventive principles, production engineers can install vacuum suction cup to separate the canned goods upon passing the sensors. Suction cup will be activated and controlled by the sensors on the conveyor belt periodically to delay the movement of cans in order to have gaps between cans [15]. Hence, the rejected cans can be diverted to the rejection gateway. The suction strength is needed to be optimized for the best solution in relation to the weight of the cans.

5 Conclusion

The ever improving technology in production line should be able to handle any sizes of products in a speedy, reliable, and cheaper cost [3-4]. In this case study, the problem can be contained and lessened by applying TRIZ tools particularly the Physical Contradiction and By-Separation tools. TRIZ helps engineers generate more feasible ideas or concepts which may lead to elegant solutions.

TRIZ requires an active cooperation between field specialists and TRIZ consultant so as to find the right root causes and derive ideas or effective solutions. It can be concluded that TRIZ is a systematic and an innovative problem solving methodology.

References

1. Sarder, MD Baniamin; Liles, Don H; Ali, M Yousuf. *Design & Implementation of production line in apparel manufacturing industry IIE Annual Conference. Proceedings* : 1-6 (2005)
2. U. Michael, G. Van, U. Elisabeth, Improving production line performance, IIE Solutions; Nov ; **32**, 11; ProQuest, pg. 36 (2000)
3. Andy Longford, Chip packaging challenges . . . a roadmap based overview, Microelectronics International, **22/2**, 17–20 (2005)
4. Lagner, P., Advantest Handling challenges of small & fine pitch devices, Advantest Technical Forum, January (2004)
5. Christine Connolly, Sensor trends in processing and packaging of foods and pharmaceuticals, **27/2** 103–108 (2007)
6. T.S. Yeoh, T.J. Yeoh, and C.L. Song, TRIZ-*Systematic Innovation in Manufacturing*, Selangor: Firstfruits (2009)
7. A.A. Guin, A.V. Kudryavtsev, V. Yu. Boubentsov, A. Seredinsky, Theory of Inventive Problem Solving, First Fruits (2009)
8. D. Mann, *Hands-On Systematic Innovation*. Belgium: CREAX Press (2003)
9. D. Mann, *Hands-On Systematic Innovation for Business and Management*, Devon: Lazarus Press (2007)
10. E. M. Smith, From Russia with TRIZ, Mechanical Engineering, **125**, 3; Academic Research Library, pp. D18, Mar (2003)
11. G.S. Altshuller, *The Innovation Algorithm, TRIZ Systematic Innovation and Technical Creativity*, Worcester, MA: Technical Innovation Centre (2000)
12. G.S. Altshuller, *40 Principles, TRIZ Keys to Technical Innovation*, Worcester, MA: Technical Innovation Centre (2002)
13. J. Terninko, A. Zusman, and B. Zlotin, *Systematic Innovation: An Introduction to TRIZ*. New York, NY: CRC Press (2010)
14. F. J. Minner, Plastic conveyor-belt fasteners eliminate scratched parts, Quality; **34**, 4; ProQuest Science Journals, pg. 138 (1995)
15. Zalud, Todd, Using suction cups to design a vacuum system, Machine Design; Dec 10; **70**, 22; pg. 114 (1998)

Research on Stiffness of the Hull Beam Model Based on ADAMS

Yanshi He, Fangzhen Song and Haiyan Shao

School of Mechanical Engineering, University of Jinan, Jinan, 250022, China

Abstract. A piecewise rigid body model of the ship is proposed, which makes the ship, a elastic body, can be analyzed in dynamics software like ADAMS and the analysis efficiency is improved. The equivalent stiffness in the model is more close to the actual situation, so it improves the precision of calculating natural frequency in the vibration characteristic analysis. Having been used in ADAMS, the modelling method turns out to be more accurate processing method than common hull beam section unit method. The processing method be used for the dynamics analysis of the similar structure.

1 Introduction

Due to the length of the ship is much larger than the high and wide of the ship, so the ship is often simplified as beam by scholar to the study of the ship vibration problem . The formula, characteristics and law of the vibration of the beam can be applied to the study of hull vibration, when the beam model is used to deal with the vibration problems of the hull [1], [2], [3]. It makes the research easier and improves the efficiency of the study.

However, the hull structure is quite complex, so it can not be fully equivalent to the beam. Therefore, when the hull simplify to hull beam model, we should first to consider that how to get a more accurate model to simulate the actual quality and stiffness of the hull. For quality, the influence of different internal hull structure is not obvious, and the change of the internal structure does not change the external hull shape. It is not difficult to get the rule of the influences of the entrained water [4], [5]. So this paper focus on the stiffness simplified method of the hull beam model. In order to facilitate the calculation, this paper uses the multi rigid body dynamics software ADAMS to assist the completion of the study [6].

For the the vibration analysis of chain structure such as hull beam, there have concentrated mass method, transfer matrix method, finite element method and so on [7,8,9]. For the method of simplified model, the literature [10,11] by using the finite element method to achieve the equivalent model of the mechanism .The transfer matrix of rigid body system connected by spring and hinge is given in literature [12].

No matter what kind of simplified method, the treatment of equivalent stiffness is an important part of the study. The accuracy of the equivalent stiffness directly affects the results of the entire study whether is correct and whether the conclusion of the research makes sense.

2 The common model of hull beam

The hull beam model for calculating the total vibration, which is a row of hull beam elements connected by nodes [13].

For each simplified beam element, the selection rules of the raw data as follows:

(1) Each simplified beam element is considered as a uniform straight beam;

(2) The arithmetic mean of the moment of inertia of the section at both ends of the beam is taken as the moment of inertia of each beam section;

(3) The quality of distribution is the sum of quality of beam section of the hull itself and the entrained water;

(4) Shear equivalent area of the beam is the average of areas on both ends of the beam section.

Generally, the longitudinal components throughout the length of the hull include bottom plate, deck, keelson, keelson, deck girder, the hatch coaming and side girder. The horizontal components include bottom plate, deck, floor board, deck beams, strong beam, hatch coaming, frame and strong frame. The transverse component is used to ensure the transverse stiffness of the hull.

Because most of the components are profiles, such as T steel and Angle steel, so its quality, the moment of inertia, cross-sectional area and section moment of inertia can be expressed by formula containing design variables. But the outside plates of irregular shape of boat, it can be obtained by the method of fitting expressions. The values of sectional moment of inertia of 20 sections respectively fill in the attribute of ADAMS.

In the solution of section parameters of beam element, the method of determining the moment of inertia and its orientation is studied [14]. X.Dong calculated the moment of inertia of car [15]. M.H.Li discussed the method for solving the moment of inertia of the cross section of the beam [16]. In this paper, taking a 54m motor cargo ship as an example, the moment of inertia of each beam element is processed. Results are shown in table 1~3:

Table 1. Sectional moment of inertia of z axis.

Num	Iz(mm4)	Num	Iz(mm4)
1	22454743151558.6	11	2997200615982.70
2	2969455466352.32	12	2997200615982.70
3	3604974664103.87	13	2997200615982.70
4	3223314900104.67	14	2997200615982.70
5	3214244418809.15	15	2997200615982.70
6	2997200615982.70	16	2997200615982.70
7	2997200615982.70	17	4608184785593.97
8	2997200615982.70	18	4340550638466.67
9	2997200615982.70	19	1836116886963.17
10	2997200615982.70	20	537280338616.40

Table 2. Sectional moment of inertia of x axis.

Num	Ix(mm4)	Num	Ix(mm4)
1	422242667234.485	11	503839201896.855
2	638843182766.61	12	503839201896.855
3	913628593998.18	13	503839201896.855
4	875216217617.815	14	503839201896.855
5	746077934134.25	15	503839201896.855
6	503839201896.855	16	503839201896.855
7	503839201896.855	17	882628554523.155
8	503839201896.855	18	1003557535370.02
9	503839201896.855	19	514515258500.11
10	503839201896.855	20	136034778370.385

Table 3. Sectional moment of inertia of y axis.

Num	Iy(mm4)	Num	Iy(mm4)
1	2232500484325.1	11	2493361414085.7
2	2330614418775.88	12	2493361414085.7
3	2691348205294.77	13	2493361414085.7
4	2348098682486.42	14	2493361414085.7
5	2468166484674.42	15	2493361414085.7
6	2493361414085.7	16	2493361414085.7
7	2493361414085.7	17	3725563864677.87
8	2493361414085.7	18	2229275264407.74
9	2493361414085.7	19	1321601628467.21
10	2493361414085.7	20	401245560250.7

Multi-rigid-body piecewise model is established in ADAMS, which is connected by beam elements. The beam model of the hull as shown in Fig. 1.

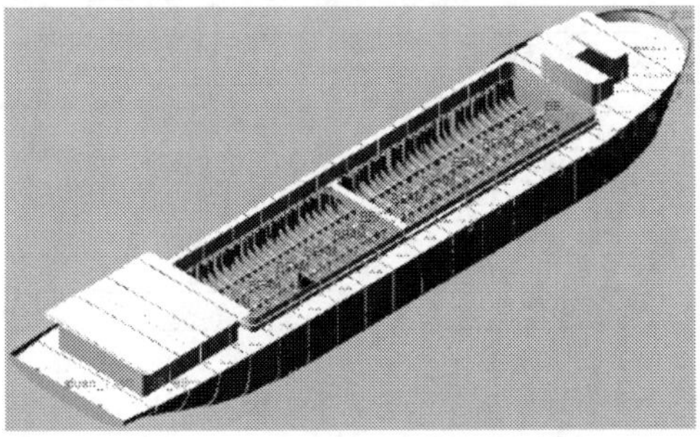

Figure 1. The model of Section beam element in ADAMS.

Results of natural frequency of the hull beam model are obtained by running the program. The error of the results and the actual natural frequency are shown in Table 4.

Table 4. Sectional moment of inertia of y axis..

Actual natural frequence (Hz)	Beam model results (Hz)	error%
0.9953	1.125	13.0312
1.3108	1.5176	15.7766
2.1445	2.634	22.8258
3.1305	3.8821	24.0089
3.3807	4.3479	28.6095
3.5282	4.6523	31.8604
3.7679	4.9875	32.3682
4.3084	5.7787	34.1264
4.8068	6.4879	34.9734
5.559	7.5946	36.6181

As can be seen from the results, the error of hull simplified by this way is great.

3 The processing of Lateral stiffness

The simplified model of the hull beam by the former way ignores the effect of the transverse components to hull stiffness. When calculate the section, transverse components such as transverse beam and ribs in the two sections of the beam are often not cut, after division of beam element. The results calculated show that the stiffness of section inertia moment will be much smaller than the actual stiffness of sectional inertia moment. And even if either cross sections or one of them is on the transverse components, the calculated results and the actual situation still differs a lot. The stiffness of the frame structure with vertical and horizontal cross is different from that of the solid structure.

So the method of calculation of section moment of inertia of y axis should be different from the x and z axis.

The method of calculation of the x and z axis remains the same, another method is chosen for y axis [17].

For any section of hull, sectional modulus can be expressed as:

$$W = \frac{Z}{G} \tag{1}$$

$$I_y = W \times y_{\max} = \frac{Z \times y_{\max}}{G} \tag{2}$$

$$\begin{aligned} Z = \sum_{i=1}^{n} \Big[& T_{Yi}(H_{Ui} - H_{Di})^3 / 12 \\ & + T_{Yi}(H_{Ui} - H_{Di})((H_{Ui} + H_{Di})/2 - G)^2 \Big] N_{Yi} \\ & + \sum_{i=1}^{m} T_{Xi} L_{Xi} (X_i - G)^2 N_{Xi} \end{aligned} \tag{3}$$

$$\begin{aligned} G = & \frac{\sum_{i=1}^{m} T_{Xi} L_{Xi} X_i N_{Xi}}{\sum_{i=1}^{m} T_{Xi} L_{Xi} N_{Xi} + \sum_{i=1}^{n} T_{Yi}(H_{Ui} - H_{Di}) N_{Yi}} \\ & + \frac{\sum_{i=1}^{n} T_{Yi}(H_{Ui} - H_{Di})(H_{Ui} + H_{Di})/2 N_{Yi}}{\sum_{i=1}^{m} T_{Xi} L_{Xi} N_{Xi} + \sum_{i=1}^{n} T_{Yi}(H_{Ui} - H_{Di}) N_{Yi}} \end{aligned} \tag{4}$$

in the formula,

G —the height from the neutral axis to the base plate when cross section is in linear elastic range, m;

T_{Xi} — The thickness of the class I components in the horizontal direction, mm;

L_{Xi} —the length of the components of I class in the horizontal direction, m;

X_i —the height from the components of I class to the base plate, m;

N_{Xi} —the number of the components of I class in the horizontal direction;

T_{Yi} —the thickness of the components of I class in the vertical direction, mm;

H_{Ui} , H_{Di} —the height from both ends of the components of I class to the base plate, m;

N_{Yi} —the number of the components of I class in the vertical direction;

m—the number kinds of horizontal components;

n—the number kinds of vertical components;

In order to achieve the purpose of high efficiency and simple, the program is prepared to calculate:

```
>>syms tx1 tx2 tx3 tx4 tx5 tx6 lx1 lx2 lx3 lx4 lx5 lx6 x1 x2 x3 x4 x5 x6 nx1 nx2 nx3 nx4 nx5 nx6 ty1 ty2 ty3 ty4 ny1 ny2 ny3 ny4 hu1 hu2 hu3 hu4 hd1 hd2 hd3 hd4;
>>G=(tx1*lx1*x1*nx1+tx2*lx2*x2*nx2+tx3*lx3*x3*nx3+tx4*lx4*x4*nx4+tx5*lx5*x5*nx5+tx6*lx6*x6*nx6+ty1*(hu1-hd1)*(hu1+hd1)/2/ny1+ty2*(hu2-hd2)*(hu2+hd2)/2/ny2+ty3*(hu3-hd3)*(hu3+hd3)/2/ny3+ty4*(hu4-hd4)*(hu4+hd4)/2/ny4)/(tx1*lx1*nx1+tx2*lx2*nx2+tx3*lx3*nx3+tx4*lx4*nx4+tx5*lx5*nx5+tx6*lx6*nx6+ty1*(hu1-hd1)*ny1+ty2*(hu2-hd2)*ny2+ty3*(hu3-hd3)*ny3+ty4*(hu4-hd4)*ny4);
>>Z=tx1*lx1*(x1-G)^2*nx1+tx2*lx2*(x2-G)^2*nx2+tx3*lx3*(x3-G)^2*nx3+tx4*lx4*(x4-G)^2*nx4+tx5*lx5*(x5-G)^2*nx5+tx6*lx6*(x6-G)^2*nx6+(ty1*(hu1-hd1)^3/12+ty1*(hu1-hd1)*((hu1+hd1)/2-G)^2)*ny1+(ty2*(hu2-hd2)^3/12+ty2*(hu2-hd2)*((hu2+hd2)/2-G)^2)*ny2+(ty3*(hu3-hd3)^3/12+ty3*(hu3-hd3)*((hu3+hd3)/2-G)^2)*ny3+(ty4*(hu4-hd4)^3/12+ty4*(hu4-hd4)*((hu4+hd4)/2-G)^2)*ny4;
>> W=Z/G;
```

Set the section moment of inertia of y axis after integration into the attributes of the software ADAMS, calculating along with the data of the x, z axis. Results as shown in table 5:

Table 5. beam model results of sectional moment of inertia of y axis after changed and the error of the actual natural frequency

Actual natural frequence (Hz)	Beam model results (Hz)	error%
0.9953	1.0252	3.0041
1.3108	1.3626	3.9518
2.1445	2.2313	4.0476
3.1305	3.2576	4.0601
3.3807	3.5724	5.6704
3.5282	3.7588	6.5359
3.7679	4.0731	8.1
4.3084	4.8248	11.9859
4.8068	5.4389	13.1501
5.559	6.6472	19.5755

From the data table can be seen, after changing the Y axis cross section moment of inertia calculation method, the accuracy of the results increased about 10 percentage points.

4 The processing of other special parts

When faced with ship with large opening, the ship is extremely prone to torsional deformation [18]. When the torsion occurs, the accuracy of simplified stiffness of Z axis is very important. It needs to make corresponding changes to calculate sectional moment of inertia of the z axis.

For analysis of hull torsion, the hull must be looked as thin-walled beam, deck and the outer plate as thin wall of hollow beam [19]. When the hull of the deck with a long opening has torsional deformation, section torsional moment of inertia:

$$I_z = \frac{1}{3}(Bt_b^3 + 2Dt_s^2 + 2Ct_d^3) \qquad (5)$$

in formula,

B—The actual width of the hull, m;

D—The actual depth of the hull, m;

C—The width of the deck plate, m;

t_b —The thickness of bottom plate, mm;

t_s —The thickness of side plate, mm;

t_d —The thickness of deck plate, mm.

This paper studied 54 m mobile carrier with the large opening, so the processing opening of section moment of inertia is considered.

After the sectional moment of inertia reprocessed and running in ADAMS, the results are as shown in table 6:

Table 6. beam model results of sectional moment of inertia of z axis after changed and the error of the actual natural frequency

Actual natural frequence (Hz)	Beam model results (Hz)	error%
0.9953	1.0157	2.0496
1.3108	1.3514	3.0973
2.1445	2.2171	3.3854
3.1305	3.2393	3.4755
3.3807	3.4565	2.2421
3.5282	3.6681	3.9652
Actual natural frequence (Hz)	**Beam model results (Hz)**	**error%**
3.7679	3.9753	5.5044
4.3084	4.6149	7.114
4.8068	5.2466	9.1495
5.559	6.2308	12.0849

It is can be seen from the results that after the processing changed, the accuracy of the sectional moment of inertia improves again. All the results of beam model and the error with actual natural frequency are all reduced by 1% or so. And they are more close to actual results.

5 Conclusions

The processing of equivalent stiffness is a very important research part in vibration analysis. The accuracy of the equivalent stiffness directly affects the results of the entire study whether is correct and whether the conclusion of the research makes sense.

In this paper, the processing method of the equivalent stiffness of the ship hull girder is studied. Using ADAMS software for auxiliary calculation obtained more accurate method than the commonly used method to deal with hull girder segment element. The equivalent stiffness is more close to the actual situation, thus improving the accuracy of the calculation of natural frequency in the study of vibration characteristics. Having been used in ADAMS, the modelling method turns out to be more accurate processing method than common hull girder section unit method.

The method is applicable for any hull which can be simplified to beam model. In this paper, the method of processing thin-walled structures with transverse and longitudinal structure can also provide a kind of support for the research of other subjects.

Acknowledgements

This work got some good people's helps. Thanks for my teacher, all my brothers and my sisters. All those who contributed directly or indirectly are thanked.

References

1. Z.Q.Li, Study on Ship Hull Vibration Model Building(Dalian University of Technology,2006)(In Chinese)
2. Y.M.Yin, Study on Modeling Method of Ship Vibration(Dalian University of Technology,2007) (In Chinese)
3. X.Z.Wang, The study of hull vibration characteristics (Dalian University of Technology,2006)(In Chinese)
4. H.Cheng,Study on the vibration of the three body ship attached to the water quality(Wuhan University of Technology, 2011) (In Chinese)
5. X.Wei,Study on Experiment and Numerical Simulation of Added Water Mass for Trimaran Vibration(Wuhan University of Technology,2012) (In Chinese)
6. Z.W.Chen, Y.L.Dong, *Multi body dynamics simulation based on MSC Adams*(China Water Conservancy and Hydropower Press, BeiJing,2012)
7. W.M.Ji, T.Fang, S.Q.Chen, *Mechanical vibration* (Science Press, Beijing,1985) (In Chinese)
8. C.Luo,C.P.Zou,Z.G.Li, Analysis of ship stern shaft bracket system modeling and vibration characteristics,Chin J Ship Res (2),55-61(2014) (In Chinese)
9. X.Peng,D.P.Luo, Modeling of ship structure and prediction of underwater vibration and radiation noise, Noise and Vibration Control **23**(06),9-12(2003) (In Chinese)
10. R.W.Che,N.L.Lu, A new method for dynamic modeling of multibody system,Turbine technology, **45**(6),364-365(2003) (In Chinese)
11. Y.M.Yin,D.Y.Zhao, Research on finite element modeling method for overall vibration of Ship Superstructure, Shipbuilding of China50(03),49-56(2009) (In Chinese)
12. X.T.Rui, W.H.Sui, Y.Z.Shao, The field transfer matrix and rigid multi-body dynamics in the application, Journal of Astronautics **4**,82-87(1993) (In Chinese)
13. X.D.Jin, J.Xia, *Ship vibration* (Shanghai Jiao Tong University press, Shanghai, 2011) (In Chinese)
14. J.Du, Research on Modeling and Simulation of simplified car body based on a passenger car(Jilin University,Jilin,2012)(In Chinese)
15. X.Dong, Calculation and FEM analysis of simplified model based on derivation of rigidity matrix(Jilin University,Jilin,2004)(In Chinese)
16. M.H.Li,Discussion on the method for solving the moment of inertia of the cross section of the beam,Gansu Science and Technology**23**(2),145-146(2007) (In Chinese)
17. Z.M.Ma,Analysis of ultimate strength of ship hull girder(Wuhan University of Technology, Wuhan, 2003) (In Chinese)
18. F.Lin, Analysis of the structural characteristics of 50000 tons large open container / bulk carrier, Journal of Marine Science and Application **1** ,36-39(1996) (In Chinese)
19. Y.H.Xie, J.G.Wu, J.L.Li, *Design of ship structure* (Publishing House of Shanghai Jiao Tong University, Shanghai,2011) (In Chinese)

Research on Goods and the Ship Interaction Based on ADAMS

Fangzhen Song , Yanshi He and Haining Liu

School of Mechanical Engineering, University of Jinan, Jinan, 250022 , China

Abstract. The equivalent method of the relative movement goods on board is discussed in details. This method is to establish dynamic model based on moving trajectory of gravity-center for goods and to take rigid body geometric model with the trajectory as constraints in ADAMS. The difference of simulation methods for the different goods in carrier rolling is compared. The interact of relative moving objects with bulk carrier is discussed by using the ADAMS model. It is verified that the ballast water can maintain the ship's stability by means of the ADAMS model.

1 Introduction

Ships often encounter all sorts of external force such as wind at anchor or voyage, so the equilibrium state of ships would destroy. Ships under external torque will be imbalanced. At this time, if the ship has proper stability, the external torque will be offset by the restoring moment which are the joint action of buoyancy and its own gravity, in order to avoid the tilt of ship continuing to expand[1]. When the external torque is eliminated, the restoring moment makes the ship (after a certain cyclical swing) to return to the original balance position. This kind of restoring ability of ship is the stability [2] problem of the ship that we study. The resilience of the ship is the problem of ship's stability that we are studying on.

The relative movement of goods on a ship can lead to the instability of the ship [3], [4]. When the ship leaves the equilibrium position and swing occurs, they intensify the oscillation amplitude of the ship and cause capsizing. It is necessary to consider relative movement of goods as the instability factors. The ship's stability is usually improved by setting ballast water tank on the ship [5], The ballast water of ballast cabin will help reduce the ship's sloshing caused by the swaying of goods. When we study the stability of ship in ADAMS[6], in order to simulate a more realistic motion of the ship and get more accurate results of the analysis, we must consider the effect of the swaying of goods and ballast water.

In this paper, grain is taken as an example of the relative movement of the goods. When grain is in natural grain piles, the nature of the flow in all directions into a cone is called grain scattered. The property of grain scattered is usually expressed as repose angle [7]. Repose angle is the angle between the cone slope that grain naturally forms falling from high and the bottom horizontal line. Repose angle is inversely proportional to scattered property. Good scattered property with small repose angle and poor scattered property with big repose angle[8].

2 The equivalent of goods

The repose angle of grains is 20°. It means when the stacked grain's surface is flat and the angle of container is more than 20 °, grains begin to flow.

The trajectory's change of center of gravity is calculated when the cargo lateral tilt changes from 20 ° to 40°. The calculation model of goods is built in Solidworks. It is shown in Fig. 1.

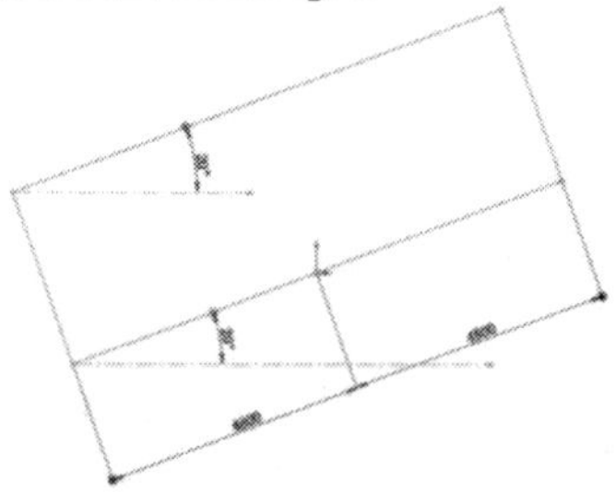

Figure 1. Calculation model of goods.

Bottom edge is h and the two lines are perpendicular to the bottom: the left line is a, the right line is b. When bottom edge is horizontal, two lines are perpendicular to the bottom are c. Cargo volume does not change, so $h \times c = h/2 \times (a+b)$, then $2c = a+b$ .It is known that h=9600mm. An angle between surface and the bottom of the cargo goods is set as θ . $c = 2750$ mm, so $2a - 9600\tan\theta = 5500$.

Every time change the angle of 2°, so the values of 10 times's change are 2917.6197, 3085.6487, 3254.5003, 3424.5960,3596.3695,3770.2715,3946.7744,4126.3779,

4309.6145, 4497.0571.

Get the change of barycentric coordinates as shown in Table 1:

Table 1. Barycentric coordinates

Nu (m)	x	y	Nu (m)	x	y
1	-606.145	1239.923	6	-1265.483	905.001
2	-740.443	1182.932	7	-1394.694	822.522
3	-873.343	1121.032	8	-1523.552	734.585
4	-1005.022	1054.141	9	-1652.345	641.023
5	-1135.667	982.167	10	-1781.398	541.642

Draw a line with the data. It is shown in figure 2:

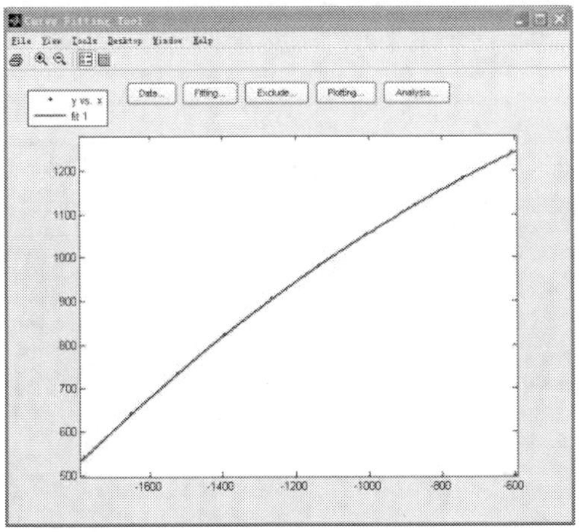

Figure 2. Barycentric coordinates curve.

Follow-up coordinate system is established. x_1 axis is horizontal to bottom line, y_1 axis is vertical to bottom line and the origin is coincidence with the one of the original coordinate system. Set the lines (attached by the point of measurement and the origin point) and x axis's angle is acute angle, the conversion formula is below:

$$\begin{cases}\cos(\theta+\alpha)\sqrt{x^2+y^2}=-x_1\\ \sin(\theta+\alpha)\sqrt{x^2+y^2}=y_1\end{cases} \tag{1}$$

The transformed coordinates are in table 2

Table 2. barycentric coordinates.

Num	α	$\theta+\alpha$	x_1	y_1
1	63.948	-41.948	-97.52406	1376.703
2	57.956	-33.956	-195.2855	1381.828
3	52.080	-26.08	-293.516	1390.428
4	46.366	-18.366	-392.504	1402.577
5	40.854	-10.854	-492.4445	1418.411
6	35.570	-3.57	-593.6175	1438.086
7	30.530	3.47	-696.3057	1461.804
8	25.741	10.259	-800.8058	1489.811
9	21.204	16.796	-907.4028	1522.424
10	16.912	23.088	-1016.473	1559.98

Draw a line with the data. It is shown in Fig. 3.

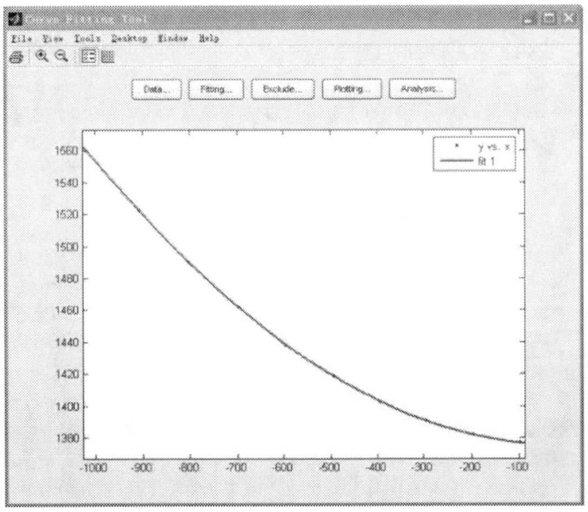

Figure 3. Barycentric coordinates curve.

The rigid body under the constraint of barycentre trajectory is shown in Fig. 4.

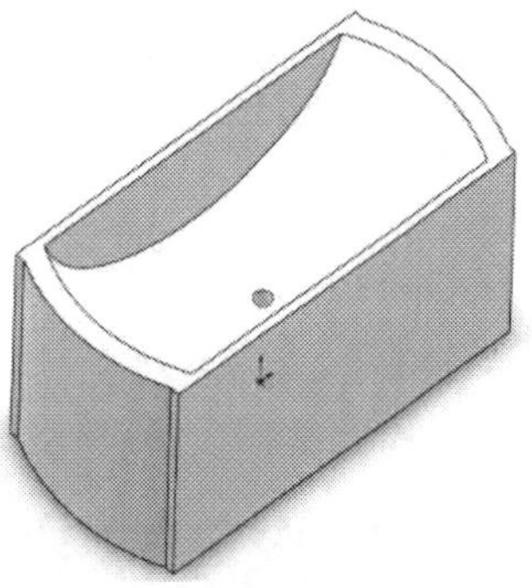

Figure 4. Rigid body under the constraint of barycentre trajectory.

The medial surface is drawn according to the calculated curve. The size of the hole which is in the middle of surface matches the spherical rigid body which is the simulation of center of gravity. When the angle of rigid body reaches 20°, center of gravity of the spherical rigid body is just through the hole on the edge in the vertical direction. It means that when angle exceeds 20 °, hole edge can no longer constraint of the spherical rigid body, the spherical rigid body begin to roll along the inside surface of the rigid body. Establish contact constraint between rigid body and spherical rigid body. The quality of the rigid body is set as a small value which affect can be ignored. The quality of spherical rigid is set as the original quality of the goods after divided into segments. Damping is associated with speed. In order to prevent the spherical rigid body due to the inertia caused reciprocating rolling, so set a larger contact constraint damping coefficient to control the velocity of the spherical rigid body.

3 The simulation of the ship in the water

In this paper, we choose a 54m motor cargo ship as the research object. In order to get a more realistic simulation of ship motion in the water, the spring is used as simulation of water surface support. Due to the spring is stretch spring, it cannot provide constraints when the force is perpendicular to the direction of the spring. The ship model will dump and then affect the simulation. The spring which is added to the bottom of the boat is changed to be added to the top of the boat to make hull be on suspension state. No matter how the hull is swinging, the hull can be restored to a state of balance due to gravity. This process has no effect on the size of force, only change the force's direction.

In order to avoid the position of gravity center and the suspension point of each section of the hull affecting the balance state of the ship, applying the spring at the four corners of the rigid deck surface of every section to ensure each rigid deck is level.

For spring stiffness, the hull support stiffness of the overall water is gotten according to the contact water area of the hull bottom, and then be divided equally to each of the spring unit of the hull support.

$F=\rho Vg$, and $F=nkx$, $\rho Sxg=nkx$, $\rho Sg=nk$, so $k=\rho Sg/n$.

ρ is the density of leaching liquid, V is the volume of leaching liquid, g is acceleration of gravity, n is the number of spring, k is spring stiffness, x is spring elongation.

In this paper, n=78, ρ=1300kg/m^3, g=9.8N/kg, S=397.1117m^2,

So each spring stiffness is

k=ρSg/n=1300×397.1117×9.8/78=64861.5777N/m

The model is established in the ADAMS is shown in Fig. 5.

Figure 5. The model in ADAMS.

4 Influence between relative moving objects and bulk carrier

Two kinds simulation way of goods are established. The model with rigid block simulation is called model 1 and the model in the center of gravity trajectory simulation called model 2. First of all, the difference of two models in kinematics is compared. Under the action of the wind and waves, the ship will leave the original balance position to one side, and under the action of the inertia, the ship will reciprocating motion. The ship with bow and stern vertical line with the line angle in the horizontal plane projection is called the roll angle [9].When the ship's stability is discussed, rolling motion occupies an important position in the stability calculation [10], [11]. So when the difference between two models is compared, therefore, the research is done from the rolling motion in the paper.

In theory, when the rolling motion occurs, the roll angle of the side is the same. So the point A at central section side is selected from the whole bulk carrier segment in the rigid body model as a measurement. Point A on model 1 along the x axis which means the lateral deviation from the origin of mobile distance changing with time curve is shown in Fig. 6.

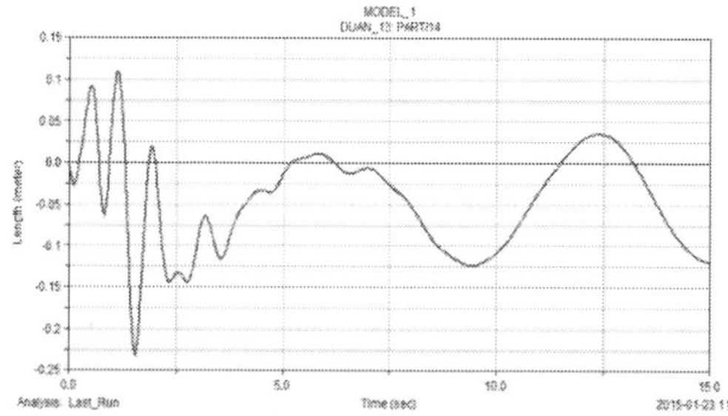

Figure 6. The lateral deviation curve of model 1.

It can be seen from the figure 6 that the maximum offset distance is 0.1249 m and the distance between the two maximum displacement points of the reciprocating swing is 0.1421 m. The curve of the roll angle changes with time is shown in Fig. 7.

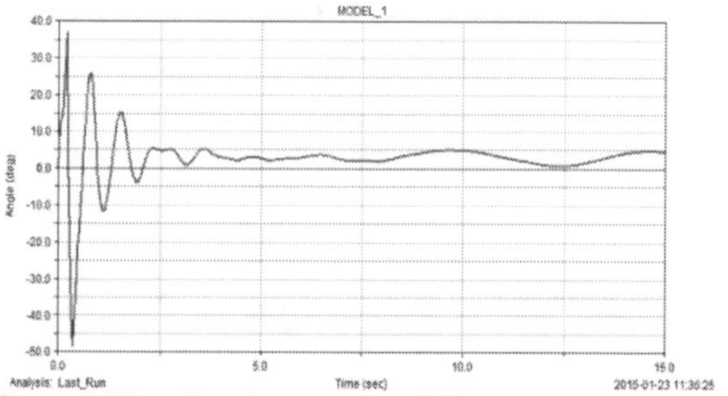

Figure 7. The roll angle curve of model 1.

It can be seen from the Fig. 7 that maximum roll angle is 5. 101°and the roll angle which the stability calculation of the original ship provides is 11.805 °. model 1 cannot accurately simulate the actual status of the ship motion. If this model is used, the result comes out to be dangerous.

Point A on model 2 along the x axis deviating from the origin's distance changing with time curve is shown in Fig. 8.

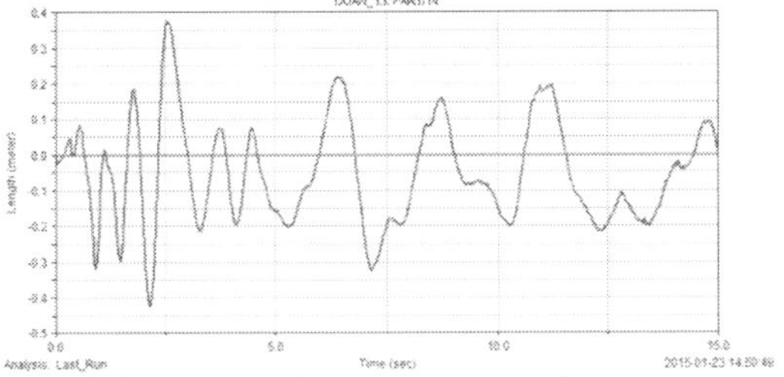

Figure 8. The lateral deviation curve of model 2.

It can be seen from the figure 8 that maximum offset distance is 0.3208m and the distance between the two maximum displacement points of the reciprocating swing is 0.5443 m. The curve of the roll angle changes with time is shown in Fig. 9.

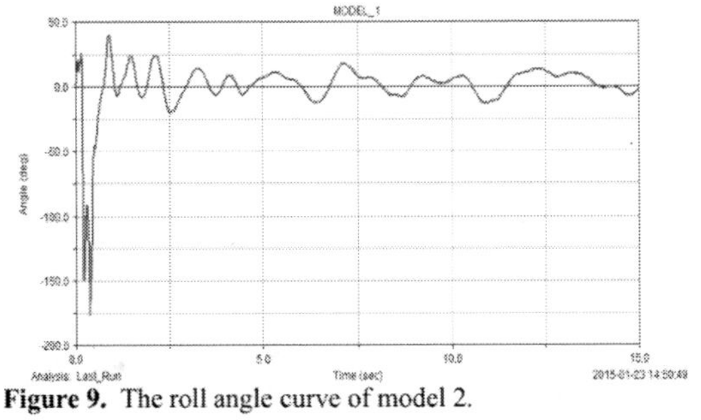

Figure 9. The roll angle curve of model 2.

It can be seen from the Fig. 9 that maximum roll angle is 17. 6866°. It can be seen that in the case of no ballast water, the ship's maximum roll angle is greater than the largest stability calculation data are given. If this model is used, the result comes to be conservative.

Add ballast water on the model 2, lateral displacement changing with time is shown in Fig. 10.

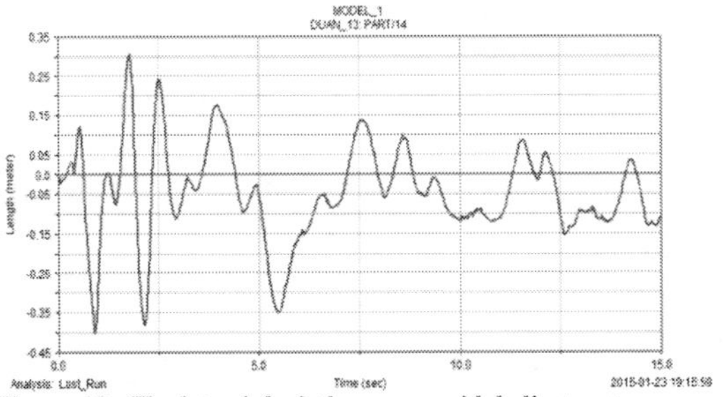

Figure 10. The lateral deviation curve with ballast water

It can be seen from the figure 10 that when at not stable state, the maximum distance is 0.4012 m and decreased by 0.0330 m compared to the distance of the displacement of the model without ballast water. Reduction is 7.60% of the model without ballast water. After reaching a stable state, the maximum distance is 0.1379 m and reduced by 57.01% compared with the model without ballast water. The maximum displacement distance is 0.2906 m when in reciprocating swing state and reduced by 46.61% compared with the model without ballast water.

Add ballast water on the model 2, roll angle changing with time is shown in Fig. 11.

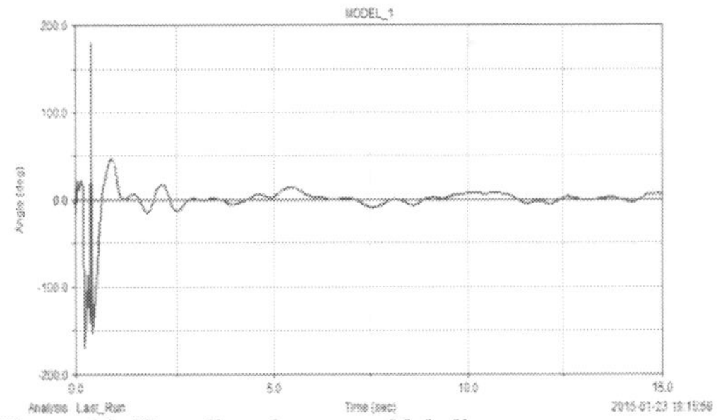

Figure 11. The roll angle curve with ballast water

It can be seen from the Fig. 11 that when at stable state, the maximum roll angel is 8.2667°and decreased by 53.26% compared to the maximum roll angle of the model without ballast water. And the results meet the specification requirements.

5 Conclusions

The equivalent method of the relative movement goods on board is discussed in details. The ADAMS dynamic model is established based on moving trajectory of gravity-center for goods and to take rigid body geometric model with the trajectory as constraints. The difference of simulation methods for the different goods in carrier rolling is compared. The interact of relative moving objects with bulk carrier is discussed by using the ADAMS model. It is verified that the ballast water can maintain the ship's stability by means of the ADAMS model.

References

1. M.Z. Qiu, Stability and Navigating Safety of a ship in Following (Harbin Engineering University, 2008)(In Chinese)
2. H.S.Yang, Re-understanding of ship stability, Marine Technology 2,21-22(2003) (In Chinese)
3. L.B.Zhou, Ship in big storm of how to prevent the movement of goods, Ship maneuvering in big waves 1 ,(2006) (In Chinese)
4. J.H .Cai, Discussion on the movement of ship cargo, Marine Technology1, 33-34 (2012) (In Chinese)
5. R.Wang, Study of Anti-heeling System Design for Ship(Harbin Engineering University,2007)(In Chinese)
6. Z.W. Chen, Y.L. Dong, *Multi body dynamics simulation based on MSC Adams*(China Water Conservancy and Hydropower Press,BeiJing,2012) (In Chinese)
7. Z.M. Wan, F.Wu, H. Li, P. Zhu, Q.T.Xu, R.G. Zhang, Study on the grain storage under different conditions, Grain Storage41 (5) ,10-15(2012) (In Chinese)

8. X.L.Zhou, H.Z.Li, S.Q.Yu, Study on static angle of natural sediment accumulation (2014 Conference on Computational Mechanics of granular materials,2014) (In Chinese)
9. P.Y. Li, T.C. Feng, Y.M.Qiu, The motion of the three body ship, Chinese shipbuilding44(01), 24-30 (2003) (In Chinese)
10. S.G.Zhang, Theoretical study on roll stability of ships in large waves, Chinese Journal Of Ship Research1(01), 50-53 (2006) (In Chinese)
11. Y.K. Hu, Study on the large rolling motion and stability of the ship in waves(Harbin Engineering University, 2011) (In Chinese)

Fuzzy Classification and Implementation Methods for Tugboat Main Engine Fault

Xiaohui Xu , Guangchen Bai

School of Automation, Beijing University of Posts and Telecommunications, 100876 Beijing, China

Abstract. The establishment of classification index system of fault in tugboat main engine was briefly discussed first. After that, the fuzzy method and procedures used in fault classification were introduced. Several implementation techniques of the fuzzy classification were explored including the knowledge expression, the hierarchical structure of knowledge library, the pre-processor and the forward inference engine. Some specific fault records of tugboat main engine were analyzed and classified for certification. And results prove that fuzzy classification has certain reference value and improves the working efficiency of tugboat personnel in equipment maintenance and management.

1 Introduction

As an important auxiliary ship for towing larger vessels in harbor, tugboat has the characteristics of flexible operation and being highly maneuverable. However, at the same time it makes the load of tugboat main engine change dramatically, where the working condition is extremely severe. When it comes to tugboat fault diagnosis, personnel often relate to unquantifiable concepts such as relatively low pressure, insufficient flow and excessive temperature rise, which have not definite boundaries, namely "fuzzy concept". In addition, a fault may result in a variety of phenomena, especially for the heavy complicated engineering equipment integrated with electromechanical and hydraulic system. Deterministic factors and random factors being staggered together, tugboat main engine fault has the distinguishing feature of changing gradually and concealment[1]. Therefore, it's necessary to explore the classification index system of tugboat fault. By researching the classification method and its implement techniques, tugboat fault can be analyzed and classified, laying the foundation for the subsequent research. Tugboat fault classification will help personnel further identify the exact location and cause of fault, predict the trend of fault and potential danger, and thus determine the appropriate measures to be taken.

2 Fault classification index system

Tugboat works in the harbor, which is highly maneuverable and has high work intensity ,therefore tugboat main engine adopt four-stroke medium speed diesel engine, with the advantages of compact structure and high power. Tugboat main engine workload's varying widely, frequent starts and stops render the mechanical load and thermal load change remarkably, leading to the extremely bad working environment, even a fault or stalling the engine. Accordingly we are supposed to make preliminary judgment on the fault phenomena timely, in an attempt to gain time for diagnosis and maintenancem.

Tugboat main engine fault system are as follows: ventilation system, fuel system, lubricating oil system, seawater system, freshwater system, starting air system, pressurization system. Then the fault phenomena having been analyzed and classified, a classification index system of tugboat main engine fault is initially built. The system classify faults with some certain index as shown in Figure 1.

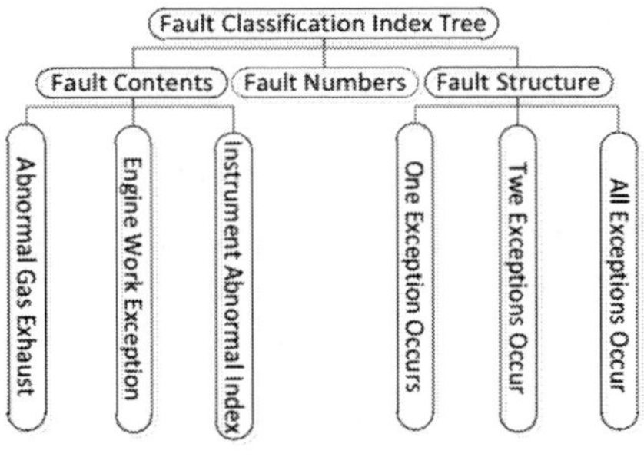

Figure 1. Fault Classification Index Tree.

3 Fuzzy methods and procedures of fault classification[2]

Fault in tugboat main engine is a relative and qualitative fuzzy concept. It's difficult to have a quantitatively accurate represent of classification index as the contents are rich and formats are variable, which renders the classification a multi-factor and multi-level fuzzy classification.

A hierarchical factor-based fuzzy classification method is used in the classification, this is also to say, to classify from the bottom to the top, as well as from a single indicator to multi-indicator considered.

1).The classification index domain of the classification object is determined, then divided into n subjects.

$$U = \bigcup_{i=1}^{n} u_i \tag{1}$$

where $u_i = \{u_{i1}, u_{i2}, \cdots, u_{ip_i}\} (i = 1, 2, \cdots, n)$ [3]

2). To each classification index subsets u_i, perform the single level and single indicator fuzzy transformation. Classification level domain $V = \{v_1, v_2, \cdots, v_q\}$ is given, where q is better to be an odd number so that we can have a middle level.

Quantify each indicator in u_i, the membership of a certain fault to all categories classified by single indicator is determined. The value of membership function can be determined using expert scoring method. So we can have the single indicator fuzzy matrix R_i of u_i.

The fuzzy weight vector of every indicator in u_i is determined as

$$A = (a_{i1}, a_{i2}, \cdots, a_{ip_i})$$

Before performing the synthesis process, A_i needs to be normalized. Ensure

$$\sum_{r=1}^{p_i} a_{ir} = 1 \tag{2}$$

Performing fuzzy transformation to A_i and R_i gives us B_i, the single level classification result vector of, which is[3]

$$\begin{cases} A_i \circ R_i = B_i \\ B_i = (b_{i1}, b_{i2}, \cdots, b_{im}) \end{cases} (i = 1, 2, \cdots, s) \tag{3}$$

The fuzzy weight vector of comprehensive index $u_i (i = 1, 2, \cdots, s)$ is given as follow

$$A = (a_1, a_2, \cdots, a_s)^{[3]}$$

3). Multi-level fuzzy classification. Repeat steps 1~2 using u_i as the comprehensive index and B_i as the single indicator classification result. We can have Multi-level fuzzy classification model.

4). Analysis of fuzzy classification results. The weighted average method is used to process fuzzy classification result vectors. The basic idea of this process is firstly to get the rank of classification result vector according to its position, which is to use $1, 2, \cdots, q$ to represent every categories, and then secondly to get the weighted sum of each components' rank of result vector, thus we can have the relative position of classified object, written as

$$E = \sum_{j=1}^{q} b_j^k j \bigg/ \sum_{j=1}^{q} b_j^k \tag{4}$$

where k is a coefficient to be determined for controlling the import of larger b_j. k is often set as 1 or 2, so we can sort several classified objects in the order of E [3].

4 Key techniques in classification

4.1 Knowledge expression

4.1.1 Production rule

Abnormal gas exhaust, engine work exception as well as instrumental abnormal index and the fault structure are identified and summarized as knowledge, based on the aforementioned classification index system. The knowledge is expressed as production rule with a basic structure below:

$$IF < A > THEN < B >$$

In which, A is the production prerequisite, and B is a set of conclusions. Both of A and B can generously be expressions composed by text, numbers and logical operators AND, OR, NOT.

Here in our research, the prerequisite A led by IF of production rule is the classification index of different levels, and conclusion B led by THEN is the corresponding membership function.

4.1.2 Frame structure

Corresponding to the hierarchical structure of classification index tree, the multiple nested frame structure, such as {Frame{Sub-Frame{Slot{Side{Sub-Side{Value...}}}}}}, is used for the knowledge expression to manage the production rules. The framework is composed by different levels. Its basic logical structure can be expressed as:

{Frame{Slot 1{Side 1{Value 1,
Value 2,
...
Value m}
Side 2{Value ...}
Slot 2{...}
...
Slot n{...}
}}

In this structure, location frame is for the classified object, such as a certain fault record. Slot is for the classification index, such as the fault content. Side is for classification sub-index, such as gas exhaust. Value is for classification indicator, such as exhaust pipe's blowing out black smoke. Compatible with the index, one side can have lots of value where production rules would be stored.

This frame-rule mixed expression of knowledge can greatly simplify the workload of the inference engine, reduce the search volume compared with a simple rule system, shorten the reasoning time and improve the efficiency of solving.

4.2 Hierarchical structure of knowledge base

A comprehensive knowledge base is necessary for the classification. Hierarchical tree structure, as shown in the Figure 2, is used to build to knowledge base, considering the contents and structure of classification index system[4].

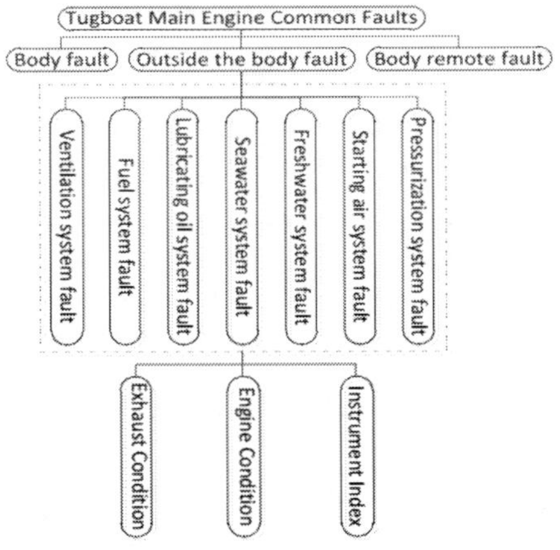

(a) Top level of knowledge base structure

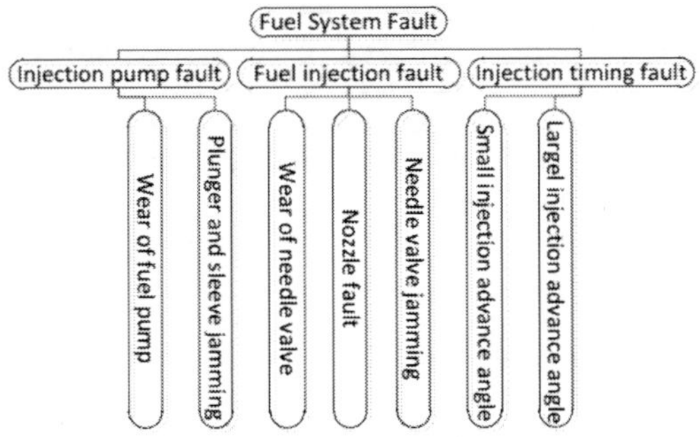

(b) Second level of knowledge base structure

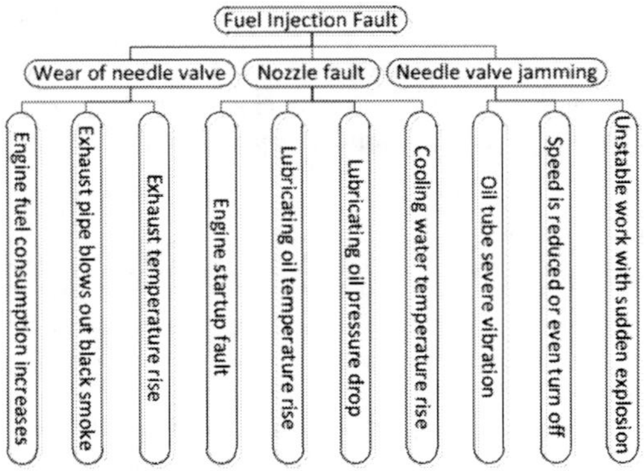

(c) Bottom level of knowledge base structure

Figure 2. Hierarchical Structure of the Knowledge Base.

The storage, management and maintain of knowledge can be easily achieved within this structure[5], through which the knowledge will also be conveniently used in inference.

4.3 Pre-process and inference

4.3.1 Pre-processor

Demanded by the classification knowledge, feature factors in a fault phenomenon must be extracted to compose the fact library. The pre-processor we used on our research is described as follows.

For a complete fault record, pre-processor firstly filtered the contents of features in it, such as many kinds of exhaust gas condition, engine condition, instrumental index, to compose the Feature Factor Sets. Then the number of all kinds of factors, which is structurally corresponding with the original text, is extracted after the filtering and counting. The results consist the fact library, and will be matched with the perquisite of rules.

A record is given here. A fault record in tugboat main engine we collected is "Exhaust pipe blows out black smoke and its temperature increases; unstable work with sudden explosion then power and speed decrease; at the same time, engine fuel consumption rises with oil tube of high temperature severe vibration; cooling water`s temperature increases; temperature of lubricating oil rises and its pressure drops".

The information extract after pre-processing is listed in Table 1.

Table 1. The pre-processor result of a fault record.

Content	Fault Name	Number
Exhaust gas condition	Blowing out black smoke	Ⅰ
	Exhaust gas`s temperature rises	Ⅱ
Engine condition	Engine power decreases	Ⅲ
	Unstable work with sudden explosion	Ⅳ
	Oil tube`s severe vibration	Ⅴ
	Engine fuel consumption rises	Ⅵ
Instrumental index	Cooling water`s temperature rises	Ⅶ
	Lubricating oil`s temperature rises	Ⅷ
	Speed decreases	Ⅸ
	Lubricating oil`s pressure drops	Ⅹ

4.3.2 Inference engine

Fitted with the knowledge expression, forward inference engine is applied in our research. By matching the fact library with the prerequisite of rules, the values of membership function can be determined through the inference engine.

Indicators in different levels of index system are filled in the hierarchical structure of knowledge base. The slot and side that will be used according to the levels of classification are firstly determined by the inference engine, and then the rules needed in solving are searched. Contents in fact library is matched with the prerequisite of the rules in the rule library one by one. The membership function contained in the conclusion of rules successfully matched will be stored as one of the row elements of membership matrix.

5 Implementation

The collected fault record in tugboat main engine are classified with the methods above. To space limitations, the fault record in 4.3 is given next as an example of the general classification.

Given the classification categories domain V.

V={Ventilation system fault (V), fuel system fault (F), lubricating oil system (L), seawater system fault (S), freshwater system fault (W), starting air system fault (A), pressurization system fault(P)}

Main indicators are chosen, according to the contents referred, to build classification indicators domain U. The general indicators domain used in classification of different fault record is transcendental, while different kinds of sub-sets of indicators can also be generally used in classification of the same type of faults. The classification result is shown in Table 2. In "categories". Ⅰ represents "Blowing out black smoke" while Ⅱ is on the behalf of "Exhasut gas`s temperature rises" and the rest are inferred in Table 1.

Table 2. The classification result of a fault record.

Categories	V	F	L	S	W	A	P
Ⅰ	0.22	0.62	0.02	0.03	0.04	0.01	0.06
Ⅱ	0.18	0.52	0.04	0.02	0.03	0.02	0.19
Ⅲ	0.22	0.31	0.05	0.03	0.21	0.02	0.16
Ⅳ	0.23	0.49	0.05	0.02	0.02	0.02	0.17
Ⅴ	0.15	0.63	0.04	0.06	0.09	0.01	0.02
Ⅵ	0.23	0.39	0.06	0.13	0.16	0.01	0.02
Ⅶ	0.21	0.24	0.02	0.21	0.27	0.02	0.03
Ⅷ	0.03	0.39	0.34	0.02	0.03	0.01	0.18
Ⅸ	0.21	0.32	0.04	0.03	0.21	0.03	0.16
Ⅹ	0.03	0.42	0.35	0.01	0.02	0.02	0.15

However, when data in Table 2 are used as the fuzzy matrix R, the value of each component of weight vector is too small to affect the classification results since there are too many categories. So we divide the data into 3 groups(the first two as a group, Ⅲ、Ⅳ、Ⅴ and Ⅵ as a group, the last four as the third group) . Fuzzy transformation is performed respectively to have

$$R = \begin{bmatrix} 0.20 & 0.57 & 0.03 & 0.02 & 0.04 & 0.02 & 0.12 \\ 0.21 & 0.46 & 0.05 & 0.06 & 0.12 & 0.01 & 0.09 \\ 0.12 & 0.35 & 0.18 & 0.07 & 0.13 & 0.02 & 0.13 \end{bmatrix}$$

Delphi method is used to determine the weights of each indicators. Here, we use the weight vector A= (0.21, 0.43, 0.36)

Fuzzy transformation is performed to have B= (0.21, 0.43, 0.18, 0.07, 0.13, 0.02, 0.13)

Weighted average method is used to analysis and process the fuzzy classification results. Set the value of each components of the resulting vector according to their positions,which means that, 1 for ventilation sysytem fault, 2 for fuel system fault, 3 for lubricating oil sysytem fault, 4 for seawater system fault, 5 for freshwater system fault, 6 for starting air system fault, 7 for pressurization system fault. The final classification results is calculated as Equation (4)(k =2, q =7).

The calculation result is $E \approx 2.45$,which shows that it is between category 2 and 3, and slightly partial to the second one. Thus it is probably a fuel system fault[6] or a lubricating oil system fault, though more likely to be the former. However, it can easily identified literally to both of those two categories. As for which one it actually belonged to, these two systems should be checked to determine the factual fault location. Nevertheless, in terms of the phenomenon of the fault record, it is difficult to distinguish whether it is a fuel system fault or a lubricating oil system fault. Therefore, this classification result is reasonable.

6 Implementation

With method and procedure above, a fault record in tugboat is classified for certification, and the result is considered reasonable and useful while the method and implementation are effective and efficient. Result shows that the classification index system, the method and the implementation technique can not only satisfy the actual need of the tugboat main engine fault classification, and also enable personnel to eliminate the fault as soon as possible, to gain time for the normal operation of tugboat.

References

1. Tikun Wang: Tugboat main engine fault diagnosis method based on the fault tree analysis (Master of Engineering, Dalian Maritime University, China 2012)
2. Xiaohui Xu, Bingshu Tong. Fuzzy evaluation and implementation techniques of product structure design for assembly. *Agricultural Machinery*. **38**,2 (2007) , p. 129-133

3. Zhongxiong He :Fuzzy mathematics and applications(Tianjin science and technology press,China,1982)
4. Qingmin Zhong: Experimental research on monitoring and diagnosing for diesel engine(Master of Science, Wuhan University of Technology, China 2008)
5. Yanyou Chai: Research on fault diagnosis of marine diesel engine based on kernel-based learning theory(Doctor of Engineering, Harbin Engineering University, China 2012)
6. Xiaojun Ma, Zhi Wang, Chunguang Xing. Research on fault diagnosis for fuel injector system of marine diesel engine. *Ship science and technology*. **30**,6 (2008) , p. 73-75,103

Comprehensive Detection Technology for Urban Subway

Jinghui Xia[1], Tao Liao[2] and Lixun Zhang[1]

[1] Zhengzhou Rail Transit Co., Ltd., China
[2] State Grid Henan Electric Power Research Institute, Zhengzhou, China

Abstract. The urban subway comprehensive detection system is a complex system, which is the collection of advanced sensor technology, computer technology, network technology, data analysis, image analysis and processing technology, artificial intelligence technology and so on. The bow net, orbit and threshold detection of basic project and detection principle for subway are expounded in this paper. And the composition and development of subway comprehensive detection system is proposed.

1 Introduction

At present, the rapid development scale and speed of China's urban subway are unprecedented. Testing the bow net, orbit and tunnel limits, and then timely discovering and dealing with potential safety hazard are the necessary measures to ensure the safe operation of the subway [1].

Urban rail transit includes three types of transportation: the subway, overhead, and light rail. The subway is in underground, do not take up floor space, plus special orbit, so compared with other ways of urban rail transit, the subway has great superiority. For example, the subway has quick speed, not traffic jam, safe, comfortable, and the advantages of small disturbance.

The function of bow net, orbit and threshold detection for electric railway integrated detection technology has been widely used in China's high-speed rail and conventional electrified railway. In the early time, on the basis of introduction of foreign technology, integrated checkout automobile has developed which is suitable for high-speed rail and electric rail.

Compared with the high speed railway and electric rail, urban subway has its characteristics in running speed, line feature parameters, tunnel length, power supply system, operation management and so on. It is not realistic that using the integrated checkout automobile for high-speed rail and electric rail simply in the urban subway. It will Faced with the problems of adaptability, cost and so on.

It is quite necessary and urgent to analysis and research the integrated detection technology that is suitable for urban subway, to research, develop and improve urban subway integrated checkout automobile [2], [3].

2 Dynamic detection of bow net

(1) The main detection project

Bow net is adopted widely in metro vehicle a current device, belonging to one of the key components of the train and its status directly affect the safe operation. Traditional inspection methods rely on artificial Govinda check, need to be done after a special power of the same inside the car depot. Train pantograph online monitoring system by using high speed image processor and sensors, with image analysis method, on the premise of not affect normal operation of the train to realize online detection of pantograph. Detection efficiency is high, the system can real-time tracking monitoring various pantograph mainly, the feature information of the running state and safety performance, and comprehensive analysis of the information processing, timely forecast affecting the safe operation of the vehicle pantograph fault, eliminate the problems.

Bow net [4] system is composed of contact system and pantograph. The main project for dynamic detection of bow net include: contact wire height (Including rigid suspension and flexible suspension), stagger value, double contact line horizontal distance and height difference, anchor segment joint contact line horizontal distance and height difference, the high change rate of high conductivity, contact pressure of bow net, offline of bow net, impact of bow net, network voltage and so on. Bow net system dynamic detection should be implemented whether in the neutral and live cases.

(2) The main project of detection principle

1) Detection of stagger value and high conductivity

The parameter of stagger value, high conductivity and so on is detected using laser scanning sensor. Laser scanning sensor is an optical measurement system which is non-contact. It installs on the roof. The way of continuous emission of laser beam is used to measure the background or the object that is under test to get their two-dimensional planar.

$$S = h/\tan\theta + C \qquad (1)$$

$$H = h + h_0 \quad (2)$$

In the equation: *S*-stagger value; *H*- conductor height; *h*-measurement height; *C*- Deflected center distance of laser scanning sensor(left is negative, right is positive); θ-angle; h_0 - height of laser scanning sensor of the rail surface.

2) Detection of bow net contact pressure

Contact wire's layout is like the Chinese character "之". The point application of contact force moves around within the scope of work in the skateboard. It is not only the position of the action of forces randomly changes, but also the direction of the contact force is changing. Contact force detection device should be able to run in the charge of cases and measure the contact force of wire in the vertical direction of the skateboard that is at any position [4], [5].

Its formula is shown in formula (3):

$$F_J = F_1 + F_2 + F_3 + F_4 \pm ma - W \quad (3)$$

In the equation: F_J -contact force of bow net in the vertical direction; m - equivalent quality of the skateboard; W - the weight of slide; a - pantograph acceleration that is measured by accelerometer installed at the bottom of the pantograph; F_1 , F_2 , F_3 , F_4 - the measured value of the force sensor that installed in the four corners of the bow head slide.

3) The detection of hard point

Hard point detection principle diagram is shown in Fig. 1. Pantograph formed the measured voltage by the sensor (the accelerometer) through the impact acceleration of the catenary hard point. The voltage is converted to frequency signal and then it will be transferred to optical communication channel by cable transmission. Optical communication channel can achieve reliable isolation between the high and low pressure. Low voltage signal is sent to the computer realize the hard point judge.

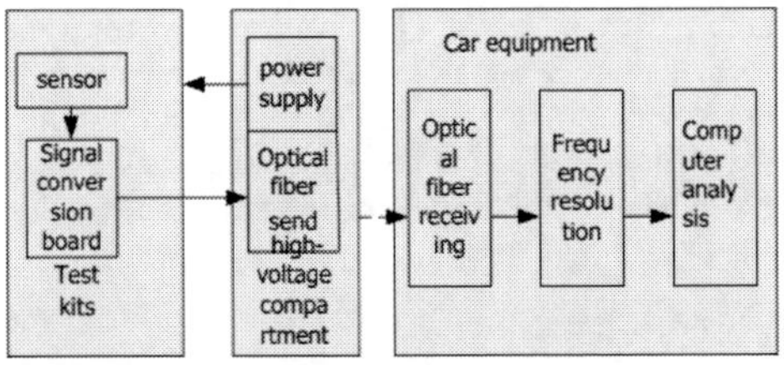

Figure 1. Schematic diagram of hard point detection

4) The detection of network voltage

Network voltage will be isolated by isolation transformation link. Then it will be converted to low voltage (0-5V) and import to acquisition channel. The network voltage measurement will be realized.

5) The dynamic compensation

Two groups of laser displacement sensor will be installed on the vehicle bogie to measure horizontal displacement of the vehicle. Two sets of laser displacement sensor and damper are installed between car body and the bogie in order to measure the relative longitudinal displacement measurement of car body and the bogie. Based on two kinds of displacement measurements, dynamic compensation of position detection quantity in level and side will be got.

Dynamic detection of track

(1) The main detection project

The main project of dynamic detection of track include: the orbit of the left and the right of discretion, the orbit of the left and the right of direction, horizontal (superelevation), warp (twist warp), track spacing and track spacing rate, curvature, vertical and lateral acceleration of car body, the top of rail, flank wear and so on.

(2) The main principle of project detection

1) Track spacing and track spacing rate

Orbital shape and the distance between the two sensors can be measured by the two-dimensional laser outline sensor which is installed on the detection beam. Then it can get the horizontal distance between the right rail and left rail which is about 16 mm under the rail top. That is the track spacing value. And then track spacing rate can be obtained.

2) The top rail and flank wear

Pattern matching and comparing will be implemented between orbital shape measured by the two-dimensional laser outline sensor and orbital shape that was set in order to get sanding top and sanding side.

3) Track curvature

Angle of rotation that corresponded to the curve track can be measured by shaking gyroscope. Then chord length of the curve track can be obtained. The ratio of the chord length and the corresponding rotation Angle is the curvature.

4) Horizontal (superelevation)

Height difference within the same cross-sectional of two rail on the top surface is called horizontal (superelevation). Its detection principle: Absolute relative angle between beam and the earth can be measured by processing measurement signal that through inclinometer and roll gyroscope. The distance of beam and rail end face can be measured by the two-dimensional laser contour sensor. And then we can get the angle between rail and beam. Though the difference of the two angles between the top of the rail and the earth we can get the absolute angle. Absolute level high value can be obtained by conversion. Then through the transformation and minus between base value and the design, we can get relative horizontal value.

5) The triangle pit

Triangle pit refers to maximum deviation (*h*) of the right and left orbit in top rail which is based on a chord length apart *L*. The basic general length is 2.2 meter.

6) Lateral acceleration and the vertical acceleration of car body

It can be measured through lateral acceleration sensor and vertical acceleration sensor installed in the car body. The measurement values can be get by removing the lateral acceleration of the centrifugal force and the initial base value of vertical acceleration.

Because of space, we escape the detection principle of orbital altitude between right rail and left rail.

7) Parts missing detection of bow net

On-line detection system using separate laser position sensor is used to identify the targeted, in order to judge whether the electric GongDuan Angle parts missing. As the train went by in the catenary at the top of the two laser position sensors for the electric GongDuan Angle position for continuous acquisition scan, will receive the reflected signal for data analysis, and judgment by electric GongDuan Angle part is complete.

After car, industrial computer through the network to read the image from the image sensor, and clear the cache initialization, such as to wait for the next shot. System reads the images after start to carry on the analysis, abrasion of pantograph slider carbon, carbon slider fracture defect, offset, tilt and missing parts such as state is analyzed, and returns the result.

3 Threshold dynamic detection, Catenary inspection technology and integrated system

1) The high-speed railway catenary system

Catenary is along the railway lines over the erection of transmission lines, is a special form of electric locomotive power supply. By the foundation and pillar, positioning device, supporting device with several contact suspension parts. Contact suspension including hanging string, the contact line, the carrier cable and connection parts. Contact suspension by supporting devices set up on the pillar, its function is to get electricity from traction substation to electric locomotive. Support device used to support contact suspension, and the load to the pillar or other buildings. According to the catenary's interval, yard and large buildings. Support device including suspension insulator string, horizontal bar, wrist, arm, rod insulator and other buildings of special support equipment. Positioning device includes locating pipe and locator, its function is fixed the position of the contact line, the line of contact within the scope of the pantograph slide running track, guarantee the contact line and pantograph not from, and will be transmitted to the level of the line of contact load. Catenary of task is to convey electricity directly from the traction substation for electric locomotive is used. Catenary and working state will directly affect the quality of the electrified railway transport capacity.

Catenary inspection technology is one of the key to the construction of high-speed railway. With the continuous acceleration of railway requirement for electrification of catenary will be higher. Will be more uncertain factors, also to test equipment requirement will be higher. Therefore, continuously improve the level of detection technology and equipment to guarantee the good condition, electrification of catenary to ensure smooth electrified railway transport.

2) Threshold detection system

Threshold detection system is used to detect the metro building line [6]. Then building outline data based on the top rail and tunnel cross section image can be output. And then the limit position of threshold needs to be got to provide accurate information for maintenance work [7].

Non-contact measurement methods, Laser radar detection, random vibration detection, data fusion, data mining, wavelet analysis, fuzzy recognition, and other advanced technology will be used. Through the application of coordinate transformation and related detection principle, holographic imaging can be measured for the equipment inside the tunnel. Then we can realize 3D image rotation and data overrun warning through special software package.

3) Integrated system

Integrated system is responsible for providing public signal, the whole system operation and management of the industrial Ethernet, and other functions for the above three detection system. The public signs provided include distance information provided by the encoder and RFID, timing signals and etc. The above three detection systems have their hardware, software, can work independently.

4 Conclusion

The urban subway comprehensive detection system is a complex system, which is the collection of advanced sensor technology, computer technology, network technology, data analysis, image analysis and processing technology, artificial intelligence technology and so on.

It is not realistic that using the integrated checkout automobile for high-speed rail and electric rail simply in the urban subway. It will Faced with the problems of adaptability, cost and so on.

So it is quite urgent to analysis and research the integrated detection technology that is suitable for urban subway, to research, develop and improve urban subway integrated checkout automobile.

Urban subway comprehensive detection system is a complex system which is the collection of advanced sensor technology, computer technology, network technology, data analysis technology, image analysis and processing technology, artificial intelligence technology. In order to ensure the operation safety of subway, based on the existing detection technology in high-speed rail and electric railway, we absorb new technologies in time. And comprehensive detection technology standards which are suitable for urban subway are studied according to the characteristics of subway [8]. It is very urgent to develop an urban subway integrated check-out car which has the characteristics of high reliability, compact structure and convenient operation maintenance. At the same time it has the technical conditions and good market for developing.

References

1. W. Hou. High-speed comprehensive detection train of NO.0. Beijing. China Railway Press, (2006)
2. W. Wang, Q. He. Dynamics of high-speed rail vehicle system. Advances in mechanics, **25**(1):134-143 (1995)
3. Ministry of railways of the People's Republic of China. Railway technology management discipline. Beijing: China Railway Press, (2010)
4. Y. Shen, L. Li, Y. Ruan. Mobile mapping technology by vehicle-borne lidar. Infrared and Laser Engineering, **38**(3):437-440 (2009)

5. W. Wang, S. Gu, C. Yang. High-speed comprehensive detection train. Railway technology innovation, (1):12-15 (2012)
6. Q. Fang, D. Zhang, L.N.Y. Wong, Shallow tunnelling method (STM) for subway station construction in soft ground. Tunnelling & Underground Space Technology, **29**(3):10-30 (2012)
7. W. Chen, Z. Long, J. Li, Research on Real-time Dynamic Monitoring of Subway Construction. Chinese Journal of Underground Space & Engineering, (2013)
8. J.F. Hou, J. Chen, R.L. Wang, Construction Monitoring Technology Analysis for Railway Tunnel. Applied Mechanics & Materials, **580-583**:1342-1346 (2014)

Considering Ergonomics in the Planning of Tugger Train Systems for Production Supply

Christopher Keuntje , Michael Kelterborn and Willibald A. Günthner

Institute for Materials Handling, Material Flow, Logistics, Technical University of Munich, 85748 Garching, Germany

Abstract. For the planning of a production supply using a tugger train system, no holistic methods are available that allow one to factor in all important criteria that influence the decision-making process. This paper investigates the possibilities for the consideration of ergonomics in the system design of tugger train systems, based on an empirical study with 42 test subjects. Three basic concepts of tugger trains used in the German automotive industry are defined and evaluated regarding ergonomics. With the aid of a software-based model for tugger trains, which takes the whole life cycle of the tugger train system into account, the economic feasibility of found ergonomic advantages is analyzed. Furthermore, it is exemplarily shown how an informed decision about technical details of a tugger train system can have a positive impact on the ergonomics and economics of the system.

1 Introduction

In the automotive industry, tugger train systems (TTS) play a vital role in the supply of the production and are used throughout the major OEMs. Tugger trains (TT) are manually operated industrial trucks driving continuously on fixed routes and – analogous to a bus in passenger transport – stop at destinations where a filled load carrier has to be supplied or an empty one has to be collected. This allows a high-frequency just-in-time supply of material and, therefore, the synchronization of the material flows and the assembly cycle. As a result, a significant improvement of key performance indicators, like the material throughput time, can be gained [1].

Despite the widespread use of TT, no standards have been implemented yet, and a large variety of techniques is used [2]. In a study conducted with partners of the German automotive industry, 26 TTS were analyzed. Based on similarities regarding additionally needed technical elements and the required handling steps for the supply of load carriers, the classification of [3] was adapted, and three basic concepts for TT were derived.

The classic TT concept is the *trailer concept*, in which the load carriers (e.g. pallet cages) are supplied to the production on the trailer itself. In contrast, in a *taxi concept*, the load carriers are supplied on additional trolleys that are transported by the wagons of the TT. The third basic concept for TT is the *roller conveyor concept*, where the load carriers are placed on roller conveyors installed on the TT wagons. To be able to supply the load carriers, permanently installed roller conveyors are additionally needed in the production. In Table 1, the introduced concepts are exemplarily illustrated by specific technical implementations of the concepts.

In addition to the introduced concepts and various technical implementations of these concepts, different operating processes and control strategies lead to a high amount of design possibilities for a TTS. In combination with various interdependencies between the TT and other elements before and after in the material flow, the planning of those systems is a complex task.

So far, no method exists that allows for a holistic planning of TTS, factoring in all criteria that should be regarded in the decision-making process [2]. In order to develop such a method, the impact of changes in the system design on these criteria has to be researched in detail to provide a significant support to the planner.

A criteria, often underrepresented in the decision-making process, is the ergonomics of a TTS. Ergonomics are a sensitive issue because, in a TTS, the manual handling of load carriers with weights of 600 kg and more can lead to a high amount of physical stress for the workers. This is especially critical considering the demographic change. An aging workforce puts high pressure on the German automotive industry, as they must be able to provide a suitable work environment [5].

On the basis of the challenge presented, this paper researches how ergonomics can be taken into account in detail in the planning of a TTS.

2 Brief literature review

In [6] and [7] the modeling of combined routing and scheduling problems is used to reduce e.g. the number of TT; the specifics of the TT and the ergonomics are not regarded.

 DOI: 10.1051/matecconf/20179511005

Table 1. Exemplary representation of technical implementations of tugger train concepts (c.f. [4]).

Concept name	Schematic representation (specific technical implementation)	Element used for production supply of the load carrier
Trailer concept		 Trailer itself
Taxi concept		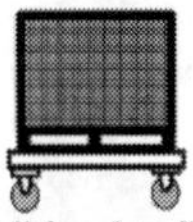 Additional trolley
Roller conveyor concept		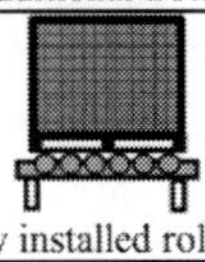 Permanently installed roller conveyor

A model for the calculation of an ideal cycle time, presented in [8], takes temporal and ergonomic restrictions into account, without regarding the technical implementation of the TT.

Regarding ergonomics, [9], [10], and [11] research the maximum acceptable weights and forces for pushing and pulling activities, focusing on the physical stress of workers; different technical designs of the handled trolleys were not analyzed in these works.

Orientating tests in companies, as described in [12], can be used for the evaluation of a specific case, but, due to various differing parameters, they can hardly be compared among themselves.

In summary, with works on TT focusing on partial aspects of the planning and works on ergonomics not factoring in alternative technical specifications, no method exists that allows for the consideration of ergonomics in detail in the system design of a TTS.

3 Ergonomic evaluation of basic tugger train concepts

In order to allow for an in-depth ergonomic evaluation of different TTS and, in this way, support the specific system design, a comprehensive test series was conducted. In this empirical study, a multi-component piezoelectric hand force measuring system was used. This allows for the measurement of the three orthogonal components of the acting force, which has to be applied by a worker in a tugger train system. Details on the methodology and experimental procedures of the study, conducted with 42 test subjects, have been published in [13]. The most significant ergonomic assessment factor, determined in the study, is the initial force – the force a worker has to apply for the pushing and pulling tasks in a TTS. The initial force is calculated based on a definition of the IFA by the 95th percentile of the data for the total force [12]. Based on the results of the study and data of another test series, performed in cooperation with the IFA and a German manufacturer of sports cars [14], the initial force is shown in Fig. 1 dependent on the weight of the load carriers for the basic TT concepts. The displayed graphs are based on mean values over various technical implementations of the concepts used in the empirical study.

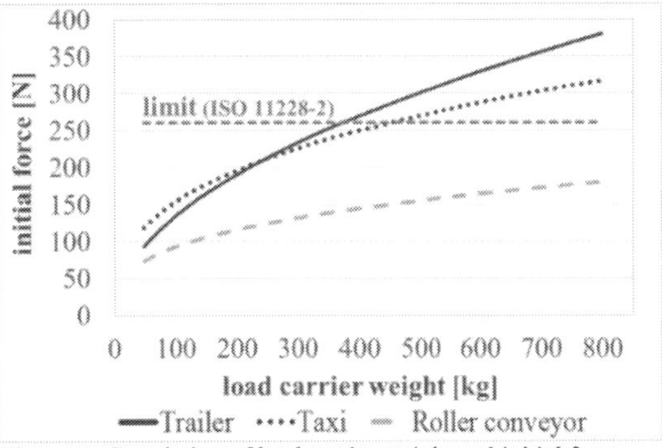

Figure 1. Correlation of load carrier weight and initial force.

The given limit of the initial force, recommended by the ISO 11228-2, takes a frequency of one pushing/pulling task every five minutes into account [15]. The stated weight only factors in the load carrier itself; additionally, the analysis takes the weight of the wagon or trolley into account.

While the roller conveyor concept can be used for weights of 800 kg and more, the recommended limit of the initial force is reached at 460 kg for the taxi concept, and 375 kg for the trailer concept.

After demonstrating the differences of the basic TT concepts regarding ergonomics, in the next section it is investigated whether the pointed out ergonomic advantages are economically feasible.

4 Economic feasibility of ergonomic findings concerning basic tugger train concepts

The economic evaluation of the basic TT concepts is done using a software-based model that, like the conducted empirical study, was developed in the research project IntegRoute [16]. Said model is an analytical model that allows for a complete dimensioning of a TTS. Therefore, the model, for example, contains a detailed calculation of the space requirement, cycle time, and personnel deployment of a TTS. On that basis, the total project costs are derived, performing a dynamic investment calculation that, for instance, takes the discounting of future cash flows and a rise in personnel costs into account.

Using the software-based model requires various input parameters. Therefore, based on the study conducted in the German automotive industry analyzing 26 TTS, an automotive production plant with characteristics close to the mean values of the examined systems was selected. As a result, the following analysis can be considered a close estimation for many production plants. In a plant with completely different characteristics, the findings should be verified using the exact input parameters.

In the selected TTS, 300 different items in load carriers are supplied to the production line, creating a throughput of 1500 load carriers per day. The weight of the load carriers is around 500 kg. The average length of a tour is 400 meters, and for every item two load carriers are present at the production line all the time. Based on said characteristics, Figure 2 shows the correlation of the project duration and the total project costs.

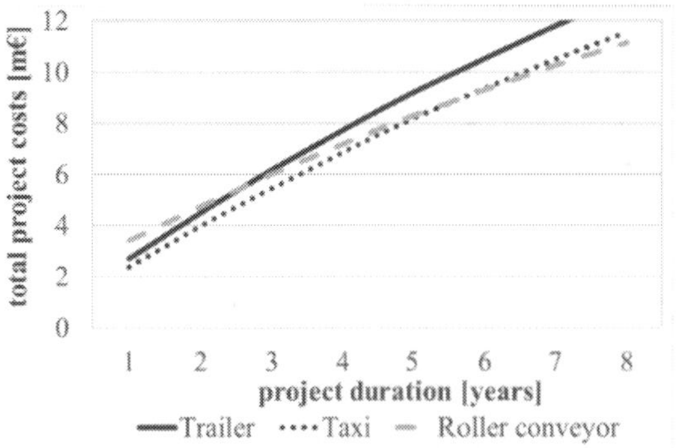

Figure 2. Correlation of project duration and total project costs.

While the highest investment for the additionally needed roller conveyors could have been expected, it is worth emphasizing that the trailer concept – which wagons are technically less sophisticated and because of that reason less expensive than the wagons in the taxi concept – leads to a higher investment than the taxi concept. While for the trailer concept a separate trailer is required to be placed at the production line for every load carrier, in the taxi concept, technically simpler trolleys are used for the placement of the load carriers. Therefore, the high amount of trailers, equipped with a clutch and costly casters made for being towed, increases the investment significantly.

Comparing the taxi concept and the roller conveyor concept in the analyzed planning scenario, after about 5.6 years the project costs equal one another. The cause is the shorter time for the exchange of load carriers in the roller conveyor concept, which leads to lower personnel costs per year.

In the automotive industry, with logistics projects being attributed to specific vehicle projects with model cycles of 6 to 8 years, the high investment for a roller conveyor concept can very well be compensated by the lower personnel costs per year.

The ergonomic evaluation of the roller conveyor concept has shown uncritical values even for load carriers with weights of 800 kg and above; therefore, there is no need for further consideration. For the taxi concept it is worth taking a closer look at different technical implementations of the concept. For this reason, the next section focuses on the taxi concept analyzing the effects of technical changes to the trolley, regarding the ergonomics and economics of the TTS.

5 Comparison of different trolleys used in a taxi concept

In this section, utilizing the data of the empirical study and the developed software-based model, different trolleys used in a taxi concept are compared for a project duration of six years. The regarded trolleys differ in configuration and material of castors used, as these factors exert the highest influence on the regarded TTS. In Table 2, the effects on ergonomic and economic key performance indicators are displayed for different trolleys. While most of the indicators are commonly used, the vibration factor β has to be explained; β was defined by the authors in [13], and is a dimensionless figure that indicates a disturbance for the worker with values above approximately 30. Based on (1), β is calculated using the values of the total force $y(n)$, which are measured in a frequency of 50 Hz.

$$\beta = \frac{MV\left|y(n) - \frac{1}{3}\sum_{i=n-1}^{n+1} y(i)\right|}{MV(y(n))} * 4000 \qquad (1)$$

while the initially regarded trolley A exceeds the recommended limit of 260 N, for the trolley B, the initial force is reduced to 253 N, due to a change in the configuration of castors. Furthermore, the existence of two fixed castors leads to a reduction of the cycle time by 0.4 minutes, which adds up to a significant decrease of investments and, as a consequence, of the total project costs.

Comparing trolley B with trolley C, a further decrease of the initial force and the cycle time is possible by changing the material of the castor from polyurethane with a Shore hardness of 96 to polyamide.

Beside the reduction of the initial force, the change of material leads to an increase of vibrations, causing a significant disturbance for the worker while pushing or pulling. Despite these contrasting ergonomic indicators, in the empirical study the castors made of polyurethane were preferred by the test subjects and only 13.5 percent favored the castors made of polyamide.

Table 2. Influence of technical specifics of trolleys on ergonomics and economics.

Factor		Trolley A	Trolley B	Trolley C
Specifics of the trolley	Configuration of castors	4 swivel castors	2 fixed castors 2 swivel castors	2 fixed castors 2 swivel castors
	Material of castors	Polyurethane (96 Shore)	Polyurethane (96 Shore)	Polyamide
Key performance indicators	Initial force [N]	272	253	231
	Vibration factor β [-]	18	19	44
	Cycle time [min]	13.6	13.2	13.1
	Investments [m€]	0.414	0.397	0.369
	Total project costs [m€]	9.26	9.10	9.06

6 Conclusion

This paper analyzes three basic concepts for tugger trains used in the German automotive industry regarding ergonomics, and it shows that – when using the roller conveyor concept – load carriers with a weight of 800 kg and more can be manipulated trouble-free; in other concepts, the weight per load carrier should not exceed 500 kg. Furthermore, it is shown that concepts with advantages regarding ergonomics do not automatically lead to higher total project costs. Therefore, the improvement of ergonomics and economics can go hand in hand.

Furthermore, the paper demonstrates the positive influence of a trolley with two fixed castors on the ergonomics and, through reduced handling time, also on the economics of a tugger train system. Regarding the material of the castors, the positive and negative characteristics of two examined materials are pointed out.

The findings of this paper show that it is possible to consider ergonomics in an early stage of the system design. A future integration of these findings in a planning method is an important step closer to a holistic planning of tugger train systems.

Acknowledgement

This paper received funding by the AiF (no. 18136 N), within the program for sponsorship by Industrial Joint Research (IGF) of the German Federal Ministry of Economic Affairs and Energy based on an enactment of the German Parliament.

References

1. J.P. Womack, D.T. Jones, *Lean Thinking*, Simon & Schuster (1996)
2. W.A. Günthner, S. Galka, E. Klenk, T. Knössl, M. Dewitz, *Results of a study*, Institute for Materials Handling, Material Flow, Logistics. Technical University of Munich (2012)
3. C. Baerwolff, *VDI-Report no. 2136* (2011)
4. The Association of German Engineers: *VDI Standard 5586* (2016)
5. C. Loch, F. Sting, N. Bauer, H. Mauermann, Harvard Business Review, **88**, 3, p. 99–102 (2010)
6. S. Emde, N. Boysen, European Journal of Operational Research **217**, p. 287–299 (2011)
7. B.S. Vaidyanathan, J.O Matson, D.M Miller, J.E. Matson, IIE Trans. **31**, 11, p. 1083–1092 (1999)
8. M. Droste, J. Deuse, Proceedings of the 4th CARV (2011)
9. S. Snook, V.M. Ciriello, Ergonomics, **34**, 9, p. 1197–1213 (1991)
10. M.J. Hoozemans, P.F. Kuijer, I. Kingma, J.H. van Dieen et al., Ergonomics **47**, 1, p. 1–18 (2004)
11. V.M. Ciriello, P.G. Dempsey, R.V. Maikala, N.V. O´Brien, Ergonomics **51**, 5, p. 593–601 (2008)
12. U. Glitsch, M. Post, I. Hermanns, DGUV Report 2/2014, p. 77-81. (2014)
13. C. Keuntje, J. Heiß, P. Gangkofner, W.A. Günthner, Proceedings of the 11. Hamburg conference on forklifts, p. 139–156 (2016)
14. U. Glitsch and R. Ellegast: *Explorative Messungen der Handkräfte beim Ziehen und Schieben von Schleppwagen*. Usage approval granted by Dr.-Ing. h.c. F. Porsche AG (2013)
15. International Organization for Standardization, *Ergonomics – Manual handling – Part 2: Pushing and pulling*. ISO 11228-2 (2007)
16. W.A. Günthner, C. Keuntje, *Research report – IntegRoute*. Institute for Materials Handling, Material Flow, Logistics, Technical University of Munich (2016).

DOI: 10.1051/matecconf/20179511006

Study on Braking Sensation Based on Urban Working Conditions

Lanie Abi[1], Tianji Feng[2], Mingyue Zhou[2], Jiajun Chen[2]
[1] Jilin University, China
[2]China Automotive Technology & Research Center, China

Abstract. In this paper we researched the vehicle braking sense in three aspects of human, vehicle and environment and analysed their impacts on brake feeling. Through the real vehicle test we analysed the relationship among pedal force, pedal travel and deceleration. We used dynamometer test method to study the brake noise question. We designed a fixture which could imitate the suspension and made the test more close to the true level. Moreover we discussed how to establish the evaluation system of vehicle braking condition. Through real vehicle test of braking, we can test and record the brake system parameters in the braking process under urban working conditions. We recorded the brake frequency, the change of brake speed and brake disc temperature. Meanwhile, based on the analysis of braking condition, we put forward the index of brake load to reflect the city's traffic conditions. Experiment show that the braking condition and brake feel are related, braking condition also provides theoretical support for the design of brake system.

1 The significance of braking sensation

With the rapid development of China's economy, the domestic auto industry is developing rapidly with the development of economy. China's car ownership was second only to the United States, ranked second in the world. Along with the development of the automobile industry, the demand of consumer is more and more high, the performance of the braking system, has been put into a more and more important position. Now the type of brake system on the market is multifarious, the different model parameters of the brake parts match assembled together to become a set of braking system, then how to judge the merits of a set of brake system. As all of us know, the automobile manufacturing is to provide a means of transport to the consumer, obviously, under the premise of satisfy the security, breaking feeling of consumers is an important index to judge the braking system of the pros and cons. According to Andrew J et al study, we should study some factors, including the pedal force, the pedal stroke, the deceleration[1,2],when we analysis braking condition, we need to focus on the characteristics of brake pedal feeling characteristics[3,4],as Jan Sendler et al study. We know that passenger car braking systems are generally hydraulic system, the pressure is very important to focus on pipeline[5]as Joel Sanford et al study. Thebrake friction coefficient[6,7,8,9] is very importantas W. Y. Loh.R. H et al study. Through this study[10,11], the automobile factory in the braking system design process, can fully take into account the braking feeling and urban conditions, clear product positioning. In the preliminary design process, the study can help to achieve the design of high efficiency, low cost and high quality products to make customer satisfy, also can let the designers know what parameters in the design process need to focus on. Study on brake noise, K Brent et al[12,13,14],shows that many factor will influence the brake noise. There are many factors we should consider[15,16,17,18,19,20] when we study noise. Urban conditions is very important for energy recovery, as Maizza V et al studies[21,22,23].When we research the energy recovery, we should also take care of the parameters of vehicle[24,25,26,27,28].

2 Test method

We recorded the data of pedal feeling, braking deceleration, wheel cylinder pressure, friction temperature changes of braking system, vacuum booster vacuum degree, vacuum booster start coordinates, brake noise, urban traffic conditions related parameters (initial braking speed, final braking speed and braking times) and so on. We will then analysis the relationship among these data. We will fully research on braking feeling influence related parameters.

2.1 Brake feeling evaluation method

Brake feel is the driver's subjective feeling in the braking process of vehicle mainly reflect through the pedal feeling and deceleration feeling, which pedal feeling refers to the braking system on the driver's foot reaction force and feeling of travel, represented by the pedal force and the pedal displacement. Feeling of moderation refers to the vehicle deceleration of the driver caused by the body inertia, visual and auditory feedback, braking deceleration characterization. In this paper we use the relationship between pedal feeling and deceleration feeling to study the characterization of vehicle braking. Their relationship is shown in Fig. 1 and Fig. 2.

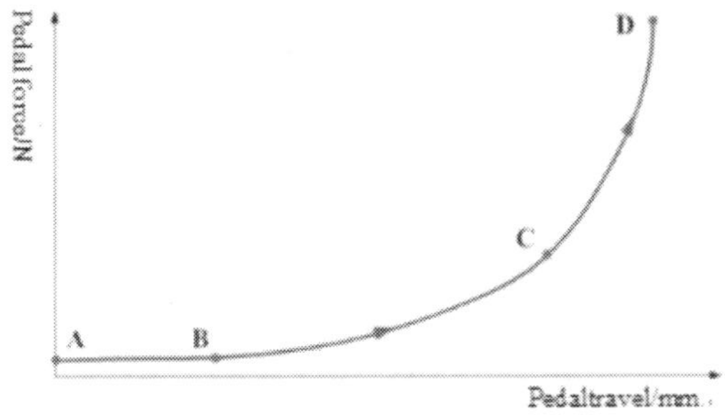

Figure1. Relationship between force and travel.

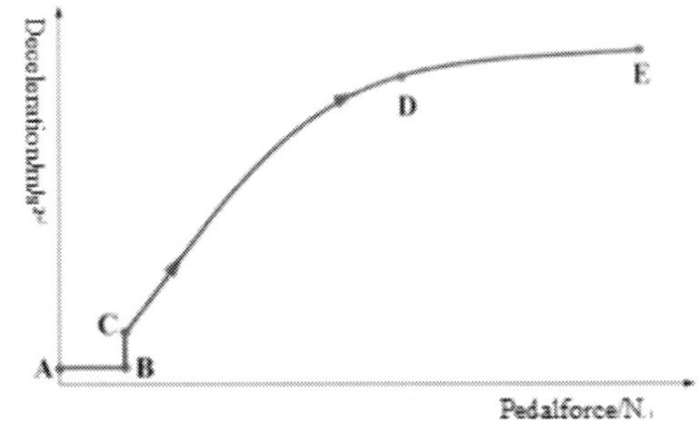

Figure2. Relationship between force and deceleration.

2.2 Brake noise test method

In the process of working of the braking system, pedal and vacuum booster have a phenomenon of air travel. This will be reflected in the test results curve, the performance is the occurrence of curve step phenomenon, the slope of the curve and so on. Here, we will study the vacuum booster step point coordinates, the reasonable range of pedal free travel.

At the same time, we will study the temperature change of the friction plate, and the working noise of the brake system. The theoretical calculation of the noise is mainly through the brake test, the friction coefficient of the test process is mainly calculated by the following formula:

$$\mu = \frac{10^6 \cdot T}{2 \cdot (P - P_T) \cdot A_P \cdot r_{eff} \cdot \eta} \qquad (1)$$

- T: output torque, the unit is N/M
- P: brake pressure, the unit is kPa
- P_T : braking actuating pressure, the unit is kPa
- A_p:piston area
- r_{eff}:effective radius
- η:efficiency

At present, there are two kinds of measuring methods of braking noise, which are vehicle measurement and bench measurement. Vehicle external environment have more noise interference on vehicle. Due to the effect of vehicle road test is affected by the weather, road conditions and driver's operation, repetition of some problem of braking conditions, especially difficult recurrence of the problems of brake noise. At this time, from the time and cost considerations, the bench test has a certain advantage. Bench test of the speed, brake pressure, temperature, humidity and other test conditions are more convenient and accurate control of the purpose. Compared with the road test, bench test has a unique advantage.

In the test of brake noise, we will compare the noise test results of the traditional fixture and the suspension clamp, so that our noise test results are more close to the real situation. Two types of fixture such as Fig. 3.

Figure3. Different types of fixtures.

2.3 Evaluation of urban conditions

We make statistics on all the relevant data of different cities in different road conditions, in order to fully reflect the working conditions of the braking system in different cities, and analysis the differences of working parameters. The processed data can provide the theoretical support for the service and the design process of the automobile factory. Particularly, in the current era, with the popularization of new energy vehicles, energy recovery become a very important research direction, in the research of urban traffic conditions we can not only on vehicle braking feeling research provide theoretical basis, and at the same time, it can provide a powerful data support for new energy vehicles energy recovery problem. The following is the calculation formula of the vehicle energy recovery rate:

$$\eta = \frac{\sum_{i=1}^{n} \int_{t_{i-1}}^{t_{i-2}} U_{i-bat} I_{i-chg} d_t}{\sum_{i=1}^{n} \left(\frac{1}{2} m v_{i-0}^2 - mgf \int v d_v - C_D A_p / 2 \int v^3 d_t \right)} \qquad (2)$$

- U_{i-bat}:Battery voltage
- I_{i-chg}:Charging current
- v_{i-0}:Velocity difference
- V: Speed of braking process

From the above formula we can see that the molecular is motor energy recovery, the denominator is Recyclable energy theory. The theory of Recyclable energy is kinetic energy loss minus the rolling resistance loss of energy, and then minus the wind age loss of energy. In the process of energy recovery, rolling resistance and wind resistance energy loss in the vehicle design already decided, this paper mainly through urban conditions of vehicle speed difference to study the kinetic energy loss and energy recovery rate. On the basis of this, we put forward the index of brake load to reflect the city's traffic conditions:

$$K = \frac{\sum_{i=1}^{n}(V_{bi}^2 - V_{ei}^2)}{2s} \tag{3}$$

where,

- V_{bi}=The initial braking speed of i (m/s)
- V_{ei}= The final braking speed of i (m/s)
- S = Travel mileage (km)

Because in each of the braking process, the loss of kinetic energy of the vehicle are not the same, in a period of time or mileage, the total amount of energy loss in one braking is not the same. To this end, we can know the braking frequency in a certain range, the initial speed of the braking process and the final speed, as well as the quality of the vehicle. The total kinetic energy of the braking loss is calculated, and the energy loss of unit mass in the unit mileage is obtained by dividing the quality and mileage of the vehicle, which is the braking load index K.

3 Data acquisition

3.1Data acquisition of pedal sensory

Pedal feel objective test is through equipment (GPS vehicle speed sensor, pedal force sensor, pull rope displacement sensor and portable data acquisition instrument) to collect brake reducing speed, braking time, pedal, pedal stroke, pipeline pressure and velocity parameters, then the collected data were processed quasi synthetic curves. We can know the change process of the pedal force, pedal travel and pipeline pressure value of the test vehicle at different characteristics by curve. The test conditions of each test vehicle are cold state, and each test is carried out by multiple groups of braking, select the best of the three sets of data, the pedal force and pedal travel to the implementation of the average value calculation. Using MATLAB to collect and process the data collection, and then compare the drawing.

3.2 Data acquisition of vehicle parameters

We will record the vacuum booster coordinates (from stepping on the pedal to the vacuum booster began to boost the role of the process, we use the coordinates which are labelled with vacuum degree and the corresponding pedal travel).In the brake noise testing process, we will use the traditional fixture test, and by using suspension clamp test method to carry on the optimization, to make the braking system noise test more close to the real work

3.3 Data acquisition of urban traffic conditions

We use LINK3802 vehicle data acquisition system to collect urban traffic working conditions. Based on the test vehicle loading 3867-DG-X4 GPS velocity sensor, 3867-DV-X1 vacuum sensor, 3867-DT-X2 brake pedal travel sensor, 3867-DF-X2 brake pedal force sensor, 3867-HT-X1 temperature and humidity sensors, friction temperature sensor, acceleration sensor and other equipment to collect the data of target cities. The experiment was divided into four conditions: urban road condition, urban road The objective test results of 7 vehicles are shown in Table 2 as follows.

Congestion condition, suburban road conditions and the conditions of highway. In the test process, each braking of vehicle will trigger data acquisition system of the preset conditions, the data acquisition system will start recording data, data acquisition respectively: reducing speed, four wheel friction temperature, number of brake, braking initial velocity and terminal velocity, environmental temperature and humidity, vacuum negative pressure pump, pedal force and stroke, and record the friction plate abrasion.

4 Data procession and test result

4.1 Comparison of subjective evaluation and objective test data

Results of the subjective evaluation of the 7 test vehicles are summarized as shown in table 1.From table 1, we can see that the subjective evaluation score of A vehicle and E vehicle is higher, and the evaluation results of E in the 7 car is the best; B and H subjective evaluation score is low, in which the H subjective evaluation results are the worst. Need to add here: the test of the 7 test vehicles are the same type of passenger cars and the price in the same competitive range

Table1. Summary of subjective evaluation results

Evaluation project		Evaluation results (10 point system)						
		A	B	C	D	E	F	G
Initial pedal setting	Pedal force	7.5	6	7	7.5	8	7	6.5
	Pedal stroke	7.5	7	6.5	7	7	6.5	6-
Vacuum booster intervention	Initial intervention point	7.5	6	7	7	7	6.5	6.5
	Ride comfort	7.5	6	6.5	7	8	6.5	6.5
Pedal stroke	Total stroke	8	7	7.5	7	8	7.5	6.5
	Stroke follow	8	7	7.5	8	8	7.5	6.5
Pedal force	Pedal force following	7	6	6.5	7	8	6.5	6.5
	Maximum pedal force	7.5	7-	7	7-	8	7.5	6.5
	Pedal force stability	7.5	6	6.5	7	7.5	6.5	6.5
Overall braking efficiency		7.5	7.5	6.5	7.5	7.5	7.5	7

The objective test results of 7 vehicles are shown in Table 2 as follows.

Table2. Objective test results

evaluating indicator		A	B	C	D	E	F	G
deceleration 1m/s2	Pedal stroke (mm)	30.7	24.9	35.4	46.2	34.7	35.9	51.8

	Pedal force (N)	30.5	44.5	40.4	33.4	34.2	42.1	39.8
deceleration 2m/s2	**Pedal stroke (mm)**	35.8	31.1	42.1	50.6	39.1	40.7	63.2
	Pedal force (N)	36.5	57.8	46.3	41.1	41.2	48.6	45.8
deceleration 3m/s2	**Pedal stroke (mm)**	41.8	38.5	48.2	57.4	45.4	45.8	75.1
	Pedal force (N)	48.7	74.9	56.7	58.3	56.9	59.7	59.5
deceleration 4m/s2	**Pedal stroke (mm)**	48.0	45.0	54.1	63.4	50.5	49.4	83.1
	Pedal force (N)	65.8	88.0	67.5	77.1	74.1	71.3	80.5
deceleration 5m/s2	**Pedal stroke (mm)**	53.0	54.4	57.9	69.0	54.7	52.7	89.9
	Pedal force (N)	81.0	109.	76.2	98.5	93.6	83.4	98.7
deceleration 6m/s2	**Pedal stroke (mm)**	58.1	61.8	62.3	74.4	58.4	56.6	96.8
	Pedal force (N)	98.5	125.	89.3	122	112	103	126

7 car comparison summary analysis diagram as shown Fig. 4 and Fig. 5.

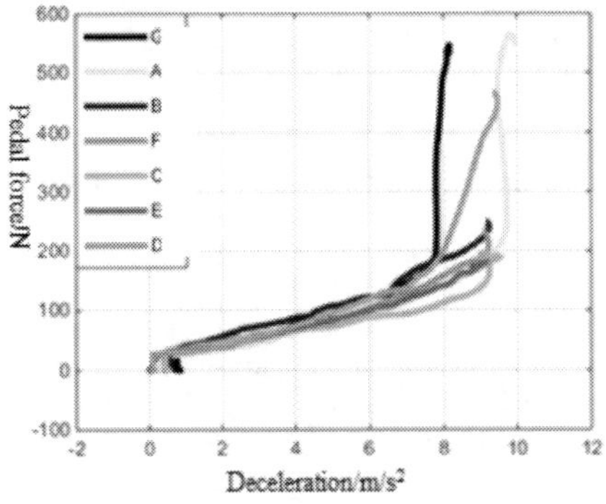

Figure 4. Relationshipamong force and ...vel

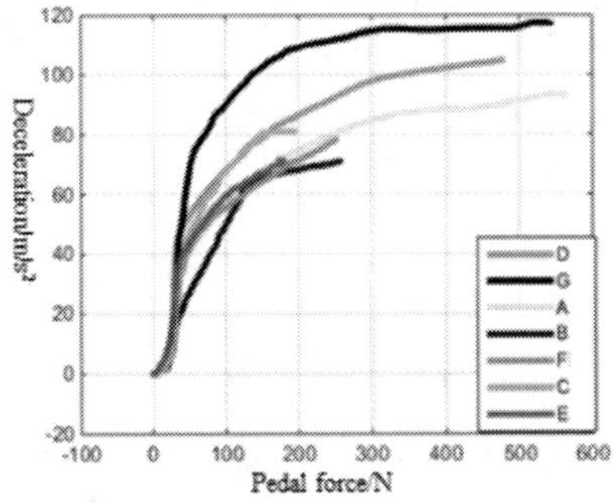

Figure 5. Relationshipamong force and deceleration

4.2 Comparison of vehicle parameter data

The key points of the objective test data are shown in Table 3.

Table 3. Objective test key point data

Key objective test data			
Objective	**Vacuum booster intervention point coordinates (a, S,F)**	**0.3g corresponding to S and F**	**0.6g corresponding to S and F**
A	(0.41, 20.4, 19.2)	(43.0, 49.6)	(58.5,96.7)
B	(0.56, 13.9,30.4)	(38.6,74.1)	(62.4, 123)
C	(0.55, 23.2, 27.3)	(49.7,56.2)	(63.2,88.3)
D	(0.09, 20.4, 26.0)	(55.3,55.9)	(72.5,120)
E	(0.43, 22.0,23.4)	(45.1,58.1)	(57.8,114)
F	(0.48, 20.4, 30.0)	(46.2, 60.1)	(56.5,105)
G	(1.21, 13.9, 16.2)	(71.2, 56.9)	(96.5,120)

According to the expert's many years of design experience and reference materials, draw as shown in Table 4, a design value range (range of values due to the vehicle type)

Table4. Range of design values

Design value range					
g-F			**g-S**		
Intervention point	**0.3g travel (mm)**	**0.6g travel (mm)**	**Intervention point**	**Pedal force of 0.3g(N)**	**Pedal force of 0.6g(N)**
(0.05-0.45, 15-25)	25-70	40-85	(0.05,-0.4,18-30)	35-90	70-150

To do the test of automobile brake system, according to the test fixture, divided into car angle type, suspension type and bridge type test. The composition of the three types of fixtures can refer to Table 5, after the two types of clamps can be preloaded to the braking angle, to ensure that the vibration of the suspension components involved, closer to the real car, but the cost is higher.

Table 5. Different fixture components

Component requirements	**Fixture level**		
	Steering knuckle	**Suspension form**	**Full shaft type**
Brake disc	Y	Y	Y
friction material	Y	Y	Y
Caliper or wheel cylinder	Y	Y	Y
Bearing	Y	Y	Y
wheel hub	Y	Y	Y
Drive shaft	Y	Y	Y
Control arm	N	Y	Y
Control arm bushing	N	Y	Y
Stabilizer bar	N	Y	Y
Shock absorber	N	Y	Y

 DOI: 10.1051/matecconf/20179511006

assembly			
Shock absorber bushing	N	Y	Y
Auxiliary frame	N	N	Y
Auxiliary frame bushing	N	N	Y

After the test, the sound pressure level and the noise frequency of the noise scream are counted, and the results are shown in Fig. 6.

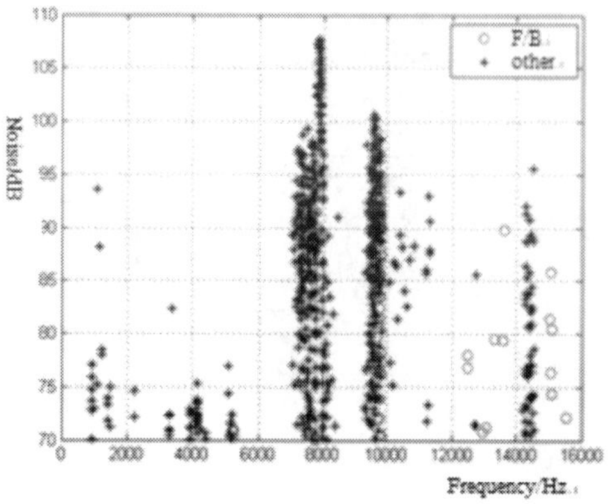

Figure 6. Relationship among frequency and noise.

Can be seen from the figure, when we use the car angle type clamp, the brake noise is mainly concentrated in the 7800Hz, 9800Hz and 14300Hz. And the same parts in the use of axle clamp, noise mainly concentrated in 2200Hz, 3300Hz and 9800Hz. This indicates that the braking noise is related to the suspension components, and the different types of fixture and suspension can cause the excitation of different frequency noises. 2200Hz and 3300Hz noise is low whistle, arises due to the mode coupling between parts, fixture impact on parts of the coupling state.

The braking test of different braking pressure on the brake system is carried out on the inertia test bench, and the test results are shown in Fig. 7.

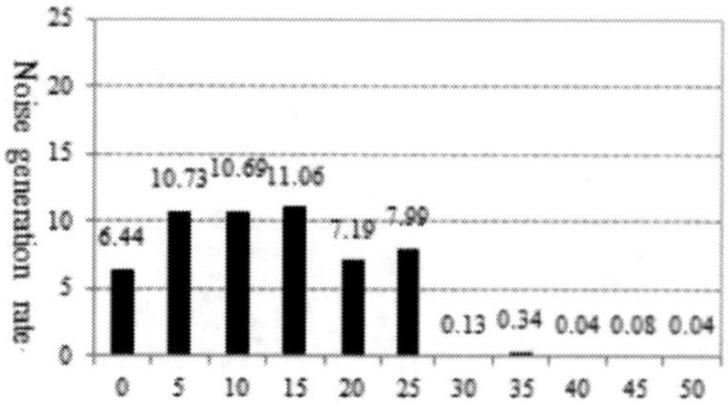

Figure7. Relationshipamongpressure and noise.

From the graph we can see that under different braking pressure, noise situation is different, the brake pressure in the range of 0-25bar is prone areas of brake noise, indicating that at low brake pressure brake noise happened tendentiousness larger. When the pressure is higher, the tendency of the brake noise is reduced, which is related to the contact stability between the friction surface and the brake disc. In the formulation of brake noise test specifications, more chapters should be set to low pressure brake.

The frequency and size of noise are different when the car is braking at high speed and low speed. This shows that the initial braking speed is also affected by the noise of the brake. In view of this situation, under the same condition to carry on the brake test to different vehicle speed, its result is shown in Fig. 8.

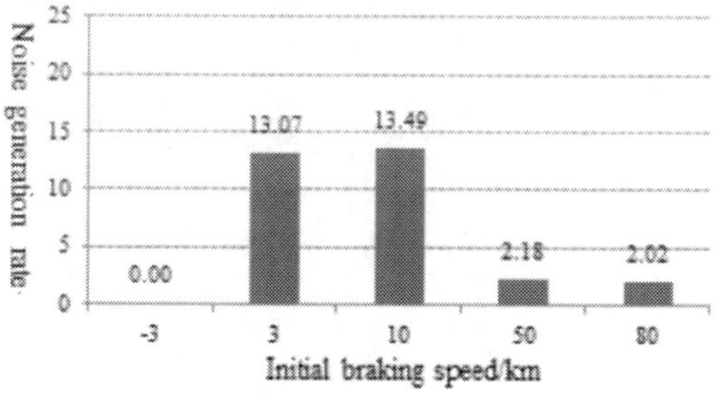

Figure 8. Relationshipamonginitial speed and noise.

When the vehicle speed is 10kph and 3kph, we have the car brake, brake noise frequency is relatively high, Vehicle braking at 50kph to 80kph, the car brake noise rate is significantly lower. From the user's point of view, brake noise is generally in low speed conditions such as traffic congestion. And working at high speed, the user brake less frequently the speed difference is not big. In the preparation of noise test specification, low speed conditions should be considered.

4.3 Comparison of urban traffic conditions

First of all, we'll process and analysis the data, then calculate these data we obtained, we will get the braking load index, according to different braking load index, we can know the different braking condition in different cities, the braking load index can reflect the loss of braking system of the city. The brake load index of different cities and the experimental data we processed are shown in Table 6.

Table 6. Data processing results of different cities

	Travel mileage	Braking times	Deceleration mean(m/s2)	Pedal force average(N)	Maximum pedal stroke(mm)	Brake load index
Changchun	602	1147	1.55	17.49	60.96	73.8
Harbin	587	1059	1.51	20.56	67.42	53.5
Dalian	598	1377	1.26	16.76	61.7	74.2
Shenyang	706	1144	1.43	17.02	63.43	66.2

Here we compare the speed variation of the braking process by Fig. 9 and Fig. 10.

 DOI: 10.1051/matecconf/20179511006

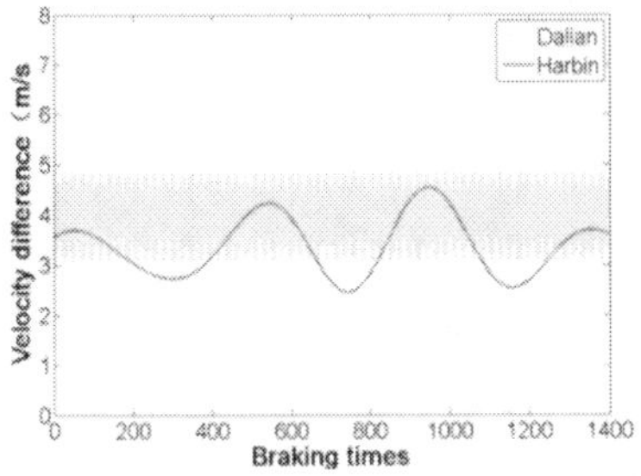

Figure 9. Velocity profile.

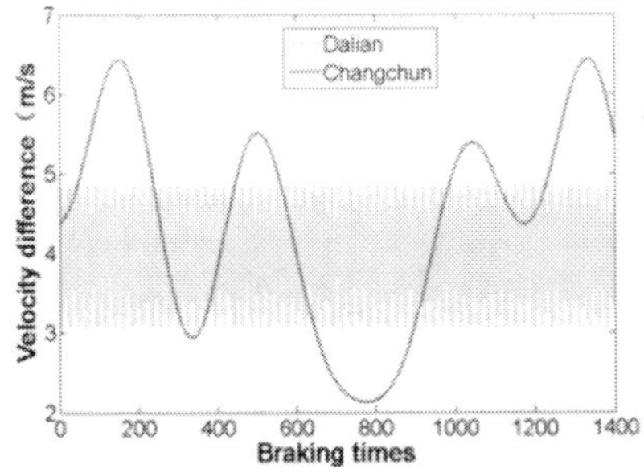

Figure 10. Velocity profile.

From Table 6 we already know that Dalian's brake load index is higher than that of Harbin, from Figure 9 we also see that the speed change of the braking process in Harbin is lower than that in Dalian. The speed change of the braking process is proportional to the energy change of the braking process, so the working strength of Dalian's braking system is higher than that of Harbin's braking system.

From Table 6 we know that the brake load index of Dalian and Changchun are very similar. Although the speed change during the braking process in Changchun is sometimes higher than in Dalian. From figure 10, by integrating the speed variation curves of the friction plate, we find that the results are similar, which indicates that the braking system of the two cities is relatively close to the working strength. But as we can see from Figure 7, the speed change in Changchun is significantly higher than that in Dalian. This shows that the driver's driving habits of Changchun significantly more intense than in Dalian. Changchun driver is more sensitive to the brake system.

The temperature change of the friction plate is shown in Fig.11 and Fig. 12.

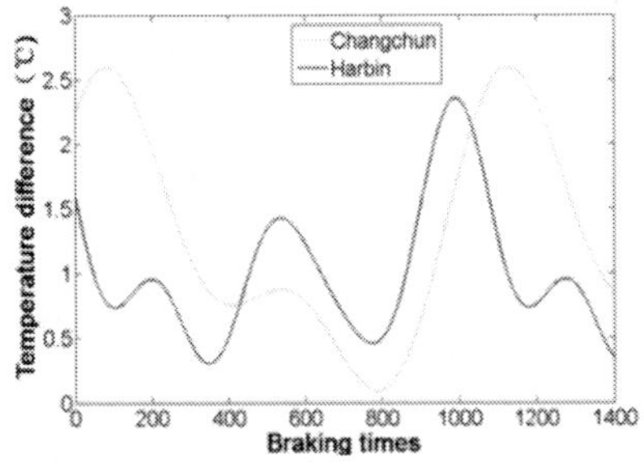

Figure11.temperature change

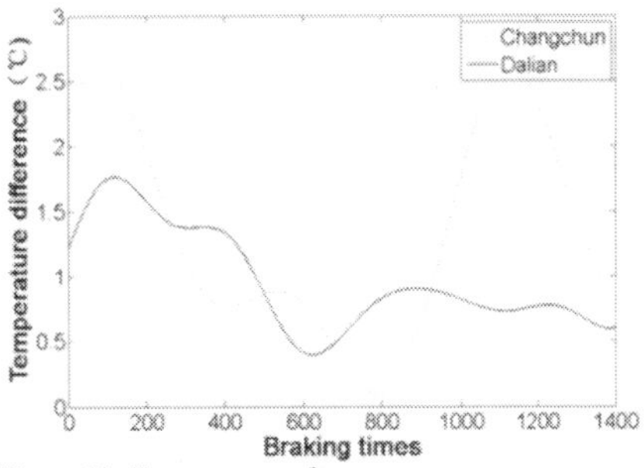

Figure12. Temperature change.

Fig. 11 is a comparison of the braking system temperature change between Changchun which has a higher braking load index and Harbin. From the figure, we can see that the temperature change of Changchun's braking system is more obvious than that of Harbin. At the same time, the temperature change of the friction plate in Harbin is more regular than that of the friction plate in Changchun. This shows that the use of the friction plate in Changchun is more intense than in Harbin, the loss of friction is greater than Harbin too. It can also be said that Changchun's driving habits are more intense than in Harbin.

Fig. 12 shows the comparison of the temperature change of the friction sheet between Changchun and Dalian, the braking load index of the two cities is close. It is not difficult to see from the diagram, the friction plate temperature of Changchun's braking system is more severe than that of Dalian (peak difference). But by integrating the temperature variation curves of the friction plate, we find that the results are similar, it shows that the energy consumption of the two cities in the braking condition is similar, brake system working strength is close. It is consistent with the results of the evaluation of the braking load index of two cities. But the difference of the peak difference shows the difference of driver's driving habits in two cities. Changchun's drivers prefer a more intense driving experience.

5 Conclusion

In the process of braking, we should collect brake feeling data and vehicle configuration parameters (pedal lever ratio, vacuum booster ratio, pipe diameter, friction, the friction coefficient) which have direct impact on vehicle braking feeling. The urban traffic condition should be considered, we will collect the traffic condition data (Including braking frequency, braking speed, friction temperature changes). Based on these data, we take brake noise tests of disc type brakes to acquire different noise under different working condition. Using different types of fixtures, brake noise performance were compared in different applying conditions. Noise occurrence was also studied in time domain. The influences of brake pressure, initial velocity and temperature on brake noise should be considered.. At the same time, the city's traffic conditions will certainly have an impact on the vehicle braking sensation. In particular, the braking condition of the vehicle also has a very important influence on the energy recovery of the new energy vehicle. Especially in the city, the speed change of the braking process directly reflects the change of the kinetic energy during the braking of the vehicle. It can be very intuitive to reflect the recycling of energy, to provide very important references for the design of new energy vehicles.

References

1. Andrew J. Day, Hon Ping Ho, Khalid Hussain. Brake System Simulation to Predict Brake Pedal Feel in a Passenger Car. SAE International, 2009, SAE2009-01-3043
2. Andrew J.Day.Break System Simulation to Predict Break Pedal Feel in Passenger Car[C].SAE 2009-01.
3. Jan Sendler, Ralf Trutschel, Klaus Augsburg. Methods of Evaluating and Developing Pedal and Brake Characteristics. ATZ Autotechnology, 2009, 111: 60-66.
4. Jan Sendler, Ralf Trutschel, Klaus Augsburg. Methods of Evaluating and Developing Pedal and Brake Characteristics. ATZ Autotechnology, 2009, 111: 60-66.
5. David Antanaitis, Mark Riefe, Joel Sanford. Automotive Brake Hose Fluid Consumption Characteristics and Its Effects on Brake System Pedal Feel. SAE Int. J. Passeng. Cars, 2010, 3(1): 113-130.
6. W. Y. Loh.R. H. Basch. Dynamic Modeling of Brake Friction Coefficients. SAE International, 2000, SAE2000-01-2753.
7. Satisfaction. Society of Automotive Engineers, 2003, SAE2003-01-3598: 1-6.
8. Rudilf Limpert. Brake Design and Safety. Warrendale: Society of Automotive Engineers, Inc, 1999.
9. Driver/Brake Pedal Under Real Road Conditions in View of Oncoming Brake-by-wire-systems. SAE International, 1999, SAE1999-01-2949: 1-9.
10. Rena Hecht Basch, Paul Sanders, Dale Hartsock. Correlation of Lining Properties with Brake Pedal Feel. SAE International, 2002, SAE2002-01-2602.
11. Andrea Fortina. Braking System Components Modelling. SAE International, 2003, SAE2003-01-3335.
12. K Brent Dunlap, Michael A Riehle, Richard E Longhouse. An Investigative Overview of Automotive Disc Brake Noise [J]. Journal of Passenger, 1999, V108-6.
13. H R Mills. Brake squeak, Technical Report 9000 B [M]. Institution of Automobile Engineers, 1938.
14. Fosberry R AC, Holubecki Z. Third report on squeal of drum brakes [R]. MIRA Report No.1957 /3.
15. R T Spurr. Atheoryofbrakesqueal, Proceedings of the Automobile Division [M]. Institution of Mechanical Engineers, 1961-1962(1) (1961) 33-52.
16. Bracken W J, Sakioka J K. Brake pad for reducing brake noise: US, US 4485898 A[P]. 1984.
17. Kusano M. Experimental study on the reduction of drum brake noise[C]. SAE851465, 1985.
18. Kamiya M, Kondo H, Sasaki S, et al. Vehicle brake system for preventing brake noise: US, US 20040212246 A1[P]. 2004.
19. Kamiya M, Kondo H, Sasaki S, et al. Vehicle brake system for reducing brake noise: US, US 20040183366 A1[P]. 2004.
20. Shin K, Brennan M J, J.-E. O H, et al. ANALYSIS OF DISC BRAKE NOISE USING A TWO-DEGREE-OF-FREEDOM MODEL[J]. Journal of Sound & Vibration, 2002, 254(5):837-848.
21. Aldo Sorniotti Virtual and Experimental Analysis of Brake Assist Systems. SAE International, 2006, SAE2006-01-0477.
22. Zheng S Y, Jin C. Research on Engine Cranking Process in Downhill Braking Condition for an Electric Drive Underground Dump Truck[J]. Applied Mechanics & Materials, 2013, 364:92-96.
23. Consonni S, Giugliano M, Massarutto A, et al. Material and energy recovery in integrated waste management systems: Project overview and main results[J]. Waste Management, 2011, 31(9–10):2057-2065.
24. Maizza V, Maizza A. Unconventional working fluids in organic Rankine-cycles for waste energy recovery systems[J]. Applied Thermal Engineering, 2001, 21(3):381-390.
25. Fernández I, Renedo C J, Pérez S F, et al. A review: Energy recovery in batch processes[J]. Renewable & Sustainable Energy Reviews, 2012, 16(4):2260-2277.
26. Fernández I, Renedo C J, Pérez S F, et al. A review: Energy recovery in batch processes[J]. Renewable & Sustainable Energy Reviews, 2012, 16(4):2260-2277.
27. Song X, Liu Z, Sun D D. Energy recovery from concentrated seawater brine by thin-film nanofiber composite pressure retarded osmosis membranes with high power density[J]. Energy & Environmental Science, 2013, 6(4):1199-1210.
28. Stragier M G. BRAKE ENERGY RECOVERY SYSTEM: US, US 20110162903 A1[P]. 2011.

Compatibility Conditions in Continua with Microstrusture

Evgenii V. Murashkin [1,2,3] and Nikita E. Stadnik [1,4]

[1] Institute for Problems in Mechanics of the Russian Academy of Sciences, Vernadsky Ave 101 Bldg 1, Moscow, 119526, Russia
[2] Bauman Moscow State Technical University, 2nd Baumanskaya Str 5/1, Moscow, 105005, Russia
[3] National Research Nuclear University T«MEPhITi, Kashirskoye shosse 31, Moscow, 115409, Russia
[4] Moscow Technological University, Vernadsky Ave 78, Moscow, 119454, Russia

Abstract. The present study is devoted to the statement of compatibility conditions on propagating wave sur- faces of strong and weak discontinuities in thermoelastic continua with microstructure. The field formalism is used to study the problem. A natural density of thermoelastic action and the corresponding variational least action principle are stated for a varying domain. A special form of the first variation of the action is employed to obtain 4-covariant boundary conditions on the wave surfaces. These are given by jumps of the Piola–Kirchhoff stress 4-tensor and the energy–momentum tensor. Problems of propagation of weak discontinuities in type-II MPTE continua are considered. Geometrical and kinematical compatibility conditions are used to study possi- ble wave surfaces of weak discontinuities. It is shown that the surfaces of weak discontinuities can propagate without weak discontinuities of the temperature displacement.

1 Introduction

Problems of micropolar continua take its origin from the classical E. and F. Cosserat's paper [1]. Micropolar (MP) continuum theories include not only translational displacements but also additional degrees of freedom. These degrees of freedom are coupled with changes in reper (three directors) associated with microvolume. Such changes may be described by a rotation vector when reper associated with microvolume are rigid rotated. In contrary to conventional elasticity a continuum with microstucture is described by the asymmetric strain and stress tensors known from many previous discussions [2, 3, 6]. Thus the asymmetric elastic theory is characterized by a comparatively large number of constitutive elastic constants need to be determined from the experimental observations. There are several phenomena (for example, additive manufacturing design [4, 5], biomechanics, nematic liquid crystals [7, 8] behavior, the anomalous piezoelectric effect in quartz, the dispersion of elastic waves, as well as a number of other experimentally observed elastic properties of the pure crystals) being beyond the scope of the conventional thermoelasticity (CTE) and piezoelectroelasticity, the finite propagation velocity of thermal waves known as second sound waves.

The phenomenon of the second sound is observed at very low temperatures in crystals of high purity. The observation of the second sound in the crystal NaF is discussed (see [9]). The second sound in the NaF is observed at a temperature of about $15K$ [10]. The existence of second sound waves was also confirmed in the NaF (see [11] and [12]); the speed of the second sound waves is $1953.1m/S$ see [13]. In the publication [14] the measurements of the second sound waves in bismuth (Bi) at a temperature of about $3K$ are described. The speed of the second sound waves in the bismuth is equal to $78 \pm 5m/S$.

The micropolar thermoelastic (MPTE) continuum may be described in terms of field formalism, for example, from positions of the Green–Naghdi thermoelasticity (GN-theory) [15, 16]. Now such mathematical frame- works of the thermoelastic behavior of solids are rapidly refined [17–20]. They are based on different modifications of the classical Fourier law of heat conduction. GN- thermoelasticity theory itself can be subdivided into the three different types. Type-I theory (GNI/CTE thermoelasticity) is based on the classical Fourier law of heat conduction with an infinite velocity of propagation of an exponentially decaying heat signal. Type-II theory (GNII, hyperbolic thermoelasticity) is characterised by the energy conservation and the finite propagation velocity of thermal waves known as second sound waves. Type-III theory (GNIII) combines both type-I and type-II thermoelasticity. Thereby type-I and type-II theories are the limiting cases of type-III theory. GNIII-theory obviously encircles a wider range of phenomena, as compared with the classical Fourier heat conduction (CTE) theory.

2 Compatibility conditions for jumps

The mathematical models are to simultaneously fulfill the following frameworks: 1.) Finiteness of the heat signal propagation velocity, and 2.) The ability of the spatial propagation of the thermoelastic waves without attenuation, and 3.) Existence of distortionless wave forms akin to the classical d'Alembert type waves.

A field theory formalism involves mathematical description of physical fields by integral action functional. A general form of action within a variable domain of 4-spacetime with the elementary volume of $d^4X=dX^1dX^2dX^3dX^4$ is

$$\Im = \int L(X^{\beta},\varphi^k,\partial_\alpha\varphi^k)d^4X, \quad (1)$$

where φ^k is the physical fields array, L is the Lagrangian density.

The least action principle states that the actual field is realized in the spacetime in a way that the action of (1) is minimum, i.e. for any admissible variations of physical fields φ^k and non-variable coordinates X^{β} are $\delta\Im = 0$. Then the classical Euler–Lagrange equations are held:

$$E_k(L) = \frac{\partial L}{\partial\varphi^k} - \partial_\beta\frac{\partial L}{\partial(\partial_\beta\varphi^k)} = 0.$$

In general, a conservation law has the following form

$$\partial_\beta J^\beta = 0,$$

where the vector J^β is the generalized vector 4-current. By finite variation $\delta^\nabla = \delta/\varepsilon$ the 4-current can be obtained as

$$J^\beta = \frac{\partial L}{\partial(\partial_\beta\varphi^k)}\delta^\nabla\varphi^k + \left(L\delta^\beta_\alpha - (\partial_\alpha\varphi^k)\frac{\partial L}{\partial(\partial_\beta\varphi^k)}\right)\delta^\nabla X^\alpha, \quad (2)$$

if the variational symmetries of the action are known from previous considerations.

The finite variations of action is (see [21])

$$\delta^\nabla\Im = \int\partial_\beta J^\beta d^4X. \quad (3)$$

Assuming the temperature field is continuous and temperature gradient of the first order can be discontinuous by passing through some bilateral surface $\Sigma^\pm$ propagating with the normal velocity G and normal unit 4-vector N_β in 4-spacetime.

We replace the integral over 4-volume in equation (3) on sum of surface integrals. Then only two surface integrals over surface Σ are remained when the variations $\delta^\nabla\varphi^k$ and $\delta^\nabla X^\alpha$ are fixed on the outer boundary of the field:

$$\delta\Im = \oint_{\Sigma^+} J^\beta N_\beta d^3\tau - \oint_{\Sigma^-} J^\beta N_\beta d^3\tau = \oint_{\Sigma}[J^\beta]N_\beta d^3\tau. \quad (4)$$

Hereafter square brackets denote the jumps.

The equation $\delta\Im = 0$ is valid for the actual field and a variations of $\delta^\nabla\varphi^k$ and $\delta^\nabla X^\alpha$ are continuous in passing through the surface Σ. Therefore following 4-covariant compatibility conditions for strong discontinuities is obtained from equation (4):

$$N_\beta\left[-\frac{\partial L}{\partial(\partial_\beta\varphi^k)}\right] = 0,$$
$$N_\beta\left[L\delta^\beta_\alpha - (\partial_\alpha\varphi^k)\frac{\partial L}{\partial(\partial_\beta\varphi^k)}\right] = 0. \quad (5)$$

One can see, that the compatibility conditions (5) on strong discontinuities surfaces contain the jumps of the energy-momentum 4-tensor and the Piola-Kirchhoff 4-tensor

$$S^{\beta\cdot}_{\cdot k} = -\frac{\partial L}{\partial(\partial_\beta\varphi^k)},$$
$$T^{\beta\cdot}_{\cdot\alpha} = L\delta^\beta_\alpha - (\partial_\alpha\varphi^k)\frac{\partial L}{\partial(\partial_\beta\varphi^k)}. \quad (6)$$

The compatibility conditions for jumps of the energy-momentum 4-tensor in view of (4) is rewritten in form

$$[T^{\beta\cdot}_{\cdot\alpha}]N_\beta = [L]N_\alpha + [\partial_\alpha\varphi^k S^{\beta\cdot}_{\cdot k}]N_\beta = 0, \quad (7)$$

Equations (7) can be transformed by the compatibility conditions of the strong discontinuities of the Piola-Kirchhoff 4-tensor into

$$[L]N_\alpha + [\partial_\alpha\varphi^k]S^{\beta\cdot}_{\cdot k}N_\beta = 0. \quad (8)$$

The only significant equation from the compatibility conditions for jumps of the energy-momentum 4-tensor by using the Hadamard-Thomas geometric first order compatibility conditions [22] $[\partial_\alpha\varphi^k] = N_\alpha F^k$ is obtained

$$[L]N_\alpha + S^{\beta\cdot}_{\cdot k}N_\beta F^k = 0. \quad (9)$$

The three-dimensional form of the compatibility conditions on the surface of strong discontinuity of the field is derived from the resulting 4-covariant form (4):

$$[S^{\beta\cdot}_{\cdot k}]n_\mu = 0, \quad (10)$$

$$[L] + (-GS^{4\cdot}_{\cdot k} + S^{\mu\cdot}_{\cdot k}n_\mu)(-G[\partial_4\varphi^k] + n_\mu[\partial_\mu\varphi^k]) = 0$$

Hereafter n_μ stand for normal unit 3-vector, $\lambda,\mu = 1,2,3$.

The compatibility conditions for strong discontinuities are complemented by well known three-dimensional geometrical and kinematic Hadamard-Thomas compatibility conditions of the first and the second order [22] valid for an arbitrary field φ^k

$$[\partial_i\partial_j\varphi^k] = C^k n_i n_j + g^{\alpha\beta}\partial_\alpha B^k(n_i\partial_\beta x_j + n_j\partial_\beta x_i) -$$
$$- g^{\alpha\beta}g^{\sigma\tau}B^k b_{\alpha\sigma}\partial_\beta x_i\partial_\tau x_j,$$
$$[\partial_i\partial_4\varphi^k] = (-C^k G + \delta_4 B^k)n_i + g^{\alpha\beta}\partial_{\alpha'}(B^k G)\partial_{\beta'}x_i,$$

$$[\partial_4\partial_4\varphi^k] = (C^k G - \delta_4 B^k)G + B^k\delta_4 G, \quad (11)$$
$$[\partial_4\varphi^k] = -B^k G, \quad [\partial_i\varphi^k] = B^k n_i,$$

where $i, j = 1,2,3$; $\alpha, \beta, \tau, \sigma = 1,2$; the prime in superscripts are denoted by the Greek indices related to surface Gaussian coordinates; $D = [\partial_i\partial_j\varphi]n^i n^j$; δ_4 is the delta-derivative operator. Those are also due to Rankine and Hugoniot [23, 24].

3 Action for micropolar continua

The theory GNII is the only thermodynamically correct theory which satisfies the principles listed in Sec. 1. The action taking into account of the polar microstructure can be adopted in the following form [21], with following field variables: X^α (α = 1, 2, 3) are the Lagrangian coordinates; x^j (j = 1, 2, 3) are the Eulerian coordinates; $\underset{a}{d}^j$ (a = 1, 2, 3) are micropolar directors associated with microvolume; ϑ is the thermal displacement field coupled with temperature θ by equation ($\theta = \dot{\vartheta}$).

Action density (1) we define in the form

$$L = \frac{1}{2}(\partial_4 x^k)\rho_{kj}(\partial_4 x^j) + \frac{1}{2}(\partial_4 \underset{a}{d}^i)\overset{ab}{I}_{ij}(\partial_4 \underset{b}{d}^j) - \\ -\psi(X^\alpha, x^j, \underset{a}{d}^j, \vartheta, \partial_4\vartheta, \partial_\alpha x^j, \partial_\alpha \underset{a}{d}^j, \partial_\alpha\vartheta). \quad (12)$$

wherein $\overset{ab}{I}_{ij}$ is the microinertion tensor; ρ_{ij} is the mass density tensor; ψ is the volume density of the Helmholtz's free energy. For tensors $\overset{ab}{I}_{ij}$ and ρ_{ij} the symmetry conditions $\overset{ab}{I}_{ij} = \overset{ab}{I}_{ji}$, $\rho_{ij} = \rho_{ji}$ are stated.

The field equations in this case read:

$$\partial_\alpha S^{\alpha\cdot}_{\cdot j} - \partial_4 P_j = -\frac{\partial L}{\partial(x^j)} \quad (\alpha = 1,2,3; j = 1,2,3),$$
$$\partial_\alpha {}^a M^{\alpha\cdot}_{\cdot j} + {}^a A_j - \partial_4 Q_j = 0 \quad (a = 1,2,3; \alpha = 1,2,3; j = 1,2,3), \quad (13)$$
$$\partial_\alpha j^\alpha_R + \partial_4 s = -\frac{\partial L}{\partial\vartheta} \quad (\alpha = 1,2,3),$$

and are supplemented by the constitutive equations:

$$\partial_\alpha S^{\alpha\cdot}_{\cdot j} = -\frac{\partial L}{\partial(\partial_\alpha x^j)}, \quad P_j = \frac{\partial L}{\partial(\partial_4 x^j)}$$
$$\partial_\alpha {}^a M^{\alpha\cdot}_{\cdot j} = -\frac{\partial L}{\partial(\partial_\alpha \underset{a}{d}^j)}, \quad {}^a A_j = \frac{\partial L}{\partial \underset{a}{d}^j}, \quad \partial_4 Q_j = \frac{\partial L}{\partial(\partial_4 \underset{a}{d}^j)}, \quad (14)$$
$$\partial_\alpha j^\alpha_R = \frac{\partial L}{\partial(\partial_\alpha\vartheta)}, \quad s = \frac{\partial L}{\partial(\partial_4\vartheta)},$$

In this notation the energy-momentum 4-tensor is rewritten in following forms

$$T^{\mu\cdot}_{\cdot\lambda} = L\delta^\mu_\lambda - S^\mu_l\partial_\lambda x^l - M^{\mu}_{\cdot j}\partial_\lambda \underset{a}{d}^l - j^\mu_R\partial_\lambda\vartheta = P^\mu_\lambda,$$
$$T^{\mu\cdot}_{4} = -S^\mu_l\partial_4 x^l - M^{\mu}_{\cdot j}\partial_4 \underset{a}{d}^l - j^\mu_R\partial_4\vartheta = \Gamma_\mu,$$
$$T^{4\cdot}_{\lambda} = -P_l\partial_\lambda x^l - Q_l\partial_\lambda \underset{a}{d}^l - s\partial_\lambda\vartheta = P_\lambda,$$
$$T^{4\cdot}_{\cdot 4} = \mathrm{L} - P_l\partial_4 x^l - Q_l\partial_4 \underset{a}{d}^l - s\partial_4\vartheta = -H,$$

Here, well known in theoretical physics characteristics, P^μ_λ is the Eshelby stress tensor, Γ_μ is the Umov–Poynting vector, P_λ is pseudo-momentum vector, H is quasidensity of the Hamiltonian.

Compatibility conditions on strong discontinuity surface propagating in MPTE-II continuum, according to eqs. (9), (10), take the forms:

$$\frac{1}{2}[(\partial_4 x^k)\rho_{kj}(\partial_4 x^j)] + \frac{1}{2}[(\partial_4 \underset{a}{d}^i)\overset{ab}{I}_{ij}(\partial_4 \underset{b}{d}^j)] - \\ -[\psi(X^\alpha, x^j, \underset{a}{d}^j, \vartheta, \partial_4\vartheta, \partial_\alpha x^j, \partial_\alpha \underset{a}{d}^j, \partial_\alpha\vartheta)] + \\ +\left(G\overset{ab}{I}_{lk}\partial_4 \underset{b}{d}^l - n_\mu {}^a M^{\mu\cdot}_{k}\right)\left(G[\partial_4 \underset{b}{d}^k] - n_\mu[\partial_\mu \underset{b}{d}^k]\right) + \\ +\left(G\rho_{lk}\partial_4 x^l - n_\mu S^{\mu\cdot}_{k}\right)\left(G[\partial_4 x^k] - n_\mu[\partial_\mu x^k]\right) + \\ +\left(Gs - n_\mu j^{\mu\cdot}_R\right)\left(G[\partial_4\vartheta] - n_\mu[\partial_\mu\vartheta]\right), \quad (15)$$
$$G\rho_{kl}[\partial_4 x^k] = n_\mu[S^{\mu\cdot}_{\cdot l}],$$
$$G\overset{ab}{I}_{kl}[\partial_4 \underset{b}{d}^k] = n_\mu[{}^a M^{\mu\cdot}_{\cdot l}],$$
$$G[s] = n_\mu[j^{\mu\cdot}_R].$$

4 Governing equations of GNII continuum. Linear case

A mathematical model of MPTE-II continuum (see [2] for details) is used throughout the paper. Such continua permit the existence of rotational degrees of freedom (a rotation vector φ) along with translational degrees of freedom (the translational displacement vector **u**). Thus the natural action density may be accepted in form

$$L = \frac{1}{2}u\cdot\boldsymbol{\rho}\cdot u + \frac{1}{2}\dot{\boldsymbol{\varphi}}\cdot \mathrm{J}\cdot\dot{\boldsymbol{\varphi}} - \psi(\boldsymbol{e}, \boldsymbol{\Gamma}, \theta, \nabla\vartheta), \quad (16)$$

where $\boldsymbol{\rho}$ is the mass density tensor, J is a tensor of inertia, ψ is the free energy per unit of volume, θ is the actual temperature, ∇ is the three-dimensional Hamiltonian differential operator (the nabla symbol), dot over a symbol denotes partial differentiation with respect to time at fixed spatial coordinates.

The asymmetric strain tensor $\boldsymbol{e}$ and bending–torsion tensor $\boldsymbol{\Gamma}$ are associated with translational displacements u and microrotations $\boldsymbol{\varphi}$ by formulae

$$\boldsymbol{e} = \nabla\otimes u - \boldsymbol{\varepsilon}\cdot\boldsymbol{\varphi}, \\ \boldsymbol{\Gamma} = \nabla\otimes\boldsymbol{\varphi}, \quad (17)$$

The equation (17) are represented in a rectangular coordinate system as

$$\begin{aligned} e_{ji} &= \partial_j u_i - \varepsilon_{jik}\phi_k, \\ \Gamma_{ji} &= \partial_j \phi_i. \end{aligned} \tag{18}$$

For MPTE-II continuum, equations of motion of micropolar medium are written (according to eq. (13)) in direct tensor representation for the case of the absence of mass forces and mass moments

$$\begin{cases} \nabla \cdot \boldsymbol{\sigma} = \rho u, \\ \nabla \cdot \boldsymbol{m} + \varepsilon \cdot\cdot \boldsymbol{\sigma} = \mathrm{J}\ddot{\varphi}. \end{cases} \tag{19}$$

where ε is the three-dimensional Levi–Civita symbol (permutation symbol, antisymmetric symbol, or alternating symbol).

Equations (19) in rectangular co-ordinate system can be deduced in form

$$\begin{cases} \partial_j \sigma_{ji} = \rho_{ij}\ddot{u}_j \\ \partial_j m_{ji} + \varepsilon_{ijk}\sigma_{jk} = \mathrm{J}_{ij}\ddot{\phi}_j \end{cases} \tag{20}$$

The entropy balance equation are taken in form

$$\dot{s} = -\nabla \cdot \boldsymbol{j}_R, \tag{21}$$

where s denotes the entropy referred to the unit of volume, $\boldsymbol{j}_R$ is the entropy flux vector.

The Clausius-Duhem inequality is satisfied by using the law of heat conduction, which states the proportionality of the heat velocity flux vector $\boldsymbol{h}$ and negative spatial temperature gradient θ

$$\dot{\boldsymbol{h}} = -\Lambda \cdot \nabla\theta,$$

where Λ is the material thermal conductivity tensor (thermal diffusion tensor). Λ is a positive definite matrix.

Since the free energy $\psi(\boldsymbol{e}, \boldsymbol{\Gamma}, \theta, \nabla\vartheta)$ is the function of the independent variables $\boldsymbol{e}$, $\boldsymbol{\Gamma}$, θ, $\nabla\vartheta$ then the constitutive equations (14) can be rewritten

$$\begin{aligned} \sigma_{ji} &= \frac{\partial\psi}{\partial e_{ji}}, & \mu_{ji} &= \frac{\partial\psi}{\partial\Gamma_{ji}}, \\ s &= -\frac{\partial\psi}{\partial\theta}, & \xi + h\cdot\frac{\nabla\theta}{\theta^2} &= 0. \end{aligned} \tag{22}$$

Let us expand the free energy ψ into the Taylor series in the vicinity of the natural state $\boldsymbol{e} = \Gamma = 0$, $\theta = 0$, disregarding the terms of higher order than the second one. The following form of the expansion is obtained for isotropic, homogeneous and centrosymmetric bodies

$$\begin{aligned} \psi &= \frac{\mu+\eta}{2}e_{ji}e_{ji} + \frac{\mu-\eta}{2}e_{ji}e_{ij} + \frac{\lambda}{2}e_{kk}e_{nn} + \\ &+ \frac{\gamma+\varepsilon}{2}\Gamma_{ji}\Gamma_{ji} + \frac{\gamma-\varepsilon}{2}\Gamma_{ji}\Gamma_{ij} + + \frac{\beta}{2}\Gamma_{kk}\Gamma_{nn} - \\ &- \alpha e_{kk}\theta - \varsigma\Gamma_{kk}\theta - \frac{\Lambda}{2}\theta^2. \end{aligned} \tag{23}$$

Therefore the constitutive equations of MPTE-II continuum can be derived according to equations (22) and (23)

$$\begin{aligned} \sigma_{ij} &= (\mu+\eta)e_{ji} + (\mu-\eta)e_{ij} + (\lambda \mathrm{tr} e_{kk} - \alpha\theta)\delta_{ij}, \\ m_{ij} &= (\gamma+\varepsilon)\Gamma_{ji} + (\gamma-\varepsilon)\Gamma_{ij} + (\beta \mathrm{tr}\Gamma_{kk} - \varsigma\theta)\delta_{ij}, \\ s &= \alpha e_{kk} + \varsigma\Gamma_{kk} + \Lambda\theta, \\ \boldsymbol{j}_{\mathrm{R}} &= \kappa\nabla\vartheta. \end{aligned} \tag{24}$$

wherein λ, μ, η, γ, β, ε are isothermal constitutive constants of type-II micropolar thermoelastic (GNII) continuum; α, ς are constitutive constants providing coupling of equations of motion and heat conduction; Λ is the material thermal conductivity (thermal diffusion); κ is the heat capacity (per unit volume) at constant (zero) strains. Constants α, ς depend not only on the mechanical properties of the continuum, but also depend on the thermal properties.

Consequently the hyperbolic heat conduction equation can be derived by transforming equations (21) and (24) as

$$\Lambda \cdot \nabla^2\vartheta - \kappa\ddot{\theta} - \alpha\nabla\cdot\dot{\boldsymbol{u}} - \varsigma\nabla\cdot\dot{\boldsymbol{\varphi}} = 0. \tag{25}$$

After substituting the stress tensors $\boldsymbol{\sigma}$ and $\boldsymbol{m}$ from the formulae (24) in the equations of motion (19) and taking account of (17) the system of coupled partial differential equations of motion and heat conduction for a linear isotropic GNII continuum in the absence of mass forces, moments, and heat sources can be written as [2, 6]:

$$\begin{cases} (\lambda+\mu-\eta)\nabla\nabla\cdot u + (\mu+\eta)\nabla^2 u + 2\eta\nabla\times\boldsymbol{\varphi} - \alpha\nabla\dot{\vartheta} - \rho u = 0, \\ (\beta+\gamma-\varepsilon)\nabla\nabla\cdot\boldsymbol{\varphi} + (\gamma+\varepsilon)\nabla^2\boldsymbol{\varphi} - 4\eta\boldsymbol{\varphi} + 2\eta\nabla\times u - \varsigma\nabla\dot{\vartheta} - \mathrm{J}\ddot{\boldsymbol{\varphi}} = 0, \\ \nabla^2\vartheta - \frac{\kappa}{\Lambda}\ddot{\vartheta} - \frac{\alpha\theta_0}{\Lambda}\nabla\cdot u - \frac{\varsigma\theta_0}{\Lambda}\nabla\cdot\dot{\boldsymbol{\varphi}} = 0. \end{cases} \tag{26}$$

Hereafter θ is the temperature increment over the referential temperature.

As pointed out the nonzero constitutive constants α, ς provide coupling of micropolar thermoelasticity equations. It is usually assumed that $\varsigma = 0$ (see [2]). This constitutive constant is kept in all further considerations for the completeness analysis.

The scalar equation in the system (26) is called a gen eralized heat conduction equation conjugate to the equations of motion (the first and the second equations in (26)).

5 Propagating weak discontinuities of thermomechanical fields in MPTE-II continuum

This section deals with the processes of propagating weak discontinuities of translational displacements **u**, microrotations φ and temperature θ in the MPTE-II continuum. System of partial differential equations (26) includes partial derivative order not higher than the second. Let a wavefront (wave surface Σ) of weak discontinuities displacements $\boldsymbol{u}$, microrotations φ and temperature θ be propagated in three-dimensional space with normal velocity G. Unit normal vector to that wave surface is denoted by $\mathbf{n}$. If surface Σ is parameterized by

$$x^i = \psi^i(y^1, y^2, t),$$

Then components of first $g^{\alpha\beta}$ and second $b_{\alpha\beta}$ fundamental forms of surface Σ are computed by formulas

$$\begin{aligned} g_{\alpha\beta} &= \frac{\partial \psi^i}{\partial y^\alpha}\frac{\partial \psi^i}{\partial y^\beta} = \partial_{\alpha'} x^i \partial_{\beta'} x^i, \\ b_{\alpha\beta} &= \frac{\partial^2 \psi^i}{\partial y^\alpha \partial y^\beta} n^i = \partial_{\alpha'}\partial_{\beta'} x^i n^i, \\ \partial_\alpha n^i &= -g^{\beta\gamma} b_{\beta\alpha} \partial_{\gamma'} x^i. \end{aligned} \tag{27}$$

Therefore, kinematical and geometrical compatibility conditions of the second and the first order according to eqs. (11) and (27) are read

$$\begin{aligned} &[\nabla \otimes \nabla \otimes \boldsymbol{u}] = \boldsymbol{n} \otimes \boldsymbol{n} \otimes \boldsymbol{A}, && [\nabla \otimes \nabla \otimes \boldsymbol{\varphi}] = \boldsymbol{n} \otimes \boldsymbol{n} \otimes \boldsymbol{S}, \\ &[\nabla \otimes \dot{\boldsymbol{u}}] = -G\boldsymbol{n} \otimes \boldsymbol{A}, && [\nabla \otimes \dot{\boldsymbol{\varphi}}] = -G\boldsymbol{n} \otimes \boldsymbol{S}, \\ &[\ddot{\boldsymbol{u}}] = G^2 \boldsymbol{A}, && [\ddot{\boldsymbol{\varphi}}] = G^2 \boldsymbol{S}, \\ &[\nabla \otimes \nabla \vartheta] = B\boldsymbol{n} \otimes \boldsymbol{n}, && [\ddot{\vartheta}] = G^2 B, \\ &[\nabla \otimes \dot{\vartheta}] = -GB\boldsymbol{n}, \end{aligned} \tag{28}$$

$B, \boldsymbol{A}, \boldsymbol{S}$ are fields defined on this surface, and the equalities $B = 0, \boldsymbol{A} = \boldsymbol{0}, \boldsymbol{S} = \boldsymbol{0}$ cannot be satisfied simultaneously at any point of the surface, if the surface is indeed the surface of weak discontinuities. Equations (26) and (28) give the following relations between the jumps of partial derivatives of the second order across the wave surface:

$$\begin{cases} (\rho G^2 - (\mu+\eta))\boldsymbol{A} - (\lambda+\mu-\eta)\boldsymbol{n}(\boldsymbol{n}\cdot\boldsymbol{A}) - \alpha GB\boldsymbol{n} = 0, \\ (\mathrm{J}G^2 - (\gamma+\varepsilon))\boldsymbol{S} - (\beta+\gamma-\varepsilon)\boldsymbol{n}(\boldsymbol{n}\cdot\boldsymbol{S}) - \varsigma GB\boldsymbol{n} = 0, \\ B - \dfrac{\kappa}{\Lambda}G^2 B + \dfrac{\alpha G}{\Lambda}\boldsymbol{n}\cdot\boldsymbol{A} + \dfrac{\varsigma G}{\Lambda}\boldsymbol{n}\cdot\boldsymbol{S} = 0. \end{cases} \tag{29}$$

The polarization vectors of the wave $\boldsymbol{A}$, $\boldsymbol{S}$ can be splited into a sum of projections onto the tangent plane and on the normal direction to the wave surface:

$$\begin{aligned} &\boldsymbol{A} = A_\perp \boldsymbol{\tau} + A_{||}\boldsymbol{n}, \quad \boldsymbol{S} = S_\perp \boldsymbol{\tau} + S_{||}\boldsymbol{n}, \\ &A_\perp = \boldsymbol{A}\cdot\boldsymbol{\tau}, \quad A_{||} = \boldsymbol{A}\cdot\boldsymbol{n}, \\ &S_\perp = \boldsymbol{S}\cdot\boldsymbol{\tau}, \quad S_{||} = \boldsymbol{S}\cdot\boldsymbol{n}. \end{aligned} \tag{30}$$

where $\boldsymbol{\tau}$ is the tangential unit vector and $\boldsymbol{n}$ is the normal unit one of the wave surface respectively. Taking account of equations (30) the system (29) is transformed into

$$\begin{cases} (\rho G^2 - (\mu+\eta))(A_\perp \boldsymbol{\tau} + A_{||}\boldsymbol{n}) - (\lambda+\mu-\eta)A_{||}\boldsymbol{n} - \alpha GB\boldsymbol{n} = 0, \\ (\mathrm{J}G^2 - (\gamma+\varepsilon))(S_\perp \boldsymbol{\tau} + S_{||}\boldsymbol{n}) - (\beta+\gamma-\varepsilon)S_{||}\boldsymbol{n} - \varsigma GB\boldsymbol{n} = 0, \\ B - \dfrac{\kappa}{\Lambda}G^2 B + \dfrac{\alpha G}{\Lambda}A_{||} + \dfrac{\varsigma G}{\Lambda}S_{||} = 0. \end{cases} \tag{31}$$

After rearrangements in equations (31) in view of (30) the following system is derived:

$$\begin{cases} (\rho G^2 - (\mu+\eta))A_\perp = 0, \\ (\rho G^2 - (\lambda+2\mu))A_{||} - \alpha GB = 0, \end{cases} \quad \begin{cases} (\mathrm{J}G^2 - (\gamma+\varepsilon))S_\perp = 0, \\ (\mathrm{J}G^2 - (\gamma+2\beta))S_{||} - \varsigma GB = 0, \end{cases} \tag{32}$$

The third equation in (31) allows us to couple a weak discontinuity of temperature displacement with a normal projection of weak discontinuities of translational displacements and microrotations by

$$B = G(\alpha A_{||} + \varsigma S_{||})\cdot(\Lambda - \kappa G^2)^{-1}. \tag{33}$$

This new equation is obtained by restrictions $\varsigma \neq 0$.

We proceed to consider the discriminated cases separately. Evidently in the case $\boldsymbol{A} = \boldsymbol{0}$, $\boldsymbol{S} = \boldsymbol{0}$ and returning to system (29) the scalar equation in (29) is satisfied identically, so the surface Σ is actually not a surface of weak discontinuities.

The propagation of transverse waves in the MPTE-II continuum is possible in the following cases:

I case — $A_\perp \neq 0$, $S_\perp = 0$;
II case — $A_\perp = 0$, $S_\perp \neq 0$;
III case — $A_\perp \neq 0$, $S_\perp \neq 0$;
IV case — $A_\perp = 0$, $S_\perp = 0$.

In the first case a weak discontinuity of translational displacements exists only on the surface propagating with the velocity

$$G = c_\perp^\mu, \tag{34}$$

where

$$c_\perp^\mu = \sqrt{\frac{\mu+\eta}{\rho}}.$$

It can be elucidated that the velocity G (34) exactly coincides with the one of transverse elastic wave if the constitutive micropolar constant η equals to zero $c_\perp = \sqrt{\mu/\rho}$.

In the second case, the third equation of the system (32) allows to compute the propagation velocity of a weak discontinuities of microrotation

$$G = c_\perp^{\mu\mu}, \tag{35}$$

where

$$c_\perp^{\mu\mu} = \sqrt{\frac{\gamma+\varepsilon}{\mathrm{J}}}.$$

Focus attention that a weak discontinuity of temperature is not associated with the tangential projections of polarization vectors of weak discontinuities $A_\perp$ and $S_\perp$.

In third case the propagation of wave surfaces of weak discontinuities displacements, microrotations and temperature is impossible if the constitutive characteristic of the type-II micropolar thermoelastic continuum does not satisfied certain limitations as determined by (32), i.e.

$$G = c_{\perp}^{\mu} = c_{\perp}^{\mu\mu}.$$

In fourth case the wave surface propagation is impossible if the normal projections of discontinuities polarization vectors are equal to zero. Thus equations (32) can be transformed into

$$\begin{cases} (\rho G^2 - (\lambda + 2\mu))A_{\|} - \alpha G B = 0, \\ (JG^2 - (\gamma + 2\beta))S_{\|} - \varsigma G B = 0, \\ B - \dfrac{\kappa}{\Lambda} G^2 B + \dfrac{\alpha G}{\Lambda} A_{\|} + \dfrac{\varsigma G}{\Lambda} S_{\|} = 0. \end{cases} \quad (36)$$

There are only two cases of transverse wave propagation in MPTE-II continuum:

I case — $A_{\|} \neq 0$, $S_{\|} \neq 0$, $B \neq 0$;

II case — $A_{\|} \neq 0$, $S_{\|} \neq 0$, $B = 0$.

In first case we can easy obtained bicubic equation for normal wave velocity as follow

$$\rho(\Lambda - \kappa G^2) + \frac{\alpha G^2}{G^2 - c_{\|}^2} + \frac{\varsigma G^2}{G^2 - c_{\|}^{\mu\mu 2}} = 0, \quad (37)$$

Here we denotes velocities by

$$c_{\|}^{\mu\mu} = \sqrt{\frac{\beta + 2\gamma}{J}}, \quad c_{\|} = \sqrt{\frac{\lambda + 2\mu}{\rho}};$$

In second case the propagation of weak discontinuities surfaces is impossible if the constitutive characteristic of the type-II micropolar thermoelastic continuum does not satisfied certain limitations as determined by (32), i.e.

$$G = c_{\|} = c_{\|}^{\mu\mu},$$

and for normal projection of discontinuities polarization vectors

$$\frac{\alpha}{\varsigma} = -\frac{S_{\|}}{A_{\|}}.$$

6 Conclusions

In the present study problems of propagating surfaces of discontinuities of displacements, microrotation and thermal displacements of the micropolar type-II thermoelastic media (GNII) has been considered. The dynamical and constitutive equations for hyperbolic thermoelastic type-II micropolar continuum have been derived by field theory technique and formalism of the variational calculus. The special form of the first variation of the action integral has been used us in order to obtained 4-covariant jump conditions on wave surfaces. The three-dimensional form of the jump conditions on the surface of a strong discontinuity of field can be derived as the result of its four-dimensional co- variant form. The possible normal velocities of the prop- agating surfaces of weak discontinuities are found. The possible normal velocities of the propagating surfaces of weak discontinuities have been found. The distinguished cases of propagating wave surfaces in MPTE-II continuum have been separately considered. Propagating surfaces of weak discontinuities have been discriminated depending on the spatial orientations of polarization vectors of dis- continuities.

Acknowledgements

The authors wish to thank Professor Yu. N. Radayev for the statement of the problem, scientific advice and neces- sary support. This work was financially supported by the Russian Science Foundation under Grant 14-19-01280.

References

1. E. Cosserat and F. Cosserat, *Theories of Deformable Bodies* (Scientific Library A. Hermann et Fils, Paris, 1909) 242
2. W. Nowacki *Theory of Asymmetric Elasticity* (Perga- mon Press, Oxford, 1986) 384
3. R. D. Mindlin and H. F. Tiersten Arch. for Rat. Mech. and Anal., **11:1** 415–448 (1962)
4. A. V. Manzhirov and D. A. Parshin, Mechanics of Solids. **50** (5) 559–570 (2015)
5. A. V. Manzhirov and D. A. Parshin, Mechanics of Solids. **50** (6) 657-675 (2015)
6. A. C. Eringen, *Microcontinuum field theories. Vol. 1. Foundations and Solids.* (Springer, Berlin; Heidelberg; N.Y., 1999) 325
7. H. Ramézani, A. El-Hraiech, and J. Jeong, Comput. Methods Appl. Mech. Engrg. **237–240** 227–243 (2012)
8. S. Chandrasekhar *Liquid Crystals* (Cambridge University Press, Cambridge, 1992) 460
9. T. F. McNelly et al. Phys. Reviews. **24:3** 100–102 (1970)
10. Jackson H. E.,Walker C. T., and McNelly T. F. Phys. Reviews. **25:1** 26–28 (1970)
11. Rogers S. J. Phys. Review B. **3:4** 1440–1457 (1971)
12. Pohl D. W. and Irniger V. Phys. Review Letters. **36:9** 480–483 (1976)
13. R.J. Hardy and S.S. Jaswal Phys. Review B. **3:12** 4385–4387 (1971)
14. Narayanamurti V. and Dynes R. C. Phys. Reviews. **28** 1461–1464 (1972)
15. V. A. Kovalev, E. V. Murashkin, and Yu. N. Radayev, Izv. Sarat. Univ. (N.S.), Ser. Mat. Mekh. Inform. **14** 1 77–87 (2014).
16. V. A. Kovalev, E. V. Murashkin, and Yu. N. Radayev, Izv. Sarat. Univ. (N.S.), Ser. Mat. Mekh. Inform. **15** 1 79–89 (2015).
17. A. E. Green and P. M. Naghdi, J. Therm. Stress, **15** 253–264 (1992)

18. A. E. Green and P. M. Naghdi, J. Elasticity.. **31** 189–208 (1993)
19. E. V. Murashkin and Y. N. Radayev, Materials Physics and Mechanics, **23** 10–13 (2015)
20. E. Murashkin and M. Polonik, On Compatibility Conditions for Propagating Surfaces with Additional Constraints. *IUTAM Symposium on Growing solids. Symposium materials*, 79–80, 2015.
21. V. A. Kovalev and Yu. N. Radayev, *Wave Problems of Field Theory and Thermomechanics* (Saratov. Univ. Pub., Saratov, 2010) 328
22. T. Y. Thomas, *Plastic Flow and Fracture in Solids* (Academic Press, N.Y., 1961) 271
23. W. J. M. Rankine, Proc. of the Royal Society of London. London: The Royal Society, **18** 80–83 (1870)
24. P. H. Hugoniot, J. Ecole Polytechnique, **CLVII** 3–98 (1887)

Mechanical Design of AM Fabricated Prismatic Rods under Torsion

Alexander V. Manzhirov [1,2,3,4]

[1]Institute for Problems in Mechanics of the Russian Academy of Sciences, Vernadsky Ave 101 Bldg 1, Moscow, 119526, Russia
[2]Bauman Moscow State Technical University, 2nd Baumanskaya Str 5/1, Moscow, 105005, Russia
[3]National Research Nuclear University T«MEPhITi, Kashirskoye shosse 31, Moscow, 115409, Russia
[4]Moscow Technological University, Vernadsky Ave 78, Moscow, 119454, Russia

Abstract. We study the stress-strain state of viscoelastic prismatic rods fabricated or repaired by additive manufacturing technologies under torsion. An adequate description of the processes involved is given by methods of a new scientific field, mechanics of growing solids. Three main stages of the deformation process (before the beginning of growth, in the course of growth, and after the termination of growth) are studied. Two versions of statement of two problems are given: (i) given the torque, find the stresses, displacements, and torsion; (ii) given the torsion, find the stresses, displacements, and torque. Solution methods using techniques of complex analysis are presented. The results can be used in mechanical and instrument engineering.

1 Statement of the torsion problem

Consider a homogeneous viscoelastic ageing rod manufactured at the initial instant of time and occupying some cylindrical domain Π_1. The lateral surface Π_1 of the body is stress-free until time $\tau_0 \geq 0$ at which forces statically equivalent to a couple with torque $M(t)$ are applied to the faces of the cylindrical body.

At time $\tau_1 \geq \tau_0$, continuous accretion of the rod by elements manufactured simultaneously with the rod begins. Let $L(t)$ be the time-varying boundary of the cross-section $\Omega(t)$, so that $L(\tau_1) = L_1$ and $\Omega(\tau_1) = \Omega_1$. The boundary $L(t)$ consists of two parts, $L(t) = L^*(t) \bigcup L_\sigma(t)$. Here $L^*(t)$ is the accretion boundary (or the growth boundary) on which the material influx occurs at time t, and $L^*(t) = L^*$ for $\tau < \tau_1$; $L_\sigma(t)$ is the stress-free boundary. We also assume that the time $\tau_0(x_1, x_2)$ at which the load is applied to the elements being added coincides with the time $\tau^*(x_1, x_2)$ at which they are added to the growing rod.

The accretion is terminated at time $\tau_2 \geq \tau_1$, and since then the rod occupies some domain $\Pi_2 = \Pi(\tau_2)$ whose cross-section $\Omega_2 = \Omega(\tau_2)$ has the boundary $L_2 = L(\tau_2)$. The boundary conditions posed on the boundary $L_2 = L(\tau_2)$ are the same as on the boundary L_1 before the beginning of growth.

Somewhat later, at time $\tau_3 \geq \tau_2$, accretion may begin again; it may well happen that the new growth boundary is not related in any way to the earlier existing one. Then one can assume that accretion terminates at time τ_4, etc., thus successively arriving at the problem of piecewise continuous accretion of a solid rod with n growth beginning instants and accordingly n growth termination instants.

We proceed to the analysis of main stages of the piecewise continuous accretion process for viscoelastic rods. From now on, we consider sufficiently slow processes, and so inertial terms in the equilibrium equations can be neglected (also see [1–7]).

2 Boundary value problem for the rod of fixed composition

Consider the stress-strain state of a viscoelastic ageing rod of fixed composition on the time interval $t \in [\tau_0, \tau_1]$. We have the boundary value problem (e.g., see [1–3]) consisting of the equilibrium equations

$$\frac{\partial \sigma_{13}}{\partial x_3} = 0, \quad \frac{\partial \sigma_{23}}{\partial x_3} = 0, \quad \frac{\partial \sigma_{13}}{\partial x_1} + \frac{\partial \sigma_{23}}{\partial x_2} = 0, \tag{1}$$

the strain–displacement relations

$$\varepsilon_{ij} = \frac{1}{2}\left(\frac{\partial u_i}{\partial x_j} + \frac{\partial u_j}{\partial x_i}\right) = 0, \ \varepsilon_{ii} = \varepsilon_{12} = 0, \ i, j = 1,2,3, \tag{2}$$

the boundary condition

$$(x_1, x_2) \in L_1: \quad \sigma_{13} n_1 + \sigma_{23} n_2 = 0; \tag{3}$$

on the lateral surface, and the constitutive equation

$$\sigma_{13} = 2G(I + N(\tau_0,t))\varepsilon_{13},$$
$$\sigma_{23} = 2G(I + N(\tau_0,t))\varepsilon_{23},$$
$$(I + N(\tau_0,t))^{-1} = (I - L(\tau_0,t)), \quad (4)$$
$$L(\tau_0,t)f(t) = \int_{\tau_0}^{t} f(\tau)K_1(t,\tau)d\tau,$$
$$K_1(t,\tau) = G(\tau)\frac{\partial}{\partial\tau}\left[G^{-1}(\tau) + \omega(t,\tau)\right],$$

where $\vec{n} = \{n_1, n_2\}$ is the unit outward normal vector on the lateral surface of the body, $G = G(t)$ is the instantaneous elastic shear modulus, $K_1(t,\tau)$ and $\omega(t,\tau)$ are the shear creep kernel and the shear creep rate kernel, respectively, and I is the identity operator. The arguments have been omitted above in a number of obvious cases. We also omit them in what follows unless their absence can lead to a misunderstanding.

Further, the following equilibrium conditions should hold for the end cross-section Ω_1 subjected to the torque $M(t)$:

$$M(t) = \iint_{\Omega_1} (x_1\sigma_{23} - x_2\sigma_{13})\,dx_1\,dx_2, \quad (5)$$
$$\iint_{\Omega_1} \sigma_{13}\,dx_1\,dx_2 = 0, \quad \iint_{\Omega_1} \sigma_{23}\,dx_1\,dx_2 = 0.$$

Relations (1)–(5) form a problem of the viscoelasticity theory for a homogeneous ageing body. The solution of this problem permits one to describe the stress-strain state of the body on the time interval $t \in [\tau_0, \tau_1]$.

Let us transform the boundary value problem for the main body. To this end, we divide the expressions containing the stresses σ_{13} and σ_{23} in (1)–(5) by G and apply the operator $(I - L(\tau_0,t))$ to them. Since the operator $(I - L(\tau_0,t))$ commutes with the divergence operator, we see that the expressions (1)–(4) acquire the form

$$\frac{\partial\sigma^0_{13}}{\partial x_3} = 0, \quad \frac{\partial\sigma^0_{23}}{\partial x_1} = 0, \quad \frac{\partial\sigma^0_{13}}{\partial x_1} + \frac{\partial\sigma^0_{23}}{\partial x_2} = 0,$$
$$\varepsilon_{ii} = \frac{\partial u_i}{\partial x_i} = 0, \varepsilon_{ij} = \frac{1}{2}\left(\frac{\partial u_i}{\partial x_j} + \frac{\partial u_j}{\partial x_i}\right) = 0, i,j = 1,2,3, \quad (6)$$
$$(x_1,x_2) \in L_1: \quad \sigma^0_{13}n_1 + \sigma^0_{23}n_2 = 0,$$
$$\sigma^0_{13} = 2\varepsilon_{13}, \quad \sigma^0_{23} = 2\varepsilon_{23},$$

where σ^0_{ij} and σ_{ij} are related by the formula

$$\sigma^0_{ij} = (I - L(\tau_0,t))\sigma_{ij}G^{-1}. \quad (7)$$

Likewise, relations (5) become

$$M^0(t) = \iint_{\Omega_1} (x_1\sigma^0_{23} - x_2\sigma^0_{13})dx_1\,dx_2,$$
$$\iint_{\Omega_1} \sigma^0_{13}dx_1\,dx_2 = 0, \quad \iint_{\Omega_1} \sigma^0_{23}dx_1\,dx_2 = 0, \quad (8)$$

where $M^0 = (I - L(\tau_0,t))MG^{-1}$.

The boundary value problem (6)–(8), in contrast to problem (1)–(5), contains time as a parameter and is mathematically equivalent to a boundary value problem of the elasticity theory with parameter t.

By substituting the strain–displacement relations into the equilibrium equations of problem (6), we obtain the equilibrium equations in terms of displacements in the form

$$\frac{\partial^2 u_1}{\partial x_3^2} + \frac{\partial^2 u_3}{\partial x_3\partial x_1} = 0, \quad \frac{\partial^2 u_2}{\partial x_3^2} + \frac{\partial^2 u_3}{\partial x_3\partial x_2} = 0,$$
$$\frac{\partial^2 u_1}{\partial x_3\partial x_1} + \frac{\partial^2 u_3}{\partial x_1^2} + \frac{\partial^2 u_2}{\partial x_3\partial x_2} + \frac{\partial^2 u_3}{\partial x_2^2} = 0. \quad (9)$$

We use the first three strain–displacement relations in (6) to obtain

$$u_i = u_i(x_2,x_3,t), \quad i = 1,2,3, \quad (10)$$

where the displacement u_i is independent of the coordinate x_i.

In view of (9) and (10), the equilibrium equations in terms of displacements become

$$\frac{\partial^2 u_1}{\partial x_3^2} = 0, \quad \frac{\partial^2 u_2}{\partial x_3^2} = 0, \quad \frac{\partial^2 u_3}{\partial x_1^2} + \frac{\partial^2 u_3}{\partial x_2^2} = 0. \quad (11)$$

It follows from the third equation in (11) that the function $u_3(x_1,x_3,t)$ is harmonic. Let us find $u_1(x_2,x_3,t)$ and $u_2(x_1,x_3,t)$. To this end, we integrate the first two equations in (11) twice with respect to x_3. As a result, we obtain

$$u_1 = f(x_2,t)x_3 + c_1(t)x_3 + f_1(x_2,t) + c_3(t),$$
$$u_2 = g(x_1,t)x_3 + c_2(t)x_3 + g_1(x_2,t) + c_4(t). \quad (12)$$

By using (12) and the fourth strain–displacement relation in (6), we obtain

$$\frac{\partial f}{\partial x_2}x_3 + \frac{\partial f_1}{\partial x_2} + \frac{\partial g}{\partial x_1}x_3 + \frac{\partial g_1}{\partial x_1} = 0. \quad (13)$$

Equation (13) permits us to conclude that

$$g = \theta(t)x_1, f = -\theta(t)x_2, g_1 = c(t)x_1, f_1 = -c(t)x_2. \quad (14)$$

By (14), for the displacements u_1 and u_2 we obtain

$$u_1 = -\theta(t)\,x_2x_3 + c_1(t)x_3 - c(t)x_2 + c_3(t),$$
$$u_2 = \theta(t)\,x_1x_3 + c_2(t)x_3 + c(t)x_1 + c_4(t). \quad (15)$$

The last terms in the expression (15) do not affect the stress state of the twisted cylindrical body. To eliminate the translational displacements of the rod along the x_1- and x_2-axes, assume that some point of the rod end is fixed, so that $x_1 = x_2 = 0$ at that point. We place the origin there and require that the additional conditions

$$\frac{\partial u_1}{\partial x_3} = \frac{\partial u_2}{\partial x_3} = \frac{\partial u_2}{\partial x_1} = 0$$

hold at that point so as to prevent the rigid rotations of the cylindrical body about the x_1-, x_2-, and x_3-axes. Then $c(t)=c_1(t)=c_2(t)=c_3(t)=c_4(t)=0$, and hence the expressions (15) for the stresses x_1 and x_2 acquire the form

$$u_1=-\theta(t)x_2x_3,\quad u_2=\theta(t)x_3x_1. \tag{16}$$

The function $\theta(t)$ is called the torsion angle (or torsion) and is related to the displacements by the formula

$$\frac{\partial}{\partial x_3}\left(\frac{\partial u_1}{\partial x_2}-\frac{\partial u_2}{\partial x_1}\right)=-2\theta(t). \tag{17}$$

By analogy with [8], set

$$u_3=\theta(t)\varphi_{\tau_1}, \tag{18}$$

where φ_{τ_1} is some function to be determined of the variables x_1, x_2, and τ_1. It is called the torsion function and characterizes the deformation of cross-sections.

Then, by (16) and (18), we have

$$u_1=-\theta(t)x_2x_3,\quad u_2=\theta(t)x_3x_1,\quad u_3=\theta(t)\varphi_{\tau_1}. \tag{19}$$

By using relations (19) and the constitutive equations in (6), we obtain

$$\sigma_{13}^0=\theta(t)\left(\frac{\partial\varphi_{\tau_1}}{\partial x_1}-x_2\right),\quad \sigma_{23}^0=\theta(t)\left(\frac{\partial\varphi_{\tau_1}}{\partial x_2}+x_1\right). \tag{20}$$

By substituting (20) into the third equilibrium equation in (3), we find that the function φ_{τ_1} satisfies the 2D Laplace equation

$$\frac{\partial^2\varphi_{\tau_1}}{\partial x_1^2}+\frac{\partial^2\varphi_{\tau_1}}{\partial x_2^2}=0; \tag{21}$$

i.e., the torsion function φ_{τ_1} is harmonic in the domain Ω_1.

Let us find the boundary condition for the function φ_{τ_1} on the contour L_1 of the cross-section Ω_1. By substituting (20) into the boundary condition (3), we obtain

$$(x_1,x_2)\in L_1:\quad \frac{\partial\varphi_{\tau_1}}{\partial n}=x_2n_1-x_1n_2. \tag{22}$$

Thus, we have reduced the original torsion problem for a prismatic body to the Neumann problem for the function φ_{τ_1} in the domain Ω_1, that is, the problem of finding a harmonic function φ_{τ_1} in the cross-section domain Ω_1 (see (1.21)) from the given values of its normal derivative on the contour L_1 (see (1.22)).

Since

$$x_1\sigma_{23}^0-x_2\sigma_{13}^0=\theta(t)\left(x_1^2+x_2^2+x_1\frac{\partial\varphi_{\tau_1}}{\partial x_2}-x_2\frac{\partial\varphi_{\tau_1}}{\partial x_1}\right), \tag{23}$$

We see that the first integral equilibrium condition in (8) acquires the form

$$M^0(t)=(I-L(\tau_0,t))M(t)G^{-1}(t)=\theta(t)D_1,$$
$$D_1=\iint_{\Omega_1}\left(x_1^2+x_2^2+x_1\frac{\partial\varphi_{\tau_1}}{\partial x_2}-x_2\frac{\partial\varphi_{\tau_1}}{\partial x_1}\right)dx_1dx_2, \tag{24}$$

and the second and third conditions are satisfied identically.

The constant D_1 is called the torsional rigidity. It only depends on the shape of the cross-section of the body.

The problem in question has two possible statements:

1. Given the torque $M(t)$, determine the stresses σ_{ij}, the displacements u_i, and the torsion angle $\theta(t)$.
2. Given the torsion angle $\theta(t)$, find σ_{ij}, u_i, and the torque $M(t)$.

The solution for the first statement can be constructed as follows: find the function φ_{τ_1} from the boundary condition, use formulas (24) to find $\theta(t)$, use formulas (19) and (20) to find u_i and σ_{ij}^0, and finally reconstruct the true stresses σ_{ij} by the formula

$$\sigma_{ij}=G(t)(\sigma_{ij}^0++\int_{\tau_0}^{t}\sigma_{ij}^0R_1(t,\tau)d\tau), \tag{25}$$

where $R_1(t,\tau)$ is the resolvent of the kernel $K_1(t,\tau)$.

For the second statement of the problem, one determines φ_{τ_1}, finds u_i and σ_{ij}^0 by formulas (19) and (20), reconstructs σ_{ij} with the use of (25), and determines $M(t)$ on the basis of (5).

Sometimes it is convenient to study the problem for the main body in terms of rates of the corresponding variables, because information about these rates is needed when solving the problem at the growth stage. To this end, one should differentiate relations (6) with respect to t. (From now on, differentiation is always understood in the sense of distributions.) Then we have

$$\frac{\partial S_{13}}{\partial x_3}=0,\quad \frac{\partial S_{23}}{\partial x_1}=0,\quad \frac{\partial S_{13}}{\partial x_1}+\frac{\partial S_{23}}{\partial x_2}=0,$$
$$D_{ii}=\frac{\partial v_i}{\partial x_i}=0,\quad D_{ij}=\frac{1}{2}\left(\frac{\partial v_i}{\partial x_j}+\frac{\partial v_j}{\partial x_i}\right)=0,\ i,j=1,2,3,$$
$$(x_1,x_2)\in L_1:\ S_{13}n_1+S_{23}n_2=0, \tag{26}$$
$$S_{13}=2D_{13},\quad S_{23}=2D_{23},$$
$$S_{ij}=\frac{\partial\sigma_{ij}^0}{\partial t},\quad v_i=\frac{\partial u_i}{\partial t},\quad D_{ij}=\frac{\partial\varepsilon_{ij}}{\partial t}.$$

Relations (26) form an initial–boundary value problem parametrically depending on time t.

By analogy with the preceding, one can show that v_i and S_{ij} have the following structure:

$$v_1 = -\theta^{\bullet}(t)x_2x_3,\ v_2 = \theta^{\bullet}(t)x_3x_1, v_3 = \theta^{\bullet}(t)\varphi_{\tau_1},$$
$$S_{13} = \theta^{\bullet}(t)\left(\frac{\partial\varphi_{\tau_1}}{\partial x_1} - x_2\right), S_{23} = \theta^{\bullet}(t)\left(\frac{\partial\varphi_{\tau_1}}{\partial x_2} + x_1\right), \quad (27)$$

where the dot stands for differentiation with respect to t (we use this notation in what follows) and $\theta^{\bullet}(t)$ is the torsion angular velocity.

By substituting (27) into the third equilibrium equation in (26), we find that the function φ_{τ_1} satisfies the 2D Laplace equation

$$\frac{\partial^2\varphi_{\tau_1}}{\partial x_1^2} + \frac{\partial^2\varphi_{\tau_1}}{\partial x_2^2} = 0; \quad (28)$$

i.e., the torsion function φ_{τ_1} is harmonic in the domain Ω_1.

The boundary condition for the function φ_{τ_1} on the contour L_1 of the cross-section Ω_1 acquires a form similar to (22),

$$(x_1, x_2) \in L(t): \frac{\partial\varphi_{\tau_1}}{\partial n} = x_2n_1 - x_1n_2. \quad (29)$$

Thus, we have reduced the original torsion problem for a prismatic body to the Neumann problem for the function φ_{τ_1} in the domain Ω_1.

The differentiated relation (24) becomes

$$M^{0\bullet}(t) = \frac{M^{\bullet}(t)}{G(t)} + \int_{\tau_0}^{t}\frac{\partial M(t)}{\partial\tau}\frac{\partial\omega(t,\tau)}{\partial t}d\tau + M(\tau_0)\frac{\partial\omega(t,\tau_0)}{\partial t},$$
$$M^{0\bullet}(t) = \theta^{\bullet}(t)D_1. \quad (30)$$

To solve the first version of the problem (where the torque $M(t)$ is given), one should use the boundary condition to find φ_{τ_1}, then determine v_i and S_{ij} by formulas (28), use formulas (30) to find the torsion angular velocity, and reconstruct the true stresses, displacement, and torsion angle by the formulas

$$\sigma_{ij}(x_1,x_2,t) = G(t)\left[\frac{\sigma_{ij}(x_1,x_2,\tau_0)}{G(\tau_0)}\left(1 + \int_{\tau_0}^{t}R(t,\tau)d\tau\right) + \right.$$
$$\left. + \int_{\tau_0}^{t}\left(S_{ij}(x_1,x_2,\tau) + \int_{\tau_0}^{\tau}S_{ij}(x_1,x_2,\varsigma)d\varsigma R(t,\tau)\right)d\tau\right],$$
$$u_i(x_1,x_2,t) = u_i(x_1,x_2,\tau_0) + \int_{\tau_0}^{t}v_i(x_1,x_2,\tau)d\tau, \quad (31)$$
$$\theta(t) = \theta(\tau_0) + \int_{\tau_0}^{t}\theta^{\bullet}(\tau)d\tau.$$

To solve the second version of the problem (where the torsion angle $\theta(t)$ is given), one should find φ_{τ_1}, determine v_i and S_{ij} by formulas (28), reconstruct u_i and σ_{ij} by formulas (31), and find the torque according to (5).

Note that the initial values of the stresses, displacements, and torsion angle in (31) are determined from the solution of the boundary value problem (1)–(5) or (6) for $t = \tau_0$.

Thus, the study of the torsion problem for a growing body at the stage preceding accretion is complete.

3 Initial–boundary value problem for a growing rod

Now consider the process of continuous accretion of a deformable body for $t \in [\tau_1, \tau_2]$. For a growing rod, we have the equilibrium equations

$$\frac{\partial\sigma_{13}}{\partial x_3} = 0,\ \frac{\partial\sigma_{23}}{\partial x_3} = 0,\ \frac{\partial\sigma_{13}}{\partial x_1} + \frac{\partial\sigma_{23}}{\partial x_2} = 0; \quad (32)$$

the strain rate–displacement rate relations

$$D_{ij} = \frac{1}{2}\left(\frac{\partial v_i}{\partial x_j} + \frac{\partial v_j}{\partial x_i}\right) = 0, D_{ii} = D_{12} = 0, i,j = 1,2,3, \quad (33)$$

the boundary condition $(t = \tau^*(x_1,x_2))$

$$(x_1,x_2) \in L^*(t): \begin{cases} \sigma_{13} = \sigma_{13}^*,\ \sigma_{23} = \sigma_{23}^*, \\ \dfrac{\partial\sigma_{13}}{\partial t}n_1 + \dfrac{\partial\sigma_{23}}{\partial t}n_2 = 0 \end{cases} \quad (34)$$

on the growth boundary, the initial–boundary condition

$$(x_1,x_2) \in L_\sigma:\ \sigma_{13}n_1 + \sigma_{23}n_2 = 0 \quad (35)$$

on the immovable boundary, and the constitutive equations

$$\sigma_{13} = 2G(I + N(\tau_0(x_1,x_2),t))\varepsilon_{13},$$
$$\sigma_{23} = 2G(I + N(\tau_0(x_1,x_2),t))\varepsilon_{23}, \quad (36)$$
$$\tau_0^*(x_1,x_2) = \begin{cases} \tau_0, (x_1,x_2) \in \Omega_1, \\ \tau^*(x_1,x_2), (x_1,x_2) \in \Omega^*(t), \end{cases}$$

where $\Omega^*(t) = \Omega(t)\setminus\Omega_1$, the operator $(I - L(\tau_0^*,t))$ and the inverse operator $(I + N(\tau_0^*t))$ can be determined from (4) by the replacement of τ_0 with τ_0^*, and $\sigma_{13}^*(x_1,x_2,\tau^*)$ and $\sigma_{23}^*(x_1,x_2,\tau^*)$ are the components of the 2D stress tensor acting on the surface accreted by the 3D body.

The equilibrium conditions for the end cross-section $\Omega(t)$ subjected to the torque $M(t)$ have the form

$$M(t) = \iint_{\Omega(t)}(x_1\sigma_{23} - x_2\sigma_{13})dx_1dx_2. \quad (37)$$

Relations (32)–(37) form the general boundary value problem for a continuously growing body. Relations (36) show that in general the process of accretion of new elements leads to constitutive relations with discontinuities on the interface between the main body and the additional bodies.

Let us transform the initial–boundary value problem for a continuously growing viscoelastic ageing body into a problem with time parameter coinciding in form with a boundary value problem of the elasticity theory. At the first stage, we transform the accretion problem for a viscoelastic body with constitutive relations (36) to the accretion problem for an elastic body described by Hooke's law.

To this end, we represent the equation of the growth boundary $L^*(t)$ for $t = \tau_1$ in the form

$$(x_1,x_2)\in L^*(t):\quad \tau_1-\tau^*=0, \tag{38}$$

where $\tau_1-\tau^*\geq 0$ for $(x_1,x_2)\in\Omega_0$ and $\tau_1-\tau^*<0$ for $(x_1,x_2)\in\Omega^*(t)$. We assume that τ^* is a sufficiently smooth function such that $\nabla\tau^*\neq 0$ for $\tau^*-\tau_1=0$. (That is, there are no singular points on the growth boundary.) We introduce the characteristic function $\theta(\tau_1-\tau^*)$, which is equal to unity if the argument of the function is nonnegative and is zero if the argument is negative. Now the operator $(I-L(\tau_0^*,t))$ can be represented in the form

$$\begin{gathered}(I-L(\tau_0^*,t))f(t)=(I-L(\tau^0,t))f(t)-\theta(\tau_1-\tau^*)L^t(\tau_0,\tau_1)f(t)\\ L^t(\tau_0,\tau_1)f(t)=\int_{\tau_0}^{\tau_1}f(\tau)K_1(t,\tau)d\tau,\\ \tau^0=\theta(\tau_1-\tau^*)(\tau_1-\tau^*(x_1,x_2))+\tau^*(x_1,x_2),\end{gathered} \tag{39}$$

where $\tau^*=\tau_1$ for $(x_1,x_2)\in L^*(\tau_1)$.

Let us divide the relations containing the stresses σ_{13} and σ_{23} in (32)–(37) by G and then apply the operator $(I-L(\tau_0^*,t))$. Then, in view of the notation $\sigma_{ij}^0=(I-L(\tau_0^*,t))\sigma_{ij}G^{-1}$, we obtain

$$\frac{\partial\sigma_{13}^0}{\partial x_3}=0,\quad \frac{\partial\sigma_{23}^0}{\partial x_3}=0,\quad \frac{\partial\sigma_{13}^0}{\partial x_1}+\frac{\partial\sigma_{23}^0}{\partial x_2}=0,$$

$$D_{ij}=\frac{1}{2}\left(\frac{\partial v_i}{\partial x_j}+\frac{\partial v_j}{\partial x_i}\right)=0,\ D_{ii}=D_{12}=0,\ i,j=1,2,3,$$

$$\sigma_{13}^0=2\varepsilon_{13},\quad \sigma_{23}^0=2\varepsilon_{23}, \tag{40}$$

$$(x_1,x_2)\in L^*(t):\ \left\{\begin{aligned}&\sigma_{13}^0=\sigma_{13}^*G^{-1},\sigma_{23}^0=\sigma_{23}^*G^{-1},\\ &\frac{\partial\sigma_{13}^0}{\partial t}n_1+\frac{\partial\sigma_{23}^0}{\partial t}n_2=0;\end{aligned}\right.$$

$$\begin{gathered}(x_1,x_2)\in L_\sigma:\ \sigma_{13}^0n_1+\sigma_{23}^0n_2=0,\\ M^0(t)=\iint_{\Omega(t)}(x_1\sigma_{23}^0-x_2\sigma_{13}^0)dx_1dx_2.\end{gathered}$$

Let us transform the initial–boundary value problem (40) to a boundary value problem for the strain rates, the displacement rates, and the operator stress rates. To this end, we differentiate the equilibrium equations, the constitutive equations, the boundary condition on the immovable boundary $L_\sigma(t)$, and the equilibrium condition for the end cross-section with respect to t and obtain the boundary value problem

$$\frac{\partial S_{13}}{\partial x_3}=0,\quad \frac{\partial S_{23}}{\partial x_3}=0,\quad \frac{\partial S_{13}}{\partial x_1}+\frac{\partial S_{23}}{\partial x_2}=0,$$

$$D_{ij}=\frac{1}{2}\left(\frac{\partial v_i}{\partial x_j}+\frac{\partial v_j}{\partial x_i}\right)=0,\ D_{ii}=D_{12}=0,\ i,j=1,2,3,$$

$$\begin{gathered}S_{13}=2D_{13},\quad S_{23}=2D_{23},\\ (x_1,x_2)\in L^*(t):\ S_{13}n_1+S_{13}n_2=0,\\ (x_1,x_2)\in L_\sigma:\ S_{13}n_1+S_{13}n_2=0;\end{gathered} \tag{41}$$

$$M^{0\bullet}(t)=\iint_{\Omega(t)}(x_1S_{23}-x_2S_{13})dx_1dx_2+\int_{L^*(t)}(x_1\sigma_{23}^0-x_2\sigma_{13}^0)s_n\,dl,$$

where the second term in the last formula is a contour integral over the boundary $L^*(t)$.

Being supplemented with the initial conditions for the main body at $t=\tau_1$, relations (40), which also contain the initial–boundary condition on the growth boundary, form an initial–boundary value problem with time parameter t.

The formulas for the variables v_i and S_{ij} can be obtained by the replacement of the function $\varphi(x_1,x_2,\tau_1)=\varphi_{\tau_1}$ by $\varphi(x_1,x_2,t)=\varphi_t$:

$$\begin{gathered}v_1=-\theta^\bullet(t)x_2x_3,\ v_2=\theta^\bullet(t)x_3x_1,\ v_3=\theta^\bullet(t)\varphi_t,\\ S_{13}=\theta^\bullet(t)\left(\frac{\partial\varphi_t}{\partial x_1}-x_2\right),S_{23}=\theta^\bullet(t)\left(\frac{\partial\varphi_t}{\partial x_2}+x_1\right).\end{gathered} \tag{42}$$

The torsion function φ_t can be found from the Neumann boundary value problem

$$\begin{gathered}\frac{\partial^2\varphi_t}{\partial x_1^2}+\frac{\partial^2\varphi_t}{\partial x_2^2}=0,\\ (x_1,x_2)\in L^*(t):\quad \frac{\partial\varphi_t}{\partial n}=x_2n_1-x_1n_2.\end{gathered} \tag{43}$$

Thus, we have reduced the original torsion problem for a prismatic body to the Neumann problem for the function φ_t in the domain $\Omega(t)$.

Obviously, in view of (42) we obtain

$$\begin{gathered}M^{0\bullet}(t)=\theta^\bullet(t)D_t+\int_{L^*(t)}(x_1\sigma_{23}^0-x_2\sigma_{13}^0)s_n\,dl,\\ D_t=\iint_{\Omega(t)}\left(x_1^2+x_2^2+x_1\frac{\partial\varphi_t}{\partial x_2}-x_2\frac{\partial\varphi_t}{\partial x_1}\right)dx_1dx_2,\end{gathered} \tag{44}$$

where D_t is the variable torsional rigidity of the growing body and φ_t is the unknown harmonic function.

At the stage of growth of the twisted body, we also have two possible versions of the statement of the problem.

To solve the first version (where the torque $M(t)$ is given), one should use the boundary condition to find φ_t, then determine v_i and S_{ij} by formulas (42), use formulas (44) to find the torsion angular velocity, and reconstruct the true stresses, displacements, and torsion angle by the formulas

$$\sigma_{ij}(x_1,x_2,t)=G(t)\left[\frac{\sigma_{ij}(x_1,x_2,\tau_0^*)}{G(\tau_0^*)}\left(1+\int_{\tau_0^*}^{t}R(t,\tau)d\tau\right)+\right.$$
$$\left.+\int_{\tau_0^*}^{t}\left(S_{ij}(x_1,x_2,\tau)+\int_{\tau_0^*}^{\tau}S_{ij}(x_1,x_2,\varsigma)d\varsigma R(t,\tau)\right)d\tau\right],$$
$$u_i(x_1,x_2,t)=u_i(x_1,x_2,\tau_0^*)+\int_{\tau_0^*}^{t}v_i(x_1,x_2,\tau)d\tau, \quad (45)$$
$$\theta(t)=\theta(\tau_0^*)+\int_{\tau_0^*}^{t}\theta^\bullet(\tau)d\tau.$$

To solve the second version of the problem (where the torsion angle $\theta(t)$ is given), one should find φ_t, determine v_i and S_{ij} by formulas (42), reconstruct u_i and σ_{ij} by formulas (45), and find the torque according to formula (5).

To solve the boundary value problem (43), one can apply methods of the theory of functions of one complex variable, which we will discuss below.

Thus, the study of the torsion problem for a growing body at the stage of continuous accretion is complete.

4 Deformation of the rod after the termination of growth

Let accretion terminate at time τ_2. At this time, the body occupies the domain Π_2 whose cross-section Ω_2 is bounded by the contour L_2. On the time interval $t\in[\tau_2,\tau_3]$ (where τ_3 is the time at which the next accretion stage begins), the problem for the nonvarying body is similar to problem (32)–(36), where the initial–boundary condition on the growth boundary $L^*(t)$ is lacking,

$$\frac{\partial\sigma_{13}}{\partial x_3}=0,\quad \frac{\partial\sigma_{23}}{\partial x_3}=0,\quad \frac{\partial\sigma_{13}}{\partial x_1}+\frac{\partial\sigma_{23}}{\partial x_2}=0,$$
$$D_{ij}=\frac{1}{2}\left(\frac{\partial v_i}{\partial x_j}+\frac{\partial v_j}{\partial x_i}\right)=0,\ D_{ii}=D_{12}=0,\ i,j=1,2,3,$$
$$\sigma_{13}=2G(I+N(\tau_0^*,t)\varepsilon_{13},\quad \sigma_{23}=2G(I+N(\tau_0^*,t)\,\varepsilon_{23},$$
$$(x_1,x_2)\in L_2:\quad \sigma_{13}n_1+\sigma_{23}n_2=0, \quad (46)$$
$$M(t)=\iint_{\Omega_2}(x_1\sigma_{23}-x_2\sigma_{13})dx_1\,dx_2.$$

By analogy with the preceding, we obtain the boundary value problem

$$\frac{\partial S_{13}}{\partial x_3}=0,\quad \frac{\partial S_{23}}{\partial x_3}=0,\quad \frac{\partial S_{13}}{\partial x_1}+\frac{\partial S_{23}}{\partial x_2}=0,\ S_{13}=2D_{13},\quad S_{23}=2D_{23},$$
$$D_{ij}=\frac{1}{2}\left(\frac{\partial v_i}{\partial x_j}+\frac{\partial v_j}{\partial x_i}\right)=0,\ D_{ii}=D_{12}=0,\ i,j=1,2,3, \quad (47)$$
$$(x_1,x_2)\in L_2:\quad S_{13}n_1+S_{13}n_2=0;$$
$$M^{0\bullet}(t)=\iint_{\Omega_2}(x_1S_{23}-x_2S_{13})\ldots dx_1\,dx_2+\int_{L_2}(x_1\sigma_{23}^0-x_2\sigma_{13}^0)s_n\,dl.$$

Problem (47) is obviously similar to problem (26). In this case, formulas (27), where the function φ_{τ_1} should be replaced by $\varphi_{\tau_2}=\varphi(x_1,x_2,\tau_2)$, apply to the variables v_i and S_{ij},

$$v_1=-\theta^\bullet(t)x_2\,x_3,\ v_2=\theta^\bullet(t)x_3\,x_1,\ v_3=\theta^\bullet(t)\varphi_{\tau_2},$$
$$S_{13}=\theta^\bullet(t)\left(\frac{\partial\varphi_{\tau_2}}{\partial x_1}-x_2\right),\ S_{23}=\theta^\bullet(t)\left(\frac{\partial\varphi_{\tau_2}}{\partial x_2}+x_1\right). \quad (48)$$

To find the function φ_{τ_2}, it suffices to solve the Neumann problem

$$\frac{\partial^2\varphi_{\tau_2}}{\partial x_1^2}+\frac{\partial^2\varphi_{\tau_2}}{\partial x_2^2}=0,\ (x_1,x_2)\in L_2:\quad \frac{\partial\varphi_{\tau_2}}{\partial n}=x_2n_1-x_1n_2, \quad (49)$$
$$M^{0\bullet}(t)=\theta^\bullet(t)D_2+\int_{L_2}(x_1\sigma_{23}^0-x_2\sigma_{13}^0)s_n\,dl,$$
$$D_2=\iint_{\Omega_2}\left(x_1^2+x_2^2+x_1\frac{\partial\varphi_{\tau_2}}{\partial x_2}-x_2\frac{\partial\varphi_{\tau_2}}{\partial x_1}\right)dx_1\,dx_2. \quad (50)$$

To find the solution of the resulting Neumann boundary value problems, one can use methods of complex analysis, which are presented, say, in [9]. The true stresses and displacements can be completely reconstructed with the use of the decoding formulas (45).

5 Conclusions

- A torsion theory of growing viscoelastic prismatic rods is constructed. Statements of the corresponding classical and nonclassical initial–boundary value problems are given.
- Solution methods for such problems are suggested based on the reduction on nonclassical problems for growing bodies to elasticity problems with a parameter. The subsequent use of complex analysis and the decoding formulas developed in the paper permits one to reconstruct the true values of stress-strain characteristics of growing rods.
- A new effect in mechanics is revealed. In the classical case, the maximum tangential stress intensity is known to be attained on the boundary of the body. In the case of a growing rod, the maximum tangential stress intensity can be attained at an arbitrary point of the grown part depending on the growth mode.
- The results obtained can serve as a basis for modeling the processes of crystal growth from melt, manufacturing or repairing machine parts by spraying, weld overlay cladding, or precipitation, and a number of other additive manufacturing processes (see also [10–16]).

Acknowledgements

This work was financially supported by the Russian Science Foundation under Grant 14-19-01280.

References

1. N. Kh. Arutyunyan, A. V. Manzhirov, and V. E. Naumov, Contact Problems in Mechanics of Growing Bodies [in Russian] (Nauka Publ., Moscow,1991)

2. A. V. Manzhirov, Journal of Applied Mathematics and Mechanics **59** (5), 805–816 (1995)
3. N. Kh. Arutyunyan and A. V Manzhirov, Contact Problems in the Theory of Creep [in Russian] (NAS Armenia, Yerevan,1999)
4. A. V. Manzhirov, Key Engineering Materials, 535–**536**, 89–93 (2013) doi: 10.4028/www.scientific.net/KEM.535-536.89
5. A. V. Manzhirov, S.A. Lychev, and N. K. Gupta, Proc. Indian Natl Sci Acad. **79** (4), 529–545 (2013) doi: 10.16943/ptinsa/2013/v79i4/47989
6. A. V. Manzhirov, Proceedings of ASME 2014 International Mechanical Engineering Congress & Exposition, Volume **9**: Mechanics of Solids, Structures and Fluids (Montreal, Quebec, Canada) 10 p. (2014) doi:10.1115/IMECE2014-36712
7. A. Manzhirov and N. Gupta, IUTAM Symposium on Growing solids. Symposium Materials (In-t Probl. Mech. RAS, Moscow, 2015) 73–76
8. N. Kh. Arutyunyan and B. L. Abramyan, Torsion of Elastic Solids [in Russian] (Fizmatgiz, Moscow, 1963)
9. N. I. Muskhelishvili, Some Basic Problems of the Mathematical Theory of Elasticity (Springer, Netherlands, 2010)
10. A. V. Manzhirov and S. A. Lychev, Lecture Notes in Engineering and Computer Science: World Congress on Engineering (London, IAENG, 2014) 1404–1409
11. S. A. Lychev and A. V. Manzhirov, Lecture Notes in Engineering and Computer Science: World Congress on Engineering (London, IAENG, 2014) 1327–1332
12. A. V. Manzhirov and S. A. Lychev, Transactions on Engineering Technologies: World Congress on Engineering 2014 (Springer, Netherlands, 2015) 99–115 doi: 10.1007/978-94-017-9804-4
13. S. A. Lychev, A. V. Manzhirov, and P. S. Bychkov, Transactions on Engineering Technologies: World Congress on Engineering 2014 (Springer, Netherlands, 2015) 239–254 doi: 10.1007/978-94-017-9804-4
14. A. V. Manzhirov and D. A. Parshin, Mechanics of Solids. **50** (5), 559–570 (2015) doi: 10.3103/S0025654415050076
15. A. V. Manzhirov and D. A. Parshin, Mechanics of Solids. **50** (6), 657–675 (2015) doi: 10.3103/S0025654415060072
16. A. V. Manzhirov, Lecture Notes in Engineering and Computer Science: Proceedings of the World Congress on Engineering 2015 (London, IAENG, 2015) 710–714

Impact of Village Resistance Values on Simulating Results of Flood Routing in Floodplains

Fengde Zhang[1] and Xiaolei Zhang[2]

[1]*Sanjiang Engineering Construction Management Bureau of Heilongjiang Province, Haerbin, China*
[2]*State Key Laboratory of Water Resources and Hydropower Engineering Science, Wuhan University, Wuhan, China*

Abstract. 2-D hydrodynamic model using unstructured triangular meshes was first established. And the model was verified against the analytical solutions and experimental data of dam-break flows in laboratory. Finally, the model was employed to preliminarily simulate inundated floods routing in Lankao and Dongming floodplains of Lower Yellow River (LYR), and the impact of roughness value in village regions on modeling results was thoroughly analyzed. The results show that the water depth for hollowing village region was largest, and the maximum value of water depth increasing was about 0.57 m. Different roughness values in village region had significant impact on flood routing near village. To be specific, the water depth of n = 0.12 was 0.25m larger than that of n = 0.06, the corresponding flow velocity decreased by 0.24 m/s.

1 Introduction

Floodplain is the important part of river, as well as habitat for human life and production. Once the overbank floods happen, the floodplain will become the natural flood detention zone, playing a role of flood detention, but it will also cause some degree of flooding, resulting in the catastrophic consequences, such as loss of people's lives and property losses. In order to assess flood risk in floodplains, the 2-D hydrodynamic model is often used to simulate the process of flood routing. The 2-D hydrodynamic from Huayuankou to Jiahetan based on the finite volume method of unstructured mesh is established, and the water-blocking buildings such as channel and road are treated with the internal boundary conditions [1]. A numerical model coupled by 1-D and 2-D numerical models based on the finite volume method is developed to simulate flood routing in Daqing River floodplains [2]. A 2-D model based on alternating direction implicit method is built, which is used to simulate flood routing in the reaches from Jiahetan to Gaocun of LYR [3]. The sediment transport and riverbed deformation process aren't generally considered by foreign researchers while simulating flood routing in floodplains, and the diffusion wave model is mostly adopted [4,5]. 2-D shallow water equations are used to simulate flood routing in Toce River [6]. The process of flood routing in the floodplains of Severn River is simulated using the diffusion wave model based on the high accuracy digital terrain elevation [7]. The above research results mostly focus on the development of mathematical model and flood routing simulation in the floodplain.

There are many villages in floodplains, and village resistance has the important effect on overbank flood routing. But now there are fewer researches on the influence of different village resistance values on flood routing process. In this paper a 2-D hydrodynamic model using unstructured triangular meshes was employed to simulate flood routing in the typical floodplains of LYR, and the impact of village resistance on modeling results was thoroughly analyzed.

2 The mathematical model

2.1 Governing equations

The process of flood routing in natural rivers and on floodplains can be described by a set of shallow water equations for flows over a horizontal plane. The governing equations used in this model can be written in the following conservative form:

$$\frac{\partial \mathbf{U}}{\partial t}+\frac{\partial \mathbf{F}}{\partial x}+\frac{\partial \mathbf{M}}{\partial y}=\frac{\partial \mathbf{F}}{\partial x}+\frac{\partial \mathbf{M}}{\partial y}+\mathbf{S} \quad (1)$$

Where **U** is vector of conserved variable; **F** and **M** is convective mass and momentum flux vector, respectively; **F** and **M** is diffusive vector related to the turbulent stress in the x and y direction, respectively; **S** is source term including bed friction and bed slope; and t is time. The above terms can be expressed in detail as:

$$\mathbf{U}=\begin{bmatrix} h \\ hu \\ hv \end{bmatrix}, \quad \mathbf{F}=\begin{bmatrix} hu \\ hu^2+\frac{1}{2}gh^2 \\ huv \end{bmatrix}, \quad \mathbf{M}=\begin{bmatrix} hv \\ huv \\ hv^2+\frac{1}{2}gh^2 \end{bmatrix}, \quad \mathbf{F}=\begin{bmatrix} 0 \\ \tau_{xx} \\ \tau_{yx} \end{bmatrix},$$

$$\mathbf{M} = \begin{bmatrix} 0 \\ \tau_{xy} \\ \tau_{yy} \end{bmatrix}, \text{ and } \mathbf{S} = \begin{bmatrix} q_s \\ hfv + gh(S_{bx} - S_{fx}) \\ -hfu + gh(S_{by} - S_{fy}) \end{bmatrix} \quad (2)$$

Where u, v is respectively depth-averaged velocity in x and y direction; h is total water depth; q_s is source (or sink) discharge per area; g is gravitational acceleration; f is Coriolis acceleration due to the Earth's rotation; S_{bx} and S_{by} is respectively bed slope in x and y direction; S_{fx} and S_{fy} is friction slope in x and y direction, respectively; and τ_{xx}, τ_{xy}, τ_{yx} and τ_{yy} are turbulent shear stresses.

2.2 Numerical solution

The governing equations are solved using the finite volume method based on an unstructured triangular mesh in the model. A cell-centred finite volume method is adopted in this model, in which the average values of the conserved variables are stored at the centre of each cell, with the three edges of each cell defining the interface of a triangular control volume. At an interface between two neighbouring cells, the calculation of flow fluxes can be treated as a locally one-dimensional problem in the direction normal to the interface, thus the fluxes can be obtained by an approximate Riemann solver. In the current model, a Roe's approximate Riemann solver, with the scheme of MUSCL [8] is employed for evaluating the normal fluxes across the cell interface, with a procedure of predictor-corrector time stepping being used to provide second-order accuracy in both time and space. Furthermore, a refined procedure for treating wetting and drying fronts is used in the model, and this method has been shown to be effective in simulating the wetting and drying processes due to rapid flooding. In addition, a small value of minimum water depth is usually used to simulate the evolution of wetting and drying fronts, and the specification of a minimum water depth also depends on the grid size, time step and the topographical features of a study area.

2.3 Model validation

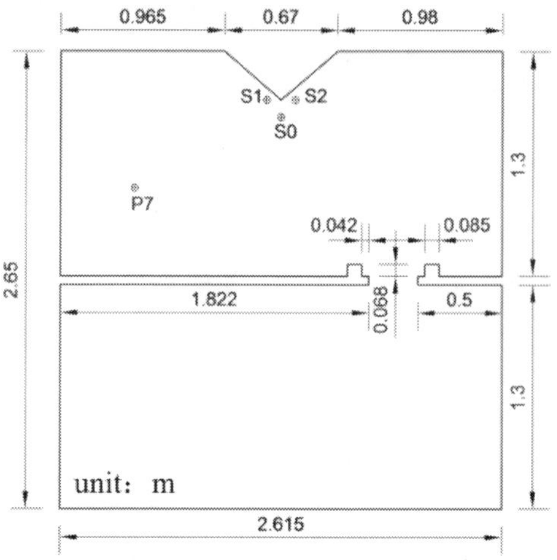

Figure 1. Sketch of a dam-break flow test.

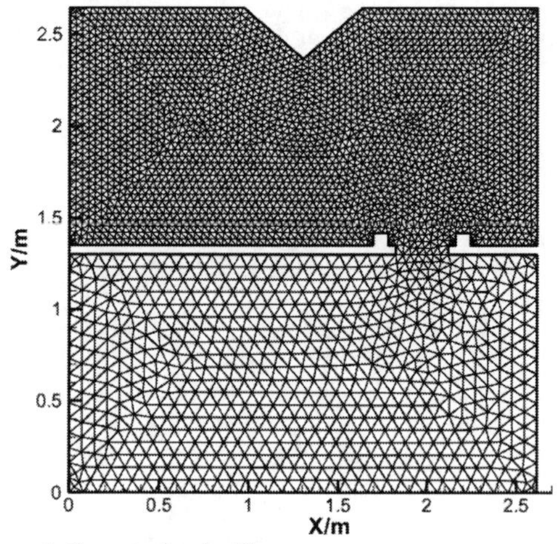

Figure 2. Computational grids.

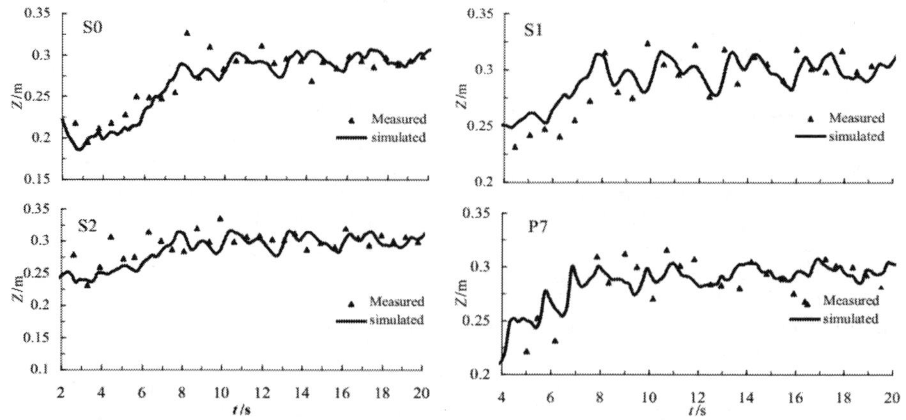

Figure 3. Comparisons between the calculated and observed water level hydrographs.

The above model is used to simulate a dam-break flow in a pool with a pyramidal obstacle, and the hydraulics model test is completed by Brufau in CITEEC laboratory of Spain [9]. The model test can be used to test the capacity of model on better capturing the shock wave in dam-break floods. This model experiment is done in a closed tank, which is divided into two parts by a solid wall, setting a sluice gate in the non-symmetric position of solid wall, and the width of gate is 0.293 m, the plane layout of model is shown in Figure 1. The closed pool's bottom is flat. The initial water depth in the pool upstream is 0.5 m, and the water depth in the pool downstream is 0.1 m. If the gate is removed suddenly, dam-break flow will happen. The whole calculation domain is composed of 24133 triangle meshes, and the computational grid is shown in Figure 2. The surrounding walls of pool and partition are concrete and solid boundary, and the corresponding roughness value is 0.018. The minimum water depth is 0.001 m, and time step is 0.025 s. The change processes of water level at different locations are given. The comparison results show that close agreement is obtained between the model predictions and the observed data. So the model can simulate the routing process of unsteady flow well in complicated boundary conditions.

3 Simulation of flood routing in typical floodplains of LYR

3.1 Simulation domain and calculation conditions

Lankao and Dongming floodplains in the LYR are chosen as the simulation domain, covering area of 193 km^2. The minimum width is 1.2 km, and the maximum width is 11.2 km. There exist lots of villages in the floodplain, which have obvious effects on overbank flood routing. Sketch of Lankao and Dongming floodplains are shown in Figure 4.

The flood of August 1982 is selected for simulation, which is the largest flood in the LYR since the Sanmenxia project has been built. The peak discharge at Jahetan station is 14500 m^3/s. In present study the process of overbank flood routing is focused, considering the recent bankfull discharge in the reaches has been restored to 7000 m^3/s, so the bankfull discharge should be subtracted from the overall discharge, and the final calculated flood process is illustrated in Figure 5. In order to reflect the effects of village resistance on simulation results, village regions are center-hollowed, and the roughness values are set to 0.06, 0.08, 0.10 and 0.12, respectively. Then the verified model is employed to simulate the overbank flood routing and the effects of different roughness values on calculation results are further analyzed.

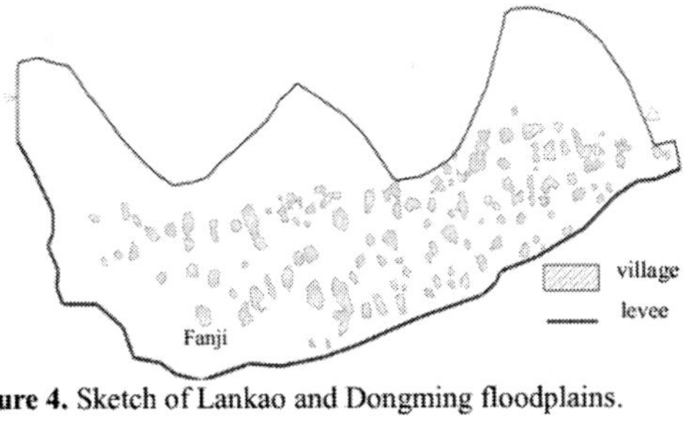

Figure 4. Sketch of Lankao and Dongming floodplains.

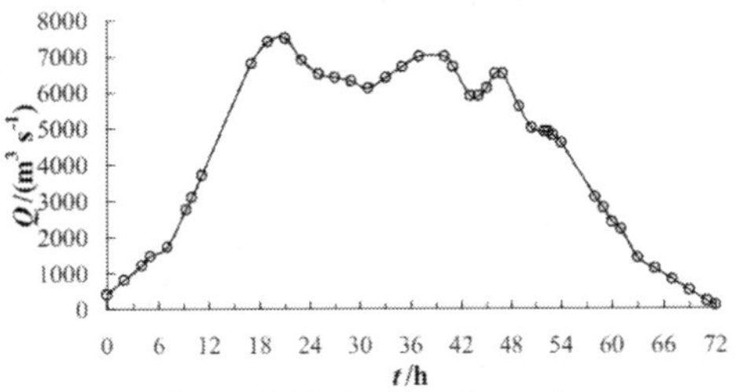

Figure 5. Discharge hydrograph.

3.2 Effect of village roughness coefficient

In order to reflect on the result of village resistance simulation, take Fanji village for an example, four representative points near the village are chosen. Local grid structure and the locations of measured points are shown in Figure 6.

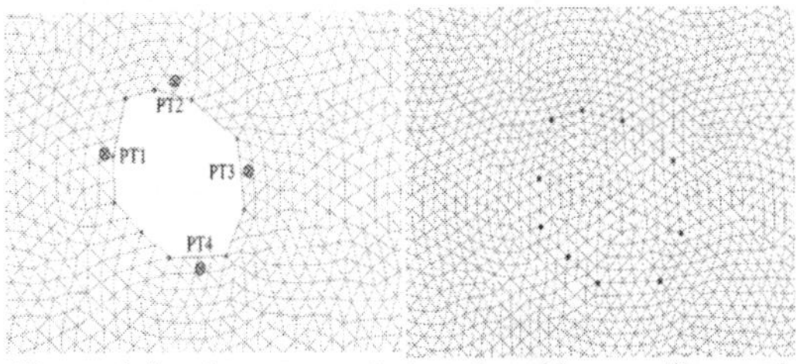

Figure 6. Local unstructured triangular grids near village and observation sites.

The analysis on results of water depth and flow velocity under different roughness value conditions at PT3 shows that the water depth under the scenario of center-hollowed village region is largest, which is 0.47 to 0.57 m larger than that under different roughness values conditions.

For different conditions of roughness values in village region, the water depth of each measuring point generally presents a gradually increasing trend with the increase of roughness value in village region, which is especially significant in the period of 18 to 36 h. Compared with the condition of n = 0.06, the corresponding maximum water depth at each measuring point increase by 0.09, 0.19, 0.25 m, respectively under the condition of n = 0.08, 0.10 and 0.12. In addition, at the same time, the flow velocity at PT1 and PT3 under the scenario of center-hollowed village region is relatively smaller, and the flow velocity at PT2 and PT4 under the scenario of center-hollowed village region is relatively larger. The flow velocity of each measuring point generally presents a gradually decreasing trend with the increase of roughness value in village region. Compared with the condition of n = 0.06, the corresponding maximum flow velocity at each measuring point decrease by 0.11, 0.19, 0.24 m/s, respectively under the condition of n = 0.08, 0.10 and 0.12. Overall, the effects of the center-hollowed village regions condition on water depth and flow velocity are very obvious. For different conditions of roughness value in village regions, the effects on water depth near the village are smaller, and the effects on flow velocity near the village are larger, especially for PT1 and PT3 in the direction of the water flow.

The water depth and velocity distribution near Fanji village with the roughness value of 0.06 and 0.12 during the flood peak are shown in Figure 7. From the figure, because of the influence of different roughness values of village regions, the water depth of village regions gradually increased with the increase of the roughness, while the flow velocity gradually decreased. When the roughness n = 0.06, the water depth is 2.81 m, and the flow velocity is 0.40 m/s; When the roughness n = 0.12, the water depth is largest and the flow velocity is smallest. The maximum water depth is 2.93 m/s, the minimum flow velocity is 0.32 m/s.

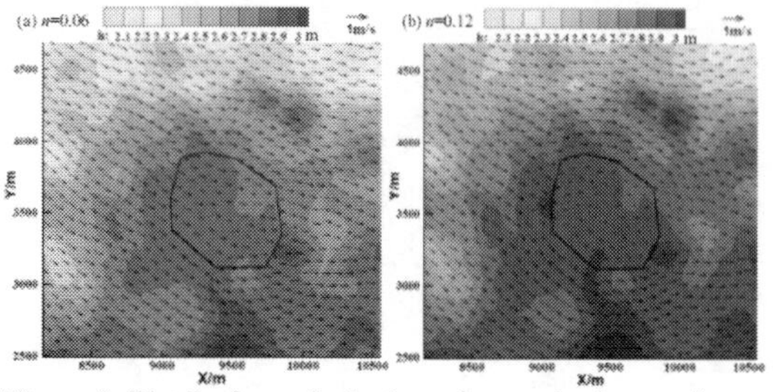

Figure 7. Distributions of velocity and water depth around a village during the flood peak.

4 Conclusions

In this paper a 2-D hydrodynamic model was set up and verified, and the process of overbank floods routing in typical floodplains of LYR was simulated. The effects of different roughness values of village regions on calculation results were analyzed. The analysis results show that center-hollowed village regions and different roughness values have significant effects on the calculation results. Specifically, the water depth under the scenario of center-hollowed village region is largest, with an increase in the maximum water depth of 0.57 m; the maximum water depth under n = 0.12 is 0.25 m greater than that under n = 0.12, with the corresponding reduction value in velocity of 0.24 m/s. In terms of practical engineering problems, the treatment method and roughness value for village regions should be considered according to the actual conditions.

Acknowledgement

This work was financially supported by the Natural Science Foundation of China (No. 51379156), the National Nonprofit Research Program of China (No. 201401038) and the Science and Technology Research Project of the Education Department of Henan Province (No. 14A570001).

References

1. H. Wang, C. Zhou, Y. Wu, Q. Wan, 2-D flood simulation of the lower Yellow River from Huayuankou to Jiahetan. Advances in Water Science, **13**, 215-222 (2002).
2. D. Li, Y. Lin, Y. Xu, Z. Zhou, Numerical model of flood propagation of rivers and flood detention basin. Journal of Tianjin University, **42**, 47-55 (2009).
3. D. Sun, X. Liao, P. Wang, X. Zhang, X. Liu, Influence of productive embankment on river flooding by 2-D numerical simulation. Journal of Hydrodynamics (Ser. A), **22**, 24-30 (2002).
4. D. Yu, S. N. Lane, Urban fluvial flood modelling using a two-dimensional diffusion-wave treatment, part 2: development of a sub-grid-scale treatment. Hydrological Processes, **20**, 1567-1583 (2006).
5. M. S. Horritt, P. D. Bates, M. J. Mattinson, Effect of mesh resolution and topographic representation in 2D finite volume models of shallow water fluvial flow. Journal of Hydrology, **329**, 306-314 (2006).
6. V. Caleffi, A. Valiani, A. Zanni, Finite volume method for simulating extreme flood events in natural channels. Journal of Hydraulic Research, **41**, 167-177 (2003).
7. P. D. Bates, M. D. Wilson, M. S. Horritt, D. C. Mason, N. Holden, A. Currie, Reach scale floodplain inundation dynamics observed using airborne Synthetic Aperture Radar imagery: Data analysis and modelling. Journal of Hydrology, **328**, 306-318 (2006).
8. J. Wang, R. Liu, A comparative study of finite volume methods on unstructured meshes for simulation of 2D shallow water wave problems. Mathematics and Computers in Simulation, **53**, 171-184 (2000).
9. P. Brufau, M. E. Vázquez-Cendón, P. García-Navarro, A numerical model for the flooding and drying of irregular domains. International Journal for Numerical Methods in Fluids, **39**, 247-275 (2002).

 DOI: 10.1051/matecconf/20179512004

Multiscale Modeling for Mechanical Properties of Cancellous Bone Based on the Schwarz Surface

ZhiQiang Huang[1,2], Yufeng Nie[1], Zihao Yang[1], YiQiang Li[1] and NiuPing Guo[3]

[1]Department of Applied Mathematics, Northwestern Polytechnical University, Xi'an, 710072, China
[2] Department of Mathematics, TaiYuan University of Science and Tcehnology, TaiYuan, 030024, China
[3] School of Statistics, Xi'an University of Finances and Economics, Xi'an, 710100, China

Abstract. Microstructure and the mechanical properties of cancellous bone were modeled by multiscale finite element method in this paper. Microstructure of the cancellous bone determines its mechanical properties and a rational geometry modeling of the cancellous bone is fundamental to predict the material properties. In this paper, we generate a new parameterized microstructural unit cell model based on triple periodic minimal surfaces (TPMS) which is established by the Schwarz surface with various volume fractions. This model truly simulates the microstructure of the cancellous bone and satisfies some biological and mechanical requirements. By using the new unit cell model, the second-order two-scale (SOTS) method is applied to predict the mechanical properties of cancellous bone. The predictive the mechanical properties of numerical results are in good agreement with the experimental data reported in literature, which demonstrate that the established unit cell model is applicable and the second-order two-scale method is valid to predict the mechanical properties of cancellous bone.

1 Introduction

The investigation of the effective properties of cancellous bone is an important topic in biomechanics which has been especially intensive as the requirement for clinical practice in recent years, for example, it may lead to the early detecting of osteoporosis[1] and the artificial orthopaedic devices, especially the complex problem of bone implant interaction emerged and recalled the importance of the interrelations between biological and mechanical properties[2] .

To understand and study the mechanical and biological functions of cancellous bone, it is essential to clarify the relationships between the mechanical properties and the structural changes at the microstructural level. Such understanding is important to analyzing implants and manufacturing synthetic bone substitutes, at the same time it can serve engineers as a guide in designs of novel artificial biomimetic materials for a wide range of engineering applications [3]. For the investigation of cancellous bone properties different approaches are developed, one of them being of experimental methods, the other ones concentrating upon developing a convenient mathematical or mechanical model. In the aspect of numerical simulation, various analytical and computational models have been proposed to predict mechanical properties of bone at these different structural scales. Sandra Ilic(2010) considers the application of multiscale finite element method to the modeling of cancellous bone as an alternative for Biot's model[4]. Hamed,Elham(2010) predict analytically the effective elastic constants of cortical bone by modeling its elastic response at these different scales[3]. A. Jaziri (2012) propose a new elastoplastic damage coupled model for the modelling of trabecular bone behavior which the damage is carried out thanks to the limit analysis based on the MCK criterion[2]. Sansalone(2014) proposes a new methodology which coupling modeling and micromechanical homogenization to estimate the material properties of bone while taking into account the uncertainties on the bone micro- and nanostructure[5] .

Due to the complicated geometry structure of cancellous bone, different mode of the unit cell can be proposed for its modeling. Most of geometric models were proposed by describing the cancellous bone using the terms "rod-like" and "plate-like" for a subjective classification of cancellous bone. Mclehaeny(1970) described the unit cell as the cubic form which relationships was predicted between bone density and Young's modulus of cancellous bone[6]. Beaupre and Hayes (1985) think the unite cell as a cell body containing spherical cavity[7]. Sandra Ilic (2010)are described for the unit cell as a cubic form with the side length which treat the solid frame as a system of thin walls whose thickness and width can be varied.[4]The other unit cell was used frequently to include Bar-Net model, Plate-Frame-Hole model, Honeybomb model, Plate-bar model etc.

However, these techniques are based on the assumption that all trabecular in the different position bones are either in the form of plates or rods, the microstructure in above models is very simplified and thus less precise results are obtained. An ideal unit cell model structure of cancellous bone should have appropriate porosity and pore-structure, to this purpose, a periodic embedded Schwarz surface (see the Fig.2)was established as the unit cell in our model, because of the Schwarz surface is more permeable, it also has a smooth surface to open the portion of the tissue which would facilitate the inflow of nutrients and the disposal of metabolic waste [8]. Most important of all, the new unit cell meet satisfy some biological and mechanical requirements of the cancellous bone.For the past few years, the second-order two-scale (SOTS) analysis method is introduced [Cui, Shin and Wang (1999)] to predict the physical and mechanical properties of periodic composites[8]. Microstructural modeling and mechanical properties computation of three-dimensional 4-directional braided composites were studied[9]. By the second-order correctors, the microscopic fluctuation of physical and mechanical behaviors inside the materials can be captured more accurately. In this paper, a new parameterized microstructural unit cell model is established, Schwarz primitive surface and it truly simulates the microstructure of cancellous bone. The parameterized design for the unit cell model is implemented by employing the TETGEN software. And then based on the established model and the mechanical parameters are predicted by using the second-order two scale method.

2 Microstructure analysis of cancellous bone

Structurally, cancellous bone is considered like a composite material with a complex structure. Form the mesoscale, ranging from hundred micrometres to several millimetres, or larger, depending on the bone size, consists of a porous network of trabeculae (Fig.1.). The pores, typically in the order of 1 mm, are filled with bone marrow, fat and bone cells. In cortical bone, this level represents randomly arranged osteons embedded in an interstitial lamella, with some resorption cavities, all surrounded by a circumferential bone.

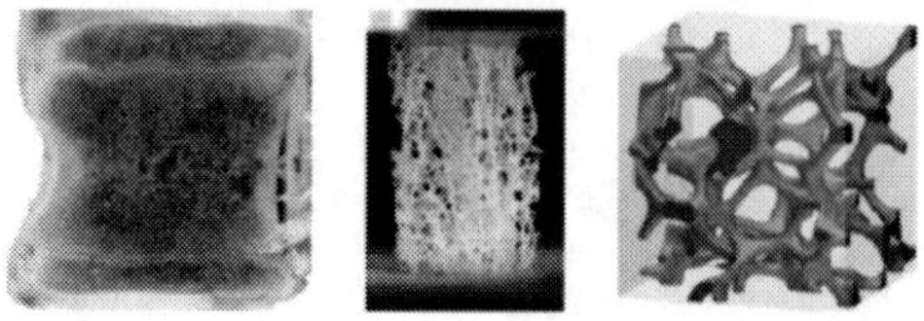

Figure 1. Specimens of trabecular bone obtained from vertebrae[10]

3 Unit cell modeling of cancellous bone

Hyperbolic surfaces have attracted the attention of physicists, chemists and biologists as they commonly exist in natural structures. Amongst various hyperbolic surfaces, minimal surfaces are the most studied, these surfaces are known as Triply Periodic Minimal Surfaces (TPMS). As the simulation of cancellous bone by the multiscale FE, it is important for unit cell of modelling the cancellous bone to have the suitable pore size, sufficient porosity, and high pore interconnectivity. The triply periodic minimal surface have been received much attentions in the recent biomaterial literatures and provide good analytic description of highly porous structures[11], however, there were few reports about the research on property of the cancellous bone simulation used these TPMS.

In a general way, a periodic surface can be generally defined as

$$\phi(\mathbf{r}) = \sum_{k=1}^{K} \mathbf{A}_k \cos[2\pi(\mathbf{h}_k \cdot \mathbf{r}) / \lambda_k + p_k] = \mathrm{C} \qquad (1)$$

An important sub-class of TPMS is those that partition space into two disjoint but intertwining regions that are bi-continuous. An example of such surfaces includes the so-called Schwartz G surface defined as

$$\begin{aligned} \phi(\mathrm{x},\mathrm{y},\mathrm{z}) &= \sin 2\pi \mathrm{x} \cos 2\pi y + \sin 2\pi y \cos 2\pi z \\ &+ \sin 2\pi z \cos 2\pi x = C \end{aligned} \qquad (2)$$

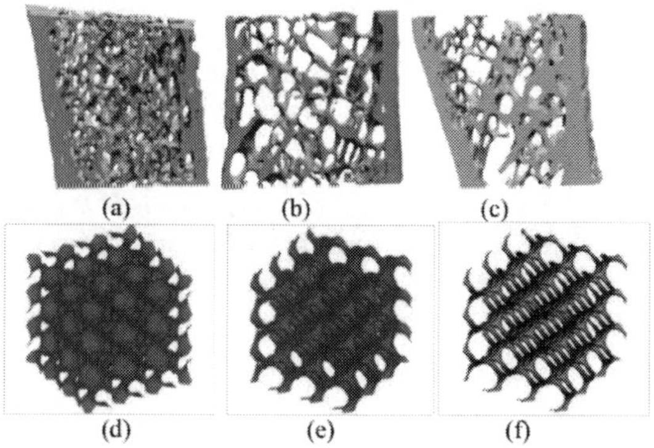

Figure 2. MicroCT imaging of iliac bone (a).In a young and healthy subject: the cortices are thick and the trabecular network is dense, (b).In a postmenopausal osteoporosis, the holes inside the network (c).In a male with idiopathic osteoporosis, note the conversion of plate into rods, although the connectivity is rather well even if trabecular are thin.[12] (d-f).The microstructure of the unit cell based on the Schwarz surface of a different porosity was used in this work.

There are the potential benefits of Schwarz surface based pore architectures which were put forward to taking advantage of these newly proposed pore geometries as unit cell to simulate the cancellous bone. Viewed from biology and biomechanics, beneficial properties can be provided by TPMS for the cancellous bone as attractive candidates for biomorphic porous architectures.

The advantages of the method utilized in the after section are satisfy some biological and mechanical requirements, such as : From the viewpoint of morphology, the unit cell based on Schwarz surface is very vividly to simulating the cancellous bone structure, and that it is important for unit cell of modeling the cancellous bone to have the suitable pore size, sufficient porosity, and high pore interconnectivity. From the viewpoint of mechanical requirements, the researched results imply that the Schwarz surface more stable structures [8] which the unit cell should consider the sufficient mechanical strength and stiffness for the cancellous bone to support the growing tissue. From the viewpoint of biological requirements, the permeability of the unit cell should be high enough to provide superior diffusion which would facilitate the inflow of nutrients and the disposal of metabolic waste. The surface of the unit cell based on Schwarz surface was sufficiently smooth which would be very beneficial to inflow and outflow of nutrients and metabolic waste. Micro-structure with a high porosity of the unit cell provides more space for cells to move into and begin to thrive, the unit cell based on Schwarz surface has a large surface area which enables much more cell attachment and growth and subsequently sufficient for tissue repair. From the viewpoint of self-optimizing of bone tissue, above these characters of the unit cell based on Schwarz surface consistent with demand of modeling cancellous bone which must satisfy some biological and mechanical requirements, at same time, which provide insights into the reason behind the natural choice by biological systems.

4 Multiscale modeling and FE analysis

4.1 Second-order two-scale formulation

From solid mechanics, the elasticity problem of the structures made from periodic composites can be expressed as follows

$$\begin{cases} \dfrac{\partial}{\partial x_j}\left(C^{\varepsilon}_{ijhk}(x)\dfrac{1}{2}\left(\dfrac{\partial u^{\varepsilon}_h(x)}{\partial x_k}+\dfrac{\partial u^{\varepsilon}_k(x)}{\partial x_h}\right)\right)=f_i(x) & x\in\Omega \\ \mathbf{u}^{\varepsilon}(x)=\bar{\mathbf{u}}(x) & x\in\Gamma_1 \\ \sigma(u)\equiv v_j C^{\varepsilon}_{ijhk}(x)\dfrac{1}{2}\left(\dfrac{\partial u^{\varepsilon}_h(x)}{\partial x_k}+\dfrac{\partial u^{\varepsilon}_k(x)}{\partial x_h}\right)=p_i(x) & x\in\Gamma_2 \\ \Gamma_1\cap\Gamma_2=\varnothing,\ \ \Gamma_1\cup\Gamma_2=\partial\Omega \end{cases} \tag{3}$$

where $i,j,k,l=1,2,3$; Ω is composed of entire ε -size cells, $\mathbf{u}^{\varepsilon}(x)$ is the displacement solution; $f_i(x)$ is a body force independent of ε ; $C^{\varepsilon}_{ijhk}(x)$ are the elastic coefficients; Γ_1 and Γ_2 denote boundaries of Ω where displacement $\bar{\mathbf{u}}(x)$ and loads $p_i(x)$ are prescribed, respectively; v_j are the normal direction cosines of Γ_2 .

Let $y=x/\varepsilon\in Y$ denotes the local coordinates on the unit cell. Then $\mathbf{u}^{\varepsilon}(x)=\mathbf{u}(x,y)$ and $C^{\varepsilon}_{ijhk}(x)=C_{ijhk}(y)$. the second-order two-scale approximate solution can be expressed as

$$\mathbf{u}^{\varepsilon}(x)=\mathbf{u}_0(x)+\sum_{l=1}^{2}\varepsilon^l\sum_{\alpha_1\cdots\alpha_l=1,2,3}\mathbf{N}_{\alpha_1\cdots\alpha_l}(y)\frac{\partial^l\mathbf{u}_0(x)}{\partial x_{\alpha_1}\cdots\partial x_{\alpha_l}} \tag{4}$$

Where $\mathbf{u}_0(x)$ is the homogenized solution that was defined on Ω , $\mathbf{N}_{\alpha_1}(y)$ and $\mathbf{N}_{\alpha_1\alpha_2}(y)$ are matrix-valued functions defined on the unit cell Y . And $\mathbf{N}_{\alpha_1}(y)$, $\mathbf{N}_{\alpha_1\alpha_2}(y)$ and $\mathbf{u}_0(x)$ are determined in the following ways

For $l=1$, $\mathbf{N}_{\alpha_1 m}(y)(\alpha_1,m=1,2,3)$ are solutions of the following problem

$$\begin{cases} -\dfrac{\partial}{\partial y_j}\left[C_{ijhk}(y)\dfrac{1}{2}\left(\dfrac{\partial\mathbf{N}_{\alpha_1 km}(y)}{\partial y_h}+\dfrac{\partial\mathbf{N}_{\alpha_1 hm}(y)}{\partial y_k}\right)\right] \\ =\dfrac{\partial}{\partial y_j}\left(C_{ijm\alpha_1}(y)\right), \quad y\in Y \\ \mathbf{N}_{\alpha_1 m}(y)=0, \quad y\in\partial Y \end{cases} \tag{5}$$

From $\mathbf{N}_{\alpha_1 m}(y)(\alpha_1,m=1,2,3)$, the homogenized parameters $\hat{C}_{ijhk}$ are calculated by the following formula

$$\hat{C}_{ijhk}=\frac{1}{|Y|}\int_Y\left(C_{ijhk}(y)+C_{ijm\alpha_1}(y)\frac{1}{2}\left(\frac{\partial\mathbf{N}_{hmk}(y)}{\partial y_{\alpha_1}}+\frac{\partial\mathbf{N}_{h\alpha_1 k}(y)}{\partial y_m}\right)\right)dy \tag{6}$$

4.2 Finite element meshes of cancellous bone

Due to the complexity of the geometry model of the cancellous bone were developed, the efficiency and precision of SOTS calculation was importantly influenced by the quality to generate meshes. In this paper, the TETGEN software package was used to generate the FE meshes of the model. The geometry model of the unit cell developed by using the MATLAB software was imported into TETGEN software, and then the model with the mineralized collagen fibrils volume fraction 21% was meshed by 112413 tetrahedron elements shown in Fig. 3.

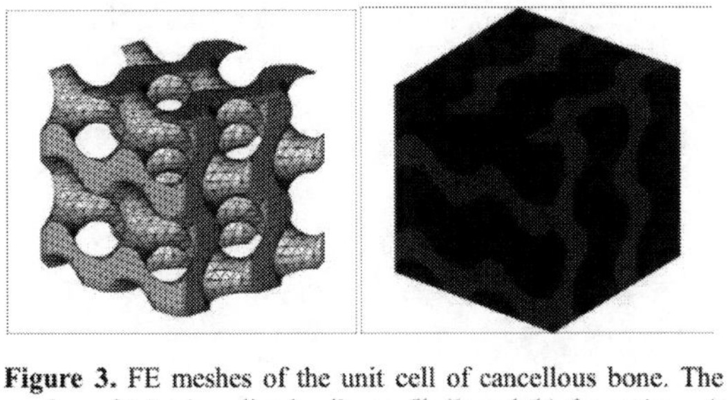

Figure 3. FE meshes of the unit cell of cancellous bone. The meshes of (a) mineralized collagen fibrils and (b) the entire unit cell

5 Numerical examples and performance analysis

In this section, the algorithm procedure of the SOTS method for predicting the mechanical properties of cancellous bone is presented, and then some numerical example are given.

5.1 Numerical examples

The finite element method was used to evaluate and simulate the mechanical parameters under different conditions such as volume fractions. We present a multiscale modeling approach for cancellous bone with an appropriate porosity and mechanical properties made from a two-phase composite of hydroxyapatite (HAp) embedded in a collagen matrix. We assume an isotropic linear elastic behavior for both collagen and hydroxyapatite, the material properties used in this paper of the collagen and the hydroxyapatite are summarized in Table 1.

Table 1. Properties of bone components selected in this work[13]

Material	Elastic modulus (GPa)	Poission's ratio	Volume fraction (%)
Collagen	1.5	0.28	90~40
Hydroxyapatite (HAp)	114	0.23	10~60

5.2 Effective material parameters

In Fig.4, the results corresponding to the unit cell with constants C in the interval 0.6–0.92 are considered. Dependence of Young's modulus and shear modulus on porosity is a smooth, monotonically decreasing function, while Poisson's ratio shows opposite behavior and it increases with increasing porosity.Young's modulus takes the values in the interval 753.21-3029.83 N/mm^2, the shear modulus 384.38-1381.32 N/mm^2 and Poisson's ratio 0.209-0.252 for the porosity from 0.4667-0.8156. These results agree well with the results obtained by Ashman and Rho (1988), that the structural elasticity modulus of cancellous bone has the values in the interval 985–2,110 N/mm^2.

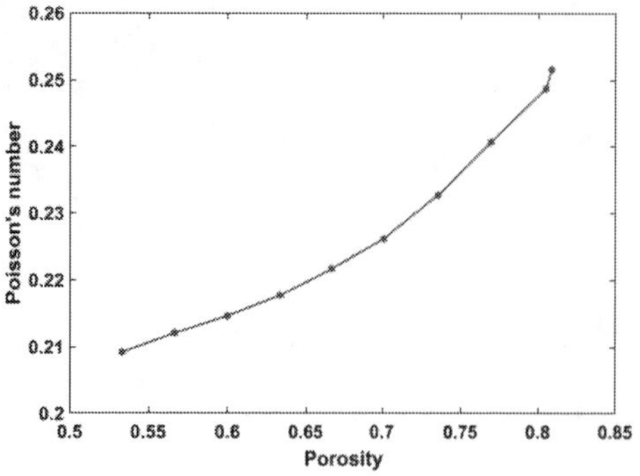

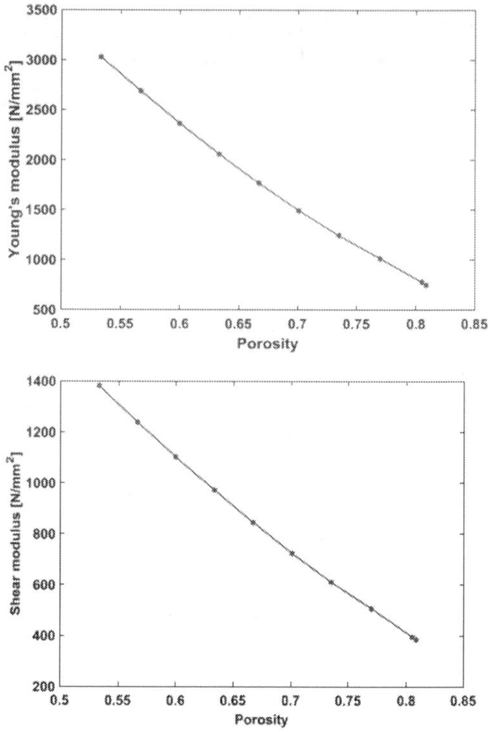

Figure 4. Change of effcetive material parameters versus porosity for the biphasic material

6 Conclusions

This study focuses on the determination of the relationship between mechanical functions and structure of bone tissues, it provides a significant improvement of the tools available for studying and understanding the mechanical functions of cancellous bone. In this paper, a new parameterized microstructural unit cell model is established and the SOTS method is applied to predict the mechanical properties of the cancellous bone. The geometry model of the unit cell developed by using the MATLAB software was imported into TETGEN software. And this model truly simulates the microstructure of the cancellous bone. From a modelling point of view, the advantages of the novel unit cell are satisfy some biological and mechanical requirements and the natural choice by biological systems. Based on these unit cell models, Poission's ratio and elastic modulus of cancellous bone are predicted by the SOTS method. The good agreement between the calculated results and experimental data and other research results, Moreover, the relations between different volume fractions and elastic constants are investigated. The results show that mechanical properties and volume fractions of cancellous bone components play an important role in its overall behavior.

Acknowledgement

This work is supported by the National Natural Science Foundation of China (Grant No.11471262), and also supported by Center for high performance computing of Northwestern Polytechnical University.

References

1. Basaruddin, K.S., N. Takano, and T. Nakano, *Stochastic multi-scale prediction on the apparent elastic moduli of trabecular bone considering uncertainties of biological apatite (BAp) crystallite orientation and image-based modelling.* Comput Methods Biomech Biomed Engin. **18**,2(2015).
2. A. Jaziri a , J.R.b., Multi-scale modelling of the trabecular bone elastoplastic behaviour under compression loading. European Journal of Computational Mechanics,**21**,3-6 (2012).
3. Hamed, E., Y. Lee, and I. Jasiuk, Multiscale modeling of elastic properties of cortical bone. Acta Mechanica, **213**,1-2(2010).
4. Ilic, S.H., Klaus Gilbert, Robert, Application of the multiscale FEM to the modeling of cancellous bone. Biomech Model Mechanobiol. **9**,1(2009).
5. Sansalone, V., S. Naili, and C. Desceliers, A stochastic homogenization approach to estimate bone elastic properties. Comptes Rendus Mécanique. **342**,5(2014).
6. Shin,J., et al., Finite Element Analysis of Schwarz P Surface Pore Geometries for Tissue-Engineered Scaffolds. Mathematical Problems in Engineering. **2012**,1-13(2012).
7. Hayes, G.S.B.W.C., Finite Element Analysis of a Three-Dimensional Open-Celled Model for Trabecular Bone. Journal of Biomechanical Engineering,**107**,3 (1985).
8. Cui, J. Z.; Shin, T. M.; Wang, Y. L. Two-scale analysis method for bodies with small periodic configuration. Structural Engineering and Mechanics, **7**, 6(1999)
9. Zihao Yang, J.C.,Yufeng Nie, Microstructural Modeling and Second-Order Two-Scale Computation for Mechanical Properties of 3D 4-Directional Braided Composites.CMC,,**38**, 3(2013)
10. Schwen, L.O.,Multiscale elasticity modeling of trabecular bone. 2013.
11. Yoo, D., New paradigms in hierarchical porous scaffold design for tissue engineering. Mater Sci Eng C Mater Biol Appl, **33**,3(2013).
12. Chappard,D.,etal.,Trabecular bone microarchitecture: a review. Morphologie. **92**,299 (2008)
13. Hamed,E.,et al., Multi-scale modelling of elastic moduli of trabecular bone. J R Soc Interfece,**9**,72(2012).

A Comparative Study on Turbulence Models for Simulation of Flow Past NACA 0015 Airfoil Using OpenFOAM

Peerakit Kekina and Chakrit Suvanjumrat

Department of Mechanical Engineering, Faculty of Engineering, Mahidol University, Salaya, Nakorn Pathom, 73170, Thailand
Laboratory of Computer Mechanics for Design (LCMD), Department of Mechanical Engineering, Faculty of Engineering, Mahidol University, Salaya, Nakorn Pathom, 73170, Thailand

Abstract. An implementation of C++ language open source code software, OpenFOAM, for simulation of flow past NACA 0015 airfoil was performed to access a suitable turbulence model. Three various turbulence models were selected which comprised of Spalart-Allmaras model, RNG k-ε model and Menter SST k-ω model, respectively. The SIMPLE algorithm and Upwind method was used to solve the governing equation to achieve flow solutions of computational fluid dynamic (CFD) models. The flow simulation obtained lift coefficient (C_L) and drag coefficient (C_D) to compare with the wind tunnel experiment data at Reynolds number (Re) of 160,000 and 360,000 with the large range airfoil angle of attack (AOA) from 0 to 20 degree. The suitable CFD model was the Menter SST k-ω model which obtained an average error of C_L and C_D less than 10.96% and 22.21%, respectively.

1 Introduction

Airfoil simulation referred to predict aerodynamics performance which was the active force acting on airfoil. Computational Fluid Dynamics (CFD) were the methodology to study phenomenon of fluid flow past airfoil with the benefit to reduce time and cost of physical experiments. The most of airfoil simulation was analyzed under the turbulent region. The precise simulation solution should be obtained when grid generation fined around wall region. The non-dimension y+ of first cell for CFD techniques was controlled into viscous sub-layer less than five [1] but some researcher had suggested one [2].

Reynolds-averaged Navier-Stokes equation (RANS) was always employed into CFD techniques which focused on the mean flow of turbulence. An extra term, Reynolds stresses, was estimated by turbulence models classified by number of additional term into the transport equation. The one transport equation, Spalart-Allmaras (S-A) model was developed to calculate kinematics eddy viscosity parameter and length scale in the term of local mean vorticity [3]-[5]. Two additional terms were used to estimate Reynolds stresses such as standard k-ε model, RNG k-ε model [6], Wilcox k-ω model and Menter shear stress transport (SST) k-ω model [7]. In an external aerodynamics problem, Spalart-Allmaras model, Wilcox k-ω model and Menter SST k-ω model were suggested for simulation. There are many researchers used commercial CFD software such as ANSYS, FLUENT and STAR-CD which contained turbulence models [8] [9]. Unfortunately they were limited on an expensive license cost. The Open Source Field Operation and Manipulation (OpenFOAM) software had been used C++ language for CFD code without license cost under GNU General Public License [10]. This research would apply turbulence models using OpenFOAM to determine an appropriate model for airfoil simulations and wind turbine blade design which was an external aerodynamics problem in a further work.

2 Turbulence models

In this research, three turbulence models comprised of Spalart-Allmaras (S-A) model, Menter SST k-ω model and RNG k-ε model were implemented to simulate flow past an airfoil.

2.1 Spalart-Allmaras model

The S-A model is form with the transport equation of the kinematic eddy viscoscity ($\tilde{v}$). The one-equation of the S-A model is written by:

$$\frac{\partial(\rho\tilde{v})}{\partial t}+div(\rho\tilde{v}U_i)=\frac{1}{\sigma_v}div\left[(\mu+\rho\tilde{v})\nabla\tilde{v}+C_{b2}\rho\frac{\partial\tilde{v}}{\partial x_k}\frac{\partial\tilde{v}}{\partial x_k}\right]$$
$$+C_{b1}\rho\tilde{v}\tilde{\Omega}-C_{w1}\rho\left(\frac{\tilde{v}}{\kappa y}\right)^2 f_w \qquad (1)$$

where $\tilde{v}$ is the kinematic eddy viscosity, f_w is the wall damping function, μ is the dynamics viscosity. The constant incluse σ_v, C_{b1}, C_{b2} and κ has value of 0.67, 0.1355, 0.622 and 0.4187, respectively.

2.2 Menter SST k-ω model

The Menter SST k-ω has been developed from the Wilcox k-ω [11] to precise simulation results on the boundary layer. Two equations of the model can be written by:

$$\frac{\partial(\rho k)}{\partial t}+div(\rho k U_i)=div\left[\left(\mu+\frac{\mu_t}{\sigma_k}\right)\nabla k\right]+P_k$$
$$-\beta^* \rho k\omega \quad (2)$$

$$\frac{\partial(\rho \omega)}{\partial t}+div(\rho \omega U_i)=div\left[\left(\mu+\frac{\mu_t}{\sigma_{\omega,1}}\right)\nabla \omega\right]$$
$$+\gamma_2\left(2\rho S_{ij}S_{ij}-\frac{2}{3}\rho\omega\frac{\partial U_i}{\partial x_j}\delta_{ij}\right)$$
$$-\beta_2\rho\omega^2+\frac{2\rho}{\sigma_{\omega,2}\omega}\frac{\partial k}{\partial x_k}\frac{\partial \omega}{\partial x_k} \quad (3)$$

where P_k is the rate of production of turbulent kinetic energy. The constant incluses σ_k, β^*, $\sigma_{\omega,1}$, γ_2, β_2 and $\sigma_{\omega 21}$ has value of 1.00, 0.09, 2.0, 0.44, 0.083 and 1.17 respectively.

2.3 RNG k-ε model

$$\frac{\partial(\rho k)}{\partial t}+div(\rho k U_i)=div\left[\alpha_k \mu_{eff}\nabla k\right]+\tau_{ij}\cdot S_{ij}-\rho\varepsilon \quad (4)$$

$$\frac{\partial(\rho \varepsilon)}{\partial t}+div(\rho \varepsilon U_i)=div\left[\alpha_\varepsilon \mu_{eff}\nabla \varepsilon\right]+C_1\frac{\varepsilon}{k}\tau_{ij}\cdot S_{ij}$$
$$-C_2\rho\frac{\varepsilon^2}{k} \quad (5)$$

where k is the turbulent kinetic energy, ε is the turbulent dissipation. The constant incluse C_μ, α_ε, C_1 and C_2 has value of 0.0845, 1.39, 1.42 and 1.68 respectively.

Figure 1. The NACA 0015 airfoil installing in the wind tunnel.

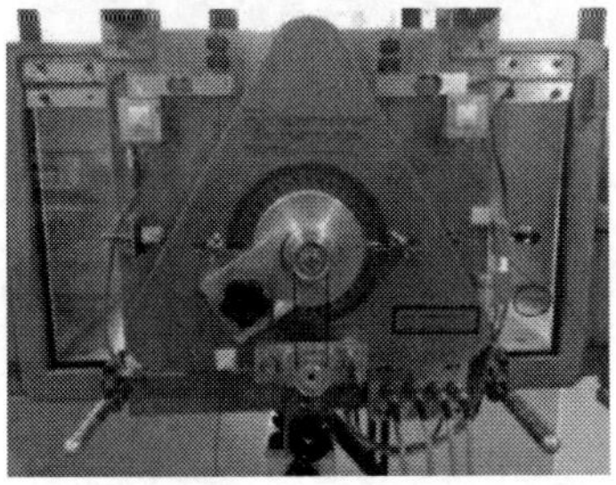

Figure 2. Triangular load cell for measuring lift force and drag force.

3 Wind tunnel experiment

The airfoil profile, NACA0015, with chorded length of 190 mm and span length of 285 mm was tested in the wind tunnel model WT300 (Fig. 1). Lift force and drag force were measured using the triangular load cell as shown in Fig. 2. The airfoil angle has been adjusted from 0 to 20 degree by using a load cell spindle. The flow velocity was controlled by an axial fan to generate Reynolds number (Re) of flow past airfoil at 160,000 and 360,000.

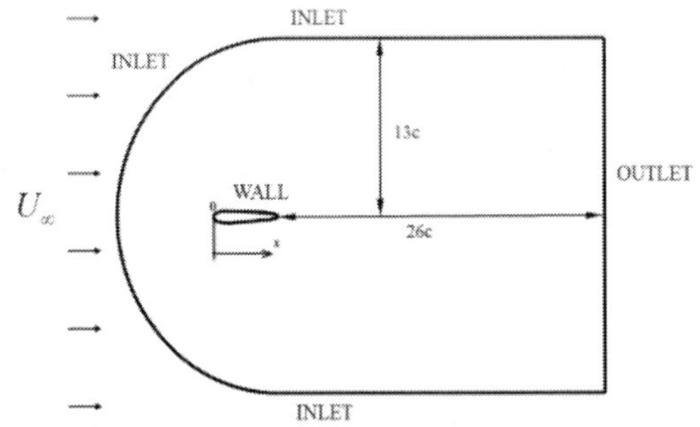

Figure 3. The C-type domain for airfoil simulation.

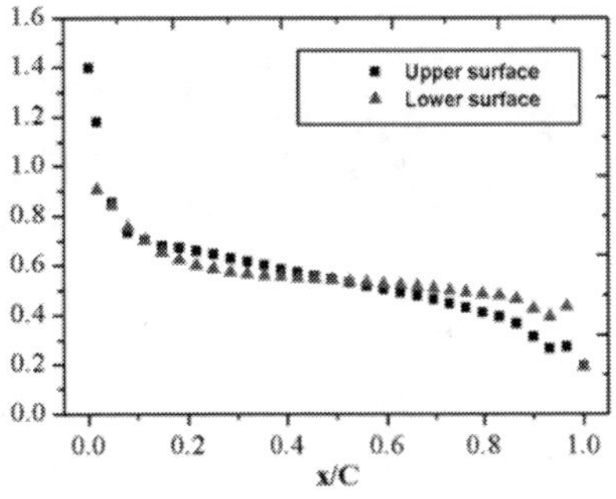

Figure 4. The y^+ distribution on airfoil surface.

4 Computational fluid dynamics

The computational domain has selected by using the C-type domain which radius and downstream length are 13 and 26 times of a chord length respectively (Fig. 3). The nearest cells or grids on airfoil were attempted to control by y+ value to be less than 1 which less than the satisfied values as equal to 11.63 [1]. The y+ value around airfoil is plotted against distance from leading edge to tailing edge as shown in Fig. 4. The cell generation is performed simultaneously after y+ controlling, then the number of nodes around the airfoil is variable from 100 to 600 which made cells being into a simulated domain from 19,818 to 100,621 respectively. The steady result of lift coefficient (C_L) and drag coefficient (C_D) of airfoil at the angle of attack (AOA) of 4 degree is the cell independent test by the node number of 300 around airfoil and the total cell of 58,454. Fig. 5 shows a cell structure of an airfoil simulation at the near wall region with y+ between 0.2 and 1.4. The SIMPLE algorithm was used to solve the governing equation of airfoil simulation, while pressure-velocity coupling problem used Upwind method.

Boundary condition was inlet flow from the left side of the CFD domain which assigned uniform velocities and outlet flow was set by an atmospheric pressure. Front and back domain boundary were set by empty type. No slip condition and no wall function were set on wall boundary. An incompressible flow was used for air with density (ρ) and a dynamics viscosity (μ) of 1.225 kg/m^3 and 1.8375×10^{-5} kg/(m s) respectively.

5 Results and discussion

The C_L and C_D achieved by calculation with simulation results of three turbulence models. The simulation and experiment result of flow past airfoil with Reynolds number (Re) of 160,000 was compared by graphs in Fig. 6 and Fig. 7. The lift force increased linearly until the AOA was 10 degree and decreased at 12 degree. On the other hand drag force was opposite to the lift force along the AOA. The RNG k-ε model had trend of the C_L graph in agreement with experimental data more than the other turbulence models. Subsequently, the SST k-ω model had C_D graph in a good agreement with experimental data more than the other turbulence models. The experimental data had shown the stall angle at 10 degree of AOA while the S-A and RNG k-ε model had shown at 14 and 12 degree respectively.

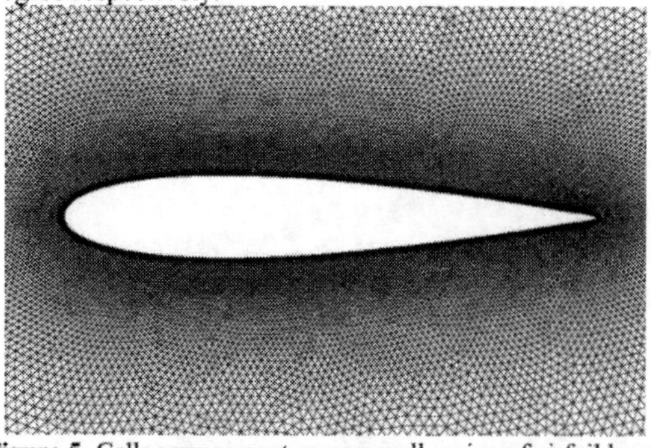

Figure 5. Cells arrangement on near wall region of airfoil by y$^+$ controlling.

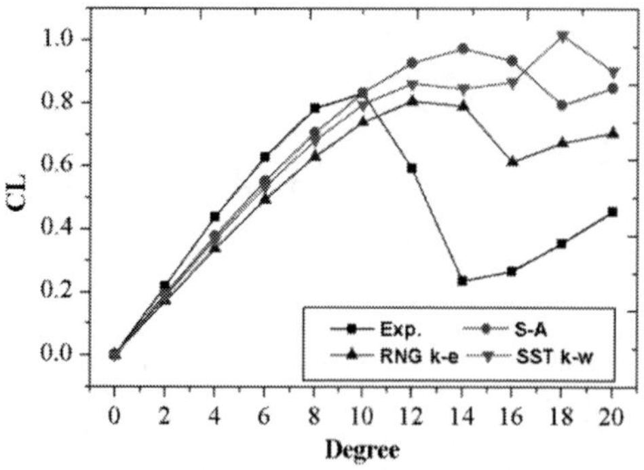

Figure 6. Lift coefficient on airfoil with various AOA by Re of 160,000.

The simulation with Re of 360,000 had predicted C_L and C_D and had been achieved the same behaviour results by using Re of 160,000. Fig. 8 and 9 show the comparison C_L and C_D graph between CFD and the physical experiment under Re of 360,000 respectively. The SST k-ω model had trend of the C_L and C_D graph in a good agreement with experimental data than the other turbulence models. The stall angle of the S-A, SST k-ω and RNG k-ε model had shown at 12 degree.

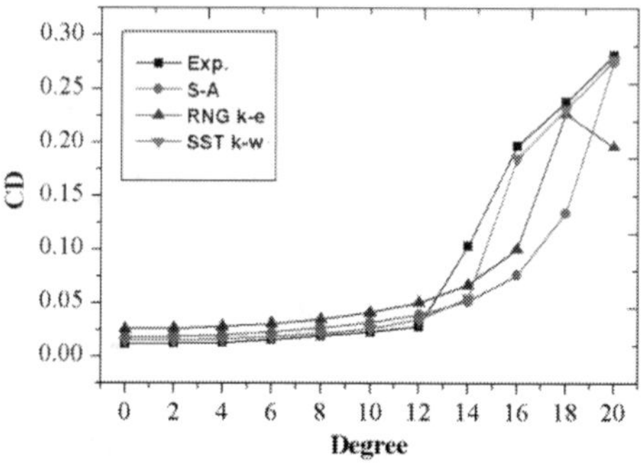

Figure 7. Drag coefficient on airfoil with various AOA by Re of 160,000.

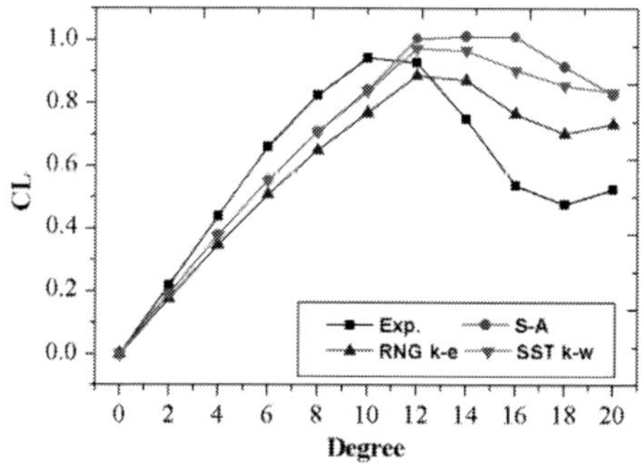

Figure 8. Lift coefficient on airfoil with various AOA by Re of 360,000.

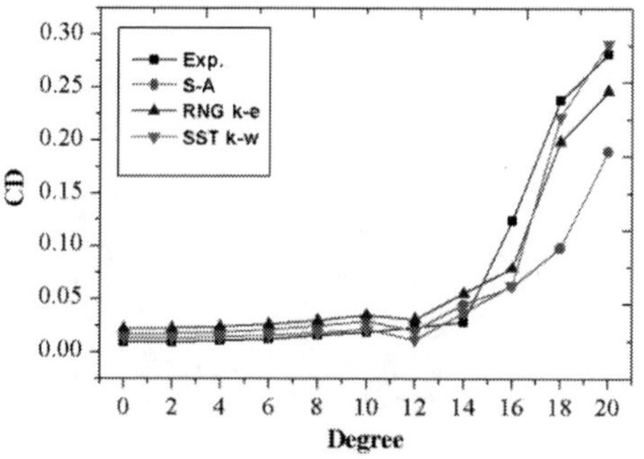

Figure 9. Drag coefficient on airfoil with various AOA by Re of 360,000.

Table 1. The average error of turbulence models in the AOA range of 0-10 degree.

Model	AOA 0-10 degree (%)			
	Re = 160,000		Re = 360,000	
	C_L	C_D	C_L	C_D
S-A	8.01	48.92	11.18	71.15
Menter SST $k-\omega$	10.61	21.25	11.30	28.28
RNG $k-\varepsilon$	16.13	102.50	17.24	114.78

The result of CFD simulation had average error when compared with experimental data that described in Table I and Table II for the AOA range of 0-10 and 11-20 degree respectively. The S-A and Menter SST k-ω model were both in good agreement with experimental data but the Menter SST k-ω model had results close to experiment data. The velocity of flow and stream line past airfoil at AOA of 12 degree under the Re of 360,000 by using the Menter SST k-ω model shows in Fig. 10. The color contour expressed the flow velocity. The maximum velocity was red and the minimum velocity was blue. The maximum velocity was about 8 m/s happened on the leading edge of airfoil when it was set AOA less than 12 degree. The solid line showed the stream line of the air flow. The wake had occurred on the tailing edge of airfoil distinctly after AOA was more than 10 degree. The stream line of flow past airfoil happened vortex on the trailing edge at AOA of 12 degree.

Table 2. The average error of turbulence models in the AOA range of 11-20 degree.

Model	AOA 11-20 degree (%)			
	Re = 160,000		Re = 360,000	
	C_L	C_D	C_L	C_D
S-A	165.61	39.64	56.06	43.02
Menter SST $k-\omega$	162.35	16.24	48.07	28.17
RNG $k-\varepsilon$	108.81	39.91	30.07	38.57

The pressure contours around airfoil by simulation at the same AOA in Fig. 10 are shown in Fig. 11. The color contour expressed the pressure around the airfoil. The maximum pressure was 80 Pa happened on lower surface contour expressed the pressure around the airfoil. The maximum pressure was 80 Pa happened on lower surface of the airfoil which depicted by red. The maximum pressure as the drag force happened at AOA of 12 degree (Fig. 11).

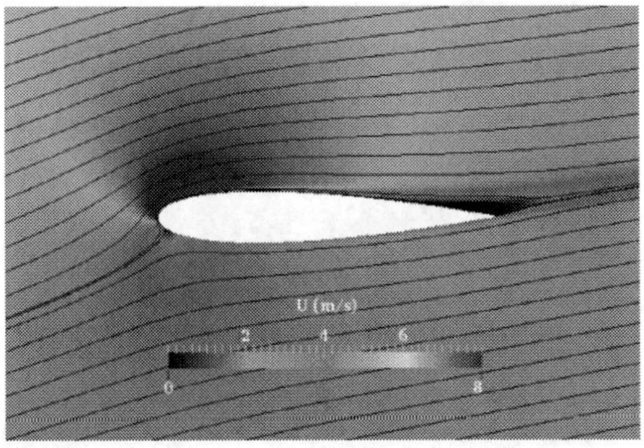

Figure 10. Streamline past airfoil at AOA of 12 degree by using the Menter SST $k-\omega$ model.

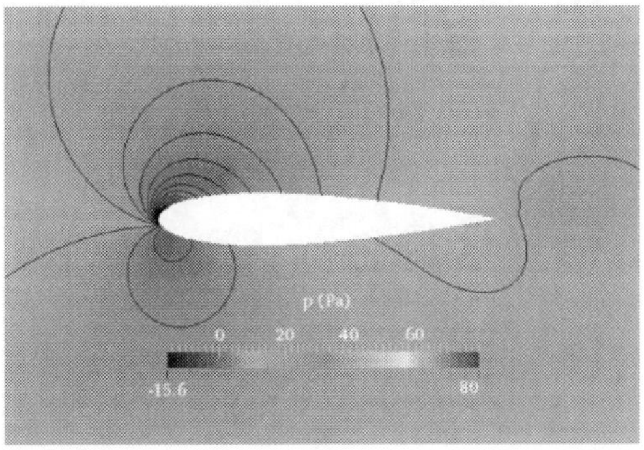

Figure 11. Pressure contour around airfoil at AOA of 12 degree by using the Menter SST $k-\omega$ model.

6 Conclusions

An implementation of turbulence models for airfoil simulation using open source software had been performed. The CFD simulation obtained results to compare with wind tunnel experiment. The lift and drag coefficient were employed to determine the suitable turbulence model. The Menter SST k-ω was a suitable model for the airfoil simulation at AOA before stall region (0-10 degree). The stream line formed to be vortex after stall region (12 degree) distinctly. For the meanwhile, the RNG k-ε model was suitable for stall region (11-20 degree). Nevertheless, the Menter SST k-ω model was recommended for studying and designing airfoil of wind turbines because the least average error was obtained in both regions.

References

1. H. K. Versteeg, W. Malalasekera, *An Introduction to Computational Fluid Dynamics: The Finite Volume Method*, second ed., Prentice Hall, England, (2007).
2. F. Villalpando, M. Reggio, A. Ilinca, Model. Simul. Eng. **2011** (2011)
3. P.R. Spalart, S.R. Allmaras, La Recherche aerospatiale. **1** (1994)
4. P.R. Spalart, S.R. Allmaras, *A one-equqtion turbulence model for aerodynamic flow*, AIAA, Washington D.C., (1992)
5. V. Yakhot, S.A. Orszag, S. Thangam, T.B. Gatski, C.G. Speziale, Phys. Fluids A. **4**, 7, (1992)
6. K. nordanger, R. Haldahl, T. Kvamsdal, A.M. Kvaving, A. Rasheed, Comput. Methods Appl. Mech. Engrg. **290** (2015)
7. F.R. Menter, AIAA J. **32**, 8, (1994)
8. R. Belamadi, A. Djemili, A. Ilica, R. Mdouki, J. Wind Eng. Ind. Areod. **112**, 146 (2016)
9. M.H. Gardiri, N.L.N. Ibrahim, M.F. Mohamed, Engineering Journal **17**, 9, (2013)
10. OpenFOAM, *The Open Source CFD Toolbox User Guide Version 2.3.0* (2014)
11. D.C. Wilcox, *Turbulence modeling for CFD*, DCW Industries, Inc. (1993)

DOI: 10.1051/matecconf/20179512006

Computational Modeling of Dynamic Stability Derivatives for Generic Airfoils

Muhammad Saleem Mumtaz, Adnan Maqsood and Salma Sherbaz

Research Center for Modeling & Simulation (RCMS), National University of Sciences and Technology (NUST), H-12 Campus, Islamabad, Pakistan

Abstract. This paper presents a method for the computation of the static and dynamic stability derivatives of generic airfoils using high fidelity Computational Fluid Dynamics. Aerodynamic coefficients are calculated for NACA 0012 airfoil and flat plate at different angles of attack. Results of lift coefficient are validated with experimental data. Static and dynamic stability derivatives are calculated by oscillating the airfoil geometry at suitable frequency. Simulations are performed at various flight conditions in terms of angles of attack, frequencies and oscillation amplitudes. The aim of the work is to decipher the behaviour of longitudinal damping derivatives used in flight mechanics through CFD. This approach enables the efficient and accurate computation of dynamic derivatives. Calculations are done for constant air velocity altering only the angle of attack. Inviscid model is tested since its results nearly match with experimental data. The simulations show that the nonlinear characteristics of the stability derivatives are captured by varying angle of attack.

1 Introduction

Design cycle of modern aircraft has reduced significantly during recent years, partly due to introduction of computers on various stages of design. Complexity in the design of manned and unmanned vehicles is gradually introduced for improved manoeuvrability and versatile mission requirements, which increases the requirement for precise guess of aerodynamic coefficients and dynamic stability derivatives during the preliminary design stage. In 1911 concept of these dynamic derivatives was given by Byran[1].Different mathematical models use these dynamic derivatives to calculate moments and aerodynamic forces of aircrafts. These aerodynamics derivatives and moments are functions of aircraft states. It is easy to predict static stability derivative than dynamic derivative. Conventionally, wind tunnel [2-9] tests were performed to measure the derivatives. For static stability derivatives wind tunnel test give good estimation but due to complex test-setup, fast design modification and time to perform these experiment, other methods such as DATCOM [10], and ESDU [11] are used. These two methods are fast compared to wind tunnel. However, these techniques may have question for accuracy. In nonlinear system [12-15] importance of dynamic derivatives has increased due to flight manoeuvres by UAVs and other flight vehicles. in recent times, Computational Fluid Dynamics (CFD) is used as an another tool to guess aerodynamic stability derivatives. Computational Fluid Dynamic has the potential to replace the wind tunnel tests and experiential methods as it is inexpensive with minimum physical and kinematical limitations. Many researchers have used CFD to calculate stability derivatives due to these advantages. In this paper sensitivity of amplitude and frequencies on dynamic stability in pitch is quantified. For validation purpose NACA 0012 airfoil is used. Initially normal force coefficient for this airfoil is calculated and is compared with experimental data. From the moment-alpha graph longitudinal derivatives for airfoil are calculated. Using the same technique, these derivatives are calculated for flat plate.

2 Geometry and meshing

For the current study, NACA0012 airfoil is chosen as the validation platform. The coordinates of the profile are produced into GAMBIT®. The chord length of the airfoil is 1m. The center of gravity was chosen as 0.25c for calculation of lift, drag and moment coefficients. For steady state analysis, airfoil remains at rest at zero degree angle of attack. The computational model with mesh for analysis is shown in Figure 1.

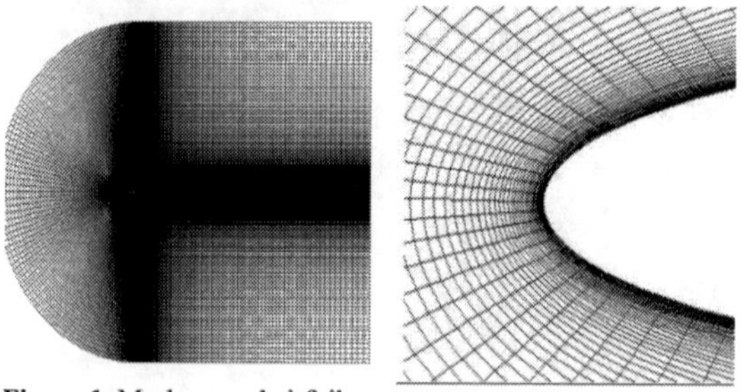

Figure 1. Mesh around airfoil.

DOI: 10.1051/matecconf/20179512006

3 NACA 0012 Airfoil studies

3.1 Steady state analysis

For steady state analysis, pre-processing is done in GAMBIT®. These files are exported in FLUENT® to solve the governing equations of fluid mechanics. For validation with experimental data, steady-state analysis is done by using different turbulence models and for different angles of attack. Parameters for this analysis are taken from AGARD 702 [16] report.

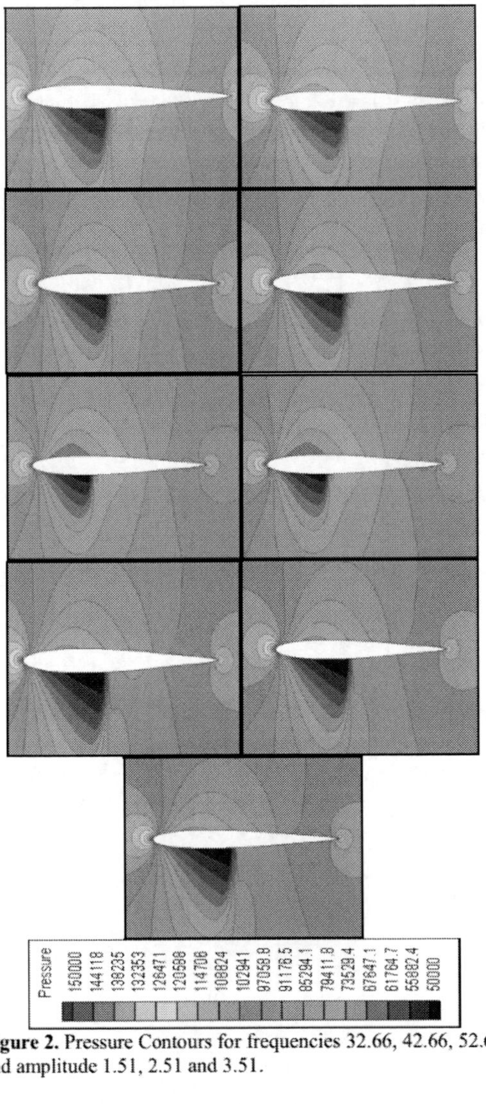

Figure 2. Pressure Contours for frequencies 32.66, 42.66, 52.66 and amplitude 1.51, 2.51 and 3.51.

When simulation for this analysis is converged, lift, drag and moment coefficients are calculated. From these coefficients, normal force coefficient is calculated and is compared with experimental data using following equation.

$$C_N = C_l \cos(\alpha) + C_d \sin(\alpha) \tag{1}$$

where C_N and α represent normal force coefficient and angle of attack respectively. Lift and drag coefficient convergence histories for steady state analysis at zero degree angle of attack using inviscid model are done. Pressure contours for NACA 0012 airfoil at different amplitudes, frequencies and at zero angle of attack are plotted in the Figure 2.

Results obtained for normal force coefficient are compared with experimental data in Figure 3.

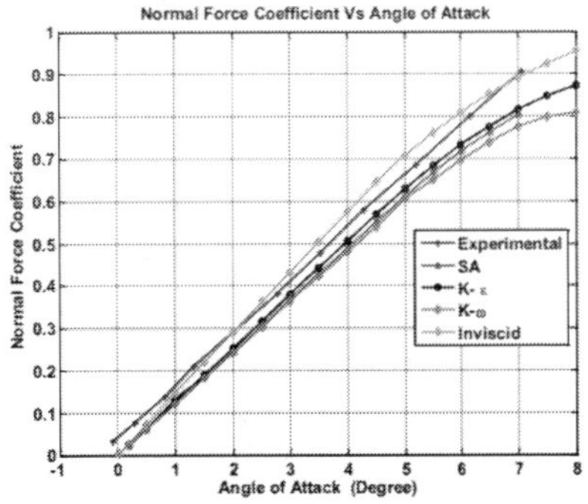

Figure 3. Normal force coefficient comparisons.

3.2 Unsteady analysis

For unsteady analysis, sinusoidal pitching motion of airfoil is achieved by applying dynamic mesh technique. Schematic of airfoil pitching motion is shown in Figure 4. A user define function is made in c language which is compiled in Fluent® that control the dynamic motion of airfoil. A domain in circular form is made around the airfoil as an interface. This interface oscillates along with airfoil with a suitable angular frequency. Motion of airfoil is along z-axis. The oscillating point of airfoil is 0.25c. The function made for oscillation is as follows

$$\alpha(t) = \alpha_0 + \alpha_l \sin(\omega t) \tag{2}$$

where α_0=Initial angle of attack, α_l=Amplitude of oscillation and ω=Angular frequency.

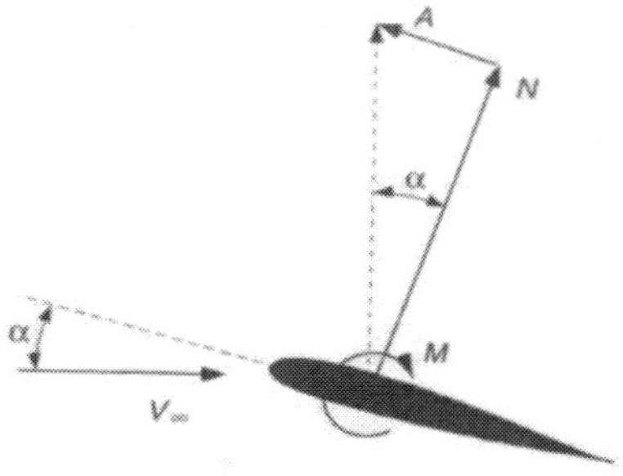

Figure 4. Airfoil pitching motion.

4 Methodology

Procedure for calculating dynamic derivative is depicted in Figure 5.

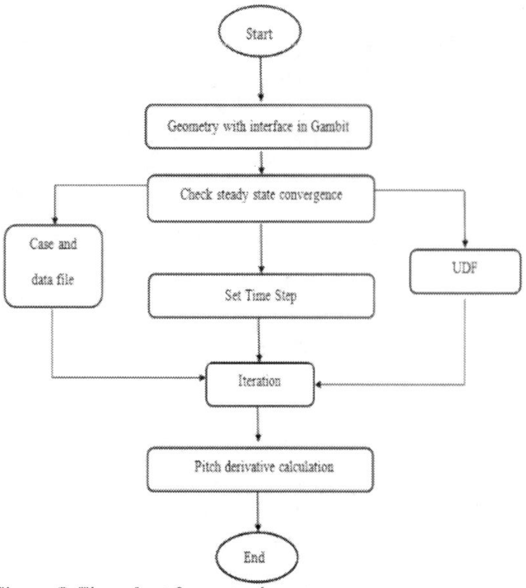

Figure 5. Flow chart for unsteady state.

Analysis parameters for unsteady analysis are also taken from AGARD 702 report and are given in the Table 1.

Table 1. Unsteady analysis parameters.

No	Parameters	Values
1	Airfoil	NACA0012
2	Mach (M)	0.755
3	Initial angle of attack	0.016 degree
4	Amplitude	2.51
5	Pitch center	0.25c
6	Reduced Frequency	0.0814
7	Angular Velocity	42.66 rad/s

Lift coefficient profiles for unsteady case are plotted below in Figure 6. Result of the current study is validated in the form of lift coefficient with experimental data which shown in Figure 7.

In aerodynamics of flight vehicles, stability plays an important role. In order to approximate dynamic stability, damping derivative calculation is necessary. These derivatives have three types, pitch, yaw and roll. Here we have calculated the pitch damping derivatives of 2-D NACA0012 airfoil. Pitch Derivatives are calculated using the following relation.

$$C_m = C_{m_0} + C_{m_\alpha}\alpha + C_{m_{\dot{\alpha}}}\dot{\alpha} + C_{m_{\ddot{\alpha}}}\ddot{\alpha} \qquad (3)$$

Neglecting higher value of derivatives, Eq (3) will reduce to

$$C_m \approx C_{m_0} + C_{m_\alpha}\alpha + C_{m_{\dot{\alpha}}}\dot{\alpha} \qquad (4)$$

Pitch derivatives for airfoil using different frequency and amplitude are calculated in the current study.In order to calculate pitch derivatives coefficients are related to oscillating parameters using time as variable. In order to determine pitch derivatives relation shown in figure 8 will helps us. The trend of the graph between the lift coefficient and the angle of attack is linear. There is shown a distinct hysteresis in the solution, as verified by the gap between the coefficient on the upstroke and down stroke of the oscillation.

This hysteresis helps us to find a dot or transient derivatives. Transient derivatives with respect to $\dot{\alpha}$ derivatives, transient derivatives represent the time lag in the development of the these coefficient resulting from a change in the flow as discussed by Etkin [17]. This idea helps us to compute the stability derivatives. To calculate stability derivatives from this solution, we compute linear least-squares fit of the output coefficient C_L with respect to α.

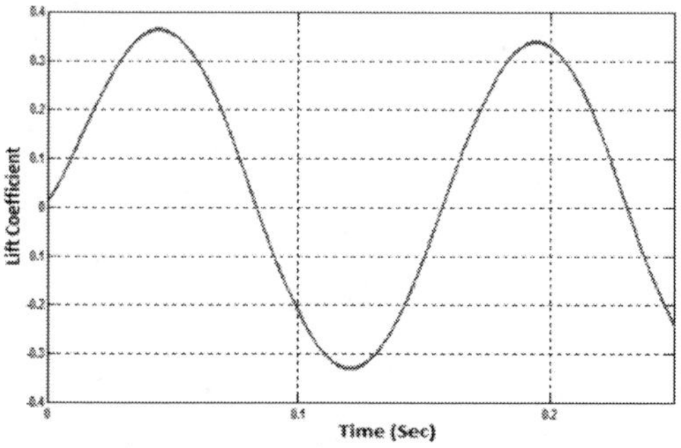

Figure 6. Lift coefficient profile.

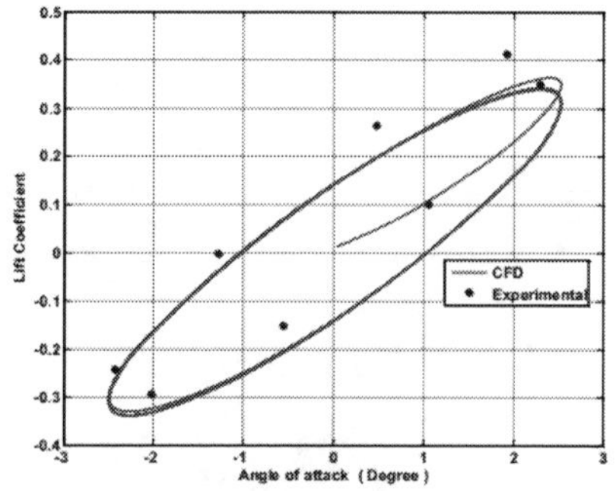

Figure 7. Comparison of lift coefficient.

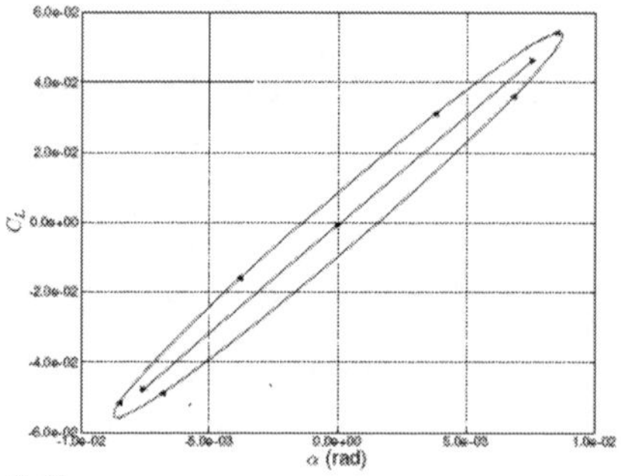

Figure 8. C_L vs α.

In the Figure 8 the stability derivative ($C_{L\alpha}$) is calculated by the slope which is obtained by least square fit method. Y-intercept of the line is the value of the lift coefficient (C_L) at the zero value of the motion perturbation $\Delta_\alpha = 0$. From this an equation of the following form is obtained

$$y = C_{i_j} x + C_{i_0} \quad (5)$$

From this hysteresis transient derivatives can be calculated. This quantity is obtained by subtracting the value of linear recession line from the time spectral solution

$$R^n_{C_i} = C^n_i - y(x^n) \quad (6)$$

From above equation another strongly linear relationship is obtained, as shown in Figure 9.

The slope of above graph gives the transient derivative $(C_{l_{\dot\alpha}})$. Using this technique Pitch damping derivative $(C_{m_{\dot\alpha}})$ is calculated for the current study. Results obtained for different amplitude and at different frequencies are given in Table 2.

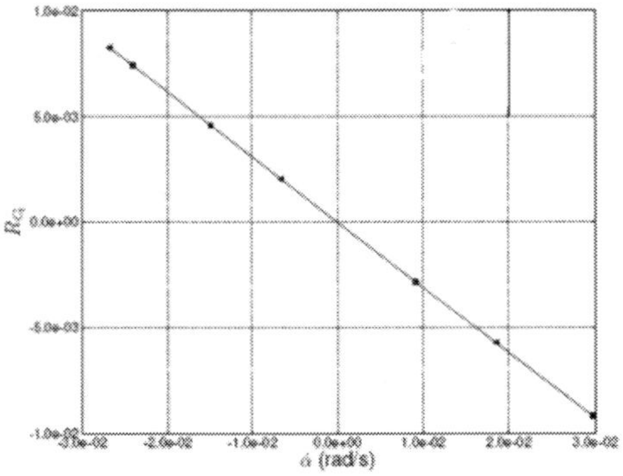

Figure 9. R_{C_i} Hysteresis vs $\dot\alpha$.

Table 2. Pitch damping derivatives.

Amplitude	Frequency					
	ω = 32.66		ω =42.66		ω = 52.66	
	Cmα	Cmα˙	Cmα	Cmα˙	Cmα	Cmα˙
1.51	0.002	-0.006	0.002	-0.006	0.002	-0.005
2.51	0.001	-0.006	0.001	-0.006	0.001	-0.0056
3.51	0	-0.005	0	-0.005	0	-0.0046

5 Flat plate geometry

Procedure for calculation of pitch derivative for NACA 0012 airfoil is also applied for flat plat e. Geometry for flat plate is shown in Figure 10.

Figure 10. Flat Plate Geometry.

Pitch derivatives are calculated at frequency (f=42.66 rad/sec) and amplitude (a=2.51 m) for different angles of attack. Results obtained are given in table 3 and are plotted in Figure 11.

Table 3. Pitch derivatives for Flat Plate.

Angle (Deg.)	Cmα	Cmα˙
0	0.002	-0.008
2	0.003	-0.010
4	-0.003	-0.004
6	-0.010	0.010
8	-0.04	-0.074

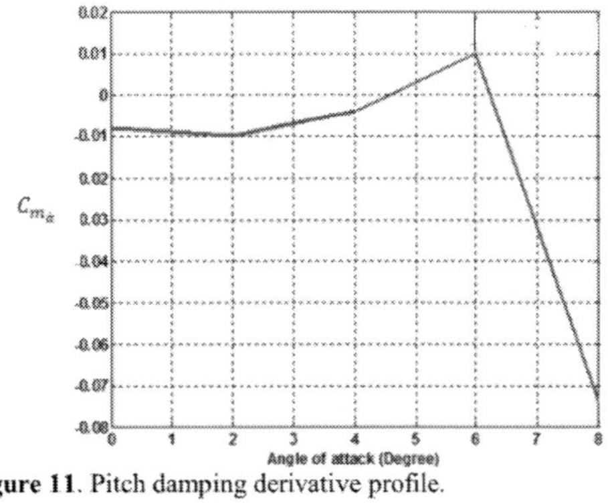

Figure 11. Pitch damping derivative profile.

6 Conclusion

The results obtained from the current simulation for NACA 0012 and for the Flat plate show good agreement with experimental data obtained from wind tunnel test and the data obtained from literature. The response of different parameters like angle of attack, angular frequency, amplitude of oscillation, time step, and turbulence model and gap variation between the flat plates was also studied. After analysis of computational data obtaining form the simulation of NACA 0012 airfoil, it was found that for the same value of angular frequencies and for increased value of oscillating amplitude the value of pitch damping derivative increases in other words we can say that dynamic stability decrease. For flat plate point of view different simulations were performed at fixed amplitude and at fixed angular frequency. From the result data it was found that by increasing the angle of attack to 6 degree pitch damping derivative increases and further increase in angle of attack its value decreases.

References

1. G. H. Bryan, Stability in aviation. London, UK: Macmillan Publishers, (1911).

DOI: 10.1051/matecconf/20179512006

2. I. Shantz and R. T. Groves, "Dynamic and static stability measurements of the basic finner at supersonic speeds," (1960).
3. L. C. MacAllister, "The Aerodynamic Properties of a Simple Non Rolling Finned Cone-Cylinder Configuration Between Mach Numbers 1.0 and 2.5," (1955).
4. G. L. Winchenbach, B. L. Uselton, and R. M. Chelekis, J. Aircr., **21**, 1, pp. 5-13, (1984).
5. B. L. Uselton, "A description of the Standard Dynamic Model (SDM)," in *56th* Supersonic Tunn. Assoc. Meet., (1981).
6. N. Alemdaroglu, I. Iyigun, M. Altun, H. Uysal, F. Quagliotti, and G. Guglieri, "Determination of dynamic stability derivatives using forced oscillation technique," in 40th AIAA Aerosp. Sci. Meet. Exhib., American Institute of Aeronautics and Astronautics, Reno, NV, USA, (2002).
7. G. Guglieri, F. Quagliotti, and P. Scarabelli, "Static and Oscillatory Experiments on the SDM at Politecnico di Torino," Nota Sci. E Tec., no. 74/93, (1993).
8. A. R. Davari and M. R. Soltani, *Iran. J. Sci. Technol.*, **31**, B1, pp. 49, (2007).
9. E. Schmidt, "Standard dynamics model experiments with the DFVLR/AVA transonic derivative balance," Institut fuer aeroelastik goettingen (1985).
10. Engineering Sciences Data Unit (ESDU sheets), www.esdu.com.
11. R. Hall, R. Biedron, D. Ball, D. Bogue, J. Chung, B. Green, M. Grismer, G. Brooks, and J. Chambers, "Computational Methods for Stability and Control (COMSAC): The Time Has Come," in AIAA Atmospheric Flight Mechanics Conference and Exhibit, American Institute of Aeronautics and Astronautics, (2005).
12. Maqsood, A. and Go, T. H., 'Multiple time scale analysis of aircraft longitudinal dynamics with aerodynamic vectoring' Nonlinear Dynamics, **69**, 3, pp. 731-742, (2012).
13. Maqsood, A. and Go, T. H., 'Longitudinal Flight Dynamic Analysis of an Agile UAV' Aircraft Engineering and Aerospace Technology, **82**, 5, pp. 288-295, (2010).
14. Go, T. H. and Maqsood, A. , 'Effect of aspect ratio on wing rock at low Reynolds number' Aerospace Science and Technology, **42**, pp. 267-273, (2015).
15. Bibi, A., Maqsood, A. and Go, T. H., 'Review on Analysis & Modeling of Dynamic Stability Characterisitics of Atmospheric Entry Vehicles' NUST Journal of Engineering Sciences, **7**, 1, pp. 15-21, (2015).
16. R. H. Landon, "Compendium of unsteady aerodynamic measurements," AGARD Rep., **702**, (1982).
17. B. Etkin, *Dynamics of Atmospheric Flight*, Dover Publications, Mineola, New York.

Analysis of Thin Walled Wavy Shell of Monge Type Surface with Parabola and Sinusoid Curves by Variational-Difference Method

Vyacheslav Ivanov and Marina Rynkovskaya

Peoples' Friendship University of Russia, Engineering Department, 117198, Miklukho-Maklaya Str., 6, Moscow, Russia

Abstract. The paper is about the stress-strain state of the thin shell in the form of Monge surface with parabola generatrix and sinusoid guide. Coordinate system of the Monge surface is a system of coordinate lines of the principal curvatures of the surface. The variational-difference method is used for analysis. Variational-difference method allows using the geometric characteristics of the middle surface of the shell, which is important in the calculation of shells of complex shape. In the finite element method, which is often used in the shells of complex shape analysis, the equation of the middle surface of the shell is used only for finite element mesh.

1 Introduction

The Monge surface is formed by the movement of a plane fixed generatrix curve which is moving in the orthogonal plane of the directrix curve, with one point of the generatrix curve is linked to the guide curve. If the guide curve is plane, the generatrix curve doesn't rotate in the orthogonal plane of the guide curve [1-6].

In paper, the variational-difference method (VDM) [2, 4, 7-13] is used for analysis of the stress-strain state of the shell. The variational-difference method is based on the Lagrange principle, the principle of minimum total potential energy [14-22]. The Lagrange principle can be performed as the following: the total potential energy reaches the absolute minimum for the displacements (among all possible displacements) which satisfies the given boundary conditions and the equilibrium equations.

When analysing shells by the VDM the middle surface of the shell is covered with a mesh of constant or variable density. The grid lines coincide with the coordinate lines of principal curvatures. The derivatives of the displacements in the strain vector are replaced with finite-difference relations. In this case the total potential energy functional becomes a function of strain nodal displacements. Then, in order to minimize the total potential energy the Ritz-Timoshenko procedure is used. Solution of the system of equations gives the nodal displacements. Using the differential equations again the tangential and bending forces can be obtained.

At the Department of Strength of Materials and Structures of the Engineering Faculty of Russian Peoples' Friendship University, a software package for the variational-difference analysis of the shells of the non-canonical form is developed [2, 4, 23, 24].

The complex includes a library of curves, which allows forming various thin-walled shells and calculating all geometric characteristics which are used in the equations of the shells.

2 Methods

2.1 Design of the shell

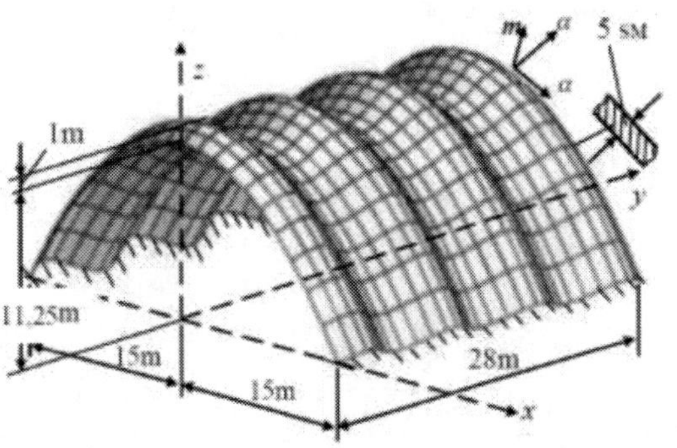

Figure 1. The wavy shell with parabola-sinusoid curves.

The guide curve is a parabola ($z = ax^2 + b$; a = -0,05; m-1; $b = 11{,}5$ m; -15 ≤ x ≤ 15 (m)) at vertical plane and generatrix curve is a sinusoid with 7 half-waves ($z = c\sin(\pi y/d) = c\sin(p\pi y/L)$; c = 1 m; L = 28 m – the length of the shell; $d = L/p$ – the length of a half-wave; p = 7 – the number of the half-waves of the shell.

If the number of the half-waves is odd the shell haves a plane of symmetry when y = L/2. The design of the shell is shown at figure 1.

At the cross-sections x = -15; x = 15 the sides of the shell are fixed; the sides y= 0, y= L are free.

The boundary conditions:

$x = \pm 15$ м – $u_1 = u_2 = u_3 =$; $\partial u_3 / \partial \alpha_1 = 0$

$y = 0$, $y = c$ – $N2 = M2 = S = H = 0$.

α_1, α_2 – the coordinates of the middle surface of the shell; u_1, u_2 – the tangential displacement, u_3 – normal displacement; N_1, N_2, S – tangential forces; M_1, M_2, H – bending forces.

Since the shell has two planes of symmetry, the analysis was performed for the shell sector

$-15 \le x \le 0$, $0 \le y \le L/2 = 14$.

At the plane of the symmetry the following conditions were realized:

$u_1 = 0$; $\partial u_3 / \partial \alpha_1 = 0$ when $\alpha_1 = 0$;

$u_2 = 0$; $\partial u_3 / \partial \alpha_2 = 0$ when $\alpha_2 = L/2$.

2.2 The results and analysis

The calculation was made with the grid of 80×140 (40 steps per one half-wave of the sinusoid). The sole weight of the shell (γ= 25kN/m3) was taken as an external load. The characteristics of the material: Young's modulus is E=35000 MPa, Poisson's ratio is ν=0.15.

The epures of the tangential force N_1 and bending M_1, M_2 moments are shown at the figures 2-4. The tangential forces N_2 are much less (5-6%) then forces N_1. The relationship for forces is $N_2 = \nu N_1$ only at the fixed edge (x= -15). That is why the epure of the force N_2 is not presented here.

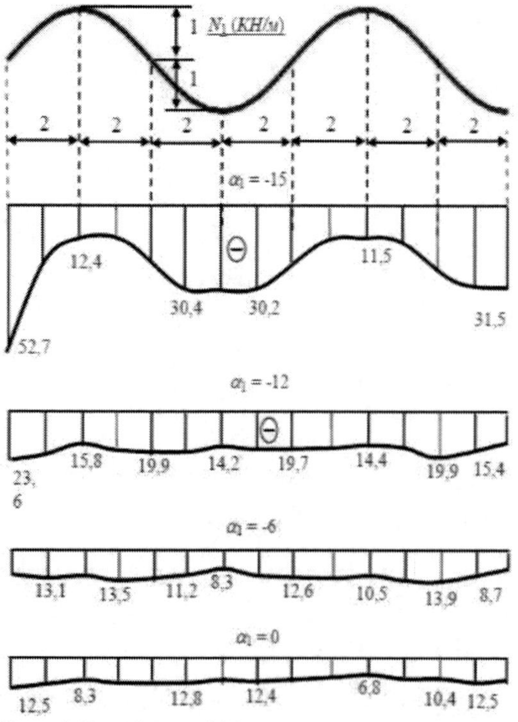

Figure 2. Normal tangential forces.

It is shown on the given graphics that the biggest compressing normal forces occur at the supporting cross-section x = -15. Those stresses decrease 2-3 times in the middle zone. It may be also seen that the internal forces have a wavy character in longitudinal direction.

The big bending moments M_1 occur only in the supporting zone which is caused by the boundary effect.

The maximal bending moments M_2 can be seen at 1/3 of the bottom part of the shell while they decrease in 2-4 times in the middle section of the shell.

The comparison of internal forces of the first three waves of 9 waves shell with internal forces in the same waves of shell with 7 waves shows that they vary slightly.

So there may be made a conclusion: for the equal (in longitudinal direction) load the first 3.5 waves (so as three last) of the many waves (7, 9 and more waves) shell may be analyzed for 3.5 waves with real boundary conditions on the left boundary and conditions of symmetry on the right side (3.5 waves).

As far as the middle waves (the forth and next) of the many-waved shell are concerned, all of them may be analyzed as 1 wave shell with the boundary conditions coinciding with conditions of the symmetry:

$u_2 = 0$; $\partial u_3 / \partial \alpha_2 = 0$ when α2=0;

$u_2 = 0$; $\partial u_3 / \partial \alpha_2 = 0$ when α2=d,

where d is a length of a half-wave of the shell.

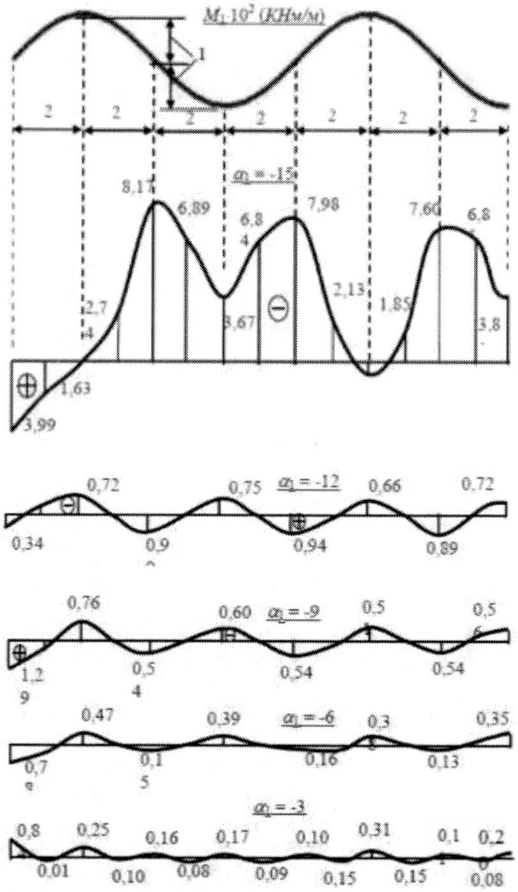

Figure 3. Bending moments $M1$.

It is possible to analyze a half part of the half-waved section, as it has a plane of symmetry.

DOI: 10.1051/matecconf/20179512007

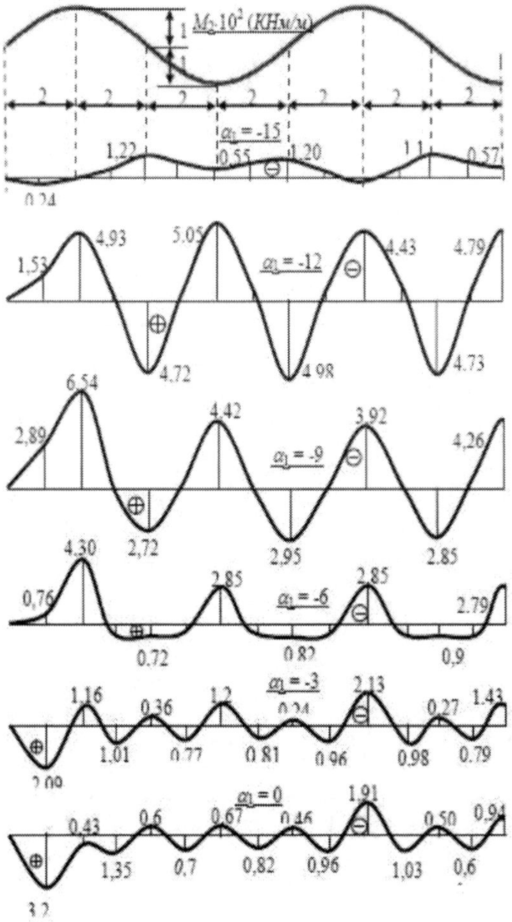

Figure 4. Bending moments M2.

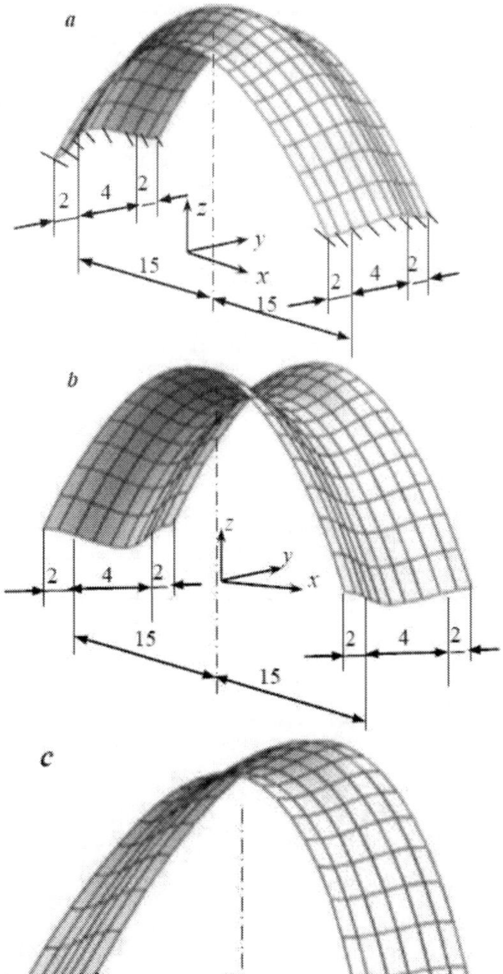

Figure 5. Middle sections of the wavy shell.

At Figure 5, the sections of the shell which may be used for analysis of the middle waves of the many-waved shell are shown.

At the Tables 1, 2, the results of calculation of the middle wave of the shell with 9 half-waves and one section (Figure 5, a) are shown.

The analysis of the middle section was made on the grid of 80×80 steps and the ¼ part of 9 half-waved surface with the grid of 80×180. This gives the equal precision of the results of calculation of both structures as they have equal steps of the grid. Generally, the difference in the strains of the regarded parts of the structure does not exceed 1-3%. Only the longitudinal strains σ_{N2} at some parts differ more. But those stresses are much less than the stresses σ_{N2} and this will have no effect on the strength of the structure.

The calculations of section which is presented at Figure 5, *b* shows mostly the same results as for sections at Figure 5, *a*.

The comparison of the stress-strain state of the section from Figure 5, a with stress-strain state of the middle (4, 5) wave of shell with 9 half-waves shows the similar results.

Table 1. Results of calculation of the middle wave of the shell.

Number of half-waves	σ_{N1} (kPA)		σ_{N2} (kPA)	
	9	отс.	9	отс.
α_1 (m)	$\alpha_2 = 0$ m			
-15	-230	-237	-34,2	-35,5
-12	-287	-290	-11,1	-13,7
-9	-272	-272	-11,2	-0,66
-6	-211	-211	-13,0	-8,62
-3	-159	-159	-16,0	-10,6
0	-138	-139	-17,7	-10,5
α_1 (m)	$\alpha_2 = 2$ m			
-15	-432	-433	-64,2	-65,0
-12	-399	-399	-3,33	-3,08
-9	-340	-340	0,60	1,09
-6	-277	-277	-1,36	-0,63
-3	-227	-227	-7,54	-6,50
0	-207	-207	-11,5	-10,3
α_1 (m)	$\alpha_2 = 4$ m			
-15	-630	-629	-94,5	-94,4
-12	-308	-306	10,2	10,5
-9	-174	-171	17,3	17,5
-6	-168	-165	15,3	15,6
-3	-213	-209	2,91	3,49
0	-238	-234	-5,88	-5,16
α_1 (m)	$\alpha_2 = 6$ m			
-15	-434	-433	-65,1	-65,0
-12	-400	-399	-31,2	-30,8
-9	-340	-340	0,88	1,29
-6	-277	-277	-1,15	-0,63
-3	-227	-227	-7,36	-6,50
0	-207	-207	-11,3	-10,3
α_1 (m)	$\alpha_2 = 8$ m			
-15	-240	-237	-35,9	-35,5
-12	-290	-290	-13,7	-13,7
-9	-271	-272	0,48	0,66
-6	-209	-211	8,23	8,62
-3	-156	-159	10,0	10,6
0	-135	-139	9,81	10,0

Table 2. Results of calculation of the middle wave of the shell.

Number of half-waves	σ_{M1} (kPA)		σ_{M2} (kPA)	
	9	отс.	9	отс.
α_1 (m)	$\alpha_2 = 0$ m			
-15	9,81	10,9	1,48	1,63
-12	-15,8	-15,6	-106	-106
-9	-12,3	-12,3	-94,2	-94,7
-6	-9,00	-9,11	-68,2	-69,6
-3	-7,49	-7,52	-50,9	-51,1
0	-7,16	-7,18	-45,6	-45,0
α_1 (m)	$\alpha_2 = 2$ m			
-15	-183	-182	-27,4	-27,3
-12	21,3	21,2	114	113
-9	12,9	12,9	68,2	68,6
-6	3,07	3,12	19,6	19,8
-3	-2,70	-2,76	-6,28	-6,16
0	-3,75	-3,90	-11,9	-11,9
α_1 (m)	$\alpha_2 = 4$ m			
-15	-91,3	-91,4	-13,7	-13,7
-12	-17,2	-17,2	-115	-115
-9	-13,4	-13,4	-102	-102
-6	-8,54	-8,61	-67,0	-67,5
-3	-4,80	-4,91	-34,4	-34,9
0	-4,03	-4,16	-22,8	-23,3
α_1 (m)	$\alpha_2 = 6$ m			
-15	-181	-182	-27,1	-27,3
-12	21,0	21,2	112	113
-9	12,9	12,9	62,8	68,6
-6	3,19	3,12	19,8	19,8
-3	-2,67	-2,76	-6,15	-6,16
0	-3,82	-3,90	-11,9	-11,9
α_1 (m)	$\alpha_2 = 8$ m			
-15	11,5	10,9	1,73	1,63
-12	-15,2	-15,6	-104	-104
-9	-12,3	-12,3	-94,4	-94,7
-6	-9,15	-9,11	-69,8	-69,6
-3	-7,59	-7,52	-51,7	-51,1
0	-7,26	-7,18	-45,7	-45,0

3 Conclusion

The analysed waved surface may be used for roofing of the buildings such as the trading, exhibition or sports centres. The variational-difference method which was used for analysing of stress-strain state of the structure allows taking into account the possibilities of the geometry of the shell. The calculation of the construction was made with different mesh density in order to receive the necessary accuracy of the result.

In paper, there are the results which were obtained for the grid of 80×140. Such approach gives the difference in results less than 5%.

The many wave shells with different boundary conditions on the left ($\alpha_2=0$) and right ($\alpha_2=L$) sides may be analyzed separately: by left and right sections with three of half waves with their boundary conditions and middle waves as well.

The boundary conditions on the left ($x=0$) and the right ($x=b$, b – the width of the shell) longitudinal sides of the shell may differ. In this case, the symmetry on coordinate x is not used.

If the load is not constant in the longitudinal direction (in direction of y) then it is necessary to analyze the whole construction.

References

1. S.N. Krivoshapko, V.N. Ivanov. Encyclopedia of Analytical Surfaces. Springer International Publishing Switzerland, 752 p. (2015)
2. V.N Ivanov, S.N. Krivoshapko. Analytical methods of analyses of shells of non canonic form. Moscow, RUDN, 542 p. (2010)
3. V.N Ivanov, M. Rizvan. Geometry of Monge surfaces and construction of the shell. Structural mechanics of s and buildings. construction: Collection of science works, Vol. **11**, ACB, p. 27-36. (2002)
4. V.N Ivanov, Younis Nasr. Analyses of the shells of complex geometry by variation differential difference method. Collection of science works, Vol. 11, ACB, p. 25-34. (2000)
5. B. Bulca, K. Arslan, "Surfaces Given with the Monge Patch in E4E4", Журн. матем. физ., анал., геом., **9**:4, p. 435–447. (2013)
6. Romain MESNIL, Yann SANTERRE, Cyril DOUTHE, Olivier BAVEREL, Bruno LEGER. Generating high node congruence in freeform structures with Monge's Surfaces. Proceedings of the International Association for Shell and Spatial Structures (IASS), (2015)
7. Ivanov V.N., Gil-oulbe Mathieu. Some Aspects of the Geometry of Surfaces with a System of Flat Coordinate Lines, International Journal of Soft Computing and Engineering (IJSCE), p. 77-82 (2014)
8. R. Lagu, A variational finite-difference method for analyzing channel waveguides with arbitrary index profiles, IEEE Journal of Quantum Electronics, Vol. **22**, Iss. 6, pp. 968-976 (1986)
9. G.D. Smit, Numerical Solution of Partial Differential Equations by Finite Difference Methods, 2nd edit., Oxford Applied Mathematics and Computing Science Series, UK, 350p. (1986)
10. S. G. Mikhlin, Variational-difference approximation, Journal of Soviet Mathematics, Vol. **10**, Iss. 5, pp. 661-787 (1978)
11. A.R. Mitchell, D.F. Griffiths, The finite difference method in partial differential equations, Wiley, 284p. (1980)11. Curant R., Variational methods for the solution of problems of equilibrium and vibrations, Bulletin of the American Mathetics Society, Vol.**49**, pp. 1-23 (1943)
12. V. A. Maksimyuk, E. A. Storozhuk I. S. Chernyshenko. Variational finite-difference methods in linear and nonlinear problems of the deformation of metallic and composite shells (review). International Applied Mechanics, Vol. **48**, Iss. 6, pp 613-687 (2012).
13. Singh J.P., Dey S.S. Variational finite difference method for free vibration of sector plates. Journal of Sound and Vibration, Vol. **136**, Iss. 1, p. 91-104 (1990)
14. Curant R., Variational methods for the solution of problems of equilibrium and vibrations, Bulletin of the American Mathetics Society, Vol.**49**, pp. 1-23 (1943)
15. Reddy J. N., Energy Principles and Variational Methods in Applied Mechanics, John Willy and Co, Inc. New Jersey, USA, 542p. (2002)
16. R. Andujar Moreno, J. Roset, V. Kilar, Variational Mechanics and Stochastic Methods Applied to Structural Design, 173p. (2014).
17. Aginam C. H., Chidolue C. A. and Ezeagu C. A., Application of direct variational method in the analysis of isotropic thin rectangular plates, ARPN Journal of Engineering and Applied Sciences, Vol. **7**, N9, pp. 1128-1138 (2012)
18. C. Lanczos, The variational principles of mechanics, University of Toronto Press, Canada, 418p. (1970)
19. J.H. Argyris, Energy theorems and structural analysis, Butterworths Scientific publications, London, UK, 85p. (1960)
20. L. Lapidus, G.F. Pinder, Numerical solution of partial differential equations in science and engineering, Wiley-Interscience, New York, 677p. (1982)
21. W. Wunderlich, W. Pilkey, Mechanics of structures. Variational and computational methods, CRC Press, pp. 852-877 (2002)
22. Ivanov V.N. Variation priciples and methods od analyses of the problems of theory of elastisityt. Moscow, RUDN, 2001, 176 p.
23. Ivanov V.N. The base of the finite elements method and the varation-difference method. Moscow, RUDN, 2008, 170 p.
24. Ivanov V.N., Kushnarenko I.V. The Variational-Difference Method for the Analisis of the Shells with Complex Geometry// International Association for Shell and Spatial Structures Proceedings of the IASS 2013 Symposium "Beyond the Limits of Man", 6 p. (2013)

 DOI: 10.1051/matecconf/20179512008

Study on Accuracy of Gough P.S and Guo Stress Formula in Calculating Propellant Bed Stress under Quasi-static Compaction

Yu-ling Zhang , Guo-qing Liu, Jin-ming Li, Guo-dong Wang

Ordnance Engineering College, 050003Shijiazhuang, China

Abstract. In order to verify Gough P.S and Guo stress formula and its accuracy in calculating propellant bed stress under quasi-static compaction, the paper adopted electronic universal testing machine to perform quasi-static compaction on propellant beds of two different diameters, and based on experimental data, the paper obtained stresses of the two beds in the process of compression. The paper used stress formula to calculate stresses of the two propellant beds respectively, based on the propellant bed feed and experimental configuration, and the results show that, during initial propellant compaction phase, the calculations are greater than experimental results, when it comes to certain compaction phase, the calculations of the formula are smaller than experimental results; the calculations of the formula are slightly smaller than experimental results.

1 Introduction

In order to accomplish tactical indicators of high muzzle velocity and gun power, the gun loading density and ballistic pressure become higher and higher, ballistic combustion and mechanical environment become more and more terrible, and recent years have seen accidents of burst barrels and barrel damages, and the safety issue of gun launch becomes more and more prominent[1-3]. Researches carried out both at home and abroad show that, the root cause for burst barrel is the propellant fragmentation and increased burning area resulted from extrusion and collision during gun launch, where the propellant bed, under the impact of ballistic gas of high pressure and high velocity within the bore, generates extrusive stress among propellant grains, causing propellant bed to crush shell bottom or bore wall at high speed. When the forces sustained by propellant grains exceed their own mechanical strengths, fragmentation or deformation occur, leading to the increased burning area of propellant charge and abrupt increase in ballistic pressure, and then burst barrel [4-5]. Hence, the quantitative description of propellant extrusion stress during gun launch will be of significance, on the one hand, predictions of extrusion stress could be used to assess whether deformation or fragmentation of propellant charge could happen, and improvement on gun or propellant charge could be made according to the predictions to avoid burst barrels; on the other hand, quantitative descriptions could be used to obtain factors affecting values of extrusion stress, and the incidence of the oversized extrusion stress could be avoided by means of altering initial feeding conditions or others. Moreover, formula of granular extrusion stress is an important substituting formula in two-phase interior ballistic equation, and its accuracy will extremely affect calculations of interior ballistic two-phase equations. If inaccurate, stress formula will cause rather big error in predictions of bore pressure and other ballistic parameters, therefore the accuracy of granular extrusion stress formula must be ensured as much as possible. Currently, the granular stress formulas used in calculating interior ballistic two-phase flow equation are mainly stress formula and stress formula, while the literature points out the approach of obtaining granular stress formula through quasi-static compaction test, and the paper focused on validating stress formula and its accuracy in calculating propellant bed stress under quasi-static compaction.

2 Test

2.1 Testing device and principle

Figure 1. Electronic universal testing machine.

This paper adopted electronic universal testing machine manufactured by Shenzhen New Sans Materials Co., Ltd, to perform quasi-static compaction on propellant bed, and the testing machine and relevant data capturing equipment are indicated in Fig. 1. In the testing process, data capturing system can record corresponding values of compaction load and compaction displacement.

Table 1. Charged condition of the two propellant beds.

propellant bed Serial No		1	2
propellant charge	quality	63.5g	103.8g
	density	1.6g/cm^3	1.6g/cm^3
	model	11/7	11/7
	height	6cm	6cm
propellant cylinder	materials	304steel	304steel
	bore	4.1cm	5.1cm
piston	materials	304steel	304steel
	diameter	4cm	5cm

2.2 Test results and analysis.

From experimental data, we obtain by direct simulation the compressive displacement-pressure curve of propellant bed, and Fig. 2 indicates compressive displacement-pressure curve of the two propellant beds.

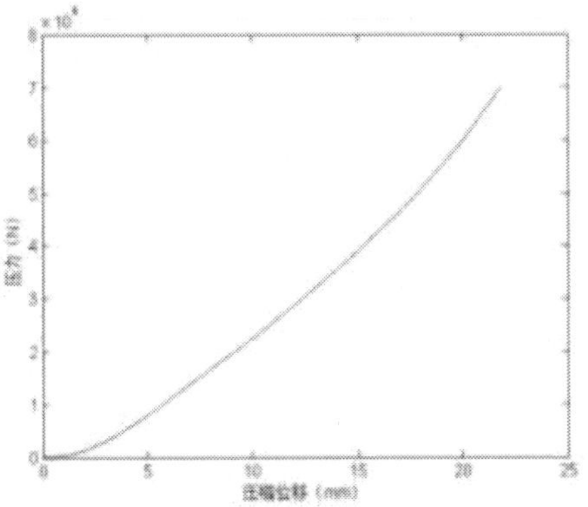

(a) First propellant bed

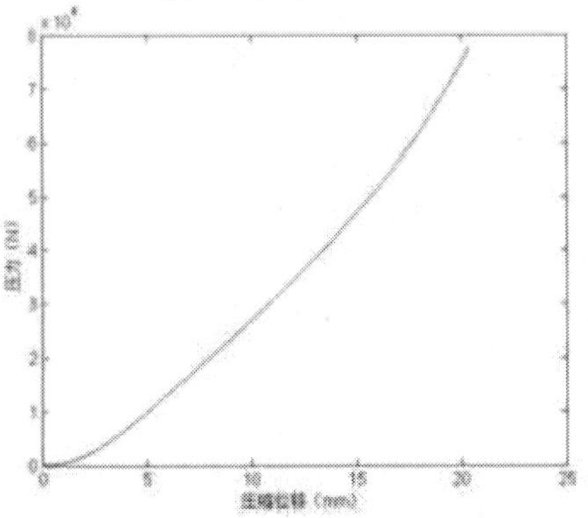

(b) Second propellant bed

Figure 2. Propellant bed Compressive displacement-stress curve.

According to definition of stress, granular extrusion stress of propellant bed at any time is given:

$$\sigma(t) = \frac{F(t)}{S(t)} \tag{1}$$

$F(t)$ is pressure at any time, $S(t)$ is extrusion area of propellant bed at any time, and assuming void ratio of propellant bed is evenly distributed, the area occupied by cross section grains of propellant bed is proportional to void ratio, i.e.:

$$S(t) = S_0(1 - \phi(t)) \tag{2}$$

where S_0 is the cross section area of propellant cylinder.

$$\phi(t) = \frac{\omega}{\rho_p S_0 (H_0 - Vt)} \tag{3}$$

where ω is the mass of propellant charge, ρ_p is propellant charge density, H_0 is initial height of propellant bed, V is compaction velocity. Substituting known data, we obtain void ratio-granular extrusion stress curve.

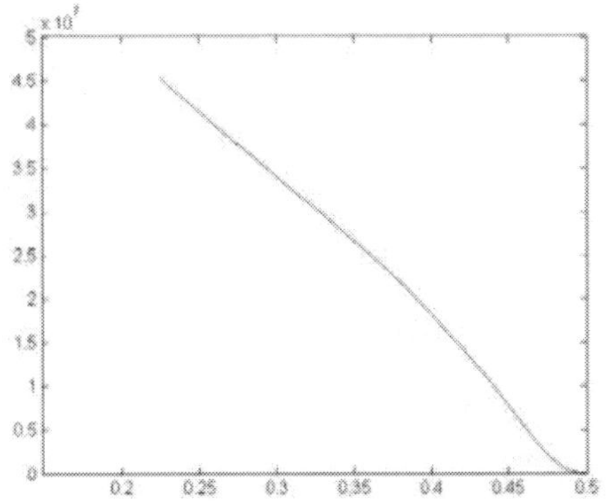

(a) First propellant bed

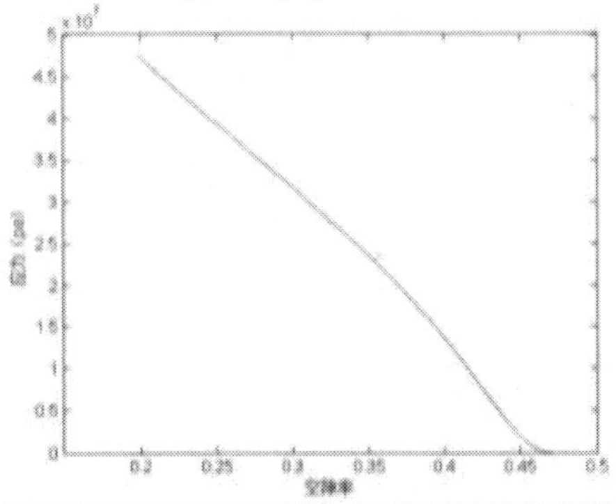

(b) Second propellant bed

Figure 3. Void ratio-stress curve of the first propellant bed.

3 Calculations of Gough P.S stress formula

3.1 Stress formula

Gough, P. S. based on the similar concept of propagation velocity of weak gas-phase perturbation, adopted equations of momentum and continuity to deduce sound speed formula of solid phase particle group.

$$a(\phi) = \sqrt{-\frac{1}{\rho_p}\frac{d[R(\phi)(1-\phi)]}{d\phi}} \quad (4)$$

The sound formula of propellant bed with different void ratios could be calculated with following formula:

$$a(\phi) = \begin{cases} a_1(\phi_0/\phi) \ldots\ldots\ldots\ldots \phi \le \phi_0 \\ a_1 \exp[-K(\phi-\phi_0)] \ldots \phi_0 < \phi \le \phi_* \\ 0 \ldots\ldots\ldots\ldots\ldots\ldots \phi \ge \phi_* \end{cases} \quad (5)$$

where ϕ_0 is compact void ratio, i.e. the void ratio value at the untamped state of propellant grains. K is stress attenuation factor, while ϕ_* represents the value when void ratio is greater than the value. The relation of ϕ_0 and ϕ_* could be given: $\phi_* = \phi_0 + 0.153$. a_1 is the value of a when $\phi = \phi_0$.

Based on the above conclusion, the granular stress formula obtained is written:

$$R = \begin{cases} -\frac{\phi-\phi_0}{1-\phi}\cdot\frac{\phi}{\phi_0}\rho_p a^2 \ldots\ldots\ldots\ldots \phi \le \phi_0 \\ \frac{1}{2K(1-\phi)}\cdot\left[1-e^{-2K(\phi-\phi_0)}\right]\rho_p a^2 \ldots \phi_0 < \phi \le \phi_* \end{cases} \quad (6)$$

where ϕ_0 is the critical void ratio of untamped propellant charge, ρ_p is propellant charge density, K is stress attenuation factor, assume $K = 45$, a_1 is solid phase sound speed when $\phi = \phi_0$, let $a_1 \approx 426.7$ m/s, where a_1 is obtained by Soper through measuring American NACO explosive, and at first he measured by certain testing means the average elastic modulus of NACO explosive, and then based on elastic wave theory, used formula $a = \sqrt{E/\rho}$ to obtain sound speed of propellant bed, where a is propellant bed velocity, and ρ is propellant bed density. At the present time, all a_1 used in formula in literature are taken approximately as $426.7m/s$.

The formula is the stress relationship obtained according to propagation theory of small-perturbation in elastic medium, and in the actual process of gun launch, only small portion of propellant charge compaction belongs to elastic deformation, as compaction continues, the propellant charge enters into stages of viscoelasticity-plastic deformation and plastic deformation.

3.2 Test results and analysis

Substituting know data into formula (6) we obtain as indicated in Fig. 4 void ratio-stress curves of the propellant beds in the test and calculation.

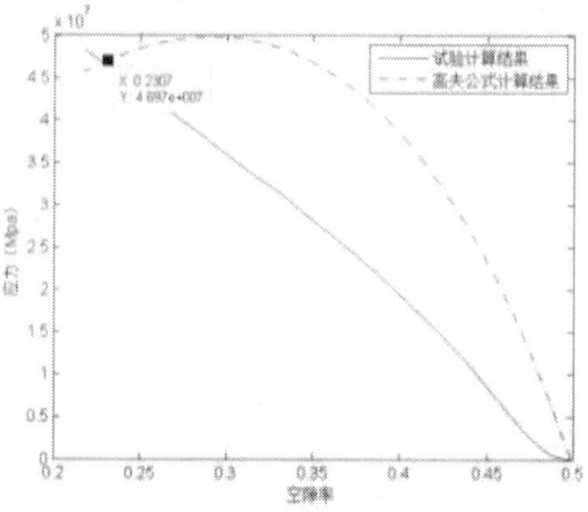

(a) First propellant bed

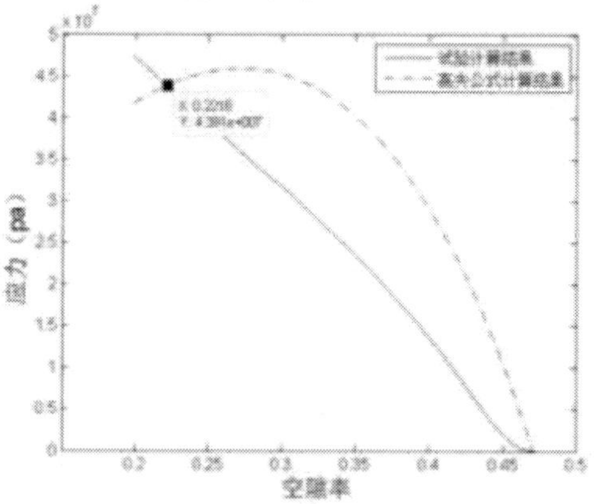

(b) Second propellant bed

Figure 4. Void ratio-stress curves of the propellant bed.

(1) Analyzing Fig. 4, we obtain that in the first propellant bed when void ratio is $(0.2307 \le \phi \le 0.499)$, formula calculations are greater than experimental results; when void ratio is 0.2307, formula calculations are equal to experimental results; within the scope of void ratio $(0.2104 \le \phi \le 0.2307)$, experimental results are greater than formula calculations. Analyzing Fig. 7, we obtain that in the second propellant bed when void ratio is $(0.2218 \le \phi \le 0.4707)$, formula calculations are greater than experimental results; when void ratio is 0.2218, the results are equal; within the scope of void ratio $(0.1968 \le \phi \le 0.2218)$, experimental results are greater than formula calculations.

(2) Analyzing Fig. 4, note that formula calculations at initial compaction stage are greater than testing results, when it comes to certain compaction stage, testing results are greater than formula calculations, which is because at the initial compaction stage the propellant bed belongs to elastic deformation, both initial void ratios of the first and

the second propellant beds are greater than 0.4, sound speed of propellant bed should be smaller than 426.7m/s, while the sound speed used in the formula is 426.7m/s, such that formula calculations are greater than experimental results; as compaction continues, the propellant bed enters into the stages of viscoelastic-plastic deformation and plastic deformation, causing formula calculations to be smaller than experimental results.

(3) Analyzing Fig. 4, obtain that formula calculations of the second propellant bed is more close to experimental results than those of the first propellant bed. One of the causes for such discrepancies between formula calculations and experimental results is sound speed variation of propellant bed under different charged conditions, currently, when applying formula (6), directly assume a_1 is 426.7m/s, which is the sound speed value obtained when void ratio of propellant bed under untamped condition is around 0.4, and if applied to propellant bed of other charged conditions, it must cause calculation errors. The bigger the difference between initial void ratio of propellant bed and 0.4, the bigger the calculation errors will be, and the void ratio of the first propellant bed is 0.4990, and that of the second propellant bed is 0.4707 which have well explained that the calculations of the second propellant bed by the formula is closer to the experimental results.

4 Calculations of Guo stress formula

4.1 Stress formula

Guo, K. K. and others adopted a device specialized for compressing propellant bed to measure its granular stress, and used a compressor with maximum load 5400N, compression rate 600N/S as an external force for the propellant bed to be loaded. The experimental results measured from propellant charges of different sizes are indicated in Fig. 2, where Ku is called compressive parameter of the dimension one, i.e.

$$Ku = \sqrt{\frac{\phi_0 - \phi}{\phi_0}}\phi_0^{0.75} \quad (7)$$

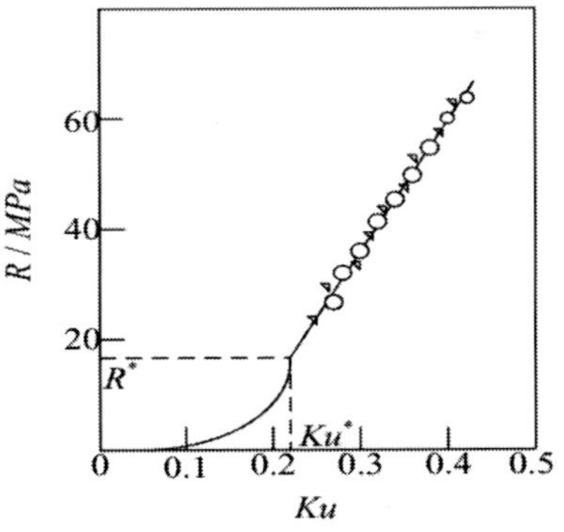

Figure 5. Relationship of Ku and granular stress.

It is related to the shape and size of propellant charges, if the size of propellant is small, the initial charged density will be big, such that ϕ_0 will be small, and conversely, it will be big. From Fig. 5, note that at $Ku = Ku^*$, there is a turning point. Soper, W. G and Eaton believed that it is caused by property variations of propellant bed in the process of compaction. After propellant grains are compressed to some extent, called even viscoelastic medium, remarkable changes have taken place in the mechanical performance of the whole propellant bed, leading to incidence of such phenomenon of turning point. As for AmericanWC-870 explosive, the measured relationship of granular stress and Ku is given as follows:

$$\overline{R} = \begin{cases} \left(\dfrac{Ku}{0.216}\right)^{3.759} & \ldots\ldots\ldots\ldots\ldots Ku \le 0.216 \\ 1 + 10.194\phi_0^{1.533}\left(\dfrac{Ku - 0.216}{\phi_0^{0.75} - Ku}\right)^{0.603} & \ldots\ldots Ku > 0.216 \end{cases} \quad (8)$$

where $\overline{R} = R/R^*$, $R^* = 14.5$ MPa, $Ku^* = 0.216$.

The formula could describe the whole process of propellant bed under quasi-static compaction, from grains filling void, elastic deformation of propellant bed as well as to viscoelasticity-plastic deformation and plastic deformation stages.

4.2 Test results and analysis

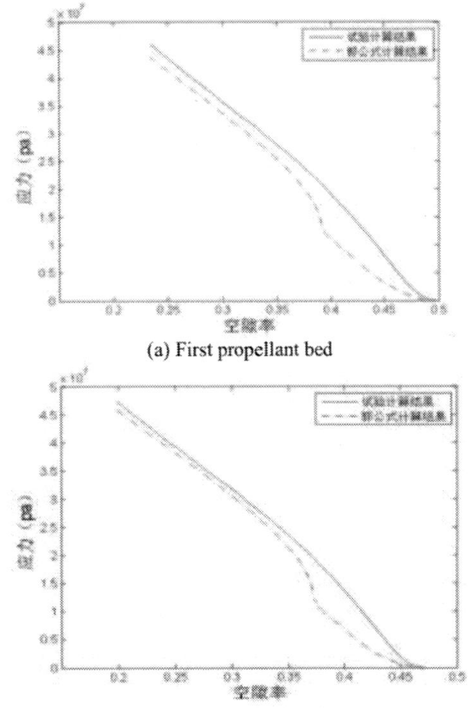

(a) First propellant bed

(b) Second propellant bed

Figure 6. The void ratio-stress curves in the test and calculation of the propellant bed.

Take formula (8) as an example for calculation, substituting the known data into formula (8) to calculate, the calculations of the propellant bed is indicated as Fig. 6.

(1) Analyzing Fig. 6, note that in the first and the second propellant beds, formula calculations are slightly smaller than the experiment results.

(2) By calculation, we obtain that formula calculations are slightly smaller than the testing results, but the difference is not too much, and it is concluded that the formula is capable of describing the quasi-static compaction process accurately, which is because the stress formula could describe accurately the whole process of propellant bed under quasi-static compaction, from stages of propellant grains filling void, elastic deformation as well as to viscoelasticity-plastic deformation and plastic deformation.

5 Conclusion

By comparing formula calculations to the experimental results, we obtain that formula calculations are greater than experimental results at the initial compressive stage, when it comes to certain compaction stage, the experimental results are greater than formula calculations, which is mainly because the formula is used to describe elastic deformation, while elastic deformation of propellant bed only happens at the initial stage, and as the compaction goes on, the propellant bed will enter into stages of viscoelasticity-plastic deformation and plastic deformation; Moreover, if the initial sound speed of the propellant bed in the formula (426.7m/s) is used for propellant beds with void ratio far greater than 0.4, it may cause some calculation errors. The formula could fail to accurately describe propellant stress under quasi-static compaction against the testing configuration.

References

1. X.T. Rui, L.F. Yun and N.S. Sha: Acta Armamentarii, Vol. **26** (2005) No.5, p.690.
2. X.T. Rui, B.B. Feng and G.P. Wang: Ordnance Industry Automation, Vol. **30** (2011) No.5, p. 56.
3. J.Y. Yang, Y.X. Yuan and X.B. Zhang: Journal of Ballistic, Vol. **11** (1999) No.4, p. 92.
4. X.T. Rui, L.F. Yun and G.P. Wang: *Direction to Launch Safety of Ammunition* (National Defense Industry Press, Beijing 2009).
5. L.F. Yun: *Study of the Method Evaluating the Launch Safety for Gun Propellant Charge* (Ph.D., Nanjing University of Science and Technology, China 2008).

FE Simulation of the Stress-Strain State during Shear-Compression Testing and Asymmetric Three-Roll Rolling Process

Alexander Pesin , Denis Pustovoytov, Veniamin Kharitonov and Aleksey Korchunov

Nosov Magnitogorsk State Technical University, 455000, Magnitogorsk, Lenin prospect, 38, Russia

Abstract. A three-roll rolling process is a significant technique in the production of wire rod, round bars and hexagonal profiles for structural applications. Better mechanical properties of wire rod, round bars and hexagonal profiles can be achieved due to large plastic deformation by the three-roll rolling process. Asymmetric rolling is a novel technique characterized by a kinematic asymmetry linked to the difference in peripheral speed of the rolls, able to introduce additional shear strains through the bar thickness. Physical simulation of shear strain, which is similar to that occurring in asymmetric three-roll rolling process, is very important for design of technology of producing ultrafine grain materials. Shear testing is complicated by the fact that a state of large shear is not easily achievable in most specimen geometries. Application of the shear-compression testing and specimen geometry to physical simulation of asymmetric three-roll rolling process is discussed in the paper. FEM simulation and comparison of the stress-strain state during shear-compression testing and asymmetric three-roll rolling process is presented. The results of investigation can be used to optimize the physical simulation of asymmetric three-roll rolling processes and for design of technology of producing ultrafine grain materials by severe plastic deformation.

1 Introduction

A three-roll rolling process is a significant technique in the production of wire rod, round bars and hexagonal profiles for structural applications. It is a rolling process in which material is shaped by three rolls in three directions, everyone of which is inclined to the others at 120° [1], [2]. Three-roll rolling process has various advantages such as easy flow control, small space requirement, and comparatively uniform deformation, etc. The restriction of the material imposed by the shape of caliber rolls in the three-roll rolling process is greater than that in the two-roll rolling process. The accuracy of the shape of products made by the three-roll rolling process is considered to be higher than that of the two-roll rolling process [3].

It is well known that efficient way to increase strength of metallic materials is to form an ultra-fine grain structure using severe plastic deformation (SPD) methods [4]. Nowadays many variants of SPD techniques are available. Asymmetric rolling process is one of them [5]-[7]. Contrary to other SPD methods such as high pressure torsion and equal channel angular pressing, the asymmetric rolling allows to provide the possibility for overcoming the limitation of producing UFG materials with large dimensions due to its continuous feature [8]-[9]. Among the asymmetric rolling processes, the differential speed rolling in which the speeds of the rolls are different, is considered to be the most effective for achieving UFG structure of materials [10]-[13]. Better mechanical properties of wire rod, round bars and hexagonal profiles can be achieved due to large plastic deformation in asymmetric three-roll rolling process. symmetric rolling results in the development of strain with strong shear components, both at the surfaces and at the center of the workpiece.

Physical simulation of shear and compression strains, which are similar to that occurring in asymmetric three-roll rolling processes, is very important for design of technology because it allows to consider a variety of technological variants and to find the best solution in relatively short time and with low costs. Shear testing is complicated by the fact that a state of large shear is not easily achievable in most specimen geometries. Problem of selection of shear testing and specimen geometry has been discussed in numerous publications.

A shear-tensile specimen was developed by Hundy and Green [14]. Another example of such a specimen is the so-called "hat specimen" which was developed by Meyer and Manwaring [15], and used thereafter by many other researchers. Another variant is the double shear specimen used by Klepaczko [16]. Very simple specimen geometry (thin foil) has been used by Clifton and Klopp [17] in their pressure-shear plate impact experiments. Peirs, et al. [18] introduced the eccentric notch shear specimen, which was designed for shear testing of sheet metals over a wide range of strain rates. A shear-compression specimen for large strain testing of materials was developed by Rittel et al. [19]. The specimen consists of a cylinder having an inclined gauge section

DOI: 10.1051/matecconf/20179512009

created by two diametrically opposed rectangular slots which are machined at 45° with respect to the longitudinal axis. The specimen was used in numerous investigations, including constitutive testing, texture evolution, and adiabatic shear banding. A miniature version of the shear-compression specimen was used for the characterization of nanomaterials. The modified shear-compression specimen was presented in [20]. In the modified specimen, the two diametrically machined gauges were semi-circular (instead of rectangular). This modification induces large strains on the mid-section of the gauge without the sharp edges and stress concentrations of the former rectangular gauges. Although many studies have been done, but there are no numerical data on the analysis of the applicability of the shear-compression testing to the modeling of the asymmetric three-roll rolling processes. The goal of this research is the numerical simulation and comparison of the stress-strain state of metal during shear-compression testing and asymmetric three-roll rolling process. The results of investigation can be used to optimize the physical simulation of asymmetric three-roll rolling processes and for design of technology of producing UFG materials by SPD.

2 Research method

DEFORM 3D, based on finite element method (FEM), was used to analyze the stress-strain state of metal during shear-compression testing and asymmetric three-roll rolling process. Simulation of the processes was performed at room temperature. Steel AISI 1010 from DEFORM 3D material library was chosen as a hardened rigid-plastic material for specimen and bar.

The specimen consisted of a parallelepiped having an inclined gauge section created by two diametrically opposed semi-circular slots which were machined at 45°. The level of strain can be controlled through adjustment of the dimensions and height reduction of the specimen during shear-compression testing. Height of the specimen was 24 mm, section dimensions were 12×12 mm, gauge thickness was 1.5 mm and gauge width was 3 mm. (Fig. 1). The specimen provides dominant shear strain in an inclined gauge-section. The specimen was meshed with 80325 tetrahedral elements. A Coulomb friction model was used between dies and specimen. The friction coefficient on the contact surfaces was 0.4. A vertical displacement of the top die was 2.4 mm. It corresponded to 10% of height reduction. Velocity of displacement was 2.0 mm/sec.

The specimen was compressed between two flat dies during shear-compression testing in accordance to the scheme (Fig. 2).

Previously [12]-[13] it was found that to obtain a high uniform shear strain through the bar thickness during asymmetric three-roll rolling process is necessary: 1) very high contact friction; 2) high elongation ratio; 3) speed ratio of rolls should be equal to elongation ratio per pass. Schematic illustration of the asymmetric three-roll rolling process in a stand with two three-roll calibers located very close to each other is shown in Fig. 3.

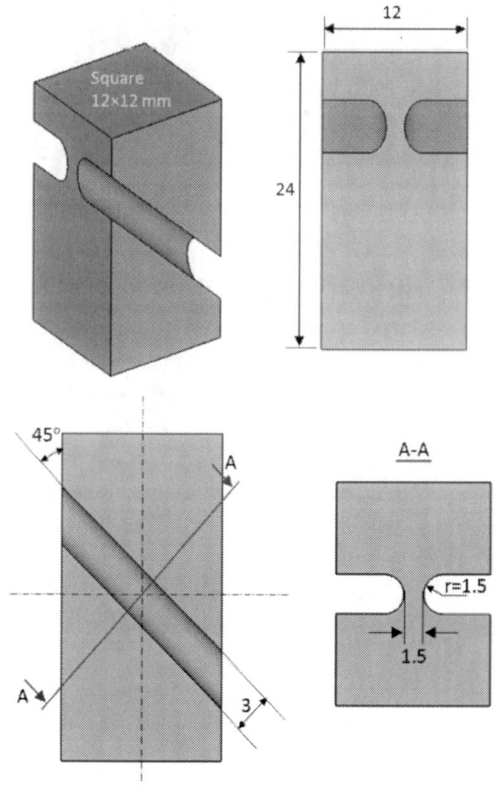

Figure 1. Geometry of the specimen.

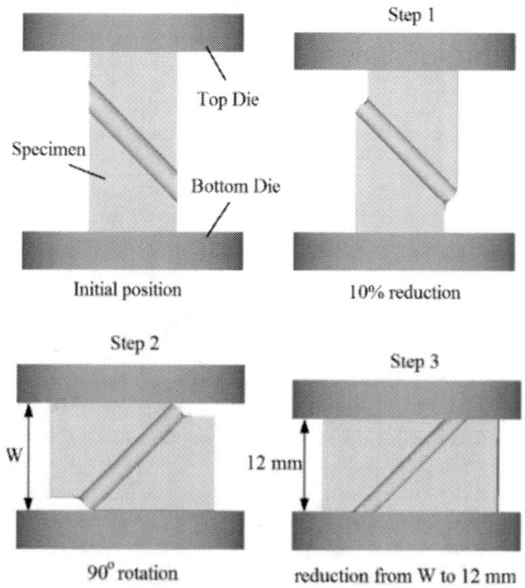

Figure 2. Schematic illustration of the shear-compression testing for one cycle.

A round bar of d=4.0 mm diameter of AISI 1010 was used. The shape of the first caliber was triangle. The shape of the second caliber was round. The initial and the final shape of the bar was a round. The modeling of the process was performed at room temperature without taking into account the increment of the metal's temperature due to the thermal effect of deformation. The isothermal condition was fulfilled due to low rolling speed. The work rolls were assumed as rigid. Peripheral speeds of the rolls were set according to the relationship (1). Peripheral speed of the first roll in all calculation variants was set constant (V_1 =0.2 m/sec). The three-roll rolling process was carried out with total elongation ratio 2. Speed asymmetry of the work rolls was equal to elongation of the bar material.

$$V_2 = V_3 = V_5 = V_6 = \frac{S_0}{S_1} V_1,\ V_4 = \frac{S_0}{S_2} V_1 \qquad (1)$$

where $V_1 \ldots V_6$ – peripheral speeds of the rolls; S_0 – the initial cross section area of the bar; S_1 – the cross section area of the bar after the first three-roll pass; S_2 – the cross section area of the bar after the second three-roll pass.

Bar was meshed with 80630 tetrahedral elements. A Coulomb friction model was used between rolls and bar. The friction coefficient on the contact surfaces was 0.4.

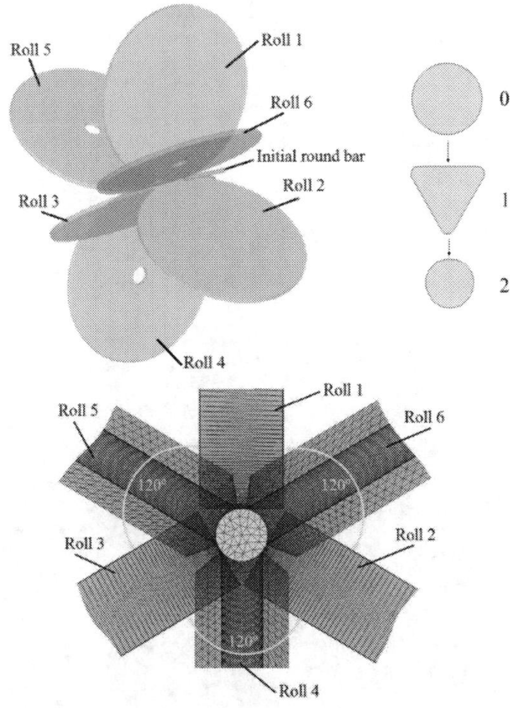

Figure 3. Schematic illustration of the asymmetric three-roll rolling process (0 – initial shape; 1 – after the first pass; 2 – after the second pass).

3 Simulation results and discussion

The deformed specimen after one cycle of the shear-compression testing is shown in Fig. 4. Strain is localized in gauge section and it is typical for large plastic deformation.

It should be noted that the dimensions of the specimen are the important factors that influence on the level of strain during shear-compression testing. With decreasing the gauge thickness and the gauge width as well as with increasing the height reduction the effective strain is increased. So the level of strain can be controlled through adjustment of the dimensions and height reduction of the specimen during shear-compression testing. The friction coefficient on the contact surfaces does not influence on the stress-strain state of metal during shear-compression testing. However the height reduction of the specimen during shear-compression testing is limited by its destruction. End effects are noticeable on both fillets of the gauge. The maximum height reduction should be established from experimental testing. The results of the stress-strain state of metal during shear-compression testing were compared with asymmetric three-roll rolling process. Distributions of the effective stress and strain along the mid-section of the gauge are shown in Fig. 4. The uniform distributions can clearly be observed during shear-compression testing of the specimen. Distributions of the effective stress and strain along the bar during asymmetric three-roll rolling process are shown in Fig. 5. It also can be observed the uniform distributions of the effective stress and strain. Effective stress is 667 MPa and effective strain is 1.0. It quite corresponds to shear-compression testing when height reduction of the specimen is 10%.

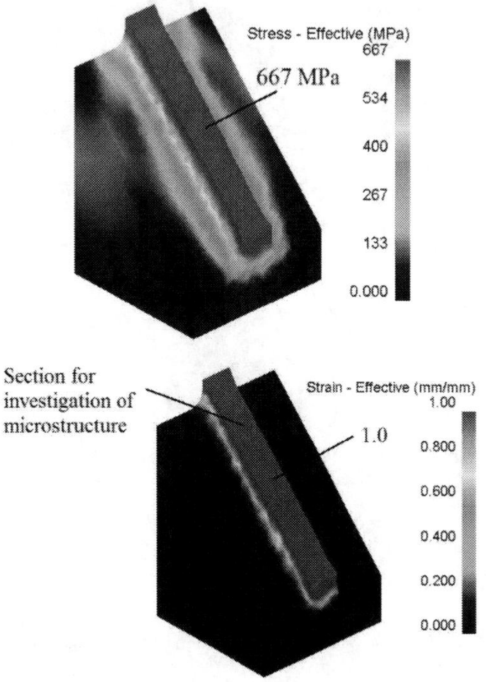

Figure 4. Fields of the effective stress and the effective strain after one cycle of the shear-compression testing.

Change in a Lagrange's mesh before and after shear-compression testing is shown in Fig. 6-Fig.7. Single cell with initial shape of square with dimensions 0.25×0.25mm became the parallelogram. The height of the square was reduced from 0.25 mm to 0.16 mm. but the length was not elongated. Lack of elongation can be explained by high metal flow in width direction.

 DOI: 10.1051/matecconf/20179512009

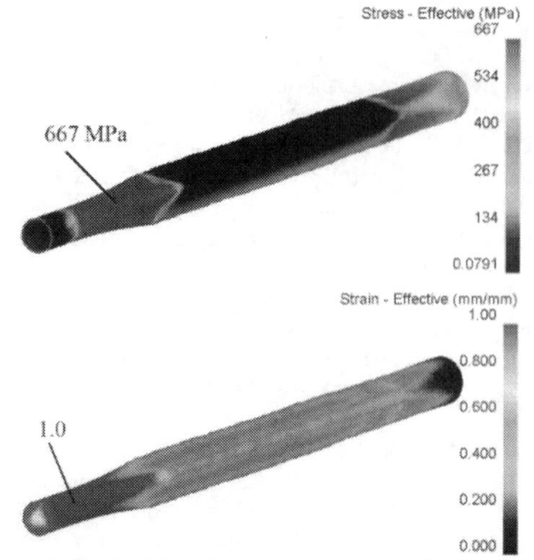

Figure 5. Fields of the effective stress and the effective strain during asymmetric three-roll rolling process with total elongation ratio 2.

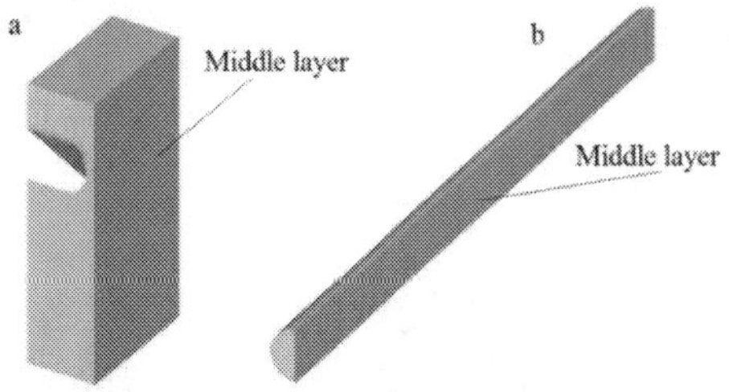

Figure 6. Position of the Lagrange's mesh: a – specimen for shear-compression testing; b – bar for asymmetric three-roll rolling process.

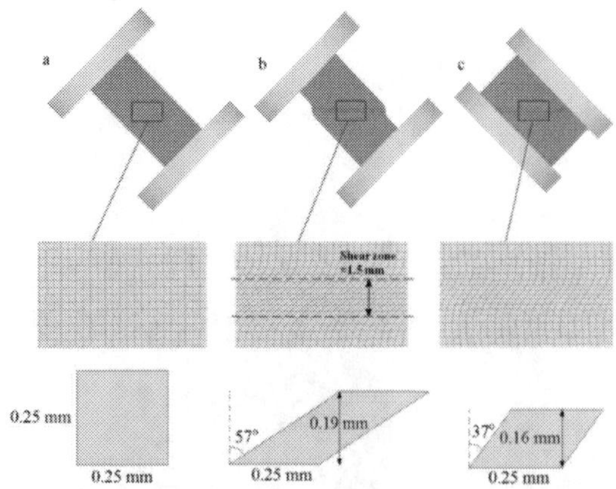

Figure 7. Change in Lagrange's mesh after one cycle of the shear-compression testing: a – initial mesh; b – mesh after step 1; c – mesh after step 3

Shear angle was increased to 57° after the first step and it was decreased to 37° after the second step (Fig. 7). Shear strain can be presented as tangent of shear angle. The shear zone in gauge section was approximately 1.5 mm after 10% of height reduction. Change in a Lagrange's mesh before and after asymmetric three-roll rolling process is shown in Fig. 8. Single cell with initial shape of square with dimensions 0.25×0.25 mm also became the parallelogram. The height of the square was reduced from 0.25 mm to 0.15 mm and the length was elongated from 0.25 to 0.50 mm. Shear angle was increased to 39° (Fig. 8). The shape of Lagrange's mesh in both cases (during shear-compression testing and asymmetric three-roll rolling process) highlights the remarkable action of the shear strain.

The accumulated strain which is equal to 1.0 can be observed during the shear-compression testing for one cycle. It corresponds to asymmetric three-roll rolling process with total elongation ratio 2. It is well known that UFG structure can be observed when accumulated strain is about 3.0…4.0. Physical simulation of the shear-compression testing for four cycles (when accumulated strain is 4.0) was performed. It quite corresponded to asymmetric three-roll rolling process with total elongation ratio 8. The microstructure of the specimen after four cycles of the shear-compression testing were examined with a scanning electron microscope (SEM, JSM-6490LV) at accelerating voltage of 20 kV. Section for investigation of microstructure is shown in Fig. 4. UFG structure was obtained (Fig. 9). Thereby the shear-compression testing can be applied to physical simulation of asymmetric three-roll rolling process. Further experimental investigation of microstructure of steel AISI 1010 processed by asymmetric three-roll rolling process is required.

Figure 8. Change in Lagrange's mesh during asymmetric three-roll rolling process: a – initial mesh; b – mesh after the second rolling pass.

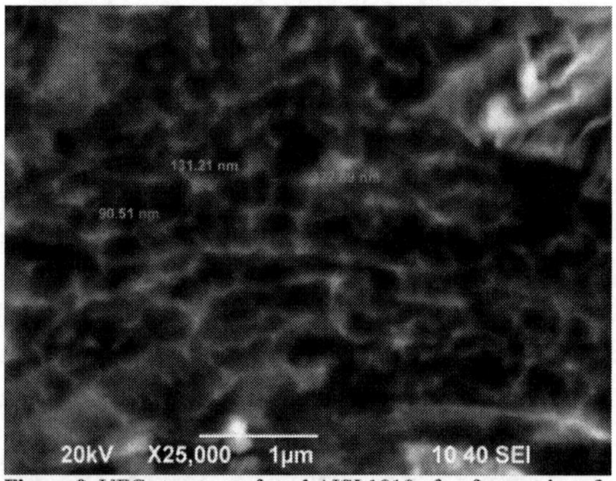

Figure 9. UFG structure of steel AISI 1010 after four cycles of the shear-compression testing (accumulated strain is 4.0).

4 Conclusions

FEM simulations and comparison of the stress-strain state of steel AISI 1010 during shear-compression testing and asymmetric three-roll rolling process were studied. The uniform distribution of the effective stress (667 MPa) and the effective strain (1.0) can be observed during asymmetric three-roll rolling process in a stand with two three-roll calibers located very close to each other. The three-roll rolling process was carried out with total elongation ratio 2. It quite corresponds to the stress-strain state of steel AISI 1010 during shear-compression testing when the specimen consists of a parallelepiped having an inclined gauge section created by two diametrically opposed semi-circular slots which are machined at 45°. Height of the specimen is 24 mm, section dimensions are 12×12 mm, height reduction is 10%, gauge thickness is 1.5 mm and gauge width is 3 mm. Thereby the shear-compression testing can be applied to physical simulation of asymmetric three-roll rolling process. The results of investigation can be used to optimize the physical simulation of asymmetric three-roll rolling processes and for design of technology of producing ultrafine grain materials by severe plastic deformation. Further experimental investigation is required.

Acknowledgement

The research work was carried out in accordance with the government order of the Ministry of Education and Science of the Russian Federation executed by Nosov Magnitogorsk State Technical University on the topic «Development of technology for manufacture of high-strength long rolled sections from materials with ultra-fine grain structure using combined processes of severe plastic deformation» (project No 11.1525K from 18.07.2014).

References

1. K. Komori, Simulation of deformation and temperature in multi-pass three-roll rolling, Journal of Materials Processing Technology. 92-93, 450-457, (1999)
2. C. Overhagen, P. Mauk, A new rolling model for three-roll rolling mills, Key Engineering Materials. 622-623, 879-886, (2014)
3. J.H. Min, H.C. Kwon, Y. Lee, J.S. Woo, Y.T. Im, Analytical model for prediction of deformed shape in three-roll rolling process, Journal of Materials Processing Technology. **140**, 471-477, (2003)
4. Laszlo S. Toth, Chengfan Gu, Ultrafine-grain metals by severe plastic deformation, Materials Characterization. **92**, 1-14, (2014)
5. Zhiming Li, Liming Fu, Bin Fu, Aidang Shan, Effects of annealing on microstructure and mechanical properties of nano-grained titanium produced by combination of asymmetric and symmetric rolling, Materials Science and Engineering: A. **558**, 309-318, (2012)
6. Y.H. Ji, J.J. Park, Development of severe plastic deformation by various asymmetric rolling processes, Materials Science and Engineering: A. **499**, 14-17, (2009)
7. Kyung-Moon Lee, Hu-Chul Lee, Grain refinement and mechanical properties of asymmetrically rolled low carbon steel, Journal of Materials Processing Technology. **210**, 1574-1579, (2010)
8. W.J. Kim, B.G. Hwang, M.J. Lee, Y.B. Park, Effect of speed-ratio on microstructure, and mechanical properties of Mg-3Al-1Zn alloy, in differential speed rolling, Journal of Alloys and Compounds. **509** 8510-8517, (2011)
9. K. Bobor, Z. Hegedus, J. Gubicza, I. Barkai, P. Pekker, G. Krallics, Microstructure and mechanical properties of Al 7075 alloy processed by differential speed rolling, Periodica Polytechnica Mechanical Engineering. **56**, 111-115, (2012).
10. Loorentz, Y.G. Ko, Effect of differential speed rolling strain on microstructure and mechanical properties of nanostructured 5052 Al alloy, Journal of Alloys and Compounds. **586**, S205-S209, (2014).
11. Q. Cui, K. Ohori, Grain refinement of high purity aluminum by asymmetric rolling, Materials Science and Technology. **16**, 1095-1101, (2000).
12. A. Pesin, M. Chukin, A. Korchunov, D. Pustovoytov, Finite element modeling of shear strain in asymmetric and symmetric rolling in multi roll calibers, Procedia Engineering, **81** 2469-2474, (2014)
13. A. Pesin, M. Chukin, A. Korchunov, D. Pustovoytov, Finite element modeling of shear strain in rolling with velocity asymmetry in multi-roll calibers, Key Engineering Materials. 622-623, 912-918, (2014)
14. B.B. Hundy, A.P. Green, A determination of plastic stress-strain relations, Journal of the Mechanics and Physics of Solids. **3**, 16-21, (1954).
15. L.W. Meyer, S. Manwaring, Metallurgical Applications of Shock-Wave and High-Strain-Rate Phenomena, Marcel Dekker, New York, 1986.
16. J.R. Klepaczko, An experimental technique for shear testing at high and very high strain rates, International Journal of Impact Engineering. **46**, 25-39, (1994).
17. R.J. Clifton, R.W. Klopp, Pressure shear plate impact testing, in: Metals Handbook, Vol. 8, 9th edition, Mechanical Testing, ASM, Metals Park, Ohio, pp. 230-239, (1985).
18. J. Peirs, P. Verleysen, J. Degrieck, Novel technique for static and dynamic shear testing of Ti6Al4V sheet, Experimental Mechanics. **52,** 729-741, (2011).
19. D. Rittel, S. Lee, G. Ravichandran, A shear-compression specimen for large strain testing, Experimental Mechanics. **42** 58-64, (2002).
20. A. Dorogoy, D. Rittel, A. Godinger, Modification of the shear-compression specimen for large strain testing, Experimental Mechanics. **55**, 1627-1639, (2015).

The Level Set Topological Optimization Method based on the Constraints of Volume and Stress

Dongyue Qu , Fanbing Liu, Xiuming Li and Chunhua Zhao

School of Harbin Engineering University, Harbin 150000, China

Abstract. This paper research the topological optimization based on level set method which based on the reaction diffusion equation. The paper propose a Level set topological optimization method which put the compliance as the objective function and take the volume as the main constraint, the element stress as the auxiliary constraint to regulate the volume constraint. This method not only can significantly speed up the optimization process, but can improve the precision of calculation and ensure the stability of the results when meet the stress constraint. Eventually, the proposed method can convenient the level set topological optimization to apply in engineering practice.

1 Introduction

At present, continuous structure topological optimization has become an important research field of engineering application. The principle of level set method is simple and clear, which won't appear the phenomenon of checkerboard, mesh dependence. So that, this method has a good development prospect.

Level set method was proposed by Osher and Sethian [1] which used to study the dynamic interface tracking as a numerical method. The reaction diffusion equation was introduced to solve the level set function method for structural topological optimization by Masaki Otomori [2]. What's more, the design of structure topological optimization which based on Flexible objective function by Gradient method was proposed by C. G. Zhang and Z. H. Xiong [3]. Moreover, X.Gu introduces some methods of stress to solve the Structural numerical optimization problem [4]. J.H. Rong and S. Ge, G etc proposed a topological optimization method of continuous structure, which combine the displacement method with the analysis of stress sensitivity [5]. ICM method is proposed to solve the problem of topological optimization that Plate and shell with stress constrains by X. D. Liu, Y. K. Sui etc [6].

2 Math model of topology optimization

The idea of level set topological optimization is that we establish higher dimension curve or surface of level set function, and then the forms of structural topology was described in the Fig. 1. During the level set topological optimization, we define ϕ as the level set function, D as design domain. The positive value of ϕ represents the material area and the negative value represents the empty domain. Besides, zero represents the boundary of structure, as shown in Eq. 1

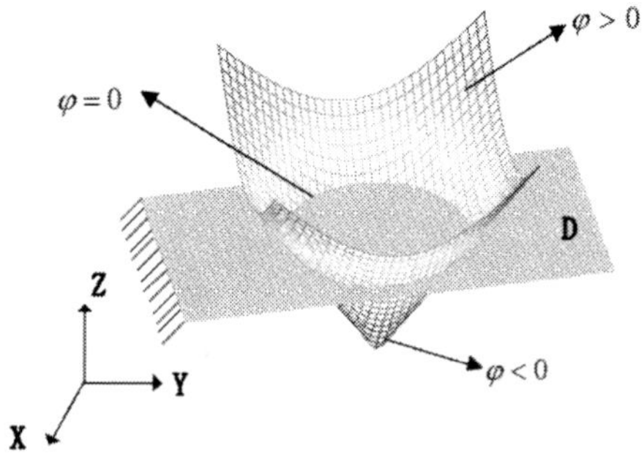

Figure 1. The level set function diagram

$$\begin{cases} \varphi(x) > 0 & if\ x \in \Omega \\ \varphi(x) = 0 & if\ x \in \partial\Omega \\ \varphi(x) < 0 & if\ x \in D/\Omega \end{cases} \quad (1)$$

2.1 The derivation formula of topological optimization

The objective function J of the design is to minimize structural compliance, and get a reasonable structure.

$$\min_{\varphi} \quad J = \int_{\Gamma} F_i u_i d\Gamma$$

$$s.t. \quad G = \int_{\Omega} d\Omega - V^* \le 0$$

$$div(E_{ijkl}u_{k.l}) = div(\sigma_{ij}) = 0 \qquad on\,\Omega. \quad (2)$$
$$u_i = 0 \qquad in\,\Gamma_u$$
$$F_i = \bar{F}_i \qquad in\,\Gamma_f$$
$$\sigma_{m,p} \leq [\sigma_s]$$

where V^* is the upper limit of volume constraint which is a shifty parameter, E_{ijkl} is the elastic tensor, u_i is displacement of unit, F_i is the external force acting at the boundary Γ_f, besides Γ_u is the boundary of fixed constraint and its value equal to 0, $div(E_{ijkl}u_{k,l}) = 0$ represent that there is no other external load on the material area. What's more, p are the iterations, $\sigma_{m,p}$ represent maximum unit stress at p times. $[\sigma_s]$ represents the allowable stress. Using Lagrange multiplier algorithm transform above formulas into a unconstrained Lagrange formula is shown as follows:

$$\bar{J} = \int_{\Gamma} F_i u_i d\Gamma + \int_{\Omega} \tilde{u}_i\, div\left(E_{ijkl}u_{k,l}\right)d\Omega + \lambda\left(\int_{\Omega} d\Omega - V^*\right) \quad (3)$$

where $\bar{J}$ represents the new objective function, $\tilde{u}_i$ and λ represents the Lagrange multipliers [2].

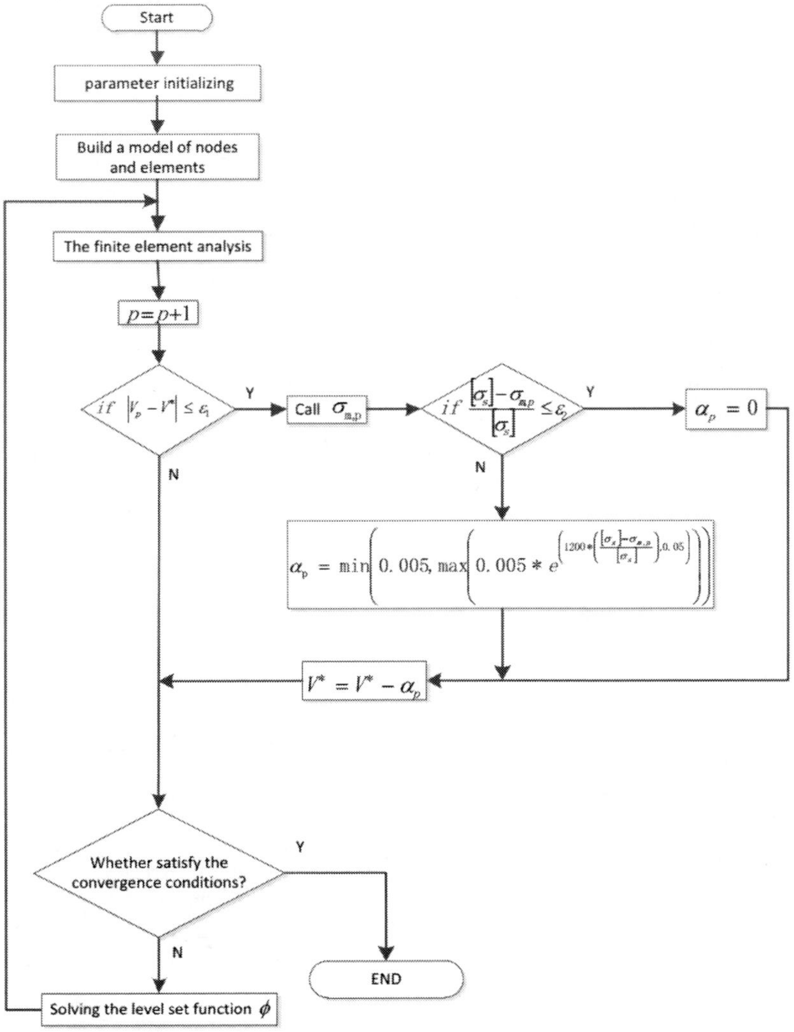

Figure 2. Flow diagram of the bidirectional improvement

2.2 Volume constraints formula

Volume constraint is applied to controls the changed volume of material. The details of solve are shown as follows:

$$V^* = V^* - \alpha_p \quad (4)$$

where, the solve of α_p as follows:

DOI: 10.1051/matecconf/20179512010

If $[\sigma_s]-\sigma_{m,p} \le \varepsilon_2[\sigma_s]$

$$\alpha_p = 0;$$

If $[\sigma_s]-\sigma_{m,p} > \varepsilon_2[\sigma_s]$ (5)

$$\alpha_p = \min\left(0.005, \max\left(0.005 * e^{\left(1200*\left(\frac{[\sigma_s]-\sigma_{m,p}}{[\sigma_s]}\right), 0.05\right)}\right)\right).$$

As shown in Eq. 5. We can see that α_p represent the value of α in p times. With the change of unit stress, the value of V^* is also changed which is the main constrain that can control the decrease of the material. Besides, the value of V^* is decided by α_p. What's more, the drop speed of volume constrain V^* will increase when the maximum stress value fully meet the located conditions. On the other hand, the drop speed of volume constrain V^* will decrease when the maximum stress value will meet the located conditions. From that, we can improve the accuracy of calculation.

2.3 The computational formula of stress

In this paper, the stress calculation formula as follows:

$$\mathbf{\sigma_i} = \mathbf{DBu_i} \quad (6)$$

where **D** represent the elastic matrix, **B** represent the strain matrix, besides, the $\mathbf{u_i}$ represent the vector of node displacement in element i.

The stress vector value of element i can be represented as $\mathbf{\sigma_i} = (\mathbf{\sigma_x}, \mathbf{\sigma_y}, \mathbf{\tau_{xy}})^{\mathbf{T}}$, and its Michelle unit stress is shown in Eq. 7.

$$\sigma_i = \sqrt{\sigma_x^2 + \sigma_y^2 - \sigma_x\sigma_y + 3\tau_{xy}^2} = \left(\mathbf{\sigma_i^T T \sigma_i}\right)^{1/2} \quad (7)$$

$$\mathbf{T} = \begin{bmatrix} 1 & -0.5 & 0 \\ -0.5 & 1 & 0 \\ 0 & 0 & 3 \end{bmatrix} \quad (8)$$

where σ_x, σ_y respectively represent the normal stress along with the direction of X and Y; $\mathbf{\tau_{xy}}$ represent the shear stress. Besides, **T** is regarded as the Coefficient matrix.

The flow diagram of stress constrain is shown as Fig. 2. In the Fig. 2, ε_1 and ε_2 are the error parameters, ε_1 is equal to 0.01, ε_2 is equal to 0.00004.

3 Numerical examples

Now, Here will prove the effectiveness and accuracy of level set topological optimization which based on the stress and volume constrains. The constrain and force is applied in a cantilever beam as shown in Fig. 3, besides the thickness value of the plate is one millimeter, and Young's modulus value of the material is $E = 2.1 * 10^{11} pa$, what's more, The value of poisson's ratio is equal to 0.3.

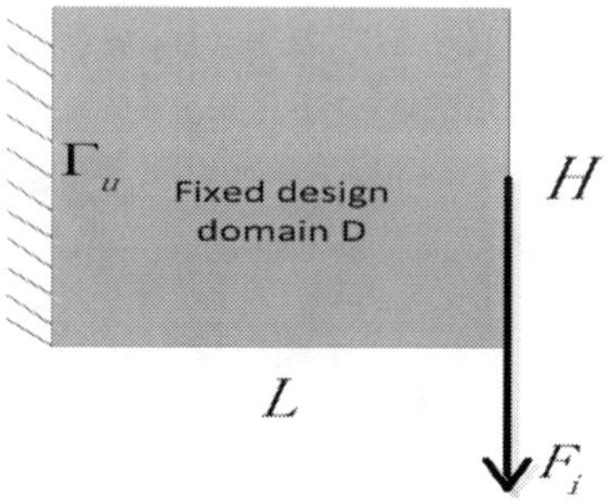

Figure 3. Structure diagram of the cantilever beam

3.1 Example 1

The wide L of the structure is equal to 80mm, high H is 64mm, and concentrated load F_i is equal to 3500N. The allowable value of stress is equal to 218MPa; the result of example 1 is shown in Fig. 4 and Fig. 5.

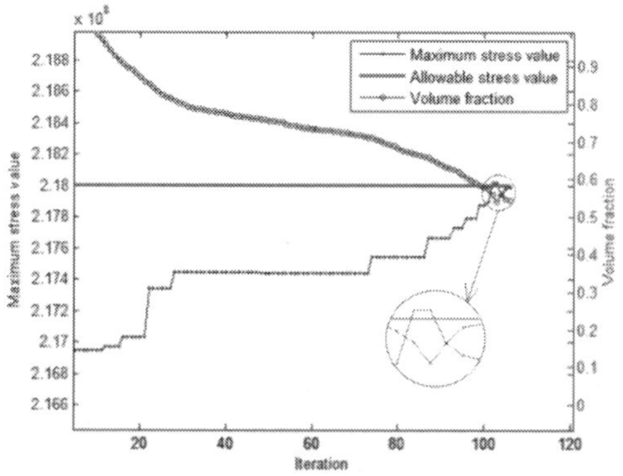

Figure 4. Variation related of parameters in example 1

Figure 5. Topological structure diagram of example 1

As shown in Fig. 4, we can see that blue line represents the maximum stress value in each iteration of the structure, red one represents the allowable stress value and the green line represents the rest of the volume fraction. Along with the iteration, the volume fraction decreased continuously, the maximum stress value of structure increased continuously and it will approach the allowable stress values.

After calculating, we can know that the volume fraction of the structure is 0.536, and the stress value is shown as $\sigma_m = 217.98MPa < [\sigma_s]$, which meet the condition that stress is not more than allowable stress.

3.2 Example 2

The wide L of the structure is equal to 40mm, high H is 80mm, and concentrated load F_1 is equal to 4000N. The allowable value of stress is equal to 249MPa; the result of example 2 is shown in Fig. 6 and Fig. 7.

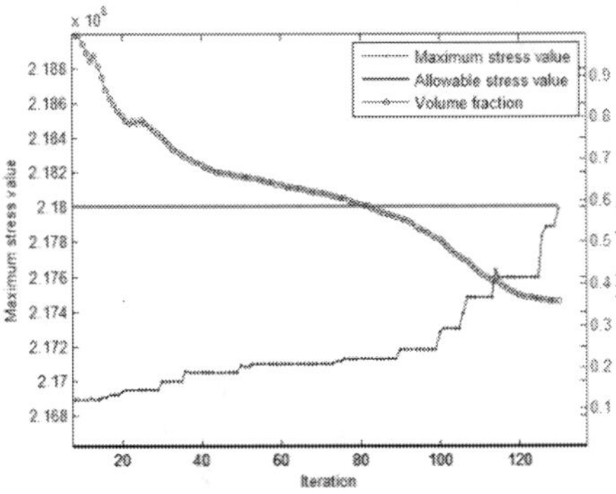

Figure 6. Variation related of parameters in example 2

Figure 7. Topological structure diagram of example 2

As shown in Fig. 6, we can see that blue line represents the maximum stress value in each iteration of the structure, red one represents the allowable stress value and the green line represents the rest of the volume fraction. Along with the iteration, the volume fraction decreased continuously, the maximum stress value of structure increased continuously and it will approach the allowable stress values.

After calculating, we can know that the volume fraction of the structure is 0.358, and the stress value is equal to $\sigma_m = 248.995MPa < [\sigma_s]$, which meet the condition that stress is not more than allowable stress.

4 Summary

This paper introduce the stress constraint to the level set method for topological optimization, and set a changed parameters as the threshold value to regulate the volume of structure, which based on the relationship between the located allowable stress and the of unit stress at every iteration. What's more, the parameters not only speed up the optimization process when the maximum stress value fully meet the located conditions, but also can slow down the optimize process, improve the accuracy of calculation, ensure stability of the results when the maximum stress value will meet the located conditions. Eventually, the proposed method can convenient the level set topological optimization to apply in engineering practice.

References

1. S. Osher and J.A. Sethian: Fronts propagating with curvature dependent speed: algorithms based on Hamilton-Jacobi formulations. J. Comput. Phys. **79** pp. 12-49, (1988).
2. M.Otomori, T. Yamada, K. Izui and S. Nishiwaki: Matlab code for a level set-based topology optimization method using a reaction diffusion equation. Struct. Multidiscip.O **51**, pp. 1159-1172, (2015).
3. C. G. Zhuang, Z. H. Xiongand D. Han:Structural Topology Optimization Based on Level Set Method and von Mises Stress, (2006).
4. W.S. Zhang, X. Guo, M.Y. Wang and P. Wei: Optimal topology design of continuum structures with stress concentration alleviation via level set method. Int. J. Numer. Methods Eng., **93**, pp. 942-959, (2013).
5. J. H. Rong, S. Ge, G. Deng, X. J. Xin and Z. J. Zhao: Structural topological optimization based on displacement and stress sensitivity analyses. Chinese Journal of Theoretical and Applied Mechanics. **4**, pp. 0518-0529, (2009).
6. X.D. Liu, Y.K. Sui and H.P. Yu: Topological Optimization of plate and shell structure with stress constraints. Mechanics in Engineering. **4**, pp. 035-039, (2012).

Entrance Region Flow Heat Transfer in Concentric Annuli with Rotating Inner Wall for Herschel-Bulkley Fluids

Srinivasa Rao Nadiminti and A. Kandasamy

Department of Mathematical and Computational Sciences, National Institute of Technology Karnataka, Surathkal, Mangalore, Karnataka - 575025, India

Abstract. A finite difference analysis of the entrance region flow heat transfer of Herschel-Bulkley fluids in concentric annuli with rotating inner wall has been carried out. The analysis is made for simultaneously developing hydrodynamic and thermal boundary layer in concentric annuli with one wall being isothermal and other one being adiabatic. The inner cylinder is assumed to be rotating with a constant angular velocity and the outer cylinder being stationary. A finite difference analysis is used to obtain the velocity distributions, pressure drop and temperature variations along the radial direction. Computational results are obtained for various values of aspect ratio, flow index, Prandtl's number and Herschel-Bulkley number. Comparison of the present results with the results available in literature for various particular cases has been done and found to be in agreement.

1 Introduction

The study of non-Newtonian laminar flow heat transfer in thermal entrance region of an annuli is of practical importance in engineering applications such as design of compact heat exchangers, axial-flow turbo machinery and polymer processing industries. Very often, optimum conditions are provided by laminar flow operations for keeping low pumping power in proportion to the heat transfer rate. In the nuclear reactor field, this happens when coolant flow rates are reduced. In case of turbulent flow, when heating starts at the duct entrance, the hydrodynamic boundary layers normally are linear near the duct entrance and the transitions to turbulence occurs at some distance downstream from the entrance. Hence, it is necessary to take into consideration the presence of such laminar entrance flow in calculating the heat transfer parameters for a duct in which the fully developed flow is turbulent. Many important industrial fluids are non Newtonian in their flow characteristics and are referred to as rheological fluids. These include blood; various suspensions such as coal water or coal-oil slurries, glues, inks, foods; polymer solutions; paints and many others. The fluid considered here is the Herschel-Bulkley model, which is of 'time independent yield stress' fluid category.

The problem of entrance region flow heat transfer in a concentric annuli for a Newtonian fluid was studied by Coney and El-Shaarawi [1]. Mishra and Kumar [2] studied the flow of the Bingham plastic in the concentric annulus and obtained the results for boundary layer thickness, centre core velocity, pressure drop. Batra and Bigyani Das [3] developed the stress-strain relation for the Casson fluid in the annular space between two coaxial rotating cylinders where the inner cylinder is at rest and outer cylinder is rotating. Maia and Gasparetto [4] applied finite difference method for the Power-law fluid in the annuli and found difference in the entrance geometries. Sayed-Ahmed and Hazem [5] have applied finite difference method to study the laminar flow of a Power-Law fluid in the concentric annuli with rotating inner wall.

The Herschel-Bulkley model represents the empirical combination of Bingham and power-law fluids. The constitutive equation for these fluids is given by Bird *et al.* [6]as

$$\tau = \tau_0 + k\left(\frac{\partial u}{\partial r}\right)^n \tag{1}$$

where τ is the shear stress, τ_0 is the yield stress, k is the consistency index, n is the flow index of the model.

Manglik and Fang [7] numerically investigated the flow of non-Newtonian fluids through annuli. Round and Yu [8] analyzed the developing flows of Herschel-Bulkley fluids through concentric annuli. Soares et al. [9] has taken up the problem of heat transfer in a fully developed flow of Herschel-Bulkley materials through annular spaces, with insulated outer wall and uniform heat flux at inner wall. Kandasamy et al. [10] investigated the entrance region flow of heat transfer in concentric annuli for Herschel Bulkley fluids and presents the velocity distributions, temperature and pressure in the entrance region. Poole and Chhabra [11] reported the results of a systematic numerical investigation of developing laminar pipe flow of yield stress fluids.

Recently, Kandasamy and Srinivasa Rao [12], [13] have investigated the entrance region flow in concentric annuli with rotating inner wall for Herschel Bulkley and Bingham fluids.

In the present work, the problem of entrance region flow heat transfer of Herschel-Bulkley fluids in concentric annuli has been investigated. The analysis has been carried out under the assumption that the inner cylinder is rotating and the outer cylinder is at rest. With Prandtl's boundary layer assumptions, the equations of conservation of mass, momentum and energy are discretized and solved using linearized implicit finite difference technique. The system of non-linear algebraic equations thus obtained has been solved by the Newton Raphson iterative method for simultaneous non-linear equations. The development of axial velocity profile, radial velocity profile, tangential velocity profile, pressure drop and temperature variations in the entrance region have been determined for different values of non-Newtonian flow characteristics and geometrical parameters. The effects of these on the temperature variations have been discussed.

2 Formulation of the problem

The Herschel-Bulkley fluid enters the horizontal concentric annuli with inner and outer radii R_1 and R_2, respectively, from a large chamber with a uniform flat velocity profile u_0 along the axial direction z and with an initial pressure p_0 and temperature t_0. The inner cylinder rotates with an angular velocity ω and the outer cylinder is at rest. The geometry of the problem is shown in Fig.1. The flow is steady, laminar, incompressible, axisymmetric with constant physical properties, having negligible viscous dissipation and no internal heat generation. Moreover, it is assumed that the axial heat diffusion is negligible as compared to the radial diffusion. We consider a cylindrical polar coordinate system with the origin at the inlet section on the central axis of the annulus, the z-axis along the axial direction and the radial direction r perpendicular to the z-axis.

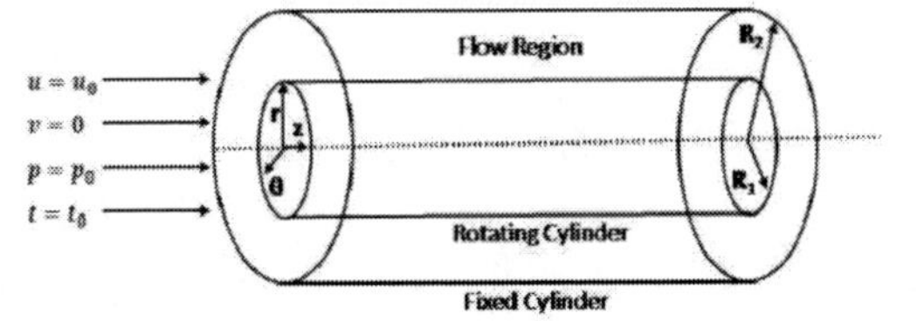

Figure 1. Geometry of the problem.

Under the above assumptions with the usual Prandtl's boundary layer assumptions [14], the governing equations in polar coordinate system (r, θ, z) for a Herschel-Bulkley fluid in the entrance region are:

Continuity Equation: $$\frac{\partial(rv)}{\partial r}+\frac{\partial(ru)}{\partial z}=0 \quad (2)$$

r-momentum Equation: $$\frac{w^2}{r}=\frac{1}{\rho}\frac{\partial p}{\partial r} \quad (3)$$

θ-momentum Equation:

$$v\frac{\partial w}{\partial r}+u\frac{\partial w}{\partial z}+\frac{vw}{r}=\frac{1}{\rho r^2}\frac{\partial}{\partial r}\left(r^2\left(\tau_0+k\left[r\frac{\partial}{\partial r}\left(\frac{w}{r}\right)\right]^n\right)\right) \quad (4)$$

z-momentum Equation:

$$v\frac{\partial u}{\partial r}+u\frac{\partial u}{\partial z}=-\frac{1}{r}\frac{\partial p}{\partial z}+\frac{1}{\rho r}\frac{\partial}{\partial r}\left(r\left[\tau_0+k\left(\frac{\partial u}{\partial r}\right)^n\right]\right) \quad (5)$$

Energy Equation:

$$v\frac{\partial t}{\partial r}+u\frac{\partial t}{\partial z}=\alpha\left[\frac{\partial^2 t}{\partial r^2}+\frac{1}{r}\frac{\partial t}{\partial r}\right] \quad (6)$$

where u, v, w are the velocity components in z, r, θ directions respectively, t is fluid temperature at any point, ρ is the density of the fluid, α is the thermal diffusivity and p is the pressure.

The boundary conditions associated with the hydrodynamic part of the problem are given by

for $z \geq 0$ and $r = R_1$, $v = u = 0$ and $w = \omega R_1$
for $z \geq 0$ and $r = R_2$, $v = u = w = 0$ (7)
for $z = 0$ and $R_1 < r < R_2$, $u = u_0$
at $z = 0$, $p = p_0$

Using the boundary conditions (7), the continuity equation (2) can be expressed in the following integral form:

$$2\int_{R_1}^{R_2} ru\,dr=(R_2^2-R_1^2)u_0 \quad (8)$$

Introducing the following dimensionless variables and parameters

$$R=\frac{r}{R_2},\ U=\frac{u}{u_0}\ \ V=\frac{\rho v R_2}{\mu_r}\ \ W=\frac{w}{\omega R_1}\ \ N=\frac{R_1}{R_2}$$

$$P=\frac{p-p_0}{\rho u_0^2},\ Z=\frac{2z(1-N)}{R_2 Re}\ \ Y_h=\frac{\tau_0}{k}\left(\frac{R_2}{u_0}\right)^n$$

$$T_a=\frac{2\omega^2\rho^2R_1^2(R_2-R_1)^3}{\mu_r^2(R_2+R_1)}\ \text{ here }\ \mu_r=k\left(\frac{\omega R_1}{R_2}\right)^n$$

$$Re=\frac{2(R_2-R_1)u_0}{\mu_r},\ T=\frac{t-t_0}{t_w-t_0}\ \ P_r=\frac{\mu C_p}{K}\left(\frac{\omega R_2}{u_0}\right)^{1-n}$$

Here Y_h is the Hershel-Bulkley number, Re Reynolds number, T_a Taylors number, μ_r is known as reference viscosity, P_r is the Prandtl's number, C_p is the specific heat at constant temperature, K is the thermal conductivity, t_w is the isothermal wall temperature and N is known as aspect ratio of the annulus.

Equations (2)-(6) and (8) in the dimensionless form are given by

$$\frac{\partial V}{\partial R}+\frac{V}{R}+\frac{\partial U}{\partial Z}=0 \tag{9}$$

$$\frac{W^2}{R}=\frac{Re^2(1-N)}{2(1+N)T_a}\frac{\partial P}{\partial R} \tag{10}$$

$$V\frac{\partial W}{\partial R}+U\frac{\partial W}{\partial Z}+\frac{VW}{R}=\frac{2}{R}\left(\frac{\partial W}{\partial R}-\frac{W}{R}\right)^n+ n\left(\frac{\partial W}{\partial R}-\frac{W}{R}\right)^{n-1}\left(\frac{\partial^2 W}{\partial R^2}-\frac{1}{R}\frac{\partial W}{\partial R}+\frac{W}{R^2}\right)+\frac{2Y_h}{R} \tag{11}$$

$$V\frac{\partial U}{\partial R}+U\frac{\partial U}{\partial Z}=-\frac{\partial P}{\partial Z}+\frac{1}{R}\left(\frac{\partial U}{\partial R}\right)^n+ n\left(\frac{\partial U}{\partial R}\right)^{n-1}\frac{\partial^2 U}{\partial R^2}+\frac{Y_h}{R} \tag{12}$$

$$V\frac{\partial T}{\partial R}+U\frac{\partial T}{\partial Z}=\frac{1}{P_r}\left[\frac{\partial^2 T}{\partial R^2}+\frac{1}{R}\frac{\partial T}{\partial R}\right] \tag{13}$$

and

$$2\int_N^1 RUdR=(1-N^2) \tag{14}$$

The boundary conditions (7) associated with the hydrodynamic part of the problem in the dimensionless form are given by

for $Z \geq 0$ and $R = N$, $V = U = 0$ and $W = 1$
for $Z \geq 0$ and $R = 1$, $V = U = W = 0$
for $Z = 0$ and $N < R < 1$, $U = 1$ (15)
at $Z = 0$, $P = 0$

For the thermal part, considering the outer cylinder to be adiabatic and the inner cylinder to be isothermal, the problem has been solved under the following boundary conditions:

for $Z \geq 0$, $T = 1$ at $R = N$
for $Z \geq 0$, $\frac{\partial T}{\partial R}=0$ at $R = 1$ (16)
for $Z = 0$ and $N < R < 1$, $T = 0$

The problem can be similarly analyzed for the case when the inner cylinder is adiabatic and the outer cylinder is isothermal. The boundary conditions for this case will be

for $Z \geq 0$, $T = 1$ at $R = 1$
for $Z \geq 0$, $\frac{\partial T}{\partial R}=0$ at $R = N$ (17)
for $Z = 0$ and $N < R < 1$, $T = 0$

3 Numerical solutions

The numerical analysis and the method of solution adopted here can be considered as an indirect extension of the work of Coney and El-Shaarawi [1]. Considering the mesh network of Fig. 2, the following difference representations are made. Here ΔR and ΔZ represent the grid size along the radial and axial directions respectively and m is the number of radial increments in the numerical mesh network.

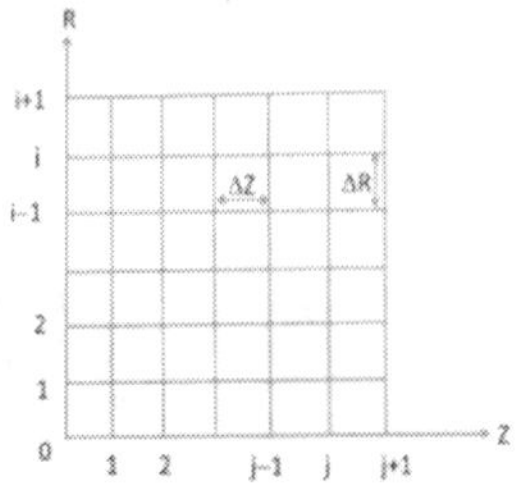

Figure 2. Grid Formation For Finite Difference Representations.

$$V_{i+1,j+1}=V_{i,j+1}\left(\frac{N+i\Delta R}{N+(i+1)\Delta R}\right)-\frac{\Delta R}{4\Delta Z}\left(\frac{2N+(2i+1)\Delta R}{N+(i+1)\Delta R}\right)* \left(U_{i+1,j+1}+U_{i,j+1}-U_{i+1,j}-U_{i,j}\right) \tag{18}$$

$$\frac{W^2{}_{i,j+1}}{N+i\Delta R}=\frac{(1-N)Re^2}{2T_a(1+N)}\frac{P_{i,j+1}-P_{i-1,j+1}}{\Delta R} \tag{19}$$

$$V_{i,j}=\left[\frac{W_{i+1,j+1}+W_{i+1,j}-W_{i-1,j}-W_{i-1,j+1}}{4\Delta R}\right]+ U_{i,j}\left[\frac{W_{i,j+1}-W_{i,j}}{\Delta Z}\right]+\frac{V_{i,j}W_{i,j}}{N+i\Delta R}=\frac{2}{N+i\Delta R}* \left[\frac{W_{i+1,j+1}+W_{i+1,j}-W_{i-1,j}-W_{i-1,j+1}}{4\Delta R}-\frac{W_{i,j}}{N+i\Delta R}\right]^n+ n\left[\frac{W_{i+1,j+1}+W_{i+1,j}-W_{i-1,j}-W_{i-1,j+1}}{4\Delta R}-\frac{W_{i,j}}{N+i\Delta R}\right]^{n-1}* \left(\frac{W_{i+1,j+1}+W_{i+1,j}-2W_{i,j+1}-2W_{i,j}+W_{i-1,j}+W_{i-1,j+1}}{2(\Delta R)^2}-\frac{W_{i+1,j+1}+W_{i+1,j}-W_{i-1,j}-W_{i-1,j+1}}{(N+i\Delta R)4\Delta R}+\frac{W_{i,j}}{(N+i\Delta R)^2}\right) +\frac{2Y_h}{N+i\Delta R} \tag{20}$$

$$P_{i,j+1}+U_{i-1,j+1}\left[-\frac{\Delta Z}{2\Delta R}V_{i,j}-\frac{n\Delta Z}{2^{n-1}(\Delta R)^{n+1}}\left(U_{i+1,j+1}-U_{i-1,j+1}\right)^{n-1}\right]+ U_{i,j+1}\left[U_{i,j}+\frac{n\Delta Z}{2^{n-2}(\Delta R)^{n+1}}\left(U_{i+1,j+1}-U_{i-1,j+1}\right)^{n-1}\right]+ U_{i+1,j+1}\left[\frac{\Delta Z}{2\Delta R}V_{i,j}-\frac{n\Delta Z}{2^{n-1}(\Delta R)^{n+1}}\left(U_{i+1,j+1}-U_{i-1,j+1}\right)^{n-1}\right]- \frac{\Delta Z}{N+i\Delta R}\left(\frac{U_{i+1,j+1}-U_{i-1,j+1}}{2\Delta R}\right)^n=P_{i,j}+U_{i,j}{}^2+\frac{Y_h\Delta Z}{N+i\Delta R} \tag{21}$$

where i=0 at R=N and i=m at R=1.

The application of trapezoidal rule to equation (14) gives

$$\frac{\Delta R}{2}\left(NU_{0,j}+U_{m,j}\right)+\Delta R\sum_{i=1}^{m-1}U_{i,j}\left(N+i\Delta R\right)=\left(\frac{1-N^2}{2}\right)$$

The boundary condition (15) gives U0, j = Um, j = 0 and the above equation reduces to

$$\Delta R\sum_{i=1}^{m-1}U_{i,j}\left(N+i\Delta R\right)=\left(\frac{1-N^2}{2}\right) \tag{22}$$

The set of difference equations (18) to (23) have been solved by the iterative procedure. Starting at the j = 0 column (annulus entrance) and applying equation (20) for $1 \le i \le m\text{-}1$, we get a system of non-linear algebraic equations. This system has been solved by using Newton Raphson method to obtain the values of the velocity component W at the second column j=1. Then applying equations (19) and (21) for $1 \le i \le m\text{-}1$ and equation (22), we get a system of non-linear equations. Again solving this system by Newton-Raphson method, we obtain the values of the velocity component U and the pressure P at the second column j=1. Finally, the values of the velocity component V at the second column j=1 are obtained from equation (18) by Gauss-Jordan method using the known values of U. Repeating this procedure, we can advance, column by column, along the axial direction of the annulus until the flow becomes axially and tangentially fully developed. With the values of V and U known, the energy equation (13) can be considered as a linear equation in T with variable coefficients. By using the implicit finite difference technique, the energy equation can be represented as

$$\begin{aligned}
&T_{i+1,j+1}\left(\frac{V_{i,j+1}+V_{i,j}}{8\Delta R}-\frac{1}{2Pr\left(\Delta R\right)^2}-\frac{1}{4(N+i\Delta R)\Pr\Delta R}\right)+\\
&T_{i,j+1}\left(\frac{U_{i,j+1}+U_{i,j}}{2\Delta Z}+\frac{1}{Pr\left(\Delta R\right)^2}\right)+\\
&T_{i-1,j+1}\left(-\frac{V_{i,j+1}+V_{i,j}}{8\Delta R}-\frac{1}{2Pr\left(\Delta R\right)^2}+\frac{1}{4(N+i\Delta R)Pr\Delta R}\right)\\
&=T_{i,j}\left(\frac{U_{i,j+1}+U_{i,j}}{2\Delta Z}-\frac{1}{Pr\left(\Delta R\right)^2}\right)+\\
&T_{i+1,j}\left(-\frac{V_{i,j+1}+V_{i,j}}{8\Delta R}+\frac{1}{2Pr\left(\Delta R\right)^2}+\frac{1}{4(N+i\Delta R)Pr\Delta R}\right)+\\
&T_{i-1,j}\left(\frac{V_{i,j+1}+V_{i,j}}{8\Delta R}+\frac{1}{2Pr\left(\Delta R\right)^2}-\frac{1}{4(N+i\Delta R)Pr\Delta R}\right)
\end{aligned} \tag{23}$$

Equation (23), with the boundary conditions (16) have been used to obtain the temperature profiles in the annular entrance region. The system of linear equations associated with each column has been solved by Gauss Jordan elimination method.

4 Results and discussion

Numerical calculations have been performed for all admissible values of aspect ratio N, flow index n, Prandtl's number and Herschel-Bulkley number. The Prandtl's number has been chosen as 7. The velocity profiles, pressure drop and temperature distribution along radial direction have been computed for N =0.3, 0.8, n=0.8, 1, 1.2 and Y_h=0, 10.

The temperature distribution along radial direction have been plotted for different axial positions with N =0.3, 0.8, n=0.8, 1, 1.2 and Y_h =0, 10. The effects of non-Newtonian characteristics and geometrical parameters on velocity profiles and pressure drop have been extensively discussed elsewhere [12]. Here the temperature distributions during the rotation of the inner wall of the annuli have been shown in Fig. 3-Fig. 8.

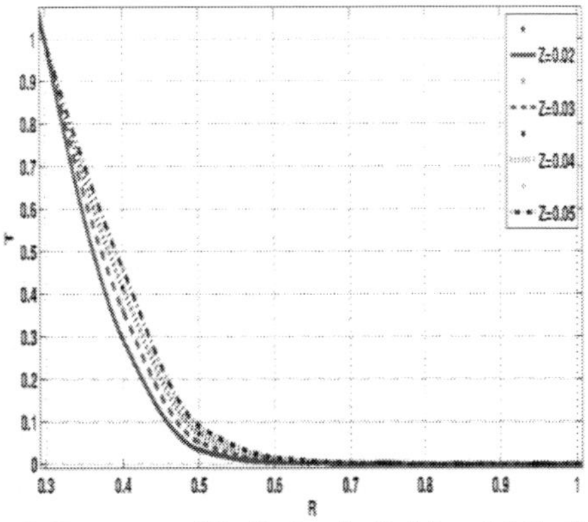

Figure 3. Temperature Distribution for N=0.3, n=0.8, Pr=7, Yh=0.

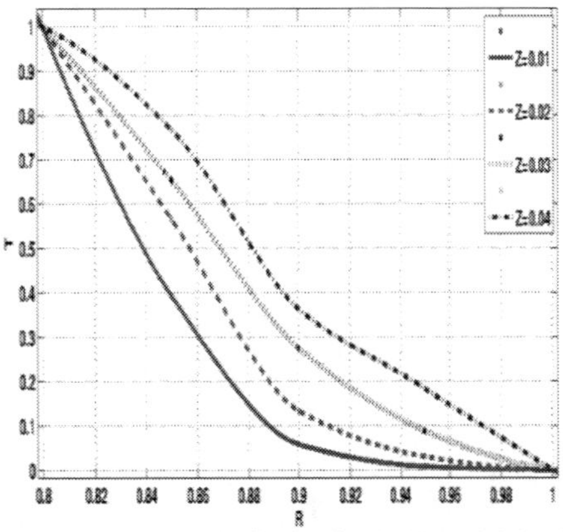

Figure 4. Temperature Distribution for N=0.8, n=0.8, Pr=7, Yh=10 .

It is observed from the results obtained, that the temperature decreases with increase of Herschel-Bulkley number for a fixed annular width. When the aspect ratio N increases, it is found that the temperature increases for a fixed Herschel-Bulkley number.

Moreover, with the increase of flow index `n' the temperature decreases for a fixed aspect ratio N and Herschel-Bulkley number. Also, it is found that with the increase of axial position the temperature also increasing for a fixed aspect ratio N, Herschel-Bulkley number.

The present results are compared with available results in literature for various particular cases and are found to be in agreement. When the Herschel-Bulkley number Y_h=0 and n=1, our results match with the results corresponded to Newtonian fluid of Coney and El-Shaarawi [1]. In the case of stationary cylinders, the results in our analysis are matching with the results of Kandasamy *et al.* [10]. Also, in the case of non-thermal part these results match with the results of Kandasamy and Srinivasa Rao [12].

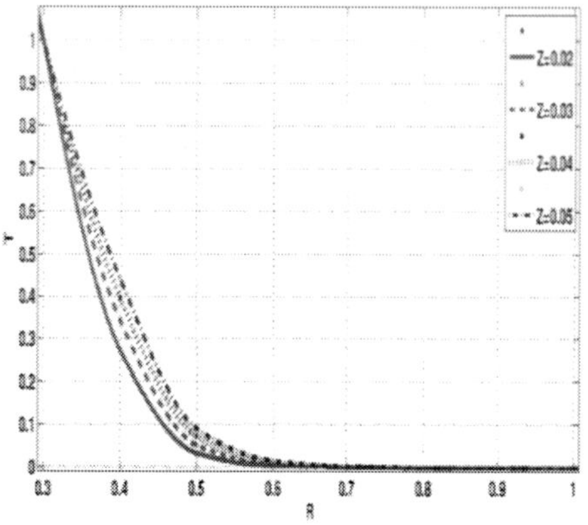

Figure 5. Temperature Distribution for N=0.3, n=1, Pr=7, Yh=0.

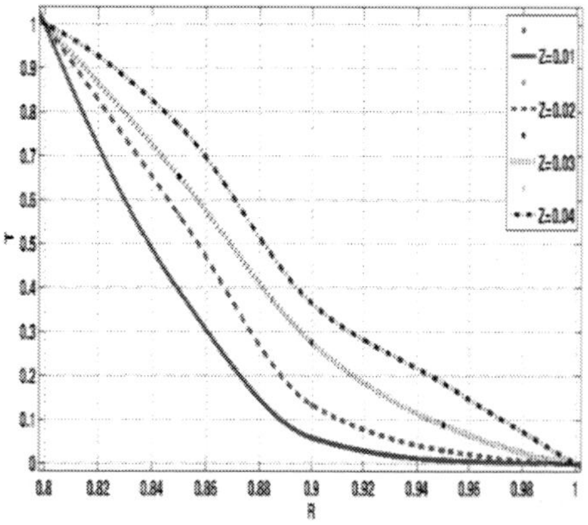

Figure 6. Temperature Distribution for N=0.8, n=1, Pr=7, Yh=10.

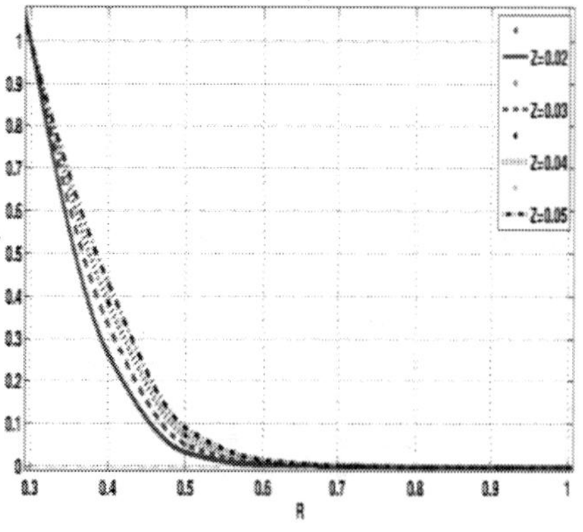

Figure 7. Temperature Distribution for N=0.3, n=1.2, Pr=7, Yh=0.

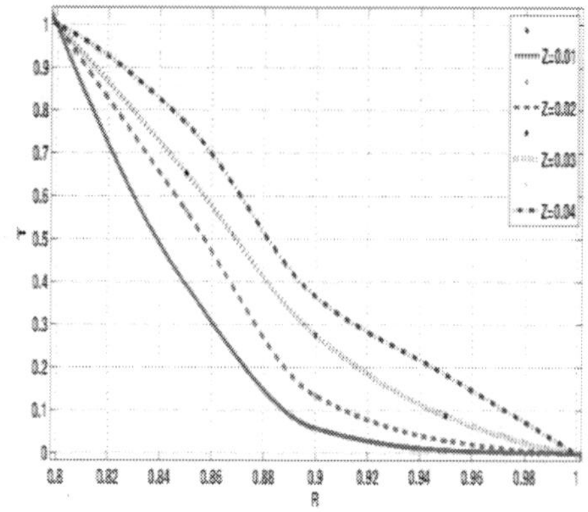

Figure 8. Temperature Distribution for N=0.8, n=1.2, Pr=7, Yh=10.

5 Conclusion

Numerical results for the entrance region flow heat transfer in concentric annuli with rotating inner wall for Herschel-Bulkley fluids were presented. The effects of the parameters aspect ratio N, Herschel-Bulkley number, flow index and Prandtl's number on the temperature distribution are studied. Numerical calculations have been performed for all admissible values of aspect ratio N, Herschel-Bulkley number, flow index and Prandtl's number. The temperature distribution along radial direction R have been presented geometrically. The present results are found in agreement with the results corresponding to various particular cases available in literature.

From this study, the following can be concluded.

1. The temperature decreases from the rotating inner wall to the stationary outer wall of the annulus.
2. When increasing the Herschel-Bulkley number, it is observed that the temperature decreases.
3. With the increase of flow index `n', the temperature decreases.
4. When aspect ratio N increases, it is found that the temperature increases.
5. With the increase of axial position Z, the temperature also increases.

References

1. J.E.R. Coney, and M.A.I. El-Shaarawi, A contribution to the numerical solution of developing laminar flow in the entrance region of concentric annuli with rotating inner walls, *J. Fluid Eng.*, **96**, 333-340 (1974)
2. I. M. Mishra, Surendra Kumar, and P Mishra, Entrance Region Flow of Bingham Plastic Fluids in Concentric Annulus, *Indian Journal of Technology*, **23**, 81-87 (1985)
3. R.L. Batra, and Bigyani Das, Flow of a Casson Fluid between Two Rotating Cylinders, *Fluid Dynamic Research*, **9**, 133-141 (1992)
4. M.C.A. Maia, and C.A. Gasparetto, A numerical solution for entrance region of non-Newtonian flow

DOI: 10.1051/matecconf/20179512011

in annuli, *Brazilian Journal of Chemical Eng.*, **20**, 201-211 (2003)

5. M.E. Sayed-Ahmed, and Hazem Sharaf-El Din, Entrance region flow of a power-law fluid in concentric annuli with rotating inner wall, *International Communications in Heat and Mass Transfer*, **33**, 654-665 (2006)
6. R.D. Bird, G.C. Dai, and B.J. Yarusso, The rheology and flow of viscoplastic materials, *Rev.Chem. Eng.*, **1**, 1-70 (1982)
7. R. Manglik, and P. Fang, Thermal processing of various non-Newtonian fluids in annular ducts, *International Journal Heat and Mass Transfer*, **45**, 803-815 (2002)
8. G.F. Round, and S. Yu, Entrance laminar flows of viscoplastic fluids in concentric annuli, *Canadian J. Chem. Eng.*, 71, 642-645 (1993)
9. E.J. Soares, M.F. Naccache, and P.R. Souza Mendes, Heat transfer to Herschel-Bulkley materials in annular flows, *Proc. 7th Brazilian Cong. Thermal Sciences*, 1146-1151 (1998)
10. A. Kandasamy, K.Karthik, and P.H.Phanidar, Entrance region flow of heat transfer in concentric annuli for Herschel-Bulkley fluids, *computational fluid dynamics journal*, **16**, 103-114 (2007)
11. R.J. Poole, and R.P.Chhabra, Development length requirements for fully developed Laminar pipe flow of yield stress fluids, *J. Fluids Eng.*, **132**, 34501-34504 (2010)
12. A. Kandasamy and Srinivasa Rao Nadiminti, Entrance region flow in concentric annuli with rotating inner wall for Herschel-Bulkley fluids, Int. Journal of Applied and Comp. Mathematics, 1, 235-249 (2015)
13. Srinivasa Rao Nadiminti and A. Kandasamy, Entrance region flow in concentric annuli with rotating inner wall for Bingham fluid, 16th International Conference on Fluid Flow Technologies, Budapest, Hungary, September 1-4, (2015).
14. H. Schlichting, and K. Gersten, Boundary Layer Theory, 8thed., Springer,(2000).

The Thermohydrodynamic Analysis of Sliding Bearing High-Speed Motorized Spindle by Rotor Dynamic

Songsheng Li , Yuan Shen and Yuxin Lu

School of Mechatronic Engineering and Automation, Shanghai Key Laboratory of Intelligent manufacturing and Robotics, Shanghai University, Shanghai University, China

Abstract. This is paper presents thermohydrodynamic characteristics of high speed motorized spindle sliding bearing rotor system. The dynamic characteristic of the oil film bearing is affected by temperature increment, thereby affecting the high-speed spindle rotor system dynamics. This study applied the hydrodynamic lubrication theory, the influence of temperature on the viscosity of lubricating oil, associated with the bearing stiffness, oil film damping and other performance parameters, is considered in generalized Reynolds equation of oil film bearing. The theoretical model of the sliding bearing rotor system is established by using the transfer matrix method to analyze the dynamic characteristic and verified by experiments. The results show the high temperature environment in the motorized spindle and the friction of the bearing lead to oil temperature rise and viscosity reduction, which influences the bearing capacity, stiffness and damping, hence impact on the critical speeds and modal shapes of the sliding bearing rotor system.

1 Introduction

With the development of science and all kinds of industrial technology, motorized spindle trends to be higher speed and larger power. Hydrodynamic sliding bearing has been widely used for its advantages of large load capacity, long service life, non contact and so on. Therefore, the hydrodynamic sliding bearing motorized spindle came into being. However, the serious friction and large heat generation of the bearing have been the key factors which restrict the development of higher speed [1]. It is a important engineering guarantee to study on the high speed hydrodynamic sliding bearing and analyze the dynamic characteristic of the high-speed motorized spindle sliding bearing rotor system.

Researches on rotating machinery occupies an important position in rotor dynamics, they are the foundation of calculating various rotor dynamics when design and optimize complex rotor system. Traditional rotor dynamics often treat bearings as accessories of the rotor neglecting the bearing itself is a important part in the rotor system [2]-[4]. For the studies of sliding bearing, since O. Reynolds derived the differential equation of oil film pressure distribution in 1886, most researches are based on the isothermal hydrodynamic lubrication model [5]. Whereas, in the actual operation process, because of the lubrication of extrusion and shear viscosity the sliding bearing will friction and generate heat, and the higher the speed, the more heat generated inside the bearing, resulting in high temperature [6], [7]. As the temperature and oil film pressure distribution of the journal bearing shown Fig. 1, the temperature rise leads to the oil viscosity decrease, the viscosity changes will also affect the dynamic pressure of oil film, bearing stiffness and damping etc. [8], [9]. So in the high temperature environment of the high-speed motorized spindle and high speed conditions, the application of traditional isothermal hydrodynamic lubrication model can not accurately reflect the dynamic characteristics of sliding bearing, and can't meet the need of practical engineering [10]-[13].

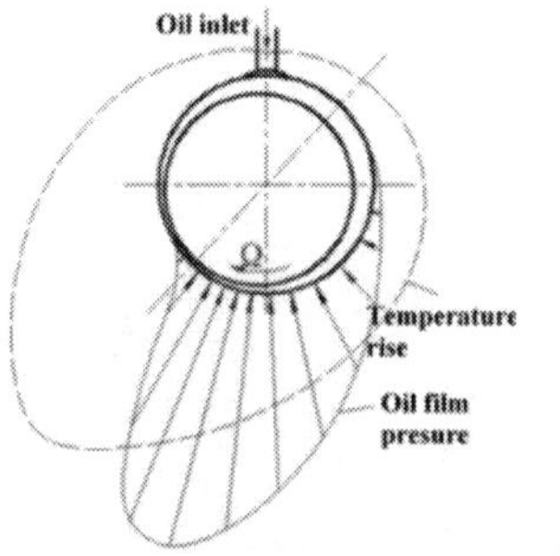

Figure 1. The temperature and oil film pressure distribution of the journal bearing.

A theoretical model, presented in this paper, applies transfer matrix method to analyze the dynamic characteristic of the sliding bearing rotor system. Based on the hydrodynamic lubrication theory, viscosity-temperature equation is considered in Reynolds equation to explore the influence of temperature on the viscosity of bearing lubricating oil film. The theoretical analysis model of high speed radial dynamic pressure sliding bearing is established combined with the energy equation. The relationship of bearing temperature, rotation speed, stiffness and damping is explored as well as their influence on the critical speed and modal shape. Simplified calculation methods of bearing total stiffness of isotropic and anisotropic sliding bearing rotor system are presented, the critical speed results of the two methods are compared and the first order critical speed is verified via experiment.

2 Dynamic analysis model of rotor system

The actual rotor is an elastic system of continuous mass distribution with infinite number of degrees of freedom. In rotor dynamics, the rotor is simplified to a multiple degrees of freedom (MDOF) model with a plurality of lumped mass in order to facilitate the calculation. That is, the mass and the moment of inertia of the rotor are lumped to a number of nodes along the center line of the shaft. These nodes are generally selected in the gear, bearing center, shaft coupling, the section of the shaft diameter changes and the shaft end and so on. The total notes N have certain requirements, present work shows that to make natural frequency error less than 1%, the number of nodes N should meet the following formula [14]:

$$N \geq 1+5.34v \quad (1)$$

where, v is the highest order of natural frequencies required to calculate.

2.1 Isotropic bearing rotor system

Excluding damping, the bending vibration of the axis deflection curve is a plane curve. The support for the rotor system is isotropic in this case, so only one plane vibration modal is needed to obtain the critical speed and corresponding modal shape. The axial direction is x direction and the radial direction is z direction, and the xoz plane is the vibration plane of rotor.

For the cross-section i of the rotor, the state vector is Z_i, which is composed of the radial displacement X_i, the torsion angle A_i, the bending moment M_i and the shear force Q_i. It can be written as:

$$Z_i = \begin{bmatrix} X & A & M & Q \end{bmatrix}_i^T \quad (2)$$

It has the following relationship with the i+1 section the state vector Z_{i+1} of the rotor:

$$Z_{i+1} = T_i Z_i \quad (3)$$

where, T_i is the transfer matrix of two sections and is shown as:

$$T_i = \begin{bmatrix} 1+\frac{l^3}{6EJ}(1-v)(m\omega^2-K_{sj}) & l+\frac{l^2}{2EJ}(I_p-I_d)\omega^2 & \frac{l^2}{2EJ} & \frac{l^3}{6EJ}(1-v) \\ \frac{l^2}{2EJ}(m\omega^2-K_{sj}) & 1+\frac{l}{EJ}(I_p-I_d)\omega^2 & \frac{l}{EJ} & \frac{l^2}{2EJ} \\ l(m\omega^2-K_{sj}) & (I_p-I_d)\omega^2 & 1 & l \\ (m\omega^2-K_{sj}) & 0 & 0 & 1 \end{bmatrix} \quad (4)$$

where, E is the elastic modulus of materials; J is the section distance; v is coefficient of shear force effect.

It can be seen that the elements of the transfer matrix are the physical parameters of the rotor components and the vortex frequency function. If the section of the rotor is not supported, or regardless the shear effect of the shaft, or regardless of the gyroscopic effect of the disk or swing effect, just make the K_{sj} or v or I_p or I_d zero respectively.

2.2 Anisotropy bearing rotor system

To consider both the motion of the rotor in horizontal and vertical two levels and the damping effect, it needs anisotropy dynamic method. The vibration quantity of the system will occur phase-difference, the magnitude of the vibration amplitude will increase or decrease with time. Therefore, the vibration quantity needs to use the complex number to express, the complex representation of the state vector is [1]:

$$Z_i = [f \vdots e]_i^T = [M_x Q_x M_y Q_y \vdots XAYB] \quad (5)$$

Besides the extended to the 8 order square matrix, it is similar to the isotropy.

3 Thermohydrodynamic lubrication model of bearing

3.1 Reynolds equation

The Reynolds equation of the bearing oil film pressure distribution at steady state is shown as follow on the assumption that the lubricating oil is laminar and incompressible, ignored the weight of the oil layer and the inertia force of the movement.

$$\frac{\partial}{\partial x}\left(\frac{h^3}{\mu}\frac{\partial P}{\partial x}\right)+\frac{\partial}{\partial z}\left(\frac{h^3}{\mu}\frac{\partial P}{\partial z}\right)=6U\frac{\partial h}{\partial x} \quad (6)$$

where, x is the coordinates along the axis neck circle direction, z is the axial coordinate, h is the radial oil film thickness, μ is oil dynamic viscosity, P is the oil film pressure, U is the linear velocity of the spindle neck.

3.2 Energy equation

Assuming all the generated heat by the bearing is carried away by the lubricating oil, that is apply the adiabatic boundary condition in the bearing and oil film interface. The equation of energy can be expressed as follows [15]:

$$\rho\left[\left(\frac{Uh}{2}-\frac{h^3}{12\mu}\frac{\partial P}{\partial x}\right)\frac{\partial(c_v T)}{\partial x}-\frac{h^3}{12\mu}\frac{\partial P}{\partial z}\frac{\partial(c_v T)}{\partial z}\right] = \mu\frac{U^2}{h}+\frac{h^3}{12\mu}\left[\left(\frac{\partial P}{\partial x}\right)^2+\left(\frac{\partial P}{\partial z}\right)^2\right] \quad (7)$$

where, ρ is the oil density, T is the oil temperature, c_v is the specific heat capacity of the oil.

Assume the specific heat capacity is constant, the results of the three-dimensional model calculations shows temperature variation is very small along the axial direction and can be neglect [4], that is $\frac{\partial(c_v T)}{\partial z}=0$. So the temperature distribution on the surface of the bearing is replaced by the temperature of z=0. By the Reynolds boundary condition, when z=0 the $\frac{\partial T}{\partial z}=0$, $\frac{\partial P}{\partial z}=0$ and the can be reduced to [16]:

$$\rho\left(\frac{Uh}{2}-\frac{h^3}{12\mu}\frac{\partial P}{\partial x}\bigg|_{z=0}\right)\frac{d(c_v T)}{dx}=\mu\frac{U^2}{h}+\frac{h^3}{12\mu}\left(\frac{\partial P}{\partial x}\bigg|_{z=0}\right)^2 \quad (8)$$

3.3 The thermohydrodynamic equation

Lubricating oil viscosity decreased with the increase of temperature quickly. The relationship can be represented by a series of measured value or calculated by empirical formula, commonly used Reynolds temperature viscosity equation is as follows [1]:

$$\mu=\mu_0 e^{-\alpha(T-T_0)} \quad (9)$$

where, T_0 is the initial temperature of lubricating oil, α is the viscidity temperature index.

The dynamic characteristics of oil film have great influence on the dynamic characteristics of the whole rotor system. Oil film will affect the natural frequency, amplitude and stability of the rotor system. In commonly used calculation, when the oil film force is obtained, the film force is regarded as a function of displacement and velocity near the equilibrium point and expand it use Taylor series. When the axial amplitude is small, the oil film can be simplified to a linear stiffness and damping characteristics. The increment of oil film force can be expressed by the following linear formula [17]:

$$\left.\begin{aligned}\Delta F_x&=k_{xx}\Delta x+k_{xy}\Delta y+c_{xx}\Delta\dot{x}+c_{xy}\Delta\dot{y}\\ \Delta F_y&=k_{yx}\Delta x+k_{yy}\Delta y+c_{yx}\Delta\dot{x}+c_{yy}\Delta\dot{y}\end{aligned}\right\} \quad (10)$$

where, k_{xx}, k_{xy}, k_{yx}, k_{yy}, c_{xx}, c_{xy}, c_{yx}, c_{yy} are the 8 oil film dynamic coefficients, the first subscript indicates the direction of the force increment, and the second subscript indicates the direction of displacement or velocity increment.

The three equations using the finite difference method for iterative solution, the oil film pressure distribution of sliding bearing can be obtained by iterative solve the formula (6), (8), (9) using the finite difference method. After obtain the distribution of film pressure, the film stiffness and damping coefficient of the bearing can be obtained by the difference method.

4 Total stiffness matrix of sliding bearing rotor system

The bearing rotor is simplified to a certain boundary conditions at the bearing node, where the bearing structure is no longer considered. The oil film stiffness (K) and damping coefficient (C) matrix of sliding bearing as follows:

$$[K]=\begin{bmatrix}k_{xx} & k_{xy}\\ k_{yx} & k_{yy}\end{bmatrix};\ [C]=\begin{bmatrix}c_{xx} & c_{xy}\\ c_{yx} & c_{yy}\end{bmatrix} \quad (11)$$

4.1 Total stiffness of isotropy sliding bearing rotor system

When the stiffness and the equivalent mass of the bearing in the x and y directions are not very different and the coupling is weak, the impact of damping can be neglected and the support is isotropic in the critical speed calculation, namely:

$$k_{xx}=k_{yy};\ k_{xy}=k_{yx}=0 \quad (12)$$

If the mass and stiffness of the bearing is considered, the oil film and the bearing house are put in series connection. Excluding the coupling of the x and y direction gives [18]:

$$K_{sj}=\frac{K(K_b-m_b\Omega^2)}{K+K_b-m_b\Omega^2} \quad (13)$$

where, K_{sj} is the j-th support of the total stiffness, K is the stiffness of the membrane, K_b is the bearing house stiffness, m_b is the vibration mass of the bearing house, Ω is the rotor whirl angular velocity

4.2 Total stiffness of anisotropy sliding bearing rotor system

When the stiffness and the equivalent mass of the bearing in the x and y directions are large and the impact of damping can't be neglected, the support is anisotropic. In this case, the total stiffness of K_s is in the matrix form, so does the stiffness, damping and mass of the bearing house [18]:

$$K_s=\begin{bmatrix}K_{sxx} & K_{sxy}\\ K_{syx} & K_{syy}\end{bmatrix}=[K_b+SC_b+S^2m_b]\cdot[K+SC+K_b+SC_b+S^2m_b]^{-1}[K+SC] \quad (14)$$

where, S is the complex frequency, $S=\lambda+i\Omega$, the real part is the attenuation index and the imaginary part for damping circular frequency, C_b is the bearing house damping coefficient matrix.

5 Analysis results and experimental verification

5.1 Results of the thermohydrodynamic model

The structure of the high-speed sliding bearing motorized spindle is shown in Fig. 2, both the front and the end of the spindle are supported by dynamic pressure sliding bearings. The spindle a simplified at a plurality of lumped mass and the bearing is considered at the model.

According to formula (1), in order to calculate the primary four orders of critical speed, the number of nodes should be divided into at least 23 nodes. In the places where diameter or material changes, or has external load, or supported by bearings and the ends of the spindle should be as nodes. Nodes for this spindle is divided into 24 nodes and is shown in Fig. 3.

DOI: 10.1051/matecconf/20179512012

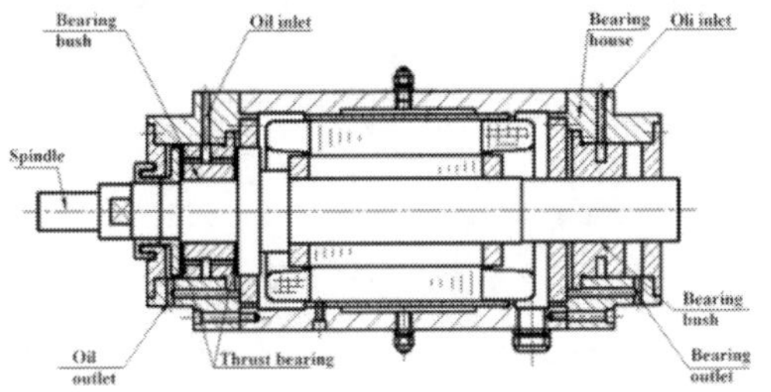

Figure 2. The structure of the sliding bearing motorized spindle.

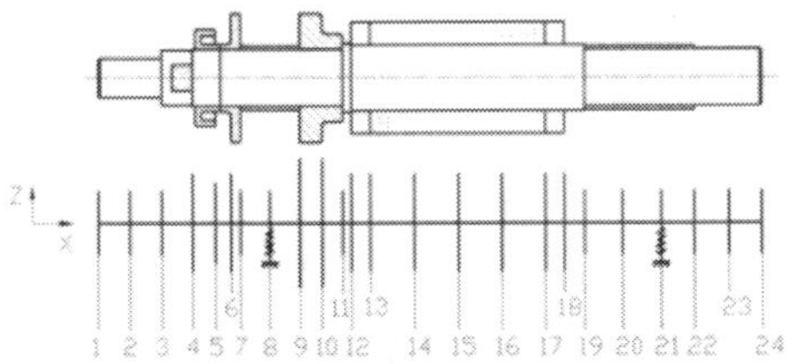

Figure 3. Lumped mass model of the spindle.

It is required to know the sliding bearings characteristics in the critical speed calculation of the sliding rotor system. The sliding bearing structure parameters is shown in Table 1. Fig. 4 shows the steady state temperature of the bearing, whose eccentricity is 0.5, changes with rotating speed. As shown in the figure, the temperature rises increases significantly when the rotating speed goes up. Since the viscosity of the lubricating oil is closely related to the oil temperature, so the influence of the temperature rise should be considered in the analysis of characteristics of high speed sliding bearing.

Table 1. Structure parameter of the sliding bearing

Parameter		Value
Diameter	d/mm	40
Contact Angle	φα/°	360
Width-diameter ratio	l/d	0.8
Clearance ratio	ψ	0.002
Average viscosity of lubricating oil	μ/Pa·s	0.02
Initial viscosity of lubricating oil	μ0/Pa·s	0.038
Initial temperature of lubricating oil	T0/℃	35
Temperature viscosity index	α	0.041

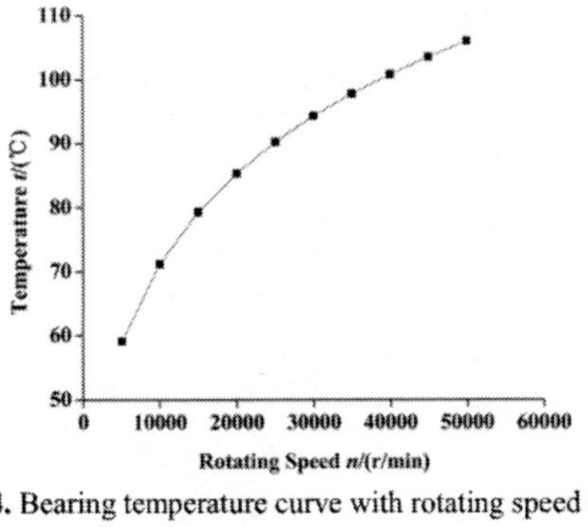

Figure 4. Bearing temperature curve with rotating speed rises.

Bearing stiffness and damping are important parameters affecting the mode of spindle. As shown in Fig. 5 and 6, under a specify load, the bearing eccentric decreases with the rotating speed increases causes the oil film stiffness varies in different trends in different directions. It also will bring lubricant flow and oil temperature increase. Fig. 7 and 8 shows the bearing eccentric increases with the load increases causes the oil film stiffness increases under a constant rotating speed. It also will bring friction of the bearing and the temperature increase. As shown in Fig. 9 and Fig. 10, if the load and the rotating speed is constant, the stiffness and damping of the bearing have different trends in different directions. Whereas, the absolute values are decrease trend. From these several figures, it is found that the effects of rotational speed, load and temperature on the bearing are mutually restricted.

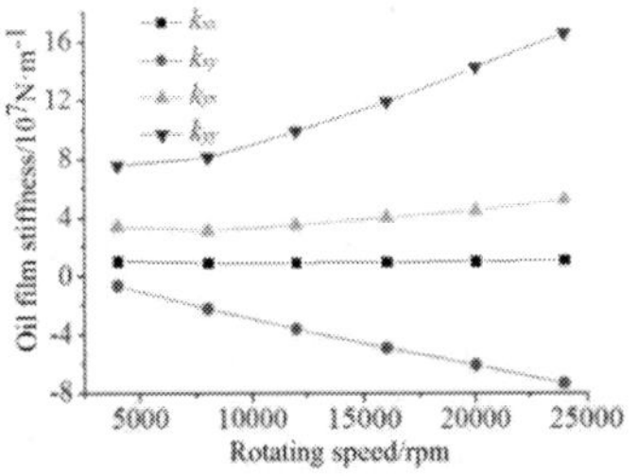

Figure 5. Effect of rotating speed on the stiffness of bearing oil film.

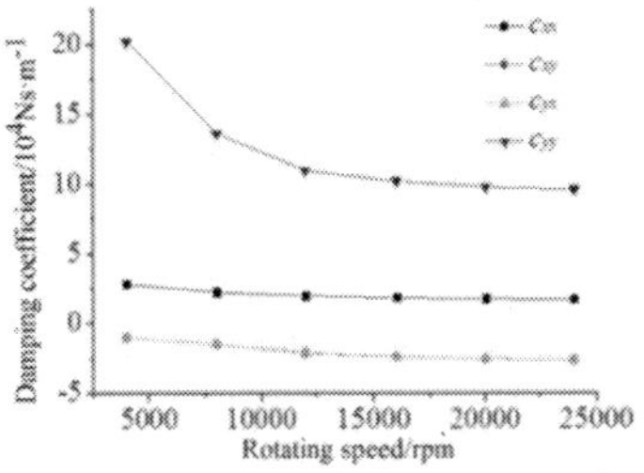

Figure 6. Effect of rotating speed on damping coefficient of bearing.

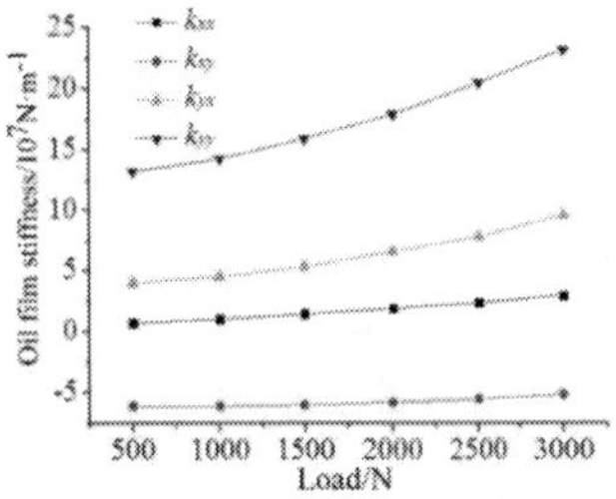

Figure 7. Effect of load on the stiffness of bearing oil film.

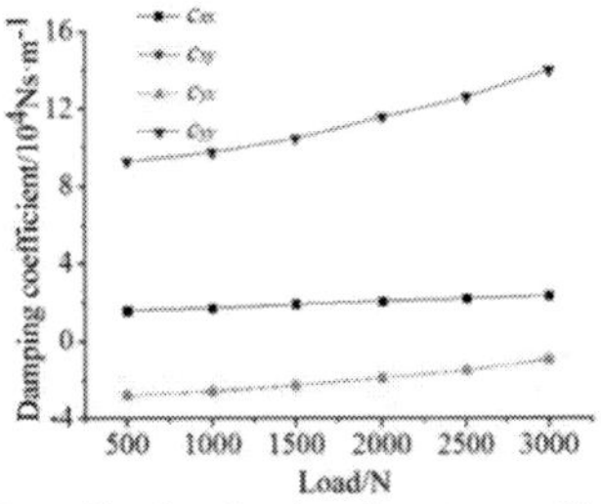

Figure 8. Effect of load on bearing damping coefficient.

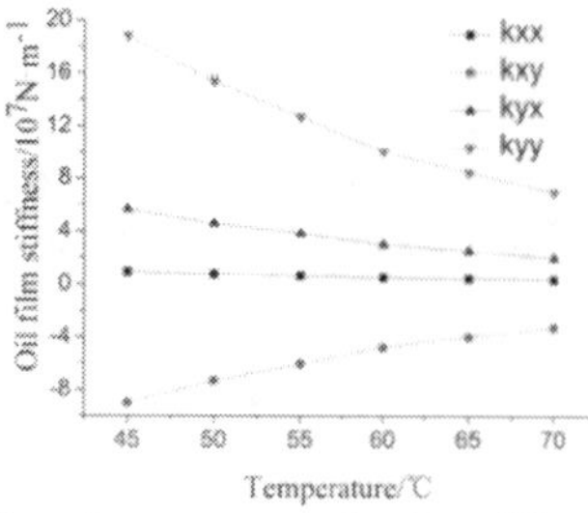

Figure 9. Effect of temperature on bearing stiffness.

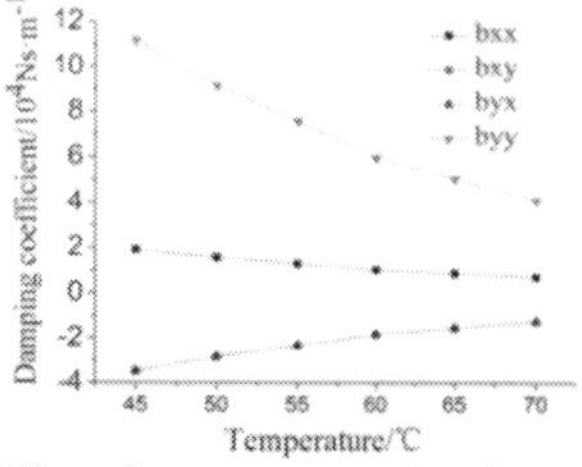

Figure 10. Effect of temperature on damping coefficient of bearing.

Fig. 11 shows the influence of bearing stiffness on the critical speed, all other things being equal. With the increase of the stiffness, the critical speed of each order shows an increasing trend, especially for the higher order. In order to improve the critical speed of the rotor system, if other conditions unchanged, it is necessary to increase the support stiffness of the rotor system. However, to make a substantial increase in the critical speed, bearing stiffness needs to increase exponentially.

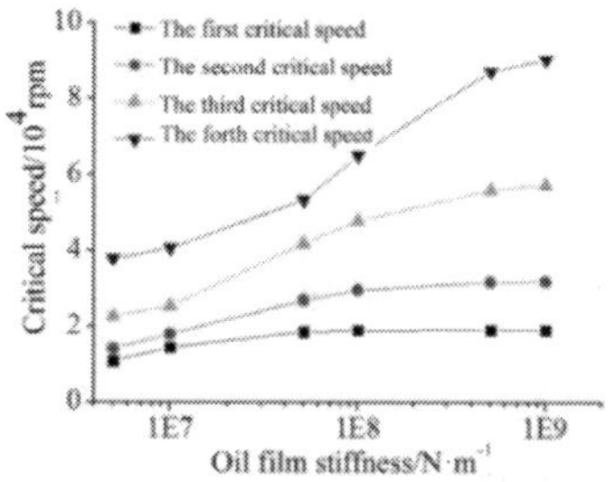

Figure 11. Effect of oil film stiffness on critical speed of rotor system.

As shown in Fig. 12, the damping coefficient of oil film also affects the critical speed. With the increase of the damping coefficient, all other things remaining unchanged, the critical speed is decreased obviously.

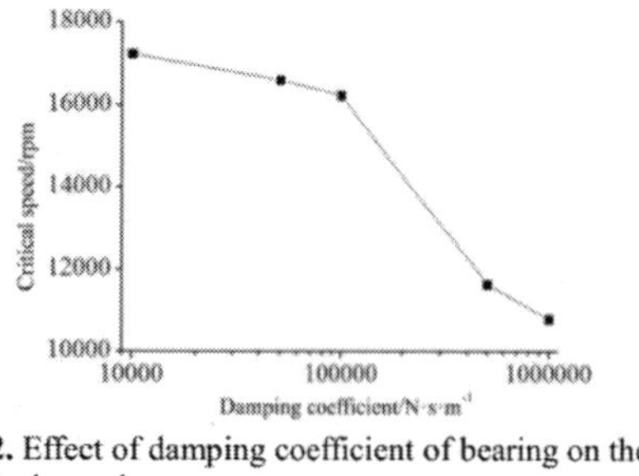

Figure 12. Effect of damping coefficient of bearing on the first order critical speed.

When calculate the rotor dynamic characteristic of the experimental motorized spindle supported by sliding bearings with 100N axial load, the influence of temperature and rotational speed on the bearing stiffness and damping coefficient is also considered. Results of the primary four orders of natural frequency and corresponding vibration mode is shown in Fig. 13.

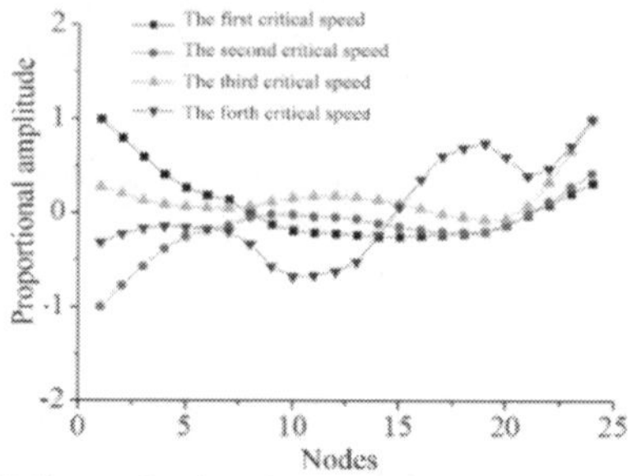

Figure 13. Rotor vibration shape at each order of critical speed.

The critical speed can also be calculated. The results of isotropy sliding bearing rotor system and anisotropy sliding bearing rotor system are shown in Table 2. As can be seen from the table, the calculation results of isotropic bearing model are larger than anisotropic bearing model. It's because of the influence of the cross stiffness and damping coefficient of the oil film are neglected in isotropic bearing model.

Table 2. The rotor system critical speed of the two model

	1st order	2nd order	3rd order	4th order
Isotropic bearing model (rpm)	19052	30651	50876	72974
Anisotropic bearing model (rpm)	17774	27764	49204	70538

5.2 Dynamic experimental verification of the sliding bearing high speed motorized spindle

Vibration of the spindle front end and motorized spindle shell are measured when the motorized spindle is operating. By comparing the amplitude of vibration under different rotating speed, the critical speed and other parameters can be obtained indirectly. Fig. 14 shows the experimental facilities.

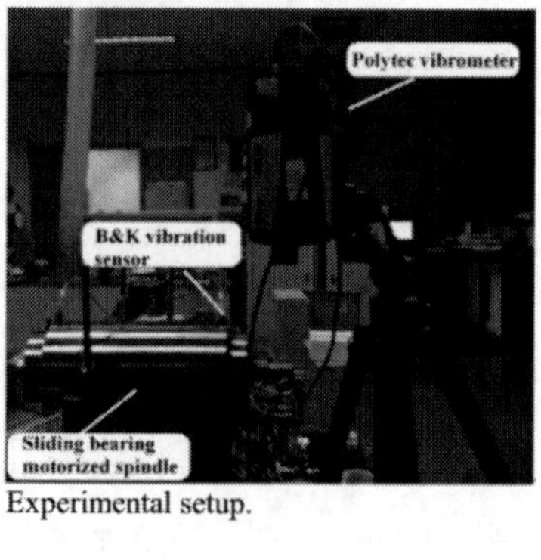

Figure 14. Experimental setup.

In the experiment, the motorized spindle runs at each speed for 90 minutes till the motorized spindle reaches thermal steady state [19]. Laser Doppler vibration meter Polytec PDV100 is used to measure the vibration value of rotor front end and vibration value of motorized spindle shell is measured by B&K vibration/noise analyze system.

Fig. 15 shows the vibration amplitude of the spindle at different speeds, Fig. 16 shows the vibration amplitude of the shell at different speeds. As shown in the figures, the amplitude appears to be the first significant maximum at 18000 rpm, which is also 300Hz. This shows that the sliding bearing system reaches the resonance at about 300Hz, that means the first critical speed of the high speed motorized spindle rotor system supported by the sliding bearing is around 18000 rpm. The anisotropy sliding bearing rotor model is more accurate. The isotropy model neglect the affect of cross stiffness and damping coefficient, they can reduce the natural frequency of the rotor system.

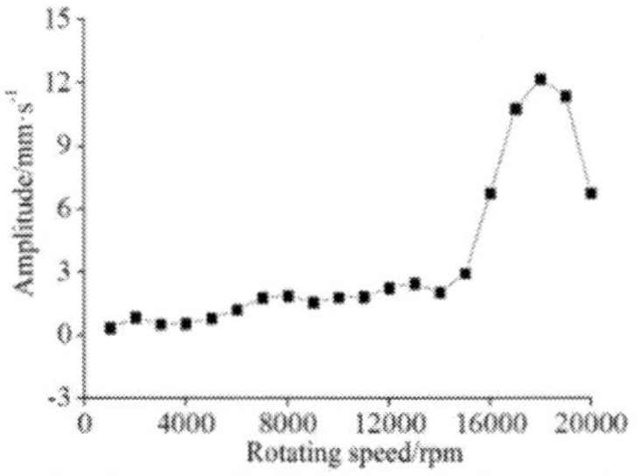

Figure 15. Vibration amplitude of the spindle at each rotating speed.

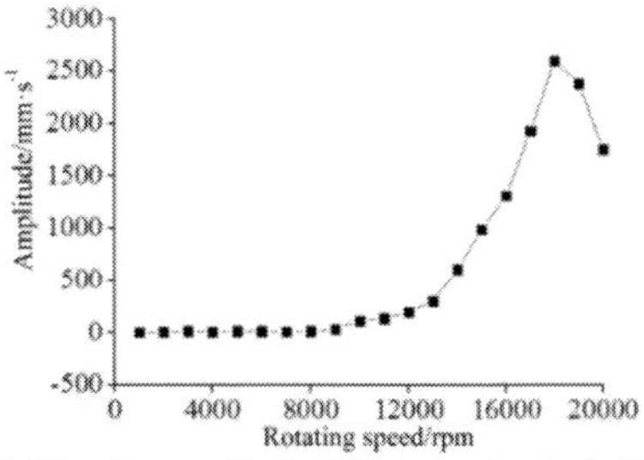

Figure 16. Vibration amplitude of the motorized spindle shell at each rotating speed.

6 Conclusions

(1) Due to the extrusion and viscous shearing of the lubricating oil, the sliding bearing will generate large amount of heat when the rotating speed is high, the oil viscosity reduces. The isothermal hydrodynamic lubrication model is no longer applicable. The thermo hydrodynamic lubrication model considering the temperature change and the viscosity temperature effect will be more appropriate in engineering.

(2) Load and rotating speed are important factors to dynamic characteristics of sliding bearing. With the increase of rotating speed, the oil film stiffness varies in different trends in different directions, and the damping coefficient decreases. The film stiffness and damping coefficient are increased with the increase of load.

(3) Experimental study shows the calculated results obtained by the anisotropic sliding bearing rotor model are better than isotropic sliding bearing rotor model. The cross stiffness and damping can reduce the natural frequency of the rotor system.

References

1. Z.Y. Zhang, Hydrodynamic lubrication theory of sliding bearing. Higher Education Press. Beijing, China, (1986)
2. P. Chang, and B. Lin, Vibration signal analysis of journal bearing supported rotor system by cyclostationarity. Shock & Vibration, **2014**(1), 1-16. (2014)
3. T.H. Machado, and K.L. Cavalca, Modeling of hydrodynamic bearing wear in rotor-bearing systems. Mechanics Research Communications, 69, 15-23, (2015)
4. K.P. Gertzos, P.G. Nikolakopoulos, A.C. Chasalevris, and C.A. Papadopoulos, Wear identification in rotor-bearing systems by measurements of dynamic bearing characteristics. Computers & Structures, **89**(12), 55-66 (2011)
5. H. Mccallion, F. Yousif, and T. Lloyd, The analysis of thermal effects in a full journal bearing. Journal of Tribology, **92**(4) (1970)
6. W. Xiong, Review on key technology of hydrodynamic and hydrostatic high-frequency motor spindles. Journal of Mechanical Engineering, **45**(9), 1-18 (2009)
7. A. Chauhan, R. Sehgal, and R.K. Sharma, Thermohydrodynamic analysis of elliptical journal bearing with different grade oils. Tribology International, **43**(11), 1970-1977 (2010)
8. S.S. Li, Y.X. Lu, M.L. Dong, J. Shao, and F. Yu, Modeling and Analysis of High-Speed Journal Bearing Based on Thermal Hydrodynamic Lubrication Theory. Applied Mechanics and Materials **380**, 82-86 (November 2013)
9. L. Daquan, T. Zheng, W. Zhang. Thermohydrodynamic study on journal bearings using finite element analysis. China Mechanical Engineering, **17**(12), 1285-1289 (2006)
10. J. Durany, J. Pereira, and F. Varas, Dynamical stability of journal-bearing devices through numerical simulation of thermohydrodynamic models. Tribology International, **43**(9), 1703-1718 (2010)
11. A.F. Clarens, P.E. Allaire, A. Younan, and S. Wang, Gas-Expanded Lubricants for Increased Energy Efficiency and Related Method and System.U.S. Patent Application No. 13/502,066 (2011)
12. S. Jerebzoff-Quintin, A thermohydrodynamic coupling model of oil aeration turbulent lubrication for journal bearing with interface effect.Journal of Tribology, **137**(4), 1718-20 (2015)
13. C. Deepthi and J. Ramathulasi, Parametric Design and Analysis of Two Station Two Spindle Rough Boring and Finish Boring SPM, International Journal of Mechanical Engineering and Robotics Research, **3**(4), 612-621, (October 2014)

14. B.C. Wen, J.C. Gu, S.B. Xia, Advanced rotor dynamics. China Machine Press. Beijing, China. (1999)
15. F. Xie, Calculation of oil film pressure distribution of journal bearing based on finite difference method. Lubrication Engineering (2012).
16. Y. Fu., X. Ma, and J. Zhu, Study on the three-dimensional temperature field of circular journal bearing. Mechanical Science & Technology, **21**(5), 711-710 (2002)
17. Y.E. Zhong, Y.Z. He, Z. Wang, The rotor dynamics. Tsinghua University Press. Beijing, China. (1987)
18. L. Yu, Y. Liu, The bearing rotor system dynamics. Xi'an Jiaotong University Press. Xi'an, China. (2001)
19. Q. He, Y. Shen, F. Ren, L. Li, and A. A. Volinsky, Numerical simulation and experimental study of the air-cooled motorized spindle. Proceedings of the Institution of Mechanical Engineers Part C Journal of Mechanical Engineering Science. (2016)

Bearings Fault Diagnosis Based on Convolutional Neural Networks with 2-D Representation of Vibration Signals as Input

Zhang Wei, Peng Gaoliang and Li Chuanhao

School of mechanics and electronics, Harbin Institute of Technology, Harbin, P. R. China

Abstract. Periodic vibration signals captured by the accelerometers carry rich information for bearing fault diagnosis. Existing methods mostly rely on hand-crafted time-consuming preprocessing of data to acquire suitable features. In this paper, we use an easy and effective method to transform the 1-D temporal vibration signal into a 2-D image. With the signal image, convolutional Neural Network (CNN) is used to train the raw vibration data. As powerful feature extractor and classifier for image recognition, CNN can learn to acquire features most suitable for the classification task by being trained. With the image format of vibration signals, the neuron in fully-connected layer of CNN can see farther and capture the periodic feature of signals. According to the results of the experiments, when fed in enough training samples, the proposed method outperforms other common methods. The proposed method can also be applied to solve intelligent diagnosis problems of other machine systems.

1. Introduction

As core components of rotating mechanism, the health conditions of rolling element bearings, for example, the fault diameters in different places under different loads could have a direct impact on the performance, stability and life span of the mechanism. To prevent possible damages, a real-time monitoring of vibration is needed during the operation of rotating mechanism. The condition signals of the rolling element bearings are collected by the sensors, and then intelligent fault diagnosis methods are implemented to recognize the fault types [1]-[3]. Most intelligent fault diagnosis methods can be divided into two steps, namely, feature extraction and classification. Features such as mean, median, kurtosis, peak to peak, minimum, maximum, standard deviation, absolute mean, skewness and root mean square of the vibration signals are common used to describe the condition of bearings [4]. In recent years, machine learning methods such as support vector machine (SVM) [5] and neural networks have become dominant methods to predict the fault types with the features extracted from the raw temporal vibration signals.

Over the last five years, with the fast development of deep learning especially the convolutional neural networks [6] (CNNs), image classification has achieved incredible success. This success came from the efficient use of GPUs, ReLUs, dropout and data augment. CNNs are now the dominant approach for almost all recognition and detection tasks and are comparable to human beings on some tasks [7]. In order to implement CNN on the 1-D signal, common method is to treat the signal as an image with height equals 1, which is already used in speech recognition [8]. However, the vibration signal is different from the speech signal. The vibration signal is periodic, which means the data in time t is not only correlated to the neighboring data but also the data collected in several cycles' intervals. In order to capture the features of periodic signals, a 2-D representation of these signals is used in this paper. This representation is first used to analyze and compress the power-quality event data [9], this 2-D transformation method is quite easy and intuitive, which will be shown in next section. With a signal image, CNN can easily extract the periodic features with small convolution and pooling kernel. We show that a segment of long periodic signals from any initialization can be classified with CNNs with the transformed image input format.

The remainder of this paper is organized as follows. The intelligent diagnosis method with new input format based on CNNs is introduced in Section 2. Some experiments are conducted to compare our methods with some common methods. Then discussions about the results of the experiments are presented in Section 3. We draw the conclusions and summarize the future work in Section 4.

2 Bearing fault intelligent diagnosis using the proposed method

2.1 A brief introduction to CNNs

The architecture of CNNs is briefly introduced in this section, more details for CNNs can be found in [6].

The convolutional neural network is a multi-stage neural network which is composed of some filter stages and one classification stage. The filter stage is designed to extract features from the inputs, which contains the convolutional layer and the pooling layer. The classification stage is a multi-layer perceptron, which is composed of several fully-connected layers. The function of each type of layer will be described as follows.

2.1.1 Convolutional layer

The convolutional layer convolves the input local regions with filter kernels, then generates the output features by the activation unit. Each filter uses the same kernel which is also known as weight sharing, to extract the local feature of the input local region. One filter corresponds to one frame in the next layer, and the number of frames is called the depth of this layer. We use $\boldsymbol{K}_i^l$ and b_i^l to denote the weights and bias of the i-th filter kernel in layer l, respectively, and use $\boldsymbol{X}^l(j)$ to denote the j-th local region in layer l. The convolutional process is described as follows:

$$y_i^{l+1}(j) = \boldsymbol{K}_i^l * \boldsymbol{X}^l(j) + b_i^l \quad (1)$$

where the notation $*$ computes the dot product of the kernel and the local regions, and $y_i^{l+1}(j)$ denotes the input of the j-th neuron in frame i of layer $l+1$.

Figure 1. Images of the vibration signals for different bear faults

After the convolutional operation, the Rectified Linear Unit (ReLU) which computes the function $f(x) = \max\{0, x\}$ is used as the activation unit of our model to accelerate the convergence of the CNNs.

2.1.2 Pooling layer

Pooling layer is usually placed after a convolutional layer in the CNN architecture. It functions as a downsampling operation which merges semantically similar features into one and reduces the parameters of the network. The most commonly used pooling layer is max-pooling layer, which performs the local max operation over the input features. It can reduce the parameters and obtain location-invariant features at the same time. The max-pooling transformation is described as follows:

$$\boldsymbol{P}_i^{l+1}(j) = \max_{\substack{(j-1)W+1<t_x\le jW \\ (j-1)H+1<t_y\le jH}} \{\boldsymbol{Q}_i^l(t)\} \quad (2)$$

where $\boldsymbol{Q}_i^l(t)$ denotes the value of t-th neuron in the i-th frame of layer l, W and H are the width and height of the pooling region respectively, and $\boldsymbol{P}_i^{l+1}(j)$ denotes the corresponding value of the neuron in layer $l+1$ of the pooling operation.

2.2 2-D representation of vibration signals

It is easy to transform the 1-D temporal signal into a 2-D image for recognition of CNN. Firstly, raw signal x is divided into n equal parts. Secondly, each part is aligned as the row of the signal image in sequence. This transformation can be described in mathematical format as follows:

$$\boldsymbol{I} = \begin{bmatrix} x(t) \cdots x(t+n-1) \\ \vdots \\ x(t+(m-1)n) \cdots x(t+mn-1) \end{bmatrix} \quad (5)$$

where $\boldsymbol{I}$ denotes the signal image and $x(t)$ is the vibration data of time t.

In order to show the image clearly, we transform a section signal with length of 40000 into a 200×200 image. The images with different faults are shown in Fig. 1. As shown in Fig. 1, the images of signals look just like texture images. People may have problem in recognizing each fault type directly from the 1-D signal wave. However, with simple transformation of each signal, we can easy classify each signal from the signal images. With this intuition, CNN may perform better on the signal image than the 1-D raw signal.

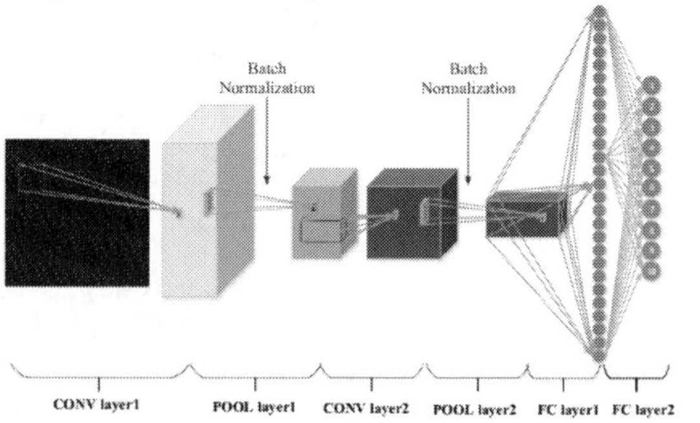

Figure 2. Architecture of the proposed CNN model

2.3 CNN-based intelligent diagnosis method

The architecture of the proposed CNN model is shown in Fig. 2. It is composed of two filter stages and one classification stage. The input of the CNN is a signal image of bearing fault vibration signal. The first convolutional layer extract features from the input image directly without any other transformation. The features of the images are extracted from two convolutional layers and two pooling layers. As shown in Fig. 2, with the increasing of the number of layers, the depth of the layers becomes larger while the width of each frame becomes smaller. The classification stage is composed of two fully-connected layers to accomplish the classification process. In the output layer, the softmax function is used to make the logits of the ten neurons accord with the probability distribution for the ten different bearing health conditions. Softmax function is:

$$f(z_j) = \frac{e^{z_j}}{\sum_k^{10} e^{z_k}} \quad (3)$$

where z_j denotes the logits of the j-th output neuron.

The loss function of our CNN model is cross-entropy between the estimated softmax output probability distribution and the target class probability distribution. Let $p(x)$ denotes the target distribution and $q(x)$ denotes estimated distribution, so the cross-entropy between $p(x)$ and $q(x)$ is:

$$H(p,q) = -\sum_x p(x)\log q(x) \quad (4)$$

In order to minimize the loss function, the Adam Stochastic optimization algorithm is applied to train our CNN models. Adam implements straightforwardly, computes efficiently and requires little memory, which is quite suitable for models with big data or many parameters. The detail of this optimization algorithm can be found in [10]. In order to make CNN more robust, Batch Normalization [11] is used to cope with problem of bad initialization. As a common way to control overfitting, dropout is used in the training process.

Table 1. Description of rolling element bearing datasets

Fault location		none	Ball			Inner race			Outer race			Load (hp)
Fault diameter(mm)			0.007	0.014	0.021	0.007	0.014	0.021	0.007	0.014	0.021	
Data A no.	Train	3000	3000	3000	3000	3000	3000	3000	3000	3000	3000	1,2,3
	Test	750	750	750	750	750	750	750	750	750	750	
Data B no.	Train	150	150	150	150	150	150	150	150	150	150	1,2,3
	Test	750	750	750	750	750	750	750	750	750	750	

3 Experiments and discussion

3.1 Data description

In order to train the CNN model sufficiently, huge numbers of training samples are prepared. The original experiments data was obtained from the accelerometers on a motor driving mechanical system at a sampling frequency of 12 kHz from the Case Western Reserve University Bearing Data center. There are four fault types of the bearing health. Namely, they are normal, ball fault, inner race fault and out race fault, and sizes of the fault diameters include 0.007mm, 0.014mm and 0.021mm, so there are all together ten conditions. In this experiment, each samples contain 2400 data points. The 2400×1 1-D signal is transformed into an image with size of 60×40. Dataset A contains 30000 training data and 7500 testing data of ten different fault conditions under three loads. In order to evaluate the necessity of large training data for the CNN, dataset B containing 1500 training is also prepared. The details of all the datasets are described in Table 1.

3.2 Experimental Setup

3.2.1 Baseline system

We compare our methods with the standard ANN system with frequency features. The ANN system have one single hidden layer. The input of the networks is the normalized 2400 Fourier coefficients transformed from the raw temporal signals using fast Fourier transformation (FFT).

3.2.2 Hyper-parameters of the proposed CNN

The architecture of the proposed CNN is composed of 2 convolutional layers, 2 pooling layers, and also one fully-connected hidden layer. The parameters of the convolutional and pooling layers are shown in Table 2. The number of the neurons in the fully-connected hidden layer is 500. The experiments were implemented using *tensorflow* toolbox of Google.

Table 2. Parameters of the convolutional and pooling layers

Layer	Kernel width	Kernel height	Kernel depth	Stride	Frames
CONV1	5	5	1	1	25
POOL1	2	2	1	2	25
CONV2	5	5	25	1	50
POOL1	2	2	1	2	50

3.3 Results

After taking into account the sample quantity of each dataset, 3000 samples and 100 samples are randomly selected as the training validation dataset of A and B, respectively. Twenty trials were implemented for the diagnosis of each dataset. Fig. 3 and Table 3 show the fault recognition results of datasets A and B, and compares this to raw 1-D signal recognition using CNN directly and the baseline system. As shown in Fig. 3(a), the diagnosis accuracies of all twenty trials are over 99.7%. Actually, except for the sixth trial, the diagnosis accuracies of rest trials on dataset A are higher than 99.9%. This means that by feeding CNN with vibration signal images, a very good result can be achieved when diagnosing the fault conditions of the rolling element bearings. The accuracy of the CNN declined when trained with dataset B, which indicates that more training data will result in higher accuracy. The reason for this would be discussed later.

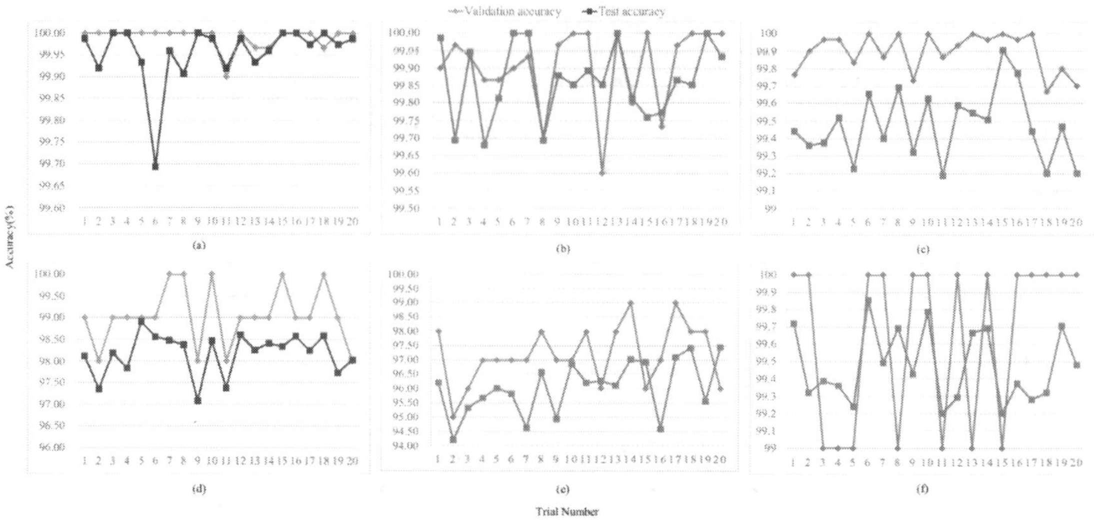

Figure 3. Fault recognition results of the proposed method and the baseline system in 20 trials. (a) and (d) are the validation and test accuracy of proposed method on dataset A and B, respectively. (b) and (e) are the validation and test accuracy of method using CNN recognition the 1-D signals, respectively. (c) and (f) are the validation and test accuracy of baseline system, respectively

Compared with other methods, the accuracies of our method outperform the others. The standard deviations of 20 trials of the proposed method on both the validation samples and test samples of dataset A are the smallest among all three methods. According to the compare above, our method is the most precise and stable one. The results on dataset B are quite different. The test accuracy of the proposed method declines near 1.8%, while the standard deviation on test samples increases by seven times. However, the results of the baseline system remain almost unchanged in terms of test accuracy and standard deviation.

Table 3. Comparison results between our method and other methods

Method	Validation accuracy	Test accuracy	Dataset
Proposed Method	99.99±0.02	99.95±0.07	A
	99.05±0.69	98.17±0.48	B
Raw 1D+CNN	99.91±0.12	99.86±0.11	A
	97.2±1.06	96.04±0.96	B
FFT+ANN	99.90±0.11	99.47±0.20	A
	99.65±0.49	99.47±0.21	B

3.4 Discussion

Some interesting points acquired from the experiments above are as follows:

The results above show that CNN performs excellently when dealing with the bearing vibration signal images. Although the performance of the FFT+ANN method is not bad, it could be quite tricky to choose between FFT or WT as the preprocessing method of the data to get a very high accuracy, because the most suitable frequency features may vary from different dataset. The advantage of our method is that it can see much further than the method which uses CNN directly on the 1-D signals. We now explore how far a neuron in the first fully-connected layer can see using the two methods. In order to ensure that the numbers of parameters remain the same, we compress the 5×5 convolution kernel into the size of 25×1 and 2×2 pooling kernel into the size of 4×1 for the CNN architecture performed on 1-D signals. The stride for convolutional layers equals 1 for both methods. Assume the first signal point the neuron in fully-connected layer can see is $x(t)$, for the raw 1-D CNN, the farthest point it can see is $x(t+136)$, while for the proposed method, the farthest point it can see is $x(t+976)$. According to the analysis, the neurons in the fully-connected layer of the proposed method see much farther than the other method. With this feature of the proposed method, CNN can better capture the periodic features than directly performing on the 1-D signals.

When the size of the dataset is large enough to train the CNN sufficiently, CNN has better and more stable performance than common methods. The major drawback of CNN is the need for huge amount of data, as can be seen from Table 3. This may result in overfitting and it could also make it hard to be applied to industries where the labeled data is hard to acquire. However, the labeled data of the rolling element bearing vibration is easy to acquire, so data augmentation is unnecessary. For example, when the sample frequency is 12 kHz, the length of the data is 2400 and the stride is 1, it takes only 1 second to acquire 9601 different labeled samples. So CNN is quite suitable for the bearing fault diagnosis since huge amount of data are available.

What's more, it is shown that it is possible to recognize the bearing fault type with periodic vibration signal images as input to the CNNs. What we want to emphasize here is that the bearing vibration signal is just a representative of the periodic signals, and this method may be good at recognizing many other kinds of periodic signals, which may provide us a new approach to deal with these kinds of signals.

4 Conclusion

In this paper, we proposed a CNN architecture with signal image as input to intelligently diagnose the bearing faults. As powerful feature extractor and classifier, CNN can learn to acquire features most suitable for the classification task by being trained. Fed with signal images, CNN can learn the periodic features of the vibration signals very well. According to the results of the experiments, the proposed method is able to mine suitable fault features adaptively from the raw bearing vibration data and classify the fault classes with high accuracy and stability. Different from the deep neural networks trained by frequency spectra, the proposed CNNs can be trained directly using raw temporal signals, which means this method can be applied to solve intelligent diagnosis problems of other machine systems.

Acknowledgment

The support of National High-tech R&D Program of China (863 Program, No. 2015AA042201), National Natural Science Foundation of China (No. 51275119) in carrying out this research is gratefully acknowledged.

References

1. Jayaswal, P., Verma, S.N. and Wadhwani, A.K., 2011. Development of EBP-Artificial neural network expert system for rolling element bearing fault diagnosis. Journal of Vibration and Control, **17**(8), pp.1131-1148.
2. Yiakopoulos, C.T., Gryllias, K.C. and Antoniadis, I.A., 2011. Rolling element bearing fault detection in industrial environments based on a K-means clustering approach. Expert Systems with Applications, **38**(3), pp.2888-2911.
3. Li, Y., Xu, M., Wei, Y. and Huang, W., 2016. A new rolling bearing fault diagnosis method based on multiscale permutation entropy and improved support vector machine based binary tree. Measurement, **77**, pp.80-94.
4. Shao, H., Jiang, H., Zhang, X. and Niu, M., 2015. Rolling bearing fault diagnosis using an optimization deep belief network. Measurement Science and Technology, **26**(11), p.115002.
5. Widodo, A. and Yang, B.S., 2007. Support vector machine in machine condition monitoring and fault diagnosis. Mechanical systems and signal processing, **21**(6), pp. 2560-2574.
6. Krizhevsky, A., Sutskever, I. and Hinton, G.E., 2012. Imagenet classification with deep convolutional neural networks. In Advances in neural information processing systems (pp. 1097-1105).
7. LeCun, Y., Bengio, Y. and Hinton, G., 2015. Deep learning. Nature, **521**(7553), pp.436-444.
8. Palaz, D., Collobert, R. and Doss, M.M., 2013. Estimating phoneme class conditional probabilities from raw speech signal using convolutional neural networks. arXiv preprint arXiv:1304.1018.
9. Gerek, O.N. and Ece, D.G., 2004. 2-D analysis and compression of power-quality event data. IEEE Transactions on Power Delivery, 19(2), pp.791-798.
10. Kingma, D. and Ba, J., Adam: A method for stochastic optimization. arXiv preprint arXiv:1412.6980 (2014).
11. Ioffe, S. and Szegedy, C., Batch normalization: Accelerating deep network training by reducing internal covariate shift. arXiv preprint arXiv:1502.03167, (2015).

Adaptive Control Algorithm of the Synchronous Generator

Victor Shevchenko , Viacheslav Pshikhopov, Mikhail Medvedev

Southern Federal University, 347922, Russia, Rostov Region, Taganrog, Shevchenko st., 2.

Abstract. The article discusses the the problem of controlling a synchronous generator, namely, maintaining the stability of the control object in the conditions of occurrence of noise and disturbances in the regulatory process. The model of a synchronous generator is represented by a system of differential equations of Park-Gorev, where state variables are computed relative to synchronously rotating d, q-axis. Management of synchronous generator is proposed to organize on the basis of the position-path control using algorithms to adapt with the reference model. Basic control law directed on the stabilizing indicators the frequency generated by the current and the required power level, which is achieved by controlling the mechanical torque on the shaft of the turbine and the value of the excitation voltage of the synchronous generator. Modification of the classic adaptation algorithm using the reference model, allowing to minimize the error of the reference regulation and the model under investigation within the prescribed limits, produced by means of the introduction of additional variables controller adaptation in the model. Carried out the mathematical modeling of control provided influence on the studied model of continuous nonlinear and unmeasured the disturbance. Simulation results confirm the high level accuracy of tracking and adaptation investigated model with respect to the reference, and the present value of the loop error depends on parameters performance of regulator.

1 Introduction

Development of the control system for the synchronous generator (SG), providing required degree of quality generated power and operation parameters of the generator, is a actual task in connection with concept of universal electrification, use energy saving and energy-efficient technologies, increase safe and sustainable functioning of the power grid. Special problem in the task of control system synthesis for synchronous generator is construction of the control, able to adapt to arising perturbations which appeared in the regulation process. [1]. Accuracy and quality of system must be invariably independently of changing object parameters. Formation a controller that can able adapt to external perturbations possibly with use adaptive control systems. Adaptation algorithms with reference model is used when object model is known and unknown parameters , but in classic execution this system unable adequately react on arising perturbation during operation [2]. This problem can be decided by introduction in object model additional parameters [3] which allow to adapt error of regulation in reference model and investigated model in set limits.

Basic control law for synchronous generator is constructed on the principles of position-trajectory control. This control successfully used in robotic mobile systems [3-10], such as control systems of autonomous marine devices [3-5] and aeronautic complex [7, 10], and for constructing regulator of stationary systems [1 - 2], such as control systems of traction electric motor [8], DC motor, synchronous generator [2], etc.

2 Synthesis of adaptive control

Mathematical model of synchronous generator is presented system differential equations by Park-Gorev. State variables in this model is currents id, iF, iq (instantaneous value), deviation of angle δ and angular velocity ω. Model of synchronous machine, operating at the system of unlimited capacity across power line, can be presented follows: [2]:

$$\frac{d\delta}{d\tau} = \omega - 1,$$

$$\frac{d\omega}{d\tau} = -\frac{L_d}{3\tau_J} i_q i_d - \frac{kM_F}{3\tau_J} i_q i_F + \frac{L_q}{3\tau_J} i_d i_q - \frac{D}{\tau_J}(\omega - 1) + \frac{M_{MX}}{\tau_J},$$

$$\frac{di_d}{d\tau} = \frac{-L_F(r_s + r_e)}{L} i_d - \omega \frac{L_F(L_q + L_e)}{L} i_q + \frac{kM_F r_F}{L} i_F - \frac{kM_F}{L} u_F + \frac{L_F U_\infty \sqrt{3}}{L} \sin(\delta) \quad (1)$$

$$\frac{di_F}{d\tau} = \frac{kM_F(r_s + r_e)}{L} i_d + \frac{-(L_d + L_e)}{L} r_F i_F + \omega \frac{kM_F(L_q + L_e)}{L} i_q + \frac{(L_d + L_e)}{L} u_F - \frac{kM_F U_\infty \sqrt{3}\sin(\delta)}{L}$$

$$\frac{di_q}{d\tau} = \frac{(L_d + L_e)}{(L_q + L_e)} \omega i_d + \frac{kM_F}{(L_q + L_e)} \omega i_F - \frac{(r_s + r_e)}{(L_q + L_e)} i_q - \frac{U_\infty \sqrt{3}}{(L_q + L_e)} \cos(\delta).$$

where $L = L_F(L_d + L_e) - kM_F^2$; L_d, L_q, L_F - own inductance of stator windings by longitudinal d and transverse q axis and inductance of rotor winding, M_F - mutual inductance of starter and rotor windings, r_s, r_F, r_e - active resistance of the stator windings, the excitation winding of the rotor and transmission lines, U_∞, u_F - voltage of infinite bus power and the excitation winding of the rotor, k – Park's constant, τ_j - mechanical time constant. All model parameters, including current, voltage and time is given in relative units.

This mathematical model of the reference and test patterns is represented by the system of equations (1), in the reference model for the state variables used, the index m.

Will make the control synthesis for the reference model, stabilizing the current frequency and power of the generator. To control these options control the mechanical torque on the turbine shaft and the excitation voltage. In accordance with the synthesis method of position-trajectory control we introduce the error control in the form:

$$\psi_{эm} = A_1 \cdot y_m - A_2 \quad (2)$$

where $y_m = \begin{bmatrix} \omega_m \\ P_{gm} \end{bmatrix}$; $T = \begin{bmatrix} T_w \\ T_p \end{bmatrix}$; $A_1 = \begin{bmatrix} A_{1w} \\ A_{1p} \end{bmatrix}$; $A_2 = \begin{bmatrix} A_{2w} \\ A_{2p} \end{bmatrix}$;

To ensure the asymptotic convergence of the estimation, we require that the error (2) subordinated to the solution of the equation:

$$\dot{\psi}_{эm} + T \cdot \psi_{эm} = 0 \quad (3)$$

The first derivative error (2) will be as follows:

$$\dot{\psi}_{эm} = A_1 \cdot \dot{y}_m \quad (4)$$

Thus, equation (3) with (2) and (4) we write as:

$$A_1 \cdot \dot{y}_m + T \cdot A_1 \cdot y_m - T \cdot A_2 = 0 \quad (5)$$

In scalar form equation (5):

$$\begin{cases} A_{1w} \cdot \dot{\omega}_m + T_w \cdot A_{1w} \cdot \omega_m - T_w \cdot A_{2w} = 0 \\ A_{1p} \cdot \dot{P}_{gm} + T_w \cdot A_{1p} \cdot P_{gm} - T_w \cdot A_{2p} = 0 \end{cases} \quad (6)$$

For ease of presentation and solution of the equations we introduce some reduction resulting from the system (1):

$$\dot{\omega}_m = f_{w\_m} + M_{mx\_m} / \tau_j \quad (7)$$

$$\dot{P}_{gm} = 6 \cdot \sqrt{3} \cdot R_n \cdot (f_{12\_m} + \frac{kM_F}{L} U_{F\_m} \cdot \sin(\delta_m)) \quad (8)$$

$$f_{12\_m} = f_{1m} - f_{2m} \quad (9)$$

$$f_{1m} = f_{iqm} \cdot \cos(\delta_m) - f_{idm} \cdot \sin(\delta_m) \quad (10)$$

$$f_{2m} = (i_{qm} \cdot \sin(\delta_m) + i_{dm} \cdot \cos(\delta_m)) \cdot (\omega_m - 1) \quad (11)$$

$$f_{iqm} = \frac{(L_d + L_e)}{(L_q + L_e)} \omega_m i_{dm} + \frac{kM_F}{(L_q + L_e)} \omega i_{Fm} - \frac{(r_s + r_e)}{(L_q + L_e)} i_{qm} - \frac{U_\infty \sqrt{3}}{(L_q + L_e)} \cos(\delta_m) \quad (12)$$

$$f_{idm} = \frac{-L_F(r_s + r_e)}{L} i_{dm} - \omega_m \frac{L_F(L_q + L_e)}{L} i_{qm} + \frac{kM_F r_F}{L} i_{Fm} + \frac{L_F U_\infty \sqrt{3}}{L} \sin(\delta_m) \quad (13)$$

Substituting expressions (7) – (13) in (7) we obtain the equations of the controllers to the reference model:

$$\begin{cases} M_{mx\_m} = (\dfrac{-T_w \cdot A_{1w} \cdot \omega_m + T_w \cdot A_{2w}}{A_{1w}} - f_{w\_m}) \cdot \tau_j \\ U_{F\_m} = (\dfrac{(-T_w \cdot A_{1p} \cdot P_{gm} + T_w \cdot A_{2p})}{A_{1p} \cdot 6 \cdot \sqrt{3} \cdot R_n} - f_{12\_m}) \cdot L / (kM_F \cdot \sin(\delta_m)) \end{cases} \quad (14)$$

Consider the synthesis of adaptive control for the investigated model, the control objective which is defined as follows:

$$\psi = A_1 \cdot y - A_2 \quad (15)$$

where $y = \begin{bmatrix} \omega \\ P_g \end{bmatrix}$.

We introduce the additional variables of the adaptive controller z:

$$\dot{z} = \begin{bmatrix} \dot{z}_1 \\ \dot{z}_2 \end{bmatrix} = \begin{bmatrix} \psi_1 - \psi_{1эm} \\ \psi_2 - \psi_{2эm} \end{bmatrix} \quad (16)$$

Error control is specified as:

$$e = \psi - \psi_{эm} + B \cdot z \quad (17)$$

where $B = \begin{bmatrix} B_1 \\ B_2 \end{bmatrix}$;

Further, similar to the procedure for the synthesis of the reference model will require the following equality:

$$\dot{e} + T \cdot e = 0 \quad (18)$$

where $e = \begin{bmatrix} e_1 \\ e_2 \end{bmatrix} = \begin{bmatrix} \psi_1 - \psi_{1эm} + B_1 \cdot z_1 \\ \psi_2 - \psi_{2эm} + B_2 \cdot z_2 \end{bmatrix}$

Define the first derivative of the control error of (17) for the regulator's synthesis with considering the equations (2), (4), (15), (16)

$$\begin{aligned} \dot{e}_1 &= A_{1p}\dot{P}_g - A_{1p}\dot{P}_{gm} + B_1(A_{1p}P_g - A_{1p}P_{gm}) \\ \dot{e}_2 &= A_{1p}\dot{P}_g - A_{1p}\dot{P}_{gm} + B_1(A_{1p}P_g - A_{1p}P_{gm}) \end{aligned} \quad (19)$$

Solving equations (18) with respect to the mechanical torque on the shaft and the excitation voltage, obtain the following expression for the adaptive controller with reference model:

$$M_{mx} = \left(\frac{k_w}{A_{1w}} - f_w\right) / \tau_j \quad (19)$$

$$U_F = (\frac{k_p}{6\sqrt{3} \cdot A_{1p}} - f_{12}) \cdot \frac{L}{k \cdot M_F \sin(\delta)} \quad (20)$$

In equations (19) and (20) used the following abbreviations:

$$k_w = A_{1w} \cdot \dot{\omega}_m - B_1 \cdot A_{1w} \cdot \omega + B_1 \cdot A_{1w} \cdot \omega_m - T_1 \cdot A_{1w} \cdot \omega + {}, + T_1 \cdot A_{1w} \cdot \omega_m - T_1 \cdot B_1 \cdot z_1$$

$$kp = A_{1p} \cdot \dot{P}_{gm} - B_2 \cdot A_{1p} \cdot P_g + B_2 \cdot A_{1p} \cdot P_{gm} - T_1 \cdot A_{1p} \cdot P_g + {}, + T_1 \cdot A_{1p} \cdot P_{gm} - T_1 \cdot B_2 \cdot z_2$$

as well as reduction as in (7)-(13).

DOI: 10.1051/matecconf/20179513002

In the General case, adaptation of the control system to respond to perturbations is due to the automatic tuning of parameters B_1 and B_2.

3 Modeling system

Fig. 1 presents the structure of an adaptive control system, which shows that the control system of a synchronous generator is constructed as a servo outputs for the reference model system. In the linear approximation adaptation to perturbations provides astatism first order. Management is built on the basis of several control loops: reference model's loop, the base management loop and adaptation loop. This structure allows you to adjust adaptation parameters B1 and B2 regardless of the settings of the base regulator.

The result of this simulation system are presented in figures 2 to 5, where solid line denotes the parameters of the model, and the dashed line indicates the parameters of the reference model.

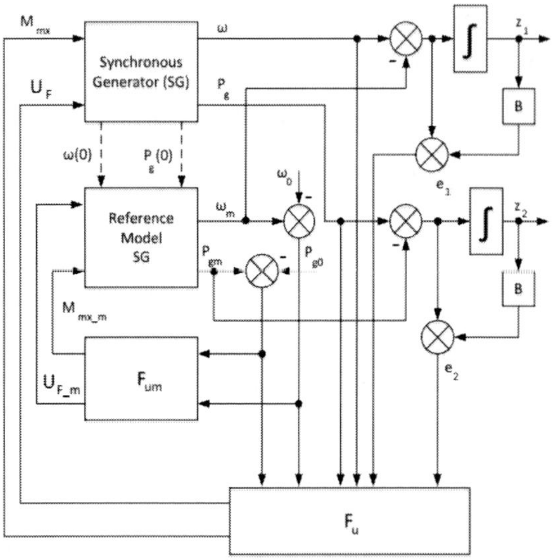

Figure 1. The structure of the adaptive control system of a SG

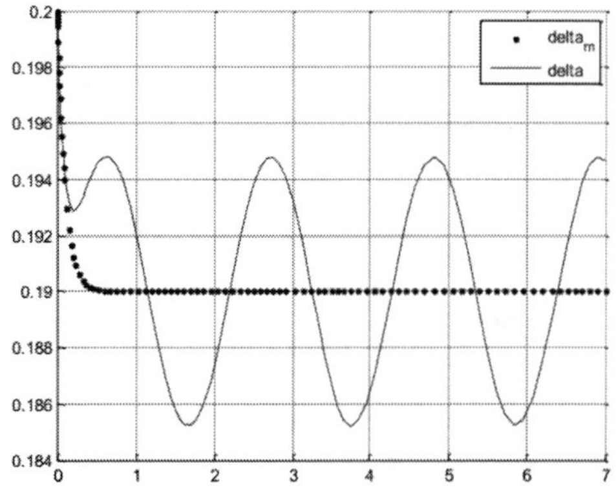

Figure 2. Angle δ.

The simulation was performed using the following system parameters: L_d=0.973; t_j=10; k=1; M_f=0.923; L_q=0.55; D=0; L_f=1.088; L=1; L_e=0.997; r_s=0.005; r_e=0.005; r_f=0.14; U_∞=1; R_n=300; A_{1w}=1; A_{1p}=1; T_{1w}=10; T_{1p}=1; A_{2w}=1; A_{2p}=1.

The disturbance acting on the reference model is zero, the perturbation value acting on the rotational speed and power 5·sin(3t).

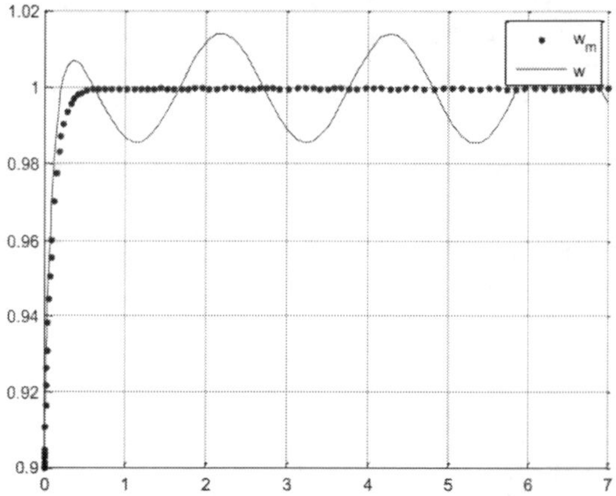

Figure 3. Shaft speed

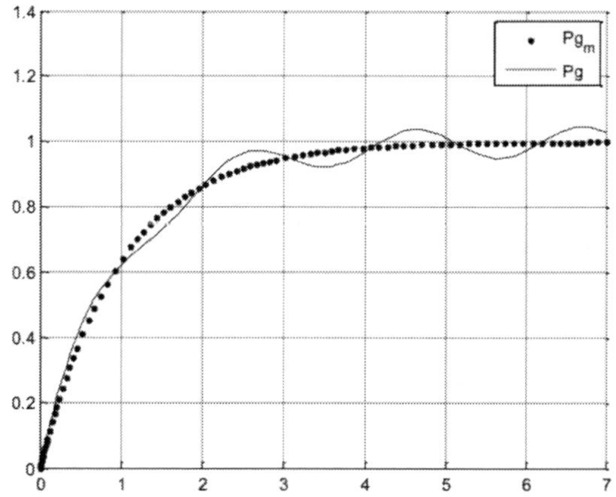

Figure 4. Generated power.

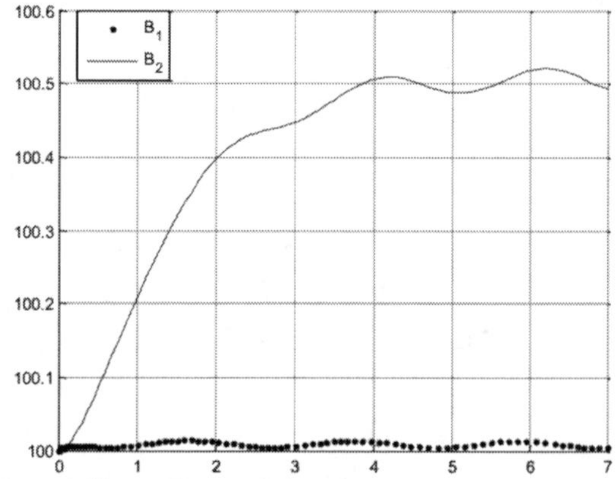

Figure 5. The coefficient of adaptation.

May be concluded from the results of simulation that the adaptive system with high precision monitors a reference model and allows you destroy the arising perturbations in the system under study and deviation values of the variables of the model in relation to the reference model depends on the speed controller parameters.

Acknowledgement

The work was supported by SFU (grant № 213.01-07/2014-01 "Theory and methods of energy saving control of distributed generation systems, transportation and consumption of electricity"). The reported study was funded by RFBR according to the research project № 16-08-00013 A.

References

1. Shevchenko, V.A., Pshikhopov, V.K., Medvedev, M.Y. Multilevel control system of the power grid // Advanced Materials Research, Volume 1049-1050, 2014, Pages 736-742.
2. Shevchenko, V.A. Development and research of system of automatic control by the synchronous generator // Proceedings of the 26th International Conference on Efficiency, Cost, Optimization, Simulation and Environmental Impact of Energy Systems, ECOS 2013.
3. Pshikhopov, V.K., Fedotov, A.A., Medvedev, M.Yu., Medvedeva, T.N., Gurenko, B.V. Position-trajectory system of direct adaptive control marine autonomous vehicles // 2014 the 4th International Workshop on Computer Science and Engineering - Summer, WCSE 2014.
4. Pshikhopov V.K., Medvedev M.Y., Gurenko B.V. Homing and docking autopilot design for autonomous underwater vehicle // Applied Mechanics and Materials. 2014. T. 490-491. C. 700-707.
5. Pshikhopov V.Kh., Medvedev M.Yu., Gaiduk A.R., Gurenko B.V. Control system design for autonomous underwater vehicle // Proceedings - 2013 IEEE Latin American Robotics Symposium, LARS 2013 2013. C. 77-82.
6. Medvedev M.Yu., Pshikhopov V.Kh. Design of robust control for block nonlinear systems by lyapunov functions method // Advanced Materials Research, Volume 1049-1050, 2014, Pages 1048-105.
7. Pshikhopov, V., Medvedev, M., Krukhmalev, V., Fedorenko, R., Gurenko, B. Method of docking for stratospheric airships of multibody transportation system // SAE Technical Papers, Volume 2014-September, Issue September, 16 September 2014.
8. Pshikhopov, V., Medvedev, M., Gaiduk, A. Control method for vehicles on base of natural energy recovery // Applied Mechanics and Materials, Volume 670-671, 2014, Pages 1330-1336
9. Neydorf, R., Krukhmalev, V., Kudinov, N., Pshikhopov, V. Methods of statistical processing of meteorological data for the tasks of trajectory planning of MAAT feeders // SAE Technical Papers, Volume 7, 2013.
10. Pshikhopov, V., Medvedev, M., Kostjukov, V., Fedorenko, R., Gurenko, B., Krukhmalev, V. Airship autopilot design // SAE Technical Papers, 2011.

Analytical Calculation and Analysis of Main Generator Inductance of Aero Brushless Three-stage Synchronous Starter/Generator System

Jixiang Wang , Weiguo Liu and Zan Zhang

School of Automation, Northwestern Polytechnical University,710072, Xi'an, China.

Abstract. Precise electro-magnetic parameters are significant in the investigation of characteristic for brushless synchronous starter/generator system, especially the inductance during the starting process.This paper presented an analytical method for inductance calculation, which analysed the influence of stator windings' parallel structure on inductance calculation and took the skewed slot structure into consideration. Due to the skewed slot structure, applicable two-dimensional coordinate systems were established .This proposed method could save time in calculation according to the simplification to some extent. The results of analytical method were presented and compared with the simulation results. The reason for the differences has been analyzed.

1 Introduction

Nowadays, more and more attention has been paid to the integrated starter/generator system which is significant to the aircraft power supply system. The three-stage starter/generator system generally consists of a pilot exciter, a main exciter and a main generator.

The main generator adopts the skewed slot structure which can improve the performance of the system by reducing the harmonics and the torque ripple. In order to analysis the motor with skewed slot structure, finite element method (FEM) or analytical calculation method are adopted. In FEM, a three-dimensional model or equivalent multiple two-dimensional models should be established, which makes the simulations complex and time-consuming [1-5]. In this paper, a two-dimensional coordinate system is proposed based on magnetic circuit method, and then on the basis of the motor structure all work was investigated in this coordinate system to calculate the parameters.

2 The influence of motor structure on inductance calculation

Considering the main generator adopts the parallel branches for stator wingdings, some simplification is made to analyse the influence. It's assumed that concentrated windings are employed and saturation is ignored. Meanwhile the MMF drop on the core of stator and rotor in the loop and the leakage flux are ignored.

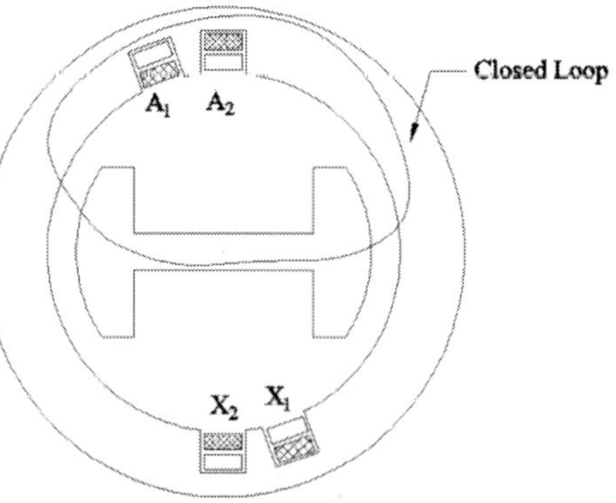

Figure 1. Simplified diagram of concentrated windings with parallel branches

As shown in Fig 1, current i is given to each parallel branch of phase A and then to phase A. On the basis of Ampere circuital theorem, the magnetic motive force (MMF) under these modes can be expressed as:

$$\begin{cases} F_{AX1} = N_1 i \\ F_{AX2} = N_2 i \\ F_A = (N_1 + N_2) i/2 = Ni \end{cases} \tag{1}$$

Due to same number of turns for each parallel branch, $N_1=N_2=N$, so that $F_{AX1}=F_{AX2}=F_A$. Meanwhile, the MMF can be written as $F=2Hl$, where H is the airgap magnetic field intensity and l is the length of the airgap. Due to the assumption, so $H_{AX1}=H_{AX2}=H_A$ which means $B_{AX1}=B_{AX2}=B_A$. The flux is equal because of the same cross-sectional area, which means $\Phi_{AX1} = \Phi_{AX2} = \Phi_A$. According to the magnetic circuit ohm's law,

$R_{AX1}=R_{AX2}=R_A$ where R is the magnetic resistance. The inductance should be described as:

$$\begin{cases} L_{AX1}=\dfrac{\varphi_{A1}}{i}=\dfrac{N_1\Phi_{AX1}}{i}=\dfrac{N_1F_{AX1}}{iR_{AX1}}=\dfrac{N_1N_1i}{iR_{AX1}}=\dfrac{N_1^2}{R_{AX1}} \\ L_{AX2}=\dfrac{\varphi_{A2}}{i}=\dfrac{N_2\Phi_{AX2}}{i}=\dfrac{N_2F_{AX2}}{iR_{AX2}}=\dfrac{N_2N_2i}{iR_{AX2}}=\dfrac{N_2^2}{R_{AX2}} \\ L_A=\dfrac{\varphi_A}{i}=\dfrac{N\Phi_A}{i}=\dfrac{NF_A}{iR_A}=\dfrac{NNi}{iR_A}=\dfrac{N^2}{R_A} \end{cases} \quad (2)$$

Finally, $L_{AX1}=L_{AX2}=L_A$ is derived from the magnetic circuit analysis above. It can be shown that the self-inductance of the phase A winding is equal to that of each parallel branch on the assumptions above which can simplify the calculation. However motor usually adopts distributed windings and the saturation can't be ignored, so they are not equal actually.

Damper wingdings are installed in the rotor of main generator to weaken the harmonics during the starting process. Since the damper windings can't be open, the inductance of armature windings can't be measured directly. It means that the measurement data needs further analysis to get the actual value[6].

3 Analytical calculation method for inductance of windings

3.1 Coordinate Systems with Skewed Slot Structure

In order to describe the MMF and permeance, appropriate coordinate systems must be established. It's assumed that the stator slots are skewed by α_c ($\alpha_c \neq 0$) in mechanical radian, the axial length of motor core is d and the length of the corresponding bevel edge is h.

The coordinate system of stator is a two-dimensional oblique coordinate system. Its abscissa axis extends along the stator circle and vertical axis extends along the skewed slots, whose origin is set by the intersection of one stator slot's symmetrical line and the abscissa axis. The coordinate system of rotor is a two-dimensional rectangular coordinate system. Its abscissa axis extends along the rotor circle and vertical axis extends along the d-axis of rotor, whose origin is set by the intersection of the d-axis and the abscissa axis, as shown in Fig 2[7].

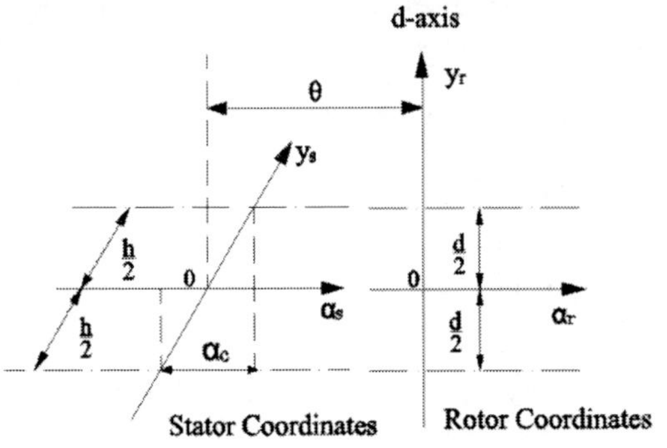

Figure 2. Coordinate systems with skewed slot structure

It is assumed that the rotor coordinates run ahead of the stator coordinates by the mechanical radian θ, as shown in Fig 2. Therefore, the conversion from one coordinate system to another can derive itself from the systems :

$$\begin{cases} \alpha_s=\theta+\alpha_r-\dfrac{\alpha_c}{d}y_r \\ y_s=\dfrac{h}{d}y_r \end{cases} \quad (3)$$

3.2 Airgap magnetic permeance

Considering the effort of the stator slot opening only, the permeance is given as follow:

$$\lambda_s(\alpha_s,y_s)=\frac{\mu_0}{\delta}\sum_{v=0}^{+\infty}a_{sv}\cos(vZ\alpha_s) \quad (4)$$

Where μ_0 is the magnetic permeance in vacuum, δ is the mean length of airgap, Z is the number of stator slots, a_{sv} is the magnitude of v-order harmonic of airgap magnetic permeance considering the effort of the stator slot opening only[8].

Considering the rotor saliency and rotational symmetry, the permeance is similarly given as follow:

$$\lambda_r(\alpha_r,y_r)=\frac{\mu_0}{\delta}\sum_{k=0}^{+\infty}b_{rk}\cos(2pk\alpha_r) \quad (5)$$

Where p is the number of pole-pairs, b_{rk} is the airgap magnetic permeance's magnitude of k-order harmonic only considering rotor saliency. The rotational symmetry determines that the permeance should be expressed as an even function.

Finally the permeance can be calculated in the superposition method:

$$\lambda=\lambda_s\lambda_r=\frac{\mu_0}{\delta}\sum_{v=0}^{+\infty}a_{sv}\cos(vZ\alpha_s)\sum_{k=0}^{+\infty}b_{rk}\cos(2pk\alpha_r) \quad (6)$$

3.3 Magnetic motive force

The fundamental harmonic of MMF for single phase of stator windings in the stator coordinate system can be expressed as follow[9]:

$$F_{si}(\alpha_s,y_s)=I_{si}F_{s1}\cos(p\alpha_s+p\alpha_{si}),\quad i\in\{a,b,c\} \quad (7)$$

3.4 Airgap flux linkage and inductance

According to (6) and (7), the airgap flux density inspired by single phase of stator windings with coordinate transformation is given as follow:

$$\begin{aligned} B_s(\alpha_s,y_s)&=\lambda F_{si}(\alpha_s,y_s) \\ &=\frac{\mu_0}{\delta}F_{si}(\alpha_s,y_s)\left\{\sum_{v=0}^{+\infty}a_{sv}\cos(vZ\alpha_s)\sum_{k=0}^{+\infty}b_{rk}\cos\left[2pk\left(\alpha_s-\theta+\frac{\alpha_c}{h}y_s\right)\right]\right\} \end{aligned} \quad (8)$$

The flux linkage of one stator winding can be expressed as follow:

$$\psi_s(\alpha_{sc1},\alpha_{sc2}) = \frac{dR_\delta N_{sc}}{h}\int_{\alpha_{sc2}}^{\alpha_{sc1}}\int_{-h/2}^{h/2} B_s(\alpha_s, y_s)dy_s d\alpha_s \quad (9)$$

Where R_δ is mean radius length of airgap, N_{sc} is the turn number of the stator winding , α_{sc1} is the upper side's coordinate value of winding , α_{sc2} is the coordinate value of lower side.

After the accumulation of flux linkage through one winding, the inductance should derive itself from the definition :

$$L_s = \sum \psi_s / I_{si} \quad (10)$$

The same procedure can be easily adapted to obtain rotor inductance parameters.

4 Verification Results

4.1 Results of Self-inductance

First, the analytical method is employed on a salient-pole synchronous motor, and then FEM simulations without damper windings are carried out. The results are shown in Fig 3.

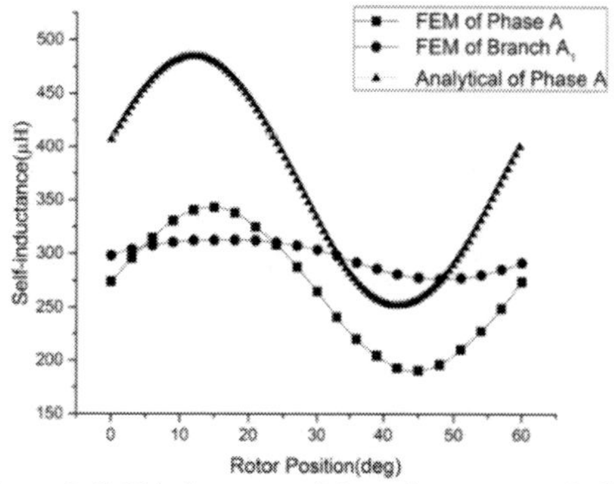

Figure 3. Self-inductance of phase A armature winding

4.2 Results of mutual-inductance

The results of mutual-inductance obtained by these two methods are shown in Fig 4.

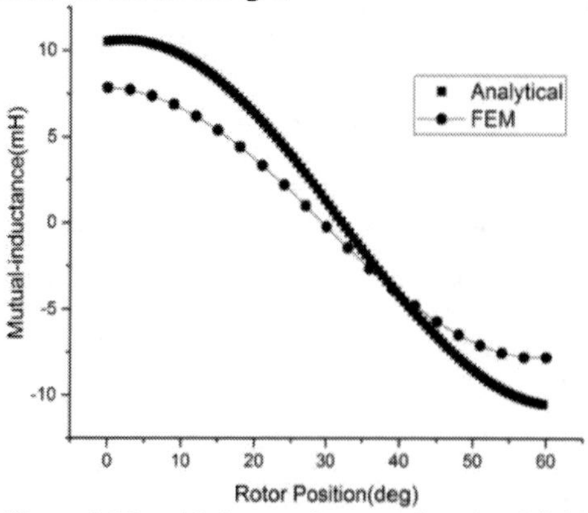

Figure 4. Mutual-inductance between phase A and field windings

4.3 FEM Simulation Results and Measured Data with Damper Windings

The self-inductance parameters are measured. Meanwhile, the FEM simulations of the established model with damper windings are carried out. The results are shown in Fig 5.

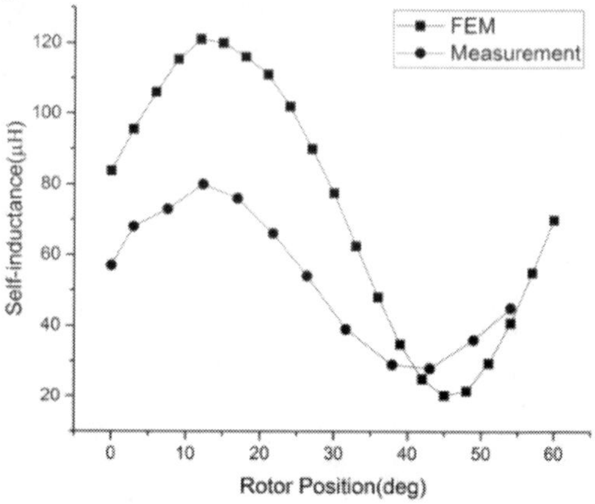

Fig 5. Self-inductance of FEM simulation and measurement

4.4 Analysis and Follow-up Work

First, the length of airgap under poles was employed in the analytical calculation instead of the mean length, so the results of analytical method are 1.36 times larger than FEM simulation results. Then, the FEM simulation results of self-inductance for phase A and branch A_1 can't match well because of different saturation resulted by different loads actually. Due to the quick and time-saving calculation, only the fundamental harmonic of MMF is employed. Considering the damper windings, the reason for the differences between the inductance parameters of FEM simulation and measurement is that the simulation models were established without skewed slots and the so-called measured self-inductance parameters are not equal to actual parameters.

In the analytical calculation, accurate parameters should be employed. At the same time, one method should be adopted to analyze the measurement data and identify the actual value of inductance.

5 Conclusion

A fast analytical calculation method of inductance based on magnetic circuit method was presented in this paper. Based on the analysis of MMF and permeance under the established two-dimensional coordinate systems, the inductance of motor with skewed slots was calculated and compared with the FEM simulation results. To some extent, the measurement data can verify the validity of FEM simulation models and the analytical method. Then more parameters of motor can be calculated such as the electromagnetic torque and analyzed next. This method can save time effectively and be applicable to other kinds of motor based on the established two-dimensional coordinate systems.

References

1. Shima K, Ide K, Takahashi M., Finite-element Calculation of Leakage Inductances of a Saturated Salient-pole Synchronous Machine with Damper Circuits[J].IEEE Transactions on Energy Conversion,2002,17（4）:463-470.
2. Chen Y. S., Zhu Z. Q., Howe D., Calculation of d- and q-Axis Inductances of PM Brushless ac Machines Accounting for Skew[J]. IEEE Transactions on Magnetics, 2005,41（10）:3940- 3942.
3. Williamson Stephen, Flack T.J., Volschenk A.F.. Representation of Skew in Time-stepped Two-dimensional Finite-element Models of Electrical Machines[J].IEEE Transactions on Industry Applications,1995,31(5):1009-1015.
4. Dziwniel P., Piriou F., Ducreux J.P., et al, A Time-stepped 2D-3D Finite Element Method for Induction Motors with Skewed Slots Modeling[J]. IEEE Transactions on Magnetics, 1999,35(3):1262-1265.
5. Akbari H.R., Sadeghi S.. Calculation of Inductances of Induction Machines Under Axial Nonuniformity Conditions[C]. IEEE International Symposium on Industrial Electronics, Vigo :IEEE,2007.1113-1118.
6. Qiao Mingzhong, Zhang Xiaofeng, Ren Xiuming, Measurement and Calculation Winding Parameters for Multi-Phase High-Power Propulsive PMSM with Damping Windings [J].Transactions of CHINA Electrotechnical Society,2003,18(6):37-41.
7. Ou Yangbin, Liu Dezhi, Zhai XIaofei, Analytical Calculation of Inductances of Windings in Electrical Machines with Slot Skew [C]. XIX International Conference on Electrical Machines - ICEM 2010, Rome.
8. Gao Jingde, Wang Xiangheng, Li Fahai, Analysis of AC Machines and Systems, 2nd ed., Beijing: Tsinghua University Press, 2005, pp. 458-462.
9. Gu Chenglin, Chen Qiaofu, Xiong Yongqian, Electrical Machinery, 3nd ed., Wuhan: Huazhong University of Science and Technology Press, 2010, pp. 392-397.

Co-simulation for the Power Generation Characteristics of an Electromagnetic Linear Actuator used in Active Suspensions

Anqi Peng, YanShi and Wenchao Yue

School of Mechanical Engineering, Nanjing University of Science and Technology, Nanjing 210094, China

Abstract. Compared to passive suspensions, active suspensions with electromagnetic linear actuator (EMLA) not only has higher performance in active force control, but also can regenerate energy form road disturbances. In this study, the characteristics of a new designed EMLA were investigated under generator mode by precisely co-simulating the quarter active suspension with single degree of freedom. This model, which consists of sprung mass, EMLA imported from Maxwell and a spring in parallel with the EMLA, is built in Simplorer. Simulation results show that the mechanical power and electric power of the EMLA are both proportional to the frequency and amplitude of excitation. The efficiency, defined as the ratio of electric energy to mechanical energy, decreases with the increase of frequency while nonlinearly changes with the increase of stroke. The yielded maximal efficiency of 94% occur sat the frequency of 2 Hzand the stroke ranges of6mm to 9mm.

1 Introduction

Under the road input, the passive suspensions absorb road vibration by converting kinetic energy into heat and dissipating it, but for the electromagnetic active suspensions, the vibration energy can be harvested into electricity and then stored in batteries. Meanwhile, due to the ability of variable active force, the vibration damping can be adjusted to adapt to the variable road input. So, electromagnetic active suspensions increasingly become attractive replacements for the currently installed passive suspensions, especially for the electric vehicles in which the high voltage batteries can be used to drive the linear actuator and store the electricity [1-3].

As the key component of the electromagnetic active suspensions, electromagnetic linear actuator (EMLA) operates in two modes, namely motor mode and generator mode. In motor mode, batteries deliver energy to EMLA to achieve desirable active force. In generator mode, EMLA feed energy by converting kinetic energy into electrical energy and charging batteries. In this paper, a newly designed EMLA is proposed to investigate the characteristics in generator mode. When the EMLA is used in the active suspension, the whole system, which involves electric, magnetic, and mechanic subsystems, is complicated. Therefore, co-simulation between Maxwell and Simplorer is adopted to model the 1/4 electromagnetic active suspension and analyze the power generation characteristic of the EMLA under different excitations.

2 1/4 Electromagnetic active suspension

2.1 Electromagnetic linear actuator

In the electromagnetic active suspensions, the EMLA is a replacement for the conventional damper, so its electromagnetic force should meet the demand of vibration damping. Therefore, as shown in Fig.1, a newly designed EMLA and its test bed are proposed, its performance parameters are listed in Table 1. The proposed EMLA is a single-phase tubular device, mainly composed of inner yoke, outer yoke, and winding coils (Fig.2). The moving part is formed by permanent magnets and inner yoke. By employing quasi-Halbach array, the permanent magnets consist of three ring magnets, of which permanent magnet 1, 3 are radially magnetized and magnet 2 is axially magnetized. Compared with a type of motor topology which has a stationary inner yoke, this structure just has one air-gap between the mover and the stator, which provides a closed return path for the flux and reduces the change of magnetic circuit caused by the mover's high frequency reciprocating motion. Meanwhile it also can greatly reduce the eddy current loss caused by the air-gap. The fixed portion consists of outer yoke and two coils in series. The inner and outer yokes are both made of DT4C, which has high magnetic permeability and small reluctance. The permanent magnets are made of NdFe45SH. The EMLA has twooperating modes, one is motor, and the other is generator. When the mover moves with surrounding vibration, the winding coils flux linkage is modulated by the linear movement of the mover, and the induced EMF is generated.

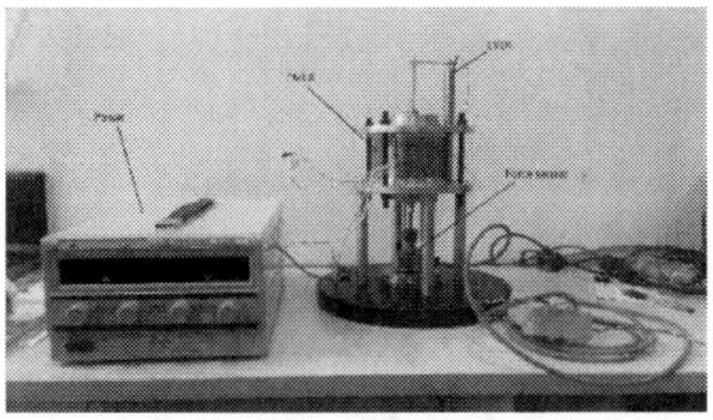

Figure 1. The test bed of EMLA

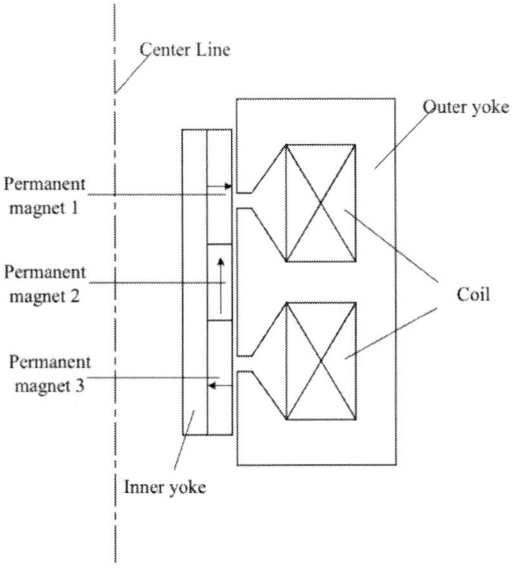

Figure 2. The structure of EMLA

Table 1. EMLA's performance parameters

Parameters	value
Resistance, $R(\Omega)$	5.4
Inductance, L(mH)	20
Motor Constant, K_i(N/A)	89
Stroke, S(mm)	± 25
Maximum Thrust, F_{peak}(N)	1000

2.2 1/4 electromagnetic active suspension system modeling

By replacing the original damper, quarter electromagnetic active suspension with single degree of freedom which are composed of vehicle body mass and EMLA paralleledwith a spring is modeled (Fig.3). Based on Newton's second law, the dynamic equation is:

$$m\ddot{z} + C(\dot{z} - \dot{q}) + K(z - q) = 0 \quad (1)$$

where m is the sprung mass, k is the spring stiffness, C is the damping coefficient of the EMLA, q is the road roughness function, and z is the vertical displacement of the sprung mass[4].

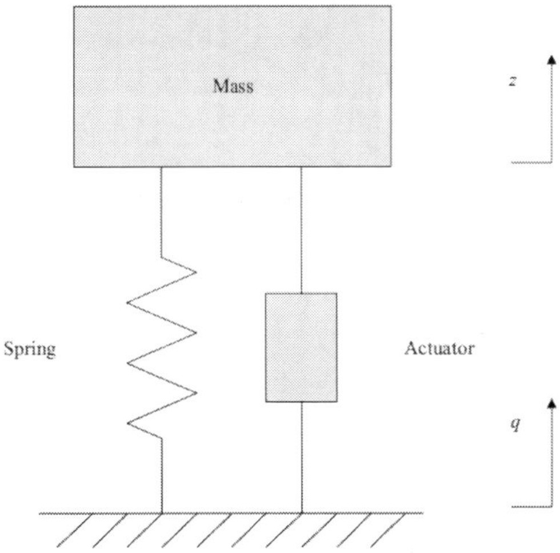

Figure 3.1/4 electromagnetic active suspension model

3 Co-simulationwith maxwell and simplorer

The characteristics of the EMLA in generator mode under variable sinusoidal excitation are accurately investigated by using a co-simulation between Maxwell and Simplorer. Maxwell is an interactive software package that uses finite element analysis (FEA) to solve three-dimensional (3D) electrostatic, magneto static, eddy current and transient problems. Simplorer is an integrated, multi-domain, mixed-signal simulator for complex technical systems [5]. The EMLA is established in Maxwell firstly by setting the initial parameters such as the structure size, material properties and coils, etc. Then, the quarter electromagnetic active suspension is set up in Simplorer, including the mechanical part and electronic part which is shown in Fig.4. The EMLA in Maxwell has been previously loaded into Simplorer through the Sub circuit. The EMLA model has 4 pins, namely 'Winding_in', 'Winding_out', 'Motion Setup_in', and 'Motion Setup_out', respectively. Road vibration is given by position excitation through pin 'Motion Setup_in'. When a Transient Co simulation Component circuit is used in a Simplorer design, Maxwell 2D or 3D transient and Simplorer transient solvers exchange data during each simulation time step. Maxwell uses a loop form of the circuit equations, while Simplorer uses a nodal form of the circuit equations. After each time step, Simplorer forms a Norton-equivalent of the drive circuit at the coupling pins between the Maxwell component and the rest of the system. Maxwell converts this to a loop matrix, solves the finite element equations, and then outputs a Thevenin-equivalent for the next Simplorer time step. Simplorer controls the time step during simulation, and also controls the linear velocity of the Maxwell component. This parameter-based coupling enhances solution accuracy and stability.

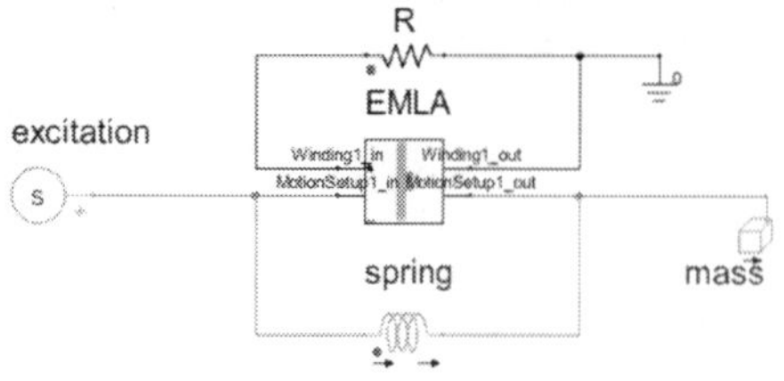

Figure 4.1/4 electromagnetic active suspension model in Simplorer

4 Results analysis

In order to observe the characteristic of the EMLAconveniently, the road disturbances are assumed as sinusoidal displacement excitation, the mechanical power, electric power, and efficiency η, defined as the ratio of electric energy to mechanical energy are given by the formulas as follows:

$$P_m = \frac{1}{t}\int_0^t Fv dt \quad (2)$$

$$P_e = \frac{1}{t}\int_0^t ui dt \quad (3)$$

$$\eta = \frac{P_m}{P_e} * 100\% \quad (4)$$

where t is the simulation time, F is the EMLA's thrust force, v is moving speed of mover of the EMLA ,u and i are voltage and current of the coils, respectively.

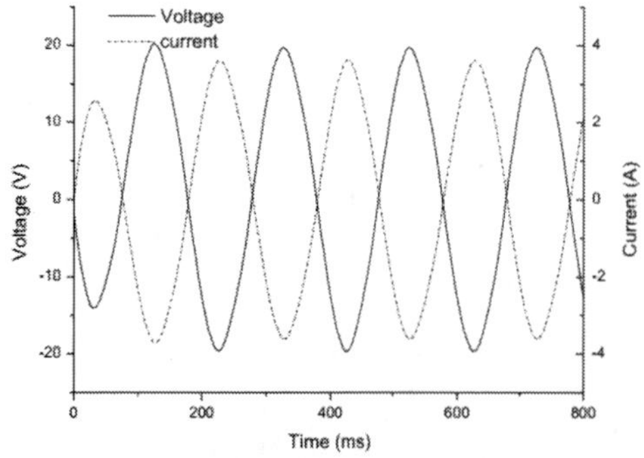

Figure 5. The voltage and current of EMLA

As shown in Fig. 5 and Fig.6, phase difference of 180° exists between voltage and the current, and this phenomenon also occurs between the force and the speed. Thus, mechanical power and electronic power are negative, which means the actuator is in generator mode. Fig.7 and Fig.8 show that the absolute value of the average of mechanical power and electric power are both proportional to the frequency and the amplitude of excitation. Fig.9 shows that the efficiency is inversely proportional to the frequency and changes nonlinearly with the increase of stroke. The higher efficiencies always occur in the stroke from 6mm to 9mm no matter how the frequency changes, and the maximum is up to 94%at the frequency of 2Hz.

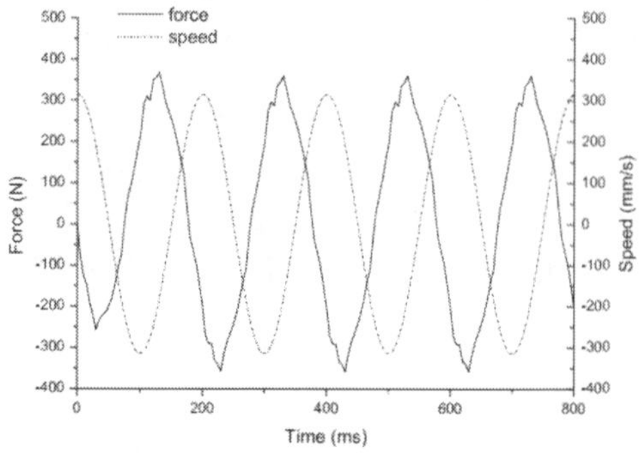

Figure 6. The force and speed of EMLA

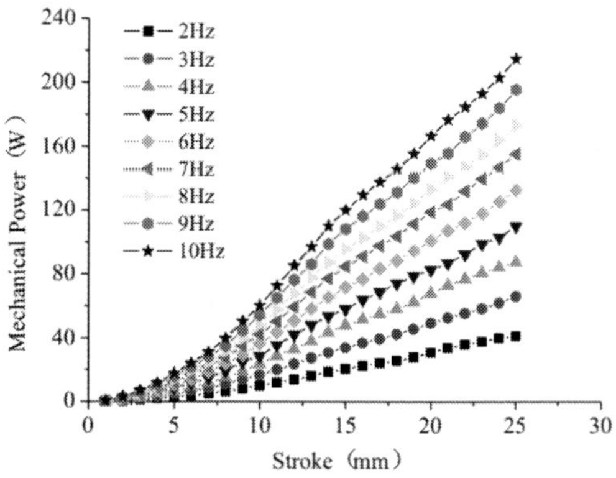

Figure 7. The absolute value of the average of mechanical power under different excitations

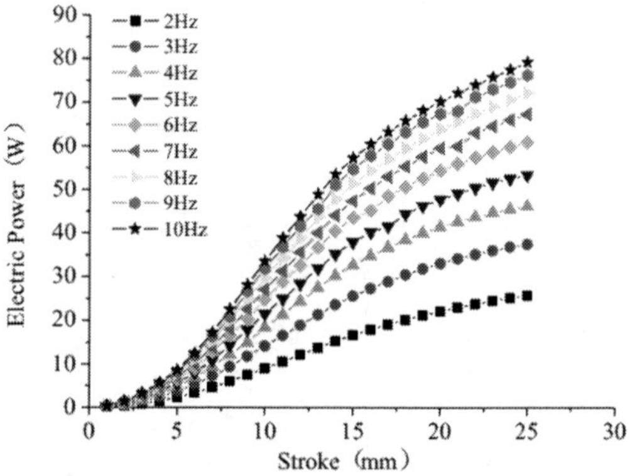

Figure 8. The absolute value of the average of electric power under different excitations

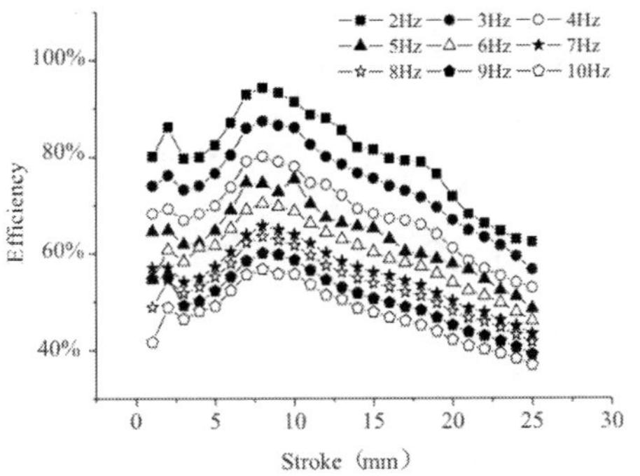

Figure 9. Efficiency of EMLA under different excitations

5 Conclusion

The EMLA, added into the active suspensions, has the benefits of altering vibration energy into electric energy while with the feature of damping road disturbances. It will first be of importance to analyze the power generation characteristics of the EMLA under different excitations. Maxwell and Simplorer co-simulation is proposed as a method to investigate how the mechanical power and the electric power change along with the excitations. The simulations show that the mechanical power and electric power are proportional to the frequency and the amplitude of excitation. The ratio of electric energy to mechanical energyvaries inversely with the increase of frequency and changes nonlinearly with the increase of stroke.

Acknowledgment

This work was supported by the Natural Science Foundation of Jiangsu Province (Grant No. BK20130757) and the Peak of the Six Talents of Jiangsu Province (Grant No. 2015-JXQC-004).

References

1. B.L.J. Gysen, J.J.H. Paulides, J.L.G.Janssen, E.A.Lomonova, IEEE. T. VEH. TECHNOL., **59**, 3 (2010)
2. J.B. Wang, W.Y Wang, K. Atallah. IEEE. T. VEH. TECHNOL.,**60**, 1 (2011)
3. M.Morteza, S. Mehdi, H. Sahar,IEEE T IND ELECTRON., **60**, **10** (2013)
4. I.Martins, J.Esteves, G.D.Marques, F. Pina da. Silva, IEEE. T. VEH. TECHNOL., **55**, 1(2006)
5. J.H.Yu,B.D.Bai, D.Z.Chen, Z.F. Hu,WCICA, **11th**, 2015.

Comparative Study between Two Schemes of Active-Control-Based Mechatronic Inerter

Lingduo He , Yaozong Liu and Silu Han

College of Mechatronic Engineering and Automation, National University of Defense Technology, Hunan, China

Abstract. Based on force-current analogy and velocity-voltage analogy in the theory of electromechanical analogy, the inerter is a device that corresponded to the capacitor completely where conquers the nature restriction of mass, what's more, it is significant to improve the ratio of the inerter's inertance to its mass for mechanical networks synthesis. And according to the principle of active-control-based mechatronic inerter, we present two implementation schemes. One was based on linear motor, and the other was based on the ball screw and rotary motor. We introduced the implementation methods and established theoretical model of the two schemes, then compared the ratio of the inerter's inertance to its mass for the two schemes. Finally, we consider the scheme is better which was based on the ball screw and rotary motor.

1 Introduction

In 2002, the concept of inerter was proposed by Malcolm C. Smith [1]. The inerter is a mechanical device which has two free and independent terminals [2], and the quantitative relation between the force acting on the two terminals and the relative acceleration between the two terminals are as shown in the formula (1).

$$\mathbf{F}=b(\dot{\mathbf{v}}_1-\dot{\mathbf{v}}_2) \tag{1}$$

in which **F** is the applied force, while $\mathbf{v}_1$ and $\mathbf{v}_2$ are the velocities of the two terminals. The constant of proportionality b is called the inertance and has units of kilograms [3]. Thus, Inerter is essentially a two terminals mass element. The concept of inerter was put forward, which solved the problem that the mass only as single terminal mass components relative to the inertial system, and the mechanical network and circuit network can be strictly corresponding to the force–current analogy [4]. It has important theory and application value for applying circuit theory to mechanical network synthesis.

The traditional inerter needs to be realized by the mechanical mechanism while enhanced the inertance needs to be increased by the transmission ratio and the mass of the flywheel [5]. However, it will inevitably lead to the increase of the mass of the inerter, and limits the application of the inerter in mass sensitive areas such as aerospace [6]. For this purpose, we put forward the concept of mechatronic inerter, and explore the feasibility of using electromagnetic force to enhance the inertance [7].

The active-control-based mechatronic inerter (hereinafter referred to as ACBMI) are composed of sensors, controllers and actuators. Its working principle is: **F** was applied to the two terminals of the actuator to make relative motion, while measured by the sensor and sent to the controller; according to set parameters and **F**, the controller control actuator always hinder the movement of **F** that the relative acceleration between the two terminals is reduced, and the inertance is enhanced[8]. The schematic diagram is shown in figure 1.

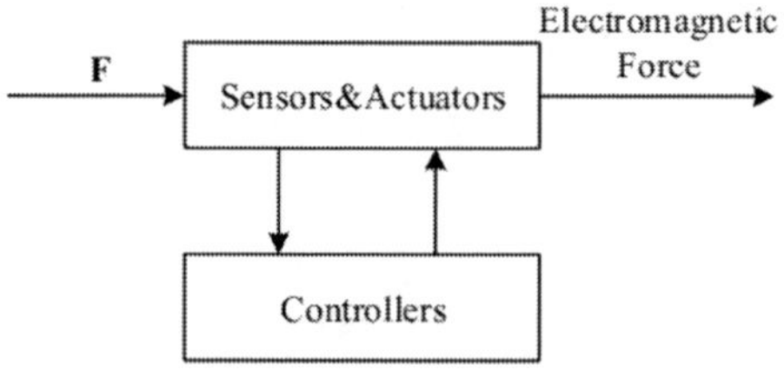

Figure 1. Principle block diagram of ACBMI.

According to different actuator implementation schemes, the ACBMIs are divided into two kinds [9]. One is the use of linear motor as the actuator (hereinafter referred to as the linear motor based ACBMI), **F** was applied to the primary and secondary of the linear motor, and linear motor directly generate electromagnetic force to prevent the linear motion caused by applied force; another is the use of ball screw and rotary motor as the actuator (hereinafter referred to as the rotary motor based ACBMI), the ball screw to achieve the conversion between linear motion and rotary motion, and the electromagnetic torque generated by the rotary motor hinders the movement of **F**.

We built the theoretical model of two kinds of the ACBMIs, and compared the relationship between the inertance and the inerter own mass [10]. The ratio between inertance and mass of inerter is defined as RMI (the ratio of mass to intance). The greater RMI of inerter, then the greater inertance can be achieved for the same mass of inerter [11]. Therefore, we can choose a better implementation scheme by comparing RMI.

2 Theoretical model of two kinds of ACBMIs

Assuming the mass of two kinds of inerters are both M, and building the theoretical model.

2.1 Theoretical model of linear motor based ACBMI

The mass of linear motor based ACBMI equal to the mass of linear motor, so the mass of linear motor is M. Among them, the mass of primary is m_1, the mass of secondary was M_1, and meet $M = m_1 + M_1$.

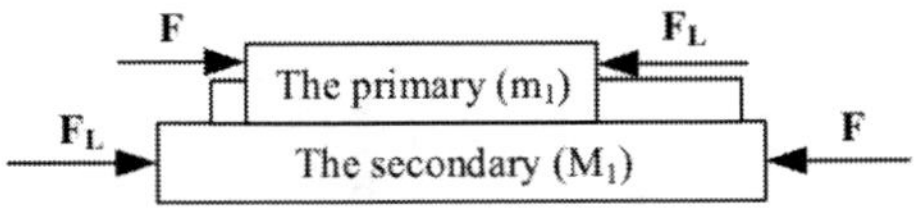

Figure 2. The force diagram of linear motor based ACBMI

As shown in Figure 2, **F** applied to the two terminals of linear motor based ACBMI, and the continuous thrust of linear motor output is $\mathbf{F}_\mathrm{L}$. Without consideration of friction, the relative acceleration of the two terminals is

$$\mathbf{a}_{\mathrm{L1}} = \frac{(\mathbf{F}-\mathbf{F}_\mathrm{L})}{m_1} \tag{2}$$

$$\mathbf{a}_{\mathrm{L2}} = \frac{(\mathbf{F}-\mathbf{F}_\mathrm{L})}{M_1} \tag{3}$$

$$\mathbf{a}_\mathrm{L} = \mathbf{a}_{\mathrm{L1}} + \mathbf{a}_{\mathrm{L2}} \tag{4}$$

$$\mathbf{a}_\mathrm{L} = \frac{(\mathbf{F}-\mathbf{F}_\mathrm{L})M}{m_1(M-m_1)} \tag{5}$$

in which $\mathbf{a}_{\mathrm{L1}}$ is the acceleration of primary, $\mathbf{a}_{\mathrm{L2}}$ is the acceleration of secondary.

The inertance b_L and the RMI g_L of the linear motor based ACBMI are respectively

$$b_L = \frac{\mathbf{F}}{\mathbf{a}_\mathrm{L}} = \frac{\mathbf{F}}{(\mathbf{F}-\mathbf{F}_\mathrm{L})} \bullet \frac{m_1(M-m_1)}{M} \tag{6}$$

$$g_L = \frac{b_L}{M} = \frac{\mathbf{F}}{(\mathbf{F}-\mathbf{F}_\mathrm{L})} \bullet \frac{m_1(M-m_1)}{M^2} \tag{7}$$

2.2 Theoretical model of rotary motor based ACBMI

The mass of rotary motor based ACBMI equal to the mass of rotary moto and the mass of ball screw. Let m_{21} be the mass of the ball screw nut, m_{22} the mass of the screw, r the radius of the screw, p the lead of the screw, J_1 the moment of inertia of rotor, M_2 the mass of the rotary motor, and $M = m_{21} + m_{22} + M_2$.

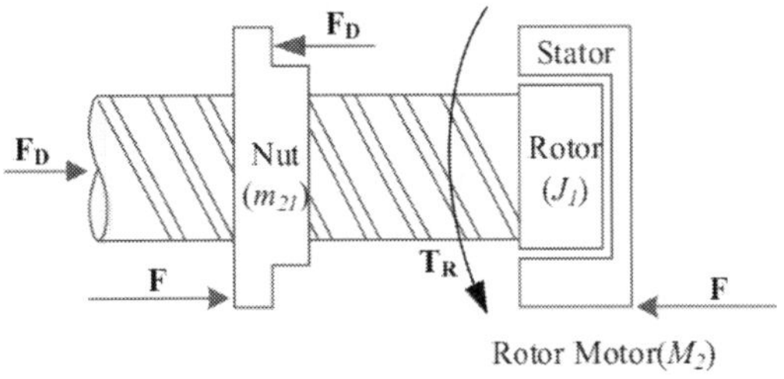

Figure 3. The force diagram of rotor motor based ACBMI

F applied to the two terminals of the linear motor based ACBMI, the output torque of the rotary motor with mass of M_2 is $\mathbf{T}_\mathrm{R}$, the force analysis of the inerter is as shown in figure 3.

Among them, $\mathbf{F}_\mathrm{D}$ is the axial force between the nut and the screw, and have the following relationship with the drive torque $\mathbf{T}_\mathrm{D}$

$$\mathbf{F}_\mathrm{D} = \frac{2\pi}{p}\mathbf{T}_\mathrm{D} \tag{8}$$

For the screw and the rotor has the following relationship

$$J_2 = \frac{1}{2}m_{22}r^2 \tag{9}$$

$$J = J_1 + J_2 = J_1 + \frac{1}{2}m_{22}r^2 \tag{10}$$

$$\mathbf{T}_\mathrm{D} - \mathbf{T}_\mathrm{R} = J\boldsymbol{\beta} \tag{11}$$

$$\mathbf{a}_{21} = \frac{\mathbf{F}-\mathbf{F}_\mathrm{D}}{m_{21}} \tag{12}$$

$$\mathbf{a}_{22} = \frac{\mathbf{F}-\mathbf{F}_\mathrm{D}}{m_{22}+M_2} \tag{13}$$

$$\frac{\mathbf{a}_{21}}{\boldsymbol{\beta}} = \frac{p}{2\pi} \tag{14}$$

$$\mathbf{a}_\mathrm{R} = \mathbf{a}_{21} + \mathbf{a}_{22} \tag{15}$$

in which J_2 is the moment of inertia of the screw, J is the moment of inertia of the screw and the rotor, $\boldsymbol{\beta}$ is the angular acceleration of rotating screw and rotor, $\mathbf{a}_{21}$ is the acceleration of nut, $\mathbf{a}_{22}$ is the acceleration of motor, $\mathbf{a}_\mathrm{R}$ is the relative acceleration of the two terminals.

Combined (8) ~ (15), then

 DOI: 10.1051/matecconf/20179513005

$$\mathbf{a}_{21} = \frac{(\mathbf{F} - \frac{2\pi}{p}\mathbf{T}_{\mathbf{R}})(m_{22} + M_2)}{\left[m_{21} + (\frac{2\pi}{p})^2 J_1 + \frac{1}{2}(\frac{2\pi r}{p})^2 m_{22}\right](M - m_{21})} \quad (16)$$

$$\mathbf{a}_{22} = \frac{(\mathbf{F} - \frac{2\pi}{p}\mathbf{T}_{\mathbf{R}})m_{21}}{\left[m_{21} + (\frac{2\pi}{p})^2 J_1 + \frac{1}{2}(\frac{2\pi r}{p})^2 m_{22}\right](M - m_{21})} \quad (17)$$

And the relative acceleration of the two terminals is

$$\mathbf{a}_{\mathbf{R}} = \frac{(\mathbf{F} - \frac{2\pi}{p}\mathbf{T}_{\mathbf{R}})M}{\left[m_{21} + (\frac{2\pi}{p})^2 J_1 + \frac{1}{2}(\frac{2\pi r}{p})^2 m_{22}\right](M - m_{21})} \quad (18)$$

The inertance b_R and the RMI g_R of the rotary motor based ACBMI are respectively

$$b_R = \frac{\left[m_{21} + (\frac{2\pi}{p})^2 J_1 + \frac{1}{2}(\frac{2\pi r}{p})^2 m_{22}\right](M - m_{21})\mathbf{F}}{(\mathbf{F} - \frac{2\pi}{p}\mathbf{T}_{\mathbf{R}})M} \quad (19)$$

$$g_R = \frac{\left[m_{21} + (\frac{2\pi}{p})^2 J_1 + \frac{1}{2}(\frac{2\pi r}{p})^2 m_{22}\right](M - m_{21})\mathbf{F}}{(\mathbf{F} - \frac{2\pi}{p}\mathbf{T}_{\mathbf{R}})M^2} \quad (20)$$

3 Comparison

3.1. $g_{Lmax}<g_{Rmax}$

For the two kinds of inerters, when the actuator output maximum continuous thrust $\mathbf{F}_{\mathbf{Lmax}}$ and maximum continuous torque $\mathbf{T}_{\mathbf{Rmax}}$, we can obtain the maximum RMI

$$g_{L\max} = \frac{\mathbf{F}m_1(M - m_1)}{(\mathbf{F} - \mathbf{F}_{\mathbf{L}\max})M^2} \quad (21)$$

$$g_{R\max} = \frac{\left[m_{21} + (\frac{2\pi}{p})^2 J_1 + \frac{1}{2}(\frac{2\pi r}{p})^2 m_{22}\right](M - m_{21})\mathbf{F}}{(\mathbf{F} - \frac{2\pi}{p}\mathbf{T}_{\mathbf{R}\max})M^2} \quad (22)$$

In order to compare g_{Lmax} and g_{Rmax}, then let

$$k_1 = \frac{\mathbf{F}}{(\mathbf{F} - \mathbf{F}_{\mathbf{Lmax}})} \quad (23)$$

$$k_2 = \frac{m_1(M - m_1)}{M^2} \quad (24)$$

$$k_3 = \frac{\mathbf{F}}{(\mathbf{F} - \frac{2\pi}{p}\mathbf{T}_{\mathbf{Rmax}})} \quad (25)$$

$$k_4 = \frac{\left[m_{21} + (\frac{2\pi}{p})^2 J_1 + \frac{1}{2}(\frac{2\pi r}{p})^2 m_{22}\right](M - m_{21})}{M^2} \quad (26)$$

And $k_1<k_3$, when

$$p < \frac{2\pi\mathbf{T}_{\mathbf{Rmax}}}{\mathbf{F}_{\mathbf{Lmax}}} \quad (27)$$

$k_{2max}=0.25$, when

$$m_1 = \frac{M}{2} \quad (28)$$

For the rotary motor inerter $m_{21} < \frac{M}{2}$, $k_2<k_4$ (the complete derivation to be given by (36) at the top of the next page) when

$$p < 2\pi\sqrt{\frac{2J_1}{M}} \quad (29)$$

So $g_{Lmax}<g_{Rmax}$, when

$$p < \frac{2\pi\mathbf{T}_{\mathbf{Rmax}}}{\mathbf{F}_{\mathbf{Lmax}}} \;\&\; p < 2\pi\sqrt{\frac{2J_1}{M}} \quad (30)$$

In order to facilitate the comparison, let

$$n_1 = \frac{\mathbf{F}_{\mathbf{Lmax}}}{M} \quad (31)$$

$$n_2 = \frac{\mathbf{T}_{\mathbf{Rmax}}}{M_2} \quad (32)$$

$$n_3 = \frac{M_2}{M} \quad (33)$$

$$n_4 = \frac{J_1}{M_2} \quad (34)$$

And

$$p < \frac{2\pi n_2 n_3}{n_1} \;\&\; p < 2\pi\sqrt{2n_3 n_4} \quad (35)$$

In order to explore whether there is the existence of p to meet the above relationship, we conducted a market research.

Table 1. Market research of linear motor.

Types	n_1
IL06	20.04
IL12	23.04
IL18	27.81
IL24	28.42
ICD05	56.69
ICD10	61.66

As shown in Table 1, with comparison for the mass of 5 kilograms of linear motor, we found that n_1 achieve the maximum value of 61.66 while selected the Kollmorgen ICD10 series linear motor. And as shown in Table 2, with comparison for the mass of 5 kilograms of rotary motor, we found that n_2 achieved the maximum value of 0.91 and n_4 achieved the maximum value of 4.42×10^{-5} while selected the Kollmorgen AKM4 series rotary motor.

$$k_4 > \frac{\left[m_{21} + \frac{M}{2} + \frac{1}{2}(\frac{2\pi r}{p})^2 m_{22}\right](M - m_{21})}{M^2} > \frac{\frac{M}{2}(M - m_{21})}{M^2} > \frac{1}{4} > k_2 \quad (36)$$

Table 2. Market research of rotary motor.

Types	n_2	n_4
AKM1X	0.53	6.11×10^{-6}
AKM2X	0.77	1.50×10^{-5}
AKM3X	0.79	2.57×10^{-5}
AKM4X	0.91	4.42×10^{-5}
CKM	0.71	3.32×10^{-5}

And through the research of the ball screw, we found that the mass of ball screw is generally below 0.6 kg and the lead of the screw are rang from 2mm to 30mm with the appropriate trip and axial load. When selecting the AKM41 rotary motor(the smallest rotary motor in AKM4 series rotary motor), n_3 gets the minimum value of 0.803.

Now we plug n_1, n_2, n_3, n_4 into formula (35), and get p<0.0529, then lead is less than 52.9mm. The RMI of the rotary motor inerter can be obtained more than the linear motor inerter by properly configuring the rotating motor and ball screw while the existing ball screw can meet this condition obviously.

3.2 Specific examples

We used inerter design by using the axis rotation of ball screw motor as the actuator, the following parameters

$$\begin{aligned} &M = 2.657kg, M_2 = 2kg, \mathbf{T}_{\mathbf{Rmax}} = 0.96N\cdot m, \\ &J_1 = 2.86\times10^{-4} kg\cdot m^2, m_{21} = 0.32kg, m_{22} = 0.337, \\ &r = 7.5mm, p = 20mm, \mathbf{F} = 400N \end{aligned} \quad (37)$$

gave a value of g_{Rmax}=39.63.

If the linear motor is used as the actuator, and the ICD10-050 linear motor with excellent performance is selected. The following parameters

$$\begin{aligned} &M = 2.8kg, \mathbf{F}_{\mathbf{Lmax}} = 171N, \mathbf{F} = 400N, \\ &m_1 = 1.9kg, M_1 = 0.9kg \end{aligned} \quad (38)$$

gave a value of g_{Lmax}=0.39.

4 Conclusions

We built the theoretical model of two kinds of inerters with different actuators, and analyzed the condition when the inertia of the rotary motor based ACBMI outweighed that of the linear motor based ACBMI. Through marker research, we found that the condition was easy to satisfy. According to specific examples, the RMI of the rotary motor inerter was much higher than that of the linear motor inerter. Therefore, the RMI of the inerters can be enhanced by applying ball screw and rotary motor as actuator.

References

1. M.C. Smith, IEEE Trans. Automatic Control, *Synthesis of Mechanical Networks: The Inerter*, **47**, 1648–1662 (2002)
2. J. Nie, X. Zhang, B. Jiang, L. Chen, Machine Design and Research, *Researh on the Inerter Structure*, **28**, 29-32 (2012)
3. D.X. Zhu, Y.Z. Liu, J.L. Wen, ICIA, *A Simulation Research of the Mechatronic Inerter Based on Active Control*, (2015)
4. M.C. Smith, F. Wang, IEEE Conference on Decision and Control, *Performance Benefits in Passive Vehicle Suspensions Employing Inerters*, **28**, 29-32 (2012)
5. C. Papageorgiou, N. Houghton, M.C. Smith, Journal of Dynamic Systems Measurement and Control, *Experimental Testing and Analysis of Inerter Devices*, **131**, 3-6 (2009)
6. M.C. Smith, F. Wang, IEEE Conference on Decision and Control, *Performance Benefits in Passive Vehicle Suspensions Employing Inerters*, **28**,29-32 (2012)
7. F. Wang, M. Hong, The Institution of Mechanical Engineers, *Designing and testing a hydraulic inerter*, **225**, 65-72 (2011)
8. L. Chen, X. Yang, R. Wang and Y. Shen, Chinese Journal of Mechanical Engineering, Analysis on Design and Performances of Vehicle Passive Inerter-spring-damper Suspension Based on Dynamic Vibration Absorbers, **28**, 30-34 (2012)
9. J. Nie, X. Zhang, B. Hu and L. Chen, Vehicle and Power Technology, A New Approach of Vehicle Passive Suspension Techniques, **128**, 59-64 (2012)
10. F. Wang and H. Chan, IEEE Conference on Decision and Control, Mechatronic Suspension Design and Its Applications to Vehicle Suspension Control, 3770 (2008)
11. S. Evangelou, D. Limebeer, J. Sharp, and M. Smith, IEEE Control Systems Magazine, Control of Motorcycle Steering Instabilities, **10**, 81-85 (2006)

DOI: 10.1051/matecconf/20179513006

Error Compensation on Probe Parameters of Articulated Arm CMM

Dateng Zheng, Gang Zhao , Zhongyue Xiao, Zhiyang Luo, Taiping Zhou and Jing Zhang

Jinggangshan University College of Mechanical & Electrical Engineering, Ji'an, 343009, China

Abstract. The error compensations on probe parameters were proposed to improve the measurement accuracy of an articulated arm coordinate measuring machines (AACMM). An AACMM's probe parameters were analyzed and a measurement model including probe parameters solely was built. A simple cone socket components were used as a calibration tool and the LM algorithm was edited to solve the probe parameters. The results show that the average measurement error dealt with the probe parameters recalibration is 0.044mm, and its standard deviation is 0.050mm. They have increased by about 31% and 33% than those without recalibration.

1 Introduction

An articulated arm coordinate measuring machine (AACMM) [1] is one kind of CMMs. Comparing with conventional orthogonal CMM, its measurement precision is lower for its complicated structure in series and others. How to improve AACMM's measurement accuracy is an objective for researchers. Wang Pingping built a measurement model with 25 parameters [2]. Wang Bin built a measurement equation with 37 errors and simplified the model with structuring recombination spherical coordinate [3]. Shang Ping analyzed the error source roundly, built the measurement equation with 60 errors [4]. Zhu Lianqing built a measurement model with 27 parameters for partly-bonded articulated arm coordinate measuring machine [5]. Hu Yi and Jorge Santolaria studied the thermal deformation error, and they built the error models including thermal deformation error using different method, respectively [6-7]. Huang Kui established a kinematic model involving 6 degrees of freedom (DOF) parameters with local product of exponentials formula[8]. Wang Wen introduced a simulated annealing algorithm to identify the eccentricity parameter s of the AACMM [9]. In our previous study, we put forward a spatial error model based on functional networks and validated the new error model and the approximate algorithm on the results of precision measurement experiment [10], and adopted the SVM method to find a way to build the different error model[11]. The existing researches help to improve AACMM's measurement accuracy, but they don't consider the factors of probe parameters separately.

As a kind of portable coordinate measurement equipment, an AACMM's probe needs to be disassembled when it is carried to another job site. The probe sometimes needs to be replaced during use. The probe parameters from measurement model are changed when it is installed in these cases. So we must find an advantageous way to recalibrate these parameters. In this study, we discuss how to build the probe parameter model, to do sample experiments simply and to calibrate the probe parameters. It is significant and helpful for primary operators because the sample method for the probe parameters is simpler than those for the whole structural parameters, and particularly the former's calibration time is shorter.

2 Probe parameter model

AACMM analogs the human's articulated structure which is mainly mechanically composed of a base, an upright column, two measuring arms, six rotary joints with circular grating encoders and a probe, as shown in Fig. 1.

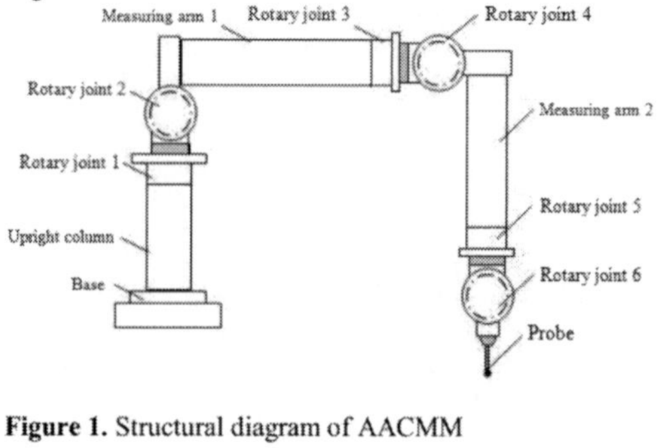

Figure 1. Structural diagram of AACMM

AACMM's measurement model is often built by the D-H (Denavit and Hartenberg) method [12], as shown in equation (1)

$$\begin{bmatrix} x_p \\ y_p \\ z_p \\ 1 \end{bmatrix} = \prod_{i=1}^{6} \begin{bmatrix} \cos(\theta_i + \theta_{i0}) & -\sin(\theta_i + \theta_{i0})\cos\alpha_i & \sin(\theta_i + \theta_{i0})\sin\alpha_i & l_i \cos(\theta_i + \theta_{i0}) \\ \sin(\theta_i + \theta_{i0}) & \cos(\theta_i + \theta_{i0})\cos\alpha_i & -\cos(\theta_i + \theta_{i0})\sin\alpha_i & l_i \sin(\theta_i + \theta_{i0}) \\ 0 & \sin\alpha_i & \cos\alpha_i & d_i \\ 0 & 0 & 0 & 1 \end{bmatrix} \begin{bmatrix} 0 \\ 0 \\ z_p \\ 1 \end{bmatrix} \tag{1}$$

Where θ_i (i=1,2,…,6) are joint variables, θ_{i0} (i=1,2,…,6) are joint zero errors; α_i(i=1,2,…,6) are torsional angles from joint axis to joint axis, including their perpendicularity errors; l_i (i=1,2,…,6) are lengths of rod pieces, including their errors; and d_i(i=1,2,…,6) are offsets of rod pieces, including their errors.

In this study, the AACMM's structural parameters are shown in table 1("–"denotes that the parameter cannot be calibrated).

Table 1. Calibration values of structural parameters

Parameter	θ_{i0} (rad)	α_i (rad)	l_i (mm)	d_i (mm)
Joint 1	—	-1.5704	29.3705	—
Joint 2	-2.7770	1.5731	-30.1459	-0.3681
Joint 3	-4.3312	-1.5711	27.1532	300.5980
Joint 4	-4.5854	1.5704	-27.0528	-0.4637
Joint 5	-2.7668	-1.5705	26.6343	278.6602
Joint 6	-0.9410	—	170.8003	-0.8118

When a probe is replaced and reinstalled, the probe parameters are changed subsequently. Apparently, the probe parameters which need to be calibrated are θ_{60}, l_6 and d_6. The Figure 2 made clear that the parameters of θ_{60}, l_6 and d_6 are before replace. The parameters of θ_{60}', l_6' and d_6' are after replace.

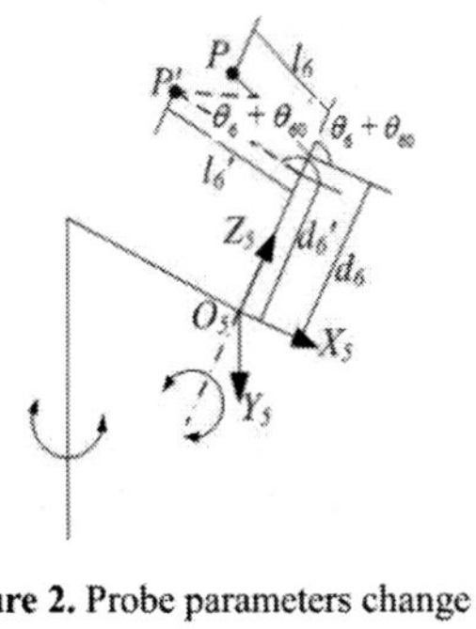

Figure 2. Probe parameters change sketch

The other parameters except the probe parameters in equation (1) are not changed only if the probe is replaced and reinstalled. So the probe parameter model is shown in equation (2).

$$\begin{bmatrix} x_p \\ y_p \\ z_p \\ 1 \end{bmatrix} = \prod_{i=1}^{5} \begin{bmatrix} \cos(\theta_i + \theta_{i0}) & -\sin(\theta_i + \theta_{i0})\cos\alpha_i & \sin(\theta_i + \theta_{i0})\sin\alpha_i & l_i \cos(\theta_i + \theta_{i0}) \\ \sin(\theta_i + \theta_{i0}) & \cos(\theta_i + \theta_{i0})\cos\alpha_i & -\cos(\theta_i + \theta_{i0})\sin\alpha_i & l_i \sin(\theta_i + \theta_{i0}) \\ 0 & \sin\alpha_i & \cos\alpha_i & d_i \\ 0 & 0 & 0 & 1 \end{bmatrix}$$
$$\times \begin{bmatrix} \cos(\theta_6 + \theta_{60}') & -\sin(\theta_6 + \theta_{60}')\cos\alpha_6 & \sin(\theta_6 + \theta_{60}')\sin\alpha_6 & l_6' \cos(\theta_6 + \theta_{60}') \\ \sin(\theta_6 + \theta_{60}') & \cos(\theta_6 + \theta_{60}')\cos\alpha_6 & -\cos(\theta_6 + \theta_{60}')\sin\alpha_6 & l_6' \sin(\theta_6 + \theta_{60}') \\ 0 & \sin\alpha_6 & \cos\alpha_6 & d_6' \\ 0 & 0 & 0 & 1 \end{bmatrix} \times \begin{bmatrix} 0 \\ 0 \\ z_p \\ 1 \end{bmatrix} \tag{2}$$

3 Calibration experiment

In order to make sample manipulate simply, we use the cone socket components as a calibration tool in this work, as shown in Fig. 3. They are mainly composed of a disc and a metallic rod. The disc has five threaded holes which are fixed the rods. One end of the rod has a standard tapered bore.

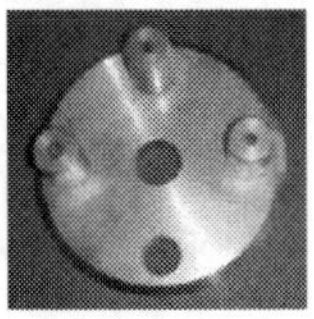

Figure 3. Cone socket components

Before sampling, we disassembled the AACMM's probe, and then assembled the same probe. We fixed the cone socket to a certain location. And then we used the AACMM's probe to detect the standard tapered bore as many postures as possible from various directions successively. Xi Si drew a conclusion that the max measurement error and standard deviation would tend to be stable when the samples reached 200 times in his master's thesis [13]. So we collected 200 sets of data to

DOI: 10.1051/matecconf/20179513006

calibrate the probe parameters (θ_{60}' , l_6' , d_6'). The calibration experiment is shown in Fig. 4.

Figure 4. Data collection in calibration experiment

4 Parameters calibration

The three probe parameters (θ_{60}', l_6', d_6') need to be calibrated to obtain precise values. We had to construct a calibration algorithm to solve the probe parameters after we collected 200 sets of samples. Many calibration algorithms had been widely studied by people all the time. CHEN Zhenyu presented a genetics-based calibration algorithm for the structural parameters of multi-joints 3D scanner [14]. WANG Xueying further presented an improved genetic algorithm for the parallel recombination simulated annealing [15]. Wang Pingping adopted the least-square parameters identification method to calibrate the AACMM's structure parameters with the standard provided by a high-accuracy orthogonal CMM [2]. GAO Guanbin presented a modified simulated annealing algorithm to identify the structural parameters of the AACMMs [16]. In terms of an algorithm's speed, effectiveness and stability, a comparative analysis was made to show that Levenberg-Marquardt(LM) algorithm as one of nonlinear least square method was the least time-consuming and produced the most stable outputs[17]. So we chose the LM calibration algorithm in this work.

One set of sample is the reading value (θ_1, θ_2, θ_3, θ_4, θ_5, θ_6) from circular grating encoders. We can calculate the coordinate value P(X, Y, Z) if the three probe parameters are calibrated. The cone socket is fixed in one location when we used the AACMM's probe to detect the standard tapered bore. So 200 coordinate values are theoretically same. The equation (3) can be built as follows.

$$\left\| P_i P_j \right\| = 0 (i \neq j) \tag{3}$$

It is easy to know that the equation (3) is not equal if the three probe parameters are not yet calibrated. But the left expression is equal to a minimal value, so the optimal object can be expressed as follows.

$$\min_{\theta_{60}', l_6', d_6' \in R} S(\boldsymbol{b}) = \sum_{i=1}^{n} \sum_{j=i}^{n} \left\| P_i P_j \right\|^2 = \left\| \boldsymbol{F}(\boldsymbol{b}) \right\|_2^2 \tag{4}$$

The LM calibration algorithm was applied in this work, the detailed solving steps were as follows.

(1) Choose an initial damping factor μ=μ00 and growth factor τ >1.

(2) Give an initial calibration parameter vector k^0, specify the accuracy ε>0, and set calculation step flag t=0.

(3) Calculate the calibration function $f^{t} = f^{(kt)}$, residual function $V^{t} = V^{(kt)}$, and $S^t = S^{(kt)} = (f^{t})$T (V^{t}), function $(f^t)^T$ is a transition matrix of f^t.

(4) Calculate Jacobian matrix A^t

$$A^t = [\frac{\partial f_u(k^t)}{\partial k^t}],$$

u=1,2,…,m, and calculate $(A^t)^T(A^t)$.

(5) Calculate

$$g^t = \nabla S(k^t) = (A^t)^T V^t.$$

(6) Solve equation to get pt:

$$((A^t)^T A^t + \mu_0^t I) p^t = -g^t.$$

(7) Calculate
$k^{t+1} = k^t + p^t$.

(8) Calculate
$V^{t+1}=V(k^{t+1})$, $S^{t+1}=S(k^{t+1})$.

(9) If $S^{t+1} \leq S^t$, then $u_0^t = u_0^t / \tau$, go to (6).

Test: if $S^{t+1} \leq S^t$, then $u_0^t = u_0^t / \tau$;

if $S^{t+1} \leq S^t$, then $u_0^t = u_0^t$.

(10) If $S^{t+1} > S^t$, then $u_{0n}^t = \tau^n u_0^t$.

Test: set n=1,2,…, in turn, and calculate u_{0n}^t respectively, go to (6).

If $S_{0n}^{t+1} \leq S^t$, then $u_0^t = u_{0n}^t$.

(11) Check termination criterion, if satisfy the criterion, then output k*=kt+1, and stop calculation; if not satisfy the criterion, then t = t+1, $k^t = k^{t+1}$, $S^t = S^{t+1}$, go to (3).

In this work, u_0^0=0.01 , τ = 3, ε = 0.01, m=200.

We edited the algorithm in MATLAB software to solve the probe parameters. The calibration results are shown in Tables 2 after 24 iterations.

Table 2. Probe parameter values before and after calibration

State	Probe parameter	value
	θ_{60}	-0.9410
Before calibration	l_6	170.8003
	d_6	-0.8118
	θ_{60}'	-0.9401
After calibration	l_6'	170.5342
	d_6'	-0.9231

5 Precision comparison

In order to verify calibration effect, we used an AACMM to measure a standard rod whose length is 100.2500 mm. We collected 100 sets of data(θ_1, θ_2, θ_3, θ_4, θ_5, θ_6). By substituting the parameters of Table 1 into equations (1) and the probe parameters (θ_{60}', l_6', d_6') of Table 2 after calibration into equations (2), we can simply compare the effect. The results are shown in Figure 5.

DOI: 10.1051/matecconf/20179513006

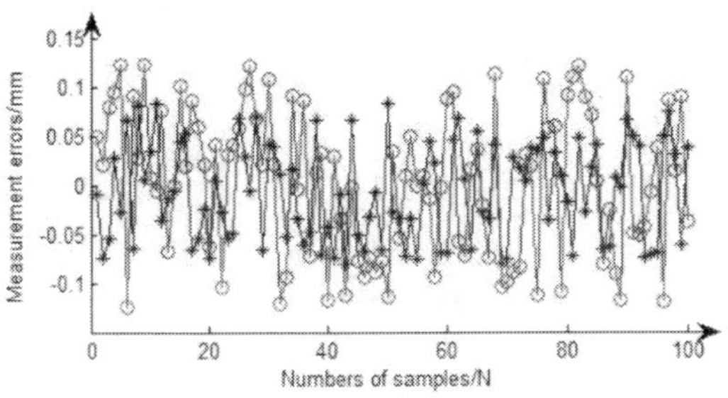

Figure 5. Measurement errors comparison

The average measurement error before the probe parameters are calibrated (red circles) is 0.064mm, and its standard deviation is 0.075mm. The average measurement error after the probe parameters are calibrated (blue stars) is 0.044mm, and its standard deviation is 0.050mm. It is clear that the measurement accuracy of AACMM dealt with the probe parameters recalibration is superior to the accuracy of no calibration.

6. Conclusions

We analyzed an AACMM's probe parameters and built its measurement model. We chose a simple cone socket components as a calibration tool and did a convenient calibration experiment. We edited the LM algorithm in MATLAB software to solve the probe parameters. The results show that the average measurement error dealt with the probe parameters recalibration is 0.044mm, and its standard deviation is 0.050mm.

Acknowledgements

We gratefully acknowledge the project 51265017 supported by National Natural Science Foundation of China and the project 20131828 by Ji'an Municipal Natural Science Foundation, which are the basis for the proposed achievements.

References

1. G.X. ZHANG. Aviation Precision Manufacturing Technology, **44**,3(2008).
2. P.P.Wang, Y.T.Fei, S.W. Lin. Journal of Xi'An Jiaotong University,**40**,3(2006).
3. B. WANG, S.G. Liu, X.Y. WANG. Measurement Technique, **5**,(2007).
4. P. Shang, Y.T.Fei. Tool Engineering, **8**,(2009).
5. L.Q. Zhu, W.X. Li, Z.K. Pan, et al. Chinese Journal of Scientific Instrument,**35**,3 (2014).
6. Y. Hu, Y.T.Fei, W.T. Cheng. JOURNAL OF MECHANICAL ENGINEERING,**47**,24(2011).
7. J. Santolaria, J.A.Yagüe, R.Jiménez, J.J.Aguilar. Precision Engineering,**33**,4(2009).
8. K. HUANG , J.H. Mo, L.H. Yu, et al. Journal of Xi'an Jiaotong University, **44**,8(2010).
9. W. WANG, K. Lin, G.B. Gao,Z.C. Chen. Optics and Precision Engineering, **18**,1(2010).
10. D.T. Zheng, Y.T.Fei. Journal of Mechnical Engineering, **46**,10(2010).
11. D.T. Zheng, Y.G. Hu. Measurement, **45**,3(2012).
12. L.D. Yu, W.T. Cheng, Y.T.Fei. Journal of University of Science and Technology of China, **39**,(2009).
13. S. Xi. Hefei University of technology,2013.
14. Z.Y. Chen, C.S. Fu, D.H. Li. Infrared and Laser Engineering, **32**,5(2003).
15. X.Y. Wang, S.G. Liu, H.T. Zhang, et al. Key Engineering Materials, **161**,(2008).
16. G.B. Gao, W. WANG, K. Lin, Z.C. Chen. Optics and Precision Engineering, **15**,8(2009).
17. W.T. Cheng, L.D. Yu, Y.T.Fei. Journal of University of Science and Technology of China, **41**,1(2011).

Temperature Insensitive Current-Mode Four Quadrant Multiplier Using Single CFCTA

Sunti Tuntrakool , Peerawut Suwanjan and Winai Jaikla

Department of Engineering Education, Faculty of Industrial Education, King Mongkut's Institute of Technology Ladkrabang, Thailand

Abstract. A four quadrant multiplier of two current input signals using active building block, namely current follower cascaded transconductance amplifier (CFCTA) is presented in this paper. The proposed multiplier consists of only single CFCTA without the use of any passive element. The presented circuit has low impedance at current input node and high impedance at current output node which is convenient for cascading in current mode circuit without the need of current buffer circuits. The output current can multiply two input currents with temperature insensitivity. Moreover, the magnitude of output current can be controlled electronically via DC bias current. With only single active building block, the presented multiplier is suitable for integrated circuit implementation for analog signal processing. Simulation results from a PSpice program are presented in order to demonstrate the multiplier proposed here.

1 Introduction

Emerging of semiconductor technology has provided a new integrated circuit (IC) for developing the new topologies for both analog and digital circuits. Multipliers are circuits that are useful in the signal processing system [1]. Also, the multiplier can be found in many fields, for examples, modulation and de-modulation system, neural networks, adaptive filter etc. The design of analog circuits has been focused in the use of active building block, because it provides the flexibility for designer in the term of minimum number of external passive elements. Especially, the realization of circuit using electronically controllable active building blocks cab avoid the use of external resistor and can be easily controlled by microcontroller or microprocessor.

The synthesis of analog circuit based on current-mode technique has been gained significant attention, because current-mode circuit offers certain advantages such as greater linearity, simple structure, better accuracy, higher-frequency operation and larger dynamic range compared with op-amps-based circuits [2]-[3]. To make the current mode circuit cascading without the use of any additional buffer, the input impedance should be low and the output impedance should be high. Many current mode multipliers using active building block have been reported in the literature for example second generation current controlled current conveyor CCCII [4]-[6], operational transconductance amplifier (OTA) [7], current controlled current differencing transconductance amplifier (CCCDTA) [8], current controlled current differencing buffered amplifier (CCCDBA) [9].

The contribution of this paper is to propose a current-mode four quadrant multiplier using CFCTA as the active element. The output current is the multiplication of two input currents with temperature sensitivity. The proposed multiplier exhibits low input impedance for all two input nodes and high output impedance. Moreover, the magnitude of output current can be electronically controlled vial DC bias current. The proposed circuit consists of single CFCTA without any passive element. Through simulation using BJT CFCTA also demonstrates improvements in terms of performance.

2 Principle of operation

2.1 Basic concept of Current Follower Cascaded Transconductance Amplifier (CFCTA)

The active building block, namely current follower cascaded transconductance amplifier (CFCTA) is used to design the proposed multiplier. The briefly review of this device is explained here. Fig. 1 shows the electrical symbol and equivalent circuit of CFCTA. The input current port, namely x port is low impedance. The output current z and z-copy (z_c) port are high impedance. Ideally, the output current i_z and i_{zc} is equal to input current i_x. The voltage at z port (v_z) is converted to be the output current at x_1 port through fist transconductance (g_{m1}) and the voltage at x_1 port is converted to be output current at x_2 port through second transconducatnce (g_{m2}). The impedance at port x_1 and x_2 is ideally high impedance. The transconductances g_{m1} and g_{m2} are electronically adjusted by DC bias current. The electrical characteristic of CFCTA is ideally written in equation (1) [10]

DOI: 10.1051/matecconf/20179514001

$$\begin{bmatrix} V_f \\ I_{z,z_c} \\ I_{x_1} \\ I_{x_2} \end{bmatrix} = \begin{bmatrix} 0 & 0 & 0 & 0 \\ 1 & 0 & 0 & 0 \\ 0 & \pm g_{m1} & 0 & 0 \\ 0 & 0 & \pm g_{m2} & 0 \end{bmatrix} \begin{bmatrix} I_f \\ V_z \\ V_{x1} \\ V_{x2} \end{bmatrix} \quad (1)$$

The g_{m1} and g_{m2} are written, respectively, as

$$g_{m1} = \frac{I_{B1}}{2V_T} \text{ and } g_{m2} = \frac{I_{B2}}{2V_T} \quad (2)$$

where V_T is the thermal voltage. I_{B1} and I_{B2} are the DC bias current which can be only positive. Fig. 2 shows the internal construction of BJT CFTA.

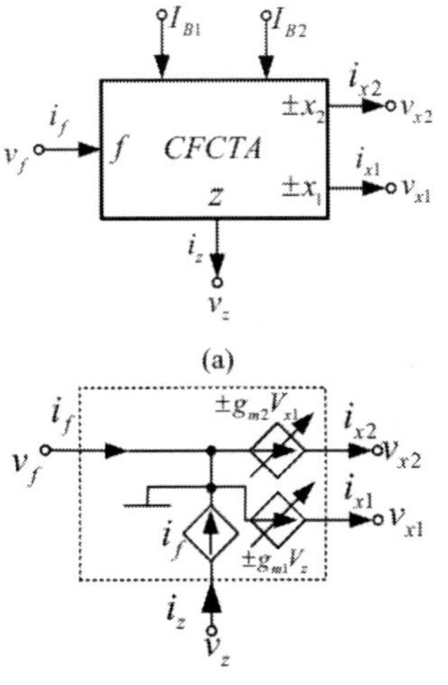

Figure 1. Circuit symbol and equivalent circuit of CFCTA.

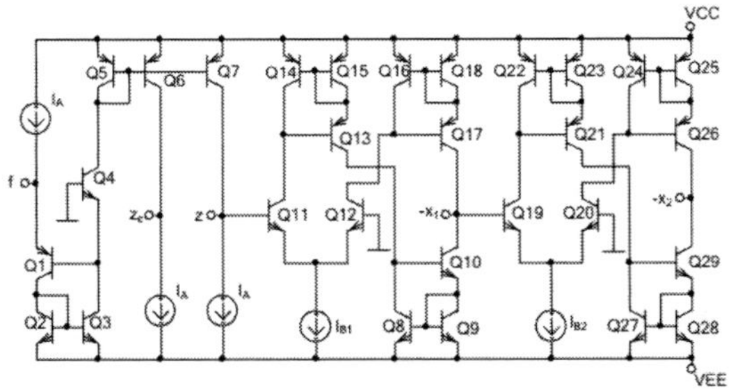

Figure 2. Internal construction of CFCTA [11]

2.2 Temperature insensitive current-mode four quadrant multiplier

Fig. 3 shows the proposed current-mode four quadrant multiplier. The proposed multiplier consists of only single CFCTA without the use of any passive element. It has low impedance at current input nodes i_{in1} and i_{in2}. Also the output current node i_{out} is high impedance which is ideal for cascade in current mode circuit without the use of buffer circuits. The DC current I_B is biased to control the g_{m1} and DC current I_B adding with i_{in2} is biased to control g_{m2}. So the g_{m1} and g_{m2} corresponding to bias current are respectively shown as

$$g_{m1} = \frac{I_B}{2V_T} \quad (3)$$

and

$$g_{m2} = \frac{I_B + i_{in2}}{2V_T} \quad (4)$$

It should be noted that i_{in2} can swing to be both positive and negative. However, the amplitude of i_{in2} should be lower or equal to DC bias current I_B. Considering the analysis of the circuit in Fig. 2, it yields the following output current

$$i_{out} = i_{in1}\left(1 - \frac{g_{m2}}{g_{m1}}\right) \quad (5)$$

Substituting Eq. (3) into Eq. (4), it yields

$$i_{out} = -\frac{i_{in1} i_{in2}}{I_B} \quad (6)$$

It should be noted from Eq. (6) that the output current i_{out} is the multiplication of i_{in1} and i_{in2} and the division by I_B with temperature insensitivity because the term of V_T is not appeared in output current. Moreover, the proposed circuit is the four quadrant multiplier because i_{in1} and i_{in2} can swing to be both positive and negative direction. Also the amplitude of the output current can be controlled electronically by means of DC bias current I_B.

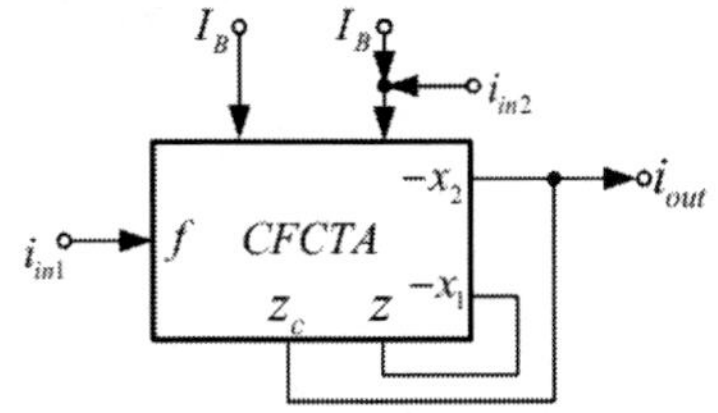

Figure 3. Proposed current mode four quadrant multiplier using single CFCTA

3 Simulation results

The proposed current multiplier was proved the performances by using the PSpice program. As illustrated in Fig. 2, the internal construction of CFCTA was constructed from BJT, the parameters of NPN and PNP transistors from the PR200N and NR200N bipolar transistors of ALA400 transistor array from AT&T were used [12]. The proposed multiplier was biased with ±2.5V supply voltages. The first simulation result shown in Fig. 4 (a) is the DC transfer characteristic of the multiplier where I_A=100μA, I_{B1}=I_{B2}=I_B=100μA. i_{in1} was sweep from -120μA to 120μA and i_{in2} was set in steps to -100μA, 50μA, -25μA, 0, 25μA, 50μA and 100μA. It is found that the input limitation of i_{in1} is not higher than value of DC bias current I_A. The DC transfer characteristic is also illustrated in Fig. 4 (b) where i_{in2} was sweep from -120μA to 120μA and i_{in1} was set in steps to -100μA, 50μA, -25μA, 0, 25μA, 50μA and

DOI: 10.1051/matecconf/20179514001

100μA. The amplitude modulation is used as application for proposed multiplier. It is found that the input limitation of i_{in2} is not higher than value of DC bias current I_B. The sinusoidal carrier signal with frequency of 10 kHz was applied to i_{in1} and the sinusoidal modulated signal with frequency of 1 kHz was applied to i_{in2}. The amplitude of both sinusoidal input signals were $160\mu A_{p\text{-}p}$. The results shown in figure 5 are the input signal and AM signal. Also, its output spectrum is shown in Fig. 6. The deviation of output current VS temperature variation is illustrated in Fig 7 where the temperature was sweep from 0-100 °C. It is found that output deviation is about -3.68% to 6.93%. This confirms the analysis in Eq. (6) that the output current slightly deviate correspondence to temperature variation. The output current for different value of I_B is shown in Fig. 8. In this case, the I_B was set to 100A, 150A and 200A. It is confirmed that the amplitude of output current can be electronically controlled by I_B as shown in Eq. (6). The deviation of total harmonic distortions (THD) against amplitude of input currents i_{in1} and i_{in2} are illustrated in Fig. 9 and 10, respectively.

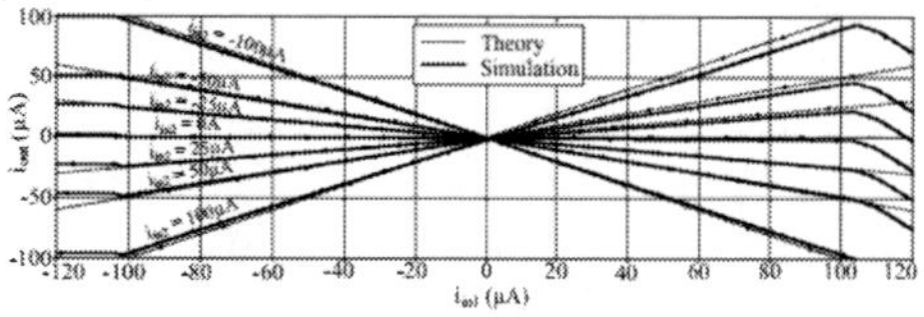

(a) i_{in1} sweep from -120μA to 120μA

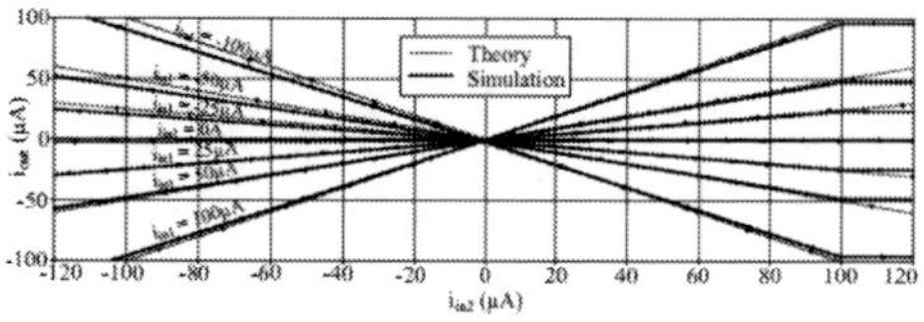

(b) i_{in2} sweep from -120μA to 120μA

Figure 4. DC transfer characteristic

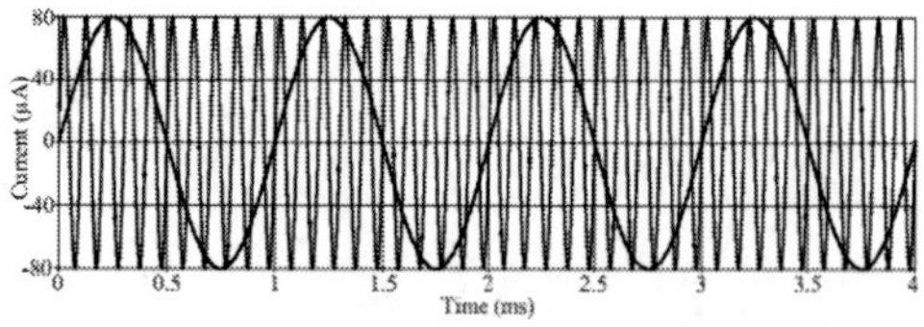

(a) Input currents of the multiplier

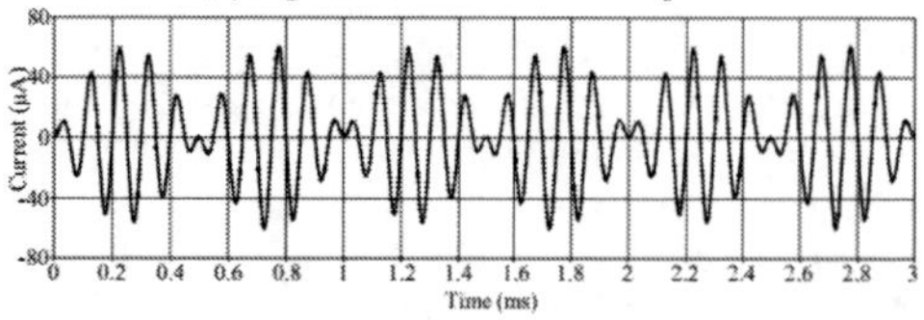

(b) AM output signal.

Figure 5. Application of proposed multiplier as amplitude modulation

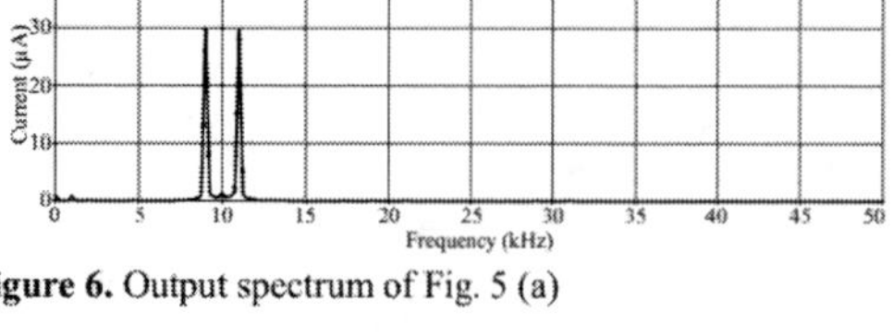

Figure 6. Output spectrum of Fig. 5 (a)

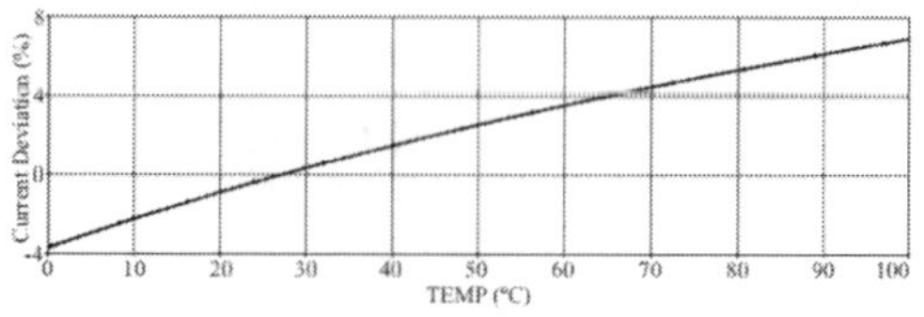

Figure 7. Output deviation VS temperature variation

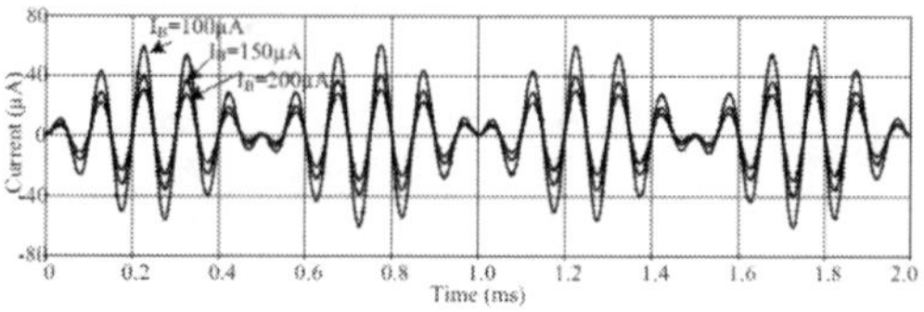

Figure 8. Output current for different value of I_B

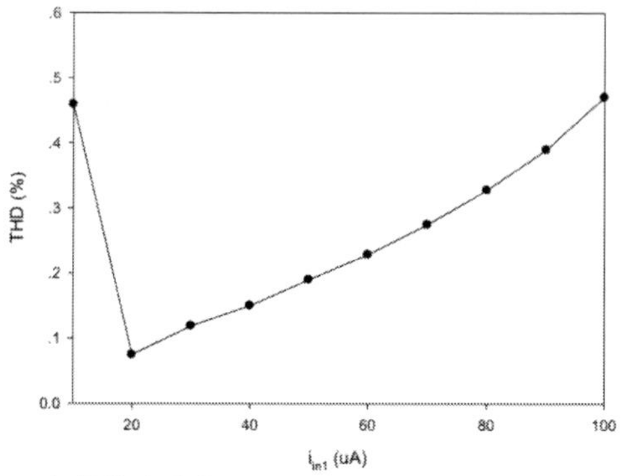

Figure 9. THD deviation against i_{in1} amplitude variation

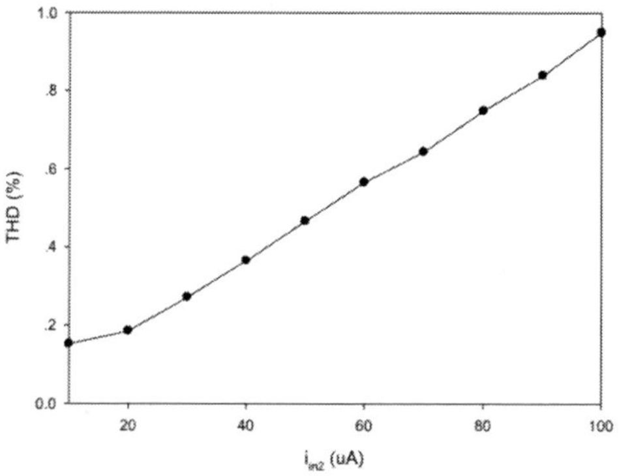

Figure 10. THD deviation against i_{in2} amplitude variation

4 Conclusion

This paper presents the current mode four quadrant multiplier. Single active building block used in this design is CFCTA. The output current is the multiplication of two input currents i_{ni1} and i_{in2} with temperature insensitivity. The output dynamic range depend on the DC bias current I_A (in CFCTA) and I_B (external bias current). The proposed multiplier doesn't require any passive element which is attractive for fabrication in chip. Moreover, the magnitude of the output current can be electronically adjusted via DC bias current I_B. The PSpice simulation results using BJT CFCTA based on parameters of PR200N and NR200N bipolar transistors of ALA400 agree well with theory expect.

Acknowledgement

Research described in this paper was financially supported from Faculty of Industrial Education, King Mongkut's Institute of Technology Ladkrabang (KMITL).

References

1. Ş.Ç. Yener, A. Uygur, H.H. Kuntman, "Ultra low-voltage ultra low-power memristor based band-pass filter design and its application to EEG signal processing," Analog Integrated Circuits and Signal Processing, (2016), Article in press
2. E. Yuce, "Current-mode electronically tunable biquadratic filters consisting of only CCCIIs and grounded capacitors," Microelectronics Journal. **40**, 719–1725, (2009).
3. G. Souliotis, "A current-mode automatic frequency tuning system for filters with current mirrors," International of Circuit Theory and Application. **38**, 591–606, (2010)
4. M. T. Abuelma'atti and M. A. Al-Qahtani, "A current-mode currentcontrolled current-conveyor-based analogue multiplier/divider," International Journal of Electronics, **85**, 71–77, (1998)
5. E. Yuce, "Design of a simple current-mode multiplier topology using a single CCCII+," IEEE Transactions on Instrumentation and Measurement. **57**, 631 - 637, (2008)
6. W. Petchakit, W. Kiranon, P. Wardkien, S. Petchakit, "A current-mode CCCII-based analog multiplier divider," International Conference on Electrical Engineering Electronics Computer Telecommunications and Information Technology (ECTI-CON), 221-224, (2010)
7. K. Kaewdang, C. Fongsamut, and W. Surakampontorn," Image processing using approximate datapath units," IEEE International Symposium on Circuits and Systems (ISCAS), 349-352, (2003)
8. M. Siripruchyanun and W. Jaikla, W. "A current-mode analog multiplier/divider based on CCCDTA," International Journal of Electronics and Communications (AEU), **62**, 223-227, (2008)
9. M. Siripruchyanun, and W. Jaikla "Current controlled current conveyor transconductance amplifier (CCCCTA): a building block for analog signal processing" Electrical Engineering, **90**, 443–453, (2008)
10. A. Chaichana and W. Jaikla, "Electronically tunable current-mode quadrature oscillator derived from first-oder allpass filter," 2014 European Modelling Symposium (EMS), 458 – 461, (2014)
11. W. Tangsrirat, "Novel current-mode and voltage-mode universal filter using single CFTA," Indian Journal of Enineering & Materials Sciences, **17**, 99-104, (2010)
12. D. R. Frey "Log-domain filtering: an approach to current-mode filtering," IEE Proc. Circuit Devices Syst., **140**, 406-416, (1993).

Universal Filter Using Single Commercially Available IC: LT1228

Surapong Siripongdee and Winai Jaikla

Department of Engineering Education, Faculty of Industrial Education, King Mongkut's Institute of Technology Ladkrabang, Thailand

Abstract. In electrical and electronic engineering, the filter is very importance circuit. A voltage-mode three inputs single output second-order multifunction filter is presented in this paper. The proposed filter is based on RLC parallel circuit. The main active element is the commercially available IC from Linear Technology Inc, LT1228. The filter consists of single LT1228, single resistor and two capacitors. Five output second-order filter responses, namely band-pass (BP), band-stop (BS), high-pass (HP), low-pass (LP) and all-pass (AP) functions are achieved. Each output response can be selected without component matching condition and the requirement of double gain amplifier. The quality factor and natural frequency can be electronically controlled. PSpice simulation results suing macro model of LT1228 are in close agreement with the theory.

1 Introduction

In electrical and electronic engineering, the biquadratic or second order filter has been important building blocks [1]. Also, this filter is the basic block to design high order filter. Especially, the second order multifunction filter which provides many filter responses in the same circuit has been gained significant attention and has become an interesting research topic. The multiple-inputs single-output (MISO) universal filter is the interesting one and has been continuously proposed. In case of voltage mode MISO, the selection of output voltage by switching on or off the input voltages should be done without the matching condition of passive and active elements. Moreover, the additional double gain amplifier should not be required.

In literature, there are a lot of MISO universal filters [2]-[15]. Unfortunately, these MISO universal filters have some drawbacks for example, need of matching condition, requirement of additional double gain amplifier, electronic uncontrollability, requirement of many active and passive element. The active elements used in some previous MISO universal filter are not commercially available.

The contribution of this paper is to propose a second order voltage-mode MISO multifunction filter using commercially available IC, LT1228 [16] as the active element. The LP, HP, BP, BS and AP responses are achieved without changing circuit topology. The matching condition of the passive and active element is not required. Also the additional double gain voltage amplifier doesn't need. The quality factor and natural frequency can be electronically tuned. It consists of single LT1228, single resistor and two capacitors. Through PSpice simulation using commercially available IC also demonstrates improvements in terms of performance.

2 Circuit configuration and analysis of the proposed filter

2.1 LT1228

Operational transconductance amplifier (OTA) has attracted considerable attention to design high performance electronics circuits. OTA is basically current mode devices and is used in high frequency applications. Mostly, its transconductance gain (g_m) can be electronically controlled which allows the OTA based circuits are controlled by microcontroller or microcomputer. Many commercial OTA ICs are available for example CA3080 from Intersil, LM13700 from National Semiconductor, MAX543 from MAXIM. The LT1228 from Linear Technology Inc is the interesting one. It is the combination of OTA and current feedback amplifier. The port name, electrical symbol and equivalent of LT1228 are illustrated in Fig. 1. The electrical characteristic is shown in Eq. (1).

$$\begin{pmatrix} I_{v+} \\ I_{v-} \\ I_y \\ V_x \\ V_w \end{pmatrix} = \begin{pmatrix} 0 & 0 & 0 & 0 \\ 0 & 0 & 0 & 0 \\ g_m & -g_m & 0 & 0 \\ 0 & 0 & 1 & 0 \\ 0 & 0 & 0 & R_T \end{pmatrix} \begin{pmatrix} V_+ \\ V_- \\ V_y \\ I_x \end{pmatrix} \quad (1)$$

where R_T is the transresistance gain. In an ideal case, R_T is typically very large and can be considered as infinite value. The g_m of LT1228 is controlled by DC bias current I_B as follows

DOI: 10.1051/matecconf/20179514002

$$g_m = 10 I_B \tag{2}$$

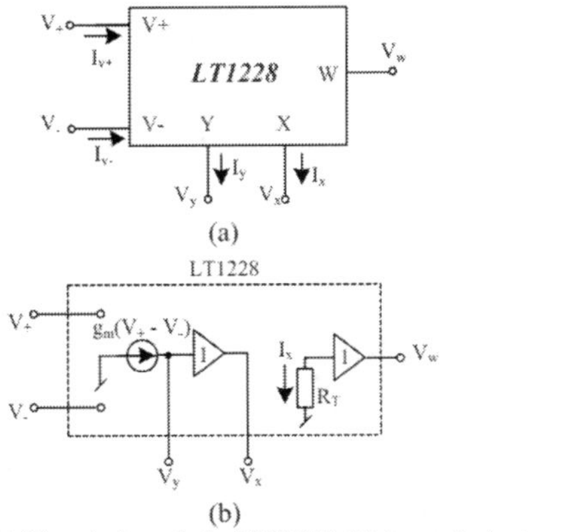

Figure 1. (a) Electrical symbol of LT1228 (b) its equivalent circuit.

2.2 Proposed three input single output voltage mode universal filter

The proposed filter is developed from passive RLC parallel circuit which consists of three input voltages, V_{in1}, V_{in2} and V_{in3} and single output voltage V_o. The filter constructed from single LT1228, single resistor and two capacitors as shown in Fig. 2. Based on RLC parallel circuit, the output voltage of the filter is obtained as

$$V_O = \frac{s^2 V_{in2} + s\frac{V_{in3}}{C_2 R} + \frac{g_m V_{in1}}{C_1 C_2 R}}{s^2 + \frac{s}{C_2 R} + \frac{g_m}{C_1 C_2 R}} \tag{3}$$

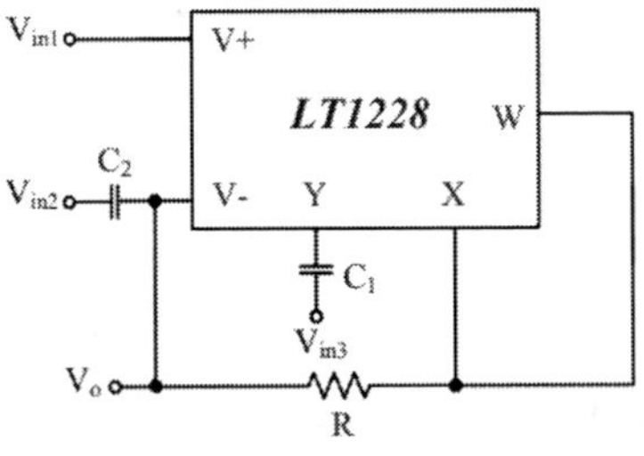

Figure 2. Three inputs single output voltage mode filter using single commercially available IC, LT1228.

From Eq. (3), a specific output filter responses can be selected by applying input voltage to node V_{in1}, V_{in2} and V_{in3} with following:

To obtain the band-pass (BP) function, the input voltage is applied at node V_{in3} while node V_{in1} and V_{in2} are grounded, then the voltage transfer function is obtained as

$$\frac{V_O}{V_{in}} = \frac{s\frac{1}{C_2 R}}{s^2 + \frac{s}{C_2 R} + \frac{g_m}{C_1 C_2 R}} \tag{4}$$

To obtain the band-stop (BS) function, the input voltage is applied at node V_{in1} and V_{in2}, while node V_{in3} is connected to ground, then the voltage transfer function is obtained as

$$\frac{V_O}{V_{in}} = \frac{s^2 + \frac{g_m}{C_1 C_2 R}}{s^2 + \frac{s}{C_2 R} + \frac{g_m}{C_1 C_2 R}} \tag{5}$$

To obtain the high-pass (HP) function, the input voltage is applied at node V_{in2}, while node V_{in1} and V_{in3} are connected to ground, then the voltage transfer function is obtained as

$$\frac{V_O}{V_{in}} = \frac{s^2}{s^2 + \frac{s}{C_2 R} + \frac{g_m}{C_1 C_2 R}} \tag{6}$$

To obtain the low-pass (LP) function, the input voltage is applied at node V_{in1}, while node V_{in2} and V_{in3} are connected to ground, then the voltage transfer function is obtained as

$$\frac{V_O}{V_{in}} = \frac{\frac{g_m}{C_1 C_2 R}}{s^2 + \frac{s}{C_2 R} + \frac{g_m}{C_1 C_2 R}} \tag{7}$$

To obtain the all-pass (AP) function, the input voltage is applied at node V_{in1}, V_{in2} and V_{in3}, however, the inverting amplifier with unit gain is required at not Vi_{n3}, then the voltage transfer function is obtained as

$$\frac{V_O}{V_{in}} = \frac{s^2 - \frac{s}{C_2 R} + \frac{g_m}{C_1 C_2 R}}{s^2 + \frac{s}{C_2 R} + \frac{g_m}{C_1 C_2 R}} \tag{8}$$

From Eqs. (4) to (8), the natural frequency (ω_0) and quality factor (Q) can be obtained as

$$\omega_0 = \sqrt{\frac{g_m}{C_1 C_2 R}} \text{ and } Q = \sqrt{\frac{g_m C_2 R}{C_1}} \tag{9}$$

If $g_m = 10I_B$ as written in Eq. (2) the natural frequency and quality factor of the proposed circuit are

$$\omega_0 = \sqrt{\frac{10 I_B}{C_1 C_2 R}} \text{ and } Q = \sqrt{\frac{10 I_B C_2 R}{C_1}} \tag{10}$$

It is found from Eq. (10) that the Q and ω_0 can be electronically adjusted. In case C_1=C_2, it is found that the natural frequency can be adjusted without affecting the quality factor as follows

$$\omega_0 = \frac{1}{C}\sqrt{\frac{10 I_B}{R}} \text{ and } Q = \sqrt{10 I_B R} \tag{11}$$

3 Simulation results

The PSpice simulation based on macro-model LT1228 IC was used to prove the performance of the filter. The proposed three-inputs single-output voltage-mode filter shown in Fig. 2 was bias with ±5 voltage supplies. It was designed with following active and passive element values: $C_1 = C_2 = 1nF$, $R=1k\Omega$ and $I_B = 100\mu A$. With this value of components, the expected value of natural frequency in Eq. (10) is 159.154k Hz, while the simulation value of natural frequency was 157.398 kHz. The deviation was about 1.103%. The error between expected and simulation value of natural frequency suffered from the parasitic elements and tracking error from input to output of LT1228. The gain and phase responses, band-pass (BP), band-stop (BS), high-pass (HP), low-pass (LP) and all-pass (AP) functions are respectively shown in Figs. 3 to 7. The selection of output responses was done as stated in section 2. This confirms that the proposed filter can provide five standard second order filter responses without changing the circuit topology. The test of transient response of the proposed band-pass response filter was done by applying the input as sinusoidal voltage signal with 80mVp and f = 160 kHz into node V_{in3}, while nodes V_{in1} and V_{in2} were grounded. The measurement of input and output voltage is shown in figure 8. The tuning of natural frequency with electronic method was tested by adjusting bias current IB as 100μA, 200μA and 400μA. With these bias current, the simulated natural frequencies were located at 157.398 kHz, 222.844 kHz and 314.775 kHz, respectively as shown in Fig. 9. It is found that not only the natural frequency is changed, but the quality factor is also changed according to the expecting in Eq. (10). However, the tuning of natural frequency can be done without affecting the quality factor by simultaneously adjusting value of C_1 and C_2 as described in Eq. (11). This confirms by the results in Fig. 10. By varying $C_1=C_2$ with different values of 0.1nF, 1nF and 10nF, the simulated natural frequencies were 769.13 kHz, 78.705 kHz and 7.88 kHz. In this case the other component values were set as $I_B=500\mu A$ and $R=20k\Omega$. It is found from Fig. 10 that the natural frequency can be tuned with constant value of quality factor.

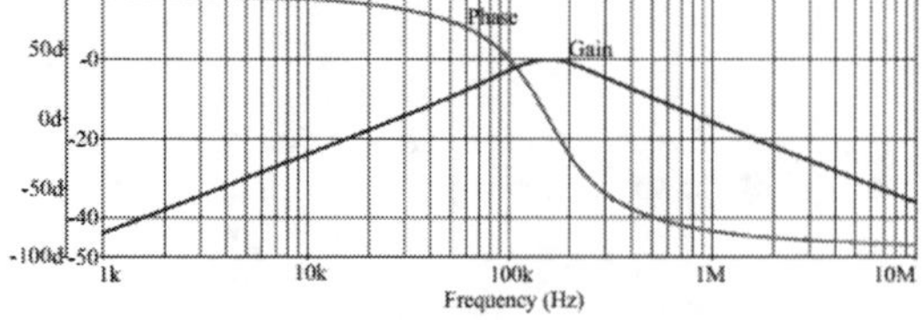

Figure 3. Simulated gain and phase responses for BP filter.

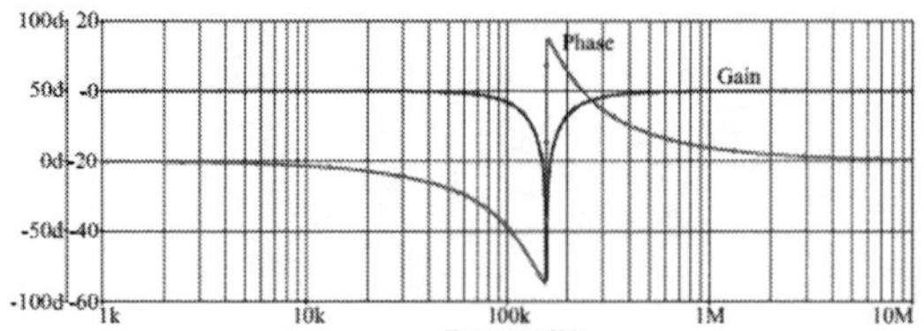

Figure 4. Simulated gain and phase responses for BS filter.

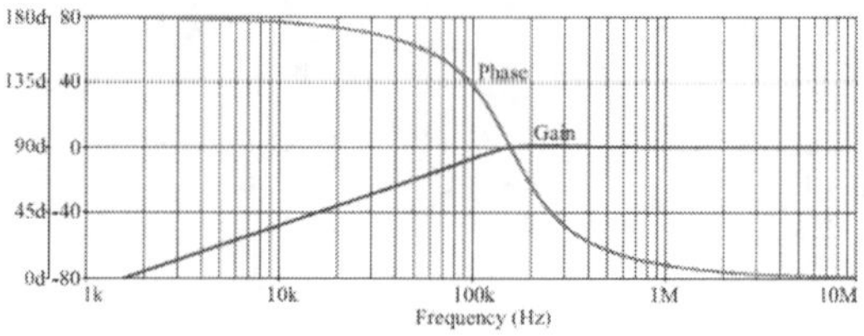

Figure 5. Simulated gain and phase responses for HP filter.

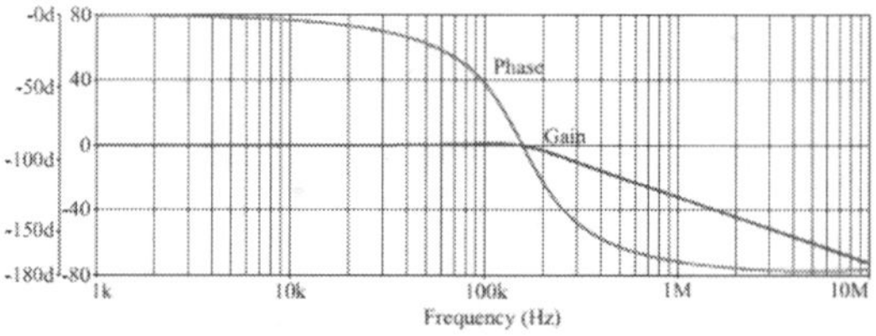

Figure 6. Simulated gain and phase responses for LP filter.

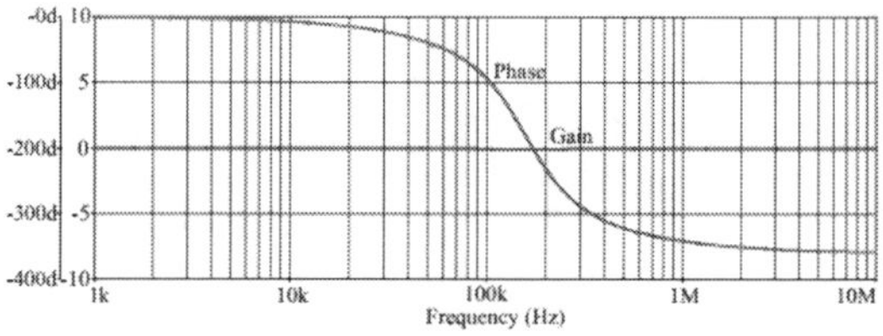

Figure 7. Simulated gain and phase responses for AP filter.

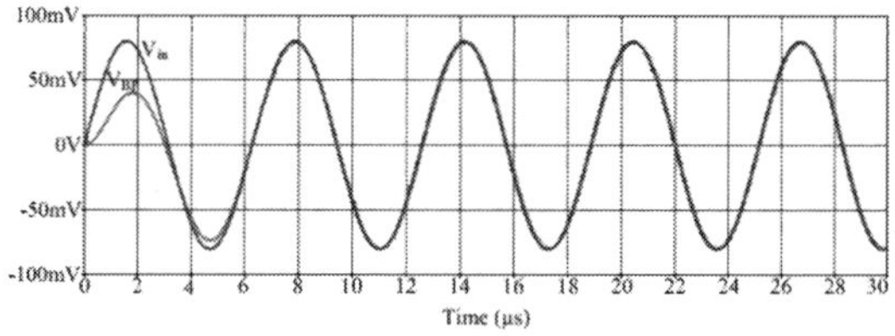

Figure 8. Transient response for BP function.

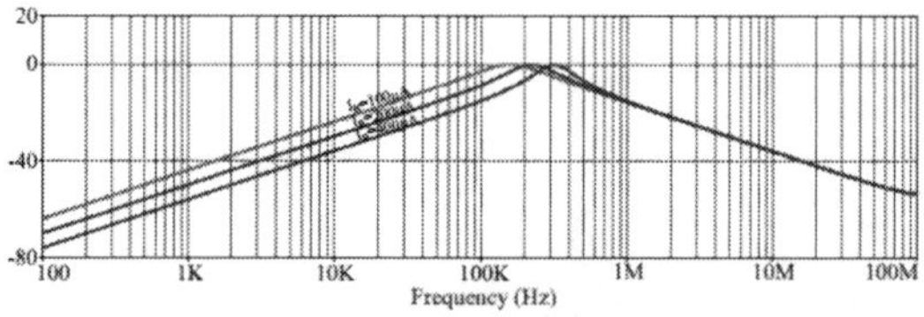

Figure 9. Simulated BP response for three different I_B values

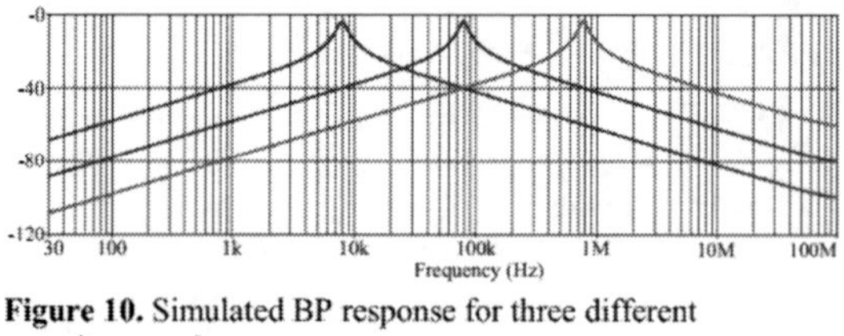

Figure 10. Simulated BP response for three different capacitance values.

4 Conclusion

In this paper, the three inputs single output voltage mode filter topology employing the commercially available IC, LT1228 is proposed. The presented filter employs only single LT1228, single resistor and two capacitors. It doesn't require passive or active component matching condition to obtain desired filter responses. Also, the selection of filter response can be done without the requirement of double gain amplifier. The natural frequency and quality factor can be tuned electronically. If value of C_1 and C_2 is simultaneously adjusted, the natural frequency can be tuned without affecting quality factor as analyzed in Eq. (10). The PSpice simulation results using macro-model of LT1228 are agree well with theory anticipation. The proposed filter is considered as simple circuit because it uses only single commercially available IC. However, the voltage buffer is required for cascading and the tune of quality factor cannot be adjusted without affecting the natural frequency.

Acknowledgement

Research described in this paper was financially supported from Faculty of Industrial Education, King Mongkut's Institute of Technology Ladkrabang (KMITL).

References

1. A.S. Sedra and K.C. Smith, "Microelectronic circuits," Florida: Holt, Rinehart and Winston 5 (2003)
2. J.W. Horng, "Voltage-mode universal biquadratic filter using two OTAs," Active and Passive Elec Comp. **27**, 85–89, (2004).
3. J.W. Horng, "High input impedance voltage-mode universal biquadratic filter using two OTAs and one CCII," Int J Electron. **90**, 185–191, (2003).
4. M. Sagbas and M. Koksal, "Voltage mode three input single output multifunction filters employing minimum number of components," Frequenz. **61**, 87–93, (2007).
5. C.M. Chang and H.P. Chen, "Single FDCCII-based tunable universal voltage-mode filter," Circuits Syst Signal Process. **24**, 221–227, (2005).
6. A. Ranjan and S.K. Paul, "Voltage mode universal biquad using CCCII," Active Passive Electron Compon. 1-5, (2011).
7. W. Tangsrirat, "Novel current-mode and voltage-mode universal biquad filters using single CFTA," Indian J Eng Mater Sci. **17**, 99–104, (2010).
8. C.M. Chang, H.P. Chen, "Universal capacitor-grounded voltage-mode filter with three inputs and a single output," Int J Electron, **90**, 401–406, (2003).
9. S. Kılınc, A.Ü. Keskin and C. Ugur, "Cascadable voltage-mode multifunction biquad employing single OTRA," Frequenz. **61**, 84–86, (2007).
10. J.K. Pathak, A.K. Singh and R. Senani, "New voltage mode universal filters using only two CDBAs," ISRN Electron, 1-6, (2013).
11. J.W. Horng, "High-input and low-output impedance voltage-mode universal biquadratic filter using DDCCs," IEEE Trans Circuits Syst II. **54**, 649–652, (2007).
12. N.A. Shah and M.A. Malik, "Voltage/current-mode universal filter using FTFN and CFA," Analog Integr Circuits Signal Process. **45**, 197–203, (2005).
13. H.P. Chen, "Voltage-mode FDCCII-based universal filters," Int J Electron Commun(AEU). **63**, 320–323, (2009).
14. W. Tangsrirat and O. Channumsin. "Voltage-mode multifunctional biquadratic filter using single DVCC and minimum number of passive elements," Indian J PureAppl Phys, **49**, 703–707, (2011).
15. W. Ninsraku, D. Biolek, W. Jaikla, S. Sriripongdee and P. Suwanjan, "Electronically controlled high input and low output impedance voltage mode multifunction filter with grounded capacitors," Int J Electron Commun(AEU), **68**, 1-8, (2014).
16. LT1228 – 100 MHz Current Feedback Amplifier with DC Gain Control, Linear Technology Corporation, www.linear.com

Multifunction Voltage-Mode Filter Using Single Voltage Differencing Differential Difference Amplifier

Amornchai Chaichana , Surasak Sangyaem and Winai Jaikla

Department of Engineering Education, Faculty of Industrial Education, King Mongkut's Institute of Technology Ladkrabang, Thailand

Abstract. In this paper, a voltage mode multifunction filter based on single voltage differencing differential difference amplifier (VDDDA) is presented. The proposed filter with three input voltages and single output voltage is constructed with single VDDDA, two capacitors and two resistors. Its quality factor can be adjusted without affecting natural frequency. Also, the natural frequency can be electronically tuned via adjusting of bias current. The filter can offer five output responses, high-pas (HP), band-pass (BP), band-reject (BR), low-pass (LP) and all-ass (AP) functions in the same circuit topology. The output response can be selected by choosing the suitable input voltage without the component matching condition and the requirement of additional double gain voltage amplifier. PSpice simulation results to confirm an operation of the proposed filter correspond to the theory.

1 Introduction

In analog image processing system, the filter is the main circuit that is used to separate the desired signal from undesired ones. Especially, the versatile filter called as multifunction filter has been gained significant attention and has become an interesting research topic. Because the multifunction filter can provide several filter responses in the same circuit topology [1]. The multiple inputs single output (MISO) multifunction filter belongs to popular filter structures. An important feature of this structure is the generation of several output transfer functions, i.e., high-pas (HP), band-pass (BP), band-reject (BR), low-pass (LP) and all-ass (AP) with single output node while the output response can be selected by suitable input signals. The popular way to select the input signal is to use the digital or electronic switch which is easily controlled by microcomputer or microcontroller. So, the MISO filter structure should not require the double input signal to avoid the addition al amplifier.

The voltage differencing differential difference amplifier (VDDDA) is the interesting active building block. This active device consists of three subtraction /addition voltage signal terminals with electronically tunable transconductance. With this feature, it can be found the VDDDA based analog circuits in the literature for examples, first order voltage mode filter [2], sinusoidal oscillator [2], inductance simulator [4], single input multiple output voltage mode filter [5]-[6], multiple input multiple output voltage mode filter [7] etc. Recently, the VDDDA based three inputs single output voltage mode multifunction filter [8] was published. This filter is attractive because it is minimum number of active element consisting of single VDDDA, single resistor and two capacitors. The natural frequency and quality factor can be electronically controlled. However, the inverting double gain voltage amplifier is required for getting the all-pass response.

In this paper, a voltage mode multifunction filter based on single voltage differencing differential difference amplifier (VDDDA) is proposed. The filter with three input voltages and single output voltage is constructed with single VDDDA, two capacitors and two resistors. Its quality factor can be adjusted without affecting natural frequency. Also, the natural frequency can be electronically tuned via adjusting of bias current. The filter can offer five output responses, high-pas (HP), band-pass (BP), band-reject (BR), low-pass (LP) and all-ass (AP) functions in the same circuit topology. The output response can be selected by choosing the suitable input voltage without the component matching condition and the requirement of additional double gain voltage amplifier. PSpice simulation results to confirm an operation of the proposed filter correspond to the theory.

2 Circuit configuration and analysis of the proposed filter

2.1 VDDDA

In order to understand the proposed filter, the brief details of the active building block, VDDDA are given in this section. The symbolic representation of VDDDA is shown in Fig. 1(a). It has five input voltage terminals, called as V+, V-, Z, P and N ports. The Z port is also the output current terminal and the W port is the output voltage. Ideally, the impedance at V+, V-, Z, P and N ports are assumed to be infinite while the impedance at W port exhibits low. The equivalent circuit is illustrated in Figure 1(b). In this design, the VDDDA is constructed from CMOS technology as illustrated in Figure 1(c). The idealization of VDDDA's parameters implies that

$$I_{V+} = I_{V-} = I_{V_N} = I_{V_P} \tag{1}$$

$$I_Z = g_m\left(V_{V+} - V_{V-}\right) \tag{2}$$

$$V_W = V_Z - V_{V_N} + V_{V_P} \tag{3}$$

The transconductance gain of VDDDA is proportional to bias current I_B and is given by

$$g_m = \sqrt{I_B \mu C_{ox} (W/L)_{15,16}} \tag{4}$$

"where μ is the mobility of the carrier for NMOS transistors, M15 and M16 in Fig. 1(c), C_{ox} is the gate oxide capacitance per unit area, W and L are the channel width and channel length, respectively." [9]

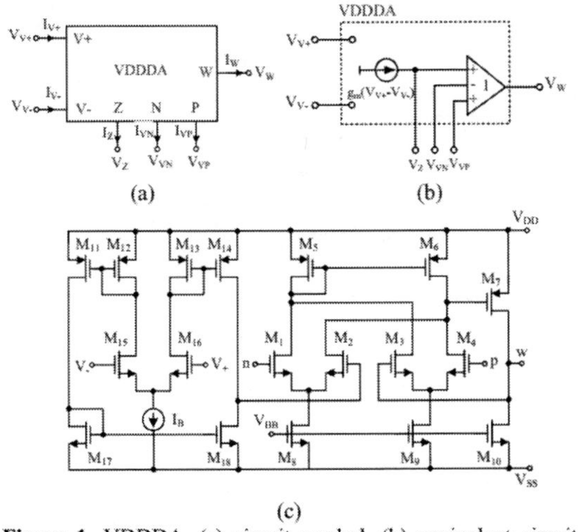

Figure 1. VDDDA, (a) circuit symbol, (b) equivalent circuit and (c) schematic diagram of CMOS VDDDA

2.2 Proposed three input single output voltage mode universal filter

Fig. 2 shows the proposed three input single output (TISO) filter. It requires only one VDDDA, two capacitors and two resistors. There are three input voltages, V_{in1}, V_{in2}, V_{in3} and single output voltage, V_o. Considering the circuit in Fig. 2, it yields the following output voltage

$$V_O = \frac{s^2 V_{in2} + s\dfrac{1}{C_1 R_1}\left(1+\dfrac{R_1}{R_2}\right)V_{in3} + \dfrac{g_m V_{in1}}{2C_1C_2R_1}}{s^2 + s\dfrac{1}{C_1 R_1}\left(1+\dfrac{R_1}{R_2}\right) + \dfrac{g_m}{2C_1C_2R_1}} \tag{5}$$

It can be noted that the proposed circuit is the unity gain filter and it can realize various filter configurations with the following input voltage selections

- The second-order LP response can be obtained if $V_{in1} = V_{in}$ (connected to the input voltage source) and $V_{in2} = V_{in3} = 0$ (grounded).
- The second-order HP response can be obtained if $V_{in2} = V_{in}$ (connected to the input voltage source) and $V_{in1} = V_{in3} = 0$ (grounded).
- The second-order BP response can be obtained if $V_{in3} = V_{in}$ (connected to the input voltage source) and $V_{in1} = V_{in2} = 0$ (grounded).
- The second-order BR response can be obtained if $V_{in1} = V_{in2} = V_{in}$ (connected to the input voltage source) and $V_{in3} = 0$ (grounded).
- The second-order AP response can be obtained if $V_{in1} = V_{in2} = -V_{in3} = V_{in}$ (connected to the input voltage source).

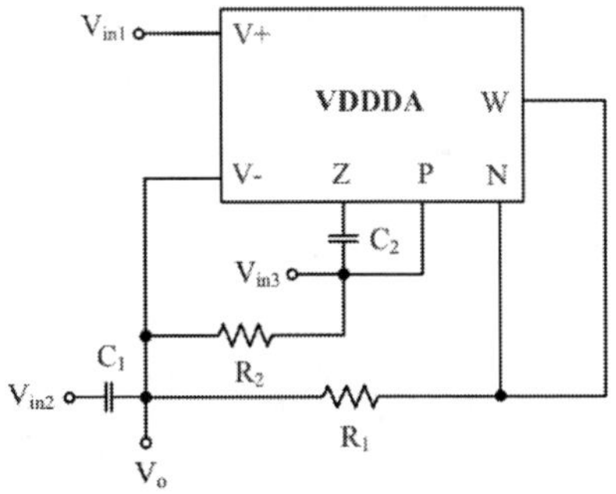

Figure 2. VDDDA based voltage mode TISO filter.

Considering to the denominator in Eq. (5), the natural frequency (ω_0) and quality factor (Q) can be give as

$$\omega_0 = \sqrt{\frac{g_m}{2C_1C_2R_1}} \text{ and } Q = \frac{1}{1+\dfrac{R_1}{R_2}}\sqrt{\frac{g_m C_1 R_1}{2C_2}} \tag{6}$$

It should be noted from Eq. (6) that the quality factor can be tuned by R_2 without affecting natural frequency. Also, the natural frequency can be electronically tuned by I_B.

2.3 Analysis of non-ideal case

The non-ideal properties of VDDDA will affect the performance of the proposed filter. In this section, the non-ideal parameters will be studies and analyzed. The first one is the voltage tracking errors from Z, N and P ports to W port. The non-ideal characteristic of VDDDA is written as

$$V_w = \beta_z V_z - \beta_n V_{vn} + \beta_p V_{vp} \tag{7}$$

where β_z, β_n and β_p are the voltage tracking errors from Z, N and P ports to W port, respectively. In this case, the output voltage will be change to

$$V_o = \frac{s^2 V_{in2} + \frac{s V_{in3}}{C_1 R_1}\left[\left(\frac{\beta_z + \beta_p}{1+\beta_n}\right) + \frac{R_1}{R_2}\right] + \frac{V_{in1} g_m \beta_z}{C_1 C_2 R_1 (1+\beta_n)}}{s^2 + \frac{s}{C_1 R_1}\left[\left(\frac{\beta_z + \beta_p}{1+\beta_n}\right) + \frac{R_1}{R_2}\right] + \frac{g_m \beta_z}{C_1 C_2 R_1 (1+\beta_n)}} \quad (8)$$

From Eq. (8), the parameters ω_0 and Q are given as

$$\omega_0 = \sqrt{\frac{g_m \beta_z}{C_1 C_2 R_1 (1+\beta_n)}} \quad (9)$$

and

$$Q = \frac{1}{\left(\frac{\beta_z + \beta_p}{1+\beta_n}\right) + \frac{R_1}{R_2}} \sqrt{\frac{\beta_z g_m C_1 R_1}{(1+\beta_n) C_2}} \quad (10)$$

Considering the nominator in Eq. (8), the voltage tracking errors will affect the gain and filter response. When consider the denominator in Eq. (8), the tracking errors will affect the natural frequency and the quality factor as shown in Eqs. (9) and (10).

3 Simulation results

The proposed filter is evaluated using PSpice simulator tool. The internal construction of VDDDA is based on CMOS technology in Fig. 1(c). The NMOS and PMOS transistor models are adopted from the Taiwan semiconductor manufacturing company (TSMC) 0.25μm CMOS process parameters at ±1.25 supply voltages. The aspect ratios of the MOS transistors are listed in Table I. The value of capacitors chosen as $C_1=C_2=80$pF. The value of resistors chosen as $R_1=2$kΩ, $R_2=10$kΩ, and the input bias currents I_B is set to 100μA. Fig. 3 depicts the gain responses of low-pass, high-pass and band-pass responses. The gain and phase responses of band-reject function are illustrated in Fig. 4. The gain and phase responses of all-pass function are illustrated in Fig. 5. The simulated natural frequency is 1.074 MHz. With the same values of resistances, capacitances and bias currents stated above, the theoretical natural frequency in Eq. (6) is 1.033 MHz. It is found that the deviation of theoretical and simulated value is about 3.96%. This deviation is caused by the voltage tracking error in VDDDA as analyzed in Section 2.3.

Table 1. Aspect ratios of mos transistors [9]

Transistor	W (μm)	L (μm)
M1-M4	1	0.25
M5-M7	15	0.25
M8-M10	3	0.25
M11-M16	5	0.25
M17-M18	3	0.25

The transient response of V_{BP} is shown in Fig. 6. In order to control the ω_0 with electronic method, the ω_0 tuning is confirmed via the BP response in Fig. 7. By using $C_1=C_2=80$pF and varying I_B with different values of 40μA, 100μA and 250μA, the simulated natural frequencies are 883.08kHz, 1.074MHz and 1.256MHz, respectively. In order to control the quality factor without affecting the ω_0, the quality factor tuning is confirmed via the BP response in Fig. 8. By using $C_1=1$nF, $C_2=50$pF, $I_B=50$μA, $R_1=2$kΩ and varying R_2 with different values of 1kΩ, 2kΩ and 4kΩ, the simulated quality factor are 1.47, 2.71 and 3.52, the theoretical natural frequency in Eq. (4) are 1.67, 2.73 and 3.75, respectively.

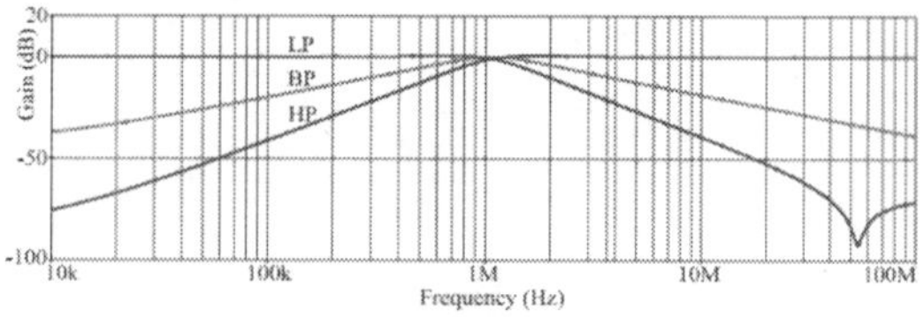

Figure 3. Gain responses for LP, HP and BP functions.

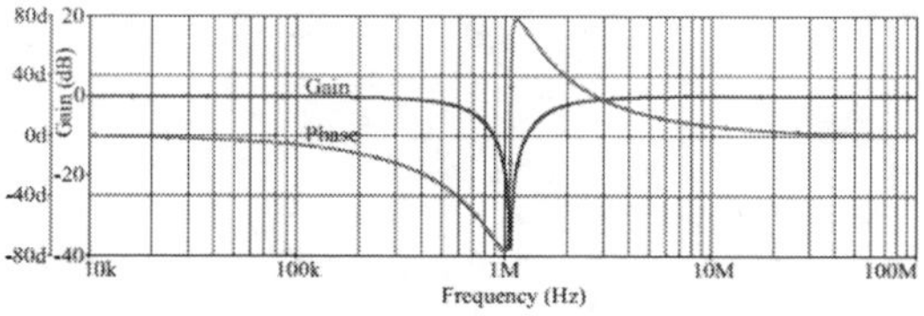

Figure 4. Gain and phase responses for BR function.

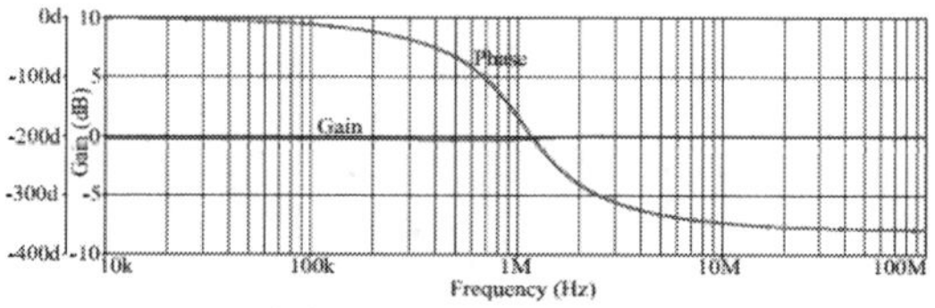

Figure 5. Gain and phase response for AP function.

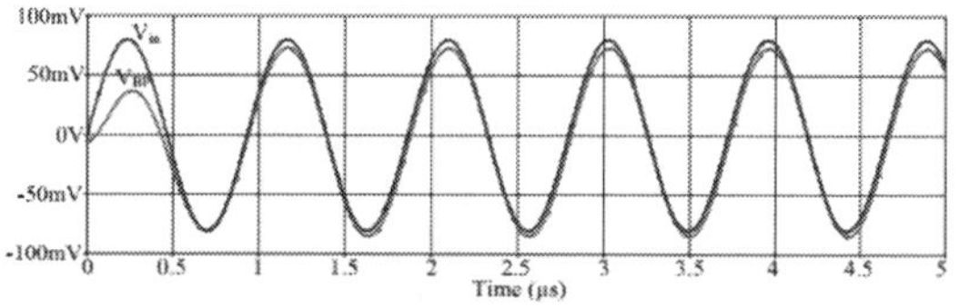

Figure 6. Transient response for BP function.

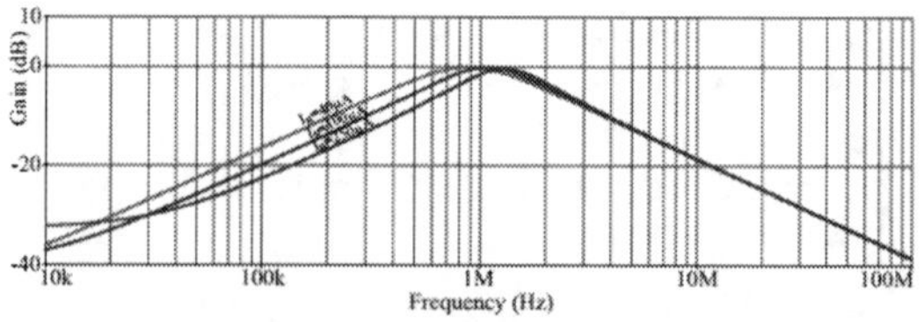

Figure 7. BP responses for different values of I_B.

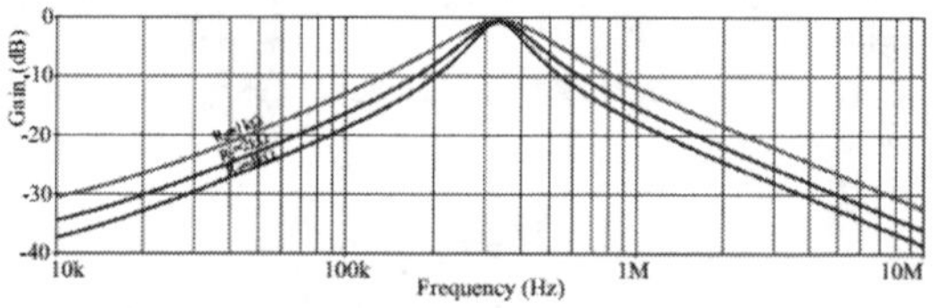

Figure 8. Quality factor responses for different values of R_2.

4 Conclusion

In this work, a multifunction filter with three-input and single-output voltage is realized using active building block, namely voltage differencing differential difference amplifier (VDDDA). The proposed filter consists of single VDDDA, two resistors and two capacitors. The proposed filter can provide five standard functions, all-pass, band-pass, high-pass, band-reject and low-pass responses without changing circuit topology. The selection of input voltages to get output filter responses can be done without the need of component matching condition and double gain amplifier. The natural frequency and quality factor can be tuned electronically. Also, the tune of quality factor can be done without affecting the natural frequency. The proposed filter is suited for fabricating into monolithic chip for low power low voltage application. To evaluate the proposed filter, we implement the circuit based on CMOS VDDDA in 0.25µm TSMC CMOS technology. The simulation results show excellent performances according theoretical expects.

Acknowledgement

Research described in this paper was financially supported from Faculty of Industrial Education, King Mongkut's Institute of Technology Ladkrabang (KMITL).

References

1. D. Biolek, R. Senani, V. Biolkova and Z. Kolka, "Active elements for analog signal processing, classification, review and new proposals," Radioengineering, **17**(4), pp. 15-32 (2008)
2. N. Herensar, R. Sotner, B. Metin, J. Koton and K. Vrba, "VDDDA - New 'voltage differencing' device for analog signal processing," Proceeding of the 8th International Conference on Electrical and Electronics Engineering. Bursa, Turkey, pp. 17-20, (2013)
3. A. S. Sedra and K. C. Smith, "Microelectronic circuit", Florida: Holt, Rinehart and Winston, (1991)
4. P. Auttaphut, "Active parallel-resistor-inductor with electronic controllability for analog signal processing", Far East Journal of Electronics and Communications, **14**(2), 93-103, (June 2015)
5. S. Chaimongkol and W. Jaikla, "A Voltage-Mode VDDDA-Based Universal Filter", International Symposium on Multimedia and Communication Technology, Sep 23 – 25, Classic Kameo Hotel, Ayutthaya, Thailand, (2015)
6. J. Koton, N. Herensar, K. Vrba and B. Metin, "Voltage-mode multifunction filter with mutually independent Q and ω0 control feature using VDDDAs," Analog Integrated Circuits and Signal Processing, **81**(1), pp 53-60 (2014)
7. S. Soisang, K. Jirasereemomkul, W. Jaikla, and K. Higuchi, "Voltage – Mode Multifunctional Biquadratic Filter Using VDDDA", Applied Mechanics and Materials, **781**, pp. 155-159, (2015)
8. N. Herensar, O. Cicekoglu, R. Sotner, J. Koton and K. Vrba, "New resistorless tunable voltage-mode universal filter using single VDIBA," Analog Integrated Circuits and Signal Processing, **76**(2), pp. 251-260 (2013)
9. S. Siripongdee and W. Jaikla, "Single VDDDA-based voltage-mode mutifunction second order filter for analog signal processing" ICIIBMS 2015, Track1: Signal Processing, Computer Networks and Telecommunications, Okinawa, Japan (2015)
10. P. Prommee, K. Angkeaw, M. Somdunyakanok, and K. Dejhan, "CMOS-based near zero-offset multiple inputs max–min circuits and its applications," Analog Integrated Circuits and Signal Processing, **61**, pp. 93–105 (2009)

 DOI: 10.1051/matecconf/20179514004

Analysis of the Functionality of the Transformer Ripple Filter for High-power DC Supplies

Lokvenc Jaroslav, Drtina Rene and Sedivy Josef

Department of Technical Subjects, Faculty of Education, University of Hradec Kralove, Rokitanskeho 62, 500 03 Hradec Kralove, Czech Repubilc

Abstract. In the field of DC power supplies stabilized classic analog circuits, which are mainly in the last twenty years has been increasingly squeezed by switching technology and digital control. Implementation of the research project stabilized linear power supply to heavy traffic conditions, which is the primary requirement for high operational reliability, low noise voltage at the output of the power supply and the high overload capacity, bring DC power supply transformer developed with us compensate for output ripple. Article presents analysis of the Functionality of the transformer ripple filter for DC power supplies.

1 Introduction

Each solution DC power supplies (analog/switching) has its advantages and disadvantages [1]. Standard analog sources usually tolerate temporary overload of 10-15% over several minutes. In industrial applications, it is then necessary for the design resources into account not only the nature of the load and possible overloading (short term and over 100%), but also the operating conditions in the environment (dust, humidity, temperature fluctuations, condensation, aggressive environment). The aim of the research project was the realization of a functional sample stabilized linear power supply to heavy operating conditions, the output current-voltage characteristic would be comparable to current-voltage characteristic of lead-acid battery.

Our proposed DC power supplies are designed for applications that do not mind a slight variation of the output voltage with the load, but the primary requirement is high operational reliability, low noise voltage at the output of the power supply and a substantial (and long term) capability. For filtering the output voltage using DC power supplies developed by us compensation transformer output ripple. DC power supplies are designed for heavy-duty industrial applications, are composed almost exclusively of passive elements and use conventional cooling.

2 Design of transformer ripple filter

Motion power filter compensation transformer are first published in connection with the construction of the single-phase linear unconventional sources [2] (Fig. 1) and then modified for three-phase linear high-power DC supply [3] (Fig. 2). The proposed solution does not use conventional technology or lower bandpass filters, but subtraction disruptive ripple from power.

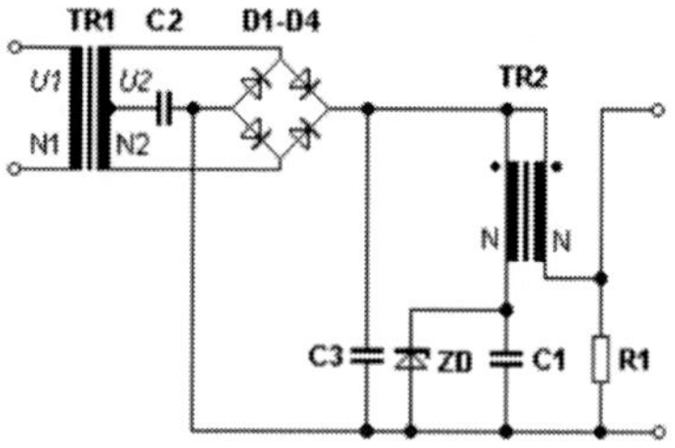

Figure 1. One-phase DC supply (by [1]).

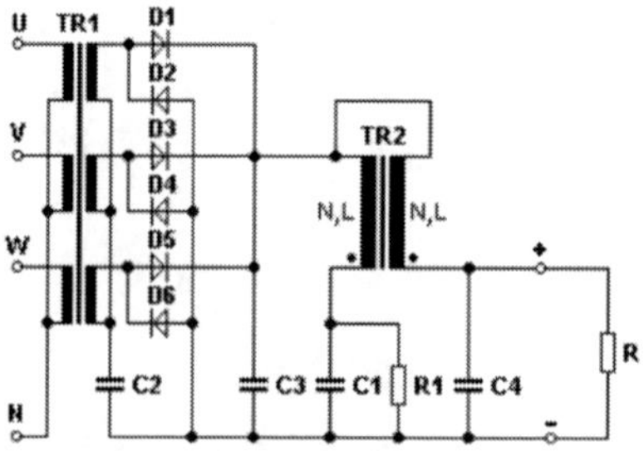

Figure 2. Tree-phase DC supply (by [2]).

The voltage at the output single-phase rectifier is by Fourier-shaped

$$U(\omega t)=\frac{4U_{\mathrm{m}}}{\pi}\cdot\left(\frac{1}{2}-\frac{1}{3}\cos 2\omega t-\frac{1}{3\cdot 5}\cos 4\omega t-\right.$$
$$\left.-\frac{1}{5\cdot 7}\cos 6\omega t-\frac{1}{7\cdot 9}\cos 8\omega t\ldots\right) \quad (1)$$

and the amplitude of the second harmonic $U_{2\mathrm{h}}$

$$U_{2hm} = \frac{4U_m}{\pi} \cdot \frac{1}{3} \quad (2)$$

The voltage at the output of a three-phase rectifier according to Fourier shaped

$$U(\omega t) = \frac{3U_m\sqrt{3}}{\pi} \cdot \left(1 - \frac{2}{5 \cdot 7}\cos 6\omega t - \frac{2}{11 \cdot 13}\cos 12\omega t - \frac{2}{17 \cdot 19}\cos 18\omega t - \ldots\right) \quad (3)$$

and the amplitude of the sixth harmonic (with respect to power frequency) U_{6hm}

$$U_{6hm} = \frac{3U_m\sqrt{3}}{\pi} \cdot \frac{2}{5 \cdot 7} \quad (4)$$

Compensation transformer is compared to the current practice designed for a closed magnetic core, which is a DC power supply output current source. Output current source I_{DC} calls the core of the transformer TR2 compensating coil with inductance L_{DC} induction sizes

$$B_{DC} = \frac{\mu_0 \mu_r N I_{DC}}{l_s} \quad (5)$$

and magnetizing current I_μ

$$I_\mu = \frac{U_{2hm}}{4\pi f L} \text{ (for single-phase supply)} \quad (6a)$$

or

$$I_\mu = \frac{U_{6hm}}{12\pi f L} \text{ (for three-phase source)} \quad (6b)$$

Aggregate peak magnetic flux density in the core must be at maximum current satisfy according the condition

$$B_{max} \geq B_{DC} + B_\mu \cdot \sqrt{2} \quad (7)$$

3 Transformer filter - analysis

Transformer operated on alternating current transmission or distribution network transfers power from the primary winding $R_1X_{1\sigma}X_h$ (from terminals 1-1') to the magnetic circuit $R_{Fe}\,X_h$ into secondary winding $R_2X_{2\sigma}X_h$ (terminals 2-2'), wherein R_1, R_2 are resistors winding and $X_{1\sigma}$, $X_{2\sigma}$ leakage inductance. For solution we use a simplified diagram of a replacement, the T-cell with $p = 1$ transfers without parasitic capacities (Fig. 3) [4]. Values R_2, $X_{2\sigma}$ not necessary to recalculate in the scheme [5].

Magnetic induction in the iron core of the transformer is in the range ± B, depending on the hysteresis loop of the material (Fig. 4), if it possible, in the linear part.

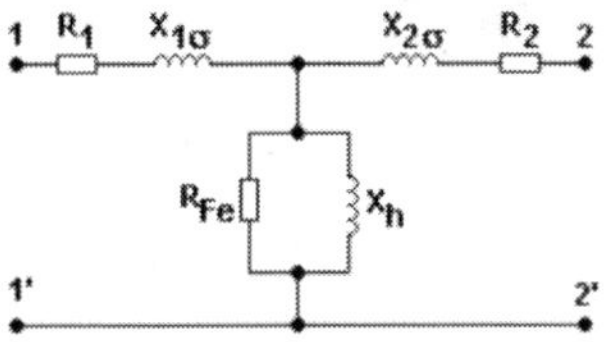

Figure 3. Equivalent diagram of transformer.

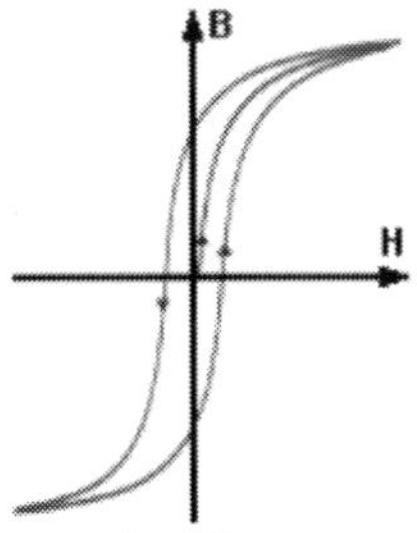

Figure 4. Hysteresis loop of transformer core.

By contrast, compensatory transfers from the transformer primary circuit (terminals 1-1' connect to the output rectifier) DC power to the secondary circuit (terminals 2-2') only through the secondary winding $R_2X_{2\sigma}X_{2h}$ (Fig. 5). The current through the secondary coil X_{2h} excite the core of transformer DC magnetic induction B_{DC} (Fig. 6) according to Eq. 5.

Primary winding $R_1X_{1\sigma}X_{1h}$ is isolated DC from the out put of the rectifier capacitor C. Windings flows through only the alternating component is superimposed to the primary coil X_{1h}, excites the core of the compensation transformer magnetic flux density $\Delta B = 2 \cdot B_\mu$ (Fig. 6) according to Eq. 5. The resulting induction in the core then varies between $B_{DC} \pm B_\mu \cdot \sqrt{2}$. Considered optimal to the working area ΔB was relatively linear part of the magnetization curve.

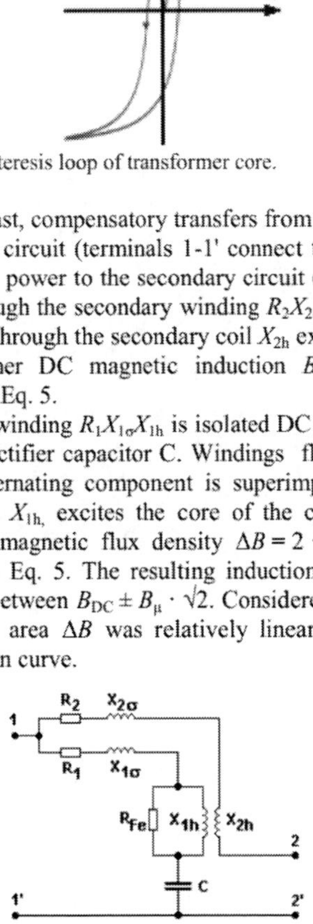

Figure 5. Equivalent diagram of compensation transformer.

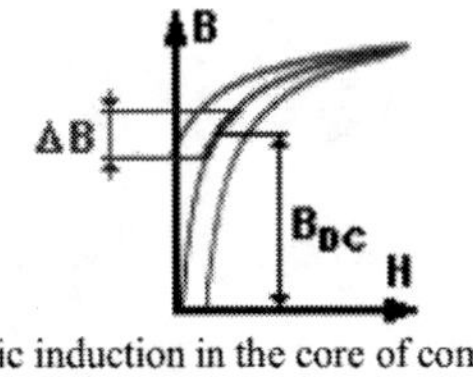

Figure 6. Magnetic induction in the core of compensation transformer.

Because the transformer ratio $p = 1$ and the windings are formed as bifilar applies $R_1 = R_2$, $X_{1\sigma} = X_{2\sigma}$, $X_{1h} = X_{2h}$. R_{Fe} are iron losses. However, they only creates alternating current component passing through the primary winding.

3.1 The influence of phase relationships of transformer operation filter

In order compensation transformer totally eliminates disruptive ripple on the output power, it would have to meet three basic requirements:

1) AC voltage at the primary and secondary side with respect to section 1 (Fig. 5) the same size;
2) the alternating voltage on the primary and secondary side are of identical shape (transformer clogged shape distortion);
3) the AC voltage at the primary and secondary side with respect to each zero phase shift.

Conditions 1) and 2) can be met relatively without much difficulty. The voltage adjust slight change the number of turns of the secondary winding, shape distortion on the secondary side DC restrict core saturation and a low level alternating magnetization, preferably within a maximum linear extent permitted magnetic induction B_{max}.

Issues voltage phase shift between the primary and secondary sides comes from the phasor diagrams loaded transformer. Zero phase shift in real magnetic circuits but virtually unattainable. Our effort is to reach a situation where $\varphi \rightarrow 0$. In practice, this means minimizing losses in the magnetic circuit. In the case of the compensation transformer, which transmits only the minimum power, it is possible to approximate this requirement. The results of simulating the impact of the phase shift between the primary and secondary voltage of the transformer compensation for the size differential voltage, which represents the spurious output are shown in Fig. 7.

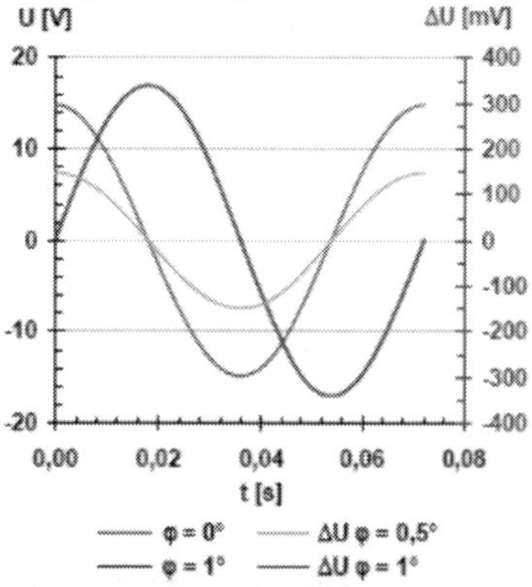

Figure 7. Simulation differential voltage.

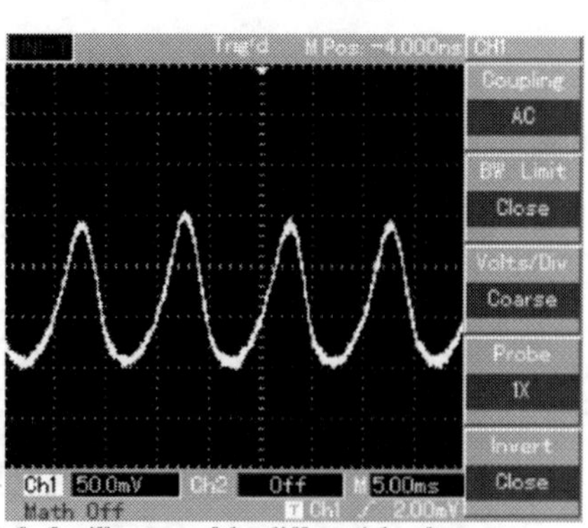

Figure 8. Oscillogram of the differential voltage.

For the amplitude of the input voltage 17 V single-phase source and the phase angle φ = 0,5° based on the residual noise voltage with amplitude of 150 mV. When the phase shift φ = 1°, then the amplitude of the disturbing voltage of 300 mV. On compensation transformer developmental sample power source we are for direct consumption I_{DC} = 10 A and measured peak to peak differential voltage 152 mV (Fig. 8), which corresponds to the effective value of the ripple 48,5 mV noise voltage at the output has a level of -50 dB.

3.2 Results of transformer filter applications in power supplies

Transformer ripple filter for high-power DC supplies have been long tested in laboratory and industrial use. Tests were run on linear single phase unstabilized supply 14,2 V/10 A and unstabilized linear three-phase source of 13,2 V/150 A. Output current-voltage characteristic (Fig. 9, Fig. 10) shows good stability of the output voltage and a very small noise voltage at the output source.

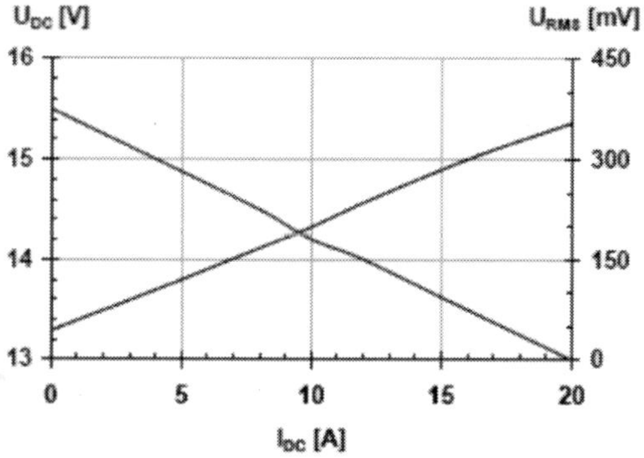

Figure 9. Output and ripple voltage (single-phase source).

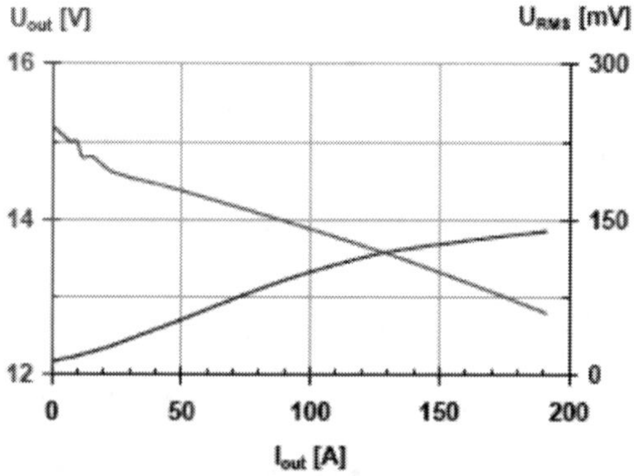

Figure 10. Output and ripple voltage (three-phase source).

Analysis of voltage drops on the individual parts of the source has proved a major impact on used rectifier output voltage drop. Used rectifier causes, depending on the load output voltage drop to 2 V. Fig. 11 shows overlapping voltage transformer for compensation at a current of 200 A, on Fig. 12 shows the ripple at the input of the compensation transformer (CH1) and the output ripple DC voltage (CH2).

 DOI: 10.1051/matecconf/20179514004

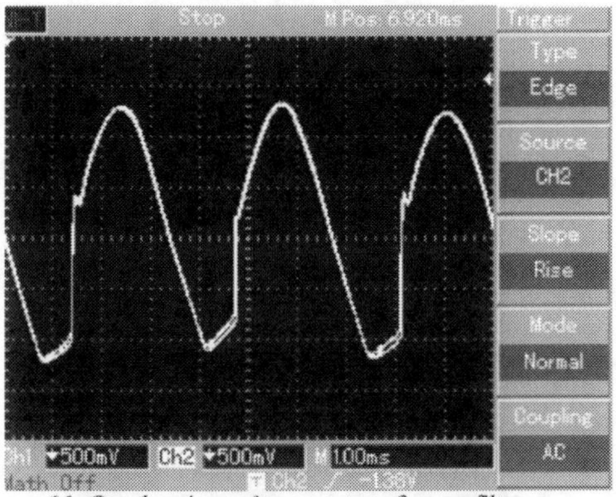

Figure 11. Overlapping voltage on transformer filter.

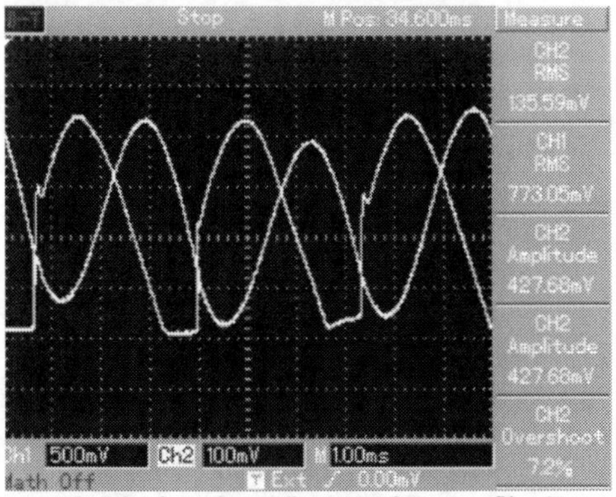

Figure 12. Ripple voltage (input and output filter).

4 Summary

Transformer ripple filter represent an effective alternative to conventional filters interference voltage. Preferably, it can be used for high-power DC supplies. Preferably, they are intended for stationary and industrial applications where higher matter weight compensation transformer filter. The advantage is the high long-term overload capability and effectively suppress interference voltages without large filter capacity. Additional capacitors can the level of disturbing voltage drop further.

The article was prepared with the support of specific research project SV PdF 2018/2012 - High-power DC supply with mid-frequency transformer compensation ripple voltage with high-overload current.

References

1. R. Drtina, J. Lokvenc and J. Andris. The Accessories for Your Laboratories, Part 9. - Switching supply for laboratory use. Media4u Magazine, **1**, pp.55-62, (2009).
2. J. Lokvenc, R. Drtina. Power supply voltage with the transformer ripple filter. In The 11th WSEAS/ IASME International Conference on Electric Power Systems, High Voltages, Electric Machines, Penang, Malaysia, pp.60-64, (2011).
3. J. Lokvenc, R. Drtina, and M. Sobek. Three-phase high overload DC power supply voltage with the mid-frequency transformer ripple filter. In The 12th WSEAS International Conference on Electric Power Systems, High Voltages, Electric Machines, Praha. pp.29-34, (2012).
4. J. H. Harlow. *Electric power transformer engineering*. BOca Raton, CRC, (2007).
5. J. J. Winders. *Power transformers: principles and applications*. New York, Dekker, (2002).

Frequency Domain Analysis for An Adaptive Windowing Parabolic Sliding Mode Filter

Shanhai Jin[1], Xiaodan Wang[1], Yonggao Jin[1] and Xiaogang Xiong[2]

[1]*School of Engineering, Yanbian University, Yanji 133002, China*
[2]*Singapore Institute Of Manufacturing Technology, 71 Nanyang Dr, Singapore 638075, Singapore*

Abstract. This paper analyses a frequency domain performance of an adaptive windowing parabolic sliding mode filter (AW-PSMF) by using Describing Function method. The analysis results show that AW-PSMF has similar gain characteristics to that of the second-order Butterworth low-pass filter (2-LPF), but AW-PSMF produces flatter gain behaviour than 2-LPF does at cutoff frequency. In addition, AW-PSMF produces smaller phase lag than 2-LPF does.

1 Introduction

In feedback control of mechatronic systems, feedback signals are usually corrupted by noise. Thus, a filter is required for removing noise from feedback signals. Linear filters are commonly applied for reducing noise because of their simplicity. However, in a linear filter, any noise component is proportionally transferred into the output. In addition, a large phase lag, which may result in the instability of feedback systems, is caused by strong noise attenuation.

Nonlinear filters have been applied for avoiding drawbacks of linear filters. For example, median filters [1] are used for removing high-frequency noise, but they are computationally expensive [2]. As another example, stochastic filters, e.g., Kalman filter [3], [4], are also applied in some applications. However, a dynamics model of the source of the signal, which is usually not available, is required. In addition, their performance depends on the model accuracy.

In the last decade, sliding mode observers based on the super-twisting algorithm [5], [6] has been attracted much attention. One advantage of these observers is that they theoretically realize finite time convergence in continuous-time analysis. However, typically with finite difference, the convergence accuracy in discrete-time implementation depends on the sampling period, as reported in [5]. Moreover, they are prone to overshoot during the convergence. Furthermore, they also require a system dynamics model.

The sliding mode filter that employs a parabolic sliding surface has been studied [7], [8]. One of major advantages of this filter is that, in the case of a constant input is provided, the output converges to the input in finite time. However, the filter is prone to overshoot. In addition, the discrete-time implementation of the filter produces high-frequency chattering due to inappropriate discretization.

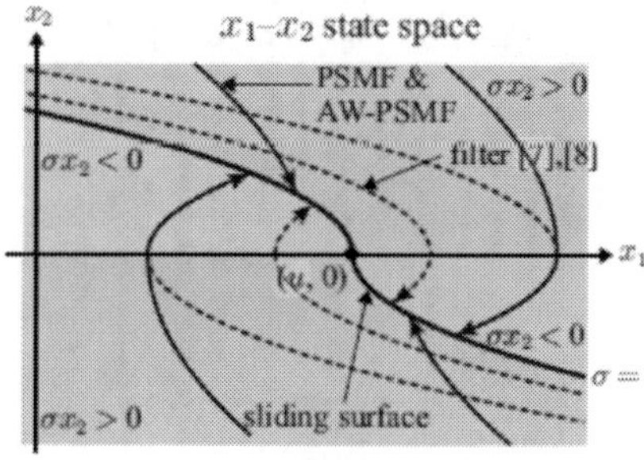

Figure 1. The parabolic-shaped sliding surface (thick solid curve) and trajectories of the state of PSMF and AW-PSMF (thin solid curve) and the filter [7], [8] (thin dotted curve).

Toward to the drawbacks of the parabolic sliding mode filter [7], [8], Jin *et al.* [9] presented a new parabolic sliding mode filter, which is referred to PSMF, for effectively removing noise in feedback control systems. It is reported in [9] that PSMF produces smaller phase lag than linear filters, and it is less prone to overshoot than the sliding mode filter [7], [8]. In addition, the algorithm of PSMF, which is derived by using the backward Euler differentiation, does not produce chattering. After that, Jin *et al.* [10] presented an adaptive windowing parabolic sliding mode filter, which is named as AW-PSMF, by extending PMSF. It is stated in [10] that AW-PSMF adaptively adjusts its window size for obtaining the largest window size that optimizes the trade-off between the filtering smoothness and the delay suppression. The effectiveness of AW-PSMF has been experimentally validated in feedback control of a mechatronic system. In [10], however, the frequency domain performance of AW-PSMF is not evaluated.

 DOI: 10.1051/matecconf/20179514005

This paper presents a frequency domain performance of AW-PSMF. It is shown that AW-PSMF has similar gain characteristics to that of the second-order Butterworth low-pass filter (2-LPF), but AW-PSMF produces smaller phase lag (maximum 140 degree) than 2-LPF (maximum 180 degree) does.

The rest of this paper is organized as follows. Section 2 provides a brief overview of parabolic sliding mode filters. Section 3 analysis the performance of AW-PSMF in frequency domain. Section 4 gives concluding remarks.

2 Parabolic sliding mode filters

In [9], Jin *et al.* presented a parabolic sliding mode filter (PSMF), of which continuous-time expression is given as follows:

$$\dot{x}_1 = x_2 \tag{1a}$$

$$\dot{x}_2 \in -\frac{F(H+1)}{2}\operatorname{sgn}(\sigma(F,u,x_1,x_2)) - \frac{F(H-1)}{2}\operatorname{sgn}(x_2), \tag{1b}$$

where

$$\sigma(F,u,x_1,x_2) \triangleq 2F(x_1-u)+|x_2|x_2 . \tag{2}$$

Here, u is the input, x_1 and x_2 are the outputs, and $F>0$ and $H>1$ are constants. In addition, sgn() is the set-valued signum function defined as follows:

$$\operatorname{sgn}(z) \triangleq \begin{cases} 1 & \text{if } z>0 \\ [-1,1] & \text{if } z=0 \\ -1 & \text{if } z<0 \end{cases} \tag{3}$$

It should be noted that $\operatorname{sgn}(z)$ returns a set instead of a single value when $z=0$.

Figure 1. illustrates the sliding surface and state trajectories of PSMF in x_1-x_2 space. It is shown that PSMF employs a parabolic-shaped sliding surface, and the state is attracted to the sliding surface from both sides, whereas as the state of the filter [7][8] moves in parallel to the sliding surface in the regions of $\sigma x_2 > 0$. Such an advantage of PSMF can be attributed to the use of $H>1$.

In [9], Jin *et al.* also provided the following PSMF's discrete-time algorithm, which is derived by using the backward Euler discretization:

Algorithm PSMF

$$x_2^*(k) := FT\Phi\left(\frac{u(k)-x_1(k-1)}{FT^2}\right) \tag{4a}$$

$$x_2(k) := \operatorname{clip}(\operatorname{clip}(x_2(k-1)-HFT, x_2(k-1)-FT, 0), \operatorname{clip}(x_2(k-1)+FT, x_2(k-1)+HFT, 0), x_2^*(k)) \tag{4b}$$

$$x_1(k) := x_1(k-1)+Tx_2(k) \tag{4c}$$

RETURN $x_1(k)$ and $x_2(k)$, (4d)

where T is the sampling period, and $\Phi()$ and clip() are functions given as follows:

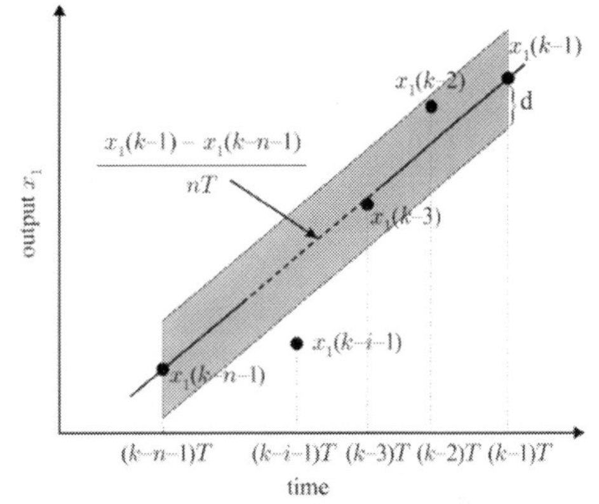

Figure 2. Windowing in AW-PSMF.

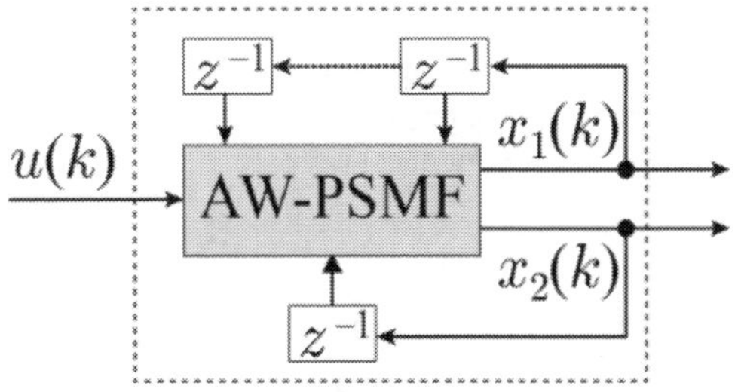

Figure 3. Block diagram of AW-PSMF.

$$\Phi(z) \triangleq \operatorname{sgn}(z)(\sqrt{1+2|z|}-1) \tag{5}$$

$$\operatorname{clip}(a,b,z) \triangleq \begin{cases} b & \text{if } z>0 \\ z & \text{if } z\in[a,b] \\ a & \text{if } z<0. \end{cases} \tag{6}$$

Here, $a<b$, and x_2^* is an intermediate variable, which is followed by x_2 under constraint of $|(x_2(k)-x_2(k-1))|/T$.

After the proposal of PSMF, in [10] Jin *et al.* presented an extension version of PSMF by introducing an adaptive window, of which windows size is adaptively changed for realizing the largest window size that optimizes the trade-off between the filtering smoothness and the delay suppression. Specifically, the complete algorithm of the modified version of PSMF, which is named as AW-PSMF, is given as follows:

Algorithm AW - PSMF

FOR $n \in \{1,\cdots,\min(k+1, N_{\max})\}$ (7a)

$a := (x_1(k-1)-x_1(k-n-1))/(nT)$ (7b)

FOR $i \in \{1,\cdots,n\}$ (7c)

$e := x_1(k-i-1)-x_1(k-1)+iTa$ (7d)

IF $|e|>C$; **EXIT LOOP** (7e)

END FOR (7f)

 DOI: 10.1051/matecconf/20179514005

IF $i < n$; **EXIT LOOP** (7g)

END FOR (7h)

$$x_2^*(k) := nFT\Phi(\frac{u(k) - x_1(k-n)}{n^2FT^2}) \quad (7i)$$

$$x_2(k) := \text{clip}(\text{clip}(x_2(k-1) - HTFT, x_2(k-1) - FT, 0), \text{clip}(x_2(k-1) + FT, x_2(k-1) + HFT, 0), x_2^*(k)) \quad (7j)$$

$$x_1(k) := x_1(k-1) + Tx_2(k) \quad (7k)$$

RETURN $x_1(k)$ and $x_2(k)$, (7l)

where N_{max} and d are parameters that should be chosen appropriately.

Figure 2. shows block diagram of AW-PSMF. In AW-PSMF, the previous outputs are used for windowing, as shown in Figure 3.. Specifically, AW-PSMF checks whether all previous outputs in the window are inside the area determined by the two end-outputs $x_1(k)$ and $x_1(k-n-1)$ and the constant d. If it is the case, the window size n is further increased for obtaining smoother output. The increase of n is continued until at least one previous output in the window lies outside the area or n reaches its maximum value N_{max}. After that, the last n value is applied for determining $x_2^*(k)$, which is an intermediate variable, and then $x_2(k)$ is determined to follow $x_2^*(k)$ under constraint of $|(x_2(k) - x_2(k-1)| / T$.

3 Frequency domain analysis of AW-PSMF

This section analysis the frequency domain behaviour of AW-PSMF by using the Bode plot. It should be mentioned that, due to the nonlinearity of AW-PSMF, the transfer function method cannot be applied here for obtaining Bode plot of AW-PSMF. Thus, as an alternative way, the section applies the Describing Function method [11], which takes into account the fundamental harmonic component of a nonlinear system's response. In this section, sampling period $T = 0.001$ s is used.

Figure 4. shows Bode plot of AW-PSMF with parameters $F = 1000$, $H = 5$, $N_{max} = 10$ and $d = 0.0001$ for the input signal $u = \sin(\omega t)$, where the frequency ω changes from 1 rad/s to 1000 rad/s. For comparison, the Bode plot of 2-LPF with cutoff frequency $\omega_{cf} = 45$ rad/s is also shown in Figure 4. Here, the term 'cutoff frequency' refers to the intersect point of low and high frequency asymptotes, and the cutoff frequency ω_{cf} of 2-LPF is chosen so that the gain plot of 2-LPF is as close as possible to that of AW-PSMF.

The gain plot shows that the high frequency asymptote of AW-PSMF is approximately -40 dB/decade, which is the same as that of 2-LPF. However, at cutoff frequency, the gain of AW-PSMF is -1.95 dB, whereas that of 2-LPF is -3 dB. It is known that the gain

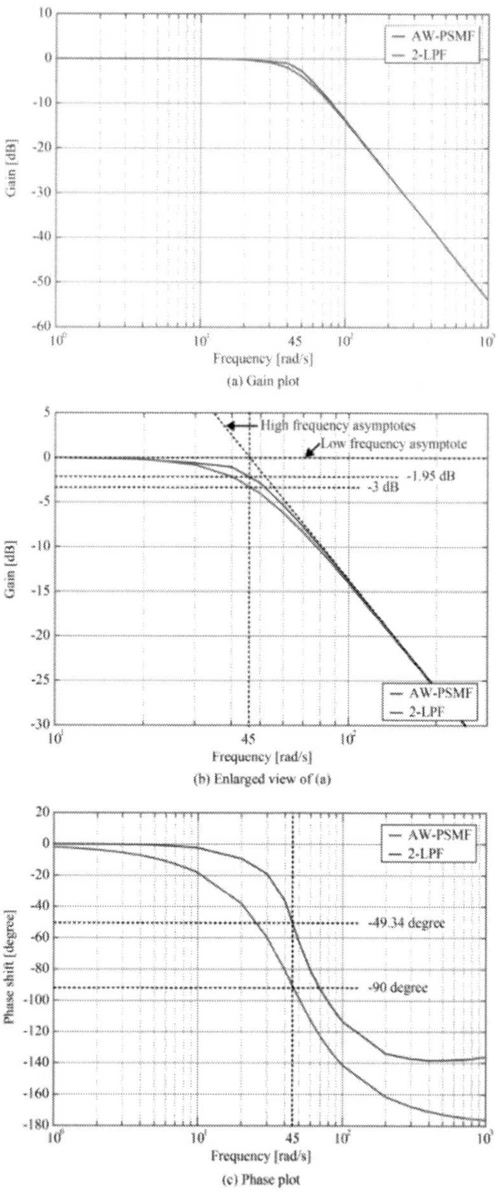

(a) Gain plot
(b) Enlarged view of (a)
(c) Phase plot

Figure 4. Bode plots of AW-PSMF with $F = 1000$, $H = 5$, $N_{max} = 10$ and $d = 0.0001$ and 2-LPF with

plots of Butterworth low-pass filters are the flattest at cutoff frequencies among all linear filters. Thus, it can be concluded that the gain characteristics of AW-PSMF is better than those of all linear filters.

The phase plot shows that the phase shift of AW-PSMF at cutoff frequency is -49.34 degree, whereas that of 2-LPF is -90 degree. In addition, one can observe that the maximum phase shifts of around -140 degree. This is smaller than that of 2-LPF, of which value is -180 degree. Thus, it can be said that the phase characteristics of AW-PSMF is advantageous over that of 2-LPF.

Figure 5. illustrates the comparison between AW-PSMF and 2-LPF in the case where the frequency ω of input $u = \sin(\omega t)$ changes from 10 rad/s to 20 rad/s at 1.5 s, and then changes again to 30 rad/s at 2.8 s. It is shown that AW-PSMF produced larger gain but smaller phase shift than 2-LPF does for all three input frequencies.

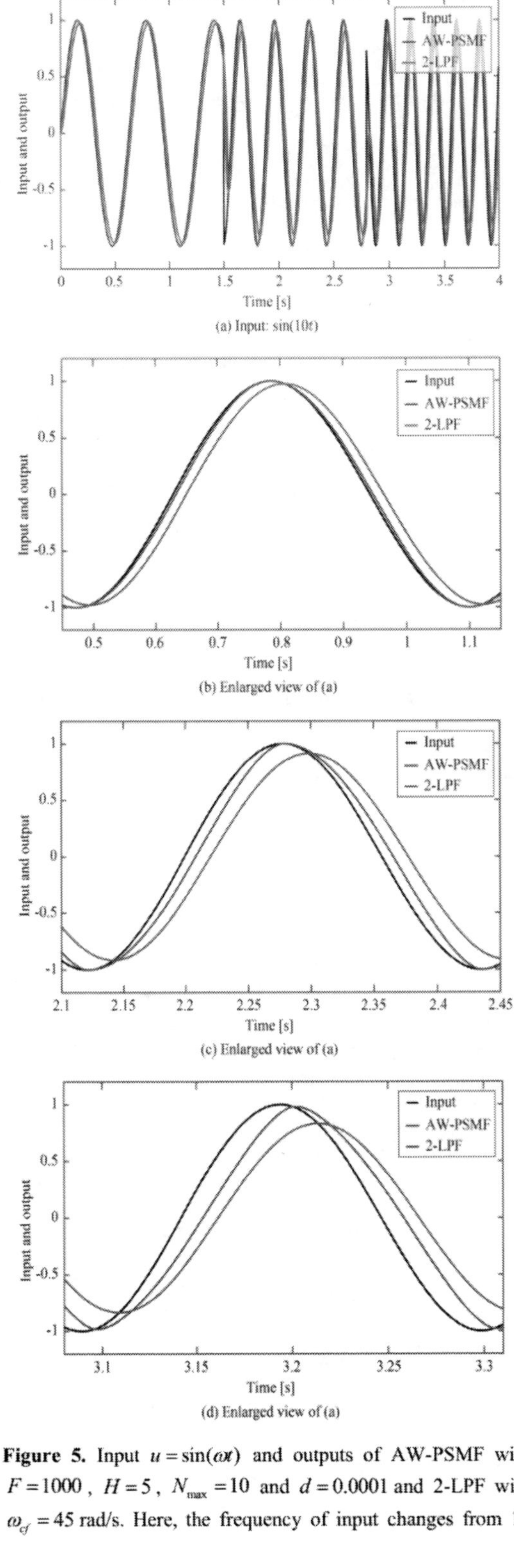

Figure 5. Input $u = \sin(\omega t)$ and outputs of AW-PSMF with $F = 1000$, $H = 5$, $N_{max} = 10$ and $d = 0.0001$ and 2-LPF with $\omega_{cf} = 45$ rad/s. Here, the frequency of input changes from 10 rad/s to 20 rad/s at 1.5 s, and then changes again to 30 rad/s at 2.8 s rad/s.

4 Conclusions

This paper has analysed a frequency domain performance of AW-PSMF by using the Describing Function method. The analysis results show that AW-PSMF has similar gain characteristics to that of 2-LPF, AW-PSMF produces flatter gain behaviour than 2-LPF does at cutoff frequency.

In addition, AW-PSMF produces smaller phase lag (-49.34 degree at cutoff frequency, maximum -140 degree) than 2-LPF (-90~degree at cutoff frequency, maximum -180 degree) does. Thus, it can be concluded that AW-PSMF has better gain and phase characteristics than 2-LPF.

References

1. Gallagher NJ, Gary LW. A theoretical analysis of the properties of median filters. IEEE Trans Acoustics, Speech and Signal Processing. 1981; 29(6):1136-1141.
2. Moshnyaga VG, Hashimoto K. An efficient implementation of 1-D median filter. In: Proc. 52nd IEEE Int. Midwest Symp. Circuits and Systems. 2009. p. 451--454.
3. Welch G, Bishop G. An introduction to the Kalman filter. Department of Computer Science, University of North Carolina. 1995. Tech Rep TR95-041.
4. Lightcap CA, Banks SA. An extended Kalman filter for real-time estimation and control of a rigid-link flexible-joint manipulator. IEEE Trans Control Systems Technology. 2010; 18(1):91-103.
5. Levant A. Robust exact differentiation via sliding mode technique. Automatica. 1998; 34(3):379-384.
6. Moreno JA, Osorio M. A Lyapunov approach to second-order sliding mode controllers and observers. In: Proc. 47th IEEE Conf. Decision and Control. 2006. p. 2856-2861.
7. Han JQ, Wang W. Nonlinear tracking-differentiator. J. System Science and Mathematical Science. 1994; 14(2):177-183. (in Chinese).
8. Emaru T, Tsuchiya T. Research on estimating smoothed value and differential value by using sliding mode system. IEEE Trans Robotics and Automation. 2003; 19(3):391-402.
9. Jin S, Kikuuwe R, Yamamoto M. A Realtime quadratic sliding mode filter for removing noise. In: Proc. 2011 Robotics and Mechatronics Conf. of The Japan Society of Mechanical Engineers. 2011. p. 1P1-F03.
10. Jin S, Kikuuwe R, Yamamoto M. Discrete-time velocity estimator basedon sliding mode and adaptive windowing. In: Proc. 2012 IEEE/SICE Int. Symposium on System Integration. 2012. p.835-841.
11. Khalil HK. Nonlinear Systems, 3rd edition. 2002.

Multiple Iteration of Weight Updates for Least Mean Square Adaptive Filter in Active Noise Control Application

Rahimie Mustafa[1], Anuar Mikdad Muad[2], Shahrizal Jelani[1] and Ahmad Nur Alifa Abdul Razap[1]

[1]*UCSI University, Cheras Malaysia*
[2]*Universiti Kebangsaan Malaysia, Bandar Baru Bangi Malaysia*

Abstract. The method of least mean square (LMS) is the commonly used algorithm in Adaptive filter due to its simplicity and robustness in implementation. In Active Noise Control application, a filtered reference signal is used prior to LMS algorithm to overcome the constraint on stability and convergence performance of the system due to the existence of the auxiliary path. This is known as Filtered-X LMS algorithm. In conventional Filtered-X LMS algorithm, each filter weight is updated once on every audio sample. This paper proposes the improved version of Filtered-X LMS algorithm with the use of multiple iteration of filter weight on every sample of audio signal. The proposed work uses field programmable gate arrays to realize real-time simulation on hardware for the noise signal of 500 Hz. Results from the real-time hardware simulations have shown much faster error convergence and better adaptation performance for different selections of learning constant μ, as compared with the conventional method.

1 Introduction

The least mean square (LMS) can be considered as the most applied algorithm for linear adaptive filter particularly in active noise control (ANC) application that required real-time processing for a successful and efficient hardware implementation [1]-[3]. The use of specialized digital signal processor (DSP) chip and with the capability of handling numerous floating point operations manage to address the real–time processing issue in narrowband attenuation of ANC headset [4]-[5], and in broadband attenuation for duct application [6] based on least mean square (LMS) algorithm. However, the performance of signal convergence in terms of the speed of filter adaptation and the excess of mean square error (MSE) is usually overlooked, which can be critical in a time time-varying environment [2], [3]. Research being carried out to improve the error convergence time of LMS-based algorithm however increased the algorithm complexity [7]-[9] and consequently, implementations are mostly evaluated successfully in a purely simulated domain [10]-[12].

The conventional digital signal processor (DSP) evaluates a signal in a sequential behaviour, where each updates on the filter weight might require several instruction cycles to complete. As a result, heavy processing for complicated LMS-based algorithm will require more instruction cycles and will invoke additional time delay to ANC system. Therefore, in general the error signal will converge much slower. To overcome this, an analysis on the LMS convergence behaviour has been carried out using field programmable gate arrays (FPGA) on ANC system for duct application [13]. It is known that FPGA has the capability of processing a signal in a parallel manner hence allowing a complete computation of weight updates within a few clock cycles. The experimental work by the author [13] also further verify the theoretical principle established from previous studies [14]-[17], that are initially based on simplifying assumptions.

In this paper, motivated by the idea presented in [13] we utilize the FPGA advantage to achieve real-time and faster convergence of error signal for LMS-based algorithm. A real-time hardware simulation is carried out to implement the modified Filtered-X LMS algorithm namely MFxLMS. Performance of the system is then compared with the conventional Filtered-X LMS algorithm also by the same hardware structure. The paper is organized as follows: Section 2 describes briefly on theory and underlying principle of LMS-based adaptive filter used, followed by our proposed modified version. In section 3, the implementation of the modified LMS-based algorithm on FPGA is described and in section 4, the real-time hardware simulation of ANC based on MFxLMS algorithm is briefly presented. The results obtained are then briefly discussed in section 5 and finally, section 6 presents the conclusion of the findings.

2 LMS-Based adaptive filter

LMS algorithm is an approximation method of steepest gradient descent that relies on the value of the instantaneous squared error signal [18], [19]. Due to this approximation, the calculation of adaptive filter weight has resulted in the simplification that is expressed as:

$$\mathbf{w}(n+1) = \mathbf{w}(n) + \mu\, \mathbf{x}(n)e(n) \quad (1)$$

where $\mathbf{w}(n) = [w_0(n), w_1(n), \ldots, w_{L-1}(n)]^T$ are the weight updates or adaptive filter coefficients vectors for L-length filter, $\mathbf{x}(n) = [x(n), x(n-1), \ldots, x(n-L+1)]^T$ correspond to the reference signal vectors, $e(n)$ is the error signal at sampling time n and μ is the learning constant.

Our implementation uses the structure of finite impulse response (FIR), a stable digital filter that has a usual canonical form of tapped-delay input as a basic element to realize the designated adaptive filter. The LMS algorithm updates the FIR filter weight to produce the output signal, $y(n)$ that can be expressed as an arithmetic sum of products:

$$y(n) = \sum_{k=0}^{N-1} w_k(n) \cdot x(n-k) \quad (2)$$

where $y(n)$ is the adaptive filter output at time n, $x(n-k)$ is the tapped-delay input reference signal and $w_k(n)$ for $k = 0,1,\ldots,N-1$ are the N-tap FIR filter coefficients. In ANC system, the residual error signal is acquired from a sensor (usually microphone) that is also corresponding to the difference between the primary noise, $d(n)$ and the adaptive filter output , $y(n)$ which is expressed as:

$$e(n) = d(n) - y(n) \quad (3)$$

In general, numerical calculation for updating the filter weights utilizing conventional DSP chip is performed one time in every sample of audio signal that is expressed as [1]:

$$w_k(n+1) = w_k(n) + \mu x(n-k)e(n) \quad (4)$$

where the error signal $e(n)$ is multiplied with each tapped-delay input of reference signal $x(n-k)$ alongside with learning constant μ before being summed up with the previous weights.

However, if we could allow the weight updates to be calculated more than once within each sample of audio signal [20], then logically we can expect faster adaptation of weight updates, hence faster convergence of error signal. In this method, the calculation of error signal, $e(n)$ from (3) has become:

$$e_j(n) = d_j(n) - y_j(n) \quad (5)$$

where j is the number of repeated calculation (iteration) of error within every sample of audio signal. Thus, by repeating the calculation of the error signal, the computation of filter weight can now be modified to:

$$w_{j,k}(n+1) = w_{j,k}(n) + \mu x(n-k)e_j(n) \quad (6)$$

3 FPGA Implementation

The FIR-based LMS adaptive filter is designed using conventional multiplier and accumulator and is implemented onto the Altera Cyclone II FPGA chip embedded on a development platform. The development platform is also featuring with a real-time audio coding and decoding (codec) that can facilitate analogue and digital conversion (DAC) of analogue audio signal at 48 kHz sampling rate along with a simple anti-aliasing and reconstruction filter. Therefore, additional design of analogue audio interface circuitry with FPGA chip is thus avoided. In digital domain, the audio signal is further down-sampled to 8 kHz to increase the number of maximum iteration possible for weight updates on every sample of audio signal. This will not affect the ANC system, since the noise frequency of interest for practical ANC application is well below 1 kHz range [2], [3]. The computation of weight updates is carried out iteratively within each sample of audio signal as depicted by the flow chart in Fig. 1.

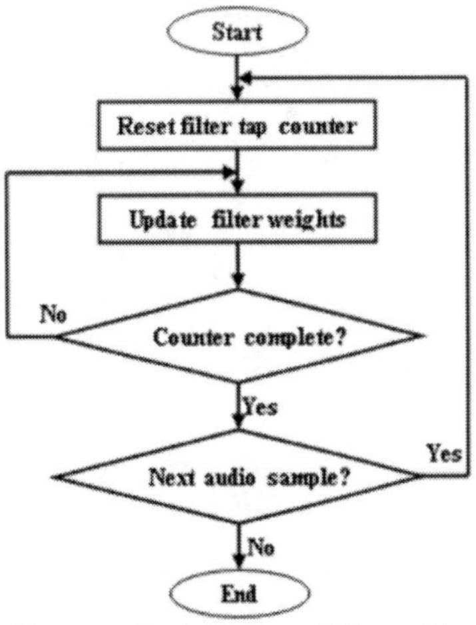

Figure 1. Flow chart for the process of filter weights updates for one sample of audio signal.

3.1. Learning constant

Due to the fixed-point data representation on FPGA, a truncation process during calculation is unavoidable especially to match the size of data at the end terminal. The truncation process also has the implication of dividing the data by the factor of 2 besides giving the effect of quantization and rounding errors. When this is implemented in the path of filter weight updates, due to this division effect, it will implicitly infers a multiplication with a learning constant, μ with the factor of reciprocal of 2 as the function of truncation size:

$$\mu = \frac{1}{2^B} \quad (7)$$

where B is a positive integer represent the size of the least significant bit that is truncated. Based on the requirement for stability and convergence of error signal established in the previous studies [1]-[3], the value of learning constant is bounded in a limited region:

$$0 < \mu < \frac{2}{NP_x} \quad (8)$$

DOI: 10.1051/matecconf/20179514006

where N is the filter length and P_x is the expectation of reference signal power. With the use of 64-taps filter length and the value of reference signal power set to be greater than one unit in average ($P_x > 1$), we obtained the practical bound of learning constant in our design, that is:

$$0 < \mu \leq 2^{-6} \tag{9}$$

The size of learning constant is 16-bit. After multiplying with the reference signal, $x(n)$ and the error signal, $e(n)$ of 24-bit, the accumulated size due to the fixed-point calculation is 64-bit in total. Since the size of the filter weight is set to 33-bit in size, then a total of 31-bit of least significant value has to be truncated. Due to this effect, we obtained the maximum inherent practical bound of the learning constant as:

$$0 < \mu \leq 2^{-31} \tag{10}$$

The maximum inherent practical bound of the learning constant is found to be within the range of the requirement as in equation (9). Thus, we can guarantee stability of the adaptation process since we can never exceed the maximum value. The choosing and selection of learning constant, μ can be made manually in between experiments by using digital switches. To simplify and to make a uniform selection (in logarithmic sense) within the range of the bounded value given in equation (10), a total number of 16 differences selection of learning constant value is used.

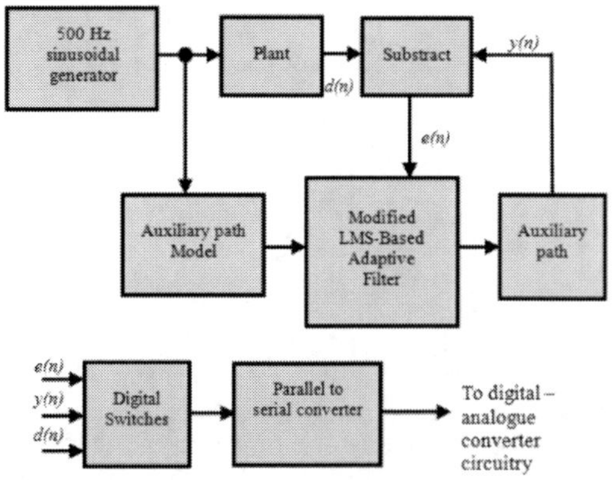

Figure 2. Block diagram of MFxLMS implemented on FPGA.

4 Real-time hardware simulation

Fig. 2 shows the block diagram of MFxLMS implemented on FPGA environment. The conventional Filtered-X LMS system also uses the same structure except the sub-block on *Modified LMS-Based Adaptive Filter* is replaced by *Conventional LMS-Based Adaptive Filter*. In Actual ANC system, the transfer function of sub-block on *Auxiliary path Model* need to adapt with the transfer function of sub-block on *Auxiliary path* by means of adaptive algorithm and need to be continuously modeled when the system is operated.

In order to conduct the experiment systematically, first the transfer function of sub-block *Auxiliary path* is fixed to two different types (pure samples delay and impulse response of all-pass filter) and the transfer function of sub-block *Auxiliary path Model* is made to follow exactly as the *Auxiliary path*. Secondly, the transfer function of sub-block on *plant* is also fixed to two different types (pure samples delay and impulse response of band-pass filter). By setting up to this configuration, there will be four differences alternative combination of *Plant* and *Auxiliary path*. In this sense, the four different combinations will gives more intuitive insight on the comparison of performance between both methods. Finally, a single 500 Hz frequency signal is used as the reference noise signal to consider the simple and basic noise for the experiment.

5 Results

The output of the results on the real-time hardware simulation is basically the analogue audio signal converted by the on-board audio codec embedded on the development platform. This audio signal is reconstructed based on the DAC circuitry that is sampled at 48 kHz sampling rate. This audio signal can be either the error signal $e(n)$, adaptive filter output $y(n)$ or the primary noise signal $d(n)$. This analogue audio signal is then captured and represented in computer terminal (PC) using Matlab for further analysis and graph presentation.

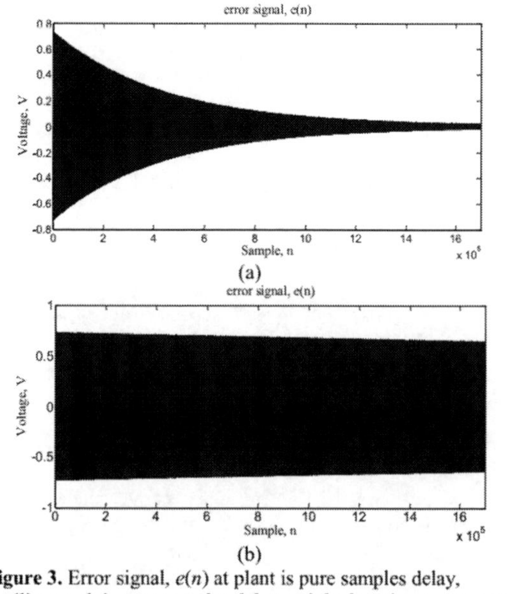

Figure 3. Error signal, $e(n)$ at plant is pure samples delay, auxiliary path is pure samples delay and the learning constant at 0.0001 x 10^{-10} (a) MFxLMS, (b) Conventional Filtered-X LMS

Fig. 3 and Fig. 4 shown two distinctive results on the experiments for MFxLMS and Conventional Filtered-X LMS for ANC system with the transfer function of *Plant* and *Auxiliary path* that is just pure samples delay. It is shown in Fig. 3 that when the learning constant is set to the lowest value which is μ = 0.0001 x 10^{-10} and when MFxLMS algorithm is applied to the ANC system, the error signal converge to its minimum value after 1600K of audio samples. This result indicates the filter weight on the adaptive filter has adapted successfully with the transfer function of the *Plant*. However, when the conventional Filtered-X LMS algorithm is applied, the error signal seems to require much more time (hence more audio sample) before it is expected to converge to its minimum value.

In Fig. 4, it is shown that when the learning constant is set to the highest value which is $\mu = 4.66 \times 10^{-10}$ and the ANC system applying the method of MFxLMS, the error signal converges to its minimum value only after 100 audio samples. Again this result indicates the filter weight on the *Plant* has been adapted successfully whereas for conventional Filtered-X LMS method, it will require more than 1500 samples of audio signal before the adaptive filter could adapt successfully with the transfer function of the *Plant*.

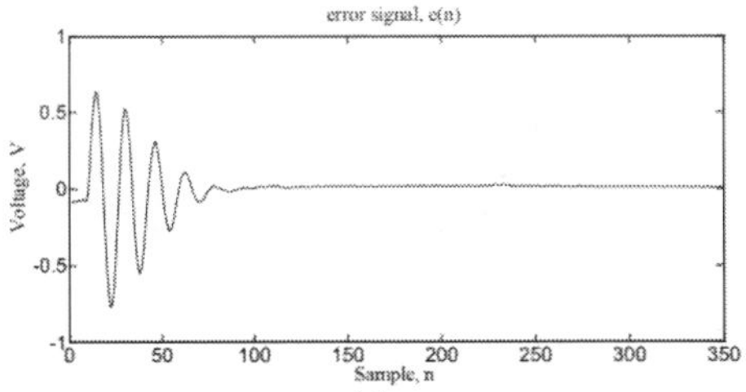

(a)

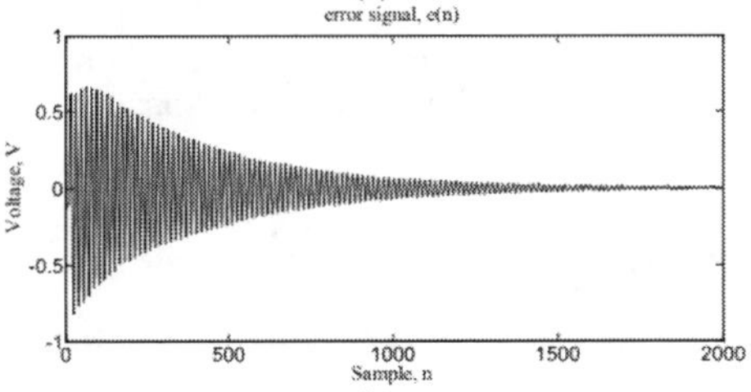

(b)

Figure 4. Error signal, $e(n)$ at plant is pure samples delay, auxiliary path is pure samples delay and the learning constant at 4.66×10^{-10} (a) MFxLMS, (b) Conventional Filtered-X LMS.

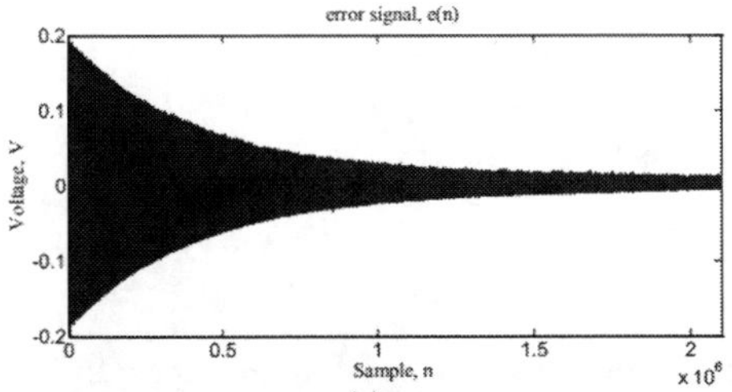

(a)

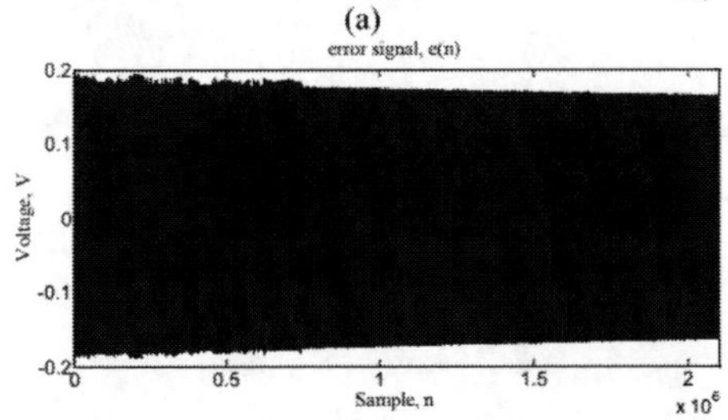

(b)

Figure 5. Error signal, $e(n)$ at plant is impulse response of band-pass filter, auxiliary path is pure samples delay and the learning constant at 0.0001×10^{-10} (a) MFxLMS, (b) Conventional Filtered-X LMS.

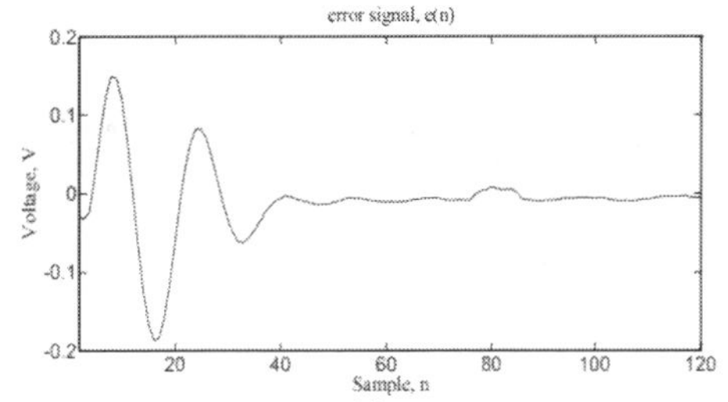

(a)

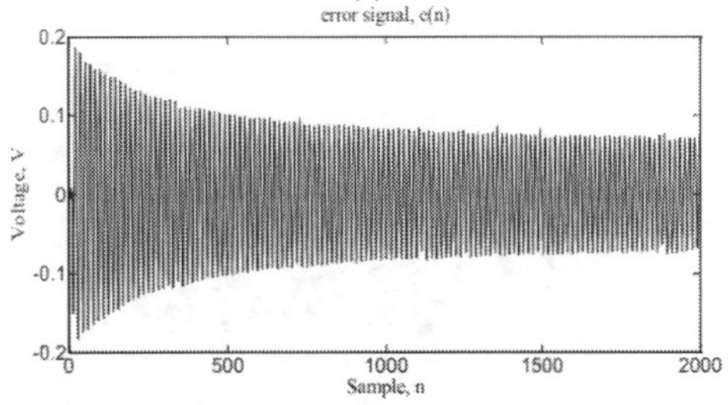

(b)

Figure 6. Error signal, $e(n)$ at plant is impulse response of band-pass filter, auxiliary path is pure samples delay and the learning constant at 4.66×10^{-10} (a) MFxLMS, (b) Conventional Filtered-X LMS.

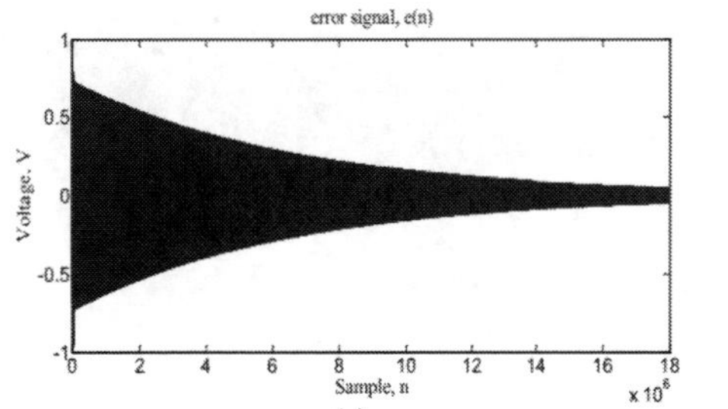

(a)

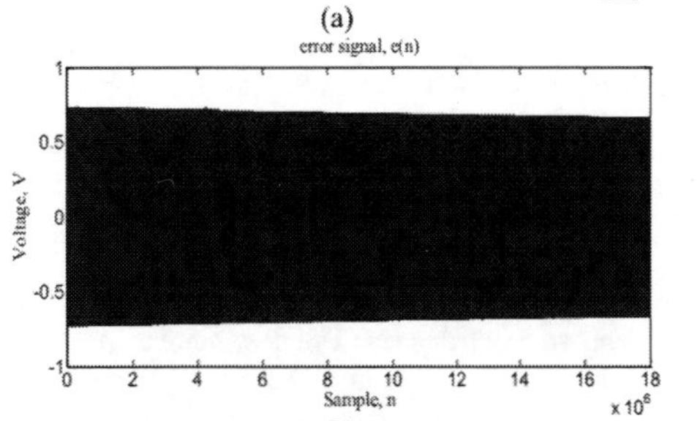

(b)

Figure 7. Error signal, $e(n)$ at plant is pure samples delay, auxiliary path is impulse response of all-pass filter and the learning constant at 0.0001×10^{-10} (a) MFxLMS, (b) Conventional Filtered-X LMS.

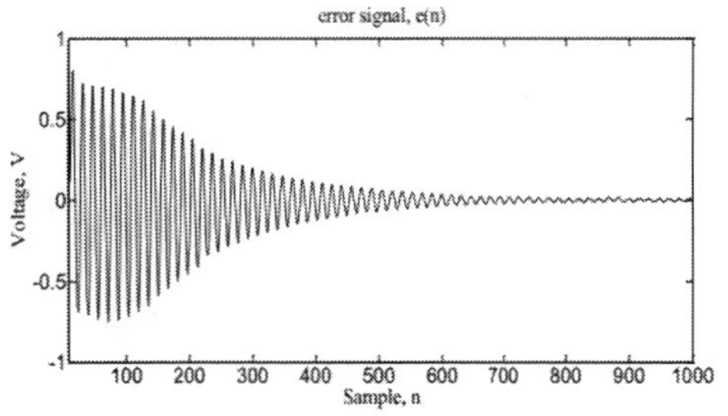

(a)

DOI: 10.1051/matecconf/20179514006

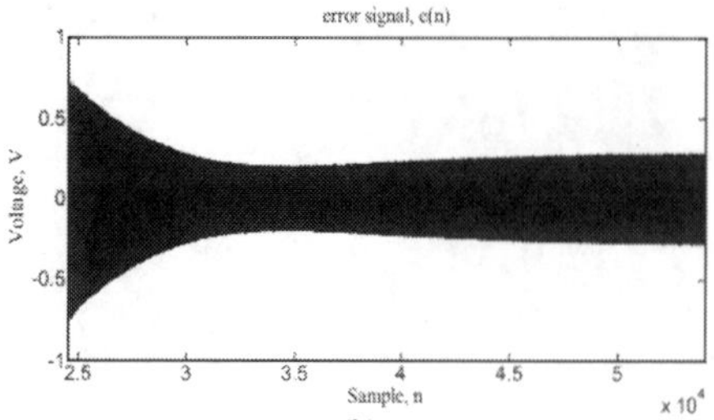

(b)

Figure 8. Error signal, $e(n)$ at plant is pure samples delay, auxiliary path is impulse response of all-pass filter and the learning constant at 4.66 x 10^{-10} (a) MFxLMS, (b) Conventional Filtered-X LMS.

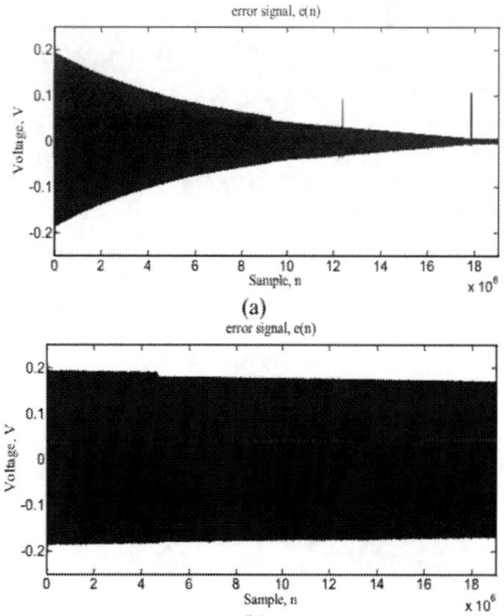

(b)

Figure 9. Error signal, $e(n)$ at plant is impulse response of band-pass filter, auxiliary path is impulse response of all-pass filter and the learning constant at 0.0001 x 10^{-10} (a) MFxLMS, (b) Conventional Filtered-X LMS.

Fig. 5 and Fig. 6 on the other hand show the results on the experiments for MFxLMS and Conventional Filtered-X LMS for ANC system when the transfer function of *Plant* is set to the impulse response of band-pass filter while the *Auxiliary path* is set to just pure samples delay. It is shown from the result that in general the MFxLMS algorithm performs much better in term of faster adaptation of filter weights. Similarly, Fig. 7 and Fig. 8 as well as Fig. 9 and Fig. 10 show the distinctive results on the comparison of error convergence performance between the method of MFxLMS and Convention Filtered-X LMS algorithm.

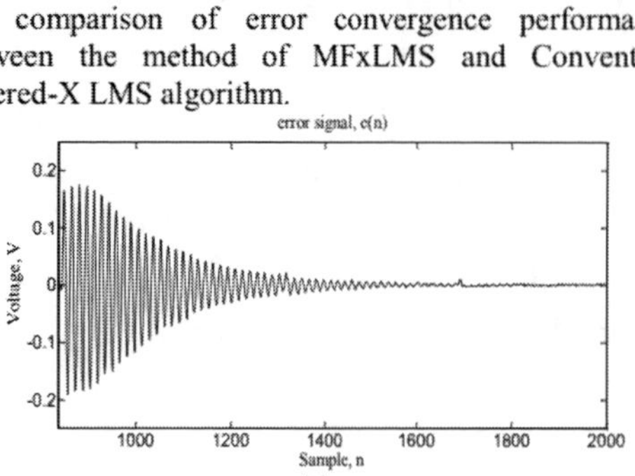

(a)

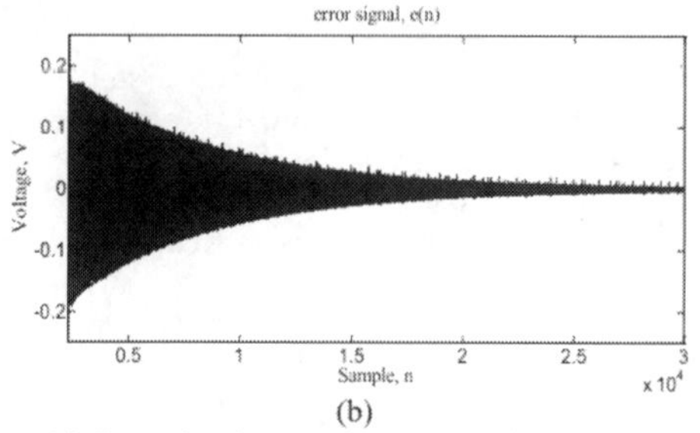

(b)

Figure 10. Error signal, $e(n)$ at plant is impulse response of band-pass filter, auxiliary path is impulse response of all-pass filter and the learning constant at 4.66 x 10^{-10} (a) MFxLMS, (b) Conventional Filtered-X LMS.

6 Conclusion

Based on the results in section 5, it is concluded that the new modified method on LMS-based adaptive algorithm that is implemented on ANC system namely MFxLMS, has shown superior performance in terms of faster adaptation of filter weights as compared to the conventional method, Filtered-X LMS.

Acknowledgment

The Author acknowledged the Faculty of Engineering and Built Environment of Universiti Kebangsaan Malaysia (UKM) for providing the laboratory facility to conduct the research.

References

1. Sen M. Kuo and Dennis R. Morgan, *Active noise control systems: Algorithms and DSP implementations*, John Wiley & Sons, Inc., New York, (1996).
2. P. A. Nelson and S. J. Elliott, *Active Control of Sound*. Academic Press, London, (1992).
3. S. J. Elliott, *Signal Processing for Active Control*. Academic Press, New York, (2001).
4. W. S. Gan, S. Mitra and S. M. Kuo "Adaptive Feedback Active Noise Control Headset: Implementation, Evaluation and Its Extensions" *IEEE Transactions on Consumer Electronics*, **51**(3), pp. 975-982, (August 2005).
5. Sen M. Kuo, S. Mitra and W. S. Gan, "Active Noise Control System for Headphone Applications" *IEEE Transactions on Control Systems Technology*, **14**(2), pp. 331-335, (March 2006).
6. C. Y. Chang "Efficient Active Noise Controller using a Fixed-point DSP" *Elsevier Signal Processing*, **89**, pp 843-850, (2008).
7. L. J. Eriksson and M. A. Allie, "Use of random noise for online transducer estimate in an adaptive active attenuation system," *Journal of Acoustical. Society of America*, **85**, pp. 797–802, (1989).
8. M. Zhang, H. Lan, and W. Ser "Cross-Updated Active Noise Control System with Online Secondary

Path Modeling" *IEEE Transactions on Speech and Audio Processing*, **9**(5), pp. 598-602, (July 2001).
9. M. T. Akhtar, M. Abe and M. Kawamata "A New Variable Step Size LMS Algorithm-Based Method for Improved Online Secondary Path Modeling in Active Noise Control Systems" *IEEE Transactions On Audio, Speech, and Language Processing*, **14**(2), pp. 720-726, (March 2006).
10. D. Zhou, V. DeBrunner, "Efficient adaptive nonlinear filters for nonlinear active noise control, *IEEE Transaction on Circuits and Systems I*, **54**(3), pp. 669–681, (March 2007).
11. P. Davari and H. Hassanpour "Designing a New Robust On-Line Secondary Path Modeling Technique for Feedforward Active Noise Control Systems" *Elsevier Signal Processing*, **89**, pp 1195-1204, (2009).
12. H. Hassanpour and P. Davari "An efficient online secondary path estimation for feedback active noise control systems" *Elsevier Digital Signal Processing*, **19**, pp. 241–249. (2009).
13. Ardekani, I. T. and Abdulla, W. H. "On the convergence of real-time active noise control systems" *Elsevier Journal on Signal Proceesing*, in Press, pp. 1-13. (2010).
14. E. Bjarnason "Analysis of the filtered-x LMS algorithm" *IEEE Transactions on Speech and Audio Processing*, **3**(6), pp. 504–514, (Nov 1995).
15. L. Vicente and E. Masgrau "Novel FxLMS convergence condition with deterministic reference" *IEEE Transactions on Signal Processing*, **54**(10), pp. 3768–3774, (Oct. 2006).
16. Y. Xiao, A. Ikuta, L. Ma and K. Khorasani "Stochastic analysis of the FXLMS-based narrowband active noise control system" *IEEE Transactions on Audio, Speech and Language Processing*, **16**(5), pp. 1000–1014, (July 2008).
17. I. Tabatabaei Ardekani and W. Abdulla "Theoretical convergence analysis of FxLMS algorithm" *Signal Processing*, **90**(12), pp. 3046–3055, (2010).
18. B. Widrow, J. R. Glover, J. M. McCool, J. Kaunitz, C. S. Williams, R. H. Hern, J. R. Zeidler, E. Dong and R. C. Goodlin, "Adaptive Noise Cancelling: Principles and Applications" *Proceeding IEEE*, **63**, pp. 1692-1716, (Dec 1975).
19. S. Haykin, *Adaptive Filter Theory*, 4th ed., Prentice-Hall, New Jersey. (2002).
20. R. Mustafa and M. A. Mohd Ali, "Fast and Efficient Least Mean Square Algorithm for Active Noise Control System Identification" *Acoustical Letter, Acoust. Sci. & Tech.* **33**(2), pp. 111-112, (2012).

Condition Evaluation of Storage Equipment Based on Improved D-S Evidence Theory

Zhang Xiao-yu and He Hua-feng

Xi'an High Tech Research Institute, 710025, Xi'an, China

Abstract. Assessment and prediction of the storage equipment's condition is always a difficult aspect in PHM technology. The current Condition evaluation of equipment lacks of the state level, and a single test data can't reflect the change of equipment's state. To solve the problem, this paper proposes an evaluation method based on improved D-S evidence theory. Firstly, use analytic hierarchy process (AHP) to establish a hierarchical structure model of equipment and divide the qualified state into 4 grades. Then respectively compare the test data with the last test value, historical test mean value and standard value. And the triangular fuzzy function to calculate the index membership degree, combined with D-S evidence theory to fuse information from multiple sources, to achieve such equipment real-time state assessment. Finally, the model is used to a servo mechanism. The result shows that this method has a good performance in condition evaluation for the storage equipment

1 Introduction

Prognostics and health management (PHM) technology is an important means to achieve maintenance and automotive support [1]. Condition assessment is an important part of the PHM design, which is a link between fault diagnosis and fault prediction o, and has been widely used. And how to make full use of test data of the storage needs a further research.

Storage equipment has the characteristics of "long-term storage, use at one time". In order to accurately grasp its condition, it's needed to test it regularly. Currently, the state assessment of equipment still applies the standard of "true or false", that is, it is considered qualified as long as index test qualified. This method has a limitation to obtain equipment's health status, which influences on arranging combat duty and maintenance solution of equipment [2-3]. According to the storage profile and performance requirements, this paper divides the storage equipment's health status level, and proposes a method based on improved evidence theory for health condition assessment.

2 Condition assessment index

2.1 Analysis of health index

Storage equipment usually is a more complex system, the need to establish a comprehensive and reasonable health index system. Referring to the idea of hierarchical analysis, this paper establishes a hierarchical structure model of equipment state assessment. This model mainly concludes goal layer, criterion layer and index layer. Goal layer represents the health state of the equipment. Criterion layer describes the health evaluation criteria and item of the equipment. Index layer is the specific test parameters of the equipment. And the hierarchical structure model of equipment state assessment is shown in Fig. 1.

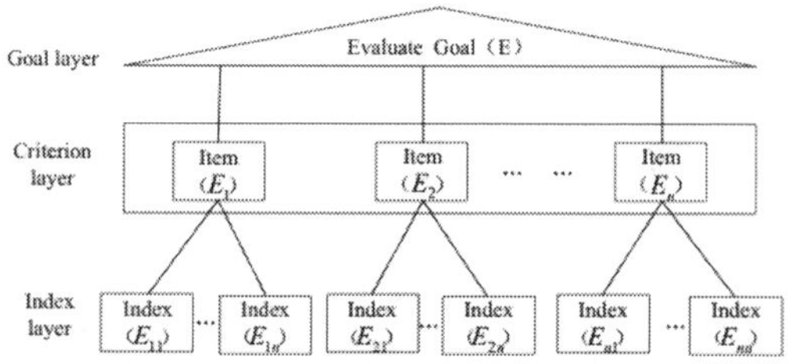

Figure 1. Hierarchical structure model of equipment state

2.2 Health status classification and membership function

It's needed to divide the equipment's state level firstly. Considering the performance and degradation degree, the qualified equipment (or the index) can be divided into good state, average state, degraded state and pseudo-fault state. There is only a fuzzy transition region instead of clear boundary between every two states. Taking actual situation of equipment's degradation and related experts' experience into account [4], this paper proposes a method of fuzzy trigonometric function to calculate state's membership degree, as shown in Fig. 2.

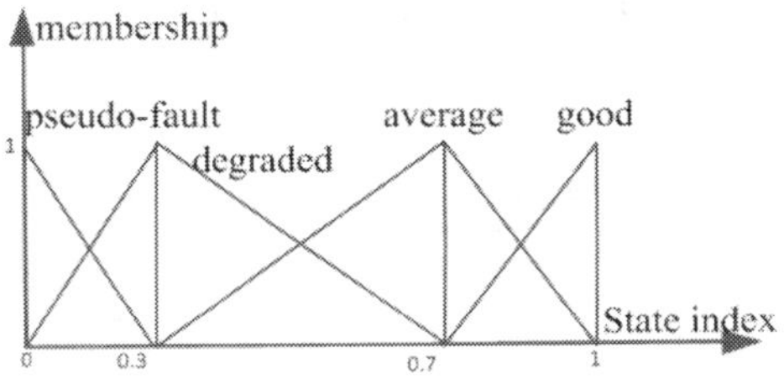

Figure 2. Fuzzy trigonometric membership function

The good state describes that the performance of the equipment is perfect, and every indicators are almost perfect. The average state represents that the equipment's state is just in a normal level. The degraded state depicts that the equipment is greatly degraded and the degradation rate speeds up. The pseudo-fault state portrays that the equipment has been in a fault season and many of indicators have already reached the critical value.

3 Test data process

It's limited to assess long-term storage equipment's health status just by once test data. Therefore, this paper proposes a quantitative method to process the test data. This method includes four kinds of data: the current data, the last test data, the historical average value and the standard value. What follows is quantization of current test data and historical average value as an example.

Firstly, calculate the absolute deviation of current test data and historical test data is calculated using formula. $\delta_h = |x - x_h|$. x refers to the current test data, and x_h represents the historical average value。

Then, calculate the normalized quantization value λ_h . The calculation formula adopts a method of half trapezoid function, as shown in following equation.

$$\lambda_h = \begin{cases} 1 & , \quad \delta_h \le 0.3\delta_0 \\ \dfrac{\delta_0 - \delta_h}{0.7\delta_0} & ,0.3\delta_0 < \delta_h \le \delta_0 \\ 0 & , \quad \delta_h > \delta_0 \end{cases} \tag{1}$$

And, the value δ_0 portrays the maximum error limit.

By the same way, λ_l represents the quantization value of current data and last data, and λ_s portrays the quantization value of current data and standard value. Considering that the importance of these three parameters are different, the weights of these three values λ_h 、λ_l and λ_s respectively are 0.2, 0.3 and 0.5. And the calculation method of state coefficient λ can be depicted as:

$$\lambda = \begin{cases} 1, \lambda_h = \lambda_l = \lambda_s = 1 \\ 0.2\lambda_h + 0.3\lambda_l + 0.5\lambda_s, \begin{matrix} \lambda_h, \lambda_l, \lambda_s \ge 0.7 \\ \textit{but not all is } 1 \end{matrix} \\ \min(0.2\lambda_h, 0.3\lambda_l, 0.5\lambda_s), 0 < \textit{any of } \lambda_h, \lambda_l, \lambda_s < 0.7 \\ 0, \quad \textit{any of } \lambda_h, \lambda_l, \lambda_s \textit{ is} 0 \end{cases} \tag{2}$$

4 Condition evaluation based on evidence fusion

4.1 D-S evidence theory

Evidence theory was proposed by Dempster [5] in 1967 and Shafer [6] made a great contribution to its development. In recent years, it is widely used in information fusion, decision analysis and target recognition [7-8], etc.

Definition1. The core of evidence theory is the method to compound evidence. Supposing $B_1, \dots, B_k$ and $C_1, \dots, C_r$ are two different focal elements , the basic compound formula of evidence theory can be shown as follow:

$$m(A) = \begin{cases} \dfrac{\sum\limits_{\substack{i,j \\ B_i \cap C_j = A}} m_1(B_i)m_2(C_j)}{1-K} & \forall A \subset U \quad A \ne \phi \\ 0 & A = \phi \end{cases} \tag{3}$$

In the formula, $K = \sum\limits_{\substack{i,j \\ B_i \cap C_j = A}} m_1(B_i)m_2(C_j) < 1$ is the conflict coefficient, and $m(A)$ is the basic probability assignment of focal element A .

4.2 The improvement of evidence theory

The compound rule is simple and practical, but when the evidence is in high conflict, the fusion result is usually contrary to the reality [9]. So far, many scholars have studied the improved method for evidence theory from different aspects. This paper combines previous research results [10-12], and proposes an improved evidence theory from two approaches of revising evidence sources and improving synthesis method, which can handle the conflict effectively.

This paper introduces a generalized Jaccard coefficient to describe the similarity between bodies of evidence to revise evidence sources [13].

The membership of every two evidences can be described as two vectors $\overrightarrow{M_i}$ and $\overrightarrow{M_j}$, and the similarity between them can be described by the generalized Jaccard coefficients as follows:

$$sim(\overrightarrow{M_i}, \overrightarrow{M_j}) = \frac{\sum\limits_{i=1}^{n} m_{ii} \bullet m_{ji}}{\sum\limits_{i=1}^{n} m_{ii}^2 + \sum\limits_{i=1}^{n} m_{ji}^2 - \sum\limits_{i=1}^{n} m_{ii} \bullet m_{ji}} \tag{4}$$

Take $s_{ij} = sim(\overrightarrow{M_i}, \overrightarrow{M_j})$ to account the similarity between bodies of evidence $\overrightarrow{M_i}$ and $\overrightarrow{M_j}$, and set up the similar matrix according to similarity of every two evidences by the equation $S(i,j) = \begin{cases} s_{ij} & i \ne j \\ 1 & i = j \end{cases}$.

Take the support as the weight of evidence, so that $w_i = \sum_{j=1,j\neq i}^{n} s_{i,j}, i = 1,2,\ldots,n$. And the weight vector of all the evidence can be described as $W = (w_1, w_2, \ldots, w_n)$. Then obtain the maximum of vector and use it to calculate the discount rate of evidence as the equation $\alpha_i = \frac{w_i}{\max\{w_1, w_2, \ldots, w_n\}}$ shows. Finally use the discount rate to revise bodies of evidence. The corrected BPA value is expressed as follows:

$$m_i^*(A) = \alpha_i m_i(A)$$
$$m_i^*(\Theta) = 1 - \sum m_i^*(A) \quad (5)$$

In the aspect of conflict management, this paper introduces the idea of local conflict distribution. The obtained modified BPA function is used to synthesize, and the synthesis formula is as follows:

$$\begin{aligned} & m(\phi) = 0 \\ & m(A) = \sum_{\substack{i,j \\ B_i \cap C_j = A}} m_1(B_i) m_2(C_j) + c(A), (A \neq \phi, A \in U) \\ & c(A) = \sum_{\substack{i \neq j \\ A \cap X = \phi}} \frac{m_i(A)}{m_i(A) + m_j(X)} m_i(A) m_j(X) \\ & m(\Theta) = 1 - \sum m(A) \end{aligned} \quad (7)$$

This synthesis formula has been proved to satisfy the good properties of D-S evidence theory, such as the exchange law and the combination law. And the value $m(\Theta)$ portrays the confidence level of the result.

4.3 Multi-index evaluation model based on evidence Fusion

The storage equipment is a complex system, with many test parameters. The state coefficient of each parameter is fuzzy and uncertain. Therefore, this paper proposes a method based on improved D-S evidence theory to evaluate the status of storage equipment. Firstly, take the indicators of index layer in part 2 as the evaluation factor set, and obtain the BPA value of each state. Then use the D-S evidence theory to fuse all the data, and finally realize the state evaluation of the storage equipment. The state evaluation model of the storage equipment based on the improved D-S evidence theory is shown in Fig. 3.

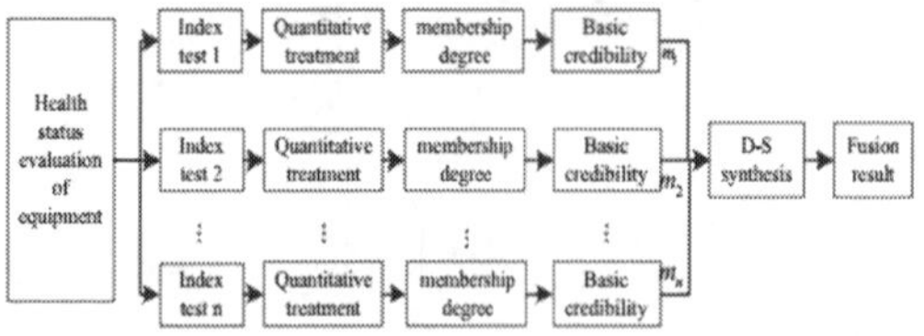

Figure 3. State assessment model based on D-S theory

As the combination rule of evidence theory satisfies the associativity and commutativity, the difference between the fusion orders has no effect on the result. Firstly, fuse various indicators under the same criterion layer, so that the state of every criterion can be got, by which it's easy to distinguish factors influencing the status of the equipment. Then, blend the BPA value of every criterion to get the real state of the equipment. Next this article will carry on the case analyses in the section 5.

5 Case Analyse

The servo mechanism is an important part of the attitude control system of launch vehicle. Most of the time, it's in a state of storage. This section takes a servo mechanism as an example and analyses its state evaluation.

5.1 State evaluation model of servo mechanism

First of all, according to the model of section 2, establish a hierarchical structure model of the servo's state. According to the test content and appraisal principle, the health index system mainly includes two aspects: static characteristic and dynamic characteristic [5], and each feature can be refined to more detailed indicators. Therefore, the establishment of the health indicators hierarchy model of the servo mechanism is shown in Fig. 4.

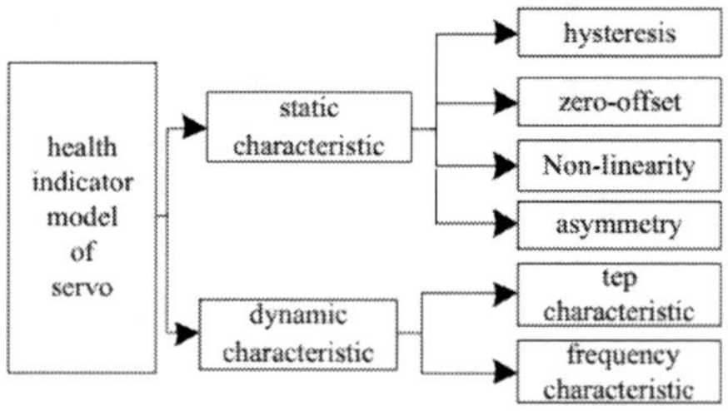

Figure 4. Health indicators hierarchy model of servo

5.2 State evaluation

According to the model established above, classify the test parameters of the servo mechanism. And also, obtain the corresponding test parameters. This servo mechanism has been in service for 5 years from 2010 to 2015. Take its test data in 2015 as the research object to check the method proposed above .Process the four key parameters of static parameter by this method. Its static characteristic test data of 2015 are shown in Table 1.

Table 1. Test data of static characteristic

No.	x	x_l	x_h	x_s	δ_0
1	0.630	0.540	0.520	0.5	0.200
2	26.80	27.50	27.12	27.00	2.500
3	48.139	46.834	46.945	47.15	3.00
4	10.528	10.983	10.346	11.00	1.000

Use the method in section 3 to process the above data and get processed data in Table 2.

Table 2. Quantization of static characteristic data

No.	λ_l	λ_h	λ_s	λ
1	0.7857	0.6429	0.500	0.6429
2	1	1	1	1
3	08071	0.86	0.9576	0.8929
4	0.7786	1	0.7543	0.8107

According to the processed data, calculate the membership degree of each parameter, and the result is shown in Table 3.

Table 3. Membership degree of each parameter

No.	good	average	degraded	pseudo-fault
1	0	0.8573	0.1427	0
2	1	0	0	0
3	0.643	0.357	0	0
4	0.369	0.631	0	0

Use weighting coefficient calculation formula in section 4 to calculate the corrected membership of each parameter, and the result is shown in Table 4.

Table 4. The revised Membership

No.	good	average	degraded	pseudo-fault	Θ
1	0	0.4934	0.0821	0	0.4245
2	0.5755	0	0	0	0.4245
3	0.638	0.3542	0	0	0.0078
4	0.369	0.631	0	0	0

In the frame of discernment ,the BPA value of the static characteristic after being synthesized is β_1=(0.5009, 0.4896, 0.0003, 0, 0.0092）.

By the same way, obtain the BPA value of the static characteristic as β_2 = (0.3048, 0.5849, 0.1, 0, 0.0103). Fuse the BPA value β_1 and β_2 to once again to get the BPA value of servo's health as β = (0.3947, 0.5872, 0.0175, 0, 0.0006). The uncertainty of the result is 0.0006, and the health status level is in average state. The comparison of the BPA value shows that the degradation of dynamic characteristics is more severe than the one of static characteristics.

5.3 Result analysis

Assess the servo's test data from 2010 to 2015 and the obtained states are good, good, good, good, average and average. And, from the result, we can know that this servo had gone to a stable storage period in 2014. The obtained states are coincided with the states experts give according to the test data basically, which verifies the validity of this method.

From the above example, different from the former method using state characteristic to assess the state of the equipment, this paper's method based on the improved evidence theory is more suitable for the equipment which has the characteristics of "long-term storage, a small amount of testing" or other equipment with no characteristic state parameters for real-time monitoring than other methods,

6 Conclusions

State assessment of the storage equipment needs consider a variety of indicators. This paper considering both performance and degradation degree of the equipment, divides its qualified state into good state, average state, degraded state and pseudo-fault state, and proposes a condition evaluation method based on improved evidence theory. In the section 5, that case verifies the proposed method's validity. This method can assess the storage equipment's real-time state, which improves the past evaluation method effectively and makes an important significance to improve the test and evaluation of other equipment.

References

1. Peng Yu, Liu Datong, Data-driven prognostics and health management: A review of recent advances [J]. Chinese Journal of Scientific Instrument, 2014, 35(3): 481-495.
2. Jin G, Matthews D E, Zhou Z B. A Bayesian framework for on-line degradation assessment and residual life prediction of secondary batteries in spacecraft [J]. Reliability Engineering and System Safety, 2013, 113: 7-20.
3. Cui J J. Rocket and missile testing technology [M]. Beijing: National Defense Industry Press, 1999:174-187.
4. Cong L H, Xu T X, Dong Q, et al. Missile condition assessment method based on improved evidence theory [J]. Science & Technology Review, 2013, 31(30):15-18.
5. Dempster A P. Upper and lower probabilities induced by a multi-valued mapping [J]. Annals Math Statist, 1967, 38(2): 325-339.
6. Shafer. Mathematical Theory of Evidence [M]. Princeton University Press, 1976.
7. Hu C H, Si X S, Shi X H. Gyroscopic drift combination forecasting model based on evidential reasoning [J]. Control and Decision, 2009, 24(2): 202-205.
8. Li P, Liu S F, Interval-valued Intuitionistic fizzy Numbers Decision-making Method Based on Grey Incidence Analysis and D-S Theory of Evidence [J]. Acta Automatica Sinica, 2011,37(8)： 993-998.
9. Zadeh L A. Review of a mathematical theory of evidence.AI Magazine, 1984, 5(3): 81−83.
10. Zhou Z, Xu X B, Wen C L, et al. An Optimal Method for Combining Conflicting Evidences [J]. Acta Automatica Sinica, 2012, 38(6):976-985.
11. Zhang Y J, Long C, Li D. Conflict Evidence Fusion Method Based on Conflict Representation [J]. PR&AI, 2013, 26(9): 853-858.

DOI: 10.1051/matecconf/20179514007

12. Guo H W, SHI W K, Liu Q K, et al. A New Combination Rule of Evidence [J]. Journal of Shanghai Jiao tong University, 2006, 40(11): 1895-1900.
13. Pan L, Lei Y L, Wang C J, et al. Method on Entity Identification Using Similarity Measure Based on Weight of Jaccard [J]. Journal of Beijing Jiao tong University, 2009, 33(6):141-143.

An Electrochemical DNA Sensor for p53 Tumor Suppressor Gene Detection

Jie Zhang[1], Long Wan[2], Hao Fan[1], Fusheng Liao[1]

[1]Department of Pharmacy, JiangXi University of Traditional Chinese Medicine, JiangXi 330004, China
[2]NanChang Teachers College, JiangXi 330004, China

Abstract. In this work, a sensitive electrochemical DNA sensor based on avidin modified electrode and DNA-functionalized Cds nanoparticle (DFNP) was developed. The DNA-Functionalized Cds nanoparticle contained two kinds of DNA, one was hairpin probe DNA with a biotin at the 3' terminal and a thiol at the 5' terminal, the other is linearity signal DNA. Without hybridized with target DNA, the loop of hairpin impeded biotin linked with avidin on electrode. However, after target hybridization, hairpin was opened and biotin was recognized by avidin resulting in DNA-functionalized Cds nanoparticle was brought on electrode surface. Electrochemical signals of methylene blue (MB) bound to the signal DNA were measured by differential pulse voltammetry (DPV). Introduction By using this new method, we demonstrate that this prototype sensor has been able to detect as low as picomolar p53 tumor suppressor gene with excellent differentiation ability for even single mismatches.

1 Introduction

The p53 tumor suppressor gene, well known as a transcription factor of cell regulation, is the most commonly mutated gene in human tumors. About 50% of all malignancies contain a mutation in p53 and aggressive growth of several types of cancer has been attributed to mutations in this gene. Moreover, p53, as the "master switch" for the control of cell proliferation, plays an important role in inducing cell cycle arrest for DNA repair or apoptosis to eliminate severely damaged cells. Sequence-specific analysis of the p53 gene could help early diagnosis of cancer development and consequently increase the success of the treatment Therefore, sensitive and rapid detection of p53 gene and the mutations in the p53 gene are of great value. [1-6]

The electrochemical DNA sensing technology has received particular attention mostly due to its high sensitivity, selectivity and easy operation with simple instrumentations. [7-10] The DNA recognition event can be detected using different strategies, including intrinsic electroactivity of the nucleic acid, enzyme labels, DNA duplex intercalators, electroactive markers, and metal nanoparticles/quantum dots.

We herein reported a non-immobilizing electrochemical DNA sensor with hybridization occurred in one homogeneous solution that employed DNA-functionalized Cds nanoparticle (DFNP) and avidin modified electrode. In the present study, we have developed a DNA-functionalized Cds nanoparticle, integrated DNA recognition; signal amplification and specific biotin- avdin link functional section, as probe for DNA detection and featured high sensitivity up to low pmtomolar. In target DNA detection, as Figure 1, when the probe DNA of DFNP is under stem-loop state, comparatively huge bulk of Cds nanoparticle and the loop of probe DNA would prevent biotin on the probe DNA be captured by the avidin on the electrode for the spatial effects. After the DFNP solution is incubated in a buffer solution containing target DNA, target DNA would combine probe DNA by forming double strand with complementary sequences, which ccdssed hairpin open and eliminate steric hindrance to biotin. Afterwards, when avidin modified electrode was immersed in a solution after the hybridization event happened, the biotin of the DFNP could have attached to avidin on the electrode and thus was captured on the electrode. The electrochemical signals of MB of signal DNA were measured by differential pulse voltammetry (DPV). Taking advantage of amplification effects of the Cds nanoparticle (CdsNP) and binding specificity of hairpin probe, this biosensor greatly simplifies the electrochemical detection method of DNA and displays higher specificity than the linear probe in DNA detection.

2 Experiment section

2.1 Apparatus

All voltammetric experiments were performed using a CHI 660 electrochemical analyzer (CHI Instrument Inc, USA). Electrochemical experiments were carried out in a 3 ml electrochemical cell at room temperature (25 °C) by using three electrode configurations. A platinum wire served as a

DOI: 10.1051/matecconf/20179514008

counter electrode and an Ag/AgCl as reference electrode with saturated KCl solution.

Unless otherwise noted, all chemicals were purchased from Dingguo Biotechnology Inc. (Shanghai, China) and of analytical reagent grade. The biotin and avidin were pur chased from Sangon Biotechology Inc. (Shanghai, China). All of the solutions were prepared with ultrapure water from a Millipore Milli-Q system.

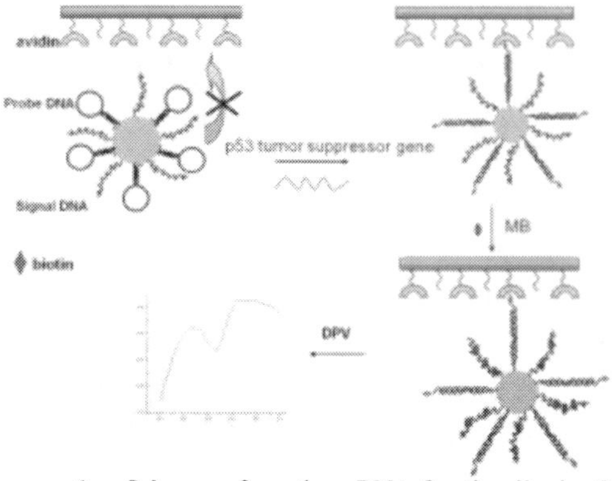

Figure 1. Scheme for the DNA-functionalized Cds nanoparticle-based electrochemical DNA sensor.

2.2 Materials

All the probes and target oligonucleotides were purchased from Takara Biotechnology Company. The oligonucleotides were purified via C18 reversed-phase HPLC and polyacrylamide gel electrophoresis (PAGE). The stem-loop probe oligonucleotide (oligo 1) has a 5′-HS label and a 3′-biotin, which will form the stem at appropriate ionic strength. The MB-labeled signal DNA (oligo 2). The sequence of the target (oligo 3) is perfectly matched to the loop sequence of the probe; oligo 4 contains a one-base mismatch, while oligo 5 has three-base mismatches. Oligo 6 is a random sequence unrelated to the probe sequence. Unless otherwise noted, all chemicals were purchased from Dingguo Biotechnology Inc. (Shanghai, China) and of analytical reagent grade. The biotin and avidin were pur chased from Sangon Biotechology Inc. (Shanghai, China). All of the solutions were prepared with ultrapure water from a Millipore Milli-Q system.

2.2 Preparation of nano CdS

$Cd(NO_3)_2$ and Na_2S solutions were filtered through a 22 m microporous membrane filter prior to use. CdS nanoparticles were prepared by using mercaptoacetic acid as the stabilizer. In brief, 9.22 μl mercaptoacetic acid was added to 50 ml 0.4 mM $Cd(NO_3)_2$ solution, and then the pH was adjusted to 7 with 0.5 M NaOH. The solution was bubbled with nitrogen for 30 min, followed by the slow addition of 1.34 mM Na2S to the mixture solution. The molar ratio of Na2S to $Cd(NO_3)_2$ was kept at 2.5. The reaction was carried out for 24 h under nitro-gen protection and then gradually a brown colloid which is the CdS nanoparticles covered with a carboxyl group was obtained. As TEM images show, the diameter of CdS nanopatricles was about 7 nm.

2.3 Preparation of the DNA-functionalized Cds nanoparticles

The process of probe DNA and signal DNA labeling was performed according to paper as follows: The mixture of 5.6 nM probe DNA and 2.8 nM signal DNA was activated with acetate buffer (pH 5.2) and 1.5 μL of 10 mM TCEP for 1 h, then added to 1 mL of freshly prepared Cds nanoparticles, and shaken gently overnight. Over the course of 16h, the DNA-CdsNP conjugates were aged in salts (0.1 M NaCl, 10 mM acetate buffer) for another 24 h. Excess reagents were removed by centrifuging at 16000 rpm for 30 min. The red precipitate was washed and centrifuged repeatly for three times. The resulting nanoparticles were dispersed into a buffer solution (0.1 M, containing 0.3 M NaCl and 2mM Mg^{2+}, pH 8.1) and stored at 4 °C.

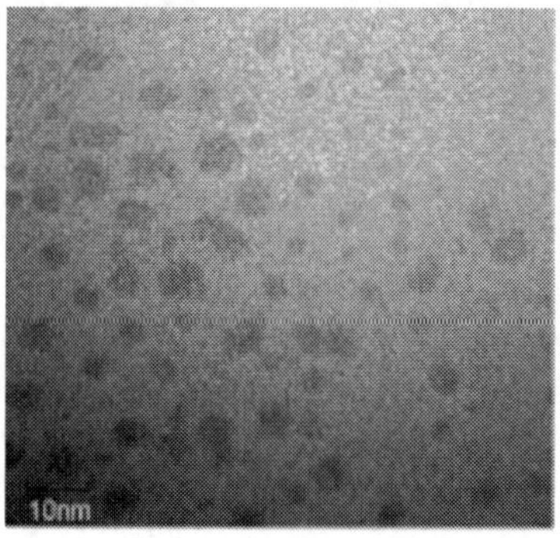

Figure 2. TEM image of the synthesized CdS nanoparticles.

2.4 Preparation of avidin-coated electrode surfaces

The gold electrode was exposed to an ethanolic 5 mM solution of 3,3'-dithiodipropionic acid for 30 min, followed by water rinsing. 5 μl of 100 mg mL-1 EDC solution was then placed on the surface, followed immediately by 5 μL of 100 mg mL-1 NHS solution. These solutions were allowed to interact with the electrode with 3,3'-dithiodipropionic acid modification for 30 min in a 100% humidity environment to prevent solution evaporation. The surface was then rinsed with water and immersed in an aqueous 0.2 mg mL-1 avidin solution for at least 120 min, after which the surface was rinsed again. The electrode was then exposed to a 1mM 2-aminoethanol solution (pH 8.0, adjusted using HCl) for 60 min, rinsed, and used in hybridization detection.

3 Results and discussion

3.1 Optimization of testing conditions

Many experimental parameters, such as hybridization temperature, hybridization time, and Mg^{2+} ion strength, capture time, can affect the detection sensitivity of the assay. A series of optimizations of these parameters were carried out to improve the sensitivity and selectivity of the DNA assay. Figure 3 shows the effect of the capture time on the signal of DNA detection.

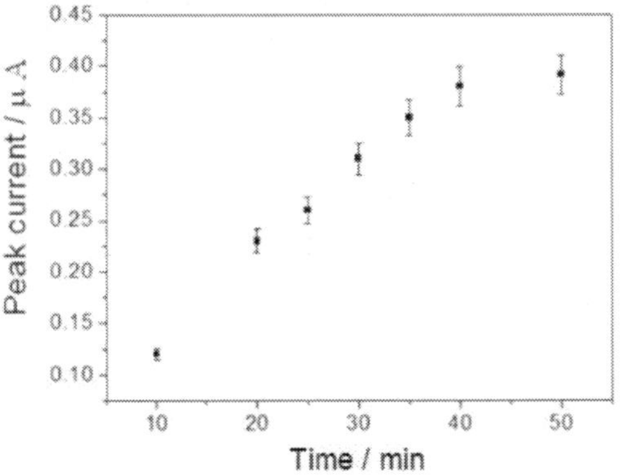

Figure 3. The DPV responds to 1.2×10^{-11} M target DNA detection with different capture time.

In this study, a 1.2×10^{-11}M complementary DNA target was incubated with DFNP for 30 min at a series of temperatures ranging from 25 to 45 °C. The highest signal is obtained at 37 °C. Therefore, a hybridization temperature of 37 °C was selected in the following experiments. The effect of the hybridization time has also been investigated. The target DNA was incubated with DFNP at 37 °C for different times from 10 to 50 min. The signal responses of the assay to the different hybridization times are shown in Supporting Information Figure S6. It can be seen from this figure that the signal increases rapidly when the hybridization time increases from 10 to 30 min and then increases slightly from 30 to 50 min, indicating that the hybridization reaches equilibriumat 30 min. As a result, a reaction time of 30 min was selected for the DNA hybridization.

The cation Mg^{2+} plays an important role in DNA hybridization. For a hairpin DNA probe-based DNA assay, there is hybridization competition between intramolecular stem part hybridization and intermolecular target DNA-probe hybridization. Therefore, Mg^{2+} concentration in the hybridization buffer should be balanced. Although a high concentration of Mg^{2+} can enhance target DNA hybridized to the loop of the structure, it may also strengthen the binding of the stem part of the probe, which may adversely affect the performance of the assay. The effect of Mg^{2+} concentration in the hybridization buffer on the response signal of the assay. The assay produces the highest signal when the Mg^{2+} concentration is 2 mM. Accordingly, a 2 mM Mg^{2+} concentration was used throughout the experiments to obtain the highest signal.

3.2 Principle of DNA detection

We employed a stem-loop DNA probe dually labeled with HS and biotin at the 5′- and the 3′- end, respectively, which could be facilely immobilized at Cds nanoparticle surfaces via the Cds-S bridge and hybridized with target DNA. The signal DNA with HS at 5′ end, which could provided electrochemical signal. Two kinds of DNA have been immobilized at Cds nanoparticle to construct DFNP.

The detection strategy is demonstrated in Figure 1. Before the hybridization, the DFNP remained in the stem-loop structure, which forced the biotin to be closed to the Cds nanoparticle. Due to the steric effect of the Cds nanoparticle, the biotin was prevented from conjugating with the avidin on the electrode and resulting in that the DFNP could not be captured by the electrode. After hybridized with the target DNA, the DFNP's loop-stem structure opened and then the biotin molecule was easily bound to avidin modified electrode and resulting in that the DFNP could be captured by the electrode and the capture efficiency was proportion with the concentration of the target DNA. The target hybridization event can be sensitively transduced via detecting the electrochemical reduction current signal of MB at the DFNP.

Indeed, in stem-loop state we only found a relatively small and stable background current (Figure 2 curve a). As we challenged the sensor with 1.1×10^{-11} M target DNA, which was expected to open the stem-loop, a large signal (Figure 4 curve b) was observed. We found that as the concentration of the complementary target DNA sequence continuously increased, which was logarithmically related to the target protein concentration from 8.3×10^{-12} to 8.3×10^{-9} M. The equation for the resulting calibration plot was calculated as $y=0.139\log x+0.163$ (x is the concentration of target DNA divided by pM, y is the DPV peak current value) with correlation coefficient of 0.9923 and detection limit of 6.2×10^{-12} M (>3 SD). This sensitivity reflected the high signal amplification of this method and the improved signal gain as a signal-on sensor used MB as indicators.

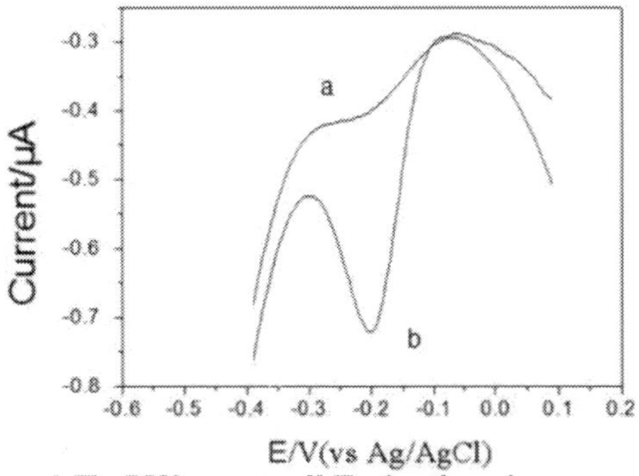

Figure 4. The DPV response of MB when the probe was incubated in phosphate buffer (a) and target DNA (b).

4 Conclusions

In summary, we completed a electrochemical DNA sensing strategy, which allowed the hybridization between DNA probe and target DNA occur in homogeneous solution. This assay protocol is simple, convenient and cost-effective by using this novel method, we could conveniently detect as few as 6.2×10^{-12} M target DNA. We propose that it might be a promising approach to perform DNA-based diagnostics where resources are limited, such as small clinics in developing countries or field detection.

DOI: 10.1051/matecconf/20179514008

Acknowledgements

This work was kindly supported by the JiangXi Science and Education Committee (Grant No. GJJ14610, GJJ14620), JiangXi University of Traditional Chinese Medicine Foundation (Grant No.JZYC14A08). JiangXi Graduate student Foundation (YC2014-S287, YC2014-S295).

References

1. K. Polyak, Y. Xia, J. L.Zweier, K.W. Kinzler and B. Vogelstein: Nature Vol. **300–305** (1997), p. 389
2. G. I. Evan, K. H. Vousden: Nature Vol. **342–348** (2001), p. 411
3. D. P. Guimaraes, P. Haincdst: Biochimie Vol. **83–93** (2002), p. 84
4. K. H.Vousden, X. Lu: Nat. Rev. Cancer Vol. **594–604** (2002), p. 2
5. J. Wang, G. Rivas, X.H. Cai, M. Chicharro, C. Parrado, N. Dontha, A. Begleiter,M. Mowat, E. Palecek and P. E. Nielsen: Anal. Chim. Acta Vol. **111–118** (1997) , p. 344,
6. D.E.J. Koshland: Science Vol. **1953** (1993), p. 262
7. J. Wang, D. Xu, A. N. Kawde, R. Polsky: Anal. Chem. Vol. **5576-5581** (2001), p. 73
8. H.Cai,; X. N.Cao, Y.Jiang, P. G.He, Y. Z. Fang:Anal. Bioanal. Chem. Vol. **287-293** (2003) p.375
9. F. Jelen, B. Yosypchuk, A. Kourilova, L. Novotny, E. Palecek: Anal. Chem. Vol. **4788-4792** (2002) p.74
10. J. Wang, A. N. Kawde, A. Erdem, M. A. Salazar: Analyst Vol. **2020-2025** (2001) p.126.

MATEC Web of Conferences **95**, 14009 (2017) DOI: 10.1051/matecconf/20179514009
ICMME 2016

Method for Detection of the Lean Meat Ratio in Pork Carcass Based on Fiber Optic Sensor

Zhichao Sun, Sheng Leng , Wantai Ma, Qing Xu and Xiaoming Xie

College of Mechanical and Electrical Engineering, Nanjing University of Aeronautics and Astronautics, No. 29, Yudao Street, Nanjing 210016, China

Abstract: One of the main evaluations in pork carcass is the lean meat ratio. A method based on Fiber optic sensor and NIR (Near Infrared) to acquire the ratio was put forward. This method acquired the reflected light intensity of the meat inside the pork carcass, by using double optical paths to reduce the errors caused by the instability of light. The fiber was lead into a probe which can be stabbed into the pork carcass, and then the reflected light was detected by a circuit named LIA (lock in amplifiers) to recognize the weak reflected light intensity effectively. The results suggested that the reflected light of the fat part of the pork was much higher than that of lean part, and through medium filter, the conditions of fluctuation was improved. The system was proved to be quick and accuracy for the detection of lean meat ration in pork carcass.

1 Introduction

Pork plays a very important role in the meat consumption in China. And with the development of society, people's requirements on the quality of pork meat becomes higher and higher. They prefer lean meat to fat meat, thus a pork's values mostly depends on the lean meat percentage. The percentage has become a very important indicator of a pork when grading the pork carcass [1]. To create a fair marketplace, many technologies had been applied into distinguishing fat meat and lean meat of pork such as ultrasonic technology, computer vision, opto-electronic, etc.

Based on Ultrasonic technology, SFK Company of Denmark developed UltraFom 300 to detect the reflected ultrasonic values of pork carcass [2]. Analyzing the different reflected values, the fat depth is got, and through collecting more than 3200 points of the pork carcass, a 3D ultrasonic image is established quickly. It can also get the estimation of marble score of pork carcass and the lean meat ratio automatically. This technology is widely used in European and American countries but seldom in China.

Computer vision is another technology used in pork carcass [3], [4]. Image-Meater is developed by CSB Company of German. It uses cameras to take photos of the pork carcass, analyzes the images and finds the relationship between image data and lean meat ratio. But the efficiency is not high as much as ultrasonic and opto-electronic.

Opto-electronic mainly uses the different reflected light values of fat and lean to detect the boundary and calculate the length of fat and lean meat. This technology is widely used in many companies like SFK, Hennessy and Destron Company.

The rapid development of the Near-infrared (NIR) Spectrum Detection technology provides a new method to detect the quality of meat nondestructively [5]. Organics or some inorganics are all made up of various molecules. In these molecules, there exists various groups such as C=C, N=C, O=C, 0=H, N=H, and these groups' movements like expansion, vibration and bending all have their own frequency. Some power of the light will be absorbed by the groups. Most of the organic compounds of meat like protein, fat and moisture have different hydrogen groups. Therefore, the reflected light values are different for fat and lean meat.

As a new sensor, FOS [6] (fiber optic sensor) is widely used in measuring sound field, electric field, pressure, temperature and velocity. It has many talent characters such as high sensitivity, low transmission loss, etc. Therefore, this article improves the opto- electronic by applying fiber optic and probe into acquiring the intensity of the reflected light of the inside meat surface, distinguishing the fat and lean meat and acquiring the length of fat and lean meat finally.

2 System design

The main components of the meat is protein (including myofibril, myosin), water, fat, etc. These substances have different absorption or reflectance of light, thus the lean meat and fat meat under the same light should own different reflectance. According to the Kubelka-Munk theory, the reflectance can be defined as:

DOI: 10.1051/matecconf/20179514009

$$\ln\frac{1}{R} = \frac{S}{K} + B \quad (1)$$

where R represents the light reflectance of pork; S and K represent the absorption coefficient and scattering coefficients respectively; B is a constant. The reflectance R is defined as:

$$R = \frac{I}{I_0} \quad (2)$$

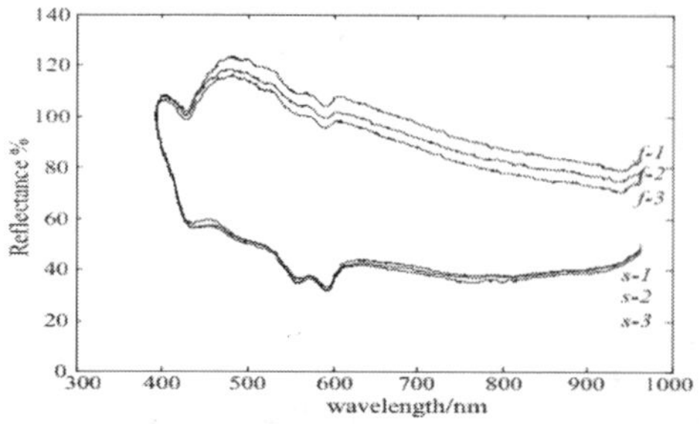

Figure 1. The different light reflectance of lean and fat meat

where I represents the intensity of reflected light and I_0 represents the intensity of input light. Obviously the lean and fat meat have different S and K coefficients and Fig. 1 shows that *f-1, f-2, f-3* are the reflectance of fat meat and *s-1, s-2, s-3* are the reflectance of lean meat. Obviously the reflectance of fat are much higher than that of lean part which gives a method to detect the fat and lean meat.

2.1 The design of FOS and Probe

To avoid the errors caused by the light source, a double optical paths system based on fiber optic is designed to detect and measure the fat and lean meat. Different from measuring the light intensity directly, double optical paths method has two signal tunnels. One is called signal arm which transmit the light from light source to the inside of pork carcass through probe, and then to the light detector after being reflected, and the other is called reference arm which means the light goes to the light detector directly without being reflected. When the light power changes, two arms will help reduce the changes. The system is shown in Fig. 2.

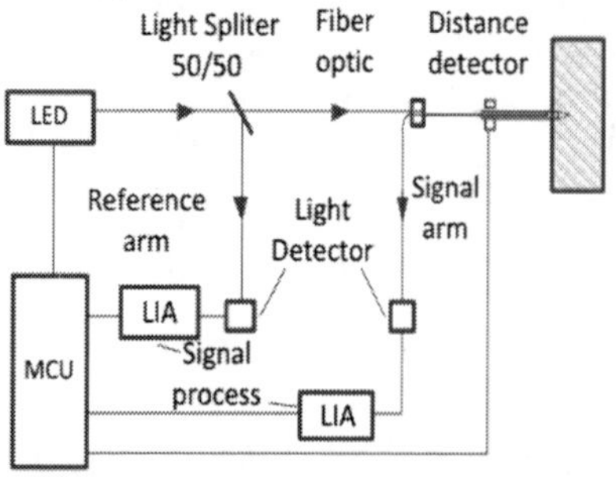

Figure 2. Double optical paths system

MCU controls the light source and receives the signals from light detectors and distance detector. The light splitter is to divide the light into two parts. One goes to the light detector directly, and the other goes to a probe which stabs into pork carcass to get reflected light from the surface of fat or lean meat and then goes back to another light detector. The reflectance then can be acquired by dividing signal arm values with reference arm values, and then the depth of probe into the pork carcass is measured by the distance detector.

The probe is designed and shown in Fig. 3. At the end of the probe, a blade is designed to stab into meat, and near the blade, there exists a small window for light. When the light arrives at the reflector glass through fiber optic, its direction changes and arrive at the surface of meat, and then get reflected and goes back to the fiber optic.

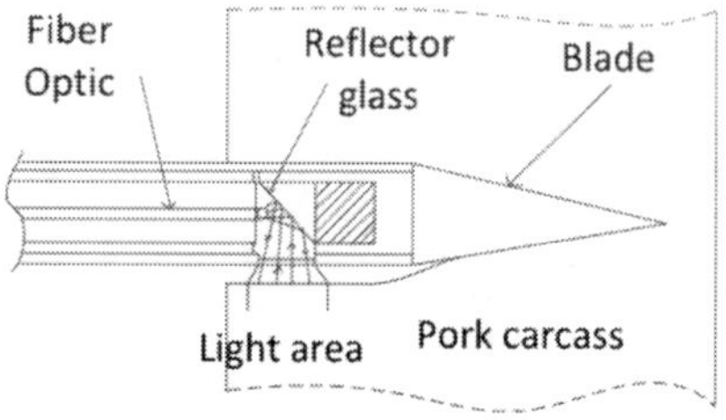

Figure 3. The design of puncture probe

2.2 The detection of light signals

The output power of the LED light is about 1.5mW which is too weak to be detected, especially for the light which is absorbed. What's more, there are many noises in circuits caused by resistors, capacity and many other active devices. These noises are mixed with useful signals and when it is directly amplified, the noise will cover the light signals. Thus, a useful technology named Lock-in-amplifiers (LIA) is taken to help detect the weak signals [7]. The light signal is modulated by a high-frequency (w) signal which can avoid $1/f$ noise. When the whole signal is amplified, a BPF (ban-pass filter) filter out most of the low frequency noise and allow the useful signal pass. And then the PSD (phase-sensitive detector) help shift the frequency of modulated signal from w to 0. Then the A/D converter will receive a DC signal after LPF (low-pass filter). The schematic diagram is shown in Fig. 4.

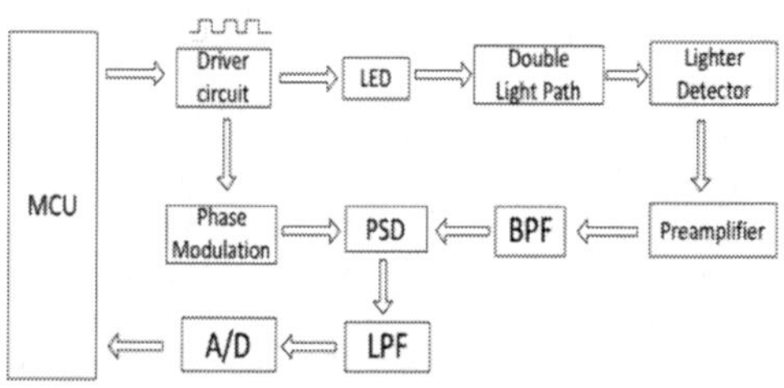

Figure 4. LIA detecting schematic

2.3 Calculation of the length of fat and lean part

When the probe fully pass through the carcass, a series of values are acquired. These values represents the different reflected light intensity. To judge the kinds of meat correctly, a threshold V is acquired.

$$V = \frac{\sum values}{Num} \quad (3)$$

It should be noticed that the values must be collected inside the pork carcass. Before calculating the length of the fat and lean part, the values in the air should be removed.

$$f(v) = \begin{cases} v \geq V & fat \\ v < V & lean \end{cases} \quad (4)$$

Eq. 2 shows how to distinguish the fat and lean part.

3 Result and discussion

Based on the theory above, a test experiment was constructed for checking the effectiveness and precision. The light source used LED (wavelength = 850nm), and the diameter of the fiber optic is 600um. And the diameter of the probe is about 5 mm. The meat samples must be fresh and have distinct boundary between the fat and the lean.

Taking different points of the pork carcass, the result is shown in Fig. 5.

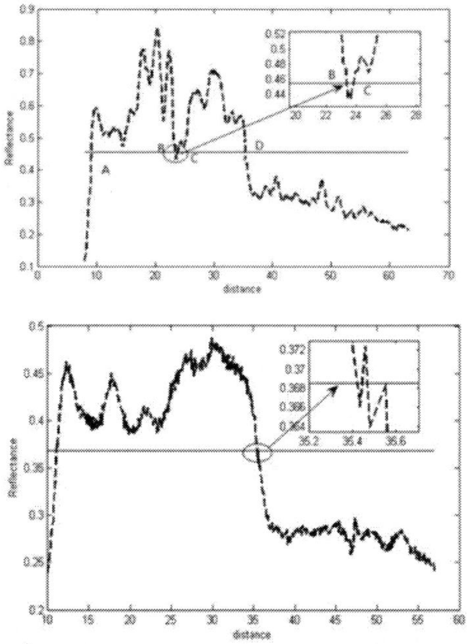

Figure 5. Result of the experiment (up: serious condition, down: ideal condition)

Apparently, the reflected light intensity of the lean is higher than that of the lean. However, the original data is difficult to calculate the lean meat ratio. The fluctuation of Fig. 5(up) is very serious, and there are some points which should be on the fat part but the values are under V line. The system will mistake this part as lean meat, then the length of the fat is L in Eq. 5.

$$L = (B.\text{distance} - A.\text{distance}) + (D.\text{distance} - C.\text{distance}) \quad (5)$$

Fig. 5(down) shows the ideal conditions that the fat and lean are easily to tell apart. But at the boundary of fat and lean meat in Fig. 5(down), there exists a slight fluctuation which will increase the work to figure the lean meat ratio.

Obviously, these situations should be avoided as much as possible. To remove the influence of the serious fluctuation of the signals, a smooth filter algorithms-medium filter is chosen.

And the result is shown in Fig. 6. The solid line is the curve after medium filter and the dashed line is the original one. It can be seen that the local fluctuation is improved and the whole curve is smoother than before. Calculate all the cross points and acquire the length of all the fat and lean part like Eq. 5.

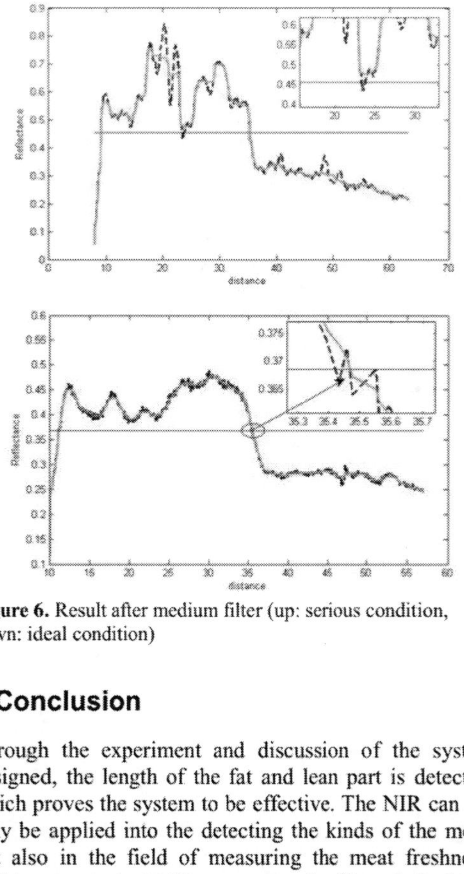

Figure 6. Result after medium filter (up: serious condition, down: ideal condition)

4 Conclusion

Through the experiment and discussion of the system designed, the length of the fat and lean part is detected, which proves the system to be effective. The NIR can not only be applied into the detecting the kinds of the meat, but also in the field of measuring the meat freshness, moisture content, etc. The average value V can help detect the kinds of pork meat correctly and the medium filter can be used in condition where the curves need to be smoothed effectively. By acquiring the fat and lean length of different parts of the pork carcass, the whole lean meat ratio of the pork then can be obtained. Through the integration of these methods, a kind of portable, low-cost and accurate solution can be applied in the food industry.

Acknowledgement

This work is being founded by the National Natural Science Foundation, China (No.51205201) and Innovation foundation for postgraduate students in

Nanjing University of Aeronautics and Astronautics (No.ykfjj20150513).

References

1. N. Zhang, G. Zhou, X. Xu, Food Ferment. Ind., **31**, 86 (2005)
2. J. Brodum, Can. J. Anim. Sci. **76**, 443 (1998)
3. J. Yin, G. H. Zhou, X. L. Xu, Food Sci. (Beijing, China), **31**, 32(2010)
4. S. W. Zhao, Y. K. Peng, H. Y. Zhang, J. Food Saf. Qual, **3**, 589(2012)
5. X. J. Yang, J. Y. Zhao, J. S. Zhao, Food Sci. Technol. (Beijing, China), **37**, 267(2012)
6. F. Liu, W. He, F. Luo, L. Zhu, Semicond. Optoelectron, **35**, 928(2014)
7. Z. W. Tian, J. Q. Xiong, J. H. Pei, Control Instrum. Chem. Ind, **5**, 509(2014)

Indoor Nearby Channel Condition Estimation System

Jia-Chyi Wu and Teng-Fu Huang

Dept. of Communications, Navigation, and Control Engr. National Taiwan Ocean University, 20242 Keelung, Taiwan

Abstract. The proposed channel estimation system is based on the collaboration with nearby users channel estimation information. Channel status condition is inaccurate while the estimating pilot sequence applied is undersized to avoid overloading the system throughput and bandwidth efficiency. In this study, we have tried to ascertain the indoor channel varying conditions by way of nearby users' channel estimated statistics. The distance of the surrounding nearby users has very high correlation to the channel frequency response profile. We have examined the channel profile performance under the proposed nearby channel condition estimations in both line-of-sight (LOS) and non-line-of sight (NLOS) indoor environments statuses.

1 Introduction

Accurate channel condition estimation system is an essential promise to gain well-enough signal transmission performance in modern communication systems. To acquire well established channel estimations information by pilot sequences before transmitting messages, the system may need overwhelming efforts to possess the necessary information, typically a much longer pilot sequence employing adequate bandwidth and more pre-processing time [1], [2]. However, the channel information is usually inaccurate estimate and the longer pilot sequence should moderate the system throughput and bandwidth efficiency.

There are two major distributed classes in channel estimation techniques. The first classification is pilot channel estimation where pilot sequences are inserted prior to the transmitted data [3], [4]. At the receiver part, the recognized pilot streams are evaluated possibly providing channel varying characteristics to gain required channel frequency response information. However, these pilots may consume loads of the transmission data flows and minimize the accessible bandwidth efficiency. The second classification is blind channel estimation scheme where channel condition estimation is acknowledged simply by the received data streams without additional pilot bits [5]. Hence, the bandwidth efficiency is promised but an additional convoluted algorithm is required to channel estimating implementation, where the resulted estimation error is higher than that with pilot channel estimation scheme.

The preliminary notion for this study is to locate sufficient channel varying characteristics at times without depriving system bandwidth efficiency [6], [7]. We then present the implementation of nearby channel condition estimation system as shown in Fig. 1. The channel information is provided by nearby users' pilot sequences estimation process. Therefore, the object (main) user does not need inserting pilot streams to perform channel estimation as that in blind channel estimation scheme.

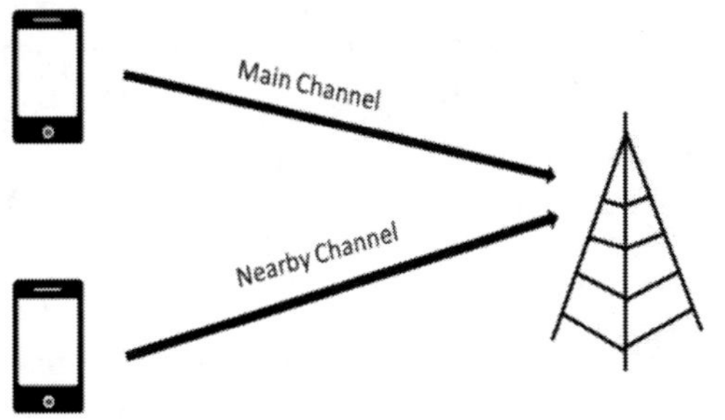

Figure 1. The object channel vs nearby channel estimation

The indoor channel estimation system configuration is described in Section 2. The indoor channel frequency response profiles estimated by nearby users compared with the object user's profiles to examine the consistency within each other are shown in Section 3. Finally, Section 4 gives some brief conclusions.

2 System configurations

Wireless communication channel has been recognized as a varying multipath channel. Data signals transmitted over wireless multipath channel come across refraction, scattering, reflection, and shadowing fading, where the signal power is distorted. The impulse response representation regarding the communication multipath channel is shown as,

$$h_l(t) = \sum_{l=0}^{L-1} a_l \delta(t-\tau_l) e^{j\theta_l(t)} \quad (1)$$

where L is the total number of the transmitted paths, a_l is the amplitude strength for the lth path, τ_l is the delay time for the lth path, and θ_l is the phase deviation for the lth path. After Fourier transformation of (1), we can see the frequency response power profile for the transmitted signal over the occupied frequency band.

$$H(f)=F[h_l(t)]=\int_{-\infty}^{\infty} h_l(t)e^{-j2\pi ft}dt=\sum_{l=0}^{L-1} a_l e^{-j[2\pi f\tau_l-\theta_l(\tau_l)]} \quad (2)$$

To be simplified in this preliminary study, we will examine the frequency response at the experimental channel paths assuming the phase deviation θ_l difference within the nearby channels can be ignored. The frequency response of the object (main) user's channel can be edited to:

$$H_M(f)=\sum_{l=0}^{L-1} a_{Ml}e^{-j2\pi f\tau_{Ml}} \quad (3)$$

The frequency response of the nearby user's channel is:

$$H_N(f)=\sum_{l=0}^{L-1} a_{Nl}e^{-j2\pi f\tau_{Nl}} \quad (4)$$

If the channel frequency response is similar, $H_N(f) \approx H_M(f)$, the nearby channel estimation can be applied to the object user's communication system.

The channel information estimated by nearby users is distributed to the object user's communication system to exclude the pilot sequence used in each transmitted block to improve the bandwidth efficiency. The required channel varying information for active object user is provided by nearby users' pilot sequences estimation process while the nearby users are in idle transmission status. Therefore, the object (main) user can exclude inserting pilot streams to perform channel estimation as that in blind channel estimation scheme. Since nearby users are assumed to be in idle communication status, they may provide longer-enough pilot sequences in each transmission block to better estimate the indoor channel condition variations around very closed areas. In this study, the nearby channel estimation layout is performed in indoor environment. The schematic diagram of the measurement is shown in Fig. 2. The experimental set-up for line-of-sight (LOS) measurement and non-line-of-sight (NLOS) measurement are shown in Fig. 3 and Fig. 4, respectively.

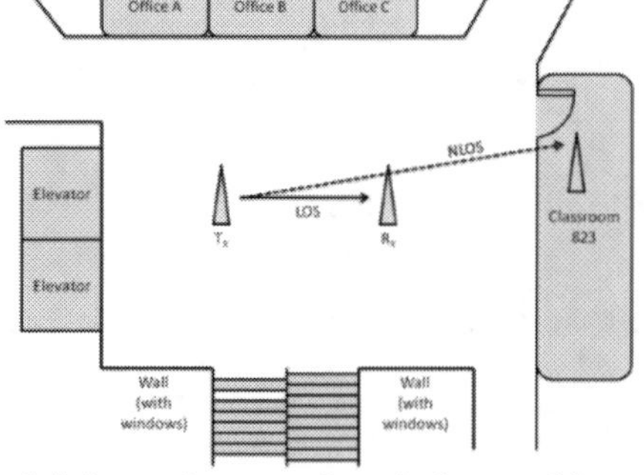

Figure 2. Indoor environment schematic diagram of the measurement

Figure 3. The operation for LOS situation measurement

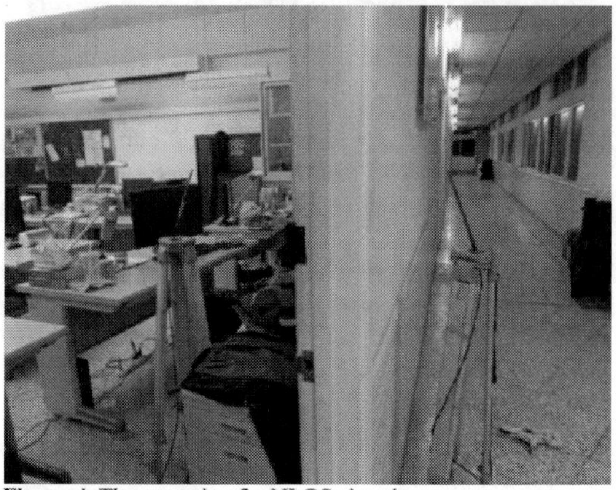
Figure 4. The operation for NLOS situation measurement

3 Results for system simulations

The indoor channel conditions estimation with line-of-sight (LOS) and non-line-of-sight (NLOS) situations are performed for distance varying from 1 meter to 5 meters, where we have found that the two nearby channels investigated have comparable channel frequency response profile (as shown in Fig. 5 to Fig. 14, for LOS/NLOS distance from 1 to 5 meters, respectively). From simulation analysis, we found that the channel frequency response is very similar when the nearby users' distance is within 1 to 3 meters for LOS/NLOS situations.

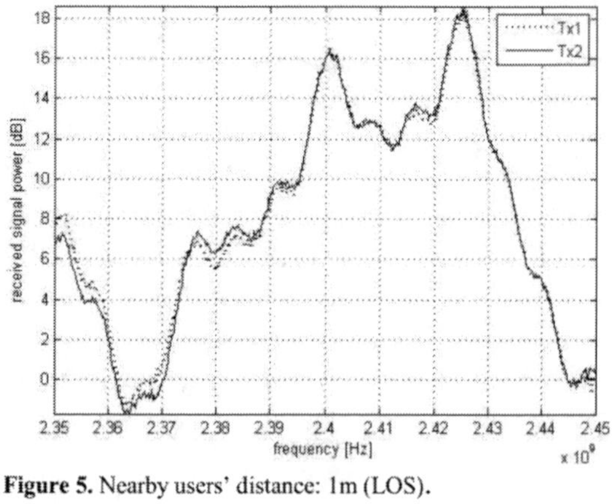

Figure 5. Nearby users' distance: 1m (LOS).

DOI: 10.1051/matecconf/20179514010

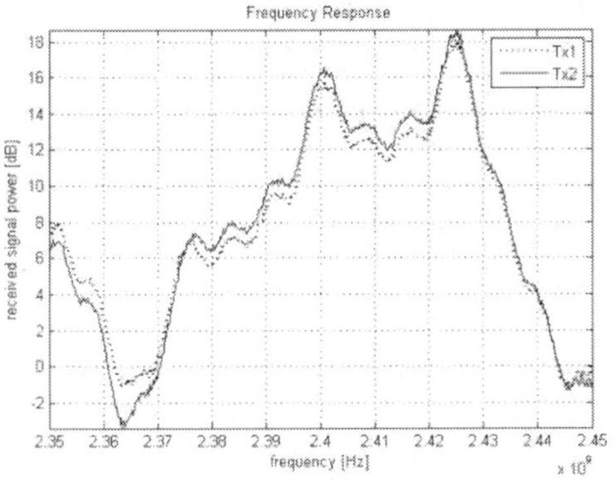

Figure 6. Nearby users' distance: 2m (LOS).

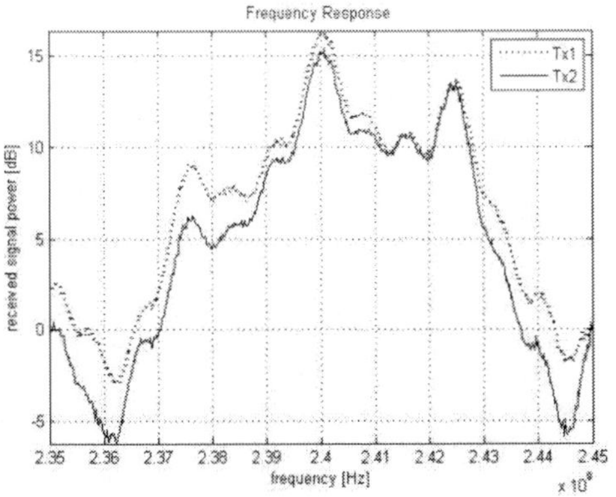

Figure 7. Nearby users' distance: 3m (LOS).

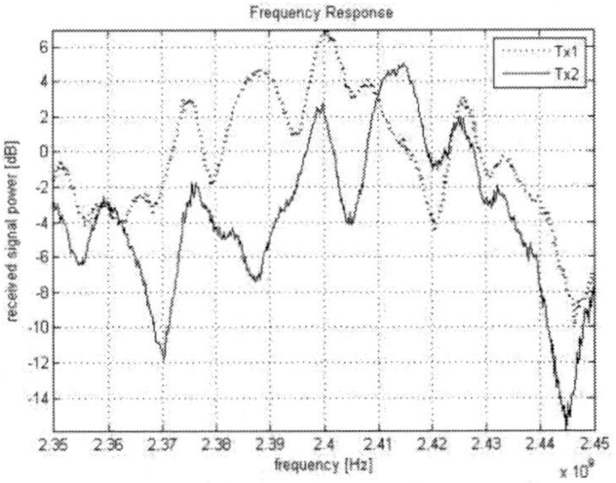

Figure 8. Nearby users' distance: 4m (LOS).

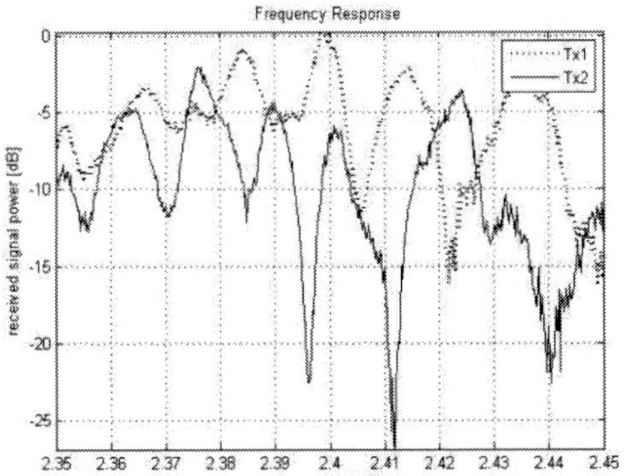

Figure 9. Nearby users' distance: 5m (LOS).

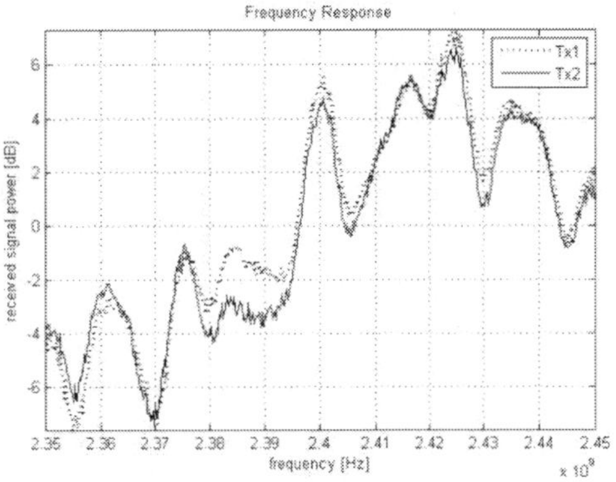

Figure 10. Nearby users' distance: 1m (NLOS).

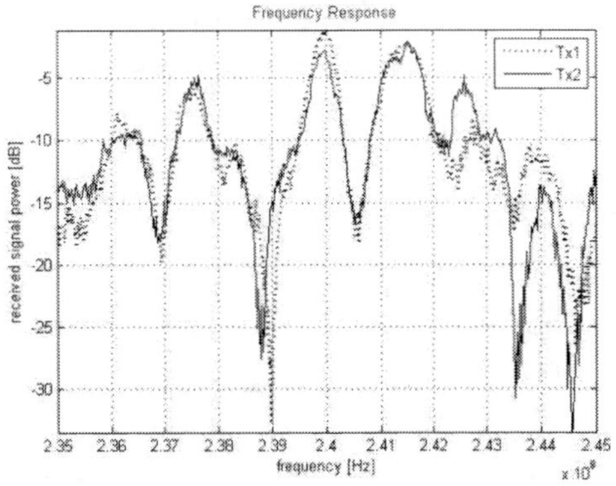

Figure 11. Nearby users' distance: 2m (NLOS).

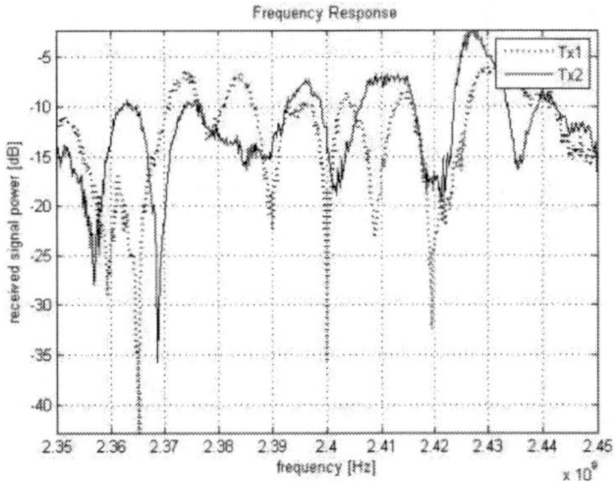

Figure 12. Nearby users' distance: 3m (NLOS).

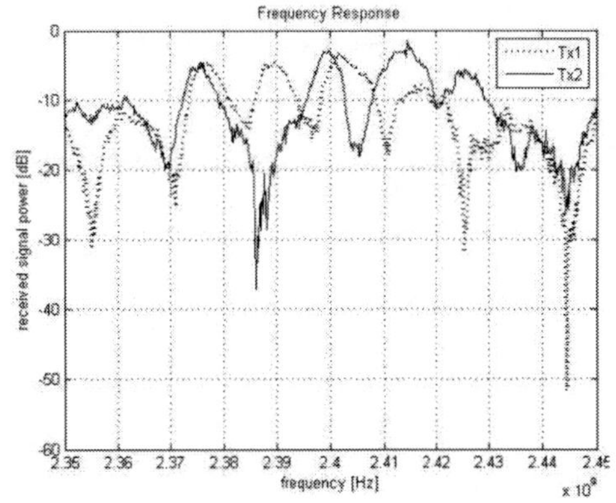

Figure 13. Nearby users' distance: 4m (NLOS).

DOI: 10.1051/matecconf/20179514010

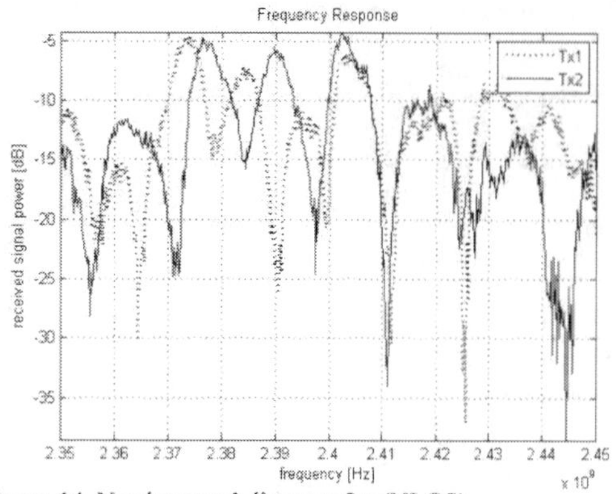

Figure 14. Nearby users' distance: 5m (NLOS).

Table 1. Cross-correlation of frequency response for real/imaginary parts between object and nearby channels (LOS).

Measurement condition	Real Part	Imaginary Part
LOS : 1m	2.2736*10^3	4.6769*10^3
LOS : 2m	1.6197*10^3	3.064*10^3
LOS : 3m	1.1913*10^3	1.8672*10^3
LOS : 4m	54.8963	91.274
LOS : 5m	14.0308	16.7565

Table 2. Cross-correlation of frequency response for real/imaginary parts between object and nearby channels (NLOS).

Measurement condition	Real Part	Imaginary Part
NLOS: 1m	237.4263	588.5835
NLOS: 2m	39.3903	64.8906
NLOS: 3m	22.4992	34.0363
NLOS: 4m	19.8203	24.8906
NLOS: 5m	12.4992	22.0363

We have observed the cross-correlation status for channel frequency response between object user's channel and nearby user's channel. Shown in Table 1 (for LOS condition) and Table 2 (for NLOS condition), are the cross-correlation values for both real and imaginary parts, separated distance from 1 meter to 5 meters. For measurement under LOS channel environments, the channel characteristics difference between antennas are quite trivial, even if the distance is about 3 meters, the real part and the imaginary part of the cross-correlation status for channel frequency response are still noteworthy. When the spacing is over 4 meters, the cross-correlation status for channel frequency response will be severe attenuated. For the test channels are NLOS environments, the cross-correlation status for channel frequency response between adjacent channels are decay much faster.

4 Conclusions

These estimated equivalent channel information from the nearby users' channel estimating profile is applicable. The nearby channel estimation system may apply to modern communication systems to gain transmission throughput with satisfactory system performance. By analyzing the real and imaginary parts of the cross-correlation function for channel frequency response of nearby users, we are able to realize the signal power strength and phase deviation over transmission channel at different separation distance. To improve the channel condition information estimation, we may collect surrounding nearby users' channel estimation data, make some statistic combination analysis. The whole transmission system can save the pilot sequence used in each transmitting block to improve bandwidth efficiency.

References

1. P. Stoica, O. Besson, Training sequence design for frequency offset and frequency-selective channel estimation. IEEE Transactions on Communications, **51**(11), 1910-1917, (2003)
2. X.W. Wang, K.J.R. Liu, Joint channel estimation and equalization in multicarrier modulation system using cyclic prefix. IEEE Transactions on Acoustics, Speech, and Signal Processing, **5**, 2733-2736, (1999)
3. M. Morelli, L. Sanguinetti, U. Mengali, Channel estimation for adaptive frequency-domain equalization. IEEE Transactions on Wireless Communications, **4**(5), 2508-2518, (2005)
4. W.A. Syafe, K. Nishijo, Y. Nagao, M. Kurosaki, H. Ochi, Adaptive Channel Estimation using Cyclic Prefix for Single Carrier Wireless System with FDE. IEEE Advanced Comm. Tec., **2**, 1032-1035, (2008)
5. F.F. Gao, A. Nallanathan, C. Tellambura, Blind channel estimation for cyclic-prefixed single-carrier systems by exploiting real symbol characteristics. IEEE Transactions on Vehicular Technology, **56**(5), 2487-2498, (2007)
6. P. Wan, M. McGuire, X.D. Dong, Near-optimal channel estimation for OFDM in fast-fading channels. IEEE Transactions on Vehicular Technology, **60**(8), 3780-3791, (2011)
7. Y.X. Li, X.J. Huang, The simulation of independent Rayleigh faders. IEEE Transactions on Communi., **50**(9), 1503-1514, (2002)

Circuit Model of Curved Ultra Wideband (UWB) Antenna

Lee Chia Ping

School of Engineering, Taylor's University, Taylor's Lakeside Campus, No. 1 Jalan Taylor's, 47500, Subang Jaya, Selangor DE, Malaysia

Abstract. This paper studies the effect of curved antenna performance on Ultra Wideband (UWB) characteristics, particularly on impedance bandwidth. A UWB antenna is designed, after which is curved and conformed onto cylindrical radius of 150 mm, 120 mm, 90 mm and 60 mm. The motivation of the study is to study the effectiveness of the antenna to be conformed onto non-planar structure applications, particularly on cylindrical body of air or land vehicles. It is observed that the degradation which occurs when the antenna is curved still adheres to the UWB characteristic of more than 500 MHz bandwidth. A circuit model is developed to understand the physics behind the antenna structure when it is curved or bent. The discontinuity of the antenna structure during the curving effect could be explained with the alteration of capacitive loading of the antenna, which is related to the electrical field of the antenna. The curved antenna proves to be suitable for UWB application and the results of its performance according to the UWB characteristics are discussed in detail.

1 Introduction

Antenna research has garnered great interest especially since the release of Ultra Wideband (UWB) frequency spectrum, which was allocated to be from 3.1 GHz to 10.6 GHz [1]. By definition, UWB refers to signals or systems with more than 500 MHz bandwidth, according to the Federal Communications Commission (FCC) [2]. Recent trend has observed more emphasis on integrating microstrip patch antenna onto curved structure of air or land vehicles because of the space constraints [3]. Related researches have been done on curved antenna. For instance, Taha A. Elwi *et al.* studied the effect of Truncated Sinusoidal Printed Circuit Antenna (TPSCA) in application for biomedical telemetry, resulting in unaffected bandwidth with change in resonant frequency [4]. Georget et al. added resonators on the dual-band antenna on flexible substrate, reporting a shift in resonant frequency, contributed by the coupling between the parasitic elements [5]. Macon et al. modelled a doubly curved conformal microstrip antenna using Green's function, which are investigated based on different orientations of probe feed positions [3]. This research aims to analyse the effect of the curved antenna on UWB characteristic, particularly on its impedance bandwidth. The antenna is designed to operate in the UWB bandwidth, after which is bent or curved on different radius. Besides, equivalent circuit is modelled for the curved antenna to study the perturbation that affects the antenna structure during bending.

2 Ultra Wideband (UWB) antenna

Fig. 1 depicts the antenna geometry, which is designed on a double-sided Roger substrate. The front part of antenna comprises the antenna radiator itself, which is made up of octagonal shape as the larger patch and fed through the feed line which is positioned in the middle of the substrate. The patch antenna is designed as such that each part of the structure would be resonant to the frequency in the return loss graph. The back part is the ground plane, comprising partial step truncated in the middle as well. The design is as such that it improves the impedance matching between the patch and ground plane so that wider bandwidth could be achieved.

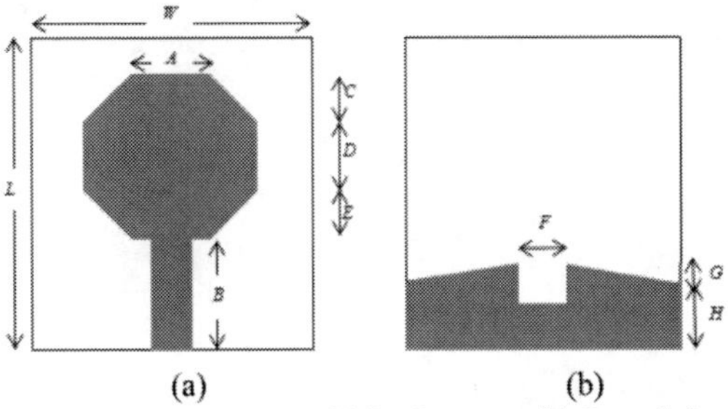

Figure 1. Antenna geometry (a) Patch antenna (b) Ground plane.

Table 1 tabulates the dimension of the antenna geometry based on Fig. 1.

Fig. 2 depicts the antenna's return loss, S11 across the UWB frequency spectrum. Based on electromagnetic simulator, the designed antenna operates with approximately 11 GHz bandwidth, covering UWB spectrum. The first resonant frequency is contributed by the antenna patch length of approximately 0.5 λ, with the S11 of -35.62 dB at 3.3 GHz, while the second and third are due to truncated step from the patch, 'E' and ground plane, respectively, thereby recording S11 of -37. 22 dB at 6.6 GHz and S11 of -22. 62 dB at 9.53 GHz. The notch at the ground plane and truncated shape of the patch antenna serve as a means to improve the impedance matching and enhancement of the S11 of the graph so that it remains below -10 dB. This is due to the matching of the capacitive coupling between the patch antenna and the ground plane. Meanwhile, the ground plane is designed partially with the notch which serves as a realization of series inductance, validated by Hoefer as a mean to compensate for the excessive capacitance due to abrupt geometry change [6].

Table 1. Antenna Dimensions.

Antenna	Variable	Dimension (mm)
Substrate	*W*	30
	L	33
Patch Antenna	*A*	8
	B	11
	C	5
	D	8
	E	5
Ground Plane	*F*	4
	G	2
	H	9

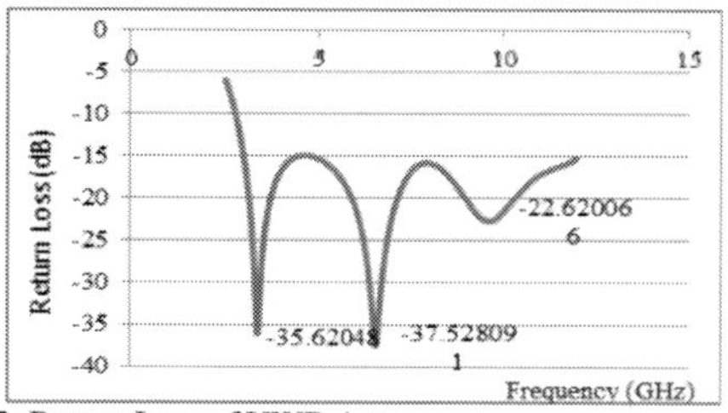

Figure 2. Return Loss of UWB Antenna.

3 Development of circuit model for antenna bending

Lumped element circuit model has been applied in various ways, such as modelling of the antenna, understanding the antenna operation and computation of antenna matching potential [7]. Literatures on such models are more extensive for narrowband antennas, especially on dipoles [8]-[10]. It has also been reported that there is a lack of study on reflection coefficient, S11 results being depicted by equivalent circuit for a wideband antenna, despite the fact that it is the primary parameter to be considered when it comes to qualifying for antenna system requirement [11]. In this paper, the circuit model features the use of lumped elements, which consists of resonant circuit represented by resistor and inductor in series with combination of capacitor and inductor in parallel, as depicted in Fig. 3.

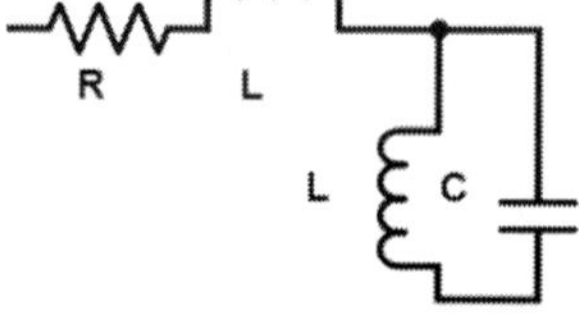

Figure 3. Basic resonant circuit

Based on Fig. 3, the series configuration represents the surface resistance, R and inductance, L, respectively, given by

$$R = \frac{1}{w\sigma_{cond}\delta}. \quad (1)$$

$$L = \frac{1}{w\sigma_{cond}\omega\delta} \quad (2)$$

where σ_{cond} represents the copper conductivity

δ represents the skin depth

The capacitance is given by the combination of Wheeler's and Schneider's model [12] :

$$C = \varepsilon_{eff}\, 2.85 \frac{1}{\ln\left\{1 + \frac{1}{2}\left(\frac{8h}{W_{eff}}\right)\left[\left(\frac{8h}{W_{eff}}\right) + \sqrt{\left(\frac{8h}{W_{eff}}\right)^2 + \pi^2}\right]\right\}} \quad (3)$$

where W_{eff} represents the effective width

ε_{eff} represents the effective permittivity

The equivalent circuit model for the designed antenna is as depicted in Fig. 4.

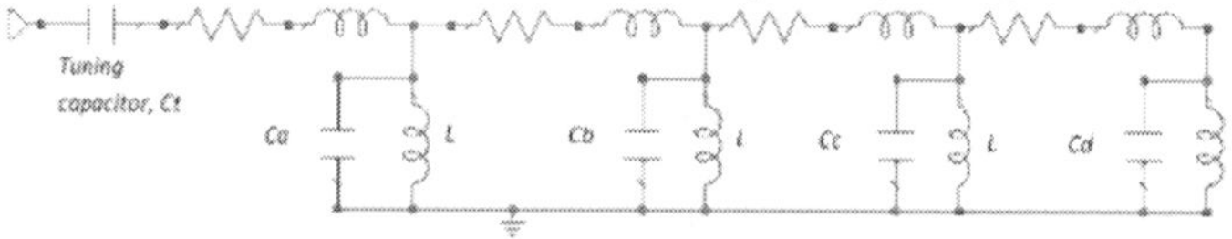

Figure 4. Lumped element circuit model of UWB antenna

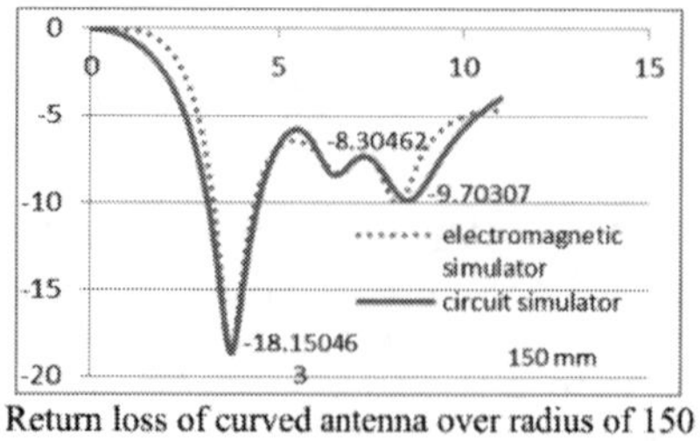

Figure 5. Return loss of curved antenna over radius of 150 mm

As depicted in Fig. 5, the bending effect of the antenna does shift its resonant frequencies, thus affecting the antenna resonant structure [13]. The first resonant frequency is shifted to the right, recording -18.15 dB at 3.73 GHz. Meanwhile, the second resonant frequency is only slightly affected with -8.3 dB at 6.67 GHz. The third resonant frequency due to the ground plane is shifted to the left, stating -9.7 dB at 8.27 GHz. The operating bandwidth is 1.09 GHz. It should be noted that the curved antenna adheres to UWB requirement of 500 MHz bandwidth with S11 below -10 dB.

Meanwhile, the radius of which the antenna is conformed onto is then set as 120 mm. The return loss is depicted in Fig. 6, which observes slight difference in the shifting of the resonant frequencies, as the first records -19.04 dB at 3.765 GHz, followed by -7.853 dB at 6.6 GHz and -9.74 dB at 8.38 GHz. The impedance bandwidth recorded is 1.065 GHz.

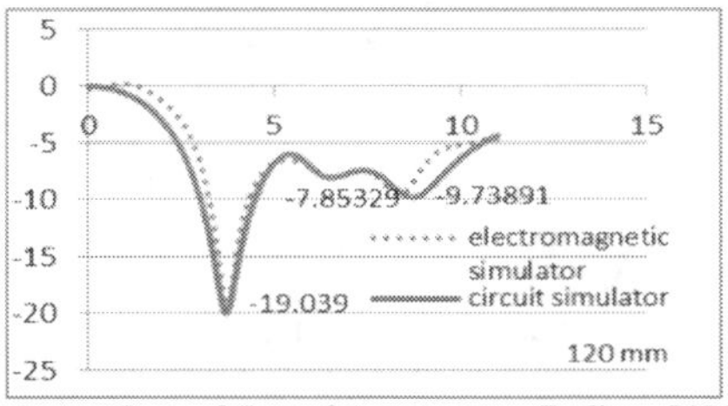

Figure 6. Return Loss of Curved Antenna over Radius of 120 mm

The same behaviour is observed in antenna being conformed onto radius of 90 mm, with minimal shifting of frequencies as depicted in Fig. 7. Recording a bandwidth of 1.11 GHz, the resonant frequencies record negligible change, but its return loss is observed to shift a little; -18.29 dB at 3.75 GHz, -8.05 dB at 6.577 GHz and -9.59 dB at 8.28 GHz.

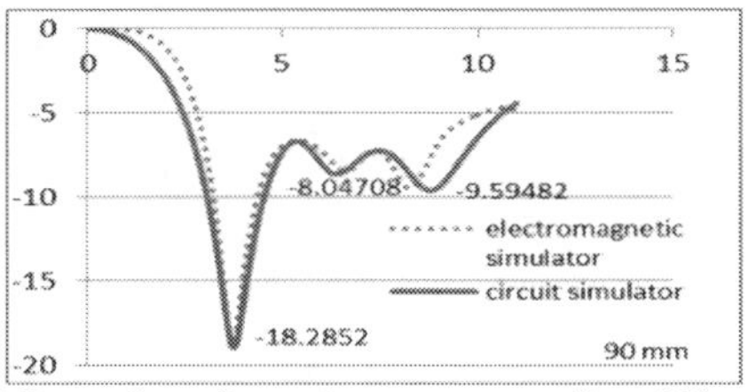

Figure 7. Return loss of curved antenna over radius of 90 mm

Similar expected result could be observed in the antenna being conformed onto radius of 60 mm, as depicted in Fig. 8. The bandwidth is of 1.095 GHz, which is similar to the previous bending. The return loss and its corresponding resonant frequencies are as follow; -18.8 dB at 3.75 GHz, -8.38 dB at 6.59 GHz and -9.28 dB at 8.25 GHz.

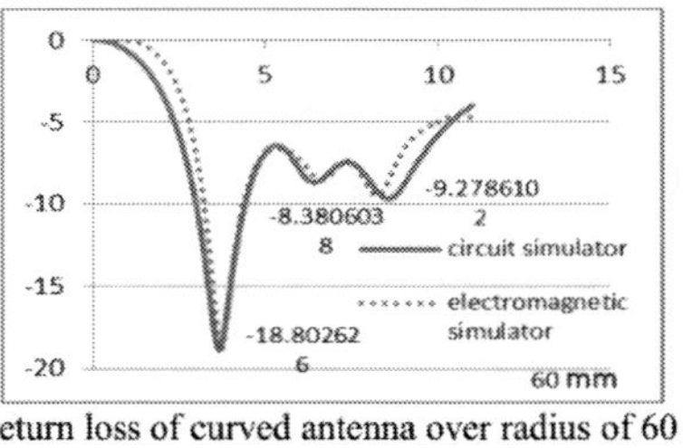

Figure 8. Return loss of curved antenna over radius of 60 mm

From the circuit analysis point of view, the bending changes the electrical field of the antenna, validated by change of the capacitance between the antenna elements [14]. The percentage of capacitance variation is depicted in Fig. 9, with each capacitance being referred to from Fig. 4. The tuning capacitor, denoted as Ct, records a decrease, indicating a decrease in terms of the coupling between the patch antenna and ground plane, therefore the reactance representation is affected. Further decreased angle for the curved antenna does not affect much on the capacitive loading. Similar behaviour could be observed in the capacitance of the patch antenna with the truncated step, as is represented by Cc. A huge spike of capacitance loading, Cb, could be observed at the patch antenna truncated part upon being curved, which then increases in value as the bending radius gets smaller. Similarly, Cd which represents the notched ground plane also records an increase. The notched ground plane structure serves as to provide a better impedance matching for the antenna. The bent antenna brings about change in which the antenna discontinuity is further deteriorated due to the curved nature. This indicates that the electrical field acts the strongest at that part. The inductive loading does not get affected much, thereby the change in magnetic field should be minimal. The development and analysis of the circuit model aid in showing the effect of parameter variation for antenna parameters, and is relatively fast, as compared to other simulators which take on transient time-domain simulations [11].

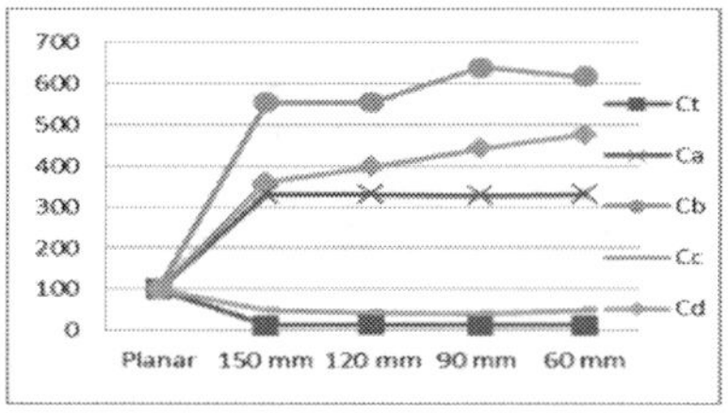

Figure 9. Percentage of capacitance variation for curved antenna

4 Conclusion

This paper analyses the effect of curved or conformal UWB antenna, particularly on the impedance bandwidth and return loss, S11. The designed UWB antenna is compared to its performance when it is conformed onto different radii over a cylindrical shape of 150 mm, 120 mm, 90 mm and 60 mm. The impedance bandwidth is affected, due to the shifting of resonant frequencies as it affects the electrical field of antenna. Nevertheless, the shifting mainly affects the resonances contributed by the patch antenna length and the ground plane. While the bandwidth is affected, it still adheres to the UWB requirement of ≥ 500 MHz.

References

1 T. Kaiser, F. Zheng: *Ultra wideband systems with MIMO,* Wiley, Chichester, Sussex. (2010)

2 FCC. *Revision of Part 15 of the commissions's rules regarding ultra-wideband transmission systems, first report and order,* ET-Docket 98-153, FCC, Washington, DC. (2002)

3 C. Macon, K. Trott, L. Kempel: A Practical Approach to Modeling Doubly Curved Conformal Microstrip Antennas. PIER, **40**, p. 295-314, (2003)

4 T. A. Elwia, H. M. Al-Rizzob, D.G. Ruckera, H.R. Khaleel: *Effects of twisting and bending on the performance of a miniaturized truncated sinusoidal printed circuit antenna for wearable biomedical telemetry devices*, IJEC, p. 217–225, (2011).

5 E. Georget, R. Abdeddaim, P. Sabouroux: *Analytical, Simulation And Measurement Studies of a Dual-Band Open-Sleeve Curved Meander Line Antenna On A Flexible Substrate*, PIER, **145**, p. 49-57, (2014).

6 W. Hoefer: *Equivalent Series Inductivity of a Narrow Transverse Slit in Microstrip*, IEEE Trans. on Microwave Theory and Techniques, **25**, p. 822-824, (1997).

7 M. Ansarizadeh, A. Ghorban: *An approach to equivalent circuit modeling of rectangular microstrip antennas*, PIERB, **8**, p. 77-86, (2008).

8 M. Hamid, R. Hamid: Equivalent circuit of dipole antenna of arbitrary length. *IEEE Trans. Antennas Propag.*, **45**, no. 11, p. 1695-1696, (1997).

9 K. Rambabu, M. Ramesh, A.T. Kalghatgi: Broadband equivalent circuit of a dipole antenna. *IEEE Proceedings Microwaves, Antennas and Propagation*, **146**, no. 6, p. 391-393, (1999).

10 J. Holopainen, R. Valkonen, O. KivekÄas, J. Ilvonen, P. Vainikainen: Broadband equivalent circuit model for capacitive coupling element-based mobile terminal antenna. *IEEE Antennas Wireless Propag. Lett.*, **9**, p. 716-719, (2010).

11 T. Tuovinen, M. Berg:*Impedance Dependency on Planar Broadband Dipole Dimensions: An Examination with Antenna Equivalent Circuits*, PIER, **144**, p. 249-260, (2014).

12 E. Bogatin: Design Rules for Microstrip Capacitance. *IEEE Transactions on Components, Hybrids, and Manufacturing Technology*, **11**, No 3, (1988).

13 P. Salonen, P.S. Oy, Y. Rahmat-Samii: Textile antennas: Effects of antenna bending on input matching and impedance bandwidth. *IEEE Aerospace and Electronics System Magazine*, **22**, p. 10-14, (2007).

14 L. C. Ping, C. K. Chakrabarty: *Circuit Modeling of Ultra Wideband (UWB) Microstrip Patch Antenna Bending Effect*, RJASET, 7, no. 11, p. 2336-2344, (2014).

A Bidding Model for a Virtual Power Plant via Robust Optimization Approach

Geng Tianxiang[1], Xiang Li[1], Ding Maosheng[1] and Li Feng[2]

[1]State Grid Ningxia Electric Power Company, 750001 Yinchuan, China
[2]Electric Power Research Institute, State Grid Ningxia Electric Power Company, 750001 Yinchuan, China

Abstract. The evolution of the energy markets has been accelerating the use of distributed energy resources (DERs) all over the world. Virtual power plant (VPP) is a new method to management this increasing two-way complexity. In this paper, a bidding model for a VPP via robust optimization in the uncertain environment of the electricity market is presented. The flexible feature embedded in the model with respect to solution accuracy and computation burden would be advantageous to the VPP. Results of a case study are provided to show the applicability of the proposed bidding model.

1 Introduction

Due to the global warming and environment concerns of coal-based power generations, the design and operation of the power system are being changed [1-2]. As the penetration of distributed energy resources (DERs) in the distribution network is increasingly grown up, for a smart power system DERs are need to supply power to the grid as much as they could. Moreover, several important issues of these generator units still remain to be solved. Virtual Power Plants (VPPs) are new methods of solving grid-integration of DERs. By integrating these units into VPPs, it can either for the purpose of trading electric energy or to provide support services to the power system.

VPPs can be seen as some conventional dispatchable units in order to compensate for the intermittency of renewable generations by generating technologies. In this sense on the one hand VPPs represents a mixture of multiple DERs and some small-scale conventional power plants. On the other hand, VPPs is operated as one unique entity to participate in the power system. According to FENIX [3], VPPs can be divided into two types, the commercial VPPs (CVPPs) and the technical VPPs (TVPPs). In smart grid, the CVPPs aims to economically optimize its dispatching schedules when trading electrical energy or to provide support services. After that, the CVPPs give the results of the optimal schedules to TVPPs as feedback signals with the consideration of the local network constraints. It should be emphasized that CVPPs only perform commercial aggregation of various kinds of DERs and do not take any network operation constraints into consideration.

At present, the challenges and opportunities in optimal scheduling and bidding strategies in electric markets of VPPs have already discussed in many literatures. In [4-5] the bid-ding strategy of VPPs with centralized control for participating in energy and spinning reserve markets are evaluated. In different cases, the results show the effectiveness and the quality of the proposed model. In [6] the impact of demand response at demand side on power system operation is assessed. Results show that a higher amount of uncontrollable capacity increase these benefits and therefore the social value of demand response. In power system, the centralize dispatch for geographically dispersed DERs will inevitably cause difficulty to the dispatch center. To deal with these problems, to maximize the profit of VPP the coordination of decision-making units embedded in DERs using novel software is proposed in [7]. An energy management system (EMS) of VPPs with a cluster of small-scale generation units, storage systems and flexible loads is proposed in [8]. With the technology of state-of-the-art communication, a multitude of DERs can be coordinated and controlled by EMS while on the other hand a VPP in reality is not a physical power plant. In EMS, DERs as VPP coalition members can be managed according to its own objectives such as maximization of the total profit of the VPP owner in conjunction with the minimization of the risk of the profit variability. An optimal bidding strategy in the day-ahead market of a microgrid consisting of distributed generation, storage, dispatchable DG and price responsive loads is proposed in [9]. The bidding problem aims to coordinate the energy production and consumption of its components and trade electricity in both day-ahead and real-time markets with the objective to minimize its daily operation cost. A bidding model for a virtual power plant consisting of a wind farm, a pumped storage power plant and three gas turbines considering uncertainties is developed in [10]. In

DOI: 10.1051/matecconf/20179515001

the model, VPP aims to maximum its profits when participate in the operation of the mid-term contract market, the day-ahead market and the balancing market. The uncertainty of the electricity price, wind power output and power unbalance penalty are considered in the model.

In conclusion, in the literature, some of worthwhile re-search works investigated specific VPPs in details and elucidates in theory possible benefits of aggregating various kinds of DERs by VPPs. Nevertheless, what is missing in the exist literature is a thorough analysis of the effect of the uncertain-ties in the bidding model for a VPP. Due to the scale of the scenario-based stochastic optimization model increases drastically when the number of the scenarios increase and it will bring huge computational burdens. On this condition, we suggest a RO-based model to deal with the uncertainties in the bidding model.

Based on the above discussion, a bidding model for a VPP via robust optimization approach is proposed. This paper is organized as follows. The VPP bidding problem in a deterministic format and a robust format is established in Section 2; Section 3 presents simulation and results based on examples of IEEE system. Finally, conclusion is drawn in Section 4.

2 Problem formulation

2.1 A deterministic bidding model

In power system, VPP works as a price taker when it sells electricity to the customers and the excess to the day-ahead market at the market price. The VPP bidding problem aims to maximize its bidding profit. The deterministic bidding model is formulated as follows:

$$\max F = \sum_{t=1}^{T} (P_t^{\mathrm{D}} \lambda_t^{\mathrm{D}} - \sum_{k \in \mathrm{GSP}} \alpha_k \{\lambda_t\} P_{k,t} - C_t) \quad (1)$$

The objective function (1) to be maximized of a profit function F computed by expected revenues minus VPP operation costs. The revenues contains two parts: the income from selling electricity to the customers and the income/cost from selling/buying electricity to the day-ahead market

The amount of hourly electricity exchange between VPP and the day-ahead market is limited to the minimum demand of customers and the capacity of interconnection with the main grid:

$$P_{k,t} \geq -\min\left\{ \max\left\{ \begin{matrix} 0, (P^{\mathrm{D,max}} - \sum_{i \in \mathrm{DG}} P_i^{\mathrm{DG,min}} \\ -\{P_t^{\mathrm{AW}}\}^{\min}) \end{matrix} \right\}, P_k^{\mathrm{Smax}} \right\} \quad (2)$$

$$P_{k,t} \leq \min\left\{ \max\left\{ \begin{matrix} 0, (\sum_{i \in \mathrm{DG}} P_i^{\mathrm{DG,max}} + \{P_t^{\mathrm{AW}}\}^{\max} \\ +\overline{e}^{\mathrm{ST}} - P^{\mathrm{D,min}}) \end{matrix} \right\}, P_k^{\mathrm{Smax}} \right\} \quad (3)$$

where P_k^{Smax} represents the rating for exchanging power with the main grid.

In VPP bidding problem, the power selling to the customers, selling/buying the excess/deficit production to/from the day-ahead market should be constrained by power balance constraint. Thus, the following constrain is considered in the bidding model:

$$\sum_i P_{i,t} + P_t^{\mathrm{W}} + \sum_{k \in \mathrm{GSP}} P_{k,t} + P_t^{\mathrm{d}} = P_t^{\mathrm{D}} + P_t^{\mathrm{c}} \quad (4)$$

The upper and lower power of a gas turbine which is con-trolled under VPP are expressed in (5) and (6) respectively.

$$P_{i,t} \leq P_i^{\max} v_{i,t} \quad (5)$$

$$P_i^{\max} v_{i,t} \leq P_{i,t} \quad (6)$$

Power output limits of the gas turbines regarding their ramp rate constraints are formulated as follows:

$$-r_i^{\mathrm{down}} \Delta t \leq P_{i,t} - P_{i,t-1} \leq r_i^{\mathrm{up}} \Delta t \quad (7)$$

Constrains (8) and (9) are necessary to model the start-up and shut-down status of the gas turbines and avoid the simultaneous commitment and decommitment of a unit.

$$y_{i,t} - z_{i,t} = v_{i,t} - v_{i,t-1} \quad (8)$$

$$y_{i,t} + z_{i,t} \leq 1 \quad (9)$$

The minimum up and down time constraints of the gas turbines are modelled in the constraint (10) to constraint (13).

$$v_{i,t} - v_{i,t-1} \leq v_{i,t+TU_{i,w}} \quad (10)$$

$$TU_{i,w} = \begin{cases} w & w \leq \mathrm{MUT}_i \\ 0 & w > \mathrm{MUT}_i \end{cases} \quad (11)$$

$$v_{i,t-1} - v_{i,t} + v_{i,t+TD_{i,w}} \leq 1 \quad (12)$$

$$TD_{i,w} = \begin{cases} w & w \leq \mathrm{MDT}_i \\ 0 & w > \mathrm{MDT}_i \end{cases} \quad (13)$$

Finally, the constraints (14) and (15) are used to denote the technical production capacity of a wind farm:

$$P_t^{\mathrm{W}} \leq \{P_t^{\mathrm{AW}}\} \quad (14)$$

$$P_t^{\mathrm{W}} \geq 0 \quad (15)$$

2.2 A robust bidding model

To deal with uncertain parameters in the optimization models, there are several methods proposed in the exist literature. These methods can be categorized into three main principal categories: probabilistic methods, possibilistic methods, and hybrid probabilistic-possibilistic [11].

Recently, robust optimization has emerged as an attractive optimization framework. Compared with other frame-work mentioned above, robust optimization can reduce the sensitivity of the optimal solution to perturbations in the parameter values. The methods which are proposed to address data uncertainty over the years can be categorized into: (1) Stochastic Programming, and (2) Robust Optimization.

As the only uncertain parameter in the deterministic bid-ding model (1)-(15) is the day-ahead market price and it just appears in the objective function, hereupon, use duality theory the robust form can be formulated as below:

$$\begin{cases} \min G = \sum_{t=1}^{T}\left[C_t + \frac{1}{2}(\bar{\lambda}_t + \underline{\lambda}_t)\sum_{k\in \mathrm{GSP}} \alpha_k P_{k,t} - P_t^{\mathrm{D}}\lambda_t^{\mathrm{D}}\right] \\ \quad +\Gamma v + \sum_{t=1}^{T}\eta_t \\ s.t.\ \text{constraints}(2\text{-}15) \\ v + \eta_t \geq \frac{1}{2}(\bar{\lambda}_t + \underline{\lambda}_t) y_t \\ -y_t \leq \sum_{k\in \mathrm{GSP}} \alpha_k P_{k,t} \leq y_t \\ v, \eta_t, y_t \geq 0 \end{cases} \tag{16}$$

3 Test examples

In order to illustrate the performance of the model pro-posed in the previous section, we present a case study using well-known IEEE 30-bus system including a VPP. The VPP is expected to aggregate and control four gas turbines at buses 26, 29 and a wind farm at bus 29 and a storage system at bus 30. For simplification of the problem, we suppose that the VPP can trade with the day-ahead market through three buses at 26, 29 and 30 (Fig. 1).

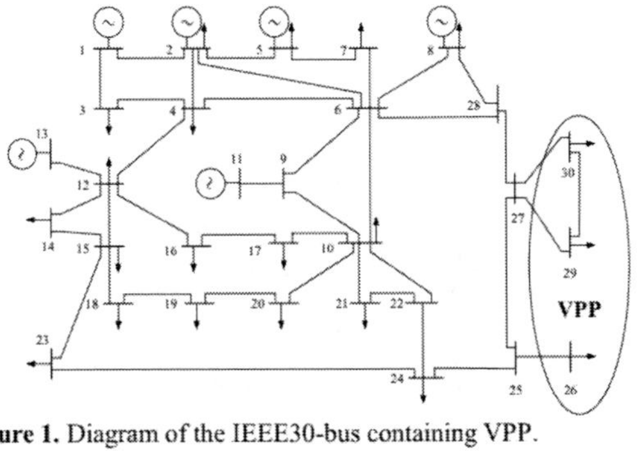

Figure 1. Diagram of the IEEE30-bus containing VPP.

Based on the economic data available in the electricity market of mainland Spain [12], the day-ahead market price forecasts is shown in Table 1. The market prices at the GSPs 26, 29 and 30 are according assumed to be 95, 100, and 105 percent of the day-ahead market prices forecasts. The wind power output forecasts is shown in Table 2.

Table 1. Day-ahead markets price forecasts.

t/h	λ /($/MWh)	*t*/h	λ *t*/($/MWh)	*t*/h	λ /($/MWh)
1	46.03	9	76.95	17	108.31
2	45.14	10	69.09	18	89.54
3	45.50	11	65.84	19	76.83
4	45.70	12	59.47	20	73.60
5	55.80	13	56.47	21	59.59
6	82.28	14	53.77	22	52.47
7	84.80	15	52.90	23	47.77
8	83.44	16	71.44	24	39.17

Table 2. Wind power output forecasts.

t/h	P^{W}_t/(MW)	*t*/h	P^{W}_t/(MW)	*t*/h	P^{W}_t/(MW)
1	39.25	9	36.91	17	34.32
2	40.73	10	34.67	18	27.85
3	36.63	11	26.36	19	22.86
4	43.77	12	27.42	20	19.47
5	39.75	13	33.12	21	27.30
6	35.10	14	38.57	22	31.35
7	30.45	15	34.56	23	30.9
8	34.43	16	34.87	24	28.17

In our case study, two values of parameter Γ , i.e., two different risk strategies are considered and compared.

Strategy A: A conservative strategy is realized when Γ=24.

Strategy B: A less conservative strategy is realized when Γ =8.

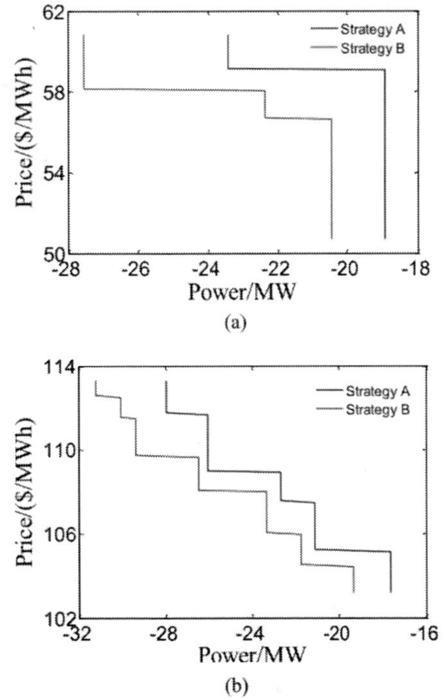

Figure 2. Bidding curve in the day-ahead market for (a) hour 5 and (b) hour 17.

Fig. 2 depicts the bidding curves in the day-ahead market during hour 5 and hour 17 for the two considered values of parameter Γ . As is shown in the figure, for a fixed value of parameter Γ , the willingness to sell energy in hour 17 is higher than in hour 5. For example, using a conservative strategy (Strategy A), the VPP decides to submit the bidding curve within the power internal [-23.4, -18.9] MW in hour 5 while in hour 17 the power internal is [-28.0, -17.6] MW. The reason is that in order to increase the bidding profit, VPP decides to sell energy in high day-ahead market price time periods.

Comparison of the results throughout the day indicates that the bidding curve covers a wider power internal in high-price hours than that in low-price hours. It is worthy to note that the bidding curve in each hour moves right by the increase of parameter Γ . In other words, adopting a less conservative strategy decreases the willing to sell energy in the day-ahead market.

VPP trading with the day-ahead market through the inter-connection grid points is given in Figure 3. It is obvious that the VPP acts as an arbitrager in some hours due to the fact that it purchases energy from the market at the cheapest bus and sells it exclusively to the market at the expensive bus.

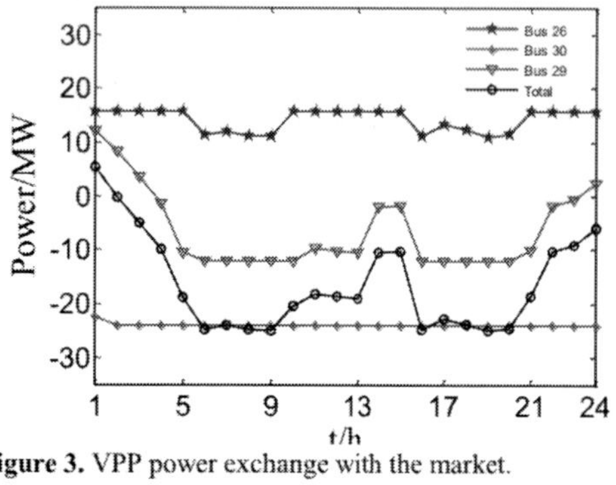

Figure 3. VPP power exchange with the market.

4 Conclusion

In this paper, a bidding model for a virtual power plant via robust optimization has been proposed. The proposed robust bidding model guarantees obtaining a maximum bid-ding profit for VPPs provided that the realized day-ahead market prices are deviated in a trust region. The advantage of the proposed approach is its flexibility in solution accuracy and computational burden. A case study demonstrates the usefulness and simplicity of the proposed model to maximize the proposed of a VPP considering a wind farm, four gas turbines and a storage system, the conclusions below are in or-der:

1) The proposed bidding model for a VPP via robust optimization allows the VPP to appropriately represent uncertain data.

2) The proposed bidding model allows the VPP to use its DERs to buy and sell energy at suitable according to its objectives.

3) The risk strategy adopted influences the bidding strategy, power traded of a VPP.

References

1. M. Shabanzadeh, M-K. Sheikh-El-Eslami and M-R. Haghifam, "The design of a risk-hedging tool for virtual power plant via robust optimization approach," Applied Energy, vol. 155, pp. 766-777, Jul 2015.
2. Z.N. Wei, S. Yu, G.Q. Sun, Y.H. Sun, Y. Yuan and D. Wang, "Concept and development of virtual power plant," Automation of Electric Power Systems, vol. 37, no. 13, pp. 1-9, Jul 2013.
3. D. Pudjianto, C. Ramsay and G. Strbac, "The FENIX vision: The virtual power plant and system integration of distributed energy resources," Contract No: SES6-518272, Deliverable 1.4.0, 2006.
4. E. Mashhour and S.M. Moghaddas-Tafreshi, "Bidding strategy of virtual power plant for participating in energy and spinning reserve markets: Part I Problem Formulation," IEEE Trans on Power Systems, vol. 26, no. 2, pp. 957-964, May 2011.
5. E. Mashhour and S.M. Moghaddas-Tafreshi, "Bidding strategy of virtual power plant for participating in energy and spinning reserve markets: Part II Numerical Analysis," IEEE Trans on Power Systems, vol. 26, no. 2, pp. 957-964, May 2011.
6. B. Dupont, K. Dietrich, C. De Jonghe, A. Ramos and R. Belmans, "Impact of residential demand response on power system operation: a Belgian case study," Applied Energy, vol. 122, pp. 1-10, Jun 2014.
7. H.M. Yang, D.X. Yi, J.H. Zhao and Z.Y. Dong, "Distributed optimal dispatch of virtual power plant via limited communication," IEEE Trans on Power Systems, vol. 28, no. 3, pp. 3511-3512, Aug 2013.
8. P. Lombaedi, M. Powalko and K. Rudion, "Optimal operation of a virtual power plant," 2009 IEEE Power & Energy Society General Meeting, pp. 1-6, 2009.
9. G.D. Liu, Y. Xu and K. Tomsovic, "Bidding strategy for microgrid in day-ahead market based on hybrid stochastic/robust optimization," IEEE Trans on Smart Grid, vol 7, no. 1, pp. 227-237, Jan 2016.
10. S.Yu, Z.N. Wei, G.Q. Sun, Y.H. Sun and D. Wang,"A bidding model for a virtual power plant considering uncertainties,"Automation of Electric Power Systems, Vol. 38,no. 22,pp. 43-49, Nov 2014.
11. A. Soroudi and T. Amraee, "Decision making under uncertainty in energy systems: State of the art," Renewable and Sustainable Energy Reviews, vol. 28, pp. 376-384, Aug 2013.
12. Iberian Electricity Market Operator, OMIP 2016, <http://www.omip.pt/>.

Comparative analysis of FACTS devices based on the comprehensive evaluation index system

Liu Jiankun[1], Chen Jing[1] and Qian Zhen[2]

[1]*State Grid Jiangsu Electric Power Company Electric Power Research Institute, Nanjing 211103, Jiangsu Province, China*
[2]*College of Energy and Electrical Engineering, Hohai University, Nanjing 211100, China*

Abstract. Flexible AC transmission system (FACTS) has the control characteristics of efficiency, flexibility and reliability. The introduction of FACTS in power system is superior to any other control methods, and different FACTS devices have different features. In this paper, a comprehensive evaluation index system is developed to study a variety of comparisons on different FACTS devices. The proposed index takes power flow controllability as its main indicator, and loss reduction, static voltage stability improvement and load shedding reduction as complementary indicators. Finally, the peak load case in 2016 of a province is adopted for case studies. The results not only show that the proposed comprehensive evaluation index is systematic, scientific, practical, but also show the superiority of unified power flow controller (UPFC).

1 Introduction

At present, the primary equipments of grid framework are mainly based on the electromagnetic and mechanical technology. Its flexibility and controllability is limited, which makes it difficult to control the power flow, voltage and others quickly and flexibly. Therefore, the technology of FACTS provides a new technical means for power grid. The concept of FACTS was initially put forward by American scholar N. G. Higorani in 1986, which injected new vitality into the traditional transmission system as a transformative technology [1]. The core idea of FACTS is to enable the system electric parameters (such as voltage, phase angle and impedance) to change fast and flexibly under the premise of keeping the security, stability and reliability of power system, thus making the most of existing resources for distributing transmission power reasonably, reducing power loss and cost and improving the efficiency of the power grid operation [2].

According to the connection mode in power system, the FACTS devices can be divided into three types: a) shunt devices like static var compensator (SVC) and static synchronous compensator (STATCOM), which are mainly used for voltage control and reactive power flow control; b) series devices like thyristor controlled series capacitor (TCSC), thyristor controlled phase shifting transformer (TCPST) and static synchronous series compensator (SSSC), which are mainly used for the control of active power flow, the improvement of the transient stability and the inhibition of power oscillator; c) comprehensive devices like unified power flow controller (UPFC), which is a combination of a) and b). UPFC is by far the most powerful FACTS and has a broad application prospect. It can not only improve the power flow distribution effectively, but also improve the stability of power system. On December 11, 2015, the demonstration project of UPFC, which has the first independent intellectual property rights of China, is officially put into operation in Nanjing. The project provides a practical basis for the theoretical analysis of UPFC operation. What's more, the demonstration project of UPFC in Suzhou is also on agenda.

Based on the above discussion, this paper proposed a new comprehensive evaluation index, which takes power flow controllability as its main indicator, and loss reduction, static voltage stability improvement and load shedding reduction as complementary indicators. This paper is organized as follows. The selection of typical FACTS devices including TCSC, TCPST, STATCOM and UPFC, and the static mathematical modeling of FACTS devices are established in Section II; In Section III, the calculation of every single index is conducted and the comparative analysis for different FACTS devices is conducted through the calculation of the comprehensive evaluation index; Section IV presents simulation and results based on the example of the peak load case in 2016 of a province. Finally, conclusion is drawn in Section V.

2 Matematical modelling of FACTS device

2.1 Static model of UPFC

Generally, the back-to-back voltage source model is adopted to account for UPFC, as illustrated in Fig. 1.

UPFC consists of a variable shunt voltage source $\dot{V}_E$ along with the impedance Z_E, and a variable series voltage source $\dot{V}_B$ with the impedance Z_B. It's assumed UPFC is installed at the s-side of line s-m. A dummy bus r can be added at the end of UPFC, rendering the UPFC to be an independent branch for calculation. It has the ability to adjust the control variables – magnitude and phase angle of the series and shunt injected voltage sources simultaneously, or independently to regulate the voltage profile and power flow in the transmission line.

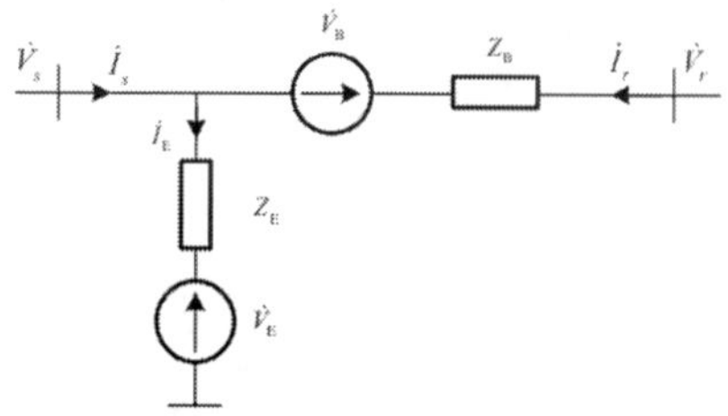

Figure 1. Dual Voltage Sources Model of UPFC.

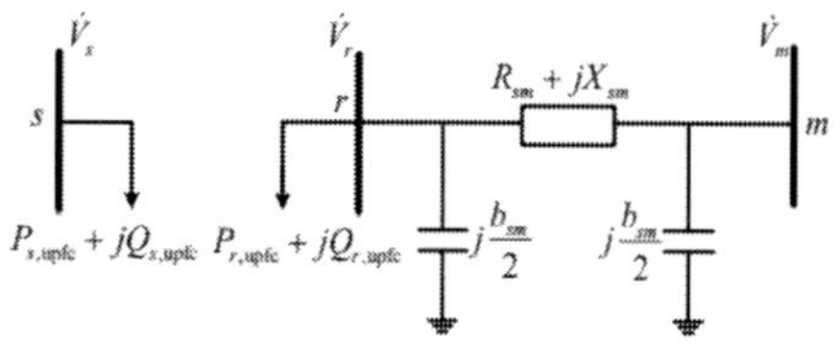

Figure 2. Diagram of UPFC branch equivalent.

When making calculations, the role of UPFC is equivalent to its injection power at each side of the branch, which are denoted as $P_{s,\text{upfc}} + jQ_{s,\text{upfc}}$ and $P_{r,\text{upfc}} + jQ_{r,\text{upfc}}$ in Fig.2. The equivalent injected power is written as:

$$\begin{cases} P_{s,\text{upfc}} + jQ_{s,\text{upfc}} = \dot{V}_s \dot{I}_s^* = \dot{V}_s \left(\dot{I}_E - \dot{I}_r \right)^* \\ P_{r,\text{upfc}} + jQ_{r,\text{upfc}} = \dot{V}_r \dot{I}_r^* \end{cases} \quad (1)$$

2.2 Static model of STATCOM

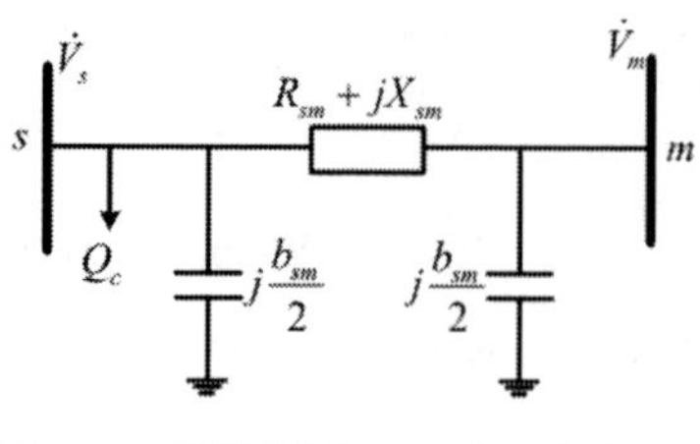

Figure 3. Diagram of STATCOM branch equivalent.

STATCOM is the type of shunt reactive compensation device, which targets at supporting the voltage magnitude profile. Hence, the STATCOM can be operated as a capacitive or inductive compensation by injecting or absorbing reactive power from the system to regulate voltage. It is regarded as a reactive power injection Q_c in Fig.3.

2.3 Static model of TCSC

TCSC can be operated to control the line flow as an expected constant within a rational operation region by changing the equivalent line reactance fastly and continuously. Figure 4 shows the model of TCSC. The series variable reactance in Fig.4, X_c, can be adjusted. The maximum value of X_c is decided by the capacity of TCSC. However, it is noted that active and reactive of the line are dependent. Therefore, active power flow is considered to be controllable.

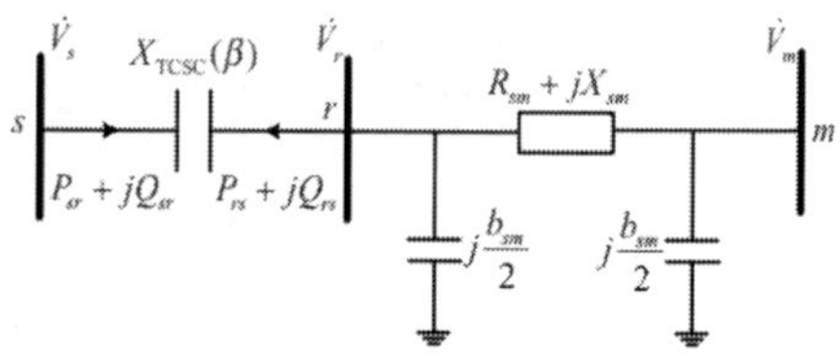

Figure 4. Diagram of TCSC branch equivalent.

2.4 Static model of TCPST

TCPST realizes rapid adjustment of the phase shifter through mechanical switch with thyristor. This fosters the wide application of TCPST. The equivalent circuit of TCPST is given in Fig. 5. It is equivalent as a series voltage source and a shunt current source. Power flow is controlled by adjusting the complex ratio. Similarly, active and reactive power cannot be controlled dependently, so active power line cab be regarded as controllable.

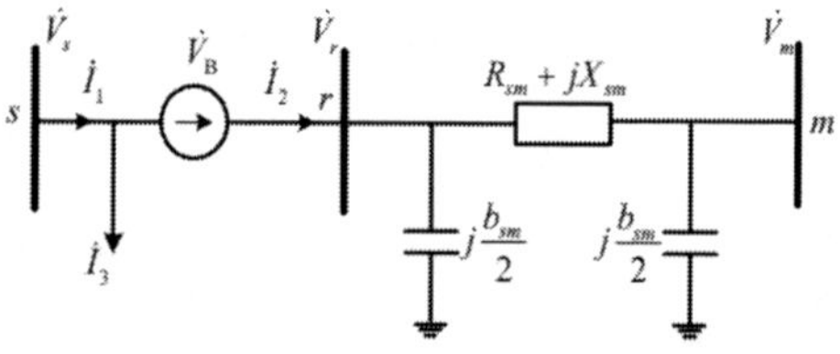

Figure 5. Diagram of TCPST branch equivalent.

3 Comprehensive evaluation index

In this part, a comprehensive evaluation index is proposed to compare the performance of the aforementioned FACTS devices. The proposed index is systematic, scientific and practical in assessing the vast series and shunt type devices. Moreover, the proposed index takes power flow controllability as its main indicator, and loss reduction, static voltage stability improvement and load shedding reduction as complementary indicators.

3.1 Indicator of control region

Three devices with the ability of series power flow control, namely TCSC, TCPST and UPFC, are compared. Figure 6 depicts the equivalent electric line installed with FACTS devices, the power flow of which is controlled by the series voltage source $\dot{U}_{pq}$. To simplify, it's assumed $\dot{U}_i = U_i\angle 0$, $\dot{U}_j = U_j\angle(-\theta)$, $Z = R + jX$.

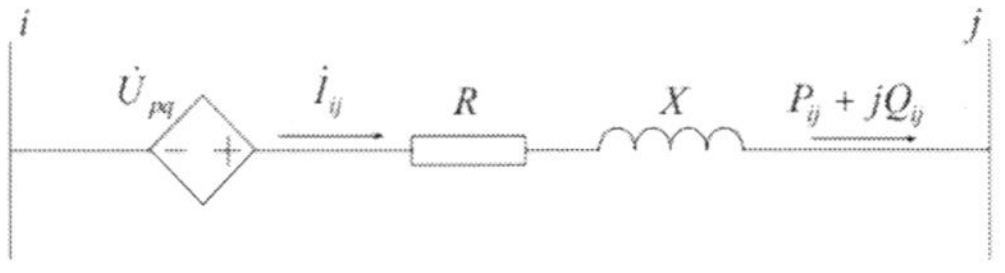

Figure 6. The equivalent electric line installed with FACTS devices.

Then the apparent line flow is written as:

$$\dot{S}_{ij} = P_{ij} + jQ_{ij} = \dot{U}_j\left[\left(\dot{U}_i + \dot{U}_{pq} - \dot{U}_j\right)/Z\right]^* \tag{2}$$

If $\dot{U}_{pq} = 0$, the electric line is not controlled by FACTS. In this case, eq. (2) can be reformulated as:

$$\dot{S}_{ij} = \dot{S}_{ij}^0 = P_{ij}^0 + jQ_{ij}^0 = \dot{U}_j\left[\left(\dot{U}_i - \dot{U}_j\right)/Z\right]^* \tag{3}$$

Equation (2) can also be written as:

$$\dot{S}_{ij} = \dot{U}_j\left[\left(\dot{U}_i - \dot{U}_j\right)/Z\right]^* + \dot{U}_j U_{pq}^* / Z^* = \dot{S}_{ij}^0 + \dot{S}_{ij}^{facts} \tag{4}$$

According to the formulations, the power flow controllable region is shown in Fig. 7, where the intersection point denotes original power flow point. The controllable region of TCSC is a section of arc going through the original point, while it is a segment (the slope is negative) going through the original point for TCPST. However, for UPFC, the controllable region is a filled circle, taking the original point as its center point.

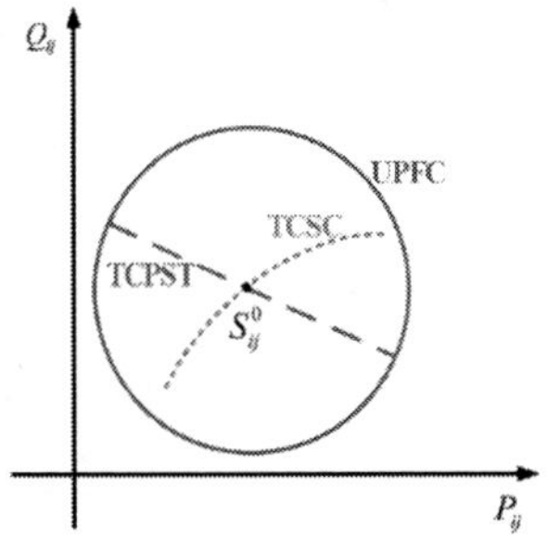

Figure 7. Diagram of power flow controllable region.

The area of the three controllable region is compared quantitatively, with the capacity of three each device being the same. It is assumed that the control area of TCSC is the area of the arc, and the rectangular area for TCPST. For UPFC, the control area is the whole circle.

3.2 Indicator of power flow losses

The reactive power optimization results are introduced to show their ability in reducing power flow losses.

Concretely, on the basis of conventional reactive power optimization model, some modifications are made in the admittance matrix and some equality constraints that are related to the buses, where FACTS devices are installed. In addition, the power balance equation of FACTS and capacity constraints are added into the model. Then, the modified optimization model is solved.

It is worth noting, no restrictions are set on control variables to enlarge the optimization space. In this way, the performance of FACTS devices in reducing power flow losses can be presented.

3.3 Indicator of voltage stability

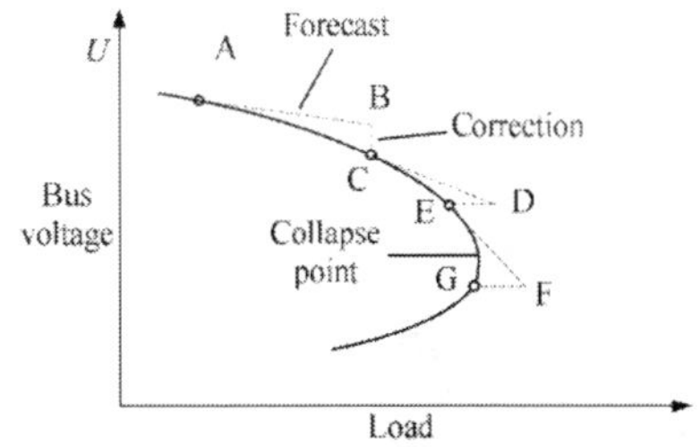

Figure 8. The calculation steps of the continuation power flow method.

The continuation power flow method is adopted in this paper to study the performance of FACTS in improving voltage stability. The calculation steps are illustrated in Fig. 8.

Loads are increased on the basis of conventional power flow as follows:

$$\begin{cases} P_{is} - U_i \sum_{j\in N} U_j (G_{ij}\cos\theta_{ij} + B_{ij}\sin\theta_{ij}) - \lambda P_{di} = 0 \quad (i = 1,2,\ldots,m) \\ Q_{is} - U_i \sum_{j\in N} U_j (G_{ij}\sin\theta_{ij} - B_{ij}\cos\theta_{ij}) - \lambda Q_{di} = 0 \quad (i = 1,2,\ldots,m) \end{cases} \tag{5}$$

Where λ is load incremental coefficient; P_{is}、Q_{is} are the predetermined active and reactive power injection at bus *i*; P_{di}、Q_{di} is the incremental direction of active and reactive load; $\dot{U}_i$ is the voltage magnitude of bus *i*; θ_{ij} is the voltage angle difference between bus *i* and bus *j*; G_{ij}、B_{ij} are the conductance and susceptance between bus *j* and bus *j*, respectively; *n* is the number of total buses; m is the number of PQ buses.

The continuation power flow method starts from the original working point and moves with the increase of load profile. The steps include prediction, correction and calculation of the complete curve. Finally, the maximum load incremental coefficient λ_{max} is obtained.

3.4 Indicator of load shedding

Load shedding reflects the insecure status of system operation and is suitable to be used as an indicator to assess the performance of FACTS devices. When the system security is endangered by an outage, a re-dispatch schedule can be obtained from optimal power flow (OPF) calculation to avoid the off-line limits. At the same time, load shedding is reduced as much as possible, if cannot be avoided. The optimal load shedding model is formulated as:

1) Objective function

$$\min \sum_{i \in n_d} C_i \quad (6)$$

2) Constraints (conventional constraints excluded)

$$\begin{cases} P_{is} - P_{Di} + C_i = 0 & i = 1,2,...,n_d \\ Q_{is} - Q_{Di} = 0 & i = 1,2,...,n_d \\ 0 \le C_i \le P_{Di} & i = 1,2,...,n_d \end{cases} \quad (7)$$

Where P_{Di}、Q_{Di} are active and reactive load at bus i; C_i is the variable for load shedding; n_d is the number of total buses.

3.5 The comprehensive evaluation index

A system of the comprehensive evaluation index is developed is this paper. The proposed index takes power flow controllability as its main indicator, and loss reduction, static voltage stability improvement and load shedding reduction as complementary indicators. The comprehensive evaluation index consists of four indicators, namely:

1)Controllable region，denoted as C_{region} , p.u.;

2)System power flow losses，denoted as P_{loss} , MW;

3)Maximum load incremental coefficient, denoted as $\lambda_{\max}$ ，dimensionless;

4)Load shedding, denoted as L_{loss} , MWh.

Denote the comprehensive evaluation index as ρ , it can be calculated as:

$$\rho=\omega_1\mu\left(C_{\text{region}}\right)+\omega_2\mu\left(P_{\text{loss}}\right)+\omega_3\mu\left(\lambda_{\max}\right)+\omega_4\mu\left(L_{\text{loss}}\right) \quad (8)$$

Where $\mu(\bullet)$ is syntropic normalization; ω_1、ω_2、ω_3、ω_4 are the weights of each indicator, and the sum of these is 1.

4 Numerical cases

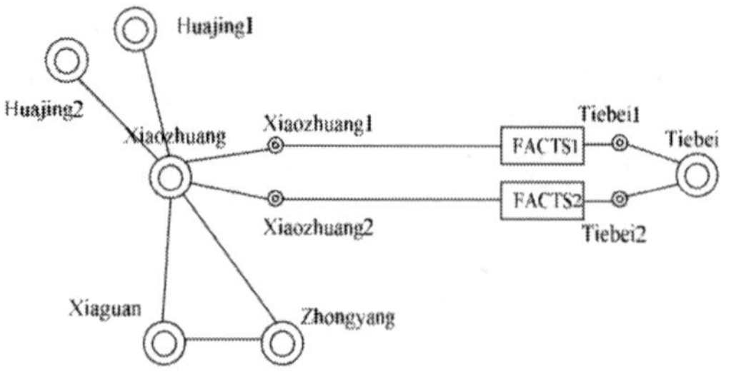

Figure 9. The system diagram installed with FACTS.

The peak load case in 2016 of a province is adopted for case studies. A comparison is made on the performance of the four FACTS devices. The system diagram installed with FACTS, including STATCOM, TCSC, TCPST and UPFC, is depicted in Fig. 9. The devices are installed at the north-side of Xiaozhuang-Tiebei double-line.

The obtained results are listed in Table 1.

Table 1. The comparative quantities of evaluation indexes of different FACTS devices.

Evaluation index	Original situation	TCSC	TCPST	STATCOM	UPFC
C_{region} /p.u.	0	0.365	1.106	0	10.401
P_{loss} /MW	1056.5	1050.9	1049.8	1037.4	1021.6
$\lambda_{\max}$	1.82	1.84	1.86	1.90	1.92
L_{loss} /MWh	9340	9024	8790	8246	7634

In Table 1, C_{region}、$\lambda_{\max}$ are positive relevant to the evaluation index while P_{loss}、L_{loss} are negative relevant. It is noted that the weighting factor can be set according to real situations, and is set to 0.25 in this paper. The evaluation results after syntropic normalization is shown in Table 2.

Table 2. The calculation results of the comprehensive evaluation index of different FACTS devices.

Comprehensive evaluation index	Original situation	TCSC	TCPST	STATCOM	UPFC
ρ	0.683	0.703	0.729	0.725	1

According to Table 2, we have following conclusions:
1) The results of FACTS's comprehensive evaluation index are bigger than the original situation, which suggests that FACTS devices can effectively improve the performance, the power supply capacity, economy, safety and reliability of system.
2) UPFC has strong ability of control, which not only can control the active power, reactive power of line flow and the voltage of shunt bus independently, but also can control the above parameters jointly. Judging from the results of the comprehensive evaluation index, UPFC is significantly better than the other FACTS devices.

5 Conclution

In this paper, a system of the comprehensive evaluation index is developed to study a variety of comparisons on different FACTS devices. The index takes power flow controllability as its main indicator, and loss reduction, static voltage stability improvement and load shedding reduction as complementary indicators. By selecting different types of FACTS devices, the calculation of the comprehensive evaluation index has been carried out, and the following conclusions could be obtained:

1) Regardless of single or comprehensive evaluation index, UPFC is significantly better than the other FACTS devices, which has important research significances to the UPFC project both in Jiangsu and Suzhou.

2) The proposed comprehensive evaluation index is a combination of the technical indicator (control region), the economic indicator (loss reduction), the safety

 DOI: 10.1051/matecconf/20179515002

indicator (static voltage stability improvement) and the reliability indicator (load shedding reduction), which is a systematic, scientific and practical evaluation index.

References

1. N. Hingorani. Future opportunities for electric power systems[J]. Power Engineering Review. IEEE, 1987, 7(10): 4–5.
2. N. G. Hingorani and L. Gyugyi, Eds. Understanding FACTS Concepts and Technology of Flexible AC Transmission Systems[M]. Wiley-IEEE Press, 2000.
3. Zhang Xiaoping. Modeling of FACTS in power flow and optimal power flow analysis[J]. Sutomation of Electric Power Systems. 2005, 29(16): 22-29.
4. C. A. Canlzares, Z. T. Faur. Analysis SVC and TCSC Controllers in Voltage Collapse[J]. IEEE Trans. Power Systems. Vol. 14, No. I, February, 1999: 158-165.
5. Nasri A, Conejo A J, Kazempour S J. Minimizing wind power spillage using an OPF with FACTS devices[J]. IEEE Transactions on Power Systems. 2014, 29(05): 2150-2159.
6. M. A. Furini and P. B. De Araujo. A comparative study of the damping oscillation function of tcsc and upfc[C]. Transmission and Distribution Conference and Exposition: Latin America, 2008 IEEE/PES. 2008: 1–6.
7. Enrigue Acha, Claudio R. Fuerte-Esquivel, Hugo Ambriz-Perez, et al. FACTS Modeling and Simulation in Power Networks[M]. John Wiley & Sons Ltd, 2004: 200-216.
8. R. Nelson, J. Bian, S. L. Williams. Transmission series power flow control[J]. IEEETransactions on Power Delivery. 1995: 437-443.
9. Mittelstaedt M, Barrios H, Schnettler A. Identification of critical states regarding voltage stability by using a Continuation Power Flow combined contingency analysis[C]// PowerTech, 2015 IEEE Eindhoven. IEEE, 2015: 256-261.
10. Pavel Etingov, Alexandre Oudalov, Nikolai Voropai. Coordinated emergency control of load shedding and FACTS devices[C]. // PowerTech, 2005 IEEE Russia. 2005: 1-8.

Optimal Allocation of Generalized Power Sources in Distribution Network Based on Multi-Objective Particle Swarm Optimization Algorithm

Li Ran[1], Yuan Jingwei[2], Song Yundong[2], Yang Yingxuan[2] and Meng Tao[3]

[1]*Shenyang Power Supply Company, Liaoning Electric Power Company Limited, Shenyang 110006, China*
[2]*Electric Power Research institute Liaoning Electric Power Company Limited, Shenyang 110006, China*
[3]*Electric Power Research institute of Jilin Electric Power Co.of State Grid, Changchun 130021, China*

Abstract. Optimal allocation of generalized power sources in distribution network is researched. A simple index of voltage stability is put forward. Considering the investment and operation benefit, the stability of voltage and the pollution emissions of generalized power sources in distribution network, a multi-objective optimization planning model is established. A multi-objective particle swarm optimization algorithm is proposed to solve the optimal model. In order to improve the global search ability, the strategies of fast non-dominated sorting, elitism and crowding distance are adopted in this algorithm. Finally, tested the model and algorithm by IEEE-33 node system to find the best configuration of GP, the computed result shows that with the generalized power reasonable access to the active distribution network, the investment benefit and the voltage stability of the system is improved, and the proposed algorithm has better global search capability.

1 Introduction

Large scale of distributed generators (DGs) supply will increase the complexity and uncertainty of planning of distribution network (DN), and even has a significant effect to this system, such as the quality of voltage deteriorate, the reliability of power supply reducing [1]. Improper allocation of DGs in active distribution network has been studied. In [2], considering the timing characteristics of the side of source and load, a mathematical model of network loss and power loss goal is established and an improved genetic algorithm is proposed to solve this model. In [3], an economic evaluation index of investment, operation, fuel and exhaust emission is set up to study DGs programming problem under different scenarios. Different methods were proposed to evaluating the planning effect of DGs in [4], however, in this methods, the multi-objective optimization problem is transformed into a single objective problem, and the choice of weighted coefficient directly affects the results of optimization, which make the considered factors of allocation schemes not comprehensive.

The concept of generalized power (GP) refers to a device that emits a certain active or reactive power, mainly DG and capacitor group. The existing research mainly puts DG and capacitor group separately and rarely considers the optimum allocation problem of GP. In [5], a multi-objective optimization model is put forward to study the optimal allocation of DG under different conditions by using the multi-objective particle swarm algorithm. The non-dominated sorting genetic algorithm is used to optimize capacitors in distribution network in [6]. At present, few literatures consider the problem of optimal allocation of GP in ADN. Therefore, optimal allocation of GP has an important significance of operation of ADN.

In view of above problems, this paper puts forward the index of voltage stability based on simple network flow. From the point of view of economic benefit of investment operation, voltage stability and pollution emissions, a multi-objective mathematical model considering GP allocation is established. A multi-objective particle swarm optimization algorithm is proposed in this paper. Finally, tested the model and algorithm by IEEE-33 node system, and study the situation of economic benefits, voltage stability and emissions under different load level to determine the best configuration of GP.

2 Multi-objective optimization model

2.1 The benefits of investment and operation

Comprehensive consideration of investment, operation cost and the marginal benefit during the process of planning, and the corresponding mathematic models for f_1:

$$f_1 = B_{\mathrm{M}} + B_{\mathrm{GP}} - C_{\mathrm{GP}} + \sum_{i=1}^{2} BR_i \tag{1}$$

where, B_M, B_{GP}, C_{GP} denote power selling benefit of microgrid, power generation benefit of GP, installation cost of GP, respectively. *BR* is the marginal benefit of installing GP, where, BR1 and BR2 denote the benefit of delay grid upgrade and the benefit of improve network loss, respectively.

2.2 Voltage stability index

Considering the operation of voltage stability of power system, and the corresponding mathematic models for f_2:

$$f_2 = VSI = \frac{1}{N}\sum_{i=1}^{N}(2V_{i+1} - V_i) \tag{2}$$

where V_i and V_{i+1} denote voltage of first node and voltage of lend node. The parameter i is referred to as the node number. N is the total number of nodes.

From (2), the voltage stability index is only concerned with voltage amplitude, and when the voltage of end node down to the half of the first node, the phenomenon of voltage instability could occur in system.

2.3 Pollution emissions index

Considering the emission index of CO_2, NO_x, SO_2, CO and other pollution gas, the corresponding mathematic model for f_3:

$$f_3 = \sum_{i=1}^{N}\sum_{j=1}^{N_{GP}}\sum_{k=1}^{N_g} S_{GP}(ij) \times w_k \times ER_{jk} \times \alpha_j \times 8760 \tag{3}$$

where, N_{GP} and N_g denote the type of GP and the type of pollution emission, respectively. $S_{GP}(ij)$ is the output power of GP from the type of j at the node of i. w_k is the weight coefficient of the pollution emission type of k. ER_{jk} is the pollution emission rate of GP. α_j is the coefficient of carbon emissions.

2.4 Configuration optimization mathematical model of GP

Considering the index of benefit of GP investment, voltage stability and pollution emission, the corresponding multi objective optimization mathematical moder as for:

$$\begin{cases} F(X) = \{f_1(X),\ f_2(X),\ f_3(X)\} \\ s.t.\quad g(X) \le 0,\ h(X) = 0 \end{cases} \tag{4}$$

where, $g(X)$ is the constraints of inequality, $h(X)$is the constraints of equality, and the parameter X is a collection of GP's capacity and location and other variables.
Current equation constraints are mainly constraint:

$$\begin{cases} P_{GP\_i} - P_{L\_i} = V_i \sum_{j=1}^{N} V_j (G_{ij}\cos\delta_{ij} + B_{ij}\sin\delta_{ij}) \\ Q_{GP\_i} - Q_{L\_i} = V_i \sum_{j=1}^{N} V_j (G_{ij}\sin\delta_{ij} + B_{ij}\cos\delta_{ij}) \end{cases} \tag{5}$$

where, P_{GP_i} and Q_{GP_i} denote active power and reactive power from GP at node i, respectively. P_{L_i} and Q_{L_i} denote active power and reactive power from load at node i, respectively. G_{ij}, B_{ij} and δ_{ij} denote admittance and voltage phase angle parameters, respectively.

Inequality constraint conditions including the installed capacity of GP, the limiting value of node voltage and branch current:

$$\begin{cases} S_{GP\_j} \le S_{GP\_j}^{\max}, i \in N \\ U_i^{\min} \le U_i \le U_i^{\max} \\ I_{ij} \le I_{ij}^{\max} \end{cases} \tag{6}$$

3 MPSO algorithm

In this paper, the multi-objective particle swarm (MPSO) algorithm is proposed based on two order particle swarm algorithm. In view of the existence problem of premature and local search on traditional algorithm, the method of linear increasing weight is used to change the weight coefficient. The formula of speed and location of MPSO as follow:

$$v_{ik}(t+1) = \omega v_{ik}(t) + c_1 r_1[p_{ik}(t) - 2x_{ik}(t) + x_{ik}(t-1)] + c_2 r_2(g_{ik}(t) - 2x_{ik}(t) + x_{ik}(t-1)) \tag{7}$$

$$x_{ik}(t+1) = x_{ik}(t) + v_{ik}(t+1) \tag{8}$$

where, $v_{ik}(t)$ and $x_{ik}(t)$ denote particle speed and position, respectively. $p_{ik}(t)$ is the individual extreme position of particle i in the t iteration under k dimension. $g_{ik}(t)$ is the global extreme position of particle i. ω is the inertia weight, and c_1, c_2 is the factor of learning. r_1 and r_2 is the random number under [0, 1].

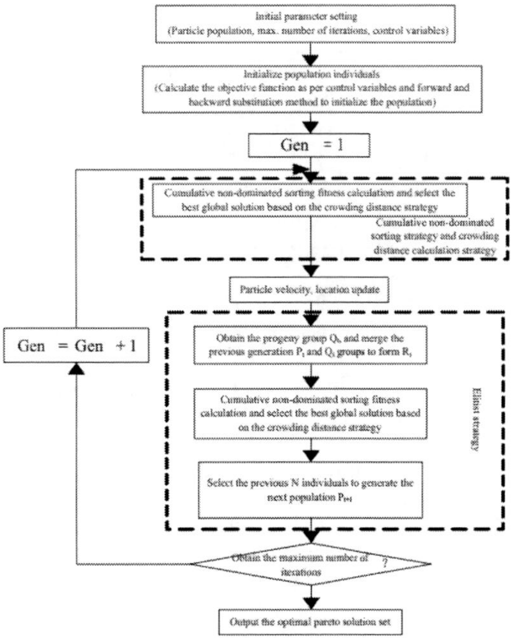

Figure 1. The process of MPSO algorithm

The strategies of cumulative rank fitness assignment, elitist and crowding distance computation are introduced during the process of calculating, and the flow chart of MPSO as shown in Fig. 1.

4 Example analisis

In this paper, the distribution system of IEEE-33 node is chose as an example and the corresponding system structure as shown in Fig. 2. GP including wind power (WT), photovoltaic (PV), fuel cell (FC), miniature turbine (MT), diesel engine (DE) and capacitor group (CB), and the related data of investment and pollution emissions are shown in Table 1. using to the multi-objective mathematical model, and use the MPSO algorithm for optimal allocation of GP, and the initial parameters of MPSO algorithm as follows: initial population of particles is set as 90, the number of iterations is set as 200, the inertia weight updating interal [0.45, 0.9], learning factor c1=c2 =1.

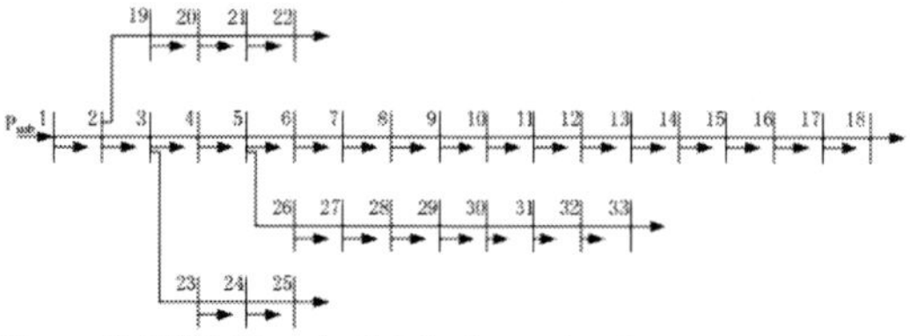

Figure 2. IEEE-33 node distribution network

Table 1. The related data of GP.

Genre	Assets cost/($/kW)	Pollution gas emissions targets(kg/kwh)				
		CO_2	$N0_x$	SO_2	CO	PM_{10}
WT	4500	0	0	0	0	0
PV	5000	0	0	0	0	0
FC	3500	0.46	0.006	0.012	0.002	0
MT	1500	0.72	0.091	0.002	0.247	0.018
DE	500	0.65	2.483	0.093	1.275	0.16
CB	90$/kvar	0	0	0	0	0

4.1 Optimal allocation of GP considering benefits of investment and voltage stability

In this section, from the angle of investment operation efficiency and voltage stability to optimize configuration of GP, and the Pareto solutions can be shown in Fig. 3.

From Fig. 3, it's not hard to see that the relationship between the benefits of investment and the index of voltage stability approximation opposite. In other words, to improve the load side voltage stability performance needs at the expense of the operation efficiency of power grid. Different schemes (A, B, C) are selected to analyze in this section, and the results can be shown in Table 2 and Fig. 4.

The Table 2 and Fig. 4 show that the level of voltage distribution is improved and the network loss is decreased with the GP reasonable access to ADN. For the more, optimal configuration of GP can improve the level of penetration of distributed power supply. Compared with un-optimization schemes, the related level of investment benefit, VSI, network loss are improved after optimization. In A, the index of VSI is higher and the efficiency of investment opration is lowest; In C, the index of VSI is the lower and the efficiency of investment opration is the highest; In B, the factors of VSI and investment benefits are considered reasonable, and the related index is more better than scheme A and scheme C, and the corresponding optimization results can be shown in Fig. 5.

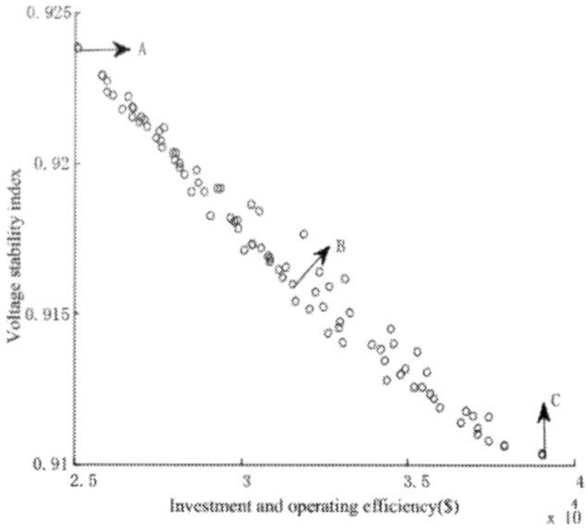

Figure 3. Pareto solution of GP optimization

Table 2. The optimal results of typical schemes.

scheme	benefit($)	VSI	Powerloss(kW)
unoptimized	—	0.8928	202.66
scheme A	25109	0.9239	134.19
scheme B	32060	0.9152	143.16
scheme C	39073	0.9104	148.63

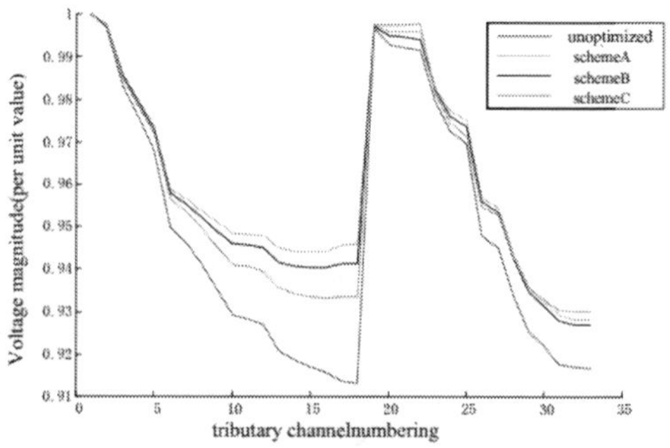

Figure 4. Voltage level under different schemes

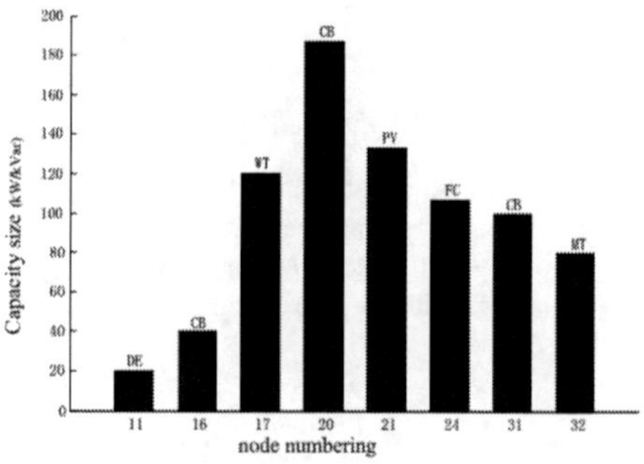

Figure 5. Optimal allocation of GP in scheme B

 DOI: 10.1051/matecconf/20179515003

4.2 Optimal allocation of GP considering benefits of investment and pollution gas emissions

From the angle of investment operation efficiency and the index of pollution gas emission to optimize configuration of GP, and the Pareto solutions can be shown in Fig. 6.

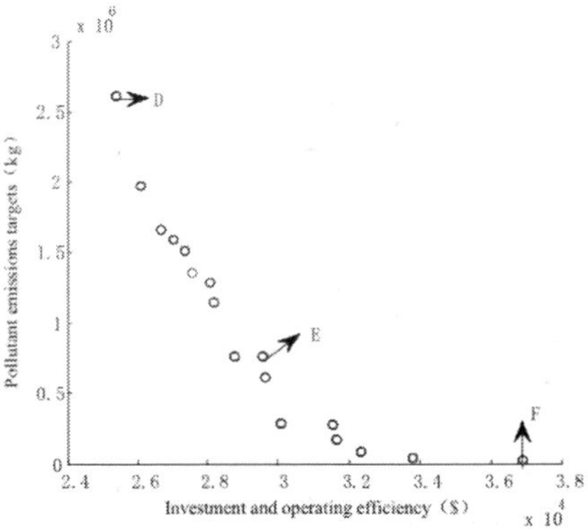

Figure 6. Pareto solution of GP optimization

From the Fig. 6, the benefits of investment is generally inverse-correlated to the index of pollution gas emission, and the reason can be explained as the capacitor groups is considered in the process of optimizing of GP. The low investment cost of capacitor groups and it has no pollution emission, therefore, it's particularly important to consider the influence of capacitor groups in the process of optimizing of GP. Different schemes (D, E, F) are analyzed in this section, and the results can be shown in Table 3 and Fig. 7.

Table 3. The optimal results of typical schemes

scheme	investment benefit ($)	gas emission (kg)	power loss (kW)
Unoptimized	—	—	202.66
scheme D	25442	2604488	117.95
scheme E	29628	760788	124.97
scheme F	36911	21024	148.75

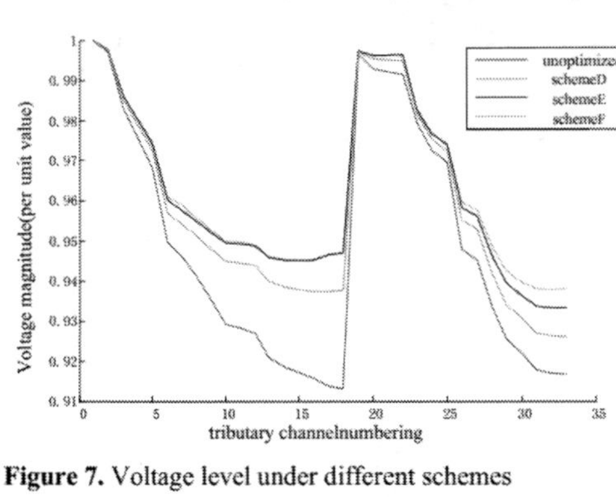

Figure 7. Voltage level under different schemes

From Table 3 and Fig. 7, the index of VSI is higher in scheme D, but, the benefit of investment is the lowest and the emission of pollution gas is the highest. In scheme F, the situation is reversed. In scheme E, the optimal allocation results is reasonablely, and the corresponding results can be shown in Fig. 8.

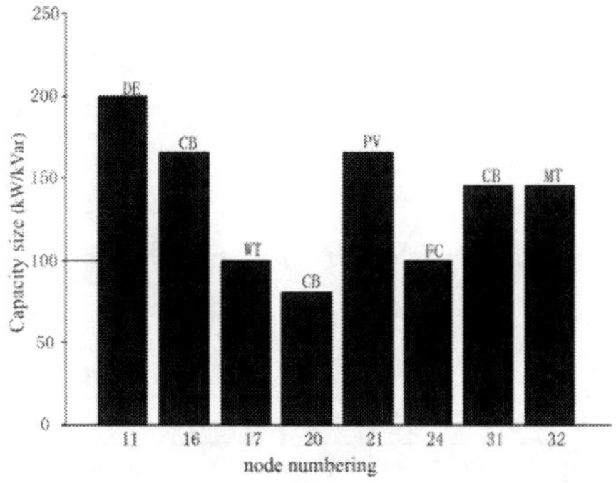

Figure 8. Optimal allocation of GP in scheme E

4.3 Optimal allocation of GP under different load levels

In this section, the IEEE-33 node is chose as examples, and the economy, technology and environmental factors are considered in the process of optimizing GP under different load levels (μ_1=0.7, μ_2=1.0, μ_3=1.3). When μ_2=1.0, the pareto solutions of different systems can be shown in Fig. 9. Different evaluation index of optimzing GP can be shown in Table 4 under different conditions.

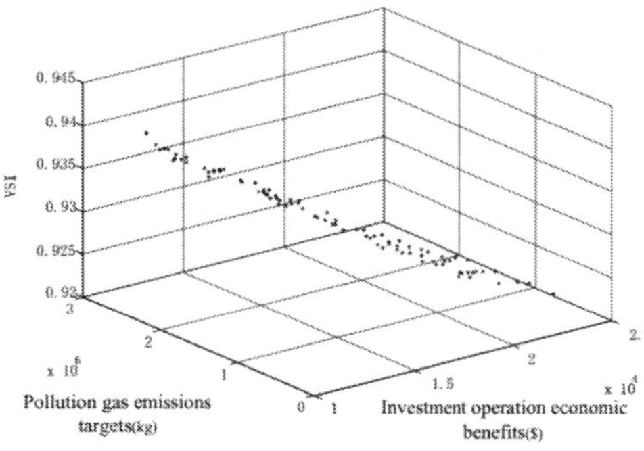

Figure 9. Pareto solution of GP in IEEE-33 node

From Fig. 9, the spatial distribution of pareto solutions is relatively uniform, which shows that the MPSO algorithm has good global search ability. At the same time, the variation patterns of key indexes are basically the same, which reflect that the MPSO algorithm has better parameter identification ability in deal with multi-objective optimization problems.

In Table 4, setting the first schemes as the base under different conditions, and the other two schemes are compared with the first scheme, and the results show that the optimize allocation schemes have obvious difference under different conditions, and the related evaluation index is approximately reversed. The situation of gas emission index is more special is that part of GP have none gas emissions.With the increase of load level, compared with original situation, the efficiency of investment operation and the stability of voltage is improved. On the other hand, compared with different optimal allocation schemes with the growth of load level, the capacity of GP is increased, but the stability of voltage is decreased, which explain that the growth trend of existing scale of GP cannot meet the long-term load demand. Therefore, it needs to expand the scale of low carbon perspective of GP.

Table 4. Optimal allocation of GP under different load rate.

μ	Plan	GP The configuration capacity (Corresponding node location)								assessment criteria		
		WT (kW)	PV (kW)	FC (kW)	MT (kW)	DE (kW)	CB1 (Kvar)	CB2 (Kvar)	CB3 (Kvar)	benefit (104$)	VSI	pollution gas(106Kg)
μ1	A1	165(1)	80(21)	200(24)	180(32)	180(11)	120(20)	120(16)	123(31)	0.4201	0.9521	2.6508
	A2	80(17)	180(21)	200(24)	160(32)	60(11)	160(20)	40(16)	200(31)	0.4717↑	0.9497↓	2.9125↑
	A3	20(17)	193(21)	20(24)	150(32)	180(11)	180(20)	130(16)	20(31)	0.9617↑	0.9391↓	0.8410↓
μ2	B1	126(1)	0(21)	100(24)	140(32)	145(11)	100(20)	160(16)	60(31)	1.2483	0.9381	2.6570
	B2	180(17)	200(21)	152(24)	40(32)	180(11)	100(20)	120(16)	180(31)	1.3012↑	0.9364↓	2.9220↑
	B3	161(17)	80(21)	60(24)	0(32)	80(11)	20(20)	40(16)	120(31)	2.0718↑	0.9256↓	1.7639↓
μ3	C1	120(17)	160(21)	20(24)	140(32)	157(11)	180(20)	120(16)	180(31)	2.4968	0.9239	2.8842
	C2	160(17)	186(21)	140(24)	117(32)	160(11)	106(20)	175(16)	132(31)	2.5595↑	0.9228↓	2.9220↑
	C3	104(17)	20(21)	190(24)	60(32)	40(11)	60(20)	150(16)	20(31)	3.8707↑	0.9104↓	1.3553↓

5 Conclusion

(1) A simple and effective index of voltage stability is put forward to assess the impact on voltage quality with GP access to DN. Rational allocation of GP can effectively improve the operation level of DN.

(2) A multi-objective optimition algorithm is proposed to optimize allocation of GP in DN, and this algorithm has better optimization performance and efficiency.

(3) With the increase of load level, rational planning of GP can improve the economic benefit for power supply service, and improve the quality of voltage for power users, and can also provide auxiliary comprehensive and reasonable decision for power dispatchers.

References

1. Y. Zhao, Y. An, Q. Ai. Research on size and location of distributed generation with vulnerable node identification in the active distribution network. IET Generation, Transmission & Distribution, **8**(11): 1801-1809 (2014)
2. C.L. T. Borges, V.F. Martins. Multistage expansion planning for active distribution networks under demand and distributed generation uncertainties. Journal of Electrical Power & Energy Systems, **36**(1): 107-116 (2012)
3. L. Li, W. Tang, M.K. Bai. Multi-objective locating and sizing of distribution generators based on time-sequence characteristics. Automation of Electric Power Systems, **37**(3): 58-64 (2013)
4. Z. Wang, B. Chen, J. Wang, et al. Robust optimization based optimal DG placement in microgrids. IEEE Transactions on Smart Grid, **5**(5): 2173-2182 (2014)
5. D. Foster, M. Berry, N. Boland, et al. Comparison of mixed-integer programming and genetic algorithm methods for distributed generation planning. IEEE Transaction on Power System, **29**(2): 833-843 (2014)
6. D.Q. Hung, N. Mithulananthan, K.Y. Lee. Determining PV penetration for distribution systems with time-varying load models. IEEE Transaction on Power System, **29**(6): 3048-3057 (2014)

Multi-Objective Low-Carbon Economic Dispatch Considering Demand Response with Wind Power Integrated Systems

Liu Wenjuan[1], Song Yundong[2], Zhang Junru[2], Liu Yang[2] and Meng Tao[3]

[1]Provincial Electric Power Co. Ltd, Shenyang 110006, China
[2]Electric Power Research institute Liaoning Electric Power Company Limited, Shenyang 110006, China
[3]Electric Power Research institute of Jilin Electric Power Co.of State Grid, Changchun 130021, China

Abstract. The generation cost, carbon emissions and customers' satisfaction are considered in this paper. On the basis of this, the multi-objective and low-carbon economic dispatch model with wind farm, this considers demand response, is established. The model user stochastic programming theory to describe the uncertainty of the wind power and converts it into an equivalent deterministic model by using distribution function of wind power output, optimizes demand side resources to adjust the next day load curve and to improve load rate and absorptive capacity of wind power, introduce customers' satisfaction to ensure that the scheduling scheme satisfies customer and integrate the resources of source and load to unify coordination wind farm access to network and to meet the requirements of energy saving and emission reduction. The search process of artificial fish school algorithm introducing Tabu search and more targeted search mechanism, an multi-objective improved artificial fish school algorithm is proposed to solve this model. Using the technique for order preference by similarity to ideal solution (TOPSIS) to sort the Pareto frontier, the optimal scheduling scheme is determined. Simulation results verify the rationality and validity of the proposed model and algorithm.

1 Introduction

As the greenhouse effect intensifying and fossil energy depletion, energy conservation and emissions reduction as the main target for the development of electric power enterprises. The carbon emissions of electric power enterprises mainly come from the large thermal power plants, thus we should change the traditional dispatching model to insure the economy and low carbon development of power system [1]. Compare with other new energy power generation, the wind power have lots of advantages: technology mature, low cost, clean and efficient. But large-scale wind power access to the power system will increase the difficulty of dispatching for the random output characteristics of wind power [2].

This problem is studied by related scholars. Literature [3] consider the randomness of wind power, and the dynamic scheduling model is established based on the stochastic programming theory. In literature [4], the dispatching question with wind power under extreme cases is change into a two-person zero-sum game between nature and decision-makes, and the robust optimization theory is used to describe the uncertainty output of wind power. On the other hand, considering the influence of the demand side resources in the process of dispatching, and unifying the resources between sources and loads to coordinate with wind power. In literature [5], the demand response resources is used as an virtual power generation to participate dispatching, and a low carbon dispatching model is established to improve the given capacity of wind power. In literature [6], the charge cars is introduced to coordination optimization dispatching, and by adjusting the weights of the indexes to implements the effective compromise between economy and carbon emissions.

In this paper, the strategy of demand response is introduced into low carbon and economic dispatching. In order to improve the given capacity of wind power, three power resources (wind power, thermal power, demand response) is considered in this paper. Firstly, application of wind power output distribution function, the multi-objective chance-constrained programming model is transformed into a deterministic model. Then, a multi-objective fish algorithm and the close to ideal solution sorting method are used to determine the optimal dispatching scheme. Lastly, by testing improved 10 machine system to verify the rationality and validity of proposed model and algorithm.

2 Stochastic model of wind power

A large amount of wind speed data analysis results show that the double parameter Weibull distribution can accurately describe the actual wind speed, and the distribution function is:

$$F(v) = 1 - \exp(-(v/c)^k) \tag{1}$$

where k, c respectively show the distribution and scale parameters; v is the wind speed.

The corresponding probability density function is:

$$f(v)=(k/c)(v/c)^{k-1}\exp[-(v/c)^k] \quad (2)$$

When know the wind speed distribution of the specific, through the analysis of the data of active output wind turbines and wind speed, the approximate function relation can be obtained, which can acquire from Fig. 1.

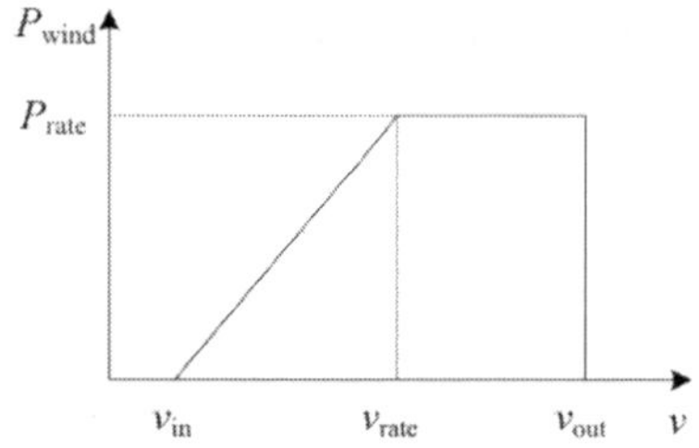

Figure 1. The active power output of wind power

From Fig. 1 we can see that the relationship of active power output of wind power (P_{wind}) and wind speed can be acquire:

$$P_{\text{wind}}=\begin{cases}0, & v\le v_{\text{in}} \text{ or } v\ge v_{\text{out}}\\ P_{\text{rate}}\dfrac{v-v_{\text{in}}}{v_{\text{rate}}-v_{\text{in}}} & v_{\text{in}}\le v\le v_{\text{rate}}\\ P_{\text{rate}} & v_{\text{rate}}\le v\le v_{\text{out}}\end{cases} \quad (3)$$

where v_{in}, v_{out}, v_{rate} respectively cut in, cut out and rated wind speed; Prate is rated active power of wind power.

According to type (3) we can acquire the distribution function of P_{wind}:

$$F(P_{\text{wind}})=\begin{cases}0, & P_{\text{wind}}<0\\ 1-\exp\{-[(1+\dfrac{v_{\text{rate}}-v_{\text{in}}}{v_{\text{in}}P_{\text{rate}}}P_{\text{wind}})\dfrac{v_{\text{in}}}{c}]^k\} & \\ +\exp[-(\dfrac{v_{\text{out}}}{c})^k], & 0\le P_{\text{wind}}\le P_{\text{rate}}\\ 1 & P_{\text{wind}}\ge P_{\text{rate}}\end{cases} \quad (4)$$

3 User satisfaction index

User satisfaction is measured by the experience of power user after the implementation of time-of-use price (TOU), which mainly include user electricity expenses satisfaction and electricity ways satisfaction index. User electricity expenses satisfaction index can be described as follow:

$$\theta=1-\frac{\sum_{t=1}^{T}(L'_j p_t-L'_{j0}p_0)}{\sum_{t=1}^{T}L'_{j0}p_0} \quad (5)$$

where θ is the electricity expenses satisfaction; P_t, L'_j respectively electricity price and load on t section after the implementation of TOU; T is the dispatching cycle; p_0, L'_{j0} respectively electricity price and load before the implementation of TOU.

User electricity expenses satisfaction index θ is higher, which shows that the electricity expenses satisfaction is greater. When θ >1, which illustrate that the electricity expenses of user is decreased.

Electricity way satisfaction index λ is described the implementation of TOU user power consumption before and after the change of time, which can be acquired as follow:

$$\lambda=1-\frac{\sum_{t=1}^{T}|(L'_j-L'_{j0})|}{\sum_{t=1}^{T}L'_{j0}} \quad (6)$$

4 Multi-objective optimization model

4.1 Multi-objective function

(1) Minimize the conventional generating units of the total cost of objective function is:

$$\min F_1=\sum_{t=1}^{T}\sum_{i=1}^{N}[I_{i,t}f_{i,t}(P_{i,t})+I_{i,t}(1-I_{i,t-1})S_{i,t}] \quad (7)$$

where F_1 is the total cost of the conventional power generation unit; T is the scheduling cycle; N is the number of conventional unit; $I_{i,t}$=0 and $I_{i,t}$=1 respectively the operation and stop station of the i th conventional unit under t time; $P_{i,t}$ is the output power of i th conventional unit; $S_{i,t}$ is the control cost of i th conventional unit.

The fuel costs of conventional units is:

$$f_{i,t}(P_{i,t})=a_iP_{i,t}^2+b_iP_{i,t}+c_i \quad (8)$$

where a_i, b_i, c_i are the energy consumption parameters of the units.

The start-stop costs of conventional units is:

$$S_{i,t}=\begin{cases}h_{si} & M_{Di}\le X_{i,t}^{\text{off}}\le M_{Di}+T_i\\ c_{si} & X_{i,t}^{\text{off}}\ge M_{Di}+T_i\end{cases} \quad (9)$$

where h_{si} and c_{si} respectively for the hot-start and cold-start costs of the ith convention units; M_{Di} is the minimum outage time of the ith convention units; $X_{i,t}^{\text{off}}$ is the continuous stoppage time of the ith convention units.

(2) Minimize the conventional generating units of the total carbon emissions of objective function is:

$$\min F_2=\sum_{t=1}^{T}\sum_{i=1}^{N}[C(P_{i,t})I_{i,t}] \quad (10)$$

where F_2 is the carbon emissions of conventional units under total scheduling cycle; $C(P_{i,t})$ is the carbon emissions, and the characteristic equation can be represented as:

$$C(P_{i,t})=\alpha_i+\beta_iP_{i,t}+\gamma_iP_{i,t} \quad (11)$$

where α_i, β_i, γ_i are the carbon emission coefficient.

4.2 Constraints

(1) Power balance constraint:

$$\sum_{i=1}^{N} P_{i,t} + P_{\mathrm{w},t} - P_{\mathrm{D},t} - P_{\mathrm{loss},t} = 0 \tag{12}$$

where Pw,t is power output of wind farm under t station; PD,t is the system load after TOU under t station; Ploss,t is network loss under t station.

(2) Wind power penetration limit constraint:

$$P_{\mathrm{w},t} \le \delta P_{\mathrm{D},t} \tag{13}$$

where δ is penetration limit coefficient, by the method of chance constrain which can be expressed as:

$$P\{\delta P_{\mathrm{D},t} \ge P_{\mathrm{w},t}\} \ge \varphi \tag{14}$$

where φ is limit constraints to satisfy the confidence level of wind farm.

(3) Output power constraint of conventional unit:

$$P_{i,\min} \le P_{i,t} \le P_{i,\max} \tag{15}$$

(4) Minimum open-stop time constraint of conventional unit:

$$\begin{cases} (1 - I_{i,t+1}) M_{\mathrm{U}i} \le X_{i,t}^{\mathrm{on}} & I_{i,t} = 1 \\ I_{i,t+1} M_{\mathrm{D}i} \le X_{i,t}^{\mathrm{off}} & I_{i,t} = 0 \end{cases} \tag{16}$$

(5) Positive and negative reserve capacity constraint of conventional unit:

$$\sum_{i=1}^{N} (P_{i,\max} - P_{i,t}) \ge P_{\mathrm{D},t}^{\mathrm{up}} \tag{17}$$

$$\sum_{i=1}^{N} (P_{i,t} - P_{i,\min}) \ge P_{\mathrm{D},t}^{\mathrm{down}} \tag{18}$$

(6) In order to guarantee the electricty ways of user does not significantly change, and the electricity way satisfaction index should be satisfied as:

$$\lambda \ge \lambda_{\min} \tag{19}$$

(7) In order to guarantee the interests of user don't damage after TOU, and the electricity expenses satisfaction index should to meet:

$$\theta \ge \theta_{\min} \tag{20}$$

(8) Open ratio should to meet:

$$Q_p / Q_v \le \omega \le 1 \tag{21}$$

5 Multi-objective optimization algorithm

5.1 Improved multi-objective artificial fish school algorithm

Artificial fish school algorithm (AFSA) and Tabu search algorithm (TS) is combined in this paper, and an improved artificial fish school algorithm (IAFSA) is proposed to improve the ability of global optimal converge. The strategies of crowding distance calculation, accumulated sort and elite reserve are introduced to make IAFSA is suitable to solve multi-objective optimization scheduling problem. Thus, the multi-objective improved artificial fish school algorithm (MIAFSA) is proposed to deal with this model, and its process can be shown in Fig. 2.

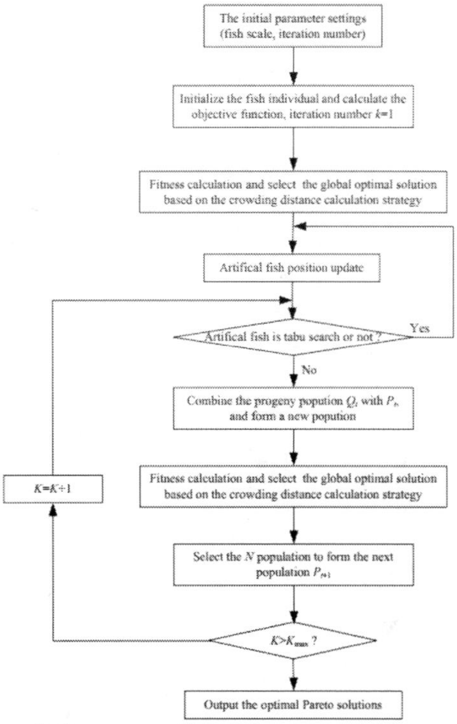

Figure 2. Ther process of MIAFSA

5.2 Multi-objective dispatching steps

(1) Input unit parameters, user response parameters, wind power prediction parameters, original electricity price and load parameters.

(2) Determine the period of electricity price, and calculate the electricity price of peak and low times.

(3) According to the principle of consumer psychology calculated system load, user electricity expenses satisfaction index and electricity way satisfaction index after TOU.

(4) Judge the related index whether meet the constraints or not, if meet in step (5), or else, turn into step (2).

(5) According to step (3) calculate the system load, and arrange the unit combination and output, and calculate the total cost and carbon emissions.

(6) Use the MIAFSA algorithm calculate the optimal Pareto solutions.

(7) Judge whether meet the maximum number of iterations or not, if meet, output the optimal Pareto solution set, or turn into step (2).

6 Example analisis

Typical 10 machibe systems with wind farms is tested in this paper. The scheduling cycle is one day, and it can be divided into 24 periods. The confidence level μ and φ respectively 0.95 and 0.85. Ordinary electricity price is 0.8 RMB/kW.h, and peak-ping-gu time period can be divided into: 9:00-15:00 and 20:00-22:00 are peak period; 6:00-9:00, 15:00-17:00, 18:00-20:00 and 22:00-23:00 are ping period; 23:00-6:00 and 17:00-18:00 are gu period. Wind farms contain 60 fan, and the related power is 2 MW, and the forecast output power in 24 hour can be shown in Fig. 3. Positive and negative rotation of spare capacity in each period are 5% of the system load, wind power penetration limit coefficient is 0.09. MIAFSA parameters setting: artificial scale is 80; iteration number is 100; crowd coefficient is 0.3; biggest foraging number is 10; the length of the Tabu table is 10. The Pareto front acquired based on MIAFSA can be shown in Fig. 4, and the diversity solutions can be shown in Table 1.

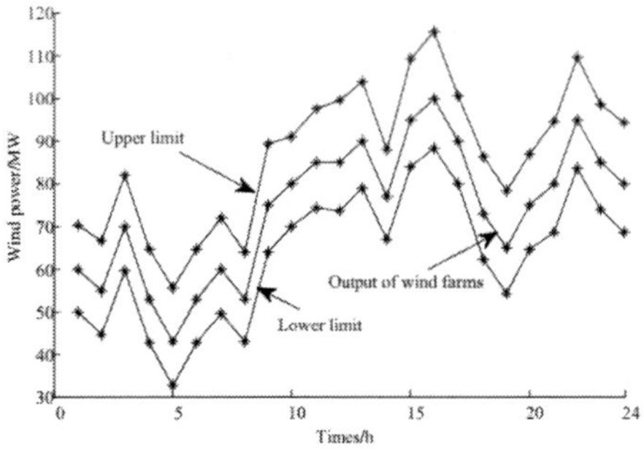

Figure 3. Ther output of wind farms and fluctuation range

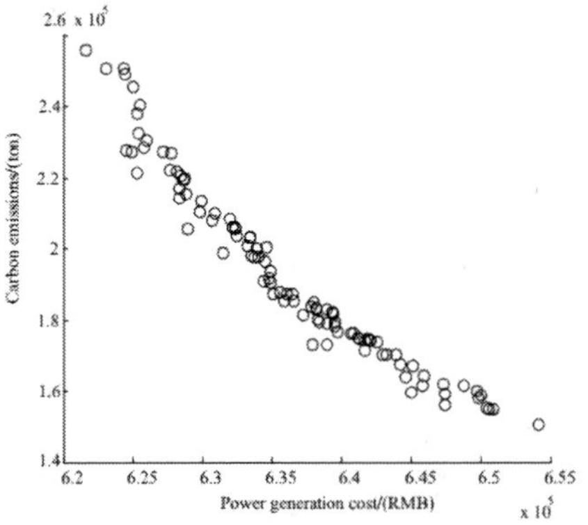

Figure 4. Ther Pareto front based on MIAFSA

From Fig. 4 we can see that the Pareto front integrity is better, and the distribution range of carbon emission is better than the power generation cost, which is mainly due to the conventional units difference is larger.

According to the data of Table 1, compared with the PSO algorithm, the power generation cost and carbon emissions of optimal solution respectively decrease 7630 RMB and 17360 ton; the electricity expense satisfaction index is approaching to 1, which indicate that the user electricity expense has decreased after TOU, but no serious damage to the power gird company's benefits. Electricity way satisfaction index is higher, which see that the daily electricity use way of user didn't upset after TOU. Opening ratio as an important index of electricity price, which directly decides the peak-valley price. The greater of the index show that the user response more fully with TOU, and peak sharpening effect is more obvious, which accord with the consumer psychology of user. The start-stop cost of units and power generation cost has no obvious relationship, but there are positively correlated with carbon emissions, which show that the frequent start-stop of units is help reduce the carbon emissions of system because of the carbon emission characteristics of small capacity unit is less. The accept ability of wind power under traditional method is 1762MW, in this paper, the accept ability is improve 7MW, 19MW, 15MW respectively under economic optimal, environment optimal and optimal solution station. Thus, compared with the traditional method, the accept ability of wind power of system is improved with demand response considered. From another hand, the peak-valley of load is decreased 334MW, which see that the load rate of system is improved with demand response considered.

The load curve with demand response considered before and after can be shown in Fig. 5. To test the authenticity of the dispatching results, the optimal solution is given in Table 2.

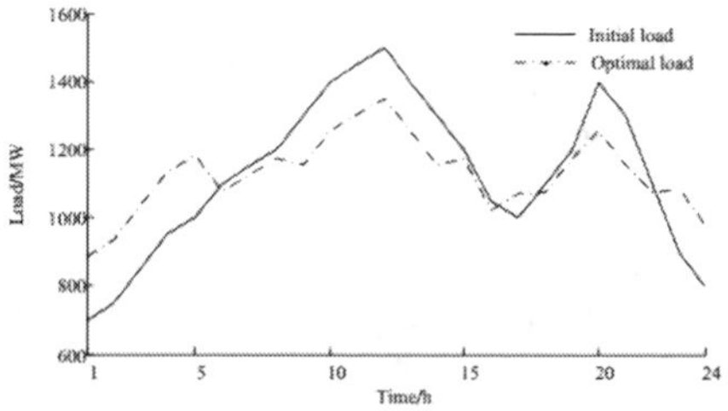

Figure 5. The load curve with demand response considered

Table 1. The diversity solutions under low carbon economic dispatching.

objects	Generation cost /(10^5RMB)	Carbon emission/(10^5ton)	electricity expenses satisfaction	Electricity way satisfaction	Control cost /(RMB)	Opening ratio	Peak valley difference /(MW)	Wind power /(MW)
Economic optimal	6.2161	2.5578	1.0102	0.8700	4200	0.8120	472	1769
Environment optimal	6.5418	1.5067	1.1424	0.9105	3970	0.6536	425	1781
Optimal solution	6.3491	1.9045	1.0841	0.8996	4190	0.7011	466	1777
PSO algorithm	6.4254	2.0781	1.0721	0.9007	4260	0.6914	455	1775

Table 2. Pareto solution of MIAFSA.

period	The output of units under different period /MW										Access ability /MW	Network loss /MW	Load /MW
	unit1	unit2	unit3	unit4	unit5	unit6	unit7	unit8	unit9	Unit10			
1	356.21	174.93	44.93	44.93	49.93	80	0	34.9	55	0	62	16.61	886.24
2	218.08	218.08	88.08	88.08	93.08	80	0	55	0	55	55	16.18	934.24
3	257.30	257.30	127.3	130	132.3	80	0	0	0	0	70	19.97	1034.2
4	319.6	257.4	130	127.4	132.4	0	85	0	55	0	53	25.52	1134.2
5	331.51	331.51	130	130	162	0	85	0	0	0	43	29.18	1183.8
6	306.73	306.73	130	0	162	0	85	0	55	0	53	24.14	1074.4
7	357.36	357.36	130	0	162	0	85	0	0	0	60	27.45	1124.2
8	414.88	403.38	71.79	0	76.79	0	76.8	55	55	0	53	31.92	1174.7
9	309.76	309.76	130	0	162	0	85	0	55	55	75	26.70	1154.8
10	359.89	359.89	130	0	162	0	85	55	55	0	80	32.13	1254.9
11	345.56	345.56	130	130	162	0	85	0	0	55	85	33.30	1304.8
12	340.63	455	130	130	162	0	85	0	0	0	85	37.31	1350.3
13	375.16	288.78	130	130	162	0	0	55	0	55	90	30.80	1255.1
14	329.67	353.50	130	130	162	0	0	0	0	0	77	27.27	1154.9
15	286.83	286.83	130	130	161.8	0	0	0	55	55	95	25.79	1174.7
16	253.41	253.41	123.4	130	128.4	0	0	0	55	0	92	18.94	1022.8
17	270.56	270.56	130	130	145.6	0	0	55	0	0	90	21.18	1070.4
18	301.28	301.28	130	130	162	0	0	0	0	0	73	23.02	1074.7
19	330.22	330.22	130	130	162	0	0	0	55	0	65	28.22	1174.2
20	339.78	339.78	130	130	162	0	0	0	55	55	75	31.72	1254.9
21	284.17	284.17	130	130	162	0	0	55	0	55	80	25.30	1155.5
22	270.38	270.38	130	130	145.3	0	0	0	0	55	95	21.42	1074.8
23	243.19	243.19	130	113.1	125.9	0	0	55	55	55	85	21.29	1084.2
24	235.06	235.06	130	105.0	162	0	0	0	0	55	80	17.94	984.55

7 Conclusion

(1) A multi-objective chance constrained program model of environment, economy and user satisfaction is established in this paper, which change the traditional economic dispatching pattern to provide a more comprehensive reference solution to decision makes.

(2) Demand response is introduced into low carbon economic dispatching can effectively cut peak-valley load difference, increase the load ability of system and the access ability of wind power.

References

1. S. Roy, Inclusion of short duration wind variations in economic load dispatch. IEEE Transactions on Sustainable Energy, **3**(2): 265-273 (2012)
2. S.W. Mei, W.T. Guo, Y.Y. Wang, et al. A game model for robust optimization of power systems and its application. Proceeding of the CSEE, **33**(19): 47-56 (2013)
3. R. Billinon, G. Baig. Generating capacity adequacy associat- ed with wind energy. IEEE Transactions on Energy Systems, **19**(3): 641-646 (2004)
4. W. Zhou, Y. Peng, H. Sun, et al. Dynamic economic dispatch in wind power integrated system. Proceeding of the CSEE, **29**(25): 13-18 (2009).
5. X. Liu, X. Ai, Q. Peng. Optimal dispatch coordinating power generation with carbon emission permit for wind farms integrated power grid considering demand response. Power System Technology, **36**(1): 213-218 (2012)
6. X. Liu, W. Xu, Economic load dispatch constrained by wind power availability: a here-and-now approach. IEEE Transactions on Sustainable Energy, **1**(1): 2-9 (2010)

MATEC Web of Conferences **95**, 15005 (2017) DOI: 10.1051/matecconf/20179515005
ICMME 2016

Analysis of the Thermal Load of Structural Elements High-Power DC Supply with the Transformer of Ripple Filter with Current Overload

Lokvenc Jaroslav, Drtina Rene and Sedivy Josef

Department of Technical Subjects, Faculty of Education, University of Hradec Kralove, Rokitanskeho 62, 500 03 Hradec Kralove, Czech Repubilc

Abstract. Design of the high-power DC supply quite logically based on the required electrical parameters and expected or defined operating conditions. A prerequisite trouble-free operation is also the correct choice of construction materials. Both in terms of mechanical strength and stability, and in terms of thermal load. The article deals with thermal conditions in the high-power DC supply with the transformer of ripple filter for long-term current overload.

1 Introduction

High-power DC supplies for industrial purposes are designed completely differently from conventional commercial devices. For industrial devices usually do not mind larger size or heavier.

The primary requirement is virtually maintenance free operation and operational reliability. A ommon requirement is too high short or long term overload. Short time overload DC supplies depend primarily on the data of the rectifier elements. Long-term overload is further more also dependent on the thermal resistance of materials used in the construction of source solution a thermal bridge and cooling space of the individual components and the entire DC supply.

Used model solution based is on theoretical knowledge of key publications [1]-[3] and the results of long-term practical tests.

2 High-power DC supply with high current overload

Tree-phase high-power DC supplies with the transformer of ripple filter for industrial purposes [4]-[6] designed for heavy-duty applications. DC supplies of this construction are primarily designed for power series motors with heavy start and, as a starting source for internal combustion engines and for power supply for small railways, trolley, electroplating, etc.

The prototype linear stabilized power supply to heavy operating conditions, the output current voltage characteristic is comparable to the current voltage characteristic of lead-acid battery was developed in the laboratories of the Department of electro technical subjects, Faculty of Education, University of Hradec Kralove. His scheme is to Fig. 1.

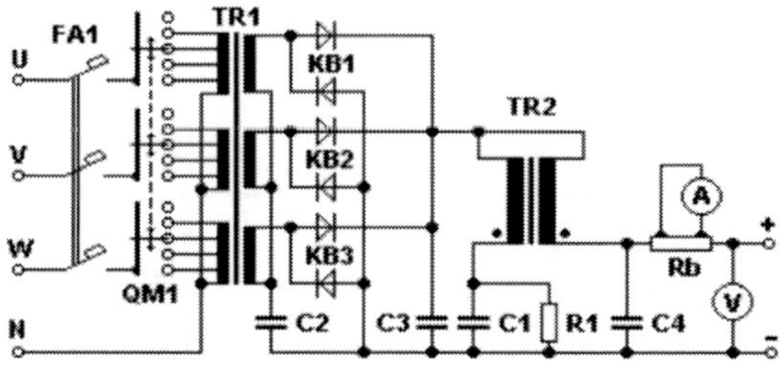

Figure 1. Scheme of the high-power DC supply 14V/140A.

Source is designed for permanent load 140A with the expected short-term overloading (in the order of minutes to tens of minutes) at 50% of rated current. Peak overload (for example, when starting production engines) is 100% for a few seconds. Overloading sources represents, among others, increased heat stress key components. The power dissipation and thereby heat generation of increases with the square of the current (Eq. 1)

$$P_{\mathrm{dissip}} = R_{\mathrm{dissip}} \cdot I^2 \qquad (1)$$

R_{dissip} represents an undesirably loss resistances in the circuit and I through the current. At 50% overload is therefore greater power dissipation of 125% (2.25×) [1, 3]. At 100% overload is then dissipated power four times higher than in normal operation. The most critical places with the greatest heat stress are a transformer TR1 [2], half-bridges rectifiers KB1, KB2, KB3 and secondary windings of the transformer TR2, through which the DC output current. Power dissipation naturally grows on the connecting wires, at the measuring resistor Rb and on the transient resistances of the output terminals (Fig. 1).

The modern half-rectifier bridges with metal base and structure have isolated when mounted correctly minimal thermal resistance and of adequate size cooler is a big factor in restrictive maximum forward current I_{FSM} and the Limiting integral I^2t.overload. Permissible operating surface temperature of modern diodes and rectifiers is normally up to 95° C.

3 Thermal resistance in transformers

Power dissipation in the transformer overloaded arises particularly in the winding, so called in losses copper. Transformers cooled by natural convection transfer heat the surface of the core and surface winding. Armature coils, insulation and covering layers represent heat bridges with large thermal resistances, which greatly limits the transmission and radiation of heat [1], [3] (Fig. 2).

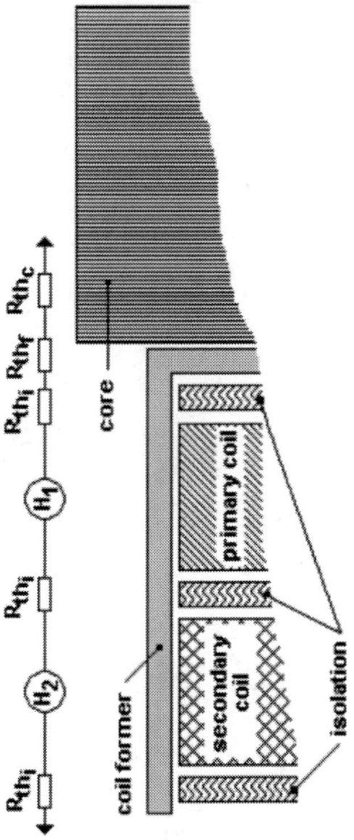

Figure 2. Simplified scheme of heat flows in the transformer.

The heat generated in the primary and secondary windings (H_1, H_2) (Fig. 2) is first accumulated in the mass of the winding. Despite the isolation (R_{thi}) and coil former (R_{thf}) only slowly penetrates to the core of it and passes into the atmosphere (R_{thc}). On the other hand, passes the heat to the atmosphere through the outer insulating layer (R_{thi}). Heat flow between H_1 and H_2 through the insulating layer (R_{thi}) is always in the direction of the temperature gradient.

From the structural design of the transformer can be inferred that the larger cooling surface of the nucleus and thus less thermal resistance of the core have the transformers with core sheathing (EI transformers). Core transformers (for example UI/UN, C-core and toroidal transformers), by contrast, have a greater cooling surface coils [2], [3]. Warming of the transformer core and windings during normal operation can be determined from Eq. 2 and Eq. 3 [3]

$$\Delta\theta_{core} = P_{core} \cdot R_{th_c} = \frac{P_{core}}{c_{core} \cdot F_{core}} \quad (2)$$

$$\Delta\theta_{coil} = P_{coil} \cdot R_{th_i} = \frac{P_{coil}}{c_{coil} \cdot F_{coil}} \quad (3)$$

where is $\Delta\theta_{core}$ warming of core, R_{thc} thermal resistance of core, P_{core} power dissipation of core, c_{core} cool down constant of core, F_{core} cooled surface of core, $\Delta\theta_{coil}$ warming of coil, R_{thi} thermal resistance of external insulation, P_{coil} power dissipation of coil, c_{coil} cool down constant of coil, F_{coil} cooled surface of coil. When the transformer overload significantly grows in power dissipation coil, power dissipation of the core remains by [3] virtually the same.

4 Thermal load elements of an overloaded high-power supply

Operational testing of a prototype high-power DC supplies with the ripple filter transformer of 14V/140A (Fig. 1) implemented by our proposal [5] include, among other things, test the temperature rise during overload. The source was burdened by a steady stream 210A, which is overloaded by 50% on nominal current. Was Measured temperature coil, coil surface temperature and core temperature on transformer T1, temperature on base half-bridges rectifiers and cooler temperature, coil temperature and core temperature on T2 transformer, shunt resistor amperemeter temperature, temperature wires and output terminals temperature.

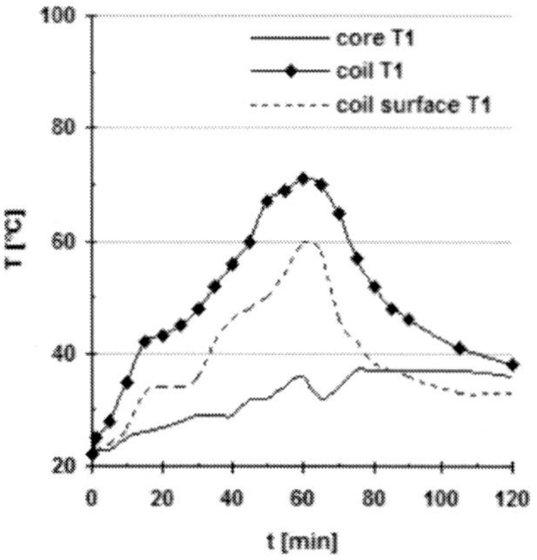

Figure 3. Temperature variation of source overload and after switch off - Transformer T1.

Overload test was limited to a maximum duration of one hour or violation of any of the temperature limit: 90°C for coil temperature, 95°C for base half-bridges rectifiers temperature, 105°C for shunt amperemeter resistor temperature and 70°C for wires temperature. We were monitored and output terminals temperature, but for the test was not limiting. Maximal temperature of output terminals is 110° C, which is much higher than the

DOI: 10.1051/matecconf/20179515005

permissible temperature of the connecting wires. After turning off the power for 1 hour monitored temperatures drop all parts source. Measurement of results is plotted in Fig. 3.

From the measured values it is clear that at ambient temperature up to 25°C are not exceeded any limits during sustained overload for 1 hour. Heat the most affected element is a three-phase source threecolumns transformer T1, which accumulates most of thermal energy. From the graph it is evident that after turning off the source of heat is accumulated in the transformer winding is transferred to the core and the core temperature rises further.

The transformer T1 due to the small cooling surface of the core and a relatively small cooling surface coils cools slowest. At 24°C was cooled to 4½ hours after the power is turned off (Fig. 3).

In contrast, the compensation transformer TR2 with single-layer winding and a large cooling surface is heated considerably more slowly and more rapidly cool (Fig. 4).

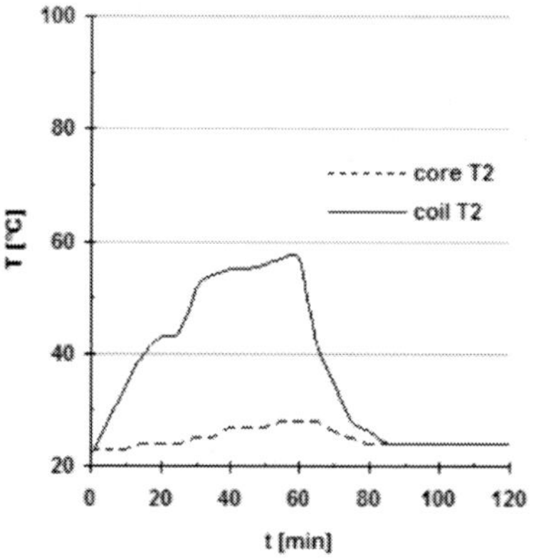

Figure 4. Temperature variation of source overload and after switch off - Transformer T2.

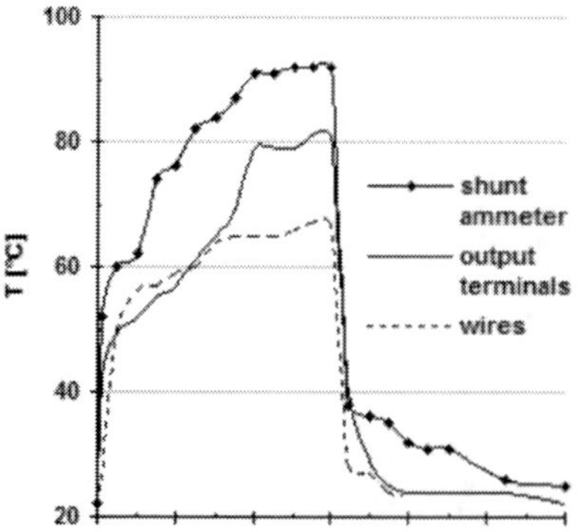

Figure 5. Temperature variation of source overload and after switch off - Shunt ammeter, internal wires, output terminals.

The graph (Fig. 5) indicates that the fastest cools amperemeter shunt resistor, wires and output terminals. Current load of the internal wires and output terminals were 8,4 A/mm^2.

The amperemeter shunt resistor is designed for a current of 250 A and therefore not overloaded. The maximum operating temperature for the amperemeter shunt is by the manufacturer determined on 155 ° C.

Temperatures half-bridges rectifiers and coolers after the power is turned off within 20 minutes aligned and cold as well (Fig. 6).

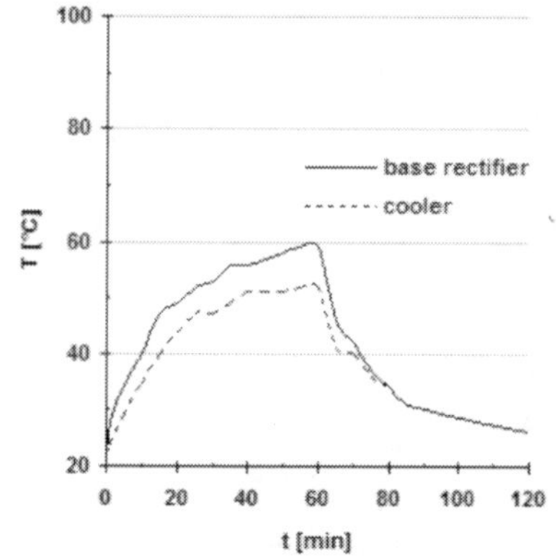

Figure 6. Temperature variation of source overload and after switch off - Transformer T2.

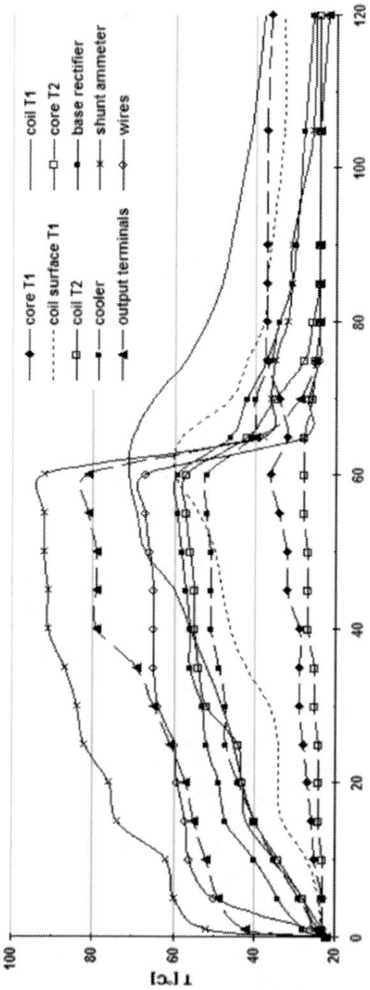

Figure 7. Temperature variation of source overload and after switch off - Summary graph.

Comprehensive summary graph of temperature measurements of the overloading power supply is on Fig. 7.

5 Summary

Of course readings, we can deduce that our proposed resource withstands significant long-term overload. Key elements of the well withstand the same overload even at ambient temperatures of 40° C, except for the connecting conductors, which would have to be dimensioned either to a higher operating temperature or at lower current density and less computational warming. For ambient temperature of 40°C can be assumed linear movement of the thermal characteristics of the transformer T1 and significantly longer cooling period.

Measurements of the prototype high-power DC supplies with the transformer of ripple filter showed that the transformer sources are also currently its substance and that even with a relatively simple design, achieving very good performance and withstand even long-term overload. Due to easy handling, high durability and high operational reliability can be easily used in automated processes.

The article was prepared with the support of specific research project SV PdF 2018/2012 - High-power DC supply with mid-frequency transformer compensation ripple voltage with high-overload current.

References

1. J. H. Harlow. *Electric power transformer engineering*. BOca Raton, CRC (2007).
2. J. J. Winders. *Power transformers: principles and applications*. New York, Dekker (2002).
3. Z. Faktor. *Transformer and coils*. Praha, BEN (1999).
4. R. Drtina, J. Lokvenc and J. Andris. The Accessories for Your Laboratories, Part 9. - Switching supply for laboratory use, Media4u Magazine, **1**, pp.55-62, (2009).
5. J. Lokvenc, R. Drtina. Power supply voltage with the transformer ripple filter, in: *The 11th WSEAS/ IASME International Conference on Electric Power Systems, High Voltages, Electric Machines*, Penang, Malaysia, pp.60-64, (2011).
6. J. Lokvenc, R. Drtina, and M. Sobek. Three-phase high overload DC power supply voltage with the mid-frequency transformer ripple filter. In *The 12th WSEAS International Conference on Electric Power Systems, High Voltages, Electric Machines*, Praha, pp.29-34, (2012).

DOI: 10.1051/matecconf/20179515006

Analysis on Crack Fatigue Growth Calculation for Nuclear Power Plant In-service Inspection

Cheng Yang , Xueliang Zhang and Peng Xia

In-service Inspection, Technical Department, Suzhou Nuclear Power Research Institute, China

Abstract. Fatigue is the common failure mode for nuclear mechanical components during operation which could induce crack initiation and growth. In order to predict the behaviour of fatigue crack propagating, fatigue calculation is introduced. A new method of crack fatigue growth calculation has been presented in this paper according to French RSE-M Rules. Regarding this method, detail analysis on key formulas selecting, essential calculation factors of occurrences combination, K-value calculation and mechanic modelling, characterization interval option has been performed. One fatigue crack growth calculation example of dissimilar weld of Reactor Pressure Vessel outlet nozzle to the safe end has been taken and the result demonstrates a good expectation.

1 Introduction

Throughout the lifetime of nuclear power plant operation, nuclear mechanical components are subject to a wide variety of loads such as pressure load, thermal load and periodic vibration load, thereby giving rise to fatigue which could induce crack initiation and growth, severely endangering the integrity of mechanical components. Therefore, during in-service inspection (ISI), once any crack has been detected through non-destructive test (NDT) which cannot be characterized by acceptance criteria, is thus required to perform crack fatigue growth calculation to predict the crack growth for the whole life of the component.

A new method of crack fatigue growth calculation has been presented in this paper according to French RSE-M Rules. Regarding this method, detail analysis on key formulas selecting, essential calculation factors of occurrences combination, K-value calculation and mechanic modelling, characterization interval option has been performed [1]. One fatigue crack calculation example of dissimilar weld of Reactor Pressure Vessel outlet nozzle to the safe end has been taken, in the hope to provide guidance on crack fatigue calculation for nuclear pressure components of M310, CPR1000, EPR and Hualong 1# reactors during ISI, or to provide reference to AP1000, Tianwan WWER-1000 reactors by ASME rules in China [2].

2 Content of crack fatigue growth calculation

The content of crack fatigue growth calculation is to calculate the crack initiation factor under combination occurrences. If the initiation factor is greater than 1, the crack should be deemed initiated.

Once crack initiated, we need to calculate the size of the crack after growing within the characterization interval. After calculating, we can obtain the fracture mechanic parameters (*Kcp* and *J* integral) and compare them to the specified limit values defined in RSE-M to verify the component structural integrity.

It is generally considered that the crack newly detected by NDT during ISI has initiated. Therefore, the initiation factor is not required to be calculated, so the theory of crack fatigue calculation in this paper is specific to the crack growth.

3 Method of crack fatigue growth calculation

The method of crack fatigue growth calculation is as follows:

(1) Calculating the stress intensity factors K_I, K_{II} and K_{III} at the crack tip in each occurrence.

K_I, K_{II} and K_{III} are the stress intensity factors of crack under three stress modes, which are known as opening mode, sliding mode and tearing mode. The maximum and minimum values of stress intensity factors in such three modes, i.e. $K_{I\ max}$ and $K_{I\ min}$; $K_{II\ max}$ and $K_{II\ min}$; $K_{III\ max}$ and $K_{III\ min}$ will be calculated.

(2) Calculating the equivalent stress intensity factor range ΔK and maximum stress intensity factor K_{max} for combination occurrences.

In case of combination of occurrences p and q, K values of stress intensity factors in such three modes can be calculated by Eq. 1, and then ΔK_I , ΔK_{II} and ΔK_{III} (difference between K values for Occurrences p and q) accordingly. This paper will calculate the equivalent stress intensity factor range and the maximum stress intensity factor using linear combination method which is simple, highly precise and easily for computer programmed, and described as follows:

$$\Delta K = \Delta K_{\rm I} + \Delta K_{\rm II} + 0.74\,\Delta K_{\rm III} \tag{1}$$

Taking the maximum value of $K_{\rm I}$, $K_{\rm II}$ and $K_{\rm III}$ for combination occurrences p and q, and then the maximum stress intensity factors K_{max} and K_{min} of such combination occurrences can be calculated.

$$K_{max} = \max\,(K_{\rm I},0) + |K_{\rm II}| + |K_{\rm III}| \tag{2}$$

$$K_{min} = K_{max} - \Delta K \tag{3}$$

Regarding different combination occurrences, it can be calculated different ΔK , K_{max} and K_{min} , and then these values could be permutated to find Maximum Value $\Delta K(m,n)$ and their corresponding $K_{max}(m,n)$, where m and n are the combination occurrences when ΔK is maximized [3].

(3) Calculating the corrected equivalent stress intensity factor range ΔK_{CP} for combination occurrences.

There is stress concentration at the crack tip and therefore existing plastic zone at the crack tip, which has changed the stress distribution of crack. So during calculation, the corrected equivalent stress intensity factor range ΔK_{CP} of the plastic zone should be considered. Here we describe the calculation method of ΔK_{CP} , taking a finite length emerging crack as an example. Suppose the dimension size of finite length emerging crack is that: Height (h) = a and Length (l) = $2c$, as shown in Figure 1.

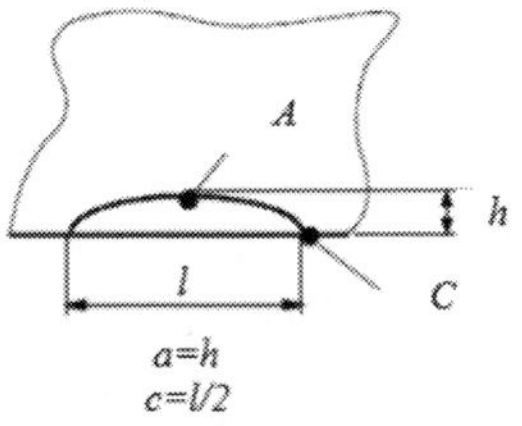

Figure 1. Dimension of finite length emerging crack.

(a) Calculating the radius of the plastic zone r_{yA}:

$$r_{yA} = \frac{1}{6\pi}\left[\frac{\Delta K_A(m,n)}{S_{yA}(m)+S_{yA}(n)}\right]^2 \tag{4}$$

where $S_{yA}(m)$ and $S_{yA}(n)$ are the yield stress values of the material at the temperatures reached at Point A, at times m and n respectively, which can be acquired by consulting the component design file. The radius of plastic zone at Point C can be calculated with the same formula above [4]0.

(b) Selecting formula :

For emerging defect, the maximum equivalent stress intensity factor $K_{\max A}^{meca}(m,n)$ due to mechanical loads under the combination occurrences at Point A should be calculated with the same method as K_{max} , while only mechanical load is used in the process of calculation.

If $K_{\max A}^{meca}(m,n) > 0.2\,K_{max}(m,n)$

Then $K_{cpA} = \Delta K_A(m,n)\sqrt{\dfrac{a+r_{yA}}{a}}$

$$\Delta K_{cpC} = \Delta K_C(m,n)\sqrt{\frac{a+r_{yC}}{a}} \tag{5}$$

Otherwise $K_{cpA} = \Delta K_A(m,n)$

$$\Delta K_{cpC} = \Delta K_C(m,n) \tag{6}$$

(4) Calculating Δa and Δc .

Fatigue crack growth is estimated by using the corrected stress intensity factor range ΔK_{CP} . The fatigue crack growth rate and the effective stress factor range has the following relational expression [5]:

$$\frac{da}{dN} = C(\Delta K_{eff})^n \tag{7}$$

$$\Delta K_{eff} = f(R)\Delta K_{CP} \tag{8}$$

C, n and $f(R)$ is related to the material properties and the value of stress intensity factor of combination occurrences, and the specific value can be determined through the recommendation table given in RSE-M, then ΔK_{eff} will be obtained by ΔK_{CP} according to Eq.8, and the defect growth rate $\frac{da}{dN}$ can be obtained, supposing the value to be μ, and then the propagating value of such crack under the combination occurrences m and n after n_{mn} times can be calculated.

$$\Delta a = \mu a n_{mn},\ \Delta c = \mu c n_{mn} \tag{9}$$

New crack size is as follows:

$$a = a + \Delta a,\ c = c + \Delta c \tag{10}$$

(5) Determination of crack size for next calculation.

Each crack tip (Point A and Point C) should be calculated, however, in each calculation, only ΔK maximum value of one tip can be found alone, if ΔK maximum value found in each combination is the value at Point A, then ΔK value at Point C for this combination may not just be the maximum value. In this case, ΔK value at Point C can be obtained according to the combination of these two occurrences and then the crack fatigue growth at Point C can be obtained, where the crack size is $a+\Delta a$ and $c+\Delta c$. If the maximum value found in each combination is the value at Point C, then ΔK at Point A can be obtained for combination of these two occurrences, finally the crack fatigue growth at Point A can be obtained, where the crack size is $a+\Delta a'$ and

$c+\Delta c'$, it is obvious that their crack sizes are different and the method of J integral is required to assess which group of crack has the most severe effect upon the integrity of component, and the size of such most severe group of crack should be used as the initial size of next crack growth calculation [6].

(6) Occurrences considered in (5) can be eliminated once new crack sizes a and c have been obtained, then in the remaining combination occurrences, a new ΔK_{CP} can be obtained by repeating the calculation by (2) and (3), and then a new crack fatigue growth size can be obtained through calculation by (4) and (5).

(7) Repeating the aforesaid steps till combination of all occurrences has been analysed and the final crack growth size a and c will be obtained.

4 Analysis on essential calculation factors

4.1. Occurrences combination

The method of occurrences combination is used in crack fatigue growth calculation and this paper will present the method on how to use occurrences combination:

(1) Suppose that in a characterization interval there are t occurrences of $a_1, a_2,\ldots,a_t$, which are set as Array X and the occur times of each occurrence is respectively $n_1,n_2,\ldots,n_t$, which are set as Array Y, then $K_{I\,max}$ and $K_{I\,min}$; $K_{II\,max}$ and $K_{II\,min}$; $K_{III\,max}$ and $K_{III\,min}$ of each occurrence can be calculated according to Eq.1 and 2 in Section 3; a group of ΔK_I , ΔK_{II} and ΔK_{III} can be obtained by combination of every two occurrences; a ΔK can be calculated per Eq.1, there are C_t^2 ΔK in such C_t^2 occurrences combinations, such C_t^2 values will then be ranked in the order of size; suppose ΔK is maximized in combination of occurrences m and n, which shall be counted as $\Delta K(m,n)$, and then maximum stress intensity factors $K_{max}(m,n)$ in combination of occurrences m and n per Eq.2. For conservative consideration, the first calculation is always given to maximum $\Delta K(m,n)$ value in the whole process, therefore, $K_{max}(m,n)$ shall be considered in the first calculation of crack fatigue growth, the occur times of occurrences m and n are respectively denoted by n_m and n_n ($n_m > n_n$), where the minimum value of such two numbers will be taken as the common occur times of the (m,n) combination and denoted by n_{mn}, which is the times of crack fatigue growth calculation used for (m,n) combination;

(2) Based on the analysis in the previous step, the occur times of occurrence n has been used up within such characterization interval, then occurrence n will be eliminated from Array X and n_n be eliminated from Array Y.So n_m-n_{mn} in occurrence m has not been used up, it should be considered in the next crack fatigue growth calculation and the new arrays are denoted by X1 and Y1, and occurrences to X1 will be recombined to obtain C_{t-1}^2 ΔK , the maximum value of which should be taken, then ΔK value for calculation of second crack fatigue growth is available, and then the number of common occur times for second calculation of crack fatigue growth is also available.

(3) Repeating the aforesaid steps till the number of occurrences combination is zeroed.

4.2. K-value calculation and mechanic modelling

In the calculation of K-value, distribution of stress σ adjacent to defect should be calculated. The defect-free mechanic model is usually used for calculation of stress distribution of the defect zone, which is detailed as follows: Defective Model (a) can be decomposed to defect-free Model (b) and singular defective Model (c), then stress σ equal in size and opposite in direction to stress in Model (a) has remained at the defect edge in Model (b), this portion of stress is generated by external load, while with regard to singular defective Model (c) where the stress is an internal one equal in size and opposite in direction, therefore, the internal stress distribution of component calculated using Model (b) is available for analysis of stress distribution adjacent to the defect zone [7].

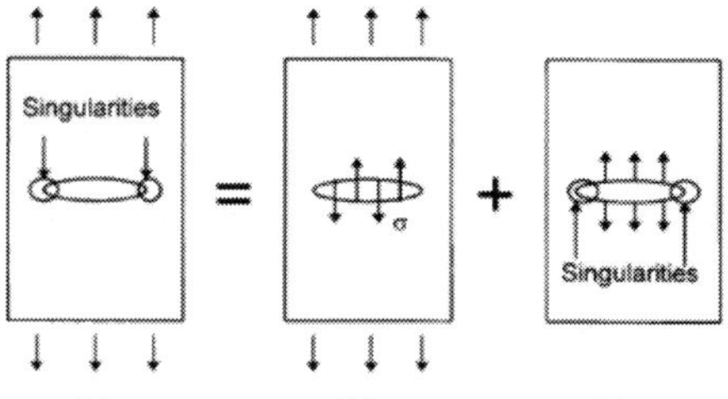

Figure 2. Equivalent analysis model for stress calculation.

Such components in the nuclear plant as the reactor pressure vessel, steam generator and pipelines are all axisymmetric models; therefore, 2D models are used to stress distribution calculation in the finite element software.

The 2D mechanic model of steam generator tube plate is shown in Fig. 3, the wall thickness of Segment AB is t, supposing x is the distance to Point A in Segment AB, and then according to RSE-M Rules, stress at the point with the distance equivalent to x from internal surface can be calculated by the following formula:

$$\sigma(x)=\begin{bmatrix}\sigma_0+\sigma_1\cdot\left(\frac{x}{L}\right)+\sigma_2\cdot\left(\frac{x}{L}\right)^2\\ +\sigma_3\cdot\left(\frac{x}{L}\right)^3+\sigma_4\cdot\left(\frac{x}{L}\right)^4\end{bmatrix} \quad (11)$$

Taking stresses at the points where their distances to the external surface are respectively equivalent to 1/5L, 2/5L, 3/5L, 4/5L and L, which can be directly obtained from the calculated result by finite element software, then establishing 5 equations per Eq.11, and then can obtain

Coefficients σ_0, σ_1, σ_2, σ_3 and σ_4. Thus, K_I can be obtained per Formula Eq.12.

$$K_I = \sqrt{\pi a}\begin{bmatrix} \sigma_0 i_0 + \sigma_1 i_1 \cdot \left(\frac{a}{L}\right) + \sigma_2 i_2 \cdot \left(\frac{a}{L}\right)^2 \\ +\sigma_3 i_3 \cdot \left(\frac{a}{L}\right)^3 + \sigma_4 i_4 \cdot \left(\frac{a}{L}\right)^4 \end{bmatrix} \quad (12)$$

where i_0, i_1, i_2, i_3 and i_4 are influence coefficients, which can be acquired by RSE-M Rules, a is the height of the crack and t is the wall thickness.

K_{II} and K_{III} are calculated as follows:

$$K_{II} = \sqrt{\pi a}\left[\tau_0 i_0 + \tau_1 i_1 \cdot \frac{a}{t}\right]$$

$$K_{III} = \sqrt{\pi a}\left[\tau_0 i_0 + \tau_1 i_1 \cdot \frac{a}{t}\right] \quad (13)$$

Once Stress σ and stress intensity factors K_I, K_{II} and K_{III} have been calculated, other parameters can be calculated by formulas in Section 3.

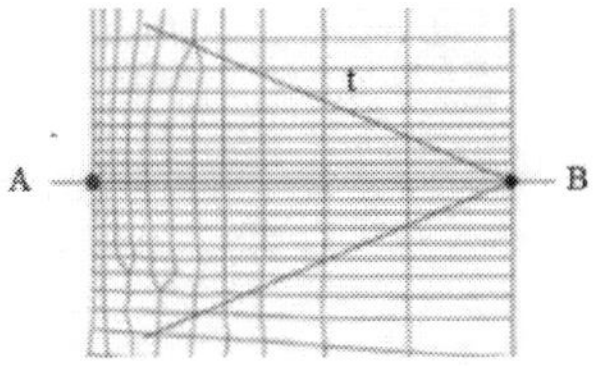

Figure 3. Diagram of 2D Mechanic Calculation Model at the Tube Plate of Steam Generator.

4.3. Option of characterization interval

The option of characterization interval has a direct effect upon the calculation of fatigue crack growth. The longer the characterization interval is, the smaller the final crack size is. This is because that the influence of the occur times of these occurrences themselves to K calculation has not been considered and thus no correction is made to K value with one characterization. Only the shorter the time interval is, the fewer the occurrences combination is, and such crack growth calculated in this interval would be taken into consideration during K calculation for the next calculation. So the shorter the time interval option is, the higher the frequency of K value correction is, then the calculation would be more precise.

The upper limit of characterization interval recommended in RSE-M Rules is 5 years and the lower limit is 1 year. Considering the refuelling period of most nuclear plants in China is 18 months, so from the perspective of in-service surveillance, the lower limit of interval is recommended to be 18 months.

Specific surveillance will be performed on any key components with defects during each outage, so during the calculation, the crack size shall be corrected according to the inspection result by NDT.

5 Crack fatigue growth calculation example

One calculation example has been given below to demonstrate this method. Suppose that there was an emerging crack whose a=5mm and $2c$=50mm on the inner surface of dissimilar weld of RPV outlet nozzle to the safe end. The operation conditions are: pressure Pc=175 MPa, temperature Tc=351°C and the thickness of the weld is 85.5mm and the material is Alloy 52.We select 6 occurrence(see Table 1 below) from the whole life time occurrences which the total number is 17 to simplify the calculation work [8].

Table 1. List of Occurrence

Occurrence	Occurrence description	Number
1	Complete plant start-up from cold shutdown to full load, with RPV open at cold shutdown	120
2	Complete plant start-up from cold shutdown to full load, with RPV closed at cold shutdown	120
3	Complete plant shutdown from full load to cold shutdown, with RPV open	120
4	Complete plant shutdown from full load to cold shutdown, with RPV closed	85
5	Unscheduled fluctuations between hot and cold shutdown	10
6	Hydraulic tests after commissioning	7

Then the mechanic model will be built in ANSYS 11 to calculate the stress intensity factors K_I, K_{II} and K_{III} of crack Tip C while not Tip A to simplify the calculation work. The 2D model is given in Figure 4.After calculating these intensity factors, according to Section 3 and 4; we use the occurrence combination to find the maximum stress intensity factors $K_{max}(m,n)$, and then ΔK_{CP} can be defined. According to RSE-M, we defined the crack fatigue growth rate calculation formula as below. So the first crack fatigue growth has been calculated used Eq. 9 and 10. Then the second occurrence combination can be defined and the crack fatigue growth also can be calculated until the number of occurrence has all been used up.

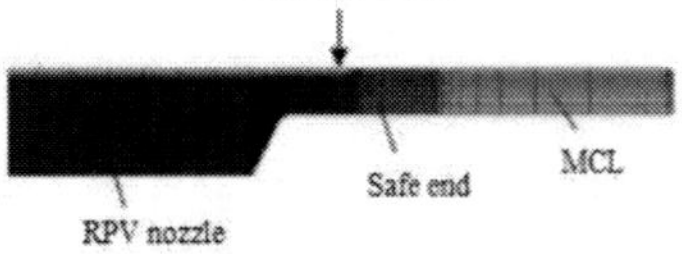

Figure 4. 2D Model of Dissimilar Weld of RPV Nozzle.

$$\frac{da}{dN} = 1.032\times10^{-12}(\Delta K_{eff})^{4.166} \quad (14)$$

$$\Delta K_{eff} = f(R)\Delta K_{CP} \quad (15)$$

DOI: 10.1051/matecconf/20179515006

$$f(R)=\frac{1}{1-\frac{R}{1.63}} \qquad R=\frac{K_{min}}{K_{max}} \tag{16}$$

After repeatedly calculating, the final crack size has been calculated (see Table 2).

Table 2. Final Crack Size

RPV outlet nozzle	Δa (mm)	$a+\Delta a$ (mm)	Δc (mm)	$c+\Delta c$ (mm)
Dissimilar weld	0.6	5.6	0.3	50.3

6 Conclusions

According to the in-depth analysis on key formulas, essential factors such as occurrences combination, K-value calculation and mechanic modelling, option of characterization interval, and the calculation example of dissimilar weld of RPV outlet nozzle to the safe end has demonstrated a good expectation, the method of crack fatigue growth calculation recommended by RSE-M could be well used to predict the crack propagating trend of nuclear power components during ISI to assess the structure integrity. Meanwhile, this method could also be used to calculate the critical crack size to define the acceptance criteria with associated safety factors for new material applied in the third generation nuclear reactors.

References

1. RSE-M, In-Service Inspection Rules for Mechanical Components of PWR Nuclear Islands, AFCEN, 701-745, (2015).
2. ASME BPVC XI, Rules for Inservice Inspection of Nuclear Power Plant Components, ASME, 437-465, (2007).
3. V. Kumar, M.D.German, *An Engineering Approach for Elastic-plastic Fracture Analysis*, EPRI, Palo Alto, 118-126, (1981).
4. Jin Cheng, Shushan Zhao, *Fracture Mechnicals*, Science Press of China, Beijing, **145**, (2006).
5. H.O.Masden, S.Krenk, N.C.Lind, *Methods of Structural Safety*, Dover Publications, New York, 65-85, (2006).
6. J. Payne, G. Welsh and J. Robert, Int J Fatigue **32**, 247 (2010).
7. Brok D, *Elementary Engineering Fracture Mechanics*, Noordhoff International Publishing, Leyden, 98~116, (1974).
8. Albahary Benoit, *EPR^{TM}-Assessment of Fatigue Risk in Most Sensitive Areas,* Areva NP, Paris, 124-139, (2013).

 DOI: 10.1051/matecconf/20179515007

Simulation Analysis for Opening Performance of Medium Voltage Vacuum Circuit Breaker Based on ADAMS and Maxwell

Ding Xianliang, Wang Zhiming , Yin Qinlin, Chen Pengwei, Zhou Shijin and Yang Yang

School of Mechanical Engineering, Nanjing University of Science and Technology, Nanjing 210094, China

Abstract. The circuit breakers play a important role in control and protect the power system and the vacuum circuit breaker has been widely used in the field of medium voltage with its excellent opening performance. Virtual prototyping technology is also became more and more popular in design and optimization of the vacuum circuit breaker.In this paper, the electromagnetic simulation software Ansoft Maxwell is used to analyze the electric repulsion of the circuit breaker in the case of open the rated short circuit breaking current. The 3D model that was built by CREO is imported into ADAMS .Then constraints, contact force, and the electric repulsion force that was analysezed in Ansoft Maxwell is added into the 3D model. Therefore, we can carry on the multi-body dynamics simulation to the 3D model. Then We can get the opening performance of the vacuum circuit breaker in the condition of open circuit rated short circuit breaking current. The simulation results show that the circuit breaker can still meet the performance requirements in the condition of open circuit rated short circuit breaking current.

1 Introduction

With the rapid development and popularization of virtual prototyping technology, more and more engineering designers begin to design and develop the products using computer. The virtual prototyping technology is more and more popular all over the world. The vacuum circuit breakers have been widely used in the medium voltage field, because it has excellent performance. The engineering and technical personnel have already begun to apply the virtual prototyping technology to design and develop the vacuum circuit breaker. For example, Tu Kai [1]and others use Pro/E to build the 3D model and then import it to the ADAMS. Finally they simulate and analyze the mechanical system of the circuit breaker, which provides a good platform for the design and optimization of circuit breaker. Zhou Yinzheng [2] and others use ADAMS software to analysis the closing bounce of vacuum circuit breaker, which to study the influence of the preload of the contact spring to the closing time. But both they only analyzed the mechanical performance of the circuit breaker under the no-load state, and not consider that there is a big electric repulsion force between the static contact and the moving contact in the condition of open the rated short-circuited breaking current, which affect the opening performance of the circuit breaker. In this paper, the model of the arc chamber is built by Ansoft Maxwell, and analyze the electric repulsion in the process of opening in the case of open the rated short circuit breaking current .The 3D model of ZN63A-12 vacuum circuit breaker was established by Creo and then is imported into ADAMS, at the same time import the electric repulsion which was analyzed in Ansoft Maxwell into ADAMS. After add constraints and contact force, we can carry on multi body dynamics simulation analysis. The finally we can get the opening performance of the vacuum circuit breaker in the case of the rated short-circuited breaking current.

2 Electric repulsion analysis and simulation

Vacuum arc extinguishing chamber is the core component of vacuum circuit breaker. And its main function is to cut off the power quickly and extinguish the arc after cutting off the power of high voltage circuit suddenly, which can avoid the occurrence of accidents. This function comes true better because of the good insulation of the vacuum. During the opening, an arc will generate between the contacts and transfer current between contacts, before the contacts contact with each other. Current changes will produce a magnetic field, which will produce electric repulsion force in the process of opening. The size of the electric repulsion relate on not only circuit configuration and contact materials loop structure, but also the size of the current. Ansoft Maxwell is a 2D/3D electromagnetic field finite element analysis software. Power、voltage and magnetic can be solved in the electrostatic field、static magnetic field、time-varying electric field、magnetic field、eddy current field、transient field and temperature field .It can be used to analyze the characteristic of static、steady state and transient state of the transformer、motor

DOI: 10.1051/matecconf/20179515007

and permanent magnet device which work in normal condition and abnormal condition and so on.

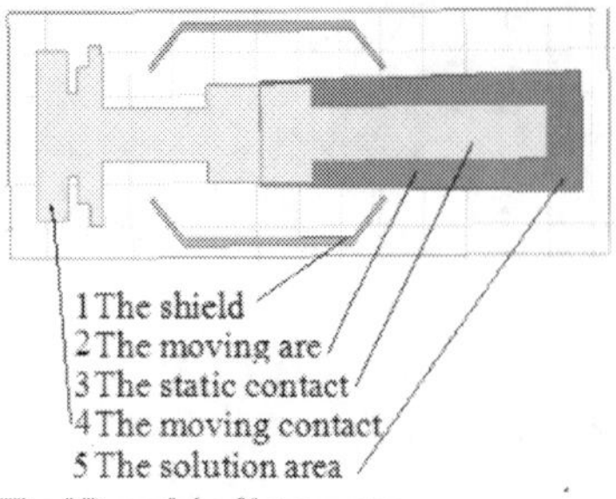

Figure 1. The 2D model of interrupter.

In this paper, the ZN63A-12 type vacuum circuit breaker is studied. As is shown in Figure 1, the 2D structure of the arc chamber is built in Ansoft Maxwell [3], which is mainly composed of the static contact、the moving contact、the shield and the solution area. Then we should define the material of each parts. The moving contact、the static contact and the shield are covered for the copper material, and others are vacuum. Then we should define the current between contacts. Because the rated short-circuited current is very complicated in the course of opening, so we import the rated short-circuited breaking current that we measured. As shown in Figure 2, it's the figure that the rated short-circuited breaking current changes with time.

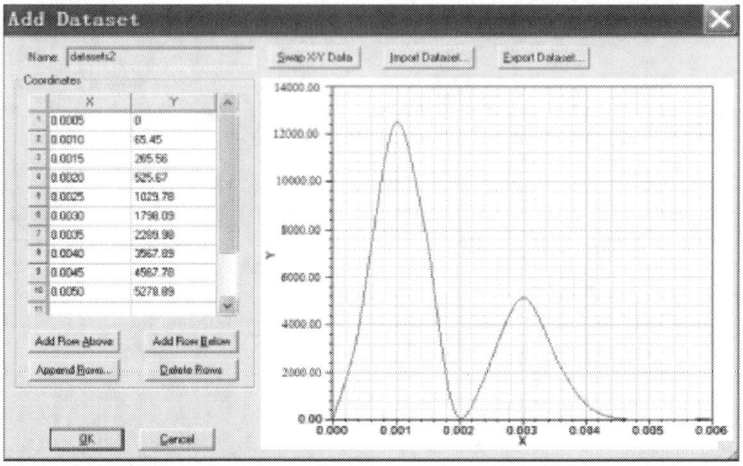

Figure 2. The rated short-circuited breaking current.

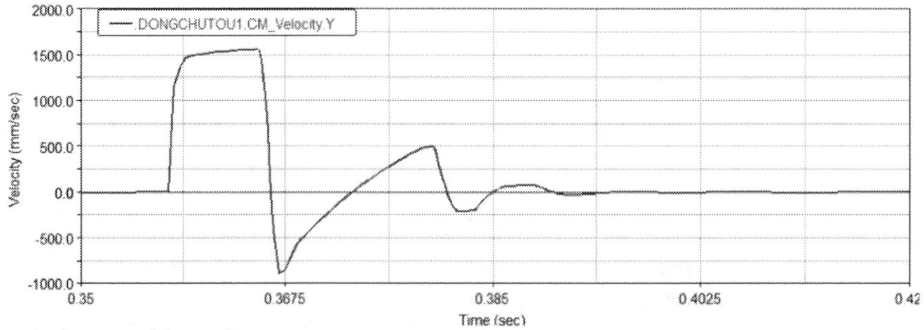

Figure 3. The velocity curve of the moving contact.

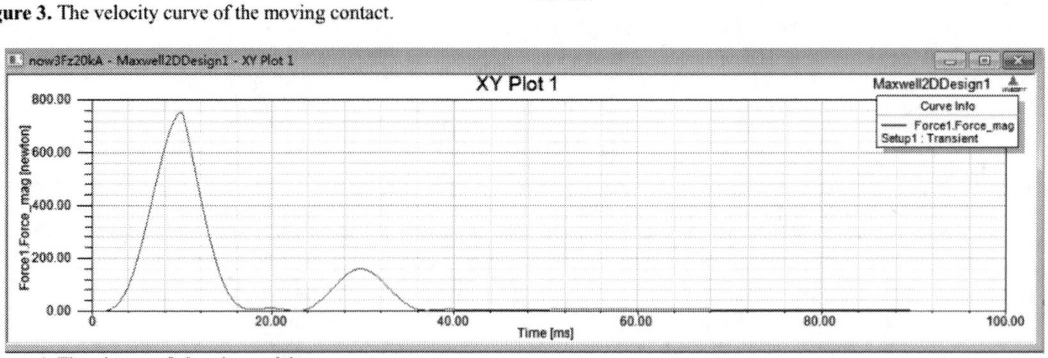

Figure 4. The picture of electric repulsion.

Next we should import the velocity of the moving contact. In order to ensure the reliability of the simulation results, we import the velocity curve of the moving contact in the ADAMS simulation under the no-load condition. As shown in Figure 3, it's the velocity curve of the moving contact in this case.

Finally after all the steps have been done, we can start the simulation analysis. And after simulate analysis, electric repulsion can be obtained like Figure 4. The maximum value of the electric repulsion is about 780N when the rated short-circuited breaking current reaches the maximum value.

3 Dynamic model establishment and simulation

The vacuum circuit breaker [4] is mainly composed of the spring operating mechanism, the vacuum arc extinguishing chamber, the transmission mechanism, the frame and so on. In order to analyze the opening performance of the vacuum circuit breaker, we need to establish the 3D model and import it into the ADAMS. Then we can carries on the dynamics simulation analysis to the 3D model.

3.1 Establish the 3D model

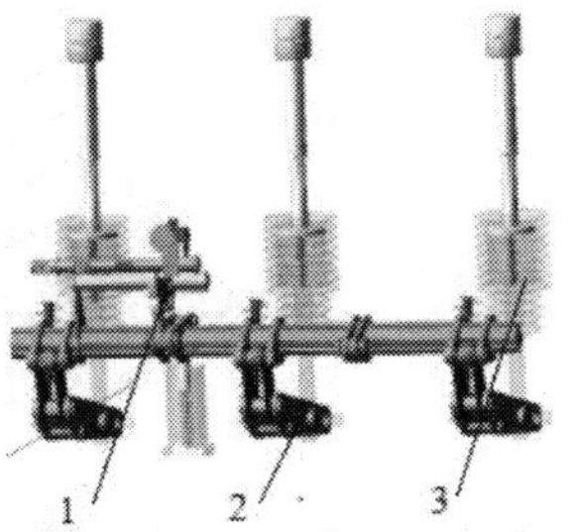

1 the spring operating mechanism

2 the transmission mechanism

3 the vacuum arc extinguishing chamber

Figure 5. The 3D model of the vacuum circuit breaker.

Each component of ZN63A-12 type vacuum circuit breaker is established by 3D modeling software CREO. In order to simplifying the model, we should reduce some holes、bolts and racks, which have no impact on the simulation analysis. It can reduce the constraints that we need add into the dynamic simulation and shorten the time to draw and simulate. Finally, the final the model is finished in the assembly module and is imported into ADAMS.As shown in Figure 5, the 3 model of the vacuum circuit breaker is mainly composed of the vacuum arc extinguishing chamber, the transmission mechanism, and the spring operating mechanism. When the power system is in trouble, the circuit breaker will receive the signal and the spring operating mechanism will respond quickly and drive the moving contact to carry on the opening operation. At last, the vacuum circuit breaker achieve opening operation, cut off the current and prevent accidents quickly. So as to ensure the safe operation of the system, The main performance parameters of the circuit breaker are as follows. Average opening velocity $1.1 \pm 0.2 m/s$; Opening time $20 \sim 50 ms$;Contact wear thickness $\leq 3mm$.

3.2 Simulation setup and analysis

ADAMS is the most outstanding mechanical system simulation software which is developed by the Mechanical Dynamic Inc. It is also the most authoritative and widely used mechanical system dynamics analysis software and has a the highest of occupancy in the global [5-6]. Firstly, we must add constraints to the various components of the 3D model. The static contact is added to the fixed constraint, the moving contact is added to the translation constraint and the appropriate constraints are added to others. Secondly, we should add the closing spring, the opening spring and the contact spring, which stiffness coefficients are 70N/mm, 20N/mm, and 100N/mm. And The stiffness coefficients and preload of all springs are shown in Table 1. Thirdly, the electric repulsion which was analyzed in the Ansoft Maxwell should be imported to ADAMS and added on the moving contact. Fourthly, we need add force to the parts which contact with each other, especially the force between the moving contact、the static contact and the brake keep pawl. The stiffness coefficient is set to 1×10^5 N/mm; the Impact index is set to 1.5; the damping coefficient is set to 50N-sec/mm and the cutting depth is set to 0.1mm. Finally, we need to apply a rotating drive on the split gate.

Table 1. The stiffness coefficients of all springs.

name	The stiffness coefficients	Preload
The contact spring	100 N/mm	2200 N
The opening spring	20 N/mm	-600 N
The buffer spring	5.0 N/mm	30 N
The closing latch spring	3 N/mm	300 N
The closing coil spring	2 N/mm	200 N

3.3 The analysis of the simulation result

We need select the appropriate simulation time and step size and begin to simulate the dynamic model. The simulation results will be processed after simulation, we can get the velocity time curve and the position time curve of the moving contact from figure 6 and figure 7. As is shown by the analysis and simulation, at the initial stage of the opening operations, the velocity of the moving contact is rapidly increased to the maximum value under the joint action of the contact spring and the opening spring, and then the velocity will slow down. When collision with the split opening buffer, the velocity will rapidly drop to zero, and then moving contact will be slowly elastic vibration until it is completely stationary. The whole process takes about 35.5ms and the time limit of the opening operation is between 20ms to 50ms,so it fully meets the time requirements.The average opening velocity is 110.3mm/s which is meet the velocity requirements of opening operations .So the vacuum circuit breaker can also meet the performance requirements when open the rated short-circuited current.

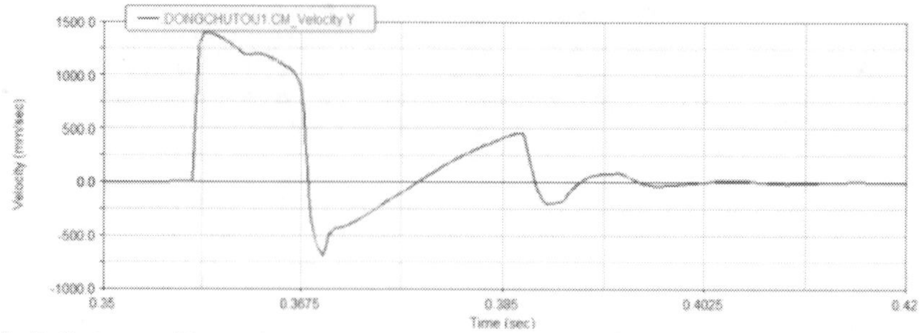

Figure 6. The velocity time curve of the moving contact.

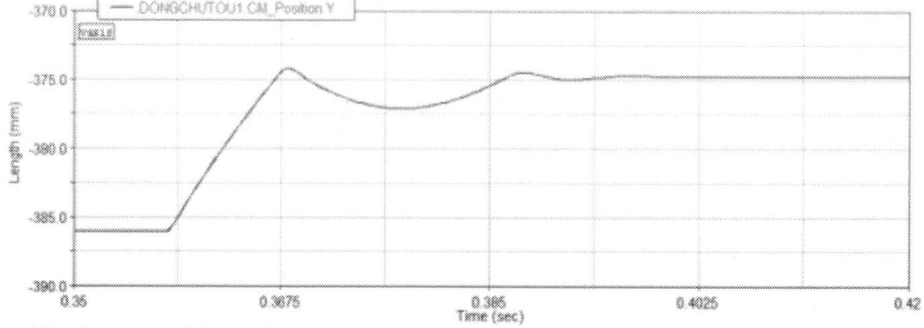

Figure 7. The position time curve of the moving contact.

4 Conclusion

With the development of the virtual prototyping technology, more and more engineering and technical personnel use the computer to carry out product design and optimization, which can not only greatly shorten the cycle of the product development and design, but also reduce the cost of development. Along with the development of CAE technology, the application of the simulation technology is more and more comprehensive and the simulation results are more and more reliable. In this paper, the 2D model of the vacuum arc extinguishing chamber is built in Ansoft Maxwell, and then we need define parameters of materials and set incentive of the model so that the electric repulsion is solved in the process of opening operation. The 3D model of ZN63A-12 vacuum circuit breaker is established by CREO and imported into ADAMS and then we need set the constraints, the contact force and the stiffness coefficients of all springs to the model and import the electric repulsion which was simulated in the Ansoft Maxwell to model. Finally we will simulate the opening process of the vacuum circuit breaker. The simulation results show that the circuit breaker can meet the requirements in the conditions of open the rated short-circuited current. The analysis results can provide reference for the further design and optimization of vacuum circuit breaker.

References

1. Tu Kai, Zhou Yi, Zhang Li. Simulation analysis of high voltage circuit breaker mechanism with ADAMS and Pro/E[A]. Machine research and Application,**49**(2009).
2. Zhou Yinzheng,Lu Min, Zhuang Huogeng. The analysis of the trip of vacuum circuit breaker based on ADAMS[A]. Low voltage electrical apparatus,**14**(2014).
3. Zhao Bo, Zhang Hongliang. Application of Ansoft 12 in engineering electromagnetic field [M]. Beijing: China Water Conservancy and Hydropower Press,38(2010).
4. Li Jianji, High voltage circuit breaker and its application ,**30**(2004).
5. Chen Fenghua. The virtual prototyping technology ADAMS 2012 from entry to the master [M]. Beijing: Tsinghua University press, **18**(2013).
6. Chen Demin, Chuang Fenghuai, Zhang Ketao. Proficient in Virtual Prototyping Technology ADAMS 2005/2007 [M]. Beijing: Chemical Industry Press, **207**(2010).

Prospects for the Use of Additional Cooling System for the Oil-Immersed Transformers with Thermoelectric Transducers

Andrey V. Dmitriev[1], Oksana S. Dmitrieva[2] and Ilnur N. Madyshev[2]

[1]Department of Theoretical Bases of Heat Engineering, Kazan State Power Engineering University, 420066 Kazan, Russia
[2]Department of Food Production Equipment, Kazan National Research Technological University, 420015 Kazan, Russia

Abstract. The main cause of the transformers failure when thermal overload, is the inter-turn short circuit inside of the winding. Therefore, the additional cooling system for the oil-immersed transformers developed, which are able to reduce the peak load and its uniform distribution throughout the operation time, due to the accumulation of cold hours corresponding to the minimum thermal load. Determined the dynamics of the ice covering on the cold surface of the thermoelectric transducer in the tank with a certain amount of water. The theoretical justification of the interaction of flows in the device presented.

1 Introduction

The oil-immersed transformers are the critical elements of electric power and electrical engineering facilities. Their reliability and occupational safety provides high quality of distributed electric energy [1]-[3]. Failure of power transformer results in a significant economic loss due to the disruption or interruption of the technological process of the industrial enterprises.

Modern cooling systems for the oil-immersed transformers do not provide a sufficiently efficient heat transfer from the winding insulations when significant short-term overloads. The result is a decrease of the service life and reliability of the transformer [4]. Therefore, as a rule, the main cause of the transformers failure when thermal overload, is the inter-turn short circuit inside of the winding [5].

In this regard, it is necessary to develop additional cooling system for the oil-immersed transformers, which are able to reduce the peak load and its uniform distribution throughout the operation time, due to the accumulation of cold hours corresponding to the minimum thermal load.

2 Additional cooling system for the oil-immersed transformers

An example of a constructive design of such system is the use of additional capacity, equipped with a cascade of semiconductor thermoelectric modules. Thermoelectric transducer is a thermoelectric refrigerator with series-connected semiconductors of p- and n-type, forming p-n and n-p-transitions between ceramic plates. Each of such transitions includes a thermal contact with one of the two radiators. As a result, transition of electric current with a certain polarity forms the temperature differential/drop between two radiators of the module: one radiator works like a fridge, another one is heated and serves for the heat dissipation [6, 7].

The lower part of the oil tank of the oil-immersed transformer includes holdover battery (Figure 1), that is designed for additional water cooling of the power transformer from the inside of a tank. For easy assembly, disassembly when capital repairs, plugs, the inlet and outlet pipes of the tank are connected by means of the pipe coil using pipes and flanges. The thermoelectric modules are fixed to one of the tank walls. Their cold side is used for the heat-transfer process of the cooling fluid inside of the tank. The heat-removing ribs are fixed to the hot side of the thermoelectric modules for the thermal discharge.

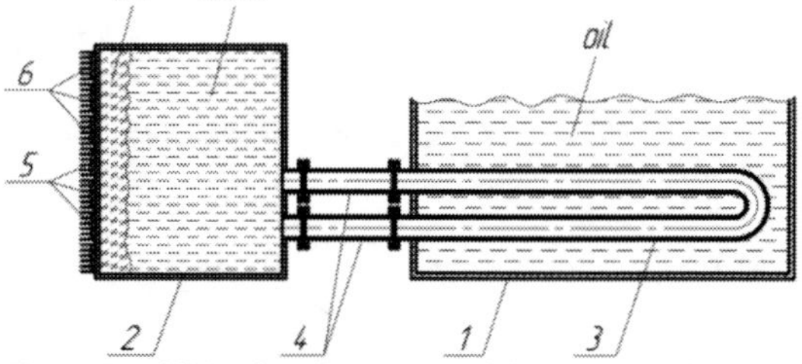

Figure 1. Additional cooling system of the power transformer: *1* – transformer tank; *2* – tank; *3* – pipe coil; *4* – communication pipes; *5* – thermoelectric modules; *6* – heat-removing ribs.

The additional cooling system for the oil-immersed transformers [8], developed by the authors of this article, works as follows. At low heat load, the coolant circulating through the pipe coil does not have time to reach high temperatures due to low heat transfer coefficients, characterizing the natural convection. Consequently, during the interval of the equipment's operational time, the cold accumulation in the tank takes place by means of thermoelectric modules. After a while on the cold side of the thermoelectric transducers, the ice forms a certain thickness. Further, with the heat load increase, which occurs mainly due to the high ambient temperatures, the pump (Figure 1 conventionally not shown) ensures forced circulation of the cooling fluid in the closed loop. The turbulence of the flow leads to higher heat transfer coefficient, thus increasing the natural circulation of the oil in the transformer tank, resulting in more effective cooling of the windings. Thus, additional liquid cooling of the bottom of the transformer tank increases the cooling efficiency of the transformer oil when maximum heat loads, preventing the failure of power transformers.

3 Calculation procedure

While design engineering process of the additional cooling systems for the oil-immersed transformers, an important task is the selection of the most efficient heat flow through the thermoelectric transducer and the selection of an optimal fluid volume, required for an adequate cooling of the power transformer in the hottest season - summer. Previously it was established that this time of a year the necessity of additional cooling for the oil-immersed transformer is about 50% of its work time per 24 hours that is about 12 hours. During this time, it is necessary to accumulate a sufficient amount of cold in the tank. In this case, as shown by the earlier numerical and experimental studies, the ice layer formation is possible on the cold side of the thermoelectric transducers 3-5 hours after shutdown of the pump, depending on the coolant temperature. The dynamic pattern of water temperature changes during this time interval can be determined by well known equations and is not of scientific interest. The greatest interest is the problem of determining the dynamics of the ice covering on the cold surface of the thermoelectric transducer in the tank with a certain amount of water.

In the studies include the assumption that heat is removed for the water cooling and its crystallization, that is, the ice formation:

$$dQ_\tau = dQ_b + dQ_i. \quad (1)$$

The summands in the equation are determined by well known equations:

$$dQ_b = m_b c_b dt_b, \quad (2)$$

$$dQ_i = r_i dm_l, \quad (3)$$

where m_b – water mass, kg; m_l – ice mass, kg; c_b – mass specific heat of water, J/(kg·K); r_i – mass specific heat of crystallization, J/kg; dt_b – water temperature change, K.

The heat amount removed from the water can be determined as follows:

$$dQ_b = \alpha_b F_b (t_b - t_n)\, d\tau, \quad (4)$$

where α_b – heat-transfer coefficient from water to the surface, W/(m^2·K); t_b – water temperature, K; t_n – surface temperature from the water side, K; $d\tau$ – time interval, s.

Equation (1) considering (3) and (4) can be transformed to:

$$r_i dm_l = dQ_\tau - \alpha_b F_b (t_b - t_n)\, d\tau. \quad (5)$$

Considering the relationship of the heat quantity with the heat flow, we may record:

$$r_i dm_l = Q d\tau - \alpha_b F_b (t_b - t_n)\, d\tau. \quad (6)$$

Earlier, the authors obtained the dependence for the calculation of the heat transfer coefficient from water when the wall temperature is equal to $t_n = 273$K:

$$\alpha_b = 13.518\, t_b^{1.827}/l^{0.248}, \quad (7)$$

where l – length of the heat-transfer area, m.

Considering this, we may record:

$$r_i dm_l = (Q - 13.518\, t_b^{1.827} F_b / l^{0.248}) d\tau. \quad (8)$$

Equation (8) can be transformed to a form suitable for the calculation of temporal mass changes:

$$dm_l = (1/r_i)(Q - 13.518\, t_b^{1.827} F_b / l^{0.248}) d\tau. \quad (9)$$

It should be noted that water temperature is a temporal changing parameter. Transforming the equation (2), we may record:

$$dt_b = \alpha_b F_b t_b d\tau / (m_b c_b), \quad (10)$$

which, taking into account the equation (7) may be recorded as follows:

$$dt_b = 13.518\, t_b^{1.827} F_b d\tau / ((m_0 - m_{li}) c_b l^{0.248}). \quad (11)$$

Finally, to calculate the temporal changes of the ice mass and temperature, we can use the following dependencies:

$$m_{li} = m_{li-1} + (1/r_i)(Q - 13.518\, t_b^{1.827} F_b / l^{0.248}) \Delta\tau, \quad (12)$$

$$t_{bi} = t_{bi-1} - 13.518\, t_{bi-1}^{1.827} F_b \Delta\tau / ((m_0 - m_{li}) c_b l^{0.248}). \quad (13)$$

The calculations were performed with different values of the heat flows through the thermoelectric transducer and the length of the heat-transfer area was 0.05 m. The initial water temperature in the tank, at which the icing started, was changed from 4.5 to 8.5°C, the water mass in the tank was estimated to be equal to 200 kg. The area of water contact with the cooling surface was 1 m^2. The results of numerical studies showed that the heat flow volume through the thermoelectric transducer affects the water cooling parameters and its crystallization only in the initial stage. Furthermore, it was found that all studied parameters, except the water temperature, do not depend on the length of the heat-transfer area.

4 Results and discussion

The Fig. 2 shows the water temperature temporal change in the tank, considering different values of the heat flow

 DOI: 10.1051/matecconf/20179515008

through the thermoelectric transducer. The diagram shows that the heat flow has a significant impact on the initial water temperature in the tank, at which the icing starts. For example, the 3 times increase of the heat flow through the thermoelectric transducer allows increasing the initial water temperature by 1.89 times. Relatively large changes are observed only in the first 3-4 hours of the thermoelectric transducer's operational time from the ice formation start. After that, the water temperature is almost independent of the heat flow values.

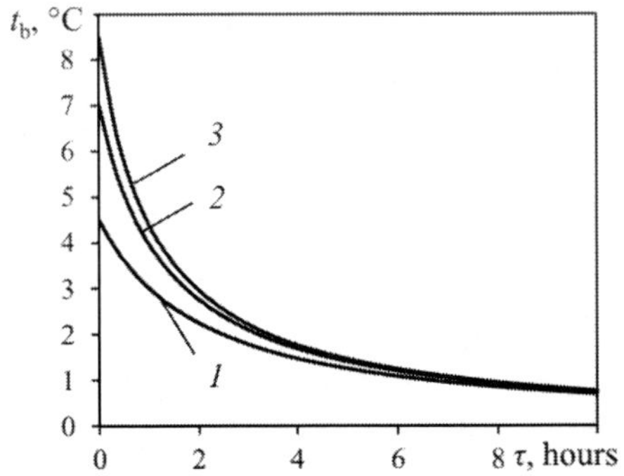

Figure 2. The water temperature temporal change in the tank, considering different values of the heat flow through the thermoelectric transducer, Q, W: *1* – 500, *2* – 1000, *3* – 1500.

The heat-transfer coefficient from the water to the surface depends on the magnitude of the heat flow only at the initial stage of cooling and liquid crystallization (Figure 3). Then, the dependence disappears and the maximum change, when τ = 4 hours, is 16.23%. The decrease of natural water convection in the tank, by reason of a substantial increase of its viscosity, results in the decrease of the heat-transfer coefficient.

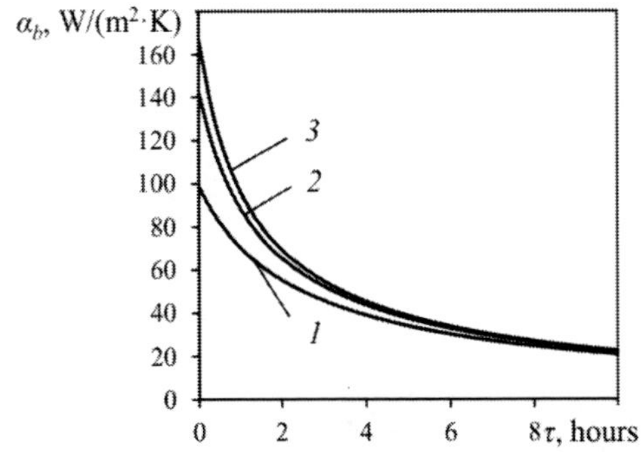

Figure 3. Temporal change of the heat-transfer coefficient from water to the surface, considering different values of the heat flow through the thermoelectric transducer, Q, W: *1* – 500, *2* – 1000, *3* – 1500.

Analyzing the diagram in Fig. 4, it can be concluded that the increase of the heat flow through the thermoelectric transducer leads to an almost proportional increase of the water fraction, converted to ice. For example, in 5 hours of the thermoelectric transducer`s operational time, the increase of the heat flow by 3 times will provide the opportunity to achieve the ice share equal to 0.343, that is, the increase will be by 3.35 times. It should be noted that the change of the water proportion , converted to ice, during the first 2 hours of operational time is characterized by power-law dependence, and in the future – linear.

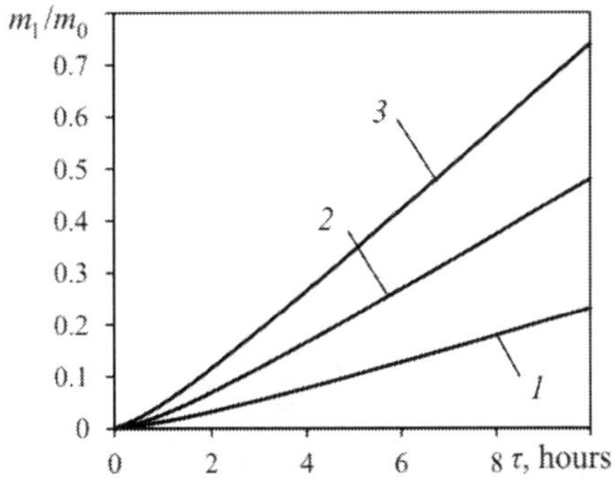

Figure 4. Temporal change of the water proportion, converted to ice, considering different values of the heat flow through the thermoelectric transducer, Q, W: *1* – 500, *2* – 1000, *3* – 1500.

The total thermal resistance increases almost in direct proportion to time (Figure 5). It is connected with the statement of the research problem, because the heat flow through the thermoelectric transducer remains constant.

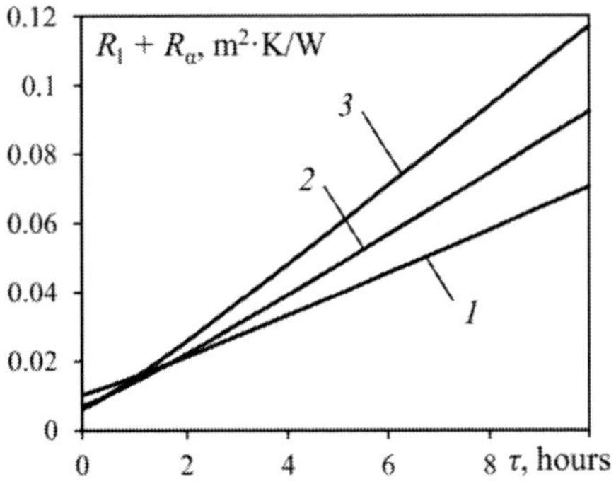

Figure 5. Temporal change of the total thermal resistance, considering different values of the heat flow through the thermoelectric transducer, Q, W: *1* – 500, *2* – 1000, *3* – 1500.

The increase of the ice thickness leads to an increase of its thermal resistance (Figure 6).

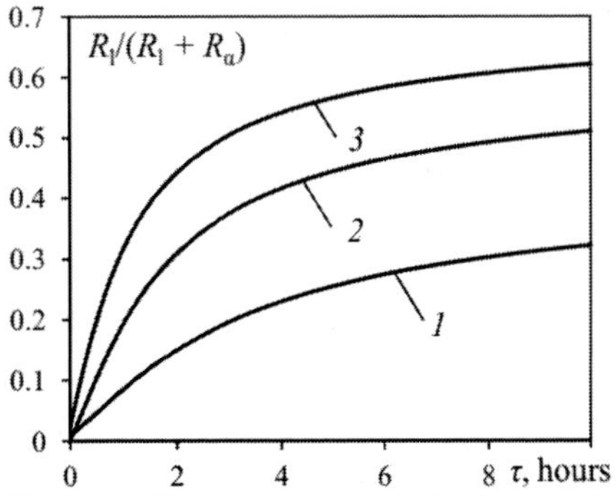

Figure 6. Temporal change of the impact of the thermal resistance of the ice layer on the total thermal resistance, considering different values of the heat flow through the thermoelectric transducer, Q, W: *1* – 500, *2* – 1000, *3* – 1500.

At the initial stage of cooling and the liquid crystallization, the thermal resistance of the ice layer provides rapid total thermal resistance. Further, when τ = 4 hours, there is a sharp decrease in the rate of change of this parameter. It should be noted that the higher heat flow through the thermoelectric transducer, the greater its thermal resistance (Figures 5, 6).

5 Conclusion

1. The numerical studies allow choosing the optimal heat flow through the thermoelectric transducer for cooling and liquid crystallization in the tank, based on the time for the cold accumulation. For example, for intensive cooling and water crystallization in a short period of time (within 3-4 hours), it is necessary to combine thermoelectric transducers with total heat flow of more than 1 kW. If the prolonged liquid cooling at low heat loads is possible, it is sufficient to use thermoelectric transducer with a heat flow of 0.5 kW.

2. The conducted research asserts that the use of the proposed additional cooling system for the oil-immersed transformers, by means of thermoelectric modules, allows to increase efficiency of their cooling when peak heat loads.

Acknowledgments

The reported study was funded by President of Russian Federation grant, according to the research project No. MK-5215.2016.8.

References

1. L. Kish, *Heat and cooling of the transformers* (Energy, Moscow, 1980). – 180 pp.
2. A.M. Golunov, N.S. Seshenko, *The cooling devices for the oil-immersed transformers* (Energy, Moscow, 1976). – 216 pp.
3. G. Gotter, *Heat and cooling of the electrical machines* (State Energy Press, Saint Petersburg, 1961). – 351 pp.
4. M.G. Bashirov, M.R. Minlibaev, A.S. Hismatullin, *Cooling efficiency increase for the oil-immersed transformers*, Neftegazovoe delo, 2, p. 347-357 (2014).
5. V.V. Sushkov, A.A. Zjabkin, *Thermal processes modeling and diagnostics of power transformers of power supply systems of oil fields*, Promyshlennaja jenergetika, 2, p. 39-42 (2013).
6. L.P. Bulat, *Thermoelectric cooling: current state and trends*, Holodil'naja tehnika, 8, p. 2-7 (2004).
7. O.S. Dmitrieva, A.V. Dmitriev, *Cooling System for the oil-immersed transformer by means of thermoelectric modules*, Izvestija vysshih uchebnyh zavedenij. Problemy jenergetiki, 11-12, p. 56-59 (2015).
8. A.V. Dmitriev, O.S. Dmitrieva, I.N. Madyshev, A.A. Sagdeev, A.N. Nikolaev, Zajavka na poleznuju model' № 2016112002 ot 30 marta 2016 goda. *A thermoelectric device for the additional cooling of oil transformer* (Moscow, 2016) (to be published)

Electromagnetic Exhaust Valve Event Optimization for Enhancing Gasoline Engine Performance

Xinyu Fan and Siqin Chang

School of Mechanical Engineering, Nanjing University of Science and Technology, 210094 Nanjing, China

Abstract. Variable exhaust valve events have the potential to further improve the engine power output, fuel economy and decrease the NO_X emissions. Based on the moving coil electromagnetic valve train applied to engine exhaust system, effects of variable exhaust valve events are analyzed in detail and the optimization approaches are carried out. Also with the fully variable intake and exhaust valve train, different internal EGR strategies can be achieved and the contrastive analyses are carried out between combustion chamber recirculation and exhaust port recirculation strategies at same operational condition. Results show that, the optimal exhaust valve opening motion can strengthen both power performance and fuel economy at engine part loads. And two principal EGR strategies are applied in a good combination under variable engine loads. At the engine speed of 2000 r/min, BMEP is about 0.3 MPa and with 30%~35% exhaust port recirculation rate, the BSFC and NO_X emissions have decrease over 10% and 85% respectively compare with initial condition.

1 Introduction

Electromagnetic valve train (EMVT) in camless engine offers large potential to maximize volumetric efficiency and minimize pumping losses, so the electromagnetic valve technology can improve engine power output and fuel economy obviously [1]. Furthermore, engine exhaust system equipped with EMVT can expand the potential of variable valve train technology, because the variable exhaust valve timing can enhance the exhaust velocity and make it easier to increase engine output power or weaken the exhaust process to reduce emissions.

E. Sher et al. researched the optimization of variable valve train for maximizing performance of an unthrottled SI engine, got conclusions that the optimal EVO and IVC for minimizing BSFC occurs in a good approximation when the engine torque maximizes in the entire range of operation condition [2]. Work by Kramer et al. demonstrated that phasing the exhaust camshaft to a later EVO and EVC at part load steady-state operation reduces hydrocarbon emissions by 10% [3]. With internal exhaust gas recirculation and variable valve timing technology, authors researched the changes of NO_X emissions, then optimization works were carried out to find the optimum exhaust valve event, it was concluded that the average reduction of NO_X emission and BSFC are 71% and 6%, respectively [4].

The high combustion pressure present in the combustion chamber would be a challenge to the operation safety and stability of electromagnetic exhaust valve train. Aiming at the large variations in combustion pressure, authors adopt corresponding close loop control strategies and resulted in a desirable performance through experiments [5]. For the moving coil EMVT, Figure 1 shows the prototype equipped on cylinder head and the structure is introduced in the reference [6]. It can be competent for the engine exhaust system with inverse system control algorithms [6-7]. Also the effect of gas pressure on the power consumption of exhaust valve and the approach to define the optimal opening motion at high load has been discussed; the results showed it get a better trade-off in terms of engine torque and exhaust valve power consumption [7]. While at the engine part load, the influence of gas pressure is inconspicuous, what we more concerned about is the way to reduce BSFC and NO_X emissions. So the objective of this paper is to extend the work described in the above, the comprehensive study on the application of variable exhaust valve events for enhancing gasoline engine performance at part load would be carried out.

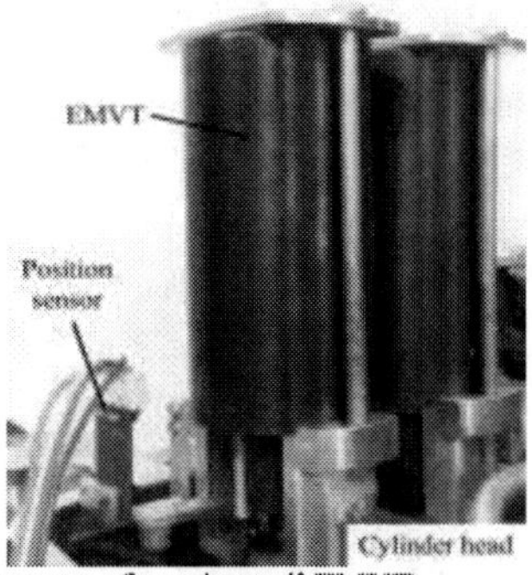

Figure 1. Prototype of moving coil EMVT.

2 Simulation modeling

The study is based on an inline four-cylinder gasoline engine and main characters are shown in table 1, only port fuel injection is used to simplify the analysis.

Table 1. Four-cylinder gasoline engine data.

Parameter	Value
Displacement	1.8 L
Bore	80 mm
Stoke	89 mm
No. of valves	16
Compression ratio	10.5
Rated power	95.7 kW (6000r/min)
Rated torque	171.3 N·m (4500r/min)

One-dimensional model of the prototype engine is carried out in AVL Boost, which can make predictions with errors less than 5% compared with test data, as shown in Figure 2. And the electromagnetic valve lift profiles are applied in the simulation model to analysis the effect of EMVT on engine performance.

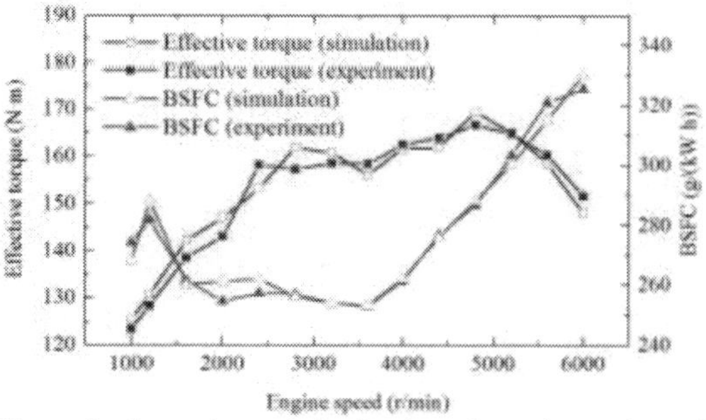

Figure 2. Comparison of simulated results and experimental results for prototype engine.

3 Exhaust valve motion control strategy

In the traditional camshaft engine, EVO timing is a limited compromise between power output, fuel consumption and complexity of mechanical actuation. With the electromagnetic exhaust valve train, the optimum compromise of expansion losses and exhaust work can be elaborated in every operating point through the variable exhaust valve motion. Later EVC timing and earlier IVO timing form valve overlap, which mainly determine the gas residual amount. While in the camless engine, internal EGR can be realized through the variable valve timing which is an effective way to control and minimize NO_X concentration in exhaust gas. For an engine with fully variable intake and exhaust valve timing, following different procedures are possible to realize internal EGR.

Intake valve partial reopening
Exhaust valve partial reopening
Early exhaust valve closing

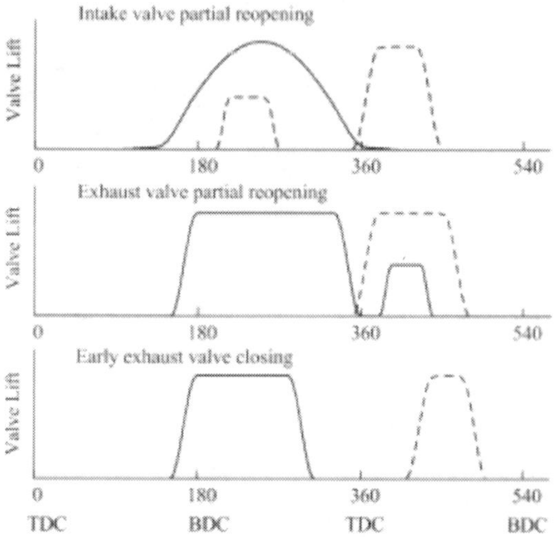

Figure 3. Internal exhaust gas recirculation strategies with variable valve events.

As shown in Figure 3, with intake valve partial reopening strategy, the intake valve opens partly during exhaust stroke. At this period, a portion of exhaust gas is pushed into intake port and would be re-pushed into combustion chamber in the next intake stroke. While the hot exhaust gas heating up fresh mixture in the intake port rapidly. This scheme to realize internal EGR has been discussed in the paper [8], it is concluded that both NO_X emissions and BSFC have a great decrease. However the research in that paper was based on the engine with fully variable intake system and camshaft exhaust system, which restricted the advantage of fully variable valve train to some extent. Also the gas pressure in combustion chamber hindered the intake valve opening during exhaust stroke, which results in an increase of intake valves' energy losses.

Exhaust valve partial reopening during intake stroke forms exhaust port recirculation (EPR), part exhaust gas flows back into combustion chamber through the exhaust valve. With the optimal exhaust and intake valve timing, pumping losses could be reduced to minimum during the period of gas exchange in camless engine.

Exhaust valve closes earlier before the piston reached to top dead center, which forms negative valve overlap and then a portion of exhaust gas is remained in cylinders, this process is called combustion chamber recirculation (CCR). The trapped exhaust gas would be compressed during the final stages of exhaust stroke and re-expand until to the intake valve opened. During the intake stroke, the fresh charge and residual exhaust gas get fully mixed. Because of the recompression processes of residual exhaust gas, pumping losses of CCR strategy are higher compared with the former two strategies. Therefore, an early exhaust valve closing event often coupled to a very late intake valve opening event, which could decrease the pumping losses effectively. With this internal EGR strategy, the temperature of fresh charge is higher than other strategies. Based on the above analysis, the optimize approach and strategy about variable exhaust valve event at engine part load would be carried out below.

4 Results and discussion

4.1 Effect of exhaust valve opening events

At the engine part load, the effect of combustion force on exhaust valve opening event is weak. So the exhaust valve opening event mainly influences the exhaust process efficiency, exhaust gas temperature and the secondary reactions of the hydrocarbon emissions in the cylinder and exhaust system. Opening the exhaust valve too early leads to higher expansion losses. Conversely, opening the exhaust valve too late leads to an increase in exhaust phase losses. Therefore, to minimize the losses in the expansion and exhaust phase, the exhaust valve opening event must be adapted to the engine load and speed.

At the engine speed of 2000r/min and BMEP is about 0.3MPa. The EVC timing is kept at top dead center and transition time (the time in which the valve moves from 5% to 95% of the travel) is 3.5ms. Also the engine load is controlled by adjusting IVC timing instead of throttle. The results are shown in Figure 4 and 5.

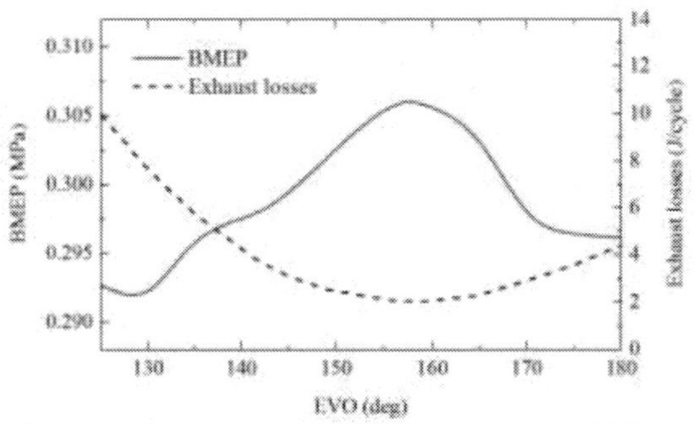

Figure 4. Effect of variable EVO timing on the BMEP and exhaust losses, at 2000 r/min and BMEP is about 0.3MPa.

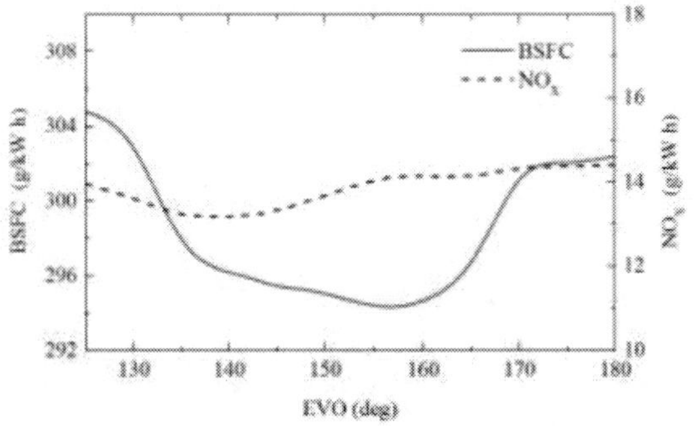

Figure 5. Effect of variable EVO timing on the BSFC and NO_X emissions, at 2000 r/min and BMEP is about 0.3MPa

For the prototype cam engine, BMEP is 0.249 MPa with the same fresh charge per cycle at 2000 r/min, the engine power has a great improvement with fully variable intake and exhaust valve train. As shown in Figure 4, the BMEP increases directly to the maximum 0.305MPa and then decreases with the delay of EVO timing. This is mainly caused by the changes of exhaust losses, the regions about exhaust losses achieve a minimum standard and BMEP attains maximum are mainly overlapped. Also we can see the BSFC reaches to lowest state at the same regions. So at the part load of camless engine, the optimal EVO timing for minimizing BSFC occurs in a good approximation when the engine exhaust losses reach to the lowest value, also it is usually late compared with conventional cam engine. While the NO_X emissions are practically insensitive to the EVO timing.

4.2 Effect of exhaust valve closing events and internal EGR strategies

The CCR formed by earlier exhaust valve closing and the EPR produced by exhaust valve partial reopening would be researched here. At the engine speed of 2000r/min and BMEP is about 0.3MPa. Only single intake valve per cylinder works for the conditions, which can improve the mixing level of fresh charge [9].

For the CCR strategy, exhaust valve opening motion remains the same and EVC timing keeps earlier ensuring larger exhaust gas stay in cylinder. While the pumping losses can be reduced to minimum coupled with later IVO timing, the choice of IVC timing is to ensure the same charge efficiency. For the EPR strategy, part exhaust gas in exhaust port flows back to cylinder with exhaust valve partial reopening. Also the result shows that changes of exhaust valve partial reopening timing in a small range has little effect on the engine performance, so a higher internal EGR rate could be carried out just through a longer opening duration time. While the choice of IVC timing is to ensure the same charge efficiency. Figure 6 (a) shows the IVO and EVC timing corresponding to variable CCR rate, figure 6 (b) shows the IVC and exhaust valve partial closing timing corresponding to variable EPR rate. Among this single/double model is applied on the exhaust valve partial reopening motion for extending the variation range of EGR rate.

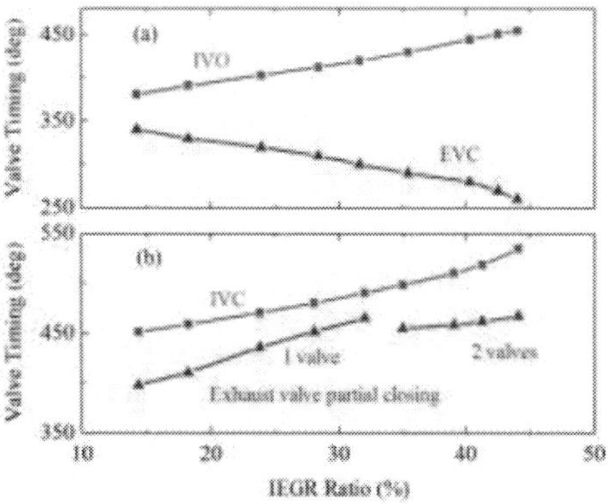

Figure 6. (a) Variable valve timing for CCR strategy. (b) Variable valve timing for EPR strategy.

Figure 7 and 8 show the thermodynamic results about two different internal EGR strategies. With the increase of internal EGR rate, BMEP increases gradually and reaches its maximum at 30%~ 35% internal EGR rate, and then dropped rapidly, also the BSFC has a small improvement but then rises sharply. The reason for this result is that the motions of intake and exhaust valve. At the part load of camless engine, the amount of fresh charge is controlled by earlier IVC timing instead of throttle, which leads to a

DOI: 10.1051/matecconf/20179515009

small pumping loss but also decreases the temperature of intake charge. This may affect the flame propagation and heat release rate at initial combustion period. So with a higher internal EGR rate, the residual exhaust gas can heat the fresh charge and improve the combustion efficiency. While the exhaust gas with high specific heat which could decrease the peak combustion temperature in cylinder and limit the production of NO_X, as shown in figure 8. With the increase of internal EGR rate further, the residuals gas rate keeps rising and results to a diluted fresh charge. Also the heating effect of exhaust gas becomes more and more obvious, which hinders the flame propagation or even results to slow-burn and partial burning. So although the NO_X emissions reduce further, yet the engine power and BSFC get worsen.

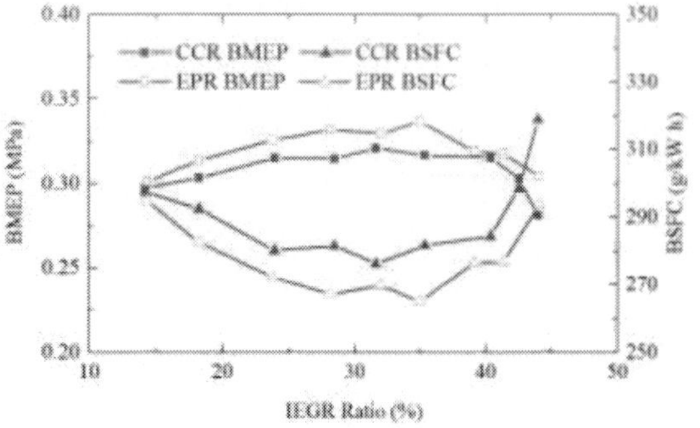

Figure 7. BMEP and BSFC with various levels of internal EGR introduction.

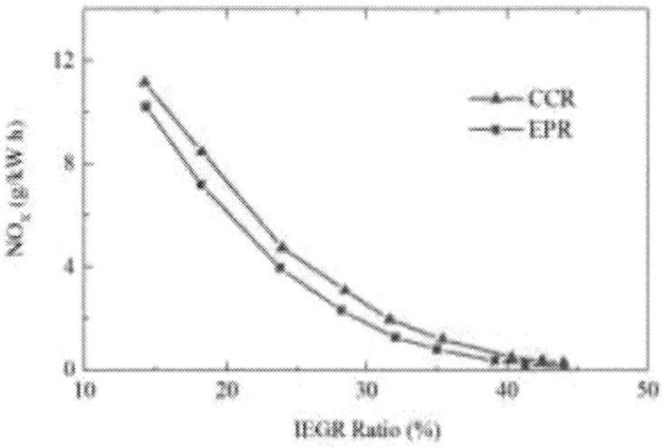

Figure 8. NO_X emissions with various levels of internal EGR introduction.

Results show the difference between CCR and EPR strategies at the same engine load. Under variable levels internal EGR rate, the BMEP and BSFC for EPR strategy are superior to the CCR strategy. Which led to this phenomenon are mainly two different aspects. Firstly, the process of CCR determines it would produce larger pumping losses than EPR inevitable even though coupled with later IVO. Figure 9 shows the P-V indicated diagram of prototype engine, 18% CCR strategy and 18% EPR strategy. Secondly, it can attribute to the effect on combustion process. As shown in Figure 10, when the piston reaches the BDC during the later stage of intake stroke, the fresh charge temperature of CCR is higher than EPR with a maximum difference about 40 K. Also it can be seen that the maximum pressure rise of CCR is higher than EPR at the same EGR rate, which means a lower quality of combustion. With the increase of internal EGR tare, the maximum pressure rise reduced significantly.

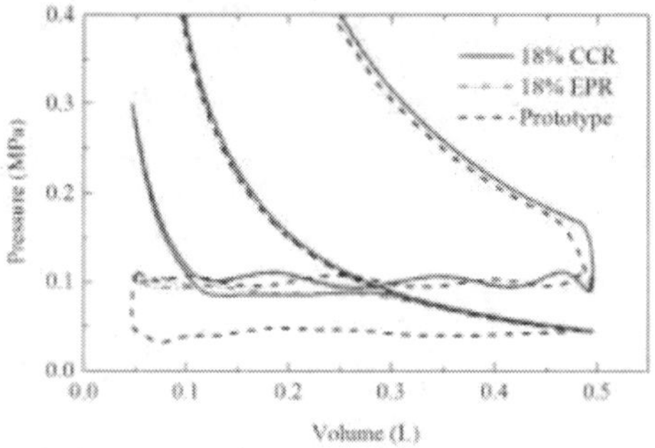

Figure 9. Pressure –Volume diagram of CCR strategy, EPR strategy and prototype.

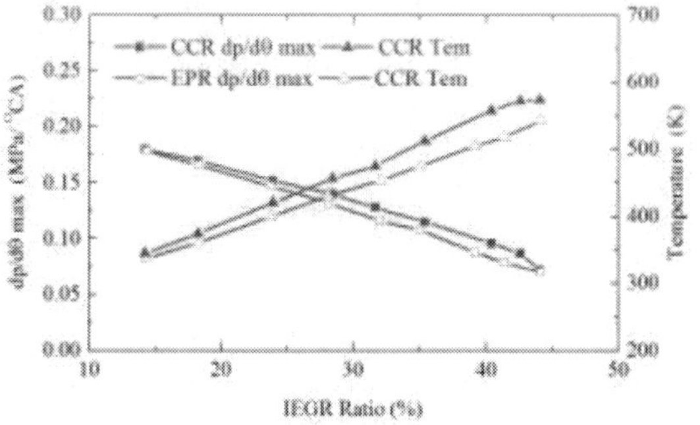

Figure 10. Maximum pressure rise and gas temperature with various levels of internal EGR introduction.

In conclusion, the maximum operational load of EPR is higher than CCR strategy. At the engine speed of 2000r/min and BMEP is about 0.3MPa, the optimal target is to maximize the engine power under invariable charge efficiency. Result shows with 30%~35% EPR rate, which get a perfect compromise between engine's fuel economy and NO_X missions. In comparison with initial condition, the BSFC and NO_X emissions have a decrease over 10% and 85% respectively.

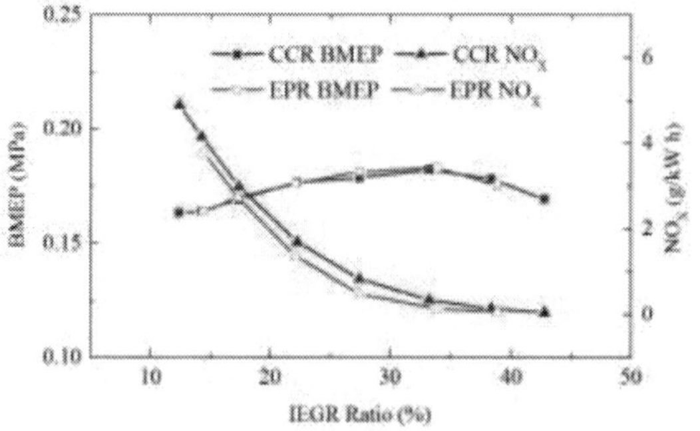

Figure 11. BMEP and NO_X missions with various levels of internal EGR introduction. At 1000 r/min and BMEP is about 0.17 MPa.

While the CCR strategy is suitable to the lower engine load conditions. Because that the camless engine does not equip with throttle system, which led to a low temperature of fresh charge and combustion efficiency. And the heating effect of exhaust gas in CCR is more obvious than others. At the engine speed of 1000r/min and BMEP is about 0.17 MPa, the results of two different EGR strategies are shown in Figure 11. It can be inferred that the difference between two EGR strategies is not significant. But from the point of view of valve motion control, the single valve motion per cycle of CCR is easier to realize than EPR. Also a low valve frequency could decrease the power consumption and heat of EMVT.

5 Map of control strategies with load/speed-range of the camless engine

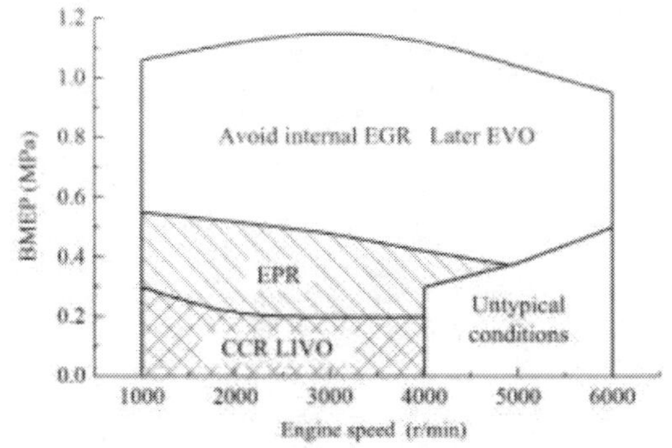

Figure 12. Map of control strategies with various operational conditions.

Based on the co-simulation models between AVL Boost and Matlab, and the genetic algorithm is used for optimization. So the map of EMVT control strategies is carried out under all operating conditions as shown in figure 12. The maximum operational region of CCR is often less than 0.2~0.3 MPa because of the prominent heating effect of exhaust gas, it also needs to couple with later IVO to avoid high pumping losses. While the application of EPR strategy makes it possible to extend the maximum load to about 0.5 MPa BMEP. With the engine load increases further, internal EGR rate should be decrease or even avoided. Because the high intake temperature caused by internal EGR would deteriorates the volumetric efficiency and combustion quality. Also later EVO should be adopted to decrease the gas pressure.

6 Conclusion

Based on the application of electromagnetic intake and exhaust valve train in camless engine, this paper mainly discusses the variable exhaust valve motions for engine performance improvement. Main results are summarized as followed:

(1) At engine part load, the optimal EVO timing for minimizing BSFC occurs in a good approximation when the engine exhaust losses reach to the lowest value. It is usually later than cam engine.

(2) EPR strategy gets a higher operational region to about 0.5 MPa BMEP compared with CCR strategy. At the engine speed of 2000r/min, BMEP is about 0.3 MPa and 30%~35% EPR, the BSFC and NO_X emissions have a decrease over 10% and 85% respectively to initial condition.

(3) CCR strategy can be applied to low load conditions with the maximum operational region less than 0.2~0.3 MPa because the prominent heating effect of exhaust gas could compensate the low intake temperature. While the pumping losses can be decreased coupled to later IVO.

Through the analysis above, the optimization approaches of exhaust valve motion and internal EGR strategies are carried out, which provide a theoretical basis for the later camless engine test.

References

1. S. Y. Tong, X. J. Li, S. Liu, et al, SAE Technical Paper 2014-01-2874 (2014)
2. E. Sher, T. Bar-Kohany, Energy **27**, 8, 757-775 (2002)
3. U. Kramer, P. Phlips, SAE Technical Paper 2002-01-1101 (2002)
4. A. M. Pourkhesalian, A. H. Shamekhi, F. Salimi, SAE Technical Paper 2010-01-1204 (2010)
5. J. F. Zhao, R. J. Seethaler, Mechatronics **20**, 4, 433-441 (2010)
6. L. Liu, S. Q. Chang, Proc IMechE, Part D: Journal of Automobile Engineering **226**, 8, 85-93 (2012)
7. X. Y. Fan, L. Liu, S. Q. Chang, et al, Int. J. Automotive Technology. (to be published)
8. B. B. Sun, S. Q. Chang, L. Liu, Automotive Engineering **36**, 2, 145-150 (2014)
9. D. E. Ramajo, N. M. Nigro, J. Eng. Gas Turbines Power **132**, 5, 052804 (2010)

Simulation of YBCO Tape and Coils in HTS Maglev System

Mengxiao Song and Jie Li

College of Mechatronic Engineering and Automation, National University of Defense Technology, Changsha 410073, P. R. China

Abstract. In the process of running high temperature superconducting maglev train, the AC(Alternating Current) loss of superconducting coil is directly related to its safe operation and operating cost. In this paper, the simulation model was built based on the finite element software COMSOL Multiphysics, and mainly simulated and calculated the AC losses of YBCO(Yttrium Barium Copper Oxide) tape and coils. In this model, as the solving object, the singular and infinite long YBCO tape and coils model was solved with H-formulation and the nonlinear characteristic (E-J constitutive law) and anisotrophy (B-J characteristic) were taken into consideration as the theoretical foundation. Then on the basis of the model under maglev suspension system, AC losses under different amplitude and frequence AC currents were calculated. The results shows that under different frequencies and dynamic components, the local maximum AC loss of YBCO tape and coils occurs when the steady-state DC(Direct Current) current is 30A . Then comparing with old maglev suspension system, the new system can greatly reduce the energy consumption and the material cost.

1 Introduction

Maglev is considered a promising way of rail transport system. It has lots of advantages which traditional wheel-rail trains do not have. Maglev trains does not contact tracks during running, so that there is no mechanical friction, which makes the noise greatly reduced. It is generally driven by linear motor, so it has great ability at climbing and small turning radius. In addition, the maglev can construct railways through viaduct, so there is less impact on the surrounding environment and lower cost[1-2].

Currently maglev train uses aluminum coils, so there are resistive losses in the suspension process which makes the electromagnets generate more heat, but also results in a waste of energy. Currently power consumption of suspension system is 0.75 kilowatts per ton. If the resistive losses can be reduced, the maglev train will save more energy[3-4].

Superconducting material can significantly reduce the resistance loss of the conventional normal conducting coils. High-temperature superconducting material is a kind of material working in the liquid nitrogen which lowers the requirements of refrigeration and insulation. High-temperature superconducting maglev is becoming the trend of future transportation development[5-6].

Research on AC loss of high-temperature superconducting maglev, obtaining loss characteristics, so we can design the HTS(High-Temperature Superconductor) suspension electromagnet, what's more, the thermal insulation structure and the power of refrigerator etc. Ultimately, assess the possibility of applying the superconducting material on electromagnetic levitation. So the primary issue of applying high-temperature superconducting material in maglev system is to study the magnitude of AC loss[7-10].

This paper is based on the E-J constitutive law of YBCO and the finite element software COMSOL Multiphysics. Mainly simulate and calculate the AC losses of YBCO tape and coils under different kinds of current inputs. The results shows that under different frequencies and dynamic components, the local maximum AC losses of YBCO tape and coils occurs when the steady-state DC current is 30A[11-13] .

2 Simulation

Fig. 1 and Fig. 3 show the simulation of YBCO tape model and coils model respectively, model parameters and expressions shown in Table 1 and Table 2.

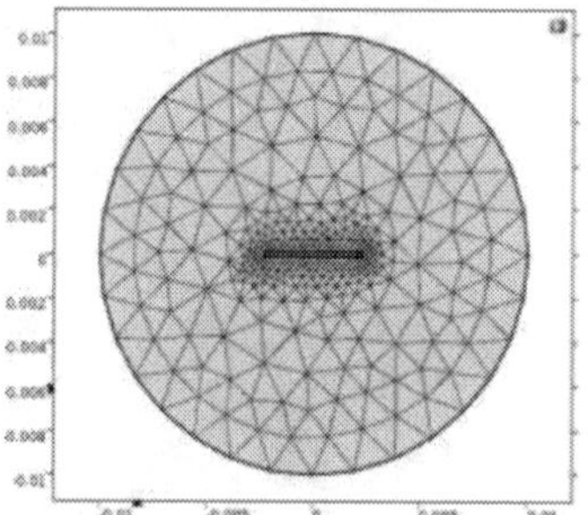

Figure 1. simulation of YBCO tape model

 DOI: 10.1051/matecconf/20179515010

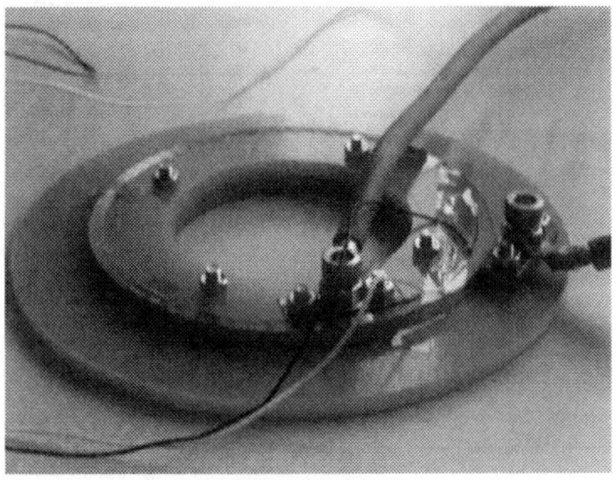

Figure 2. the photo of YBCO coils

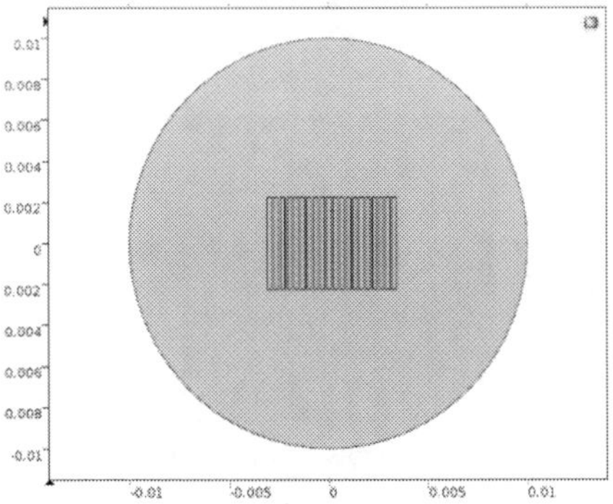

Figure 3. simulation of YBCO coils model

Table 1. Primary parameters

tape width	4.5 mm
tape thickness	0.23 mm
critical current	120 A
critical current density	1.17e8 A/m^2
simulation time	0.1s
distance between turns	0.1mm

Table 2. Primary variables and expressions

variables	expressions	description
J	$\frac{\partial H_y}{\partial x}-\frac{\partial H_x}{\partial y}$	Current density (A/m^2)
E	$\begin{cases} 0 & J<J_c \\ E_c(\frac{J-J_c}{J_c})^n & J\geq J_c \end{cases}$	Electric field strength (V/m) E-J constitutive law
Q	$\iint_{t\,s} E\times J$	AC loss (W/m)
I	$I_d+I_0\sin(wt)$	current applied (A)

This simulation is based on different types of given current to calculate the AC losses of the tape and coils.

3 Results and analysis

3.1. The tape simulation

Current given is $I=I_d+I_a\sin(2\pi ft)$.

AC losses under various parameters are shown below. Among them I_d changes from 10A to 70A with the step of 10A. I_a changes from 5A to 20A with the step of 5A. f changes from 10Hz to 500Hz.

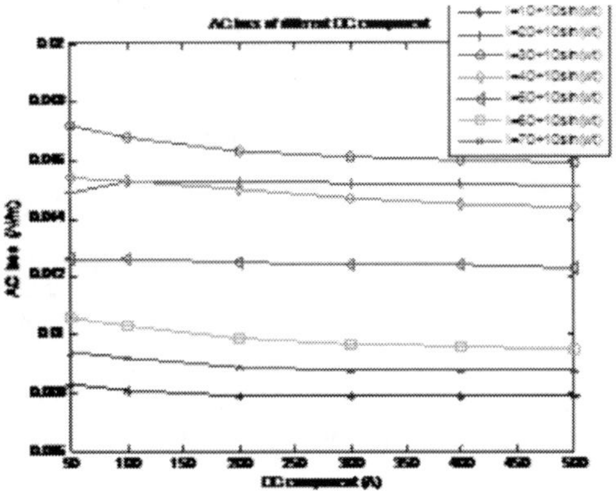

Figure 4. AC losses of different DC components under different frequencies

As can be seen from Fig. 4, with I_a being 10A, the change of frequency makes no significant effect on AC losses in different

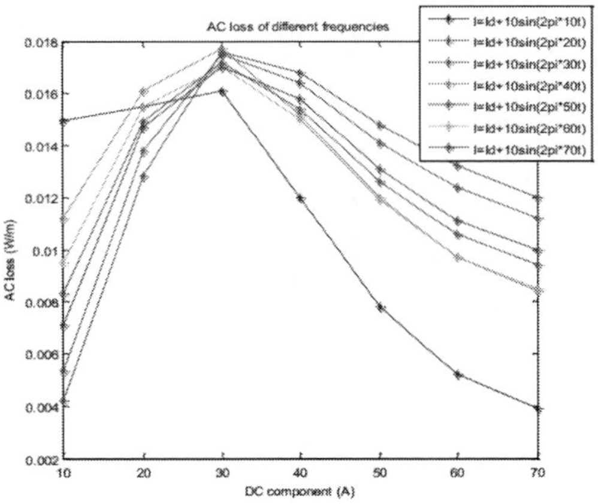

Figure 5. AC losses of 10~70Hz under different DC components

DOI: 10.1051/matecconf/20179515010

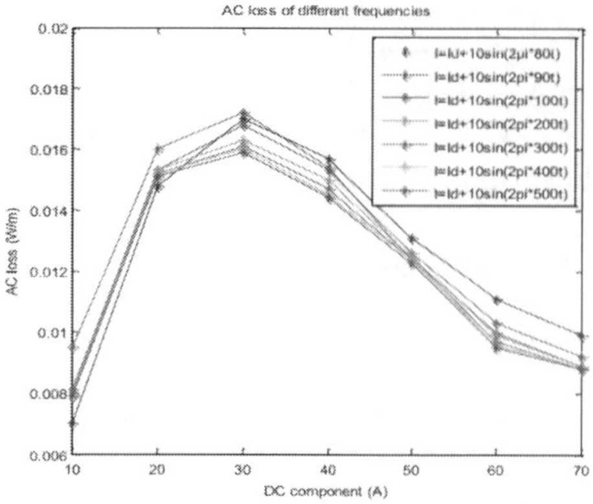

Figure 6. AC losses of 80~500Hz under different DC components

As can be seen from Fig. 5 and Fig. 6, with I_a being 10A, at each frequency, the AC loss occurs local maximum when the DC component is 30A. Thus AC losses can be reduced if avoiding 30A and its neighborhood.

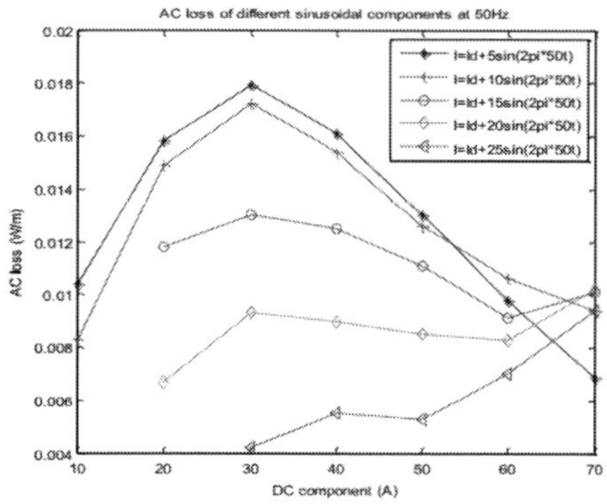

Figure 7. AC losses of different sinusoidal components under different DC components at 50Hz

As can be seen from Figure 7, at 50Hz, for different sinusoidal components, the AC loss occurs local maximum when the DC component is 30A. Although there is a rally after 60A, AC losses can be reduced if avoiding 30A and its neighborhood.

3.2 The coils

From above, select a set of representative parameters to calculate AC loss of YBCO coils model.

Current given is $I = I_d + I_a \sin(2\pi ft)$

AC losses under various parameters are shown below. Among them I_d changes from 10A to 70A with the step of 10A. I_a = 5A and f = 20Hz.

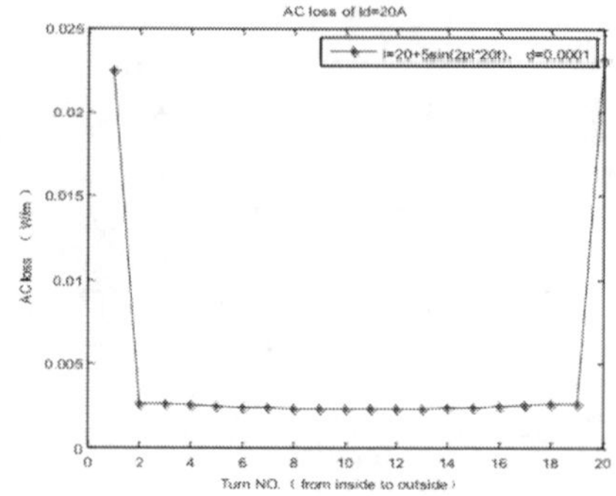

(a)

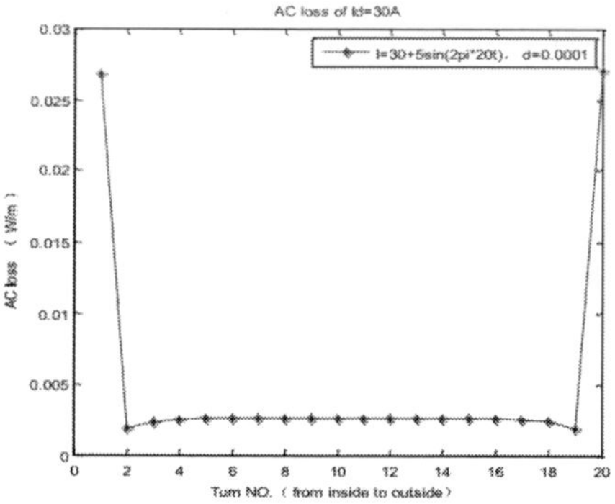

(b)

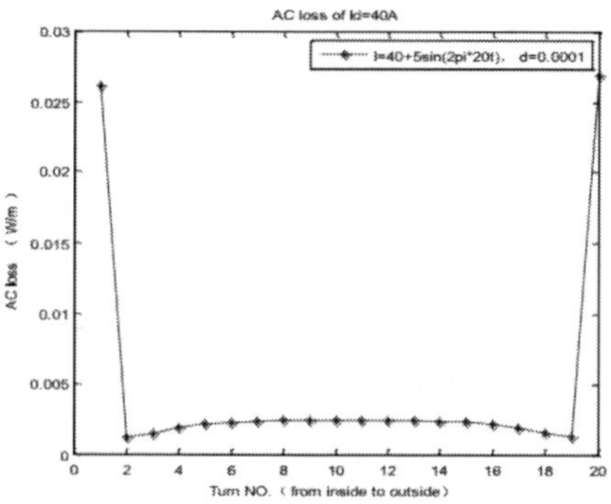

(c)

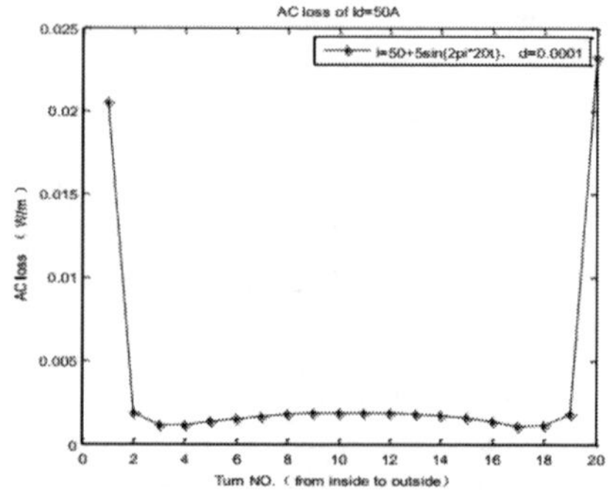

(d)

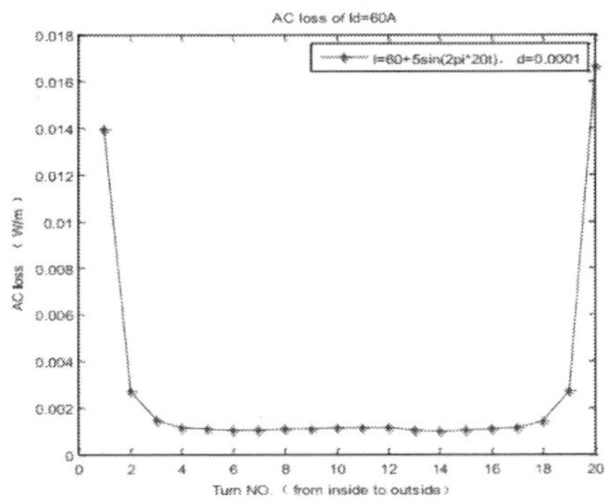

(e)

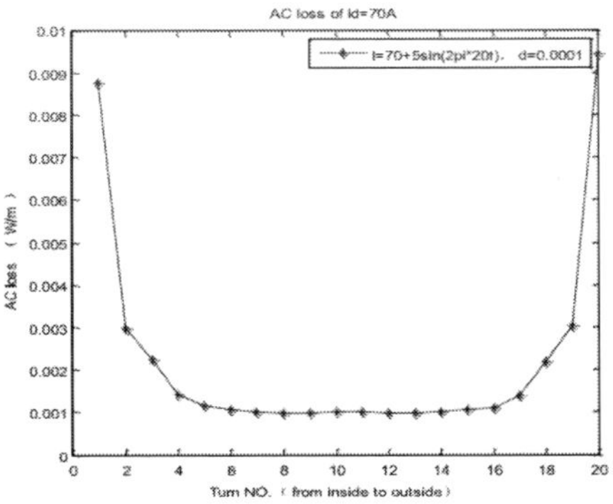

(f)

Figures 8. (a~f) AC losses of 20Hz and I_a = 5A under different DC components from 20A to 70A

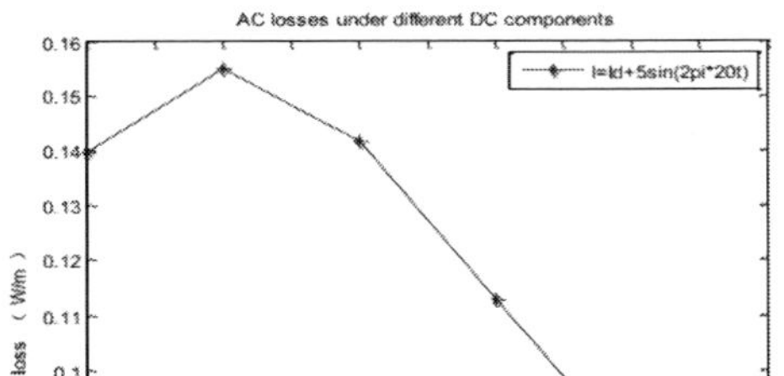

Figure 9. normalized AC losses under different DC components

This normalization means multiplying the corresponding length of every turn of the coils then add them all to obtain the total AC loss.

As can be seen from the simulation above, with I_a being 10A, f being 20Hz, the AC loss occurs local maximum when the DC component is 30A. This result is the same as the simulation of YBCO tape.

4 Conclusion

Equations should be centred and should be numbered with the number on the right-hand side.As for maglev system, saving the cost and reducing energy consumption have always been the objectives to pursue.

Based on the above conclusions, if we improve the DC components of work current from 30A to 50A, the AC loss will drop from 0.0172W/m to 0.0126W/m, a decrease of 26.74%. Meanwhile, with the same magnetic motive force, the coil turns required will reduce by 40%. What' s more, the total reduction of cost will be more than 40% due to the winding accumulation and the low success rate of winding coils. For example, when the DC component of current input is 30A, we need about 400m tape with the market price about 400 yuan / m. Therefore, this improvement will bring very substantial cost savings. Then we turn back to see the energy consumption.

$$E_{new} = (1-26.74\%)\times(1-40\%)E_{old} \quad = 44.0\% E_{old}$$

This improvement will reduce the energy consumption by more than 50%. The new one is only 44% of the old one.

The above conclusions are on the contrary to the traditional cognition that the greater the current is, the greater the AC loss will be. The author was very surprised when firstly found this conclusion, after constantly repeating and verifying this simulation, still got this conclusion. Finally decide to propose this result by this paper to have a discussion with the readers.

References

1. Yan L G. IEEE Transactions on Applied Superconductivity, 2006, 16(2):1138-1141.
2. Yan L G. IEEE Transactions on Applied Superconductivity, 2008, 18(2), 92-99.
3. Salih E, Lachowicz S, Bass O, et al. International Conference on Green Energy. 2014:267-272.
4. Higashikawa K, Nakamura T, Hoshino T. Physica C Superconductivity & Its Applications, 2005, 419(s 3–4):129-140.
5. Zhang M, Kvitkovic J, Kim C H, et al. Journal of Applied Physics, 2013, 114(4):043901-043901-8.
6. Yuan W, Campbell A M, Hong Z, et al. Superconductor Science & Technology, 2010, 23(8):32-32.
7. Zhang, Min. Study of second-generation high-temperature superconducting coils[J]. 2013(October).
8. Zermeno V M R, Abrahamsen A B, Mijatovic N, et al. Journal of Applied Physics, 2013, 114(17):173901-173901-9.
9. Nguyen D N, Sastry P V P S S, Schwartz J. Journal of Applied Physics, 2007, 101(5):053905-053905-9.
10. Grilli F, Ashworth S P. Superconductor Science & Technology, 2007, 20(8):794-799.
11. Ritesh Kumar and Amit Kumar, International Journal of Mechanical Engineering and Robotics Research, Vol. 3, No. 1, pp. 42-49, January 2014.
12. Boyd A. Nicholds and John P. T. Mo, International Journal of Mechanical Engineering and Robotics Research, Vol. 5, No. 2, pp. 138-143, April 2016. DOI: 10.18178/ijmerr.5.2.138-143

13. Ngoctan Tran, Yaw-Jen Chang, Jyh-tong Teng, and Thanhtrung Dang, International Journal of Mechanical Engineering and Robotics Research, Vol.4, No. 3, pp. 204-208, July 2015. DOI: 10.18178/ijmerr.4.3.204-208

MATEC Web of Conferences **95**, 15011 (2017) DOI: 10.1051/matecconf/20179515011
ICMME 2016

Magnetic Flux Density Feedback Control for Permanent Magnetic-Electromagnetic Hybrid Suspension System

Qiang CHEN, Jie LI, Guanchun LI, Siyang ZHOU

College of Mechatronics Engineering and Automation, National University of Defense Technology, 410073 Changsha, China

Abstract. Permanent magnetic-electromagnetic hybrid suspension system can effectively reduce energy consumption and heat release of the system, but also increase the difficulty of suspension control because of the existence of permanent magnets. The traditional current feedback control method is not conducive to the stability of the system and is difficult to debug. In this paper, the models of permanent magnetic-electromagnetic hybrid suspension system based on current feedback and magnetic flux density feedback are established. The effects of current feedback and magnetic flux density feedback on the stability of the system are analyzed in theory and the advantages of flux density feedback are pointed out. The model of magnet flux feedback is simple and it can overcome the disadvantages of current feedback, which is beneficial to the stability of the system. The magnetic flux density feedback control of permanent magnetic-electromagnetic hybrid suspension system is realized by simulation and experiment. Control system performs well and is easy to debug.

1 Introduction

Maglev transportation has advantages of safe, comfort, low noise, small turning radius and strong climbing ability [1]. It is a promising way of rail transit, which has received extensive attention in recent years. Countries such as Germany, Japan, China, South Korea and America have carried out relevant research on maglev technology and all have made great progress [2, 3]. In permanent magnetic-electromagnetic (PM-EM) hybrid suspension system [4], the permanent magnets provide static suspension force, which electromagnets provide dynamic electromagnetic force required to adjust. It also has the advantages of low energy consumption, low heat release and small installed capacity of vehicle power. While at the same time, the difficulty of the suspension control becomes harder because of the existence of permanent magnets.

Suspension control is the core technique of PM-EM hybrid suspension system. Currently the major method of suspension control is gap-current double loops cascade control [5]. The gap outer-loop realizes stable suspension of the system and tracks the change of the gap. The current inner-loop realizes the quick response of the system. However, for the PM-EM hybrid suspension system, the relationship of magnetic force, suspension gap and current is complex and overlapping. The inter-loop using current feedback control is not beneficial to the design of suspension controller and the development of suspension performance. Considering that for the PM-EM hybrid suspension system, the magnetic force is decided by the magnetic flux density in the suspension gap only, therefore using magnetic flux density feedback control inter-loop can make the design of the suspension controller easier. Besides, it will be more convenient to debug the control system [6, 7].

In this paper, relevant theoretical research is carried out. By comparing with the current feedback control, the effects of current feedback and magnetic flux density feedback on the stability of the system are analyzed and the advantages of magnetic flux density feedback control are pointed out. Using the magnetic flux density feedback control method realized the PM-EM hybrid suspension system stable suspension in simulation and experiment.

2 System modelling

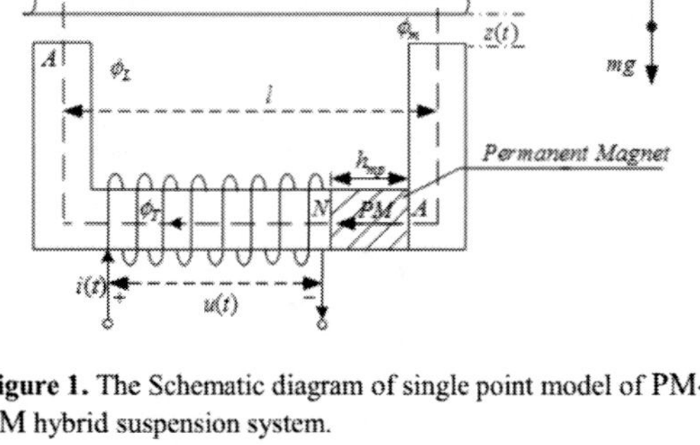

Figure 1. The Schematic diagram of single point model of PM-EM hybrid suspension system.

The PM-EM hybrid suspension system is made up of permanent magnets and electronic magnets. In PM-EM hybrid suspension system, each bogie has two suspension modules. Each module has two suspension points. By mechanical and electrical decoupling, each suspension point is controlled separately. Thus the hybrid suspension system can be decomposed and considered as single point suspension system. Single point model of permanent magnetic-electromagnetic (PM-EM) hybrid suspension system is shown in Fig. 1.

The following assumptions are made for the single point suspension system:

(1) The leakage flux of coil winding, permanent magnet and the magnetic saturation characteristics of materials are ignored.

(2) The reluctance of magnet in core and guide rail is ignored, which means that the magnetic potential drops on the permanent magnet and the gap uniformly.

(3) The elastic vibration or dynamic deformation of the track is not considered, which means that the related stiffness coefficient of the track and the electronic magnet is infinite.

Considering the gap and the current as state variables, the single point model of the PM-EM hybrid suspension system can be described as

$$\begin{cases} m\ddot{z}(t) = -F(t) + mg \\ F(t) = \dfrac{\mu_{mp}\mu_r A[Ni(t) + H_c h_{mp}]^2}{[2\mu_r z(t) + h_{mp}]^2} \\ u(t) = Ri(t) + \dfrac{\mu_{mp} A N^2}{2\mu_r z(t) + h_{mp}}\dot{i}(t) - \dfrac{2NA\mu_{mp}\mu_r[Ni(t) + H_c h_{mp}]}{[2\mu_r z(t) + h_{mp}]^2}\dot{z}(t) \end{cases} \quad (1)$$

Considering the gap and the magnetic flux density as the state variables, the single point model of the PM-EM hybrid suspension system can be described as

$$\begin{cases} m\ddot{z}(t) = -F(t) + mg \\ F(t) = \dfrac{A}{\mu_0} B^2(t) \\ u(t) = \dfrac{R}{\mu_{mp} N}\{[2\mu_r z(t) + h_{mp}]B(t) - \mu_{mp} H_c h_{mp}\} + NA\dot{B}(t) \end{cases} \quad (2)$$

The meanings of symbols are illustrated in table 1.

Table 1. Meanings of symbols for the PM-EM hybrid suspension system.

Symbol	Meaning
ϕ_m	Gap flux
ϕ_T	Main pole flux
ϕ_L	Leakage flux
$B(t)$	magnetic flux density
$u(t)$	Electromagnet coil voltage
$i(t)$	Electromagnet coil current
R	Electromagnet coil resistance
N	Electromagnet coil turns
A	cross-sectional area of iron core
H_c	Coercive force of permanent magnet
h_{mp}	Permanent magnet thickness
μ_0	Air permeability
μ_{mp}	Permeability of permanent magnet
μ_r	Relative permeability of permanent magnet
$z(t)$	Gap between magnet and track
$F(t)$	electromagnetic force
m	The total mass
g	Gravitational acceleration

The magnetic flux density in the suspension gap of the PM-EM hybrid suspension system can be expressed as

$$B(t) = \frac{\mu_{mp}[Ni(t) + H_c h_{mp}]}{2\mu_r z(t) + h_{mp}} \quad (3)$$

Therefore formula 1 and formula 2 are equal, and it can be easily proved that the PM-EM hybrid suspension system is a non-linear and unstable system. But if current is defined as the state variable, from formula 1, the magnetic force is correlated with many variables such as current and gaps, and the change of current will cause the change of gaps. The relationship of them is complex and coupling. However, using magnet flux density as state variable, from formula 2, it can be seen that magnetic force is directly related with magnet flux density. The equation is simple. Therefore, using the magnetic flux density feedback inter-loop control, the suspension controller is easier to design and debug.

3 Comparison of magnetic flux density feedback and current feedback

3.1 Current feedback

Linearizing formula 1 at the equilibrium point (z_0, i_0)

$$\begin{cases} \Delta\ddot{z} = -\dfrac{1}{m}\Delta F \\ \Delta F = \dfrac{2NA\mu_{mp}\mu_r(Ni_0 + H_c h_{mp})}{(2\mu_r z_0 + h_{mp})^2}\Delta i - \dfrac{4A\mu_{mp}\mu_r^2(Ni_0 + H_c h_{mp})^2}{(2\mu_r z_0 + h_{mp})^3}\Delta z \\ \Delta u = R\Delta i + \dfrac{\mu_{mp} A N^2}{2\mu_r z_0 + h_{mp}}\Delta\dot{i} - \dfrac{2NA\mu_{mp}\mu_r(Ni_0 + H_c h_{mp})}{(2\mu_r z_0 + h_{mp})^2}\Delta\dot{z} \end{cases} \quad (4)$$

Let

$$K_{IZ} = \frac{2NA\mu_{mp}\mu_r(Ni_0 + H_c h_{mp})}{(2\mu_r z_0 + h_{mp})^2} \tag{5}$$

$$K_{ZZ} = \frac{4A\mu_{mp}\mu_r^2(Ni_0 + H_c h_{mp})^2}{(2\mu_r z_0 + h_{mp})^3} \tag{6}$$

$$K_{UI} = \frac{2\mu_r z_0 + h_{mp}}{\mu_{mp}AN^2} \tag{7}$$

$$K_{ZI} = \frac{2\mu_r(Ni_0 + H_c h_{mp})}{(2\mu_r z_0 + h_{mp})N} \tag{8}$$

Making $(\Delta z, \Delta\dot{z}, \Delta i)^T$ as state variable, the state-space equation is

$$\begin{bmatrix} \Delta\dot{z} \\ \Delta\ddot{z} \\ \Delta\dot{i} \end{bmatrix} = \begin{bmatrix} 0 & 1 & 0 \\ \frac{1}{m}K_{ZZ} & 0 & -\frac{1}{m}K_{IZ} \\ 0 & K_{ZI} & -RK_{UI} \end{bmatrix} \begin{bmatrix} \Delta z \\ \Delta\dot{z} \\ \Delta i \end{bmatrix} + \begin{bmatrix} 0 \\ 0 \\ K_{UI} \end{bmatrix} \Delta u \tag{9}$$

The block diagram of PM-EM hybrid suspension control system based on current feedback inter-loop is shown in Fig. 2. Using gap-current double loops cascade control. Gap outer-loop is gap feedback controller $C_z(s)$, which realizes the system levitate stably and track the change of gaps. Current inter-loop is current feedback controller $C_i(s)$, which makes the system response quickly.

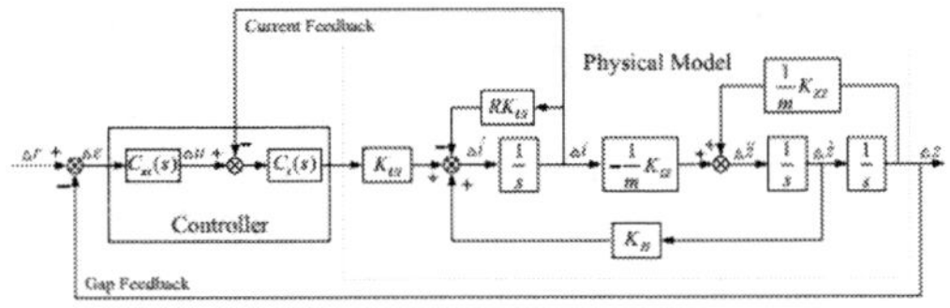

Figure 2. The block diagram of PM-EM hybrid suspension control system based on current feedback inter-loop.

The transfer function of current loop is

$$G_i(s) = \frac{\Delta i(s)}{\Delta u(s)} = \frac{K_{UI}C_i(s)}{s + RK_{UI} + K_{UI}C_i(s)} \tag{10}$$

The current loop uses proportional control, $C_i(s) = k_i$ and the current loop is a first order system. Time constant is

$$\tau_i = \frac{1}{RK_{UI} + K_{UI}k_i} \tag{11}$$

If there is no current inter-loop, the transfer function from control voltage Δu to electromagnetic current Δi is

$$G_i'(s) = \frac{K_{UI}}{s + RK_{UI}} \tag{12}$$

Time constant will be

$$\tau_i' = \frac{1}{RK_{UI}} \tag{13}$$

So current feedback makes the time constant of current loop smaller, and the current of electromagnets can be adjusted to track the change of control voltage in very short time, the response velocity will be quicker.

In order to investigate the effect of current feedback on the whole system, do not consider the gap feedback at first and the transfer function from Δu to Δz is

$$\frac{\Delta z(s)}{\Delta u(s)} = -\frac{k_i K_{UI} K_{IZ}}{ms^3 + (k_i + R)K_{UI}ms^2 + (K_{IZ}K_{ZI} - K_{ZZ})s - (k_i + R)K_{UI}K_{ZZ}} \tag{14}$$

The root locus of the transfer function 14 with the change of current feedback coefficient k_i is shown in Fig. 3. With the increase of k_i, the unstable pole in the right half s-plane will increase, making the system diverge faster. So current feedback cannot solve the stable problem, nor it will make the suspension system more unstable.

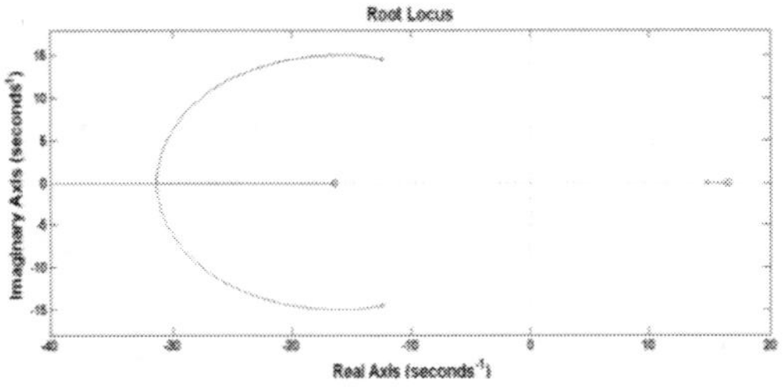

Figure 3. The root locus of current feedback control.

3.2 Magnet flux density feedback

Linearizing formula 2 at equilibrium point (z_0, B_0).

$$\begin{cases} \Delta\ddot{z}(t) = -\frac{1}{m}\Delta F \\ \Delta F = \frac{2AB_0}{\mu_0}\Delta B \\ \Delta u = NA\Delta\dot{B} + \frac{(2\mu_r z_0 + h_{mp})R}{\mu_{mp}N}\Delta B + \frac{2\mu_r RB_0}{\mu_{mp}N}\Delta z \end{cases} \tag{15}$$

Let

$$K_{BZ} = \frac{2AB_0}{\mu_0} \tag{16}$$

$$K_{BB} = \frac{(2\mu_r z_0 + h_{mp})R}{\mu_{mp}AN^2} \tag{17}$$

$$K_{ZB} = \frac{2\mu_r RB_0}{\mu_{mp}AN^2} \tag{18}$$

Making $(\Delta z, \Delta\dot{z}, \Delta B)^T$ as state variable, the state-space equation is

$$\begin{bmatrix} \Delta\dot{z} \\ \Delta\ddot{z} \\ \Delta\dot{B} \end{bmatrix} = \begin{bmatrix} 0 & 1 & 0 \\ 0 & 0 & -\frac{1}{m}K_{BZ} \\ -K_{ZB} & 0 & -K_{BB} \end{bmatrix} \begin{bmatrix} \Delta z \\ \Delta\dot{z} \\ \Delta B \end{bmatrix} + \begin{bmatrix} 0 \\ 0 \\ \frac{1}{NA} \end{bmatrix} \Delta u \quad (19)$$

The block diagram of the PM-EM hybrid suspension control based on the feedback inter-loop of the magnetic flux density is shown in Fig. 4. Using gap-magnet flux density double loops cascade control. Gap outer-loop control realizes the system levitate stably and track the change of the gap. The magnet flux density inter-loop realizes the quick response of the system.

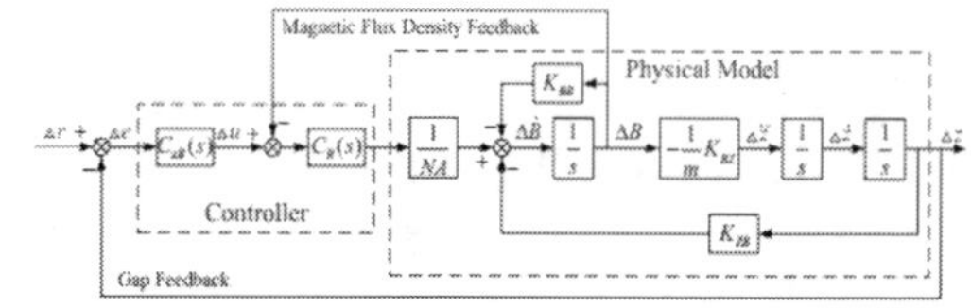

Figure 4. The block diagram of the PM-EM hybrid suspension control based on magnetic flux density feedback inter-loop.

The transfer function of magnet flux density loops is

$$G_B(s) = \frac{\Delta B(s)}{\Delta u(s)} = \frac{C_B(s)}{NAs + NAK_{BB} + C_B(s)} \quad (20)$$

The magnet flux density inter-loop uses proportional control, $C_B(s) = k_B$, and the magnet flux density loop is first order system. Time constant is

$$\tau_B = \frac{NA}{NAK_{BB} + k_B} \quad (21)$$

If there is no magnet flux density inter-loop, the transfer function from control voltage Δu to gap magnet flux density ΔB is

$$G'_B(s) = \frac{1}{NAs + NAK_{BB}} \quad (22)$$

Time constant will be

$$\tau'_B = \frac{1}{K_{BB}} \quad (23)$$

So magnet flux density feedback makes the time constant of magnet flux density loop smaller, and the magnet flux density can be adjusted to track the change of control voltage in very short time, the velocity of the response will be quicker. At the same time, it is conducive to system control because magnetic force is decided by magnet flux density directly.

In order to investigate the effect of magnet flux density feedback on the whole system, do not consider the gap feedback at first and the transfer function from Δu to Δz is

$$\frac{\Delta z(s)}{\Delta u(s)} = -\frac{k_B K_{BZ}}{mNAs^3 + (k_B + NAK_{BB})ms^2 - NAK_{BZ}K_{ZB}} \quad (24)$$

The root locus of the transfer function 24 with the change of magnet flux density feedback coefficient k_B is shown in Fig. 5. With the increase of k_B, the unstable pole p goes near the virtual axis. While $k_B \to \infty$, $p = 0$, the system will be critical stable. It can be seen that the magnet flux density feedback is beneficial to the stabilization of the system.

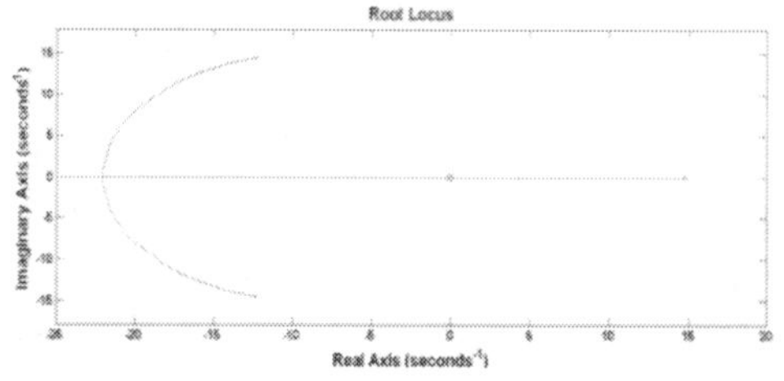

Figure 5. The root locus of magnet flux density feedback.

4 Simulation and experiment

Adapting the control system in Fig. 4, proportional control is used in magnet flux density inter-loop. $C_B(s) = k_B$, PID controller is used in the gap outer-loop. $C_{zi}(s) = k_{zip} + k_{zip}\frac{1}{s} + k_{zid}s$, the air gap is 9mm, tracking amplitude 1mm square wave. The change of the gap in simulation is shown in Fig. 6.

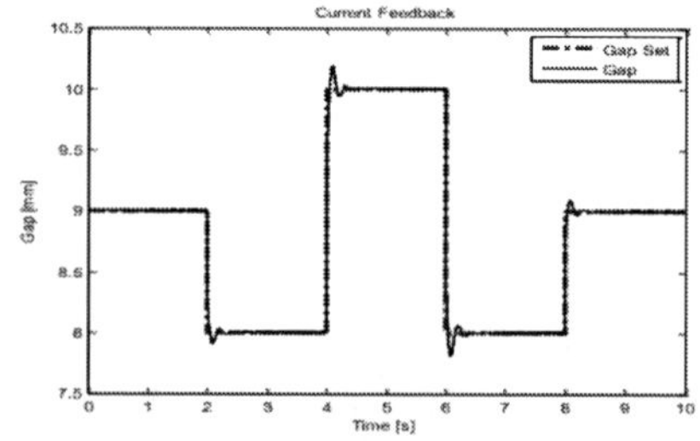

Figure 6. simulation results of tracking square wave.

Conducting suspension control experiment on the single bogie experimental platform shown in Fig. 7, using the same control method mentioned above. The air gap is 9mm, tracking amplitude 1mm square wave. The change of the gap is shown in Fig. 8. Magnet flux density curve is shown in Fig. 9.

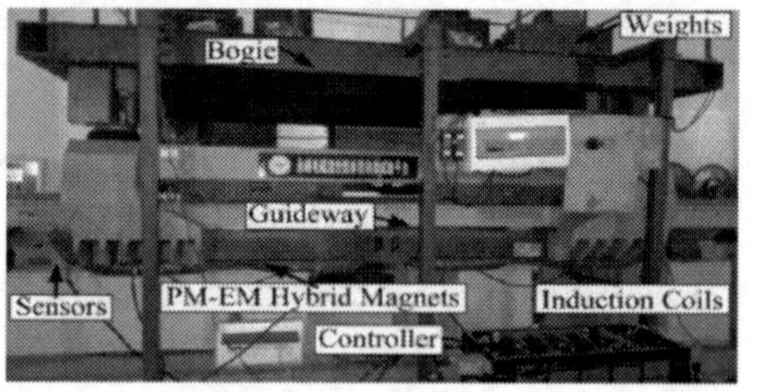

Figure 7. Single bogie experimental platform of the PM-EM hybrid suspension system.

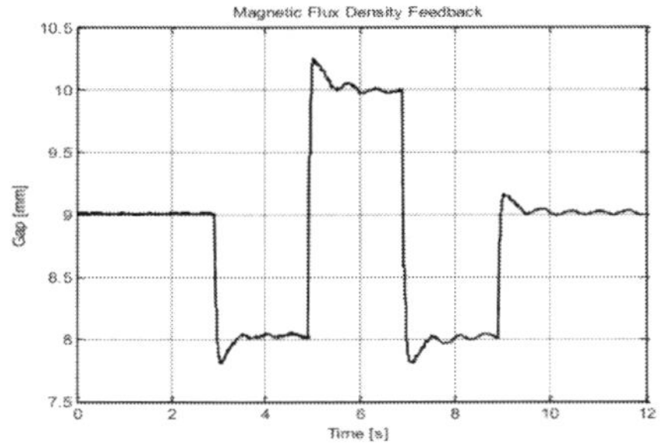

Figure 8. Experiment results of tracking square wave.

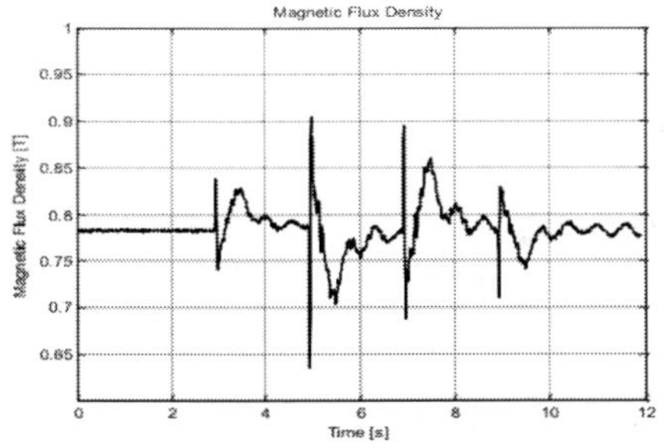

Figure 9. Magnet flux density curve.

From the simulation and experiment, because of the effect of gap outer-loops, the hybrid suspension system can be stable and achieve good control performance. However, the method of magnet flux density feedback is easier and is convenient to debug.

5 Conclusion

The influence of the current feedback inter-loop and the magnet flux feedback inter-loop on the PM-EM hybrid suspension control system is analyzed in theory based on the system model. Current feedback inter-loop is not conducive to the stability of the system and the performance of the outer-loop will be affected, what's worse, the system may be diverged. The model of magnet flux feedback is simple and it can overcome the disadvantage of current feedback, which is beneficial to the stability of the system. The results of the simulation and experiment show that the magnet flux feedback control system performs well, and it is easy to debug.

References

1. Lee H W, Kim K C, and Lee J. Review of Maglev Train Technologies. IEEE Transactions on Magnetics, 2006, 42(7):1917~1925.
2. Yan L G. Development and Application of the maglev transportation System. IEEE Transactions on Applied Superconductivity, 2008, 18(2):92-99.
3. Yan L G. Progress of the Maglev Transportation in China. IEEE Transactions on Applied Superconductivity, 2006, 16(2):1138~1141.
4. Cho H W, Han H S, Lee J M, Kim B S, Sung S Y. Design Considerations of EM-PM Hybrid Suspension and Propulsion Device for Magnetically Levitated Vehicle. IEEE Transactions on Magnetics, 2009, 45(10): 4632~4635.
5. LI Jin-hui, Zhou Dan-feng, LI Jie, Zhang Geng, Modeling and simulation of CMS04 Maglev Train with Active Controler, Journal of Central South University of Technology, 2013, 4 (22) : 1366-1377.
6. Wenqing Zhang, Jie Li, Kun Zhang, Peng Cui. Measurement and Control of Magnetic Flux Signal in Maglev System. Asian Journal of Control, 2014, 16(6), 1-11.
7. Wenqing Zhang, Jie Li, Kun Zhang, Peng Cui. Design of Magnetic Flux Feedback Controller in Hybrid Suspension System. Mathematical Problems in Engineering, 2013.

A Hybrid Controller for Position Control of a Pneumatic Actuator under Variable Loading Conditions

Sukhdeep Singh Dhami

Mechanical Engineering Department, National Institute of Technical Teachers Training and Research, Chandigarh, India

Abstract. Pneumatic actuators are preferred for low to medium power requirements due to their desirable features like large power to weight ratio, low cost and safe working medium. Position control of pneumatic actuators is however difficult due to inherent nonlinearity-contributing factors. Therefore, position control of pneumatic actuators have been found to be a challenging task using fixed gain controllers. Sliding mode control can be used to overcome this difficulty. Further, the nonlinear response of pneumatic actuators under variable loading conditions makes it a fit case for using heuristic control based on the knowledge of the system behaviour. This may be done by using fuzzy logic based control approach. In the present work, a hybrid controller was synthesized using fuzzy logic control and sliding mode control for position control of a pneumatic actuation system. The quantitative results emphasized that hybrid FLC-SMC controller exhibited effective position control for pneumatic actuation system as fuzzy logic control was effective for combinations of lower payloads and smaller piston displacements, whereas sliding mode controller was effective for combinations of higher payloads and larger piston displacements.

1 Introduction

Pneumatic actuators are preferred for low to medium power requirements due to their desirable features like large power to weight ratio, low cost and safe working medium. Position control of pneumatic actuators is however difficult as compared to hydraulic and electromechanical systems due to nonlinearity-contributing factors like air compressibility, airflow through orifice and friction etc. A multitude of control techniques have been reported in literature for minimizing the positional error in pneumatic servo systems. Such techniques include PID control, modified PID control, fuzzy logic based control and sliding mode control etc.

2 Literature Review

Nguyen et al. [1] proposed a sliding mode approach for positioning a 2 kg mass using a double-acting pneumatic motion control system. The tuning of the controller parameter was done experimentally. The results for 20 mm amplitude and 0.5 Hz frequency sinusoidal and square wave trajectories, exhibited a tracking error of ±2 mm range. Bone and Ning [2] developed two sliding mode controllers, one based on linearized plant model, labeled Sliding Mode Control Linear (SMCL) and the other based on nonlinear plant model, labeled Sliding Mode Control Nonlinear (SMCN), for positioning 1.9 kg, 5.8 kg and 10.8 kg payloads. The controller parameters were tuned manually for the nominal load of 5.8 kg, in both cases. The results showed that the SMCL exhibited superior control of 10.8 kg payload and inferior control of 1.9 kg payload, whereas the opposite was true for SMCN. Taghizadeh et al. [3] designed and implemented a multimodel controller for position control of a variably loaded pneumatic actuator. A multimodel controller was designed for a number of constant loads and corresponding to each load, a linear model was identified and a PD-controller was tuned for the system. A switching algorithm was applied which determined the best model and selected the corresponding controller in any load condition. Experimental results indicated good performances of the multimodel controller under variable load conditions. Taleb et al. [4] presented a new second-order sliding-mode algorithm for position control of a pneumatic actuator. Due to the dynamic adaptation of the gains, the controller design did not require complete information on the bounds of uncertainties and perturbations. It automatically decreased the gains and oscillations due to a too large discontinuous-control magnitude. Salim et al. [5] proposed a self-regulating nonlinear PID controller for a pneumatic positioning system for a variably loaded pneumatic actuator. Simulation and experimental tests conducted with different types of inputs showed improved transient and steady state response vis-à-vis previously reported nonlinear PID controllers.

3 Objective

The objective of present work is to synthesize a hybrid controller, using fuzzy logic control and sliding mode control, for position control of a pneumatic actuation system shown in Figure 1. The nomenclature and values of system parameters are given in the Appendix.

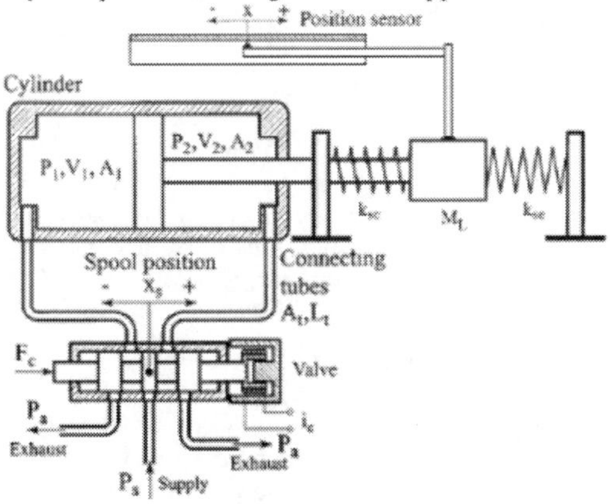

Figure 1. Schematics of pneumatic servo system [6]

The variable loading conditions, in terms of combinations of piston payload and desired piston displacement, considered in the present work are shown in Table 1.

Table 1: Payload (M_L) and desired piston displacement (x_d) combinations

0.1 kg, ±30 mm	0.4 kg, ±30 mm	0.7 kg, ±30 mm	1.0 kg, ±30 mm
0.1 kg, ±20 mm	0.4 kg, ±20 mm	0.7 kg, ±20 mm	1.0 kg, ±20 mm
0.1 kg, ±10 mm	0.4 kg, ±10 mm	0.7 kg, ±10 mm	1.0 kg, ±10 mm

4 Methodology

MATLAB and Simulink application software were used for modeling and simulating the pneumatic actuation system as well as for designing fuzzy logic and sliding mode controllers for the system. Modeling and simulation of the pneumatic actuation system was done using its mathematical and virtual models developed by Dhami et al. [7]

4.1 Modeling, Simulation and Control of Pneumatic Servo System

The mathematical model of the pneumatic actuation system consists of dynamics of control valve, air flow dynamics through control valve ports and connecting tubes, pressure dynamics in cylinder chambers and dynamics of piston and load. Virtual model of the pneumatic actuation system was developed in MATLAB/Simulink and used for simulating the system.

For designing Fuzzy Logic Controller for the system, the positional error, rate of change of error and the applied external load were taken as input variables and the control signal as the output variable. Trapezoidal membership functions were used for defining the fuzzy sets of input parameters and linear membership functions for output variable. A Sugeno type fuzzy logic controller was designed using the input and output data of dynamic response of pneumatic actuation system obtained by simulations.

The surface of sliding mode controller for pneumatic servo system is given by

$$S = (\ddot{x} - \ddot{x}_d) + 2\lambda(\dot{x} - \dot{x}_d) + \lambda^2(x - x_d) \qquad (1)$$

where λ is a strictly positive constant. The dynamics in sliding mode are given by

$$\dot{S} = 0 \qquad (2)$$

Solving equation (2), the equivalent control signal 'u_{eq}' was obtained and combined with a saturation term for satisfying sliding mode condition. The resulting control action 'u' was given by:

$$u = u_{eq} - k\, sat\left(\frac{S}{\phi}\right) \qquad (3)$$

where 'sat' is the saturation function defined as

$$\left.\begin{aligned} sat(y) &= -1, if\ y < -1 \\ sat(y) &= y, if\ -1 \le y \le 1 \\ sat(y) &= +1, if\ y > 1 \end{aligned}\right\} \qquad (4)$$

The controller parameters - λ, k and ϕ were manually tuned. It was found that satisfactory positional response was not possible for different combinations of payloads and linear displacements, for a single set of controller parameter values. The most dominant parameter 'λ' was varied and tuned for different operative conditions. The tuned values of k and ϕ were found to be 10 and 100 respectively. Values of λ tuned for different operative conditions are given in Table 2.

Table 2. Values of λ for different payload and piston displacement combinations

load → / displacement ↓	0.1 kg	0.4 kg	0.7 kg	1.0 kg
- 30 mm	220	210	200	170
- 20 mm	220	180	170	150
- 10 mm	210	170	150	120
+ 10 mm	210	110	100	90
+ 20 mm	210	160	130	110
+ 30 mm	200	180	170	150

4.2 Performance evaluation of fuzzy logic and sliding mode controllers

The positional response of the pneumatic actuation system was simulated for different operative conditions using FLC and SMC independently. The transient and steady state positional responses of pneumatic actuation system were quantified in terms of rise time, percent overshoot, settling time, steady state error and integral of absolute error. The first four parameters were computed using Control System Toolbox of MATLAB. Integral of Absolute Error (IAE) is a quantitative measure of the performance of dynamic systems and it is particularly useful for computer simulation studies [8]. IAE is mathematically defined as

$$IAE = \int_0^T |e(t)|\, dt \qquad (5)$$

where e(t) is the instantaneous error
T is the simulation time duration

The IAE performance index gives a high value for both excessively overdamped and underdamped systems. A smaller value of IAE signifies improved transient response.

Table 3 shows the comparative values of transient and steady state parameters obtained using fuzzy logic control (FLC) and sliding mode control (SMC) independently. The units of Payload, Reference Position, Rise Time, Overshoot, Settling Time, Steady State Error and IAE are kg, mm, second, percentage, second, mm and mm respectively. Further, the settling time was computed using ±5% criterion. For steady state error more than ±5%, the settling time is shown as NaN.

The following inferences were drawn from positional response and data in Table 2:

- Rise time was smaller for sliding mode control for all payload-displacement conditions.
- Overshoot was lower for sliding mode control for payloads of 0.4 kg, 0.7 kg and 1.0 kg but it was higher for payload of 0.1 kg.
- Settling time was less for sliding mode control for 1.0 kg payload and ±30 mm piston displacement, but it was higher for most of other payload-displacement conditions.
- Steady state error was less for sliding mode control in case of higher payload and higher piston displacement combinations.
- IAE was remarkably lower for sliding mode control in case of combinations of 0.7 kg and 1.0 kg payloads and ±20 mm and ±30 mm piston displacements.
- It was observed from positional response that Sliding Mode Control resulted in excessive piston dithering at steady state for lower payloads (0.1 kg and 0.4 kg) and all reference positions. For higher payloads and smaller reference positions, marginal dithering of piston was observed at steady state.

4.3 Synthesis of Hybrid Controller

Based on the quantitative analysis of transient and steady state piston positional response, a hybrid-control matrix was formulated for position control of pneumatic actuation system, Table 4. The matrix represents hybrid control approach as it employs both Fuzzy Control and Sliding Mode Control approaches for relevant combinations of piston payload and piston position.

Simulink model of the synthesized hybrid model is shown in Figure 2. The exploded view of the hybrid controller is shown in Figure 3, in which the 'Control Signal Selection Criteria' block incorporates Table 3.

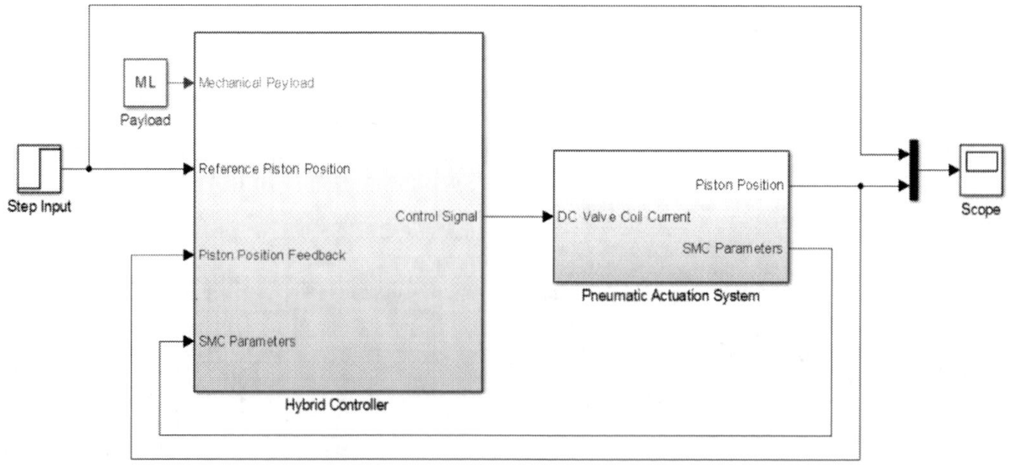

Figure 2. Simulink model of synthesized hybrid controller

5 Results and discussions

The positional response of the pneumatic actuation system using hybrid controller for different payload-displacement combinations is presented in Figure 4. Hybrid controller selects sliding mode control for payload-displacement combinations of (0.7 kg, ±30 mm), (1.0 kg, ±20 mm) and (1.0 kg, ±30 mm), whereas for other combinations fuzzy control was employed.

For 0.1 kg payload, fuzzy control results in very good transient as well as steady state response. For 0.4 kg payload, fuzzy control results in slightly increased transient oscillations, but both transient and steady state response is satisfactory as indicated by IAE, settling time, steady state error. For payload-displacement combinations of (0.7 kg, ±20 mm), (0.7 kg, ±10 mm) and (1.0 kg, ±10 mm), fuzzy control results in high overshoot, but it is preferred as settling time and steady state error are less than those in case of sliding mode control. For the remaining combinations of payload and piston displacement, sliding mode control resulted in piston dithering, but it was quite small and the transient as well as steady state responses were very good as indicated by performance parameters for these combinations.

The steady state positional error and integral of absolute error obtained with hybrid control of pneumatic actuation system under different operative conditions are shown in Figure 5. The maximum steady state error was 1.66 mm (5.5%) for a desired piston position of 30 mm with 0.1 kg payload. The average steady state error for lower payloads (0.1 kg and 0.4 kg) is almost same and higher than that obtained for other two payloads. The maximum IAE was 7.25 for a desired piston position of 30 mm with 0.4 kg payload. The average IAE of 6.74 for payloads of 0.1 kg, 0.4 kg and 0.7 kg was more than double for that obtained for 1.0 kg payload. Higher IAE for payloads of 0.1 kg and 0.4 kg was due to higher steady state positional error at these payloads for all reference positions. Higher IAE for 0.7 kg payload was due to increased transient oscillations for ±20 mm reference piston positions.

Table 3. Transient and Steady State Performance Indicators

Pay Load	Ref. Pos.	Rise Time		Overshoot		Settling Time		Steady State Error		Integral of Absolute Error	
		FLC	SMC	FLC	SMC	FLC	SMC	FLC	SMC	FLC	SMC
0.1	-30	0.03	0.02	0	9.03	NaN	2	1.62	-1.26	5.65	3.68
	-20	0.06	0.02	6.18	12.44	0.22	NaN	-0.54	-1.99	4.28	4.55
	-10	0.02	0.01	57.93	23.48	0.26	NaN	0.38	-2.91	1.69	6.46
	+10	0.02	0.01	54.61	58.73	0.21	NaN	-0.15	-0.85	1.32	4.32
	+20	0.06	0.01	6.68	22.21	0.24	2	0.58	-1.34	4.74	3.44
	+30	0.09	0.01	0	14.75	NaN	NaN	1.66	-2.41	5.82	5.11
0.4	-30	0.04	0.02	8.33	2.8	NaN	1.68	-1.56	-0.31	7.24	1.89
	-20	0.06	0.02	29.56	3.36	0.67	1.99	0.84	-0.9	6.21	2.41
	-10	0.03	0.03	77.21	4.42	NaN	NaN	0.76	-1.14	3.08	2.54
	+10	0.02	0.02	77.03	20.27	NaN	NaN	-0.78	-0.86	3.16	1.86
	+20	0.06	0.01	24.64	8.84	0.55	NaN	-0.27	-0.6	5.68	1.63
	+30	0.04	0.01	8.85	11.44	0.69	0.6	1.3	-0.76	7.25	2.07
0.7	-30	0.04	0.02	19.76	1.83	0.95	0.06	-1.41	-0.33	7.78	1.69
	-20	0.06	0.02	39.97	2.83	0.92	NaN	0.75	-0.66	7.1	1.94
	-10	0.03	0.02	64.29	0	0.53	NaN	-0.16	-0.8	2.75	1.86
	+10	0.03	0.04	61.53	13.41	0.53	NaN	0.37	-0.39	3.1	1.09
	+20	0.06	0.02	41.06	5.59	0.91	1.79	-0.5	-0.53	7.16	1.41
	+30	0.04	0.04	19.78	19.78	0.96	0.96	1.13	-0.38	7.58	1.56
1.0	-30	0.06	0.03	12.47	0.61	1.23	0.06	-1.35	-0.41	9.66	1.9
	-20	0.1	0.03	35.84	0.93	1.39	1.95	0.82	-0.56	10.31	1.78
	-10	0.04	0.02	69.87	0.03	0.61	NaN	-0.06	-0.58	3.1	1.43
	+10	0.04	0.06	74.81	5.89	0.7	1.99	-0.16	-0.3	3.31	0.9
	+20	0.1	0.02	36.03	4.83	1.04	0.04	-0.65	-0.37	10.23	1.23
	+30	0.06	0.02	14.58	19.34	1.13	0.06	0.94	-0.35	9.4	1.59

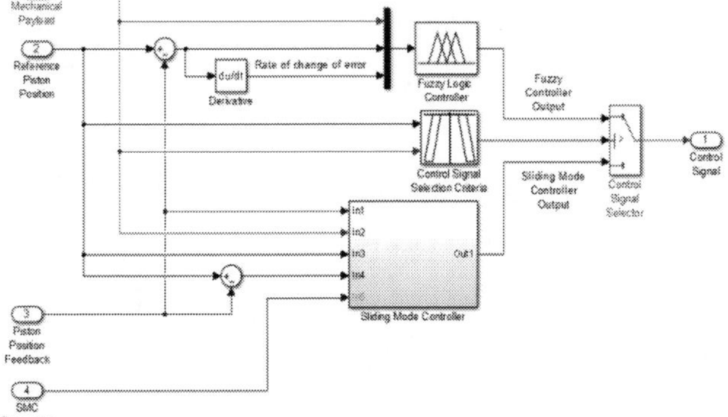

Figure 3. Exploded view of hybrid controller

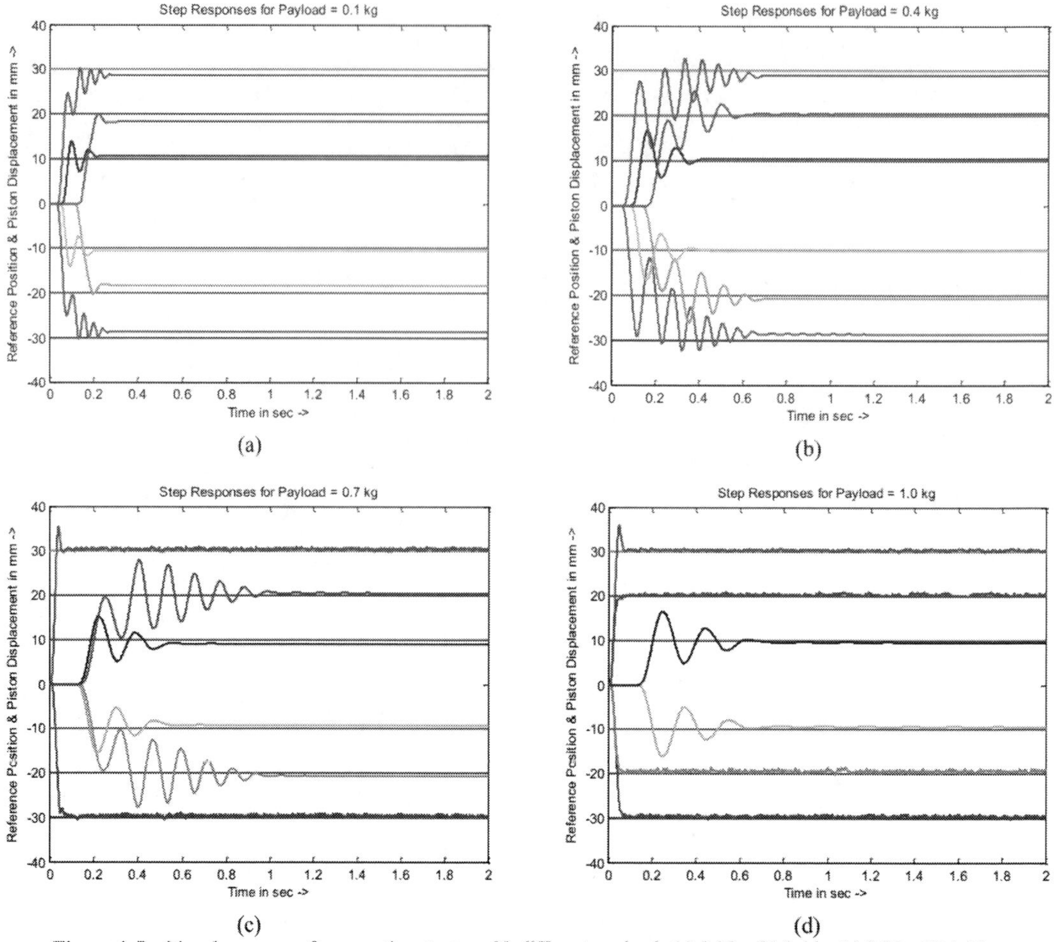

Figure 4. Positional response of pneumatic actuator with different payloads (a) 0.1 kg (b) 0.4 kg (c) 0.7 kg (d) 1.0 kg

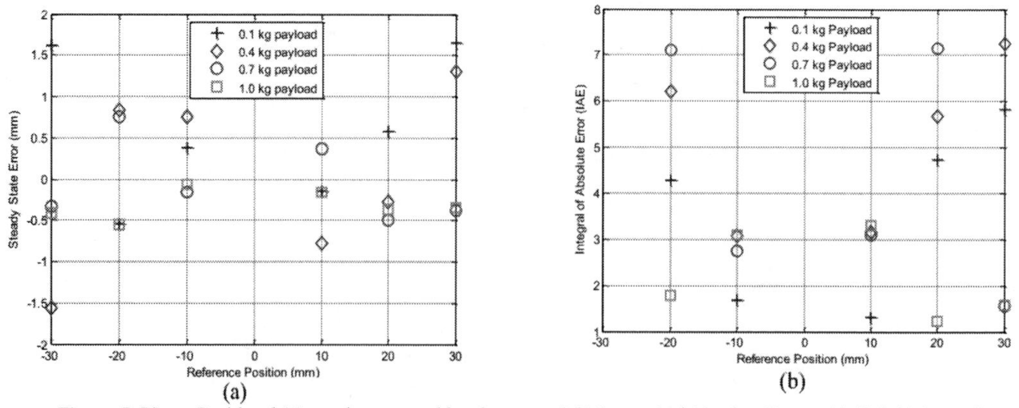

Figure 5. Piston Positional (a) steady state positional error and (b) Integral of Absolute Error with Hybrid Controller

Table 4. Control matrix for payload and piston displacement combinations

Ref. Pos. (mm)↓ / Pay Load (kg)→	0.1	0.4	0.7	1.0
-30	FLC	FLC	SMC	SMC
-20	FLC	FLC	FLC	SMC
-10	FLC	FLC	FLC	FLC
+10	FLC	FLC	FLC	FLC
+20	FLC	FLC	FLC	SMC
+30	FLC	FLC	SMC	SMC

DOI: 10.1051/matecconf/20179516001

6 Conclusions

A hybrid controller was synthesized using fuzzy logic control and sliding mode control for position control of a pneumatic actuation system. Simulation was employed for obtaining the transient and steady state response of the pneumatic actuator for various combinations of applied payloads and piston displacements. The quantitative results emphasized that hybrid FLC-SMC controller exhibited effective position control for pneumatic actuation system as fuzzy logic control was effective for combinations of lower payloads and smaller piston displacements, whereas sliding mode controller was effective for combinations of higher payloads and larger piston displacements.

References

1. T. Nguyen, J. Leavitt, F. Jabbari, J. E. Bobrow, IEEE/ASME Transactions on Mechatronics, 12 (2), pp. 216-219, (2007)
2. G. M. Bone, S. Ning, IEEE/ASME Transactions on Mechatronics, 12 (5), pp. 557-561, (2007)
3. M. Taghizadeh, F. Najafi, A. Ghaffari, International Journal of Advanced Manufacturing Technology, 48(5), pp. 655 – 662, (2010)
4. M. Taleb, A. Levant, F. Plestan, Control Engineering Practice, 21(5), pp. 727 – 736, (2013)
5. S. N. S. Salim, M. F. Rahmat, A. A. M. Faudzi, Z. H. Ismail, N. Sunar, Mathematical Problems in Engineering, http://dx.doi.org/10.1155/2014/957041, (2014)
6. Edmond Richer and Yildirim Hurmuzlu, ASME J. Dynamic Systems, Measurement and Control, **122**, pp. 416-425, (2000)
7. S. S. Dhami, S. S. Bhasin, P. B. Mahapatra, ASME International Mechanical Engineering Congress and Exposition, November 13-19, Lake Buena Vista, Florida, USA, (2009)
8. R. C. Dorf, R. H. Bishop, *Modern Control Systems*, Pearson Education, Delhi, India, (2005)

Appendix

A_1: Piston area on cap end side (2.85 x 10^{-4} m2)
A_2: Piston area on rod end side (2.53 x 10^{-4} m2)
A_r: Piston rod area (3.17 x 10^{-5} m2)
C_f: Valve orifice discharge coefficient (0.25)
c_s: Viscous friction coefficient of control valve (7.5 N-s/m)
D: Connecting tube diameter (0.0032 m)
F_{df}: Kinetic friction force for cylinder (0.486 N)
F_{sf}: Static friction force for cylinder (3.8 N)
K_{fc}: Valve coil force coefficient (2.78 N/m)
k_s: Spring stiffness coefficient of control valve (975 N/m)
k_{se}: Spring stiffness coefficient of load (205 N/m)
L: Total cylinder stroke (0.0762 m)
L_t: Length of connecting tube (0.5 m)
M_p: Piston and rod combined mass (0.055 kg)
n: Polytropic constant (1.3)
P_a: Atmospheric pressure (101.3 x 10^3 Pascal (absolute))
P_s: Supply pressure (506.5 x 10^3 Pascal (absolute))
T_a: Air temperature (293 K)
V_{01}: Dead volume of cap end side cylinder-chamber (2.17 x 10^{-6} m^2)
V_{02}: Dead volume of rod end side cylinder-chamber (1.93 x 10^{-6} m^2)
β: Viscous friction coefficient for cylinder (4.47 N-s/m)
x_d : Desired piston position
x: piston displacement measured from the mid-stroke

Analytical Formulations in the Domain Decomposition Method using Cylindrical Control Surface

Bo Tian , Hui-long Ren

College of ShipBuliding Engineering, Harbin Engineering University, Harbin, Heilongjiang, China

Abstract. The seakeeping problem is very important for marine structures, but present boundary element methods have still unresolved difficulties. In this work, the Green function method (Green function satisfying the linear boundary condition at free surface) is combined with the Rankine panel method so that two methods keep their advantages and brings important benefits. The fluid domain is divided into two subdomains by a control surface of cylindrical form. In the interior one the ship is arbitrary, the Rankine source method is then applied. In the exterior one the shape of the control surface is analytically known while the velocity potential and its normal derivative are expressed by expansions of orthogonal functions. Across the control surface two additional conditions must be satisfied: both the velocity potential and its normal derivative are to be continuous. In order to improve the computational accuracy, velocity potential is decomposed by using the Fourier series in polar direction and Laguerre function in vertical direction on the cylindrical control surface. The matrix elements of boundary integral equations consist of analytical integrations in wavenumber. Furthermore, the numerical results of the integrals are presented for seakeeping problems.

1 Introduction

The Rankine panel method is one of the well-known methods used for sea-keeping problems. It uses the simple Green function ($1/r$) for unbounded fluid, where r is the distance between the field point and the source point. On the other hand, to satisfy the free surface and body boundary conditions, the integrations must be performed over all surfaces. Thus, a large amount of panels are necessary due to panelizing the free surface. In addition, to avoid the reflected wave from the sides of a numerical fluid domain a damping zone has to be introduced [1], [2].

As it is known, the implementation of the Kelvin source method using the Green function for a moving and oscillating source arises to several well-known problems concerning its wave component. In particular, it behaves with singularities and high oscillations when both the field and source points tend to the free surface. Also, it is the slow convergent integral defined along the dispersion curves. In every case, for each body shape this wave component should be evaluated very carefully [1], [2].

In this paper, we modify the Green function method (Green function satisfies the free surface boundary condition) by combining it with the Rankine panel method. In this work, we suggest to divide the fluid domain into two subdomains by a control infinitely-long cylindrical surface. This surface separates the problem into two problems: the interior one in which the ship is of any form, the Green function is Rankine source Green function; the exterior one in which the shape of the control surface is known and velocity potential is assumed to be known. Across the control surface two additional conditions must be satisfied: both the velocity potential and its normal derivative are to be continuous. The second problem provides us the Dirichlet-Neumann map which is used to solve the first problem, and as result the original one by Rankine panel method.

The combination of these two methods keeps their advantages and brings important benefits: area to be discretized becomes smaller; no need to introduce the damping zone, the solution of the problem is that for unbound fluid domain. The calculation of the Green function for ship motion may be done only once for large set of the different ships for the one particular velocity U, incoming wave heading and frequency, and an additional parameter which describes the control surface.

2 Problem statement

A ship or a body of general shape floats in unbounded fluid. The fluid is assumed inviscid, incompressible and the flow is irrotational. The motions of the ship are assumed to be small so that the linear theory can be applied. The coordinate system is situated above the center of the mass of the ship on the level of the undisturbed free surface $(z=0)$, see Fig. 1. The problem is described by the following boundary value problems for the velocity potential $\phi_j(x,y,z)$ [1].

$$\Delta\phi_j(x,y,z)=0, \qquad (x,y,z)\in\Omega \tag{1}$$

$$\frac{\partial}{\partial n}\phi_j(x,y,z)=n_j(x,y,z), \qquad (x,y,z)\in B \tag{2}$$

$$\frac{\partial}{\partial z}\phi_j(x,y,z)-\frac{\omega^2}{g}\phi_j(x,y,z)=0, \quad (x,y,z)\in F \tag{3}$$

where Ω is the fluid domain, B is the submerged part of the body hull, F is the free surface, $\vec{n}=(n_1,n_2,n_3)$ is the normal vector on the body directed toward the fluid, $(n_4,n_5,n_6)=\vec{r}\times\vec{n}$, $\vec{r}$ is a vector from a reference point to the point (x,y,z) on the body, g is acceleration due to the gravity, ω is an incoming wave frequency.

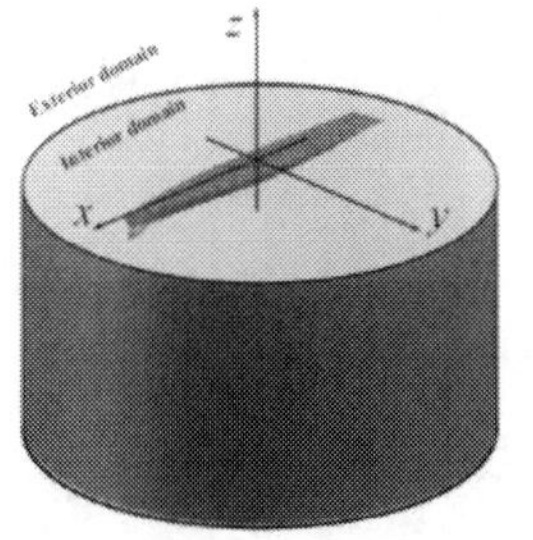

Figure 1. The coordinate system and the domain decomposition.

As it is shown in Fig.1, the cylindrical control surface C divides the fluid domain into two sub-domains: the exterior domain Ω_e and the interior domain Ω_i. The solution in the interior domain is solved using the constant Rankine panel method while in the exterior one Kelvin source method is applied. To couple both solutions two additional conditions should be used:

$$\phi_j^e(x,y,z)=\phi_j^i(x,y,z),\ \frac{\partial}{\partial n}\phi_j^e(x,y,z)=\frac{\partial}{\partial n}\phi_j^i(x,y,z),$$
$$(x,y,z)\in C \tag{4}$$

which are the conditions of continuous of the velocity potential and its normal derivative across the control surface.

3 Mathematical formulations

3.1 External Problem

The cylindrical control surface is presented by distribution of singularities for the Green function is well known and is given as the following[1]:

$$G(P,Q)=-\frac{1}{4\pi r(P,Q)}-\frac{1}{4\pi r^*(P,Q)}-\frac{k_0}{2\pi}\int_0^\infty\frac{e^{-k|z+\zeta|}}{k-k_0}J_0(kR)dk \tag{5}$$

where $k_0=\omega^2/g$, g is acceleration due to the gravity, ω is an incoming wave frequency. $P(x,y,z)$ is a field point, $Q(\xi,\eta,\zeta)$ is a source point, J_0 is the Bessel function, $r(P,Q)$ and $r^*(P,Q)$ are the distances between the source and the field points and between the filed point and mirror of the source point with respect to the mean free surface $z=0$, $R^2=(x-\xi)^2+(y-\eta)^2$. This Green function satisfies the free surface condition.

Since the shape of control surface in this paper, it is convenient to introduce the cylindrical coordinate system in the following way and make the z-axis downwards in Fig. 2

$$x=\rho\cos\psi,\quad y=\rho\sin\psi,\quad z=z \tag{6}$$

here ρ is a radius of the cylindrical surface which is given, $\psi\in[-\pi,\pi]$ varies in the plane xOy.

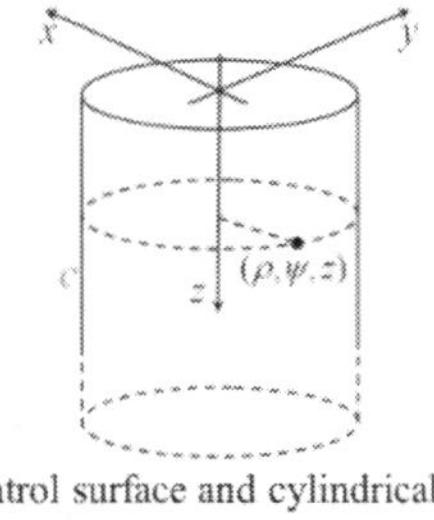

Figure 2. The control surface and cylindrical coordinate system.

Application of the Green's second identity provides

$$\iint_C[\phi_n(Q)G(P,Q)-\phi(Q)G_n(P,Q)]ds=\begin{cases}0, & P\notin\Omega_e\\ \phi(P)/2, & P\in\partial\Omega_e\\ \phi(P), & P\in\Omega_e\setminus\partial\Omega_e\end{cases} \tag{7}$$

here n is the normal vector toward the fluid.

Any function of variables ψ and z can be presented in the form of infinite series using complete sets of base functions for each variable. For the convenience we chose harmonic functions $e^{il\psi}$ and Laguerre functions $\ell_k(z)$ [3]:

$$f(x,y,z)=\sum_{l=-\infty}^{\infty}\sum_{k=0}^{\infty}A_{kl}\ell_k(z)e^{il\psi} \tag{8}$$

where

$$\ell_k(z)=e^{-\frac{z}{2}}L_k(z),\ \int_0^\infty\ell_n(z)\ell_k(z)dz=\int_0^\infty L_n(z)L_k(z)e^{-z}dz=\delta_{kn} \tag{9}$$

here $L_k(z)$ is the Laguerre polynomials, δ_{kn} is the Kronecker delta. The coefficients A_{kl} are found as[3]

$$A_{kl}=\frac{1}{2\pi}\int_0^\infty\int_{-\pi}^{\pi}f(x,y,z)\ell_k(z)e^{-il\psi}d\psi dz,\ (k,l)\in\mathbb{N}\times\mathbb{Z} \tag{10}$$

Thus, according to (8)-(10), the velocity potential ϕ and its normal derivative $\varphi\equiv\phi_n$ can be rewritten in the form

$$\phi(x,y,z)=\sum_{l=-\infty}^{\infty}\sum_{k=0}^{\infty}\phi_{kl}\ell_k(z)e^{il\psi},\quad \varphi(x,y,z)=\sum_{l=-\infty}^{\infty}\sum_{k=0}^{\infty}\varphi_{kl}\ell_k(z)e^{il\psi} \tag{11}$$

where the coefficients φ_{kl} are unknown and ϕ_{kl} are assumed to be know.

After substituting (11) into (7), we obtain

$$\frac{1}{2}\sum_{l=-\infty}^{\infty}\sum_{k=0}^{\infty}\phi_{kl}\ell_k(z)e^{il\psi}=\iint_C G(P,Q)\left[\sum_{l=-\infty}^{\infty}\sum_{k=0}^{\infty}\varphi_{kl}\ell_k(z')e^{il\psi'}\right]ds$$
$$-\iint_C G_n(P,Q)\left[\sum_{l=-\infty}^{\infty}\sum_{k=0}^{\infty}\phi_{kl}\ell_k(z')e^{il\psi'}\right]ds \tag{12}$$

Here, we assumed that both points are on the same surface C and we denoted $(\psi,z)=(x,y,z)\in C$ and $(\psi',z')=(\xi,\eta,\zeta)\in C$. Multiplying the both sides of (12) by $\frac{1}{2\pi}\ell_n(z)e^{-im\psi}$, then integrating the obtained product with respect to $(\psi,z)\in C[-\pi,\pi]\times[0,+\infty)$ yield

$$\frac{1}{2}\phi_{nm}=\frac{1}{2\pi}\sum_{l=-\infty}^{\infty}\sum_{k=0}^{\infty}\varphi_{kl}\int_0^{\infty}\int_{-\pi}^{\pi}\iint_C G(P,Q)\ell_k(z')e^{il\psi'}\ell_n(z)e^{-im\psi}dsd\psi dz$$
$$-\frac{1}{2\pi}\sum_{l=-\infty}^{\infty}\sum_{k=0}^{\infty}\phi_{kl}\int_0^{\infty}\int_{-\pi}^{\pi}\iint_C G_n(P,Q)\ell_k(z')e^{il\psi'}\ell_n(z)e^{-im\psi}dsd\psi dz \quad (13)$$

As it is shown in Fig.3, the integral infinitesimal of the control surface

$$ds=dldz=\rho d\psi dz \quad (14)$$

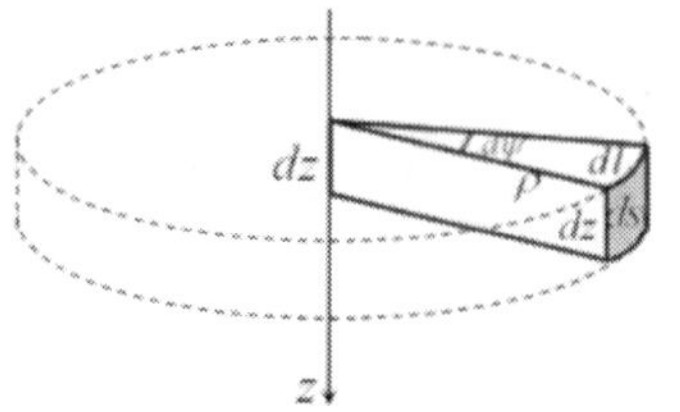

Figure 3. The integral infinitesimal.

hence,

$$\iint_C f(r,\psi,z)ds=\iint_{r=\rho}f(\rho,\psi,z)ds=\int_0^{+\infty}\int_{-\pi}^{\pi}f(\psi,z)\rho d\psi dz \quad (15)$$

then, (13) is

$$\frac{1}{2}\phi_{nm}=\frac{\rho}{2\pi}\sum_{l=-\infty}^{\infty}\sum_{k=0}^{\infty}\varphi_{kl}\int_0^{\infty}\int_{-\pi}^{\pi}\int_0^{\infty}\int_{-\pi}^{\pi}G(P,Q)\ell_k(z')e^{il\psi'}\ell_n(z)e^{-im\psi}d\psi'dz'd\psi dz$$
$$-\frac{\rho}{2\pi}\sum_{l=-\infty}^{\infty}\sum_{k=0}^{\infty}\phi_{kl}\int_0^{\infty}\int_{-\pi}^{\pi}\int_0^{\infty}\int_{-\pi}^{\pi}G_n(P,Q)\ell_k(z')e^{il\psi'}\ell_n(z)e^{-im\psi}d\psi'dz'd\psi dz \quad (16)$$

3.2 Internal Problem

Application of the Green's second identity provides

$$\iint_{B+F+C}[\phi_n(Q)G(P,Q)-\phi(Q)G_n(P,Q)]ds=c(P)\phi(P),\ P\in B\cup F\cup C \quad (17)$$

where

$$c(P)=\begin{cases}1/2, & P\in B\cup F\\ -1/2, & P\in C\end{cases} \quad (18)$$

and n is a normal vector directed inward the fluid domain on the body B and free surface F and outward (that is why $c(P)=-1/2$) on the control surface C,

$$G(P,Q)=-\frac{1}{4\pi r(P,Q)} \quad (19)$$

is the Rankine Green function.

The both body hull and free surface are divided by and sub-surfaces (panels), respectively. Thus, (17) becomes

$$\left(\sum_{\alpha=1}^{N_B}\iint_{B^\alpha}+\sum_{\beta=1}^{N_F}\iint_{F^\beta}+\iint_C\right)[\phi_n(Q)G(P,Q)-\phi(Q)G_n(P,Q)]ds=c(P)\phi(P) \quad (20)$$

According to the boundary conditions, the normal derivative on the body is known function and on the free surface, due to the free surface boundary condition (3), it is

$$\frac{\partial\phi}{\partial n}=-\frac{\partial\phi}{\partial z}=-k\phi, \qquad z=0, \quad (21)$$

Hence, (20) can be rewritten as the following

$$c(P)\phi(P)=\iint_C[\phi_n(Q)G(P,Q)-\phi(Q)G_n(P,Q)]ds$$
$$+\sum_{\alpha=1}^{N_B}\iint_{B^\alpha}\phi_n(Q)G(P,Q)ds-\sum_{\alpha=1}^{N_B}\iint_{B^\alpha}\phi(Q)G_n(P,Q)ds \quad (22)$$
$$-\sum_{\beta=1}^{N_F}\iint_{F^\beta}\phi(Q)[kG(P,Q)+G_n(P,Q)]ds$$

On each panel the velocity potential and its normal derivative are assumed to be a constant and we may write

$$c(P)\phi(P)=\iint_C[\phi_n(Q)G(P,Q)-\phi(Q)G_n(P,Q)]ds$$
$$+\sum_{\alpha=1}^{N_B}\varphi_\alpha\iint_{B^\alpha}G(P,Q)ds-\sum_{\alpha=1}^{N_B}\phi_\alpha\iint_{B^\alpha}G_n(P,Q)ds \quad (23)$$
$$-\sum_{\beta=1}^{N_F}\phi_\beta\iint_{F^\beta}[kG(P,Q)+G_n(P,Q)]ds$$

On the surface C the velocity potential and its normal derivative are defined as:

$$\phi(\xi,\eta,\zeta)=\sum_{l=-\infty}^{\infty}\sum_{k=0}^{\infty}\phi_{kl}\ell_k(z')e^{il\psi'},\quad \varphi(\xi,\eta,\zeta)=\sum_{l=-\infty}^{\infty}\sum_{k=0}^{\infty}\varphi_{kl}\ell_k(z')e^{il\psi'} \quad (24)$$

The substitution (24) into (23) yields

$$c(P)\phi(P)=\sum_{l=-\infty}^{\infty}\sum_{k=0}^{\infty}\varphi_{kl}\iint_C\ell_k(z')e^{il\psi'}G(P,Q)ds$$
$$-\sum_{l=-\infty}^{\infty}\sum_{k=0}^{\infty}\phi_{kl}\iint_C\ell_k(z')e^{il\psi'}G_n(P,Q)ds$$
$$+\sum_{\alpha=1}^{N_B}\varphi_\alpha\iint_{B^\alpha}G(P,Q)ds-\sum_{\alpha=1}^{N_B}\phi_\alpha\iint_{B^\alpha}G_n(P,Q)ds \quad (25)$$
$$-\sum_{\beta=1}^{N_F}\phi_\beta\iint_{F^\beta}[kG(P,Q)+G_n(P,Q)]ds$$

Depending on the position of the point P, different cases should be considered:

When $P\in C$, According to (11), we can get

$$-\frac{1}{2}\phi_{nm}=\sum_{l=-\infty}^{\infty}\sum_{k=0}^{\infty}\varphi_{kl}\frac{1}{2\pi}\int_0^{\infty}\int_{-\pi}^{\pi}\iint_C\ell_k(z')e^{il\psi'}G(P,Q)\ell_n(z)e^{-im\psi}dsd\psi dz$$
$$-\sum_{l=-\infty}^{\infty}\sum_{k=0}^{\infty}\phi_{kl}\frac{1}{2\pi}\int_0^{\infty}\int_{-\pi}^{\pi}\iint_C\ell_k(z')e^{il\psi'}G_n(P,Q)\ell_n(z)e^{-im\psi}dsd\psi dz$$
$$+\sum_{\alpha=1}^{N_B}\varphi_\alpha\frac{1}{2\pi}\int_0^{\infty}\int_{-\pi}^{\pi}\iint_{B^\alpha}G(P,Q)\ell_n(z)e^{-im\psi}dsd\psi dz \quad (26)$$
$$-\sum_{\alpha=1}^{N_B}\phi_\alpha\frac{1}{2\pi}\int_0^{\infty}\int_{-\pi}^{\pi}\iint_{B^\alpha}G_n(P,Q)\ell_n(z)e^{-im\psi}dsd\psi dz$$
$$-\sum_{\beta=1}^{N_F}\phi_\beta\frac{1}{2\pi}\int_0^{\infty}\int_{-\pi}^{\pi}\iint_{F^\beta}[kG(P,Q)+G_n(P,Q)]\ell_n(z)e^{-im\psi}dsd\psi dz$$

When $P_\gamma=P\in B^\gamma$ or $P_\gamma=P\in F^\gamma$,where γ is the number of the panel, then (25) becomes

$$c(P_\gamma)\phi_\gamma=\sum_{l=-\infty}^{\infty}\sum_{k=0}^{\infty}\varphi_{kl}\iint_C\ell_k(z')e^{il\psi'}G(P_\gamma,Q)ds$$
$$-\sum_{l=-\infty}^{\infty}\sum_{k=0}^{\infty}\phi_{kl}\iint_C\ell_k(z')e^{il\psi'}G_n(P_\gamma,Q)ds$$
$$+\sum_{\alpha=1}^{N_B}\varphi_\alpha\iint_{B^\alpha}G(P_\gamma,Q)ds-\sum_{\alpha=1}^{N_B}\phi_\alpha\iint_{B^\alpha}G_n(P_\gamma,Q)ds \quad (27)$$
$$-\sum_{\beta=1}^{N_F}\phi_\beta\iint_{F^\beta}[kG(P_\gamma,Q)+G_n(P_\gamma,Q)]ds$$

(26) and (27) from the system of linear algebraic equations, which in the matrix form can be written as

DOI: 10.1051/matecconf/20179516002

$$\mathbf{M\Phi}=\mathbf{F}, \quad \mathbf{\Phi}=\left(\phi_\alpha,\phi_\beta,\phi_{kl}\right)^T \quad \mathbf{F}=\left(F_1,F_2,F_3\right)^T \tag{28}$$

where vector $\mathbf{\Phi}$ is unknown. The vector $\mathbf{F}$ is

$$F_1=\left[\left\{\sum_{\alpha=1}^{N_B}\varphi_\alpha\iint_{B^\alpha}G(P_\gamma,Q)ds\right\}\right] \qquad (P_\gamma\in B^\gamma)$$

$$F_2=\left[\left\{\sum_{\alpha=1}^{N_B}\varphi_\alpha\iint_{B^\alpha}G(P_\gamma,Q)ds\right\}\right] \qquad (P_\gamma\in F^\gamma)$$

$$F_3=\left[\left\{\sum_{\alpha=1}^{N_B}\varphi_\alpha A_{B,nm,\alpha}\right\}\right] \qquad (P\in C)$$

The structure of the matrix $\mathbf{M}$ is showed as follow:

$$\mathbf{M}=\begin{bmatrix}M_{11} & M_{12} & M_{13}\\ M_{21} & M_{22} & M_{23}\\ M_{31} & M_{32} & M_{33}\end{bmatrix}=\begin{bmatrix}B\leftrightarrow B & B\leftrightarrow F & B\leftrightarrow C\\ F\leftrightarrow B & F\leftrightarrow F & F\leftrightarrow C\\ C\leftrightarrow B & C\leftrightarrow F & C\leftrightarrow C\end{bmatrix} \tag{29}$$

4 Numerical results

In order to calculate the see-keeping of a sphere as Fig. 4 showed, a hemisphere is calculated as Fig. 5 showed. Parameters of the calculation are given in Table 1. In numerical model all series are represented by the finite series

$$\sum_{l=-\infty}^{\infty}\sum_{k=0}^{\infty}\phi_{kl}\ell_k(z)e^{il\psi}\rightarrow\sum_{l=-L_{\max}}^{L_{\max}}\sum_{k=0}^{K_{\max}}\phi_{kl}\ell_k(z)e^{il\psi} \tag{30}$$

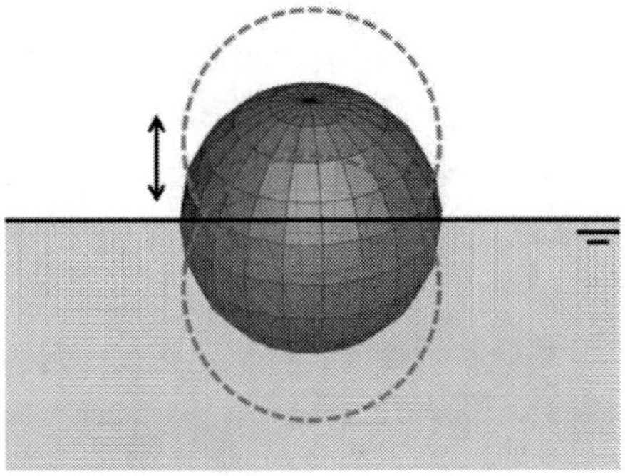
Figure 4. The heaving sphere.

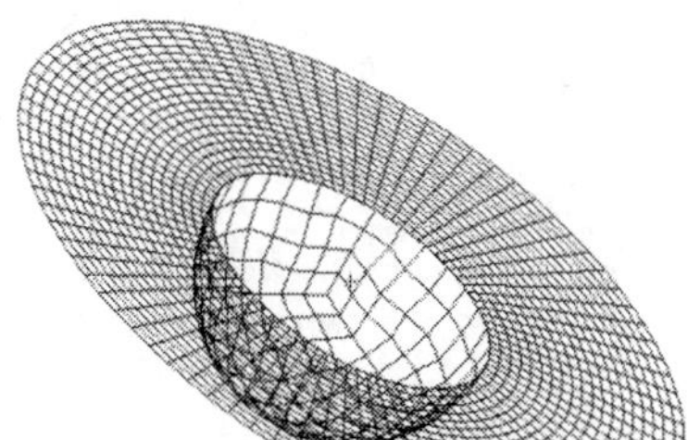
Figure 5. The floating sphere and the free-surface.

The results of velocity potential on the body panels with different frequency for comparing between the current method and the Kelvin source method are shown in the Fig. 6 to 11, ϕ_1 represents the velocity potential caused by the motion of surge, and ϕ_3 represents the motion of heave.

Table 1. The parameters of floating sphere.

Parameter's name	Notation	Values
Radius	r	25m
Number of panels of the body	N_B	192
Number of panels of the free-surface	N_F	1710
Radius of the control surface	R_C	60m
Number of Laguerre functions	$K_{\max}$	10
Number of harmonic functions	$L_{\max}$	10

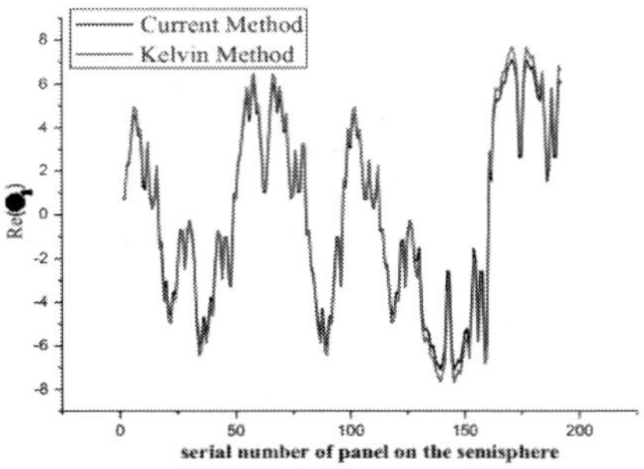

Figure 6. ϕ_1 on each panel when the wave number $k=1.5$.

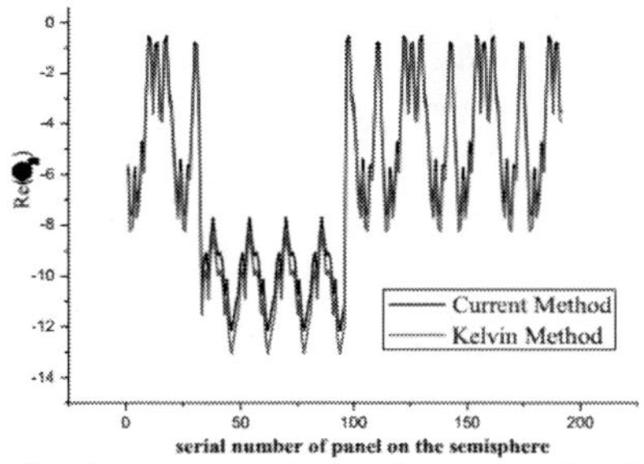

Figure 7. ϕ_3 on each panel when the wave number $k=1.5$.

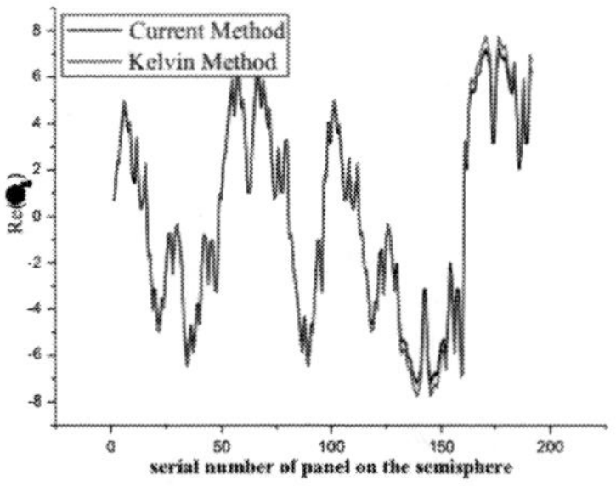

Figure 8. ϕ_1 on each panel when the wave number $k=2.5$.

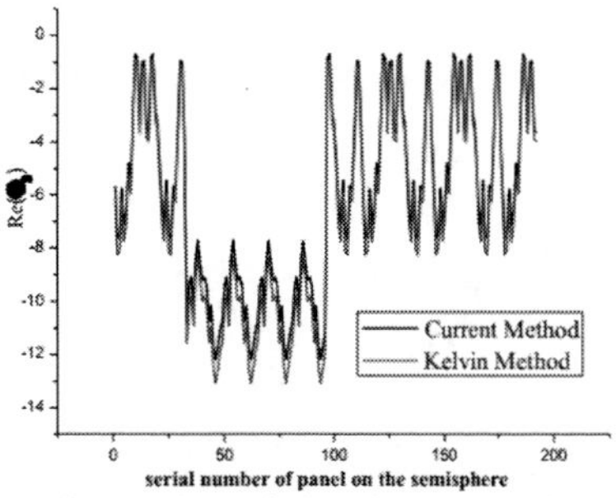

Figure 9. ϕ_3 on each panel when the wave number $k=2.5$.

DOI: 10.1051/matecconf/20179516002

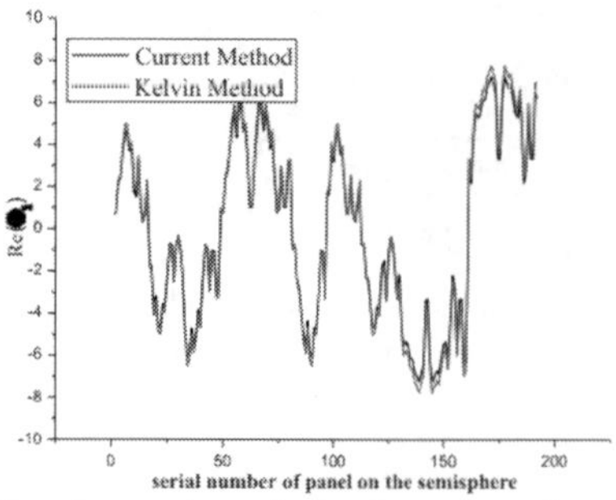

Figure 10. ϕ_1 on each panel when the wave number $k = 3.5$.

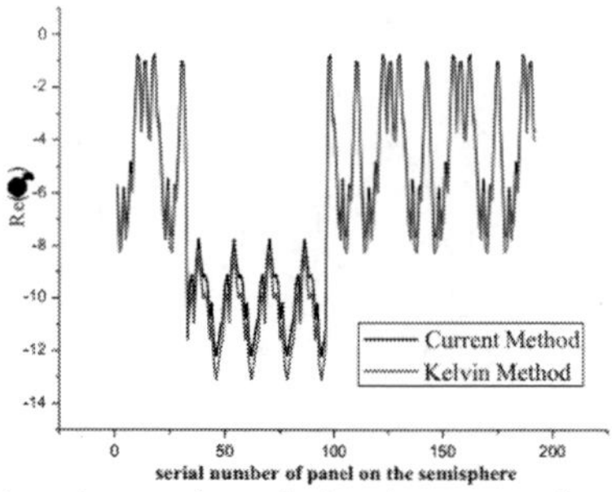

Figure 11. ϕ_3 on each panel when the wave number $k = 3.5$.

In order to study the effect of the control surface radius R to calculation accuracy, this paper choses different R/r values where the r is the radius of hemisphere, the results are compared with the analytical values [4] as the Fig.12 to Fig.15 showed.

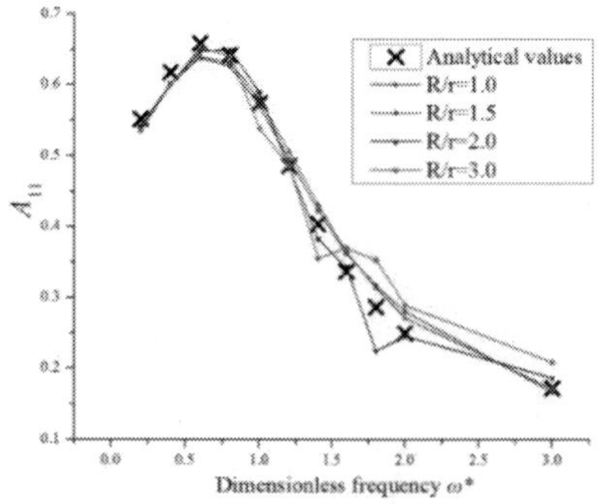

Figure 12. Surge added mass A_{11} of hemisphere.

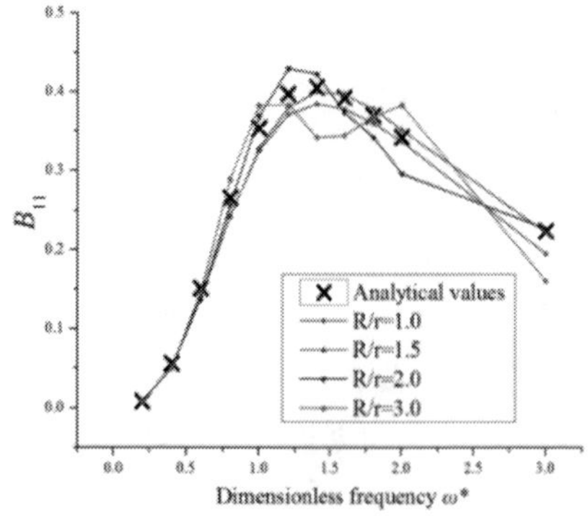

Figure 13. Surge damping coefficient B_{11} of hemisphere.

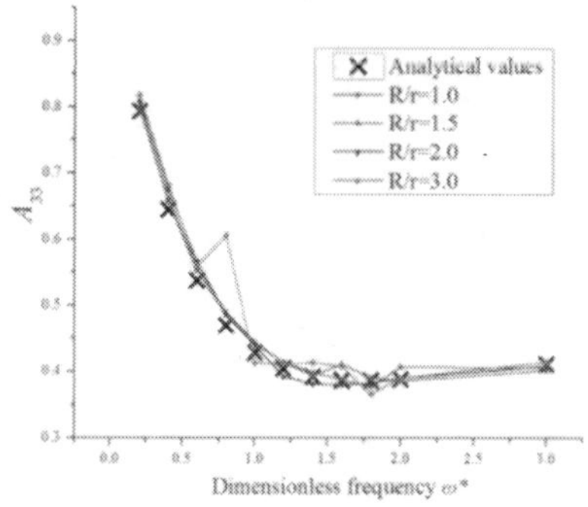

Figure 14. Heave added mass A_{33} of hemisphere.

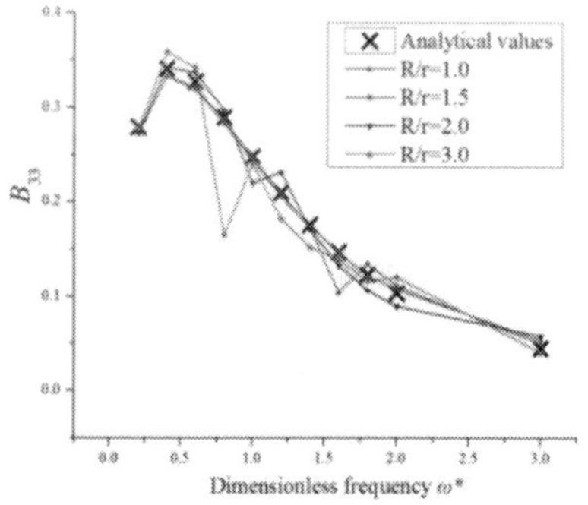

Figure 15. Heave damping coefficient B_{33} of hemisphere.

The results show that with the control surface radius becoming larger, the deviation between numerical results and the analytical values of added mass and damping coefficient becomes larger. Therefore, choosing smaller control surface radius, not only ensures high accuracy, but also reduces the number of free-surface units to improve computational efficiency.

5 Conclusions

In this paper, we have considered the seakeeping problem for a hemisphere with zero speed using the hybrid method. The fluid domain is divided into two subdomains by a control surface of cylindrical form. In the interior one the Rankine source method is applied, and in the exterior one the Kelvin source method is applied. Unlike the classical BEM method, the control surface is not divided into panels, but uses the analytical expressions. Using the cylindrical surface, we can decouple the vertical and horizontal direction. The results of the comparison between the current method and the Kelvin method show that the velocity potentials on the panels of the ship fit well each other.

References

1. X.B.Chen, F.Dias and W.Y. Duan "Introduction of dissipation in potential flows," Proc. of 7th International Workshop on Ship Hydrodynamics, Shanghai, China, **213-322**(2011)
2. I. Ten and X.B. Chen "A coupled Rankine-Green function method applied to the forward-speed

seakeeping problem," Proc. 25th IWWWFB, Harbin, China, **165-168**(2010)
3. A. Milton and A.S. Irene "Handbook of Mathematical functions, "Dover publications Inc, New York. (1964).
4. A.Hulme, "The wave forces acting on a floating hemisphere undergoing forced periodic oscillations", J. Fluid Mech., 121:**443-463**(1982)
5. OZDEMIR Y H, BARLAS B, YILMAZ T, BAYRAKTAR S. "Numerical and experimental study of turbulent free surface flow for a fast ship model". BRODOGRADNJA/SHIPBUILDING 65(1):**39-54**, (2014).

Research and Application of an Automatic Clam Collecting Device

Bin Wang[1], Kaijie Hu[1,2] and Ping Wang[1]

[1]Yancheng Institute of Technology, School of Mechanical Engineering, 224051 Yancheng, China
[2]Jiangsu University, School of Mechanical Engineering, 212013 Zhenjiang, China

Abstract. To collect clams automatically and effectively on coastal beach, an automatic clam collecting device was designed. The device consists of a connecting device, a shovelling device, a conveying device, and a filtering device. The mechanical device is designed based on some of the presented devices, such as a blade with slope, a pipelined conveyor belt, planar linkage mechanisms and a ski mechanism. The connecting device is connected to the device body by bolts. The shovelling device adopts a blade with slope, which can reduce the resistance between sandy soil and the device. The transmission device adopts a conveyor belt with a two-stage reducer, which can effectively control the speed of the transmission and avoid the splash of mud. A mesh structure is used for soil filtering, which is designed with a certain slope, thus sandy soil and other impurities fall from the mesh for its own weight. The designed device for clam collecting will improve efficiency and decrease cost effectively.

1 Introduction

The development of shellfish collecting machine has more than 70 years of history. There have been many kinds of shellfish catching machines used to seed, shallow sea resources and collect shellfish and dead shells. A number of collecting machines have been analysed, which mainly have the following characteristics [1]. 1) Collecting clams, mussels, oysters and thin shell of vermiculite. 2) Working on tidal flats or collecting shellfish in shallow sea. 3) Can collect various kinds of shellfish on the surface of the mud, sand, sand or gravel. 4) Collecting machine is divided into types of semi mechanical devices, which are pushed by hand and types of devices with different degree of mechanization.

FletcherHanks catching machines are widely used in many countries for seashell collecting, which is improved on different purposes based on the structure of Mya arenaria Linnaean's collecting machine [2]. In recent years, the design of a novel clam collecting machine adopting mechanical lifting mechanism has been completed, which has been manufactured and tested in Virginia State.

A type of hydraulic collecting machine has been developed in Canada [3], which is pushed by hand. While the machine is working normally, a man pushes the machine forward slowly, and the other person collects the clam in the ditch. This machine is especially suitable for operation on gravel or shallow water.

A high efficiency hydraulic clam collecting machine was developed in the UK [4]. Sediment and clams on the tidal flat is dug up by high pressure water, which is spurted from high pressure water pump. And another auxiliary pump sucks gravel, clams and other impurities in to the shipboard. At last, the young shellfishes and other impurities are filtered out through the metal net.

Marine fishery headquarters and Rudong County [5] jointly developed an automatic clam collecting device. They do a test in a high yield area where hourly clam yield can reached 900 kg to 1000 kg. Although in theory this device is reasonable, the actual use has a large gap with the imagination. Because the machine and the ship are not matching, and the operation is not convenient, so the machine still needs to be improved. On this basis, Yellow Sea Fisheries Research Institute [6] developed the type of wcs-6 shellfish collecting machine. Experiments were carried out in the Laizhou Bay. Derived from the experiments, the device can capture clams about 150-200 kg per hour, and the damage rate is relatively low, which is not more than 4-5%.

Overall, the current situation is that clam collection of the demand for labor is relatively large with high cost. So in view of the current problems, a collection device for clams in beach is designed.

2 Innovative Design

The innovative device adopts the principle of similar ploughing device to dig the clams on the beach. Under the driving of the motor, clams and sediment which have been spaded are uplifted through the role of the common conveyor and speed reducer [7]. The final collecting of clam is completed through the implementation of the sediment filter.

Since the device is required to work on the beach, it cannot adopt ordinary wheels. Therefore, the device adopts track wheel, which can increase the contact area

DOI: 10.1051/matecconf/20179516003

with sand [8]. By adopting this design, car is not easy to fall into the sediment, and normal running of the car on the surface of the sand can be ensured. In addition, the design of the support of the device adopts sliding plate structure, which can reduces drag and saves the cost of the device. The significance of this device is that it can liberate the dependence on labor and accelerate efficiency of collection.

Because of the innovative design mentioned in this paper, the device has the following main advantages.

- It can achieve the goal of massive cultivation, liberate the dependence on the labor, and accelerate the efficiency of the clams' collection.
- It can save the resource efficiently.
- It can promote the development of aquaculture.

3 The structure and working principle of the collecting device

3.1 Device structure

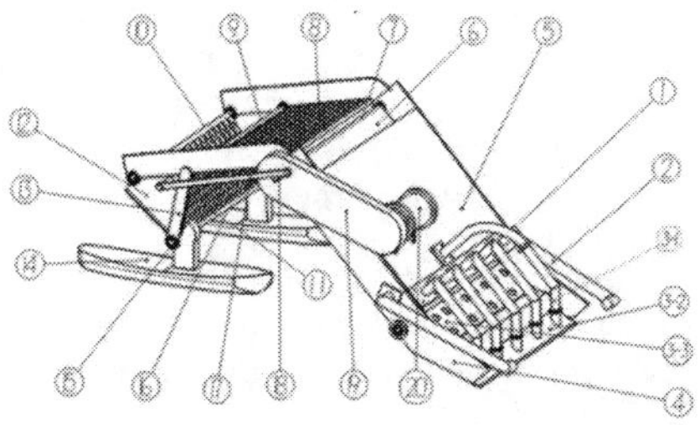

Figure 1. Device structure diagram.

The machine is consisted of main connecting support 1, connecting support 2 on two sides, blade assembly 3, blade linkage 3-1, blade 3-2, filter plate with holes 3-3, baffles on both sides 4, upper baffle 5, transmission belt 6, roll bar 7, filter screen 8, mesh swash-plate at front 9, mesh swash-plate at back 10, V-shape cage 11, fan-shaped rocker 12, connecting support 13, ship form support 14, bearing 15 the roll bar under the V-shape cage 16, connecting rod 17, crank 18, secondary expansion helical gear reducer 19, motor 20, transmission belt's roll bar bellow.

The device is connected with the body of tractor through connecting support 1 and connecting support 2 on two sides. When the tractor is moving, it drives clams collecting device to move on the beach through connecting support 1 and connecting support 2 on two sides. The device adopted support, which is ship type. Small contact area with sediment and small sliding resistance can be achieved by this way.

The device has blade assembly 3 at front. The blade assembly is consisted of blade linkage 3-1, blade 3-2 and filter plate with holes 3-3. One side of blade linkage 3-1 is connected with the shank of blade 3-2 by bolts. Another side of blade linkage 3-1 is fixed on upper baffle 5. The blade 3-2 has some curves.

The drive assembly is consisted of motor 22, big belt-pulley 21, small belt-pulley 19, belt 20, roll bar 7, transmission belt 6 and baffle 5. The motor 22 is fixed on baffle 5. The power of motor 22 is directly output on big belt-pulley 21.

One side of roll bar 7 is connected with crank 18 by spline and drives crank 18 to rotate around its axle. Crank 18 is connected with connecting rod 17 by bears and pins. Connecting rod 17 covers around cylindrical shank of fan-shaped rocker 12 by bears.

3.2 Shovel device design

As the device is moving, the blade 3-2 will dig and uplift sediment, it will also dredge the clams out. While the device keeps moving, the sediment and clams will be uplifted by the blade 3-2, and bilateral baffles 4 keep the sediment and clams in the device from moving out. After the sediment and clams move through the blade 3-2 with curves, they will be conveyed to filter plate with holes 3-3 and some of the sediment will be filtered.

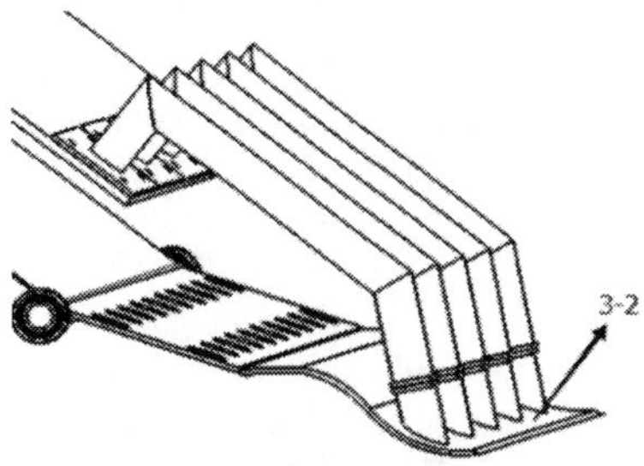

Figure 2. Schematic diagram of shovel.

3.3 Sand filter design

Big belt-pulley output power on small belt-pulley 19 by belt 20 and drives small belt-pulley to rotate. The small belt-pulley is connected with roll bar 7 by spline. The rotation of small belt-pulley drives the rotation of roll bar 7. Roll bar 7 nested on baffle 4 by bear. The transmission belt 6 is to transmit clams and sediment.

There is a round hole at V-shaped bottom of fan-shaped rocker 12. The round hole covers around V-shape cage 16 by bears [9]. Connecting rob 17 drives fan-shaped rocker 12 swing around V-shape cage 16. The swing can shake off the sediment.

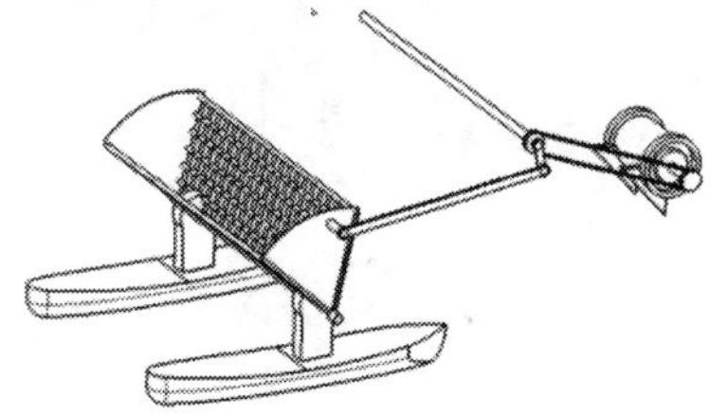

Figure 3. Schematic diagram of sand filter.

When the transmission belt 6 transmits clams and sediment to filter screen 8, the filter screen will filter the sediment. Clams will fall into V-shape cage 11. There are baffles on both sides 4 over the V-shape cage 11, which are designed to prevent clams from getting out. Mesh swash-plate at front 9, mesh swash-plate at back 10 are fixed in the hole of baffle 4. They can also prevent clams from getting out.

4 Detail design

4.1 Shovel device

Shovelling blade is made of wear resistance and impact resistance steel. It needs to be heat-treated. Because the friction between the shovelling blade and sandy soil is pretty heavy and always results in abrasion of the blade.

4.1.1 Resistance calculation of the blade

Literature [10] proposed an equation for the calculation of shovelling blade's cutting resistance, which can be used for the resistance calculation of this paper. The equation is listed as follows.

$$K_b = 10^{-6}\frac{p_0}{bh}(MPa) \tag{1}$$

Where P_0 is the cutting resistance on shovelling blade (N); b is the width of shovelling blade (m); h is the cutting depth (m).

We can get cutting force of the blade is:

$$P_0 = K_b l_1 h_{cp} \times 10^6 \tag{2}$$

Where K_b is cutting rate resistance (MPa); l_1 is the length of blade's cutting edge (m); h_{cp} is average cutting thickness (m).

Now combine the dimension parameter, we can get the cutting resistance, which is 2.5 kN.

4.2 Power train design

The driving power of the device comes from the engine of the tractor which drives the tractor to move ahead. The rotational speed of the tractor is too high to be directly used for device. Therefore a helical gear reducer is designed to decrease the input speed of the device. The sketch drawing of the two-stage reducer is illustrated in Fig. 4.

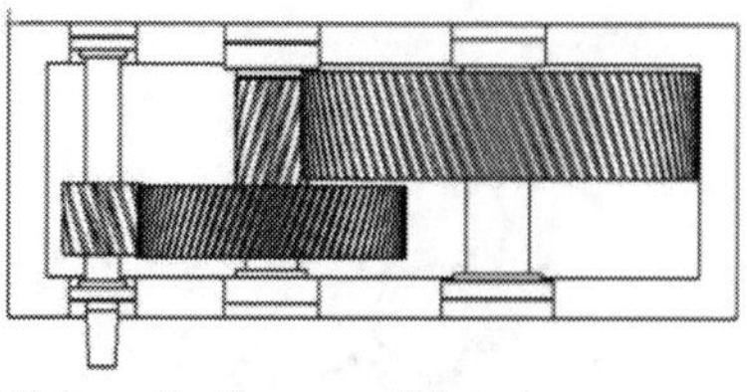

Figure 4. Schematic diagram of shovel.

4.2.1 Design parameter

According to the working condition of the device, the parameters of the transmission system of the device should be well designed in advance. Considering the friction of sandy soil and shovel structure, the parameters are listed in the following table.

Under the working condition of continuous unidirectional rotation, the load of the device has fluctuation. Therefore, Allowable error of the conveyer belt's working speed is between -6% and +6%.

Table 1. Value of parameters.

Parameter	Value
Conveyer belt's Working pull F (N)/N	2800
Conveyer belt's working speed V/m/s	1.5
Drum's diameter D/mm	320

4.2.2 Transmission scheme

Due to the vibration when machine is working, the material of conveyor belt should be selected according to the effect of vibration reduction. Reducer adopted two-stage expanded helical gear reducer [12]. But the Gear and bearing is not on the same line, so the shaft should have relatively large stiffness. In addition, the layout of the high speed gear also have requirements. In order to reduce the phenomenon of the uneven distribution of loading, so the high speed gear is required to be put away from input terminal.

According to the requirements, the transmission scheme are as follows.

- Outer transmission is V-belt drive.
- Reducer is two-stage expanded helical gear reducer.
- Schematic diagram is shown as follows.

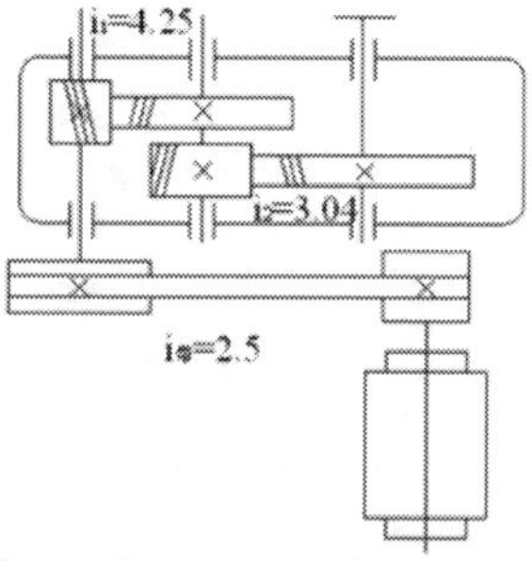

Figure 5. Diagram of two stage reducer.

4.3 Calibration for sand filter

This paper adopts the planar linkage mechanism as sand filter device. This planar linkage has a variety of applications. Advantages are as follows.

- Between the various components of the device is a surface contact, which can withstand greater load. Also, this way is very suitable for lubrication, which reduce wear and tear between components of the device effectively.
- Because the device adopts the plane contact. Therefore, the device is relatively easy to be manufactured and processed. Also, the device can get high accuracy.

 DOI: 10.1051/matecconf/20179516003

- The device can change the length of each bar, so as to realize a variety of motion.

4.4 Design and calculation of V-belt drive

4.4.1 The choice of V-belt

This device is used for clam collecting in shallow beach, so applicable load and the diameter of the belt pulley should not be too large. According to the design requirement and users' feedback, ordinary V-belt was chose. The reason is that the common V belt is cord structure, which can resist tension.

4.4.2 Design of pulley

Because the power of the motor is low, so aluminium was chosen as the material of pulley. In order to make the belt wheel do not shake in the drive, the pulley needs to be corrected by a dynamic balance way. Based on comprehensive thinking, we decided to choose wheel-spoke pulley. Because it fits in multiple environment. In high rational speed condition, we chose cast steel. In low power condition, we chose cast aluminium.

4.4.3 Contented conditions needed

To prevent relative displacement when pulley is working, the peripheral force is needed to be ensure that it is smaller than maximum effective tension. To extend the life of V-belt, we need to ensure that belt's maximum stress is smaller than its allowable stress.

5 Program benefits

The plan adopted crank-rocker mechanism to achieve the swing of V-shape cage and high degree of filtration of sediment. The power of crank-rocker mechanism and transmission belt comes from roll bar, which can ensure that crank-rocker mechanism and transmission belt rotate in same speed.

There are two mesh swash-plates, which is installed on the top of the v shape cage. As the V-shape cage is swing, mesh swash-plate at front and mesh swash-plate at back hold contact with the V-shaped hole of the V-shape cage from beginning to end, so there will be no spaces between V-shape cage and filter screen.

This device adopted sliding plate structure as the brace to increase the measure of support surface and prevent it from sliding into the sediment.

It adopted two-stage reducer, which can control the rate easily.

6 Conclusion

The invention is applicable to the large-scale acquisition of mudflat shellfish such as clams with the high degree of automation, low cost, high efficiency, and low leakage rate of mining. The device can not only effectively dug out clams and other shellfish, transport clams and other shellfish, but also can greatly reduce the resistance, so as to realize the collection of the clam of tidal flats.

Also the device has some problems, because of the clam survival environment will change, and the device can realize the acquisition of clams on the tidal flats, unable to adapt to some other environment, so the device still needs to be improved. In the future, the device should be improved in this direction, the design of a waterproof device may be better to adapt to changes in the environment.

Acknowledgment

The authors gratefully acknowledge the financial support of Jiangsu Provincial Science and Technology Program under Grant No. BY2015057-38 and Yancheng Agricultural Science and Technology Innovation Special Guiding Program under Grant No. YK2015023.

References

1. B. Massoud. European Journal of Operational Research,1999,112:258-272.
2. Z. J. Yu, L. Zhang, W. P. Shao. Mechanical design problem analysis and problem solving. Shen Yang. Press 2000.
3. W. Q. Yan. Manufacture Information Engineering of China, 2004,33(5):69-70
4. S. M. Liu, X. Q. Gong. *Sichuan steam technology*, 2000,(4):10-14.
5. W. C. Ye. Concise design handbook of mechanical engineering and automation (volume one). China Machine Press 2001.
6. J. X. Hu. Mechanical parts design and practical manual [M] China Machine Press 1999.
7. Q. G. Wang, B. W. Li, J. X. Huang. Mining machinery and supporting equipment. Xu Zhou. China University of Mining and Technology Press 2006.
8. X. D. Wang, X. T. Lan, T. S. Yu. Hebei Journal of Industrial Science and Technology, 2006.
9. Z. H. Sun, W. Fu. The process design of production management and working standard. Beijing. PT Press 2006: 128.
10. Y. C. Ye, L. H. Ke, D. Y. Hunag. Technology and application of comprehensive evaluation system. Beijing metallurgical industry.
11. Y. M. Li. Mechanical manufacturing process design manual. China Machine Press 1995
12. W. H. Liang, K. Liu, Y. H. Cui. Calculation and Experimental Verification of Helical Gear Churning Power Loss Basseed on Finite Element Analysis. Chinese Mechanical Engineering Society, 2014,18:2510-2515.

Structural Reliability Analysis Using Orthogonalizable Power Polynomial Basis Vector

Yejun Li and Bin Huang

School of Civil Engineering and Architecture, Wuhan University of Technology, Wuhan, China

Abstract. A new method for structural reliability analysis using orthogonalizable power polynomial basis vector is presented. Firstly, a power polynomial basis vector is adopted to express the initial series solution of structural response, which is determined by a series of deterministic recursive equation based on perturbation technique, and then transferred to be a set of orthogonalizable power polynomial basis vector using the orthogonalization technique. By conducting Garlekin projection, an accelerating factor vector of the orthogonalizable power polynomial expansion is determined by solving small scale algebraic equations. Numerical results of a continuous bridge structure on reliability analysis shows that the proposed method can achieve the accuracy of the Direct Monte Carlo method and can save a lot of computation time at the same time, it is both accurate and efficient, and is very competitive to be used in structural reliability analysis.

1 Introduction

Bridge is the hinge of transportation system and the important component of lifeline project, and the safety of the bridges is directly related to the national economy and people's livelihood. Thus, conducting structural reliability analysis is in general an important task, which has aroused the interest of many scholars in recent years.

The failure probability is not only a key indicator to measure structural safety, but also a basic research area in structural reliability analyses. Since multiple probability factors existed in the safety analysis of structural component or system, a multi-fold integral needs to be calculated to obtain the failure probability of the structure. However, even if the performance function is explicit, the nonlinearity of the function and the non-Gaussian distribution of random variables always make it impossible to obtain the exact solution of this multi-fold integral [1], [2]. Various methods have been used to analyze the failure probability of structures subjected to random material parameters, such as the first- and second- order reliability methods (FORM/SORM) [3], [4], the simulation methods [5]-[7], the traditional perturbation stochastic finite element methods (PSFEM) [8], [9], the response surface method [10], [11], and so on. However, more or less problems exist in the above mentioned methods. For example, FORM/SORM are ineffective when response sensitivities are not available, the limit state function is highly nonlinear or multiple MPPs exsit. The most representative and effective simulation method is the Direct Monte Carlo simulation (DMC) [12], which is usually used to verify the accuracy of the other methods, but it is also very time-consuming which restricts its application in structural safety analysis, especially in large and complex structures. PSFEM is widely used for stochastic series expansion, but the random responses of structures may be unstable if the uncertainty of some random parameters is large. The response surface method is a good choice when some specific types of multivariate polynomial are assumed to represent the response surface of structure, but further research is needed to ensure the correctness of the assumed form of the response surface.

In this study, an innovative method is presented to analyze the failure probability of structures which is mainly based on an improvement strategy of high order perturbation stochastic finite element method [13], 14]. Afterwards, in terms of an orthogonalization technique, the initial series solution is transferred to be a set of orthogonalizable power polynomial basis vectors. Through the usage of Galerkin projection scheme, an accelerating factor vector of the orthogonalizable power polynomial expansion is determined by solving small scale algebraic equations. Due to the acquisition of the explicit expression of random displacement response, the DMC method is then used to evaluate the failure probability directly and quickly. Finally, a numerical example is provided to illustrate the efficiency and accuracy of the proposed method (OSFEM-P).

2 The OSFEM-P Method

2.1 Random static equilibrium equation

When small deformation of a linear elastic structural system occurs under an load, the equation can be expressed as the following

$$\mathbf{Kd} = \mathbf{f} \tag{1}$$

where $\mathbf{K}$ is the structural stiffness matrix; $\mathbf{f}$ is equivalent nodal load vector.

The uncertain stiffness matrix can be expressed as

$$\mathbf{K} = \mathbf{K}_0 + \sum_{i=1}^{n_1} \alpha_i \mathbf{K}_i \tag{2}$$

where $\mathbf{K}_0$ is a $N \times N$ dimensional deterministic matrix with respect to deterministic mean parameters. $\mathbf{K}_i$ is a $N \times N$ dimensional matrix.

When external load includes l random variables, it can be written as

$$\mathbf{f} = \mathbf{f}_0 + \sum_{i=n_1+1}^{n} \mathbf{f}_i \alpha_i \quad (n = n_1 + l) \tag{3}$$

where $\mathbf{f}_0$ is a $N \times 1$ dimensional deterministic vector with respect to deterministic mean parameters and $\mathbf{f}_i$ is a $N \times 1$ dimensional vector.

The random response vector $\mathbf{d}$ is used by the following multivariate power series expansion as

$$\begin{aligned}\mathbf{d}(\mathbf{X},\boldsymbol{\alpha}) &= \mathbf{d}_0(\mathbf{X})\varphi_0 + \sum_{i=1}^{\infty} \mathbf{d}_i(\mathbf{X})\varphi_1(\boldsymbol{\alpha}) + \\ &\sum_{i=1}^{\infty}\sum_{j=1}^{i} \mathbf{d}_{ij}(\mathbf{X})\varphi_2(\boldsymbol{\alpha}) + \sum_{i=1}^{\infty}\sum_{j=1}^{i}\sum_{k=1}^{j} \mathbf{d}_{ijk}(\mathbf{X})\varphi_3(\boldsymbol{\alpha}) + \cdots\end{aligned} \tag{4}$$

where $\varphi_i(\boldsymbol{\alpha})$ denotes the multidimensional power series, the first four terms can be written as

$$\varphi_0(\boldsymbol{\alpha}) = 1;\ \varphi_1(\boldsymbol{\alpha}) = \alpha_i;\ \varphi_2(\boldsymbol{\alpha}) = \alpha_i\alpha_j;\ \varphi_3(\boldsymbol{\alpha}) = \alpha_i\alpha_j\alpha_k \tag{5}$$

2.2 Perturbation technique

Substituting (2-4) into (1) results in the following

$$(\mathbf{K}_0 + \sum_{i=1}^{n} \alpha_i \mathbf{K}_i)[\mathbf{d}_0\varphi_0 + \sum_{i=1}^{n} \mathbf{d}_i\varphi_1 + \cdots] = \mathbf{f}_0 + \sum_{i=1}^{n} \mathbf{f}_i\alpha_i \tag{6}$$

where $\mathbf{d}$ is defined as a truncated series which includes n random variables at the same time, and

$$\mathbf{K}_i = 0\,(i = n_1 + 1, \cdots n);\quad \mathbf{f}_i = 0\,(i = 1, \cdots n_1) \tag{7}$$

Since every sample of random vector $\boldsymbol{\alpha}$ should satisfy (6), then the sum of coefficients of the same order terms based on power series expansion should be equal to zero. Therefore, the coefficient vectors of the power series expansion can be determined by perturbation technique.

2.3 Orthogonalizable technology and Galerkin projection scheme

In general, the convergent rate for the power series expansion is not often satisfied, thus, Galerkin projection is taken as an improvement measure.

Before utilizing that scheme, $\mathbf{d}$ can be transferred into an orthogonal polynomial expansion $\bar{\mathbf{d}}$ by using a transformation matrix in order to use Galerkin projection scheme more easily, then

$$\bar{\mathbf{d}} = \bar{\mathbf{d}}_0\chi_0 + \sum_{i=1}^{n} \bar{\mathbf{d}}_i\chi_1 + \sum_{i=1}^{n}\sum_{j=1}^{i} \bar{\mathbf{d}}_{ij}\chi_2 + \cdots \tag{8}$$

where χ_0, χ_1, $\chi_2 \cdots$ are orthogonal basis vectors.

If correction factors are employed in each term in (8), then

$$\bar{\mathbf{d}} = \beta_0\bar{\mathbf{d}}_0\chi_0 + \beta_1\sum_{i=1}^{n} \bar{\mathbf{d}}_i\chi_1 + \beta_2\sum_{i=1}^{n}\sum_{j=1}^{i} \bar{\mathbf{d}}_{ij}\chi_2 + \cdots = \sum_{i=0}^{m} \beta_i\psi_i \tag{9}$$

where $\psi_0 = \bar{\mathbf{d}}_0\chi_0$; $\psi_1 = \sum_{i=1}^{n} \bar{\mathbf{d}}_i\chi_1$; $\psi_2 = \sum_{i=1}^{n}\sum_{j=1}^{i} \bar{\mathbf{d}}_{ij}\chi_2$ and so on, β_0, $\beta_1 \cdots \beta_m$ are correction factors.

By using Galerkin projection scheme, (9) can be written as following

$$\sum_{j=0}^{m}\sum_{i=0}^{n} < \alpha_i(\psi_k)^T \mathbf{K}_i\psi_j > \beta_j = \sum_{i=0}^{n} < \alpha_i(\psi_k)^T \mathbf{f}_i > \tag{10}$$

where $k = 0, \cdots m$, $< \cdot >$ denotes the operator of mathematical expectation.

By solving the small scale (m+1) × (m+1) linear algebraic equations, the unknown coefficients β_0, $\beta_1 \cdots \beta_m$ can be easily determined. Then, the random response vector is capable of being got.

3 Reliability analysis

3.1. Failure probability

According to the definition of the failure probability P_F, it can be determined by direct integration over the failure domain as follows

$$\begin{aligned} P_F &= P(\Psi(\boldsymbol{\alpha}) < 0) \\ &= \int_{\Psi(\boldsymbol{\alpha})<0} f_p(\alpha_1, \alpha_2, \cdots, \alpha_n) d\alpha_1 d\alpha_2 \cdots d\alpha_n \end{aligned} \tag{11}$$

where $f_p(\alpha_1, \alpha_2, \cdots, \alpha_n)$ is the joint probability distribution function of the random variables $\{\alpha_1, \alpha_2, \cdots, \alpha_n\}$ and $(\Psi(\boldsymbol{\alpha}) < 0)$ defines structural failure.

3.2 Calculation of multifold Integral

Generally, even if the explicit expressions of random structural responses are determined, it is still difficult to calculate the exact failure probability due to the reason that the failure domain of numerical interval is difficult to obtain or the explicit performance function is complex.

If the partial derivative of performance function is given, FORM, SORM and other similar methods can be used to calculate the failure probability. However, results based on FORM/SORM should be interpreted with caution for highly nonlinear performance functions or multiple design point, which exists in many structural systems.

Since the random response vector is obtained, so the response surface is apparent.

$$P(\boldsymbol{\alpha}^{(i)}) = \Theta - \mathbf{d}(\boldsymbol{\alpha}^{(i)}) \quad (12)$$

where $\boldsymbol{\alpha}^{(i)}$ is the ith sample of the random vector $\boldsymbol{\alpha}$; Θ denotes the threshold value of displacement at some point of the structural system; $\mathbf{d}(\boldsymbol{\alpha})$ is the displacement response vector of the structural system under external loads.

The Monte Carlo simulation is used to calculate the failure probability directly.

$$P_F = \frac{1}{N}\sum_{i=1}^{N} \mathrm{H}[P(\boldsymbol{\alpha}^{(i)}) < 0] \quad (13)$$

where N is the sample size; H is defined as counting function; if $P(\boldsymbol{\alpha}^{(i)})$ is in failure area, then $\mathrm{H}=1$.

4 Numerical example

Considering a continuous beam type of bridge structure with two spans, which can be simplified into a one-dimensional continuous beam subjected to a concentrated force, as shown in Fig. 1. The whole length of the beam is $8m$, and the beam is divided into 16 elements. Assuming that the bending rigidity EI of the beam and the external load F are both random variables. The means of EI and F are $3.4\times10^4 kN\cdot m^2$ and $40kN$ respectively.

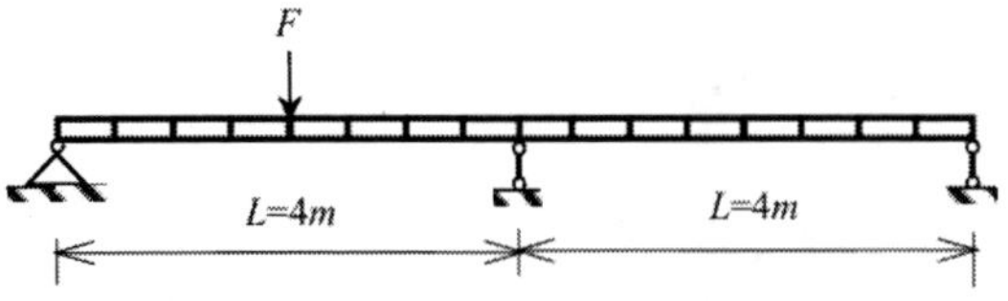

Figure 1. Random continuous beam

Assuming that EI and F are both of Beta distribution, thus we have

$$EI = EI_0(1+\sqrt{5}\cdot\alpha_1\cdot\delta_1);\ F = F_0(1+\sqrt{5}\cdot\alpha_2\cdot\delta_2) \quad (14)$$

where EI_0 and F_0 denote the means of EI and F respectively; δ_{EI}, δ_F are the coefficients of variation (COVs) of EI and F respectively. In addition, $\alpha_i (i=1,2)$ are independent random variables with Beta distribution, of which the probability density function is written as

$$f(\alpha) = \frac{3}{4}(1-\alpha^2) \quad \alpha \in (-1,\ 1) \quad (15)$$

Here $\mathbf{d}(\boldsymbol{\alpha}^{(i)})$ denotes the vertical displacement at the middle of the left span of the beam. The failure probability of the random beam is computed by four methods, i.e., FORM, SORM, OSFEM-P, DMC. the results of failure probability calculated by these four methods are listed in Table 1.

Here it is assumed that δ_F is equal to 0.1 and δ_{EI} is supposed to be 0.1, 0.15, 0.2, 0.25 and 0.3 respectively. The threshold value, Θ, of the vertical displacement at the middle of the left span of the beam is assigned as $1.702\times10^{-3} m$.

Table 1 shows the results of OSFEM-P are the most close to that of DMC, and SORM is more precise than FORM, especially when the δ_{EI} is small.

Table 1. The failure probability of the beam from different methods with $\delta_F = 0.1$

Methods	δ_{EI}				
	0.1	0.15	0.2	0.25	0.3
FORM	8.40e-4	1.80e-2	6.37e-2	1.16e-1	1.63e-1
SORM	5.62e-4	1.33e-2	5.60e-2	1.10e-1	1.59e-1
OSFEM-P	1.00e-4	1.31e-2	5.43e-2	1.09e-1	1.60e-1
DMC	1.00e-4	1.31e-2	5.41e-2	1.09e-1	1.60e-1

The spent computation time of the four methods are also recorded and listed in Table 2, it can be seen that the DMC method, the highest accurate method, is the most time-consuming method, which is hundreds times of the proposed method and the widely used FORM and SORM. The time cost of the proposed method is very close to the FORM and SORM, but it can significantly increase the computation accuracy. Thus, the proposed method is both accurate and efficient in structural reliability analysis.

Table 2. CPU time spent of different methods.

Methods	DMC	OSFEM-P	FORM	SORM
CPU times (s)	1326.4	3.2	1.2	1.5

To fully illustrate the effects of the random inputs on the statistical properties of random response, the curves of probability density function (PDF) of the vertical displacement at the middle of the left span of the beam with the proposed method is shown in Fig. 2 and Fig. 3. It can be seen that the PDF curves of the vertical displacement at the middle of the left span of the beam vary with different δ_{EI} or δ_F, herein, we choose the two cases as following:

Case 1: smallest failure probability in Table 1, $\delta_{EI} = 0.1$ with $\delta_F = 0.1$

Case 2: largest failure probability in Table 1, $\delta_{EI} = 0.3$ with $\delta_F = 0.1$

DOI: 10.1051/matecconf/20179517001

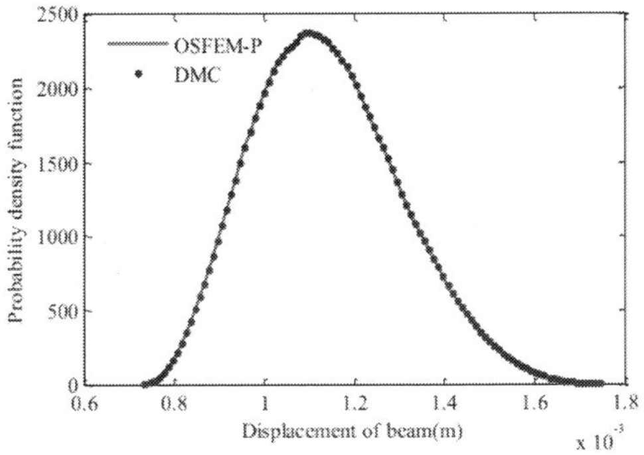

Figure 2. PDF from DMC and proposed method (case 1).

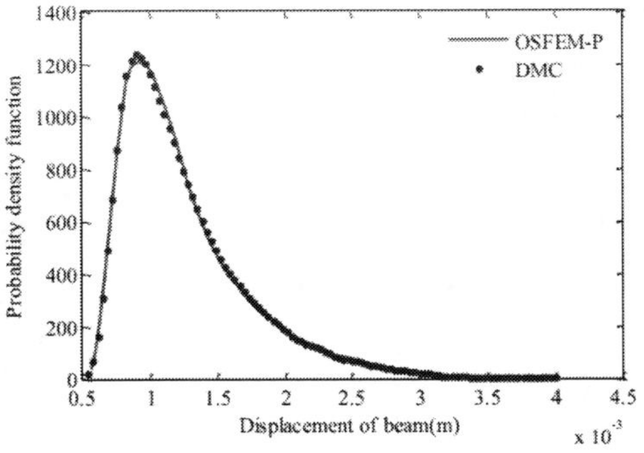

Figure 3. PDF from DMC and proposed method (case 2).

Fig. 2 and Fig. 3 show that the output displacements of the beam are diverse from each other for different δ_{EI}, and the PDF curves of OSFEM-P are very close to that of DMC no matter δ_{EI} is small or large, which illustrates the accuracy of the proposed OSFEM-P is very good.

5 Conclusions

In this study, a new method, i.e. OSFEM-P, for structural reliability analysis based on orthogonalizable power polynomial basis vector is presented. Firstly, a power polynomial basis vector is adopted to express the initial series solution of structural response, which is determined by a series of deterministic recursive equation based on perturbation technique, and then transferred to be a set of orthogonalizable power polynomial basis vector using the orthogonalization technique. By conducting Garlekin projection, an accelerating factor vector of the orthogonal power polynomial expansion is determined by solving small scale algebraic equations. Numerical results of a continuous bridge structure shows that the proposed method is both accurate and efficient, it can significantly improve the calculation accuracy compared with the widely used FORM/SORM and save a lot of computation time compared with the DMC method.

Financial support from the National Nature Science Foundation of China under Project No. 51378407 and 51578431 is gratefully acknowledged.

References

1. K. Breitung. Asymptotic approximations for multinormal integrals. J Eng Mech, **110**(3): 357-366 (1984).
2. G.Q. Cai and I. Elishakoff. Refined second-order reliability analysis. STRUCT SAF, **14**(4): 267-276 (1994).
3. M.A. Valdebenito, H.J. Pradlwarter, G.I. Schuëller. The role of the design point for calculating failure probabilities in view of dimensionality and structural nonlinearities. Struct SAF, **32**(2):101-111 (2010)
4. R. Rackwitz. Reliability analysis—a review and some perspectives. Struct SAF, **23**(4): 365-395 (2001).
5. G.I. Schuëller, H.J. Pradlwarter, P.S. Koutsourelakis, A critical appraisal of reliability estimation procedures for high dimensions. Probabilist Eng Mech. **19**(4): 463-474 (2004).
6. S.K. Au and J.L. Beck. Estimation of small failure probabilities in high dimensions by subset simulation. Probabilist Eng Mech, **16**(4): 263-277 (2001).
7. G.I. Schuëller, H.J. Pradlwarter and P.S.Koutsourelakis. A critical appraisal of reliability estimation procedures for high dimensions. Probabilist Eng Mech, **19**(4): 463-474 (2004).
8. J.D. Arregui-Mena, L. Margetts, P.M. Mummery. Practical Application of the Stochastic Finite Element Method. Arch Comput Method E, **23**(1):171-190 (2014).
9. B. Huang, Q.S. Li and W.H. Shi, Z. Wu. Eigenvalues of structures with uncertain elastic boundary restraints. Appl Acoust, **68**(3): 350-363 (2007).
10. N. Balakrishnan, M.H. Ling. Gamma lifetimes and one-shot device testing analysis. Reliab Eng Syst Safe, **126**(6), 54-64 (2014).
11. X.L. Guan and R.E.Melchers. Effect of response surface parameter variation on structural reliability estimates. Struct Saf, **23**(4): 429-444 (2001).
12. R.Y. Rubinstein. *Simulation and the Monte Carlo method*. Wiley, New York, America (1981).
13. B. Huang, Q.S. Li, A.Y. Tuan and H. Zhu. Recursive approach for random response analysis using non-orthogonal polynomial expansion. Comput Mech, **44**(3): 309-320 (2009).
14. B. Huang, R.F. Seresh and L. Zhu. Statistical analysis of basic dynamic characteristics of large span cable-stayed bridge based on high order perturbation stochastic FEM. Adv Struct Eng, **16**(9): 1499-1512 (2013).

Study on the Thermal Environment and Indoor Comfort of Half Glass Curtain Buildings – a Case of Kouhu Visitor Center in Yunlin County, Taiwan

Ta-Ching Liang[1] and Szu-Hsien Peng[2]

[1]Department of Leisure and Recreation, National Formosa University, Yunlin 632, Taiwan, R.O.C.
[2]Department of Spatial Design, Chienkuo Technology University, Changhua 500, Taiwan, R.O.C.

Abstract. Located in the subtropics, the high temperature in summer affects indoor living environments in Taiwan. The use of perspective glass material for modern public buildings increases the transparency of landscape views; however, the solar radiation heat also increases the heat in indoor environment. In particular, visitor centers which offer service functions for large quantities of visitors should satisfy the demands for indoor environment comfort and energy saving. "Numerical simulation" and "field data" are therefore utilized in this study for mutual comparison and the analyses of duration of sunshine, temperature distribution, heat balance, and space comfort. The research result reveals that different building envelope materials, with distinct thermal resistance and thermal conductivity coefficients, as well as sunshine direction and duration, affect the heat transmitting indoors, change the indoor temperature and further influence the comfort of the indoor environment. In the future indoor venue comfort will affect visitors' recreational behaviours.

1 Introduction

To match public service buildings with landscapes and neighboring environments, high-transparency glass wall design is often introduced. Nonetheless, located in the subtropics, the high temperatures in summer influence the comfort of indoor environments in Taiwan. Although glass surfaces are specially treated to enhance the proportion of reflection and diffusion and reduce the transmission material, the indoor temperature is still high. Whenever the outdoor temperature rises in the summer, the external heat enters indoors through conduction, radiation, and convection, thus, increasing the temperature of the indoor environment [1,2]. In order to maintain the basic indoor living comfort, air conditioning systems have become essential equipment in such public spheres. Moreover, the power consumption of air-conditioners increases with rising outdoor temperature; in particular, this increases the energy consumption in large spheres offering public services. The call for green buildings offering energy saving, carbon reduction and environmental protection is increasing globally in recent years [3]. However, this kind of practice presents difficulties because of the physical factors of global warming in thermal environment and indoor comfort, such as heat radiation, air movement, and temperature & humidity distribution. These four factors must be properly controlled to achieve better environments and comfort [4]-[6].

Taking the Kouhu Visitor Center in the Southwest Coast National Scenic Area as the experimental target, "numerical simulation" and "field data" are applied to collect data and carry out mutual verification at the same time. Under simulated weather conditions, the duration of sunshine, temperature distribution, heat balance, and space comfort between the measurement and the numerical simulation are mutually compared to analyze the effects of building envelope materials, with distinct thermal resistance and thermal conductivity coefficients, and sunshine direction & duration. The effect of indoor venue comfort on visitors' recreational behaviors is examined. The findings can serve as a reference for relevant maintenance and management sectors.

2 Material and method

2.1 Research area

Kouhu Visitor Center in the Southwest Coast National Scenic Area Administration of the Tourism Bureau, M.O.T.C. was selected as the experimental area. The building plan features landscape architecture, and two irregular masses interlocking the image of a bird spreading its wings before flight. The building body is mainly built by reinforced concrete and large glass walls matched with wooden gratings and the roof is designed as a slope extending from a scenic green roof trail to the ground. In addition to the shadow of the green roof, there are no building or tree shelters around the building. Although the glass wall is designed with wooden gratings, there are no large window openings (Fig. 1). The indoor space of the building is mainly divided into three zones (A, B, C), as shown in Fig. 2.

Figure 1. Building appearance of the Kouhu Visitor Center in the Southwest Coast National Scenic Area.

Figure 2. Indoor space zones of the building.

2.2 In situ meteorology

The meteorological data reveal high temperatures and humidity. The climatic zone is located 20km north of the Tropic of Cancer, in the Subtropics. The mean annual temperature is about 22.6°C, with up to 220 days with a mean temperature above 22°C per year, and the changes in four seasons are not apparent. In regard to the temperature change cycle, the temperature gradually rises up to 20°C in March, and the highest temperatures in a year occur from June to September. The highest temperature in summer reaches 28 °C and even higher. The temperature in October and November gradually drops, and the lowest temperature appears between December and February, when the lowest winter temperature drops down below 20°C. The mean relative humidity is around 80%-90%.

2.3 Numerical simulation

Ecotect, the environment numerical simulation software [7], is used in this study for the numerical modeling based on the completed drawing size and boundary conditions of the numerical simulation, including the meteorological data (organized from the Data Bank for Atmospheric & Hydrologic Research and Yunlin Weather Station of the Central Weather Bureau), the elevation coordinates (+4m above sea level), and the geographic location (East longitude 120.1877° and North latitude 23.5460°).

2.4 Field data

An infrared thermal imager (a contactless and high-sensitivity infrared thermal imager with the measuring range of -20°C~250°C) is used for irradiating the building surface and measuring the surface temperature to further deduce the actual temperature change and distribution of the building envelope. Meanwhile, a multi-functional thermometer was placed inside the visitor center for the continuous monitoring from February - April, 2013. Under the conditions of natural ventilation without use of the air-conditioning system, the long-term observed data were recorded by monitoring the temperature, humidity, and solar radiation.

3 Result and discussion

According to the analyses with an infrared thermal imager and numerical simulation, the hourly, daily temperatures in the visitor center are shown in Fig. 3 and Fig. 4. The outdoor temperature in zones A, B, and C obviously rises after 6 a.m. because of the duration of sunshine. The highest temperature appears at 11 a.m., and the temperature becomes stable after 6 p.m. The obvious temperature difference between day and night corresponds the findings of previous works [8]. As seen in Fig. 3, the building materials in zones A and B show higher thermal resistance; although the temperature rises slowly, the high thermal resistance causes higher indoor temperatures compared to the outdoor temperature during the day. In this case, the indoor temperature, under the same external climate conditions, apparently changes with distinct building orientation, window design, and building materials (zone B reveals higher indoor temperature).

According to the calculation with Ecotect [9], the hourly heat analysis is used for simulating the thermal environment in this study. Moreover, it is assumed that the internal temperature of any building approaches the local outdoor temperature and generates delay or restraint caused by the heat capacity as well as the shelter of the building body. As a result, the indoor temperature changes are similar to the changes of the outdoor temperature. As a result, when the total heat loss is equal to the total heat gain, the indoor temperature is stable. Meanwhile, the heat gain and loss results based on the HVAV load, structural conduction, temperature (SOL-AIR), solar radiation heat, natural ventilation, internal personnel and equipment, and inter-zonal are also analyzed.

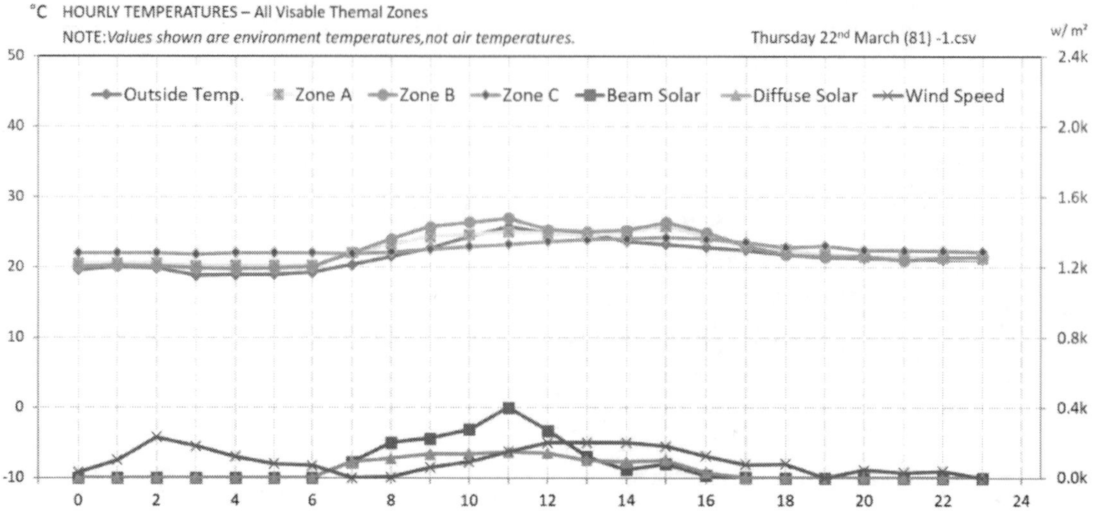

Figure 3. Indoor/outdoor hourly temperature.

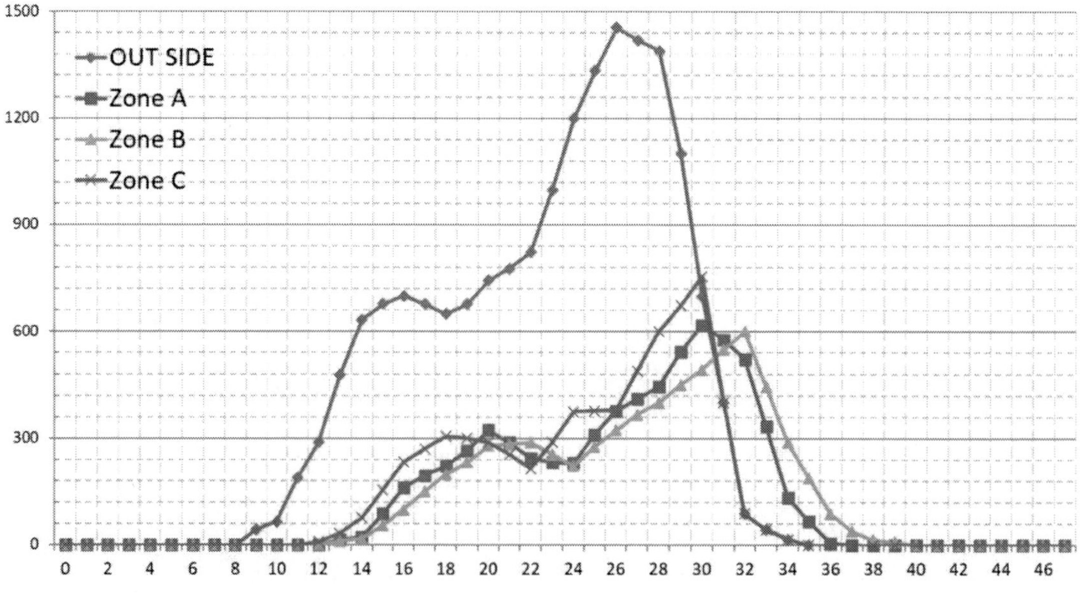

Figure 4. Outdoor and research area temperature distribution.

The calculation shows that the solar radiation accounts for the majority of heat gain in the summer, about 31.0%, followed by internal personnel and equipment, 21.4%, building structural conduction, 21.0%, and comprehensive temperature heat, 16.9%. Ventilation and inter-zonal heat gain and loss are the least, at about 8.0% and 1.8%, respectively. The duration of sunshine is longer and the mean temperature is high in summer. The building body is radiated by the sun for a long time and the heat from the sun conducts through the structure as the building body absorbs large amounts of heat; the heat is also conducted to the building through the structure; and, the radiation heat and the outdoor heat blown in by the wind enter indoors through doors and windows, adding to the heat from indoor personnel and equipment.

Regarding the three zones in the indoor space, numerical simulation is used for accumulating the number of annual temperature hours. People perceive more serious discomfort in zones A and B than in zone C. Discomfort in zone B appears to be even higher; visitors explained that they felt uncomfortable in zone B, and rarely expressed that they felt chilly in the building. Apparently, the indoor temperature of the visitor center is high in summer. As a result, visitors perceived higher discomfort.

4 Conclusions

Taiwan, with its hot-humid climate, was selected as the research subject. Aiming at the open building environment of a visitor center, the envelope surface temperature, physical environment measurements, and numerical simulation are analyzed. The research result reveals that sunshine orientation affects the change of indoor temperature. Moreover, longer duration of sunshine continuously increases the temperature, and the heat is trapped indoors and slowly distributed when the thermal resistance is high and the thermal conductivity coefficient is low, resulting in sultriness. The heat gain in the visitor center throughout the year results from solar radiation, 31%, personnel and equipment, 21.4%, and structural conduction, 21%. Under the same weather conditions, the shadow caused by the building orientation affects the indoor temperature. Various data reveal that the indoor temperature in zone B is obviously higher with greater discomfort than in other zones in summer. However, the temperature and discomfort are lower in this zone in the winter. These results could serve as a reference for future improvement or the design of other buildings.

References

1. R.L. Hwang, M.J. Cheng, The Open Construction and Building Technology Journal **1** (2007).
2. C.A. Roulet, Journal of the Energy and Environment **33**, 3 (2000).
3. Construction & Planning Agency, *Building Energy-Saving Design Specifications*, Ministry of the Interior, Taiwan, (2011).
4. ASHRAE, ANSI: Standard 55-2004, *Thermal Environmental Conditions for Human Occupancy, Atlanta: American Society of Heating* (Refrigerating, and Air-conditioning Engineers. Inc., USA, 2004).
5. C. Bouden, Journal of the Renewable Energy **32**, 1 (2007).
6. Y. Etzion, E. Erell, Building & Environment **35**, 5 (2000).
7. Ecotect, *Ecotect Tutorial. Ver. 5.20*, Joondalup, WA, Square One research Pty Ltd., (2003).
8. T.C. Liang, W.C. Hsu, Journal of National Huwei University of Science & Technology **32**, 1 (2014).
9. G. Mihalakakou, A. Ferrante, Energy Conversion & Management **41** (2000).

DOI: 10.1051/matecconf/20179517003

Study for the Pile Foundation on Steep Slope Applied in UHV Transmission projects

Jin Li and He Chuan

Hubei Electric Power Design Institute, Wuhan, China

Abstract. When designing a pile on the steep slope, the calculation is based on a hypothesis of horizontal ground. Obviously, it is difficult to estimate the difference between the calculation result and the actual loading behaviour of the pile foundation. For a UHV (Ultra High Voltage) power transmission project in China, results are compared by two kinds of methods. One is traditional theory; the other is numerical simulation method. It shows that quite different deformation of ground foundation occurs near two sides of the pile cap. The traditional calculating method for the pile foundation can consider the effect by increasing the part of the pile above the ground surface. Reasonability of the increased value can be assessed by the numerical method. The result of numerical simulation shows that the maximum deformation appears in the middle of the pile, but it occurs in the top of the pile by the traditional method. And deformation is evidently larger by the traditional method. Deformation curve is different between the traditional method and the numerical method which can consider frictional resistance around the pile. Quite different are vertical stress values of the pile obtained by two methods.

1 Introduction

With the implementation of Chinese "West-East electricity transmission project", "UHV (Ultra High Voltage) power transmission" and other significant energy strategies, the technology of ± 800kV HVDC power transmission has been widely used for the output of hydropower from the Jinsha River and Yalong River. Some projects have been put into operation, such as Xiangjiaba-Shanghai ± 800kV HVDC power transmission project, Jinping - Southern Jiangsu ± 800KV HVDC power transmission project. With the construction of these important projects, power transmission towers are higher, so the vertical and the lateral load applied on one foundation are getting larger and larger [1].

As is known to all, China has rich hydropower resources in Sichuan, Yunnan, Tibet, and so on. There are more high mountains, steep slopes in these regions. So the pile foundation on steep slope is very common [2-3]. Currently, it is based on the hypothesis of flat ground layer when designing a pile foundation subjected to horizontal load, which is suggested by the code. So it is difficult for engineers either designing the pile foundation on steep slope or locating the tower near a steep slope. Thus, it is valuable in theory and practice to study loading behaviour of piles on steep slope and its calculation method.

Foreign researchers have made some study earlier about the pile foundation bearing horizontal load. Achievements have been obtained in the field of pile groups, deep foundations, pile on slope or Rigid soil, etc [4-7]. Recently, with the construction of highways, hydro projects, and UHV power transmission projects in China, it have made some researches about design methods for pile foundation on slope and the interaction between pile and soil. These have produced many findings [8-11].

However, few of results have referred to the loading behaviour of pile foundation pulled up on a steep slope [8]. For power transmission projects, uplift loads often play a decisive role in the design of pile foundation, which is typically different from piles in other projects.

Based on the available achievements, theoretical analysis and numerical calculation are both conducted for a pile foundation on steep slope in a Chinese UHV power transmission project. It shows distinctly the loading behaviour of pile foundation on steep slope either pulling up or pressing down, which can offer data support for designing the pile foundation on steep slope in transmission projects.

2 Project background

Figure 1. Location of the power transmission towe

A ± 800kV HVDC power transmission project passes through five provinces in China, the flight distance of which is over 2000km. The elevation difference of this project is nearly 2900m. Almost 60% of the line is in steep mountains, involving a large number of pile foundations on steep slope.

In this project a tower has a rotation of turning left 0° 37'. The location is in the ridge slope of a mountain along the Yangtze River, and it is a cypress grove (shown in Fig. 1). The front slope occurrence is 220° ∠15° ~20° , and the left slope occurrence is 140° ∠37° ~ 43° , and the right slope occurrence is 310° ∠ 35° ~42° . Dolomitic limestone appears sparsely in this area, the occurrence of which is 347° ∠18° .

Bedrock under the tower is dolomitic limestone, covered with a thin layer. The cover is of Quaternary eluvial hard plastic cohesive soil which has a thickness of 0.3~2.5m. Quaternary cover layer's thickness is generally 0.0 ~ 1.8m around the tower. In the water-eroded groove, some is relatively thicker, and the thickness is up to 3.8m. The mechanical parameters of each layer are shown in Table 1.

Table 1. Mechanical parameters of the rock and soil

Stratigraphic position	Name	Weight (kN/m^3)	Elastic modulus (MPa)	Poisson ratio	Cohesion (kPa)	Internal friction Angle (°)	Tensile strength (kPa)
top	silty clay containing rock fragments	19.0	5.5	0.42	40	10	10
middle	Weathered dolomitic limestone	22.5	26	0.30	60	22	60
bottom	Weak weathered dolomitic limestone	22.5	2.8E4	0.26	130	30	200

Table 2. Design parameters of the pile

diameter (m)	Depth under ground (m)	Depth above ground (m)
2.6	9.0	6.0

Table 3. Mechanical parameters of the pile

Weight (kN/m^3)	Elastic Modulus (GPa)	Poisson ratio
24.0	28.0	0.30

Table 4. Load cases

Case number		Load in direction X (kN)	Load in direction Y (kN)	Load in direction Z (kN)
Pressing down	standard value 1	591.11	-2868.15	523.70
	Design value1	798	-3872	707
Pulling up	standard value2	477.78	1951.85	377.04
	Design value2	645	2635	509

3 Theoretical analysis

Theoretical calculation adopts the method recommended in "Technical code for building pile foundations" [12], [13] (hereinafter referred to as "the Code"). Design parameters of the pile are listed in Table 2 and Table 3. Load cases are shown in Table 4. Cases listed in table 4 are calculated according to the Code. Calculation diagram is shown in Fig. 2.

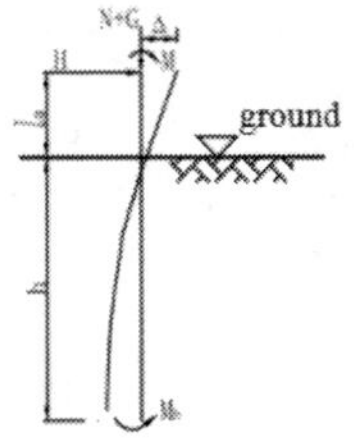

Figure 2. Calculation diagram of the pile

Assuming the cross-section of the pile is a plane anytime, at the ground surface, the horizontal displacement x_0 and rotation angle φ_0 should be

$$\begin{cases} x_0 = H_0\delta_{HH} + M_0\delta_{HM} \\ \varphi_0 = -\left(H_0\delta_{MH} + M_0\delta_{MM}\right) \end{cases} \quad (1)$$

Considering as a single pile, then $M_0 = M + Hl_0$, $H_0 = H$, coefficients of δ_{HH}, δ_{MH}, δ_{HM} and δ_{MM} are determined from the table in the code.

Horizontal displacement of the pile top is

$$\Delta = x_0 - \varphi_0 l_0 + \Delta_0 \quad (2)$$

In which $\Delta_0 = \frac{Hl_0^3}{3nEI} + \frac{Ml_0^2}{2nEI}$, EI is bending stiffness of the pile.

Results calculated for each load case are shown in Tables 5 and Table 6. The calculation increases the depth above ground to consider the effect of the actual steep slope. It can be found that the top of the pile has a displacement either in the case of pulling up or in the case of pressing down. Compared calculation results of two load cases, the ratio of the value of the pile top displacement is proportional to that of the absolute value of the resultant force of horizontal loads(X and Z direction). Besides, the position of the maximum bending moment is at the same depth in two cases, and the maximum stress occurs at the bottom of the pile. Due to the assumption of flat ground layer, it is difficult to consider the missing soil in the outside of the steep slope. Usually it is a simple and approximate method to increase the part of the pile above the ground surface. Obviously the results by this method need to be checked and compared by other methods.

DOI: 10.1051/matecconf/20179517003

Table 5. Calculation results of deformation and internal force

		Design value1	Design value2	Standard value1	Standard value2
At the ground surface	Horizontal displacement(mm)	4.01	3.09	2.97	2.29
	Rotation angle(FI) (rad)	-0.001039	-0.000801	-0.000770	-0.000593
	Axial force (kN)	-4636.538	1870.462	-3632.688	1187.312
	Shear (kN)	1066.139	821.648	789.730	608.632
	Moment (kN·m)	6396.836	4929.890	4738.377	3651.792
Pile top displacement (mm)		10.25	7.90	7.59	5.85

Table 6. Calculation of bending moment, maximum stress (depth is 0 at the design ground surface)

		Design value1	Design value2	Standard value1	Standard value2
Moment	Maximum bending moment (kN·m)	7893.379	6083.241	5846.923	4506.131
	Axial force at the maximum bending moment (kN)	-4922.84	1584.16	-3918.99	901.01
	Corresponding depth (m)	2.247	2.247	2.247	2.247
Stress	Maximum stress (KPa)	-400.133	-308.373	-296.394	-228.426
	Corresponding depth (m)	9	9	9	9

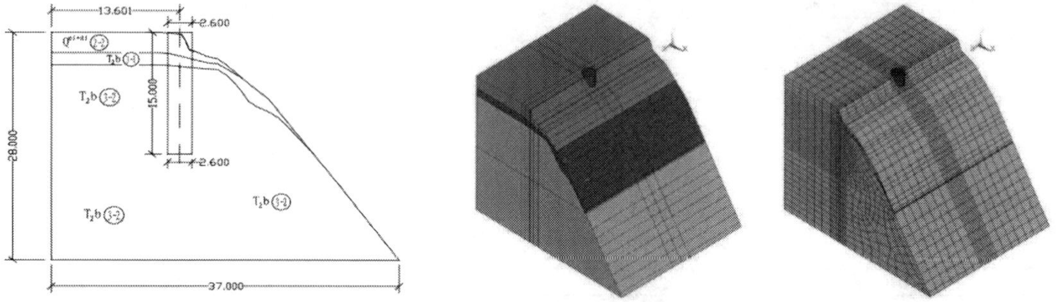

(a) design section (unit: m) (b) material zoning (b) meshing

Figure 3. The model of pile foundation on the steep slope

4 Setting of numerical analysis

4.1 Model and parameters

Numerical calculation is carried out by the method of FFM. The design section and calculation model are shown in Figure 3. The whole calculation model is divided into 26660 elements, 28689 nodes. According to the geological survey data, the parameters of the rock and soil are identified in Table 1 based on practical conditions. And the physical and mechanical parameters of the pile can be found in table 3.

4.2 Simulation method of pile-soil interaction

In order to fully evaluate the pile-soil interaction, it is necessary to determine stress and strain distribution of the pile and the soil respectively, and the data of displacement and stress distribution on the contact area. Then a comprehensive study can be made about factors influencing the pile-soil interaction. This problem is one of contact problems between different kinds of media in solid mechanics. It is a typical nonlinear problem, which could contain nonlinear material, or nonlinear contact, etc. In the calculation, the interaction between the pile and the soil is simulated by setting contact surfaces between the pile and the soil. That is to say, Surface-to-Surface contact is built. Then the interaction effect between the two kinds of media is represented by the contact stiffness, $F=k\Delta$ (k is the contact stiffness). This is shown in Fig. 4.

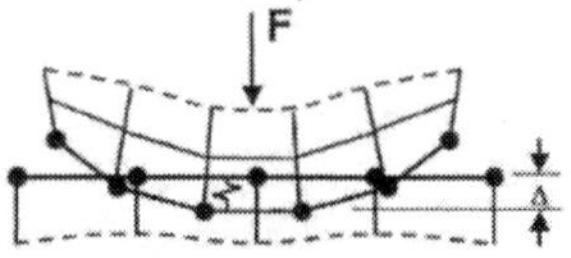

Figure 4. Schematic diagram of pile-soil contact

Eight-node isoparametric element is used in rock and soil (SOLID185). When analyzing the contact of the pile and soil, surfaces of the pile are of rigid target surface (TARGE170), and surfaces of the rock and soil are of contact surface (CONTA174). The target surface and the contact surface are combined together to form a contact pair.

5 Results and discussion

5.1 Deformation behavior of the pile

In view of the horizontal displacement of the pile, at the depth of 6m from the pile top, in two cases the result of the pile axis by numerical simulation is close to that of standard values of loads by theoretical calculation according to the code (see Table 7).

Table 7. deformation at the section of 6m depth of the pile

Method	Horizontal displacement (mm)	
	in pressing case	in pulling case
Theoretical method in the code	2.97	2.29
Numerical method	2.46	2.41

Based on the traditional theory suggested by the code, the result is reliable using a larger height value (6.0m) of the pile above ground surface than the actual height. The horizontal distance is 7.0m from the centre of the section at the depth of 6.0m to the slope surface. It is about 2.7 times the diameter of the pile. Overall, numerical results show that the maximum deformation occurs in the middle of the pile and theoretical results is at the top of the pile (see Table 5 and Fig. 5). Considering a free height of 6.0m, the theoretical deformation is significantly larger. In fact the height of 6.0m is surrounded by soil and rock on the steep slope.

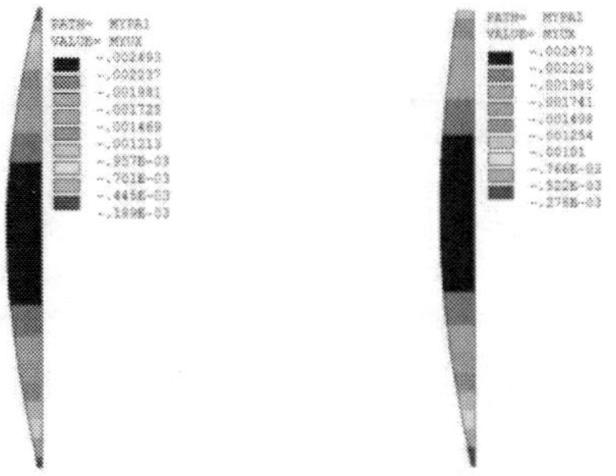

(a) in pressing case (b) in pulling case
Figure 5. Horizontal displacement (Ux) at the pile axis (Unit: m)

In view of vertical displacement, the tension side obviously has a difference with the compression side of the pile (see Fig. 6), which is consistent with the characteristics of bending deformation. This could not be considered by the theoretical method.

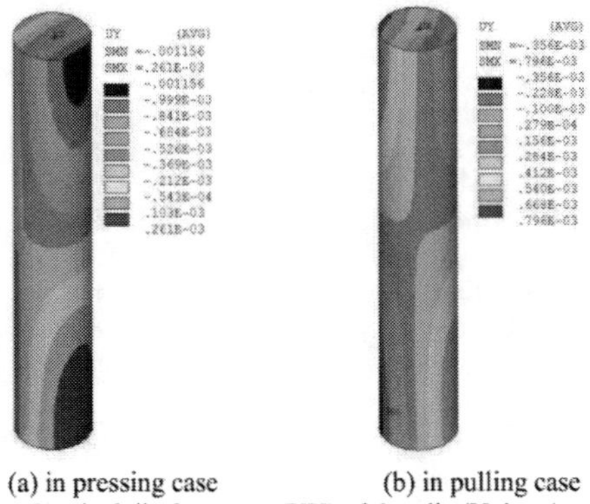

(a) in pressing case (b) in pulling case
Figure 6. Vertical displacement (UY) of the pile (Unit: m)

Based on the traditional theory suggested by the code, the result is reliable using a larger height value (6.0m) of the pile above ground surface than the actual height. The horizontal distance is 7.0m from the centre of the section at the depth of 6.0m to the slope surface. It is about 2.7 times the diameter of the pile. Overall, numerical results show that the maximum deformation occurs in the middle of the pile and theoretical results is at the top of the pile (see Table 5 and Fig. 5). Considering a free height of 6.0m, the theoretical deformation is significantly larger. In fact the height of 6.0m is surrounded by soil and rock on the steep slope.

In view of vertical displacement, the tension side obviously has a difference with the compression side of the pile (see Fig. 6), which is consistent with the characteristics of bending deformation. This could not be considered by the theoretical method.

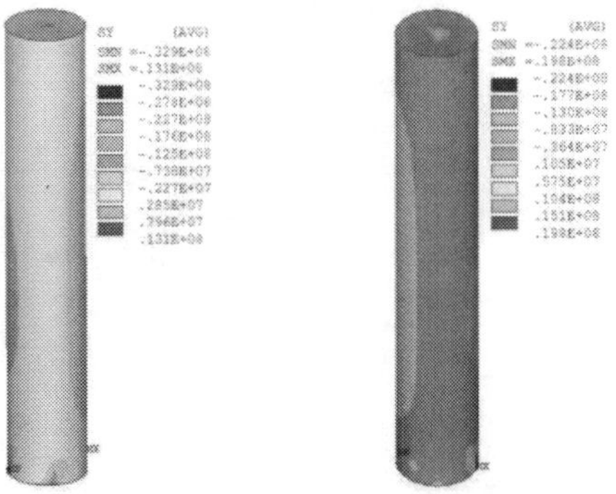

(a) in pressing case (b) in pulling case
Figure 7. Vertical stress (SY) of the pile (unit: Pa)

5.2 Vertical stress (Sy) of the pile

In view of vertical stress (see Fig. 7), in two cases the traditional method or the numerical method both show that maximum vertical stress emerged at the bottom of the pile. Two methods also reveal that the position of maximum moment is basically same, which is located on the middle of the pile.

From the view of stress value, it is quite different between theoretical calculation and numerical simulation. The theoretical calculation assumes that there is no lateral frictional force, and adopts the equation of deflection curve. However, the numerical simulation results, which have considered the frictional resistance, show that there are some differences in the curves of deformation, which leads to the difference of the value of stress in Y direction. In addition, numerical results indicate stress concentration at the bottom of the pile.

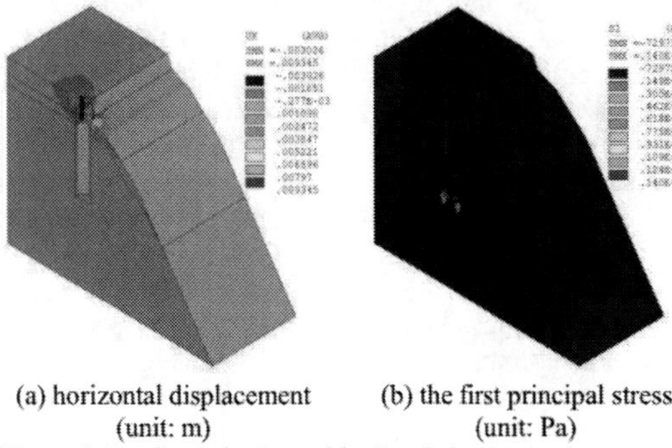

(a) horizontal displacement (unit: m) (b) the first principal stress (unit: Pa)
Figure 8. Loading behaviour of the foundation in pressing case

5.3 Foundation deformation and stress

Fig. 8 shows the horizontal displacement and the first principal stress distribution in pressing case. Fig. 9 shows them in pulling case. Due to the landform of steep slope, displacement down slope is large. In two cases it could see obvious deformation of the covering soil layer near the pile cap, but it is not obvious for the horizontal displacement of the bedrock. Compared to results of the pile foundation in flat ground, the pile foundation on steep slope presents an asymmetry image as to the ground deformation near the pile head. This is because of the difference of in-situ stress between flat ground and steep slope.

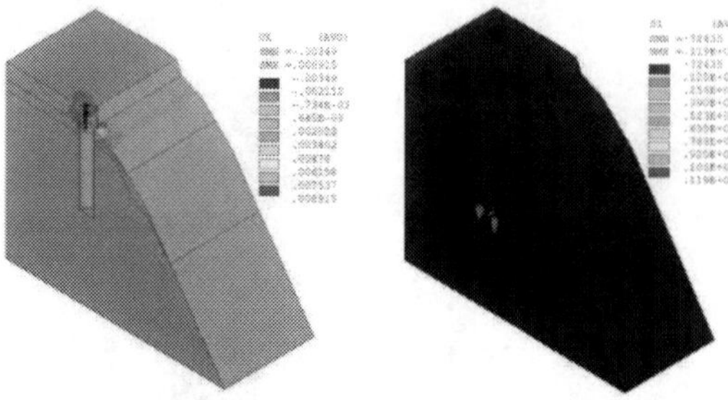

(a) horizontal displacement (unit: m) (b) the first principal stress (unit: Pa)

Figure 9. Loading behaviour of the foundation in pulling case

Stress concentration also can be seen for the first principal stress at the bottom of the pile in two load cases. This is because of the rotation of the pile occurs in the foundation under horizontal loads, when considering pile-soil interaction by contact surfaces.

6 Conclusions

Based on the above analysis, conclusions can be made as follows:

(1) For the pile on steep slope, the traditional calculation models could basically meet the need of UHV transmission projects, but the effect of steep slope can never be ignored. It is a simple and effective method to increase the part of the pile above the ground surface. The value increased is suggested by a formulation of $(2.5\sim3.0)\times D\times \tan\alpha$, in which D is the diameter of the pile , and α is the dip angle of the steep slope.

(2) The result of numerical simulation shows that the maximum deformation appears in the middle of the pile which has been supposed to appear on the head of the pile according to theoretical calculation. It is considered unconstrained for the part above the ground surface increased factitiously, but this part is surrounded by soil and rock on the steep slope. The theoretical deformation is obviously larger than the actual value. When subjected to horizontal load, the theoretical calculation assumes that there is no lateral frictional force around the pile, and adopts the equation of deflection curve. However, the numerical simulation results, which have considered the frictional resistance, show that there are some differences in the deformation curves. Two methods have showed the same position of the maximum moment, which is located in the middle of the pile.

(3) For the pile foundation on steep slope, numerical simulation presents an asymmetry image as to the ground deformation near the pile head. This is because of the difference of in-situ stress between flat ground and steep slope. Due to the effect of horizontal loads, stress concentration is obvious at the bottom of the pile foundation on steep slope either in pressing case or in pulling case.

Overall, the topography of steep slope will result to obvious differences of the deformation and stress distribution on both sides of the pile foundation. When considering surface frictional resistance of the pile, the deformation of the pile is more complicated. So the method of numerical simulation can be used for checking results by the theoretical method according to the code when designing pile foundation on steep slope for UHV transmission projects. And further studies should be made to modify the traditional theory considering the effect of a steep slope.

References

1. Shu Yinbiao, Liu globe, Gaoli Ying, et al. 800kV 6400MW UHVDC transmission project design. Power system technology, **30** (1):1-8, (2006).
2. Zhang Yongjie, Li Youjun, Zhao Minghua, et al. Design and calculation method of pile groups under high and steep slope. Chinese Journal of highway and transport, **27** (10).84-92, (2014).
3. Zhao Ming-hua, Yin Ping-bao, Zhang Yong-jie, et al. The Design and Calculation Method of Pile-column Bridge Pile Foundation in High and Steep Slope. Engineering Mechanics, **30** (3):106-111, (2013).
4. Alizadeh, M., and Davisson, M. T. Lateral load test on piles-Arkansas River Project. J. Soil Mech. and Found. Div., ASCE, (96), 1583-1603, (1970).
5. ANG L M, MCVAY M C, HAN S J, et al. Effects of dead loads on the lateral response of battered pile groups. Canadian Geotechnical Journal, (39): 561-575, (2002).
6. Reese LC, Welch RC. Lateral loading of deep foundations in stiff clay. Journal of Geotechnical Engineering Division, ASCE, **101**(GT7):633-649, (1975).
7. Gabr MA, Borden RH. Lateral analysis of piers constructed on slopes. Journal of Geotechnical Engineering. **116**(12):1831-1850, (1990).
8. He Jian. Testing study on anti-draw bearing properties of cast-in-place rock-socked pile . Chinese Journal of Rock Mechanics and Engineering, **23** (2): 315-319, (2004).
9. Rao Pingping, Li Jingpei, Zhan Yue. The adjacent slopes sink pile squeezing effect of soil particle flow numerical simulation. Journal of Water Resources & Water Engineering, **24** (4):1-5, (2013).
10. Gong Xiarrbing, Yang Minghui, Zhao Minhua, et al. Load-bearing Mechanism Model Test for Bridge Pile Foundation in High-steep Transverse Slope. China Journal of Highway and Transport, **26**(2):56-62, (2013).

DOI: 10.1051/matecconf/20179517003

11. Cheng Liu-yong, Xu Xi-chang, Chen Shan-xiong, et al. Model test and numerical simulation of horizontal bearing capacity and impact factors for foundation piles in slope . Rock and Soil Mechanics, **35**(9): 2685－2691, (2014).
12. Electric Power Planning & Engineering Institute. DL/T 5219–2014 Technical code for design of foundation of overhead transmission line. Beijing: China Planning Press, (2014).
13. The Professional Standards Compilation Group of People's Republic of China. JGJ94-2008 Technical Code for Building Pile Foundations. Beijing：China Architecture & Building Press, (2008).

Study on the Influence of Piloti Ratio on Residential Block's Wind Environment by Coupled Simulation Method

Tianyu Xi[1], Jianhua Ding[2] and Hong Jini[1]

[1]*Architecture School of Harbin Institute of Technology, Harbin City, 150006, China*
[2]*School of Architecture, Tsinghua University, Peking City, 100084, China*

Abstract. The environmental degradation results in not only the increase of energy consumption in cities but also healthy problem of people, and many researches regarding to outdoor environment and thermal comfort are carried out. Wind velocity is considered to be one of the thermal comfort influential factors, and there are few work studied the comprehensive influence of piloti on wind environment by coupled simulation method, which offers high precision CFD simulation results. This paper taking the city of Guangzhou, China, as an example, adopting coupled simulation method, studied the wind environment of residential blocks under various piloti ratio (0%, 40%, 60%, 80 and 100%). Both of the cumulative distribution and average wind velocity under piloti and out of piloti area are simulated and analysed. This work is expected to offer reference for architects and civil engineers in outdoor environment and thermal comfort design, and contribute to further study in sustainable and healthy building environment research field.

1 Introduction

The degradation of outdoor environment leads to not only energy consumption but also people's health problems. of Urban heat island is now regarded as one of the most serious environmental problems, and to optimize the outdoor thermal environment and thermal comfort, a lot of research were carried out [1]-[4], including plant, high reflective capability materials of building envelope, green building walls and roofs, improvement of city ventilation, use of sea wind and decrease of artificial heat, etc.

To get good outdoor ventilation and create shadow areas, piloti is commonly used in tropical and subtropical climate zones (Fig. 1), but there are few studies reveal the comprehensive influence of piloti on wind environment in residential blocks, and furthermore, a coupled simulation method, which offers high precision prediction result, is seldom taken into consideration in piloti research field. Xi et al. published field measurement results of various human built elements in Guangzhou [5]-[8], including piloti, but no systematic analysis by simulation was used in those research. The influence of piloti on mean radiant temperature simulated by 3-D unsteady state heat balance radiation calculation method was published by Xi, Hong Jin et al. but no wind environment was mentioned [9], [10]. Xi, Qiong Li et al. studied on the effects of semi-open space (100 percent ratio piloti) on the outdoor thermal environment of residential communities, by adopting coupled simulation method, but no different piloti ratio cases were compared [11].

This research takes residential blocks in Guangzhou, China as an example, aiming at study on the comprehensive influence of piloti ratio on the wind environment. A coupled simulation method, which combines the commercial STAR-cd software and homemade program (coded in Tokyo University) was adopted in this study, and this method is expected to get high precision prediction result.

Figure 1. Building piloti design in subtropical climate zones.

2 Methodology

2.1 Coupled simulation method

Firstly, non-isothermal CFD analysis is carried out by STAR-cd, using data from local meteorological bureau (wind velocity and prevailing wind direction, and air temperature) and observed data (ground and building surface temperature). In this step, the surface temperature is set to be mean value as traditional method, which cannot reflect the real surface situation and will cause buoyance calculation omit in the velocity simulation.

Secondly, ground and building surface temperatures are recalculated in step 2 by homemade program (coded in Tokyo University), which is based on unsteady state heat balance calculation including 3-dimentional radiation and 1-dimentional conduction calculations.

Based on outcomes in above 2 steps, CFD analysis is carried out one more time in STAR-cd and more accurate results will be obtained (Fig. 2).

2.2 Time selected in simulation

Based on TMY (typical meteorological year) data, 24-hour meteorological data are averaged per day to get mean daily data. Consequently, the whole summer mean data are obtained by averaging the cumulated mean daily data. Finally, analysis date is determined by choosing the day whose data are most close to the whole summer mean data. Therefore, 14th July in Guangzhou is selected as a typical day of summer. Analysis time is selected based on at the time when maximum temperature occurred during analysis date (15:00).

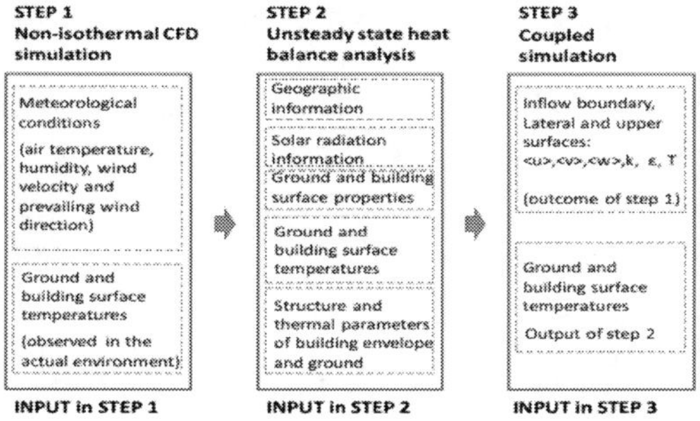

Figure 2. Coupled simulation chart flow.

2.3 Cases design

The domain size, wind direction and building orientation are shown in Fig. 3, the distance between buildings and inflow and wall boundary is set to be 5 times of building height, and the distance between buildings and outlet boundary is set to be 15 times of building height, both of which are set due to AIJ guidelines. The evaluated area is set in the middle of the community to avoid the lateral influence of CFD simulation. The piloti ratio of each case is set to be 0, 40, 60, 80 and 100 percent. To avoid the lateral influence, the standard area was chosen in the centre of simulated cases (Fig. 4).

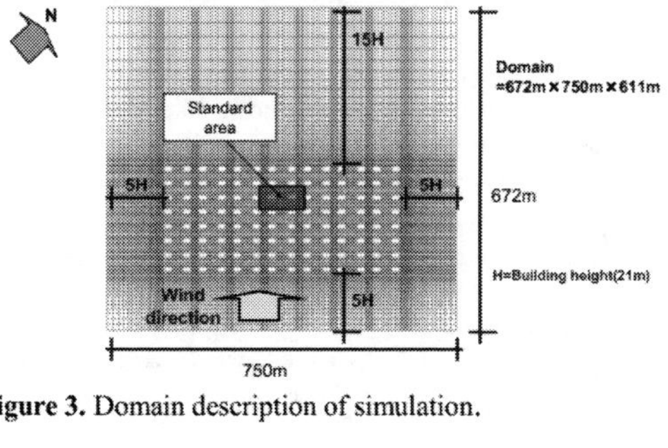

Figure 3. Domain description of simulation.

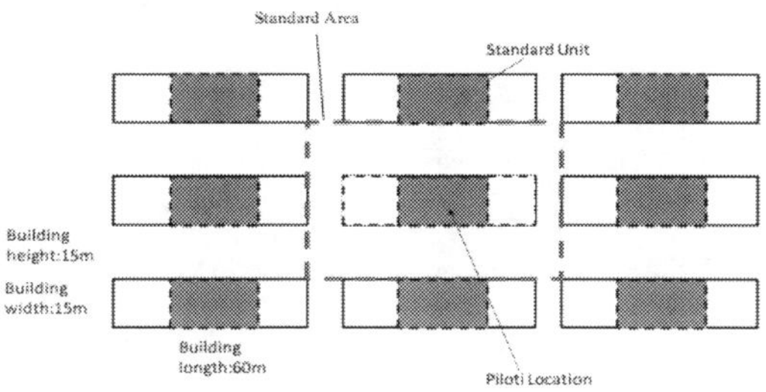

Figure 4. Standard unit and standard area.

3 Results and analysis

3.1 Wind environment of non-piloti area

The cumulative distribution of wind velocity of those non-piloti areas is shown in Fig. 6. It shows that when piloti ratio is lower than 40 percent, the average wind velocity is clearly divided into 2 levels, and the blank interval is vanished when piloti ratio reaches 60 percent, which should be because the non-piloti area includes wind channel area and space before and after buildings. When piloti ratio is 100 percent, no low wind velocity exists anymore, and the lowest wind velocity is over 0.9m/s.

The average wind velocity of non-piloti area is plotted in Fig. 5, and it shows that 40 percent piloti ratio cannot optimize the wind environment efficiently, and the slop becomes shaper when piloti ratio increases from 60 to 80 percent, which seemed to be the inflection point of non-piloti space wind velocity.

3.2 Wind environment of piloti area

The cumulative distribution of wind velocity of piloti areas is shown in Fig. 7, and the average wind velocity of non-piloti area is plotted in Fig. 8. When piloti ratio equals to 40%, the wind velocity is very low, and no velocity is beyond 0.3m/s. Each 20 percent piloti ratio increase can highly improve the wind velocity, and both of the lowest and highest wind velocity value keeps increasing along with the piloti ratio's change from 40% to 100%. 80 percent and 100 percent piloti ratio can highly optimize the wind environment under piloti area, leaving blank in low velocity intervals (Fig.7). The increase slop changed at the 80% piloti ratio, and before that the increasing trend seemed to be constant.

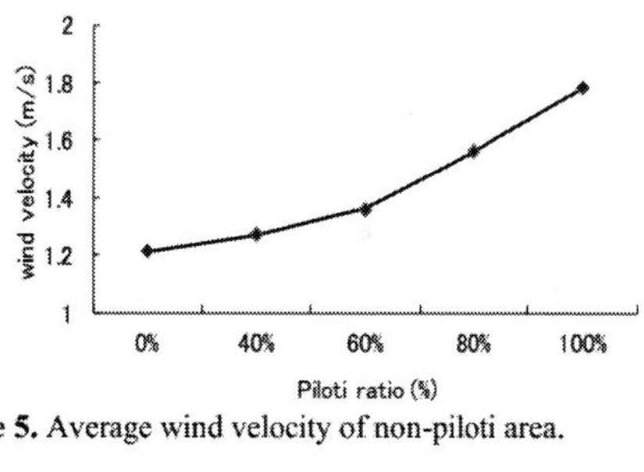

Figure 5. Average wind velocity of non-piloti area.

DOI: 10.1051/matecconf/20179517004

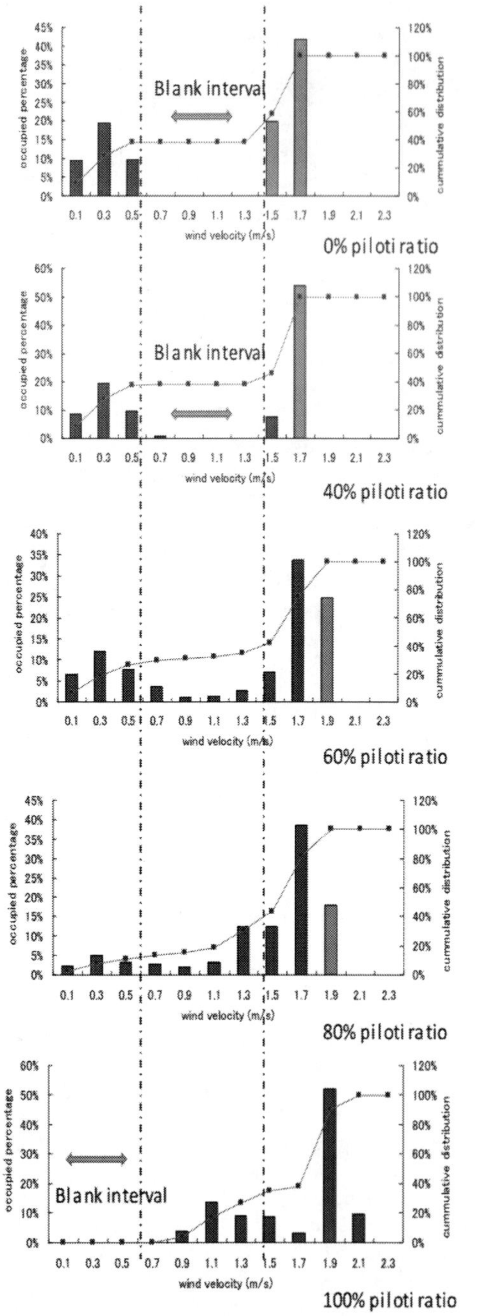

Figure 6. Wind velocity cumulative distribution of non-piloti area.

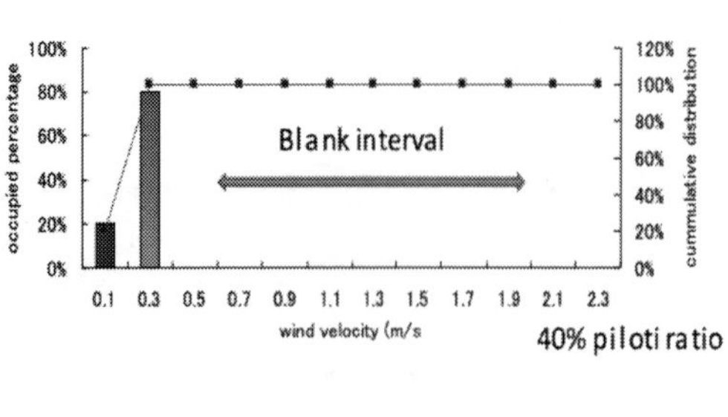

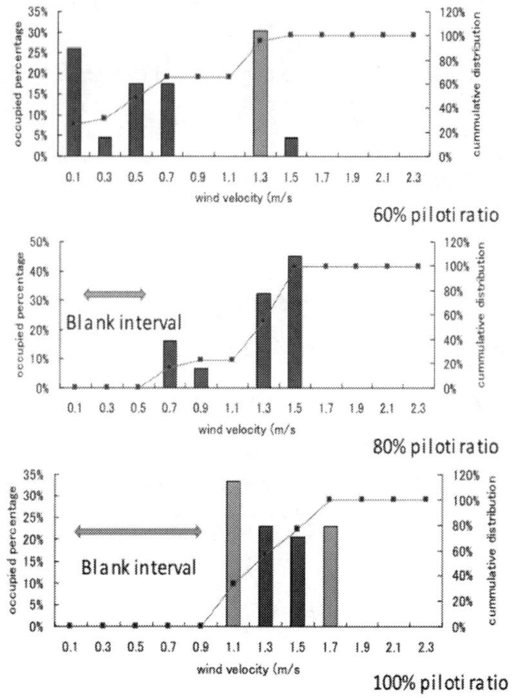

Figure 7. Wind velocity cumulative distribution of piloti area.

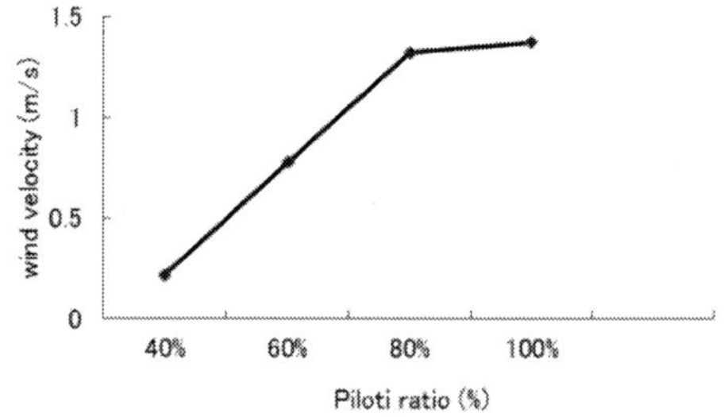

Figure 8. Average wind velocity of piloti area.

4 Conclusions

This work studied the influence of piloti ratio on residential block's wind environment by coupled simulation method, and to avoid the lateral influence, the standard area was chosen in the centre of simulated cases, which was divided into 2 parts to be analysed, the area under piloti and the space out of piloti.

For the non-piloti space, high wind velocity area (wind channel between buildings) and low wind area (before and after the buildings) are clear to be observed when piloti ratio is lower than 80 percent, and When piloti ratio is 100 percent, no low wind velocity exists anymore, and the lowest wind velocity is over 0.9m/s. For the average wind velocity, the slop becomes shaper when piloti ratio increases from 60 to 80 percent, which seemed to be the inflection point of non-piloti space wind velocity.

For area under piloti, the optimization on wind environment of 40 percent piloti ratio is very limited, and 80 and 100 percent piloti ratio is recommended in subtropical climate cities, whose lowest wind velocity is 0.7m/s and 1.1m/s independently. The increase of velocity is constant before 80 percent, and only minus increase was observed when piloti ratio changed from 80 to 100 percent.

Acknowledgement

This work was financially supported by the National Natural Science Foundation of China (Grant No. 51408160), China Postdoctoral Science Foundation (2015M571419), and the Project-sponsored by SRF for ROCS, SEM. Work partially supported by the Fundamental Research Fund for the Central Universities (HIT.NSRIF.2014 075), Harbin Science and Technology Bureau (2015RQQXJ068) and Heilongjiang Province Postdoctoral Science Foundation(LBH-Z15084).

References

1. Tetsu Kubota, Supian Ahmad, Journal of Asian Architecture and Building Engineering, 199-206, (2006)
2. Uehara Kiyoshi, Ymamao Yukio, Oikawa Susumu and Mochida Akashi, Japan Society for Atmospheric Environment, **42** (5), (2007)
3. Takanobu Moriizumi, Toru Kawai, Atsushi Inagaki, Manabu Kanda, Paper Collection of Hydro Science and Hydraulic Engineering, **52**, (2008)
4. Yoshida Shinji, Murakami Shuzo, Mochida Akashi et al. Paper Collection of Planning System of Annual Meeting of Architectural Institute of Japan, No. 529, 77-84, (2003)
5. Tianyu Xi, Qiong Li, Akashi Mochida, Qinglin Meng, Journal of Habitat Engineering, **3**(2), 253 – 260, (2011)
6. Tianyu Xi, Qiong Li, Akashi Mochida, Qinglin Meng, Building and Environment, **52**, 162-170, (2012)
7. Tianyu Xi, Qiong Li and Akashi Mochida, Summaries of Technical Papers of Annual Meeting of Architectural Institute of Japan, Environmental Engineering, 805-806, (2011)
8. Tianyu Xi, Qiong Li, Akashi Mochida and Qinglin Meng, 4th International Conference on Human-Environment System, ICHES2011 in Sapporo, Japan, 85-90, (2011)
9. Tianyu Xi, Jianhua Ding and Hongjin, Journal of Harbin Institute of Technology, **20** (4), 91-95 (2014)
10. Tianyu Xi, Hong Jin and Jianhua Ding, International Conference on Materials Engineering and Environment Science, MEES2015 in Wuhan, China, 478-485, (2015)
11. Tianyu Xi, Yingli Xuan and Akashi Mochida, Technical papers of Annual Meeting of Society of Heating, Air-conditioning and Sanitary Engineers of Japan, 405-408, (2009)

The maximum sloshing wave height evaluation in cylindrical metallic tanks by numerical means

Walid Samir Manser[1], Mokhtar Touati[1] and Rui CarneiroBarros[2]

[1] *Faculty of Civil Engineering, USTHB University, Structural and Materials Department, 16111,Bab Ezzouare, Algiers*
[2] *Faculty of Engineering, FEUP University, Civil Engineering Department, 4200-465, Porto, Portugal*

Abstract. The metallic cylindrical storage tanks are very common structures in the field of civil engineering; These facilities are especially used in the industry in which they are used to store all kinds of products-which are for the most toxic or flammable. The tanks are also used in the storing of drinking water. When earthquakes, these structures must be strictly maintained in order to avoid that they lose their precious contents causing reactions that can cause more damage than the earthquake itself. In this study, the effects of the liquid height, the geometric parameters of tanks in the variation of the maximum sloshing wave height are studied: For this purpose, the software ANSYS V11.0 is used for modelling the tanks, the results found are compared with thus given in the Euro code 8

1 Introduction

Seismic events hitting Japan, the USA and Turkey have shown that the storage tanks are highly vulnerable to large seismic movements; failure modes such as: sliding, buckling of the walls and thero of damage (Figure 1) can lead to the outbreak of fire, that is why important parameters such as sloshing of fluid, basis upliftand the interaction: liquid-shell shall be considered in the dynamic analysis.

Figure 1. Damage on the roof of a tank after excessive sloshing liquid content.

The importance of LST seismic analysis is most seen their complex behaviour during vibration, a behaviour that has been proven during major earthquakes in areas containing tanks **(Longbeach 1933, Chile 1960, Alaska 1964, Turkey1999, Japan 2011)** where accidents occurs: the tanks containing flammable or toxic products are real bombs, and can sometimes cause more damages than the earthquake itself. During the **Kocaeli earthquake in Turkey in1999**, that hit the refinery **Turpas - 7thEuropean refinery**-and caused more than **US $500 millions** of damages and losses. Duringthis earthquake, , covered tanks have suffered excessive sloshing and their roof rubbing with the walls and then created instantly sparks igniting the liquid .The fire then spread to crude oil tanks damaging 30to45 tanks covered by a roof (Figure 2).In addition to economic losses, large quantities of toxic material have been released in the environment. After every earthquake, the observations of damage to facilities are made in order assess the losses but also verify the validity of design standards, and improve them if necessary.

Figure 2. Activation of a fire in a tank in Turpas raffinery after Koaceli's Earthquake 1999 (Turkey).

2 Importance of sloshing and the MSWH (maximum sloshing wave height)

The sloshing phenomenon is the oscillation of the liquid surface due to excitation of the container. In fact, during an earthquake, the fluid in a tank oscillates with a rather special way: The upper part move with large periods of time, this part is usually called: the convective part. While the rest of the liquid (the lower part) called the impulsive part has a similar movement to the rigid shell of the tank (Figure 3).

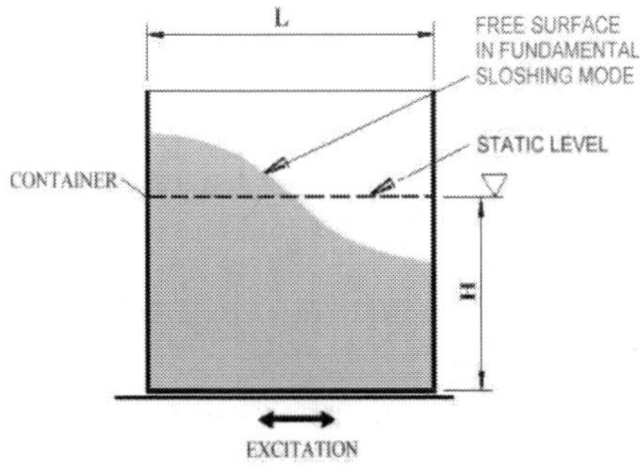

Figure 3. Schematic presentation of the sloshing.

Since the vertical movement of the convective fluid can lead to the loss of the liquid contained in the tank oralthough a damage in the tank's roof, the sloshing height must be held into account during the design and calculation of tanks.

3 The sloshing according to standards

The most important aspect for the tanks is the sloshing behaviour and the dynamic response of the tanks under seismic excitation.

Sloshing waves with large amplitudes cause damage to the roofs of the tanks and make them temporarily unusable, therefore, the liquid in this area can cause fires. To avoid such disasters, the sloshing of the liquid in the tanks must be considered to determine the minimum **"Free-Board"** required .The **"Free- Board"** is given by the expressions available in the various regulations covering the tanks:

EC8

$$d_{max}=0,84\ R\ S_e(T_{c1}) \tag{1}$$

where: (Tc1) is the appropriate elastic spectrum of the acceleration in g.

When an impulsive acceleration act son a liquid, this can produce a pressure significant hydrodynamic pressures produced by the free surface on the tank walls, the estimation methods of these impacts and the hydrodynamic pressures are not yet well developed and are often determined by the experimental means.

The tank diameter **(D)**, the height**(H)**and the liquid level**(HL)**are the dominant factors for evaluating the sloshing of the fluid.

4 Purpose of the study

Given the importance of sloshing in the seismic response of metallic cylindrical tanks to an excitation, the study of the evolution of the maximum sloshing height according to various parameters such as the geometric properties of the tank is done. For this purpose, 45tanks constrained at the base with different ratios (H/R) were treated with dynamic on- linear analyzes, these tanks are simulated in ANSYS v11.0 Software program using the finite element method for 3Dmodelling of the fluid-shell system. Material properties were assumed homogeneous. The shell has been modelled by SHELL63 elements and the fluid contained in the tank is modelled by FLUID 80elements.The results are then compared with the values prescribed in EUROCODE8.

For the analysis , a recording of Tabas earthquake with PGA 0,328g was used , only excitation in the horizontal direction UX has been taken into account in the time-history analysis to study the variation of the MSWH's values during the earthquake (see Fig. 4, 5).

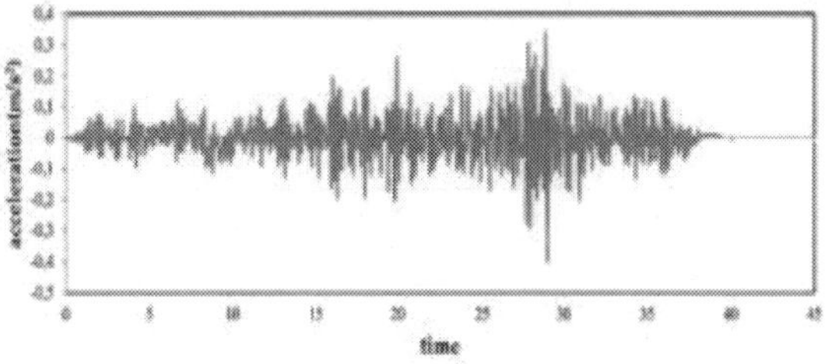

Figure 4. Horizontal component of the recording of Tabas's Earthquake.

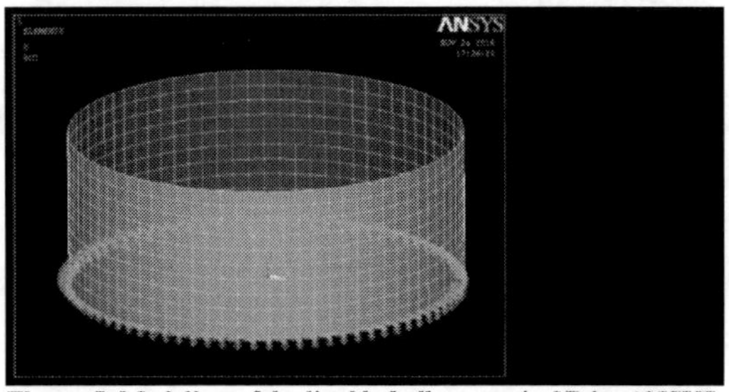

Figure 5. Modeling of the liquid-shell system in 3D by ANSYS Software.

5 Specifaication of the studied tanks

$$T_d=\frac{4,9\ D\ (Hl-0,3)G}{Sd}+CA \tag{2}$$

$$T_t = \frac{4,9\, D\, (Hl-0,3)}{St} \quad (3)$$

Where:

D: Tank diameter.

H_l: The liquid's height

G: the specific density of the liquid content.

CA: the corrosion factor.

S_d: The allowable stress for calculation conditions in MPa.

S_t: The allowable stress for the hydrostatic test in MPa.

The results obtained are summarized in the following tables:

Table 1. Tanks's dimensions of the 1st Part.

Part I					
n	Tank height H(m)	Liquid height H_l (m)	Tank radius R (m)	Thikness t (mm)	Thikness adopted t_{adop} (mm)
1	5	2,5	10	1,58	10
2	6	3	10	1,93	10
3	7	3,5	10	2,29	10
4	8	4	10	2,65	10
5	9	4,5	10	3	10
6	10	5	10	3,36	10
7	11	5,5	10	3,72	10
8	12	6	10	4,08	10
9	13	6,5	10	4,44	10
10	14	7	10	4,8	10
11	15	7,5	10	5,16	10
12	16	8	10	5,52	10
13	17	8,5	10	5,88	10
14	18	9	10	6,24	10
15	19	9,5	10	6,6	10

Table 2. Tanks's dimensions of the 2nd Part.

Part II					
n	Tank height H(m)	Liquid height H_l (m)	Tank radius R (m)	Thikness t (mm)	Thikness adopted t_{adop} (mm)
16	20	12	10	8,36	30
17	20	12	12	10,04	30
18	20	12	14	11,72	30
19	20	12	16	13,39	30
20	20	12	18	15,07	30
21	20	12	20	16,74	30
22	20	12	22	18,41	30
23	20	12	24	20,09	30
24	20	12	26	21,76	30
25	20	12	28	23,44	30
26	20	12	30	25,11	30
27	20	12	32	26,78	30
28	20	12	34	28,46	30
29	20	12	36	30,13	30
30	20	12	38	31,81	30

Table 3. Tanks's dimensions of the 3rd Part.

Part III					
n	Tank height H(m)	Liquid height H_l (m)	Tank radius R (m)	Thikness t (mm)	Thikness adopted t_{adop}(mm)
31	10	2,5	10	1,57	10
32	10	3	10	1,93	10
33	10	3,5	10	2,29	10
34	10	4	10	2,65	10
35	10	4,5	10	3	10
36	10	5	10	3,36	10
37	10	5,5	10	3,72	10
38	10	6	10	4,08	10
39	10	6,5	10	4,44	10
40	10	7	10	4,8	10
41	10	7,5	10	5,16	10
42	10	8	10	5,52	10
43	10	8,5	10	5,88	10
44	10	9	10	6,24	10
45	10	9,5	10	6,6	10

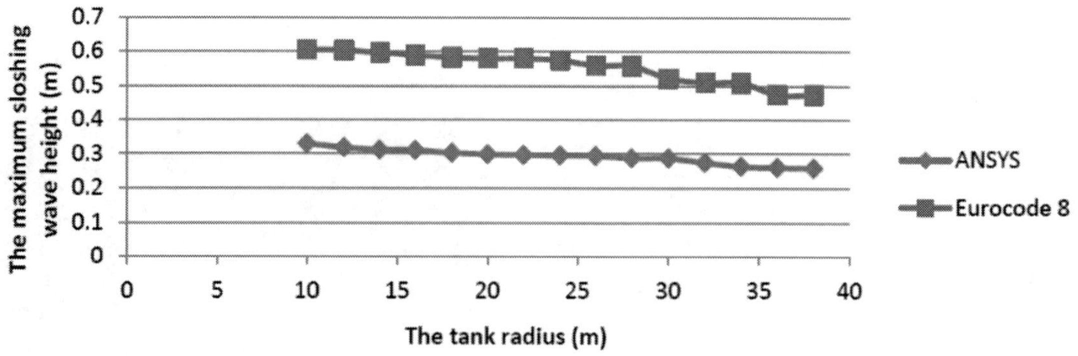

Figure 6. The MSWH's evolution versus the tank Height.

Generally, there are two methods to design the thickness of the tank shell plates: The first one is called "the variable-design point method that is mainly used in tanks with radius higher than 60 m, while the second method – used in our case - is used for tanks with small diameters, it is called "the 1 foot method". In the latter method, the minimum thickness to be used is the greater of the two following values [3] (see Fig. 6-Fig. 8).

DOI: 10.1051/matecconf/20179517005

6 Numerical analysis

To study the effects of different parameters on the MSWH, the tanks are analyzed under the following conditions:

1.Tanks with height's variation only (Part I) .
2.Tanks with radius variation only (Part II).
3.Tanks with variation in the liquid's heights (Part III).

6.1.Tanks with variation in height only

In this case, the tanks have a constant radius of 10m while the changing their heights. The dimensions of each tank and the results are summarized in the table below:

Table 4. Results for the 1st Part.

Part I			
n	H(m)	MSWH ANSYS (m)	MSWH EC8 (m)
1	5	0,247	0,63
2	6	0,252	0,63
3	7	0,254	0,672
4	8	0,266	0,714
5	9	0,274	0,756
6	10	0,294	0,808
7	11	0,305	0,839
8	12	0,308	0,86
9	13	0,32	0,845
10	14	0,313	0,84
11	15	0,308	0,814
12	16	0,303	0,809
13	17	0,293	0,804
14	18	0,288	0,798
15	19	0,278	0,798

6.2.Tanks with variation in radius only:

At present, we are interested on the effect of the radius on the MSWH. For this purpose, we studied the variation of the MSWH depending different radius. The height of the tanks being set 20m.

Table 5. Results of the 2nd Part.

Part II			
n	R(m)	MSWH ANSYS (m)	MSWH EC8 (m)
16	10	0,33	0,606
17	12	0,319	0,604
18	14	0,312	0,597
19	16	0,311	0,59
20	18	0,303	0,583
21	20	0,299	0,58
22	22	0,297	0,58
23	24	0,296	0,575
24	26	0,295	0,56
25	28	0,288	0,558
26	30	0,288	0,522
27	32	0,275	0,51
28	34	0,264	0,509
29	36	0,261	0,4743
30	38	0,259	0,474

6.3.Tanks with variation in liquid's height :

In this part, a container having a height of 10m and a radius of 10 m was selected. The liquid level was changed each time while visualizing the variation of the MSWH's values .The results found are shown in Table 3:

Table 6. Results for the 3rd Part

Part III			
n	H(m)	MSWH ANSYS (m)	MSWH EC8 (m)
31	2,5	0,243	0,63
32	3	0,255	0,63
33	3,5	0,263	0,672
34	4	0,269	0,714
35	4,5	0,274	0,756
36	5	0,294	0,808
37	5,5	0,287	0,839
38	6	0,292	0,86
39	6,5	0,309	0,845
40	7	0,323	0,84
41	7,5	0,339	0,814
42	8	0,326	0,809
43	8,5	0,317	0,804
44	9	0,315	0,798
45	9,5	0,314	0,798

7 Conclusion

In this work, we studied the effects of different geometrical properties of metal cylindrical tanks on maximum heights of sloshing waves occurring during a possible excitation in the base. At first, the accuracy of the finite element method has been validated, then the tanks were modelled and subsequently analyzed. the conclusions to be drawn from this study are summarized in the following points:

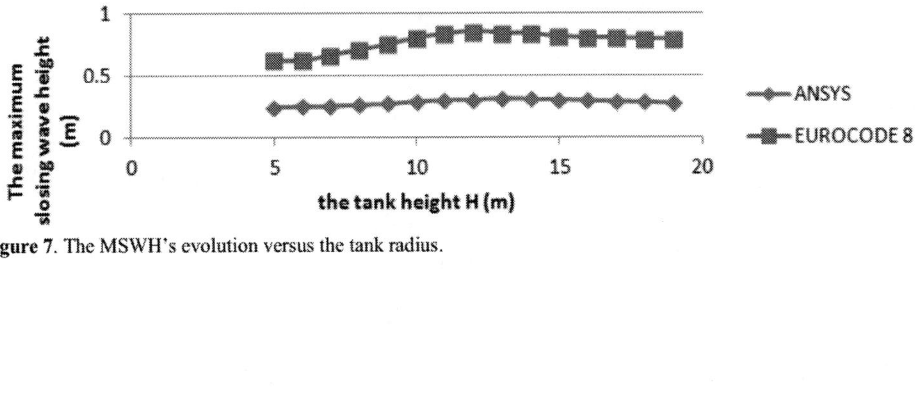

Figure 7. The MSWH's evolution versus the tank radius.

DOI: 10.1051/matecconf/20179517005

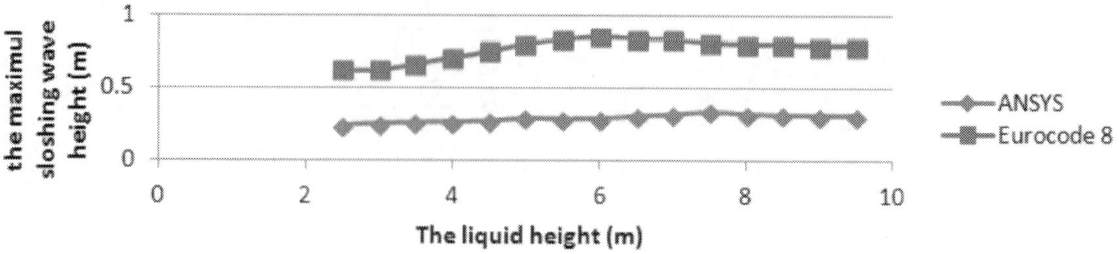

Figure 8. The MSWH's evolution versus the Liquid's Height.

-By varying the height only, it was observed that at the beginning the height of sloshing increase with increasing the height but after reaching a critical value, the values of the wave's height decrease, this decreasing is due to increasing of the flexibility of the tank because of sloshing.

From tanks analyzed with change of the radius only , it was observed that the maximum height of sloshing wave decrease with increasing the radius : indeed for the the same height , a tank having a larger radius is more rigid that a further having a smaller diameter .

For analysis where the liquid's height is varied, it was seen that the maximum height of the sloshing wave increases by increasing the height of the liquid .But after a critical value , the values of the MSWH decrease as the liquid-shell syste becomes more flexible. So there is a point to wherein the security of the tanks can be ensured.

8 Recommendations

From the evolution of the maximum wave sloshing values based on the heights of the tank and their radius, we can see that the critical values -where the waves are maximals- are for tanks having radius lower than their heights.

From the evolution of the maximum values waves sloshing based on liquid heights, we can see that there is a gap for the liquid height between 0.5 H and 0.7 Hwhere the waves values are critical.

In the light of the results that either Eurocode 8 or digital experimentation upon the Ansys software, we recommend performing cylindrical metallic tanks for storage of small heights and large diameters, In that case the maximum heights of sloshing waves are most optimistic.

References

1. ANSYS, ICEM CFD, 11.0 User's Manual, ANSYS (2009).
2. R.C. Burros, R.F. Alves, WIT *Transactions on the Built Environment*, **57**, 733-743 (2001).
3. Standard, British. *Eurocode 8: Design of structures for earthquake resistance* (2005).
4. D. Goulmot,*Nouvelles méthodes numériques pour l'analyse sismique des réservoirs de liquide* (2012).
5. M. Gradinscak, PhD dissertation, *Victoria University Melbourne*, Australia (2009).
6. ASCE, *Technical Council on Lifeline Earthquake Engineering*, New York (1984).
7. M.A. Haroun,G.W. Housner, ASCE, *Journal of technical councils*, **107**, 1,191-207 (1981).
8. E.J. Hopfinger, V.Baumbach,EDP Sciences,*Progress in Propulsion Physics*,**1**, 279-292 (2009).
9. G.W. Housner, *Bulletin of the Seismological Society of America*, **53**, 2, 381-387 (1963).
10. J. B. Frandsen, *Journal of Computational Physics*, **196**, 1, 53-87 (2004).
11. O.R. Jaiswal, S. Kulkarni, P. Pathak,*Proceedings of the 14th World Conference on Earthquake Engineering*, Beijing China, 12-17, October (2008).
12. H. Kazem, S. Mehrpouya, Journal of Structural Engineering and Geo-Techniques, **2**, 1, 55-59 (2012).
13. R. Livaoglu,*Soil dynamics and Earthquake Engineering*,**28**, 2, 132-146 (2008).
14. M. J. N. Priestley, et al. *Recommendations of a Study Group of the New Zealand National Society for Earthquake Engineering*, **17** (1986).
15. A.N. Oskouei, E.N. Naghani,*Journal of current research in Science*, **2** , 5, 664-672 (2014).
16. R.S. Wozniak, W.W. Mitchell, API*Refining Department 43rd Midyear Meeting*, Toronto,Canada (1978).

 DOI: 10.1051/matecconf/20179517006

Carved Monge Surfaces as New Forms in the Architecture

Jevgenija Filipova , Marina Rynkovskaya

Peoples' Friendship University of Russia, Engineering Department, 117198, Miklukho-Maklaya Str., 6, Moscow, Russia

Abstract. On the example of today's existing buildings and cities the tendency of using thin construction in modern world can be clearly noticed. To create new ideas and images engineers have to study new geometrical forms and its calculation methods. The article deals with carved Monge surface. The possibilities of its usage in the architecture, its geometry and form making are well examined in the article, and are the basic problems considered in the article. As well as the basic calculations of Monge surface were made in this article using moment less theory and finite element method, and we can clearly see the advantages of using this shell. The main aim of this article is to show the relevance of carved Monge surface.

1 Introduction

Nowadays for development and creation of the new urbanistic cities designers need to pay more and more attention to such tendencies in the construction which allow us to use more expressive architectural images, to cut the consumption of materials, labor input of production and installation of designs, as well as to solve important functional problems. In this regard the efficiency of space structures' usage is indisputable. But one shouldn't pay attention only to architectural expressiveness of shell of difficult geometry [1]. It should be noted that the shells allow to achieve the reliable designs with big stiffness, stability and durability with rather small consumption of materials. Changing various geometrical parameters of a shell, it is possible to vary over a wide range of fields of movements and tensions of construction [2]. Therefore, research of thin structures of uncanonical forms is in demand and can have a practical usage [3].

Nowadays different shell structures are used in different industries (shipbuilding, aircraft industry, automotive engineering, etc.), in civil engineering and architecture [4].

2 Main features of carved Monge surface

The usage of shape structures in different buildings is possible due to well-studied geometry, the existing theories of calculations, usage of computers and software. However, the majority of shape structures used nowadays are limited to number of surfaces only: circular cylindrical, conical, spherical and some other traditional shapes. These surfaces are just a small part of geometrical forms available by geometry [1, 3-12].

One of the directions in structural theory of thin constructions is a detailed study of geometry, new calculation methods of shells of uncanonical forms and research of the correlation between thin structure shape and its strength properties [13]. Therefore we would like to touch the less studied surfaces which have no less architectural expressiveness and other advances than different well known shell structures.

Great opportunities in creation of new architectural forms are given by carved Monge shell. These surfaces are rather technological and allow simplifying construction process due to ability of their middle surface to wrap on the surfaces without folds and gaps [14].

Concerning a shell structure it is necessary to announce several subjects of research: form making, geometrical characteristics, architectural features, possibility of its usage in modern buildings, and also methods of strength calculation.

2.1 Architectural features of carved Monge surface

Speaking about relevance of ruled carved Monge surface usage, it is necessary to examine architectural aspect. This surface can be used in solution of the most modern tendencies and questions of design today. First of all, shell structure, as an esthetic form, serves as a spectacular example of modern architecture. Its smooth and flexible forms, reflects to today's world tendencies. Secondly, the structure is an effective from the economic point of view. As the shell structure is a complete construction and works like an integrated construction, it does not require massive bearing walls. Thirdly, it is ability to seal big areas, and also the lack of the additional columns, cutting space.

If we speak about practical usage of this shell structure, it should be noted, that it is possible to use it as an independent bearing construction (showrooms, places for events with a big flow of people and others), and also as an additional form for different buildings integral characters creation (canopy, architectural details and others) [15].

Speaking about the usage of carved Monge surfac in the modern architecture, they are at the initial stage of introduction in modern architecture, because they are not so well studied. Nevertheless, it is possible to carry some existing buildings and constructions to carved Monge surface. As the example: Hydroelectric Tidal House – project made by Serbian architect (Fig. 1).

Figure 1. Hydroelectric Tidal House

2.2 Geometry of carved Monge surface

Speaking about form making, it is necessary to discuss the main methods of creation and also the mathematical description of this surface [16].

Gaspard Monge has given the definition of carved surfaces as the surfaces formed by the movement of the flat curve lying in the plane roll without sliding on some cylindrical surface (Fig. 2).

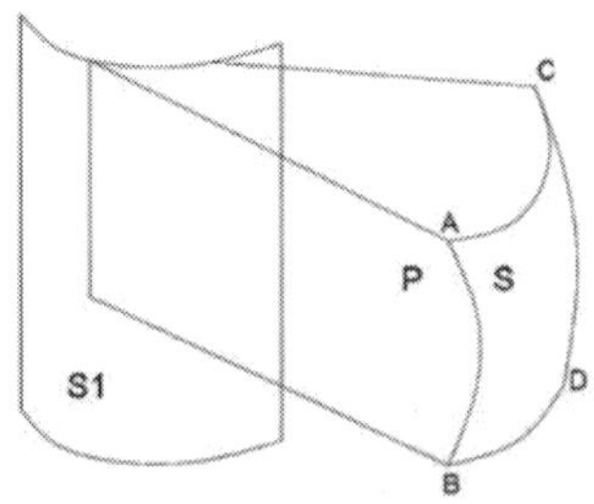

Figure 2. Form making method of carved Monge surface.

Speaking about geometry, we would like to begin with defining the equations of surface.

The general vector equation of a carved surface looks as follows:

$$\bar{r} = \bar{\rho}(\alpha) + \varphi(\beta)\bar{p}(\alpha) + \psi(\beta)\bar{q}(\alpha), \quad (1)$$

where $\varphi(\beta)$, $\psi(\beta)$ are the parametrical equations of the forming curve in mobile Cartesian reference system, $\bar{p}$, $\bar{q}$ are unitary vectors of mobile Cartesian reference system in the normal plane of the directing line $\bar{\rho}(\alpha)$.

There are more than 10 types of carved surfaces [17]. Depending on change of various parameters, for example the form of a moving line or a guiding line, the appearance of a surface will change (figs. 3- 5). Let us have a look on a specific case: ruled carved Monge surface (fig. 5). In this example, we accept an involute of circle as a parallel and an intercept as the meridian (fig. 6).

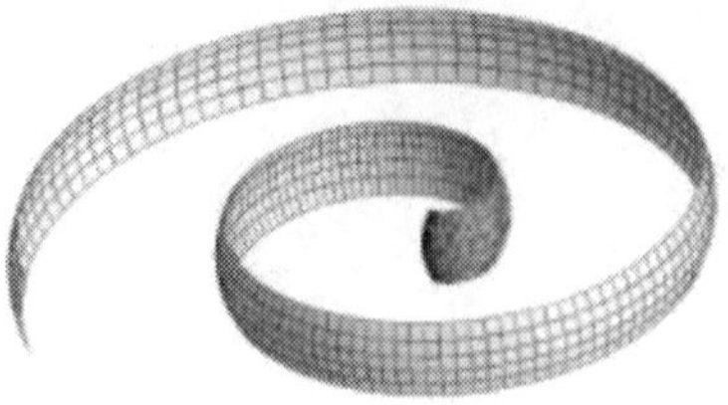

Figure 3. Carved Monge shell with cylindrical guide surface and parabola as a longitude.

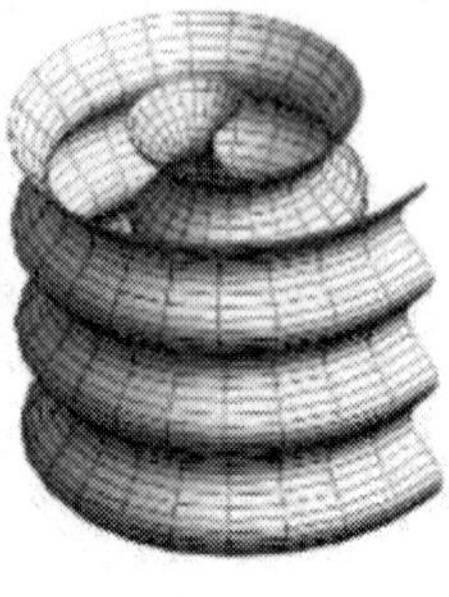

Figure 4. Carved Monge shell with cylindrical guide surface and sinusoid as a longitude.

Figure 5. Carved ruled Monge shell with circular-cylindrical guide surface.

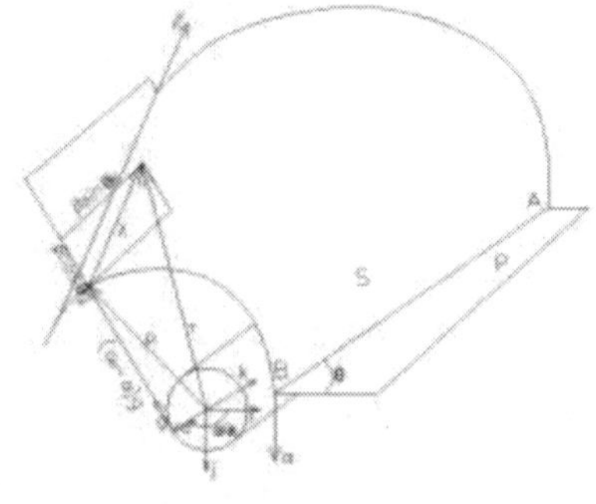

Figure 6. Carved ruled Monge shell assignment.

Vector equation (1) of a special case of ruled carved Monge surface can be presented :

$$r(\alpha, \beta) = a \cdot \bar{e} + (\alpha_0 - \alpha + \beta \cdot \cos(\theta)) \cdot \bar{g} + \beta \cdot \sin(\theta) \cdot \bar{k} \quad (2)$$

where, e, g are vector circular functions; $\rho = a \cdot e + (\alpha_0 - \alpha) \cdot g$ is the equation of a parallel, which represents itself the involute of circle with radius $r = a$; α is natural parameter; $\beta = |\lambda|$ is binormal of a guide line, θ is a corner between the vectors g and λ; α_0 is initial point unwrapping of a surface, $\lambda = \beta \cdot \cos(\theta) \cdot g + \beta \cdot \sin(\theta) \cdot k$ is the meridian equation in any situation.

Let us change a vector way of equations representation to a parametrical one.

Formulas (2) give the chance to receive the parametrical equations of the considered surface:

$$X = a \cdot \cos\left(\frac{\alpha}{a}\right) + \alpha_0 \cdot \sin\left(\frac{\alpha}{a}\right) - \alpha \cdot \sin\left(\frac{\alpha}{a}\right) + \cos(\theta) \cdot \sin\left(\frac{\alpha}{a}\right) \cdot \beta,$$
$$Y = -a \cdot \sin\left(\frac{\alpha}{a}\right) + \alpha_0 \cdot \cos\left(\frac{\alpha}{a}\right) - \alpha \cdot \cos\left(\frac{\alpha}{a}\right) + \cos(\theta) \cdot \cos\left(\frac{\alpha}{a}\right) \cdot \beta,$$
$$Z = \sin(\theta) \cdot \beta. \quad (3)$$

Curvilinear coordinate α coincides with a surface parallels, and β-lines – its lineal moving line. Then the main quadratic forms coefficients of a surface and main curvatures are determined by formulas:

$$A = \frac{(\alpha_0 - \alpha + \beta \cdot \cos(\theta))}{a} = C_0 + C_1 \cdot \alpha + C_2 \cdot \beta; \quad (4)$$
$$B = 1;$$
$$L = \frac{-A \cdot \sin(\theta)}{a};$$
$$F = M = N = 0;$$
$$k_1 = \frac{\sin(\theta)}{a \cdot A}; \; k_2 = 0,$$

where $C_0 = \frac{\alpha_0}{a}$; $C_1 = -\frac{1}{a}$; $C_2 = \frac{\cos(\theta)}{a}$.

3 Methods of strength analysis

Let us look at the main initial methods of strength analysis for checking their correctness in relation to this shell structure.

The momentless theory represents simplified option, where influence of bending moment, torque moment and transverse force on the stress strain behavior is not considered (Fig.7). Nowadays momentless theory is not so popular, but still there are some works which take momentless theory as the basic calculation method, such as [18-20].

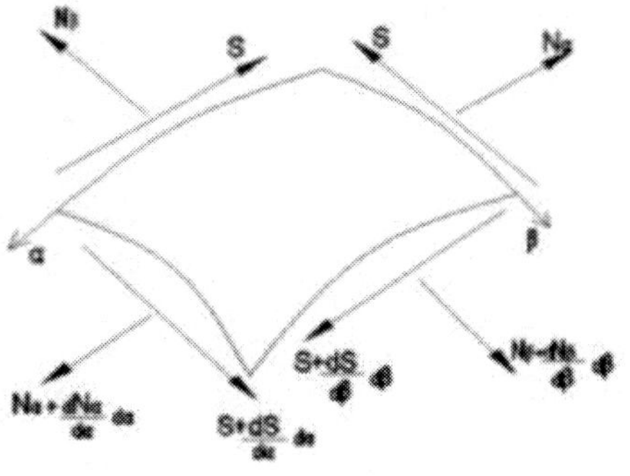

Figure 7. Efforts in the momentless theory

The equations of the momentless theory for a carved Monge shells with circular-cylindrical guide surface will be:

$$\frac{\partial N_\alpha}{\partial \alpha} + \frac{1}{A} \cdot \frac{\partial}{\partial \alpha}(A^2 \cdot S) + A \cdot X =;$$
$$\frac{\partial}{\partial \beta}(A \cdot N_\beta) - \frac{\partial A}{\partial \beta} \cdot N_\alpha + \frac{\partial S}{\partial \beta} + A \cdot Y = 0;$$
$$\frac{N_\alpha}{R1} - Z = 0 \quad (5)$$

where, N_α is normal force in the direction α, N_β is normal force in the direction β, S is shearing force, X,Y,Z are projections of the spreaded load on the directions α, β, z.

Will enter new constants [18]:

$$m = \frac{a}{\sin(\theta)}; \; C_0 = \frac{\alpha_0}{a};$$
$$C_1 = -\frac{1}{a}; \; C_2 = \frac{\cos(\theta)}{a}.$$

Integrating the equation (5) we receive:

$$N_\alpha = m \cdot A \cdot Z;$$
$$S = -\frac{1}{A^2}\int_{\beta 0}^{\beta} A \cdot \left(A \cdot X + \frac{\partial N_\alpha}{\partial \alpha}\right)\partial\beta + \frac{x1(\alpha)}{A^2};$$
$$N_\beta = -\frac{1}{A}\int_{\beta 0}^{\beta}\left(A \cdot Y - \cdot C_2 \cdot N_\alpha + \frac{\partial S}{\partial \alpha}\right)\partial\beta + \frac{x2(\alpha)}{A}. \quad (6)$$

Let's consider an example of a dead load action.

So X=0; $Y = q \cdot \sin(\theta)$, $Z = -q \cdot \cos(\theta)$.

Integrated equations (6) will look:

$$N_\alpha = -\frac{a}{sin(\theta)} \cdot A \cdot q \cdot cos(\theta),$$
$$S = -\frac{1}{A^2} \cdot \left(\frac{1}{3}C_2^2 \cdot X \cdot \beta^3 + \frac{1}{2} \cdot C_2 \cdot (2 \cdot)\right) + \frac{f1(\alpha)}{A^2} \quad (7)$$

In the results we will receive the following diagrams of normal efforts N_β and N_α, and shearing force S (Fig. 8-10).

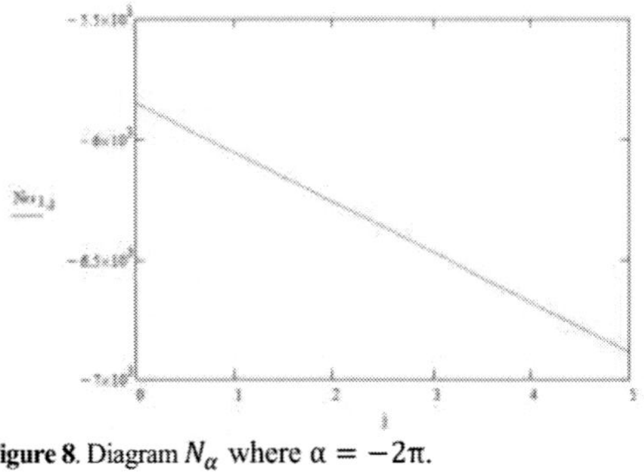

Figure 8. Diagram N_α where $\alpha = -2\pi$.

Now we will consider the turned-out results of the calculation of a carved Monge surface by finite element method. For calculation purposes the design model has been created using computer program LIRA 2013R (fig. 11). Nowadays different computer programs are the most popular and easiest way of calculation [21, 22], but analytical methods are still in usage too [23]. All the parameters were accepted the same as for the momentless theory. The accepted shell thickness is 50 mm. Examine the same option

 DOI: 10.1051/matecconf/20179517006

of a dead loads similar to the momentless theory. We will receive izofields of internal forces and movements (Fig. 12).

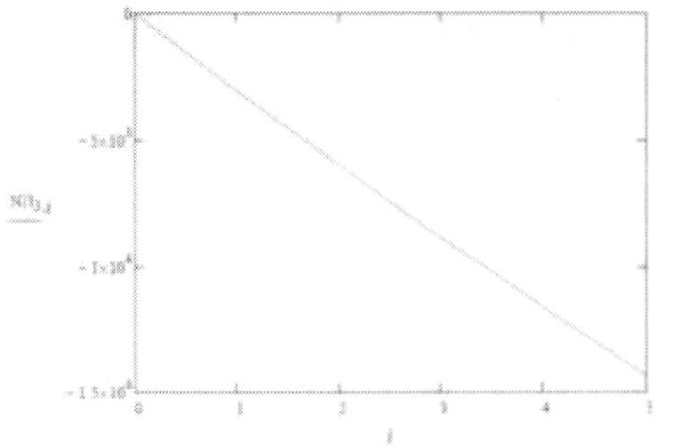

Figure 9. Diagram N_β where $\alpha = -2\pi$.

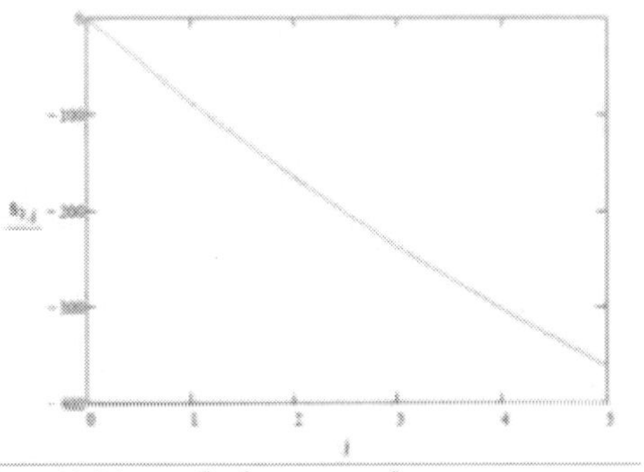

Figure 10. Diagram S where $\alpha = -2\pi$.

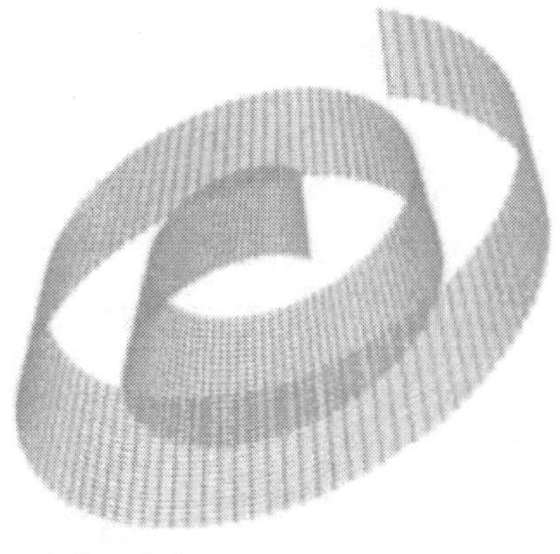

Figure 11. General view

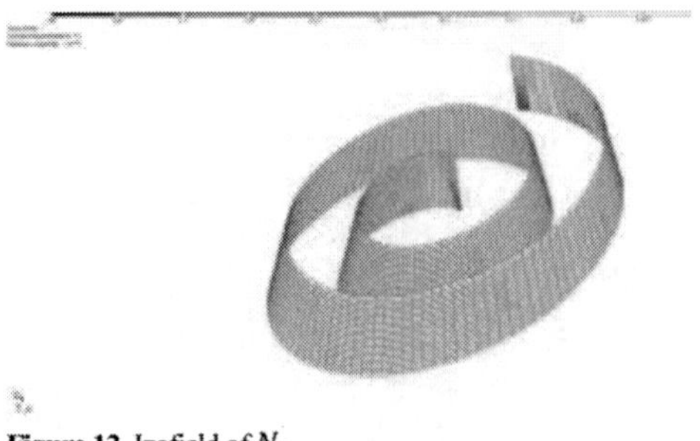

Figure 12. Izofield of N_β

To compare with the momentless theory we will analyze normal efforts N_y ,N_x and shearing force S. For the analysis of the calculations made, we will make the table of normal efforts N_β, N_α and the shearing force counted by a momentless method and by finite element method where α=-2π and calculate differences in the results (table 1).

Table 1. Comparison of two calculation methods.

	α=-2π		
Nβ	momentless kg/m	FEM kg/m	difference %
β=-6	0.00	-0.40	-
β=-4.5	-3.80	-4.12	7.8
β=-3	-7.44	-8.09	8.1
β=-1.5	-10.95	-11.75	6.8
β=0	-14.34	-14.53	1.3
Nα	momentless kg/m	FEM kg/m	difference %
β=-6	-5.85	-6.28	6.9
β=-4.5	-6.11	-6.72	9.1
β=-3	-6.37	-6.90	7.7
β=-1.5	-6.63	-7.37	10.1
β=0	-6.89	-7.42	7.2
S	momentless kg/m	FEM kg/m	difference %
β=-6	0.00	-11.27	-
β=-4.5	-107.85	-110.60	2.5
β=-3	-202.76	-215.42	6.2
β=-1.5	-286.71	-278.51	2.9
β=0	-361.34	-316.92	12.3

As we can see in results, the difference between two methods of calculation is not so big. An average difference is between 5%-10% (percent). It is a positive result because the finite element method includes such important things as bending moment, torque moment and transverse force, which influence the construction work.

Conclusion

Carved Monge surfaces, as it was shown earlier, have the following advantages as:

1. architectural expressiveness,
2. the reliable design with high stiffness and durability,
3. rather small consumption of materials (massive bearing walls aren't required),
4. technological effectiveness of construction that allows to carry out construction process directly on a building site,
5. possibility of rational use of space,
6. presence of smooth lines pleasant for eye perception,
7. an opportunity to use shell for buildings of different functions.

Also the analysis of two calculation methods was made, and the results show us that calculations are correct, as the average difference in two methods is in limits of error.

References

1. Roberto Cipolla, Peter Giblin. Visual Motion if Curves and Surfaces, Cambridge University Press, UK, 187 p (2001).
2. Deserno M. Notes on differential geometry with special emphasis on surfaces in R3, UCLE, Los Angeles, 2004. – p. 64.
3. Ivanov V.N., Rynkovskaya M.I. Application of circular surfaces to the architecture of the buildings, structures and products. Bulletin of PFUR: Engineering Researches, **3**, p.111-119 (2015).
4. Rynkovskaya M. On application and analysis of helicoidal shells in architecture and civil engineering, Bulletin of PFUR, engineering, **4**, p. 84-90 (2012).
5. S.N. Krivoshapko, Geometry and strength of general helicoidal shells, **5**, p. 161-175, (1999).
6. S.N. Krivoshapko, V.N. Ivanov, Encyclopedia of Analytical Surfaces, Springer International Publishing Switzerland, 752 p (2015).
7. C.A. Bock hyeng., S.N. Krivoshapko Umbrella-type surfaces in architecture of spatial structures, IOSR JOURNAL OF ENGINEERING, **3**, p. 43-53 (2013).
8. R. Bradshaw, D. Campbell, M. Gargari, A. Mirmiran, P. Tripeny, Special structures. Past, present, and future, Journal of Structural Engineering, p. 691-701 (2002).
9. S.N. Krivoshapko, Cable-stayed structures, Structural Mechanics of Engineering Constructions and Buildings, **1**, p. 9-22, (2016).
10. V.N. Ivanov. The problems of the geometry and the architectural design of shells based on cyclic surfaces, Spatial structures in new and renovation projects of buildings and construction, Proceedings international congress ICSS-98, 2, - p. 539-546 (1998).
11. Hyeng, A. Christian Bock, E.B. Yamb, Aplication of cyclic shells in the architecture, machine design, andbiones, Int. Journal of Modern Engineering Researches **12**, 799-806, (2012).
12. S.N. Krivoshapko, K.P. Pjatikristovsky. On history of building of wooden shells and their opportunities at present and in the future, Structural Mechanics of Engineering Constructions and Buildings, **1**, p. 3-18, (2014).
13. S.N. Krivoshapko. Static analysis of shells with developable middle surfaces, Applied Mechanics Reviews, **51**, No12, Part 1, p. 731-746 (1998).
14. Barony S.Y., Tottenham H. The analysis of rotational shells using a curved ring element and the mixed variational formulation II Int. J. Numer Meth. Eng.**10**. N4, p.861-872 (1976).
15. V.N. Ivanov, R. Muxammad, Geometry of carved Monge surface and shell structuring, Structural Mechanics of Engineering Constructions and Buildings, **1**, p. 27-36, (2002).
16. Olodo E.T., Adjovi E.C., Krivoshapko S.N. Geometrical modeling of a composite folded membrane by a developable membrane with parabolic guidelines of any order// International Journal of Scientific & Engineering Research, **4**. p. 338-343, (2013).
17. Luther Pfahler Eienhart. A Treatise on the Differential Geometry of Curves and Surfaces, Dover Publicans, New York, 479 p (2004).
18. Arbelaes, Juxanio Marulanda, Analysis of Shells in the Form of Ruled Monge surface, PhD Diss. In technic., Moscow, UDN, p. 154, (1970).
19. M.J. Fares, Momentless theory of carved Monge surface of double curvature, p.118 (1974).
20. Krivoshapko S.N., Gbaguidi Aïssè Gerard. Two methods of analysis of thin elastic open helicoidal shells// International Journal of Research and Reviews in Applied Sciences, **12**, p. 382-390, (2012).
21. Bathe K.J. Numerical Methods in finite element analysis-prentice-hall, Englewood Cliffs, New Jersey, p. 631, (1976).
22. Bushel 1 D. Computer analysis of complex shell structures II Pres. At AIAA eight Aerospace sci. Meet.N.Y.AJAA paper, **70**, p. 138, (1970).
23. Rynkovskaya M.I. On problem of strength analysis of thin linear helicoidal shells, Structural Mechanics of Engineering Constructions and Buildings, **6**, p. 13-15, (2015)

Presenting a Model Based on Fuzzy Application to Optimize the Time of IBS Projects in Gas Refineries

Abbas Naderpour[1, 2], Javad Majrouhi Sardroud[3] and Masood Mofid[4]

[1]*Ph.D. Student, Faculty of Engineering, Islamic Azad University, Central Tehran Branch, Iran.*
[2]*Head of Structural Engineering, Management of Coordination and Production Supervision, National Iranian Gas Company, Iran.*
[3]*Assistant Professor, Faculty of Engineering, Islamic Azad University, Central Tehran Branch, Iran.*
[4]*Professor, Department of Civil Engineering, Sharif University of Technology, Iran.*

Abstract. Nowadays, the construction industry has started to embrace IBS as a method of attaining better construction quality and productivity and reducing risks related to occupational safety and health. The built of pre-fabricated component in factories reduces many problems related to lack of purposing uncertainty in scheduling calculation and time management of projects. In the case of using IBS method for managing time in projects, former studies such as Allan Tay's research, indicates that this method can save up at least 29% of overall completion period versus the conventional method. But beside mentioned advantages of this technical method, the projects could be optimized more and more in scheduling calculations. This issue is critical in gas refineries, since special parameters such as risk of spreading poison H_2S gas and mandatory of performing projects in short time period events such as maintenance overhauls demands to perform projects in optimum time. Custom scheduling calculation of project planning uses the Critical Path Method (CPM) as a tool for Planning Project's activities. The researches of this paper's authors indicated that Fuzzy Critical Path Method (FCPM) is the best technique to manage the uncertainty in project scheduling and can save up the construction project's time versus the custom methods. This paper aims to present a model based on fuzzy application in CPM calculations to optimize the time of Industrial Building System.

1 Introduction

IBS can be defined as a construction system which components are manufactured in a factory, on or off-site, Positioned and assemble in to structures with minimal additional site work [1]. The typical classifications is shown in Fig. 1, [2].

Figure 1. Industrialized building system types (Box-Frame, Beam-Panel system and panel systems)

Former studies indicate that IBS can save up about 29% of overall completion time versus the conventional method [3].

Beside mentioned advantages of these technical methods, the projects could be optimized more and more in scheduling calculations. The researches of this paper's authors indicated that Fuzzy Critical Path Method (FCPM) is the best technique to manage the uncertainty in project scheduling and can save up the construction project's time versus the custom methods.

Table 1. Building system classification according to relative weight of component

No	General System	Conventional Method	Production Material	Comparisons
1	Overall Construction Period	24 months + 17 days	19 months + 12 days	Saving up to 135 days or 29% overall completion period.
2	Main Building Structure	464 days	329 days	
3	Foundation Work	83 days (usual piling method)	50 days (raft foundation)	Saving up to 33 days

2 Fuzzy critical path method (FCPM)

FCPM is an extension of Critical Path Method by fuzzy

approach. In this method, fuzzy function defines the time of project activities to manage their uncertainties. For understanding the main concepts of FCPM, consider the project network indicated in Fig. 3. This project network is a part of a small one-story Industrial building System project that contained of excavation, concreting and installing the building structures. The resistance system against the earthquake is A.D.A.S bracing.

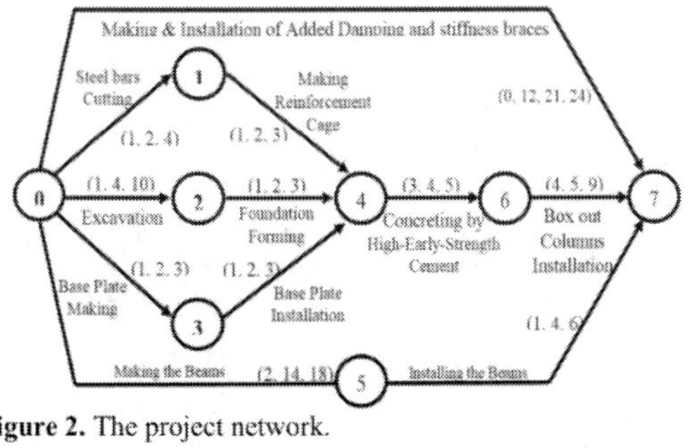

Figure 2. The project network.

As it could be seen from Fig. 2, 10 of whole 11 activity times have triangular fuzzy type format and only the remaining one is in trapezoidal form. In order to determine the project time According to CPM calculation, it is necessary to rank the fuzzy numbers and select the maximum. For ranking the fuzzy numbers brings from CPM calculations, 5 methods available. The result shows in Fig. 3 [4].

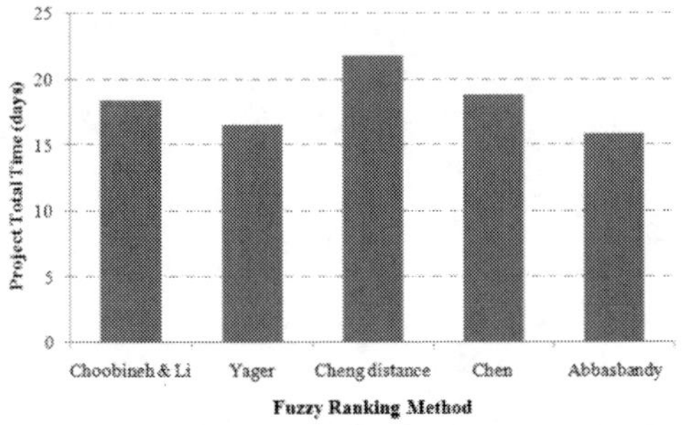

Figure 3. Project total time calculated based on different fuzzy ranking methods

In FCPM calculations, as the results indicate, selecting the fuzzy ranking method for each project scheduling has main influence on determination of project total time and critical path. Also suitable fuzzy ranking method, compatible to project nature should be considered by project management team. The scheduling calculation of this paper's authors indicated that Fuzzy Critical Path Method (FCPM) can save up project total time versus the custom project management methods.

3 Presenting a fuzzy base model to optimize the time of (IBS) projects

For implementing a model based on Fuzzy technique In order to optimize the time of (IBS) Projects various membership functions and linguistic variables have been defined by judgment of experts. Linguistic variables, such as Site Organization, weather, Labor Skills and Quality of equipment, need to be translated in to mathematical measures. The value of these considered variables can be classified in to 5 types: Very Poor, Poor, Medium, High and Very High. These are not clearly defined but are meaningful classifications.

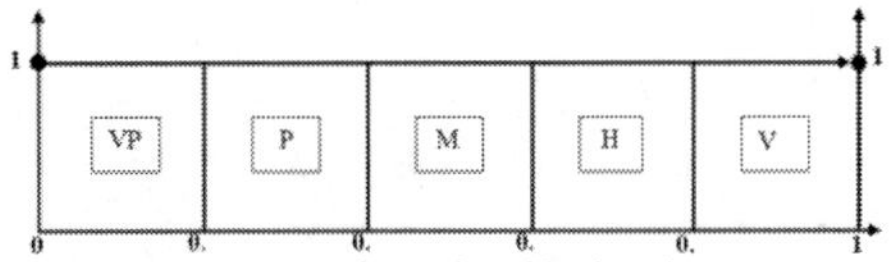

Figure 4. Categorization of membership functions

So in first step the entire project activities will be converted to any of categories mentioned above. The values of membership function may also change with respect to activity and the results obtained from the simulation of different graphs are thus interpreted and modified as per requirements of duration change. Second step is the schedule calculations and analysis.

As described in the example of section 2 (the project network indicated in Fig. 2) project network calculation is a very time consuming process. So we need to use software tools for the optimizer model. The research of this paper indicated that, fuzzy toolbox of MATLAB is the best tools for our goal.

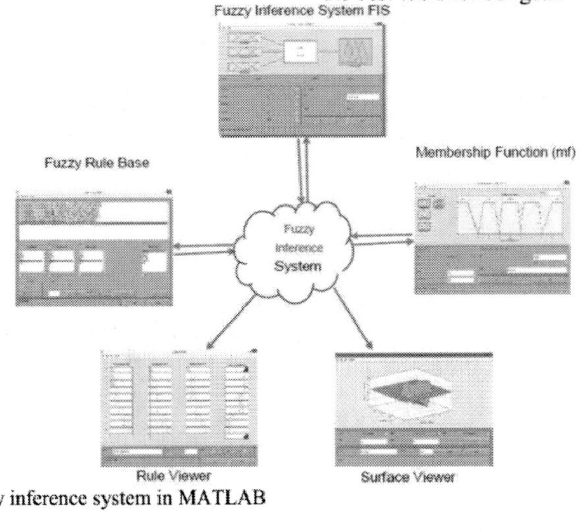

Figure 5. Pictorial view of fuzzy inference system in MATLAB

Output is a crisp result which is obtained after undergoing the process of: Pre-Processing (1), Fuzzification (2), Rule base (3), Inference Engine (4) and Defuzzification (5) [5]-[12]. The output can be decided by user to obtain required result after the whole inference process, which can be desired result of whole designed fuzzy system. In the first step the scenario of analysis is designed.

4 The result of implementation of proposed model in a gas refinery

Corresponding author of this paper has implemented the Recommended model to manage the project of a specific Gas Refinery in east-north of Iran in statistical society of 30 selected projects among all of industrialized projects (Selected according the Cochran formula).

Table 2. Configuration of Selected Projects for Research Calculation and Analysis

Project ID	Title of project
P001	Construction of Pardis Staff dormitory
P002	Construction of Oily Water Separator
P003	Degassing of Granulation Unit
P004	Construction of Housing Services center
P005	Construction of Ware House Administrative Building
P006	Movement of Gonbazli Dehydration sole
P007	Extending of Central Restaurant
P008	Construction of Sculpture Platform
P009	Optimization of Shahid Mohajer Pool
P010	Construction of Gas station
P011	Construction of Shahid Torshizi Sewage
P012	Construction of TPL Fencing
P013	Construction of oil Loading Pavement
P014	Construction of Transportation Sole
P015	Performing of Refinery F & G System
P016	Construction of Sculpture Unit Road
P017	Construction of Senior Operator Room
P018	Construction of HSE Energy Chanel
P019	Performing P.F Wall in Torshizi Residential
P020	Performing of Pardis Complex Gas line
P021	Performing of Pardis Complex Data line
P022	Performing of Pardis Complex Water line
P023	Performing of Pardis Complex 6Kv Elec line
P024	Performing of Pardis Complex Sewage line
P025	Construction of Contractor Building
P026	Performing of O.W.S Supports
P027	Zero Flaring
P028	Construction of CMF Pipe Line
P029	Performing of General Civil Maintenance
P030	Restaurant's Cold and Mechanical Room

Table 2 indicates the configuration of selected projects. The members of expert team in this research were Chosen from contractors, consultant and supervisors of owner and the time of research took about two years between 2014 up to 2016. Table 3 indicates the identification of research expert team and the result of analysis is shown in Fig. 6.

The result of analysis indicates the implementation of the recommended model, lead to reducing the total time of selected project (as a minimum 7% of project total time) versus the custom methods. The rate of this reduction estimated by Corresponding Author, and the result is according to Fig. 6.

Table 3. The identification of expert team of research

Member No.	Job Title - Duration of Experience
1	Supervisor - Civil Engineer - 14
2	Supervisor - Civil Technician - 17
3	Consultant - Civil Engineer - 16
4	Contractor- Civil Engineer - 12
5	Supervisor- Mechanical Eng -14
6	Supervisor- Mechanical Eng -17
7	Contractor- Mechanical Tech -16
8	Consultant- Mechanical Eng -8
9	Contractor- Mechanical Tech -15
10	Supervisor- Electrical Engineer-17
11	Consultant- Electrical Engineer-16
12	Contractor- Electrical Tech -15
13	Contractor- Electrical Tech -14
14	Contract Administration Staff- 17
15	Contract Administration Staff- 16
16	Budget Expert Staff - 12
17	Budget Expert Staff - 16
18	Business Administration Staff - 12

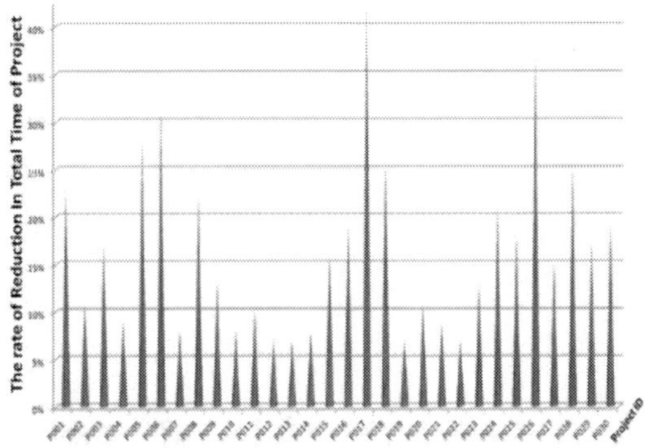

Figure 6. Research Results (Project ID - rate of Reduction in Total Time)

5 Conclusions

In the field of using new technical method in construction industry such as IBS, many studies indicate that these new technical methods can save up at least 29% of overall completion period versus the conventional method. So when these new technical methods are merged by intelligent project planning approaches, the results can improve more and more. Custom scheduling calculation of project planning uses the Critical Path Method (CPM) as a tool for measuring activities time progress.

Whereas CPM has the main weak point which is related to lack of purposing uncertainty in scheduling calculation. The research of this paper's authors indicated that Fuzzy Critical Path Method (FCPM) that added Fuzzy approach to CPM is the best technique to manage the uncertainty in project scheduling and it can save up as a minimum 7% of project total time versus the custom project management methods.

So when this project planning approach is merged with new technical method in construction industry such as IBS, we can save up more than 36 percent in total time of project. Finally, we can save up one third of project total time.

References

1. M. R. Abdul Kadir, W.P. Lee, M.S. Jaffar, S.M. Sapuan, A.A.A. Ali, Construction Performance Comparison between Conventional and Industrialized Building Systems in Malaysia, Structural Survey, **24**, 5, pp. 412-424 (2006).
2. N.A. Ahmad Bari, R. Yussuff, N. Ismail, A. Jaapar, N. Ahmad, Factors Influencing the Construction Cost of Industrialised Building System (IBS) Projects, Procedia-social and Behavioral Science, Elsevier, **35**, pp. 689-696 (2012).
3. A. Tay Eng Mian, Industrialised Building System Formation Scheduling for Public Buildings, Master Thesis, University of Technology Malaysia Publishing (2006).
4. A. Naderpour, M. Mofid, J. Majrouhi Sardrood, State-of-the-Art Review on Project's Scheduling Network and Fuzzy Critical Path Method Calculation (FCPM), The Second International Conference of Science, Engineering and Technology, Istanbul, Turkey (2015).
5. S. Rao Rachakondra, Fuzzy logic integration into construction management planning, Master Thesis, Chapter 3, Polytechnic University of Catalonia Publishing (2015).
6. A. Ozta, O. Oktem, in: Evolutionary A CPM-Based Scheduling Method for Construction Projects with Fuzzy Sets and Fuzzy Operations, **56**, pp. 2-9 (2014).
7. F. Choobineh, H. Li, in: An index for ordering fuzzy numbers, Journal of Fuzzy Sets and Systems, **5**, pp. 1-19 (1985).
8. A. Naderpour, M. Mofid, in: Improving construction management of an educational centre by applying earned value technique. Procedia engineering, **14**, pp. 1945-1952 (2011).
9. C.H. Cheng, in: A new approach for ranking fuzzy numbers by distance method, Journal of Fuzzy Sets and Systems, 95, pp. 307-317 (1998).
10. C. Chen, H.Tang, in: Ranking nonnormal pnorm Trapezodial fuzzy numbers with integral value, Journal of Computers and Mathematics with Applications, **56**, pp. 2340-2346 (2008).
11. S. Abbasbandy, B. Asady, in: Ranking of fuzzy numbers by sign distance, Journal of Information Science, **176**, pp. 2405-2416 (2006).
12. S. Sathish, K.Ganessan, in: A simple approach to fuzzy critical path analysis in project networks, International Journal of Scientific & Engineering Research, **2**, pp. 1-6 (2011).

Application Status and Trends of Bolting-Shotcreting-Mesh Supporting Technology in Underground Engineering and Its Application in Ivory Coast

Kouame Joseph Arthur Kouame , Fuxing Jiang, Yu Feng and Sitao Zhu

School of Civil and Environmental Engineering, University of Science and Technology Beijing, Beijing, China

Abstract. Many gold mines deposits have been discovered recently in Ivory Coast. The current government tries all the possible ways to develop the mining sector. However, the artisanal gold mine leads the operation in all the major parts of the country. The bolting-shotcreting-mesh supporting technology, which serves as an anchorage technology, has mainly been a part of traditional civil engineering. In case we refer to the historical development of the bolting-shotcreting-mesh supporting technology in geotechnical engineering, some of its application status aspects include standard construction, design, anchoring material and construction technology. For this reason, this paper focuses on the problems of the anchorage technology and how the development trend of the anchorage technology can be useful in Ivory Coast gold's mining industry.

1 Introduction

According to its definition, Bolting-shotcreting-mesh supporting technology as an anchorage technology is a technical method with the superiority of technology and economy. So far, Bolting-shotcreting-mesh supporting technology is not only widely used in the world's major coal countries (such as China and Australia), but it can also be applied to the projects of metallurgy, water conservancy and hydropower, railway, highway, constructions and so on. This is more especially in the underground geotechnical engineering, which studies the rock masses and soil in order to develop and use the underground space resources including underground buildings, subways, highway tunnels, and so on ([1]-[4]).

For bolting-shotcreting-mesh supporting technology, anchor rods and sprayed concrete together form load-carrying structures. Fig. 1 below shows how the effect of an anchor rod and sprayed concrete would effectively limit the free flow of surrounding rock deformation, and prevent rock mass from scattering and falling.

According to historical records, the bolting-shotcreting-mesh supporting technology's historical background lie in the fact that bolts were firstly used in the supporting of rock roadways in 1911 in the US,. People first used bolt cables in the West Lycia mines in 1918. In Africa, more precisely, in Algeria, a pre-stressed bolt with the bearing capacity of 10,000kN was firstly used in order to realize the stability of the dam engineering.

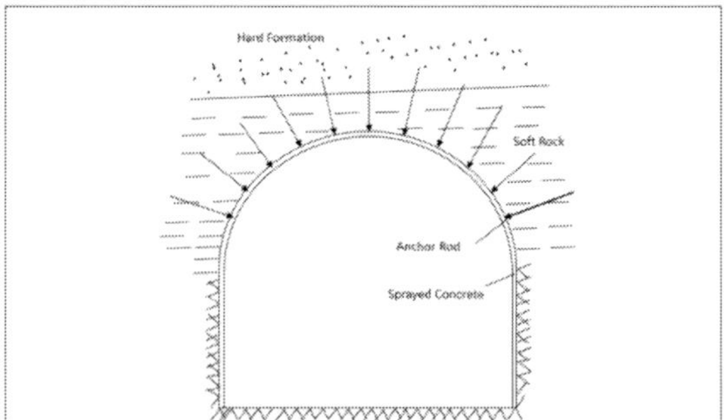

Figure 1. The bolting-shotcreting-mesh supporting schematic diagram

In 1989, Australia employed a new anchor bearing the capacity of 16500kN, which was made up of 65 sticks of steel strand. China, especially in Beijing, will also develop some big foundation pit projects, by using pre-stressed underground anchors to support pile wall structures in Beijing city such as Beijing city building, Wangfu hotel, and so on ([5], [6]).

The recent deep exploration in Ivory Coast showed that the country possesses rich mineral resources. Though the government has paid more attention to mining development than before, the management of artisanal gold mines remains a key problem in the Ivory Coast mining industry. In addition, many of the mines are unexplored and many known deposits are undeveloped which has hindered the rapid growth of economy. Therefore, we face enormous problems in its level of development. The problems are of an environmental, technological, economic, and local society nature issue.

Nowadays, to face the underground engineering in many aspects, the bolting-shotcreting- mesh supporting technology will be used and spread deeply, in order to enhance mining development. Some countries such as Australia, Britain, USA, FIB, Japan and China, have successively promulgated the technical manual of an underground anchor. Even if in this decade, many researchers and scholars have obtained many research achievements in the supporting technology and theory, there is still a need for more deep researches working to reach the goal. Hence, the importance to find the development trend of the supporting technology according to its applied status is required. That is why this paper strongly suggests the use of Bolting-shotcreting-mesh supporting technology in the Ivory Coast mining industry, in order to reduce the risks of artisanal mining in the country ([7]-[11]).

2 Application statuses

2.1 Standard construction

The government of Ivory Coast can, in order to design and construct the geotechnical anchorage technology economically and safely, use bolt shotcrete supporting technology specification, soil anchor design and construction specifications, bolt shotcrete supporting technology specification, rock and soil anchor rod technical procedures, which will reflect the new achievement in the Ivory Coast anchorage technology field as it has been done in China several decades ago.

Ivory Coast has the biggest hydraulic which produces and provides electricity to many neighbouring countries such Burkina Faso and Mali. Hence, the department of hydraulic and electric engineering and construction department can also be the main corresponding industrial standard, including hydraulic pre-stressed anchorage design specification, construction of foundation pit supporting technology procedures, water and electricity engineering pre-stressed anchorage design specification, hydropower project construction specification support by rock bolt shotcrete and reinforced cement-soil pile anchors, to support technology procedures.

2.2 Design method

The types of projects differ from one to another, so some big differences exist among the design methods and analytical theories in the geotechnical anchorage technology. The anchorage design methods applied in underground geotechnical engineering are: analytical method, including elastic-plastic analysis and ultimate analysis, empirical method including engineering analogy method and expert experience method, and new Austrian tunneling method. For sure those methods can be useful in Ivory Coast as it has been successful in China.

2.3 Anchorage material

In the case of Ivory Coast, we also will increase the varieties of anchors as it had been done recently in order to improve the adaptability of anchorage material in different working conditions, and increase its economy. Normally the ordinary anchors which are applicable in engineering practices include the grouting type and mechanical type pre-stressed anchor, tensile type and compressive type pre-stressed anchor, loading-distributed anchor, whole- bond type anchor, hollow grouting anchor, friction anchor bolt and many others.

It is notable that the Ivory Coast economy still weak, and not strong as countries such as Australia, USA or China. As such, applying the early strength cement anchor for the binding material of the rock bolt remains the proper method, and it is a fast and cheap method. The pulling resistance of the cement volume anchor can bear up to 145kN after 3 hours installation in Ivory Coast.

3 Development trend

3.1 Technical study

The theoretical researches of anchorage technology in geotechnical engineering are as follows:

(1) The research on the force-transferring mechanism should consider the fact that the bond stress is not uniform, and proposes a calculation method for the real bearing capacity of anchor.

(2) According to the half theory and half experience design philosophy, a calculation method of the bolt supporting system that considers the anchor group effect should be proposed.

(3) Studying the anchorage mechanism including the effect of pre-stress on the stress distribution in rock or soil mass, the effect of anchorage solid on the physical and mechanical properties of rock or soil mass, the interaction between the anchor and the rock or soil mass.

(4) Study and propose the reasonable simulation model, and the force-transferring rule in tensile type, compressive type and shear type anchorage solid.

(5) Study the simulation method of anchorage effect and mechanical effect of the anchor.

(6) Study the working mechanism of composite soil nailing wall and its design method.

(7) Study the theoretical analysis of an anchorage system and its computing method.

(8) Study the relationship between the bonding property of the bonding part of the anchor, and the bonding length, and study the factors affect the anchorage effect.

(9) Study the effect of pre-stress of an anchor on the stress redistribution of the rock and earth mass, and its mechanical properties.

(10) Study the anchor properties and design method under special conditions like earthquakes, impacts, dynamic loads, high temperatures.

3.2 Engineering application

The applied research of anchorage technology in geotechnical engineering is as follows:

(1) Research, development and application of high strength anchor.

(2) Research and development of light, high efficient, rapid and multi-functional drilling rig and its equipment; the development of the comprehensive anchoring machine; research and development of the comprehensive pre-stressed bolt cable.

To apply to large geotechnical anchorage engineering, recently, Ivory Coast will introduce all kinds of crawler-type hydraulic machines and some new type of drilling machines from some countries such as USA, German, Italia' China and Japan. Some drilling rigs made in China such as CM351 type, KQJ-100B type, QZ-100K type rock bolt drilling rig and 811L type, YTM87 type, KGM5 type soil anchor drilling machines, can be useful in the Ivory Coast because they can protect the casing well, and drill up 60 m ([11]-[14]).

(3) Develop new kinds of anchors and new technology; improve industrialized production of the anchor and its devices. Additionally, develop removable core bolts and enlarged-top bolts, self-drilling bolt, pressure-relief anchor, high strength pre-stressed long bolt cable.

To improve the adaptability of anchorage material in different working conditions and increase its economy, the varieties of anchors have been increased in recent years. The friction type rock bolt, including slot-tube anchor and water expansion anchor, has many advantages. For example, it can make the surrounding rock under three dimensional pre-stress status after installed, have good ductility. Its anchorage force could be increased substantially when the surrounding rock moves or is shaken. As such, it is especially applicable for soft rock surroundings or blasting effect underground projects. So far, this kind of friction type anchor has been applied into more than 100 underground coal mine projects. Anchors with resin bonding and anchors with early hardening cement has the characteristics of early strength, enough supporting resistance timely, which are used more and more. Compared with the normal anchor, pressure-relief anchor or retractable type anchor can bear bigger deformation. And its application can supply a more effective supporting method for the roadway with big deformation or affected the dynamic loading.

(4) Improve the long-term performance of anchor and safety assessment of anchorage projects, improve the construction quality control and project reliability detection, test the grouting density, and study real-time monitoring and network transmission test technology on the basis of network information.

(5) Study the pre-stress loss control and anticorrosion, control of mechanical stability and chemical stability of soil bolt anchor, pre-stressed bolt cable stability under the conditions of formation creep, steel relaxation, temperature change, impact force, and earthquake.

(6) Develop an expert system for the design of bolt-shotcrete supporting technology under different geological conditions.

(7) Discover how to recycle the anchor. According to the development history of the recyclable anchor structure, there have been many kinds of recyclable anchors including the slant wedge, double wedge, expanding shell bolt, invert sphenoid bolt, cement bolt, and so on.

4 Conclusions

The technology geotechnical anchorage technology still cannot satisfy the requirement of engineering practices, even if it has made much progress in the past several years. For this reason theoretical research and engineering practices should be improved and innovated in order to influence the outcome of the engineering practices requirements. And its applied range and status will be spread and developed with the improvement of the technology level. As new trends come in the Ivory Coast mining industry, the country needs to develop and apply electrical isolation bolts technology in order to test the anticorrosion property, which can become anchor standard in Ivory Coast and Africa. Ivory Coast can also develop soil anchors in order to change the force-transferring mechanism and decrease the fixed length. The artisanal gold mines over the country will greatly be reduced.

Acknowledgement

Foremost, I would like to express my sincere gratitude to my advisor Prof. Fuxing Jiang for the continuous support of my Ph.D. study and research, for his patience, motivation, enthusiasm, and immense knowledge. His guidance helped me in all the time of research and writing of this thesis. I could not have imagined having a better advisor and mentor for this paper.

Besides my supervisor, I would like to thank the rest of my others Professors for their encouragement, insightful comments, and hard questions on my work.

My sincere thanks also goes to my classmates and labmates for offering me the opportunities in their groups and leading me working on diverse exciting projects.

Last but not the least, I would like to thank God, my family and my friends for supporting me spiritually throughout my life.

References

1. Z. Fang and H.Q. Zhang, Experimental Research on Interfacial Bond Performance in Rock Anchor with High-Performance Grouting Medium. In Key Engineering Materials **517**, pp. 932-938 (July 2012)
2. L. Cheng, J. Han, and P. Zhang, Long-term performance and safety assessment of anchorage in geotechnical engineering, *Chinese Journal of Rock Mechanics and Engineering*, **5**, 003 (2008)
3. X.Y. Xie, K. Zhu, J.Y. Hua, H.W. Ying, and K.F. Liu, Behaviour and In Situ Tests of Self-Drilling Prestressed Anchors, In Applied Mechanics and Materials **130**, pp. 3536-3542 (January 2012)
4. Y.G. Cheng, Slope Anchoring Technology in China: State of the Art Report and its Prospect, In *Applied Mechanics and Materials* **256**, pp. 26-33 (December 2012)
5. Z.Q. Yue, W. Gao, J. Chen, and C.F. Lee, Drilling process monitoring for a wealth of extra factual data from drillhole site investigation, In Proceedings of the 10th international congress of the International

Association of Engineering Geology (IAEG2006 Engineering Geology for Tomorrow's Cities), Nottingham, UK, pp. 6-10 (September 2006)

6. K. Zhang, Z. Fang, A. Nanni, J. Hu, and G. Chen, Experimental Study of a Large- Scale Ground Anchor System with FRP Tendon and RPC Grout Medium, Journal of Composites for Construction, **19**(4), 04014073 (2014)
7. B. Leyland, Small hydroelectric engineering practice, CRC Press, (2014)
8. L.Y. Guo, M. Xiao, and Y.C. Xiao, Rock bolt mechanical analysis and its application to engineering. *Journal of Coal Science and Engineering (China)*, **14**, 58-61 (2008)
9. Z.O.U. Jun, Prestressed anchoring technology for the high slope of Three-gorge Lock, *The Chinese Journal of Geological Hazard and Control*, **2**, 021 (2004)
10. W.J. Lei, and G.X. Xu, Experimental investigation on the anchorage mechanics of anchorage cable of the late model with the pressure uniform, In *Advanced Materials Research* **163**, pp. 3798-3803 (January 2011)
11. R.J. Zhang, J.J. Zheng, P.Y. Li, J. Zhang, and S. Yu, A method for predicting mechanical behaviour of HPJG–Anchors–Part I: Mechanical characteristics and load transfer models. Computers and Geotechnics, **45**, 62-73 (2012)
12. X.Y. Luo, X.T. Xie, K.L. Li, and Y.C. Kuang, Test on Corrosion Resistance of GFRP Anchor Bolt, In *Advanced Materials Research*, **535**, pp. 1927-1935, (2012)
13. T. Nishikawa and S. Fujita, *Nanoimprint Biosensors: The Fusion of Nanofabrication, Nanophotonics and Nanobiology*, CRC Press, (2015)
14. Z. Tan, Y. Yu, M. Wang, and J. Yang, Experimental research on bolt anchorage effect on large-section deep-buried tunnel in loess, *Chinese Journal of Rock Mechanics and Engineering*, **8**, 013 (2008)

A Study on Landslide Risk Management by Applying Fault Tree Logics

Danish Kazmi[1], Sadaf Qasim[1], I.S.H Harahap[2], Syed Baharom[2], Arjumend Masood[1] and Muhammad Imran[2]

[1]*Department of Civil Engineering, NED University of Engineering & Technology, Karachi, Pakistan - 75270*
[2]*Department of Civil & Environmental Engineering, Universiti Teknologi Petronas, Perak, Malaysia - 32610*

Abstract. Slope stability is one of the focal areas of curiosity to geotechnical designers and also appears logical for the application of probabilistic approaches since the analysis lead to a "probability of failure". Assessment of the existing slopes in relation with risks seems to be more meaningful when concerning with landslides. Probabilistic slope stability analysis (PSSA) is the best option in covering the landslides events. The intent here is to bid a probabilistic framework for quantified risk analysis with human uncertainties. In this regard, Fault Tree Analysis is utilized and for prediction of risk levels, consequences of the failures of the reference landslides have been taken. It is concluded that logics of fault trees is best fit, to clinch additional categories of uncertainty; like human, organizational, and knowledge related. In actual, the approach has been used in bringing together engineering and management performances and personnel, to produce reliability in slope engineering practices.

1 Introduction

Learning from failures has long been a vital part of geotechnical engineering. Although there are many types of geotechnical failures, dam failures are especially significant and lean to be explored vigilantly. The dam failures at South Fork, Saint Francis, Malpasset, Vaiont, and the slope failure at Aberfan exhibit the consequences of failure, to tackle geotechnical and geological issues satisfactorily [1]. However, the failures also refers that an inadequate safety culture, overconfidence, and confused management are often more imperative than purely technical issues [1]. The problem of landslides in Malaysia is also dominated by these issues. This is not new; the catastrophic failure of Kwun Lung Lau landslide in Hong Kong is the outcome of human uncertainty. In this connection, this document involves current state of art in risk analysis to overcome or minimize the effects of slope failures. As in last few decades, the need to identify and quantify human uncertainties has not been accepted or catered fully by the geotechnical industry [2], [3], [4] and [5].

2 Objectives of the study

The objective of this study is to substantiate a need of human reliability analysis in geotechnical industry by putting numerous devastating slope failures into consideration. Secondarily, to overcome or minimize the effects of slope failures consequences, this study has developed fault tree analysis of significant scenarios responsible for slope failures

Table 1. Classes for probability of occurrence [8].

Qualitative Evaluation	Quantitative Evaluation	Value
Certain	Every time	1.0
Very High	One in a ten	10^{-1}
High	One in a hundred	10^{-2}
Moderate	One in a thousand	10^{-3}
Low	One in ten thousand	10^{-4}
Very Low	One in hundred thousand	10^{-5}
Extremely Low	One in a million	10^{-6}
Practically Zero	One in ten million	10^{-7}

3 Methodology of the study

The goal here is to bid a probabilistic framework for quantified risk analysis in relation with human uncertainties. As mentioned before, that Fault Tree Analysis (FTA) approach is employed by this study to overcome the likelihood situation of a catastrophe. The first step in quantifying a fault tree is to allot initial probabilities to the basic events. This step is taken by assembling information from a focus group decipherable with particular situations. The attached Table 1 is used as a channel for collecting this qualitative information. The group includes evaluation from registered professional engineers (civil and geotechnical), Contractors: from Pusat Khidmat Kontraktor (PKK), construction industry development officials and slope engineering division officials of Jabatan Kerja Raya Malaysia. Their opinions further justified by Aggregated Individual Method [6] however at few places Consensus Group Method [6] is also followed for the purpose of validation.

After having one concluded parameter of risk, "probability of failure", another parameter has been established bytaking selected case studies of Malaysian region. In case of following Santamarina and associates [7] approach, having in hand quantities of causalities and economic losses of the reference landslides can give an idea about risk levels to access whether the risk is acceptable or marginally acceptable.

4 Results and discussions

The above mentioned fault tree analysis technique is used to determine the chances of failure of the specific design, construction and maintenance subtasks (Refer Table 2) by fault tree logics.

Table 2. Probability of Failures of Selected Tasks/Scenarios.

Scenarios	Probability of Failure
Development plans	0.75
Testing/selection of shear strength parameters	0.124
Drainage planning/design	0.001
Installation/construction of drains	0.001
Retaining wall construction	0.069
Soil nail wall construction	0.068
Ground anchor maintenance	0.025
Other slope strengthening works	0.194

With reference to a fault tree for selecting/testing the shear strength parameters, one of the fact discovered through expert analysis is that improper experimental setup has also been observed in many cases. According to the study of Tan and Gue [9], selection of appropriate shear strength parameters is very crucial in design of cut slopes. It is significant that, materials like residual soils have discontinuities which the small scale tests may sometimes unable to detect. On the other hand, large-sized particles, effect cannot be quantitatively assessed and usually under-estimate the shear strength parameters of the in-situ material mass. Hence special care has to be taken in the selection of representative soil strength for slope stability analysis [9].

In reference to developing plans, "perception lacking" and "trend to follow thumb rules/patent methods" is quite high; according to quantitative estimation, it is one in ten. In fact, many of the problems identified in developing plans revolve around poor accountability towards the projects, no right person in making decision of the project due to lack of knowledge and experiences, no close coordination among stakeholders and lack of communication among project participants [10]. According to Oakes [11], the benefits of a good project development plans must include earlier identification of risks and issues; adoption of good practice; availability of skills and experience; improved communication; improved ability to allocate resources; improved predictability of project delivery; and greater confidence to take risks.

Regarding the flaws in construction of structures such as soil nail walls / retaining walls, no clear guidelines regarding installation and fixing of the bearing plates (in soil nail walls) exist. According to Chin and Meng [12], the designer should clearly indicate the constraints/steps in the relevant documentation, and the constructor should strictly obey the instructions of designer. It is recommended that stages of work for soil nailing to be included in construction drawings and relevant work specifications [12].

Among other slopes strengthening structures, "prestress ground anchors maintenance" involves unexpected diversion of water, blockage, seismic forces, soil mass movements improper coating on the anchors, sustained loading, environmental effects, seepage of water and material deficiency. Between the above mentioned basic events, subjective probability of seepage is reported quite high. It is approximately one in ten according to the given scale in Table 1. For the sake of maintenance, Malaysian local body Jabatan Kerja Raya (JKR) has also set some criteria and standards for monitoring and inspection for complex structures of ground anchors, surface subsurface drains, soil nailing etc. [13]. In slope strengthening structures, basic events of improper fixing and layout along with unknown cause plays a significant role which give rise to further event of obstruction in drainage. Water damming effect is also one of the events responsible to produce the above-mentioned event among three primary events. The remaining two includes biased results through inclinometers and making an assumption that structures are maintenance free. This third primary event is totally illogical but as mentioned by Gue and Fong [14], this illogical approach also exists.

As already mentioned in the previous studies [15], [5], and [16], that consequence analysis requires some previous landslides for a reference. Consequences can be in terms of causalities (non-payable), property (payable) or environmental (long lasting). The risk level parameters can be established by taking reference of those landslides which have been occurred due to human errors and taking the empirical rates of failure, for civil engineering facilities estimated previously has been taken into account [17]. The reported landslides (Table 3), as investigated by the geologists and experts are the cases of design flaws, improper sequencing of construction and overall poor or non-maintenance factors. Specifically talking about Highland Tower collapse, inadequate and poorly maintained drainage is the major contributor. Design flaws are also prevailing but the collapse of the Highland Tower has been taken place after 15 years of construction. In case of the next two case histories selected in this study, Bukit Antarabangsa (1999) and (2008), one common feature among all three is the poor or inadequate drainage. Clogged drains or even no sign of berms drain construction noted in Bukit Antarabangsa 1999.

Table 3. Prominent human error cases of landslides

Year	Location	Causes	Lives	RM(Million)
1993	Highland Tower	Improper rubble walls [18] Maintenance [19],[20],[21]	48	184.9
1996	Pos Dipang	Design error [22],[20]	44	69.0
1996	Keningau	Design error [20]	302	458.9
1999	Kg. Gelam	Design error [20]	17	29.5
1999	Bukit Antarabangsa	Design /construction Flaws [20],[22],[23]	-	5.5
2002	Simunjan Sarawak	High infiltration [20]	16	28
2002	Taman Hillview	Inadequate design of adjacent slope [21],[20],[24]	8	17.4
2003	Bukit Lanjan	Adverse estimation of geological conditions [22],[20]	-	836
2004	Taman Harmonis	Design error [20]	1	-
2006	Taman Zooview	Poor drainage [24],[20]	4	20.7
2008	Bukit Antarabangsa	Non maintenance drainage [25]	5	10.3

Table 4. Parameters for risk levels of major selected landslides.

Location	Year	Pf	F	EC(RM)
Highland Towers	1993	0.194	48	184.9
Pos Dipang	1996	0.001	44	69.0
Keningau	1996	0.001	302	458.9
Kg. Gelam, Sandakan	1999	0.001	17	29.5
Bukit Antarabangsa	1999	0.75	-	5.5
Simunjan Sarawak	2002	0.75	16	28
Taman Hillview	2002	0.001	8	17.4
Bukit Lanjan	2003	0.75	-	836
Harmonis, Gombak	2004	0.001	1	-
Bukit Antarabangsa	2008	0.194	4	20.7

According to literature review, consequences of failure cannot be benchmarked [16]. As indicated for Bukit Antarabangsa 1999, the probability of failure is 0.75 (Table 4) but its original consequences reflect very minimal losses as compared to other landslides of Highland Towers and Bukit Antarabangsa 2008 (Table 4). The landslides of Bukit Lanjan Selangor and Simunjan Sarawak also carry a probability of failure of 0.75 but the losses occurred from all the three landslides are totally different with each other. This is now certain that, for risk evaluation of slope failures, probability of failure is a better option.

5 Conclusion

The development of fault trees is to clinch additional categories of uncertainty like human, organizational, and knowledge related. The results of fault tree logics are used to combine, engineering and management performances to furnish quality and reliability in slope engineering practices. The intent here is to propose a probabilistic framework for quantified risk analysis with human uncertainties. Formally, QRA emphasizes on measurement of probabilities and losses of the past scenarios which have been focused to predict the risk's severity.

References

1. Christian, J. T., "Failures and Reliability of Large Projects" Metropolitan Section, Geo -Institute Chapter ASCE, (2013).
2. Baecher, G. B. and Christian, J. T., *Reliability and* Statistics *in Geotechnical Engineering: John Wile y & Sons Inc*, (200)3.
3. Morgenstern, N. R., "Managing risk in geotechnical engineering," in *10th Pan American Conference on Soil Mechanics and Foundation Engineering*, (1995).
4. Stewart, R. A., "Dam Risk Management," in *International Conference on Geotechnical and Geological Engineering (GeoEng2000)*, Melbourne, Australia, pp. 721–748, (2000).
5. Ho, K., et al., "Quantitative risk assessment: application, myths and future direction," 2000, pp. 19 –24.
6. Kirwan, B., "A Guide to Practical Human Reliability Assessment." Taylor & Francis London, (1994).
7. Santamarina, J., et al., "Reliability of Slopes: incorporating Qualitative Information," Transportation Research Record (1992).
8. Kirsten, H., "Workshop on Evaluation of Risk as Decision Making with the Criterion," Denver, CO, (1999).

9. Chin, T. Y. and Sew, G. S., "The Determination of Shear Strength in Residual Soils for Slope Stability Analysis," presented at the Seminar Cerun Kebangsaan, Cameron Highlands (2001).
10. Too, E., et al., "Project Governance in Malaysia Hillside Developments," in Sixth International Conference on Construction in the 21st Century (CITC-VI) *"Construction Challenges in the New Decade"*, Kuala Lumpur, Malaysia, (2011).
11. Oakes, G., "Project Reviews, Assurance and Governance" Aldershot, England; Burlington, VT: Gower, 2008.
12. Chin, T. Y. and Meng, C. C., "Slope Stabilization using Soil Nails Design Assumptions and Construction Realities," presented at the Malaysia-Japan Symposium on *Geohazards and Geoenvironmental Engineering*, Bangi, Malaysia, (2004).
13. Salahuddin, J. S., "Guidelines on Slope Maintenance in Malaysia," Slope Engineering Branch, Public Works Department Malaysia, (2006).
14. Gue, S. S. and Fong, C. C., "Slope safety: Factors and common misconceptions," Buletin Ingenieur, vol. **19**, (2003).
15. Uzielli, M., et al., "A conceptual framework for quantitative estimation of physical vulnerability to landslides," Engineering Geology, vol. **102**, pp. 251-256, (2008).
16. Ellignwood, B., "Assessment and Mitigation of Risk from Competing Low- Probability High Consequences Hazards," Symposium on *Emerging developments in Multi-Hazard Engineering*, (2007).
17. Baecher, G. B., "Geotechnical Risk Analysis Users Guide," FHWA/RD -87-011 Federal Highway Administration McLean, (1987).
18. Aun, O. T., "Mitigation And Rehabilitation of Natural Disasters In Malaysia," The Ingenieur, vol. **39**, (2008).
19. Gue, S. S. and Tan, Y. C., "Landslides: Abuses of The Prescriptive Method," *International Conference on Slope 2006 Kuala Lumpur*, (2007).
20. Raya, J. K., "National Slope Master Plan" Sectoral Report Research and Development, vol. Jabatan Kerja Raya Malaysia, (2009).
21. Samah, F. A., "Landslide in hillside development in the Hulu Kelang, Klang Valley," presented at the Post -Graduate Seminar Universiti Teknologi Malaysia, (2007).
22. Jaapar, A. R. B., "A framework of a national slope safety system for Malaysia," University of Hong Kong, (2006).
23. Gue, S. S. and Liong, C. H. (2007). Is the Ground in Ulu Klang Unstable? www.gnpgroup.com
24. Aiun, O. T. and Miin, O. H., "Geotechnical Failures/Issues, Dispute Resolution and Mitigation," The Journal of the Institute of Engineers Malaysia, vol. **71**, (2010).
25. JKR, "Jabatan Kerja Raya Final Investigation Report Investigation of Slope Failure at Taman Bukit Mewah, Bukit Antarabangsa Hulu Klang Salengor," (2009).

The Strengthening of Geological Infrastructure, Research and Data Acquisition - Using Gis in Ivory Coast Gold Mines

Kouame Joseph Arthur Kouame , Fuxing Jiang, Yu Feng and Sitao Zhu

School of Civil and Environmental Engineering, University of Science and Technology Beijing, Beijing, China

Abstract. The artisanal gold mining in Ivory Coast has become a key problem in the mining sector. A diverse group of people in Ivory Coast, including the young and the old, are all engaged in these activities that are reportedly better than agricultural inputs. . However, it is still a high-risk activity that leads to pollution, environmental degradation and the loss of human life. About ten people die each year in the gold mines. This paper focuses on gold mine safety by using the Geographic Information System (GIS) as a major solution to solve the artisanal gold mines problem, and also seeks to promote the mining industry in Ivory Coast.

1 Introduction

1.1 Definition of artisanal mining

The term Artisanal and Small-scale mining broadly refers to a mining practiced by individuals, groups, or communities, often occurring informally in developing nations. A common definition of this sector has not been adopted yet as a legal status, and local definitions vary from country to country. Due to this reason, it is difficult to estimate the extent of artisanal mining in developing contexts due to the lack of a common definition, its use of seasonal workers, and the lack of official statistics [1] (Hilson, G., 2002).

Artisanal mining is a thorny issue for both governments and large-scale mining areas in developing countries due to the various associated benefits and risks. In the last ten to fifteen years, the introduction of technology that has made this mining process more efficient. The problem however is that artisanal mining is not all that risk free. This paper focuses on artisanal mining in the Ivory Coast.

1.2 Artisanal mining in the Ivory Coast

Ivory Coast in the western part of Africa has a high fertility rate that is expected to reach nearly 80% in 2018. The increasing poverty rate of the population is mainly due to the presence of the youth, men, and women in the artisanal gold mines. Many children who became orphans during the rebellion in 2002 and the post electoral crisis; view working in the artisanal gold mines as the perfect job to provide for their daily bread and to support their relatives. A majority of the breadwinners of a family in Africa are children. ([2], [3]) (African Economic Outlook, 2013 and Carisch, 2012). The country has a large portion of the Birimian geological formations that make West Africa a mining region par excellence. In the case of the Ivory Coast, the issue of mining became rampant during the period of turmoil between 2003 and 2011, mainly due to the absence of a state authority, particularly in the north that was occupied by rebels at the time. Many efforts have been made to the promotion and transfer of artisanal mining in the so-called "small scale" mine by West African states, through projects and multi-sectorial funding programs that aim at optimizing national investment and promoting socio-economic welfare through the fight against poverty, assistance in communities settlement, protection of the environment, and quality of life improvement, among others [4] (Sogal, 2012).

Recent research has shown that artisanal miners have increased in number, and artisanal mining activities have become a privileged destination for local people and foreigners in the western part of Africa such as Ivory Coast, Guinea, Mali, and Ghana. This is mainly due to their changed livelihood standards, as a result of this activity ([5]-[7]) (Hilson et al., 2013, Bryceson and Jonsson 2009, Nyame et al, 2010).

According to SODEMI, the unique national company in Ivory Coast 24 out of 31 regions in the country are being illegally exploited by more than 500,000 miners which are mostly foreigners and indigenous people [8] (Assie, 2009). But we cannot determine the real number of the people who venture into illegal gold mine activities in the country, due to the fact that local people sometimes connive with foreigners for these activities. Despite the current government of Ivory Coast taking a decision to close several illegal mines around some cities like Ity,

 DOI: 10.1051/matecconf/20179518001

Djekenou Zanzan region, Zouan-Hounien, and other regions, illegal miners always back to the gold exploitation on the same sites after few months ([9]-[11] (Randgold Resources Ltd., 2012, Economie, 2015, AZUBUIKE.OKEH, 2014).

The vast majority of artisanal mining is illegal. Miners usually have to apply for a license to mine from the local authorities before they start digging, but the local industry estimates that 95% of all the artisanal mining goes ahead without such regulations. Regulation, rather than banning artisanal mining altogether is the only sustainable solution.

In the 1990's, the liberalization of the gold mining industry meant a downward shift in terms of environmental, human rights as well as transparency standards in many of the West African states. This is the case because the goal at the time was to lure foreign investors. However, governments at present need to come together and to harmonize the standards upwards. There has been some progress in this regard, due to the efforts of regional bodies such s the Economic Community of West African States (ECOWAS), which is working on a new mining code to apply to all its members. It is expected that this law would be deliberated on and passed in the following years, after heads of state within the region passed a directive on high level commitment to mining. With artisanal mining growing steadily and industrial scale mining set to significantly increase between now and 2020, networks and industries need to work hard and keep tabs on the industry at all levels.

However, Ivory Coast is now becoming a trustful country in the world after 10 years of political unrest. The current government attaches a great importance to the safety of persons and property. The poverty rate averages slightly, and the country's GDP grew from 7% in 2011 to 13% in 2013. The country can borrow funds from the World Bank and also from other financial institutions such as the Inter-American Development Bank (IDB) to acquire new technologies and invest fully in the mining sector. This is the case because recent research has shown satisfactory results that have discovered a huge mining potential in many parts of the national territory. (Fig. 1).

In the gold mine or coal mine production process, safety is a very important part of the disaster prevention in terms of mine gas disasters, coal dust hazards, and fire hazards. Unfortunately, Ivory Coast at present has a lot of artisanal gold mines such that the safety information management level is still relatively backward and has not completely desorbed from the traditional manual operation. The presence of low work efficiency, staff duplication, difficult data found in long series of problems in the information transfer cycle, and untimely safety information and accident reporting treatment, leads to vicious consequences of property damage and casualties. Therefore, carrying out a research of mining safety information on the safety management of gold production in Ivory Coast is of important significance.

The geospatial information technologies (GIS) and geospatial integrated information technology (IT) are increasingly used in the field of the mining industry. This new technology continues to improve the views of all their utilities in management of mining resources, planning and good management infrastructure in localisation and assessments of mining sites. These modern advanced technologies are also important in the health field of education and administration.

The majority of the mining deposit in Ivory Coast is mainly found in a gold mine that is located in the northern, western, central and eastern part of the country, as the most explored targets for gold deposits. The gold deposit in Ivory Coast has been previously interpreted since the independence, but today young gold has been discovered in Angovia and bonikro. Today several foreign mining companies are heading to the country for further deep explorations and exploitations for those already had the exploration permits. This is due to the fact that the government of the Ivory Coast has begun to take an interest in the mining sector in order to diversify its economy that is primarily based on agriculture. However some mining investors have met with problems such conflict over land and insecurity due to the presence of thousands of artisanal gold miners. For this reason, this paper is mainly focussed on offering GIS as a tool to establish a real predictive model for mapping gold mineralization in all the regions in the country to preventing such outcomes.

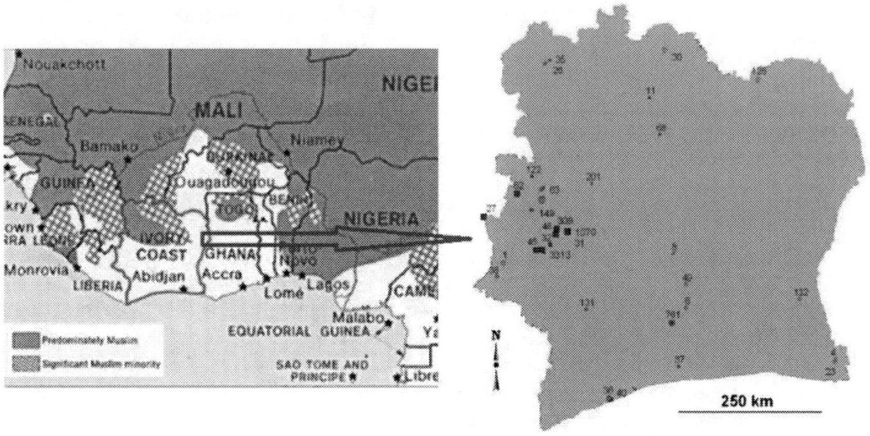

Figure 1. Legend and ore deposit name in Ivory Coast

2 System design

2.1 Functions of GIS system

The geographic information system (GIS) will exist to elaborate the map information management and to deal with monitoring information system management in Ivory Coast (Fig. 2).

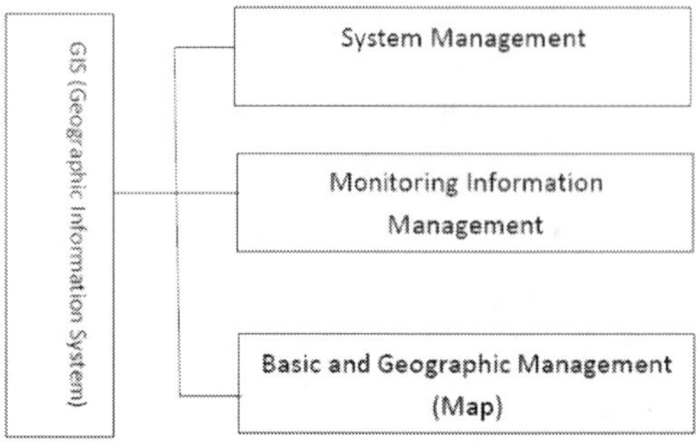

Figure 2. System functional

2.2 Elaboration of the map information management

GIS-based graphics leads to achieve the map to zoom, pan, layer control and other editing functions. At the same time, it realizes maps and attributes peer review. Ivory Coast is today a privilege destination for artisanal miners. Thousands foreigners and citizens are involved in artisanal and small-scale mining (ASM) in Ivory Coast due to the high rate of poverty and unemployment, which are the direct consequences of the political unrest from 1999 to 2011 [12] (UN, 2014).

The adoption of the new technology will play a major role in the Ivory Coast mining industry, by elaborating maps which will locate all the mining sites of the national territory first, and then identify all the artisans involved in the artisanal gold mining sector who illegally exploit the natural resources of countries. The Government of Ivory Coast will as such be able to control the quasi totality of the mining sector and take necessary steps towards better management of natural resources. This is the case because the lower system operating environment requirements provides a friendly user interface, which is easy for non-GIS personnel to operate the application.

2.3 Monitoring information system management

The real development of the mining sector in the Ivory Coast necessitates the understanding of knowledge of the mineral potential of each region and locality of the country. The new advanced technologies will be useful in obtaining accurate information on natural resources in Ivory Coast. Data based information will be acquired, processed and analyzed according to the geographical information of each locality and area of the Ivory Coast. Due to the dynamics of new advanced techniques and also their capacity expansion, the advanced technologies can collect geological data, resulting in many problems such as soil degradation and restoration of the environment being resolved. Land used jointly by mining, communities and epidemiology will be evaluated. Additionally, the deaths and flooding of the mine site will also be attenuated. As a result, the possibility to record all mineral production from the Ivory Coast and provide transparent and based information to the mining administration through the use of new technologies will be a great asset to the country's economy. Ivory Coast will experience a rapid economic development, epidemiologically secures evaluation, and disaster recovery. This is due the fact that the mine safety monitoring system with real-time data exchange, data exchange files for safety monitoring system, in accordance with a given data format specification, development of the data read and display algorithm, can read data files in real-time monitoring systems, the underground monitoring basic information and its operational status information about the device is displayed on the map.

3 The main function

The GIS system will be important in the mining industry in Ivory Coast in the following way:

(1) The introduction of the GIS in Ivory Coast mining operation will lead to the safety of local people and miners due to the medical system management in the operational mining areas. Numerous diseases such as lungs, heart and skin diseases, and also the spreading of HIV / AIDS are very often recorded on mining sites.

It is clear that GIS is now an important element in the medical service. The medical sector actors use GIS to invest in the geographical effects on humans. The Ivory Coast must be endowed with GIS for environmental preservation. Thus the risk of epidemic and infectious diseases proliferation will be greatly reduced on mining sites, and the local population will be free from diseases. In our advanced case, the use of GIS is essential because it can detect danger on future mining operation sites. The government of Ivory Coast will certainly invest huge funds for the acquisition of new geologic technologies, but the rewarding of that will greatly benefit the reduction of the WHO liabilities, which include the construction of hospitals and treatment of those suffering from HIV / AIDS and, tuberculosis. The proceeds will be used to solve some immediate problems of the population such as unemployment, securities, and poverty.

(2) GIS is a system which has good interactive capabilities, automatic connection and a two-way electronic map and database queries, and functions according to the need for thematic map output. Compared to the previous cut graphics and data established mutual monitoring software, this system has successfully realized the data and graphics correspondence and sync output. Many mining sites in Ivory Coast remain undiscovered or undeveloped, and some exploited sites even have often been abandoned by their operators due to the political unrest or land conflict between the mining companies and local people. The existence of many undeveloped mining sites is thus due to the lack of the high technology in the

mining sector in Ivory Coast. Some of the major strengths of a Geographic Information System (GIS) include its ability to plan a reclamation activity. Additionally, potential problems in operation can be remotely monitored and diagnosed by experts, who can propose effective solutions and provide technical support directly through online consultation

The active use of GIS applications will lead to the restoration of most of the abandoned mining sites in Ivory Coast. The minerals (gold, diamond, iron..) which are currently exploited by illegal miners in the northern, western, central and even in the southern part of the country can be controlled by the government and attributed to some reliable foreign investors such as Amara mining, the British mining company; which is expected to run the biggest gold mine deposit in 2017. The mining industry will be the second largest sector of the country's economy beside cocoa ([13-15]) (United Nations. Resolution 1528 (2011); United Nations. Resolution 2045 (2011); Ogwang Tom, 2011).

4 Conclusions

GIS-based integrated safety management is a system that holds a great and high level of visual, interactive and good man-machine interface coordination. This will assist in creating a real stable and fast response, with a strong security comprehension due to the smart work of GIS and its applications. The system plays an impressive role in safety management, as well as controlling and reducing the mine disasters during the operation. A good use of GIS leads to the development of the mining sector and enhances better management levels. Besides the technical support of GIS system in the advance sustainable development of gold mine production safety, GIS will play an increasingly large role in the mining industry in Ivory Coast. As such, the development of GIS – based graphic safety technology with an integrated visual information management system, which aims at enhancing the level of mine safety management in the Ivory Coast, is essential.

References

1. G. Hilson, Small-scale mining and its socio-economic impact in developing countries, *In Natural Resources Forum*, Blackwell Publishing Ltd. **26**(1): 3-13 (February 2002)
2. African Economic Outlook, Promoting youth employment in Africa, Why an African economic outlook on youth employment, (2013)
3. E. Carisch, Conflict gold to criminal gold: The new face of artisanal gold mining in Congo, South Africa: Southern Africa Resource Watch, (2012).
4. S.G. Sogah, *The effects of differences in agro-ecosystems on the diversity and distribution of avifauna in selected areas in the Eastern Region of Ghana*, Doctoral dissertation, (2012).
5. G. Hilson and C. Garforth, Everyone now is concentrating on the mining: drivers and implications of rural economic transition in the eastern region of Ghana, Journal of Development Studies, **49**(3):348–364 (2013)
6. J.B. Jønsson and D.F. Bryceson, Rushing for Gold: Mobility and Small-Scale Mining in East Africa. Development and Change, **40**(2): 249-279 (2009)
7. F.K. Nyame and J. Blocher, Influence of land tenure practices on artisanal mining activity in Ghana, Resources Policy, **35**(1), 47-53 (2010)
8. K.E. Assie, Lode gold mineralization in the Paleoproterozoic (Birimian) volco-sedimentary sequence of Afema gold district, southeastern Cote d'Ivoire, PhD thesis, Technical University of Clausthal, Germany, (2009)
9. Randgold resources Ltd., 2011 annual report: London, United Kingdom, Randgold resources Ltd., p. 185, (2012)
10. Economie. (December 21, 2015). *Centre-Ouest ivoirien: La mine d'or de Gamina exploitée illégalement un mois après sa fermeture (REPORTAGE)*. Retrieved Decembre 24, 2015, from Abidjan.net: http://news.abidjan.net/h/576676.html
11. AZUBUIKE.OKEH. (October 23, 2014). *Cote d'Ivoire closes 150 illegal gold mines*. Retrieved November 12, 2014, from NEWS AGENCY OF NIFERIA: http://nannewsnigeria.com/cote-d%E2%80%99ivoire-closes-150-illegal-gold-mines
12. United Nations Security Council, 2011a, Letter dated 17 October 2011 from the Chair of the Security Council Committee established pursuant to resolution 1572 (2004) concerning Côte d'Ivoire addressed to the President of the Security Council: United Nations Security Council report S/2011/642, p. 34
13. UN Security Council Security Council members unanimously adopt resolution 2153 (2014), 2014 Reed, Jared, 29.04.2014 http://www.dw.de/un-lifts-diamond-export-embargo-for-ivory-coast/a-17601628
14. United Nations Security Council, 2011b, Letter dated 20 April 2011 from the Chair of the Security Council Committee established pursuant to resolution 1572 (2004) concerning Côte d'Ivoire addressed to the President of the Security Council: United Nations Security Council report S/2011/272, p. 80
15. T. Ogwang, The root causes of the conflict in Ivory Coast, *Backgrounder*, (5):1, (2011)

Performance of Grass Filter Strip in Copper and Zinc Removal in Surface and Subsurface Runoff

Weijie Huo[1], Huaidong Zhou[1], Linghua Liu[1], Gaofeng Zhao[1], Zhaohui Yang[2] and Yilei Yu[3]

[1]Department of water environment, China Institute of Water Resources and Hydropower Research, Beijing, China, 100038
[2]State Key Laboratory of Simulation and Regulation of River Basin Water Cycle, China Institute of Water Resources and Hydropower Research, Beijing, China, 100038
[3]Beijing Key Laboratory of Wetland Services and Restoration , Institute of Wetland Research, Chinese Academy of Forestry, Beijing, China, 100091

Abstract. Three filter strips were conducted on self-designed soil bins. Taking a filter strip with no vegetation as contrast, the effectiveness of vegetation and soil conditions on heavy metals (including copper and zinc) removal efficiencies were investigated by simulated runoff experiment. The results showed that the adsorbed state is the main existing form of heavy metal. For surface runoff, most of total copper and total zinc are trapped in first 4m and it is ineffective to increase the distance beyond 4m for removal. Vegetation has no significant effect on total copper and total zinc removal, while the soil with higher content of organic matter is contributing to total Zn interception. For subsurface runoff, the removal efficiencies of total copper and total zinc can reach to above 95.38% and both vegetation and soil conditions have no significant effects. Vegetation is contributing to copper ion and zinc ion removal significantly. Soil condition is only a significant factor to zinc ion, with higher content of organic matter as a contributing factor.

1 Introduction

Even Intensive agriculture has sustained the world development, but some agricultural activities are contributing to the environmental health decreases at local, regional, and global scales. In China, the contaminants discharged in key river basins from agriculture are much more than from industry [1]. The excess sediment, nutrient and heavy metals from agricultural runoff have been a major cause to water quality deterioration.

Vegetative filter strip here is defined as a band of soil planted with permanent vegetation, and when the cropland runoff flows across the area, it undergoes a decrease of pollutant concentration and volume. Over the past decades, numerous studies have shown that it is a proposing way to attenuate sediment and nutrient in cropland run off [2-6].

Vegetative filter strip can disperse current and reduce the velocity of runoff and promote sedimentation, infiltration and adsorption. Vegetation plays an important role, such as soil conservation, biological utilization, improving soil permeability. The vegetation types used in studies include perennial grasses [7-11], shrubs [12-14], forests [15,16] and combinations of those above.

Compared to forest and shrub, grass filer strip is still economical alternative to reduce non-point source pollutants, and particularly those native grasses with tall, erect and stiff stems and strong root systems, is highly valued for removing particulate contaminants. Fescue is a common choice to construct grass filter strip, which is a kind of tufted grasses with strong fibrous root system, and widely used for buffer zones. In a field experiment, Daniels found that the fescue filter strip reduced runoff load by 50~80%, TSS by 80%, TP by 50% and retained NH4+-N by 20~50% [17]. Under simulated rainfall conditions, Magette found that the Kentucky-31 fescue filter strips were effective to remove nutrient in runoff [7]. Another study compared three practices, and reported that the fescue strip with switchgrass barrier was the most effective to reduce losses of organic N, NO_3^--N, NH_4^+-N, PP and PO_4^{3-}-P [8]. These studies have shown the effect of tall fescue filter strip on contaminants removal successfully.

Besides vegetation, soil condition is also an important factor to the effect of filter strip. The component, porosity, and texture of soil are usually positively correlated with precipitation volume and removal efficiency of pollutant.

Heavy metal pollution is long lasting, irreversible and it is hard to give rid of its influence once it is introduced into environment. It can harm human health and ecosystem safety through food chain. However, the studies on heavy metal attenuation by vegetative filter strips have been fewer.

The objectives of this study were : (i) to compare copper (Cu) and zinc (Zn) removal efficiencies by

DOI: 10.1051/matecconf/20179518002

contrast filter strip, tall fescue filter strips with different soils,(ii) to estimate the effects of vegetation and soil on total heavy metals and heavy metal ions removal in surface runoff and subsurface runoff.

2 Materials and methods

2.1 Construction of experimental system

Experimental system is consisted of PE bucket and soil bin system. The volume of PE bucket is 800L, and water is controlled by flow meter. Soil bin system is consisted of drainage channel and soil bin. Drainage channel is 7.5×10^{-2} m^3 (Length 0.3m * width 0.5m * high 0.5m), while soil bin is 1.5 m^3 (long 6.0m * width 0.5m * high 0.5m). Water from PE bucket flows firstly into the drainage channel, and then enters into soil bin by horizontal overflow. There are three surface holes along the bin of 2m, 4m and 6m (No.01, 03 and 05), and three bottom holes (No.02, 04 and 06) corresponding to surface holes vertically. The soil bin is placed on the bracket with adjustable slope, and the structure is shown in Figure 1.

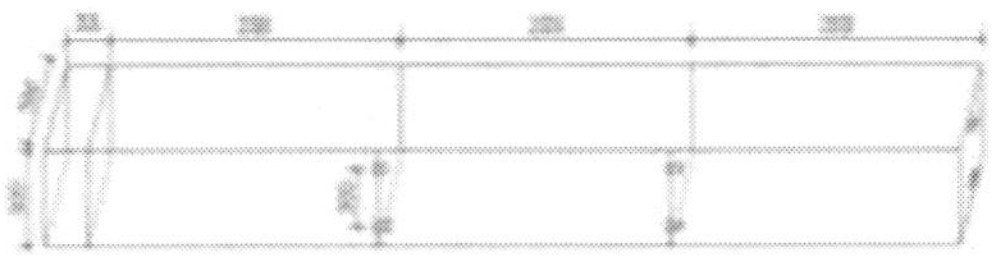

Figure 1. Sketch Map of Soil Bin System (unit: mm).

Soil bin is filled with soil of 0.35m depth. There are three filter strips constructed on three soil bins. The set is as follows: C is a contrast filter strip with no plant in it, and 1# soil is filled in. A is a tall fescue filter strip with 1# soil filled in. The tall fescue grows with coverage 90%, average root length of 35cm, average tissue height aboveground of 42cm, fresh weight of 4.11 kg m^{-2}, dry weight of 0.721 kgm^{-2}. B is also a tall fescue filter strip with 2# soil filled in. The tall fescue grows with coverage around 90%, average root length of 36cm, average tissue height aboveground of 40cm, fresh weight of 4.76 kg m^{-2}, dry weight of 0.913 kg m^{-2}. The main physical and chemical properties of 1# and 2# soil are listed in Table 1. According to the soil mechanical composition data and international standard of soil texture classification, 1# soil is specified as silt loam, while 2# soil is specified as clay loam [18].

Table 1. The main physical and chemical properties of the tested soil.

Soil No.	organic matter	Cu	Zn	CEC	particle composition%		
	g Kg^{-1}	mg Kg^{-1}		cmol $^{+}Kg^{-1}$	sand	powder	clay
1#	5.64	25.1	156	3.41	70.96	26.00	3.04
2#	18.25	18.2	59.0	12.5	61.83	20.76	17.41

Note: The unit of CED is $cmol^{+}$ Kg^{-1}, which is positive charge per kilogram.

2.2 Experimental scheme

Experimental devices are located on the bank of Kunyu River, near Yu Yuan Tan park, in Beijing, China. Simulated runoff experiments are applied. Experimental water is extracted from river into bucket (800L), and soil particles and concentrated solutions of heavy metal (Cu, Zn) are accurately added into bucket according to the experimental design, which is to simulate the sand and heavy metal in cropland runoff.

Grouped comparison is applied. Vegetation effect is investigated by contrast filter strip versus tall fescue filter strip(C versus A).Soil condition effect is investigated by soil 1# versus soil 2#(A versus B).

Since the start of the simulation runoff, system inlet water samples from outlet of bucket are collected by every 10-15 minutes. Surface runoff samples from surface holes along the bin of 2m, 4m and 6m are collected. Subsurface runoff samples from bottom holes along the bin of 2m, 4m, 6m are also collected. The concentrations of heavy metals are determined immediately after collection.

Heavy metal determined in this study is divided into total amount of heavy metals and heavy metal ions. The concentration of total amount was determined by total water sample acid digestion. Briefly, a volume of 15mL of total water sample was weighed. About 1.5mL concentrated HNO_3 was added to sample and the sample was heated to concentrate into 2-5ml.Then 1.5ml HNO_3 and 0.2 mL H_2O_2 were added into that concentrated solution. Then it continues to concentrate into 1ml. Lastly, ionized water is added to sample until the volume to 25mL. Determination of heavy metal ions is directly to use the water sample that is filtered by 0.45μm membrane. All the heavy metal concentrations determination is completed by ICP-MS(Perkin-Elmer, USA).HNO_3 and H_2O_2 used in experiment are guarantee reagent. Parallel sample determination should be done for ensure the precision of the experiment. The relative standard deviations of heavy metals were less than 5%.

3 Results and analysis

3.1 Analysis of interception characteristics of heavy metals in surface runoff

Heavy metal existing in runoff water and sediment is generally divided into particulate and dissolved fractions by filter through a 0.45 μ m membrane. Heavy metals ion may be combined with the active groups of soil particles by adsorption, and may get into water by desorption. Heavy metal ion exists in the balance of adsorption and desorption. Cu and Zn are chosen as target pollutants and their interception features are investigated by surface runoff and subsurface runoff.

Heavy metal determination in surface runoff is total amount determination, including adsorbed heavy metals combined with suspended matter, as well as heavy metal ion with reactive activity that dissolved in water. Total Cu removal efficiencies in surface runoff are shown in Figure 2.

DOI: 10.1051/matecconf/20179518002

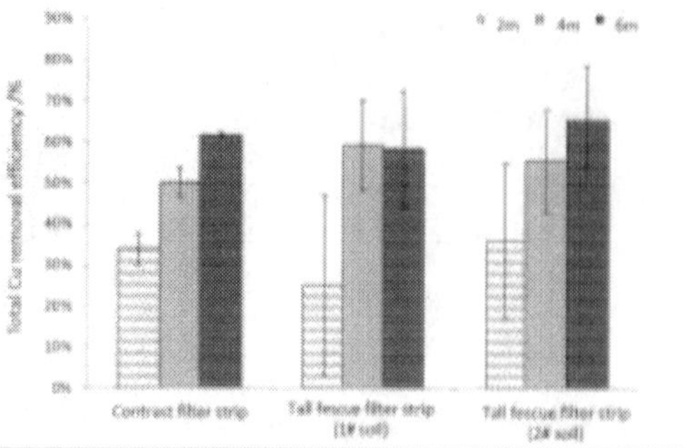

Figure 2. Total Cu removal efficiencies in surface runoff along 2m, 4m, and 6m of strip.

Total Cu removal efficiencies of contrast filter strip with 2m, 4m and 6m are 33.89%, 50.13% and 61.61% successively. Total Cu removal efficiencies of tall fescue filter (1# soil) are 25.09%, 59.19% and 58.30% successively. Total Cu removal efficiencies of tall fescue filter (2# soil) are 36.00%, 55.47% and 65.42%. The result is shown that the total Cu removal efficiencies of all filter strips have the trend of increasing with the filter width increasing. Comparing the total Cu removal efficiencies along 2m, 4m, and 6m of filter strip, that of 4m is significantly higher than that of 2m, and the removal is slowing down with width increasing from 4m to 6m.Therefore, 0-4m is considered to be the main strip width of Cu interception.

Adopting the method of spss17.0 one-way analysis of variance to compare the Cu removal efficiency of 6m filter strip, it is found that there is no significant difference($p=0.753$) between contrast filter strip and tall fescue filter strip (A with 1# soil). And there is no significant difference($p=0.562$) between tall fescue filter strip with 1# soil (A) and tall fescue filter strip with 2# soil (B). Therefore, the vegetation condition and soil condition are not significant factors to Cu removal efficiency.

Total Zn removal efficiencies in surface runoff along 2m, 4m, and 6m are shown in Figure 3. Total Zn removal efficiencies of contrast filter strip with 2m, 4m and 6m are24.78%, 36.29% and35.97% successively. Total Zn removal efficiencies of tall fescue filter (1# soil) are 35.28%, 30.78% and 31.48% successively. Total Zn removal efficiencies of tall fescue filter (2# soil) are 24.21%, 44.35% and 51.71% successively.

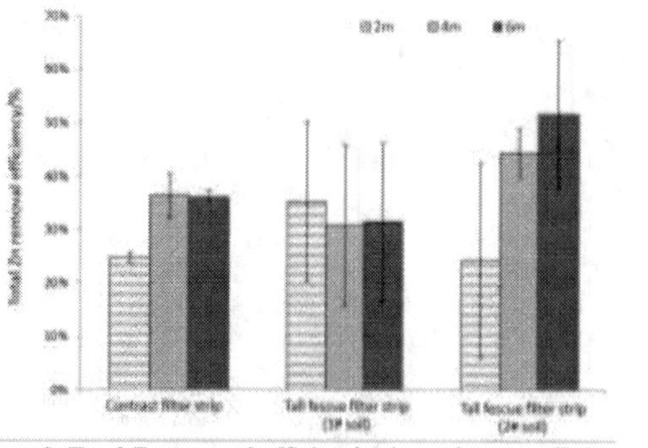

Figure 3. Total Zn removal efficiencies in surface runoff along 2m, 4m, and 6m of strip.

The result is shown that the total Zn removal efficiencies of contrast filter strip and tall fescue filter strip (with 2# soil) have the trend of increasing with the filter width increasing. But the tall fescue filter strip with 1# soil does have the trend. It is same as total Cu interception that 0-4m is considered to be the main range of strip width for interception.

Adopting the method of spss17.0 one-way analysis of variance to compare the Zn removal efficiency of 6m filter strip, it is found that there is no significant difference ($p=0.167$) between contrast filter strip and tall fescue filter strip (A with 1# soil). But the removal of tall fescue filter strip with 2# soil is significantly better than tall fescue filter strip with 1# soil ($p<0.05$). It can be drawn that vegetation condition is not a significant factor to total Zn removal efficiency, but soil condition is a significant factor. The soil with higher organic matter content has the higher removal efficiency. In addition, Zn content of 2# soil is 59 mg Kg-1, which is lower than that of 1# soil (156 mg Kg^{-1}). The soil owning lower Zn content has the greater adsorption capacity, which might be one of the reasons that affect the removal efficiency of filter strip.

The total content of heavy metals in surface runoff is determined, which includes adsorbed heavy metals and heavy metal ion. The inflow concentration of total Cu is 68.76～99.23μg L^{-1} and that of total Zn is 267.95～496.00μg L^{-1}.The inflow concentration of Cu^{2+} is 4.310～9.737μg L^{-1} and that of Zn^{2+} is 24.885～81.901μg L^{-1}. By comparison of Cu and Zn, it is found that ion component has smaller proportion in total content, with Cu^{2+} accounting for 6.27%～9.81% and Zn^{2+}accounting for 9.29%～16.51%. Therefore, the heavy metal adsorbed with sediment is the main existence in inflow water.

When water flows through the filter system consisting of soil and vegetation, the pollutants get preliminarily intercepted by physical and chemical processes like deposition, filtration, adsorption within a short retention time. And the intercepted pollutants will be transformed further by plant absorption, microbial metabolism and chemical reaction [19-20].

The vegetation is a necessary element of filter strip system, which can block the surface runoff and reduce the speed of runoff. So vegetation can promote particles deposition and the adsorbed pollutant retention. The retention time will be extended, which promotes the interception between pollutants and adsorption sites [21-22]. In addition, the plant tissue can contact with runoff, which can produce adsorption to heavy metal. The growth of vegetation can increase the content of organic matter in soil, which is conductive to the adsorption reaction.

Comparing the effects of soil conditions, it can be concluded that Zn in tall fescue filter strip (2# soil) is significantly higher than tall fescue filter strip (1# soil) , while the soil has no significant effect on removal efficiency of Cu. The organic matter content of 1# soil was 5.86～6.90 g Kg^{-1} and that of 2# soil is 11.62～13.92 g Kg^{-1}. The content of organic matter of soil 2# is significantly higher than that of soil 1# ($p<0.01$).Except organic matters, the caption exchange capacity (CEC) and clay proportion of 2# soil are also higher than those of soil 1#. And the condition mentioned above is conductive to the adsorption reaction [23-24].

3.2 Analysis of interception characteristics of heavy metals in subsurface runoff

The grass filter strip system constructed between pollution source and receiving water can change the hydraulic characteristics of surface runoff. Vegetation barrier and soil percolation can slow surface runoff, and the hydraulic retention time is extended. All the above can promote surface runoff to infiltrate. Therefore, soil percolation is an important way for filter strip to intercept pollutants. In this study, seepage water samples are collected from bottom holes, and the interception effect of heavy metal during seepage through 0.35m of soil layer is studied.

The interception effect of total amount heavy metal and heavy metal ion are investigated in this section. The removal efficiencies of total Cu and Cu^{2+} in water of 6m outlet is shown in Figure 4. Total Cu removal efficiencies of contrast filter strip, tall fescue filter strip (1# soil) and tall fescue filter strip (2# soil) are 96.16%, 96.50% and 95.62% successively. And those of Cu^{2+} are 56.31%, 76.23% and 74.55% successively. Adopting the spss17.0 one-way analysis of variance to compare the total Cu removal efficiencies, it can be drawn that there is no significant difference (p=0.825) between contrast filter strip and tall fescue filter strip. But Cu^{2+} removal efficiency of tall fescue filter strip is significantly higher than contrast filter strip ($p < 0.05$. There is no significant difference of total Cu and Cu^{2+} removal efficiencies in both tall fescue filter strips (p=0.845 and p=0.727).

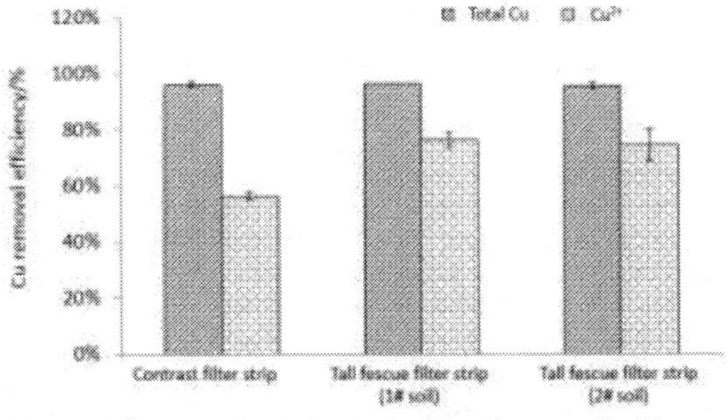

Figure 4. Cu removal efficiencies in subsurface runoff of 6m strip.

The removal efficiencies of total Zn and Zn^{2+} are shown in Figure 5. Total Zn removal efficiencies of contrast filter strip, tall fescue filter strip (1# soil) and tall fescue filter strip (2# soil) are 99.70%, 96.57% and 95.38% successively. And the removal efficiencies of $Zn^{2+}$are 62.68%, 81.43% and 86.42% successively.

Adopting the spss17.0 one-way analysis of variance to compare the total Zn removal efficiencies, it can be drawn that there is no significant difference (p=0.869) between contrast filter strip and tall fescue filter strip. But Zn^{2+} removal efficiency of tall fescue filter strip is significantly higher than contrast filter strip ($p < 0.05$). There is no significant difference of total Zn removal efficiency in both tall fescue filter strip (p=902), but Zn^{2+} removal efficiency of strip with 2# soil is significantly higher than that of strip with 1# soil ($p < 0.05$).

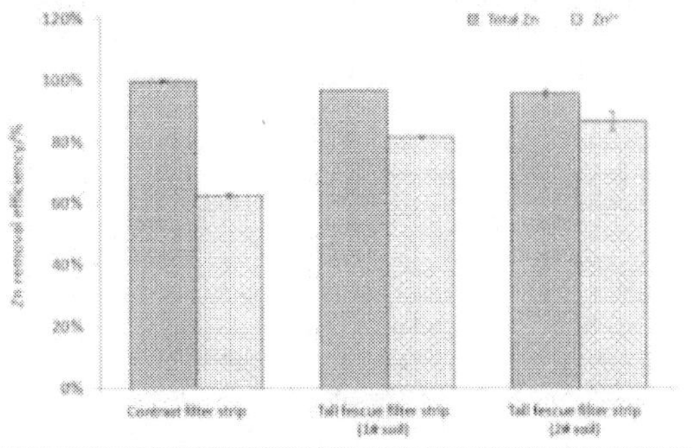

Figure 5. Zn removal efficiencies in subsurface runoff of 6m strip.

Based on the analysis above, it can be drawn that the interception of Cu and Zn in 0.35m of soil layer are similar. For the total amount of heavy metal, the vegetation condition and soil condition are not significant factors. For heavy metal ion, vegetation condition is an significant factor, and the existence of vegetation is advantageous to the interception of Cu and Zn ions. The soil condition is a significant factor to Zn^{2+}, and 2# soil is more advantageous to the interception of Zn^{2+}.

Total heavy metal includes heavy metal ion and the absorbed heavy metal, and the absorbed one takes on a greater proportion. During the process of water infiltration, most of suspended matters can not seepage through the soil pore, and has been intercepted at the soil layer. Therefore, there is no significant difference for total heavy metal interception in different conditions of vegetation and soil. And the total heavy metal removal efficiency can reach to above 95.38%, which is significantly higher than corresponding ion. The interception process of heavy metal ions through the soil infiltration is more complicated, which including the physical and chemical processes like soil adsorption and desorption, vegetation adsorption and microbial metabolism [25-28]. But within a short term of hydraulic retention, the adsorption and desorption should be a more important process. The water electrical conductivity of surface runoff is around 585.5～655.5μs cm^{-1}, while the seepage water electrical conductivity is 424.0～526.5 μ s cm^{-1}, and there is no significant difference (p=0.856) between different vegetation and soil conditions. The change of water electrical conductivity shows the process mentioned above are less, such as mineralization of organic matter, biodegradation and caption exchange, that can increase the number of ions in solution. While the adsorption that can decrease the ion number may be the main process to reduce electrical conductivity, which is also verify that adsorption effect is the primary mechanism in penetration process.

4 Conclusion

(1) The interception characteristics of total heavy metal are investigated in surface runoff. Adsorbed state is the main existence form of heavy metal in surface runoff, and their interceptions are occurred mainly in the length of 0-4m. The removal is slowing down with length increasing from 4m to 6m.The conditions of vegetation and soil have different effects on total Cu and total Zn removal. Vegetation has no significant effect on both heavy metals removal, but the soil with higher content of organic matter is contributing to the interception of total Zn.

(2) The interception characteristics of total heavy metal and heavy metal ions are investigated in subsurface runoff. By the seepage of 0.35m of soil layer, the removal efficiencies of total Cu and total Zn can reach to above 95.38%, which is significantly higher than corresponding ions, and both vegetation and soil conditions have no significant effects. For ions, vegetation is contributing to Cu^{2+} and Zn^{2+} removal significantly. Soil condition is only a significant factor to Zn^{2+}, with higher content of organic matter as contributing factor.

Acknowledgements

This research was supported by the National Water Pollution Control and Treatment Science and Technology Major Project, Grant NO. 2013ZX07104-004.

References

1. Xiong Deng :J. ACTA SCIENTIARUM NATURALIUM UNIVERSITATIS SUNYATSENI. **146**, 244-247, (2007) (In Chinese)
2. SergiSabater, Andrea Butturini, Jean-Christophe Clement, et al. J. Ecosystems, **6**, 20-30, (2003)
3. T.J.Schmitt,M.G.Dosskey, K.D.Hoagland. J. Journal of Environmental Quality, 28, 1479-1489 (1999)
4. Jon E. Schoonover, Karl W.J. Williard, James J. Zaczek, et al.J. Agroforestry Systems, **64**, 169-180, (2005)
5. Patty, L., B.Rheal, et al. J. Pesticide Science, **49**, 243-251 (1997)
6. Dillaha,T.A., R.B. Reneau, *et al.* J. USEPA, Chesapeake Bay Liaison Office, Annapolis,MD,(1987)
7. Magette, W.L., R.B. Brinsfield, R.E. Palmer, and J.D. Wood.J. Trans. ASAE. **32**. 663–667, (1989)
8. Humberto Blanco-Canqui, C. J. Gantzer, S. H. Anderson, E. E. Alberts, and A. L. Thompson . J. Soil Sci. Soc. Am. **68**,1670–1678, (2004)
9. Lee, K.H., T.M. Isenhart, R.C. Schultz, and S.K. Mickelson. J. Trans. ASAE. **39**, 2231–2239, (1999)
10. Coyne, M.S., R.A. Gilfillen, R.W. Rhodes, and R.L. Blevins. J. Soil Water Conserv. **50**,405–408 (1995)
11. Eghball, B., J.E. Gilley, L.A. Kramer, and T.B. J. Soil Water Conserv. **55**, 172–176, (2000)
12. Monica Vianello, CostantinoVischetti, Luciano Scarponi, Giuseppe Zanin. J. Chemosphere. **61**, 717–725, (2005)
13. Mankin KR, Ngandu DM, Barden CJ, Hutchinson SL, Geyer WA. J. Journal of the American Water Resources Association. **43**, 1108–1116, (2007)
14. Schultz, R.C., Colletti, J.P., Isenhart, T.M., Simpkins, W.W., Mize, C.W.,Thompson, M.L., J. Agroforestry Syst. **29**, 201–226, (1995)
15. Lowrance, R., G. Vellidis, R. D. Wauchope, P. Gay, and D. D.Bosch. J. Trans. ASAE **40**, 1047–1057, (1997)
16. Doyle, R. C., Stanton, G. C., and Wolfe, D. C. J. American Society of Agricultural Engineers. **9**, 2451-2501, (1977)
17. R.B.Daniels,J.W.Gilliam. J. Soil Science. **60**, 246–251, (1993)
18. Dayi Lin. Soil experimental guidance. M.China Forestry Press. **37**, (2004), (In Chinese)
19. Songlin Zhang, QingshiDong,XibinZhou,et al. J. Journal of Lanzhou University(Natural Sciences). **44**, 47-50,(2008)
20. Bailey Ann Sullivan.. Kansas State University, 2006
21. Toshiro Yamada, Sally D.Logsdon, Mark D.Tomer, Michael R.Burkart.J. Environment. **385**, 297-309, (2007)
22. Dosskey,Dosskey, M.G., Helmers, M.J., Eisenhauer, D.E., Frantia, T.G. &Hoagland, K.D. J.Journal of Soil and Water Conservation, **57**, 336–343, (2002)
23. Lowrance R., Dabney S. and Schultz R. J. Soil Water Conserv. **57**, 36–43, (2002)
24. Standard examination methods for water and wastewater (the 4th edition). M.China (2002)(In Chinese)
25. Lena B.-M., Gilles P , Ann F . J. Landscape Urban Plan. **31**, 323 – 331, (1995)
26. Majed Abu-Zreig,Ramesh P Rudra, Hugh R. Whiteley. J. Journal of Environmental Quality. **32**, 613-619, (2003)
27. Dillaha,T.A.Reneau, R.B. J. Transactions of the American Society of Agricultural Engineers. **32**, 513-519 (1989)
28. Broersma K., Robertson J.H. and Chanasyk D.S. J. Soil Sci Plant Anal. **26**, 1795–1811, (1995)

Stochastic Lot-Sizing under Carbon Emission Control for Profit Optimisation in MTO Manufacturing

A. Qiao, S.H. Choi , X.J. Wang and Y.C. Zhao

Department of Industrial and Manufacturing Systems Engineering, the University of Hong Kong, Hong Kong

Abstract. Aggravating global warming has heightened the imminent need by the world to step up forceful efforts on curbing emission of greenhouse gases. Although manufacturing is a major resource of carbon emission, few research works have studied the impacts of carbon constraints on manufacturing, leading to environmentally unsustainable production strategies and operations. This paper incorporates carbon emission management into production planning for make-to-order (MTO) manufacturing. This paper proposes a model that solves lot-sizing problems to maximise profits under carbon emission caps. The model adopts stochastic interarrival times for customer orders to enhance the practicality of the results for real-world manufacturing. Numerical experiments show that reducing carbon emission undercuts short-term profits of a company. However, it is conducive to the company's market image as being socially responsible which would attract more customers who concern about environmental protection. Hence, reducing carbon emission in manufacturing is beneficial to long-term profitability and sustainability. The results provide managerial insights into manufacture operations for balancing profitability and carbon control.

1 Introduction

In recent decades, global warming has been aggravated due to excessive carbon emission from human activities. World governments are under huge pressure to mitigate the harmful impacts on the environment by tightening regulations on carbon emission. One of the most significant programs for carbon emission control is the United Nations Framework Convention on Climate Change held by 154 nations in 1992. In 2015, the Paris Agreement was concluded in the 21st Session of the Conference of the Parties, which elevated the international corporation in carbon emission management to a new stage [1].

A major source of carbon emission is industrial production, since it consumes huge amounts of materials and energy and releases much wastes. Besides the traditional make-to-stock (MTO) production mode that maintains inventory for incoming orders, more companies start to focus on the make-to-order (MTO) mode that produces products only after receiving customer orders, which is considered conducive to carbon emission management by applying suitable emission constraints.

Many researchers have contributed to the study of carbon emission management. Sarkis et al. conducted a review of organizational theoretic literature, while Brandenburg et al. conducted a review about the quantitative models that have been used in sustainable supply chain management [2], [3]. Accenture examined supply chain information about the risks and opportunities related to climate change in its supply chain report [4]. Besides, Govindan et al. studied the recent status and opportunities for the green supply chain management [5].

Although manufacturing is a major source of GHGs, only some researchers have attempted to incorporate carbon control into optimization of production operations. Absi et al. provided production planning considering carbon emission policies and Chen et al. worked on the economic order quantity (EOQ) model considering the carbon emission in production cost. [6], [7]. Among the above works reviewed, few have focused on MTO.

To mitigate the research gap, this paper incorporates carbon control in stochastic lot-sizing optimization for MTO manufacturing. It proposes a model for optimal lot size that maximises the profit without emitting carbon more than a given cap during the production process. To enhance its practicality for real-world manufacturing, the model adopts interarrival times with stochastic distributions for customer orders. Numerical experiments are conducted and the result helps manufacturers balance short-term profitability and carbon control, which is vital to long-term profitability and business sustainability.

In the remainder of this paper, Section 2 elaborates the production planning problems and the mathematical formulas in detail. Section 3 conducts a numerical case study and presents the analytical results, while section 4 draws conclusions.

2 Production model formulation

2.1 The stochastic MTO production system

Fig. 1 shows the production system for single-product MTO manufacturing, in which independent customer orders arrive at stochastic interarrival times. When orders accumulate to a given size Q, they are gathered and transferred in a batch to the setup stage for the initial processing. These batches are then moved to the processing stage to be manufactured one by one. When the product for an order is completed, it is directly delivered to the end customer without waiting for other orders in the same batch. The lot size Q is crucial to the performance of such a production system. While a small lot size causes frequent setups and hence shop floor nervousness, a large one leads to high work-in-process (WIP) inventory which increases not only costs but also carbon emission during storage.

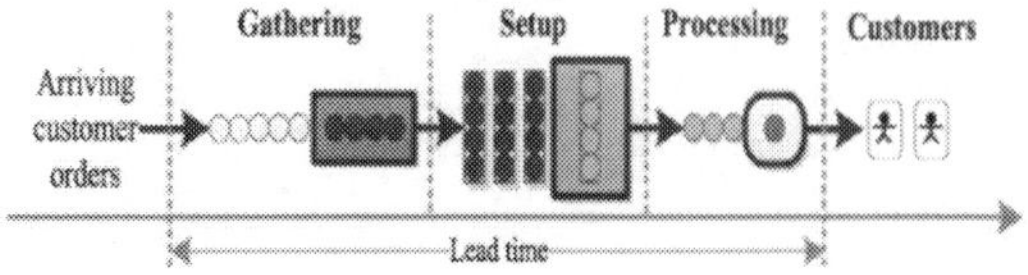

Figure 1. Stochastic MTO production system

2.2 The proposed model

The proposed model therefore aims to solve for an optimal lot size that maximises the profit of products while limiting the total carbon emission during production below a given carbon cap, which is deemed by governments as an effective way to control carbon emission by individual manufacturers. Governments generally set carbon caps for manufacturers smaller than their annual emissions. If a manufacturer does not exceed its carbon cap, it pays no cost; otherwise, it has to pay extra tax or purchase carbon quota for its excessive emission. Hence, the proposed model supposes that the manufacturer will adjust its lot size Q to reduce carbon emission to avoid extra cost due to excessive emission. Other assumptions made for the model are:

(1) Orders are independent so that all orders come in stochastically with an independent and stable increment.
(2) All machines are independent, and machine breakdown is not considered.
(3) Orders are processed based on the first-come-first-serve principle.
(4) Each order includes only one single product.
(5) The product price is exogenous, and the manufacturer has no influence on the product price.
(6) The decision variables and stated indexes do not influence the stochastic process.

Notations

Q	Lot size
R	Operational profit
D	Uncertain marketing demand
γ	Unit sales price
C_V	Variable costs
C_F	Fixed costs
s	Unit setup cost
h	Unit WIP carrying cost
W	Total work flow time
X_i	Interarrival time of the ith order, all X_i 's are independent and identically distributed (IID)
Y_i	Setup time of the ith batch, all Y_i 's are IID
Z_i	Processing time of the ith order, all Z_i 's are IID
X^b	Interarrival times of batches for the assumed GI/G/1 queuing model
T_b	Service time for the assumed GI/G/1 queuing
τ	Mean setup time
μ	Mean processing rate
e_m	Carbon emission from manufacturing
e_{WIP}	Carbon emission from WIP inventories
M	Emission cap

2.2.1 Benchmark model without carbon emission constraint

This research first derives the proposed model for MTO production without considering carbon emission as a benchmark against production under carbon emission constraint. The operational profit R is formulated as:

$$R = D\gamma - C_V - C_F \qquad (1)$$

with

$$C_V = D(s/Q + E(W)h + \omega) \qquad (2)$$

where ω represents all the unit variable costs irrelevant to the product lot size and the expected leading time, and $E(W)$ means the mean WIP inventory time.

To solve $E(W)$, all the involved random variables are characterized based on the expected values and variances without assumptions on the distribution. Based on Fig. 1, the total work flow time is:

$$E(W) = E(W_{qg}) + E(W_g) + E(W_{qs}) + E(W_s) + E(W_{qp}) + E(W_p) \qquad (3)$$

When an order is received, it is immediately gathered into batches without waiting, so that queuing time for gathering $E(W_{qg})= 0$. The gathering time for the ith order is the time that it waits for the remaining Q-i orders to be gathered with itself into a batch, i.e.,

$$E(W_g) = E_i(E(W_g|i) = (Q-1)/D \qquad (4)$$

To calculate queuing time for setup $E(W_{qs})$, the setup and processing stages may be simplified as a GI/G/1 queuing model. It is similar to the production situation, except the processing time is overestimated, which will not affect the W_{qs}. Based on the equations for GI/G/1 model,

$$E(W_{qs}) = [Var(X^b) + Var(T^b)]/2[E(X^b) - E(T^b)] \qquad (5)$$

$$\begin{aligned} E(X^b) &= Q/D, Var(X^b) = Q\sigma_X^2, \\ E(T^b) &= \tau + Q/\mu, \\ Var(T^b) &= \sigma_Y^2 + Q\sigma_Z^2 \end{aligned} \qquad (6)$$

With $\rho = E(T^b)/E(X^b)$, where $\sigma_x^2, \sigma_y^2, \sigma_z^2$ represent the variance of interarrival times of orders, setup and processing correspondingly.

The queuing time for processing is:

$$E(W_{qp} \mid i) = E(Z_1 + Z_2 + \ldots Z_{i-1} \mid i) = (i-1)/\mu \quad (7)$$

$$E(W_{qp}) = E_i(E(W_{qp} \mid i)) = (Q-1)/2\mu \quad (8)$$

Besides, it has already been obtained that setup and process time are:

$$E(W_s) = \tau, E(W_p) = 1/\mu \quad (9)$$

By substituting $E(W_{qg})$= 0, (4), (5), (8), and (9) into (3) yields

$$E(W) = \frac{Q-1}{2D} + \frac{Q(\sigma_X^2 + \sigma_Z^2) + \sigma_Y^2}{2\left[Q\left(\frac{1}{D} - \frac{1}{\mu}\right) - \tau\right]} + \frac{Q+1}{2\mu} + \tau \quad (10)$$

In the real production situation, the lot size Q may not be less than one and the traffic intensity ρ has to be less than 100% for a realistic queuing case, so that:

$$Q \geq 1, \quad \rho < 100\% \quad (11)$$

2.2.2 Model with carbon emission constraint

A carbon cap is used as a constraint in the proposed model to restrict the total amount of carbon emission in the production process below the carbon cap when maximizing the profit. The production process in Figure 1 includes two major carbon emission sources, namely the manufacturing process and the WIP inventory.

Carbon emission from manufacturing is mainly due to the fossil fuel and electricity used to operate vehicles and machines and carbon emission from WIP inventory is incurred during storage and transfer of semi-finished products. Besides, WIP inventory is related to demand, as well as to the holding time $E(W)$. Hence the total carbon emission is:

$$e = e_m + e_{WIP} = k_0 + Dk_1 + g_0 + g_1 DE(W) \quad (12)$$

where k_0 means the fixed carbon emission factor for the carbon emission by vehicles driven empty and machines run without production, while k_1 means the variable carbon emission factor, calculating the surplus marginal carbon emission for each one more product produced. g_0 and g_1 are the fixed and variable carbon emission factors for the WIP inventory correspondingly.

Consequently, the total carbon emission e must be no more than the allowable emission cap M, i.e.

$$e = e_m + e_{WIP} \leq M \quad (13)$$

2.2.3 Propositions

Based on the above formulations, propositions are summarized, but due to the space limitation, the propositions are listed without detailed illustration as below.

Proposition 1. The operational profit R in the benchmark model is concave in its domain.

Proposition 2. The optimal lot size Q^* that maximises the operational profit R for the benchmark model must satisfy the following quartic equation:

$$\frac{AB^2h}{2}Q^4 - ABh\tau Q^3 + \left(\frac{Ah\tau^2}{2} - \frac{Ch}{2} - sB^2\right)Q^2 + 2Bs\tau Q - s\tau^2 = 0 \quad (14)$$

with $A = \frac{1}{D} + \frac{1}{\mu}, B = \frac{1}{D} - \frac{1}{\mu}$, and $C = (\sigma_X^2 + \sigma_Z^2)\tau + \left(\frac{1}{D} - \frac{1}{\mu}\right)\sigma_Y^2$

Proposition 3. There exists only one optimal lot size Q^* for the benchmark model.

Proposition 4. For the benchmark production model the profit curve is (1) positively sloped in the interval $[Q_{min}, Q^*]$; (2) negatively sloped in the interval $[Q^*, +\infty]$, where $Q_{\min} = \max\{\tau/(1/D - 1/\mu), 1\}$.

Proposition 5. The carbon emission function e is convex in terms of Q in its domain.

Proposition 6. Under the carbon emission constraint, there are three different situations for Q^*, as follows:

a. when $Q^{e,1} \leq Q^* \leq Q^{e,2}$, the optimal lot sizing solution is still Q^*

b. when $Q^* \leq Q^{e,1}$, the optimal solution is $Q^{e,1}$ with profit decreased

c. when $Q^* \geq Q^{e,2}$, the optimal solution is $Q^{e,2}$ with profit decreased

3 Numerical case study

Numerical experiments are conducted in this section to validate the proposed MTO manufacturing optimization model. The theoretical propositions proven in the last section are used to determine the globally optimal solution. In these numerical examples, the data for the random variables involved are based on the heat treatment industry case study by Lambrecht, Chen, and Vandaele [8], as follows:

Demand ratio $1/D$=1.0000 minutes, variance of interarrival times of orders σ_X^2 =0.5000, mean setup time τ =10.0000 minutes, variance of setup times σ_Y^2 =10.0000, mean process time $1/\mu$=0.5000 minutes, variance of processing times σ_Z^2 =0.0625.

All the other parameters pertinent to carbon emission, costs, and assets can be collected from the manufacturer's managerial accounting systems [9]. Hence these data are:

Fixed carbon emission factor in manufacturing k_0=2 ton/year, variable carbon emission factor in manufacturing k_1=0.02 kg, fixed carbon emission factor in WIP inventories g_0=3 ton/year, variable carbon emission factor in WIP inventories g_1=0.01 kg/min, sum of other variable costs ω=\$2.5, unit setup cost s=\$1200, unit WIP cost h=\$1.5, fixed cost C_F=\$2.0 million, unit sales price γ =\$150.

Since the carbon cap is usually annually based, the case study is set as a 2-year production plan to observe the influence of carbon cap onto the production performance. The production line operates 12 months per year, 20 days per month, and 8 hours per day, so that the total demand for 2 years is

$$D_{Total} = D \times 60 \times 8 \times 20 \times 12 \times 2 = 230400 .$$

3.1 Benchmark model

By substituting the above data into (19), the optimal lot size Q^* for the benchmark mode is 34.9502, the expected total work time is 37.9645 mins, the maximal profit is \$5.4 million per year, and the total carbon emission is 51 tons per year.

Based on the Fig. 2, it can be observed that the minimal value for $E(W)$ is 33.38, which limits the value of carbon emission cap M to be no less than 46 tons to get the real number solution for $Q^{e,1}$ and $Q^{e,2}$.

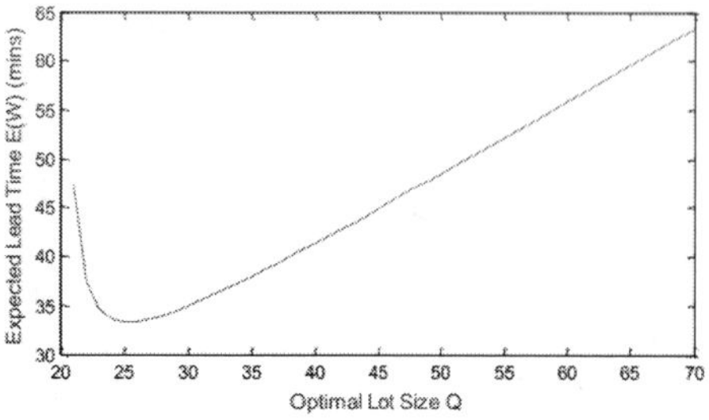

Figure 2. The expected leading time function of lot size

3.2 Model under carbon constraint

The carbon cap given by the government is usually for the total amount of annually carbon emission by the company. The manager therefore needs to determine the carbon emission quota released to different departments or production lines based on their carbon emission amounts. To study the influence of carbon cap on the production performance, the carbon emission cap for the proposed production system is set as between 46 and 65 tons to study situations with different carbon constraints.

By changing the carbon emission cap, the lot sizing interval will vary correspondingly, leading to different cases as discussed in Proposition 6. Table 1 shows the lot size interval, optimal lot size and profit of proposed model under the carbon cap constraints with comparison of optimal lot size of benchmark model.

Table 1. Effects of carbon constraint on lot sizing optimisation

Cap M (ton)	$[Q^{e,1},Q^{e,2}]$	Q^*	Optimal Lot size Q	Profit R (mil)	Carbon emission (ton)
46	[24.09, 26.94]	34.95	26.94	5.06	46
47	[23.12, 29.07]	34.95	29.07	5.28	47
48	[22.67,30.68]	34.95	30.68	5.38	48
49	[22.33, 32.18]	34.95	32.18	5.44	49
50	[22.09, 33.57]	34.95	33.57	5.47	50
51	[21.90, 34.93]	34.95	34.93	5.48	51
52	[21.75, 36.23]	34.95	34.95	5.48	51
54	[21.51, 38.78]	34.95	34.95	5.48	51
56	[21.34, 41.27]	34.95	34.95	5.48	51
58	[21.20, 43.73]	34.95	34.95	5.48	51
60	[21.09, 46.15]	34.95	34.95	5.48	51
65	[20.03, 52.89]	34.95	34.95	5.48	51

Based on the Table 1, the function curves are drawn as below. Fig. 3.1 shows that when the carbon cap is smaller than the carbon emission value of benchmark model ($M \leq 51$ tons), the carbon emission amount is restricted by the carbon cap and it is always equal to the carbon cap. Fig. 3.2 implies that a smaller lot size leads to lower carbon emission when $M \leq 51$ tons. When $M > 51$ tons, the lot size remains at the value that achieve maximal profit. On the other hand, Fig. 3.3 indicates the impact of carbon cap on the profit, in that the profit increases with the increase in carbon cap until it flattens out at $M > 51$ tons. Fig. 3.4 illustrate the relationship between the lot size and the profit. A large lot size increases profit but leads to more carbon emission. The case study shows that in response to being imposed of a small carbon cap, a firm can reduce its lot size for production to reduce carbon emission.

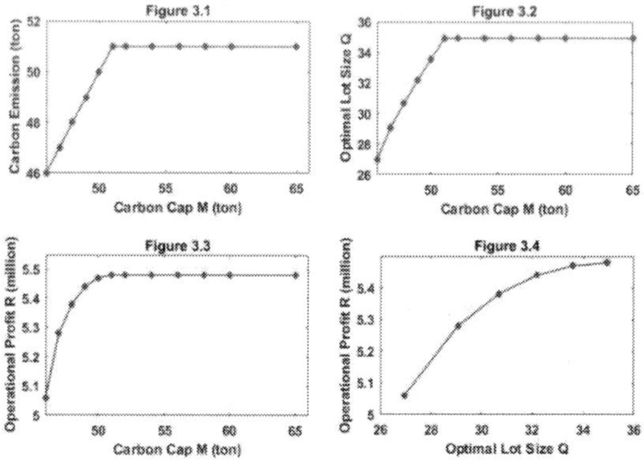

Figure 3. The function curves

4 Conclusion

This paper deals with the stochastic lot sizing problem for profit maximization in MTO manufacturing under carbon emission caps. It aims to optimise the lot size for profit maximization while limiting the total carbon emission below the carbon cap. Numerical experiments are carried out to validate the propositions and analytical insights of the model. The results show that a lower carbon cap tends to motivate a company to set small lot sizes, which can reduce carbon emission with slightly decrease in profit. Although sacrifice of short-term profits may be unavoidable, controlling carbon emission help enhance the company's market image and reputation as being socially responsible which would attract more customers who concern about environmental protection. Hence, manufacturers are encouraged to adjust production plans or use new technologies to control carbon emission accordingly.

References

1. United-Nations, *Paris Agreement: Framework Convention on Climate Change*, **2** (2015)
2. J. Sarkis, Q. Zhu, and K.-h. Lai, *An organizational theoretic review of green supply chain management literature*, Int. J. Prod. Econ., **130**, pp. 1-15 (2011)
3. M. Brandenburg, K. Govindan, J. Sarkis, and S. Seuring, *Quantitative models for sustainable supply*

chain management: Developments and directions, Eur. J. Oper. Res., **233**, 299-312 (2014)
4. Accenture, *Collaborative Action on Climate Risk-- CDP Supply Chain Report* (2014)
5. K. Govindan, J. Sarkis, C. J. C. Jabbour, Q. Zhu, and Y. Geng, *Eco-efficiency based green supply chain management: Current status and opportunities*, Eur. J. Oper. Res., **2**, 293-298 (2014)
6. N. Absi, S. Dauzère-Pérès, S. Kedad-Sidhoum, B. Penz, and C. Rapine, *Lot sizing with carbon emission constraints*, Eur. J. Oper. Res., **227**, 55-61 (2013)
7. X. Chen, S. Benjaafar, and A. Elomri, *The carbon-constrained EOQ*, Oper. Res. Lett., **41.2**, 172-179 (2013)
8. M. R. Lambrecht and N. J. Vandaele, *A general approximation for the single product lot sizing model with queueing delays*, Eur. J. Oper. Res., **95**, 73-88 (1996)
9. R. H. Garrison and E. W. Noreen, *Managerial Accounting*. USA: Irwin McGraw-Hill (2000).

Estimating Household Willingness to Pay for Improved Solid Waste Management: A Case Study of Thu Dau Mot City, Binh Duong

Pham Thi Thuy Trang[1], Dinh Quang Toan[1,2] and Nguyen Thị Xuan Hanh[1]
[1]Falcuty of Resources and Environment, Thu Dau Mot University, Binh Duong, Viet Nam
[2]College of Natural Resources and Environment, Northwest A&F University, Yangling, Shaanxi 712100, China

Abstract. The main objective of this paper is to investigate the household willingness to pay (WTP) for improved solid waste management in Thu Dau Mot city, Binh Duong by using Dichotomons choice Contingent Valuation Method. A questionnaire was designed to draw a sample of 330 respondents for the study. The results show that mean household willingness to pay for solid waste management system in Thu Dau Mot is about 24 thousand VND/ month/ household, higher than the current fee (20 thousand VND month/ household).The paper used a logistic regression to establish the factor affecting to the household willingness to pay for solid waste management system. The logit results show that those with a higher income and more educated is willing to pay more for improvement in management of their solid waste. Also, more environmental aware the person the more they are willing to pay. Moreover, females are less likely to give a positive response to WTP compare to male.

1. Introduction

Thu Dau Mot is one of the economic, financial and cultural centres of Vietnam with 28 industrial parks. The continuous growth of economic and population has brought a new challenge to the local government especially in dealing with the issue of municipal solid waste disposal, particularly domestic household wastes. Currently, there are 120 tons domestic household wastes generated per day, with 70% is collected and 30% is illegally dumped. Also, further with increase horizontal expansion of Industry Park, it has become very difficult to find adequate waste disposal sites in the future. The city is now facing with many problems regarding pollution and health because the local waste management system cannot cope with the volumes of wastes being generated. One of the approaches has been now considered by local authority is to increase frequency of collecting waste and number of vehicles of collection.

Actually, improving solid waste management is recognized as critical to improve health and promote environmental protection in the developing countries. To establish an effective management project, it is necessary to be contributed by communities who are the primary producers and suffer the effects of unmanaged solid waste. However, currently in Vietnam, policy makers mainly focus on local authorities and private provider while there is still a lack of data on households' demands and their willingness to pay to improve solid waste management system. Therefore, the key research question is how much household are willing to pay for cost-effective delivery of household solid waste management to residential areas and what are main factors influencing their willingness to pay. To estimate the willingness to pay of individuals, one of the techniques used is contingent valuation method (CVM).The contingent valuation method involves asking to a sample of the population about their willingness to pay for the provision of a given good or service [1] [2]. It is found that CVM can be fruitfully used to support the design and implementation of new solid waste management [3].

The purpose of this paper is to apply the CVM approach to find out the monetary value that households willing to attach for an improved solid waste management service in Thu Dau Mot city and determine variables affecting to their willingness to pay.

According to requirement of CVM study, firstly, the respondent should be given detailed information about the current situation of solid waste management to be valued and the hypothetical scenario under which it is improved. Second, a method which elicits respondent's willingness to pay is applied. Third, household demographic and socio-economic characteristics such as age, gender, income, education level are needed to estimate the valuation function for the environmental service [4]. Finally, regression model is used to analyse obtained data to estimate how willingness to pay are related to the respective variables based on theory. The study result of this paper is essential for policy makers to improve solid waste management in Thu Dau Mot city in particular and other city in Vietnam in general.

2. Methodology

2.1. Questionnaire design

a questionnaire that is the main instrument for collecting data for CVM study is designed with four sections [2].

-*Section 1*: Information on the household demographic and socio-economic characteristics (age, gender, education level, environmental awareness, family size, monthly income) is interviewed to generate data that may influence the household willingness to pay.

-*Section 2:* Information about respondents' environmental awareness and challenges regarding current household waste management service in Thu Dau Mot.

-*Section 3:* Demonstrate general problems of current management service and propose detailed hypothetical scenario of the solid waste management system improved.

-*Section 4:* Contingent Valuation question designed to estimate willingness to pay for improved solid waste management system in Thu Dau Mot city. Normally, there are three approaches of CVM that willingness to pay can be elicited [1]. In this paper, authors used dichotomous choice CVM approach which individuals were asked the amount of money, called bid level, to state whether they were willing to pay or not by responding "YES" or "NO". Their choice is motivated by discussion in previous sections.

2.2. Sampling Strategy

More than 300 samples should be represented for huge population [5]. Therefore, with total 45.000 households of Thu Dau Mot, in total 330 households were selected to interview by using systematic sampling technique which every 10th building in the area was sampled. The next one in the list will be sampling if every 10th building is vacant.

In-person interviews were conducted with one respondent per household. With dichotomous choice (DC) question used in section 3 of the questionnaire, it is expected that the individual will answer "YES" when their WTP is greater than the suggested bid amount. In this study, different sub sample respondents are asked with different bid level. Based on DWEABS model, the authors establish five different bid levels (23000VNĐ, 26000VNĐ, 28000VNĐ, 30000VNĐ and 35000VNĐ) [6]. To avoid estimation bias, each sub sample with same number of respondents should be interviewed for each level of bid, so we have a total of 330 observations divided in five sub samples, with approximately the same number of 66 individuals in each sub.

2.3. Model specification

Data analysis was done using Stata statistical analysis software programs. With dichotomous structure of the dependent variable WTP, the model can be estimated through a non-linear probability model and the most common one is the logistic regression model. The logistic regression model allows estimate the probability (P) that an event will occur or not through prediction of a binary dependent outcome[7], coded to take the values 1 or 0 for 'yes' and 'no' answer, respectively, from a set of independent variables:

$$P = \frac{1}{1+e^{-(\beta_0+\beta_1 X1+..+\beta k Xk)}} \quad (1)$$

$$\mathbf{WTP} = \text{Logit} = \text{Ln}\left(\frac{p}{1-p}\right)$$
$$= \ln\left(\frac{\Pr(\text{yes})}{1-\text{pr}(\text{yes})}\right) = \beta_0 + \beta_1 bid + .. + \beta_k X_k. \quad (2)$$

This paper analyses the effect of bid level asked and six selected variables that are explained in Table 1 on household willingness to pay, so a logistic model is constructed to check the validity and explanatory power of the relevant factors is as follows:

$$\mathbf{WTP} = \text{Logit} = \text{Ln}\left(\frac{\text{prob(yes)}}{1-\text{prob(yes)}}\right) = \beta_0 + \beta_1 bid + \beta_2 gender + \beta_3 age + \beta_4 edu + \beta_5 env + \beta_6 income + \beta_7 size + \varepsilon. \quad (3)$$

where:
WTP = the dependent variable obtained from respondents in the form of "yes" or "no" answer;
WTP=1 if the respondent answers yes and 0 otherwise,
ε= random disturbance term;
β_o is the constant term;
β_1 is the bid amount coefficient.

Table 1. Description of explanatory variables.

Variable name	Description	Type of data
Age	Age of Respondents(number of year)	Interval data
Gender	Gender of respondent (1 = female; 0 otherwise)	Nominal data
Edu	Education of Respondent (Number of years spent in the school)	Interval data
Env	Environmental awareness of the respondent. Based on answer in section 2 of questionnaire. (0 = not aware; 1 = fairly aware; 2 = much aware).	Ordinal data
Size	Number of member in the Household	Interval data
Income	Monthly Income of the Household (VND)	Interval data

Mean household willingness to pay was calculated using the formula given as [7]: Only influence variables are included in the formula:

$$\text{Mean (WTP)} = -\frac{\alpha Z}{\beta} \quad (4)$$

$$= \frac{-\beta_0 + \beta_2 * \text{mean}(X2) + \beta_3 * \text{mean}(X3) + \cdots + \beta_k * \text{mean}(Xk)}{\beta_1}$$

3. Results and discussion

3.1. Demographic and socio-economic factors

The demographic and socio-economic characteristics of 330 sample respondents in Thu Dau Mot city are showed in Table II.

The female accounted for 59 % while 41% were male. It is reasonable because women often stay at home more than men due to responsible for household job. The distribution of household size indicates that 91% of the household have between 1 to 5 members while only 9 % represent those that between 6 to10 members. The mean household size is about 4 members. The average age of respondents was 48 years. The age range with the highest frequency is 41 – 50 years which accounted for 27% of the respondents. This implies that respondents are in their active age and therefore can work to earn, more income which can affect their decision to pay for improved waste services. The majority of respondents (46%) had education of high school. The mean years of education in the study area is 8 years which is still not high enough. This may be one of reasons explaining why the majority of respondents (61%) in Thu Dau Mot are not aware environment. The average monthly income of household respondents was 9.5 million VND.

Table 2. Demographic and socio-economic characteristics

Demographic factors	Percentage	Mean
Age		
≤ 30	8.79	
31- 40	20.01	
41 -50	27.28	48.09394
51- 60	23.65	
above 60	20.27	
Gender		
Male=0	40.91	0.5909091
Female =1	59.09	
Edu		
1-6	38.78	8.469697
7-12	45.57	
13- 16	15.65	
Env		
0 = not aware;	61.21	0.4909091
1 = fairly aware;	28.48	
2 = much aware	10.30	
Size		
1-5	91	4.075758
6-10	9	
Income		
<10.000.000	51.51	9578788
10.000.000-20.000.000	46.07	
Above 20.000.000	2.42	

3.2. Logistic regression analysis

The logistic regression results are presented in Table III. We found that multicollinearity is not a serious problem in our dataset. Testing for heteroskedasticity shows no problem. In order to test for the goodness of fit, the Pseudo R2 and linktest was used. The pseudo R^2 is 67.5, which implies that 67.5 % of the variation in the probability of being willing to pay or not is explained by the variables included in the model.

3.3. Determinants of willingness to pay

An important aspect to check when using contingent valuation data is that individuals should be sensible to the bid amount, so we expect that as the bid amount goes up the proportion of individuals that give a positive answer goes down. As expected, our data show that the proportion of positive answers goes down as the bid amount goes up, presented in Table IV.

Table 3. The logistic regression results

Variables	Coefficient	Standard error	Significant
Age	0.0126626	0.0180913	0.484
Gender	-1.883937	0.4669733	0.000*
Edu	0.4234707	0.0677565	0.000*
Env	2.519017	0.4205125	0.000*
Size	-0.0134297	0.1554821	0.931
Income	2.14e-07	6.28e-08	0.001*
Bid	-0.0003129	0.0000657	0.000*
Cons	1.796039	2.069628	
Statistic Summary No. of Obs 330 Logi-likelihood -72.135473Pseudo R2 67.5% LR Chi2 299.92Prob> Chi2 0.0000			Significance Level *Significance at 1%

Table 4. Total number of respondents in each level of bid

WTP	23	26	28	30	35	Total
Yes	37	28	27	27	13	132
No	29	38	39	39	53	198
Respondents	66	66	66	66	66	330

The logit results showed that age of respondents and household size do not significantly influence the probability of answering "YES" of respondents to the bid price asked. This relationship was not expected by many researches because people with significant age are often likely to take care for their environment [4] [8]. And in some cases, the more number of people in the household, the more willing the household will appreciate a clean environment.

On the other hand, monthly income, education, environmental awareness and gender of households are statistically significant at 1%.Particularly, the positive sign of the coefficient on income indicate that the relative probability of a 'yes' increases with income, which is consistent with economic theory that indicates that income is positively related with demand in general and the same with environmental demand [8]. This is in conformity with the work other researches [4].

Also, educational level is positively related to willingness to pay to improve waste management services. It means that as level of education increases the tendencies to agree and pay for improved services will also go up. The higher the education attained, the higher the probability of the individual's willingness to pay for improved waste disposal services. The same story happens to the environmental awareness which show that individuals who are aware with environment have a greater position towards the willingness to pay for improved solid waste management than those who are not aware. While, the negative sign of gender indicate that females are less likely to give a positive response.

DOI: 10.1051/matecconf/20179518004

3.4. WTP responses

From logit result, all explainable variables are statistically influence willingness to pay at significant of 1%, except for age of respondents and household size. So, the mean WTP will be calculated without age and household size variables. Using mean values of variables from Table II, we have:

$$\text{Mean (WTP)} = -\frac{\alpha Z}{\beta}$$

$$= \frac{-\beta_0 + \beta_2 * \text{mean}(gender) + \beta_4 * \text{mean}(edu) + \beta_5 * \text{mean}(env) + \beta_6 * \text{mean(income)}}{\beta_1}$$

= 24.141 (VND).

4. Conclusion

By using dichotomous choice CVM, the mean household willingness to pay for improved solid waste management in Thu Dau Mot is estimated about 24.000 VND/ household. This level is higher compare to the current management fees (20.000 VND). Household's educational level, environmental awareness, gender and income were noted to significantly influence their willingness to pay. In general, the household are eager for improved solid waste management, so there is plenty of room to increase the fee and acquire sufficient funds to substantially improve waste management in Thu Dau Mot. This result will add to a small but growing body of knowledge to policy maker in determining the socially optimal charges for solid waste services where households would receive planned and properly scheduled waste management services. It is also recommended that the educational campaigns of the various metropolitan, municipal and district assemblies on waste management should be intensified to increase the awareness of households.

References

1. A.L. Feldma, Introduction to contingent valuation using Stata, Munich Personal RePEc Archive Paper no.41018 (2012)
2. R.C. Mitchell and R.T. Carson, *Using Surveys to Value Public Goods: The Contingent Valuation Method*, Washington, DC: Resources for the Future (1989)
3. W. Fonta, H.E. Ichoku, K.K. Ogujiuba and J.O. Chukwu, Using a contingent valuation approach for improved solid waste management facility: evidence from Enugu state, Nigeria, J Afr Econ. **17**, 2 (2008)
4. T.R. Ashish and D. Uttam, Households Willingness to Pay for Improved Waste Management In Silchar Municipal Area: A Case Study In Cachar District, Assam, Int. J. Human. Soc. Sci. Res. **6**, 5 (2013)
5. R. Worthington and T. Whittaker, Scale Development Research A Content Analysis and Recommendations for Best Practices, Couns Psychol **36**, 6 (2006)
6. J. Cooper, Optimal Bid Selection for Dichotomous Choice Contingent Valuation Surveys, J. Environ. Econ. Manage. **24**, 1 (1993)
7. J. Loomis, T. Brown, B. Lucero and G. Peterson, Evaluating the Validity of the Dichotomous Choice Question Format in Contingent Valuation, Environ. Resour. Econ **10**, 2 (1997)
8. N. Aggrey and O.G. Douglason, Determinants of Willingness to Pay for Solid Waste Management in Kampala City, J. Econ Theory **2**, 3(2010)

DOI: 10.1051/matecconf/20179518005

Analysis of the changing Solar Radiation Angle on Hainan Island

Zhiwu Ge[1], JingJingHuang[2], HongxiaLi[2] and HuizhenWang[2]

[1]*Hainan normal University, Physics Department, No. 99, Long Kun south road, Haikou, China*
[2]*Hainan normal University, Electronic Information and Technology Department, No. 99, Long Kun south road, Haikou, China*

Abstract. As the only tropical provinces in China, Hainan province has advantageous geographical location, and abundant solar energy resources. But because of Local ideas and habits, especially the lack of theoretical research on local solar resources, development and application of solar energy in Hainan is almost blank. In this paper, we studied the variation regularity of sunlight angle on Hainan tropical island, analyzed the revolution and rotation of the earth, and the change rule of sunlight angle caused by the sun's movement between the tropic of cancer and the tropic of capricorn, deduced the change rule of sunlight angle in the spring equinox, the autumnal equinox, summer solstice and winter solstice day, and got the movement rules of solar elevation angle throughout the year. Theoretic analysis is consistent with field measurement results. These rules are of importance and can effectively guide the local People's daily life and production, such as the reasonable layout of the buildings, floor distance between different heights of buildings, the direction of the lighting windows of tall buildings, installation angle of photovoltaic panels, and other similar solar energy absorbing and conversion equipment.

1 Preface

With the growing development of science and technology, people's living standard is increasing day by day, the material demand is growing also. The resource consumption will greatly increase. But as the earth's non-renewable resources is becoming less and less, it is possible to exhaust the non-renewable resources in the near feature. In the exploration process of new energy, it is found that the light energy from the sun is almost inexhaustible, and it is the first choice to take the place of the non-renewable resources [1, 2, 3].

In recent years, under national policy support, solar energy has become a fast growth energy application form all over the world. Our country also spare no effort in the development and applications of solar energy [1, 2, 3].

In order to understand and use the solar energy fully, it is necessary to research and grasp the change rules of the solar radiation angle, and to improve the efficiency of the use of solar energy.

Due to the different latitude, the sunlight angle are also different in the same time, and this paper study the changing rules of the solar inclination angle in Hainan tropical island.

2 The relevant laws of the movement of sunlight angle

Every day, in the morning, sunlight angle of the sun is the smallest, with the rotation of the earth, solar altitude angle changes from the original small to large, At noon to reach the maximum, which is the maximum altitude angle of the Sun. Subsequently, the solar altitude angle is gradually reduced from noon to dusk. At dusk reach the minimum as shown in Figure 1.

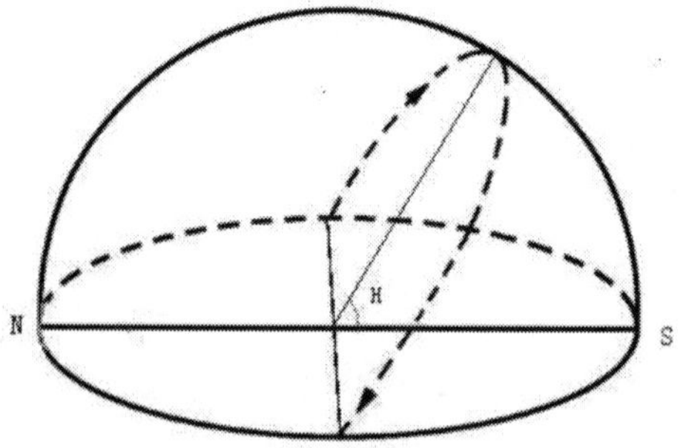

Figure 1. The trajectory of the sun.

The direct point of the sun is at 12 o'clock at noon in the local and moves along the trajectory as shown in Figure 2 [4].

The vernal equinox day just before and after March 21st, the direct point of sun near the equator (0 °), to move from north to the tropic of cancer (23 ° 26'N), this time is summer solstice before and after June 22nd. Then from the North to the South, moving to near the equator (0 °) .This time is autumnal Equinox before and after September 23rd. Finally moved to near Capricorn (23 ° 26'S), before and after the December 22nd. Since then, continued move circularly from South to North. The movement of the direct point of the sun brings the change of the year round, while the length of day and night changes, such as from long to short or from short to long.

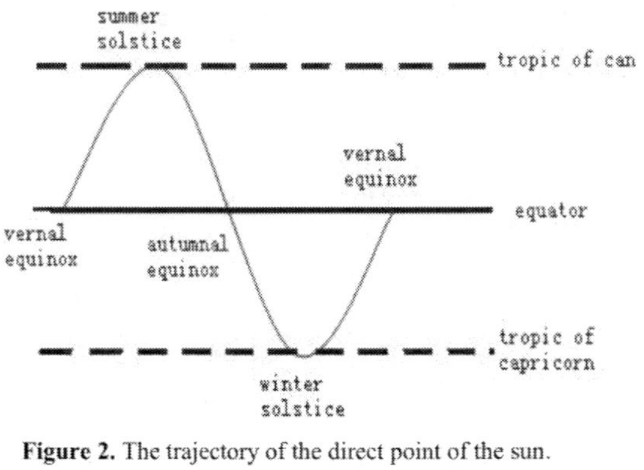

Figure 2. The trajectory of the direct point of the sun.

When the solar altitude angle is equal to 0 degrees in the terminator line, Half of the solar altitude angle of the earth Day is more than 0 degrees, Half of the solar altitude angle of the earth night is less than 0 degrees, can't see the Sun. When the parallel. When direct point of the sun located is 90 degrees, reaching half of the maximum solar altitude angle of the day earth. The solar altitude angle is decreasing along the north and south, with the change of latitude, in the same hemisphere, the latitude changes 1 degree, the angle reduced by 1 degree, just like different hemisphere. Therefore, we can know the latitude of the sunlit point of the sun at a certain time, then it can be inferred that the solar altitude angle at the same time in any place. Thus we can deduce the formula (1)[4].

$$H = 90^\circ - |\theta - \delta| \quad (1)$$

H is the sunlight angle, the latitude which direct point of the sun located, the local latitude we desired. 90 degrees subtract the latitude difference then reserve the absolute value, that is the desired local solar altitude angle. What needs to pay attention is that if the direct point of the sun and the desire place in the same hemisphere of the earth, to do subtraction between the two places to get the latitude difference then reserve the absolute value. If the direct point and the desire place in the different hemisphere of the earth, figure out the sum of the two places. For example, when the direct point of the sun at 40 degrees north latitude, figure out the solar altitude angle at 60 degrees north latitude. Apply the formula 2-1, 10 degrees north latitude, for the local latitude is 20°north latitude, the difference between the two is -20°. Due to The absolute value is 20 degrees in the northern hemisphere Then the solar altitude angle at 60 degrees north latitude is H=90°-70°=20°. If this time, the direct point of the solar altitude angle at 10 degrees south latitude, asking for figuring out the solar altitude angle at 50° north latitude. Also apply the formula 2-1, 10 degrees south latitude, for the local latitude is 50°north latitude. Because the two places in the different hemispheres of earth, adding together: 10 °+60° =70°. Then the solar altitude angle at 60 degrees north latitude is H=90°-70°=20°, of course, even don't know the latitude of direct point of the sun at this time , we still able to calculate the latitude that the sun located by the date, then figure out the solar altitude angle of the place, then deduced the solar altitude angle at the place, as shown in formula (2) [4].

$$ED = 0.3723 + 23.2567\sin\theta + 0.1149\sin 2\theta - 0.1712\sin 3\theta - 0.758\cos\theta + 0.3656\cos 2\theta + 0.0201\cos 3\theta \quad (2)$$

In formula(2), ED is the solar declination angle that we required, expressed in degrees. The day angle is θ , $\theta = 2\pi t/365.2422$, the angle of π is 180 degrees. The time that the earth makes one revolution is 365.2422 days. The calculation formula of it is shown as in formula (3).

$$t = N - N_0 \quad (3)$$

Annual accumulation day refers to the total number of days in a year. For example, January 1st's annual accumulation day is 1, December 30st's annual accumulation day is 364. Of course, December 31st's annual accumulation day is 365 in a common year, December 31st's annual accumulation day is 366 if it is a leap year, and so on. Derivation of formulas is shown as in formula (4)[4].

$$N_0 = 79.6764 + 0.2422*(X - 1985) - \text{int}[(X - 1985)/4] \quad (4)$$

Among them, X in the formula (4) is the year of that time, then put formula (4) into the formula(3) obtained formula (5).

$$t = N - 79.6764 - 0.2422*(X - 1985) + \text{int}[(X - 1985)/4] \quad (5)$$

3 Analysis of the movement laws of sunlight angle in Hainan province

Hainan Province is located at the southernmost of China. The Qiongzhou Strait is to its north delimiting Hainan and Guangzhou. It faces Democratic Republic of Vietnam across the Beibu Gulf in the west, Taiwan province across the South China Sea in the east. In the southeast and south, it is adjacent with Malaysia, Brunei and Philippines in the South China Sea. It's the China's largest province including Hainan island, Xisan island, Zhongsha island, Nansha island and its sea. The province's total land area is 35.4 thousand square kilometres and its sea area is 2 million square kilometres. Hainan island is between latitude 18° 10′ ~20° 10′ to north and longitude 108° 37′ ~111° 03′ to east. Its outline is like to a oval and big pear. Its major axis is 290 kilometres long across northeast to southwest and is 180 kilometres wide across northwest to southeast. Hainan island's area is 33.9 thousand square kilometres.

Hainan is the second largest island that is only smaller than Taiwan island.

Next, detail analysis of changing solar radiation angle on Hainan island will be given in the vernal and autumnal equinoxes, the summer solstice and winter solstice.

3.1 Winter Solstice

December 22nd is winter solstice. When the solar elevation angle locates at tropic of capricorn（23° 26′ S）(see Figure.2).In the winter solstice of north hemisphere, the sun irradiates directly and the antarctic circle appears polar day. The sun of equator's area rises from due east by south 23° 26′ vertically. It goes down from due west by south 23° 26′ vertically. The north hemisphere is in winter solstice. However, south hemisphere is in summer solstice. The daytime of north hemisphere is the shortest throughout the year. The length of objects' shadow reaches the longest. The noon's solar elevation angle is smallest. At this moment of south hemisphere, the daytime reaches the longest, the shadows of objects become the shortest and noon's solar elevation angle is the biggest. Hainan island's daytime is shortest that is about 11 hours, and night time is the longest. The solar elevation angle of noon is the smallest and the shadows of objects is longest. The sun irradiates the tropic of capricorn directly and the solar elevation angle is 90 degree. We can get the solar elevation angle of Hainan at noon by using formula 2-1.The angle of subsolar point is 23° 26′ S. The geographical location of Hainan is latitude 18° 10′ N to 20° 10′ N that is and it is between 18° 10′ and 20° 10′ .Using formula 2-1, because of the subsolar point is located at south hemisphere. But Hainan is at 18° 10′ N to 20° 10′ N which is in north hemisphere. Due to locating at different hemispheres, we plus two latitudes that is 41° 36′ N to 43° 36′ N. So H=90° -（41° 36′ ~43° 36′）=（48° 24′ ~46° 24′）. Midday solar elevation angle is about 48° 24′ to 46° 24′ on December 21st .

3.2 Spring Equinox

The sun moves from the tropic of capricorn（23° 26′ S） to north. On March 21st which is spring equinox, subsolar point locates at equator. Different regions have the same way that the positions of sunrise and sunset are due east, due west respectively(everywhere but south pole).And with latitude going up, the dihedral angle between solar irradiation and the horizon changes from 90°to 0°.

The time of day and night is equal in the south and north hemisphere, but seasons are opposite. The north hemisphere is in spring equinox, relatively, The south hemisphere is in autumn equinox. In north hemisphere, The time of day and night changes to that day is longer than night. With the daytime increasing, the objects' shadow decreases and the midday solar elevation angle becomes bigger. But south hemisphere is complete opposite. Hainan is located at north hemisphere. Using formula 2-1, we can calculate Hainan's midday solar elevation angle. The solar elevation angle of subsolar point is 0° . The geographical location of Hainan is latitude 18° 10′ ~20° 10′ to north that is =（18° 10′ ~20° 10′）. Using formula 2-1,because subsolar point is at equator and Hainan is located at north hemisphere, we get the two latitudes substracted and then get the absolute value that is H=90° - 0-（18° 10′ ~ 20° 10′）|=（71° 50′ ~69° 50′）.Hainan's midday solar elevation angle is about 71° 50′ to 69° 50′ . At this moment, the time of day and night changes to that the time of day is equal to night. With the time going by, daytime is shorter and the object's shadow of night are shorter, too.

3.3 Summer Solstice

The sun moves northwards from equator（0° ）to the tropic of cancer（23° 26′ N）.On June 22nd, subsolar point is at the tropic of cancer. On north hemisphere's summer solstice, the sun irradiates the tropic of cancer directly. Within the Antarctic circle, there appears the polar night. In equator region, the sun rises from due east by north 23° 26′ vertically and it goes down from due west by north vertically. The higher the latitude is, the larger the angle which deviates northeast is. In polar night, the position of sunrise and sunset becomes one point, locating at due north. The daytime of north hemisphere is the longest and the night time of south hemisphere is the longest. The north hemisphere is in summer and the south hemisphere is in winter. The daytime of north hemisphere reaches the longest throughout the year.

In the north of tropic of cancer, midday solar elevation angle is the largest in the year. And objects' shadow is the shortest. At latitude 23° 26′ to north, shadow coincides with body. The night time of south hemisphere is the longest. The midday solar elevation angle of south hemisphere is the smallest and the shadow reach the shortest. At this time, the daytime of Hainan is longest. On June 22nd, 2014, the sun rises at 06:09:04.It is highest at 12:49:58 and it goes down at 19:30:15.The daytime of that day reaches 13 hours 21 minutes 11 seconds that is the longest. The change of solar elevation angle experiences two periods. Firstly, it moves northwards from equator to latitude 18° 10′ ~20° 10′ to north. Hainan's solar elevation angle becomes larger and larger. The daytime changes from short to long. The shadow is shorter. Until May 13th to May 22nd, the midday solar elevation angle of Hainan is max that is 90 degrees and the shadow coincides with body. Secondly, the sun moves northwards from 20°north latitude to the tropic of cancer （23° 26′ N）. In the process, the midday solar elevation angle of Hainan changes from big to small, the daytime becomes longer unceasingly and the shadow becomes shorter. Finally, when subsolar point is at the tropic of cancer（23° 26′ N）, daytime reach the longest in Hainan. At the same time, midday solar elevation angle can be calculated by formula 2-1.We all know that the solar elevation angle of subsolar point equals 23° 26′ N. The geographical location of Hainan is latitude 18° 10′ ~20° 10′ to north that is = (18° 10′ ~ 20° 10′) .Using formula 2-1, because of subsolar point is at the tropic of cancer and Hainan is located at north hemisphere, we get the two latitude substracted and then get the absolute value that is H=90° -|23° 26′ - (18° 10′ ~20° 10′) |= (84° 44′ ~ 86° 44′) .Midday solar elevation angle is about 84° 44′ to 86° 44′ in Hainan. At this moment, daytime is the longest and is going to be shorter. Meanwhile, the night time changes from short to long. The midday solar elevation angle changes from small to big and the shadow changes long to short.

3.4 Autumn Equinox

The sun moves from tropic of capricorn (23° 26′ S) to south. On September 23rd, which is spring equinox, subsolar point locates at equator. The day and night is divided equally .Different regions have the same way that the position of sunrise and sunset are due east, due west respectively(everywhere but south pole). And with latitude going up, the dihedral angle between solar irradiation and the horizon changes from 90°to 0°.That is to say solar elevation angle changes from 90°to 0°.The time of day and night is equal in the south and north hemisphere, but seasons are opposite. The north hemisphere is in autumn equinox, relatively, the south hemisphere is in spring equinox. In north hemisphere, the time of day and night changes to that night is longer than day. With the night time increasing, the objects' shadow increases and midday solar elevation angle becomes smaller. But south hemisphere is complete opposite. The change of solar elevation angle experience two periods. Firstly, it moves southwards from the tropic of cancer to 18° 10′ to 20° 10′ north latitude whose time is from July 23rd to August 1st .And the midday solar elevation angle of Hainan is max that is 90 degrees and the shadow coincides with body. With the time going by, the sun moves to south unceasingly, solar elevation angle becomes smaller and smaller in Hainan. The daytime becomes shorter and shorter and the night time increases. The shadow is longer and longer. Secondly, subsolar point is at equator (0°) in autumn equinox. According to formula 2-1, we can calculate the midday solar elevation angle of Hainan. The solar elevation angle of subsolar point is 0° . The geographical location of Hainan is latitude 18° 10′ ~20° 10′ to north that is = (18° 10′ ~20° 10′) . Using formula 2-1, because of subsolar point is at the equator and Hainan is located at north hemisphere, we get the two latitudes substracted and then get the absolute value that is H=90° -|0° - (18° 10′ ~20° 10′) |= (71° 40′ ~69° 40′) .Hainan' s midday solar elevation angle is about 71° 40′ to 69° 40′ . At this moment, Hainan's daytime is the longest and is going to be equal with night time. With the time going by, the daytime is shorter and the night time is longer. The shadow changes from short to long.

3.5 Summery

Subsolar point moves from the tropic capricorn (23° 26′ S) of winter solstice to the equator (0°) of spring equinox in Hainan, and then moves to the tropic of cancer (23° 26′ N) , and then moves to the equator (0°) of autumn equinox, finally returns to the tropic of capricorn (23° 26′ S) . It takes one year for a cycle. It's winter solstice when the sun is at the tropic of capricorn (23° 26′ S) . Hainan is at 18° 10′ to 20° 10′ north latitude. Its daytime is the shortest. The sun rises from southeast and the shadow is toward northwest. The midday solar elevation angle is 48° 24′ to 46° 24′ . The subsolar point moves from the tropic of capricorn (23° 26′ S) to equator (0°) . The daytime of Hainan is longer,the solar elevation angle is bigger and the shadow is shorter. Firstly, The subsolar point is at equator in spring equinox that is March 21st. Hainan is at 18° 10′ to 20° 10′ north latitude. The midday solar elevation angle is 71° 50′ to 69° 50′ . The daytime is equal with the night time theoretically. The position of sunrise and sunset are due east, due west respectively. When the subsolar point moves northwards to the tropic of cancer (23° 26′ N) , the daytime of Hainan is longer and longer. The solar elevation angle of Hainan changes from small to large and it reaches the biggest in March 13th to March 22nd .Secondly, the subsolar point moves to north unceasingly. The daytime of Hainan becomes longer and longer. In June 22nd ,the subsolar point reaches the tropic of cancer (23° 26′ N) . At this moment, the daytime of Hainan is the max. Hainan is at latitude north 18° 10′ to 20° 10′ . The midday solar elevation angle is 84° 44′ to 86° 44′ . The daytime is the longest throughout the year. The sun rises from northeast and goes down from southwest. Thirdly, the subsolar point moves southwards from the tropic of cancer (23° 26′ N) .And it reaches equator (0°) in September 23rd .Meanwhile, the midday solar elevation angle in Hainan changes from small to big. And then it is the max in July 23rd to August 1st .With the subsolar point moving southwards, the solar elevation angle changes from big to small. In September 23rd the solar elevation angle is 71° 50′ to 69° 50′ .The daytime is shorter and the shadow is longer. Finally, the subsolar point moves from equator (0°) and reach the tropic of capricorn (23° 26′ S) in December 22nd .The midday solar elevation angle changes from big to small, the daytime becomes shorter and shorter and the shadow changes from short to long.

The movement rules of solar elevation angle throughout the year in Hainan is shown in figure 3.

In winter solstice that is December 22nd , the midday solar elevation angle is min (48° 24′ ~46° 24′) throughout the year. In spring equinox that is March 21st, it rises at 71° 50′ to 69° 50′ and reaches the max (90°) in May 13th. In June 22nd that is summer solstice, it goes down at 84° 44′ to 86° 44′ . With the subsolar moving southwards, the solar elevation angle of Hainan reaches the max again around July 23rd. Thus it can be seen that there are two times which the midday solar elevation angle of Hainan reaches the max. And then, with the subsolar point moving southwards, the midday solar elevation angle of Hainan decreases at 71° 50′ to 69° 50′ in September 23rd .In December 22nd ,the subsolar point returns to the tropic of capricorn23° 26′ S) and the solar elevation angle of Hainan is the min (48° 24′ ~46° 24′). By analysing the change of the midday solar elevation angle throughout the year, then we look how the midday solar elevation angle changes in a month.

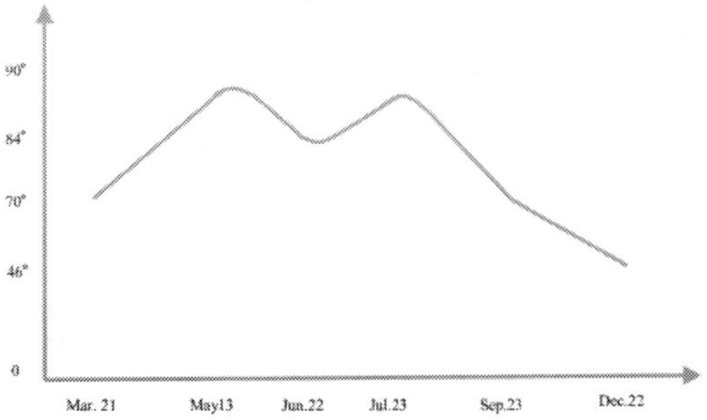

Figure 3. Hainan's solar elevation angle throughout the year.

For example, In November 1st, the change of subsolar point has its specific track in the year. Using formula 2-2,we can calculate that the subsolar point of that day is at 14° 14′ S. That day, the subsolar point is at south hemisphere. According to formula 2-1,we can calculate that the midday solar elevation angle of Hainan (18° 10′ ~20° 10′) is about 57° 36′ to 56° 36′ .On this account, we can calculate midday solar elevation angle of Hainan everyday .The result is shown in table 1.

Table 1. Hainan midday sun angle in November.

Date	Angle	Date	Angle	Date	Angle
1	57°36′	11	54°35′	21	52°03′
2	57°16′	12	54°18′	22	51°50′
3	56°57′	13	54°02′	23	51°37′
4	56°39′	14	53°46′	24	51°24′
5	56°20′	15	53°03′	25	51°12′
6	56°02′	16	53°15′	26	51°
7	55°44′	17	53°	27	50°49′
8	55°26′	18	52°45′	28	50°38′
9	55°09′	19	52°31′	29	50°27′
10	54°52′	20	52°17′	30	50°17′

As shown in table1, Hainan' sun Angle decreases in November, the day time get shorter and shorter.

4 Conclusion

The special movement laws of sunlight angle in Hainan province has been analyzed, the results showed the different characteristic of solar energy resources in Hainan compared with that in inland. According to the above analysis, solar elevation angle throughout the year changes along unique bimodal curve on Hainan island. It's easy to verify the results, set a known length of a stick at noon, measure the length of its shadow, and the sun angle can be calculated out. Above analysis results are also in accordance with the curve got according to the local meteorological data [5].

The results can guide daily life and production of local residents effectively, and provide direction for economically and reasonably using solar energy. For example, options of building orientation, orientation of heat acquisition plate or photovoltaic panels, etc., are all closely related with the angle of the sun.

Acknowledgement

Thanks for the funding of Hainan province natural science foundation (Grant Number 20165198).

References

1. http://xueshu.baidu.com/s?wd=paperuri:(c8a0b8d51f472cede3d00a617f9c147c)&filter=sc_long_sign&sc_ks_para=q%3D%E6%B5%B7%E5%8D%97%E7%9C%81%E5%A4%AA%E9%98%B3%E8%83%BD%E5%88%A9%E7%94%A8%E7%8A%B6%E5%86%B5%E8%B0%83%E6%9F%A5&tn=SE_baiduxueshu_c1gjeupa&ie=utf-8&sc_us=7327178283693853975
2. Xufei. zhang. Renpeng. Xiao, Science and Technology and Equipment, **12**, 35-37, (2010).
3. Yi .shen, Thesis of Lanzhou University, 1-13,(2014)
4. Hanyun. Huang, Principle of solar photovoltaic power generation applications, 9-15, (2012)
5. Changan. wang. Guojie. wang, Journal of Hainan Normal University(Natural Science), 168-173, (2011)

An Experimental Study on the Inner and Outer Glass Cover Temperatures of Solar Still

Pankaj K. srivastava[1], AshutoshDwivedi[2], Mihir Kumar Pandey [3], Abhay Agrawal[4] and R.S. Rana[5]

[1] Mechanical Engineering Department, Rewa Engineering College, Rewa (M.P.), India
[2] Mechanical Engineering Department, Vindhya Institute of Technology and Science, Satna (M.P.), India
[3] Mechanical Engineering Department, Jai Narayan College of Technology, Rewa (M.P.), India
[4] Mechanical Engineering Department, Rewa Engineering College, Rewa (M.P.), India
[5] Mechanical EngineeringDepartmen,tMaulana Azad National Institute of Technology, Bhopal (M.P.) India

Abstract. The glass cover of a solar still plays a major role in its performance andefficiency since the output depends directly on the water and glass cover temperature difference. In most of the experimental studies, the outer glass cover temperatureswererecorded and considered as the glass cover temperature for calculations, however the condensation on the glass cover surface dependsmore on the inner surface temperature than the outer. An experimental study was conducted in order to study the difference between the inner and the outer glass cover temperature difference in a conventional basin type solar still and a low thermal inertia solar still obtained by modifying the conventional still. It was observed that for theconventional still, the temperature difference was maximum upto7.14% at noon and insignificant in the morning and evening hours. For the low thermal inertia still, the difference was more upto 10.2% at around 11:00 hrs and zero during the early morning and the evening hours. A significant difference in the glass cover temperature curves for the two different types of solar stills can be seen .

1 Introduction

In most of the remote and rural regions of the world.electric power supply and potable water availability has always been a major problem and urgent necessity of the natives. To cope with such problems, equipments and devices which do not require electricity for converting brakish or saline water into potable or drinking water are desired. Solar stills are simple devices which can provide potable water by using the thermal energy of the sun.Despite of being simpleand economical in construction and working, solar stills suffer from a poor thermal efficiency and hence low distillate output.

A lot of research work has been conducted by several researchers on solar stills all over the world. Heat and mass transfer relations were developed by Dunkle [1]. For given ambient and basin water temperatures, the glass cover temperature could be calculated by trial and error method. The performance of a solar still improves with the reduced air gap, which is due to the fact that, at reduced air gap, the vapor length is small and the still can operate by diffusion mass transfer as investigated by Sodha et al. [2]. Furthermore, due to reduced mass the air gets saturated more quickly. Sharma and Mullick [3] developed an empirical relation for the calculation of glass cover temperature with the help of which the heat transfer coefficients, upward heat flow and the evaporation in a solar still could be estimated.

It was established by Tiwari et al. [4] that the inclination of the still cover and the direction of its inclination depend upon the latitude of the location. Besides, in case of the cover angle equal to the latitude of the location, the sun rays are received by the still nearly normal throughout the year. It was also investigated that the cover temperature significantly affects the productivity of the solar still. A decrease in the cover temperature increases the productivity of the still, since a larger cover and basin water temperature difference increases both the convective and evaporative heat transfer in the solar still. Minasian and Al-Karaghouli [5] developed a wick-basin type solar still with improved productivity. An experimental study was conducted by Madani and Zaki [6] on the solar still with porous basin and an improvement in the performance was observed. A.A. El-Sebaii [7] investigated that the higher wind velocity increases the convective heat transfer from cover to atmosphere, which in turn increases the rate of condensation and evaporation, thus increasing the productivity. Bassam et al. [8] conducted an experimental study by placing sponge cubes on basin water and obtained an increase in the productivity. Karaghouli and Alnaser [9] conducted an experimental study to compare the performance of single and double slope solar stills. Shukla and Sorayan [10] developed a mathematical model based on regression analysis to determine the values C and n with the help of the experimental data. Expressions for internal heat transfer coefficients were determined by using both inner and outer glass cover temperatures. Expressions for water and glass temperatures, hourly yield and instantaneous efficiency for double slope distillation systems were derived by Shukla and Rai [11]. In this analysis a thermal model was developed to predict the performance of the solar still based on the inner and outer glass cover temperatures.

Kabeel et al [12] conducted an experimental study in Kafrelsheikh, Egypt in which two types of solar distillation systems were tested. It was observed that the stepped solar still gave higher distillate production than the conventional still. For 5mm water depth about 57% higher output was obtained in the stepped still. Srivastava and Agrawal [13] conducted an experimental and theoretical analysis of a basin type solar still with multiple floating porous absorbers. They obtained an increase in the distillate output by around 68% over the conventional basin type solar still. Srivastava and Agrawal [14] developed an extended porous fin type still by placing jute cloth porous fins in the basin water in such a way that the fins were partially dipped and partially extended above the basin water surface. The performance of the improved still in the winter and summer season has been reported. An appreciable improvement in the productivity along with nocturnal production was observed due to porous fins. Srivastava and Agrawal [15] conducted an experimental study on the floating absorber type solar still and observed a highest of 74.6% more output in the month of January and lowest 35.5% more in August as compared to that of the conventional still.BadushaanfArjunan [16] usedcomputationalFluid Dynamics (CFD) method, to develop a two phase threedimensionalmodel for simulation of a solarstill. Simulation resultwerecomparedwithactualexperimental data of solarstill. A good agreement wasfoundbetweenexperimental data and CFD data of fresh water productivity, water temperature and heattransfer coefficients. Kumar and Tiwari [17] estimated convective heat transfer in a solar distillation system.Hijleh [18] performed water film cooling of the glass cover and obtained an increase in the performance of the solar still. Sebaiiet. Al. [19] investigated the effect of wind speed on some designs of solar stills.Velmuruganet. Al. [20] experimented by integrating fins on the basin plate with the help of which they achieved an increase in the distillate production in the solar still. Tanaka [21] conducted experimental study with internal and external reflecters. It was observed that a slight inclination of the relector in the winter season gave an increase in the distillate yield by 70 to 80 percent. Eldalil and Khalid [22] applied vibratory harmonic effect to a basin type solar still. A productivity increase upto 5,85l/m^2 and average daily efficiency of 60% was observed. Samee et. al. [23] conductedan study on the design and performance of a solar still.

In the present work, an experimental study on solar still glass cover temperatures conducted during the winter season in central Indian location is presented.

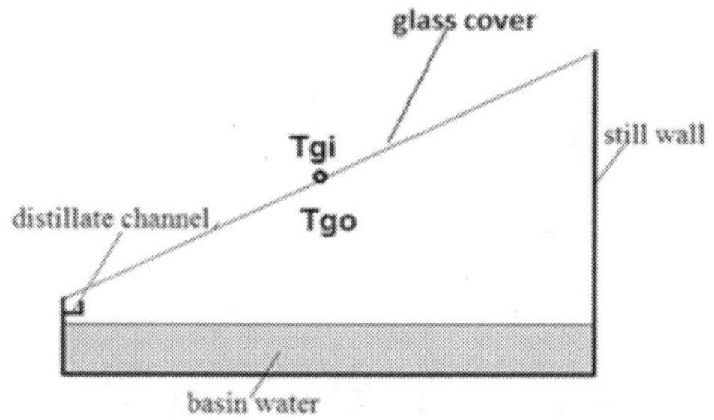

Figure 1.Conventional basin type solar still.

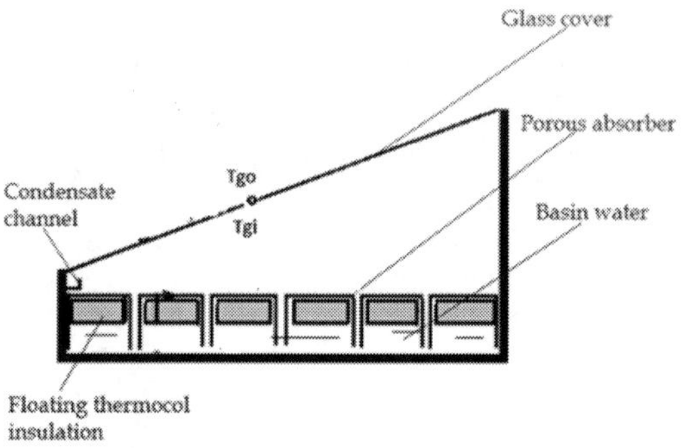

Figure 2. Low thermal inertia solar still.

2 Experimental procedure

The experimental study was conducted during the winter month of November in a Central Indian location. The experimental setup consisted of two identical single sloped basin type solar stills made of wood with effective basin area of 0.5m^2 each. One of the stills was modified into low thermal inertia solar still by floating blackened jute cloth absorbers with the help of 20 mm thick thermocol planks on the basin water. The 4mm thick glass covers in both the stills were kept at an angle of 25° which is nearly equal to the latitude of the place of experimental study. Beside, at the cover angle equal to that of the latitude of the place, the solar still receives the solar radiation nearly perpendicular to it all round the year.The basin water level in the two stills was kept as 50mm.Vapour tightness of the glass covers was ensured by applying rubber lining at the glass edges. Thermocouple wires for recording the inner and outer glass cover temperaturesTgi and Tgorespectively were attached on the lower and upper sides respectively at the centres of each glass covers. The multi-point digital temperature indicator gave the temperatures of all the desired points for both the solar stills.

The experiments were conducted from 08:00 hrs to 17:00 hours every day for five consecutive days. The temperatures were recorded at hourly intervals. The water level in the stills was kept constant at 50mm by topping up the basin water at regular intervals. The best observations of the five days experimentation are presented.

3 Thermal relations

The amount of distillate per hour per square metrefor the solar still can be obtained from the relation,

$$m_w = \frac{q_{ewg}}{h_{fg}} \quad (1)$$

In the equation no. (1) qewg is the evaporative heat transfer from the water surface to the glass cover inner surface and h_{fg} is the latent heat of vaporization of water.

The evaporative heat transfer is given as

$$q_{ewg} = h_{ewg}(T_w - T_g) \quad (2)$$

Where h_{ewg} is the evaporative heat transfer coefficient as per Dunkle [1], Tw and Tg are the water surface and glass cover temperatures.

The evaporative heat transfer coefficient between the water surface and the glass cover depends upon the convective heat transfer and hence it is given as,

$$h_{ewg} = \frac{0.01623 * h_{cwg}(P_w - P_g)}{(T_w - T_g)} \quad (3)$$

The convective heattransfer coefficient isdependentupon the difference in the water and the glass covertemperatures and the partial pressures at the temperaturesof the water and glass cover surface.Henceitisrepresentedas,

$$h_{cwg} = 0.884\left[(T_w - T_g) + \frac{(P_w - P_g)T_w}{(268.9 * 10^3 - P_w)}\right]^{1/3} (4)$$

In the above equation P_w and P_g are the partial presures at water and glass cover temperatures respectively.

4 Results and discussion

Fig. 3 shows the hourly variations of the inner and the outer glass cover temperatutres of the conventional basin type solar still.It can be seen that the maximum cover temperatures were reached at around 01:00 hrs and the maximum temperature difference was obtained at 11:00 hrs. In the morning hours from 08:00 to 10:00, the temperature difference is negligible. This is due to the fact that the basin water of the conventional solar still has high thermal inertia, which causes a time delay in reaching the maximum water temperature and hence the glass cover temperature. The glass cover receives the major amount of latent heat from the water vapours during the day hours. Although the ambient temperature is also high during this duration, but the heat flux from the basin water is high enough to create a significant temperature difference between the inner and the outer surface. Fig. 4 represents the hourly variation of inner and the outer glass cover temperatures for the low thermal inertia solar still. This type of solar still has a higher evaporation rate as higher absorber surface and glass cover temperatures are reached. Therefore the temperature differences in this type of still is also higher.The maximum temperature difference was observed upto 5°C in the low thermal inertia type still, whereas it was upto 3°C for the conventional solar still as can be seen from the Fig. 5 In this case also the temperature difference between the inner and outer glass surface is large whereas negligible during the morning and the evening hours. Figure 6gives an idea of the variation of the percentage difference between the inner and the outer glass cover temperatures for the two types of stills. The difference is higher for the low thermal inertia type than the conventional. For the former the maximum % temperature difference observed was 10.2% at about 11:00 hrs, whereas for the later it was 7.14% around 12:00 hrs. The irregular variations in between can be caused due to random wind or cloud shading effect. The thermal relation for the distillate output indicates that the distillate production depends upon the evaporative heat transfer from the water surface to the glass cover. This mode of heat and mass transfer is dependent upon the tempearture difference between the water and glass cover surfaces. Since the water vapours directly come in contact with the inner glass surface, therefore the inner surface temperature is very much significant for determining the heat transfer coefficients and the distillate production of the solar still, wether it may be conventional type or low thermal inertia type solar still.

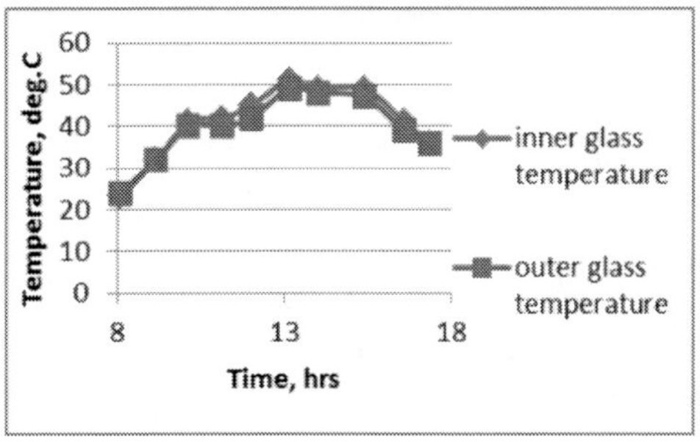

Figure 3. Glass cover temperatures of conventional solar still.

5 Conclusion

The study reveals a significant difference in the glass cover temperature profiles for the two different types of solar stills. In case of the low thermal inertia type solar still, the temperature difference is even more significant. However in both the cases the temperature difference between the inner and the outer glass cover surfaces is very small or even nedligible. The study can be helpful for the researchers in having a prelimnary idea of the inner and the outer glass cover temperature differences.

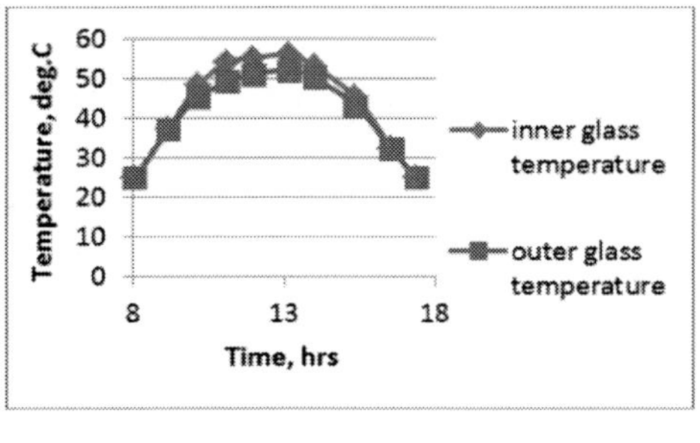

Figure 4. Glass cover temperatures of low thermal inertia solar still.

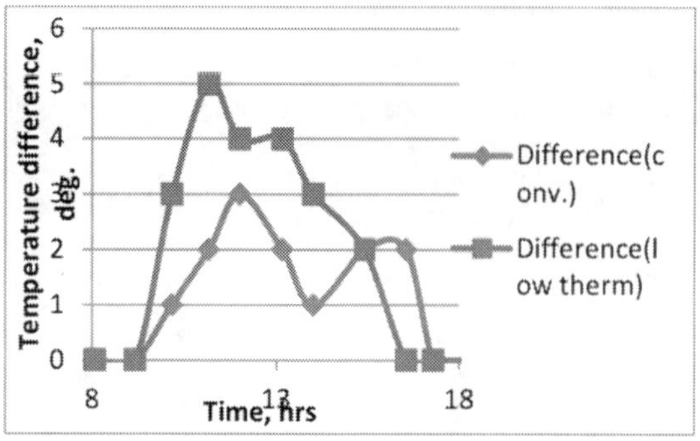

Figure 5. Difference of glass cover inner and outer temperatures.

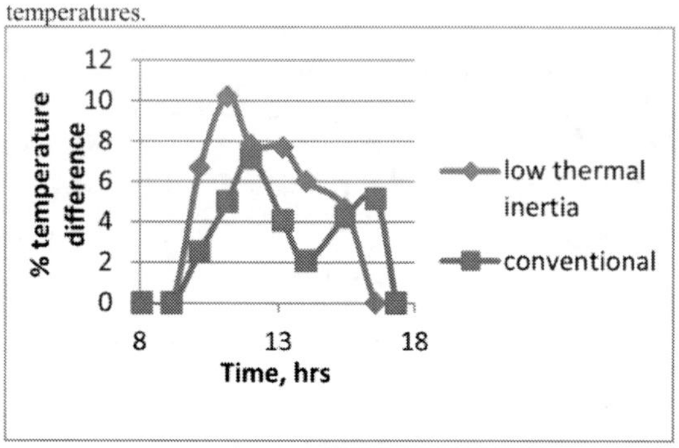

Figure 6. Percentage temperature difference for both types of solar stills.

References

1. R.V. Dunkle, International Developments in Heat Transfer, Proceedings of International Heat Transfer Conference, University of Colorado ASME, 895–902, (1961).
2. M.S. Sodha, A. Kumar, G.N. Tiwari, R.C. Tyagi, Solar Energy,**26** 127-131(1981).
3. V.B. Sharma, S.C. Mullick, , J. Solar energy Trans. of ASME, **115,** 231-236(1993).
4. G.N. Tiwari, J.M. Thomas, Emran Khan, Heat Rec. Syst. CHP **14,** 447–455(1994).
5. A.N. Minasian, A.A. Al-Karaghouli, En.Conv. Mgmt. **36,** 213–217 (1995).
6. A.A. Madani, G.M. Zaki, App.En.**52**, 273–281(1995).
7. A.A. El-Sebaii, J. En.Conv. Mgmt. **41**, 523–538(2000).
8. A. Bassam, K. Abu –Hijleh, Hamzeh M. Rababa'h, J.En.Conv. Mgmt**44,** 1411–1418 (2003).
9. A.A. Al-Karaghouli, W.E. Alnaser, Appl. En. **77**, 317–325(2004).
10. S.K. Shukla, V.P.S. Sorayan, Ren. En.**30,** (5) 683–699(2005).
11. S.K. Shukla, A.K Rai, Therm. Sc. **12,** (3) 139–152(2008).
12. Kabeel A.E., Khalil A.,Omara Z.M., Younes M.M., Desalination**289,** 12-20(2012).
13. Pankaj K. Srivastava, S.K. Agrawal, Desalination **311**, 198–205(2013).
14. Pankaj K. Srivastava, S.K. Agrawal, Desalination **319**, 73–78(2013).
15. Pankaj K. Srivastava, S.K. Agrawal, Int. J. Ren. En.Tech.**5** (2), 159 – 175(2014).
16. RegilBadusha A,T V Arjunan, IJMERR. **2** (4), 74-81 (2013).
17. .S. Kumar, G,N, Tiwari, Solar Energy, **57**, 459-464 (1996).
18. Abu- Hijleh, B.A.K,Desalination, **107**, 235-244 (1996).
19. .A.A. Sebaii, A.A. Al-Ghamdi, F.S. Al-Hazmi, Abdel S. Fidah, App. En.,**86**, 1187-1195 (2009).
20. V. Velmurugan, M. Gopalakrishnan, R. Raghu K. SritharEn. Con.Mgmt**49**, 2602-2608 (2008).
21. Hiroshi Tanaka, Desalination, **30**,1-15, (2009).
22. M.S. Eldalil, M.S. Khaled, Desalination,**251**, 3-11, (2010).
23. M.A. Samee, U.K. Mirza, T. Majeed, N. Ahmad, Ren. Sus.En.Rev., **11**, 543-549 (2007).

Appendix:

h_{fg}= 3044205.5 -1679.1109T_w– 1.14258T_w^2

$$p = 165960.72X10^{-[x(a+bx+cx^3)/T(1+dx)]}$$

Where, $x = 647.27 - T,\ a = 3.2437814,$

$$b = 5.86826\,X\,10^{-3}, \quad c = 1.1702379\,X\,10^{-8}$$

$$and\ d = 2.1878462\,X\,10^{-3}$$

MATEC Web of Conferences **95**, 18007 (2017) DOI: 10.1051/matecconf/20179518007
ICMME 2016

Experiment on Measurement of Interfacial Tension for Subsurface Conditions of Light Oil from Thailand

Jiramet Jiravivitpanya, Kreangkrai Maneeintr and Thitisak Boonpramote

Carbon Capture, Storage and Utilization Research Laboratory, Department of Mining and Petroleum Engineering, Faculty of Engineering, Chulalongkorn University, Thailand

Abstract. One of enhanced oil recovery techniques to increase oil production is surfactant flooding. Surfactants are considered as effective chemical agents used in oilfield in Thailand. It is used to reduce the interfacial tension (IFT) of two fluids and to make them flow easier in the reservoir. In this study, Monoethanolamide (MEA) commonly used for carbon dioxide capture, is applied as a surfactant to reduce IFT between oil and brine. Therefore, the aim of this work is to investigate and measure the IFT based on the conditions of subsurface at the oilfield in Thailand. These parameters such as temperature, pressure, salinity as well as the concentration of surfactant are adjusted to investigate the effects on IFT reduction. From the results, it is reported that pressure from 1000 to 2000 psi and temperature varied from 70°C to 90°C can reduce IFT insignificantly. However, salinity and surfactant concentration are the main parameters that impact on the IFT reduction. It can greatly decrease IFT up to 87.13% for surfactant concentration and up to 74.06% for salinity. Finally, the results can be applied to use in the real field for enhanced oil production in Thailand.

1 Introduction

Nowadays, the world economic growth is developing continuously. Therefore, energy consumption is growing rapidly as well. Oil production needs to be enhanced to serve this increasing demand. Enhanced oil recovery (EOR) technology as tertiary recovery is getting more attention because of ineffective primary and secondary recoveries [1]. Surfactants are considered as one of the effective chemical flooding agents because it can significantly reduce the interfacial tension (IFT) between water and oil interfaces and increase values of both sweep efficiency and displacement efficiency, as well as decrease residual oil saturation [2].

However, there are some challenges that limit the prediction of efficient surfactant flood [3]. The effective in-situ IFT on surfactant flooding is controlled by several factors: (1) reservoirs conditions; (2) partitioning of surfactant between trapped oil and brine phases; (3) degree of surfactant dilution; and (4) surfactant adsorption onto the rock surface. These causes are difficult to know how IFT exactly changes under reservoir conditions, but there are researches [4] reporting some of concerning chemical parameters that effect on IFT. For this study, the research is focusing on the effect between aqueous surfactant solution and hydrocarbon phase such as pressure, temperature, salinity and surfactant concentration.

The mechanism for IFT reduction in aqueous solution can be described in term of colloidal aggregates of surfactant molecules which are called micelles. When surfactant is put in aqueous solution, the dissolved surfactant molecules are dispersed on the interface as monomers and the surface free energy has been decreased. When the surfactant concentration is added more and the specific surfactant concentration called critical micelle concentration (CMC) is reached, the surfactant starts aggregating into micelles [1]. At above critical micelle concentration, the additional of surfactants will be insignificant on IFT reduction and also excess amount of surfactant in aqueous solution. The CMC values are used in process industry surfactant applications such as mineral processing to determinate the personal care food, to separate the different types of plastic and to bring the new surfactant into proper form because the greatest surfactant effect will be accomplished when significant micelles concentration is propose [5]. Therefore, the CMC is very important to concern when discussing in the formulation of micelles and applied in many process surfactant industries.

In this study, Monoethanolamide (MEA) is adopted as surfactant solution which is widely used in many chemical processes such as acid gas removal because it is cheap and can be easily found in the market. The molecular structure of MEA is presented in Fig. 1. From the figure, it is clear that MEA structure has 2 main parts, a hydrophilic part (hydroxyl and amino groups) which can dissolve in aqueous phase, and another part is called hydrophobic part (hydrocarbon chain) that can dissolve in oil phase Therefore, the aim of this work is to investigate and measure the IFT at oil-water interfaces based on the conditions of subsurface at the oilfield in Thailand.

DOI: 10.1051/matecconf/20179518007

Figure 1. Molecular structure of MEA

2 Experiment

2.1 Chemicals

Oil sample is obtained from the oilfield in the North of Thailand with the density of 0.84 g/cm^3. MEA is purchased from Sigma-Aldrich mixed with simulated brine. Brine is prepared from distilled water and sodium chloride purchased from Ajax at various salinity depending on the sources in the oilfield.

2.2 Equipment

The interfacial tension meter in Fig. 2 (Model 700, Vinci Technology) used for measuring the interfacial tension (IFT) between liquid-liquid and liquid-gas interface has chamber volume of 1×10^{-5} m^3. The maximum working pressure and temperature are 69 MPa (or 10,000 psi) and 180°C (or 350°F), respectively. Then, a computer connected with camera will analyze the IFT value and also record the shape of the drop with Drop Analysis Software (DAS) provided by Vinci Technology.

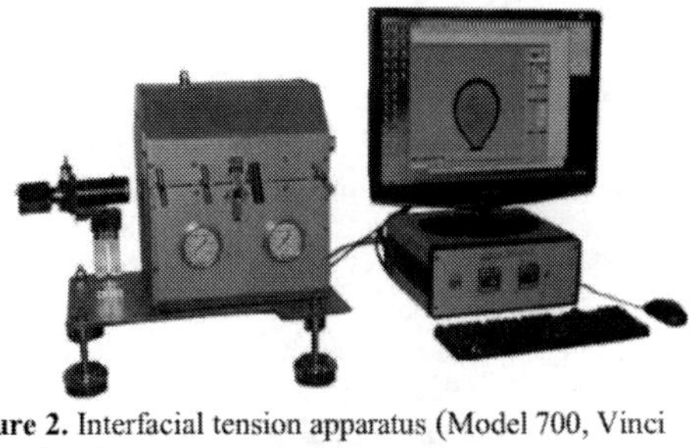

Figure 2. Interfacial tension apparatus (Model 700, Vinci Technology)

Figure 3. Rising oil drop

2.3 Experimental procedures

The IFT 700 is set up to run the experiment. Before measuring the IFT, software requires the density of oil and surfactant solution. The density can be measured by density meter. The running process would be cleaned by using acetone and distilled water. Then, both oil sample and surfactant solution are fed to sample cylinder and heated up to the desired temperature. The surfactant solution as well as the oil would be injected to the chamber of IFT 700 at designed pressure. The camera connected to computer will detect the generated drop which is created by calibrated capillary in the chamber as presented in Fig. 3. The results will then be recorded by integrated software.

The parameters will be varied to measure IFT as shown in Table 1. First, the surfactant concentrations are prepared at 0.5, 1.0, 2.0 and 4.0 wt%. Second, the salinity is set at 7,500, 10,000, 15,000 and 17,500 ppm. Then, the temperature is also varied at 70, 80 and 90°C. Lastly, pressure would be tested for 3 values which are 1,000, 1,500, 2,000 psi. The operating conditions for this study are presented in Table 1.

Table 1 Operating conditions for IFT measurement

Parameters	**Operating conditions**
Type of surfactant	Monoethanolamine (MEA)
Solution concentration (wt%)	0.0, 0.5, 1.0, 2.0, 4.0
Temperature (°C)	70, 80 and 90
Pressure (psi)	1,000 1,500 and 2,000
Salinity (ppm)	0, 7,500, 10,000, 15,000, and 17,500

3 Results and discussion

3.1 Effect of pressure on IFT

As shown in Fig. 4, when pressure changes, the IFT just changes a small extent. Basically pressure has more effect on gas but the liquid phase system has intermolecular force higher than gas phase system. Therefore, pressure has less effect on IFT with liquid-liquid system. According to D.W. Green and G.P. Willhite [6], it shows that the pressure on phase behavior at constant temperature would have a small effect on IFT.

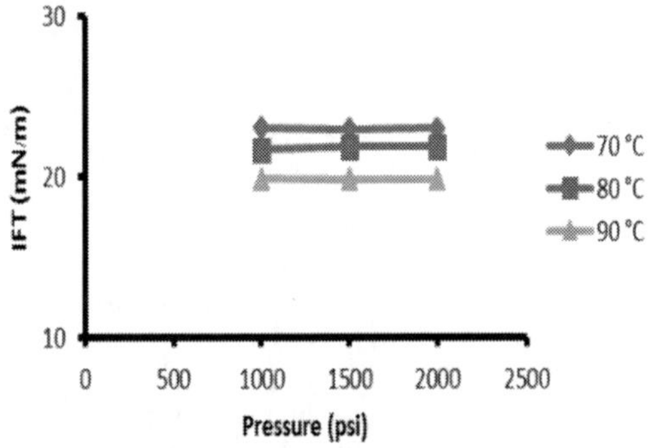

Figure 4. Effect of pressure on the IFT at MEA concentration 0.05wt% and salinity 15,000 ppm

3.2 Effect of temperature on IFT

The range of temperature in this study will cover the temperature of the conditions in the oilfield from 70°C to 90°C. Fig. 5 and 6 show the effect of temperature and salinity for the solutions without and with surfactant, respectively. Both cases have an opposite effect of temperature on IFT in that IFT will be decreased as temperature increases. When the temperature is higher,

DOI: 10.1051/matecconf/20179518007

the free energy between oil and solution would reduce thus resulting in the enhanced mobility of oil and water and also decreasing the IFT [7]. One more reason is that the increment of temperature can effect on surfactant solution because of the weakening of intermolecular forces at the oil-water interface [8]. From Fig. 5, the results indicate that the percentage of IFT changes in case of 15,000 ppm of salinity without surfactant, temperature from 70°C to 80°C and 80°C to 90°C are 2.0% and 0.7% respectively. Also, in Fig. 6 the rate of IFT change in case of 15,000 ppm with 1wt.% MEA, temperature from 70°C to 80°C and 80°C to 90°C are 12.2% and 7.2% respectively. The temperature from 80°C to 90°C can describe that interface between oil and water is less space for occupying of surfactant. Therefore, the rate of declining IFT trends to be decreased.

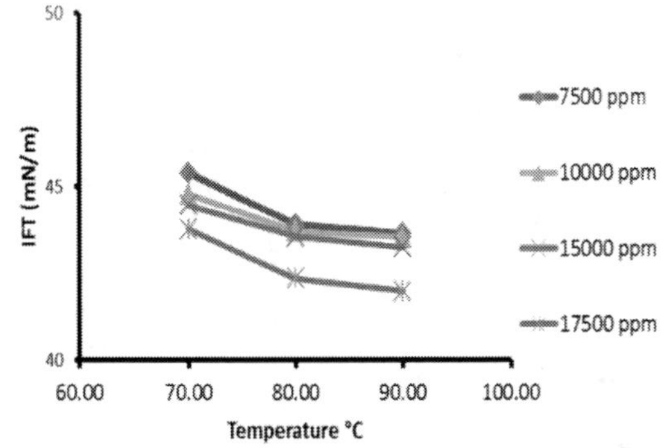

Figure 5. Effect of temperature on the IFT at various salinities without surfactant and pressure at 1,500 psi

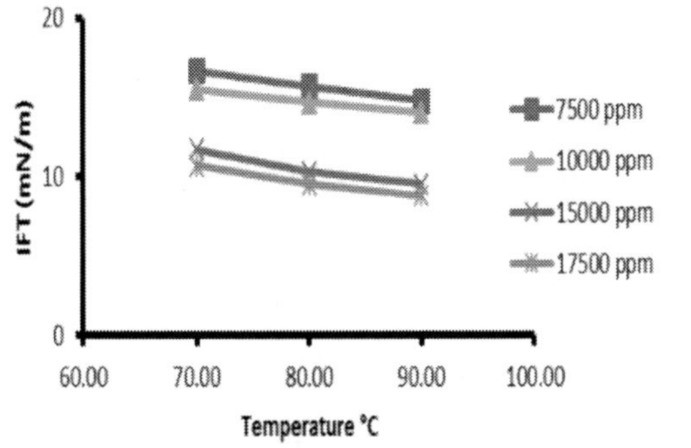

Figure 6. Effect of temperature on the IFT at MEA concentration 1wt.% and pressure at 1,500 psi

3.3 Effect of salinity on IFT

The water produced from the oilfield is composed mainly of the various ions in the reservoir. Most of them are sodium and chloride ions accounting for 96%. Fig. 7 indicates the IFT trends to be lower as the salinity of brine increases. The reason for these changes is that the ionic solution pushes the surfactant to oil-water interface. Besides, the incremental of salinity concentration can enhance the absorption of surfactant (MEA) at the oil-water interface that causes the lowering of IFT [9]. However, for the salinity ranging from 15,000 to 17,500 ppm, the IFT slightly decreases. Particularly, for case of concentration at 4.00 wt%, salinity has less significant on IFT reduction because high salt changes try to drive amount of in-situ surfactant to the oil-water interface, but the space in there is not sufficient for occupying by surfactant.

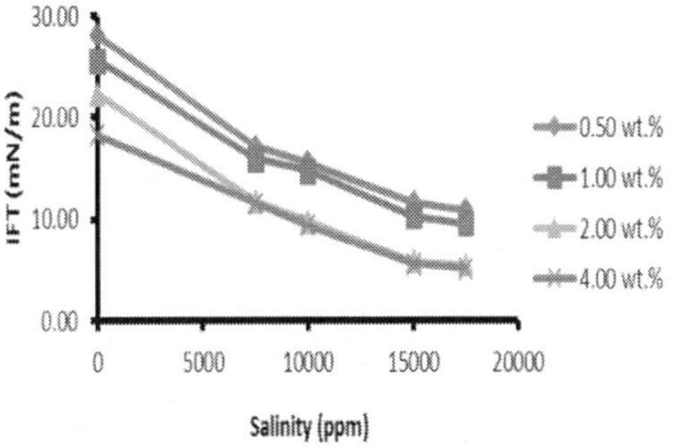

Figure 7. Effect of Salinity of brine on the IFT at temperature at 80°C and pressure at 1,500 psi

3.4 Effect of surfactant concentration on IFT

As shown in Fig. 8, the effect of surfactant concentration is presented at various brine concentrations with 1,500 psi and 80°C. The result shows that at any salinity concentration, the reduction in IFT in all cases is caused by chemical characteristic of MEA. There are two parts of MEA's molecular structure. One part is called hydrophilic part (hydroxyl and amino groups) which can dissolve in aqueous phase, and another part is called hydrophobic part (hydrocarbon chain) that can dissolve in oil phase. The surface is covered by the surfactants molecules as monomers and the surface free energy (surface tension) has decreased. Thus, an increase in surfactant concentration can dissolve more both oil and water thus making the reduction of IFT corresponding to Wu et al [10]. In addition, the low concentration varied from 0%wt. to 2.0 %wt can greatly decrease the IFT more than 74%. Nevertheless, when the surfactant concentration is higher than 2.0 wt%, the IFT value becomes stable or a little bit change in interfacial tension. The reason is that the surfactant is added in aqueous solution excessively resulting in the surfactant aggregating into micelles which is called critical micelle concentration (CMC). Consequently, the amount of MEA is difficult to possess in oil-water interface.

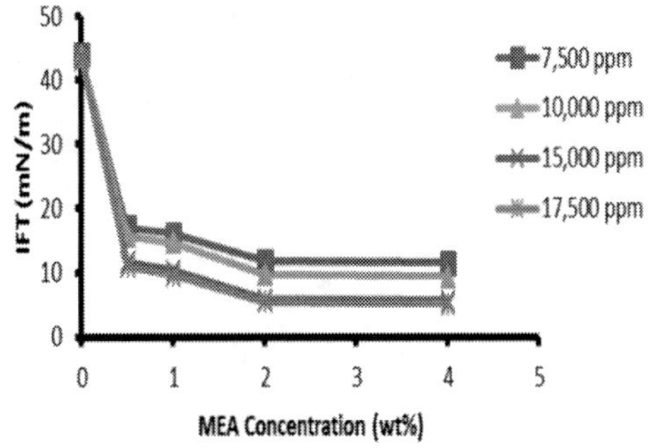

Figure 8. Effect of surfactant concentration on the IFT temperature at 80°C and pressure at 1,500 psi

DOI: 10.1051/matecconf/20179518007

4 Summary

Chemical enhanced oil recovery techniques are used to increase oil production by using surfactant flooding. Surfactants are considered as effective chemical agents used in oilfield in Thailand. It is used to reduce the interfacial tension of two fluids and to make them flow easier in the reservoir. In this study, Monoethanolamide (MEA) commonly is applied as a surfactant to reduce IFT between oil and brine based on the conditions at the subsurface from the oilfield in Thailand. From this research, the results report the IFT data at various conditions as well as the effects of each parameter such as pressure, temperature, surfactant concentration and salinity on IFT reduction at reservoir conditions in Thailand. Pressure ranging from 1,000 psi to 2,000 psi at constant temperature has less significant effect on IFT reduction and can be ignored. Also, temperature has less effect on IFT reduction and the rate of declining IFT trends to be decreased with the increase of temperature. However, the IFT can be reduced by increasing salinity from 0 to 15,000 ppm accounting up to 74.06%. Also, IFT becomes more stable when salinity is higher than 15,000 ppm. The IFT can greatly decrease up to 87.13% as surfactant concentration increases from 0% to 2 wt%, but it will be relatively constant at the concentration higher 2.0 wt%. These investigated effects of each parameter will be useful to understand and used as fundamental data to apply for oil recovery with reservoir conditions in the oilfield in Thailand.

Acknowledgement

The authors would like to thank the 90th Anniversary of Chulalongkorn University, Rachadapisak Sompote Fund and Graduate School Thesis Grant and Overseas Academic Presentation Scholarship for Graduate Students, Chulalongkorn University for financial support. Moreover, Defense Energy Department at Fang district, Chiang Mai, Thailand is acknowledged for light oil sample.

References

1. J. J. Sheng, in: *Modern Chemical Enhanced Oil Recovery: theory and practice*, Gulf Professional Publishing (2011)
2. M. Ali Ahmadi, Y. Arabsahebi , S. Reza Shadizadeh, S. Shokrollahzadeh Behbahani. Preliminary evaluation of mulberry leaf-derived surfactant on interfacial tension in an oil-aqueous system: EOR application, Fuel 117, 749–755 (2014)
3. Y. P. Zhang, S.G. Sayegh and S. Huang, Effect of Oil/Brine Ratio on Interfacial Tension in Surfactant Flooding, Canadian International Petroleum Conference, 12-14 June, Calgary, Alberta, Canada, (2007)
4. J. Buckley and T. Fan, Crude Oil/Brine Interfacial tension, The Society of Core Analysts Symposium **48,** (2005)
5. L. Schramm, N. Stasiuk and D. Gerrard Marangoni, Surfactants and their applications, Annual Reports Section "C' (Physical Chemistry) **99,** 10 (2003).
6. D.W. Green, G.P. Willhite, *Enhanced oil recovery SPE Textbook Series* **6**, USA, 262 (1998)
7. X. Wei, in: Experimental Investigation of Dynamic Interfacial Interactions at Reservoir Conditions, Louisiana State University and Agricultural and Mechanical College, (2005).
8. W. Karnanda, M.S. Benzagouta, A. AlQuraishi and M.M. Amro: Arab J. Geosci. **6**, 3535 (2013)
9. A.J. Prosser and E.I. Franses: Colloids and Surfaces A: *Physicochemical and Engineering Aspects* **178,** 1 (2001).
10. Z. Wu, X. Yue, T. Cheng, J. Yu and H. Yang: Journal of Petroleum Exploration and Production Technology **4**, 9 (2014).

Prediction of Shanghai Index based on Additive Legendre Neural Network

Bin Yang and Haifeng Wang

School of Information Science and Engineering, Zaozhuang University, Zaozhuang, China

Abstract. In this paper, a novel Legendre neural network model is proposed, namely additive Legendre neural network (ALNN). A new hybrid evolutionary method besed on binary particle swarm optimization (BPSO) algorithm and firefly algorithm is proposed to optimize the structure and parameters of ALNN model. Shanghai stock exchange composite index is used to evaluate the performance of ALNN. Results reveal that ALNN performs better than LNN model.

1 Introduction

Artificial neural networks (ANNs) are powerful mathematical methods that can be used to learn complex linear and non-linear continuous functions, and have been successfully applied to many areas in the past decades [1]. Due to that traditional neural network has some disadvantages such as low efficiency, long learning time and easy to fall into the local minimum solution, Legendre neural network (LNN) was proposed. LNN modle has no hidden layer and could add dimensionality of the input layer with a set of nonlinear functions. Patra et al. proposed a Legendre neural network model for equalization of nonlinear communication channels with 4-QAM signal constellation [2]. Patra et al. also propose a computationally efficient Legendre neural network for identification of nonlinear dynamic systems [3]. Pei et al. forecastedand investigated the stock prices of the financial model by an improved Legendre neural network–Legendre neural network with random time strength function [4].

The structure of Legendre neural network is very simple and learning speed is fast. But the number of input Legendre polynomials is large due to the fact that each input variable has *n* order Legendre polynomials. The stucture of LNN is fixed and only task is to optimize the paramters of LNN. To reduce the optimization complexity and improve efficiency, in this paper, a novel Legendre neural network model is proposed, namely additive Legendre neural network (ALNN). Binary particle swarm optimization (BPSO) algorithm is proposed to select proper input Legendre polynomials in order to construct proper structure. Firefly algorithm is used to optimize the parameters of ALNN. Shanghai stock exchange composite index is used to evaluate the performance of ALNN.

2 Method

2.1 Structure of ALNN

Legendre neural network (LNN) was first proposed by Yang and Tseng for function approximation in 1996 [5]. LNN has less parameter and does not have hidden layer, which uses Legendre orthogonal polynomials as the activation functions of hidden layer neurons. Due to the absence of hidden layer, LNN provides computational advantage over the MLP.

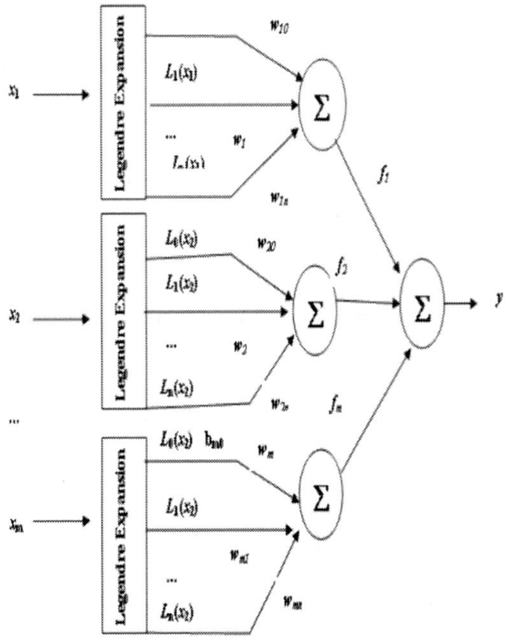

Figure 1. The structure of ALNN.

In order to select proper Legendre polynomials for specific problems, additive Legendre neural network (ALNN) is proposed. The structure of an ALNN is shown in Figure. 1. Suppose that input vector [$x_1, x_2, \ldots x_m$] and n order Legendre polynomials.

The Legendre polynomials of each input variable are described as followed.

$$\begin{cases} L_0(x) = 1 \\ L_1(x) = x \\ L_2(x) = \frac{1}{2}(3x^2 - 1) \\ \vdots \\ L_n(x) = \frac{1}{n}[(2n-1)xL_{n-1}(x) - nL_{n-2}(x)] \end{cases} \tag{1}$$

The output of f_i is defined as followed.

$$f_i = \tanh(\sum_{k=0}^{n} L_0(x_k) \times b_{ik} \times w_{ik}) \tag{2}$$

where $\tanh()$ is Hyperbolic tangent, b_{ik} is boolean value (0 or 1). When b_{ik} is equal to 1, Legendre polynomial $L_0(x_k)$ is selected as input data and weight w_{ik} is assigned.

The final output y is defined as followed.

$$y = \sum_{k=1}^{m} f_k \tag{3}$$

2.2 Structure optimization of ALNN

Additive Legendre neural network could not allow all Legendre polynomials as input data. It uses evolutionary method to select proper Legendre polynomials. In this paper, binary particle swarm optimizaiton (BPSO) is used to gain the optimal vector $[b_{10,}b_{11}, \ldots, b_{1n}, \ldots, b_{m0}b_{m1}, \ldots, b_{mn}]$.

In BPSO, the moving trajectory and velocity of each particle is defined in term of probability. The moving trajectory represents changes of probabilities of a certain value. The moving velocity is defined as probability of a state or another state. Thus each bit $x_i(t)$ of one particle is restricted to 0 or 1. Suppose that ALNN has m input variables and n order Legendre polynomials. The length of each particle is m*(n+1). Each $v_i(t)$ represents the probability of bit $x_i(t)$ taking the value 1. A new velocity $v_i(t)$ for particle i is updated as same as PSO, which is defined as followed.

$$\begin{aligned} v_i(t+1) = w * v_i(t) + c_1 r_1 (Pbest_i - x_i(t)) + \\ c_2 r_2 (Gbest - x_i(t)). \end{aligned} \tag{4}$$

where w is the inertia weight, c_1 and c_2 are positive constants and r_1 and r_2 are uniformly distributed random number. $Pbest_i$ is the best fitness of particle i and $Gbest$ is the best position among all particles.

$xi(t)$ is calculated as followed [6].

$$x_i(t) = \begin{cases} 1, & r < Sig(v_i(t)) \\ 0, & other. \end{cases} \tag{5}$$

where r si created randomly from range [0,1], and the function Sig is defined as followed.

$$Sig(v_i(t)) = \frac{1}{1 + e^{-v_i(t)}}. \tag{6}$$

2.3 Parameters optimization of ALNN

According to the optimal structure of ALNN, tally the number (p) of 1 in the optimal particle $[x_1, x_2, \ldots, x_{m*(n+1)}]$ of BPSO. p parameters $[w_1, w_2, \ldots, w_p]$ need be optimized. Firefly algorithm (FA) is an efficient optimization algorithm which was proposed by Xin-She Yang in 2009 [7]. It is very simple, has few parameters and easy to apply and implement, so this paper uses firefly algorithm to optimize the parameters of Legendre neural network.

Firefly algorithm is the random optimization method of simulating luminescence behavior of firefly in the nature. The firefly could search the partners and move to the position of better firefly according to brightness property. A firefly represents a potential solution. In order to solve optimization problem, initialize a firefly vector $[x_1, x_2, \ldots, x_n]$ (n is the number of fireflies). As attractiveness is directly proportional to the brightness property of the fireflies, so always the less bright firefly will be attracted by the brightest firefly.

The brightness of firefly i is computed as

$$B_i = B_{i0} * e^{-\gamma r_{ij}} \tag{7}$$

where B_{i0} represents maximum brightness of firefly *i* by the fitness function as $B_{i0} = f(x_i)$. γ is coefficient of light absorption, and r_{ij} is the distance factor between the two corresponding fireflies *i* and *j*.

The movement of the less bright firefly toward the brighter firefly is computed by

$$x_i(t+1) = x_i(t) + \beta_i(x_j(t) - x_i(t)) + \alpha\varepsilon_i \qquad (8)$$

where α is step size randomly created in the range [0, 1], and ε_i is gaussian distribution random number.

2.4 Fitness function

Root mean squared error (*RMSE*) is used to search the optimal ALNN according to actual data and predicted data.

$$RMSE = \sqrt{\frac{1}{N}\sum_{i=1}^{N}(y^i_{actual} - y^i_{predicted})^2} \qquad (9)$$

Where N is the number of data points, y^i_{actual} is the actual output of i-th data point and $y^i_{predicted}$ is the predicting output of i-th data point.

3 Experiments

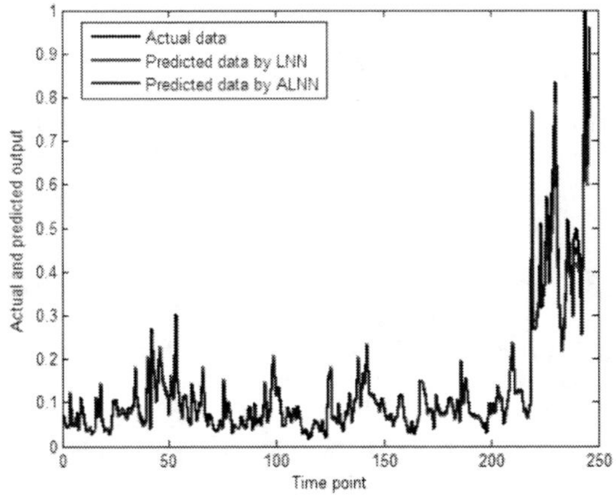

Figure 2. Actual data and predicted data by LNN and ALNN

We choose the Shanghai stock exchange composite index (Shanghai index, stock id: 000002) from 04 January, 2011 to 01 January, 2015 as the samples for validating the model.

We make the comparison between ALNN and LNN models. Through several runs, the predicted data by two methods are described in Figure 2. The predicted errors are illustrated in Figure 3. From two figures, we can see that ALNN and LNN models could predict stock index accurately and the error of ALNN is less than LNN.

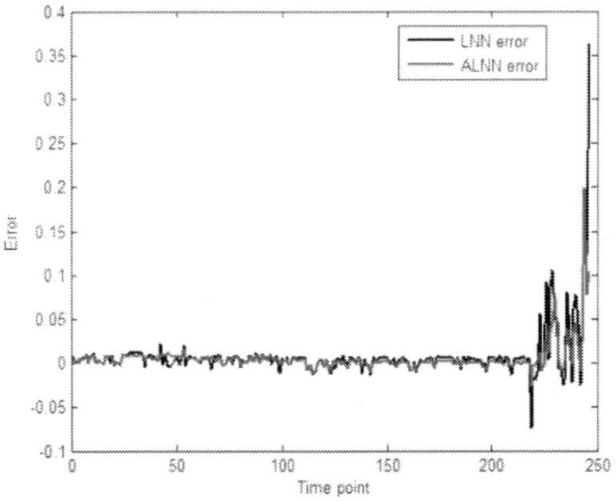

Figure 3. Predicted error by LNN and ALNN

4 Conclusion

In order to resolve the disadvantages of LNN, this paper propose a novel Legendre neural network model, namely additive Legendre neural network (ALNN). ALNN utlizes binary particle swarm optimization algorithm to select proper input Legendre polynomials in order to determine the struture of ALNN. Firefly algorithm is used to optimize the parameters of ALNN model. Shanghai stock exchange composite index is used to test the performance of ALNN. Results reveal that ALNN predicts more accurately than LNN model.

Acknowledgement

This work was supported by the PhD research startup foundation of Zaozhuang University (No.2014BS13), Zaozhuang University Foundation (No. 2015YY02), and Shandong Provincial Natural Science Foundation, China (No. ZR2015PF007).

References

1. P.J. Angeline, G.M. Saunders, J.B. Pollack, IEEE Trans. Neural Networks **5** (1994).
2. J.C. Patra, P.K. Meher, G. Chakraborty, Signal Processing **89(11)** (2009)
3. J.C. Patra, C. Bornand, IJCNN 2010 (2010)
4. A.Q. Pei, J. Wang, W. Fang, Soft Computing (2015)
5. S.S. Yang, C.S. Tseng, IEEE Trans. Syst. Man Cybern. Part-B **26 (5)** (1996)
6. H. Nezamabadi-pour, M. Rostami, The Journal of Computer Society of Iran (CSI) On Computer Science and Engineering (JCSE) **6(1-A)** (2008)
7. X.S. Yang, Lecture Notes in Computer Sciences **5792** (2009).

An Improved Brain Storm Optimization with Dynamic Clustering Strategy

Zijian Cao , Xiaofeng Rong, Zhiqiang Du

School of Computer Science and Engineering, Xi'an Technological University, Xi'an 710021, China

Abstract: Intelligence algorithms play an increasingly important role in the field of intelligent control. Brain storm optimization (BSO) is a new kind of swarm intelligence algorithm inspired by emulating the collective behavior of human beings in the problem solving process. To improve the performance of the original BSO, many variants of BSO are proposed. In this paper, an improved BSO algorithm with dynamic clustering strategy (BSO-DCS) is proposed as a variant of BSO for global optimization problems. The basic framework of BSO is firstly introduced. Then to reduce the time complexity of the original BSO, a new grouping method named dynamic clustering strategy (DCS) is proposed to improve the clustering method in the original BSO. To verify the effectiveness of the proposed BSO-DCS, it is tested on 12 benchmark functions of CEC 2005 with 30 dimensions. Experimental results show that DCS is an effective strategy to reduce the time complexity, and the improved BSO-DCS performs greatly better than the original BSO algorithm.

1 Introduction

In the past few decades, a lot of swarm intelligence optimization algorithms, such as particle swarm optimization (PSO) [1], ant colony optimization (ACO) [2], bee colony optimization (BCO) [3], firefly optimization algorithms (FFO) [4], bacterial forging optimization (BFO) [5], artificial raindrop algorithm (ARA) [6], have been proposed to tackle increasingly complex real-world optimization problems in the field of intelligent control [7-9].

Brain storm optimization (BSO) is a new kind of swarm intelligence algorithm inspired by emulating the collective behavior of human beings in the problem solving process [10, 11]. Like other swarm intelligence algorithms, BSO has achieved successful applications in areas such as optimal satellite formation reconfiguration [12], the design of DC Brushless Motor [13], economic dispatch considering wind power [14], and multi-objective optimization problems (MOPs) [15-16].

However, in swarm intelligence research, there have always been attempts to further improve the performance of any given findings. In this paper, an improved BSO algorithm with dynamic clustering strategy (BSO-DCS) is proposed as a variant of BSO for global optimization problems. After we firstly introduce the basic framework of BSO, in order to reduce the time complexity of the original BSO, a new grouping method named dynamic clustering strategy (DCS) is proposed to improve the clustering method in the original BSO. To verify the effectiveness of the proposed DCS, it is tested on 12 benchmark functions of CEC 2005 with 30 dimensions.

The rest of this paper is organized as follows. Section 2 introduces briefly the BSO algorithm. Section 3 describes the BSO with a dynamic clustering strategy (BSO-DCS). Section 4 presents the test benchmark functions, the experimental setting for each algorithm and the experimental results. Section 5 discusses the differences between the BSO-DCS and the original BSO algorithm, and conclusions are also given in this Section.

2 Brain storm optimization

The BSO algorithm is motivated by the philosophy of brainstorming. Brainstorming is a widely used tool for increasing creativity in organizations which has achieved wide acceptance as a means of facilitating creative thinking [17]. Similar to other swarm intelligence algorithms, BSO is also a population-based stochastic optimization technique. A potential solution in the fitness landscape is named an idea in BSO. BSO sticks to the rules of interchange of ideas by a team and uses clustering, replacing and creating operators to produce global optimum generation by generation. As presented in [10-11], the process of BSO can be described as follows.

Firstly, N ideas are randomly initialized within the searching space. Then each idea is evaluated and its fitness value is calculated. Next m points of cluster center are also randomly initialized like N ideas, where m is lesser than N.

Next, like other evolutionary algorithms and swarm intelligence algorithms, the main operations will be divided into converging operation and diverging operation.

2.1 Converging operation

The converging operation primarily is the clustering of ideas. Clustering is a process of grouping similar objects together, and during each generation, all the ideas are clustered into *m* clusters according to idea individual features such as the distance. The best idea in each center is chosen as the cluster center, and the clustering operation can refine a search area. In BSO, the basic method of clustering is *k*-means.

2.2 Diverging operation

The diverging operation mainly includes disrupting cluster center and creating individuals.

2.2.1 Cluster center disrupting

Cluster center disrupting operation randomly chooses an idea in a cluster center and replaces it with a newly-generated idea with a probability of *p_replace*, which is also named as the replacing operation. The value of *p_replace* is utilized to control the probability to replace a cluster center by a randomly generated solution. This can avoid the premature convergence, and help idea individuals "jump out" of the local optima.

2.2.2 Creating individuals operation

To maintain the diversity of population, a new idea individual can be generated based on one idea or two in one cluster or two, respectively. In the creating operation, BSO first randomly chooses one cluster or two according to a probability of *p_one*. Then in the basis of choosing one cluster or two, an idea of cluster center or a random idea is selected with a probability of *p_one_center* and *p_two_center*. The selecting operation is defined below as

$$x_{select} = \begin{cases} x_{ij}, & \text{one cluster} \\ rand * x_{i1,j} + (1-rand) * x_{i2,j}, & \text{two clusters} \end{cases} \tag{1}$$

where *rand* is a random value between 0 and 1. After choosing one idea or two, the selected idea(s) is updated according to as

$$X_{new} = X_{select} + \xi * normrnd(0,1) \tag{2}$$

where *normrnd* is the Gaussian random with mean 0 and variance 1, ξ is an adjusting factor slowing the convergence speed down as the evolution goes, which is expressed as

$$\xi = rand * \log sig\left(\frac{0.5 * max\_iteration - current\_iteration}{k}\right) \tag{3}$$

where *rand* is a random value between 0 and 1. The *max_iteration* and *current_iteration* denote the maximum number of iteration and current number of iteration respectively. The *logsig* is a logarithmic sigmoid transfer function, and such form is beneficial to global search ability at the beginning of the evolution and enhances local search ability when the process is approaching to the end. *k* is a predefined parameter for changing slopes of the *logsig* function. The new created idea is evaluated, and if the fitness value is better than the current idea, the old idea will be replaced by the new one.

3 BSO with dynamic clustering strategy

3.1 Dynamic clustering strategy

In the original BSO, a *k*-means clustering method was adopted to group similar ideas into serval groups in the converging operation. As we all known, the *k*-means clustering method is time consuming and needs heavier time computational burden. During the evolutionary process, BSO executes the *k*-means clustering in every generation to group ideas. However, it is not necessary to use *k*-means clustering method to group the ideas into different groups in every generation.

In our proposed algorithm, the dynamic clustering strategy is used to improve the *k*-means clustering method. The main thinking of the dynamic clustering strategy is that we periodically execute the *k*-means clustering after a certain number of generations (re-clustering period), so that the exchange of information covers all ideas in the clusters to achieve proper exploration ability.

Hence, in our dynamic clustering strategy, the key point is the size of re-clustering period. If the re-clustering period is higher, the algorithm will get a high degree of exploitation and a high convergence rate, and the time complexity of the algorithm will be correspondingly higher. On the contrary, if the re-clustering period is lower, the algorithm will achieve more exploration and diversity, and the run time of the algorithm will be reduced in proportion. So, an appropriate re-clustering period helps to balance the exploitation and exploration of the algorithm, and will achieve better performance than the original BSO algorithm.

In this paper, we use a probabilistic parameter *p_dynamic* to denote the re-clustering period. *p_dynamic* is a value between 0 to 1. According to the above analysis, the pseudo-code for the implementation of dynamic clustering strategy is summarized in Algorithm 1.

Algorithm1: pseudo-code of Dynamic Clustering

```
Begin
  if rand()<p_dynamic
    execute k-means method
  end
end
```

The setting of the parameter *p_dynamic* will be introduced in detail in Section 4.2.

3.2 Dynamic step size parameter control

For an optimization algorithm, the good balance between exploration and exploitation is comparatively necessary. Exploration refers to the ability of the algorithm to explore or search different areas of the feasible search space whereas exploitation means the ability of convergence of all the individuals to the near optimal solutions as fast as possible. Too much stress on exploration would result in pure random search, whereas too much stress on exploitation would result in pure local search.

To adjust the convergence speed as the evolution goes in idea generation, the original BSO algorithm defines an adjusting factor ξ described in Formula (3). Through numerical experiments, we find that at first the adjusting factor keeps around 1, while when half the number of generations has been reached, it rapidly turns to near 0. This method to control the size of step can also balance exploration and exploitation at different searching generations. However, it just takes effect only for very short interval. Hence, we introduce a simple dynamic step size strategy. The dynamic function is described as the following

$$\xi = rand * \exp\left(1 - \frac{max\_iteration}{max\_iteration - current\_iteration + 1}\right) \quad (4)$$

where *rand* is a random value between 0 and 1. The *max_iteration* and *current_iteration* also denote the maximum number of iterations and current number of iteration, respectively.

3.3 Pseudo-code of the BSO-DCS

As described above, the pseudo-code of the improved BSO-DCS is summarized in Algorithm 2.

Algorithm2: BSO-DCS( )

```
Begin
 Randomly initialize N ideas and evaluate their fitness
 Initialize m cluster centers (m<N)
 while(stopping condition not met)
  Execute Algorithm 1: Dynamic Clustering( )
  for (i=1 to N)
   if rand()<p_one
     Randomly select a cluster
     if rand()<p_one_center
       Give the cluster center idea to X_selected
     else
       Randomly select an idea in a cluster to X_selected
     endif
   else
     Randomly select two clusters
     if rand()<p_two_center
       Combine the two clusters' center ideas to X_selected
     else
       Combine two random ideas to X_selected
     endif
   endif
   Create X_new using X_selected with formula (2) and (4)
   Accept X_new if f(X_new) is better than f(Xi )
  endfor
 endwhile
end
```

In Algorithm 2, the main process of BSO-DCS is similar with the original BSO. In converging operation, we improve the *k*-means clustering method with the dynamic clustering strategy, and in creating individuals operation, we adopt a new dynamic step size strategy described in Section 3.2. As a whole, we do not change the main framework of the original BSO for a fair comparison with BSO.

4 Benchmark tests and experimental results

4.1 Benchmark functions

In this paper, we choose 12 generally known shifted and rotated benchmark functions in CEC2005 given in Table 3[18]. All functions are tested on 30 dimensions. The searching range and theory optima for all functions are also given in the Table 1. Among 12 benchmark functions, F1 to F5 are shifted unimodal functions, F6 to F12 are shifted or rotated multimodal functions.

Table 1. Numerical benchmark function definition.

Fun.	Function Name	Range	Optima
F1	Shifted sphere	[-100 100]	0
F2	Shifted Schwefel	[-100 100]	0
F3	Shifted rotated high conditioned elliptic	[-100 100]	0
F4	Shifted Schwefels problem 1.2 with noise	[-100 100]	0
F5	Schwefels problem 2.6	[-100 100]	0
F6	Shifted Rosenbrock	[-100 100]	0
F7	Shifted rotated Griewank	[0 600]	0
F8	Shifted rotated Ackley	[-32 32]	0
F9	Shifted Rastrigin	[-5 5]	0
F10	Shifted rotated Rastrigin	[-5 5]	0
F11	Shifted rotated Weierstrass	[-0.5 0.5]	0
F12	Schwefels problem 2.13	[-100 100]	0

4.2 Parameter settings of comparative algorithms

For a fair comparison, all the experiments are conducted on the same machine with an Intel 3.4 GHz CPU, 4GB memory. The operating system is Windows 7 with MATLAB 8.0 (R2012b).

Compared with the original BSO, Our BSO-DCS algorithm requires a new parameter (p_dynamic) called re-clustering period. Fig.1 shows the variation of convergence performance under different p_dynamic which of values are 0.1, 0.3, 0.5, 0.7 and 0.9 for function F1.

From the Fig.1, we can observe that for function F1, higher re-clustering period leads to higher accuracy. However, the higher the re-clustering period, the more time complexity is required. We set *p_dynamic*=0.5 to get a trade-off between solution accuracy and time complexity.

DOI: 10.1051/matecconf/20179519002

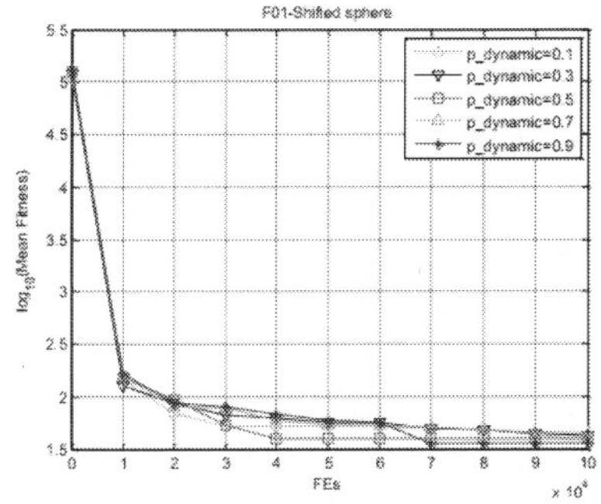

Figure 1. Convergence performance under different *p_dynamic*

Because the BSO algorithm is analogy with the PSO algorithm, we specially compare the performance of BSO-DCS with the PSO. To eliminate influences of statistical errors, each problem function is independently simulated 25 times which is a prescribed evaluation criterion in CEC2005 [18]. For all algorithms, the population size is set to 30. The same stopping criterion is used in all algorithms, i.e. reaching certain number of iterations or function evaluations (FEs).

In our experiments, we have run all algorithms on the benchmark functions using the same FEs of 10000*D for a fair comparison, where D is the size of the problem dimension. The parameter settings of all the algorithms are given in Table 2.

Table 2. The parameter settings.

NO.	Algorithm	Parameter Setting
1	BSO-DCS	M=5,p_one=0.8,p_one_center=0.4, p_two_center=0.5,p_dynamic=0.5
2	BSO	M=5,p_replace=0.2, p_one=0.8, p_one_center=0.4, p_two_center=0.5,K=20
3	PSO	ω: 0.9~0.4, c1= c2=1.49445, global version

4.3 Comparison results

4.3.1 Comparisons on solution accuracy

The results of solution accuracy are given in Table 3 in terms of the mean optimum solution and the standard deviation of the solutions obtained in the 25 independent runs by each algorithm over 300,000 FEs on 12 benchmark functions. In all experiments, the dimensions of all problems are 30. In each row of the table, the mean values are listed in the first part, and the standard deviations are listed in the last part, and the two parts are divided with a symbol "±". The best results among the algorithms are shown in boldface.

Table 3. The results of solution accuracy.

Fun.	BSO-DCS	BSO	PSO
F1	**8.79E-14±9.36E-14**	2.65E-03±3.03E-03	4.67E+03±2.86E+03
F2	**3.50E-04±2.58E-04**	2.76E+00±8.16E-01	6.51E+03±8.15E+03
F3	**8.57E+05±3.69E+05**	1.33E+06±3.58E+05	2.29E+07±4.93E+07
F4	**1.48E-01±1.00E-01**	1.29E+04±9.06E+03	4.56E+03±6.54E+03
F5	**3.03E+04±4.55E+02**	3.12E+04±6.02E+02	3.43E+04±2.28E+03
F6	**4.20E+02±4.92E+02**	1.52E+03±2.71E+03	8.15E+08±9.30E+08
F7	**5.12E+03±1.30E+02**	5.35E+03±2.32E+02	5.36E+03±5.34E+02
F8	**2.02E+01±6.10E-02**	2.05E+01±1.17E-01	2.09E+01±6.56E-02
F9	**3.11E+01±7.30E+00**	4.97E+01±1.02E+01	6.90E+01±2.28E+01
F10	**2.89E+01±6.82E+00**	4.59E+01±1.19E+01	1.26E+02±3.82E+01
F11	**9.99E+00±2.36E+00**	2.43E+01±2.58E+00	2.02E+01±2.59E+00
F12	4.95E+03±4.32E+03	3.16E+04±1.70E+04	**4.83E+03±5.14E+03**
w/t/l	-	12/0/0	11/0/1

The comparison results among BSO-DCS and other algorithms are summarized as "*w/t/l*" in the last row of the Table, which means that BSO-DCS wins in *w* functions, ties in *t* function and loses in *l* functions. From the Table 5 it can be observed that the mean value and the standard deviation value of the BSO-DCS performs better for all 12 functions than the original BSO and PSO. We can conclude that the BSO-DCS algorithm remains a better performance of the solution accuracy for complex shifted and rotated benchmark functions than the original BSO and PSO.

4.3.2 The Comparison results of convergence speed

To compare the convergence speed of different algorithms, the mean CPU times of all algorithms in one run are shown in Table 4.

Table 4. The mean CPU times of 3 algorithms.

Fun.	BSO-DCS	BSO	PSO
F1	3.31E+02	3.93E+02	3.48E+02
F2	3.46E+02	4.16E+02	3.52E+02
F3	4.76E+02	5.72E+02	4.97E+02
F4	3.59E+02	4.10E+02	3.67E+02
F5	3.40E+02	3.90E+02	3.55E+02
F6	3.34E+02	3.82E+02	3.41E+02
F7	4.75E+02	5.39E+02	5.14E+02
F8	4.76E+02	5.90E+02	5.05E+02
F9	3.33E+02	4.70E+02	3.38E+02
F10	4.75E+02	6.56E+02	4.93E+02
F11	6.19E+02	8.15E+02	6.42E+02
F12	3.83E+02	5.05E+02	3.91E+02
Total	**4.95E+03**	6.14E+03	5.14E+03
Rank	1	3	2

Table 6 shows that the mean CPU times of BSO-DCS are shorter than BSO and PSO for functions F1 to F12.

Fig.2 presents the convergence graphs in terms of the mean fitness values achieved by each of 3 algorithms for 25 runs. From the Fig.2 we can observe that BSO-DCS has fast or similar convergence speed than the other two algorithms.

DOI: 10.1051/matecconf/20179519002

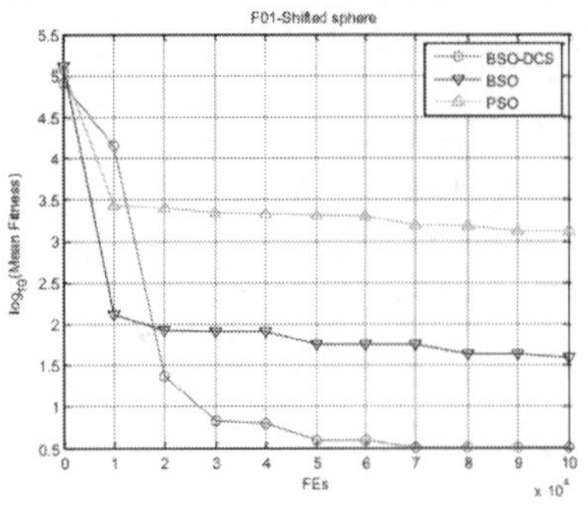

(1) F01-Shifted sphere

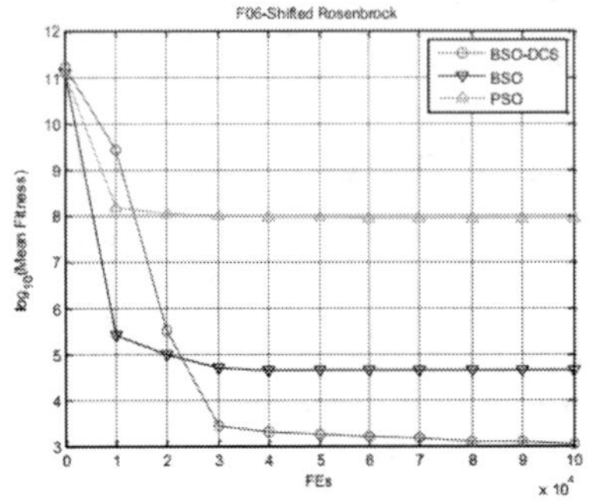

(2) F06-Shifted Rosenbrock

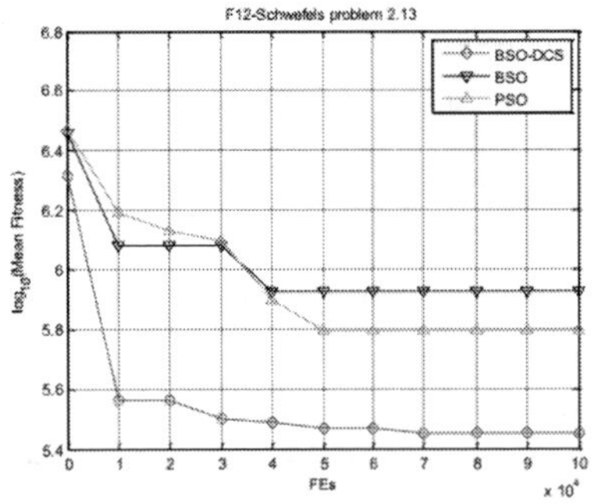

(3) F12-Schwefels problem 2.13

Figure 2. Convergence performance on part of functions

5 Discussions and conclusions

There are mainly two differences between the improved BSO-DCS and the original BSO.

Firstly, to reduce the time complexity of the algorithm, the proposed BSO-DCS adopted the dynamic clustering strategy to adjust the *k*-means clustering method. The experimental results show that the mean CPU times of BSO-DCS is lower significantly than the original BSO algorithm.

Besides, we modified the creating individuals' operator with a dynamic step size parameter control strategy. The step size can dynamically adjust the convergence speed as the evolution goes in idea generation, and may overcome the shortage of the original adjusting factor which updates in very short interval. On the other hand, the dynamic step size parameter control strategy also reduces the load of parameter settings, since it is no longer required the parameter *k*.

In this paper, an improved BSO with dynamic clustering strategy named BSO-DCS is proposed for solving complex global optimization problems. Experiments on the 12 chosen test problems were carried out. From the results of experiments, it is observed that the dynamic clustering strategy can greatly reduce the run time of the BSO-DCS algorithm, and with the aid of dynamic step size parameter control, the solution accuracy of the BSO-DCS also is better than the original BSO. From the result analysis, we can conclude that BSO-DCS significantly improves the performance of the original BSO.

We expect that BSO-DCS will be further applied to constrained, uncertain and dynamic, large scale optimization domains. Moreover, the algorithm may be used to solve some real-world intelligent control problems, such as the fuzzy model predictive control of non-linear processes, chaotic time series prediction, etc.

References

1 J. Kennedy, R. C. Eberhart, Particle swarm optimization, Proceedings of IEEE International Conference on Neural Network, pp. 1942-1948, 1995.
2 M. Dorigo, V. Maniezzo, A. Colorni, Ant system: Optimization by a colony of cooperating agents. IEEE Transactions on Systems, Man. Cybernetics B, vol. 26, no. 2, pp. 9-41, Feb. 1996.
3 D. Karaboga, B. Basturk, A powerful and efficient algorithm for numerical function optimization: Artificial bee colony (ABC) algorithm, Journal of Global Optimization, vol. 39, no. 3, pp. 459-471, Nov. 2007.
4 X. Yang, Nature-inspired metaheuristic algorithms, Beckington, UK: Luniver Press, 2008.
5 K. M. Passino, Bacterial foraging optimization, International Journal of Swarm Intelligence Research, Vol.1, no.1, pp.1-16, 2010.
6 Q. Y. Jiang, L. Wang, X. H. Hei, et al, Optimal appro ximation of stable linear systems with a novel and eff icient optimizaion algorithm, IEEE Congress on Evol utionary Computation, pp. 840-844, Jul. 2014.
7 H. Sarimveis, G. Bafas. Fuzzy model predictive control of non-linear processes using genetic algorithms. Fuzzy Sets and Systems. 2003, 139: 59-80.
8 J. Wang, D. Chi, J. Wu, H. Y. Lu . Chaotic time series method combined with particle swarm optimization and trend adjustment for electricity demand forecasting. Expert Systems with Applications, 38(7): 8419-8429, 2011
9 V. A. Gromov, A. N. Shulga. Chaotic time series prediction with employment of ant colony optimization. Expert Systems with Applications, 39(9): 8474-8478, 2012
10 Y. H. Shi, Brain storm optimization algorithm, The second International Conference on Swarm Intelligence, pp. 303–309, 2011.

11 Y. H. Shi, An optimization algorithm based on brainstorming process, International Journal of Swarm Intelligence Research, vol. 2, no. 4, pp. 35–62, Oct. 2011.
12 C. H. Sun, H. B. Duan, Y. H. Shi, Optimal satellite formation reconfiguration based on closed-loop brain storm optimization, IEEE Computational Intelligence Magazine, vol. 8, no. 4, pp. 39-51, 2013.
13 H. B. Duan, S. T. Li, Y. H. Shi, Predator-prey based brain storm optimization for DC brushless motor. IEEE Transactions on Magnetics, vol. 49, no. 10, pp. 5336-5340, 2013.
14 H. T. Jadhav, U. Sharma, J. Patel, et al, Brain storm optimization algorithm based economic dispatch considering wind power. IEEE International Conference on Power and Energy, pp. 588-593, 2012.
15 Y. H. Shi, Multi-objective optimization based on brain storm optimization algorithm, International Journal of Swarm Intelligence Research, vol. 4, no. 3, pp. 1-21, Jul. 2013.
16 J. Q. Xue, Y. L. Wu, Y. H. Shi, S. Cheng, Brain storm optimization algorithm for multiobjective optimization problems, The third International Conference on Swarm Intelligence, pp. 513-519, 2012.
17 A. F. Osborn, Applied imagination: Principles and procedures of creative problem solving. Charles Scribner's Son, New York, 1963.
18 P. N. Suganthan, N. Hansen, J. J. Liang, et al, Problem definitions and evaluation criteria for the CEC 2005 special session on real-parameter optimization, Nanyang Technological University, Singapore, Technical Report, 2005.

AUTHOR INDEX

AUTHOR INDEX

AUTHOR INDEX

AUTHOR INDEX

AUTHOR INDEX

AUTHOR INDEX